# Molecular Biology and Biotechnology

# Editorial Board

# Molecular Biology and Biotechnology

## A Comprehensive Desk Reference

*Edited by*
Robert A. Meyers

Robert A. Meyers, Ph.D.
3715 Gleneagles Drive
Tarzana, CA 91356, USA

*Management Supervised by:*
Chernow Editorial Services, Inc.,
1133 Broadway, Suite 721, New York, NY, USA

*Cover design by:* G & H SOHO, Inc.

Cover art courtesy of Dr. Alexander Wlodawer from Figure 1 of his article, "AIDS, Inhibitor Complexes of HIV-1 Protease in."
Art prepared by Dr. Jacek Lubkowski.

**Library of Congress Cataloging-in-Publication Data**

Molecular biology and biotechnology : a comprehensive desk reference / Robert A. Meyers, editor.
p. cm.
Includes bibliographical references and index.
ISBN 1–56081–569–8 (alk. paper). — ISBN 1–56081–925–1 (pbk.: alk. paper)
1. Molecular biology—Encyclopedias. 2. Biotechnology—Encyclopedias.
I. Meyers, Robert A. (Robert Allen), 1936–
QH506.M66155 1995
574.8'8'03—dc20 95-9063
CIP



Printed in the United States of America

ISBN 1-56081-569-8 (hardcover)

Printing History:
10 9 8 7 6 5 4 3 2

ISBN 1-56081-925-1 (softcover)

Printing History:
10 9 8 7 6 5 4 3 2

Published jointly by:

VCH Publishers, Inc.
220 East 23rd Street
New York, NY 10010
USA

VCH Verlagsgesellschaft mbH
P.O. Box 10 11 61
D-6940 Weinheim
Federal Republic of Germany

VCH Publishers (UK) Ltd.
8 Wellington Court
Cambridge CB1 1HW
United Kingdom

Fax: (212) 481-0897
E-mail address: order@vch.com

# Dedication

During 1994 a unique Nobel Laureate passes away: Linus Pauling. He occupies a special position among the laureates, since he remains the only person ever to have received two undivided Nobel Prizes: the Chemistry Prize in 1954 and the Peace Prize in 1962 (presented in 1963).

Pauling is widely regarded as one of the most important chemists of the 20th century. He described the nature of chemical bonds and did epoch-making work on the structure of proteins. His research linked physics with chemistry, and chemistry with biology and medicine. He was a colorful person with unconventional working methods. To solve difficult problems, he used his intuition. His close associates maintained that the depth of Paulings's intuitive understanding of a question was such that he often knew the solution to a problem before having developed a theory to explain it.

Aside from his scientific work, Pauling pursued an indefatigable campaign against nuclear weapons testing, proliferation of nuclear weapons, and war in general as a method of resolving international conflicts. Pauling's intuition undoubtedly also played a role when he raised the issue of the effects of radioactive fallout on human genes. In a speech to the students of Washington University, he said: "I believe that no human being should be sacrificed to the project of perfecting nuclear weapons that could kill hundreds of millions of human beings and devastate this beautiful world in which we live."

This was the speech in which he outlined the main principles of his appeal against nuclear weapons testing, which attracted huge publicity. More than 11,000 scientists from some fifty countries signed the appeal, which Pauling handed to the Secretary General of the United Nations, Dag Hammarskjöld.

*Bengt Samuelsson,*
*Professor and Chairman of the Board of the Nobel Foundation*
*(By permission, from a translation of his speech*
*given at the 1994 Prize Award Ceremony)*

# Contents

List of Articles . . . . . . . . . . . . . . . . . . . . . . . . . . . . . . . . . ix

Preface. . . . . . . . . . . . . . . . . . . . . . . . . . . . . . . . . . . . . . . . xv

Acknowledgments . . . . . . . . . . . . . . . . . . . . . . . . . . . . . . . xvii

How to Use This Book . . . . . . . . . . . . . . . . . . . . . . . . . . . . xxi

Contributors . . . . . . . . . . . . . . . . . . . . . . . . . . . . . . . . . . xxiii

Articles . . . . . . . . . . . . . . . . . . . . . . . . . . . . . . . . . . . . . . . 1

Glossary of Basic Terms . . . . . . . . . . . . . . . . . . . . . . . . . . 985

Index. . . . . . . . . . . . . . . . . . . . . . . . . . . . . . . . . . . . . . . . 987

# List of Articles


Achilles' Cleavage
*Michael D. Koob*

Aging, Genetic Mutations in
*James E. Fleming*

AIDS, Inhibitor Complexes of HIV-1 Protease in
*Alexander Wlodawer*

AIDS, Transcriptional Regulation of HIV in
*Melanie R. Adams and Boris Matija Peterlin*

AIDS HIV Enzymes, Three-Dimensional Structure of
*Zuzana Hostomska and Zdenek Hostomsky*

AIDS Therapeutics, Biochemistry of
*Christine Debouck and Martin Rosenberg*

Alzheimer's Disease
*Frank Ashall and Alison M. Goate*

Amino Acid Synthesis
*Robert M. Williams*

Annexins
*Carl E. Creutz*

Antibody Molecules, Genetic Engineering of
*Sherie L. Morrison*

Antisense Oligonucleotides, Structure and Function of
*Eugen Uhlmann and Anusch Peyman*

Arabidopsis Genome
*Elliot M. Meyerowitz*

Autoantibodies and Autoimmunity
*K. Michael Pollard, Edward K. L. Chan, and Eng M. Tan*

Automation in Genome Research
*Edward Theil*

Bacterial Growth and Division
*Stephen Cooper*

Bacterial Pathogenesis
*Luciano Passador, Kenneth D. Tucker, Barbara H. Iglewski*

Bacteriorhodopsin
*Janos K. Lanyi*

Bacteriorhodopsin-Based Artificial Photoreceptor
*Tsutomu Miyasaka*

Biochemical Genetics, Human
*Frits A. Hommes*

Biodegradation of Organic Wastes
*Duane Graves*

Bioenergetics of the Cell
*Stephen E. Specter and Barbara A. Horwitz*

Bioinorganic Chemistry
*R. Bruce Martin*

Biomaterials for Organ Regeneration
*Susan L Ishaug, Robert C. Thomson, Antonios G. Mikos, and Robert Langer*

Biomimetic Materials
*Robert J. Campbell*

Biomolecular Electronics and Applications
*Elias Greenbaum*

Bioorganic Chemistry
*Simon H. Friedman and George L. Kenyon*

Bioprocess Engineering
*Kimberly L. Ogden and Milan Bier*

Biosensors
*Anthony E. G. Cass*

Biotechnology, Governmental Regulation of
*David T. Kingsbury*

Biotransformations of Drugs and Chemicals
*Paul R. Ortiz de Montellano*

Body Expression Map of the Human Genome
*Kousaku Okubo and Kenichi Matsubara*

Brain Aromatic Amino Acid Hydroxylases
*Seymour Kaufman and Paula Ribeiro*

Breast Cancer, Genetic Analysis of
*Patricia Margaritte-Jeannin*

Calcium Biochemistry
*Joachim Krebs*

Cancer
*Thomas G. Pretlow and Theresa P. Pretlow*

Carbohydrate Analysis
*Martin F. Chaplin*

Carbohydrate Antigens
*David R. Bundle*

Carbohydrates, Industrial
*Margaret A. Clarke*

Cardiovascular Diseases
*Todd Leff and Peter J. Gruber*

Cell-Cell Interactions
*Geoffrey M. W. Cook*

Cell Death and Aging, Molecular Mechanisms of
*Michael Hengartner*

Chaperones, Molecular
*Elizabeth A. Craig*

Chemiluminescence and Bioluminescence, Analysis by
*Larry J. Kricka*

Chirality in Biology
*Ronald Bentley*

Chlamydomonas
*Jean-David Rochaix*

Chromatin Formation and Structure
*Timothy J. Richmond and Fritz Thoma*

Circular Dichroism in Protein Analysis
*Nicholas C. Price*

Coenzymes, Biochemistry of
*Donald B. McCormick*

Colon Cancer
*Joanna Groden*

Combinatorial Phage Antibody Libraries
*R. Anthony Williamson and Dennis R. Burton*

Computing, Biomolecular
*Felix T. Hong*

Cytochrome P450
*Michael R. Waterman*

Cytokines
*Alan G. Morris*

Cytoskeleton-Plasma Membrane Interactions
*Fredrick M. Pavalko*

Denaturation of DNA
*Richard D. Blake*

Diabetes Insipidus
*Walter Rosenthal, Anita Seibold, Daniel G. Bichet, and Mariel Birnbaumer*

Diabetes Mellitus
*David Jenkins, Catherine H. Mijovic, and Anthony H. Barnett*

DNA Damage and Repair
*George J. Kantor*

DNA Fingerprint Analysis
*Lorne T. Kirby, Ron M. Fourney, and Bruce Budowle*

DNA in Neoplastic Disease Diagnosis
*David Sidransky*

DNA Markers, Cloned
*Eugene R. Zabarovsky*

DNA Repair in Aging and Sex
*Carol Bernstein and Harris Bernstein*

DNA Replication and Transcription
*Yusaku Nakabeppu, Hisaji Maki, and Mutsuo Sekiguchi*

DNA Sequencing in Ultrathin Gels, High Speed
*Robert L. Brumley, Jr. and Lloyd M. Smith*

DNA Structure
*Olga Kennard and S. A. Salisbury*

Down's Syndrome, Molecular Genetics of
*Charles J. Epstein*

Drosophila Genome
*John Merriam*

Drug Addiction and Alcoholism, Genetic Basis of
*Judith E. Grisel and John C. Crabbe*

Drug Synthesis
*Daniel Lednicer*

Drug Targeting and Delivery, Molecular Principles of
*Alfred Stracher*

E. coli Genome
*Hirotada Mori and Takashi Yura*

Electric and Magnetic Field Reception
*Eberhard Neumann*

Electron Microscopy of Biomolecules
*John Sommerville and Hanswalter Zentgraf*

Electron Transfer, Biological
*Christopher C. Moser, Ramy S. Farid, and P. Leslie Dutton*

Endocrinology, Molecular
*Franklyn F. Bolander, Jr.*

Environmental Stress, Genomic Responses to
*John G. Scandalios*

Enzyme Assays
*Robert Eisenthal*

Enzyme Mechanisms, Transient State Kinetics of
*Kenneth A. Johnson*

Enzymes
*Jon A. Christopher, Stephen W. Raso, Miriam M. Ziegler, and Thomas O. Baldwin*

Enzymes, High Temperature
*Michael W. W. Adams*

Enzymology, Nonaqueous
*Alan J. Russell, Sudipta Chatterjee, and Darrell Williams*

Expression Systems for DNA Processes
*Ka-Yiu San and George N. Bennett*

Extracellular Matrix
*Linda J. Sandell*

Fluorescence Spectroscopy of Biomolecules
*Joseph R. Lakowicz*

Food Proteins and Interactions
*Nicholas Parris and Charles Onwulata*

Fragile X Linked Mental Retardation
*A. J. M. H. Verkerk and B. A. Oostra*

Free Radicals in Biochemistry and Medicine
*Barry Halliwell*

Fuel Production, Biological
*Lee R. Lynd*

Fungal Biotechnology
*Brian McNeil and Linda M. Harvey*

Gaucher Disease
*Ernest Beutler*

Gel Electrophoresis of Proteins, Two-Dimensional Polyacrylamide
*Peter J. Wirth*

Gene Expression, Regulation of
*Göran Akusjärvi*

Gene Mapping by Fluorescence in Situ Hybridization
*Amanda C. Heppell-Parton*

Gene Order by FISH and FACS
*Malcolm A. Ferguson-Smith*

Genetic Analysis of Populations
*A. Rus Hoelzel*

Genetic Diversity in Microorganisms
*Werner Arber*

Genetic Immunization
*Stephen A. Johnston and De-chu Tang*

Genetics
*D. Peter Snustad*

Genetic Testing
*Frank K. Fujimura*

Genomic Imprinting
*Wolf Reik and Nicholas D. Allen*

Glycobiology
*Akira Kobata*

Glycogen
*Mathieu Bollen and Willy Stalmans*

Glycoproteins, Secretory
*Alistair G. C. Renwick*

Growth Factors
*Antony W. Burgess*

Heart Failure, Genetic Basis of
*Frank Kee and Alun E. Evans*

Heat Shock Response in E. coli
*Takashi Yura and Hirotada Mori*

Hemoglobin
*Gino Amiconi and Maurizio Brunori*

Hemoglobin, Genetic Engineering of
*Timm-H. Jessen*

Hemophilia
*Francesco Giannelli*

Histones
*Gary S. Stein, Janet L. Stein, and André J. van Wijnen*

HPLC of Biological Macromolecules
*Karen M. Gooding*

Human Chromosomes, Physical Maps of
*Cassandra L. Smith, Denan Wang, and Dietmar Grothues*

Human Disease Gene Mapping
*Scott R. Diehl*

Human Genetic Predisposition to Disease
*Belinda J. F. Rossiter and C. Thomas Caskey*

Human Linkage Maps
*Jean Weissenbach*

Human Repetitive Elements
*Jerzy Jurka*

Hybridization for Sequencing of DNA
*William Bains*

Hydrogen Bonding in Biological Structures
*George A. Jeffrey*

Immunology
*Jiri Novotny and Anthony Nicholls*

Immuno-PCR
*Takeshi Sano, Cassandra L. Smith, and Charles R. Cantor*

Immunotoxicological Mechanisms
*Rodney R. Dietert*

Infectious Disease Testing by Ligase Chain Reaction
*Helen H. Lee and Gregor W. Leckie*

Inorganic Solids, Biomolecules in the Synthesis of
*Trevor Douglas and Stephen Mann*

Insecticides, Recombinant Protein
*James Y. Bradfield and Beverly J. Burden*

Interferons
*Michael J. Clemens*

Interleukins
*Michael J. Clemens*

Ionizing Radiation Damage to DNA
*Clemens von Sonntag and Heinz-Peter Schuchmann*

Isoenzymes
*Adrian O. Vladutiu and Georgirene D. Vladutiu*

Kallikrein-Kininogen-Kinin System
*Michael E. Rusiniak and Nathan Back*

Ligation Assays
*Clair Delahunty, Pui-Yan Kwok, and Deborah Nickerson*

Lipid Metabolism and Transport
*Clive R. Pullinger and John P. Kane*

Lipids, Microbial
*Colin Ratledge*

Lipids, Structure and Biochemistry of
*Donald M. Small*

Lipopolysaccharides
*Ilkka M. Helander, P. Helena Mäkelä, Otto Westphal, and Ernst T. Rietschel*

Lipoprotein Analysis
*E. Roy Skinner*

Liposomal Vectors
*Roger R. C. New*

Liver Cancer, Molecular Biology of
*Mehmet Ozturk*

Lung Cancer, Molecular Biology of
*Jack A. Roth*

Major Histocompatibility Complex
*James Driscoll*

Mammalian Genome
*John Schimenti*

Mass Spectrometry of Biomolecules
*Raymond E. Kaiser, Jr.*

Medicinal Chemistry
*David J. Triggle*

Membrane Fusion, Molecular Mechanism of
*Koert N. J. Burger*

Memory and Learning, Molecular Basis of
*Timothy E. Kennedy and Eric R. Kandel*

Metalloenzymes
*Walther R. Ellis, Jr. and Gregory M. Raner*

Mitochondrial DNA
*Ulf Gyllensten*

Mitochondrial DNA, Evolution of Human
*Linda Vigilant*

Molecular Genetic Medicine
*Bernice E. Morrow and Raju Kucherlapati*

Motor Neuron Disease
*Michael Sendtner*

Motor Proteins
*Ravindhra G. Elluru, Janet L. Cyr, and Scott T. Brady*

Mouse Genome
*Stephen D. M. Brown*

Muscle, Molecular Genetics of Human
*Prem Mohini Sharma*

Mycobacteria
*Johnjoe McFadden and Neil Stoker*

Nematodes, Neurobiology and Development of
*Darryl MacGregor, Ian A. Hope, and R. Elwyn Isaac*

Neurofibromatosis
*David H. Viskochil and Roger K. Wolff*

Neuropeptides, Insect
*Edward P. Masler*

Neuropsychiatric Diseases
*F. Owen*

Nitric Oxide in Biochemistry and Drug Design
*Larry K. Keefer*

Nuclear Magnetic Resonance of Biomolecules in Solution
*Betty J. Gaffney and Brendan C. Maguire*

Nucleic Acid Hybrids, Formation and Structure of
*James G. Wetmur*

Nucleic Acid Sequencing Techniques
*Christopher J. Howe*

Nutrition
*C. J. K. Henry*

Oligonucleotide Analogues, Phosphate-Free
*Rajender S. Varma*

Oligonucleotides
*Fritz Eckstein, Olaf Heidenreich, Mabel Ng, and Tom Tuschl*

Oncogenes
*Anthony Byrne and Desmond N. Carney*

Origins of Life, Molecular Biology of
*James P. Ferris*

Paleontology, Molecular
*Rob DeSalle*

Partial Denaturation Mapping
*Ross B. Inman*

PCR Technology
*Henry A. Erlich*

Peptides
*Tomi K. Sawyer, Wayne L. Cody, Daniele M. Leonard, and Mac E. Hadley*

Peptides, Synthetic
*Gregory A. Grant*

Peptides and Mimics, Design of Conformationally Constrained
*Victor J. Hruby and Lakmal W. Boteju*

Peptide Synthesis, Solid-Phase
*R. C. Sheppard*

Pesticide-Producing Bacteria
*Jeffrey L. Kelly, Michael David, and Peter S. Carlson*

Pharmaceutical Analysis, Chromatography in
*Satinder Ahuja*

Pharmacogenetics
*Ann K. Daly and Jeffrey R. Idle*

Phenylketonuria, Molecular Genetics of
*Randy C. Eisensmith and Savio L. C. Woo*

Phospholipases
*Suzanne E. Barbour and Edward A. Dennis*

Phospholipids
*Dennis E. Vance*

Photosynthetic Energy Transduction
*Neil R. Baker*

Plant Cells, Genetic Manipulation of
*Roland Bilang and Martin Schrott*

Plant Cell Transformation, Physical Methods for
*James Oard*

Plant Gene Expression Regulation
*Eric Lam*

Plant Pathology, Molecular
*Sarah Jane Gurr*

Plasma Lipoproteins
*Thomas L. Innerarity*

Plasmids
*Kimber Hardy*

Polymers, Genetic Engineering of
*Maurille J. Fournier, Thomas L. Mason, and David A. Tirrell*

Polymers for Biological Systems
*Robert C. Thomson, Susan L. Ishaug, Antonios G. Mikos, and Robert Langer*

Population-Specific Genetic Markers and Disease
*L. B. Jorde*

Protein Aggregation
*Jeannine M. Yon*

Protein Analysis by Integrated Sample Preparation, Chemistry, and Mass Spectrometry
*Kenneth R. Williams and Steven A. Carr*

Protein Analysis by Raman Spectroscopy
*George J. Thomas, Jr.*

Protein Analysis by X-Ray Crystallography
*Gordon V. Louie*

Protein Designs for the Specific Recognition of DNA
*Aaron Klug*

Protein Folding
*Thomas E. Creighton*

Protein Modeling
*Thomas R. Defay and Fred E. Cohen*

Protein Phosphorylation
*Clay W Scott*

Protein Purification
*Murray P. Deutscher*

Proteins and Peptides, Isolation for Sequence Analysis of
*Larry D. Ward and Richard J. Simpson*

Protein Sequencing Techniques
*Jeffrey N. Keen and John B. C. Findlay*

Protein Targeting
*Brian Austen and David Stephens*

Radioisotopes in Molecular Biology
*Robert James Slater*

RecA Protein, Structure and Function of
*Michael M. Cox*

Receptor Biochemistry
*Tatsuya Haga*

Recognition and Immobilization, Molecular
*Irwin M. Chaiken*

Recombination, Molecular Biology of
*Hannah L. Klein*

Renal System
*Samir S. El-Dahr, R. Ariel Gomez, and L. Gabriel Navar*

Repressor-Operator Recognition
*Peter G. Stockley and Simon E. V. Phillips*

Restriction Endonucleases and DNA Modification Methyltransferases for the Manipulation of DNA
*Eric W. Fisher and Richard I. Gumport*

Restriction Landmark Genomic Scanning Method
*Yoshihide Hayashizaki, Shinji Hirotsune, Yasushi Okazaki, Masami Muramatsu, and Jun-ichi Asakawa*

Retinoblastoma, Molecular Genetics of
*Brenda L. Gallie and John Wu*

Retinoids
*Robert R. Rando*

Ribosome Preparations and Protein Synthesis Techniques
*Gary Spedding*

Ribozyme Chemistry
*John M. Burke*

RNA Replication
*Claude A. Villee*

RNA Secondary Structures
*John A. Jaeger*

RNA Structure, Nonhelical
*Brian Wimberly*

RNA Three-Dimensional Structures, Computer Modeling of
*François Major*

Scanning Tunneling Microscopy in Sequencing of DNA
*Thomas L. Ferrell, D. P. Allison, T. Thundat, and R. J. Warmack*

Scleroderma Diagnosis with Recombinant Protein
*Yoshinao Muro, Kenji Sugimoto, Masaru Ohashi, and Michio Himeno*

Sequence Alignment of Proteins and Nucleic Acids
*William R. Taylor*

Sequence Analysis
*Martin Bishop*

Sequence Divergence Estimation
*Takashi Gojobori and Kazuho Ikeo*

Sex Determination
*Stephen S. Wachtel*

Steroid Hormones and Receptors
*Robin Leake*

Superantigens
*Monique Lafon*

Synapses
*Hui-Quan Han and Robert A. Nichols*

Synthetic Peptide Libraries
*Kit S. Lam*

Taxol® and Taxane Production by Cell Culture
*Arthur G. Fett-Neto and Frank DiCosmo*

Theoretical Molecular Biology
*Andrzej K. Konopka*

Tomatoes, Gene Alterations of
*Belinda Martineau and William R. Hiatt*

Trace Element Micronutrients
*Robert J. Cousins*

Transgenic Animal Modeling
*Carl A. Pinkert, Michael H. Irwin, and R. Jeffrey Moffatt*

Transgenic Animal Patents
*W. Lesser*

Transgenic Fish
*Thomas T. Chen, J. K. Lu, and Kathy Kight*

Translation of RNA to Protein
*Robert A. Cox and Henry R. V. Arnstein*

Transport Proteins
*Milton H. Saier, Jr.*

Transposons in the Human Genome
*Abram Gabriel*

Tumor Suppressor Genes
*Prem Mohini Sharma and S. R. Dev Sharma*

Ultraviolet Radiation Damage to DNA
*David L. Mitchell*

Vaccine Biotechnology
*Tilahun D. Yilma*

Viral Envelope Assembly and Budding
*Milton J. Schlesinger*

Viruses, DNA Packaging of
*Philip Serwer*

Viruses, Photodynamic Inactivation of
*James L. Matthews and Millard M. Judy*

Viruses, RNA Packaging of
*Dennis H. Bamford*

Vitamins, Structure and Function of
*Donald B. McCormick*

Whole Chromosome Complementary Probe Fluorescence Staining
*Heinz-Ulrich G. Weier, Daniel Pinkel, and Joe W. Gray*

X-Ray Diffraction of Biomolecules
*K. Ravi Acharya and Anthony R. Rees*

Yeast Genetics
*Iain Campbell*

Zinc Finger DNA Binding Motifs
*John W. R. Schwabe and Louise Fairall*

# Preface

The overall objective of this publication is to provide a professional-level reference work with comprehensive coverage of the molecular basis of life and the application of that knowledge in genetics, medicine, and agriculture. The result is a comprehensive desk reference of the following:

1. life processes at a molecular level;
2. genetic disease diagnosis and genetic therapy;
3. the theory and techniques for understanding, manipulating, and synthesizing biological molecules and their aggregates; and
4. the application of biological processes to make or modify products to improve plants or animals or to develop microorganisms for specific uses.

Our purpose is to provide a single-volume desk reference for the expanding numbers of scientists who will contribute to this field, many of whom, as in the past, will enter molecular biology research from majors or careers in physics, chemistry, mathematics, and engineering, as well as animal, plant, and cell biology, and medicine. It is also our purpose to provide an easily accessible reference work for clinical physicians who will employ the advances in genetic medicine in their practice. Teachers and professors in schools and universities will use this volume for course preparation, and members of the press will find useful background information on new developments in biotechnology and genetic medicine.

Each subject is presented on a first-principles basis, including detailed figures, tables, and drawings to elucidate atomic or structural features. The authors present a concise treatment of their field of expertise at a level useful to both colleagues and researchers who are experts in related fields, as well as to university students and physicians requiring an introduction to a specific molecular biology discipline.

The more than 250 articles that comprise the book cover the following major areas of modern molecular biology and biotechnology: Genetics and Nucleic Acids Structure and Processes; Human Genome Project; Molecular Biology of Specific Organisms and Organ Systems; Molecular Biology of Specific Disease; Biotechnology Techniques, Applications and Products; Proteins, Peptides and Amino Acids; Lipids, Carbohydrates and Relation to Molecular Biology; Immunology and Biomolecular Interactions; Biochemistry and Relation to Molecular Biology; and Pharmacology and Relation to Molecular Biology.

It must be admitted that in some cases readers at the beginning of the educational hierarchy, e.g., university students with a year of college chemistry, physics, and biology, may need to refer to a dictionary of scientific terms to access a few of the articles. These users will find the extra effort well worthwhile. Also, readers at the top of the hierarchy, i.e., molecular biologists and bioengineers, who are experts in one or more discipline within molecular biology, may find some of the articles in their field of research to be less detailed than desired. However, these scientists will find the articles in related disciplines to be useful in planning experimentation and interpreting results within their line of research.

Robert A. Meyers
Tarzana, CA, USA

**VCH Publishers, Inc., New York, NY**

Frank B. Cermak, *President*
Barbara M. Goldman, *Ph.D., Editorial Director*
Camille Pecoul, *Director, Production and Manufacturing*

**VCH Verlagsgesellschaft mbH, Weinheim, Germany**

H.-J. Kraus, *Ph.D., Executive Editor, Life Sciences*

Michael Beaubien, *Managing Editor, Chernow Editorial Services, New York, NY*

Brenda Griffing, *Copy Editor*
Virginia Hobbs, *Indexer*

# Acknowledgments


## PUBLISHER'S ACKNOWLEDGMENTS

The publisher gratefully acknowledges the contributions of the following people to the development of this reference work.

Abul K. Abbas
*Harvard Medical School*

Reid G. Adler
*National Institutes of Health*

W. L. Alworth
*Tulane University*

Michael Ashburner
*Cambridge University*

Joe Bentz
*Drexel University*

Paul Berg
*Stanford University Medical School*

James L. Bittle
*Scripps Research Institute*

Pamela Bjorkman
*California Institute of Technology*

Kristina Bostrom
*University of California, Los Angeles, School of Medicine*

Lisa Brannon-Peppas
*Eli Lilly and Company*

H. Franklin Bunn
*Harvard Medical School*

Ranajit Chakraborty
*University of Texas*

Robert S. Coleman
*University of South Carolina*

Robert W. Colman
*Temple University School of Medicine*

Lisa Conigho
*Cornell University*

David Dantzker
*Long Island Jewish Medical Center*

Kay E. Davies
*John Radcliffe Hospital*

Thomas Devlin
*Hahnemann University School of Medicine*

H. Dintzis
*Johns Hopkins School of Medicine*

G. A. Dover
*University of Leicester*

William A. Eaton
*National Institutes of Health*

Richard H. Ebright
*Rutgers University*

M. A. El-Sayed
*University of California, Los Angeles*

S. Fakan
*University of Lausanne*

Lawrence A. Frohman
*University of Illinois at Chicago College of Medicine*

Irving Geis
*Scientific Illustrator, New York*

Nicholas W. Gillham
*Duke University*

E. M. Golenberg
*Wayne State University*

Paul Goodfellow
*Washington University*

Jay A. Gottfried
*New York University*

J. Greer
*Abbott Laboratories*

Jack Griffith
*University of North Carolina*

Carol A. Gross
*University of California, San Francisco*

George Guilbault
*University of New Orleans*

Christine Guthrie
*University of California, San Francisco*

Barry Hall
*University of Rochester*

Mike Harrington
*California Institute of Technology*

Patrick Hearing
*State University of New York, Stony Brook*

M. T. W. Hearn
*Monash University*

Roger Hendrix
*University of Pittsburgh*

R. A. Herbert
*University of Dundee*

John B. Hibbs, Jr.
*University of Utah School of Medicine*

Ralph F. Hirschmann
*University of Pennsylvania*

H. L. Holland
*Brock University*

Leaf Huang
*University of Pittsburgh School of Medicine*

Iain Hunter
*University of Glasgow*

George Alan Jeffrey
*University of Pittsburgh*

John B. Jenkins
*Swarthmore College*

Ramachandra V. Joshi
*Western Michigan University*

Dick Junghans
*Deaconess Hospital*

David Kaplan
*Natick Research, Development, and Engineering Center*

Claudia Kappen
*Mayo Clinic Scottsdale*

Richard A. King
*University of Minnesota*

Claude B. Klee
*National Cancer Institute*

George Klein
*Karolinska Institute*

Ted Klein
*E. I. duPont de Nemours & Co.*

Michael Lerman
*National Cancer Institute*

Beatriz Levy
*Gladstone Institutes*

Yuhua Li
*Clemson University*

Jane B. Lian
*University of Massachusetts Medical Center*

Tim Lohman
*Washington University School of Medicine*

Mike Luster
*National Institutes of Health*

J. P. Luzio
*Cambridge University*

Coling T. Mant
*University of Alberta*

Koichiro Matsuno
*Nagaoka University of Technology*

L. McLaughlin
*Boston College*

William C. Merrick
*Case Western Reserve University School of Medicine*

Alfred H. Merrill, Jr.
*Emory University*

Peter B. Moore
*Yale University*

Krishna Murthy
*Temple University Medical School*

Leslie E. Orgel
*The Salk Institute*

R. H. Pain
*Jozf Stefan Institute*

Richard Porter
*State University of New York, Stony Brook*

George Poste
*SmithKline Beecham Pharmaceuticals*

C. G. Proud
*University of Bristol*

Thomas Quinn
*Johns Hopkins University*

Francesco Ramirez
*Mount Sinai Hospital*

J. B. Rattner
*University of Calgary*

Eugene Rearick
*Amalgamated Sugar Co.*

D. C. Rees
*California Institute of Technology*

John Saari
*University of Washington*

Takeshi Sano
*Boston University*

D. Schlessinger
*Washington University School of Medicine*

Charles Schwartz
*Greenwood Genetic Center*

Charles R. Scriver
*Montreal Children's Hospital*

Jonathan Smith
*Rockefeller University*

Robert F. Steiner
*University of Maryland, Baltimore County*

John T. Stults
*Genentech, Inc.*

Michael F. Summers
*University of Maryland*

Lorraine Symington
*Columbia University*

James P. Tam
*Vanderbilt University*

M. Thompson
*University of Toronto*

Paul Todd
*University of Colorado*

Nicholas J. Turro
*Columbia University*

Jan Vilcek
*New York University Medical Center*

Christine A. White
*Scripps Memorial Hospital*

Emily S. Winn-Dean
*Applied Biosystems Inc.*

William B. Wood
*University of Colorado*

June Hsieh Wu
*Chang-Gung Medical College*

Charles Wyman
*National Renewable Energy Laboratory*

## EDITOR'S ACKNOWLEDGMENTS

I would like to acknowledge Robert A. Meyers, Jr. and Tamara H. Cochrane, who maintained the encyclopedia article and topical outline database that I used to monitor author manuscript scheduling and to correlate the contents to assure completeness of coverage. I would also like to acknowledge my wife, Ilene Meyers, who assisted in this effort in the areas of project planning and scheduling. And finally, I thank Madeleine Lewis, who so ably assisted me with author recruitment during the concluding year of manuscript preparation.

# How to Use This Book

The more than 250 articles in *Molecular Biology and Biotechnology: A Comprehensive Desk Reference* are designed as *self-contained treatments* of the important topics in molecular biology. Each article begins with a *key word section,* including definitions, to assist the scientist or student who is unfamiliar with the specific subject area. The book includes more than 1,600 key words, each of which is defined within the context of the particular scientific field covered by the article, since the same word may have different meanings in different fields.

In addition to the key word definitions, the *Glossary of Basic Terms* found at the back of the book defines 40 of the most commonly used terms in molecular biology.

These definitions, along with the reference materials printed on the inside front and back covers *(the genetic code, the common amino acids with their one- and three-letter abbreviations, and the structures of the deoxyribonucleotides)* should allow most readers to understand articles in the *Desk Reference* without referring to a dictionary, textbook, or other reference work.

The detailed *subject index* provides the reader with an additional tool for locating topics of particular interest.

A concise definition of the subject and its importance begins each article, followed by the body of the article and a bibliography of references for further reading. Because of the self-contained nature of each article, some overlap between articles on related topics exists. Cross-references to related articles are provided to help the reader expand his or her range of inquiry.

The articles in the *Desk Reference* may be categorized as follows: (1) *core articles* that give perspective on the major disciplines, e.g., DNA structure, DNA replication and transcription, translation of RNA to protein, molecular basis of genetics, human genetic predisposition to disease, molecular genetic medicine, peptides, enzymes, bioprocess engineering, immunology, medicinal chemistry, cancer; (2) *satellite core articles* that give perspective on particularly active areas of research and importance, e.g., the core article on cancer is supported by satellite core articles on oncogenes and tumor suppressor genes; and (3) *specific subject articles*, e.g., on liver cancer, breast cancer, and colon cancer. Since the subjects are advancing so rapidly, sometimes a specific field is covered within core or satellite core articles, rather than in an individual subject article. The hope is that the reader will find as complete and authoritative a coverage of the molecular basis of life as can be provided in a single volume.

# Contributors

K. Ravi Acharya
School of Biology and Biochemistry
University of Bath
Bath BA2 7AY, United Kingdom
*X-Ray Diffraction of Biomolecules*

Melanie R. Adams
Scientific Services
Irwin Memorial Blood Centers
San Francisco, CA 94118, USA
*AIDS, Transcriptional Regulation of HIV in*

Michael W. W. Adams
Department of Biochemistry
University of Georgia
Athens, GA 30602, USA
*Enzymes, High Temperature*

Satinder Ahuja
CIBA-GEIGY Corporation
Pharmaceuticals Division
Suffern, NY 10901, USA
*Pharmaceutical Analysis, Chromatography in*

Göran Akusjärvi
Department of Medical Immunology and Microbiology
Uppsala University
S-751 23 Uppsala, Sweden
*Gene Expression, Regulation of*

Nicholas D. Allen
Laboratory of Developmental Genetics and Imprinting
The Babraham Institute
Cambridge CB2 4AT, United Kingdom
*Genomic Imprinting*

D. P. Allison
Oak Ridge National Laboratory
Oak Ridge, TN 37831, USA
*Scanning Tunneling Microscopy in Sequencing of DNA*

Gino Amiconi
Department of Biochemical Sciences
University of Rome La Sapienza
00185 Rome, Italy
*Hemoglobin*

Werner Arber
Department of Microbiology
Biozentrum
University of Basel
CH-4056 Basel, Switzerland
*Genetic Diversity in Microorganisms*

Henry R. V. Arnstein
King's College, London
London WC2R 2LS, United Kingdom
*Translation of RNA to Protein*

Jun-ichi Asakawa
Department of Genetics
Radiation Effects Research Foundation
5–2 Hijiyama Park
Minami-ku
Hiroshima 732, Japan
*Restriction Landmark Genomic Scanning Method*

Frank Ashall
Department of Psychiatry
Washington University School of Medicine
St. Louis, MO 63110, USA
*Alzheimer's Disease*

Brian Austen
Department of Surgery
St. George's Hospital Medical School
London SW17 0RE, United Kingdom
*Protein Targeting*

Nathan Back
Department of Biochemical Pharmacology
School of Pharmacy
State University of New York
Buffalo, NY 14260, USA
*Kallikrein-Kininogen-Kinin System*

William Bains
PA Consulting Group
Cambridge Laboratory
Royston, Hertfordshire SG8 6DP, United Kingdom
*Hybridization for Sequencing of DNA*

Neil R. Baker
Department of Biology
University of Essex
Colchester CO4 3SQ, United Kingdom
*Photosynthetic Energy Transduction*

Thomas O. Baldwin
Department of Biochemistry and Biophysics
Texas A&M University
College Station, TX 77843, USA
*Enzymes*

Dennis H. Bamford
Institute of Biotechnology
and
Department of Biosciences Division Genetics
Biocentre
University of Helsinki
FIN-00014 Helsinki, Finland
*Viruses, RNA Packaging of*

Suzanne E. Barbour
Department of Chemistry and Biochemistry
University of California, San Diego
La Jolla, CA 92093, USA
*Phospholipases*

Anthony H. Barnett
Department of Medicine
University of Birmingham
Queen Elizabeth Hospital
Birmingham B15 2TH, United Kingdom
*Diabetes Mellitus*

George N. Bennett
Department of Biochemistry and Cell Biology
Rice University
Houston, TX 77251, USA
*Expression Systems for DNA Processes*

Ronald Bentley
Department of Biological Sciences
University of Pittsburgh
Pittsburgh, PA 15260, USA
*Chirality in Biology*

Carol Bernstein
Department of Microbiology and Immunology
University of Arizona
College of Medicine
Tucson, AZ 85724, USA
*DNA Repair in Aging and Sex*

Harris Bernstein
Department of Microbiology and Immunology
University of Arizona
College of Medicine
Tucson, AZ 85724, USA
*DNA Repair in Aging and Sex*

Ernest Beutler
Department of Molecular and Experimental Medicine
The Scripps Research Institute
La Jolla, CA 92037, USA
*Gaucher Disease*

Daniel G. Bichet
Department of Medicine
Centre de Recherche
Hôpital du Sacré-Coeur de Montréal
Montreal, Quebec H4J 1C5
Canada
*Diabetes Insipidus*

Milan Bier
Center for Separation Sciences
University of Arizona
Tucson, AZ 85721, USA
*Bioprocess Engineering*

Roland Bilang
Institute of Plant Sciences
Swiss Federal Institute of Technology
ETH-Zentrum
CH-8092 Zurich, Switzerland
*Plant Cells, Genetic Manipulation of*

Mariel Birnbaumer
Department of Cell Biology
Baylor College of Medicine
Houston, TX 77030, USA
*Diabetes Insipidus*

Martin Bishop
UK HGMP Resource Centre
Cambridgeshire CB10 7KQ, United Kingdom
*Sequence Analysis*

Richard D. Blake
Department of Biochemistry, Microbiology, and Molecular Biology
University of Maine
Orono, ME 04469, USA
*Denaturation of DNA*

Franklyn F. Bolander, Jr.
Department of Biological Sciences
University of South Carolina
Columbia, SC 29208, USA
*Endocrinology, Molecular*

Mathieu Bollen
Afdeling Biochemie
Katholicke Universiteit Leuven
Campus Gasthuisberg
B-3000 Leuven, Belgium
*Glycogen*

Lakmal W. Boteju
Department of Chemistry
University of Arizona
Tucson, AZ 85721, USA
*Peptides and Mimics, Design of Conformationally Constrained*

James Y. Bradfield
Department of Entomology
Texas A&M University
College Station, TX 77843, USA
*Insecticides, Recombinant Protein*

Scott T. Brady
Department of Cell Biology and Neuroscience
University of Texas Southwestern Medical Center
Dallas, TX 75235, USA
*Motor Proteins*

Stephen D. M. Brown
Department of Biochemistry and Molecular Genetics
St. Mary's Hospital Medical School
London W2 1PG, United Kingdom
*Mouse Genome*

Robert L. Brumley, Jr.
Department of Chemistry
University of Wisconsin
Madison, WI 53706, USA
*DNA Sequencing in Ultrathin Gels, High Speed*

Maurizio Brunori
Department of Biochemical Sciences
University of Rome La Sapienza
00185 Rome, Italy
*Hemoglobin*

Bruce Budowle
Laboratory Division
Forensic Science Research and Training Center
Federal Bureau of Investigation Academy
Quantico, VA 22135, USA
*DNA Fingerprint Analysis*

David R. Bundle
Department of Chemistry
University of Alberta
Edmonton, Alberta T6G 2G2
Canada
*Carbohydrate Antigens*

Beverly J. Burden
Department of Biological Sciences
Louisiana State University
Shreveport, LA 71115, USA
*Insecticides, Recombinant Protein*

Koert N. J. Burger
Department of Cell Biology
University of Utrecht
Medical School AZU
3584 CX Utrecht, The Netherlands
*Membrane Fusion, Molecular Mechanism of*

Antony W. Burgess
Ludwig Institute for Cancer Research
Royal Melbourne Hospital
Victoria 3050, Australia
*Growth Factors*

John M. Burke
Department of Microbiology and Molecular Genetics
Markey Center for Molecular Genetics
University of Vermont
Burlington, VT 05405, USA
*Ribozyme Chemistry*

Dennis R. Burton
Departments of Immunology and Molecular Biology
The Scripps Research Institute
La Jolla, CA 92037, USA
*Combinatorial Phage Antibody Libraries*

Anthony Byrne
Department of Clinical Pharmacology
Royal College of Surgeons in Ireland
Dublin 2, Ireland
*Oncogenes*

Iain Campbell
International Centre for Brewing and Distilling
Heriot-Watt University
Edinburgh EH14 4AS, United Kingdom
*Yeast Genetics*

Robert J. Campbell
Chemistry and Biological Sciences Division
U.S. Army Research Office
Research Triangle Park, NC 27709, USA
*Biomimetic Materials*

Charles R. Cantor
Center for Advanced Biotechnology
and
Departments of Biomedical Engineering and Pharmacology
Boston University
Boston, MA 02215, USA
*Immuno-PCR*

Peter S. Carlson
Crop Genetics International
10150 Old Columbia Road
Columbia, MD 21046, USA
*Pesticide-Producing Bacteria*

Desmond N. Carney
Department of Medical Oncology
Mater Misericordiae Hospital
Dublin 7, Ireland
*Oncogenes*

Steven A. Carr
Department of Physical and Structural Chemistry
SmithKline Beecham Pharmaceuticals
King of Prussia, PA 19406, USA
*Protein Analysis by Integrated Sample Preparation, Chemistry, and Mass Spectrometry*

C. Thomas Caskey
Department of Molecular and Human Genetics
and
Howard Hughes Medical Institute
Baylor College of Medicine
Houston, TX 77030, USA
*Human Genetic Predisposition to Disease*

Anthony E. G. Cass
Center for Biotechnology
Imperial College of Science, Technology and Medicine
South Kensington
London SW7 2AZ, United Kingdom
*Biosensors*

Irwin M. Chaiken
Department of Molecular Genetics
SmithKline Beecham Pharmaceuticals
King of Prussia, PA 19406, USA
*Recognition and Immobilization, Molecular*

Edward K. L. Chan
Department of Molecular and Experimental Medicine
W. M. Keck Autoimmune Disease Center
The Scripps Research Institute
La Jolla, CA 92037, USA
*Autoantibodies and Autoimmunity*

Martin F. Chaplin
School of Applied Science
South Bank University
London SE1 0AA, United Kingdom
*Carbohydrate Analysis*

Sudipta Chatterjee
Department of Chemical Engineering
Center for Biotechnology and Bioengineering
University of Pittsburgh
Pittsburgh, PA 15261, USA
*Enzymology, Nonaqueous*

Thomas T. Chen
Center for Marine Biotechnology
Maryland Biotechnology Institute
Baltimore, MD 21202, USA
*Transgenic Fish*

Jon A. Christopher
Department of Chemistry
Texas A&M University
College Station, TX 77843, USA
*Enzymes*

Margaret A. Clarke
Sugar Processing Research Institute, Inc.
New Orleans, LA 70124, USA
*Carbohydrates, Industrial*

Michael J. Clemens
Division of Biochemistry
Department of Cellular and Molecular Sciences
St. George's Hospital Medical School
London SW17 0RE, United Kingdom
*Interferons; Interleukins*

Wayne L. Cody
Department of Chemistry
Parke-Davis Pharmaceutical Research
Ann Arbor, MI 48206, USA
*Peptides*

Fred E. Cohen
Department of Pharmaceutical Chemistry
University of California
School of Pharmacy
San Francisco, CA 94143, USA
*Protein Modeling*

Geoffrey M. W. Cook
Department of Anatomy
Cambridge University
Cambridge CB2 3DY, United Kingdom
*Cell-Cell Interactions*

Stephen Cooper
Department of Microbiology and Immunology
University of Michigan Medical School
Ann Arbor, MI 48109, USA
*Bacterial Growth and Division*

Robert J. Cousins
Food Science and Human Nutrition Department
Center for Nutritional Sciences
University of Florida
Gainesville, FL 32611, USA
*Trace Element Micronutrients*

Michael M. Cox
Department of Biochemistry
University of Wisconsin
Madison, WI 53706, USA
*RecA Protein, Structure and Function of*

Robert A. Cox
National Institute for Medical Research
London NW7 1AA, United Kingdom
*Translation of RNA to Protein*

John C. Crabbe
Veterans Affairs Medical Center
and
Department of Medical Psychology
Oregon Health Sciences University
Portland, OR 97201, USA
*Drug Addiction and Alcoholism, Genetic Basis of*

Elizabeth A. Craig
Department of Biomolecular Chemistry
University of Wisconsin
Madison, WI 53706, USA
*Chaperones, Molecular*

Thomas E. Creighton
European Molecular Biology Laboratory
D-690 Heidelberg, Germany
*Protein Folding*

Carl E. Creutz
Department of Pharmacology
Health Sciences Center
University of Virginia
Charlottesville, VA 22908, USA
*Annexins*

Janet L. Cyr
Department of Cell Biology and Neuroscience
University of Texas Southwestern Medical Center
Dallas, TX 75235, USA
*Motor Proteins*

Ann K. Daly
Department of Pharmacological Sciences
University of Newcastle upon Tyne
The Medical School
Pharmacogenetics Research Unit
Newcastle upon Tyne, NE2 4HH, United Kingdom
*Pharmacogenetics*

Michael David
Crop Genetics International
10150 Old Columbia Road
Columbia, MD 21046, USA
*Pesticide-Producing Bacteria*

Christine Debouck
Biopharmaceutical Research and Development
SmithKline Beecham Pharmaceuticals
King of Prussia, PA 19406, USA
*AIDS Therapeutics, Biochemistry of*

Thomas R. Defay
Graduate Group in Biophysics
University of California
Davis, CA 94143, USA
*Protein Modeling*

Claire Delahunty
Department of Molecular Biotechnology
University of Washington
Seattle, WA 98195, USA
*Ligation Assay*

Edward A. Dennis
Department of Chemistry and Biochemistry
University of California, San Diego
La Jolla, CA 92093, USA
*Phospholipases*

Rob DeSalle
Department of Entomology
American Museum of Natural History
New York, NY 10024, USA
*Paleontology, Molecular*

Murray P. Deutscher
Department of Biochemistry
University of Connecticut Health Center
Farmington, CT 06032, USA
*Protein Purification*

Frank DiCosmo
Department of Botany
University of Toronto
Toronto, Ontario M5S 3B2
Canada
*Taxol® and Taxane Production by Cell Culture*

Scott R. Diehl
Molecular Epidemiology and Disease Indicators Branch
National Institute of Dental Research
National Institutes of Health
Bethesda, MD 20892, USA
*Human Disease Gene Mapping*

Rodney R. Dietert
Institute for Comparative and Environmental Toxicology
Cornell University
Ithaca, NY 14853, USA
*Immunotoxicological Mechanisms*

Trevor Douglas
School of Chemistry
University of Bath
Bath BA2 7AY, United Kingdom
*Inorganic Solids, Biomolecules in the Synthesis of*

James Driscoll
Department of Molecular Biology
Harvard Medical School
Boston, MA 02115, USA
*Major Histocompatibility Complex*

P. Leslie Dutton
Department of Biochemistry and Biophysics
Johnson Research Foundation
University of Pennsylvania
Philadelphia, PA 19104, USA
*Electron Transfer, Biological*

Fritz Eckstein
Max-Planck-Institut für Experimentelle Medizin
D-37075 Göttingen, Germany
*Oligonucleotides*

Randy C. Eisensmith
Department of Cell Biology and Molecular Genetics
Baylor College of Medicine
Houston, TX 77030, USA
*Phenylketonuria, Molecular Genetics of*

Robert Eisenthal
Department of Biochemistry
University of Bath
Bath BA2 7AY, United Kingdom
*Enzyme Assays*

Samir S. El-Dahr
Department of Pediatrics
Division of Pediatric Nephrology
Tulane University School of Medicine
New Orleans, LA 70112, USA
*Renal System*

Walther R. Ellis, Jr.
Department of Chemistry
University of Utah
Salt Lake City, UT 84112, USA
*Metalloenzymes*

Ravindhra G. Elluru
Department of Cell Biology and Neuroscience
University of Texas Southwestern Medical Center
Dallas, TX 75235, USA
*Motor Proteins*

Charles J. Epstein
Department of Pediatrics
University of California
San Francisco, CA 94143, USA
*Down's Syndrome, Molecular Genetics of*

Henry A. Erlich
Human Genetics Department
Roche Molecular Systems, Inc.
Alameda, CA 94501, USA
*PCR Technology*

Alun E. Evans
Department of Epidemiology and Public Health
The Queens University of Belfast
Belfast VT12 6BJ, Northern Ireland
*Heart Failure, Genetic Basis of*

Louise Fairall
Medical Research Council
Laboratory of Molecular Biology
Cambridge CB2 2QH, United Kingdom
*Zinc Finger DNA Binding Motifs*

Ramy S. Farid
Department of Biochemistry and Biophysics
Johnson Research Foundation
University of Pennsylvania
Philadelphia, PA 19104, USA
*Electron Transfer, Biological*

Malcolm A. Ferguson-Smith
Department of Pathology
Cambridge University
Cambridge CB2 1QP, United Kingdom
*Gene Order by FISH and FACS*

Thomas L. Ferrell
Oak Ridge National Laboratory
Oak Ridge, TN 37831, USA
*Scanning Tunneling Microscopy in Sequencing of DNA*

James P. Ferris
Department of Chemistry
Rensselaer Polytechnic Institute
Troy, NY 12180, USA
*Origins of Life, Molecular Biology of*

Arthur G. Fett-Neto
Centre for Plant Biotechnology
Department of Botany
University of Toronto
Toronto, Ontario M5S 3B2
Canada
*Taxol® and Taxane Production by Cell Culture*

John B. C. Findlay
Department of Biochemistry and Molecular Biology
University of Leeds
Leeds LS2 9JT, United Kingdom
*Protein Sequencing Techniques*

Eric W. Fisher
Department of Biochemistry
College of Medicine and School of Chemical Sciences
University of Illinois at Urbana-Champaign
600 South Mathews Avenue
Urbana, IL 61801, USA
*Restriction Endonucleases and DNA Modification Methyltransferases for the Manipulation of DNA*

James E. Fleming
Department of Biology
Eastern Washington University
Cheney, WA 99004, USA
*Aging, Genetic Mutations in*

Ron M. Fourney
Biology Research and Development Section
Central Forensic Laboratory
Royal Canadian Mounted Police
Ottawa, Ontario K1G 2M3
Canada
*DNA Fingerprint Analysis*

Maurille J. Fournier
Department of Biochemistry and Molecular Biology
University of Massachusetts
Amherst, MA 01003, USA
*Polymers, Genetic Engineering of*

Simon H. Friedman
Department of Pharmaceutical Chemistry
School of Pharmacy
University of California
San Francisco, CA 94143, USA
*Bioorganic Chemistry*

Frank K. Fujimura
Myriad Diagnostic Services, Inc.
Salt Lake City, UT 84108, USA
*Genetic Testing*

Abram Gabriel
Department of Molecular Biology and Biochemistry
Rutgers University
Piscataway, NJ 08854, USA
*Transposons in the Human Genome*

Betty J. Gaffney
Department of Chemistry
Johns Hopkins University
Baltimore, MD 21218, USA
*Nuclear Magnetic Resonance of Biomolecules in Solution*

Brenda L. Gallie
Department of Ophthalmology and Division of Immunology and Cancer Research Institute
The Hospital for Sick Children
and
Department of Medical Genetics and Ophthalmology
University of Toronto
Toronto, Ontario M5G 1X8
Canada
*Retinoblastoma, Molecular Genetics of*

Francesco Giannelli
Division of Medical and Molecular Genetics
United Medical and Dental Schools
Guy's and St. Thomas's Hospitals
London SE1 9RT, United Kingdom
*Hemophilia*

Alison M. Goate
Department of Psychiatry
Washington University School of Medicine
St. Louis, MO 63110, USA
*Alzheimer's Disease*

Takashi Gojobori
DNA Research Center
National Institute of Genetics
Mishima 411, Japan
*Sequence Divergence Estimation*

R. Ariel Gomez
Department of Pediatrics
Division of Pediatric Nephrology
University of Virginia Health Sciences Center
Charlottesville, VA 22908, USA
*Renal System*

Karen M. Gooding
SynChrom, Inc.
Lafayette, IN 47903, USA
*HPLC of Biological Macromolecules*

Gregory A. Grant
Department of Molecular Biology and Pharmacology
Washington University School of Medicine
St. Louis, MO 63110, USA
*Peptides, Synthetic*

Duane Graves
Biotechnology Applications Center
IT Corporation
Knoxville, TN 37923, USA
*Biodegradation of Organic Wastes*

Joe W. Gray
Division of Molecular Cytometry
Department of Laboratory Medicine
University of California
San Francisco, CA 94143, USA
*Whole Chromosome Complementary Probe Fluorescence Staining*

Elias Greenbaum
Chemical Technology Division
Oak Ridge National Laboratory
Oak Ridge, TN 37831, USA
*Biomolecular Electronics and Applications*

Judith E. Grisel
Veterans Affairs Medical Center
and
Department of Medical Psychology
Oregon Health Sciences University
Portland, OR 97201, USA
*Drug Addiction and Alcoholism, Genetic Basis of*

Joanna Groden
Department of Molecular Genetics, Biochemistry, and Microbiology
University of Cincinnati College of Medicine
Cincinnati, OH 45276, USA
*Colon Cancer*

Dietmar Grothues
Biochemical Instrumentation
EMBL
6900 Heidelberg, Germany
*Human Chromosomes, Physical Maps of*

Peter J. Gruber
Department of Surgery
Johns Hopkins Hospital
Baltimore, MD 21287, USA
*Cardiovascular Diseases*

Richard I. Gumport
Department of Biochemistry
College of Medicine and School of Chemical Sciences
University of Illinois at Urbana-Champaign
600 South Mathews Avenue
Urbana, IL 61801, USA
*Restriction Endonucleases and DNA Modification Methyltransferases for the Manipulation of DNA*

Sarah Jane Gurr
Department of Plant Sciences
Oxford University
Oxford OX1 3RB, United Kingdom
*Plant Pathology, Molecular*

Ulf Gyllensten
Department of Medical Genetics
University of Uppsala Biomedical Center
S-751–23 Uppsala, Sweden
*Mitochondrial DNA*

Mac E. Hadley
Department of Cell Biology and Anatomy
University of Arizona
Tucson, AZ 85724, USA
*Peptides*

Tatsuya Haga
Department of Biochemistry
Institute for Brain Research
University of Tokyo
7-3-1 Hongo
Tokyo 113, Japan
*Receptor Biochemistry*

Barry Halliwell
Pharmacology Group
University of London
King's College
London SW3 6LX, United Kingdom
*Free Radicals in Biochemistry and Medicine*

Hui-Quan Han
Laboratory of Molecular and Cellular Neuroscience
The Rockefeller University
New York, NY 10021, USA
*Synapses*

Kimber Hardy
Glaxo Institute for Molecular Biology
Geneva, Switzerland
*Plasmids*

Linda M. Harvey
Department of Bioscience and Biotechnology
University of Strathclyde
Glasgow G1 1XW, United Kingdom
*Fungal Biotechnology*

Yoshihide Hayashizaki
Genome Science Laboratory
Tsukuba Life Science Center
The Institute of Physical and Chemical Research (RIKEN)
3-1-1 Koyadai Tsukuba Ibaraki 305, Japan
*Restriction Landmark Genomic Scanning Method*

Olaf Heidenreich
Max-Planck-Institut für Experimentelle Medizin
D-37075 Göttingen, Germany
*Oligonucleotides*

Ilkka M. Helander
Department of Bacterial Vaccine Research and Molecular Biology
National Public Health Institute
FIN-00300 Helsinki, Finland
*Lipopolysaccharides*

Michael Hengartner
Cold Spring Harbor Laboratory
Cold Spring Harbor, NY 11724, USA
*Cell Death and Aging, Molecular Mechanisms of*

C. J. K. Henry
School of Ciological and Molecular Science
Oxford Brookes University
Gipsy Lane Campus, Headington
Oxford OX3 0BP, United Kingdom
*Nutrition*

Amanda C. Heppell-Parton
Clinical Oncology and Radiotherapeutics Unit
Medical Research Council Centre
Cambridge CB2 2QH, United Kingdom
*Gene Mapping by Fluorescence in Situ Hybridization*

William R. Hiatt
Calgene, Inc.
Davis, CA 95616, USA
*Tomatoes, Gene Alterations of*

Michio Himeno
Department of Applied Biological Chemistry
College of Agriculture
University of Osaka
Sakai, Osaka 593, Japan
*Scleroderma Diagnosis with Recombinant Protein*

Shinji Hirotsune
Genome Science Laboratory
Tsukuba Life Science Center
The Institute of Physical and Chemical Research (RIKEN)
3-1-1 Koyadai Tsukuba Ibaraki 305, Japan
*Restriction Landmark Genomic Scanning Method*

A. Rus Hoelzel
Laboratory of Viral Carcinogenesis
National Cancer Institute
Frederick, MD 21702, USA
*Genetic Analysis of Populations*

Frits A. Hommes
Human Genetics Program
Department of Pediatrics
New York University Medical Center
New York, NY 10016, USA
*Biochemical Genetics, Human*

Felix T. Hong
Department of Physiology
Wayne State University
Detroit, MI 48201, USA
*Computing, Biomolecular*

Ian A. Hope
Departments of Pure and Applied
Biology and Genetics
University of Leeds
Leeds LS2 9JT, United Kingdom
*Nematodes, Neurobiology and Development of*

Barbara A. Horwitz
Division of Biological Sciences
Section of Neurobiology, Physiology,
and Behavior
University of California
Davis, CA 95616, USA
*Bioenergetics of the Cell*

Zuzana Hostomska
Agouron Pharmaceuticals, Inc.
San Diego, CA 92121, USA
*AIDS HIV Enzymes, Three-Dimensional Structure of*

Zdenek Hostomsky
Agouron Pharmaceuticals, Inc.
San Diego, CA 92121, USA
*AIDS HIV Enzymes, Three-Dimensional Structure of*

Christopher J. Howe
Department of Biochemistry
Cambridge University
Cambridge CB2 1QW, United Kingdom
*Nucleic Acid Sequencing Techniques*

Victor J. Hruby
Department of Chemistry
University of Arizona
Tucson, AZ 85721, USA
*Peptides and Mimics, Design of Conformationally Constrained*

Jeffrey R. Idle
Department of Pharmacological Sciences
University of Newcastle upon Tyne
The Medical School
Pharmacogenetics Research Unit
Newcastle upon Tyne, NE2 4HH,
United Kingdom
*Pharmacogenetics*

Barbara H. Iglewski
Department of Microbiology and
Immunology
University of Rochester School of
Medicine
Rochester, NY 14642, USA
*Bacterial Pathogenesis*

Kazuho Ikeo
DNA Research Center
National Institute of Genetics
Mishima 411, Japan
*Sequence Divergence Estimation*

Ross B. Inman
Institute for Molecular Virology
and
Department of Biochemistry
University of Wisconsin
Madison, WI 53706, USA
*Partial Denaturation Mapping*

Thomas L. Innerarity
The Gladstone Institute of
Cardiovascular Disease
Cardiovascular Research Institute
University of California
San Francisco, CA 94141, USA
*Plasma Lipoproteins*

Michael H. Irwin
Department of Comparative Medicine
University of Alabama
Birmingham, AL 35294, USA
*Transgenic Animal Modeling*

R. Elwyn Isaac
Department of Pure and Applied Biology
University of Leeds
Leeds LS2 9JT, United Kingdom
*Nematodes, Neurobiology and Development of*

Susan L. Ishaug
Institute of Biosciences and
Bioengineering
Rice University
Houston, TX 77251, USA
*Biomaterials for Organ Regeneration; Polymers for Biological Systems*

John A. Jaeger
Genta, Inc.
San Diego, CA 92122, USA
*RNA Secondary Structures*

George A. Jeffrey
Department of Crystallography
University of Pittsburgh
Pittsburgh, PA 15260, USA
*Hydrogen Bonding in Biological Structures*

David Jenkins
Department of Medicine
University of Birmingham
Queen Elizabeth Hospital
Birmingham B15 2TH, United Kingdom
*Diabetes Mellitus*

Timm-H. Jessen
Hoechst-AG
General Pharma Research
D-6230 Frankfurt 80, Germany
*Hemoglobin, Genetic Engineering of*

Kenneth A. Johnson
Department of Biochemistry and
Molecular Biology
Pennsylvania State University
University Park, PA 16802, USA
*Enzyme Mechanisms, Transient State Kinetics of*

Stephen A. Johnston
Department of Biochemistry
University of Texas
Southwestern Medical Center
Dallas, TX 75235, USA
*Genetic Immunization*

L. B. Jorde
Department of Human Genetics
University of Utah Health Sciences
Center
Salt Lake City, UT 84112, USA
*Population-Specific Genetic Markers and Disease*

Millard M. Judy
Baylor Research Institute
Baylor University Medical Center
Dallas, TX 75226, USA
*Viruses, Photodynamic Inactivation of*

Jerzy Jurka
Linus Pauling Institute
Palo Alto, CA 94306, USA
*Human Repetitive Elements*

Raymond E. Kaiser, Jr.
Lilly Research Laboratories
Eli Lilly & Company
Indianapolis, IN 46285, USA
*Mass Spectrometry of Biomolecules*

Eric R. Kandel
Center for Neurobiology and Behavior
Howard Hughes Medical Institute
College of Physicians and Surgeons of Columbia University
New York, NY 10032, USA
*Memory and Learning, Molecular Basis of*

John P. Kane
Cardiovascular Research Institute
University of California
San Francisco, CA 94143, USA
*Lipid Metabolism and Transport*

George J. Kantor
Department of Biological Sciences
Wright State University
Dayton, OH 45435, USA
*DNA Damage and Repair*

Seymour Kaufman
Laboratory of Neurochemistry
National Institute of Mental Health
Bethesda, MD 20892, USA
*Brain Aromatic Amino Acid Hydroxylases*

Frank Kee
Department of Epidemiology and Public Health
The Queens University of Belfast
Belfast VT12 6BJ, Northern Ireland
*Heart Failure, Genetic Basis of*

Larry K. Keefer
National Cancer Institute
Frederick Cancer Research and Development Center
Frederick, MD 21702, USA
*Nitric Oxide in Biochemistry and Drug Design*

Jeffrey N. Keen
Department of Biochemistry and Molecular Biology
University of Leeds
Leeds LS2 9JT, United Kingdom
*Protein Sequencing Techniques*

Jeffrey L. Kelly
Crop Genetics International
1050 Old Columbia Road
Columbia, MD 21046, USA
*Pesticide-Producing Bacteria*

Olga Kennard
Cambridge Crystallographic Data Centre
Cambridge CB2 IEZ, United Kingdom
*DNA Structure*

Timothy E. Kennedy
Department of Anatomy
School of Medicine
University of California
San Francisco, CA 94143, USA
*Memory and Learning, Molecular Basis of*

George Kenyon
Department of Pharmaceutical Chemistry
School of Pharmacy
University of California
San Francisco, CA 94143, USA
*Bioorganic Chemistry*

Kathy Kight
Center for Marine Biotechnology
Maryland Biotechnology Institute
Baltimore, MD 21202, USA
*Transgenic Fish*

David T. Kingsbury
Division of Biomedical Information Sciences
Johns Hopkins School of Medicine
Baltimore, MD 21205, USA
*Biotechnology, Governmental Regulation of*

Lorne T. Kirby
Department of Pathology
University of British Columbia
Vancouver, British Columbia V6H 3V4
Canada
*DNA Fingerprint Analysis*

Hannah L. Klein
Department of Biochemistry
New York University Medical Center
New York, NY 10016, USA
*Recombination, Molecular Biology of*

Aaron Klug
MRC Laboratory of Molecular Biology
Cambridge CB2 2QH, United Kingdom
*Protein Designs for the Specific Recognition of DNA*

Akira Kobata
Tokyo Metropolitan Institute of Gerontology
25–3 Sakaecho, Itabashi-Ku
Tokyo-173, Japan
*Glycobiology*

Andrzej K. Konopka
BioLingua Research
Frederick, MD 21702, USA
*Theoretical Molecular Biology*

Michael D. Koob
Department of Neurology
University of Minnesota Medical School
Minneapolis, MN 55455, USA
*Achilles' Cleavage*

Pui-Yan Kowk
Division of Dermatology
Washington University School of Medicine
St. Louis, MO 63110, USA
*Ligation Assay*

Joachim Krebs
Laboratory of Biochemistry
Swiss Federal Institute of Technology (ETH)
CH-8092 Zurich, Switzerland
*Calcium Biochemistry*

Larry J. Kricka
Department of Pathology and Laboratory Medicine
University of Pennsylvania
Philadelphia, PA 19104, USA
*Chemiluminescence and Bioluminescence, Analysis by*

Raju Kucherlapati
Department of Molecular Genetics
Albert Einstein College of Medicine
Bronx, NY 10461, USA
*Molecular Genetic Medicine*

Monique Lafon
Departement de Virologie
Institut Pasteur
75724 Paris, France
*Superantigens*

Joseph R. Lakowicz
Department of Biological Chemistry
University of Maryland
School of Medicine
Baltimore, MD 21201, USA
*Fluorescence Spectroscopy of Biomolecules*

Eric Lam
Department of Plant Sciences
Waksman Institute of Molecular Biology
Rutgers University
Piscataway, NJ 08854, USA
*Plant Gene Expression Regulation*

Kit S. Lam
Arizona Cancer Center
Department of Medicine
College of Medicine
University of Arizona
Tucson, AZ 85724, USA
*Synthetic Peptide Libraries*

Robert Langer
Department of Chemical Engineering
Massachusetts Institute of Technology
Cambridge, MA 02139, USA
*Biomaterials for Organ Regeneration; Polymers for Biological Systems*

Janos K. Lanyi
Department of Physiology
University of California
Irvine, CA 92717, USA
*Bacteriorhodopsin*

Robin Leake
Department of Biochemistry
University of Glasgow
Glasgow G12 8QQ, United Kingdom
*Steroid Hormones and Receptors*

Gregor W. Leckie
Probe Diagnostics Business Unit
Abbott Laboratories
Abbott Park, IL 60064, USA
*Infectious Disease Testing by Liagse Chain Reaction*

Daniel Lednicer
National Cancer Institute
National Institutes of Health
Bethesda, MD 20892, USA
*Drug Synthesis*

Helen H. Lee
Probe Diagnostics Business Unit
Abbott Laboratories
Abbott Park, IL 60064, USA
*Infectious Disease Testing by Liagse Chain Reaction*

Todd Leff
Department of Biotechnology
Parke-Davis Research
and
Department of Biological Chemistry
University of Michogan
Ann Arbor, MI 48105, USA
*Cardiovascular Diseases*

Daniele M. Leonard
Department of Chemistry
Parke-Davis Pharmaceutical Research
Ann Arbor, MI 48206, USA
*Peptides*

W. Lesser
Department of Agricultural Economics
Cornell University
Ithaca, NY 14853, USA
*Transgenic Animal Patents*

Gordon V. Louie
Department of Crystallography
Birbeck College
University of London
London WC1E 7HX, United Kingdom
*Protein Analysis by X-Ray Crystallography*

J. K. Lu
Center for Marine Biotechnology
Maryland Biotechnology Institute
Baltimore, MD 21202, USA
*Transgenic Fish*

Lee R. Lynd
Thayer School of Engineering
Dartmouth College
Hanover, NH 03755, USA
*Fuel Production, Biological*

Darryl MacGregor
Department of Pure and Applied Biology
University of Leeds
Leeds LS2 9JT, United Kingdom
*Nematodes, Neurobiology and Development of*

Brendan C. Maguire
Department of Chemistry
Johns Hopkins University
Baltimore, MD 21218, USA
*Nuclear Magnetic Resonance of Biomolecules in Solution*

François Major
Département d'Informatique et de Recherche Opérationnelle
Université de Montréal
Montreal, Quebec H3C 3J7
Canada
*RNA Three-Dimensional Structures, Computer Modeling of*

P. Helena Mäkelä
Ilkka Helander
Department of Special Bacterial Pathogens
National Public Health Institute
FIN-00300 Helsinki, Finland
*Lipopolysaccharides*

Hisaji Maki
Department of Prokaryotic Molecular Genetics
Graduate School of Biosciences
Nara Institute of Science and Technology
Ikoma, Nara 630–01, Japan
*DNA Replication and Transcription*

Stephen Mann
School of Chemistry
University of Bath
Bath BA2 7AY, United Kingdom
*Inorganic Solids, Biomolecules in the Synthesis of*

Patricia Margaritte-Jeannin
INSERM U155
Unité de Recherches d'Epidémiologie Génétique
Château de Longchamp
75016 Paris, France
*Breast Cancer, Genetic Analysis of*

R. Bruce Martin
Department of Chemistry
University of Virginia
Charlottesville, VA 22901, USA
*Bioinorganic Chemistry*

Belinda Martineau
Calgene, Inc.
Davis, CA 95616, USA
*Tomatoes, Gene Alterations of*

Edward P. Masler
U.S. Department of Agriculture
Agricultural Research Service
Plant Sciences Institute
Insect Neurobiology and Hormone Laboratory
Beltsville, MD 20705, USA
*Neuropeptides, Insect*

Thomas L. Mason
Department of Biochemistry and Molecular Biology
University of Massachusetts
Amherst, MA 01003, USA
*Polymers, Genetic Engineering of*

Kenichi Matsubara
Institute for Molecular and Cellular Biology
Osaka University
Osaka 565, Japan
*Body Expression Map of the Human Genome*

James L. Matthews
Baylor Research Institute
Baylor University Medical Center
Dallas, TX 75226, USA
*Viruses, Photodynamic Inactivation of*

Donald B. McCormick
Department of Biochemistry
Emory University School of Medicine
Atlanta, GA 30322, USA
*Coenzymes, Biochemistry of; Vitamins, Structure and Function of*

Johnjoe McFadden
School of Biological Sciences
University of Surrey
Guildford, Surrey GU2, 5XH, United Kingdom
*Mycobacteria*

Brian McNeil
Department of Bioscience and Biotechnology
University of Strathclyde
Glasgow G1 1XW, United Kingdom
*Fungal Biotechnology*

John Merriam
Department of Biology
University of California
Los Angeles, CA 90024, USA
*Drosophila Genome*

Elliot M. Meyerowitz
Division of Biology
California Institute of Technology
Pasadena, CA 91125, USA
*Arabidopsis Genome*

Catherine H. Mijovic
Department of Medicine
University of Birmingham
Queen Elizabeth Hospital
Birmingham B15 2TH, United Kingdom
*Diabetes Mellitus*

Antonios G. Mikos
Institute of Biosciences and Bioengineering
Rice University
Houston, TX 77251, USA
*Biomaterials for Organ Regeneration; Polymers for Biological Systems*

David L. Mitchell
University of Texas M. D. Anderson Cancer Center
Science Park—Research Division
Smithville, TX 78957, USA
*Ultraviolet Radiation Damage to DNA*

Tsutomu Miyasaka
Fuji Photo Film Co., Ltd.
Ashigara Research Laboratories
Minamiashigara
Kanagawaken 250–01, Japan
*Bacteriorhodopsin-Based Artificial Photoreceptor*

R. Jeffrey Moffatt
Department of Comparative Medicine and The Transgenic Animal/ES Cell Resource
University of Alabama
Birmingham, AL 35294, USA
*Transgenic Animal Modeling*

Hirotada Mori
Nara Institute of Science and Technology
Ikoma, Nara 630–01, Japan
*E. coli Genome; Heat Shock Response in E. coli*

Alan G. Morris
Department of Biological Sciences
University of Warwick
Coventry CV4 7AL, United Kingdom
*Cytokines*

Sherie L. Morrison
Department of Microbiology and Molecular Genetics
The Molecular Biology Institute
University of California
Los Angeles, CA 90095, USA
*Antibody Molecules, Genetic Engineering of*

Bernice E. Morrow
Department of Molecular Genetics
Albert Einstein College of Medicine
Bronx, NY 10461, USA
*Molecular Genetic Medicine*

Christopher C. Moser
Department of Biochemistry and Biophysics
Johnson Research Foundation
University of Pennsylvania
Philadelphia, PA 19104, USA
*Electron Transfer, Biological*

Masami Muramatsu
Genome Science Laboratory
Tsukuba Life Science Center
The Institute of Physical and Chemical Research (RIKEN)
3-1-1 Koyadai Tsukuba Ibaraki 305, Japan
*Restriction Landmark Genomic Scanning Method*

Yoshinao Muro
Department of Dermatology
Nagoya University School of Medicine
Nagoya 466, Japan
*Scleroderma Diagnosis with Recombinant Protein*

Yusaku Nakebeppu
Department of Biochemistry
Medical Institute of Bioregulation
Kyushu University
Fukuoka 812, Japan
*DNA Replication and Transcription*

L. Gabriel Navar
Department of Physiology
Tulane University School of Medicine
New Orleans, LA 70112, USA
*Renal System*

Eberhard Neumann
Faculty of Chemistry
University of Bielefeld
D-33501 Bielefeld, Germany
*Electric and Magnetic Field Reception*

Roger R. C. New
Cortecs Research Laboratory
London School of Pharmacy
London WC1N 1AX, United Kingdom
*Liposomal Vectors*

Mabel Ng
Max-Planck-Institut für Experimentelle Medizin
D-37075 Göttingen, Germany
*Oligonucleotides*

Anthony Nicholls
Department of Biochemistry and Molecular Biophysics
Columbia University
New York, NY 10032, USA
*Immunology*

Robert A. Nichols
Departments of Pharmacology and Anatomy & Neurobiology
Medical College of Pennsylvania
Philadelphia, PA 19219, USA
*Synapses*

Deborah Nickerson
Department of Molecular Biotechnology
University of Washington
Seattle, WA 98195, USA
*Ligation Assays*

Jiri Novotny
Department of Macromolecular Structure
Bristol-Myers Squibb Research Institute
Princton, NJ 08543, USA
*Immunology*

James Oard
Department of Agronomy
Louisiana State University
Baton Rouge, LA 70803, USA
*Plant Cell Transformation, Physical Methods for*

Kimberly L. Ogden
Center for Separation Sciences
University of Arizona
Tucson, AZ 85721, USA
*Bioprocess Engineering*

Masaru Ohashi
Department of Dermatology
Nagoya University School of Medicine
Nagoya 466, Japan
*Scleroderma Diagnosis with Recombinant Protein*

Yasushi Okazaki
Genome Science laboratory
Tsukuba Life Science Center
The Institute of Physical and Chemical Research (RIKEN)
3-1-1 Koyadai Tsukuba Ibaraki 305, Japan
*Restriction Landmark Genomic Scanning Method*

Kousaku Okubo
Institute for Molecular and Cellular Biology
Osaka University
Osaka 565, Japan
*Body Expression Map of the Human Genome*

Charles Onwulata
U.S. Department of Agriculture
Agricultural Research Service
Eastern Regional Research Center
Philadelphia, PA 19118, USA
*Food Proteins and Interactions*

B. A. Oostra
Department of Clinical Genetics
Erasmus University
3000 DR Rotterdam, The Netherlands
*Fragile X Linked Mental Retardation*

Paul R. Ortiz de Montellano
Department of Pharmaceutical Chemistry
School of Pharmacy
University of California
San Francisco, CA 94143, USA
*Biotransformations of Drugs and Chemicals*

F. Owen
Division of Neuroscience
School of Biological Sciences
University of Manchester
Manchester M13 9PT, United Kingdom
*Neuropsychiatric Diseases*

School of Biology and Biochemistry
University of Bath
Bath BA2 7AY, United Kingdom
*X-Ray Diffraction of Biomolecules*

Mehmet Ozturk
Laboratoire d'Oncologie Moléculaire
INSERM CJF 9302
Centre Leon Berard
69373 Lyon, France
*Liver Cancer, Molecular Biology of*

Nicholas Parris
U.S. Department of Agriculture
Agricultural Research Service
Eastern Regional Research Center
Philadelphia, PA 19118, USA
*Food Proteins and Interactions*

Luciano Passador
Department of Microbiology and Immunology
University of Rochester School of Medicine
Rochester, NY 14642, USA
*Bacterial Pathogenesis*

Fredrick M. Pavalko
Department of Physiology and Biophysics
Indiana University School of Medicine
Indianapolis, IN 46202, USA
*Cytoskeleton-Plasma Membrane Interactions*

Boris Matija Peterlin
Department of Medicine
Howard Hughes Medical Institute
University of California
San Francisco, CA 94143, USA
*AIDS, Transcriptional Regulation of HIV in*

Anusch Peyman
Hoechst Aktiengesellschaft
General Pharma Research G838
D-65926 Frankfurt am Main, Germany
*Antisense Oligonucleotides, Structure and Function of*

Simon E. V. Phillips
Department of Biochemistry and Molecular Biology
University of Leeds
Leeds LS2 9JT, United Kingdom
*Repressor-Operator Recognition*

Daniel Pinkel
Division of Molecular Cytometry
Department of Laboratory Medicine
University of California
San Francisco, CA 94143, USA
*Whole Chromosome Complementary Probe Fluorescence Staining*

Carl A. Pinkert
Department of Comparative Medicine and The Transgenic Animal/ES Cell Resource
University of Alabama
Birmingham, AL 35294, USA
*Transgenic Animal Modeling*

K. Michael Pollard
Department of Molecular and Experimental Medicine
W. M. Keck Autoimmune Disease Center
The Scripps Research Institute
La Jolla, CA 92037, USA
*Autoantibodies and Autoimmunity*

Theresa P. Pretlow
Department of Pathology
Case Western Reserve University School of Medicine
Cleveland, OH 44106, USA
*Cancer*

Thomas G. Pretlow
Department of Pathology
Case Western Reserve University School of Medicine
Cleveland, OH 44106, USA
*Cancer*

Nicholas C. Price
Department of Biological and Molecular Sciences
University of Stirling
Stirling FK9 4LA, United Kingdom
*Circular Dichroism in Protein Analysis*

Clive R. Pullinger
Cardiovascular Research Institute
University of California
San Francisco, CA 94143, USA
*Lipid Metabolism and Transport*

Robert R. Rando
Department of Biological Chemistry and Molecular Pharmacology
Harvard Medical School
Boston, MA 02115, USA
*Retinoids*

Gregory M. Raner
Department of Chemistry
University of Utah
Salt Lake City, UT 84112, USA
*Metalloenzymes*

Stephen W. Raso
Department of Chemistry
Texas A&M University
College Station, TX 77843, USA
*Enzymes*

Colin Ratledge
Department of Applied Biology
University of Hull
Hull HU6 7RX, United Kingdom
*Lipids, Microbial*

Anthony R. Rees
School of Biology and Biochemistry
University of Bath
Bath BA2 7AY, United Kingdom
*X-Ray Diffraction of Biomolecules*

Wolf Reik
Laboratory of Developmental Genetics and Imprinting
The Babraham Institute
Cambridge CB2 4AT, United Kingdom
*Genomic Imprinting*

Alistair G. C. Renwick
The International Medical College
21 Jalan Selangor
46050 Petaling Jaya
Selangor Darul Ehsan, Malaysia
*Glycoproteins, Secretory*

Paul Ribeiro
Institute of Parasitology
McGill University
Ste. Anne-de-Bellevue, Quebec H9X 3V9
Canada
*Brain Aromatic Amino Acid Hydroxylases*

Timothy J. Richmond
Institute for Molecular Biology and Biophysics
Swiss Federal Institute of Technology (ETH)
CH-8093 Zurich, Switzerland
*Chromatin Formation and Structure*

Ernst T. Rietschel
Forschungsinstitut Borstel
Institut für Experimentelle Biologie und Medizin
D-23845 Borstel, Germany
*Lipopolysaccharides*

Jean-David Rochaix
Departments of Molecular Biology and Plant Biology
University of Geneva
CH-1221 Geneva 4, Switzerland
*Chlamydomonas*

Martin Rosenberg
Biopharmaceutical Research and Development
SmithKline Beecham Pharmaceuticals
King of Prussia, PA 19406, USA
*AIDS Therapeutics, Biochemistry of*

Walter Rosenthal
Rudolf-Buchheim-Instut für Pharmakologie
Justus-Liebig-Universität Giessen
35392 Giessen, Germany
*Diabetes Insipidus*

Belinda J. F. Rossiter
Department of Molecular and Human Genetics
Baylor College of Medicine
Houston, TX 77030, USA
*Human Genetic Predisposition to Disease*

Jack A. Roth
M.D. Anderson Cancer Center
Department of Thoracic and Cardiovascular Surgery
University of Texas
Houston, TX 77030, USA
*Lung Cancer, Molecular Biology of*

Michael E. Rusiniak
Departments of Molecular and Cellular Biology
Roswell Park Cancer Institute
Buffalo, NY 14263, USA
*Kallikrein-Kininogen-Kinin System*

Alan J. Russell
Department of Chemical Engineering
Center for Biotechnology and Bioengineering
University of Pittsburgh
Pittsburgh, PA 15261, USA
*Enzymology, Nonaqueous*

Milton H. Saier, Jr.
Department of Biology
University of California, San Diego
La Jolla, CA 92093, USA
*Transport Proteins*

S. A. Salisbury
Cambridge Crystallographic Data Centre
Cambridge CB2 IEZ, United Kingdom
*DNA Structure*

Ka-Yiu San
Department of Chemical Engineering
Rice University
Houston, TX 77251, USA
*Expression Systems for DNA Processes*

Linda J. Sandell
Departments of Orthopaedics and Biochemistry
Veterans Affairs Medical Center
Seattle, WA 98108, USA
*Extracellular Matrix*

Takeshi Sano
Center for Advanced Biotechnology and
Department of Biomedical Engineering
Boston University
Boston, MA 02215, USA
*Immuno-PCR*

Tomi K. Sawyer
Department of Chemistry
Parke-Davis Pharmaceutical Research
Ann Arbor, MI 48206, USA
*Peptides*

John G. Scandalios
Department of Genetics
North Carolina State University
Raleigh, NC 27695, USA
*Environmental Stress, Genomic Responses to*

John Schimenti
Jackson Laboratory
Bar Harbor, ME 04609, USA
*Mammalian Genome*

Milton J. Schlesinger
Department of Molecular Microbiology
Washington University School of Medicine
St. Louis, MO 63110, USA
*Viral Envelope Assembly and Budding*

Martin Schrott
Institute of Plant Sciences
Swiss Federal Institute of Technology
ETH-Zentrum
CH-8092 Zurich, Switzerland
*Plant Cells, Genetic Manipulation of*

Heinz-Peter Schuchmann
Max-Planck-Institut für Strahlenchemie
D-45470 Mülheim an der Ruhr, Germany
*Ionizing Radiation Damage to DNA*

John W. R. Schwabe
Medical Research Council
Laboratory of Molecular Biology
Cambridge CB2 2QH, United Kingdom
*Zinc Finger DNA Binding Motifs*

Clay W Scott
Pharmacology Department
Zeneca Pharmaceuticals
Wilmington, DE 19897, USA
*Protein Phosphorylation*

Anita Seibold
University of Texas Health Science Center
Graduate School of Biomedical Sciences
Houston, TX 74030, USA
*Diabetes Insipidus*

Mutsuo Sekiguchi
Department of Biochemistry
Medical Institute of Bioregulation
Kyushu University
Fukuoka 812, Japan
*DNA Replication and Transcription*

Michael Sendtner
Department of Neurology
University of Würzburg
D-97080 Würzburg, Germany
*Motor Neuron Disease*

Philip Serwer
Department of Biochemistry
The University of Texas Health Science Center
San Antonio, TX 78284, USA
*Viruses, DNA Packaging of*

Prem Mohini Sharma
Molecular Biology and Virology Laboratory
The Salk Institute for Biological Studies
San Diego, CA 92186, USA
*Muscle, Molecular Genetics of Human; Tumor Suppressor Genes*

S. R. Dev Sharma
PrimeX Corporation
Carlsbad, CA 92008, USA
*Tumor Suppressor Genes*

R. C. Sheppard
Medical Research Council
Molecular Biology Laboratory
Cambridge CB2 2QH, United Kingdom
Present Address:
15 Kinnaird Way
Cambridge CB1 4SN, United Kingdom
*Peptide Synthesis, Solid-Phase*

David Sidransky
Department of Otolaryngology—Head and Neck Surgery
Head and Neck Cancer Research Division
Johns Hopkins University School of Medicine
Baltimore, MD 21205, USA
*DNA in Neoplastic Disease Diagnosis*

Richard J. Simpson
Joint Protein Structure Laboratory
Ludwig Institute for Cancer Research (Melbourne Branch)
and
The Walter and Eliza Hall Institute of Medical Research
Parkville, Vic 3050, Australia
*Proteins and Peptides, Isolation for Sequence Analysis of*

E. Roy Skinner
Department of Molecular and Cell Biology
University of Aberdeen
Marischal College
Aberdeen AB9 1AS, United Kingdom
*Lipoprotein Analysis*

Robert James Slater
Division of Biosciences
University of Hertfordshire
Hatfield, Hertfordshire AL10 9AB, United Kingdom
*Radioisotopes in Molecular Biology*

Donald M. Small
Department of Biophysics
Boston University Medical Center
Boston, MA 02118, USA
*Lipids, Structure and Biochemistry of*

Cassandra L. Smith
Center for Advanced Biotechnology and
Departments of Biomedical Engineering, Biology, and Pharmacology
Boston University
Boston, MA 02215, USA
*Human Chromosomes, Physical Maps of; Immuno-PCR*

Lloyd M. Smith
Department of Chemistry
University of Wisconsin
Madison, WI 53706, USA
*DNA Sequencing in Ultrathin Gels, High Speed*

D. Peter Snustad
Department of Genetics and Cell Biology
University of Minnesota
St. Paul, MN 55108, USA
*Genetics*

John Sommerville
School of Biological and Medical Sciences
University of St. Andrews
St. Andrews, Fife KY16 9TS, United Kingdom
*Electron Microscopy of Biomolecules*

Clemens von Sonntag
Max-Planck-Institut für Strahlenchemie
D-45470 Mülheim an der Ruhr, Germany
*Ionizing Radiation Damage to DNA*

Stephen E. Specter
Department of Nutrition
University of California
Davis, CA 95616, USA
*Bioenergetics of the Cell*

Gary Spedding
Department of Chemistry
Butler University
Indianapolis, IN 46208, USA
*Ribosome Preparations and Protein Synthesis Techniques*

Willy Stalmans
Afdeling Biochemie
Katholicke Universiteit Leuven
Campus Gasthuisberg
B-3000 Leuven, Belgium
*Glycogen*

Gary S. Stein
Department of Cell Biology
University of Massachusetts Medical Center
Worcester, MA 01655, USA
*Histones*

Janet L. Stein
Department of Cell Biology
University of Massachusetts Medical Center
Worcester, MA 01655, USA
*Histones*

David Stephens
Department of Surgery
St. George's Hospital Medical School
London SW17 0RE, United Kingdom
*Protein Targeting*

Peter G. Stockley
Department of Genetics
University of Leeds
Leeds LS2 9JT, United Kingdom
*Repressor-Operator Recognition*

Neil Stoker
Department of Clinical Sciences
London School of Hygiene & Tropical Medicine
London WC1E 7HT, United Kingdom
*Mycobacteria*

Alfred Stracher
Department of Biochemistry
State University of New York
Health Science Center at Brooklyn
Brooklyn, NY 11203, USA
*Drug Targeting and Delivery, Molecular Principles of*

Kenji Sugimoto
Laboratory of Applied Molecular Biology
Department of Applied Biological Chemistry
College of Agriculture
University of Osaka
Sakai, Osaka 593, Japan
*Scleroderma Diagnosis with Recombinant Protein*

Eng M. Tan
Department of Molecular and Experimental Medicine
W. M. Keck Autoimmune Disease Center
The Scripps Research Institute
La Jolla, CA 92037, USA
*Autoantibodies and Autoimmunity*

De-chu Tang
Departments of Internal Medicine and Biochemistry
University of Texas
Southwestern Medical Center
Dallas, TX 75235, USA
*Genetic Immunization*

William R. Taylor
Division of Mathematical Biology
National Institute of Medical Research
London NW7 1AA, United Kingdom
*Sequence Alignment of Proteins and Nucleic Acids*

Edward Theil
Human Genome Center and Engineering Division
Lawrence Berkeley Laboratory
University of California
Berkeley, CA 94720, USA
*Automation in Genome Research*

Fritz Thoma
Institute for Cell Biology
Swiss Federal Institute of Technology (ETH)
CH-8093 Zurich, Switzerland
*Chromatin Formation and Structure*

George J. Thomas, Jr.
Division of Cell Biology and Biophysics
School of Life Sciences
University of Missouri
Kansas City, MO 64110, USA
*Protein Analysis by Raman Spectroscopy*

Robert C. Thomson
Institute of Biosciences and Bioengineering
Rice University
Houston, TX 77251, USA
*Biomaterials for Organ Regeneration; Polymers for Biological Systems*

T. Thundat
Oak Ridge National Laboratory
Oak Ridge, TN 37831, USA
*Scanning Tunneling Microscopy in Sequencing of DNA*

David A. Tirrell
Department of Biochemistry and Molecular Biology
University of Massachusetts
Amherst, MA 01003, USA
*Polymers, Genetic Engineering of*

David J. Triggle
School of Pharmacy
State University of New York
Buffalo, NY 14260, USA
*Medicinal Chemistry*

Kenneth D. Tucker
Department of Microbiology and Immunology
University of Rochester School of Medicine
Rochester, NY 14642, USA
*Bacterial Pathogenesis*

Tom Tuschl
Max-Planck-Institut für Experimentelle Medizin
D-37075 Göttingen, Germany
*Oligonucleotides*

Eugen Uhlmann
Hoechst Aktiengesellschaft
General Pharma Research G838
D-65926 Frankfurt am Main, Germany
*Antisense Oligonucleotides, Structure and Function of*

Dennis E. Vance
Department of Biochemistry and Lipid and Lipoprotein Research Group
University of Alberta
Edmonton, Alberta T6G 2S3
Canada
*Phospholipids*

André J. van Wijnen
Department of Cell Biology
University of Massachusetts Medical Center
Worcester, MA 01655, USA
*Histones*

Rajender S. Varma
DNA Technology Laboratory
Houston Advanced Research Center
The Woodlands, TX 77381, USA
*Oligonucleotide Analogues, Phosphate-Free*

A. J. M. H. Verkerk
Department of Clinical Genetics
Erasmus University
3000 DR Rotterdam, The Netherlands
*Fragile X Linked Mental Retardation*

Linda Vigilant
Department of Anthropology
Pennsylvania State University
University Park, PA 16802, USA
*Mitochondrial DNA, Evolution of Human*

Claude A. Villee
Department of Biological Chemistry
Harvard Medical School
Boston, MA 02115, USA
*RNA Replication*

David H. Viskochil
Department of Pediatrics
Division of Medical Genetics
University of Utah School of Medicine
Salt Lake City, UT 84112, USA
*Neurofibromatosis*

Adrian O. Vladutiu
Departments of Pathology, Microbiology, and Medicine
State University of New York
School of Medicine and Biomedical Sciences
Buffalo General Hospital
Buffalo, NY 14203, USA
*Isoenzymes*

Georgirene D. Vladutiu
Department of Pediatrics
State University of New York
School of Medicine and Biomedical Sciences
Children's Hospital of Buffalo
Buffalo, NY 14209, USA
*Isoenzymes*

Stephen S. Wachtel
Division of Reproductive Genetics
Department of Obstetrics and Gynecology
University of Tennessee
Memphis, TN 38103, USA
*Sex Determination*

Denan Wang
Departments of Biomedical Engineering, Biology, and Pharmacology
Center for Advanced Biotechnology
Boston University
Boston, MA 02215, USA
*Human Chromosomes, Physical Maps of*

Larry D. Ward
Biopharmaceutical Research Laboratory
AMRAD Operation Pty Ltd
Hawthorn, Vic 3122, Australia
*Proteins and Peptides, Isolation for Sequence Analysis of*

R. J. Warmack
Oak Ridge National Laboratory
Oak Ridge, TN 37831, USA
*Scanning Tunneling Microscopy in Sequencing of DNA*

Michael R. Waterman
Department of Biochemistry
Vanderbilt University
School of Medicine
Nashville, TN 37232, USA
*Cytochrome P450*

Heinz-Ultich Weier
Division of Molecular Cytometry
Department of Laboratory Medicine
University of California
San Francisco, CA 94143, USA
*Whole Chromosome Complementary Probe Fluorescence Staining*

Jean Weissenbach
CNRS-URA 1922, Généthon
91002 Evry, France
*Human Linkage Maps*

Otto Westphal
Chemin de Ballalaz 18
CH-1820 Montreux, Switzerland
*Lipopolysaccharides*

James G. Wetmur
Department of Microbiology
Mount Sinai School of Medicine
New York, NY 10029, USA
*Nucleic Acid Hybrids, Formation and Structure of*

Darrell Williams
Department of Chemical Engineering
Center for Biotechnology and Bioengineering
University of Pittsburgh
Pittsburgh, PA 15261, USA
*Enzymology, Nonaqueous*

Kenneth R. Williams
W.M. Keck Foundation Biotechnology Resource Laboratory
Howard Hughes Medical Institute
Yale University
New Haven, CT 06536, USA
*Protein Analysis by Integrated Sample Preparation, Chemistry, and Mass Spectrometry*

Robert M. Williams
Department of Chemistry
Colorado State University
Fort Collins, CO 80523, USA
*Amino Acid Synthesis*

R. Anthony Williamson
Departments of Immunology and Molecular Biology
Scripps Research Institute
La Jolla, CA 92037, USA
*Combinatorial Phage Antibody Libraries*

Brian Wimberly
Department of Molecular Biology
The Scripps Research Institute
La Jolla, CA 92037, USA
*RNA Structure, Nonhelical*

Peter J. Wirth
Laboratory of Experimental Carcinogenesis
National Cancer Institute
National Institutes of Health
Bethesda, MD 20892, USA
*Gel Eletrophoresis of Proteins, Two-Dimensional Polyacrylamide*

Alexander Wlodawer
Macromolecular Structure Laboratory
National Cancer Institute
Frederick Cancer Research and Development Center
Frederick, MD 21702, USA
*AIDS, Inhibitor Complexes of HIV-1 Protease in*

Roger K. Wolff
Mercator Genetics, Inc.
Menlo Park, CA 94025, USA
*Neurofibromatosis*

Savio L. C. Woo
Department of Cell Biology and Molecular Genetics
Baylor College of Medicine
Houston, TX 77030, USA
*Phenylketonuria, Molecular Genetics of*

John Wu
Division of Hematology-Oncology
Department of Pediatrics
University of British Columbia
and
British Columbia's Children's Hospital
Vancouver, British Columbia V6H 3V4
Canada
*Retinoblastoma, Molecular Genetics of*

Tilahun D. Yilma
Department of Veterinary Pathology, Microbiology, and Immunology
International Laboratory of Molecular Biology for Tropical Disease Agents
University of California
Davis, CA 95616, USA
*Vaccine Biotechnology*

Jeannine M. Yon
Laboratoire d'Enzymologie Physicochimique et Moléculaire
Unité de Recherche du Centre National de la Recherche Scientifique
Université de Paris-Sud
Orsay 91405F, France
*Protein Aggregation*

Takashi Yura
HSP Research Institute
Kyoto Research Park
Kyoto 600, Japan
*E. coli Genome; Heat Shock Response in E. coli*

Eugene R. Zabarovsky
Microbiology and Tumor Biology Center
Karolinska Institute
Stockholm S-171 77, Sweden
*DNA Markers, Cloned*

Hanswalter Zentgraf
Deutsches Krebsforschungszentrum KS
D-69009 Heidelberg, Germany
*Electron Microscopy of Biomolecules*

Miriam M. Ziegler
Department of Biochemistry and Biophysics
Texas A&M University
College Station, TX 77843, USA
*Enzymes*

# Achilles' Cleavage

*Michael D. Koob*

*Key Words*

**Genome** The entire DNA sequence content of an organism.

**Methyltransferase** Enzyme that makes the recognition sequence of its corresponding restriction endonuclease resistant to cleavage by adding a methyl group to one of the bases within that sequence.

**Recognition Sequence** Sequence of nucleotides in a DNA molecule that is specifically recognized by a DNA-binding molecule (e.g., a repressor) or a DNA-modifying enzyme (e.g., a restriction endonuclease or methyltransferase).

**Restriction Endonuclease** Enzyme that produces a double-stranded cleavage of DNA wherever its recognition sequence appears in that DNA molecule.

Achilles' cleavage (AC) is a general method for combining the cleavage specificity of a restriction endonuclease with the binding specificity of other DNA-binding molecules. Restriction enzymes, which typically have recognition sequences 4–8 base pairs long, cleave unmodified DNA from complex genomes at thousands or even millions of sites. By limiting the activity of these enzymes to a specific subset of these sites, the AC process makes it possible to cleave even the human genome at only a single site. This capability allows researchers both to determine the location of a particular AC site relative to other mapped sites and to specifically isolate the DNA between any two AC sites.

## 1 PRINCIPLES

The principle behind AC is outlined in Figure 1. DNA treated with a methyltransferase is completely resistant to cleavage by its corresponding restriction endonuclease. In the AC procedure, a sequence-specific DNA-binding molecule or complex is added to unmodified DNA to specifically protect one or more of the restriction recognition sequences from methylation. The methyltransferase is then added, and all the unprotected sites are modified and thus made resistant to cleavage. Finally the methyltransferase and the protecting complex are removed and the DNA is cleaved by the restriction enzyme exclusively at the unmodified restriction site(s). Because cleavage occurs only at sites at which the recognition sequence for the protecting group and the restriction enzyme overlap, the site specificity of this reaction is significantly greater than that of either of these two molecules alone.

Methylation protection in the AC reaction can be mediated by any DNA-binding molecule or group of molecules that form sequence-specific complexes capable of excluding a methyltransferase. The binding characteristics of the protecting group used in a given AC protocol dictate the number and type of restriction sites that are ultimately cleaved and therefore determine the types of application for which that AC reaction is suited. Specific examples of AC reactions and their uses are given in Section 4.

The experimental goal when developing or applying an AC procedure is to obtain efficient DNA cleavage that is highly specific for the targeted AC site(s). Incomplete methylation of the unprotected restriction sites results in partial cleavage by the restriction enzyme at these sites and therefore lowers the specificity of the AC reaction. Poor binding of the protecting molecule, on the other hand, allows partial methylation at the AC site and decreases the efficiency of the Achilles' cleavage. For these reasons, AC protocols are optimized to obtain complete methylation of the unprotected restriction sites under conditions suitable for stable, specific binding of the protecting group.

## 2 NOMENCLATURE

The name ''Achilles' cleavage'' was given to this technique because it calls to mind a well-known Greek myth. Thetis, the mother of Achilles, bathed her son in the river Styx to make him invulnerable. The heel by which she held him was not affected by the water, however, and remained susceptible. By analogy, ''bathing'' DNA with the methyltransferase makes it impervious to attack by the restriction endonuclease except at the site at which it is ''held.'' This remaining restriction site is referred to as the *Achilles' cleavage site* (AC site). The name of a particular AC reaction reflects the name of the molecule used to protect a restriction site from methylation by referring to ''*X*-mediated AC'' (*X*-AC), where ''*X*'' is the proper name of the protecting molecule (e.g., an AC reaction using the RecA protein would be ''RecA-mediated AC'' or ''RecA-AC'').

## 3 BACKGROUND

The principle of AC was first proposed and experimentally demonstrated on two model systems in a paper published in 1988. In this work, repressor proteins from the *lac* operon of *Escherichia coli* (LacI) and from the bacteriophage lambda (λ cI) were shown to specifically and completely protect from methylation restriction sites within their operator binding sites. The other similar restriction sites on the plasmid DNA were methylated and thus rendered uncleavable. Restriction enzymes that would otherwise recognize and cleave sequences four and five base pairs in length were then used to cleave the plasmids exclusively at the AC recognition sequence of 20 base pairs (LacI binding site) and 17 base pairs in length (λ cI binding site), respectively.

This work clearly demonstrated that the cleavage specificity of restriction enzymes could be easily combined with the much larger

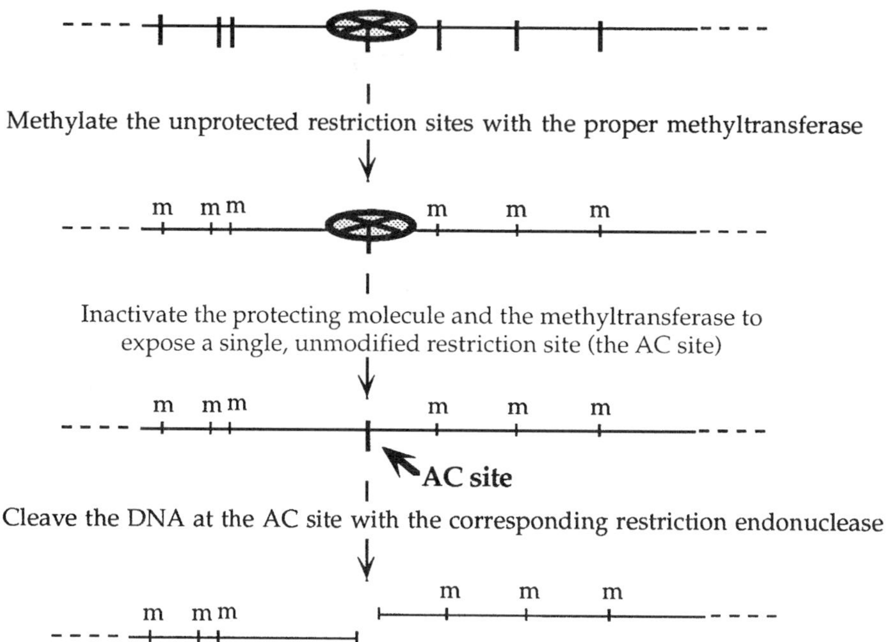
Add a DNA-binding molecule to purified DNA to cover a restriction site in that molecule's recognition sequence
Methylate the unprotected restriction sites with the proper methyltransferase
m mm m m m
Inactivate the protecting molecule and the methyltransferase to expose a single, unmodified restriction site (the AC site)
AC site
Cleave the DNA at the AC site with the corresponding restriction endonuclease

**Figure 1.** The Achilles' cleavage procedure for generating new DNA cleavage specificities: long, heavy bars, cleavable restriction sites; short bars marked "m," methylated, uncleavable restriction sites.

binding specificity of other DNA-binding molecules when the AC process was performed on relatively small (approximately 4000 base pairs) plasmid DNA. Experiments published in 1990 showed that an appropriately modified LacI-AC process could also efficiently cleave the genomes of the bacteriophage lambda (44,000 base pairs), the bacterium *E. coli* (4.7 megabase pairs), and the yeast *Saccharomyces cerevisiae* (15 megabase pairs) exclusively at inserted *lac* operator sequences. At this point, the feasibility of using AC to physically map and precisely dissect large genomes was established.

## 4 TYPES OF AC AND THEIR USES

The binding specificity of the protecting group used in a given AC protocol dictates the number and type of restriction sites that are protected from methylation. The molecule chosen to mediate the AC reaction therefore determines the applications for which that particular AC protocol is suited. AC restriction sites that can be generated fall into four broad categories: (1) large sites that do not naturally occur in the genome and must first be introduced at the location of interest, (2) sites that are rare or unique but occur by chance within the sequence of genome, (3) biologically significant sites that have evolved to occur throughout the genome, and (4) any restriction site located within a known DNA sequence. Examples of protecting molecules that generate each of these types of AC site (see Table 1) as well as some applications for each category of AC are given in Sections 4.1 to 4.4.

### 4.1 Unique AC Sites: Cleavage at a Single Inserted Site

A recognition sequence 20 base pairs long should occur by chance only once in every $1.1 \times 10^{12}$ bases of sequence and so does not appear within the sequence of most genomes (the human genome is approximately $3 \times 10^9$ bases). AC reactions mediated by DNA-

**Table 1** Examples of DNA-Binding Molecules and Their AC Sites

| DNA-Binding Molecule | Type of Binding | Example AC Recognition Sequence[a] |
|---|---|---|
| LacI | Very specific | AATTGTGAGCGCTCACAATT |
| IHF | Relaxed | ATGCAGTCACTATGAATCAACTACTTA |
| Oligonucleotide | Triple helix (polypurine sequence) | GAAGAAAAGAAGAAAGAAAAAGAA(TTC) |
| GCN4 | Gene-specific | TGGATGACTCATTTTTTT |
| RecA | Binds single-stranded DNA to homologous sequence | (Any unique sequence) |

[a]Restriction recognition sequence underscored; nucleotides in parentheses outside binding sequence.

binding proteins with large, highly specific recognition sequences, such as LacI or λ cI, therefore produce cleavage only at AC sites that have been specifically introduced into a genome. These unique AC sites can be inserted into a genome either randomly on mobile genetic elements or, in many organisms, at the location of a cloned genomic fragment using the cellular process of homologous recombination. The physical distance of that DNA fragment or mobile genetic element from either the end of the chromosome or a second AC site can then be determined by cleaving the chromosome at the inserted AC site(s) and measuring the size of the resulting DNA cleavage fragments.

### 4.2 Rare but Random AC Sites: Cleavage at Naturally Occurring Sites

Proteins having recognition sequences that are either comparatively short or "relaxed" (i.e., the protein recognizes many different related sequences) bind to sites that occur at random in most large genomes. An example of this type of protein is the IHF protein from *E. coli,* which recognizes and binds to sequences that are rich in adenine/thymine (A/T) base pairs. Although IHF binds to and protects approximately 27 base pairs of DNA, the high degree of flexibility in the recognition sequence gives rise to a binding pattern more typical of a protein that recognizes only 8 base pairs. Potential AC sites occur where an IHF binding site overlaps with a restriction recognition sequence. The number of IHF-AC sites cleaved in a genome depends on the size and base pair composition of the recognition sequence for the restriction endonuclease/methyltransferase pair used. Short restriction sites (4 base pairs) that are high in A/T content would arise more often in the A/T-rich IHF binding sites, whereas long restriction sites with a low A/T content would rarely if ever overlap with an IHF binding site. IHF-AC is used in the same manner and serves the same purpose as other rare-cutting restriction digests.

Another DNA-binding molecule that can recognize multiple sequences within a genome is a particular type of synthetic oligodeoxyribonucleotide that is capable, under the proper conditions, of pairing with a region of double-stranded chromosomal DNA to form a triple-stranded structure. Unlike IHF, however, these molecules can be designed to bind to only one of their many potential binding sites and thus are able to produce a unique AC site. The binding sites for these oligonucleotides are stretches of purines (adenine or guanine) on one of the DNA strands that are 15–25 bases long, interrupted by no more than two thymines. These sequences are thought to partially overlap with restriction sites as often as every 10,000–30,000 bases of genomic sequence. Triple helix mediated AC is therefore capable of targeting a unique cleavage event to the general region of interest within even a very complex genome.

### 4.3 Biologically Relevant AC Sites: Gene-Specific Cleavage

Rather than using a DNA-binding protein from one organism and relying on the fortuitous presence of binding sites in the genome of another, proteins that have specifically evolved to interact with sites throughout the genome of interest can be utilized in the AC process. An example of such a protein is the transcription regulatory protein GCN4 from *S. cerevisiae.* The typical recognition sequence for GCN4 contains a restriction recognition sequence and normally appears at several sites near the genes that are regulated by this protein. GCN-4 mediated AC of the *S. cerevisiae* genome therefore results in a sort of "gene-specific" cleavage at most of the GCN4-regulated genes. Analysis of the location and degree of this cleavage reveals both the physical location of and the relative strength of binding to that subset of GCN4 binding sites that contains the restriction site. Since GCN4-related transcription factors and binding sites have been highly conserved in evolution, similar AC procedures are possible in most eukaryotic organisms.

### 4.4 Designer AC Sites: Cleavage at Any Predetermined Restriction Site

The RecA protein from *E. coli* is a DNA repair protein that promotes the exchange of a single-stranded DNA fragment with a homologous strand in double-stranded DNA. In this process, the RecA polymerizes on the single-stranded DNA to form a nucleoprotein filament that in turn binds to the homologous sequence in the duplex DNA. If one of the cofactors in this reaction is replaced with a synthetic analogue, the process is frozen at this binding stage and strand exchange does not occur. Such "frozen" RecA nucleoprotein filaments can be used as the protecting agent in an AC reaction. Since the binding specificity for the RecA filament is derived entirely from the single-stranded DNA moiety, RecA-AC allows any restriction site within a unique DNA sequence to be converted into a unique cleavage site.

Genetic and physical maps for the chromosomes of many organisms are currently being generated in which the basic landmarks are short tracts of unique sequence known as sequence-tagged sites (STSs). Since it is likely that essentially all STSs will contain a restriction site that can serve as a RecA-AC site, cleavage at adjacent STSs offers a means of both determining the precise physical distance between these landmarks and specifically isolating the intervening region of the chromosome. This ability to isolate a particular DNA fragment gives researchers immediate access to chromosomal regions of interest to facilitate further mapping, cloning and sequencing studies.

Achilles' cleavage is still an emerging technology and will undoubtedly be subject to significant improvements. In particular, the quality and number of available methyltransferases will inevitably increase, and new protecting agents, both natural and synthetic, will continue to be added to the repertoire. The applications for this technique will also evolve as these new reagents and improved protocols become available.

*See also* RecA Protein, Structure and Function of; Restriction Endonucleases and DNA Modification Methyltransferases for the Manipulation of DNA.

## *Bibliography*

Koob, M. (1992) Conferring new cleavage specificities on restriction endonucleases. In *Methods in Enzymology,* Vol. 216, *Recombinant DNA,* Part G, R. Wu, Ed., pp. 321–329. Academic Press, San Diego, CA.

Strobel, S. A., and Dervan P. B. (1992) Triple helix-mediated single-site enzymatic cleavage of megabase genomic DNA. In *Methods in Enzymology,* Vol. 216, *Recombinant DNA,* Part G, R. Wu, Ed., pp. 309–321. Academic Press, San Diego, CA.

## Acquired Immunodeficiency Syndrome: *see* articles beginning with AIDS.

# Aging, Genetic Mutations in

*James E. Fleming*

*Key Words*

**Aging** A deteriorative process that occurs in organisms, leading to an increase in vulnerability and a decrease in vitality. Used synonymously with senescence. In evolution, a persistent decline in the age-specific fitness components of an organism.

**Antagonistic Pleiotropy** Evolutionary hypothesis that genes selected on the basis of advantages to fitness early in life have negative effects on other characteristics, particularly at late ages.

**Error Catastrophe Theory** The theory that aging is caused by the self-catalyzing propagation of errors in the synthesis of proteins or nucleic acids.

**Mutation Accumulation** Evolutionary process that arises from the accumulation of mutations that are deleterious to an organism only in late life. Such mutations originally occur in the germ line.

**Somatic Mutation Theory** Hypothesis that aging is caused by the accumulation of deleterious mutations in the somatic cells of an organism. Somatic cells are all cells other than the reproductive cells; thus somatic mutations are not heritable.

The concept of "mutation" can enter discussions of organismal aging in three ways: (1) the hypothesis that mutations occur in the somatic cells during the course of aging; (2) evolutionary theories of aging, which predict mutation accumulation over the germ lines of successive generations; and (3) the identification of genes either by mutation, transformation, or selection that play a major role in regulating the life span of animals.

In this entry, the word "aging" is defined as a deteriorative process that occurs in organisms, giving rise to an increase in vulnerability and a decrease in vitality. In this context, aging is synonymous with senescence, since we are concerned with the degenerative aspects of the process. In the light of evolutionary discussions of aging, one is forced to think of decreases in both age-specific survival rates and age-specific reproductive rates. In this case, aging is a persistent decline in the age-specific fitness components of an organism due to internal physiological deterioration.

Mutation is a fixed permanent change in the genetic material of a cell. Somatic mutations occur in the nonreproductive cells that comprise an organism, and germ line mutations occur in the reproductive cells or their precursors and can be transmitted to the organism's descendants. This distinction is important for a complete understanding of the role of mutations in the aging process. The accumulation of age-dependent somatic mutation is a hypothesis that explains why individual organisms grow old and die, whereas the process of mutation accumulation in the germ line is used by evolutionary biologists to explain why different species have different life spans.

## 1 MUTATIONAL THEORIES OF AGING

### 1.1 Somatic Mutation

The origin of the somatic mutation theory of aging can be traced back to early studies of radiation biology. Doses of radiation that are sublethal usually serve to reduce the life span of mammals. For example, irradiated mice show evidence of premature aging and shortened life spans. It was reasoned that since radiation, which shortens life span, induces mutations in somatic cells, perhaps it accelerates the natural accumulation of mutations during aging. The somatic mutation theory predicts that age-dependent permanent genetic changes occur in dividing cells and/or postmitotic cells, leading to a reduction in the efficiency of tissues to carry out their functions. This concept represented one of the first attempts to explain the aging process at the molecular level. It was a very popular idea and has thus undergone a great deal of scrutiny. Recent refinements of the concept suggest that the mutations occur in specific genes that regulate such processes as DNA replication, transcription, or translation.

One early test of the theory was carried out in an elegant study on haploid and diploid strains of the wasp *Habrobracon juglandis*. As expected, the life span of haploid male wasps was significantly more reduced than that of diploid males following X-ray treatment. X-irradiation induces significantly more lethal mutations in haploids, since they are devoid of the extra allele found in the diploid strains. However, in the absence of radiation, the life spans of the haploid and the diploid strains were identical. This result argues strongly against the accumulation of point mutations as the cause of aging. More recently, molecular methods have been employed to test the somatic mutation theory of aging. For example, a specific mutation in human hemoglobin was examined in which isoleucine is substituted for valine. An increase in this mutation was detected in individuals exposed to radiation; however, no significant age-related substitution has been observed in the absence of radiation. Interestingly, the conclusions of the two studies are identical: radiation induces mutations but normal aging apparently does not. Other laboratories have investigated the response of various species, with differing life spans, to DNA damaging agents. A statistically significant correlation between the life span of a species and the capacity to repair UV-damaged DNA has been reported but does not necessarily imply that mutations are accumulating in senescent cells. So far, most of the published evidence provides a convincing refutation of the theory. Yet, recent techniques (e.g., two-dimensional DNA typing, two-dimensional protein electrophoresis, the use of shuttle vector systems to rescue and clone specific sequences from transgenic mice) may allow one to definitively determine the level of somatic point mutation frequency in vivo. Although no convincing support has been provided for the role of somatic mutations in the aging process, chromosomal aberrations have been consistently observed in senescent cells. One area of research that deserves considerable attention in the near future is the need for a molecular explanation for age-dependent chromosomal abnormalities.

### 1.2 Error Theories

Another group of related theories that involve alterations in the fidelity of information flow during aging can be categorized generally as error theories. The prototype for this type of theory, the "error catastrophe" theory, does not predict primary damage to DNA as do the "mutational" ones. Rather, it postulates that self-catalyzing errors will accumulate in the biosynthesis of proteins or nucleic acids. The decrease in the accuracy of protein synthesis brought on by an accumulation of such errors would be self-amplifying. When the accuracy of the translation process is reduced catastrophically, the cell should no longer function. This idea presented the experimental gerontologist with an attractive and testable hypothesis. One prediction is that measurable levels of such errors would be expressed as misincorporation of amino acids during protein synthesis. Misincorporation of amino acids into proteins that catalyze the biosynthesis of nucleic acids or other proteins could eventually lead to the catastrophic event. Most experimental tests of this theory have been negative. No compelling evidence of decreases of the fidelity of translation in old cells has been observed. In fact, most studies have provided overwhelming proof that the accuracy of protein synthesis is preserved in old age. For example, two-dimensional protein electrophoresis of the soluble proteins from the common fruit fly, *Drosophila melanogaster,* have shown that there are no detectable age-dependent changes in either the charge or the size of polypeptides under normal metabolic conditions. Direct evidence against errors in the enzymes involved in replication or protein synthesis as predicted by the error catastrophe theory was recently upheld using a clever molecular technique. In this study, errors in mammalian DNA polymerase were examined by reversion of amber mutants in bacteriophage $\phi$ X 174. DNA polymerase $\alpha$ or $\beta$ was isolated from regenerating livers of young and old mice. It was shown that the age of the mouse did not have any effect on the fidelity of copying the bacteriophage template. Most likely, any age-related change observed in protein structure or function during aging is the result of posttranslational modification and not the effect of alteration in the primary structure of the polypeptide.

### 1.3 Nuclear Versus Mitochondrial Genomes

The preceding sections have suggested that there is no compelling evidence for the occurrence of age-dependent mutational events. However, this conclusion is based on experimental work that has largely concerned surveys of the nuclear genome. One other genome that is essential to consider in the analysis of organismal aging is the mitochondrial genome, which has unique features that render it a likely target of age-related damage. In mammals, it is a small (ca. 16.5 kilobases) circular genome that is not protected by histones and apparently lacks DNA repair mechanisms comparable to those found in the nucleus. The mitochondrial genome experiences a much higher mutation frequency than nuclear DNA in human populations. Moreover, mitochondrial DNA is exposed to a higher degree of oxidative damage than the nuclear DNA. This latter observation may be related to the high rate of DNA-damaging oxyradicals generated in the mitochondria during respiration. Thus the mitochondrial genome may be more vulnerable to age-related damage than the nuclear genome. Very recently, several laboratories have reported that deletions in mitochondrial DNA occur in an age-dependent manner, and some of these deletions have been linked to specific pathologies in human populations (e.g., Kearnes–Sayre syndrome). In one sense, mitochondrial mutations may represent a new and refined version of the somatic mutation theory. The analysis of mitochondrial genetics will be an important area of research on the molecular biology of aging in the next few years.

## 2 EVOLUTIONARY ASPECTS OF GENE ACTION

Any discussion on the role of mutations in the aging process must consider the evolutionary contribution of mutation accumulation. The evolutionary theory of aging suggests that the force of natural selection greatly declines at late ages. In this context, the evolutionary theory of aging is an outgrowth of basic concepts of population genetics and thus lends itself to experimental examination. Two of the major mechanisms that may explain the genetic aspects of the evolution of aging are antagonistic pleiotropy and mutation accumulation.

### 2.1 Antagonistic Pleiotropy

Antagonistic pleiotropy is simply the idea of evolutionary tradeoffs—that is, a long life is sacrificed to win increased reproduction. As such, this mechanism is driven by natural selection, which tends to strengthen the mean fitness of a population. Experimental evidence supporting the idea of antagonistic pleiotropy as a mechanistic force driving the evolution of aging can be summed up as follows. There should be a negative genetic correlation between early and late life-history traits and antagonistic correlated responses of life-history characters to selection on other life-history characters. Some of the results from *Drosophila* support the antagonistic pleiotropy concept; however, the results are not so clear for other organisms.

### 2.2 Mutation Accumulation

The driving force of the mutation accumulation theory is of course mutation. Basically, the theory posits the accumulation of late-acting mutations that are not deleterious until late in life. Selection declines at late ages, and mutation rates then dominate gene frequency determination. Some *Drosophila* selection experiments support the idea that mutation accumulation is occurring in the evolution of senescence. In these experiments, where genes with late-acting deleterious effects are allowed to accumulate, later fitness components are subject to the accumulation of these age-specific alleles.

Understanding how evolution has shaped the life spans of populations should eventually provide important insight into the mechanisms active in senescence. It is worth noting that these two mechanisms, antagonistic pleiotropy and mutation accumulation, may be operating simultaneously to regulate the life spans of organisms, since one process can be acting on one set of loci while the other is acting at another set of loci.

## 3 LONGEVITY GENES

One of the current goals in the study of the biology of aging is to identify the genes that are associated with long life. Such genes are thought to play a major role in promoting the longevity of an organism and thus delaying the onset of senescence. Several experimental strategies are available for identifying longevity genes in experimental animals. One may mutate particular genes by the use of mutagens or site-directed mutagenesis and then test the effects of their expression, or lack of expression, on life span. Also, an experimentalist could employ genetic transformation techniques to generate transgenic animals expressing extra copies of genes that delay the onset of senescence. Alternatively, one may select for postponed aging in a population of animals and then identify the genes selected.

### 3.1 Mutation

Early studies on the genetics of aging suggested that the analysis of mutant strains with reduced life spans might lead to the discovery of genes that are critical to the maintenance of long life; but this approach has provided very little information relevant to the mechanism of senescence. Most mutations shorten the life span of experimental animals, and many of these genes clearly do not have any bearing on regulating normal senescent processes. On the other hand, development of mutant strains with extended life spans should furnish experimental gerontologists with a well-founded understanding of the molecular genetic mechanisms operating during senescence. Recently, mutagenesis was employed to develop a mutant strain of the nematode *Caenorhabditis elegans,* which displays delayed senescence. The function of this gene, designated *Age*-1, presently remains undetermined, however, the identification of this and similar ''gerontogenes'' represent important goals in gerontological research.

### 3.2 Genetic Transformation

Genetic transformation, a method for introducing foreign genes into the genomes of experimental animals, permits the gerontologist to analyze the contribution of specific genes to the aging process. The goal of such studies is to introduce genes that delay the onset of senescence. So far, several model systems have been employed to test the role of selected genes on the rate of aging. For example, the genes for elongation factor 1 alpha, CuZn superoxide dismutase, and catalase have been introduced into *Drosophila melanogaster* using p-element mediated transformation. Several of these studies have reported statistically significant increases in the mean life span, whereas others have noted no increase in longevity associated with the expression of the added genes. These studies are clearly pioneering efforts, of various adequacy with respect to replication and analysis of gene expression. However, they establish that p-element transformation is a useful strategy for postponing aging in *Drosophila* and should eventually yield the identification of genes that are important to controlling longevity. Site-directed p-element mutagenesis can also be used to disrupt the function of selected genes involved in setting limits to the life span of *Drosophila.*

### 3.3 Genetic Selection

As pointed out earlier, the development of experimental animals in which aging is postponed represents a significant step in the quest for uncovering the genes that control the aging process. Artificial genetic selection experiments have been used to develop stocks of *Drosophila* with extended life spans. This model system readily lends itself to an investigation of the physiological and molecular mechanisms that are responsible for the increased life span. Presently, the genes that were selected for in these long-lived strains have not been identified. One way in which these stocks can be analyzed is to make biochemical comparisons between the controls and those with postponed aging. The identification of the specific loci associated with the long-lived phenotype may be determined by two-dimensional protein electrophoresis, restriction fragment length polymorphism mapping combined with standard genetic crosses, or the use of differential molecular genetic screening techniques. Once the genes that are associated with postponed aging in experimental animals have been identified, they can be cloned and introduced into other strains using genetic transformation techniques to directly test their effects on the aging process.

*See also* DNA Repair in Aging and Sex.

### *Bibliography*

Comfort, A. (1979) *The Biology of Senescence.* Elsevier, New York.

Finch, C. E. (1990) *Longevity, Senescence, and the Genome.* University of Chicago Press, Chicago.

Fleming, J. E., Quattrocki, E., Latter, G., Miquel, J., Marcuson, R., Zuckerkandl, E., and Bensch, K. G. (1986) Age-Dependent Changes in Proteins of *Drosophila melanogaster. Science,* 231:1157–1159.

Joenje, H., Ed. (1992) *Molecular Basis of Aging: Mitochondrial Degeneration and Oxidative Damage.* Special issue of *Mutat Res,* 275(3–6).

Johnson, T. E. and Lithgow, G. J. (1992) The search for the the genetic basis of aging: The identification of gerontogenes in the nematode caenorhabditis elegans. *J. Am. Ger. Soc.,* 40:936–945.

Rose, M. R. (1991) *Evolutionary Biology of Aging.* Oxford University Press, New York.

Warner, H. R., Butler, R. N., Sprott, R. L., and Schneider, E. L., Eds. (1987) *Modern Biological Theories of Aging.* Raven Press, New York.

# AIDS, Inhibitor Complexes of HIV-1 Protease in

*Alexander Wlodawer*

### *Key Words*

**AIDS** Acquired immunodeficiency syndrome; a disease caused by the human immunodeficiency virus (HIV). At the present time, AIDS cannot be completely reversed once the infection has taken place.

**HIV Protease** An enzyme, encoded by the genome of the human immunodeficiency virus, which is necessary for proper maturation of the newly synthesized viral particles.

**Rational Drug Design** An approach to drug design in which the molecular structure of the target is used to design and synthesize putative drugs, the structures of complexes of the drug(s) are evaluated, and the compounds are modified to improve their properties.

---

AIDS is a presently incurable disease caused by infection with the human immunodeficiency virus (HIV). Its progress can be arrested, but not reversed, by treatment with nucleoside analogues such as AZT, ddI, and ddC. These compounds, together with a number of newer nonnucleoside drugs, are inhibitors of reverse transcriptase, one of four enzymes encoded in the HIV genome. However, their side use usually leads to serious side effects and to development of drug resistance, and thus new anti-AIDS drugs need to be discovered and developed. HIV protease is potentially an excellent drug design target, since this enzyme is necessary for virus maturation, and its inhibition should lead to arresting of viral infection. Crystal structure of HIV protease has been determined both by itself and in complexes with many inhibitors, which were either discovered in random drug screens, adapted from work on antihypertensive drugs that had been targeted at renin, or developed specifically for this target. Crystallographic results were used to suggest chemical changes that could increase the potency or selectivity of inhibitors, and this technique has been shown to be very useful in rational drug design. The development of some protease inhibitors has reached the stage of phase III clinical trials for the treatment of AIDS in humans.

## 1 HIV PROTEASE AS A TARGET FOR DRUG DESIGN

### 1.1 Rational Design of Drugs Against AIDS

Human immunodeficiency virus, the causative agent of AIDS, encodes only four enzymes in its genome. In addition to reverse transcriptase and integrase, it encodes an aspartic protease (HIV PR), whose function is essential for proper virion assembly and maturation. Inactivation of HIV PR by either mutation or chemical inhibition leads to the production of immature, noninfectious viral particles. HIV PR inhibitors could become a new class of therapeu-

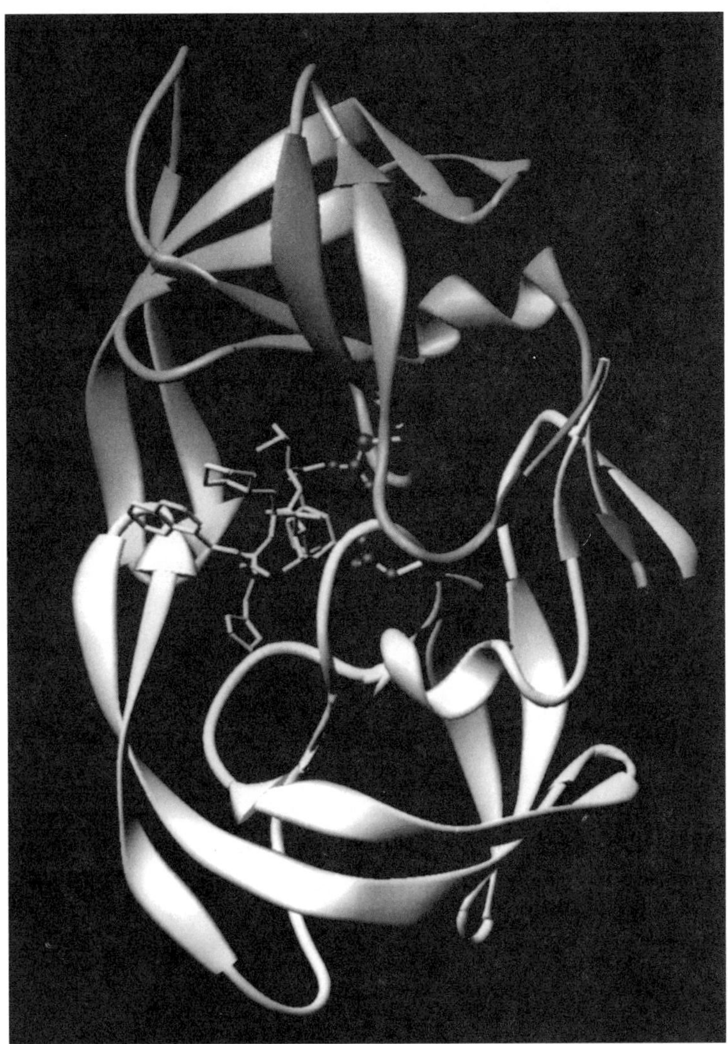

**Figure 1.** Computer graphics image of the HIV protease complexed with an inhibitor, U-75875. The chains of the enzyme are in two different colors, the inhibitor is shown in bond representation, and the active site aspartates are shown in the ball-and-stick format. (See color plate 1.)

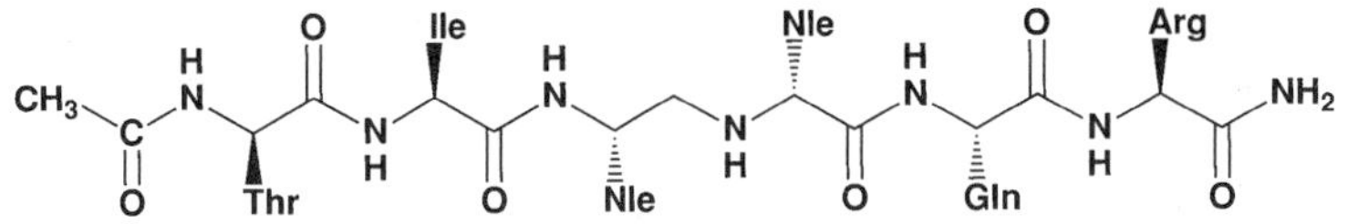

MVT-101

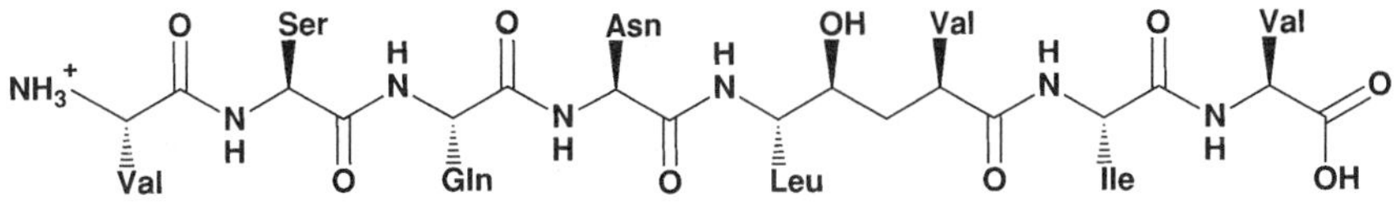

U-85548e

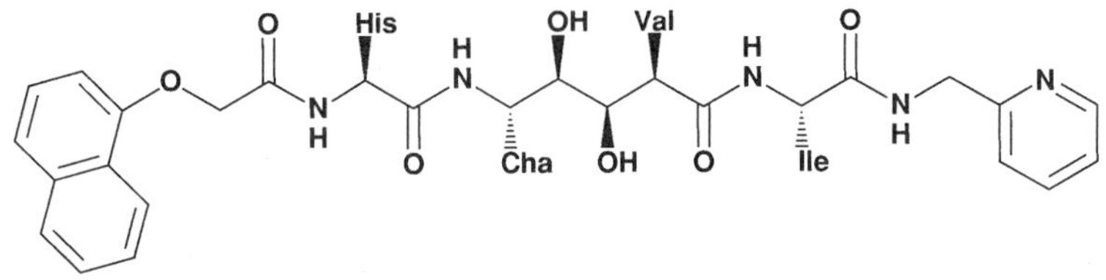

U-75875

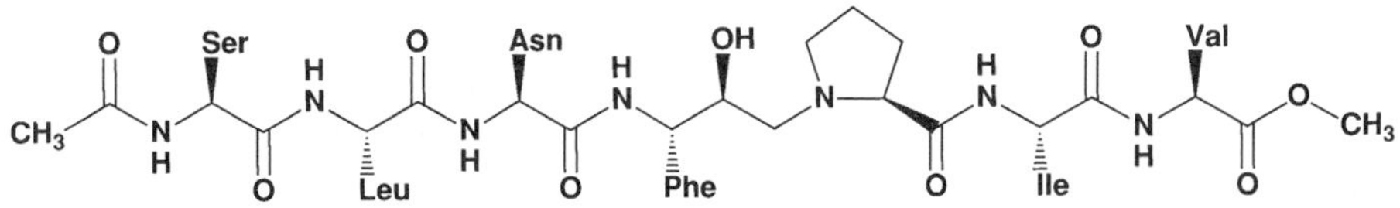

JG-365

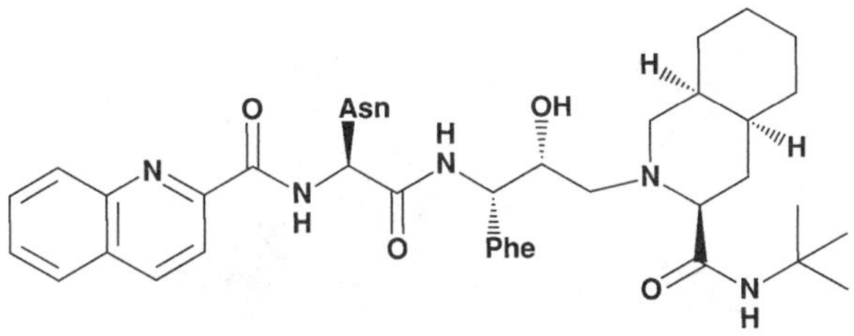

Ro-31-8959

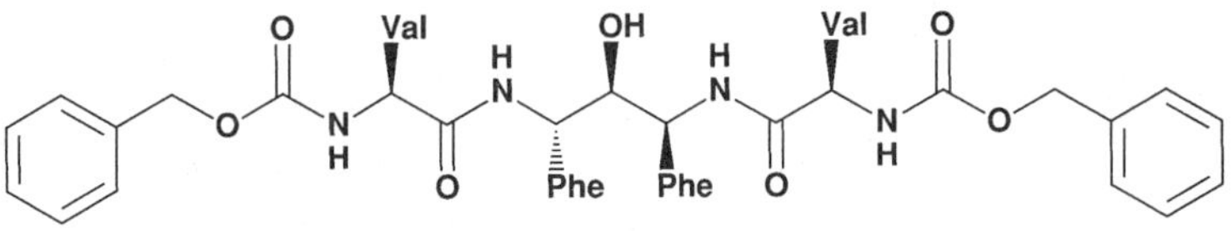

A-74704

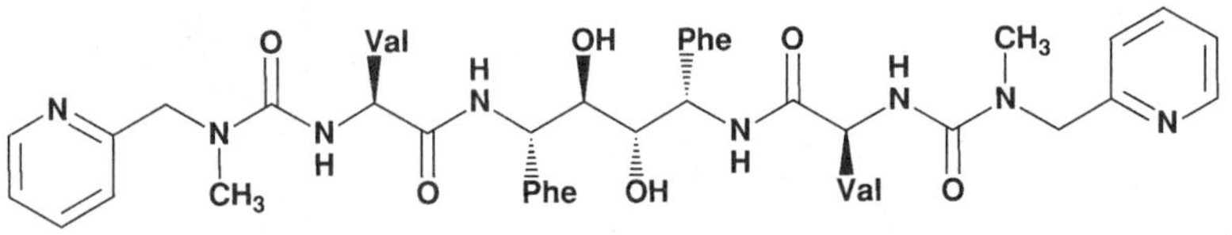

A-77003

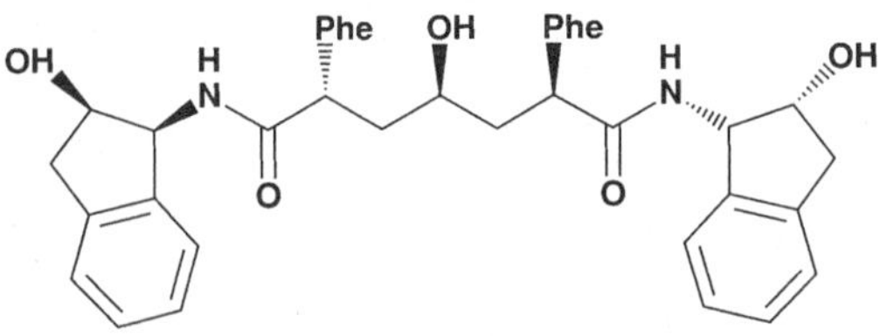

L-700,417

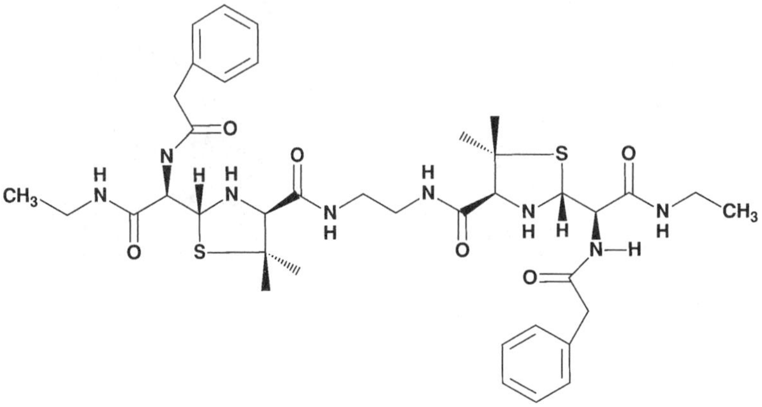

GR-116624X

**Figure 2.** Chemical structures of some inhibitors for HIV PR/inhibitor complexes whose structures have been solved using X-ray crystallography and reported. The top five inhibitors are peptidomimetics; next three are symmetric and based on peptide chemistry; the last compound was found in a random screen.

tic agents complementing existing approaches to antiviral therapy that target the HIV reverse transcriptase with drugs such as AZT, ddI, and ddC.

Rational drug design combines chemical synthesis of compounds with structure determination methods, such as protein crystallography, nuclear magnitude resonance spectroscopy, and computational biochemistry. Protein crystallography has been particularly useful in helping to elucidate precise interactions between inhibitors and their target enzymes. This information can, in turn, be used to help explain the basis of effective inhibition, and to improve the potency and specificity of newly synthesized compounds. In the past few years there has been a virtual explosion of new X-ray crystal structures of HIV PR/inhibitor complexes. Indeed, the level of involvement of crystallographers in this area is unprecedented in the history of the field: at least 300 structures have been solved in 21 laboratories between 1989 and 1994, yielding invaluable information about the interactions between the enzyme and its ligands, and helping to design and discover novel compounds with antiviral activity.

### 1.2 Structural and Biological Properties of HIV PR

HIV PR is a homodimer of two 99 amino acid chains. Since only small amounts of HIV PR could be isolated from viral sources, the enzyme was cloned in a variety of vectors, and it was also prepared by total chemical synthesis. The experimental crystal structure of recombinant native HIV-1 PR was independently verified using the synthetic enzyme. As predicted, the topology of each monomer is similar to that of a single domain of such aspartic proteases as pepsin, chymosin, and renin. The carboxylate groups of Asp25 from both chains in the single active site of the enzyme are nearly coplanar and in close contact (Figure 1; see also color plate 1).

## 2 INHIBITORS OF HIV PR

### 2.1 Asymmetric Peptidic Inhibitors

The smallest polypeptide substrate of HIV PR consists of at least six amino acids. The side chain of each amino acid occupies a clearly defined subsite, named Pn—Pn′, with the cleaved bond residing between P1 and P1′. Replacement of the potentially scissile peptide bond by a nonscissile analogue can turn a substrate into an inhibitor. Most of the medicinal chemistry efforts employed to design HIV PR inhibitors have been based on classical substrate or transition state analogue-based approaches. This strategy has been highly successful in the design of peptidomimetic inhibitors for the related aspartic protease, human renin. The scissile P1—P1′ amide bond can be replaced by a nonhydrolyzable isostere such as reduced peptide bonds, hydroxyethylene, dihydroxyethylene, phosphinate, statine, and norstatine. This knowledge has resulted in a variety of structurally diverse and potent HIV PR inhibitors, many of which are effective inhibitors of virus replication in vitro. Representatives of different classes of peptidic inhibitors are listed in Figure 2.

The identification of amino acid context and of preferred side chains in particular positions in protein and peptide substrates has led to predictable trends for peptidomimetic inhibitor design. Several laboratories have demonstrated that a high degree of inhibitor potency can be achieved by incorporating into peptide mimics hydrophobic blocking groups and amino acid side chains that do not occur naturally. Incorporation of the hydroxyethylene isostere into peptide mimics resulted in potent antiviral agents when the amide was incorporated into a cyclic structure such as prolyl or piperidinyl group.

### 2.2 Symmetric Peptidic Inhibitors

The development of peptide-based inhibitors into effective drugs has been hampered by the inherently poor pharmacological properties of peptides. For this reason, several groups have departed from traditional peptidomimetic designs and have embarked on what may be termed structure-based inhibitor design. One of the approaches utilized the concept of active site symmetry to design twofold ($C_2$) symmetric or pseudo-$C_2$ symmetric inhibitors. The observation that the active site of HIV PR was $C_2$ symmetric led to the design of simple lead structures that would mimic this symmetry, and also satisfy the hydrogen bonding and subsite preferences known to be important for binding. A potent inhibitor, A-74704, was synthesized, and its structure was then used to guide the design of pharmacological improvements by focusing synthetic efforts on the peripheral blocking groups. This led to synthesis of A-77003, a potent inhibitor that was also water-soluble and exhibited achievable antiviral blood levels in monkeys and dogs upon oral administration. Another pseudo-$C_2$ symmetric hydroxyethylene-based inhibitor of HIV PR, L-700,417, has been reported. The foregoing examples illustrate how knowledge of enzyme structure may be used for the conceptualization of novel lead compounds.

### 2.3 Nonpeptidic Inhibitors

HIV PR structure was used directly to search structural databases for compounds that contained shapes complementary to the enzyme active site. The search procedure identified accessible surfaces or cavities on macromolecules that were used to search a structural database. Using this technique, the known compound bromperidol was identified and a closely related analogue, haloperidol, was shown to weakly inhibit the enzyme. This series of compounds is still under development.

Not all potent inhibitors of HIV PR have resulted from rational substrate- or structure-based design. Large-scale screening of chemical databases has been employed in many laboratories with some notable successes. In particular, screening of a number of penicillin analogues has led to the discovery of a crude sample of a penicillin dimer that had considerable inhibitor activity. Further purification of this sample led serendipitously to the breakage of β-lactam rings and the production of a $C_2$ symmetric diester. Synthetic elaboration of this compound resulted in GR-116624X (Figure 2) which is a potent inhibitor of HIV PR.

## 3 PROSPECTS FOR DEVELOPMENT OF CLINICALLY USEFUL DRUGS

Comparison of the crystal structures of numerous HIV PR inhibitor complexes with structurally diverse compounds reveals several general features important for future design strategies. The ability of peptidomimetic inhibitors to bind in the nearly symmetric substrate binding cleft of the enzyme is due to the flexibility of the inhibitor backbones, which facilitate the optimization of both polar backbone–backbone interactions and the largely nonpolar inhibitor side chain–subsite interactions. The design of more reasonable, small, rigid, nonpeptide inhibitors for HIV PR will likely require that even more attention be given to symmetry to achieve high potency.

A number of problems related to the development of practical drugs that would be active by inhibiting HIV PR have not yet been solved. In particular, it is still very difficult to correlate the in vitro inhibition properties with in vivo activity, both in cell culture and in organism. This is a general problem encountered in drug design and is not unique to HIV protease. In addition, the ways in which such drugs could reach their targets are still not well understood. However, a number of compounds have shown sufficient promise to enter clinical studies. In particular, the inhibitor Ro-31-8959 is currently undergoing phase III clinical trials using oral delivery. It can be reasonably hoped that this or other compounds will finally lead to practical utilization of this completely new class of anti-AIDS drugs.

*See also* AIDS, Transcriptional Regulation of HIV in; AIDS HIV Enzymes, Three-Dimensional Structure of; AIDS Therapeutics, Biochemistry of.

### *Bibliography*

Huff, J. R. (1991) *J. Med. Chem.* 34:2305–2314.

Meek, T. D. (1992) *J. Enzyme Inhibition,* 6:65–98.

Mitsuya, H., Yarchoan, R., and Broder, S. (1990) *Science,* 249:1533–1544.

Tomasselli, A. G., Howe, W. J., Sawyer, T. K., Wlodawer, A., and Heinrikson, R. L. (1991) *Chim. Oggi,* 9:6–27.

Wlodawer, A., and Erickson, J. W. (1993) *Annu. Rev. Biochem.* 62:543–585.

# AIDS, Transcriptional Regulation of HIV in

*Melanie R. Adams and Boris Matija Peterlin*

### *Key Words*

**AIDS** Acquired immune deficiency syndrome: a systemic immunodeficiency characterized by decreasing $CD4^+$ T lymphocytes and increasing susceptibility to opportunistic infection.

**Latency** A concept describing (1) an asymptomatic clinical condition, (2) the state of viral activity within a population of cells, or (3) the downregulation or absence of gene expression within an infected cell.

**LTR** The long terminal repeat (LTR), a 713 base pair DNA sequence repeated at the 5′ and 3′ ends of the genome, consisting of the enhancer and promoter regions for gene expression, the RNA start site, and untranslated RNA sequences such as the genomic repeat and polyadenylation sites.

**Nonprocessive Transcription** Initiation with inefficient elongation; transcription complexes pause and release, leading to an abundance of short, nonpolyadenylated RNA and only rare full-length mRNAs.

**Processive Transcription** Efficient elongation of transcripts leading to high levels of polyadenylated mRNA.

**Promoter/Enhancer Module** Multiple, sometimes overlapping binding sites for modulatory proteins that in some cases bind cooperatively and substitute for each other when multimerized, a redundancy of possible evolutionary importance.

**TAR** The transactivation response (TAR) element, the target for Tat binding, is an RNA stem-loop positioned immediately 3′ of the transcription start site.

**Tat** The virally encoded transactivating protein.

---

The human immunodeficiency virus (HIV) is a human lentivirus capable of both establishing long-term latency and rapidly destroying its host. Its replicative versatility may be determined by the unique mechanism by which it regulates transcription. The HIV long terminal repeat (LTR) promoter directs the assembly of transcription complexes that initiate transcription but are incapable of efficient elongation in the absence of the virally encoded transactivator, Tat. Tat is unique among known transactivators insofar as it binds the stem-loop formed by the leader sequence of the nascent RNA, a region called the transactivation response (TAR) element. Basal transcription by an initiated yet unstable transcription complex and dependence on cellular activation for the synthesis of the stabilizing factor Tat are phenomena that precisely regulate the viral life cycle, allowing the virus to remain transcriptionally poised, yet silent, until cellular activation signals trigger the transition to full viral replication. This entry focuses on HIV-1, reviewing the cellular transcription factors involved in basal and activated rates of LTR-directed transcription, Tat and its interaction with TAR, and the consequences of these interactions as they are reflected in disease progression.

## 1 HIV LTR

### 1.1 Enhancer Region

The enhancer/modulatory region of the HIV LTR (positions −454 to −78 relative to the RNA cap site) contains binding sites for at least seven cellular transcription factors important in the regulation of cell activation and growth (Figure 1). Individually, these upstream modules appear to play only a small role in LTR transcriptional regulation, but functional redundancy may be necessary in the dynamic in vivo environment. For example, while it seems probable that the module for nuclear factor of activated T cells (NFAT) responds to the increased levels of NFAT-1 found in activated T lymphocytes, recent studies indicate that mutation of the motif does not change the LTR response to activation.

The most extensively studied LTR enhancing element is the nuclear factor kappa B (NF-κB). The two NF-κB motifs bind dimers of the *rel* proto-oncogene family. In resting T cells, NF-κB dimers are complexed with the inhibitory factor I-κB and are retained in the cytoplasm. Upon cell activation, I-κB is phosphorylated and dissociates, allowing NF-κB translocation to the nucleus and interaction with the HIV LTR. There are reports that NF-κB participates in basal and activated HIV gene expression, although, similar to the other regulatory modules, there are in vivo experimental systems in which NF-κB binding sites are dispensable.

### 1.2 Promoter Region

The basal rate of transcription initiation is regulated at the HIV-LTR promoter (−78 to −1) through the binding of the constitutively

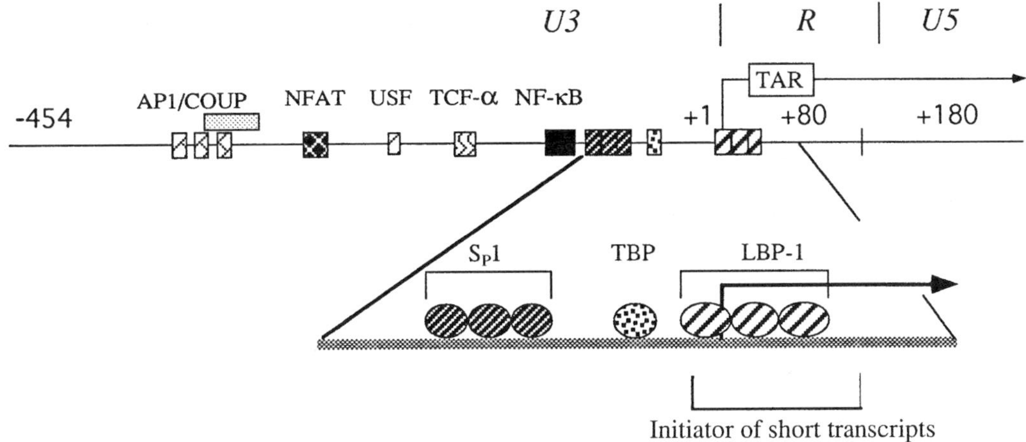


**Figure 1.** Cellular transcription factor modules on the HIV-1 LTR. The full LTR consists of 713 nucleotides and is divided into the U3 (untranslated 3′), R (genomic repeat), and U5 (untranslated 5′) regions. The enhancer/modulatory region includes binding sites for activator protein 1 (AP1), chicken ovalbumin upstream promoter (COUP), nuclear factor of activated T cells (NFAT-1), upstream stimulatory factor (USF), T-cell factor 1a (TCF-1a), and nuclear factor kappa B (NF-κB). The promoter region includes the Sp1 module and the TATA box, which binds the TATA-binding protein (TBP). The sequences from positions −5 to +82, termed the initiator of short transcripts (IST), include the binding sites for untranslated binding protein-1 (UBP-1 or LBP-1).

expressed protein Sp1. Sp1 is also strongly supportive of Tat-mediated transactivation, although other transcriptional activators can substitute for Sp1 without loss of Tat response. In contrast, the specific LTR TATA sequence is critical to Tat transactivation; alternative TATA sequences, while maintaining the basal rate of initiation, abolish the Tat response.

Other factors recognize DNA modules that overlap the TATA and flanking regions. LBP-1/UBP-1, HIP116, and TFII-I vary in affinity for their binding sites, and their relative contribution to basal promoter activity requires further study. A region further downstream (−5 to +82) has been termed the initiator of short transcripts (IST) region, in analogy to the U1 and U2 small nuclear RNA gene promoters, which direct the assembly of transcription complexes that are responsive to specific downstream elongation blocks not recognized by other complexes (Figure 1). Therefore, the LTR TATA box and the sequences extending through the TAR DNA, may direct the assembly of a class of transcription complexes that attenuate transcription prematurely and depend on accessary proteins such as Tat for efficient elongation.

### 1.3 TAR

The TAR region of the HIV LTR encodes a regulatory structure unique among those known in eukaryotic promoter regulation: a stem-loop RNA that binds a transactivating factor. The unique interaction of Tat and TAR is dependent on the shape and position of TAR: a 59-nucleotide hairpin positioned immediately 3′ of the LTR cap site (Figure 2). The minimal functional region extends from positions +19 to +42; the lower stem is dispensable for transactivation. In vitro binding studies, with site-directed and compensatory TAR mutations, have revealed that Tat recognizes the trinucleotide bulge and the specific base pairs of the stem adjacent to the bulge. More detailed NMR spectroscopy suggests the bulge changes conformation upon Arginine binding, with U23 of the bulge and A27-U38 of the stem stabilizing the spacing of the backbone phosphates (the arginine fork) and also stabilizing the Arg-hydrogen bonding to G26 and the phosphates.

TAR loop mutation studies predict the presence of cellular TAR binding factors involved in Tat function for, while Tat does not bind to the TAR loop, loop mutations destroy Tat transactivation. Three TAR-binding nuclear proteins have been discovered that confer specific transactivating activity (Figure 2). The HeLa nuclear protein p68 binds to the loop and increases Tat transactivation. Two other TAR binding proteins, TRP-1 (TAR binding protein 1, also called TRP-185) and TRP-2 (TAR binding protein 2), may assist in Tat binding or form a transactivating complex with Tat. Further evidence of TAR-binding cellular factors comes from the observation that Tat transactivation is species specific: it is high in primate cells, low in rodent cells. Complementation of Chinese hamster ovary (CHO) cells with human chromosome 12 reconstitutes Tat function. Recently, Tat chimeric proteins with heterologous RNA tethering mechanisms were found to return Tat activity in CHO cells, suggesting that rodent cells lack a TAR binding protein that helps Tat to bind TAR. Numerous Tat chimeric protein studies have been performed in which TAR is replaced by a heterologous RNA binding site without loss of ability to transactivate, implying that cellular TAR-binding proteins are expendable when Tat is tethered near the promoter.

### 1.4 Transcription in the Absence of Tat

RNase protection and nuclear run-on assays first revealed that LTR-directed transcription in the absence of Tat is predominantly nonprocessive (Figure 3). Cellular transcription complexes assemble on the promoter, initiate transcription, and then terminate prematurely. This leads to a gradient of transcriptional activity along the viral DNA: the greater the distance from the promoter, the fewer the transcripts. In addition, short, nonpolyadenylated transcripts corresponding to the TAR stem-loop accumulate in the cytoplasm of *tat*-mutant transfected and infected cell lines, representing either stable RNase-resistant products of randomly aborted nonpolyadenylated transcripts or the products of pauses signaled by the TAR stem-loop. For fully efficient transcription to take place. Tat must be present.

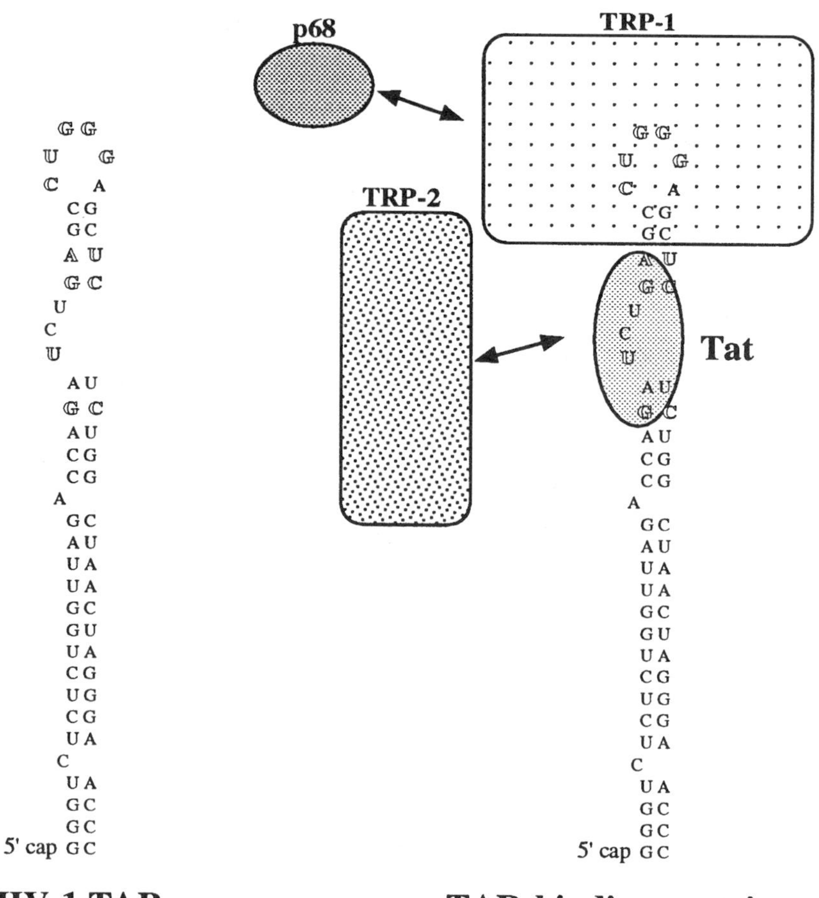


Figure 2. HIV-1 TAR RNA sequence and structure with regions targeted by Tat and nuclear TAR-binding proteins TRP-1, TRP-2, and p68.

## 2 TAT TRANSACTIVATOR

### 2.1 Tat Structure

HIV-1 Tat is a small protein of 82 to 101 amino acids, depending viral strain, that localizes to the nucleus. Functional Tat is translated from three multiply spliced mRNAs with a fourth spliced form, Tev or Tnv, joining the first exon of *tat* to a portion of *env* and *rev* (Figure 4). The first exon of *tat* encodes a 72 amino acid protein that functions efficiently in vitro, although the full-length protein may be required for most efficient transactivation in vivo. As shown in Figure 4, these 72 amino acids are divided into five domains: the N-terminal, cysteine-rich, core, basic, and C-terminal domains (regions I–V, respectively). The cysteine-rich, core, and basic domains are conserved between species and are essential for Tat activity.

Similar to other transcriptional activators, Tat's domains serve two functions: binding and activation. Mutation and chimeric protein studies have shown that the activating region (residues 1–48) is composed of the cysteine-rich and core domains with the N-terminal region augmenting transactivation efficiency. The cysteine residues in the cysteine-rich region are critical for function, as is lysine 41. The cysteines contribute to protein folding and may form a $Zn^{2+}$ finger or binding pockets for divalent cations. It has been suggested that divalent cations may be a link between homodimers; however, Tat homodimers have not been observed in cells. The amino acid sequence in the core region cannot be altered. The activation domain can be exchanged between HIV-1, HIV-2, Simian Immunodeficiency Virus (SIV) and Equine Infectious Anemia Virus (EIAV) (a region of only 26 cysteine-rich and core amino acids in EIAV vs. 48 in HIV-1); the specificity of the different Tat species is confirmed by the binding domain.

Experiments in which Tat was fused to RNA-binding proteins demonstrated that Tat will transactivate from a nascent RNA chain containing the appropriate RNA binding site. RNA binding and nuclear transport have been localized to the basic domain (residues 48–60). Domain substitutions between the RNA-binding regions of HTLV-I Rex protein, the N protein of bacteriophage lambda, and HIV-1 have identified the consensus binding sequence: Arg/Lys-X-X Arg-Arg-X-Arg-Arg. However, random substitutions of Arg and Lys, including nine Lys with one Arg, seem to bind equally well when presented to TAR RNA as a short peptide. The C-terminal amino acids, together with the basic domain, may also be required for nuclear localization.

Tats lacking the basic domain can inhibit transactivation by wild-type Tat, although they never reach the nucleus. This suggests there is a cytoplasmic protein important for viral transactivation that associates with a nonbinding domain of Tat. The protein could modify Tat directly, or perhaps interact with TAR or the transcription complex once transported to the nucleus. A protein of 36 kDa has been isolated by Tat affinity chromatography that increased levels of transactivation by Tat, possibly by stabilizing its interac-

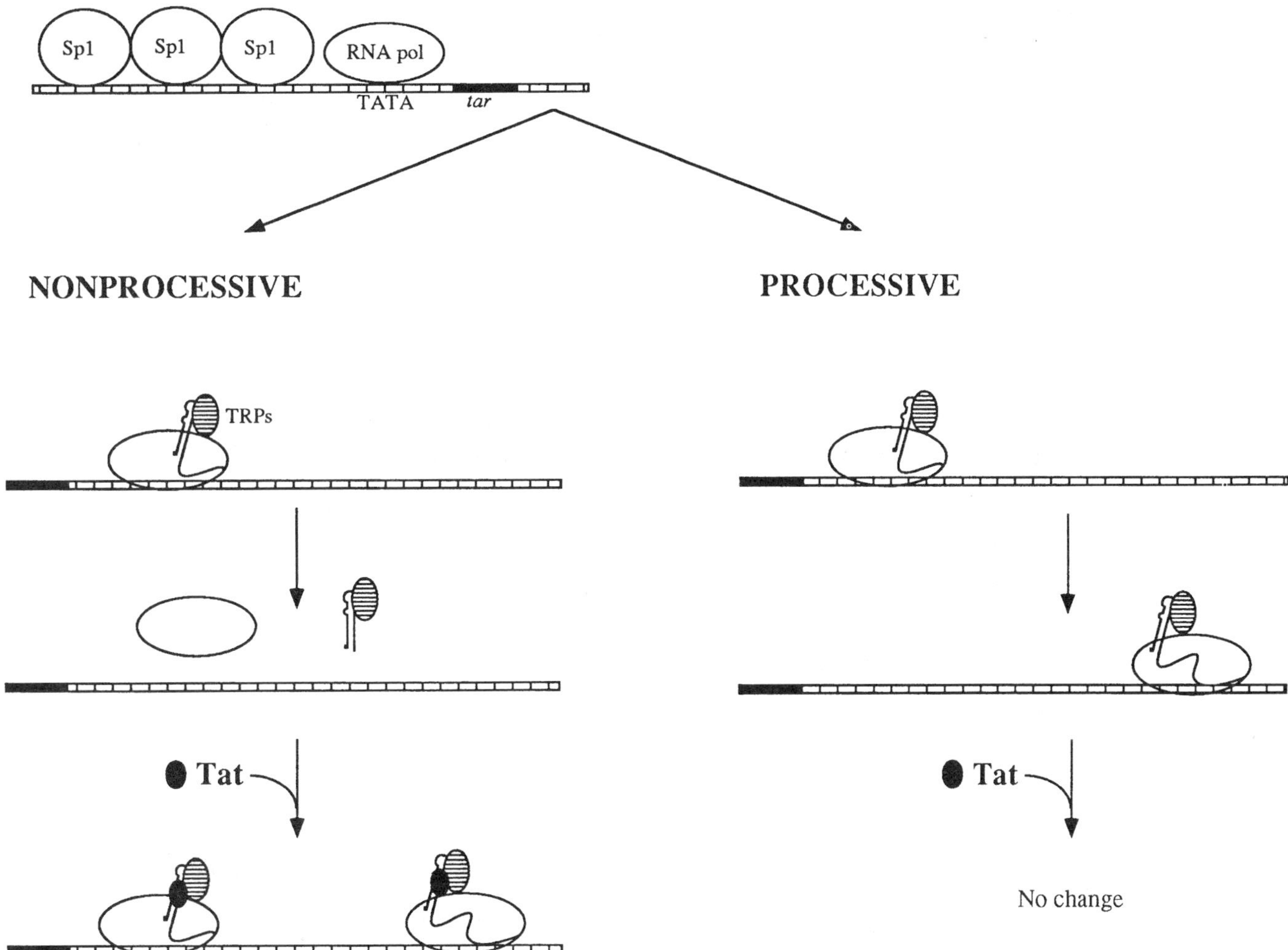


**Figure 3.** A model for HIV transcription. Transcription complexes initiate at the promoter region in the absence of Tat. These complexes are predominantly nonprocessive, releasing prematurely attenuated transcripts, which are then digested by RNases to the stable TAR stem-loop structure. Tat acts on these nonprocessive complexes, greatly increasing elongation efficiency. It is unknown whether two discrete classes of complex form at the promoter, one processive the other nonprocessive, or all are nonprocessive but capable of rare elongation completion in the absence of Tat. RNA-binding proteins (TRPs) may complex with TAR throughout the process.

tion with TAR RNA. Recently, a human protein termed TBP-1 has been isolated that decreases Tat transactivation. It remains to be determined whether any of these proteins bind Tat in cells or whether there are other proteins that play more important roles in transactivation.

### 2.2 Tat Function

Tat stimulates efficient transcription and abundant full-length polyadenylated RNA collects in its presence. The Tat binding site has been found to be the nascent RNA chain; the site and mechanism of Tat action remains the subject of intense investigation, however, and has yet to be defined. Research is complicated by the possibility that Tat's role in the transcription process may vary depending on experimental conditions and cellular environment.

#### 2.2.1 Tat as Processivity Factor

In vivo analyses of *tat*-mutant transfected and infected cell lines indicate a strong processive role for Tat (Figure 3). Utilizing RNase protection and nuclear run-on analysis, these studies showed no detectable increase in total viral RNA following addition of Tat, only a change in the pattern of transcript elongation. In vitro cell-free transciption experiments confirmed that Tat promotes transcript elongation, stabilizing the RNA polymerase complex as it transcribes the full length of the genome. The bacteriophage λ antiterminator, N protein, provides a prokarytoic model of this type of RNA-directed elongation regulation; the N protein binds an RNA sequence, *nut B,* and facilitates transcription elongation through defined termination sites.

#### 2.2.2 Tat as Initiation Factor

It is has been postulated that Tat facilitates complex assembly or transcription initiation. In systems in which the basal level of transcriptional initiation was low, Tat was found to increase RNA polymerase density near the promoter as well as along the full genome. In contrast, increasing the basal rate of initiation by the addition of the general transcriptional activators E1A or PMA

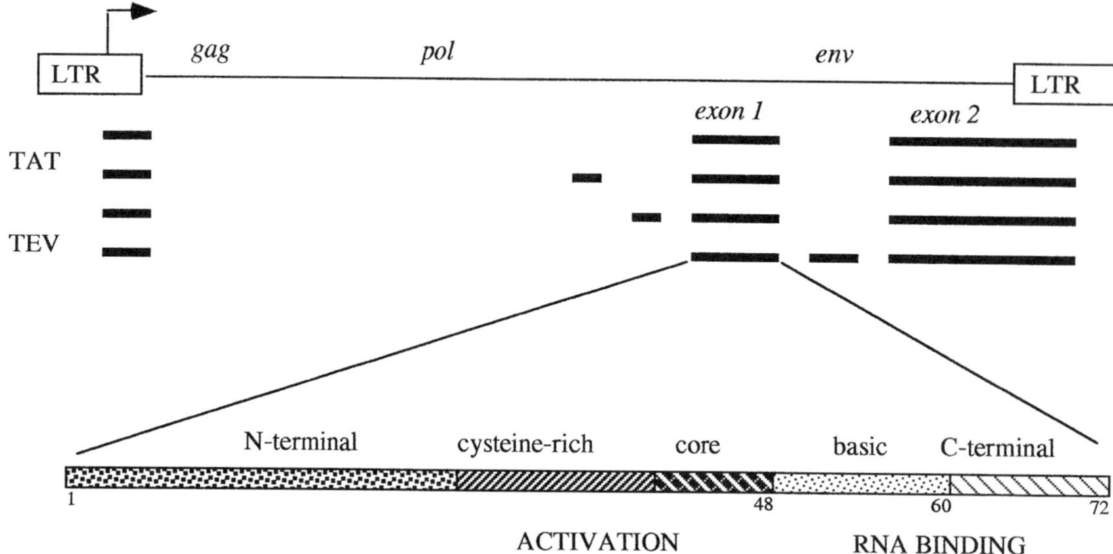


**Figure 4.** The multiply spliced transcripts that code for Tat and the functional domains of the Tat protein. Exon 1 codes for the region required for optimal activity; the first 60 amino acids are essential for transactivation.

resulted in a strongly synergistic increase in processivity with only an additive effect on initiation. Therefore, it was suggested that the role Tat manifests is determined by the level of basal initiation, a rate that may vary with experimental system and cell type. A dual function for Tat would be analogous to the action of a cellular transcription factor for polymerase II: this transcription factor, TFIIF, a complex of a 74 kDa and a 30 kDa protein, is capable of increasing both initiation and elongation of RNA polymerase II transcription. In an in vitro reconstruction experiment, purified TFIIF increased activity from the LTR promoter to such a degree that Tat effect was undetectable.

Tat also appears to function in the manner of an enhancer when tethered to upstream DNA. It has been demonstrated that a Tat-GAL4 fusion protein can stimulate transcription when bound via six GAL4 DNA binding sites upstream of a minimal promoter containing the TATA box alone. These results suggested that Tat could function in a manner similar to an enhancer where increasing the number of binding sites has a synergistic effect. While confirming that Tat interacts with the transcription complex in the proximity of the promoter, these results do not directly address the question of the transcriptional process most affected by Tat.

### 2.2.3 Tat and Translation

In some experimental systems the Tat-induced levels of viral mRNA may not change after a certain point, although levels of reporter proteins increase. Therefore, Tat may be functioning to increase the efficiency of both transcription and translation. Since TAR can be inhibitory to translation in rabbit reticulocyte lysates and in the *Xenopus* oocyte, it was proposed that Tat increases the efficiency of polysomal loading onto mRNA by masking or modifying the inhibitory TAR structure. TAR is not inhibitory in primate cells, however, and Tat appears to exert its effect soley during transcription.

## 3 CELL–VIRUS INTERACTION

### 3.1 The Viral Life Cycle

The life cycle of the virus is linked to that of the host cell by its LTR. Stimulation of viral transcription parallels cellular activation by antigens, mitogens, and UV light. Therefore, detailed understanding the viral life cycle requires examination of the LTR in the context of a changing cellular milieu. In the resting cell, transcription is both processive and nonprocessive, the rate of each determined by the individual cells' production of factors such as NF-κB and Sp1. Following cell activation, levels of basal transcription factors, including factors such as TFIIF and possible cellular TAR-binding factors, may rise, along with rising levels of enhancing proteins. A cellular factor as potent as Tat is unlikely to be expressed during this period, inasmuch as wild-type Tat must be present for elevation of p24 *gag* levels following PMA stimulation. Stimulation increases both processive and nonprocessive transcription from the LTR, leading to the increased production of both Tat protein and Tat responsive target. A threshold level of Tat may be necessary before it can initiate the subsequent cascade of viral production.

### 3.2 Latency

Because positive control of viral transcription plays such a prominent role in viral replication, negative control of transcription may play a role in latency. Negative as well as positive control of transcription may also be part of the virus's armamentarium, possibly conferring evolutionary advantage to the virus.

Models for negative control of transcription include nonspecific inhibition of the viral LTR via inhibition of the infected cell itself. There is evidence that $CD8^+$ suppressor cells will specifically repress activation of infected $CD4^+$ T cells. Other $CD4^+$ cell types, such as monocytes and dendritic cells, may differ in the level or type of basal and activating factors they express, leading to variation in the expression of viral genes. Alternatively, some cells may possess proteins that block HIV transcription by hypermethylation of the LTR after integration.

The viral regulatory proteins Tat and Rev may be inhibited by cellular or viral factors. These viral products are present in some in vivo models of latency, but at levels limited by unknown factors. Studies of the T and monocytic cell line models of latency, ACH2 and U1, respectively, reveal that basal processive transcription can occur during latency with spliced transcripts coding the regulatory proteins predominating. Both *tat-* and *rev*-mutant virus will give a similar Northern blot pattern in vivo, leading to the suggestion that in some cases latency is sustained by the maintenance of a subcritical level of viral regulatory proteins. The switch to productive infection would then occur after a threshold level of Tat or Rev has been reached.

Although the role Tat may play in the establishment, maintenance, and termination of latency is speculative, it is clear that controlling Tat levels would efficiently contribute to a latent state. If Tat protein is produced during latency as described in the cell line models, some form of Tat inactivation must take place. In this context it is striking that basal nonprocessive transcription leads to the cytoplasmic accumulation of TAR RNA. It has been shown that high level expression of TAR-containing constructs (TAR decoys) will almost completely suppress Tat transactivation and HIV replication, probably by sequestering both Tat and cellular TAR-binding proteins. Depending on the ratio of initiation to elongation in an individual cell, there might be sufficient endogenous TAR decoys present to sequester the few Tat proteins also produced and effect a prolonged state of latency. A threshold level of free Tat may be required before an increase in viral transcription becomes irreversible, even following cellular activation and the synthesis of increased levels of Tat. In such a scenario, it is the equilibrium established between the production of Tat and the production of its inhibitors that determines whether a cell will be latent, persistent, or actively productive of virus.

## 4 CONCLUSION

The study of HIV has revealed that nonprocessive transcription is a powerful regulatory mechanism when coupled to an RNA-targeted transactivator. The functional role of nonprocessive transcription in human genes is unknown, although premature termination of transcription has been described for the human c-*myc,* c-*fos,* and c-*myb* genes. *Drosophilia* heat shock gene synthesizes short (25-nucleotide) transcripts until heat induction, when full-length transcripts predominate. Nonprocessive transcription allows for gene silence in the context of continuous transcriptional initiation, creating abundant targets for RNA-directed transactivation and rapid response to activation. That activation of the HIV LTR is triggered by activation of the host cell ensures that the immune system will eventually be unable to mount a reaction to any organism, including HIV itself. However, understanding the regulatory mechanisms behind HIV transcription may allow us to break this linkage and influence the course of the disease. Tat antagonists, transdominant or inhibitory Tats, and TAR decoys hold promise as future antiviral agents.

*See also* AIDS HIV Enzymes, Three-Dimensional Structure of; Viruses, DNA Packaging of.

### *Bibliography*

Cullen, B. R. (1990) The HIV-1 Tat protein: An RNA sequence-specific processivity factor? *Cell,* 63:655–657.

———, (1991) Human immunodeficiency virus as a prototypic complex retrovirus. *J. Virol.* 65:1053–1056.

Gaynor, R. (1992) Cellular transcription factors involved in the regulation of HIV-1 gene expression. *AIDS,* 6:347–363.

Jones, K. A. (1989) HIV trans-activation and transcription control mechanisms. *New Biol.* 1:127–135.

Haseltine, W. A. (1989) Development of antiviral drugs for the treatment of AIDS: Strategies and prospects. *J. Acquired Immune Defic. Syndrome,* 2:311–334.

Peterlin, B. M., Adams, M., Alonso, A., Bauer, A., Ghosh, S., Hu, X., and Huo, Y. (1993) Tat transactivator. In *The Molecular Biology of Human Retroviruses,* B. Cullen, Ed., pp. 75–100. Oxford University Press, Oxford.

Proudfoot, N. J. (1989) How RNA polymerase II terminates transcription in higher eukaryotes. *Trends Biochem. Sci.* 14:105–110.

Sharp, P. A., and Marciniak, R. A. (1989) HIV TAR: An RNA enhancer? *Cell,* 59:229–230.

# AIDS HIV Enzymes, Three-Dimensional Structure of

*Zuzana Hostomska and Zdenek Hostomsky*

### *Key Words*

**Domain** Polypeptide chains fold into globular units called domains, which may be joined by flexible segments. Domains can be further divided into subdomains.

**Human Immunodeficiency Virus (HIV)** Retrovirus associated with acquired immune deficiency syndrome (AIDS). Two types of human immunodeficiency virus are designated HIV-1 and HIV-2.

***pol* Gene** Large open reading frame located near the center of the viral genome. All essential retroviral enzymes—protease, reverse transcriptase, ribonuclease H, and integrase—are encoded by the *pol* gene.

**Retrovirus** Class of viruses containing a (+) RNA genome inside an icosahedral shell. The virion particles of all retroviruses contain the enzyme reverse transcriptase, required for the synthesis of a DNA copy of the viral RNA genome.

**Subunit** Each polypeptide chain in a protein is called subunit.

**X-ray Crystallography** Technique that can reveal the precise three-dimensional arrangement of most of the atoms in a protein molecule based on the analysis of the X-ray diffraction of a single protein crystal.

**α-Helix and β-Sheet** Periodic structural motifs found in proteins. In contrast to the tightly coiled main chain of an α-helix, which forms a rodlike structure, the chain in a β-sheet is almost fully extended.

---

X-ray crystallography has provided insights into the structures of HIV enzymes, which recently were elucidated at high resolution. These structures promise to provide a basis for the rational design of new antiviral agents.

## 1 INTRODUCTION

Since its discovery as the causative agent of AIDS, human immunodeficiency virus (HIV) has been the focus of intensive worldwide scientific attention. The need to identify agents to treat the disease has stimulated research into the structure and function of HIV genes and their products. This entry summarizes current knowledge of the atomic structures of enzymes encoded by human immunodeficiency virus. As a result of the extensive efforts currently devoted to structure determination of HIV components, three-dimensional structures of HIV protease, ribonuclease H, and reverse transcriptase have been elucidated by means of X-ray crystallography, and it is likely that structures of other major components of HIV will be available within several years. Knowledge of retroviral structures will allow additional novel insights and strategies for antiviral intervention.

## 2 CRYSTAL STRUCTURES OF HIV ENZYMES

### 2.1 HIV-1 Protease

In the course of viral replication, retroviral structural proteins and enzymes are initially synthesized as polyproteins, which undergo enzymatic cleavage to generate functional proteins. Following the autocatalytic release from the *gag-pol* polyprotein precursor, virally encoded protease is responsible for the further processing of the structural proteins of the viral capsid as well as of the reverse transcriptase–RNase H complex and integrase. The HIV protease (PR) has been shown to belong to the class of aspartyl proteases on the basis of a variety of criteria, including distant sequence homology with cellular proteases and inhibition studies, and by mutational analysis of the active site residue. In recent years there has been an explosion of X-ray structures from numerous laboratories aimed at the characterization of PR and PR complexed with inhibitors on atomic level. With only 99 amino acid residues, HIV-1 PR is the smallest known retroviral protease, much smaller than the related mammalian aspartyl proteases. Cellular pepsin-like enzymes typically consist of a single polypeptide chain with homologous N- and C-terminal domains that may have evolved from a gene duplication event. The retroviral enzymes are dimeric, composed of two identical polypeptide chains, so that the resulting overall arrangement is reminiscent of that found in mammalian aspartyl proteases. The tertiary structure of PR contains primarily β-sheet, turn, and extended polypeptide structural elements. The chain adopts a fold in which the N- and C-terminal strands are organized together in a four-stranded β-sheet, which forms the dimer interface. A helix precedes the C-terminal strand. PR is unusual in that

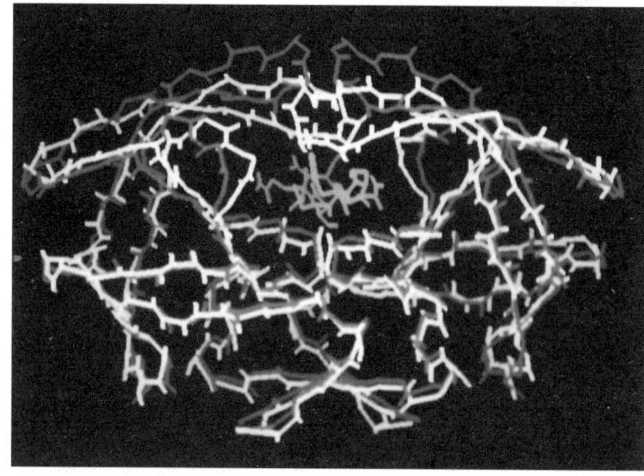

**Figure 1.** The α-carbon backbone of the homodimer of HIV-1 protease. Apoenzyme is shown in blue. The inhibited enzyme is in yellow complexed with an inhibitor in red. (Courtesy of Dr. K. Appelt.) (See color plate 2.)

it is a dimer made up of two identical subunits but with only one active site. Side chains of the conserved Asp-Thr-Gly tripeptide form the active site at the dimer interface. The active site is located at the base of a cleft in which the peptide to be processed is positioned prior to proteolytic cleavage. Each monomer contributes one of the two aspartyl residues within the Asp-Thr-Gly sequences. The comparison of the free PR with PR complexed with an inhibitor (Figure 1; see also color plate 2) shows that large conformational changes occur upon ligand binding. PR contains two identical "flap" regions, which are part the substrate/inhibitor binding sites. These flaps move as much as 7 Å upon binding of an inhibitor. The movement results in the rearrangement of β-strands in the flap region. The large changes probably reflect the need to recognize the wide range of sequences processed by the HIV-1 PR. Analysis of details of the movements in the structure upon inhibitor binding is essential for design of better PR inhibitors.

The crystal structures of PR–inhibitor complexes are of particular interest for drug design. More than 100 structures of PR with different ligands have been elucidated. This unprecedented effort led to the detailed structural characterization of PR complexed with analogues mimicking the geometry of the tetrahedral intermediate in peptide hydrolysis; in such analogues, the scissile bond of the substrate is replaced with stable isosteres (e.g., statine, hydroxyethylene, hydroxyethylamine, reduced amide). These inhibitors bind in extended conformations. A network of hydrogen bonds and van der Waals contacts is formed between the enzyme and functional groups of an inhibitor. A typical feature of these complexes is a tightly bound water molecule, which is hydrogen-bonded between two flaps and the inhibitor. The homodimeric nature of PR was explored for the design of symmetric compounds that inhibit PR at the nanomolar concentrations. The PR structure was also used for structure-based discovery of novel nonpeptidic inhibitors. Searching of chemical database for steric complementarity with the PR active site led to identification of haloperidol as a weak inhibitor. While haloperidol may not be useful for treatment of HIV infection, it validates the approach of using structural databases as sources of lead compounds for drug design.

### 2.2 HIV-1 Reverse Transcriptase

Reverse transcriptase (RT) is an enzyme responsible for copying the single-stranded RNA viral genome into a double-stranded proviral DNA form for integration into the host genome. This process involves three distinct enzymatic activities of RT: copying of the RNA template to generate complementary minus-strand DNA; selective hydrolytic degradation of the RNA portion of the intermediate DNA/RNA hybrid; and copying of minus-strand DNA to complete DNA formation. The overall process is complex and involves a series of template–primer strand transfer reactions. RT protein is a heterodimer composed of two subunits of 66 and 51 kD a (p66/p51). The p66 subunit has a N-terminal polymerase and a C-terminal RNase H domain, while the p51 subunit has only the polymerase domain, since the RNase H region has been proteolytically removed. The crystal structure of RT complexed with Nevirapine, a drug with potential for treatment of AIDS, was elucidated at 3.5 Å resolution. A highly asymmetric dimer was revealed. The two subunits in the heterodimer (p66 and p51) are not related by a simple rotation axis. The most astonishing trait in the RT structure consists of the very different folding patterns adopted by its two polymerase domains, even though they have the same amino acid sequences. The polymerase domain of p66 consist of four subdomains. Its anatomical resemblance to a right hand led to naming three subdomains as fingers, palm, and thumb. The fourth subdomain lies between the polymerase and the RNase H domain, and it is called the connection subdomain. The tertiary structures of the four corresponding subdomains in p51 are largely the same as in p66, but dramatic differences are seen in both subunits in relative positions of subdomains and interdomain packing contacts.

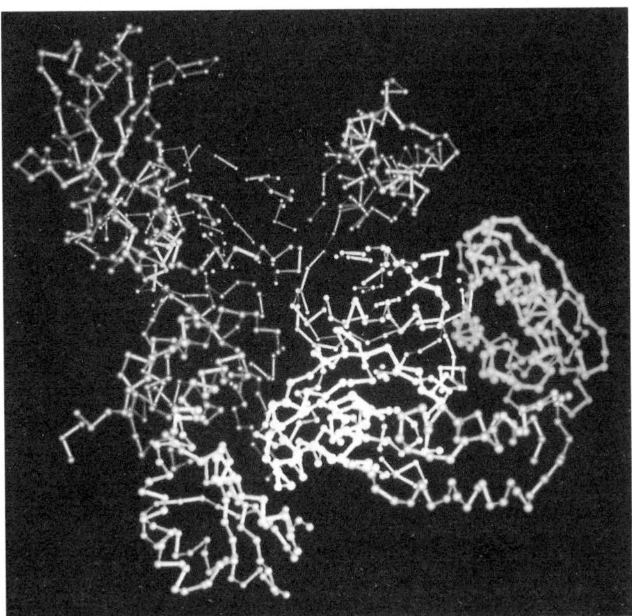

**Figure 2.** The α-carbon backbone of the heterodimer of HIV-1 reverse transcriptase. Polymerase domains of p66 and p51 are folded differently into four separate subdomains: in blue, terminal subdomain—finger; in purple, catalytic subdomain—palm; in yellow, connecting region between polymerase and RNase H domains; in green, flexible region between connecting and catalytic subdomains—thumb. The RNase H domain is orange. [Reprinted with permission from L. A. Kohlstaedt et al., *Science*, 256:1783–1790 (1992). Copyright 1992 by the American Association for the Advancement of Science.] (See color plate 3.)

In p66 the connection domain (see Figure 2; see also color plate 3) contacts the RNase H thumb domain and the connection of the p51 subunit, whereas in p51 it contacts all three of the other p51 subdomains as well as the connection of p66. The thumb consists of a four-helix bundle, and the fingers are formed by a mixed β-sheet/α-helix structure. The palm subdomain, containing the polymerase active site, and the connection subdomain are formed mostly from a five-stranded β sheet.

The determined structure of RT contains Nevirapine, a nonnucleoside, noncompetitive inhibitor of RT, which shows specificity different from that of 3′-azido-2′,3′-dideoxythymidine (AZT). The drug binds in a deep pocket that lies between the β-sheets of the palm and the base of the thumb subdomains. The side chains of Tyr 181 and Tyr 188 are in contact with the hydrophobic inhibitor. The mode of action of this drug is not yet known. Unfortunately, as in the case of AZT, frequent genomic mutations affecting these two residues decrease the affinity of RT for Nevirapine and reduce the effectiveness of the drug.

Independently, the crystal structure of RT complexed with double-stranded DNA template–primer, and a noninhibitory monoclonal antibody Fab fragment has been determined at 3.0 Å. The structure provides detailed information on the interactions of the 19/18 base duplex with p66/p51 RT. This complex shows RT in a catalytically relevant DNA-dependent DNA polymerization mode. The contacts between RT and DNA involve the sugar phosphate backbone of the duplex and amino acid residues in the palm, thumb, and finger subdomains of p66. Comparison of RT structures with and without the DNA duplex shows a significant movement of the p66 thumb subdomain from the template–primer binding groove upon nucleic acid binding. The conserved residues in the palm region together with those in the p66 thumb position the DNA correctly to the active site.

The relative locations of polymerase and RNase H active sites with respect to the template–primer were determined. The distance between the active sites is 17 or 18 nucleotides. The primer 3′-hydroxyl is close to the polymerase active site, in the vicinity of catalytic residues Asp 110, Asp 185, and Asp 186. The template–primer duplex is in the A-form DNA near the polymerase active site. A significant bend separates A-form DNA from a region of B-form DNA, which is close to the RNase H active site. The bend of 40–45° is distributed over four nucleotides. The structural information revealed in complexes of RT with the template–primer or Nevirapine will help in the design of new RT inhibitors and in the rationalization of emerging resistant forms of RT.

### 2.3 HIV-1 Ribonuclease H

The HIV ribonuclease H (RNase H) activity of RT is indispensable at several stages of reverse transcription: it degrades RNA template during synthesis of the plus-strand DNA, and it specifically removes primers by an endonucleolytic mechanism. The RNase H cleaves the RNA strand of a DNA/RNA hybrid in a divalent metal activated reaction by hydrolyzing a phosphodiester bond. The RNase H constitutes the carboxy-terminal domain of the p66 subunit of the p66/p51 heterodimer.

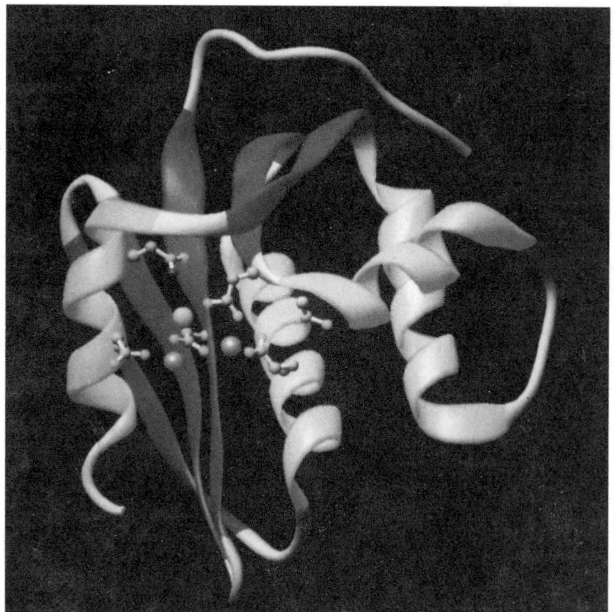

**Figure 3.** Ribbon representation of the overall fold of the RNase H domain of HIV RT: β-strands are in blue and helices in yellow. Manganese binding sites (silver balls) are shown with respect to the seven invariant residues, which cluster near the catalytic site. Proposed phosphate site is shown as a yellow ball. (Courtesy of Dr. D. Matthews.) (See color plate 4.)

The crystal structure of the isolated RNase H domain of RT has been determined at a resolution of 2.4 Å. The molecule contains a mixed five-stranded β-sheet flanked on one side by a single helix and three helices on the other side (Figure 3; see also color plate 4). The structure of RNase H domain is remarkably similar to the structure of active bacterial RNase H from *Escherichia coli.* Although the level of amino acid sequence similarity is relatively low, key amino acid residues are conserved, and folds of the two enzymes are closely related. The major difference between the enzymes is the connection between the second and fourth helices, which forms helical loop protrusion of 20 residues in the *E. coli* enzyme but only a 5-residue loop in HIV RNase H. This region is rich in basic amino acids and could be involved in binding the DNA/RNA hybrid substrate. It seems that lack of productive binding of the substrate to the RNase H domain could be the reason for lack of activity of the isolated domain. To reconstitute the activity in vitro, the RNase H domain must be combined with the p51 polymerase domain.

RNase H requires divalent metals for activity. Metal binding was studied by soaking RNase H crystals in $MnCl_2$. High resolution data allowed the determination of the positions of the side chains of individual amino acid residues and showed two tightly bound $Mn^{2+}$ ions in close proximity to four acidic residues (Asp 443, Glu 478, Asp 498, and Asp 549: Figure 3), which are conserved in all known bacterial and retroviral RNase H sequences. The geometry of the active site appears to be the same as in the *E. coli* enzyme. The presence of two metals suggests that RNase H works via two metal ion mechanism.

The structure offers insights concerning how precursor p66/p66 of RT might be processed by HIV protease. To generate the p66/p51 heterodimer, proteolysis must occur in only one of two p66 subunits. An unexpected finding was that the site of processing by HIV PR between the polymerase and the RNase H domains is completely inaccessible in the structure of RNase H. This suggests that one of the RNase H domains in the p66/p66 precursor is partially unfolded. A consequence of this unusual processing strategy is the release of the C-terminal fragment, which probably is further degraded because it has lost residues needed to form the central β-sheet.

## 3 PERSPECTIVES

How will information of detailed three-dimensional structures help discovery of new drugs? The availability of the crystallographic structures provides the opportunity for molecular events to be correlated with structural features of proteins, and this is a prerequisite for rational inhibitor design. Therefore, to facilitate the design of more potent inhibitors interacting at sites that must be conserved for the enzymatic activity, it is of importance to determine further structural details of the interaction of HIV enzymes with their substrates.

Traditional drug screening technologies will be combined with computer-aided drug design based on knowledge of the structure of the target enzyme. Useful lead compounds that are identified will be modified on the basis of structural considerations, and this process will lead to the identification of derivatives that are substantially better inhibitors.

*See also* AIDS, Inhibitor Complexes of HIV-1 Protease in; AIDS, Transcriptional Regulation of HIV in; Protein Analysis by X-Ray Crystallography.

### *Bibliography*

Creighton, T. E. (1993) *Proteins: Structures and Molecular Properties.* Freeman, New York.

De Clercq, E., Ed. (1990) *Design of Anti-AIDS Drugs.* Elsevier Science Publishing, Amsterdam.

Gallo, R. C., and Jay, G., Eds. (1991) *The Human Retroviruses.* Academic Press, San Diego, CA.

Gluskar, J. P., and Trueblood, K. N. (1985) *Crystal Structure Analysis.* Oxford University Press, New York.

Laver, W. G., and Air, G. M., Eds. (1990) *Use of X-Ray Crystallography in the Design of Antiviral Agents.* Academic Press, San Diego, CA.

Skalka, A. M., and Goff, S. P., Eds. (1993) *Reverse Transcriptase.* Cold Spring Harbor Laboratory Press, Plainview, NY.

# AIDS Therapeutics, Biochemistry of

*Christine Debouck and Martin Rosenberg*

### *Key Words*

**Polyprotein Precursor** Large precursor protein that is typically inactive and is proteolytically cleaved to yield several smaller, functional proteins.

**Protease** Generic term for an enzyme that cleaves a peptide bond within a polypeptide chain. Proteases are classified into four groups according to their catalytic mechanism: serine, metallo, thiol, and aspartyl proteases. For example, in aspartyl proteases, two aspartic acid residues located in the active site participate in catalysis.

**Retroviruses** Family of enveloped viruses whose genome consists of single-stranded positive-sense RNA that is replicated by a unique, virally encoded enzyme called reverse transcriptase. These viruses occur widely among vertebrates and are associated with various diseases.

**Reverse Transcriptase** Enzyme encoded by retroviruses that produces a DNA copy of a primed RNA molecule. The enzyme catalyzes three consecutive reactions: RNA-dependent DNA synthesis, ribonuclease digestion of the RNA in the resulting RNA-DNA duplex, and DNA-dependent DNA synthesis.

**Transactivation** Positive regulation of gene expression consisting of a significant increase in the level of transcription and/or translation. This upregulation typically results from the action of factors interacting with the core cellular transcriptional or translational machinery.

The etiologic agent of acquired immune deficiency syndrome (AIDS) is the human immunodeficiency virus type 1 (HIV-1), a member of the retrovirus family. HIV-1 infection in man results in serious impairment of the immune system that ultimately leads to death. The time between infection and appearance of disease symptoms is usually long and can vary from a few months to years (as many as 10 and even longer). Infected individuals may be chronic asymptomatic or symptomatic carriers of the virus, and they can transmit HIV-1 through sexual activity, by blood-to-blood contact, and transplacentally. There is no effective vaccine to prevent HIV-1 infection, and it remains unclear when or if such a vaccine will ever be available. Thus, to date, HIV-1 infection has been managed almost exclusively through therapeutic intervention.

## 1 MOLECULAR TARGET STRATEGIES

The basic tools of molecular and cellular biology have allowed an amazingly rapid and efficient dissection of the genetic and structural

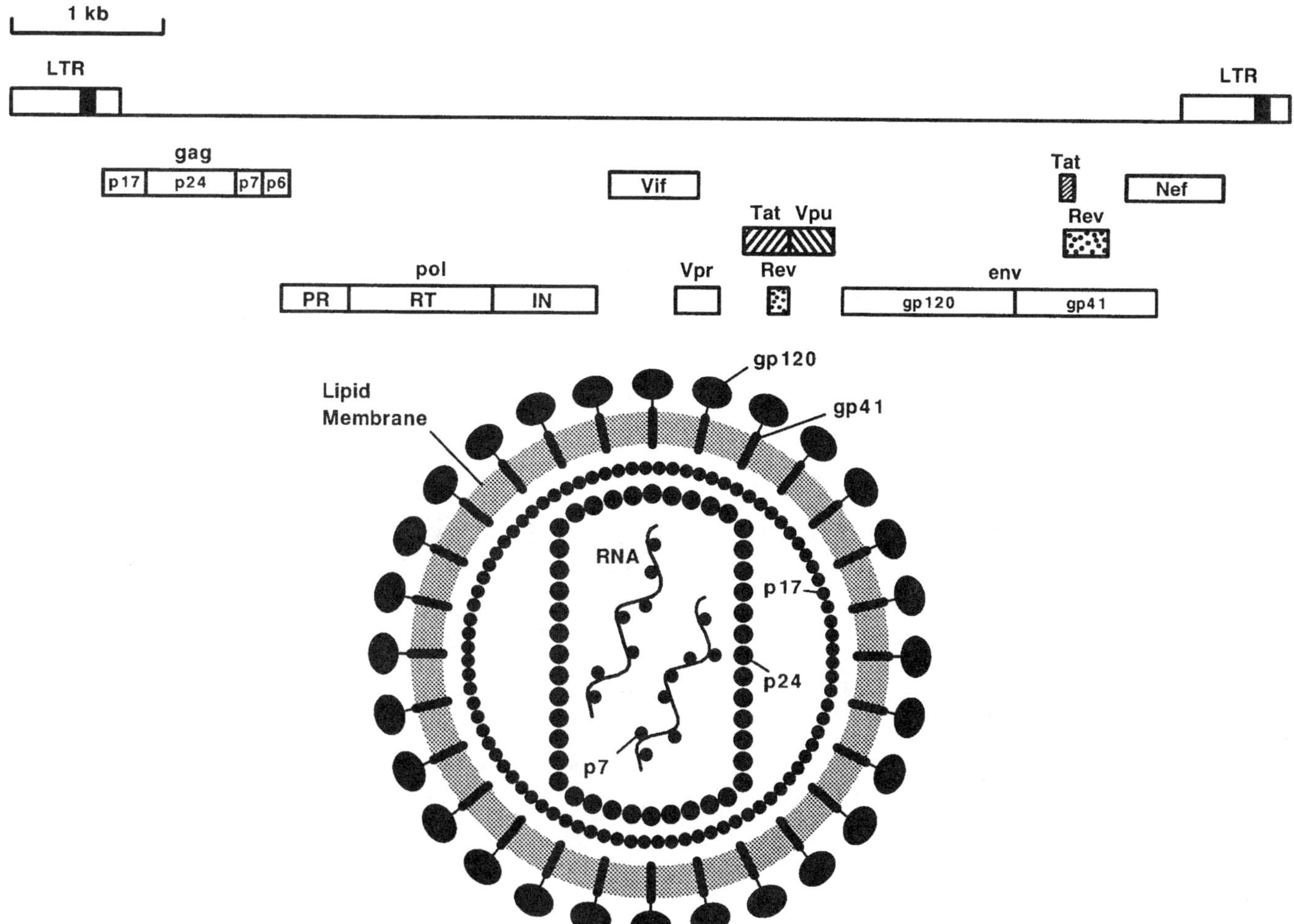


**Figure 1.** Genetic and structural organization of HIV-1. Like other retroviruses, HIV-1 is characterized by a relatively short (ca. 10,000 bases) single-stranded RNA genome that contains three major open reading frames, *gag, pol,* and *env*. *gag* comprises the structural proteins of the virion (p17, p24, and p7), *pol* encodes the viral enzymes, protease (PR), reverse transcriptase (RT), and integrase (IN), and *env* produces the envelope glycoproteins, gp120 and gp41. HIV-1 also encodes six small proteins: two essential regulatory proteins, Tat and Rev, and proteins with accessory, ill-defined function (Vif, Vpu, Vpr, Nef).

The genome of HIV-1 is shown schematically with the long terminal repeats (LTR) at each end and the position, size, and designation of all coding regions derived from the nucleotide sequence. The 1 kb scale represents 1000 nucleotides. The structure of the HIV-1 virion is also shown schematically.

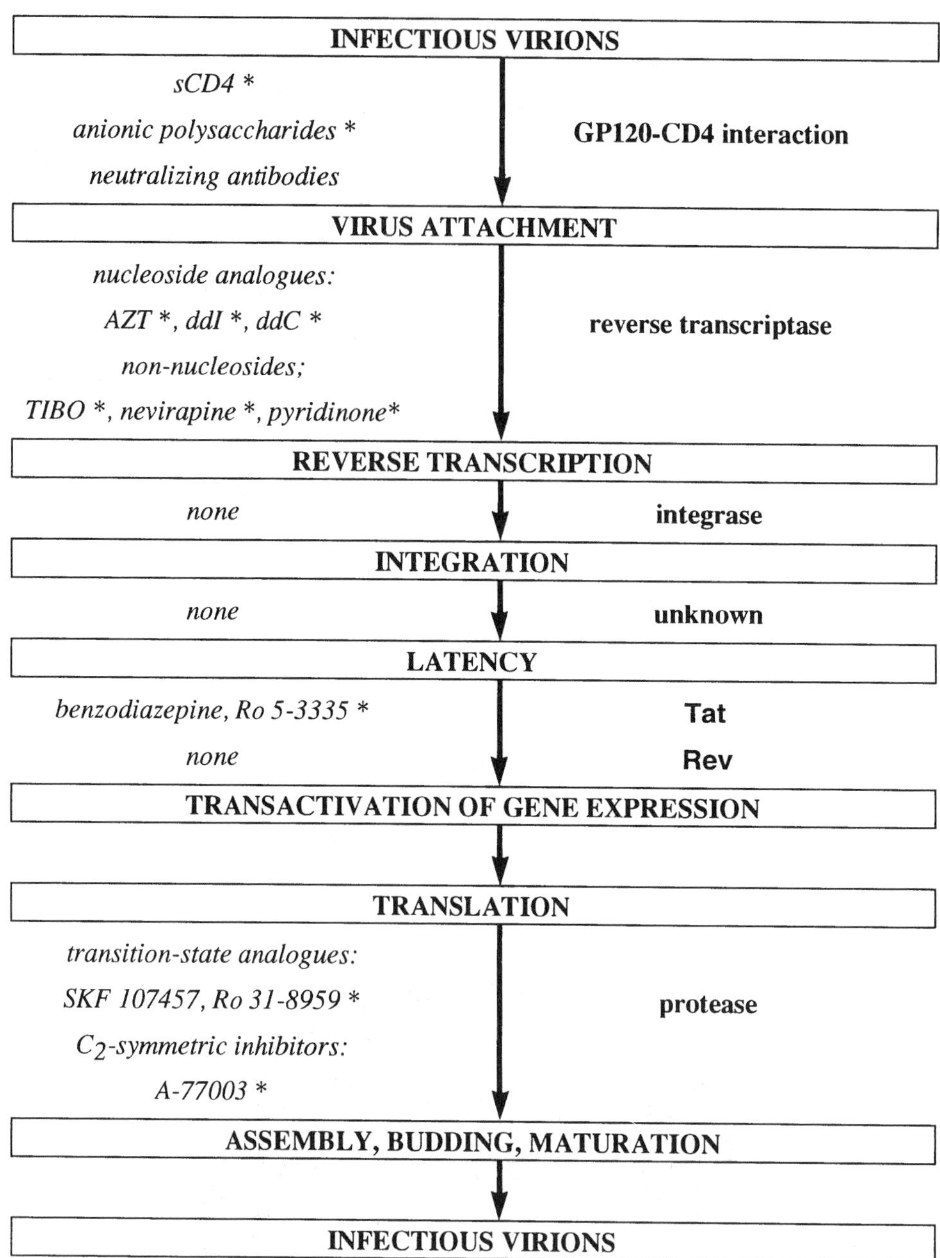


**Figure 2.** Sites for therapeutic intervention in the life cycle of HIV-1. The salient features of the HIV-1 life cycle are shown schematically: the viral proteins required for the completion of these steps appear in the right-hand column, and examples of inhibitors for several of these viral functions are shown on the left. An asterisk (*) indicates a compound that has been administered to man. The outer envelope protein, gp120, mediates the initial attachment of the virion to the target cell through specific interaction with the human CD4 cell surface receptor. The next critical step involves conversion of the viral RNA genome into double-stranded DNA, a unique reaction that is carried out by viral reverse transcriptase. The DNA then migrates to the nucleus, where it becomes integrated in the host cell chromosome by the action of viral integrase. This step is responsible for the long-term chronic nature of retroviral infection. Once integrated, the provirus can remain latent for extended periods of time. Upon emergence from latency by means of mechanisms sitll poorly understood, the expression of viral RNA and proteins is dramatically enhanced by the action of the two viral regulatory proteins, Tat and Rev. The translation of the *gag* and *pol* regions results in large polyprotein precursors that must be processed by virally encoded protease to yield mature, infectious virions and complete the cycle.

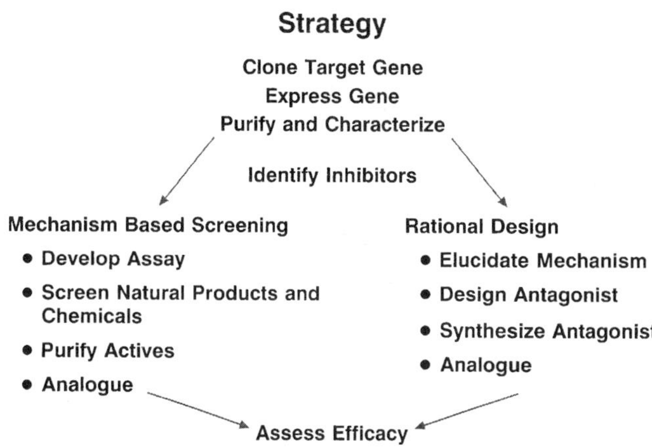


**Figure 3.** Drug discovery strategies: the mechanism-based screening and rational design approaches. Increasingly, the first step in either pathway has come to consist of the molecular cloning, expression, and characterization of the therapeutic target of interest.

organization of HIV-1 (Figure 1). This information has resulted in the development of molecular strategies for identifying inhibitors of specific viral functions critical to the HIV-1 life cycle. Several stages of this life cycle have been exploited as targets of therapeutic intervention (see Figure 2). For example, viral replication, gene expression, and maturation are dependent on the function of three virally encoded proteins. These are, respectively, reverse transcriptase, necessary for RNA-dependent DNA synthesis prior to integration of the viral genome into the host cell DNA; Tat, the transactivating effector of HIV-1 gene expression; and protease, responsible for the proteolytic maturation of the Gag-Pol polyprotein precursors. In addition, viral entry into human cells and cell-to-cell spread of the virus depend on the selective interaction between the viral envelope protein and the human CD4 receptor found on cells involved in host immune function. The function of each of these molecular targets absolutely depends on a specific viral gene product to allow viral growth and development.

Two distinct strategies are typically used to identify inhibitors of selected molecular targets (Figure 3). Common to both strategies is the initial use of recombinant DNA technologies to isolate and characterize the target gene and its protein product. These reagents and the information generated from their characterization are then used as the molecular building blocks to form the basis of the two drug discovery strategies.

In one strategy, a mechanism-based screening approach, the recombinant gene product is used to develop a relatively simple cell-free assay that is applied to high throughput screening of both natural products and synthetic chemical compounds. A novel ''lead'' activity may be identified that selectively inhibits the target function. Such lead structure then serves to guide the chemists in further synthetic work aimed at maximizing the potency of the lead and related compounds. Compounds showing sufficient activity in the cell-free assay are then tested for antiviral activity in cell-based assays.

The other strategy for identifying inhibitors relies on a rational design approach. In this case, the intention is to generate sufficient information on the mechanism of action of the target protein to permit the de novo design and synthesis of selective antagonists. The designed antagonist can be protein, peptide, or chemical in nature and is initially tested for selective inhibitory activity in the cell-free assay. Again, analogues are synthesized to enhance potency. These compounds are then tested for efficacy and potential cell toxicity in the same cell-based assays used to assess compounds identified by screening.

At this point, compounds identified by either approach typically undergo additional modifications in an attempt to improve such properties as pharmacokinetics, bioavailability, toxicity, and formulation characteristics. In the final analysis, the antiviral efficacy profile of the compound relative to its toxicity profile and its overall acceptability for administration to humans determines its potential for clinical testing.

## 2 INHIBITION OF VIRUS ATTACHMENT

HIV-1 binds to its primary target cells through a specific, high affinity interaction between the viral envelope glycoprotein and the CD4 receptor present on the surface of a subset of human lymphocytes and certain other human cells. A recombinant, soluble form of this receptor, called sCD4, was the first candidate antiviral agent to result from a rational design strategy. Soluble CD4 exhibited potent antiviral activity in vitro by competing with the cell surface receptor for binding of various laboratory-adapted strains of HIV-1. Unfortunately, it was found later that clinical isolates of HIV-1 required significantly higher concentrations of sCD4 for neutralization than did the laboratory strains, thereby possibly explaining the lack of efficacy of this therapeutic candidate in humans. Modified derivatives of sCD4, exhibiting either significantly better pharmacokinetic properties (longer serum half-life in man) or increased potency in cell culture experiments, have also been designed and clinical trials initiated in humans. To date, none of these compounds have progressed into efficacy trials, and some of them have exhibited significantly greater dose-limiting toxicities in man than did the original sCD4 candidate.

Other molecules that interfere with the viral envelope–CD4 interaction have been identified by screening approaches. For example, polyanionic compounds, such as dextran sulfate, were shown to behave as potent inhibitors of HIV-1 in vitro, but they have failed to exhibit efficacy in the clinic. Another example is *N*-butyldeoxynojirimycin, a compound that interferes with glycosylation of the envelope protein, a posttranslational modification that is essential for the ability of HIV-1 to infect cells. This compound has been shown to decrease HIV-1 infectivity in vitro and is currently undergoing clinical evaluation.

Neutralizing antibodies targeted to the HIV-1 envelope also represent an attractive rational approach to interfere with the initial virus-binding step. Efforts to identify and characterize neutralization epitopes on the viral envelope have proceeded with difficulty because antibodies against the principal neutralization epitope derived from one HIV-1 strain often show little if any cross-neutralization of other HIV-1 strains, presumably as a result of the extensive genetic variation exhibited by the virus. Recently, several antibodies that appear to be effective against a broader range of laboratory and clinical strains of HIV-1 have been described. It is unclear whether these will be useful clinically. If effective, such antibodies could be very useful in immediate postexposure treatment (e.g., for needle sticks). Moreover, the definition of broadly neutralizing antibodies and their epitopes could have tremendous impact on HIV-1 vaccine research efforts.

## HIV-1 reverse transcriptase inhibitors

*nucleoside analogues*

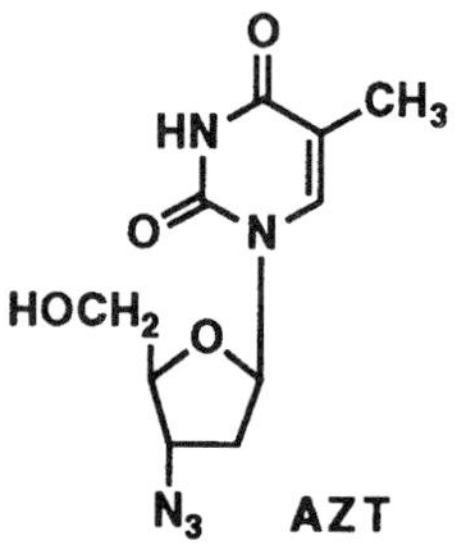

AZT 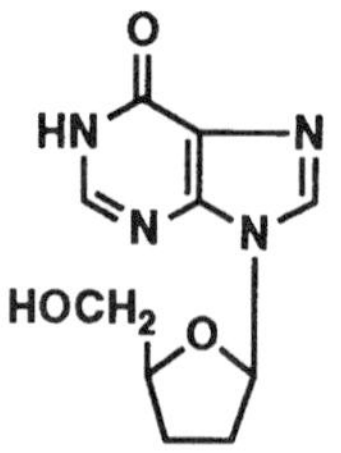 DDI 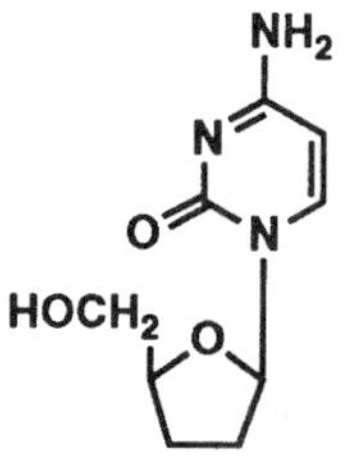 DDC

*nonnucleosides*

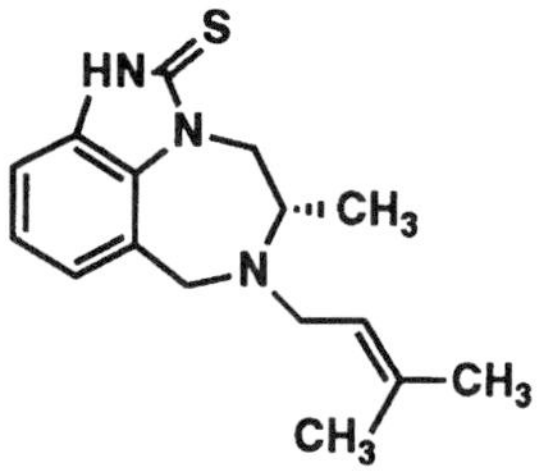

TIBO (R82150) 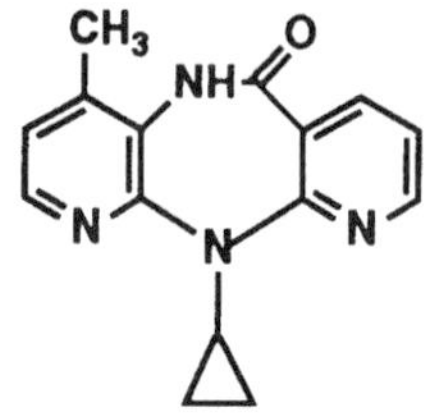 Nevirapine (BI-RG-587) 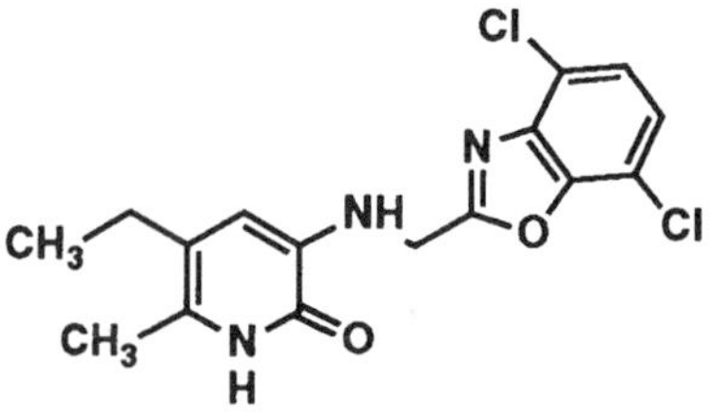 Pyridinone (L,697,661)

## HIV-1 Tat inhibitors

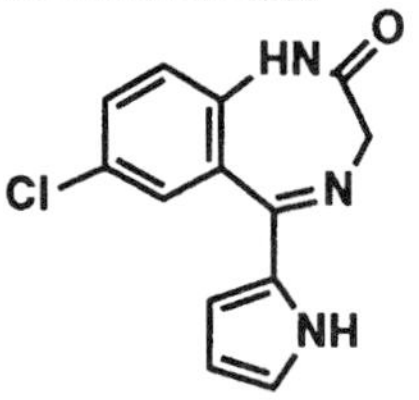

Ro 5-3335

## HIV-1 protease inhibitors

*transition-state analogues*

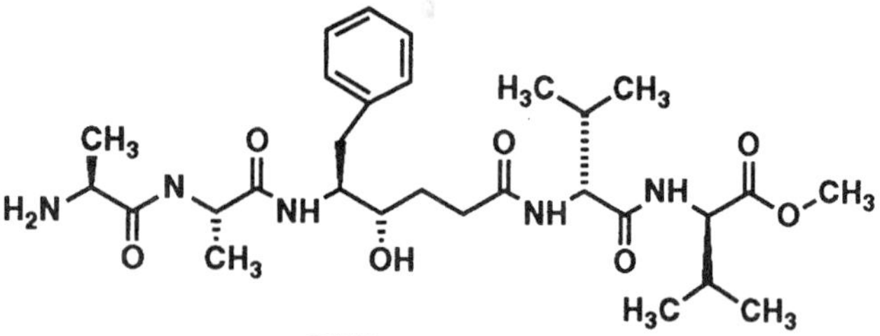

SKF 107457 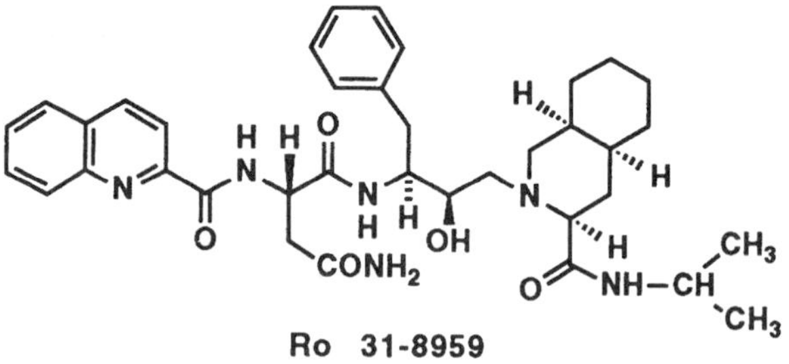 Ro 31-8959

*$C_2$-symmetric inhibitors*

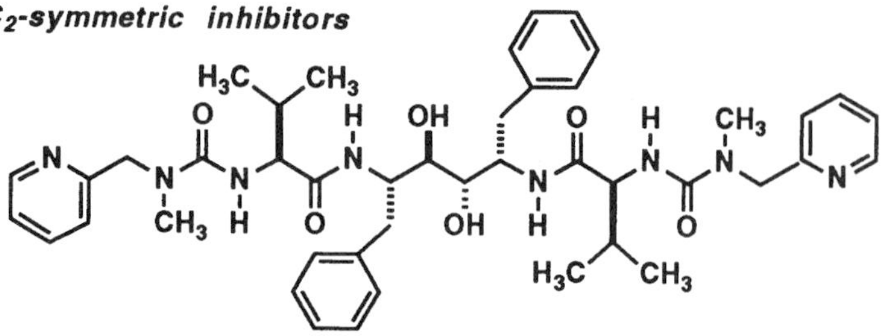

A-77003

Figure 4. Chemical structure of inhibitors of HIV-1 protein functions. The structures of selected potent inhibitors of two HIV-1 enzymes (reverse transcriptase and protease) and one HIV-1 regulatory protein (Tat) are shown with name and/or compound number.

## 3 INHIBITION OF HIV-1 REVERSE TRANSCRIPTASE

Reverse transcriptase is the viral replicative polymerase and performs unique enzymatic processes that cannot be carried out by host cellular polymerases: RNA-dependent DNA synthesis and ribonuclease digestion of the RNA in the resulting RNA:DNA hybrid. Reverse transcriptase is the target of inhibition by the first antiviral agents reported to have anti-HIV-1 activity in vitro. Indeed, a combined rational and screening approach led to the identification of several nucleoside analogues as inhibitors of HIV-1 replication in 1985, less than two years after the discovery of the virus. These compounds act by direct competitive inhibition of reverse transcriptase and also behave as chain terminators by blocking the elongation of nascent complementary DNA. Nucleoside analogues represent the first class of antiviral drugs to be tested in man against HIV-1 and, to date, the only drugs to be approved by the Food and Drug Administration (FDA).

More specifically, AZT (3′-azido-2′,3′-dideoxythymidine, zidovudine, Retrovir, Figure 4) was shown to suppress HIV-1 replication in patients, resulting in longer survival, improved quality of life, and even delayed progression to AIDS when administered to asymptomatic HIV-1-seropositive individuals. The in vitro and in vivo antiviral properties of AZT stimulated the design and synthesis of novel nucleoside analogues as well as the screening of existing nucleoside analogues, two of which have recently been approved by the FDA: ddI (2′,3′-dideoxyinosine) and ddC (2′,3′-dideoxycytidine)(Figure 4). Despite these highly promising results, nucleoside analogues have two clinical shortcomings. First, their imperfect selectivity for reverse transcriptase translates into toxic effects that often preclude their long-term use in patients (e.g., bone marrow suppression, anemia, neuropathy, pancreatitis). Second, HIV-1 variants resistant to nucleoside analogues have been isolated from patients within 6 to 12 months of treatment and could result in the resurgence of clinical symptoms.

A totally different class of nonnucleoside reverse transcriptase inhibitors was identified more recently via a screening-based strategy. Several groups extensively screened existing synthetic compound banks for their ability to inhibit HIV-1 replication in vitro or to directly inhibit reverse transcriptase enzymatic activity. The inhibitors identified by this approach were very potent against HIV-1 in vitro and exhibited much better selectivity for reverse transcriptase than nucleoside analogues, although their inhibitory activity did not extend to reverse transcriptases from other retroviruses such as HIV-2. These inhibitors are derivatives of benzodiazepines (R82150), dipyridodiazepinones (BI-RG-587), pyridinones (L,697-661) (Figure 4) and bisheteroarylpiperazines. Although these compounds differ significantly in chemical structure, they are believed to inhibit reverse transcriptase by binding to the same allosteric site on the enzyme. The clinical evaluation of these compounds is ongoing, although it has already met one potential hurdle caused by the rapid emergence of HIV-1 variants highly resistant to the compounds.

## 4 INHIBITION OF HIV-1 TAT TRANSACTIVATOR

Tat is a small regulatory protein containing 86 amino acids; its function is absolutely required for the replication and propagation of HIV-1. Tat is responsible for dramatically increasing the amount of viral RNA and all viral proteins by acting at the transcriptional and/or posttranscriptional level, a phenomenon referred to as "transactivation." The precise mechanism of action of Tat remains ill-defined, making a rational design approach to inhibition of its function impossible at this time. However, a variety of cell-based assays have been developed to mimic Tat-mediated transactivation, and these assays have been applied to the screening of synthetic compounds from chemical banks. One inhibitor identified by this approach, the benzodiazepine Ro 5-3335 (Figure 4), has been shown to exhibit potent antiviral activity in vitro. Ro 5-3335 is currently being tested in phase I/II dose-escalating pharmacokinetic and safety studies in symptomatic HIV-1 seropositives.

## 5 INHIBITION OF HIV-1 PROTEASE

HIV-1 produces a small, dimeric aspartyl protease that specifically cleaves the polyprotein precursors encoding the structural proteins (Gag) and enzymes (Pol) of the virus (Figure 1). This proteolytic activity is absolutely required for the production of mature, infectious virions and is therefore an attractive target for therapeutic intervention. The identification of HIV-1 protease inhibitors has been primarily driven by rational drug design, made possible by important contributions resulting from a unique collaboration between molecular and structural biology, medicinal chemistry, and biochemistry. First, HIV-1 protease produced in its authentic, mature form using recombinant DNA technology was extensively characterized enzymologically. High throughput biochemical assays were then developed in which purified recombinant HIV-1 protease cleaved oligopeptide substrates. Armed with the knowledge of competent peptide substrates and of the classification of HIV-1 protease as an aspartyl protease, medicinal chemists undertook to design and synthesize inhibitors of this critical viral enzyme using two different rational design approaches.

Most laboratories applied the concept of transition state analogue to the design of HIV-1 protease inhibitors. This was done by preparing minimal peptide substrates in which the peptide bond normally cleaved by HIV-1 protease was replaced by a nonhydrolyzable surrogate, mimicking the tetrahedral transition state motif known to be utilized by this enzyme class. Inhibitors of this type were shown to block purified HIV-1 protease in vitro very effectively. The transition state surrogates that have been used in HIV-1 protease inhibitors include, in increasing order of potency against the enzyme, aminomethylene isosteres, statine analogues, phosphinic acid isosteres, $\alpha,\alpha$-difluoroketones, dihydroxyethylene isosteres, hydroxyethylene isosteres (e.g., SKF 107457, Figure 4), and hydroxyethylamine isosteres (e.g., Ro 31-8959, Figure 4).

The synthesis of another class of rationally designed inhibitors of HIV-1 protease was prompted by the elucidation of the three-dimensional structure of the enzyme by X-ray diffraction crystallography. The twofold rotational symmetry of HIV-1 protease and, more specifically, of its active site, inspired some medicinal chemists to design protease inhibitors that possess a $C_2$ axis of symmetry positioned on or near the carbonyl carbon of the scissile amide bond and superimposable on the axis of symmetry of the enzyme. Inhibitors of this class have also exhibited highly potent inhibition of HIV-1 protease in vitro (e.g., A-77003, Figure 4). It is important that the three-dimensional structures of protease–inhibitor complexes continue to provide guidance for the design of new, more potent inhibitors of either class of inhibitors (peptide-based or symmetric).

Protease inhibitors of both classes were shown to effectively block the protease within HIV-1 infected cells and to block virus propagation in vitro in both human lymphocytes and monocyte-macrophages, two cell types known to be relevant to infection in humans. One of the most attractive antiviral properties of these protease inhibitors resides in their ability to block HIV-1 infection not only in acutely infected cells, but also, and perhaps more importantly, in chronically infected cells, a state that more appropriately mimics the situation in infected humans. This may be a major advantage over compounds like AZT that are incapable of blocking HIV-1 in chronically infected cells. The first HIV-1 protease inhibitors that were identified displayed rather short serum half-life, rapid biliary clearance, and low oral availability. These shortcomings appear to have been overcome, and certain protease inhibitors are currently in late stage preclinical or early clinical evaluation.

## 6 PERSPECTIVES

The demonstrated albeit imperfect clinical efficacy of nucleoside analogues clearly indicates that the search for new, potent, specific inhibitors of HIV-1 in vitro is a valid approach to therapeutic intervention. On the one hand, this search has been facilitated by the genetic, functional, and structural characterization of several critical HIV-1 enzymes and regulatory proteins. On the other hand, the development of promising compounds has been greatly hampered by the lack of appropriate animal models for HIV-1 infection and disease. Thus, compounds showing promise in vitro are taken directly to man for efficacy studies without the advantage of an intermediate animal model. This experimental circumstance accounts in part for the large number of drug candidate failures in the clinic, since preclinical testing in animals can address only potential toxicity.

Since the in vivo efficacy of nucleoside analogues has been relatively limited and hindered by both their associated toxicities and the rapid emergence of drug-resistant HIV-1 variants, the search for additional potent and specific inhibitors of HIV-1 replication must continue. Novel compounds with activity against the viral targets discussed here must be sought using rational design and screening strategies. Antiviral approaches based on antitemplate strategies (antisense oligonucleotides, ribozymes) are also being pursued to interfere with HIV-1 replication. Although these approaches have resulted in antiviral effects in vitro, significant problems of delivery, pharmacokinetics, and cost-effective production will have to be overcome before such agents can achieve their full therapeutic potential in vivo.

Other HIV-1 gene functions are also beginning to receive more attention as targets for therapeutic intervention. These include the viral integrase, which is responsible for integration of the proviral DNA into the host cell genome; the regulatory protein Rev, which is a posttranscriptional transactivation factor required for high level expression of HIV-1 Gag, Pol, and Env proteins; and ribonuclease H, which carries out the essential degradation of the viral RNA template during its conversion into DNA. Assays are currently being developed for these viral functions and their mechanism of action is being elucidated, thereby allowing these molecular targets to be utilized in the identification of new drug candidates.

It is hoped that many of the problems of toxicity and drug resistance associated with current and future therapies may be circumvented or diminished by the administration of drugs in combination. Combination therapies using either concomitant or alternating drug regimens are most likely to evolve as the means to effectively block a virus such as HIV-1, which infects multiple cell types latently, chronically, or acutely and undergoes rapid adaptive changes as a result of its unusual ability to mutate.

Last, it is important to point out that clinical investigation of new antiviral drugs has focused largely on demonstrating drug efficacy in later stages of the disease, after the chronic infection has existed for years in an asymptomatic state. This rationale has been driven by the necessity of defining clinically relevant end points for measuring drug efficacy, as well as maximizing the perceived benefit-to-risk ratio associated with experimental drug treatment in diseased versus asymptomatic patients. By its very nature, this rationale minimizes the chances for achieving effective therapy, since the viral infection has persisted and damage has accumulated for many years in these patients. Hence, a key objective for any HIV-1 therapeutic agent will be its use as early as possible in the infection process, to maximize the potential for slowing viral progression and for delaying the shift from the asymptomatic to the diseased state. Achieving this goal will require the identification of appropriate surrogate markers for measuring viral progression in the asymptomatic state, as well as developing the associated technologies needed for reproducibly monitoring these surrogate markers. These considerations indicate that safe, therapeutic agents with effective anti-HIV activities continue to be urgently needed for the potential treatment of AIDS, and, most importantly, early viral infection.

*See also* AIDS, Inhibitor Complexes of HIV-1 Protease in; AIDS, Transcriptional Regulation of HIV in; AIDS HIV Enzymes, Three-Dimensional Structure of; Molecular Genetic Medicine.

### *Bibliography*

Debouck, C. (1992) The HIV-1 protease as a therapeutic target for AIDS. *AIDS Res. Hum. Retroviruses,* 8:153–164.

De Clercq, E. (1992) HIV inhibitors targeted at the reverse transcriptase. *AIDS Res. Hum. Retroviruses,* 8:119–134.

Haseltine, W. A. (1989) Development of antiviral drugs for the treatment of AIDS: Strategies and prospects. *J. AIDS,* 2:311–334.

Mitsuya, H., Yarchoan, R., Kageyama, S., and Broder, S. (1991) Targeted therapy of human immunodeficiency virus-related disease. *FASEB J.* 5:2369–2381.

Mohan, P. (1992) Anti-AIDS drug development: Challenges and strategies. *Pharm. Res.* 9:703–714.

Rosen, C. A. (1992) HIV regulatory proteins: Potential targets for therapeutic intervention. *AIDS Res. Hum. Retroviruses,* 8:175–181.

Schinazi, R. F., Mead, J. R., and Feorino, P. M. (1992) Insights into HIV chemotherapy. *AIDS Res. Hum. Retroviruses,* 8:963–990.

Sweet, R. W., Truneh, A., and Hendrickson, W. A. (1991) CD4: Its structure, role in immune function and AIDS pathogenesis, and potential as a pharmacological target. *Curr. Opinion Biotech.* 2:622–633.

## Algae: *see* Chlamydomonas.

# Alzheimer's Disease

*Frank Ashall and Alison M. Goate*

### *Key Words*

**Autosomal Dominant Trait** A phenotype that is expressed in a heterozygote and maps to a chromosome other than the sex chromosomes.

**Differential RNA Splicing** Conversion of a single hnRNA molecule into more than one possible mRNA product, each product encoding a different protein.

**DNA Polymorphism** A DNA sequence that occurs in the population in two or more variants, each with a significant frequency (> 1%).

**Posttranslational Event** Any modification of a protein, (chemical, physical, or altered levels) that occurs after the protein has been made in a cell.

---

Progress in understanding the molecular basis of Alzheimer's disease (AD) has been rapid in recent years, particularly as a result of identification of the chemical constituents of plaques and tangles in brains of AD patients and identification of genetic mutations that cause familial forms of the disease. The amyloid precursor protein (APP) has a key role in AD, since it gives rise to β-amyloid, the major constituent of plaques. Mutations in the gene encoding APP are known that predispose individuals to early onset AD. In addition, the APP gene is encoded by chromosome 21; people with Down's syndrome have three copies of chromosome 21 and also develop AD in their thirties. Knowledge of the biochemical basis of β-amyloid production from APP is critical for understanding Alzheimer's disease. APP is processed by proteases by at least two routes; one involves secretion of APP without β-amyloid formation, while the other pathway involves lysosomal proteases and may involve β-amyloid production. Cells also secrete a soluble form of β-amyloid, which may be important in plaque formation.

# 1 ALZHEIMER'S DISEASE

## 1.1 Pathology and Epidemiology

Alzheimer's disease, which was first described in 1907 by Alois Alzheimer in a 55-year-old patient, is the commonest cause of dementia and the fourth leading cause of death in the United States. The disease is confined mainly to the aged, and it is progressive. It affects 1% of people in developed nations and is likely to become

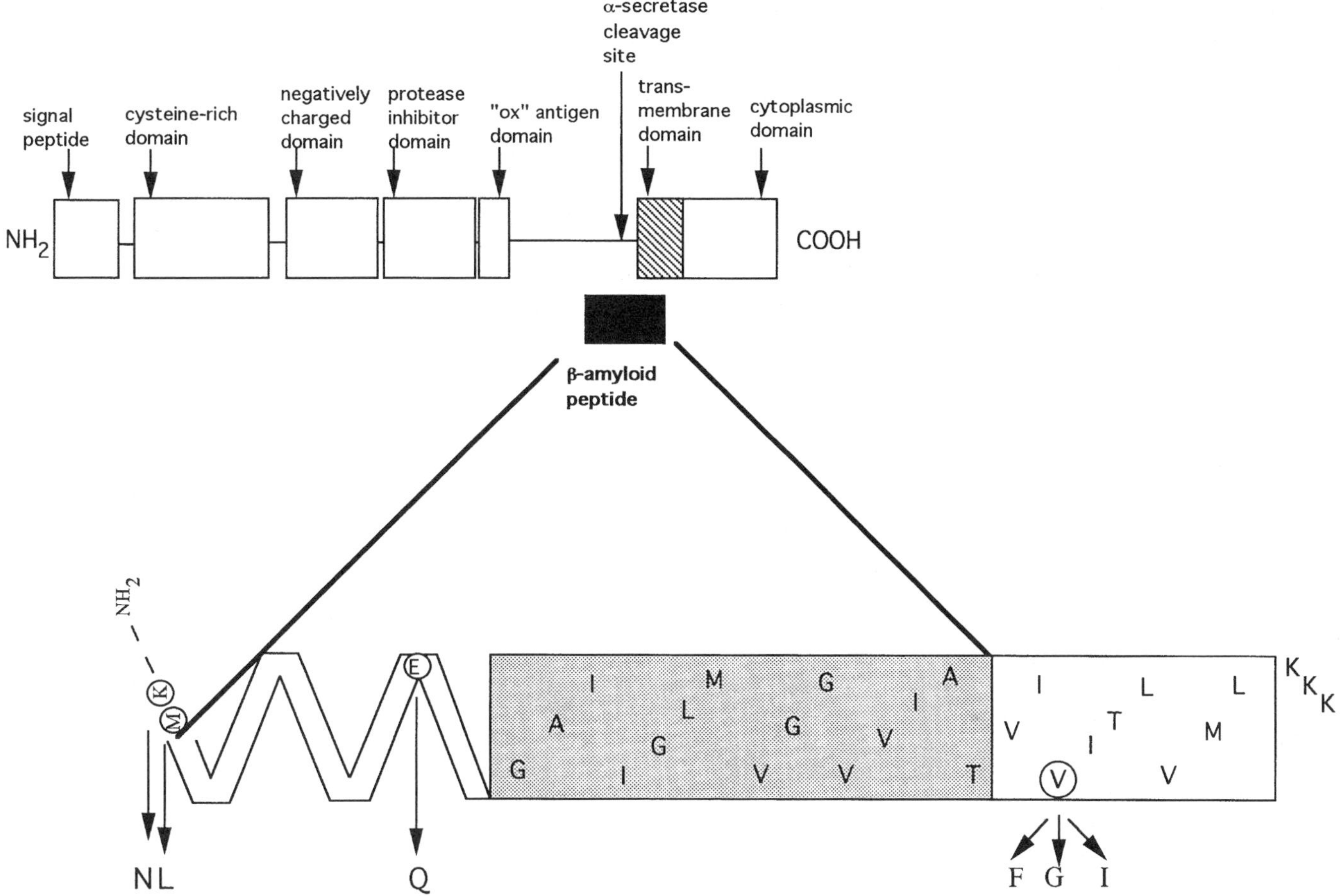


**Figure 1.** The amyloid precursor protein (APP) contains numerous domains, including a transmembrane region and a protease inhibitor domain. The β-amyloid peptide spans part of the membrane domain and part of the extracellular portion of APP. Amino acid changes caused by mutations in the APP gene can lead to familial forms of Alzheimer's disease. A double mutation (Lys-Met [K-M]) at the N-terminal region of the β-amyloid peptide region, and mutation of valine (V) to phenylalanine (F), glycine (G) or isoleucine (I) a few amino acids on the carboxyl side of the β-amyloid region predispose to familial AD. Another mutation, changing a glutamic acid (E) to a glutamine (Q) within the extracellular portion of the β-amyloid peptide region is associated with Dutch-type amyloidosis. **(HCHWA-D)**

a major problem in developing nations as the proportion of elderly persons increases in these populations. The incidence of AD rises with increasing age: 1 to 5% of those aged 65 years are affected, while 10 to 20% of those over the age of 80 years have AD. The first clinical manifestation is usually a loss of short-term memory, and this is followed, over the next 6 to 20 years, by progressive loss of memory, reasoning, judgment, and orientation. Death occurs, on average, about 10 years after onset of symptoms. Many epidemiological surveys have been carried out to identify risk factors. There is a clear correlation between AD and a positive family history (in about one-third of cases); and most, if not all, individuals with Down's syndrome develop AD by middle age. Severe head trauma has been identified as a risk factor in some but not all studies. Others have suggested links with dietary aluminum intake and thyroid disease, but these remain uncertain risk factors.

Alzheimer's disease is characterized neuropathologically by the presence of large numbers of extracellular neuritic plaques and intracellular neurofibrillary tangles within the brain cortex. In addition, there is massive neuronal cell loss, and amyloid deposits are found in the walls of blood vessels of the meninges, cerebral cortex, and hippocampus. While plaques and tangles do occur in normal aging brains, they are more numerous and more widely distributed in AD brains. The major protein component of the plaque both in the parenchyma and in the blood vessels is a 39 to 43 amino acid peptide called β-amyloid. Subsequent cloning of the cDNA encoding this peptide demonstrated that it was derived from a much larger precursor protein termed the amyloid precursor protein (APP). This protein has properties characteristic of a transmembrane protein (see Figure 1) and was found to be expressed in every tissue studied and not specifically in the brain.

### 1.2 Association with Other Diseases

β-Amyloid deposition has also been observed in association with several other pathological conditions. Individuals with Down's syndrome (trisomy 21) who survive to their thirties develop plaques, tangles, and other pathological features that are indistinguishable from AD. Temporal studies in Down's syndrome suggest that β-amyloid deposition is the earliest of the pathological markers to appear and that neurofibrillary tangles and neuronal cell loss occur decades after the first β-amyloid deposition. Dementia is observed in these individuals 10 to 20 years after tangle formation. Cerebrovascular deposition of β-amyloid is also observed in hereditary cerebral hemorrhage with amyloidosis, Dutch type (HCHWA-D). Individuals with this disorder do not dement but have frequent and fatal cerebral hemorrhages in early middle age. β-amyloid is found deposited in the blood vessels of the brain and as diffuse plaques (not neuritic plaques) in the brain parenchyma. Diffuse β-amyloid plaques (but not neuritic plaques) and neurofibrillary tangles are observed in another rare disorder, *dementia pugilistica,* which is thought to be caused by repeated severe head trauma and is observed in boxers and battered wives.

## 2 GENETICS OF ALZHEIMER'S DISEASE

### 2.1 Familial Alzheimer's Disease

The most striking evidence suggesting the importance of genetic factors in the etiology of AD is the existence of pedigrees in which the illness appears to segregate as a fully penetrant autosomal dominant trait. The disease in these families characteristically has an early age of onset (< 65 years) but does not appear atypical in other ways. In one population, the Volga Germans, there is a high incidence of familial AD. It appears likely that this is due to a "founder effect." About a thousand Germans settled along the Volga in two villages (Walter and Frank) during the eighteenth century. It is thought that one of those settlers had a genetic form of AD and that subsequent intermarriage in a restricted population led to an increase in the incidence of the disease among these Volga Germans. Most cases of AD occur in elderly people. Although many cases of AD in the elderly appear to be familial, this correlation may be due merely to the high frequency of the disease. AD is not a Mendelian trait in the elderly, and any genetic influence is likely to be the result of interaction of several genes with each other or between genes and the environment. Such a model is very similar to that proposed for heart disease and cancer, where rare inherited forms of the diseases occur at an earlier age than the more common forms that occur later on and for which there is no clear pattern of inheritance.

### 2.2 Mutations in the Gene Encoding the Amyloid Precursor Protein

The existence of large pedigrees with an inherited form of AD has enabled the use of a genetic linkage strategy to localize the disease genes. The association between AD and Down's syndrome led researchers to analyze the inheritance of genes on chromosome 21 in these families. Some but not all families showed linkage between AD and polymorphic DNA markers on chromosome 21; that is, the disease cosegregated with one allele of the polymorphic marker in all affected but no unaffected individuals, indicating that the gene causing the disease in these families was located close by.

Localization of the APP gene had shown that it mapped to the central region of the long arm of chromosome 21 (the same region as the genetic linkage to AD in these families). Linkage studies between a DNA polymorphism within the APP gene and AD families originally excluded the APP gene as the site of the defect. However, those early studies were based on the false premise that AD is genetically homogeneous. We now know that this is incorrect and that the APP gene is specifically mutated in some, but not all, familial forms of AD. Indeed, analysis of several families in which there was clear linkage to markers on chromosome 21 demonstrated no recombinants between the disease and the APP gene, suggesting that this gene could be the site of the genetic defect in these families.

APP is encoded by a large gene covering 400 kb of DNA divided into 18 exons. Since exon 17 encodes part of the β-amyloid region and is the site of the HCHWA-D mutation, it was the first exon to be sequenced in individuals from the chromosome 21 linked families. The first mutation to be reported in an individual with AD was a valine-to-isoleucine substitution at codon 717 of APP (APP717), within the transmembrane segment, two amino acid residues beyond the C-terminus of the β-amyloid peptide. The mutation cosegregates with the disease in some AD families but has not been found in any unaffected individuals. No genetic or genealogical evidence exists to link any of these families (some are Japanese; the others are Caucasian), suggesting that the APP717 mutation occurred independently in these families. The APP717 valine→isoleucine mutation seems to be associated with disease onset between 50 and 60 years of age, with a typical clinical picture involving severe AD pathology.

Subsequent investigations have identified within the same APP717 codon two further mutations that cause familial AD. In one British family with disease onset of 59 ± 4 years, sequencing of exon 17 revealed a T-to-G transition at base pair 2150 (the middle base pair of codon 717), changing valine to glycine. Sequencing of a U.S. family with early onset AD (age of onset 40 years) revealed a C-to-T change at bp 2149, changing the valine to a phenylalanine. These mutations have not yet been identified in any other cases of AD. The identification of three different mutations in the same codon of the gene suggests that this codon may be particularly important for processing of APP. More recently a Swedish pedigree has been described in which individuals with AD carry an unusual double point mutation, which leads to the substitution of two consecutive amino acids in exon 16 of APP (APP670/671).

Several other mutations within exon 17 of the APP gene that lead to a different phenotype have been reported. In addition to the original glutamic acid→glutamine substitution at APP693, which was reported in affected individuals from a family with HCHWA-D, a second mutation has been reported at the adjacent codon (APP692). This substitution causes a disease similar to HCHWA-D in some family members and AD in others, suggesting that additional factors may influence the expression of this mutation.

### 2.3 Other Genetic Loci Predisposing to Alzheimer's Disease

Mutations in the APP gene account for only a small proportion of familial early onset Alzheimer's disease cases, and both linkage analysis and DNA sequencing suggest that other loci are causative for most other cases of familial AD. Recent evidence shows that a second genetic locus predisposing to early onset AD is located on the long arm of chromosome 14. Work is now underway to localize and clone this gene. There must be at least one other familial AD locus in the genome, since families of a Volga German origin show no linkage between Alzheimer's disease and markers on chromosome 21 or chromosome 14.

Genetic linkage studies in families with late onset AD provided weak evidence for genes on chromosome 19 that are predisposed to AD. Binding studies demonstrated that β-amyloid binds to apolipoprotein E *in vitro* and that both proteins are found in AD plaques. Since the gene coding for apolipoprotein E (ApoE) maps to the region of chromosome 19 showing weak evidence for linkage, it was a candidate gene. Using association methods, many groups have now demonstrated an association between the ApoE-ε4 allele and AD. This association is observed in cases with and without a family history of AD and in both early and late onset cases. The risk of developing AD appears to be dosage dependent; that is, individuals with two ApoE-ε4 alleles have the highest risk and those with no ApoE-ε4 alleles have the lowest risk, with one ApoE-ε4 allele conferring an intermediate risk. Preliminary evidence suggests that the ApoE-ε2 allele may have a protective effect. Unlike the APP mutations, ApoE-ε4 alleles are not sufficient to cause disease. They merely increase an individual's chance of developing AD. It has not yet been conclusively shown that the ApoE-ε4 allele is the biological effector. An alternative explanation for the present data is that the ApoE-ε4 sequence variant lies very close to another sequence variant that causes the increased risk.

## 3 BIOCHEMISTRY OF β-AMYLOID FORMATION

### 3.1 β-Amyloid Precursor Protein (APP)

Because β-amyloid is the major constituent of plaques in AD brains and because mutations in APP clearly predispose individuals to early onset AD, the biochemistry and cell biology of APP expression and the mechanisms by which it is converted to β-amyloid are crucial to understanding the pathogenesis of Alzheimer's disease. APP expression is complex, involving at least five differentially spliced messenger RNAs, which give rise to different forms of APP (APP563, APP695, APP714, APP751, and APP770) with different combinations of exons and different molecular sizes. There is some degree of tissue specificity of expression of the different forms of APP: all mammalian cell types so far examined express APP, and levels of different transcripts as well as levels of the different protein products vary from one type to another. APP695 is particularly abundant in neurons.

In addition to differential RNA splicing, APP undergoes posttranslational modification, including phosphorylation, N- and O-linked glycosylation, sulfation, and proteolytic processing. Moreover, the relative rates of these modifications may differ between tissues and may vary according to changes in environmental conditions. It is not clear how these different events control APP expression, particularly formation of β-amyloid fragments; a detailed understanding of these posttranslational events will undoubtedly entail information relevant to amyloid plaque formation and to elucidating the biological function(s) of APP and β-amyloid.

The APP molecule consists of numerous domains (see Figure 1). One of these domains is homologous to regions found in proteinaceous protease inhibitors of the Kunitz type. Indeed, the secreted forms of APP containing the Kunitz protease inhibitor domain are identical to protease nexin-2, a serine protease inhibitor known to be involved in regulation of extracellular protease activity. The β-amyloid region of APP encompasses a portion of the extracellular domain of APP and a portion of the membrane-spanning domain.

A major constituent of neurofibrillary tangles is a protein called tau, which is a microtubule-associated protein. In AD the tau present in tangles is hyperphosphorylated compared with normal tau, and it is possible that this modification of the protein leads to tangle formation. Exactly how plaque and tangle formation are related to each other and to neuronal cell death is unclear. One possibility is that β-amyloid deposition or excessive levels of extracellular β-amyloid are toxic to neurons and cause tangles to appear in neurons as well as death of neurons.

### 3.2 Proteolytic Processing of APP

Proteolytic cleavage of APP is an area of research that is under particularly intensive study, especially because β-amyloid is a product of this process. Proteolytic processing of APP occurs by at least two routes in normal cells. Whether processing occurs along an abnormal route in AD cells remains to be determined, although β-amyloid production does occur in normal cells, suggesting that excessive β-amyloid production in AD may be due to enhanced processing of APP by normal pathways that generate β-amyloid. One route of APP processing is the so-called α-secretase pathway, in which cell surface membrane-bound APP is cleaved to release a large amino-terminal soluble fragment of APP into the cell's environment and a small membrane-associated C-terminal fragment, which is subsequently degraded in the cell. Secretion of APP by this route does not involve production of β-amyloid. Neither the secreted APP nor the membrane-bound C-terminal fragment can be converted to β-amyloid because neither contain the full-length β-amyloid region: the α-secretase route involves cleavage of membrane APP within the β-amyloid region.

A second enzyme activity has been detected that cleaves APP on the N-terminal side of β-amyloid. Although the enzyme has not been purified the activity has been called β-secretase. Cleavage at this site is compatible with β-amyloid production and is thought to be increased in individuals with the APP670/671 mutation. The C-terminal fragment produced by this cleavage can be detected in lysosomes but β-amyloid has never been detected inside lysosomes, suggesting that production of soluble β-amyloid probably occurs in another cellular compartment or at the cell surface. APP shows cell-specific processing. Neurons have high levels of β-secretase activity compared with other cell types studied. This may, in part, explain the restriction of AD pathology to the brain even though APP is expressed in all cell types.

A soluble form of β-amyloid is also secreted by normal cells, and levels of it have been reported to be elevated in cells transfected with the APP gene from familial AD containing a double mutation at codons 670/671. Soluble β-amyloid is also present in cerebrospinal fluid and blood of normal humans. The mechanism by which secreted soluble β-amyloid is produced from APP is unclear: it may be derived from the lysosomal route, in which case cells would release lysosomal β-amyloid or precursors of it into their environment; or it may be formed by a third processing pathway, for example, by cleavage of membrane-bound or secreted full-length APP outside the β-amyloid region to generate β-amyloid or precursors of it.

β-amyloid shows both C- and N-terminal heterogeneity. The most soluble β-amyloid is β-amyloid (1-40), i.e. amino acids one to forty. A minor species in soluble β-amyloid is two amino acids longer (β-amyloid 1-42). However, β-amyloid (1-42) is the major species present in plaques. It is much less soluble than β-amyloid (1-40). Recent experiments have demonstrated that the APP717 mutations cause an increase in the proportion of β-amyloid (1-42) produced. Given the decreased solubility of this form of β-amyloid, this may explain the massive amyloid accumulation seen in these patients and the early onset of disease.

Tremendous progress has been made in understanding the molecular pathogenesis of some forms of AD. During the next few years, it is anticipated that we will learn more about how these early changes cause the characteristic pathological changes we call AD and why these changes occur in distinct neuronal populations.

*See also* Neuropsychiatric Disease.

### *Bibliography*

Ashall, F. and Goate, A. M. (1994) The Role of β-amyloid precursor protein in AD. *Trends in Biochemical Sciences* 19:42–46.

Roth, M., and Iversen, L. L. Eds. (1986) Alzheimer's disease and related disorders. *Br. Med. Bull,* 42, No. 1.

Selkoe, D. (1992) The molecular pathology of Alzheimer's disease. *Neuron* 6:487–498.

# Amino Acid Synthesis

*Robert M. Williams*

### *Key Words*

**Chiral** Describing a specific geometric property of a molecule that is nonsuperimposable on its mirror image.

**Diastereomer** Stereoisomers that are not related by an enantiomeric (mirror image) relationship.

**Diastereoselective** Describing a reaction that favors the formation of one diastereomer rather than other possible diastereomers. Diastereoselectivity is expressed as the ratio of one diastereomer to another(s).

**Enantiomeric Excess** Expressed as %ee = [R] − [S]/[R] + [S] × 100, where [R] and [S] are the amounts of the individual enantiomers produced.

**Enantiomorphic** Usually synonymous with enantiomeric. A compound that is enantiomorphic to another will be a nonsuperimposable mirror image of that compound.

**Optical Purity** Expressed as % op = $[\alpha]_{observed}/[\alpha]_{maximum} \times 100$, where $[\alpha]_{observed}$ is the specific optical rotation of the sample under consideration and $[\alpha]_{maximum}$ is the specific optical rotation of the optically pure substance (i.e., specific rotation of a sample of a single enantiomer uncontaminated by the other), recorded under identical conditions.

**Prochiral** Describing an achiral compound that has two enantiotopic atoms or groups. If reaction (i.e., chemical transformation) of one of these enantiotopic groups as opposed to the other produces enantiomeric (i.e., chiral) products, the whole molecule is said to be prochiral.

**Racemic** Describing a sample comprised of equal amounts of two enantiomers of a given compound.

**Zwitterionic** A molecule containing nonadjacent plus (+) and minus (−) charges.

The importance of the roughly 20 proteinogenic amino acids that comprise the building blocks of proteins has been augmented by the discovery and utilization of naturally occurring, semisynthetic, and totally synthetic α-amino acids. Many of these nonproteinogenic amino acids have demonstrated significant biological function. Chemists and biologists have extensively utilized synthetic chemical and biochemical techniques to incorporate nonproteinogenic amino acids into biologically important peptides and proteins. Since the large majority of these amino acids are chiral molecules, that is, capable of existing in either the L- (*S*-) or D- (*R*-) absolute configuration, the development of reliable, practical methods to access the optically pure enantiomers is a very significant area of research. Glycine is the only natural proteinogenic α-amino acid that is achiral.

Each amino acid, as a result of the nature of the functionality in the side chain, presents a unique set of synthetic chemical challenges. The experimentalist will undoubtedly be confused and bewildered in selecting *the most appropriate* synthetic method to prepare the amino acid of interest. Of all the methodologies extant, there is no single *best* method that a laboratory may acquire for solving *every* amino acid synthesis problem that might be encountered. The task then, is to match the most appropriate methodology to the problem at hand. This entry deals specifically with the chemical synthesis of chiral α-amino acids. The existing conceptual approaches to preparing stereochemically defined α-amino acids are briefly considered, and the methods presented represent tried and true techniques for synthesizing amino acids in a standard organic chemistry laboratory setting. The literature is vast, and the works cited were chosen for their demonstrated practicality, reliability, and versatility.

## 1 INTRODUCTION

With the advent of a variety of sophisticated spectroscopic and computational methods to elucidate the relationships between amino acid sequence, protein conformation, and corresponding chemical, physical, and biological properties, a tremendous level of interest has been generated in the de novo design and synthesis of unnatural amino acids for the purposes of imparting enzyme-inhibitory, antimetabolite, protease resistance, and unique conformation-inducing properties to peptides and derivatives. In addition, techniques have recently been developed to incorporate unnatural amino acids into proteins that have shown promise in probing and altering enzymic mechanism and function. As a consequence, the development of versatile new methodology for the preparation of proteinogenic, natural, and unnatural amino acids in optically pure form has emerged as a highly significant and challenging synthetic endeavor (see Bibliography). The diverse nature of functional groups found or desired in the amino acid α-substituent ("R") and the obligate importance of accessing either the "L" (*S*) or "D" (*R*) absolute configuration requires the conception and development of numerous strategic solutions to this problem. Extensive reviews on this subject are listed in the Bibliography.

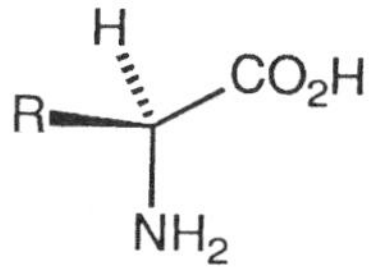

**L** 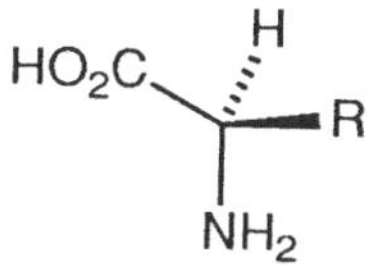 **D**

The established methods for the asymmetric synthesis of amino acids can be divided into roughly six categories in accordance with the system of Williams (1989). Methods in the first group accomplish the highly stereoselective hydrogenation of chiral, nonracemic dehydro amino acid derivatives or the asymmetric hydrogenation of prochiral dehydro amino acid derivatives. Chiral glycine equivalents serve as useful α-amino acid templates undergoing homologation via carbon–carbon bond formation at the α-position through nucleophilic carbanion alkylation (the second approach) or electrophilic carbocation substitution (the third). The fourth and fifth approaches, quite recently developed, consist of nucleophilic amination and electrophilic amination of optically active carbonyl derivatives. Finally, enzymatic and whole-cell-based syntheses have recently become more attractive in terms of substrate versatility, cost, and scale. All these methods have their relative strengths and weaknesses; the optimum methods for individual applications must still be chosen on a case-by-case basis with respect to functionality, quantity desired, cost, and time. Sections 2 to 8 describe very briefly the major, useful methods to prepare optically active amino acids.

## 2 ASYMMETRIC DERIVATIZATION OF GLYCINE

Several chiral, nonracemic glycine templates have been developed that have demonstrated utility in the preparation of unnatural amino acids. The three templates shown below (**1**, **2**, and **3**) have enjoyed the widest utility, as evidenced by the number of primary papers from the principal authors and significantly, the number of papers by other groups successfully utilizing these systems for amino acid preparation. All three of these systems are now commercially available.

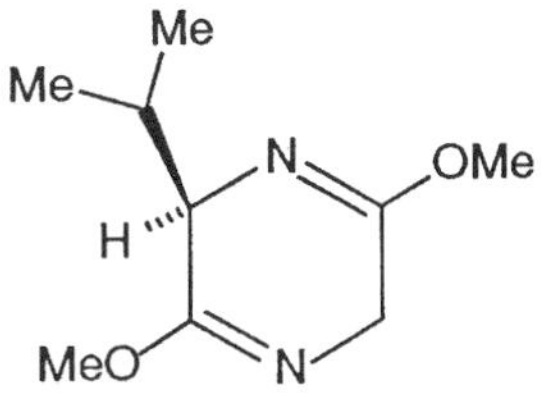

**1a** 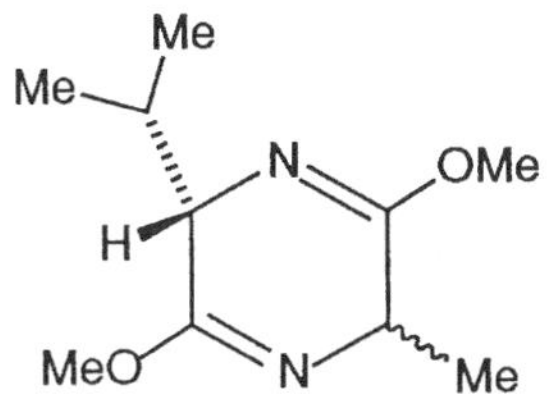 **1b**

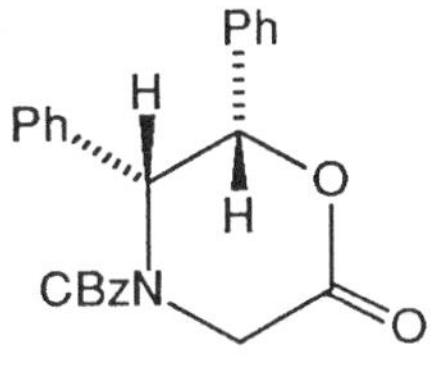

**2a** 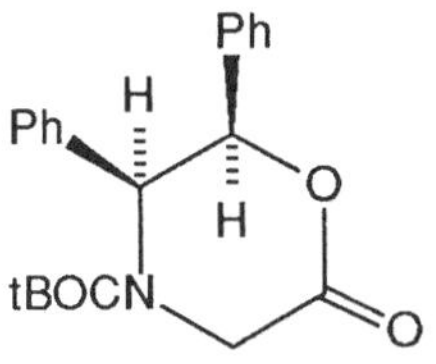 **2b**

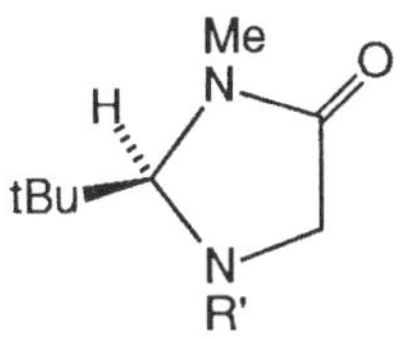

**3a**, R' = COPh
**3b**, R' = CBz
**3c**, R' = t-BOC

The bislactim ethers (**1**), developed by Schollkopf and collaborators, have been the most extensively studied glycine enolate equivalents. Both templates (**1a** and **1b**) are commercially available (Merck, Darmstadt) in both enantiomeric forms providing access to D- and L-configured α-amino acids. Metallation of the bislactim ethers (**1**) with *n*-butyllithium in tetrahydrofuran (*n*-BuLi in THF) at low temperature followed by quenching the enolate with a reactive electrophile (e.g., alkyl halides, aldehydes, α,β-unsaturated carbonyl derivatives, acid chlorides epoxides, ketones) provides the functionalized bislactim ether. Mild acid hydrolysis of these substances provides the α-amino acid methyl esters, which can be further hydrolyzed to the free, zwitterionic amino acids. The alkylations are generally highly diastereoselective giving the amino acid products in great enantiomeric excess (typically $> 95\%$ ee).

The diphenyl oxazinones (**2**) developed by Williams et al. are also available commercially (Aldrich) in both enantiomorphic forms, as either the *N*-benzyloxycarbonyl (*N*-CBz, **2a**) or *tert*-butyloxycarbonyl (*N*-*t*-BOC, **2b**) substrates. Like the systems already described, these can be converted into their corresponding enolates by treatment with lithium, sodium, or potassium hexamethyldisilazane in THF at low temperature; subsequent alkylation with a reactive alkyl bromide or iodide provides the corresponding α-functionalized lactones. Alternatively, this system has proven very useful as a glycine cation equivalent for reaction with organometallic reagents. Treatment of **2** with *N*-bromosuccinimide (NBS) in warm carbon tetrachloride gives the corresponding α-bromodiphenyloxazinone; subsequent reaction with a carbon nucleophile (organometallic) also provides highly diastereoselective access to the corresponding α-functionalized lactones. Both the enolate reactions and the bromide displacements proceed with generally high diastereoselectivity giving amino acids with typically 96% ee. Williams et al. have developed both oxidative and reductive conditions for cleaving the corresponding α-functionalized lactones into their corresponding α-amino acid derivatives; the choice of cleavage conditions depends on the nature of the functionality present in the "R" group. Where possible, dissolving metal reduction of the *N*-*t*-BOC series gives direct access to the corresponding *N*-*t*-BOC-α-amino acids.

The imidazolidinones (**3**) developed by Seebach and co-workers are also available commercially in both enantiomorphic forms (Aldrich) and have been extensively studied for enolate alkylations and aldol condensations. Vigorous acid hydrolysis provides the corresponding α-amino acids in high optical purity.

All three systems (**1**, **2**, and **3**) have been successfully applied to the problem of making congested α-methyl and other α-alkyl-α-amino acids. These systems are best suited for providing access to gram or multigram quantities of unusual amino acid derivatives for basic research applications and are not really practical for large industrial (multikilo scale) syntheses.

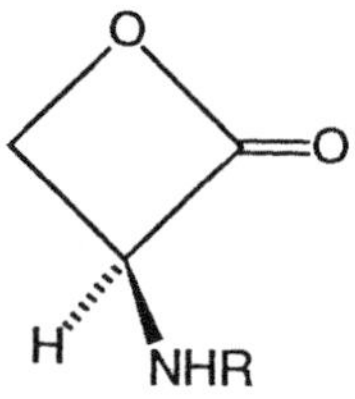

## 3 HOMOLOGATION OF THE β-CARBON

Derivatization of the β-carbon can formally considered to be the derivatization of the alanine β-carbon (or serine/cysteine). The most practical approaches appear to be that developed by Vederas, Baldwin, Nakajima, and Garner (structures 4–8).

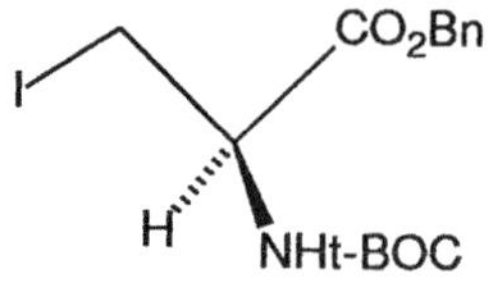

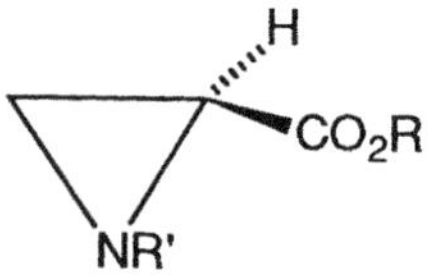

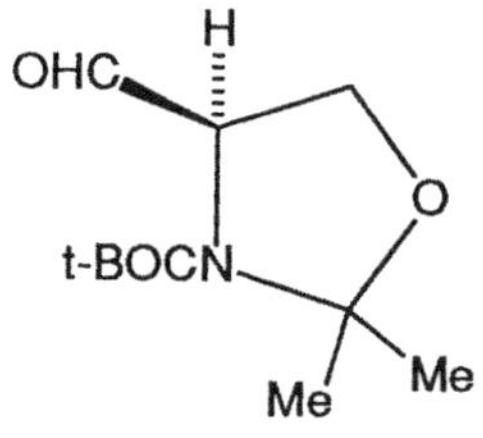

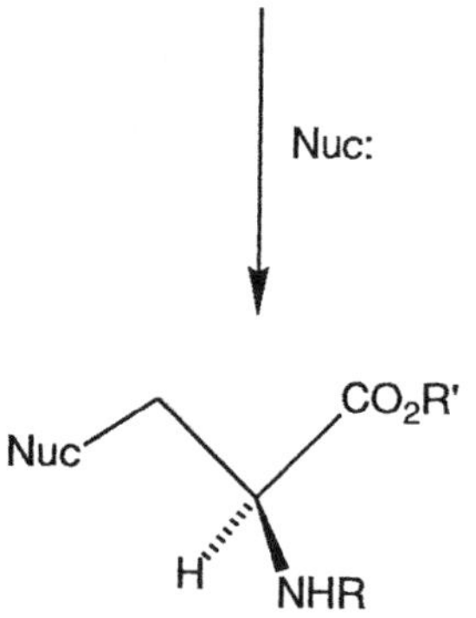

Vederas has converted serine into the corresponding β-lactones, **4**. The β-lactones are activated for nucleophilic ring-opening reactions with either organometallic reagents (forming a C—C bond) or with heteroatom nucleophiles under both Lewis acidic and basic conditions. These systems offer the advantage that the N-acylated (CBz or *t*-BOC) amino acid derivatives (**8**) are directly accessed.

NHR
**4a**, R = CBz
**4b**, R = t-BOC

$CO_2Bn$
NHt-BOC
**5**

$CO_2R$
NR'
**6**

OHC
t-BOCN
Me Me
**7**

Nuc:

Nuc
$CO_2R'$
NHR
**8**

Baldwin and associates have examined various reactions of β-iodoalanine derivatives, **5.** Free radical reactions derived from the reaction of allylstannanes gives highly functionalized amino acids **8**.

Aziridines (**6**) derived from serine and threonine have been extensively examined by Nakajima, Okawa, and associates. In these earlier studies, ring opening of the aziridine was primarily studied with heteroatom nucleophiles; more recently these workers and others have examined various carbon nucleophiles for providing access to β-substituted amino acids.

Garner and co-workers have reported the conversion of serine into the serinal derivative **7**. Many papers have appeared on various methods to functionalize the aldehyde of this very versatile and useful synthetic template. With this approach, it is necessary to oxidize the serine-derived hydroxymethyl group up to the carboxylic acid oxidation state after the aldehyde (derived from the serine carboxyl) has been derivatized. To date, none of the templates mentioned in this section are commercially available, hence they must be prepared by the investigator.

## 4 ELECTROPHILIC AMINATION OF ENOLATES

The reaction of various enolates with electrophilic sources of nitrogen is a relatively new and sparsely studied approach. The relative lack of activity in this potentially very versatile area is a manifestation of the paucity of recognized, reliable sources of electropositive nitrogen transfer reagents. Three systems are worthy of note.

Four groups (Evans, Gennari, Vederas, Oppolzer) simultaneously discovered the utility of diazodicarboxylates as useful nitrogen transfer reagents to ester and amide enolates. The overall approach involves treating the appropriate enolate species with a diazodicarboxylate, affording the corresponding α-hydrazido adducts. Depending on the nature of the specific chiral auxiliary, the carboxyl groups and the chiral auxiliary are removed to give the unusual α-hydrazido-α-amino acids. The N–N bond of the hydrazide can then be cleaved reductively to give the corresponding α-amino acid. This method allows for the preparation of the interesting α-hydrazido-α-amino acids, which have just started to become recognized as potentially biologically interesting amino acid surrogates.

A much more useful approach to electrophilic amination has been described by Evans and associates that involves the azidation of enolates with trisyl azide (2,4,6-triisopropylbenzenesulfonyl azide). A number of complex natural products by this group and others have been successfully prepared by this approach.

## 5 NUCLEOPHILIC AMINATION OF α-SUBSTITUTED ACIDS

The complement to the approach just described is the displacement of a leaving group alpha to the carboxyl residue with a nucleophilic

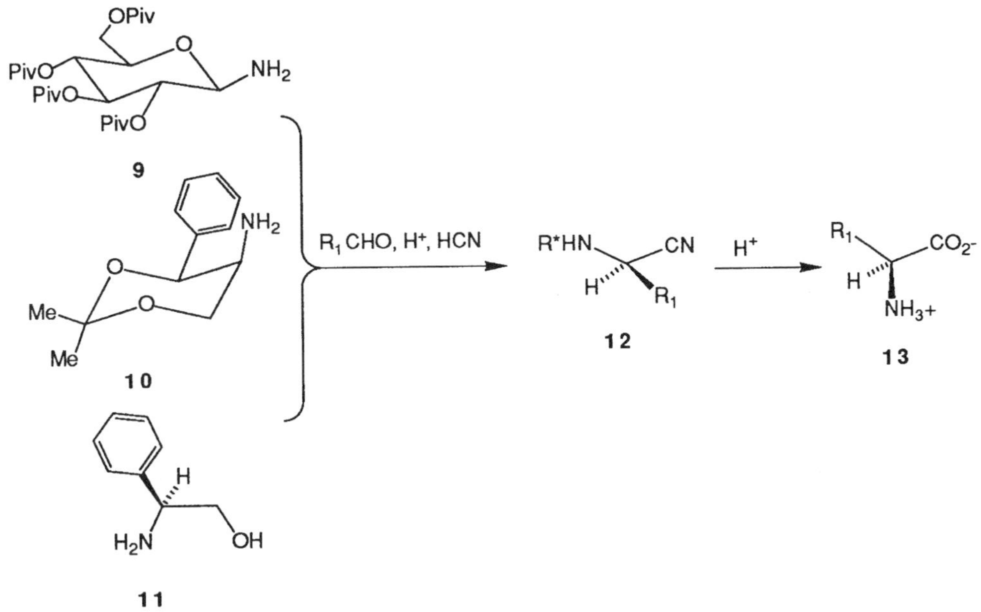

OPiv PivO PivO PivO NH2 **9**

NH2 O O Me Me **10**

H H2N OH **11**

$R_1CHO$, $H^+$, HCN

R*HN CN H R1 **12**

$H^+$

R1 CO2- H NH3+ **13**

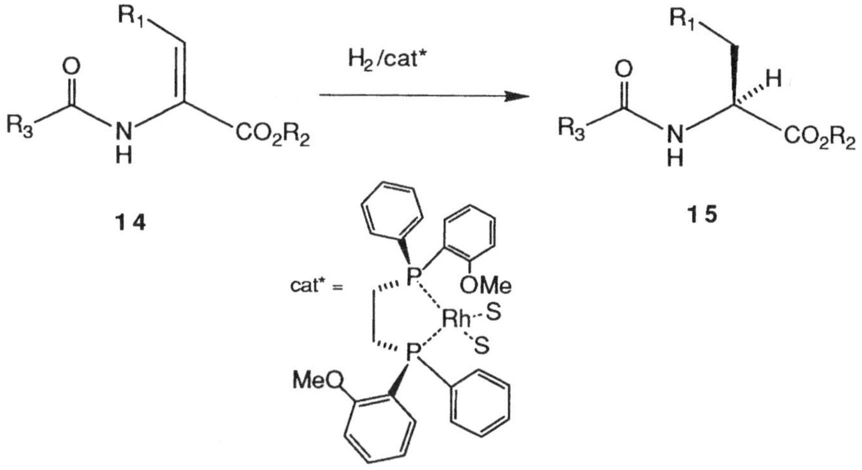

R1 O R3 N H CO2R2 **14**

$H_2$/cat*

R1 O H R3 N H CO2R2 **15**

cat* = P OMe Rh S S P MeO

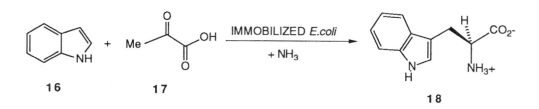

source of nitrogen. Here again, many approaches have been utilized (see Williams, 1989); the most useful general approach appears to be the displacement of a halogen with azide. Several excellent chiral auxiliaries have been utilized to direct the halogenation of a substituted acid.

## 6 ASYMMETRIC STRECKER SYNTHESES

The classical Strecker synthesis, known since 1850, is perhaps one of the most versatile methods to prepare amino acids in racemic form. Recent interest in asymmetric and practical versions of this reaction promise that the method will enjoy increased utility in the future. Many researchers have examined a variety of optically active amines to induce transfer of chirality to the newly created stereogenic center. All these methods share the classic Strecker protocol involving the condensation of an aldehyde (or in some cases, ketone) with the amine in the presence of cyanide or a cyanide equivalent. The intermediate α-amino nitrile (or equivalent) is subsequently hydrolyzed to the corresponding amino acid. The limitation of this method resides in the availability of the aldehyde component and the stability of the ''R'' group to the obligatory hydrolytic processing of the α-amino nitrile derivative. Several relatively useful chiral, nonracemic amines for this application have been used (see structures **9–13**).

## 7 ASYMMETRIC HYDROGENATION OF DEHYDRO AMINO ACIDS

The asymmetric hydrogenation of dehydro amino acids has been a very useful method for the preparation of amino acids, particularly those in the phenylalanine structural manifold. The approach illustrated on page 31 involves the preparation of a (typically) *Z*-dehydro amino acid (**14**) by the condensation of an aldehyde with a phosphorylglycine derivative. The dehydro amino acid is then subjected to catalytic hydrogenation with a chiral, cationic rhodium catalyst. Although many chiral, nonracemic phosphine ligands have been studied for this reaction, the bis-phosphine (DIPAMP) system shown in structures **14** and **15** is the most generally reliable system. This method is not applicable to the synthesis of α-alkylated amino acids or substances with reductively labile side chain functionality.

## 8 ENZYMATIC SYNTHESES OF α-AMINO ACIDS

Synthesis of many different classes of amino acids with purified, immobilized enzymes, cell-free enzyme preparations, and whole-cell reactors has become an important method for the industrial-scale preparation of amino acids. Two major classes of reaction types have been employed: enzymatic resolution of racemic amino acid derivatives (generally prepared by a classic Strecker protocol) and asymmetric bond-forming reactions on prochiral substrates catalyzed by various classes of enzymes.

Enzymatic resolutions have been studied with amidases, acylases, esterases, and nitrilases and nitrile hydratases. Many of these enzymes are now commercially available. Many variations on this general theme have been reported, including the efficient kinetic resolution of amino amides with L-specific amidases, the enantioselective hydrolysis of *N*-acylated amino acid derivatives with lipase and acylase, enantioselective alkaline protease and papain cleavage of amino acid esters, chymotrypsin-mediated hydrolysis of α-nitro carboxylic acids, and cleavage of α-amino nitriles by means of nitrile hydratase.

An example of the second general class, (asymmetric bond-forming reactions on prochiral substrates catalyzed by an enzymatic system) is given in structures **16–18**. In this system, immobilized *E. coli* cells effect the oxidative enzymatic coupling of indole (**16**) to pyruvic acid (**17**), yielding an incipient α-keto acid which is subsequently transaminated (enzymatically) to give L-tyrosine (**18**). Numerous other examples of this type of process include the following conversions: phenylpyruvic acid to phenylalanine, α-ketoadipic acid into L-α-aminoadipate by glutamate dehydrogenase, *trans*-cinnamic acid into L-phenylalanine by phenylalanine ammonia–lyase, and fumaric acid into aspartic acid by aspartase. Most of these processes have been engineered toward the preparation of (primarily) L-amino acids for large, industrial-scale synthesis. Some inroads have been made in this field with respect to making unnatural amino acids and amino acids with the D-configuration; however, for research quantities of unusual amino acids, the methods outlined in Sections 2, 3, and 5 probably provide the easiest access for most research labs.

*See also* CHIRALITY IN BIOLOGY.

## *Bibliography*


Bosnich, B., Ed. (1986) *Asymmetric Catalysis.* Nijhoff, Dordrecht.

Drauz, K., Kleeman, A., and Martens, J. (1982) Induction of asymmetry by amino acids. *Angew. Chem. Int. Ed. Engl.* 21:584.

Kochetkov, K. A., and Belikov, V. M. (1987) Modern asymmetric synthesis of α-amino acids. *Russian Chem. Rev.* 56:1045.

Martens, J. (1984) Asymmetric synthesis with amino acids. *Topics Curr. Chem.* 125:167.

Morrison, J. D., Ed. (1983, 1984) *Stereodifferentiating Addition Reactions,* Vols. 2 and 3 in *Asymmetric Synthesis.* Academic Press, Orlando, FL.

———, Ed. (1985) *Chiral Catalysts,* Vol. 5 in *Asymmetric Synthesis.* Academic Press, Orlando, FL.

———, and Mosher, H. S. (1971) *Asymmetric Organic Reactions.* American Chemical Society, Washington, DC.

O'Donnell, M. J., Ed. (1988) α-Amino acid synthesis (*Tetrahedron* Symposium-in-Print). *Tetrahedron* 44:5253–5614.

Valentine, D., and Scott, J. W. (1978) Asymmetric synthesis. In *Synthesis* 329.

Wagner, I., and Musso, H. (1983) New naturally occurring amino acids. *Angew. Chem. Int. Ed. Engl.* 22:816.

Williams, R. M. (1989) *Synthesis of Optically Active α-Amino Acids.* J. E. Baldwin, Ed. Pergamon Press, Oxford.

———. (1992) *Aldrichim. Acta* 25:11.

# Annexins

*Carl E. Creutz*

## *Key Words*

**Exocytosis** Release of secretory product from a cell by fusion of an intracellular vesicle containing the product with the cell surface membrane.

**Membrane Fusion** Merger of two biological membranes to form a single membrane, as when the membrane of a secretory vesicle fuses with the surface membrane of the cell.

**Phospholipid** Common constituent of biological membranes composed of two fatty acids esterified to a glycerol backbone with a head group containing a phosphate ester.

**Phosphorylation** Covalent attachment of phosphate, as in the enzyme-catalyzed phosphorylation of serine or tyrosine amino acid side chains in a protein.

The annexins are a family of structurally related, calcium-dependent, phospholipid-binding proteins that have been postulated to mediate calcium-dependent activities at membrane surfaces such as membrane fusion, lipid metabolism and reorganization, and ion permeation. The basic annexin structure consists of four homologous 70 amino acid repeats and a unique N-terminal domain. These repeats do not contain sequences found in other intracellular calcium-binding proteins; therefore the annexins represent a novel class of calcium-binding proteins. The "core" domains of all the annexins, comprised of the four 70 amino acid repeats, are 40 to 60% identical in sequence. One annexin family member, (annexin VI) has been formed as a result of gene duplication and consists of eight of the 70 amino acid repeats. Another type of duplication has occurred with annexin II (calpactin), in which two 36 kDa molecules, the "heavy chains," each containing four repeats, bind to a dimer of an 10 kDa protein, the "light chains," to form a tetramer.

In contrast to other lipid-binding proteins such as protein kinase C or phospholipase $A_2$, the annexins are unique in that they are

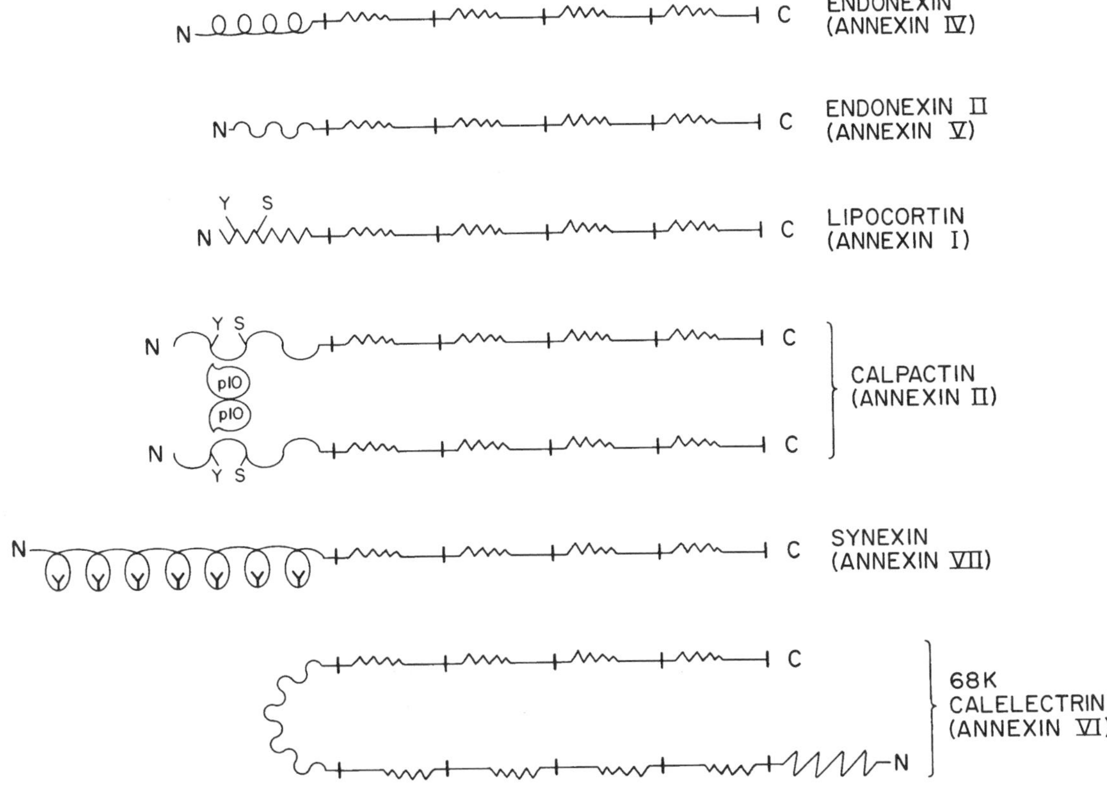


**Figure 1.** Schematic illustration of the primary structures of six annexins. Each of the four (or eight) homologous domains contains the 17 amino acid "endonexin fold" sequence represented by a sawtooth line (sequence: KGhGTDExxLIpILApR: h, hydrophobic residue; p, polar residue; x, variable residue). The unique N-terminal structures are on the left (or right in the case of annexin VI); Y and S represent phosphorylation sites in the tails of annexins I and II. The annexin II tetramer is drawn showing the association of the N-termini of the heavy chains with the light chain (p10) dimer. The *Y*'s inside the loops in the tail of synexin represent a pro-β helix. [Reproduced from Dedman and Smith (1990), with permission.]

bivalent. That is, they can attach to two membranes, rather than just one, and as a consequence draw them together.

## 1 DIVERSITY AND FUNCTIONS

Members of the annexin family of proteins, of which there are now 10 recognized mammalian variants, have been independently discovered in a number of contexts. As a consequence, a variety names have been used for the different family members, such as *synexins, chromobindins, lipocortins, calcimedins, calphobindins,* and *calpactins.* However, a standard reference nomenclature has been adopted involving the term *annexin* followed by a Roman numeral (Fig. 1). Representative compounds have been isolated from animals, plants, and the slime mold *Dictyostelium,* attesting to the universality of the functions performed by annexins.

Because of the bivalent activity of the annexins, one of their postulated roles is in the promotion of membrane fusion in exocytosis. After secretory vesicle (chromaffin granule) membranes have been aggregated by an annexin, they undergo fusion if exposed to cis-unsaturated fatty acids. In a cell, such fusogenic fatty acids might be made available through activation of lipases. Alternatively, the annexins might function in concert with other proteins that promote membrane fusion and remodeling after the membranes have been brought into close apposition by the annexin.

In addition to a possible role in exocytosis and membrane trafficking, some of the annexins have been suggested to be mediators of the anti-inflammatory effects of steroids, components of the submembranous cytoskeleton, inhibitors of blood coagulation, transducers of signals generated by tyrosine kinases at the cell membrane, mediators of bone formation and remodeling, mediators of cell–matrix interactions, voltage-dependent ion channels, actin bundling proteins, regulators of calcium release channels in muscle, enzymes involved in the metabolism of inositol phosphates, regulators of DNA polymerase activity, and mediators of phagocytosis.

This diversity of hypotheses attests to the ubiquitous and abundant nature of these proteins. In most cases, the data supporting a role for the annexins in these various processes is derived from in vitro systems involving calcium and lipid membranes, and it has been difficult to relate these activities to true cellular functions. However, it seems likely that this family of proteins has radiated to perform a variety of cellular functions, since so many are coexpressed in single cells and their individual differences have been highly conserved during evolution. In that sense, the annexins may be similar to the other major class of calcium-binding proteins, the "EF-hand" family, which includes calmodulin and troponin C, and is responsible for a variety of nonoverlapping cellular phenomena.

## 2 STRUCTURE AND MECHANISM OF ACTION

The annexins, in general, have proven to be relatively easy to isolate in high purity, and this advantage has led to several successful attempts to crystallize members of the family. The structure of annexin V reveals that each of the four 70 amino acid repeats forms a compact bundle of α-helices, and these four bundles are arranged in a plane (Figure 2; see color plate 5). Viewed from the side, the molecule displays a convex face, thought to be the membrane-binding face, and a concave face, thought to face the cytoplasm. Each homologous domain forms one or two potential calcium-binding sites, which are complex in that they involve convergent loops from different parts of the domain. In addition, the positions of sulfate ions in the loops (derived from the crystallization medium) suggest that the phosphate of a lipid head group may also sit in the binding pocket and coordinate the calcium ion. This geometry would explain the interdependence of calcium and lipid binding by the annexins.

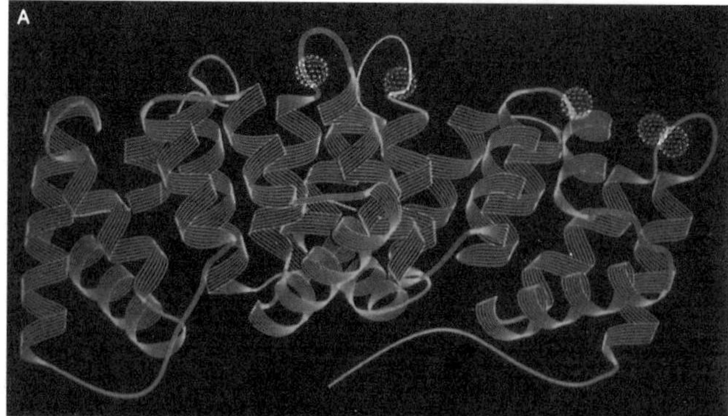


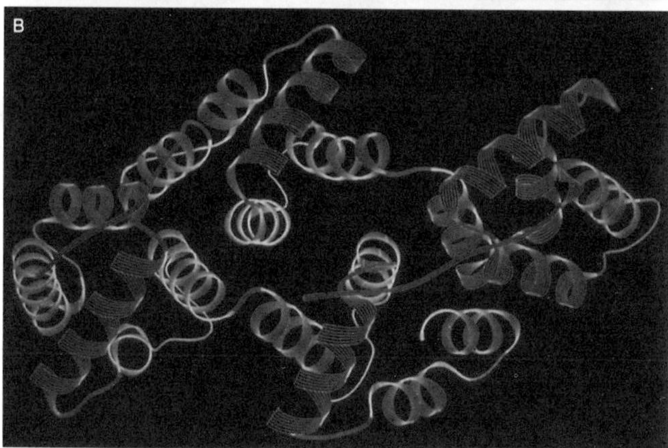


**Figure 2.** Ribbon plots of the structure of annexin V as determined by X-ray diffraction. (A) "Side" view of the molecule, with the calcium-binding sites and the membrane-binding face at the top. Calcium ions are represented by the red spheres. The high affinity sites are represented by the first, second, and fifth spheres from left to right; the third and fourth spheres indicate low affinity ion-binding sites that were identified by lanthanide binding. The extended N-terminus is seen at the bottom. (B) View of the "cytoplasmic" side of the molecule and the N-terminus. The calcium-binding sites are on the opposite face. Note the potential channel structure in the center of the molecule. [Creutz (1992), with permission. Original figure kindly provided by A. Burger and R. Huber.] (See color plate 5.)

Since all the lipid-binding sites of the annexin core domains face a single membrane, they may be able to integrate information about membrane lipid composition. In general, all the annexins bind with higher affinity (i.e., at lower levels of calcium) to acidic phospholipids. However, there is some variation reported in the literature for different annexins with regard to order of preference for different acidic lipids. In addition, the four homologous repeats may have different specificities, since the repeats are only 40 to 50% identical in sequence.

The crystal structure of annexin V has also been interpreted in terms of the ion channel forming properties of the protein. The molecule has a hydrophilic pore perpendicular to the face of the membrane (Figure 2). Because the molecule has no hydrophobic external surfaces, it is assumed it cannot actually enter the bilayer.

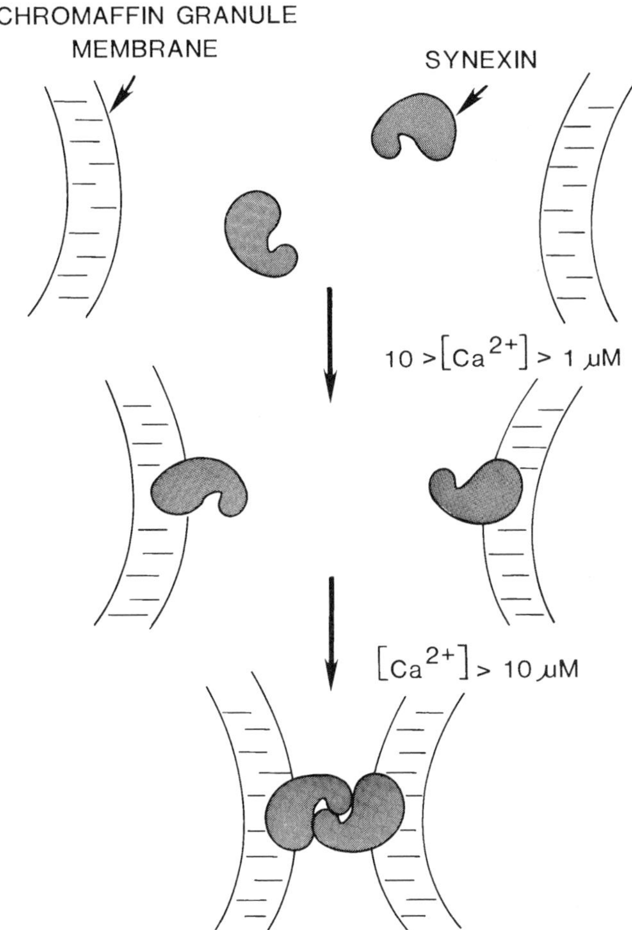


**Figure 3.** Hypothesis for the mechanism of action of synexin (annexin VII) in promoting intermembrane contacts in a cell. In the resting cell (top) the $Ca^{2+}$ concentration is less than 1 μ*M* and synexin is freely soluble. In the stimulated cell (middle), as the calcium concentration rises to between 1 and 10 μ*M*, synexin attaches to membranes. When the calcium concentration further increases above 10 μ*M* (perhaps near the plasma membrane), synexin molecules on different membranes self-associate to bring the membranes together. [Reproduced from Creutz and Sterner (1983), with permission.]

However, it has a calculated external electrical field similar to the field strength necessary to punch a hole through a membrane by electroporation. Therefore, it has been suggested that the molecule sits on the surface of the membrane and leads to a disordered state in the lipids immediately below it. In this disordered state the membrane conducts ions that must also pass through the hydrophilic pore in the molecule. The molecule thus provides the ion selectivity of the overall transmembrane channel. A similar mechanism of membrane permeabilization has been suggested for the action of cholera toxin.

Since the calcium/lipid-binding sites all appear to be on one side of the annexin molecule, it is not obvious how the annexins may aggregate membranes. Since, however, there is some flexing of the molecule about a central cleft when calcium is bound, it is not inconceivable that the molecule could completely double over and expose lipid-binding sites on opposite sides. It seems more likely, though, that a self-association of annexin molecules attached to different membranes is required for membrane aggregation (Figure 3). The concave faces of the molecules that face the cytoplasm might interlock during this self-association event. In the case of synexin acting in vitro on chromaffin granules, binding to membranes occurs at low levels of calcium (< 10 μ*M*), but membrane aggregation and fusion depend on higher levels of calcium (> 100 μ*M*). The calcium dependence of granule aggregation correlates exactly with the calcium dependence of synexin self-association in the absence of membranes, suggesting a mechanism for membrane aggregation whereby membrane-bound synexin molecules undergo self-association to bring the two membranes together. If this mechanism is correct, then the "bottleneck" in the system is the high level of calcium needed to promote annexin self-association.

The annexin II (calpactin) tetramer promotes the aggregation and fatty acid dependent fusion of chromaffin granules at the lowest level of calcium of any annexin (~ 1 μ*M*). This may occur because the tetramer represents a permanently self-associated annexin (Figure 1). The two heavy chains might be associated through the light chain dimer in such a way that each heavy chain can bind a different membrane and thus pull them together. Therefore, the processes of overall membrane aggregation and fusion are catalyzed by the low levels of calcium needed simply to promote membrane binding.

## 3 REGULATION

The N-terminal domains of the annexins are the major sites of sequence divergence in the family (Figure 1). They may be the site of interaction with other proteins, as in the case of the annexin II heavy chain, which associates with p10, and annexin XI, which associates with the small calcium-binding protein, calcyclin. The N-terminal domains are also potential regulatory sites for the rest of the molecule. For example, cleavage of the N-terminus of the isolated annexin II heavy chain is necessary for this protein to aggregate chromaffin granules at low levels of calcium. Furthermore, the calcium sensitivity of membrane aggregation by such cleaved molecules is strongly affected by the exact site of cleavage. In addition, annexins I and II are phosphorylated in the N-terminal domain by tyrosine- or serine/threonine-specific kinases. Phosphorylation of the N-terminus of annexin I by protein kinase C *blocks* the ability of this protein to aggregate chromaffin granules, while slightly *enhancing* its ability to bind the granule membrane.

These biochemical observations on the importance of the annexin N-terminal structure can be interpreted in terms of the crystal structure of annexin V. Since the N-terminus is located on the "cytoplasmic" side of the membrane-bound annexin, it may be in a position to participate in annexin–annexin self-association essential to membrane aggregation. This may explain why alterations of the tails of other annexins strongly affect membrane aggregation but not membrane binding. Such alterations, resulting from either proteolysis or phosphorylation, may provide additional mechanisms for cellular control of these calcium-dependent proteins.

*See also* CALCIUM BIOCHEMISTRY; MEMBRANE FUSION, MOLECULAR MECHANISM OF; PHOSPHOLIPIDS.

### *Bibliography*

Creutz, C. E. (1992) The annexins and exocytosis. *Science* 258:924–931.
———. (1993) Calcium-dependent membrane-binding proteins in cell-

free models for exocytotic membrane fusion. *Methods Enzymol.* 221:190–203.

Creutz, C. E., and Sterner, D. C. (1983) Calcium dependence of the binding of synexin to isolated chromaffin granules. *Biochem. Biophys. Res. Commun.* 114:355–364.

Dedman, J. R., and Smith, V. L., Eds. (1990) *Stimulus–Response Coupling: The Role of Intracellular Calcium-Binding Proteins.* CRC Press, Boston.

Moss, S. E., Ed. (1992) *The Annexins.* Portland Press, London.

## Antibodies: *see* Antibody Molecules, Genetic Engineering of; Immunology.

# Antibody Molecules, Genetic Engineering of

*Sherie L. Morrison*

### *Key Words*

**Bacteriophage** Bacteriophage are viruses that infect bacteria. Bacteriophage λ is a temperate phage that can grow lytically where it lyses the bacteria; it will form a clear plaque on a lawn of bacteria. Infection with filamentous phages such as M13 is not lethal and the host bacteria do not lyse. Instead, their rate of growth slows and they form turbid plaques on the bacterial lawn.

**Domain** An independently folded structural unit within a protein.

**Fab** A monovalent antigen-binding fragment of an antibody that consists of one light chain and part of one heavy chain (the variable region and first constant region domain). It can be obtained by digestion of intact antibody with papain or by genetic engineering techniques.

**Fc** A non-antigen-binding fragment of an antibody that consists of the carboxyl-terminal portion of both heavy chains. It can be obtained by papain digestion of an intact antibody. Two Fab fragments and one Fc constitute a complete IgG antibody.

**Fv** A monovalent antigen-binding fragment of an antibody composed of the variable regions from the heavy and light chains.

**Monoclonal Antibody** Homogenous antibodies of the same antigenic specificity representing the product of a single clone of antibody-producing cells.

The antibody molecule is a key component of the immune response because of such properties as the ability to recognize a vast array of different foreign substates and the capacity to interact with and activate the host effector systems. The value of antibodies with defined specificities is clear, and with the development of hybridoma technology, it is now possible to produce such antibodies. Protein engineering can be used to further improve antibody production and to produce antibodies with novel structure and functional properties. Genetically engineered mice are also available which produce specific antibodies that are totally human in sequence. Bacteriophage expression systems hold the promise of being able to produce specific antibodies in the absence of animal immunization.

## 1 Antibody Structure and Antibody Engineering

Antibodies are composed of light and heavy chains joined by disulfide linkages (Figure 1). It is the heavy chain that determines the biologic properties of the antibody and several classes, or isotypes, of antibodies with different heavy chains are found in mammals. The light and heavy chains fold into functional domains, two for the light chain, three or four for the heavy chain depending on the isotype. The N-terminal domain from each chain forms the variable regions, which constitute the antigen-binding sites. The other domains contribute to the activation of host effector mechanisms. The domain structure of the molecule is favorable to protein engineering, facilitating the exchange between molecules of functional domains carrying antigen-binding activities (Fabs or Fvs) or effector functions (Fc). The structure of the antibody also makes it easy to produce antibodies with the antigen recognition capacity

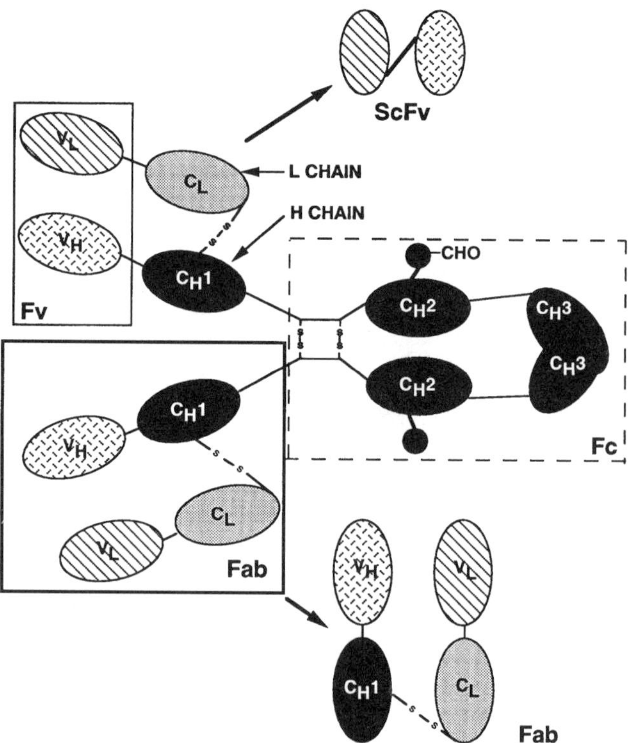


Figure 1. Schematic diagram of the antibody molecule. Distinct domains are represented by ellipses. The constant region domains of the heavy chain, which determine its isotype, are indicated as $C_H1$, $C_H2$, and $C_H3$; the constant region domains of the light chain are indicated by $C_L$. The carbohydrate (CHO) present on the $C_H2$ domain of the heavy chain is indicated as a solid circle. The $V_H$ and $V_L$ (variable domains), which form the antibody-binding site, are also shown. Functional antibody fragments are indicated both as part of the complete antibody and as separate fragments such as would be synthesized by bacteria. The disulfide bonds (S—S) critical for the covalent structure of the antibody are indicated.

joined to molecules such as toxins, lymphokines, or growth factors. Antibody molecules have proven to be surprisingly amenable to such modifications.

## 2 EXPRESSION SYSTEMS FOR ANTIBODY PRODUCTION

Expression systems are available for antibody production in bacteria, yeast, plants, baculovirus, and mammalian cells; each expression system has advantages and limitations.

Cultured mammalian cells are an excellent system for the production of complete antibody molecules because of their ability to correctly process, posttranslationally modify, and secrete the antibody. Bacteria have been used most effectively to produce antibody fragments with the expression of complete functional antibodies in bacteria presenting problems of glycosylation, assembly, and disulfide bond formation. While cytoplasmic expression of antibody in *E. coli* results in the formation of insoluble and inactive protein aggregates (inclusion bodies), which require refolding and renaturation to be active, antibody fragments supplied with leader sequences can be expressed and secreted in fully functional form.

It has recently become possible to make mice that produce antigen-specific antibodies that are totally human. This has been accomplished by inserting elements of the human heavy and light chain loci into mice in which production of endogenous murine heavy and light chains was disrupted. The resulting mice can synthesize human antibodies specific for many antigens, including antigens of human origin, and can be used to produce hybridomas making human antibodies. To date, these engineered mice can produce only some of the human isotypes, but further genetic modifications promise to expand their potential.

## 3 IN VITRO LIBRARIES FOR ANTIBODY IDENTIFICATION

The technique of expressing immunoglobulin genes in bacteriophage has been developed as a means for obtaining antibodies with the desired binding specificities. Expression systems based on bacteriophage λ and, more recently, filamentous phage have been developed. The bacteriophage expression systems can be designed to allow heavy and light chains to form random combinations, which are tested for their ability to bind the desired antigen.

The high packaging efficiency of phage λ and the ability to detect protein expression at the level of single-phage plaques using plaque lifts make it an attractive system. Libraries of about $10^6$ clones can be efficiently screened and are adequate to isolate specific antibodies from an immunized host. Since, however, the screening procedure tends to identify only antibodies with high affinity for antigen, it is difficult to isolate specific antibodies from unimmunized donors where the antibodies are of predominantly low affinity.

To obtain larger libraries, methods have recently been developed to display Fv and Fab fragments on the surface of filamentous bacteriophage (f1, M13, and fd) as either single or multiple copies of the antibody of interest depending on the phage protein used for fusion. Expression of multiple copies facilitates the identification and isolation of low affinity antibodies. The resulting phage bind specifically to antigen, and rare phage with specificity for a given antigen can be isolated and expanded after affinity chromatography. Multiple rounds of enrichment consisting of binding to immobilized antigen, expanding the bound phage, and further enriching by again binding to immobilized antigen can yield specific phage even if the desired specificity was present on less than 1 in $10^6$ of the original phage in the library. Filamentous phage therefore offer an excellent system for screening large numbers of clones and selecting rare specificities.

While initial studies isolated specific clones from a library of heavy and light chain variable regions prepared using immunized donors, more recently experiments have isolated specific antibodies from libraries prepared from peripheral blood lymphocytes (PBLs) of unimmunized donors. Further modifications of techniques have made it possible to produce phage libraries of very large diversity, which should permit the isolation of high affinity antibodies specific for virtually any antigen. The variable regions in the phage vectors can be of murine or human origin. Use of human variable regions makes it possible to create antibodies entirely encoded by human sequences.

Binding specificities can be expressed as either single-chain fusions or as Fabs. However, the use of Fab fragments offers advantages for the construction of combinatorial libraries, since heavy and light chain libraries can be produced independently and resorted. By introducing an amber mutation between the antibody chain and the coat protein, it is possible to either display the antibody on phage using suppressor strains of bacteria, or to produce soluble fragments using nonsuppressor strains. It is possible to use the phage system to mutate and select antibodies in vitro, thereby acquiring antibodies with improved binding properties.

## 4 CHIMERIC ANTIBODIES

Although clinical trials with murine monoclonal antibodies have shown promise, murine antibodies are rapidly cleared from circulation, are ineffective in interacting with the host immune effector systems, and in most cases elicit an immune response to the murine protein (human antimouse response, or HAMA). To address these problems, chimeric antibodies have been produced with murine variable regions joined to human effector regions. Antibody molecules are particularly well suited to such manipulations because of their domain structure: by exchanging domains or entire isotypes between antibodies, it is possible to produce functional molecules. For the most part, these antibodies retain their target specificity and show reduced HAMA responses.

Depending on the application, antibodies would have different biologic properties. For example, the ideal diagnostic antibody may be target-specific, radiolabeled, and rapidly cleared from circulation. In contrast, a therapeutic antibody might be designed to engage a localized immune response. Chimeric antibody construction provides the means to produce and compare antibodies with different structures, to ensure the rational selection and design of the antibody best suited for the desired purpose.

Genetic engineering has also proved useful in the isolation and characterization of immunoglobulin from species for which myeloma lines are unavailable. Murine variable regions have been expressed with rabbit constant regions. Similarly Ig isotypes have been isolated from bovine serum. Genetic engineering can also be used to produce isotypes such as IgD, which are normally available in limited quantities.

## 5 HUMANIZED ANTIBODIES

In chimeric antibodies the variable region remains completely murine. However, the structure of the antibody domain makes it possible to produce variable regions of comparable specificity which are predominantly human in origin. The antigen-combining site of an antibody is formed from the six complementarity-determining regions (CDRs) of the variable portion of the heavy and light chains. Each antibody domain consists of seven antiparallel β-strands forming a β-barrel with loops connecting the β-strands. Among the loops are the CDR regions. It is feasible to move the CDRs and their associated specificity from one scaffolding β-barrel to another. However, it is rarely sufficient merely to combine the CDRs from a murine antibody with a completely human framework. Usually the mouse residues that make key contacts with the CDRs must be also introduced into the human framework along with the CDRs themselves. If it is found that the CDR-grafted antibody has reduced or no binding activity, new constructs can be made that incorporate additional mouse residues near the CDRs until binding is restored. A major impetus behind CDR grafting has been the need to produce antibodies that are less immunogenic, hence should function more effectively as therapeutic agents. However the CDR regions are associated with the idiotypic determinants and the humanized Mabs may evoke an anti-idiotypic response.

## 6 FUSION PROTEINS

Fusion proteins with antibodies can be achieved in several different ways. The variable regions from the antibody can be fused to nonantibody proteins, endowing them with the binding specificity inherent in the antibody molecule. Depending on the position of the substitution, different antibody-related effector functions and biologic properties will be retained. An example is the fusion of the catalytic β-chain of tissue-type plasminogen activator (t-PA) to an antifibrin antibody in which the combination of high affinity fibrin binding and plasminogen activation results in a thrombolytic agent that is more specific and more potent than t-PA alone. Alternatively, nonimmunoglobulin sequences can be substituted for the variable region of the antibody; examples are CD4, interleukin 2, and tumor necrosis factor receptor. These molecules have been called ''immunoadhesins'' because they contain an adhesive molecule linked to the immunoglobulin Fc effector domains. In these proteins the fused moiety acquires antibody-associated properties such as effector functions or improved pharmacokinetics.

## 7 ANTIBODY FRAGMENTS

Fab, Fv, and single chain Fv (scFv) fragments with $V_H$ and $V_L$ joined by a polypeptide linker exhibit specificities and affinities for antigen similar to the original monoclonal antibodies (Figure 1). The scFv fusion proteins can be produced with a nonantibody molecule attached to either the amino or carboxy terminus. In these molecules the Fv can be used for specific targeting of the attached molecule to a cell expressing the appropriate antigen. Bifunctional antibodies can also be created by engineering two different binding specificities into a single antibody chain.

## 8 CATALYTIC ANTIBODIES

Catalytic antibodies represent a novel extension of classical notions of protein structure and function. Although both enzymes and antibodies exhibit binding specificity, enzymes bind and stabilize the transition state of a reaction while antibodies normally bind a compound in its ground state. However an antibody specific for the transition state should accelerate a reaction by binding and stabilizing the transition state, and the use of transition state analogues as haptens has been effective for in vitro production of monoclonal catalytic antibodies. While most of the catalytic antibodies described to date have been produced by standard hybridoma technology, the bacteriophage λ system has been used in attempts to increase the repertoire of catalytic antibodies. The advantage of this system is that antibody production and screening can be performed more rapidly, and with many more candidates, than is possible with hybridoma production. As discussed earlier, the bacteriophage λ combinatorial libraries are limited in that the screening systems can identify only antibodies that strongly bind antigen. These are not necessarily the best catalytic antibodies; it is often preferable that the antibody bind the transition state analogue weakly. Possibly the filamentous phage libraries will provide better catalytic antibodies.

## 9 CONCLUSIONS

Rapid progress has been made in producing genetically engineered antibodies with novel characteristics. Technologies for expressing foreign DNA in a variety of host cells have made it possible to produce antibodies in sufficiently large quantities for study, therapy, or other applications. Developments such as the bacteriophage expression systems may be possible to obtain a greater variety of combining specificities than were available earlier. The challenge for the future is to exploit the available information and production systems for the design and synthesis of antibodies with the desired combination of binding specificities and biologic properties.

*See also* COMBINATORIAL PHAGE ANTIBODY LIBRARIES; IMMUNOLOGY.

### *Bibliography*


Adair, J. R. (1992) Engineering antibodies for therapy. *Immunol. Rev.* 130:5–40.

Burton, D. R., and Woof, J. M. (1992) *Human Antibody Effector Function*, Vol. 51, pp. 1–84. Academic Press, San Diego, CA.

Hoogenboom, H. R., Marks, J. D., Griffiths, A. D., and Winter, G. (1992) Building antibodies from their genes. *Immunol. Rev.* 130:41–68.

Pluckthun A. (1992) Mono- and bivalent antibody fragments produced in *Escherichia coli:* engineering, folding and antigen binding. *Immunol. Rev.* 130:151–188.

Winter, G., and Milstein, C. (1991) Man-made antibodies. *Nature*, 349:293–299.

Wright, A., Shin, S.-U., and Morrison, S. L. (1992) Genetically engineered antibodies: Progress and prospects. *Crit. Rev. Immunol.* 23:301–321.


# ANTISENSE OLIGONUCLEOTIDES, STRUCTURE AND FUNCTION OF

*Eugen Uhlmann and Anusch Peyman*

### *Key Words*

**Antisense Oligonucleotide** Synthetic single-strained nucleic acid, built up of about 12 to 25 mononucleotides, which binds *via* Watson-Crick base-pairing to certain complementary ("sense") regions on messenger RNA resulting in inhibition of protein synthesis.

**Antisense Oligonucleotide Analogue** An antisense compound whose chemical nature has been altered relative to the naturally occurring oligodeoxyribonucleotide structure aiming at improved biophysical or biological properties.

**Binding Affinity** Measure for the strength of binding of an oligonucleotide to its target nucleic acid, usually expressed as the experimentally determined melting temperature $T_m$, the temperature at which 50% of the double strand has dissociated into its single strands.

**Cellular Uptake** Time-dependent process by which an oligonucleotide penetrates into the cell during its incubation in tissue culture.

**Nuclease Stability** Stability of an oligonucleotide against various nucleic acid degrading enzymes present in serum or within cells.

**RNase H** An ubiquitous cellular enzyme which recognizes a hetero-duplex between DNA and RNA, resulting in the cleavage of the RNA strand of said duplex.

---

In recent years, a noval class of potential pharmaceuticals of high interest has been developed which bind in a predictable way to certain nucleic acid target sequences aiming at selective inhibition of expression of disease-causing genes. The chemical structure of these so-called antisense oligonucleotide compounds must be altered relative to their natural models to render them stable under in vivo conditions and to allow their penetration to the site of action inside cells. In this chapter, the principle of antisense oligonucleotide function, the structure of antisense oligonucleotide analogues, the mechanism of action, the different strategies for improvement of their biological potency, and selected reports on successful in vitro and in vivo studies using these antisense oligonucleotides are summarized.

## 1 PRINCIPLES

Rational drug design is the ultimate goal in pharmaceutical research. Although many of today's drugs act by direct inhibition of enzymes or other proteins, the exact nature of such interactions is known only in a few cases. Moreover, a therapeutic intervention at the messenger RNA or DNA level often appears to be advantageous. On active transcription, every gene is transcribed into $10^2$ to $10^4$ copies of mRNA, which in turn are translated into $10^4$ to $10^6$ copies of protein molecules (Figure 1).

Molecules inhibiting transcription or translation can be designed on a highly rational basis. Zamecnik and Stephenson in 1978 were the first to propose the use of oligonucleotides directed against complementary viral nucleic acid sequences for inhibition of virus replication. Since then a great deal of work has been devoted to this new therapeutic principle.

*Antisense oligonucleotides* are synthetic oligonucleotides that bind via Watson–Crick base pairing to certain complementary regions on the mRNA, thereby inhibiting protein biosynthesis. In contrast, *triple helix forming oligonucleotides* bind via Hoogsteen base pairing to the major groove of double-stranded DNA, resulting in inhibition of transcription (Figure 2).

Specificity and rational design are the main attributes of these two classes of potential nucleic acid therapeutics: because of the general base-pairing rules, therapeutic intervention using oligonucleotide antagonists is a universal approach, which is applicable to many diseases whose causative agents or targets, such as viruses, oncogenes, receptors, ion channels, and immunomodulators, have been characterized on the DNA level. The rational design of oligonucleotide antagonists aimed at the inhibition of gene expression is simply based on the target nucleic acid base sequence. Their specificity is determined by the occurrence of a given sequence. It has been calculated that a 17-mer oligonucleotide sequence appears statistically just once in the human genome ($4 \times 10^9$ base pairs). This consideration, supported by in vitro hybridization studies, suggests that oligonucleotide antagonists should indeed act highly specifically. Finally, it ought to be mentioned that the antisense principle is also used in nature (antisense RNA) to regulate gene expression.

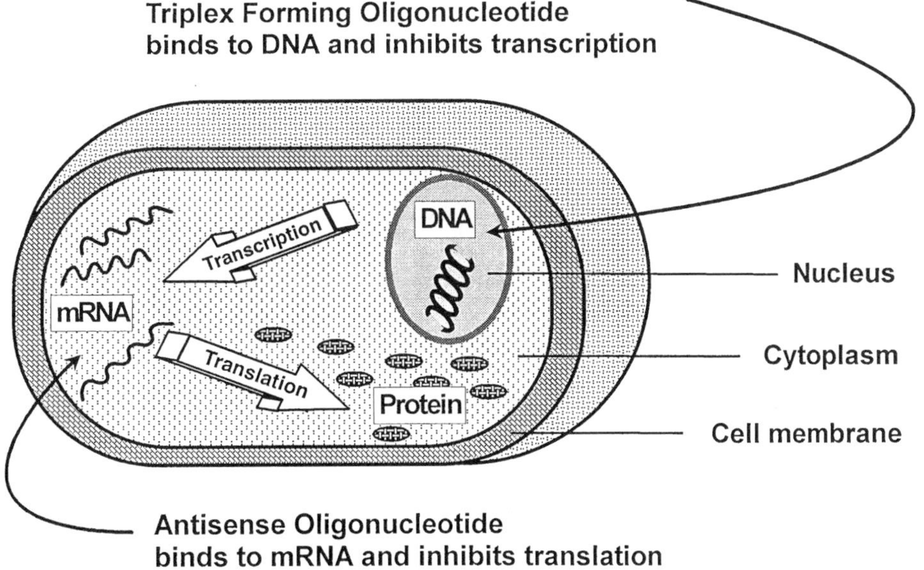


**Figure 1.** Principle of action of antisense oligonucleotides and triple helix forming oligonucleotides.

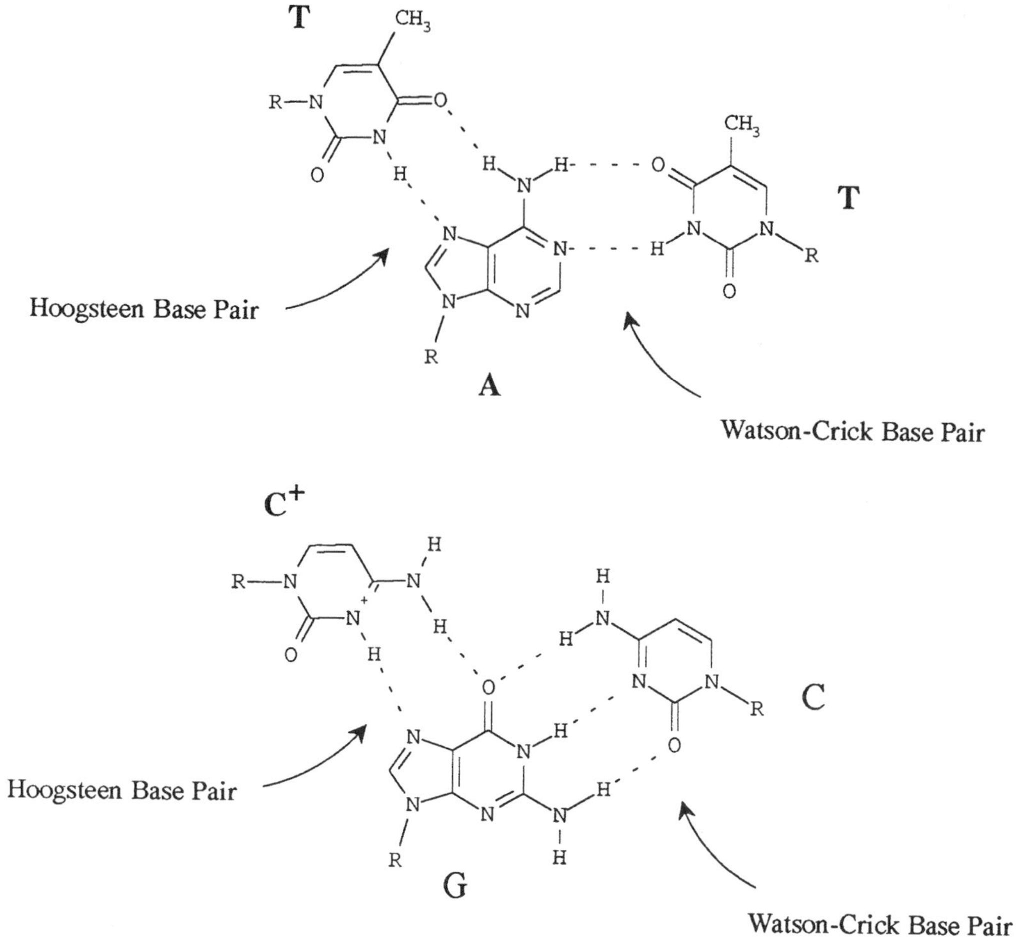

**Figure 2.** Base triplets TAT and $C^+GC$. The third strand is bound in the major groove by Hoogsteen hydrogen bonding to the purine in the Watson–Crick duplex. Whereas duplex formation is a highly general principle simply based on Watson–Crick base pairing, triple-helix formation by Hoogsteen base pairing is limited to purine-rich stretches of DNA.

## 2 STRUCTURE AND PROPERTIES OF ANTISENSE OLIGONUCLEOTIDES

Antisense and triple helix forming oligonucleotides can be used as tools for mechanistic studies in molecular biology or as gene expression inhibiting therapeutics. From a practical point of view, the development of antisense oligonucleotides and triplex-forming oligonucleotides as drugs is still at an early stage. To put the antisense or triple-helix principle into practice, the oligonucleotide analogues must have the following properties:

1. They must be sufficiently stable in serum and inside the cell to be able to carry out the desired inhibitory effect.
2. They must be able to penetrate cellular membranes, and they should enter the various organs of the body to reach their side of action.
3. They should form stable Watson–Crick or Hoogsteen complexes with complementary target sequences under physiological conditions.
4. The interaction of the oligonucleotide with its target sequence must be sequence specific.

### 2.1 Structure of Oligonucleotides and Their Analogues

To meet the four requirements just listed, numerous chemical modification of ''natural'' oligonucleotides have been made, including modifications or replacement of the phosphodiester backbone, base and sugar modifications, and conjugate formation at different sites of the oligonucleotide. It is beyond the scope of this entry to discuss the whole range of chemical variations that have been realized so far, and for which in many cases relevant biological data have not yet been reported. We restrict ourselves to the most common and best investigated modifications, which are those of the phosphate backbone, especially phosphorothioates, phosphoramidates, and methylphosphonates, and the blocking or removal of the hydroxyl group at the 3′ terminus (Figure 3). Recently, ''dephospho'' oligonucleotides in which the phosphodiester group has been replaced completely have been described, examples are formacetals, thioformacetals, methylhydroxylamines, and many others (Figure 4). In even more drastic changes such as the ''peptide nucleic acids'' or the ''morpholino nucleoside'' oligomers (Figure 4), the whole phosphate sugar backbone was replaced by a peptide chain or a carbamate-linked morpholino chain, respectively. Although some

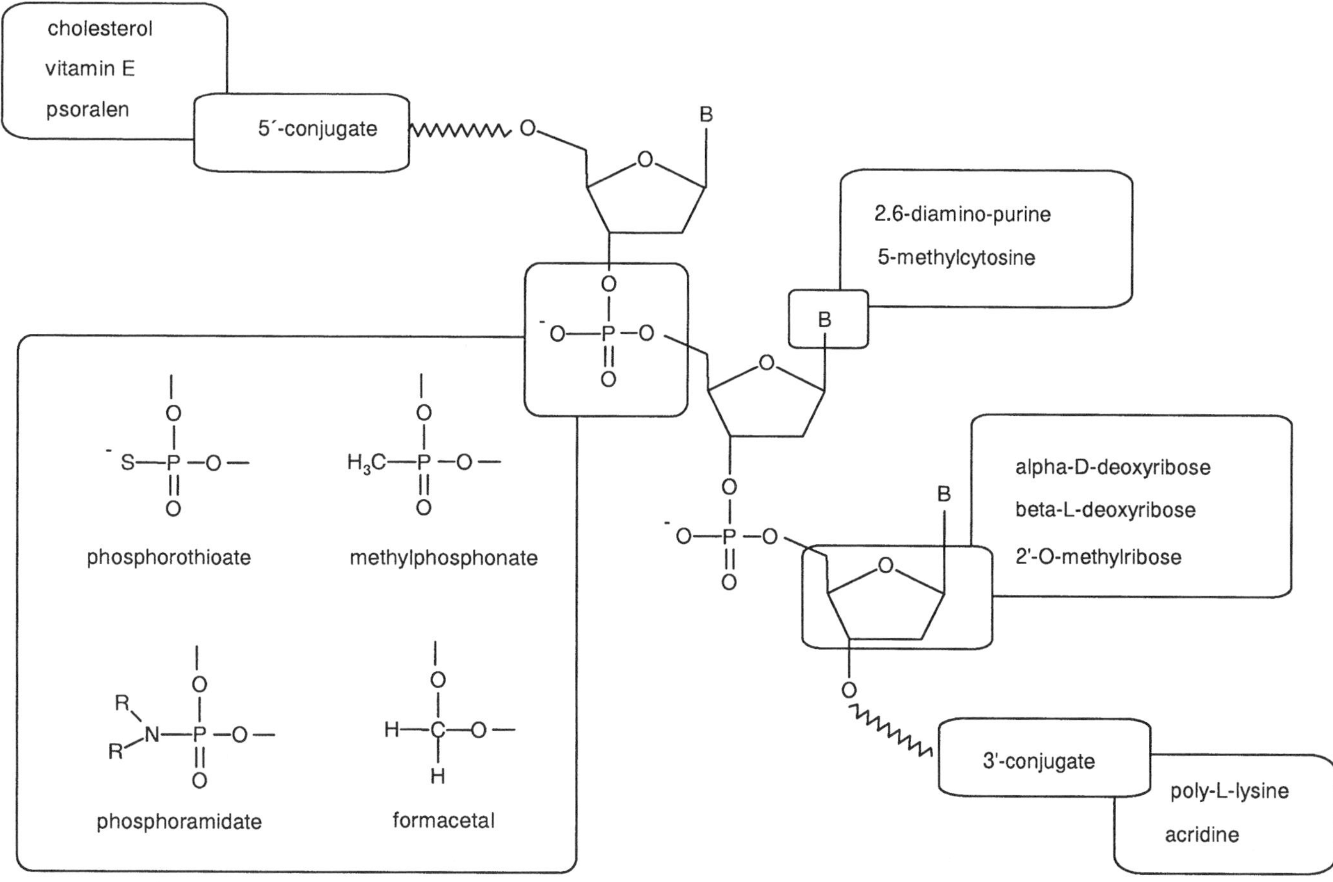


**Figure 3.** Possibilities for the modification of oligonucleotides.

of these "dephospho" oligonucleotides exhibit excellent hybridization properties and nuclease stability, there are only limited data available on their biological activity including cellular uptake.

## 2.2 Stability of Oligonucleotides Against Nucleases

Unmodified oligonucleotides with a natural phosphodiester backbone are degraded in serum within a few hours, mainly by the action of 3′-exonucleases. Removal or blockage of the 3′-hydroxyl group of oligonucleotides as well as modification of the last few 3′-terminal internucleoside phosphate linkages enhances considerably their stability against 3′-exonucleases. In some tissues a significant 5′-exonucleolytic activity is observed. Therefore, a quite popular modification is exchange of the last two to five phosphate residues at the 3′ and 5′ ends of oligonucleotides for phosphorothioate or methylphosphonate linkages, respectively. This exchange enhances nuclease stability by at least an order of magnitude while retaining the sequence specificity of the oligonucleotides. These "end-capped" oligonucleotides are not perfectly stable under in vivo conditions because they are subject to degradation by endonucleases. However, modification of the whole oligonucleotide backbone can strongly influence other properties of the corresponding oligonucleotide analogues, such as lipophilicity, solubility, or binding affinity to complementary sequences, as discussed in the following sections. Degradation of oligonucleotides in biological systems also can be slowed down by many other derivatizations, such as sugar modifications (α-deoxyribose, 2′-modified ribose) or certain base modifications. In addition, protection against nucleases has been achieved by packaging of oligonucleotides into liposomes or into nanoparticles, which also can be considered to be convenient carriers for in vivo applications.

## 2.3 Cellular Uptake of Oligonucleotides

As with nuclease stability, the membrane transport and cellular distribution of oligonucleotides is largely a function of the chemical modifications introduced. Unmodified oligonucleotides are typically 12 to 25 nucleotides long and can be characterized as highly hydrophilic anionic compounds. Nonetheless they are readily taken up by living cells by endocytosis, which most likely is receptor-mediated. However, the concentration of free oligonucleotide in the cytoplasm is considerably lower because the oligomer is packaged into lysosomes or endosomes. Interestingly, once the oligonucleotide has been set free from this reservoir or the oligonucleotide has been microinjected into the cytoplasm, it rapidly penetrates to the cell nucleus. Improved cellular uptake can be achieved by lipophilic derivatization of the oligonucleotides (e.g., as lipophilic conjugates at the 5′ terminus), by conjugation to poly-L-lysine, or by packaging into antibody-targeted liposomes. It is still not clear whether uncharged and lipophilic methylphosphonates are taken up by passive diffusion or by an active uptake mechanism. Phosphorothioates at high concentration can compete with normal phospho-

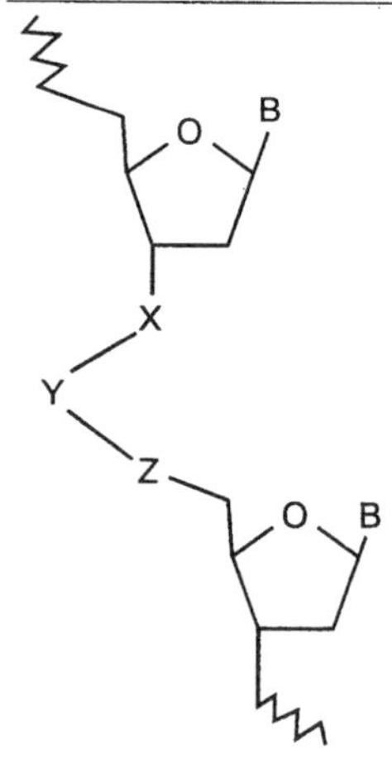

| X | Y | Z | Compound | Literature |
|---|---|---|---|---|
| O | $CH_2$ | O | Formacetal | M. Matteucci, *Tetrahedron. Lett.*, 1990 |
| O | $CH_2$ | S | Thioformacetal | M. Matteucci, *J. Am. Chem. Soc.*, 1991 |
| $CH_2$ | N—$CH_3$ | O | Methylhydroxylamine | J.J. Vasseur, *J. Am. Chem. Soc.*, 1992 |
| CH | =N | O | Oxime | J.J. Vasseur, *J. Am. Chem. Soc.*, 1992 |
| $CH_2$ | N—$CH_3$ | N—$CH_3$ | Methylenedimethylhydrazo | J.J. Vasseur, *J. Am. Chem. Soc.*, 1992 |
| $CH_2$ | $SO_2$ | $CH_2$ | Dimethylenesulfone | S. Benner, *J. Org. Chem.*, 1991 |
| O | $Si(CH_3)_2$ | O | Silyl | H. Seliger, *Nucleotides Nucleosides*, 1987 |

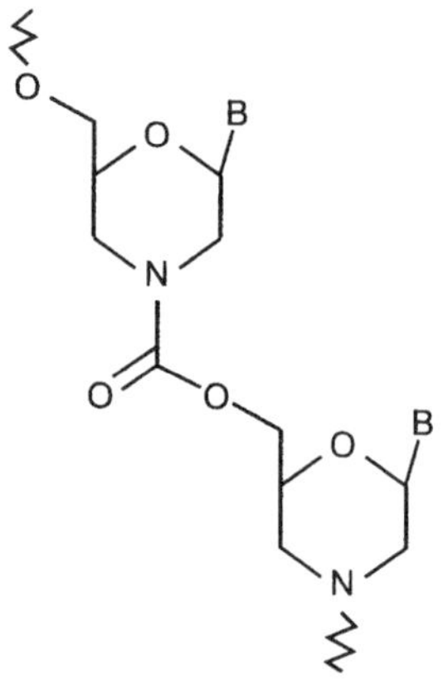

Morpholino nucleoside oligomers
E.P. Stirchak et al., *Nucleic Acids Res.*, 1989

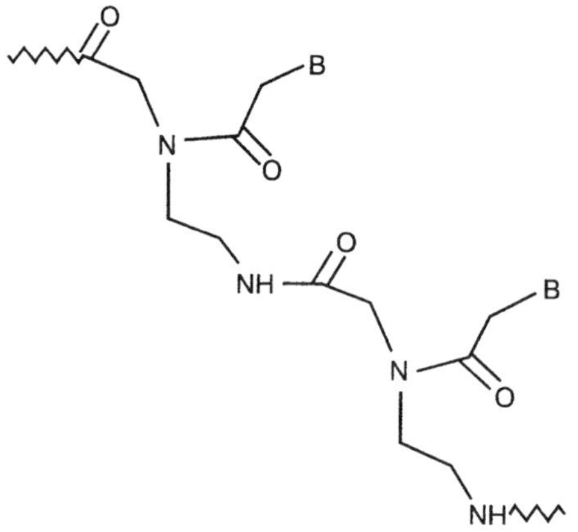

Peptide nucleic acids
P.E. Nielsen et al., *Science*, 1991

**Figure 4.** Examples of "dephospho" oligonucleotides.

diester oligonucleotides for uptake, and they need more time than normal oligonucleotides to reach an intracellular equilibrium. Peptide nucleic acids appear to penetrate very poorly into cells. In general, cellular uptake of oligonucleotides depends on time, energy, and the cell type used for the study. Since there exists also an efflux of oligonucleotides from the cells, recycling of oligonucleotides must be taken into account. To obtain predictions of biological efficacy, care is necessary when comparing different chemical derivatizations for their uptake. As with other therapeutic compounds, even of low molecular weight, the cell uptake of oligonucleotides in tissue culture may not necessarily correlate with their pharmacological efficacy in animal studies.

## 2.4 Binding Affinity of Oligonucleotide Analogues

Modification of an oligonucleotide is usually accompanied by a change in binding affinity to the complementary sequence. If the internucleotide bond is altered, the factors that determine affinity are charge, sterical requirements, and absolute stereochemical arrangement. The latter problem is especially important for the most common modifications: replacement of one oxygen atom in the phosphate moiety by sulfur, methyl, or alkylamino substituents creates a center of chirality at the phosphorus. Therefore oligonucleotides having $n$ internucleotidic phosphate groups replaced by methylphosphonate moieties will result after standard oligonucleotide synthesis in a mixture of $2^n$ diastereoisomers. The binding affinity of random methylphosphonates is roughly the same as or slightly lower than that of unmodified oligonucleotides, probably as a result of reduced charge repulsion, whereas methylphosphonates with regular stereochemistry (*R* configuration at the phosphorus) have a much higher affinity. Introduction of phosphorothioate moieties (*R, S* mixture) into oligonucleotides usually lowers their melting temperature by approximately 0.4–1.0°C per altered residue. The melting temperature $T_m$ is the temperature at which 50% of the double strand has dissociated into its single strands. It should be pointed out that the $T_m$ values often differ significantly when

RNA, not DNA, serves as the complementary strand. Unfortunately, there is still no suitable stereospecific synthesis available for methylphosphonates or for phosphorothioates. The absolute stereochemistry of phosphorothioates seems to have less influence on binding. Introduction of α-nucleotides or 2′-*O*-alkylribonucleotides into oligonucleotides enhances their binding affinity to RNA. Peptide nucleic acids also bind more strongly than natural oligonucleotides to RNA and DNA. Other approaches to the improvement of binding affinity are conjugation with intercalating or cross-linking agents such as acridine or psoralen. For triple helix forming oligonucleotides, improvement in binding is brought about by the use of modified bases, such as 5-methylcytosine. Interestingly, cyclic oligonucleotides with improved nuclease stability have been reported which bind with high affinity through triple-helix formation to appropriate single-stranded target sequences. Thereby, the antisense part of the cyclic oligonucleotide forms a duplex via Watson–Crick base pairing and the triplex-forming part of the cyclo-oligonucleotide completes the triple stranded structure via Hoogsteen base pairing.

## 3 SYNTHESIS OF OLIGONUCLEOTIDES

Small-scale synthesis of oligonucleotides is preferably performed using solid phase chemistry in combination with the phosphoramidite method. Commercially available DNA synthesizers allow synthesis at 0.2 μM to 2 mM scale, thus providing gram amounts of oligonucleotides for animal testing. The challenge for the development of oligonucleotides as drugs is certainly the achievement of synthesis on a large scale. There is an ongoing scientific discussion about what type of chemistry (solution or solid phase synthesis, phosphoramidite, *H*-phosphonate, or phosphotriester chemistry) should be used for this purpose; A detailed analysis of the vast field of chemistry involved, however, is far beyond the scope of this entry. A more detailed description of the synthesis of oligonucleotides and some analogues is given under Oligonucleotides. We restrict ourselves to mentioning innovative approaches such as the use of soluble polymeric supports to combine the advantages of solid phase and solution chemistry, or the use of unprotected monomeric building blocks, a great simplifying and cost-reducing measure. Other recent important developments for pharmaceutical production in the analysis of oligonucleotides include electrospray mass spectrometry and capillary electrophoresis.

## 4 MECHANISM OF ACTION

Triplex-forming oligonucleotides are designed for transcriptional inactivation and antisense oligonucleotides for translational inactivation, respectively. The mechanism of the latter largely depends on the selected target site. Several processes may be interrupted: splicing, polyadenylation, correct RNA folding, translocation and initiation of translation of mRNA, or ribosome movement along the mRNA. Complications may arise because completely different mechanisms of action can dominate effects seen in cell culture experiments, namely, blockade of cellular receptors or direct inhibition of viral enzymes such as DNA polymerase or reverse transcriptase. In antiviral assay systems, blockade of virus adsorption or induction of the interferon system may also account for the biological effect. The question of the extent to which the antisense effect is supported through cleavage of the mRNA by cellular RNase H remains to be answered completely. RNase H recognizes DNA-RNA hybrid molecules and cleaves the RNA strand. It has been suggested that DNA-mRNA hybrids spanning only a few base pairs are substrates of RNase H, leading to unspecific breakdown of mRNA. This is why the action of RNase H could be deleterious to the antisense approach when natural phosphodiesters or phosphorothioate oligonucleotides are used. However, many oligonucleotide derivatives, such as methylphosphonates, α-anomeric oligonucleotides, 2′-*O*-alkylribose derivatives and dephospho-oligonucleotide analogues, do not induce RNase H cleavage. Consequently, hybrid molecules bearing stretches of these non-RNase H inducing modifications at the 3′ and 5′ termini, while having only small regions (''windows'') of RNase H inducing nucleotides, are under investigation as potential inhibitors of increased specificity.

## 5 APPLICATIONS

### 5.1 In Vitro Activity

A number of papers have been published in recent years describing the efficacy of antisense oligonucleotides directed against various targets in cell culture experiments. Table 1 summarizes some representative examples of in vitro activities of antisense and triplex-forming oligonucleotides against different viruses, oncoproteins, receptors, cytokines, and growth factors. In most instances, the effective dose for inhibition of expression is in the range of 0.1 to 50 μM of oligonucleotide. Unmodified oligonucleotides are active only in the absence of serum; phosphorothioate oligonucleotides seem to be more active than methylphosphonate oligonucleotides. Conjugation of psoralen to the latter can enhance significantly their activity under irradiation. In summary, oligonucleotide antagonists have demonstrated a broad palette of sequence-specific activities against different targets: viruses, receptors, enzymes, oncogenes, growth factors, and other host gene products.

### 5.2 In Vivo Activity, Acute Toxicity, and Organ Distribution of Antisense Oligonucleotides

Efficacy of antisense oligonucleotides in vivo against different targets, including viral infections, has been reported recently. Topical application of methylphosphonate oligonucleotides to mice was shown to be effective against herpes simplex virus 1 (HSV-1), and in another study topical application of phosphorothioate antisense oligonucleotides to the cornea of mice infected with HSV-1 inhibited viral growth and cured the infection. In this model antiviral activity of the antisense oligonucleotide was equivalent to that of trifluorothymidine without concomitant occurrence of local or systemic toxicity. Oligonucleotides with alkylating groups indicated protection against tick-borne encephalitis virus infection in mice. Successful in vivo inhibition of duck hepatitis B virus replication by phosphorothioates has demonstrated antiviral efficacy of antisense oligonucleotides when administered systemically. Specific inhibition of neurotransmitter receptor expression in the living brain by antisense oligonucleotides has been shown to modulate anxiety. Only recently, c-*myb* antisense oligonucleotides administered as pluronic solution were shown to inhibit intimal arterial smooth muscle cell accumulation in damaged neck arteries of laboratory rats. However, the mechanism of action of the oligonucleotide antagonists has still to be evaluated. Another animal study

**Table 1** Examples of in Vitro Activities of Antisense and Triplex-Forming Oligonucleotides

| Targets | Type of Cell | Type of Oligomer[a] | Effective Dose (μM) | Literature |
|---|---|---|---|---|
| **Viruses** | | | | |
| Rous sarcoma | Fibroblasts | P-O, P-S | 10 | P. Zamecnic, *Proc. Natl. Acad. Sci. U.S.A.*, 1978 |
| HIV | H T Cells | P-S | 0.5 | M. Matsukura, *Proc. Natl. Acad. Sci. U.S.A.*, 1987 |
| HSV | Vero | P-Me | 50–100 | C. C. Smith, *Proc. Natl. Acad. Sci. U.S.A.*, 1986 |
| HSV | Vero | P-Me-(psoralen) | 5 | C. C. Smith, *Proc. Natl. Acad. Sci. U.S.A.*, 1986 |
| VSV | L 929 | P-O-(poly-L-Lys) | 0.1 | L. Lemaitre, *Proc. Natl. Acad. Sci. U.S.A.*, 1987 |
| Influenza | MDK | P-S | 1.25 | J. Leiter, *Proc. Natl. Acad. Sci. U.S.A.*, 1990 |
| Hepatitis B | PLC/PRF/5 | P-O, P-S | 0.3–17 | G. Goodarzi, *J. Gen. Virol.*, 1990 |
| SV 40 | CV-1 | P-O-(acridine) [tfo] | 15–30 | F. Birg, *Nucleic Acids Res.*, 1990 |
| **Nuclear Oncoproteins** | | | | |
| c-*myb* | HL-60 | P-O-(TF-poly-Lys) | 50 | G. Citro, Proc. Natl. Acad. Sci. U.S.A., 1992 |
| c-*myb* | BC3H1 | P-S | 10–25 | M. Simons, *Circ. Res.*, 1992 |
| c-*myc* | HL-60 | P-O, P-S | 10 | E. L. Wickstrom, *Cell Dev. Biol.*, 1989 |
| c-*myc* | Burkitt cells | P-O | 100 | M. E. McManaway, *Lancet*, 1990 |
| c-*myc* | T lymphocytes | P-O | 30 | R. Heikkila, *Nature*, 1987 |
| **Cytoplasmic/Membrane Oncoproteins** | | | | |
| Ha-*ras* | T24 | P-O-Acr | 1–10 | E. L. Wickstrom, *Cell Dev. Biol.*, 1989 |
| *bcl*-2 | Pre-B ALL | P-O, P-S | 25–150 | J. C. Reed, *Cancer Res.*, 1990 |
| **Cellular Receptors** | | | | |
| ICAM-1 | A 549 | P-S | 0.5–1 | M.-Y. Chiang, *J. Biol. Chem.*, 1991 |
| IGF-1R | BALB/c3T3 | P-O | 15 | E. Surmacz, *Exp. Cell. Res.*, 1992 |
| IL2Rα | PB lymphocytes | P-O-(amino) [tfo] | 10 | F. M. Orson, *Nucleic Acids Res.*, 1991 |
| **Cytokines/Growth Factors** | | | | |
| IL-2 | T cells | P-O | 5 | A. Harel-Bellan, *J. Exp. Med.*, 1988 |
| IL-1β | Monocytes | P-S | 0.5–2.5 | J. Manson, *Lymphokine Res.*, 1990 |
| IL-1α | HUVEC | P-O | 10 | J. A. Maier, *Science*, 1990 |
| CSF-1 | Monocytes (FL) | P-O | 5–10 | R. M. Birchenall, *J. Immunol.*, 1990 |

[a]P-O, phosphodiester; P-S, phosphorothioate; P-Me, methylphosphonate; -(psoralen), psoralen conjugate; -(poly-L-Lys), poly-L-lysine conjugate, -(TF-poly-Lys), transferrin-poly-L-lysine conjugate; -(acridine), acridine conjugate; -(amino), aminopropandiolphosphate; tfo, triplex-forming oligonucleotide.

suggests sequence-specific inhibition of gene expression by antisense oligonucleotides administered locally by continuous perfusion, resulting in decreased tumor growth. Treatment of human leukemia in a severe combined immunodeficiency (SCID) mouse model could be simulated using phosphorothioate oligonucleotides against c-*myb*. Furthermore, ablation of transplanted HTLV-I tax-transformed tumors in mice by antisense inhibition of nuclear factor NF-κB has been observed by intraperitoneal injection of 3′-end-capped phosphorothioates. In the first human trial, no response of the leukemia was observed in a patient after a 10-day infusion with a phosphorothioate antisense oligonucleotide (0.05 mg/kg/h) directed against p53. However, this study showed that systemic administration to human beings is possible without major toxicity. This conclusion is supported by the few animal studies reported so far, in which no serious acute toxicity has been detected. In continuous infusion experiments on adult male rats, applying 5–150 mg of 27-mer phosphorothioate oligonucleotide via osmotic pumps, no significant decreases in body weight or organ weights were observed. In another study in which phosphorothioate oligonucleotides were administered in a single dose either intravenously or intraperitoneally, the oligomer was found for up to 48 hours in most of the tissues. Daily administration of 100 mg of oligonucleotide per kilogram of body weight for 14 days did not cause observable toxicity in mice. Organ distribution of a 3′-end-modified oligonucleotide 4 hours after intravenous or intraperitoneal injection has been reported to be in the following rank order: kidney > liver > spleen > heart, lung > muscle, ear >> brain.

## 6 PERSPECTIVES

In summary, antisense and triple helix forming oligonucleotides are valuable tools in pharmacology and promising candidates for pharmaceutical development: chemical modifications render oligonucleotide derivatives that are stable enough to survive in serum for a pharmacologically relevant time. New oligonucleotide analogues containing dephospho internucleoside linkages are still being investigated for enhanced properties. In tissue culture experiments, antisense as well as triple helix forming oligonucleotides inhibit gene expression, usually in the range of 0.1 to 50 μM, depending mainly on the target sequence and the chemical modification of the oligonucleotide. Finally, preliminary human data indicate no acute toxicity and the most advanced compounds have entered phase III clinical studies.

*Bibliography*

Agrawal, S., Temsamani, J., and Tang, J. Y. Pharmacokinetics, biodistribution and stability of oligonucleotide phosphorothioates in mice. *Proc. Natl. Acad. Sci. U.S.A.* 88:7595 (1991).

Bayever, E., Iversen, P., Smith, L., Spinolo, J., and Zon, G. Systemic human antisense therapy begins. *Antisense Res. Dev.* 2:109 (1992).

Beaucage, S., and Iyer, R. Advances in the synthesis of oligonucleotides by the phosphoramidite approach. *Tetrahedron,* 48:2223 (1992).

Cook, P. D. Medicinal chemistry of antisense oligonucleotides—Future opportunities. *Anti-Cancer Drug Design,* 6:585 (1991).

(a) Crooke, S. T. Therapeutic applications of oligonucleotides. *Bio/Technology,* 10:882–885 (1992). (b) Crooke, S. T. Therapeutic applications of oligonucleotides. *Annu. Rev. Pharmacol. Toxicol.* 32:329 (1992).

Eguchi, Y., Itoh, T., and Tomizawa, J. Antisense RNA. *Annu. Rev. Biochem.* 60:631 (1991).

Engels, J., and Uhlmann, E. Gensynthese. *Angew. Chem.* 101:733–752 (1989); *Angew. Chem., Int. Ed. Engl.* 28:716 (1989).

Iversen, P. In vivo studies with phosphorothioate oligonucleotides: Pharmacokinetics prologue. *Anti-Cancer Drugs Design,* 6:531 (1991).

Kitajima, I., Shinohara, T., Bilakovics, J., Brown, D. A., Xu, X., and Nerenberg, M. Ablation of transplanted HTLV-I *tax*-transformed tumors in mice by antisense inhibition of NF-κB. *Science,* 258:1792 (1992).

Kool, E. T. Molecular recognition by circular oligonucleotides: Increasing the selectivity of DNA binding. *J. Am. Chem. Soc.* 113:6265 (1991).

Miller, P., and Ts'o, P.O.P. A new approach to chemotherapy based on molecular biology and nucleic acid chemistry: Matagen (masking tape for gene expression). *Anti-Cancer Drug Design,* 2:117 (1987).

Neckers, L., Whitesell, L., Rosolen, A., and Geselowitz, D. A. Antisense inhibition of oncogen expression. *Crit. Rev. Oncogenesis,* 3:175 (1992).

Offensperger, W.-B., Offensperger, S., Walter, E., Teubner, K., Igloi, G., Blum, H. E., and Gerok, W. In vivo inhibition of duck hepatitis B virus replication and gene expression by phosphorothioate modified antisense oligodeoxynucleotides. *EMBO J.* 12:1257 (1993).

Ratajczak, M., Kant, J. A., Luger, S. M., Hijiya, N., Zhang, J., Zon, G., and Gewirtz, A. M. In vivo treatment of human leukemia in a scid mouse model with c-*myb* antisense oligonucleotides. *Proc. Natl. Acad. Sci. U.S.A.* 89:11823 (1992).

Simons, M., Edelman, E., DeKeyser, J., Langer, R., and Rosenberg, R. Antisense c-*myb* oligonucleotides inhibit intimal arterial smooth muscle cell accumulation in vivo. *Nature,* 359:67 (1992).

Uhlmann, E., and Peyman A. Antisense oligonucleotides: A new therapeutic principle. *Chem. Rev.* 90:543 (1990).

Uhlmann, E., and Peyman, A. *Oligonucleotide analogues containing dephospho internucleoside linkages. In Methods in Molecular Biology: Oligonucleotide Synthesis Protocols,* S. Agrawal, Ed. Humana Press, Totowa, NJ, (1993).

Wahlestedt, C., Pich, E. M., Koob, G. F., Yee, F., and Heilig, M. Modulation of anxiety and neuropeptide Y-Y1 receptors by antisense oligonucleotides. *Science,* 259:528 (1993).

Whitesell, L., Rosolen, A., and Neckers, L. M. In vivo modulation of N-*myc* expression by continuous perfusion with an antisense oligonucleotide. *Antisense Res. Dev.* 1:343 (1991).

Zamecnik, P., and Stephenson, M. Inhibition of Rous sarcoma virus replication and cell transformation by a specific oligodeoxynucleotide. *Proc. Natl. Acad. Sci. U.S.A.* 75:280 (1978).

Zendegui, J. G., Vasquez, K. M., Tinsley, J., Kessler, D. J., and Hogan, M. E. In vivo stability and kinetics of absorption and disposition of 3′-phosphopropyl amine oligonucleotides. *Nucleic Acids Res.* 20:307 (1992).


# *Arabidopsis* Genome

*Elliot M. Meyerowitz*

## *Key Words*

**Angiosperm** A flowering plant.

***Arabidopsis thaliana*** A small plant in the mustard family (Brassicaceae) of flowering plants used as a laboratory organism in plant genetics and molecular biology.

**Genome** The total of nucleic acid sequences of an organism.

**Nuclear Genome** The total of DNA sequences found in the nucleus of a eukaryotic organism.

**YAC** Yeast artificial chromosome: a type of molecular cloning vector that allows for the molecular cloning of genomic fragments several hundred kilobase pairs long.

The genome of the flowering plant *Arabidopsis thaliana* is of interest because the small amount of nuclear DNA and simple repetitive sequence structure of the genome has allowed several new approaches to the molecular cloning of higher plant genes. Because of the convenience of *A. thaliana* for molecular cloning experiments, and because of the rapid growth, short generation time, small size, and amenability to the methods of classical genetics of this member of the mustard family, *Arabidopsis* has become a favorite laboratory organism of plant molecular biologists and of plant geneticists.

## 1 GENOME SIZE

### 1.1 Size Measurements

That *Arabidopsis thaliana* is among the higher plants with the smallest genomes has been known for some time: Sparrow and Miksche showed in 1961 that radiation sensitivity and DNA content are related in plants, and that *Arabidopsis* is highly resistant to ionizing radiation. In later work, Sparrow and collaborators showed a correlation of nuclear volume and genome size in plants and found *A. thaliana* to have the smallest nuclear volume among the angiosperms examined. Subsequent measurements are more readily converted to quantitative estimates of nuclear genome size: Fuelgen microspectrophotometry indicates a haploid nuclear genome size of around 200 Mb, DNA reassociation analysis shows a size of around 80 Mb, quantitative gel blot hybridization gives a size of 50 Mb, and electron microscopic measurements of chromosome volume indicate a haploid nuclear genome of about 100 Mb. Flow cytometry measurements give an *A. thaliana* haploid genome size of 86 Mb when the yeast *Saccharomyces cerevisiae* is used as a size standard, and a size of 145 Mb when chicken red blood cells are used as a size standard. While this information does not give an exact genomic size for *Arabidopsis,* it indicates a small range around 100 Mb, which can for the present be taken as the best estimate of haploid genome size for the plant.

### 1.2 Coding Content

The nuclear genomes of angiosperms typically are 5–100 times larger than that of *Arabidopsis*, which has the smallest genome known among this group of plants. Estimates of the total number and average size of distinct transcripts in higher plants range from 15,000 transcripts averaging 1.2 kb (in parsley and tobacco), to 60,000 transcripts of average size 1.24 kb (in tobacco). There is no reason to think that *Arabidopsis* has fewer or different genes from other, related plants; its morphology, anatomy, growth, development, and environmental responses are all typical of flowering plants, and it is known that closely related plants can have very different genome sizes. If *Arabidopsis* falls within the range of gene numbers estimated for parsley and tobacco, around one- to three-quarters of the total *A. thaliana* nuclear genome is genic and transcribed, and a gene should be found on average in every 2 to 7 kb of the nuclear DNA.

## 2 SEQUENCE ORGANIZATION

Nonetheless, there is a considerable portion of the *Arabidopsis* nuclear genome that is not genic. Reassociation kinetic studies have shown approximately 10% of the nuclear genome to consist of highly repeated sequences. Only 20% of these have been described at the molecular level, and these constitute three different families of tandem repeats. These are a 180 bp sequence repeated around 5000 times in largely tandem arrays, a 500 bp repeat present in approximately 500 copies, also largely in tandem, and a 160 bp repeat present in around 1500 copies, again, largely in tandem arrays. These repeats have been shown by in situ hybridization to occur, in the main, in centromeric heterochromatin. There are also other repeated sequences in the *Arabidopsis* nuclear genome. Ribosomal DNA repeats around 10 kb long are repeated approximately 600 times per haploid genome, thus comprising 6% of the total nuclear DNA. The ribosomal DNA repeats have been shown by in situ hybridization to cluster on two of the five chromosomes of the haploid genome, chromosomes 2 and 4. There are also dispersed repeats that have sequence similarity to retrotransposons, though each known type of these is present in no more than a few copies per haploid genome. Additional dispersed, moderately repetitive elements have been described, but an analysis of genomic lambda clones representing almost 1% of the nuclear genome revealed only five of these, which seem to exist with an average spacing of one every 125 kb.

## 3 APPLICATIONS TO CLONING METHODS

The pattern of repeated DNA organization described in Section 2 is, so far, unique among the flowering plants. The typical case is for the predominant classes of DNA to be repetitive, with only very small regions of single-copy or low-copy DNA separated by large amounts of repetitive sequence. While the evolutionary advantage or disadvantage of the *Arabidopsis* type of organization is unknown, there is a practical advantage: it is possible to perform chromosome walking in *Arabidopsis;* that is, overlapping cloned fragments can be isolated, which enables the investigator to proceed from a given starting point to any gene of interest. Repetitive DNA makes such experiments difficult, if not impossible, in other plant species. In fact, several genes have now been cloned by chromosome walking in *Arabidopsis.*

The small genome size has also potentiated several other methods of gene cloning. One is subtractive hybridization, in which deletion mutations are obtained and the DNA deleted is identified by hybridizing an excess of mutant plant DNA to wild-type sequences, whereupon the wild-type fraction not hybridized by the mutant genome is recovered. By this method, the *GA1* gene (involved in gibberellin biosynthesis) has been cloned. Another is T-DNA tagging, in which *Agrobacterium*-mediated transformation results in a collection of plants with different locations of integrated, foreign DNA. Because the *Arabidopsis* genome is largely genic, a high proportion of such transformants, when made genetically homozygous for the inserted foreign DNA, show new mutations. These are often caused by the introduced DNA, allowing molecular cloning of the mutated gene by use of probes to the foreign DNA.

## 4 ADDITIONAL GENOMIC PROPERTIES

Other characteristics of the *Arabidopsis thaliana* nuclear genome that have been reported include a haploid chromosome number of 5 and a total genetic map distance of around 500 to 600 centimorgans. The G+C content of the DNA has been estimated at 41.4%, a typical figure for a member of the mustard family. Cytosine methylation has also been reported, amounting to approximately 6% of nuclear cytosine on a molar basis, a rather low figure for an angiosperm. *Arabidopsis thaliana* telomeres have been cloned and sequenced, and each consists of a total of about 350 tandem repeats of the simple sequence 5-CCCTAAA-3′.

## 5 RFLP AND PHYSICAL MAPPING

The small size of the *Arabidopsis* genome has made approachable the goals of obtaining very high physical density maps of restriction fragment length polymorphism (RFLP), as well as a complete physical map of the genome. The present *Arabidopsis* RFLP map has more than 300 molecular markers, which is on average one genetically mapped fragment every 300 to 350 kb. Starting points are thus available for chromosome walks to virtually every genomic region, and the walk to any gene is on average only one yeast artificial chromosome (YAC) clone. Since YAC libraries of *Arabidopsis* DNA are generally available, any component laboratory can perform map-based cloning in this species. Work toward providing a complete physical map of the *A. thaliana* genome is in progress, with the genome presently available as 750 groups of contiguous cosmid clones that represent around 90 to 95% of the genomic DNA, and also as a collection of 296 ordered YAC clones (obtained by finding YAC clones that cross-hybridize with 125 genetically mapped RFLP markers), which include 30% of the nuclear genome. Rapid progress in physical mapping has been possible with this species because of the small genome and generally single-copy DNA of *Arabidopsis.*

## 6 GENOMIC SEQUENCING

More traditional means of gene cloning have also been applied to *Arabidopsis,* in particular cloning by homology to known genes, cloning from cDNA libraries, and cloning from protein sequences. These have resulted in a collection of 2351 separate sequence entries in the GenBank DNA sequence database (release 78), which represent more than 1600 kb of DNA sequence. Given its small size, it seems likely that the *Arabidopsis* genome will be the first plant nuclear genome whose DNA sequence will be completely known. An international effort has been organized to achieve the goals of complete physical mapping and complete genomic sequencing of the genome of *Arabidopsis thaliana.*

*See also* PLANT CELLS, GENETIC MANIPULATION OF; PLANT GENE EXPRESSION REGULATION.

### *Bibliography*

Koncz, C., Chua, N., and Schell, J., Eds. (1992) *Methods in* Arabidopsis *Research.* World Scientific, Singapore.

Meyerowitz, E. M. (1987) *Arabidopsis thaliana. Annu. Rev. Genet.* 21:93–111.

———, and Pruitt, R. E. (1985) *Arabidopsis thaliana* and plant molecular genetics. *Science.* 229:1214–1218. [Reprinted in *Biotechnology: The Renewable Frontier,* D. E. Koshland, Jr., Ed. American Association for The Advancement of Science, Washington, DC, 1986, pp. 311–320.]

National Science Foundation (190) A long-range plan for the Multinational Coordinated *Arabidopsis thaliana* Genome Research Project. NSF 90-80, Washington, DC, 14 pp. [Progress Report, Year 1, NSF 91-60, 17 pp; Progress Report, Year 2, NSF 92-112, 27 pp; Progress Report, Year 3, NSF 93-173, 71 pp.]

## Aromatic Amino Acid Hydroxylases of the Brain: *see* Brain Aromatic Amino Acid Hydroxylases.

# AUTOANTIBODIES AND AUTOIMMUNITY

*K. Michael Pollard, Edward K. L. Chan, and Eng M. Tan*

### *Key Words*

**Antibody** A protein product of B cells composed of heavy and light chains and able to combine with a specific molecular target called an antigen.

**Antigen** Substance that interacts with antibodies.

**Indirect Immunofluorescence (IIF)** A technique whereby an antibody is overlaid onto an antigen-containing cellular substrate and the antigen–antibody complex formed is detected by a fluorescently labeled anti-antibody.

Autoimmunity is an immunological reaction against constituents of an organism that normally are tolerated by the immune system of that organism. Autoimmune reactions can be either cell- or antibody-mediated. Autoantibodies are therefore the antibodies that recognize normally tolerated cell and tissue constituents (or autoantigens). The antigenic specificity of an autoantibody can be a useful aid in clinical diagnosis. Autoantibodies are either cell (or tissue) specific, as found in organ-specific autoimmune disease such as autoimmune thyroiditis, or non–organ specific, as found in multisystem autoimmune disease such as the systemic rheumatic diseases. The latter group includes autoantibodies that recognize macromolecular complexes of nucleic acids and/or proteins such as small nuclear ribonucleoprotein particles (snRNPs), nucleosomal and subnucleosomal structures, and tRNA synthetases, which are intrinsic components of all nucleated cell types present in an organism. Autoantibodies also recognize components of subcellular structures including mitochondria, ribosomes, Golgi apparatus, nuclear membrane, chromosomes, and substructures within the nucleus and nucleolus. The ability of autoantibodies to recognize components of the cellular machinery of replication, transcription, RNA processing, RNA translation, and protein processing has made them important reagents for the cDNA cloning of proteins involved in these cellular processes and for probing the relationship between molecular and cellular structure and function. The evolutionarily conserved nature of many autoantigens allows autoantibodies to be used to identify their target antigens in diverse species, ranging in some cases from human to lower eukaryotes such as yeast. Autoantibodies have been used to inhibit the biological function of autoantigens and/or to recognize autoantigens in a defined functional state. Techniques for assessing the ability of autoantibodies to influence function have included microinjection of antibody into living cells and addition of antibody to in vitro functional assay systems.

## 1 AUTOANTIBODIES AND AUTOIMMUNITY

An autoimmune response is an attack by the immune system on the host itself. The responsible effector mechanisms, which appear to be no different from those used to combat exogenous infective agents, include soluble products such as antibodies (humoral immunity) as well as direct cell-to-cell contact resulting in specific cell lysis (cell-mediated immunity). No single mechanism has been described that can account for the diversity of autoimmune responses, or the production of autoantibodies. Common features of theoretical models of autoantibody elicitation include a genetic predisposition, an unknown but possibly environmental trigger, disregulation of the immune response, and the availability of autoantigen to drive the autoimmune response. In only a few diseases are autoantibodies thought to be the causative agents of pathogenesis (e.g., anti–acetylcholine receptor autoantibodies in myasthenia gravis, and anti–thyroid stimulating hormone receptor autoantibodies in Graves' disease). In some autoimmune diseases autoantibodies have been shown to participate in pathogenetic events by way of complexing with their cognate antigens to cause immune complex mediated inflammation.

## 2 AUTOANTIBODIES AS DIAGNOSTIC MARKERS

The diseases associated with autoantibodies can be divided into two broad groups: the organ-specific autoimmune diseases, in which autoantibodies react with a particular organ or tissue, and the multisystem autoimmune diseases, in which autoantibodies react with common cellular components and appear to bear little resemblance to the underlying clinical picture. In both cases particular autoantibody specificities can serve as diagnostic markers (Table 1). Other autoantibody specificities may occur at different frequencies in a variety of diseases, and the resultant profile consisting of distinct groups of autoantibodies in different diseases can have diagnostic use. In some cases the grouping of autoantibody specificities, such as the preponderance of antinucleolar autoantibodies

**Table 1** Examples of Clinical Diagnostic Specificity of Autoantibodies[a]

| Autoantibody Specificity | Molecular Specificity | Clinical Association |
|---|---|---|
| *Multisystem Autoimmune Diseases* | | |
| Anti-ds DNA[b] | B form of DNA | SLE |
| Anti Sm[b] | B, B′, D, and E proteins of U1, U2, U4–U6 snRNP | SLE |
| Anti-nRNP | 70 kDa, A and C proteins of U1 snRNP | MCTD, SLE |
| Anti-SS-A/Ro | 60 and 52 kDa proteins associated with hY1-Y5 RNP complex | SS, neonatal lupus, SLE |
| Anti-SS-B/La | 47 kDa phosphoprotein complexed with RNA pol III transcripts | SS, neonatal lupus, SLE |
| Anti-Jo-1[b] | Histidyl tRNA synthetase | PM |
| Antifibrillarin[b] | 34 kDa protein of U3 snoRNP | Scleroderma |
| Anti–RNA polymerase I[b] | Subunits of RNA polymerase I complex | Scleroderma |
| Anti-DNA topoisomerase I (anti-Scl-70)[b] | 100 kDa DNA topoisomerase I | Scleroderma |
| Anticentromere[b] | Centromeric proteins CENP-A, B, C | CREST (subset of scleroderma) |
| cANCA | Serine proteinase (proteinase 3) | Wegener's vasculitis |
| *Organ-Specific Autoimmune Diseases* | | |
| Anti–acetylcholine receptor[b] | Acetylcholine receptor | Myasthenia gravis |
| Anti–TSH receptor[b] | TSH receptor | Graves' disease |
| Antithyroglobulin[b] | Thyroglobulin | Chronic thyroiditis |
| Anti–thyroid peroxidase[b] | Thyroid peroxidase | Chronic thyroiditis |
| Antimitochondria[b] | Pyruvate dehydrogenase complex | Primary biliary cirrhosis |

[a]Abbreviations: cANCA, cytoplasmic antineutrophil cytoplasmic antibody; MCTD, mixed connective tissue disease; PM, polymyositis; SLE, systemic lupus erythematosus; SS, Sjøgren's syndrome; TSH, thyroid-stimulating hormone.
[b]Disease-specific diagnostic marker autoantibody.

in scleroderma (Table 1), provides provocative but as yet little understood relationships with clinical diagnosis.

## 3 AUTOANTIBODIES AS MOLECULAR AND CELLULAR PROBES

The most visually impressive demonstration of the usefulness of autoantibodies as biological probes is the indirect immunofluorescence (IIF) test. This technique is being used to identify an increasing number of autoantibody specificities that recognize cellular substructures and domains (Table 2). One such example is the coiled body. This subnuclear structure, initially described from observations made using light microscopy by the Spanish cytologist Ramon y Cajal in 1903, can now be more easily identified using human autoimmune sera. Features of these subcellular structures such as size, shape, and distribution can be studied by IIF during the cell cycle, viral infection, mitogenesis, and many other cellular responses. For example, autoantibodies have been used to localize nuclear membrane, nucleolar proteins, and nuclear proteins to the periphery of chromosomes during mitosis. Comparative studies using human autoantibodies and nonhuman antibodies raised by immunization against specific cellular proteins detected by different fluorochromes can be useful in localizing the specific protein. Immunolocalization of the non–small nuclear ribonucleoprotein particle spliceosome component SC-35 was achieved in this way by comparison of anti-SC-35 antisera with the IIF pattern of autoimmune anti-Sm sera, which recognize protein components of the spliceosome.

One feature of autoantibodies that distinguishes them from antibodies raised by specific immunization, and underscores their uniqueness, is the ability to recognize a target antigen not only from the host but from a variety of species. The extent of this species cross-reactivity is dependent on the evolutionary conservation of the autoantigen and is related to the conservation of protein sequence. One example is the 34 kDa protein fibrillarin, a component of the U3 small nucleolar ribonucleoprotein (snoRNP) particle and a target of autoantibodies in human scleroderma. Using autoantibodies in a variety of immunological techniques, including IIF, this protein can be recognized from species as diverse as *Homo sapiens* and the unicellular yeast *Saccharomyces cerevisiae*. cDNA cloning of fibrillarin has confirmed the expected high degree of conservation of the protein sequence.

The reactivity of autoantibodies with conserved sequence and conformational protein elements has made them useful reagents in the cloning of cDNAs of expressed proteins from cDNA libraries from a variety of species. Because of their reactivity with the human protein, however, they have found most use in the cloning and characterization of the primary structure of numerous human cellular proteins. The diversity of the targets that can be exploited by this approach is clearly illustrated in Tables 1 and 2.

Many of the autoantibodies from systemic autoimmune diseases target autoantigens within the nucleus of the cell. Elucidation of the structure of these autoantigens has revealed that many are functional macromolecular complexes involved in nucleic acid synthesis and processing (Table 3). A distinguishing feature of many of these complexes of nucleic acid and/or proteins is that autoantibodies do not recognize all the components of the complex. Nonetheless the use of autoantibodies that identify specific components of such complexes has aided in identifying other subunits of these complexes with profound consequences. Thus the initial identification of anti-Sm (Smith) and anti-nRNP autoantibodies in systemic lupus erythymatosus (SLE) led to the observation that these particles recognize some of the protein components of the uridine-rich small nuclear ribonucleoprotein (U snRNP) particles and fueled subsequent studies that showed the U snRNPs as components of the spliceosome complex that functions in pre-mRNA splicing.

As the functional associations of autoantigens have become known, attempts to uncover the role of the autoantigen itself have

**Table 2** Examples of Subcellular Structures and Domains Recognized by Autoantibodies[a]

| Subcellular Structure | Molecular Specificity | Autoantibody |
|---|---|---|
| *Cytosolic Components* | | |
| Mitochondria | Pyruvate dehydrogenase complex | Antimitochondria |
| Ribosomes | Ribosomal P proteins | Antiribosome |
| Golgi apparatus | 95 and 160 kDa golgins | Anti-Golgi |
| Microsomal | Cytochrome P450 superfamily | Antimicrosomal |
| Lysosomes | Serine proteinase (proteinase 3) | cANCA |
| Midbody | 38 kDa protein | Antimidbody |
| Centriole/centrosome | 48 kDa protein | Anticentrosome/centriole |
| *Nuclear Components* | | |
| Chromatin | Nucleosomal and subnucleosomal complexes of histones and DNA | Antichromatin |
| Nuclear pore | 210 kDa glycoprotein (gp210) | Antinuclear pore |
| Nuclear lamina | Nuclear lamins A, B, C | Antilamin |
| Centromere | CENP-A, B, C, F | Anticentromere |
| Coiled body | p80-coilin | Anti-p80 coilin |
| PIKA | p23-25 kDa proteins | Anti-PIKA |
| Mitotic spindle apparatus | 238 kDa protein | Anti-NuMA |
| *Nucleolar Components* | | |
| Dense fibrillar component | 34 kDa fibrillarin | Antifibrillarin |
| Fibrillar center | RNA polymerase I | Anti-RNA polymerase I |
| Granular component | 75 and 100 kDa proteins of PM-Scl complex | Anti-PM-Scl |
| NOR | 90 kDa doublet of hUBF | Anti-NOR 90 |

[a]Abbreviations not given in Table 1. hUBF, human upstream binding factor; NOR, nucleolar organizer region; NuMA, nuclear mitotic apparatus; PIKA; polymorphic interphase karyosomal association; PM-Scl, polymyositis-scleroderma.

revealed that autoantibodies can directly inhibit the function of their target autoantigen (Table 3). Although it remains to be determined, it seems likely that such inhibition reflects the involvement of conserved protein sequence or structure in functional activity. An increasing number of autoantibodies, many of unknown molecular specificity, recognize their autoantigen only in a particular functional state or phase of the cell cycle. Of the several examples known, the best characterized is proliferating cell nuclear antigen (PCNA), which is recognized by autoantibodies only during S phase even though PCNA is present throughout the cell cycle. These intriguing features of some autoantibodies have added new dimensions to their biological usefulness and have suggested that functionally active macromolecular complexes may play a role in the elicitation of the autoantibody response.

## 4 PERSPECTIVES

Initially used as aids to clinical diagnosis, autoantibodies have become increasingly useful ''reporter'' molecules in the identification of structure–function relationships. New autoantigens continue to be discovered, while many described autoantigens remain to be characterized both structurally and functionally. Autoantibodies will figure prominently in these characterization studies. As the molecular structures of the interaction between autoantigen and autoantibody become known, it should be possible to design peptide configurations capable of perturbing the functional activity of numerous cellular processes.

*See also* COMBINATORIAL PHAGE ANTIBODY LIBRARIES; IMMUNOLOGY.

### *Bibliography*

Earnshaw, W. C., and Rattner, J. B. (1991) The use of autoantibodies in the study of nuclear and chromosomal organization. In *Methods in Cell Biology*, Vol. 35, B. A. Hamkalo and S.C.R. Elgin, Eds. Academic Press, New York.

Fu, X.-D., and Maniatis, T. (1990) Factor required for mammalian spliceosome assembly is localized to discrete regions in the nucleus. *Nature* 343:437–441.

McCarty, G. A., Valencia, D. W., and Fritzler, M. A. (1984) Antinuclear antibodies: Contemporary Techniques and Clinical Applications to Connective Tissue Diseases. Oxford University Press, New York.

Naparstek, Y., and Plotz, P. H. (1993) The role of autoantibodies in autoimmune disease. *Annu. Rev. Immunol.* 11:79–104.

Pollard, K. M., Reimer, G., and Tan, E. M. (1989) Autoantibodies in scleroderma. *Clin. Exp. Rheumatol.* 7/S-3:57–62.

Rose, N. R., and Mackay, I. R., Eds. (1992) *The Autoimmune Diseases*, Vol. 2. Academic Press, New York.

Tan, E. M. (1989) Antinuclear antibodies: Diagnostic markers for autoimmune diseases and probes for cell biology. *Adv. Immunol.* 44:93–151.

**Table 3** Examples of the Functions of Nuclear Autoantigens and the Effect of Autoantibody on Antigen Function

| Autoantigen | Function | Autoantibody Effect[a] |
|---|---|---|
| *Known Function* | | |
| Sm/nRNP (U1,2,4-6 snRNP) | Pre-mRNA splicing | Inhibition of pre-mRNA splicing |
| PCNA (DNA polymerase δ auxiliary protein) | DNA replication | Inhibition of DNA replication and repair |
| RNA polymerase I | Transcription of rRNA | Inhibition of rRNA transcription |
| tRNA synthetase | Aminoacylation of tRNA | Inhibition of charging of tRNA |
| Ribosomal RNP | mRNA translation | Inhibition of protein synthesis |
| Centromere/kinetochore | Microtubule-based chromosome movement during mitosis | Inhibition of centromere formation and function |
| *Probable Function* | | |
| Fibrillarin (U3 snoRNP) | Processing and methylation of pre-rRNA | Not tested |
| NOR-90 | RNA polymerase I transcription factor | Not tested |

[a]Inhibition of function has been demonstrated in vitro or following injection of autoantibody into living cells.

———. (1989) Interactions between autoimmunity and molecular and cell biology: Bridges between clinical and basic sciences. *J. Clin. Invest.* 84:1–6.

Todd, J. A., and Steinman, L., Eds. (1992) Autoimmunity. *Curr. Opinion Immunol.* 4:699–778.

# Automation in Genome Research

*Edward Theil*

## *Key Words*

**Automated Sequencer** Any of several devices by which the actual sequence of nucleotides of single-stranded DNA may be understood without a high level of manual intervention. Typically, an automated sequencer will consist of an electrophoresis system combined with a laser scanning device to image the gel and specialized signal-processing software to find and interpret the individual bands of DNA as they appear on the image.

**Bottom-Up Design** A design principle in which individual autonomous modules are developed without necessarily planning in detail how the final system will eventually be constructed (see System Integration). The risk is that short-term goals may be accomplished at the expense of the long-term success of a project.

**Contigs** A contig (from "contiguous") is an ordered collection of clones. Although the ordering may have been accomplished in any of several ways, the result is a set of clones in which the sequence of each one partially overlaps that of its neighbors, so that the result represents a contiguous part of a genome.

**Electrophoresis** A technique to separate molecules of different sizes, typically of DNA or proteins. The electrically charged sample is placed in a semisolid matrix, called a gel, and exposed to an electric field. The DNA migrates through the gel in response to the field, and, when this motion occurs, smaller particles travel faster and move further than larger ones.

**Mass Spectrometry** A technique in which streams of charged particles are separated according to each particle's respective mass, as determined by the spectrum of the ensemble. This technique, perfected in physics and physical chemistry, is gradually being applied to large organic molecules like DNA, although the ratio of mass to charge makes accurate detection a difficult research problem at this time.

**Multiplex** Usually reserved to describe a data communications link that carries more than one message simultaneously, the term is used here to describe processes and instruments that operate on more than one sample at a time.

**PCR (polymerase chain reaction)** PCR is an in vitro method for enzymatic amplification of specific DNA sequences, using short DNA primers derived from knowledge of the sequence surrounding the region to be amplified, and Taq polymerase to effect the reactions. By repeated duplication of a selected sequence of DNA through a process based on DNA replication, geometric increase in the amount of that sequence in the range of $10^5$ to $10^7$ is achievable.

**System Integration** The process whereby various independent modules or components are formed into a larger, intercommunicating entity. The modules discussed here are usually electromechanical, and the integration takes place through data communication protocols and software.

The word *automation* is used in a broad sense in this article to mean not only a variety of automated instruments that may or may not be under computer control, but also the important associated idea of system integration. Without system integration of automatic instruments, high throughput of genomic markers and DNA sequence will not be achieved because inevitably there will be bottlenecks where the smooth flow of data is interrupted.

It can be argued that without automation at a level previously unnecessary in molecular biology, the goals of the Human Genome Project (and other genome projects) cannot be obtained. Thus, the role of automation in large-scale sequencing and mapping, for example, is critical. Furthermore, it promises to play a growing part in the commercialization of technologies now being developed through the impetus of the various Genome Projects.

## 1 INTRODUCTION

### 1.1 Background

As the Human Genome Project (HGP) matures, it is increasingly clear that automation methods are beginning to play the major role originally foreseen for them. This new technology, some of which is described here, is gradually changing the way much of molecular biology is done. For our purposes, robotics and automation form a part of bioengineering; specifically, the application of engineering principles and design to problems in molecular biology. In such applications, the usual considerations of engineering projects arise: cost estimates and tradeoffs, design factors, life cycle, ease of use, and so on. The Human Genome Project has provided some of the challenges and wherewithal to encourage this approach to molecular biology. Nevertheless, development and operating costs are still significant factors. A rough rule of thumb useful for budding biotechnologists is that for any significant idea in the field, it will take four to five times as long to produce a solution that works "in our hands" as it did to accomplish the proof of principle. Then, it takes four to five times longer still to create a robust instrument or procedure that others can also use effectively. Finally, support and enhancement of a new device is generally at least 50 to 80% of the total project time and money. Given this very significant level of effort, the payoff must clearly be worth the price before development of a new instrument or automation system can proceed very far. It is highly unlikely the HGP will achieve its goals without appropriate new instrumentation, but the cost, while substantial, will result in new techniques for biological research that extend beyond the field of genomics.

For the sake of specificity, a hypothetical sequencing project is assumed. The system approach to genetic or physical mapping would be quite similar. Sequencing projects provide good prototypes for automation because of the number of different operations and their repetitive nature. In Section 1.2, some rudimentary value engineering reveals interesting notions about cost and throughput. Section 2 discusses the notion of system integration and suggests three criteria that an automation system for genomics should satisfy. Section 3 describes several state-of-the-art devices now in use at various laboratories. These devices are prototypical of the kind that can form building blocks for the general system approach outlined in Section 4. Informally and rather briefly, we will show how to design a plan in which robots, instruments, and computer control are integrated to monitor data flow, capture results, and reduce human error. First, however, it is worth examining some numbers that help to determine required throughput.

### 1.2 Some Useful Arithmetic

In view of the attendant disadvantages of high cost and lengthy development time noted at the outset, why is it that robotics and automation are nevertheless becoming increasingly important in genome research? Some simple arithmetic helps to provide an answer and also serves as a first step in system analysis.

The figure $3 \times 10^9$ is frequently cited as the number of nucleotide pairs in the human genome, but that's too large a number to work with easily. Instead, imagine a single large-scale sequencing project 3 to 5 years hence, designed to produce 10 million bases of sequence each year at an initial cost of $1/base. (This is probably half to one-third of current costs.) Assuming that both strands of DNA are to be sequenced, at least $2 \times 10^7$ bases will actually be run through the sequencing machines yearly. (In practice, the amount of redundant sequencing increases that number by 50–300%.) Given 250 working days per year, at least 80,000 bases must be sequenced *per day* at this one center! By genome standards, this proposal seems modest: 1/300 of the genome per year. Even so, the rate exceeds the best current efforts of well-funded, well-staffed laboratories by a factor of at least 10.

Furthermore, it is critical to determine how the $10 million annual budget should be spent. If, for example, most of it goes into the salaries of, say, 70 to 100 staff members who perform a great deal of repetitive labor, there will be little opportunity to lower costs in future years merely by hiring more people, as opposed to improving the technology.

Now, let us also examine the throughput of one popular automated sequencer, the Applied Biosystems Model 373A. The 373A can perform a sequencing run in about 10 to 12 hours and runs 24 lanes in one gel. Assuming an average length of 350 to 400 nucleotides per lane, a single machine can produce up to 9600 bases per run. Even if we assume only one run per day, eight to ten machines should be adequate to meet our hypothetical project's goal. Although the machines are expensive, their costs can be amortized over several years. Thus, it becomes clear immediately that *the sequencing operation itself is not a primary bottleneck.*

### 1.3 The Need for Automation

If the sequencing operation itself is not the principal problem, what is? As has been emphasized by several writers, the relatively slow production of finished sequence is due to a number of compounding factors involved in the multistep sequencing process. Some of these are upstream of the actual sequencing step, such as the preparation and manipulation of clone libraries and the purification of the source DNA. Others are downstream, such as the assembly of short sequenced fragments into larger units called "contigs." The upstream steps are typically labor intensive, repetitive, and prone to human error (e.g., placing two clones in the same tube or the same clone in different tubes). The downstream steps (e.g., sequence assembly, contig reconstruction, analysis) tend to be computer intensive but currently in a highly interactive way, which does not lend itself directly to further improvements in throughput. Thus, barring striking new advances in this field (which would inevitably involve their own kind of automation), it seems clear that the only way to achieve the required throughput is by the introduction of laboratory protocols that are designed to take advantage of the parallel (or *multiplexed*) operations on samples possible with robotics, automatic instruments, and computerized acquisition and management of data. The numbers of manipulations and the sheer volume of data (with its consequent acquisition, interpretation, and representation) dictate that significant use of automatic instruments, computers, and specialized software is inevitable, if the Human Genome Project is to achieve its goal of elucidating the genetic code during the next 10 to 20 years.

## 2 SYSTEM REQUIREMENTS

### 2.1 Controlling the Data Flow: The Long-Range Objective

While individual devices such as automated sequencers and synthesizers are crucial for large-scale experiments in molecular biology, they will not be sufficient. The reason is that the stream of data is rate-limited by its slowest link. To take an extreme case for the purposes of illustration, there is little point in increasing the output of raw DNA sequence by a factor of 100 if the assembly of that data increases by a factor of only 50. Inevitably, the amount of data backed up will grow exponentially over time. The situation is analogous to increasing the number of lanes on a crowded expressway, only to decrease them again at the only exit point.

What is missing is the "glue" that ties the instruments together. This glue—a hardware and software infrastructure that facilitates data and information flow—can be implemented in many ways, but it is important to recognize that one critical element is reliable data acquisition and tracking. Keeping track of materials, results, intermediate data (e.g., PCR results) and final data (e.g., information about where the sequence resides on a physical map of the genome) becomes as important as the production of the sequence itself.

A frequent requirement is that the data be captured in computer-readable form at virtually all stages of the experiment; this includes the experimental parameters as well as raw and derived results, and general information about the experiment (time, date, name, etc.). The use of bar codes on laboratory equipment such as microwell plates is a common partial solution to this problem, but much more is required: for example, a database designed for the purpose, and a model of the experimental process. This point is discussed further in Section 4.

### 2.2 Three Desirable Attributes of an Automation System

What should be the goals of an automation system for a large-scale project in sequencing or mapping? Each one will have its

**Figure 1.** Device for picking and arraying colonies from clone libraries, designed and built at Lawrence Berkeley Laboratory. Twelve tungsten needles are mounted on a carousel; there are two computer-controlled, movable tables, a sterilizing tank at the rear, an imaging station (used separately), and computer hardware and software. The device picks 40 colonies per minute.

own strategy and requirements, but the following seem desirable for almost any large project.

1. The system must take advantage of economies of scale. That is, doubling the output should not double the cost.
2. The system should permit early starts. That is, it must be possible to accomplish useful work early in the life cycle, not only when the entire package is available. All too commonly, software systems, including many that are apparently well designed, don't work at all until they're completely finished.
3. The system should allow modification during its life cycle to accommodate new technology. Inflexibility is a particular pitfall in a field in which innovation is a principal characteristic.

### 2.3 Bottom-Up Evolutionary Design

A design approach that has the potential of achieving all three of the goals just listed consists of introducing individual modules optimized to specific tasks that satisfy obvious needs, then gradually incorporating and refining them while they serve as nodes in an integrated system. This approach amounts to a *bottom-up design* that evolves as the system grows and becomes better understood, with the additional requirement that the individual modules serve useful purposes at all stages of their evolution. By recognizing early in the life cycle of the project that the details of the system almost certainly will evolve over time (as is common in research projects), it is possible to identify and avoid critical bottlenecks while the system grows. Another term for this style of gradual development is *stepwise refinement.* Of course, it is important in this approach that users and builders (i.e., biologists and engineers) share a common set of realistic expectations.

The advantages to this approach include the following:

1. Modules can be replaced by newer technology as it becomes available, without major impact on the overall system. An example that may become practical in the future is the use of mass spectroscopy in place of electrophoresis to size individual fragments of DNA.
2. Individual modules can be made available for other projects, even if the system within which they currently operate is not itself adaptable to such use.
3. The system yields useful results even from its earliest stages.

## 3 EXAMPLES OF MODULES AVAILABLE TODAY

### 3.1 Automatic Colony Picking and Library Arraying

Two groups, one at Lawrence Berkeley Laboratory and another at the Medical Research Council in Cambridge, England, have

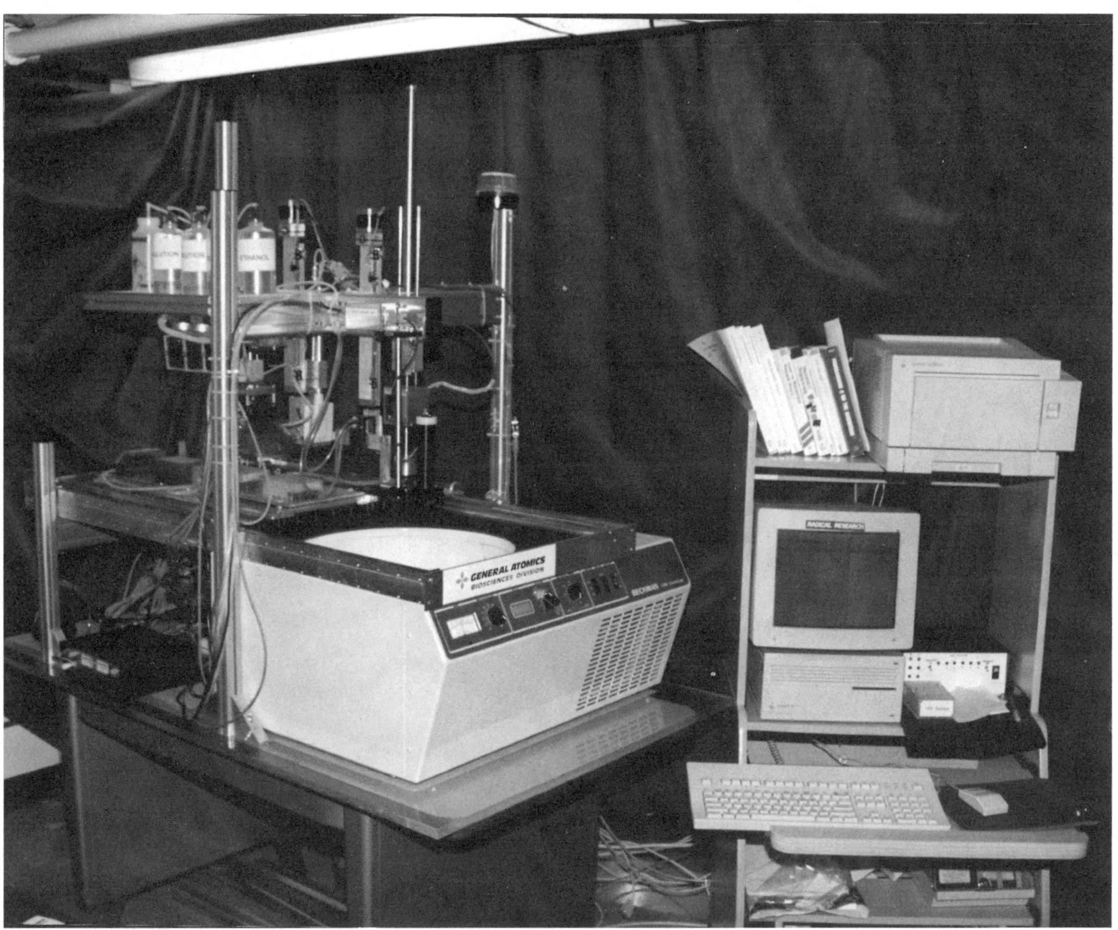

**Figure 2.** Prepper, Ph.D., is a stand-alone automated (mini prep) system for DNA purification and extraction developed at General Atomics. It can prep yeast artificial chromosome (YAC), plasmid, and cosmid DNA at the rate of 48 samples per hour. Samples are processed in batches, 96 at a time, in deep-well microtiter plates; after processing, they are ready for cycle sequencing or as templates for PCR screening. Centrifugation, fluid dispensing, and transfer pipetting are all done under computer control. Software options permit modifications of the protocols; for example, RNAse and proteinase K can be optionally added.

developed machines with approximately the same functionality. We briefly describe the LBL version here.

Machine vision is supplied by first obtaining a digitized image of a petri dish on which 100 to 200 bacterial or yeast colonies are growing on agar. This dish, held in a custom fixture, provides a constant reference frame. The image is captured by a frame grabber on a personal computer, using commercial hardware and software. A customized application of a standard image analysis software package automatically locates the coordinates of the colonies against the lighter agar background.

The petri dish and the coordinate information are passed directly to the colony picker and its control computer respectively. Twelve tungsten needles rotate on a circular carousel that forms part of the picker (Figure 1). As each one reaches the ''picking position,'' a small movable table controlled by the computer positions the petri dish so that the next colony is ready to be picked. The needles then rotate to the target microtiter plate, located on an identical table on the other side of the picker. Since this table also has moved, the next available well of the microtiter plate is positioned to receive some of the colony. Later in the rotation cycle, the needle passes through a sterilizing tank filled with hot water and alcohol. When a plate is filled or a dish is completely picked, the operation automatically pauses until the next dish or plate is in place (replacements are supplied either manually or by robot). Picking, depositing, and sterilization all occur at the same time with different needles. This device can pick and array approximately 40 colonies per minute exclusive of imaging time and has routinely picked 10,000 colonies in a single day without the handling errors common to manual picking.

### 3.2 Multiplexed Thermal Cyclers

The polymerase chain reaction (PCR) technique is rapidly becoming as ubiquitous in genome research as electrophoresis. Several genome centers are designing major projects in which PCR plays a critical role in the screening of libraries or the development of high resolution physical maps prior to the actual sequencing steps.

Thermal cyclers that handle up to 96 samples at a time are currently available, but an increase of 5 to 10 times that rate will be required for major sequencing projects. Different approaches have been taken by engineering groups to increase throughput. At the Whitehead Institute at MIT, an industrial collaboration with

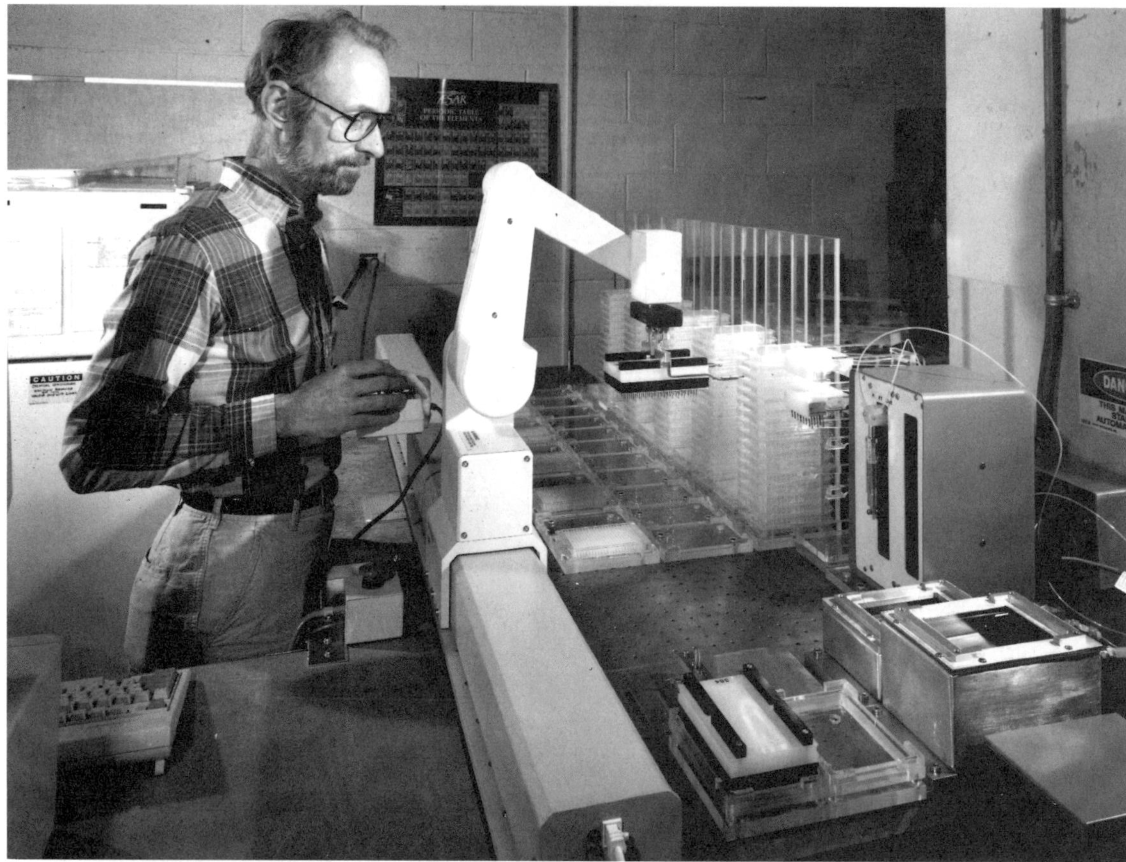

**Figure 3.** Human operator monitor, Hewlett Packard ORCA robot, adapted for use in the Human Genome Project at Lawrence Berkeley Laboratory. As discussed in the test, most general-purpose laboratory robots require custom peripherals and software. Here, the robot is holding a 96-pin tool, designed to deliver material to all the wells of a conventional microtiter plate in one action. Also visible are a sterilizing tank for the tool (right foreground) and custom holders and stackers for the plates.

Intelligent Automation Corporation has produced a thermal cycler that resembles a waffle iron, where microwell plates take the place of waffles. The unit, which accommodates up to sixteen 96-well plates, heats the plate electrically and cools them by circulating water through channels built into the device.

At LBL's Human Genome Center, rapid heating and cooling (desirable to suppress the amplification of unwanted fragments) is achieved by surrounding the plates with water baths at different temperatures. The water is provided by four reservoirs maintained at different temperatures. Because the temperatures in the reservoirs are already established, there is no chance of overshoot in either heating or cooling. Both this arrangement and the rapid transition between temperature levels are features that appear to enhance the biochemical mechanisms, resulting in fewer undesired amplifications and cleaner data.

### 3.3 Automatic Preparation and Purification of DNA

A number of biotechnology companies have marketed machines for the automatic purification and extraction of DNA. General Atomics has developed a high throughput device that can handle up to four microtiter plates at a time (Figure 2). It performs centrifugation-based DNA separations using alkaline lysis preps. The machine is controlled at the low level by stepper motors and a pneumatic actuator driven gripper. All high level control is performed by a Macintosh computer.

### 3.4 Robots

At most genome laboratories, the emphasis has been less on building robots than on using them effectively. On the whole, robots have been less useful for performing highly specialized tasks than for conveying materials from one special device to another while assisting in the automatic tracking of data. Robots such as the Beckman Biomek 1000 and 2000 and the Hewlett Packard ORCA system (Figure 3) are being adopted with increasing frequency; nevertheless, the development of any nontrivial application not foreseen by the manufacturers requires a significant investment of time and effort, often including the design and fabrication of special-purpose auxiliary tools. Unlike their science fiction counterparts, the robots are severely limited in functionality, and it is therefore not surprising that custom hardware and software frequently must be added. In important respects, the robots are the physical analogues of computer software: effective at repetitive tasks, potentially but not immediately flexible, and subject to total failure due to small errors.

A system proposed at General Atomics integrates a custom robot with a number of routine tasks required for screening large clone

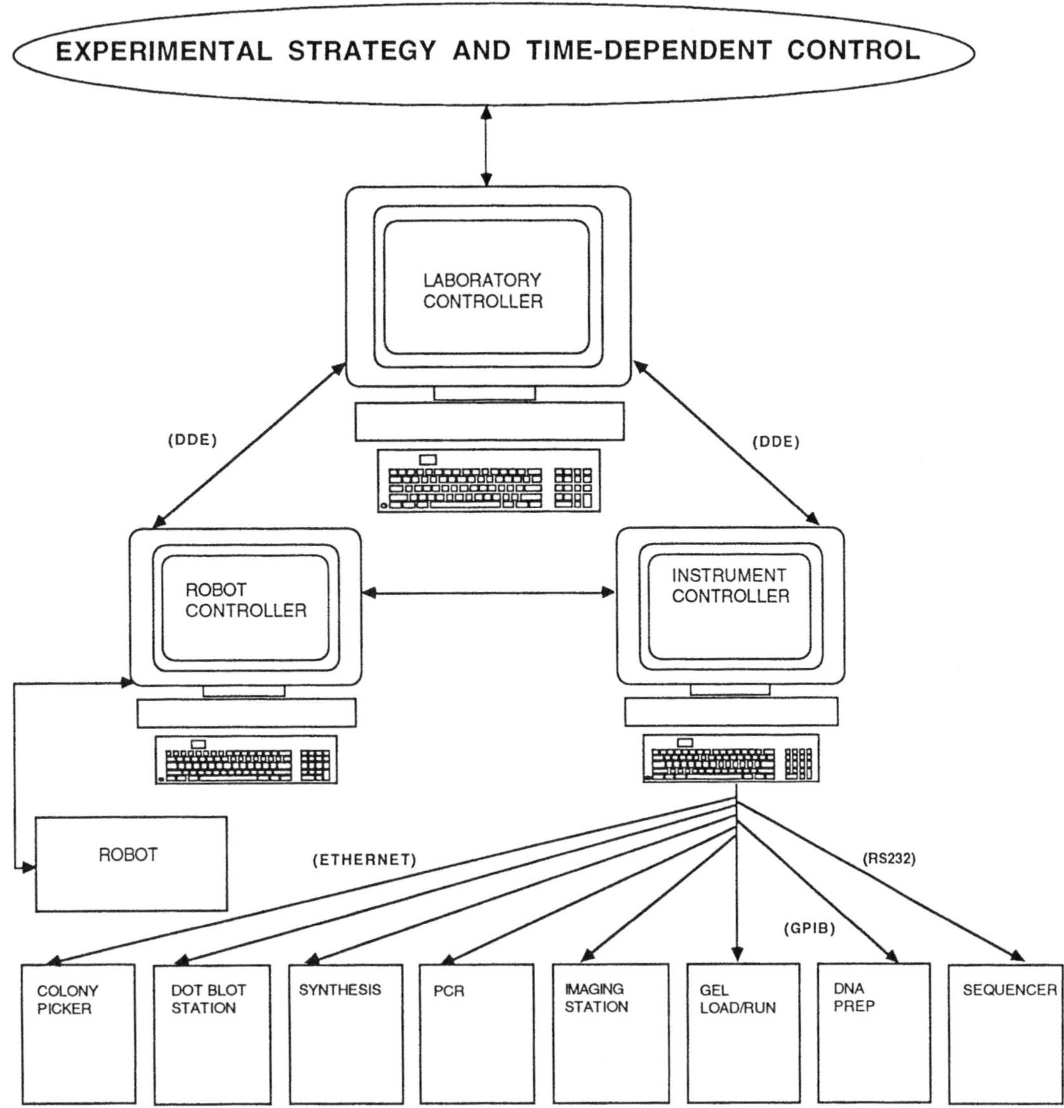


**Figure 4.** A conceptual architecture for an automation system for molecular biology. Labels appearing on the lines connecting computers and devices are intended to suggest some possible communication protocols that might be used in a particular implementation, such as RS232, Ethernet, and Microsoft's Dynamic Data Exchange (DDE), even though these standards apply to rather different levels of communication, ranging from simple physical connections to data exchange at the level of software applications. There are many possible variations on such a system. Nevertheless, one key feature is the hierarchical approach to control and data acquisition: the lowest level consists of the individual modules and their controller(s). At the next level is a laboratory controller, with software that controls the sequence of operations and a database for experimental parameters and data capture. Above that are other computers (only suggested in the bubble) for viewing the reduced data, dealing with time-dependent processes, and monitoring the experiment.

libraries. The system will perform pipetting, centrifugation, and thermal cycling, and it also has the ability to retrieve material from freezers. Samples are held in 96-well or 864-well microtiter plates.

Robots have been used extensively for preparing high density filters for hybridization experiments, for loading PCR machines and gels, for replicating clone libraries, and for many other common tasks. The genome center at Tsukuba University in Japan has constructed an automation and robotics system that handles most of the major steps in sequencing, including DNA extraction, sequencing reactions, fluorescent sequencing, and sequence assembly. The system, however, is not commercially available at this time.

Perhaps the most important use of robots, however, goes beyond any of these: because their control software forms part of the flow of information and data through the experiment, they can act as integrating agents in the construction of an automation system, as described in Section 4.

## 4 DESIGNING THE OVERALL SYSTEM

### 4.1 A General Structure

The modular instruments described in Section 3 meet the general criteria proposed in Section 2.2. Nevertheless, a system view is still lacking. At first, it might seem very difficult to generalize about the construction of an automation system, since such systems would seem to depend so strongly on the strategies and approaches of a particular genome center. Fortunately, even without recourse to a specific experimental strategy, it is still possible to present an architecture that is sufficiently flexible to support a variety of functions and experimental strategies. Figure 4, for example, is loosely based on a system designed by J. Jaklevic, J. Meng, and co-workers at LBL, but it shares design elements with similar projects at other genome laboratories, such as those at the Medical Research Council in Cambridge, England, and General Atomics in La Jolla, California, even though the approaches and specific goals at these centers differ considerably in detail.

### 4.2 Hardware Architecture

In this general scheme, individual modules are associated with computers called instrument controllers, including one for robots. Each of these computers knows about the particular device it controls, but not the overall experimental strategy or other modules. (Although Figure 4 shows only one instrument controller, in practice it is simpler to allocate one computer per instrument.) The larger strategy is allocated to a laboratory control computer (''lab controller'') and to computers that will maintain databases and exercise time-dependent control. For example, if the results from one day's operation affect what is to be done on the next day, that information is hidden from the lower level computers, which monitor and control the operation of the individual modules, acquire processed data and the results of individual analyses, and pass that information up to the lab controller. This is a hierarchical architecture common in a variety of control applications, such as the control of large-particle accelerators in high energy physics. As a design principle, it encourages modularity and evolutionary development.

### 4.3 Software Architecture

A specific large project, such as production sequencing or mapping, can be broken down into a series of protocols, which in turn are broken down into a sequence of operations using individual modules. In that way, the complex operations that constitute the experiment are dispersed among the individual controllers, and the software design is simplified.

At LBL, for example, experiments will be tracked by a set of electronic ''notebooks.'' Each notebook contains information on a single subject, with each page containing a unit of information created by filling out a form on a computer screen. Some of the information is entered manually at the beginning of an experiment or an experimental run; other data are recorded automatically as the experiment proceeds. The various notebook types are conceptually organized as follows.

The *experiment* notebook, each page of which defines an experiment by naming a protocol and a set of source material.

The *protocol* notebook, each page of which specifies a system module and, if necessary, its operating mode. Material or data coming from the module is labeled during this step of the protocol as part of the input to the next step.

The *work order* notebook, each page of which contains a work order for a system module. Each order is tagged with the identity of the material going into the module.

Other notebooks keep track of labels and act as logbooks. Pages of each notebook are collected into chapters. Each experimenter has his or her own chapters, and the pages of the chapters define the individual experiment. Collectively, the notebooks are implemented as a single database on the laboratory controller.

Precise identification of the materials of the experiment is essential. Each experiment must be uniquely labeled, as must its source material (e.g., bacterial colonies, raw DNA) and any generated material and derived data (e.g., PCR amplifications, results of gel analysis, DNA sequence). Bar codes are a useful tool for this purpose. The labels permit the laboratory controller to maintain an unambiguous association between the experiment and all its parts. Typically, this information is formally structured in a database that is written to by the instrument controllers. The database is one mechanism by which information is passed from one level of the hierarchy to another.

In addition to the organization of data, control and scheduling information (e.g., the temporal sequence of operations) must be modeled within the lab controller. A *Dispatcher* software module is responsible for this critical part of the overall operation. The Dispatcher polls the notebooks and modules, looking for active experiments and responding to status messages from the instrument controllers. It responds by creating work orders that are then downloaded to appropriate modules at the instrument controller level.

Even this highly simplified description should suggest the value of evolved system growth and modular design. Hierarchical hardware and software architectures provide one path toward that goal, but there are others. By working toward a system that continually responds to the needs of biologists and by developing an interdisciplinary collaboration with a common set of expectations, laboratory automation can provide the three desirable characteristics discussed in Section 2.2: flexibility, early impact, and incremental growth.

*See also* Sequence Analysis.

## *Bibliography*

Endo, I., Soeda, E., Nishi, K., and Murakame, Y. (1991) Construction of a fully automated line system for sequencing the human genome. Presented at Human Genome III, San Diego, CA.

Jaklevic, J., Hansen, A.D.A., Theil, E., and Uber, D. C. (1991) Application of robotics and automation in a genomic laboratory. *Lab. Rob. Autom.* 3:161–168.

Jones, P., Watson, A., Davies, M., and Stubbings, S. (1992) Integration of image analysis and robotics into a fully automated colony picking and plate handling system. *Nucleic Acids Res.* 20(17):4599–4606.

Landegren, U., Kaiser, R., Caskey, C., and Hood, L. (1991) Large-scale and automated DNA sequence determination. *Science,* 254:59–67.

Martin, W., and Walmsley, R. M. (1992) Vision assisted robotics and tape technology in the life science laboratory: Applications to genome analysis. *Bio/Technology,* 8:1258–1262.

National Research Council. (1988) *Mapping and Sequencing the Human Genome.* National Academy Press, Washington, DC.

U.S. Department of Energy, Office of Energy Research, Office of Health and Environmental Research. (1992) *Human Genome 1991–92 Program Report.* DOE/ER-0544P.

# B

## Bacterial Growth and Division

*Stephen Cooper*

### *Key Words*

**Bidirectional Replication** The replication of DNA with two replication forks proceeding in opposite directions away from an origin of replication.

**Cell Membrane** In bacteria, the lipid-containing structures that lie adjacent to the peptidoglycan layer. Some cells have a single membrane layer, while others have an outer and an inner membrane which are different.

**C Period** The time needed for a round of DNA replication, from initiation at the origin to completion at the terminus. In *Escherichia coli* the C period, for cells growing between 20 and 60 minute doubling times, is a constant of approximately 40 minutes.

**Cytoplasm** The collective name for the numerous components of the cell that are not associated with either the cell wall or the genome. Cytoplasm is composed of the small molecules such as ions, metabolites, and cofactors; macromolecules such as soluble enzymes, tRNAs, and mRNAs; and higher complexes such as ribosomes.

**D Period** The time between termination of DNA replication and cell division. In *Escherichia coli* the D period, for cells growing between 20 and 60 minute doubling times, is a constant of approximately 20 minutes.

**Multiple Forks** The result of initiation of DNA replication on a molecule that has not finished the preceding round of replication; there are forks upon forked material.

**Origin** A sequence on a chromosome or a plasmid at which normal replication of DNA initiates.

**Peptidoglycan** The stress-bearing, presumably shape-maintaining, layer of the bacterial cell wall, composed of glycan chains cross-linked with amino acids.

**Segregation** The distribution of cell material from the dividing mother cell into the two daughter cells. Different cell components may have different segregation patterns.

**Terminus** A point on a chromosome or plasmid at which a normal round of replication terminates.

A dividing bacterial cell must, on average, have precisely twice as much of everything in a newborn cell. How does a cell duplicate its components between divisions, and how does a cell ensure that it does not divide before all its components have been duplicated? These questions have been investigated by studying biosynthesis during the division cycle.

Cell cytoplasm increases uniformly and exponentially during the division cycle. There do not appear to be any cell-cycle-specific syntheses of any cytoplasmic components during the division cycle. DNA replication is initiated in the division cycle at a specific time: when the cell mass per chromosome origin reaches a particular value. Once initiated, DNA replicates at a constant rate such that the time between initiation and termination of replication is relatively independent of the growth rate. The time between termination of replication and cell division is also relatively constant. High-copy plasmids replicate throughout the division cycle and low-copy plasmids replicate at a particular time during the division cycle according to rules similar to that for initiation of chromosome replication. Cell surface grows in response to the increase in cell mass so that the turgor pressure inside the cell, and the cell density, are constant during the division cycle. At division, the components of the cell are segregated to the two daughter cells. The cytoplasm segregates randomly, the DNA segregates stochastically but nonrandomly, and the peptidoglycan segregates in a manner consistent with the fixed location of the synthesized peptidoglycan.

The only "events" during the division cycle are the initiation and termination of DNA replication and the initiation and termination of pole formation. There do not appear to be any other cell-cycle-specific events or syntheses during the division cycle that are related to the regulation of the division cycle.

## 1 THE BACTERIAL DIVISION CYCLE

### 1.1 Cell Age and the Age Distribution

A newborn cell has an age of zero and grows during the division cycle to divide at age 1.0. The cell ages during the division cycle are thus fractions indicating where, in the division cycle, a cell is located. A cell half-way between birth and division is age 0.5. Cell age, by definition, increases linearly during the division cycle. Because of the variability in the absolute time required for a division cycle, however, cells at the same absolute time after birth are only approximately the same cell cycle age.

In a culture, the distribution of cell ages is not uniform. There are twice as many newborn cells as dividing cells. The age distribution is given by $2^{1-x}$, where $x$ is the cell age between zero and 1.0. If all cells grew with exactly the same interdivision time, this formula would give the age distribution exactly. But because of cycle variability there is a smoothing of the function, and the actual distribution is only an approximation of the theoretical distribution. The age distribution means that the properties of an exponentially growing culture are independent of time, and all patterns of synthesis within the division cycle give an exponentially increasing amount of that material in a growing culture. Thus, whether something is made linearly, exponentially, or at an instant during the division cycle, the amounts of all cell components in an exponentially growing culture increase exponentially and in parallel. This is referred to as balanced growth.

Given that between the birth of a cell and its subsequent division, all of the cell components double, we can ask, At what rate is material synthesized during the division cycle, and how does the cell ensure that there is a precise doubling of cell components by division? Let us first analyze which components must be dealt with in terms of the division cycle.

### 1.2 Aggregation Theory and the Control of Synthesis During the Division Cycle

In economics, the aggregation problem is defined in terms of combining various sectors of an economy. Should the figures for the production of capital machinery be combined with those for the production of consumer goods? Is paper produced for boxes in the same economic category as stationery? It is difficult to treat each item in an economy individually; to understand the whole system, some aggregation is necessary. Now consider the aggregation problem for the analysis of the bacterial division cycle.

How should one aggregate the different parts of the cell to achieve an understanding of the biochemistry of growth and division? Is there a unique pattern of synthesis during the division cycle for each enzyme, or are there a limited number of patterns with different enzymes or molecules synthesized according to any one of these patterns? Are there ways of grouping proteins or RNA molecules so that one can consider classes of molecules rather than individual molecular species? Should we consider the cell membrane to be a different category from that of peptidoglycan? There are approximately a thousand proteins in the growing cell, and if each protein had a unique cell cycle synthetic pattern, or if there were only a few enzymes exhibiting any particular pattern, we would have an insuperable task describing the biosynthesis of the cell during the division cycle.

Fortunately, one need consider only three categories of molecules, each of which is synthesized with a unique pattern. The growth pattern of the cell is the sum of these three biosynthetic patterns. The first category is the cytoplasm, which is the entire accumulation of proteins, RNA molecules, ribosomes, small molecules, water, and ions making up the bulk of the bacterial cell. It is the material enclosed within the cell surface that is not the genome. The second category is the genome, the one-dimensional linear DNA structure. The third category is the cell surface, which encloses the cytoplasm and the genome. The surface is composed of peptidoglycan, membranes, and membrane-associated proteins and polysaccharides. Everything in the cell fits into one of the three categories, and each category has a different pattern of synthesis during the division cycle. These three patterns are simple to understand because they can be derived from our current knowledge of the principles involved in the biosynthesis of cytoplasm, genome, and cell surface.

## 2 CYTOPLASM SYNTHESIS DURING THE DIVISION CYCLE

Consider a unit volume of cytoplasm. It contains enzymes required for the breakdown of nutrients, the biosynthesis of low molecular weight precursors of macromolecules, and the synthesis of macromolecules. Each unit of cytoplasm produces a small amount of new cytoplasm over a small interval of time. If the new cytoplasm is indistinguishable from the old, and if the new cytoplasm acts to synthesize more cytoplasm immediately, then the pattern of cytoplasm synthesis is exponential. This is because after each interval of synthesis, the new increment of material added to the cytoplasm results in an increase in the rate of synthesis. This exponential pattern is illustrated in Figure 1, where both the rate of synthesis and the amount of material increase exponentially during the division cycle.

The pattern for cytoplasm as a whole is reflected in the synthesis of the individual cytoplasmic components. There are no changes in the specific rate of synthesis or pattern of cytoplasm synthesis that are related to any particular cell cycle event. No particular molecule of the cytoplasm is made differently from any other molecule of the cytoplasm during the division cycle. In addition, there is no variation, during the division cycle, in the relative concentration of any component of the cytoplasm. All parts of the cytoplasm accumulate exponentially during the division cycle, and this pattern is unchanged even when the cell is dividing. At the instant of division, the combined rate of cytoplasm synthesis in the two new daughter cells is precisely equal to the synthetic rate in the dividing mother cell. (If DNA is sequestered in the membrane at different times during the cell cycle, it may be that transcription from these genes could cease for a short period of time; but this process should not detract from the basic rule that the synthesis is not regulated in any necessary way with regard to the cell cycle.)

The evidence for a smooth, continuous, and exponential increase in cytoplasm comes from experiments that use the membrane elution method, colloquially referred to as the ''baby machine'' (Section 12). The pattern of DNA synthesis during the division cycle, which has been shown to be correct by flow cytometric and other analyses, was actually discovered using the membrane elution method. Thus, the results from a method capable of producing an accurate analysis of biosynthesis during the division cycle have confirmed the exponential synthesis of cytoplasm.

In addition to this experimental support, an evolutionary argument can be made for an invariant rate of accumulation of cytoplasm during the division cycle. If the synthetic rate of an enzyme changed abruptly during the division cycle, there would be a relative excess or deficiency of that enzyme at some point during the division cycle; this would be an inefficient use of resources. In the optimal allocation of resources, each component would be at a constant concentration during the division cycle. Cells would not evolve controls making them less efficient in cell production. The ideal pattern of cytoplasm synthesis is an invariant cytoplasm composition during the division cycle.

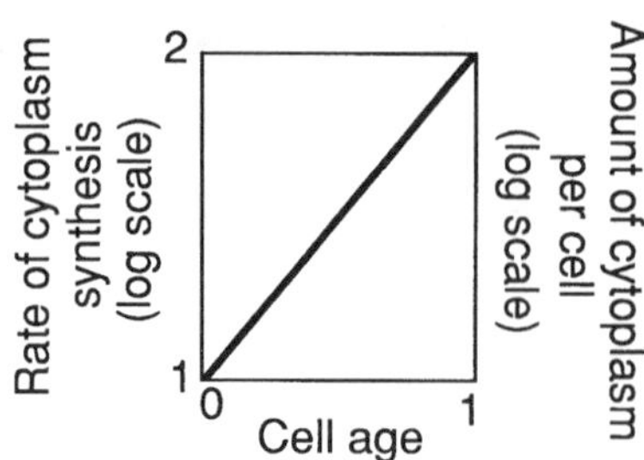


**Figure 1.** Cytoplasm synthesis during the division cycle. Both the rate of cytoplasm synthesis and the pattern of accumulation of cytoplasm are exponential during the division cycle.

A third argument supporting exponential cytoplasm synthesis is that only this pattern is explained in known biochemical terms. The cycle-independent exponential synthesis of cytoplasm can be derived from our current understanding of the biochemistry of macromolecule synthesis. Enzymes make RNA and protein, which make ribosomes, which then lead to more protein synthesis. More proteins mean more RNA polymerases, more catabolic and anabolic enzymes, and the continuously increasing ability of the cell to make more and more cytoplasm. In contrast, if newly synthesized material were *not* activated for synthesis when made, but instead were recruited for biosynthesis only at the instant of division or at some specific cell cycle age, one would have linear cytoplasm synthesis during the division cycle. There are no known mechanisms enabling the cell to distinguish between newly synthesized cytoplasm and old cytoplasm, or which lead to cell-cycle-specific variation in the synthetic rate of any particular molecule during the division cycle. We therefore conclude that in theory as well as practice, cytoplasm increases uneventfully and exponentially during the division cycle.

## 3 DNA SYNTHESIS DURING THE DIVISION CYCLE

One of the most important generalizations regarding the regulation of linear macromolecule synthesis is that the rate of synthesis is regulated at the point of initiation of polymerization. This principle is exhibited most clearly in DNA replication, where the process of initiation of replication can be dissociated from the continued replication of DNA that has already initiated replication. If one inhibits initiation of replication, rounds of DNA replication in progress will continue until their normal termination.

### 3.1 Pattttern of DNA Synthesis During the Division Cycle

The DNA synthesis pattern during the bacterial division cycle (specifically for *Escherichia coli* and related bacteria) is comprised of one or more periods of constant rates of DNA accumulation (Figure 2). In cells with a 60-minute interdivision time, there is a constant rate of DNA synthesis for the first 40 minutes, with one bidirectional pair of replication forks, followed by a zero rate of synthesis during the last 20 minutes. In cells with a 30-minute interdivision time, there is a constant rate of DNA synthesis for the first 10 minutes, which falls to a constant rate two-thirds of the initial rate for the last 20 minutes. In cells with a 20-minute interdivision time, there is a constant rate of DNA synthesis throughout the division cycle. Other growth rates have similar particular patterns of DNA synthesis.

The constant rates of DNA synthesis can be understood in terms of the molecular aspects of DNA synthesis. DNA is synthesized by the movement of a replication point along the parental double helix, leaving two double helices in its wake. The rate of movement of a growing point appears to be invariant and independent of its location in the genome. This means that the rate of DNA synthesis is proportional to the extant number of replication points and that the number of replication points is constant for any period of the division cycle. Thus the rate of DNA synthesis during any period of the division cycle is also constant. The pattern of constant rates of synthesis is derived from, and consistent with, our understanding of the biochemistry of DNA synthesis.

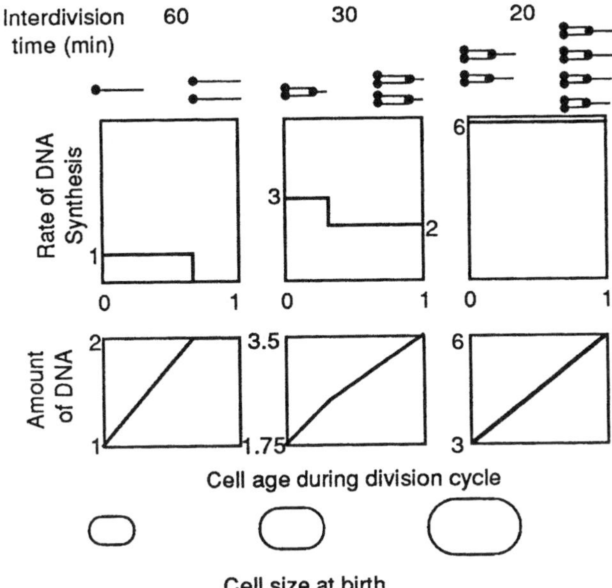


**Figure 2.** DNA synthesis during the division cycle. Three different patterns are illustrated for cells growing with 20-, 30-, and 60-minute doubling-times. For each growth rate, the time from initiation to termination of DNA replication is 40 minutes. The time between termination of replication and cell division is 20 minutes. The proposed chromosome patterns at the start and finish of the division cycle are illustrated above the graphs. In addition to the rate of DNA synthesis during the division cycle, the pattern of accumulation of DNA during the division cycle is presented. Accumulation of DNA is composed of periods of linear synthesis. The rates during these periods are proportional to the existing number of growing points. The graph of the rate of synthesis is the differential of the accumulation plot. A representation of the cell size expected for cells at the start of the division cycle is presented below each synthetic pattern. The sizes of the newborn cells are in the ratio of 1:2:4 in the three cultures illustrated here. The new initiations at the start of each division cycle also occur in the ratio of 1:2:4.

### 3.2 Regulation of Initiation of DNA Replication

But why is replication initiated at these particular times during the division cycle? How is the initiation process regulated? Rounds of replication are initiated in a bacterial cell when the amount of cell mass or something proportional to cell mass (the ''initiator'') is present at a certain amount per origin. Initiation of new rounds of replication occurs once per division cycle when the mass (or ''initiator'') per origin reaches a fixed value. In the 60-minute cell, initiation occurs at one origin with a cell of unit size (size 1.0). In the 30-minute cells there are two origins in the newborn cell, and the size of the newborn cell is twice that of a 60-minute cell. And the 20-minute cell has four origins initiated in the newborn cell and the cell size is four times the newborn 60-minute cell. Thus, not only does the DNA per cell increase with growth rate, but the average cell size increases with growth rate as well. In the cells shown in Figure 2, the relative sizes of the newborn cells, as well as the average cell in the culture, are in the ratio of 1:2:4.

The actual nature of the initiator is not known, but a number of candidate molecules have been identified. Among these the DnaA protein appears as a good candidate for the regulator of DNA initiation.

Alternative models of initiation have been proposed, such as the sudden accumulation of an inhibitor of initiation that is diluted out by cell growth. This model does not accommodate the constant synthesis of cell cytoplasmic components. In all cases it is indistinguishable from the initiator accumulation model, and there is no evidence that an inhibitor is synthesized during the division cycle.

## 4 SURFACE SYNTHESIS DURING THE DIVISION CYCLE

The cell surface is made to perfectly enclose, without excess or deficit, the cytoplasm synthesized by the cell. Cell cytoplasm increases continuously during the division cycle, and therefore cell surface is made continuously. How is the bacterial cell surface made and duplicated during the division cycle? How can we describe the rate and topography of cell surface synthesis during the division cycle?

Consider an imaginary cell in which the cytoplasm is enclosed in a cell surface tube that is open at each end; the cytoplasm remains within the bounds of the tube. The cytoplasm in the newborn cell is encased in the cylinder of cell surface, which is made up of membrane and peptidoglycan. As the cytoplasm increases exponentially, the tube length increases to exactly enclose the newly synthesized cytoplasm. In this imaginary case, the cell surface increases exponentially in the same manner as the cytoplasm. When the cell cytoplasm has doubled, the tube divides into two new cells, and the cycle repeats. In this imaginary cell, the cytoplasm increases exponentially, the internal volume of the cell increases exponentially, the surface area increases exponentially, and the density of the cell (i.e., the total weight of the cell per cell volume) is constant during the division cycle.

But a real rod-shaped cell does have ends, and therefore the pattern of cell surface synthesis during the division cycle is not simply exponential. If the cell surface were synthesized exponentially, the cell volume could not increase exponentially. Since the cytoplasm increases exponentially, there would have to be a change in cell density. Cell density, however, is constant during the division cycle. A proposal for cell surface synthesis that allows an exponential increase in cell volume, and therefore a constant cell density, is presented in Figure 3. Before invagination, the cell grows only in the cylindrical sidewall. After invagination, the cell grows in the pole area and the sidewall. Any volume required by the increase in cell cytoplasm that is not accommodated by pole growth is accommodated by an increase in sidewall. The cell is considered to be a pressure vessel, and when the pressure in the cell increases, there is a corresponding increase in cell surface area. The pole is preferentially synthesized, and any residual pressure due to new cell cytoplasm that is not accommodated by pole growth in relieved by an increase in cylindrical sidewall area.

The resulting pattern of synthesis is approximately exponential. The formula describing surface synthesis during the division cycle is a complex one, including terms for the shape of the newborn cell, the cell age at which invagination starts, the pattern of pole synthesis, and the age of the cell. One consequence of this model of surface synthesis is that at no time is surface synthesis exponential, since the rate of surface synthesis is not proportional to the amount of surface present over any period of time.

As with cytoplasm and DNA synthesis, we can derive the cell cycle pattern of surface synthesis from our understanding of the molecular aspects of peptidoglycan synthesis. The peptidoglycan sacculus of the cell is comprised of glycan strands encompassing the cell perpendicular to the long axis, as shown in Figure 4. These strands are cross-linked by peptide chains. Because the cross-linking in the load-bearing layer is not complete, one can have new strands in place, below the taut load-bearing layer, before the bonds linking old strands are cut. As shown in Figure 4, the load-bearing layer is stretched by the turgor pressure in the cell. An infinitesimal increase in cytoplasm leads to an infinitesimal increase in the turgor pressure of the cell. This increase in turgor pressure places a stress on all the peptidoglycan bonds, which reduces the energy of activation for bond hydrolysis. The result is an increase in the cutting of stressed bonds between the glycan chains. When a series of cuts is made, allowing the insertion of a single new strand into the load-bearing layer, there is an infinitesimal increase in cell volume. This increase in volume leads to a reduction of the stress on the remaining bonds. By a continuous repetitive series of cytoplasm increases, surface stresses, enzymatic cuts, and volume increases, an increased cell volume that precisely accommodates the increase in cell cytoplasm is obtained.

An inside-to-outside mode of peptidoglycan growth, similar to that for *Bacillus subtilis,* has been proposed for *E. coli.* This proposal is based on the observed recycling of murein, the calculated amount of peptidoglycan per cell, and the existence of trimeric

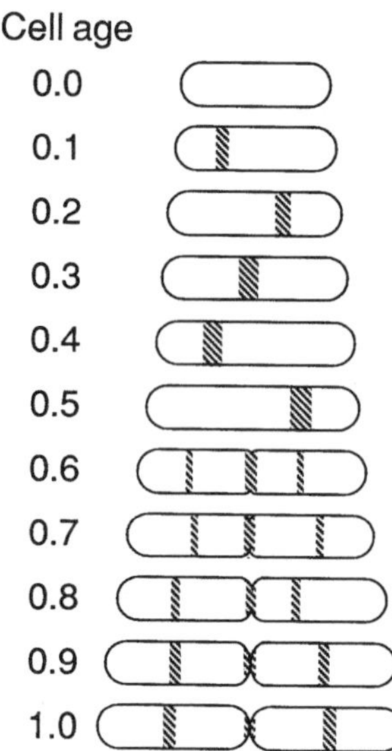


**Figure 3.** Cell surface synthesis during the division cycle. Before invagination, the cell grows only by cylinder extension. Each cell is drawn to scale, with the volume of the cells increasing exponentially during the division cycle. The shaded regions of the cell indicate the amount and location of wall growth (whether in pole or sidewall) each tenth of a division cycle. The width of the shaded area is drawn to scale. Cell surface growth actually occurs throughout the sidewall, not in a narrow contiguous zone. The variable locations of wall growth in this figure have no specific meaning; rather, they indicate that synthesis is occurring all over the sidewall during the division cycle. Before invagination, the ratio of the rate of surface increase to the rate of volume increase is constant. When pole synthesis starts, at age 0.5 in this example, there is an increase in the ratio of the rate of surface increase to the rate of volume increase. Any volume not accommodated by pole growth is accommodated by cylinder growth. At the start of pole growth, there is a reduction in the rate of surface growth in the cylinder. As the pole continues to grow, there is a decrease in the volume accommodated by the pole, and an increase in the rate of growth in the sidewall.

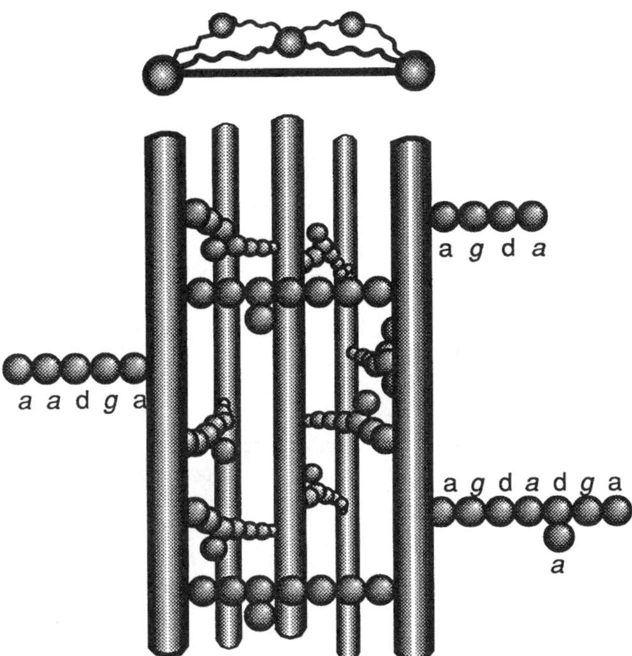


**Figure 4.** Idealized three-dimensional representation of peptidoglycan structure as seen from the outer surface. The thick bars represent glycan chains at the outside of the peptidoglycan layer. The thinner bars represent glycan chains below the outer layer. The stretched chains of circles represent amino acids cross-linking the glycan chains. The chains below the stretched surface of the cell rise to the outer layer when the taut layers of the peptidoglycan are hydrolyzed. Above is a cross-sectional view through the glycan chains illustrating the taut outer layer and the more loosely inserted inner material. The letters indicate the amino acids composing the cross-links: a, alanine; d, diaminopimelic acid; g, glutamic acid; italicized letters are present in the D-configuration.

and tetrameric fragments that are consistent with a multilayered peptidoglycan structure. The insertion of new peptidoglycan strands in an unstressed configuration prior to their movement into the load-bearing layer of the peptidoglycan can explain all these observations. The recycling of peptidoglycan may be a strain-specific result, as there is no major release or recycling of peptidoglycan in *Salmonella typhimurium* or in *E. coli* B/r. At this time, the inside-to-outside mode of surface growth cannot be excluded.

Although this discussion of the rate of surface synthesis during the division cycle has dealt primarily with peptidoglycan, it applies equally to membranes and other surface-associated elements. The cell membrane grows in response to the increase in peptidoglycan surface, and it coats the peptidoglycan without stretching or buckling. The area of the membrane increases in the same way that bacterial peptidoglycan increases.

One must make distinctions among cell components with regard to their location in the cell, rather than with regard to their chemical properties. Cell proteins can be divided into two categories, those associated with cytoplasm and those associated with surface. Proteins associated with the membranes are included in the surface category of synthesis during the division cycle. If a bacterial cell had a histonelike protein associated with DNA, there would be a third category of protein synthetic pattern, that which is synthesized during the division cycle with constant rates of synthesis. Proteins are synthesized during the division cycle with a pattern that is presumably consistent with their final location or category. Proteins are not a monolithic group, nor are they divided into a large number of groups. The three-category system proposed here allows us to conceptualize the possible patterns of protein synthesis observed during the division cycle.

## 5 PLASMID REPLICATION DURING THE DIVISION CYCLE

### 5.1 High-Copy Plasmids

If a cell has a large number (20–100) of plasmids per newborn cell, these plasmids can be considered to replicate during the division cycle in proportion to those present (i.e., exponentially), thereafter assorting themselves randomly into the newborn cells. Because the plasmid number is high, there is little chance of one of the daughter cells ending up with no plasmids as all of the plasmids segregate to the other daughter cell. This pattern appears to be the case for naturally occurring, high-copy number plasmids. One possible exception is the artificial plasmid made by cloning the origin (*oriC*) region of the bacterial chromosome. This appears to have a random assortment mechanism, but it is synthesized at a precise time during the division cycle that coincides with the normal time at which the chromosome initiates replication.

### 5.2 Low-Copy Plasmids

Low-copy plasmids are usually larger than high-copy plasmids and are present on the order of 1 to 2 per cell. Examples of such plasmids are the F-factor and P1 plasmid, which replicate in a precise way at a particular time during the division cycle. The rules of their replication are similar to that regulating chromosome replication, with initiation of plasmid replication occurring when a fixed amount of cell mass is present per plasmid origin.

## 6 SEGREGATION OF CELL COMPONENTS AT DIVISION

### 6.1 Segregation of Cytoplasm

The cytoplasmic components of the cell appear to segregate randomly at division. There does not appear to be any compartmentalization or hindrance to segregation of cytoplasm (Figure 5).

### 6.2 Segregation of Wall

The peptidoglycan is a large macromolecule that grows by intercalation of new material between old material. Since all peptidoglycan is covalently linked in one macromolecule, there is no ability for the peptidoglycan components to move with respect to one another. The cell wall segregation is therefore determined by the location of the cell wall. Once in place, the wall segregates into the newborn cells so that the old poles go to each cell, and the sidewall is segregated to the newborn cells (Figure 5).

### 6.3 Segregation of DNA

Two pieces of quantitative data describe the nonrandom segregation of DNA in *E. coli.* When segregation was analyzed using the membrane elution method, nonrandom segregation was constant over a wide range of growth rates. When cells were analyzed by

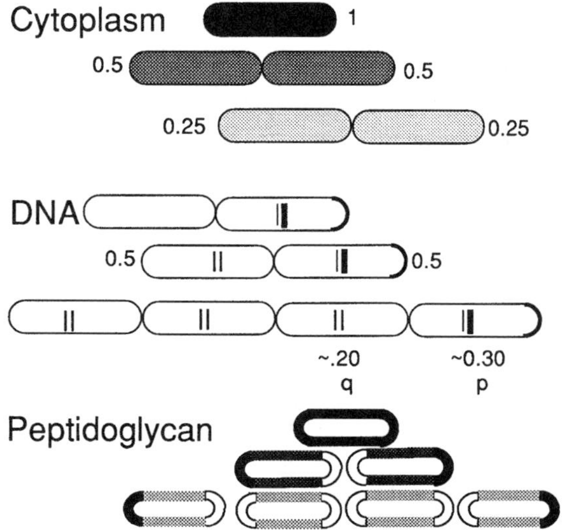


**Figure 5.** Segregation of cytoplasm, surface, and DNA at division. The cytoplasm of the cell is segregated randomly at division. Each daughter cell shares equally in the parental cytoplasm (ribosomes, soluble proteins, ions, etc.), which can be considered to be randomly distributed throughout the dividing cell. The DNA is segregated nonrandomly, but each daughter cell gets a complete and equal complement of DNA from the dividing cell. The nonrandom segregation is at the level of a single strand. Since each DNA duplex is made up of an ''older'' strand and a new complement made on the older strand, and each cell is composed of an ''older'' pole and a pole just made at the preceding division, one can see whether strands of one age go preferentially toward one or the other age pole. Experimental analysis indicates that the DNA strands segregate as though the older DNA strand goes preferentially toward the pole that it went to at the preceding division. This pattern may be phrased as follows: the newer strand goes toward the newer pole. The degree of nonrandom segregation is not perfect, but it is stochastic, and it is approximately 60:40 rather than 50:50. The surface of the cell segregates to the daughter cells at division such that the poles are conserved, and the sidewall segregates with no apparent conservation of sidewall peptidoglycan. It is not clear whether insertion of new wall material over the side is perfectly uniform or whether there is some nonuniform pattern of insertion.

the ''presegregation Methocel'' method, segregation was found to be nonrandom at slow growth rates, becoming more random as the growth rate increased. Although these two results seem contradictory, both are accommodated by a model proposing that DNA segregates as though the new strands were attached to new wall material. The apparent discrepancy between the two methods reflects differences between the approaches: the membrane-elution method examines the nonrandomness of segregation at a particular cell division, while the presegregation Methocel method measures segregation in the total cell population. The constancy of nonrandom segregation using the membrane elution technique means that the shape of the cell is constant at all growth rates. A constant shape means that the fraction of the cell devoted to pole is constant at all growth rates.

## 7 ARE THERE EVENTS DURING THE DIVISION CYCLE?

There are only four discrete events during the division cycle of a growing rod-shaped bacterial cell: (1) the initiation of DNA replication, (2) the termination of DNA replication, (3) the initiation of pole formation, and (4) the completion of pole formation. Other aspects of cell growth are too continuous to be considered to be cell-cycle-specific events. For example, the addition of new nucleotides at position 43,567 on the chain of DNA may be considered to be a unique event, and one that occurs at a particular time in the division cycle; because of cell cycle variability, however, the time of this synthetic occurrence can never be precisely known. The time of insertion of a particular nucleotide pair has no meaning for the cell cycle; only the initiation and the termination of DNA replication are important. With regard to the division cycle, the DNA is an amorphous material with no biochemical specificity along the strand.

Similar considerations apply to the cell surface and cytoplasm. No aspect of peptidoglycan strand insertion is unique in time during the division cycle. A new strand is inserted at random sites in response to the stretching of the cell surface. This may occur in one cell at position 0.376 of the cell length; at the same time, in another cell, the new strand may be inserted at position 0.549. For an analysis of the cell cycle, we should consider pole and sidewall growth as uneventful extensions of cell surface. Regarding the cytoplasm, one may call the increase in ribosome content from 37,411 to 37,412 an event; but such an individual event is unmeasurable and without meaning in the cell cycle. Thus, cytoplasm synthesis, DNA synthesis, and cell surface synthesis are uneventful during the division cycle.

In contrast to our understanding of the principles of cytoplasm, cell surface, and DNA synthesis, the biochemical mechanisms of the four events occurring during the division cycle—the initiation and termination of DNA synthesis and the initiation and cessation of pole formation—are still unknown. While a great deal is known about the biochemistry of initiation of DNA replication, the mechanism by which this initiation is regulated is less well understood. In addition, very little is known of the biochemical principles involved in either the termination of DNA replication or the start and cessation of pole formation. A cell surface structure, the periseptal annulus, has been proposed as a possible first step in the formation of a new pole. If the periseptal annulus is the start of pole formation, the important question is whether there are definable steps between the formation of the periseptal annulus and the start of invagination.

## 8 VARIATION OF THE BACTERIAL DIVISION CYCLE

Division cycles in a culture are not precisely timed. There is a great deal of variability in division cycles: a culture having a 60-minute doubling time probably has cells with interdivision times between 40 and 100 minutes. The variability of the division cycle can be visualized as due to the variability of the separate components that make up the division cycle. Thus, two identical newborn cells can divide at different times because of:

Variability in mass synthesis in the two cells leading to variation in time of initiation of DNA replication
Variation in the rate of DNA replication in two cells
Variation in the time between termination and division
Variation in the equality of division (which leads to variation in the next division cycle)

## 9 MODELS OF DIVISION CYCLE CONTROL

The driving force of the division cycle is cytoplasm synthesis. Some part of the energy used by the cell to make cytoplasm drives the biosynthesis of the cell surface by causing stresses along the cell surface, and this stress leads to the breaking of peptidoglycan bonds. The insertion of new peptidoglycan then leads to the increase in cell size. Cytoplasm increase also regulates DNA synthesis. Cell size at birth is greater at faster growth rates (Figure 2) because the initiation of new rounds of DNA replication occurs when the cell has a unit (or critical) amount of cytoplasm per origin of DNA. Cell size at initiation is proportional to the number of replication points in the cell, and that is why cells are larger at faster growth rates—at faster growth rates, there are more origins per cell at initiation. The cell ''titrates'' the amount of cytoplasm, or some specific molecule that is a constant fraction of the cytoplasm. Thus DNA synthesis is initiated at all available origins when the amount of cytoplasm per origin reaches a particular value.

The cell cytoplasm increases as fast as it can, given the external nutrient conditions. The synthesis of DNA and cell surface cannot outpace cytoplasm synthesis or lag behind it. DNA synthesis cannot go faster because it is waiting to initiate new rounds of DNA replication in response to cytoplasm synthesis, nor can it go slower, because the cytoplasm would increase without initiating new rounds of DNA replication. Cell surface is made to just enclose the newly synthesized cytoplasm. The regulation of surface and DNA synthesis by cytoplasm thus explains why there is no dissociation of these syntheses during the division cycle, and during the growth of a culture. Furthermore, the failure to complete DNA replication (so that a cell has only one genome) prevents cell division; the cell thus ensures that there will be enough material (genome, cytoplasm, and cell surface) to allow the production and survival of two new daughter cells.

## 10 THE BACTERIAL GROWTH LAW DURING THE DIVISION CYCLE

Some believe that there is a ''bacterial growth law'' that can be discovered by sensitive methods of analysis. Does the cell grows linearly, bilinearly, or exponentially, or perhaps follow some other, yet undiscovered, growth law? If there were a general law of cell growth, independent of the biosynthetic patterns of the three categories that comprise the cell, the individual categories of biosynthesis would have to accommodate themselves to this overall growth law.

There is no simple mathematical growth law regulating or describing bacterial growth during the division cycle. Bacterial growth during the division cycle is the simple weighted sum of the biosynthetic processes that are described by the three categories proposed here. From Table 1 it can be seen that the components of the cytoplasm of the cell are about 80% of the total cell weight. This means that the growth of the cell is approximately exponential. A slight deviation from exponential growth is due to the contribution of cell surface and DNA synthesis. Cell growth, then, is the result of a large number of individual reactions, regulated by local conditions, and not conforming to any overriding growth law. Thus the growth law is simply the sum of the individual synthetic patterns of the three categories of cell material.

**Table 1** Composition of *Escherichia coli,* a Typical Gram-Negative, Rod-Shaped Bacterium

| Cell Component | Molecules per cell | Dry Weight (%) |
|---|---|---|
| Cytoplasm | | |
| Protein | 2,350,000 | 55.0 |
| RNA | 255,480 | 20.5 |
| Polyamines | 6,700,000 | 1.1 |
| Glycogen | 4,300 | 3.5 |
| Metabolites, ions, cofactors | | 2.5 |
| Surface | | |
| Peptidoglycan | 1 | 2.5 |
| Lipid | 22,000,000 | 9.1 |
| Lipopolysaccharide | 1,430,000 | 2.4 |
| Genome | | |
| DNA | 2.1 | 3.1 |

## 11 BACTERIAL GROWTH AT DIFFERENT GROWTH RATES

When bacteria are grown in different media (at a given temperature), a regular pattern of cell composition is observed. The faster

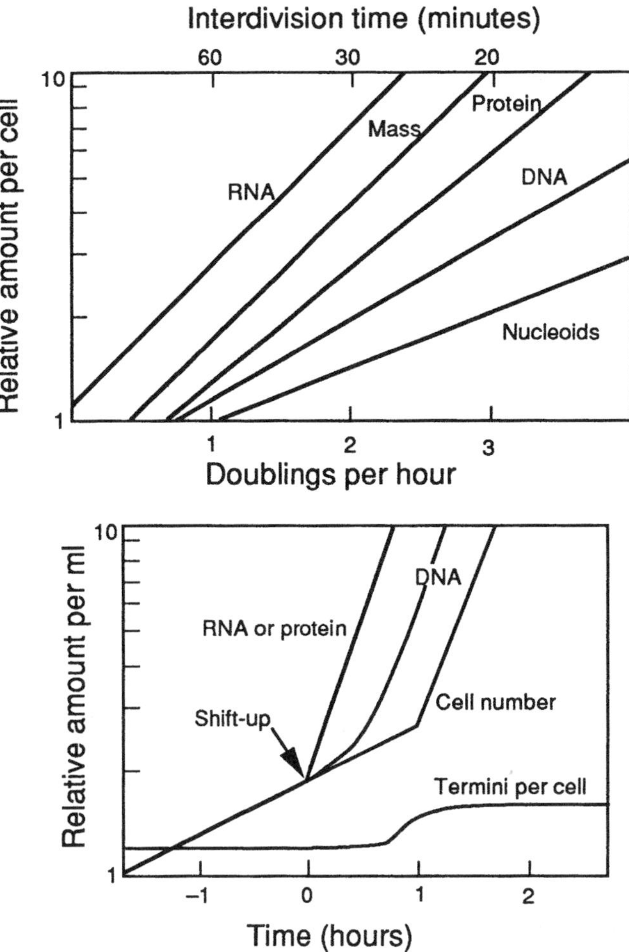


**Figure 6.** Composition of bacteria as a function of growth rate (top). Synthesis of cell components during a shift-up from minimal medium to faster growth in richer medium (bottom).

a cell grows (richer medium), the larger the cell and the more DNA, RNA, protein, and other components per cell (Figure 6). When the components are accurately measured, it is found that the material per cell increases exponentially as a function of the reciprocal of the doubling time (i.e., growth rate). This experimental observation is now known to be theoretically expected on the basis of the constancy of the C and D periods. While the amount of everything increases with increasing growth rate, the slopes of these increases are different.

During a shift-up from a slow growth medium (minimal medium) to rapid growth in a rich medium, there is a regular pattern of change. There is a very rapid change in the rate of RNA and protein synthesis to the new rate of increase, a slower change in the rate of DNA replication, and finally a sharp change in the rate of cell increase from the old rate to the new rate at 60 minutes after the shift-up (Figure 6). The shift-up results are now understood to be expected in the light of the constant C and D periods.

## 12 EXPERIMENTAL ANALYSIS OF THE BACTERIAL DIVISION CYCLE

How can one study the pattern of synthesis of cell components during the division cycle? At first thought it appears that a synchronized bacterial culture (i.e., a culture composed of cells all of the same age—for example, newborn cells) might be prepared, and measurements taken of the cell components at different times during the division cycle. It turns out that this method has rarely if ever led to any concrete results, presumably because of the perturbations and artifacts introduced when cells are synchronized. The artificial process, in turn, is necessary because no naturally synchronized populations are available.

The methods that have given results are nonperturbing, nonsynchrony methods. One is the membrane elution method (or "baby machine"), which analyzes the division cycle of unperturbed cells. In this method a label is added to a growing culture, the labeled cells are filtered onto a membrane, the cells bind as though bound at one end or pole, and when medium is pumped through the inverted membrane, newborn cells are eluted from the bound cells. Since the newborn cells arise from the bound cells as a function of the original cell age, the first newborn cells arise from the oldest cells at the time of labeling, and with time the newborn cells arise from cells labeled at younger and younger cells ages. By measuring the amount of label per cell during elution, one can determine the pattern of synthesis during the division cycle. This has been done successfully for DNA, plasmids, protein, RNA, peptidoglycan, and cell membrane.

Another method for cell cycle analysis is flow cytometry, where the amount of material per cell in a cell population is determined for individual cells by a complex system of liquid handling and laser excitation engineering. The quantitative distribution of cell material can give the relative rate of synthesis during the division cycle. This method has been successful for DNA replication during the division cycle and has confirmed the initial results obtained with the membrane elution method.

## 13 *DIE EINHEIT DER ZELLBIOLOGIE*

Are the interrelationships of cell cycle regulation, proposed here for *Escherichia coli,* applicable to other organisms? Other regulatory mechanisms have been proposed in other cell types. For example, a circular regulatory system was proposed for *Caulobacter* and a stochastic mechanism as well as a rate-controlled model for animal cells. Is the mechanism of cell cycle regulation subject to historical accident? Or is there a unity of mechanisms regulating the division cycle of all types of cells? I conjecture that the *E. coli* model of cell cycle regulation is generally applicable to other organisms, and that the optimal design of the cell cycle is exhibited by the *E. coli* cell cycle.

In the third decade of this century, Kluyver and his colleagues explicitly proposed the principle of the unity of biochemistry, or as originally proposed, *die Einheit der Biochemie.* Although the original proposal dealt primarily with the nature of oxidation–reduction reactions, the basic principle has evolved to encompass even more unifying principles. The nature of the gene, the genetic code, the mechanism of protein synthesis, the use of enzymes for biochemical changes, the patterns of enzyme regulation, and all the fundamental aspects of biochemistry appear to be similar throughout the different organisms on earth. While there are exceptions and unique aspects to many organisms, the essential principle that there is a unified biochemistry has been supported.

Here we must now add, with regard to the cell cycle, the unity of cell biology—*die Einheit der Zellbiologie*—which is analogous to the unity of biochemistry. Phenomena suggesting the existence of other modes of cell cycle regulation in animal cells have been shown to be consistent with the model for the bacterial division cycle proposed here. For example, the constancy of C and D periods is mimicked by the constancy of their analogues in animal cells, the S and G2 periods. The variation in cell size with growth rate and the rules regarding size determination also appear to hold for all organisms. The rules of the cell division cycle—that is, the logic and design principles of the division cycle—are the same for *Escherichia* and escargot, *Salmonella* and salmon. Although there are many detailed differences between eukaryotic and prokaryotic division cycles, at the deepest level the principle of the unity of cell biology proposes that the division cycle is ultimately regulated by the continuous accumulation of some molecule that is titrated against a fixed amount of another cell component. From this viewpoint any cell-cycle-specific events identified in eukaryotic cells are the result or symptom of a deeper regulatory principle that is produced in a cell-cycle-independent manner. The question for future analysis is whether there is some deeper rule that ensures this common pattern in all cells. I conjecture that this is indeed the case, and I hope that a search for the underlying meaning of size control in all cells is now the new goal of cell cycle research.

*See also* BACTERIAL PATHOGENESIS; *E. COLI* GENOME; HEAT SHOCK RESPONSE IN *E. COLI*; LIPIDS, MICROBIAL; MYCOBACTERIA.

## *Bibliography*

Cooper, S. (1984) The continuum model as a unified description of the division cycle of eukaryotes and prokaryotes. In *The Microbial Cell Cycle,* P. Nurse and E. Streiblova, Eds. pp. 7–18. CRC Press, Boca Raton, FL.

———. (1991) *Bacterial Growth and Division: Biochemistry and Regulation of Prokaryotic and Eukaryotic Division Cycles.* Academic Press, San Diego, CA.

———. (1991) Synthesis of the cell surface during the division cycle of rod-shaped, Gram-negative bacteria. *Microbiol. Rev.* 55:649–674.

Okuda, A., and Cooper, S. (1989) The continuum model: An experimental and theoretical challenge to the G1-model of cell cycle regulation. *Exp. Cell Res.* 185:1–7.

# Bacterial Pathogenesis

*Luciano Passador, Kenneth D. Tucker, and Barbara H. Iglewski*

*Key Words*

**Adhesin** A microbial component, such as pili, which allows a pathogen to adhere to its target cell or tissue.

**Capsule** A compact or loose layer of extracellular polysaccharide that surrounds some bacteria and may play a role in the evasion of host defenses.

**Clonality** The ability of a microbial clone to be more virulent than another of the same species.

**Pathogenicity** The ability of an organism to cause damage (disease) to a host, which it infects.

**Virulence** The degree of pathogenicity exhibited by an organism.

The ability of bacteria to be pathogenic is directly associated with the ability of those organisms to produce various cell-associated or extracellular components often termed virulence factors. These components are often involved in processes such as adherence to host cells (as in the case of pili) and the evasion of host defense mechanisms (as in the case of various proteases and toxins). The presence of these factors allows the bacterium to gain an advantage over the host and results in various pathologies.

The ability to clone and express the gene(s) involved in the synthesis and expression of virulence factors has been pivotal in gaining an understanding of how bacteria cause disease. The application of molecular biological techniques to the study of bacterial pathogenesis has proven to be a powerful approach by which to study the involvement and interaction of the various factors in the development of disease. The knowledge gleaned through the use of these techniques also provides a means by which novel therapeutic approaches may be designed. This contribution offers a general overview of the molecular basis of bacterial pathogenesis of the organisms that must colonize the host.

## 1 INTRODUCTION

Bacteria cover and colonize most of the exterior surface of the body including mucosal surfaces. While the majority of these organisms usually serve a beneficial role to the host, or at least do not appear to be detrimental, a small percentage of bacteria are pathogenic. To survive, any organism must be able to respond and adapt to its changing environment. Pathogens are no exception in that they must adapt to changes when being transmitted to a new host or to changes in the new host as a result of their presence. For example, certain pathogens can evade the host immune system while other pathogens are virulent only in immunocompromised hosts. Some organisms use a combination of tactics to cause disease, while others rely on a single approach. Pathogenesis may be caused by factors produced after the bacteria have colonized the host or by factors produced exogenous to the host. The latter case requires the host to contact preformed toxins produced by the bacteria and is primarily limited to certain forms of food poisoning not covered in this contribution.

## 2 TRANSMISSION OF PATHOGENS TO THE HOST

There is very little known about the requirements for the effective transmission from one host to another. Some pathogens, such as *Bordetella pertussis,* the causative organism of whooping cough, can be transmitted via aerosols created by sneezing. Others, including *Treponema pallidum,* which causes syphilis, are transmitted via sexual contact. Still others, such as *Vibrio cholerae,* are internalized by ingestion of contaminated matter or via the fecal–oral route. No matter the route of entry, all the pathogens share the common difficulty of having to move from one environment to another, often drastically different, environment.

## 3 ENTRY INTO THE HOST

The primary barrier to infection is the epithelium. Bacteria may breach this protective layer in many ways. One approach is to enter through breaks in the skin caused by trauma including cuts, burns, and insect bites. For example, *Yersinia pestis,* the causative organism of the bubonic plague, can be transmitted by arthropod vectors. Additional routes of entry include mucosal surfaces that are involved in nutrient uptake or gas exchange (e.g., gastrointestinal tract and respiratory tract) and are in constant exchange with the environment. This tissue is a thin barrier between the host and its environment and is the site of entry of many bacteria that invade the host through intact epithelia.

## 4 ADHERENCE TO HOST CELLS

Some pathogens remain on the surface of the cell and multiply at that location, while others make use of their ability to invade the tissue in search of a more suitable location. Nonetheless, even invasive organisms must initially attach to a host cell. Specific components (adhesins) determine the pathogen's ability to adhere to receptor sites on the host cell. The types of adhesin are varied, as indicated by the use of surface proteins (as in the case of the pili of enteropathogenic *Escherichia coli*) and lipoteichoic acids (as demonstrated in the group A streptococci). Most pathogens maintain several alternative means of adhering to cells which may be used under specific environmental conditions or to allow attachment to specific cell types. Although pathogens theoretically could interact with a myriad of host surface molecules, only a few molecules have actually been identified as receptors. One of the most studied, fibronectin, is a large matrix glycoprotein that is used as an adherence molecule by eukaryotic cells and serves as a receptor for the lipoteichoic acid of *Streptococcus pyogenes.*

## 5 EVASION OF THE HOST IMMUNE RESPONSE

After attachment, the bacteria must survive in their new environment. Toward this end, the bacteria must compete for nutrients—not only with the host but with the indigenous microbial flora. In addition, the organisms must be able to evade the components of the host immune response. Antibodies and complement are the major components of the humoral response, and when these proteins encounter an organism or its soluble components, the antibodies bind to form an antibody–antigen complex. In the case of whole organisms, a series of complement proteins binds the complex; the result is the lysis and death of the pathogen. In the case of soluble factors, the formation of the complex facilitates phagocytosis and removal of the dead pathogen by specific cells. However, the formation of the immune complexes themselves may cause pathology, as seen in complex-mediated hypersensitivity in which the host immune system is turned against the host itself. An excellent example is the case of poststreptococcal glomerulonephritis. The classic view of this disease postulates that lysis products of *Streptococcus pyogenes,* which result from an attempt by the host inflammatory response to clear a throat infection, eventually enter the bloodstream and initiate an antibody response. The high levels of antibodies produced by the host then interact with antigens present in the bloodstream, resulting in the formation of antibody–antigen complexes that may then accumulate in the kidney. The complexes themselves may elicit an inflammatory response, resulting in damage to kidney tissue and eventually impair normal kidney function. The same organism is also responsible for rheumatic heart disease in which antibodies directed against a number of *S. pyogenes* products include those against M protein, a cell surface protein of *S. pyogenes* that shares epitopes with proteins on the surface of cardiac tissue. Hence the formation of antibodies against M protein results in the production of a set of antibodies that may also recognize the host tissue, leading to an immune response against the host.

Alternatively, the interaction of pathogens with antibodies may increase the susceptibility of the former to phagocytosis. The process of coating cells with antibodies, termed opsonization, leads to their recognition by phagocytic cells and other cells of the immune system. This interaction may also lead to hypersensitivity reactions, since many of these cells will release molecules that result in an inflammatory response. Some pathogens (e.g., *Pseudomonas aeruginosa*) directly attack the immune response by producing proteolytic enzymes that can degrade certain classes of antibodies and complement components. Others (e.g., *Klebsiella pneumoniae*) evade the immune response by cloaking themselves in exopolysaccharide capsules, which are believed to impair the ability of antibodies to bind and to prevent phagocytosis, or by periodically changing their surface antigens to render existing antibodies ineffective (antigenic variation). *Neisseria gonorrheae,* the causative organism of gonorrhea, provides a good example for antigenic variation. This organism is capable of using a variety of methods to vary the amount and composition of the pili on its surface. In this way it presents an ever-changing surface to the host response, resulting in decreased ability of the host to clear the organism. Still other organisms utilize a process of molecular mimicry in which they present a surface marker that is recognized by the host as "self," which means that the host does not respond with a normal immune response. Good examples of this mimicry are the hyaluronic acid capsule of *S. pyogenes* or the sialic acid capsule of *Neisseria meningiditis.* The importance of evading the immune response is illustrated by opportunistic pathogens, which are limited to immunocompromised hosts; that is, they cannot survive in a host whose immune system is intact.

## 6 INVASION OF HOST CELLS

Some pathogens have evolved the mechanisms required to invade cells, a development that has several advantages. First, it sequesters the organism from the potentially lethal effects of the host immune system; second, it places the organism in a nutrient-rich environment; and third, it positions the organism in a location in which competition from other organisms is almost nonexistent. However, invasion requires a set of biological mechanisms complex enough to allow the organism a means not only to enter the cells, but to also survive, multiply, and eventually disseminate from the target tissue. Little is known about the adaptations for survival within the cell, but novel techniques should lead to a rapid appreciation of the complexity required. The importance of host cell receptors in invasion is suggested by the observation that some invasive pathogens can invade an extremely diverse population of cells while others are restricted to specific cell types. Integrins, which play a role in host–cell interactions including attachment, phagocytosis, and the binding of proteins present in the extracellular matrix, have been identified as receptors. *Yersinia pseudotuberculosis, Mycobacterium tuberculosis,* and *Legionella pneumophila* have all been shown to utilize members of the integrin family of proteins as receptors.

## 7 DISSEMINATION

Many pathogens remain localized to the initial site of attachment, while others are capable of dissemination within the host. The ability to disseminate allows the pathogen to move to environments of greater nutrient availability as well as to increase the probability of locating a preferred niche. Dissemination may involve movement through cells or between cells in the extracellular matrix. Many pathogens are capable of eroding tissue through the secretion of extracellular products. *P. aeruginosa* and *Staphylococcus aureus* may secrete a large number of extracellular products, including toxins (which can have an exfoliative effect) and collagenases and elastases (which degrade major connective tissue components).

## 8 REGULATION OF VIRULENCE

Given that a pathogen must be able to sense the condition of its host, it is not surprising that many of the signals that initiate expression of virulence factors are environmental. For example, low temperature and calcium limitation have been well documented as signals for virulence of *Yersinia* species, as has the availability of iron on the expression of virulence factors of many organisms. All the species of *Yersinia* produce a series of proteins known as Yops, which are excreted but may also be membrane associated. The function of the Yops is not known in all cases, but those whose function has been determined appear to be involved in the inhibition of phagocytosis of the organism. The Yop proteins appear to be maximally expressed at 37°C, but only in the absence of calcium. The presence of calcium even in millimolar amounts dramatically reduces the expression of the Yops. Thermal and calcium regulation

**Table 1** Some Representative Virulence Factors/Mechanisms and Their Functions

| Virulence Factor/ Mechanism | Organism(s) | Virulence Function |
|---|---|---|
| Motility | *Pseudomonas aeruginosa* | Enhanced ability of motile organism to disseminate to other tissues |
| Altered lipopoly-saccharide side chains | *Salmonella typhimurium* | Inhibition of cell lysis by inhibiting the formation of the activated complement complex |
| Proteases | *P. aeruginosa* | Destruction of tissue for dissemination, provision of nutrients |
| | *Streptococcus pyogenes* | Destruction of complement, antibodies, and other host cell defense proteins |
| Antigenic variation | *Neisseria gonorrheae* | Variation of surface proteins to evade host antibody response |
| Exotoxins | A-B types *Corynebacterium diphtheriae*, *P. aeruginosa*, *Clostridium tetani*, *Vibrio cholerae* | Killing of host immune response cells (phagocytes, leukocytes, etc.); Destruction of tissue to enhance nutrient availability and possible dissemination to other sites |
| | Membrane disrupting *Listeria monocytogenes*, *Clostridium perfringens* | |
| Pili or fimbriae | *P. aeruginosa*, *Escherichia coli*, *V. cholerae* | Adhesion to host cells and mucosal tissue |
| Nonpilus adhesins | *P. aeruginosa*, *S. pyogenes*, *V. cholerae* | Adhesion to cells and mucosal tissue |
| Secreted (siderophores) and surface proteins that bind iron | *E. coli*, *P. aeruginosa*, *Yersinia pestis* | Acquisition of iron from host sources |
| Capsules | *Streptococcus pneumoniae* | Inhibit phagocytosis; Decrease activation of complement |
| Survival within phagocytes | *S. typhimurium* | Evasion of host immune response |

processes involve a large number of other protein products, some of which respond to temperature and others to calcium levels.

## 9 CLONALITY

With the large number of microorganisms in existence, it is intriguing that relatively few are capable of causing disease. Even within a given species, only a small number of strains may actually become pathogenic. However, the acquisition of virulence is not a random event. The current evidence suggests that strains and species that carry specific virulence determinants have evolved. The reason for this clonality is not clear, but the phenomenon itself underscores the complexity of what determines the pathogenicity of an organism.

## 10 CLOSING REMARKS

Bacterial pathogenesis involves a large, complex system that serves to allow pathogens to inhabit their hosts and cause disease. This interplay between host and bacteria has resulted in many pathogens becoming specific for certain host species as they evolve to proficiently colonize and invade the host. Pathogens adapt to changes in their environment by being able to sense changes in their hosts and to respond by producing virulence factors, which aid in their survival. As one might imagine, these adaptations require an intricate balance of the expression of certain genes. Details regarding the mechanisms and genetics have been elucidated for some of the virulence factors (see Table 1). In many cases the genetic regulation is poorly understood. While we have made great progress in characterizing the molecular basis of pathogenesis, we cannot define the full complement of factors required for pathogenicity. This remains a future goal.

*See also* Bacterial Growth and Division; Mycobacteria.

### *Bibliography*

Ewald, P. W. (1993) The evolution of virulence. *Sci. Am.* 268:86–93.

Finlay, B. B., and Falkow, S. (1989) Common themes in microbial pathogenicity. *Microbiol. Rev.* 53:210–230.

Groisman, E. A., and Saier, M. H., Jr. (1990) *Salmonella* virulence: New clues to intramacrophage survival. *Trends Biochem. Sci.* 15:30–33.

Mekalanos, J. J. (1992) Environmental signals controlling expression of virulence determinants in bacteria. *J. Bacteriol.* 174:1–7.

Moxon, E. R., and Kroll, J. S. (1990) The role of bacterial polysaccharide capsules as virulence factors. *Curr. Top. Microbiol. Immunol.* 150:65–85.

# Bacteriorhodopsin

*Janos K. Lanyi*

*Key Words*

**Bacteriorhodopsin** A retinal protein in the cytoplasmic membrane of halobacteria that functions as a light-driven proton pump.

**Ion Pump** A membrane protein that utilizes ATP hydrolysis, redox reactions, or retinal isomerization to drive the uphill transport of a cation or anion.

**Protonmotive Force** Difference of the electrochemical potential for protons across a membrane.

**Time-Resolved Spectroscopy** Measurement of absorption spectra during a single turnover of a catalytic cycle, revealing the rise and decay of intermediate states.

The light-driven proton pump bacteriorhodopsin is the simplest active (uphill) ion transport system. It couples the isomerization of its retinal chromophore to directed changes in the proton affinity of strategically arranged groups between the two membrane surfaces. The mechanism of proton translocation in this system provides clues to how active transport is achieved by pumps in general.

## 1 STRUCTURE OF THE PROTEIN

Bacteriorhodopsin is a small integral membrane protein; under anaerobic conditions it is the most abundant component of the cell envelopes of some strains of extremely halophilic bacteria. It consists of seven transmembrane helical segments, A–G, which span the width of the lipid bilayer. Lys-216, near the middle of helix G, binds an all-*trans*-15-*anti*-retinal via a Schiff base linkage; this and the surrounding protein residues form the purple chromophore. The protein is assembled into extended two-dimensional crystalline arrays on the surface of the cells. Figure 1 shows the approximate structure of a monomer, and the locations of the retinal and some residues with functional roles. Absorption of light causes isomerization of the retinal to 13-*cis* and sets off a sequence of reactions ("photocycle"), in which the initial state is regained within a few tens of milliseconds and a proton is translocated from one side of the membrane to the other. Thus, upon illumination, bacteriorhodopsin functions in the cytoplasmic membrane of the halobacteria as an electrogenic pump for protons and contributes to the creation of protonmotive force that energizes ATP synthe-

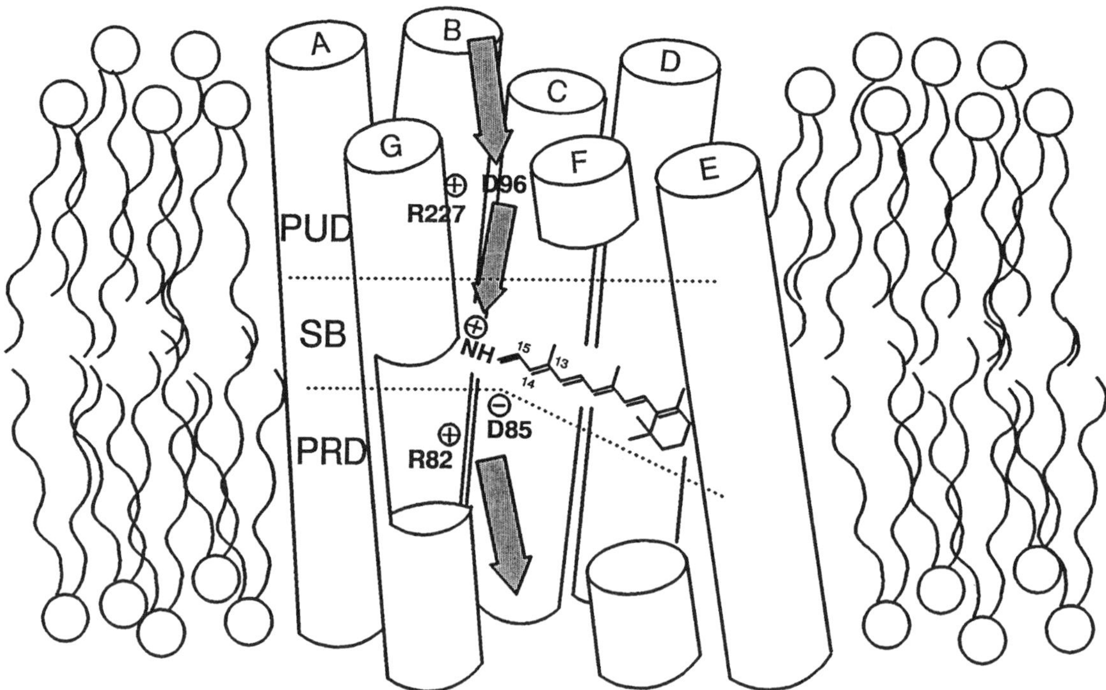


**Figure 1.** Sketch of bacteriorhodopsin in the lipid bilayer, with its seven transmembrane helices labeled A–G. These monomers assemble into the extended two-dimensional lattice in the plane of the bilayer called the "purple membrane." Parts of helices F and G are cut away for better view. The approximate positions of the retinal and some residues that play roles in the transport are indicated. The protein may be divided into three domains: PUD (proton uptake domain), SB (Schiff base), and PRD (proton release domain).

sis and active transport of other ions and nutrients. With sunlight, the halobacteria can grow photoautotrophically and will survive without respiratory activity—for example, under anaerobic conditions.

## 2 THE PHOTOCYCLE

It has been the hope of a large number of investigators that studies of bacteriorhodopsin will lead to an understanding of the mechanism of ionic pumps in general. Understanding the bacteriorhodopsin photocycle would permit the description of the reactions of the retinal chromophore as followed by a wide variety of spectroscopic measurements; understanding how proton is transported would permit the description of the internal and external proton transfer reactions, which move the proton across the width of the membrane. The photocycle is studied mainly by visible and Raman spectroscopy, the protonation reactions in the protein mainly by infrared spectroscopy, and the accompanying charge transfers by photoelectric measurements of oriented bacteriorhodopsin films or membranes. Site-specific mutagenesis has produced a large variety of partly or wholly defective proteins. Their properties have shed light on the interactions of protein residues with one another and with the retinal. These investigations together have produced a kinetics scheme for the photocycle and a mechanistic model for the functioning of this proton pump.

The chromophore reactions after the initial photoinduced isomerization of the retinal can be described by a linear sequence of intermediate states termed J, K, L, M, N, and O, and some of their substates. These states are identified by their absorption maxima in the visible and by distinct bands in resonance Raman and Fourier transform infrared difference spectra. As the retinal skeleton passes through the J and K states, it undergoes a series of torsions and rotations that result in a relaxed 13-*cis*-15-*anti* configuration in the L state. The Schiff base loses its proton in the M state, and regains it in N. Crystallographic studies of M have revealed structural changes originating most likely from tilting of helices F and G. A late M state and the N intermediate are characterized also by a changed protein backbone conformation, evident from changed amide bands in the infrared. The retinal regains its all-*trans* configuration in O, but recovers fully only as the back-reaction (BR) state is finally repopulated.

The rise and decay of these chromophoric and protein states contain more time constants than would be expected on basis of a unidirectional reactions sequence. These complications have led to a large variety of proposed photocycle schemes. It is now recognized by many investigators that the additional time constants originate from equilibration (i.e., back-reactions of rates comparable to the forward reactions) in many of the interconversions between successive photocycle intermediates. As a first approximation, the photocycle is described by the minimal scheme

$$BR \xrightarrow{h\mu} K \leftrightarrow L \leftrightarrow M_1 \rightarrow M_2 \leftrightarrow N \leftrightarrow O \rightarrow BR$$

The existence of two M substates connected by a unidirectional reaction is required by the decay kinetics of L and is supported by the fact that under some conditions an early and a late M intermediate, exhibiting different absorption maxima, can be distinguished.

## 3 PROTON TRANSFER REACTIONS IN THE PROTEIN

The molecular events that underlie these spectroscopically detected reactions involve the retinal Schiff base and several key protein residues. In the L-to-$M_1$ reaction, the proton of the Schiff base is transferred to Asp-85, a residue that lies toward the extracellular surface. At pH values exceeding 6, this is followed by release of a proton to the extracellular aqueous phase from a complex of residues termed XH in which not only Asp-85 but also apparently the neighboring Arg-82, Tyr-57, and a bound water play essential roles. At pH values less than 6, proton release does not occur at this time because below its p*K*a the protonated form of XH is stabilized. The subsequent transition from $M_1$ to $M_2$ occurs independently of the release of the proton at this time. This step, referred to as the reprotonation switch, changes access of the Schiff base from the extracellular to the cytoplasmic side. Its nature is not yet understood, but it may be linked to a retinal single-bond rotation, a protein conformational change, or both. Once the connection of the Schiff base to the extracellular side has been broken, it is reprotonated from the cytoplasmic side by Asp-96 in the M-to-N reaction via what appears to be a hydrogen-bonded chain over about 12 Å, involving a few bound water molecules. Two events follow: proton uptake by a complex of Asp-96, Arg-227, Ser-226, and Thr-46, and probably bound water, and reisomerization of the retinal to all-*trans*. The latter is observed as the N-to-O chromophore reaction. The complex pH dependence of the kinetics of these reactions is not well understood. It may originate from the interdependence of proton uptake and retinal reisomerization. At pH values exceeding 6, proton uptake precedes the reisomerization, resulting in the accumulation primarily of the N state before the final return to BR. In the low pH pathway reisomerization precedes the proton uptake, and it is primarily the O state that accumulates. The initial deprotonated state of the Asp-85 residue is regained in the final O-to-BR reaction. At high pH this consists of proton transfer to the complex group XH. At low pH, where XH remained protonated, this reaction consists apparently of proton release directly to the extracellular bulk.

## 4 TRANSPORT MECHANISM

The protein consists of three domains layered between the cytoplasmic and extracellular surfaces of the membrane (Figure 1): the proton uptake domain (PUD), the Schiff base (SB) domain, and the proton release domain (PRD). The proton pump functions by coupling internal proton transfers, which redistribute protons in the three domains to proton exchange with the aqueous phase. During the photocycle, the central domain containing the Schiff base communicates alternately with extracellular and cytoplasmic domains. Initially each domain contains one mobile proton. Through the specific proton transfers described earlier, a proton excess develops in the extracellular domain (PRD) and a deficit in the cytoplasmic domain (PUD). Proton release on the extracellular surface and uptake on the cytoplasmic surface correct this imbalance and result in the net translocation of a proton across the three domains.

*See also* Bacteriorhodopsin-Based Artificial Photoreceptor; Bioenergetics of the Cell; Retinoids.

*Bibliography*

Ebrey, T. G. (1993) Light energy transduction in bacteriorhodopsin. In *Thermodynamics of Membranes, Receptors and Channels,* M. Jackson, Ed., pp. 353–387. CRC Press, New York.

Lanyi J. K. (1992) Proton transfer and energy coupling in the bacteriorhodopsin photocycle. *J. Bioenerg. Biomembranes,* 24:169–179.

Mathies, R. A., Lin, S. W., Ames, J. B., and Pollard, W. T. (1991) From femtoseconds to biology: Mechanism of bacteriorhodopsin's light-driven proton pump. *Annu. Rev. Biophys. Biophys. Chem.* 20:491–518.

Oesterhelt, D., Tittor J., and Bamberg, E. (1992) A unifying concept for ion translocation by retinal proteins. *J. Bioenerg. Biomembranes,* 24:181–191.

Rothschild, K. J. (1992) FTIR difference spectroscopy of bacteriorhodopsin: Toward a molecular model. *J. Bioenerg. Biomembranes,* 24:147–167.

# Bacteriorhodopsin-Based Artificial Photoreceptor

*Tsutomu Miyasaka*

*Key Words*

**Charge Displacement** Spatial displacement of a charge within a molecule due to energetic excitation.

**Langmuir–Blodgett (LB) Film** Stack of monomolecular layers built up on a solid surface by transferring a monolayer of an organic surfactant formed at the interface of air and a liquid phase (normally water).

**Photoreceptor** The light-receiving system (cell) of the eye, corresponding to the retina.

**Proton Pumping** Active unidirectional transport (translocation) of protons across a cellular membrane to develop a membrane potential by a pH gradient.

**Purple Membrane** A cellular membrane, isolated from halobacteria, comprising a two-dimensional crystalline array of bacteriorhodopsin molecules.

**Visual Information Processing** Processing of optical information for pattern recognition, modeled on human visual perception and neural networks.

Bacteriorhodopsin (bR), an analogue of the visual pigment rhodopsin, is the only protein of the purple membrane (PM), a functionally specialized domain in the cell membrane of *Halobacterium salinarium (halobium).* It enables these organisms, which live in the highly concentrated brine of salt lakes, to use light as respiratory energy when the oxygen concentration in the brine becomes too low. Illumination of bR drives a rapid, cyclic series of conformational changes in the retinal chromophore and protein; as a result, protons are transported across the membrane and generate a membrane potential (proton gradient). The conformational changes and proton transport can be detected by external electrodes in vitro. Purified PM, capable of "proton pumping," can easily be obtained in large amounts from cultures of *H. salinarium.*

Molecular electronic devices based on bR have been constructed for potential applications in optical computers and light-sensing systems. Of particular interest are visionlike properties of bR. An artificial photoreceptor fabricated from bR photocells is capable of image sensing and can be applied for pattern recognition in real-time visual information processing systems.

## 1 PHOTOCYCLE OF BACTERIORHODOPSIN AND PHOTOELECTRIC BEHAVIOR

The chromophore all-*trans*-retinal is bound as a protonated Schiff base to a lysine residue of the 248 amino acid in bR. Light excitation causes a rapid *trans*-to-*cis* double-bond isomerization of the retinal, which drives a cyclic series of conformational changes (photocycle) of the protein. During the photocycle, unidirectional proton pumping is achieved which generates the electrochemical proton gradient needed to the ATP synthesis for respiration. Figure 1 shows a simplified scheme of the cycle. The first shown, thermal intermediate state K, arises in less than 5 ps; proton transport occurs during the following intermediate states L—O, and the cycle is completed within 10 to 15 ms after the regeneration of *trans*-retinal. Note the drastic chromism in the formation of each intermediate. The quantum efficiency for cycling is 0.7, similar to the visual pigment. In contrast, however, bacteriorhodopsin (bR) does not require metabolism for regeneration of the initial state and is remarkably stable under continuous illumination. This advantage, as well as its unique optical functions, hold great promise for potential applications of this protein in optical memories including holographic devices.

The vectoral displacement of charges and changes in dipole moment during the photocycle can be detected as photocurrents with electrodes attached to oriented assemblies of purple membrane (PM). Dry oriented PM films have been extensively studied for photoelectric measurement since 1978, mainly using electrodeposited PM films. The extremely short time constant for the bR-to-K conversion and its electric response, surpassing results obtained with solid state devices, are significant advantages for molecular devices using this retinal protein.

## 2 DEVISING bR-BASED PHOTOCELLS

The conventional technique to elicit a bR-induced photoelectric response employs sandwich-type solid junction photocells (electrode/dry bR film/electrode) that produces transient and continuous photovoltages corresponding to changes in light intensity. We have found that a charge displacement photocurrent of bR takes place in the perfect time-differential mode against a change in light intensity when a thin molecular assembly of bR is immobilized at the interface of an electrode and an aqueous electrolyte. The photocell having such responsivity is fabricated by immobilizing a thin Langmuir–Blodgett (LB) film of purple membranes on a conductive electrode of tin oxide or indium oxide and placing the film into contact with an aqueous gel containing a concentrated salt. A typical cell has a junction structure of $SnO_2$/PM/electrolyte gel/Au, where the electrolyte gel is inserted as a thin layer (thickness $< 300$ μm) and comprises an aqueous mixture of 4% carboxymethylchitin and 2 M KCl. Employing an electrolyte wherein bR is immersed in water maintains the highest activity of the photocycle and, simultaneously, provides ample electric conductivity to the medium connecting the bR molecules and two electrodes.

Light irradiation of the photocell just described produces a transient stroke of photocurrent with a peak value and direction depen-

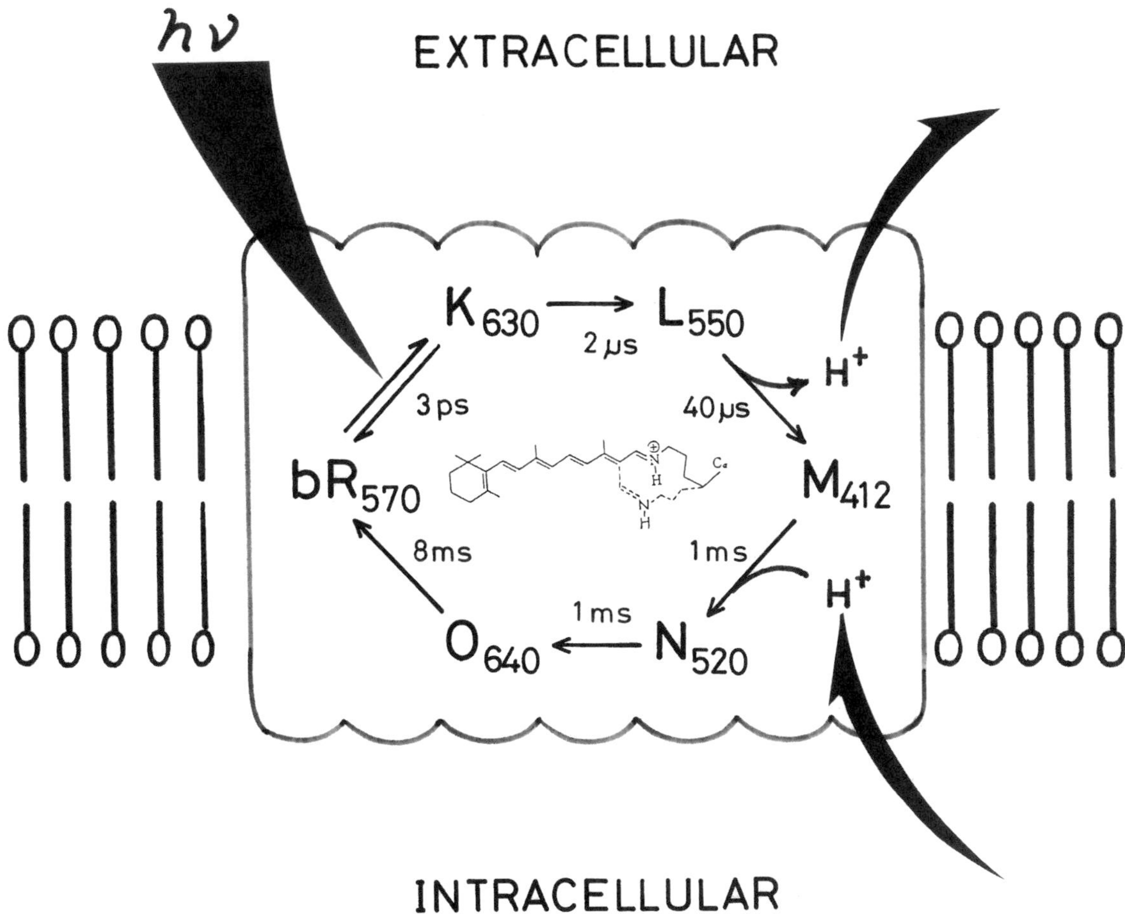


**Figure 1.** Photocycle of bR comprising a series of intermediates (K, L, M, N, O). Subscripts denote the peak absorption wavelengths of intermediates. Inset: structures are *trans* (solid line) and *cis* (dashed line) forms of retinal.

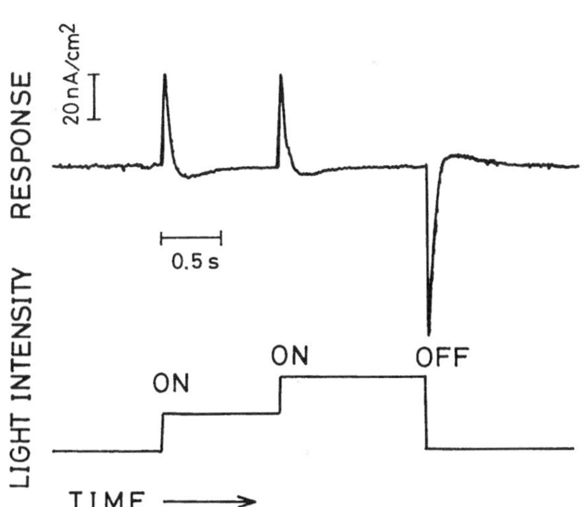


**Figure 2.** Differential response profile of bR-induced photocurrent. Reference signal (light intensity) of incident light, monitored by a photodiode, is given below. Data were obtained with six PM layers coated on an $SnO_2$ electrode under irradiation of green light through a band-pass filter (540 ± 130 nm).

dent on the differential of the change in light intensity. A typical response profile is shown in Figure 2. The bR-induced differential photocurrents are always strongly rectified in the cathodic direction: that is, electrons flow from the electrode to the electrolyte side. The action spectrum of the photocurrent matches the optical absorption of bR (Figure 3), peaking at around 560 nm. The photocurrent exhibits optimum characteristics at a thickness of about 10 layers of PM film (about 40–50 nm), demonstrating that the bR molecules present close to the electrode surface are the chief contributors to the generation of displacement photocurrents.

Displacement current is a nonfaradaic current, which involves no carrier (electron) transfer at the electrode–species (here, bR) interface. It is electrostatically induced when the electrode receives a transient electric field generated by a charge displacement in bR molecules. Positive and negative strokes of photocurrent correspond to forward and return processes of charge displacement, respectively, between which no current occurs unless further displacement is caused by a change in light intensity.

The light-sensing capability of the photocell is guaranteed by the excellent linearity in its output-versus-intensity characteristics, as plotted in Figure 4. With this linearity, which covers a wide intensity region of more than three orders of magnitude, light quanta (photons) are converted to a flow of electrons at an estimated quantum efficiency of $10^{-2}$. Speed of light detection examined by

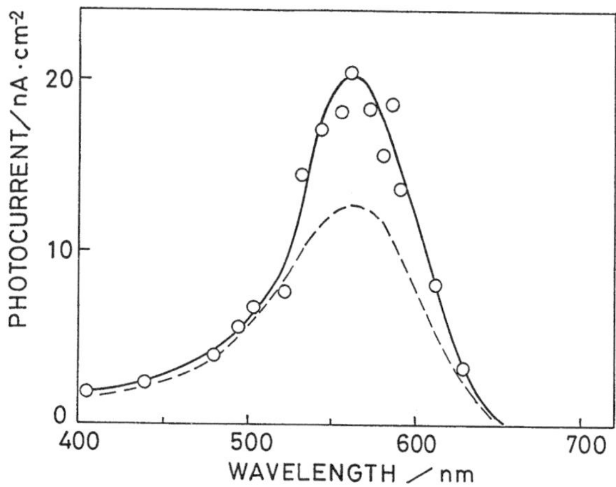


Figure 3. Wavelength dependence of bR-induced photocurrent. Data obtained with 2 PM layers coated on an $SnO_2$ electrode; dashed line represents the absorption spectrum of bR.

pulse excitation is on the order of $10^{-4}$ s and is apparently affected by the *RC* time constant of the cell and measurement circuits.

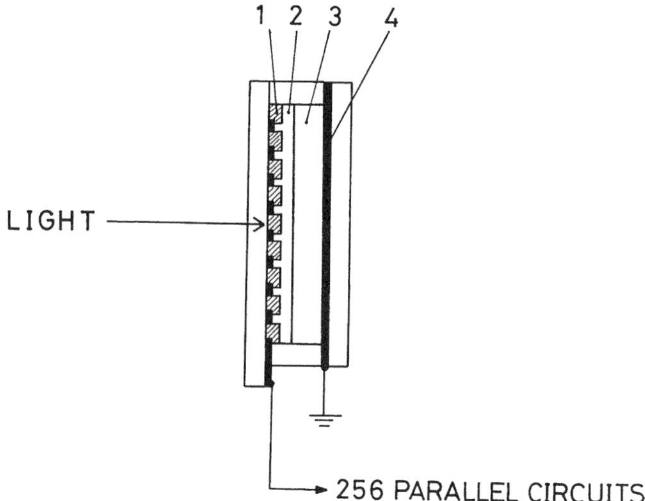


Figure 5. Structure of bR-imobilized artificial photoreceptor: 1, 256-pixel ITO electrode (100 nm thick patterned ITO layer); 2, PM LB film (10–14 PM layers); 3, polymer electrolyte gel (300 μm thick); 4, Gold counter-electrode coated on a glass. The light-receiving area of the photoreceptor is 31 × 31 mm.

## 3 IMAGE SENSING AND PROCESSING BY A bR-BASED PHOTORECEPTOR

Rapid and clear differential responsivity of the bR-based photocell provides a simple model for the visual photoreceptors in humans and animals. In visual cells, optical information received by rhodopsin undergoes preliminary processing through a network of ganglion cells that output a differential signal useful for edge detection and motion sensing. Preprocessing of visual information (image) is thus performed at the material level of the retina before higher processing (pattern recognition, etc.) occurs in the neural networks of the cerebrum. Such an image-processing function can be realized in an artificial photoreceptor that incorporates bR as a photosensitive material.

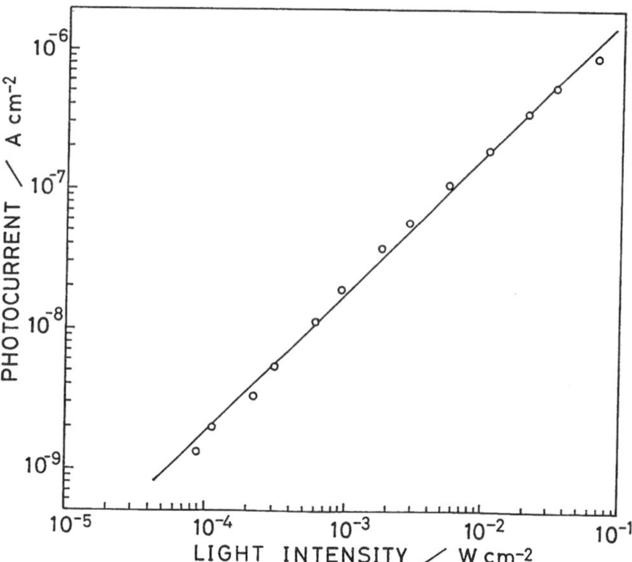


Figure 4. A bR-induced peak photocurrent as a function of light intensity. The experimental conditions are the same as in Figure 2.

An artificial photoreceptor has been fabricated by implementing a 256-pixel "composite eye" made of a two-dimensional array of bR-based sandwich photocells. The photoreceptor structure is sketched in Figure 5. An indium–tin oxide (ITO) transparent electrode (3 cm × 3 cm) bearing a network pattern of 256 square pixels (pixel size: 1.3 mm × 1.3 mm) and lead wires is coated with a PM LB film and formed into a junction with the chitin gel layer and a gold counterelectrode. The photoreceptor is connected to a signal amplification circuit that converts small photocurrents from the pixel array into voltage signals, with final output to a display panel made of a 256-pixel array of light-emitting diodes (LEDs). Using this system, optical images captured by the photoreceptor are simultaneously displayed through parallel transaction of two-dimensional signals.

A unique sensing function of the photoreceptor is presented in its motion detection capability. When a spot of light emitted from a pencil lamp illuminates the photoreceptor, an image of a ring or spot appears on the display panel; the image disappears the instant the light stops moving. Using a commercial video projector, an attempt was made to sense moving images of everyday objects. The projector was optically coupled with the photoreceptor, and hand motions captured by a video camera were directly projected to the photoreceptor. Figure 6 exhibits an image of a hand motion detected by the 256-pixel photoreceptor. Apparently, the photoreceptor is detecting a moving hand alone (b), ignoring all still objects when the hand movement stops (a). Another important behavior, edge detection of a moving object, can be seen in Figure 6. The photoreceptor is capable of extracting an edge profile from an image simultaneously with motion detection.

Edge detection, or contrast enhancement, is a significant step in performing pattern recognition. We have carried out a pattern-sensing experiment using a alphabetical letter image as a target. By performing spatial oscillation of the optical axis incident to the photoreceptor (i.e., mechanically oscillating the lens system or the

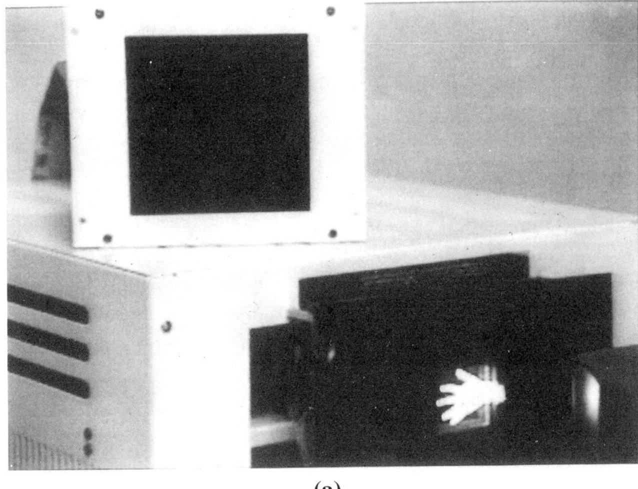

(a)

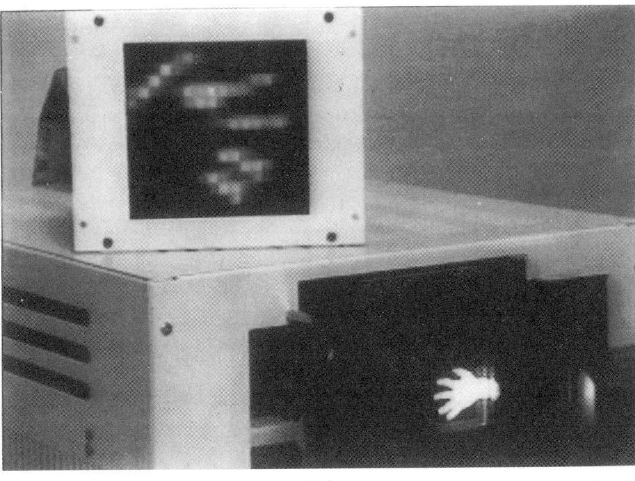

(b)

**Figure 6.** Detection of a moving hand by a 256-pixel bR-based photoreceptor. A hand motion monitored by a video camera and projected onto the photoreceptor (lower right) is displayed on the LED panel (upper left). The photoreceptor is not sensitive to still images when the hand stops its motion (a), but it detects the edge profile of the hand at the moment the hand starts jiggling (b).

photoreceptor itself), it was possible to extract selectively an edge component of the letter image in a desired direction (horizontal, vertical, etc.) where fluctuation in light intensity occurred. Edge patterns thus extracted in the vectorial mode are typically utilized for letter recognition systems using computations based on neural networks.

The oscillation modulation of an image incident to the bR-based photoreceptor corresponds to the eye movement phenomenon in visual perception, by which the human eye is able to perceive still objects. The bR-based artificial photoreceptor further functions as a direction sensor by distinguishing the edge lines of a moving image depending on the direction of movement.

Systems for visual information processing have been developed based on solid state devices. Such systems, however, need processing and computation circuitry in addition to photodetectors. Material level, real-time processing of information can be achieved only when an intelligent material that has processing ability is employed as a light detector. Bacteriorhodopsin is a typical example of such a material.

## 4 PERSPECTIVES

Because of its versatility in photochemistry and photoelectronics, bacteriorhodopsin holds enormous potential in the design of intelligent optical and optoelectronic devices that possess multiple functions at the material level. Real-time processing of visual information can be implemented by the bR-based pixelized photoreceptor, which is most promising for future image compression technologies. The high stability of bR against light has already been established experimentally, clearly confirming that the responsivity of the bR photoreceptor is maintained for years under ambient conditions including long-term exposure to sunlight.

*See also* BACTERIORHODOPSIN; BIOMOLECULAR ELECTRONICS AND APPLICATIONS; BIOSENSORS.

### *Bibliography*

Birge, R. R. (1990) Photophysics and molecular electronic applications of the rhodopsins. *Annu. Rev. Phys. Chem.* 41:683–733.

Koyama, K., Yamaguchi, N., and Miyasaka, T. (1994) Antibody-mediated bacteriorhodopsin orientation for molecular device architectures. *Science* 265:762–765.

Miyasaka, T., Koyama, K., and Itoh, I. (1992) Quantum conversion and image detection by a bacteriorhodopsin-based artificial photoreceptor. *Science* 255:342–344.

———, and ———. (1992).Rectified photocurrents from purple membrane Langmuir–Blodgett films at the electrode–electrolyte interface. *Thin Solid Films* 210/211:146–149.

———, and ———. (1993). Image sensing and processing by a bacteriorhodopsin-based artificial photoreceptor. *Appl. Opt.* 32:6371–6379.

Oesterhelt, D., Bräuchle, C., and Hampp, N. (1991) Bacteriorhodopsin: A biological material for information processing. *Q. Rev. Biophys.* 24:425–478.

Skulachev, V. P. (1988) *Membrane Bioenergetics.* Springer-Verlag, Berlin.

Trissl, H. -W. (1990) Photoelectric measurements of purple membranes. *Photochem. Photobiol.* 51:793–818.

# BIOCHEMICAL GENETICS, HUMAN

*Frits A. Hommes*

### *Key Words*

**Enzyme** A biocatalyst.

**Heteropolymer** A macromolecule made up of unequal subunits.

**Phenotype** Structural and functional presentation of an organism as the result of the collective effects of the entire genome and environmental factors.

**Splicing** Cutting and rejoining of a linear biopolymer.

**Transcription** The process of copying a strand of DNA to yield a complementary strand of RNA.

Human biochemical genetics is concerned with the genetics of enzyme function, its variability and its metabolic consequences as an explanation for human biological individuality and pathology.

## 1 INTRODUCTION

Human biochemical genetics started in 1908 with the publication in *The Lancet* of "The incidence of alkaptonuria, a study in chemical individuality," in which Garrod made the observation that individualities in metabolism are examples of variations of chemical behavior, the extremes of which result in overt pathology, such as alkaptonuria. Garrod concluded, furthermore, that such extremes had a genetic basis. Six years later, in his Croonian lectures, likewise published in *The Lancet,* he further elaborated the concept into "inborn errors of metabolism": that is, a disturbance in metabolism due to a decreased activity or the absence of an enzyme, the decrease in activity or absence being under genetic control. This proved to be an extremely productive concept, the importance of which can hardly be overestimated, the more so since it was not until 1926 that Sumner demonstrated that enzymes are proteins, and not until 1945 that Beadle formulated the one gene–one enzyme concept. Although in the strictest sense these paradigms are no longer valid—some RNA molecules have catalytic properties for example, and some enzymes consist of heteropolymers requiring more than one gene—the basic concept defined as one gene–one gene product approaches more closely reality. Even that definition, however, does not do justice to current concepts such as alternative splicing of mRNA, giving rise to different gene products, or to DNA rearrangements prior to transcription, giving rise to different immunoglobulins.

The one gene–one gene product concept is, however, the basis for the understanding of the inborn errors of metabolism, of biological variability, and of biochemical genetics. Implicit in this concept is the idea that all biochemical reactions are under genetic control and that each biochemical reaction is controlled by a different, single gene. If a mutation occurs in such a gene or its promoter, leading to altered activity of the gene product or to the absence of the gene product, the flux of the metabolic pathway in which that gene product functions will be altered. If the change in activity is sufficiently severe, it may lead to pathology. A mutation leading to a less dramatic change in functional activity of the gene product contributes to individuality and biological variability. It does not necessarily have a deleterious effect on the well-being of its bearer, but it does determine phenotypic variability. In some cases it may predispose the organism to the development of disease, especially when a number of such mutations—each in itself potentially inconsequential—cooperate to influence the flux of a metabolic pathway to such an extent that pathology can develop.

## 2 DEFINITION

Biochemical genetics is concerned with the study of the identification of altered enzyme function and the consequences for the organism when this altered enzyme function has a genetic basis. When applied to humans, biochemical genetics gains special significance because of its potential for explaining human variability and pathology.

## 3 METHODOLOGY

The identification of altered enzyme function has in the past been pursued mainly by measuring enzyme activity in selected tissues and comparing it with that tissue from phenotypically normal controls. Extensive clinical and chemical investigations, including detailed analyses of metabolites, usually precede such enzyme studies to pinpoint as precisely as possible the candidate enzyme to assay. More than 350 such conditions are presently known, with the biochemical defect of the majority of the genetic diseases still to be uncovered. With increasing sophistication and refinement of analytical techniques for the detection of metabolites, biochemical genetics will continue to contribute to the identification of altered enzyme function. The power of this approach is, however, limited. It relies heavily on the specificity and sensitivity of the analytical techniques used for the identification and quantitation of metabolites in physiological fluids. Despite considerable progress in this area, there is still an urgent need for more refined analytical procedures. This is especially true for genetic diseases involving the central nervous system, the more so since the central nervous system is so frequently affected. An additional complicating factor for the central nervous system is accessibility. Biological variation is another interfering issue. Even if detection and quantitation can be performed in a reliable way, small deviations from the mean or range of values are sometimes difficult to interpret because phenotypically normal individuals may display a wide range of values. Gross abnormalities due to severe changes in gene product activity will be identifiable, but more subtle mutations leading to less dramatic changes in metabolite levels may be buried in the noise of normal biological variation.

A significant contribution to the identification of mutations will therefore be made by reverse genetic techniques and other molecular methods, as the elucidation of the basic defects of Duchenne muscular dystrophy and cystic fibrosis have demonstrated. Yet routine diagnosis of these conditions currently is done by immunological methods and measurement of chloride in sweat, respectively: that is, by biochemical techniques rather than by molecular biology techniques.

Details of the consequences of altered gene product function—the other area of concern in biochemical genetics—are incompletely understood. This is in part due to the aftereffects of the one gene–one gene product concept and its implication that each biochemical reaction is under the ultimate control of a single gene. Although not necessarily incorrect, this concept has emphasized the idea of a rate-limiting enzyme for a metabolic pathway. Metabolite pathways are not necessarily controlled by one rate-limiting enzyme, although some may be. When a mutational change causes the activity of an enzyme to decrease below a certain level, that enzyme may become rate limiting for the metabolic pathway in which it functions. This will be the case in particular when the residual activity is very low. All pathology must then be related to the low activity of that gene product. The control of the flux through a metabolic pathway is, however, in many cases shared by all enzymes of the pathway, with each enzyme having its own specific contribution. Mutations leading to more subtle changes in the activity of the gene product change the flux of the metabolic pathway in a more delicate way. This forms the basis of biological variability. Theories of metabolic control to analyze such changes have been developed but have as yet had few applications in the areas of genetic diseases and biological variability. It will nevertheless provide the answer to questions such as why the same mutation leads to significantly different phenotypical presentations and how severe mutations must be in multifactorial conditions before changes in the flux of the pathway give rise to pathology.

*See also* ENZYMES; PHENYLKETONURIA, MOLECULAR GENETICS OF.

## *Bibliography*


Beaudet, A. L., Scriver, C. R., Sly, W. S., McKusick, V. A., Stanbury, J. B., Wyngaarden, J. B., Frederickson, D. S., Goldstein, J. L., and Brown, J. S. (1990) *Introduction to Biochemical and Molecular Genetics.* McGraw-Hill, New York.

Cornish-Bowden, A., and Cárdenos, M. L. (1990) *Control of Metabolic Processes.* Plenum Press, New York.

Harris, H. (1980) *The Principal of Human Biochemical Genetics,* 3rd ed., Elsevier/North Holland Biomedical Press, Amsterdam.

Hommes, F. A. (1991) *Techniques in Diagnostic Human Biochemical Genetics. A Laboratory Manual.* Wiley-Liss, New York.


# BIODEGRADATION OF ORGANIC WASTES

*Duane Graves*

## *Key Words*

**Bioremediation** The application of biodegradative processes to the treatment of soil, sediment, sludge, air, or water contaminated with wastes or pollutants.

**Cometabolism** The metabolism by a microorganism of a chemical (cosubstrate) that the organism cannot use as a nutrient or source of carbon or energy and that is metabolized only when the microorganism is using another chemical (substrate) for growth.

**Halogenated Organic Compound** Any organic compound that contains a covalently bonded halogen atom (F, Cl, Br, I) as part of its molecular formula. Chlorinated and brominated compounds are more common environmental contaminants.

**Reductive Dehalogenation** The removal of a halogen atom from a molecule with the concurrent addition of electrons to the molecule.

Biodegradation is essential for cycling carbon, nitrogen, phosphorus, oxygen, and several other elements in the biosphere. Decay is a natural, biologically driven process that facilitates the decomposition of complex organic molecules into simple compounds that can be reassimilated into biomass. This cyclical process can also be applied to the decomposition and detoxification of anthropogenic wastes. Bioremediation is the application of biodegradative processes to the treatment of undesirable wastes for the purpose of restoring, reclaiming, or remediating soil, sediment, air, and water that has been negatively impacted by the presence of wastes. The biological processes involved, the microbial population dynamics in the environment, and the ability of microorganisms to biodegrade specific compounds are issues central to the application of biodegradation for environmental remediation.

## 1 COMMON ORGANIC WASTES AND THEIR ORIGIN

The Industrial Revolution radically changed mankind's lifestyle and accelerated the rate at which collective knowledge was gathered, disseminated, and applied. A significant side effect of the industrialization of the world has been the disposal of various manufacturing by-products. Additionally, the clustering of jobs in industrialized areas created significant new problems for the disposal and treatment of domestic wastes and sewage as more people lived in closer proximity.

Open dumping, land-filling, burial, discharge into rivers and lakes, and emission of volatile compounds into the air are common historic avenues for the disposal of industrial and domestic wastes, including organic wastes. Recycling, waste minimization, and recovery and reuse were generally considered to be uneconomical and were not practiced. Many of the wastes released into the environment are hazardous to human health and wildlife. Many are considered to be hazardous at very low levels, and they often degrade or dissipate to nontoxic levels at a very low rate, causing chronic environmental contamination. Regulatory restrictions on waste disposal have encouraged industries to become more conscientious in handling their by-products and have forced the remediation of sites impacted by the intentional or accidental release of hazardous materials.

Wastes may be of several general types, including organic compounds, metals, acids, bases, and salts. Many wastes may be subject to biological treatment when appropriate conditions are applied; however, biodegradation is generally limited to the biological oxidation, reduction, or transformation of organic compounds. Domestic sewage has been treated biologically for more than a century, but the application of biological treatment processes to more complex organic wastes in environmental matrices has been actively employed for less than two decades.

Industrial organic wastes encompass a wide range of organic compounds including aliphatic and aromatic hydrocarbons derived from petroleum, coal, and wood, as well as natural products, halogenated and oxygenated solvents, pesticides, herbicides, and explosives. Table 1 lists several common organic contaminants. Single compounds and complex mixtures of several compounds can be found in soil and groundwater at many sites across all industrialized nations. The remediation of soil, sludge, sediment, air, and water impacted with organic wastes is a difficult and complex process. Because microorganisms (bacteria and fungi) are essentially ubiquitous and capable of assimilating a wide range of organic compounds, their metabolism has come under close scrutiny as a possible route to the treatment of organic wastes present in the environment.

The biodegradation of organic wastes is a useful side effect of microbial metabolism. Thus the fundamental principles of biodegradation are integrally linked to microbial physiology. The diversity of bacteria found in the environment, their genetic promiscuity (biodegradative pathways are often carried on plasmids), their rapid growth rate, and their ability to use a wide range of organic compounds as sources of carbon and energy permits them to biograde a variety of organic compounds. Several general types of metabolism are known to promote the biodegradation of organic wastes. Obligate chemolithotrophs and photoautotrophs are, in general, not prominent biodegraders of organic wastes.

Microbes surviving in natural environments need, at a minimum, inorganic nutrients including but not limited to fixed nitrogen, preferably as ammonium, phosphate, potassium, and trace levels of several metals; an electron acceptor such as oxygen, nitrate, sulfate, or carbon dioxide; and a source of organic carbon, which may include such compounds as petroleum, petrochemicals, solvents, alcohols, and xenobiotics (human-made chemicals that do not occur naturally).

Bacteria typically live in a ''feast or famine'' situation. Adaptation to this type of existence has led to the evolution of several mechanisms that permit bacteria to survive in inhospitable environments until conditions improve. The goal of bioremediation is to enhance the local environment to provide favorable conditions for bacterial metabolism. The consequence of this action is consumption of organic carbon by bacteria present in the treated matrix.

**Table 1** Examples of Common Organic Wastes

| General Classification | Examples |
|---|---|
| Petroleum | Gasoline |
| | Diesel |
| | Jet fuel |
| | Crude oil |
| Volatile organic compounds (VOC) | Benzene |
| | Toluene |
| | Ethylene benzene |
| | *o*-, *m*-, and *p*-Xylene |
| | Trichloroethylene |
| | Vinyl chloride |
| Polynuclear aromatic hydrocarbons (PAH) | Naphthalene |
| | Phenanthrene |
| | Anthracene |
| | Fluorene |
| | Benz(*a*)pyrene |
| Aliphatic compounds | Decane |
| | Paraffins |
| | Octane |
| | Cyclohexane |
| Aromatic compounds | Benzene |
| | Phenol |
| | Dioctyl phthalate |
| | Naphthalene |
| | 2,4-Dinitrotoluene |
| Solvents | Trichloroethene (TCE) |
| | Tetrachloroethene (PCE) |
| | Methyl ethyl ketone (MEK) |
| | Acetone |
| | Ethanol |
| Pesticides/Herbicides | 2,4-Dichlorophenoxyacetic acid |
| | Lindane |
| | Dieldrin |
| | Aldrin |
| | Atrazine |
| Explosives | Trinitrotoluene (TNT) |
| | Hexahydro-1,3,5-trinitro-1,3,5-triazine (RDX) |
| | Octahydro-1,3,5,7-tetranitro-1,3,5,7-tetraazocine (HMX) |
| | Nitroglycerin |
| Halogenated compounds | Trichloroethene |
| | Tetrachloroethene |
| | Chlorobenzene |
| | 1,2-Dichloroethane |
| | Dichloromethane (methylene chloride) |
| | Polychlorinated biphenyl (PCB) |

## 2 BACTERIAL METABOLISM AND BIODEGRADATION

### 2.1 Aerobic Biodegradation

Aerobic bacteria require the presence of molecular oxygen. They represent a very diverse group of bacteria that exhibit a wide range of metabolic capabilities. The biodegradation of many organic compounds proceeds most rapidly under aerobic conditions. Within the general grouping of aerobic bacteria, heterotrophy and methanotrophy represent the two most widely studied types of metabolism.

Heterotrophs are bacteria that use organic compounds for carbon and energy. Many species of heterotrophic bacteria have been observed to biodegrade organic compounds. Several different pathways have been elucidated or partially elucidated for the aerobic biodegradation of chemical contaminants. A common first step is the addition of one or two oxygen atoms to the compound being biodegraded.

The aerobic biodegradation of petroleum hydrocarbons proceeds by several different biochemical pathways. The typical first step for the biodegradation of linear alkanes is an oxidation of the terminal carbon to produce a primary fatty alcohol. This alcohol is then converted via an aldehyde to a fatty acid. Some organisms convert the carbons on both ends of the alkane to a carboxylic acid. Figure 1A illustrates the production of a monocarboxylic fatty acid from a linear alkane. Other bacteria are able to oxidize subterminal carbons, often with the production of acetic acid and a fatty acid two carbons shorter than the original alkane. Once a fatty acid has been generated, it can be further biodegraded by the sequential removal of two carbon units until the entire molecule has been decomposed. Cyclic alkanes are biodegraded by oxidative processes that lead to linearization, usually resulting in a dicarboxylic fatty acid, which is further biodegraded.

Simple aromatic compounds such as benzene, ethylbenzene, toluene, and xylenes are common environmental contaminants. The well-documented biodegradation of benzene proceeds by the formation of catechol with subsequent ring cleavage—via orthofission resulting in muconic acid or via metafission resulting in 2-hydroxymuconic semialdehyde (Figure 1B). Both these degradation intermediates are readily biodegraded. The first steps in the biodegradation of other simple aromatic compounds result in the production of catechol or substituted catechols. Similarly, some chlorinated aromatic compounds are biodegraded with the intermediate production of chlorinated catechols (Figure 1C).

Polynuclear aromatic hydrocarbons (PAHs) are also hydroxylated on one of the rings, yielding a dihydroxy PAH, which is susceptible to ring cleavage and further degradation. High molecular weight, extremely insoluble PAHs are difficult to biodegrade because of their limited concentration in the aqueous phase, and their size may prohibit attack by microbial enzymes. Additionally, the microbe must expend considerable energy to generate compounds it can use. Thus biodegradation of large compounds is not energetically favorable for bacteria, although its occurrence at relatively slow rates has been documented.

As hydrocarbons undergo the various oxidative steps that ultimately lead to biodegradation, the physical characteristics of the intermediates can vary considerably compared to the parent compounds. Many of the intermediates are more soluble in water than the parent hydrocarbon. Microbes may also facilitate the solubilization of hydrophobic compounds by exuding biosurfactants. The purpose of this response is to move more organic carbon into the aqueous phase. Since microbes exist in an aqueous environment, the availability of carbon sources is limited by solubility; the ability to drive hydrophobic compounds into the aqueous phase provides an obvious advantage for microbes competing for limited resources.

Methanotrophs (or methane-oxidizers) are obligate aerobes that use methane as their sole carbon and energy source. A similar group of bacteria, known as propane oxidizers, use propane as their sole carbon and energy source. These two groups of bacteria are important because they express a monooxygenase that can cometabolize trichloroethylene (TCE: a common industrial solvent). Because of this feature of their metabolism, they have received much attention for the bioremediation of soil and groundwater contaminated with TCE.

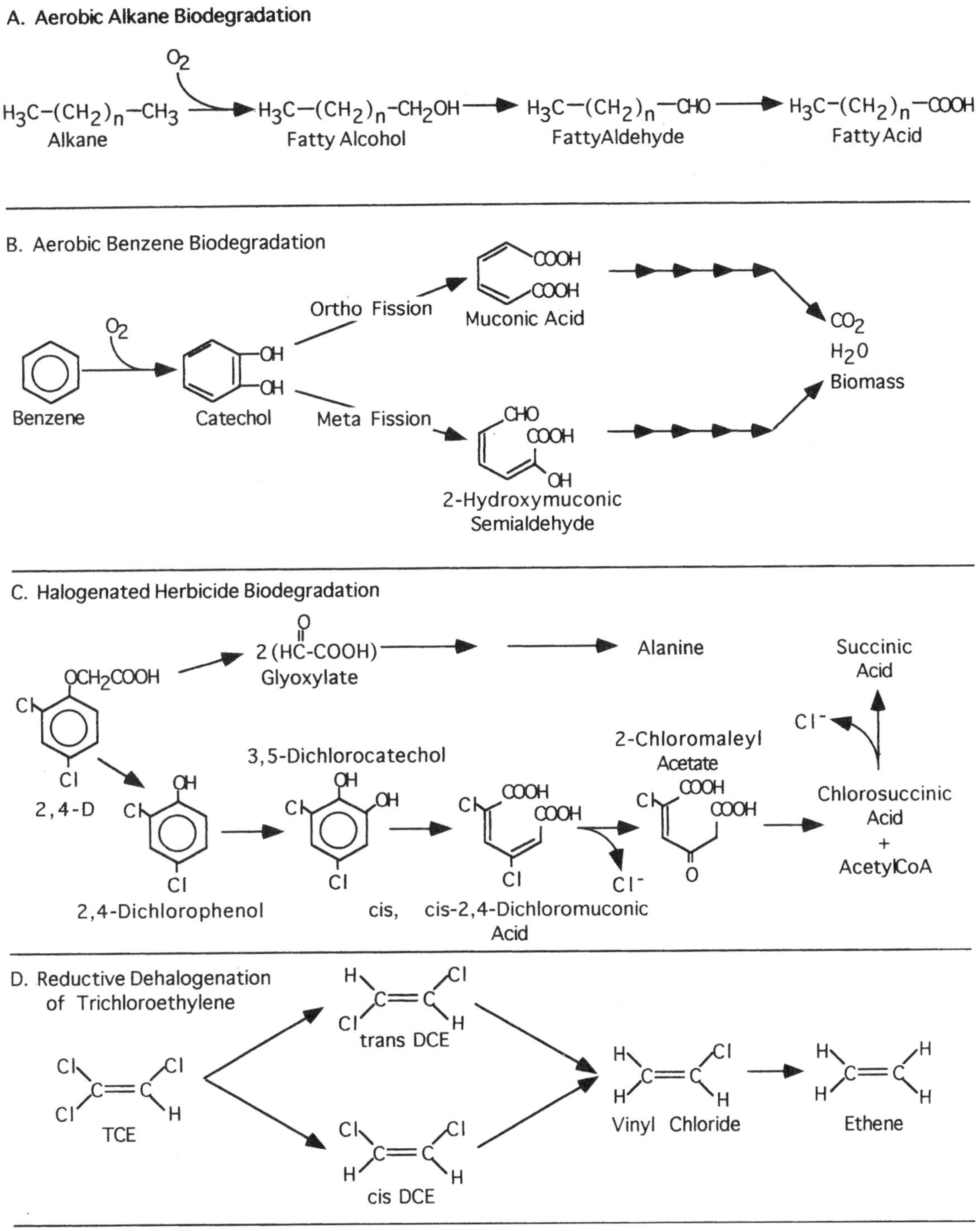

Figure 1. Common biodegradative pathways.

### 2.2 Cometabolism

Cometabolism is the process whereby a chemical is metabolized by a bacterium while the bacterium is obtaining carbon and energy from another organic compound. The aerobic biodegradation of TCE is a common example of cometabolism. Bacteria of several types express monooxygenase enzymes that will cometabolize TCE when they are induced to metabolize other compounds. Bacteria that biodegrade toluene or phenol using monooxygenases will also initiate the first step in the biooxidation of TCE, which appears to be the production of an epoxide. Methanotrophic bacteria also express a monooxygenase in response to methane and propane which will gratuitously cometabolize TCE. Nitrifying bacteria express ammonia monooxygenase, which can cometabolize TCE.

TCE is not capable of inducing the expression of monooxygenases; therefore, the cometabolism of TCE is integrally linked to the presence of the primary metabolite. The primary metabolite has a kinetic advantage in binding at the monooxygenase's active site and is thus more readily biodegraded than the cometabolite. In the case of ammonia monooxygenase, TCE inactivates the enzyme. Recently, mutants have been isolated which constitutively express an aromatic monooxygenase. These organisms have the potential of biodegrading TCE in the environment in the absence of an inducer or primary metabolite, provided a viable population can be established and maintained.

### 2.3 Anaerobic Metabolism

Aerobic bacteria are generally considered to be the most efficient at biodegrading hydrocarbons; however, anaerobic modes of metabolism have been observed to support biodegradation. Anaerobic bacteria use compounds other than oxygen as the terminal electron acceptor in their respiratory system. Nitrate, sulfate, iron ($Fe^{3+}$), manganese ($MN^{4+}$), selenate, carbon dioxide, and phosphate may serve as electron acceptors for anaerobic bacteria. Bacterial metabolism driven by each of these types of respiration may degrade organic wastes. However, the most commonly observed types of anaerobic respiration associated with biodegradation of organic wastes are nitrate reduction, sulfate reduction, and methanogenesis.

Anaerobic bacteria capable of utilizing nitrate as the terminal electron acceptor have been demonstrated to degrade hydrocarbons; however, degradation rates are generally less than the rates measured for aerobes. The range of compounds observed to be biodegraded under nitrate reducing conditions is similar in scope to the range of those biodegraded by aerobes.

A few instances of hydrocarbon biodegradation have been reported for anaerobic sulfate-reducing bacteria, but their metabolism is thought to depend primarily on the availability of small, partially oxidized carbon compounds such as simple alcohols and carbohydrates. A principal by-product of their metabolism is hydrogen sulfide. Because sulfide is highly reduced, the production of sulfide has been used to reduce oxidized metals such as water-soluble $Cr^{6+}$ to insoluble $Cr^{3+}$.

Methanogens use carbon dioxide as their terminal electron acceptor, and they usually utilize simple carbon compounds such as acetate and sugars as their sources of carbon and energy. Methanogens are known to biodegrade or biotransform petroleum compounds with the production of methane; however, the range of substrates and the degradation pathways used by methanogens are not well defined.

### 2.4 Reductive Dehalogenation

Reductive dehalogenation is an important process that usually occurs under anaerobic conditions, although aerobic reductive dehalogenation has been documented. Reductive dehalogenation is a common mechanism by which halogenated organic compounds are attacked by bacteria. Many pesticides, herbicides, and solvents are halogenated hydrocarbons, as are the insulating oils, polychlorinated biphenyls (PCBs). The first step in the biodegradation of these compounds is often the removal by anaerobic bacteria of one or more chlorine atoms.

Two general mechanisms are employed to achieve halogen removal. Hydrogenolysis is the displacement of the halogen atom with a proton. This process may occur with alkyl and aryl compounds; however, bacteria display substrate specificity, with aryl dehalogenators being less substrate specific and more widespread. Alternatively, the halogen atoms on adjacent carbons may be replaced with two electrons, resulting in a double bond between carbon atoms. This process is known as vicinal reduction.

The extent to which reductive dehalogenation will remove chlorine atoms is not certain and depends on the microbial population structure and activity, and on environmental conditions. Most research on reductive dehalogenation has been conducted using mixed microbial populations because the isolation of pure strains has proven to be very difficult. Communities of bacteria that can reductively dehalogenate chlorinated compounds may exist under codependent or syntrophic relationships, making the survival of one member of the population dependent on the survival and activity of others. It is generally assumed that individual strains cannot completely dehalogenate polychlorinated compounds but rather, the removal of multiple chlorine atoms is the result of the concerted action of several bacterial species.

The metabolic benefit of reductive dehalogenation is not well characterized; however, in at least one case, dehalogenation is thought to be coupled to ATP synthesis through a proton ATPase. Reductive dehalogenation is most commonly observed in anaerobic bacteria, including methanogens, sulfate reducers, and denitrifiers. Aryl dehalogenation has also been documented in some aerobic bacteria and facultative anaerobes.

The removal of chlorine or other halogens from organic compounds facilitates the further biodegradation of the compound. In general, the less halogenated a compound, the more likely it is to be further biodegraded. Therefore, dehalogenating bacteria play an important role in initiating the biodegradation of halogenated compounds.

Methanogens dehalogenate TCE via vinyl chloride to ethene. Polychlorinated biphenyls are also reductively dehalogenated to biphenyls with a lower level of chlorination. Dehalogenation makes PCBs more susceptible to aerobic biodegradation. Chlorobenzenes, chlorophenoxyacetates, pentachlorophenol, and other halogenated aromatic hydrocarbons are also reductively dehalogenated. An example of reductive dehalogenation is shown in Figure 1D.

## 3 BIODEGRADATION OF ORGANIC WASTES BY FUNGI

White rot fungi (*Phanerocheate* spp.) are capable of biodegrading lignin, an extremely high molecular weight plant product that is insoluble and difficult to biodegrade. The white rot fungi express extracellular peroxidases that oxidize lignin, leading to the fragmentation and solubilization of high molecular weight polymers into smaller subunits, which are more susceptible to further assimilation. The capacity to biodegrade complex organic molecules expressed by the white rot fungi has been applied with varying degrees of success to the biodegradation of a variety of organic wastes. In general, the fungi require complex growth media, and conclusive demonstration of biodegradation is often difficult to obtain due to limitations in analytical methods.

## 4 GENETIC ENGINEERING, MOLECULAR BIOLOGY AND BIOLOGICAL WASTE TREATMENT

Key pathways and genes have been identified for several important biodegradative processes. It is possible to engineer bacteria that overexpress, have altered regulatory processes, carry multiple biodegradative operons, or are otherwise modified to be more effective biodegraders of specific contaminants. However, current state and federal regulations generally prohibit the introduction of engineered bacteria into the environment. This restriction has suppressed the development of strains designed to biodegrade contaminants. Additionally, the development and characterization of useful strains is laborious, the long-term stability of the engineered strains must be tested, and the ability of the strains to survive and function in a complex environment already populated with bacteria has to be evaluated for each new strain.

There are several cases of genetically engineered bacteria that serve as indicators of biodegradative processes. One common example is placement of the lux operon, which contains genes for bioluminescence, under the control of an operon encoding a biodegradative pathway. Such engineered organisms produce light when their biodegradative pathway is being expressed. Bioluminescence is used to indicate the level of biodegradative activity stimulated by various treatments and environmental conditions.

Isolated and cloned genes specific for a particular biodegradative pathway are being used to determine gene frequency within a bacterial population and to determine the presence of bacteria capable of biodegrading a specific compound in a contaminated environment. Methods are also being developed to use messenger RNA to quantify the level of expression for critical genes in a degradative pathway.

Many other potential applications for genetically engineered microorganisms exist. However, the most important current applications for genetic engineering and molecular biology are to facilitate understanding of the cellular mechanisms that result in contaminant biodegradation, and of the population dynamics, regulatory mechanisms, and level of activity occurring during the biodegradation of a target compound.

*See also* Bacterial Growth and Division; Biotechnology, Governmental Regulation of; Fuel Production, Biological.

### *Bibliography*

Atlas, R. M., Ed. (1984) *Petroleum Microbiology*. Macmillan, New York.

Chaudhry, G. S., and Chapalamadugu, S. (1991) Biodegradation of halogenated organic compounds, *Microbiol. Rev.* 55(1):59–79.

Hinchee, R. E., and Olfenbuttel, R. F. (1991) *In Situ Bioreclamation*. Butterworth-Heinemann, Stoneham.

Mohn, W. W., and Tiedje, J. M. (1992) Microbial reductive dehalogenation, *Microbiol. Rev.* 56(3):482–507.

Rochkind-Dubinsky, M. L., Sayler, G. S., and Blackburn, J. W. (1987) *Microbiological Decomposition of Chlorinated Aromatic Compounds*. Dekker, New York.

**Biodiversity:** *see* **Genetic Diversity in Microorganisms.**

# Bioenergetics of the Cell

*Stephen E. Specter and Barbara A. Horwitz*

### *Key Words*

**Bioenergetics** The biochemical reactions involved in energy changes within a living system. That is, the mechanisms by which the energy made available by the oxidation of substrates, or by the absorption of light, are coupled to energy-requiring reactions such as the synthesis of ATP or the accumulation of ions across a membrane.

**Electron Transport Chain** The series of electron transfers taking place on the cell membrane of bacteria, the thylakoid membrane of plant cell chloroplasts, and the inner mitochondrial membrane of eukaryotic cells. Electrons derived from the cofactors reduced during nutrient metabolism are transferred in a stepwise fashion to a lower energy state until they are taken up by oxygen, the final electron acceptor in the chain.

**Oxidative Phosphorylation** The synthesis of ATP from energy released as electrons are transferred "down" the electron transport chain. As electrons pass along the respiratory chain, the released energy is harnessed to pump protons across a membrane, establishing an electrochemical proton gradient. This gradient in turn drives protons back through an enzyme complex in the membrane, causing the enzyme (ATP synthetase) to drive the phosphorylation of ADP.

**Photophosphorylation** The synthesis of ATP from light energy. In plants cells and cyanobacteria, the energy available from the absorption of quanta of visible light is used to generate a proton electrochemical gradient. Electrons derived from water are subsequently able to be driven "up" an energy gradient to acceptors such as $NADP^+$ ($NADP^+ + 2e^- \rightarrow NADP_2$). Thus, light energy is ultimately used to generate ATP that photosynthetic cells use to convert inorganic carbon sources (e.g., $CO_2$) to glucose and other organic molecules.

No study of living systems would be complete without taking time to understand the fundamentals of bioenergetics. The regulated use of energy to ensure survival is intrinsic to every form of life, from bacteria and green plants to higher animals. All biochemical reactions involve energy changes, and all living systems share common strategies for using the energy obtained from their surroundings to function and ultimately flourish. Macronutrient energy must be converted into a physiologically useful form, then stored or employed directly to carry out metabolic work. In both simple and complex organisms, this conversion of energy to work is carried out within the cell. Thus, information derived from the study of energy metabolism at the level of the individual cell can be applied to understanding metabolism over the entire range of living systems.

## 1 METABOLISM IN LIVING SYSTEMS

### 1.1 Introduction

Living systems require energy to maintain homeostasis, to grow, to reproduce, and to perform work. Regulated use of energy to support life is a fundamental property of metabolism in simple and complex organisms. Metabolism, which is characterized by continual exchange of matter and energy, encompasses the physiological changes involved in the use of energy from dietary or stored substrates to drive complex anabolic (e.g., biosynthetic) reactions. Both simple and complex organisms use the pathways of intermediary metabolism to meet their energy needs. While coordination among specialized tissues allows more complex forms of life to cope with a varying nutrient supply, the energy requirement of any living organism is in reality a response to metabolic demands that originate in individual cells. That is, microorganisms, plants, and higher animals share a common link in the regulation of energy balance. Thus, the bioenergetics of all living systems can be simplified and perhaps best understood by examining their metabolism at the cellular level.

### 1.2 Prokaryotic Versus Eukaryotic Cells

Contemporary views regarding the regulation of energy metabolism are built upon early studies of prokaryotic (Greek: *pro,* before; *karyon,* nucleus) cells. Prokaryotes comprise different families of unicellular microorganisms and are characterized by their lack of internal membranes. Eukaryotic (Greek: *eu,* good or true) cells, which evolved perhaps a billion years after the prokaryotes, have their genetic material enclosed within a nucleus and contain other subcellular organelles, such as mitochondria and chloroplasts.

Prokaryote metabolism must by and large conform to the nutrient supply available in the external environment at any given time. Higher organisms, by contrast, have evolved specialized tissues to store energy, allowing eukaryotes to regulate metabolism independent of the need for an immediately available exogenous energy source. Even so, the fundamental mechanisms for harnessing energy from such disparate sources as light and glucose are virtually the same in all cells.

## 2 THE THERMODYNAMICS OF BIOLOGICAL ENERGY TRANSFER

### 2.1 Energy and Cellular Work

Energy is the capacity to do work. Work is the energy change accomplished by ordered or coherent molecular movement: for example, matter moving across a concentration gradient (osmotic work) or electrons moving between two different oxidation potentials (electrochemical work). Free energy, which in biological reactions is conserved primarily in the form of phosphoanhydride bonds, such as those found in adenosine triphosphate (ATP), is the major form of energy that cells can use to do work. Cellular work (e.g., muscle contraction, nerve impulse transmission, maintenance of ion gradients, protein synthesis, cell division) is accomplished at the expense of stored bond energy.

Biochemical energy transformations are governed by the laws of thermodynamics. Energy is classically described as occurring in two forms: kinetic energy (KE), the energy of molecules in motion, and "bound" or potential energy (PE). Potential energy can be either converted into KE or released to do work. During a biological reaction, a significant portion of PE is released as heat into the surrounding environment, reflecting the fundamental inefficiency of energetic conversions. Because inefficiency and heat loss inescapably accompany any physiological conversion, free energy must continually be supplied to the cell from its environment in order to support life.

### 2.2 Gibbs Free Energy

Gibbs free energy ($\Delta G$) is the maximum amount of work that can be obtained from a given chemical reaction under isothermal conditions. It is the free energy of the products minus the free energy of the reactants in a chemical exchange. The value for $\Delta G$ reflects the maximum amount of energy that has been liberated and can be passed on during the course of a reaction occurring under defined conditions. The energy released can perform work; it can be dissipated as heat or transferred as chemical bond energy into the components of another reaction to which it is coupled. If a system is at equilibrium and equivalent concentrations of reactants and products are present, $\Delta G = 0$. Under these conditions, no work can be performed.

A significant portion of the energy derived from organic substrates and liberated during oxidative stages of intermediary metabolism is conserved by the cell and transiently stored as chemical energy capable of supporting cellular work. The concept of "high energy" phosphate bonds was introduced in 1941 by Fritz Lipmann, who first proposed ATP as the intermediary between endergonic (energy-requiring) and exergonic (energy-releasing) processes. Conservation of energy in the form of high energy phosphate bonds is one of the unifying components of energy transfer in all living organisms.

Reactions with a $+\Delta G$ are endergonic and must be coupled to exergonic reactions. A common strategy of cells is to couple energetically unfavorable reactions (with a relatively high $+\Delta G$) to the splitting of ATP into adenosine diphosphate (ADP) and inorganic phosphate ($H_2PO_4^-$), a reaction that has a relatively large $-\Delta G$, as for example, in the first step of glycolysis: glucose + ATP $\rightarrow$ glucose -6-phosphate + ADP. Synthesis of complex macromolecules, including proteins, lipids, nucleic acids, and carbohydrates, also is endergonic and usually is coupled to ATP hydrolysis. In contrast, photosynthesis (the light-driven synthesis of organic molecules) is endergonic: it results in increased free energy ($+\Delta G$) of a system (e.g., plant cells) at the expense of radiant energy.

## 3 ENERGY METABOLISM IN AUTOTROPHS AND HETEROTROPHS

The ultimate energy source for all living organisms is the sun. However, only autotrophic organisms can directly utilize light energy to carry out biosynthesis of organic molecules like glucose and amino acids from inorganic precursors such as carbon dioxide ($CO_2$), water ($H_2O$), and ammonia ($NH_3$). Autotrophs include the green leaf cells of plants and photosynthetic bacteria. Heterotrophic organisms, which encompass the entire animal kingdom, nonphotosynthesizing plants, and most microorganisms, must obtain carbon (and nitrogen) from their environment in the form of relatively complex organic nutrients. Heterotrophs obtain chemical energy from the stepwise combustion of macronutrients via intermediary metabolism; $CO_2$, $H_2O$, and heat are regenerated and ultimately returned to the environment.

In aerobic heterotrophs, enzyme cofactors (e.g., $NAD^+$, FAD) that are reduced during nutrient catabolism can be reoxidized by transferring their electrons to a series of proteins in the electron transport chain (ETS). This electron transfer is accompanied by the release of chemical energy and is used to drive ATP production by a process known as oxidative phosphorylation. While autotrophic organisms are similarly capable of carrying out oxidative phosphorylation, they can also produce ATP via the light-driven reactions of photophosphorylation. It is via photosynthesis that autotrophs convert light energy into $NADPH_2$ and ATP, "fixing" (i.e., covalently linking) inorganic carbon sources (e.g., $CO_2$) and subsequently storing them as glucose and other organic molecules. The ETS and the enzymes involved in oxidative phosphorylation are ordered within the plasma membrane (in prokaryotes) or the inner mitochondrial membrane (in eukaryotes). In photosynthetic cells, the analogous enzymes are located in the chloroplasts on specialized infoldings termed thylakoid membranes.

## 4 ATP SYNTHESIS VIA OXIDATIVE PHOSPHORYLATION

### 4.1 Early Hypotheses

It was initially believed that the flow of electrons down the ETS might be coupled to ATP synthesis by one or more diffusible or localized energy-rich intermediate(s) via a mechanism similar to substrate level phosphorylation.*

Attempts to uncover the hypothesized intermediates have been unsuccessful, however, and the idea of a "phosphorylating respiratory chain" has, for the most part, been abandoned. A related concept was the "conformational hypothesis," which linked energy conservation and transfer to hypothesized conformational changes taking place in membrane-localized electron transport enzymes. Despite evidence that membrane proteins do in fact undergo conformational changes during ATP formation, this theory was never widely accepted. In any event, it fails to adequately describe a mechanism for the transfer of energy between electron transport and ADP phosphorylation.

* Although mitochondrial oxidative phosphorylation is the chief source of ATP in eukaryotic cells, ATP (or GTP, its metabolic equivalent) can be directly generated during both glycolysis and the tricarboxylic acid (TCA) cycle via substrate level phosphorylation. Energy-coupling reactions are referred to as substrate level phosphorylations when a metabolic intermediate is used to generate ATP directly. For example, the following reactions occur in the glycolytic pathway:

glyceraldehyde 3-phosphate + $P_i$ + $NAD^+$
1,3-bisphosphoglycerate + NADH + $H^+$
1,3-bisphosphoglycerate + ADP ⇌ 3-phosphoglycerate + ATP

Energy is released with the oxidation of an aldehyde to a carboxyl group and conserved by the coupled formation of ATP from ADP and inorganic phosphate. A similar mechanism occurs in the TCA cycle during the oxidation of succinyl coenzyme A (CoA) to succinate.

### 4.2 The Transmembrane Proton Electrochemical Gradient

The possibility that the "intermediate" linking cellular respiration (i.e., substrate oxidation and the flow of electrons down the ETS) and ATP synthesis might involve a proton gradient across a membrane was first suggested by the British biochemist Peter Mitchell in 1961. According to Mitchell's "chemiosmotic hypothesis," the transport of electrons down the ETS is coupled to ATP synthesis by a transmembrane proton electrochemical gradient. This gradient is the sum of two components: one is the difference in voltage across the membrane or the membrane electrical potential, $\Delta\psi$; the other is the difference in hydrogen ion concentration across the membrane, ΔpH. The proton electrochemical gradient is currently termed the "protonmotive force," Δp.

The value of Δp is a thermodynamic measure of the extent to which the proton gradient is removed from equilibrium. In mitochondria, $\Delta\psi$ is the dominant component, and the pH gradient is small (~0.5 pH unit), consistent with the fact that enzymes on both sides of the inner mitochondrial membrane operate optimally near neutral pH. In chloroplasts, by contrast, no enzymes are contained within the thylakoid space, and the hydrogen ion concentration gradient may exceed 3 pH units.

Movement of electrons "down" the thermodynamic gradient of the ETS is tightly linked to the translocation of protons out of the mitochondrial matrix. Thus, Δp is established by the oxidation of reducing equivalents (or cofactors) in mitochondria during the flow of electrons down the ETS, or by photon capture in chloroplasts. Proton movement in the reverse direction (*back into* the mitochondrial matrix) is also tightly linked or "coupled" to ATP synthesis. Since the inner mitochondrial membrane is impermeable to ions in general and to protons in particular, damage to the membrane or the presence of lipid-soluble proton conductors results in destruction of the electrochemical potential, dissociation of oxidative phosphorylation from electron transport (a phenomenon known as "uncoupling"), and release of the energy of oxidation as heat.*

The protonmotive force generated during electron transfer drives a membrane-bound ATPase backward (i.e., in the direction of ATP formation). This reversible ATPase is an universal feature of energy-transducing membranes in mitochondria, chloroplasts, and photosynthetic bacteria. Thus, Mitchell's suggestion that ATP synthesis is driven (at least in part) by an ion gradient is consistent with experimental evidence from numerous biological systems.

### 4.3 The Electron Transport Chain

All the enzymatic steps in the oxidative degradation of organic substrates in aerobic cells converge into a final common pathway,

* Although mitochondrial uncoupling is generally considered to be pathological, brown adipocytes, which are found in most terrestrial, placental mammals at some stage of life, are physiologically adapted to convert large amounts of energy to heat via uncoupled oxidative phosphorylation. The thermogenesis of brown adipose tissue, subsequent to activation by the sympathetic nervous system, is a physiological strategy used to generate warmth in response to cold (nonshivering thermogenesis) or, alternatively, to expend excess energy intake (diet-induced thermogenesis).

the electron transport chain. This stage is commonly referred to as ''respiration'' because electrons, originally derived from reduced cofactors (e.g., NADH + $H^+$, $FADH_2$) flow down an energy gradient to oxygen, the final electron acceptor. (The term ''reduced cofactor'' or ''reducing equivalent'' commonly is used to designate a single electron equivalent participating in a redox reaction.) Electron-donating reducing equivalents are generated as organic substrates are catabolized during earlier stages of metabolism. For example, reduced cofactors are generated by conversion of glucose to pyruvate during glycolysis, by β-oxidation of fatty acids to acetyl-CoA, and by passage of acetyl-CoA through the TCA cycle. $NAD^+$ collects pairs of reducing equivalents from many different substrates in one molecular form, NADH + $H^+$.

With one ''turn'' of the TCA cycle, four pairs of hydrogen atoms are removed from metabolic intermediates by the action of specific dehydrogenase enzymes. Most of these combine with $NAD^+$ to form NADH, which enters the ETS at the first enzyme complex. A second entry point accepts reduced flavin cofactors ($FADH_2$) generated by the TCA cycle (i.e., succinate + FAD → fumarate + $FADH_2$) and during β-oxidation. Electrons entering the ETS pass through a structured sequence of more than 20 different electron carriers organized into one of four enzyme complexes anchored to the inner mitochondrial membrane by transmembrane proteins. The movement of electrons down the energy gradient involves transfers of high structural and chemical specificity. Each successive carrier has greater affinity than the preceding one for the incoming electron, and oxygen, with the greatest relative affinity, serves as the final electron acceptor.

An ''electrical potential drop'' occurs as electrons move between complexes, releasing sufficient free energy to drive protons out of the inner mitochondrial matrix and into the intermembrane space. This movement decreases the pH outside the mitochondrial matrix, and the intermembrane space becomes more negative relative to the matrix. The resultant pH and voltage gradients drive the return of protons into the mitochondrial matrix. The flow of protons *down* their gradient is the motive force for ATP synthetase to catalyze the phosphorylation of ADP. This is what is meant by the chemiosmotic coupling of electron transport to ATP synthesis.

## 5 ATP SYNTHESIS IN PHOTOSYNTHETIC ORGANISMS

Heterotrophic organisms require an exogenous supply of chemical energy and $O_2$ from their environment to transduce the potential energy stored in organic nutrients into ATP. Thus, heterotrophs are dependent on the unique ability of autotrophic organisms to carry out photosynthesis: that is, to harness the energy of solar radiation via photophosphorylation, using it to drive the conversion of atmospheric $CO_2$ into glucose and more complex organic molecules. Photosynthesis occurs in green plants, eukaryotic organisms such as algae, and the cyanobacteria (green and purple sulfur bacteria).

Photosynthetic organisms can be divided into two classes, those that produce oxygen and those that do not. Green leaf plant cells, for example, yield molecular oxygen during photosynthesis, with water as the hydrogen (or electron) donor, as described by the following general equation:

$$nH_2O + nCO_2 \xrightarrow[\text{chloroplasts}]{h\nu} (CH_2O)_n + nO_2 \qquad (1)$$

where $CH_2O$ refers to general organic material and $n$ is typically assigned a value of 6 to correspond to synthesis of glucose ($C_6H_{12}O_6$). With the exception of cyanobacteria, photosynthetic bacteria are strictly anaerobic and do not generate oxygen; however, they can utilize a variety of organic and inorganic hydrogen donors.

Photosynthetic mechanisms are further categorized according to whether they involve light. The nonphotochemical pathways, where $CO_2$ is fixed and glucose is synthesized, are usually referred to as ''dark reactions'' because they do not require light to proceed. A separate set of radiant-energy-dependent ''light reactions'' comprises two processes: (1) the conversion NADP → $NADPH_2$ (photoreduction) and (2) the production of ATP (photophosphorylation), both of which are driven by photosynthetic electron transfer. Production of glucose from inorganic precursors is primarily active in the light because of its ongoing requirement for concomitant regeneration of $NADPH_2$ and ATP.

Both oxidative and photophosphorylation are coupled to the flow of energy-rich reducing equivalents down an electron transport chain, with a portion of the original energy ultimately conserved as ATP. Carrier-mediated transfer of electrons down an energy gradient in thylakoid membranes shares many features of respiration-driven electron transport in the inner mitochondrial membrane. Water, cleaved by light in the presence of chlorophyll, serves as the sole electron donor in chloroplasts; in mitochondria, electrons enter via reduced cofactors, generated chiefly via β-oxidation of fatty acids and oxidation of TCA cycle intermediates; in microbes, electrons are chiefly supplied by inorganic matter such as $H_2S$.

Protons are translocated outside the matrix space in mitochondria, across the plasma membrane in bacteria, and from outside *to inside* the thylakoid space in photosynthetic cells. The movement of protons across a membrane to establish a transmembrane electrochemical potential difference is a universal strategy employed by mitochondria, chloroplasts, and bacterial cells to drive the phosphorylation of ADP. In respiring heterotrophs, however, the Δp across the inner mitochondrial membrane is generated by oxidative metabolism, while in autotrophic organisms, the Δp is driven by photochemical events. The final coupling factor, the enzyme ATP synthetase, catalyzes the terminal electrogenic exchange between protons and the production of ATP:

$$ADP + H_2PO_4^- \xrightarrow[\substack{\text{mitochondria}\\ \text{chloroplasts}\\ \text{bacteria}}]{\Delta p} ATP + H_2O \qquad (2)$$

## 6 CONTROLLING THE RATE OF CELLULAR RESPIRATION AND ATP SYNTHESIS

Intermediary metabolism (via pathways involving nutrient catabolism or biosynthesis) is highly sensitive to changes in energy balance. Critical steps in a pathway (i.e., a single reaction) can respond to multiple internal signals. A number of theories have been advanced to explain the regulation of cellular respiration and ATP synthesis. Briefly, Chance and Williams described ATP production as governed by the demands of cytosolic energy-utilizing reactions. In their view, first published in 1955, ATP hydrolysis and regeneration of the substrates ADP and $P_i$ were the rate-limiting steps in the ATP synthetase reaction. Therefore, as cytosolic ATP was used up, more ADP would move into the mitochondrial matrix to be phosphorylated. Atkinson later reasoned that since adenine nucleotides act as allosteric effectors of glycolytic enzymes, TCA cycle dehydrogenases, and ATP-utilizing pathways, energy metabolism would be regulated by the relative amount of ATP available, as described by the adenylate charge:

$$\text{adenylate charge} = \frac{[\text{ATP}] + 0.5[\text{ADP}]}{[\text{ATP}] + [\text{ADP}] + [\text{AMP}]} \tag{3}$$

According to the "near-equilibrium" hypothesis of Erecinska and Wilson, published in 1982, if the ETS and oxidative phosphorylation operate near equilibrium (with the exception of cytochrome $aa_3$), mitochondrial respiration and ATP synthesis are chiefly controlled by (1) the mitochondrial NADH/$NAD^+$ ratio, (2) the ATP/ADP ratio or phosphorylation potential (i.e., ratio of ATP to ADP + $P_i$) and (3) effectors of cytochrome $aa_3$ (e.g., pH or $[O_2]$). Three years later, in separate publications, Denton and McCormack and Hansford suggested that calcium ions ($Ca^{2+}$) regulate the rate of oxidative phosphorylation through their effect on key mitochondrial dehydrogenases and possibly through direct stimulation of pathways requiring ATP.

There is considerable evidence that the rate of oxidative phosphorylation is in fact influenced by a number of factors, including the rate of ATP hydrolysis, the delivery of reducing equivalents to the respiratory chain, and the flux of key effectors, such as calcium ions. This complex multistep process probably entails several signals or factors in the regulation of energy balance in general and cellular respiration in particular.

## 7 SUMMARY

Living organisms continually exchange matter and energy with their surroundings. Prokaryotes have a limited capacity to store energy and must depend on their immediate environment for available nutrients. The energy needs of eukaryotic cells are more likely to be met via a coordinated exchange of information and substrates among multiple specialized tissues. Chemical energy stored in higher animals (as triacylglycerol) or in plants (as starch) is released as fatty acids and glucose and taken up by individual cells according to their demand for metabolic work.

A cell functions as an individual bioenergetic unit. ATP is the currency of internal energy transfer, and each cell is obligated to furnish its own supply. To do this, the organism (and then the cell) must take in energy or catabolize a source of stored energy. Autotrophic organisms can synthesize their own supply of organic molecules, utilizing photic energy to fix atmospheric $CO_2$, while heterotrophs must obtain chemical energy from preformed sources of essential nutrients. The potential energy stored in organic nutrients must somehow be transferred to a metabolic intermediate capable of coupling the release of its free energy to the complex and often highly specialized endergonic reactions within the cell.

Cells accomplish this task by coupling electron transport to ATP synthesis via oxidative and photophosphorylation. The passage of electrons down an energy gradient generates sufficient energy to create a transmembrane electrochemical potential, driving the phosphorylation of ADP by ATP synthetase. This gradient can be established across the inner mitochondrial, thylakoid, and bacterial plasma membranes. Nearly all cells take advantage of the mechanism of chemiosmotic coupling of electron transport to ATP synthesis to meet their immediate energy needs. The overall energy requirement for any living system is reflected by specific demands in specialized tissues which, in turn, are governed by physiological conditions at the cellular level.

*See also* ELECTRON TRANSFER, BIOLOGICAL; NUTRITION; PROTEIN PHOSPHORYLATION.

### *Bibliography*

Alberts, B., Bray, D., Lewis, J., Raff, M., Roberts, K., and Watson, J. D. (1989) *Molecular Biology of the Cell,* 3rd ed. Garland, New York.

Balaban, R. S. (1990) Regulation of oxidative phosphorylation in the mammalian cell. *Am. J. Physiol.* 258:C377–C389.

Becker, W. M. (1977) *Energy and the Living Cell.* Lippincott, Philadelphia.

Brown, G. C. (1992) Control of respiration and ATP synthesis in mammalian mitochondria and cells. *Biochem. J.* 284:1–13.

Nicholls, D. G., and Ferguson, S. J. (1992) *Bioenergetics II.* Academic Press, San Diego, CA.

# BIOINORGANIC CHEMISTRY

*R. Bruce Martin*

### *Key Words*

**Chelate** (from Greek claw) Multiple bonding of two or more atoms of a single molecule to a metal ion, often to form five- or six-membered rings.

**Ligand** A molecule or ion with a donor atom possessing a lone pair of electrons that interacts with a metal ion.

**pH = −log ($H^+$)** The negative decade logarithm of the hydrogen ion activity or concentration. Almost all chemists and biochemists calibrate a pH meter with buffers that yield a pH scale based closely on hydrogen ion activity. Many coordination chemists calibrate a pH meter using known concentrations of hydrogen ion. In the region of 0.1 to 0.2 ionic strength, the concentration scale yields acidity constant logarithms ($pK_a$ values) about 0.12 log unit less than the more common activity-based scale. Stability constants of metal ions with ligands are unaffected by the choice of pH scale.

Bioinorganic chemistry focuses on the roles of noncarbon elements in life processes. Yet, bioinorganic chemistry is inseparable from the general chemistry of life. Carbon itself cycles among the many bioorganic compounds and inorganic carbon dioxide and carbonates. More than 80% of all the carbon in the earth's crust occurs as $CaCO_3$. About 30% of all enzymes contain metal ion cofactors. The most common metal ion, zinc, appears in more than 100 enzymes; iron and copper, in a substantial number; manganese, cobalt, and molybdenum, in a few cases. Selenium appears in the enzyme glutathione peroxidase. Metal ions stabilize nucleic acid polymers, which bear a negative charge on each residue. Though blood is loaded with an array of organic molecules, its main constituent is NaCl; intracellular fluids are composed chiefly of KCl. Even among the vitamins, a word representing "amines essential to life," the action of vitamin $B_{12}$ depends on a cobalt ion. Metal ions and nonmetals other than carbon are intimately and inseparably involved in life processes.

## 1 ESSENTIALITY

Twenty-one elements are essential to humans. A nutrient is essential if a deficiency results in an impairment in function that is relieved only by administration of that substance. Vitamins by definition and some minerals are essential. The significance of essentiality may be illustrated by burlesquing an old adage.

> For want of a nail the shoe is lost, for want of a shoe the horse is lost, for want of a horse the rider is lost.
> George Herbert, *Jacula Prudentum* (Outlandish Proverbs), 1640

> For want of a nutrient the enzyme is lost, for want of an enzyme the function is lost, for want of a function the life is lost.
> Bruce Martin, *Summa Veritatis* (Lofty Truth), 1989

Four essential elements (H, O, C, and N) comprise more than 99 atom % and about 96 wt % of the human body. These 4, plus 14 other essential elements, occur among the first 30 elements (through zinc) of the periodic table. Three heavier trace elements (Se, Mo, and I) are also essential in humans. For 17 essential elements, Table 1 shows the predominant elemental form at pH 7, typical adult concentrations in the blood plasma or serum, the approximate amount found in a 70 kg adult, and a recommended daily allowance for adults. In addition to the basic four elements, the essential elements include two alkali metal ions, two alkaline earth metal ions, seven transition metals (the most common, iron, contributes less than 0.01% of body weight), phosphorus, sulfur, selenium, and three halogens. Table 1 shows that most of the remaining 4% of body weight consists of calcium and phosphorus, two elements found in bone. Many of the elements do not exist predominantly in their pH 7 forms in the serum because they are combined with other components. For example, $Fe^{3+}$ does not precipitate as the hydroxide but is retained by tightly chelating ligands. There is little free iodide; iodine occurs as part of the thyroid hormones. For sulfur Table 1 lists the total serum concentration, most of which appears in proteins; there is only about 1 mM nonprotein sulfur. Sulfur is not important as an inorganic element but only as part of the amino acids cysteine and essential methionine. An additional four elements not included in Table 1 are essential for other organisms (B, Si, V, and Ni).

Essentiality is not the only criterion for inclusion of elements in a survey of bioinorganic chemistry. Organisms may accumulate elements that are not essential. Some elements such as arsenic and antimony have been used therapeutically, and bismuth still is. Others, such as aluminum and the heavy metals cadmium, mercury, thallium, and lead, are toxic and prevalent, and their interactions with life processes are matters of concern.

## 2 STABILITY SEQUENCES

From the many studies on stability constants, one finds the order of metal ion stabilities dependent on the ligand. Increasing metal ion stabilities follow the orders:

$$\text{Glycine: Ca, Mg} << \text{Mn} < \text{Fe, Cd, Pb,} < \text{Co, Zn} < \text{Ni} << CH_3Hg^+, \text{Cu} << \text{Hg}$$

$$\text{1.2-Diaminoethane: Mg} << \text{Mn} << \text{Fe} < \text{Pb, Cd, Co, Zn} << \text{Ni} < CH_3Hg^+ << \text{Cu} <<<< \text{Hg}$$

The ion $Ca^{2+}$ does not form stable amine complexes. Except for methylmercury, $CH_3Hg^+$, all the metal ions carry two positive charges. Owing to the strongly chelating bidentate ligands, glycine (gly) and 1,2-diaminoethane (en), $CH_3Hg^+$ with only a single strong binding site is at a competitive disadvantage in the series just listed. To a unidentate ligand, $CH_3Hg^+$ binds more strongly than all the foregoing metal ions except $Hg^{2+}$.

In the foregoing series each inequality sign stands for an approximate 10-fold increase in stability constant. The two series gly and en are similar, their major difference being a stability constant span from $Mg^{2+}$ to $Hg^{2+}$ of $10^9$ for glycine and $10^{14}$ for en. Generally the increment between metal ions increases on passing from O < N < S donor atoms. The presence of a sulfur donor promotes $Cd^{2+}$

**Table 1** Essential Elements in Humans[a]

| Element | pH 7 Form | Serum Concentration | Amount in Humans | Daily Allowance |
|---|---|---|---|---|
| Na | $Na^+$ | 140 mM | 70 g | 1–2 g |
| K | $K^+$ | 4 mM | 130 g | 2–5 g |
| Mg | $Mg^{2+}$ | 0.8 mM | 22 g | 0.3 g |
| Ca | $Ca^{2+}$ | 2.4 mM | 1100 g | 0.8 g |
| Mn | $Mn^{2+}$ | 10 nM | 12 mg | 3 mg |
| Fe | $Fe(OH)_3\downarrow$ | 17 μM | 4 g | 10–20 mg |
| Co | $Co^{2+}$ | 2 nM | 1 mg | 3 μg vit. $B_{12}$ |
| Cu | $Cu^{2+}$ | 17 μM | 80 mg | 3 mg |
| Zn | $Zn^{2+}$ | 14 μM | 2.3 g | 15 mg |
| Cr | $Cr(OH)_2^+$ | 3 nM | 6 mg | 0.1 mg |
| Mo | $MoO_4^{2-}$ | 6 nM | 5 mg | 0.2 mg |
| Cl | $Cl^-$ | 104 mM | 80 g | 2–4 g |
| P | $HPO_4^{2-}$ | 1.1 mM | 600 g | 1 g |
| S | $SO_4^{2-}$ | 24 mM | 120 g | 0.7 g Met[b] |
| Se | $HSeO_3^-$ | 1 μM | 5 mg | 0.1 mg |
| F | $F^-$ | 2 μM | 2.5 g | 2 mg |
| I | $I^-$ | 0.4 μM | 30 mg | 0.15 mg |

[a]In addition to H, C, N, and O.
[b]Essential amino acid methionine.

and $Pb^{2+}$ to higher positions than they occupy in the foregoing series. The order of increasing sulfur binding strengths is $Zn^{2+} < Cd^{2+} < Pb^{2+} < CH_3Hg^{+} < Hg^{2+}$. Sulfhydryl group interactions are the main mode of toxicity of the heavy metal ions. These general bonding features are expressed in the concept of the stability ruler.

## 3 STABILITY RULER

For a single ligand the Irving–Williams stability sequence of dipositive metal ions is Mg < Mn < Fe < Co < Ni < Cu < Zn, invariant of ligand. The uniformly progressive part of this sequence from $Mg^{2+}$ to $Cu^{2+}$ defines a *stability ruler* to which more variable metal ions may be compared. The stability ruler appears across the top of Table 2. For several ligand–donor sets, Table 2 shows the relative binding strengths of the variable dipositive metal ions $Zn^{2+}$, $Cd^{2+}$, $Pb^{2+}$, and $Hg^{2+}$. Their placement in Table 2 corresponds to their stability constants compared with those of the metal ions that define the ruler. The entries for glycine and 1,2-diaminoethane agree with the series in Section 2. With most of the metal ions in Table 2, histidine will be tridentate. The increment between metal ions, hence the length of the stability ruler, increases with the substitutions O < N < S. The ruler length appears in the second to last column of Table 2 in log *K* units for binding to each ligand. The longer the ruler length, the more discriminating the ligand in selecting among metal ions. Thus ligands with oxygen donor atoms are least discriminating among metal ions, nitrogen donors intermediate, and sulfur donors most discriminating.

Not only do $Co^{2+}$ and $Zn^{2+}$ display nearly identical radii, they also exhibit similar stabilities for all ligands in Table 2, except for those involving sulfur and hydroxide. The similarities allow the facile and useful substitution of $Co^{2+}$ for $Zn^{2+}$ in many enzymes. When a sulfhydryl group is present, as in 2-mercaptoethylamine, $Zn^{2+}$ binding strengthens, equaling that for $Ni^{2+}$. Table 2 also shows the relative strengthening of $Cd^{2+}$ and $Pb^{2+}$ binding with the sulfhydryl donor in 2-mercaptoethylamine.

For all donor sets in Table 2, $Hg^{2+}$ binds so strongly that it is off the end of the ruler scale. The numbers under Hg (last column of Table 2) refer to the relative distance by which the length of the whole log stability constant scale from $Mg^{2+}$ to $Cu^{2+}$ must be extended to reach the value for $Hg^{2+}$. A telling contrast appears between the length of the extension for most bidentate compared to unidentate ligands. The scale extension for $Hg^{2+}$ amounts to 0.2 to 0.4 log unit for bidentate ligands and to 1.1 to 1.2 log units for the three unidentate ligands at the end of Table 2. The difference arises because $Hg^{2+}$ prefers linear, two-coordination and binds the second donor atom in small chelate rings much more weakly than the first donor atom.

We may generalize the results and conclusions of the stability ruler by noting that alkali metal and alkaline earth metal ions, lanthanides, and $Al^{3+}$ prefer oxygen donors; transition metal ions, oxygen and nitrogen donors; and the heavy metal ions, nitrogen and sulfur donors.

## 4 METAL ION BINDING CHARACTERISTICS

### 4.1 Amino Acids

Amino acids bind transition metal ions more avidly with increasing pH, since the amino group suffers less competition from the proton. The potentially tridentate amino acids histidine and cysteine are especially strong transition metal ion binders. Alkali and alkaline earth metal ions, $Al^{3+}$, and lanthanides bind only weakly to amino acids. Owing to loss of basic carboxylate and amino groups, peptides usually bind metal ions more weakly than amino acids unless the metal ion deprotonates the peptide nitrogen.

### 4.2 Proteins

Metal ions interact in numerous ways with proteins, from the weak fairly, nonspecific interactions of many metal ions with proteins such as serum albumin, to the highly specific protein sites made for exclusive binding of a single kind of metal ion. For example, the iron-transporting protein of the plasma, transferrin, binds two $Fe^{3+}$ ions under blood plasma conditions with conditional dissociation constants of $10^{-22}$ M. Such a minuscule value is necessary because the solubility of goethite, FeO(OH), only allows $10^{-21}$ M $Fe^{3+}$ at pH 7.4. Individual cases are the subject of other entries in this volume.

### 4.3 Nucleosides

Let us designate the nucleic bases with their usual alphabetical symbols, A, C, G, T, and U. The order of decreasing basicity is given by

$$H^{+}: T3 > U3 > G1 >> C3 > A1 > G7 > A7$$

Table 2 Stability Ruler

| | | Dipositive Cations | | | | | | | | | | | | |
|---|---|---|---|---|---|---|---|---|---|---|---|---|---|---|
| Ligand | Donors | Mg | < | Mn | < | Fe | < | Co | < | Ni | < | Cu | Length (log *K*)[a] | Hg[b] |
| Hydroxide | $OH^-$ | | | | Cd | | | | | | Zn | Pb | 3.7 | 1.2 |
| Acetate | $O^-$ | | | | | | | | | Zn | Cd | Pb | 1.3 | 1.2 |
| Imidazole | =N | | | | | | | Zn | Cd | | | | 4.0 | 1.1 |
| Ammonia | $NH_3$ | | | | | | Pb | | Zn | Cd | | | 4.0 | 1.1 |
| Oxalate | $O^-,O^-$ | | | | Cd | | | Zn | Pb | | | | 2.1 | |
| Glycine | $N,O^-$ | | | | | Cd | Pb | Zn | | | | | 6.1 | 0.4 |
| Histidine | N,=N | | | | | Cd | Pb | Zn | | | | | 8.2 | 0.2 |
| 1,2-Diaminoethane | N,N | | | | | Pb | Cd | Zn | | | | | 10.1 | 0.4 |
| 2-Mercaptoethylamine | $N,S^-$ | | | | | | | | | Zn | Cd | Pb | 12 (est.) | 0.2 |

[a]Ruler length in difference between log stability constants of $Cu^{2+}$ and $Mg^{2+}$ complexes.

In contrast, for a heavy metal ion in neutral solutions, the order of decreasing stability is given by

Heavy metal ion: G7 > A7 > C3 > A1 > G1 > U3 > T3

Comparison of the two series reveals a marked promotion of the G7 and A7 sites on passing from the first to second series. Moreover, in the DNA double helix the N7 sites are not involved in interstrand hydrogen bonding and are consequently relatively exposed. Thus the antitumor agent *cis*-diaminodichloroplatinum (II) binds mainly at the guanosine N7 site of DNA. Though proposed many times as metal ion binding sites, primary amino ($—NH_2$) groups located at C-4 in cytidine, C-6 in adenosine, and C-2 in guanosine are neither proton nor metal ion binding sites in neutral solutions. The sugar moiety binds metal ions very weakly and insignificantly.

### 4.4 Nucleic Acids

With the introduction of more than one basic phosphate group, the di- and trinucleotides ($pK_a$ 6.5) become strong metal ion binders. Virtually all reactions of ATP require an $Mg^{2+}$ cofactor, and ATP occurs as an $Mg^{2+}$ complex in cells. Nucleic acid chemists were slow to recognize the role of ambient metal ions in stabilizing polymeric nucleic acid structures. With a negative charge on the phosphate of each residue, only a random coil form exists at low salt concentrations. The phosphate groups in the polymers are not basic ($pK_a \approx 1$) and bind metal ions very weakly. Required for structures such as the double helix are nonspecifically bound alkali metal ions and $Mg^{2+}$ serving as counterions to offset the negative charges on the phosphates. $Mg^{2+}$ is also important in stabilizing various RNA structures.

*See also* Calcium Biochemistry; Cytochrome P450; Hemoglobin; Metalloenzymes; Trace Element Micronutrients; Zinc Finger DNA Binding Motifs.

### *Bibliography*

Berthon, G., Ed. (1995) *Handbook on Metal–Ligand Interactions in Biological Fluids:* Vol. 1, *Chemistry of Metal–Ligand Interactions in Biological Fluids.* Dekker, New York.

da Silva, J.J.R.F., and Williams, R.J.P. (1991) *The Biological Chemistry of the Elements.* Clarendon Press, Oxford.

Martin, R. B. (1979) Interactions between metal ions and nucleic bases, nucleosides, and nucleotides in solution. *Metal Ions Biol. Syst.* 8:57.

Ochiai, E. (1987) *General Principles of Biochemistry of the Elements.* Plenum Press, New York.

Sigel, H., Ed. (1973–1995) *Metal Ions in Biological Systems,* Vols. 1–28. Dekker, New York.

## Biological Electron Transfer: *see* Electron Transfer, Biological.

## Biological Fuel Production: *see* Fuel Production, Biological.

## Bioluminescence: *see* Chemiluminescence and Bioluminescence, Analysis by.

# Biomaterials for Organ Regeneration

*Susan L. Ishaug, Robert C. Thomson, Antonios G. Mikos, and Robert Langer*

### *Key Words*

**Allogeneic** Of the same species but genetically different.

**Angiogenesis** Development and formation of new blood vessels; vascularization.

**Autogeneic** Derived from the same individual.

**Macroencapsulation** The envelopment of a large mass of cells or tissue in biomaterial membranes for immune protection.

**Microencapsulation** The envelopment of single cells, islets, or a small number thereof, within biomaterial membranes for immune protection.

**Scaffold** Biomaterial framework or structure used to organize and direct the growth of cells.

**Xenogeneic** Denoting individuals or tissues from different species.

Cell-based artificial organs and therapies can have a great impact in medicine for the treatment of a variety of acquired or inherited diseases. Cell transplantation can provide an alternative treatment to whole-organ transplantation for failing or malfunctioning organs. Because many isolated cell populations can be expanded in vitro using cell culture techniques, only a very small number of donor cells may be needed to prepare an implant. Consequently, the living donor need not sacrifice an entire organ; thus the donor pool is expanded significantly. Novel biocompatible and biodegradable materials, which incorporate specific peptide sequences to improve cell adhesion and promote differentiated cell growth by releasing growth factors, angiogenesis factors, and other bioactive molecules, can provide temporary scaffolding to transplanted cells and by so doing allow the cells to secrete extracellular matrix (ECM), enabling a completely natural tissue replacement to occur. Until recently, most research in the field has focused on minimizing biological fluid and tissue interactions with biomaterials in an effort to prevent fibrous encapsulation brought on by foreign body reaction or clotting in blood that has contact with artificial devices. In short, much biomaterials research has focused on making materials invisible to the body. Innovations that use the inverse approach—programmed extensive interaction of the material with biological tissue—will give biomaterials research a new focus.

## 1 PRACTICES

The human body is a complex biological machine; when it is functioning properly, its various parts interrelate and work together smoothly. In the absence of such balanced relationships, however, an entire organ or a major part thereof fails. Often the only way to correct this problem is by whole-organ transplantation. Many lives are saved or improved by this procedure, but multitudes are lost owing to the tremendous scarcity of donor organs. Donor scarcity is the major limitation to the whole-organ transplantation procedure. For example, chronic liver disease and cirrhosis was the tenth leading cause of death in the United States in 1990, corresponding to approximately 30,000 deaths. Moreover, the number of organ donors falls far short of the need: fewer than 3000 human livers are available annually.

In addition to donor scarcity, other complications with whole-organ transplantation have compelled researchers to search for alternative methods for replacing or repairing the failed or weakly functioning organs. Such problems include foreign tissue rejection and patients' continual need for immunosuppressive drugs, difficulty in preservation and transportation of the donor organs, and the complicated, severe, and expensive surgery involved in organ transplants. Biomaterials can be utilized to replace or repair tissue function, induce cell proliferation, or provide a matrix for cell attachment and tissue ingrowth. Not only nondegradable synthetic materials but also bodily tissues and biodegradable synthetic polymers can be used to replace or repair the defective organ.

Long-lived biomaterials used in artificial organs such as the plastics that make up artificial hearts or the metals in pins used to join fractured bones are designed to reside in the body for extensive periods of time. These either completely replace the organ or are intended to facilitate the recovery of a failed organ or tissue. Since these materials are foreign to the body, there are usually problems with biocompatability, including rejection and failure of the material with the eventual need for replacement with a fresh implant. Cases associated with the rupturing of silicone breast implants are notable examples.

Since the functions of most organs are carried out by the parenchymal cells, some techniques focus on utilizing only these cells for regenerating organ function. This alternative approach to whole- or partial-organ transplantation may reduce the chance of rejection. By using the separated parenchymal cells, thus eliminating other cell types that might contain added rejection potential, the immune responses may be reduced. Since many cell populations can grow in vitro, cells from a small biopsy of organ tissue can be cultured to produce a significant mass to be used for implantation. The biopsy tissue may be obtained either from the patient or from a close relative. This approach would greatly increase the donor pool and diminish the need to wait for cadaver organs, a time during which some transplant candidates die. It would also avoid additional surgical trauma for the patients and result in reduced medical costs.

Parenchymal cells alone have been injected into bodily cavities in hopes that organ tissue will form. However, no organ structure has been observed for any cells transplanted in this manner. Most normal cells are anchorage dependent, whereas to survive, proliferate, and function properly, they need to attach to some type of matrix. Therefore, another technique involves attaching cells to biomaterials and transplanting these assemblies into the desired surgical site. It has been observed that the cells need to be in close proximity to interact and function as a tissue. They may need to be close to blood vessels for the unhindered exchange of nutrients and waste material essential to their survival. With these factors in mind, bodily extracellular matrix (ECM) proteins such as collagen and glycosaminoglycans have been formed into matrices and implanted, for the purpose of recruiting cells from healthy tissue or for use as a template for transplanted cells to regenerate nerves, skin, blood vessels, peridontal tissue, cartilage, and bone. Yet, these protein templates are often difficult to produce or to produce in the preferred shapes with the desired mechanical properties, and their introduction can lead to proliferation of cell types different from those desired.

More recently, biomaterials studied for organ regeneration involve biocompatible and biodegradable synthetic polymers. These materials usually function as a scaffold for the growth and organization of implanted organ cells. Parenchymal cells are isolated from the tissue and seeded into the polymer, and the cell–polymer structure is implanted (Figure 1). Basically, the scaffold degrades simultaneously as the cells proliferate and excrete their ECM substances. The growing cells, ECM, and nutrient-supplying vascular tissue continually replace the void spaces of the disappearing scaffold until eventually the scaffold/cell implant has been replaced by natural organ tissue.

The biodegradable polymers must be fabricated to allow for cell adhesion and for the retention of cell function, and to enhance the growth rate. The majority of the polymers used in cell transplantation are polyesters and polyanhydrides. They have been produced

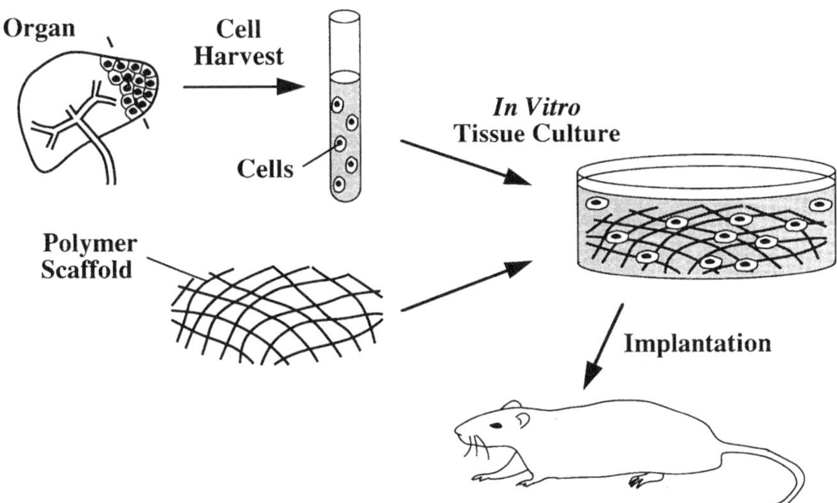


Figure 1. Schematic representation of organ regeneration methodology by cell transplantation using biodegradable polymer scaffold.

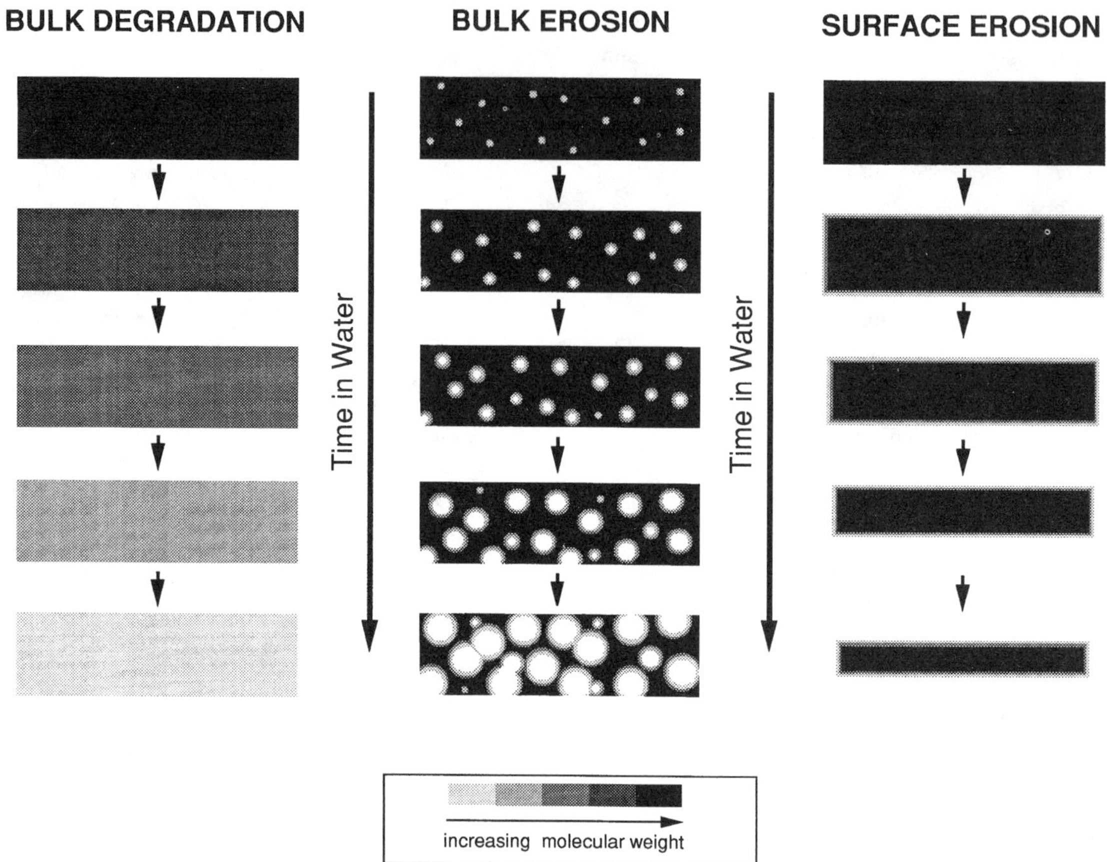


**Figure 2.** Mechanistic comparison of polymer matrices undergoing bulk degradation, bulk erosion, and surface erosion.

in sheets, in porous scaffolds, and in fibrous strands that can be woven into a mesh. The degradation of the material can occur by bulk degradation, bulk erosion, or surface erosion. For "ideal" bulk degradation, the polymer chains are cleaved uniformly throughout the entire material volume without significant loss of matter. The molecular weight of the entire polymeric material continually decreases until some critical point is reached in which the chains are small enough and dissolve. For "ideal" bulk erosion, the polymer breaks down, and matter is removed by dissolution, nonuniformly from the entire material volume. For "ideal" surface erosion, polymer chains are cleaved at, and matter is depleted from, the exterior surface, while the interior of the material remains unchanged (Figure 2).

An important class of degradable polyester biomaterials include polylactic acid, polyglycolic acid, and their copolymers, which are among the few synthetic polymers approved for human clinical use. They are also referred to as bioresorbable, in that their degradation products are low molecular weight compounds, such as lactic acid and glycolic acid, which enter into normal metabolic pathways. These synthetic polymers can be fabricated in specific shapes needed for implants, their pore size and surface area can be altered for increased cell attachment, their mechanical properties can be engineered, and their degradation rates can be tailored to vary from a few days to years. They degrade by bulk degradation, the rate of which depends on the copolymer ratio, pH, and temperature. The degradation mechanism involves the hydrolysis of the ester linkage (Figure 3). The rate of hydrolysis, which is minimal for homopolymers of lactic acid or glycolic acid, becomes maximum for poly(lactic-*co*-glycolic) acid with a 50:50 copolymer ratio (Figure 4). The degradation rate also increases with temperature and is enhanced in both acidic and basic environments.

Biodegradable polymer foams can be prevascularized to allow for sufficient nutrient exchange for the transplantation of cells and tissues where survival and function depend on their adequate nourishment. A variety of bioactive molecules also can be embedded in the polymer to inhibit inflammatory responses and increase cell attachment. Such molecules include growth factors, growth inhibitors, angiogenic agents, immunosuppression agents, and differentiation agents, which are slowly released with the gradual degradation of the polymer to aid in the growth and differentiation of the tissue. Biomolecules can be physically introduced into a polymer matrix or chemically bonded to a polymer backbone (Figure 5).

For some organs, the primary cells cannot be cultured *in vitro* to provide the critical mass required for cell transplantation and functional replacement. In this case, xenografts (tissue from other species) can be used to replace the tissue function. When the main function of the transplanted cell population is metabolic regulation and activity, xenogeneic cells or tissues can be encapsulated in membranes and implanted. These membranes are designed to isolate the cells from the body, thereby avoiding the immune responses the foreign cells could initiate, and also to allow the diffusion in and out of the membrane of the desired metabolites, such as insulin and glucose for pancreatic cells.

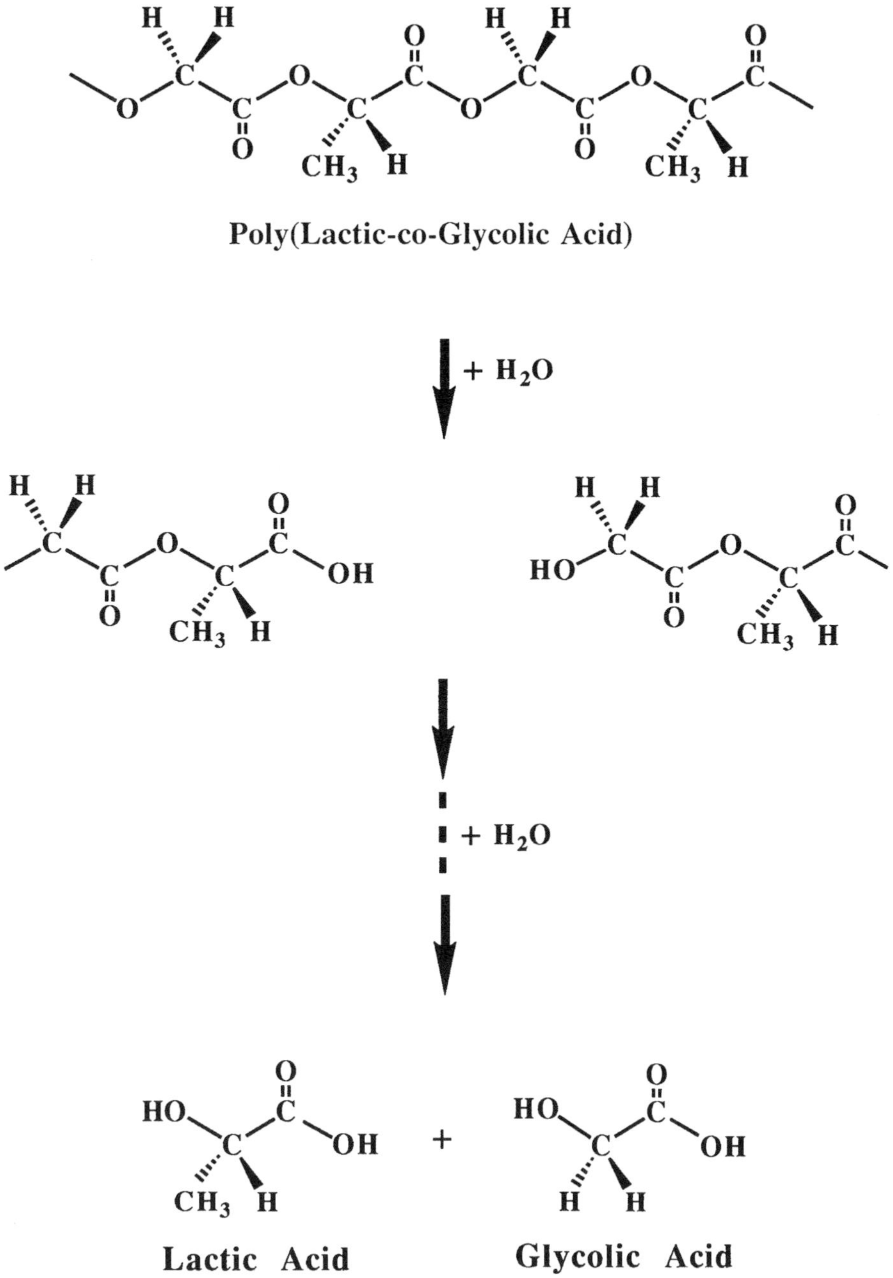

Poly(Lactic-co-Glycolic Acid)

$+ H_2O$

$+ H_2O$

Lactic Acid + Glycolic Acid

**Figure 3.** Schematic representation of the hydrolytic degradation of poly(lactic-*co*-glycolic acid) copolymers.

The cells of each organ have unique functions and thus usually require different types of biomaterials for their regeneration. The specific properties and function of biomaterials utilized in the regeneration of pancreas, liver, bone, and cartilage are discussed in Sections 1.1 to 1.4. Nevertheless, the use of biomaterials is by no means limited to these organs and tissues.

### 1.1 PANCREAS

Approximately 1 in 35 people in the United States are diagnosed with diabetes. Personal costs incurred by those afflicted by the disease are high, owing to the strict dietary regime and the need for insulin injections. The monetary costs to the nation are also high. One estimate sets the total annual cost in the United States in 1987 at $20 billion. Currently, the only proven alternative to exogenous insulin injection is a whole-organ pancreatic transplant, but the scarcity of organ donors has led to a situation of demand far outweighing supply.

Allogeneic cell transplantation is not a viable means of increasing the donor pool because currently it is not possible to grow mature pancreatic islets *in vitro*. Therefore, only xenogeneic transplantation of pancreatic islets would expand the donor pool significantly. Such a procedure entails the use of biomaterials to isolate the transplanted cells from the recipient's immune system, hence preventing graft rejection. It is hypothesized that the requisite isolation may be achieved by enveloping the islets in a polymer membrane

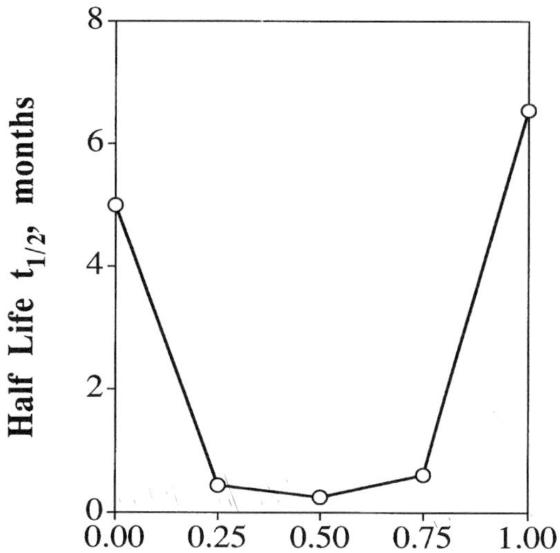


**Figure 4.** Degradation half-life of poly(lactic-*co*-glycolic acid) copolymers as a function of the lactic acid copolymer ratio.

that allows the transport of small molecules, including those that are essential to the correct functioning of the islets (e.g., glucose, insulin, oxygen) while at the same time precluding the passage of the larger molecules of the immune system (e.g., immunoglobulins).

To develop the polymer membranes that would be required for successful xenogeneic transplantation, initial studies in this area focused on allogeneic transplants. Microencapsulation of individual rat pancreatic islets within a hydrophilic alginate gel, followed by further polymer coatings to form a size-selective membrane, is a technique that has been widely studied. There have been reports of improved biocompatibility achieved by forming microcapsules with an alginate outer coating. In allograft experiments the use of these microcapsules resulted in normoglycemia (i.e., normal glucose level in the blood) in diabetic rats for 10 months. Precipitates of polyelectrolyte complexes can also be used, and the ease with which islets can be encapsulated within such biomaterials is a factor in their desirability. Although microencapsulation is an attractive system in terms of the large surface area for mass transfer, which is imparted by the small size of the microencapsulants, there are still problems associated with the diffusion of glucose and insulin across the capsule membrane.

In a normal pancreas, there is a network of capillaries that penetrate even into the islet cell aggregates, to provide an adequate supply of oxygen to every cell. That is, a minimum diffusion distance exists for metabolites between the vascular system and the cells. With encapsulated cells, even under optimal conditions for vascularization, there will still be an additional diffusion resistance associated with both the outer membrane and the alginate gel. The increased diffusion resistance also causes a delay in the dynamic insulin secretory response of islets to glucose stimulation. In addition, aerobic metabolic processes within each cell will cause a diminution of the concentration of, for example, oxygen (between the cells on the periphery of the islet and those toward its center). The successful use of biomaterials for islet encapsulation therefore depends on minimizing these problems of diffusion to ensure adequate mass transfer rates to and from the cells.

In addition to microcapsules, other immunoprotection devices are used which encapsulate large numbers of islets in one compart-

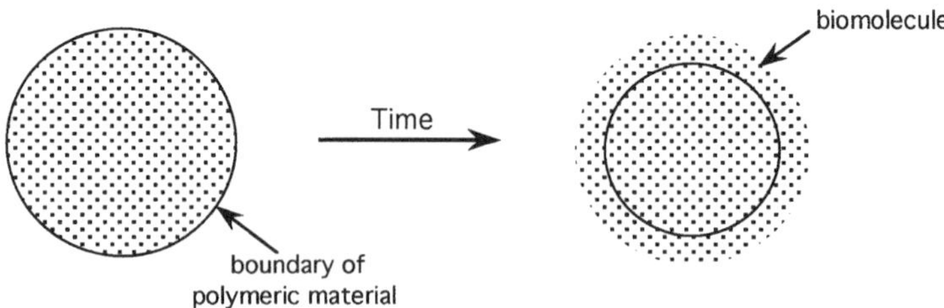


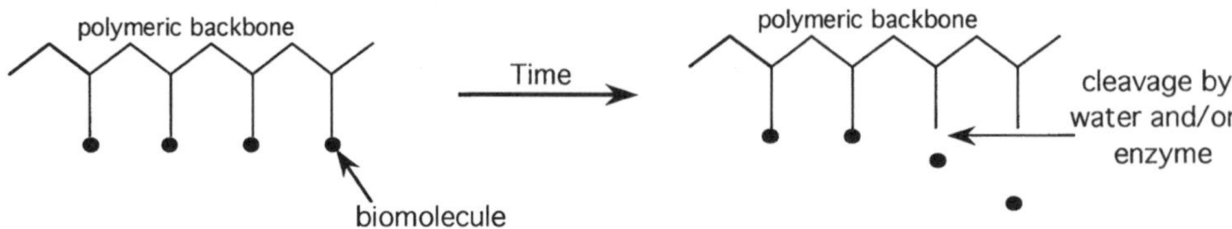


**Figure 5.** Schematic representation of the release of biomolecules physically or chemically embedded in polymer matrices.

ment. This approach is referred to as macroencapsulation. Such devices may be intravascular or extravascular and are constructed of the same type of semipermeable membrane used in microencapsulants. Intravascular devices consist of a double-walled tube that is connected directly to the vascular system. Blood flows through the center tube, and small molecules pass through the membrane wall into the islet compartment. The major disadvantages of this device are its relatively small mass transfer area and its greater membrane thickness in comparison to microcapsules, which results in increased diffusion resistance. The mass transfer area can be increased by using bundles of small diameter tubes, but this results in the formation of blood clots due to increased plasma shear stresses. Extravascular devices are simply larger versions of microcapsules and have the same diffusion problems associated with intravascular devices. The use of macroencapsulated individual cells, as opposed to islets, has been reported and may improve the mass transfer of molecules from the inside of the encapsulant to the inside of the cell.

It seems that xenogeneic transplantation, using polymer membranes for graft isolation, is a promising means by which to expand the donor pool. The biomaterials used as membranes and elsewhere in encapsulation devices must be biocompatible. In addition, the membranes must be tailored to prevent the transport of large molecules, such as immunoglobulins, without unduly increasing the diffusion resistance to small molecules.

### 1.2 Liver

Unlike pancreatic islets, hepatocytes can be cultured *in vitro.* As a result, autogeneic and allogeneic cell transplantation may be used to replace lost liver function. The first biomaterials utilized for liver regeneration were naturally occurring extracellular matrix proteins such as laminin. These proteins, which provided an environment for the transplanted hepatocytes similar to that found in liver, proved to be essential for long-term differentiated function. The major drawback of ECM proteins is that they have insufficient structural strength to be used as practical implantation devices. Synthetic biomaterials have therefore been developed with a view to mimicking the three-dimensional ECM framework and, at the same time, providing adequate strength. Liver is a highly vascularized organ with an organized cellular architecture; hence the design of an artificial scaffold should provide for these requirements.

Collagen-coated dextran microcarriers have been used successfully as substrates for hepatocyte attachment. The collagen provides the surface characteristics required for cell attachment, while the properties of structural strength and high surface are imparted by the dextran microspheres. When implanted into rats, these microcarriers form aggregates that provide a template for new tissue formation and neovascularization. This technique has been used, with rats, both to correct genetic enzyme deficiencies (e.g., analbuminemia) and to provide temporary support of liver function in partially hepatectomized animals.

Biodegradable polymers, such as polygalactin, polyorthoesters, polyanhydrides, and poly($\alpha$-hydroxy esters), as well as nondegradable ones like polyvinyl alcohol, have been used as substrates for transplanted hepatocytes. Their major advantage over naturally occurring ECM proteins, however, is the ease with which their properties can be tailored to the needs of a particular organ. For example, they can be processed with specific pore morphologies to maximize vascularization prior to cell implantation.

Microencapsulation of hepatocytes has also been proposed with a view to xenogeneic transplantation. Initial studies have been carried out *in vitro* using rat hepatocytes. However, it remains to be seen whether the diffusion problems associated with this method (see Section 1.1) will allow successful *in vivo* culture. Given that small numbers of donor cells can be multiplied *in vitro,* albeit at a rate lower than that observed *in vivo,* it appears unlikely that xenogeneic transplantation will ever be required to increase the size of the donor pool.

To achieve successful liver cell transplantation, biomaterials must provide a prevascularized site for cell seeding, a substrate for cell attachment that results in differentiated cell function, and a matrix to direct the organization of cells, newly formed tissue, and capillaries.

### 1.3 Bone

Biomaterials can be utilized to repair bone loss due to injuries (e.g., car accidents), to aid in the regeneration of bone parts excised as a result of developing tumors, to reconstruct physical deformities, and to reinforce tissue that is aged or diseased (e.g., due to osteoporosis). In some cases the entire bone requires replacement and in others the damaged bone needs only reinforcement to enhance the function of its self-repair mechanism.

A very common bone replacement is the hip endoprosthesis. Artificial hips have been made from cast metal alloys and fibrous compound materials. Since the stem or acetabular cup often loosens and other malfunctions often occur, these implants must be replaced by a second or third surgical procedure. Pins, which are used to reinforce the healing of damaged bones, are generally made of metal; bone cement, used to fill in irregular shaped defects, consists primarily of polymethyl methacrylate. Metals, fibrous compounds, and synthetic nondegradable polymers of bone implants are much stiffer than bone and deprive the bone of normal stresses, causing the long-term growth of an inferior bone structure adjacent to the implant. These nondegradable synthetic implants not only produce this stress shielding effect, they encounter problems with interface adhesions and scar tissue formation.

Bone reconstructive surgery often requires implants, which are usually performed with autogenous bones taken from the ribs or iliac crests. These implants are effective for filling defects, but the method requires an additional surgical procedure to isolate the donor bone, imposing additional postoperative discomfort on the patient. When a large bone graft is needed, it is usually obtained from a bone bank. Yet, the supply of this material is limited. Also, immune responses and transmission of diseases are inevitably encountered with both allogeneic and xenogeneic bone grafts. Studies have been undertaken to alter the allografts to decrease the immunogenicity or stimulate the formation of new bone. Freeze-drying and defatting allografts results in loss of immunogenicity. Decalcification and demineralization of the allogeneic bone grafts also seems to increase bone growth into these implants.

As an alternative to the replacement by bone grafts, scaffolds have been utilized to recruit bone cells and tissues for the purpose of regenerating bone. The scaffold needs to be biologically compatible, readily available, and easily made into the desired shapes and sizes; in addition, replacement of the scaffold by natural bone tissue should occur readily. Two bone implant materials, sea coral and hydroxyapatite (dense ceramic coralline material) meet the foregoing criteria and have been used with some success. Other porous composites of aluminum, calcium and phosphorus oxides, and tricalcium phosphate, have also been used as bone graft materials. The coralline substances are cheap and easily sterilized, and since they are inert, are accepted by bone and the body without signs of infectious complications. Coral is biodegradable, being 99% calcium carbonate; it reossifies well and eventually is replaced by natural bone. These substances, however, are naturally brittle, and because of their large pore size, their mechanical strength is low.

Bone pins, plates, and intramedullary rods may be replaced by the use of degradable polymers of the poly($\alpha$-hydroxy ester) family. These synthetic polymers can be manufactured with various pore sizes, and their mechanical strength can be altered by varying the monomer ratio of the copolymer. Osteoblasts also can be seeded within these polymers to facilitate the proliferation and differentiation of the bone tissue. Degradable polymers cause less inflammation; they are replaced by natural bone tissue, thus not requiring replacement with a fresh implant, and when completely replaced by new bone, they may function like the original bone with respect to environmental stresses. Work still needs to be done to improve the strength and enhance cell growth within these polymers and to ensure retention of biomechanical requirements during the course of polymer degradation.

### 1.4 Cartilage

Cartilage replacement or regeneration is utilized mainly in reconstructive surgery, such as the nose and ears, and orthopedic surgery for the correction of joint problems, such as those caused by arthritis. The nondegradable synthetic compounds usually used to replace cartilage are silicone rubber, porous polyethylene, polytetrafluoroethylene, various ceramics, and metals.

Silicone implants are sometimes used to replace cartilage. Implants made of this seemingly inert material usually do not change in shape and volume, may be produced in desired shapes, and do not support bacterial growth. Yet, in some cases a high rate of extrusion for the implants can be found. Porous polyethylene and polytetrafluoroethylene show good macroscopic incorporation into the joint surface. They are even incorporated better than silicon, showing that these newly investigated implant materials have potential as adequate alternative cartilage implants. Ceramics and metals are usually used to replace cartilage that experiences higher stresses, such as that in the femoral head and hip joint. Even though these materials are usually biocompatible and nondegradable, they lead to damage and sometimes complete loss of the opposite facing joint cartilage with which the implant is in contact.

The materials just named do not regenerate the cartilage and are used only as replacements. The materials that play a major role in the regeneration of cartilage are autogeneic or allogeneic cartilage grafts, and also biodegradable natural or synthetic biomaterials, which are utilized as templates for cell transplantation. Autogeneic cartilage grafts are usually obtained from the ribs. Although these implants are readily incorporated into the desired region, they impose additional surgical trauma and discomfort to the patient in the retrieval of the donor tissue and cannot be used when large cartilage sections are needed.

Allogeneic cartilage is used when autogeneic grafts cannot be obtained. These can be devitalized before implantation by irradiation, formalin, glutaraldehyde, and alcohol. The devitalization reduces the immune response, thereby enabling some of the grafts to be immunologically inert. Since, however, there are no viable chondrocytes to maintain the matrix, the devitalized allogeneic grafts eventually undergo resorption. If adequate growth of new cartilage within these degrading grafts does not occur, the strength of the structure may be insufficient for the transplant site.

Since chondrocyte culture is relatively easy, the chondrocytes derived from a small biopsy sample of cartilage can be expanded for seeding into biodegradable natural or synthetic biomaterials. As earlier mentioned, natural biomaterials such as collagen and glycosaminoglycans can be used as templates for cartilage regeneration, but problems are encountered in their production into desired shapes, and cell types different from those desired tend to proliferate.

Synthetic biocompatible and biodegradable polymers can be seeded with chondrocytes and implanted to regenerate cartilage. The continuously degrading polymer scaffold is progressively replaced by cartilage that maintains the exact three-dimensional shape of the original polymer/cell scaffold. Therefore it seems that synthetic biocompatible biodegradable materials could be a useful alternative as a template for cartilage regeneration. These materials can be altered in many different ways if required and are eventually completely replaced by natural cartilage. Further investigations are required into the production of polymeric devices with anatomical shapes, the efficient seeding of chondrocytes throughout the scaffold volume, and the retention of the differentiated function of the attached chondrocytes.

## 2 FUTURE PERSPECTIVES

In addition to offering a new approach to solve the devastating problem of organ loss, cell transplantation can also provide a new route for *in situ* drug administration. Missing genes can be introduced to different cell types, which are transplanted back to the host for the production and local release of proteins and other therapeutic drugs. The treatment of coronary and peripheral diseases using genetically altered endothelial cells seeded onto degradable or nondegradable stents is one example of the enormous impact cell therapies could have in the future. Transplanted genetically modified endothelial cells can also provide a new means of systemic delivery of drugs (such as tissue-type plasminogen activator), since their secretions enter directly into the bloodstream. Other cell types that are potential candidates for therapeutic gene transfer include hepatocytes (for treatment of low density lipoprotein receptor deficiency), myoblasts (for treatment of muscular dystrophy), hematopoetic stem cells (for treatment of hemoglobinopathies or congenital white cell deficiencies), fibroblasts, and keratinocytes.

*See also* Biomimetic Materials; Polymers, Genetic Engineering of.

### *Bibliography*

Davies, J. E., Ed. (1991) *The Bone–Biomaterial Interface*. University of Toronto Press, Toronto.

Folkman, J., and Klagsbrun, M. (1987) Angiogenic factors. *Science* 235:442–447.

Langer, R. (1990) New methods of drug delivery. *Science* 249:1527–1533.

Langer, R., and Vacanti, J. P. (1993) Tissue engineering. *Science* 260:920–926.

Mikos, A. G., Papadaki, M. G., Kouvroukoglou, S., Ishaug, S. L., and Thomson, R. C. (1994) Islet transplantation to create a bioartificial pancreas. *Biotechnol. Bioeng.* 43:673–677.

Mikos, A. G., Sarakinos, G., Lyman, M. D., Ingber, D. E., Vacanti, J. P., and Langer, R. (1993) Prevascularization of porous biodegradable polymers. *Biotechnol. Bioeng.* 42:716–723.

Miller, R. A., Brady, J. M., and Cutright, D. E. (1977) Degradation rates of oral resorbable implants (polylactates and polyglycolates): Rate modification with changes in PLA/PGA copolymer ratios. *J. Biomed. Mater. Res.* 11:711–719.

Peppas, N. A., and Langer, R. (1994) New challenges in biomaterials. *Science* 263:1715–1720.

Tamada, J. A., and Langer, R. (1993) Erosion kinetics of hydrolytically degradable polymers. *Proc. Natl. Acad. Sci. U.S.A.* 90:552–556.

# Biomimetic Materials

*Robert J. Campbell*

## *Key Words*

**Ceramic** Relating to the manufacture of functional or ornamental products from inorganic, nonmetallic mineral, hardened to a stage characterized by an ability to resist wear and tear. Ceramics do not corrode or disintegrate.

**Codon** The triplet nucleotide sequence of messenger RNA that represents a particular amino acid in the genetic code and is translatable into that amino acid at the ribosome.

**Composite** Structural material made up of two or more diverse substances.

**Liquid Crystal** Substance exhibiting optical or other physical properties of an intermediate state, resembling ordered crystal with different values measured along axes in different directions, yet having viscosity low enough to behave mechanically as a liquid.

**Mimetic** Characterized by, or exhibiting a tendency toward, imitation, mimicry, or simulation.

**Photodynamic Protein** Protein capable of transducing radiant energy (as light) into a change of color (photochromic), an electromotive force (photovoltaic), or a change in wavelength (nonlinear optical), or in some other way using a photon to energize a process (e.g., information storage).

**Polymer** Large molecule comprising a chain of smaller molecules linked together by covalent chemical bonds. For example, a peptide polymer (polypeptide or protein) is a chain of amino acids joined by peptide bonds.

Biomimetic materials may be rather broadly defined as materials produced through direct manipulation of a synthetic process of biological origin, or by engineered chemical synthesis deriving its product design or function from a naturally occurring biological system. To successfully mimic or copy a biological process or product, we must first understand what mechanisms and pathways are involved in that process or what materials and structures comprise that product, or better yet, be able to characterize both process and product.

Creation of a biomimetic material then, requires us to incorporate into the research enterprise at least one of three attributes: a lesson from nature, mimicry of a biological process, or mimicry of the biological product. Recent advances in physical and structural analysis, in the case of materials sciences, and in biochemical and molecular genetic manipulation, in the case of chemistry and biological sciences, are generating breakthroughs in our ability to design and synthesize along the lines of one or more of these three themes, with the aim of obtaining materials with tailor-made properties optimized for a particular purpose.

## 1 THE SCOPE OF BIOMIMETIC MATERIALS

### 1.1 Lessons and Examples from Nature

Nature provides both general lessons and specific examples of how biomolecular structures, cellular product architectures, and fabrication strategies have satisfied form- and function-specific requirements in what appears to be optimal fashion, given a defined set of building blocks and adaptive pressures played out over extended time scales. In the realm of biology, there exist almost unlimited examples of the organization of molecules into structures that effect the transfer of mass, charge, and energy over performance durations ranging from picoseconds to seconds. Likewise, examples abound for biological production of hierarchical supportive and protective structures over the course of minutes to days or more. These examples of everyday activities of living cells, and the material results of these activities, represent but a piece of the wealth of biological information applicable to development of biomimetic materials. With an eye toward the goal of eventual conceptualization as ''food for thought'' blueprints for biomimetic materials, the object here is to become familiar with what is available in nature's searchable storehouse of ideas. Exploring and understanding what is on offer is the key—describing the possibilities via analytical studies of naturally occurring structures, materials, processes, and systems from the biological world.

### 1.2 Mimicking the Process

A substantial number of pharmaceuticals are now made partially or totally with the aid of biosynthetic capabilities of, for example, microorganisms. In much the same way, and with newly defined modifications allowing greater freedom of expression with these and more highly evolved cellular systems, the manipulation and exploitation of biological processes for novel synthesis of advanced materials offers great promise as it emerges to complement and enhance the repertoire of classical methods of materials synthesis. Biological cells operating as manufacturing plants have elaborate compartmentalized and sequentially regulated routes of synthesis that are difficult, if not impossible, to duplicate today, even using the most rigorous chemical synthetic techniques. They can, however, be mimicked by applying them wholly or in part to nonbiological molecules and structures. These biosynthetic pathways, whether functioning in nature in the wild-type cell, in cells genetically engineered for mimetic purposes, or in cell-free biomimetic systems containing minimal components, do so in an energetically favorable and environmentally benign manner, requiring much less in the way of physicochemical extremes of temperature, pressure, and so on. Nevertheless they get the job done with higher degrees of selectivity, specificity, and hierarchical complexity than are found to be possible with purely chemical synthetic routes.

### 1.3 Mimicking the Product

Naturally occurring multicomponent biological materials are distinguished from current synthetic materials by the sophisticated structure–function relationships exhibited at all levels of organization. This is true not only for the many supramolecular complexes serving a vital role in, for example, energy transduction, but also for specialized hard or soft structural materials of a composite nature. As such, each of the diverse examples of these intricate biological materials inspires design and synthesis of man-made materials that mimic their most useful characteristics. Indeed, the potential of natural systems for providing innovative structural and functional concepts for improved or altogether new materials is enormous. A thorough appreciation of biological structures at various levels of scale can be used for design and fabrication of mimetic structures from substances never before available, and with properties for which a requirement has never existed in natural development, but for which the need and the precursors now exist in the material world. The routes to getting there, to mimicking the product, are as varied as the synthetic pathways found in cells, or in a laboratory sourcebook of synthetic protocols.

Table 1 The Scope of Biomimetic Materials

| Topic | Ref. |
|---|---|
| Mimicking biological polymerization | |
| Engineered ribosomal biosynthesis | Ellman et al. |
| Peptide polymers | Tirrell et al. |
| Degradable thermoplastic biopolyesters | Steinbüchel |
| Biomimetic molecular electronics, photonics, and mechanics | |
| Photodynamic protein mimetics for optical processing | Oesterhelt et al. |
| Mimicking electron transfer in proteins | Boxer |
| Mechanically active macromolecules | Cross and Kenrick-Jones |
| Organized biomolecular arrays | |
| Membrane pores and channels | Bayley |
| Self-assembling supramolecular structures | Whitesides et al. |
| Biomimetic patterning elements for nanometer-scale fabrication | Sleyter et al. |
| Mimicking biological reaction specificities | |
| Biomolecular recognition components for sensors | Nakamura et al. |
| Bioengineered catalysis | |
| Enzymes by design | Clark and Estell |
| Catalytic antibodies | Benkovic |
| Protein substructure mimetics | Rizo and Gierasch |
| Hierarchically ordered composite materials | |
| Nonmineralized systems for rigidity or compliance | Aksay et al. |
| Bioceramic structures | Aksay et al. |

### 1.4 The Best of Both Worlds

Various approaches toward progress in this exciting area of biomimetic materials are under way. These range from purely bioanalytical methods, where the objective is to understand the candidate system well enough to mimic it, to purely chemical techniques that have as their goal materials taking their cue from the biological world, but drawing their components from conventional chemical building blocks and their fabrication from traditional synthetic methods. Some very innovative approaches lie somewhere in between these two extremes and are beginning to attract the attention they deserve for their potential contributions to successful application of biotechnology to production of useful materials not otherwise available, or available only at a competitive disadvantage. These approaches capitalize not only on the traditional methods where appropriate, but also on adaptation of design principles and self-organizing properties found in nature, as well as the creative engineering of molecular biological tools and processes to achieve product and production superiority over conventionally synthesized special-purpose materials. Examples of investigative efforts in the area of biomimetic materials are too numerous to describe here in detail. Suffice it to say that examples of research directions enabling potential development of biomimetic materials are as diverse as the topics noted in Table 1. Specific instances of three representative biomimetic approaches to materials synthesis are more fully described in Section 2.

## 2 SOME EXAMPLES OF BIOMIMETIC MATERIALS

### 2.1 Photodynamic Protein Mimetics for Optical Processing

Bacteriorhodopsin (bR) is a retinal-containing, seven-helix membrane protein that functions as a light-driven proton pump in the purple membrane of the archaebacterium *Halobacterium halobium* (*salinarium*). As such, it appears to have been optimized for its photophysical properties in its biological development over the course of time, and to have evolved complete with excellent chemical and thermal stability and resistance to photochemical degradation.

This photodynamic protein complex has been studied extensively, not only for what it offers scientifically in the way of unique biological structure–function relationships, but also for what it may lead to technologically, in the form of novel photonic materials development. Present in its native state (purple membrane) as one of the few naturally occurring two-dimensional protein crystals, bR's characteristic structure of both hydrophilic and hydrophobic domains makes it feasible to fabricate oriented bR derivatives into thin films via biomimetic self-organizing assembly. New families of biomimetic materials based on this complex biological macromolecule are under continuing study in a number of laboratories.

In addressing the need for advanced photonic materials, investigators have recently shown that bR has promising photochromic, photovoltaic, and nonlinear optical properties with high potential for device application. Because of such photophysical properties as high extinction coefficient, high quantum efficiency, and rapid photochromic shifts in absorption, the specifications of devices employing biomimetic material derived from bR include high sensitivity and extremely rapid response. Durability of such materials would likely be high also, owing to the photochemical properties

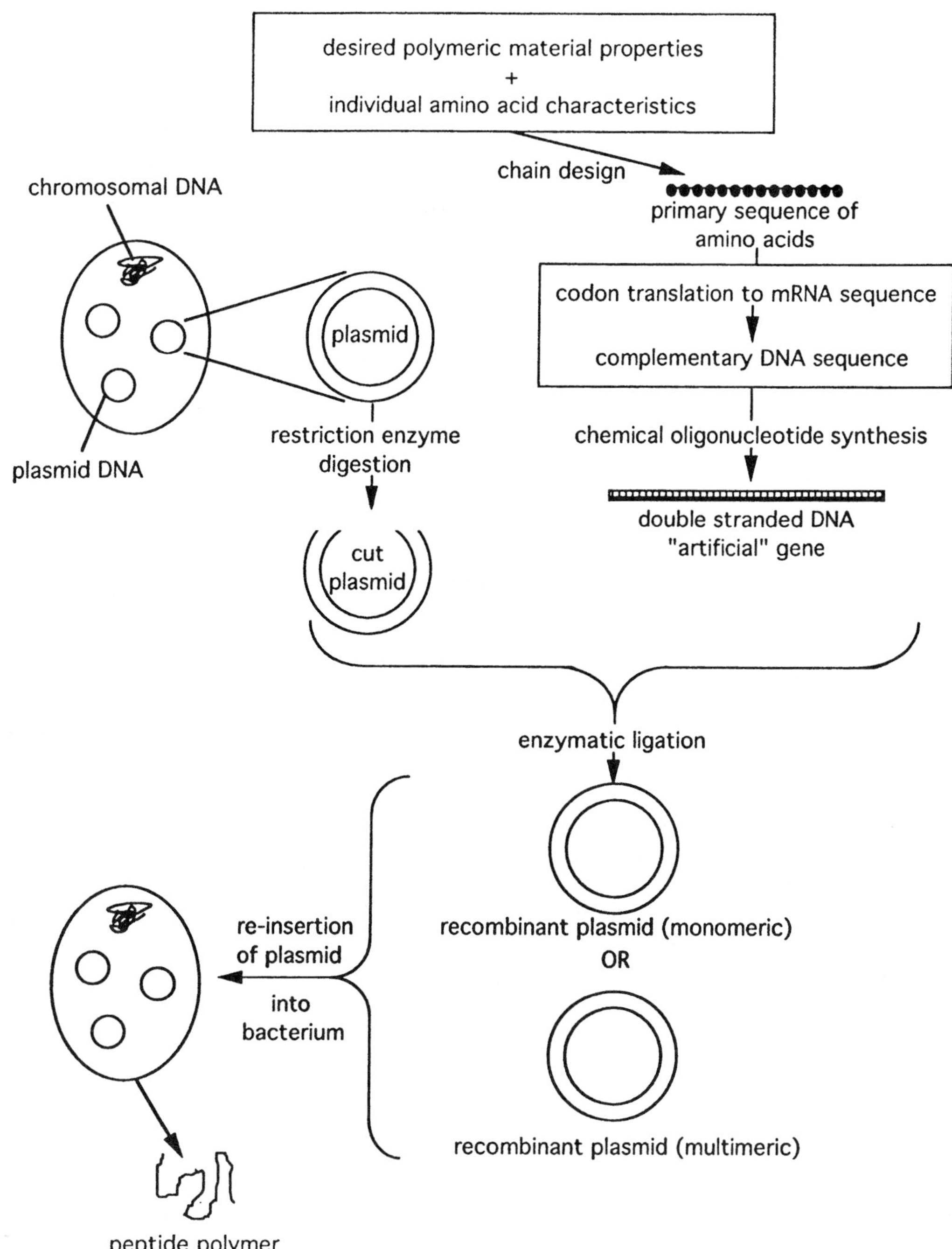


**Figure 1.** Recombinant DNA technology applied toward peptide polymer production in a bacterium.

of bR, which unlike many other optical materials does not exhibit photodegradation (i.e., fatigue) when it is light-cycled. In the class of biologically derived material, its chemical stability is indeed noteworthy; under conditions of polymerization into rubber or plastic, thermoreversible photochemistry persists.

Among the means of generating bR variants with chromophore system structures appropriate for desired characteristics of a potential device component are (1) replacement of the retinal in the binding site with structures analogous to the retinal, and (2) genetic modification of the bR protein primary structure. Regarding the former, there are practically unlimited possibilities of variation of retinal substitution. Hundreds of biomimetic bRs containing hundreds of different retinal replacements have been produced already, some with very interesting properties. One particular substitution results in a material that might be useful as an optical switch in the 10 ps range. Molecular genetic manipulations aimed at alteration of the photodynamic behavior of bR take the form of both random and site-directed mutagenesis. Recent development of a halobacterial transformation system enables expression of the stable purple membrane conformation in *H. halobium* and allows for production of very large quantities of high quality, mutated bR. By genetically modifying amino acids that surround the retinal

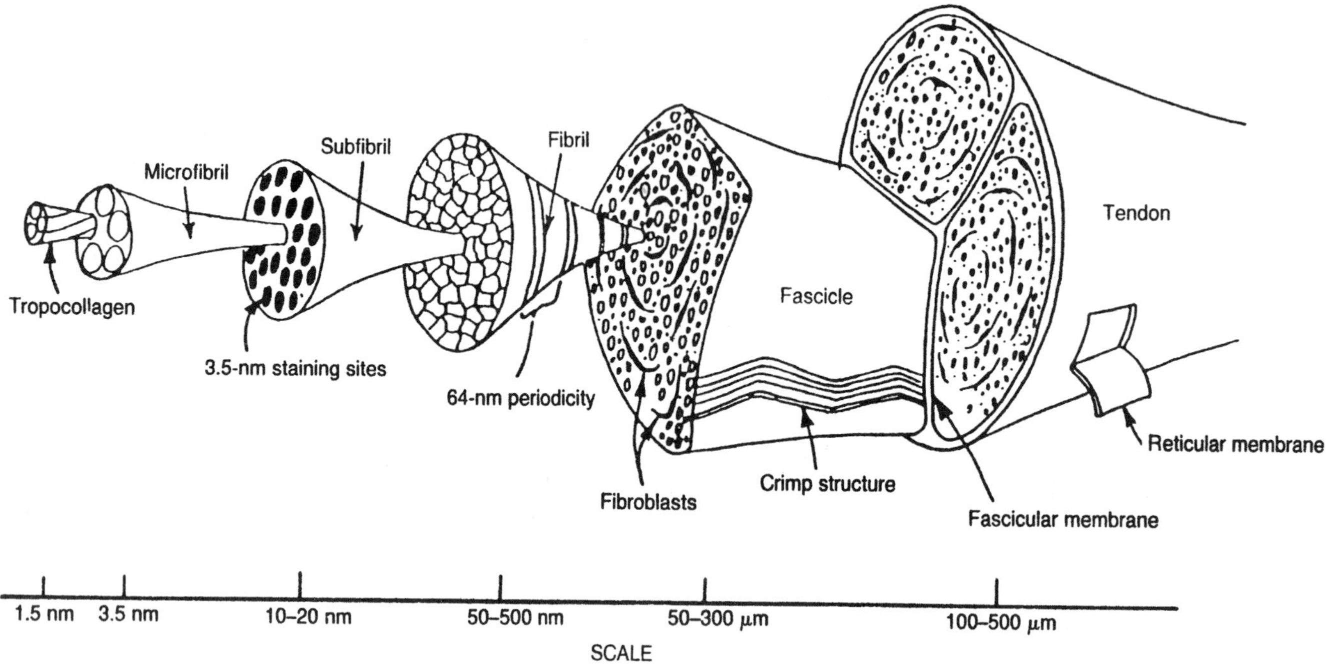


**Figure 2.** Tendon. Composed mostly of collagen embedded in a noncollagenous matrix, a tendon has an architecture that exhibits a hierarchical organization of subunits distinguishable structurally at six levels of scale. Starting with the three-membered helical chain of tropocollagen protein molecules, this structural hierarchy builds through two progressively larger combinations at nanometer scale (microfibril, subfibril) to a fourth level, the fibril, which shows a characteristic crimping pattern useful in tension damping. Groups of these fibrils, in aggregate with other cellular material, make up the fascicle which, in turn, is packaged with one or two other fascicles within a reticular membrane to form a tendon. This hierarchical organization, and the subcomponent interactions it engenders, is the basis of the tendon's toughness. (Reprinted with permission from E. Baer et al., *Physics Today,* October 1992.)

and play a key role in Schiff base protonation and electronic excited states of the retinal, it is possible to create novel photonic properties. Biomimetic bR materials resulting so far, from changing amino acid residues of wild-type purple membrane bR, possess properties suitable for optically addressed spatial light modulators, real-time interferometry, holographic recording films and pattern recognition devices, optical neural net synaptic elements, and devices for nonlinear optical phase conjugation and second-harmonic generation.

## 2.2 Peptide Polymers

It seems highly likely that future development in the area of polymeric materials requiring uniform, well-defined chain populations will involve a substantial transition from synthesis and fabrication using conventional methodology to more sophisticated biomimetic synthetic techniques. Standard chemical synthesis carries with it the inherent limitation of statistical distribution of product structural variables such as molecular weight, composition, sequence, and stereochemistry. In contrast, biomimetic synthesis allows precise control over these important variables, yielding a purer polymeric material of desired characteristics.

Among a number of classes of naturally occurring biological polymers, the proteins appear to be the most versatile in terms of form and function. Much progress has been made over the past decade in establishing the merits of further study and development of materials based on periodic repeats of polypeptide units, along the lines of those that exist in proteins such as silk, elastin, and collagen. In nature's cellular production process for these peptide polymers, control of macromolecular properties is readily accomplished at the molecular level by virtue of the fidelity of RNA and DNA replicative mechanisms for subunit order and specificity, as well as the unique ability of the ribosomal protein synthetic machinery to exert exquisite control over initiation, sequence read-through, and termination of the amino acid polymerization. Via recombinant DNA technology, and the continuing advances in related areas of molecular biology, that same control over uniformity of composition, length, and sequence is now afforded the polymer chemist or materials scientist willing to engage in the synthesis and expression of natural or tailor-made genes for biomimetic peptide polymers. By inserting appropriate DNA sequences encoding polypeptide repeats into microorganisms, where they are stably maintained and expressed, investigators have learned how to induce these cells to reproducibly synthesize polymers they do not ordinarily produce, in quantities sufficient for further study and eventual application in the form of novel materials.

A generalized recombinant DNA strategy using an artificial gene for production of a peptide polymer in a bacterial host system is shown in Figure 1. Based on the desired macromolecular material properties and on the known characteristics of individual amino acid behavior in chain structure, a sequence of amino acids with greatest potential to confer these properties in the resulting polymer is designed. In accordance with the genetic code, appropriate codons translated from the amino acid sequence dictate the messenger RNA template, which in turn is used to determine the complementary DNA sequence for the artificial gene. Double-stranded oligonucleotide encoding a single unit or multiple repeats of the designed polypeptide is synthesized, purified, and, through standard restriction enzyme and ligation methodology, inserted as a complete coding sequence into an appropriate DNA vector. Recombinant plasmids containing the synthetic insert are identified in bacterial cell transformants, then isolated and purified. Cloned monomeric oligonucleotide is excised and self-ligated to generate a distribution of multimeric oligonucleotides, each of which is inserted into transfer vectors for cloning into bacterial host expression systems for production of peptide polymers of varying, but defined, sequence, length, and composition.

By means of molecular genetic techniques similar to those just mentioned, innovative material design strategies allow the preparation of a number of interesting sequence monomers and assembly of these monomers singly into peptide homopolymers or, in combination, into block copolymers. Oligonucleotide synthesis derived gene segments in the range of 100 to 500 bases encoding reasonably large amino acid sequence monomers are possible. With multimerization of the monomeric gene segments, production of peptide polymers on the order of 100 kDa molecular weight can be achieved. Some of those produced to date exhibit liquid crystalline properties; others self-assemble into ordered monolayer films. Still others show additional properties of interest. Given the high feasibility of chemical manipulation of functional groups of the polymer amino acid components before or after synthesis, the potential impact of the biomimetic polymers for application in the areas of electronic, optical, mechanical and medical materials is far-reaching indeed.

## 2.3 Hierarchically Ordered Composite Materials

A composite may be defined as a multiphase material comprising two or more components of different composition which, in combination, exhibit more useful structural or functional properties, or both, relative to performance of any of the constituents individually. Many structural materials in the biological world contain such combinations of distinct phases. Nature is very good at processing composite materials, and at building composite structures, ordered hierarchically through all levels, ranging from molecular through macroscopic scale. Likewise, nature is very good at specifying where and how these composites are used to achieve appropriate combinations of strength, rigidity, and toughness, or other physical properties required for the purpose intended. Proteins, polysaccharides, minerals, and water may not offer much alone in the way of optimization for balance of whichever of the latter properties are needed. But, uniquely combined as constituent phases of a highly ordered composite material, they meet or exceed the particular performance demands for which they have evolved in that form. Thus, for one use, protein fibers might be embedded in a polysaccharide matrix; for another use, polysaccharide fibers might be surrounded by a proteinaceous matrix; for yet another purpose, reinforcement in the form of minerals might be added. Interactions between these phases during their growth are thought to influence the size, orientation, and rate of formation of embedded fibers or particles and consequently to play a key role in determining mechanical properties of the resulting biocomposite structure. These mechanical properties are expressed in complex natural materials as varied as seashells, teeth, bone, wood, and insect exoskeleton. Each of these natural composite systems provides evidence of fine structural control at all levels of synthesis. In that way, they appear to be much more sophisticated than any man-made composite systems. Indeed, they serve as object lessons for scientists interested in improving the versatility and performance capability of synthetic materials and structures by applying some of the same design principles and materials utilization concepts to biomimetic composite fabrication techniques.

### 2.3.1 Nonmineralized Systems for Rigidity or Compliance

To build materials with the proper combination of strength and stiffness, nature has been able to devise the requisite protein–carbohydrate molecular organization and polymer structure in composite systems observed in both the animal and plant kingdoms. Biocomposites containing collagen, the fibrous protein found in animal connective tissue, or cellulose, the structural polysaccharide common in plants, have reached a stage of design at which they are able to withstand complex stresses without suffering damage. In both cases, the organization of inherently weaker subunits into more complex assemblies allows otherwise flexible individual macromolecules to reach the level of intermolecular interaction necessary to sustain rigidity and tissue structural integrity under displacement forces. Variations in composition and conformation even within the same type of composite are widespread because of differences in functional requirements, physical location, and chemical environment. Thus, collagens in different parts of an animal's body are different. Substantial research has been conducted on these various forms of collagen. Structure–function relationships have been thoroughly analyzed in terms of hierarchical levels of organization (Figure 2). Take-home lessons from these studies provide polymer and materials scientists with novel biomimetic concepts for the fabrication of high performance composite polymeric materials.

Proteins other than collagen, and polysaccharides other than cellulose, are used in nature when other composite physical properties are needed. For example, when more extensibility is required, perhaps for mechanical energy storage, rubber- or elasticlike proteins such as resilin or elastin are utilized in natural products. Both these materials are under study for what they may offer in the way of mimicry of their most useful characteristics. Spider silk is another biological material of great interest—its high tensile strength is one of a number of versatile properties. Investigations are under way to thoroughly characterize the molecular biology and genetic engineering of this silk and other silklike proteins and, where possible at this stage, to define biochemically the structural features responsible for the unique physical properties exhibited by these exceptional biological macromolecules. Further studies of these materials, and of some extraordinary protein-based biopolymeric adhesives and coatings as well, are sure to provide insight for future development of biomimetic composite materials in terms of both process and product.

Where mechanical demands call for more protective and supportive roles, a shift is seen in the composition of biological composites, from a relatively pliant matrix toward a stiffer fiber-reinforced matrix. An example of the latter is the insect outer skeleton, or cuticle. This fibrous polymeric composite possesses a very wide range of useful properties, providing stiffness where needed for protection, springiness for powerful jumping relative to weight, and an ability to be stretched without damage. Cuticle is found in extreme abundance in the insect world, evidently because it has been optimized so well for those purposes. Composed of a very stiff polysaccharide fiber, chitin, embedded in a protein matrix in varying concentration, this biological fiber-reinforced composite is structurally very similar to man-made fiber-reinforced composites. As such, it is now the subject of active structural analysis for similarities in design. Perhaps more important, the significant differences that are found in cuticle composite design are now being examined in detail for hints about the design methods nature uses to fabricate such an excellent composite. These include multiangle lamination of unidirectionally oriented sheets, the presence of a continuously variable gradient of matrix cross-link density, and variation of fiber cross-sectional size and shape across composite thickness. Indeed, the fiber–matrix interface in insect cuticle has been found to be superior to synthetic composites in terms of shear strength and toughness, most likely as a result of the totality of microfiber bridging architecture. It is hoped that continued study of the relevance of these and other structural characteristics to mechanical properties will yield fundamental information that can be applied in biomimetic fashion to the synthesis of artificial composites for improved performance precisely in the areas of strength and toughness at low weight.

### 2.3.2 Bioceramic Structures

To construct very stiff protective or supportive materials, or shaped hard tools, which better resist bending or compressive loads, nature introduces a mineral phase into its composite matrix. Biomineralization, the process of incorporation of inorganic substances into composites to form ceramic structures, finds wide distribution in the biological world. Organisms ranging from microorganisms to higher animals make all sorts of bioceramics, either crystalline or amorphous. Examples of intricate shapes abound. The most common amorphous biomineral is silica, found in plants and lower animals. The predominant types of crystalline biomineralization product are calcium phosphate, found in bones and teeth, and calcium carbonate, found in outer protective coverings of marine organisms (e.g., shells and corals). Other bioceramics that occur in some abundance are the oxides and sulfides of manganese, iron, and other heavy metals found primarily in prokaryotes.

Mineralized biocomposite materials are distinguishable from man-made composites by the complexity of their essentially flawless microstructure and by the level of sophistication achieved in their macroscopic morphology. Their structure and organization attest not only to a high degree of spatial regulation of the nucleation and growth of mineral but also to fine control of phase composition and of development of composite microarchitecture. In contrast to techniques for preparation of synthetic ceramic composites, where mineral particles are blended in random dispersive fashion into bulk polymeric resin, the prescribed shape, size, density, and orientation of mineral in the biocomposite matrix all point to the existence of an optimized surface-interactive process whereby ceramic structures form by growth of the inorganic mineral phase in the polymeric matrix under organic substrate control. From the biologically controlled production of structured magnetite particle arrays in magnetotactic bacteria, to the bones of vertebrates and the shells of invertebrates, nature serves us up a feast of examples of how, utilizing simple starting materials, the necessary processes have apparently been optimized to achieve truly unique composite architectures for particular purposes. The question of how nature is able to produce such remarkable ceramic architectures is of great interest. It would be of value to understand better the nature of the interactions between organic and inorganic phases by which the cell regulates the rate and organizes the structure of biomineral growth, and to be able to mimic that process, where feasible, for synthesis of man-made ceramics.

*See also* Biomaterials for Organ Regeneration; Biosensors; Inorganic Solids, Biomolecules in the Synthesis of; Polymers, Genetic Engineering of.

### *Bibliography*


Aksay, I. A., Baer, E., Sarikaya, M., and Tirrell, D. A., Eds. (1992) *Hierarchically Structured Materials,* Vol. 255 in *Materials Research Society Symposium Proceedings.*

Bayley, H. (1991) Monolayers from genetically engineered protein pores. In *Materials Synthesis Based on Biological Processes,* M. Alper, R. Calvert, R. Frankel, P. Rieke, and D. Tirrell, Eds. *Materials Research Society Symposium Proceedings,* 218:69–74.

Benkovic, S. J. (1992) Catalytic antibodies. *Annu. Rev. Biochem.* 61:29–54.

Boxer, S. G. (1990) Mechanisms of long distance electron transfer in proteins: Lessons from photosynthetic reaction centers. *Annu. Rev. Biophys. Biophys. Chem.* 19:277–299.

Clark, D. S. and Estell, D. A., Eds. (1992) *Enzyme Engineering XI.* New York Academy of Sciences, New York.

Cross, R. A., and Kendrick-Jones, J., Eds. (1990) *Motor Proteins. A Volume Based on the EMBO Workshops.* The Company of Biologists Limited, Cambridge.

Ellman, J., Mendel, D., Anthony-Cahill, S., Noren, C. J., and Schultz, P. G. (1991) Biosynthetic method for introducing unnatural amino acids site-specifically into proteins. In *Proteins and Peptides: Principles and Methods,* Vol. 202, in *Methods in Enzymology.* Academic Press, San Diego, CA, pp. 301–337.

Nakamura, R. M., Kasahara, Y., and Rechnitz, G. A., Eds. (1992) *Immunochemical Assays and Biosensor Technology for the 1990s.* American Society for Microbiology, Washington, DC.

Oesterhelt, D., Bräuchle, C., and Hampp, N. (1991) Bacteriorhodopsin: A biological material for information processing. *Q. Rev. Biophys.* 24(4):425–478.

Rizo, J., and Gierasch, L. M. (1992) Constrained peptides: Models of bioactive peptides and protein substructures. *Annu. Rev. Biochem.* 61:387–418.

Sleytr, U. B., Prim, D., Sára, M., and Messner, P. (1992) Two-dimensional protein crystals as patterning elements in molecular nanotechnology. In *Molecular Electronics—Science and Technology,* A. Aviram, Ed., pp. 167–177.

Steinbüchel, A. (1991) Polyhydroxyalkanoic acids. In *Biomaterials: Novel Materials from Biological Sources,* D. Byrom, Ed. Macmillan, Basingstoke, pp. 123–213.

Tirrell, D. A., Fournier, M. J., and Mason, T. L. (1991) Protein engineering for materials applications. *Curr. Opinion Struct. Biol.* 1:638–641.

Whitesides, G. M., Mathias, J. P., and Seto, C. T. (1991) Molecular self-assembly and nanochemistry: A chemical strategy for synthesis of nanostructures. *Science* 254:1312–1319.


**Biomolecular Computing:** ***see*** **Computing, Biomolecular**

# Biomolecular Electronics and Applications

*Elias Greenbaum*

### *Key Words*

**Bacteriorhodopsin** The pigment protein complex of the bacterium *Halobacterium halobium.*

**Bioassay** Analytical chemistry procedure for determining biomolecules.

**Biphotonic** Requiring two photons.

**Lumen** The inner region of photosynthetic membranes.

**Photoisomerization** Isomerization triggered by light. Isomerization, the internal rearrangement of atoms within a molecule, changes the structure but not the chemical formula of the molecule.

**Photosynthesis** The biological process in which water, carbon dioxide, and light are converted into stored chemical energy of plant matter.

**Photosystem** The system that converts light energy into chemical energy by photosynthesis; the two photosystems in higher plants are photosystems I and II.

**Porphyrin** A light-sensitive molecule commonly found in biological systems.

**Reaction Center** The locus of conversion of light energy into chemical energy in photosynthesis. ''Reaction center'' and ''photosystem'' are sometimes used interchangeably.

**Redox Enzyme** Enzyme that performs its function by cycling through reduction and oxidation states, which involves accepting and donating electrons.

**Stroma** The outer region of photosynthetic membranes.

***trans*-Retinal** The initial configuration of bacteriorhodopsin (prior to light absorption).

**Tunneling** According to quantum mechanics, the process in which an electron, for example, can move from one energy state to another of equal or lower energy by penetrating a barrier of higher energy. Such a process is forbidden according to classical (Newtonian) physics.

---

This contribution provides a brief introduction and overview of the principles of biomolecular electronics, techniques and applications, and future prospects.

## 1 INTRODUCTION

The term ''biomolecular electronics'' evokes two powerful images of twentieth-century science and technology. First, the ''bio'' prefix explicitly acknowledges the molecular basis and understanding of living-state systems. It is this understanding that comprises the foundation for molecular biology, immunology, and genetic engineering. Second, ''electronics,'' in the context of technology, is understood as electronic devices, starting from vacuum tubes and progressing through transistors, integrated circuits, and the semiconductor electronics industry comprising communications and consumer electronics.

The conflation of these two terms into ''biomolecular electronics'' implies a radical concept: the construction of practical electronic devices from biomolecular components. Whereas the desired device-producing properties of semiconductors are derived from the *bulk* properties of appropriately doped material, in biomolecular electronics, the imputed device-producing property and the potential for high density packing are intrinsic characteristics of individual molecules or ensembles of molecules.

## 2 PRINCIPLES OF BIOMOLECULAR ELECTRONICS

The principles of biomolecular electronics can be illustrated by two examples: photosynthesis and bacteriorhodopsin. Although photosynthesis is generally understood to be the process in which light energy is converted into chemical energy for plant growth, modern research has produced a precise understanding of the molecular architecture of the pigment protein complexes that comprise the photosynthetic reaction center. In fact, for the photosynthetic bacterium *Rhodopseudomonas viridis,* the reaction center has been crystallized and the three-dimensional structure is known. Higher plant photosynthesis consists of two reaction centers, photosystems I and II, connected in series by an electron transport chain. Figure 1 is a schematic illustration of the structure of photosystem I. The reaction centers of photosynthesis are paradigms of biomolecular optoelectronic devices. For each photon that is absorbed in the reaction center, an electron is ejected from the lumen side to the stroma side, thereby creating a voltage difference across the membrane. The magnitude of this voltage is approximately 1 V. The characteristic response time of the forward electron transfer is $\approx$ 5–10 ps, and, as can be seen in Figure 1, the characteristic linear dimension is $\approx$ 5 nm. This is a remarkable structure from a molecular electronics perspective: extremely small, extremely fast, and possessing formidable electrical properties.

A second biomolecular optoelectronic system is the pigment protein complex bacteriorhodopsin, which has potential in protein-based optical computing and optical memories. Bacteriorhodopsin (Figure 2) is the light-harvesting protein in the membrane of *Halobacterium halobium.* Unlike the photosynthetic reaction center, which generates a voltage by an electron transfer charge separation reaction, bacteriorhodopsin pumps a proton across the membrane, thereby creating a chemical and osmotic potential which, in the living organism, serves as a source of energy under low-oxygen conditions. Bacteriorhodopsin in its resting state is all-*trans*-retinal. It is bound to the protein matrix by a linkage to a lysine residue. Upon absorption of a photon, a shift in the distribution of electron density occurs, triggering a rotation around the double bond between C-13 and C-14. This process, photoisomerization, causes a vectorial shift in positive charge perpendicular to the membrane and generates a signal in $\approx$ 5 ps. Biomolecular electronic devices using bacteriorhodopsin have potential applications as spatial light modulators (optical computers) and memory devices. Two-dimensional optical processing systems are capable of performing mathematics and information processing for such practical applications as pattern recognition, image processing, equation solving, and linear algebra. Thin films of bacteriorhodopsin have also been proposed as the photoactive components in memory devices. From the foregoing examples, it is clear that biomolecules can perform specific devicelike electronic functions and can be thought of as intelligent materials and microstructures.

## 3 TECHNIQUES AND APPLICATIONS

In the fabrication of biomolecular electronic devices, it is essential to make direct electrical contact with a specific electroactive structure. One method of making direct electrical contact with the electron-emerging region of the photosystem I reaction center involves the precipitation of metallic platinum at the reducing end of photosystem I. The key advance in this work is the achievement of

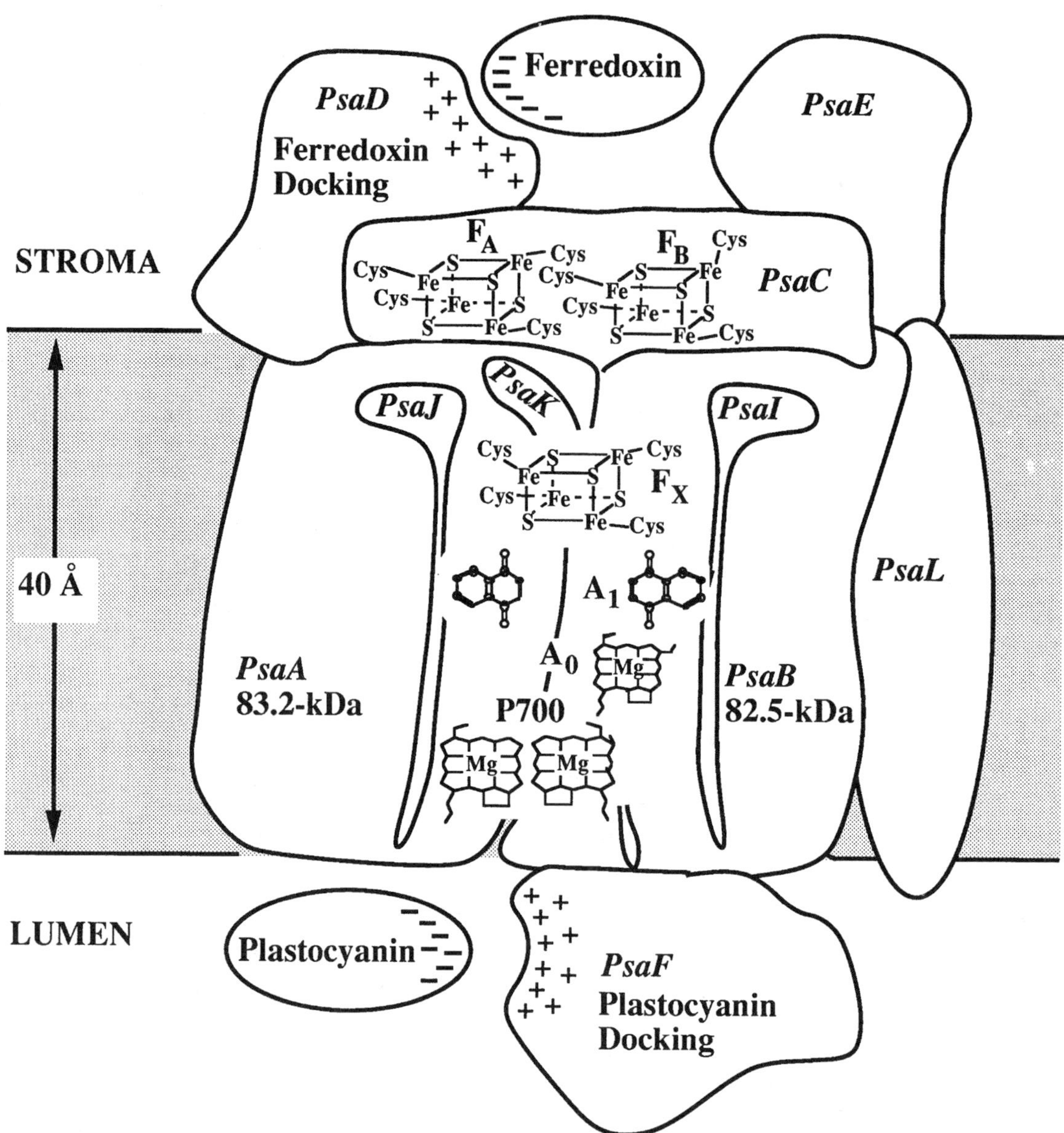


**Figure 1.** Schematic illustration of the molecular architecture of the photosystem I reaction center of photosynthesis (linear dimension in Ångstroms. 1nm = 10Å). Electrical contact with platinum metal contact has been achieved at the ferredoxin docking site. Similar contact is potentially achievable at the plastocyanin docking site. (After J. H. Golbeck, *Annu. Rev. Plant Physiol. Plant Mol. Biol.* 43:293. Copyright 1992, Annual Reviews, Inc. Used with permission.)

metal contact at pH 7 and room temperature, experimental parameters that are compatible with the preservation of biological function. This result has been experimentally verified. A schematic illustration of the stroma–platinum contact adjacent to the ferredoxin docking site is illustrated in Figure 1. Also illustrated is a potential platinum contact at the plastocyanin site. When this contact is achieved, complete electrical communication with isolated photosystem I reaction centers will have been obtained.

Molecular electronic devices should be connected with molecular wires. The construction of a picosecond optical switch based on biphotonic excitation of an electron donor–acceptor–donor molecule illustrates an approach to the construction of molecular wires and spacers. As illustrated in Figure 3, this optical switch consists of two porphyrin donors rigidly attached to the two-electron acceptor *N,N′*-diphenyl-3,4,9,10-perylenebis (dicarboximide). In this structure, the porphyrin molecules at each end are the light receptors. Excitation with subpicosecond laser pulses results in the transfer of one or two electrons from the porphyrins to the electron acceptor, depending on the light intensity. A single electron donated to the electron acceptor produces a strong optical absorbance at 713 nm, whereas two electrons produce the change at 546 nm.

For redox enzymes to function in biosensor biomolecular electronic devices, they must be appropriately wired. Because for most enzymes the redox active centers are buried in the protein, hence are not easily accessible, one possible approach is to chemically modify the enzyme so that it retains its catalytic properties while acquiring the ability to conduct electrons to the electrode.

Both scanning tunneling microscopy and atomic force microscopy play important roles in the construction as well as characterization of molecular and biomolecular electronic devices, although,

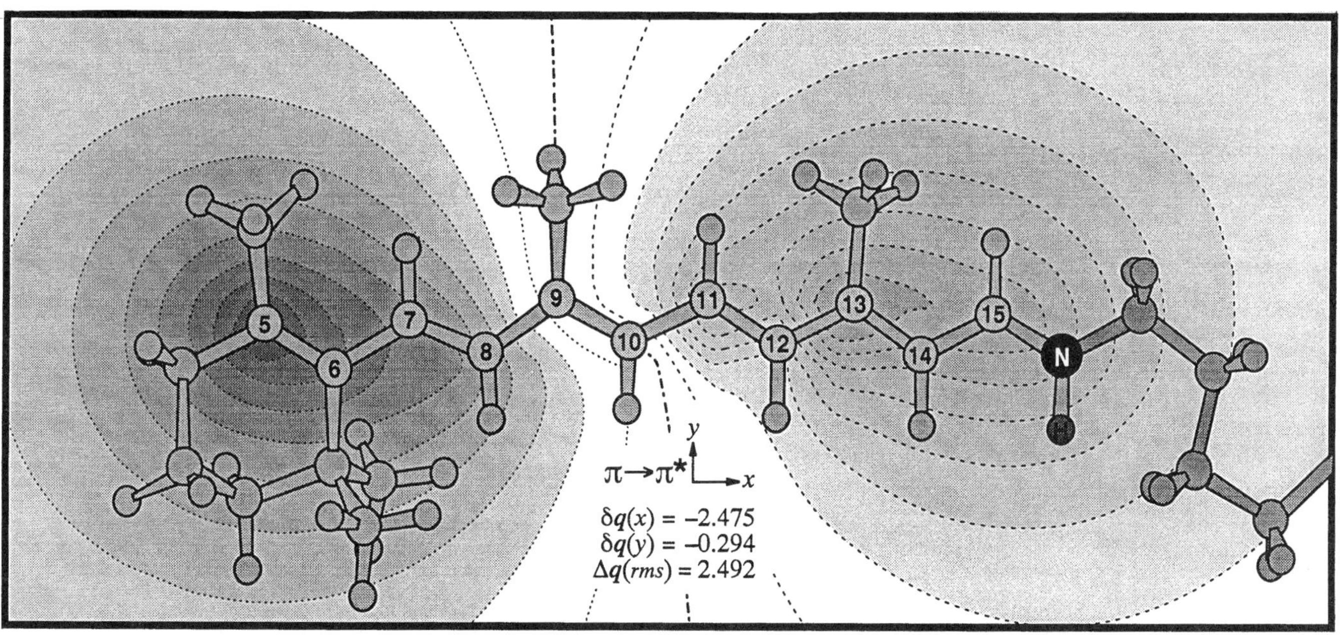


**Figure 2.** Molecular model of the retinal chromophore in light-adapted bacteriorhodopsin. The contour lines indicate the shift in charge distribution following electronic excitation. (R. R. Birge, *Computer* 25:56. Copyright 1992, Institute of Electrical and Electronics Engineers, Inc. Used with permission.)

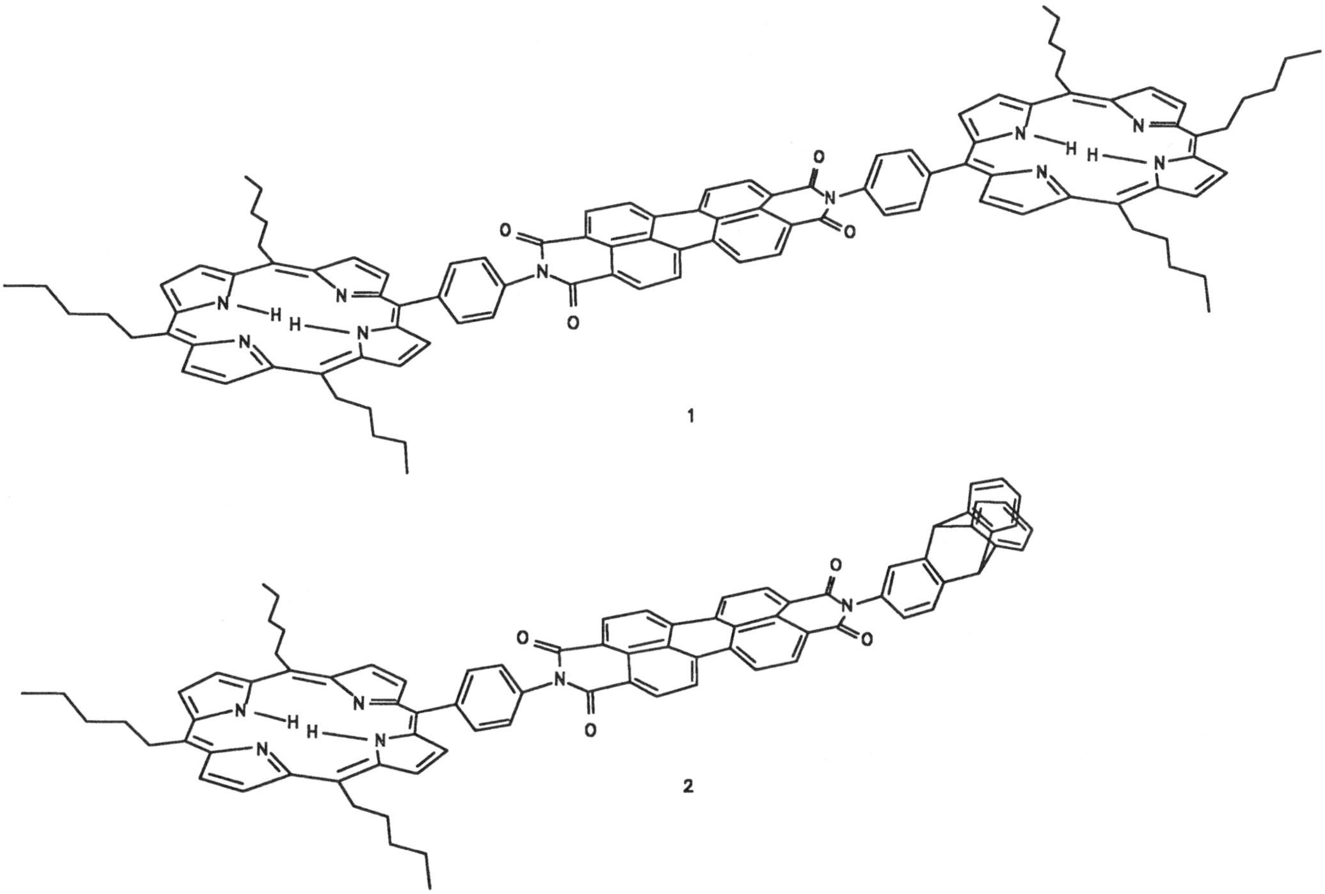

1

2

**Figure 3.** Molecular model of an electron donor–acceptor–donor molecule. The porphyrins at each end of the molecule are light-activated electron donors that donate electrons to the central electron acceptor. (M. P. O'Neil et al., *Science,* 257:63. Copyright 1992, American Association for the Advancement of Science. Used with permission.)

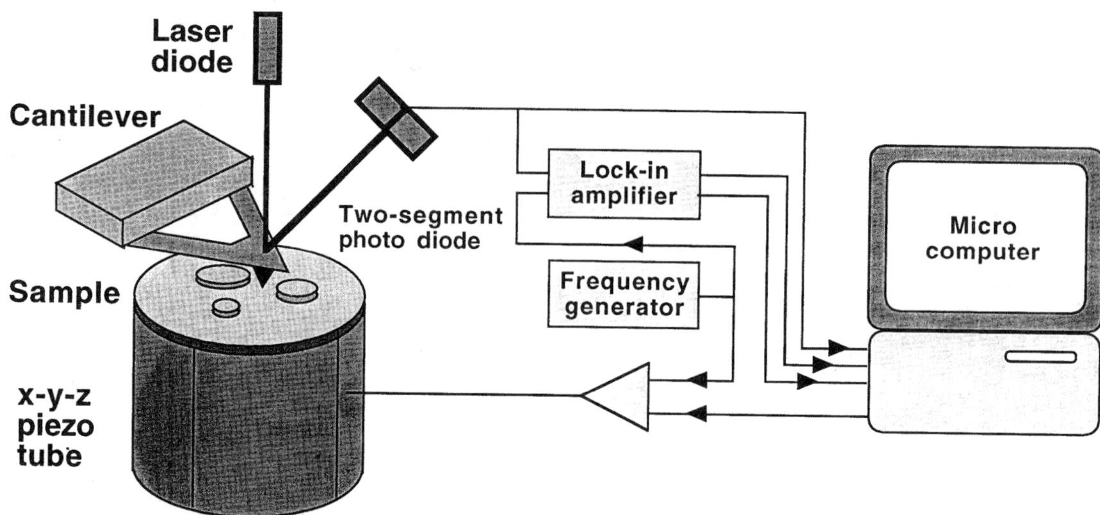


Figure 4. Schematic illustration of an atomic force microscope, which can be used to image surfaces with molecular dimension resolution. (M. Radmacher et al., *Science*, 257:1900. Copyright 1992, American Association for the Advancement of Science. Used with permission.)

depending on the masses of the biomolecules involved, construction capability may be restricted. The optoelectronic devices just described require no batterylike power source because the absorption of light triggers the requisite reactions. However, some devices may be powered by nanobatteries that are 70 nm diameter. One of these silver–copper devices is about the size of a common cold virus. The nanobattery was developed in the course of research using scanning tunneling microscopy to build ultrathin metallic structures.

Unlike conventional microscopes, which use light or the wavelike quantum mechanical properties of electrons and focusing optics to produce visual information on the structures under study, scanning tunneling microscopy and the related atomic force microscopy techniques produce surface replications of the sample based on the phenomenon of quantum mechanical tunneling. This technology can be used to image soft samples such as molecules and cells (Figure 4). A refinement of this idea leads to the concept of a molecular microscope that combines atomic force microscopy and magnetic resonance imaging techniques to image individual molecules in their normal environment (Figure 5). The proposed technique combines the atomic scale resolution of atomic force microscopy with three-dimensional magnetic resonance imaging structural information on the types and positions of nuclei within a single molecule.

One of the most popular techniques for molecular insertion and orientation is the use of Langmuir–Blodgett films (Figure 6). If an insoluble monolayer of hydrocarbon is spread on the surface of water, it can be lifted off the surface and deposited onto a solid

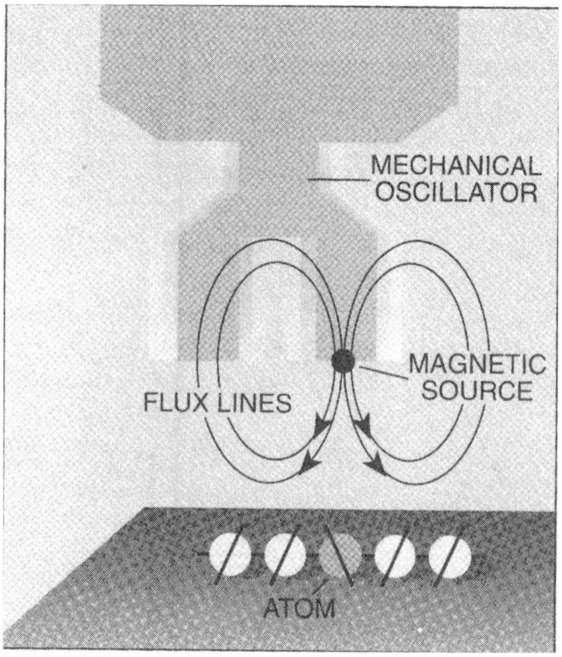

Figure 5. Conceptual illustration of a "molecular microscope." This idea combines the resolution of atomic force microscopy with the structural information obtained with magnetic resonance imaging. (J. Sidles, *Science*, 257:750. Copyright 1992, American Association for the Advancement of Science. Used with permission.)

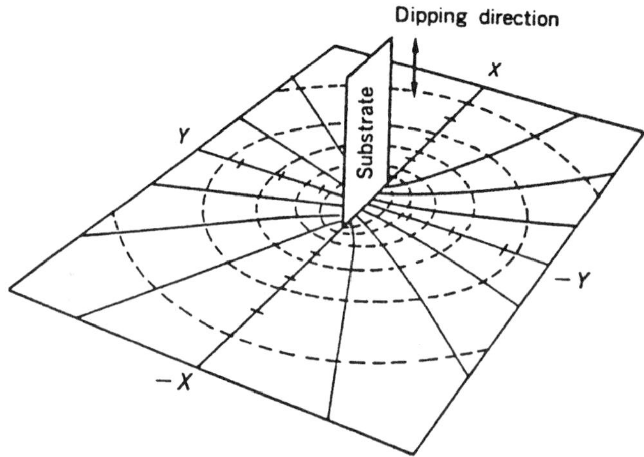


Figure 6. Schematic representation of the Langmuir–Blodgett deposition process. The equipotential and stream lines are illustrated by dashed and solid lines, respectively. (M. Sugi, *FED J.* 2(suppl.):43. Copyright 1992, Research and Development Association for Future Electron Devices. Used with permission.)

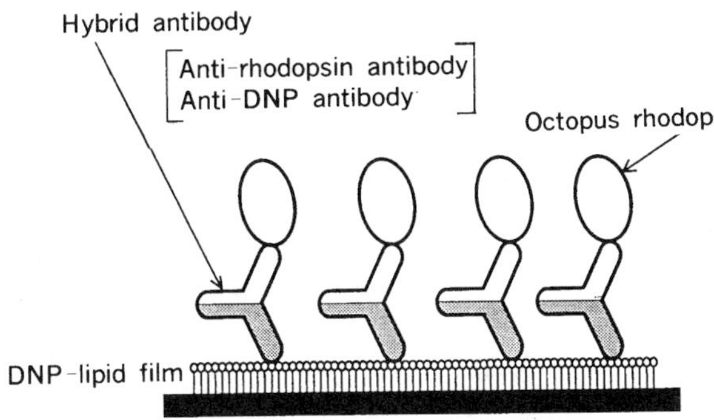


Figure 7. Example of the Langmuir–Blodgett deposition process: ordered rhodopsin molecules using hybrid antibodies; DNP is deoxyribonucleic protein. (T. Ishibashi, *FED J.* 2(suppl.):53. Copyright 1992, Research and Development Association for Future Electron Devices. Used with permission.)

substrate. One popular substrate that is used in this technique is glass. By incorporating appropriate biomolecular-active species in the organic film, two-dimensional-array devices can be constructed. One such example is a two-dimensional protein film using hybrid antibodies (Figure 7). Such "biochips" can, in principle, be used for information processing. The first step in fabricating such systems is developing a method for two-dimensional ordering of functional biomolecules on a substrate. This can be achieved by constructing thin protein films using the Langmuir–Blodgett technique followed, for example, by antigen–antibody reactions. Figure 7 illustrates the ordering of rhodopsin molecules using this technique.

## 4 FUTURE PROSPECTS

It can be seen that biomolecular electronics is a scientific and technological discipline that is in its infancy. However, significant prospects exist for interesting and important applications, of which light detection, optical switching, information storage processing, optical computing, spatial addressing, molecular rectifiers, image detection, and tactile and olfactory devices are but a few.

*See also* BACTERIORHODOPSIN-BASED ARTIFICIAL PHOTORECEPTOR; BIOSENSORS; COMPUTING, BIOMOLECULAR.

### *Bibliography*

*Advanced Materials for Optics and Electronics* (formerly *Journal of Molecular Electronics*. Wiley, Chichester.

Aviram, A. (1992) *Molecular Electronics—Science and Technology.* American Institute of Physics, New York.

Birge, R. R., Ed. (1993) *Molecular and Biomolecular Electronics,* ACS Advances in Chemistry Series. American Chemical Society, Washington, DC.

Bloor, D., Bryce, M., and Petty, M., Eds. (1994) *Introduction to Molecular Electronics.* Edward Arnold, Dunton Green, England.

Carter, F. L., Ed. (1982) *Molecular Electronic Devices.* Dekker, New York.

Carter, F. L., et al., Eds. (1988) *Molecular Electronic Devices.* Elsevier, New York.

*FED Journal,* Vol. 2, Supplement (1992). Shuku Maeda for the Research and Development Association for Future Electron Devices, Tokyo, Japan.

Kuhn, H. (1967) *Naturwissenschaften,* 54: 429.

## Biomolecules in the Synthesis of Inorganic Solids: *see* Inorganic Solids, Biomolecules in the Synthesis of.



# BIOORGANIC CHEMISTRY

*Simon H. Friedman and George L. Kenyon*

### *Key Words*

**Catalytic Antibodies** Antibodies that have been generated to catalyze specific chemical reactions.

**Enzyme Inhibitors** Molecules that are able to prevent enzyme catalytic action.

**Phosphorothioate-Containing DNA** A nuclease-resistant form of DNA in which one of the two unesterified phosphate oxygens is replaced with a sulfur atom.

**Solid Phase Oligonucleotide Synthesis** The primary technique used for chemical synthesis of oligonucleotides in which the oligonucleotide is tethered at one end to an insoluble resin.

**Transition State Analogues** Molecules that closely resemble the shape and charge of a reaction's transition state (more accurately known as the activated complex).

Bioorganic chemistry is the study of biological systems utilizing the tools of the organic chemist, usually including a synthetic component. Bioorganic chemists pursue a wide range of activities including the study of biomechanisms, enzyme models, biosynthesis, biomimetic synthesis, molecular recognition, enzymology, peptide chemistry, nucleic acid chemistry, immunology, and the design and synthesis of therapeutic agents. Within the past decade, the classical distinctions among scientific disciplines have gradually faded, and frequently one finds the synthetic organic chemist making compounds of biological interest and, conversely, the biochemist using generally simpler synthetic techniques to aid in answering biological questions. This presentation of recent highlights of bioorganic chemistry, ranging from the more organic to the more biochemical, emphasizes areas of greatest interest to molecular biologists.

## 1 NUCLEOTIDE INTERACTIONS

A wide variety of bioorganic investigations deal with molecules that interact with DNA. The purposes behind these investigations are manyfold, ranging from the development of therapeutics to the understanding of the forces required for molecular recognition. Section 1.1 deals with synthetic oligonucleotides interacting with DNA; Section 1.2 discusses small molecules that interact with DNA.

### 1.1 OLIGONUCLEOTIDES

The revolution in recombinant DNA relies critically on the ability to make a variety of oligonucleotides quickly and cheaply. The development of solid phase oligonucleotide synthesis has made

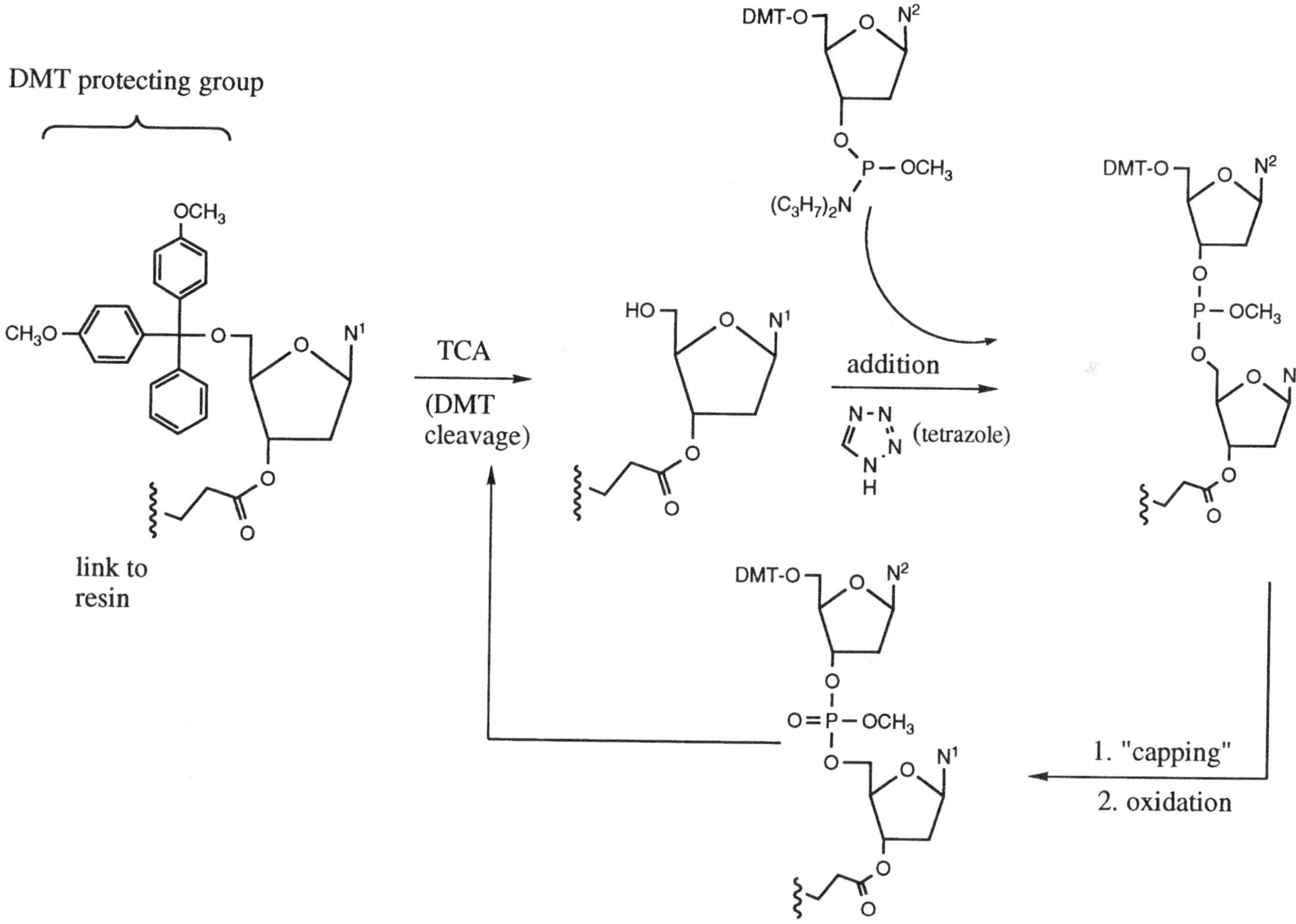


**Figure 1.** Oligonucleotide synthesis usually begins with the tethering of the first monomer to an insoluble resin via an ester linkage to its 3′ hydroxyl group. To prevent esterification of the 5′ hydroxyl, it is blocked with a dimethoxytrityl (DMT) group. After the ester linkage to resin has been formed, the DMT group is cleaved with trichloroacetic acid (TCA). The next monomer (also 5′-hydroxyl DMT protected) contains a phosphoramidite group attached to its 3′ hydroxyl. The displacement of the phosphoramidite amine by the deprotected 5′ hydroxyl is catalyzed by the addition of the mild acid tetrazole. "Capping" or acetylation of the 5′ hydroxyl group of any monomer that did not react with the phosphoramidite prevents errors from compounding. The capping prevents the unreacted, and therefore incorrect, oligomer from undergoing further elongation. After the coupling between monomer has taken place, the phosph*ite* triester is oxidized to form the phosph*ate* triester using either hydrogen peroxide or iodine. The DMT group is cleaved, thereby releasing the 5′ hydroxyl group, which can then participate in the next round of monomer addition. When the oligomer is complete, a final step of deprotection removes the final DMT group and any protecting groups on the nucleotide bases and converts the phosphate triester into the naturally occurring diester.

this process relatively simple. In addition to the many uses of these oligonucleotides in recombinant DNA manipulation [ranging from use as polymerase chain reaction (PCR) primers to the generation of site-directed mutants], they have potential as therapeutics in the field of antisense therapy. Antisense therapy involves the inhibition of gene expression by treatment with oligonucleotides complementary to mRNAs or DNA. Paralleling that used in solid phase peptide synthesis, a common synthetic strategy that is used in oligonucleotide synthesis is depicted in Figure 1.

There are many variations on this chemistry, not only for the production of natural DNA but also for the production of unnatural polynucleotides, which have alternative backbone and nucleic acid structures. An example of the latter is phosphorothioate-containing DNA, in which one of the two unesterified phosphate oxygens is replaced with a sulfur. This exchange may be achieved by replacing the hydrogen peroxide or iodine in the phosphite oxidation step shown in Figure 1 with a solution of elemental sulfur. Phosphorothioate DNA has been found to have greater resistance to nucleases (enzymes that cleave the diester backbone of DNA), which makes it more suitable for antisense therapy than naturally occurring DNA.

### 1.2 Small Molecules

Literally hundreds of small molecules have been found to have interesting or useful effects that depend on their binding to DNA. Uses of these compounds include antiviral and antineoplastic (anticancer) chemotherapies (e.g., bleomycin), nucleotide stains (e.g., ethidium bromide), and biochemical probes for nucleic acid structure (e.g., phenanthroline complexes). The major interaction modes are binding to the minor and major grooves and intercalation between stacked base pairs, as well as a host of potential covalent modifications (including bond cleavage and alkylation) of the nucleotides, sugars, and phosphate backbone. A well-studied DNA-binding small molecule is the natural product distamycin (Figure

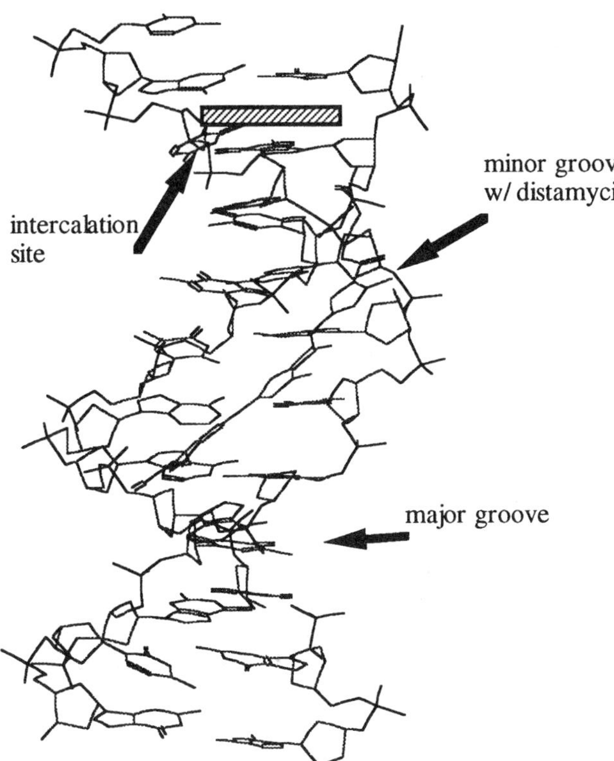


**Figure 2.** Major sites of small molecule interaction with nucleotides.

2), which has been shown to bind noncovalently and selectively to the minor groove of AT-rich portions of DNA. In addition to having antiviral properties, which probably are linked to its ability to inhibit the enzyme DNA polymerase, distamycin has found use as a template for designing molecules with altered sequence specificity and sequence-specific cleavage.

## 2 PROTEIN STUDIES

### 2.1 Enzyme Inhibition

The study of enzyme inhibition is a major focus of biological chemistry for two major reasons. First, enzyme inhibitors are useful tools for understanding enzyme structure and mechanism of action. Second, enzyme inhibition has proven to be a successful strategy for the design of therapeutics. Examples of drugs that act by inhibiting enzymes include methotrexate, Deprenyl, and Enalapril. Inhibitors can block enzyme function by preventing access of the natural substrate to the active site or by stopping subsequent reaction from taking place by, for example, chemically blocking a key catalytic group on the enzyme.

There are two broad classes of inhibitors, covalent and noncovalent, the distinction being whether a covalent chemical bond does or does not form between enzyme and inhibitor. These classifications may be further divided into categories based on kinetic and mechanistic criteria. Major subclasses include transition state analogues, multisubstrate analogues, affinity labels, and mechanism-based inhibitors.

The strength of the binding of inhibitor with enzyme, indeed the interaction of any two molecules, depends primarily on the sum of four main types of force: (1) hydrophobic interactions, which entail the tendency of nonpolar surfaces to associate in aqueous solution, (2) van der Waals interactions (i.e., attractive forces between closely associated but uncharged groups caused by momentary fluctuations in charge density), (3) hydrogen bonding, a principally electrostatic interaction between hydrogen atoms that are bonded to electronegative atoms such as nitrogen and oxygen and a lone electron pair that resides on another electronegative atom, and (4) electrostatic or charge–charge interactions. Water plays a crucial role in modulating the strength of all these forces. For example, to separate a closely associated carboxylate anion and ammonium cation in vacuo (i.e., with no solvent) requires ~60 kcal/mol, whereas to separate the two while in contact with water requires ~5% of that. The reason for this is that while the anion and cation have a strong mutual charge–charge electrostatic attraction, the ions also have an almost equally strong charge–dipole interaction with water.

### 2.2 Catalytic Antibodies and Transition State Analogues

Catalytic antibodies, also known as abzymes, are antibodies capable of catalyzing specific chemical reactions. They are able to perform this task because they have been elicited against special antigens known as transition state analogues. These are molecules that closely resemble in shape and charge the reaction's transition state (i.e., the highest energy species in the reaction pathway, more accurately, but less widely known, as an activated complex). Enzymes are observed to bind tightly not to the substrates and the products of the reaction they catalyze, but rather to the transition state, the highest energy species that forms during the reaction. Enzymes have evolved to use this strategy for catalysis. By forming an active site that is complementary in shape and charge to the transition state, enzymes stabilize or lower the energy of this species. The effect of this is to accelerate the reaction because the activation barrier is more easily overcome. This complementarity in charge and shape to the transition state is readily demonstrated by the effectiveness of transition state analogues as enzyme inhibitors. A classic example of a transition state analogue is a phosphonate-containing molecule that mimics the transition state of the hydrolysis of the corresponding acyl derivative (Figure 3). When an ester is hydrolyzed, the central carbonyl group changes from being planar ($sp^2$ hybridization of the carbonyl carbon) to being tetrahedral ($sp^3$ hybridization). The phosphonate-containing analogue is able to reproduce the shape of the transition state as well as the partially negatively charged oxygens. In addition, it is chemically stable, whereas the actual transition state exists only fleetingly. In this example, an enzyme that catalyzed the hydrolysis of this ester would also bind tightly to the phosphonate analogue.

The general strategy behind the production of a catalytic antibody is to (1) design and synthesize a molecule whose shape and charge closely resemble those of the transition state of the reaction one wishes to catalyze, (2) tether this molecule to a larger molecule, (3) elicit an immune response to this conjugate, and (4) screen the resultant monoclonal antibodies for catalytic activity of the type desired. Antibodies that are elicited against the transition state analogue and therefore have the ability to bind them will be chemi-

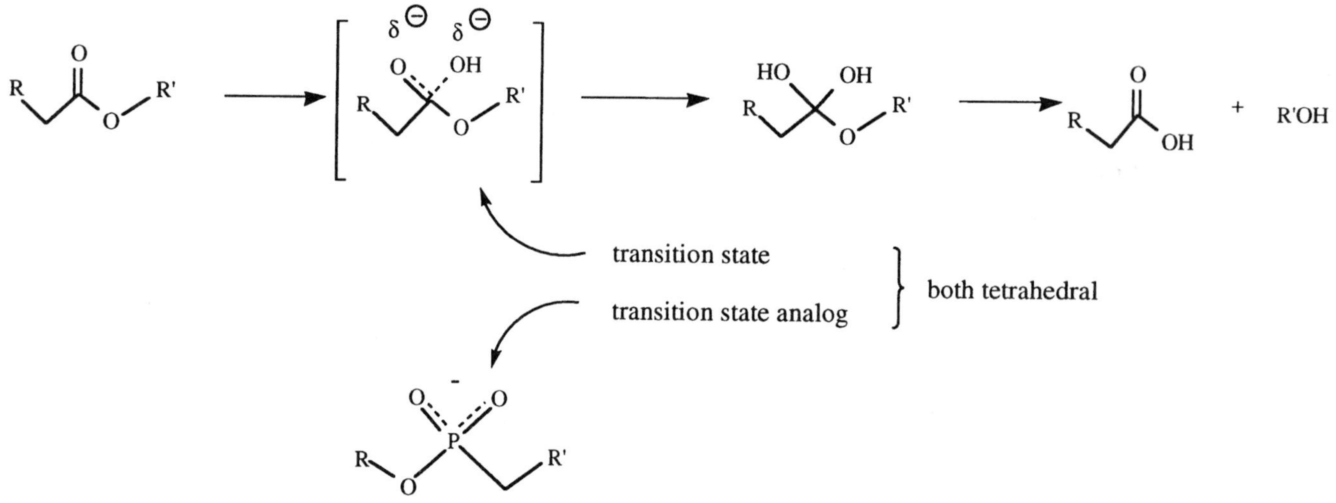

Figure 3. Comparison of transition state structure and transition state analogue structure for an ester hydrolysis.

cally and sterically complementary to the transition state and will therefore be potentially capable of catalyzing the reaction.

## 3 SUMMARY

We have briefly highlighted areas of bioorganic chemistry of general importance and of special interest to the molecular biologist. Bioorganic chemistry makes connections with a very broad range of specialties, ranging from physical chemistry to biochemistry. Obviously, we have just scratched the surface of these topics in this vital and rapidly expanding discipline.

*See also* Drug Synthesis; Oligonucleotide Analogs, Phosphate-Free; Receptor Biochemistry.

### *Bibliography*


Benkovic, S. J. (1992) Catalytic antibodies. *Annu. Rev. Biochem.* 61:29–54.

Breslow, R. Ed. (1986) *International Symposium on Bioorganic Chemistry* New York Academy of Sciences, New York.

Cohen, J. S. (1991) Oligonucleotides as therapeutic agents. *Pharmacol. Ther.* 52:211–225.

Dugas, H. (1989) *Bioorganic Chemistry: A Chemical Approach to Enzyme Action.* Springer-Verlag, New York.

Muscate, A. Levinson, C. L., and Kenyon, G. L. (1993) Enzyme inhibitors. In *Encyclopedia of Chemical Technology,* J. I. Kroschwitz, Ed. Wiley, New York.

Nielsen, P. E. (1990) Chemical and photochemical probing of DNA complexes. *J. Mol. Recognition,* 3:1–25.

Silverman, R. B. (1992) *The Organic Chemistry of Drug Design and Drug Action* Academic Press, San Diego, CA.


# Bioprocess Engineering

*Kimberly L. Ogden and Milan Bier*

### *Key Words*

**Bioprocessing** The engineering component of the commercial exploitation of biological materials, living organisms, and their activities.

**Hybridoma** Cell line produced by fusing a myeloma (cancer) cell with a lymphocyte; used to produce antibodies.

**Recombinant DNA** A general term for the result of laboratory manipulation in which DNA molecules or fragments from various sources are severed, combined enzymatically, and reinserted into host organisms.

**Transgenic** Describes animals created by introducing new DNA sequences into the germ line via addition to the egg.

---

Bioprocessing has been defined as the engineering component of the commercial exploitation of biological materials, living organisms, and their activities. It is a diverse field requiring the collaboration of scientists, physicians, and engineers working in the areas of biology, biochemistry, genetics, medicine, agriculture, materials science, chemical and electrical engineering, and more recently, environmental sciences. The biotechnology industry is as diverse as the disciplines that are involved in the fundamental research. Current biotechnology products include pharmaceuticals, fine chemicals, and agricultural and food products. In addition, new technologies being developed in the areas of environmental remediation, energy, biomaterials, and bioelectronics will one day increase the scope of biotechnology products even further. The current worth of the biotechnology industry is estimated to between \$7 billion and \$15 billion, and this amount is expected to increase to at least \$50 billion by the year 2000.

## 1 INTRODUCTION

The diverse roots of bioprocess engineering are found in the food, tobacco, tea, leather, and other ancient industries. Of particular relevance is the fermentation industry, represented not only by beer brewing, but also by the production of antibiotics, citric acid, various amino acids, and other biomolecules. The art of yeast and bacterial fermentation is therefore well established. Current frontiers are in the culture of mammalian, insect, and plant cells.

Another important precursor of present-day bioprocessing is the fractionation of human plasma proteins for the production of its components: mainly serum albumin, immunoglobulins, fibrinogen, and various clotting factors. Usage of purified specific proteins is a far more efficient utilization of human plasma—an expensive national resource—than total plasma transfusion. The industry originated during World War II and is based on fractional precipitation of plasma component proteins by the addition of ethanol under controlled conditions of temperature, pH, and ionic strength. The ethanol process was developed by Cohn and his numerous associates and is used on a worldwide basis with remarkably few changes since its introduction decades ago.

Modern purification of therapeutic proteins that are produced through recombinant DNA technology is not based on ethanol precipitation, but mainly relies on chromatography. Nevertheless, protein purification benefited enormously from the older plasma fractionation industry, which developed such essential ancillary equipment and processes as centrifuges, filters, freeze-dryers, sterile operation, pyrogen testing, and quality control. Plasma fractionation still processes much larger volumes of proteins and at a lower cost than any other therapeutic protein processing.

The biotechnology industry currently relies on several complementary technologies. The backbone of the industry is the use of recombinant proteins, obtained by inducing a host organism to express foreign proteins. To this end, a gene regulating the production of this protein is inserted into a plasmid, an autonomous, self-replicating piece of DNA that is inserted into a living cell. The host organism can be a bacterium, a yeast, or a mammalian or insect cell, as well as an intact plant or higher animal. Transgenic plants and animals may eventually prove to be the most economical source of recombinant proteins. The plasmid can be constructed to cause the cell to manufacture a great quantity of the foreign protein (up to 50 g of protein per liter of cell medium). The excess protein is in many cases expressed in a denatured form as a densely aggregated particle. This step facilitates the purification of the protein, but it is then necessary to solubilize and refold the protein in the biologically active form—an often daunting problem, making the choice of both plasmid and host of utmost importance. The formulation of the original gene (i.e., the sequence of the nucleotides forming the DNA) is greatly facilitated by the polymerase chain reaction (PCR) method, first developed in 1984–1985 by the Cetus Corporation and commercialized by Perkin-Elmer. PCR is an enzymatic process that is carried out in a discrete cycle of amplification, each cycle doubling the amount of available DNA. Thus, minute quantities of a rare target sequence can be amplified to produce adequate quantities with excellent faithfulness. Specific sequences of DNA can also be chemically synthesized using solid phase procedures.

A second important technology is the production of monoclonal antibodies. Sensitized antibody-producing spleen cells are hybridized with stable cancer cell lines; the so-formed hybrid has the longevity of typical cancer cells in tissue culture and the specificity of the original splenocyte. The reproducible character of the monoclonal antibodies has permitted the development of new diagnostic procedures of exquisite specificity and also has the potential of therapeutic use. While recombinant proteins can be expressed in bacteria and other host organisms, the production of hybridoma cells requires more stringent—and far more expensive—mammalian tissue culture procedures.

A third major branch of biotechnology is the direct chemical synthesis of peptides using the solid phase methodology first developed by Bruce Merrifield, a Nobel laureate. The first amino acid of a desired peptide sequence is coupled to a latex particle, and subsequent amino acids are coupled to it, step by step, in highly automated instruments. There is a lively competition between direct peptide synthesis and recombinant methods for the more economic approach to peptide production. The direct synthesis yields an easier approach for the synthesis of families of peptides differing only in a single amino acid—a point mutation. A newly developed computer-controlled apparatus permits simultaneous production of 12 individually programmed peptide sequences. These are used to determine the essential sequence or part of a sequence or part of a sequence of a biologically active peptide.

Detailed descriptions of all these and other current processes is beyond the scope of this brief review. Therefore, the key features of only four aspects of bioprocessing are presented: enzyme engineering, bioreactors, receptor and cell transport, and bioseparations or downstream processing.

## 2 FOUR ASPECTS OF BIOPROCESSING

### 2.1 Enzyme Engineering

Enzymes are an alternative to live cells for synthesis of biological products. The applications vary from alternative fuel synthesis to environmental remediation to biosensors to food processing. The most common reactor configuration is a packed or fluidized immobilized enzyme reactor. This system has the advantage of being a continuous process able to handle high throughputs of process streams. Enzymes are reasonably protected in this environment, as well. Cellulase is an example of an enzyme used in an immobilized configuration for the conversion of biomass to feedstock for ethanol in the production of alternative fuels. In addition, these systems are beginning to prove useful in the environmental area for the selective removal of heavy metals or the transformation of hazardous substances, such as organics, into nontoxic compounds.

Enzyme technology is exploited in the area of biosensors. Enzyme-coated electrodes provide a highly selective and sensitive method for determining the amount of a given substrate. Examples of electrodes include sensors for common substrates like glucose, urea, and nitrate as well as sensors for fermentation products and intermediates like amino acids, lactic acid, penicillin, and alcohols. The future use of these specific electrodes will facilitate control of bioprocesses, and thus, higher productivity in all aspects of bioprocessing.

A more recent development is the use of ribonucleic acid (RNA) as a catalyst inhibitor or "antisense" blocker of gene expression. Immobilized enzyme and template technology is being studied to facilitate the transcription in vitro of RNA in a continuous reactor. Start-up biotechnology companies such as Nexagen in Boulder, Colorado, are attempting to use this technology to make therapeutic proteins and vaccines that will combat AIDS. Larger companies like US Biochemicals are working with Nobel laureate Thomas Cech to develop RNA pharmaceutical enzyme technology.

At the level of everyday use, enzymes and other bioproducts are currently employed in the food industry as preservatives, thickeners, coloring and flavoring agents, and emulsifiers. Enzymes are able to replace or minimize the number of additives used in the food processing industry. Enzymes are also used to increase the shelf life of many fruits and vegetables. The issue becomes one of how to process or cheaply make large quantities of the enzymes.

This leads us into the next area of the bioprocessing industry, live-cell bioreactors.

## 2.2 Whole-Cell Bioreactors

Living cells can be viewed as small biochemical reactors of great complexity. They are utilized in most aspects of biotechnology and are currently the primary step in most biotechnology processes. Despite the advances in enzyme technology, it is important to keep in mind that enzymes come from cells, thus whole-cell bioreactor technology will always be an important part of bioprocessing. The roots of fermentation technology can be traced back many centuries. The earliest bioreactors or fermentors were used in the brewing and wine making industries. During the last 50 years, fungal fermentations have been used to make antibiotics, beginning with penicillin. Now, fermentation is also extensively used in the developing biotechnological/pharmaceutical industry. The first major products made by recombinant organisms were insulin (Eli Lilly), tissue plasminogen activator (Genetech), and erythropoietin (Amgen). The most common organism used in recombinant fermentations is *Escherichia coli,* a genetically well-characterized bacterium.

Optimization of production of recombinant proteins is a continuing objective of industrial and academic research. Engineering and science work together to reach this objective. For instance, a common problem is genetic instability (segregational, structural, host cell gene mutation, and/or growth rate dominated), since the overproduction of foreign proteins is always detrimental to cell growth and survival. Novel reactor strategies that have been developed to alleviate segregational instability include recycle reactors that selectively recycle plasmid-bearing bacterial cells to a fermentor through flocculation and size separation, while nonproductive, rapidly growing, segregant cells are removed from the reactor. Thus, a productive, continuous reactor can exist "in theory" for infinite time. Cells have been genetically manipulated to alleviate this problem, as well, by inserting antibiotic resistance genes into the plasmids and supplying antibiotics to the medium; in another approach, the survival of plasmid-bearing cells alone is ensured by inserting a gene necessary for cell survival, such as the SSB (single-strand binding protein, needed for cell replication) gene, into the plasmid of a host that lacks this gene. More current areas of research include limitation of protein synthesis by the amount of mRNA in the cells, decreasing protease activity in cells to reduce cellular protein degradation, manipulation of cells to inhibit inclusion body formation, and development of new vector–host systems for the overexpression of proteins.

In addition to work aimed at overcoming genetic instability, metabolic engineering is being investigated as a means of further enhancing productivity. Metabolic engineering involves endowing an organism with new or augmented pathways, and/or selecting culture conditions that favor product synthesis over other metabolic activities. The new organism may be easier to culture in a fermentor, or it may be able to degrade two hazardous chemicals simultaneously rather than just one. The potential applications are the production of antibiotics and vitamins by fermentation, the development of bacteria with enhanced biodegradative capabilities for waste detoxification, and the design of improved strains and processes for the production of bulk chemicals such as organic solvents and acids. Metabolic engineering is studied for all types of cells, including bacterial, fungal, plant, and animal cells.

Heretofore, only bacterial cells have been discussed in detail. Plant, mammalian, and insect cells hold greater potential for the biotechnology industry; however, their complexity has slowed development of industrial processes. Plants nevertheless offer the advantage of diversity. As many as 25% of today's pharmaceuticals are extracted from plants naturally grown. Current research is directed at the propagation of plant cell lines in vitro, obviating the need for intact plants. Only a few plant cell systems are used commercially in Japan and Germany. A most promising area for plant reactors in the production of Taxol, an anticancer drug recently approved by the U.S. Food and Drug Administration. Currently, Taxol is extracted from yew tree bark, but the demand is greater than the amount currently available from the natural source.

Cell cultures derived from insects, a relative newcomer to biotechnology, are more complex than plant cells but easier to establish than mammalian-derived cell lines. A successful expression system is the baculovirus system. It consists of a strong promoter that allows a cell to make up to 40% of its protein as the target or product protein with practically no cell multiplication.

Many proteins of mammalian origin are not simply products of gene expression; they require glycosylation and sometimes other secondary enzymatic processes for biological activity. Since bacteria cannot carry out these conversions, production of such proteins requires mammalian or at least eukaryotic cell lines. Mammalian cells grow much more slowly than bacteria, are more demanding in terms of media composition, and may require anchorage to solid supports. They are also very shear sensitive and may be suppressed by their own metabolites. It is relatively easier to establish long-lasting or nearly permanent mammalian culture cells using hybridoma cells for the production of monoclonal antibodies. Hybridoma cells do not require anchorage and have, as a rule, the longevity of their parent cancer cell lines.

A technology that is still in its infancy is the use of transgenic plants and animals to produce proteins. The living plant or animal becomes the "bioreactor." This technology involves inserting new genetic information in the embryo and having the nontoxic protein expressed by the mature animal. One strategy is to have a protein secreted into an animal's milk. Then the product is simply separated from the rest of the milk constituents.

To optimize production, fermentation processes are usually performed under conditions of controlled pH, dissolved oxygen, agitation, and temperature. More complicated control strategies involve regulating protein secretion or substrate addition. New biosensors are being developed to aid in determining the amount of a substrate or product. These sensors include enzymes (used to detect specific products or substrates), bacteria (to determine biological oxygen demand), *Desulfovibrio desulfuricans* (to determine sulfur content), and minced carnation petals (to determine urea concentration). Better sensors lead to improved bioreactor control. One of the important parameters involved in sensing and controlling reactors is understanding the transport of cells and chemicals, which is the next topic.

## 2.3 Transport and Adhesion of Cells

Cellular and viral transport and adhesion are of importance in immobilized reactors, development for new drugs, and in situ bioremediation efforts. Cells of all types—blood, bacterial, tumor, and endothelial—respond to environmental stimuli through receptors found on the cell surface. For example, a cell will have a receptor on its surface for epidermal growth factor; binding and internalization of the growth factor then stimulate DNA synthesis. Cell receptors are responsible for cell adhesion to surfaces such as tissues, and for self-agglutination. White blood cells will adhere to ligands on vessel walls, or mammalian cells will adhere to surfaces coated with ligands through integrins. Cell receptors are also responsible for chemotaxis—the migration of some cells in response to a specific chemical gradient. Chemotaxis is of particular importance in development and repair of inflammation. Understanding the mechanisms of intra- and intercellular transport is key to the development of site-directed or target-specific drugs. For instance, monoclonal antibodies may be used to deliver drugs to cancer cells only, through specific ligand–receptor binding.

Understanding cellular transport can also aid in environmental remediation. Once microorganisms have been injected into the environment, they must move from the point of injection to the contamination site. Cell receptors known as pili, located on the surface of the cells, are responsible for some cell adhesion to the soil, minerals, and other contaminants found in the soil. If the adhesion protein synthesis and binding mechanism is understood, then the efficiency of biodegradation to remediate soil and water can be determined through process engineering analysis and compared to other existing technologies, such as chemical oxidation and incineration. When cells move slowly toward a chemical contaminant as they degrade it, chemotaxis is useful in bioremediation efforts.

## 2.4 Bioseparations

Although a large quantity of work has been done on fermentation and other upstream processing, the main cost of making a biotechnological product is in the downstream processing or separation steps. Many of the separation techniques employed are simple scale-ups of analytical and micropreparative techniques pioneered in life sciences laboratories, and improved methods are needed.

After a product has been made by a cell, it must be separated from the rest of the by-products. The standard industrial techniques for separating cells from spent medium are centrifugation and cross-flow filtration. If the cell product is an intracellular product, the cells have to be lysed through homogenization or osmotic shock. Bacterial cells like *E. coli* typically do not secrete protein products, whereas mammalian cell typically will. The expensive part is then the further purification of the product, not the initial separation of cells from product. Precipitation of all cellular protein is a common first step. A number of technologies exist for further product separations, including differential precipitation, affinity and other types of chromatography, electrophoresis, liquid–liquid extraction, and use of specialized membranes.

As essential step in downstream processing is quality control, the assessment not only of sterility, absence of genetic materials such as viruses or nucleic acids, and absence of pyrogens, but also of the purity of the final product, such as the desired protein. Various chromatographic and electrophoretic procedures are routinely utilized, often combined with immunochemical reagents for further product identification. While electrophoresis is generally recognized as having the highest resolution for protein isoforms, high performance liquid chromatography (HPLC) is also widely used, in part because of the simplicity of operations and the similarity to the bulk purification process. Electrophoresis has not yet made significant inroads into protein downstream processing, although two research-scale instruments are commercially available. Both are based on isoelectric focusing in free solution, and neither requires gels or other supporting matrices. Preparative isoelectric focusing does not call for expensive buffers and could easily be scaled up for production purposes.

Chromatography is defined as the separation of various substances by the selective binding of components to porous, solid, sorptive media. Various types of chromatographic separation are employed, including absorption, liquid–liquid partition, ion exchange, gel filtration or molecular sieving, affinity, hydrophobic, and HPLC. Typically a combination of chromatographic processes is used to purify proteins. For example, an ion exchange column can be used to separate negatively charged products from positively charged and neutral products, followed by an affinity column to isolate the desired product. Affinity chromatography can be a particularly versatile purification method, often displaying exquisite selectivity. Various affinity systems can be used: antigen–antibody, lectin–carbohydrate moiety, enzyme–substrate, immobilized metal ion–histidine-containing protein, receptor–hormone, and so on. In addition, affinity chromatography may be used for environmental remediation by using ligands (chelators or macromolecules) that can specifically bind metal ions. Several approaches are currently being pursued to reduce the cost and accelerate the chromatographic process: membrane affinity chromatography, perfusion chromatography, and the use of magnetically stabilized fluidized beds. Attachment of affinity ligands to membranes combines the low cost, high throughput characteristic of filtration with the selectivity of membranes. A similar objective is the aim of perfusion chromatography, based on the recent development of highly porous particulate chromatographic media, in which process fluid perfuses through the media interstices, thereby accelerating the absorption–desorption process. Finally, magnetically stabilized fluidized beds may eliminate the occasional problem of fluid flow channeling within a chromatographic column. A mixture of magnetic supports and nonmagnetic affinity beads, such as modified Sephadex, are utilized, and a magnetic field is applied to the system to stabilize the bed. The stability problem is also addressed in expanded-bed chromatography, in which media are gently suspended by the upward plug flow of the eluent from below.

Liquid–liquid extraction originally used immiscible liquids such as an organic and an aqueous phase. For example, an aqueous DNA solution is often extracted with phenol and ether to remove lipids and other cell debris. Extraction has also been used to separate inhibitory end products such as alcohols from fermentation broth. Organic solvents are not compatible with many proteins. Instead, aqueous two-phase extraction has been utilized to separate proteins. Typical aqueous phases contain water-soluble but mutually incompatible polymers such as polyethylene glycol and dextran. The latter polymer, or any other bottom phase polymer, such as a cellulose derivative, may be substituted by salt solutions at high concentration.

Current research focuses on fundamental understanding of the principles of extraction by applying thermodynamic and fluid dynamic principles to the separation processes. The goal is to be able to predict in advance whether bioproducts can be separated using this technique. Heretofore, trial-and-error experiments have been performed for each individual system to determine the feasibility of extraction as a separation process for the product of interest.

Membranes are used to concentrate proteins and to clarify solutions, but they are not widely used in the bioprocess industry to purify individual proteins. They are used to purify water through reverse osmosis. The major drawback of membranes is biofouling. Layers of proteins build up on the protein surface and clog the membranes. Cross-flow or tangential flow filtration decreases fouling but does not eliminate it. Current research involves understanding fouling through theory and experiment, and the development of nonfouling membranes.

## 3 CONCLUDING REMARKS

Biotechnology is already a multibillion dollar industry and will continue to grow. It involves the collaboration of scientists and engineers from an emerging tricornered coalition of government agencies, industry, and universities. The current thrust is the production of high value human and veterinary therapeutics. Of greater future socioeconomic impact may be agricultural products enhanced with genetic modification of plants and animals to maximize desirable features and minimize undesirable ones. Examples include the development of insect-, drought-, or salt-resistant plants, and modification of the quantity and level of unsaturation of oils. Finally, one may expect the biological production of low cost, high tonnage commodity chemicals, such as pesticides and food components. Advances in current bioprocess engineering techniques will be essential if the cost of production is to be reduced. The competition is international, and many countries are trying to position themselves on the forefront of bioprocessing technology.

*See also* Antibody Molecules, Genetic Engineering of; Biodegradation of Organic Wastes; Expression Systems for DNA Processes; Fuel Production, Biological; Taxol® and Taxane Production by Cell Culture.

### *Bibliography*

Bailey, J. E. and Ollis, D. F. (1986) *Biochemical Engineering Fundamentals.* McGraw-Hill, New York.

Belter, P. A., Cussler, E. L., and Hu, W.-S. (1988) *Bioseparations, Downstream Processing for Biotechnology.* Wiley, New York.

Chmiel, H., Hammes, W. P., and Bailey, J. E., Eds. (1987) *Biochemical Engineering, A Challenge for Interdisciplinary Cooperation.* Gustav Fischer Verlag, Stuttgart.

Georgiou, G. (1988) Optimizing the production of recombinant proteins in microorganisms. *AIChE J.* 34:1233–1248.

Ghose, T. K., Ed. (1989) *Bioprocess Engineering, The First Generation.* Ellis Horwood, Chichester.

National Academy of Sciences/National Research Council. (1992) *Putting Biotechnology to Work: Bioprocess Engineering.* National Academy Press, Washington DC.

Shuler, M. L., and Kargi, F. (1992) *Bioprocess Engineering, Basic Concepts.* Prentice Hall, Englewood Cliffs, NJ.

Todd, P., Sikdar, S. K., and Bier, M., Eds. (1992) *Frontiers in Bioprocessing II.* American Chemical Society, Washington DC.

# Biosensors

*Anthony E.G. Cass*

### *Key Words*

**Amperometric** Describing measurement of the current arising from an electrode reaction when a potential is applied.

**Analyte** The substance being measured.

**Analyte Matrix** The material in which the analyte is found.

**Electrode** An electrically conducting material at which changes in current or potential can occur.

**Fluorescence** The emission of light from a molecule that has been illuminated at short wavelength.

**Piezoelectric** Describing a material that undergoes a mechanical distortion in response to an applied potential.

**Potentiometric** Describing measurement of the potential difference under circumstances of no current flowing.

**Transducer** Device that converts a physical or chemical change into an electrical signal.

---

Biosensors are a class of analytical device in which the molecular recognition and catalytic properties of biological materials are combined with the signal-processing features of microelectronic instrumentation. This close integration of highly specific detection methods with electronic measurement and display technologies offers the opportunity to develop small, simple-to-use, and intelligent analyzers that can be employed at locations remote from central laboratories. Such devices are also well suited to continuous monitoring, and they offer novel approaches to the measurement of biologically important parameters.

## 1 INTRODUCTION

Biosensors are analytical devices that have as their central feature the close physical and functional coupling of biological materials to optically or electronically active surfaces in such a way that the biomolecular recognition reaction is converted or transduced into an electrical signal. The close integration of the biochemical events with the generation of a signal offers the potential for fabricating compact and easy-to-use analytical tools of high sensitivity and specificity. A large number of choices for both the biological component and the signal-generating surface have been exploited, and some examples are shown in Table 1. We can think of biosensors as converting a chemical information flow into an electrical information flow through the following series of events (Figure 1):

1. Diffusion of the anlayte molecule from the solution to the biosensor surface.
2. Reaction of the analyte with the biological material at the surface.
3. Alteration of the physicochemical properties of the surface in response to the biorecognition.
4. Detection of that altered physicochemical state by a change in some property of the surface that alters its optical or electronic properties.
5. Measurement, amplification, and display of the resulting electrical signal.

To ensure an unambiguous relationship between the signal and the analyte concentration, it is important that the remaining components of the analyte matrix not interfere with either the biorecognition event or the subsequent change in the surface properties. In this respect biosensors share features with all other chemical and physical sensors and indeed with all other analytical techniques.

## 2 GENERAL FEATURES OF BIOSENSORS

Although there are many classification schemes for biosensors, one of the most useful entails separating devices that act through the

Table 1 Some Biological Materials and Optically or Electronically Active Devices Commonly Used in Biosensors

| Biological Material | Device |
|---|---|
| Enzymes | Potentiometric electrodes |
| Nucleic acids | Amperometric electrodes |
| Antibodies | Wave guides |
| Lectins | Grating couplers |
| Cells | Acoustic wave sensors |
| Organs | Conductimetric sensors |
| Tissue slices | Thermometric sensors |

biological material converting the analyte into a different chemical species (catalytic sensors) from those that act through simply binding the analyte (affinity sensors). In the context of familiar laboratory equipment, biosensors of the first type are analogous to devices such as the polarographic (Clark) oxygen electrode, while those of the second type more resemble the glass pH electrode. Catalytic sensors are kinetic devices; the continued generation of a signal is dependent on the transport of fresh analyte to the sensor's surface, where it is converted to product. Affinity sensors tend to be equilibrium devices (i.e., the differences in the surface before and after binding are compared), although it is also possible to operate affinity sensors in a kinetic mode (i.e., the rate of approach to equilibrium is measured). These two different types of sensor therefore tend to have their performance determined by different characteristics; catalytic sensors depend largely on the kinetic properties of the biological material ($K_m$ and $k_{cat}$), while affinity sensors depend on the dissociation constant of the complex of the biological material and the analyte ($K_D$). Both devices will also give a response that depends on the total amount of biological material on the sensor's surface, and the lifetime of the sensor may be determined by the stability of the biological material.

The foregoing features tend to imply that in ideal circumstances the characteristics of the biosensor could be predicted simply from a knowledge of the basic properties of the biological components. Certain factors, however, complicate such a simple analysis. The first factor, and one that is particularly relevant for catalytic sensors, is that the conversion of analyte by the biological component is not the only kinetic step. As mentioned earlier, the generation of the signal occurs as analyte is converted to product, and to sustain this signal, fresh analyte molecules must be transported to the sensor surface to replenish those lost through reaction. There is thus a transport step as well. Hence the rate of signal generation has at least two components, and the one that is rate-limiting step will determine the overall performance of the biosensor. When the catalytic reaction is rate limiting, the response characteristics of the biosensor will to a great extent be governed by the properties of the biological component; when transport is rate limiting, however, the characteristics of the sensor will be distinctly different from those expected from simply considering the nature of the biological material. This distinction can be appreciated by considering the concentration profile of substrate away from the sensor surface. When the catalytic activity at the surface is high, the instantaneous concentration of analyte there is essentially zero; that is, substrate molecules are converted to product as soon as they arrive. Under these conditions of catalytic excess there are enzyme molecules at the surface which are not taking part in the reaction; hence the rate of substrate conversion is less than would be expected, were all the catalysts involved. Moreover the concentration of substrate experienced by the catalyst is lower than the bulk concentration, and therefore the enzyme is less saturated than would be expected. Finally the excess of enzyme activity and the transport-limited rate of signal generation mean that enzyme molecules can become inactivated or can begin to operate under less than optimal conditions without a change in the signal occurring. Under transport-limited conditions, therefore, the biosen-

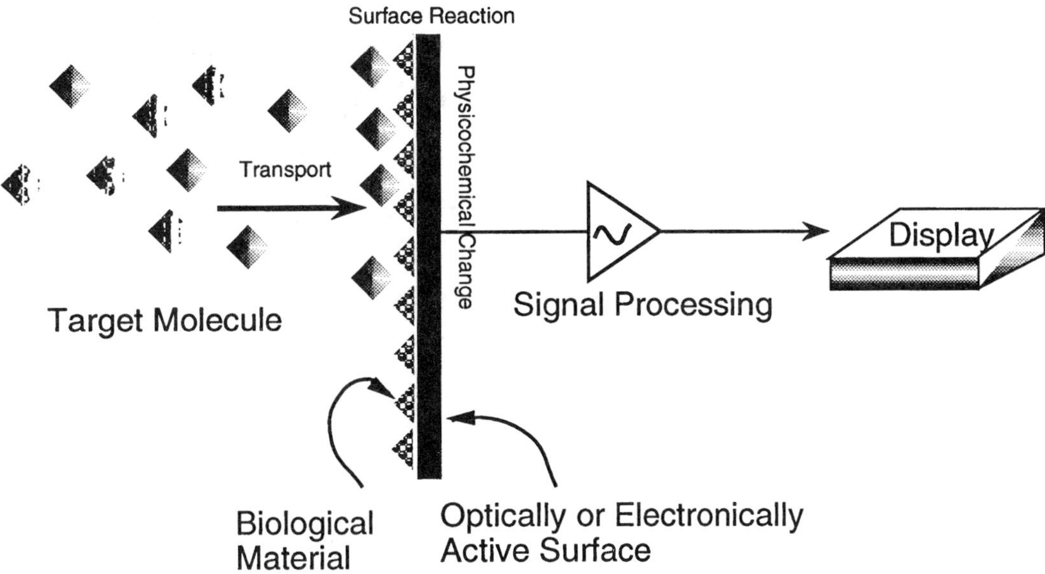


Figure 1. Schematic diagram of a generic biosensor showing the major components and the processes that contribute to signal generation.

sor will tend to exhibit a reduced signal, an extended linear range, and an increased lifetime.

Affinity sensors when operated as equilibrium devices do not exhibit these transport-dependent properties. They do however need to be designed to avoid an alteration in the $K_D$ value as a consequence of the methods used to immobilize the biological material. Moreover the nonspecific binding (absorption) directly to the optical or electronic surface of other compounds in the matrix can give rise to a spurious signal.

## 3 ELECTROCHEMICAL BIOSENSORS

Enzyme electrodes were among the earliest biosensors to be described in the literature and are still probably the most intensively studied. They can be classified as either potentiometric or amperometric, depending on whether a potential difference or a current is produced, and they are almost exclusively associated with catalytic sensors. In the case of potentiometric sensors, the surface tends to be that of an ion-selective or gas-sensing electrode, and the catalytic reaction changes the surface concentration of either a specific ion or a gas. One of the attractions of potentiometric biosensors is that very small units can be made in large quantities by semiconductor fabrication methods, particularly in the guise of ion-selective field effect transistors (ISFETS). As with all potentiometric devices, the potential is related to the analyte concentration through a logarithmic function, where under ideal (or Nernstian) conditions for a singly charged ion there is a 57 mV shift in potential for each order of magnitude change in concentration.

A classic example of a potentiometric enzyme electrode is that for urea, where the enzyme urease catalyzes the reaction:

$$CO(NH_2)_2 + 2H_2O + H^+ \rightarrow 2NH_4^+ + HCO_3^- \qquad (1)$$

Immobilization of urease at the surface of either a pH-sensitive or an ammonium ion sensitive electrode or at the surface of an electrode that is sensitive to ammonia or carbon dioxide gas offers possible configurations for a urea-sensitive electrode. In practice, potentiometric enzyme electrodes based on pH-sensitive devices tend to be the most common, although both the background pH and the buffer capacity must be carefully controlled if high precision measurements are to be made.

Other potentiometric enzyme electrodes include those for organophosphorus pesticides, based on the inhibition of acetylcholinesterase, which catalyzes the reaction:

$$H_2O + (CH_3)_3N^+CH_2OCOCH_3 \rightarrow (CH_3)_3N^+CH_2OH + CH_3COO^- + H^+ \qquad (2)$$

The turnover of acetylcholinesterase generates protons, which are measured with a pH electrode, and in the presence of organophosphorus pesticides the enzyme is strongly inhibited, preventing the acidification at the electrode surface. In these types of enzyme sensor, where the analytical information arises from the inhibition of catalytic activity, it is important to ensure that the electrode's response is not transport limited. Otherwise the sensitivity toward the inhibitor will be poor.

Amperometric enzyme electrodes exploit the catalysis of the oxidation or reduction of the substrate under the influence of an applied potential to generate a current whose size is proportional to the substrate concentration. Amperometric enzyme electrodes may be based on systems in which the enzyme catalyzes the redox reaction between the analyte ($SH_2$) and a freely diffusing cofactor [often $NAD(P)^+$]:

$$SH_2 + NAD(P)^+ \rightarrow S + NAD(P)H + H^+ \qquad (3)$$

The reaction cycle is completed by the reoxidation of NAD(P)H:

$$NAD(P)H \rightarrow NAD(P)^+ + H^+ + 2e^- \qquad (4)$$

Alternatively a protein-bound redox center (E·P)may be first reduced by the analyte ($SH_2$), then oxidized by an electron acceptor (A):

$$E{\cdot}P_{ox} + SH_2 \rightarrow E{\cdot}P_{red} + S \qquad (5)$$

$$E{\cdot}P_{red} + A \rightarrow E{\cdot}P_{ox} + AH_2 \qquad (6)$$

At a suitable electrode surface, the reduced acceptor is reoxidized:

$$AH_2 \rightarrow A + 2H^+ + 2e^- \qquad (7)$$

In the case of the first group of enzymes the reoxidation of NAD(P)H is thermodynamically favored but in early work tended to require at high overpotentials. New chemically modified or organic conducting salt electrodes have largely overcome this problem, however. In the case of the second group of enzymes, the natural electron acceptor (usually oxygen) can be employed and the product (hydrogen peroxide) oxidized, although this again occurs at high overpotentials, leading to possible difficulties due to the simultaneous oxidation of endogenous molecules. An alternative approach is to take advantage of the observation that since reaction (6) is relatively nonspecific for A, a range of synthetic molecules (collectively referred to a mediators) can be used. These molecules have advantages in terms of their lower oxidation potentials, rapid reactions with both enzyme and electrode, and versatile synthetic chemistries.

An example of the first type of enzyme electrode is a biosensor for ethanol in which $SH_2$ is ethanol and NADH can be reoxidized either at a chemically modified electrode comprising a carbon paste mixed with phenothiazine or phenoxazine derivatives or, alternatively, at a conducting organic salt electrode made of *N*-methylphenazinium tetracyanoquinodimethide ($NMP^+TCNQ^-$). The second type of electrode is represented by a glucose biosensor; here the flavoenzyme glucose oxidase has its FAD prosthetic group reduced by β-D-glucose and reoxidized by a mediator such as the ferricenium ion.

A final type of enzyme electrode is that based on conductimetry —that is, measurement of the conductance of the solution between a pair of electrodes subject to an alternating voltage. Traditional conductimetry usually employs a pair of large electrodes immersed in solution. However the techniques of lithography and screen printing make it possible to place the two electrodes close together on a planar plastic or ceramic base, usually in an interdigitated format, facilitating measurements in very thin layers of liquid. In this type of biosensor, enzymes that give rise to charged products cause a local change in conductivity proportional to the initial substrate concentration. Thus the example of urease-catalyzed hydrolysis of urea described in equation (1) could also be applied to a conductimetric sensor.

## 4 OPTICAL BIOSENSORS

Optical sensors have been developed that employ a wide range of phenomena to measure both catalytic and affinity reactions, either taking by advantage of some spectroscopic property (such as fluorescence or absorbance) or by measuring the change in the surface loading of a dielectric material (such as a protein) that affects the intrinsic optical properties of the surface. Various optical components have been employed including waveguides (fibers and planar slabs) and dispersion elements (such as gratings and prisms).

Opt(r)odes are the optical equivalent of electrodes. At their simplest, they use the fibers to carry light from source to sample and then onto the detector. The light can be used to measure either absorbance or fluorescence, and the biosensors have been used both with enzymes that convert a colorless or nonfluorescent substrate into a colored or fluorescent one and with optically sensitive pH- or oxygen-responsive compounds in a manner analogous to pH or oxygen electrodes. Optical sensors offer advantages over electrochemical ones in being more immune to noise, in not needing a reference electrode, and in avoiding a direct electrical connection to the system being measured. This last feature is particularly relevant if in vivo measurements are being made. Optical fibers have also been used in competitive binding assays, where their numerical aperture has been used to distinguish between the fluorescence coming from a labeled ligand in front of the fiber's tip and the fluorescence from the labeled ligand bound to a coaxial membrane protruding beyond the end of the fiber.

Although fibers can act as simple light pipes to transfer radiation by total internal reflection, both fibers and planar waveguides show the phenomenon of an evanescent wave just beyond the surface. During the transmission of radiation by total internal reflection, the reflected light penetrates just beyond the surface of the waveguide in the form of an evanescent field; the penetration is dependent on a number of factors but is typically of the order of a few hundred nanometers. If there is nothing within this evanescent layer for the radiation to interact with, it couples back into the waveguide and is transmitted without loss. The presence of absorbing or fluorescing molecules within this layer, however, will result in an optical signal arising from within this very thin layer. In the context of an antibody binding its antigen, then, in a competition assay characterized by fluorescently labeled and unlabeled antigen competing for a limited amount of surface-bound antibody, the evanescent field is able to discriminate between the fluorescent label bound to the surface of the waveguide and the (excess) unbound label in the solution, offering the opportunity of separation-free immunoassays without the need for washing steps. A further advantage of the waveguide as a basis for immunosensing is the ability to incorporate standard and control measurements onto a single waveguide along with the samples.

This type of fluorescence binding assay is not restricted to antibody–antigen reactions, and any pair of complementary molecules could be used in a similar fashion—for example, in DNA hybridization or carbohydrate–lectin binding assays.

Spectroscopic sensors, whether based on light guides or on evanescent excitation, rely on the use of labels to provide a signature for measurement. In a second class of optical sensor the signal is generated through a change in the mass loading of protein (or other dielectric material) on the surface. This second group of methods does not need a label but is capable of direct sensing and is exemplified by surface plasmon resonance (SPR). In SPR a dispersion element (prism or grating) is coated with a thin metal film (a layer of gold or silver about 5 A thick) and when monochromatic light from a laser is incident on the grating at a specific angle, the momentum of the photons is transferred to the electrons in the metal film, setting up an electron wave known as a plasmon. The loss of photon momentum at the critical angle means that the intensity of the light reflected is much reduced and the value of the critical angle is dependent on (among other factors) the refractive index of the layer above the metal surface, which in turn depends on the composition of this layer and its thickness. When antibody molecules are bound to the metal surface (or confined within 100 nm of it), the binding of a macromolecular antigen changes the refractive index of the layer (the refractive index of water is 1.33 and of protein about 1.5), hence shifts the critical angle. Measurement of this angular shift can then be related to the amount of antigen present. Moreover by measuring the rate of change of angle under conditions of controlled transport, one can determine the rates of binding and dissociation, thus allowing a complete kinetic and thermodynamic analysis of the interaction.

This principle provides a potentially powerful tool for characterizing novel therapeutic agents because it can be readily extended to any macomolecular interaction (e.g., of receptors with their ligands), or with agonists and antagonists.

SPR is not the only optical method that takes advantage of changes in refractive index. The efficiency of gratings in coupling light into and out of optical fibers also depends on the refractive index at the grating surface, and this too has been used to measure binding reactions at their surface.

## 5 ACOUSTIC WAVE BIOSENSORS

Similarly, optical methods are not the only ones that are sensitive to mass loading, hence can monitor binding reactions without recourse to labeling. A number of piezoelectric devices show a sensitivity in their frequency of vibration to the surface mass. These devices are essentially acoustic and rely on establishing a vibrational wave in the material under the influence of an alternating voltage applied to electrodes coated on the surface. Depending on the nature of the acoustic wave, which in turn depends on the structure of the device, these are known variously as the quartz crystal microbalance (thickness–shear mode), SAW (surface acoustic wave), Lamb wave (flexural plate wave), or acoustic plate mode devices. Coating the surface of these devices with antibodies and then exposing them to the complementary antigen results in an increase in the surface mass loading, hence a decrease in their vibrational frequency proportional to the mass increase.

## 6 CONCLUSIONS

There is certainly no shortage of elegant and sophisticated biosensor designs, and many laboratory prototypes have been demonstrated. Future considerations of the feasibility of their widespread adoption must therefore be concerned with the large-scale manufacture of reliable, cheap, and effective devices that not only satisfy the requirements of their users but also meet legislative and economic criteria.

*See also* Bacteriorhodopsin-Based Artificial Photoreceptor; Biomolecular Electronics and Applications.

*Bibliography*

Buck, R. P., Hatfield, W. E., Umana, M., and Bowden, E. F., Eds. (1990) *Biosensor Technology.* Dekker, New York.

Cass, A.E.G., Ed. (1990) *Biosensors. A Practical Approach.* IRL Press, Oxford.

Hall, E. A. (1990) *Biosensors.* Open University Press, Milton Keynes.

Scheller, F., and Schubert, F. (1992) *Biosensors.* Elsevier, Amsterdam.

Turner, A.P.F., Karube, I., and Wilson, G. S., Eds. (1987) *Biosensors. Fundamentals and Applications.* Oxford University Press, Oxford.

# Biotechnology, Governmental Regulation of

*David T. Kingsbury*

*Key Words*

**EPA** Environmental Protection Agency

**FDA** U.S. Food and Drug Administration.

**NIH Guidelines** National Institutes of Health Guidelines for Research Involving Recombinant DNA.

**USDA** U.S. Department of Agriculture.

Biotechnology is an *enabling* technology; it has broad applications in many diverse aspects of basic research, industry, and commerce. Because of these broad applications, cooperative regulation based on a consensus between different agency viewpoints and approaches is essential in this field where generic review standards are not well established and products tend to be reviewed on a case-by-case basis. The U.S. government's 1986 Coordinated Framework marked the beginning of current regulatory policy, with the stated intention that the policies would evolve with the field. The Coordinated Framework was a broad and complex coordinated policy statement that explained the application of existing statutes to the regulation of biotechnology and outlined an approach to interagency coordination, which is so vital to this field.

The document was presented in two parts in the *Federal Register* on November 14, 1985 (Vol. 50, pp. 47174–47195) and on June 26, 1986 (Vol. 51, pp. 23302–23393). The second part consisted of six elements, the preamble and statements of policy from the Food and Drug Administration, the Environmental Protection Agency, the U.S. Department of Agriculture, the Occupational Safety and Health Administration, and the National Institutes of Health. These two *Federal Register* notices illustrate the intent of the separate agencies to form a coordinated group. Each of the agency statements stood alone, each describing the agency's separate and respective policy, within the context of the applicable laws. Since the initial publication, several agency notices of rule making and policy refinement have appeared.

One critical element in the coordinated regulatory framework was the use of common definitions for the products subject to particular types of regulatory oversight. Common definitions are essential because government regulation is structured around products, developed by various technologies for specific intended purposes. The principal focus of the new policy was environmental release (deliberate release) of new organisms. The Coordinated Framework and its subsequent clarifications explain the application of certain statutes to genetically modified organisms and in some cases impose an abbreviated review on unmodified organisms applied to environmental uses.

## 1 HISTORY OF BIOTECHNOLOGY REGULATION

It is generally accepted that U.S. regulatory agencies focus regulation on products to be introduced into the marketplace, not the means by which those products have been produced. Therefore, a food additive is judged to be safe based on its properties, not on whether it was derived by synthetic organic chemistry or extraction from a natural source. In some instances a manufacturing element is considered, usually in the context of quality assurance and manufacturing reproducibility. Why is it, then, that one of the continuing issues in the regulation of biotechnology is the regulation of the "process" of biotechnology, not the *products?* To understand this issue, it is necessary to step back and take a brief look at the origins of biotechnology regulation.

In 1973 Herbert Boyer and Stanley Cohen first described their development of recombinant DNA technology, and during that year, concerns began to appear regarding the potential hazards of this new technology, even though recombinantlike processes were already recognized in nature. The discussion expanded throughout 1973 and continued to grow over the next few years. In the well-known Asilomar meeting in 1975, a group of scientists called for the NIH to develop "guidelines" for the use of this technology, including a set of safety procedures. In response, a series of containment levels was developed and experiments were categorized by their *predicted* level of hazard. Some experiments were banned based on their *perceived* danger. The debate over the safety of recombinant DNA technology became extremely emotional and politically charged, and several cities banned all recombinant DNA work within their jurisdictions. A large number of environmental groups and social activist organizations joined in the debate.

The release of the NIH Guidelines in 1975 helped to stabilize the scientific environment but failed to silence the ongoing debate. Therefore, one price the scientific community paid for an element of stability was the establishment of the concept, which remains in the public policy debates even today, that the *technology* needed to be regulated rather than the specific results of the technology. This issue continues to dominate discussions of the regulation of products to be introduced into the environment as living organisms, even though it was very clear to most scientists that the hypothesized hazards were not real, and no problems have arisen as a result of recombinant DNA research.

In the mid-1980s the government moved to reexamine the regulatory environment, especially as it related to commercial development, to replace the NIH Guidelines with a regulatory policy more in keeping with the statutory authorities backing the several regulatory agencies, and to extract the NIH from acting as a de facto regulatory body. The eventual result of that effort was the Coordinated Framework for the Regulation of Biotechnology, which appeared in June 1986.

## 2 U.S. REGULATORY POLICY

### 2.1 General Policy

The Coordinated Framework released in 1986 and the 1990 Report on National Biotechnology Policy published by the Executive Office of the President describe the basic tenets of U.S. regulatory policy. The combination defines a broad and complex policy, which explains the application of existing statutes to the regulation of biotechnology and outlines the approach to interagency coordination, which is so vital to this field. Furthermore, it seeks to restrict regulation to instances of unaddressed and unreasonable risks.

Government regulation is not organized around technological processes. Rather, it tends to be structured around products, developed by various technologies for specific purposes. Therefore, one critical element in a coordinated regulatory framework is a widely accepted definition of the nature of the products subject to particular types of regulatory oversight.

The principal focus of "biotechnology" regulatory policy is environmental release of new organisms. There has been general acceptance of the regulation of nonliving products of biotechnology, such as pharmaceuticals, and the 1986 policy affirmed that the past regulatory practices would be maintained. The new policy explained the application of certain statutes over genetically modified organisms, and in some cases even imposed an abbreviated review over unmodified organisms when applied to environmental uses. A clear policy was established requiring review of genetically engineered microorganisms prior to release into the environment, with some organisms subject to an abbreviated review.

The regulation of biotechnology products in the United States falls within the responsibility of several government agencies, principally the Food and Drug Administration, the Department of Agriculture, and the Environmental Protection Agency. The underlying regulatory philosophy and the mechanism of interagency coordination was outlined in the Coordinated Framework and further clarified in a 1992 publication, Exercise of Federal Oversight Within Scope of Statutory Authority: Planned Introduction of Biotechnology Products into the Environment. These publications provide a road map for the movement of products within the regulatory system and incorporate the principle that products are to be regulated on the basis of their characteristics and risk, not the process used in production. These principles are to be applied to both living and nonliving products. Most agencies have refined their initial position through additional rule making and policy clarification. Likewise, the NIH Guidelines have been revised many times, and the stringent oversight of the initial guidelines is no longer in place. Approval for environmental introductions on noncommercial research generally must be sought from the appropriate funding agency, or, in the case of non-government-funded work from the agency responsible for similar commercial products (e.g., pesticides to the EPA, plants to the USDA).

### 2.2 The Food and Drug Administration

The Food and Drug Administration has consistently taken the position that no new procedures or requirements are necessary for the oversight of products resulting from the new biotechnology. The FDA has steadfastly regarded the techniques of the new biotechnology as simply extensions or refinements of older forms of genetic manipulation, and the products have been subject to the same regulatory requirements as other products.

Although there are no statutory provisions or regulations that address biotechnology specifically, the laws and regulations under which the agency approves products place the burden of proof of safety as well as effectiveness of products on the manufacturer. The agency possesses extensive experience with these regulatory mechanisms and applies them to the products of biotechnological processes.

In May 1992 FDA published a statement of policy regarding the oversight of new varieties of food plants, again stating the principle that the techniques of their construction do not constitute the trigger for special review. The FDA has identified scientific and regulatory issues that may require a consultation between the developer of a new variety and the FDA. These issues are related to characteristics of foods that raise safety questions and would trigger a higher level FDA review. The issues identified in the FDA statement included the presence in the new variety of a substance that is completely new to the food supply, the presence of an allergen in an unusual or unexpected setting, changes in levels of a major nutrient, and an increase in the level of a toxin normally found in food. New varieties without these characteristics are subject to lower level scrutiny. The technique employed in the development of the new variety does not in itself determine the need for, or the level of, review.

### 2.3 The Department of Agriculture

The Department of Agriculture outlined its approach in the 1986 Coordinated Framework and has subsequently published several statements of clarification and rule making. The central office within the department having jurisdiction over biotechnology products is the Animal and Plant Health Inspection Service (APHIS). In 1987 rule making APHIS established procedures for the issuance of permits for the field testing of new plant varieties derived from recombinant DNA technology. Subsequently hundreds of trials have been safely performed on a number of plant varieties. In light of this extensive experience with field trials, in November 1992 APHIS published a proposal that would require only *notification* to APHIS, rather than prior approval, for field trials of transgenic plants that meet criteria designed to ensure that the plants are not plant pests. These proposals are still in the public comment stage and have not been adopted as final rules. If they are adopted, however, the APHIS regulations would be in concert with the risk-based process outlined in the Report on National Biotechnology Policy and "scope" announcement, and currently practiced by the Food and Drug Administration.

The Research and Education division of the Department of Agriculture had intended to develop its own "guidelines" for field research. Since, however, the procedures then in effect had proved workable, this intention was abandoned. Left in place was a mechanism of oversight for research with field trials subject to the jurisdiction of various agencies, depending on the characteristics of the organism in question or its intended use, not its source of funding. For example, veterinary vaccines and plant pests would be subject to USDA/APHIS oversight, human vaccines to the FDA, and microbial pesticides to the Environmental Protection Agency.

### 2.4 The Environmental Protection Agency

The Environmental Protection Agency's statutory authority for regulation of microbial products falls under two federal statutes, the Federal Insecticide, Fungicide, and Rodenticide Act (FIFRA) and the Toxic Substances Control Act (TSCA). These statutes have differing requirements for regulatory oversight, and differing procedures regarding notification and permitting. Neither statute was specifically written to address the problems of biotechnology, and without additional rule making the EPA felt that the procedures were not sufficiently explicit. In the 1986 Coordinated Framework, EPA outlined a policy and announced the intention to release "a significant new rule" to cover this new group of organisms. Because of the difficulties associated with making the appropriate definitions based on process rather than biological characteristics, the EPA has been unable to come forward with acceptable new rules. The agency is continuing to work on these rules, and publication should follow.

At present two EPA requirements are in place as outlined in the 1986 notice. First is a notification and reporting requirement for small-scale field tests, and the experimental use permit and registration requirements under FIFRA; second, there are the premanufacture notice requirements under TSCA for "new" microorganisms, as defined in the EPA section of the Coordinated Framework. While it is prudent for anyone with a product subject to federal regulation to make a preliminary contact with the appropriate agency, vagaries of the FIFRA and TSCA language and its interpretation for enforcement purposes make this step even more essential for products subject to EPA regulation.

## 3 INTERNATIONAL COORDINATION

International coordination of regulation has been a goal since the beginning of the regulatory process, and initially the NIH Guidelines were reflective of the international viewpoint. However, as commercial applications developed, international coordination became more difficult. Just recently implementation of the European Community policies has been achieved and is embodied in two directives, 90/219/EEC and 90/220/EEC. However, the regulatory approach taken by the EC is likely to make coordination with the United States an even greater problem. The following analysis was recently published by the Office of Technology Assessment of the U.S. Congress:

> In enacting directives that specifically regulate genetically modified organisms, the EC has established a regulatory procedure that is significantly different from that of the United States. In the EC, regulation is explicitly based on the method by which the organism has been produced, rather than the intended use of the product. This implies that the products of biotechnology are inherently risky, a view that has been rejected by regulatory authorities in the United States. In addition, manufacturers are concerned that their new biotechnology-driven products may face additional barriers before they can be marketed, for the product may also be subject to further regulations based on its intended use.

Experience with continued development of biotechnology products and their introduction into the marketplace will be the test of these regulatory approaches. It is clear that the EC policy makers were dealing with both scientific information and cultural diversity while enacting their directives, and the provision of a consistent policy across Europe was essential.

Over the past 6 years there has been a concerted attempt to coordinate the regulatory policies of the countries of the Americas. This goal has not been fully realized. However, there are in place recommendations made by the collective bodies of the Pan American Health Organization (PAHO), the Organization of American States (OAS), and the Inter-American Institute for Cooperation in Agriculture (IICA). The United States, followed closely by Canada, has been the leader in the development of biotechnology in the Americas and in the development of regulatory policy and practice as new biotechnology products came to market. The other countries in the Americas have accepted a derivative version of the NIH Guidelines as a working base, and this has provided an environment for international trust and cooperation.

In contrast, Japan has adopted a *process-based* regulatory approach to products of the new biotechnology, which, while annoying to the Japanese pharmaceutical industry, has not retarded development. This is not true in the areas of agriculture and gene therapy. There has not been a single clinical trial of gene therapy in Japan, nor is there any evidence that such trials are contemplated in the foreseeable future, despite the sophistication of the Japanese medical system. Likewise, there has been only one field trial of a recombinant DNA manipulated plant, and none of recombinant microorganisms. The Japanese government has not provided any risk-based regulatory guidance to those contemplating such trials, and furthermore, the Japanese Ministry of Health and Welfare has imposed a strict regulatory regime specific to foods and food additives manufactured with recombinant DNA techniques.

## 4 UNCERTAINTY IN THE FUTURE

Guidance for future refinement in the regulatory environment was provided by the President's Council on Competitiveness in the publication *Report on National Biotechnology Policy* issued in February 1991, referred to earlier. In that report the president's advisers decried the still present inconsistencies in regulatory oversight, especially in cases of living organisms to be used in the open environment, and urged agencies to reaffirm a "risk-based," not a "process-based," approval process. The *Report* outlined the following principles of regulatory review:

1. Federal government regulatory oversight should focus on the characteristics and risks of the biotechnology product—not the process by which it is created.
2. For biotechnology products that require review, regulatory review should be designed to minimize regulatory burden while assuring protection of public health and welfare.
3. Regulatory programs should be designed to accommodate the rapid advances in biotechnology. Performance-based standards are, therefore, generally preferred over design standards.
4. In order to create opportunities for the application of innovative new biotechnology products, all regulation in environmental and health areas—whether or not they address biotechnology—should use performance standards rather than specifying rigid controls or specific designs for compliance.

However, since the issuance of this report no rule making has occurred, and the U.S. elections of 1992 changed the political climate for regulatory policy. The Council on Competitiveness has been eliminated, and no clear picture has emerged regarding the coordinated formation of regulatory policy. It is noted, however, that the vice president, while serving in the Senate, encouraged regulation in this area, and his domestic affairs adviser was the author of the congressional bill for the comprehensive regulation of field research with *recombinant DNA manipulated organisms,* which was defeated in 1990. Biotechnology is rapidly developing as a tool in so many areas of potential benefit to the environment that the administration cannot afford to interfere with one of the strongest elements of U.S. technology-driven industry.

*See also* TRANSGENIC ANIMAL PATENTS.

### *Bibliography*

European Community (1990) Council Directive on the contained use of genetically modified microorganisms (90/219/EEC). *Offic. J Eur. Commun.* L117:1–14.

———. (1991) Council Directive on the deliberate release into the environment of genetically modified organisms (91/220/EEC). *Offic. J Eur. Commun.* L239:23–26.

Exercise of federal oversight within scope of statutory authority: Planned introduction of biotechnology products into the environment. (1992) *Fed. Regist.* 57:6753–6762.

Report on National Biotechnology Policy. (1991) The President's Council on Competitiveness, Executive Office of the President, Washington, DC.

U.S. Congress, Office of Technology Assessment. (1991) Biotechnology in a global economy, OTA-BA-494. U.S. Government Printing Office, Washington, DC.

# BIOTRANSFORMATIONS OF DRUGS AND CHEMICALS

*Paul R. Ortiz de Montellano*

### *Key Words*

**Electrophile** A compound, group, or function that is electron deficient and therefore preferentially reacts with electron-rich moieties.

**Nucleophile** A compound, group, or function that is electron rich and therefore preferentially reacts with electron-deficient moieties.

**Xenobiotics** Compounds foreign to an organism.

Life is based on the segregation of biological systems into compartments by means of lipid bilayer membranes. This applies to subcellular organelles such as the nucleus and mitochondria, to the cells themselves, and through the assembly of cells into surfaces, to organs and larger entities. The membrane barriers prevent the passage of charged or highly polar molecules, with the exception of small molecules that diffuse through pores in the membrane barriers, or molecules for which active transport systems exist. Lipophilic molecules can cross the membrane barriers relatively unimpeded. In contrast, mammals and other complex organisms efficiently excrete polar or ionic compounds only because lipophilic compounds are sequestered in membranes and are readily reabsorbed from excretory compartments such as the kidney. Xenobiotics that are readily taken up by biological systems are therefore the very same compounds that cannot be readily excreted.

The general purpose of xenobiotic metabolism is to convert lipophilic compounds to polar materials that can be eliminated. For this reason, the highest concentrations of enzymes of drug metabolism are found at sites that are portals of entry into the organism, such as the lungs, intestine, and liver of mammals. It is instructive in this context to consider the case of compounds, mostly man-made polyhalogenated chemicals, that are resistant to the enzymes evolved for the metabolism of natural products. The polyhalogenated biphenyls, for example, accumulate in adipose tissues and may remain there for the lifetime of the individual. This accumulation of lipophilic, poorly metabolized compounds in fatty tissues is the basis for the well-known concentration of xenobiotics in natural food chains.

## 1 CLASSIFICATION AND PROPERTIES OF ENZYMES OF XENOBIOTIC METABOLISM

The majority of the enzymes of drug metabolism catalyze oxidation/reduction, hydrolysis, and conjugation reactions, although there is some ambiguity in assigning enzymes to the drug metabolism system. For organizational purposes, the redox and hydrolytic enzymes are considered to be responsible for early stages of xenobiotic metabolism and are known as phase I enzymes. The conjugative enzymes are similarly thought to be responsible for the latter stages of metabolism and are known as phase II enzymes. This classification is artificial because phase II conjugation reactions are often not preceded by phase I reactions, although it is true that phase I redox or hydrolytic reactions usually do not occur after phase II conjugations. Phase I xenobiotic metabolism is predominantly catalyzed by cytochrome P450 and flavoprotein monooxygenases, monoamine oxidases, alcohol and aldehyde dehydrogenases, esterases, amidases, and epoxide hydrolases. The enzymes involved in phase II metabolism are primarily the glucuronyltransferases, sulfotransferases, *N*-acyltransferases, and glutathione transferases. Despite their involvement in drug metabolism, some members of both the phase I and phase II classes of enzymes are primarily or exclusively involved in processing of endogenous substrates. Furthermore, many drugs resemble endogenous substances and are therefore metabolized by more specialized enzymes. A case in point is provided by fatty acid β-oxidation, a pathway for the catabolism of endogenous fatty acids that also degrades xenobiotic alkyl chains that terminate in carboxyl groups.

Some generalizations can be made concerning the properties of enzymes that are primarily involved in xenobiotic metabolism. In contrast to most enzymes, the enzymes of drug metabolism are not highly substrate specific and are usually stereoselective rather than stereospecific. There are not many classes of drug-metabolizing enzymes, but each class encompasses multiple enzyme forms. This combination of loose substrate specificity and enzyme multiplicity provides the flexibility required to metabolize the diversity of natural and man-made substances to which organisms are routinely exposed. The ability to deal with specific xenobiotics is enhanced by the transient increase in the concentrations of some of the enzymes, most notably individual cytochrome P450 monooxygenases, upon exposure to their substrates. This selective induction of

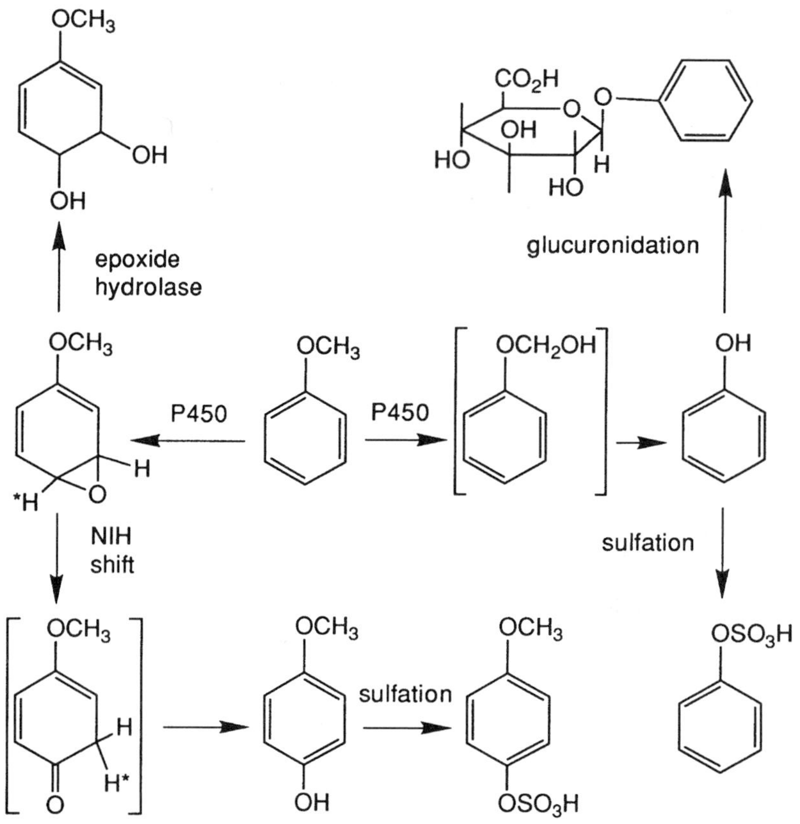

Figure 1. Selected pathways for the metabolism of anisole.

enzyme forms required to deal with particular xenobiotics enhances the ability of the system to deal with xenobiotic exposure. The mechanism of enzyme induction has been shown in the case of one cytochrome P450 enzyme to involve binding of inducers to a soluble receptor, which translocates to the nucleus and enhances expression of the appropriate gene. The extent to which this mechanism applies to the induction of other enzymes, even other cytochrome P450 enzymes, is not yet known.

## 2 PHASE I ENZYMES

### 2.1 Cytochrome P450 and Flavoprotein Monooxygenases

The cytochrome P450 monooxygenases are hemoproteins that insert one atom of molecular oxygen into their substrates while reducing the other to water. The electrons required for oxygen reduction are provided by reduced pyridine nucleotides via a flavoprotein known as cytochrome P450 reductase. The catalytically activated oxygen is inserted into C–H or N–H bonds to give hydroxy derivatives, added to $\pi$ bonds to give epoxides, or transferred to nitrogen or sulfur electron pairs to give *N*-oxides or sulfoxides. Other oxidative outcomes (e.g., dehydrogenation of alkyl chains, deformylation of aldehydes) are known but are much less common. Under conditions of low oxygen tension, cytochrome P450 enzymes also catalyze reductive reactions, including nitro and azo reduction and alkane dehalogenation.

The initial metabolites often decompose to more stable products, thereby greatly enhancing the diversity of products formed by cytochrome P450. For example, the hydroxymethyl product generated by cytochrome P450 catalyzed hydroxylation of the methoxy group of anisole (methoxybenzene) readily decomposes to give phenol and formaldehyde, and the unstable epoxide of benzene produced by the enzyme rearranges to phenol by what is known as the NIH shift mechanism (Figure 1). Further product diversity is introduced by the presence of multiple forms of cytochrome P450 that differ in substrate specificity and, for substrates common to more than one form of the enzyme, in regio- and stereoselectivity.

The flavoprotein monooxygenases catalyze the NADPH- and oxygen-dependent oxidation of electron-rich nitrogen and sulfur atoms. These enzymes therefore have a more restricted metabolic scope than cytochrome P450 enzymes but can play a major role in the metabolism of drugs and xenobiotics containing nitrogen or sulfur.

### 2.2 Monoamine Oxidase

Monoamine oxidase is an oxygen-dependent, flavin-containing enzyme that oxidizes the amine functions of endogenous neurotransmitters and related xenobiotics to the corresponding imines. The imines are then hydrolyzed to the carbonyl derivatives, as shown for 2-phenylethylamine:

$$PhCH_2CH_2NH_2 \rightarrow PhCH_2CH{=}NH \rightarrow PhCH_2CHO \qquad (1)$$

### 2.3 Alcohol and Aldehyde Dehydrogenases

Alcohol dehydrogenases catalyze the oxidation of alcohols to aldehydes or ketones as well as the reverse reaction. The direction of the reaction depends on the redox balance of the pyridine nucleotide pool and whether the alcohol or carbonyl compound is the substrate. Aldehyde dehydrogenases, however, unidirectionally catalyze the oxidation of aldehydes to acids. The simplest drug oxidized by these enzymes is ethanol, which is converted to acetaldehyde and subsequently to acetic acid by the sequential action of these two enzymes.

### 2.4 Esterases and Amidases

The hydrolysis of xenobiotic ester and amide bonds is catalyzed by a ubiquitous but still poorly defined family of enzymes. As is true chemically, ester hydrolysis occurs more readily than amide hydrolysis. The products of these reactions are the corresponding acid and alcohol or amine.

### 2.5 Epoxide Hydrolases

Epoxide hydrolysis is catalyzed by both microsomal and cytosolic epoxide hydrolases. The reaction involves backside addition of water to the epoxide to give the *trans*-diol. Epoxide functionalities are not generally found in drugs or xenobiotics because of their chemical reactivity, but epoxide hydrolysis is a very important process because epoxides are formed in situ by the cytochrome P450 catalyzed oxidation of double bonds (Figure 1).

## 3 PHASE II ENZYMES

### 3.1 Glucuronyltransferases

Glucuronidation is a common, quantitatively very important pathway of xenobiotic metabolism. The glucuronyltransferases are membrane-bound enzymes that transfer the glucuronic acid moiety from uridine-5′-diphospho-α-D-glucuronic acid to a nucleophilic functional group on the substrate. The reaction involves backside displacement of the phosphate leaving group by the substrate nucleophile. Functional groups bearing an OH, NH, or SH moiety, including phenols, alcohols, carboxylic acids, hydroxylamines, amines, and mercaptans, readily undergo glucuronidation (Figure 1). Although less common, glucuronidation of trialkylamines to give quaternary ammonium glucuronides is also observed, as is the glucuronidation of carbons that are ionized at physiological pH. With few exceptions, glucuronidation occurs only once and is the last step in a metabolic sequence.

### 3.2 Sulfotransferases

Sulfation of xenobiotics occurs nearly as frequently as glucuronidation. The sulfate donor pool is limited; therefore, unlike the case of glucuronidation, the quantitative importance of sulfation usually decreases as the dose of the drug or xenobiotic increases. Sulfotransferases are soluble enzymes that transfer the sulfate from 3-phosphoadenosine-5′-phosphosulfate to nucleophilic functions on the xenobiotic or its phase I metabolites (Figure 1). The range of functionalities that undergo sulfation is similar to, but more limited than, the range that undergoes glucuronidation. Phenol, alcohol, and amine groups are readily sulfated, but carboxyl and sulfhydryl groups and trialkylamines are not detectably sulfated.

### 3.3 *N*-Acyltransferases

*N*-acyltransferases are involved in xenobiotic metabolism in three ways. Xenobiotics with a carboxyl group can be conjugated to amino acids, commonly glycine or glutamine, by a process that involves activation of the acid to its CoA ester followed by *N*-acyltransferase catalyzed condensation with the amino acid. The conversion of benzoic to hippuric acid is an example of amino acid conjugation:

$$PhCO_2H \rightarrow PhCONHCH_2CO_2H \quad (2)$$

A second reaction mode involves acetylation of xenobiotic amino groups by *N*-acyltransferase catalyzed reaction with acetyl CoA, as illustrated for phenelzine:

$$PhCH_2CH_2NHNH_2 \rightarrow PhCH_2CH_2NHNHCOCH_3 \quad (3)$$

The third important reaction catalyzed by *N*-acyltransferases is the acetylation of cysteine conjugates produced by the glutathione pathway (see Section 3.5). Acetylation is one of the two general reactions that decrease rather than increase polarity and therefore can occur at any stage in the metabolic process.

### 3.4 Methyltransferases

The second general reaction that decreases substrate polarity is the methylation of phenols, thiols, and selected amines by *S*-adenosyl-L-methionine-dependent methyltransferases. O-Methylation of one of the hydroxyl groups in dopamine and N-methylation of the imidazole in histamine are examples of this biotransformation.

### 3.5 Glutathione Transferases

The glutathione conjugation system provides the major line of defense against electrophilic, chemically reactive xenobiotics and xenobiotic metabolites. Glutathione is a tripeptide in which the carboxyl group of cysteine in linked to a glycine, and the amino group to the γ-carboxyl group of glutamic acid (Figure 2). The key function in the tripeptide is the cysteine sulfhydryl. Glutathione conjugation involves addition of the sulfhydryl to electrophilic agents to give thioether or thioester products. Sufficiently reactive xenobiotics or xenobiotic metabolites can be trapped by simple chemical reaction with the 5 to 10 mM concentrations of glutathione normally found in tissues, but a family of glutathione transferases is available to catalyze the reaction with less reactive electrophiles. Essentially all electrophilic functionalities are subject to reaction with glutathione, but the most common include alkyl and acyl halides, α,β-unsaturated carbonyl functions, and epoxides.

The formation of glutathione conjugates is only the first step of the glutathione metabolic pathway (Figure 2). The thioether or thioester conjugates are stripped in the kidney of the glycine and glutamic acid moieties, yielding cysteine adducts that are then N-acetylated (see Section 3.3). The resulting *N*-acetylcysteine adducts, known as mercapturic acids, are excreted in the urine. The cysteine adducts can also be cleaved by a β-lyase to give a derivative of the xenobiotic bearing a sulfhydryl group at the reaction site. The sulfhydryl is usually methylated and oxidized to the sulfoxide.

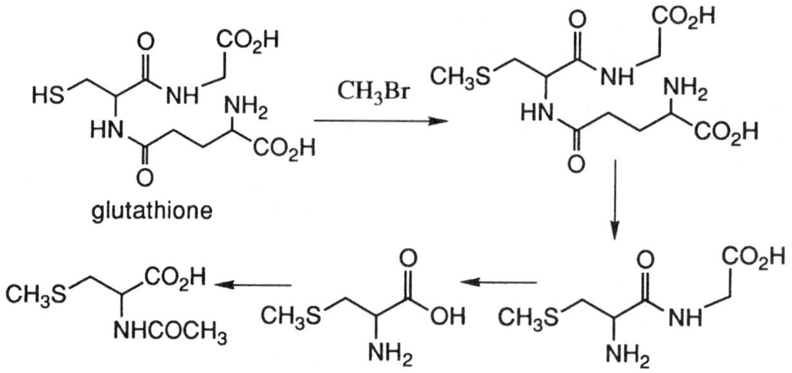

**Figure 2.** Metabolism of methyl bromide by the glutathione pathway. The final *N*-acetylcysteine adduct (''mercapturic acid'') is the excreted product.

## 4 XENOBIOTIC METABOLISM AND TOXICITY

The low reaction control implicit in the broad specificity of drug-metabolizing enzymes and the unpredictable diversity of xenobiotic structures make drug metabolism a major contributor to the toxicity and carcinogenicity of xenobiotics. For example, the cytochrome P450 catalyzed oxidation of aflatoxin and polycyclic aromatic hydrocarbons to epoxides that bind covalently to DNA is directly responsible for the carcinogenic properties of these substances. The analogous covalent binding of reactive metabolites to proteins is responsible for the toxic properties of many xenobiotics. As noted in Section 3.5, the principal role of the glutathione system is to prevent tissue damage by trapping metabolically produced reactive species. The relationship between the glutathione system and toxicity has been most clearly demonstrated in the case of acetaminophen, which is oxidized in part to a reactive iminoquinone. The iminoquinone is nontoxic at low doses because it is trapped by glutathione, but it causes hepatic necrosis at doses that deplete the glutathione supply. Thus, the metabolism of lipophilic compounds is essential for their elimination but frequently produces species of potentially higher toxicity than the parent structure.

## 5 THERAPEUTIC ASPECTS OF DRUG METABOLISM

The degree and duration of drug action is controlled by the extent and rate of absorption of the drug, the rate at which it is delivered to the target site, the extent and rate of its metabolism, and the rate at which it is cleared from the body. Factors that influence the role of metabolism in determining drug action include exposure to other drugs or xenobiotics, genetic makeup, gender, diet, age, and alterations in physiological status. Concomitant exposure to secondary agents can increase the metabolism of xenobiotics (by enzyme induction) or decrease xenobiotic metabolism (by enzyme inhibition). For example, when both phenytoin and phenobarbital are used in the treatment of epilepsy, the latter increases the metabolism of the former by inducing cytochrome P450 isozymes that oxidize phenytoin. The genetic makeup of the individual is important because the levels of individual drug-metabolizing enzymes in the human population are genetically determined. Gender and physiological status are important because hormonal factors alter, among other parameters, the levels and types of drug-metabolizing enzymes. The profile of drug metabolism is thus determined by a combination of heredity and environment and varies from individual to individual. It is therefore not surprising that there are also major differences in the metabolism of xenobiotics by different species. The species dependence of drug metabolism is a major concern in the extrapolation to man of kinetic and toxicological data obtained with animals.

*See also* CYTOCHROME P450; ENZYMES.

### *Bibliography*

Gibson, G. G., and Skett, P. (1986) *Introduction to Drug Metabolism.* Chapman & Hall, London and New York.

Guengerich, F. P., Ed. (1987) *Mammalian Cytochromes P450.* CRC Press, Boca Raton, FL.

Mulder, G. J., Ed. (1990) *Conjugation Reactions in Drug Metabolism. An Integrated Approach.* Taylor & Francis, London and New York.

Ortiz de Montellano, P. R., Ed. (1986) *Cytochrome P450: Structure, Mechanism, and Biochemistry.* Plenum Press, New York.

Testa, B., and Jenner, P. (1976) *Drug Metabolism: Chemical and Biochemical Aspects.* Dekker, New York.

# BODY EXPRESSION MAP OF THE HUMAN GENOME

*Kousaku Okubo and Kenichi Matsubara*

### *Key Words*

**Poly(A)-Tailed mRNA** In the majority of cases, eukaryotic messenger RNA has a sequence of polyadenylic acid at the 3′ end of the molecule. Deoxyoligothymidylic acid (oligo dT), complementary to the polyadenylated tail, is widely used to initiate the synthesis of first-strand cDNA, copying the messenger RNA from its 3′ terminus.

**Randomly Primed cDNA Library** A cDNA library can be constructed by using a set of randomly designed oligodeoxynucleotides as the primer of the first-strand cDNA synthesis. In contrast to the oligo dT-primed method, most of the cDNAs are copies of the internal sequences of the mRNAs, making the randomly primed library suited for representing the coding region of messenger RNA.

**3′-Directed cDNA Library** A duplex DNA (cDNA) is synthesized by using in vitro reverse transcription to copy the sequence of RNA. A collection of clones carrying cDNA in a cloning vector is called cDNA library. A cDNA library that is designed specifically to represent the 3′ sequence of mRNA using oligo dT primer is called a 3′-directed cDNA library. Similarly, a 5′-directed library is designed to represent the 5′ sequence of mRNA, but details of the mechanism are different.

---

Body mapping refers to the process of generating a gene expression map of the body by compiling data telling what genes are active to what extent in a given cell or tissue of the body. This database is constructed by compiling the expression profile of active genes in a cell or tissue. The profile consists of the names of active genes, known or novel, and the frequency of their transcripts in the mRNA population. Although the number of genes and the number of different cells or tissues in the database are far from complete at present, such expression profiles are being collected with as many cells or tissues of the body as are available, through use of high throughput sequencing and computing systems. Thus, in the near future, a body map will become a useful database for investigations of expression control of the 100,000 or so human genes, with their tissue specificity, frequency of expression, and changes in expression control reflecting cell physiology or development. Upon completion, the body map will be useful for describing the normal or aberrant physiology of cells, but more importantly, generation of this database is complementary to efforts now under way to map the human genome and sequence its DNA. In addition, it will serve as a powerful tool for discovering novel genes of interest such as cell-specific or ubiquitous genes, based on their expression pattern. The body map is different from the so-called expressed sequence tag (EST) approach, which aims at discovering genes of interest by partially sequencing cDNA clones and looking for those genes that have interesting amino acid motifs.

## 1 BACKGROUND

The human body consists of about 60 trillion cells of various shapes and functions, whose morphological, biochemical, biophysical, and developmental properties, as well as interactions, have been studied intensively. Despite the wide variety in cell phenotypes, the genomic DNA carried in each cell of a person's body is essentially identical, being a faithful copy of that originally carried in the fertilized egg. Each cell in the human body is capable of expressing a set of selected genes, probably around 10,000–20,000, under regulation by one or more mechanisms established by the cell lineage and the positional information exerted by direct or indirect cell–cell interactions. The set of expressing genes may differ from cell to cell; the details have not been elucidated so far. Body mapping aims at assigning individual genes in the human genome to the site(s) in the body where each one is active. It is a new effort in molecular biology, promoted by rapidly developing DNA technology.

In making a body map, we first generate a "gene expression profile" in the cells or tissues by (1) identifying and registering the active genes, and (2) assaying the level of their transcripts. This is done through sequencing randomly selected clones from the nonbiased 3′-directed cDNA library prepared from the appropriate cell or tissue. Using the sequence itself for identifying the active gene, and the frequency of appearance of the same sequence, the expression profiles of thousands of genes have been obtained. The number of genes that can be surveyed in the first-phase studies is in the order of thousands. We then extend these efforts to as many different cell types as possible and compile the expression profiles. Through these efforts, a significant fraction of total genes in the human genome (estimated to be about $10^5$) will be identified and mapped to the sites of expression in the body. The human body consists of some 60 trillion cells. However, because they can be grouped into only approximately 200 different types, we should be able to complete the first-phase study of gene expression profiles in the not-too-distant future. Perhaps then we can proceed to the next phase, that is, the investigation of the minority cell types, cells affected by diseases, cells in the process of differentiation, and cells that appear only during a particular stage of development, such as in the embryo and early fetus.

## 2 PRINCIPLE

### 2.1 Housekeeping Genes, Tissue-Specific Genes, and Their Abundance

In 1974 Bishop and co-workers made pioneer attempts to gain insight into global gene expression in eukaryotic cells. Upon analysis of the reassociation kinetics of messenger RNA and cDNA reverse-transcribed from mRNA, they obtained a rough estimation of the nature and number of expressing genes. Their findings can be summarized as follows:

1. In most somatic tissues of animals, about 10,000–20,000 species of genes are expressed. This means that 10–20% of the total genes in the genome are active in a given cell.
2. About one-quarter to one-half of the messenger RNA population, comprising about 10,000 species, are expressed at a low level, and about 90% of the gene products are shared by many, perhaps all, cell types. They may be called "housekeeping genes," since they are responsible for basic and common cellular activities, such as energy production, low molecular mass substrate production, and macromolecule synthesis.
3. The remaining mass of the messenger RNAs consists of abundant classes, comprised of a relatively limited number of species. Most of them code for tissue-specific proteins and often are the major products of the cell. Although such "tissue-specific" genes are very informative, we did not attempt to elucidate the nature of the active genes.

Recent advances in rapid DNA sequencing technology combined with computerized data processing have made possible the analysis at the sequence level of thousands of cDNAs. Thus, the objective of describing all the genes that are active in each of the different cell types has been realized. The relevant elements in the technology include quantitative conversion of a population of messenger RNA into a cDNA library, amplification of each cDNA by the polymerase chain reaction (PCR) or by plasmid preparations, large-scale automated cDNA sequencing, and high throughput data processing and analysis.

### 2.2 Gene Signatures, Expression Profiles, and Body Mapping

To study the expression profile of a gene in a given cell or tissue with the currently available technology, a 3′-directed cDNA library is prepared, then randomly selected clones are sequenced on a large scale. The 3′ region of mRNA, just upstream of the polyadenylated stretch in mRNA, can be quantitatively converted to cDNA, and therefore, the 3′-directed library is a faithful representative of the molar composition of the mRNA. Since the sequences at the 3′ region are unique, sequencing data of about 150–300 nucleotides are sufficient to characterize the gene. This short sequence is called the "gene signature." Since identifying the primary structure of encoded proteins is not the aim of sequencing, the gene signature sequence does not have to cover the coding region.

Recently, several groups have initiated efforts to collect partial cDNA sequences, using randomly primed cDNA libraries or sets of clones called 5′-directed cDNA libraries. These analyses have been powerful tools for discovering active genes whose products are similar to proteins of biological or commercial interest, or, at least whose products have some interesting amino acid sequence motifs (e.g., secretion signal, DNA-binding domain, metal-binding domain). Such partial cDNA sequences, called expressed sequence tags, may also be used for registering active genes in the tissue. However, ESTs cannot be used to describe the molar composition of the mRNA because a randomly primed cDNA library does not tell the investigator whether two different partial sequences are from different genes, or from different parts of the same transcript. The currently available 5′-directed cDNA libraries or size-selected ones are also inadequate for quantitative studies.

A homology search of gene signatures in the DNA data bank, in which about $5 \times 10^5$ sequences have been installed, reveals that about 20% of the gene signatures thus collected can be identified as known genes; thus 80% are novel genes. The abundance of the mRNA can be measured by counting the number of appearances of the same "signature." With the help of automated sequencers, the mRNA population of a given cell or tissue can yield thousands of gene signatures, with their frequency profiles. The resulting data are referred to as the "gene expression profile" of the tissue. By comparing the expression profiles of various materials, tissue-specific and housekeeping genes can be identified. Through the stepwise collection of expression profiles of as many tissues or cells as possible, we can approach the "body map of expressing genes," a comprehensive collection containing each gene in the human genome at the assigned site(s) of expression in the body. In other words, in body mapping genes are assigned not to chromosomes, but to the tissues in which they are active.

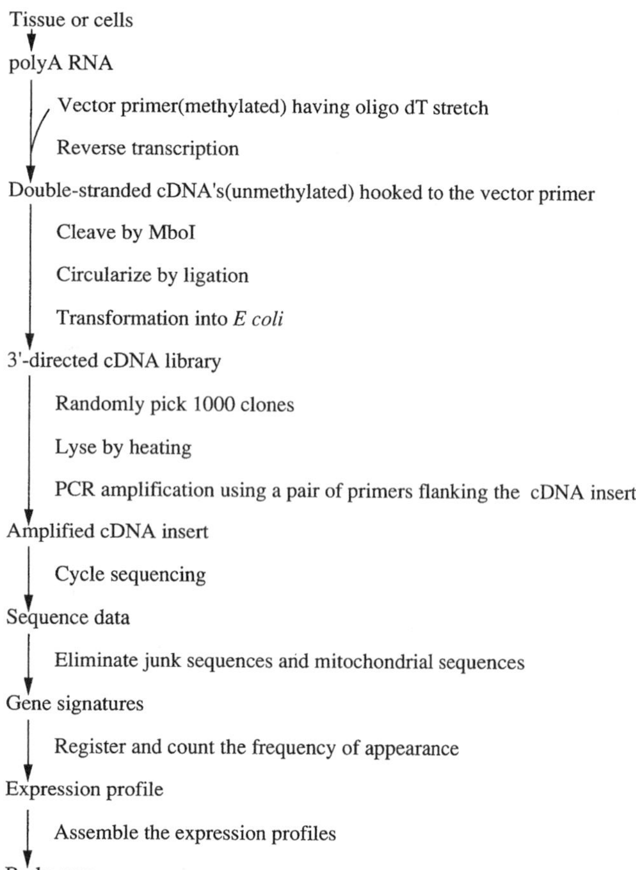


**Figure 1.** Flowchart for constructing the body map of expressed genes.

## 3 DATA COLLECTION

Figure 1 shows the flow chart of data collection. Instructions are given in Sections 3.1 and 3.2.

### 3.1 The 3′-Directed cDNA Library

Select any fresh tissue or a cell line. To reduce the complexity of the data, the sample should be as homogeneous as possible. Prepare poly(A)-RNA and use it as template for the reverse transcriptase synthesis of cDNA. Use a vector primer that is dam-methylated at the *Mbo* I(GATC) sites and has a 3′ protrusion of homo dT stretch at one end. Cleave the double-stranded cDNA moiety with *Mbo* I. Then the vector that is attached with the 3′-cDNA [from poly(A) to the nearest *Mbo* I site] will be circularized by ligation and transformed into *E. coli*. Pick the resulting transformants randomly, lyse individually by brief boiling, and use the lysate as a template for PCR amplification. Use the resulting short double-stranded DNA for the cycle sequencing reaction, and analyze using an autosequencer. High throughput sequencing analysis is crucial to this type of study.

### 3.2 Data Analysis

Collect the resulting sequence data that are longer than 20 nucleotides, and subject them to survey by a computer. The mitochondria-coded gene transcripts are eliminated first because this study is focused on genes on the chromosomes. The remaining sequences are the gene signatures. Even if the sequence contains some repetitive elements such as Alu or GT repeats, these can serve as signatures as long as the rest of the sequence is unique. Register them and count the frequency of appearance to obtain the expression profile; then subject them to a similarity search in a DNA data bank. The gene expression profile of the tissue consists of identified and novel genes, but no discrimination should be made.

As a rule, about 1000 clones are analyzed in a given cell or tissue, and the majority of the abundant transcripts is expected to be included among these. To go beyond this number, elimination of the abundant clones is recommended. Finally, the frequency profiles of gene expression are compared among different tissues for body mapping.

## 4 EXAMPLES

The abundance of each gene signature is in parallel with the gene's relative expression level in the tissue and represents tissue-specific

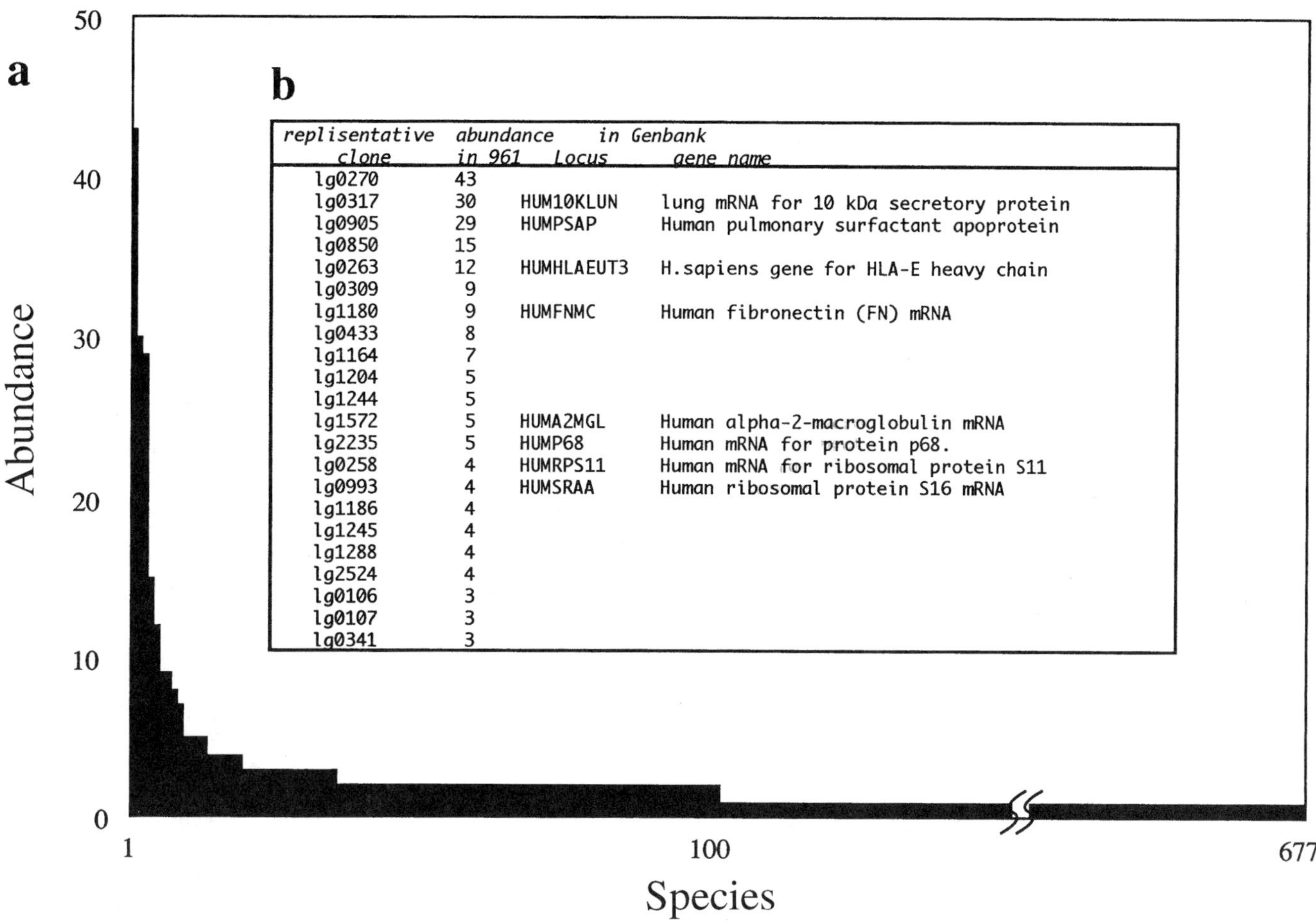


| replisentative clone | abundance in 961 | in Genbank Locus | gene name |
|---|---|---|---|
| lg0270 | 43 | | |
| lg0317 | 30 | HUM10KLUN | lung mRNA for 10 kDa secretory protein |
| lg0905 | 29 | HUMPSAP | Human pulmonary surfactant apoprotein |
| lg0850 | 15 | | |
| lg0263 | 12 | HUMHLAEUT3 | H.sapiens gene for HLA-E heavy chain |
| lg0309 | 9 | | |
| lg1180 | 9 | HUMFNMC | Human fibronectin (FN) mRNA |
| lg0433 | 8 | | |
| lg1164 | 7 | | |
| lg1204 | 5 | | |
| lg1244 | 5 | | |
| lg1572 | 5 | HUMA2MGL | Human alpha-2-macroglobulin mRNA |
| lg2235 | 5 | HUMP68 | Human mRNA for protein p68. |
| lg0258 | 4 | HUMRPS11 | Human mRNA for ribosomal protein S11 |
| lg0993 | 4 | HUMSRAA | Human ribosomal protein S16 mRNA |
| lg1186 | 4 | | |
| lg1245 | 4 | | |
| lg1288 | 4 | | |
| lg2524 | 4 | | |
| lg0106 | 3 | | |
| lg0107 | 3 | | |
| lg0341 | 3 | | |

**Figure 2.** (a) Expression profile of genes in the adult human lungs: the frequency of appearance of each gene signature in 961 randomly sequenced clones. (b) Table of data bank identities of the most abundant 22 signatures.

**Table 1** Body Map of Human Genes[a]

| Gene Signatures | HepG2 Cells | Liver 19w | Liver 40w | H160 Cells | Macro-phages | Granulo-cytes | Adipose | Aorta | Lung | Colon |
|---|---|---|---|---|---|---|---|---|---|---|
| 001 | 0 | 0 | 0 | 0 | 0 | 0 | 0 | 0 | 43 | 0 |
| 002 | 0 | 0 | 0 | 0 | 0 | 0 | 0 | 0 | 30 | 0 |
| 003 | 0 | 0 | 0 | 0 | 0 | 0 | 0 | 0 | 29 | 0 |
| 004 | 0 | 0 | 7 | 2 | 2 | 0 | 15 | 2 | 15 | 44 |
| 005 | 0 | 0 | 0 | 1 | 0 | 5 | 0 | 6 | 12 | 0 |
| 006 | 5 | 2 | 3 | 7 | 2 | 1 | 92 | 6 | 9 | 27 |
| 007 | 0 | 4 | 6 | 0 | 0 | 0 | 0 | 31 | 9 | 0 |
| 008 | 0 | 0 | 17 | 1 | 4 | 6 | 23 | 13 | 8 | 44 |
| 009 | 0 | 0 | 16 | 1 | 2 | 0 | 8 | 0 | 7 | 20 |
| 010 | 8 | 4 | 0 | 23 | 10 | 2 | 16 | 9 | 5 | 5 |
| 011 | 0 | 0 | 4 | 0 | 0 | 1 | 13 | 0 | 5 | 20 |
| 012 | 0 | 5 | 3 | 0 | 0 | 0 | 0 | 0 | 5 | 1 |
| 013 | 0 | 0 | 0 | 0 | 0 | 2 | 7 | 2 | 5 | 0 |
| 014 | 2 | 1 | 2 | 11 | 0 | 1 | 5 | 4 | 4 | 0 |
| 015 | 1 | 1 | 0 | 2 | 3 | 1 | 1 | 1 | 4 | 0 |
| 016 | 0 | 0 | 0 | 0 | 1 | 1 | 1 | 5 | 4 | 2 |
| 017 | 0 | 1 | 0 | 0 | 1 | 0 | 0 | 0 | 4 | 0 |
| 018 | 0 | 0 | 0 | 0 | 0 | 0 | 0 | 0 | 4 | 8 |

[a]Expression levels of the 18 most active human genes in the adult lungs are listed. The numbers in each column represent the frequency of appearance of gene signatures in each library among 1000 randomly selected clones.

expression control of each gene. Figure 2 shows an example of the expression profile of the adult human lung. This profile contains the results of sequence analysis of about 961 cDNA clones. Table 1 summarizes the assembly of such data with tissue or cell lines. For simplicity, only the top 18 genes are shown. For example, genes 1 through 3 are expressed heavily in the adult human lungs but not in other tissues. On the contrary, genes 4 through 6 are expressed ubiquitously in human tissues. All numbers in Table 1 are relative expression levels represented by the frequency of appearance in about 1000 randomly selected cDNA clones. Although no amino acid sequence information is shown, it is clear that gene 1 is related to a lung-specific function, whereas gene 4 is related to housekeeping function. At present such information exists for about 5000 human genes.

## 5 PERSPECTIVES

Today, large-scale fragmentary cDNA sequencing is being performed to collect expressed sequence tags, mostly to discover novel genes that have interesting amino acid sequence motifs for possible commercial use. In contrast, the gene signatures are collected primarily for expression profiles and body mapping. When data of this type have been comprehensively collected into a body map, a significant fraction of human genes can be mapped to the tissue at which they are active, representing the functional aspects of the genome. Both the gene signatures and ESTs can be used for mapping genes along the genomic DNA, which represents the structural aspects of the genome. Relating these two maps will then lead to the decoding of all genetic information carried in the human genome.

Expression profiles are powerful tools that are used in biological sciences. By comparing the expression profiles of a cell undergoing differentiation, or by comparing normal and disease-affected cells, it is easy to find up-regulating or down-regulating gene expression. The discovery of genes acting exclusively in a specific tissue will profoundly affect the pharmaceutical sciences. It is not too far-fetched to imagine that diagnosis, designing of drugs, gene therapy, and other medical sciences and technologies will all be affected by these studies, along with basic biology.

There is increasing demand that the gene signature contain full size sequence information with which the function of genes can be predicted. Analyses of small samples, or a much larger number of samples, or analysis at a higher speed, is also awaited. The rapid progress in DNA technology has made us optimistic about our ability to cope with these demands.

*See also* DNA Markers, Cloned; Gene Expression, Regulation of; Genomic Imprinting; Human Disease Gene Mapping.

### *Bibliography*

Adams, M. D., Kelley, J. M., et al. (1991) Complementary DNA sequencing: Expressed sequence tags and the Human Genome Project. *Science,* 252:1651–1656.

Alberts, B., Bray D., et al. (1989) *Molecular Biology of the Cell,* 3rd ed. New York, Garland Publishing.

Bishop, J. O. (1974) The gene numbers game. *Cell,* 2:81–86.

Lewin, B. (1980) *Gene Expression 2.* Wiley-Interscience, New York.

Matsubara, K., and Okubo, K. (1993) Identification of new genes by systematic analysis of cDNA and database construction approach. *Curr. Opin. Biotechnol.* 4:672–677.

Okubo, K., Hori, N., et al. (1992) Large scale cDNA sequencing for analysis of quantitative and qualitative aspects of gene expression. *Nature Genet.* 2:173–179.

Southern, E. M. (1992) Genome mapping: cDNA approaches. *Curr. Opin. Genet. Dev.* 2(3):412–416.

# Brain Aromatic Amino Acid Hydroxylases

*Seymour Kaufman and Paula Ribeiro*

### *Key Words*

**$BH_4$** (6*R*)-5,6,7,8-Tetrahydrobiopterin; a naturally occurring pterin cofactor of the aromatic amino acid hydroxylases.

**CaM Kinase II** $Ca^{2+}$/calmodulin-dependent protein kinase.

**Catecholamines** A class of compounds that are synthesized from tyrosine, including the neurotransmitters dopamine and noradrenaline, and the neurohormone adrenaline.

**PKA** cAMP-dependent protein kinase.

**PKC** $Ca^{2+}$/phospholipid-dependent protein kinase.

Tyrosine and tryptophan hydroxylase, the aromatic amino acid hydroxylases of the mammalian nervous system, catalyze the rate-limiting steps in the biosynthesis of catecholamines and serotonin, respectively, which are among the most widely distributed and functionally diverse of mammalian neurotransmitters. The two neuronal enzymes are members of a family of pterin-dependent aromatic amino acid hydroxylases whose prototype is phenylalanine hydroxylase of liver. The three hydroxylases share important structural and mechanistic properties and are almost certainly derived from the same family of proteins. On the other hand, each hydroxylase is responsive to specific physiological requirements and is therefore subject to unique mechanisms of regulation.

The purpose of this chapter is to discuss some of the salient properties of the neuronal aromatic amino acid hydroxylases. Rather than attempt to review everything that is known about these enzymes, we focus attention on their structural properties and on the molecular mechanisms that regulate their activities.

## 1 REACTIONS OF TYROSINE AND TRYPTOPHAN HYDROXYLASE

An understanding of the processes that mediate the hydroxylation of aromatic amino acids came from studies of a nonneuronal enzyme, hepatic phenylalanine hydroxylase. This enzyme is part of a multicomponent system that catalyzes the hydroxylation of phenylalanine to tyrosine. In addition to the hydroxylase, the system consists of two other essential components, a nonprotein coenzyme that was identified as tetrahydrobiopterin ($BH_4$) and a second enzyme, identified as dihydropteridine reductase. $BH_4$ and the reductase were later shown to be essential components of the tyrosine and tryptophan hydroxylating systems. The details of these reactions are summarized schematically in Figure 1. Tryptophan hydroxylase catalyzes the hydroxylation of tryptophan to 5-hydroxytryptophan (5-HTP), a short-lived intermediate that is rapidly decarboxylated to serotonin. On the other hand, tyrosine hydroxylase catalyzes the hydroxylation of tyrosine to 3,4-dihydroxyphenylalanine (dopa),

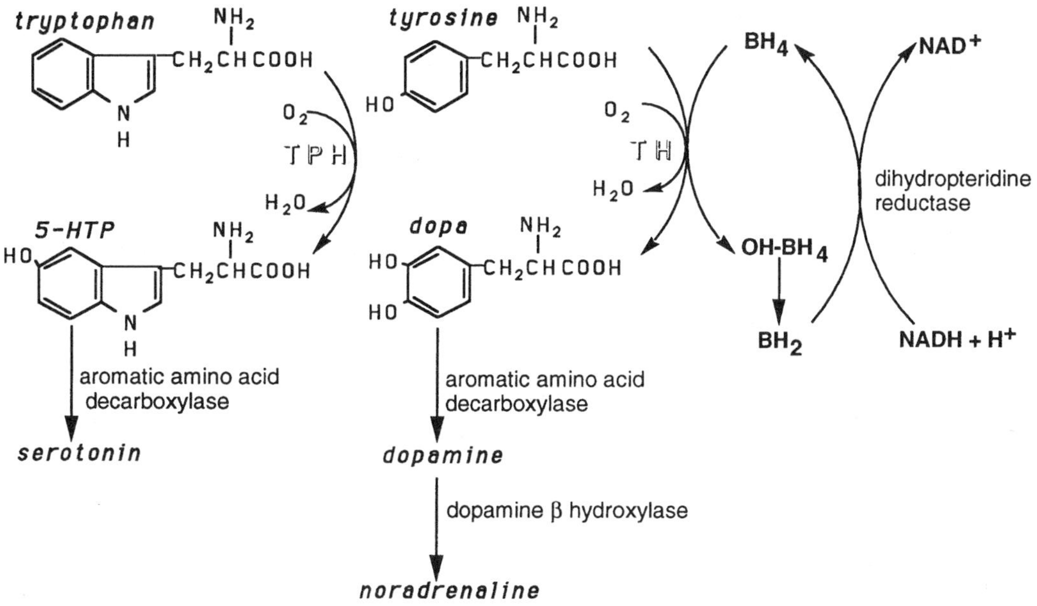

**Figure 1.** Scheme for the reactions of tyrosine hydroxylase (TH) and tryptophan hydroxylase (TPH) and the pathways leading to the biosynthesis of catecholamines and serotonin, respectively.

the first in a series of reactions that leads to the synthesis of dopamine and noradrenaline. Each of these hydroxylation reactions requires molecular oxygen and $BH_4$; the latter compound is converted to 4a-hydroxytetrahydrobiopterin (OH-$BH_4$), which is formed in equal amounts with the hydroxylated amino acid product during the tyrosine hydroxylase catalyzed reaction and, in all likelihood, during the reaction catalyzed by tryptophan hydroxylase as well. OH-$BH_4$ is converted to quinonoid dihydropterin ($BH_2$) by a dehydratase; $BH_2$ is then reduced to the tetrahydro level by the action of dihydropteridine reductase, which utilizes NADH as the electron donor. This reaction serves to regenerate the active form of the pterin which, in turn, allows the coenzyme to function catalytically.

## 2 TYROSINE HYDROXYLASE

### 2.1 Physical Properties

Tyrosine hydroxylase is present in catecholaminergic cells of the central nervous system (CNS) and periphery. In the brain, the enzyme is enriched in the areas associated with the dopaminergic nigrostriatal system and the noradrenergic neurons of the locus coeruleus. Outside the CNS, tyrosine hydroxylase is found in the sympathetic nervous system, in the adrenal medulla, and in adrenomedullary tumors (pheochromocytoma). The hydroxylase has been purified to homogeneity from brain, adrenal medulla, and pheochromocytoma. The pure enzyme is a tetramer ($M_r \approx 240$ kDa) composed of four identical subunits. Like phenylalanine hydroxylase, tyrosine hydroxylase requires iron for catalytic activity; the purified enzyme contains amounts of iron that approach 1 mol per mol of enzyme subunit. Although the form of the enzyme-bound iron that functions in catalysis is believed to be in the ferrous state, some of the iron in tyrosine hydroxylase is ferric, Fe(III), and appears to be coordinated to low levels of catecholamines that are copurified with the enzyme. This complexation with catecholamines has important consequences for the regulation of enzyme activity (see Section 2.3).

The amino acid sequence of tyrosine hydroxylase has been deduced from a complementary DNA nucleotide sequence that is highly conserved in a large number of mammalian species. Based on the analysis of a full-length cDNA derived from rat pheochromocytoma, each subunit of tyrosine hydroxylase contains 498 amino acids with a predicted molecular weight of 55,903, which is consistent with the estimated molecular weight of the tetramer. In lower mammals (i.e., nonprimates) there is but one species of tyrosine hydroxylase messenger RNA that codes for a single type of enzyme. In primates, however, and particularly in humans, the primary mRNA transcript undergoes alternative splicing to generate multiple tyrosine hydroxylase mRNA species. Man has four such mRNAs that express four types of enzyme (types 1–4), with slightly different numbers of amino acids and catalytic activities. Type 1 has the highest specific activity and shows the greatest homology with rat tyrosine hydroxylase. Types 1 and 2 are also the most abundant and are widely distributed in the human central nervous system and periphery. By comparison, human types 3 and 4 are relatively rare, particularly in the brain, where they represent less than 1% of all the tyrosine hydroxylase mRNA.

A structural model for tyrosine hydroxylase is briefly outlined in Figure 2. The model is derived primarily from studies of tyrosine hydroxylase from rat pheochromocytoma, but the main components of the model are conserved in other species, including man. In this model, each subunit of tyrosine hydroxylase is organized into a regulatory domain and a catalytic center. The catalytic center binds the essential components of the tyrosine hydroxylase reaction, the substrates (tyrosine and oxygen), the cofactor ($BH_4$), and iron. Its location has been defined on the basis of proteolytic and mutagenesis data and by analyses of sequence homology with other aromatic amino acid hydroxylases. Based on these results, the amino end of the catalytic domain has been tentatively placed between amino acid residues 158 and 184, and the carboxyl end at or prior to position 455. There are at least four cysteine residues in the catalytic domain, which are conserved across species and in the sequences of the other aromatic amino acid hydroxylases. These sites are thought to be situated at the active site, where they may play a

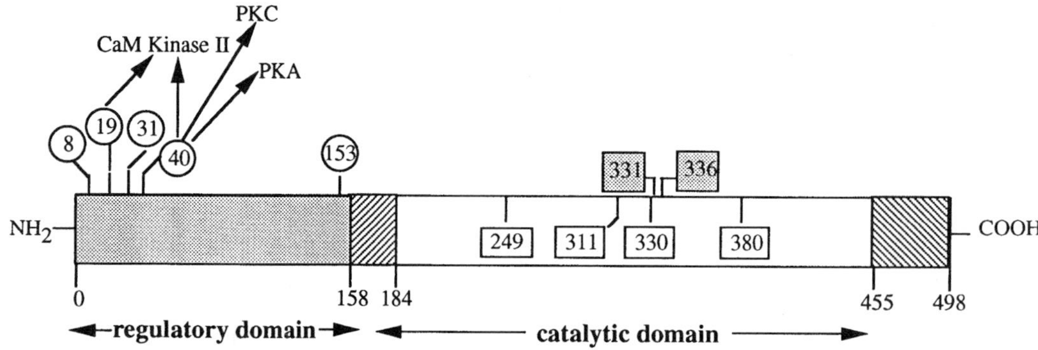


**Figure 2.** Schematic representation of the regulatory and catalytic domains of rat pheochromocytoma tyrosine hydroxylase: [After Abate and Joh, *J. Mol. Neurosci.* **2**: 203 (1991), Ribeiro *et al. J. Mol. Neurosci.* 4:125 (1993), Gibbs *et al J. Biol. Chem.* 268:8046 (1993) and Zigmond *et al Annu. Rev. Neurosci.* 12:415 (1989). Circles indicate the position of the phosphorylation sites ($Ser^{8}$, $Ser^{19}$, $Ser^{31}$, $Ser^{40}$, and $Ser^{153}$) with their corresponding protein kinases; open squares depict the position of conserved cysteine residues ($Cys^{249}$, $Cys^{311}$, $Cys^{330}$, and $Cys^{380}$); shaded squares give the position of conserved histidine residues ($His^{331}$ and $His^{336}$). Details of this model are described in the text.

role in maintaining proper folding through the formation of disulfide bonds. In addition, there are two highly conserved histidine residues which may function as nonheme binding sites for iron. Beyond residue 455 is a region of 43 amino acids that is thought to be involved in the folding and/or oligomerization of the enzyme.

In addition to the catalytic region, Figure 2 shows the regulatory domain of tyrosine hydroxylase, located within the first N-terminal 157 amino acids. The removal of this region either by proteolysis or deletion mutagenesis causes a substantial increase in catalytic activity. This suggests that the N-terminus imposes a constraint on the active site and may mediate the down regulation of the enzyme. The N-terminus contains all the phosphorylation sites for tyrosine hydroxylase. Each subunit has multiple phosphorylation sites: all serine residues located at positions 8, 19, 31, 40, and 153. The main sites of phosphorylation, $Ser^{19}$ and $Ser^{40}$, are conserved in all mammalian species, including man. $Ser^{40}$ is phosphorylated by a multiplicity of protein kinases, including cAMP-dependent protein kinase (PKA), $Ca^{2+}$/phospholipid-dependent protein kinase (PKC), and $Ca^{2+}$/calmodulin-dependent protein kinase (CaM kinase II), although it is unclear whether the last kinase phosphorylates $Ser^{40}$ in vivo. On the other hand, CaM kinase II is the only protein kinase that phosphorylates $Ser^{19}$ residues *in vitro* as well as in intact nerve endings. The third highly conserved residue, $Ser^{31}$, is phosphorylated in vitro and in rat PC12 cells by ERK1 and ERK2, two myelin basic protein- and microtubule-associated protein kinases, whose activities are regulated by the phosphorylation of tyrosine and threonine residues in the kinase. Finally, $Ser^{153}$ is a substrate for PKA, and $Ser^{8}$ is phosphorylated both by a protein kinase that copurifies with tyrosine hydroxylase (endogenous kinase) and by a proline-directed protein kinase, so named because it recognizes the sequence Xaa-Ser/Thr-Pro-Xaa. These two sites are not conserved in a number of mammalian species, however, and their phosphorylation may or may not have any significance in vivo.

## 2.2 Activation by Phosphorylation

There is increasing evidence that the phosphorylation of specific sites of tyrosine hydroxylase mediates the acute regulation of enzyme activity in response to neuronal stimulation. For example, depolarization of catecholaminergic nerve terminals leads to an increase in tyrosine hydroxylase activity that correlates with phosphorylation of the enzyme at $Ser^{19}$. The second messenger in this pathway appears to be calcium, which stimulates endogenous CaM kinase II activity and consequently increases the phosphorylation of $Ser^{19}$ residues. This same mechanism appears to be involved in the heteroregulation of neuronal tyrosine hydroxylase activity by nicotinic agonists. On the other hand, the regulation by certain peptide neurotransmitters and possibly dopamine, acting via presynaptic autoreceptors, is mediated by the modulation of PKA activity and the subsequent phosphorylation of $Ser^{40}$. Thus, the multiplicity of phosphorylation sites in tyrosine hydroxylase and their specificities for different protein kinases allow for subtle regulation of enzyme activity by a variety of neuroeffectors.

A greater understanding of how phosphorylation affects the properties of tyrosine hydroxylase has come from *in vitro* studies of the purified enzymes. These studies have shown that the phosphorylation of purified tyrosine hydroxylase by PKA activates the hydroxylase (Table 1). Among the kinetic changes caused by phosphorylation are an upward shift in the pH optimum toward 7 to 7.5, a modest decrease in the $K_m$ for the pterin cofactor, and an increase in $V_{max}$ at pH 7.2. Thus, the overall effect of phosphorylation by PKA is to increase activity at physiological pH and to render the enzyme somewhat more responsive to $BH_4$.

Phosphorylation of tyrosine hydroxylase by other protein kinases also leads to kinetic activation, but the underlying mechanisms appear to be different. Thus, phosphorylation by PKC causes a decrease in the $K_m$ for $BH_4$ but does not increase $V_{max}$ significantly, even at neutral pH. On the other hand, phosphorylation by CaM kinase II leads to an increase in $V_{max}$ without an apparent change in the $K_m$ for $BH_4$. Phosphorylation by this kinase does not activate

**Table 1** Effects of Phosphorylation with PKA on the Kinetic Profile of Purified Rat Pheochromocytoma Tyrosine Hydroxylase

| | At pH 7.2 | | |
|---|---|---|---|
| Hydroxylase | $K_m$ $BH_4$ (μM) | $V_{max}$ (μmol/min/mg) | Optimum pH |
| Nonphosphorylated | 65 | 0.3 | 6.3 |
| Phosphorylated | 20 | 0.9 | 7–7.5 |

tyrosine hydroxylase directly, however; rather, it makes the enzyme susceptible to activation by another component of the CaM kinase II phosphorylation system, namely, the "activator protein." The activator is not necessary for the phosphorylation of $Ser^{40}$ or $Ser^{19}$, but its presence is required for the expression of enhanced activity of the phosphorylated enzyme.

### 2.3 Substrate and Product Inhibition

Tyrosine hydroxylase shares with the other pterin-dependent enzymes the property of being inhibited by excessive concentrations of either of its substrates, tyrosine and oxygen. Tyrosine hydroxylase is also sensitive to inhibition by its catechol products, dopa, and, particularly, catecholamines, which are believed to be competitive with $BH_4$. Substrate inhibition is probably of physiological significance, since the enzyme is about 50% inhibited at concentrations of oxygen and tyrosine only twice their normal tissue levels. Similarly, the $K_i$ for catecholamine inhibition is thought to be within the range of the intracellular levels of these neurotransmitters and, thus, end-product inhibition is also likely to be operative under physiological conditions.

Phosphorylation of tyrosine hydroxylase by PKA has no apparent effect on substrate binding or inhibition. On the other hand, phosphorylation markedly decreases the sensitivity to product inhibition, suggesting that the interaction between these two mechanisms may be of importance for the regulation of the enzyme. A possible scheme for this dual regulation has been recently suggested. One of the foundations of this model is the finding that pure tyrosine hydroxylase contains low levels of tightly bound catecholamines, which are believed to maintain the enzyme in a state of low activity. On phosphorylation, these inhibiting catecholamines are thought to dissociate, leading to activation of the enzyme.

## 3 TRYPTOPHAN HYDROXYLASE

Tryptophan hydroxylase is present in brain, within the serotonergic neurons of the dorsal raphe nuclei, and in several peripheral tissues, including the endocrine pineal gland, mast cells and small intestine, and in carcinoid and mastocytoma tumors.

Considerably less is known about tryptophan hydroxylase than about the other two hydroxylases. This is principally because the former enzyme is difficult to purify and tends to be unstable after purification. Nonetheless, highly purified tryptophan hydroxylase has been obtained from rat and rabbit brain and mouse mastocytoma; the enzyme is a tetramer ($M_r \approx$ 220–280 kDa) composed of one type of subunit with an approximate relative molecular mass of 53 to 64 kDa. In contrast to tyrosine hydroxylase, which has the same properties in brain and in the periphery, tryptophan hydroxylase from brain and the pineal gland appear to have different properties and may be distinct enzymes.

There are strong indications that like its cognate enzymes, phenylalanine and tyrosine hydroxylase, tryptophan hydroxylase is also an iron enzyme. In addition, just as tyrosine hydroxylase is sensitive to inhibition by excessive concentrations of its substrates, tryptophan hydroxylase is inhibited by high levels of tryptophan and oxygen. A noteworthy difference between these two enzymes, however, is that tryptophan hydroxylase is much less sensitive to inhibition by its products, 5-hydroxytryptophan or serotonin. Furthermore, it remains to be established whether product inhibition plays a role in the *in vivo* regulation of tryptophan hydroxylase.

Much of the information available on the primary structure of tryptophan hydroxylase comes from sequence analyses of full-length tryptophan hydroxylase cDNAs that have been isolated from rat brain and from both rat and rabbit pineal gland. In analogy with the homologous sequences of tyrosine and phenylalanine hydroxylase, the catalytic domain of tryptophan hydroxylase has been assigned to the central and C-terminal regions of the enzyme. On the other hand, the N-terminal end is nonhomologous with the other hydroxylases and therefore probably encodes the regulatory domain. The N-terminal region of rat pineal tryptophan hydroxylase has the consensus sequence for a PKA phosphorylation site at position 58. Outside the N-terminal region, two serines at positions 260 and 443 are possible candidates for phosphorylation by CaM kinase II. At present there is no direct evidence that these sites are indeed phosphorylated. Nonetheless, the sequence analyses are supported by biochemical evidence that purified rat brain tryptophan hydroxylase is activated by exposure to PKA or CaM kinase II under phosphorylating conditions. The activation is observed only in the presence of activator protein, however, even when the added kinase is PKA. The nature of the activation is the same irrespective of the kinase, and typically leads to a small increase in specific activity and a modest decrease in the $K_m$ for the cofactor. These findings suggest that tryptophan hydroxylase may be regulated *in vivo* by cAMP- and $Ca^{2+}$-mediated second-messenger systems that involve PKA and CaM kinase II, respectively. Indeed, there is evidence from studies of crude brain extracts that both cAMP and $Ca^{2+}$ can stimulate tryptophan hydroxylase activity. It remains to be seen, however, whether this type of regulation occurs *in vivo* and whether the activation is mediated by the phosphorylation of specific sites.

*See also* Enzymes.

## *Bibliography*


Anderson, K. K., Cox, D. D., Que L., Jr., Flatmark, T., and Haavik, J. (1988) *J. Biol. Chem.* 263:18621–18626.

Dahlström, A., Belmaker, R. H., and Sandler, M., Eds. (1988) Progress in catecholamine research. In *Neurology and Neurobiology,* Vol. 42A. Liss, New York.

Darmon, M. C., Guibert, B., Leviel, V., Ehret, M., Maitre, M., and Mallett, J. (1988) *J. Neurochem.* 51:312–316.

Fujisawa, H., and Okuno, S. (1988) *Adv. Enzyme Regul.* 28:93–110.

Gibbs, B. S., Wojchowski, D., and Benkovic, S. J. (1993) *J. Biol. Chem.* 268:8046–8052.

Goldstein, M., and Greene, L. A. (1987) Activation of tyrosine hydroxylase by phosphorylation. In *Psychopharmacology: The Third Generation of Progress,* H. Meltzer, Ed., pp. 75–80. Raven Press, New York.

Haavik, J., Martinez, A., and Flatmark, T. (1990) *FEBS,* 262:363–365.

Haycock, J. W., Ahn, N. G., Cobb, M. H., and Krebs, E. G. (1992) *Proc. Natl. Acad. Sci. U.S.A.* 89:2365–2369.

Kaufman, S. (1987) Aromatic amino acid hydroxylases. In *The Enzymes,* Vol. 18, pp. 217–282. Academic Press, Orlando, FL.

Makita, Y., Okuno, S., and Fujisawa, H. (1990) *FEBS,* 268:185–188.

Nagatsu, T., and Ichinose, H. (1990) *Comp. Biochem. Physiol.* 98C:203–210.

Tong, H. Joh, Ed. (1990) Catecholamine genes. In *Neurology and Neurobiology,* Vol. 57. Wiley-Liss, New York.

Zigmond, R. E., Schwarzschild, M. A., and Rittenhouse, A. R. (1989) *Annu. Rev. Neurosci.* 12:415–461.

# Breast Cancer, Genetic Analysis of

*Patricia Margaritte-Jeannin*

*Key Words*

**Genetic Distance** A function of the proportion of parental gametes that have recombined during meiosis. The unit of measurement is the morgan: 0.01 morgan, or 1 centimorgan (cM) corresponds to a proportion of recombined gametes of 1%.

**Genetic Marker** Polymorphic DNA sequence with a known location on the genome.

**Linkage** Describes the tendency of genes to be inherited together as a result of their location on the same chromosome. Linkage is measured by the genetic distance between loci.

**Penetrance** Probability that an individual with a given genotype will develop the disease.

**Segregation** The way in which a trait (e.g., a disease) is distributed among family members.

In spite of improvements in health care and disease prevention, breast cancer remains an important cause of female mortality and a major public health problem. It is the most frequent female cancer and occurs only rarely in men. Given the current life expectancy, the risk that a woman will develop such a cancer is close to 10%. For a long time it has appeared that certain families are more affected than others by this disease, suggesting that certain cases might be due to hereditary factors. For this reason, several research groups have performed familial analyses of breast cancer.

Genetic analysis includes several steps. The first step involves segregation analysis, to describe the distribution (the "segregation") of the disease in a sample of families ascertained through affected probands. In particular, we wish to determine whether the observed segregation can be explained for a family subsample by the transmission of a mutation that predetermines the development of the disease. The second step, linkage analysis, is aimed by locating the mutated gene with the help of genetic markers. The ultimate aim is the determination of the sequence of the mutated gene and the associated protein deficit, in the hope of treating the disease. In an intermediate step, information on markers linked to the mutation can be used to provide families with genetic counseling.

## 1 SEGREGATION ANALYSIS

### 1.1 Principle

Segregation analysis is the attempt to determine the genetic model that best fits the disease segregation observed in a large number of patients and their families. If the effect of a major gene is found, we can estimate the frequency of the disease allele and the penetrance—the probability that carriers of this allele and noncarriers will develop the disease. Penetrance may be a function of the age of the individuals.

### 1.2 Results for Breast Cancer

The most recent segregation analyses of families of women with breast cancer have concluded that two distributions coexist:

1. One corresponds to the presence of a low frequency mutation with dominant autosomal transmission and almost complete penetrance at 80 years of age in women who carry this allele ("genetic cases"). In other words, the women who carry this allele have a probability of nearly 100% of developing breast cancer if they live beyond the age of 80.
2. The other distribution corresponds to "sporadic cases" (i.e., those that appear randomly and are not genetically predetermined). The probability that a woman without the disease allele will develop the disease is approximately 10% for a life span of 80 years.

According to recent analyses, 5% of the cases are "genetic" and 95% are "sporadic."

Penetrances have been estimated for different age groups of carriers and noncarriers of the mutated gene. Women who have inherited the mutation are generally diagnosed with cancer when young, while the other women, if they develop cancer, do so at a later age.

Although estimations for the parameters (frequency of the mutation and penetrances) vary slightly from one study to another, the above-mentioned conclusions are common to all the studies.

### 1.3 Difficulties in the Analysis

Certain studies do not find both distributions. This lack of power is probably due to the high frequency of sporadic cases in the general population. In a genetically susceptible family, not all affected women necessarily carry the disease allele. Similarly, breast cancer can appear sporadically in two close relatives.

## 2 LINKAGE ANALYSIS

### 2.1 Principle

When a mutated gene is associated with the disease, linkage analysis tries to determine its location using genetic markers, polymorphic DNA sequences whose location on the genome is known. Studying the transmission of the marker and the appearance of the disease in families with multiple cases can reveal whether the pathological and the marker alleles have loci on the same pair of chromosomes. If they do, we can then estimate the genetic distance between them.

### 2.2 Results for Breast Cancer

In 1990 a research group in the United States located, on the long arm of chromosome 17, a mutated gene associated with the development of breast cancer. Later the gene, called *BRCA1,* was more precisely located near the marker D17S579.

### 2.3 Difficulties in the Analysis

A mutation on *BRCA1* gene, however, does not explain all so-called genetic cases. Certain cases seem to be predetermined by a mutation located elsewhere on the genome.

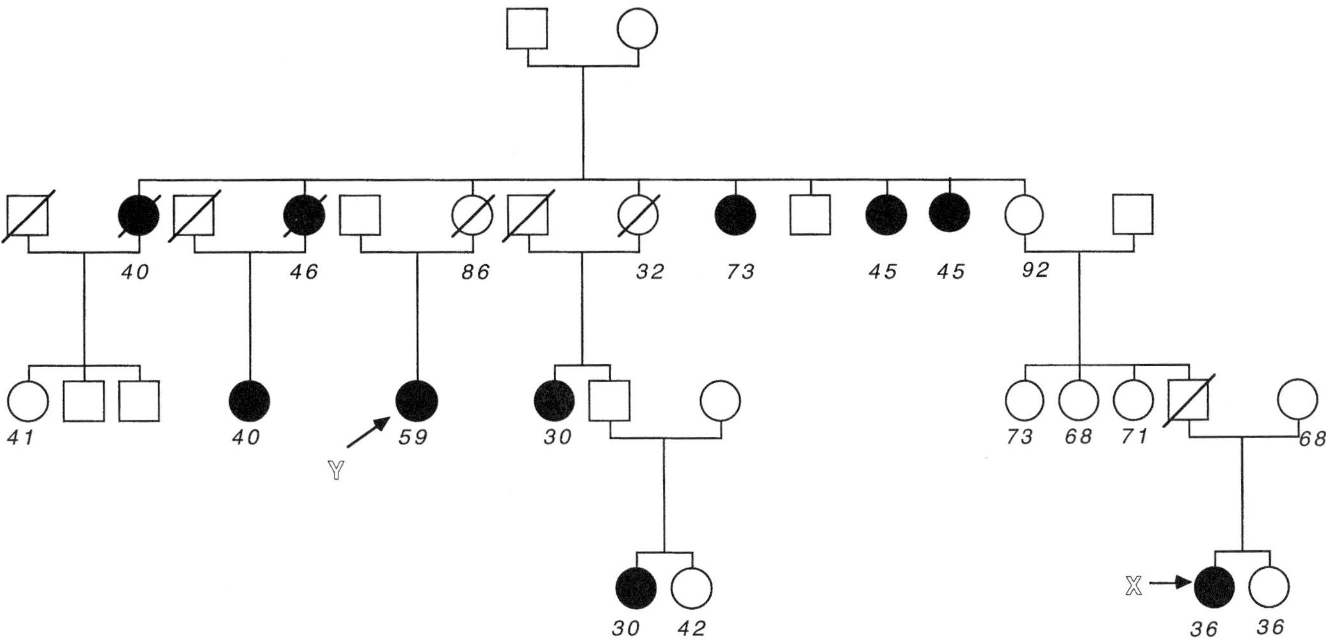


**Figure 1.** Presence of breast cancer in members of family F: squares, men; circles, women; open symbols, unaffected; solid symbols, affected; diagonal slash through symbol, deceased. Numbers indicate current age or age at death (unaffected) or age at diagnosis (affected).

Among the group of families studied by the Breast Cancer Linkage Consortium (1992), the proportion of families in which a mutation on *BRCA1* is present has been estimated at 62% and, among women having both breast and ovarian cancers, at 100%. The estimation of both the genetic distance between the mutated gene and the marker and the percentage of families with a mutation on *BRCA1* depends on the estimation of the frequency of the mutation and the penetrances obtained by segregation analysis. The values for these parameters are also used in genetic counseling.

## 3 GENETIC COUNSELING

The purpose of genetic counseling is the determination of a woman's risk of developing breast cancer. This entails the calculation of her probability of inheriting this morbid mutation. This probability depends on several factors:

Her age and her status: for example, a woman without breast cancer at 90 years of age has a probability of carrying the mutation that is close to zero.

The age and the status of her relatives: the risk of inheriting the mutation will increase, if, for example, her mother is affected, and it will increase even more if that cancer appeared at a young age.

Family F (Figure 1) includes 10 affected women, most of whom developed the disease when young. It is very likely that most of the cases in this family are ''genetic''; but given the frequency of sporadic cases in the general population, some of these women may have developed the disease without being carriers of the mutation. To calculate the probability that a woman is a carrier of the disease allele, one must use, in addition to information on the family (status and age of all individuals), penetrance values and a value for the frequency of the mutation, both obtained by segregation analysis.

We performed these calculations using values from a recent analysis of a large number of families (hereafter S1): the risk for all affected women in family F was found to exceed 90%, except for members X and Y. For them, the probability is 6 and 4%, respectively. When we look more carefully at this family, however, we find that X is linked to the other affected women by a grandmother not affected at 92 years of age. For Y, the link is her mother, not affected at 86 years of age. It therefore seems unlikely that X and Y have inherited the mutation; they are probably sporadic cases.

We repeated the calculations using values for frequencies and penetrances estimated by a different segregation analysis (hereafter S2). With S2, the probability that X is a carrier is 54%. This difference of 48% with respect to the S1 results shows how sensitive risk calculations are to estimated values for mutation frequency and penetrances.

If we have a genetic marker linked to the pathological locus, we can obtain additional information: with the segregation of the marker, we can indirectly follow that of the mutation. If, for example, we consider the alleles of the marker D17S579 carried by the affected women in family F, we find that all of them, except X and Y, have one allele in common. In this example, the marker gives us additional information, which confirms the information provided by the age and the status of the members of the family: X and Y do not carry the mutation. Interestingly, an estimation of the probability of carrying the mutated gene in which this marker is considered makes that probability less sensitive to the parameters (frequency and penetrances). Thus if we use parameter values estimated by S1 and S2 but take the marker into account, the probability of X carrying the mutation is 5 and 11%, respectively. These calculations, however, require that we provide a value for the genetic distance between the mutation and the marker. This distance is also sensitive to the parameter values used.

Thus, those engaged in genetic counseling must be aware of the problem posed by the sensitivity to the parameter values used. This is especially important in breast cancer because of its high frequency in the general population. In the families in which a mutated allele is segregating, women who have not inherited this allele may still develop the disease.

*See also* CANCER; HUMAN GENETIC PREDISPOSITION TO DISEASE; HUMAN LINKAGE MAPS; ONCOGENES; TUMOR SUPPRESSOR GENES.

## *Bibliography*

Elson, R. C. (1986) Modern methods of segregation analysis. In *Modern Statistical Methods in Chronic Disease Epidemiology,* S. H. Moolgavkar and R. L. Prentice, Eds., pp. 213–224. Wiley, New York.

Morton, N. E. (1982) *Outline of Genetic Epidemiology.* Karger, Basel and New York.

Ott, J. (1991) *Analysis of Human Genetic Linkage.* Johns Hopkins University Press, Baltimore and London.

# Calcium Biochemistry

*Joachim Krebs*

## *Key Words*

**Calcium** The fifth most abundant element of the human body, calcium has a stabilizing function in shells, bones, and teeth (bound as hydroxyapatite) and in proteins in the extracellular fluid. It also is a second messenger inside cells, interacting with specific $Ca^{2+}$-binding proteins (epg., calmodulin).

**Calmodulin** Highly conserved, intracellular $Ca^{2+}$-binding protein involved in triggering $Ca^{2+}$ signals, containing four helix–loop–helix or EF-hand $Ca^{2+}$-binding domains.

**Channel** Structure composed of integral membrane proteins responsible for transport across membranes; opening and closing (i.e., gating) can be voltage- or ligand-operated.

**EF-Hand Proteins** An expression coined by R. H. Kretsinger to describe the helix–loop–helix calcium-binding domains of specific proteins. The highly conserved motif (first described on the basis of the crystal structure of parvalbumin containing six helices, A–F) in which certain amino acids are invariant consists of two helices enclosing the $Ca^{2+}$-binding loop. As a model, the forefinger and the thumb of the right hand can resemble the two helices (epg., E and F of the second $Ca^{2+}$-binding domain of parvalbumin) and the bent midfinger the enclosed loop (i.e., EF hand).

**Exchanger** Transmembrane protein using the downhill gradient of one ion as an energy source to transport another ion against its gradient across the membrane (epg., $Na^+/Ca^{2+}$).

**Pump** Transmembrane protein using ATP or another nucleotide triphosphate as an energy source to transport an ion against its gradient across the membrane.

**Second Messenger** Intracellular small molecules (epg., cyclic nucleotides or inositol polyphosphates) or ions such as $Ca^{2+}$ indispensable for the transduction of signals converting extracellular stimuli—for example, by hormones (= primary messengers)—into intracellular responses.

Calcium is of pivotal importance for many biological processes. It may have a quite static, structure-stabilizing role, or it may participate as one of the second messengers of the cell in signal transduction pathways, fulfilling a more dynamic function. This activity is made possible by such specific properties of the $Ca^{2+}$ ion as high dehydration rate, great flexibility in coordinating ligands, and a largely irregular coordination sphere geometry.

The control of calcium homeostasis is of central importance for the organism, involving an exchange of the mineral between the skeleton as the major calcium reservoir, the intestine and the kidney as the organs for calcium absorption or reuptake, respectively, and the extracellular fluid and intracellular calcium. This highly integrated process consists of a number of hormonally controlled feedback loops and an elaborate system of channels, exchangers, and pumps to control $Ca^{2+}$ fluxes into and out of cells. This contribution describes the different roles of calcium in the regulation of biological functions and the proteins involved in these processes.

## 1 INTRODUCTION

Calcium, one of the most common elements on earth, is the fifth most abundant element in the human body. In addition to its central role in cellular functions as one of the second messengers, calcium is a major constituent of the skeleton. It has a stabilizing function in shells, bones, and teeth and therefore is an old component of organisms, the record being $2 \times 10^9$ years (blue-green algae). In minerals as well as in solution, calcium occurs predominantly in a complex form, mostly as the phosphate [epg., hydroxyapatite, $Ca_{10}(PO_4)_6(OH)_2$], which makes up 60% of the weight of human skeleton. That is, the skeleton of a man contains 1.0–1.3 kg of calcium, that of a woman 0.8–0.9 kg, comprising 99% of the total calcium of a human body. Compared with this amount, the calcium found in the extracellular fluid (ECF) and intracellularly in the cytosol or in other intracellular compartments is almost negligible.

In the ECF or in the lumen of intracellular reticular systems, the calcium concentration is millimolar (2–5 m*M*, of which about 50% is unbound, i.e., corresponds to free $Ca^{2+}$ concentration), whereas the cytosolic free $Ca^{2+}$ concentration of a resting cell is about 100–300 n*M*. This discrepancy results in a steep concentration gradient of ionized $Ca^{2+}$ across cellular membranes which is regulated by a variety of channels, pumps, and other transporting systems controlling the fluxes of $Ca^{2+}$ into and out of the cell and between the various intracellular compartments.

On the other hand, calcium homeostasis of the ECF is maintained through a highly integrated and complex endocrine system. This involves the interplay between two antagonistic polypeptide hormones, parathyroid hormone (PTH) and calcitonin (CT), and a vitamin D metabolite, 1,25-$(OH)_2D_3$. These substances regulate the flow of calcium into and out of the ECF through acting on target cells of intestine, kidney, and bone. Normally PTH prevents calcium of the ECF from falling below a threshold level, whereas CT prevents abnormal increases of serum calcium. PTH is also responsible for the formation of 1,25-$(OH)_2D_3$, which acts on specific receptors in the intestine to promote absorption of $Ca^{2+}$.

This contribution describes in detail the different roles of calcium in extracellular and intracellular compartments and discusses the interrelation of the different functions.

## 2 CALCIUM LIGATION

The ligation of $Ca^{2+}$ usually occurs via carboxylates (mono- or bidentate) or neutral oxygen donors, since the calcium ion overwhelmingly prefers oxygen as donor groups of ligands. As a result of calcium's flexibility in coordination (coordination numbers usually 6–8, but up to 12 are possible) and a largely irregular geometry both in bond length and in bond angles, calcium is superior in its ability to bind to proteins (compared to $Mg^{2+}$, which has a smaller ionic radius than calcium, 0.64 vs. 0.97 , hence requires a fixed geometry of an octahedron with six coordinating oxygen atoms and a fixed ionic bond distance). This difference in flexibility of complexation geometry affords calcium a greater versatility in coordinating ligands, leading to a higher exchange rate. This is reflected in a three-order-of-magnitude difference in dehydration rate between the two ions, which makes $Ca^{2+}$ much more suitable as a signal transducing factor, especially in the presence of a high excess of $Mg^{2+}$, the concentration of which is in the millimolar range on both sides of the cellular membrane.

## 3 CALCIUM IN THE EXTRACELLULAR SPACE

Calcium in the ECF is tightly controlled to maintain its concentration in a range of 2–5 m*M*. One of the important functions of $Ca^{2+}$ in the ECF is to stabilize the structure of proteins and to mediate cell–cell or cell–matrix interactions. One important difference in comparing the $Ca^{2+}$-binding sites of extracellular and intracellular proteins (epg., EF-hand proteins, see Section 5.1) is the spatial arrangement of the ligating groups. As discussed in detail later, the intracellular $Ca^{2+}$-binding proteins show a sequential arrangement of their ligating residues, whereas in the extracellular proteins these ligands usually are located at distant positions in the amino acid sequence, resulting in a preformed, rigid cavity of $Ca^{2+}$-binding sites of low affinity. But because of the relatively high concentration of $Ca^{2+}$ in the ECF, these proteins occur in their $Ca^{2+}$-bound form and are therefore protected against proteolytic cleavage.

Bone-forming cells (i.e., osteoblasts) synthesize and secrete a number of noncollagenous $Ca^{2+}$-binding proteins such as osteocalcin, osteopontin, and osteonectin, which bind to bone minerals such as hydroxyapatite in a calcium-dependent manner. Osteocalcin belongs to the class of $Ca^{2+}$-binding proteins rich in the unusual amino acid $\gamma$-carboxylglutamic acid (mediating the $Ca^{2+}$-binding property), whereas osteopontin, a glycoprotein, is rich in sequences of aspartic or glutamic acid residues (responsible for $Ca^{2+}$-binding). Osteonectin, on the other hand, is an extracellular $Ca^{2+}$-binding protein that can contain high affinity $Ca^{2+}$-binding sites of at least one EF-hand type; normally these sites are typical for intracellular $Ca^{2+}$-binding proteins. This EF hand is stabilized by an S—S bridge that normally does not occur in EF-hand-type proteins (see Section 5.1), an interesting abnormality.

## 4 SIGNAL TRANSDUCTION PRINCIPLES

### 4.1 General Properties

Cell surface receptors can recognize extracellular signals ("primary messengers") and multiply them into a cascade of intracellular events using a limited number of intracellular signal transducers, so-called second messengers. The most common intracellular signal multiplication system makes use of the phosphorylation/dephosphorylation of proteins or enzymes by activating a group of different kinases/phosphatases. Four classes of intracellular messengers are known:

1. Cyclic nucleotides [adenosine cyclic 3′,5′-monophosphate (cyclic AMP); guanosine cyclic 3′,5′-monophosphate (cyclic GMP)].
2. Derivatives of phosphatidylinositol [inositol polyphosphates, epg., inositol 1,4,5-triphosphate ($IP_3$), inositol 1,3,4,5-tetraphosphate ($IP_4$), and diacylglycerol (DAG), stemming from the same precursor phosphatidylinositol 4,5-diphosphate ($PIP_2$)].
3. Free $Ca^{2+}$ ions.
4. Gases such as nitric oxide (NO); probably also carbon monoxide (CO).

The primary event in all signal-transducing pathways is the reception of an external signal by a specific receptor in the cellular membrane, activating a chain of reactions that will finally result in an intracellular response. That is, the cellular or plasma membrane is the main information barrier.

$Ca^{2+}$ is able to perform its second messenger function because changes in the distribution at the two sides of the membrane modulate its messenger function, in contrast to the metabolic synthesis and degradation of the other second messenger molecules. If the intracellular $Ca^{2+}$ rises because $Ca^{2+}$ channels have opened upon receiving an extracellular signal, the ion will bind to specific proteins. This event will result in a conformational change, providing the triggering device to multiply the incoming signal.

The different second messengers are central components of intracellular control mechanisms. They are interconnected in their action through a complex network of feedback mechanisms. One of the earliest responses upon binding of growth factors to their receptors as a primary signal is a rapid increase of cytosolic free $Ca^{2+}$ due to receptor-activated, membrane-bound phospholipase C. This enzyme specifically cleaves phosphatidylinositol 4,5-diphosphate ($PIP_2$) to release $IP_3$ and DAG, the former releasing $Ca^{2+}$ from intracellular stores, whereas DAG activates the $Ca^{2+}$- and phospholipid-dependent protein kinase C (PKC).

### 4.2 Calcium and the Cell Cycle

Calcium is one of the control elements of the cell division cycle (cdc) during proliferation of cells. Usually, there are three control points in a cell cycle: START, mitosis ENTRY, and mitosis EXIT. At all three points a transient rise of the intracellular $Ca^{2+}$ can be observed. The probable mediator of these $Ca^{2+}$-dependent cell cycle controls is calmodulin, a multifunctional intracellular calcium-binding protein described in Section 5.1. It has been shown that overexpression of calmodulin leads to a reduction of the $G_1$ phase in the cell cycle, whereas reduced calmodulin synthesis leads to arrest in $G_1$ and mitosis EXIT.

## 5 INTRACELLULAR CALCIUM-BINDING PROTEINS

### 5.1 The EF-Hand Protein Family

As pointed out before, the concentration of the intracellular $Ca^{2+}$ in eukaryotic cells is closely regulated to remain below $5 \times 10^{-7}$ *M* in a resting cell, whereas outside cells the concentration is $10^{-3}$ *M*, resulting in a steep gradient across the plasma membrane. To function as a second messenger, the $Ca^{2+}$-signal must be transduced by a variety of $Ca^{2+}$-binding proteins. In contrast to the extracellular, low affinity $Ca^{2+}$-binding proteins, the intracellular proteins have a sequential arrangement of the amino acids ligating calcium in a loop flanked by two helical segments, which serves to bind

$Ca^{2+}$ with high affinity. This common helix–loop–helix $Ca^{2+}$-binding motif, also known as the EF hand (Kretsinger's term), is an important entity of the intracellular $Ca^{2+}$ receptor proteins able to trigger cellular responses. The helix–loop–helix motif can be present in several copies in these proteins. The number of the latter is steadily increasing (to date more than 200 have been described, but for most of them a precise function is not known yet). The most important members of this protein family are calmodulin, troponin C, parvalbumin, and calbindin.

The residues forming the ligands to $Ca^{2+}$ are highly conserved within a contiguous sequence of 12 residues spanning the loop and the beginning of the second helix. Loop residues in positions 1, 3, 5, and 12 contribute monodentate (1, 3, 5) or bidentate (12) $Ca^{2+}$ ligands through side chain oxygens; residue 7 acts through its backbone carbonyl oxygen. Therefore an invariant glycine residue is in position 6 to permit the sharp bend necessary to ligate $Ca^{2+}$ through the oxygen of a side chain (5) and a backbone carbonyl (7). Residue 9 provides an additional ligand, either directly through an oxygen of its side chain or indirectly via a water molecule. Usually an aspartate is located in position 1 of the loop, whereas in position 12 glutamic acid is invariant.

The $Ca^{2+}$-binding domains usually occur in pairs stabilized by hydrogen bond bridges between the central residues of adjacent loops which form a minimal antiparallel β-sheet. These pair-forming $Ca^{2+}$-binding domains serve to enhance the $Ca^{2+}$ affinity to these sites and the cooperativity of binding. Calmodulin and troponin C contain four $Ca^{2+}$-binding sites. Both proteins display $Ca^{2+}$-binding characteristics compatible with a "pair of pairs" model of EF hands; that is, the two globular domains at the N- and C-terminal sites each contain a pair of EF hands, and the domains are connected by a long central helix, providing a dumbbell-shaped appearance as revealed by crystal structure determination. The N- and C-terminal domains bind calcium cooperatively but, in the absence of a target, independently. Recent multinuclear magnetic resonance experiments of the $Ca^{2+}$-bound form of calmodulin corroborated the general validity of the dumbbell-shaped structure in solution but indicated a high degree of flexibility of the central part of the helix. This flexibility in the center of the molecule is an important property for the interaction with targets: it is because of this that interaction the extended, dumbbell-shaped calmodulin molecule is transformed, by bending the central helix and bringing the two domains closer together, in a more compact form. This has been demonstrated recently by NMR and X-ray studies, respectively, on complexes between calmodulin and peptides corresponding to the binding domain of skeletal or smooth muscle myosin light chain kinase (MLCK). These studies revealed an unusual binding mode sequestering the bound peptides into a hydrophobic channel formed by the two domains of calmodulin. A key finding was the observation that the peptides were anchored to the two N- and C-terminal domains of calmodulin by two hydrophobic residues separated by a minimum length in the amino acid sequence of the peptide.

### 5.2 The Annexins

In addition to the family of the EF hand containing $Ca^{2+}$-binding proteins, another intracellular $Ca^{2+}$-binding protein family has become known in recent years. These soluble, amphipathic proteins bind to membranes containing negatively charged phospholipids in a $Ca^{2+}$-dependent manner and are therefore called annexins. They are widespread in the animal and plant kingdoms and are said to be involved, for example, in membrane fusion, interaction with the cytoskeleton, anticoagulation, signal transduction, and phospholipase inhibition. These cellular functions may be attributable to the proteins' ability to act as gated calcium channels, as suggested for annexin V on the basis of its crystal structure. In contrast to the EF hand containing $Ca^{2+}$-binding proteins, the ligands coordinating calcium in the three binding sites of the annexins are not adjacent in sequence. The ligating oxygens mainly stem from peptide carbonyls or water molecules.

## 6 SYSTEMS CONTROLLING INTRACELLULAR $CA^{2+}$ CONCENTRATIONS: STRUCTURAL AND FUNCTIONAL PRINCIPLES

A number of different transmembrane $Ca^{2+}$-transporting systems participate in controlling the free $Ca^{2+}$ concentration in the cell. Most of these systems are located in the plasma membrane ($Ca^{2+}$ channel, ATP-dependent transporting system, $Na^+/Ca^{2+}$ exchanger), in the sarco(endo)plasmic reticular system (ATPase, $Ca^{2+}$-release channel), or in mitochondria (an electrophoretic uptake system, a $Na^+$-dependent $Ca^{2+}$ exchanger).

### 6.1 Calcium Transport Systems of the Plasma Membrane

#### 6.1.1 The Calcium Channel

Transient changes of the intracellular free $Ca^{2+}$ can be due to calcium deriving either from intracellular stores (see Section 6.1.2) or from the extracellular fluid (by passing through specifically regulated, i.e., gated, channels in the plasma membrane down the electrochemical gradient). Electrophysiological and pharmacological properties have been used to characterize different subtypes of the calcium channels. To date, four different channel types have been identified. On the basis of their electrophysiological properties as determined using the well-established patch-clamp technique, it can be shown that these channels differ in their tissue distribution, their opening kinetics, their conductance, and their pharmacological characteristics.

#### 6.1.2 The $Ca^{2+}$ Pump

The $Ca^{2+}$ pump of plasma membranes is an important enzyme involved in the fine-tuning of the intracellular free $Ca^{2+}$ concentration. Since its discovery by Schatzmann in 1966, it has attracted investigators to study the structural and functional aspects of this protein. It belongs to the P-type ion-motive ATPases; that is, it forms an aspartyl phosphate intermediate to transport $Ca^{2+}$ against its concentration gradient across the plasma membrane at the expense of ATP. It is a protein of low abundance, but ubiquitous in plasma membranes of all cells. In contrast to the $Ca^{2+}$ pump of the sarco(endo)plasmic reticulum (see Section 6.2.3) calmodulin (CaM) interacts directly with the $Ca^{2+}$ pump of plasma membranes, thereby lowering the $K_M$ for $Ca^{2+}$ by an order of magnitude and increasing the $V_{max}$ two- to threefold. In addition to being stimulated by CaM, the plasma membrane $Ca^{2+}$ pump can be activated by alternative treatments, including acidic phospholipids, fatty acids, phosphorylation by different protein kinases (PKA or PKC), oligomerization, and controlled proteolytic treatment. The latter procedure helped to identify a number of functional domains of the enzyme. Additional properties, at least in a reconstituted system, concern the 1 : 1 $Ca^{2+}$/ATP stoichiometry and its electroneutral operation: that is, $Ca^{2+}$ is exchanged for $H^+$ in the ratio of 1-2.

The $Ca^{2+}$ pump of plasma membranes consists of a single polypeptide (ca. 1200 amino acids ≅ 135 kDa, depending on the isoform) spanning the membrane an even number of times (probably 10 times); that is, the N- and the C-termini are located on the same side of the membrane. The cytosolic part of the enzyme can be divided into three different units:

1. *Transduction domain,* which is proposed to couple ATP hydrolysis to ion transport.
2. *Catalytic domain,* which contains the aspartyl phosphate site, the ATP-binding site as identified by FITC (fluoresceinisothyocyanate) and the so-called hinge region. The latter, a highly conserved amino acid sequence among ion-pumping ATPases, is supposed to be responsible for bringing the phosphorylation site near to the bound ATP. This unit further contains a receptor for the "autoinhibitory" calmodulin-binding domain, a property that seems to be typical for CaM-dependent enzymes.
3. *Regulatory domain,* which contains the sequence protruding into the cytosol after the last transmembrane domain up to the C-terminus. It comprises several sites important for the regulation of the $Ca^{2+}$-pump: the CaM-binding domain and consensus sequences of two protein kinases, PKA and PKC.

To date, four different genes of the plasma membrane $Ca^{2+}$ pump have been identified in mammalian species. Additional variability of this multigene family is produced by alternative splicing (two different sites have been identified in the primary structure at which alternative splicing mainly could occur). These splicing isoforms could give rise to differences in tissue-specific expression. The four genes have been localized on human chromosomes: *PMCA1* on chromosome 12, *PMCA2* on chromosome 3, *PMCA3* on chromosome X and *PMCA4* on chromosome 1.

### 6.1.3 $Na^+/Ca^{2+}$ Exchanger

In addition to the $Ca^{2+}$ pump, there exists an $Na^+$-dependent $Ca^{2+}$ exchanger in the plasma membranes of many cells, especially in excitable tissues. This $Na^+/Ca^{2+}$ exchanger has a low affinity for $Ca^{2+}$ but a much higher transport capacity. It is an electrogenic system; that is, it transports three $Na^+$ ions for one of $Ca^{2+}$. The direction of transport depends on the ionic transmembrane gradients or the transmembrane electrical potential (i.e., the operation of the exchanger is fully reversible). Recently, the exchanger has been cloned from cardiac tissues indicating a molecular weight of 120 kDa. The protein is very polar, but analysis of the primary structure revealed 12 possible transmembrane regions. It further contains a number of consensus phosphorylation sites, which probably contribute to the regulation of the function of the protein. Almost no sequence homology to other protein structures is known, even to the recently cloned and sequenced $Na^+/Ca^{2+}$, $K^+$ exchanger from plasma membranes of photoreceptor cells.

## 6.2 $Ca^{2+}$ Transport Systems of the Reticulum

The $Ca^{2+}$-transporting systems located in the plasma membrane are supplemented by intracellular organelles involved in controlling the free $Ca^{2+}$ ion concentration of the cell. These reticular systems have best been studied in skeletal, smooth, and cardiac muscle cells (i.e., the sarcoplasmic reticulum SR), but recently the related endoplasmic reticulum (ER) of other cells has gained importance, especially since it became known that the $Ca^{2+}$ content of these stores can be released into the cytosol by inositol 1,4,5-phosphate ($IP_3$). This latter second messenger links plasma membrane receptor activation response to extracellular hormonal stimuli to $Ca^{2+}$ mobilization from intracellular stores.

### 6.2.1 The $IP_3$ Receptor

Inositol 1,4,5-phosphate is produced by receptor-activated phospholipase C–dependent hydrolysis of phosphatidylinositol diphosphate ($PIP_2$). Subsequently $IP_3$ binds to a specific receptor located in the membranes of the ER, thereby releasing calcium into the cytosol through the $Ca^{2+}$ ion channel of the receptor. The $IP_3$ receptor, which is highly concentrated in Purkinje cells of the cerebellum, has been identified and purified to a single protein band corresponding to 260 kDa. However, the molecular weight of the native receptor is about 1 million daltons, indicating that the receptor is a homotetramer. It is a very conserved protein, demonstrating a high degree of similarity to the ryanodine-binding $Ca^{2+}$ release channel of the SR with which it shares functional similarity.

$IP_3$ receptors can be phosphorylated by different kinases (PKA, PKC, CaM kinase II) in a stoichiometric manner resulting in a reduced potency of $IP_3$ in releasing $Ca^{2+}$ from the ER. Thus regulation of the $IP_3$-induced $Ca^{2+}$ release by phosphorylation of the receptor is a means of "communication" between the different second messenger systems.

### 6.2.2 The $Ca^{2+}$ Release Channel

The sarcoplasmic reticulum of striated muscles is an important feature in the regulation of the excitation–contraction coupling of muscles. It is composed basically of two elements: the longitudinal tubules surrounding the myofibrils and the terminal cisternae. The latter are in contact with the transverse tubular system (T tubules), a periodic inflection from the plasma membrane (the sarcolemma) forming a junctional gap, which is crossed by periodic "footlike" structures. These structures, originally described as the ryanodine receptor, have been identified as the calcium release channel of the SR, forming tetramers as functional units. The amino acid sequence of the monomer has been deduced from the cDNA, consisting of more than 5000 amino acids. An unusual feature of these proteins is the existence of only four potential transmembrane domains right at the C-terminal end, indicating that more than 90% of the protein protrudes into the cytosol. This characteristic, on the other hand would provide an attractive morphological explanation for the existence of these "foot" structures spanning the 150 Å gap between the SR and the T tubules and, furthermore, would provide the possibility for a physical interaction between the $Ca^{2+}$ release channel of the SR and the voltage-dependent $Ca^{2+}$ channel concentrated in the T tubules and identified as dihydropyridine receptors (see earlier). These features would provide a rational explanation for a direct triggering of $Ca^{2+}$ release by these $Ca^{2+}$ channels located in the T tubules.

### 6.2.3 The $Ca^{2+}$ Pump

The principal protein component of the SR membranes is an ATPase pumping $Ca^{2+}$ from the cytosol into the lumen of the reticulum against the concentration gradient across the membrane. This protein can represent as much as 90% of the membrane protein (in

SR of skeletal muscles), but even in the SR of heart cells it can still make up to 50%. Similar to the plasma membrane $Ca^{2+}$ pump, the protein consists of a single polypeptide chain of about 100 kDa, which is significantly lower than the homologous protein of the plasma membrane described earlier. The properties of the SR $Ca^{2+}$ pump can be summarized as follows:

1. The hydrophobic portion is made up of 10 putative transmembrane helices.
2. Five predicted α helices contiguous with the proposed transmembrane helices make up the stalk sector, rich in polar amino acids.
3. The major protein mass protruding into the cytosol is divided into a transduction and a catalytic domain, the latter consisting of the phosphorylation- (aspartyl phosphate) and the ATP-binding domain, and the hinge region.
4. One of the significant differences between the two pumps concerns the C-terminal part. In contrast to the plasma membrane $Ca^{2+}$ pump, the SR protein does *not* directly interact with calmodulin; that is, it lacks the corresponding regulatory domain. Some isoforms, however, can be regulated by CaM-kinase through phosphorylation of phospholamban.
5. The SR $Ca^{2+}$ pump of cardiac or smooth muscles (i.e., slow-twitch muscles) is regulated by a highly hydrophobic, phosphorylatable protein called *phospholamban,* a homopentamer with a subunit of 6 kDa. The protein can be phosphorylated by several different kinases (PKA, PKC and a CaM-dependent kinase), thereby leading to a significant stimulation (up to fivefold) of the SR $Ca^{2+}$ pump. In this way phospholamban exerts an effect on the SR pump of slow-twitch muscles similar to that of CaM on the homologous enzyme of the plasma membrane: it increases the $V_{max}$ of the $Ca^{2+}$ translocation reaction and induces an increase in the affinity of the enzyme for $Ca^{2+}$. In reverse, the unphosphorylated form of phospholamban can be viewed as an endogenous inhibitor of the SR $Ca^{2+}$ pump, being an autoinhibitory domain like the CaM-binding domain of the $Ca^{2+}$ pump of plasma membranes.

As noted for the plasma membrane $Ca^{2+}$ pump, there also exist at least two different genes for the SR (ER) $Ca^{2+}$ pump differentiating between a fast-twitch and a slow-twitch muscle form, the latter including those for cardiac SR and the nonmuscle (i.e., ER) isoform. As a result of alternative splicing each of these genes gives rise to several isoforms (differences between adult and neonatal isoforms).

### 6.3 The Mitochondrial Calcium-Transporting Systems

Like the reticular systems, mitochondria provide intracellular calcium-transporting systems sequestering $Ca^{2+}$ by means of energy-dependent processes. In contrast to the former, the mitochondrial calcium transporters probably play a minor role in the constant regulation of the cytosolic $Ca^{2+}$ concentration, since the mitochondrial $Ca^{2+}$ uptake rate is about 10-fold slower than that of the SR. Under more pathological conditions, however, the mitochondrial $Ca^{2+}$ transporters may have a major role, since mitochondria possess by far the highest calcium storage capacity in cells. Different $Ca^{2+}$ uptake and release systems have been characterized electrophysiologically, but a detailed description on a molecular basis is lacking.

### 6.4 Calcium in the Nucleus

From electron probe analysis of rapidly frozen cells, it is known that the nucleus contains about 1 mmol of calcium per kilogram dry weight, corresponding to $10^{-}$s3–$10^{-4}$ $M$ $Ca^{2+}$, of which only $10^{-7}$ $M$ is free. That is, most of the nuclear calcium is bound. But even if the function of the nuclear $Ca^{2+}$ is only poorly understood, it has become evident in recent years that changes in free nuclear $Ca^{2+}$ do not follow passively the changes occurring in the cytosol; rather, nuclear $Ca^{2+}$ must be actively regulated. The possibility for a functional role of nuclear $Ca^{2+}$ became evident with the finding that calmodulin can shuttle between the cytosol and the nucleus, and, most importantly, that some of the key nuclear functions (epg., DNA replication, DNA repair) seem to depend on $Ca^{2+}$, and probably on calmodulin as well.

## 7 CONCLUSIONS

In summarizing, the following general points can be made.

1. As one of the oldest components of organisms, calcium plays a central role in biological systems. It can fulfill a more static function by stabilizing structures or a more dynamic function by participating in signal transduction pathways as a second messenger.
2. Calcium homeostasis in an organism is carefully controlled, and such regulation involves a variety of systems in the skeleton, the extracellular fluid, and inside cells. Depending on its function, calcium can be complexed in different forms: by hydroxyapatite in a skeleton; by acidic, low affinity proteins in the extracellular fluid; or by the high affinity EF-hand proteins inside cells. To fulfill its triggering function, calcium has advantages over, for example, $Mg^{2+}$: it can be dehydrated much faster, and it demonstrates a higher flexibility in its coordination properties.
3. Extracellular and intracellular concentrations of calcium differ by several orders of magnitude. Therefore cells are exposed to a steep $Ca^{2+}$ gradient across their membranes, which makes it possible for even small changes of membrane permeability to lead to substantial changes in the concentration of intracellular free $Ca^{2+}$. The ability to accommodate a wide range of concentrations is a prerequisite to the use of $Ca^{2+}$ as a second messenger in the transmission of signals from the extracellular space to be amplified inside the cell using different signal transduction pathways. The control of cellular calcium is maintained by an elaborate system of channels, exchangers, and pumps located both in the plasma membrane and in intracellular membranes.
4. The so-called EF-hand or helix–loop–helix proteins play a pivotal role in permitting $Ca^{2+}$ to function as a second messenger. These proteins bind $Ca^{2+}$ with high affinity, selectivity, and cooperativity. The binding of calcium permits them to undergo conformational changes, which in turn permit interactions with targets. More than 200 different proteins of this class have developed during evolution to fulfill the different tasks of calcium-dependent mechanisms such as glycogen metabolism, muscle contraction, excitation–secretion coupling, cell cycle control, and mineralization.

*See also* Bioinorganic Chemistry.

## Bibliography

Carafoli, E. (1992) The $Ca^{2+}$ pump of the plasma membrane. *J. Biol. Chem.* 267:2115–2118.

———, Krebs, J., and Chiesi, M. (1988) Calmodulin in the transport of calcium across biomembranes, in *Calmodulin,* C. B. Klee and P. Cohen, Eds., pp. 297–312. Elsevier, Amsterdam.

Clore, G. M., Bax, A., Ikura, M. and Gronenborn, A. M. (1993) Structure of calmodulin-target peptide complexes. *Curr. Op. Struct. Biol.* 3: 838–845.

Krebs, J. (1991) Calcium, biochemistry, in *Encyclopedia of Human Biology,* Vol. 2, R. Dulbecco, Ed., pp. 89–99. Academic Press, San Diego, CA.

Krebs, J. and Guerini, D. (1995) *The Calcium-Pump of Plasma Membranes in Biomembranes* Vol. 6, A. G. Lee, Ed. JAI Press Connecticut.

Kretsinger, R. H. (1987) Calcium coordination and the calmodulin fold: Divergent versus convergent evolution. *Cold Spring Harbor Symp. Quant. Biol.* 52:499–510.

MacLennan, D. H. (1990) Molecular tools to elucidate problems in excitation–contraction coupling. *Biophys. J.* 58:1355–1365.

McPhalen, C. A., Strynadka, N. C. J., and James, M. N. G. (1991) Calcium binding sites in proteins: A structural perspective. *Adv. Protein Chem.* 42:77–144.


# Cancer


*Thomas G. Pretlow and Theresa P. Pretlow*


## Key Words

**Cancer** Term for many different cellular tumors that, when not treated, grow, invade other tissues, usually spread, and usually are fatal.

**Lyon Hypothesis** The genes on only one X chromosome (selected randomly) are expressed in each mammalian cell.

**Monoclonal** Referring to the progeny of a single cell.

**Neoplasm** A new growth. Generally includes both benign tumors and cancer.

**Philadelphia Chromosome** Chromosomal aberration, involving chromosomes 9 and 22, that is observed in most cases of chronic myelogenous leukemia.

**Tumor** Any mass. Often used loosely synonymously with "cancer."

"Cancer" is a term that is applied to a wide variety of very different diseases. With the discovery that most cancers contain nonrandom chromosomal aberrations, the tools of molecular and cell biology became very relevant to the dissection of the specific molecular changes that are observed in cancers at the molecular genetic level. Many of these changes complement the usual histopathological definitions of these diseases and facilitate more precise predictions of prognosis and responses to specific forms of therapy. There are increasing efforts to use the techniques of molecular biology to understand and to attack the growing number of known specific molecular defects in cancers.

## 1 NATURE OF CANCER

### 1.1 Definition

The term "cancer" is applied to a large and diverse group of diseases, as different from one another as a wart from a fracture. The definitions of "cancer" in general and specific cancers in particular (prostate cancer, colon cancer, etc.) are based on morphological criteria as assessed by an experienced histopathologist. Basal cell cancers of the skin are very common cancers that occur primarily in sun-exposed areas in older people. Basal cell cancers may invade tissues locally, but they metastasize to distant locations only very rarely even when, untreated, they are permitted to grow and invade other tissues locally. In contrast to basal cell cancer, another kind of skin cancer, malignant melanoma of the skin, usually will metastasize if left without treatment and allowed to grow large. Bronchogenic cancer of the lung, the most common cause of cancer deaths in the United States, grows and metastasizes rapidly. Often it is composed of squamous cells and may appear histopathologically similar to squamous cell cancer of the skin, a cancer that metastasizes to distant locations only in a small minority of cases. In contrast to squamous cell cancer of the skin, squamous cell bronchogenic cancer of the lung kills most of those who contract it in less than 6 months without treatment; with the best treatment available, the disease usually is fatal in less than 2 years. The molecular biology, genetics, pathogenesis, biochemistry, and prevention of different cancers are as different as their respective biological behaviors. Discussion of these areas becomes more instructive when it is focused on a specific disease (see Colon Cancer, Liver Cancer, Breast Cancer, etc.).

Knowledge of the pathobiology of most cancers is still very phenomenological. To the extent that we know anything of this area, there is an extensive overlap and an intermixture of knowledge that might be considered to be related to the biochemistry, molecular biology, genetics, pathology, and cell biology of cancer.

### 1.2 Monoclonality

In contrast to many tumors that are caused by viruses in lower animals, most human tumors are monoclonal (i.e., derived from single cells). Exceptions to this generalization include some lymphomas that develop in patients who are immunosuppressed with cyclosporin to prevent rejection of transplanted organs, and some Kaposi sarcomas, tumors that develop with increased frequencies in homosexual men both with and without detectable infections with the human immunodeficiency disease virus. Early demonstrations of the monoclonality of most human tumors depended on the demonstration of the expression of a single isoenzyme of glucose 6-phosphate dehydrogenase by the neoplastic cells in cancers of women who are heterozygous for this enzyme, the gene for which is located on the X chromosome. In contrast to cancers, the normal tissues in these women are a mosaic of cells, approximately half of which express one of the isoenzymes and half of which express the other, consistent with the Lyon hypothesis. More recently, the monoclonality of most human cancers has been demonstrated by a variety of approaches, including the use of other markers on the X chromosome of heterozygous females, demonstration of the clonal expression of single immunoglobulin molecules in B-lymphocyte lymphomas, the expression of identical T-lymphocyte receptors on the neoplastic T lymphocytes in T-cell lymphomas, the expression of mutations of identical oncogene (see Oncogenes) or suppressor genes (see Tumor Suppressor Genes) in the neoplastic cells of a tumor, the presence of clonal gene rearrangements (e.g., the translocation of chromosomes 9 and 22 in chronic myelogenous leukemia, with the production of the Philadelphia chromosome), and the demonstration of clonally aberrant karyotypes in the neoplastic cells of cancers.

It has been widely assumed that the monoclonal nature of most human cancers suggests that cancers arise as the result of changes that occur in a single cell. An alternative explanation for the monoclonality of most cancers is that even if the cancer arises in multiple, parallel neoplastic cells in the same part of the same tissue, one of the clones may grow faster than the others and may replace and destroy the others in a fashion similar to the destruction of the normal tissues that are invaded and destroyed by cancers.

## 2 HISTORICAL KNOWLEDGE OF MOLECULAR DEFECTS

In the 1960s and 1970s, an enormous effort was made to prove that viruses cause most human cancers. While it is certain that many kinds of cancer in laboratory animals and in wild animals are caused by viruses, there is no conclusive evidence that a virus is a direct, independent, and sufficient cause of any human cancer; and only a few of the many of hundreds of precisely defined human cancers are known to be frequently associated with specific viral infections. There is good reason to believe that some of the human cancers that are often associated with virus infections show modified patterns of growth because of the viral infections. Despite extreme efforts to establish a viral etiology of human cancer, viruses do not play a prominent role among the many carcinogens that are well established cancer-causing agents in humans. Those who Klein (in *Cancer Medicine*) has called the ''proponents of the 'panvirological' view of tumor origin'' postulated in the ''first oncogene theory of Heubner and Todaro'' that oncogenic viruses are naturally endowed with ''oncogenes'' that serve no purpose for the viruses except to transform the cells of their hosts to a neoplastic condition that lacks survival value both for the host and for the virus. The study of these ''viral oncogenes'' led to what Klein has termed the ''unexpected [by the viral oncologists] gift of viral oncology to cellular and cancer biology'' of the discovery of genes that control growth under normal circumstances and control aberrant growth in cancers.

## 3 GENETIC INSTABILITY

Cancers have been shown to contain genomic aberrations that are (1) not random and (2), to a variable degree, characteristic of particular kinds of cancers. For example, the Philadelphia chromosome is seen in most cases in chronic myelogenous leukemia; loss of function of the *Rb* gene is observed in retinoblastoma; and alterations of several genes are observed commonly in colon cancers; as noted in this section.

The complexity of the genetic and/or chromosomal aberrations observed in human cancers varies enormously. When cancers are familial and associated with germ line mutations, the known genetic alterations may be as simple as the mutations of both *Rb* genes in retinoblastoma; and a particular oncogene or suppressor gene may affect more than one cancer. For example, patients who have familial retinoblastoma and are cured of their retinoblastomas develop osteosarcomas with increased frequency later in life.

Many of the common cancers that develop later in life, with or without familial patterns of occurrence, involve complex genetic and cytogenetic alterations. The most thoroughly studied of these neoplastic diseases is colon cancer (see Colon Cancer). No gene has been identified that is mutated in all cases of colon cancer; however, there are several genes that are commonly mutated in colon cancer. These include the suppressor gene p53 on chromosome 17, the DCC (deleted in colon cancer) gene on chromosome 18, and the *fap* (APC) and MCC genes on chromosome 5. In addition, the oncogene c-Ki-*ras* is mutated in approximately half of colon cancers. In addition to these most frequent, identified mutations in oncogenes and suppressor genes, surveys of all chromosomes in colon cancers by a variety of techniques have shown that segments of any chromosome may be lost with a frequency that suggests a high level of genomic instability. Although it has been postulated that some of this genomic instability may result from defects in methylation of DNA in cancer that make chromosomes stick together and fragment during mitosis, we do not have a good understanding of the detailed mechanism(s) responsible for the widespread chromosomal and genomic instability in many common cancers.

Characteristic cytogenetic and molecular genetic aberrations have proved to be complementary to conventional histopathological and immunohistochemical techniques for the classification of some kinds of tumor. While still in its infancy, the use of molecular genetic and cytogenetic information for the identification of subpopulations of tumors that are different in prognosis and/or in response to therapy is becoming more common as our knowledge of these areas expands. For many years, we have known that more aggressive tumors commonly exhibit biochemical characteristics different from those observed in less aggressive tumors. For example, the enzymology of prostatic carcinoma permits one to distinguish more aggressive prostatic carcinomas from those that are less aggressive. While it is unclear whether the new genetic information about tumors will be more valuable than the biochemical information that has accumulated for many years, many of the clinical facilities in which patients are treated are better equipped and staffed to carry out cytogenetic and molecular genetic studies than to extract tumors for enzymatic studies.

## 4 IMPACT OF MOLECULAR KNOWLEDGE ON THERAPY

In addition to complementing traditional methods for the more accurate classifications of patients into subgroups with different prognoses and different responses to therapy, the discovery of basic molecular defects in cancers has suggested new forms of therapy. For example, Weber took advantage of the elevated activities of inosine monophosphate dehydrogenase in neoplastic cells of many kinds and successfully treated cancers in experimental animals with tiazofurin, an inhibitor of this dehydrogenase. This drug has been studied in humans and has demonstrated activity as an antineoplastic agent for the treatment of some leukemias.

In the area of molecular biology, new approaches to therapy were suggested by the discovery that some tumors have oncogenes or suppressor genes that are altered either by mutation or by translocation, with the apposition of genetic components that are not normally associated. In some cases, new gene products have resulted from the joining of pieces of two different chromosomes; and there is hope that these new gene products may confer unique vulnerability to appropriately designed antineoplastic therapeutic agents. This line of reasoning is developmentally young and still fraught with assumptions and misunderstandings; however, it is quite promising.

There are many examples of the reversal of malignant neoplastic phenotypes by the introduction of ''corrected'' genetic information into neoplastic cells in culture. For example, some cancer cells that lack specific suppressor genes can be induced to cease behaving like cancer cells by introducing specific tumor suppressor genes into them. While the correction of the neoplastic phenotype has not been possible in vivo in experimental animals or patients with established solid tumors, there are examples of the successful treatment of established experimental tumors by the introduction into the host, at a site remote from these tumors, of genetically altered neoplastic cells of the same strain. Probably the best studied of these systems is a transplantable glioma, a kind of brain tumor, in rats. This glioma differs from most fatal tumors in man in that it does not metastasize; however, it does kill by continued local growth. The glioma makes insulinlike growth factor. When antisense DNA for insulinlike growth factor is introduced into glioma cells that are then injected into animals bearing advanced transplanted gliomas, not only do the cells with the transfected antisense nucleic acid fail to form large tumors, the previously well established gliomas in the same host regress and disappear. The mechanism for this regression is not well defined but is believed by many to be immunological.

There are still many problems to be solved before ''gene therapy'' can be tested in patients. Probably the most important problem is the difficulty that may be encountered in introducing altered genetic information into well-established tumors in vivo. In the same vein, if antineoplastic therapy with the tools of molecular biology is to depend frequently on immunological mechanisms, the effectiveness of therapy will depend on the degree to which immunological defenses are specific for neoplastic cells to the exclusion of immunological attacks on normal tissues. These are difficult problems; however, the available techniques are becoming more sophisticated, and it seems likely that these technical issues can be approached.

*See also* Breast Cancer, Genetic Analysis of; Colon Cancer; DNA in Neoplastic Disease Diagnosis; Liver Cancer, Molecular Biology of; Oncogenes; Retinoblastoma.

### *Bibliography*

Holland, J. F., Frei, E., Bast, R. C., Kufe, D. W., Morton, D. L., and Weichselbaum, R. R., Eds. (1993) *Cancer Medicine,* 3rd ed. Lea & Febiger, Philadelphia.

Pretlow, T. G., and Pretlow, T. P., Eds. (1991, 1994) *Biochemical and Molecular Aspects of Selected Cancers.* Academic Press, San Diego, CA.

Weber, G. (1977) Enzymology of cancer cells. *N. Engl. J. Med.* 296:486–493, 541–551.

———. (1983) Biochemical strategy of cancer cells and the design of chemotherapy: G.H.A. Clowes Memorial Lecture. *Cancer Res.* 43:3466–3492.

———. (1988) Enzyme-pattern-targeted chemotherapy with tiazofurin and allopurinol in human leukemia. *Adv. Enzyme Regul.* 27:405–433.

# Carbohydrate Analysis

*Martin F. Chaplin*

### *Key Words*

**FAB-MS** Fast atom bombardment mass spectrometry. Underivatized biomolecules, dissolved in a nonvolatile liquid under vacuum, are bombarded with a beam of accelerated atoms. This causes the molecules to be desorbed from the solution and ionized. Primary ions fragment to form secondary ions in a predictable manner. The masses of the desorbed primary and secondary ions allow the molecular weights and structures of molecules to be determined.

**GLC** Gas–liquid chromatography

**HPAEC-PAD** High performance anion exchange chromatography with pulsed amperometric detection.

**HPLC** High performance liquid chromatography.

**ICUMSA** The International Commission for Uniform Methods of Sugar Analysis (c/o British Sugar plc, Technical Centre, Colney, Norwich, England, U.K.)

**NMR** Nuclear magnetic resonance.

**TLC** Thin-layer chromatography.

Carbohydrates form the most diverse of all the groups of biological compounds. They possess a wide variety of structural types with very different chemical and physical characteristics. There is no single method applicable to their analysis. Even for simple carbohydrates like glucose, a number of methods are used, depending on the purpose of the analysis. Complex carbohydrates may consist of several monosaccharide moieties combined in differing ways and proportions. They may be covalently attached to proteins, lipids, inorganic anions, or other molecules. In addition, the more complex carbohydrates are rarely found to exist as a single structural entity; heterogeneity is common, and a given substance often is found as a family of molecules with closely related structures. The principal classes of carbohydrates are listed in Table 1, together with the most commonly encountered analytical methods.

## 1 MONOSACCHARIDES

There are well-established and standardized methods, validated by the International Commission for Uniform Methods of Sugar Analysis (ICUMSA), for analyzing low molecular weight carbohydrates in foodstuffs. These make use of colorimetric assays, the reduction of cupric ions (Table 2), or physical methods such as optical rotation and specific gravity. Enzymatic methods of analysis are becoming more commonly used. Monosaccharides, free or derived from more complex carbohydrates by hydrolysis with dilute acid, are analyzed using gas–liquid chromatography (GLC) or high performance liquid chromatography (HPLC) by comparison with known standards. A currently popular and efficient separatory method is high performance anion exchange chromatography with pulsed amperometric detection (HPAEC-PAD) using quaternary ammonium resins in the hydroxide form.

## 2 OLIGOSACCHARIDES

Oligosaccharides may be purified by means of high performance thin-layer chromatography (TLC) or, more commonly, HPLC. An important method for HPLC involves ion-moderated partition on sulfonated polymeric resins in the calcium form, using water as the eluant. Known carbohydrate oligomers are usually quantified, as part of their separation and purification procedure, by comparison with standards. Confirmation of oligosaccharide structures often can be achieved by the use of specific enzymic hydrolysis (Table 2). The structures of pure, but unknown, oligosaccharides

Table 1 The Important Analytical Methods for Carbohydrates

| Carbohydrate | Usual Analytical Method(s) |
|---|---|
| **Monosaccharides and disaccharides** | |
| In syrups | ICUMSA methods |
| In food | Enzymic methods |
| Derived from more complex carbohydrates | HPLC, GLC |
| **Oligosaccharides** | HPLC, TLC |
| **Neutral polysaccharides and proteoglycans** | |
| | Separation and extractive techniques, component analysis, methylation analysis, enzymic hydrolysis, FAB-MS, $^{13}C$ and $^{1}H$ NMR |
| **Glycoproteins and glycolipids** | |
| | Cleavage from protein–lipid, lectin binding, component analysis, methylation analysis, enzymic hydrolysis, FAB-MS, $^{13}C$ and $^{1}H$ NMR |

of NMR and mass spectrometry (MS; see Section 6). The size distribution of oligosaccharide mixtures is determined using gel permeation chromatography (GPC).

## 3 POLYSACCHARIDES

Polysaccharides are high molecular weight polymers (> 5000 daltons) of monosaccharides. They may be linear or highly branched and may contain just one type of monosaccharide or several. Most polysaccharides possess repeating units of relatively uniform structure. It is this structural unit, together with the molecular weight, which is generally sought rather than an accurate picture of a complete molecule. The type of polysaccharide may be concluded from the component analysis obtained after acid hydrolysis. The glycosidic linkage positions may be determined by means of methylation analysis. The anomeric configuration of these links may be discovered by means of their susceptibility to hydrolysis using various specific glycosidases (Table 2).

## 4 PROTEOGLYCANS

Proteoglycans consist of a number of highly charged polyanionic glycosaminoglycan chains attached to a central protein core via serine or threonine residues. These glycans are distinguished from those occurring in glycoproteins in that the former are high molecular weight molecules, essentially unbranched chains of repeating disaccharide units. They contain carboxyl groups and/or are substituted with sulfate ester groups. Analysis of proteoglycans involves the determination of the protein's structure and the linkage sites, in addition to the structural determination of the attached glycans.

Glycosaminoglycan chain fragments may be cleaved from the proteoglycans by enzymic hydrolysis or β-elimination with base. They are purified by anion exchange HPLC and may then be analyzed by, for example, $^{13}C$ NMR spectrometry. Although the class of the glycosaminoglycans can be relatively easily deduced on the basis of a component analysis and chemical and enzymic degradation, the discovery of detailed fine structure is much more arduous.

## 5 GLYCOPROTEINS

Generally, the carbohydrate-containing molecules most important to molecular biologists are the glycoproteins, where the carbohydrate groups often confer the specificity inherent in many molecular and cellular interactions. They possess diverse structures in which

Table 2 Nonchromatographic Analytical Methods for Carbohydrates

| Carbohydrate Group | Assay | Mechanism of Action |
|---|---|---|
| Reducing sugars | Lane and Eynon titration | Reduction of alkaline cupric tartrate to cuprous oxide |
| b1 | Luff–Schoorl method | Iodometric determination after the reduction of alkaline cupric citrate |
| Hexose (e.g., glucose, free or combined) | Phenol sulfuric acid assay | Condensation of phenol with dehydrated hexoses to yield a brown solution |
| Uronic acid (free or combined) | Carbazole assay | Condensation of carbazole with dehydrated uronic acids to yield a pink solution |
| Glucose | Enzymatic | Production of NADPH, which absorbs UV light of wavelength 340 nm<br>$\text{glucose} + \text{ATP} \xrightarrow{\text{hexokinase}} \text{glucose 6-phosphate} + \text{ADP}$<br>$\text{glucose 6-phosphate} + \text{NADP}^+ \xrightarrow{\text{glucose 6-phosphate dehydrogenase}} \text{6-phosphogluconolactone} + \text{NADPH} + \text{H}^+$ |
| Sucrose | Enzymatic | Production of glucose, then as with preceding glucose reactions<br>$\text{sucrose} + H_2O \xrightarrow{\text{invertase}} \text{fructose} + \text{glucose}$ |
| Lactose | Enzymatic | Production of glucose, then as for sucrose<br>$\text{lactose} + H_2O \xrightarrow{\beta\text{-galactosidase}} \text{glucose} + \text{galactose}$ |
| Starch | Enzymatic | Production of glucose, then as for sucrose<br>$\text{starch} + (n-1)\,H_2O \xrightarrow{\text{amyloglucosidase}} n\text{-glucose}$ |

carbohydrate moieties, known as glycans, are covalently attached to the proteins. A number of different carbohydrates may be attached in different ways, with varying anomeric configurations, linkage positions, and locations within the sequence. If the complete analysis of the glycoprotein is to be achieved, all these parameters must be determined. Usually the glycans are of relatively low molecular weight, containing between 1 and 20 monosaccharides of between 1 and 5 different types, often involved in highly branched structures. The same type of monosaccharide may be connected in different ways even within a single glycan group. It is frequently found that several different types of glycan are attached to one protein molecule (see Figure 1). Glycan heterogeneity is commonly found, the result of incomplete glycan biosynthesis or prior partial hydrolysis. Such differences in structure often do not significantly influence the chromatographic behavior. A number of distinct, but related, glycans may be present and coupled to the same amino acid residue, in different molecules, even in ''purified'' glycoprotein preparations. The analysis of such structures is a serious undertaking, since there is no single, generally applicable, analytical technique. Methods may vary between laboratories even for similar glycoproteins. There are, however, stages through which any complete analysis must pass:

1. *Isolation and purification.* This stage involves a number of chromatographic steps such as fast protein liquid chromatography (FPLC), HPLC, GPC, and lectin affinity chromatography.
2. *Determination of the constituent carbohydrates.* Monosaccharides may be liberated from the glycoproteins by hydrolysis using dilute aqueous or methanolic hydrogen chloride. Once liberated, and after any necessary derivatization, they are separated and determined in one step by either GLC or HPLC. This component analysis usually gives definite pointers to the type of glycans, hence, to the glycan–protein linkages involved.
3. *Cleavage of the intact glycans from the protein moieties.* The different families of glycoprotein glycans require different methods for their cleavage (see Figure 1). As an alternative to the use of enzymes, an automated chemical cleavage method using hydrazinolysis is available which releases either O-linked or O- and N-linked intact oligosaccharides. The glycosylation sites may be determined from glycoprotein peptide mapping. The cleaved glycans are often labeled with a radioactive tracer to facilitate following them throughout the purification.
4. *Glycan isolation and purification.* A combination of GPC and HPLC, using HPAEC-PAD, is commonly used for the purification of glycans. Lectin affinity chromatography is a powerful technique used for the separation of different classes of glycans.

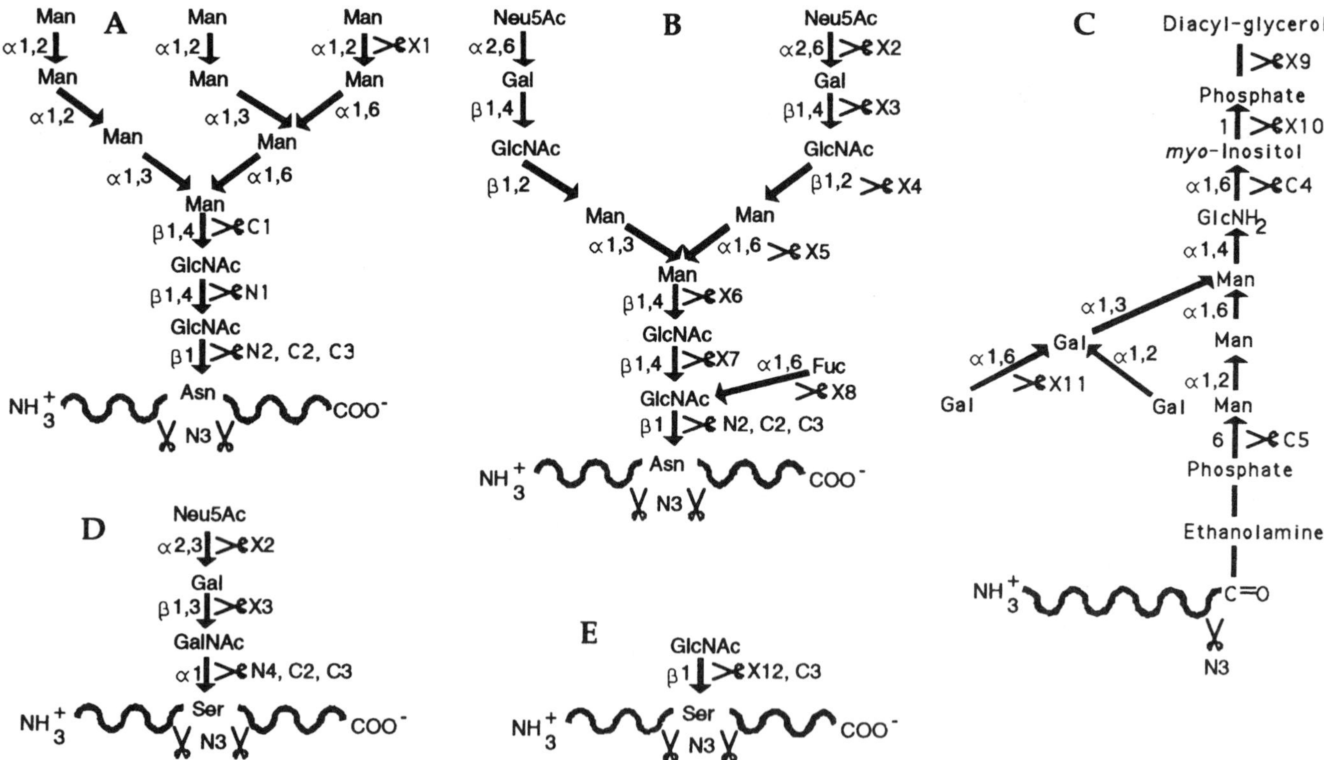


Figure 1. Examples of the major classes of glycoprotein glycans, showing some methods for their cleavage: (A) N-linked high mannose triantennary, (B) N-linked complex biantennary, (C) glycosyl-phosphatidylinositol anchor, (D) O-linked complex, (E) O-linked *N*-acetylglucosamine. The cleavages may be achieved by the *exo*-glycosidases: X1, α-mannosidase; X2, neuraminidase; X3, β-galactosidase after X2; X4, β-*N*-acetylglucosaminidase after X3; X5, α-mannosidase after X4; X6, β-mannosidase after X5; X7, β-*N*-acetylglucosaminidase after X6; X8, α-fucosidase; X9, phospholipase C; X10, GPI-phospholipase D; X11, α-galactosidase; X12, β-*N*-acetylglucosaminidase; by the *endo*-glycosidases: N1, *endo*-glycosidase H; N2, peptide-*N*-glycosidase F; N3, pronase; N4, *O*-glycosidase after X2; and by the chemical methods: C1, trifluoroacetolysis; C2, hydrazinolysis; C3, alkaline sodium borohydride; C4, nitrous acid; C5, cold hydrofluoric acid. Asn, asparagine; Fuc, fucose; Gal, galactose; GalNAc, *N*-acetylgalactosamine; GlcNAc, *N*-acetylglucosamine; $GlcNH_2$, glucosamine; Man, mannose; Neu5Ac, *N*-acetylneuraminic acid; Ser, serine (or threonine).

5. *Structural studies on the intact glycan.* Much analytical information may be obtained by the use of $^{13}C$ NMR and two-dimensional $^{1}H$ NMR spectrometry at high field strengths corresponding to $^{1}H$ resonance frequencies of 400 to 600 MHz, and FAB-MS of the intact glycopeptides. FAB-MS is a sophisticated technique that enables oligosaccharides, and even complex glycan structures, to be determined relatively rapidly using small samples of materials. Further confirmatory information can be obtained from the FAB-MS data of the fully methylated glycans. Mass spectrometry using other ionization techniques such as laser desorption (UV-LDI MS) are also in use. Although a powerful tool for determining the presence, sequence, and arrangement of carbohydrates in a glycan chain of up to about 40 residues, MS is generally not able to distinguish the anomeric type or the linkage positions. NMR spectrometry, on the other hand, provides a complementary tool that can provide detailed information on the anomeric type of the linkages present and some information on the linkage positions, despite the lack of an easy way to determine sequences. Lectin binding is a powerful technique for discriminating between glycan types. In some cases, lectin binding is able to differentiate between structures that are otherwise difficult to distinguish: for example, α2-6- and α2-3-linked terminal sialic acids.
6. *Complete glycan analysis.* To be sure of any conclusions drawn from studies on an intact glycan, it is necessary to know its complete component composition. In addition, it is usual to confirm the sequence of carbohydrates and their anomeric linkages through the use of specific *endo-* and *exo*-glycosidases (see Figure 1 for examples). The linkage positions of the glycan components are determined by means of methylation analysis. In this technique all the free hydroxyl groups in the intact glycan are methylated. The fully methylated glycan is then hydrolyzed. The linkage positions are deduced from the positions of the methyl and hydroxyl groups in the resultant partially methylated monosaccharides, as determined using GLC.

## 6 GLYCOLIPIDS

Glycolipids form a diverse group of biological molecules containing carbohydrate glycans covalently bound to a wide variety of lipid molecules. They are generally purified by HPLC. The purified but intact glycolipid molecules may be analyzed by a combination of NMR and FAB-MS, or by other techniques similar to those applicable to glycoprotein glycans.

*See also* GLYCOBIOLOGY; HPLC OF BIOLOGICAL MACROMOLECULES; LIPOPOLYSACCHARIDES.

### *Bibliography*

Chaplin, M. F., and Kennedy, J. F., Eds. (1994) *Carbohydrate Analysis: A Practical Approach,* 2nd ed. Oxford University Press, London and New York.

Dell, A., and Rogers, M. E. (1989) *Trends Anal. Chem.* 8:375–378.

Pareth, R. B., and Patel, T. P. (1992) *Trends Biotechnol.* 10:276–280.

Welply, J. K. (1989) *Trends Biotechnol* 7:5–10.

White, C. A., Ed. (1991) *Advances in Carbohydrate Analysis,* Vol. 1. JAI Press, London.

# CARBOHYDRATE ANTIGENS

*David R. Bundle*

### *Key Words*

**Epitope** Antigenic determinant of known structure.

**Glycoconjugates** Naturally occurring glycolipids, glycoproteins, and glycopeptides, including synthetic oligosaccharides, covalently linked to carrier molecules such as proteins or lipids.

***N*-Acetylneuraminic Acid** Multifunctional nine-carbon monosaccharide bearing a carboxyl group at C-1.

**Tumor-Associated Carbohydrate Antigens** Oligosaccharide epitopes of glycolipids and glycoproteins that are virtually absent in progenitor cells but accumulate in tumor cells as the result of blocked or enhanced synthesis of particular carbohydrate chains.

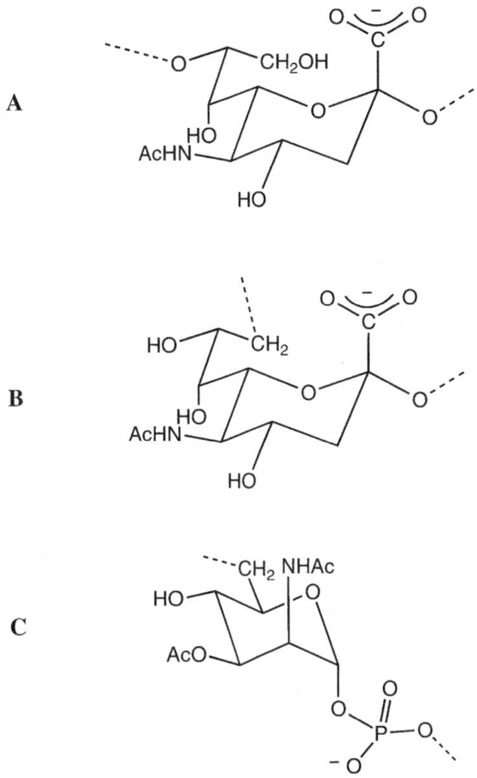

A

O, O, C, O, $CH_2OH$, O, O, HO, AcHN, HO

B

O, O, C, HO, $CH_2$, O, O, HO, AcHN, HO

C

$CH_2$ NHAc, HO, O, AcO, O, O, P, O, O

**Figure 1.** (A) Monomeric repeating unit of *N. meningitidis* group B capsular antigen, a 2,8-linked homopolysaccharide of α-D-Neu*p*5Ac. It is also the structure of the *E. coli* K1 capsule, sometimes referred to as colominic acid. (B) An identical monosaccharide repeating unit occurs for the *N. meningitidis* group C antigen, but the glycosidic linkage connects the 2 and 9 positions of adjacent residues, rather than the 2 and 8 positions of the group B antigen. The result is a structurally and immunologically distinct antigen. (C) *N. meningitidis* group A antigen is a homopolysaccharide of 2-acetamido-2-deoxy-α-D-mannopyranose phosphate. The phosphodiester linkage bridges the C-1 and C-6 positions of adjacent residues. Random but regioselective acetylation at the O-3 position also occurs with, on average, 60–70% of the repeating units carrying this noncarbohydrate substituent.

**Table 1** Structures of the Capsular Polysaccharides of Group B *Streptococcus*

| Group | Structure |
|---|---|
| Ia | →4)βD–Glc*p*–(1→4)–βD–Gal*p*–(1→<br>3<br>↑<br>1<br>αD–Neu*p*NAc–(2→3)–βD–Gal*p*(1→4)Glc*p*NAc |
| Ib | →4)–βD–Glc*p*–(1→4)–βD–Gal*p*–(1→<br>3<br>↑<br>1<br>αD–Neu*p*NAc–(2→3)–βD–Gal*p*(1→3)βD–Glc*p*NAc |
| II | →4)–βD–Glc*p*NAc–(1→3)–βD–Gal*p*(1→4)–βD–Glc*p*–(1→3)–βD–Glc*p*–(1→2)–βD–Gal*p*–(1→<br>6 3<br>↑ ↑<br>1 2<br>βD–Gal*p* αD–Neu*p*NAc |
| III | →4)–βD–Glc*p*–(1→6)–βD–Glc*p*NAc–(1→3)–βD–Gal*p*–(1→<br>4<br>↑<br>1<br>αD–Neu*p*NAc–(2→3)–βD–Gal*p* |
| A[a] | αD–GalNAc(1→3)βD–Gal*p*(1→3/4)βD–Glc*p*NAc<br>2<br>↑<br>1<br>αL–Fuc*p* |

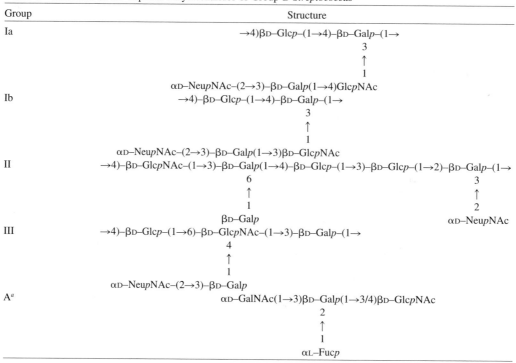

Carbohydrate antigens are cell surface polysaccharides or glycoconjugates that evoke an oligosaccharide-specific antibody response or react with antibodies. When pure, they are T-independent antigens that exhibit severe limitations in their ability to generate a carbohydrate-specific T-cell response because of their inability to form an immunogenic complex with restriction elements of a major histocompatibility complex. Capsular and cell wall polysaccharides of bacteria and fungi are prominent and exposed components of these organisms, hence induce an often dominant mammalian immune response during infections. Mammalian glycoconjugates (glycolipids or glycoproteins) are important human blood transfusion and organ transplantation antigens of tissues, leukocytes, and red blood cells. Strictly self antigens, they react with natural antibodies, and in cancers, immune antibodies directed toward tumor-associated carbohydrate antigens (TACA: carbohydrate epitopes carried on glycolipids or glycoproteins) are frequently encountered. Mammalian glycoconjugates also function in cell–cell interactions, and there is a strong likelihood that in both cases the relevant carbohydrate structures are developmentally regulated and the control of tissue specific expression is regulated at the level of glycosyltransferase gene expression.

## 1 BACTERIAL ANTIGENS

Capsular and cell wall polysaccharides function as virulence factors by protecting bacteria from nonspecific host defenses such as complement and phagocytes. As type-specific antigens, they also serve as markers of infection and aids to serodiagnosis. Since anticapsular antibodies can neutralize the shielding effect of capsules, there has been a resurgence of interest in polysaccharide vaccines, heightened by their lack of serious side effects. In addition to the capsular and lipopolysaccharide antigens discussed here, many bacteria also produce other carbohydrate antigens such as teichoic acids, *S*-layer glycoproteins, and glycolipids.

### 1.1 Capsular Antigens

Capsular polysaccharides are high molecular weight antigens ($10^5$–$10^6$ daltons) built from several hundred oligosaccharide repeating units; typically these repeating units contain from one to six monosaccharide residues. Fungi and bacteria are capable of assembling a diverse array (several hundred) of linear and branched polysaccharide structures that often incorporate rare monosaccharides, as well as noncarbohydrate substituents such as acetate or phosphate esters and pyruvate ketals. The simplest capsular polysaccharides are homopolymers (polymers composed of a single monosaccharide) such as the *Neisseria meningitidis* groups A, B and C antigens. The B and C antigens are homopolymers of glycosidically linked sialic acid residues (Figure 1A, 1B), while the A antigen typifies many capsular antigens in that it utilizes a phosphodiester to connect the repeating units, in this case a monosaccharide (Figure 1C).

Antibodies are able to discriminate between capsular antigens with only fine structural differences (cf. the *N. meningitidis* group B and C antigens). Consequently antibodies raised in rabbits and goats to capsular and cell wall polysaccharides are widely used in conventional bacterial serotyping. Conversely, the detection of immune antibodies in human or animal sera by their reactions with purified carbohydate antigens can be used to confirm the group and serotype of infecting bacteria. Since these antibodies are often protective against subsequent infection, several modern bacterial

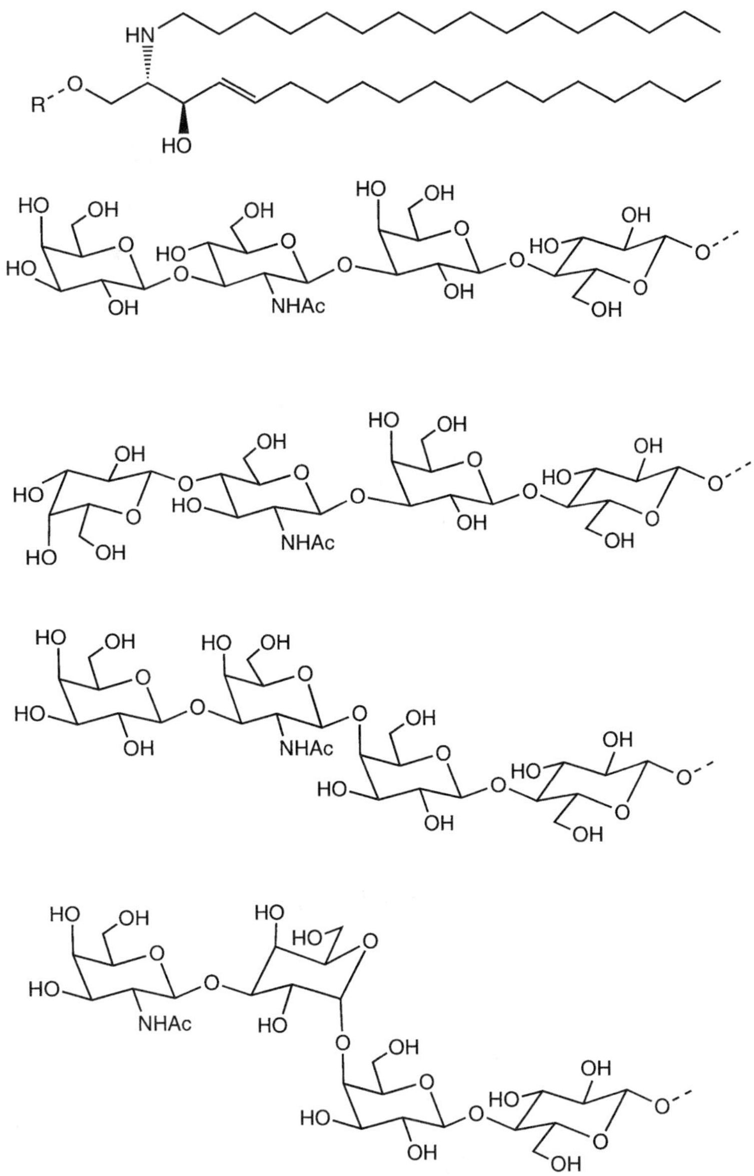

**Figure 2.** (A) The ceramide portion of glycosphingolipids is *N*-acylated by fatty acids of various lengths. Glucose is always β-linked to the primary hydroxyl group of the lipid. (B) Tetrasaccharide core sequences found in glycosphingolipids (i) lactotetraosylceramide, (ii) neolactotetraosylceramide, (iii) gangliotetraosylceramide, and (iv) globotetraosylceramide. All possess a common lactose element.

vaccines are composed of purified bacterial polysaccharides, either alone or covalently linked to a carrier protein. Purified capsular polysaccharides provide type-specific protection in human adults; for example, vaccination with its polysaccharide antigen provides immunity against infection by *Streptococcus pneumoniae* type 3 but not to infection by *S. pneumoniae* of other types.

A general characteristic of all purified, protein-free polysaccharides is membership in the T-cell-independent class of antigens. This means they do not evoke a T-cell response, with the result that there is no immunological memory nor class switch with significantly elevated antibody levels on subsequent immunization. Polysaccharide vaccines are effective in adults because, unlike mice, humans respond to a single injection with a persistent level of IgM or IgG antibody. Infants are unresponsive to such vaccines, however, and efforts to conjugate polysaccharides to proteins, to enable a T-cell response, are under way.

The capsular antigens of *N. meningitidis* and group B *Streptococcus* closely resemble carbohydrate epitopes that occur on mammalian glycolipids and glycoproteins. For example, the polysialosyl antigen of group B meningococcus (Figure 1A) resembles the glycan chains present on the fetal neural cell adhesion molecule [E]N-CAM. Oligosaccharide sequences that occur in the Group B streptococcal polysaccharides exhibit structural homology with human serum glycoproteins and glycolipids, since the capsular antigens of type Ia, Ib, II, and III possess the αDNeu5Ac(2→3)βD-Gal(1→4)βDGlcNAc epitope (Table 1), found as terminal structures in glycopeptides and glycolipids. This mimicry enables the bacteria to evade the host's immunological system.

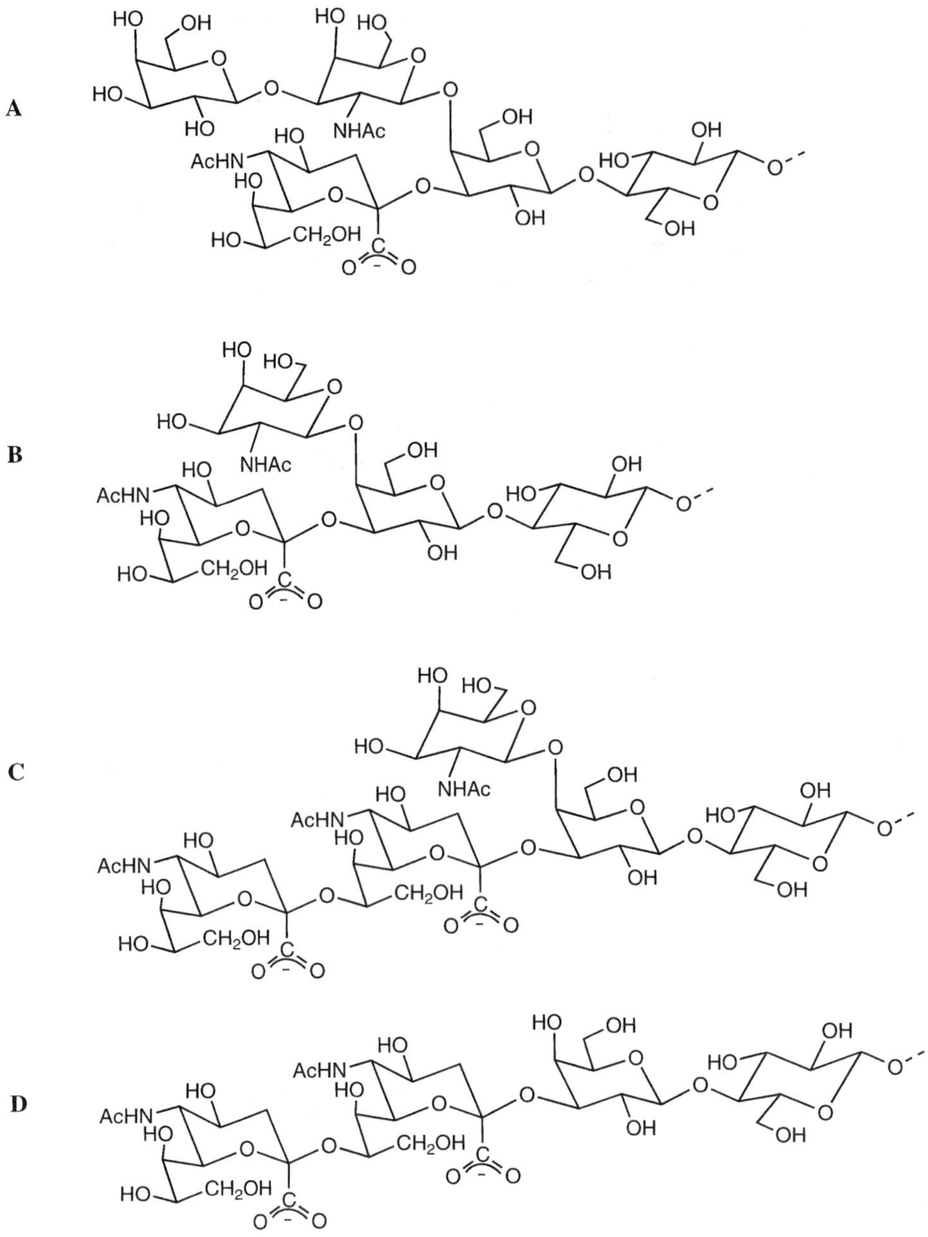

A

B

C

D

**Figure 3.** (A) Ganglioside $GM_1$; α-D-Neup5Ac, with 2,3-linkage to the lactose element of the gangliotetraosyl ceramide. (B) Ganglioside $GM_2$ is a truncated homologue of $GM_1$. (C) Ganglioside $GD_2$, a TACA, is structurally related to $GM_2$ but contains an α-D-Neup5Ac(2→8)α-D-NeupAc disaccharide rather than a single α-D-Neu5Ac residue. (D) Ganglioside $GD_3$, a melanoma TACA, consists of the α-D-Neup5Ac(2→8)α-D-Neup5Ac disaccharide 2,3-linked to lactose.

### 1.2 O-Polysaccharide Antigens

Lipopolysaccharides, particularly the polysaccharide O-antigen and oligosaccharide core components, are major antigens of Gram-negative bacteria. The O-antigens resemble capsular antigens in having a repeating unit motif, but they possess even greater structural diversity.

The immune response to Gram-negative bacteria is usually dominated by *O*-polysaccharide-specific antibodies. These antibodies are protective for and diagnostic of the infecting bacteria. When a single monosaccharide can inhibit the binding of antibody to the antigen, it is said to be immunodominant, as in the case of 3,6-dideoxyhexoses, which occur in the *Salmonella* serogroup A, B, and D antigens. A crystal structure of an antibody Fab complexed with a *S. typhimurium* epitope shows that the immunodominant activity arises because the 3,6-dideoxyhexose residue is completely buried in the antibody combining site, where it coordinates a network of hydrogen bonds between the monosaccharide, water, and the protein surface.

Certain bacteria such as *Haemophilus influenzae, Neisseria meningitidis,* and *N. gonorrhoeae* produce rough-type lipopolysaccharides (devoid of O-polysaccharide) only, and in these circum-

**Table 2** Structures of the Human ABH and Lewis Antigenic Determinants

| Antigen | Structure |
|---|---|
| Lewis a | βD–Gal*p*(1→3)βD–Glc*p*NAc<br>4<br>↑<br>1<br>αL–Fuc*p* |
| Lewis b | βD–Gal*p*(1→3)βD–Glc*p*NAc<br>2 4<br>↑ ↑<br>1 1<br>αL–Fuc*p* αL–Fuc*p* |
| Lewis x | βD–Gal*p*(1→4)βD–Glc*p*NAc<br>3<br>↑<br>1<br>αL–Fuc*p* |
| H type 1 | αL–Fuc*p*(1→2)–βD–Gal*p*(1→3)βD–Glc*p*NAc |
| H type 2 | αL–Fuc*p*(1→2)–βD–Gal*p*(1→4)βD–Glc*p*NAc |
| B[a] | αD–Gal*p*(1→3)–βD–Gal*p*(1→3/4)β–D–Glc*p*NAc<br>2<br>↑<br>1<br>αL–Fuc*p* |
| A[a] | αD–GalNAc(1→3)βD–Gal*p*(1→3/4)βD–Glc*p*NAc<br>2<br>↑<br>1<br>αL–Fuc*p* |

[a]Terminal trisaccharide epitope may be carried on either a type 1 or a type 2 core structure.

stances the outer and inner core regions of the molecule assume greater importance as antigenic components.

## 2 MAMMALIAN GLYCOCONJUGATES

Cell surface oligosaccharide epitopes of mammalian cells are covalently attached to lipids or proteins. These macromolecules, which cover large areas of the cell surface, are located at the cell membrane with the carbohydrates protruding from it. Interactions between the carbohydrate, the lipid/protein, and the membrane surface may modulate the antigenic characteristics of certain oligosaccharide epitopes. The functions of cell surface oligosaccharide antigens are just beginning to be understood, and even for structures as thoroughly investigated as the human blood group system, their precise roles are only poorly appreciated. It has been demonstrated that some pathogenic bacteria possess oligosaccharide-specific receptors that enable them to adhere to epithelial cells.

### 2.1 Glycolipids

Glycosphingolipids possess the generalized structure shown in Figure 2A. The hydrophilic oligosaccharide is covalently linked via a β-D-glucopyranosyl residue to ceramide, the hydrophobic lipid that anchors the molecule to the membrane. Glycosphingolipids are complex structures that defy classification into a simple systematic nomenclature, but it is possible to define four characteristic core tetrasaccharide elements (Figure 2B) from which most glycosphingolipids are derived by elaboration or truncation. Gangliosides are

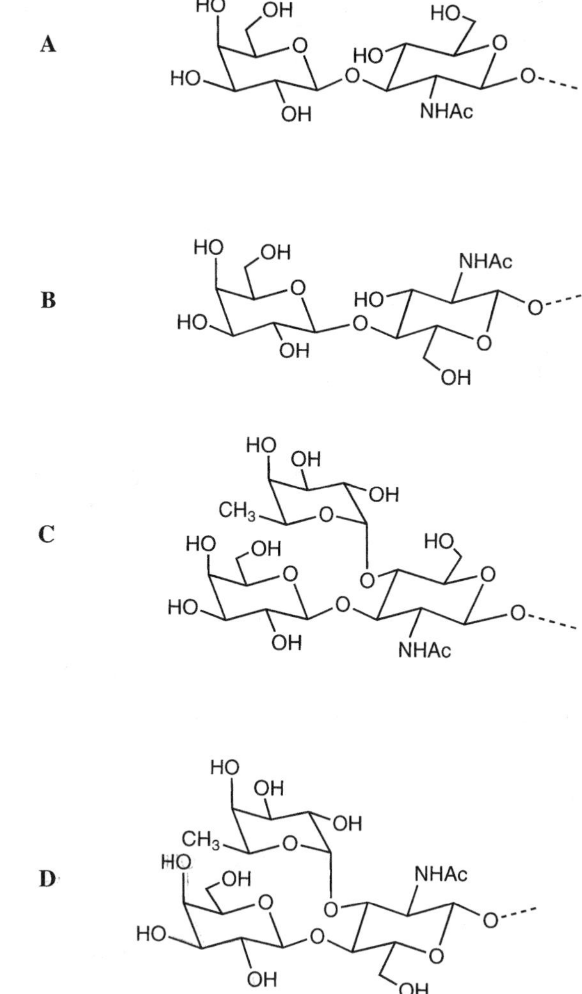


**Figure 4.** (A) Type 1 chain present in blood group ABH and Lewis antigens. This element is part of the lactotetraosylceramide structure, Figure 2B(i). (B) Type 2 disaccharide is the major chain type of the ABH glycolipids present on erythrocytes (cf. neolactotetraosylceramide). (C) The Lewis-a epitope formed by introduction of an α-L-fucose residue at GlcNAc O-4 of the type 1 disaccharide. (D) Lewis-x trisaccharide is closely related to the $Le^a$ structure but unique to type 2 chains. The galactose and fucose residues of $Le^a$ and $Le^x$ occupy similar three-dimensional space and differ principally by the relative orientation of the acetamido (NHAc) and hydroxymethyl ($CH_2OH$) groups.

a special class of glycosphingolipids that contain one or more residues of *N*-acetylneuraminic acid (Neu5Ac) (Figure 3).

Human blood group ABO(H) antigens (Table 2) are carried on red cell membranes linked to type 2 oligosaccharide core structures present in glycosphingolipids and glycoproteins. In contrast, the Lewis antigens ($Le^a$ and $Le^b$) of erythrocytes are not synthesized in hematopoietic tissues or mature erythrocytes but instead are incorporated into the red cell membrane from serum. The terminal structures that define the AB(H) determinants may be present on either type of precursor core structures, the type 1 or type 2 chains (Figure 4A,B). However, the Lewis-a and Lewis-b antigens may be synthesized only on type 1 chains (cf. lactotetraosylceramide), since for type 2 chains the site of the galactose linkage to the 4

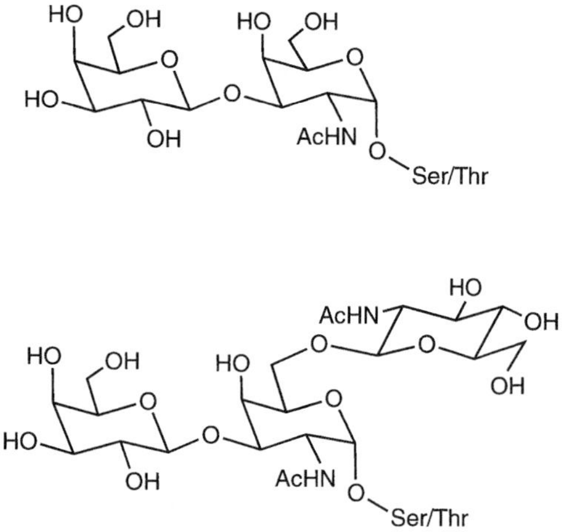

Type 1

Type 2

**Figure 5.** Mucin type 1 core structure provides short linear oligosaccharide chains of the type seen in the T antigen. Type 2 mucin core may carry short or long glycan chains on the branching GlcNAc residue.

position of GlcNAc precludes the formation of the $Le^a$ structure (Figure 4C). The Lewis-x antigen results when L-fucose is added to the *O*-3 atom of the GlcNAc residue of a type 2 chain (Figure 4D). Blood group antigens may also be elaborated on the globo core structure, but this combination is relatively rare on erythrocytes.

Certain gangliosides are important in pathogenic processes. Some, such as $GD_3$ and $GD_2$, occur as tumor-associated carbohydrate antigens, and fucosylated structures related to the Lewis antigens have been identified as oncofetal antigens (e.g., di- and trimeric forms of $Le^x$). The $Le^x$ structure bearing a Neu5Ac residue at the *O*-3 atom of galactose, sialyl Lewis x, is a specific ligand for members of the selectin family of cell adhesion molecules, which are involved in inflammation. Cholera toxin from *Vibrio cholerae* binds to $GM_1$, while tetanus and botulinum toxins, both of which block neurotransmitter release, bind to the major neuronal gangliosides $GT_{1b}$ and $GD_{1b}$.

### 2.2 Glycoproteins

The two general classes of glycoproteins possess either *N*-linked mannose-containing sugar chains, linked via 2-acetamido-2-deoxy-β-D-glucose to the amide group of asparagine, or mannose-free oligosaccharides, *O*-linked via 2-acetamido-2-deoxy-α-D-galactose to the hydroxyl group of serine or threonine. Frequently N-linked glycoproteins possess extensions to the branched mannose core structures, and these may serve as substrates for glycosyltransferases that elaborate blood group epitopes and sialylated structures. Structures of this type have been isolated from red cell membranes.

More importantly, some *O*-linked glycoproteins, the mucins, form prominent glycoprotein antigens and TACAs. They have high apparent molecular weights and carbohydrate contents exceeding 50%, and they are the major glycoproteins of mucus secretions. Two different classes of *O*-linked oligosaccharides are commonly found in cell surface mucins, characterized by type 1 or type 2 core structures (Figure 5, not to be confused with the glycolipid type 1 and 2 chains). Polylactosamine chains occur on the β-D-GlcNAc arm of the type 2 core structure. These mucins are important markers for cancer diagnosis and potential targets for treatments that attempt to induce T-cell responses. Their epitopes include the Tn, sialyl Tn, and T antigens of epiglycan, a mucin. The epitopes must be presented as peptide conjugates or synthetic glycoconjugates to obtain T-cell responses.

*See also* Carbohydrate Analysis; Carbohydrates, Industrial.

## *Bibliography*

Ajl, S. J., Weinbaum, G., and Kadis, S., Eds. (1971) *Microbial Toxins,* Vol. 4. Academic Press, New York.

Aspinall, G.O., Ed. (1983) *The Polysaccharides,* Vol. 2. Academic Press, New York.

Germanier, R. (1984) *Bacterial Vaccines.* Academic Press, New York.

Hakomori, S. (1989) Aberrant glycosylation in tumors and tumor-associated carbohydrate antigens. *Adv. Cancer Res. 52:257–331.*

Harris, H., and Hirschhorn, K. (1980) *Advances in Human Genetics,* Vol. 10. Plenum Press, New York.

Jann, K. and Jann, B., Eds. (1990) *Bacterial Capsules,* Vol. 150 in *Current Topics in Microbiology and Immunology.* Springer-Verlag, Berlin.

Korhonen et al., Eds. (1992) *Molecular Recognition in Host–Parasite Interactions.* Plenum Press, New York.

Lasky, L. A. (1992) Selectins: Interpreters of cell-specific carbohydrate information during inflammation. *Science* 258:964–969.

Paul, W. E., Ed. (1989) *Fundamental Immunology,* 2nd ed. Raven Press, New York.

Sharon, N., and Lis, H. (1993) Carbohydrates in cell recognition. *Sci. Am.* 82–89.

# Carbohydrates, Industrial

*Margaret A. Clarke*

## *Key Words*

**Disaccharide** A molecule composed of two monosaccharide units chemically joined by a glycosidic bond.

**Food Ingredient** A component of naturally occurring or processed foods.

**Gums** Nonscientific nomenclature for polysaccharides.

**Monosaccharide** Classically defined as a simple sugar, $C_nH_{2n}O_{n-1}$, usually in five- or six-membered ring form (occasionally 7) with an oxygen atom as part of the ring.

**Oligosaccharide** A short polymer of up to 10 monosaccharide units.

**Polysaccharide** A polymer of 11 or more monosaccharide units.

**Starch** A polymer of more than 20 glucose units with a majority of α-(1 → 4) linkages.

**Sugar** In common usage, the dissacharide sucrose; in chemical terms, a saccharide of one to four monosaccharide units.

**Sweetener** A compound, usually soluble in water, that in certain concentration ranges elicits the organoleptic response "sweet."

Carbohydrates, found in plants, animals, and microorganisms, are the most abundant organic compounds on earth. Carbohydrates in industrial use can be classified as mono-, di-, and trisaccharides (used primarily as food ingredients for sweetness) and higher oligo- and polysaccharides, starch, cellulose, hemicellulose, and gums (used as food ingredients for caloric supply and bulking agents). Individual physical and chemical properties give particular roles to each class and compound. Cellulose, starch, their common component glucose, and sucrose have other industrial roles as starting materials for the synthesis of organic chemicals, from ethanol to plastics, but these uses are not discussed here.

## 1 MONOSACCHARIDES

### 1.1 Glucose

In the aldohexose form α-D-glucose ($C_6H_{12}O_6$), glucose, known commercially as dextrose, is the major product from starch hydrolyzed by acid and/or enzymes (see Section 4.2). The major starch source in the United States and Japan is corn *(Zea mays);* in Europe, wheat and potato are the main sources. There is some starch and starch hydrolysate production from cassava in the tropics. Glucose, which has sweetness of 70–80 relative to sucrose at 100, is sold as anydrous dextrose, more commonly as dextrose monohydrate, as glucose syrup or corn syrup (epg., Karo syrup), or as dried corn syrup. Total U.S. production of glucose and glucose syrup was 3.5 million metric tons (mmt) in 1992–1993. The combined sweetness and colligative properties of glucose find it many uses in the confectionery industry, in the making of preserves, and in frozen food manufacture.

### 1.2 Fructose

Fructose in the ketohexose form β-D-fructose ($C_6H_{12}O_6$) is produced from glucose by the enzyme isomerase, which converts glucose to fructose; subsequently, the fructose fraction (equilibrium conversion is about 50%) is enriched, or isolation of fructose is followed by crystallization. Products are high fructose corn syrup, the most widely used monosaccharide sweetener, at 42, 55, and 90% fructose (with glucose the other major component) and crystalline fructose. Fructose sweetness ranges from 120 to 160, varying from solid through various concentrations in aqueous solution. Total U.S. production of fructose syrups and crystalline fructose in 1992–1993 was 6.2 mmt. Major uses are as sweetener in carbonated soft drinks and as sweetener/preservative/phase stabilizer in canned and frozen foods and preserves. Crystalline fructose, the most expensive sweetener in the group, appears in reduced caloric formulations and dietetic foods because its sweetness/caloric ratio is higher than that of sucrose.

## 2 DISACCHARIDES

### 2.1 Sucrose

Sucrose, α-D-glucopyranosyl-β-D-fructofuranoside, is the traditional sweetener, table sugar. It is highly soluble and is present in most fruits, some root vegetables, many trees, and grasses. It is most concentrated in sugarcane, *Saccharum officinarum,* and the sugar beet, *Beta vulgaris,* the former grown in tropical and subtropical areas, the latter in temperate zones. From either crop, juice is extracted, purified, and concentrated to syrup. Sucrose is crystallized from the syrup, in serial crystallization; the final residual viscous syrup is molasses. From both crops come a range of white sugar products: solid-granulated, powdered, cubes, liquid sucrose, and invert syrups (blends of sucrose, glucose, and fructose). Primarily from sugarcane come brown (soft) sugars, edible molasses, golden syrups, and dark cane syrups. In 1992–1993, of 111.3 mmt of sugar produced worldwide, 77.1 mmt came from sugarcane and 38.8 mmt from sugar beet; U.S. production was 3.36 mmt from cane and 4.34 from beet. Sugar beets are processed to final product in one stage (one factory). Sugarcane processing usually requires two stages: a raw sugar, produced where cane is grown, is shipped to areas of consumption, where it is refined to stable white and brown sugars. Sugarcane cannot be stored once harvested because sucrose decomposes rapidly after the cane has been cut. Sugar beet can be stored at cool temperatures for several weeks, or for months, preferably in frozen condition in piles or in temperature-controlled sheds.

Sucrose is used as a table sweetener, in baking, in brewing, in confectionery, and in bread making, as well as in canned and frozen foods, dry mixes, preserves, liqueurs, soft drinks, and alcohol production. In the United States, use of sucrose in soft drinks and in cheaper grades of preserves, confectionery products, and canned and frozen foods, has largely been replaced by the cheaper corn syrups since the early 1980s.

### 2.2 Lactose

Lactose, 4-(β-D-galactoside)-D-glucose, or milk sugar, is found only in milk and is isolated from milk as crystalline, or spray-dried, α-lactose or α-lactose monohydrate. It is not noticeably sweet (30 vs. sucrose at 100) and is frequently used as a carrier or bulking agent for nonnutritive sweeteners in table sweeteners or pharmaceuticals. The alkaline isomerization product of lactose, lactulose (4-*O*-β-D-galactopyranosyl-D-fructose), is about twice as sweet as lactose. It can be isolated from waste whey. It does not cause lactose intolerance and thus finds use in infant and health foods.

### 2.3 Leucrose

Leucrose, like lactulose a biotechnology product, has the same components as sucrose but, as 5-α-D-glucopyranosyl-D-fructopyranose, it is a reducing sugar, 50% as sweet as sucrose. Leucrose is noncariogenic and has no laxative effect. It is produced by *Leuconostoc mesenteroides* from sucrose in the presence of fructose. Since it crystallizes readily, leucrose is finding application in the manufacture of confectionery and preserves in Europe. It is not yet available in the United States.

### 2.4 Maltose

Maltose, α-D-glucopyranosyl-α-D-glucopyranose, is another product of starch hydrolysis, sold in syrup form or as crystalline maltose. It has about one-third the sweetness of sucrose and is used as a bulking agent, preservative against staling of baked goods, and filler and stabilizer for frozen foods.

"Malt syrup" and "malt extract" contain some maltose but get their name from the same source as maltose: the malting process, or germination of barley grain to produce a solution of amylolytic enzymes.

### 2.5 Palatinose

Palatinose, or isomaltulose, 5-*O*-α-D-glucopyranosyl-D-fructofuranose, is made by the action of α-glucosyltransferase on sucrose. Its main use is as a starting material for hydrogenation to Palatinit, or isomaltitol, a lowered calorie, non-cariogenic sweetener sold in Europe. Palatinose is about half as sweet as sucrose and is noncariogenic.

### 2.6 Honey

Honey, produced by honeybees (*Apis mellifera* and *Apis dorsata*), is a liquid product, about 80% solids, containing a mixture of simple carbohydrates: 24–45% fructose, 25–37% glucose, 2–12% maltose, and 0.5–3% sucrose, with traces of many other sugars depending on the bees' floral source. In 1992–1993, 100,000 metric tons of honey was produced in the United States and another 50,000 metric tons imported. Honey is available in liquid form, in the comb (as bees store it), and in dry granulated form; it is used as a sweetener in dairy and processed foods, in frozen foods and, in dry form, in dry mixes.

### 2.7 Maple Sweeteners

Maple syrup and sugar, made from sap of the sugar maple tree (*Acer saccharum*) through concentration and crystallization, are, like honey, mixtures of simple sugars. Freshly made maple syrup, generally about 65% solids by weight, contains 50–63% sucrose, and from 0 to 8% combined glucose and fructose, but no maltose. As maple syrup is stored, the sucrose will invert to glucose and fructose. Because of their delicate flavors and their expense, maple products generally are consumed directly as sweets or as pancake syrup, or used in home baking, but there is some industrial consumption in confectionery and baked goods. In 1992–1993, 1.64 million gallons of maple syrup was produced in the United States. About 0.5% of this was crystallized to maple sugar.

## 3 OLIGOSACCHARIDES

The oligosaccharides are a group of carbohydrates containing two to twelve monosaccharide units linked together. The science of glycobiology is based on certain of these compounds. Other oligosaccharides of industrial importance are discussed here.

### 3.1 Dextrins

Dextrins, products of the hydrolysis of starch when the conversion process is stopped at product size of 3 to about 30 glucose units, have dextrose equivalent (D.E.) values below 20. Corn syrup solids are the small dextrins. Slightly sweet, they contain three to six glucose units and are used as fillers and hygroscopic agents in foods and dry mixes. The larger oligosaccharides, called maltodextrins, are usually spray-dried. They are less hygroscopic and find a wide range of use as bulking agents, dispersants, and nutritive fillers. A specific group of maltodextrins has emerged as fat substitutes in processed foods because of their emulsifying and mouthfeel properties.

Cyclodextrins are rings of glucose units, usually six, seven, or eight (α, β, or γ) in number, which have a strong potential for inclusion complex formation. Widely used as stabilizing agents for controlled-release pharmaceuticals and agricultural chemicals, the cyclodextrins are finding application as flavor and odor compound stabilizers.

### 3.2 Fructosyloligosaccharides

Fructosyloligosaccharides (FOS, or neosugars) are fairly recent market entries. These mixtures contain varying quantities of fructosyl sucrose oligomers: 1-kestose ($GF_2$, where G represents glucose and F fructose), nystose ($GF_3$ and $GF_4$, or 1′-β-fructofuranosyl nystose, where the fructoses are linked by β-(2→1) bonds. These mixtures are prepared by isolation from inulin-containing plants (Jerusalem artichoke and chicory) in Europe, and by the action of fructosyl transferase (usually fungal α-fructofuranosidase, from *Aspergillus niger*) in the United States and Japan. Recent work has shown a thermal route to this preparation.

The mixtures are sold as noncariogenic and nonnutritive sweeteners in Japan, where they are also used in confectionery and soft drinks. The U.S. product can be sold "as is" but has not received clearance from the Food and Drug Administration as a food ingredient: it is useful as a bulking agent, or as a carrier for nonnutritive sweeteners, and it shows a positive effect on human digestion by increasing bifidobacteria in the small intestine. FOS neosugars also find use as animal feed additives, for weight gain. A special product containing small inulin hydrolysates, from chicory inulin, is sold in Europe as a fat substitute.

## 4 POLYSACCHARIDES

Some 90% of nature's carbohydrates are polysaccharides; cellulose, the most abundant, is a structural element in all plants. Industrial carbohydrates, other than as sweeteners or food ingredients, are in the polysaccharide group.

### 4.1 Cellulose

Cellulose, a high molecular weight polymer of β-D-glucose units in β-(1→4) linkage, is the principal cell wall component of higher plants and the major component of wood and cotton, hence of paper. The cellulose that is insoluble in 18% alkali is termed α-cellulose; β-cellulose is soluble in 18% alkali but precipitates upon neutralization, whereas γ-cellulose remains in solution. Wood is the main source for industrial cellulose for, in addition to paper, production of rayons and nonwoven materials. Viscose and cellophane are made through cellulose xanthate intermediate. Other industrially useful cellulose esters are acetate (almost 400,000 tons/yr) for fiber; acetate–butyrate mixture, a tough plastic; and nitrate, or gun cotton. Esters of cellulose also have many applications: ethyl cellulose is a thermoplastic and a binder; hydroxyethyl cellulose and similarly substituted celluloses are finding uses in coatings and films; both microcrystalline cellulose and, increasingly, carboxymethyl-cellulose (now 40,000 tons/yr), are finding roles as fillers and binders in processed foods. The last group can also be classified as gums or hydrocolloids.

### 4.2 Starch

Starch, a polymer of α-(1→4)-linked α-D-glucose units, is the common form of energy storage in all green plants. Starch has two major components: amylose and amylopectin. Amylose, the linear polymer, 0–40% content varying with plant type, has a molecular weight up to $5 \times 10^6$ Da. Amylopectin, the branched-chain polymer, branched at α-(1→6) linkages, has a molecular weight from 50 to 200 × $10^6$ Da. Common starches of commerce are corn (maize), wheat, potato, rice, and barley. The major uses for unmodified starch, other than as food in its plant sources, are as a raw material for starch hydrolysate sweeteners and dextrins, as a thickener in food mixtures (epg., puddings), and in laundry preparations.

Modified starches have a multitude of uses. Acid-modified starches are used in textile finishing; cross-linked starches are used as thickeners in processed foods; cationic starches (substituted ammonium starch ethers) are used in papermaking and finishing, as are other starch ethers (epg., hydroxyethyl starch). Copolymers of starch—with, for example, acrylonitrile—are used in coatings and additives for extensive water absorbency (e.g., diapers, hospital pads).

Most starches must be heated to gelatinize (i.e. to hydrate and solubilize, giving them their thickener properties), but a new modification process has produced starches that will solubilize and swell in cold water. Members of another new group of modified starches function as fat substitutes in processed foods. These starches come from corn in U.S. markets, from wheat or potato in Europe, and from maize, rice, or cassava in South America and the Far East. Some 20 million tons of starches and derivatives is used industrially each year, worldwide.

### 4.3 Gums and Hydrocolloids

Gums include hetero- and homopolysaccharides that dissolve in water to form viscous solutions or gels. Except for heparin and hyaluronic acid, two hydrocolloids important in warm-blooded animals, they are of plant or microbial origin. Several gums with greatest industrial use are listed, with tonnages where available. Some hemicelluloses (e.g., gum arabic) are also classed as gums.

Algins, salts of polymers of mannuronic and guluronic acid, are extracted from algae in seaweed and are used at levels of some 25,000 tons per year, mostly as food gels.

Carrageenans (30,000 tons/yr) are components of seaweed and extracted therefrom, with galactose as their main structural component. These substances, which also form gels, have found a widely publicized use as a fat substitute in meat products.

Dextrans, microbial products of *Leuconostoc mesenteroides* on sucrose, are used clinically as a blood plasma replacement (1000 tons/yr) and in laboratory and industrial separations as a chromatographic support (2000 tons/yr). Dextrans are α-(1→6)-linked polymers of D-glucose.

Guar gum and gum arabic are natural products of semitropical trees; guar has a β-(1→4)-linked D-mannose backbone with galactose side chains, and gum arabic has a β-galactose backbone with side chains that can contain galactose, arabinose, rhamnose, and glucose. Both have wide application in food processing, at rates of some 60,000 tons/yr for guar and 25,000 for gum arabic. Both are increasingly being replaced by gellan gum and welan gum, bacterial polysaccharides developed for this purpose. Bacterial polysaccharides have more reproducible structures, and more predictable sources of supply, than natural gums.

Pectins (15,000 tons/yr), with a backbone of α-(1→4)-linked D-galacturonic acid and side chains of rhamnose, xylose, and glucose according to source, are made from citrus, apples, and sugar beet. Their major use is as gelling agents in high viscosity, high sugar gels (high methoxyl pectins) or low viscosity, low sugar gels (low methoxyl pectins).

Xanthan gum, one of the first bacterial polysaccharides to find wide industrial application (25,000 tons/yr), has a celluloselike backbone of α-(1→4)-linked D-glucose units, with alternate units substituted on O-3 by trisaccharides of mannose, glucose, and 6-*O*-acetyl mannose. Xanthan gum, which forms a heat-stable gel, is used to thicken dressings, puddings, and various processed foods.

### 4.4 Chitosan

Chitosan is the deacetylation product of chitin, the major carbohydrate in shellfish and arthropod shells. Shells (crab, shrimp, etc.) at seafood processing plants are the main source, although chitin is, after cellulose, the most abundant polymer in nature. Chitosan is a polymer, having a molecular weight of up to several million daltons, of 2-amino-2-deoxy-D-glucose. It has only recently (mid-1980s) become available in tonnage amounts. Its cationic properties and nontoxic biocompatibility are leading to many applications in water and beverage treatment, cosmetic, pharmaceutical and medical areas, agricultural uses (seed coatings), and food and feed processing.

*See also* Carbohydrate Analysis.

## Bibliography

Aspinall, G. O., Ed. (1982, 1983, 1985) *The Polysaccharides,* Vols.1–3. Academic Press, New York.

Clarke, M. A., Ed. (1992) *Carbohydrates in Industrial Synthesis.* Verlag Dr. A. Bartens, Berlin.

Lichtenthaler, F. W., Ed. (1991) *Carbohydrates as Organic Raw Materials.* VCH, Weinheim.

Marie, S., and Piggott, J. R. (1991) *Handbook of Sweeteners.* Blackie, London.

Pancoast, H. M., and Junk, W. R., Eds. (1980) *Handbook of Sugars,* 2nd ed. AVI, Westport, CT.

Pigman, W., and Horton, D., Eds. (1970) *The Carbohydrates: Chemistry and Biochemistry,* Vol. IIA, Academic Press, New York.

———, and ———, Eds. (1972) *The Carbohydrates: Chemistry and Biochemistry,* Vol. IA, 2nd ed. Academic Press, New York.

U.S. Department of Agriculture, Economic Research Service. (1994) *Sugar and Sweetener Situation and Outlook Reports.* ERSS-NASS, Herndon, VA.

Whistler, R. L., and BeMiller, J. N., Eds. (1992) *Industrial Gums,* 3rd ed. Academic Press, San Diego, CA.

Yalpani, M. (1988) *Polysaccharides.* Elsevier, Amsterdam.

———, Ed. (1988) *Industrial Polysaccharides.* Elsevier, Amsterdam.

# Cardiovascular Diseases

*Todd Leff and Peter J. Gruber*

## Key Words

**Atherosclerosis** Disease that leads to the formation in an artery of a lesion that eventually impedes or blocks the flow of blood.

**Coronary Arteries** The arteries that supply blood to the heart muscle.

**Gene Expression** The multistep process in which a gene sequence is converted into a functional protein. The main steps in this process are transcription of a DNA sequence into RNA and translation of RNA into protein.

**Lipoproteins** Spherical protein-coated particles that carry cholesterol and other lipids in the circulation.

**Promoter** The region of DNA that is at or near the 5′ end of a gene and controls the rate of transcription of the gene.

**Transgenic** An animal that carries a foreign gene integrated into its genetic material, which has been inserted by in vitro methods.

---

The hallmark of atherosclerosis, the major cause of cardiovascular disease, is the formation of lipid-laden cellular lesions in one or more of the coronary arteries that supply the heart muscle with blood. Such a lesion can constrict the lumen of the artery, reducing the flow of blood and starving a portion of the heart for oxygen. Ultimately, the lesion may rupture, inducing the formation of a blood clot and completely blocking the flow of blood to a portion of the heart. This event—the classic heart attack, or myocardial infarction—can have a variety of effects on the heart muscle depending on the length and severity of the blood flow restriction. These effects range from mild reversible damage to the heart to severe permanent injuries like tissue necrosis and scarring. In the most extreme cases, the result of a myocardial infarction is cardiac arrest and the death of the patient.

This brief review excludes many pathologies that are subjects of molecular biology research, including the effects on the cardiovascular system of viral illness, systemic hypertension, diabetes, kidney failure, and substance abuse, as well as congenital abnormalities of the heart. As in atherosclerotic disease, these conditions can, under some circumstances, damage the heart muscle. This loss of myocardial function induces an adaptive compensatory response that initially allows the heart to maintain normal function, but eventually initiates a program of physiological changes that weakens the heart and ultimately leads to cardiac failure and death. This set of events, triggered by an initial insult and subsequent compensation, is known generically as heart failure and is a leading cause of death in many countries. Although the molecular events involved in heart failure are not fully understood, progress has been made in recent years in the establishment of model systems amenable to molecular approaches. Using these systems, some genes have been identified that display altered patterns of expression during one or more stages of heart failure. The role these genes play in the development of heart failure is a topic of current investigation.

## 1 ATHEROSCLEROSIS

The formation and development of an atherosclerotic lesion depends on the presence of one or more physiological conditions or risk factors. Primary among these is an abnormal profile of lipid-carrying lipoprotein particles. Specifically, high levels of low density lipoprotein (LDL) have been shown to be associated with an increased risk of developing atherosclerosis. The mechanism by which abnormal lipoprotein levels induce or enhance the formation of the atherosclerotic lesion is not understood. However, some of the key events in the formation of the lesion can be described.

The event that triggers the formation of a lesion is thought to be an injury to the layer of endothelial cells that line the inside of the artery. This injury initiates a repair response characterized by the infiltration of monocytes and T lymphocytes into the space under the endothelial cell layer. In the presence of high levels of LDL, this repair process is accompanied by an infiltration of cholesterol-rich lipoprotein particles. The amount of lipid that accumulates in the subendothelial space is related to the amount of LDL in the circulation. Some fraction of this lipoprotein may become damaged or modified by partial oxidation. Oxidized LDL appears to serve as another chemoattractant that draws additional monocytes and lymphocytes into the subendothelial space. This prolongs and accentuates the initial response to the damaged vessel wall.

Physiological conditions that increase the level of LDL oxidation probably accelerate the development of the lesion. Diabetes, a known risk factor for atherosclerosis, may have its effect by increasing the oxidative damage of LDL. Many important risk factors that contribute to the formation of atherosclerotic heart disease are not discussed in this entry, including risk factors that are themselves disease states (e.g., diabetes, hypertension, obesity), as well as specific elements of the blood clotting and immunological systems. Section 2 focuses on the molecular biology of the lipoprotein system, and specifically on the genes involved in determining the level of LDL in the blood.

## 2 LIPOPROTEIN METABOLISM

To appreciate the contribution of molecular biology to our understanding of lipoprotein metabolism, it is necessary to review the basic features of this complex system. Lipoproteins are spherical particles composed of a polar surface of protein, phospholipid, and free cholesterol, plus a nonpolar core containing cholesterol esters and triglyceride. The primary function of these particles is to transport lipids, mainly cholesterol and triglyceride, between organs and tissues. There are several types of lipoprotein, characterized by distinct functions and containing specific sets of specialized proteins called apolipoproteins (Tables 1 and 2).

Lipids absorbed in the diet (mainly triglyceride and cholesterol) are packaged by the intestine into lipoprotein particles called chylomicrons. The lipids contained in chylomicrons are delivered to peripheral tissues, mainly muscle and adipose, and also to the liver. The liver produces another important substance, very low density lipoprotein (VLDL), which transports triglyceride and cholesterol from the liver to other tissues in the body. As triglyceride is removed from VLDL it is gradually converted into LDL, the major vehicle for cholesterol transport between tissues. LDL particles were removed from the circulation by the LDL receptor, which is present on the surface of all cells in the body (Table 2). The LDL receptor binds strongly to LDL particles by recognizing specific apolipoproteins (either apo B or in some cases apo E) that are present on the surface of the LDL particle.

Another important member of the lipoprotein family is high density lipoprotein (HDL), is a cholesterol-rich lipoprotein particle that is formed in plasma and is believed to carry out a process called reverse cholesterol transport. This process apparently clears excess cholesterol from peripheral tissues and transports it to the liver, where it is excreted in the bile. High levels of HDL in the

**Table 1** Major Classes of Lipoproteins

| Class | Primary Function | Apolipoproteins |
|---|---|---|
| Chylomicron | Transports dietary triglyceride from the gut to peripheral tissues | $B_{48}$, E, C-I, C-II, C-III |
| VLDL (very low density lipoprotein) | Delivers hepatic derived triglyceride to peripheral tissues | $B_{100}$, E, C-I, C-II, C-III |
| LDL (low density lipoprotein) | Transports cholesterol from the liver to peripheral tissues | B |
| HDL (high density lipoprotein) | Reverse cholesterol transport from peripheral tissue back to the liver | A-I, AI-V, A-II |
| Lp(a) (lipoprotein a) | Modified version of LDL, unknown function | B, apo(a) |

circulation are believed to be protective against the development of atherosclerosis. Although the mechanism of this protection is unknown, it is clear that in the general human population, there is an inverse correlation between HDL levels and the incidence of atherosclerosis.

Most of the key proteins involved in lipoprotein metabolism were purified and biochemically characterized during the last 20 years. The initial contribution of molecular biology to the study of atherosclerosis was the cloning and characterization of the genes encoding these proteins. Once these genes were known, it was possible to identify and characterize naturally occurring mutations that provided insights into the specific physiological roles of these proteins. It was also possible, using transgenic and gene targeting techniques, to produce genetically altered strains of mice that were either missing specific genes or overexpressing specific gene products. Finally, it has been possible to explore the regulation of the expression of these genes and to determine what role gene expression plays in normal and pathological states.

**Table 2** Genes Encoding Proteins Involved in Lipoprotein Metabolism

| Gene | Site of Expression | Site of Action | Protein Function |
|---|---|---|---|
| **Apolipoproteins** | | | |
| —apo A-I | Liver and small intestine | Serum | Activates lecithin cholesterol acyltransferase, a component of HDL and chylomicrons |
| —apo A-II | Liver and small intestine | Serum | Unknown function, component of HDL and chylomicrons |
| —apo A-IV | Liver and small intestine | Serum | Unknown function, component of HDL and chylomicrons |
| —apo B | Liver and small intestine | Serum | B100: ligand for the LDL receptor, synthesis and secretion of VLDL, component of VLDL and LDL<br>B48: synthesis and secretion of chylomicrons, component of chylomicrons |
| —apo C-I | Liver and small intestine | Serum | Modulates uptake of triglyceride-rich lipoproteins, component of HDL, VLDL and chylomicrons |
| —apo C-II | Liver and small intestine | Serum | Activates lipoprotein lipase, component of HDL, VLDL and chylomicrons |
| —apo C-III | Liver and small intestine | Serum | Modulates uptake of triglyceride-rich lipoproteins, component of HDL, VLDL, and chylomicrons |
| —apo E | Liver, small intestine, macrophages, and many other tissues | Serum | Ligand for the LDL and chylomicron remnant receptors; component of HDL, VLDL, and chylomicrons |
| —apo(a) | Liver | Serum | Attaches to the apo B protein of LDL creating Lp(a), unknown function |
| **Lipoprotein Receptors** | | | |
| LDL (B/E) receptor | All cells | Plasma membrane of all cells | Removal of LDL from plasma |
| Scavenger receptor | Mainly macrophages | Plasma membrane of macrophages | Removal of oxidized lipoprotein from plasma and from atherosclerotic lesions |
| Remnant receptor (not confirmed) | Liver | Plasma membrane of hepatocytes | Removal of VLDL remnants from plasma |

## 3 GENETIC ABNORMALITIES

Naturally occurring genetic mutations are extremely useful tools for exploring the function of a specific gene and for identifying key genes involved in a particular physiological system. A great deal of information about the function of individual gene products can be obtained by correlating an exact molecular genetic defect with a specific physiological phenotype. As described earlier, one of the more important risk factors for the development of atherosclerosis is the level of LDL in the circulation. For this reason, some of the first studies on genetic abnormalities affecting atherosclerosis were focused on genes involved in LDL metabolism. Particular attention has been paid to the genes encoding the LDL receptor and the apolipoprotein ligands for the LDL receptor (apo B and apo E).

An excellent example of how molecular analysis of a genetic defect has provided a great deal of information about a disease is in the study of the severe genetic abnormality called familial hypercholesterolemia (FH). Patients with this disorder have extremely high LDL levels and develop severe premature atherosclerosis, frequently suffering heart attacks as teenagers. The pioneering work of Michael Brown and Joseph Goldstein demonstrated that these patients had inherited two defective alleles of the LDL receptor gene and did not have functional LDL receptors. Since these receptors normally function to remove LDL from the circulation, LDL accumulates to high levels in the blood of these patients and induces the development of atherosclerosis. Heterozygous FH patients who have inherited one defective allele are quite common in the population, with a frequency of about 1 in 500. These patients have about twice the normal level of LDL and have an increased risk of developing atherosclerosis.

Analysis of these mutations at the molecular level has revealed that the population contains a heterogeneous set of mutant alleles. Several classes of LDL receptor mutations have been characterized. Some mutations cause a complete failure of the protein to be synthesized, while others produce a protein that is either not transported to the cell surface, or is incapable of recognizing its ligand and therefore fails to bind LDL. The characterization of these various classes of mutations has provided a great deal of information about the LDL receptor and how it works. The FH syndrome constitutes a vivid experiment of nature, providing detailed information about the crucial role of the LDL receptor in the regulation of plasma LDL levels. In addition, since these patients frequently have no other risk factors, the syndrome directly demonstrates the key role of elevated LDL levels in the development of atherosclerotic disease.

Inherited disorders of lipoprotein metabolism have been mapped to many other genes in the lipoprotein system. These include several apolipoprotein genes, as well as genes for various enzymes involved in the metabolism of lipoprotein particles. Analysis of these mutations and their physiological effects has been invaluable in determining the function of many of the components of the lipoprotein system.

## 4 TRANSGENIC ANIMALS

Perhaps the most dramatic application of molecular biology to the study of the lipoprotein system and its role in the development of cardiovascular disease has been the use of transgenic animals. This technology allows the introduction of a foreign gene into the germ line of a mouse. The gene can be a duplicate copy of a normal endogenous mouse gene, or it can be a foreign gene from another species. The transgene can be unaltered (wild type), or it can be modified to carry specific mutations designed to provide information about the function of particular regions of the gene or domains of the protein. Finally, the expression of the transgene can be under the control of its own promoter (regulatory region) or under a foreign promoter with the desired regulatory characteristics. Transgenic animals are made by injecting a purified DNA into the pronucleus of a fertilized mouse egg. Injected eggs are then implanted into the uterus of a foster mother, where some fraction of them will develop normally. The incorporation of the foreign DNA into the mouse genome is scored in newborn pups either by analysis of tail tip DNA or by observation of the presence of the transgenic product in the blood. Modification of the mouse genome to overexpress genes involved in lipoprotein metabolism has provided a great deal of insight into the functional roles of specific gene products.

The first gene involved in lipoprotein metabolism to be overexpressed in a transgenic animal was the human LDL receptor gene. As noted earlier, the absence of this gene in patients causes severe hypercholesterolemia and premature atherosclerosis. Transgenic mice overexpressing the receptor gene showed an increased rate of LDL clearance and dramatically lower cholesterol levels. Another transgenic line that showed an alteration of LDL metabolism was a line expressing the rat apo E gene. Overexpression of *apo E,* which is a ligand for the LDL receptor, also increased the rate of LDL clearance. An excellent example of the use of transgenic technology to clarify a metabolic role for a specific gene product was the study of the *apo*CIII gene. In vitro evidence and epidemiological studies had suggested that *apo*CIII inhibited the clearance of the triglyceride-rich lipoproteins VLDL and chylomicrons. This was confirmed when transgenic lines overexpressing the *apo*CIII gene were found to have extremely high triglyceride levels and reduced rates of VLDL clearance.

The hypothesis that HDL plays a protective role in the development of atherosclerotic heart disease was tested by using transgenic technology to generate a mouse with increased levels of HDL. This was accomplished by inserting multiple copies of the human gene encoding the major apolipoprotein component of HDL (*apo*AI). The increased production of *apo*AI caused an increase in HDL levels. Some of the transgenic mice with elevated HDL levels were found to be resistant to the development of atherosclerotic lesions, supporting the hypothesis that HDL plays a protective role.

The complement of the transgenic technology, which generates an animal overexpressing a specific protein, is the gene knockout technique, which results in a mouse line in which a specific gene has been inactivated. In this method, the gene of interest is inactivated by homologous recombination between one of the endogenous copies of the gene and an exogenous cassette containing a mutated version of the same gene. The recombination event takes place in cultured embryonic stem cells. Once the cells containing the recombined, or mutated, gene have been isolated, they are introduced into a developing mouse blastocyst. Some of the mice that develop from these chimeric blastocysts may be composed of cells containing the mutated gene. If the germ line cells are derived from the mutant cells, the inactivated gene will be inherited by subsequent generations. This technique has been used with dramatic results to demonstrate the crucial role of the apo E gene in lipopro-

Table 3 Genes Encoding Key Enzymes Involved in Cellular Cholesterol Metabolism

| Gene | Protein Function | Regulation |
|---|---|---|
| LDL receptor | Adsorption of LDL, increases cellular cholesterol content | Transcription |
| HMG CoA synthetase | Biosynthesis of cholesterol | Transcription |
| HMG CoA reductase | Biosynthesis of cholesterol | Transcription, translation, protein stability, enzymatic activity |
| Acyl-CoA cholesterol acetyltransferase (ACAT) | Conversion of free cholesterol to cholesterol ester for storage | Enzymatic activity |
| 17α-Hydroxylase | Degradation of cholesterol, production of bile acids | Transcription, enzyme activity |

tein metabolism and atherosclerotic heart disease. Mice that were homozygous for a defective allele of the apo E gene and had no circulating *apo E* showed very high LDL cholesterol levels and rapidly developed atherosclerosis. This result is especially remarkable because rodents are normally resistant to atherosclerosis, even when fed a high fat diet.

## 5 GENE EXPRESSION

In some circumstances, abnormalities in the lipoprotein system may be due to dysfunction in the expression of specific genes. A great deal of work has been done on the molecular mechanisms that regulate the expression of genes involved in lipoprotein metabolism. The regulatory regions of many key genes have been characterized, and the protein factors that control their transcriptional activities have been identified. A clear example of the key role of transcriptional regulation is the control of the genes involved in maintaining cellular cholesterol homeostasis. The transcriptional regulation of this set of genes is a key feature of cellular cholesterol metabolism and plays an important role in modulating plasma cholesterol levels.

The level of cholesterol inside the cell is maintained within narrow limits. This is accomplished by balancing the rate of cholesterol removal with the rates of cholesterol synthesis and cholesterol absorption (from circulating LDL via the LDL receptor). When the cholesterol level rises, the metabolic machinery of the cell responds by decreasing the activity of the cholesterol biosynthetic pathway and by reducing the number of LDL receptors on the surface of the cell. This results in lower rates of cholesterol biosynthesis and reduced uptake of cholesterol from the circulation. These effects are accomplished, in part, by reducing the transcriptional activity of two of the key genes in the cholesterol biosynthetic pathway: the LDL receptor gene and a pair of 3-hydroxy-3-methylglutaryl coenzyme A compounds, HMG CoA reductase and HMG CoA synthetase (Table 3). The promoter region of each of these genes contains one or more copies of an eight nucleotide long sequence element called a sterol response element (SRE). The regulation of gene expression by cholesterol is dependent on the presence of this element. If the SRE is removed or mutated, the transcriptional activity of these genes no longer responds to cellular cholesterol levels. The activity of the SRE is mediated by a transcription factor that either directly or indirectly interacts with cholesterol (or one of its metabolic derivatives), binds to the SRE, and modulates the transcriptional activity of the gene. HMG-CoA reductase is one of the most highly regulated enzymes in the cell. In addition to transcriptional regulation of the gene, the translation of the mRNA, the stability of the protein, and the activity of the enzyme are regulated by cholesterol.

The regulation of gene expression may play an important role in many aspects of arthrosclerotic disease, including lesion development, lipoprotein metabolism, and, as just described, cholesterol metabolism. In addition, aberrant gene regulation may be a key feature of heart failure, another important cardiovascular disease. Analysis of the transcriptional regulatory mechanisms of key genes involved in the cardiovascular diseases is an active area of investigation.

*See also* Heart Failure, Genetic Basis of; Lipid Metabolism andTransport; Plasma Lipoproteins.

### *Bibliography*

Braunwald, E., Ed. (1992) *Heart Disease: A Textbook of Cardiovascular Medicine,* 4th ed. Saunders, Philadelphia.

Goldstein, J. L., and Brown, M. S. (1990) Regulation of the mevalonate pathway. *Nature* 343:425–430.

Havel, R. P., and Kane, J. P. (1989) Structure and metabolism of plasma lipoproteins. In *The Metabolic Basis of Inherited Disease,* 6th ed., C. R. Sriver, A. L. Beaudet, W. S. Sly, and D. Valle, Eds., pp. 1129–1138. McGraw-Hill, New York.

Ross, R. (1993) The pathogenesis of atherosclerosis: A perspective for the 1990s. *Nature* 362:801–809.

# Cell–Cell Interactions

*Geoffrey M. W. Cook*

### *Key Words*

**Cadherins** A family of cell surface glycoproteins that mediate $Ca^{2+}$-dependent cell–cell adhesion in vertebrates.

**CAMs** Cell adhesion molecules: a collective term for cell surface glycoproteins that regulate cell–cell adhesion. CAMs may be divided into two classes; those that express their adhesive function only in the presence of $Ca^{2+}$ and those that function independently of $Ca^{2+}$. The term **primary CAM** is used to describe CAMs that appear early in embryogenesis on derivatives of all three germ layers. **Secondary CAMs** appear somewhat later on a more restricted set of tissues.

**Fab Fragment** The fragment obtained by papain-catalyzed hydrolysis of the immunoglobulin molecule. It consists of one light chain linked to the N-terminal half of the contiguous heavy chain. Two Fab fragments are obtained from each immunoglobulin molecule and, since each contains one antibody combining site, it can combine with antigen as a univalent antibody.

**Glycoprotein** Protein containing oligosaccharides covalently attached to selected amino acid residues. A common linkage involves C-1 of *N*-acetylglucosamine and the amide group of asparagine to produce *N*-glycans. Alternatively C-1 of *N*-acetylgalactosamine may be joined by glycosidic bond to the hydroxylated side chain of serine or threonine to form an *O*-glycan. Both types of linkage can occur in the same glycoprotein.

**Glycosyltransferases** Collective term for the enzymes directly involved in the biosynthesis of *N*- and *O*-glycans of glycoconjugates. They catalyze the transfer of sugar from a sugar nucleotide to a suitable acceptor, which may be an appropriate amino acid residue in a polypeptide or a carbohydrate sequence in an incomplete glycan. When present at the cell surface, these enzymes are termed ectoglycosyltransferases.

**Immunoglobulin Domain** The homology regions of immunoglobulin, consisting of disulfide bonded domains containing approximately 100 amino acids, which are organized into two parallel β sheets stabilized by disulphide bonds.

**Lectin** Term used to describe a class of proteins of nonimmunological origin that bind carbohydrates.

**Sialic Acid** Collective term to describe all acylated neuraminic acids. Neuraminic acid is a nine-carbon sugar acid, with an amino group in the molecule.

---

The ability of cells to become associated with one another to form a tissue is the target of intense investigation at the cellular, biochemical, and genetic level. In dealing with ''cell–cell interactions,'' particular attention has focused on ''cell adhesion'' as an essential feature of normal development, as well as pathological processes such as the spread of tumors. While ''cell adhesion'' may be described as the ability of cells to remain in association, ''adhesiveness'' is usually defined operationally in terms of the method of measurement employed. For example, ''adhesiveness'' can be a measure of the ease with which cells can be separated. Alternatively the ability of single cells in suspension to come together to form aggregates or to attach to a previously formed cellular aggregate or to a monolayer of cells may be used as a means of assessing ''cell adhesion.'' Though these forms of measurement are ideal for laboratory investigations, they may not replicate all the conditions prevailing in vivo. Nonetheless, as a result of applying such assay methods, together with an immunological approach, investigators have identified a number of cell adhesion molecules (CAMs), which are of considerable biological importance.

More recently attention is also being given to the mechanisms by which cells avoid certain other cell types, a process that has been termed ''cell–cell repulsion.'' Both ''cell adhesion'' and ''cell repulsion'' can be considered as different extremes of ''cell–cell interactions''; in vivo, both processes may be operative but to different extents at any one time.

Cells interact not only with one another but also with components of the extracellular matrix (ECM). A number of substrate adhesion molecules (SAMs)—that is, molecules found in the ECM that bind to particular receptors on cells—have been identified but are not considered here.

## 1 CELL ADHESION

### 1.1 Historical Aspects

The study of cell–cell adhesion dates from the early part of this century. An American scientist, Wilson, found that a suspension of single cells could be obtained by pressing fragments of sponges through a fine sieve with seawater. These cells retained their ability to combine to form clusters and ultimately small sponges. Interestingly, not only did the sponge cells possess adhesive properties but, by using mixtures of cells derived from various species of sponge distinguishable by their different colors, Wilson found that re-formed sponges were from one species only, providing an experimental basis for studying ''adhesive selectivity.''

The modern investigation of vertebrate embryonic cell adhesion is usually regarded as stemming from the work of Holtfreter, another American investigator, and subsequently from work with his student Townes. Holtfreter found that when tissue segments taken from different germ layers of an amphibian were combined in vitro, they would adhere to form an aggregated mass in which distinct groups of cells sorted out from one another in a way that was reminiscent of their topographical deposition in the embryo and was also in keeping with the timing of morphogenetic movements in vivo. This work demonstrated that the adhesive interactions between cells constitute a highly selective process that is crucial to normal morphogenesis. Finding that amphibian embryos can be dissociated in alkaline solution, Townes and Holtfreter repeated ''sorting-out'' experiments with single cells taken from different regions of the embryo. Cells from different regions, when mixed and allowed to form aggregates, could ''sort out'' and become segregated to form groups and layers reminiscent of the normal histological organization of the embryo, again illustrating that adhesion is important in morphogenetic movements.

The work just described raised the question of whether the differences in affinities between groups of cells could be explained in qualitative or quantitative terms. A number of scientists have tried to determine this matter; the studies of Steinberg and Moscona and their colleagues are especially noteworthy. Steinberg, who recognized the importance of calcium ions in cell–cell interactions, originated the ''differential adhesion hypothesis,'' which placed weight on quantitative differences in adhesive affinities to explain cell sorting, Moscona, on the other hand, emphasized the importance of qualitatively distinct cell surface proteins in cell adhesion and sorting.

Moscona contributed an important methodological advance to the study of cell–cell adhesion. The earlier studies, such as those of Holtfreter, had to rely on random collisions between cells in stationary cultures, Moscona placed his cultures on a rotating platform so that cells could be brought into contact by centripetal forces, thus eliminating the need for cell migration.

## 1.2 CAMs: Immunological Approach to an Analysis of Cell Adhesion

Considerable advances in the identification and characterization of CAMs have taken place over the last two decades. This progress has resulted from an immunological approach to the problem of solving what molecules are involved in cell adhesion. The first group to exploit antibodies in a sustained manner for analyzing cell–cell interactions and identifying putative adhesion molecules was a German team led by Gerisch, using the slime mold *Dictyostelium* as a model system. Appreciating that bivalent antibodies are liable to agglutinate cells, these investigators pioneered the use of univalent Fab fragments for inhibiting cell adhesion.

The immunological approach to vertebrate embryonic cell adhesion was first applied by Edelman and his colleagues. They studied the adhesive interactions between chick embryo neural retinal cells which, for the purposes of their experiments, had been released as single cells by trypsin treatment. Adhesion was measured by the decrease in the number of particles present in cultures that had been subjected to the rotating procedure devised by Moscona (Section 1.1). This adhesion was blocked by the use of monovalent Fab fragments. Since the Fab fragments were derived from rabbit serum, which was likely to contain several antibodies, it was necessary to identify the molecule relevant to the adhesive process. This was achieved by fractionating tissue culture supernatant fluid by a combination of ion exchange chromatography, gel filtration, and gel electrophoresis to isolate the relevant neutralizing antigen. Once the antigen had been identified, this purified material was used to raise specific antibodies to the adhesion molecule. To recover an intact adhesion molecule, the neural retina was solubilized in detergent and the extract subjected to affinity chromatography using the specific antibody. The CAM isolated by this procedure from neural tissue was designated N-CAM. N-CAM has been isolated from both adult and embryonic tissue. Material purified from embryos consists of glycoproteins of molecular weight 250,000 to 200,000 while that from adult sources has a lower molecular weight (160,000–110,000). These differences in molecular weight arise as a result of differences in the covalently attached carbohydrate, as discussed later.

From the knowledge of the partial amino acid sequence of isolated N-CAM and using specific antibodies reacting with this CAM, the cDNA clone of N-CAM has been isolated and its complete structure elucidated. Interestingly N-CAM and a number of related adhesion molecules have in the extracellular part of their polypeptide chains a high degree of homology with the immunoglobulins (see Figure 1); as a result, N-CAM has been classed as a member of what is termed the *immunoglobulin superfamily*. Members of this superfamily are characterized by possessing one or more immunoglobulinlike domains. The domains are typically about 100 amino acids long, with two cysteines separated by approximately 50 amino acids, and with many other conserved amino acids, particularly around the cysteine residues. Each of these domains is folded

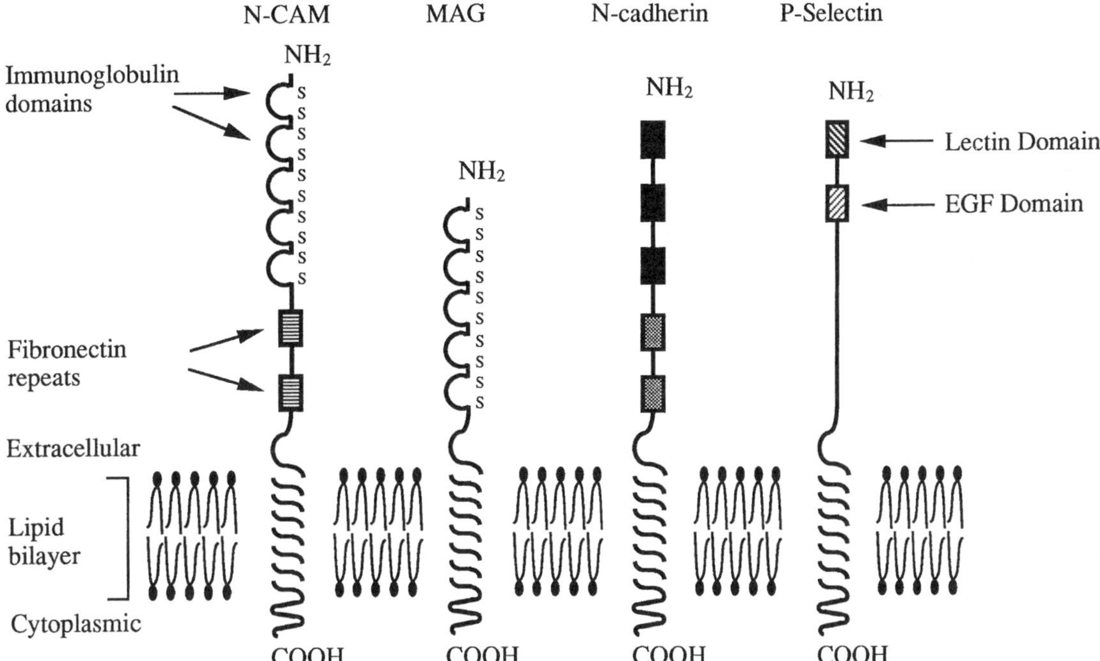


**Figure 1.** Schematic depiction of cell surface molecules involved in cell adhesion oriented in a plasma membrane. N-CAM and MAG (myelin-associated glycoprotein) are examples of members of the immunoglobulin superfamily, characterized by multiple repeats of disulfide-linked loops homologous to domains first characterized in the immunoglobulin molecule and depicted as half-circles. N-CAM also contains domains similar to the type III repeat in the substrate adhesion molecule, fibronectin. Although depicted here as a transmembrane form, N-CAM also exists as a glycosylphosphatidylinositol-anchored form and in a secreted state. There are five sites of possible N-linked carbohydrate attachment: one just before the third domain (from the N-terminus), one on the fourth immunoglobulin domain, and three in the fifth immunoglobulin domain. The primary structure of the cadherin family of $Ca^{2+}$-dependent adhesion molecules is illustrated with N-cadherin. The extracellular part of the molecule contains five ectodomains: three with internal homology (solid symbols) and two less homologous repeats (lighter symbols, below). The primary structure of the selectins is illustrated with P-selectin. After the "C-type" lectin and EGF domains, there is a domain of variable length containing a series of repeating regions that are related to complement–regulatory proteins. A transmembrane region of approximately 25 amino acids is followed by a short cytoplasmic C-terminal domain.

into a characteristic sandwich structure made up of two β-sheets, each consisting of three to four antiparallel β-strands of five to ten amino acids each, stabilized by intrachain disulfide bonding between the conserved cysteine residues. Three types of immunoglobulin domain have been proposed: V, C1, and C2; the domains found in the CAMs are of the C2 type. N-CAM mediates cell–cell adhesion by a calcium-independent mechanism involving homophilic binding (i.e., like molecule to like molecule); the interaction of the immunoglobulinlike domains is considered to play an important role in this association.

In addition to the immunoglobulinlike domains, some CAMs (e.g., N-CAM) possess a second type of tandem protein repeat, the fibronectin type III domain, in the extracellular part of the molecule. This motif of some 90 amino acids was first identified as one of the major repeat domains in mammalian fibronectin. Though exhibiting less extensive homology from one to another than do the immunoglobulin domains, they have several amino acids that are well conserved. The functional significance of fibronectin III domains in the CAMs is still unknown. As a glycoprotein, N-CAM has carbohydrate attached at three sites. Unique among membrane glycoproteins, the carbohydrate is distinguished by possessing polysialic acid: that is, *N*-acetylneuraminic acid residues linked to one another by $\alpha_{2-8}$-glycosidic bonds. In the embryonic form of N-CAM the sialic acid content is of the order of 30 g per 100 g of polypeptide and in the adult form this decreases to 10 g per 100 g of polypeptide; this E(embryonic) to A(adult) conversion is accompanied by a fourfold increase in the rates of binding. The polysialic acid, because of its highly charged nature, exerts a modulatory effect on the homophilic binding of N-CAM.

N-CAM, an example of what has been termed a primary CAM, exists in several forms with regard to its attachment to the membrane and extent of the cytoplasmic domain. Each is encoded by a distinct messenger RNA, the different mRNAs being generated by alternative splicing of an RNA transcript produced from a single large gene. In addition to forms that traverse the lipid bilayer and possess a cytoplasmic domain of different proportions, N-CAM may be attached to the plasma membrane via a glycosylphosphatidylinositol residue.

A number of the members of the immunoglobulin superfamily have been characterized in detail. For example, L1 is involved in neuron–neuron adhesion, a process termed *neurite fasciculation.* This molecule is structurally related to Ng-CAM, which had been described as a secondary CAM.

CAMs that are members of the immunoglobulin superfamily are not confined to nerve cells. Some of the molecules involved in lymphoid cell adhesion, such as CD4, a cell surface marker expressed in a subpopulation of T cells, are also members of the immunoglobulin superfamily.

### 1.3 The Cadherins: Calcium and Cell Adhesion

It has long been recognized that calcium ions play an important role in cell–cell interactions, and a number of investigators have noted that it is easier to dissociate embryonic tissue in calcium-free medium. Calcium, it is suggested, plays a protective role against proteolytic degradation, and the removal of this divalent cation from tissues may render them more susceptible to tryptic digestion. A Japanese scientist, Takeichi, put the subject of calcium in cell adhesion on a firm biochemical basis when he showed that vertebrate cells can possess two independent adhesion systems, one calcium dependent and the other calcium independent. He found that if trypsin is used in the presence of $Ca^{2+}$, the calcium-independent site is destroyed but the calcium-dependent site is protected from proteolytic degradation. Monoclonal antibodies can be produced that inhibit the $Ca^{2+}$-dependent adhesion of a number of cell types. These antibodies define distinct members of what are termed the cadherin family. (Cadherins are given a letter designation: E cadherin, also called L-CAM, is found on epithelial cells, N on nerve cells, and P in the placenta.) The direct role of these molecules in cell adhesion has been demonstrated by gene transfer experiments. For example L fibroblasts, which do not form tight intercellular connections, may be transfected with E-cadherin cDNA, after which the resultant cells come to adhere tightly to each other. Cadherins appear to play an important role in cell condensation, for example, in compaction of blastomeres, in mesenchyme-to-epithelium transitions, and in the folding of epithelial sheets, as occurs during neural tube formation. By way of example, when in the chick embryo the neural tube forms and pinches off from the overlying ectoderm, it loses E-cadherin and acquires N-cadherin (as well as N-CAM). That cadherins associate with the same cadherin but not significantly with those of different subclasses may be the basis for the selective cell adhesion seen in the earlier classical experiments on in vitro aggregation.

The prototype $Ca^{2+}$-dependent adhesion molecule is L-CAM, which Edelman and his colleagues have studied in chicken liver cells by the antibody inhibition method discussed earlier. Again using the antibody method, it has been possible to identify uvomorulin, the $Ca^{2+}$-dependent adhesion molecule involved in the compaction of early mouse embryos, which is the mouse homologue of L-CAM. Structural information on the cadherins has been achieved by the analysis of primary sequences obtained following the cloning of the respective cDNA. There is a remarkable degree of homology of around 50% between the different cadherins, justifying their classification as a separate superfamily of molecules. The polypeptide chains of most members can be roughly divided into three subdomains; a large extracellular domain, a single hydrophobic transmembrane region, and a cytoplasmic domain. Analysis of the large extracellular domain of the cadherins reveals three major interhomologous ectodomains and one or two less homologous repeats more proximal to the membrane (see Figure 1). The first two and the fourth of these domains are characterized by the conservation of LDRE sequences. All the first four N-terminal ectodomains contain the DXNDNXP sequence, and four cysteine residues located in all the cadherins. The cytoplasmic domain, whose function is most probably to connect with the cytoskeleton, is the most highly conserved region, and antibodies generated against peptide sequences contained in this portion of the molecule have been used to isolate novel cadherins. The truncated (T) cadherins deviate from this pattern in that they lack most of the cytoplasmic domain and are thought to be anchored to the plasma membrane via a glycosylphosphatidylinositol moiety. The cadherins are considered to mediate cell adhesion by homophilic binding, though this is not an absolute requirement. It appears that the homophilic binding site and the specificity-determining region reside in the N-terminal part of the molecule. The sequence HAV found in the first ectodomain appears to be important for cadherin action, and the amino acids flanking this sequence play a role in determining cadherin specificity. With regard to the $Ca^{2+}$-binding region, the DXNDNXP is one putative site, and other aspartic acid stretches also have been implicated.

## 2 SURFACE CARBOHYDRATES AND CELL–CELL INTERACTIONS

Carbohydrates are ubiquitous components of the cell periphery, being present, on glycoproteins and glycolipids, in the outer leaflet of the lipid bilayer of the plasma membrane. The biological function of surface carbohydrate has long been viewed by workers in this area as providing important recognition structures in cell–cell interactions. This view was driven by the knowledge that within oligosaccharides there is an enormous potential for diversity of structure. Monosaccharides, like amino acids, can be linked together in different sequences. However, sugars have the additional advantage that the glycosidic bond can be either in the $\alpha$ or $\beta$ configuration; also, sugars carry a number of free hydroxyl groups, allowing branch structures to be formed between the monosaccharide units in oligosaccharides. An example of the potential diversity of structure possible with sugars can be given: two identical amino acids can form only one dipeptide, while two identical monosaccharides potentially can form 11 different disaccharides. With different monosaccharides, the potential for diversity of structure expands enormously. For example, a recent review points out that four different monosaccharides can generate 35,560 unique tetrasaccharides. While the biological role of surface carbohydrates as recognition determinants has long been attractive, conclusive evidence that these molecules are involved in cell–cell interactions has proved more elusive.

### 2.1 Ectoglycosyltransferases

One way in which carbohydrate at the cell surface could be recognized was proposed in 1970 by Saul Roseman, who speculated that glycosyltransferases normally found in the endoplasmic reticulum and in the Golgi apparatus might also be present at the cell periphery (as ectoglycosyltransferases), where they could play a role in cell–cell interactions. It was envisaged that enzyme–substrate recognition, by the ectoglycosyltransferase on one cell surface interacting with the appropriate substrate on an apposing cell surface, would result in adhesion. In the presence of sugar nucleotide donor, transglycosylation would then take place and the adherent cells would dissociate, thus explaining the observed mutable adhesions between cells. This intriguing idea has been quite difficult to prove because products of enzymatic degradation of radiolabeled sugar nucleotide donor tend to be taken up by intact cells, giving a false impression of transglycosylation by an ectoglycosyltransferase. However, four pieces of information suggest that ectoglycosyltransferases should be considered as potential candidates for modulating cell–cell interactions. First, enzymatic activity can be demonstrated with whole, intact cells in the absence of any "leaked" glycosyltransferase activity. Second, isolated plasma membranes have been shown to possess glycosyltransferase activity. A third piece of evidence comes from the use of UDP-aldehyde, which does not get incorporated into cells but can inhibit the glycosyltransferase reaction with intact cells. Fourth, antibodies prepared against purified glycosyltransferases bind to the plasma membrane. The evidence to date suggests that ectoglycosyltransferases may be involved in cellular movements of gastrulation and morphogenesis.

### 2.2 The Selectin Family of Cell Adhesion Molecules

A boost to an understanding of how sugars might be important in cell recognition came nearly 20 years ago when Ashwell and Morell found that hepatocytes contain a $Ca^{2+}$-dependent carbohydrate-binding protein or lectin that interacts with serum glycoproteins that have lost their terminal sialic acid residues. Animal lectins may be classed as "C type" ($Ca^{2+}$ dependent), of which the hepatic asialoglycoprotein receptor is the prototype, or "S type" (thiol dependent). The "C-type" lectins possess a $Ca^{2+}$-dependent carbohydrate recognition domain consisting of some 130 amino acid residues, within which 18 invariant residues are found in a conserved pattern; particularly striking is the highly conserved placement of several cysteine residues, which appear to be involved in disulfide bonds. "S-type" lectins, which in the main have a specificity for $\beta$-galactosides, have no invariant cysteine residues. Over the last two or three years molecular cloning has revealed a family of structurally homologous adhesion proteins known as the selectins. Each of these proteins has an amino-terminal $Ca^{2+}$-dependent "C-type" motif that share homology with other "C-type" lectins. To date, three selectins have been described; L-selectin which is found in lymphocytes, E-selectin (ELAM-1) in endothelial cells, and P-selectin (GMP140; PADGEM) in platelets. In the case of these three lectins, five distinguishable domains have been found. The lectin domain at the N-terminus is followed by an EGF domain, then a domain of variable length containing a series of repeating regions some 60 amino acids in size that are related to complement–regulatory proteins. Next there is a transmembrane region followed by a short cytoplasmic, C-terminal domain (see Figure 1). P-Selectin has been shown to have two high affinity binding sites for $Ca^{2+}$, and there is evidence that the lectin domain has, in addition to carbohydrate binding, $Ca^{2+}$-binding activity; it is possible that the coordination sites for $Ca^{2+}$ are near to the carbohydrate-binding site in all selectins. In view of their structural organization, these proteins were termed LEC-CAM (lectin/EGF complement binding–Cell Adhesion Molecule). More recently the term "selectin" has been adopted to denote this family of selective lectin-dependent cell adhesion molecules.

These proteins appear to be important in lymphocyte recirculation and in inflammatory responses, and there is much interest in the therapeutic potential of compounds that inhibit the binding of adhesive proteins involved in the pathology of inflammation. L-Selectin is involved in lymphocyte adhesion to high endothelial venules in the lymph node. E-Selectin is expressed transiently in endothelial cells, within a few hours of being activated by IL-1 and other inflammatory cytokines, where it may act to arrest neutrophils that are rolling along the inner lining of blood vessels. However, endothelial cells also have an internal stock of P-selectin, as do platelets, which can be mobilized to the cell surface within minutes after an infection begins. Besides having a role in inflammation, selectins may play a role in other disease states (e.g., the spread of cancer cells throughout the body).

The nature of the carbohydrate structure interacting with the selectins is under active investigation. Interest has centered on fucosylated oligosaccharides; in particular, the sialyl Lewis$^x$ determinant is regarded as the specific ligand for E-selectin. In these studies molecular biology techniques such as transfecting cells with appropriate glycosyltransferase genes to generate appropriate ligands are proving valuable.

## 3 CELL–CELL REPULSION

A growing body of evidence suggests that cell–cell repulsion may be as important a phenomenon as adhesion in cell–cell interactions, especially in the field of developmental neurobiology. In tissue culture experiments it has been found that the growth cone, the amoeboid tip of each developing nerve cell, which is involved in crawling through surrounding tissue, can under various circumstances undergo a dramatic change from a spread morphology to a bulletlike shape; this phenomenon is called growth cone collapse. Retinal growth cones in culture undergo growth cone collapse upon making contact with sympathetic axons. The same phenomenon takes place when sympathetic growth cones make contact with retinal axons. Cell–cell repulsion is now considered important in explaining how peripheral spinal nerves become segmented (by avoiding the posterior half of somites in a developing embryo) and the way in which retinal axons are guided in the developing brain.

A number of in vitro methods to monitor cell–cell repulsion have been devised, including an assay based on the growth cone collapse phenomenon. A number of inhibitory glycoproteins have been identified, though at the moment amino acid sequence data are available for only one of these proteins. Clearly it will be interesting to see whether these proteins, identified by a functional assay in vitro, are novel molecules or whether they represent additional members of existing gene families. Though the molecules mediating cell–cell repulsion have yet to be fully characterized they may be placed at one extreme of a spectrum of cell–cell interactions, with cell–cell adhesion at one end and cell–cell repulsion at the other. Cell–cell repulsion, however, may be explained in terms of interactions between a specific ligand and a surface receptor, which in the case of the growth cone conveys the appropriate information to the cytoskeleton via an appropriate second-messenger system.

*See also* CALCIUM BIOCHEMISTRY; CYTOSKELETON-PLASMA MEMBRANE INTERACTIONS; GLYCOBIOLOGY.

### *Bibliography*

Cummings, R. D., and Smith, D. F. (1992) The selectin family of carbohydrate-binding proteins: Structure and importance of carbohydrates for cell adhesion. *BioEssays,* 14:849–856.

Curtis, A.S.G., and Lackie, J. M., Eds. (1991) *Measuring Cell Adhesion.* Wiley, Chichester.

Edelman, G. M. (1989) *Topobiology: An Introduction to Molecular Embryology.* Basic Books, New York.

Geiger, B., and Ayalon, O. (1992) Cadherins. *Annu. Rev. Cell Biol.* 8:307–322.

Keynes, R. J., and Cook, G.M.W. (1990) Cell–cell repulsion: Clues from the growth cone? *Cell,* 62:609–610.

Pigott, R., and Power, C. (1993) *The Adhesion Molecule: Facts Book.* Academic Press. London.

Rathjen, F. G., Ed. (1991) *Neural Cell Contact and Recognition,* Vol. 3 in *Seminars in the Neurosciences.* Saunders, Philadelphia.

Schwab, M. E., Kapfhammer, J. P., and Bandtlow, C. E. (1993) Inhibitors of neurite growth. *Annu. Rev. Neurosci.* 16:565–595.

Sharon N., and Lis, H. (1993) Carbohydrates in cell recognition. *Sci. Am.* 268(1):74–81.

# CELL DEATH AND AGING, MOLECULAR MECHANISMS OF

*Michael Hengartner*

### *Key Words*

**Apoptosis** Initially used to describe a subset of programmed cell deaths sharing a particular set of morphological features, which include membrane blebbing, shrinkage of cytoplasm, chromatin condensation, and formation of a "DNA ladder." Sometimes used to refer to programmed cell deaths of all types. The word is derived from the Greek word for the shedding of leaves from trees.

**Programmed Cell Death** Type of cell death in which a cell, in response to specific physiological or developmental signals, undergoes a regulated series of events that will lead to its death and removal from the organism.

**Senescence** Age-related changes that adversely affect the normal functioning of a cell or organism. One distinguishes organismal senescence, which results in increased rates of mortality as a function of time, and cellular senescence, which results in a decreased capacity to proliferate.

Naturally occurring or programmed cell deaths, which play important roles in animal development and homeostasis, are observed in a wide variety of tissues in both vertebrates and invertebrates. Such deaths remove cells that might be harmful, are not needed, or have served their purpose. In many tissues, cell death and cell proliferation are precisely balanced, to maintain the proper number and types of cells. Disruption of this balance can result in disease. Both cell death and aging appear to be under genetic control. Identification of the genes involved in these processes promises to greatly enhance our understanding of these phenomena.

## 1 CELL DEATH

### 1.1 PROGRAMMED CELL DEATH VERSUS PATHOLOGICAL CELL DEATH

Programmed cell death is distinguishable, both morphologically and functionally, from another type of cell death called necrosis. Programmed cell death is a natural form of death that organisms use to get rid of cells. Cells dying by programmed cell death usually shrink, rarely lyse, and are efficiently removed from the organism without the appearance of inflammation (Figure 1). Necrosis, on the other hand, is a pathological type of cell death observed following physical or chemical injury, exposure to toxins, or ischemia (lack of oxygen). Necrosis is characterized by swelling, rupture of the plasma membrane and cellular organelles, and release of the cellular content into the surrounding tissue, leading in vivo to inflammation.

Programmed cell deaths, as the name suggests, are generally believed to be brought on through activation of an endogenous

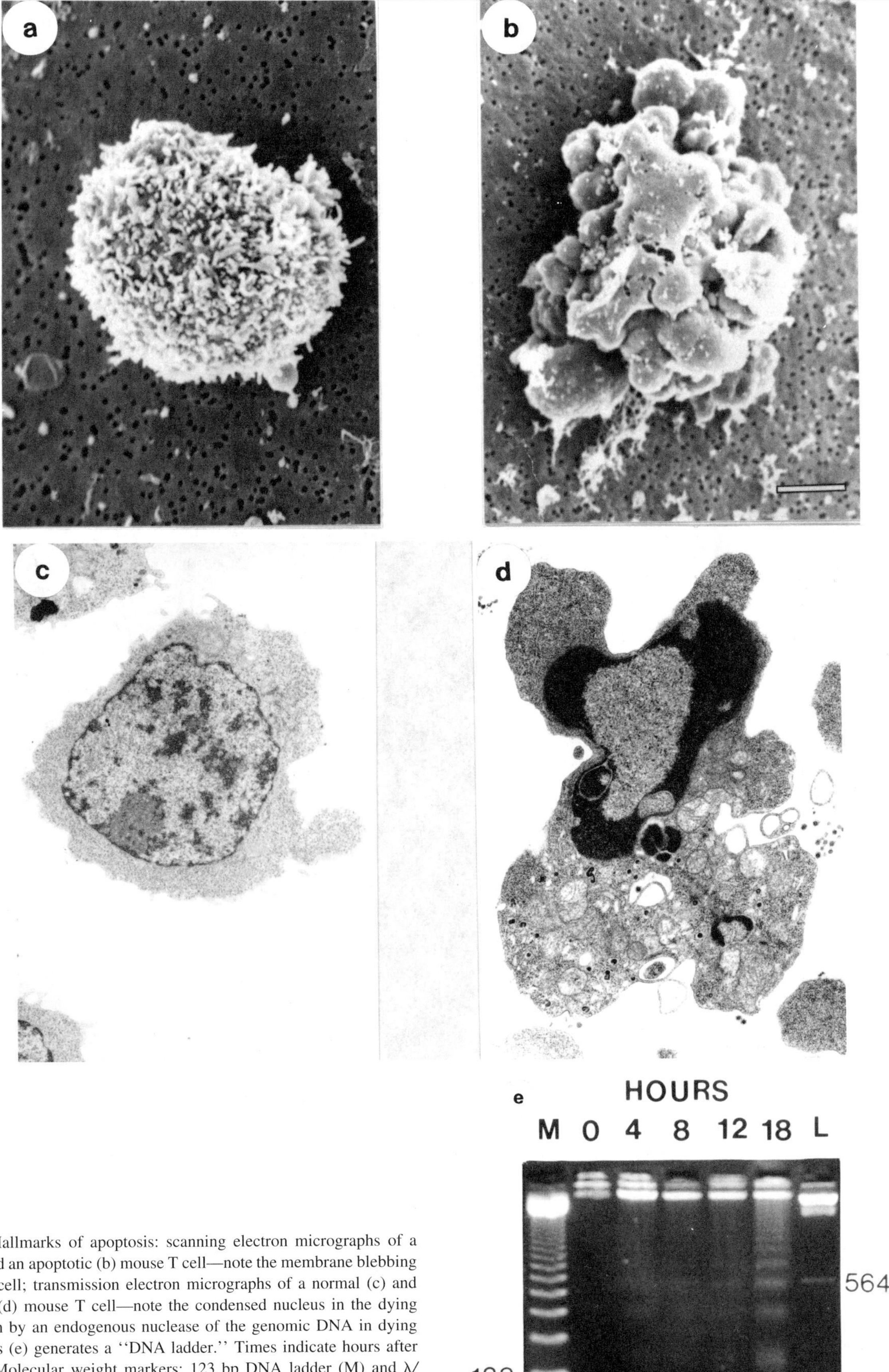


**Figure 1.** Hallmarks of apoptosis: scanning electron micrographs of a normal (a) and an apoptotic (b) mouse T cell—note the membrane blebbing in the dying cell; transmission electron micrographs of a normal (c) and an apoptotic (d) mouse T cell—note the condensed nucleus in the dying cell; digestion by an endogenous nuclease of the genomic DNA in dying mouse T cells (e) generates a ''DNA ladder.'' Times indicate hours after stimulation. Molecular weight markers: 123 bp DNA ladder (M) and λ/ *Hin* dIII (L). Bar represents ~ 2 μm. (Photo: Barbara Osborne, Sallie Smith, and Larry Schwartz).

genetic program. The existence of such a program has been supported by the observation that in many experimental systems, induction of cell death can be prevented through addition of inhibitors of RNA or protein synthesis, such as actinomycin D or cycloheximide, suggesting that for the cell to die, new polypeptides must be synthesized. Further evidence of a genetic program comes from studies of cell death in *Caenorhabditis elegans,* where a number of genes involved in programmed cell death have been identified genetically and ordered into a pathway.

## 1.2 Programmed Cell Death in the Nematode *Caenorhabditis elegans*

Of the 1090 somatic cells generated during the development of the *C. elegans* hermaphrodite, 131 undergo programmed cell death. As with most of *C. elegans* development, these deaths are highly reproducible from animal to animal: the same cells always die, and each cell dies at its own characteristic point in development. This reproducibility allows study of cell death at single-cell resolution. Morphologically, programmed cell deaths in *C. elegans* are quite similar to apoptotic cell death in mammals: as a cell dies, its volume shrinks and its nucleus condenses. The nuclear membrane breaks down, and membranous whorls appear in the cytoplasm. Subsequently, a neighboring cell engulfs the dying cell and degrades it. Genetic studies have led to the identification of mutations that affect this process; these mutations define a genetic pathway for programmed cell death in *C. elegans* (Figure 2).

A number of genes have been identified that affect the decision of individual cell types to live or to die. For example, in the wild-type animal two specific neurons in the pharynx called the NSM sister cells undergo programmed cell death. Mutations in *ces*-1 (*ce*ll death *s*pecification) and *ces*-2 allow these cells to survive and differentiate instead of dying as in the wild type. Thus, the *ces*-1 and *ces*-2 genes might identify components involved in the cell-specific control of genes involved in all cell deaths, such as *ced*-9, *ced*-3, and *ced*-4.

The activities of two genes, *ced*-3 and *ced*-4 (*ce*ll *d*eath abnormal), are necessary for programmed cell deaths to occur: mutations that inactivate either *ced*-3 or *ced*-4 result in the survival of almost all cells that normally die during development. Both genes probably function within the cells that die, suggesting that the programmed cell deaths in *C. elegans* are not ''murders,'' but rather ''suicides'' and that the dying cell plays a central role in bringing about its own demise.

A third gene, *ced*-9, acts to protect cells from programmed cell death. A mutation that abnormally activates *ced*-9 prevents the programmed cell deaths that occur during normal *C. elegans* development. Conversely, mutations that inactivate *ced*-9 cause cells that normally live to undergo programmed cell death; these mutations result in embryonic lethality, indicating that *ced*-9 function is essential for development. *ced*-9 appears to function by negatively regulating the activities of the *ced*-3 and *ced*-4 genes, keeping the cell death program off in cells that are scheduled to live. These results indicate that many, if not all, cells in *C. elegans* carry the information and mechanisms to undergo programmed cell death, although it appears that the program is usually suppressed through the activity of the *ced*-9 gene.

Seven genes (*ced*-1, 2, 5, 6, 7, 8, 10) have been identified that function in the engulfment of dying cells by their neighbors. Whereas in wild-type animals, dead cells are rapidly and efficiently engulfed, in these mutants, unengulfed, undegraded cell corpses accumulate throughout the animal. These genes could be required for the generation of an engulfment-stimulating signal by the dying cell, for the reception of such a signal, or for the actual engulfment step itself.

The gene *nuc*-1 (*nuc*lease) functions in the last step of the pathway: *nuc*-1 mutants lack a particular endonuclease activity that is required to degrade the DNA of dead cells. Cells still die in *nuc*-1 worms, indicating that this nuclease activity is not required for cell death to occur in *C. elegans.*

## 1.3 Programmed Cell Death in Mammals

In mammalian cells, the decision to live or to die often is determined by signals from other cells or from the environment. Both death-

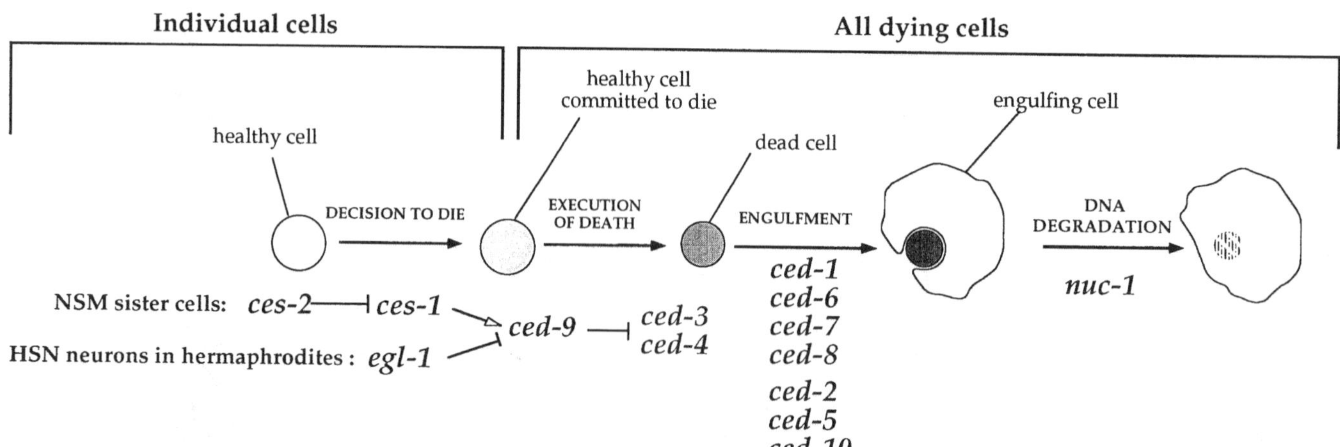


**Figure 2.** Genetic pathway for programmed cell death in *C. elegans.* Fourteen genes have been identified that act in programmed cell death in this nematode species. The genes define four distinct steps: decision of individual cells to live or die, actual execution of the death sentence, engulfment of the dying cell by a neighbor, and degradation of the engulfed cell. The arrows represent the deduced genetic interactions between the various genes in the pathway. For example, *ces*-1 is thought to be a positive regulator (→) of *ced*-9, which itself negatively regulates (—|) the *ced*-3 and *ced*-4 genes. Dominant mutations in the *egl*-1 (*egg*-laying abnormal) gene lead to the programmed death of the NSM neurons in hermaphrodites. The other genes are described in the text. For additional information, see R. Ellis, J. Yuan, and H. R. Horvitz, *Annu. Rev. Cell Biol.* 7:663–699 (1991).

preventing and death-inducing stimuli exist. For example, many mammalian cells are dependent for their continued survival both *in vivo* and in culture on the presence of trophic (i.e., survival) factors. The two best known classes of trophic factors are the neurotrophin family (comprised of the related molecules NGF, BDNF, and NT3), required by various classes of neurons (e.g., sympathetic neurons require NGF), and a subset of the interleukins, which act on various lymphocyte populations. If the trophic factor is withdrawn, the factor-dependent cells rapidly (often within 24 hours) undergo programmed cell death. These cells therefore require the continual presence of a signal that prevents them from undergoing programmed cell death.

Signals also exist that induce programmed cell death. An interesting example is provided by the Fas/APO-1 cell surface antigen, which is found on activated B and T cells. Activation (through cross-linking) of this receptor induces the cell to undergo apoptosis. *lpr* mice, which carry a mutation in the mouse *Fas/APO-1* gene, develop autoimmune disease, suggesting that Fas/APO-1 plays an important role in eliminating autoreactive T cells. T cells also can be induced to undergo apoptosis by exposing them to a variety of other signals, including, for example, glucocorticoids.

How can one identify the actual genes involved in killing cells? One approach has been to search for genes whose expression patterns change following induction of cell death, hoping that genes that show increased or decreased expression will play a functional role in this process. A second approach has been to start with a cell line that can be induced to die, and to isolate mutants that now are resistant to the death-inducing signal. A third approach would be to identify diseases caused by the disruption of the normal cell death pattern and then clone the gene responsible for each disease. For example, the survival of cells that normally should die can lead to cancer or autoimmune disease. On the other hand, many human disorders, such as neurodegenerative diseases, are characterized by extensive cell deaths; these could possibly be the result of an aberrant activation of programmed cell death in cells that should normally live.

An oncogene for *bcl*-2 (*B*-*c*ell *l*ymphoma) provides a good example of a cell death gene identified in the third way just described. *bcl*-2 was cloned based on its involvement in a t(14;18) translocation that is found in the majority of follicular lymphomas diagnosed in the United States. The translocation fuses *bcl*-2 to the immunoglobulin heavy chain locus and results in overexpression of *bcl*-2; this overexpression protects these cells from undergoing apoptosis. Overexpression of *bcl*-2 can also protect other cells from apoptosis: for example, sympathetic neurons expressing high levels of *bcl*-2 do not die following NGF withdrawal. *bcl*-2 might therefore be a universal antidote to programmed cell death.

One of the hallmarks of apoptosis is the digestion of the genomic DNA of the dying cell into small fragments of about 180 bp or multiples thereof (Figure 1). These fragments are generated by the action of a $Ca^{2+}$-dependent endonuclease that is activated upon induction of apoptosis, and it has been suggested (though not proved) that activation of this nuclease is required for the cell to die. The exact nature of the endonuclease and the mechanism of activation (e.g., increased expression, proteolytic activation, increase in intracellular calcium levels, access to the nucleus) have yet to be determined.

*trans*-Glutaminase activity is also increased upon initiation of cell death. This enzyme cross-links proteins by creating peptide bonds through ε-(γ-glutamyl)lysine bonds. The result of this activity is to transform the normally fluid cytoplasm into a tight network, a step thought to be important for maintaining the integrity of the dying cell and preventing inflammation.

Mammalian cells undergoing programmed cell death are rapidly recognized and engulfed by macrophages. This process is so efficient that even in tissues experiencing extensive cell death, only a small number of dying cells can be observed at any given time (explaining at least in part why the role of cell death during development has long been underestimated). The recognition of a dying cell by a macrophage appears to be mediated through several, partially redundant pathways. A similar observation has been made in *C. elegans.* Once engulfed, the cellular debris fuses with the phagocytes' lysosomal compartments and is completely digested.

### 1.4 Evolutionary Conservation of the Molecular Pathway for Programmed Cell Death

Recent advances in the molecular characterization of the *C. elegans* cell death genes revealed that *ced*-3 and *ced*-9, two key genes involved in the control and execution of the death sentence, are similar in sequence and function to mammalian cell death genes. For example, the CED-9 protein shows significant similarity to mammalian Bcl-2, which like *ced*-9 is a negative regulator of programmed cell death (see section 1.2 above). Furthermore, overexpression of *bcl*-2 can prevent programmed cell death in *C. elegans.* Similarly, CED-3 is homologous to the mammalian cysteine protease interleukin-1β converting enzyme (ICE) and the product of the mouse *nedd*-2 gene, and overexpression of either ICE or CED-3 in mammalian cells induces apoptotic cell death, suggesting that a CED-3-like protease also mediates programmed cell death in mammals. The involvement of members of the *ced*-9/*bcl*-2 and *ced*-3/ICE gene families in mediating programmed cell death in both *C. elegans* and humans—and the observation that these structurally similar genes are functionally interchangeable—strongly suggest that nematodes and mammals share a common molecular pathway for programmed cell death. This common genetic program presumably predates the evolutionary separation of nematodes and vertebrates and thus seems likely to be of ancient origins.

## 2 Aging

### 2.1 Organismal Aging

The many theories that have been advanced to explain why organisms age can be broadly divided into two groups. Theories in the first group suggest that aging is the result of the accumulation of damage due to a constant onslaught of external insults. This damage would then result in a decreased function of the affected tissues or cells, ultimately leading to a "system failure" and death. One theory in this group suggests that aging is the result of oxidative damage produced by activated oxygen species such as oxygen superoxide anion ($O_2^{\cdot -}$) and hydrogen peroxide ($H_2O_2$). These molecules, often by-products of metabolic activities such as mitochondrial respiration, react with and modify intracellular molecules, including proteins, lipids, and DNA. The oxygen radical theory is supported by several observations. For example, this theory suggests that longevity will be directly proportional to the degree of antioxidant activity observed in the organism. Indeed, long-lived species in general possess higher levels of enzymes that protect from oxygen radicals than do shorter lived ones.

Theories in the second group postulate that aging is built into the basic biology of a species and is therefore in a sense "programmed." The idea that aging is a genetically determined phenomenon has been supported by the identification of long-lived mutants in organisms as diverse as *Drosophila, C. elegans,* and the bread mold *Neurospora crassa.* Aging mutants also exist in humans: two rare genetic diseases, progeria and Werner's syndrome, cause what appears to be accelerated aging in connective tissues.

The two groups of models are not mutually exclusive; for example, the level of antioxidant activity, and therefore longevity, could be genetically determined in each species. Indeed, the long-lived strains of *Drosophila* do have higher level of protective enzymes than their wild-type counterparts.

### 2.2 Cellular Aging

Almost all differentiated eurkaryotic cells have a limited capacity to proliferate: after a finite number of cell divisions, their capacity to go through another cycle gradually decreases and eventually stops altogether. Several observations suggest that this phenomenon, known as cellular senescence, might be a reflection at the cellular level of organismal aging. For example, when primary cultures of fibroblasts from various species were tested for their proliferative capacity, the number of cell divisions a given fibroblast culture could go through before entering senescence was directly proportional to the maximum life span of the species and inversely proportional to the age of the donor. This observation suggests that there is a species-specific number of cell divisions a cell can go through before ceasing to divide.

While most cells in a senescent culture will eventually cease to proliferate, a very small proportion (varying from 1 in $10^{-6}$ in rodents to about 1 in $10^{-12}$ in humans) of the cells somehow escape senescence and become immortal. These cells have acquired one or more mutations that allow them to now grow continuously in culture, and they show characteristics of transformed (i.e., tumorous) cells. Interestingly, cell fusion experiments showed that the senescence phenotype is dominant over the immortality phenotype. This result, in conjunction with the low frequency of appearance of immortal cells, suggests that immortality arises as a result of the loss of activity in one or a few dominantly acting senescence gene(s). Complementation tests between various immortal cell lines revealed that there are at least four complementation groups for the immortality phenotype. Identification of the genes involved in the control of senescence promises to further our understanding of the molecular basis of cellular aging.

*See also* Aging, Genetic Mutations in; DNA Damage and Repair.

### *Bibliography*

Arking, R. (1991) *Biology of Aging: Observations and Principles.* Prentice-Hall, Englewood Cliffs, NJ.

Ellis, R., Yuan, J., and Horvitz, H. R. (1991) Mechanisms and functions of cell death. *Annu. Rev. Cell Biol.* 7:663–698.

Finch, C. E. (1990) *Longevity, Senescence, and the Genome.* University of Chicago Press, Chicago and London.

Tomei, L. D., and Cope, F. O., Eds. (1991) *Apoptosis: The Molecular Basis of Cell Death.* Cold Spring Harbor Laboratory Press, Plainview, NY.

# Chaperones, Molecular

*Elizabeth A. Craig*

### *Key Words*

**Cytoplasm** The substance of a cell minus the nucleus.

**Cytosol** The cytoplasm minus the mitochondria, endoplasmic reticulum, and other organelles.

**Endoplasmic Reticulum (ER)** A system of membrane-enclosed compartments in the cytoplasm through which proteins to be secreted pass.

**Mitochondrium** An organelle of the cytoplasm constructed of an outer and inner membrane; the site of aerobic respiration.

**Protein Folding** The formation of an native (active) state of a protein with a well-defined secondary and tertiary structure.

**Protein Translocation** The vectorial movement of proteins synthesized in the cytosol across membranes (e.g., into a mitochondrium or the endoplasmic reticulum).

Molecular chaperones are ubiquitous proteins that play a critical role in the cellular processes of protein folding and translocation of proteins across membranes into organelles. Chaperones act by binding a wide array of proteins, with a preference for those in an unfolded conformation. Chaperones bind to nascent chains during synthesis on the ribosome and to newly completed polypeptides. Correct folding of many polypeptides in vivo requires chaperone action. In addition, chaperones within the endoplasmic reticulum (ER) and mitochondria are critical for the translocation of proteins into these organelles from the cytosol, acting by binding to the translocating polypeptide as it emerges from the membrane surrounding the organelles.

## 1 GENERAL ROLES OF CHAPERONES

Chaperones derive their name from their function: they prevent illicit interactions between proteins by interacting with many proteins from the time of synthesis on the ribosome until they are in their final location in the cell and properly folded. However, chaperones were first identified as proteins whose synthesis was induced upon a temperature upshift and have therefore been called heat shock proteins (hsps) for many years. The hsps (chaperones) are encoded by families of genes; most of the genes are expressed under optimal growth conditions and are essential for cell viability. The two most abundant and best characterized chaperones are members of the hsp70 and hsp60 (also called chaperonin) multigene families. These families are discussed in more detail in Sections 2.1 and 2.2

Binding of chaperones to their denatured or partially unfolded substrate proteins can have two different effects: (1) maintaining an unfolded conformation, thus preventing folding into a more compact structure, and (2) facilitating the folding of proteins into their native conformation. Although much remains to be learned

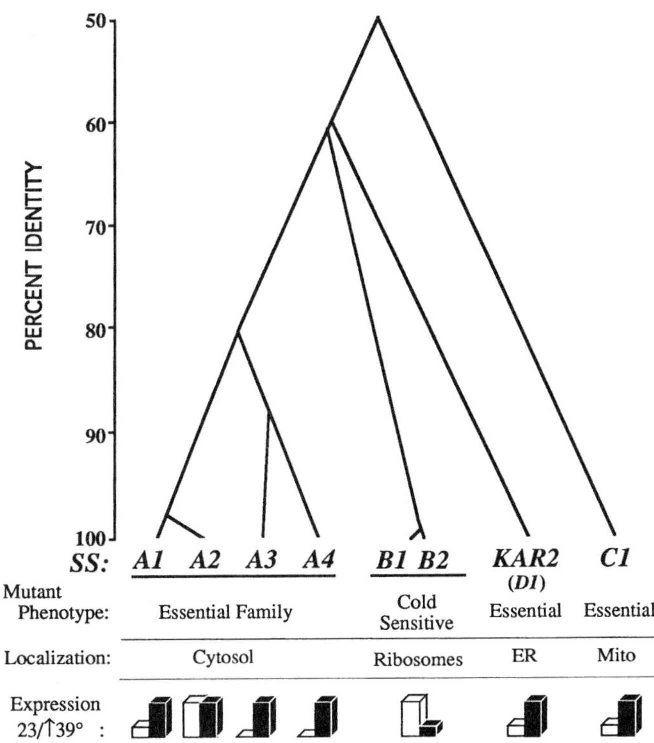


**Figure 1.** The hsp70 multigene family of *S. cerevisiae.* (From E. A. Craig, The heat shock response, in *The Molecular Biology of the Yeast* Saccharomyces cerevisiae, (E. Jones, J. Pringle, and J. Broach, Eds.) Cold Spring Harbor Press, Plainview, NY, 19xx, pp. 501–538.

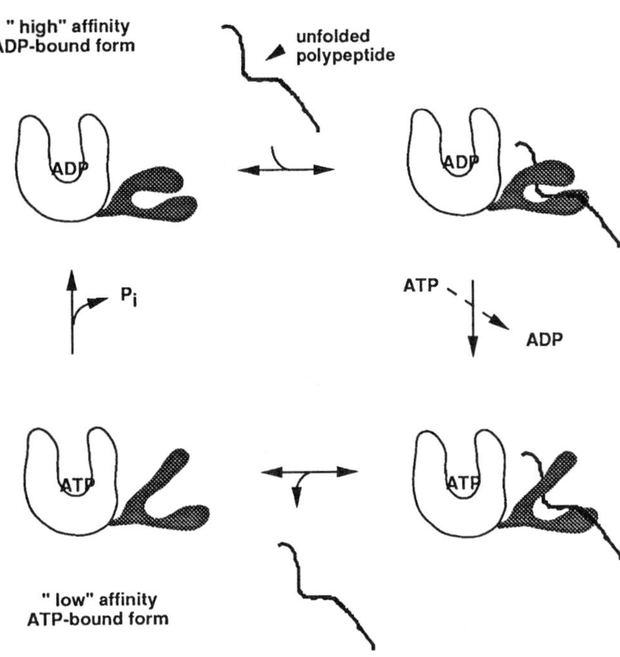


**Figure 2.** Model for nucleotide-dependent cycle of interaction of an hsp70 with substrate polypeptides. hsp70 is depicted as a bilobed structure with an NH2 terminal adenosine nucleotide binding domain and a COOH-terminal peptide binding domain. The structure of the nucleotide binding domain has been determined using X-ray crystallographic methods; the structure of the peptide-binding domain is unknown, but speculated to be related to the well-characterized major histocompatibility class I peptide binding domain. The ADP-bound form favors binding; the ATP-bound form favors release.

about chaperone action, one class of chaperones, hsp70s, seems to act to prevent the folding of proteins, while another class, chaperonins (hsp60s), facilitates protein folding.

## 2 THE MAJOR CHAPERONES: hsp70 AND hsp60

### 2.1 hsp 70s

The hsp70s have been found in all organisms examined, including bacteria, plants, and animals. Eukaryotes contain multigene families. Proteins encoded by different members are found in organelles as well as in the cytosol. For example, in the yeast *Saccharomyces cerevisiae* (Figure 1), one hsp70 is found in the matrix of the mitochondria, one in the lumen of the ER, and several in the cytosol. Similar complex families are found in other eukaryotes including mammals.

The hsp70 proteins within one organism, as well as between organisms, are between 50 and 90% identical in amino acid sequence. The conserved amino-terminal two-thirds of hsp70s is more highly conserved than the carboxy-terminal third. The conserved amino-terminal portion contains an ATP binding site; the carboxyl-terminal region contains the peptide binding domain. hsp70s have a weak ATPase activity essential for chaperone action. The ADP-bound form favors peptide binding; the ATP-bound form favors peptide release. Therefore cycles of binding and release from peptide substrates is achieved by cycles of ATP binding and hydrolysis and exchange of bound ADP for ATP (Figure 2). The hsp70s bind short peptide segments of about seven amino acids. The rules of hsp70 binding to substrate peptides have not yet been established; however there seems to be a preference for peptides of hydrophobic character and a bias against highly charged residues. Different hsp70s have different binding specificities, binding different polypeptides with very different affinities.

The hsp70s of the cytosol appear to bind to nascent polypeptides as they are synthesized on the ribosome, presumably to maintain the polypeptide chain in a relatively unfolded conformation, preventing aggregation and misfolding. Cytosolic hsp70s are also necessary for normal translocation of proteins from the cytosol into either the ER or mitochondria. Based on knowledge of the interaction of hsp70s with unfolded proteins and the requirement for proteins to be in an unfolded conformation to cross membranes, it has been assumed that binding of hsp70s to proteins destined for organelles maintains these proteins in an unfolded, translocation-competent conformation. Mitochondrial hsp70 binds to translocating polypeptides once a portion has entered the matrix. This binding is required for translocation of the entire protein across both the inner and outer membranes (Figure 3). The hsp70 is located in the lumen of the ER (often called BiP) acts in a similar manner, interacting with proteins crossing the membrane of the endoplasmic reticulum. It has been proposed that the organellar hsp70s provide the driving force for translocation of proteins into these organelles.

The hsp70s do not act alone, but with other proteins. These factors have been best studied in *E. coli,* where two proteins (DnaJ

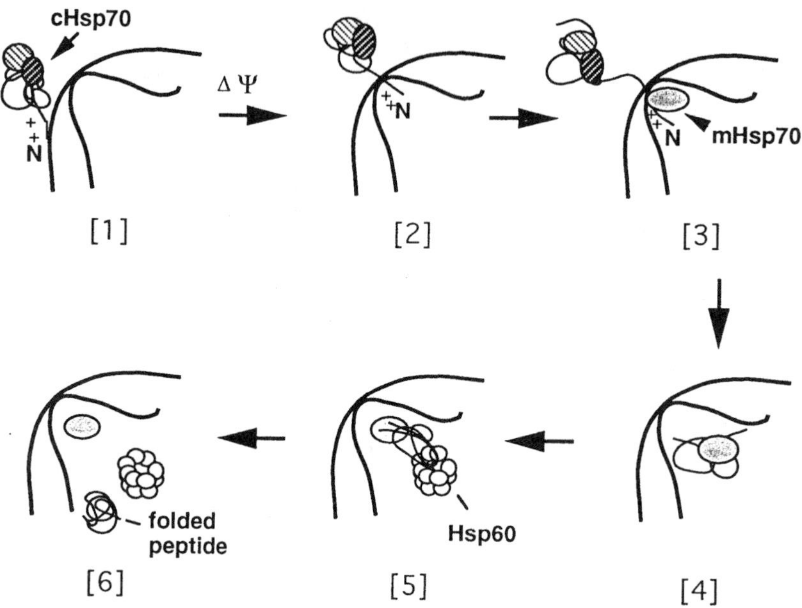


**Figure 3.** A model of the role of mitochondrial hsp70 and hsp60 in protein translocation into mitochondria: 1, precursor, bound to cytosolic hsp70s (cHsp70) and perhaps additional factors that aid in maintaining a relatively unfolded conformation, binds to a receptor on the cytosolic surface of the outer mitochondrial membrane; 2, The N-terminal presequence is inserted into the outer membrane and, because of the membrane potential ($\Delta\Psi$) across the inner membrane, is translocated across the membranes; 3,4, mitochondrial hsp70 (mHsp70) binds tightly to the precursor protein in the matrix, perhaps at a number of sites as translocation progresses, preventing movement back toward the cytoplasm; 5, Ssc1p ''passes off'' the protein to hsp60, where folding and/or assembly occurs; 6, The folded protein is released. (Adapted from E. A. Craig, The heat shock response, in *The Molecular Biology of the Yeast* Saccharomyces cerevisiae, E. Jones, J. Pringle, and J. Broach, Eds. Cold Spring Harbor Press, Plainview, NY, 19xx, pp. 501–538.

and GrpE) have been shown to physically interact with hsp70 and to affect nucleotide binding and hydrolysis.

### 2.2 hsp 60

hsp60 was first identified in *E. coli* and subsequently in chloroplasts and mitochondria. hsp60s, also called chaperonins, are large, approximately ribosome-sized structures composed of 14 subunits of about 60 kDa each arranged in two seven-membered rings, which lie on top of each other. Chaperonin-60 acts with a 10 kDa protein called chaperonin-10, which also forms a seven-membered ring structure. Like hsp70s, chaperonins bind a wide array of polypeptides and their release depends on ATP hydrolysis. However, while hsp70s generally bind extended unstructured polypeptides, maintaining an unfolded conformation, chaperonins bind polypeptides that already have nativelike secondary structure, α-helices, and β-sheets, but a nonorganized tertiary globular structure, facilitating proper folding. In studies with purified components, chaperonins have been shown to allow the folding of proteins, which alone are unable to fold properly.

Chaperonins appear to act by decreasing the rate of nonproductive off-pathway reactions, thus preventing aggregation and the formation of inactive forms of the polypeptides. They appear to do so by binding the polypeptide inside a central cavity such that this space is surrounded by the individual members of the chaperonin complex, each of which may have an individual binding site. The bound polypeptide is released from the binding sites, in a reaction dependent on ATP hydrolysis, and can begin to fold while still associated with the chaperonin. There is no evidence that chaperonins actually increase the rate of the on-pathway reactions of protein folding. Rather, proteins contain within their amino acid sequence the information needed for proper folding into an active conformation; chaperonins act by allowing folding to occur more efficiently. In the cell, chaperonins may also be required for the assembly of multimeric protein structures.

Recently a protein (called Tcp1) related to hsp60s has been identified in the eukaryotic cytosol. Preliminary results indicate that the essential Tcp1 protein facilitates the folding of at least some proteins of the cytosol. Therefore, chaperonins may well be required for protein folding in all cellular compartments.

### 2.3 hsp70s and hsp60 Interactions

The hsp70s and the chaperonins act in a pathway for protein folding. For example, in the mitochondria the hsp70 in the matrix interacts with proteins as they enter; the protein is then ''passed'' to hsp60, where folding occurs (Figure 3). The details of these interactions remain to be worked out, but a similar pathway likely exists in the cytoplasm, where hsp70s bind nascent and newly completed polypeptides.

## 3 OTHER CHAPERONES

### 3.1 hsp90

hsp90 is a highly conserved class of chaperones found in all prokaryotes and eukaryotes that have been tested. The best characterized chaperone of this class, hsp90 from mammalian cells, is found bound to inactive steroid hormone receptor. Steroid hormone receptors are positively acting transcriptional activators which, upon interaction with steroid hormones, bind DNA upstream of steroid-responsive genes. Binding of steroid hormone to the receptor is accomplished by release of hsp90, unmasking the DNA binding site. In addition, binding of hsp90 to the receptor may be required for the folding of the receptor into a conformation capable of DNA binding. hsp90 associates with a number of other proteins including *src* protein kinase.

### 3.2 SecB

SecB, a chaperone of *E. coli,* interacts with a subset of proteins destined for export into the periplasmic space. Binding of SecB appears to prevent premature folding of the polypeptide, stabilizing a translocation-competent conformation. The general role of SecB is reminiscent of the role of hsp70s in protein translocation in the cytosol of eukaryotic cells, although there is no sequence similarity between the two proteins.

*See also* Protein Aggregation; Protein Folding; Translation of RNA to Protein.

### *Bibliography*

Craig, E. A., Gambill, B. D., and Nelson, R. J. (1993) *Microbiol. Rev.* 57:402–414.

Hendrick, J. P. and Hartl, F.-U. (1993) Molecular chaperone functions of heat-shock proteins. *Annu Rev. Biochem.* 62:349–384.

Jakob, U. and Buchner, J. (1994) Assisting Spontaneity: the role of hsp90 and small hsps as molecular chaperones. *Trends Biochem. Sci.* 19:205–211.

Lorimer, G. (1992) Role of accessory proteins in protein folding. *Curr. Opinion Struct. Biol.* 2:26–34.

Morimoto, R. I., Tissieres, A., and Georgopoulos, C., Eds. (1994) *The Biology of Heat Shock Proteins and Molecular Chaperones.* Cold Spring Harbor Laboratory Press, Cold Spring Harbor, NY.

Nelson, R. J., and Craig, E. A. (1992) Tcp1—A molecular chaperonin of the cytoplasm. *Curr. Biol.* 2:487–490.

# Chemiluminescence and Bioluminescence, Analysis by

*Larry J. Kricka*

## *Key Words*

**Bioluminescence** Light emission from a living organism.

**Chemiluminescence** Light emission that occurs in chemical reactions as a result of the decay of chemiexcited species.

**Homogeneous Assay** Immunoassay or nucleic acid hybridization assay that does not require a separation step.

**Immunoassay** Assay that uses specific antibodies or labeled antibodies to detect an antigen (the analyte).

**Luciferase** Generic name for an enzyme that catalyzes a bioluminescent reaction.

**Luciferin** Generic name for the substrate of a luciferase.

**Luminometer** Instrument for detecting and measuring light emission.

**Nucleic Acid Hybridization Assay** Assay that utilizes the specific recognition of a nucleic acid probe or labeled nucleic acid probe for a complementary sequence on target nucleic acid.

**Photoprotein** Tightly bound protein–luciferin complex that can be activated to produce light.

---

Bioluminescent (BL) and chemiluminescent (CL) reactions are used as analytical tools in immunoassay, in protein and nucleic acid blotting, in nucleic acid sequencing and hybridization assays, and in reporter gene studies. In a BL or CL assay the intensity or the total light emission is measured and related to the concentration of the unknown analyte. Light can be measured quantitatively using a luminometer (photomultiplier tube as the detector) or qualitatively by means of photographic or X-ray film. The main advantages of these reactions are their simplicity and analytical sensitivity: for example, attomole to zeptomole ($10^{-18}$–$10^{-21}$ mol) amounts of analyte. In immunoassay and nucleic acid hybridization tests, these reactions or their components are used in two ways, either as labels or as detection systems for other types of labels (e.g., enzymes). Acridinium esters, acridinium sulfonyl carboxamides, isoluminol derivatives, the photoprotein aequorin, and luciferases from the firefly, marine bacteria, and *Renilla* are the principal luminescent labels. CL and BL detection schemes have been developed for assaying alkaline phosphatase, glucose oxidase, glucose 6-phosphate dehydrogenase, horseradish peroxidase, and xanthine oxidase labels. In addition, simple homogeneous assays are possible with CL acridinium ester and isoluminol labels. CL and BL reactions can be adapted for the analysis of enzymes, substrates, cofactors, inhibitors, and metal ions, and for the study of cellular processes (e.g., phagocytosis). Luciferase genes (*luc, lux*) are now used as reporter genes, and the gene expression is quantitated by measuring the expressed luciferase in a BL assay.

## 1 CHEMILUMINESCENCE

Chemiluminescence is the light emission produced in certain chemical reactions. Excited singlet state molecules are formed via the decomposition of high energy intermediates such as 1,2-dioxetanes, dioxetanones, peroxides, or *endo*-peroxides. These decay to the ground state and some of the energy is released as light. The majority of CL results from oxidation reactions that can provide sufficiently energetic intermediates for light emission (e.g., 63.5 kcal/mol for blue light at 450 nm). One of the best known CL reactions is the oxidation of luminol (5-amino-2,3-dihydro-1,4-phthalazinedione) to produce a blue CL at 425 nm, due to the decay of singlet excited state 3-aminophthalate molecules. Generally CL reactions are inefficient in aqueous media (CL quantum yield for luminol = 0.01) but are improved in aprotic solvents (e.g., dimethyl sulfoxide).

## 2 BIOLUMINESCENCE

Bioluminescence (''living light'') is the light emitted from living organisms. Many thousands of BL species have been identified (666 genera from 13 phylla), and the firefly (*Photinus pyralis*) is the best known example. Light emission (562 nm) is produced in an ATP-dependent oxidation of firefly luciferin, catalyzed by firefly luciferase (EC 1.13.12.7). This highly efficient reaction has a quantum yield approaching unity. However, the majority of BL reactions are considerably less efficient than the firefly reaction.

$$\text{firefly luciferin} + \text{ATP} \xrightarrow{\text{firefly luciferase, } Mg^{2+}} \text{AMP} + CO_2 + PP_i + \text{oxyluciferin} + \text{light} \quad (1)$$

where $PP_i$ is an organic pyrophosphorus compound.

## 3 APPLICATIONS

Assays simply involve mixing the sample with the appropriate CL or BL reagents and measuring or detecting the light emission using a luminometer, charge-coupled device (CCD) camera, or film (photographic or X-ray). CL and BL reactions are sensitive and simple to perform; they utilize nonhazardous reagents and have a wide dynamic range (several orders of magnitude), as well as diverse applications (assays for proteins, DNA, RNA, enzymes, substrates, inhibitors, metal ions, and biomass).

### 3.1 Immunoassay, Blotting, and Nucleic Acid Hybridization Assays

The extreme sensitivity of CL and BL assays (detection limits in the attomole to zeptomole range) has made them ideal for the subject applications. CL and BL substances are effective alternatives to radioactive labels ($^{32}P$, $^{125}I$), and generally, CL and BL reactions are superior to colorimetric and fluorescent assays for conventional enzyme labels.

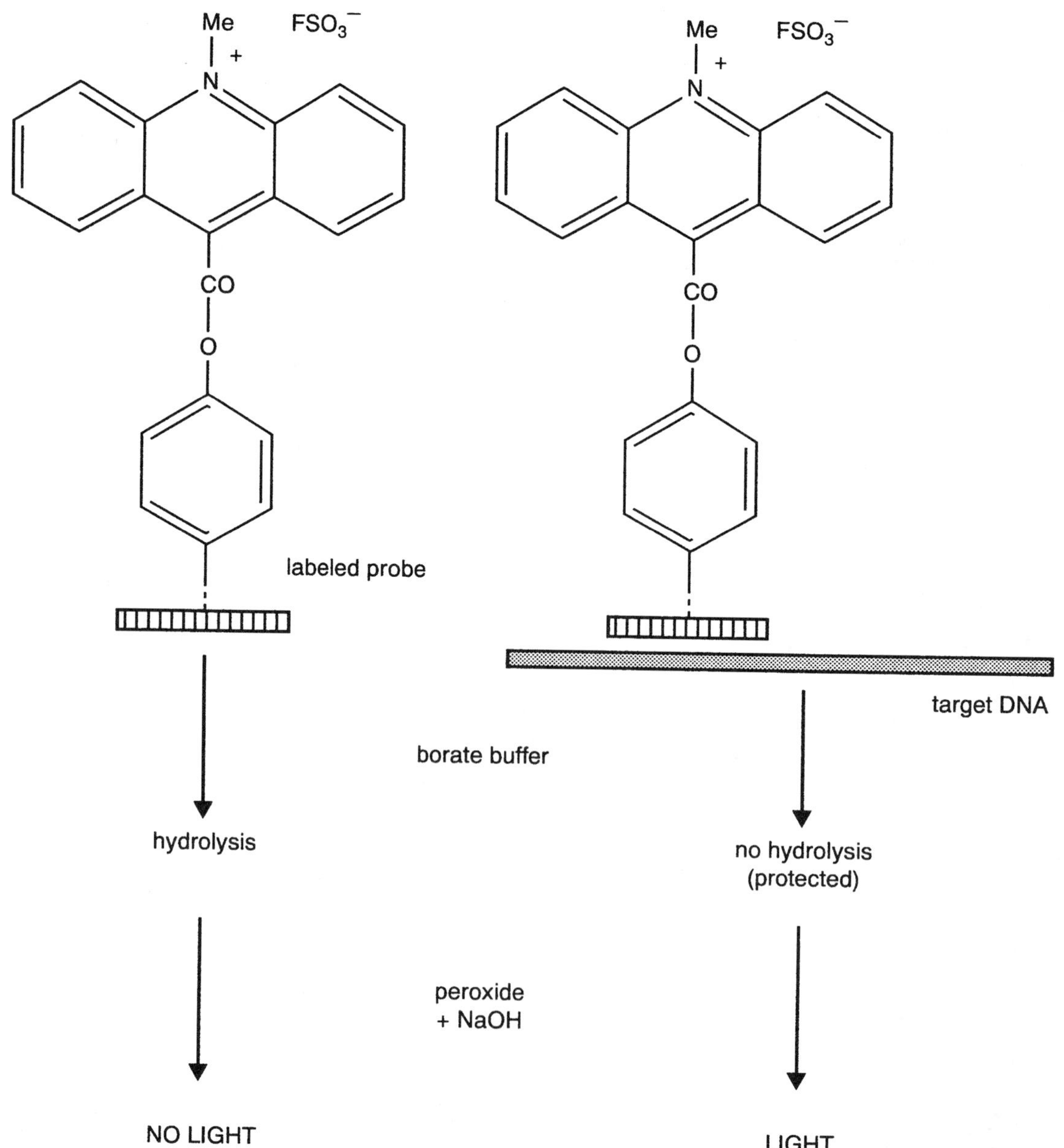


**Figure 1.** Hybridization protection assay for detecting target nucleic acid (DNA) using a chemiluminescent acridinium ester label.

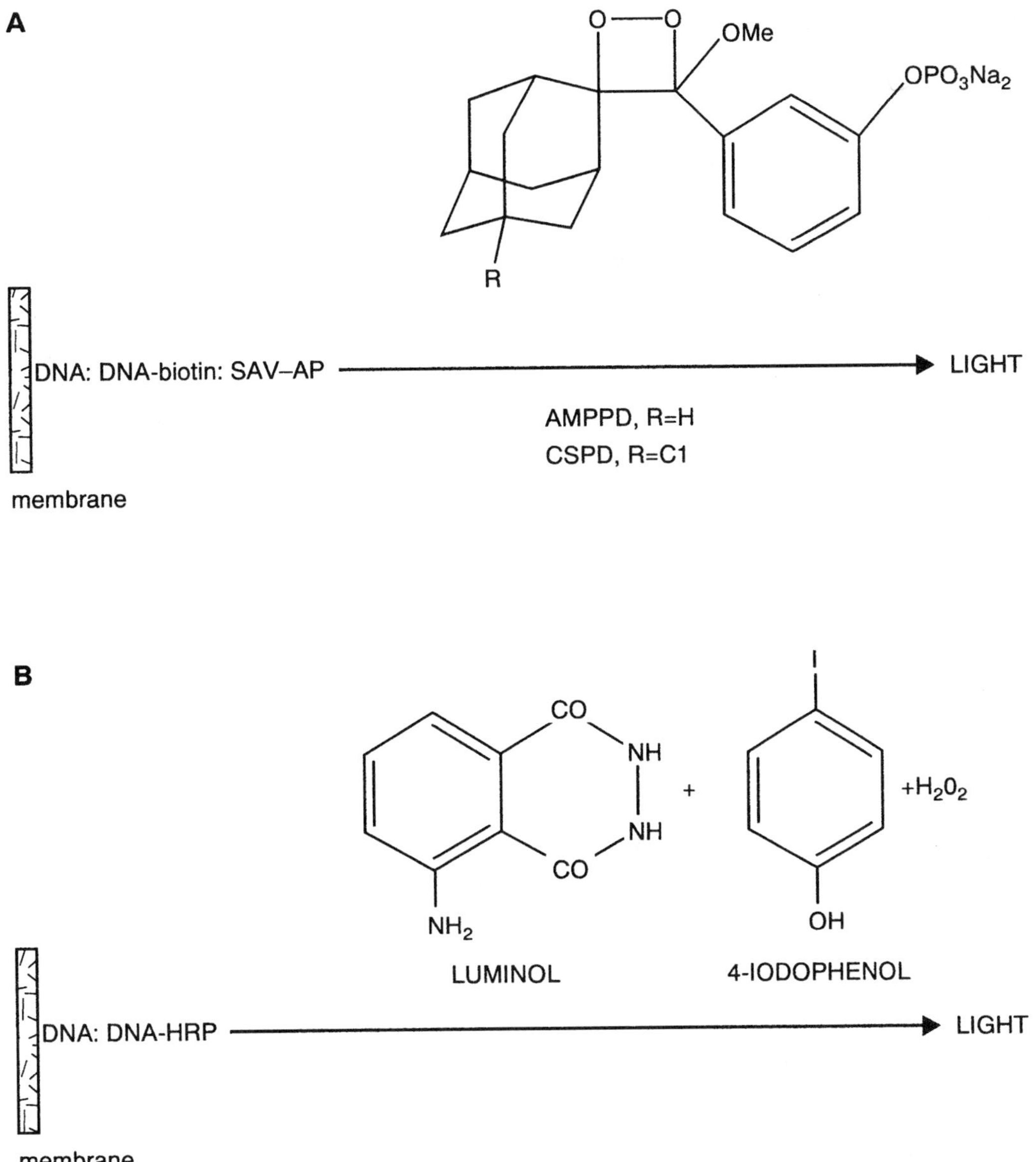


**Figure 2.** (A) Chemiluminescent detection of an alkaline phosphatase label using an adamantyl 1,2-dioxetane aryl phosphate (AMPPD, CSPD): AP, alkaline phosphatase; HRP, horseradish peroxidase; SAV, streptavidin. (B) Detection of a peroxidase label using the luminol-based enhanced chemiluminescent reaction.

### 3.1.1 Labels

Luminol, isoluminol, acridinium ester, acridinium sulfonyl carboxamide, firefly luciferase, marine bacterial luciferase, *Renilla* luciferase, *Vargulla* luciferase and aequorin have all been tested as labels. Acridinium esters are one of the principal types of CL label in current use in immunoassay and hybridization assays. A rapid flash of light is produced by treating the label with $H_2O_2$ + NaOH. Nonseparation (homogeneous) nucleic acid assays can be formatted with an acridinium ester label, as in the ''hybridization protection assay.'' In this three-step assay, first the sample and an acridinium ester labeled probe are incubated together; next the label on unhybridized probe is hydrolyzed to a nonchemiluminescent product (probe bound to target nucleic acid is protected from hydrolysis); and finally a chemiluminescent signal is obtained from the acridinium ester label on the bound probe by treatment with oxidant in basic conditions (Figure 1). The recombinant photoprotein aequorin is the most promising BL label. Light emission (rapid flash) is triggered by reacting the aequorin with calcium ions.

### 3.1.2 Detection Reactions

The two most popular enzyme labels, alkaline phosphatase and horseradish peroxidase, can be quantitated by a range of CL and BL reactions. For alkaline phosphatase, the CL assay using adamantyl 1,2-dioxetane aryl phosphate substrates (e.g., AMPPD, CSPD) (Figure 2) is convenient and very sensitive (detection limit, 1 zeptomole). For horseradish peroxidase the enhanced chemiluminescent assay, which uses luminol and an enhancer molecule (e.g., 4-iodophenol or 4-hydroxycinnamic acid), is the most sensitive assay (detection limit, 5000 amol) (Figure 2). In addition, sensitive CL or BL detection reactions of varying complexity are available for glucose oxidase, glucose 6-phosphate dehydrogenase, xanthine oxidase, and β-galactosidase labels.

### 3.2 Enzymes, Substrates, and Cofactors

The luminol, firefly luciferase, and marine bacterial luciferase reactions are indicator reactions for the production or consumption of peroxide, ATP, and NAD(P)H, respectively. They can be coupled to other reactions involving oxidases, kinases, and dehydrogenases and used to measure any component of the coupled reaction (enzyme, substrate, cofactor)—for example, assays for alcohol and glucose, as in equations (2) and (3). Coimmobilization of the luciferases with other enzyme reagents improves the efficiency of the reaction by channeling of the intermediates produced by successive coupled enzymes.

$$\text{alcohol} + \text{NAD} \xrightarrow{\text{alcohol dehydrogenase}} \text{acetaldehyde} + \text{NADH}$$

$$\text{NADH} + \text{FMN} \xrightarrow{\text{NAD(P)H:FMH oxidoreductase}} \text{NAD} + \text{FMNH}_2$$

$$\text{FMNH}_2 + \text{decanal} + \text{O}_2 \xrightarrow{\text{marine bacterial luciferase}} \text{FMN} + \text{decanoic acid} + \text{light} \quad (2)$$

$$\text{glucose} + \text{O}_2 \xrightarrow{\text{glucose oxidase}} \text{gluconolactone} + \text{H}_2\text{O}_2$$

$$\text{H}_2\text{O}_2 + \text{luminol} \xrightarrow{\text{peroxidase}} \text{3-aminophthalic acid} + \text{N}_2 + 2\text{H}_2\text{O} + \text{light} \quad (3)$$

### 3.3 Reporter Genes

The gene for firefly luciferase has been cloned; this gene, *luc*, is an effective reporter gene for studying transcriptional activity of cloned genomic sequences. Luciferase activity in transfected cells is measured by lysing the cells with a detergent (e.g., Triton X-100), adding ATP + $Mg^{2+}$, then injecting firefly luciferin and measuring the flash of light (detection limit 1–3 $\times$ $10^5$ molecules). The gene for *Vargulla* luciferase, apoaquorin, and the fused marine bacterial luciferase *lux* A and *lux* B genes, are also useful as reporter genes. Other alternatives include the CL assay (adamantyl 1,2-dioxetane aryl galactoside substrate) of β-galactosidase expressed by the *lac* Z gene and the BL assay (firefly luciferin-*O*-phosphate substrate) of enzyme expressed by the alkaline phosphatase gene.

### 3.4 Rapid Microbiology

ATP is present in all living cells, and measurement of ATP using the firefly luciferase reaction is a convenient means of enumerating cells, offering a sensitivity of $10^5$ colony-forming units per milliliter (cfu/mL).

### 3.5 Cellular Luminescence

Cellular CL is widely used for the study of phagocytosis. Activation of polymorphonuclear neutrophils by phagocytosable particles leads to the production of reactive oxygen species and a weak CL that is enhanced in the presence of luminol or lucigenin (*N,N'*-dimethyl-9,9'-biacridinium dinitrate).

*See also* Fluorescence Spectroscopy of Biomolecules; Hybridization for Sequencing of DNA.

## *Bibliography*

Alam, J., and Cook, J. L. (1990) Reporter genes: Application to the study of mammalian gene transcription. *Anal. Biochem.* 188:245–254.

Bronstein, I., Edwards B., and Voyta, J. C. (1989). 1,2-Dioxetanes: Novel chemiluminescent enzyme substrates. Applications to immunoassays. *J. Biolumin. Chemilumin.* 4:99–111.

Campbell, A. K. (1988) *Chemiluminescence,* Ellis Horwood, Chichester.

Kricka, L. J. (1991) Chemiluminescent and bioluminescent techniques. *Clin. Chem.* 37:1472–1481.

———, Ed. (1992) *Nonisotopic DNA Probe Techniques.* Academic Press, San Diego, CA.

Matthews, J. A., and Kricka, L. J. (1988) Analytical strategies for the use of DNA probes. *Anal. Biochem.* 169:1–25.

Pazzagli, M., Cadenas, E., Kricka, L. J., Roda, A., and Stanley, P. E. (1989) *Bioluminescence and Chemiluminescence: Studies and Applications in Biology and Medicine.* Wiley, Chichester.

Stanley, P. E., and Kricka, L. J., Eds. (1991) *Bioluminescence and Chemiluminescence: Current Status.* Wiley, Chichester.

# Chirality in Biology

*Ronald Bentley*

## *Key Words*

**Chiral** Describing an object or molecule whose image in a plane mirror cannot be superposed on the original; hence, chirality.

**Configuration** The three-dimensional arrangement of atoms or groups of atoms at a center of chirality; arrangements resulting from rotation about single bonds are excluded.

**Diastereoisomer** Stereoisomers that are not in an object–mirror image relationship.

**Enantiomer** Stereoisomers that are in an object–mirror image relationship.

**Isomer** Compounds having the same molecular formula but differing in the nature or sequence of bonding of their atoms, or in the spatial arrangement of their atoms.

**Optical Activity** The property, possessed by some substances (or solutions of some substances) of rotating the plane of polarization of plane-polarized light.

**Prochiral** Describing a molecule that yields a chiral structure if one of two chemically like groups is replaced by a different group; thus, prochiral Caabc, by replacement of one a by d, yields chiral Cabcd.

**Racemic** Describing a mixture of equal amounts of the two enantiomers of a compound; such a mixture is optically inactive.

**Racemization** The conversion of an enantiomer to a racemic mixture characterized by the time-dependent loss of optical activity.

**Stereoisomers** Isomers differing only in the spatial arrangement of their atoms.

Chiral objects and molecules are those for which the image in a plane mirror cannot be superposed on the original. The two human hands are chiral; the mirror image of the left hand is the right hand. Many chiral structures or chiral forms of growth occur in nature. Most metabolites are chiral, and enzymes usually exhibit high degrees of specificity toward molecules of opposite chirality. Chiral molecules, related as left and right hands (enantiomers), can have very different physiological effects. The recognition of chiral molecules is a central event in biology.

## 1 DEFINITION

If two objects are superposable by only rotations and/or translations, they are identical. Objects for which the reflected image in a plane mirror cannot be superposed on the original by rotational and/or translational operations are described as chiral (Greek, χειρ, hand). The mirror image of the left hand cannot be superposed on itself; it is a right hand. Chirality in biology has ramifications for gross and molecular structure.

## 2 CHIRALITY IN NATURAL OBJECTS

Humans and most animals show an approximate bilateral symmetry. A middle plane at right angles to the body gives a left side that is the mirror image of the right side. This left–right symmetry does not extend to the internal organs; the heart occurs in the left side of the chest cavity and the major liver lobe is on the right side of the abdomen. The rare condition *situs inversus* (with organ positions reversed), occurs in about 1 out of 10,000 births.

Asymmetric structures in living organisms include the following.

*Microorganisms:* Some bacteria will grow as double-stranded filaments formed by association of many single cells. These helical arrangements can be either left- or right-handed. Mutants exist with only one helical sense. In a unique case of specialization, entomogenous fungi of the genus *Laboulbenia* parasitize only the left hind leg of certain beetles—never the right.

*Plants:* In tobacco plants, leaves alternate along stems in a helical pattern. In a large population, half the plants are left-handed, the other half right-handed. Many vines climb with either left- or right-handed helicity; the honeysuckle twines in a left-handed helix and the morning glory to the right.

*Animals:* Male fiddler crabs have one extraordinary large claw; in half of the animals in a large population the claw is on the left side of the body, in the other half on the right. The narwhal has a large ivory tusk up to 10 feet long, which usually pierces the upper lip on the left-hand side; it has a helical striation from right to left. Very rarely, tusked females are encountered, as are males with two tusks. One bird, the New Zealand wrybilled plover, has a beak that is always bent to the right-hand side of the bird's body. The direction of mouth opening in some fish is either left- or right-handed (with 50% of each type in a large population); in flat fishes the eyes are either on the right (e.g., sole, halibut), or on the left (e.g., turbot). Many mollusk shells have beautiful helical structures, usually right-handed; some whelks (e.g., the Florida lightning whelk) are left-handed. Rarely, left-handed, "sinistral" shells are found for normally right-handed types.

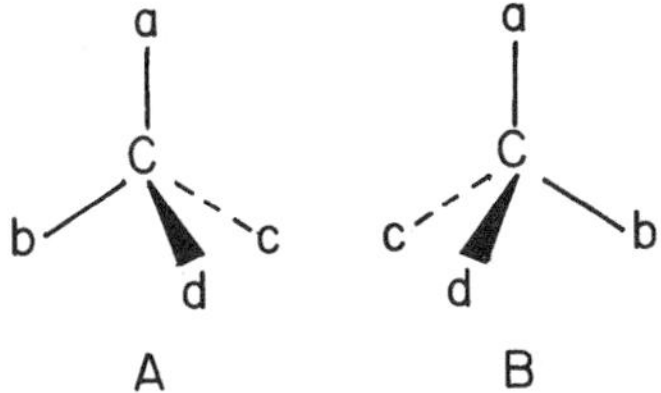


**Figure 1.** The two stereoisomers of Cabcd. (A) and (B) are enantiomers and are in an object–mirror image relationship. Chirality is not implied in the letters (a, b, c, d) used to represent atoms or groups of atoms.

## 3 CHIRALITY AT THE MOLECULAR LEVEL

Many naturally occurring "essential oils" were found to be optically active, as were solutions of some compounds such as cane sugar (in water) and camphor (in alcohol). In organic compounds, optical activity was associated with the presence of an "asymmetric" carbon atom, Cabcd, where a, b, c, and d are different atoms or groups of atoms. The tetrahedral disposition of the valence bonds of carbon, allows two three-dimensional models for Cabcd (Figure 1), related as object and mirror image. One stereoisomer rotates the plane of polarized light to the right (+), the other equally to the left (−). The two stereoisomers of chiral compounds that are related as object and mirror image are known as enantiomers (Figure 1A, B). Stereoisomers not in an object–mirror image relationship are diastereoisomers. Optical activity also occurs in allenes, abC=C=Cab. These structures with an element of symmetry—a twofold axis of rotation—were once termed "dissymmetric." Both cases, Cabcd and abC=C=Cab, are now simply described as chiral. The Cabcd type has a center of chirality; the allene type, an axis of chirality.

The three-dimensional arrangement of atoms at a chiral center is termed the configuration; this term generally excludes possible arrangements differing only after rotation about single bonds. The arrangements shown in Figure 1 are the configurations of Cabcd. They may be depicted in three dimensions (Figure 1) or as Fischer projection formulas in two dimensions (Figure 2). For most biologically interesting compounds, the actual "absolute configurations" are known. A high degree of specificity in chemical reactions characterizes living systems. This specificity is not absolute and there are many exceptions.

Chemists specify enantiomer configurations by the unambiguous *R,S* (Cahn–Ingold–Prelog) system. In biology, it is traditional to apply a "local" system for related compounds. A configurational descriptor, D or L, is based on the orientation of a selected functional

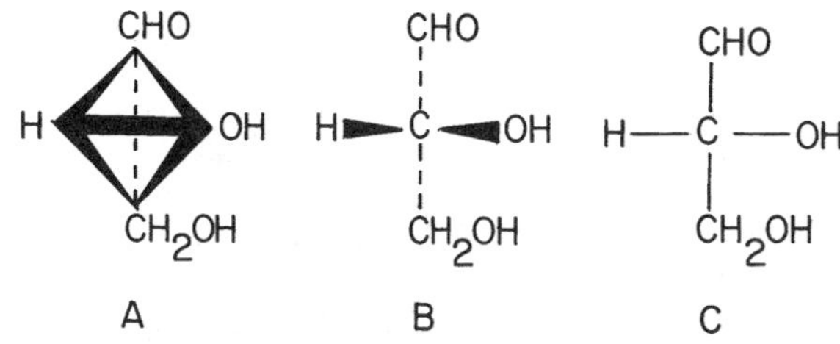


**Figure 2.** Fischer projection formulas. The compound used for illustration is (+)-D-glyceraldehyde. It is crucial that the three-dimensional model be oriented exactly as shown in (A). This model is simplified in (B), and the two-dimensional Fischer structure is (C). Structure C always implies the three-dimensional arrangement of A and B.

group at the chiral center in a Fischer projection formula. If this group is to the left, the descriptor is L, to the right, D. Two configurational reference standards are serine for amino acids and glyceraldehyde (see also Figure 2) for carbohydrates.

```
     COOH                           CHO
      |                              |
H2N—C—H    (−)-L-Serine    H—C—OH   (+)-D-Glyceraldehyde
      |                              |
     CH2OH                          CH2OH
```

Chiral compounds are *generally* isolated from living systems as single enantiomers, while laboratory synthesis *generally* leads to an optically inactive 1:1 mixture of both enantiomers. Such an equimolar mixture of two enantiomers is said to be racemic. The time-dependent conversion of one enantiomer to a racemic mixture is termed racemization; it may be spontaneous (usually slow), or it may be promoted by catalysts. "Racemase" enzymes are known, for instance, for some amino and hydroxy acids.

Optically inactive compounds containing opposed centers of chirality are described as meso structures. Thus, tartaric acid has two enantiomers, one optically inactive meso structure and a racemic mixture of the two enantiomers; the subscript g indicates reference to the glyceraldehyde standard.

```
     COOH                 COOH                 COOH
      |                    |                    |
 H—C—OH (R)          HO—C—H (S)           HO—C—H (S)
      |                    |                    |
HO—C—H (R)            H—C—OH (S)          HO—C—H (R)
      |                    |                    |
     COOH                 COOH                 COOH
```

(+)-$L_g$-Tartaric acid (−)-$D_g$-Tartaric acid *meso*-Tartaric acid

### 3.1 Amino Acids

Of the 20 amino acids used for ribosomal protein synthesis, Gly is achiral (but prochiral, see later), Thr and Ile contain two chiral centers, and the rest have one chiral center. The "charging" of tRNA molecules with the appropriate amino acid by the aminoacyl tRNA synthetases is highly specific not only for individual amino acids, but also for those with the L configuration. Proteins initially synthesized by ribosomes only contain L-amino acids. Nevertheless, not all the aminoacyl tRNA synthetases have absolute chiral specificity. Thus, the *Escherichia coli* enzyme charging L-Tyr can also use D-Tyr. Moreover, D-Leu and D-Val inhibit reaction with the L forms, so presumably they are bound by the enzyme. In other cases the specificity is absolute; D-Arg and D-Pro are not activated by, and do not inhibit, the respective synthetases.

Some mammalian proteins, particularly those of long life, contain D-amino acids derived by spontaneous, posttranslational racemization. D-Asp and D-Ser occur in lens proteins, in tooth dentine and enamel, and in brain myelin; D-Asp occurs in human erythrocytes. Posttranslational racemization yields D-amino acids (e.g., D-Ala) present in opioid peptides of certain frogs (e.g., dermorphin). This enzyme-catalyzed process probably requires dehydrogenation of L-Ala to dehydro alanine, followed by rehydrogenation to the D-enantiomer. Ribosomally produced lantibiotics (e.g., nisin) contain lanthionine (with D-Ala) and β-methyllanthionine (with D-aminobutyric acid).

Mammals have a very active enzyme, D-amino acid oxidase (EC 1.4.3.3), as well as a less active L-amino acid oxidase (EC 1.4.3.2). The precise function of the D-enzyme is enigmatic; it may be involved in the catabolism of D-amino acids from plant dietary sources or from microorganisms. A D-aspartate oxidase (EC 1.4.3.1) is also present. Free D-Asp occurs in brain and other tissues of rodents, and in human blood. D-Ala and D-Ser occur commonly in the free state in insects, and D-Ser is present in earthworms both free and in combined forms.

Microbial secondary metabolites often contain D-amino acids. There is frequently an association between antibiotic activity and the presence of the "unnatural" amino acid as in penicillin. Of the three chiral centers in benzylpenicillin (penicillin G), two derive from the precursor δ(L-α-aminoadipyl)-L-cysteinyl-D-valine. D-Amino acids occur in lantibiotics, gramicidins, and bacitracin and in the immunosuppressive agent, cyclosporin A.

Complex peptidoglycan structures in bacterial cell walls contain tetrapeptides; the overall structure is variable but always includes a D-amino acid. *Staphylococcus aureus* contains L-Ala-D-isoGlu-L-Lys-D-Ala, and *E. coli* has L-Ala-D-isoGlu-*meso*-diaminopimelate-D-Ala. Some bacterial capsules contain a D-Glu polymer.

### 3.2 Carbohydrates

Carbohydrates usually contain more than one chiral center. Membership in the D- or L-configurational series is determined by the orientation of the OH group at the chiral carbon with the highest number in the Fischer projection formula. In the following list, the configuration-defining OH group is in bold type:

```
    CHO      1        CH2OH    1
     |                  |
 H—C—OH     2          CO      2           CHO      1
     |                  |                   |
HO—C—H      3       HO—C—H     3        H—C—OH     2
     |                  |                   |
 H—C—OH     4        H—C—OH    4       HO—C—H      3
     |                  |                   |
 H—C—OH     5        H—C—OH    5       HO—C—H      4
     |                  |                   |
    CH2OH    6        CH2OH    6          CH2OH     5
  D-Glucose          D-Fructose         L-Arabinose
```

The most abundant carbohydrate on earth is D-glucose. L-Glucose either does not occur or is extremely rare; it is, however, oxidized by some microorganisms. *N*-Methyl-L-glucosamine is a component of streptomycin. Most carbohydrates occur in the D configuration—for instance, D-ribose and 2-deoxy-D-ribose, in the nucleic acids. For galactose, the D form predominates. In agarose, a major polysaccharide of agar-agar, D-galactopyranose and 3,6-anhydro-L-galactopyranose alternate. Other polysaccharides may contain L-guluronate or L-iduronate. Arabinose, fucose, and rhamnose occur mainly as the L forms. Some rare 3,6-dideoxyhexoses have separate names for each enantiomer (e.g., abequose and tyvelose, and colitose and ascarylose).

### 3.3 Other Components

#### 3.3.1 Hydroxy Acids

Anaerobic glycolysis in muscle yields (+)-L-lactic acid; this enantiomer occurs in bacteria, and racemic (±)-lactic acid is formed by *Lactobacillus plantarum*. (−)-D-Lactic acid is a component of bacterial cell wall peptidoglycan, which contains *N*-acetylmuramic acid (the 3-*O*-D-lactyl ether of D-glucosamine). (+)-L-Tartaric acid is common in plants (especially in grape juice) and some microorganisms; (−)-D-Tartaric acid and *meso*-tartaric acid occur in other plants. The racemic mixture, originally known as paratartaric acid, occurs irregularly during the manufacture of wine. It was the first such mixture to be resolved into its enantiomers (Pasteur, 1848).

### 3.3.2 Terpenes and Alkaloids

The occurrence of both enantiomers, sometimes as racemic mixtures, is particularly common among the terpenes (e.g., borneol, camphor, carvone, and limonene). Many alkaloids and other secondary metabolites occur as both enantiomers. The alkaloidal drug atropine is a racemic mixture of hyoscyamine enantiomers. The naturally occurring and physiologically active form is (−)-hyoscyamine. Partial racemization occurs during drug isolation and is completed subsequently by treatment of the mixture with dilute alkali.

## 4 PHYSIOLOGICAL RESPONSES TO ENANTIOMERS

Responses of humans, animals, and most living systems to the enantiomeric forms of a given material cover a wide range. For instance, taste differences between amino acid enantiomers have long been known and are exploited commercially. L-Glu has a "meaty" flavor and is widely used as monosodium L-glutamate (MSG). D-Glu, however, is "flat" or tasteless. Sweet tastes are associated with D-Asn, D-His, D-Leu, D-Phe, D-Trp, D-Tyr, and L-Ala; their enantiomers are either tasteless, flat, or bitter. Moreover, the methyl ester of L-Asp-L-Phe has a remarkably sweet taste and is widely used in the food and beverage industries as "Aspartame." Enantioaspartame (DD) and the diastereoisomers, DL and LD, have bitter tastes. Taste differences among carbohydrate enantiomers are less clear-cut.

Significantly different odor responses to enantiomers are known. *R*-Carvone (from peppermint oil) has a spearmint odor, and *S*-carvone a caraway odor. In other cases [e.g., (+)-and (−)-camphor], there are no odor differences. Considerable individual variation in olfactory responses has hampered investigation; about 50% of a given population may have an anosmia to a particular odorant.

Drug enantiomers may have widely different pharmacological actions in humans. A tragic situation occurred when racemic thalidomide was used as a sedative by pregnant women; many children were born with missing or abnormal limbs (phocomelia). Although the situation is still not completely clear, it is likely that the teratogenic activity resides in the L(*S*) enantiomer, while the D(*R*) form is sedative. There is now considerable emphasis on the desirability of using drugs in pure enantiomeric forms. In at least one case, two enantiomers are marketed under separate, palindromic names—Darvon, (+), is a narcotic and analgesic, and Novrad, (−) is an antitussive. Some carcinogenic materials also show enantiomeric specificity.

For chiral plant growth regulators, one enantiomer is generally more potent than the other. Thus, the herbicidal activity of α-phenoxypropionic acid is strongest with the R enantiomer. Chiral insect pheromones show all the following specificities:

1. Only one enantiomer is active; the other enantiomer either inhibits or does not inhibit.
2. Both enantiomers are active.
3. Different enantiomers used by different species.
4. Pheromone action requires both enantiomers.

## 5 CHIRAL RECOGNITION

The recognition of one molecule by another is a central event in biology; if the event involves an enantiomer, it is described as "chiral recognition." Chiral recognition eventually leads to specific physiological effects (see Section 4). In living organisms, the recognition usually involves enzymes, and receptors for drugs and other materials. The specific requirements for chiral recognition include two necessary factors. For the reaction of each enantiomer with an enzyme or receptor, the two pathways must be of different energetics as a result of diastereoisomeric geometries in the two cases. Such diastereoisomeric geometries arise when chiral molecules interact with environments that are also chiral. This is a necessary but not sufficient condition. There must also be specific "short-range" interactions between the enantiomer and enzyme (or receptor), either of attraction or repulsion.

1. Consider the potential reaction of two enantiomers (+)A and (−)A with (+)B. For differentiation to occur, at some point in a reaction intermediate or overall pathway leading to the AB products, there must be a diastereoisomeric relationship between the two possible pathways, that is, (+)A(+)B and (−)A(+)B; this relationship cannot be that of two enantiomers, (+)A(+)B and (−)A(−)B.
2. In the absence of specific constraints, a total of four interactions (involving eight atomic centers, with four different, nonplanar centers on each molecule) is the minimum required. Under certain conditions (where the diastereoisomers adopt the same relative orientation), the minimum number of interactions is three. This situation is generally implied for enzymes and various receptors; the substrate can approach only from one direction (the "surface"), not from the other (the "interior" or "backside"). In many cases of enzyme–substrate or drug–receptor interaction, chiral recognition may be rationalized by a "three-point attachment" hypothesis.

## 6 ENZYME SPECIFICITY

Enzymes generally show a high degree of specificity in interactions with a chiral compound; there are, however, many exceptions to this statement. One highly specific enzyme, glucose oxidase (EC 1.1.3.4), finds applications in industry and in medical practice; it has been tested against more than 50 carbohydrates. If the D-glucose activity is set at 100%, only 2-deoxy-D-glucose (at 25%) shows more than 3% of this activity. In particular, L-glucose does not react. A very few other materials with the D configuration have activity in the range of 1–3%. A further specificity concerns the anomeric position, C-1. The enzyme is specific for β-D-glucopyranose (this yields the 100% activity); the very low activity level with α-D-glucopyranose is attributed to spontaneous interconversion of the two anomers (mutarotation) during the reaction.

Other examples of enzymes regarded as highly specific are D-amino acid oxidase (EC 1.4.3.3) and L-amino acid oxidase (EC 1.4.3.2). At least 20 other enzyme pairs, one specific for the D substrate and the other for the L substrate, are known.

Many highly purified enzymes utilize both substrate enantiomers in reactions at the chiral center itself (e.g., 2-hydroxy-4-ketoglutarate aldolase) or at a remote site (e.g., glutamine synthetase, nicotine oxidase). Chiral specificity may depend on the enzyme source. Horse liver alcohol dehydrogenase oxidizes both enantiomers of butan-2-ol and octan-2-ol; yeast alcohol dehydrogenase does not oxidize the *R* enantiomers. Racemase enzymes for amino and hydroxy acids clearly utilize both substrate enantiomers. In some cases, one enantiomer, though not a substrate, binds to the enzyme and is inhibitory. Thus, phenylalanine ammonia-lyase, which converts L-Phe to *trans*-cinnamic acid, is inhibited by D-Phe. Inhibition of some aminoacyl-tRNA synthetase enzymes by D-amino acids was noted earlier.

## 7 PROCHIRALITY

Enzymes can distinguish chemically like atoms or groups of atoms that are in different stereochemical environments in the same molecule; these differentiations can include the faces of trigonal atoms. Only a few particularly important cases can be illustrated from the very large literature. The requirements for differentiation of groups or faces in a prochiral molecule are essentially the same as those for chiral recognition.

The achiral structure, Caabc, is often described as prochiral; replacement of one group, a, at the prochiral center, by a different group, d, gives the chiral molecule Cabcd. Glycerol is an example. In the glycerol kinase reaction only one of the chemically like —$CH_2OH$ groups undergoes phosphorylation (P = $PO_3H_2$).

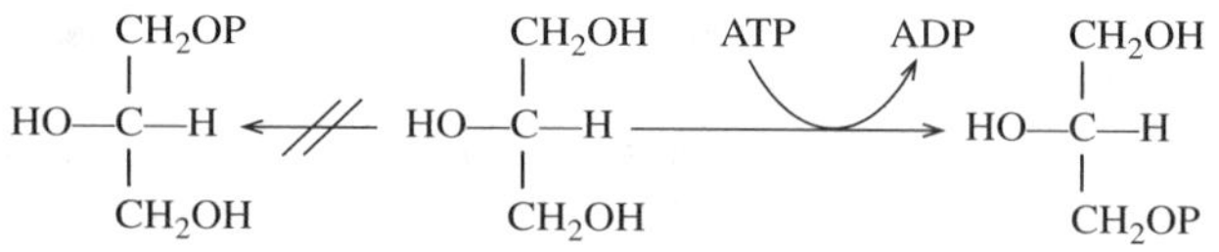

$CH_2OP$ | HO—C—H | $CH_2OH$ ⇸ $CH_2OH$ | HO—C—H | $CH_2OH$ —(ATP → ADP)→ $CH_2OH$ | HO—C—H | $CH_2OP$

In the action of fumarase, only one of the two hydrogens at a methylene group is removed. The stereochemistry of such reactions (i.e., the fate of individual atoms) is investigated by use of tracer isotope techniques, either with low (tracer) levels of $^3H$ or with $^2H$ (up to 100% enrichment).

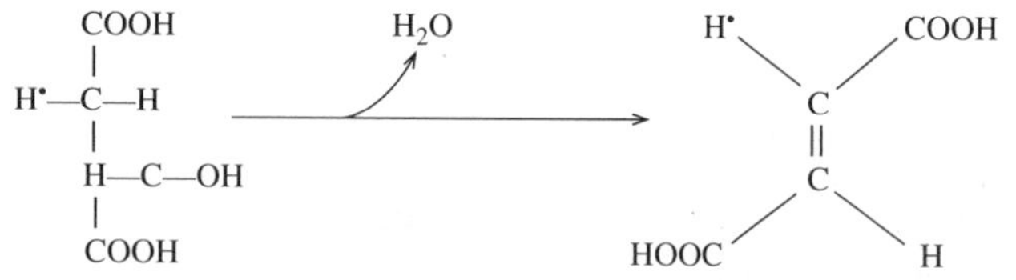

COOH | H•—C—H | H—C—OH | COOH —($H_2O$)→ (H•)(COOH)C=C(HOOC)(H)

The two stereochemically different "faces" of the carbonyl group in abC=O can be differentiated by enzymes—for example, in the reduction of acetaldehyde by alcohol dehydrogenase (Figure 3). Not only is a hydrogen added to one specific face of the carbonyl; in addition, only one of the two hydrogens at C-4 in NADH is utilized and the hydrogen is placed into only one (prochiral) position in the product, ethanol.

Similar differentiation of "faces" occurs in some structures with a C=C bond. In the fumarase reaction, in the direction fumarate → L-malate, only one of the two stereochemically distinct faces of fumarate is selected (Figure 4).

Finally, since $^2H$ and $^{18}O$ can be obtained at 100% enrichment, enzyme stereochemistry involving $^3H^2H^1H$—C—X and $^{18}O^{17}O^{16}O$—P—X stereocenters can be studied.

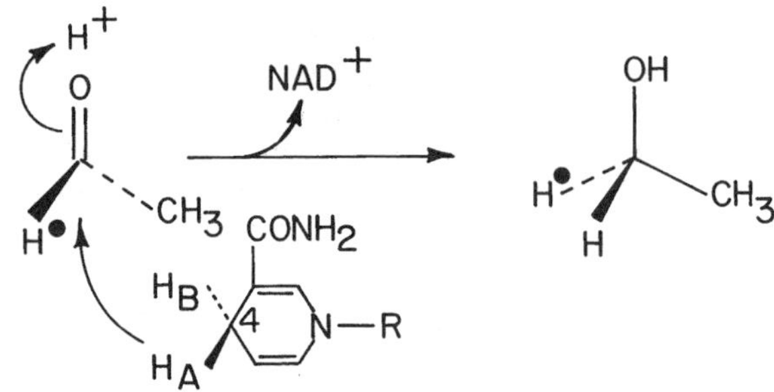


**Figure 3.** Stereospecificity of alcohol dehydrogenase. The two (prochiral) hydrogens at C-4 of NADH are traditionally termed $H_A$ and $H_B$ in the biochemical literature. They are formally described as $H_R$ and $H_S$, respectively. Only $H_A$ is used in this reaction and only one "face" of acetaldehyde is attacked. If H• were $^2H$, the ethanol would be *S* and would be observably levorotatory. (In this figure, *R* indicates the rest of the NADH molecule.)

## 8 THE ORIGIN OF CHIRALITY ON EARTH

Organic compounds associated with living organisms clearly existed on the prebiotic earth. However, the products of experiments simulating prebiotic conditions are all racemic mixtures if a chiral center is involved. A mechanism must have existed to account for the selectivity observed today—the carbohydrate components of the nucleic acids are exclusively D and, for the most part, the protein amino acids are L. Despite much speculation and debate, there is little agreement. Two general possibilities are that the first significant material was already formed with one chiral sense in place, or that racemic mixtures were produced initially and were subsequently enriched in one chiral sense. Possible mechanisms for which there is experimental support, include interactions with polarized β-rays from nuclear decay, interactions with circularly polarized light, or selective adsorption and/or polymerization of one enantiomer on quartz crystals or clay crystals (template catalysis). Using circularly polarized light, all the following processes have

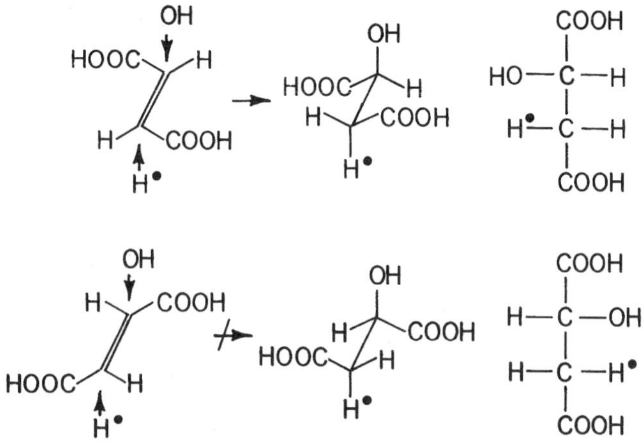

(Top: fumarate with OH added from above and H• from below → malic acid, three-dimensional and Fischer projection: COOH | HO—C—H | H•—C—H | COOH)

(Bottom, not formed: COOH | H—C—OH | H—C—H• | COOH)

**Figure 4.** Stereospecificity of fumarase. The addition of the elements of water to fumaric acid proceeds only as shown on the top line of structures; the "face" of fumarate, diagrammed at the bottom, is not attacked. The product, malic acid, is shown first as a three-dimensional representation, then as a Fischer projection formula. If H• were $^2H$, the fumarase reaction would yield (3*R*,3-$^2H$)-L-malic acid.

been shown in photochemical operations: asymmetric photodestruction, partial photoresolution, and asymmetric syntheses.

The possibility that optically active "seeds" or components reached earth from elsewhere in the universe has also been considered. The "Murchison meteorite" contains a number of amino acids, but it is not clear whether the amino acids were present as racemic mixtures. No clear-cut decision as to the terrestial origin of molecular chirality is now possible. However, the idea that molecular asymmetry on earth reflects the fundamental structure of matter itself, and particularly the nonconservation of parity in the weak interactions, is favored.

*See also* Amino Acid Synthesis; Origins of Life, Molecular Biology of; Receptor Biochemistry; Recognition and Immobilization, Molecular.

### *Bibliography*

Alworth, W. L. (1972) *Stereochemistry and Its Application in Biochemistry.* Wiley-Interscience, New York.

Bentley, R. (1969, 1970) *Molecular Asymmetry in Biology,* Vols. I and II. Academic Press, New York.

Buckingham, J., and Hill, R. A. (1987) *Atlas of Stereochemistry,* two vols. and suppl., 2nd ed. Chapman & Hall, London.

Eliel, E. L., Wilen, S. H., and Mander, L. N. (1994) *Stereochemistry of Organic Compounds,* Wiley & Sons, New York.

Gardner, M. (1990) *The New Ambidextrous Universe,* 3rd rev. ed. Freeman, New York.

Jacques, J. (1993) *The Molecule and Its Double.* McGraw-Hill, New York.

Mislow, K. (1965) *Introduction to Stereochemistry.* Benjamin, New York.

Neville, A. C. (1976) *Animal Asymmetry.* Edward Arnold, London.

Rétey, J., and Robinson, J. A. (1982) *Stereospecificity in Organic Chemistry and Enzymology.* Verlag Chemie, Weinheim.

Tamm, C., Ed. (1982) *Stereochemistry. New Comprehensive Biochemistry,* Vol. 3. Elsevier Biomedical Press, Amsterdam.

Testa, B. (1979) *Principles of Organic Stereochemistry.* Dekker, New York.

# Chlamydomonas

*Jean-David Rochaix*

### *Key Words*

**Cosmid** A plasmid that contains phage lambda cos sites and can therefore be packaged in vitro into phage coats.

**DNA Complexity** The total length of different DNA sequences contained in a given DNA preparation; the DNA complexity is usually given in thousands of base pairs (kbp).

**Intron** A segment of DNA within a gene that is transcribed but removed from the transcript by splicing together the flanking coding sequences (exons).

**Linkage Group** All loci that are linked together by genetic criteria and correspond to a chromosome.

**Open Reading Frame (ORF)** DNA sequence that potentially encodes a protein; an ORF consists of a series of triplets (codons) coding for amino acids.

**Splicing** Process of removing introns from a precursor RNA and joining flanking exons.

The highly diversified, polyphyletic genus *Chlamydomonas* comprises more than 450 species of unicellular flagellated photosynthetic algae. These organisms are especially attractive for several areas of research in cell and molecular biology because of their small size, fast growth rate, and short sexual cycle. However, only few species have been actively used for molecular genetic studies. These include the heterothallic and interfertile species *C. reinhardtii* and *C. smithii,* which are on the line of descent leading to the multicellular species *Volvox;* the sibling species pair *C. eugametos* and *C. moewusii,* both very distinct from *C. reinhardtii;* and the homothallic species *C. monoica.* This contribution focuses on *C. reinhardtii* because this alga has emerged as the organism of choice for most investigations. The alga is amenable to extensive genetic analysis, and its sexual cycle is well characterized. Major technical advances with *C. reinhardtii* include the development of reliable transformation methods for the nuclear, chloroplast, and mitochondrial genomes. Gene tagging and nuclear gene rescue with genomic libraries are also feasible. *C. reinhardtii* has therefore become a powerful model system, especially in the areas of photosynthesis, organellar biogenesis, and flagellar function and assembly.

## 1 THE ORGANISM

Characteristic features of *Chlamydomonas* are its unique cup-shaped chloroplast, which occupies nearly half the cell volume, and its two flagella. In this organism photosynthetic function is dispensable, provided a reduced carbon source such as acetate is included in the growth medium. The alga can therefore be grown under three different conditions: phototrophic growth with $CO_2$ assimilated through photosynthesis as unique carbon source, heterotrophic growth in the dark with acetate, and mixotrophic growth with acetate in light.

Haploid vegetative cells of *C. reinhardtii* exist as mating-type + or −, determined by alternative alleles of a nuclear gene, and can be propagated indefinitely through mitotic divisions. Upon transfer to a medium lacking nitrogen, vegetative cells differentiate into gametes. After gametes of opposite mating type have been mixed, their flagella adhere rapidly, initiating a series of complex reactions that ultimately lead to the fusion of gametes and their nuclei, chloroplasts, and probably mitochondria. The resulting zygote matures into a thick-walled zygospore, which can be induced to undergo meiosis and produce a tetrad of four haploid spores. Vegetative diploids can also be selected from matings or cell fusions mediated by polyethylene glycol. They divide mitotically and allow one to determine whether a mutation is dominant or recessive.

## 2 THREE GENETIC SYSTEMS

As higher plants, the *Chlamydomonas* species contain three genetic systems, located in the nucleocytosol, the chloroplast, and the mitochondria. In contrast to nuclear genes, which are transmitted to the offspring in a Mendelian fashion, chloroplast and mitochondrial DNAs of *C. reinhardtii* usually are transmitted uniparentally from the mating-type + and mating-type − parent, respectively. In *C. eugametos* and *C. moewusii,* both organellar genomes are transmitted uniparentally from the mating-type + parent (Table 1).

The complexity of the nuclear genome of *C. reinhardtii* has been estimated at $1 \times 10^5$ kbp. The GC content of the nuclear DNA is 64%, and there is a marked bias in the codon usage of

**Table 1** Three Genetic Systems in *Chlamydomonas*

| System | Complexity of genome: Species[a] | Complexity of genome: kbp | Number of Linkage Groups | Copy Number | Inheritance[b] |
|---|---|---|---|---|---|
| Nucleus | *C.r.* | $10^5$ | 17 | | me |
| Chloroplast | *C.r.* | 196 | 1 | 80 | $up^+$ |
| | *C.e.* | 243 | 1 | | $up^+$ |
| | *C.m.* | 292 | 1 | | $up^+$ |
| Mitochondria | *C.r.* | 15.8 | 1 | 50 | $up^-$ |
| | *C.e.* | 20.2 | 1 | | $up^+$ |
| | *C.m.* | 18.5 | 1 | | $up^+$ |

[a]*C.r., C. reinhardtii; C.e., C. eugametos; C.m., C. moewusii.*
[b]me, mendelian inheritance; $up^+$, $up^-$, uniparentally inherited from the mating type + and −, respectively.

most nuclear genes that have been sequenced. At least 17 nuclear linkage groups have been identified with numerous markers. These include auxotrophic mutations, drug-resistant markers, and mutations that directly or indirectly affect photosynthesis, flagellar function, and mating. Efforts to establish restriction fragment length polymorphism (RFLP) maps for several of the nuclear linkage groups are under way.

The chloroplast DNA of *C. reinhardtii* consists of 196 kbp circular molecules, which are arranged into 8–10 nucleoids within each chloroplast. The chloroplast genome of this alga is, therefore, larger than the plastid genomes of higher plants, which range between 120 and 160 kbp. Although the informational content of chloroplast DNA is low (about 0.2% of the cell DNA complexity), it constitutes 10 to 15% of the cellular DNA mass because there are approximately 80 copies per cell. The chloroplast genome of *C. reinhardtii* contains nearly the same set of photosynthetic and ribosomal protein genes, as well as ribosomal RNA and tRNA genes that are found in the plastid genomes of higher plants. However, although chloroplast gene arrangement is highly conserved in most higher and lower plants, this conservation does not extend to *Chlamydomonas,* where chloroplast gene order differs markedly even between different species.

Several chloroplast genes of *Chlamydomonas* are discontinuous and contain either group I or group II introns, which have also been found in a large number of organellar genomes. Many of the *Chlamydomonas* group I introns are able to self-splice in vitro and some of them are mobile at the DNA level, provided a suitable integration site is available. This process, called intron homing, has been extensively studied for the intron of the chloroplast 23*S* ribosomal RNA gene. This intron contains an open reading frame encoding a protein with double-stranded DNA endonuclease activity that cleaves the intronless allele of the 23*S* ribosomal RNA gene at the site of intron insertion. Cleavage triggers transposition of an intron copy to this target site and leads to efficient spreading of the intron to all target sites available. In particular, when an intron-containing strain is crossed to a strain lacking that particular intron, all the progeny inherit the intron. This has been found for the intron of the mitochondrial cytochrome *b* gene of *C. smithii* that is absent from *C. reinhardtii* upon crossing these two species.

Besides cis-splicing, trans-splicing has been found in the chloroplast of *C. reinhardtii.* The chloroplast *psa A* gene, encoding one of the reaction center subunits of photosystem I, consists of three exons that are widely scattered on the genome. Each exon is flanked by sequences characteristic of group II introns and maturation of the *psa A* message depends on two trans-splicing reactions. Surprisingly, intron 1 itself has a tripartite structure: the middle part of this intron is encoded by a locus that is distant from both exons 1 and 2. Another remarkable feature is that at least 14 nuclear loci are involved in the first trans-splicing reaction. Whether the unusual split structure of the *psa A* gene reflects an ancient primitive gene structure or whether it was created through DNA rearrangements remains an open question.

The mitochondrial DNA of *C. reinhardtii* consists of 15.8 kbp linear molecules which have been entirely sequenced. This genome displays several unusual features. It encodes a few polypeptides of the mitochondrial respiratory chain and only three tRNA genes, suggesting that the other mitochondrial tRNAs are encoded by nuclear genes and imported into the organelle. The genes of the two mitochondrial ribosomal RNAs are fragmented into smaller gene pieces encoding specific ribosomal RNA domains, which are intermingled with each other and with protein and tRNA genes on the mitochondrial chromosome. Whereas both mitochondrial genomes of *C. eugametos* and *C. moewusii* are circular and collinear with each other, they are considerably rearranged compared to the mitochondrial genome of *C. reinhardtii.*

## 3 RECENT TECHNICAL ADVANCES

A major breakthrough for molecular studies on *Chlamydomonas* was the establishment of reliable methods for transformation of the nuclear, chloroplast, and mitochondrial genomes. Nuclear transformation can be easily performed with cell-wall-deficient *C. reinhardtii* mutants by using the particle gun to vortex the cells with glass beads and DNA or with walled strains. Three host strains can be used: *arg 7,* deficient in argininosuccinate lyase; *nit-1,* deficient in nitrate reductase; and *Fud44,* deficient in photosynthesis. It is also possible to introduce additional genes into the nuclear genome by cotransformation. Selection is performed on medium lacking arginine, ammonium, or acetate, respectively, by transforming with the corresponding wild-type cloned genes. In most cases transformation occurs through nonhomologous recombination as the transforming DNA appears to integrate randomly into the nuclear genome. This property has been used successfully for tagging genes. In this approach new mutations are induced through the integration of the transforming DNA into nuclear genes. The bacterial vector sequences in the transforming DNA can then be used as a probe for isolating the mutated gene. Two mobile elements of *C. reinhardtii* Gulliver, which resembles classical transposable elements, and *TOC1,* which is related to retrotransposons, have been characterized and can also be used for nuclear gene tagging. The high nuclear transformation yield has also allowed gene rescue by complementing nuclear mutations with a genomic cosmid library of *C. reinhardtii.*

While nuclear transformation works rather efficiently when cloned *C. reinhardtii* nuclear genes are used as transforming DNA, it has not yet been possible to express foreign genes efficiently in this organism, even when they are fused to *C. reinhardtii* promoters and polyadenylation sites. It is not yet clear whether this difficulty is due to the biased codon usage of nuclear genes of *C. reinhardtii* or to other factors.

Chloroplast transformation of *C. reinhardtii* can be achieved with a particle gun in which cells are bombarded with DNA-coated tungsten particles. Chloroplast mutants carrying defective photosynthetic genes are usually transformed with the corresponding wild-type genes, which integrate into the chloroplast genome by homologous recombination. Alternatively, nonphotosynthetic markers can be used for selection, such as ribosomal RNA genes with mutations conferring resistance to streptomycin and spectinomycin. Resistance against these antibiotics can also be obtained with the bacterial *aad A* (aminoglycoside adenyl transferase) gene fused to a chloroplast promoter and 5′ leader region. These tools have opened the door for genetic engineering of the chloroplast genome. It is now possible to perform chloroplast gene disruptions and site-directed mutagenesis and to insert and express foreign genes (e.g., chimeric *GUS* constructs) at specific sites in the chloroplast genome.

## 4 *CHLAMYDOMONAS* AS MODEL SYSTEM

### 4.1 Function and Assembly of the Photosynthetic Apparatus

As in higher plants, the biosynthesis of the photosynthetic apparatus in *C. reinhardtii* occurs through the concerted action of genetic systems located in the nucleus and chloroplast, respectively. Several subunits of the photosynthetic complexes are encoded by the chloroplast genome and translated on 70*S* chloroplast ribosomes. The remaining subunits are encoded by nuclear genes, translated on cytosolic 80*S* ribosomes, and imported into the chloroplast compartment, where all the subunits are assembled to form functional complexes.

Because photosynthetic function is dispensable when *C. reinhardtii* cells are grown on medium containing acetate, it has been possible to isolate numerous nuclear and chloroplast mutants deficient in photosynthetic activity. These mutations fall in two major classes. The first includes mutations within genes encoding components of the photosynthetic system. The second class, which consists primarily of nuclear mutations, affects genes whose products are required for the proper expression of chloroplast genes. These factors appear to act at several posttranscriptional levels, such as chloroplast RNA stability, RNA processing and splicing, translation and, most likely, at the level of assembly of the photosynthetic complexes. Surprisingly, the number of these nuclear loci required for chloroplast gene expression is quite large, and most of their products appear to act in a gene-specific manner.

### 4.2 Function and Assembly of the Flagellar Apparatus

*C. reinhardtii* possesses two flagella located at the anterior end of the cell. They are assembled on basal bodies that are also involved in microtubule organization within the cell and correspond to the centrioles of higher eukaryotic cells. The flagellar system of *Chlamydomonas* has proven to be particularly well suited for studying microtubule assembly and function, and cell motility. This is because flagellar biosynthesis can be readily synchronized and numerous mutants affected in the function and assembly of the flagellar apparatus have been isolated. These mutants can be ordered into two major classes: those with abnormal or no motility, and those with defects in flagellar assembly.

While most of these mutations map to various linkage groups, several are linked together on the *uni* linkage group, named after the *uni* mutation that affects basal body assembly and leads to the formation of uniflagellated cells. The high degree of clustering of functionally related genes on the *uni* linkage group is unusual for eukaryotic chromosomes. Conflicting results on the intracellular location of the *uni* linkage group have been reported. It is not yet clear whether the *uni* linkage group is associated with the basal bodies, the nucleus, or both organelles.

## 5 PERSPECTIVES

*Chlamydomonas* has emerged as an attractive model system for studies of the molecular and cellular biology of eukaryotic photosynthetic cells. This alga can be manipulated with relative ease at the biochemical, molecular, and genetic levels. *C. reinhardtii* is, at present, the only organism in which nuclear, chloroplast, and mitochondrial transformation is feasible. *Chlamydomonas* will remain valuable for studies on flagellar function, assembly, and cell motility. Because the photosynthetic apparatus of *C. reinhardtii* is similar to its homologue in higher plants, this alga is also an excellent system for studying the biogenesis and function of the photosynthetic complexes and for investigating in general terms the interactions between the nuclear, chloroplast, and mitochondrial genomes. Other areas of research for which *Chlamydomonas* is uniquely suited include phototoxis, cell wall synthesis, mating reactions and gametogenesis, and the metabolism of carbon, nitrogen, and sulfur.

*See also* Genetic Analysis of Populations; Mitochondrial DNA; Origins of Life, Molecular Biology of.

### *Bibliography*

Boynton, J. E., Gillham, N. W., Newman, S. M., and Harris, E. H. (1992) Organelle genetics and transformation of *Chlamydomonas*. In *Plant Gene Research, Cell Organelles*, R. G. Herrmann, Ed., pp. 3–64. Springer-Verlag, New York.

Dutcher, S. K. (1989) Linkage group XIX in *Chlamydomonas reinhardtii* (Chlorophyceae): Genetic analysis of basal body function and assembly. In *ALgae as Experimental Systems*, pp. 39–53. Liss, New York.

Harris, E. H. (1989) *The* Chlamydomonas *Sourcebook*. Academic Press, San Diego, CA.

Lefebvre, P. A., and Rosenbaum, J. L. (1986) Regulation of the synthesis and assembly of flagellar proteins during regeneration. *Annu. Rev. Cell Biol.* 2:517–546.

Luck, D. J. (1984) Genetic and biochemical dissection of the eucaryotic flagellum. *J. Cell Biol.* 98:789–794.

Rochaix, J. D. (1992) Post-transcriptional steps in the expression of the chloroplast genome of *Chlamydomonas reinhardtii*. *Annu. Rev. Cell Biol.* 8:1–28.

# Chromatin Formation and Structure

*Timothy J. Richmond and Fritz Thoma*

### *Key Words*

**Chromatin Fiber** The higher order structure formed by the condensation of a nucleosomal filament.

**Chromatosome** A particle containing the entire histone protein complement for one nucleosome and 166 DNA base pairs arranged in two complete turns of superhelix, but lacking the DNA that links two successive nucleosomes in the nucleosomal filament.

**Histone Octamer** The protein core of a nucleosome, composed of two each of the four different core histone proteins: H2A,

H2B, H3, and H4 organized as a single $(H3)_2(H4)_2$ tetramer and two (H2A)(H2B) dimers.

**Linker Histone** The histone protein H1 or variant, which binds to the outside of the DNA superhelix of the nucleosome.

**Nucleosome** The fundamental, nucleoprotein repeating unit of chromatin, composed of the histone octamer, a linker histone, and 170 to 240 DNA base pairs.

**Nucleosome Core Particle** The greater part of the nucleosome, containing 146 DNA base pairs wrapped around the histone octamer in 1.8 superhelical turns.

---

DNA, the genetic material of higher organisms from yeast to mammals, is organized in the cell nucleus as a nucleoprotein complex called chromatin. If all the DNA molecules in the human genome were laid in a line with the Watson–Crick B-form structure, they would span a distance exceeding 1 meter. Nevertheless, the two complete genomes of a diploid cell, each with $3.9 \times 10^9$ nucleotide base pairs, fit into a cell nucleus of only 1 to 2 μm in diameter. This packaging of DNA into chromatin is remarkable not only because of the $10^6$-fold compaction that takes place, but because it also allows access to the DNA double helix for use in gene expression, replication, recombination, and repair.

Chromatin is organized in several structural levels: nucleosomes as elemental subunits, nucleosomal filaments condensed into chromatin fibers, and ultimately, during cell division, the fibers assembled into recognizable chromosomes. Chromatin contains roughly equal masses of DNA and protein, where the histone proteins account for more than 90% of the total protein and are the principal component responsible for giving chromatin its structure. Considering the near universal coverage of the genomic DNA in nucleosomes and their essential role in gene activity, nucleosome position with respect to DNA sequence appears to be of central importance. The chromatin-associated, nonhistone proteins are numerous and diverse in structure and function. Some, such as the nonspecific high mobility group and gene-specific regulatory proteins, temper the chromatin fiber to prepare it for participation in various life functions, while others, including RNA and DNA polymerases, conduct the required processes.

## 1 HISTONE PROTEINS

The histone proteins are relatively small, basic proteins with molecular masses of 10 to 21 kDa. Each of the four types of core histones, H2A, H2B, H3, H4, contains a region of its amino acid sequence containing 4 α-helices, flanked at the N-terminus, and for H2A also at the C-terminus, by an apparently unstructured "tail" sequence. Pairs of histone subunits interact to form stable complexes as H2AH2B dimers and H3H4 tetramers. Under conditions of high salt concentration, two copies of the lysine-rich H2AH2B dimer and one of the H3H4 tetramer assemble into a histone octamer or core. The unstructured terminal regions are not required for these stable interactions, but they do contain the sites of posttranslational modifications found in vivo, including methylated and acetylated lysine, phosphorylated serine, and a lysine linkage to the small protein ubiquitin. Histone protein variants expressed from species and cell-type-specific genes also show their greatest differences in these tail regions. While the core histones are extremely highly conserved throughout eukaryotic species, the fifth histone H1 is much more highly variant, to the extent that it has not been observed in some of the lower eukaryotes such as yeast. Histone H1 consists of a short N-terminal tail of about 20 amino acids, a central globular domain of about 80 amino acids, and a highly positively charged C-terminal tail, which contains approximately 100 amino acids and also undergoes modification. The globular domain of H1 is highly homologous in its three-dimensional structure to the sequence-specific, DNA-binding catabolic activator protein from bacteria as well as to a liver-specific transcription factor. H1 does not interact stably with the core histones in the absence of DNA.

## 2 NUCLEOSOMES

Histone octamers are associated repeatedly with every 160 to 240 bp of DNA in chromatin to form arrays of nucleosomes. The exact periodicity, which depends on species and cell type, represents an average over all DNA sequences for which local deviations can be significant. The mass of the nucleosome is nearly equally divided between DNA and protein, and can be viewed as a structure assembled in several steps.

1. One histone H3H4 tetramer binds to the central 6 turns of the DNA double helix.
2. Two histone H2AH2B dimers bind on opposite faces of the tetramer–DNA complex, and each organizes the next 4 turns of DNA. This particle, which can be isolated as a nucleosome core particle from intact chromatin by DNA cleavage with the enzyme micrococcal nuclease, contains 146 base pairs of DNA organized in 1.8 left-handed superhelical turns. The core histone proteins form a helical ramp with the order H2A-H2B-H4-H3-H3-H4-H2B-H2A, which guides the path of the DNA.
3. The next 10 base pairs of DNA at each end of the DNA superhelix, beyond that seen in the core particle, converge to complete 2 full turns of superhelix. These DNA extensions, presumably in conjunction with the middle of the central superhelix turn, form the binding pocket for the globular domain of histone H1. This particle, the chromatosome, can also be isolated after micrococcal nuclease digestion. Tail regions of histones H2A and H3 may also bind to these segments of DNA.
4. The remaining DNA links the nucleosome to its neighbors in the nucleosome filament and perhaps is associated with the N-terminal and long C-terminal tail regions of histone H1.

Overall, the nucleosome is a disk like structure approximately 57 A in height along the superhelix axis and 110 Å in diameter, with the linker DNA entering and exiting at the point of histone H1 binding on the periphery (Figure 1; see also color plate 6). The nucleosome core particle can, in principle, express an overall two-fold molecular symmetry, as has been seen for the histone octamer, but it may also show inherent or DNA sequence-dependent asymmetry. The addition of a single histone H1 molecule necessarily makes the nucleosome an asymmetric particle.

Because the DNA is spooled around the outside of the histone octamer, the DNA has an inaccessible inner surface facing the histone core and adjacent turn of DNA superhelix, and an outer surface accessible from solution. The inner and outer surfaces running along the length of the DNA define its rotational setting in the particle. While the outer surface is accessible to proteins, such as nucleases or regulatory factors, the inner surface is protected

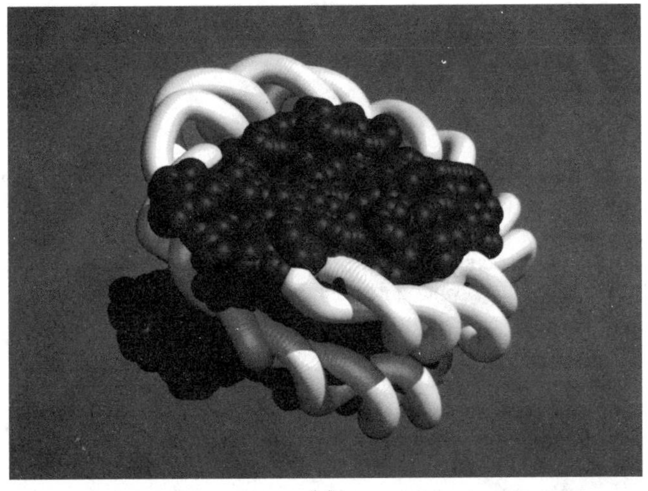

(a)

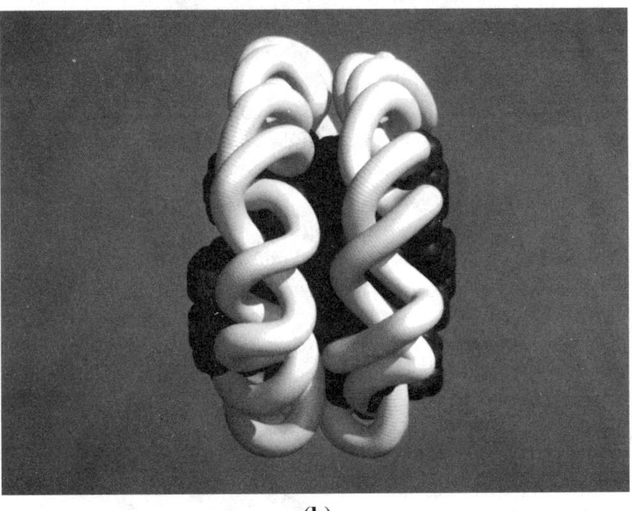

(b)

**Figure 1.** The X-ray crystal structure of the nucleosome core particle at 7 Å resolution, showing that the 146 bp of DNA in the particle are wrapped in 1.8 turns of a flat superhelix around the histone octamer. The electron density of the core histone proteins is represented by overlapping green spheres. The paths of the phosphodiester chains of the DNA double helix are shown as intertwined gold tubes. (a) View oblique to both the DNA superhelix axis and the molecular dyad axis. The center of nucleosomal DNA, the binding site of the linker histone, is surrounded above and below by protrusions from the histone core. (b) View down the molecular pseudodyad axis, passing between the gyres of the DNA superhelix. [Adapted from T. J. Richmond, J. T. Finch, B. Rushton, D. Rhodes, and A. Klug, *Nature* 311:532 (1984).] (See color plate 6.)

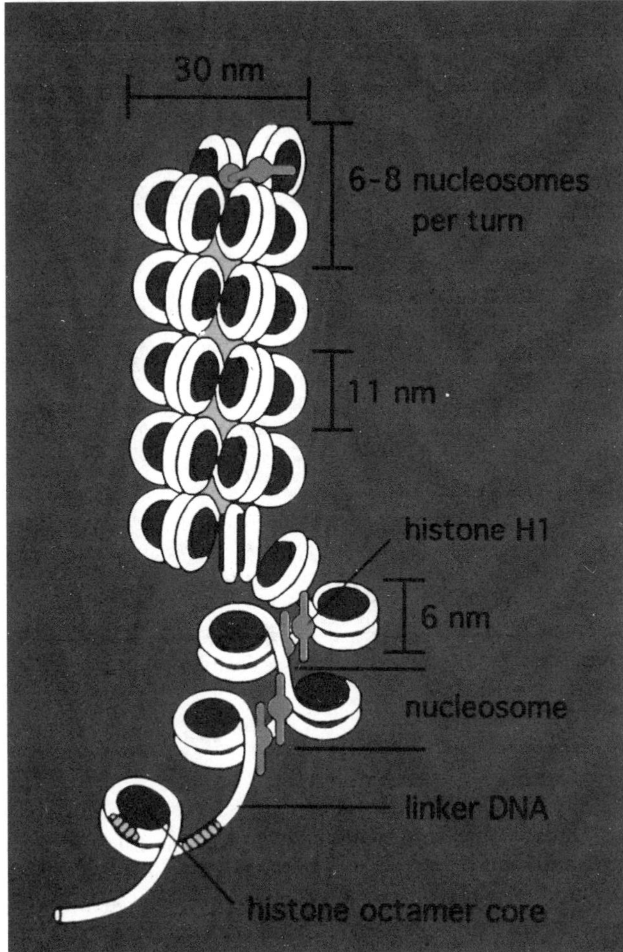


**Figure 2.** Model of the chomatin fiber built from nucleosomes in a higher order helix. The disklike nucleosomes associate to form a higher order helix with 6 to 8 nucleosome units per turn. The colors of the DNA and core histone proteins are the same as in Figure 1. The interactions stabilizing the fiber come from the histone H1, shown in red, thought to be at the center of helix, and possibly also from the histone tail regions. The accessibility of sites (striped patches) for other DNA-binding proteins depends on the different levels of chromatin organization. [Adapted from F. Thoma, T. Koller, and A. Klug, *J. Cell Biol.* 83:403 (1979).] (See color plate 7.)

by interactions with the histone proteins. This property is illustrated by the pattern of DNase I attack, which demonstrates that the enzyme can cleave the DNA phosphodiester backbone only when it faces outward, producing cuts in increments of approximately 10.4 base pairs. The intrinsic curvature and bendability of a particular DNA sequence appear to govern the rotational aspect of the DNA orientation. Given a single rotational orientation, the translation position of the histone octamer along the concave inner face of the DNA may show multiple settings, occurring in steps of 10 base pairs or one turn of the double helix. The DNA minor groove always faces outward at the center of the nucleosomal DNA, at the site where the apparent dyad symmetry axis passes through the double helix. The DNA in the nucleosome core particle is not smoothly folded in the superhelix, but displays four points of tight bending, about 1 and 4 helix turns from the center, where the radius of curvature is about 20 Å compared to that of 46 Å over the whole particle. In addition, the DNA is generally overwound compared to its nominal value of 10.5 bp in solution to 10.0 bp per turn in the nucleosome, with the exception of the central 2 to 3 turns, which are underwound to 10.7 bp.

## 3 CHROMATIN FIBER

The nucleosome filament, which at low ionic strength can be visualized in the electron microscope as a string of beads connected in a "zigzag" conformation, is condensed to a "30 nanometer" diameter fiber at physiological salt concentration. This nucleosome higher order structure or chromatin fiber is the dominant structure of chromatin in nuclei and the substrate on which all DNA-dependent reactions are initiated. Studies using a variety of techniques indicate that the isolated nucleosome core particle is very similar structurally to the nucleosome core of the nucleosome in these fibers. In the simplest model, the higher order structure is a low pitch, single start helix with 6 to 8 nucleosomes per 10 nm of fiber length; the superhelix axis of each nucleosome is oriented nearly perpendicular to the axis of the fiber, but parallel to the path of the higher order helix in the fiber (Figure 2; see also color plate 7). The connectivity and location of the linker DNA in the 30 nm fiber is unknown, but is envisioned to occupy the central cylinder along the fiber axis. This model also places in the central channel the H1 linker histone, which is requisite for fiber formation, and its variants. Additional stability may be provided by the core histone tails, which although they are not required for nucleosome formation, could interact with adjacent nucleosomes and linker DNA. While the structure of the nucleosome core particle is highly conserved in evolution, the linker DNA varies between individual nucleosome pairs along the same filament. The higher order structure must accommodate this variation and may contribute to the observed irregularity of the structure.

## 4 CHROMATIN ASSEMBLY

During DNA replication and prior to segregation of homologous chromosomes in cell division, new chromatin in assembled. DNA replication begins by DNA strand separation at replication origins, which are defined sequences in yeast and some viruses but less easily characterized in higher eukaryotes. The replication machinery generates a "DNA replication fork" that divides the DNA into leading and lagging strands, which show continuous and discontinuous DNA synthesis, respectively, and result in the creation of two double-stranded daughter molecules. Histone octamers most likely dissociate from the chromatin near the replication fork. New sets of histone proteins are synthesized in the cytoplasm and enter the nuclei, mixing with the free histone proteins already there. Reassembly of nucleosomes along the daughter molecules occurs within a few hundred base pairs behind the replication fork. New and old histone proteins appear to distribute randomly between leading and lagging strands, although at least the H3H4 tetramer associations are maintained. Several chromatin assembly factors that do not remain in association in the final product may play a role in the generation of chromatin. For example, chaperonin proteins exist which can combine with the histone proteins while they are displaced from DNA. Specifically, N1 combines with the H3H4 tetramer and nucleoplasmin with the H2AH2B dimer. During chromatin assembly, other proteins may compete with the histone proteins for DNA binding and thereby establish specific chromatin structures required for initiation of transcription or replication.

Nucleosomes can be assembled in vitro by mixing core histones with DNA in high salt and slowly reducing the ionic strength by various protocols. Histones H3 and H4 bind first as a tetramer while two H2AH2B dimers complete the formation of the nucleosome core. While in vitro reconstitution of nucleosome core particles on single particle length DNA is highly efficient, reassembly of histones on longer DNA fails to generate the natural spacing of nucleosomes. By using nuclear extracts that include chromatin assembly factors, a more natural spacing of nucleosomes along the DNA can be obtained.

## 5 NUCLEOSOME POSITIONING

While early studies suggested that nucleosomes might be arranged randomly with respect to DNA sequence, more recent evidence has established that nucleosomes can occupy preferred or precise positions, particularly in gene promoter regions. The parameters that affect nucleosome positioning have been studied by in vitro reconstitution of the histone octamer with defined sequence DNA, by sequence analysis of nucleosomes purified from cell nuclei, and by analysis of chromatin formed on DNA plasmid constructs introduced into yeast. So far, three parameters have been found to contribute to nucleosome positioning.

1. DNA sequences that most facilely accommodate the structural distortions in the path of the superhelix induced by the histone octamer are superior in nucleosome formation and lead to positioned nucleosomes in vitro and in living cells. Hence, nucleosome position and stability will be affected by sequence-dependent properties of DNA, such as its intrinsic curvature and bendability, and the arrangement of segments of DNA with differing properties over the length of DNA in the nucleosome.
2. Nucleosomes can be positioned by boundary effects such as the presence of tightly bound, sequence-specific DNA-binding proteins. These nonhistone proteins may localize nucleosomes by direct recognition of a histone protein or simply by excluding nucleosome formation from a particular sequence. The presence of boundaries may have long-distance effects on the arrangement of nucleosomes: either the nucleosomes are laid down during replication on a next-available-site basis or nucleosomes pack between barriers to maximally fill the available space.
3. The formation of nucleosomal higher order structure or chromatin fibers can also modulate the position of nucleosomes along the DNA sequence. Chromatin in living cells is a dynamic structure that depends on multiple variables to adopt the forms that accommodate its functions.

## 6 NUCLEOSOMES AND TRANSCRIPTION

In eukaryotic cells, three different RNA polymerases and a multitude of transcription factor proteins execute the process of DNA transcription to RNA. RNA polymerases I and III make transfer RNA and ribosomal RNA, and RNA polymerase II synthesizes the messenger RNA molecules that are translated into proteins. For "*Pol* II" genes, a set of general transcription factor proteins assemble on the promoter sequence adjacent to and upstream of the polymerization initiation site and direct RNA polymerase II to start basal level transcription. Activation and repression of *Pol* II genes is regulated by additional transcription factor proteins that bind specifically to enhancer sequences in, adjacent to, or even far away from the gene promoter region. Hence, to control gene expression, gene regulatory factors are able to communicate over long distances in the DNA sequence with the general transcription factor complex assembled near the site of transcription initiation. It appears that both gene specific and gene general factors bind stably to their respective sites, and therefore DNA loop formation must accompany their interaction. In this case, the chromatin structure assembled on the intervening DNA segment has the potential to facilitate or hinder gene activation.

It is firmly established that histones act as nonspecific repressors of transcription and that nucleosome formation is a crucial determinant of basal transcription levels. General repression by histone proteins, which cover most of the genomic DNA and severely restrict access to nonhistone factors, is not surprising. Nucleosomes may be involved in specific gene repression as well. The histone octamer may bind to certain promoter sequences essential for gene activation (e.g., the TATA box) and cover them under the guidance of repressor proteins. For gene activation, it was commonly envisioned that nucleosomes would hand over their DNA to transcription factors when challenged, and this step was thought to explain the hypersensitivity to nuclease digestion at various regulatory sequences. However, the details of such mechanisms as they emerge are increasingly intricate, and in some cases transcription factors apparently collaborate with the histone proteins to cause activation. Alternatively, some sites may be preselected by binding of the appropriate nonhistone factors during DNA replication, but the details of such processes are yet to be elucidated.

During the elongation of the RNA transcript by RNA polymerase in vivo, chromatin unfolds and the histone proteins in front of the polymerase dissociate. Behind the polymerase, the histone proteins reassemble on the bare DNA. This process may be enhanced by the structure of the transcription complex itself. The self-assembly of nucleosomes and the self-assembly of the chromatin fiber are apparently in competition with transcription, so that, depending on the number and activity of RNA polymerase molecules bound to the template, the chromatin will be at various stages of completeness. Posttranslational modifications, particularly acetylation, are likely to weaken the association of the histone proteins with DNA, making the DNA more accessible to polymerase.

*See also* DNA Replication and Transcription; Histones; Protein Folding.

### *Bibliography*


Grunstein, M. (1990) Histone function in transcription. *Annu. Rev. Cell Biol.* 6:643.

McGhee, J. D., and Felsenfeld, G. (1980) Nucleosome structure. *Annu. Rev. Biochem.* :1115.

Pederson, D. S., Thoma, F., and Simpson, R. T. (1986) Core particle, fiber, and transcriptionally active chromatin structure. *Annu. Rev. Cell Biol.* 2:49117.

van Holde, K. E. (1988) *Chromatin.* Springer-Verlag, New York.


## Chromatography: *see* type, e.g., HPLC of Biological Macromolecules; Pharmaceutical Analysis, Chromatography in.

# Circular Dichroism in Protein Analysis

*Nicholas C. Price*

### *Key Words*

**Chirality** The asymmetry (optical activity) of a molecule which is not superimposable on its mirror image.

**Circular Dichroism (CD)** The differential absorption of the left and right circularly polarized components of plane-polarized radiation by a chromophore that is associated with a chiral center.

**Ellipticity** A measure of the difference in absorption of the left and right circularly polarized components of plane-polarized radiation; defined as $\tan^{-1}(b/a)$, where $b$ and $a$ are the minor and major axes, respectively, of the resulting elliptically polarized light.

**Mean Residue Weight** The average molecular mass (weight) of the repeating unit of a polymer.

**Secondary Structure Analysis** The analysis of the CD spectrum of a protein in the far UV to estimate the proportions of different types of secondary structure present.

Circular dichroism (CD) refers to the differential absorption of the left and right circularly polarized components of plane-polarized radiation. CD is exhibited by chromophores that either are intrinsically chiral (optically active) or are placed in an asymmetric environment. The principal types of secondary structure adopted by proteins give rise to characteristically different CD spectra in the far UV region (< 240 nm); this allows one to make rapid quantitative estimates of the secondary structure contents of proteins with very modest amounts of material. The CD spectrum in the near UV (260–320 nm) can give information on the tertiary structure of a protein. CD is also widely used to study the folding, stability, and binding properties of proteins.

## 1 INTRODUCTION

### 1.1 What Is Circular Dichroism?

Circular dichroism (CD) refers to the differential absorption of the left and right circularly polarized components of plane-polarized radiation. This effect is observed when the chromophore is chiral (optically active), either intrinsically or because it has been placed in an asymmetric environment.

### 1.2 Origin of the CD Effect

As shown in Figure 1a (**I**), plane-polarized radiation can be considered as being composed of the two circularly polarized components. In practice, the separation can be achieved by passing the radiation through a *modulator* [either an electro-optic crystal such as ammonium dihydrogen phosphate (Pockels cell) or a piezoelastic crystal such as quartz] subjected to an alternating (50 kHz) electric field. The modulator will transmit the two circularly polarized components of the radiation in turn. When the two components are combined without being altered, radiation polarized in the original plane is regenerated. If, however, one of the components is absorbed to a greater extent than the other by passage through the sample, the resulting radiation will now be elliptically polarized; that is, the resultant will trace out an ellipse, as in **II** (Figure 1a).

The CD instrument (or spectropolarimeter) consists essentially of a light source (usually a xenon arc), a prism (to produce the plane-polarized radiation), a monochromator (to select the chosen

(a)

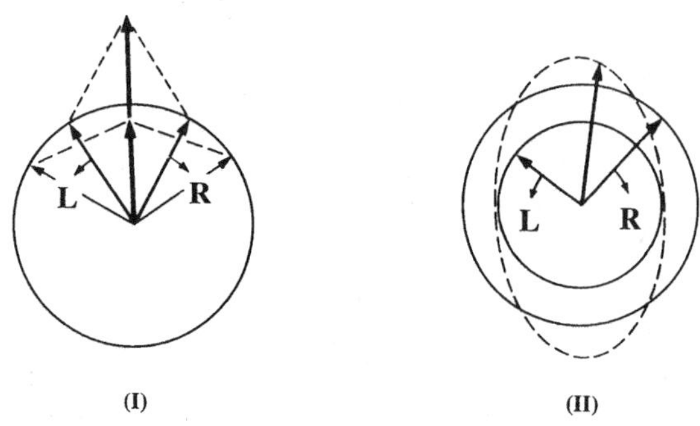


(b)

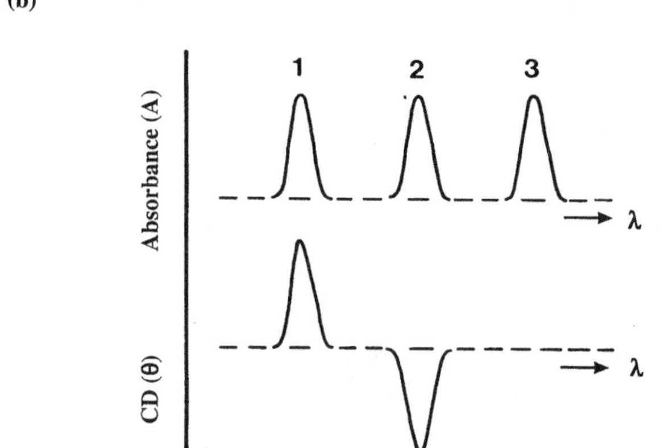


**Figure 1.** Origin of the CD effect. (a) The left (L) and right (R) circularly polarized components of plane-polarized radiation: **(I)** the two components have the same amplitude and when combined regenerate plane-polarized radiation; **(II)** the components are of different amplitudes and the resultant (dashed line) is elliptically polarized. (b) The relationship between absorption and CD spectra. Band 1 has a positive CD peak with L absorbed more than R; band 2 has a negative CD peak; band 3 is not chiral.

wavelength), a modulator (to resolve the components of the polarized radiation), and a photomultiplier (to detect the radiation transmitted after passage through the sample). The resulting dichroism will be displayed as either the difference in absorbance ($\Delta A = A_L - A_R$) or as the ellipticity in degrees ($\theta$); $\theta = \tan^{-1}(b/a)$, where $b$ and $a$ are the minor and major axes of the resulting ellipse (**II** in Figure 1a). There is a simple numerical relationship between $\Delta A$ and $\theta$, namely:

$$\theta = 32.98\Delta A$$

where $\theta$ is in degrees.

In practice in most biological work, the observed dichroism is very small, with ellipticities of the order of 10 millidegrees, corresponding to differences in absorbance ($A_L - A_R$) of the order of $3 \times 10^{-4}$ absorbance unit. Considerable care must therefore be taken to ensure that meaningful data are obtained (Section 2.1). A CD spectrum is obtained when the dichroism is measured as a function of wavelength. Figure 1b shows the relationship between the absorption and CD spectra. Further background information can be found in the reviews by Adler et al., Bayley, and Timasheff.

### 1.3 Chromophores that Can Contribute to the CD Spectrum of Proteins

Proteins possess many different chromophores. The aromatic amino acid side chains (Phe, Tyr, and Trp) absorb in the range 250–290 nm; the side chain of His has a broad absorption band around 245 nm and the disulfide bond of cystine (if present) absorbs over the range 240–290 nm. By far the strongest absorption in the UV, however, arises from the amide (peptide) bonds. These absorb in the range from 180–240 nm (involving a weak but broad $n \rightarrow \pi^*$ transition centered around 210 nm, and an intense $\pi \rightarrow \pi^*$ transition at about 190 nm). In addition, nonprotein chromophores such as flavin or heme may be present; each of these may give rise to CD signals in the appropriate region of the spectrum corresponding to its absorption, and such signals can be used to monitor the precise environment of the chromophore.

### 1.4 Type of Information Available from CD

From the far-UV CD spectrum ($\leq$ 240–190 nm) it is possible in most cases to obtain quantitative estimates of the secondary structure content of proteins. This is because each of the different regular forms of secondary structure found in proteins ($\alpha$-helix, $\beta$-sheet, etc.) generates a distinct pattern of optical activities associated with the peptide bond resulting in characteristic CD spectra (Section 3.1).

The overall folding of the polypeptide chain (i.e., the tertiary structure) places aromatic amino acid side chains in optically active environments; these will give rise to signals in the near UV (260–320 nm). The CD spectrum from the near UV can serve as a sensitive fingerprint of the structure and can be of great value, for example, in screening the structures of a range of mutants produced by site-directed mutagenesis (Section 3.2).

Changes in the CD signals can be used to monitor the rates of structural transitions in proteins. Considerable success has been achieved in recent years studying the rates of processes occurring on the time scale of 5 ms or more, using stopped-flow CD methods, and this work has been of great value in examining the early stages of protein folding (Section 3.3).

The CD signals can be used to monitor perturbations in the structure caused, for example, by structure-disrupting agents (e.g., guanidinium chloride, urea) or by the binding of ligands, allowing estimates to be made of the stability of the native conformation and the binding parameters, respectively (Sections 3.4 and 3.5).

### 1.5 Strengths and Weaknesses of CD as a Structural Technique

X-Ray crystallography and NMR spectrometry are capable of yielding highly detailed structural information on proteins. By contrast, CD can give only overall descriptions of structure, or information about the environment of a particular chromophore (e.g., a heme group in a protein). CD, however, is much less demanding than the other two techniques in terms of time and amount of sample required, and this is its principal advantage. A good far-UV CD spectrum can be obtained within 30 minutes using as little as about 100 μg of protein (Section 2.3) whereas NMR spectrometry requires several hours if not days of instrument and data analysis time and tens of milligrams of sample in a concentrated solution; X-ray work is even more demanding. CD is thus likely to play a very important role when the amounts of time and sample are limited or when it is desired to select particular samples for more detailed study. CD also offers the possibility of examining the extent and rate of structural changes; this is much more difficult using the high resolution techniques. Thus CD should be seen as complementary to, rather than competitive with, the more detailed structural techniques.

## 2 PRACTICAL ASPECTS

### 2.1 How to Obtain Reliable Data

The major problems in obtaining reliable data in CD can be divided into those that are due to the instrument and those that arise from the nature of the sample.

#### 2.1.1 The Instrument

The instrument should be regularly calibrated as a check on both the wavelength and the amplitude of the signal. The most commonly used substance is (1*S*)-(+)-10-camphorsulfonic acid (CSA), the CD spectrum of which shows a maximum at 290.5 nm ($\Delta\varepsilon = 2.3$ $cm^{-1}$ $M^{-1}$) and a minimum at 192.5 nm ($\Delta\varepsilon = -4.72$ $cm^{-1}$ $M^{-1}$). Thus the ellipticities for a 2.59 m*M* solution of CSA in a cell of pathlength 10 mm would be 201.8 and −403.5 mdeg at 290.5 and 192.5 nm, respectively.

In view of the small size of the dichroism signal for most biological samples, various instrumental parameters, including the time constant, scan rate, number of scans, and spectral bandwidth, should be optimized. The *time constant* is kept relatively high to allow the accumulation of several hundreds or thousands of measurements of the dichroism. Typical values are 0.25 or 2 seconds when signals of the order of 100 or 10 mdeg, respectively, are being measured. (Of course, considerably shorter time constants—e.g., 0.25 ms—are used in stopped-flow CD work; this leads to high levels of noise.) The *scan rate* should not be too fast, otherwise distortions in the position and size of bands will result. Generally the product of the scan rate and the time constant should not exceed 0.5 nm: that is, for a time constant of 2 seconds, the scan rate should not exceed 0.25 nm/s (i.e., 15 nm/min). A *number of scans* can be accumulated to improve the signal-to-noise ratio (the ratio is proportional to the square root of the number of scans). Increasing the *spectral bandwidth* by increasing the slit width will allow more radiation to fall on the photomultiplier and thus reduce noise. To avoid unreasonable loss of wavelength resolution, however, the bandwidth should be kept low (generally < 2 nm).

Correction for any signals due to buffers and/or cuvettes should be made by running appropriate blanks.

#### 2.1.2 The Sample

For meaningful work, samples of proteins should be homogeneous and solutions should be clarified by centrifugation or passage through a 0.2 μm filter. Significant quantitative errors in calculating the molar values of the dichroism (Section 2.3) can arise from uncertainties in determining the protein concentration. Many of the routinely used procedures (Lowry, dye binding, etc.) rely on comparisons with standard proteins (e.g., bovine serum albumin) and may thus be inaccurate. The most reliable methods are those that can be calibrated against absolute values (e.g., $A_{280}$ measurements calibrated with reference to dry weight or to amino acid analysis data).

Problems arising from the sample are usually due to excessive absorbance from other components in the sample. Ideally CD signals should be measured when the total absorbance of the sample is less than about 1 absorbance unit, to avoid excessive noise and consequent artifactual readings, particularly at low wavelengths. The signal-to-noise ratio will be improved if the absorbance of the component of interest is as high as possible compared with that of the other components. For work in the far UV, it is essential to flush the sample compartment with $N_2$ (≥ 10 L/min) to remove $O_2$, which absorbs strongly below 200 nm. Phosphate, borate, and low molarity (20 m*M*) Tris buffers have relatively low absorbances above 190 nm in cells of pathlength 0.1 cm or less. However, most buffers that are appropriate at lower pH values—for instance, citrate or acetate or zwitterionic buffers (MOPS, MES, etc.)—have strong absorbances below 200 nm unless present at low concentrations (≥ 10 m*M*). Chloride ions absorb strongly below 195 nm; thus NaCl is not recommended as a component of a buffer for CD work; $Na_2SO_4$ is much more suitable to maintain ionic strength. Components such as EDTA and DTT, which are often added as stabilizers, do not cause problems at low (≥ 1 m*M*) concentrations. Denaturants such as urea or guanidinium chloride (even of the highest purity available) absorb strongly below 210 nm when present at the concentrations usually employed (6–8 *M*).

*In all cases the best procedure is to run a solvent blank (i.e., not containing the peptide or protein) first, to check that absorbance is not excessive. If necessary, the solvent or buffer may need to be diluted, or a shorter pathlength cell used.*

### 2.2 Amounts of Sample Required

Typical amounts of protein required can be judged from the need to keep the absorbance less than about one unit. Typical pathlengths used are in the range 0.01–0.05 cm and protein concentrations in the range 0.2–1 mg/mL. The volume of sample required can range from 1 mL down to 0.05 mL depending on the type of cell being used, though the smaller volumes used in demountable cells generally are not recoverable. Thus the amount of protein required to obtain a far-UV CD spectrum can be as little as 10 μg, but generally 100–500 μg is preferred, to explore optimum conditions.

The signals in other parts of the spectrum (near UV, visible, near IR) are comparatively much weaker than those in the far UV. For such measurements, it would be fairly typical to use a protein concentration of 0.5–2 mg/mL and a pathlength of 0.5–2 cm. The amounts of protein required are thus likely to be of the order of several milligrams.

### 2.3 Units

As mentioned in Section 1.2, CD data can be reported in units corresponding to absorbance or to ellipticity. For small, monomeric molecules, the molecular weight (mass) would be used in expressing the results in molar terms. In the cases of biological polymers, molar values usually refer to the molarity of the repeating unit, using the mean residue weight (MRW)—that is, the molecular weight (mass) divided by the number of repeating units. For proteins the MRW is usually about 110, but it can deviate significantly from this value if the amino acid composition is very unusual or if the protein undergoes extensive posttranslational modification (e.g., a high degree of glycosylation). The molar values of the dichroism are used in the analysis of secondary structure (Section 3.1).

When CD data are expressed in terms of absorbance, the units will be those of the *difference in molar absorbance* ($\Delta\varepsilon = \varepsilon_L - \varepsilon_R$), that is, reciprocal centimeters divided by concentration ($cm^{-1}\ M^{-1}$).

In terms of ellipticity, the *molar ellipticity* ($[\theta]_\lambda$) at a wavelength $\lambda$ is quoted in units of degree-square centimeters per decimole ($deg.cm^2.dmol^{-1}$) and is given by:

$$[\theta]_\lambda = \frac{\text{MRW}\ \theta_\lambda}{10dc}$$

where $\theta_\lambda$ is the observed ellipticity, $d$ is the pathlength (cm), and $c$ is the concentration (g/mL). For small, monomeric molecules, the molecular weight (mass) would be used in place of MRW.

The numerical relationship between values in the two sets of units is:

$$[\theta] = 3298\Delta\varepsilon$$

## 3 APPLICATIONS OF CD TO THE ANALYSIS OF PROTEINS

### 3.1 Protein and Peptide Secondary Structure

As shown in Figure 2, the different forms of regular secondary structure found in proteins and peptides exhibit distinct CD patterns in the far UV. Most recent approaches to the problem of establishing the contributions of the various structures from the CD spectrum have been based on sophisticated fitting procedures involving data sets of spectra of well-characterized proteins of known secondary structure. The various algorithms used in data analysis are available from the respective authors. One of the most widely used methods of analysis, the CONTIN procedure of Provencher and Glöckner, involves the direct analysis of a CD spectrum as a linear combination of the CD spectra of 16 proteins whose secondary structures have been determined by X-ray crystallography to high resolution. When CD data over the range of 190–240 nm were used, the α-helix and β-sheet contents could be estimated fairly reliably; however the contributions of β-turns and remainder (random structure) were much less well estimated. Although more structural information is available if the CD spectrum can be recorded down to 178 nm, it is often extremely difficult in practice to obtain reliable data below 190 nm because of interference by other components. It is however possible to make fairly reliable estimates of the α-helix contents of proteins and peptides from the sizes of the CD signals at 208 or 222 nm:

$$\%\ \alpha\text{-helix} = \frac{[\theta]_{208} + 4000}{-29{,}000} \times 100\%$$

$$\%\ \alpha\text{-helix} = \frac{[\theta]_{222} - 3000}{-39{,}000} \times 100\%$$

Complications can arise in the analysis of far-UV CD data when aromatic side chains and disulfide bonds make significant contributions in this spectral region, as in the case of lysozyme. Sections 3.1.1 and 3.1.2 present examples of the analysis of secondary structure of peptides and proteins.

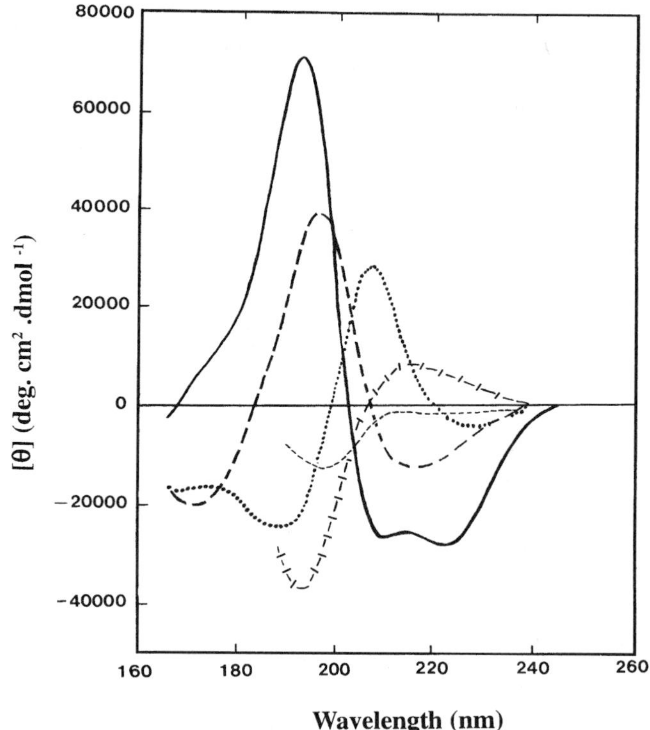


Figure 2. Far-UV CD spectra associated with various types of secondary structure: ———, α-helix; — — — —, antiparallel β-sheet; · · · ·, type I β-turn; -|-|- left-handed extended $3_{10}$ helix; - - - -, irregular structure.

#### 3.1.1 Amyloid Peptides

The A4 or β-peptides (39–43 amino acids in length) are the principal proteinaceous component of amyloid deposits in Alzheimer's disease. CD has been used to study the structure of synthetic amyloid β-peptides (1–39 and 1–42) as well as fragments (1–28 and 29–42) under a variety of experimental conditions (pH, solvent composition, temperature, concentration). In aqueous solution both 1–39 and 1–42 form mainly β-sheet structures, but in the presence of trifluoroethanol or hexafluoroisopropanol, α-helical structure is favored. Adoption of the β-structure, which is particularly pronounced at high concentrations of the full length peptides at pH values between 4 and 7 and in the case of the 29–42 peptide under all conditions, leads to the formation of aggregates. The hydrophobic region between 29 and 42 may thus represent the site of nucleation of the β-pleated sheet structure found in the amyloid deposits in the diseased state.

#### 3.1.2 Alcohol Dehydrogenase

CD has been used to explore the secondary structural relationships between the two main classes of alcohol dehydrogenases (ADHs), which appear to be unrelated at the amino acid sequence level. Long chain ADHs (350 amino acids) exhibit a preference for primary alcohols, have a requirement for $Zn^{2+}$ ions, and show 4-*pro-R* stereospecificity of hydride transfer. By contrast, the short chain ADHs (250 amino acids) prefer secondary alcohols, have no requirement for metal ions, and display 4-*pro-S* stereospecificity. Some long-chain enzymes (e.g., horse liver ADH) have been characterized in great structural and kinetic detail. Short chain enzymes such as the ADH from *Drosophila* are much less well explored.

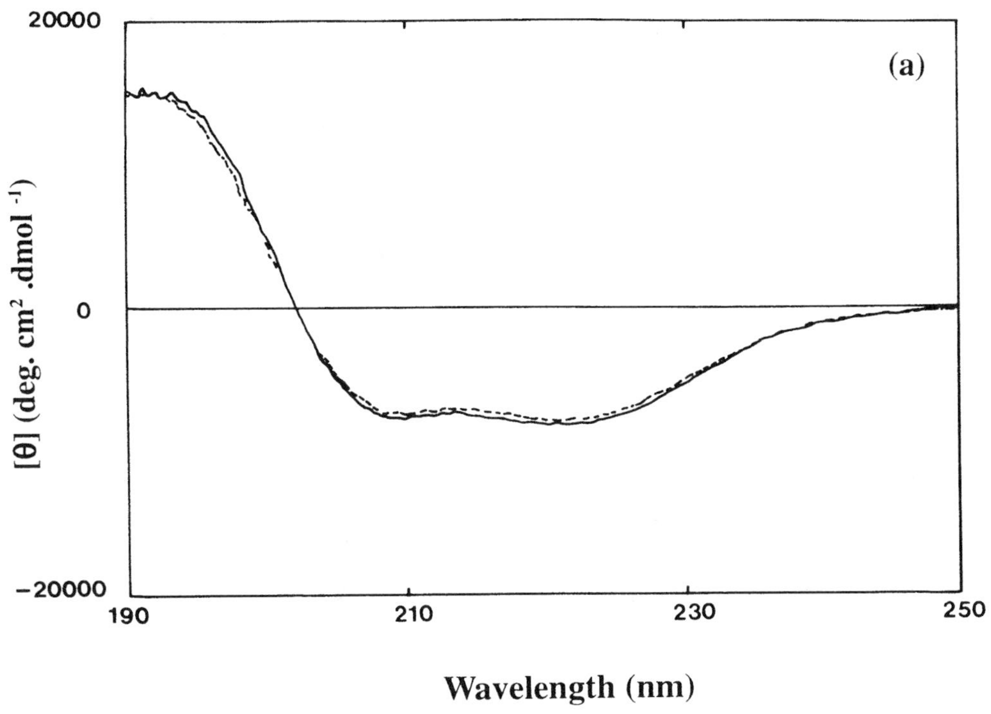


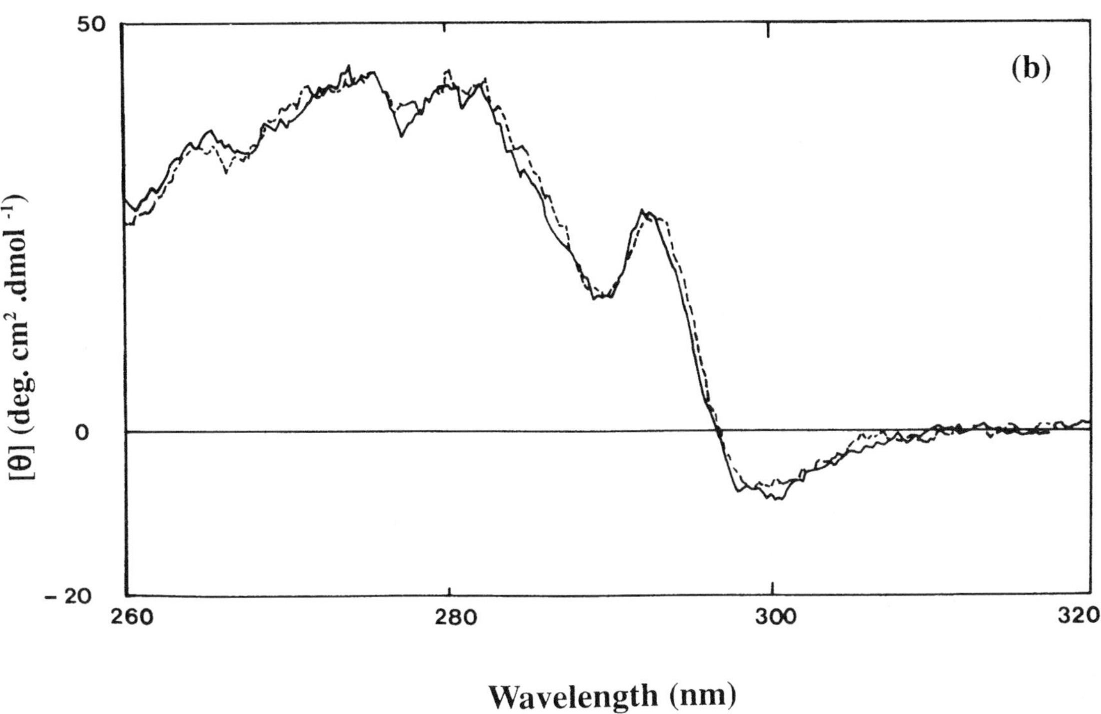


Figure 3. CD spectra of isocitrate lyase from *E. coli:* (a) far-UV CD spectra and (b) near-UV CD spectra. The solid and dashed lines depict the spectra of wild-type and mutant enzyme, respectively. In the mutant the Cys 195 at the active site has been replaced by Ala. The spectra of the mutant with Cys replaced by Ser are superimposable on those of the wild-type enzyme.

In the case of the horse liver enzyme, analysis of the far-UV CD spectrum by the CONTIN procedure gives values for α-helix and β-sheet (28 and 35%, respectively) that are in good agreement with those derived from X-ray (28 and 34%). The CD analysis of the *D. melanogaster* enzyme reveals that the α-helix and β-sheet contents are similar to those of the horse liver enzyme (29 and 31%, respectively); this result is also consistent with the pattern of helices and sheets predicted from the amino acid sequence. The similar content of helix and sheet in the two enzymes suggests that a typical $NAD^+$-binding domain (helix–sheet motif) also exists in the *Drosophila* enzyme.

### 3.2 Analysis of the Tertiary Structure of Proteins

The near-UV CD spectrum of a protein arises from the environments of each aromatic amino acid side chain as well as contributions from disulfide bonds or nonprotein cofactors that may be present. In favorable cases, such as that of interleukin 1β, assignment of bands to particular side chains has been undertaken by the systematic replacement of the aromatic side chains using site-directed mutagenesis.

The near-UV CD spectrum of a protein can be regarded as providing a detailed "fingerprint" of the tertiary structure of proteins containing a number of aromatic amino acid side chains or other contributory chromophores. This has been used, for example, to show that mutants of isocitrate lyase from *E. coli* in which the active site Cys 195 has been replaced by Ser or Ala have essentially identical structures to the wild-type enzyme (Figure 3). Analysis of the far-UV CD spectrum of the wild-type enzyme by the CONTIN procedure gives values for the α-helix and β-sheet contents of 23 and 33%, respectively. CD provides a rapid way of focusing attention on mutant proteins that may possess different structures from the wild type and will thus justify a more detailed analysis by more precise structural techniques (NMR or X-ray).

### 3.3 Analysis of Protein Folding

Because CD has the ability to yield structural information on a reasonably rapid time scale with modest amounts of material, the technique has been widely used in studies of the processes of protein folding. CD has been used to assess the ability of putative folding domains of proteins to adopt stable structures. Stopped-flow CD (both in the far and near UV) has been used in conjunction with other techniques, such as fluorescence and two-dimensional NMR spectrometry to explore the early events in the folding of lysozyme from a denatured state. In the very early parts of the process, secondary structural elements are formed, as shown by the far-UV CD spectra; the subsequent acquisition of the correct tertiary structure (as revealed by near-UV CD) is a considerably slower process.

### 3.4 Analysis of Protein Stability

CD can be used to measure the stability of the native protein structure toward various perturbations such as those brought about by heat or agents such as guanidinium chloride (GdnHCl). The GdnHCl-induced unfolding of dehydroquinase from *E. coli* could be monitored by the loss of the CD signal at 225 nm. When the enzyme is treated with substrate and then reduced with $NaBH_4$, there is a marked increase in stability, with the mid-point for denaturation increased from 1.2 *M* to 4.2 *M* GdnHCl. Analysis of the unfolding curves shows that the stability of the folded state (relative to the unfolded state) has increased from 20.5 kJ/mol to 40.5 kJ/mol as a result of the modification.

### 3.5 Binding of Ligands

Changes in the CD spectrum of a protein can be used to monitor the binding of ligands, hence to deduce the characteristics (stoichiometry, affinity, etc.) of the interaction. Thus addition of *S*-adenosylmethionine to various DNA methylases has been shown to lead to a marked decrease in the size of the CD signal in the 200–230 nm region, consistent with a decline in the α-helix content. From the titration curve, the binding parameters can be evaluated.

The binding of ligands to proteins can lead to induced asymmetry, providing an additional means of monitoring structural changes in proteins. For instance, addition to carbonic anhydrase of the inhibitor Neoprontosil (an azosulfonamide) leads to the production of large CD signals in the near-UV and visible regions. Since the azosulfonamide is intrinsically optically inactive, this result must imply that it has become highly immobilized in an asymmetric environment in the protein. The CD signals from the bound inhibitor have been used to highlight subtle structural differences between different isoforms of the enzyme.

## 4 SUMMARY AND PERSPECTIVES

It has been shown how CD can be used to analyze proteins in a variety of ways, to give information on secondary and tertiary structures as well as changes in these structures. The preferred analysis of far-UV CD spectra uses a set of real proteins as models rather than relying on reference spectra of model poly-L-amino acids. There is still, however, the limitation that any particular protein may not be comparable to the range of reference proteins in the data set. It is questionable whether any type of analysis of CD spectra will ever yield estimates of secondary structure content as precise as those afforded by X-ray studies. However, the strengths of CD (especially speed and economy of sample: Section 1.5) are particularly important in the early stages of structural investigations and may well render CD the only realistic approach in cases of proteins or peptides that cannot be crystallized in a form suitable for X-ray work or are too large or insufficiently soluble to be studied by NMR techniques. Developments in stopped-flow CD and in rapid (diode array) detection make it likely that events on the millisecond time scale will be increasingly amenable to study, ensuring that CD will play an ever more important part in the study of ligand-induced transitions and protein folding.

*See also* Chirality in Biology; Protein Modeling.

## *Bibliography*

Adler, A. J., Greenfield, N. J., and Fasman, G. D. (1973) *Methods Enzymol.* 27:675–735.

Bayley, P. (1980) In *An Introduction to Spectroscopy for Biochemists*, S. B. Brown, Ed., pp. 148–234. Academic Press, London.
Chaffotte, A. F., Guillou, Y., and Goldberg, M. E. (1992) *Biochemistry*, 31:9694–9702.
Johnson, W. C., Jr. (1990) *Proteins: Struct. Function Genet.* 7:205–214.
Kahn, P. C. (1979) *Methods Enzymol.* 61:339–378.
Pflumm, M., Luchins, J., and Beychok, S. (1986) *Methods Enzymol.* 130:519–534.
Provencher, S. W., and Glöckner, J. (1981) *Biochemistry*, 20:33–37.
Strickland, E. H. (1974) *CRC Crit. Rev. Biochem.* 2:113–175.
Timasheff, S. N. (1970) *Enzymes* Vol. 2, 3rd ed. pp. 371–443. Academic Press, New York.

# Coenzymes, Biochemistry of

*Donald B. McCormick*

*Key Words*

**Apoenzyme** Protein moiety of an enzyme that requires a coenzyme.

**Cofactor** Natural reactant, usually either a metal ion or coenzyme, required in an enzyme-catalyzed reaction.

**Holoenzyme** Catalytically active enzyme constituted by coenzyme bound to apoenzyme.

**Vitamin** Essential organic micronutrient that must be supplied exogenously and in many cases is the precursor to a metabolically derived coenzyme.

A coenzyme is an organic molecule that can bind within the enzymes that require its function to catalyze a biochemical reaction. Hence, coenzymes are organic cofactors that augment the diversity of reactions that otherwise would be more limited to chemical properties of side chain substituents from amino acid residues within protein enzymes. Coenzymes bind to apoenzymes to generate catalytically competent holoenzymes. With tight binding, coenzymes may be referred to as prosthetic groups; with loose binding, they may be called cosubstrates.

## 1 GENERAL HISTORY

Most work on the chemical nature of coenzymes began in the 1930s, when the structures of some vitamins in the B complex were being elucidated. In 1932 Auhagen had succeeded in dissociating a heat-stable component of the yeast "carboxylase," which he called "cocarboxylase." By 1937, Lohmann and Schuster had determined its structure as thiamine pyrophosphate (TPP). During the same period Warburg and Christian were active in elucidating the nature of the oxidation–reduction coenzymes that contain riboflavin (FMN, FAD) and nicotinamide (NAD, NADP). By the mid-1940s, the coenzyme forms of vitamin $B_6$ (PLP, PMP) were recognized by Gunsalus, Snell, and others. The discovery by Lipmann in 1947 of a coenzyme of acetylation (CoA), which contained pantothenate, led to full structure elucidation by Baddiley et al. in 1953. Advances also made in the 1950s by a number of laboratories led to recognition of coenzyme forms of folacin, and by the end of the decade, Lynen et al. had characterized covalent 1′-*N*-carboxybiotin and Barker and his associates had identified a coenzyme form of vitamin $B_{12}$. Prerequisite to determining structures of coenzymes was recognition and purification of enzyme systems that depend on the catalytic participation of these organic cofactors. Continued examinations of diverse enzymes are extending the list of newer coenzymes, some of which do not derive from vitamins.

## 2 INDIVIDUAL COENZYMES

### 2.1 Vitamin-Derived Coenzymes

Many coenzymes are the metabolic result of converting ingested vitamins, especially those of the B complex, to forms suitable for binding and function in enzyme systems. Of the 13 vitamins presently known to be required in the diet of humans and most animals, at least eight (thiamine, riboflavin, niacin, vitamin $B_6$, folacin, vitamin $B_{12}$, biotin, and pantothenate) are simply the essential precursors for coenzymic forms made in the body. These coenzymes and their functions are listed in Table 1.

### 2.2 Other Coenzymes

A growing number of coenzymes are not formed from vitamins; hence they can be biosynthesized by at least most of the organisms that require them. Several additional coenzyme types are described briefly.

#### 2.2.1 Pyruvoyl Functions

A simpler, but less frequently encountered, variation on one way in which pyridoxal 5′-phosphate operates is provided by the pyruvoyl N-terminus of some enzymes. In these electrophilic centers, the amino function of a substrate amino acid condenses to form a ketimine, which facilitates loss of a carboxyl group from the attached amino acid moiety. For example, *S*-adenosylmethionine decarboxylases from mammals as well as *Escherichia coli* use such a system to form spermidine from putresine and methionine.

#### 2.2.2 Lipoyl Functions

α-Lipoic (thioctic) acid occurs naturally in amide linkage to the ε-amino group of lysyl residues within transacylases that are core protein subunits of α-keto acid dehydrogenase complexes. In such cases, the functional dithiolane ring is on an extended flexible arm. The lipoyl function mediates the transfer of the acyl group from α-hydroxyalkyl-TPP to CoA in a cyclic system that transiently generates the dihydrolipoyl residue.

#### 2.2.3 Ubiquinones

The ubiquinones (coenzyme Q's) are a group of "ubiquitous" substituted benzoquinones with variable-length terpenoid chains. In eukaryotes, they are found mainly in the mitochondrial inner membrane, where they function to accept electrons from several different dehydrogenases and relay them to the cytochrome system. In this process, the quinone function can be cyclically reduced and reoxidized. Recently evidence has been found for a function of CoQ in plasma membrane electron transport, which influences mammalian cell growth.

**Table 1** Coenzymatic Forms and Functions of B Vitamins

| Coenzyme Forms | Vitamin Enzyme Systems | Functions |
|---|---|---|
| | *Thiamine ($B_1$)* | |
| Thiamine pyrophosphate (perhaps triphosphate) | α-Keto acid decarboxylases, *trans*ketolase | Decarboxylations of α-keto acids from metabolism of carbohydrates and amino acids; glycolaldehyde transfers; nerve membrane ion transport |
| | *Riboflavin ($B_2$)* | |
| Flavin mononucleotide, flavin-adenine dinucleotide (covalent and noncovalent) | Oxidases, dehydrogenases | Hydrogen and electron transfers from substrates to oxygen and to cytochromes |
| | *Niacin* | |
| Nicotinamide adenine dinucleotide and the 2′-phosphate | Reductase, dehydrogenases | Hydride ion transfer |
| | *Vitamin $B_6$* | |
| Pyridoxal 5′-phosphate, pyridoxamine 5′-phosphate | Aminotransferases, amino acid decarboxylases, amino acid dehydratases, cystathionine synthetase, phosphorylase, 5-aminolevulinate synthetase, etc. | Transaminations, decarboxylations, and side-chain cleavages of amino acids; breakdown of glycogen; biosynthesis of heme, phospholipid, etc. |
| | *Folacin* | |
| Tetrahydrofolate and γ-glutamyl conjugates | Transformylases, thymidylate synthase hydroxymethyltransferases, formiminotransferase, etc. | One-carbon activation and transfer as in purine and pyrimidine biosynthesis, catabolism of serine and histidine, etc. |
| | *Vitamin $B_{12}$* | |
| 5′-Deoxyadenosylcobalamin | L-Methylmalonyl-CoA mutase | Isomerization of L-methylmalonyl-CoA to succinyl-CoA |
| Methylcobalamin | 5-Methyltetrahydrofolate-homocysteine methyltransferase | Methyl group transfer as in methionine biosynthesis |
| | *Biotin* | |
| 1′-*N*-Carboxybiotin as amide-linked to ε-lysyl residues | Acyl-CoA carboxylases and other bicarbonate-dependent carboxylases | $Co_2$ activation and transfer in formation of acids |
| | *Pantothenic Acid* | |
| CoA | Acyltransferases, etc. | Fatty acid metabolism, etc. |
| 4′-Phosphopantetheine | Acyl carrier protein of fatty acid synthetase | Fatty acid biosynthesis |

### 2.2.4 Pyrroloquinoline Quinone and Quinoproteins

The coenzyme pyrroloquinoline quinone (PQQ) originally was isolated from bacteria possessing a methanol dehydrogenase. PQQ functions as a redox component of holoenzymes in which the coenzyme can be reduced in some cases by successive one-electron steps to the radical and dihydro forms and in other cases directly to the dihydro level by two-electron processes. More recent detailed examinations of several quinoproteins have elucidated other quinone-hydroquinone-related coenzymes as side chain altered prosthetic groups. Examples include the 6-hydroxydopa-(trihydroxyphenylalanyl) residue at the active site of serum amine oxidase and the tryptophan tryptophylquinone in a bacterial methylamine dehydrogenase.

### 2.2.5 Methanogen Coenzymes

There are a diverse group of coenzymes required to convert carbon dioxide to methane—rather novel compounds that vary from a complex nickel-porphyrin-like $F_{430}$ to the phenyl ether of methanofuran, the unusual pterin of tetrahydromethanopterin, the deazaflavin of $F_{420}$, to the somewhat simpler sulfur-containing mercaptoheptanoylthreonine and coenzyme M. These coenzymes are used by highly specialized archaebacteria (known as methanogens) that can accomplish the reduction of $CO_2$ through formyl, methyl, methylene, and methyl stages of $CH_4$. The importance of the coenzymes that participate can be appreciated by the quantitative and global aspects of carbon cycling through methane-forming systems, which are not only free-living microbes but also are found in the human intestinal tract.

*See also* ENZYMES; VITAMINS, STRUCTURE AND FUNCTION OF.

## *Bibliography*

Aurbach, G. D., and McCormick, D. B., Eds. (1983–1991) *Vitamins and Hormones,* Vols. 40–46. Academic Press, New York.

Blakley, R. L., and Benkovic, S. J. (Vol. 1) and Blakley, R. L., and Whitehead, V. M. (Vol. 3) (1984) *Folates and Pterins.* Wiley, New York.

Chytil, F., and McCormick, D. B., Eds. (1986) *Vitamins and Coenzymes,* Parts G and H, in *Methods in Enzymology.* Academic Press, New York.

Curti, B., Ronchi, S., and Zanetti, G., Eds. (1991) *Flavins and Flavoproteins.* de Gruyter, Berlin and New York.

Frey, P. A. (1988) Structure and function of coenzymes, in *Biochemistry,* G. Zubay (Coord. Au.). Macmillan, New York.

Fukui, T., Kagamiyama, H., Soda, K., and Wada, H., Eds. (1991) *Enzymes Dependent on Pyridoxal Phosphate and Other Carbonyl Compounds as Cofactors.* Pergamon Press, Oxford.

McCormick, D. B. (1991) Coenzymes, biochemistry, in *Encyclopedia of Human Biology,* Vol. 2, R. Dulbecco, Ed.-in-Chief. Academic Press, San Diego, CA.

Rouviere, P. E., and Wolfe, R. E. (1988) Novel biochemistry of methanogenesis (minireview). *J. Biol. Chem.* 263:7913–7916.

# COLON CANCER

*Joanna Groden*

## *Key Words*

**Adenoma** A benign tumor with a glandular structure; of epithelial cell origin.

**Allelotyping** A process of genotyping polymorphic loci to look for loss of alleles in a tumor.

**Carcinoma** An invasive tumor of epithelial cell origin.

**Dominant** Describing an allele or gene that confers an effect or action that is expressed in favor of another allele or gene at the same locus; the opposite of recessive.

**Dominant Negative** An inactivating mutation of a gene whose mutant protein product interferes with the function of the normal gene product from the other allele.

**Familial** Describing the appearance of a particular trait within a family.

**Frameshift** A type of mutation resulting from the insertion or deletion of nucleotides in a DNA sequence that disrupts the open reading frame.

**Gene Targeting** The introduction of a DNA sequence into, or modification of, the sequence of a gene, based on a technology that is specifically designed to select that gene or sequence for modification.

**Heptad Repeat** A repeat of seven amino acids that is apolar-X-X-apolar-X-X-X; a protein domain that may confer the ability to interact with other proteins through the formation of a helical structure.

**Linkage Analysis** The statistical method used to determine the physical proximity of two loci based on the coinheritance of particular alleles at these loci in families.

**Loss of Heterozygosity (LOH)** The absence of one of two polymorphic alleles at a locus known originally to be heterozygous; LOH is a marker for loci that encode tumor suppressors inactivated by mutation.

**Methylation** The addition of methyl groups to DNA; generally associated with the inactivation of a gene or genes on a chromosome.

**Mutation** A nucleotide change that results in a heritable alteration of the function or sequence of a DNA segment or gene.

**Oncogene** Gene that functions to produce a tumor.

**Penetrance** The ability of a gene to express its effect.

**Phenotype** The physical characteristics that result as a consequence of the interaction of the environment and the genotype of a cell or an organism.

**Point Mutation** Type of mutation that results from the alteration of a single nucleotide in a sequence. A missense mutation changes the amino acid at that position in the protein; a nonsense mutation introduces a stop codon into the open reading frame.

**Recessive** Describing an allele or gene that confers an effect or action that is not expressed in favor of another allele or gene at the same locus, unless that allele or gene is also recessive; the opposite of dominant.

**Tumor Suppressor Gene** A gene that functions to suppress growth or tumorigenicity of a cell. Such genes are often mutated in tumors and are believed to control checkpoints in cell growth of normal cells; when aberrant, they do not work to control these checkpoints, so that cells grow in an unrestricted manner.

---

Colon carcinoma develops through a benign intermediate tumor known as an adenoma. Therefore, the genetic pathway that takes normal epithelium to carcinoma can be analyzed by applying the techniques of molecular biology to this series of tumors. Inherited syndromes of colon cancer predisposition also enable researchers to identify and study genes involved in tumor formation. Consequently, the genes that function in the control of cell growth and differentiation have been discovered. The study of these genes will allow better diagnosis and treatment of cancer and other human diseases.

## 1 INTRODUCTION

Colonic tumors provide an ideal system for the identification of genes that are affected in mutational pathways. Tumors of the colon, both adenomas and carcinomas, are common in the general population, are accessible at a variety of stages, and can be found at an increased incidence in some inherited syndromes. Also, most colon carcinomas develop through a benign intermediate, the adenoma or adenomatous polyp. Genes involved in this sequence can be identified by analyses of sporadic tumors and by studies of families in which the incidence of colon cancer is unusually high. Work of this kind has illuminated important aspects of how colon cancers occur in the general population and how these tumors might be prevented and treated.

## 2 WAYS OF DEFINING STEPS IN FORMATION OF COLON CANCERS

### 2.1 Study of Genetic Materials

#### 2.1.1 Allelic Loss and Loss of Heterozygosity

One approach to defining the genetic steps involved in the formation of colonic tumors is the detection in tumor cells of allelic loss, or "allelotyping," to pinpoint regions of the genome that have undergone significant genetic changes during tumor formation. Models that propose the involvement of "tumor suppressor" genes in the control of cell growth also propose that disruption of these genes can occur by deletional mechanisms that result in hemi- or homozygosity for a recessive mutation. Therefore, losses of genetic information by allelic deletion in tumors can mark the physical locations of specific genes that normally function in the suppression of tumorigenicity.

Colonic tumors frequently show and loss of heterozygosity affecting chromosome arms 5q, 8p, 17p, and 18q. Other chromosome arms also lose heterozygosity, but not as often. Increased loss of heterozygosity in a tumor correlates with poor clinical prognosis, even when the tumor is similar in size and clinical stage to tumors with fewer allelic losses.

Loss of heterozygosity affecting chromosome 5q has occurred in about 30% of adenomas and about 50% of carcinomas studied. These observations imply that mutations affecting one homologue of chromosome 5 occur early in the establishment of a tumor and that loss of the normal allele occurs later in tumor progression. The mapping of an inherited colon cancer syndrome, familial adenomatous polyposis coli (APC), to the same region of 5q also implies that this part of the genome contains a gene or genes involved in an early and perhaps rate-limiting step in the mutational pathway to colon cancer. When adenomas from APC patients are examined, allelic loss in the 5q region is observed in some adenomas and carcinomas; this loss has affected the allele inherited from the normal parent.

Losses of heterozygosity on chromosomes 17p and 18q occur later in tumor formation; both occur at high frequencies, 75 and 70% respectively. These losses seem to correlate with the transition from the adenoma stage to that of the carcinoma, as they are usually found only in carcinomas.

Regions of the genome with significant loss of alleles have been examined more closely by mutational studies of tumors and by the screening of candidate genes in these regions for mutations. The tumor suppressor genes known to be involved in the formation of colon tumors, first mapped to chromosomes 5, 17, and 18 by allelotyping and by linkage analysis in APC families, are *APC, p53,* and *DCC,* respectively. Each was identified by the detection of small deletions, chain-terminating nucleotide alterations, or missense mutations in tumors. As yet, only APC and *p53* have been shown directly to function as suppressors of tumorigenicity in colon carcinoma cells. However, transfer of whole normal chromosomes 5, 17, or 18 into carcinoma-derived cell lines has corrected tumorigenicity in the cultured cells.

#### 2.1.2 Study of the *APC* Gene

Direct DNA sequencing shows that a majority of colon carcinomas and adenomas, even small adenomas, contain at least one mutated *APC* gene and that the frequency of such events does not increase as tumors progress. Most mutations that have been identified in this gene are chain-terminating mutations. These data confirm that *APC* mutation plays a major if not essential role in the early development of colonic tumors. One study reported that inactivation of both *APC* alleles had occurred in 10 of 15 adenomas from an APC patient in whom the germline mutation had already been identified. The second somatic event was identified as loss of heterozygosity. This result also suggests that inactivation of both alleles of *APC* may be necessary for the development of early adenomas.

The predicted amino acid sequence of the *APC* gene product reveals no significant homology to known proteins; its function is not yet known. The amino-terminal domain contains heptad repeats, suggesting that this protein could form homo- or heterodimers. Interestingly, an alternative splice sometimes removes two heptad repeats and inserts a novel, single heptad repeat in its place. Preliminary work suggests that the first heptad repeat in the amino terminus of the protein functions in the establishment of homodimers and that truncated proteins are produced from some of the mutant *APC* alleles in tumors and in APC patients. Whether these mutant proteins affect normal function of the full-length protein is unknown, although some evidence suggests that there may be a dominant-negative function derived from some truncated *APC* alleles.

Alternative splicing also occurs at the 5′ end of the *APC* gene, as four novel exons appear in different transcripts. Interestingly, each of these splice forms has two other forms: one that includes exon 1 and one that does not. Exon 1 encodes most of the first heptad repeat shown to confer homodimerization ability, so this phenomenon has intriguing consequences for functional regulation.

#### 2.1.3 Study of the *p53* Gene

The *p53* gene encodes a protein with transcriptional activation ability that in a tetrameric structure binds a unique DNA sequence.

This gene plays an important role in controlling cell growth and cell cycle progression. Its product is found at a low concentration in cells under normal conditions, but when DNA damage occurs, its expression is increased and the cell cycle is arrested. Expression of *p53* remains high until the DNA damage is repaired. Wild-type *p53,* when overexpressed, blocks transition from G1 to S phase in the cell cycle. Targets for p53 protein include genes that regulate genomic stability, response of the cell to DNA damage, cell cycle progression, and the induction of apoptosis by radiation or chemotherapeutic DNA-damaging drugs.

Inactivation of p53 occurs by either of two mechanisms: mutation of the gene, or binding of the protein to other oncogenic proteins such as MDM2 (the *mdm2* gene often is amplified in human sarcomas), the *E1B* gene product of adenovirus, the *E6* gene product of human papilloma virus, or the large T antigen of the simian virus SV40. Mutation or inactivation of *p53* is the most frequently observed molecular alteration in all human tumors, not only colonic tumors. These mutations are primarily missense mutations that occur within evolutionarily conserved regions of amino acids; the majority occur in one of four hot spots. Such mutations usually result in the inability of the protein to bind target sites on DNA; many change the overall conformation of the protein molecule. In tumors, both *p53* alleles usually are inactivated by allelic loss or mutation, suggesting that complete loss of wild-type function leads to tumor formation. However, some mutant forms of *p53* also acquire a gain of function and appear to stimulate cell growth; alterations of these types can act in the heterozygous state and are known as dominant-negative mutations.

Germline mutations of *p53* have been described in kindreds with the Li–Fraumeni cancer syndrome, in patients with multiple sarcomas or multiple primary tumors, and in some patients with strong family histories of cancer. Interestingly, colon cancers are not commonly found in Li–Fraumeni families, again suggesting that mutation of *p53* is not an early event in colon tumor formation.

### 2.1.4 Study of the *DCC, Ki-ras,* and *MCC,* Genes

The *DCC* (deleted in colon cancer) gene on chromosome 18 encodes a protein with significant homology to neural cell adhesion molecules and related cell surface glycoproteins. The gene is normally expressed in colorectal mucosa, but its expression is absent or reduced in colonic tumors. Mutation of both *DCC* alleles may play a role in tumor development by disrupting normal cell–cell contacts or contacts between cells and the extracellular matrix. Antisense constructs to *DCC* induce formation of transformed foci in RAT-1 cells, strengthening the evidence for the role of *DCC* in the control of cell growth.

Almost 50% of large adenomas and colon carcinomas carry *Ki-ras* mutations. Only 10% of small adenomas carry such mutations. These are dominant mutations that lead to a gain of function; the majority occur in codon 12 or 13 of the gene. Although *ras* mutations occur early in the pathway from adenoma to carcinoma, they are not the initiating event. Interestingly, mutations in the *NF1* gene also are found in some colonic tumors. The *NF1* gene product is neurofibromin, a protein that contains a GTPase-activating domain involved in the catalysis of RAS-GTP to RAS-GDP. Normally, proteins like neurofibromin, along with others such as GAP, the GTPase-activating protein, work together in signal transduction pathways to regulate cell growth. The disruption of any of the genes involved could result in deregulation of growth or differentiation in the colonic epithelium; this sequence has been demonstrated by cell culture experiments in which the tumorigenicity of colon cancer cell lines has been eliminated by removing the activated copy of the *Ki-ras* allele. One diagnostic test for adenomas and carcinomas, in fact, is based on the presence of activated *Ki-ras* alleles in DNA from stool samples.

Mutational analyses of tumors have shown that the *MCC* (mutated in colon cancer) gene, adjacent to *APC* on chromosome 5q, also contains missense mutations in about 10% of colon carcinomas. Such mutations may be merely the result of a high mutational background in tumor cells, or this effect may indeed be a significant, although not rate-limiting, event. The function of the *MCC* gene product is unknown. The predicted protein sequence contains heptad repeats, suggesting that MCC may form protein–protein complexes.

Other molecular alterations observed in colonic tumors include generalized hypomethylation of DNA and a higher than normal expression of the *c-myc* oncogene. The cause or significance of these changes is not well understood.

## 2.2 Study of Individuals or Families

A second approach to the delineation of genes involved in the adenoma–carcinoma sequence is the study of individuals or families in which the incidence of one of these tumor types is higher than that of the general population. Such inherited disorders can be divided into two broad categories: hereditary polyposis colorectal cancer syndromes and hereditary nonpolyposis colorectal cancer syndromes. The first group can be further divided into adenomatous and hamartomatous syndromes.

### 2.2.1 Adenomatous Syndromes

The adenomatous syndromes of colorectal cancer predisposition include adenomatous polyposis coli, Gardner syndrome, and Turcot syndrome. APC is an autosomal dominant disorder characterized by the presence, at an early age, of hundreds to thousands of adenomatous polyps in the colon. The likelihood that colon cancer will develop from one or more of these polyps is almost 100%. Gardner syndrome is a variant of APC that is characterized by extracolonic manifestations such as osteomas, desmoids, and fibromas, as well as by polyposis. Both syndromes map to chromosome 5q21 and are associated with disruption of the *APC* gene. As yet, the cause for this variation in expressivity of extracolonic symptoms is unknown. It is not related to the position or the type of mutation present within the *APC* gene.

A large number of disease-causing *APC* mutations have been observed; with the exception of four, all are chain-terminating mutations and include frameshifts, point mutations that introduce premature stops into the open reading frame of the gene, and mutations that affect splice junctions. Most APC families carry unique mutations. Only two mutations are known to have occurred in more than a few kindreds; each is a five-base-pair deletion that disrupts the open reading frame. Together, they account for less than 10% of the known disease-causing *APC* mutations. Mutations have been found throughout the *APC* gene from exon 3 through exon 15.

Attenuated APC, or AAPC, is characterized by a lower number of adenomatous polyps, usually fewer than 100, and the development of colon cancer at a later age than is the case in classical APC. This disorder also maps to the *APC* locus. *APC* mutations have been observed in DNA from affected individuals. Interestingly, in three AAPC kindreds, the mutations were more 5′ than mutations in APC or Gardner's syndrome. All were found in, or affected the splicing of, exons 3 and 4 of the *APC* gene. These observations lend support to the dominant-negative hypothesis: perhaps the mutant alleles in AAPC families produce truncated peptides that are either too short or too unstable to form a peptide long enough to interfere with the normal protein dimers or complexes. One argument against the dominant-negative hypothesis, however, has come from the discovery of APC patients with cytogenetically visible deletions of the entire locus.

The incidence of tumors outside the colon and rectum in APC patients is elevated. Carcinoma of the upper gastrointestinal tract is the most common cause of death following colorectal cancer in APC individuals. Other malignancies include papillary carcinoma of the thyroid, carcinoid tumors of the ileum, and brain tumors of different histological types. This observation may account for the designation of Turcot syndrome for individuals in kindreds where brain tumors and polyposis occur together. Both dominant and recessive modes of inheritance have been used to explain the segregation of these two phenotypes in families. Recently, germline mutations of *APC* have been found in patients with Turcot Syndrome.

### 2.2.2 Hamartomatous Syndromes

The hamartomatous syndromes include Cowden syndrome, familial juvenile polyposis, and Peutz–Jegher syndrome. Each of these rare inherited disorders is characterized by multiple hamartomatous polyps in the gastrointestinal tract, as well as other features. Hamartomas in themselves are not considered to have malignant potential; yet affected individuals carry a significant risk of colonic cancer.

Hereditary nonpolyposis colon cancer (HNPCC) syndromes are split into two groups: both are affected by an increased risk of (usually right-sided) colon cancer, but one, also known as cancer family syndrome, is characterized by the occurrence of other tumors as well. Diagnostic criteria include the occurrence of colon cancer in at least three first-degree relatives representing at least two generations, and the development of cancer in one of these individuals before age 50.

Linkage analysis in two large HNPCC kindreds first mapped a disease allele to chromosome 2p, but analysis of other HNPCC kindreds excluded linkage to that chromosomal region. Therefore, the disease is heterogeneous in its underlying genetic basis. It was shown subsequently that germline mutations in any one of four homologues of bacterial DNA mismatch repair genes (*hMLH1, hMSH2, hPMS1,* of *hPMS2*) resulted in HNPCC. Disruption of these genes, therefore, causes an increase in genomic instability, as stretches of tandemly repeated dinucleotide and trinucleotide sequences have expanded or contracted in some sporadic right-sided colonic tumors and in tumors from HNPCC patients. This instability may then result in the accumulation of mutations that lead to tumor formation.

Last, studies show that a person who has even one relative with either a colon cancer or an adenomatous polyp may carry an elevated risk for colon cancer. Therefore, aberrant forms of other genes or variants of those already isolated may predispose some individuals to colon cancer.

## 3 PERSPECTIVE

The definition of the genetic changes that accompany the transition of normal colonic epithelium through the intermediate of the adenoma to a carcinoma have yielded important insights into how cells grow and differentiate. The coordinated study of tumors and individuals at risk for a particular type of cancer have resulted in an outline of the tumorigenic process, at least for colon cancer. Further study will no doubt lead to the discovery of even more valuable information.

*See also* CANCER; HUMAN GENETIC PREDISPOSITION TO DISEASE; ONCOGENES; TUMOR SUPPRESSOR GENES.

### *Bibliography*

Fearon, E. R., and Vogelstein, B. (1990) A genetic model for colorectal tumorigenesis. *Cell* 61:759–767.

Steele, G., Burt, R. W., Winawer, S. J., and Karr, J. P., Eds. (1988) *Basic and Clinical Perspectives of Colorectal Polyps and Cancer,* Vol. 249 in *Progress in Clinical and Biological Research.* Liss, New York.

Yamada, T., Alpers, D. H., Owyang, C., Powell, D. W., and Silverstein, F. E., Eds. (1991) *Textbook of Gastroenterology.* Lippincott, Philadelphia.

# COMBINATORIAL PHAGE ANTIBODY LIBRARIES

*R. Anthony Williamson and Dennis R. Burton*

### *Key Words*

**Antibody Repertoire** The array of different antibody molecules that may be synthesized by the immune system; thought to be in the region of $10^8$ in humans.

**Combinatorial Antibody Library** A large ensemble of antibody molecules expressed in a given vector, generally derived by the independent cloning of single heavy and light chain genes in the vector.

**Panning** An affinity-based selection procedure for the isolation of antibody–phage with specificity for a desired antigen or epitope.

**Phage Display** The expression of proteins or peptides on the surface of filamentous bacteriophage.

Combinatorial antibody library technology has evolved as a means of capturing and selecting from antibody repertoires. By expressing antibody libraries on the surface of filamentous bacteriophage, it is possible to select from a library those phage displaying, and containing the information encoding, antibody with specificity for a particular antigen. Once isolated, antibodies may be readily subjected to further manipulation to modify specificity and affinity. Combinatorial libraries may be constructed from a number of species, from both immunized and nonimmunized sources, and from synthetic or partially synthetic genes. One of their most important applications probably lies in the generation of previously inaccessible human antibodies, which may be expected to be valuable reagents for the prevention, therapy, and diagnosis of human disease.

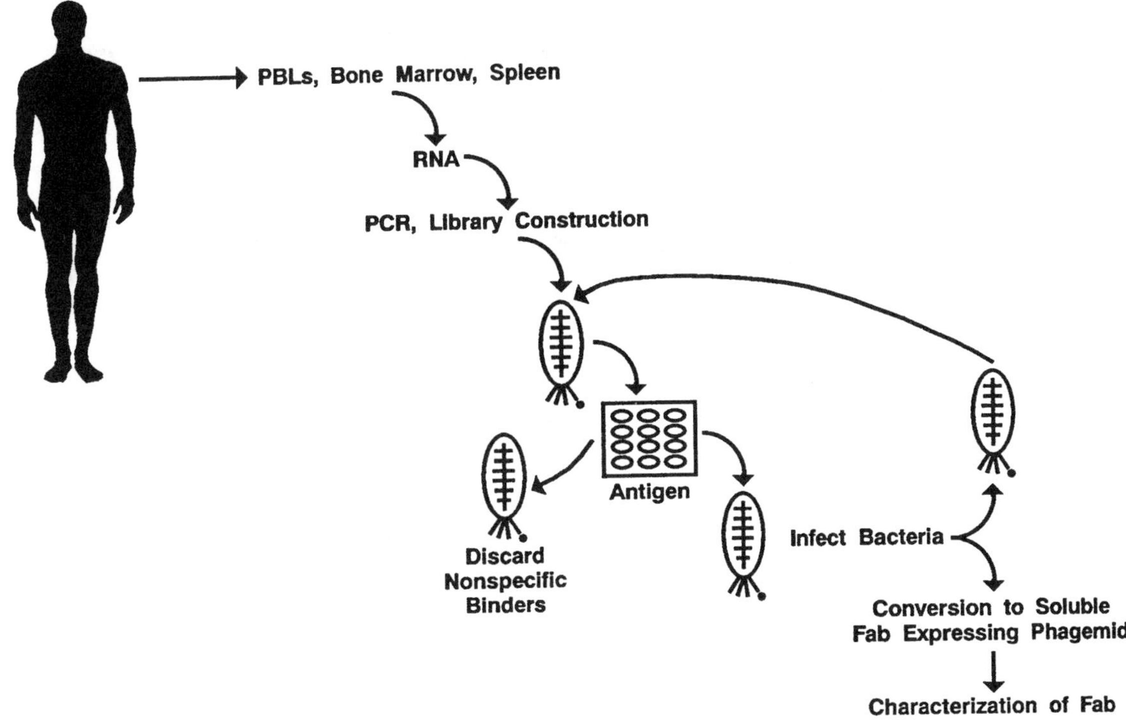


**Figure 1.** Strategy for cloning human monoclonal Fab fragments from phage display combinatorial libraries. RNA prepared from antibody-producing tissue sources [e.g., bone marrow and peripheral blood lymphocytes (PBLs)] is reverse transcribed, and the light chain and Fd portion of the heavy chain are amplified using PCR. The amplified genes are then cloned sequentially into a phagemid vector and "rescued" to a phage display library, in which each phage expresses Fab on its surface. The library is then "panned" over an immobilized antigen and specific phage–Fabs converted to a soluble Fab expressing system for further characterization. [Adapted from D. R. Burton and C. F. Barbas, *Chem. Immunol.* 56:112–126 (1993).]

## 1 CONSTRUCTING ANTIBODY LIBRARIES ON PHAGE

Generating human monoclonal antibodies by application of hybridoma and Epstein–Barr virus transformation methodologies has yielded largely disappointing results. In the search for an alternative approach to monoclonal antibody production, two developments were critical. First, the expression of antigen-binding fragments of antibody (Fv and Fab) in bacteria, second the use of the polymerase chain reaction (PCR) and a family of oligonucleotide primers to amplify antibody genes from a mixed population of antibody producing cells. Taken together, these techniques meant that one could, by using an appropriate vector, express in bacteria the antigen-binding domains of an array of antibodies from any animal for which one had antibody sequence for the construction of appropriate PCR primers.

The first antibody libraries were constructed in phage lambda. This system, however, was hampered by an inefficient screening methodology, which restricted the number of clones that could be examined. Subsequently, several groups developed systems to display antibody fragments on the surface of M13 phage. For the purposes of this contribution, we will concentrate on the pComb3 phagemid vector system.

Figure 1 outlines the overall strategy for cloning monoclonal antibodies from combinatorial libraries on the surface of phage. The pComb3 vector is illustrated and described in Figure 2. During library construction, PCR-amplified heavy and light chain DNAs are cloned sequentially into the vector and electrotransformed into bacteria. It is important at this point to note that inherent in the construction of combinatorial libraries is the loss of the original in vivo heavy–light chain combinations. However chain promiscuity—the ability of a particular heavy chain to accept a number of different light chains while retaining antigen binding—determines that the frequency of productive combinations is higher than might have been envisioned a priori.

The vector is designed to fuse the antibody heavy chain to the carboxy-terminal domain of the phage coat protein III (cpIII). Following translation, the heavy chain is transported under the direction of a *pel* B leader sequence to the periplasm of the bacterium, where it is tethered via cpIII to the inner membrane. Here the light chain assembles onto the heavy chain template, forming a functional antibody Fab. Inclusion of the f1 origin of replication in pComb3 means that upon coinfection with M13 helper phage, it is possible to rescue and package the phagemid DNA carrying the antibody genes, to a phage display library (detailed in Figure 3). This process also serves to amplify the initial library, yielding $10^4$–$10^5$ copies of each clone.

## 2 AFFINITY-BASED SELECTION FROM PHAGE DISPLAY LIBRARIES

Following amplification, the antibody phage may be screened or panned against antigen. The panning methodology is illustrated in Figure 4. Antigen can be immobilized (e.g., by binding to the wells of an ELISA plate or to synthetic beads), or may be part of an intact cell surface. Phage are incubated with antigen and nonbound

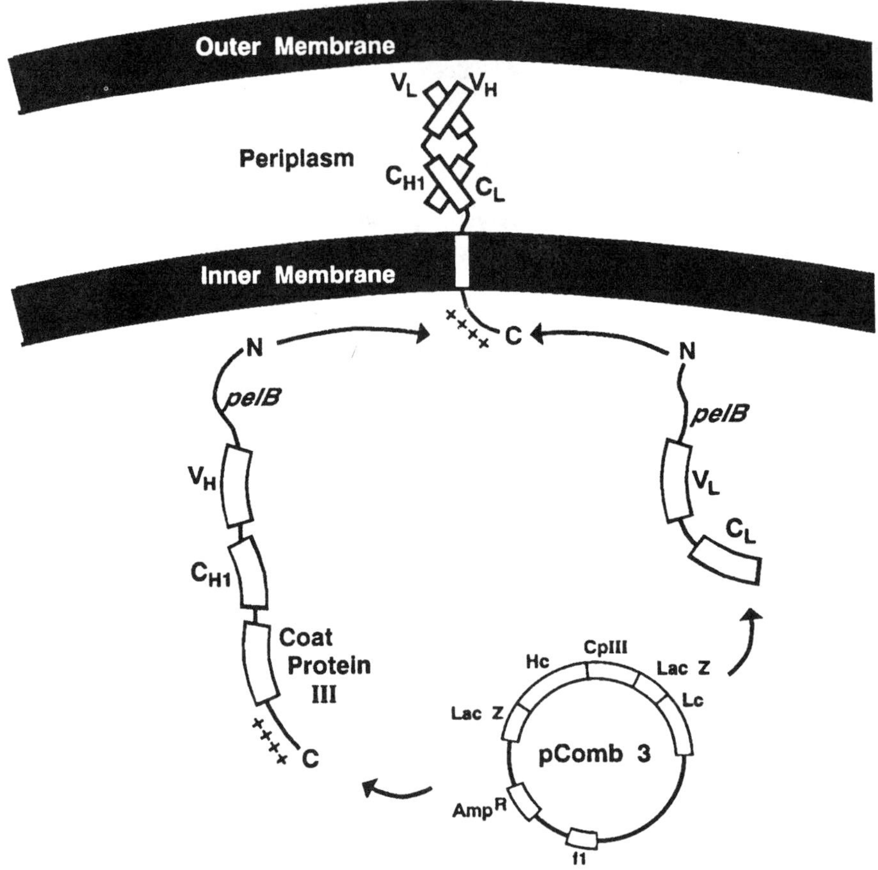


Figure 2. The proposed pathway of Fab assembly in the bacterial periplasm. Expression of cloned Fd (Hc, heavy chain)/cpIII fusion and light chain (Lc) is controlled by *lac* promoter/operator sequences. The translated antibody is directed to the periplasmic space by *pel*B leader sequences. Here the light chain assembles onto the heavy chain template, which is anchored to the inner membrane by the cpIII fusion. (Reproduced from Barbas et al., in *Methods, Vol. 2, A Companion to Methods in Enzymology,* pp. 119–124. Academic Press, Orlando, FL, 1991.)

phage removed by repeated washing. The remaining phage are then eluted in low pH and reamplified, and the panning is repeated, usually three or four times. Enrichments during each panning round (1) will favor those antibody–phage with the highest affinity for antigen and (2) can allow the isolation of specific phage occurring at a frequency as low as 1 in $10^7$ in the original library. Phage eluted from the final round of panning are converted to plasmids expressing soluble Fab by excision of the cpIII gene from the vector. Fab, either as a crude cell extract or following affinity purification, may then be assessed for reactivity with antigen.

## 3 ANTIBODIES FROM IMMUNE AND NONIMMUNE LIBRARIES

Recombinant antibodies were first isolated from combinatorial libraries derived from immunized mice and humans. Specific Fab have also been prepared from humans who have not been actively immunized but have at some time had contact with antigen, typically an infectious pathogen. We have shown that if polyclonal antibodies with specificities for a number of antigens can be detected in the donor serum, then these specificities may be isolated at a monoclonal level from the corresponding library. Several of these antibodies, as Fab fragments, have displayed powerful antiviral activities both in vitro and in vivo.

Isolating antibodies from an immune source has a particular advantage—the molecules will already have been selected by the host. However, there are many molecules of importance (epg., self antigens) against which no high affinity antibodies will be produced in the normal human host. Libraries containing antibody genes derived from so-called unimmunized and/or semisynthetic sources provide an opportunity to create these specificities. Unimmunized or "naive" antibody repertoires may be prepared using the IgD and IgM antibody classes, as these antibody gene pools are less biased by the immune history of the host than those encoding IgG. Antibodies selected from these libraries will normally be of lower affinity than those typically obtained from IgG libraries. However, because the genetic information is at hand, it is possible to increase the affinity of these clones through the random mutagenesis of selected complementarity-determining regions (CDRs) of the heavy chain, followed by reselection against antigen. Additionally, or alternatively, one can take a particular heavy or light chain and "cross" or reclone it into a light or heavy chain library, respectively, in the hope that it will pair with a more suitable partner. These strategies can often improve binding constants by more than two orders of magnitude.

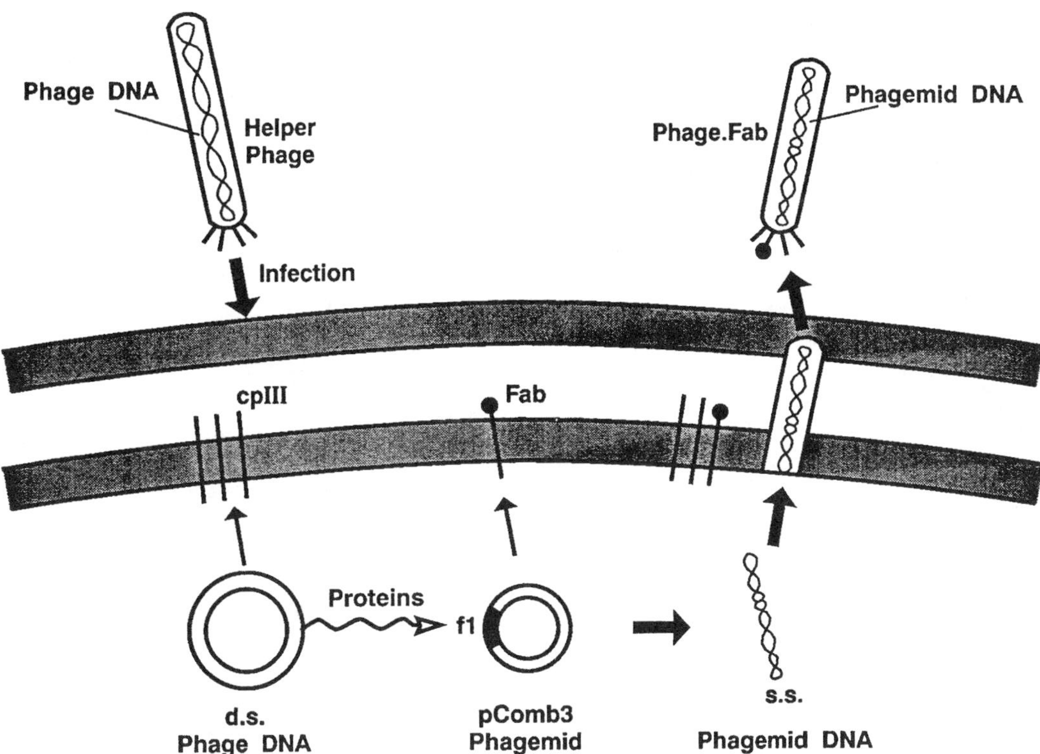


Figure 3. Helper phage rescue of pComb3 phagemid yielding a phage display library. Incorporation of the f1 origin of replication into pComb3 phagemid means that upon coinfection with M13 helper phage, single-stranded pComb3 DNA will be packaged and extruded from the bacterial cell as an assembled phage. As the assembling phage is extruded from the cell, it is "capped" by both native cpIII provided by the helper phage and cpIII-Fab complex derived from pComb3. The resulting phage has Fab displayed primarily monovalently on its surface and contains the corresponding antibody genes inside. [Adapted from D. R. Burton and C. F. Barbas, *Immunomethods,* 3:155–163 (1993).]

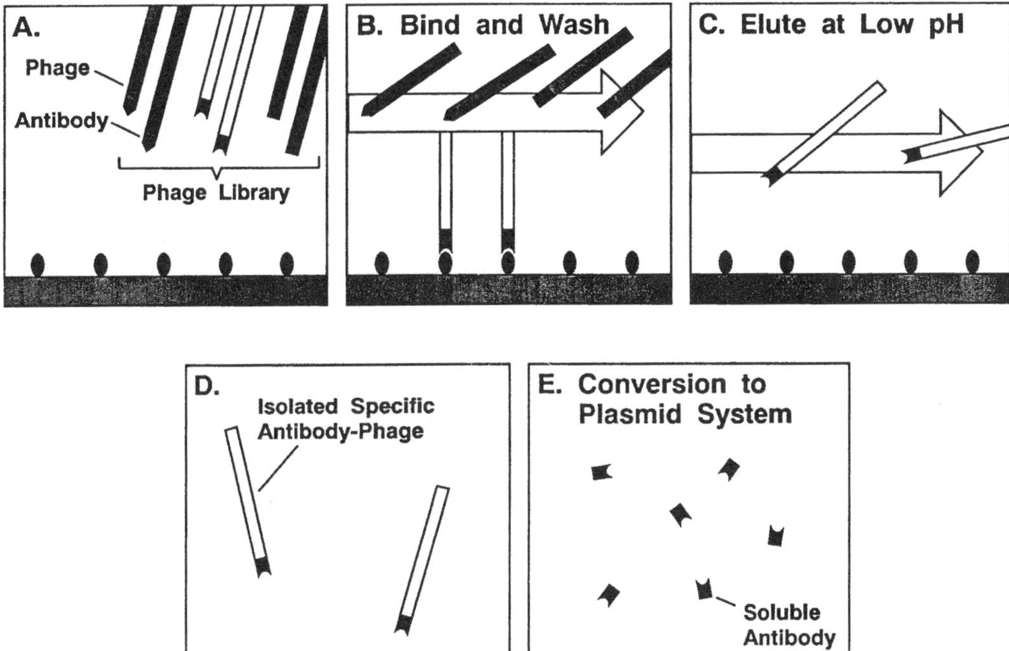


Figure 4. Panning of the combinatorial phage display library. Phage are incubated with immobilized antigen (A), and nonbound phage removed by washing (B). The remaining phage are then eluted in low pH or in the presence of excess soluble antigen (C, D). Eluted phage are reinfected into bacteria, reamplified, and reapplied to antigen. Repeating steps A–D three or four times results in a progressive enrichment for antigen specific phage-Fab. Phage from D are converted to the phagemid form of the vector, the DNA prepared and the cpIII gene excised. Religation then yields a reconstructed phagemid, which may be used to transform bacteria for the production of soluble Fab. [Adapted from D. R. Burton, *Hosp Pract.* 27:67–74 (1992).]

Semisynthetic libraries are constructed from naturally occurring antibody genes in which some or all of the CDRs are composed of synthetic gene segments. This aspect of library technology offers many possibilities, but is constrained by the difficulty of generating libraries large enough to sample a significant proportion of the diversity that may be generated. For example, the complete randomization of 16 amino acids in a single heavy chain CDR will alone require a library of more than $10^{20}$ clones to fully represent it. This figure is much greater than the transformation efficiency of *E. coli* ($10^{8}$–$10^{9}$) clones per microgram of vector DNA), although systems in which Fab are created via in vivo recombination between heavy and light chain libraries on different replicons may increase library size by two to three orders of magnitude. However, how immunogenic semisynthetic antibodies will prove to be in therapeutic applications remains to be seen.

*See also* Antibody Molecules, Genetic Engineering of; Immunology.

### *Bibliography*

Barbas, C. F., III, Kang, A. S., Lerner, R. A., and Benkovic, S. J. (1991) Assembly of combinatorial antibody libraries on phage surfaces: The gene III site. *Proc. Natl. Acad. Sci. U.S.A.* 88:7978–7982.

Burton, D. R., and Barbas, C. F., III (1993) Human antibodies from combinatorial libraries. In *Protein Engineering of Antibody Molecules for Prophylactic and Therapeutic Applications in Man,* M. Clark, Ed., pp. 65–82. Academic Titles, Nottingham, U.K.

Hoogenboom, H. R., Marks, J. D., Griffiths, A. D., and Winter, G. (1992) Building antibodies from their genes. In *Immunological Reviews,* Vol. 130, G. Møller, Ed., pp. 41–68. Munksgård, Copenhagen.

# Computing, Biomolecular

*Felix T. Hong*

### *Key Words*

**ATP (Adenosine Triphosphate)** A high energy compound that functions in many biological systems. It can be hydrolyzed either to AMP (adenosine monophosphate) or ADP (adenosine diphosphate).

**Cytochrome** A hemoprotein that contains an iron–porphyrin complex as a prosthetic group functioning as an electron carrier by virtue of the reversible valance change of the heme iron atom.

**Rhodopsin** A visual pigment that has an absorption maximum at 500 nm.

The notion of biomolecular computing differs from the conventional one as seen in the operation of a digital computer. Biomolecular computing is not mere number-crunching or performing logic operations; rather, it encompasses pattern recognition, process control, adaptation to environmental changes, optimization for long-term survival, and various other aspects of problem solving and information processing.

As described in elementary biology, information processing and process control are mediated by the nervous and the endocrine systems. These two systems have much in common at the membrane and molecular levels. In fact, both systems share many schemes on which their respective functional processes are carried out. This contribution examines the salient features of biomolecular computing from the point of view of a biophysicist and bioengineer.

## 1 FUNDAMENTAL FEATURES OF BIOMOLECULAR COMPUTING

In 1972 Michael Conrad and his colleagues were among the first to propose the use of biological information processing as an alternative architectural paradigm for future computers. The motivation stemmed from the apparent deficiency of modern digital computers in sophisticated decision-making processes such as pattern recognition, creative learning, and other cognitive processes. They pointed out that the predominant mode of biological information processing is not simple logic operations based on a finite set of switches, but a subtle form of information processing involving molecular recognition much like the matching of a key and a lock (shape-based information processing). A large variety of biomolecules distributed in a large number of different cell types constitute the infrastructure on which biological information processing is being carried out.

## 2 HIERARCHY OF THE DYNAMICS IN BIOCOMPUTING

There is a fundamental difference between conventional digital computing and biomolecular computing. The internal dynamics of a conventional computer is completely suppressed in the sense that the outcome of each constituent computing element is completely predictable. The digital computer is therefore under complete control of the software. In contrast, the internal dynamics of a biological organism is actually enlisted for information processing. Conrad and his colleagues characterized the organization of computational dynamics in biology as the Macro–micro (M-m) scheme of molecular computing. The signaling between neurons or other cells can be considered as macroscopic events. Beneath these macroscopic events, we can consider the intracellular molecular dynamics exhibited by chemical reactions and diffusion of both reactants and products. Thus, intracellular molecular information can be characterized as spatial and temporal patterns of various molecular species in the cell. The coupling between the macroscopic and the microscopic events is mediated by the so-called mesoscopic events at the membrane level, or rather signal transduction at the membrane level. Mesoscopic events occur on a spatial scale too small to be regarded as macroscopic but too large to be characterized as microscopic. The mesoscopic scale is comparable to the dimension of macromolecules and to the membrane thickness. Conrad and his colleagues pointed out the important role played by second messengers, such as cyclic AMP, cyclic GMP, and calcium ions. The action of these second messengers depends on the integrity of the plasma membrane.

## 3 MEMBRANE AS THE MESOSCOPIC SUBSTRATE

The importance of the plasma membrane cannot be overemphasized. The plasma membrane serves to maintain concentration gradients of many biologically important small ions, on which many macroscopic signaling processes are based. It also serves as an intermediary in the mesoscopic signaling processes that link the macroscopic to the microscopic dynamics. The plasma membrane is truly the mesoscopic substrate for biological information processing. Many important events occur in the mesoscopic range at the membrane surface and the membrane interior.

It is well known that biological membranes are formed by two molecular layers of phospholipid, of which the hydrophobic hydrocarbon domains form the core of the membrane, whereas the hydrophilic polar head domains border the two aqueous phases. While the phospholipid bilayer often is conveniently viewed as an inert supporting structure in which many integral proteins are embedded, the electrical events taking place on its two surfaces and its interior have an enormous impact on the mesoscopic events.

The plasma membrane often carries surface charges. These charges are located either at the polar head region of phospholipids or at the two hydrophilic domains of many transmembrane integral proteins. The surface charges confer surface potentials to the membrane. As a consequence, an electric field may appear at the membrane surfaces and/or inside the membrane phase. The electric potential must decrease (in magnitude) from its value at the membrane surface (i.e, surface potential) to the reference value (usually zero) at the remote bulk region. The remote bulk region is actually only a short distance away from the membrane surface, as indicated by the Debye length (< 1 Å in physiological solutions). The electric field, which is the negative slope of the potential profile, can be enormous within this range (known as the diffuse electrical double layer). As a consequence of a process known as charge screening, this electric field is highly localized because it is of a shorter range than that prescribed by the law of inverse squares.

Charge screening may be incomplete, and the electric field may extend to the opposite side of the membrane and thus confer a surface potential there even if the opposite membrane surface carries no surface charge. Any slight inequality of the two surface potentials will be translated into a significant intramembrane electric field primarily because of the small membrane thickness. For example, a modest difference of 60 mV between the two surface potentials leads to an intramembrane electric field of $10^5$ V/cm.

A similar situation occurs at the surface and the interior of a macromolecule. Many macromolecules contain charged groups at the exposed hydrophilic domain. These surface charges confer a zeta potential to the macromolecules and an intense but short-ranged electric field at the surface of the macromolecules. Likewise, surface charges or charged groups buried in the hydrophobic domain exert electric forces on the interior of the macromolecule and may affect its overall conformation. This intramolecular electric field is also responsible for the shift, to a longer wavelength, of the absorption band of the chromophore, retinal, in the visual pigment rhodopsin (a single-chain polypeptide with 348 amino acid residue, with 11-*cis*-retinal covalently bound to lysine residue 296).

These short-ranged electric fields play an important role in molecular recognition as well as in mesoscopic events associated with the plasma membrane. We shall refer to them collectively as electrostatic interactions. The adjective ''electrostatic'' is applicable even to when there is a dynamic change of charge distribution because the redistribution of ions in water is very fast compared to the speed of many conformation-mediated biological events. Thus, the electrostatic equilibrium always prevails in the mesoscopic molecular dynamics as well as in the majority of internal molecular dynamics.

## 4 ELECTROSTATIC INTERACTIONS AS CONTROL FACTORS OF THE INTERNAL DYNAMICS

Superficially, the intracellular dynamics appears to be rather stochastic because diffusion is a random process and chemical reactions in solutions are mediated by random collisions. One may wonder how coupling of reactions and diffusion can lead to meaningful events characteristic of intelligence in life. An examination of the factors that govern the intracellular dynamics is in order.

First of all, the cell is highly compartmentalized, and this compartmentalization is made possible by an extensive intracellular membrane system. Distributed among various compartments and often linking various membranous components is an intricate network of cytoskeleton, formed by a large variety of macromolecules that can undergo rapid polymerization and depolymerization in response to some mesoscopic factors. Thus, desirable reactants tend to be grouped together by transport processes, some of which are mediated by membranes and others by the cytoskeleton. Additional factors are also at work to make desirable chemical reactions less random and more deterministic. For example, consider the conversion of ADP and inorganic phosphate to ATP in the inner surface of mitochondrial inner membranes. The membrane potential across the inner membrane has a polarity that keeps ADP inside the matrix space, where ATP is being synthesized or hydrolyzed, and ATP outside. The reaction condition is thus tipped in the direction that favors the forward reaction (i.e., synthesis rather than hydrolysis of ATP).

Another important factor is the electrostatic interaction, mentioned earlier. Reactant molecules find each other by random collisions, but they are just as easily deflected from each other before they have time to initiate the desired reaction. Electrostatic interactions in the form of salt bridges tend to stabilize the reaction complex and to ensure the consummation of the desired reaction. In addition, charged groups placed at strategic locations have the effect of making molecular recognition more specific than mere shape fitting. An interesting example is provided by cytochrome *c,* which is a small membrane-bound redox molecule. Its prosthetic group, heme, can undergo redox reactions and vary its valence from +3 to +2 and vice versa. Its functional role is to serve in the mitochondrial membrane as a mobile electron carrier, shuttling electrons from the reducing end of the cytochrome $b$-$c_1$ complex to the oxidizing end of cytochrome oxidase. Cytochrome $c$ utilizes virtually the same charged groups to ''dock'' with either complex. Specificity conferred by matching pairs of surface charges also guarantees that the matching macromolecules have the correct relative orientation for the electron transfer to take place.

When a mesoscopic event takes place in the vicinity of a membrane, electrostatic interactions further make the event more efficient than otherwise. An example is provided by the distribution of negative surface charges at the mouth of the sodium ion channel. By virtue of the intense local electric field at the mouth region, the local (surface) concentration of sodium ions can be raised easily two orders of magnitude over the bulk concentration in the remote region, and thus the permeability is greatly enhanced without having to raise the bulk $Na^+$ concentration.

Electrostatic interactions can be used for switching purposes if a mechanism to vary surface charges exists. The surface charge density may be regulated by changing the pH, which alters the ionization states of acidic groups or basic groups. It can also be regulated by bringing about a change of the dissociation constant of those ionizable groups as a consequence of a conformational change: negative charges can be added or removed from macromolecules by means of phosphorylation or dephosphorylation of tyrosine, serine, or threonine residues of a protein. Since a phosphate group carries multiple negative charges, phosphorylation adds fixed negative charges to the exposed hydrophilic domain of a protein molecule whereas dephosphorylation removes them.

Many enzymes are activated by phosphorylation. We suspect that electrostatic interactions may play a prominent role in enzyme activation. Through judicious activations of selected enzymes, nature is able to preferentially channel reactions in desirable directions and to maintain optimal concentrations of various molecules. By virtue of a concerted regulatory regime, a high degree of determinacy can be achieved.

## 5 CONCEPT OF INTELLIGENT MATERIALS

The intricate control mechanism of the internal dynamics of biocomputing leads to a highly efficient and distinctly purposeful computing action of the organism, which is manifested as intelligence. How the organism acquired the features through evolutionary learning is stunning, but the process offers important lessons for investigators devoted to the development of an intelligent molecular machine. The apparent harmony of the internal dynamics from the macroscopic through the mesoscopic and down to the microscopic level can actually be extended to the submolecular level. In other words, many biomolecules possess a rudimentary form of intelligence.

The concept of intelligent materials was proposed by the U.S. Army Research Office and was first applied to the materials science of ceramics and composites. Intelligent or smart materials are composite materials in which sensors and actuators are embedded in the structural components. Japanese investigators extended the concept to include single molecules: all the sensor, actuator, and even processor capabilities are included and compacted into a single molecule. According to the Science and Technology Agency (STA) of Japan, intelligent materials are "materials with the ability to respond to environmental conditions intelligently and [to] manifest their functions."

Intelligent materials distinguish themselves from ordinary materials in that they are able to perform a host of individual tasks in a purposeful manner; moreover, they seem to be able to deliver more than conventional materials. An example illustrating this point is the function of hemoglobin, the oxygen carrier molecule of the blood. Hemoglobin consists of four subunits and can bind up to four molecules of oxygen. The oxygen-binding ability of hemoglobin is actually enhanced when more and more oxygen molecules are bound (cooperativity). This leads to a sigmoid-shaped oxygen-binding curve, which is also pH dependent (Bohr effect). The implication is that hemoglobin, under these conditions, actually binds more oxygen at the lung where the oxygen tension is high and the blood pH is low, and releases more oxygen in tissues where the opposite conditions occur.

Technical applications of intelligent biomaterials often call for modification of raw biomaterials because the users' purposes often differ from what was originally intended by nature. In the past, molecular engineering or protein engineering was a difficult and labor-intensive task. With the advent of modern recombinant DNA technology, it is possible to subcontract the task to bacteria. One need only provide the instruction in the form of the desired genetic code.

## 6 ERRORS, GRADUALISM, AND EVOLUTIONARY PROGRAMMING

As mentioned earlier, the internal dynamics of a conventional digital computer is completely suppressed to assure complete determinacy in the outcome of computation. In contrast, biomolecular computing actually exploits the internal dynamics and, as we shall see, also takes advantage of the errors that are inherent in the internal dynamics. Whereas errors in digital computing (known as "bugs") often cause a program to crash, errors in biomolecular computing are more subtle. This is because proteins, which make up the bulk of biomaterials, have sufficient built-in, mechanistically superfluous redundancy to provide stability, a unique feature known as gradualism. Thus, single point mutations usually do not generate a change drastic enough to disrupt stable folding of the protein, yet they provide variability on which evolutionary selection can work to improve its functional features. Thus there is an increase in the dimensionality of the adaptive surface on which evolution is depicted as moving from one peak of maximum novelty to another. As a result, it is often possible to find the combination of dimensions that will permit the investigator to avoid being thwarted by an unfathomable (untraversable) abyss separating the two peaks. This is what Conrad called extradimensional bypass.

## 7 FUTURE PERSPECTIVES

A major advantage of conventional digital computing lies in its programmability, both at the hardware and the software levels; that is, it is both structurally programmable and task programmable under complete human control. To achieve this high degree of programmability and, thus, a high degree of human control, the internal dynamics of the computing elements (which appear as logic gates, flipflops, etc.) must be completely suppressed. The input/output relationship is highly deterministic. However, such a rigid system completely lacks the ability to evolve both structurally and operationally. A single "point mutation" either at the hardware or the software level frequently paralyzes the computer. With the exception of the kind of software that modifies itself during its execution, the normal mode of operation of a digital computer requires a fixed layout of computational networks to operate reliably. The fixed architecture is largely responsible for the low resource utilization and low computational efficiency. Compounding this problem, a sequential (von Neumann type) digital computer further reduces resource utilization by having a "one-track mind." Most computing elements are idle most of the time. Biocomputing is just the opposite: biocomputing possesses high computational efficiency and high evolutionary learning capability but low structural programmability. This feature was formulated by Conrad and his colleagues as a *tradeoff principle,* stated as follows. "A computing system cannot have all of the following three properties: structural programmability, high computational efficiency, and high evolutionary adaptability."

Apparently, nature decided that the tradeoff is worthwhile. A biosystem is less deterministic, but the built-in redundancy makes the system fault tolerant. The seemingly chaotic reactions and diffusion comprising the microscopic dynamics are made more deterministic through compartmentalization, cytoskeletal dynamics, gene activation, enzyme catalysis, and so on. The biocomputing system is a highly versatile one. Sometimes it utilizes switches, sometimes it utilizes a more subtle mode of computation based on ''shape fitting'' between macromolecules. The moving charge carriers can be electrons, protons, or ions. On top of these motions, the atomic nuclei also undergo limited relative motion known as the conformational change of the protein. Last but not least, entire macromolecules can move from one subcellular compartment to another.

In the preceding discussion, we singled out electrostatic interactions as decisive forces that operate at the various levels of the internal dynamics. Electrostatic interactions often carry out the switching function, which makes the networklike connections rather dynamic. This switching function pervades all levels of the internal dynamics. At the submolecular level, electrostatic interactions are important in controlling the protein conformations. At the microscopic level, electrostatic interactions restructure the networklike reaction–diffusion regime.

At the mesoscopic level, the electrostatic interactions apparently play an important role in many membrane-related phenomena. An outstanding example is provided by the state 1–state 2 transition in the photosynthetic membrane of the green plant chloroplast. The light-regulated cycle of phosphorylation and dephosphorylation of a key integral protein complex is the major control for dissociation and association of photosystem I and photosystem II—a process tantamount to forming and breaking of a network.

Electrostatic interactions may also be important in establishing new neuronal network connections. This is achieved through phosphorylation and dephosphorylation of synapsin I and a host of related phosphoproteins. Synapsin I is an integral protein component of synaptic vesicle membranes. Its phosphorylation state profoundly affects the efficiency of neurotransmitter release and is believed to play an important role in neural plasticity.

Through the structural control of networking at the various levels, the biocomputing system is able to achieve high computational efficiency. This is accomplished through self-organizing and self-assembly of the components. External forces, such as human intervention, have little control. If humans had total external programming control, death would not be inevitable.

Throughout the various levels of interactions, the entire biosystem works in a concerted manner as if each part knows what the other part expects and acts accordingly in harmony. The result is an enhanced functional performance suggestive of intelligence. This intelligence can be traced from the macroscopic level all the way down to the submolecular level. The concept of intelligent materials serves as an inspiration and a guiding principle for technology-minded investigators who attempt to engineer biomaterials for molecular device construction.

The intelligence unique to biosystems is the product of evolutionary learning. The errors, which appear as a by-product of unleashing the internal dynamics in biocomputing, actually fuel evolutionary learning. By the same token, we suspect that the human frailty of making occasional or not so occasional errors may turn out to be the source of human creativity. After all, a completely deterministic machine can only reflect the intelligence of its creator.

Although it is risky to second-guess nature, it appears that molecular materials were chosen instead of inorganic materials for construction of living organisms because of the rich repertoire of chemical reactions unique to organic materials. Whereas conventional computing exploits the physical properties of bulk inorganic materials, biomolecular computing includes both physical and chemical properties of organic materials as the ''substrates'' for building the internal dynamics.

Molecular machines are perceived as being fragile, and yet nature seems to choose molecular materials as the building blocks. The reconciliation of this puzzle lies in the observation that a biosystem is capable of self-repair. Self-repair and self-assembly are the keys to the success of using molecular materials for device construction by nature and so will be the keys if humans attempt to build molecular machines from molecular materials.

*See also* BACTERIORHODOPSIN-BASED ARTIFICIAL PHOTORECEPTOR; BIOMOLECULAR ELECTRONICS AND APPLICATIONS; BIOSENSORS.

### *Bibliography*

Alberts, B., Bray, D., Lewis, J., Raff, M., Roberts, K., and Watson, J. D. (1989) *Molecular Biology of the Cell,* 2nd ed. Garland, New York and London.

Blank, M., and Vodyanoy, I., Eds. (1994) *Biomembrane Electrochemistry,* ACS Advances in Chemistry Series No. 235. American Chemical Society, Washington, DC.

Conrad, M. (1990) Molecular computing. In *Advances in Computers,* Vol. 31, M. C. Yovits, Ed., pp. 235–324. Academic Press, Boston, San Diego, New York, London, Sydney, Tokyo, Toronto.

———, Ed. (1992) Special issue on molecular computing. *Computer (IEEE)* 25(11).

Hameroff, S. R. (1987) *Ultimate Computing: Biomolecular Consciousness and Nanotechnology.* Elsevier/North Holland, Amsterdam.

Hong, F. T. (1992) Intelligent materials and intelligent microstructures in photobiology. *Nanobiology,* 1:39–60.

———. (1992) Do biomolecules process information different than synthetic organic molecules?, *BioSystems,* 27:189–194.

———, Ed. (1994) Special issue on molecular electronics. *IEEE Eng. Med. Biol.* EMB-13(1).

Kaminuma, T., and Matsumoto, G., Eds. (1991) *Biocomputers: The Next Generation from Japan.* Chapman & Hill, London, New York, Tokyo, Melbourne, and Madras. (Translated by N. D. Cook.)

# CYTOCHROME P450

*Michael R. Waterman*

### *Key Words*

**Endogenous Substrates** Substrates for enzymatic (P450) reactions which are natural compounds synthesized within the organism harboring the enzyme.

**Exogenous Substrates** Substrates for enzymatic (P450) reactions which are derived from the environment surrounding the organism harboring the enzyme.

**Hemoprotein** Protein containing the prosthetic group heme.

**Superfamily** Group of proteins that are evolutionarily related as determined by their primary amino acid sequence.

Cytochrome P450 is the generic name applied to a large superfamily of hemoprotein, mixed-function oxidases that metabolize a structurally diverse group of exogenous and endogenous organic substrates. The name is derived from the prominent absorption band observed at 450 nm following reduction of the heme iron and its coordination with carbon monoxide. These enzymes are widely distributed among microorganisms, plants, insects, fishes, and animals, and they catalyze a reaction of the general nature:

$$O_2 + \underset{\text{biological reducing equivalents}}{NAD(P)H} + \underset{\text{organic substrate}}{AH} \xrightarrow{P450} \underset{\text{hydroxylated organic product}}{AOH} + H_2O + NAD(P)$$

In the case of exogenous substrates, this reaction most frequently is involved in an organism's effort to detoxify foreign compounds (xenobiotics) derived from the environment. In the case of endogenous substrates, the reaction is generally involved in the production of biologically more active compounds (e.g., steroid hormones) from less active compounds (e.g., cholesterol).

## 1 CHARACTERISTICS OF P450s

Ryo Sato and Tsuneo Omura described in 1962 an unusual pigment in a subcellular fraction of rabbit liver, the endoplasmic reticulum, which upon reduction and coordination with carbon monoxide showed an intense absorbance band at 450 nm. Hence the name, *pigment 450 nm* or *P450.* This substance was subsequently clearly established to be a protoheme-containing protein, the same prosthetic group associated with hemoglobin. However, P450 and hemoglobin are quite distinct in that when reduced and coordinated with CO, hemoglobin has an intense absorbance band at 420 nm. It is now apparent that P450 has a thiolate ligand provided by cysteine in the fifth coordination position of the heme iron, while hemoglobin contains an imidazole ligand from histidine at this coordination position. This difference is the basis of the spectral difference between these two heme proteins. Functionally, hemoglobin serves to reversibly bind and transport oxygen and therefore does not reduce $O_2$, while P450 function hinges on the ability to reduce oxygen during the mixed-function oxidation reaction. The thiolate ligand participates by lowering the redox potential of the heme iron in P450, thereby facilitating oxygen reduction.

In bacteria, cytochromes P450 are soluble in the cytoplasm, while in all higher species including plants, P450s are firmly anchored in membranes. Most forms of P450 are found in the endoplasmic reticulum, where they face the cytosol. A few forms of cytochrome P450 are localized to the inner mitochondrial membrane, where they face the matrix.

Cytochromes P450 usually cannot function on their own. They require transfer of reducing equivalents from reduced pyridine nucleotides (NADH or NADPH) to the P450 heme iron. In bacteria, virtually all forms utilize electrons from NADH, which are transferred via an electron transport chain consisting of a flavoprotein that extracts electrons from NADH and passes them on to an iron–sulfur protein, which in turn reduces the P450. Mitochondrial P450s represent a variation on this theme whereby electrons are transferred from NADPH via a flavoprotein and iron–sulfur protein located in the mitochondrial matrix. P450s in endoplasmic reticulum (microsomes) derive their reducing equivalents from NADPH via a ubiquitous flavoprotein known as NADPH cytochrome P450 reductase. This flavoprotein is also firmly embedded in the membrane of the endoplasmic reticulum. In animals, there appears to be a single form of P450 reductase, which services a rather large number of different forms of microsomal P450.

Thus one way of classifying P450s is based on the manner in which electrons are transferred.

Bacterial and Mitochondrial P450s:

$$NAD(P)H \rightarrow \underset{\text{(FAD-containing)}}{\text{flavoprotein}} \rightarrow \underset{\text{(2Fe-2S protein)}}{\text{iron–sulfur protein}} \rightarrow P450$$

Microsomal P450s:

$$NADPH \rightarrow \underset{\text{(FAD + FMN-containing)}}{\text{flavoprotein}} \rightarrow P450$$

Recently an interesting variation of this microsomal P450 system has been characterized as a soluble protein in bacteria. A P450 protein is extended at its carboxyl terminus by a domain having most of the characteristics of the microsomal P450 reductase. This fusion protein is the sole known bacterial P450 to resemble the microsomal P450 system rather than the mitochondrial system and is the single form of P450 able to function on its own with only the presence of NADPH.

## 2 ACTIVITIES OF P450s

Virtually every animal cell contains one or more forms of P450. Another way of classifying P450s has been based on their enzymatic properties. In the broadest sense, P450s can be grouped into forms that metabolize endogenous substrates and forms that metabolize exogenous substrates.

The forms involved in endogenous substrate metabolism tend to be localized to specific functional sites such as steroidogenic tissues, (adrenal, gonads, placenta) or a specific organ such as the kidney or liver, while the forms involved in exogenous substrate metabolism are sometimes widely distributed. Sites of exogenous metabolism are found in every organ including the skin, with highest concentration and numbers of these forms of P450 being found in the liver. Recently is has been learned that unique P450s can exist in only a limited number of cells such as the nasal epithelium, and it is imagined that many forms of P450 yet undiscovered reside in such limited locations.

P450-mediated endogenous substrate metabolism includes steps in cholesterol biosynthesis (14-demethylase), biosynthesis of steroid hormones from cholesterol (progesterone, aldosterone, cortisol, testosterone, estrogen), bile acid biosynthesis, fatty acid hydroxylation, and arachidonic acid metabolism. These reactions lead to production of key regulators of biological processes such as reproduction and vascular activity. Other potential roles for P450s include biosynthesis of neurosteroids and activation or inactivation of important compounds in growth and development, including derivatives of vitamin A. In the best studied of the endogenous systems, that involved in biosynthesis of steroid hormones, it is clearly established that amino acid mutations within the P450 polypeptide chain lead to genetic diseases that alter the homeostasis and even the phenotype of the individual. It can be presumed that genetic diseases will be found to be associated with most or all of the forms of P450 metabolizing endogenous compounds.

While the endogenous activities just listed clearly define the biological significance of the P450 systems, P450-mediated metabolism of xenobiotic compounds is equally important. This too, represents a broad scope of activities ranging from use of environmental compounds as the sole carbon source in bacteria to the biotransformation and detoxification of xenobiotics in animals. As a general consideration, these forms of P450 may metabolize structurally diverse chemicals and demonstrate lower substrate specificity than do the forms involved in metabolism of endogenous compounds. Types of biotransformations include N-, S-, and C-hydroxylations and dehalogenations, as well as deaminations, dealkylations, and reductions of countless drugs, chemical carcinogens, mitogens, and other environmental contaminants. Included in xenobiotic biotransformation is drug metabolism, an issue of considerable interest to the pharmaceutical industry, particularly in light of the variability in the levels of different P450s in different people. In studies with laboratory animals, inbred strains are generally used in which all individuals contain relatively constant levels of the different forms of P450. Humans are, of course, outbred, and each individual can be considered to have his or her own unique P450 profile. Consequently the effectiveness of drug therapy may be dependent on this profile, the individual P450 pattern being dependent on genetic (sex, age) as well as environmental (nutrition, exposure to foreign compounds) factors. In some cases different forms of P450 metabolize the same chemical substrate by different reactions, leading to different patterns of products. Thus individual variation in the levels of different P450s can significantly affect drug metabolism, particularly in instances of combined drug therapy. Clear examples of these individual variations are seen in the polymorphisms that exist within the human population with respect to P450-dependent metabolism of certain drugs, such that specific individuals will be defined as poor metabolizers (i.e., they do not clear the drug as efficiently as the general population, with resultant unwanted effects reminiscent of drug overdose).

While the detoxification of foreign compounds is generally beneficial to animals, in some cases reactive intermediates produced by P450 metabolism can be toxic, mutagenic, or carcinogenic. For example, benzo[*a*]pyrene, a polycyclic aromatic hydrocarbon in tobacco smoke, is metabolized by a specific form of P450 into a carcinogen that is thought to play a key role in lung cancer.

## 3 HOW MANY P450s ARE THERE?

The P450 superfamily is both large and evolutionarily ancient. While subcellular location and enzymatic activity are important characteristics of the P450s, these properties do not provide the most useful approach to classifying this large number of proteins. Rather, a systematic classification for P450s has been developed based on amino acid sequence. This scheme leads to classification into gene families and subfamilies based on primary sequence relatedness. Thus the 30 known human P450s are distributed among 12 gene families scattered among 9 different chromosomes. In addition to the mammalian P450s, unique gene families have been found in insects, snails, yeast, fungi, plants, and bacteria. The total number of known P450s in biology exceeds 240 and is rapidly increasing as new forms are identified.

In humans, the most abundant forms of P450 have already been identified. Nevertheless, we can expect that a substantial number of forms that are localized in specific cell types in the brain, intestine, and other organs remain to be discovered. While these may be of relatively low abundance compared to many of the forms already known, they will be surely found to play very important roles. For example, in the nasal epithelium unique forms of P450 have been found which presumably play important roles in detoxification and protection from desensitization of the sensory system involved in smell. With the application of the powerful techniques of molecular biology to the identification of P450s, the sequences of proteins are easily determined by cloning, but the substrate specificity and therefore the function of these proteins cannot be so easily elucidated. Thus we now know the sequence of several P450s for which the function remains unknown. It may be that there will be as many as 200 different forms of P450 identified in humans. Even though P450s have related amino acid sequences and, in fact, contain diagnostic sequence motifs, universal P450 probes (either immunological or oligonucleotides) have not yet been developed. Thus we can expect that it will take several more years before the majority of the rare human P450s are identified.

Since eukaryotic forms of P450 are integral membrane proteins, they are difficult to release from their membrane environment and subsequently purify. Thus it has been difficult to obtain sufficient quantities of these P450s for biophysical investigation of their structure–function relationships. Accordingly, the only detailed structural information available on members of the P450 superfamily is from soluble, bacterial P450s. While this information provides a general picture of P450 structure, until the detailed structure of eukaryotic forms of P450 is determined we will not know for certain the common structural features associated with all P450s. Nevertheless, the ability to carry out site-directed mutagenesis based on predictions from bacterial P450 structural analysis, coupled with heterologous expression of cDNAs in systems having low or absent background levels of P450, is leading to detailed information on how P450s work.

## 4 REGULATION OF P450s

Regulation of gene expression is one of the key biological processes controlling development, tissue specificity, and homeostasis. The developmental roles of the P450 superfamily are not yet well defined. In a few instances it is clear that the timely expression of specific genes encoding P450s is essential; for example, formation of male secondary sex characteristics requires expression of P450s involved in testosterone synthesis at a specific moment in early fetal life. Certainly it will be found that various forms of P450 play important roles in different aspects of growth and development, including metabolism of endogenous compounds that serve as signals for different growth processes. Likewise, the molecular basis of tissue-specific expression of P450s is not yet well understood, but many forms are found only in certain cell types, indicating that complex regulatory processes are important at this level as well. The maintenance of P450 activities associated with endogenous substrate metabolism throughout adult life is known to be controlled by other endogenous compounds—for example, peptide hormones and steroid hormones. However, the detailed biochemistry of regulation of these genes is not yet understood.

One of the intriguing aspects of the P450 system is that many forms involved in the metabolism of exogenous compounds are induced (increased) by exogenous compounds. Often substrates for specific forms of P450 will induce these forms; for example, the forms of P450 that metabolize polycyclic aromatic hydrocarbons are induced by polycyclic aromatic hydrocarbons. Thus the environment regulates the drug-metabolizing profile in individuals. In some of these instances a particular P450 is undetectable until the individual is challenged by specific xenobiotics, while in other cases lower levels are increased to higher levels by such challenge. In summary, the regulation of the P450 profile in individuals will prove to be a very complex process involving developmental, tissue-specific, endogenous, and exogenous factors.

## 5 FUTURE DIRECTIONS

Many questions remain to be answered concerning the P450 superfamily in relation to the number of P450s, their structure–function relationships, their roles in key biological process such as growth and development, and the basis on which expression of P450 genes is regulated. In addition, the application of the unique P450-dependent chemistry to the production of fine chemicals or to the removal of environmental contaminants using microorganisms and engineered P450s in heterologous systems can be anticipated. Finally, we can foresee a time when P450 profiles will be analyzed in a noninvasive fashion and drug regimens for treatment of diseases will be tailor-made for the individual. The P450 superfamily will surely continue to be a major focus of biomedical research, yielding intriguing new discoveries well into the twenty-first century.

*See also* METALLOENZYMES.

### *Bibliography*

Gonzalez, F. J. (1988) The molecular biology of cytochrome P450s. *Pharmacol. Rev.* 40:243–488.

Guengerich, F. P., Ed. (1992) Cytochrome P450: Advances and prospects. *FASEB J. (thematic issue),* 6(2).

Nebert, D. W., and Gonzalez, F. J. (1987) P450 genes: Structure, evolution and regulation. *Annu. Rev. Biochem.* 56:945–993.

Nelson, D. R., Kamataki, T., Waxman, D. J., Guengerich, F. P., Estabrook, R.W., Feyereisen, R., Gonzalez, F. J., Coon, M. J., Gunsalus, I. C., Gotoh, O., Okuda, K., and Nebert, D. W. (1993) The P450 superfamily: Update on new sequences, gene mapping, accession numbers, early trivial names of enzymes, and nomenclature. *DNA Cell Biol.* 1:1–51.

Waterman, M. R., and Johnson, E. F., Eds. (1991) *Cytochrome P450,* Vol. 206 in *Methods in Enzymology.* Academic Press, Orlando, FL.

# CYTOKINES

*Alan G. Morris*

### *Key Words*

**Cytokine** A polypeptide factor produced transiently by a range of cell types, acting usually locally, altering the physiology of target cells by binding to cell surface receptors and activating the expression of specific genes.

**Growth Factor** Cytokine controlling proliferation or survival of cells.

**Interleukin** Cytokine produced by leukocytes and controlling leukocyte function.

**Lymphokine and Monokine** Cytokines produced by lymphocytes and monocytes, respectively: terms now becoming obsolete.

About 40 polypeptide factors currently are regarded as cytokines. Examples include the interleukins, the interferons, and the tumor necrosis factors, which are involved in control of growth and differentiation of many cell types. A large group of cytokines is important in hematopoiesis and immune or inflammatory responses. Usually, cytokines are produced in response to more or less specific stimuli, and their production ceases when the stimulus is removed. Growth factors, a set of factors often considered in connection with the cytokines, control the growth or survival of nonhematopoietic cells in both adult and embryo, and these are produced constitutively. Some cytokines are important in the therapy of disease, and others may become so.

## 1 DEFINITION AND GENERAL PROPERTIES OF CYTOKINES

As with many biological phenomena, it is impossible to provide precise definitions of what a cytokine is: the best that can be done is to build up an idea of what they are by describing their general properties. Cytokines are a large and disparate group of agents, however, and all generalizations about their properties are subject to exception.

All factors regarded as cytokines are proteins with molecular weights ranging from a few thousands to a few tens of thousands. They may be thought of as short-range hormones, and it is not easy to justify their separate categorization inasmuch as they share many properties with polypeptide hormones. Classical hormones such as insulin are produced by specialized cells within particular organs and act at a distance via the circulation. Cytokines on the other hand tend to be produced locally, often by nonspecialized cells (i.e., cells that are not devoted exclusively to the production of one particular cytokine). They often function locally (at the site of production), and their half-life in the circulation is generally short: a clear exception to this generalization is erythropoietin, which is produced by the kidney and stimulates terminal erythrocyte differentiation in the bone marrow. The production of cytokines is often transient; this is especially true of cytokines involved in immune responses produced by activated lymphocytes or macrophages. However, growth factors—including hematopoietic growth factors—are produced constitutively by certain cell types (e.g., bone marrow stromal cells, fibroblasts).

Many cytokines play a role in hematopoiesis or the immune response, and these factors were originally described in this connection, but it is wrong to think of them exclusively as effector molecules of the immune system: cytokines also have nonimmunological functions (e.g., in wound healing). Equally, it is wrong to suppose that other diffusible factors, not regarded as cytokines, have no immunological or hematological effects: both prostaglandins and some classical hormones—steroids, for example—can have immunological effects. These different systems overlap and interact.

Cytokines act on their target cells via cell surface receptors, since as proteins they are unable to diffuse into cells and act directly on intracellular structures, as steroid hormones do. Often but not exclusively the receptors are specific for a particular cytokine, but there are examples of related cytokines binding the same receptor. Cells lacking a receptor for a particular cytokine will not respond to that cytokine. The binding of a cytokine to its receptor activates physiological changes in the target cell. These changes usually depend on the activation of cellular genes through intracellular signaling systems targeted on transcription factors (and so take some time), but there are also examples of cytokines activating preexisting enzyme systems within cells (and so acting much more quickly). The receptors for cytokines are now in many cases cloned and so defined in molecular terms: they fall into a number of gene families. Except for growth factor and interferon signaling, however, signaling pathways are not yet fully defined. Growth factors function via activation of a cascade of protein kinases, ultimately activating nuclear transcription factors: interferons function via activation of a multimeric cytoplasmic factor, which translocates to the nucleus and there acts as a transcription factor.

It is characteristic of cytokines that they are often *redundant* in their effects (i.e., different cytokines may have the same effect); moreover, they often have *multiple* effects on the same cell, and they are often *pleiotropic* in their actions (i.e., they have different effects on different cells). The multiplicity of actions is due to the activation of different subsets of genes in different cells, presumably—since there is no evidence that different cell types have different receptors for individual cytokines—as a consequence of different signaling pathways downstream of the receptor. Different cytokines may *interact*, either synergistically or antagonistically or merely in an additive manner; this feature gives rise to the idea of a complex *cytokine network.*

In the early days of cytokine research, the multiplicity of cytokine effects led to great confusion about just how many distinct proteins were involved—whether the different activities were due to one or several proteins in a given crude preparation. It was far from certain, for example, that antiviral effects *and* cell growth inhibitory effects *and* activation of histocompatibility antigens ascribed to interferon were due to one and the same protein. The resolution of this confusion came through purification—no easy task, since cytokines generally have very high specific activities (i.e., number of units of biological activity per milligram of pure protein), hence are present at very low concentrations in crude preparations. The use of monospecific antibodies in cross-neutralization experiments can establish the molecular identity of two different activities, but interactive effects may nevertheless confuse the issue. Finally, even though gene cloning and expression allow fairly definitive allocation of functions to molecules, one caveat remains: since cytokines can induce other cytokines, the *apparent* activity of cytokine A *may* be due to induction of cytokine B, and this possibility may be hard to exclude.

The foregoing properties of cytokines—redundancy, pleiotropy, and interaction—make it extremely difficult to identify their physiological roles in vivo, and indeed these roles have not been established in detail. It seems unlikely that *all* the many properties described in vitro for a given cytokine are important, or at any rate equally important, in vivo. One experimental approach is the administration of cytokines to observe their in vivo effects on, for example, the development of disease. This simple route has many disadvantages, not least that if administered via the circulation, the cytokine may not reach the appropriate site of action: more subtly, the administration of a cytokine may make no difference where that particular cytokine is already perhaps present in excess through local production. A better design is to eliminate a cytokine from an animal and determine how this change affects the subject's physiology or response to disease. For example, antibodies can be used to deplete cytokines from tissues. The disadvantage here is that the antibody administered into the circulation again may not reach the site of cytokine production and action. More interesting is the gene knockout approach, where a cytokine gene is eliminated by homologous recombination and transgenic animals lacking the cytokine gene are produced. Derangements in the subjects' physiology or response to disease may be very informative. A good example is the deletion of the gene for transforming growth factor β in mice: the animals die within a few days of birth with massive inflammatory infiltrates in many tissues, implying a major role for this cytokine in the control of inflammation.

## 2 NOMENCLATURE

The nomenclature of cytokines is still chaotic and reflects the piecemeal development of our understanding of these factors. That the same factor can have multiple effects has led to the same factor having multiple names. These are listed in Table 1, together with the usual abbreviations.

An attempt to systematize nomenclature by the use of the term *interleukin* (IL) with a number has been incompletely successful. The term was introduced to indicate that the factor in question was involved in interactions between leukocytes: however some ILs may be produced by nonleukocytes and plainly act on nonleukocytic cells. Other factors that clearly are important in leukocyte interactions retain other names. At present, however, new leukocyte products are always assigned IL numbers upon the establishment of a gene sequence. IL-13 is the most recent.

The terms *lymphokine* and *monokine,* to indicate cytokines of lymphocytic and monocytic origin, respectively, are dropping out of use. The meaning of the term *growth factor* is obvious except that at least one of the cytokines so designated (TGF-β) is growth inhibitory; many of the ILs also function as growth factors. The *colony-stimulating factors* are also growth factors but acquired their name through the original assay for their activity, which was by the development of colonies of hematopoietic cells in soft agar cultures. The *interferons* are named from the phenomenon of viral interference in which infection with one virus may prevent superinfection with another. Several other factors are named specifically on a functional basis, reflecting usually the phenomenon by which each was first recognized (*tumor necrosis factor, leukemia inhibitory factor,* etc.).

## 3 CYTOKINE FUNCTIONS

Table 1 lists the cytokines according to the most commonly used name, with a selection of synonyms, loosely grouped according to nomenclature rather than function. What may be the major functions are indicated: we have already warned that it is difficult to be definitive about function. Cross-references identify cases of factor names that may be misleading in terms of function (e.g., some hematopoietic growth factors are listed in the interleukin section). Molecular weights are given because they *may* be helpful for identification (e.g., on Western blots); otherwise, however, they are of little interest and indeed may be misleading owing to variation in glycosylation. Typical cell sources are given, but there is no guarantee that the cell type mentioned is the primary producer in vivo.

## 4 CYTOKINES IN PATHOLOGY AND THERAPY OF DISEASE

Although cytokines are obviously important as protective responses to disease, in some situations they play a part in *pathogenesis.* This is probably most obvious in autoimmune and allergic diseases, where the inflammatory effects of inappropriately produced cytokines damage normal tissue. But inappropriate cytokine production

**Table 1** Listing of Cytokines

| Usual Abbreviation | Other Names in Common Use[a] | Molecular Weights (kDa)[b] | Common Cell Source[c] | Accepted Major Functions |
|---|---|---|---|---|
| | | The Interleukins (ILs) | | |
| IL-1[d] -α, -β | Lymphocyte activating factor (LAF): endogenous pyrogen (EP) | 17.5 | *Macrophage,* keratinocyte, astrocyte, etc. | Accessory factor in T-cell activation: stimulates acute phase response |
| IL-1 receptor antagonist, IL-1ra | | 25 | *Macrophage* etc. | Blocks binding of IL-1 to its receptor *without* activating target cell, hence is antagonist |
| IL-2 | T-cell growth factor (TCGF) | 15.5 | T cell ($T_h1$ or Tc) | Supports proliferation and activation of T, B, and NK cells |
| IL-3 | Multi-colony-stimulating factor (multi-CSF) | 15–25 | T cell ($T_h1$ & 2) | Hematopoietic growth factor |
| IL-4 | B-cell growth (or stimulatory) factor I (BCGF I or BSF I) | 25 | T cell ($T_h2$) | Activates B cells |
| IL-5 | Eosinophil differentiation factor (EDF) or B-cell growth factor (BCGF) II | 40–50 (dimer) | T cell ($T_h2$) | Promotes eosinophil growth and differentiation: accessory factor in B-cell activation |
| IL-6[e] | B-cell stimulatory factor (BSF) II; hepatocyte-stimulating factor (HSF) | 26 | T cell ($T_h2$), macrophage fibroblast, endothelial cell, etc. | Stimulates acute phase response: accessory factor for B-cell activation |
| IL-7 | Lymphopoietin | 20–25 | Bone marrow stroma | Hematopoietic growth factor |
| IL-8 and many related factors[f] | Monocyte-derived neutrophil chemotactic factor (MDNCF); neutrophil-activating factor (or peptide) (NAF or NAP), etc. | 8–9 | *Macrophage* etc. | Chemotactic factor for neutrophils |
| IL-9 | | 52 | T cell ($T_h2$) | Mast cell, T-cell growth factor |
| IL-10 | Cytokine synthesis inhibitory factor (CSIF) | 30–35 | T cell ($T_h2$) | Inhibitor of $T_h1$ cell cytokine production and macrophage function |
| IL-11 | | 23 | Fibroblasts, trophoblasts | Hematopoietic growth factor |
| IL-12 | Natural killer cell stimulatory factor (NKSCF) | 75 (hetero-dimer) | Macrophage B cell | Activates NK cells: controls $T_h1/T_h2$ cell differentiation |
| IL-13 | | 17 | T cells | Inhibits macrophage activation: anti-inflammatory |
| | Hematopoietic Growth Factors: Colony-stimulating factors (CSFs), etc.; See also IL-3, IL-7, IL-11 | | | |
| Monocyte (M-CSF) | CSF-1 | 44–60 (dimer) | T cells, bone marrow stroma, etc. | Stimulates growth and differentiation of monocyte precursor |
| Granulocyte/ monocyte (GM-CSF) | | 18–24 | T cells, ($T_h2$), bone marrow stroma, etc. | Stimulates growth and differentiation of myeloid precursor: macrophage activator |
| Granulocyte G-CSF[e] | | 20 | Macrophage, bone marrow stroma, etc. | Stimulates growth and differentiation of granulocyte precursor: neutrophil activator |
| Stem cell factor (SCF) | C-*Kit* ligand, mast cell growth factor, steel factor | 53 | Bone marrow stroma, hepatocyte | Acts synergistically with other factors on early hematopoietic precursors |
| Erythropoietin (EO) | | 30 | Kidney | Stimulates growth and differentiation of erythroid precursor |

| Nonhematopoietic Growth Factors (GFs) and Other Growth Regulatory Cytokines | | | | |
|---|---|---|---|---|
| Fibroblast GFs 1–7 (FGF-1 to FGF-7)[f] | FGF-1: acidic (aFGF) FGF-2: basic (bFGF) FGF-7: keratinocyte (kFGF) | 16–18 | Central nervous system, etc. | Mitogenic for many cell types, angiogenic |
| Nerve (NGF) | | 26.5 (dimer) | Central nervous system, etc. | Mitogen |
| Platelet-derived (PDGF) | | 30 (dimer) | *Platelets,* fibroblasts, etc. | Mitogen |
| Insulin-like (IGF-1, IGF-II) | | | Liver | Mitogen |
| Epidermal (EGF) | | 6 | Many | Mitogen |
| Transforming: -α, TGF-α | | 6 | Many | Mitogen |
| TGF-β[g] | | 25 | Many | Growth INHIBITOR: stimulates wound healing; anti-inflammatory; angiogenic |
| Hepatocyte growth factor (HGF) | Scatter factor | 82 | *Platelets,* etc. | Mitogen, stimulates movement |
| Oncostatin-M (OSM) | | 28 | T cells, other leukocytes | Growth INHIBITOR for tumor cells: stimulates acute phase response |
| Many others | | | | |
| Antiviral Cytokines: The Interferons (IFNs) | | | | |
| IFN-α[f] | Type 1: leukocyte IFN | 20 | Many | Antiviral: activates class I: MHC activates NK cells |
| IFN-β | Type 1: fibroblast IFN | 20 | Many | Antiviral: activates class I MHC: activates NK cells |
| IFN-γ | Type 2, immune IFN, macrophage activating factor (MAF) | 20–25 | T cells ($T_h1$ or Tc) | Antiviral: activates class I and II MHC; activates macrophages |
| Tumor Necrosis Factors (TNFs) | | | | |
| TNF-α | Cachectin | 45 trimer | Macrophage | Cytotoxic: activates cytokine and MHC antigen production |
| TNF-β | Lymphotoxin (LT) | 60–75 trimer | T cell ($T_h1$ or Tc) | Cytotoxic activates phagocytes |
| Cytokines Involved in Pregnancy and Proliferation of Embryonic Cells | | | | |
| Leukemia inhibitory factor (LIF)[e] | Hepatocyte-stimulating factor (HSF); differentiation-inducing factor (D-factor, DIF, DIA, etc.) | 32–67 | Uterus | Growth factor for embryonic stem cells; growth INHIBITOR for tumor cells; hemotopoietic growth factor: stimulates acute phase response |
| Ovine and bovine trophoblast protein-1[h] | Antiluteolytic protein | | Trophoblast | |
| Many others, including growth factors listed above | | | | |

[a]Not an exhaustive list—many other terms are in use, especially in the older literature.

[b]Approximate: variation will occur, usually because of differing degrees of glycosylation. Where the normal form is (usually!) multimeric, this is indicated.

[c]Not exhaustive: where several cell types produce a cytokine, the probable major source is italicized.

[d]There are two genes and so two proteins, with about 25% identity at the amino acid level and with essentially the same biological activities.

[e]Supergene family with others marked.

[f]Supergene family.

[g]Supergene family unrelated to TGF-α.

[h]Molecularly related to IFN-α.

or function probably is involved in the development of a range of diseases, including atherosclerosis, AIDS, cancer (through angiogenesis), neural degenerative diseases, and toxic shock syndrome, as well. Indeed the symptoms of viral infections (headache, fever, malaise, etc.) are attributed to cytokine action.

The driving force for the study of cytokines is of course their therapeutic potential, recognized since the very early days of interferon research. By and large, cytokines are not established in clinical practice on any significant scale, with the important exceptions of EO for anemia and INF-α for hepatitis-B. The use of colony-stimulating factors to correct leukopenia due for example to cancer chemotherapy is in its infancy, as is the use of these factors for treatment of leukemia. In general, the use of cytokines in cancer therapy has been very disappointing, except for one or two rare diseases (e.g., hairy cell leukemia and renal cell carcinoma).

Where cytokines are involved in pathogenesis, a potential therapy consists of their depletion or neutralization using either specific antibodies, soluble receptors, or specific antagonists (natural or synthetic). An example of this is the treatment of toxic shock with antibodies to TNF-α.

*See also* GROWTH FACTORS; INTERFERONS; INTERLEUKINS.

### *Bibliography*


Aggarwal, B. B., and Gutterman, J. U., Eds. (1993) *Human Cytokines: Handbook for Basic and Clinical Research.* Blackwell Scientific, Oxford.

Anonymous. (1994) Minireviews in R & D Systems Catalog. R & D Systems, Minneapolis, MN.

Balkwill, F., Ed. (1991) *Cytokines: A Practical Approach.* IRL Press at Oxford University Press, Oxford.

Callard, R., and Gearing, A. (1994) *The Cytokine Factsbook.* Academic Press, San Diego, CA.

Miyajima, A., Kitamura, T., Harada, N., Yokota, T., and Arai, K. (1992) Cytokine receptors and signal transduction. *Annu. Rev. Immunol.* 10:295–331.

Moller, G., Ed. (1992) *Immunological Reviews,* Vol. 127. *Cytokines in Infectious Disease.* Munksgaard, Copenhagen.

Thomson, A., Ed. (1991) *The Cytokine Handbook.* Academic Press, London.


# CYTOSKELETON–PLASMA MEMBRANE INTERACTIONS

*Fredrick M. Pavalko*

### *Key Words*

**α-Actinin** A protein that functions as both an actin-bundling protein and a linker between actin filaments and integrins.

**Cytoskeleton** Filamentous structures in cells, including actin microfilaments, intermediate filaments, and microtubules.

**Focal Adhesions** Specialized regions of the plasma membrane in adherent cells that serve as anchorage sites for microfilaments.

**Integrins** A family of cell surface glycoproteins that function as cell–extracellular matrix and cell–cell adhesion molecules.

There is ample evidence that the cytoskeleton attaches to the membrane via the cytoplasmic domains of cell surface adhesion molecules. Actin filaments associate with adhesion molecules of several types, including integrins and cadherins, at adhesive junctions in a variety of cells types, and with the glycoprotein Ib/IX complex in platelets. Intermediate filaments associate with cadherins in desmosomes at sites of cell–cell association. The physical associations between cytoskeletal filaments and transmembrane glycoproteins are indirect, and the molecular mechanisms of these attachments have only recently become better understood. The importance of interactions between the cytoskeleton and the cytoplasmic domains of adhesion molecules are evident from the severe effects of cytoplasmic domain deletion and mutagenesis studies on normal cell function in a number of experimental systems. Transfection studies with mutant and chimeric adhesion molecules along with protein binding studies are beginning to clarify some of the mechanisms that mediate signaling events across the cell membrane and regulate attachments between the cytoskeleton and the plasma membrane.

## 1 INTRODUCTION

Interactions between cytoskeletal filaments and the plasma membrane occur at discrete domains of the membrane. Sites of cytoskeletal filament anchorage are usually sites of adhesion between cells or between the cell and the extracellular matrix. These attachment sites are found in both isolated cells and in tissues (Table 1). Several classes of anchoring junctions have been distinguished. The actin filaments of cells grown in culture interact with the cytoplasmic face of the plasma membrane at discrete sites of tight attachments between the ventral plasma membrane and the substrate. Actin does not interact directly with the membrane at these sites, known as focal adhesions or focal contacts. Instead, several cytoskeleton-associated proteins, which are concentrated in focal adhesions, function as links between actin filaments and the cytoplasmic domains of transmembrane adhesion molecules called integrins. Recent efforts have also focused attention on the role of integrin-mediated phosphorylation of cytoskeleton-associated proteins in the organization of actin microfilaments.

Another class of cytoskeleton-associated junctions contain members of the family of transmembrane glycoprotein cell adhesion molecules called cadherins. Interactions between the cytoskeleton and cadherins occur in the zonula adherens and desmosomes of epithelial cells. Actin microfilaments are associated with the ad-

**Table 1**

| Site of Cytoskeleton–Membrane Interaction | Type |
|---|---|
| Focal adhesions (focal contact, adhesion plaque) | Cell–ECM |
| Activated platelets | Cell–cell and cell–ECM |
| Myotendinous junctions | Muscle–tendon |
| Neuromuscular junctions | Nerve–muscle |
| Smooth muscle dense plaques | Cell–cell (smooth muscle) |
| Intercalated disks | Cell–cell (cardiac muscle) |
| Z-lines | Cell–cell (striated muscle) |
| Epithelial zonula adherens (adhesion belt) | Cell–cell (epithelia) |
| Epithelial hemidesmosomes | Epithelial–basal lamina |
| Desmosomes (intermediate filaments) | Cell–cell (many cell types) |

herens-type junctions, while intermediate filaments are associated with desmosomes. Two common features of junctional sites of different types include involvement of multiple linker proteins and a requirement for the cytoplasmic domains of adhesion molecules to mediate attachments with cytoskeletal filaments. In platelets, one of the major mechanisms of attachment of the membrane skeleton to the plasma membrane occurs through the linkage of actin filaments to the membrane glycoprotein Ib/IX complex (GP Ib/IX). Evidence from *Dictyostelium* suggests that the primary mechanism for actin–plasma membrane interaction is via the protein ponticulin. One of the best characterized models of cytoskeletal–membrane interactions is in erythrocytes, where spectrin, band 4.1, akyrin, and actin form the membrane skeleton that links to band 3 (the anion transporter) and glycophorin in the membrane. Nonerythorocyte spectrins may also link actin to the membrane in other cell types. This entry focuses on recent progress toward understanding the molecular interactions necessary for establishing and maintaining interactions between the cytoskeleton and receptors in the plasma membrane.

## 2 CYTOSKELETAL INTERACTIONS WITH INTEGRINS

### 2.1 In Vitro Interactions Between the Cytoskeleton and Integrins

Talin is one of several cytoskeleton-associated proteins that are concentrated in focal adhesions. Using a gel filtration assay designed to detect low affinity interactions between proteins in vitro, purified talin was shown to bind integrins that had been isolated from membranes. The recent demonstration that talin can bind to actin filaments suggests that talin can function as a direct link between actin and the membrane. Another focal adhesion protein, vinculin, can also bind both talin and α-actinin, suggesting that this protein may participate in linking actin filaments to integrins as part of a multiprotein chain between integrins and actin. The interaction between talin and integrin is, at least in vitro, of relatively low affinity. A higher affinity interaction between the cytoplasmic domain of the integrin $\beta_1$ and $\beta_2$ subunits and the protein α-actinin has been demonstrated using cytoplasmic domain peptide affinity chromatography and solid phase binding assays. The binding site for integrin is contained within the rod domain of α-actinin and is preserved in a proteolytic fragment of α-actinin that is distinct from the actin-binding domains.

### 2.2 Focal Adhesions

Cells grown in tissue culture adhere tightly to the substratum through specialized regions of the membrane called focal adhesions or focal contacts. Members of the integrin family of transmembrane receptors for extracellular matrix are concentrated in the focal adhesions of cells grown in culture and mediate transmembrane communication between the actin cytoskeleton and the extracellular matrix. Bundles of actin filaments, known as stress fibers, terminate at the cytoplasmic face of the membrane in focal adhesions. Several cytoplasmic proteins are found at focal adhesions, two of which, talin and α-actinin, have been shown to bind directly to the cytoplasmic domain of the $\beta_1$ integrin subunit in vitro and may form direct links between actin filaments and integrins in focal adhesions. Adhesion of cells to certain extracellular matrix (ECM) proteins also activate signals that lead to an increase in phosphorylation of at least two focal adhesion proteins, tensin and paxillin, which may be involved in the coordinated reorganization of the actin cytoskeleton during cell spreading.

Evidence that α-actinin plays a role in linking actin stress fibers to the membrane at focal adhesions in living cells comes from microinjection studies using the integrin-binding domain of α-actinin. Microinjection of high concentrations of the fluorescent-labeled rod domain of α-actinin into living cells results in its rapid colocalization with integrin in focal adhesions. Localization of the α-actinin rod domain in focal adhesions is quickly followed by a loss of endogenous α-actinin from these sites and the detachment of actin stress fibers from the membrane. This result argues that the cells' endogenous, intact α-actinin molecules, which are completed from focal adhesions by the integrin-binding fragment of α-actinin, are necessary for the attachment of actin filaments to focal adhesions in vivo. The importance of the cytoplasmic domains to normal integrin function and cytoskeletal organization is indicated by studies in which mutated integrins with modified cytoplasmic domains were transfected into cells, resulting in abnormal integrin localization, altered cytoskeletal interactions, and reduced ligand-binding activity.

### 2.3 Cytoskeletal–Membrane Interactions in Platelets

A dramatic reorganization of the actin cytoskeleton appears to be crucial to the function of activated platelets in vivo. A primary mechanism for attachment of actin filaments in platelets is the heterotrimeric membrane complex, GP Ib/IX. This nonintegrin receptor binds to von Willebrand factor and mediates the adhesion of platelets to injured blood vessels. One of the first demonstrations of a direct linkage between actin filaments and the plasma membrane came with the finding that platelet "actin-binding protein" (ABP) mediates actin membrane attachment by linking actin filaments to the cytoplasmic domain of the α-chain of GPIb. Actin filaments in platelets may also associate with the integrin glycoprotein IIb/IIIa via the cytoplasmic domain of GPIIIa, which is the integrin $\beta_3$ subunit. Purified GPIIb/IIIa incorporated into phospholipid vesicles associated with purified α-actinin suggesting a link to actin filaments. Also, talin has been shown to redistribute to the subplasma membrane region upon activation of platelets. The cytoplasmic domain of GPIIIa shares extensive homology with both the integrin $\beta_1$ and $\beta_2$ subunits in the amino-terminal region, which is a consensus site for binding to α-actinin.

### 2.4 Cytoskeletal–Membrane Interactions in Leukocytes

The identification of interactions between purified proteins provides important clues about the mechanisms that may attach actin filaments to the membrane in living cells. Confirmation that these interactions are relevant in vivo is more difficult. Recent studies using human neutrophils indicate that an interaction between α-actinin and the integrin $\beta_2$ subunit is induced upon activation of neutrophils with chemotactic peptides. In lymphocytes, activation via the T-cell receptor may induce a similar association of $\beta_2$ integrins with the actin cytoskeleton involving α-actinin. Taken together with the microinjection studies using the integrin-binding domain of α-actinin already described, current evidence strongly suggests that α-actinin serves a physiologically relevant role linking actin filaments to the membrane in vivo.

## 3 CYTOSKELETAL INTERACTIONS WITH CADHERINS

### 3.1 Adherin-Type Junctions

A second category of adhesive junctions, distinct from those involving integrins, are cell–cell adhesions that require calcium and are mediated by the family of adhesion molecules called cadherins. Within this category, two types of filament anchorage at the cytoplasmic face of the membrane can be distinguished. One involves the attachment of actin filaments at sites such as the zonula adherens of epithelial cells. These sites contain the transmembrane glycoprotein E-cadherin and a number of cytoplasmic "plaque" proteins, which may attach actin filaments to cadherins. Transfection of cells with mutated cadherin cDNA, which code for a molecule lacking the cytoplasmic domain, demonstrate that these regions are necessary for cytoskeletal–cadherin associations. Several of the cytoplasmic proteins have been identified including α-, β- and τ-catenins, α-actinin, vinculin, and radixin. Both the colocalization of cytoskeletal proteins with cadherins and the demonstration that nonionic detergents fail to remove cadherins from adhesive junctions support the existence of a physical association of cadherins with the cytoskeleton.

### 3.2 Desmosomes

A second type of cadherin-containing adhesive junction provides attachment sites for intermediate filaments composed of cytokeratins, desmin, or vimentin. These sites, called desmosomes, contain members of a complex subfamily of transmembrane cadherins called desmogleins and desmocollins. At least three distinct desmogleins and three desmocollins have been identified. Desmosomal plaques also contain the proteins desmoplakin and plakoglobin at the cytoplasmic face of the membrane, as well as several other cell-type-specific proteins. Several studies using chimeric molecules transfected into cells lacking endogenous cadherins have shown that the cytoplasmic tails of E-cadherin and of desmosomal cadherins, particularly the highly conserved carboxy-terminal domains, contain sufficient information to direct the recruitment of plaque-associated proteins and of microfilaments or intermediate filaments that insert at the plaques. The specific molecular interactions that link cytoskeletal filaments to cadherins have not been determined. With regard to this question, it is noteworthy that α-catenin shows homology to the focal adhesion protein vinculin, which appears to be involved in linking actin filaments to integrins at these sites.

## 4 REGULATION OF CYTOSKELETON–MEMBRANE INTERACTIONS

Transformed cells exhibit decreased adhesion and a reduced number of actin-containing stress fibers and focal adhesions. Changes in adhesion can be largely explained by decreased amounts of the ECM protein fibronectin and a loss of high affinity fibronectin receptor ($\alpha_5/\beta_1$ integrins) on the cell surface. Enhanced tyrosine kinase activity in the focal adhesions of transformed cells mediated by the tyrosine kinase $pp60^{v-src}$ suggests that phosphorylation of actin–integrin linker proteins may be involved in altered cytoskeletal organization. Several focal adhesion proteins, including talin, vinculin, paxillin, and integrin, contain slightly elevated levels of phosphotyrosine in RSV transformed cells, although the functional consequences on actin–membrane interactions are unclear. Several proteins have been shown to interact with tyrosine phosphorylated proteins via regions of *src* homology (SH2). The focal adhesion protein tensin has been shown to contain multiple actin-binding domains and an SH2 domain, suggesting a role in binding to tyrosine-phosphorylated focal adhesion proteins.

Tensin and paxillin are phosphorylated by the focal adhesion kinase ($pp125^{FAK}$), a kinase that is activated by certain integrin–ligand interactions. This observation further supports the possibility that phosphorylation on tyrosine may be part of a signal transduction pathway through integrins that regulates cytoskeletal–membrane interactions. Another molecule that appears to play a role in the reorganization of the actin cytoskeleton that is mediated by phosphorylation is the MARCKS protein. MARCKS is a substrate for the $Ca^{2+}$-dependent serine–threonine protein kinase C (PKC) and is localized to sites of actin–membrane interaction in a phosphorylation-dependent manner. MARCKS can cross-link actin filaments and bind directly to phospholipids, suggesting a potentially complex regulation of actin–plasma membrane interaction in vivo.

*See also* Cell-Cell Interactions.

## *Bibliography*

Andrews, R. K., and Fox, J.E.B. (1990) Platelet receptors in hemostasis. *Curr. Opinion Cell Biol.* 2:894–901.

Burridge, K., Fath, K., Kelly, T., Nuckolls, G., and Turner, C. (1987) Focal adhesions: Transmembrane junctions between the extracellular matrix and the cytoskeleton. *Annu. Rev. Cell Biol.* 4:487–525.

Burridge, K., Nuckolls, G., Otey, C., Pavalko, F., Simon, K., and Turner, C. (1990) Actin–membrane interaction in focal adhesions. *Cell Diff. Dev.* 32:337–342.

Geiger, B., and Ayalon, O. (1992) Cadherins. *Annu. Rev. Cell Biol.* 8:307–332.

Palek, J., and Sahr, K. E. (1992) Mutations of the red blood cell membrane proteins: From clinical evaluation to detection of the underlying genetic defect. *Blood,* 80:308–330.

Pardi, R., Inverardi, L., Rugarli, C., and Bender, J. R. (1992) Antigen–receptor complex stimulation triggers protein kinase C-dependent CD11a/CD18–cytoskeleton association in T lymphocytes. *J. Cell Biol.* 116:1211–1220.

Pavalko, F. M., and Otey, C. A. (1994) Role of adhesion molecule cytoplasmic domains in mediating interactions with the cytoskeleton. *Proc. Soc. Exp. Biol. Med.* 205:282–293.

Pavalko, F. M., Otey, C. A., Simon, K. O., and Burridge, K. (1991) α-Actinin: A direct link between actin and integrins. *Biochem. Soc. Trans.* 19:1065–1069.

Schwarz, M., Duden, R., Cowin, P., and Franke, W. W. (1990) Desmosomes and hemidesmosomes: Constitutive molecular components. *Annu. Rev. Cell Biol.* 6:461–491.

Turner, C. E., and Burridge, K. (1991) Transmembrane molecular assemblies in cell–extracellular matrix interactions. *Curr. Opinion Cell Biol.* 3:849–853.

Zachary, I., and Rozengurt, E. (1992) Focal adhesion kinase ($pp125^{FAK}$): A point of convergence in the action of neuropeptides, integrins and oncogenes. *Cell,* 71:891–894.

# Denaturation of DNA

*Richard D. Blake*

## Key Words

**Conformation (DNA)** The three-dimensional arrangement of the polynucleotide chains that form the double helix; resulting from local configurational tendencies of the chains and various noncovalent interactions.

**Degradation (DNA)** Breakdown of covalent bonds supporting the polynucleotide chains, caused by exposure to certain chemical agents or extremes in environmental conditions. The process is generally irreversible.

**Melting Curve (DNA)** The variation with temperature of any property (e.g., absorbance at 260 nm) sensitive to the native structure of DNA.

**Stacking Forces (DNA)** The forces between stacked base pairs.

**$T_m$ (DNA)** The melting temperature of DNA, determined from the midpoint in the change in the dependent variable (e.g., absorbance at 260 nm) with temperature.

The *denaturation of DNA* is brought about by any of the many artificial means of disturbing the noncovalent interactions supporting the native conformation, causing either alterations at the level of local structure or the total collapse of the double helix. The process is generally reversible.

## 1 Denaturation and Degradation

The native helical conformation of DNA is supported by noncovalent forces that are substantially weaker than the covalent forces that hold the polynucleotide chains together. For the most part, this noncovalent support consists of a short-range dispersion forces sensitive to temperature. Watson–Crick hydrogen bonds between bases maintain the alignment of the two strands, but the vertical "stacking" forces between nearest-neighbor base pairs are the major sources of support to the helix. Stacking forces vary with the sequence of stacked pairs in the duplex e.g.,

$$\begin{matrix}5'\\3'\end{matrix} < \begin{matrix}(\text{G}\cdot\text{C})\\(\text{G}\cdot\text{C})\end{matrix} > \begin{matrix}3'\\5'\end{matrix}, \quad < \begin{matrix}(\text{G}\cdot\text{C})\\(\text{C}\cdot\text{G})\end{matrix} >, \quad < \begin{matrix}(\text{G}\cdot\text{C})\\(\text{A}\cdot\text{T})\end{matrix} >, \ldots$$

There are 10 unique nearest-stacked-neighbors, each with a different stacking energy. These differences lead to variations in local stability of the DNA as well as a strong dependence of the structural state of each pair on the state of its neighbors. Indeed, the dependence extends well beyond next neighbors. The energy needed to free a pair from the stacked array is so great that when provided, it leads to the cooperative collapse of entire stretches that may involve as many as 250 to 500 contiguous pairs. Factors used to break these bonds usually bring it about over a very narrow range of conditions and affect large stretches of the DNA.

In addition to stacking and hydrogen bonds, a number of weaker bonds contribute to the support of the duplex, primarily from solvation effects, the occasional water bridge, and hydrogen bonds associated with certain sequence arrangements. Also, the phosphodiester groups ($\cdots$—O—P[$O^-_2$]—O—$\cdots$) that join nucleotides in the backbone have a negative charge; therefore the polynucleotide chain is a polyanion with strong repulsive forces between opposing chains, counteracting the stabilizing forces from stacking and hydrogen bonds. The high charge density of the helix causes counterions (e.g., $Na^+$, $K^+$, $Mg^{2+}$) to condense on the helix in high concentration, effectively reducing the net charge on the phosphates and screening the repulsion between negative charges.

If these weak stabilizing forces are broken artificially, the double helix collapses and the DNA is *denatured.* Denaturation leads to nonhelical structures and, if sufficiently severe, to total separation of the polynucleotide chains. If *all* noncovalent bonds are broken, the chains assume the conformation of random coils. However, denaturation is rarely if ever so complete; rather, the DNA assumes a complex disordered state where residual noncovalent bonds are mostly misdirected. If DNA is forced into an altered structure, one that is not everywhere helical or everywhere recognizable by the various enzymes and proteins that maintain or interact specifically with the native structure, it is still said to be *denatured.* Noncovalent bonds can be broken through one of four means, involving exposure of the DNA to high temperatures, extreme pH values, low counterion concentrations, and various denaturing solvents. Denaturation is distinguished from the more severe process of *degradation,* during which covalent bonds are broken. Thus to avoid degradation, denaturation of DNA by high temperatures for studies requiring the integrity of the sequence be maintained (e.g., heteroduplex formation, sequencing, and the polymerase chain reaction) should involve a minimum exposure of the single strands to temperatures much above 50°C. Single-stranded DNA is particularly prone to depurination at high temperature and low pH. Many procedures for denaturing DNA call for heating in boiling water for 2 to more than 10 minutes. The incubation time need only ensure that thermal equilibrium has been achieved, since denaturation is almost instantaneous.

The structure of native single-stranded DNA is more complex than that of duplex DNA, with many regions of single-strand stacking, double-strand hairpins, and segments of random and dynamic conformation. Single-stranded DNA adopts these local structural features according to the sequence. Double-stranded DNA collapses into a similar but metastable, quasi-stochastic structure if quick-cooled from high temperature. Complete single-strand stacking does not occur until well below room temperature. Also, since the conformational states of stacked bases in single-stranded DNA

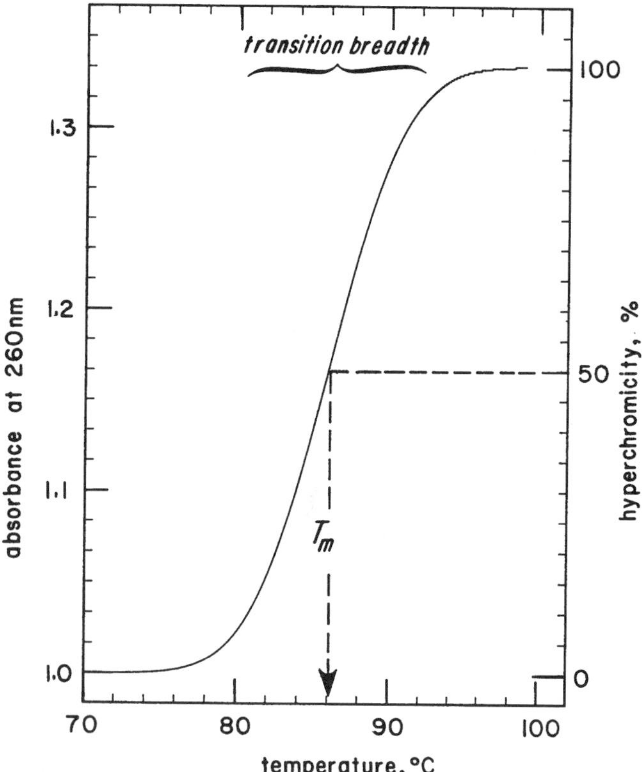


**Figure 1.** Typical thermal denaturation (melting) curve for DNA, showing the variation in optical absorbance at 260 nm with increasing temperature. The $T_m$, usually defined as the temperature corresponding to half the overall change in absorbance, has physical significance only for heterogeneous DNAs of considerable size.

are essentially independent of those of their neighbors denaturation occurs over a wide range of conditions. Melting is gradual, occurring over a range of 80 to 100°C, whereas duplex DNA typically melts within 12 to 16°. Indeed, if the sequence is sufficiently uniform, duplex DNA may melt within 0.1°C. Most physical and chemical properties of single-stranded DNA are like those of RNA.

## 2 ANALYTICAL METHODS FOR FOLLOWING THE DENATURATION PROCESS

Methods for monitoring the denaturation of DNA can be classified as short-range methods (those that depend on specific short-range physical properties), long-range methods (which reflect alterations in the macromolecular properties), chemical methods (which probe structural alteration by changes in reactivity) and biological methods (which rely on sensitivities to various biological probes such as nuclease sensitivity). *Differential scanning calorimetric* and various spectroscopic methods probe short-range effects, while hydrodynamic and electrophoretic methods monitor the long-range effects. The most common method of monitoring the denaturation process is by *absorbance spectroscopy* in the ultraviolet region of 200 to 290 nm. The mean absorption coefficient for $\pi \rightarrow \pi^*$ electronic transitions of the heterocyclic bases is unusually large over this region. In the helical state these transitions are neighbor-perturbed in the close-ordered packing of the bases. The molar absorption coefficient for the helix, $\varepsilon_{260nm}(P)$, is approximately $7 \times 10^3$L/mol·cm under physiological solvent conditions. During denaturation, the bases unstack and the molar absorption coefficient increases to $\sim 10^4$L/mol·cm, close to that for the free nucleotide monomers. The increase at 260 nm is approximately 38%, which means that denaturation can be monitored with considerable precision on nucleotide residue concentrations in the micromolar range.

A typical thermal denaturation or melting curve of DNA is shown in Figure 1. A melting temperature $T_m$, determined from the midpoint of the transition, is generally taken as a measure of the thermal stability of the DNA; while the transition breadth is taken as a measure of the variation in (G+C) content. Fine detail is evident in the high resolution, first-derivative melting profile of Figure 2, which shows the melting of pN/MCS-6, a recombinant plasmid of 4750 base pairs. Denaturation takes place in saltatory fashion, by domains of increasing stability with increasing temperature. A domain in this instance is a particular stretch of DNA that dissociates or melts in a cooperative, all-or-none fashion; it is seen in profiles such as this one as sharp peaks representing subtransitions.

Methods have been devised for assigning subtransitions to specific sequences. The $T_m$ determined from the midpoint of integration in Figure 2 is now seen to have no particular significance for the stability of specific sequences. The subtransition for denaturation of a 330 bp (A·T)-rich insert can be seen clearly at 76.40°C. The temperatures at which different domains melt depend on a number of factors, including the frequencies of the 10 unique nearest neighbors and the number of changes in helix–coil boundaries separating domains. The insert domain of Figure 2 melts as an internal loop, generating two new boundaries. The $T_m$ for the insert subtransition is only 75.11°C when the insert melts without generating new boundaries (i.e., when it is located at the end of the DNA).

*Circular dichroism,* also effective in the region of 200 to 290 nm, measures the difference in molar extinction of the left and right circularly polarized components of monochromatic light. As such it is used to detect the spatial asymmetry and helical handedness as the chromophoric base pairs twist one way or the other around the major axis. Data from *infrared* and *Raman spectroscopy* in the 500 to 1800 $cm^{-1}$ region reflect changes in *vibrational* and *rotational* modes affected by changes in hydrogen bonding and stacking interactions. Changes involving polar groups give the most intense absorptions and are useful for following the dissociation of hydrogen bonding. There is a superabundance of spectral detail in these spectra, since the number of transitions for all possible modes approaches $3n - 6$, where $n$ is the number of atoms. The downside is that extinction coefficients are typically several orders of magnitude smaller than in the electronic region, with the result that DNA concentrations need to be higher. Working concentrations are usually 0.01 to 1.0 mg of DNA per milliliter, depending on the transition and sensitivity of the facility.

*Nuclear magnetic resonance,* the measure of circular motion of magnetic nuclei due to their interaction with the magnetic component of electromagnetic radiation in the radio frequency range ($\sim 10^{10}$ Hz), is the most powerful method for investigating dynamic structural changes during denaturation. One-dimensional spectra in the range of 9.5 to 0.5 ppm are used mainly for studies of the exchangeable imino and amino resonances, important indicators of the hydrogen-bonded states of the bases. Phase-sensitive, two-dimensional NMR in $D_2O$ is used for tracing single-step connectivities and for measuring coupling constants. Since there are only the four bases in DNA, it is difficult to assign spectral features to specific bases in a sequence of any length. To reduce or eliminate ambiguity in assignments, specimens usually are oligomeric, and one must contend with end effects.

Hydrodynamic methods reflect the reduction in frictional coefficient during denaturation when DNA changes from a rigid rod

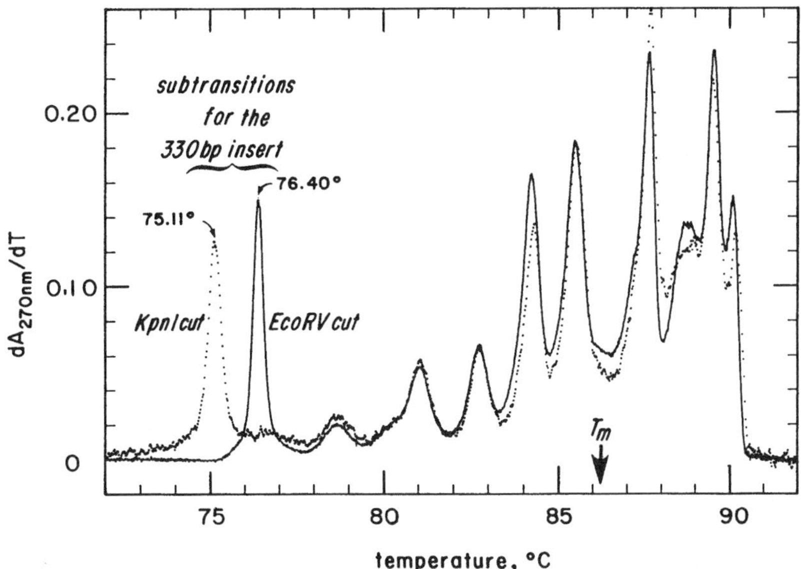


**Figure 2.** The example of a melting curve of a homogeneous DNA of discrete length (4750 bp), obtained and plotted as the derivative of the change in absorbance with temperature, $dA_{270nm}/dT$. At 270 nm the change in absorbance for dissociation of A·T base pairs is equal to the change for G·C pairs. The specimen is the plasmid pN/MCS-6, a pBR322 derivative with a multiple cloning sequence (MCS) of 54 bp at the unique *Nru* I cleavage site, containing an insert at the *Sma* I locus of the MCS. The *Nru* I locus divides two very (G+C)-rich regions that serve as strong helical boundaries for inserts. The insert is 330 bp and has the sequence $[AAGTTGAACAAT]_{27}$AAGTTG [25% (G+C)] (prepared by S. G. Delcourt). The melting subtransition for the insert is seen as an isolated peak at 76.40 ± 0.03°C. The plasmid was linearized prior to melting by cutting at the unique *Eco*RV locus, a considerable distance from the insert, so that the insert melts as a loop. When the plasmid is linearized by cutting at the unique *Kpn* I immediately adjacent to the insert, the insert melts at 75.11 ± 0.03°C. The concentration of $Na^+$ is 0.075 M, and the rate of heating was 6.00°C/h.

into flexible chains. *Viscometry, sedimentation,* and *dynamic light scattering* are generally more difficult and cumbersome than spectral methods, requiring greater amounts of monodisperse material. The long-range effects of denaturation also have been studied visually by *electron microscopy.* The methodology involves fixing partially denatured regions at intermediate temperatures with glyoxal. This bifunctional reagent reacts with imino and amino groups, preventing re-formation of the helix. *Denaturing gradient gel electrophoresis* (DGGE), a new technique that also monitors changes in conformation during denaturation, has been used to detect minor sequence variations and mutations. The sensitivity of this technique to small thermodynamic differences between specimens is approximately the same as that obtainable by electronic absorption spectroscopy (Figure 2). However, in this instance the sensitivity is most effective in detecting small differences *between* specimens. In *parallel* DGGE a gradient of denaturing conditions is established between the electrodes at the two ends of a (vertical) slab gel. The gradient is approximately 30 to 70% (v/v) formamide plus 7 M urea, and the run is carried out at 60°C. During electrophoresis, the mobility of the specimen DNA remains constant until one of the domains denatures. When the DNA enters a region of the gradient that leads to denaturation of the weakest domain, the mobility of the DNA abruptly decreases. Two fragments with a limited number of distinct melting domains, differing by a single base pair, have been clearly resolved by DGGE.

## 3 PHYSICAL AND CHEMICAL FACTORS AFFECTING THE STABILITY OF DNA

A number of factors from within and without the DNA affect its sensitivity to denaturation:

1. The (G+C) content, or more precisely the distribution of the 10 nearest-neighbor frequencies in the DNA.
2. The local sequence and conformational state of neighboring domains.
3. The presence of wobble or other mispairs.
4. The DNA length.
5. The counterion or cation type and concentration.
6. pH.
7. Temperature.
8. Various salts and nonaqueous solvents.
9. Supercoiling stresses.

The first denaturation experiments were performed by titration with acid and alkali. Protonation and deprotonation of the bases at the upper and lower extremes of hydrogen ion concentration alter the hydrogen bond donor–acceptor relationships between pairs and place charges on the bases. DNA denatures below about pH 4 and above about pH 11—somewhat higher and lower, respectively, than the p$K$ values of the free bases. The difference between p$K$s for the free bases and p$H_m$ for denaturation reflects the free energy of isolating the ionizable groups by Watson–Crick hydrogen bonding. There are practical reasons for avoiding pH values below about 5, however: protonation leads to precipitation of intact polymeric DNA, while still lower pHs increase the rate of depurination and subsequent chain cleavage. Denaturation at high pH, on the other hand, appears to be less problematic and represents a method for separating the helical strands. Provided the strands differ from unity in the ratio of purines to pyrimidines, they can be separated by CsCl buoyant density gradient centrifugation at about pH 11, where thymine and guanine are titrated.

The effect of temperature has been described by example (Figure 1). The $T_m$ shows a strong dependence on the mole fraction of guanine plus cytosine residues, $F_{GC}$, as well as on the cation concentration. These variables can be wed into a single empirical relationship, which, updated and refined for the conditions $0.3 < F_{GC} < 0.7$

and $0.01 \lesssim [Na^+] \lesssim 0.4$ M, is given in degrees Celsius by the Marmur–Schildkraut–Doty equation:

$$T_m = 193.67 - (3.09 - F_{GC})(34.64 - 2.83 \log_{10}[Na^+])$$

As noted, the stability of DNA derives mainly from 10 neighbor-stacking interactions; not just from (A·T) and (G·C) pairs. Therefore this expression represents an approximation of the more complex dependence on stacking energies; as such, it is applicable only to DNAs with suitably large and random distributions of neighbor pairs. The expression also disguises the effects of neighboring domains on denaturation, as seen in the example of Figure 2.

## 4 UTILITY OF DENATURATION STUDIES

Studies involving the deliberate denaturation of DNA are useful for many reasons, both biological and structural.

First, denaturation is necessary for producing hybrids of complementary polynucleotide chains from different source material and for forming heteroduplexes for Southern blotting or for the detection of evolutionary or mutational differences.

Second, knowledge of the precise conditions of denaturation is needed to establish a level of stringency, to ensure that chains of lesser sequence similarity are excluded from kinetically competing with complementary regions of sequence and slowing up the hybridization process.

Third, repeated cycles of denaturation and synthesis form the basis for the amplification of minute quantities of DNA by the *Taq*-polymerase chain reaction, a simple alternative to cloning.

Fourth, the thermal denaturation of heteroduplexes is an easy and sensitive means of measuring sequence divergence, since mispairs contribute to greater configurational entropy and lower stability. The reduction of stability (i.e., the drop in melting temperature) is proportional to the frequency of mispairs in the heteroduplex.

Fifth, denaturation is also a sensitive method for resolving distributions of sequence families of different base contents in the overall thermal denaturation profiles of total genomic DNAs.

Sixth, denaturation is required for primer bonding, an important first step in the sequencing of DNA.

Seventh, denaturing gradient gel electrophoresis is useful for the detection of point mutations and sequence variations in mixed populations of DNAs.

Eighth, denaturation of total genomic DNAs followed by careful measurements of the kinetics of renaturation is used to measure different levels of sequence complexity: the distributions of single-copy, middle, and highly repetitive sequence families.

Finally, the thermal denaturation of DNA is an important adjunct of structural studies, providing the means for identifying various molecular sources of stability in the helix and for quantitating the thermodynamic characteristics of those sources.

*See also* HYDROGEN BONDING IN BIOLOGICAL STRUCTURES; PARTIAL DENATURATION MAPPING.

### *Bibliography*

Anderson, C. F., and Record, M. T., Jr. (1990) *Annu. Rev. Biophys. and Biophys. Chem.* 19:423.

Bloomfield, V. A., Crothers, D. M., and Tinoco, I., Jr. (1974) *Physical Chemistry of Nucleic Acids.* Harper & Row, New York.

Cantor, C. R., and Schimmel, P. R. (1980) *Biophysical Chemistry,* Part III: *The Behavior of Biological Macromolecules.* Freeman, San Francisco.

Erlich, H. A. (1989) *PCR Technology.* Stockton Press, New York.

Hames, B.D., and Higgins, S. J. (1985). *Nucleic Acid Hybridization.* IRL Press, Washington, DC.

Howe, C. J., and Ward, E. S. (1989) *Nucleic Acid Sequencing.* IRL Press, New York.

Lilley, D. M. J. and Dahlberg, J. E. (1992) *Methods Enzymol. 212b.*

Manning, G. S. (1978) *Q. Rev. Biophys.* 11:179.

Poland, D. (1974) *Biopolymers,* **13:**1859–1871. Wartell, R. M., and Benight, A. S. (1985) *Phys. Rep.* 126:67.

Record, M. T., Jr., Anderson, C. F., and Lohman, T. (1978) *Q. Rev. Biophys.* 11:103.

Wetmur, J. G. (1991) *Crit. Rev. Biochem. Mol. Biol.* 26:227–259.

# DIABETES INSIPIDUS

*Walter Rosenthal, Anita Seibold, Daniel G. Bichet, and Mariel Birnbaumer*

### *Key Words*

**Central Diabetes Insipidus (CDI)** Form of diabetes insipidus caused by inadequate or no release of vasopressin from the posterior pituitary.

**Diabetes Insipidus** An acquired or inherited disease characterized by the two key symptoms: polyuria and polydipsia.

**Nephrogenic Diabetes Insipidus (NDI)** Form of diabetes insipidus caused by a resistance of the kidneys to vasopressin.

**Vasopressin Receptors** Cell surface receptors mediating cellular responses to vasopressin. Whereas the antidiuretic response is mediated by $V_2$ receptors, other responses are mediated by $V_1$ receptors.

**Vasopressins** Nonapeptides synthesized in hypothalamic nuclei and released from the posterior pituitary. Their main function is the conservation of water. Vasopressins also contract vascular smooth muscle, stimulate glycogenolysis in liver cells, and enhance the release of adrenocorticotropic hormone (ACTH) from the anterior pituitary. *Synonym:* antidiuretic hormone (ADH).

The nonapeptide vasopressin acts via two different cell surface receptors. Whereas the antidiuretic response is mediated by $V_2$ receptors, other responses are mediated by $V_1$ receptors. Diabetes insipidus is a disease characterized by two key symptoms: polyuria and polydipsia. Central diabetes insipidus (CDI) is caused by insufficient or no release of vasopressin from the posterior pituitary. Nephrogenic diabetes insipidus (NDI) is caused by resistance of the kidney (lack of responsiveness) toward vasopressin. Acquired and inherited forms of CDI and NDI have been described. Recent work shows that autosomal-dominant CDI is caused by defects in the vasopressin precursor gene, and that the major cause of X-chromosomal recessive NDI is a defect in the $V_2$ receptor gene. Identification of the genes responsible for these diseases facilitates carrier identification and prenatal diagnosis; it is also a step toward an improved treatment of CDI and NDI patients.

## 1 VASOPRESSIN AND VASOPRESSIN RECEPTORS

### 1.1 Vasopressin Structure, Synthesis, and Release

Vasopressins (also referred to as antidiuretic hormones or ADH) are nonapeptides characterized by an amidated C-terminal glycine residue and a disulfide bridge between cysteine residues at positions 1 and 6. In all mammals except swine, the nonapeptide 8-arginine vasopressin (AVP) is found; the porcine variant is 8-lysine vasopressin (LVP). The inactive vasopressin precursor is synthesized in neurons of the hypothalamic supraoptic and paraventricular nuclei. The gene encoding the precursor (vasopressin gene) contains three exons (1–3) (Figure 1), which encode the functional domains of the vasopressin preprohormone consisting of a signal peptide, the hormone, a peptide essential for the axonal transport of the hormone (neurophysin II), and a C-terminal glycopeptide of unknown function. Cleavage of the precursor occurs in the endoplasmic reticulum and in neurosecretory granules during axonal transport to the posterior pituitary gland from which vasopressin and neurophysin II are released by an exocytotic pathway. In man, the release is triggered by minute (2%) increases in plasma osmolality sensed by peripheral and central osmoreceptors. The release of vasopressin is also stimulated by a more pronounced (10%) decrease in extracellular fluid volume, sensed by cardiovascular baroreceptors.

### 1.2 Vasopressin Receptors and Transmembrane Signaling

Vasopressin acts via two types of cell surface receptor, the $V_1$ and the $V_2$ receptors. Hydropathy profiles based on cDNA-derived amino acid sequences are consistent with seven transmembrane domains (Figure 2), a characteristic feature of the large group of G-protein-coupled receptors. Within this category, the two vasopressin receptors together with the oxytocin receptor form a subfamily; well-conserved regions include the transmembrane domains and the extracellular loops (E2, E3, E4: see Figure 2). The human gene for the $V_2$ receptor contains three expressed exons framing two introns (positions indicated by arrowheads in Figure 2). The first exon contains the start codon; it is separated from the second exon by an intron 360 base pairs long, which interrupts codon 9. The second exon encodes approximately 80% of receptor sequence. The second intron interrupts codon 304, located at the junction of the third extracellular loop and the seventh transmembrane domain. Exon 3 encodes the seventh transmembrane domain and the C-terminus. The gene structure of the $V_1$ receptor is not known.

$V_1$ and $V_2$ receptors activate different signal transduction pathways. The $V_1$ receptor stimulates phospholipase C via a G protein of the $G_q$ family. Phospholipase C induced formation of inositol-1,4,5-triphosphate and diacylglycerol causes a release of $Ca^{2+}$ from intracellular stores (increase in cytosolic $Ca^{2+}$) and activation of protein kinase C, respectively. Pharmacological evidence suggests the existence of two types of $V_1$ receptor. $V_{1a}$ receptors are found on hepatocytes and vascular smooth muscle cells, mediating vasopressin-induced stimulation of glycogenolysis and vasoconstriction, respectively. $VG_{1b}$ receptors promote the release of adrenocorticotropic hormone (ACTH) from corticotropic cells of the anterior pituitary. $V_2$ receptors couple to adenylyl cyclase via the stimulatory G protein, $G_s$. Receptor activation leads to an increase in cellular cyclic AMP and subsequent activation of cAMP-dependent protein kinase. In mammals, the $V_2$ receptor is found in the contraluminal membrane of epithelial cells lining the renal collecting duct and also—to varying degrees from species to species—in epithelial cells lining the medullary thick ascending loop of Henle. The final event following the activation of the $V_2$ receptor in the collecting duct is the fusion of water channel-containing vesicles with the apical membrane. The insertion of these water channels into the plasma membrane dramatically increases the water permeability of epithelial cells and allows the transfer of fluid from the tubular lumen to the interstitium. This is the major mechanism by which vasopressin exerts its antidiuretic effect. Stimulation of $V_2$ receptors located in the ascending loop of Henle causes an increase in the activity of the $Na^+$:$K^+$:$2Cl^-$ cotransporter and thereby an increase in salt transport across the epithelium. As a consequence, the tubular fluid is diluted, and the osmolality of the medullary interstitium is increased.

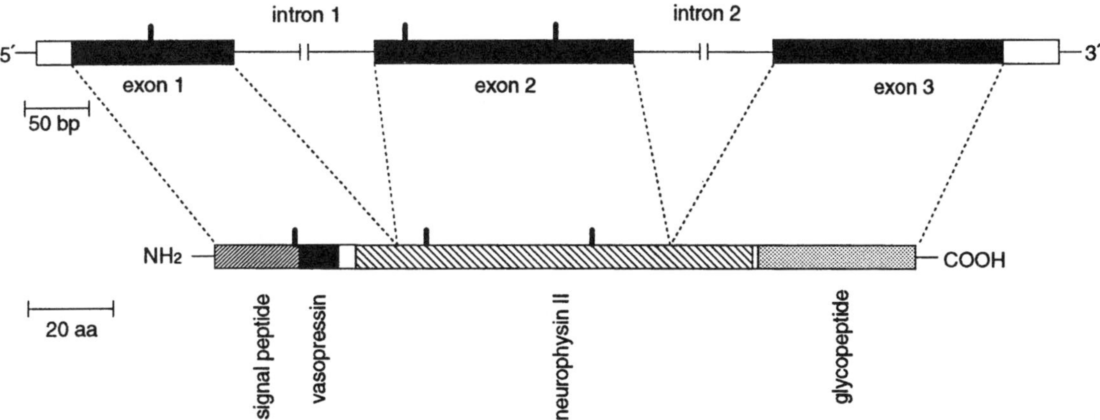


**Figure 1.** Structure of the vasopressin precursor gene and the vasopressin precursor. Translated and untranslated portions of exons are shown as solid or open bars, respectively. Introns are depicted as interrupted horizontal lines. The length of intron 1 is 1374 bp, and that of intron 2 is 165 bp. The various filled portions of the precursor indicate the different peptides; open portions are spacer peptides. Corresponding regions of the gene and the peptide are assigned to each other by dotted lines. The sites of the three mutations found in patients with autosomal dominant CDI are indicated in the gene and the peptide (bold vertical lines).

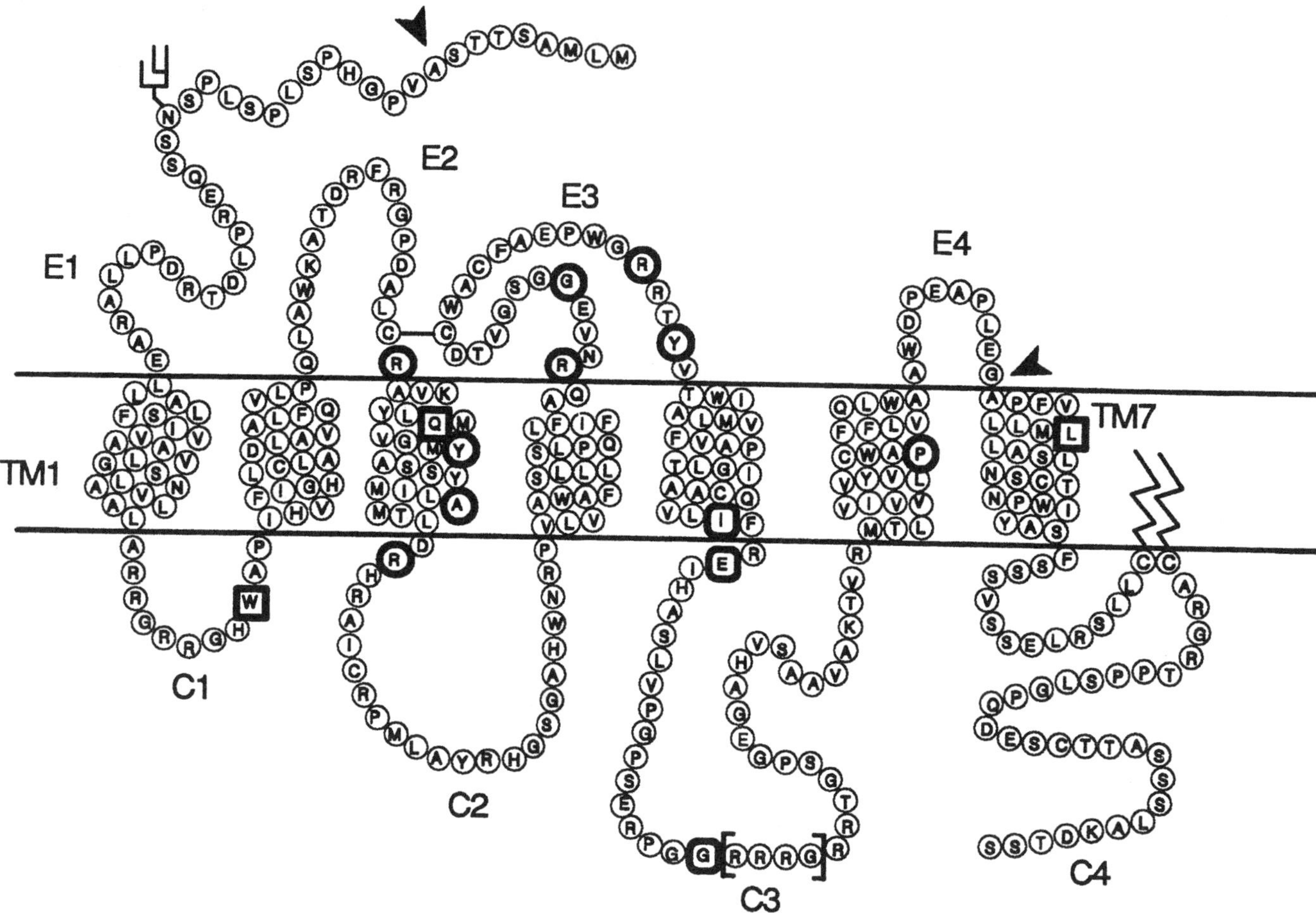


Figure 2. A secondary structure model of the human $V_2$ vasopressin receptor. The depicted membrane topology is consistent with the hydropathy profile of the polypeptide chain of the $V_2$ receptor and is assumed to be analogous to that of bacteriorhodopsin, a bacterial light-driven proton pump, and to that of the vertebrate photoreceptor pigment rhodopsin. Similar models have been proposed for the $V_1$ receptor and the oxytocin receptor. Predicted amino acids are given in the one-letter code: E1–E4, extracellular domains; C1–C4, intracellular domains. Transmembrane regions (TM1–TM7) are counted from left to right. Arrowheads indicate exon–intron junctions. Based on the presence of consensus regions for modifications in the amino acid sequence, the model depicts a sugar residue at asparagine 22 of the extracellular C-terminus, a disulfide bridge between cysteines 142 and 192 of the first (E2) and second extracellular loop (E3), respectively, and fatty acid residues at cysteines 341 and 342, attaching the intracellular C-terminus to the plasma membrane. Mutations found in patients with X-linked NDI are indicated by bold circles (missense mutations), bold squares (nonsense mutations), rounded squares (deletion or insertion causing a frameshift and premature stop), and brackets (in frame deletion). For more details on mutations see Section 4.3 and Table 1.

Evidence for the existence of extrarenal $V_2$ receptors is provided by observations made with a synthetic analogue of vasopressin, desmopressin (1-deamino[8-D-arginine]vasopressin), a selective $V_2$ receptor agonist:

1. In various species including man, rat, dog, and monkey, desmopressin (or the combination of vasopressin plus a selective $V_1$ receptor antagonist) causes a decrease in blood pressure and peripheral resistance (vasodilatory response).
2. In humans, dogs, and monkeys, desmopressin increases circulating levels of coagulation factor VIIIc, which is synthesized in hepatocytes, and of von Willebrand factor, which is synthesized in epithelial cells and facilitates platelet adhesion and serves as the plasma carrier for factor VIIIc (coagulation response).

## 2 CLINICAL MANIFESTATIONS OF DIABETES INSIPIDUS

The main symptoms of diabetes insipidus are polyuria and polydipsia. Insufficient fluid intake causes dehydration and a high serum osmolality (hypernatremia). Affected human adults pass large volumes (2.5–30 L/day) of hypoosmotic urine (osmolality $< 290$ mmol/kg).

## 3 CENTRAL DIABETES INSIPIDUS (CDI)

### 3.1 Pathophysiology and Etiology

CDI, also referred to as neurogenic or neurophyseal diabetes insipidus, is caused by inadequate or no release of vasopressin, which is low or not detectable in peripheral blood even after water deprivation. However patients respond normally to administered vasopressin or vasopressin analogues.

Frequent causes of the disease are brain tumors, pituitary or hypothalamic surgery, and severe head injuries. CDI of unknown cause (idiopathic CDI) is also common. Familial CDI is a rare autosomal-dominant trait. Similar to idiopathic CDI, it does not manifest itself immediately after birth. In fact, patients may remain free of symptoms for months or years.

### 3.2 Molecular Genetic Defects in Autosomal Dominant CDI

Recently, missense mutations were found in the vasopressin gene of CDI patients (see Figure 1). In two affected members of a Japanese family, an A → G transition was found, which results in a substitution of Gly 57 of neurophysin II for Ser. In a Dutch family, a G → T transversion, converting Gly 17 of neurophysin II to Val, cosegregates with the disease. Moreover, a G → A transition was identified in patients but not in healthy members of a Caucasian (presumably North American) family; the mutation replaces the C-terminal amino acid of the signal peptide, Ala 19, with Thr. The same mutation was found in a Danish and a Japanese CDI kindred.

At present it is not clear how the genetic defect in one allele causes CDI. As the patients are heterozygous for the mutation, one expects that the wild-type precursor is also formed in their hypothalamic nuclei. It seems plausible that the mutant precursors are not properly processed, and, as a consequence, accumulate in the hormone-producing neurons. The intracellular protein deposits may impair cell function and eventually cause cell death.

## 4 NEPHROGENIC DIABETES INSIPIDUS (NDI)

### 4.1 Pathophysiology and Etiology

Patients with NDI show normal or elevated plasma levels of vasopressin and an adequate increase of vasopressin levels upon dehy-

**Table 1** Mutations in the Human $V_2$ Receptor Gene Associated with X-Chromosomal Recessive Nephrogenic Diabetes Insipidus

| $V_2$ Receptor Mutant[a–d] | Type of Mutation | Change in Nucleotides[b] | Predicted Change in Amino Acid | Predicted Protein Domain | Reference[e] |
|---|---|---|---|---|---|
| W71X* | Nonsense | G → A at 284 | Trp → stop at 71 | C1 | Bichet et al., 1993 |
| R113W* | Missense | C → T at 408 | Arg → Trp at 113 | E2 | Bichet et al., 1993; Holtzman et al., 1993 |
| Q119X* | Nonsense | C → T at 426 | Gln → stop at 119 | TM3 | Pan et al., 1992 |
| Y128S* | Missense | A → C at 454 | Tyr → Ser at 128 | TM3 | Pan et al., 1992 |
| A132D[†] | Missense | C → A at 466 | Ala → Asp at 132 | TM3 | Rosenthal et al., 1992 |
| R137H* | Missense | C → A at 481 | Arg → His at 137 | C2 | Bichet et al., 1993; Rosenthal et al., 1993 |
| (1) R181C*<br>(2) 810del12* | (1) Missense<br>(2) In-frame deletion | (1) C → T at 612<br>(2) deletion of 12 bp 3′ to 810 | (1) Arg → Cys at 181<br>(2) Deletion of Arg247-Arg248-Arg249-Gly250 | (1) E3<br>(2) C3 | Pan et al., 1992 |
| G185C[‡] | Missense | G → T at 624 | Gly → Cys at 185 | E3 | van den Ouweland et al., 1992 |
| R202C[‡] | Missense | C → T at 675 | Arg → Cys at 202 | E3 | van den Ouweland et al., 1992 |
| Y205C[‡] | Missense | A → G at 685 | Tyr → Cys at 205 | E3 | van den Ouweland et al., 1992 |
| 755insC[§] | Frameshift | Insertion of C in codon 228 (nucleotides 753-755) | Frameshift 3′ to codon 228; codon 258 → stop | TM5 and C3 | Merendino et al., 1993 |
| 763delA* | Frameshift | Deletion of A at 763 | Frameshift 3′ to codon 231, codon 270 → stop | C3 | Pan et al., 1993 |
| 804dX* | Frameshift | Deletion of G between 804 and 809 | Frameshift 3′ to codon 247; codon 270 → stop | C3 | Rosenthal et al., 1992 |
| P286R[‖] | Missense | C → G at 928 | Pro → Arg at 286 | TM6 | Pan et al., 1992 |
| L312X* | Nonsense | T → A 1006 | Leu → stop at 312 | TM6 | Bichet et al., 1993 |

[a]In case of single-base substitution, mutants are named according to the amino acid change (one-letter code) at the indicated codon. X: termination at the indicated codon. In case of a deletion (del), the nucleotide number is given first. Following "del" is the total number of deleted nucleotides or—in case of a single-base deletion—the base. In case of the single-base insertion, (755insC), nucleotide 755 (C) may precede or follow the inserted base (C).
[b]Nucleotide numbers are given according to the sequence numbering of GenBank entry Z11687 in which Met 1 corresponds to nucleotides 72–74.
[c]Origin of families: *, North America; [†], Iran; [‡], Netherlands; [§], Lithuania; [‖], El Salvador.
[d]Mutations preceded by numbers in parentheses were found in the same patient.
[e]*References:*

Bichet, D. G., Arthus, M.-F., Lonergan, M., Hendy, G. N., Paradis, A. J., Fujiwara, T. M., Morgan, K., Gregory, M. C., Rosenthal, W., Antaramian, A., Didwania,, A., and Birnbaumer, M. (1993) X-linked nephrogenic diabetes insipidus in North America and the Hopewell hypothesis. *J. Clin. Invest.*, 92:1262–1268; Holtzman, E. J., Harris, H. W., Kolakowski, L. F., Guay-Woodford, L. M., Botelho, B., and Ausiello, D. A. (1993) Brief report: A molecular defect in the vasopressin $V_2$-receptor gene causing nephrogenic diabetes insipidus. *New Engl. J. Med.* 328:1534–1537.

Merendino, J. J., Spiegel, A. M., Crawford, J. D., O'Carroll, A.-M., Brownstein, M. J., and Lolait, S. L. (1993) Brief report: A mutation in the vasopressin V2-receptor gene in a kindred with X-linked nephrogenic diabetes insipidus. *New Engl. J. Med.* 328:1538–1541.

Pan, Y., Metzenberg, A., Das, S., Jing, B., and Gitschier, J.: (1992) Mutations in the $V_2$ vasopressin receptor gene are associated with X-linked nephrogenic diabetes insipidus. *Nature Genet.* 2:103–106.

Rosenthal, W., Antaramian, A., Gilbert, S., and Birnbaumer, M. (1993) Nephrogenic diabetes insipidus: A V2 vasopressin receptor unable to stimulate adenylyl cyclase. *J. Biol. Chem.* 268:13030–13033.

Rosenthal, W., Seibold, A., Antaramian, A., Lonergan, M., Arthus, M.-F., Hendy, G. N., Birnbaumer, M., and Bichet, D. G. (1992) Molecular identification of the gene responsible for congenital nephrogenic diabetes insipidus. *Nature,* 359:233–235.

van den Ouweland, A. M. W., Dresen, J. C. F. M., Verdijk, M., Knoers, N. V. A. M., Monnens, L. A. H., Rocchi, M., and van Oost, B. A. (1992) Mutations in the vasopressin type 2 receptor gene (AVPR2) associated with nephrogenic diabetes insipidus. *Nature Gen.* 2:99–102.

dration. However the kidney fails to respond to the endogenous or to administered hormone.

A common cause for acquired NDI is $Li^+$, used for the treatment of manic–depressive (bipolar) illness. Other drugs that antagonize the antidiuretic actions of vasopressin include certain antibiotics and volatile anesthetics. Acquired NDI may also occur in association with systemic diseases. Hereditary NDI is a rare, typically X-linked disease.

### 4.2 Prevalence and Specific Features of X-Chromosomal Recessive NDI

X-chromosomal recessive NDI appears to occur worldwide. In Quebec, the prevalence of the disease is estimated to be 3.7 per million males. However in other defined regions in North America, the prevalence is considerably higher. Although the concentration defect of the kidney can be demonstrated shortly after birth, polyuria and polydipsia often remain undiagnosed in babies. As a consequence, repeated episodes of severe dehydration, especially if they occur during the first years of life, frequently lead to mental retardation, hypocaloric dwarfism, or even death. Most heterozygous females do not present with symptoms, and only a few are severely affected. Patients lack not only renal but also extrarenal responses to desmopressin.

### 4.3 Molecular Genetic Defects in X-Chromosomal Recessive NDI

Based on the analysis of restriction fragment length polymorphism (RFLP) haplotypes, the gene responsible for NDI has been assigned to the subtelomeric region of the long arm of the human X chromosome (Xq28). More recently, the human $V_2$ receptor gene was cloned and mapped to the same region. Subsequently, mutations in the $V_2$ receptor gene, segregating with the clinical phenotype, were found in NDI families from North and South America, several European countries, and Iran (Table 1). The data indicate that familial NDI is most often ascribable to a defect in the $V_2$ receptor gene.

Mutations are scattered throughout the coding region of the $V_2$ receptor gene. Most common are single base substitutions giving rise to a missense mutation. Expression of the R137H mutant cDNA in mammalian cells showed that the mutant receptor is unable to stimulate the $G_s$/adenylyl cyclase system in response to even high doses of vasopressin, although its affinity to vasopressin is comparable to that of the wild type. Other mutants exhibit a reduced affinity for AVP or a decreased expression on the cell surface. Nonsense mutations lead to the expression of a truncated protein, which is very unlikely to be active. An extreme example is the W71X mutation. The predicted mutant protein consists of the extracellular N-terminus, transmembrane domain 1 (TM1), and the first cytosolic loop (C1), comprising together just 20% of the amino acids of the wild-type receptor. Except for one in-frame deletion (810del12), the identified deletions or insertions cause a frameshift and introduce a premature stop codon. The predicted mutant proteins contain nonreceptor sequences 3′ of the site of mutation and are truncated to varying degrees. Similar to nonsense mutations, frameshift mutations normally abolish the function of a protein.

Of particular interest is the diversity of mutations found in North American NDI families (see Table 1). Although the W71X mutation causes NDI in the Hopewell family, the largest NDI kindred in North America, it does not explain the origin of the disease in the other North American families. Thus the "Hopewell hypothesis," proposing a common ancestor for most North American NDI patients, needs to be revised.

## 5 OUTLOOK

The identification of genes responsible for familial forms of CDI and NDI will facilitate early diagnosis in patients, carrier identification and prenatal diagnosis; it is also the first step toward an improved treatment of CDI and NDI patients. In particular, the expression and biochemical characterization of $V_2$ receptor mutants associated with familial NDI may help to identify patients suitable for drug therapy.

Since submission of the manuscript, additional mutants associated with X-linked NDI have been identified and functionally characterized (e.g. Bichet et al., 1994, *Am. J. Hum. Genet.* 55:278–286; Birnbaumer et al., 1994, *Mol. Endocrinol.* 8:886–894). It has also been shown that an atypical form of congenital NDI with an autosomal recessive inheritance and preserved extrarenal responses to desmopression is caused by defects in the recently identified AVP-regulated water channel, aquaporin-CD (Deen et al., 1994, *Science* 264:92–95).

*See also* Endocrinology, Molecular; Human Genetic Predisposition to Disease; Receptor Biochemistry.

### *Bibliography*

Bichet, D. G. (1992) Nephrogenic Diabetes insipidus. In *Oxford Textbook of Clinical Nephrology,* S. Cameron, A. M. Davison, J.-P. Grünfeld, D. Kerr, and E. Ritz, Eds., pp. 789–800. Oxford University Press, Oxford.

Birnbaumer, L., Abramowitz, J., and Brown, A. (1990) Receptor–effector-coupling by G-proteins. *Biochim. Biophys. Acta,* 1031:163–224.

Czernichow, P., and Robertson, A. G., Eds. (1985) *Frontiers in Hormone Research,* Vol. 13, *Diabetes Insipidus in Man.* Karger, Basel.

Moses, A. M., and Streeten, D. H. P. (1991) Disorders of the neurohypophysis. In *Harrison's Principles of Internal Medicine,* 12th ed., J. D. Wilson, E. Braunwald, K. J. Isselbacher, R. G. Petersdorf, J. B. Martin, A. S. Fauci, and R. K. Root, Eds., pp. 1682–1691. McGraw-Hill, New York.

Reeves, W. B., and Andreoli, T. E. (1989) Nephrogenic diabetes insipidus. In *The Metabolic Basis of Inherited Disease,* 6th ed., R. Scriver, A. L. Beaudet, W. S. Sly, and D. Valle, Eds., pp. 1985–2011. McGraw-Hill, New York.

# Diabetes Mellitus

*David Jenkins, Catherine H. Mijovic, and Anthony H. Barnett*

### *Key Words*

**Diabetes** Heterogeneous disorder characterized by persistent hyperglycemia.

**Human Leukocyte Antigen (HLA) Class II** Heterodimeric cell surface protein found on antigen-presenting cells that presents antigen to CD4 T-cell receptors.

**Insulin** Hormone with several metabolic effects. Deficiency of insulin results in diabetes.

**Major Histocompatibility Complex (MHC)** Cluster of genes that include several loci, many of which are involved in regulation of the immune response, particularly allograft rejection.

**Maturity Onset Diabetes of the Young (MODY)** Subset of non-insulin-dependent diabetes with autosomal-dominant inheritance.

---

Diabetes mellitus is a heterogeneous disorder. Insulin-dependent diabetes mellitus (IDDM) is determined by both genetic and environmental factors. Several genes are involved, some of which occur within the major histocompatibility complex. Linkage between IDDM and the insulin locus has also been shown in certain subsets of patients. None of the genes have been precisely identified. Non-insulin-dependent diabetes mellitus (NIDDM) also has a complex etiology. It is more difficult to investigate because of disease heterogeneity and lack of sufficient families. Abnormalities of the glucokinase gene have been identified as causes of some cases of maturity onset diabetes of the young, a subset of NIDDM. It seems likely that recent advances in molecular genetics will further our knowledge of these complex disorders.

## 1 INTRODUCTION

Diabetes mellitus is a heterogeneous condition of complex etiology characterized by persistent hyperglycemia. Insulin-dependent diabetes (IDDM) is an autoimmune, pancreatic β-cell-specific disease distinct from non-insulin-dependent diabetes (NIDDM) in several respects. Predisposition to both diseases appears to be determined by multiple genes.

Analysis of genetic predisposition to diabetes has confirmed that IDDM and NIDDM are different. IDDM has a lower concordance rate among identical twins than NIDDM. Subjects with IDDM are also less likely to have a diabetic relative, suggesting that the genetic predisposition to NIDDM is greater than that for IDDM.

## 2 METHODS OF IDENTIFYING DISEASE SUSCEPTIBILITY GENES

Certain genes are polymorphic, different forms (alleles) occurring in different individuals. Any allele increased in frequency among diabetic patients compared with healthy control subjects may directly predispose to diabetes, or (as is more likely) be in linkage disequilibrium with a disease susceptibility allele.

This approach to gene mapping was limited by the few known polymorphic genes. The polymerase chain reaction (PCR) has revolutionized genetics and has led to the discovery of a huge variety of polymorphic loci throughout the human genome. Most are noncoding repeated nucleotide sequences (minisatellites and microsatellites) of unknown function. Associations between such marker loci and diabetes suggest linkage disequilibrium, indicating possible locations of disease susceptibility genes.

Family studies provide complementary methods of gene mapping. Conventional linkage analysis is useful for studying monogenic disorders with known mode of inheritance. NIDDM and IDDM are determined by several genes of uncertain mode of inheritance. Linkage analysis has, therefore, been of little value in these disorders, with the notable exception of maturity-onset diabetes of the young (MODY), a subset of NIDDM. Analysis of allele sharing by affected sib-pairs can be applied to polygenic disorders. This approach has been useful in IDDM, but large numbers of affected sib-pairs are required. In NIDDM the late onset of the disease together with its excess mortality seriously hinders the identification of sufficient informative families for sib-pair analysis. Heterogeneity of NIDDM also confounds comparison of findings between different studies, particularly when different ethnic groups have been used.

Animal models for IDDM and NIDDM may clarify the genetics of both diseases. Identification of susceptibility genes in animal models produces candidate genes for the human disease.

## 3 IDDM AND THE MHC

### 3.1 HLA Associations

The human leukocyte antigen (HLA) genes associated with IDDM occur in the human major histocompatibility complex (MHC), on chromosome 6 (Figure 1). The disease is most strongly associated with alleles of the class II genes, particularly DR and DQ. Class II A and B genes encode α and β genes, respectively. These combine to form cell surface molecules on antigen-presenting cells which restrict the CD4 T-cell response to foreign antigen. DRB1 alleles determine the various DR antigens such as DR3 and DR4, which are positively associated with IDDM, and DR2, which is negatively associated with this form of diabetes. It is unclear whether the DRB1 alleles themselves affect disease predisposition. Mode of inheritance studies indicate that the susceptibility alleles associated with DR3 and DR4 appear distinct. The DQA1 and DQB1 genes are also polymorphic. DR and DQ alleles are now typed precisely using DNA sequencing and allele-specific gene probing of PCR-amplified DNA.

In Caucasian populations, IDDM is strongly associated with the DQ alleles DQA1*0301, DQB1*0302, and DQB1*0201, and negatively associated with DQB1*0602. It was hypothesized that susceptibility is strongly influenced by an arginine residue at position 52 (Arg 52) of the DQα chain and absence of aspartate at position 57 (non-Asp 57) of the DQβ chain. It was also suggested that individuals able to encode DQ heterodimers comprising both Arg 52 DQα and non-Asp 57 DQβ chains were strongly predisposed to IDDM. This was supported by analysis of DQ genotypes in Caucasian populations. These associations may occur, however, because of linkage disequilibrium between DQ and other genes, which encode susceptibility.

### 3.2 Trans-Racial Studies in Mapping Susceptibility to IDDM

Improved mapping of disease susceptibility has come from studying disease associations in different races. Although linkage disequilibrium is strong between MHC genes, recombination during evolution has generated combinations of MHC alleles (extended haplotypes) that vary between populations. This has caused disease-predisposing alleles to be in linkage disequilibrium with different marker alleles in different races. Any allele that is consistently associated with a disease in all races, irrespective of the extended haplotype on which it occurs, is a candidate disease susceptibility determinant. This assumes that IDDM has the same genetic basis in all races. Comparison of HLA associations with IDDM in different demo-

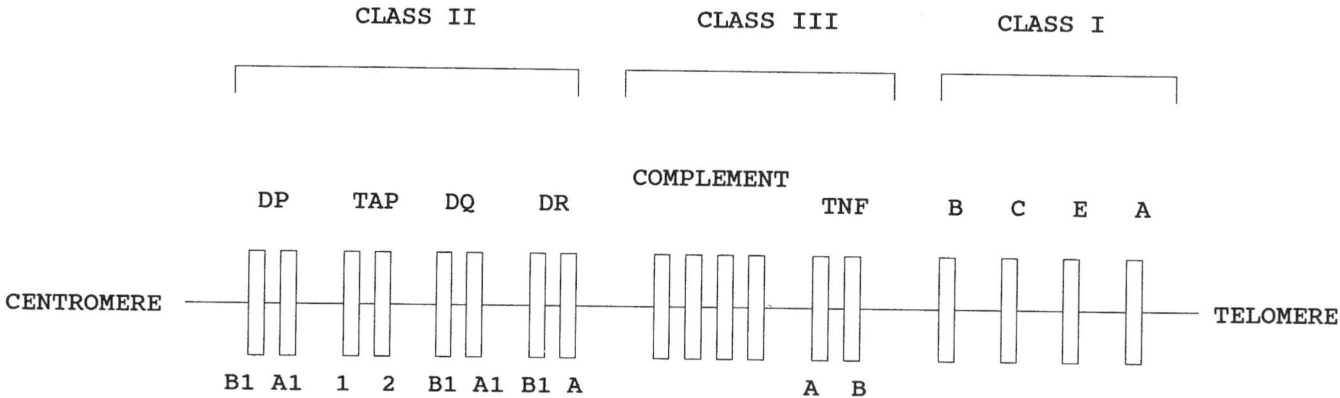


**Figure 1.** Simplified diagram of the MHC (approximately 4 megabases in length) on the short arm of chromosome 6. The diagram is not to scale. Pseudogenes and several recently discovered genes have been omitted.

graphically defined races (transracial studies) therefore require rigorous application of diagnostic criteria for disease, and careful matching of disease and control populations for ethnic origin.

IDDM is positively associated with DQB1*0201 in all races with one exception; this allele is very rare in the Japanese. (The rarity of this disease-associated allele may contribute to the low prevalence of IDDM in Japanese subjects.) DQB1*0201 is in linkage disequilibrium with DR3 and may direct determine DR3-associated susceptibility to IDDM. The DR2-associated allele DQB1*0602 and the very similar DQB1*0603 are negatively associated with IDDM in all races and may directly protect against the disease. DQA1*0301 (in linkage disequilibrium with DR4) predisposes to IDDM in all races except the Chinese. DQA1*0301, therefore, may not predispose to disease directly but may be in linkage disequilibrium with a predisposing allele at another locus. Alternatively, any effect of DQA1*0301 is modified by another susceptibility gene. (There are no data to suggest that IDDM in Chinese subjects is distinct from the disease in other races.) It seems likely, therefore, that DQ alleles alone cannot explain DR4-associated susceptibility to IDDM.

The role of non-Asp 57 DQβ chains in IDDM is questioned by the positive association between the Asp 57-encoding allele DQB1*0401 and IDDM in Japanese subjects. It seems likely that a non-Asp 57 DQβ chain is simply a useful marker of susceptibility in Caucasians.

If DQ genes do directly encode IDDM susceptibility, disease-predisposing DQ molecules may present β-cell antigens to CD4 cells, triggering β-cell destruction. The identity of such an antigen is being intensively investigated.

### 3.3 The NOD Mouse

The role of MHC class II genes in diabetes susceptibility has been studied further in the nonobese diabetic (NOD) mouse. This animal spontaneously develops IDDM. I-Aβ (the murine analogue of DQβ) is non-Asp 57 in NOD mice, supporting the proposed protective role of Asp 57. Insertion of both Asp 57 and certain non-Asp 57 I-Aβ genes into transgenic NOD mice, however, can protect against diabetes. Although these data implicate DQ in IDDM susceptibility, they call into further question the Asp 57 hypothesis.

## 4 OTHER POSSIBLE DETERMINANTS OF SUSCEPTIBILITY

The MHC contains several other genes that might affect susceptibility to IDDM. Some studies have suggested that alleles of HLA-B, complement, tumor necrosis factor (TNF), and transporter-associated peptide (TAP) genes may alter disease predisposition. These reports require confirmation. The continuing discovery of more genes within the MHC may identify other candidate determinants of susceptibility.

Although the major component of inherited susceptibility to IDDM appears to be MHC-encoded, non-MHC genes are also implicated. Candidates include genes encoding insulin (INS), the constant portions of the heavy ($G_m$) immunoglobulin chains, and the constant portions of the T-cell receptor (TCR). IDDM is positively associated with INS polymorphism. Linkage between INS and IDDM has recently been demonstrated in certain patient subsets. This suggests that a second gene encoding IDDM susceptibility lies within the INS region.

Positive interactions between $G_m$ alleles, HLA alleles, and IDDM have also been reported but require confirmation. Weak associations between IDDM and TCR polymorphism are not supported by linkage analysis.

## 5 INHERITED SUSCEPTIBILITY TO NIDDM

### 5.1 Identification of Candidate Genes

Attempts to map susceptibility to NIDDM have relied on identifying polymorphic candidate genes. NIDDM is characterized by both insulin resistance and abnormal insulin secretion. Many studies have searched for associations between NIDDM and genes that might alter insulin secretion and action, including the genes for insulin, the insulin receptor, the various glucose transporters (GLUT), and amylin. These studies have been disappointingly negative or inconsistent. There is, however, one candidate gene that has yielded an unequivocally positive result.

### 5.2 Maturity Onset Diabetes of the Young

MODY is a well-defined subset of NIDDM with an early age of onset and an autosomal-dominant mode of inheritance. It contrasts, therefore, with the majority of NIDDM. The existence of large MODY pedigrees has made this disorder amenable to conventional linkage analysis. It was postulated that glucokinase, which acts as a glucose sensor on human B cells, might be a candidate gene for MODY. Linkage between MODY and glucokinase has now been found in a number of families, and abnormalities of the glucokinase gene have been identified in the affected individuals. It should be noted, however, that linkage with glucokinase was not found in all MODY families, indicating that MODY itself is heterogeneous. No consistent association between glucokinase polymorphism and the majority of individuals with NIDDM has been found. Perhaps the main benefit of identifying glucokinase as the determinant of some cases of MODY is the clear demonstration of NIDDM heterogeneity. Distinction of other subsets of NIDDM may allow identification of other genes.

## 6 FUTURE STUDIES

The genetics of both IDDM and NIDDM remain incompletely understood. Major determinants of susceptibility to IDDM have been mapped to the DQ subregion of the MHC. The INS region is also implicated. The exact identities have yet to be established, as have the mechanisms by which these genes predispose to pancreatic β-cell destruction. Other genes are likely to be involved. The genetic defect responsible for a small subset of NIDDM has now been established.

The candidate gene approach, which proved disappointing for so long, has started to yield results. An alternative method is to screen systematically gene markers distributed throughout the genome for linkage with diabetes. This approach may prove useful in IDDM, for which large numbers of multiplex families are now available. No such resource is available for NIDDM. At present the best approach for NIDDM seems to be to characterize the metabolic defects that identify subsets of NIDDM and then to study candidate genes that seem appropriate. Elucidation of the molecular genetics of diabetic diseases should clarify their pathogenesis and may lead to improvements in treatment and prevention.

*See also* Diabetes Insipidus; Immunology; Major Histocompatibility Complex.

### *Bibliography*

Cox, N. J., and Bell, G. I. (1989) Disease associations: Chance, artifact, or susceptibility genes? *Diabetes,* 38:947–950.

Field, L. L. (1991) Non-HLA region genes in insulin-dependent diabetes mellitus. In *Bailliere's Clinical Endocrinology and Metabolism,* Vol. II, *Genetics of Diabetes,* L. C. Harrison and B. D. Tait, Eds., Bailliere Tindall, London.

Jenkins, D., Mijovic, C., Fletcher, J., Jacobs, K. H., Bradwell, A. R., and Barnett, A. H. (1990) Identification of susceptibility loci for type 1 (insulin-dependent) diabetes by transracial gene mapping. *Diabetologia,* 33:387–395.

O'Rahilly, S., Wainscoat, J. S., and Turner, R. C. (1988) Type 2 (non-insulin-dependent) diabetes mellitus: New genetics for old nightmares. *Diabetologia,* 31:407–414.

Parham, P. (1990) A diversity of diabetes. *Nature,* 345:662–664.

Randle, P. J. (1993) Glucokinase and candidate genes for type 2 (non-insulin-dependent) diabetes mellitus. *Diabetologia,* 36:269–275.

Rich, S. S. (1990) Mapping genes in diabetes. *Diabetes,* 39:1315–1319.

Thomson, G., Robinson, W. P., Kuhner, M. K., Joe, S., MacDonald, M. J., Gottschall, J. L., et al. (1988) Genetic heterogeneity, modes of inheritance, and risk estimates for a joint study of Caucasians with insulin-dependent diabetes mellitus. *Am. J. Hum. Genet.* 43:799–816.

Todd, J. A., and Bain, S. C. (1992) A practical approach to identification of susceptibility genes for IDDM. *Diabetes,* 41:1029–1034.

# DNA Damage and Repair

*George J. Kantor*

### *Key Words*

**DNA Damage** Any modification of the structure of the DNA polymer.

**DNA Repair** A restoration of damaged DNA to its original structure, usually accomplished through a multistep enzymatic process.

**Endogenous Agents** Physical and chemical agents of intracellular origin that are part of the normal intracellular environment.

**Exogenous Agents** Physical and chemical agents of extracellular origin that can enter the intracellular environment.

**Mutation** Any change in the sequence of nucleotides in genomic DNA.

The genetic molecule DNA is continually exposed to a wide variety of chemical and physical agents that can change its structure. These changes interfere with replication and transcription of DNA and are thus referred to as DNA damage. Biological consequences include cell death and mutations, events that may cause cancers, mental retardation, and reduced growth and development. Cells defend against the effects of DNA damage by using enzymatic repair to restore the DNA to its original structure. If repair is timely, the consequences of DNA damage are eliminated. Obtaining an understanding of the nature of DNA damage, its biological consequences, and its elimination by repair is an important step toward resolving the many health problems associated with exposure to DNA-damaging agents.

## 1 DEFINITION OF DNA DAMAGE

Cellular modifications of DNA are induced by several agents, including electromagnetic radiation, heat, pressure, and some chemicals. Several forms of damage have been identified, but only in a few cases has their biological effectiveness been estimated.

Early measures of the biological effects of radiation showed that nucleic acids were target molecules for the induction of mutations and cell death caused by ultraviolet light. Scientists thus decided to examine the genetic molecule DNA for the chemical changes responsible. One of the first forms of DNA damage defined as biologically significant was the cyclobutyl pyrimidine dimer, the result of a covalent joining of two adjacent pyrimidines in one of the DNA strands. This structure is illustrated in the representation of DNA in Figure 1. Other examples of exogenously induced damage to DNA include breaks in the backbone structure of the

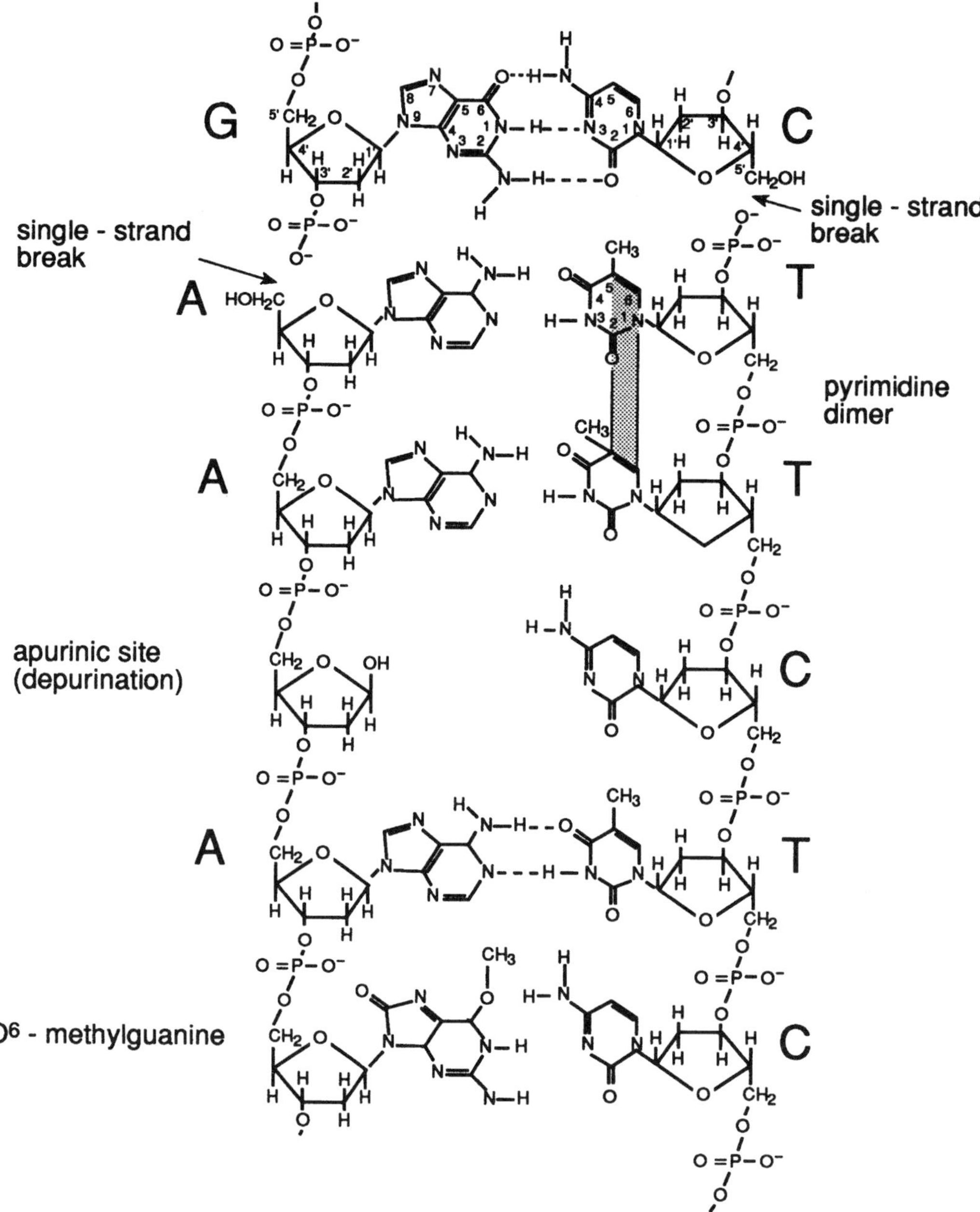


**Figure 1.** Examples of modifications that have been identified as biologically relevant DNA damage. A portion of a double-stranded DNA molecule is represented using structural formulae. Complementary base pairings achieved through hydrogen bonding (- - - -) of adenine (A) to thymine (T) and guanine (G) to cytosine (C) are illustrated. The numbers in the ring structures identify specific atoms in the nucleotides. Examples of DNA damage include the UV-induced pyrimidine dimer, the alkylation product O$^6$-methylguanine, an apurinic (AP site), and two nearby single strand breaks that create a double strand break. Modifications in the bases interrupt the standard hydrogen bonding between bases.

DNA polymer, caused by ionizing radiation such as X-rays and by some chemicals. Heat promotes the removal of purines and pyrimidines, creating apurinic or apyrimidinic (AP) sites. Some chemicals can alkylate the purine and pyrimidine rings of DNA. Highly reactive radicals formed by ionizing radiation react with the purines and pyrimidines to create several oxidative products. DNA strand cross-links, which are covalent bonds that hold the two complementary strands of DNA in an inseparable state, can be created by radiation and by the action of some chemicals. Some typical forms of the kinds of DNA damage discussed are illustrated in Figure 1.

Damage to DNA induced spontaneously by endogenous agents may account for the greatest fraction of damage experienced by most animal cells. Human body temperatures (37°C) can break the glycosidic bond between a sugar and a purine or a pyrimidine. Such action results in absent bases at sites along the DNA strands (AP sites). Oxidants produced in cells as by-products of normal metabolism induce several DNA structural modifications. It has been estimated that these natural reactions result in $10^4$ apurinic sites and an equal number of oxidant-induced damaged sites per human cell per day. These numbers represent a small fraction (about 0.0004%) of the DNA bases per diploid cell. However, each damaged site can cause a mutation or cell death. The recognition that normal intracellular events cause a significant level of DNA damage is a major advance in our understanding of the role of DNA damage in the health of individual cells.

## 2 DEALING WITH DNA DAMAGE

### 2.1 Biological Indications of Repair

Scientists have discovered mutant cells that are much more sensitive than normal cells to the lethal action of DNA-damaging agents. This extrasensitivity is a readily observable indicator of a defect in dealing with DNA damage. Normal cells can restore damaged DNA to its original structure; sensitive cells usually lack this capability.

### 2.2 Mechanisms of DNA Repair

Scientists have also discovered many different enzymatic mechanisms for repairing damaged DNA. These include damage-elimination by cutting it out and restoring the cut-away section and direct reversal of damage. The most common repair mechanism, referred to as excision repair, employs enzymes to remove a short section of the DNA strand containing the damage. The opposite complementary strand is left intact. Subsequent DNA synthesis and rejoining steps fill in the resulting single-strand gap, thereby restoring the molecule to its original structure. Photoreactivation is an example of a direct mechanism for reversing UV-induced pyrimidine dimers. A specific enzyme complexed with a dimer can absorb a photon of blue light and split the cyclobutyl structure. The enzyme–DNA complex then dissolves, leaving the DNA in its original unaltered structure.

## 3 DNA REPAIR IN HUMAN CELLS

### 3.1 Repair in Normal Human Cells

Most DNA damage is removed rapidly in an error-free manner in human cells. For example, cultured human cells remove about 80% of UV-induced pyrimidine dimers in 24 hours. Skin cells in humans may repair the same amount of damage even faster, in 2 to 3 hours. The mechanisms for this repair are poorly understood but probably involve processes of excision and DNA resynthesis.

### 3.2 Hereditary Diseases Associated with Defects in Repair of DNA

Normal excision repair is defective in several human hereditary diseases, including xeroderma pigmentosum (XP), ataxia telangiectasia, Cockayne's syndrome, Fanconi anemia, and Bloom's syndrome. Defective repair may also cause the symptoms associated with these diseases. For example, XP patients have a high incidence of skin cancer on parts of their body that are exposed to sunlight. Their individual cells do not repair most of the DNA damage induced by the UV portion of sunlight. This association represents some of the best evidence that DNA damage, if not repaired in a timely fashion, can induce human malignant lesions and other developmental problems.

*See also* Environmental Stress, Genomic Responses to; Ionizing Radiation Damage to DNA; Ultraviolet Radiation Damage to DNA.

### *Bibliography*


Ames, B. N., and Gold, L. S. (1991) Endogenous mutagens and the causes of aging and cancer. *Mutat. Res.* 250:3–16.

Bohr, V. A., Phillips, D. H., and Hanawalt, P. C. (1987) Heterogeneous DNA damage and repair in the mammalian genome. *Cancer Res* 47:6426–6436.

Cleaver, J. E., and Kraemer, K. H. (1989) Xeroderma pigmentosum, in *The Metabolic Basis of Inherited Disease,* C. R. Scriver, A. L. Beaudet, W. S. Sly, and D. Valle, Eds., pp. 2949–2971. McGraw-Hill Information Sciences, New York.

Friedberg, E. C. (1985) *DNA Repair.* Freeman, New York.

Mullaart, E., Lohman, P. H. M., Berends, F., and Vijg, J. (1990) DNA damage metabolism and aging. *Mutat. Res.* 237:189–210.

Piette, J. (1991) Biological consequences associated with DNA oxidation mediated by singlet oxygen. *J. Photochem. Photobiol. B: Biol.* 11:241–260.

Setlow, R. B. (1982) DNA repair, aging and cancer. *Natl. Cancer Inst. Monogr.* 60:249–255.

Sutherland, B. M., and Woodhead, A. D. Eds. (1990) *DNA Damage and Repair in Human Tissue.* Plenum Press, New York and London.


# DNA Fingerprint Analysis

*Lorne T. Kirby, Ron M. Fourney, and Bruce Budowle*

### *Key Words*

**AMP-FLP (Amplified Fragment Length Polymorphism)** PCR-amplified fragment lengths consisting of VNTRs.

**AP-PCR (Arbitrary Primer—PCR).** See RAPD.

**Chain of Custody (Continuity)** A record of the custody and handling from the time evidence material is first obtained until it is entered in a court of law.

**DAF, DNA Amplification Fingerprinting** See RAPD.

**Exclusion** The elimination of the possibility of a crime suspect or putative parent being the source of a specimen or the biological father.

**Hypervariable Region** A segment of DNA characterized by considerable variation in the number of tandem repeats or sequence array.

**Inclusion** The result of a determination that the crime suspect or putative parent cannot be excluded as the source of a specimen or as the biological father.

**Minisatellites** Regions of 16-50 base pair tandem repeats in the genome.

**MVR-PCR, (Minisatellite Variant Repeat—PCR)** A PCR procedure based not only on tandem repeat copy number but also on the different interspersion pattern that exists among the alleles. Sequence variation is measured along minisatellite alleles consisting of two types (a and t) of repeat units that differ at one polymorphic site. The DNA profiles of the alleles at the same locus in a diploid organism are superimposed to generate a ternary code where each repeat unit site is aa, tt, or at(ta).

**PCR** Polymerase chain reaction. An in vitro DNA amplification procedure.

**RAPD (Random Amplified Polymorphic DNA)** RAPD, AP-PCR, and DAF are PCR procedures that differ in the length of primers used, the amplification conditions, and the resolution and visualization of the products; usually only one arbitrary sequence primer per reaction is used for the generation of DNA fingerprints.

**STR (Short Tandem Tepeat)** Also referred to as SSLP (simple sequence length polymorphism). STRs are tandem repeat regions 2 to 7 base pairs long, scattered throughout genomes. PCR procedures are used to amplify these regions for the generation of DNA profiles.

**Tandem Repeat** The end-to-end duplication of a series of identical or almost identical motifs (usually 2–80 base pairs each) of DNA scattered throughout genomes.

**VNTR (Variable Number of Tandem Repeats)** The variable number of tandem repeats forming alleles at a locus.

---

DNA fingerprint analysis (fingerprinting or typing) is the process of preparing and interpreting bar-code-like profiles of DNA segments for individual identification. Nanogram-to-microgram quantities of genomic DNA are isolated and polymorphic segments are directly analyzed or amplified by means of the polymerase chain reaction (PCR) and analyzed. With the exception of identical twins and clones, profiles of the segments (usually separated and characterized according to length or by sequence differences) from humans, other animals, microorganisms, fungi, and plants are unique to each individual (Figures 1 and 2).

The analysis of DNA is revolutionizing the field of identification. The need for only minute quantities of tissue, the stability of DNA, and the high degree of assay accuracy and precision have contributed to the universal application of DNA typing. The use of this procedure is rapidly altering the manner in which genetic analyses are carried out in human parentage, rape, and homicide cases; in animal poaching, parentage, breeding, and population studies; in medical analysis; and in plant patent disputes, clone identity investigations, parentage testing, and gene bank management programs.

## 1 PRINCIPLES

### 1.1 Specimen Processing

DNA is the molecule of heredity. It is an integral component of all living matter with the exception of RNA viruses. Any living or nonliving organic matter containing relatively intact DNA fragments can be used for DNA typing analysis. Common sources include human or other animal blood, semen and solid tissues, and plant foliage and seeds.

A document detailing chain of custody (continuity) is required for a legal case specimen. The drawing and labeling (name ID, date, location) of blood samples in paternity and immigration cases must be witnessed, as must details of specimen collection and custody in rape and homicide cases.

Specimens are stored under clean, dry, cool conditions to reduce contamination by microorganisms and DNA degradation caused by cell lysis and DNAase activity. Successful analyses have been carried out on DNA extracted from dried blood stains, dried museum specimens, and frozen tissue preserved for many decades. Indeed, these studies have formed the basis of recent new theories in molecular evolution and anthropology.

A good yield of high molecular weight DNA devoid of organic and inorganic contaminants can be extracted from tissues. The enzyme proteinase K is generally used to assist with cell lysis and to digest proteins. The detergent sodium dodecyl sulfate facilitates the separation of residual protein from DNA. Proteinaceous materials are removed by phenol extraction (or a nonorganic NaCl or

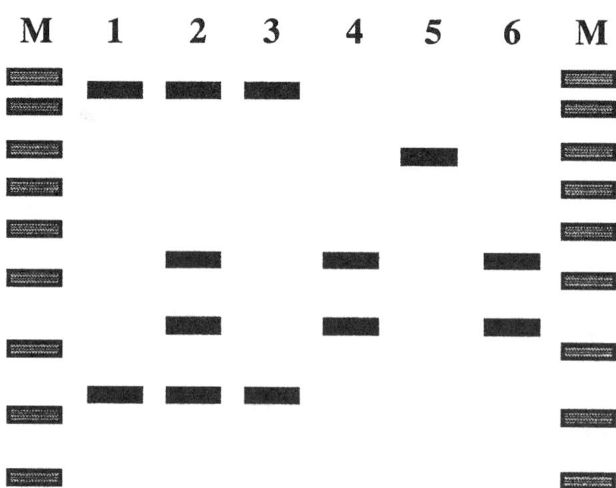


Figure 1. DNA typing in a sexual assault case.

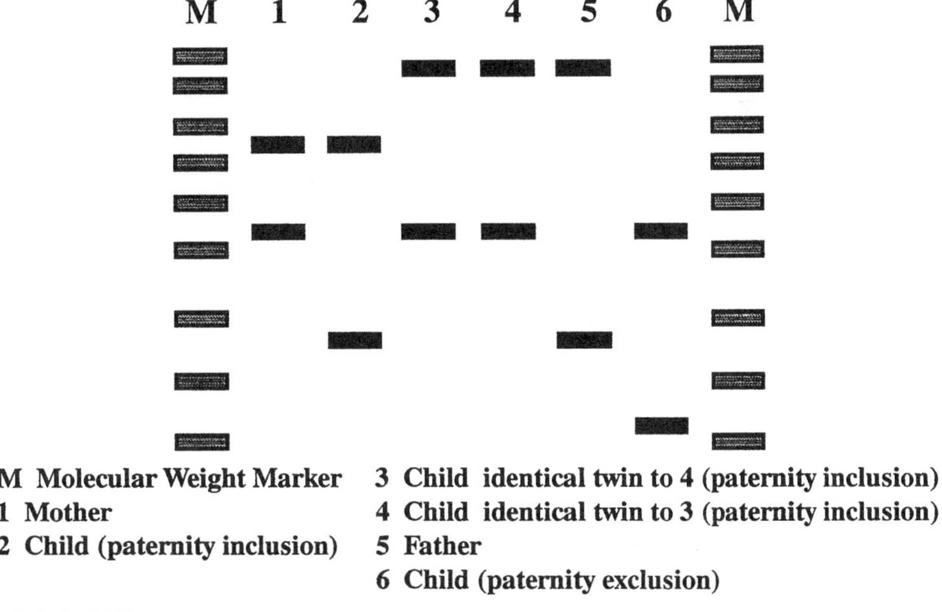


**Figure 2.** Paternity analysis by DNA.

LiCl precipitation), and chloroform is added to remove traces of phenol. Most of the contaminating RNA can be eliminated, provided the phenol is appropriately equilibrated. High molecular weight DNA is recovered in a relatively pure form by ethanol precipitation in the presence of salt. The extraction and purification steps can, under certain circumstances, be bypassed after cellular lysis. Fragments of DNA can be directly amplified by PCR and the amplified fraction used to construct a DNA fingerprint profile. If this approach is followed, possible inhibitors of the PCR reaction can be removed by means of a simple nonorganic technique using an ion exchange resin such as Chelex 100 or by washing through a Centricon 100 filter.

The analyst may be presented with mixed specimens. It is possible to separate mixtures such as sperm and female cells often present in vaginal swabs from rape victims. The female cells are lysed in extraction buffer devoid of dithiothreitol (DTT) and the DNA is isolated from the supernatant. The sperm pellet is then lysed by adding extraction buffer containing DTT and increased detergent.

DNA quality can be evaluated by ethidium bromide staining or by southern blotting and hybridizing with a species specific probe. The degree of degradation is estimated by observing, under UV light, ethidium bromide stained fragment lengths subjected to agarose gel electrophoresis. High molecular weight DNA appears as a single large band, whereas partially degraded DNA forms a long smear of large to small fragments ($>$ 20,000 base pairs to a few hundred base pairs for human DNA). Ethidium bromide staining cannot differentiate DNA from different organisms—species-specific probes are used for this purpose.

Quantitation is the final step in preparing DNA for analysis. At least three different procedures are available. In the first technique, the absorbance of a DNA solution is measured in a spectrophotometer at 260 nm wavelength. Second, the DNA and a series of standards of known concentration are subjected to agarose gel electrophoresis and stained with ethidium bromide. Last, an aliquot of the DNA is fixed on a nylon membrane and is hybridized with, for example, a labeled, highly repetitive, primate-specific $\alpha$-satellite probe p17H8 that detects locus D17Z1.

### 1.2 Specimen Analysis

The analysis objective is the isolation of DNA segments that will facilitate identity discrimination among individuals. Profile patterns from tissues from the same individual should be identical.

Many polymorphic genetic markers (segments) consist of tandem repetitive sequences. Depending on the locus, each repeat can consist of a minimum of 2 to perhaps 80 base pairs and may contain possible single base pair polymorphic sites. The number of tandem repeats varies from one to hundreds.

A locus may, therefore, be polymorphic for the number of tandem repeats [referred to as variable number (VNTRs) or short (STRs)] and for the base pair composition at specific sites within each repeat [minisatellite variant repeat (MVR)]. This variation, when analyzed at one or more loci by RFLP, AMP-FLP, or MVR-PCR techniques is the basis of DNA profile construction (Figure 3).

Polymorphic sites that may or may not be located in tandem repeat units provide another form of variation useful in identity analysis. The RAPD, AP-PCR, or DAF system is based on the annealing of arbitrary sequence PCR primers to template sequences sufficiently close together on complementary DNA strands to facilitate amplification. Systems differ in their specifics: the length of the primers used, the amplification conditions, and the resolution and visualization of the products. Amplification will not occur if the template sequence differs significantly from the primer sequence or perhaps if competition occurs between the primer and hairpin loop structures in the template.

The final DNA pattern or profile can be visualized with color dyes, fluorescent dyes, radioisotopes, or other stains such as silver. There are three analytical approaches used for DNA fingerprinting: VNTR analysis, dot blot analysis of sequence polymorphisms, and direct sequence analysis. For VNTR analysis, the restriction-

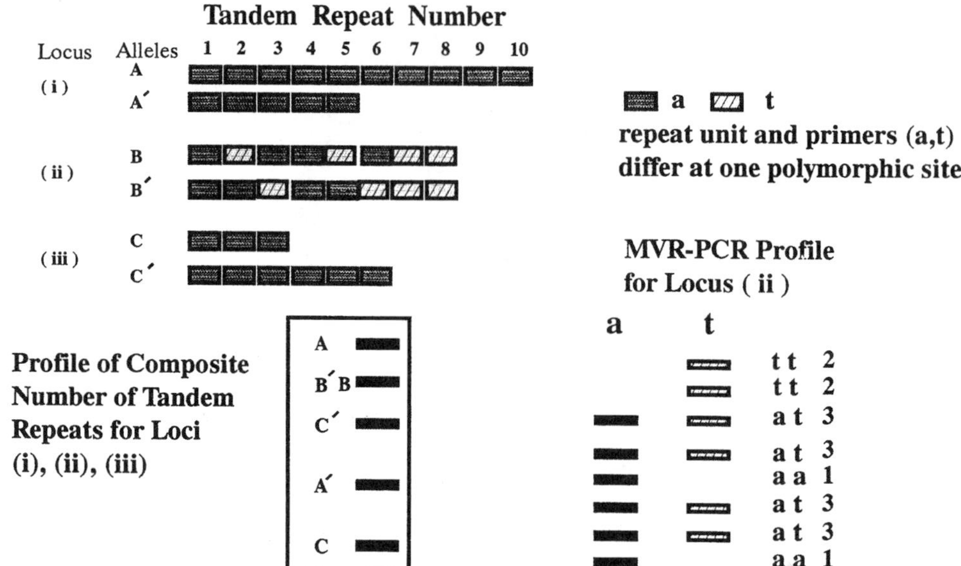


Figure 3. Variable number of tandem repeats and MVR-PCR profiles.

digested or -amplified DNA fragments are separated, generally based on size, by electrophoresis in a sieving medium. The relative positions of the bands after electrophoretic separation are determined with reference to size standards. In dot blot analysis, the target DNA sequence is amplified by PCR and the amplified products are fixed to membranes. The DNA is then incubated with allele-specific probes. If hybridization occurs, the dot will be marked; if the DNA does not contain the particular allele, the spot will remain blank. Alternatively, dot blots can be configured where the allele-specific probes are fixed to the membrane and the amplified DNA is allowed to hybridize to the fixed probes. This reverse dot blot approach is used for HLA-DQα typing. Finally, the sequence of DNA bases can be read by direct sequencing.

A number of analytical approaches can, therefore, be used to isolate and identify profile patterns. Isolated and purified DNA or nonextracted DNA may be used. DNA from each of a number of loci may be analyzed separately and the results combined to form a profile, or DNA consisting of similar base pair sequences from multiple loci may be analyzed coincidentally. The DNA may be analyzed directly, or it may be amplified and then analyzed.

## 1.3 Data Processing

Genetic profiles are presented as (1) the presence or absence of dots, (2) the position or size of DNA fragments separated by an electrophoretic approach, or (3) direct sequence information (e.g., AATCGTACCTGATCC). The results produced by any of these methods can be evaluated visually. However, as data become more abundant and complex, computer-assisted analyses are needed to aid in pattern interpretation, data storage and manipulation.

For dot blot profiles, most interpretations are performed by "eye," and the data are generally transcribed manually into a computer storage bank. However, since most dots are colored or visible on X-ray film, densitometric scanning systems can be used to automatically score and store the data for future analysis.

For most VNTR analyses, which are based on size or position, the first generation of data acquisition used to determine the relative size of unknown DNA fragments was processed by measurement with a ruler and comparison with known size standards (on autoradiograms). Later, semiautomated computer-assisted image capture systems (which are essentially electronic rulers) replaced the manual method of data analysis. With the use of fluorescent or UV detection systems, labeled DNA fragments can be detected in real-time analysis and automatically transcribed to the computer. Both VNTR typing and sequencing can be automated using this approach.

The main advantages of nonmanual data acquisition are the reduction of human transcriptional errors and, hopefully, a reduction in labor. Once the data have been stored in the computer, various simple and sophisticated statistical analyses can be performed for determinations, such as the likelihood of occurrence of DNA profiles for identity typing purposes and for linkage analysis.

## 1.4 Quality Assurance

Laboratory quality assurance is the documented verification that proper procedures have been carried out by skilled and highly trained personnel to ensure that valid and reliable results can be obtained. The validity of DNA typing results centers on correctly identifying true nonmatches and matches from a prospective DNA fingerprint profile. Test reliability measures reproducibility under defined conditions of use and should transcend different laboratories and practitioners. Quality assurance must encompass all significant aspects of the DNA typing process, which include personnel education and training, documentation of records, data analysis, quality control of reagants and equipment, technical controls, proficiency testing, reporting of results, and auditing of the laboratory procedures. Quality assurance and appropriate standards for DNA typing evolved initially from the experience of clinical laboratories; more recently, however, they have been carefully defined through consensus from forensic laboratories represented in the North American nonregulatory federal initiative, the Technical Working Group on DNA Analysis Methods (TWGDAM). It is important to note that quality assurance guidelines in forensic science must retain enough flexibility to accommodate the unique nature of forensic samples as well as future advances in recombinant DNA technology and molecular biology. Good quality control management is paramount for obtaining quality results. It is also a key factor in establishing uniform reliability between laboratories.

## 2 APPLICATIONS

DNA fingerprinting is applicable to any identification problem for which a minimal quantity of relatively intact DNA is available. The requirement for intact DNA depends on the method of analysis. Typing by means of restriction fragment length polymorphisms (RFLP) generally requires high molecular weight DNA, whereas most genetic marker typing based on PCR can make use of much smaller, even somewhat degraded, fragments.

### 2.1 Human

Suspect or victim tissue found at the scene of a rape or a homicide provides a source of forensic evidence for determining whether a suspect is the source of the crime scene specimens (Figure 1). Accident or homicide victims unidentifiable from their physical features can be identified provided typable DNA is isolated from the victim's remains and is matched with DNA, perhaps from hair roots obtained from his or her hair brush. If the victim's putative parents or other relatives are available, parentage (or other biological relationship) testing also may be carried out.

Parentage can be determined for child custody and immigration cases, and for counseling about genetic diseases. The bands in the offspring's DNA profile must have been inherited from the biological parents. Barring mutations, the presence of bands not found in either parental profile is indicative of nonparentage (Figure 2).

Applications in medicine include genetic counseling, tracing the percentage of donor versus recipient cells in bone marrow transplants, determination of possible "contamination" of fetal chorionic villi sample (CVS) tissue with maternal tissue, tissue culture cell line identification, and the confirmation of twin zygosity.

Identification monitors are feasible for security purposes in the armed forces and for identification in mass disasters or war. For these applications, DNA profiles are recorded for future match comparisons.

### 2.2 Animal

Parentage, poaching, identification in cases of theft or loss, population studies, the detection of trait markers, and breeding programs are areas of application in animal husbandry.

Proof of parentage may be important for pedigree registration and for establishing sale value.

A DNA profile from animal remains can be compared with a profile from a frozen steak or from a trophy specimen in a poaching case.

The evolution of populations can be traced, for example, by analyzing DNA from dried museum specimens or from insects housed in amber for millions of years. Base pair sequences and profile bands can be compared with those from modern relatives.

Profiling is used in breeding programs for populations of endangered species with small gene pools. Animals exhibiting the greatest number of DNA profile differences and, therefore, probably the greatest genetic diversity, are chosen for breeding. Also, biological relationships can be confirmed in artificial insemination and embryo transfer programs for domesticated animals.

### 2.3 Microorganisms

DNA fingerprinting provides a powerful tool for microorganism identification. Different bacterial strains, for example, have unique profiles.

Profiling also provides a basis for issuing patents for microorganisms such as yeast used in the brewing industry.

### 2.4 Plants

The field of plant identification for patent, parentage, theft, and trait marker purposes has recently received considerable impetus because of DNA analysis, especially using the RAPD-type systems.

The infringement of breeders' rights relating to the origin of cloned material such as cuttings, parentage of plants with desirable traits, the theft of expensive trees, and gene bank management for identification and the measurement of variation are examples of applications.

## 3 PERSPECTIVES

### 3.1 DNA Fingerprinting Versus Other Identification Techniques

Many of the conventional biochemical identification techniques such as isozyme electrophoresis are being replaced by DNA typing, especially where more information is required. Although conventional methods such as blood typing are rapid, simple, and definitive for exclusions, they are much less powerful than DNA analysis for the probability of inclusion. If a mother and a putative father both have blood type O, their offspring cannot be type A, B, or AB. The putative father is readily excluded as the biological father in this example. If the offspring is type O, possible exclusion of the putative father is considerably more difficult using conventional methods.

Additional key factors in choosing DNA for analysis include the availability of the hereditary material, as well as its highly polymorphic nature, continuity within the same organism, stability relative to that of enzymes, and ability to amplify segments in vitro when a minimal quantity of tissue is available or the DNA is moderately degraded.

### 3.2 Future Directions

Automation is a primary objective. Current methodology is labor intensive because of the requirement of a number of isolated steps, including combinations of extraction, amplification, endonuclease digestion, electrophoresis, labeling, hybridization, blot scanning, profile comparison matching, data reduction, and probability calculation. Automation is especially desirable for direct sequencing of highly polymorphic regions of the genome because it will result in the elimination of most of the tedious current techniques.

Although fingerprinting approaches using RAPD (DAF, AP-PCR) systems circumvent the need for probes, a limitation arises in the calculation of probabilities. It is extremely difficult to determine population frequencies of the amplified fragments. Without these data, the calculation of the probability of profile matches is not readily feasible. Probes (together with their population allele frequencies) specific to each species are, therefore, urgently needed.

DNA fingerprinting began in 1985 with multilocus analysis of sperm and blood in a double rape–homicide case in Britain. In less than a decade, applications have reached into every taxonomic kingdom—Monera, Protista, Fungi, Animalia, and Planta. It appears that the potential uses of molecular biology techniques for identity testing have only begun to be realized.

*See also* Human Repetitive Elements; PCR Technology; Restriction Endonucleases and DNA Modification Methyltransferases for the Manipulation of DNA.

### *Bibliography*

Erlich, H. A., Ed. (1989) *PCR Technology: Principles and Applications for DNA Amplification.* Stockton Press, New York.

Gresshoff, P. M., Ed. (1992) *Plant Biotechnology and Development.* CRC Press, Boca Raton, FL.

Jeffreys, A. J., Wilson, V., and Thein, S. L. (1985) Hypervariable "minisatellite" regions in human DNA. *Nature,* 314:67–72.

Kearney, J. J., et al. Technical Working Group on DNA Analysis Methods (TWGDAM), and California Association of Criminalists Ad Hoc Committee on DNA Quality Assurance. (1991) Guidelines for quality assurance program for DNA analysis. *Crime Lab. Dig.* 18(2). U.S. Department of Justice, Federal Bureau of Investigation, Quantico, VA.

Kirby, L. T. (1990) DNA *Fingerprinting: An Introduction.* Stockton Press, New York.

Reynolds, R. Sensabaugh, G., and Blake, E. (1991) Analysis of genetic markers in forensic DNA samples using the polymerase chain reaction. *Anal. Chem.* 63:1–15.

Sambrook, J., Fritsch, E. F., and Maniatis, T. (1989) *Molecular Cloning: A Laboratory Manual,* Vols. 1, 2, and 3, 2nd ed. Cold Spring Harbor, Cold Spring Harbor, NY.

Weir, B. S. (1992) Population genetics in forensic DNA debate. *Proc. Natl. Acad. Sci. U.S.A.* 89:11654–11659.

# DNA in Neoplastic Disease Diagnosis

*David Sidransky*

### *Key Words*

**Allelic Loss** Loss of a paternal or maternal chromosomal allele from a tumor cell; such an event is detected by loss of a band on a DNA blot.

**Cytology** Microscopic examination of cells obtained from bodily fluids or tissue for pathologic analysis.

**Linkage** The likelihood that a disease allele and a chromosomal marker (usually nearby) will be inherited together.

**Oncogene** Cellular gene altered in cancer progression.

**Polymerase Chain Reaction (PCR)** Enzymatic amplification of DNA molecules through the use of specific primers in vitro.

**Proto-oncogene** Cellular gene capable of dominant transforming function in cancer when one copy is altered or mutated (i.e., activated).

**Translocation** Physical movement of genetic material from one chromosome to another.

**Tumor Suppressor Gene** Cellular gene whose normal wild-type suppressor function is lost when both copies are altered or mutated (i.e., inactivated).

DNA is the genetic code upon which all life is sustained. Much of human disease is due to errors within DNA that is either inherited (germ line) or acquired after birth (somatic). Neoplasms are now known to arise through a series of changes in specific oncogenes. Because these changes are intimately involved in tumor progression, they provide novel markers for cancer detection. The ability to detect these genetic changes within DNA allows identification of individuals at risk for the development of certain diseases such as cancer. Furthermore, the polymerase chain reaction has allowed amplification of small quantities of DNA, leading to a revolution in molecular diagnosis. Detection of DNA alterations in blood and cytologic samples can allow rapid and sensitive diagnosis of many neoplasms and has revolutionized the approach to cancer diagnosis.

## 1 MUTATIONS IN DNA

DNA, the double-stranded helix composed of polynucleotides, is the basis of all life processes. To ensure healthy, viable offspring, all creatures—from the fly to man—must successfully replicate DNA that is free of errors. Despite the existence of a variety of mechanisms to prevent the formation of nucleotide substitutions, however, excessive mutagens or defective repair enzymes may lead to mutation fixation. Although many DNA mutations are silent, a particular change within a regulatory or coding region of a gene may lead to human disease.

For a variety of human diseases, a specific genetic mutation has already been described. The pace of new discoveries is rapidly accelerating as new genes are found and new mutations within these genes are characterized. In particular, cancer is at the forefront of these discoveries, as genes involved in both cancer susceptibility and spontaneous tumor formation are rapidly being identified. Recent discoveries include the genes responsible for a variety of cancer predisposition syndromes (Table 1). All these diseases share a high susceptibility to particular forms of cancer, with the majority of affected patients developing tumors at an early age. Furthermore, many spontaneous (noninherited) tumors have been shown to arise from somatic mutation and inactivation of these genes.

## 2 GENETICS OF CANCER

Inactivation of tumor suppressor genes was predicted by a model based on a hypothesis of Alfred G. Knudson, Jr. Knudson reasoned, after studying the kinetics of childhood tumors, that genetic susceptibility to cancer could be explained by inactivation of one allele in the germ line of affected patients. The other allele was then mutated or deleted in a somatic cell. The concept of tumor suppressor genes has spread to spontaneous tumors, where cancers are

**Table 1** Inherited Cancer Susceptibility Syndromes in Which the Candidate Gene has been Identified

| Disease | Gene | Clinical Manifestations | Type of Cancer[a] |
|---|---|---|---|
| Familial adenomatous polyposis | *APC* | Colonic polyps | Colon |
| Neurofibromatosis-1 | *NF2* | Neurofibromas | Neurofibrosarcoma |
| Neurofibromatosis-2 | *NF1* | Loss of hearing | Acoustic neuroma, meningioma |
| Retinoblastoma | *RB1* | Loss of vision | Retinoblastoma |
| Von Hippel–Lindau disease | *VHL* | Renal defects | Kidney |
| Li–Fraumeni syndrome | p53 | Multiple tumors | Sarcoma, breast |
| Wilms' tumor | *WT1* | Renal mass | Nephroblastoma |
| Hereditary nonpolyposis coli | *MSH2, MLH1* | Multiple tumors | Colon, uterus |
| Multiple endocrine neoplasia[b] | *RET* | Hypocalcemia | Medullary thyroid cancer |
| Hereditary breast cancer | *BRCA-1* | Breast cancer | Breast |
| Xeroderma pigmentosum | *ERCC* | Photosensitivity | Skin cancer |
| Familial Melanoma | p16 *(CDKN2)* | Multiple nevi | Melanoma |

[a]Most common tumors in this syndrome.
[b]The only proto-oncogene implicated in these syndromes.

often found to have inactivated both copies of a particular gene during tumor progression. Often, one allele (gene copy) is mutated and the other allele is lost by deletion or recombination. Thus, p53 mutations are usually accompanied by loss of heterozygosity on chromosome 17p, inactivating the second copy of this tumor suppressor gene. Because these inactivation events take time, affected patients usually present with single spontaneous tumors much later in life.

The other major class of oncogenes, termed proto-oncogenes, were originally discovered in tumor viruses. Subsequently, these genes were found to be derived from normal cellular genes picked up by the viruses that then were reintroduced in activated (mutated) form during transformation of normal cells to cancer. In tumors, these genes are commonly activated by translocation or point mutation during cancer progression. Only one proto-oncogene has been found to be an inherited cause of cancer susceptibility. However, both classes of oncogenes contain genetic alterations (Table 2), and their detection facilitates DNA diagnosis. Thus, alterations in oncogenes are useful targets for assessing patients at risk for developing cancer and for detecting the presence of cancer cells.

## 3 TUMOR PROGRESSION MODELS

Tumors are now known to arise through a series of genetic steps during progression involving specific activation of proto-oncogenes or inactivation of tumor suppressor genes. To understand genetic progression, a histopathologic progression model for particular tumor type serves to outline the progressive development of a normal cell into a cancer cell. Pathologic analysis has detailed various histopathologic steps for colorectal carcinoma, from adenoma to carcinoma. These steps have been correlated with specific genetic events that drive the progression pathway (Figure 1). Careful review of this model reveals that activation of proto-oncogenes such as k-*ras* and inactivation of a tumor suppressor gene such as *APC* on chromosome 5q occur early. Late events in progression include loss of chromosome 17p (and p53 mutations), as well as additional loss of other chromosomal arms. It is immediately obvious from this progression model that early steps involved in progression such as k-*ras* and *APC* are useful targets for cancer diagnosis. Analysis of individual tumors revealed that the accumulation of these genetic changes, not necessarily their order of progression, leads to tumor outgrowth. However, the general order of progression can provide insights into the best targets for initial screening. Patients who might harbor inherited mutations of *APC* could be screened for cancer susceptibility. Furthermore, identification of specific mutations of k-*ras* or *APC* in stool could lead to detection of colorectal cancer.

**Table 2** Proto-oncogenes and Tumor Suppressor Genes Found to Be Altered in the Progression of Common Cancers[a]

| Cancer | Tumor Suppressor Genes | Proto-oncogenes | Chromosomal Loss |
|---|---|---|---|
| Lung | p53, *Rb* | k-*ras*, *erb* B | 3p, 5q, 9p, 13q, 17p |
| Head and neck | p53 | cyclin $D_1$ | 3p, 9p, 13q, 17p |
| Breast | p53 | cyclin $D_1$, *Her2/neu* | 1q, 3p, 11p, 13q, 17 |
| Colon | p53, *APC* | k-*ras* | 5q, 8p, 17p, 18q |
| Prostate | p53, *Rb* | | 8p, 11p, 13q, 16q, 17p |
| Pancreas | p53, p16 (*CDKN2*) | k-*ras* | 17p, 18q |

[a]Areas of allelic loss on chromosomal arms represent regions thought to contain inactivated tumor suppressor genes; many have not been identified.

## 4 DIAGNOSTICS

### 4.1 Inherited Susceptibility

Although a variety of tumor suppressor genes have been shown to control susceptibility to the formation of certain tumors, only a small fraction of cancers have been linked to a particular genetic susceptibility. While cancers may "run" in a certain family, often no specific inheritance pattern can be designated. This absence of pattern may be partly attributable to the tendency of many cancers to arise spontaneously instead of being inherited. Moreover, cancer is a complex disease that entails some background susceptibility, and exposure to environmental factors, as evidenced by a mouse model demonstrating variable susceptibility to lung cancer linked to a gene on mouse chromosome 4. In families where particular cancers are common and a candidate tumor suppressor gene is known or localized to a specific chromosomal region, however, the ability to test these family members is of great importance.

In cases where the gene is not known, linkage has been the hallmark of screening strategies. Linkage involves the use of polymorphic (informative) markers from chromosomal regions that are closely associated with the diseased gene. Even if the exact gene

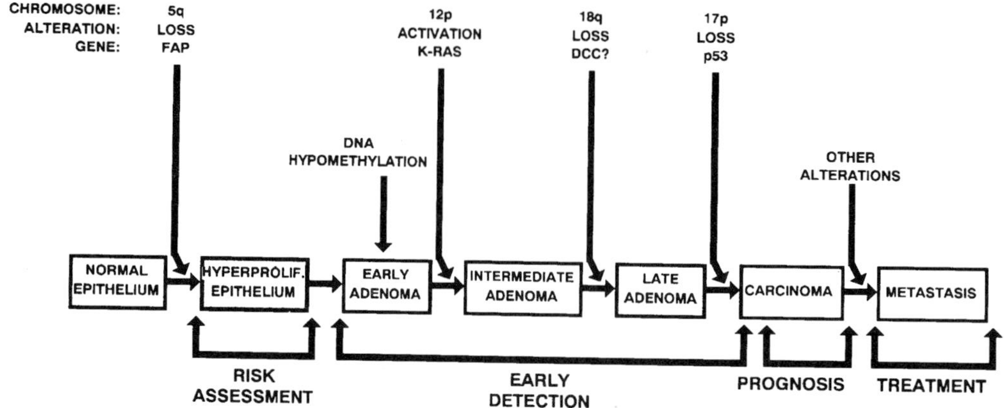


**Figure 1.** Genetic progression of colorectal cancer: the accumulation of genetic changes, not necessarily the order, allows progression. (Reprinted with kind permission from Fearon, E. R., Vogelstein, B., *Cell,* 61:759–767, 1990. Copyright *Cell Press,* 1990.)

has not been isolated, these markers allow rapid identification of anonymous alleles. If the patient is known to have inherited a nearby allele from an affected parent, then it is quite likely that the patient carries the abnormal gene and will develop the disease. If, however, the patient does not appear to carry this nearby allele, he or she is quite likely to be disease free. Depending on the number of markers used and the proximity of the markers to the gene, such linkage studies can be nearly definitive for either the absence or presence of the abnormal gene.

Several new DNA diagnostic tests have been developed to test for cancer susceptibility when the responsible gene has been identified. In retinoblastoma, the polymerase chain reaction (PCR) allows amplification of target DNA from the retinoblastoma gene in blood, followed by specific hybridization with probes recognizing the most common mutations. The ability to detect these specific mutations may be useful in preventive screening (e.g., to help test individuals prior to the development of a tumor). Alternatively, if these tumors are found to be sporadic and not due to an inherited gene mutation, other family members can be assured that their risk of developing this eye tumor is negligibly small.

Additionally, novel screening techniques have been developed for patients with familial adenomatous polyposis (FAP): symptoms include hundreds of polyps, and the likelihood of colon cancer approaches 100%. These tests are based on the fact that *APC* mutations are responsible for the disease and often lead to termination codons, which in turn produce truncated protein. Patients who carry mutations will have short protein products, which can be identified by PCR of the *APC* gene, followed by in vitro translation, and separation of protein products on a gel.

Other attempts have been made to facilitate detection of mutations of p53 in patients with the Li–Fraumeni syndrome. These PCR-based techniques involve functional tests of PCR products obtained from the target p53 gene. The PCR products are cloned into vectors and introduced into bacteria or yeast, then identified as normal or mutant by a color assay. However, for most cancer susceptibility syndromes, the mutations either are not known or are so diverse that direct tests cannot be done. In these cases, only a rigorous and laborious sequencing of the entire gene will identify the causative mutation, making the diagnosis expensive and not readily available outside the research setting.

### 4.2 Tumor Diagnosis: Hematologic Malignancies

Specific genetic alterations that activate proto-oncogenes or inactivate tumor suppressor genes have become the basis of novel diagnostic assays for cancer. In the great majority of cancer patients, no specific inherited gene mutations have been identified. Therefore, diagnosing affected patients early and accurately can lead to definitive and often life-saving therapy. Additionally, the ability to identify patients with minimal residual disease (e.g., residual cancer) as determined by novel PCR-based techniques has led to an important application for patients with hematologic disorders.

Perhaps the best-known activating mutation in hematologic malignancies involves the translocation between chromosomes 9 and 22 resulting in the Philadelphia (Ph) chromosome (Table 3). This translocation produces a *BCR/Abl* rearrangement resulting in a chimeric protein with transforming activity. Rearrangement occurs in 95% of adults with chronic myelogenous leukemia (CML) and can also be identified in 20–35% of adult patients with acute lymphoid leukemia (ALL). The presence of this translocation is a poor prognostic indicator for patients with ALL. Initially performed predominantly by Southern blot techniques, fluorescence in situ hybridization (FISH) and novel PCR analysis can allow rapid and easy diagnosis of this critical rearrangement. FISH allows hybrid-

**Table 3** Some Common Translocations in Leukemia and Lymphomas and the Putative Proto-oncogenes Identified: Many Translocations and Chimeric Gene Products Are Variable

| Leukemia[a] | Cytogenetics | Proto-oncogene |
|---|---|---|
| CML, ALL | t(9, 22) | *BCR/Abl*[b] |
| APML | t(15, 17) | *PML/RARα*[b] |
| ALL/AML | t(4, 11) | *ALL1/AF4*[b] |
| AML | t(8, 21) | *AML1/ETO*[b] |
| Burkitt's lymphoma | t(8, 14) | c-*Myc* |
| Centrocytic lymphoma | t(11, 14) | *Bcl*-1 |
| Nodular lymphoma | t(14, 18) | *Bcl*-2 |
| Lymphocytic lymphoma | t(3, 14) | *Bcl*-6 |

[a]CML, chronic myelogenous leukemia; ALL, acute lymphoid leukemia; AML, acute myeloid leukemia; APML, acute promyelocytic leukemia.
[b]Chimeric (fusion) gene products.

ization of genomic clones derived from the chromosomal regions near these breakpoints, followed by fluorescent detection of their immediate proximity due to the translocation. Most PCR techniques are based on identification of chimeric RNA transcripts from critical translocations (Table 3). Following isolation of RNA, cDNA is formed by reverse transcriptase followed by PCR amplification (RT-PCR), utilizing primers derived from each of the genes normally located on different chromosomes. The presence of a chimeric product immediately identifies the abnormal juxtaposition of the two genes caused by the translocation, providing a very sensitive and specific diagnostic method. Additionally, detection of the *BCR/Abl* gene rearrangement can be used to monitor CML and ALL patients following bone marrow transplants. Although there is some disagreement over the value of finding residual disease by PCR, relapses identified by PCR may precede cytogenetic or actual clinical relapses by 6–8 months.

Another major rearrangement is the reciprocal translocation between chromosomes 14 and 18 which brings together the immunoglobulin heavy chain region and the critical *Bcl*-2 gene in a large percentage of nodular and follicular lymphomas and in a smaller percentage of diffuse large cell lymphomas. The *Bcl*-2 gene is involved in cellular apoptosis (or programmed cell death), and altered expression allows cells to remain resistant to this phenomenon. Detection of rearrangements of *Bcl*-2 by Southern blot assessment is useful for determining clonality as well for differential diagnosis (of follicular lymphoma vs. a reactive lymph node). There may be additional prognostic information in that *Bcl*-2-positive patients may be less likely to enter complete remission. Recent PCR-based techniques are significantly superior to conventional diagnostic methods. As in the other PCR-based techniques, assessment of minimal residual disease can be done following bone marrow transplantation.

Additional translocations involve rearrangement of chromosomes 8 and 14 leading to overexpression of c-*myc,* a critical nuclear transcription factor, in Burkitt's lymphoma and some diffuse lymphomas. The novel *RAR-α/myl* rearrangement between chromosomes 15 and 17 is found in acute promyelocytic leukemia. This translocation may have prognostic importance because of the induction of complete remissions in 80% of patients with this rearrangement following treatment with retinoic acid. Additionally, there are new breakpoints constantly being cloned, present in a smaller percentage of leukemias, which can be detected by Southern blot analysis, FISH, or PCR-based techniques. These may help in establishing definitive diagnosis in some of these leukemias, and all may play a role in assessment of minimal residual disease as described. Many of these novel translocations are listed in Table 3, with the corresponding gene rearrangements.

Another molecular application for hematological malignancies lies in the assessment of clonality. Clonal rearrangement of the immunoglobulin (IG) heavy chain region indicates the presence a clonal population of B cells. Almost all mature B-cell cancers such as CLL, follicular lymphomas, and multiple myeloma contain rearrangements of both the heavy chain and adjoining regions. Detection of these rearrangements establishes the presence of clonality (and therefore malignancy) in difficult-to-diagnose proliferative lymphocytic disorders. Rearrangement is usually done by Southern blot analysis, although progress is being made in PCR-based techniques. Residual presence of a clonal population of cells can be used in a fashion similar to that described for the *BCR/Abl* rearrangement. Furthermore, rearrangements of the β chain in the T-cell receptor can be used in similar strategies for T-cell hematologic malignancies.

### 4.3 Tumor Diagnosis: Solid Tumors

Although proto-oncogenes are occasionally involved in the development of solid tumors, tumor suppressor genes are commonly inactivated in the progression of most epithelial cancers. Again, because many patients do not have inherited gene mutations, it is difficult to identify high risk patients by family history alone. Therefore, diagnosing patients with early tumors while they can still be successfully treated is of paramount importance. Recently, novel PCR-based assays were developed to identify rare cancer cells among an excess background of normal cells. These assays depend on molecular genetic techniques of increased precision, able to detect point mutations or genetic alterations within cancer-associated genes or loci that are specific for different tumor types. Cytologic samples can be analyzed by PCR both to obtain the diagnosis and, potentially, to provide prognostic information.

Initial demonstration of a PCR-based assay was used in the detection of p53 mutations in the urine from patients with bladder cancer. Investigators used the PCR to amplify a portion of the p53 gene followed by cloning to separate PCR molecules and then probing with specific oligomers able to recognize one mutant copy among 10,000 normal cells. Even though these samples were cytologically negative, 1–7% of the cells were found to be positive by this assay. This work was quickly followed by similar detection of *ras* gene mutations in the stool of patients' colorectal cancers. Moreover, *ras* and p53 genetic changes were detected in sputum months before some patients developed lung cancer. Importantly, these patients were amenable to surgical resection and could have undergone complete removal of the tumors. Other investigators have used techniques such as enriched PCR or allele-specific amplification, to further enrich for the mutant molecules from the cancer cells for accurate diagnosis. Additional techniques such as the ligase chain reaction can also provide information from the samples with a minimal number of cells. As promising as these techniques appear, they will have to be validated in prospective trials to test the sensitivity and specificity of the various assays. Identification of the high risk populations (e.g., patients who smoke for identification of lung cancer) will help validate results and eventually lead to the application of these tests for the population at large.

In addition to diagnostic information, knowing the specific genetic changes involved in various tumors can provide important prognostic information. For example, inactivation of p53 by immunohistochemical analysis or direct sequencing can provide physicians with the knowledge that many patients exhibiting these mutations will die as a result. This outcome was best demonstrated with respect to bladder cancer and appears to be probable for many other tumor types such as lung cancer. It is well known that inactivation of the retinoblastoma gene in bladder cancer also may provide a poor prognosis. As other genetic changes are identified in various tumors, they should provide important prognostic information for clinical outcome. Clinicians can then use the identification of these genetic changes to appropriately put patients into high risk groups for intensive therapy. Investigators have also used specific genetic changes to identify infiltrating tumors cells in apparently normal histologic margins and lymph nodes. PCR-based assays similar to those described detect these rare cells by light microscopy in apparently normal margins or tissue. The identification of these infiltrating tumor cells may have important implications for prognosis and therapeutic intervention for the patients.

## 5 FUTURE IMPACT

Molecular biology has provided an entirely new approach to cancer diagnosis. Since DNA is intimately involved in every aspect of normal cell life, the ability to detect abnormal or mutated DNA has become the basis of a new rational approach. The ability to detect patients susceptible to different types of cancer has profound implications for affected patients and their families. These tests will continue to aid in the diagnose of patients with hematologic disorders and in the precise tracking of their disease. Furthermore, the ability to identify tumors early, when they are still surgically resectable, can have a significant effect on the general population.

It has been mentioned that tumor suppressor genes are often found to exist in areas of chromosomal loss. Many such genes may yet be identified on these chromosomes (Table 2), and their mode of mutation is expected to be similar to that of genes previously identified. These new genes could then serve as additional markers for cancer diagnosis. New genes may be found to be the cause of other cancer susceptibility syndromes, and/or they may be implicated in the progression of many other tumors. Detection of a combination of gene alterations can then lead to improved diagnostic tests. Thus, a new era of diagnostics is at hand, based on the molecular biology of cancer and the ability to detect the fundamental DNA alterations of this deadly disease.

*See also* CANCER; DNA MARKERS, CLONED; GENETIC TESTING; ONCOGENES; TUMOR SUPPRESSOR GENES.

### *Bibliography*

Ambinder, R. F., and Griffin, C. A. (1991) Biology of the lymphomas: Cytogenetics, molecular biology, and virology. *Curr. Opin. Oncol.* 3(5):806–812.

Bishop, J. M. (1991) Molecular themes in oncogenesis. *Cell,* 64:235–248.

Fearon, E. R., and Vogelstein, B. (1990) A genetic model for colorectal tumorigenesis. *Cell,* 61:759–767.

Knudson, A. G., Jr. (1985) Hereditary cancer, oncogenes, and anti-oncogenes. *Cancer Res.* 45:1437–1443.

Landegren, U., Kaiser, R., Caskey, C. T., and Hood L. (1988) DNA diagnostics—Molecular techniques and automation. *Science,* 242:229–237.

Sidransky, D., Tokino, T., Hamilton, S. R., et al. (1992) Identification of *ras* oncogene mutations in the stool of patients with curable colorectal tumor. *Science,* 256:102–105.

# DNA MARKERS, CLONED

*Eugene R. Zabarovsky*

### *Key Words*

**Blue–White Selection** A misnomer: the process described is really color identification. Usually recombinants form white plaques or colonies and parental vectors produce blue plaques (colonies) in the presence of X-gal.

**Genetic Selection** In cloning, usually, selection against parental, nonrecombinant molecules in favor of recombinant. For lambda-based vectors used for construction of genomic libraries, the two most common types are *spi-* and *supF* selection.

**Polylinker** Short DNA fragment (in the vector) that contains recognition sites for many restriction enzymes.

**Restriction Enzyme** Enzyme that recognizes a specific sequence in DNA and can cut at or near this sequence.

---

It is possible to clone (isolate) any DNA fragment in the genome of an organism and, after reverse transcription, any transcribed gene in the form of a cDNA. The cloning procedure involves the insertion of the DNA fragment into a vector, capable of replication in a microorganism; it is then possible to produce a large quantity of the DNA fragment for physical or biological analysis. Determination of the location in the genome from which the particular DNA fragment was derived confers on that fragment the property of a DNA marker. Such DNA markers are a prerequisite for physical and genetic mapping of the genome of the organism. DNA markers are also of importance for the diagnosis of genetic diseases. DNA markers can be divided into several different classes. Vectors and clone libraries of different types can be used.

## 1 PRINCIPLES

In molecular biology, cloning is the insertion of DNA with interesting information into a specific vector that allows replication and transfer of the cloned DNA from one host to another. The clone containing the inserted DNA is called a ''recombinant vector'' to distinguish it from its parental vector, which does not contain any inserted foreign DNA. The main idea of cloning is to obtain the interesting piece of DNA in a large quantity convenient for analysis and further experiments. Now there are many different types of vectors and many strategies used for cloning. This entry concentrates on the widely used lambda-based vectors and the construction of different types of genomic libraries.

A genomic library is a collection of recombinants containing DNA fragments that represent the genome of a particular organism. Genomic libraries can be either general, containing DNA fragments covering the whole genome, or special, containing only specific genome fragments, which differ in certain parameters (see Section 3). Cloned DNA fragments can be located to a specific site of a chromosome, whereupon they can serve as markers for physical and genetic mapping. Anonymous markers represent randomly cloned DNA fragments whose functions or specific features are not known. Other DNA markers can contain a known gene, recognition sites for rare cutting restriction enzymes, and so on. Such markers can be polymorphic, that is, having different structures in different individuals (they are usually distinguished on the basis of differing mobility in gel electrophoresis).

Polymorphic markers come in three common types. Restriction fragment length polymorphism (RFLP) markers recognize genomic fragments that contain polymorphic recognition sites for a particular restriction endonuclease (epg., *Taq* I, *Msp* I). The same chromosomal regions in different individuals contain or lack this recognition site.

Another form of DNA polymorphism results from variation in the number of tandemly repeated (VNTR) DNA sequences in a particular locus. Usually the VNTR sequences are divided in two classes: mini- and microsatellites. Minisatellites are DNA fragments 0.2 to 2 kilobases long that contain many copies of 15 to 60 bp repeats. Microsatellites are relatively short DNA fragments (usually <100 bp) that contain runs of tandemly repeated DNA with repeat unit 1 to 5 bp.

## 2 TECHNIQUES

### 2.1 General Characteristics of Lambda-Based Vectors Used for Construction of Genomic Libraries

Extensive modifications of lambda phage vectors were developed which combine the features of different vector systems. Cosmids are essentially plasmids that contain the *cos* region of phage lambda responsible for packaging of DNA into the phage particle. Since the plasmid body is usually small (3–6 kb), big DNA molecules (46–49 kb) can be cloned in these vectors. Phagemids are lambda phages that have an inserted plasmid. Diphagemids are the vectors that combine the advantages of phages (λ and M13) and plasmids.

### 2.2 Construction of General Genomic Libraries

Representativity is one of the most important features of a genomic library. In a representative genomic library, every genomic DNA fragments will be present in at least one of the recombinant phages of the library.

Two common ways to construct genomic libraries are shown in Figure 1. The first method includes generation of sheared genomic DNA fragments of a particular size. The vector DNA is digested with two restriction enzymes. The stuffer piece and both arms will now have different sticky ends, preventing re-creation of the original vector molecule during subsequent ligation with genomic DNA fragments. The ''partial filling-in'' method avoids fractionation steps. Phage arms are prepared as described before, and the sticky ends produced are partially filled in with the Klenow fragment of DNA polymerase I. Genomic DNA partially digested with *Sau*3AI is also partially filled in, but in the presence dATP and dGTP. Under the preceding condition, self-ligation of vector arms or genomic DNA is impossible.

## 3 APPLICATIONS AND PERSPECTIVES

### 3.1 Cloning DNA Markers Specific for a Particular Chromosome

One approach to cloning DNA markers that are specific for a given chromosome is based on the use of fluorescence-activated cell sorters (FACS). DNA isolated from sorted chromosomes can be used for construction of chromosome-specific genomic libraries. Another approach is based on the use of hybrid cell lines. To obtain such somatic cell hybrids, human cells are fused with rodent cells. When the resulting hybrid cells are grown in culture, there is a progressive loss of human chromosomes until only one or a few of them are left. Human-specific clones can be isolated from the library by hybridization to total human DNA.

### 3.2 *Alu*-PCR as a Tool to Clone Markers from Specific Regions of the Chromosomes

*Alu* repeats are the most abundant repetitive sequences in the human genome. Related repeats exist also in other mammals. It is possible to find conserved sequences that are species specific. These conserved sequences can be used as primers in the polymerase chain reaction (PCR), to specifically amplify human sequences in the presence of nonhuman DNA. These features are the basis for using *Alu* repeats for isolation (and cloning) of human chromosome specific sequences from hybrid cell lines containing human and nonhuman DNA sequences.

### 3.3 Use of Microdissection to Construct Region-Specific Libraries

Another approach to obtaining region-specific libraries is chromosome microdissection. This approach makes it possible to physically remove the chromosomal region of interest and to clone minute quantities of microdissected DNA by a microcloning procedure. DNA obtained from only a few (2–20) chromosomes is enough for PCR amplification and construction of a region-specific library.

### 3.4 CpG Islands as Powerful Markers for Genome Mapping

Stably unmethylated sequences (about 1% of the genome) have been observed in human chromosomal DNA. They are usually called CpG (rich) islands because they contain more than 50% of C+G. It is now clear that the majority if not all CpG islands are associated with genes. It has been shown that recognition sites for many of the rare cutting enzymes are closely associated with CpG islands (e.g., 82% of all *Not* I sites are located in the CpG islands).

### 3.5 Linking and Jumping Libraries

The *Not* I jumping library and the ''general'' jumping (hopping) library are the two best-known kinds; only the former is discussed here. The basic principle of both methods is to clone only the ends of large DNA fragment rather than continuous DNA segments, as in yeast artificial chromosome (YAC) clones. Internal DNA is deleted by controlled biochemical techniques. Jumping clones contain DNA sequences adjacent to neighboring *Not* I sites, and linking clones contain DNA sequences surrounding the same restriction site.

There are two main approaches for the construction of *Not* I jumping and linking libraries. In the first method (Figure 2A), jumping and linking libraries are constructed using selective genetic marker. An integrated approach for construction of jumping and linking libraries is outlined in Figure 2B.

### 3.6 *Alu*-PCR and Subtractive Procedures to Clone CpG Islands from Defined Regions of the Chromosomes

The disadvantage of the *Alu*-PCR approach described in Section 3.2 is that the DNA fragments cloned are small and not linked with genes or markers of other kinds. According to another scheme, linking libraries are constructed from different hybrid cell lines containing either whole or deleted human chromosomes. Then total DNA isolated from these libraries can be used for *Alu*-PCR. Genomic subtractive methods represent potentially powerful tools for identification of deleted sequences and cloning region-specific markers. But the great complexity of the human genome has generated serious problems. These problems can be overcome by

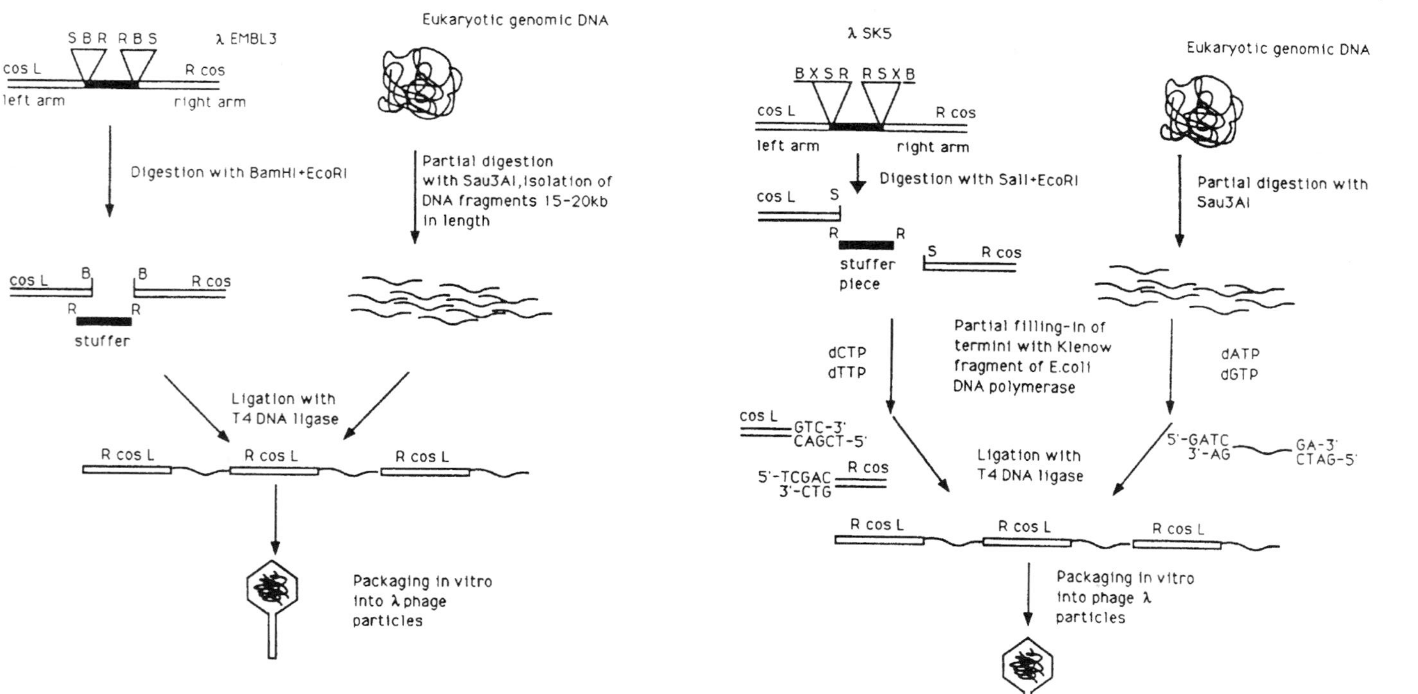


**Figure 1.** Two approaches for constructing genomic libraries: (A) classical method and (B) partial filling-in method. Left and right parts of the *cos* site designated cos L and cos R, respectively; B, *Bam*HI; R, *Eco*RI; S, *Sal* I; X, *Xma* III.

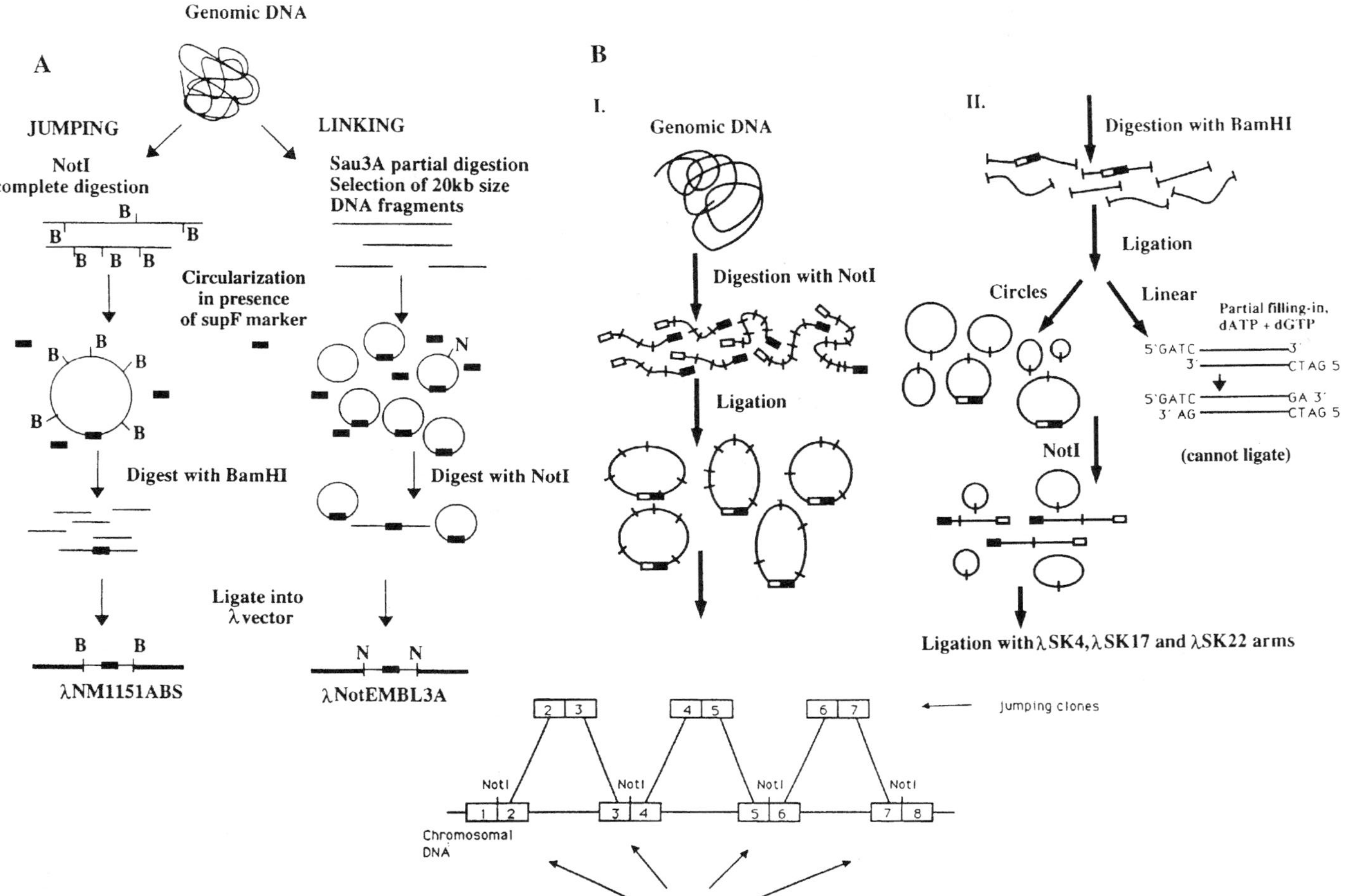


**Figure 2.** Jumping and linking libraries for genome mapping constructed using *supF* marker (A) and partial filling-in method (B). The most important feature here is that the same vectors and protocol are used for construction of both libraries. No selective genetic markers are exploited in this scheme. Solid boxes designate *supF* marker in (A) and *Bam*HI in (B). Solid and open bars in (B) denote *Not* I sites, and vertical slashes represent *Bam*HI sites. Construction of the jumping library represented in (BI) and (BII). In the construction of the linking library (BI), digestion of the genomic DNA with *Bam*HI is the first step. Diagram at bottom shows long-range mapping using jumping and linking libraries (see Section 3.7).

reducing the complexity of the human genomic sequences. Two approaches have been suggested to achieve this aim. In one (representational difference analysis), only a subset of genomic sequences (epg., *Bam*HI fragments *1 kb) is used for subtractive procedures, and this approach will result in the cloning of random sequences. In the other, Not* I or *Xho* I (*Sal* I) linking libraries are used instead of whole genomic DNA. It is important to note that linking clones are associated with genes.

### 3.7 Use of Linking and Jumping Clones to Construct a Physical Map of Chromosomes

When linking clones are used to probe a genomic blot (*Not* I digested genomic DNA) produced by pulsed field gradient electrophoresis, each clone should reveal two DNA fragments, which are adjacent in the genome. Thus, in principle, one should be able to order the rare cutting sites with just a single library and one digest, although it seldom is possible to distinguish between two fragments of the same size. The use of jumping and linking libraries in a complementary fashion simplifies this approach (Figure 2).

*See also* Genetic Testing; Human Chromosomes, Physical Maps of; Nucleic Acid Sequencing Techniques; PCR Technology.

### *Bibliography*

Collins, F. S. (1988) Chromosome jumping. In *Genome Analysis: A Practical Approach,* K. E. Davis, Ed., pp. 73–94. IRL Press, Oxford.

Poustka, A., and Lehrach, H. (1988) Chromosome jumping: A long-range cloning technique. In *Genetic Engineering Principles and Methods,* Vol. 10, J. K. Setlow, Ed., pp. 169–193. Brookhaven National Laboratory, Upton, NY; Plenum Press, New York and London.

Sambrook, J., Fritsch, E. F., and Maniatis, T. (1989) *Molecular Cloning: A Laboratory Manual,* 2nd ed. Cold Spring Harbor Laboratory Press, Cold Spring Harbor, NY.

Watson, J. D., Gilman, M., Witkovski, J., and Zoller, M. (1992) *Recombinant DNA,* 2nd ed. Scientific American Books, New York.

# DNA Repair in Aging and Sex

*Carol Bernstein and Harris Bernstein*

### *Key Words*

**Aging** The progressive impairments of function experienced by many organisms throughout their life span.

**Complementation** The masking of the expression of mutant genes by corresponding wild-type genes when two homologous chromosomes share a common cytoplasm

**DNA Damage** A DNA alteration characterized by an abnormal structure that cannot itself be replicated when the DNA is replicated, although it may be repaired.

**DNA Repair** The process of removing damage from DNA and restoring the DNA structure.

**Mutation** Change in the sequence of DNA base pairs, which may be replicated and thus inherited.

**Sex** The process by which genetic material (usually DNA) from two separate parents is brought together in a common cytoplasm where recombination of the genetic material ordinarily occurs, followed by the passage of the recombined genome(s) to progeny.

---

A number of theories have been proposed to account for the biological phenomena of aging and of sexual reproduction (sex). These theories can be found in volumes by Arking (aging) and by Michod and Levin (sex) listed in the Bibliography. An emerging unified theory, which accounts for much of the data relating to both aging and sex, is presented here. Although this theory is still somewhat controversial, it is consistent with and encompasses most other theories of aging and at least one other theory of sex (see volume by Bernstein and Bernstein listed in the Bibliography).

## 1 OVERVIEW OF THE UNIFIED THEORY

Both aging and sexual reproduction (sex) appear to be consequences of a universal property of life: the vulnerability of genetic material to ''noise'' (DNA damage and mutation). DNA damages occur very frequently, and organisms use enzyme-mediated repair processes to cope with them. In any cell, however, some DNA damage may remain unrepaired despite the existence of repair processes. Aging seems to be due to the accumulation of unrepaired DNA damages in somatic cells, especially in nondividing cells such as those in mammalian brain and muscle. On the other hand, the primary function of sex appears to be repair of damages in germ cell DNA, through efficient recombinational repair when chromosomes pair during the sexual process. This mechanism allows an undamaged genome to be passed on to the next generation. In addition, in diploid organisms, sex allows chromosomes from genetically unrelated individuals (parents) to come together in a common cytoplasm (that of progeny). Since genetically unrelated parents ordinarily would not have common mutations, the chromosomes present in their progeny should complement each other, masking expression of any deleterious recessive mutations that might be present. Thus it appears that in general, aging reflects the accumulation of DNA damage and sex reflects the removal of DNA damage plus, in diploid organisms, the masking of mutations by complementation.

## 2 THE DNA DAMAGE THEORY OF AGING

Considerable evidence indicates that DNA damage occurs frequently in somatic cells. If unrepaired, DNA damages interfere with RNA transcription and DNA replication. The result of interference, caused by accumulating DNA damage, can be progressive impairment of cell function and cell death.

Nondividing or slowly dividing cells, such as long-lived neurons and differentiated muscle cells, have been shown to accumulate DNA damage with age. In mammalian brain, mRNA synthesis and protein synthesis decline, neuron loss occurs, tissue function is reduced, and functional impairments directly related to the central processes of aging (e.g., cognitive dysfunction and decline in homeostatic regulation) occur. Similarly, in muscle cells, mRNA and protein synthesis decline, cellular structure deteriorates, and cells die, and these processes are accompanied by a reduction in muscle strength and speed of contraction. Thus, for mammalian brain and muscle, there seems to be a good correlation between the accumulation of DNA damage and major features of aging.

In other cell populations where cell division is infrequent, such as those of liver and lymphocytes, there is also evidence that DNA damage leads to functional declines with age. In contrast to nondividing or slowly dividing cells, at least some types of rapidly dividing cell populations seem to cope with unrepaired damages by replacing lethally damaged cells through the duplication of undamaged ones. Such cell populations, which show little or no sign of aging, include colon and duodenum epithelial cells and hemopoietic cells of the bone marrow.

Normal mammalian cellular metabolism produces oxidative free radicals from molecular oxygen, and these reactive molecules appear to cause large numbers of DNA damages per average cell per day. It is estimated that the total number of all types of oxidative DNA damage per cell per day may be about 10,000 in man and about 100,000 in rat. Although repair processes can remove most of these damages, a significant fraction remain unrepaired in some tissues and accumulate with age. In comparisons with human, monkey, rat, and mouse, shorter life span correlates with a higher incidence of oxidatively altered DNA bases. Thus DNA damage caused by oxidative reactions may be a major cause of aging in mammals.

Support for the DNA damage theory of aging comes from some additional types of evidence. Among numerous mammalian species, life span correlates with DNA repair capacity. Also, treatment of mammals with exogenous DNA damaging agents, such as X-rays, induces life-shortening with some similarity to normal aging. Furthermore, there are at least 10 genetic syndromes of humans in which individuals appear to have both an increased incidence of DNA damage and some features of premature aging.

## 3 REPAIR OF DNA DAMAGE

The number of DNA damages in a cell is determined by the balance between production of damages and their repair. DNA damages can be recognized by repair enzymes because of their abnormal DNA structure. This is in contrast to mutations, which cannot be recognized because they are structurally normal. There are two basic types of DNA damage: single-strand and double-strand damages. Single-strand damages, which alter only one of the two DNA strands, can be repaired by a number of different repair processes. Excision repair, however, appears to be the predominant type used in mammalian cells. Thus excision repair may be the most important type in resisting aging. All enzymatic pathways of excision repair involve the removal of the damaged section from one DNA strand and its accurate replacement by copying from the undamaged complementary strand.

When the two strands of DNA are structurally altered at or near the same position, as occurs in a double-strand break or an interstrand cross-link, the damage is referred to as double-strand damage. This type of damage probably cannot be repaired by excision repair, because the process depends on the ability of the DNA strand opposite the damaged one to act as an intact template for repair synthesis. Double-strand damages are formed by oxidative reactions and heat, although not as frequently as single-strand damages. Since double-strand damages are more difficult to repair than single-strand damages, they may contribute substantially to mammalian aging despite their low frequency. To repair a double-strand damage in one DNA molecule, it is necessary to obtain intact information from a second homologous DNA molecule. Such repair involves replacement of the lost information by means of a transfer of a single-stranded section of the intact DNA to the damaged DNA. This process is referred to as recombinational repair. Whether recombinational repair is important in resisting mammalian aging is unknown, but the process appears to be important in sex.

## 4 THE DNA REPAIR (AND COMPLEMENTATION) THEORY OF SEX

Sexual reproduction, the most common form of reproduction in higher animals and plants, occurs among many microorganisms as well. Only about 0.1% of higher animal species and about 8% of higher plant species entirely lack sexual reproduction. Sex, however, is a very costly process from a number of perspectives, including the time and energy females and males must spend in making contact with the opposite sex. Furthermore, sexual reproduction, in contrast to female parthenogenesis, allows males to pass on about half their genes to each progeny while frequently avoiding the burden of nurturing these progeny. Thus a parthenogenetic female, with no male partner, can in principle often transmit her genes twice as effciently as a sexual one. If sex is adaptive in a Darwinian sense, it must provide a large benefit to compensate for its large costs. Two adaptive benefits of sex appear to be the promotion of recombinational repair of DNA damage in the germ line, and the masking of mutations by complementation.

Recombinational repair has been shown to be efficient at overcoming DNA damages in microorganisms, especially double-strand damages. It also has been shown to be versatile (able to handle a wide variety of different types of damage) and prevalent in nature (occurring in a broad spectrum of organisms).

A key stage of the sexual cycle in eukaryotes is meiosis, the process by which germ cells, such as egg and sperm, are formed. Recombination is promoted in the phase of meiosis during which the homologous chromosomes derived from each parent line up intimately in pairs. Gene functions necessary for meiotic recombination are required for DNA repair, and the pathway of meiotic recombination involves DNA repair steps. Meiotic recombination appears to occur by the double-strand break repair mechanism, in which a double-strand break in one chromosome is extended to a double-strand gap. The chromosome containing the gap then receives a single-strand segment of DNA from the homologous chromosome with which it is paired. The succeeding steps involve completion of the exchange process, resulting in two intact chromosomes. This process appears to be designed to remove DNA damage in a simple, direct fashion, by creating a gap followed by double-strand break repair at the site of the damage. Thus meiotic recombination may reflect a general mechanism for repairing all types of DNA damage. If this view is correct, then meiosis can be regarded as an adaptation for clearing damages from the DNA passed on to progeny via germ cells. This can provide a strong selective advantage to compensate for some of the cost of sex.

Outcrossing is a second key attribute of sex. This appears to be an adaptation for dealing with the problem of deleterious mutation rather than DNA damage. For instance, humans are estimated to carry (in a heterozygous state) an average of about two deleterious recessive mutations per genome. Such mutations can impose a large physiological penalty on inbreeding. That is, incestuous matings in humans often result in genetically defective progeny. In an organism having a diploid stage in its life cycle, wild-type (functional) genes in one copy can complement (mask) mutated genes in the other copy present in the same cell. However complementation is most beneficial if the mating partners are unrelated, since their genomes are unlikely to carry the same deleterious recessive mutations. Thus complementation that occurs upon outcrossing provides a further strong selective advantage in diploids that could also compensate for some of the high cost of sex.

In conclusion, sex seems to be maintained in diploid organisms both by the advantage of recombinational repair during meiosis and the advantage of outcrossing, which facilitates complementation of recessive deleterious mutations. It appears that the need to cope with genetic noise (DNA damage and mutation) in the transmission of genetic information has generated a continuous selective pressure for sex.

In humans, the availability of efficient repair during meiosis may underlie the potential immortality of the germ line. In contrast, the less efficient DNA repair available in the somatic cells may account for the aging and mortality of the individual.

*See also* Aging, Genetic Mutations in; Cell Death and Aging, Molecular Mechanisms of; DNA Damage and Repair.

### *Bibliography*


Ames, B. N., and Gold, L. S. (1991) Endogenous mutagens and the causes of aging and cancer. *Mutat. Res.* 250:3–16.

Arking, R. (1991) *Biology of Aging; Observations and Principles.* Prentice Hall, Englewood Cliffs, NJ.

Bernstein, C., and Bernstein, H. (1991) *Aging, Sex and DNA Repair.* Academic Press, San Diego, CA.

Bernstein, H., Hopf, F. A., and Michod, R. E. (1987) The molecular basis of the evolution of sex. *Adv. Genet.* 24:323–370.

Holmes, G. E., Bernstein, C., and Bernstein, H. (1992) Oxidative and other DNA damages as the basis of aging: A review. *Mutat. Res.* 275:305–315.

Michod, R. E., and Levin, B. R., Eds. (1988) *The Evolution of Sex; An Examination of Current Ideas.* Sinauer, Sunderland, MA.

Mullaart, E., Lohman, P. H. M., Berends, F., and Vijg, J. (1990) DNA damage metabolism and aging. *Mutat. Res.* 237:189–210.


# DNA Replication and Transcription

*Yusaku Nakabeppu, Hisaji Maki, and Mutsuo Sekiguchi*

### *Key Words*

**DNA Polymerase** Enzyme that performs elongation of the DNA chain by adding deoxyribonucleotides to the 3′-hydroxyl end of a preexisting DNA strand.

**Eukaryote** Organism whose cell(s) contains a nucleus in which the genome DNA is organized as chromosome(s).

**Holoenzyme** Active form of enzyme that consists of multiple subunits.

**Origin (*ori*)** Unique site on the chromosome from which DNA replication starts in one or both directions.

**Prokaryote** Single-cell organism in which the genome DNA is not enclosed in a separate organelle (e.g., nucleus).

**Promoter** Oriented DNA sequence recognized by the RNA polymerase holoenzyme to initiate transcription.

**Replication Fork** Moving front of DNA replication at which a double-stranded DNA helix is separated into two newly replicated helices.

**RNA Polymerase** Multisubunit enzyme that synthesizes RNA complementary to the DNA template.

**Transcription Factor** Class of proteins that bind to a promoter or to a nearby sequence of DNA to facilitate or prevent transcription initiation.

---

Since an organism's genetic information is stored in its DNA as a nucleotide sequence array, the cell must replicate its chromosomal DNA very precisely. The self-complementary nature of DNA is the basis for accurate replication. DNA polymerase adds deoxyribonucleotides to preexisting DNA chains in the 5′ →3′ direction. One strand is synthesized continuously (the leading strand), while the other is synthesized discontinuously (the lagging strand). To extract information from the DNA, a limited region of DNA, corresponding to a gene or gene cluster, is transcribed into RNA. This process is catalyzed by RNA polymerase, a multisubunit enzyme that adds ribonucleotides to the growing RNA chain in the 5′→3′ direction, along the template DNA. Both starting point and the frequency of transcription are determined by the promoter, a specific sequence of DNA to which RNA polymerase and transcription factors bind.

## 1 DNA Replication

Replication of DNA is a key event in cell propagation. A matching set of chromosomal DNA must be transferred precisely into each of the daughter cells. This is achieved only when chromosomal DNA is accurately replicated, providing two copies of the entire genome, with one copy faithfully distributed into each daughter cell. To this end, the cell has special mechanisms to keep the fidelity of DNA replication very high, to segregate the replicated DNA, and to tightly coordinate DNA replication and cell division within the cell cycle.

### 1.1 Structural Aspects of DNA Replication

DNA replication starts at a particular site on the DNA molecule called the origin of replication. In prokaryotic cells (e.g., bacteria), the origin is located at a unique site on the circular chromosome, and DNA replication proceeds bidirectionally from the origin to a terminus site (Figure 1, left). On the other hand, eukaryotic cells have many replication origin sites on each chromosome (Figure 1, right). In both prokaryotes and eukaryotes DNA synthesis is catalyzed only at the replication fork. In duplex DNA, the replication fork advances in a semidiscontinuous manner. One strand, the leading strand, is synthesized continuously and the other, the lagging strand, is synthesized discontinuously.

### 1.2 Biochemistry of DNA Replication

Many proteins are involved in DNA synthesis at the replication fork. In *Escherichia coli,* more than 20 different proteins participate in this process. Among them, DNA polymerase III holoenzyme, which consists of 10 distinctive polypeptides, is responsible for elongation of the DNA chain in both leading and lagging strand replication. This enzyme extends the DNA chain with a high processivity and a high catalytic efficiency, in contrast with the other *E. coli* polymerases, DNA polymerases I and II. A 3′→5′ exonuclease

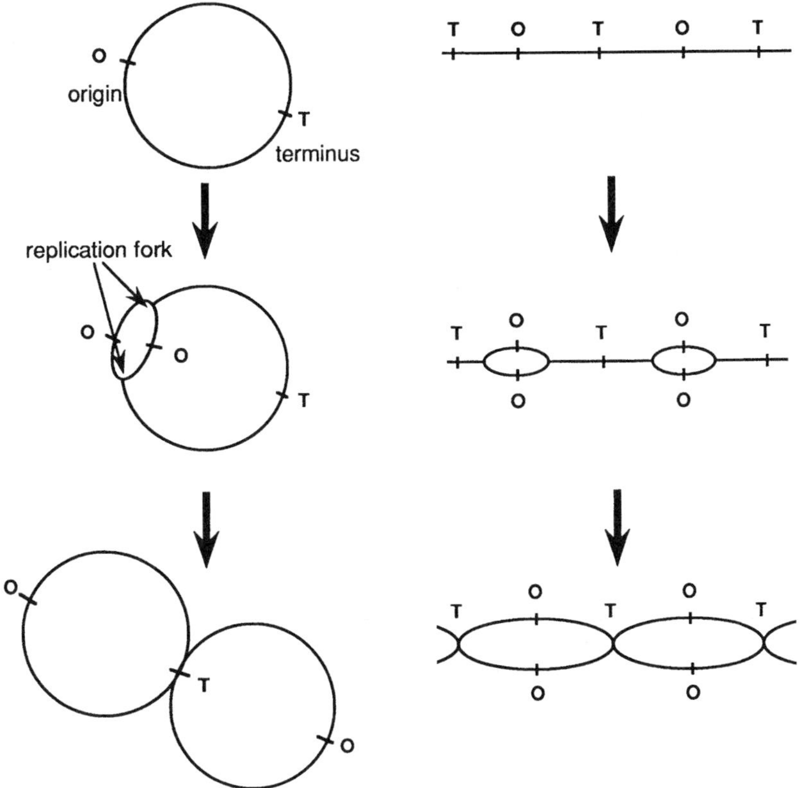


**Figure 1.** Units of DNA replication in prokaryotic and eukaryotic cells: O, replication origin; T, replication terminus.

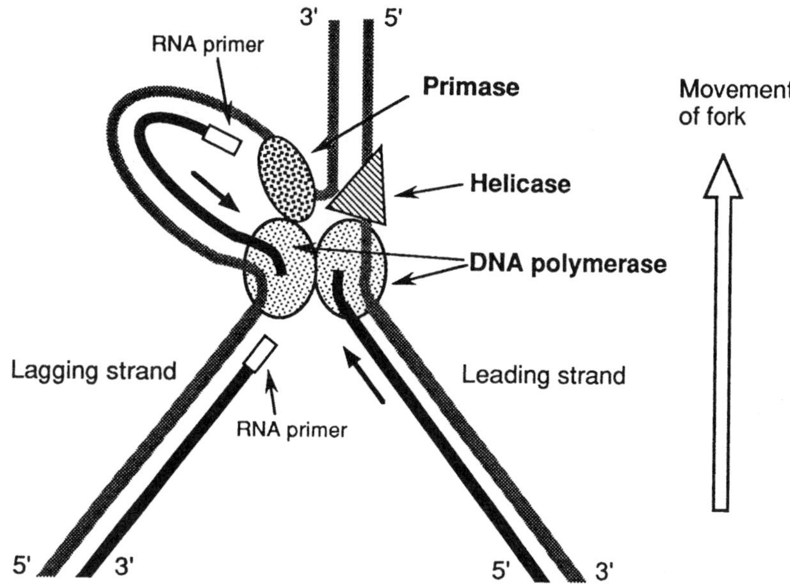


**Figure 2.** Simplified structure of the replication apparatus of *Escherichia coli.*

activity is associated with polymerase III and provides the enzyme with a proofreading capacity to correct replicational errors. DNA polymerases δ and ε, replicative enzymes recently found in eukaryotic cells, possess similar characteristics.

The DNA polymerase requires a single-stranded region of DNA as a template, and a piece of RNA or DNA annealed to the template as a primer. For this reason, two other enzymes are essential to double-stranded DNA replication. DNA helicase opens the duplex DNA to provide a single-stranded region at the replication fork, while primase synthesizes a short RNA transcript that acts as a primer for DNA chain elongation. Since chain elongation by DNA polymerase occurs only in the $5' \rightarrow 3'$ direction, lagging-strand DNA synthesis proceeds opposite to replication fork movement. This property requires a repeated synthesis of primer RNA for initiating synthesis of the nascent short fragment (Okazaki fragment) on the lagging strand. A concurrent DNA synthesis of both leading and lagging strands is achieved by the formation of a complex called the replisome (Figure 2).

In addition to the components of the replisome, several auxiliary proteins are needed for DNA replication. Single-stranded DNA-binding protein (SSB) stabilizes the structure of the replication fork, while DNA topoisomerase, acting as a swivel, releases torsional stress produced ahead of the replication fork. To complete the discontinuous DNA synthesis, RNase H (for removing primer RNA), DNA polymerase I (for filling the gap), and DNA ligase (for connecting the nascent fragments) are required.

### 1.3 Regulation of DNA Replication

DNA replication is principally regulated at the step of initiation of synthesis. The initial signal that turns on DNA replication involves growth factors in eukaryotic cells, or nutrition in prokaryotic cells, but it is largely unknown how these signals are transduced to the initiation event.

As revealed with an *E. coli* reconstituted system, at least eight proteins are required for replication of plasmid DNA carrying *ori* C, the unique 245 bp chromosomal origin of *E. coli.* A key event in initiation is formation of a complex in which the initiator protein (DnaA protein) tightly binds to the 9 bp dnaA boxes in the *ori C* sequence. Following this step, duplex DNA is opened in an ATP-dependent manner, and a helicase and SSB are settled in the single-stranded region to form a prepriming complex. Primase and DNA polymerase III holoenzyme then start DNA synthesis. A similar mechanism has been found in the initiation of SV40 (a eukaryotic virus) DNA replication in vitro; large T antigen acts as a helicase as well as an initiator protein specifically bound to the SV40 origin. Although DNA replication origins and proteins binding to them have been found in yeast *Saccharomyces cerevisiae,* the initiation mechanisms in eukaryotes are largely unknown.

## 2 TRANSCRIPTION

Transcription is the process of synthesizing an RNA molecule complementary to its template DNA, whereby the transfer of information from DNA to protein begins. DNA-dependent RNA polymerase binds to a specific DNA sequence called the promoter; then it unwinds the duplex DNA for about one turn of the helix, to expose a short stretch of single-stranded DNA so that complementary base pairing can be made with incoming ribonucleotides. The enzyme joins two of the ribonucleoside triphosphate monomers and then moves along the DNA strand, extending the growing RNA chain in the $5' \rightarrow 3'$ direction until it encounters a second special sequence, called the terminator, which signals where RNA synthesis should stop. After transcription has been completed, each RNA chain is released from the DNA template as a free, single-stranded RNA molecule.

**Table 1** Subunit Composition of the *E. coli* RNA Polymerase

| Subunit | Gene | Number of Amino Acids | Molecular Weight | Function |
|---|---|---|---|---|
| α | rpo *A* | 329 | 36,512 | Connecting ββ′-subunits |
| β | rpo *B* | 1342 | 150,618 | Catalyzing RNA synthesis |
| β′ | rpo *C* | 1407 | 155,163 | Template binding and association with σ-subunit |
| $\sigma^{70}$ | rpo *D* | 613 | 70,263 | Recognition of general promoters |

### 2.1 RNA Polymerase

Prokaryotic cells are equipped with a single type of RNA polymerase for transcription of genes, beside DnaG primase for synthesis of RNA primer required for initiation of DNA replication. The RNA polymerase of *E. coli* is composed of multiple subunits, with a total mass of about 450 kDa. The holoenzyme, the catalytically active form with promoter selectivity, consists of five subunits: two α-subunits, and one each of β, β′, and σ (Table 1). The σ-subunit, which ensures that the holoenzyme binds to the proper site in the promoter, is released once eight or nine nucleotides have been joined. There are multiple forms of sigma, each responsible for recognizing a particular class of promoters. The predominant form, serving for general promoters, is $\sigma^{70}$.

In eukaryotic cells, transcription is accomplished by three distinct types of RNA polymerase. RNA polymerase I (pol I) makes ribosomal RNA (18 and 28S rRNA) and pol II mostly makes messenger RNA, while pol III makes transfer RNA, 5S ribosomal RNA, and a number of other small RNA molecules (Table 2). All these enzymes are large and complex, with molecular weights of about half a million. For example, RNA polymerase II contains two large subunits (~200 and ~140 kDa) and a collection of smaller polypeptides, each less than 50 kDa. The largest subunit shares amino acid sequence homology with the β-subunit of the *E. coli* enzyme, while the second largest is related to β′.

No eukaryotic RNA polymerase can initiate transcription by itself, and the initiation at proper sites requires a group of general initiation factors. For pol II, at least six general transcription factors (TFIIA, TFIIB, TFIID, TFIIE, TFIIF, and TFII-I) are required. The polymerase enzyme together with the general factors constitute a basic transcription apparatus that can initiate transcription at the proper sites.

### 2.2 Transcription Units

Transcription takes place within a limited region of the genomic DNA, and generally only one of the two DNA strands is used as a template. The promoter determines the starting point of transcription as well as which of the two strands is copied. A transcription unit extends from the promoter to the terminator, where the RNA

Table 2 Eukaryotic RNA Polymerases[a]

| Polymerase | Location | Copies per Cell | Products | Polymerase Activity of Cells (%) |
|---|---|---|---|---|
| pol I | Nucleolus | ~40,000 | 35S and 47S pre-ribosomal RNA | 50–70 |
| pol II | Nucleoplasm | ~40,000 | Heterogeneous RNA (pre-messenger RNA)<br>Small nuclear RNA (U1, U2, U4, U5) | 20–40 |
| pol III | Nucleoplasm | ~20,000 | Transfer RNA, 5S RNA, 7S RNA, snRNA (U6), other small RNA molecules | ~10 |

[a]Mitochondria and chloroplasts contain distinctive RNA polymerases.

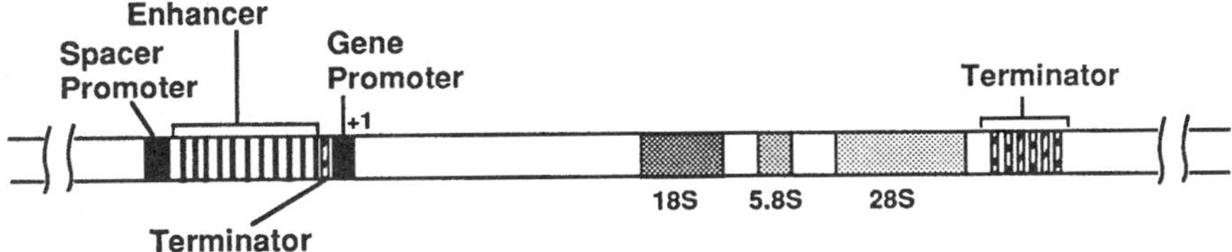


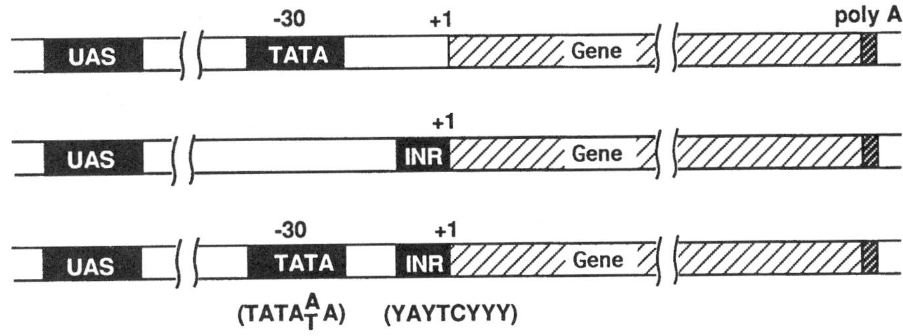


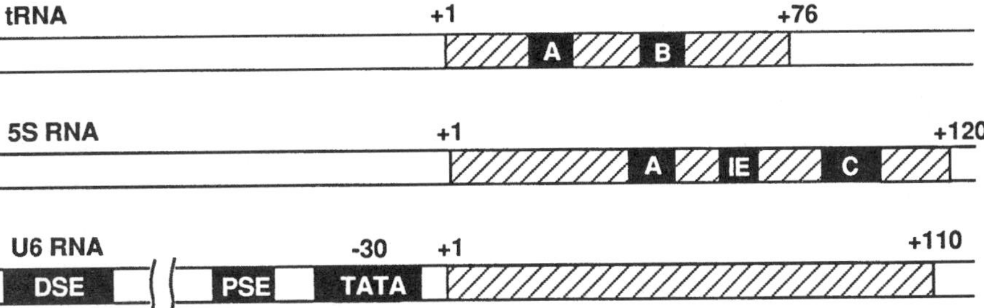

Figure 3. Transcriptional units in prokaryotic and eukaryotic cells: UAS, upstream activating sequence; TATA, TATA box; INR, initiator; A, B, C, and IE, intragenic promoter elements for RNA polymerase III; DSE, distal sequence element; PSE, proximal sequence element; Y, pyrimidine.

polymerase stops adding nucleotides to the growing RNA chain. A critical feature of the transcription unit is that it constitutes a stretch of DNA transcribed as a single RNA molecule, as shown in Figure 3. A transcription unit may include only one gene or several.

Bacterial promoters are identified by two short conserved sequences, centered at −35 and −10 nucleotide relative to the startpoint. Promoters whose −35 and −10 regions closely approximate the consensus sequences (TTGACA for -35 and TATAAT for −10 in *E. coli*) are strong promoters, whereas promoters whose sequences differ significantly from the consensus sequences are weak. Bacterial RNA polymerase terminates transcription at sites of two types. Factor-independent sites contain a GC-rich hairpin followed by a run of U residues, while factor-dependent sites require rho factor.

In eukaryotes, the promoters recognized by each RNA polymerase are distinct. The promoters for RNA polymerases I and II are mostly upstream of the starting point, whereas the promoter for RNA polymerase III most often lies downstream. Promoter elements that are necessary and sufficient for specific initiation by RNA polymerase II with its general factors are referred to as minimal, or core, promoter elements. The most common core elements are the TATA box found about 30 nucleotides upstream of the transcription start site and less well-characterized initiator (INR) elements at the start site. Specific genes may contain either or both of these elements, which may function together with other cis elements. Little is known about the termination process for eukaryotic RNA polymerases, even their termination signals.

### 2.3 Regulation of Transcription

Initiation of RNA synthesis is a primary control point in the regulation of differential gene expression. Cells respond to intra- and extracellular signals by turning certain genes on or off and modulating the extent of transcription of active genes. In bacterial cells, there are several different mechanisms by which modulation of transcription initiation is achieved: exchange of sigma factors, inactivation of repressor protein, activation of transactivator, and conversion of the DNA topology by topoisomerase.

In eukaryotic cells, an assortment of regulatory elements for RNA polymerase II transcription can be scattered both upstream and downstream of the transcription start site for a gene. Each gene has a particular combination of positive and negative regulatory cis elements that are uniquely arranged with respect to number, type, and spatial array. These elements are binding sites for sequence-specific transcription factors that activate or repress transcription of the gene. The DNA binding proteins also interact with activator proteins that also interact with the core transcription apparatus (e.g., TAFs). Usually, cis elements are arrayed within several hundred base pairs of the initiation site, but some elements can exert their control over much greater distances (1–30 kb). The control region in the vicinity of a transcription start site is called the promoter as described earlier, and regions that regulate the promoter activity from a distance and in an orientation-independent fashion are called enhancers. In addition, transcription can be regulated by the premature termination of elongation, or attenuation.

*See also* Chromatin Formation and Structure; DNA Structure.

*Bibliography*

Alberts, B., Bray, D., Lewis, J., Raff, M., Roberts, K., and Watson, J. D. (1989) *Molecular Biology of the Cell,* 2nd ed. Garland, New York.

Kornberg, A., and Baker, T. A. (1992) *DNA Replication,* 2nd ed. Freeman, New York.

Lewin, B. (1990) *Gene IV.* Cell Press, Cambridge, MA.

Moses, R. E., and Summers, W. C. (1988) *DNA Replication and Mutagenesis.* American Society for Microbiology, Washington, DC.

Reznikoff, W. S., Burgess, R. R., Dahlberg, J. E., Cross, C. A., Record, M. T., and Wickens, M. P. (1987) *RNA Polymerase and the Regulation of Transcription: A Steenbock Symposium.* Elsevier, New York.

# DNA Sequencing in Ultrathin Gels, High Speed

*Robert L. Brumley, Jr. and Lloyd M. Smith*

*Key Words*

**Anode** In an electrophoresis cell, the terminal toward which negatively charged species migrate.

**Cathode** In an electrophoresis cell, the terminal toward which positively charged species migrate.

**Electrophoresis** The movement of small ions and/or macromolecules (such as DNA) through a matrix or solution under the influence of an electric field.

**Polyacrylamide** Polymer matrix formed by the copolymerization of acrylamide and bisacrylamide.

DNA sequencing by horizontal ultrathin gel electrophoresis (HUGE) is a method of separating DNA in thin (usually < 0.1 mm) polyacrylamide gels so that the relative size of the DNA fragments (and thus, the sequence) can be determined. When used with conventional radioisotope-based sequencing protocols, HUGE can resolve up to 350–400 bases of sequence information in less than 30 minutes of electrophoresis, instead of the several hours usually required by traditional vertical electrophoresis. HUGE is also compatible with automated DNA sequencing methods and reduces the time of analysis by about a factor of 10. DNA separation by HUGE is a relatively new technique that is gaining acceptance as a practical way to significantly decrease the time required for the electrophoresis step of DNA sequencing.

## 1 PRINCIPLES

At present, almost all DNA sequencing is based on the separation of DNA fragments in polyacrylamide gels. One approach to reducing the time required for this step is to increase the speed of the separations. Several researchers have shown that capillary gel electrophoresis permits the time required for the electrophoretic separation of DNA fragments to be greatly reduced (see Bibliography for review by Luckey et al.). Capillary gels are only 50 μm in diameter and are very efficient at dissipating the heat generated

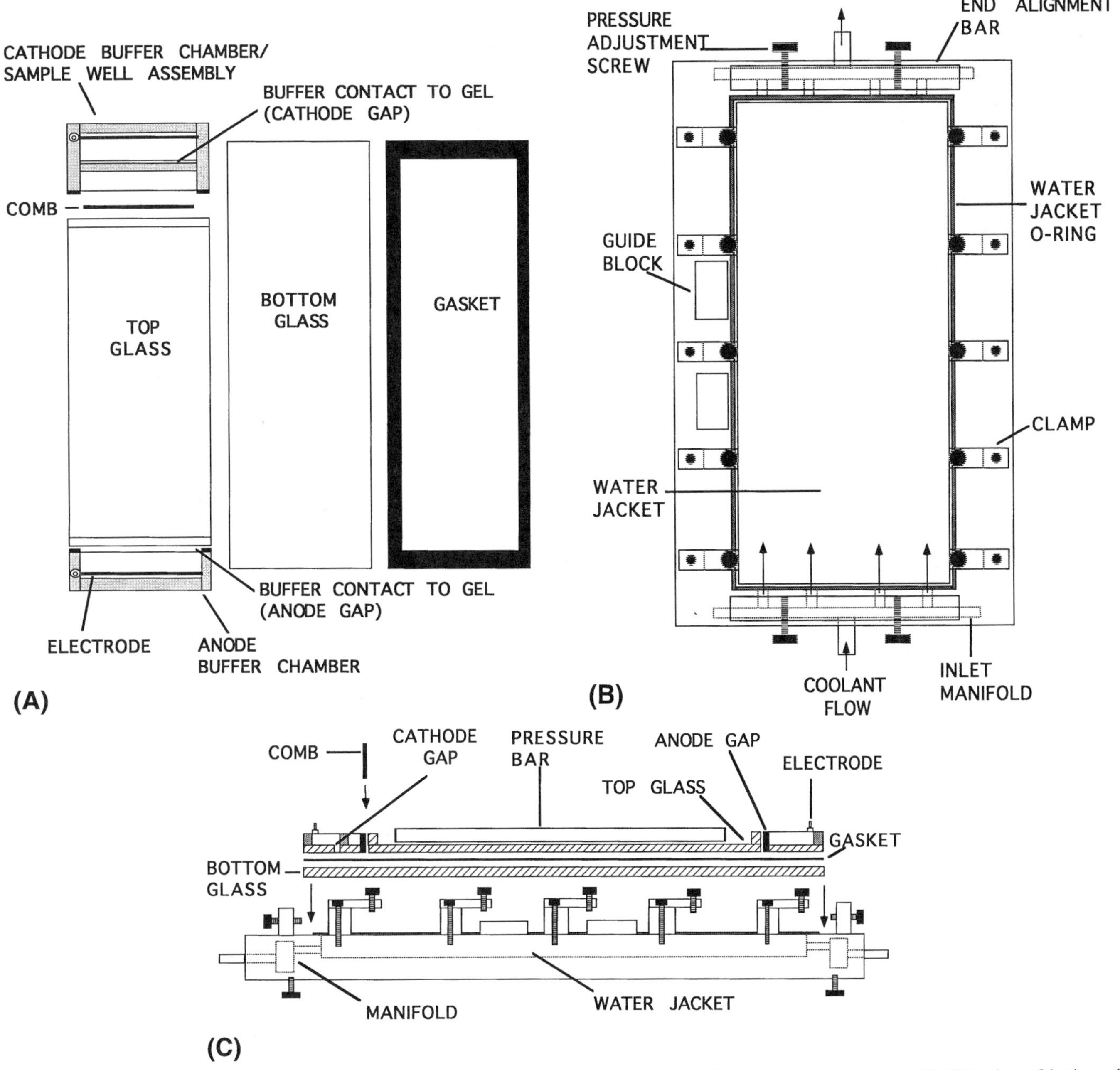


**Figure 1.** The HUGE setup. (A) Top view of glass and plastic components. (B) Top view of horizontal apparatus base. (C) Side view of horizontal apparatus base and glass components. (Reprinted by permission of Oxford University Press from R. L. Brumley, Jr., and L. M. Smith. Copyright 1991, *Nucleic Acids Research.*)

in electrophoresis. Thus, electric field strengths as high as 400 V/cm can be used for DNA separation in capillary systems without excessive Joule heating. The use of high electric fields can result in separation speeds increased as much as 26-fold over conventional electrophoresis. However, for each DNA sample that is analyzed in parallel by capillary gel electrophoresis, one gel-filled capillary must be prepared. Our approach to achieving rapid separations in polyacrylamide gels has been to develop a method of introducing samples onto 50 μm slab gels. We describe here the instrumentation and methods used for high speed DNA sequencing by horizontal ultrathin gel electrophoresis (HUGE).

## 2 APPARATUS DESIGN

Schematic diagrams of the apparatus are shown in Figure 1. The two major subdivisions of the horizontal apparatus are the glass and plastic "components" (Figure 1A) and the apparatus "base" (Figure 1B). The "components," when assembled, define the gel dimensions, sample wells, and electrophoresis buffer chambers. The "base" provides a water jacket, guide blocks, and clamps to hold the "components" in place. The temperature of the gel is regulated during electrophoresis by circulating coolant through the water jacket under the glass plates. The manifolds (Figure 1C) help

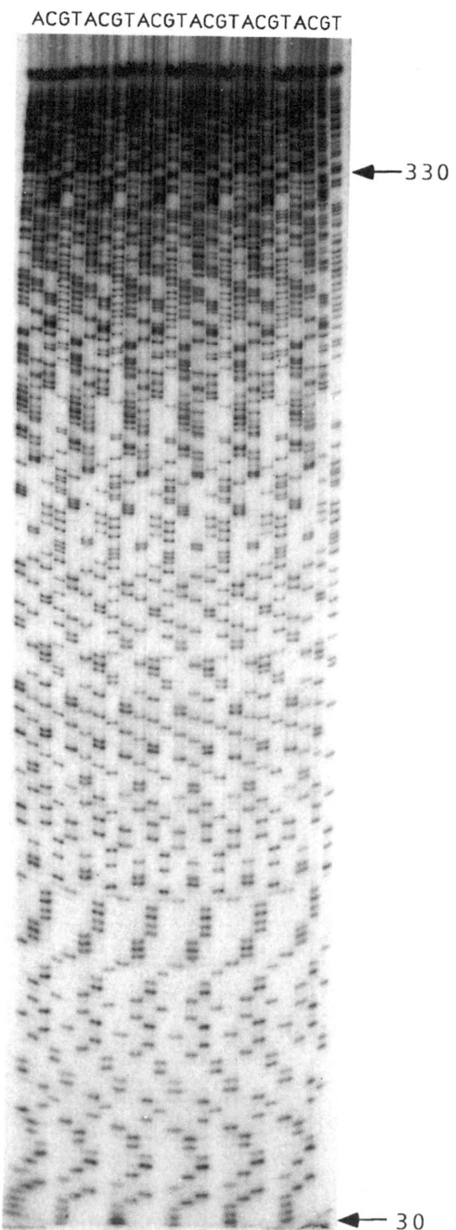


**Figure 2.** Autoradiograph obtained by HUGE: six identical sets of sequencing reactions. The gel was electrophoresed for 17 minutes at 40 W constant power (128–213 V/cm). (Reprinted by permission of Oxford University Press from R. L. Brumley, Jr., and L. M. Smith. Copyright 1991, *Nucleic Acids Research*.)

to disperse the fluid across the width of the glass and aid in the uniform distribution of temperature across the gel.

## 3 TECHNIQUES

### 3.1 Gel Preparation and Assembly

Prior to gel preparation, all glass components are cleaned and the bottom glass is treated so that the gel will bond to the glass when the "components" are disassembled after electrophoresis. The gel is cast directly on the apparatus by assembling the "components" as shown in Figure 1C. When these units have been assembled and clamped into place, the polyacrylamide solution is poured into one of the buffer chambers and the gel flows down the glass within the inside border of the gel gasket. The three vertical "gaps" (comb, anode, and cathode) are filled with gel solution, and the comb is placed in the slot between the cathode chamber and top glass plate. The gel is allowed to polymerize for about 60 minutes before use.

### 3.2 Sample Loading and Electrophoresis

About 15 minutes before samples are to be loaded, the water in the circulating bath is preheated to the desired temperature (usually about 35–40°C for sequencing gels). After preheating, the water is allowed to circulate through the water jacket via the inlet and outlet manifolds, to prewarm the glass plates. The comb is then removed, the sample wells are rinsed with distilled water, and sequencing reaction products are dispensed into the wells with a 1.0 μL syringe. The electrodes are connected from the power supply to the apparatus, and electrophoresis buffer is added to the anode and cathode chambers. Gels are usually electrophoresed at 40–50 W constant power, which results in a field strength of 150–250 V/cm. Electrophoresis time varies, depending on the amount and location of the sequence of interest. If the first 300 bases of sequence are desired, electrophoresis is stopped after about 15–20 minutes, when the bromophenol blue tracking dye reaches the anode buffer chamber (this dye is usually included in the "stop" solution supplied with many commercial radioactive sequencing kits). When electrophoresis is complete, the buffer is removed from the chambers and properly discarded (the buffer in the anode chamber is radioactive). The water circulator is turned off and the water jacket is drained of fluid. The "components" are removed from the apparatus base and disassembled such that the gel adheres to the bottom glass plate. The gel can then be "fixed" with an alcohol/acid solution and dried or wrapped in plastic wrap and exposed to X-ray film to visualize the separated fragments. Figure 2 is an example of a typical sequencing "ladder" produced by the Sanger dideoxy radioactive sequencing protocol.

## 4 AUTOMATED DNA SEQUENCING

The speed of DNA sequencing can be increased further by using fluorescent-labeled DNA instead of radioactive labels. The rationale and methodology of this approach have been reviewed extensively (see Bibliography). We have developed a detection system (Figure 3) that is compatible with HUGE in which 450 bases of sequencing information can be obtained from each of 18 samples

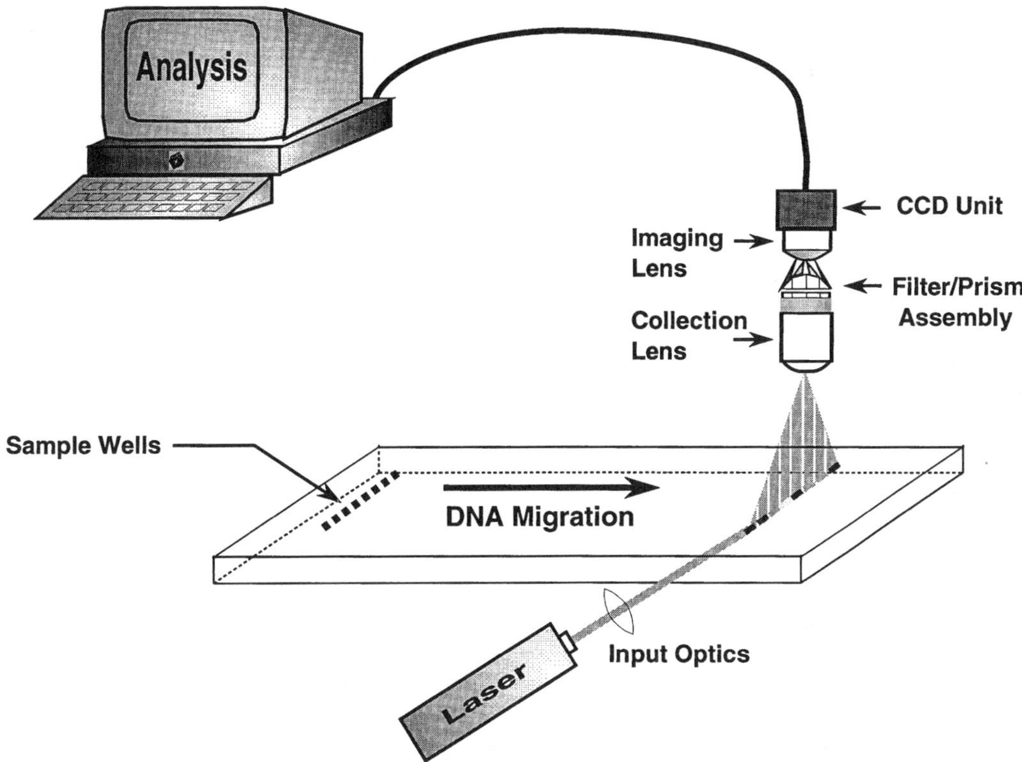


**Figure 3.** Schematic diagram of the high speed automated DNA sequencing instrument.

in about one hour. This system consists of a laser and optical components to provide a line of fluorescence excitation across the gel, collection optics, and a charge-coupled device (CCD) detector to measure the fluorescence emitted from the DNA samples. The detection system is connected to a computer that controls the CCD camera functions and stores and analyzes the data. The output of data appears on the computer screen as a series of peaks, each peak corresponding to a fragment of DNA (Figure 4). This automated system is currently under development to optimize its performance and promises to substantially increase the speed and reduce the cost of DNA sequence analysis.

*See also* AUTOMATION IN GENOME RESEARCH; NUCLEIC ACID SEQUENCING TECHNIQUES.

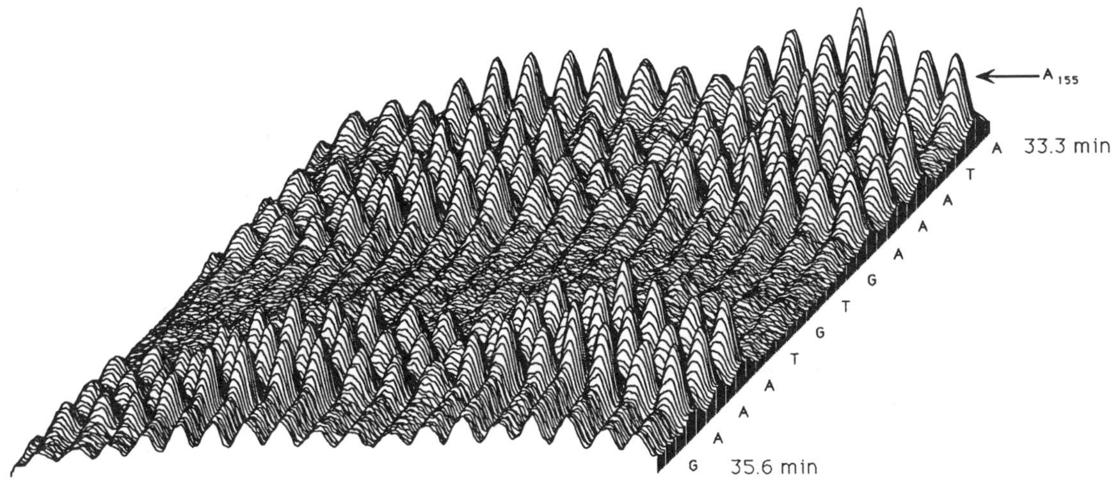


**Figure 4.** Automated sequencing output obtained from one of four wavelengths in the analysis of 18 samples in parallel. (Reprinted with permission from A. J. Kostichka, M. L. Marchbanks, R. L. Brumley, Jr., H. Drossman, and L. M. Smith. Copyright 1992, *Bio/Technology.*)

*Bibliography*

Brumley, R. L. Jr., and Smith, L. M. (1991) Rapid DNA sequencing by horizontal ultrathin gel electrophoresis. *Nucleic Acids Research,* 19(15):4121–4126.

Hood, L. E., Hunkapiller, M. W., and Smith, L. M. (1987) Automated DNA sequencing and analysis of the human genome. *Genomics,* 1:201–212.

Kostichka, A. J., Marchbanks, M. L., Brumley, R. L. Jr., Drossman, H., and Smith, L. M. (1992) High speed automated DNA sequencing in ultrathin slab gels. *Bio/Technology,* 10(1):78–81.

Luckey, J. A., and Smith, L. M. (1992) Automated methods in DNA sequence analysis. *Lab. Robotics Autom.* 3:175–180.

Luckey, J. A., Drossman, H., Kostichka, T., and Smith, L. M. (1993) High-speed DNA sequencing by capillary gel electrophoresis. *Methods Enzymol.* 218:154–172.

Smith, L. M. (1989) DNA sequence analysis: Past, present, and future. *Am. Biotechnol. Lab.* 7:10–25.

Smith, L. M. (1991) High-speed DNA sequencing by capillary gel electrophoresis. *Nature* (product review section), 349:812–813.

# DNA Structure

*Olga Kennard and S. A. Salisbury*

## *Key Words*

**Base Pair** The base portions of two nucleotides associated by hydrogen bonds.

**Duplex** Two nucleic acid chains linked through base pairs.

**Mismatch** A base pair other than those between the Watson–Crick partners: A and T; G and C.

**Oligonucleotide** General reference to a fragment of nucleic acid, commonly of two to 20-nucleotide residues, but with no rigid upper limit.

***R* Factor** A measure of correspondence between experimental X-ray diffraction data and those calculated for a given molecular model and, therefore, the reliability of the model.

DNA is central to the chemistry of life; its involvement extends beyond the storage and duplication of genetic information to include participation in the regulation of its own expression and evolutionary development. Interaction with the many other cellular components involved in these processes depends essentially on nonbonded interactions, and these are modulated by the spatial relationships of atoms imposed by the three-dimensional structure of the macromolecule. The conformational study of the DNA molecule provides an opportunity to observe structural detail at a level often not accessible in models that are more functionally complete but unworkably complex. Investigations on simple analogous systems are potentially, therefore, keys to understanding, in molecular terms, the operation of these important biological mechanisms.

## 1 TECHNIQUES

Natural DNA is classically a high polymer of four mononucleotide units. The superficial regularity of this arrangement in many cases hinders conformational study at the level of resolution required to describe the properties of relatively short sequences of specialized function. Chemical synthesis of oligonucleotides has in recent years made an important contribution here, enabling pioneering work on DNA isolated from natural sources to be extended to fragments of known sequence. Wide-ranging physicochemical and biochemical techniques have been applied to provide, in varying degrees of completeness, different and often complementary views of DNA structure. This is true not only of the parameters that are accessible by different methods, but also of the restrictions in molecular environment imposed by individual experimental systems.

In vivo, DNA forms only one component of the apparatus employed in its storage and use: it is associated with proteins, other nucleic acids, and species of low molecular weight, all of which can be expected to influence its structure in some degree. Such interactions are paralleled by those in solutions or crystals whose contents are designed to model the biological environment.

### 1.1 X-Ray Diffraction

Native DNA precipitates from aqueous solutions as a gel that can be drawn into fibers. Individual molecules are aligned both with the fiber axis and in register with each other according to the regularly repeating structure of the sugar–phosphate backbone. X-ray diffraction patterns of such fibers reflect the periodicity of regions of high electron density such as base planes viewed edge-on, or the crossing points of the backbone. Although these patterns alone are inadequate to define the three-dimensional structure in detail, they can provide constraints from which to deduce generalized models such as the familiar A and B forms of DNA.

Single crystals are essential for high resolution structural studies. Fragments of natural or enzymatically polymerized DNA produced by mechanical shearing and size fractionation can form single crystals, but they are insufficiently ordered for structure determination at high resolution. Synthetic oligonucleotides, by contrast, commonly form highly ordered single crystals that diffract at or near atomic resolution (2.5–0.9 Å). The molecular structure can be derived directly from the experimental data using mathematical methods at high resolution (up to 1.3Å). Otherwise the structure can be solved only if a suitable starting model is found which can be refined to fit the experimental data to a reliability factor (*R*) of 15–25%. In such an analysis the conformational details, albeit not the individual atomic positions, are accurately determined.

The X-ray data also contain information about intermolecular interactions, either between DNA oligomers themselves, or with proteins and drugs in mixed crystals, all of which are valuable in understanding in vivo behavior. Oligonucleotide crystals contain approximately 50% by weight of water. Only a small proportion of these water molecules are sufficiently ordered for their positions in the crystal to be defined precisely, but those close to available H-bond donors and acceptors in the nucleic acid can be identified, providing information about the contribution of hydration to the stability of secondary structure.

The chief limitation of the method is the need for single crystals. The chances of crystallizing a random sequence are small but are improved by the selection of sequences compatible with known packing motifs. At a practical level, already determined structures are used as templates whose limited modification can allow other sequences to be examined, or the effects of modified nucleotides or noncomplementary bases observed. A few structures have also been solved that are not recognizable as fragments of a conventional extended duplex.

### 1.2 NMR Spectroscopy

When studying DNA in solution, nuclear magnetic resonance spectroscopy is the method of choice for obtaining comprehensive structural information. The method takes advantage of the abundance of hydrogen atoms in nucleotides: diverse magnetic environments both within the individual monomer units and those superimposed by the secondary structure cause the proton spectra often to be well dispersed.

Conformational mobility of oligonucleotides in solution can result in broadening of spectral lines, but a range of spectroscopic data is accessible when signals are sharp. In particular, three-bond spin–spin couplings and the rates of nuclear Overhauser effect buildup between protons provide information on torsion angles and interproton distances, respectively. These data can be used as constraints to derive reliable computer-generated three-dimensional structures.

Ring current effects of adjacent heterocyclic bases within the secondary structure give rise to substantial variation in the chemical shifts of signals of otherwise equivalent protons. Duplex formation is associated with generally upfield shifts in the resonances of base protons, when increased stacking places the protons close to the faces of adjacent aromatic rings. Precise structural interpretation is not possible, but distinctively shifted signals are indicators of structural anomaly.

Present capabilities owe much to the development of two-dimensional acquisition techniques, which greatly enhance the effective spectral resolution. In favorable cases data obtainable from spectrometers operating in the range of 400–600 MHz for protons can define the conformation of an oligonucleotide containing up to 15 unique nucleotide residues. Some specific conformational investigations are possible on longer sequences. Only a few milligrams of sample is required at these magnetic field strengths.

### 1.3 Vibrational Spectra

Bands in infrared and Raman spectra correspond to differences between the vibrational energy levels of bonds. An infrared absorption line is observed if the relevant bending or stretching mode is associated with a change in dipole moment, while for Raman scattering a change in polarizability is required. The spectra may be seen as complementary; but in practice, for complex molecules such as DNA, reality lies between these extremes, and infrared and Raman spectra are often qualitatively similar.

Individual lines in the spectra, though they may be due predominantly to the behavior of one particular bond, are combination bands of all deformations in the vicinity and so contain conformational information. They refer necessarily to localized features; however, some of these have in turn been shown to correlate with global properties of the double helix and so are of more general significance as indicators of three-dimensional structure.

Vibrational frequencies such as those due to carbonyl stretching are directly interpretable—for example, in terms of bond strength changes due to hydrogen bonding; but in other cases only an empirical approach is possible. By comparing spectra of polynucleotides with known and different conformations, investigators have identified groups of bands associated with specific structural features—for example, in distinguishing 2′-*endo* and 3′-*endo*-furanose conformations typical of B- and A-DNA, respectively, and from which more general conclusions can be drawn.

A particular advantage of the method is its applicability to a range of solid and solution samples, including mixtures. It enables the comparison of structures in a number of physical environments, including those for which no other methods for extracting structural information are available.

### 1.4 Electronic Spectra

Nucleic acid bases possess strong ultraviolet chromophores with extinction coefficients of around $10^4$. Spectra of oligo- and polynucleotides resemble the sum of their component nucleotides, but the molecular environment within primary, secondary, and tertiary structures, where the bases are in close proximity, modifies the resultant spectra. The hypochromic effect observed at wavelengths close to the absorption maximum (ca. 260 nm) upon association of complementary strands into a duplex is commonly applied to monitor the temperature-dependent helix–random coil transition, or melting of DNA. At a more detailed level, changes in the relative intensities of component bands of the absorption envelope cause its shape to change as a result of secondary structure in the polymer.

The chiral environments of secondary structures, such as the various families of double helix, give rise to distinctive circular dichroism (CD) spectra. The potential for obtaining structural information combined with its sensitivity make CD a simple and attractive technique for conformational study. It is, however, a comparative method that can detect changes in conformation or identifies a stereotype by reference to standard spectra. Structures must first be identified using X-ray diffraction or some other definitive technique. The observation that A, B, and Z forms of DNA produce characteristic CD spectra has enabled extensive investigation of their interrelations in terms of nucleotide sequence or solution composition.

### 1.5 Electron Microscopy

Images of DNA can be obtained using the electron microscope. Early studies, which required specimens to be treated with heavy atom stains, contributed much to our knowledge of higher order DNA structure and topology. More recently, developments in scanning tunneling microscopy have made possible the visualization of the general form of the double helix such that its component residues can be discerned and its dimensions and periodicity measured.

With so alluringly visual a technique, it is especially important to bear in mind the experimental environment and sample preparation involved. Reassuringly, the results obtained are in general accord with the established view.

### 1.6 Biochemical Methods

Precise molecular recognition, which is a notable feature of many biological systems, provides a means of identifying certain structural characteristics of DNA. When interactions are essentially independent of nucleotide sequence, they can be exploited as structural probes. Examples of this approach are the location of single-stranded regions in looped structures by cleavage using a specific nuclease and the histological visualization of Z-DNA by antibodies raised against synthetic polynucleotides adopting that conformation under physiological conditions.

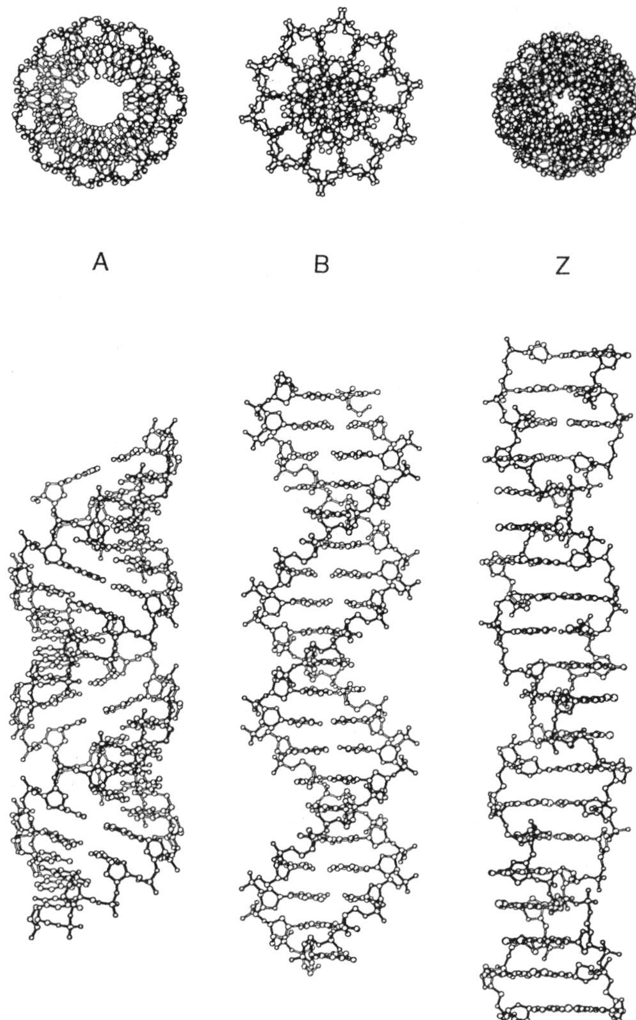


**Figure 1.** Skeletal representations of 20 base pair fragments of A-, B-, and Z-DNA according to idealized coordinates derived from X-ray diffraction data obtained from fibers (A- and B-) and single crystals of $d(CG)_3$ (Z-). Plotted using the program Molscript [P. J. Kraulis, *J. Appl. Crystogr.* **24:**946–950 (1991)].

Though providing limited direct information on three-dimensional structure, such methods can be deployed with high sensitivity when physicochemical methods are not practicable. They are, therefore, important among the array of techniques available for increasing our understanding of the complex behavior of DNA in living things.

## 2 DNA STRUCTURE

Fiber diffraction patterns defined the classical double helix. These sequence-averaged data imply a necessarily regular structure within which the base pairs are interchangeable. High resolution studies, chiefly using single crystal X-ray analysis, have penetrated this superficial regularity to expose sequence-linked variations consistent with the inherent flexibility of the sugar–phosphate backbone.

The A, B, and Z forms of DNA are well established. Early studies showed that DNA-B was generally associated with higher water activity and lower ionic strength than the A form. DNA-B was, therefore, presumed to be prevalent in living systems. Results with fibers are paralleled in solutions of polynucleotides where the interconversions can be monitored using CD spectroscopy. The A form is stabilized by the presence of organic cosolvents and high salt concentrations, but the transition conditions are also dependent on base composition and sequence. Alternating copolymers, notably poly[d(C-G)] and its analogues, adopt the Z form under conditions of high salt or torsional stress. This left-handed double helix was discovered by high resolution single crystal X-ray analysis of $d(CG)_3$ but has subsequently been observed by a variety of techniques.

The three families are morphologically distinct (Figure 1), though within especially the A and B types a number of minor variations have been described. General features are two clefts bounded by helical ridges formed by the sugar–phosphate backbone. These grooves, denoted "major" and "minor," reflect the asymmetric disposition of glycosidic bonds in Watson–Crick base pairs (Figure 2). They contain the hydrogen bond donors and acceptors of the heterocyclic bases and are sources of recognition information, since the arrangement of functional groups is unique to the particular nucleotide sequence. Groove geometry is important to the accessibility of these groups and varies greatly between the different families, while the relative disposition of the features depends on other conformational characteristics of the double helix.

In B-DNA the base pairs are roughly perpendicular to and intersect with the helix axis, giving an approximately 10-fold repeat. With this arrangement the major and minor grooves are respectively 10 and 5 Å wide and have similar depths. The A form appears hollow when viewed end-on. The base pairs are displaced from the helix axis and inclined to give a repeat of about 11. Though of similar widths to those of the B form, the minor groove is shallower and the major groove much deeper. Z-DNA is a left-handed double helix with the base pairs perpendicular to the helix axis. The helical repeat is approximately 12 base pairs but, in contrast to the right-handed forms, individual nucleotide units are not interchangeable. The true repeating unit is a dinucleotide, since the glycosidic conformations of successive nucleotides are alternately syn- and anti-, and in consequence the backbone follows a zigzag course. Slight displacement of the base pairs from the center of the helix produces a major groove that is very shallow, but the minor groove is deep and extends to the helix axis. The Z conformation is principally associated with alternating purine–pyrimidine sequences in which the purine has a syn-glycosidic conformation. Characteristic dimensions of the helical types are listed in Table 1. Key conformational features have been rigorously defined according to the schemes in Figures 3, 4, and 5.

The three families of duplex DNA are the result of subtle equilibria between cohesive and disruptive forces within the secondary structure and its envelope of solvent molecules and counterions. These include stacking between the heterocyclic bases, hydrogen bonds, and ionic interactions around the phosphodiester groups and other centers of localized charge. Within basic structural outlines, variations are seen that may serve to shape molecular surfaces suited to specific recognition events. Correlations between certain backbone torsion angles (see, e.g., Figure 3) are observed experimentally and appear to result from accommodation that preserves the overall helical shape. Too few structures have, however, been determined firmly to establish more comprehensive descriptions of sequence dependence.

Unless balanced by the presence of positive ions, repulsion between the numerous centers of negative charge, principally phosphate and carbonyl oxygen atoms, is a major obstacle to formation of multiple-stranded helices. This is also the case for higher order

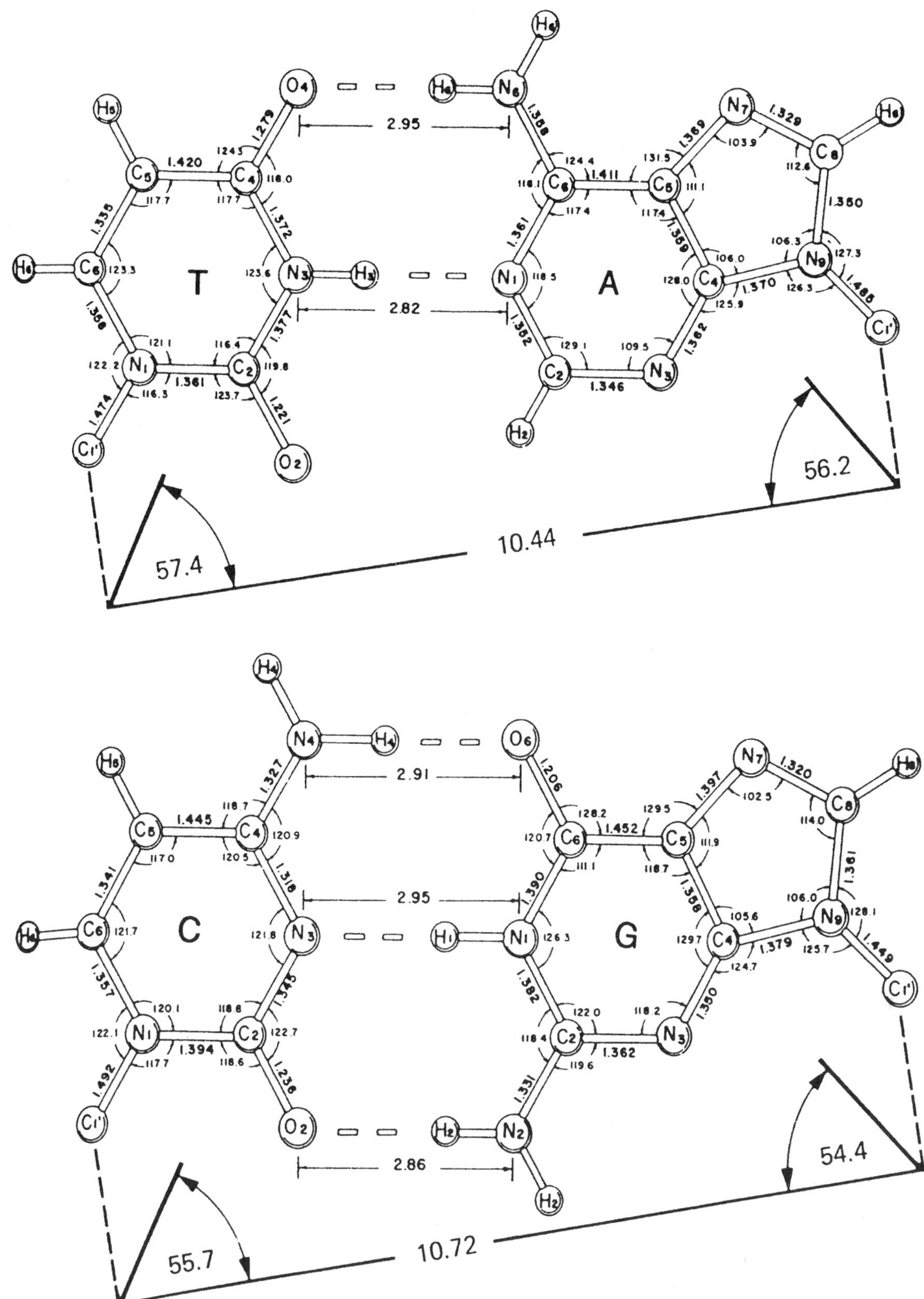


**Figure 2.** Watson–Crick base pairs formed in the accurately determined X-ray crystal structures of r(ApU) and r(GpC), which illustrate the geometry of the glycosidic bonds and other characteristic dimensions applicable to analogous pairs in DNA.

structure. In vivo, these cations are provided by proteins: histones in eukaryotic chromatin form a core on which the double helix is wrapped before further compaction by subsequent levels of coiling. In the absence of proteins, DNA condenses into higher order structures when polyamines such as spermine are added. The intermolecular interactions in these assemblies may be similar to those in oligonucleotide crystals, where close contacts between duplexes are observed.

**Table 1** Typical Parameters Associated with the Three DNA Families[a]

| Parameter | A-DNA | B-DNA | Z-DNA |
|---|---|---|---|
| Helix sense | Right handed | Right handed | Left handed |
| Base pairs per turn | 11 | 10 | 12 |
| Helical twist | 33° | 36° | 10°, −50° |
| Rise per base pair | 2.9Å | 3.4Å | 3.7Å |
| Helix pitch | 32Å | 34Å | 35Å |
| Tilt | 13° | 0° | −7° |
| Roll° | 6.0 ± 4.7° | −1 ± 5.5° | 3.4 ± 2.1° |
| Glycosidic conformation | *anti-* | *anti-* | *anti-, syn-* |
| Propeller twist | 15.4 ± 6.2° | 11.7 ± 4.8° | 4.4 ± 2.8° |

[a]Data based on single crystal X-ray structures.

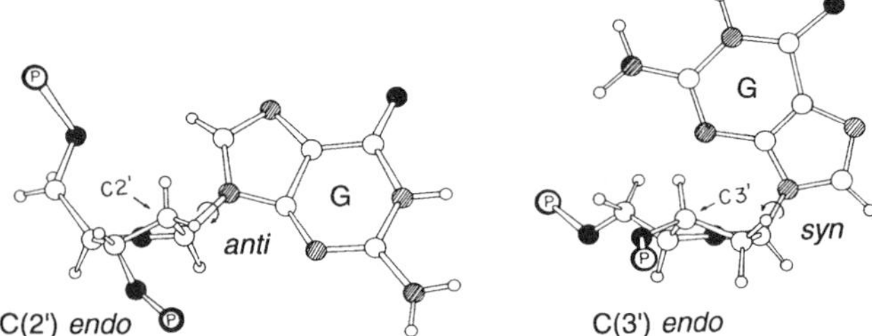


**Figure 4.** Principal conformations of nucleoside units in DNA illustrated by guanosine in B-DNA and Z-DNA, respectively; both the sugar pucker and orientation about the glycosidic bond are close to the extreme values for these parameters.

Certain DNA sequences can associate into three- and four-stranded assemblies using the abundance of hydrogen-bonding sites of the four DNA bases. Triplexes composed of purine·pyrimidine·purine (A·T·A, G·C·G) and pyrimidine·purine·pyrimidine (T·A·T, C·G·CH+) base triples have been characterized by chemical, biochemical, and spectroscopic techniques that establish the direction of the third strand bound in the major groove relative to the other two, which form a Watson–Crick double helix. Particular sequences may relate to special biological functions. Characteristic guanine repeats are found, for example, in telomeres at the ends of eukaryotic chromosomes. Fragments of related sequence have been shown to associate into quadruplex structures made up of guanine tetrads and have been extensively studied by NMR and X-ray analysis. Deoxycytidine oligomers also form four-stranded structures in acidic solution (pH 5) in which interdigitation of the photonated C·CH$^+$ pairs produces the ''I-motif'' quadruplex. Examples of these two structures are illustrated in Figure 6.

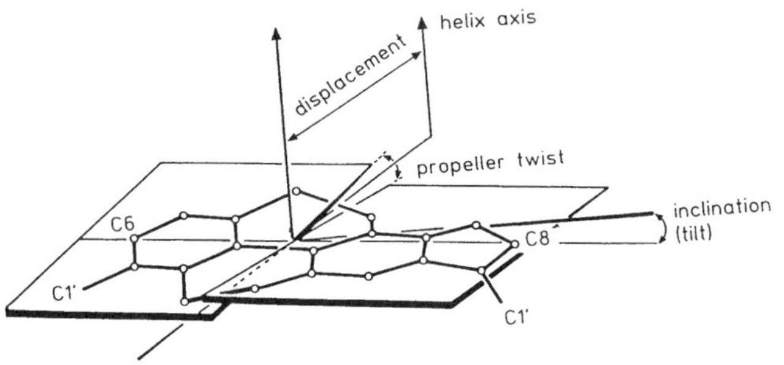


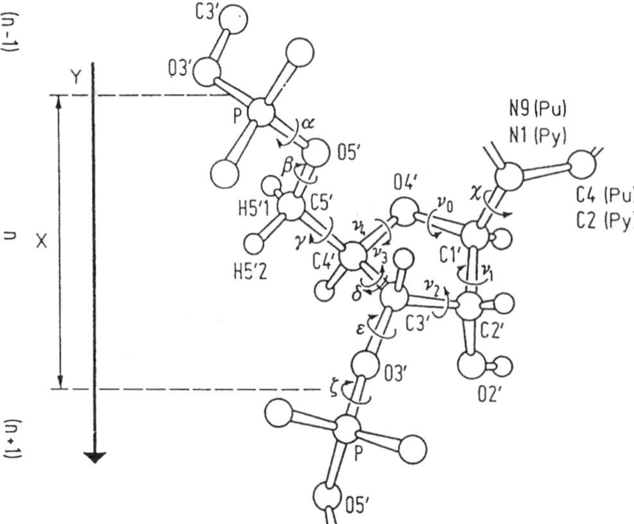


**Figure 3.** A deoxyribose–phosphate unit in the backbone, showing atom numberings and the notation used to identify backbone torsion angles.

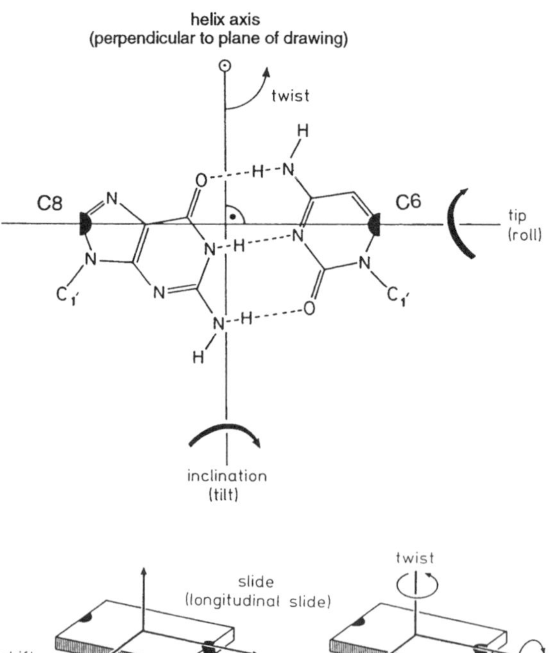


**Figure 5.** Schematic definitions of the numerous parameters devised to describe the geometry of individual base pairs and the relationships between successive steps of the double helix. Alternatives are given in parentheses.

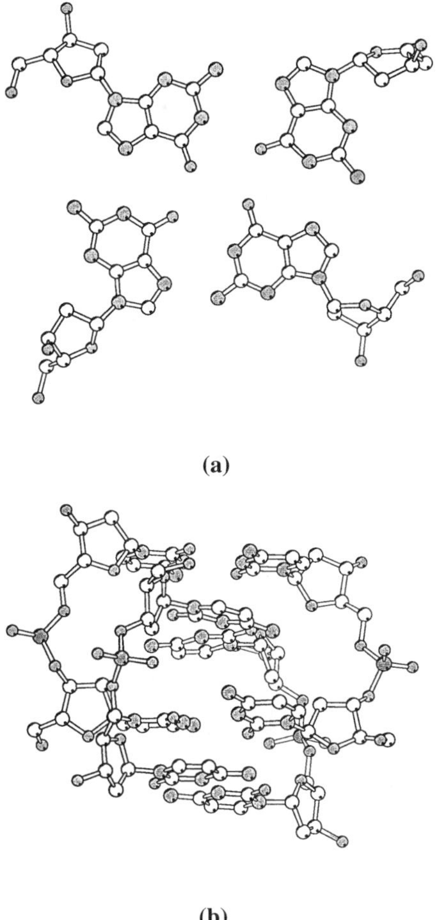

(a)

(b)

**Figure 6.** (a) One of the G tetrads in the X-ray crystal structure d(GGGGTTTTGGGG), where pairwise association of the looped oligonucleotide generates fragments of quadruplex [C.-H. Kang, X. Zhang, R. Ratliffe, R. Moyzis, and A. Rich, *Nature*, 346:126–131 (1992)]. The base tetrad itself is a common feature of oligonucleotide models of G quadruplexes, both in solution and the solid state, although different orientations of the sugar–phosphate backbone are observed. (b) The i-motif portion of the solution structure of d(TCC) determined by high resolution NMR spectroscopy (M. Guéron, personal communication). The quadruplex is composed of two interlocking duplexes, each formed by the self-pairing typical of hemiprotonated polycytidylic acids.

## 3 PERSPECTIVES

The double helix was discovered 40 years ago by constructing molecular models consistent with X-ray diffraction patterns obtained from DNA fibers. This view of general conformation remains essentially unchanged, but our increasing appreciation of the complexities of genetic organization and regulation has propelled the search for functionally relevant, local structural features encoded in the nucleotide sequence. Detailed analysis has been made possible by technological developments both in instrumentation for studying the shapes of large molecules and in the synthetic chemistry and enzymology that make available fragments of DNA of defined sequence. The intrinsic flexibility of the linear primary structure is potentially a source of great conformational variety that is best explored by analogy with simpler model compounds. Such studies have illustrated a number of sometimes unexpected structural possibilities available to nucleic acids that may be exploited in vivo. The approach can supply detailed views of fragments of the double helix that contain potentially mutagenic lesions resulting from errant base pairs between unmodified or chemically altered bases. Similarly, complexes of oligonucleotides with drugs and with proteins display details of the molecular recognition events involved.

The ultimate goal is to achieve at the structural level a predictive understanding of the behavior of DNA in living organisms. Here the molar quantities of material present and the complicating effects of other substances in the environment preclude direct study. Different model systems and analytical techniques provide alternative perspectives of this problem, supplying complementary information at various levels of detail. Interpretation of the data as a whole offers the possibility that the influences of individual experimental conditions can be identified and that the finer detail supplied by simpler models relate to more elaborate systems that are closer representations of DNA in vivo.

*See also* ELECTRON MICROSCOPY OF BIOMOLECULES; NUCLEAR MAGNETIC RESONANCE OF BIOMOLECULES IN SOLUTION; NUCLEIC ACID HYBRIDS, FORMATION AND STRUCTURE OF; X-RAY DIFFRACTION OF BIOMOLECULES.

### *Bibliography*

Cantor, C. R., and Schimmel, P. R. (1980) *Biophysical Chemistry*. Freeman, San Francisco.

Dickerson, R. E. (1992) *Methods Enzymol.* 211:67–111.

Kennard, O., and Hunter, W. N. (1991) *Angew. Chem. Int. Ed. Engl.* 30:1254–1277.

Rich, A., Nordheim, A., and Wang, A. H.-J. (1984) *Annu. Rev. Biochem.* 53:791–846.

Sanger, W. (1984) *Principles of Nucleic Acid Structure*. Springer, New York.

Watson, J. D., and Crick, F. H. C. (1954) *Nature,* 171:737–738.

Wütrich, K. (1986) *NMR of Proteins and Nucleic Acids*. Wiley, New York.

# DOWN'S SYNDROME, MOLECULAR GENETICS OF

*Charles J. Epstein*

### *Key Words*

**Acrocentric** Describing a chromosome in which the centromere is very close to one end.

**Autosomes** All the chromosomes in the genome except for the sex chromosomes (X and Y).

**Centromere** The structure within each chromosome at which the fibers required to move the chromosomes during meiosis or cell division (mitosis) attach.

**Meiosis** The process within the germ cells during which genetic recombination occurs and the number of chromosomes within the egg or sperm is reduced from two homologous sets found in somatic cells to one.

**Phenotype** The characteristics or traits that result from a genetic mutation or other genetic alteration.

**Trisomy** The presence in the genome of three rather than two copies of a specific chromosome.

---

Down's syndrome is the commonest of the genetically caused forms of mental retardation.It occurs with a frequency of approximately one per 800–1000 live births, and its incidence increases with increasing maternal age. Down's syndrome is caused by the presence of an extra chromosome 21 within the genome, which, in turn, results in a 50% increase in the expression of the genes contained on the chromosome. By mechanisms currently undefined, the increased expression of several genes on human chromosome 21 results in a syndrome characterized by mental and growth retardation, a distinctive set of major and minor congenital malformations, a variety of cellular abnormalities, and, later in life, by the development of Alzheimer's disease.

## 1 PHENOTYPE OF DOWN'S SYNDROME

The most immediately apparent, if not the most serious, manifestations of Down's syndrome (DS) are the many minor dysmorphic features that collectively constitute its distinctive physical phenotype. Salient among these are upslanting palpebral fissures, epicanthic folds, flat nasal bridge, brachycephaly, short broad hands, incurved fifth fingers, loose skin of the nape of the neck, open mouth with protruding tongue, and transverse palmar creases. Although any individual with DS will have many of the characteristic features and can be easily recognized as having the disorder, none of these features is present in all persons with DS.

Down's syndrome is manifested in the nervous system in three principal ways: hypotonia, which occurs in virtually all affected newborns and infants; delayed psychomotor development in infancy and mental retardation throughout life; and neuronal degeneration during the adult years. The latter process, which is pathologically identical to Alzheimer's disease (presenile and/or senile dementia), results in significant pathologic changes in the brain and may further compromise the already impaired mental functioning.

Down's syndrome is associated with two types of major congenital malformation. Most frequent (about 40%) is congenital heart disease, usually of the endocardial cushion type or one of its variants. Gastrointestinal tract abnormalities occur in about 4.5% of individuals, more than half of whom have duodenal stenosis or atresia. Structural abnormalities of the thymus and functional defects in T-cell function leading to an increased susceptibility to infection are present, and there is a 10–18 times normal incidence of childhood leukemia, with the frequent occurrence of acute megakaryoblastic leukemia.

## 2 CYTOGENETICS

Down's syndrome is the phenotypic manifestation of trisomy 21. As such, it occurs when a third copy of chromosome 21 is present in the genome, either as a free chromosome or as part of a Robertsonian fusion chromosome (in which the long arms of two acrocentric chromosomes are joined at the centromeres). Except in the 2–4% of cases in which there is mosaicism—with two populations of cells, one diploid and one trisomic—all cells of the body are trisomic. Although these cytogenetic abnormalities involve most or all of chromosome 21, in rare cases translocations occur in which only part of the long arm of the chromosome is triplicated. Depending on the region of the chromosome that is involved, such cases may or may not express the classical DS phenotype.

Analyses using DNA markers have shown that maternal nondisjunction, the failure of paired chromosomes to separate properly, accounts for 95% of all cases of trisomy 21, with 69% percent of the informative cases occurring at meiosis I and 28% at meiosis II, and 3% at mitosis. Furthermore, there appears to be a reduced rate of recombination in the maternal meioses that gave rise to trisomic offspring, findings that are consistent with the hypothesis that absence of pairing and/or reduced chiasma frequency and recombination predispose to nondisjunction.

## 3 STRUCTURE OF CHROMOSOME 21

Chromosome 21 is the smallest of the human autosomes, constituting approximately 1.7% of the length of the haploid genome. It consists of approximately $51 \times 10^6$ base pairs, with a sex-equal genetic length of about 56 centimorgans (cM). The correspondence between DNA and genetic lengths is reasonable, since it is assumed that 1 cM is roughly equivalent to $10^6$ base pairs. In physical terms, chromosome 21 is an acrocentric chromosome; the centromere is very close to one end, and the short arm is very small. The short arm (21p) terminates in a satellite region, which varies in size from molecule to molecule. Proximal to the satellite is the stalk (secondary constriction) which, as the nucleolar organizer region (NOR), contains multiple copies of the ribosomal RNA genes *(RNR4)* and stains characteristically with silver. The degree of silver staining appears to be a representation of the molecular activity of the ribosomal RNA genes that the chromosome contains. The region of 21p adjacent to the centromere contains highly repeated DNA sequences, which consist of the satellite (including alphoid) and the "724" families of sequences. None of these gene families is unique to chromosome 21. It is believed that these families of repeated gene sequences may be involved in the juxtaposition or association of the satellite regions (satellite association) of the acrocentric chromosomes during mitotic metaphase and with formation of the nucleolus during interphase.

The major part of chromosome 21 is the long arm (21q), which has a characteristic banding pattern consisting of three or four bands at low resolution and as many as 11 dark and light bands resolvable by prometaphase banding. All genes of known function (other than for ribosomal RNA) are located on this arm of chromosome 21, and only this arm is essential for normal development and function. The presence of a Robertsonian fusion chromosome in which the short arms of two acrocentric chromosomes (sometimes both chromosomes 21) are deleted does not cause detectable abnormalities if the genome is otherwise balanced.

## 4 MAPS

A list of the loci of known function or with known products mapped to human chromosome 21 is presented in the legend to Figure 1. In addition to these genes, the database of the Human Genome Project contains more than 400 anonymous DNA marker segments derived from chromosome 21 and more have been described since the generation of the maps in Figure 1. While the functions, if any, of the regions from which these segments are derived are unknown, their sequences provide valuable probes for mapping the chromosome and for following the segregation of loci at meiosis.

Two types of chromosome 21 map are now under construction. The first is a *physical* map on which the physical locations of

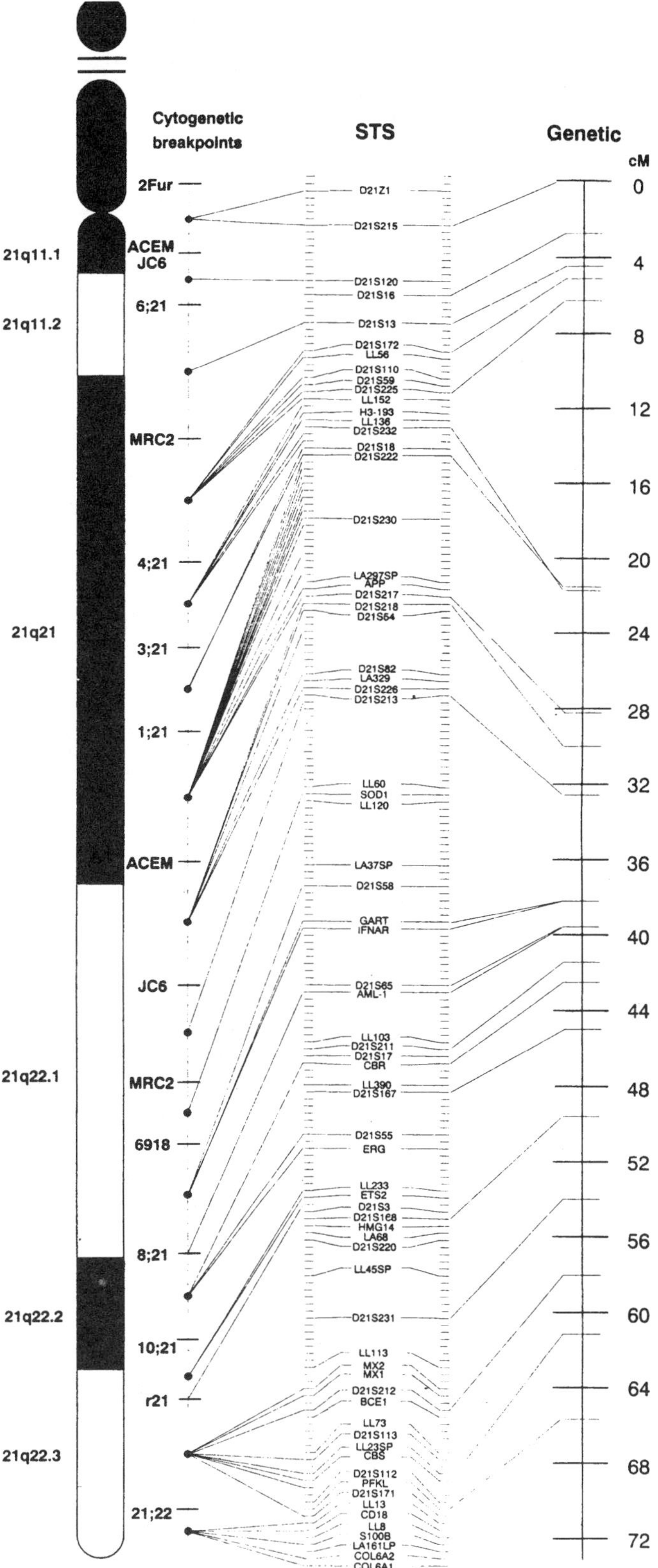


individual loci are placed, using features of the banding pattern of the chromosome as specific landmarks. Construction of a physical map is based on the use of a variety of tools, including somatic cell hybrids, DNA dosage studies of individuals with chromosome 21 duplications or deletions, in situ hybridization of labeled DNA probes to metaphase chromosomes, sequence tagged sites (STS) that are detectable by means of the polymerase chain reaction, and yeast artificial chromosomes. It has been possible with these techniques to map regionally many of the defined and anonymous loci on chromosome 21; the results of these efforts are contained in Figure 1.

The second type of chromosome 21 map is a *genetic* map, which shows the linkage relationships between loci and the inferred genetic distances of loci from one another. This type of map is derived from studies of the segregation of polymorphic loci within families. Because of intensive efforts devoted to the mapping of chromosome 21, the genetic map has rapidly evolved, and a recent version of the map is presented in Figure 1: the majority of the genetic distance, in physical terms, clearly appears to be located in the distal tenth of the chromosome.

Cases in which only part of chromosome 21 is triplicated have been intensively studied to arrive at a phenotypic map that will permit the correlation of particular phenotypic features of DS with specific regions or loci of the chromosome. On the basis of both molecular and cytogenetic analyses, it appears that many of the phenotypic features of DS are caused by triplication of the region around *D21S55* (which has been localized to band 21q22.2). The full extent of the region, however, remains undefined. More proximal triplication may result in some degree of mental retardation without the visible phenotype of DS. The region associated with the development of Alzheimer's disease has not been defined.

## 5 PATHOGENESIS

In searching for mechanisms to relate the components of the DS phenotype to specific genes on chromosome 21, it is important to realize that the immediate consequence of having a third copy of all or part of a chromosome is a gene dosage effect for each of

**Figure 1.** Physical and genetic maps of human chromosome 21. *Left:* a low resolution G-banding pattern next to which is a physical map of a large number of sequence tagged sites (STS) located with regard to a series of chromosomal breakpoints defined by translocations. *Right:* genetic map showing the locations of several of the STSs; the loci of known function in the STS list are APP, amyloid-β (A4) precursor protein; SOD1, superoxide dismutase 1, soluble (CuZnSOD), the same locus as ALS (amyotrophic lateral sclerosis); GART, a trifunctional enzyme with phosphoribosylglycinamide synthetase, phosphoribosylaminoimidazole synthetase, and phosphoribosylglycinamide formyltransferase activities; IFNAR, interferon alfa (and beta) receptor; AML1, acute myeloid leukemia 1 oncogene; CBR, carbonyl reductase (NADPH); ERG, avian erythroblastosis virus E26 (v-*ets*) oncogene-related sequence; ETS2, avian erythroblastosis virus E26 (v-*ets*) oncogene homologue 2; HMG 14, high mobility group protein 14; MX2, MX1, homology to murine myxovirus (influenza) resistance 2 and 1; BCEI, estrogen-inducible sequence expressed in breast cancer; CBS, cystathionine-β-synthetase; PFKL, phosphofructokinase, liver type; CD18, lymphocyte function-associated antigen, β subunit; S100B, S100 protein, β subunit; COL6A2, COL6A1, collagen, type VI, α-2 and α-1. Not shown are GLUR5, glutamate receptor subunit 5 (near SOD1); IFNGT1, interferon gamma transducer 1 (near FNAR); EPM1, progressive, myoclonic epilepsy (Unverricht–Lundborg type); and CRYA1, crystallin, α-polypeptide 1 (between CBS and PFKL). (Modified and reprinted with permission from *Nature, 359:*380–287. Copyright 1992 Macmillan Magazines Limited.)

the loci on the triplicated chromosome or chromosome segment. Such gene dosage effects, in which the concentration of gene products is 1.5 times normal, as would be expected from the presence of three rather than two copies of each gene, occur for all but one of the eight tested chromosome 21 loci. The only significant exception is in the levels of the APP (amyloid precursor protein) mRNA in brains from fetuses with DS, in which increases to about four times normal levels have been found. It thus appears that some type of dysregulation of the expression of the APP locus is superimposed on the gene dosage effect.

On the basis of studies in animal models (see Section 6), decreased platelet serotonin uptake and prostaglandin synthesis, both of which occur in DS, have been ascribed to increased activity of CuZn-superoxide dismutase (SOD1). With these exceptions, it has not been possible to relate any component of the DS phenotype to overexpression of specific loci. However, it is possible that the development of Alzheimer's disease in DS is related to the overproduction of APP, perhaps superimposed on an intrinsically defective nervous system.

## 6 ANIMAL MODELS

Many of the consequences of aneuploidy in humans arise during the period of morphogenesis, placing a special stumbling block in the way of their investigation. Research on events occurring during gestation, especially early gestation, is both technically impractical and, at the present time and for the foreseeable future, ethically and legally impossible. For these reasons, interest has turned to the development of animal models that will duplicate the human condition—in developmental and functional terms—as closely as possible. The mouse has been the model animal of choice for several reasons: ease of manipulation, genetic control, and similarities to the human in the processes of morphogenesis and probably of central nervous system function, in neurobiological if not psychological terms.

Three types of animal models for DS now exist. In one type, the effects of increased concentrations of specific gene products are analyzed in transgenic mice carrying one or more copies of individual human chromosome 21 genes. Strains of transgenic animals carrying the genes for human CuZnSOD, APP, and S100β have been made, and some features of DS were observed in the CuZnSOD-transgenic animals. At the other end of the extreme, mice with trisomy 16 have been bred. Mouse chromosome 16 carries many of the genes present on human chromosome 21q (as far distal as the loci for the Mx proteins), in addition to a large number of unrelated genes. Trisomy 16 mice, which survive only until the end of gestation, have many features of interest, including congenital heart disease (endocardial cushion defects), abnormalities of the thymus and brain, dysregulation of APP expression, and atrophy of transplanted neurons. Postnatally viable mice with an extra copy of just the region of mouse chromosome 16 that is homologous to 21q have also been generated. Because they reproduce the genetic imbalance found in the cases of partial trisomy 21 that exhibit the DS phenotype, these animals should be particularly valuable for studies of the pathogenesis of DS.

*See also* ALZHEIMER'S DISEASE; HUMAN GENETIC PREDISPOSITION TO DISEASE.

### *Bibliography*

Epstein, C. J. (1993) Down syndrome, in *The Metabolic Basis of Inherited Disease,* 7th ed., C. R. Scriver, A. L. Beaudet, W. S. Sly, and D. Valle, Eds. McGraw-Hill, New York.

———, Ed. (1993) *The Phenotypic Mapping of Down Syndrome and Other Aneuploid Conditions.* Wiley-Liss, New York.

Human Gene Mapping 11 (1991). *Cytogenet. Cell Genet.* 58:1–2200.

# DROSOPHILA GENOME

*John Merriam*

### *Key Words*

**Chromosome Walks** A method of selecting and aligning cloned DNA from a genomic library whereby consecutive hybridizations of probes originating from one end of a clone are used to identify the next clone in line.

**Enhancer Trap** A transposon containing the *lacZ* gene of *E. coli* that is transcribed only in response to tissue and stage acting enhancer elements adjacent to the chromosome site where the transposon is inserted (also called "enhancer sniffer," or "enhancer detector").

**In Situ Hybridization** Location of labeled RNA or DNA probes to chromosome sites by annealing to a slide of chromosome squashes denatured to be single stranded while retaining recognizable chromosome morphology.

**Polytene Chromosomes** "Many strands" of parallel, slightly condensed identical chromatids forming a visibly thick and long interphase chromosome after rounds of DNA replication without division.

**Transposable Element** Any member of a number of different families of viruslike gene segments characterized by repeated end sequences and located at chromosome sites that differ between unrelated lines.

The fruit fly *Drosophila melanogaster* has long been the geneticists' friend; the species is easy (and cheap) to grow, with a rapid 2-week generation time, and mutations occur readily. Recent molecular technological advances have shifted emphasis to questions on the normal gene structure and function after a period of exclusive consideration of mutant alleles. Through conserved gene sequences, *Drosophila* has become a model organism for determining the role of specific genes in multicellular development and behavior. Experimental molecular manipulations include loss of gene function, gain of function, controlled expression, sequence changes in the gene product, and simultaneous changes in other genes. Effects are measured on expression of gene products and on phenotypes of cells, tissues, or behaviors.

## 1 CHROMOSOME MORPHOLOGY

*Drosophila melanogaster* is by far the best studied of all the many species of fruit flies. As such, a description of the *melanogaster* genome is considered to be representative of all the *Drosophila* species. One example of similarities (and differences) is the microscopic description of chromosome shapes, sizes, and numbers. All *Drosophila* species can be seen to be different arrangements of five standard arms plus a tiny ''dot'' chromosome (Figure 1B), with the haploid number ranging from four to six. The other apparent differences are found in the sizes of the centric heterochromatic blocks and in the number of DNA sequences in the genome that are repeated (Figure 1A). Each chromosome is a single DNA double helix (''unineme'') without interruption from the telomere at one end through the centromere to the telomere at the other end.

### 1.1 Euchromatin

Euchromatin is classically the chromosome section that is invisible during interphase and the last to coil for mitotic divisions. Functionally it contains most of the Mendelian genes and transcription units. The amount of euchromatin is fairly constant between species, although it may vary as the proportion of a species' total genome. About two-thirds of the metaphase chromosome lengths in *melanogaster* are euchromatic.

### 1.2 Heterochromatin

Heterochromatin is classically the chromosome section that is slow to uncoil after mitosis (and may remain in a ''relic coil''). Functionally it contains few Mendelian genes or transcription units. Heterochromatic metaphase sections are marked by constrictions at the centromeres and at the nucleolar organizer regions (NORs, coding for the rRNAs). Heterochromatic regions are late replicating; they are generally transcriptionally inert. The Y chromosome is entirely heterochromatic. In the other chromosomes heterochromatin is differentiated between the majority $\alpha$-heterochromatin, with the classic properties, and a smaller $\beta$-heterochromatin, with properties such as its location, which is intermediate between euchromatin and $\alpha$-heterochromatin.

## 2 DNA CONTENT

### 2.1 The Nuclear Genome

The haploid genome size is about 165 million base pairs (165 Mb) with an overall GC content of 43%. About 110 Mb are present in the euchromatin. This consists mostly of ''unique'' sequences present only as single copies in the genome. About 15% of the DNA sequences are present in a few to several dozen copies each (''moderately repetitive''). These sequences, which make up most of the $\beta$-heterochromatin, include transposable elements (about 50 families known) and the tandemly repeated genes coding for histones and ribosomal RNAs. About 20% of the DNA is present as several families of highly repeated simple sequences that can be isolated as peaks of AT-rich satellite DNA. These sequences are entirely located in the pericentric $\alpha$-heterochromatin and are not amplified in polytene chromosomes. Earlier, the sizes of the repeated DNA classes were measured indirectly, by physical means; this approach is giving way to cloning the different sequences and the direct measurement of their abundances and locations in the genome.

Almost all the euchromatic genome is available in clones that have been traced to known cytological locations. Genomic walks, single clones, and cDNAs in plasmid or lambda vectors cover about 30%. Cosmid, P1, and YAC vector-based libraries have been screened for individually larger clones to localize larger fractions of the genome in fewer fragments. Clones in cosmid vectors cover about 40% of the genome; clones in the P1 vector show almost ''one-hit'' coverage, with at least 50% covered; yeast artificial chromosome clones cover almost 90%.

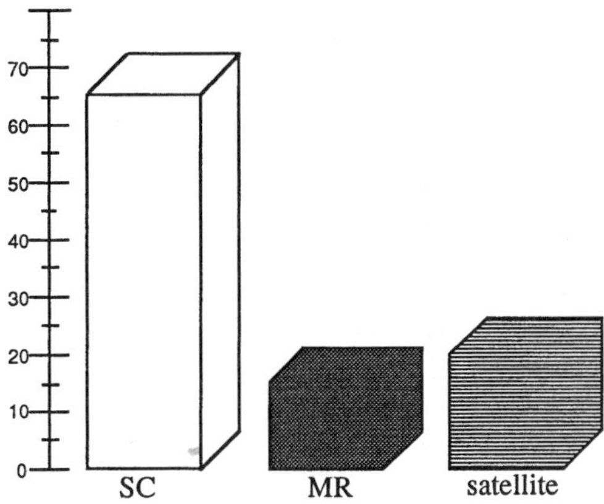


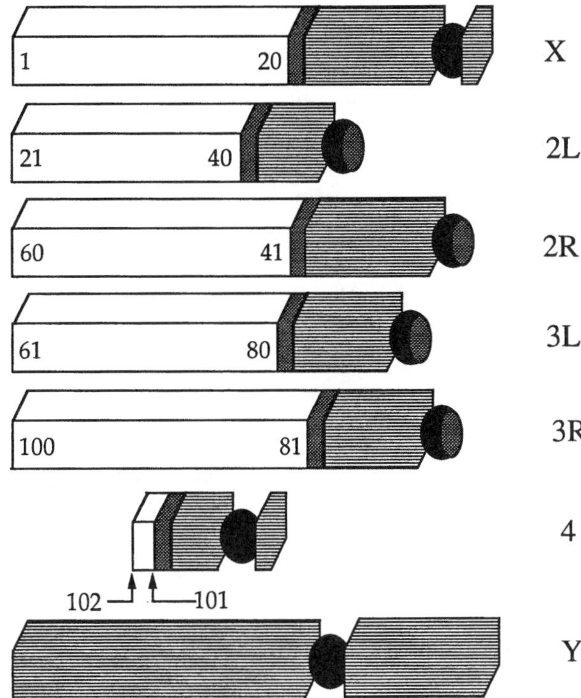


Figure 1. (A) The proportion of *Drosophila melanogaster* DNA sequence classes from cot curve estimates: SC, single copy; MR, middle repetitive; satellite, most of the highly repeated DNA classes. (B) The five ''elemental'' mitotic chromosome arms plus the ''dot'' chromosome of *D. melanogaster* showing the approximate locations and sizes of the DNA sequence classes: open bars, euchromatin; stippled bars, $\beta$-heterochromatin; striped bars, $\alpha$-heterochromatin. The numbers on each euchromatic section (e.g., 1–20) relate polytene chromosome sections to locations on the mitotic chromosomes. The X, Y, and fourth chromosomes are pictured with their subterminal centromere locations. For the other arms, telomeres are located to the left and centromeres to the right.

### 2.2 Transcribed Regions

The number of genes in *D. melanogaster* is estimated at 15,000, and about 1100 have been identified in clones so far (Table 1). Transcripts have been identified for most cloned genes. The average mature transcript size is 3.1 kb. The smallest known is 0.3 kb and the largest 13 kb, much the same as found in mammals. The initial transcript unit (the length of DNA transcribed into the primary transcript) is larger, of course, because of introns; but neither the average transcript unit size nor the extent of the euchromatic genome that is transcribed is known. Intron sizes are either small (63 nucleotides modal among those sequenced) or exceptionally large (e.g., 45 kb in the *Ubx* gene). The average intron number is not known but is likely to be low. Many genes are described as having one to a few introns. Alternative transcript starts and terminations are common, as is the alternative splicing of exons. A degree of complexity exists in overlapping transcription units (i.e., exons of gene 1 within introns of gene 2), including examples mostly transcribed in opposite directions, but also including an example of different transcripts from the same strand.

**Table 1** Current Knowledge about the *Drosophila melanogaster* Genome

Chromosome Descriptions

Haploid number = 4
Euchromatin: 110 Mb in approximately 5012 polytene bands
Heterochromatin: 55 Mb in 60 mitotic blocks
Recombination: 290 map units

Nuclear Genes

Size: approximately 165 Mb
Composition: 43% GC
Portion euchromatic genome covered by localized clones (approximate)
  Plasmid or lambda clones: 30%
  Cosmid clones: 40%
  P1 clones: 50%
  YACs: 90%

DNA Sequence Classes

Cot curve estimates: 65% single copy, 15% middle repetitive, 20% highly repeated
Isolated satellite species: 20%
Transposable elements: 10%

Transcribed Regions

Mature transcript—average size: 3.1 kb, range: 0.3–13 kb
Intron—modal size (sequenced): 63 nucleotides, range: 51–5392 nucleotides
Primary transcript—range: 0.3–80 kb

Gene Number Data

Estimated number of genes: 15,000
Estimated number of lethal complementation groups: 5000
Number of identified genes: 4000
Number of cloned genes: 1100
Number of sequenced genes: 900
Number of identified transcript units: 800

Mitochondrial Genes

Size: 18.4–19.5 kb
Composition: 21% of GC

### 2.3 The Mitochondrial Genome

The maternally inherited mitochondrial DNA molecule is a closed circle that varies from 18.4 to 19.5 kb in length with 21% GC content. The variation is at the extremely AT-rich block containing the replication origin. The entire molecule has been sequenced. It codes for 13 polypeptides, plus rRNAs, and 22 tRNAs.

## 3 MAPPING TECHNIQUES

The biological feature that most distinguishes the *Drosophila* genome is the polytene, or ''giant,'' chromosomes of the larval salivary gland cells. Without any stretching, each euchromatic chromosome arm is longer than 0.2 mm and wider than 0.01 mm (Figure 2). This unusual size results because the euchromatin of these interphase (elongated) chromosomes has undergone up to 10 rounds of DNA replication without division. Moreover, the sister chromatids are held together along their lengths in parallel, like a thick rope. Local areas of coiling along the length give a ''banded'' appearance that serves to uniquely identify every chromosome region. Detailed maps are available showing the standard coordinate positions along the chromosome. The total genome has been divided into 102 primary units, 20 per arm plus 2 on the fourth, on the basis of predominant bands. Each of these has been subdivided further into six lettered (A–F) units. Within each lettered unit the bands are numbered sequentially. The approximately 5012 polytene chromosome bands provided the first high quality physical map of any genome.

### 3.1 In Situ Hybridization

Determining the location of a gene through hybridization of a labeled DNA or RNA probe to polytene chromosomes is now standard procedure (Figure 2). Because of the multiple target number through polyteny, even probes of single-copy genes can be located with ease. With biotin as the label, probes frequently can be resolved to the average band size of 25 kb, and sometimes as low as 10 kb.

### 3.2 Chromosome Rearrangements

In *Drosophila*, the salivary gland polytene chromosomes also show tight association of homologous chromosomes along their euchromatic length. Effectively this is the visual equivalent of the meiotic gene-by-gene synapsis at early prophase, as inferred from recombination studies. This phenomenon was recognized because inversion heterozygotes show the synapsed homologues with a loop extending from the first to the second of the chromosome breakpoints characterizing the inversion limits. Any chromosome rearrangement above a minimum size, and assuming heterozygosity, can be located in this way to precise positions along the chromosome map. Rearrangement breakpoints resulting in mutations through interrupting normal gene sequences (inversions, translocations), or through absence of up to 1000 kb of the normal gene sequence (deletions = deficiencies) have been instrumental in mapping more than 3000 genes to small sections of the chromosomes.

### 3.3 Recombination

Mapping a new trait or genetic variant is most often done by using recombination studies with known markers and the new trait. The

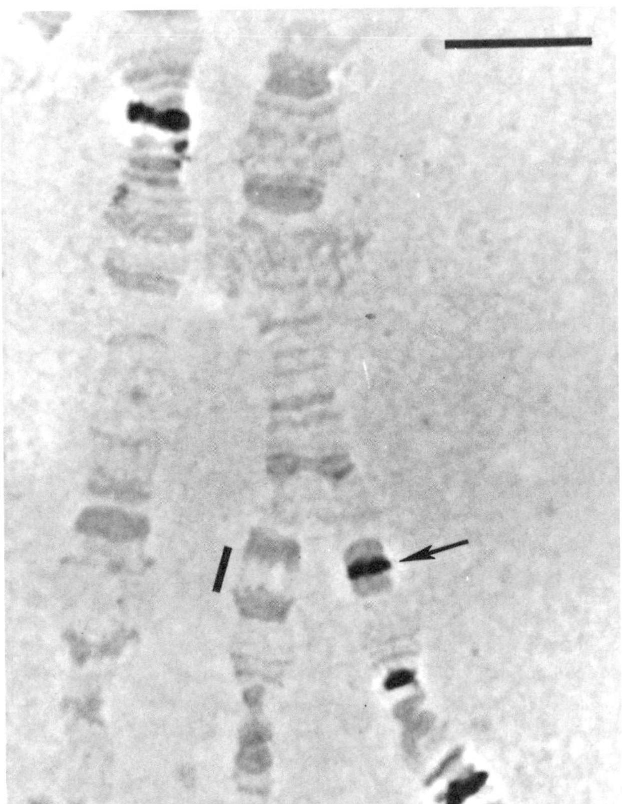

**Figure 2.** An enlarged view of partially stretched and asynapsed polytene X chromosomes that have been hybridized with a biotinylated probe consisting of the *Hobo* transposable element. The chromosomes came from a female heterozygous for a deficiency, *Df (1)3C7; 3D,* of about 200 kb. The normal chromosome is shown with a short bar on the left of the two homologous chromosomes. The deficiency breakpoints with *Hobo* end points are indicated by the arrow to the right-hand chromosome. Other biotin-labeled sites are not identified. The size bar at the top of the figure indicates 0.01 mm. The two homologous chromosomes are normally synapsed toward the tip, above the section marking the *Hobo* inserts and the deficiency. A section of chromosome 2L is present at the extreme left. (Photo micrograph courtesy of Johng Lim, University of Wisconsin, Eau Claire.)

chromosome and recombination maps are collinear, although recombination per unit DNA is not completely uniform. Over most of the chromosome, one map unit (1% recombination = 1 centimorgan) is about 300 kb. At the chromosome tips, and especially near the centromeres, recombination is suppressed, with the result that one map unit covers more DNA. The size of the X chromosome is about 65 map units; the second and third chromosomes are slightly more than 100 map units apiece. Neither the fourth nor the Y chromosomes, nor heterochromatic regions, recombine. Recombination also does not occur in the male. Recombination mapping of traits is reemerging in importance as the initial low resolution mapping pointing to chromosome sections for more exact studies using rearrangements.

## 4 GENETIC ANALYSIS

### 4.1 Mutant Screens

A goal of genetic analysis is to identify all genes and their functions through mutations that lead to altered phenotypic traits. Techniques for making large numbers of stocks homozygous for mutagenized chromosomes, or hemizygous for sections, have identified more than 3000 genes with allelic variants. This is a large fraction of the overall estimated 5000 genes considered vital because they can be mutated to lethality (as homozygotes; the mutant alleles are maintained as heterozygotes with balancer chromosome stocks). Mutations in the other 10,000 genes will be more difficult to identify through loss of function because the traits are slight or are masked by related genes with similar functions, or because the genes do not mutate easily. Improvement of stock-keeping techniques by freezing will be important for genetic analysis on a larger scale.

### 4.2 Transformation Rescue

No development within the past decade has had more effect than gaining the ability to transform genes into the genome. This breakthrough came by inserting genes into the P transposable element as a vector. One result of the flood of research that followed has been confirmation of which genes correspond to a trait by determining whether the transformed wild-type gene sequence suppresses the mutant trait. Another result has been the expression of experimental sequences at defined stages and tissues, which makes it possible to assay their effects. Transformation is the basis for direct determination of genetic potential by manipulating the genome one bit at a time.

### 4.3 Enhancer Traps

Transposable element inserts coupled with a reporter gene are now in common use to detect cis-acting enhancer regulatory sites. Mutagenesis with transposable elements moving to different chromosome sites now includes the *lacZ* gene from *E. coli* in the transposon. Tissue- and stage-specific expression of galactosidase from the *lacZ* gene is triggered by local signals from enhancer elements adjacent to the insert site. Genes normally influenced by these enhancer elements can be identified on the basis of simply staining the β-galactosidase activity with X-gal. More than 1500 P element transposons mapped to different chromosome sites are maintained at stock centers as a resource for defining adjacent genes.

### 4.4 Genetic Mosaics

Lethality of homozygous mutants at an early stage precludes examination of the mutant phenotype at later stages. A way around this problem in the genetic analysis of function through phenotype entails the creation of genetic mosaics as chimeras with only some cells homozygous for the mutant allele. This approach does not interfere with the animal's viability and does allow phenotype to be determined in small patches. Such determinations are accomplished by induced mitotic chromosome recombination or loss during development, to create cell homozygous for mutant alleles carried heterozygously in the surrounding cells.

## 5 PERSPECTIVES

### 5.1 The Most Asked Questions

The most asked question is, What mutant gene (trait) corresponds to a DNA sequence cloned by a laboratory? The answer, if any, has to come from locating the DNA clone and genetic variants to the same site on the polytene chromosome physical and molecular maps. The second most asked question is, How can mutations be created in the DNA sequence identified by a clone? The answer, if any, has to come from locating the DNA clone to sites that can be targeted by chromosome rearrangements for mutant selection.

### 5.2 FlyBase: A Database Project

FlyBase is a computer-based project working to interactively display information about genes and/or chromosome locations. Preliminary (flat file) versions may be accessed electronically by gopher or file transfer protocol to IUBIO.BIO.INDIANA.EDU. Biannual reports are published as Drosophila Information Service volumes.

*See also* Chromatin Formation and Structure; Genetics; Hybridization for Sequencing of DNA; Recombination, Molecular Biology of; Sequence Analysis.

### *Bibliography*

Ashburner, M. (1989) *Drosophila: A Laboratory Handbook.* Cold Spring Harbor Press, Cold Spring Harbor, NY.

Lindsley, D. L., and Zimm, G. (1992) *The Genome of Drosophila melanogaster.* Academic Press, San Diego, CA.

Merriam, J., Ashburner, M., Hartl, D., and Kafatos, F. (1991) Toward cloning and mapping the genome of *Drosophila. Science,* 254:221–225.

Roberts, D., Ed. (1986) *Drosophila: A Practical Approach.* IRL Press, Oxford.

Sorza, V. (1988) *Chromosome Maps of Drosophila.* CRC Press, Boca Raton, FL.

# Drug Addiction and Alcoholism, Genetic Basis of

*Judith E. Grisel and John C. Crabbe*

### *Key Words*

**Dependence** Characterized by frequent and regular drug use and craving, the hallmark of addiction; also accompanied by tolerance and withdrawal.

**Genotype** The genetic constitution of an organism.

**Inbred Strain** A group of animals mated with close relatives (usually siblings) for at least 20 generations, making them genetically identical (i.e., like identical twin) except for gender.

**Phenotype** The observable properties or response of an organism, resulting from the interaction of genotype and environment.

**Psychoactive Drug** A substance that changes how people think or feel.

**Selected Line** Animals genetically bred for a particular phenotype.

**Tolerance** Decrease in drug effect with repeated use, or conversely, a need to increase drug dose to obtain the same effect.

**Withdrawal** Abstinence syndrome, characterized by effects opposite to those induced by the drug.

Although a genetic influence on the etiology of alcoholism has been suspected for centuries and empirically determined for several decades, its precise neurobiological underpinnings are only recently beginning to be elucidated. This brief overview presents some of the evidence supporting the existence of a molecular genetic predisposition to abuse alcohol and drugs.

Recent progress in mapping strategies suggests that a bridge is being forged, effectively linking molecular techniques with the use of genetic animal models. This merger, as applied to the study of drug abuse, is likely to result in significant advances in our understanding of the specific relationship between genetics and behavior. Addiction researchers working at the level of behavior and those whose focus has been on the genome are beginning to find common ground in this interdisciplinary neurobiology, enhancing our understanding of the addictive process.

## 1 INTRODUCTION

People have been known to use distilled spirits, opiates (e.g., heroin), and other natural or synthetic drugs since the beginning of recorded history. But despite the use of psychoactive substances to varying degrees in many societies, incidences of addiction occur among only a small percentage of those who expose themselves to these substances. As yet there is no definitive biological marker for addiction, so the condition is defined behaviorally. Broadly speaking, behavior is labeled addictive when it is excessive, compulsive, and destructive psychologically or physically. The United States has the highest level of psychoactive drug use of any industrialized society, and according to the U.S. Surgeon General, 30% of all deaths in the country are premature because of alcohol and tobacco use.

The tendency to abuse alcohol seems to be codetermined with the tendency to abuse other drugs. Determinants of substance abuse include an interaction of biological and environmental factors. For instance, drug availability, a family history of drug abuse, and low economic status have all been associated with the prevalence of addiction. Furthermore, alcoholism runs in families. One-third of alcoholics have at least one alcoholic parent, and children of alcoholics are several times more likely to become alcoholic than children of nonalcoholics, even if they are adopted by nonalcoholic parents.

## 2 PREALCOHOLISM AND BIOMARKERS OF RISK

Several lines of research indicate that those who develop problems with alcohol and other drugs differ from other people even before their drug problems begin. This is a critical matter because it indicates that some of the abnormalities found in alcoholics and addicts are not due solely to the effects of chronic drug use but may predate drug exposure. This subset of characteristics, which can be used as markers indicating a risk for future development of abuse, may help target individuals for preventive intervention. Psychological studies show that prealcoholics (people tested before drinking begins, who subsequently develop alcoholism) are more independent, impulsive, undercontrolled, and nonconformist than normal control subjects. This tendency toward what is called "antisocial personality" has also been shown to be heritable, and in addition to a positive family history is a major predictor of subsequent alcoholism.

Some biomarkers of risk are also suspected. Brain electrophysiological evidence suggests that a specific, genetically determined anomaly exists in both alcoholics and their nondrinking offspring. When exposed to novel stimuli, about 35% of 7–13 year-old sons of alcoholic fathers, who presumably have not had alcohol themselves, have evoked potentials (a particular pattern of brain waves) like those of chronic alcoholics. It is not known whether this neurological marker is linked to chemical abnormalities that generate alcohol craving, is associated with emotional or behavioral problems that lead to compulsive drinking, or relates to the behavioral changes associated with alcoholism in some other way. In addition, there is evidence that the cellular response to ethanol differs in alcoholic and nonalcoholic subjects. When lymphocytes from alcoholic subjects are grown in a culture dish in the absence of alcohol, these blood cells show enhanced cyclic adenosine monophosphate (cAMP) signal transduction compared to lymphocytes from nonalcoholic subjects. As a second messenger, cAMP plays a critical role in the transmission of information in the brain by activating protein kinases, thereby leading to changes in the membrane potential of nerve cells. These results indicate that the regulation of cAMP signal transduction may be altered in subjects at risk for alcoholism. Although it is not clear how these in vitro findings might relate to a predisposition for drug abuse, the differences just noted may constitute useful markers for alcoholism liability.

## 3 ANIMAL MODELS

The use of animal models has promoted studies of the genetic basis of drug and alcohol abuse. Genetic animal models offer several advantages over studies using human subjects. In particular, the experimenter is in control of the subject's genotype. Whereas among human subjects only monozygotic twins have identical genotypes, numerous stable genotypes of rat and mouse are available. This broad pool of subjects is propitious for the sharing of information between laboratories.

Studies employing genetic animal models can be divided into three general classes. The first evaluates sensitivity to the acute effects of a drug. Examples in this area include such responses as ethanol-induced hypothermia or loss of righting reflex, and the motor-stimulating effects of amphetamine or cocaine. Studies in a second category investigate the response to chronic drug exposure. The adaptation of the nervous system underlying tolerance or sensitization, as well as phenomena such as dependence or withdrawal, are examples in this category. Finally, models have been developed that assess drug reward (or aversion), reflecting the reinforcing properties of the drug.

The first step is to study these special genetic strains of animal (usually rats and mice) to see how they differ in response to drugs. Once strain differences in behavior have been determined, investigators try to find out which gene products are influencing behavior and how they do it. In other words, what proteins are being coded, and by which mechanisms do these products alter behavior? Animal studies of specific phenotypes that may influence genetic risk are likely to yield identification of genetic markers that can predict drug abuse susceptibility in humans. Hundreds of such studies have been conducted since genetic animal models were first employed in the 1950s, and reviews of this literature are available. For our purposes, representative examples from each of these types of studies are discussed, to give a general indication of the methodology, progress, and application in this field of research.

### 3.1 Effects of Amphetamine and Cocaine on Mice

Amphetamine and cocaine greatly increase the activity of rodents, and this effect is known to depend in part on genetic determinants. One reason researchers are interested in the stimulating effects of these drugs is that they are thought to be related to the substances' rewarding effects (and thus the abuse potential). Inbred strains of mice vary widely in their response to an injection of amphetamine; some strains (e.g., C57BL/6) exhibit hyperactivity, and others (e.g., BALB/c mice) show amphetamine-induced inhibition of locomotion. $F_1$ hybrids derived from these progenitor strains have also been evaluated for their locomotor response to an acute dose of amphetamine. In this case, offspring behaved similarly to their C57BL/6 parent, and so the gene(s) responsible for an excitatory response to amphetamine appear to be dominant with respect to those leading to an inhibitory effect. It has been determined that the locomotor response to amphetamine is polygenic (i.e., determined by multiple genes). Similar differences exist between C57BL/6 and BALB/c mice with regard to the acute effects of cocaine on locomotor activity. Since, however, in this instance, $F_1$ hybrid strains exhibit an intermediate response to cocaine, it is clear that the response is incompletely dominant. Examination of locomotor activity induced by cocaine in other inbred strains revealed that this phenotype is also polygenic. However, different strains are sensitive to the two drugs, and different strains resistant, indicating that the genes underlying the responses to amphetamine and cocaine are not the same.

### 3.2 Effects of Alcohol and Other Depressants on Mice

Seizure susceptibility following withdrawal from a number of drugs that depress brain function has also been intensively studied using genetic animal models. Selected lines of Withdrawal Seizure-Prone (WSP) and Withdrawal Seizure-Resistant (WSR) mice have been bred to exhibit a specific reaction following ethanol dependence and withdrawal. WSP mice have much more severe alcohol withdrawal than WSR mice because most of the genes leading to severe withdrawal have been fixed in the former line. If WSP and WSR mice differ in other phenotypes, it can be assumed that alcohol withdrawal genes are responsible for these effects as well. WSP and WSR lines show similar differences when withdrawn from benzodiazepines (e.g., valium), barbiturates (e.g., phenobarbital, a general anesthetic) and nitrous oxide (laughing gas), indicating the likelihood of a common brain substrate for susceptibility to withdrawal-induced seizure upon cessation of ingestion of depressants. This finding may relate to the tendency to see multiple addictions in human subjects.

The neural changes associated with chronic exposure to ethanol have also been studied in WSR and WSP mice. Following chronic ethanol treatment, there are differences in the number of one kind of receptor on the surface of brain cells (calcium channels), as well as in the amount of zinc found in the hippocampus, a brain region associated with the control of seizures. These changes may contribute to the difference in neural excitability seen following drug withdrawal. Interestingly, these lines do not differ with respect to many other drug effects such as psychomotor stimulation and hypothermic tolerance, suggesting that the mechanisms underlying sensitivity, tolerance, and dependence are at least partially genetically independent.

Alcohol may be differentially regulating gene transcription in different populations. Some evidence for this hypothesis comes from studies on Long-Sleep (LS) and Short-Sleep (SS) mice. The lines have been selectively bred based on the duration of ethanol-induced ''sleep time'' or loss of righting reflex. These lines differ markedly in their response to ethanol and other depressant drugs (e.g., benzodiazepines, barbiturates), all of which act at the γ-aminobutyric acid (GABA) receptor. Specifically, they enhance activity at the GABA receptor, which modulates inhibition in the brain by decreasing the excitability of nerve cells. There is evidence that LS and SS mice differ in their response to alcohol (and presumably to other depressant drugs) because they differ in the molecular structure and function of GABA receptors. Exposure to ethanol increases the activity of GABA receptors more in LS than in SS mice. Furthermore, when messenger RNA (mRNA) for GABA is expressed in *Xenopus* oocytes, and those cells are then tested for ethanol sensitivity, they respond differently depending on whether the mRNA came from LS or SS mice. Presumably, this is because genes code for different GABA receptors in the two selected lines. Ethanol more readily inhibits neurol activity, through its action at GABA receptors, when these receptors are translated from the mRNA of LS mice. These findings indicate that the phenotypic selection of LS and SS mice produces genotypes that differ in GABA receptor function.

### 3.3 Self-Administration Studies

A common method for studying the ''abuse liability'' for a particular substance in humans is to use self-administration experiments in animals. Self-administration models for evaluating genetic differences measure either the amount of drug-adulterated water an animal chooses to consume compared to plain water or the number of times an animal emits a response (e.g., bar pressing) that results in drug delivery. These studies came out of work involving electrical self-stimulation of particular brain regions. From the initial experiments, in which animals were shown to act purposefully to electrically stimulate their own brains, the ''neurobiology of craving'' has been worked out. Most drugs that have addictive potential activate the neurotransmitter (brain chemical messenger) dopamine in an area of the brain called the mesolimbic pathway. Opioids, stimulants, nicotine, barbiturates, and alcohol are among the drugs having high abuse liability in humans that release dopamine in this midbrain area when self-administered by animals.

Of particular research interest are the neural substrates that determine differences in the likelihood that an individual will self-administer a particular drug. More than 30 years ago it was found that certain inbred mouse strains (e.g., DBA/2 and BALB/c) tend to avoid alcohol while others (C57BL/6 in particular) readily drink an ethanol solution in preference to water. Although the precise causes are not currently known, these behaviors appear to be determined by several genes and to involve differences in the salience of sensory stimuli (such as taste and odor), neurotransmitter levels in particular brain regions, and endogenous opioid levels.

Endogenous opioids, a class of internal morphinelike substances, play a role in many different behaviors including reinforcement, learning, and feeding. Endogenous levels of β-endorphin (one such opioid) increase in a dose-dependent manner after ethanol administration and do so to a greater degree in C57BL/6J mice than in DBA/2 mice. These differences were also present in human subjects who were either at high risk (three-generation history of alcoholism) or low risk for future development of alcoholism. Individuals in the high risk group had lower basal levels of β-endorphin circulating in the blood than individuals of the low risk group. Furthermore, in response to a low dose of ethanol, only subjects in the high risk group showed an increase in β-endorphin. This observation is potentially important because one of the general theories of alcoholism is that alcoholics self-medicate to compensate for unusually low levels of endogenous opioids. Using another tactic, a number of selected rodent lines have been developed to prefer or reject alcohol solutions; these lines include Preferring (P) and Nonpreferring (NP) and High (HAD) and Low (LAD) Alcohol Drinking rats. Differences in endogenous opioids as well as in monoamine levels (including dopamine and serotonin) in the mesolimbic pathway have all been related to the drinking differences in these lines.

## 4 QUANTITATIVE TRAIT LOCI MAPPING

Now we not only know that drug addiction is genetically transmitted, but we have some idea of ways in which drug responsiveness may differ between subjects who are and are not abuse-prone. Molecular techniques have recently been brought to bear on these findings and the precise genetic influences are beginning to be uncovered. One useful technique involves quantitative trait loci (QTL) mapping, which has led to exciting new discoveries in the field of drug addiction.

Traits such as those just discussed can be effectively studied in recombinant inbred (RI) strains, and genetic loci contributing to the phenotypic expression are currently being identified. The RI strains are inbred from the $F_2$ cross of two inbred parent strains. Since they are inbred, RIs are genetically identical and, in this case, have a fixed mixture of genes from either of the two progenitor strains. Drug responses are tested in both parent strains and a battery of their RIs. Under controlled conditions these mice will differ according to their unique genotypes, and strain means can then be compared to previously typed genetic maps. When a relationship between strain mean patterns and a pattern for a previously mapped gene is determined, the QTL is assumed to be linked to the previously identified gene. That is, it is located very near to the known gene, on the same chromosome. This means that the previously mapped gene may then be used as a marker for the QTL to predict whether individuals are likely to have the QTL. It also facilitates the use of molecular biological techniques to find the QTL itself so that its function can be studied.

The BXD RI series, derived from the cross of C57BL/6 and DBA/2 parent strains, is presently one of the most extensively studied. More than a thousand loci have been mapped from these RI strains, making QTL analysis particularly productive. For example, it was known that amphetamine-induced hyperthermia was much greater in DBA/2J than in C57BL/6J mice. A QTL analysis demonstrated a 0.96 correlation with the *Lamb*-2 locus on mouse chromosome 1 with this trait, suggesting that the thermic response to amphetamine is largely determined by a gene (QTL) near the *Lamb*-2 locus.

This finding is not prototypic, inasmuch as most drug responses appear to be codetermined by several QTL. Ethanol acceptance, a measure of how readily an ethanol solution will be consumed, was found to be associated with 17 distinct QTL. These QTL, located on several regions of several chromosomes, account for 95% of the variability associated with ethanol acceptance. Other phenotypes have been related to these particular QTL, and so some areas of a chromosome appear to be associated with multiple traits. In other cases, multiple QTL appear to codetermine responses to different drugs. For example, the QTL markers *Car-2* and *Ly-9* (found on mouse chromosomes 3 and 1, respectively) are correlated with both ethanol withdrawal severity and amphetamine hyperthermia. In some cases the correlation between a QTL and a phenotype is so high that this technique may be identifying the precise loci of influence. When evaluated across 20 inbred strains, one of the QTL for ethanol acceptance is apparently identical or very closely linked to the gene *Ltw-4,* on chromosome 1. The product of this gene is a very abundant brain protein, although its function is not yet understood.

Combined with other advances in molecular biology, these techniques show promise for even greater application. For example, several QTL candidates that may influence the severity of ethanol withdrawal have been identified. One of these sites, **Pmv-7** on mouse chromosome 2, accounts for about 40% of the variance in acute ethanol withdrawal. Chronic ethanol withdrawal severity and nitrous oxide withdrawal are also linked to the same QTL, indicating that genes at this locus confer susceptibility to withdrawal from multiple drugs of abuse. Nearby candidate genes include *Gad-1,* which is related to neural excitability in that it codes for glutamic acid decarboxylase, the rate-limiting enzyme catalyzing synthesis of the inhibitory neurotransmitter, GABA. Other candidate genes near this locus are those that encode the α-subunit of voltage-dependent sodium channels in the brain, which also are related to neural excitability and possibly to withdrawal severity.

*See also* Mouse Genome; Neuropsychiatric Diseases.

### *Bibliography*

Crabbe, J. C., and Harris, R. A., Eds. (1991) *The Genetic Basis of Alcohol and Drug Actions.* Plenum Press, New York.

———, Belknap, J. K., and Buck, K. J. (1994) Genetic animal models of alcohol and drug abuse. *Science,* 264:1715–1723.

Kalivas, P. W., and Samson, H. H., Eds. (1992) The Neurobiology of Drug and Alcohol Addiction. *Annals of the New York Academy of Sciences,* Vol. 264.

Nestler, E. T., Hope, B. T., and Widnell, K. L. (1993) Drug addiction: A model for the molecular basis of neural plasticity. *Neuron,* 11: 995–1006.

*Trends in Pharmacological Sciences.* (1992) Whole issue, 13(5).

# Drug Synthesis

*Daniel Lednicer*

### *Key Words*

**Adrenergic** Branch of the involuntary nervous system.

**Agonist** Substance that elicits the same response as the endogenous hormone.

**Analgetic** More commonly *analgesic.* Obtunding pain.

**Antagonist (sometimes Blocker)** Substance that prevents the response of an endogenous hormone.

**Beta-Adrenergic** Subdivision of the adrenergic system that influences such responses as heart rate, blood pressure, and bronchial constriction.

**Drug Metabolism** Chemical modification of drugs by the host after administration. May increase or decrease activity and sometimes toxicity.

**Hormones** Chemical substances synthesized by various structures in the body that serve to elicit biological responses at either a proximate or remote site. Consist of both simple molecules (e.g., adrenaline) and complex peptides (e.g., growth hormone).

**Lead** Usually the first compound to display activity on a bioassay. Basis for the design of all subsequent related compounds.

**Medicinal Chemistry** Area of chemistry that deals with drugs; includes the design and preparation of drugs as well as related areas such as the study of their disposition and metabolism.

**Receptors** Specific structures on cells that bind hormones and/or drugs; binding event results in biological response. Binding of antagonist blocks response to hormone.

**SAR** Structure–activity relationship; study of the effect on biological activity of modification of chemical structure within a series of active compounds. Used extensively in optimizing the structures of leads (q.v.).

**Screen** Bioassay used to test compounds to determine initial biological activity; designed as a model for the desired therapeutic effect. May be carried out either in vitro (e.g., antibacterial agents) or in small animals (e.g., analgetics). Many initial screens are designed for high throughput.

---

Drug synthesis encompasses the organic chemical manipulations involved in the preparation of pure compounds intended to have useful biological activity for the treatment of disease states. The design of target molecules involves the closely allied area often called drug design. Drug synthesis is closely associated with and often regarded as a segment of medicinal chemistry.

## 1 INTRODUCTION

The use of chemical compounds for treating disease predates chemistry by at least a millennium. Empirical ingestion of plants and plant extracts as potential cures and palliations for mankind's ills led quite early on to a sizable pharmacopoeia. With the rise of technology came attempts to improve on the specificity of those botanicals by use of various purification techniques. The isolation from opium of the pure organic compound morphine in 1803 thus predates Wohler's synthesis of urea (1828), an event often held to mark the beginning of organic chemistry.

With the rise of organic chemistry, it became recognized that most if not all items in the pharmacopoeia owed their activity to constituents that were usually organic compounds. The shortcomings of the early drugs combined with the development of techniques for the manipulation of the structure of organic compounds led to the nascence of drug synthesis as a technique for developing effective therapeutic agents.

## 2 LEADS FOR SYNTHETIC TARGETS

### 2.1 Natural Products

Several overlapping strategies have been used over the years to develop useful synthetic drugs. One of the earliest and still prominent approaches starts with a natural product that has demonstrated biological activity. That observation of activity may range all the way from folkloric usage (as in the case of opium) to complex in vitro or in vivo screens or assays (as in the case of many antibacterial compounds). The first-identified natural product more often than not suffers some serious shortcoming, which prevents its use as a therapeutic agent: it may not work when administered orally; or there may be undesirable side effects, too short duration of action due to rapid metabolism, or even serious supply problems. In a typical example, the structure of the active natural product is used as a starting point in designing simplified structures to attempt to find the minimal portion of the molecule which retains activity.

The natural product morphine **(1)** found early and widespread application in medicine because of its potent analgetic (i.e., pain-killing) activity. The serious shortcomings of this drug, most prominently addiction potential and respiratory depression, provided some of the impetus for synthesis programs aimed at better-tolerated drugs. Analgetic activity, it was found, was retained in the face of deletion of major parts of the molecule, as in the totally synthetic compound levallorphan **(2)**; activity is retained even when yet another ring is deleted as in benzomorphan A **(3)**. The simple phenylpiperidine, meperidine **(5)**, actually prepared a good many years prior to the two bridged polycyclic compounds, represents almost the ultimate simplification from morphine. All the simplified molecules show the same shortcomings as the parent; replacement of the *N*-methyl group by *N*-dimethylallyl in a benzomorphan leads to pentazocine **(4)**, an analgetic with reduced addiction potential.

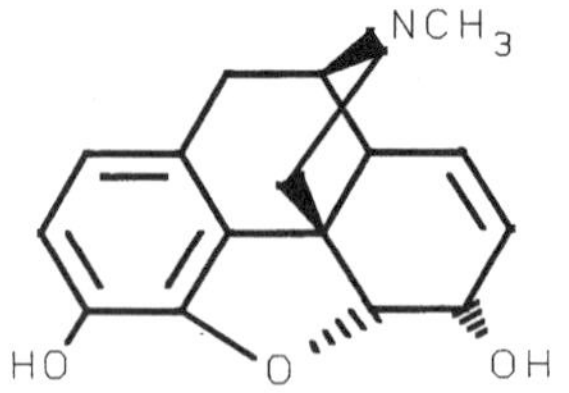

**1** Morphine

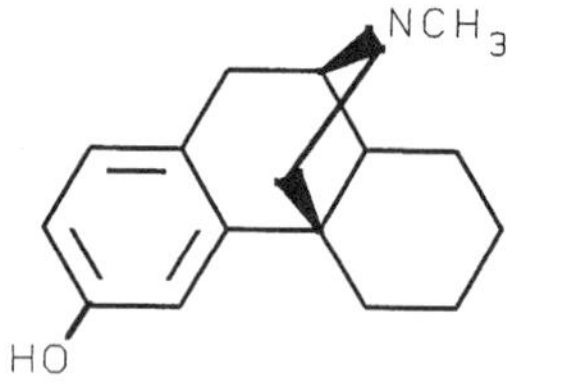

**2** Levalorphan

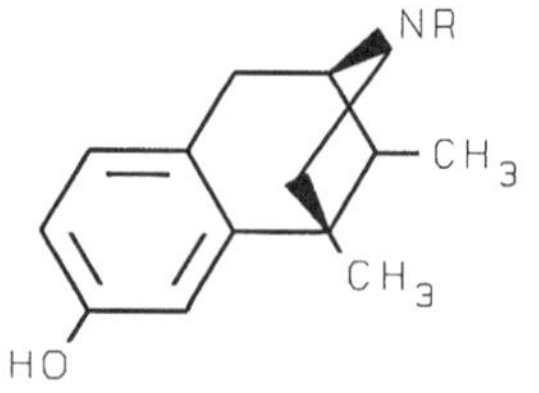

**3** Benzomorphan A, R = $CH_3$
**4** Pentazocine, R = $CH_2CH{=}C(CH_3)_3$

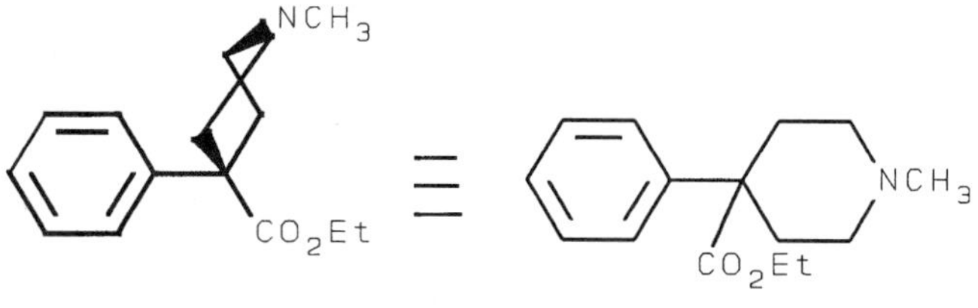

**5** Meperidine

### 2.2 Endogenous Hormones

A closely related approach starts with the structure of endogenous hormones or messenger substances. The targets of synthetic manipulations in such cases include compounds that will actually block the action of the endogenous substance and/or show improved absorption on oral administration. The familiar hormones epinephrine **(6)** (adrenaline) and norepinephrine **(7)** (noradrenaline) are intimately involved in the regulation of the smooth muscles involved in such basic functions as heart rate, blood pressure, and bronchial function. These endogenous hormones are not suitable for use as drugs per se because the multiplicity of end points each affects, as well as their short duration of action.

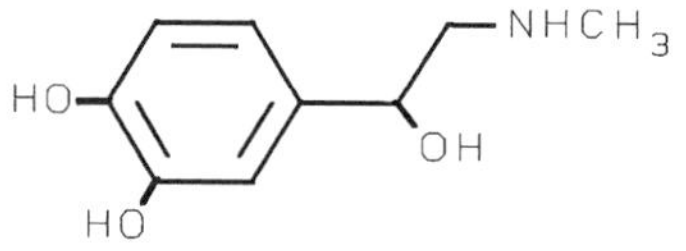

**6** Epinephrine

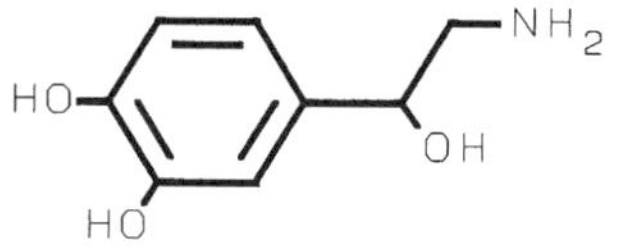

**7** Norepinephrine

Synthetic programs based on these α-amino alcohols succeeded in providing structurally related subseries, which showed quite discrete activities. This result was later attributed to structural modifications, which led to changed affinities to subsets of adrenergic receptors. The availability of the compounds was in fact essential to the recognition of those subsets. Replacement of the two phenolic hydroxyls in epinephrine by a *p*-acetamido group, insertion of an oxymethylene in the side chain, and replacement of *N*-methyl by *N*-isopropyl gives the β-adrenergic blocking agent atenolol **(8)**, a very widely used antihypertensive agent. On the other hand, a β-adrenergic agonist is produced by interposing a methylene group between oxygen and carbon in the meta phenol and replacing the *N*-methyl by a *tertiary* butyl group. This compound, albuterol **(9)**, is used as a bronchodilator in the treatment of asthma.

Emerging structure–activity relationships (SARs), which reveal the dependence of activity and/or potency on the presence or absence of various structural features, relative solubility in oil versus water (lipophilicity ratio), electronic distribution, or shape, help

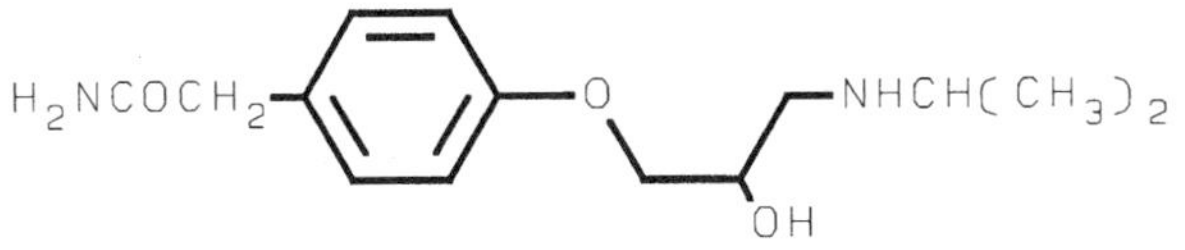

8 Atenolol

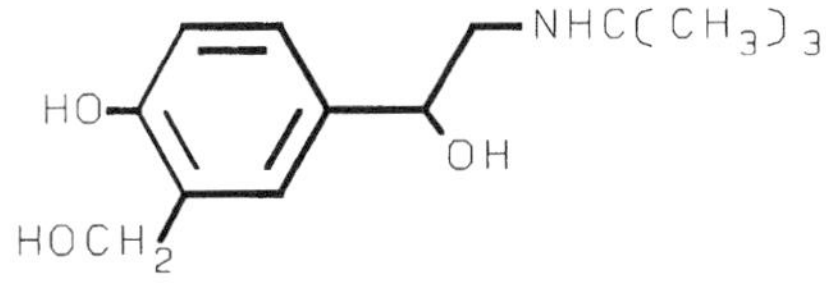

9 Albuterol

guide further synthesis. Systematic application of a study of those factors (quantitative SAR: QSAR) may provide added insight. The recent wide availability of microcomputers has had a profound effect on drug design by facilitating SAR and QSAR studies. The development of software for modeling organic compounds has allowed the expansion of SAR studies to three dimensions by providing a tool for readily studying the overlay of active compounds.

### 2.3 Random Screening

One very common and undeniably successful approach relies initially on random screening. This strategy starts with the development of an assay, historically performed in vivo but now increasingly done in vitro, which is intended to mimic some disease state. Since typically compounds known to be active against that disease do not exist, the screen will be used to test the widest possible structural array of compounds. Test candidates usually are obtained from a number of sources including, most prominently, collections maintained by the laboratory doing the screening. Once an active compound has been found, related structures are synthesized, to systematically modify the structure of the initial active lead. The SAR developed by this procedure is used much as just described, to optimize the activity.

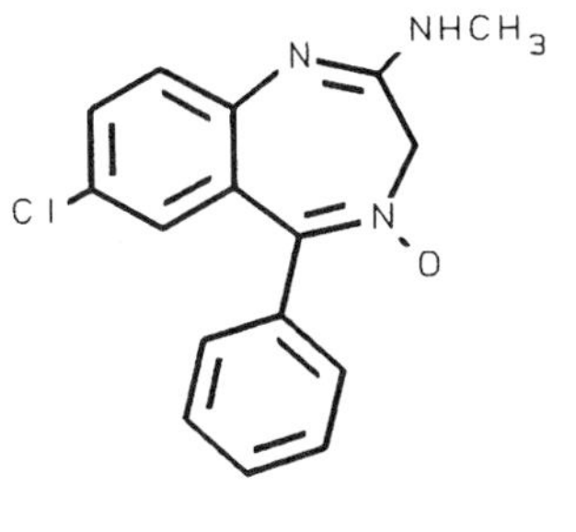

**10** Chlordiazepoxide

One of the first compounds to show activity in a bioassay intended to identify anxiolytic agents, then known as minor tranquilizers, was the benzodiazepine chlordiazepoxide **(10)**.

The finding that this compound, under the trade name of Librium, was quite active as an anxiolytic agent in humans occasioned an enormous amount of work on analogues, both in the laboratory that made the discovery and among its competitors. The number of benzodiazepines prepared ranks in the tens of thousands, some dozens of which have found clinical use. The spectrum of activity of a typical benzodiazepine is quite complex and includes hypnotic and muscle relaxant properties.

A significant number of drugs have been developed historically by modifying the structures of active leads to emphasize one particular aspect of a lead compound's side effects. Sulfonamide antibiotics were among the first compounds that had documented therapeutic activity in the treatment of bacterial infections; these agents pointed the way to drugs that were toxic to bacteria but relatively innocuous to a mammalian host. Some of these were found to exhibit diuretic effects in the clinic, while others seemed to lower blood sugar in patients. Systematic modification of those compounds led to the structurally related thiazide diuretic drugs and oral sulfonylurea antidiabetic agents.

Corresponding structural modifications of benzodiazepines provided drugs that emphasized specific facets of the lead compound's spectrum of activities. Triazolam **(11)**, for example, is used mainly for its hypnotic activity, while its analogue midazolam **(12)** finds use mainly as an injectable anesthetic. Deletion of the pendant ring

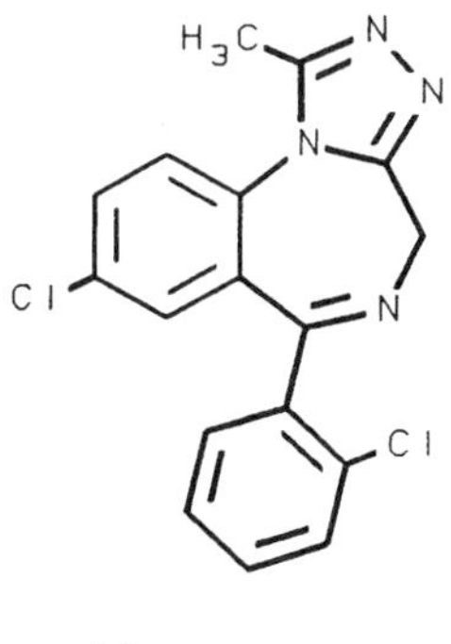

**11** Triazolam 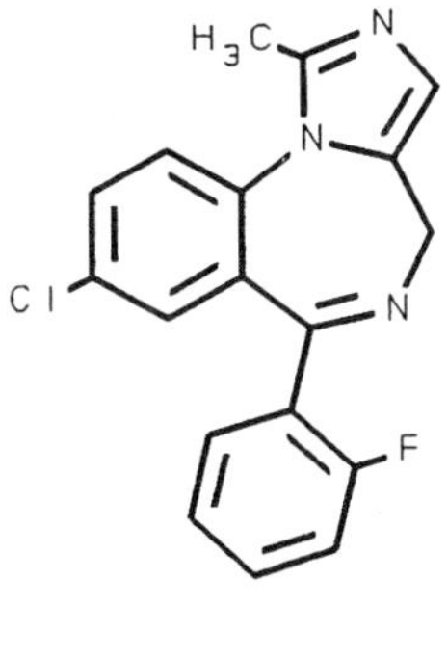 **12** Midazolam

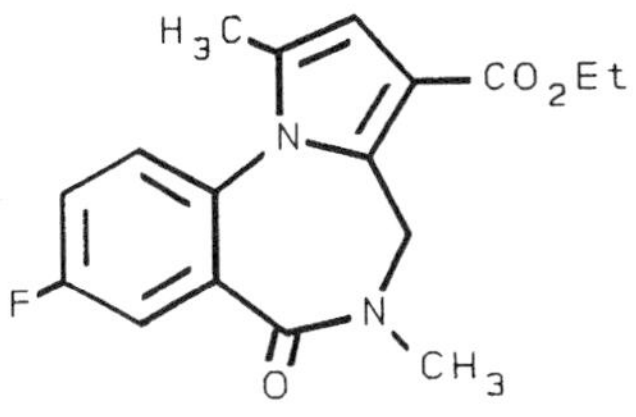

**13** Flumezanil

typical of the classical benzodiazepine led to flumezanil **(13)**; this compound is actually a benzodiazepine antagonist with potential utility in treating drug overdoses.

## 3 DRUG TARGET BASED DESIGN

### 3.1 Structures of Drug Receptors: New Tools in Drug Design

Many drugs owe their effect to the binding of drug molecules to specific structures on cells called receptors or, alternatively, to enzymes. Detailed knowledge of the three-dimensional structure of a receptor should in theory allow the design of better drugs by optimizing the fit of the drug molecule to its receptor through appropriate structural modification. Increased understanding of the molecular biology and biochemistry associated with cell function today often allows the identification and isolation of enzymes or cell receptors that are presumed drug targets.

Detailed three-dimensional chemical structures for many such receptors and enzymes are now available from data provided by X-ray crystallographic studies combined with computerized molecular modeling programs. Models derived from those studies have in several cases provided detailed structures of the drug–receptor interaction site. The use of this potentially powerful tool for designing de novo molecules that will fit those sites, and thus, ideally, new drugs, is being pursued quite intensively.

### 3.2 Computer Modeling

Many biopolymers that form receptors cannot be obtained in a form suitable for detailed structure determination. Information on those receptor sites must be gleaned by using an alternative strategy. Computer modeling of overlays of drugs that are known to interact with a common receptor will often reveal common structural features among the various compounds. It is assumed that those shared groups define interaction sites with the putative receptor. New compounds can then be designed which will position the critical functions at the interaction sites. This approach to drug design is also the focus of much current attention.

*See also* Biotransformations of Drugs and Chemicals; Medicinal Chemistry.

## *Bibliography*

Albert, A. (1985) *Selective Toxicity.* Chapman and Hall, New York.

Hansch, C., Sammes, P. G., and Taylor, J. B. (1990) *Comprehensive Medicinal Chemistry.* Pergamon Press, New York.

Lednicer, D., and Mitscher, L. A. Vol. 1 (1977), Vol. 2, (1980), Vol. 3 (1984), Vol. 4 (1990) *Organic Chemistry of Drug Synthesis.* Wiley, New York.

Wilson, C. O., Gisvold O., Delgado, J. N., and Remers, W. A. (1991) *Wilson and Gisvold's Textbook of Organic and Pharmaceutical Chemistry.* Lippincott, Philadelphia.

Wolf, M., Ed. (1980) *Burger's Medicinal Chemistry,* 4th. Ed. Wiley, New York.

# Drug Targeting and Delivery, Molecular Principles of

*Alfred Stracher*

## *Key Words*

**Bioavailability** That portion of an administered drug which becomes available to a cell for biological action.

**Drug Delivery** The means, either chemical or mechanical, by which drugs are sent to their intended destination.

**Drug Targeting** The use of carrier systems to deliver a drug to its intended area of action.

**Epitope** An antigenic determinant or a group recognized by an antibody.

**Liposomes** Microparticulate colloidal systems that are capable of encapsulating a variety of drugs.

**Monoclonal Antibodies** Antibodies synthesized by a population of identical antibody-producing cells (clones).

**Parenteral** Describing the administration of a drug by means other than by way of the intestines, such as intramuscular, intravenous, or subcutaneous.

**Permeation Enhancers** Small molecules, usually lipophilic, that enhance the penetrability of proteins or polypeptides across mucosal membranes.

**Polymers** Large molecules composed of individual monomers linked in long chains and held together by covalent or noncovalent forces. Can be either naturally occurring (e.g., starch) or synthetic (e.g., nylon).

**Toxins** A colloidal proteinaceous poisonous substance that is produced by a lower organism; usually very unstable and toxic to tissue.

---

Conventional drug delivery technology, which in the past has concentrated on improvements in mechanical devices such as implants or pumps to achieve more sustained release of drugs, is now advancing on a molecular level. Recombinant technology has produced a variety of new potential therapeutics in the form of peptides and proteins, and these successes have spurred the search for newer delivery and targeting methods. The development of new polymeric materials, the discovery of monoclonal antibodies, and the achievement of greater knowledge of cellular receptors have resulted in new attempts to deliver and target drugs via conventional as well as unconventional routes of administration. Delivery and/or targeting across the blood–brain barrier remains an area of intense interest. Microencapsulation of drugs within biodegradable polymers and liposomes has already achieved nominal successes in improving the pharmacodynamics of a variety of drugs such as antibiotics and chemotherapeutic agents. Challenges in the delivery of large molecules such as proteins and genes lie ahead, but already some progress has been made in these areas as well. Although drug delivery technology is still in its infancy, and drug targeting even more so, the burgeoning interest in this field provides ample evidence that further progress is forthcoming.

## 1 INTRODUCTION

### 1.1 General Principles of Drug Targeting/Delivery

When a pharmaceutical agent is administered to a patient, either orally or by injection, the drug distributes itself in most of the whole-body water and tissues, while only a small portion of the administered dose goes to the diseased area where it is expected to have its curative effect. Not only is this wasteful of expensive drugs, since larger doses have to be given to ensure an effect in only a part of the body, but the remainder of the medication, now in general circulation, can produce severe undesirable effects in organs for which it was not intended. Thus, the means by which a drug reaches its target site takes on increasing significance, as well as its delivery at the right moment and frequency.

Recent developments have fueled an increased intensity in research aimed at creating new drug delivery systems. Much of this interest has stemmed from the new advances in biotechnology and immunology, which have resulted in the creation of a new class of peptide and protein drugs. Concurrent attempts to overcome the barriers that limit the availability of these macromolecules has led to an exploration of nonparenteral routes, with the goals of achieving systemic delivery as well as finding means to overcome the enzymatic and absorption barriers for the purpose of increasing their bioavailability. Although for conventional drugs the oral route has been the most convenient and popular, most peptide and protein drugs have low oral uptake as a result of proteolytic degradation in the gastrointestinal tract and poor permeability of the intestinal mucosa to high molecular weight substances. Several approaches to overcome these obstacles are now under intense investigation: (1) inhibiting proteolytic degradation, (2) increasing the permeability across the relevant membrane, (3) achieving structural modification to improve resistance to breakdown or to enhance permeability, and (4) creating specific pharmaceutic formulation to prolong the retention time of drugs at the site of administration using so-called controlled-delivery systems.

### 1.2 Controlled-Release Systems

A number of combinations and variations on the foregoing themes have also been investigated. These include linkage of drugs to monoclonal antibodies (Mabs), encapsulation of drugs within liposomes, modification of the liposome surface to alter the pharmacokinetics, coating of proteins and/or liposomes with polymers or polysaccharides, and fusion of toxins to antibodies via recombinant technology. All these modifications are designed to accelerate and control the transport of pharmacologically active agents from sites of administration to organs and tissues by increasing residence time, bioavailability, and penetrability.

### 1.3 Site-Specific Delivery (Targeting)

The alterations in drug structure just listed are not limited entirely to enhancing the stability of the drug but are also designed to improve the targeting of the drug to a specific organ or tissue. By taking advantage of a feature on the cell membrane that becomes a focal point for incorporating a specific carrier into the design of the drug to carry it to its designated goal, targeting or site-specific delivery can be improved. The carriers generally utilized to the present have been monoclonal antibodies to specific cell membrane epitopes or receptors. A greater understanding of membrane-specific features, however, might enable one to design small molecular carriers attached to drugs for enhanced uptake.

Thus, new drugs in the form of peptides, proteins, oligonucleotides, and genes are now on the horizon. Our limitations at this juncture may be summarized by a question: How do we deliver the drugs, intact, to preferred sites within the cell, to achieve maximal physiologic effectiveness?

## 2 GENERAL METHODOLOGY

### 2.1 Routes of Administration

Classically, drugs have been administered orally or parenterally, with the oral route being the most convenient and popular. To achieve maximal effectiveness, most of the new proteinaceous therapeutics have been delivered parenterally: intramuscularly, intravenously, or subcutaneously. Low patient compliance for long-term treatments, however, has led to investigation of nonparenteral routes for the systemic delivery of peptide and protein drugs. Thus, there have been studies of administration by other routes (nasal, buccal, rectal, ocular, pulmonary, vaginal), using a variety of so-called permeation enhancers, to increase bioavailability. The major barriers to be overcome are significant proteolytic activities and the presence of various epithelia at different locations. Since the skin is relatively impermeable, the nasal route appears to be the most attractive alternative to parenteral administration for the delivery of polypeptides, such as insulin.

The epithelial membranes, which cover the organ tissues that are used in nonparenteral drug administration, constitute a highly efficient barrier to drug absorption. Drugs must penetrate the barrier by active transport, vesicle transport, or concentration-dependent diffusion. Alternatively, the drug must navigate through the tight junctions between cells. Studies have indicated that the pore radius cannot accommodate molecules of a size greater than a tripeptide. Absorption enhancers have been used to increase pore size, however, enabling molecules the size of insulin to be transported across the mucosal membrane. Some commonly used enhancers are sodium deoxycholate, sodium glycocholate, dimethyl-β-cyclodextrin, lauroyl-1-lysophosphatidylcholine, to mention but a few. In clinical applications, however, chronic use of these additives has been accompanied by changes in the mucosal surface, and new enhancers are being evaluated to overcome such difficulties. The mechanism by which enhancers exert their effect is not well known, although it is thought that they function by disrupting the ordered membrane phospholipid domain, thereby lowering barrier function and increasing permeability. Most effective enhancers have structural similarities to the phospholipid domains of the mucosal membrane.

### 2.2 Delivery and Targeting of Drugs

At the same time that new routes of administration are under investigation, other studies to deliver drugs by conventional routes, in a more selective manner, continue. Still dominating current delivery methodology are mechanical methods (pumps, patches, osmotic devices, etc.) to improve release over longer periods, as well as the older "sustained release" technology, which includes formulating drugs in suspensions, emulsions, slowly dissolving coatings, and compressed pills. As a result of the development of new biomaterials, as well as new advances in molecular biology, immunology, and membrane biology, however, novel approaches to drug delivery and targeting are already having an impact on the pharmaceutical industry and will continue to do so in the years ahead. The most advanced of these methods incorporate the use of liposomes, polymers, and monoclonal antibodies.

Liposomes, which are capable of encapsulating a variety of drugs and delivering them via the circulatory system, are microparticulate colloidal systems composed of lipids. They are the most widely studied vesicles to date, and they can be formulated with a variety of lipid types and compositions, which can alter their stability, pharmacokinetics and biodistribution. A major disadvantage of liposomes as delivery systems is their size, which prevents them from crossing most normal membrane barriers and limits their administration to the intravenous route. In addition, their tissue selectivity is limited to the reticuloendothelial cells, which recognize them as foreign microparticulates and are concentrated in tissues such as the liver and spleen. Nevertheless, their biodistribution can be altered to some degree by attaching various ligands to the surface, such as monoclonal antibodies. Other modifications, such as the simultaneous attachment of polyethylene glycol and monoclonal antibodies, the so-called stealth liposomes, have led to improved circulation time as well as improved site-directed delivery. More recently, so-called cationic liposomes, constructed from positively charged lipids or by the attachment of cationic ligands to the surface, have been used for constructing DNA/liposome ternary complexes for the purpose of delivering genes, leading to increased transfection of cells for purposes of gene therapy.

Polymers have also been used as drug delivery systems. They generally release drugs by (1) polymeric degradation or chemical cleavage of the drug from the polymer, (2) swelling of the polymer to release drug trapped within the polymeric chains, (3) osmotic pressure effects, which create pores that release a drug, which is dispersed within a polymeric network, and (4) simple diffusion of the drug from within the polymeric matrix to the surrounding medium.

Polyesters are the most widely studied biodegradable synthetic polymers for drug delivery. They degrade by hydrolysis of the ester bonds to form nontoxic organic alcohols and acids. Because of their low toxicity and biocompatibility, polyesters of lactic acid and glycolic acid are the most commonly used materials. They are now being used for the slow release of large molecules such as proteins, polysaccharides, and oligonucleotides. Release usually occurs by one of the mechanisms just enumerated. Release rates are affected by polymer size and composition as well as by macromolecule size, solubility, and stability. Polymer systems are now being used to deliver many large molecules, including insulin, growth factors, and oligonucleotides, as well as smaller drugs such as anticancer agents, and steroids.

Monoclonal antibodies have offered the greatest potential for selective targeting to specific cells. Thus, bioactive agents such as drugs, radioisotopes, or toxins have been chemically linked to the desired antibody with the intent of targeting these agents to selective cells for treatment. Toxins such as ricin have been used in this manner in cancer chemotherapy in the hope that the toxin will accumulate at the tumor site and kill the cancer cells selectively without damaging surrounding tissue. However, this technology is limited by the low availability of antibodies selective for a given tumor cell and by the changing surface configuration of the tumor cell.

Nevertheless this attractive approach is still under intense investigation. A number of variations on this theme have been used as well; for example, monoclonal antibodies have been attached to liposomes in an attempt to target these vesicles to a desired site. Additional attachment of polyethylene glycol to MAb-modified liposomes has given rise to the term ''stealth'' liposomes, which are more stable and have longer circulation times than unmodified forms. It is likely that these types of modification will achieve increasing clinical applicability in the near term.

*See also* Antibody Molecules, Genetic Engineering of; Biotransformations of Drugs and Chemicals; Drug Synthesis; Medicinal Chemistry; Molecular Genetic Medicine.

## *Bibliography*

Gregoriadis, G. (1989) *Targeting of Drugs: Implications in Medicine.* Wiley, New York.

Langer, R. (1990) New method of drug delivery. *Science,* 249:1527–1533.

Pardridge, W. M. (1991) *Peptide Drug Delivery to the Brain.* Raven Press, New York.

## *E. coli,* Heat Shock Response in: *see* Heat Shock Response in *E. coli.*

# *E. coli* GENOME

*Hirotada Mori and Takashi Yura*

*Key Words*

**DNA Databank** Databank for experimentally determined DNA sequences.

**Genetic Map** Linear or circular diagram showing the relative position of genes in a chromosome.

**Genome** The total complement of genetic material in a cell or individual.

**Protein Databank** Databank for amino acid sequences of proteins determined experimentally or deduced from DNA sequences.

*Escherichia coli* is one of the best studied organisms and has contributed extensively to understanding the fundamentals of molecular biology. *E. coli* strain K12 is most widely used as a host–vector (EK) system for recombinant DNA and other experiments. Thus knowledge of the complete nucleotide sequence of its genome is important not only to gain insights into the structural and functional organization and evolution of the genome, but to make the best use of *E. coli* as an experimental system.

## 1 THE *E. coli* MOLECULE

The chromosome of *E. coli* is a single circular DNA molecule that consists of about 4700 kilobases (kb), in which some 1500 genes have been identified. The number of entries in databanks in the United States, Europe, and Japan of the *E. coli* DNA sequence is about 2300, amounting to 4000 kb (January 1993); this includes data from diverse strains and mutants. When all repeats and overlaps are eliminated from the total entries specifically for *E. coli* K12, the actual segments of chromosomal DNA that have been sequenced account for more than 40% of the genome (Figure 1).

## 2 MAPPING THE GENOME

The construction of a physical map of an entire genome depends on the availability of ordered clones of chromosomal segments in the form of recombinant plasmids or phages. The current physical map of the *E. coli* genome was established particularly by such clones of lambda phage (''Kohara clones'') that carry the overlapping DNA segments covering virtually the entire genome. This makes the *E. coli* genome an ideal subject for systematic sequencing. Such a ''genome project'' consists of three phases of operation: (1) compilation of DNA sequence data currently available at the databank, (2) systematic DNA sequencing, and (3) theoretical and experimental analyses of DNA sequence data.

### 2.1 Compilation of DNA Sequence Data

The genetic linkage map of *E. coli* K12 with all known genes, together with information on gene products and comprehensive literature, has been compiled and updated by continuing efforts of the *E. coli* Genetic Stock Center. Attempts to incorporate the genetic data into the DNA sequence data from databanks have been made, and this result along with the *E. coli* protein index are open to the public through the European Molecular Biology Library as ''ECD'' (*E. coli* database). Novel algorithms were developed to correlate and integrate the *E. coli* genetic map with the physical map, including DNA sequence data, thus integrating adjacent DNA sequences into contiguous structures and eliminating redundancies. Furthermore, a relational database, GeneScape, became available on personal computers (Macintosh) to permit easy inspection and manipulation of genomic DNA sequences.

### 2.2 Systematic DNA Sequencing

With recent advances in sequencing strategies and technical improvements including the use of automatic sequencers and the polymerase chain reaction (PCR), goals of determining complete genome sequences have been set for bacteria and lower eukaryotes. In *E. coli,* a set of 476 lambda recombinant phage clones that covers more than 99% of the genome provides most useful material for such endeavors. Each clone carries a DNA fragment of 15 to 20 kb, which was prepared by digesting DNA with *Sau*3AI or *Eco*RI, and alignment was achieved by analyzing restriction fragments obtained by each of eight restriction enzymes (Figure 2). These and other recombinant clones have been used for systematic sequencing of the *E. coli* genome. Starting from 0 minutes of the genetic map, sequences for 111.4 kb (0–2.4 min) have been determined (see Figure 2). Other regions that have been sequenced include the 84.5–86.5 minute region (near the chromosomal replication origin, *ori C*) of 91.4 kb and a 150 kb sequence around the replication terminus *(ter C)*. The map of about 90 genes and putative genes predicted from the contiguous sequence data for the 0–2.4 minute region is shown in Figure 3.

### 2.3 Analysis of Sequence Data

In general, novel nucleotide sequences are first subjected to analysis for open reading frames (ORFs) with searches in three phases and for both orientations. Possible ORFs thus obtained are analyzed

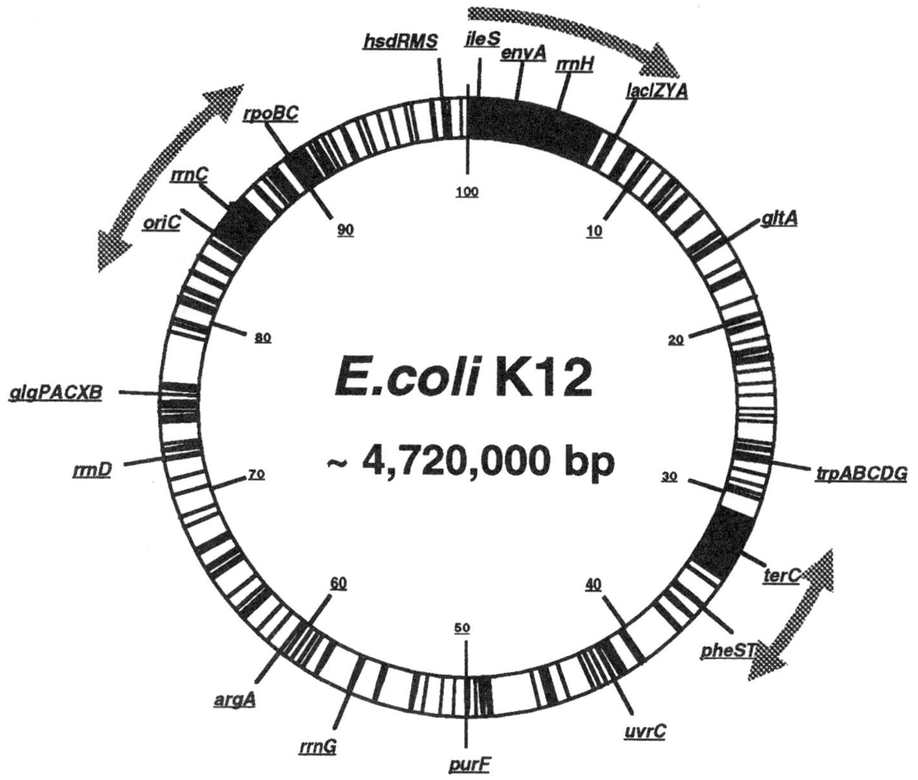


Figure 1. Sequenced areas of the *E. coli* K12 chromosome. Arrows indicate positions and direction of systematic sequencing projects currently in progress. Black bars on the genetic map indicate regions that have been sequenced.

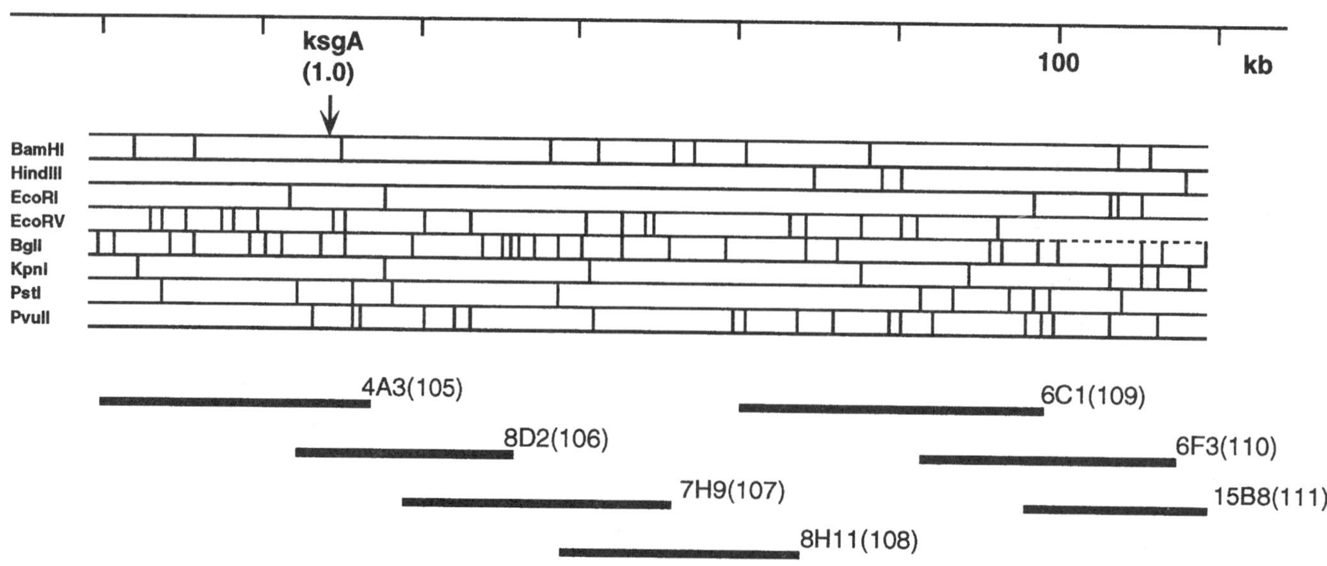


Figure 2. Portion of physical map of *E. coli* K12 and Kohara clones. The top bar shows scales expressed in kilobase coordinates; the position of *thr A* at 0 minutes on the genetic map is taken as 0 kb. Horizontal bars under the restriction map represent individual clones.

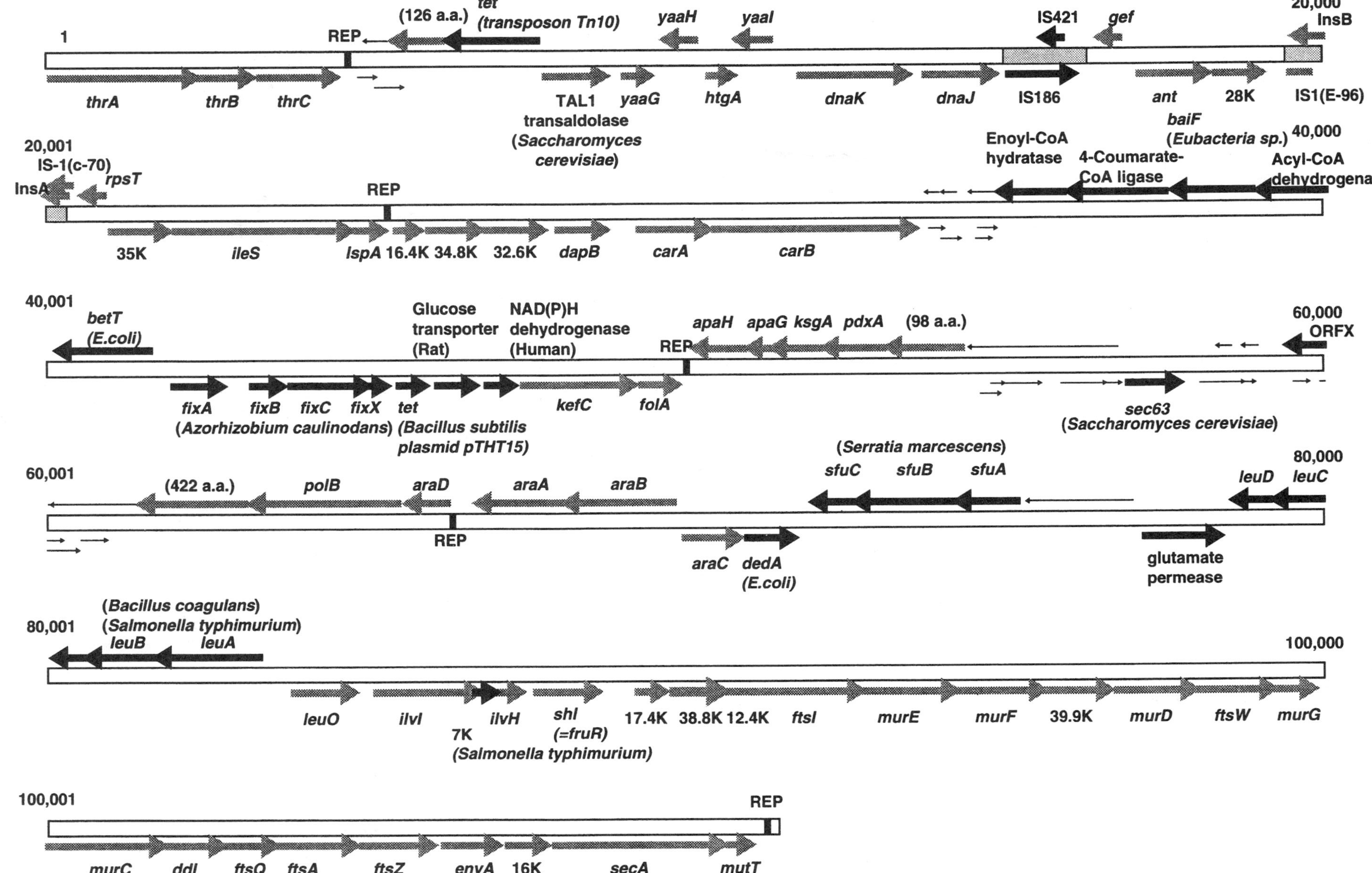


Figure 3. Physical map of ORFs on the 0–2.4 minute region of the *E. coli* chromosome: thick, black arrows, putative genes predicted by homology analysis; thin black arrows, uncharacterized ORFs; lighter arrows, previously sequenced genes.

for sequence similarity against a large collection of protein sequence data (e.g., the database of the Protein Identification Resource in the United States). Through such analyses, the possible nature or functions of putative ORFs can be predicted and the results can then be utilized for further investigation, either theoretical or experimental. For example, analysis of the 0–2.4 minute region led to the prediction of the occurrence of several novel clusters of genes similar to genes of other organisms that are involved in fatty acid metabolism, iron transport, and symbiotic nitrogen fixation (Figure 3). In view of the tight arrangements of genes on the chromosome with little spacing between, it seems that about 4000 genes may constitute the genome of *E. coli*.

Extensive analyses of *E. coli* chromosomal DNA, combined with currently available sequence data, revealed some interesting features, such as repeated sequences, gene families, functional organization of genes, and clues to the evolution of the genome. It is anticipated that an increasing number of important novel features that are essential for deeper understanding of the genome will be unraveled as contiguous sequence data accumulate.

*See also* BACTERIAL GROWTH AND DIVISION; EXPRESSION SYSTEMS FOR DNA PROCESSES; HEAT SHOCK RESPONSE IN *E. COLI*; NUCLEIC ACID SEQUENCING TECHNIQUES.

## *Bibliography*

Bouffard, G., Ostell, J., and Rudd, K. E. (1992) *Comput. Appl. Biosci.* 8:563–567.

Daniels, D. L., Plunkett III, G., Burland, V., and Blattner, F. R. (1992) *Science,* 257:771–778.

Drlica, K., and Riley, M., Eds. (1990) *The Bacterial Chromosome.* American Society for Microbiology, Washington, DC.

Kröger, M., Wahl, R., Schachtel, G., and Rice, P. (1992) *Nucleic Acids Res.* 20:2119–2144 (Suppl.).

Neidhardt, F. C., Ingram, J. L, Low, B. K., Magasanik, B., Schaechter, M., and Umbarger, H. E., Eds. (1987) Escherichia coli *and* Salmonella typhimurium: *Cellular and Molecular Biology.* American Society for Microbiology, Washington, DC.

Yura, T., Mori, H., Nagai, H., Nagata, T., Ishihama, A., Fujita, N., Isono, K., Mizobuchi, K., and Nakata, A. (1992) *Nucleic Acids Res.* 20:3305–3308.

# ELECTRIC AND MAGNETIC FIELD RECEPTION

*Eberhard Neumann*

## *Key Words*

**Chemical Electric and Magnetic Field Effects** The consequences of the primary field effects on ions and dipoles, leading to the dependencies of the rate and equilibrium constants of reactive chemical processes and phase transitions on the electric field strength and, to a weaker extent, on the magnetic flux density.

**Cooperativity** A mechanism for the simultaneous state transition of a cluster or domain of strongly coupled molecules in a narrow field strength range.

**Electric and Magnetic Field Strength** Measures of the force on electric charges and electric and magnetic dipoles, respectively. The targets are either freely mobile or fixed ionic and magnetic groups of larger macromolecules.

**Interfacial Electric Polarization** Field-induced ion accumulation at interfaces of media with different dielectric constants, leading to ionic charging of membranes and to large induced transmembrane voltages by small external fields.

**Magnetic Induction of Electric Fields** The main effect of time-varying magnetic fields, causing ionic (ac) currents and dipolar displacement currents, both of which interfere with chemical reactivity.

**Magnetosomes** Small elongated magnetite ($Fe_3O_4$) organelles which in magnetobacteria serve for geomagnetic field sensing and direction finding. Also found in cells of higher organisms.

Electric ($E$) and magnetic ($B$) field reception comprise the mutual interaction of external fields with the relatively high internal fields of the ionic and the (electric and magnetic) dipolar components inclusively the cellular magnetosomes, involved in sensory processes. Physicochemically, fields primarily act on charges, changing their positions and motional states. As a consequence, the extent and the rate of chemical reactions, conformational changes, and phase transitions of the field-receiving molecules are dependent on the strength of the $E$ and $B$ fields. Although there are *always* finite chemical shifts in rate and equilibrium constants, weak external fields require amplification processes—for instance, interfacial polarization of larger particles and cooperativity in domain structures.

The natural $E$ fields are particularly apparent in the electrophysiological function of nerves. Man-made electric and magnetic implements have become indispensable tools of daily life, although it is not yet clear whether power transmission lines and household devices bear potential health risks even under normal safety conditions. To estimate beneficial and negative field effects, it is necessary to understand the physical and chemical interaction mechanisms of electric and magnetic field reception.

## 1 CONCEPTS

### 1.1 NATURAL FIELDS

All life processes have evolved in the natural electric and magnetic force fields of the earth. Moreover, even in its lowest forms, life makes use of electric and magnetic field effects, most obviously in the rapid signal transmission on the microscopic level of neurons, involving in particular the gated transport of $Na^+$, $K^+$, $Ca^{2+}$ and $Cl^-$ ions.

Generally, electric and magnetic fields act as forces on charges: electrolyte ions and fixed ionic groups as well as free dipolar molecules and fixed dipolar groups of macromolecules, including membrane lipids and cellular magnetosomes. As a consequence of these primary field effects, reactive chemical interactions and phase transitions are dependent on the electric field strength $E$ (V/m) and, to a much weaker extent, on the magnetic field strength $H$ (A/m).

### 1.2 Definition of Field Reception

In molecular biology, the term "field reception" appears to be meaningful only if it incorporates more than just the direct interactions of the force fields (and their spatial gradients) with molecules. Direct field interaction primarily causes changes in the positional and motional states of the targets. An adequate definition of field reception must specify at least the next biochemical or biophysical process that follows the direct force effects. Thus field reception indispensably includes field effects on the reactive interactions, the molecular orientations, and the transport processes of the "field-receiving" ions and electric and magnetic dipoles.

Briefly, the effects of external $E$ and $B$ fields may be summarized as a cascade of mutual interaction steps:

$$(E,B)_{\text{ext}} \leftrightarrow (E,B)_{\text{int}} \leftrightarrow \begin{Bmatrix} \Delta c, \Delta\phi_m \\ \Delta K/K_0, \Delta\beta/\beta_0 \\ \Delta k/k_0 \end{Bmatrix}$$

The primary step in field reception (and emission) is the mutual interference of the external $E$ and $B$ fields, $(E,B)_{\text{ext}}$, with the local (body and cell) internal fields $(E,B)_{\text{int}}$, provided the external static and low frequency $E$ fields can penetrate the systems—for instance, via direct (metal) electrode contact. The subsequent steps may occur in parallel:

1. Mutual changes in electrolyte flows (resulting in local ion concentration changes $\Delta c$) and in the internal fields, including the relatively very strong $E$ fields of the cell membranes ($E_m \approx 10^3$–$10^4$ kV/m) corresponding to electric membrane potential differences in the order of $|\Delta\phi_m| \approx 0.01$ to 0.2 V.
2. Field-induced changes of ligand binding processes, of conformational transitions, and of domain phase transitions—for instance, in membranes. Theoretically the field-induced changes may be quantified as relative changes ($\Delta K/K_0$, $\Delta\beta/\beta_0$, and $\Delta k/k_0$) in the equilibrium constants ($K$), in the extents of ligand binding and of transitions ($\beta$), and in the rate coefficients ($k$).

## 2 EQUILIBRIUM AND RATE CONSTANTS

### 2.1 Dipole Reactions and Transitions

Field effects on chemical processes involving dipolar educts $J_r$ and products $J_p$ according to

$$\Sigma_r|\nu_r|J_r \underset{\overleftarrow{k}}{\overset{\overrightarrow{k}}{\rightleftharpoons}} \Sigma_p\nu_p J_p \tag{1}$$

are characterized by the molar reaction dipole moment $\Delta M = M_p - M_r$ as the difference between the orientational averages of the molar (electric and magnetic) dipole moments of $J_r$ and $J_p$ and by the stoichiometric coefficients of the reactants $\nu_r$ and products $\nu_p$.

Introducing now $F_{(E,B)} = F_E = |\mathbf{E}|$ or, in the other case, $F_{(E,B)} = F_B = |\mathbf{B}|$ for the amount of the electric field strength vector $\mathbf{E}$ and that of the magnetic flux density vector $\mathbf{B}$, respectively, the general expression of the (isothermal, isobaric) field-induced change in the equilibrium concentration ratio or equilibrium constant $K$ is given by

$$\left(\frac{\partial \ln|K|}{\partial F_{(E,B)}}\right)_{p,T} = \frac{\Delta M}{RT} \tag{2}$$

where $R = N_A k_B$, with $N_A$ the Avogadro constant and $k_B$ the Boltzmann constant, and $T$ is the thermodynamic (Kelvin) temperature. For $F = E$, $\Delta M$ is the electric reaction moment; for $F = B$, $\Delta M$ refers to the molar difference in the magnetic moments. Note that $K = \Pi_p c_p^{\nu_p}/\Pi_r c_r^{|\nu_r|} = \overrightarrow{k}/\overleftarrow{k}$, where $\overrightarrow{k}$ and $\overleftarrow{k}$ are the rate constants for the forward and reverse reaction steps, respectively, and $c_p$ and $c_r$ are concentrations of the products and reactants, respectively.

Integration of equation (2) yields

$$K = K_0\, e^X \tag{3}$$

where $K_0$ is the $K$ value at zero field. The "field effect exponent" $X$ is given by the ratio of the electric or magnetic free energy $\Delta\hat{G}_{(E,B)}$, respectively, to the thermal energy ($RT$):

$$X = \frac{-\Delta\hat{G}_{(E,B)}}{RT} = \frac{\int \Delta M dF_{(E,B)}}{RT} \tag{4}$$

The integration boundaries are $F_{(E,B)}$ and $F_{(E,B)} = 0$, respectively.

### 2.2 The Weak-Field Approximation

If $\Delta\hat{G}_{(E,B)} << RT$, then $X << 1$ and $K = K_0\, e^X$ takes the form $K = K_0(1 + X)$.

For practical purposes a relative change $\Delta K/K_0$ is introduced, which readily permits the expression of field-induced changes in percent.

$$\frac{\Delta K}{K_0} = \frac{K - K_0}{K_0} = e^X - 1 \tag{5}$$

For ($X << 1$), we obtain the weak-field approximation

$$\frac{\Delta K}{K_0} = X \tag{6}$$

In a random ensemble of reaction partners, we have $\Delta M \propto F_{(E,B)}$ and thus, in the weak-field limit, we see that both $X$ and $\Delta K/K_0$ are proportional to the square of $F$ (see Figure 1).

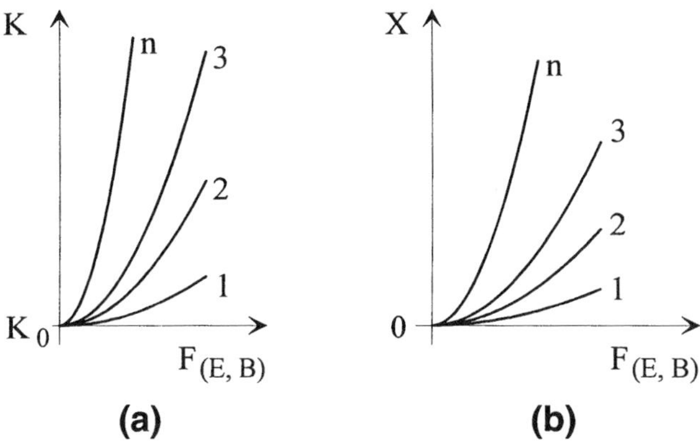


**Figure 1.** Field dependences of (a) the distribution (equilibrium) constant $K = K_0\, e^X$ and (b) the field effect exponent $X = \int \Delta M\, dF_{(E,B)}/(RT) \propto F^2_{(E,B)}$. The steepness terms $(\partial K/\partial F)_{p,T} = K \cdot \Delta M/RT$) and $(\partial X/\partial F_{p,T} = \Delta M/RT$ increase with $\Delta M$ and with the domain size $n$. Since the relative change $\Delta\beta/\beta_0$ in the advancement of a chemical process is proportional to $X \propto F^2_{(E,B)}$ we have $\Delta\beta/\beta_0 \propto F^2_{(E,B)}$. The rate constants $k$ exhibit qualitatively the same field dependences as $K$; that is, $\Delta k/k_0 \propto F^2_{(E,B)}$.

The expressions for the rate coefficients $k$ are qualitatively the same as for the $K$ terms. For instance,

$$k = k_0\, e^{Xa}$$

$$x_a = \frac{\int M_a dF_{(E,B)}}{\mathrm{RT}}$$

where $M_a$ is the activation dipole moment of the reaction step considered. Here, too, for $x_a << 1$, we have the weak-field approximation $\Delta k/k_0 = x_a$. Theory provides the explicit expressions for the elementary reactions such as $LB = L + B$ and $B = B'$ for dipolar and ionic reaction partners: $K = [L][B]/[LB]$ and $K = [B']/[B]$, respectively.

From equations (4) and (6) we see that no matter how small the differences $\langle \Delta M \rangle$ in the reaction dipole moments of the interaction partners, there is *always a finite change* caused by a field. Of course, the inevitable thermal motion ($k_B T$) leads to local activity fluctuations $\delta a_j = y_j \delta c_j + c_j \delta y_j$. The question is how relevant are the small, field-induced changes $\Delta c_j$ and $\Delta y_j$ in concentration and activity coefficient, respectively, compared with the random thermal fluctuations $\pm\ \delta c_j$ and $\pm\ \delta y_j$?

## 3 AMPLIFICATION MECHANISMS

Enhancement mechanisms for both small external fields and locally small chemical shifts ($\Delta K/K_0$, $\Delta k/k_0$) are important features of the molecular characterization of field reception. A very efficient field enhancement is due to interfacial polarization of single cells or organelles, or clusters of cells and tissue species, but the phenomenon also is observed in a string of cells connected by conducting gap junctions (Figure 2). The field amplification factors are geometry and size dependent and may reach enhancement values of $10^3$ to $10^5$.

Another efficient enhancement is encountered when whole domains of many molecules cooperate (i.e., undergo changes simultaneously). If the cooperative domain size $n >> 1$, then $\Delta M_n = n\Delta M$, and thus $X_n$ and $\Delta K(n)/K_0$ are very large (Figure 1).

A particular amplification mode resides directly in the field dependence of equilibrium constants $K$. This enhancement mode appears to be especially important for all *membrane processes* occurring in the strong natural transmembrane fields. From equations (3) and (4) it is seen that the steepness of the $K(X)$ relationship, given by

$$\left(\frac{\partial K}{\partial F_{(E,B)}}\right)_{p,T} = \frac{K(E)\Delta M}{RT} \tag{7}$$

depends not only on $\Delta M$ but also on $K(E)$. The nonlinearity of the $K(X)$ function provides a mechanism not only for amplification but also for asymmetric concentration changes.

(a)

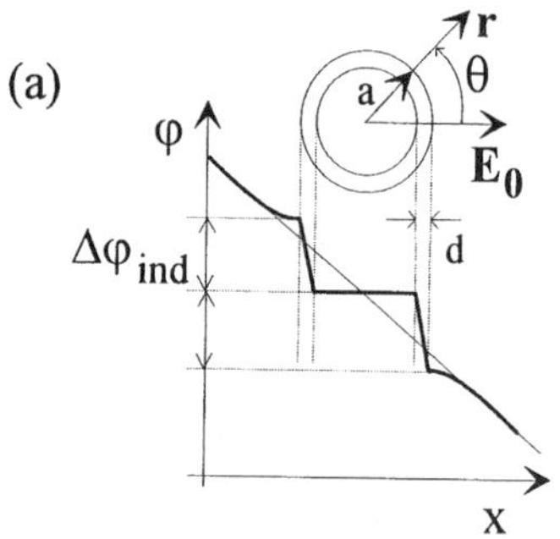


(b)

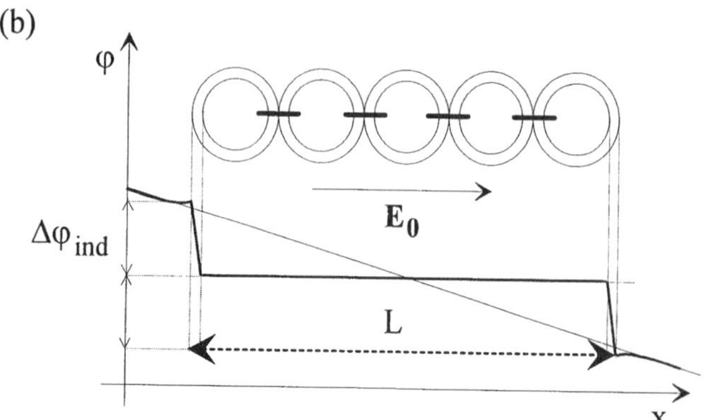


**Figure 2.** Field amplification by interfacial polarization through dc and low frequency **E** fields. Electric potential profile $\varphi(x)$ in the $x$ direction of the external field $\mathbf{E}_0$ vector through the pole caps. (a) Single cell sphere of radius $a$, membrane thickness $d$, and radius vector $\mathbf{r}$; the steep potential drop across the membranes in the direction of $\mathbf{E}_0$ is given by $\Delta\varphi_{\text{ind}} = -1.5\ |\mathbf{E}_0|a|\cos\theta|$ if the conductivity of the membrane is very small compared to the cell inside and outside. The field amplification factor at the pole caps is $f(a) = (E_{\text{ind}}/\mathbf{E}_0) = 1.5\ a/d$. (b) Array of cells connected by conducting gap junctions. The field amplification factor is $f(L) = (E_{\text{ind}}/\mathbf{E}_0) = 0.5\ L/d$.

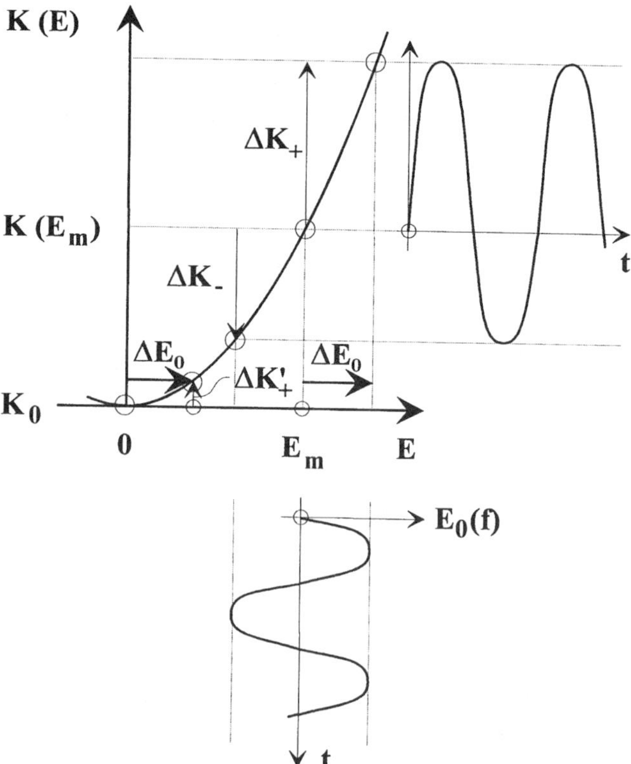


**Figure 3.** Amplification at high field and asymmetric field effect for oscillatory external fields (ac). Application of an external field, inducing $E_{\text{ind}} = \Delta|\mathbf{E}_0|$ at $E = 0$, causes only a small change $\Delta K'_+$. If applied at $E = E_m$ (e.g., at the resting potential $\Delta\varphi \approx -100$ mV), the same external field causes a much larger change $\Delta K_+$ because $\Delta K/\Delta E = K(E)\Delta M/(RT)$ depends on $K(E)$. The asymmetry of the $K(E)$ dependence leads to an asymmetric amplification of $K$: $\Delta K_+ > \Delta K_-$, if the symmetric external field $\mathbf{E}_0(f) = \hat{\mathbf{E}}_0 \cdot \sin(2\pi f \cdot t)$ operates at $\mathbf{E}_m$. At $E = 0$, $\Delta K_+ = \Delta K_-$. In a similar way, the rate constant $k$ is asymmetrically dependent on $\mathbf{E_0}(f)$.

If the external force is an oscillatory (ac) field, say $\mathbf{E}_0(f) = \hat{\mathbf{E}}_0 \sin(2\pi f t)$, where $\hat{\mathbf{E}}_0$ is the peak amplitude, the enforced oscillatory change $\Delta K \pm (f)/K(E_m)$ in the equilibrium constant of a reaction is asymmetric: $\Delta K_+/K(E_m) \neq \Delta K_-/K(E_m)$ (see Figure 3). For instance, the induced concentration changes $\Delta C_+$ are larger for one ac field direction ($+\Delta\mathbf{E}_0$) compared with the $\Delta c_-$ values of the other direction ($-\Delta\mathbf{E}_0$). In a similar way, the change in the forward rate constant is different from that of the reversed rate constant.

The interpretation of experimental data on weak-field effects is still very controversial. The formalism indicated in this short digression may help (1) in the design of experiments that also use as variables, for instance, species and sample size, and (2) in the analysis of data in terms of thermodynamic and kinetic parameters consistent with the first principles of chemical reactivity in electric and magnetic fields.

*See also* FREE RADICALS IN BIOCHEMISTRY AND MEDICINE; SYNAPSES.

### *Bibliography*

Neumann, E. Chemical electric field effects in biological macromolecules. *Prog. Biophys. Mol. Biol.* 47:197–231 (1986).

Polk, C., and Postow, E., Eds. *Handbook of Biological Effects of Electromagnetic Fields.* CRC Press, Boca Raton, FL, 1986, 503 pp.

# ELECTRON MICROSCOPY OF BIOMOLECULES

*John Sommerville and Hanswalter Zentgraf*

### *Key Words*

**Cryoelectron Microscopy** Imaging of unfixed, unstained biomolecules in a hydrated state after rapid freezing to low temperature (−160°C).

**Heavy Metal Shadowing** Evaporation of a film of heavy metal (e.g., platinum–palladium or tungsten–tantalum) onto a dehydrated preparation. The deposit of metal particles around the biomolecules improves contrast and gives a shadowed, three-dimensional appearance.

**High Resolution Autoradiography (EM ARG)** Detection of radiolabeled molecules by coating a stained or shadowed preparation with a film of photographic emulsion. Radioactive emissions hit silver halide crystals in the emulsion, which can be developed into silver grains.

**Immunoelectron Microscopy** The application of antibodies to map specific sites on the biomolecules. The antibody molecules may be visible after shadowing or negative staining, although detection can be improved by binding of a secondary antibody linked to an electron-dense tag (e.g., ferritin or colloidal gold particles).

**Miller Spread** Deposition of dispersed chromatin onto a carbon-coated grid by centrifugation through a dense phase containing sucrose and formaldehyde.

**Negative Staining** Instead of staining the biomolecules themselves (positive staining), a solution of heavy metal salt (e.g., uranyl acetate or sodium phosphotungstate) is deposited in the hydrated spaces around and within the molecules.

**Replica Casting** Adsorption of biomolecules to a mica surface followed by heavy metal shadowing and coating of the preparation with a film of carbon. The carbon–metal replica (minus the biomolecules) is then removed from the mica and mounted on an EM grid for viewing.

**Support Film** A thin film of plastic or carbon, which is attached to an EM grid and provides a substrate onto which biomolecules can be adsorbed.

Electron microscopy (EM) is a method appropriate for examining details of the sizes and shapes of biological macromolecules and is particularly useful in studying isolated or reconstituted macromolecular assemblies. Small amounts of material (often less than 1 μg) can be used and prepared in a state suitable for viewing in as little as several minutes. The basic procedure involves adsorption of the biomolecules onto a support film, followed by staining or shadowing of the preparation and viewing of the dehydrated molecules in vacuo in an electron beam. A resolution of 1–2 nm is routine with conventional microscopes. The basic procedure can be adapted to give information about sites of specific epitopes, location of newly synthesized components, internal structures, and atomic composition.

Structures most suitable for EM analysis are nucleoprotein complexes including ribosomes, nucleosomes, and virus particles. DNA and RNA molecules are also suitable, and their lengths can be directly related to the number of base pairs or nucleotide residues determined biochemically. In general, small proteins ($<$ 50,000 Da) are difficult to visualize, but good detail can be obtained if they are isolated as multimeric complexes or can be induced to form filaments or crystalline arrays.

A wide range of applications is available using EM techniques, including virus identification, mapping of hybridized regions in heteroduplexes, detailing of macromolecular interaction, and analysis of the organization of molecular components in replication, transcription, splicing, and translation complexes.

## 1 PRINCIPLES

The transmission electron microscope (TEM) consists of a metal column from which air is evacuated and through which a linear beam of electrons is accelerated and focused by electromagnetic lenses. The biomolecules are adsorbed onto a support film, stained and dehydrated (or frozen in an aqueous film on the grid), and introduced into the electron beam through an air lock. Whereas some of the electrons collide with atoms in the specimen, lose energy, and are scattered, the remaining electrons pass through the preparation and are focused to form an image on a phosphorescent screen (for direct viewing) or on a photographic plate (for later examination). Under ideal conditions, a resolution of 0.1–0.2 nm can be obtained; however limitations are imposed by the naturally low masses of atoms (primarily hydrogen, carbon, and oxygen) contained in biomolecules, by distortions and artifacts created during sample preparation, and by radiation damage. Techniques are designed with the following objectives:

1. To minimize distortion by immobilization of the molecules, in their native state, onto an appropriate support
2. To stabilize molecular complexes by suitable chemical fixation
3. To improve contrast by staining the preparation with heavy metal salt or by shadowing with heavy metal

Irrespective of the method adopted, data derived from EM examination of biomolecules should be entirely consistent with their known biochemical and biophysical properties. Whenever possible, apparent sizes should be checked by independent measurement of sedimentation rate, electrophoretic mobility, or gel filtration elution. Also, features revealed by electron microscopy of isolated molecules should be compared with observations made on them in situ, by EM examination of cell or tissue sections.

## 2 TECHNIQUES

### 2.1 General

Similar general principles of sample preparation (Figure 1) apply for nucleic acids, proteins, and nucleoprotein complexes, since most of these are less than 20 nm thick and can be adsorbed directly from solution onto the support matrix. A suitably high concentration of molecules is required so that several examples can be viewed together in the one field. The efficiency of uptake onto the support is not always predictable, and a range of initial concentrations should be tried. It is important that the biomolecules be held in a solution known to maintain the proper structural features directly prior to applying to the support.

The grid consists of a fine meshwork, usually of copper, which must be coated with a thin support film of plastic (e.g., collodion) or carbon. Carbon films are preferable because they are thin (down to 2 nm) and contribute little to the image. However, they are frequently hydrophobic and should be subjected to ionizing gases (by "glow discharging") or treated chemically (e.g., with Alcian blue 8GX) to render them hydrophilic before use.

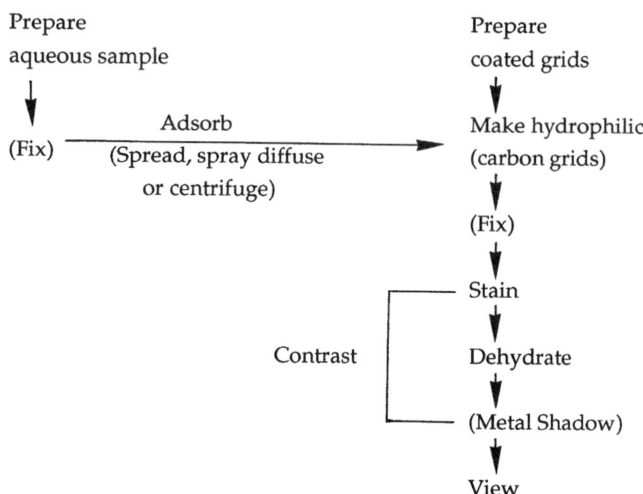


**Figure 1.** Flow diagram of the basic procedure. Steps bracketed are not always used.

### 2.2 Nucleic Acids

Nucleic acids are used at a concentration of 1–5 μg/mL and as little as 1 μg is required. The molecules can be native DNA or RNA, cloned DNA, single-stranded or double-stranded forms, partially denatured duplexes, or denatured and hybridized structures. A key step in their preparation for electron microscopy is the spreading out of the molecules to permit their adsorption in an extended, nonaggregated form on the support film. The most successful method of spreading is that devised originally by Kleinschmidt and Zahn. The negatively charged nucleic acid is mixed with a basic protein (usually cytochrome *c*) in a solution appropriate for maintaining the required conformation. This spreading mix, or hyperphase, is spread, via a glass ramp, onto the surface of a second solution, the hypophase, and a molecular monolayer of nucleic acid and protein is formed at the liquid–air interface.

Alternatively, the molecular monolayer can be formed by diffusion to the surface of the spreading mix contained in a small vessel or even in a droplet sitting on a hydrophobic surface. The nucleic acid–protein complex is adsorbed through brief contact with the support film, stained with ethanolic uranyl acetate, rinsed in ethanol, and air-dried. To improve contrast, the preparation is rotary-shadowed, usually with platinum–palladium, at an angle of 3–10° from the plane of the metal vapor. Metal shadowing greatly exaggerates the thickness of the molecules, but double-stranded and single-stranded regions can still be differentiated (Figure 2A).

The use of cytochrome *c* in spreading obscures fine detail of nucleic acid structure and must be avoided if experimentally bound proteins are being studied. Protein-free spreading methods have been devised to permit detection of proteins bound to specific regions of nucleic acid molecules. These methods employ low molecular weight substances (e.g., benzyldimethylalkylammonium chloride or ethidium bromide) in place of cytochrome *c*. In working with protein molecules bound to nucleic acids, care must be taken to stabilize the complexes by using glutaraldehyde or formaldehyde to cross-link the molecules in solution, prior to spreading. Good detail of the association can be obtained by making a carbon–metal replica of the molecules bound to the surface of freshly cleaved mica. The use of platinum–carbon or tungsten for shadowing reveals finer detail. Complexes formed between *E. coli* RNA polymerase and bacteriophage T7 DNA are shown in Figure 2B.

### 2.3 Proteins

The most commonly used method for contrast enhancement of proteins is negative staining. This extremely rapid method involves drying of an aqueous solution of heavy metal salt (e.g., uranyl acetate) with the protein molecules on the coated grid. The metal salt occupies hydrated regions in and around the biomolecules and, upon drying, forms a dense cast. Since the biomolecules themselves are not stained, but the background is stained, a negative image is created. Negatively stained microtubules are shown in Figure 2C.

An alternative to the adsorption of proteins from droplets is to spray them, in aerosols, onto the surface of carbon-coated grids. The spraying is most easily achieved by touching the tip of a glass capillary tube, containing both sample and stain, into a stream of nitrogen. Elongate, or rod-shaped proteins can be mixed with glycerol (and no stain) and sprayed onto the surface of a piece of freshly cleaved mica. The deposit is then shadowed finely and carbon-coated to produce a replica cast of the protein molecules.

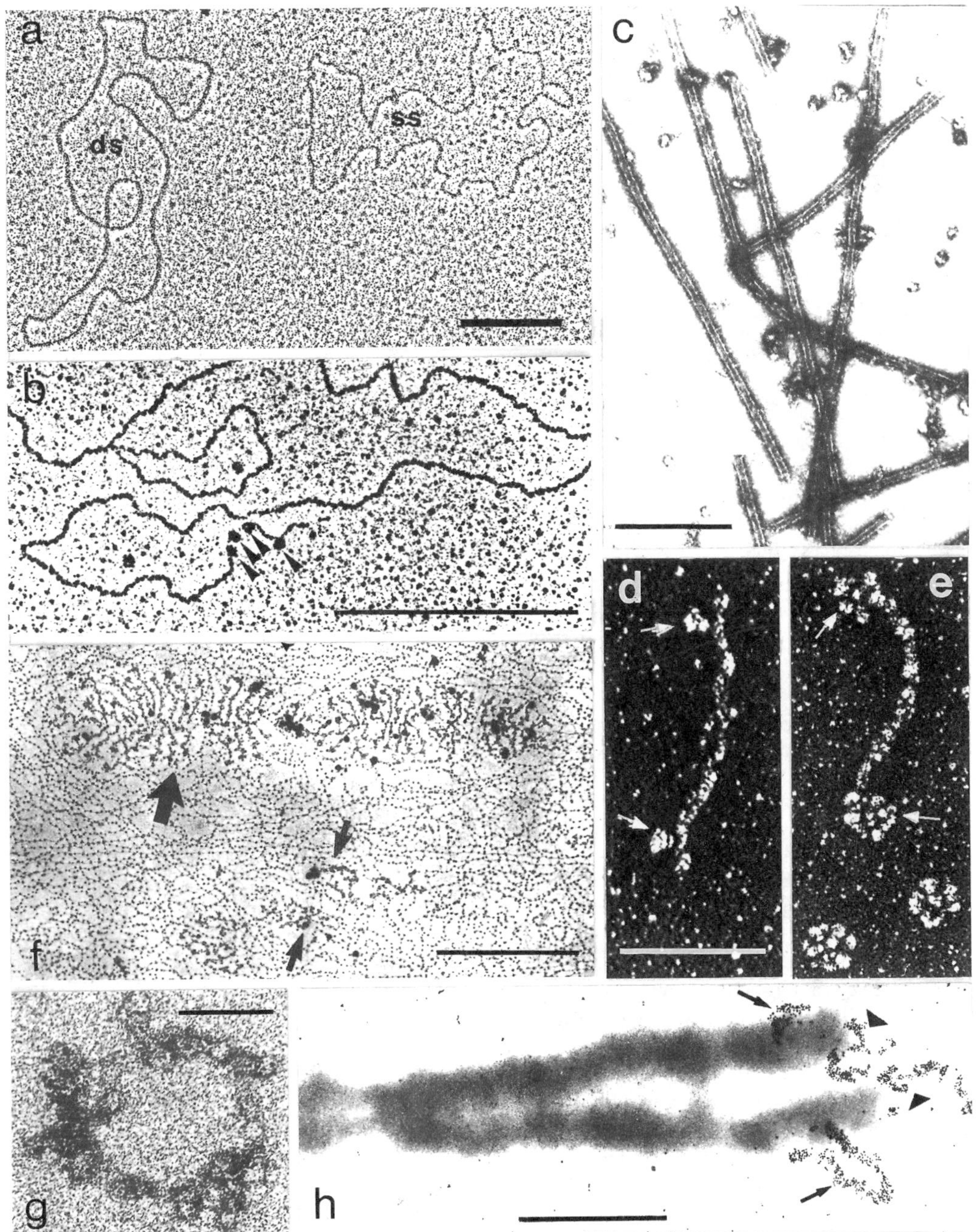


**Figure 2.** Electron micrographs of selected biomolecules. (a) Double-stranded (ds) and single-stranded (ss) DNA of bacteriophage PM2 spread in 40% formamide, 0.01 M EDTA, 0.1 M Tris-HCl, pH 8.5. The preparation was stained with uranyl acetate and then rotary-shadowed with platinum–palladium. Note that dsDNA is wider and has smoother contours than ssDNA. Bar represents 0.25 μm. (Courtesy of L. W. Coggins.) (b) *E. coli* RNA polymerase molecules (arrows) bound to binding sites I–IV at the genetically left end of the T7 DNA molecule. The preparation was adsorbed to mica, fixed in 0.1% glutaraldehyde, and used to make the platinum–carbon replica shown. Bar represents 0.5 μm. (Courtesy of J. Sogo.) (c) Tubulin assembled *in vitro* into microtubules in the presence of microtubule-associated proteins. The preparation was negatively stained with 1% uranyl acetate. Bar represents 0.5 μm. (Courtesy of E. Spiess.) (d) and (e) Molecules of the protein spectrin reacted with monoclonal antibodies. The molecule consists of two heterodimers, which self-associate head to head to give rise to the double-stranded tetramers shown. The antibody molecules (arrows) react with only one of the subunits and are seen bound equidistant from the two ends. Note the structural difference between IgG (d) and IgM (e) molecules. The preparations were sprayed onto mica, fine-shadowed with tantalum–tungsten, and used to make the carbon replicas shown. The micrographs are printed in reverse contrast. Bar represents 100 nm. (Courtesy of J. R. Glenney.) (f) Autoradiograph of chromatin from mouse P815 cells spread by the Miller technique after 5 minutes of incubation with [$^3$H]uridine. Note the dense silver grains lying over the transcription complexes of pre-rRNA (large arrow) and of nonnucleolar RNA (small arrows). The background consists of many strands of nucleosomal chromatin. The preparation was stained with phosphotungstic acid, rotary-shadowed with platinum, and exposed with Ilford L4 emulsion. Bar represents 1 μm. (Courtesy of S. Fakan and J. Fakan.) (g) The SV40 minichromosome showing its nucleosomal configuration. The preparation was spread from a droplet of low ionic strength buffer (2 mM EDTA, 1 mM Tris-HCl, pH 8.4) and negatively stained with 2% uranyl acetate. Bar represents 100 nm. (h) Whole chromosome mount from *Xenopus* culture cell after in situ hybridization with a mixture of biotinylated DNA encloding tRNA and oocyte-specific 5*S* RNA. The chromosomes are partially unfolded and the sites of hybridization are detected using secondary antibodies tagged with colloidal gold. The gold particles are seen to decorate DNA loops containing tRNA genes (arrows) and 5*S* RNA genes (arrowheads). The bar represents 2 μm. (Courtesy of S. Narayanswami and B. A. Hamkalo.) All micrographs printed with permission from Oxford University Press.

Molecules of spectrin have been prepared in this way (Figure 2D, E).

### 2.4 Chromatin

Chromatin spreading techniques are mostly adapted from the procedure devised by Miller and co-workers. Large chromatin units are obtained by lysing cells or nuclei in a low salt concentration, high pH buffer (often referred to as ''pH 9 water''). The chromatin is left to disperse and is then centrifuged at low speed through a denser solution containing fixative (formaldehyde) onto the surface of a carbon-coated grid. The grid is dipped into a solution of a commercial wetting agent (Kodak Photoflo 200) and air-dried. The preparations can be positively stained (with ethanolic phosphotungstic acid), negatively stained (with aqueous uranyl acetate), or metal-shadowed. Chromatin spread from mouse cells shows nucleosomal chromatin, a nucleolar transcription unit, and nonnucleolar transcription complexes (Figure 2F).

For small chromatin units, such as chromatin fragments and viral chromatin (minichromosomes), high speed centrifugation is required to deposit the chromatin onto the coated grid. Alternatively, small chromatin units can be adsorbed directly by floating the grid, carbon film down, on the surface of a droplet containing the sample. The nucleosomal configuration of simian virus 40 chromatin (the SV40 minichromosome) is shown in Figure 2G.

### 2.5 Macromolecular Assemblies

Multimeric enzyme complexes, ribosomes, viruses, and other particles can be treated in a way similar to that described for protein filaments or for small chromatin units. Particles in the size range of 10–200 nm diameter are excellent targets for negative staining, which often reveals considerable detail of surface structure. Nevertheless, the size of the particles of dried stain (or of the metal particles after shadowing) limits the resolution of structural detail. A superior technique, cryoelectron microscopy, involves the rapid freezing of a thin (100 nm) aqueous film containing the specimen particles. Below −143°C, water can be held in a vitrified state, which resembles the liquid state and avoids formation of ice crystals. In this condition macromolecules and macromolecular assemblies can be viewed free of artifacts from fixation, dehydration, and staining. Underfocus phase contrast is used to produce a high resolution image.

## 3 APPLICATIONS

### 3.1 Use of Antibodies

Antibody molecules (IgG and IgM) are large enough to be resolved in metal-shadowed or negatively stained preparations (see Figure 2D,E). Monoclonal antibodies can be used to map sites on nucleic acid or protein molecules and to identify individual components within macromolecular assembles. Antibodies can also be used to build up denser structures around small proteins that are themselves difficult to resolve. Secondary antibodies tagged with electron-dense markers such as ferritin or colloidal gold particles are routinely used to improve detection. A specific example is the use of antibiotin to detect biotinylated nucleic acid probes, which have been hybridized to complementary sequences in chromatin preparations or in situ in isolated chromosomes. In this example, the antibody binding site is enhanced for EM detection by addition of a secondary antibody tagged with colloidal gold particles. Thus specific genes can be detected within chromatin masses (see Figure 2H).

### 3.2 Radiolabeled Molecules

Although high resolution autoradiography (EM ARG) is applied mostly to the detection of newly synthesized (radiolabeled) components in situ at the cellular level, application is possible with spread molecules and molecular complexes. However, EM ARG is limited to the use of radioisotopes emitting soft β-particles, usually $^3H$. Molecules that have been radiolabeled either in vivo or in vitro are prepared for electron microscopy by the method most appropriate for the sample type. After staining, the preparation is covered, first with a thin protective coat of carbon and then with a layer of photographic emulsion. Ideally, the emulsion should consist of a homogeneous monolayer of the silver halide crystals. The emulsion is placed in the dark long enough to allow a sufficient amount of radioactive decay; then it is developed, leaving silver grains above the sites of radiolabeling in the molecules. EM ARG is subject to two severe limitations:

1. The accuracy with which the developed silver grain can be located to the source of radiation in the specimen molecule. This depends on the thickness of the preparation and the diameter of the silver halide crystal in the emulsion and results in scattering of grains with a half-distance of about 100 nm.
2. The efficiency in detecting radioactive disintegrations. Even molecules labeled to high specific activities require exposure times of generally more than a few weeks.

In spite of these limitations, EM ARG has been used successfully in studying sites of replication in isolated DNA molecules and the location of active transcription complexes in spread chromatin (Figure 2F).

### 3.3 Electron Spectroscopic Imaging (ESI)

By adapting the EM to include an imaging electron energy filter, positional information can be recovered from electrons that have lost a specific and characteristic amount of energy in colliding with atoms in the specimen molecules. This approach has been used successfully in the location of phosphorus atoms, particularly those in the phosphodiester bonds of nucleic acids. The resulting ESI has been used to delineate the path of the sugar–phosphate backbone in nucleosomes and the configuration of 7*SL* RNA within the signal recognition (ribonucleoprotein) particle. This approach, as well as related techniques in element mapping within biomolecules and their complexes, is undergoing constant sophistication and promises many diverse applications.

### 3.4 Image Reconstruction

Individual molecules, under optimum conditions of spreading and staining, give weak and ill-defined images. To improve on the amount of structural detail, information from many molecules can be combined to smooth out random variations between images. To do this it is necessary to use molecules (e.g., protein filaments) that contain regular, repeating arrays of subunits. Alternatively, the biomolecules can be induced to form crystalline arrays of regular, tightly packed, oriented units. Electron micrographs of these types of array can then be used for image processing to produce enhanced structural detail.

*See also* DNA STRUCTURE; IMMUNOLOGY; NUCLEIC ACID HYBRIDS, FORMATION AND STRUCTURE OF; RNA THREE-DIMENSIONAL STRUCTURES, COMPUTER MODELING OF.

### *Bibliography*

Dubochet, J., Adrian, M., Chang, J., Homo, J.-C., Lepault, J., McDowall, A. W., and Schultz, P. Cryoelectron microscopy of vitrified specimens. *Q. Rev. Biophys.* 21:129–228 (1988).

Harris, J. R., Ed. *Electron Microscopy in Biology: A Practical Approach.* Oxford University Press, London and New York, 1991.

Ottensmeyer, F. P. Elemental mapping by energy filtration: Advantages, limitations and compromises. *Annals N.Y. Acad. Sci.* 483:339–353 (1986).

Sommerville, J., and Scheer, U., Eds. *Electron Microscopy in Molecular Biology: A Practical Approach.* Oxford University Press, London and New York, 1987.

# ELECTRON TRANSFER, BIOLOGICAL

*Christopher C. Moser, Ramy S. Farid, and P. Leslie Dutton*

### *Key Words*

**Franck–Condon Factors** The dependence of the electron transfer rate on the overlap of the reactant and product nuclear wave functions.

**Reorganization Energy** The energy required to distort the geometry of the electron donor–acceptor pair to resemble the equilibrium geometry after electron transfer.

Respiratory electron transfer involves the guided stepwise transfer of electrons from a source of reducing equivalents to a sink of oxidizing equivalents. Photosythetic electron transfer adds to this framework light-pumped cyclic electron transfer by virtue of chlorophyll's ability to act both as an electron donor (in the excited state) and as an electron acceptor (in the oxidized ground state). It is because bioenergetic electron transfer proteins are intimately associated with a closed membrane that this redox energy can be converted and stored in other forms. The chemiosmotic model describes the principles by which electron transfer reactions are oriented across the membrane and coupled to the absorption and release of protons, whereupon they generate transmembrane electric and proton gradients.

This entry explores the underlying principles by which proteins guide electron transfer to assure reactions that are productive within the chemiosmotic framework and also maintain engineering efficiency. We will show that the fundamental role of protein is to provide a scaffolding that separate redox centers by controlled distances in specific directions. Nearly as important is the influence of the protein environment on the free energy of reactions and types of relaxation that occur concurrent with the shift of charge.

## 1 ELECTRON TRANSFER THEORY

### 1.1 TUNNELING

Fundamentally, all biological electron transfers are tunneling reactions. Although redox reactions involving mobile redox carriers such as water-soluble cytochrome *c* or membrane-bound ubiquinone may be rate-limited by diffusive reactions, and critical reactions that couple electron transfer to proton uptake or release may be dependent on the dynamics of proton donors or acceptors, every electron transfer involves the quantum mechanical tunneling of an electron from donor to acceptor through the barrier of the intervening protein medium. The quantum mechanical nature of this reaction is most obvious in intraprotein electron transfer, especially in the photosynthetic reaction center in which light-initiated electron transfer is nearly independent of temperature down to liquid helium temperatures. This relative temperature independence contrasts with a classical transition state description in which a reactant absorbs thermal energy to surmount an activation energy barrier and decays into products.

A simple description of the rate of intraprotein electron transfer is provided by Fermi's Golden Rule, derived from quantum mechanical perturbation theory applied to well-separated (weakly coupled) redox centers. This equation is written to separate the influence of the electronic and nuclear wavefunctions of the reactant and product. The electron transfer rate ($k_{et}$) is proportional to the electronic coupling, which is in turn proportional to the square of the overlap of the electronic wave functions of the reactant and product ($V^2{}_R$). The rate is also proportional to the Franck–Condon factors ($FC$), which describe the overlap of the reactant and product nuclear wave functions:

$$k_{et} = \frac{2\pi}{\hbar} V_R^2 \, FC \tag{1}$$

The electronic wave functions of the transferred electron have large values only immediately around the atoms of the redox centers themselves. In the case of such small centers as FeS clusters, almost all the transferred electron density is within a sphere of a few angstroms, while for a porphyrin-derived redox center, such as hemes or chlorophylls, electron density will be delocalized over a 10 A disk. Electron density in the intervening medium where donor and acceptor wavefunctions overlap will be quite small. Perhaps the simplest way to approximate the wavefunctions in this critical area is to model the redox centers as relatively narrow electron potential wells surrounded by barrier medium of constant potential. In such a case the electronic wave functions will decay exponentially with the edge-to-edge distance between redox centers, and their decay rate will depend on the height of the potential barrier of the intervening medium. These assumptions lead to the relation:

$$V_R^2 = V_0^2 \, e^{-\beta R} \tag{2}$$

where $R$ is the edge-to-edge distance between the redox centers, $\beta$ the effect of the barrier height on the wave function decay, and $V^2{}_0$ the electronic coupling that is extrapolated to a condition where the redox center edges overlap. This relation is a simple one, which ignores the atomic heterogeneity of the protein medium between redox centers and the possibility that electron transfer may occur along specific low potential paths following the chemical bonds of amino acids.

### 1.2 FRANCK–CONDON FACTORS

Marcus has provided a simple description of the essential elements of the nuclei-dependent Franck–Condon factors. As in the classical transition state description, reactant and product states, representing

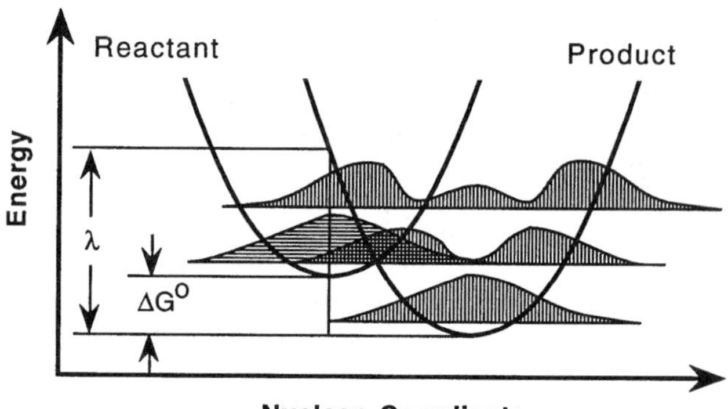


**Figure 1.** Parabolic harmonic oscillator potential surfaces representing the energy of the reactant and product states before and after electron transfer. The minimum of each potential well represents the equilibrium position of the nuclei when the electron is being transferred in on the donor or acceptor. The energy required to distort the nuclei of the reactant to the equilibrium geometry of the product before electron transfer defines the reorganization energy (λ). Superimposed on the classical potential surfaces are vibrational quantum energy levels and representations of the density of harmonic oscillator wave functions. The Franck–Condon factors will be proportional to the overlap of reactant and product nuclear wave functions, shown schematically as shaded areas.

the electron localized on the donor and acceptor, lie at the bottom of different potential energy wells (see Figure 1). In the Marcus description these potential energy surfaces are represented by simple parabolas because the nuclei of the reactants and products are treated as a generalized harmonic oscillator. For simplicity, the frequency of the reactant and product harmonic oscillators are considered unchanged. Only the geometry of the nuclei and the overall free energy change upon electron transfer. As the geometry of the reactant is distorted from the lowest energy equilibrium configuration, the energy of the reactant state will rise. The energy required to distort the geometry of the reactant to the equilibrium geometry of the product defines the reorganization energy (λ). The greatest value of the reactant nuclear wave function will tend to be near the equilibrium nuclear geometry. However, as the temperature is raised, higher vibrational energy levels will be populated in a Boltzmann distribution, with the effect of spreading the wave functions and generally increasing the overlap with the product wave functions. The calculated overlap of the nuclear wave functions is given by the Gaussian expression:

$$FC = \sqrt{4\pi\lambda k_B T}\exp -\frac{(-\Delta G^\circ - \lambda)^2}{4\lambda k_B T} \tag{3}$$

This expression in combination with Fermi's Golden Rule predicts a Gaussian dependence of the electron transfer rate on the free energy of the reaction at a given fixed distance. The maximum rate is found when the free energy of the reaction matches the reorganization energy, a situation that corresponds to the intersection of the reactant and product potential surfaces at the minimum of the reactant parabola. The Marcus relation predicts the surprising behavior that increasing the free energy of a reaction beyond the reorganization energy (into what is called the inverted region) actually decreases the expected rate of the reaction.

The intersection of the reactant and product parabolas in the Marcus picture occurs at a free energy $(-\Delta G^\circ - \lambda)^2/4\lambda$ above the bottom of the reactant well. This energy formally corresponds to a classical activation energy and leads to an Arrhenius-type temperature dependence. Yet a number of biological electron transfer reactions are known to become essentially temperature independent at low temperatures, suggesting that a quantum mechanical modification of the Franck–Condon factors must be considered. A semiclassical relation presented by Hopfield maintains the overall Gaussian dependence of the FC factors on free energy, but replaces the Marcus variance of $4\lambda kT$ with a new variance

$$\sigma^2 = \lambda\hbar\omega \coth\left(\frac{\hbar\omega}{2k_B T}\right) \tag{4}$$

introducing the quantum vibrational energy of the oscillator as $\hbar\omega$. At high temperatures (thermal energy significantly greater than half the quantum vibrational energy) the coth term reduces to $2kT/\hbar\omega$ *and the variance to* $2\lambda kT$. This is the same variance found in the Marcus expression, predicting the same Arrhenius temperature dependence. At low temperature, the coth term reduces to 1, the variance to $\lambda\hbar\omega$, and the FC factors become nearly independent of temperature.

A completely quantum mechanical expression for the FC factors, using quantum harmonic oscillator wave functions, has been described by DeVault. The form of the expression is more difficult to follow intuitively because a term rising exponentially in free energy competes with a simultaneously falling modified Bessel function $I_P$.

$$FC = \frac{1}{\hbar\omega}\exp\left\{-S(2n+1)\left(\frac{n+1}{n}\right)^{P/2} I_P[2S\sqrt{n(n+1)}]\right\} \tag{5a}$$

$$S = \frac{\lambda}{\hbar\omega};\ P = \frac{-\Delta G^\circ}{\hbar\omega};\ n = \frac{1}{e^{\hbar\omega} - 1} \tag{5b}$$

The net effect of the opposing functions is a rising and falling of the electron transfer rate with free energy that is nearly Gaussian for relatively small $\hbar\omega$ but becomes asymmetric as the reorganization energy approaches $\hbar\omega$.

Equations (5) strictly define an FC factor only for free energies corresponding to exact multiples of the quantum vibrational energy. However, all these expressions can be modified to accommodate the coupling of electron transfer to more than one nuclear vibration by convoluting together independent expressions for each vibrational frequency. This tends to smooth out the discrete behavior of equations (5). The convolution of two Gaussians is another Gaussian with a summed variance. For expression (4), the variance becomes a reorganization energy weighted sum of all vibrational frequencies, with vibrations smaller than $2kT$ contributing an effective vibrational energy of $2kT$. In other words, in the case of a spectrum of vibrations coupled to electron transfer, it becomes meaningful to consider a total reorganization energy and an overall characteristic frequency.

## 2 EMPIRICAL RESULTS

### 2.1 Free Energy Dependence of Intraprotein Electron Transfer

Systems in which the parameters of the foregoing theoretical expressions are accurately measured and systematically varied but still relatively rare in both chemistry and biology. However, several systems of covalently linked redox centers display a conspicuous

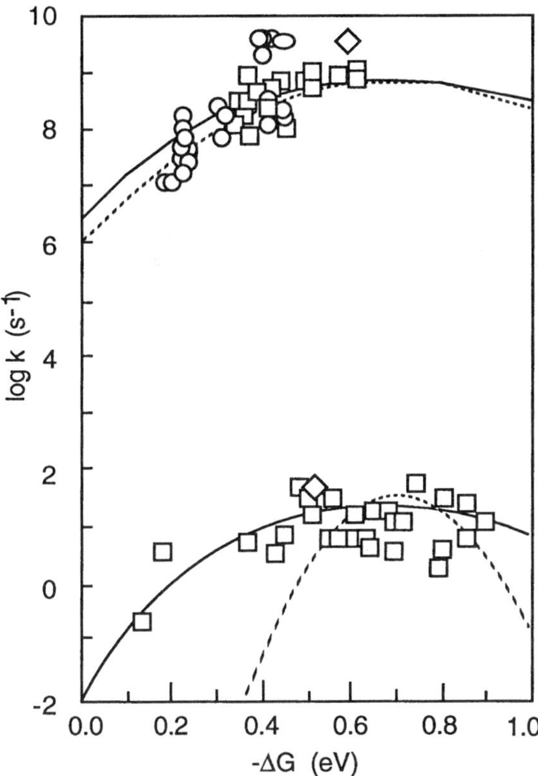


**Figure 2.** The free energy dependence of the rate of a number of intraprotein electron transfers appears to be approximately Gaussian (parabolic on a log plot). Examples of the free energy dependence of the rate of physiologically productive electron transfer over 10 (edge to edge) from photoreduced bacteriopheophytin (BPh-) to primary quinone $Q_A$, and physiologically unproductive charge recombination over 22.5 from $Q_A$ to the photooxidized bacteriochlorophyll dimer ($BChl_2$). The data for the latter reaction are shown at a temperature of 35 K. The maximum of the parabola defines the reorganization energy and the free energy optimal rate. Superimposed on these data are the theoretical relationships of equation (4) (solid lines) and (3) (dashed lines). At low temperature the classical relation of equation (3) narrows considerably.

Gaussian-like rate versus free energy behavior predicted by the Marcus relation. In bacterial photosynthetic reaction centers, X-ray diffraction provides well-defined structures, while quinone substitution reveals a similarly Gaussian-like behavior within an approximately order of magnitude experimental scatter (see Figure 2). The rate maxima suggests a reorganization energy of about 700 meV and the breadth suggests a characteristic frequency of about 70 meV. Six other electron transfer reactions between reaction center redox components have been described, although with a less extensive free energy dependence. Approximately Gaussian relationships with a rate maximum at about 1100 meV have been found in various heme proteins in which surface histidine residues have been ruthenium-modified to create light-activatable redox centers. A similar reorganization energy is anticipated for pulse radiolysis induced electron transfer from a disulfide bridge to the Cu center in azurin.

The vast majority of experimental intraprotein electron transfer systems have limited free energy variation and uncertain reorganization energies and distances. However, we note that reorganization energy for various biological reactions generally tends toward moderate values of 500 to 700 meV for reactions that take place within the relatively nonpolar protein interior, increasing toward 1000 to 1200 meV for reactions that involve redox centers at the relatively polar exteriors. Unusually low reorganization energy values seem to be associated with ultrafast picosecond or sub picosecond electron transfer between chlorins in the initial charge separation, perhaps because of the delocalization of the electrons over the chlorin ring structure.

We also note that a characteristic frequency of about 70 meV is adequate to model the width of the Gaussian rate–free energy relationships within the scatter of the data available. This frequency generates curves that are very similar to the classical Marcus relation at room temperature and do not narrow as dramatically at low temperatures (see Figure 2). Since lowering characteristic frequencies below 50 meV should have no effect on electron transfer rates at physiological temperatures, natural selection is not expected to favor low values. On the other hand, on the basis of the data sets we have examined, it is clear that characteristic frequencies higher than about 100 meV have not been selected. It appears that the typical vibrational frequency spectrum of a condensed organic medium such as protein provides the basis for a characteristic frequency in this range and that natural selection cannot effectively modify this parameter.

### 2.2 Distance Dependence of Optimal Electron Transfer Rate

Comparing the free energy optimized rate of various intraprotein electron transfer reactions both minimizes the variations in electron transfer rate due to FC factors and reveals the dependence of the rate on the electronic wavefunction overlap found in equation (1). If the logarithms of the optimal rates just described are plotted versus edge-to-edge distance, a surprisingly linear relationship is found over 12 orders of magnitude (see Figure 3). This result suggests that a model of simple exponential decay of electronic wave functions through a medium barrier of constant average height ($\beta = 1.4\ A^{-1}$) is usefully accurate in intraprotein electron transfer.

The same $\beta$ value does not seem to apply to chemically synthesized covalently linked systems. It appears that most synthetic systems provide a more direct through- (covalent) bond link between redox centers that has the effect of more efficiently propagating the electronic wave functions of the donor and acceptor ($\beta \sim 0.7\ A^{-1}$). It appears that the relatively large size of typical biological redox centers (e.g., porphyrins) and the typical geometry of amino acid packing in a protein has led to multiple paths for wave function propagation and average $\beta$ values similar to those found in an organic solvent glass. Furthermore, it is clear from the reaction center examples presented that natural selection has not modified $\beta$ of protein medium to systematically favor physiologically productive electron transfer over energy-wasting charge recombination reactions.

In some cases it appears to be possible to construct an electron transfer system that operates at a rate significantly different from that predicted by the average protein $\beta$. Figure 3 shows that a cytochrome *c* ruthenated at histidine 62 gives an electron transfer rate two orders of magnitude less than that anticipated for the distance given the expected free energy and reorganization energy, while the azurin reactions are two orders of magnitude faster. (Generally, edge-to-edge distance includes within redox centers all atoms that make up the conjugated $\pi$-ring system of porphyrin-

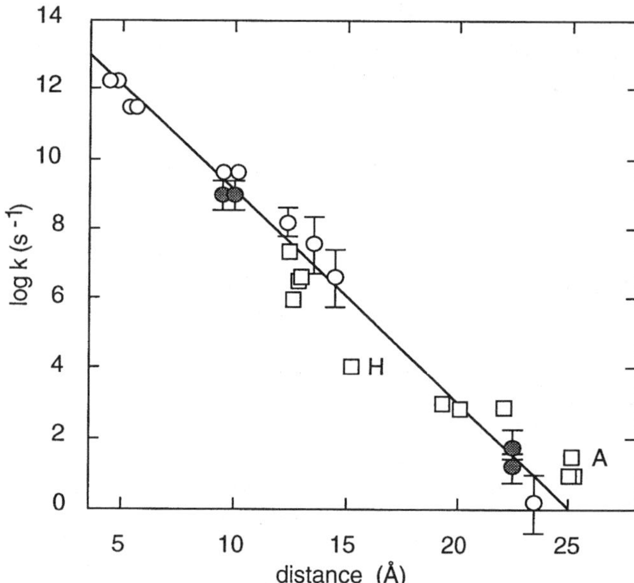


**Figure 3.** A plot of the log of the free energy optimal rate as a function of the edge-to-edge distance between redox centers in proteins generates a surprisingly linear relationship over 12 orders of magnitude of rate. Rates of charge separating and recombination reactions in two species of photosynthetic reaction centers are shown as circles; solid circles are not shown in Figure 2. Rates of nonphysiological reactions in modified electron transfer proteins are shown as squares. Two rates with obvious deviations above and below the average β are labeled H (His-62 ruthenated cytochrome *c*) and A (disulfide bridge to Cu center in azurin).

and quinone-based redox centers, and ligands of the smaller metal atom centers.) Indeed, a detailed bond-by-bond tunneling pathway analysis can be successfully applied to these examples. The medium between redox centers in cyt *c*-Ru-(His 62) is found to be unusually free of covalent links, while the beta barrel structure of azurin leads to an unusually large number of covalent links.

There may be several factors in biologically relevant reactions that disfavor the appearance of highly linked or unlinked paths and favor an average β. Molecular dynamics and the large size of common biological redox centers can increase the averaging effect of multiple competing pathways between redox centers. Other concerns such as protein assembly and stability may dominate long-range structural choices that affect connectivity, while changes in free energy and reorganization energy can effect rate changes without significant structural change. While rate modulation greater than two orders of magnitude by connectivity changes may be physiologically possible, it remains to be seen whether an expanded study of natural electron transfer systems will uncover examples of β modification driven by natural selection.

## 3 ELECTRON TRANSFER PROTEIN DESIGN

So far we have considered optimal electron transfer rates in what is essentially "solid state," nonadiabatic intraprotein electron transfer. In these systems the only first-order parameters of electron transfer theory that appear to be modified to set electron transfer rates are distance, free energy, and reorganization energy. A first-order numerical estimate for the upper limit of the electron transfer rate is

$$\log k_{et} = 15 - 0.6R - 3.1\frac{(\Delta G^\circ - \lambda)^2}{\lambda} \quad (6)$$

where the electron transfer rate is expressed in reciprocal seconds, the edge-to-edge distance in angstrom units, and the free energy and reorganization energy in electron volts.

Nevertheless, many electron transfer rates are slowed by rate-limiting diffusion (e.g., with the redox carriers cytochrome *c* or ubiquinone) or coupling to other events such as proton binding or release. Biological electron transfer systems seem generally tolerant of coupling reactions in the micro- and millisecond range as long as the overall rate exceeds $10^3\ s^{-1}$. For a typical small free energy respiratory reaction, this rate can be met, provided the redox centers are separated by an edge-to-edge distance of less than 17 A. Because the low dielectric membrane thickness is 35 A, this means that redox proteins in a chemiosmotic scheme must have at least two electron transfer reactions or three redox centers to complete transmembrane electron transfer. This 17 A distance also represents the amount of protein insulation that is required to prevent electron transfer to adventitious redox centers that may encounter the redox protein.

Nanosecond and faster reactions are essential in the face of rapid decay mechanisms, such as are found in light reactions of photosynthesis. Efficient light trapping by photosynthetic systems requires electron transfer to be significantly faster than the roughly nanosecond fluorescence decay of chlorophyll. In the bacterial reaction centers this speed of electron transfer appears to be assured by placing the chlorins of the first two electron transfer reactions at approximately 5 A edge-to-edge spacing. In addition, the greater than 1 eV energies associated with the creation of an excited state pose a special problem for photosynthetic systems in that energy dumping charge recombination reactions to ground state will have a very large free energy that is potentially competitive with productive charge separation. The Marcus "inverted region" discussed in Section 2.2 provides a means to assure that such large free energy reactions will be slow, provided the reorganization energy for the reaction is less than half the total available energy. Indeed, the reorganization energies of the initial charge separation reactions appear to be among the lowest observed. The failure to satisfy these criteria often results in relatively poor quantum efficiency and charge separation stability in synthetic redox systems.

The general first-order description of intraprotein electron transfer described here should prove useful for systems less well characterized than the bacterial photosynthetic reaction centers. In the many bioenergetic proteins for which an X-ray crystal structure does not exist, equation (6) in combination with rough estimates of the free energy and reorganization energy of the reactions will predict edge-to-edge distances within a few angstroms. This relation will also prove useful in the design of de novo protein electron transfer systems. Indeed, if one is armed with an awareness of the importance of a low reorganization energy for initial charge separation, equation (6) can provide a description of the distances and free energies to construct charge-separating proteins that match the reaction center in quantum efficiency and exceed it in overall efficiency of energy conversion.

*See also* BIOENERGETICS OF THE CELL.

*Bibliography*

Beratan, D. N., Betts, J. N., and Onuchic, J. N. (1991) Protein electron transfer rates set by the bridging secondary and tertiary structure. *Science,* 252:1285–1288.

Bolton, J., McLendon, G. L., and Mataga, N. Eds. (1991) *Electron Transfer in Inorganic, Organic, and Biological Systems.* American Chemical Society, Washington, DC.

Kirmaier, C., and Holten, D. (1987) Primary photochemistry of reaction centers from the photosynthetic purple bacteria. *Photosynth. Res.* 13:225–260.

Marcus, R. A., and Sutin, N. (1985) Electron transfers in chemistry and biology. *Biochim. Biophys. Acta,* 811:265–322.

Moser, C. C., and Dutton, P. L. (1992) Engineering protein structure for electron transfer function in photosynthetic reaction centers. *Biochim. Biophys. Acta,* 1101:171–176.

———, Keske, J. M., Warncke, K., Farid, R. S., and Dutton, P. L. (1992) The nature of biological electron transfer. *Nature,* 355:796–802.

# Endocrinology, Molecular

*Franklyn F. Bolander, Jr.*

*Key Words*

**Adaptor** A small molecule consisting only of phosphotyrosine (SH2) and polyproline helix (SH3) binding domains and used to associate proteins.

**G Protein** A GTP-binding protein with intrinsic GTPase activity; it acts as a molecular switch (which is ''on'' when bound to GTP) with a built-in timer.

**Hormone** A chemical, nonnutrient, intercellular messenger that is effective at very low concentrations.

**Hormone Response Element (HRE)** A DNA sequence recognized by a transcription factor that is primarily regulated by hormones.

**Oncogene** A gene that usually encodes a component of the growth factor pathway and has undergone a mutation resulting in constitutive activity and tumor formation.

Endocrinology is the study of hormones, chemicals that cells use to communicate with one another. Traditionally, endocrinology was studied at the organismal level: the effects of hormones on tissue or animal growth, reproduction, or metabolism. Molecular endocrinology seeks to determine the molecular mechanisms for these gross effects: the actions of hormones, direct or through mediators, on enzymes, transport processes, the cytoskeleton, and transcription factors. Such information is vital to understanding how complex metazoans coordinate cellular functions. In addition, this knowledge can be useful in the evaluation and treatment of various endocrine diseases that have genetic bases. In addition, many tumors arise when the genes for various growth-promoting hormones or for components of their signaling pathway undergo mutations that render them constitutively active.

## 1 INTRODUCTION

Multicellular organisms have two major coordinating systems: the nervous and the endocrine systems; both utilize chemical messengers. In the endocrine system, these molecules are called hormones and are usually secreted into a circulatory system for general distribution. In the nervous system, the molecules are called neurotransmitters and are released into synapses for more precise effects. The distinction is not always this clear; many of the chemicals are used by each system and their mechanisms of action are identical. In addition, some hormones are made and act locally in a very defined area, while some neurotransmitters can leak out into the general circulation. Thus, there is a general tendency to consider all chemical messengers as a single functional group.

## 2 HORMONES

Structurally, hormones are extremely diverse; nearly every organic group is represented. Proteins and peptides are the largest group. In addition, there are hormones that are derivatives of amino acids, sterols, fatty acids, phospholipids, nucleotides, and carbohydrates. There are even several gaseous hormones: ethylene and nitric oxide.

However, one property divides all hormones into two major groups and will determine their mechanisms of action: solubility in water. Hydrophobic ligands have no problem crossing the plasma membrane; as such, these hormones have a direct mechanism of action. In particular, they migrate to the nucleus, where they interact with transcription factors. Since their mechanism is primarily genomic, their effects tend to be delayed and long-term; they are often involved in developmental processes. Hydrophilic hormones cannot cross the plasma membrane and must interact with binding proteins, the receptors. Since these receptors are integral membrane proteins, they must generate another signal on the cytosolic side; if the hormone is the primary messenger, this subsequent factor becomes a second messenger. These latter messengers often activate kinases, which may have acute effects on metabolism and cell structure or may phosphorylate transcription factors for more long-term effects.

## 3 RECEPTORS

### 3.1 Nuclear Receptors

As noted in Section 2, the receptors for lipophilic hormones are ligand-regulated transcription factors. These receptors are all homologous: the amino terminus possesses a transcription activation domain (TAD), the center has a DNA-binding region, and the carboxy terminus, which binds the hormone and heat shock proteins (hsps), contains a dimerization domain and a second TAD. The hsps are necessary to maintain the receptors in a conformation required for ligand binding; once the hormone has bound, the hsps dissociate (Figure 1). Ligand binding also induces receptor dimerization, phosphorylation, nuclear translocation (although some receptors are constitutively located in the nucleus), DNA binding, and transcription activation. After the hormone has dissociated, the receptor is dephosphorylated and recycled.

The nuclear receptors can be divided into three families based on their structures and the DNA sequences to which they bind. The glucocorticoid family is the most recently evolved group and

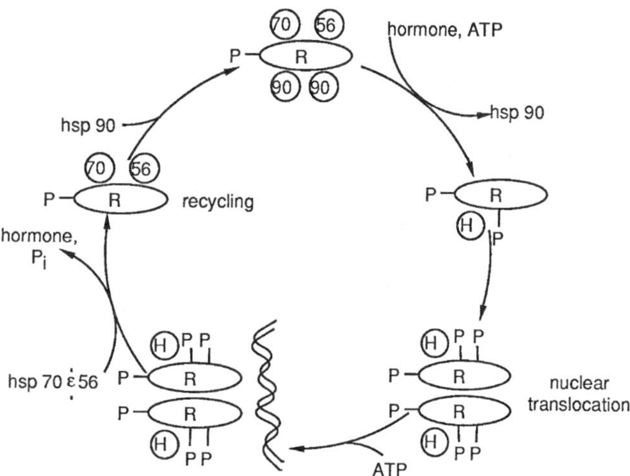


**Figure 1.** Activation and recycling of a nuclear receptor: hsp 56, 70, 90, heat shock proteins; H, hormone; P, phosphorylation site; R, receptor.

contains the cortisol, aldosterone, androgen, and progesterone receptors. The members of this family are basically homodimers; they require hsp 90 and bind inverted repeats of the hormone response element (HRE) TGTTCT. The thyroid hormone family is the oldest and most diverse group and includes receptors for the thyroid hormone, vitamins A and D, ecdysone, and arachidonic acid. These substances are most active as heterodimers, do not require hsp 90, and can bind either direct or inverted repeats of TGACC. The estrogen family contains only the estrogen receptor and a few related receptors whose ligands are still unidentified. Its properties lie between the two other groups: it binds the thyroid hormone HRE but only as inverted repeats; in addition, it forms homodimers and requires hsp 90, like the glucocorticoid family.

### 3.2 Membrane Receptors

Membrane receptors are considerably more diverse. Four basic superfamilies are recognized: the enzyme-linked, fibronectinlike, serpentine, and ion channel receptors (Figure 2). The enzyme-linked receptors are the simplest: the basic structure consists of a single protein that traverses the plasmalemma once, via an α-helix. The amino-terminal extracellular domain binds the hormone, while the carboxy-terminal cytosolic domain possesses a catalytic site. Ligand binding induces dimerization and enzyme activation. Three types of enzymatic activity have been identified: the tyrosine kinases are represented by four families (the epidermal growth factor, insulin, platelet-derived growth factor, and nerve growth factor groups); the serine–threonine kinases are found only in the transforming growth factor β-family; and the guanylate cyclases generate cyclic GMP (cGMP) in response to atrial natriuretic factors.

In the case of the receptor tyrosine kinases (RTKs), the major substrates are themselves. Phosphorylated tyrosines are recognized by SH2 domains, found in many enzymes and adaptors; these latter proteins will bind the autophosphorylated receptors and then mediate many of the biological activities of RTKs (see Section 4).

The fibronectinlike receptors have the same structure as the enzyme-linked binding proteins except that there is no recognizable catalytic site in the cytosolic domain. The extracellular region is composed of two modified units first identified in fibronectin, a matrix protein. In the cytokine (class 1) receptors, these units form two seven-stranded β-sheets that join at right angles to create

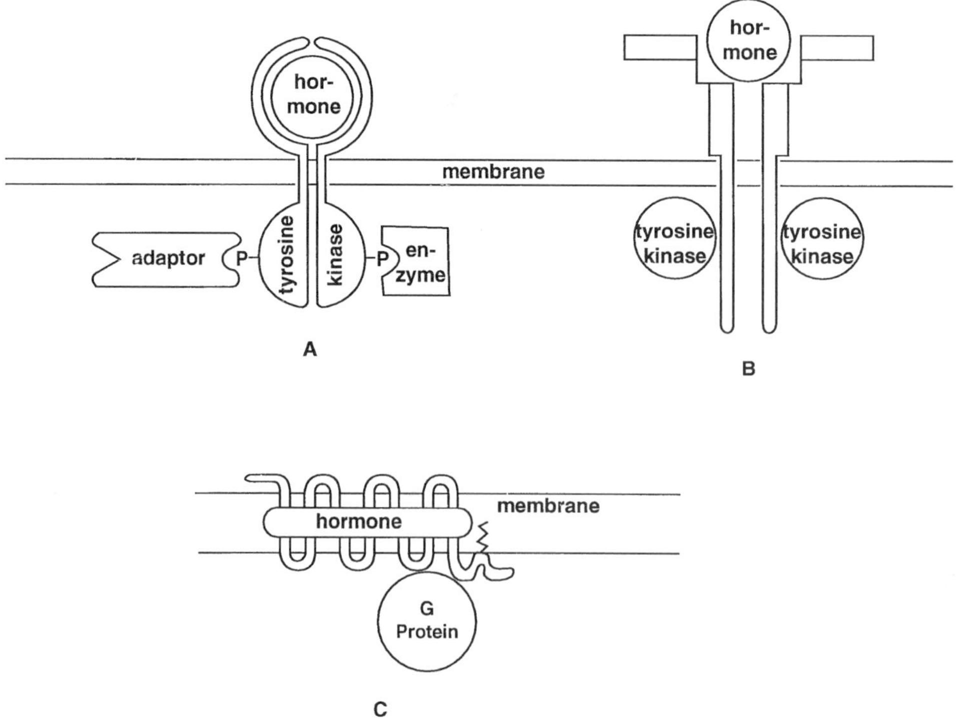


**Figure 2.** Schematic representation of three of the classes of membrane receptors: (A) receptor tyrosine kinase, (B) cytokine receptor, and (C) serpentine receptor.

a ligand-binding pocket (Figure 2B). This core may be further embellished by immunoglobulin loops and/or additional fibronectin domains. The class 2 receptors form repeats of five-stranded β-sheets that extend over the hormone like fingers.

In the cytosolic juxtamembrane region of the fibronectinlike receptors is a conserved proline-rich region that constitutively binds soluble tyrosine kinases. Like RTKs, these kinases are activated by receptor aggregation induced by ligand binding. As such, the fibronectinlike receptors can be considered to be RTKs whose catalytic site is located on a separate subunit.

The next superfamily has many names: G-protein coupled, seven-transmembrane segment, and serpentine receptors. We will use the latter term because of its simplicity. The serpentine receptors, probably the oldest and most diverse of the membrane receptors, mediate both sensory and endocrine transduction. The protein traverses the plasma membrane seven times; several of these transmembrane α-helices have conserved prolines that kink the helices, resulting in the formation of a ligand-binding pocket (Figure 2C). As such, most of the hormones using these receptors are small (the catecholamines, histamine, the prostaglandins, etc.). Receptors with larger ligands often have extended amino termini to aid in hormone binding. Several of the intracellular loops bind and activate G proteins (see Section 4.1), especially the third loop and the juxtamembrane portion of the carboxy terminus.

The last superfamily consists of the ion channel receptors. The ligand-gated channels are pentamers of homologous subunits; each subunit contributes an α-helix toward forming the wall of the channel. The hormone appears to bind between subunits and to open the channel either by pushing the subunits further apart or by twisting the subunits. Acetylcholine activates sodium channels; glycine and γ-aminoisobutyric acid activate chloride channels; and serotonin and glutamate activate calcium channels.

The voltage-gated channels are homotetramers; each subunit has six transmembrane helices. The pore is formed by a β-loop between the last two helices; together, the four subunits create an eight-stranded β-barrel channel. Two families from this group have become regulated by second messengers: first, the cyclic nucleotide gated channels, are cation pores opened by cAMP or cGMP. The second family, which gates channels that release internal stores of calcium, includes the inositol 1,4,5-trisphosphate ($IP_3$) and the cyclic ADP–ribose receptors (see Section 4.2).

## 4 SECOND MESSENGERS

### 4.1 G Proteins and Cyclic Nucleotides

G proteins are molecular switches: they are active when GTP is bound to them and inactive when GDP is bound. In addition, they have intrinsic, although weak, GTPase activity; as such, they eventually turn themselves off when they hydrolyze the bound GTP to GDP. In the active state, they can stimulate enzymes and ion channels, and affect the cytoskeleton and vesicular trafficking. There are several G-protein families, but only two that have been closely associated with hormone action are discussed here: a small G protein (Ras) and a large G protein (the αβγ trimer).

For a hormone to "flip" this switch on, it must facilitate the exchange of GDP for GTP. For example, the conversion of Ras·GDP to Ras·GTP is accomplished by a Ras GDP dissociation stimulator (RasGDS) (Figure 3), which is under hormone regulation. In fact, there are several different types of RasGDS, each with its own mechanism of control. In immune cells, one RasGDS (called Vav) is directly phosphorylated and activated by RTKs. Another RasGDS (called mammalian SOS) is stimulated indirectly: receptor or soluble tyrosine kinases phosphorylate either themselves or some substrate. An adaptor (Grb2) binds to the phosphotyrosine via its SH2 domain; Grb2 also has an SH3 domain that binds polyproline helices. Such a helix is present in RasGDS, so that a complex among a tyrosine phosphorylated protein, Grb2, and RasGDS is formed. This complex activates RasGDS, which then stimulates nucleotide exchange on Ras. Once formed, Ras·GTP binds to Raf, a serine–threonine kinase, and brings it to the plasmalemma, where it is stimulated by a still unknown process. Finally, Raf initiates a protein kinase cascade that eventually activates MAP kinase, a critical kinase in transcription regulation (see Section 5.2).

The GTPase activity of small G proteins is so weak that accessory proteins, the GTPase-activating proteins (GAPs) are required. This is another potential site for hormone input: RasGAP has an SH2 domain that binds phosphorylated tyrosines. Upon autophosphorylation, RTKs can bind and sequester RasGAP, thereby prolonging the activated state of Ras·GTP.

The large G proteins occur as heterotrimers: the α-subunit is the GTPase, while the βγ dimer is involved with membrane localization and protein association. Although the GTPase activity of α is slow, the large G proteins do not need GAPs. There are four subfamilies: $G_s$ stimulates adenylate cyclase, $G_i$ inhibits adenylate cyclase, $G_q$ stimulates phospholipase Cβ (PLCβ), and $G_{12/13}$ activates the ($Na^+$,$H^+$) exchanger, NHE-1. In addition $G_s$ and $G_i$ directly stimulate several ion channels. The hormone–receptor complex directly accelerates nucleotide exchange, resulting in the activation of $\alpha_s$ and the dissociation of βγ (Figure 4). As long as GTP remains bound to $\alpha_s$, adenylate cyclase will be stimulated. However, $\alpha_s$ will eventually hydrolyze GTP and reassociate with βγ. The $G_i$ proteins represent a counterregulatory mechanism that operates in a parallel manner except that both $\alpha_i$ and βγ participate in the inhibition of adenylate cyclase.

The ultimate output to this pathway is cAMP, which can activate several enzymes and ion channels; however, the major effector for cAMP is a serine–threonine kinase (protein kinase A, PKA). This tetramer has two catalytic subunits and two inhibitory subunits (Table 1). cAMP binds to the regulatory subunits and causes them to dissociate, thereby removing the inhibition.

cGMP is another cyclic nucleotide used as a mediator of hormone action; it may be generated by two different pathways. The first way, discussed in Section 3.2, involves a hormone directly binding and activating a membrane-bound guanylate cyclase. cGMP may also be synthesized by a soluble cyclase; this enzyme is activated indirectly by hormones that elevate calcium (see Section 4.2). Like cAMP, cGMP stimulates both ion channels and a homologous kinase, protein kinase G (PKG). The regulatory and catalytic subunits in PKG are fused into a single protein.

### 4.2 Calcium, Calmodulin, and Phospholipids

Calcium is an abundant cation in extracellular fluids; in addition, it is concentrated within several cellular organelles, such as mitochondria and elements of the smooth endoplasmic reticulum. However, cytosolic levels are kept extremely low, because many cellular processes are dramatically affected by calcium. Thus hormones

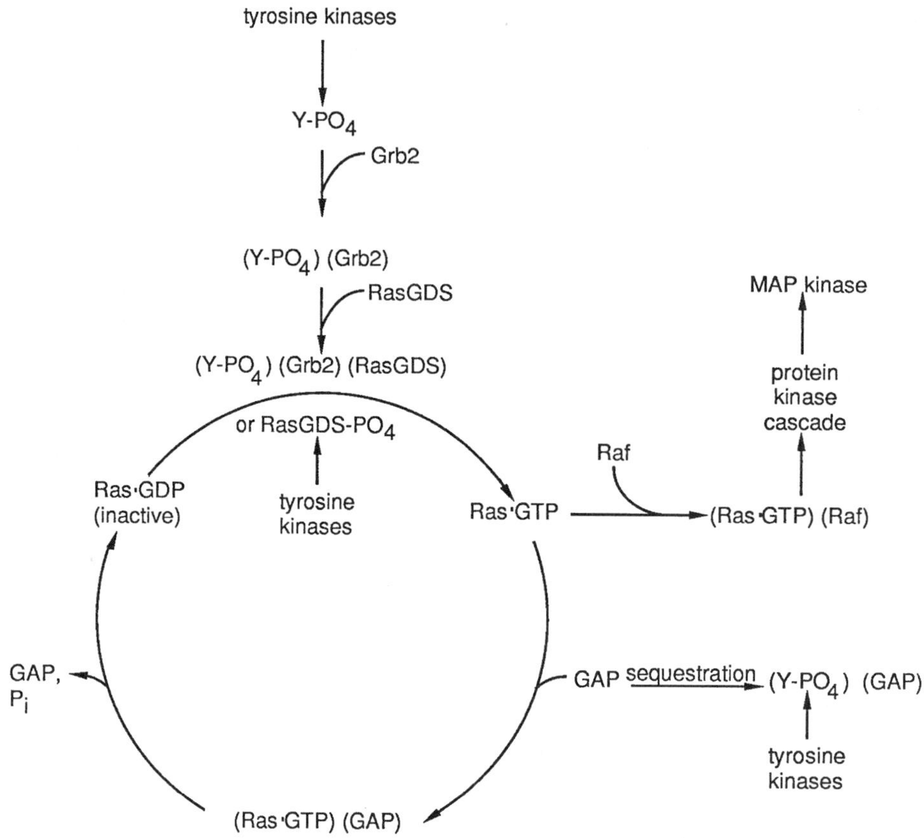


**Figure 3.** Activation–inactivation cycle for Ras, a small G protein: GAP, GTPase activating protein; GDS, GDP dissociation stimulator; Grb2, an adaptor; Raf, a protein kinase; Ras, a small G protein; Y-$PO_4$, tyrosine phosphorylation site.

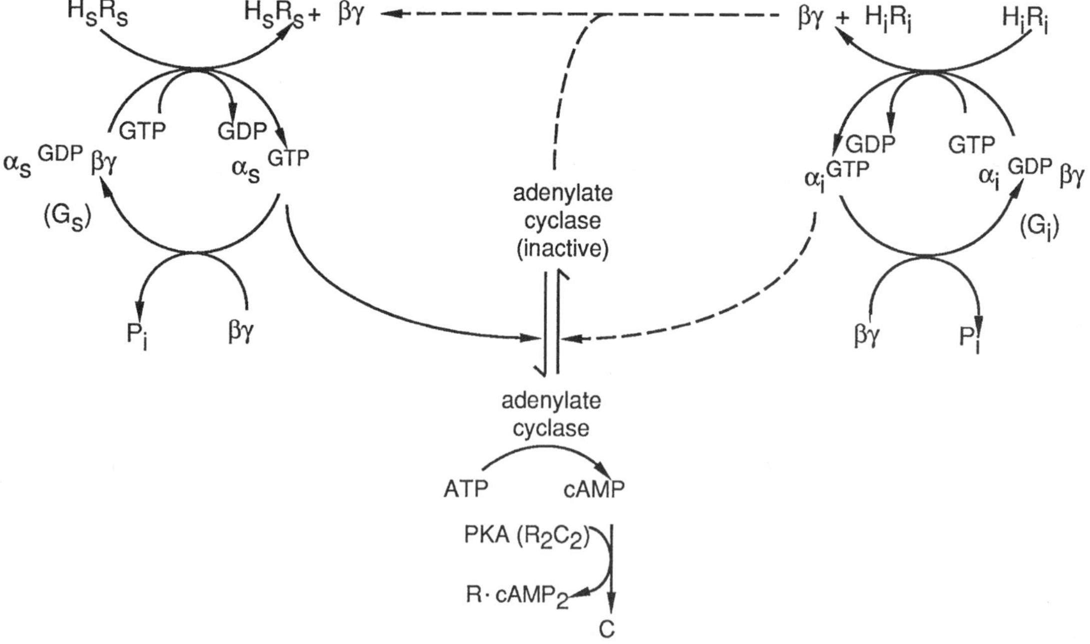


**Figure 4.** Regulation of adenylate cyclase by several G-protein trimers; inhibitory effects are depicted by dashed lines: $H_iR_i$, adenylate cyclase inhibiting hormone and its receptor; $H_sR_s$, adenylate cyclase stimulating hormone and its receptor; $R_2C_2$, PKA tetramer containing two regulatory (R) and two catalytic (C) subunits.

Table 1 Some Multifunctional Serine–Threonine Kinases Involved with Hormone Action

| Kinase | Abbreviation | Structure | Activators |
|---|---|---|---|
| Protein kinase A | PKA | Tetramer ($R_2C_2$) | cAMP binds and dissociates inhibitory R subunits |
| Protein kinase C | PKC | Monomer | |
| Conventional | cPKC | | Calcium and diacylglycerol |
| Novel | nPKC | | Diacylglycerol |
| Atypical | aPKC | | Ceramide; arachidonic acid; polyphosphoinositides |
| Protein kinase G | PKG | Homodimer | cGMP binding |
| Calmodulin-dependent protein kinase II | CaMKII | Homodecamer or homo-octamer | Calmodulin masks autoinhibitory domain |
| Casein kinase II | CKII | αβ dimer | Polyamines and phosphorylation |
| Mitogen-activated protein kinase | MAPK | Monomer | Ras and PKC via a kinase cascade |

can regulate these cellular functions by controlling the cytoplasmic calcium concentrations. The most direct mechanism for elevating calcium would be for hormones to activate ligand-gated calcium channels, such as the glutamate receptor; by merely opening these channels, hormones would cause external calcium to flood the cytoplasm.

However, a major source of hormonally released calcium is internal (Figure 5). Briefly, hormones stimulate a phospholipase (PLC) to hydrolyze a phospholipid (polyphosphoinositide) into diacylglycerol (DG) and the former head group ($IP_3$). There are several PLC groups distinguished both by their structure and by their hormone regulation. PLCγ possesses two SH2 domains through which it binds to autophosphorylated RTKs; this association brings the enzyme into close proximity to its substrate and allows RTKs to phosphorylate and stimulate the PLCγ. On the other hand, PLCβ, and probably PLCδ, are activated by G proteins.

Once released, $IP_3$ diffuses through the cytosol to its receptor, a calcium channel on the endoplasmic reticulum. $IP_3$ binding opens the channel and allows the internally stored calcium to enter the cytoplasm. The DG remains in the plasma membrane, where it stimulates a calcium-activated, phospholipid-dependent protein kinase, protein kinase C. Actually, there are three PKC groups that differ in their calcium and DG requirements (Table 1).

Phospholipids may give rise to many other second messengers; for example, phospholipase $A_2$ can liberate arachidonic acid, which may stimulate enzymes or may activate a nuclear receptor (Table 2). Arachidonic acid can also be converted to prostaglandins or

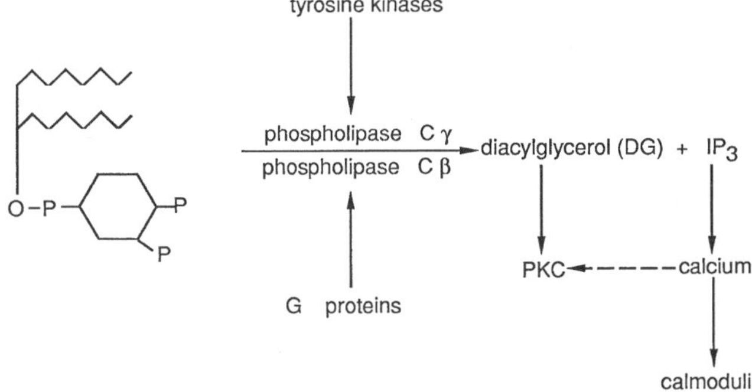


Figure 5. Brief summary of the polyphosphoinositide pathway: $IP_3$, inositol 1,4,5-trisphosphate.

other eicosanoids. These fatty acid derivatives are hormones in their own right: they are secreted; they bind to specific membrane receptors; and they stimulate various second-messenger pathways. However, they usually act locally. Such molecules, which are called parahormones, may adjust a cell's response to other hormones or to local conditions. A lysophospholipid is what remains after the arachidonic acid has been removed, and it is also biologically active. Finally, another membrane lipid, sphingomyelin, can be the source of several mediators of hormone action. Sphingomyelin is structurally similar to phosphatidylcholine except that it has a serine rather than a glycerol backbone. Its hydrolysis can generate ceramide and sphingosine 1-phosphate, which can then affect protein phosphorylation and transcription.

Once elevated, calcium alone can directly affect many enzymes, the cytoskeleton, and other biological processes. However, it frequently acts through a calcium-binding protein, calmodulin (CaM). CaM is a small dumbbell-shaped peptide: two globular ends separated by an α-helix. Calcium binding to the globular ends allows the groove found in each end to wrap around an α-helix in a target protein. For example, the CaM-dependent protein kinase II, a general-purpose serine–threonine kinase, has a CaM-binding site that blocks the ATP-binding site and inhibits kinase activity. In the presence of calcium, CaM binds this site, prevents it from interfering with ATP binding, and activates the kinase. There are many other CaM-responsive enzymes that operate on this same principle.

Another enzyme activated by CaM is nitric oxide synthase; this enzyme generates nitric oxide, which then activates a soluble guanylate cyclase. This represents an alternate pathway for the production of cGMP.

## 5 BIOLOGICAL EFFECTS

### 5.1 Nongenomic

There are two major mechanisms by which hormones can affect cellular processes: allosterism and phosphorylation. For example, G proteins can directly bind and alter the activity of both enzymes and transporters. Similarly, cyclic nucleotides can stimulate nucleotide-gated channels and protein kinases. Lipophilic hormones, which have direct access to the cellular interior, have also been shown to bind directly several enzymes, but the physiological relevance of these observations is still controversial.

**Table 2** Some Transcription Factors Mediating Hormone Action

| Transcription Factor | Abbreviation | Primary Activator(s) |
|---|---|---|
| cAMP response element binding protein | CREB | PKA phosphorylation |
| CCAAT/enhancer-binding protein β | C/EBPβ | CaMKII and MAPK phosphorylation |
| Jun-Fos | AP-1 | PKC phosphorylation (indirect) |
| Myc-Max | | CKII and MAPK phosphorylation |
| Nuclear factor of activated T lymphocytes | NFAT | Dephosphorylation by PP-2B, a CaM-regulated phosphatase |
| Nuclear factor for κ genes of B lymphocytes | NF-κB | PKC phosphorylation |
| Nuclear receptors | GR, ER, TR, etc. | Ligand binding by steroids, retinoids, thyroid hormones, etc. |
| Peroxisome proliferator activated receptor | PPAR | Fatty acid binding |
| Serum response factor | SRF | CKII and MAPK phosphorylation |
| Signal transducers and activators of transcription | STAT | Direct phosphorylation by receptor or soluble tyrosine kinases |

However, the broadest effects are generally achieved by phosphorylation. Indeed, virtually every known second-messenger pathway activates at least one protein kinase and several also affect protein phosphatases (Table 1): the extent of this modification is controlled by a balance between these two enzyme groups. Phosphorylation is a major regulatory mechanism in metabolism; glycogen metabolism is one of the best-known examples. Hormones that stimulate glycogen breakdown activate PKA, which initiates a protein kinase cascade leading to the phosphorylation and stimulation of glycogen phosphorylase. In addition, PKA and other kinases modify and inhibit glycogen synthetase; that is, glycogen breakdown is activated, while its synthesis is blocked. On the other hand, insulin favors glycogen synthesis; it does so by activating the phosphatases that reverse these phosphorylations. Many other metabolic cycles are also regulated by this modification.

In addition, phosphorylation can affect other cellular processes: it can alter the function of various channels and transporters and trigger the breakdown of the cytoskeleton.

### 5.2 Genomic

Hormones and their mediators can also affect gene expression by the same mechanisms just described (Table 2). For example, nuclear receptors are actually transcription factors that are allosterically regulated by their ligands, usually hormones but also second messengers like arachidonic acid. Many other transcription factors are controlled by phosphorylation. The simplest mechanism would be for an RTK or a cytokine receptor with associated soluble tyrosine kinase to phosphorylate directly a transcription factor that would then migrate to the nucleus and activate gene expression. The STAT family is activated by such a mechanism (Figure 6A).

However, most transcription factors are phosphorylated via second-messengers pathways. For example, hormones whose activities are mediated by cAMP first activate $G_s$; the α-subunit then stimulates adenylate cyclase, and the resulting accumulation of cAMP activates PKA. Finally, PKA modifies a cAMP response element binding protein (CREB); in particular, the phosphorylation occurs in a TAD and renders CREB a more efficient transcription activator (Figure 6B).

Phosphorylation can also occur on accessory proteins: NF-κB is a transcription factor frequently associated with defense genes. It is held in the cytoplasm by an inhibitory subunit, IκB. After being modified by PKC, IκB dissociates, allowing NF-κB to translocate to the nucleus, bind its HRE, and activate transcription (Figure 6C, top). Phosphorylation is not always stimulatory: NFAT, another transcription factor associated with defense genes, is tonically inhibited by phosphorylation. Activation occurs when hormones elevate calcium concentrations that stimulate protein phosphatase 2B (PP-2B), leading to the dephosphorylation of NFAT (Figure 6C, bottom).

The preceding discussion is an oversimplification. In fact, many transcription factors may be modified by several different kinases. This cross-talk allows for the integration of multiple signals, although it may at times appear confusing. Therefore, this general overview has concentrated on the dominant regulators of several major pathways.

## 6 CLINICAL APPLICATIONS

Nowhere have the benefits of molecular biology been more obvious than in the diagnosis and management of endocrine diseases. Hormones are so active that their serum concentrations are in the nanomolar range or less; their receptors rarely exceed more than a few thousand per cell. With such small numbers, conventional protein purification and characterization techniques are inadequate. Most defects in hormones, receptors, and transducers are identified by nucleotide sequencing after amplification by the polymerase chain reaction. Such techniques have been so successful that we cannot begin to cover all the endocrinopathies whose molecular bases have been determined; however, selected examples illustrating many of the major categories are presented.

### 6.1 Endocrine Deficiencies

Endocrine deficiencies can occur if there are mutations in the hormone, its receptor, or its transducer; *diabetes insipidus* offers an excellent example, inasmuch as defects have been found in all three. Diabetes insipidus is a functional impairment of the antidiuretic hormone (ADH) pathway. ADH binds a serpentine receptor in the kidney, activates adenylate cyclase, and translocates a water channel (aquaporin-2) from the cellular interior to the plasma membrane. This sequence of events is essential for the body to resorb water from the urine; in diabetes insipidus, this pathway is nonfunctional and the patient excretes large quantities of dilute urine.

ADH is actually synthesized as a polyprotein whose amino terminus is ADH and whose carboxy terminus is a transport protein. In *familial neurohypophyseal diabetes insipidus,* the carboxy terminus harbors a mutation that impairs ADH packaging and transport, leading to ADH deficiency, although ADH itself is normal. In one form of *hereditary nephrogenic diabetes insipidus,* the defect is in the renal isoform of the ADH receptor. In another form, aquaporin-2 is mutated. Inactivating mutations in $G_s$ are also known, but because so many hormones utilize $G_s$, the effects involve multiple endocrine systems.

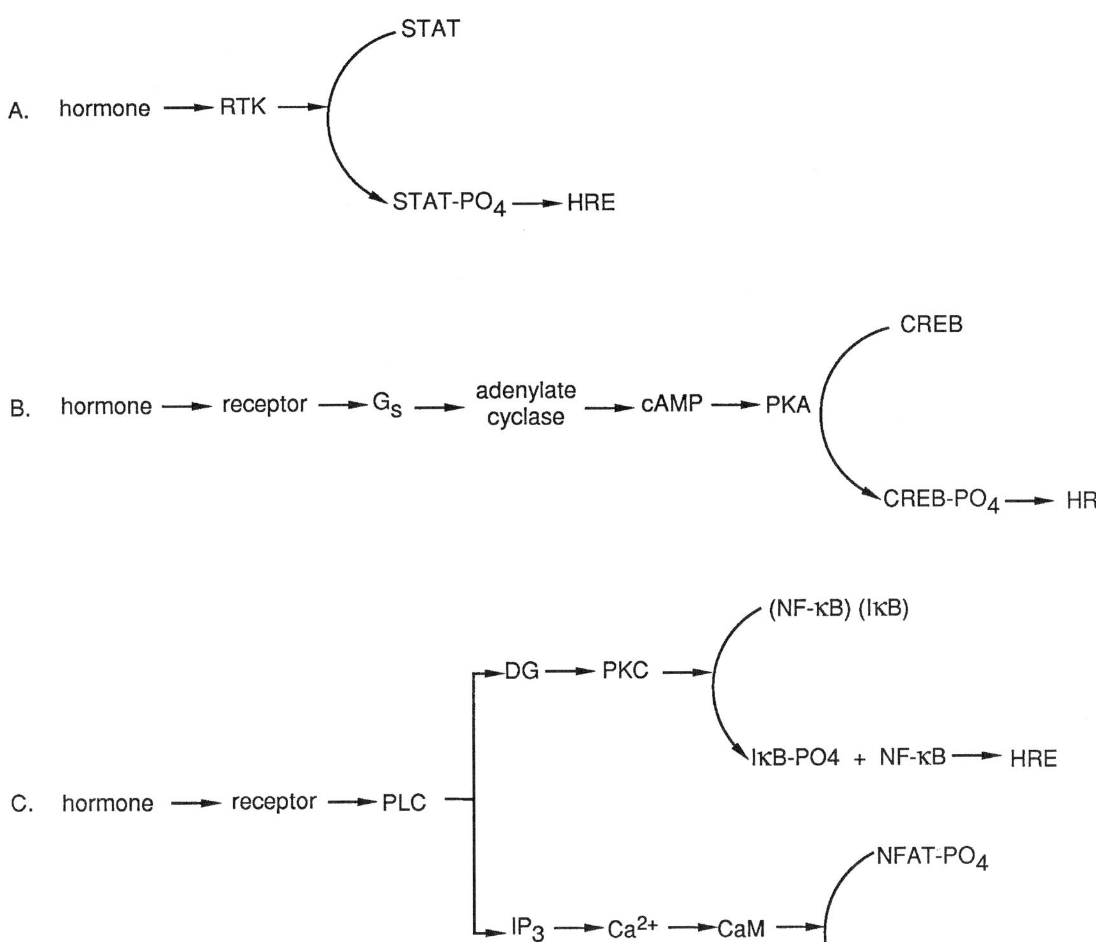


**Figure 6.** Activation of several transcription factors by various second-messenger pathways: CaM, calmodulin; DG, diacylglycerol; HRE, hormone response element; PLC, phospholipase C; RTK, receptor tyrosine kinase. See Tables 1 and 2 for other abbreviations.

### 6.2 Endocrine Excesses and Oncogenesis

Perhaps more fascinating than loss-of-function mutations are those that lead to overactivity. No mutations that increase the intrinsic activity of a hormone have been reported, but one mutation in angiotensinogen appears to increase its half-life in serum, allowing it to accumulate. Angiotensin is a very potent vasoconstrictor, and this mutation has been associated with some types of *essential hypertension.* Impaired catabolism and an altered gene regulatory region are the bases for two forms of hyperaldosteronism: *11β-hydroxysteroid dehydrogenase deficiency* and *glucocorticoid-remediable aldosteronism,* respectively.

Finally, hormone secretion can be elevated by faulty negative feedback. Parathormone (PTH) maintains serum calcium levels by resorbing this mineral from bone and recovering it from the urine. As calcium levels rise, PTH secretion shuts off. A mutant calcium sensor blocks this negative feedback and results in the persistent secretion of PTH, which in turn elevates calcium. The heterozygous state is called *familial hypocalciuric hypercalcemia,* while the homozygous condition is called *neonatal severe hyperparathyroidism.*

Mutations can also render receptors constitutively active in the absence of their natural ligand. Several activating mutations have been reported in serpentine receptors; they usually occur in either the transmembrane helices or the cytosolic loops. The former are thought to mimick the ligand occupied state, while the latter improve coupling to G proteins. Luteinizing hormone (LH) is a pituitary hormone that stimulates the gonads to produce sex steroids. Activating mutations in its receptor can stimulate steroidogenesis prematurely, giving rise to *familial male precocious puberty.* Because women require a second hormone for estrogen production, mutations in the LH receptor alone do not lead to any symptoms.

Finally, activating mutations have been identified in a transducer. The protein $G_s$ inactivates itself when it hydrolyzes GTP to GDP (see Section 4.1); mutations that adversely affect its GTPase activity without altering its ability to interact with its effectors would result in the constitutive activation of $G_s$. Because $G_s$ is integral to so many pathways, this type of mutation would be lethal if it were genomic; however, somatic mutations would produce a genetic mosaic in which only some cells would possess the defective gene. Individuals with this type of $G_s$ mosaicism have *McCune–Albright syndrome,* and their symptoms reflect overstimulation from all the hormones that utilize cAMP as a second messenger. These hormones include those that stimulate pigment cells (café au lait spots), the gonads (precocious puberty), the thyroid gland (hyperthyroidism), the adrenal glands (adrenal hyperplasia), and bone (dysplasias).

When these activating mutations occur in the growth factor pathway (e.g., in RTKs, cytokine receptors, Ras, or related transcription factors), tumors often result. Such altered genes, called *oncogenes,* may arise spontaneously within the organism's genome, or they may be introduced by oncogenic viruses. In the latter case, it is believed that these viruses originally acquired proto-oncogenes from their hosts and later converted them to oncogenes because in this altered form they conferred some reproductive advantage on the virus. For example, a rapidly dividing cell may have an abundance of replicative machinery that the virus can then commandeer for itself.

### 6.3 Tumor Resistance to Therapy

In addition to carcinogenic mechanisms, molecular endocrinology can help to explain the development of tumor resistance to certain forms of therapy. When hormone target tissues become malignant, they often retain their endocrine sensitivity. Initially such tumors can be treated by hormone therapy: breast cancer may response to antiestrogens, prostate cancer to antiandrogens; and leukemias and lymphomas to glucocorticoids. Hormone therapy is not only effective but also among the most easily tolerated treatments, because of its relative lack of undesirable effects. The benefits of this therapy are very often short-lived, however, as the tumor eventually develops resistance. Quantitative and structural studies of nuclear receptors can help to determine which tumors will initially respond to hormone therapy and can also explain why many of these growths later become resistant.

Many resistant tumors have inactive receptors. The simplest case is leukemias and lymphomas that undergo apoptosis in response to glucocorticoids. Resistant tumors either lack glucocorticoid receptors or possess defective receptors and, therefore, can no longer respond to these steroids. It is felt that these deletions and mutations arise spontaneously; those that inactivate the receptor offer a selective advantage by allowing the cells that have them to evade the cytotoxic effects of glucocorticoids. Such cells survive and reproduce to create a cancer that is resistant to these steroids.

On the other hand, breast and prostate cancers are dependent on sex steroids for survival; as such, they are treated by hormone ablation or antagonists. Therefore, simple deletion or inactivation of the receptor would prove detrimental to the survival of the cancer. Two types of mutation are observed in these cases. First, some mutations render the receptor constitutively active regardless of the presence or absence of hormone agonists or antagonists. Second, mutations in the ligand-binding domain may reduce the binding specificity of the receptor, making it possible for other steroids, or even steroid antagonists, to activate the receptor. In either instance, the tumors have either reduced or eliminated their hormone dependence.

## 7 PERSPECTIVES

To be able to survive in a competitive and dangerous environment, single cells require careful regulation of metabolism, growth, and reproduction. The development of multicellularity introduced the additional problem of intercellular communications to coordinate these processes. Nature's frugal solution was simply to couple external chemical signals to these preexisting intracellular regulators, which then became second messengers. Therefore, knowledge about molecular endocrinology is knowledge about the very essence of how cells control all their internal functions. In addition, this information provides valuable insights into ourselves via the various disease states that can now be explained by molecular biology. Thus molecular endocrinology affords a critical understanding of many basic physiological processes of life.

*See also* Human Genetic Predisposition to Disease; Oncogenes; Receptor Biochemistry; Steroid Hormones and Receptors.

### *Bibliography*

Blumer, K. J., and Johnson, G. L. (1994) Diversity in function and regulation of MAP kinase pathways. *Trends Biochem. Sci.* 19:236–240.

Bolander, F. F. (1994) *Molecular Endocrinology,* 2nd ed. Academic Press, San Diego, CA.

Cohen, P., and Foulkes, J. G., Eds. (1991) *The Hormonal Control of Gene Transcription.* Elsevier, Amsterdam.

Exton, J. H. (1994) Phosphoinositide phospholipases and G proteins in hormone action. *Annu. Rev. Physiol.* 56:349–369.

McPhaul, M. J., Marcelli, M., Zoppi, S., Griffin, J. E., and Wilson, J. D. (1993) The spectrum of mutations in the androgen receptor gene that causes androgen resistance. *J. Clin. Endocrinol. Metab.* 76:17–23.

Moudgil, V. K., Ed. (1994) *Steroid Hormone Receptors: Basic and Clinical Aspects.* Birkhäuser, Boston.

Petersen, O. H., Petersen, C.C.H., and Kasai, H. (1994) Calcium and hormone action. *Annu. Rev. Physiol.* 56:297–319.

Raymond, J. R. (1994) Hereditary and acquired defects in signaling through the hormone-receptor–G protein complex. *Am. J. Physiol.* 266: F163–F174.

Weintraub, B. D., Ed. (1994) *Molecular Endocrinology: Basic Concepts and Clinical Correlations.* Raven Press, New York.

# Environmental Stress, Genomic Responses to

*John G. Scandalios*

### *Key Words*

**Genome** The totality of a cell's genetic information, including genes and other DNA sequences.

**Genomic Fluidity** The capacity of the genome to reorganize rapidly in response to a given stimulus or signal.

**Oxidative Stress (Oxystress)** An elevation in the steady-state concentration of reactive oxygen species. Occurs when the balance between the mechanisms triggering oxidative conditions and cellular antioxidant defenses is impaired.

**Promoter** A sequence of nucleotides on DNA that is required for the initiation of transcription by RNA polymerase.

**Reactive Oxygen Species (including "Free Radicals")** Toxic by-products of reduced oxygen. These include hydrogen peroxide, superoxide anion, hydroxyl radical, and singlet oxygen that can cause damage by initiating oxidation of various macromolecules.

**Transposable Elements (also, Mobile Elements; "Jumping Genes")** DNA segments that can move from one place to another in a genome.

---

All living organisms are affected by the environment in which they exist. Differences among individual organisms in response to environmental stresses are common, whether the stressing factors be infectious agents, natural environmental variations, environmental chemicals, or any other natural or anthropogenic environmental variables. During their evolution, organisms have evolved a variety of ways of adapting to environmental changes. However, the underlying mechanisms by which cells or organisms perceive environmental adversity and mobilize their defenses to it are far from understood. Such an understanding is essential in any future attempts to engineer organisms for greater tolerance or resistance to more frequent and rapid environmental changes, be they due to natural or anthropogenic causes. As a consequence of recent developments in molecular biosciences and dissection of the genome, a deeper appreciation of the mechanisms by which genes may perceive environmental signals and start a cascade of biochemical events to effect a response to a given signal is now emerging.

## 1 INTRODUCTION

Every living organism is affected by its environment. The environment, whether internal or external to the organism, is continually changing and for the organism to survive it must adapt.

Environmental changes, irrespective of source, cause a variety of "stresses" or "shocks" that a cell must face repeatedly, and to which its genome must respond in a programmed manner for the organism to survive. Examples are responses to oxidative stress, pathogenicity, wounding, anaerobiosis, thermal shock, and the "SOS" response in microorganisms. For cells of any organism to respond to external cues, they must be able to perceive these cues or signals and transduce that perception into the appropriate response. Some sensing mechanism(s) must be present to alert the cell to imminent danger, and to trigger the orderly sequence of events that will mitigate this danger. In addition, there are genomic responses to unanticipated, unprogrammed challenges for which the genome is unprepared, but to which it responds to discernible though initially unforeseen and unpredictable ways.

Many, though not all, signals are perceived at the cell surface by plasma membrane receptors. Activation of such receptors by mechanisms such as ligand binding may lead to alterations in other cellular components, ultimately resulting in alterations in cell shape, ion conductivity, gene activity, and other cellular functions. Identification and isolation of mutants that are unable to respond, or that respond abnormally to a particular signal, may provide ways to decipher the mechanisms by which a particular signal is transduced into a given response.

Long before humans began manipulating and altering their environment, organisms from the simplest to the most complex had started evolving methods to cope with stressful stimuli. Consequently, most living cells possess an amazing capacity to cope with a wide diversity of environmental challenges, including natural and synthetic toxins, extreme temperatures, high metal levels, and radiation. Numerous studies in the past have demonstrated clear "cause-effect" relationships upon exposure of a given organism or cell to a particular environmental factor or stressor. But, it was only recently that certain environmental insults have been shown to elicit specific genomic responses. At present, little is known of the underlying molecular mechanisms by which the genome perceives environmental signals and mobilizes the organism to respond. Such information is not only interesting in and of itself but is also essential in any future attempts to engineer organisms for increased tolerance to environmental adversity.

The recent dramatic advances in molecular biology have made it possible to investigate the underlying mechanisms utilized by organisms to cope with environmental stresses. Investigations of genomic responses to challenge are beginning to shed some light on unique DNA sequences capable of perceiving stress signals, thus allowing the cell or organism to mobilize its defenses.

## 2 GENE RESPONSES

Terminally differentiated cells express an array of genes required for their stable functioning and precise metabolic roles. A genome can respond in a rapid and specific manner by selectively decreasing or increasing the expression of specific genes. Genes whose expression is increased during times of stress presumably are critical to the organism's survival under adverse conditions. Examination and study of such "stress-responsive" genes has implications for human health and well-being, agricultural productivity, and for furthering basic biological knowledge. In addition to aiding the organism under stress, genomes that are modified by stress can be utilized to study the molecular events that occur during periods of increased or decreased gene expression. The mechanisms by which an organism recognizes a signal to alter gene expression and responds to fill that need, are important physiologically and render possible the examination of gene regulation under various environmental regimens.

The mechanisms of induction of stress-response genes are similar among various organisms examined. Similarities in stress-induced changes in gene expression have been observed for a variety of stressors. Some which have been studied in some detail include thermal shock, pathogenic infections, anaerobiosis, photostress, oxystress, water stress, and heavy metals. In all cases, specific changes in transcript and/or protein expression have been observed in various organisms subjected to such challenges. Some of these are discussed below.

## 3 THERMAL STRESS

### 3.1 Heat Shock

Organisms have evolved a variety of ways to adapt to fluctuations in temperature. The most readily discernible response to thermal stress in most organisms is the "heat shock" response. Cells from virtually every kind of organism react to hyperthermic shock by activating a small number of genes, thus inducing the synthesis of a set of proteins, the "heat-shock proteins (HSPs)," that protect the cell from thermal damage. These genes were initially recognized in the fruitfly, *Drosophila melanogaster* embryos as "puffs" in polytene chromosomes arising shortly after subjecting the embryos to a heat shock; i.e., shifting them from their normal growth temperature (25°C) to an increased temperature (37°C). Subsequently, the protein (HSP) products of these genes were identified and characterized. This gene activation is rapid; HSPs appear within a few minutes after heat shock initiation, and it is reversible. A transition from hardly detectable levels of transcription at the normal temperature to extremely high transcription rates occurs during heat shock, leading to rapid accumulation of high levels of HSPs.

### 3.2 Heat-Shock Genes and Proteins

Three major types of HSPs are found in most organisms: (1) the large HSPs ranging in subunit molecular weights from 80K–100K; (2) the intermediate size group (65K–75K); and (3) the small HSP group (15K–30K). The structures of these major HSPs are strongly conserved among animals, plants, yeast, and bacteria. The activation mechanism, and its components, is virtually identical among higher eukaryotes, is similar to lower eukaryotes, and is remotely similar to prokaryotes. For example, the maize (corn) *hsp70* gene has a 75% sequence homology to the *Drosophila hsp70* gene. Detailed promoter analysis of the *hsp70* gene from different species also shows commonality in the presence of a short DNA sequence upstream of the TATA box that is required for heat inducibility. A palindromic consensus regulatory sequence (CT-GAA--TTC-AG) was shown to be sufficient for conferring heat inducibility on a heterologous gene. This sequence is referred to as the ''heat-shock regulatory element (HSE)'' and can be found within the first 400bp upstream of every *hsp* gene, from every higher eukaryote, which has been sequenced. In addition, there are several secondary heat-shock consensus elements located further upstream.

### 3.3 Alternative Inducers of HSPs

Many of the molecular events accompanying heat-shock are also apparent on subjecting cells to other kinds of stress. Many of the stressors seem to interfere with oxidative phosphorylation, or energy production in general. Such agents as metal poisons, sulfhydryl oxidants, and amino acid analogues induce proteins that are identical to the HSPs. In plants, the synthesis of certain HSPs can readily be induced, in addition to heat-shock, by such factors as osmotic stress, arsenite, anaerobiosis, high concentrations of growth factors such as abscisic acid, auxin, and ethylene, and high salt. In addition, some of the *hsp* genes are activated and expressed during normal development. For example, low molecular mass HSPs, particularly HSP70, are transiently expressed during the normal cell cycle. Such findings implicate these proteins in basic metabolic activities of the cell.

### 3.4 Functions of HSPs

Many studies have clearly demonstrated that a pre-shock treatment can render a biological system more resistant to a subsequent thermal stress (thermotolerance) and that this protective effect may be transient. Thermotolerance appears to be important for survival under stress conditions. For example, thermotolerant maize (corn) plants are able to survive the lethal temperature, and to outgrow the control plants when returned to normal temperatures. It has also been demonstrated that HSPs accumulate in field grown, heat-stressed plants, suggesting that these proteins may be critical for plants to cope with natural stress adversity. Other functions attributed to HSPs include a role (HSP70) in translocation of proteins into various eukaryotic organelles, and their participation in the folding and assembly of polypeptides. It is apparent that the *hsp* genes have pleiotropic effects, and that other roles, in addition to conferring thermotolerance will be uncovered as further studies are executed.

### 3.5 Cold Acclimation

Much less is currently known about cold acclimation than is the case with heat adaptation. What is presently known is based primarily on recent work with plants. Perhaps the most dramatic manifestation of cold acclimation (cold hardiness) is the increased freezing tolerance that occurs in many plant species. Extreme examples of cold adapted perennial species include the dogwood and birch. Species not acclimated to cold are killed by temperatures of about −10°C, whereas some cold-hardy species can survive experimental temperatures of −196°C.

Some biochemical alterations found to accompany cold acclimation include changes in lipids, increases in soluble protein and sugars, expression of new isozymes, and changes in mRNA populations. It was demonstrated as early as 1912, by wheat breeding experiments, that frost hardiness had a complex quantitative genetic basis. More direct evidence has recently been obtained from molecular investigations leading to the isolation and characterization of specific ''cold-regulated *(cor)* genes'' from various plant species. Although correlations between *cor* gene expression and freezing tolerance have been found, the exact regulation, expression and role of these genes await further resolution.

The more thoroughly investigated heat-shock response is quite different than the cold acclimation response. Unlike heat-shock, the changes in *cor* gene expression that accompany cold acclimation are relatively mild, are not transient, and the genes expressed at normal temperatures continue to be expressed at the low temperatures. Whereas, expression of *hsp* genes is generally accompanied by suppression of preexisting mRNAs. Thus, cold acclimation and heat-shock appear to be distinct responses.

## 4 OXIDATIVE STRESS

Oxygen is essential for life on Earth. In its ground state (its normal configuration, $O_2$) oxygen is relatively unreactive. However, during normal metabolism, and as a consequence of various environmental perturbations and pollutants (e.g., radiations, drought, air pollutants, cigarette smoke, temperature stress, herbicides, etc.) oxygen ($O_2$) gives rise to various highly toxic and lethal intermediates. These intermediates, referred to as active oxygen species and free radicals include the superoxide radical ($\cdot O_2^-$), hydrogen peroxide ($H_2O_2$), and the hydroxyl radical ($\cdot OH$). These and the physiologically energized form of dioxygen, singlet oxygen ($^1O_2$), are the biologically most important active oxygen species. All of these are extremely reactive and cytotoxic to all organisms. For example, $\cdot OH$, one of the most potent oxidizing agents known, reacts with most macromolecules (DNA, proteins, lipids, etc.) to cause severe cellular damage leading to physiological dysfunction and cell death. Some of the biological consequences of oxidative damage include peroxidation of membrane lipids, loss of organelle function, mutations, enzyme inactivation, reduced metabolic efficiency, and reduced carbon fixation leading to impaired photosynthetic capacity in plants. Free radicals and derivatives have been implicated as causative agents in numerous human diseases including the aging process, cancer, emphysema, and immunologic impairments. Thus, the effective and rapid elimination of active oxygen species is essential to the proper functioning and survival of all living organisms.

### 4.1 Protective Antioxidant Defenses

Nature has thus presented us with the "oxygen paradox." For life to be sustained, oxygen is required; yet in its reduced state this sustainer of life becomes a deterrent to life. As a consequence of this paradox, organisms evolved antioxidant defense mechanisms to protect themselves. Such defenses include both non-enzymatic as well as enzymatic mechanisms. Among the former are included such substances as β-carotene, vitamins C and E, flavonoids, and hydroxyquinones. Enzymatic defenses include enzymes capable of removing, neutralizing, or scavenging free radicals and oxy-intermediates. Without such defenses, plants could not efficiently convert solar into chemical energy, and life on Earth, as we know it, would not be possible.

Examples of enzymatic antioxidant defenses include ascorbate peroxidase and glutathione reductase, believed to scavenge hydrogen peroxide in chloroplast and mitochondria, respectively; catalase (CAT) and peroxidases that efficiently remove hydrogen peroxide from cells, and superoxide dismutase (SOD) that scavenges the superioxide anion. The CAT and SOD enzymes are the most efficient antioxidant enzymes, whose combined action converts the superoxide radical and hydrogen peroxide to water and molecular oxygen, thus abating the formation of the most toxic and highly reactive hydroxyl radical, averting cellular damage.

Although numerous defenses to cope with oxidative stress exist in all aerobic organisms, it is not yet clearly understood how these organisms mobilize their defenses to respond at the appropriate time. Increases in oxystress often lead to correlative increases in some antioxidant defenses. However, little is currently understood as to how the genome perceives oxidative insult and mobilizes a response to it. Such information is essential in any future efforts to engineer organisms for increased tolerance to environmental oxidative stress. To understand the underlying mechanisms, a great deal of effort has been expanded in recent years at elucidating the responsive antioxidant defense genes. Using state of the art molecular techniques, their structure, regulation, and expression is being resolved giving some insight as to how such genes act to mobilize cellular defenses. Some recent findings are discussed below.

### 4.2 Enzymatic Antioxidant Defenses

The genes encoding various catalases and superoxide dismutases have been isolated and characterized, and found to be evolutionarily conserved, from numerous organisms. These genes and their products have been effectively utilized in some novel experiments which have clearly demonstrated that these genes indeed play major protective roles against oxidative damage. For example, there is a yeast mutant that lacks the gene responsible for producing the manganese superoxide dismutase (MnSOD) that translocates to its mitochondria where it normally functions. This mutation is conditionally lethal in that this yeast is unable to grow in an oxygen environment. In recent experiments, the *MnSod (Sod3)* gene of maize was isolated and successfully transferred to the MnSOD-deficient yeast cells. The maize gene was properly transcribed, translated, and its protein product (the MnSOD maize protein) was properly imported and processed into the yeast's mitochondria. Most significantly, the corn MnSOD protein functioned correctly resulting in the rescue of the transgenic yeast cells which now grew normally in an oxygen environment, and were able to cope with imposed oxidative stress. Whether such approaches will prove successful in engineering various other organisms for greater tolerance to oxystress remains to be seen. However, the prospects appear fruitful.

Isolation of such genes provides the opportunity to determine intron-exon boundaries, start and stop sites of transcription, *cis*-regulatory regions, and to identify sequences responsive to oxidative stress signals that might indicate how these genes are regulated to respond and protect cells against oxystress.

### 4.3 Antioxidant Gene Responses

Oxidative stress in bacteria has been shown to activate the transcriptional regulator *oxyR,* which in turn induces genes whose products protect cells from oxidative damage. The OxyR protein is directly activated by the metabolic stimulus, an oxidant, to become the transcriptional activator. Thus, OxyR protein is both the sensor of oxidative stress and the mediator of enhanced transcription of genes whose products are components of the protective response. For example, $H_2O_2$ added to bacterial growth medium, interacts with OxyR to change its oxidation state, causing a conformational change that affects the way the protein interacts with its target promoters. The OxyR protein is encoded in a single dominant regulatory gene *(oxyR).* The several antioxidant defense structural genes scattered around the genomes of *E. coli* and *S. typhimurium* and controlled by *oxyR* constitute a "regulon."

Regulons parallel to the bacterial "OxyR regulon" have not been identified in eukaryotic cells. However, as the structure of eukaryotic antioxidant defense genes is being unraveled, other signal recognition factors are being encountered.

For example an eleven base pair motif (5′-puGTGACNNNGC-3′) referred to as the antioxidant response element (ARE) has been found in the promoter region of several eukaryotic antioxidant defense genes, including the maize *Cat1* gene. There exists convincing evidence from rat studies that the ARE might represent a *cis*-acting element which activates genes that protect eukaryotic cells against oxidative stress.

It is becoming clear that as more and more genes responsive to environmental stimuli are isolated and characterized, motifs or elements, are identified which may be responsible for perceiving the stimuli and initiating the appropriate defensive response. Thus, a fruitful research direction involves the identification of both *cis*-acting elements and *trans*-acting substances (like the OxyR protein) relative to environmental stimulus responsive gene systems. It is also interesting that genes responsive to environmental stimuli are often found to be highly polymorphic, and to have additional functions such as roles in development. It is likely that polymorphism in genes and systems that participate in stress defense strategies is essential for the organism to survive varied, and continually changing environmental stresses.

## 5 RAPID GENOMIC RESPONSES

There are programmed responses to threats that are initiated within the genome itself, that can lead to new and irreversible genomic modifications. Thus, in many organisms, the genome may reorganize itself on facing an adversity for which it is unprepared in order to ensure the organism's survival. For example, cells are able to sense the presence in their nuclei of ruptured ends of chromosomes and then to activate mechanisms that will bring together and then unite these ends, one with another. This is a most revealing example of the sensitivity of cells to what is occurring within them.

Although the capacity of the genome to reorganize and respond rapidly (''genomic fluidity'') has been considered for a long period, it was not fully appreciated until the pioneering work of Barbara McClintock, demonstrating that discrete genetic loci could transpose in the genome of maize, was fully accepted and shown to be universal among all organisms studied.

### 5.1 Transposable Elements

McClintock demonstrated that transposable elements regulate gene expression following their insertion into a given locus (she used the term ''controlling elements''). Transposable elements provide a regulated disruption of the chromosomes in cells in which they are active. Activation of transposable elements is a recognizable consequence of the cell's response to trauma. The mobility of these elements allows them to enter different loci and to take control of the action of the gene in which the element is inserted. In addition to modifying gene action, transposable elements can restructure the genome at various levels from changes involving a few nucleotides to gross modifications of large chromosome segments.

## 6 ADDITIONAL GENOMIC RESPONSES TO STRESS

### 6.1 Plant Galls

Numerous cases are known in nature where one organism is forced to produce a wholly new structure for the benefit of another. Plant galls provide numerous such examples. The exact structural organization of a given gall gives it a uniqueness that begins with an initial stimulus, characteristic of each invasive species. The galls on legume roots associated with nitrogen-fixing bacteria provide a clear example of reprogramming of the plant (legume) genome by a stimulus received from foreign organisms (bacteria). In other examples, a single plant may have on its leaves several distinctly different galls, each catering to a different insect species. The stimulus for placing the insect egg into the leaf initiates reprogramming of the plant's genome, forcing it to make a unique structure adapted to the needs of the developing insect. Such genomic reprogrammings by a variety of organisms, including bacteria, fungi, and insects, are not a requisite response of the plant host genome during its normal developmental cycle.

### 6.2 Tissue Culture Induced Changes

An interesting example of how a genome may modify itself when confronted with unusual conditions is provided by culturing cells, particularly plant cells, in artificial conditions. When cells are removed from their normal locations and placed in tissue culture, a variety of changes are observed. It is common to observe novel phenotypes emerge in many species when plants have been regenerated from protoplast or tissue cultures. Much of the variation observed in cultures is a result of genomic change (''somaclonal variation''). The passage from cell isolation, callus formation, and the ensuing production of whole plants, inflicts on the cells a succession of traumatic stresses. These stresses may result in the abnormal reprogramming of the genome in a way that does not follow the same orderly sequence that occurs in natural conditions. Thus, giving rise to a wide range of unexpected altered phenotypes, which may or may not be heritable. Somaclonal variation provides fertile ground for selection, by breeders, of useful traits. It thus appears that tissue culture represents a form of stress on cultured cells, which in turn respond by decidedly restructuring and resetting the genome.

### 6.3 Photoresponses

Light is the most important environmental stimulus to which plants react. Light plays an indispensable role throughout the life of higher plants. Light provides the energy for photosynthesis, so essential for the existence of life on Earth, and it has profound effects on gene expression by serving as a trigger and modulator of complex developmental and regulatory mechanisms. Fluctuations in light quality and quantity lead to alterations in the activity of specific genes, which culminate in a variety of developmental responses, ranging from seed germination, differentiation, and flowering. Light responses are mediated by at least four photoreceptors: protochlorophylide, phytochrome, blue light receptors, and an ultraviolet B receptor. These photoreceptors, probably in conjunction with different proteins, give signals to regulatory sequences associated with light-response genes. In fact, numerous genes are now known whose amount of mRNA transcribed changes directly in response to light. Several such genes have been shown to have a consensus ''light response element (LRE)'' within their promoter regions. The *trans*-acting factors that interact with such elements to effect the appropriate gene response to light are yet to be identified.

Although essential, light imposes considerable stress. Consequently, plants and other organisms have incorporated considerable light signals in their developmental pathways to cope with a photodependent lifestyle.

The exquisite precision by which light governs the growth, development and aging in plants, is just now beginning to be understood and appreciated. Unraveling the underlying mechanisms of the light-induced responses, and how they are modulated or modified by other factors will prove challenging, and will provide a deeper understanding of the regulatory networks within cells. It will also provide a better understanding of how cells are able to selectively derive the beneficial and minimize the detrimental aspects of the photoenvironment.

## 7 CONCLUSIONS AND SPECULATIONS

Genomic flexibility is an extraordinary adaptability of organisms to their environment. Although the capacity of the genome to reorganize in the face of adversity had been considered since Mendel's laws of heredity were formulated, it was the seminal contributions of Barbara McClintock that set the stage for our current understanding of genomic fluidity.

In the course of evolution, those organisms that had the capacity to adapt to new environments survived. This capacity was due to the ability of each organism or species that survived, to possess in its biological repertoire the means to cope with environmental adversity. In situations where multiple environments existed, those organisms possessing motility (e.g., animals) could exchange one environment for another; those with a stationary lifestyle (e.g., plants) had to rely on alternate defenses. However, in adverse environments (natural or anthropogenic) from which organisms cannot escape, they must rely on diverse and unique sets of responses encoded within, or resulting from the flexibility of their genome. Recent work, as discussed above, has demonstrated that environmental stress itself can instigate genome modifications by mobilizing cell mechanisms that can restructure genomes in various ways. Such genomic reorganization allows the organism to cope with stress, and in the long term may provide the bases for evolution of new species.

As the genomes of various organisms are being investigated with state of the art molecular techniques, more knowledge is being gained as to structural components involved in the response of specific genes to environmental insult. However, we know virtually nothing about how the cell perceives danger and initiates the truly remarkable observed responses to it. Once such information is available, and a better understanding of the defense response is attained, the engineering of organisms for resistance or sustained tolerance to adverse environmental stresses will be possible.

*See also* FREE RADICALS IN BIOCHEMISTRY AND MEDICINE; GENETIC DIVERSITY IN MICROORGANISMS; HEAT SHOCK RESPONSE IN *E. COLI*.

### *Bibliography*

Bienz, M. (1985) Transient and developmental activation of heat-shock genes. *Trends Biochem. Sci.* 10:157–161.

McClintock, B. (1984) The significance of responses of the genome to challenge. *Science,* 226:792–801.

Scandalios, J. G., Ed. (1987) *Molecular Genetics of Development.* Academic Press, San Diego, CA.

Scandalios, J. G., Ed. (1990) *Genomic Responses to Environmental Stress.* Academic Press, San Diego, CA.

Scandalios, J. G., Ed. (1992) *Molecular Biology of Free Radical Scavenging Systems.* Cold Spring Harbor Laboratory Press, Cold Spring Harbor, NY.

# ENZYME ASSAYS

*Robert Eisenthal*

### *Key Words*

**Initial Rate** Rate of conversion of substrate to product in an enzyme-catalyzed reaction, measured during the steady state.

$K_m$ **(Michaelis Constant)** Substrate concentration at which the rate of an enzyme-catalyzed reaction is half the maximum velocity, $V_{max}$.

**Product Inhibition** Inhibition of an enzyme-catalyzed reaction caused by product(s) binding to enzyme species.

**Steady State** Period of an enzyme-catalyzed reaction during which the concentrations of reaction intermediates are effectively constant.

The assay of enzyme activity is one of the most frequently performed experimental procedures in biochemistry. Enzyme assays are essential for assessing the activity of genetically engineered enzymes, for determining kinetic parameters, for estimating the amount of enzyme present in a cell or tissue, and for following the progress of an enzyme purification procedure. This entry describes the principles of designing and carrying out enzyme assays, but it is not a compilation of assays for individual enzymes. Such information may be found in appropriate volumes of *Methods in Enzymology* and in *Methods of Enzymatic Analysis.*

## 1 UNITS OF ENZYME ACTIVITY

The most frequently used definition of enzyme activity is the *International Unit* (IU; also enzyme unit), defined by the Enzyme Commission of the International Union of Biochemistry as the amount of enzyme that catalyzes the transformation of substrate to product at a rate of *one micromole per minute.* It should be noted that the enzyme unit refers to an *amount* of substrate transformed, and not to a *concentration.* Since the rate of change in an enzyme assay is often measured as a concentration change, it is important to take into account the volume of the assay mixture when interconverting measured rates and enzyme units.

The introduction of SI units brought about the proposal that the katal be used as the unit of enzyme activity. The *katal* is defined as the amount of enzyme catalyzing the conversion of substrate to product at a rate of *one mole per second.* This is an inconveniently large unit in terms of amounts of enzyme activity actually measured *in vitro* (or indeed occurring *in vivo* in any single organism) and is less often used. The conversion factors for international units and katals are as follows:

$$1 \text{ Kat} = 6 \times 10^7 \text{ IU}$$
$$1 \text{ IU} = 16.67 \text{ nKat}$$

*Specific activity,* a useful term for describing the purity and activity of an enzyme, is usually expressed as *enzyme units per milligram of protein.* The *molecular activity* of an enzyme is defined as *enzyme units per micromole of enzyme.* If the molecular weight of the enzyme and the purity of the preparation are known, the molecular activity may be calculated from the specific activity. Molecular activity is expressed in units of reciprocal time, and is equivalent to the catalytic constant, $k_{cat}$. It is the number of molecules of substrate transformed to product per molecule of enzyme per unit time (conventionally per second), and is a measure of the catalytic power of the enzyme. *Turnover number,* a term still occasionally used, is equivalent to the molecular activity, but is sometimes used to refer to the catalytic site activity, which is the molecular activity divided by the number of catalytic sites per enzyme molecule.

When reporting the results of activity assays, information on the pH and composition (including ionic strength) of the assay buffer, and the temperature, should always be stated, as all these affect enzyme activity.

## 2 INITIAL RATES

It is essential for proper design of an enzyme assay that the measured activity be proportional to the total concentration of enzyme present. For this to be true, the *initial rate* (i.e., the steady state rate) must be determined. Most enzyme assays are conducted at near-ambient temperatures, with the initial substrate concentration $S_0$ greatly exceeding the total enzyme concentration $E_0$. The condition $S_0 >> E_0$ ensures the validity of the steady state assumption and enables the rate of substrate disappearance to be equated with the rate of product appearance, so that either may be followed to assess the rate of the reaction. Under these conditions, the steady state is usually attained well within one second of mixing the enzyme with the substrate(s). From this point on, the rate (the change of product concentration per unit time) will usually decrease. Provided that an accurate estimate of the rate can be made at "zero" time, the rate may be termed an initial rate and will be proportional to the enzyme concentration. Another reason for using the initial rate is that $S_0$ is most accurately known (at zero time). It is usually desirable to keep $S_0 \geqslant 10 \times K_m$ (but beware substrate inhibition, see Section 3.1). This condition not only makes the rate relatively insensitive to errors in $S_0$ but also increases the time over which the initial rate maintains near-constancy. If the progressive decrease in rate as the reaction proceeds occurs to a sufficient extent to make the estimate of initial rate inaccurate, the measured rate will not be proportional to the enzyme concentration. For this reason, continuous assays (see Section 4.1) are to be preferred to discontinuous assays, since the former, deviations of the progress curve from linearity will be easy to identify. Suitable methods may then be applied to correct for the curvature, or the assay method itself may be altered to improve the linearity of the progress curve.

## 3 WHY DOES THE RATE DECREASE?

The most common causes for downward curvature in the reaction time course (or *progress curve*) are described in Sections 3.1 to 3.5.

### 3.1 Substrate Depletion

Unless a method exists for maintaining the substrate concentration at a constant value (e.g., by recycling the product), substrate depletion is an inevitable consequence of the progress of the reaction. For this reason the initial substrate concentration is usually set at a value greater than $K_m$, the Michaelis constant of the enzyme (ideally $\geqslant 10 \times K_m$), so that the rate will change little as substrate is used up. However one should always take care not to use unnecessarily high initial substrate concentrations, which could lead to substrate inhibition, a not uncommon phenomenon. This in turn could give rise to progress curves in which the rate actually increases as the substrate is used up. High substrate concentrations may also result in the depletion of cofactors such as metal ions that may be essential for enzyme activity.

The effect of substrate depletion may be minimized by observing the rate of reaction over a period of time such that the fraction of the total substrate concentration consumed is very small. This may be accomplished by increasing the sensitivity of the method of detection to allow the determination of reaction rate over a shorter time period or by decreasing the amount of enzyme in the assay. A serious case of substrate depletion results if the substrate is chemically unstable under the assay conditions. Although it may be possible to correct for such instability by analyzing the kinetics of the decomposition reaction, it is preferable to alter the assay conditions so that the substrate is stable. For example, assays involving $NAD(P)^+$-dependent dehydrogenases are rarely carried out at pH values much below 7 because NAD(P)H is unstable under acid conditions.

### 3.2 Approach to Equilibrium

If the catalyzed reaction is reversible, the contribution of the back-reaction to the net rate will increase as the reaction approaches equilibrium. This effect can be minimized by running the assay under conditions that shift the equilibrium in favor of product formation. Thus, enzymes that catalyze reactions producing hydrogen ions, such as those involving the conversion of $NAD(P)^+$ to NAD(P)H, are often assayed at pH > 9.

A related approach is to use trapping reagents that remove the product as it is formed. An example of this is found in the assay of alcohol dehydrogenase, which catalyzes the oxidation of ethanol by $NAD^+$. This assay is often run in the presence of semicarbazide, which reacts with the acetaldehyde produced to form the semicarbazone and thus removes the product from the reaction mixture. If the purpose of the assay is simply to assess the amount of enzyme activity, it may be helpful to run the assay in reverse [i.e., to interchange substrate(s) and product(s)]. For example, at pH values below 9, the equilibrium of the reaction catalyzed by alcohol dehydrogenase favors the formation of alcohol; thus the progress curve is linear for longer periods of time when assays are conducted in this direction.

### 3.3 Product Inhibition

Curvature of the progress curve can result from product inhibition even when the reaction is effectively irreversible. This effect is sometimes so severe that it is necessary to sacrifice the convenience of a continuous assay for the accuracy of a more sensitive discontinuous one. The only way to eliminate this phenomenon (i.e., product inhibition) is to remove the product from the assay as rapidly as it is formed. In the assay of catechol-*O*-methyltransferase (COMT), the enzyme is powerfully inhibited by the reaction product, *S*-adenosylhomocysteine. This can be removed by inclusion in the assay of adenosine deaminase (ADA), which converts the inhibitory product into *S*-inosylhomocysteine, which does not inhibit COMT.

### 3.4 Inactivation of Enzyme

Inactivation of the enzyme during the assay will also cause a decrease in rate and curvature of the time course. However, the remedies suggested in Sections 3.1 to 3.3 are unlikely to remove the cause. Indeed some of the measures suggested (e.g., dilution of the enzyme) may well exacerbate the problem if, for example, the cause of the inactivation is irreversible dissociation of the enzyme into inactive subunits, or adsorption on glass. A simple graphical test for enzyme inactivation during an assay is available and should be applied whenever this is suspected. If inactivation is found to be occurring, one or several approaches may be tried.

Proteins are frequently stabilized by the presence of other proteins; for this reason serum albumin is often included (at 0.1–1.0 mg/mL) in assay mixtures. If the cause of enzyme inactivation is oxidation of —SH groups, addition of thiol reagents such as dithiothreitol or mercaptoethanol may help, although in extreme cases it may be necessary to use degassed solutions and to exclude oxygen from the assay. Trace heavy metals in assay components may also cause loss of enzyme activity; for this reason chelating agents, such as EDTA, are often included in assay mixtures. The presence of glycerol (5–20% v/v) has also been reported to stabilize enzymes in dilute solution. Treatment of glass vessels with siliconizing agents or the use of plastic vessels may help if inactivation is due to adsorption.

### 3.5 Artifactual Causes

A well-designed assay should not be plagued by artifacts, but one should be on guard for the presence of, for example, inadequate buffering in a reaction involving production or consumption of hydrogen ions, inadequate temperature control of the assay or addition of cold assay components, and nonlinearity of the instrumental method (frequently spectrophotometric) used for the assay.

## 4 TYPES OF ASSAYS

Assays can be continuous or discontinuous, direct or indirect.

### 4.1 Continuous or Discontinuous Assay?

Continuous assays may be used whenever a sufficient difference exists between a directly detectable property of a substrate and product. Most commonly, continuous assays utilize changes that may be followed photometrically, fluorimetrically, polarographically, polarimetrically, or through changes in pH that occur as the reaction proceeds. A great advantage of such assays is that the shape of the progress curve is revealed, and any irregularities in curvature are readily detected. Such information not only simplifies initial rate estimation, but where anomalous behavior is observed, it may disclose interesting properties of the enzyme. A possible disadvantage is that continuous assays must be carried out sequentially. Automated instrumentation is commercially available and may be useful where many samples must be processed under identical assay conditions. For example, enzyme-linked immunosorbent assay (ELISA) plate readers, which can read 96 samples almost simultaneously and repeatedly, enable a large number of continuous assays to be processed in one operation.

Discontinuous assays, sometimes termed stopped assays or sampling assays, are carried out by terminating the reaction by inactivating or removing the enzyme and/or stopping the reaction, often by raising or lowering the pH of the assay mixture or of a sample withdrawn from the assay. Substrate and product are then separated, or the quenched reaction mixture is treated to produce a chemical change in either substrate or product (e.g., a color reaction), allowing one or the other to be detected. Discontinuous assays have two disadvantages compared with continuous ones. First, the shape of the progress curve is not readily apparent (unless samples are taken at many time points). Also, timing and volume inaccuracies may be introduced, associated with the termination of the reaction and withdrawal of samples at fixed times. Thus it is especially important to ensure that rates estimated using such assays are truly proportional to enzyme concentration. However, discontinuous methods do allow many assays to be run at the same time. They are also inherent in the use of sensitive and selective assay techniques that require separation of product and substrate either prior to (as in radioactivity measurement) or as an integral part [as in high performance liquid chromatographic (HPLC) techniques] of the measurement.

### 4.2 Direct Assays

The spectrophotometric assay of xanthine oxidase is a good example of a direct continuous assay. Here the substrate, xanthine, is converted into uric acid. Both substrate and product absorb light in the UV region but have different absorbance maxima. The difference between the absorption coefficients of substrate and product is greatest at 292 nm—a convenient wavelength to use for the assay. Many HPLC and radiochemical assays are examples of direct discontinuous methods (see Section 4.1).

### 4.3 Indirect Assays

Many assays are indirect, in that the change in product or substrate concentration is measured after a subsequent chemical reaction has produced an observable phenomenon (e.g., a color change). Coupled assays are an important example of indirect assays.

### 4.4 Coupled Assays

Although it is usually wise to keep experimental protocols as simple as possible, coupled enzyme assays are often used. In this procedure, a reaction of interest is linked to one or more subsequent reactions. Coupled assays may be continuous or discontinuous. As an example of the former, one may use the often-employed pyruvate kinase–lactate dehydrogenase (PK-LDH) system. This enzyme couple is used to assay enzyme-catalyzed reactions that produce ADP [e.g., glycerol kinase (GK)]:

$$\text{glycerol} + \text{ATP} \xrightarrow{\text{GK}} \text{glycerophosphate} + \text{ADP}$$

$$\text{ADP} + \text{phosphoenolpyruvate} \xrightarrow{\text{PK}} \text{ATP} + \text{pyruvate}$$

$$\text{pyruvate} + \text{NADH} \xrightarrow{\text{LDH}} \text{lactate} + \text{NAD}^+$$

This system allows GK to be assayed continuously by observing the decrease in absorbance at 340 nm as NADH is converted to $NAD^+$. For this, the assay mixture would contain phosphoenolpyruvate, PK, LDH, and NADH, in addition to GK, glycerol, and ATP. The disadvantages of increasing the complexity of the assay are outweighed by the simplicity of the continuous spectrophotometric method. As a bonus, the substrate, ATP, is recycled. An example of a discontinuous coupled assay is that described for COMT (see Section 3.3). Here the disadvantage of introducing an additional component (ADA) is more than compensated by the removal from the system of the strongly inhibiting product.

In a coupled assay, the rate actually measured should equal the rate of the reaction being catalyzed by the enzyme under investigation. Thus it is essential that the coupling enzyme(s) be present in sufficient excess (in activity terms) over the enzyme under study to prevent the rate of the coupling reactions from becoming limiting. There is also an inevitable lag in such assays as the substrate(s) of the coupling reaction(s) build up to their steady state levels in the reaction mixture. Increasing the amount of coupling enzyme in the assay will reduce this lag. However, it is unwise simply to add a ''vast excess'' of coupling enzyme. In addition to being wasteful and costly, this practice may lead to unexpected complications if the coupling enzyme contains contaminating enzyme activities that interfere with the overall assay. Methods are available for the calculation of the amounts of coupling enzyme necessary to give desired lag times and ratios of measured to true activity.

## 5 ESTIMATION OF INITIAL RATE

If a progress curve is sufficiently linear, the initial rate may be estimated by eye and a ruler. When fitting by eye, care must be taken to avoid subjective bias. For a reasonably accurate estimate, the extent of reaction followed should be no more than 5%, and the substrate concentration should be no greater than $5 \times K_m$. In some circumstances (e.g., the equilibrium is highly unfavorable, or the substrate concentration necessarily is less than $K_m$), it may be impossible to avoid highly curved time courses. In these cases it is best to use an objective analytical method, such as fitting the time course to a polynomial. A somewhat less empirical approach based on the integrated Michaelis–Menten equation is also available. Such methods are equally applicable to discontinuous and continuous assays (provided a sufficient number of points are obtained). However, when many discontinuous assays are to be performed under varying conditions (e.g., of substrate or inhibitor concentrations), determination of a progress curve of several points for each assay may involve an impracticably large number of experiments. In such cases it is usually worthwhile to try to linearize the initial portion of the progress curve by altering the fixed conditions of the system. Such investigations should be carried out under conditions of the variable(s) giving the highest and lowest rates in the experimental series. If the progress curves are linear for these rates, it is usually safe to assume that they will be linear over that time period for all assays measuring intermediate rates. As a further check, proportionality of rate with enzyme concentration should be determined for conditions giving the highest and lowest rates.

Lags or bursts in progress curves, which may cause problems in initial rate estimation, can arise for artifactual reasons (instrumental response, dust particles, temperature equilibration, etc.) or from factors inherent in the assay or the enzyme, such as product activation, pre-steady-state transients, or hysteretic effects. It is obviously essential to determine the cause of such behavior, to ensure that the correct portion of the progress curve is used to estimate the initial rate. Once a reliable method for initial rate estimation has been established for a given set of assay conditions, it is usually safe to assume that it will be valid for all smaller rates measured.

## 6 LINEARITY OF MEASURED RATE WITH ENZYME CONCENTRATION

Provided the considerations discussed above have been met, a plot of activity versus enzyme concentration will be a straight line going through the origin. If the activity is expressed as a molar concentration change, if the substrate concentration is saturating, and if the molar concentration of the enzyme is used, the slope of this line will be the turnover number. However deviations from linearity may occur, and it is important to realize when these are due to some property of the enzyme are a result of an experimental artifact.

Failure to subtract a blank rate will result in a straight line that intersects the ordinate axis, usually above the origin. Blank rates (i.e., nonenzymic rates) may arise from several causes. The most obvious is the occurrence of the uncatalyzed reaction. However instability of the detected substrate, e.g., resulting from the presence of NADH oxidase in the GK assay system (Section 4.4), would also result in a blank rate. Whenever assay conditions are altered, such rates should be determined and subtracted from the observed enzymic rate. Any artifactual situation that affects the shape of progress curves, such as instrumental limitations, may also result in deviation from linearity of the relationship of rate to enzyme concentration. It should be realized that such behavior may be a result of a property of the enzyme itself. If the enzyme dissociates into subunits that are inactive, the plot of velocity versus $E_0$ may show upward curvature, since the fraction of enzyme in the active form will increase with increasing $E_0$. The converse will obtain if it is the dissociated form is active and the aggregated form that is inactive. Be warned that similar results may be obtained if the enzyme stock solution is contaminated with a reversibly dissociating activator (upward curvature) or inhibitor (downward curvature).

## 7 SUMMARY

The assay of enzyme activity, which is essentially a kinetic measurement, presents many pitfalls for the unwary. If the caveats described here are heeded, enzyme assays give valuable information and allow reliable conclusions to be drawn from kinetic data.

*See also* ENZYME MECHANISMS, TRANSIENT STATE KINETICS OF; ENZYMES.

### *Bibliography*

Bergmeyer, H. U., Ed. (1974 and 1983) *Methods in Enzymatic Analysis,* Vols 1–4, 2nd ed. and 1–10, 3rd ed. Verlag Chemie, Weinheim, and Academic Press, New York.

Colowick, S. P., Kaplan, N. O., et al., Eds, (1955–) *Methods in Enzymology,* Vol. 1 *ff.* Academic Press, New York.

Cornish-Bowden, A. (1979) *Fundamentals of Enzyme Kinetics,* Chapters 2 and 3. Butterworths, London.

Tipton, K. F. (1992) Chapter 1 in *Enzyme Assays,* R. Eisenthal and M. J. Danson, Eds. Oxford University Press, Oxford and New York.

Wharton, C., and Eisenthal, R. (1981) *Molecular Enzymology,* Chapter 10. Blackie & Son, Glasgow.

# ENZYME MECHANISMS, TRANSIENT STATE KINETICS OF

*Kenneth A. Johnson*

### *Key Words*

**Chemical-Quench-Flow** A method of observing rapid reactions by first mixing two solutions to initiate the reaction followed by mixing with a third solution to quench the reaction. Products of the reaction are then separated to allow quantitation.

**Enzyme Kinetics** The study of the rates of enzyme-catalyzed reactions with the goal of establishing the mechanism of reaction.

**Stopped-Flow** A method of observing a reaction by mixing two solutions and then following the reaction time course by optical methods after the flow into an observation flow cell is stopped. Reaction rates up to 500 to 1000 $s^{-1}$ can be observed.

**Transient State** The rapid analysis of chemical or enzymatic reactions in the early phase of a reaction.

Transient state kinetic analysis involves following the time course of a reaction to completion for two reasons: to determine the rate of the reaction and to use that information to establish mechanism. The term *transient state* usually refers to the rapid (millisecond time scale) analysis of chemical or enzymatic reactions in the early phase of the reaction preceding the more commonly studied *steady state.* However, most of the kinetic analysis of biological reactions relies on transient state kinetic methods because these methods pertain to single molecular events. For example, the binding of a ligand to a cell surface receptor and the binding of a repressor to DNA fall under the realm of transient kinetics because both involve a single reaction proceeding to completion, as opposed to the

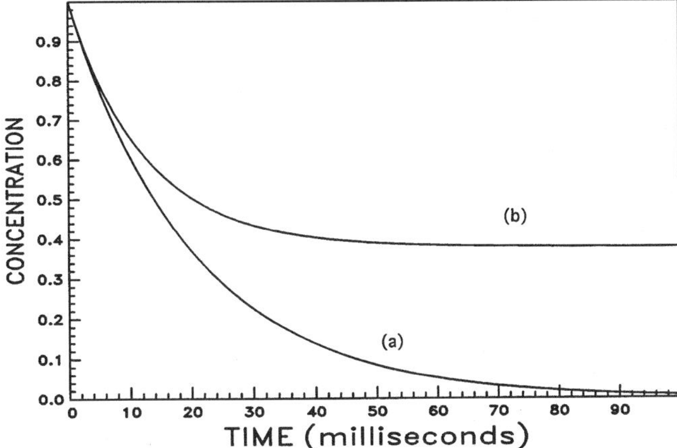


**Figure 1.** The time courses two hypothetical enzyme–substrate binding reactions: (a) irreversible and (b) reversible. The curves were calculated by numerical integration of the simple binding mechanism described in the text with [E] = 1 μ$M$, [S] = 50 μ$M$, $k_1 = 1 \times 10^6\ M^{-1}\ s^{-1}$ and $k_1 = 30\ s^{-1}$ in curve b or 0 in curve a.

multiple turnovers that characterize the conceptual framework of steady state enzyme kinetic analysis.

The contribution describes the fundamentals of transient state kinetic analysis of bimolecular and unimolecular reactions in biological systems.

## 1 THEORETICAL ASPECTS

A reaction involving the binding of a substrate S to an enzyme E is described by the kinetic equation:

$$E + S \underset{k_{-1}}{\overset{k_1}{\rightleftharpoons}} ES$$

The constant $k_1$ defines the *second-order rate constant* for substrate binding to the enzyme site, while the value of $k_{-1}$ defines the *first-order rate constant* for dissociation of substrate from the enzyme site. A first-order rate constant gives the rate of spontaneous reaction in units of events per second ($s^{-1}$). A second-order rate constant includes the concentration dependence in units of events per second divided by concentration ($M^{-1}\ s^{-1}$). For most biological reactions, second-order binding rates fall in the range of $10^4$–$10^9\ M^{-1}\ s^{-1}$, where the upper limit is a function of the diffusional collision frequency. Dissociation rates are typically on the order of $10^{-4}$–$10^3$ $s^{-1}$ (half-lives of 2 to < 1 m).

If the binding of E to S is irreversible ($k_{-1} = 0$), and the concentration of S is greater than the concentration of E, the reaction will proceed to completion at a rate of $k_{obs} = k_1[S]$ as shown in Figure 1. The time course of disappearance of E will follow a *single exponential* equation, $[E] = [E]_0 \exp(-k_1[S]t)$. Analysis of the reaction time course by nonlinear regression would give the value of $k_{obs}$ ($s^{-1}$). Division of the observed rate by the concentration of substrate gives the value for the second-order rate constant ($M^{-1}\ s^{-1}$).

In practice, reactions in biological systems are rarely irreversible, so we must consider the kinetics of the reversible reaction, where the magnitude of $k_{-1}$ is comparable to $k_1[S]$. Under these conditions, the time course still follows a single exponential, but the observed rate constant $k_{obs}$ is $k_1[S] + k_1$. Because the data analysis involves fitting to a single exponential, the observed rate of approach to equilibrium is equal to the sum of the forward and back rates. The time course of reaction is given by the equation:

$$[E] - [E]_\infty = ([E]_0 - [E]_\infty)\exp(-k_{obs}\,t)$$

While the rate of the observed reaction is increased by the rate of the reverse reaction, the amplitude of the observed reaction is reduced by the extent of reaction at equilibrium, where $[E]_\infty$ is the concentration of free enzyme at infinity.

This elementary kinetic analysis provides the fundamental conceptual basis for examination of all biological reactions. Additional complexities arise as a result of the addition of multiple steps along a reaction sequence, and each step in a pathway contributes a single exponential to the equation describing the process. For example, a conformational change may occur in the ES complex leading to tighter binding (E′S complex):

$$E + S \underset{k_{-1}}{\overset{k_1}{\rightleftharpoons}} ES \underset{k_{-2}}{\overset{k_2}{\rightleftharpoons}} E'S$$

The time dependence of the two-step binding reaction follows an equation involving the sum of two exponential terms. In the most simple case involving two irreversible reactions, the rates of the two phases of the reaction equal the rates of the two steps ($k_1$ and $k_2$). However, in the general equation allowing for reversible reaction steps, the magnitude of each of the two observed rates will be a function of all four rate constants. Different results can be obtained, depending on the signal that is being measured and the relative rates of each reaction.

Data analysis can be accomplished by fitting data to integrated rate equations by nonlinear regression. However, the solution of rate equations for complicated pathways is tedious and often requires the use of simplifying assumptions. The use of computer simulation by numerical integration of the rate equations overcomes the limitations of conventional data analysis and allows complicated reaction pathways to be analyzed rigorously with no simplifying assumptions.

## 2 METHODS

Transient kinetic analysis in enzymology, which entails the study of events occurring at the enzyme active site, requires specialized instrumentation to measure reactions over the millisecond time scale. A *stopped-flow instrument* allows enzyme and substrate to be rapidly mixed and the progress of reaction monitored by following an optical signal (e.g., absorbance or fluorescence). A computer is triggered to measure the intensity of light as a function of time after mixing for periods ranging from a few milliseconds to tens or hundreds of seconds. A *chemical quench-flow* set up allows the mixing of enzyme and substrate to initiate a reaction followed, and after a defined time period, by mixing with a quenching agent (acid, base, or other denaturant) to prevent further reaction. The amount of product formed during the reaction must then be quantified by methods usually involving a chromatographic separation of reactants from products.

Modern stopped-flow and chemical quench-flow instruments employ a computer-controlled motor to drive syringes at a precise speed and distance to force the mixing of a defined volume of solution either into an observation cell, in the case of stopped-flow; or through a length of tubing before mixing with the quench-

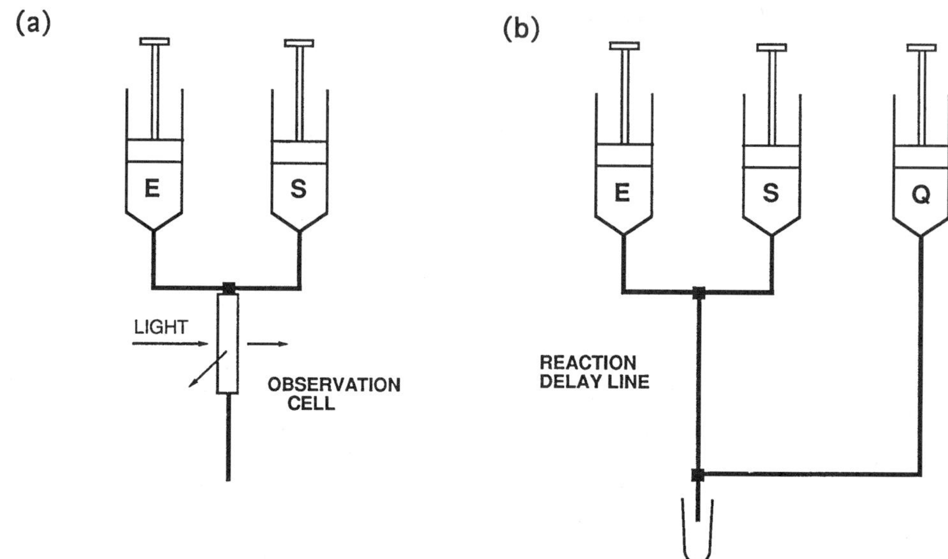


**Figure 2.** The arrangement of syringes used to rapidly mix solutions in (a) stopped-flow and (b) chemical quench-flow instruments. (a) The rapid mixing of two solutions into a small quartz cell allows observation of a reaction within a millisecond after mixing. (b) Enzyme is mixed sequentially with substrate to initiate reaction; the reacting solutions flow through the "reaction delay line" and then mix with quench solution to prevent further reaction. The time of reaction is determined by the rate of flow and the volume of solution between the two points of mixing; the shortest possible reaction times is 3 ms.

ing agent in a quench-flow instrument (see Figure 2). The volumes of solution required for each reaction can be quite small (20–40 μL) in the best of the modern instruments, opening up these methods to almost any enzyme system under investigation.

## 3 EXAMPLES

In the field of enzymology, transient state kinetic analysis has had a profound effect on our understanding of enzyme-catalyzed reactions. *Steady state kinetic* methods provide information pertaining only to the binding of substrates and to the magnitude of the rate-limiting step for catalysts in the form of $k_{cat}$ and $K_m$. Information defining the events occurring after substrate binding and before product release is simply not available by analysis in the steady state. Transient kinetic analysis allows the direct measurement of events occurring during catalysis, in addition to providing direct quantification of binding steps. For example, transient state kinetic analysis has afforded resolution of enzyme-bound intermediates, the kinetics of ribozyme binding, and the mechanism of DNA polymerization.

The enzyme EPSP synthetase provides the best example of the importance of a complete kinetic analysis of an enzyme reaction pathway by transient state kinetics. This enzyme, which is the target of the herbicide glyphosate, catalyzes the transfer of an enolpyruvoyl group from phosphoenol pyruvate to shikimate 3-phosphate in a reaction that proceeds by an addition–elimination mechanism via a tetrahedral intermediate. It was the direct observation by rapid quench kinetic methods of the kinetics of formation and decay of the tetrahedral intermediate that led to its isolation and structure proof by NMR spectrometry. Estimates were obtained for the rate and equilibrium constants governing all six steps in the pathway.

Analysis of the mechanism of DNA polymerization has been most profoundly affected by the application of transient kinetic methods. A comprehensive mechanistic analysis by rapid quench kinetic methods has shown that high fidelity of DNA replication is due to a two-step nucleotide-binding sequence. In the first step, the binding of a deoxynucleotide triphosphate (dNTP) to an enzyme–DNA complex is largely a function of ground state base-pairing free energy and contributes a factor of 200 toward the overall fidelity in discriminating against an incorrect base pair. The rate-limiting step for polymerization is a conformational change in the enzyme–DNA–dNTP complex which is a function of the base-pairing geometry and contributes a factor of 2000 to the overall fidelity. The key steps in the reaction responsible for the extraordinary fidelity of DNA replication are transparent to steady state kinetics measurements, which are limited by the slow rate of release of DNA from the enzyme.

*See also* Enzymes; Gene Expression, Regulation of; Zinc Finger DNA Binding Motifs.

### *Bibliography*

Anderson, K. S., and Johnson, K. A. (1990) *Chem. Rev.* 90:1131–1149.
———, Sikorski, J. A., and Johnson, K. A. (1988) *Biochemistry,* 27:7395–7406.
———, ———, and ———. (1988) *J. Am. Chem. Soc.* 110:6577–6579.
———, Miles, E. W., and Johnson, K. A. (1991) *J. Biol. Chem.* 266:8020–8033.
Barshop, B. A., Wrenn, R. F., and Frieden, C. (1983) *Anal. Biochem.* 130:134.
Bevilacqua, P. C., Kierzek, R., Johnson, K. A., and Turner, D. H. (1992) *Science,* 258:1355–1358.
Johnson, K. A. (1992) *Enzymes,* XX:1–61.
———, (1993) *Annu. Rev. Biochem.* 62:685–713.
Kati, W. M., Johnson, K. A., Jerva, L. F., and Anderson, K. S. (1992) *J. Biol. Chem.* 267:25988–25997.
Zimmerle, C. J., and Frieden, C. (1989) *Biochem. J.* 258:381–387.

# Enzymes

*Jon A. Christopher, Stephen W. Raso, Miriam M. Ziegler, and Thomas O. Baldwin*

*Key Words*

**Activation Energy ($\Delta G^{\ddagger}$)** The energy barrier that must be overcome for a chemical reaction to occur; formally, the free energy needed to promote a molecule (or molecules) from the ground state to the transition state.

**Active Site** The region of an enzyme at which the substrate binds and chemistry occurs.

**Allostery** Means of enzyme regulation in which a regulatory molecule acts on the enzyme by binding to it at a site distant from the active site; usually an induced conformational change is responsible for the change in enzyme activity.

**Equilibrium** The state of a chemical reaction in which the forward rate is equal to the rate of the back-reaction; there is no net change in product or reactant concentrations.

**Equilibrium Constant ($K_{eq}$)** The equilibrium ratio of product concentrations to reactant concentrations under a given reaction for a given set of conditions (e.g., temperature); $K_{eq}$ is independent of *initial* concentrations of products and reactants.

**Gibbs Free Energy** The measure of energy available to do work in a chemical system.

**Substrate** The specific compound on which an enzyme acts; the reactant.

**Subunit** A single polypeptide chain in a protein.

**Transition State** The short-lived, unstable (high energy) species of a molecule (or molecules) that forms transiently during a chemical reaction, in which chemical bonds are partially formed and/or broken; transition states cannot be isolated.

**$V_{max}$** The theoretical maximum velocity (or rate) of an enzymatic reaction asymptotically approached as the substrate concentration increases.

Enzymes are the biological catalysts (usually proteins) that help carry out nearly all chemical reactions in living systems; indeed, life as we know it would not be possible without them. Enzymes are true catalysts: that is, they increase the rate of a reaction but do not participate in it; they are returned to their original form at the end of the reaction cycle. Enzymes are special as catalysts only because their specificity and rate enhancements are unparalleled by any man-made or chemically developed catalysts. Enzymatic rate enhancements can range from $10^3$ to $10^{16}$ times the rates of uncatalyzed reactions. This ability allows otherwise exceedingly slow chemical reactions to occur on a time scale that is biologically meaningful. Like other catalysts, enzymes lower the activation energy of a reaction, making the product kinetically accessible.

The science of enzymology has made remarkable strides in the last 50 years, largely because of advances in recombinant DNA techniques, X-ray crystallography, various types of spectroscopy, and the ready availability of isotopes for kinetic and structural studies. Our understanding of enzymatic mechanisms of catalysis is increasing every day. Continued study of enzymes may lead to the development of supercatalysts for use in industry, in bioremediation of toxic waste, and as therapeutic agents for the treatment of disease. Many diseases are the result of a missing or defective enzyme; research in this area is bringing us closer to the ability to repair or create substitutes for these essential molecules.

## 1 INTRODUCTION/HISTORY

Enzymes are responsible for carrying out the highly complex chemical reactions necessary to sustain life. Like chemical catalysts of nonbiological origin, enzymes only increase the rate of a reaction that would naturally occur; they cannot induce a reaction that would not ordinarily happen. For example, in the absence of enzymes, the starch in last night's baked potato would break down into simpler sugars, but the process would take hundreds if not thousands of years. In the presence of some of your digestive enzymes, the job is done by the time of this morning's drive to work—an enzyme has increased the rate of a spontaneous reaction. No enzyme, however, could catalyze the conversion of simple sugars to starch (in a potato, for example) without the input of additional energy; this is because the process would not occur naturally. Another feature that enzymes have in common with nonbiological catalysts is the regeneration of the catalyst at the end of the reaction; enzymes are not consumed in the course of a reaction and are returned to their initial form by the end of the catalytic cycle.

Enzymes may be *highly* specific for the reactions they catalyze (see Figure 1), selectively binding their substrates and efficiently converting them to different chemical forms, the products. Often an enzyme will selectively bind one substrate out of the many thousands of chemicals in a cell, and increase the rate of one specific reaction by millions of times. Other enzymes are much less specific in their *binding,* but the reaction catalyzed is still highly specific. For example, the mammalian cytochrome P450 binds a variety of substrates, but in each case the chemical reaction is the addition of a hydroxyl group (—OH) to the substrate.

Specificity is a major feature that distinguishes enzymes from nonbiological catalysts. Another feature that distinguishes enzymes from chemical catalysts is regulation. Enzymes may be regulated in a variety of ways that may increase or decrease their catalytic efficiency. Finally, enzymes may be distinguished from most chemical catalysts by the phenomenon of saturation. Enzymes have a

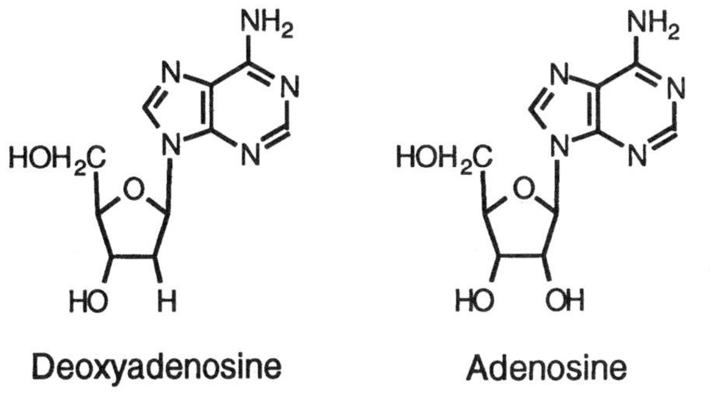

Deoxyadenosine Adenosine

**Figure 1.** Enzymes can easily distinguish between these two very similar molecules.

**Table 1** A Brief History of Enzymology

| Date | Investigators | Contribution |
|---|---|---|
| 1833 | Payen and Persoz | First observation of enzyme activity in a test tube |
| 1878 | Kühne | Introduces the word "enzyme" (in yeast) |
| 1898 | Duclaux | First suggestion for enzyme nomenclature |
| 1890s | Fischer | "Lock and key" model of enzyme action |
| 1913 | Michaelis and Menten | Mathematical model of enzyme action |
| 1920s | Willstätter | First pure enzymes |
| 1926 | Sumner | First crystallization of an enzyme (urease) |
| 1948 | Pauling | Proposed transition state theory of enzyme action |
| 1951 | Pauling and Corey | Discovery of α-helix and β-sheet structures in enzymes |
| 1953 | Sanger | First determination of the amino acid sequence of a protein (insulin) |
| 1961 | Perutz and Kendrew | X-Ray structure of myoglobin (not an enzyme) |
| 1986 | Cech | Discovery of catalytic RNA |
| 1986 | Lerner and Schultz | Catalytic antibodies developed |

Several of the investigators above have received the Nobel Prize for their work in the fields of protein chemistry and enzymology. Sumner, Pauling, Corey, Perutz and Kendrew received the Nobel Prize for work in protein structure and enzymology, and Cech for the description of catalytic RNA molecules. Sanger has received two Nobel Prizes, one for determination of the covalent structure of insulin and one for development of the methodology that is used to determine the sequence of bases in DNA. Pauling also received two Nobel Prizes, one for his scientific contribution and the second (the Nobel Peace Prize) for his efforts to limit nuclear proliferation.

maximum rate with respect to substrate concentration, whereas the majority of chemical catalysts do not.

Table 1 outlines some of the landmark events in the history of enzymology. Throughout much of this history, all enzymes were believed to be proteins—polymers of amino acids, the so-called building blocks of life. However, it has recently been shown that ribonucleic acids (nonprotein biopolymers) can catalyze certain reactions as well, and as biological catalysts, these RNA molecules fit the old definition of an enzyme. Most biochemists now have narrowed the definition of the term "enzyme" to refer only to proteins and have accepted the term "ribozyme" to describe catalytic RNA molecules. Another recent development has been the discovery of the means to produce antibodies that can catalyze specific reactions. These catalytic antibodies can also be regarded as enzymes, although not all the reactions they catalyze are of biological significance.

## 2 CLASSIFICATION AND NOMENCLATURE

Enzymes are given names that help to classify them. The agency responsible for naming enzymes is the Commission on Biochemical Nomenclature of the International Union of Biochemistry, which periodically publishes updates to a manuscript entitled *Enzyme Nomenclature*. There are three acceptable ways to refer to an enzyme: its Enzyme Commission code number, its systematic name, and its recommended name.

The EC code number provides a rigid systematic method for classifying enzymes. It takes on the general form of four numbers, separated by periods, following the letters "EC." For example, EC 1.1.1.1 represents alcohol dehydrogenase. The number in the first position of the code number denotes the class of the enzyme. This information indicates the general type of reaction the enzyme carries out. The second number is the enzyme subclass. The third indicates the sub-subclass. It is important to realize that the subclass and sub-subclass designations mean different things in different classes. The last number is the enzyme's serial number. Enzymes within a given sub-subclass are assigned sequential serial numbers, giving each enzyme a unique code.

## 3 CHEMICAL ENERGETICS

### 3.1 Thermodynamics

To understand how enzymes work, we must first examine the energetics of chemical reactions. There are two major sets of properties common to all chemical reactions: kinetic and thermodynamic. In a simple sense, the kinetic parameters of a reaction describe how fast the reaction (or chemical steps within the reaction) will take place. Thermodynamics, on the other hand, indicates the extent to which reactants will be converted to products. Thermodynamic predictions rely on the relative stabilities of products and reactants.

Enzymes affect *only* the *kinetics* of the reaction, since they, like all other chemical catalysts, cannot alter the equilibrium ratio of products to reactants. The most useful thermodynamic value for determination of whether a reaction will occur is the change in Gibbs free energy ($\Delta G$), named for the American chemist J. W. Gibbs (1839–1903). The general relationship between the change in Gibbs free energy and other thermodynamic parameters for any chemical reaction is as follows:

$$\Delta G = \Delta H - T\Delta S \tag{1}$$

where $\Delta G$ is the change in free energy, $\Delta H$ is the change in enthalpy, $T$ is the temperature in degrees kelvin, and $\Delta S$ is the change in entropy. Enthalpy is a measure of stored energy and entropy, in a broad sense, is the degree of disorder of a system.

All chemical reactions are theoretically reversible, such that reactants are constantly being re-formed from products. The rates of the forward and backward reactions depend on the concentrations of the reactants and products as well as on "rate constants" for the forward and back-reactions, which are different for every reaction. The higher the concentration of reactants, the faster the forward reaction will go; the higher the concentration of products, the faster the reverse reaction will go. Since the reactants are depleted as they become transformed into products during the course of a chemical reaction, the rate of the forward reaction is decreasing and the rate of the reverse reaction is increasing.

Eventually, the reaction will reach a state of equilibrium; at equilibrium there is no *net* formation of products or reactants—products are converted to reactants and reactants are converted to products at equal rates.

To determine whether a reaction will occur, it is useful to look at the equilibrium constant ($K_{eq}$) for the reaction. For the chemical reaction shown in equation (2),

$$A + B \rightleftharpoons C + D \tag{2}$$

the equilibrium constant is defined as follows

$$K_{eq} = \frac{[C][D]}{[A][B]} \tag{3}$$

where [X] indicates molar concentration of X at *equilibrium.* If the ratio of the initial concentrations of products to reactants do not satisfy the ratio defined by $K_{eq}$, the reaction will proceed spontaneously in whichever direction is necessary to achieve the equilibrium ratio $K_{eq}$. It is important to realize that a change in any of the initial concentrations will alter the concentrations of reactants and products needed at equilibrium to insure that their ratio is still equal to $K_{eq}$.

Now that we have defined the equilibrium constant, its relationship to the Gibbs free energy can be shown:

$$\Delta G = \Delta G^\circ + RT \ln \left(\frac{[C][D]}{[A][B]}\right) \tag{4}$$

where $\Delta G^\circ$ is the standard free energy change, a constant for a given chemical reaction; $R$ is the ideal gas constant, and $T$ is the temperature in degrees kelvin.

By definition $\Delta G = 0$ at equilibrium, so the relationship becomes:

$$\Delta G^\circ = -RT \ln K_{eq} \tag{5}$$

The importance of $\Delta G$, specifically $\Delta G^\circ$, is evident. Since $\Delta G^\circ$ is directly related to the equilibrium constant, it can be shown that any reaction that will proceed in the forward direction will have a negative $\Delta G$. Armed with the initial concentrations of reactants and products and with the value of the constant $\Delta G^\circ$ (or with the value of $K_{eq}$, which can be used to calculate $\Delta G^\circ$), it becomes an easy task to predict the direction of the reaction (i.e., whether more C and D will be formed or more A and B). The value of $\Delta G^\circ$ is proportional to the negative log of the equilibrium constant (equation 5), and therefore indicates the extent to which the reaction will favor products over reactants at equilibrium.

Two points should be noted:

1. The $\Delta G$ for a reaction changes with the concentrations of reactants and products. A positive $\Delta G$ for a reaction mixture does not mean that no reaction will occur; the positive sign simply means that the *forward* reaction will not go spontaneously under that set of conditions. In fact, a positive $\Delta G$ value means that under those conditions, there will be a net *back*-reaction. A change in the initial concentrations or temperature may make the forward reaction favorable.
2. The value of $\Delta G$ indicates nothing about how fast a reaction will proceed. The $\Delta G$ can provide information only about the direction and extent of the reaction that will bring the reaction to equilibrium under a certain set of conditions. Even though the products of a reaction may be greatly favored at equilibrium (large $K_{eq}$), it could take many, many years for that state to be reached.

## 3.2 Kinetics

We now turn our attention to the rates of chemical reactions. The rate of most chemical reactions is dependent on the concentrations of the substrates (reactants). For the reaction shown in equation (6),

$$A + B \rightleftharpoons C + D \tag{6}$$

the forward rate ($v$) can be expressed as the product of the concentrations of A and B times a rate constant $k$:

$$v = k\,[A][B] \tag{7}$$

As stated in Section 3.1, thermodynamics does not provide information about the rate of a reaction. By using transition state theory, however, we can obtain the rate law for this reaction. The most convenient way to apply the properties of chemical reactions just described to the explanation of rates and catalysis is to construct a reaction coordinate diagram. Figure 2 shows reaction coordinate diagrams for an uncatalyzed and an enzyme-catalyzed reaction. Since the final products of the reaction possess less free energy (lower free energy is a more stable, thermodynamically favored state), thermodynamics tells us that the product is favored at equilibrium. However, the reactants must first assume a very unstable, high energy state called the transition state. The free energy needed to promote a molecule (or molecules) from the ground state to the transition state is called the activation energy (denoted $\Delta G^{\ddagger}$). As the reaction proceeds from left to right in the uncatalyzed instance, a large activation energy must be overcome.

Transition state theory assumes that the reactants (A and B) in a chemical process are in equilibrium with the transition state ($AB^{\ddagger}$). Because of this equilibrium, we may apply the formula for Gibbs free energy to the activation energy, giving:

$$\Delta G^{\ddagger} = -RT \ln\left(\frac{[AB]^{\ddagger}}{[A][B]}\right) \tag{8}$$

Statistical thermodynamics allows us to calculate the probability that a molecule will have a particular energy. If we take $\Delta G^{\ddagger}$ as that energy, we obtain the probability of a molecule having enough energy to reach the transition state:

$$p = \frac{\kappa T}{h} \exp\left(-\frac{\Delta G^{\ddagger}}{RT}\right) \tag{9}$$

where the new parameters are $\kappa$, the Boltzmann constant, and $h$, Planck's constant.

It is obvious that the rate of a reaction is governed by the number of molecules that have enough energy to reach the transition state of the *slowest* step in the reaction, the rate-determining step. Assuming that once the transition state has been reached, half the molecules will go forward to products and half will go back to reactants, we can express the rate law of the reaction as follows:

$$v = \frac{\kappa T}{2h} \exp\left(-\frac{\Delta G^{\ddagger}}{RT}\right) [A][B] \tag{10}$$

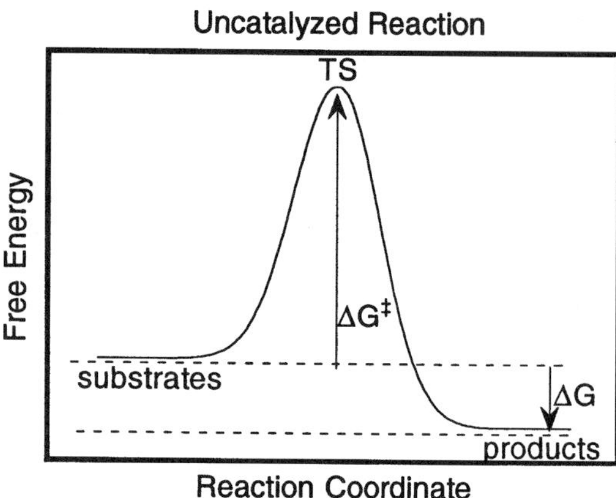


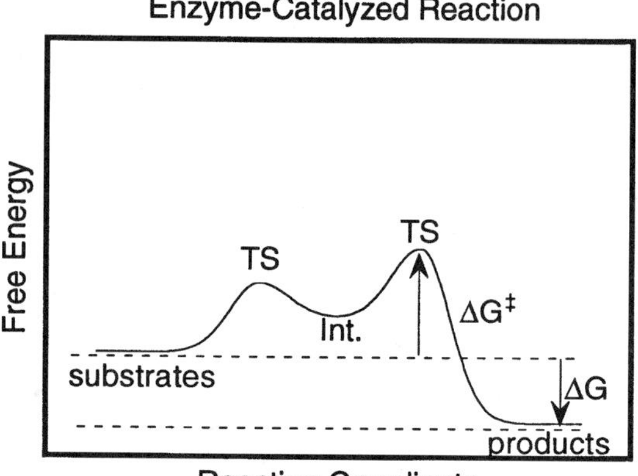


**Figure 2.** Reaction coordinate diagrams for uncatalyzed and catalyzed reactions. This uncatalyzed reaction has a single transition state, but some reactions may have more than one. The catalyzed reaction has two transition states, indicating that catalyzed reactions need not have the same mechanism as uncatalyzed ones. TS: transition state. ΔG: free energy of the reaction. $\Delta G^{\ddagger}$: activation energy of the reaction. Int.: an intermediate in the reaction.

and therefore the rate constant is given by

$$k = \frac{\kappa T}{2h} \exp\left(-\frac{\Delta G^{\ddagger}}{RT}\right) \tag{11}$$

which depends only on $\Delta G^{\ddagger}$ and the temperature.

Since very few molecules possess the energy needed to attain the transition state, the magnitude of $\Delta G^{\ddagger}$ provides the major kinetic obstacle in any reaction. The reaction coordinate diagram of an enzyme-catalyzed reaction shows the alternate chemical pathway with a more accessible, lower free energy transition state (and thus a larger rate constant). It should be noted that the relative stability of products and substrates remains unchanged, and the rate of the reverse reaction is also enhanced by the same factor as the forward reaction.

## 4 HOW ENZYMES WORK

Enzymes enhance reaction rates by factors of $10^3$ to $10^{16}$ relative to the rates of the uncatalyzed reactions. Some enzymes (e.g., carbonic anhydrase, triosephosphate isomerase, catalase) are considered to be ''catalytically perfect'' because the rate of the reaction is limited only by the rate at which the substrates diffuse through water to encounter the catalyst. Enzymes achieve these phenomenal rate enhancements by lowering the reaction's activation energy. To accomplish this, enzymes use a variety of strategies, including stabilization of the transition state, catalysis by approximation, acid–base and redox catalysis, and covalent catalysis. These strategies are discussed in more detail in Sections 4.1 to 4.4.

### 4.1 Transition State Stabilization

The most important aspect of enzymatic catalysis is the ability of the enzyme to stabilize the transition state of a reaction and thereby lower the activation energy of the overall reaction. Enzymes are able to selectively bind the transition state more strongly than the ground state of a molecule because they contain a highly evolved binding site that is complementary in structure to the transition state of the reaction rather than the ground state of the reactants. Since the enzyme is designed to bind the transition state, when the substrate binds the substrate is ''bent'' into a conformation that is closer to the transition state and thereby lowers the energy necessary to attain the transition state. The enzyme may be complementary to the transition state in size, shape, and charge distribution. The transition state fits into the enzyme as a key fits into a lock, and binds tightly. For efficient catalysis, however, the enzyme needs to be able to ''turn over'' (i.e., to complete one reaction cycle) as quickly as possible; thus the enzyme should not bind the *product* very tightly.

### 4.2 Catalysis by Approximation

For molecules to react in the uncatalyzed reaction, they must collide with one another in a very precise orientation. In contrast, when a molecule is bound to an enzyme, it is held in an optimal orientation and proximity for reaction with another chemical group. The randomness, or entropy, involved in obtaining the correct reactive conformation is greatly reduced when the reactants are bound to the enzyme, resulting in a smaller $\Delta G^{\ddagger}$.

### 4.3 Covalent Catalysis

The chemical side chains of an enzyme's constituent amino acids present in the active site may actually take part in the catalyzed reaction, and be restored to their original state by the end of the reaction cycle. A commonly observed mechanism is one in which an amino acid side chain displaces a chemical group of a substrate molecule to form a covalently bound enzyme–substrate intermediate, which is more reactive than the original substrate. A second substrate may then react with the intermediate complex to yield product(s) and the enzyme in its original, catalytically active form. This strategy is well precedented in hydrolysis and other transfer reactions. In these cases, the enzyme provides an alternative reaction mechanism with a smaller $\Delta G^{\ddagger}$.

### 4.4 Acid–Base and Redox Catalysis

Amino acid side chains may also cause a chemical reaction to occur more readily by acting as acids (donors of protons) or bases (acceptors of protons). It is important to note that other (nonprotein) acids or bases can also act as catalysts, but in enzymes, these

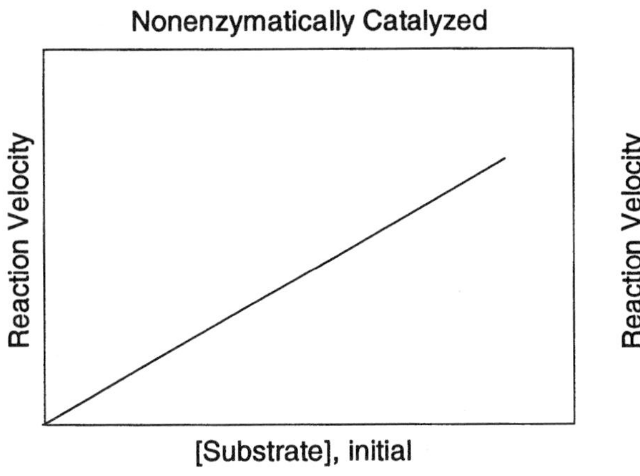


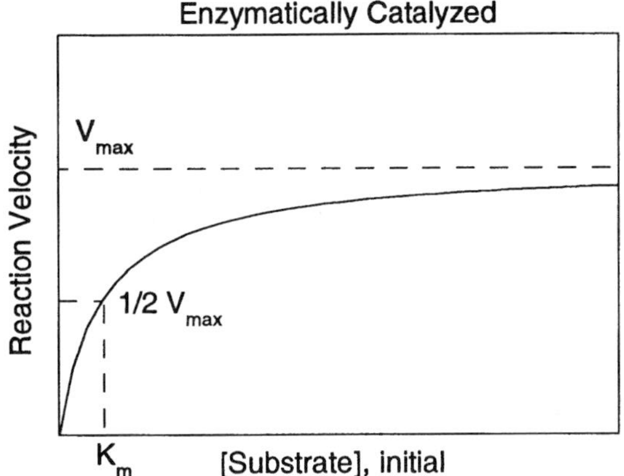


**Figure 3.** Effect of substrate concentration on the initial velocity of reactions catalyzed nonenzymatically and enzymatically.

reactive groups are oriented in the active site so that the necessary proton transfer may occur most efficiently.

Electrons may also be transferred during the course of a reaction; such reactions are called *redox* (reduction–oxidation) reactions. Reduction occurs when a molecule gains electrons and oxidation occurs when electrons are lost. Enzymes most commonly accomplish these electron transfers by incorporating transition metals or specialized organic molecules called *coenzymes.* Metals and coenzymes are generally referred to as *cofactors* and may also take part in acid–base chemistry and/or in covalent catalysis. Again, there are nonenzymatic redox catalysts, but the active site of an enzyme is situated so that everything is in the correct orientation to optimize catalytic efficiency.

## 5 ENZYME KINETICS

One of the characteristics distinguishing enzymes from most nonenzymatic catalysts is the phenomenon of *saturation.* As shown in Figure 3, a plot of the initial velocity of an enzyme-catalyzed reaction versus the initial concentration of the substrate indicates that the rate of the reaction varies linearly with low substrate concentrations; the reaction is *first-order* (varying linearly) with respect to substrate concentration. However the rate soon asymptotically approaches a limiting value, called $V_{max}$, becoming *zero-order* with respect to substrate concentration (not varying with concentration). By comparison, a similar plot typical for a reaction catalyzed nonenzymatically shows no such limit; it is first-order with respect to substrate concentration over the entire range of substrate concentrations that can be obtained.

### 5.1 Michaelis–Menten Equation

Henri, Michaelis, and Menten developed a mathematical treatment to explain substrate saturation, which was extended by Briggs and Haldane. Consider a simple enzyme-catalyzed reaction as depicted in equation (12) in which substrate (S) binds to enzyme (E) in a reversible equilibrium and is then converted into a product (P) with the regeneration of the enzyme.

$$\mathrm{E} + \mathrm{S} \underset{k_1}{\overset{k_1}{\rightleftharpoons}} \mathrm{ES} \overset{k_2}{\longrightarrow} \mathrm{E} + \mathrm{P} \tag{12}$$

Here we define the concentration of the enzyme–substrate complex to be [ES], the concentration of the free enzyme to be [E], and the total enzyme concentration to be $[E_0]$ (= [E] + [ES]). The concentrations of substrate and product are represented by [S] and [P], respectively. The rate constants for the two forward reactions are indicated by $k_1$ and $k_2$, while the reverse reaction is indicated by $k_{-1}$.

Mathematically, a first-order rate is expressed as a rate constant times a concentration. For example, the rate expression for the conversion of ES to E and P would be written as $k_2$[ES], the rate constant times the concentration of ES. The rate of *change* of the concentration of the intermediate (*d*[ES]/*dt*) is equal to the rate of its formation from E and S, minus its rate of breakdown. There are two modes of breakdown, one to E and S, the other to E and P. The overall rate of change of the concentration of ES is therefore given by:

$$\frac{d[\mathrm{ES}]}{dt} = k_1[\mathrm{E}][\mathrm{S}] - k_{-1}[\mathrm{ES}] - k_2[\mathrm{ES}] \tag{13}$$

Briggs and Haldane introduced the concept of the *steady state.* The steady state assumption postulates that although initially the rate of production of ES would be positive, because no ES is initially in solution, very shortly after E and S have been mixed, the rate of formation of ES equals the rate of its breakdown. Therefore *d*[ES]/*dt* = 0, implying a steady state, so that

$$k_1[\mathrm{E}][\mathrm{S}] - k_{-1}[\mathrm{ES}] - k_2[\mathrm{ES}] = 0 \tag{14}$$
$$k_1\ \{[\mathrm{E_0}] - [\mathrm{ES}]\}[\mathrm{S}] - k_{-1}[\mathrm{ES}] - k_2[\mathrm{ES}] = 0$$

Solving for [ES], one obtains:

$$[\mathrm{ES}] = \frac{k_1[\mathrm{E_0}][\mathrm{S}]}{k_{-1} + k_2 + k_1[\mathrm{S}]} \tag{15}$$

Since the rate of the overall reaction is the rate of formation of P from ES, the rate expression is $v = k_2$[ES]. The rate expression then becomes:

$$v = \frac{k_1 k_2[\mathrm{E_0}][\mathrm{S}]}{k_{-1} + k_2 + k_1[\mathrm{S}]} \tag{16}$$

Finally, substituting $k_0$ for $k_2$, where $k_0$ may be a combination of several rate constants, not simply a single step as indicated in our example, and setting $K_m = (k_{-1} + k_2)/k_1$, one arrives at the Michaelis–Menten equation:

$$v = \frac{k_0[E_0][S]}{K_m + [S]} \quad (17)$$

This equation, which describes a rectangular hyperbola, explains the substrate saturation behavior characteristic of enzyme-catalyzed reactions. One notices that as the substrate concentration [S] becomes very large ($\gg K_m$) the value of $K_m$ + [S] may be approximated by [S]. The rate equation then reduces to:

$$v = k_0[E_0] \quad (18)$$

explaining the zero-order behavior at high concentrations of substrate. The maximal velocity, $V_{max}$, is thus mathematically equal to $k_0[E_0]$.

At [S] equal to $K_m$, the rate equation becomes:

$$v = \frac{V_{max}}{2} \quad (19)$$

Therefore the $K_m$ value is not only a ratio of rate constants as defined earlier; it is also the substrate concentration necessary to obtain half of $V_{max}$.

### 5.2 Regulation

Another distinguishing feature of enzymatic catalysis is that it can be regulated. An enzyme may respond to changing conditions within a cell by slowing down or speeding up its activity as necessary. This is usually accomplished by the binding of a small molecule to the enzyme. This molecule changes the conformation of the enzyme in such a way as to make it either more or less efficient, depending on the needs of the cell. Enzymes that are regulated in this manner still follow Michaelis–Menten kinetics, albeit slightly modified.

A second form of regulation of enzyme activity is the phenomenon of *allostery,* meaning "other site."Allosteric enzymes are frequently composed of multiple subunits. The essence of allostery is that ligand binding to one subunit induces conformational changes in that subunit and also in the other *subunits,* altering the enzymatic properties of *the other subunits.* A "ligand" in this case may be a substrate or an allosteric effector (either an activator or inhibitor). The kinetic behavior of allosteric enzymes is frequently so complex that it cannot be described by the Michaelis–Menten equation. Some allosteric enzymes are regulated by binding of the allosteric ligand(s), altering the apparent affinity of the enzyme for substrate, a property that is reflected in the $K_m$. Allosteric regulation of other enzymes occurs at the level of catalytic efficiency and results in a change in $V_{max}$. Binding of allosteric ligands to some enzymes will alter both $V_{max}$ and $K_m$.

## 6 NONTRADITIONAL ENZYMES

For many years it was thought that all enzymes were the highly evolved proteins described in the preceding discussion. In 1986 reports were published of two new types of biological catalyst: catalytic antibodies and catalytic RNA.

The catalysis of reactions by antibodies is not a natural occurrence. Antibodies are protein molecules that bind foreign substances as part the immune response. Antibodies are designed to bind the ground states of molecules. Scientists were able to produce catalytic antibodies by taking advantage of transition state theory, using a molecule similar in structure to the proposed transition state of the desired reaction to elicit an antibody that can catalyze the reaction. The production of catalytic antibodies by this method is a major validation of the hypothesis that transition state stabilization contributes significantly to reaction rate enhancement by enzymes.

Catalytic RNA molecules, or *ribozymes,* discovered by Thomas Cech, are the first known example of nonprotein biological catalysts. Ribonucleic acid is another kind of biological polymer that is heavily involved in the synthesis of proteins using the genetic information stored in a cell's DNA. RNA and DNA have different characteristics; the most important difference for catalysis is the ability of RNA to assume complex three-dimensional structures. Biological catalysts must possess the ability to bind a specific substrate; the three-dimensional folds in an enzyme or an RNA molecule allow selective substrate binding. Catalytic RNA molecules are able to cut and splice themselves into a form that can then act catalytically on other RNA molecules. To date, the only ribozyme-catalyzed reactions known are cleavages of RNA and DNA. Ribozymes are true catalysts because they enhance reaction rates without any net change to themselves. They are capable of turnover (recycling) and show kinetics typical of enzymes.

A very interesting aspect of the discovery of catalytic RNA lies in the evolutionary implication. It is conceivable that RNA molecules were in existence before DNA or proteins. It has been proposed by Walter Gilbert that RNA molecules first catalyzed their own replication and then developed other catalytic abilities. It is not difficult to envision that RNA molecules eventually developed the ability to synthesize proteins. Proteins, which were able to catalyze many more reactions with greater efficiency than ribozymes, became the primary biological catalysts. It is speculated that DNA was first synthesized by reverse transcription of the genetic information on RNA and that DNA became the carrier of genetic information because it is more stable than RNA.

## 7 PERSPECTIVES

Enzymology has a long and exciting history. Largely because of development of recombinant DNA technology, however, we are now able to isolate in pure form large amounts of numerous enzymes that could not be isolated from the native biological source, and to alter the catalytic activity of natural enzymes. Recent improvements in physical methods, including NMR spectroscopy and X-ray diffraction from crystals of large molecules, are allowing dramatic advances to be made in our understanding of the mechanisms and modes of communication between enzymes and other biological macromolecules. It would appear that the glory days of enzymology are not behind us, but before us.

*See also* Enzyme Mechanisms, Transient State Kinetics of; Protein Folding.

### *Bibliography*

Baldwin, T. O., Raushel, F. M., and Scott, A. I., Eds. (1991) *Chemical Aspects of Enzyme Biotechnology—Fundamentals.* Plenum Press, New York.

Cornish-Bowden, A., and Wharton, C. W. (1988) *Enzyme Kinetics.* IRL Press, Washington, DC.

Denbigh, K. (1992) *The Principles of Chemical Equilibrium,* 4th ed. Cambridge University Press, Cambridge.

Dixon, M. D., and Webb, E. C. (1979) *Enzymes.* Academic Press, New York.
Fersht, A. (1985) *Enzyme Structure and Mechanism,* 2nd ed. Freeman, New York.
Jencks, W. P. (1969) *Catalysis in Chemistry and Enzymology.* McGraw-Hill, New York.
Stryer, L. (1988) *Biochemistry,* 3rd ed. Freeman, New York.
Walsh, C. (1979) *Enzymatic Reaction Mechanisms.* Freeman, New York.

# ENZYMES, HIGH TEMPERATURE

*Michael W. W. Adams*

## *Key Words*

**Amylase** Enzyme that degrades starch.

**Archaea** Formerly Archaebacteria, one of the three domains into which all life forms have been classified. The others are Bacteria, which includes the majority of microorganisms, and Eucarya, which consists of all higher life forms.

**Dehydrogenase** Enzyme that catalyzes a reaction involving electron transfer to or from an external electron carrier, which is typically NAD or NADP.

**Hydrogenase** Enzyme that catalyzes the reversible activation of hydrogen gas.

**Hydrothermal Vent** Superheated, mineral-rich water that emanates at temperatures approaching 400°C from deep-sea volcanic sites.

**Hyperthermophile** Microorganism with an optimum growth temperature of at least 80°C and a maximum growth temperature of 90°C or above.

**Oxidoreductase** Enzyme that catalyzes a reaction involving electron transfer to or from an external electron carrier, which is typically a redox protein.

**Protease** Enzyme that catalyzes the hydrolysis of (specific) peptide bonds within proteins.

**Universal Ancestor** The original life form(s) from which all present-day organisms evolved.

In the last few years, microorganisms have been discovered that grow optimally at temperatures near and above 100°C, the normal boiling point of water. Most of these so-called hyperthermophiles have been found in marine volcanic environments, which include deep-sea hydrothermal vents. They are the most ancient organisms known. The majority are strict anaerobes and depend on the reduction of elemental sulfur ($S^0$) to hydrogen sulfide ($H_2S$) for optimal growth. Most of them utilize proteins, peptides, and sugars as growth substrates, which are converted to simple organic acids and gases such as hydrogen ($H_2$) and carbon dioxide ($CO_2$). These potentially rich sources of a variety of "hyperthermostable" enzymes are able to catalyze reactions at extreme temperatures. So far about 30 enzymes have been purified and characterized from the sulfur-reducing species. All have optimal temperature for catalysis above 100°C and some have half-lives at this temperature of several days.

## 1 INTRODUCTION

A revelation of great significance in the field of microbiology, with profound implications in microbial metabolism, biochemistry, and biotechnology, occurred in 1982 when microorganisms that grew at temperatures exceeding 100°C were isolated from shallow marine volcanic vents. Although we know little about the novel biochemistry that must be required to sustain these organisms under such conditions, they are potential sources of a range of "hyperthermostable" or high temperature enzymes. This contribution discusses what is known about hyperthermophilic organisms and the enzymes that have been characterized from them, with an emphasis on the enzymes from the "sulfur-dependent" species, the predominant type of hyperthermophile.

## 2 SOURCES OF HIGH TEMPERATURE ENZYMES

Hyperthermophiles are defined here as organisms able to grow at 90°C and above with an optimal growth temperature of at least 80°C. They have been isolated in just the last few years, and only two of the 20 or so hyperthermophilic genera currently known are conventional bacteria. The majority of them are classified as Archaea (formerly Archaebacteria). Archaea were recognized as the third kingdom of life in the late 1970s on the basis of molecular (rRNA) sequence analyses. These studies also indicate that hyperthermophilic organisms are the most ancient of life forms, the first to have diverged from some universal ancestor, which suggests that the rest of biology is the result of evolutionary pressures to adapt to low (< 100°C) temperatures.

The known hyperthermophilic genera are listed in Table 1. All have been isolated from geothermally heated environments, which include deep-sea hydrothermal vents located up to 4000 m below sea level. The majority of the hyperthermophiles are referred to as sulfur-dependent organisms, since their growth depends to a greater or lesser extent on the reduction of elemental sulfur ($S^0$) to hydrogen sulfide ($H_2S$). Almost all grow only under strictly anaerobic conditions and are strict organotrophs that use complex organic mixtures as the carbon and nitrogen sources, including yeast and meat extracts, tryptone, peptone, and casein. A few are also able to use carbohydrates as additional or primary C sources. They are potential sources of a variety of hydrolytic-type enzymes, such as proteases and amylases, which are of some industrial importance.

## 3 PROPERTIES OF HIGH TEMPERATURE ENZYMES

The enzymes that have been purified from the $S^0$-dependent hyperthermophilic genera are listed in Table 2. All exhibit maximal catalytic activity above 100°C. Protease activity has been detected in several hyperthermophiles, but these organisms have been purified from only two species. The serine-type protease of *Pyrococcus furiosus* was isolated by boiling cell-free extracts with the detergent sodium dodecyl sulfate (SDS, 1%) for 24 hours, a process that destroyed virtually all other cellular proteins. Several carbohydrate-metabolizing enzymes able to utilize starch, maltose, cellulose, and/or xylan as substrates have been characterized. These represent the most stable enzymes yet characterized from any organism, with half-lives of one to two days near 100°C. The activities of

Table 1 Hyperthermophilic Genera

| Genus | $T_{max}$ (°C)[a] | Growth Substrates[b] | Habitat[c] |
|---|---|---|---|
| **$S^0$-Dependent Archaea** | | | |
| *Thermoproteus* | 92 | Pep, CBH, $H_2$, $S^0$ | c |
| *Staphylothermus* | 98 | Pep, $S^0$ | d/m |
| *Desulfurococcus* | 90 | Pep, $S^0$ | d/c/m |
| *Thermofilum* | 100 | Pep, $S^0$ | c |
| *Pyrobaculum* | 102 | Pep, $H_2$, $S^0$ | c |
| *Acidianus* | 96 | $S^0$, $H_2/S^0$, $O_2$ | c/m |
| *Pyrodictium* | 110 | Pep, CBH, $H_2$, $S^0$ | d/m |
| *Thermodiscus* | 98 | Pep, $S^0$ | m |
| *Pyrococcus* | 105 | Pep, CBH, $\pm$ $S^0$ | d/m |
| *Thermococcus* | 97 | Pep, CBH, $\pm$ $S^0$ | d/m |
| *Hyperthermus* | 110 | Pep, $H_2$, $\pm$ $S^0$ | m |
| "ES-1" | 91 | Pep, CBH, $S^0$ | d |
| "ES-4" | 108 | Pep, CBH, $S^0$ | d |
| "GB-D" | 103 | Pep, $S^0$ | d |
| "GE-5" | 102 | Pep, $S^0$ | d |
| "JDF-3" | 108 | Pep, CBH, $S^0$ | d |
| **Sulfate-reducing Archaea** | | | |
| *Archaeoglobus* | 95 | CBH, $H_2$, $SO_4$ | d/m |
| **Methanogenic Archaea** | | | |
| *Methanococcus* | 91 | $H_2$, $CO_2$ | d/m |
| *Methanothermus* | 97 | $H_2$, $CO_2$ | c |
| *Methanopyrus* | 110 | $H_2$, $CO_2$ | d/m |
| **Bacteria** | | | |
| *Thermotoga* | 90 | Pep, CBH, $\pm$ $S^0$ | d/m |
| *Aquifex* | 95 | $S^0$, $H_2$, $O_2$, $NO_3$ | m |

[a]Maximum growth temperature.
[b]Growth substrates utilized including peptides and proteins (Pep), carbohydrates (CBH) or hydrogen gas ($H_2$), elemental sulfur ($S^0$), oxygen ($O_2$), nitrate ($NO_3$), sulfate ($SO_4$); $\pm$ $S^0$ indicates can grow without $S^0$.
[c]Indicates whether species have been isolated from shallow marine vents (m), deep-sea hydrothermal vents (d), and/or from continental hot springs (c).

*Pyrococcus woesei* amylase, *P. furiosus* sucrose α-glucohydrolase, and ES-4 amylopullulanase have been measured at 130, 117, and 135°C, respectively. Similarly, the immobilized xylanase from a *Thermotoga* strain retained 25% of its activity after 1 hour at 130°C in molten sorbitol.

Many of the hyperthermophiles also metabolize hydrogen gas ($H_2$), and hydrogenase is the enzyme responsible for catalyzing its production and activation. Hydrogenases usually utilize as an electron carrier a low molecular weight redox protein known as ferredoxin, which also has been purified from these organisms. Another low molecular weight redox protein known as rubredoxin has been purified from *P. furiosus* as well, although its function in the cell is not known. The rubredoxin is of some importance, since it is the only hyperthermophilic protein for which a three-dimensional structure is available (determined independently by NMR spectroscopy, X-ray crystallography, and molecular modeling). Surprisingly, the overall folding patterns of the mesophilic and hyperthermophilic rubredoxins are remarkably similar, and enhanced stability appears to arise from rather minor changes.

Several ferredoxin-dependent oxidoreductases have also been purified. These oxidize various substrates and transfer electron to ferredoxin ultimately for $H_2$ production or $S^0$ reduction. The aldehyde oxidoreductase of *P. furiosus* converts aldehydes to the corres-

Table 2 Properties of Enzymes Purified from $S^0$-Reducing Hyperthermophiles

| Enzyme | $t_{50\%}$ (h/°C)[a] | Catalytic Activity |
|---|---|---|
| ***Pyrococcus furiosus*** | | |
| Protease | 33/98 | Peptide hydrolysis |
| Amylase | 2/120 | Starch hydrolysis |
| α-Glucosidase | 48/98 | Maltose hydrolysis |
| Sucrose α-glucohydrolase | 48/95 | Sucrose hydrolysis |
| Hydrogenase | 2/100 | $H_2$ production (Fd)[b] |
| Ferredoxin | >24/95 | Electron transfer |
| Rubredoxin | >24/95 | Electron transfer |
| Aldehyde oxidoreductase | 6/80 | Aldehydes to acids (Fd)[b] |
| Pyruvate oxidoreductase | 0.3/90 | Pyruvate to acetyl CoA (Fd)[b] |
| Glutamate dehydrogenase | 10/100 | Glutamate to 2-OG[c] (NAD/P)[b] |
| DNA polymerase | 20/95 | DNA replication |
| ***Thermococcus litoralis*** | | |
| Ferredoxin | >24/95 | Electron transfer |
| Formaldehyde oxidoreductase | 2/80 | Formaldehyde to formate (Fd)[b] |
| DNA polymerase | 7/95 | DNA replication |
| ***Pyrococcus woesei*** | | |
| Amylase | 6/100 | Starch hydrolysis |
| GAPDH[c] | 0.7/100 | GAP to BPG[c] (NAD/P)[b] |
| ***Thermoproteus tenax*** | | |
| GAPDH[c] | 0.3/100 | GAP to BPG[c] (NAD)[b] |
| GAPDH[c] | 0.5/100 | GAP to BPG[c] (NADP)[b] |
| ***ES-4*** | | |
| Amylopullulanase | 20/98 | Starch hydrolysis |
| ***Pyrodictium brockii*** | | |
| Hydrogenase | 1/98 | $H_2$ oxidation (unknown)[b] |
| ***Desulfurococcus mucosus*** | | |
| Protease | 1.5/95 | Peptide hydrolysis |
| ***Thermotoga maritima*** | | |
| 4-α-Glucanotransferase | 3/80 | Starch hydrolysis |
| Hydrogenase | 1/90 | $H_2$ production (unknown) |
| GAPDH[c] | 2/95 | GAP to BPG[c] |
| Lactate dehydrogenase | 1.5/90 | Pyruvate to lactate (NAD)[b] |
| ***Thermotoga* sp. strain FjSS3-B.1** | | |
| Xylanase | 1.5/95 | Xylan hydrolysis |
| Cellobiohydrolase | 1.1/108 | Cellulose hydrolysis |

[a]Time required to lose 50% of catalytic activity after incubation at the indicated temperature.
[b]Indicates whether the enzyme utilizes ferredoxin (Fd), NAD, NADP, or both (NAD/P) as an electron carrier, or if the physiological carrier is unknown.
[c]2-OG, 2-oxoglutarate; GAPGH, glyceraldehyde 3-phosphate dehydrogenase; GAP, glyceraldehyde 3-phosphate; BPG, 1,3-bisphosphoglycerate.

ponding acid and is part of a novel pathway for carbohydrate oxidation in this organism. The formaldehyde oxidoreductase of *Thermococcus litoralis* converts formaldehyde to formate, but this is thought to be involved in the oxidation of amino acids inside the cell. These two enzymes are unusual in that their catalytic sites contain tungsten, an element seldom found in biological systems.

Dehydrogenase-type enzymes utilize NAD or NADP as electron carriers rather than ferredoxin. However, these cofactors have half-lives at 100°C of only a few minutes and it is not known how they are stabilized inside the cell. Interestingly, the half-life of *Thermotoga* lactate dehydrogenase at 90°C increased from 2 minutes to 150 minutes when NAD was present, suggesting that the cofactor and the enzyme may stabilize each other at the growth temperature of the organism. Similarly, the substrates for glyceraldehyde 3-phosphate dehydrogenase (GAPDH) are very unstable, and accurate assays are not possible much above 70°C.

High temperature DNA polymerases are of particular interest because of their use in the polymerase chain reaction (PCR), now a routine tool in molecular biology. Three DNA polymerases are commercially available, and the properties of those from *T. litoralis* and *P. furiosus* have been described. A thermostable DNA polymerase is essential in PCR work.

## 4 STRUCTURAL INFORMATION ON HIGH TEMPERATURE ENZYMES

So far there is only limited information on the enzymes and proteins that have been purified and characterized from hyperthermophilic $S^0$-reducing organisms. From the Archaea, five complete amino acid sequences have been reported: three derived from proteins (ferredoxin and rubredoxin from *P. furiosus,* and *T. litoralis* ferredoxin) and two from gene sequences (*P. woesei* GAPDH and *T. litoralis* DNA polymerase). In addition, four complete amino acid sequences have been reported for *Thermotoga maritima* proteins: for lactate dehydrogenase (from the protein), GAPDH, glutamine synthetase, and EF-Tu (from gene sequences). So far the genes for three hyperthermophilic proteins have been expressed in *E. coli* (*P. woesei* GAPDH, and glutamine synthetase and EF-Tu from *Thermotoga maritima*). The three-dimensional structure of one hyperthermophilic protein has been determined, namely *P. furiosus* rubredoxin.

*See also* Bacterial Growth and Division; Enzymes; Heat Shock Response in *E. coli*.

### *Bibliography*

Adams, M. W. W. (1990) The metabolism of hydrogen by extremely thermophilic, sulfur-dependent bacteria. *FEMS Microbiol. Rev.* 75: 219–238.

———. (1993) Enzymes and proteins from microorganisms that grow near and above 100°C. *Annu. Rev. Microbiol.* 47:627–658.

———, and Kelly, R. M., Eds. (1992) *Biocatalysis at Extreme Temperatures: Enzyme Systems Near and Above 100°C,* ACS Advances in Chemistry Series No. 498. American Chemical Society, Washington, DC.

Stetter, K. O., Fiala, G., Huber, G., Huber, R., and Segerer, G. (1990) Hyperthermophilic microorganisms. *FEMS Microbiol. Rev.* 75: 117–124.

# Enzymology, Nonaqueous

*Alan J. Russell, Sudipta Chatterjee, and Darrell Williams*

### *Key Words*

**Lyophilized Enzyme Powder** Freeze-dried preparations of biological catalysts, which, if desired, can be suspended directly into a water-free organic solvent to promote activity.

**Polyethylene Glycol Modification** The covalent attachment of a hydrophobic polymer to the surface of an enzyme, enabling the protein to be solubilized in a variety of organic solvents.

**Reversed Micelles** Aggregates of amphiphilic molecules in an organic solvent, which can encapsulate a water pool.

**Supercritical Fluid** A material above its critical temperature and pressure; its physical properties lie between those of liquids and gases.

Nonaqueous enzymology is concerned with the utilization and understanding of enzymes in essentially organic environments. The development of structure-function-environment relationships is central to our understanding of how enzymes work as catalysts in vivo and in vitro. This article introduces the overall subject, and presents briefly some of the advances made to date.

## 1 INTRODUCTION

Living organisms synthesize a wide array of enzymes, which catalyze a myriad of reactions both inside and outside the cell. The environment of the enzymes not associated with a membrane is essentially aqueous. For this reason, conventional biocatalysis is carried out in aqueous media, and it is not surprising that most methods developed to study enzyme performance are water based. Recently, however, interest has been focused on using enzymes to catalyze reactions in organic media. Industrial enzymatic processes offer the advantage of mild reaction conditions, extreme selectivity, and environmental acceptability. If an enzyme could function in an essentially organic environment, increased ease of product recovery, increased hydrophobic reactant solubility, and reduced microbial contamination would be properties contributing to wide applicability. This entry focuses on another remarkable feature of enzyme catalysis in solvents other than water: namely, that altering the solvent can have a dramatic effect on enzyme function, enabling the production of catalysts by design via "solvent engineering."

Modern-day nonaqueous enzymology evolved in a number of distinct phases. Initially, water-miscible organic solvents, such as acetone and ethanol, were added to aqueous enzyme solutions to determine the maximal concentrations of solvents the enzymes could tolerate (~50% for most conventional enzymes). Next, biphasic mixtures in which an aqueous solution of enzyme was emulsified in such water-immiscible solvents as chloroform and ethyl acetate were evaluated. This approach was further developed by the use of reversed micelles to stabilize the enzyme in such a mixture. In 1966 Dastoli and Price demonstrated that freeze-dried enzyme powders retained significant activity when suspended in a wide variety of organic solvents. In the early 1980s, Klibanov revolutionized the field of nonaqueous enzymology by demonstrating that enzymes could catalyze novel reactions in organic media, and even exhibit enhanced thermostability. Klibanov's work led to the development of a new approach for enzyme studies: not only can we investigate how an enzyme works, but we can also determine whether it can be fine-tuned by altering its environment.

Catalysis is dependent on the intact three-dimensional structure of a protein. Water plays a critical role in all noncovalent interactions that maintain an enzyme in its catalytically active form. It was assumed therefore that replacement of water by a nonaqueous medium would result in the loss of enzyme activity. However, an enzyme freeze-dried (lyophilized) from aqueous medium retains a small amount of water (~15% by weight, and less than a monolayer per molecule of protein). Thus a dry enzyme powder, suspended in a nonaqueous dispersant, can be protected from bulk solvent by a shell of water, enabling the enzyme to function. Because of the water associated with enzymes suspended in anhydrous solvents, the systems are often termed "essentially anhydrous."

Lyophilized enzyme powders suspended in organic solvents have been used to perform a variety of important catalytic functions such as esterifications, transesterifications, aminolyses, oximolyses, racemate resolutions, oxidations, reductions, peptide syntheses, polymerizations, and depolymerizations. The insolubility of enzymes in organic solvents, resulting in a biphasic system, also results in such advantages as ease of enzyme recovery by filtration. The absence of water leads to improved enzyme thermostability (water is a reactant in many processes that irreversibly denature enzymes), and the reduction of undesirable side reactions that require water as a substrate. Since heterogeneous systems can be diffusionally limited, mixing and sonication are usually employed. A recent development in nonaqueous enzymology is the use of polyethylene glycol (PEG) modified enzymes, which are soluble in organic solvents. These enzymes form homogeneous reaction systems and thereby eliminate the problems of diffusion. Figure 1 summarizes the various approaches that can be utilized to enable enzymes to function in organic solvents. While the given methods involve the general principles of using enzymes in connection with an organic solvent or compressed gas (supercritical fluid), each system offers several advantages and disadvantages that dictate the choice of system.

## 2 SOLVENT ENGINEERING VERSUS PROTEIN ENGINEERING

The most remarkable feature about the effect of a solvent on an enzyme is that the environment can be used to alter the "natural" activity and specificity of the catalyst. This approach to catalyst design has been termed "solvent engineering." Both solvent and protein engineering (where one changes the catalyst itself) have the same ultimate aim, that is, the manipulation of an enzyme's functional properties. Table 1 lists the goals of solvent and protein engineering, many of which have been achieved using solvent engineering, although the field is still in its infancy. In this discussion we focus on the use of dry enzyme powders in organic solvents.

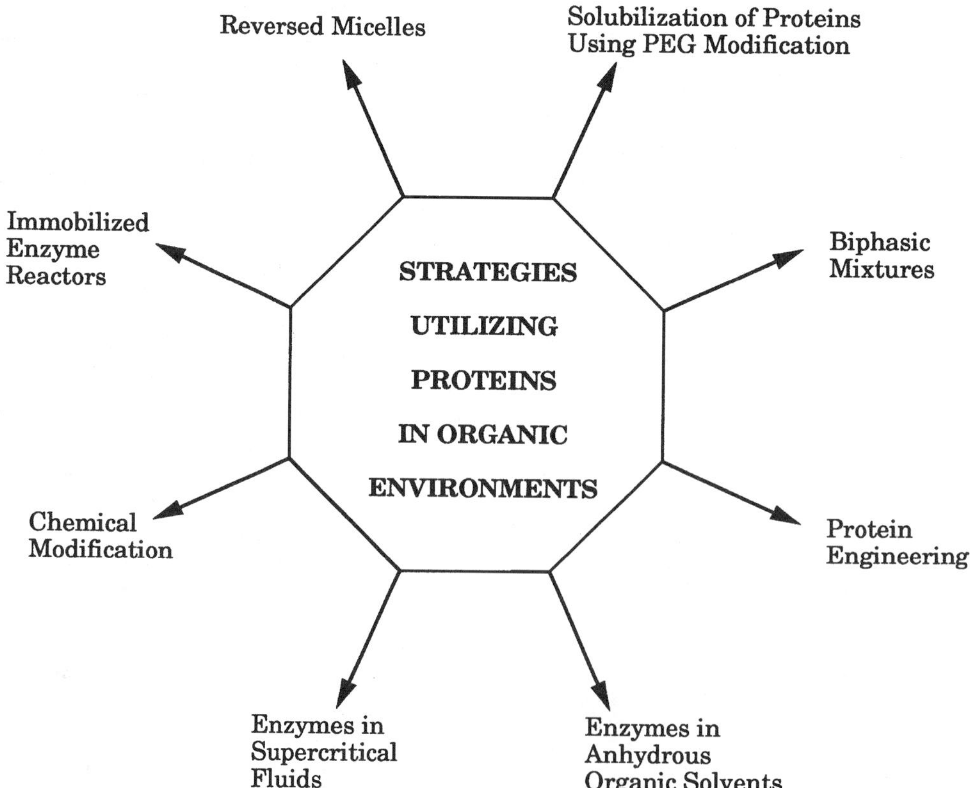


Figure 1. Strategies that enable the utilization of enzymes in the presence of significant concentrations of nonaqueous media.

Table 1 Typical Goals of Solvent and Protein Engineering

| Goal | Success Using Solvent Engineering | Success Using Protein Engineering |
|---|---|---|
| Modification of catalytic activity | X | X |
| Modification of enzyme selectivity | X | X |
| Control of enzyme enantioselectivity | X | X |
| Increase of enzyme thermostability | X | X |
| Modification of enzyme pH dependence | X | X |
| Control of allosteric function | | X |
| Utilization of enzymes in non-native reactions | X | X |
| Stabilization of enzymes against denaturants | | X |

### 2.1 Effect of Solvent on the Physical Properties of an Enzyme

A complete characterization of enzyme function requires a detailed study of the effect of environment on protein structure, activity, and specificity. Morphological studies on protein powders suspended in organic solvents have shown considerable changes in the powder as a function of the water associated with the preparation. In general, most protein powders are reported to be flaky or spherical, with an average diameter of 100 μm. Structural studies on proteins suspended in organic solvents have been performed using electron paramagnetic resonance spectroscopy, nuclear magnetic resonance spectroscopy, infrared spectroscopy, and circular dichroism. All studies conclude that the structural integrity of enzymes is not altered in the presence of a hydrophobic bulk solvent. Organic solvents can, however, affect enzyme structure and function by directly interacting with water molecules bound to the enzyme. Hydrophilic solvents can easily "strip" the "essential" water from the surface of enzyme molecules, thus removing the protective layer from the enzyme and enabling direct interaction between nonaqueous solvent and protein. These direct interactions may cause fluctuations in enzyme structure, inactivating the protein.

The effect of solvent on the mobility of proteins has also been investigated, with interesting results. Proteins in water are like a "Swiss cheese"; the holes are full of water, which lubricates the protein. Proteins are by their nature very flexible, and the activity of the molecule is in large part determined by the flexibility of the molecule. An enzyme suspended in a hydrophobic solvent will usually retain enough water, and thus flexibility, to maintain function. In hydrophilic solvents, however, the loss of water from the powder to bulk solvent results in rigidity and decreased activity. This limitation can be overcome to some degree by simply adding extra water to the solvent.

### 2.2 Engineering Enzyme Activity and Substrate Specificity

Electrostatic forces such as ion pairing, hydrogen bonding, and hydrophobic and electrostatic interactions influence substrate binding and catalysis. Conventional protein engineering uses such methods as site-specific mutagenesis to add or remove residues at or near the active site, thus controlling enzyme specificity and stereoselectivity. Solvent engineering allows control of all these properties by controlling the reaction medium.

The nature of the reaction medium affects enzyme activity by influencing kinetic parameters such as catalytic turnover, enzyme–substrate binding, and the overall efficiency of the enzyme. Upon transition from aqueous to nonaqueous media, turnover decreases and binding increases. Substrate and solvent hydrophobicity often determine the extent of the effect of the medium on enzyme efficiency. As substrate hydrophobicity increases, the substrate will have a greater tendency to partition into the solvent rather than into the now relatively hydrophilic enzyme active site, and more substrate will be required to saturate the enzyme.

It has also been suggested that solvent can modulate specificity. Enzymes like chymotrypsin and subtilisin, whose substrate-binding pockets are lined by hydrophobic amino acid residues, show a preference for hydrophobic nonpolar substrates in water. However, in organic media the same enzymes show a preference for hydrophilic substrates. This is because polar substrates can easily partition into the enzymes' active sites, which are less hydrophobic than the reaction medium. Hence, enzymes in organic media often show altered substrate specificity compared to that in water.

In general, enzymes retain their stereoselectivity in organic media. It is hypothesized that L-isomers bind "correctly" and D-isomers "incorrectly" to the active site. The consequent release of water from the active site, due to the binding of L-isomers, is therefore, an energetically favored process in water rather than in organic solvents. As a result, the transition from aqueous to nonaqueous medium depresses the catalytic efficiency of L-isomers more than it does that of D-isomers. Hence, in some cases a nonaqueous medium "relaxes" an enzyme's enantioselectivity. In addition, solvents can be used to fine-tune the enantioselectivity to a predetermined degree.

Two novel features of enzymes in nonaqueous media are inhibitor-induced enzyme memory and pH memory. When an enzyme is lyophilized in the presence of a competitive inhibitor, the high rigidity of enzymes in organic media allows the enzyme to maintain an activated form even after the ligand has been removed. The phenomenon of "pH memory" refers to the ability of charged groups on the enzyme to retain the ionization state corresponding to the pH of the aqueous solution from which it was lyophilized. These two phenomena show how enzyme pretreatment offers a means to regulate enzyme activity and substrate specificity in nonaqueous environments.

### 2.3 Engineering Protein Stability

A relatively common approach in the design of improved biocatalysts is to alter the protein itself, using either chemical modification or genetic engineering. Improvements in stability toward denaturants, oxidants, proteolysis, and high temperatures are frequently the desired end results of such strategies. Interestingly, enzymes in organic solvents often show enhanced thermostability (enzymes are stable at temperatures of up to 150°C when suspended in anhydrous organic solvents under optimized conditions). Irreversible protein denaturation is often limited by the rates of disulfide bond interchange, deamidation of glutamine and asparagine residues, and hydrolysis of peptide bonds. Each of these reactions requires water as a substrate. The absence of water in nonaqueous media, therefore, positively affects enzyme stability.

## 3 PROTEIN ENGINEERING FOR NONAQUEOUS MEDIA

A recent trend in nonaqueous enzymology has been the use of protein engineering techniques to design enzymes with improved stability and activity in organic media. Rules for redesigning proteins are formulated via a deep understanding of protein–solvent interactions, as well as the effect of the solvent on the noncovalent forces that affect protein structure and function. Efforts at redesigning proteins have been directed at improving the conformational stability and the compatibility of the protein surface with the organic media. It has been suggested that the former may be accomplished by the introduction of residues that can form new disulfide bridges, increased van der Waals and electrostatic interactions, and improved hydrogen bond formation. Replacement of both charged surface amino acids, which require water for solvation, and surface hydrogen bonding sites, which cannot be satisfied by nonaqueous media, have been demonstrated to improve protein–solvent compatibility.

α-Lytic protease and subtilisin E are two enzymes that have been redesigned for nonaqueous biocatalysis. Both enzymes have been tested in mixtures of water and polar organic solvents for the effectiveness of the mutations. Substitution of surface charge residues by uncharged residues and the addition of residues with enhanced hydrogen bonding abilities had a positive effect on the enzymes' stability. Interestingly, the redesigned proteins, which performed so well in mixtures of water and organic solvents, have also been shown to function with improved efficiency in essentially anhydrous systems.

## 4 GOALS FOR THE FUTURE

The opportunities presented by nonaqueous biocatalysis are limited only by the imagination. In just one decade, this area of research has become one of the most active in science and engineering. While the technology has already been commercialized, full-scale adoption of nonaqueous enzymology will depend on the generation of a predictive model to correlate the effect of solvent properties to the binding and catalytic steps of an enzyme-catalyzed reaction. The design of bioreactors, and the development of upstream and downstream processing units for nonaqueous enzymology, are also central to further utilization of this core technology. The clear need for biocatalysts in industry should not obscure what the use of enzymes in extreme environments can teach us about the molecular basis of enzyme catalysis. Research is already demonstrating how the delicate interaction between environment and biological molecules serves to modulate function.

*See also* CHIRALITY IN BIOLOGY; ENZYME ASSAYS; ENZYMES.

### *Bibliography*

Dordick, J. S., Ed. (1991) *Biocatalysts for Industry.* Plenum Press, New York.

Klibanov, A. M. (1989) Enzymic catalysis in anhydrous organic solvents. *Trends Biochem. Sci. (Pers. Ed.)* 14:141.

Laane, C., Boeren, S., Hilhorst, R., and Veeger, C. (1987) In *Biocatalysis in Organic Media*, C. Laane, et al., Eds., pp. 65–87. Elsevier, Amsterdam.

Russell, A. J., Chatterjee, S., Rapanovich, I., and Goodwin, J. (1992) Mechanistic enzymology in non-aqueous solvents. In *Biomolecules in Organic Solvents,* A. Gomez-Puyou, Ed., pp.92–109. CRC Press, Boca Raton, FL.

Zaks, A., and Russell, A. J. (1988) Enzymes in organic solvents: Properties and applications. *J. Biotechnol.* 8:259–269.

**Evolution:** *see* **Genetic Analysis of Populations; Genetic Diversity in Microorganisms; Mitochondrial DNA, Evolution of Human; Paleontology, Molecular.**

# EXPRESSION SYSTEMS FOR DNA PROCESSES

*Ka-Yiu San and George N. Bennett*

### *Key Words*

**Antibiotic Resistance** The ability of the cell to grow in the presence of a chemical (antibiotic) that normally inhibits an essential cell function.

**Bacteriophage** A virus that infects bacteria.

**Coding Region** The nucleic acid segment that contains the linear arrangement of codons specifying the order of amino acids in the protein.

**Codon** A three-nucleotide unit in a molecule of mRNA that specifies the particular amino acid or stop signal in the mRNA as it undergoes translation.

**Fusion Protein** A genetic construct that attaches a gene encoding a protein or a portion of the gene to another coding sequence such that the protein formed during translation is a linear combination of the two proteins.

**Partition** The usual segregation of the genetic material of the dividing cell into the two daughter cells.

**Promoter** The DNA site at which the RNA polymerase specifically binds to initiate transcription.

**Protease** An enzyme that cleaves the peptide bonds between amino acids comprising the protein.

**Ribosome Binding Site** The region of the mRNA near the site of translation initiation that is important for ribosome recognition.

**RNA Secondary Structure** A structure formed within an RNA molecule by bending the molecule such that a series of specific base pairs can form between nearby complementary nucleotides, forming a small region that has double-stranded character.

**Translation** The process of protein synthesis in which the order of codons on the individual mRNA are used to specify the order of linkage of the amino acids of that particular protein as it is made on the ribosome.

**Virus** An agent naturally able to infect a host cell and bearing genes allowing its complete reproduction within a specific host cell.

---

The production of a specific protein by recombinant DNA technology entails cloning the gene that encodes the protein from the desired organism, forming a suitable genetic construct such that the gene can be expressed in a host organism, and introduction and maintenance of the construct in the host to allow the production of an adequate yield of functional protein. The ability to take a gene from any organism and express it in adequate quantity in a system permitting the analysis of the structure and function of the protein encoded by the gene has great importance in the research laboratory. Physical and enzymatic features of the protein can be studied, and its functional location in cells or tissues determined, in an effort to better define the biological function of a newly discovered gene. Industrial applications include the production of therapeutic and diagnostic proteins having great impact in the field of medicine, and the production of stable bulk enzymes for the food and specialty chemical industries.

## 1 INTRODUCTION TO HOST–VECTOR SYSTEMS

Before a specific RNA or protein product can be produced in a particular host cell, a suitable DNA construct must be prepared. This entry discusses the essential features of expression systems that are widely used for overproduction of proteins for a variety of purposes. The expression system is composed of an expression vector and a specific host cell. Not only is the selection of the ideal expression vector important, but the choice of the appropriate host can affect production efficiency considerably. First the expression vectors is considered, and then the important attributes of the host are discussed. A number of organisms have been used as hosts for expression of foreign proteins. Since, however, the most complete picture is available for *Escherichia coli,* that prokaryotic organism serves as an example and is discussed in more detail.

Expression vectors usually consist of small, circular plasmids specifically designed with several key features that allow a foreign gene inserted into the plasmid to be expressed in the host cell. Important elements of the plasmid include (1) an origin of replication, which allows the plasmid to be replicated in the host, (2) a selectable genetic marker, which allows cells bearing the plasmid to preferentially grow on a specifically composed medium, (3) transcription and translation signals recognized by the host cell, and (4) a suitable unique restriction site located appropriately with regard to the transcription and translation signals, where the plasmid can be cleaved and the foreign DNA can be introduced. In Sections 2 and 3, the generalized structure typical of many expression vectors is illustrated, and the variation and importance of each of these features in the overall production system is explained.

## 2 COMPONENTS OF PLASMID VECTORS

### 2.1 Origin of Replication

The origin of replication of the plasmid is a specific DNA sequence that defines the beginning of the replication of the circular plasmid DNA molecule. Because of its essential function in this event, its structure also limits the frequency with which replication is initiated. Replication frequency defines the copy number of the plasmid (i.e., the number of individual plasmid molecules present in each cell). This parameter, in turn, is important because of the gene dosage effect. If each plasmid carries a copy of the gene, and each copy can give rise to a certain maximal amount of product per unit time, the more copies of the plasmid present in the cell, the greater will be the level of production that can be attained per cell. Of course, there is a limit to this effect, and above a certain level the cell machinery becomes saturated and the increased metabolic burden of maintaining the plasmid overcomes the advantage of the gene dosage in the production process.

The most widely used expression vectors for *E. coli* are derived from that of the ColE1 plasmid and have copy numbers of between 10 and 100. The useful properties of high gene dosage have been used advantageously through the construction of temperature-sensitive copy number control systems in which the copy number can be raised dramatically by increasing the temperature. A typical prokaryotic expression vector is represented in Figure 1.

Another feature specified by the origin of replication is the compatibility of the plasmid. Since it is sometimes desirable to maintain two distinct plasmids within the same cell, two different origin of replication types would be employed on the two different plasmids. If the two plasmids bear the same type of origin, one eventually will dominate without special selection and the other will be lost from the cell as the cells grow and divide. This phenomenon is called incompatibility, and plasmids with the same origin type are defined as being in the same incompatibility group.

### 2.2 Selectable Markers

The ability to select the vector and maintain the plasmid in the host cell is in important aspect of the expression system, especially in production regimens that involve a longer time course or continuous culture fermentations. The most common type of selectable marker incorporated on the plasmid is an antibiotic resistance element allowing selection with ampicillin, kanamycin, or tetracycline. Other approaches place an essential gene on the vector and use a special strain in which that gene is deleted from the chromosome (e.g., the valine tRNA synthetase, *val S* or the single-stranded DNA binding protein, *ssb*). If the plasmid is lost from the cell, the host cell alone is unable to grow. In such a system, addition of an antibiotic is not needed to select for the plasmid. The exploitation of naturally occurring partition systems that maintain copy number and select against plasmid loss has also been successful. An example is the incorporation of the *par B* locus on vectors. This locus selects strongly against plasmid loss by killing the cells that no longer carry the *par B* locus. This system does not require any specially constructed host strain.

### 2.3 Transcription and Translation Signals

The transcription and translation signals allow the coding region of the gene to be efficiently recognized and expressed by the host cell machinery (i.e., RNA polymerase and ribosomes) that decodes the nucleotide sequence of the DNA of the gene into the appropriate sequence of amino acids in the protein. The signal specifying where transcription starts is called the promoter. The more efficient the promoter, the greater the synthesis of messenger RNA of the gene under its control. Certain expression vectors employ a relatively strong promoter, which is unregulated and is called a constitutive

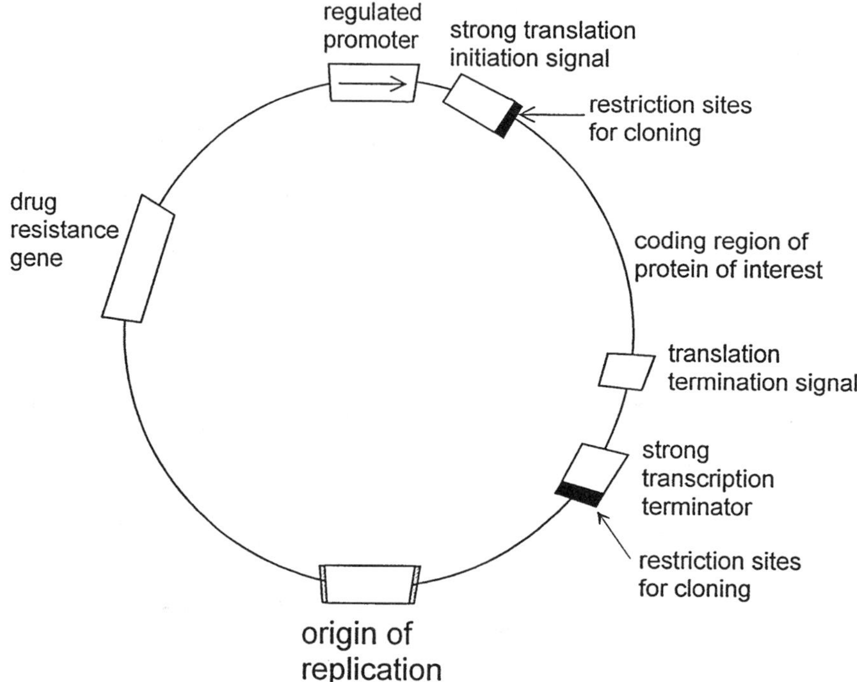


**Figure 1.** Typical prokaryotic expression vector.

promoter. This approach is feasible if the protein to be made is very stable in the cell and does not cause any toxic effects or retard growth. More generally, however, a regulated promoter that can allow production of the desired protein at a suitable time is used. This approach has two advantages: a short period of protein synthesis and induction in the growth of the batch culture at the most opportune time (i.e., the period during which the optimal combination of cell density and specific protein production can be achieved).

The most widely used control systems for transcription initiation are those regulated by temperature, or by the addition of a chemical agent or metabolite. A strong promoter from the bacteriophage lambda is normally prevented by a repressor from binding RNA polymerase. If an altered form of the repressor is used that is unstable at high temperature, an increase in temperature will release this negative control of the promoter, allowing large quantities of the desired messenger RNA to be synthesized. This promoter system can also be released from control of the repressor by addition of DNA damaging agents. This method is sometimes used instead of an increase in temperature.

Another widely used system is based on the repressor control of the lactose promoter. The DNA sequence bound by the lactose repressor can be combined with a variety of other stronger promoters (e.g., the tryptophan operon or lipoprotein gene promoter), which then become controlled by the lactose repressor. The addition of an analogue of lactose, isopropylthiogalactoside, is most commonly used to induce expression from this system.

A number of other regulation systems have been advocated for process control. Among these are phosphate, oxygen, and acid-induced promoters. Some promoters are regulated by specific activators that stimulate transcription rather than by a negatively acting repressor system. An extension of the idea of regulating transcription initiation is the production of the desired messenger RNA through the use of a special RNA polymerase not normally found in the cell, in combination with its unique promoter sequence. A system based on the bacteriophage T7 RNA polymerase and the T7 promoter it recognizes can yield very high specific production of the desired product. In this case, the T7 promoter would be located on the expression vector and the T7 RNA polymerase gene would be located on the chromosome under the control of a regulated system, typically of the type just mentioned. Thus induction in this system first produces the T7 RNA polymerase, which in turn synthesizes the messenger corresponding to the gene of interest.

Once a suitable means of producing the messenger RNA encoding the protein of interest has been defined, the next consideration is to translate the messenger RNA with optimal efficiency. Several factors affect this efficiency. One is the stability of the message. If the messenger RNA has a long half-life in the cell, it can be translated proportionally more often before it is degraded. The presence of hairpin secondary structures, where the nucleotide sequence allows intramolecular base pairing, particularly near the end of the message, seems to help stabilize the messenger RNA from degradation. The hairpin sequences act as part of the transcription termination signal and are normally included in the expression vector construct at the 3′ end of the message beyond the protein coding sequence.

The major factor usually limiting translation is the ability of the ribosome to initiate translation on the messenger RNA. The ability of the ribosome to bind effectively to the messenger RNA can be impeded by secondary structures in the message which block access to the translation initiation codon. The binding of the ribosome to the messenger requires interaction between the ribosome, specifically a segment of the ribosomal RNA, and a region of the messenger RNA about 10 nucleotides 5′ from the translation initiation codon. Optimal sequences for this ribosome binding region have been analyzed for a number of genes, and a highly efficient ribosome binding site is often included in the vector just preceding the position at which the protein coding sequence is to be placed. Nucleotide sequences frequently placed in vectors for their efficient ribosome sites include chemically synthesized optimal consensus sequences, the T7 g10-L region, or those of other highly translated genes. While these standard translation initiation systems are suitable for adequate expression of most proteins, the sequence in this area has to be optimized to gain the highest level of protein production. Translational limitations can also occur as a result of the specific nature of the protein coding region (e.g., unusual composition, unique codon usage, or special secondary structures). These problems cannot be addressed at the expression vector level.

### 2.4 Restriction Sites for Cloning

Careful selection of the location of the restriction endonuclease cleavage site at which the protein coding sequence is to be introduced into the plasmid can allow the expression of the protein in its most useful form. In many cases it is desired to produce the complete intact natural protein, beginning with its N-terminal amino acid and ending with the native C-terminal amino acid. This frequently is the case when the exact structure of a mature, natural, pharmaceutically active protein is needed. To position the coding sequence correctly with respect to the ribosome binding site, a specific restriction endonuclease site incorporating an ATG sequence is used to place the translation initiation codon appropriately in the vector. Two commonly used sites that contain an ATG and rarely exist in any given DNA sequence are *Nco* I and *Nde* I. A second unique restriction site is often used where the C-terminal end of the gene is joined to the vector to orient the protein coding segment during cloning. Such a construct would then enable translation of the complete cloned protein coding region.

Although not all mature proteins have an N-terminal methionine, this is the primary product formed in bacterial systems. The extra amino acid can be removed from the protein after synthesis by in vitro methods, or the desired protein can initially be made as part of a longer peptide chain with subsequent cleavage by a specific protease to generate the N-terminus of the mature protein product. A variety of N-terminal protein carrier elements have been used to make protein fusion products. In the formation of a fusion protein, the vector carries the appropriate signals for translation initiation, as well as a coding region of the N-terminal component of the protein and a restriction endonuclease cleavage site where the DNA fragment bearing the foreign gene can be placed in the proper reading frame. The properties of the N-terminal component of the protein fusion can be used to assist in purification of the fusion protein. For example, if this segment encodes an easily purified protein such as the maltose binding protein or glutathione *S*-transferase, the combined fusion product can be readily isolated using affinity chromatography. The addition of tracts of basic, acidic, or metal-binding amino acids in the carrier protein can also be used to aid purification. This protein fusion concept can be utilized to allow for protein processing in vivo. For example, if the N-terminal segment encodes a localization signal, the protein can be directed to the periplasmic space of *E. coli* or, in some cases, into the extracellular medium. This type of export can aid in the purification and stabilization of the protein. A vector employing protein A as the carrier is one of this type.

In some applications the expression of the entire protein is not necessary. For raising antibodies or for the detection of an antigen by antibodies, only a recognizable epitope is necessary, so in these cases expression of only a short protein segment is sufficient. The cloned segment comprising the antigenic site can be attached to a protein carrier segment that will allow effective exposure of the foreign peptide. Placement of the foreign segment at the C-terminal end of β-galactosidase has been used effectively for this purpose. In some cases a suitable surface protein can be used as the carrier portion to ensure that the foreign protein segment is localized to the surface of a cell or bacteriophage particle, thus presenting the immunologically active antigen or antibody species in a way that allows detection, isolation, or use of the living cell or virus particle.

## 3 HOST CELL FEATURES

### 3.1 *Escherichia coli*

The use of the correct host can have significant impact on the final yield of the desired protein product. Factors generally useful to the host include the ability to grow rapidly and to a high density, the ability to be transformed in a manner that facilitates the introduction of the DNA construct, and the possession of a low recombination and mutagenic rate, to ensure that the plasmid is not frequently lost, damaged, or otherwise inactivated. Stability to degradation is a major problem in the production of certain proteins. Therefore, host cells with reduced protease levels have been used to enhance the in vivo stability of the foreign protein of interest. Commonly used mutations in *E. coli* that have reduced protein degradation rates because of the inactivation of proteases are *lon, rpo H,* and *clp.* In some cases the protein stability can be addressed at the expression vector level by using a fusion protein construct. The presence of the longer carrier protein often will effectively stabilize a small foreign peptide segment. Translation of large proteins is also a limitation in many strains; however other *E. coli* strains may have an increased ability to translate the particular product.

### 3.2 Other Hosts

While *E. coli* continues to be the most widespread expression system, other host–vector systems have certain advantages. The strengths of the *E. coli* system are the variety of vectors and specifically altered hosts available and the well-studied methods for manipulating this organism. High levels of production can be attained. However protease problems, formation of inactive inclusion bodies containing the product, and the lack of a eukaryotic glycosylation system limit production of a number of proteins from mammalian sources. Other bacteria have received some attention owing to their ability to grow on particular compounds, their potential for secretion of the protein, or their industrial potential.

Yeasts are a suitable production system for a number of processes. Not only are they well studied like *E. coli,* and amenable to scale-up, but they are able to carry out some posttranslational modification of eukaryotic proteins.

Fungi can produce high yields of commercial proteins and are reasonably capable of glycosylation and secretion of proteins into the medium. Methods have not been so completely developed, and the use of organisms of this class has not been widespread.

Viruses that infect insect cells (baculoviruses) have gained attention as a system for producing glycosylated proteins at reasonably high levels but at less expense than is incurred using mammalian cells. The vaccinia virus has been used to express foreign antigens in whole animals, demonstrating its potential for use as a vaccine vector system. Although mammalian cells have been studied extensively and have the advantage of producing truly identical processed mammalian protein, the costly, difficult scale-up has limited commercial production with this system.

## 4 PERSPECTIVES

The ability to clone and express high levels of proteins in other organisms has led to great advances in the speed and detail with which biological systems can be analyzed. With the expansion of this technology to other organisms and the construction of more complicated and sophisticated derivatives of currently studied systems, the impact of this area will increase. Recombinant protein production has now begun to see use in applications to industrial microorganisms, the formation and use of transgenic plants and animals, and the analysis of unknown proteins associated with genetic disorders.

*See also* DNA Markers, Cloned; *E. coli* Genome; Fungal Biotechnology; Gene Expression, Regulation of; Plasmids; Yeast Genetics.

### *Bibliography*

Alitalo, K. K., Huhtala, M.-L., Knowles, J., and Vaheri, A., Eds. (1990) *Recombinant Systems in Protein Expression.* Elsevier, Amsterdam.

Barr, P. J., Brake, A. J., and Valenzuela, P., Eds. (1989) *Yeast Genetic Engineering.* Butterworths, Boston.

Goeddel, D. V., Ed. (1990) *Gene Expression Technology,* Vol. 185 in *Methods in Enzymology.* Academic Press, San Diego, CA.

Henninghausen, L., Ruiz, L., and Wall, R. (1990) Transgenic animals—Production of foreign proteins in milk. *Curr. Opin. Biotechnol.* 1:74–78.

Hruby, D. E. (1990) Vaccinia virus vectors: New strategies for producing recombinant vaccines. *Clin. Microbiol. Rev.* 3:153–170.

Kriegler, M. (1991) *Gene Transfer and Expression: A Laboratory Manual.* Freeman, New York.

Old, R. W., and Primrose, S. B. (1990) *Principles of Gene Manipulation: An Introduction to Genetic Engineering,* 4th ed. Blackwell Scientific Publications, Oxford.

O'Reilly, D. R., Luckow, V. A., and Miller, L. K. (1992) *Baculovirus Expression Vectors—A Laboratory Manual.* Freeman, New York.

Reznikoff, W., and Gold, L., Eds. (1986) *Maximizing Gene Expression.* Butterworths, Boston.

Ridgeway, A. A. (1988) Mammalian expression vectors. *Biotechniques* 10:467–492.

Vasil, I. L. (1990) The realities and challenges of plant biotechnology. *Bio/Technology* 8:296–301.

Wang, L.-F., and Doi, R. H. (1992) Heterologous gene expression in *Bacillus subtilis.* In *Biology of Bacilli: Applications to Industry,* R. H. Doi and M. McGloughlin, Eds., pp. 63–104. Butterworth-Heinemann, Boston.

# Extracellular Matrix

*Linda J. Sandell*

### *Key Words*

**Exon** DNA Of the gene that is represented in the mature RNA product.

**Extracellular Matrix** Material lying adjacent to and between cells.

**Gene Structure** Organization of the gene, including the promoter, exons, and introns.

**Intron** DNA of the gene that is removed from the mature RNA product.

**mRNA Splicing** The removal of introns from the pre-mRNA and joining of exons in mature RNA; thus introns are spliced out, while exons are spliced together.

---

The extracellular matrix is necessary for development and normal functioning of all cell types in an organism: it is made up of secreted proteins and glycoproteins whose complex interactions determine the matrix properties. The composition of the extracellular matrix is controlled at the level of gene expression by providing the quantity of mRNA sufficient to produce adequate proteins and by alternative splicing of pre-mRNA to generate proteins with the proper functional domains.

## 1 INTRODUCTION

The extracellular matrix (ECM) of all tissues is a complex mixture of secreted proteins that collectively play a critical role in determining and maintaining tissue function. The ECM is located primarily around connective tissue cells and under epithelia. ECM proteins range from the multifunctional fibronectins and thrombospondins to the large families of collagen types, proteoglycans, and laminins, among others. Fibrillar collagens types I, II, and III are the principal ECM proteins that confer the structural characteristics typical of tissues such as bone, skin, blood vessels, and cartilage. Other ECM proteins such as the fibronectins, laminins, and tenascin play critical roles in cell–cell interactions, cell migration, and cytoskeletal organization. The structure and function of extracellular matrix proteins have been reviewed recently in two very informative books by Hay (1993) and Kreis and Vale (1993). While not inclusive, this contribution discusses the molecular biology of the major categories of ECM components.

### 1.1 Regulation of ECM Gene Expression

It is becoming increasingly clear that the regulation of ECM gene expression, both transcriptionally (to regulate quantity) and post-transcriptionally (using alternative splicing of mRNA to include or remove functional domains), is of crucial importance during morphogenesis and cell differentiation, cell migration and proliferation, wound healing, and disease processes such as fibroses and arthroses. ECM expression is regulated by a wide variety of growth factors and cytokines, being generally stimulated by, for example, transforming growth factor beta, and insulin-like growth factors I and II, and inhibited by interleukin 1 and interferon gamma. Interestingly, as more information accumulates regarding regulation of ECM expression, it is apparent that ECM molecules are quite independently controlled in their expression and that regulation is dependent on cell type. Table 1 shows some examples of regulation of ECM molecules by cytokines. This is a rapidly emerging field and more information is published every day. It is now known that certain ECM molecules can effect the expression of themselves or other ECM molecules: for example, the amino-propeptide of type I collagen is thought to down-regulate collagen synthesis, while the small proteoglycan decorin may be a component of feedback regulation of cell growth.

### 1.2 Gene Structure

Matrix proteins tend to be quite large, and often undergo extensive posttranslation modification. The genes are large and generally are composed of exons averaging 100–300 base pairs. Over the last 10 years, a considerable amount of information has accumulated, describing the structure, synthesis, and turnover of extracellular matrix proteins and, in particular, the genes coding for these proteins. Without exception, all these vertebrate ECM genes are single-copy, multiexon structures which, in many examples, represent among the most complex genomic structures yet described. The human α1(XIII) collagen, for example, is approximately 140 kb in size and contains 39 exons; yet only 1.8% of this large gene is used to code for the 2300 bp of the α1(XIII) mRNA.

### 1.3 Alternative Splicing of Pre-mRNA

Recently, several examples of alternate usage of particular exons were observed in a number of different ECM genes. This process of alternate exon usage, also referred to as alternative splicing, has been shown to account for isoforms of fibronectin and several proteoglycans. In addition, several investigators have demonstrated alternate exon usage to be a feature within the genes coding for several collagen types. In some instances, alternate exon usage is clearly responsible for the production of functionally distinct isoforms of an ECM protein. In other examples, the functional significance of alternative splicing of a pre-mRNA coding for a particular ECM protein is not yet clear. It does seem, however, that the mechanism of alternative splicing is useful for generating subtle differences within these large proteins while keeping the common parts of the protein identical.

## 2 COLLAGENS

### 2.1 Function

The collagens are a large family of genetically and structurally distinct proteins that collectively make up a major component of the ECM of most multicellular organisms. This family of triple-helical connective tissue proteins plays a critical role in the development and maintenance of a variety of functions including tissue architecture, tissue strength, and cell–cell interactions. Work over the last few years has clearly shown the important role that collagens play in diseases ranging from inherited skeletal dysplasias to tumor metastasis. This work has also led to the suggestion that collagen defects may be involved in such common diseases as osteoarthritis, osteoporosis, and disorders of the cardiovascular system. To date, 18 members of the vertebrate collagen family of proteins have been identified, coded for by at least 30 different genes. The collagens are the largest gene family and have been studied in most detail; consequently, the molecular biology of the collagens serves as our focus.

### 2.2 Gene Structure

The collagens are divided into two major groups: fibrillar and nonfibrillar. The fibrillar collagens, sometimes also called interstitial, include types I, II, III, V, and XI. The nonfibrillar include types VI, IX, X, XI, XII, XIV, and XIII. A subgroup of the non-fibril-forming collagens are classified as FACIT (fibril-associated collagens with interrupted triple helices), as they are found in close association with fibrillar collagens. The interstitial collagen chains are composed of three distinct or identical chains arranged in a ''collagen'' triple helix having the amino acid sequence Gly-X-Y, where X is often proline and Y, hydroxyproline. The collagens are initially synthesized as procollagens with globular propeptides at each end. The non-fiber-forming collagens are similar in overall structure, but they have noncollagenous, globular domains interspersed between Gly-X-Y domains. The protein structure is uniquely reflected in the gene structure. In general, the Gly-X-Y domains are encoded by very regularly sized exons, while the globular propeptides reflect gene structure observed in other genes. For example, the *COL2A1* gene is made up of 54 exons and 53 introns. One exon (exon 2) is known to be alternatively spliced. This exon–intron structure is striking and represents a consistent pattern of all the fibrillar collagen genes (including types I, II, III, V, and XI). With very few exceptions, the conservation of exon structure is independent of species or type of fibrillar collagen. All the Gly-X-Y coding exons are 54 bp, 45 bp, or multiples of 54 or 45 (90, 108, and 162 bp). The translation of this gene structure into protein is shown diagrammatically in Figure 1. At each end of the triple-helical encoding exons are junction exons, which contain some Gly-X-Y coding domain and part of the nonhelical domain. The junction exons vary to a small extent and have even evolved to add Gly-X-Y coding triplets to type III collagen. In the propeptide-encoding domains, there are features of structural conservation within the fibrillar collagen genes, although not to the same extent as observed in the collagenous coding domain. For example, the penultimate exon, which contains a highly conserved coding sequence around a carbohydrate attachment site, is 243 bp in all the fibrillar genes where DNA sequence in this region is available. Equally, the coding region of the last exon, which con-

**Table 1** Some ECM Molecules Regulated by Cytokines

| | Molecules | | | | | | |
|---|---|---|---|---|---|---|---|
| Cytokine | FN | COLI | COLII | Decorin | Biglycan | Versican | Aggrecan |
| TGF-β | Up | Up | Up | No change | Up | Up | Up |
| IL-1 | Down | Up | Down | Up | n.d. | Down | Down |
| Retinoic acid | Up | Up | Down | Up | Down | Up | Down |
| IFN-γ | Down | Down | Down | n.d. | n.d. | n.d. | n.d. |

*Abbreviations:* FN, fibronectin; COLI, II, collagens I and II; TGF-β, transforming growth factor beta; IL-1, interleukin I; IFN-γ, interferon gamma; n.d., not determined.

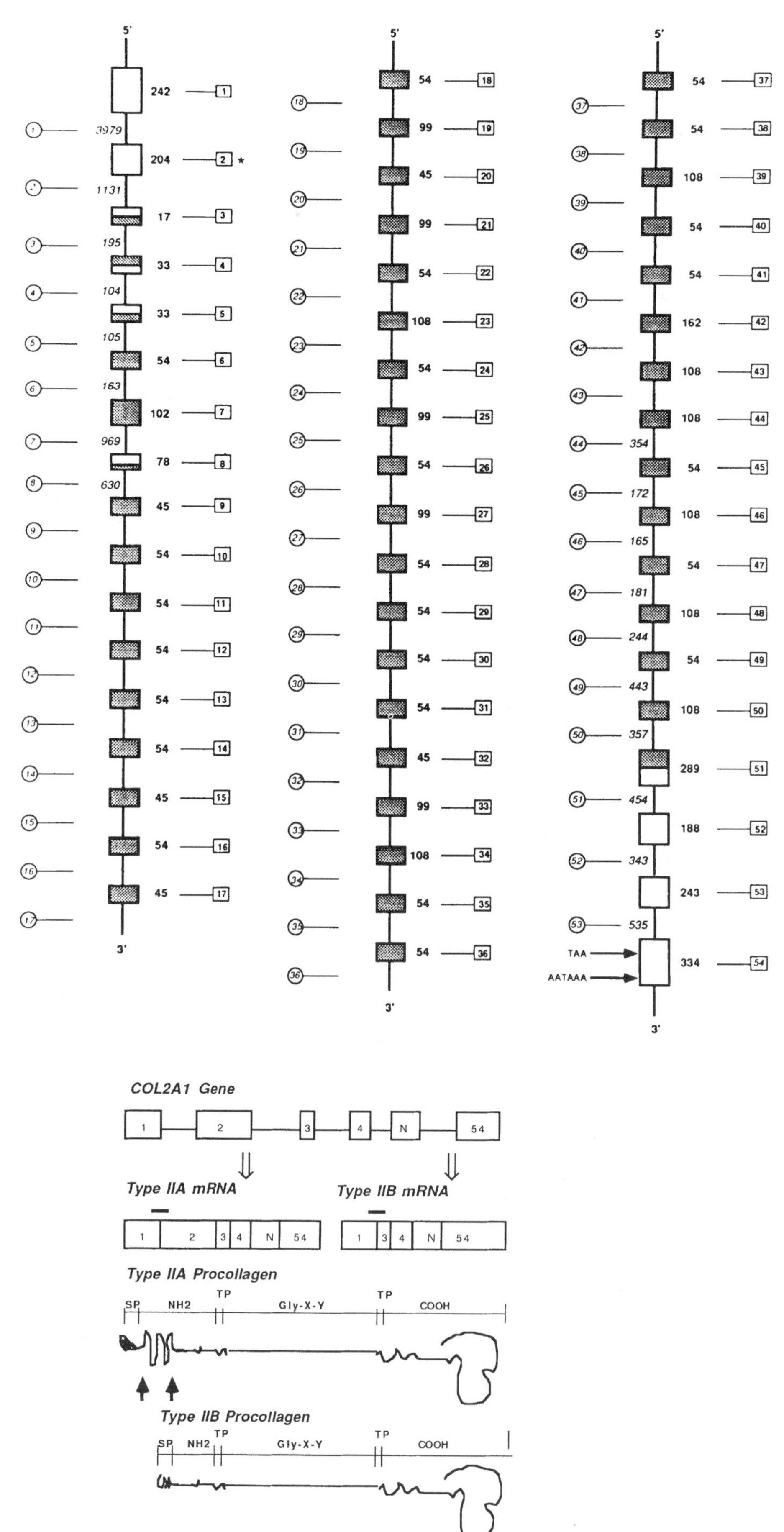

5'
242
3979
204
1131
17
195
33
104
33
105
54
163
102
969
78
630
45
54
54
54
54
54
45
54
45
3'
5'
54
99
45
99
54
108
54
99
54
99
54
54
54
54
45
99
108
54
54
3'
5'
54
54
108
54
54
162
108
108
354
54
172
108
165
54
181
108
244
54
443
108
357
289
454
188
343
243
535
TAA
AATAAA
334
3'
COL2A1 Gene
1
2
3
4
N
54
Type IIA mRNA
Type IIB mRNA
1
2
3
4
N
54
1
3
4
N
54
Type IIA Procollagen
SP
NH2
TP
Gly-X-Y
TP
COOH
Type IIB Procollagen
SP
NH2
TP
Gly-X-Y
TP
COOH

tains the C-terminus of the C-propeptide in all phylogenetic species and all fibrillar collagen genes, is 144 bp in length. The other two exons that make up the N-terminus of the C-propeptide and the C-telopeptide vary in size to a small extent. The high degree of conservation of the C-propeptide between collagen genes argues for a similar and critical function of the propeptide in all collagens—perhaps the alignment of the three chains and initiation of triple-helix formation. The C-propeptide has been isolated from the extracellular matrix of cartilage (chondrocalcin) and may have an additional function, possibly in mineralization. On the other hand, the 5′ region of the fibrillar collagen genes, coding for the N-propeptide domain, clearly shows the highest degree of structural and sequence divergence, indicating that the function of the amino propeptide differs somewhat between collagen types.

### 2.3 Alternative Splicing

In the family of collagen genes, there are a number of examples of alternative splicing of pre-mRNA, thus providing further diversity to an already large gene family. The simplest alternative splicing occurs in the *COL2A1* gene shown in Figure 1A. Here, exon 2, which comprises the large (69 amino acid), cysteine-rich domain of the N-propeptide, is either included (type IIA) or excluded (type IIB) from the mature mRNA. These two collagen mRNAs are tissue specific, with the type IIB mRNA found in chondrocytes resident in cartilage and the type IIA mRNA found in chondroprogenitor cells, bone precursors, skin precursor cells in avians, and in a variety of epithelial–mesenchymal junctions of the developing embryo. It is known that the type IIB procollagen provides structural integrity to the cartilage ECM. The type IIA procollagen may be involved in induction of skeletal structures or have some other function during differentiation.

Expression of specific domains of type VI and type XIII collagens also is regulated by alternative splicing of exons. While some of the alternative expression is tissue specific and may be developmentally regulated, there is no clear pattern, nor any proposed function for alternative splicing in these genes, at the present time.

### 2.4 Alternative Use of Gene Promoters

The collagens also provide examples of alternative promoter use. The α1(IX) and α2(I) collagen genes also generate diversity by differential usage of exons. However, the mechanism is not by alternative splicing of the pre-mRNA, but by alternative start sites of transcription. These events eliminate most of the 5′ exons. In both the chicken and human α1(IX), the choice of transcription start site is tissue specific: in cartilage, transcription begins at exon 1 and contains a large N-terminal non-triple-helical domain of 266 amino acids (NC4); in primary corneal stroma, transcription begins in the intron between exons 6 and 7, eliminating the expression of the NC4 domain. In the chicken gene, these promoters are separated by 20 kb of DNA. The function of the alternatively spliced region is unknown; however, it has an estimated isoelectric point of pI 9.7 and 10.55, in the chicken and human, respectively, and is believed to bind to proteoglycans.

Transcription of the α2(I) collagen gene is also tissue specific; however, the reading frame of the protein is changed by the alternative use of promoter, producing a different and noncollagenous protein. In most type I collagen synthesizing tissues, all the previously described exons are used whole; in chondrocytes, transcription begins at a new site within intron 5 and includes an additional exon. The use of the cartilage transcription start site replaces exons 1 and 2 with a 96 bp exon contained within intron 2. The resulting transcripts contain a reading frame that is out of frame with the collagen coding sequence. The putative protein product of this mRNA is consequently not collagenous and may encode a protein of regulatory function. Interestingly, this anomaly provides a mechanism for inhibition of type I procollagen synthesis by chondrocytes. In prechondrogenic limb mesenchyme, transcription of the α2(I) collagen initiates at the previously described bone/tendon promoter, while during differentiation into chondrocytes, the promoter utilization is switched to the cartilage promoter.

## 3 ELASTIN

Tropoelastin is the soluble precursor to elastin, the major protein component of elastic fibers. Several isoforms of tropoelastin have been identified in a variety of elastic tissues; however, unlike the other connective tissue genes, there is no elastin family of genes, only a single gene. Isoforms are generated solely through alternative splicing of pre-mRNA. Tropoelastins from all vertebrates studies are approximately 70 kDa in size. The complete, derived amino acid sequence from overlapping tropoelastin cDNA and genomic DNA clones from several vertebrate species has confirmed earlier peptide sequence data suggesting the existence of several functional domains in tropoelastin—including alanine-rich, lysine-containing regions necessary for the formation of desmosine cross-links. The resilient properties of tropoelastin are conferred by multiple hydrophobic domains rich in valine, proline, and leucine. In addition to a signal sequence at the amino-terminal end of tropoelastin, there are several other domains, including the only cysteine-containing region located at the carboxy terminus of the protein. Most of these domains are encoded by separate exons within the tropoelastin gene.

The complete genomic sequence coding for tropoelastin has been analyzed in rats, humans, and cows. In all species, the tropoelastin gene is approximately 40 kb in length and is composed of about 36 exons. The majority of these exons code for either hydrophobic or cross-link domains. By DNA sequencing and S1 nuclease protection analysis, it is clear that a total of approximately 40% of these exons are subject to alternate usage in elastic tissues such as nuchal ligament, aorta, lung, and skin from varying developmental ages of rats, cows, or humans. Figure 2 summarizes all known exons or domains within exons in the vertebrate tropoelastin gene that are subject to alternative splicing. Any clear correlation with development or tissue-specific expression of the tropocollagen isoforms

◄

**Figure Figure 1.** Diagram of the *COL2A1* gene and procollagen α-chain: SP, signal peptide; $NH_2$, the amino-terminal propeptide; TP, the telopeptide; Gly-X-Y, the triple-helical domain; COOH, the carboxy-terminal propeptide. Regions indicated by a straight line are triple helical; curves lines indicate regions that are globular portions of the protein. Arrows indicate the differentially spliced domain of the type IIA procollagen. [From Sandell et al, 1991,

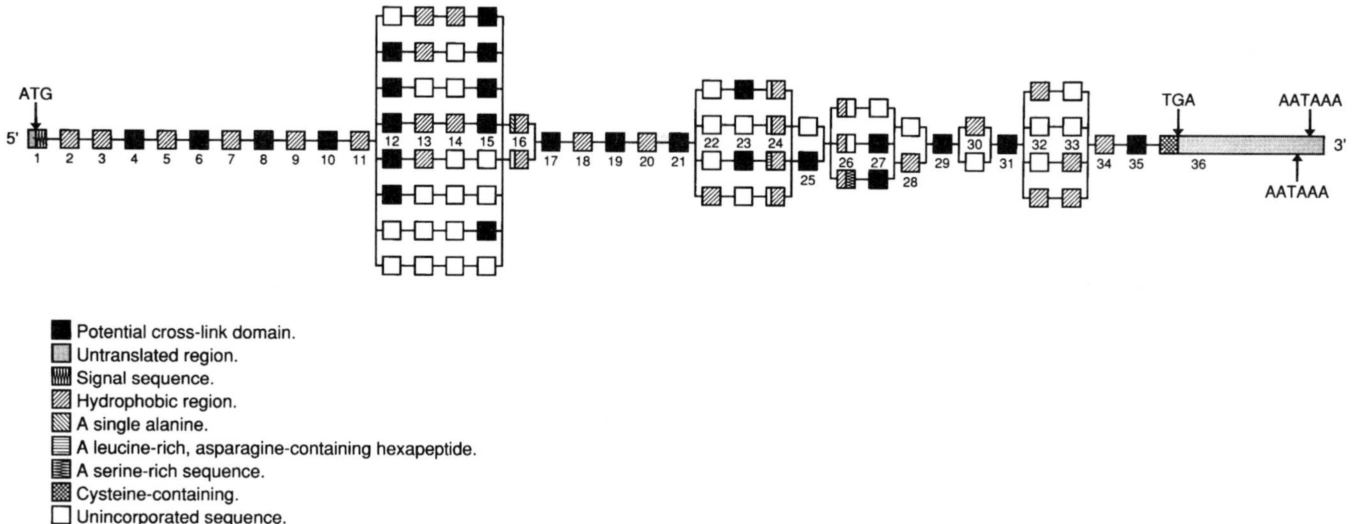


**Figure 2.** A summary of exons alternatively spliced in the vertebrate tropoelastin gene. Exons are numbered from the 5′ end of the gene and the various domains encoded by exons are identified by variable shading explained in the legend. Known permutations of alternatively spliced exons are illustrated. The initiation and termination codons and polyadenylation signals are also indicated. The relative size of exons and introns are not indicated [From Boyd et al. (1993).]

cannot be observed, and the function of these multiple forms is unknown.

## 4 FIBRONECTIN

Fibronectins (FN) are large, extracellular matrix glycoproteins that function as both plasma and ECM proteins. FN contains binding domains for fibrin, cells, heparin and proteoglycans, and collagen. The gene structure is shown in Figure 3. The plasma and cellular forms of FN have been isolated and extensively studied. The difference between these two forms and other forms with more subtle differences can be accounted for by alternative splicing.

Alternative splicing of fibronectin pre-mRNA was the first example of alternative exon usage as a means of generating ECM protein diversity. As a major cell adhesion molecule, FN has multiple sites for interaction with cell surface receptors and is an essential structural component of the extracellular network. The binding domains are composed of sets of repeating modules called types I, II, and III repeats. Single exons encode each type I and type II repeat and pairs of exons combine to code for each type III repeat (see Figure 3). The FN gene consists of 48 exons and 47 introns that are spliced to give an mRNA about 8 kb long. Within this framework, the FN mRNA varies at three sites via two different patterns of alternative splicing. The first of these regions to be identified, the variable (V) region, exemplifies exon subdivision in which a single 467-base exon containing alternate and constitutive coding sequences is partially or completely included. In rodents and cows, the 360-base V region is either entirely included or entirely excluded, or the first 75 bases are omitted during splicing. The number of splice variants at this site differs among species, with five in humans and two in chickens and frogs. The exon is never skipped because 107 bases encoding half the flanking type III repeat are constitutively included. The resulting polypeptides contain additional 120 or 95 amino acid segments or no extra residues, thus generating three types of FN subunits V120, V95, and V0, respectively. This segment has a unique sequence; unlike the remainder of FN, it is not part of the repeating structure.

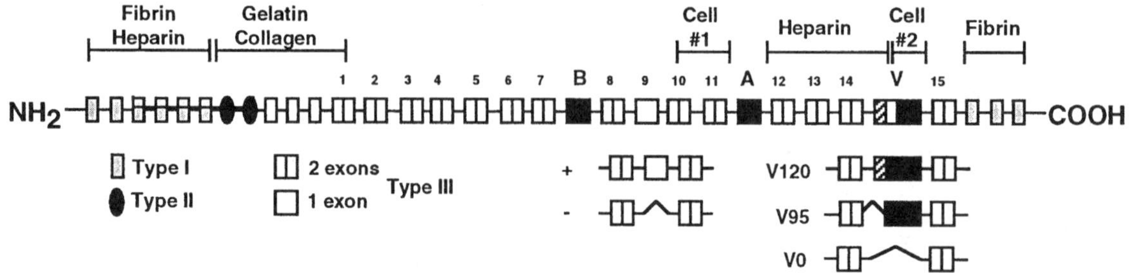


**Figure 3.** FN subunit variants. The repeat structure of FN is shown and consists of type I, type II, and type III repeats as indicated. Each type I and II repeat is contained within a single exon in the gene. All but three of the 17 type III repeats are encoded by two exons; the exceptions are the alternatively spliced repeats EIIIA (A) and EIIIB(B) and one constitutive repeat, exon 9. Inclusion (+) or omission (−) of EIIIA and/or EIIIB results in the structures shown. Exon subdivision of the V region generates three variants differing in the number of extra amino acids (V120, V95, V0). The function-binding sites are indicated above the gene structure. The two cell-binding sites are recognized by different integrin receptors: 1, an RGD site, is recognized by α5β1; 2, a EILDV site, is recognized by α4β1 integrin. [Modification of a figure from Boyd et al (1993).]

The second pattern of alternative splicing is exon skipping. At two sites, the single exons EIIIA and EIIIB, encoding single type III repeats, are included or skipped independently, generating mRNAs with all combinations of these two exons. Mechanistically, the EIIIB splicing event appears to involve formation of a specific protein RNA complex on the intervening sequence 5′ of this exon.

All FN alternate exon usage is regulated in a tissue-specific manner. Liver splicing of FN is the most simple and distinctive. Liver synthesizes circulating plasma FN, a heterodimer of one V+ subunit (containing either the V120 or V95 segment) and one V0 subunit and lacking both EIIIA and EIIIB. Most, if not all, other tissues express the cellular form of FN, which is composed of V95 or V120 (V+) FNs with varying amounts of EIIIA and/or EIIIB.

The most striking shifts in the splicing repertoire occur during embryogenesis, wound healing, and oncogenesis. All three situations show an increase in the amount of EIIIA+ and/or EIIIB+ mRNA relative to normal adult tissues. This is most clearly illustrated by in situ hybridization of cutaneous wound sections. Adjacent normal tissue cells express V+ FN, while fibroblasts within the granulation tissue of the wound bed are also very positive for EIIIA+ and EIIIB+ mRNAs. In addition, quantitation of EIIIA and EIIIB mRNAs identified by RNase protection analysis demonstrates a dramatic drop in their levels at birth and increased inclusion of some oncogenically transformed cells. Thus, changes in the combination and proportions of EIIIA+ and EIIIB+ RNAs at different stages of tissue development, remodeling and tumor formation could have significant effects on the interactions and functions of FN.

## 5 PROTEOGLYCANS

Structurally, proteoglycans are distinguished from other glycoproteins in possessing glycosaminoglycan chains on a protein core. The glycosaminoglycan chains themselves differ in number (from 1 to 100) and glycan composition: chondroitin sulfate, dermatan sulfate, heparan sulfate, and keratan sulfate. Proteoglycans are present in a variety of basement membranes, ECMs of most connective tissue cells, and cell membranes. While functionally disperse, the proteoglycan core proteins can be classified into a number of families: the large aggregating proteoglycans (aggrecan, versican, and CD44) and the small leucine-rich proteoglycans (decorin, biglycan, fibromodulin, and lumican). The large proteoglycans of this group bind to hyaluronic acid and have between 15 (versicon) and 100 (aggrecan) glycosaminoglycan chains. Functions for the small proteoglycans include both collagen and fibronectin binding.

Other proteoglycans include serglycin (in intracellular storage granules), perlecan (the membrane heparan–sulfate proteoglycan), syndecan and fibroglycan (transmembrane proteoglycans), and betaglycan (a TGF-β binding protein). The syndecan family (four members) is important in embryogenesis and organ development, while perlecan is the largest and most common proteoglycan of basement membranes.

The large proteoglycan mRNAs are up to 10–12 kb, and the genes can be quite large. Aggrecan, the related glycoprotein link protein CD44, and the large proteoglycan versican are all alternatively spliced. One of the domains that is alternatively spliced is an epidermal growth factor–like module. A member of the small proteoglycan family, decorin, exhibits usage of alternative promoters.

## 6 OTHER INTERACTIVE GLYCOPROTEINS: LAMININS, TENASCIN, THROMBOSPONDIN

Laminin is a member of a family of basement membrane proteins including merosin, *S*-laminin, and *S*-merosin. This family is expanding rapidly and is now known to include epiligrin. All the laminins are made up of three distinct gene products; A, B1, and B2 subunits (recently renamed α, β and γ, respectively), with a total molecular weight of 800 kDa. The laminins function in adhesion, spreading, migration, differentiation, and growth of cells. Many of the genes have been isolated, and there may be alternative splicing of some members of the family. The amino acid sequence of the laminin B1 and B2 chains indicates that they are the products of closely related but distinct genes, which are likely to be derived from a common ancestor gene.

Tenascin (also called hexabrachion, J1 glycoproteins, and cytotactin) is a large disulfide-linked hexmeric ECM glycoprotein with many repeated structural units such as heptad, EGF-like, and fibronectin type III repeats, as well as a homology to the globular domain of β- and γ-fibrinogen. cDNA sequences are available for chicken, mouse, and human. Interestingly, a tenascinlike transcript is encoded on the opposite strand of the human steroid 21-hydroxylase/complement C4 locus on chromosome 6p21. Tenascin is expressed during embryogenesis, during regeneration, and in tumors; it can be alternatively spliced in a tissue-specific manner.

Thrombospondin is a 420 kDa adhesive glycoprotein composed of three subunits of equal molecular weight. It is expressed in developing heart, muscle, bone, and brain, and in response to injury and inflammation in adult tissue. Thus far there are four members of the thrombospondin gene family including a protein found primarily in cartilage, cartilage oligomeric protein (COMP).

*See also* CELL-CELL INTERACTIONS; CYTOSKELETON-PLASMA MEMBRANE INTERACTIONS; GLYCOPROTEINS, SECRETORY.

### *Bibliography*

Boyd, C. D., Pierce, R. A., Schwarzbauer, J. E., Doege, K., and Sandell, L. J. (1993) Alternative exon usage in matrix genes. *Matrix* 13:457–469.

Burgeson, R. E., Chiquet, M., Duetzmann, R., Ekbloom, P., Engel, J., Kleinman, H., Martin, G. R., Meneguzzi, G., Paulsson, M., Saues, J., Timpl, R., Trygguason, K., Yamada, Y., and Yukchenco, P. D. (1994) A new nomenclature for the laminins. *Matris Biol.* 14:209–211.

Hay, E. D. (1993) *Cell Biology of Extracellular Matrix*, 2nd ed. Plenum Press, New York.

Kreis, T., and Vale, R. (1993) *Guidebook to the Extracellular Matrix and Adhesion Proteins*. Sambrook and Tooze, Oxford University Press, Oxford.

Ramirez, F., and Di Liberto, M. (1990) Complex and diversified regulatory programs control the expression of vertebrate collagen genes. *FABSEB J.* 4:1616–1623.

Sandell, L. J., and Boyd, C. D. (1990) Conserved and divergent sequence and functional elements within collagen genes. In *Extracellular Matrix Genes*, pp. 1–56. Academic Press, Orlando, FL.

Fish, Transgenic: *see* Transgenic Fish.

# Fluorescence Spectroscopy of Biomolecules

*Joseph R. Lakowicz*

*Key Words*

**Fluorescence** The emission of light from molecules in excited electronic states.

**Fluorophores** Fluorescent molecules, which can be intrinsic or added to biological molecules; also called probes.

Fluorescence methods are widely used in the biological sciences, including biochemistry, biophysics, and cell biology. Fluorescence is the emission of light from molecules in excited electronic states. This phenomenon is often observed in everyday life as the green or orange glow from new antifreeze or from wall posters under black lights. Fluorescence can be detected with high sensitivity because the light is seen against a dark background, just as stars are more easily seen at night.

Many biological molecules display intrinsic fluorescence, or can be labeled with extrinsic fluorophores. The emissions of these probes can now be used to reveal the structure and function of biological macromolecules. Additionally, fluorescence is used in DNA sequencing, for cell identification and sorting in flow cytometry, in clinical chemistry, and to reveal the localization and movement of intracellular substances using fluorescence microscopy. Advances in the technology of laser light sources and detectors is providing increased resolution of biological structures and advanced biomedical diagnostics.

## 1 PHENOMENON OF FLUORESCENCE

### 1.1 Typical Fluorescent Substances

Fluorescence molecules are typically aromatic species that display light absorption in the ultraviolet (UV: 250–400 nm) to visible (400–700 nm) regions of the spectrum. A wide variety of fluorescent substances are known. Some of these occur naturally, such as tryptophan in proteins (Figure 1). The indole residues are the dominant source of UV absorbance and emission in proteins.

In the case of biological membranes, there are few natural (intrinsic) fluorophores. Hence it is common to add extrinsic fluorescence labels, usually called probes, to obtain a useful signal. The probes can simply partition into the membranes, as occurs for diphenylhydantoin (DPH), or they can resemble the lipids, like a rhodamine (Rh) fatty acid (Figure 1). While the nitrogenous bases of DNA absorb UV light, they do not display useful fluorescence. However, a wide variety of dyes bind spontaneously to DNA, such as acridines, ethidium bromide, and other planar cationic species. There are a few naturally fluorescent bases, such as the Y-base, which occurs in the anticodon region of a particular tRNA (Figure 1).

## 2 MOLECULAR INFORMATION FROM FLUORESCENCE

### 2.1 Emission Spectra and the Stokes Shift

The most dramatic aspect of fluorescence is that it occurs at wavelengths longer than absorption, as seen for all the fluorophores in Figure 1. These Stokes shifts, which are most dramatic for polar fluorophores in polar solvents, are due to interactions between the fluorophore and its immediate environment. Indole is one such solvent-sensitive fluorophore. Consequently, the emission spectra can reveal the location of tryptophan residues in proteins. The emission from an exposed surface residue will occur at longer wavelengths than for a tryptophan residue in the protein's interior (Figure 1).

### 2.2 Quenching of Fluorescence

A wide variety of small molecules or ions act as quenchers of fluorescence; that is, they decrease the intensity of the emission. These substances include iodide ($I^-$), oxygen, chlorinated hydrocarbons, amines, and disulfide groups. The accessibility of fluorophores to quenchers is widely used to determine the location of probes on macromolecules, or the porosity of proteins and membranes to the quenchers. This is illustrated in Figure 2, which shows the intensity of protein- and membrane-bound fluorophores in the presence of dissolved water-soluble quenchers [Q]. The emission intensity of a tryptophan on the protein's surface (right), or on surface of a cell membrane, will be decreased in the presence of a water-soluble quencher. The intensity of a buried tryptophan residue or of an interior probe in the membrane will be less affected by the dissolved quencher (left). One can also add lipid-soluble quenchers, such as brominated fatty acids, to study the interior acyl side chain region of membranes.

### 2.3 Fluorescence Polarization or Anisotropy

Fluorescence probes usually remain in the excited state from 1 to 100 ns, a duration called the fluorescence lifetime. Because rotational diffusion of proteins also occurs in 1–100 ns, fluorescence lifetimes are a favorable time scale for studies of the associative and/or rotational behavior of macromolecules. By means of fluorescence, and in particular, measurements of fluorescence polarization (or anisotropy), it is possible to measure rotational motions of proteins in dilute (physiologically relevant) solutions. This possibil-

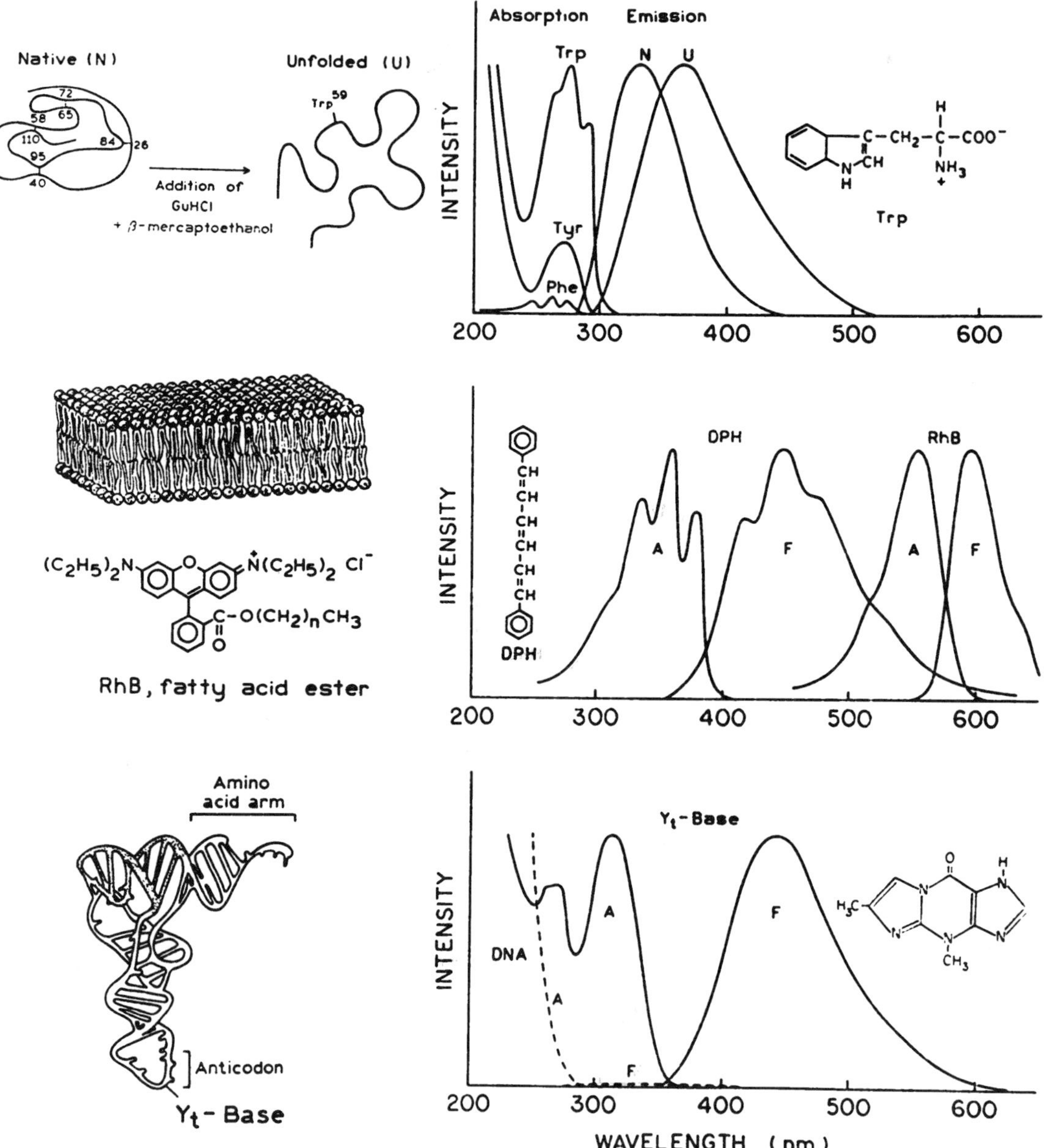


**Figure 1.** Absorption and emission spectra of biomolecules: top, tryptophan emission from proteins; middle, spectra of extrinsic membrane probes; bottom, spectrum of $Y_t$-base, a naturally occurring fluorescent base. DNA itself (dashes) displays very weak emission.

ity exists because of the polarization properties of light, and the dependence of light absorption on the alignment of the fluorophores with the electric vector of the incident light. If a sample is illuminated with vertically polarized light (Figure 3), the emission can also be polarized. However, rotational motions of the protein during the lifetime of the excited state randomize the excited state population. Faster rotational diffusion results in lower polarization of the emitted light.

The phenomena of fluorescence polarization can be used to measure the apparent volume (or molecular weight) of proteins. This measurement is possible because larger proteins rotate more slowly. Hence, if a protein binds to another protein, the rotational rate decreases, and the anisotropy(s) increases (Figure 3).

Fluorescence polarization measurements have also been used to determine the apparent viscosity of the side chain region (center) of membranes. Such measurements of microviscosity are typically performed using a hydrophobic probe like DPH (Figure 1), which partitions into the membrane. Fortunately, DPH does not fluoresce in the aqueous phase, simplifying the measurements—since all the detected light is from the membrane-bound probe.

## 2.4 Fluorescence Resonance Energy Transfer

Fluorescence resonance energy transfer (FRET) provides an opportunity to measure the distances between sites on macromolecules. This phenomenon of FRET is a predictable through-space interac-

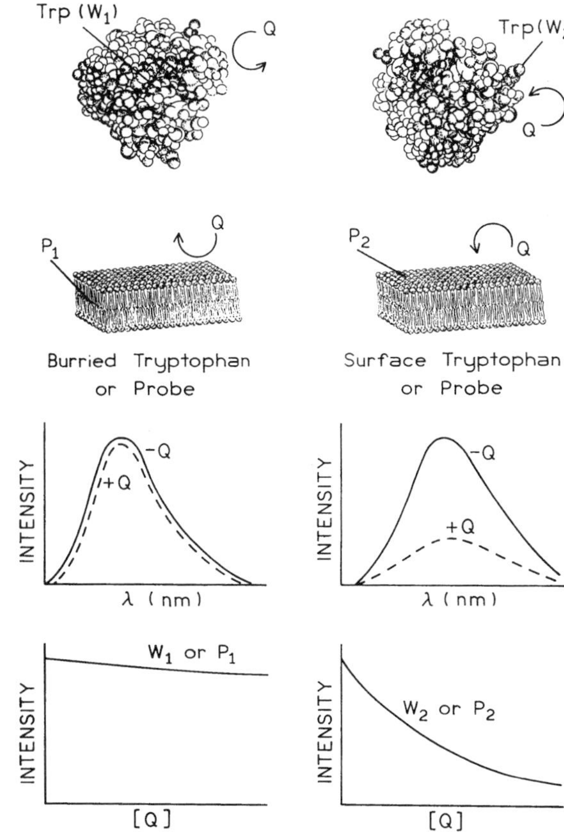


Figure 2. Accessibility of fluorophores to quenchers.

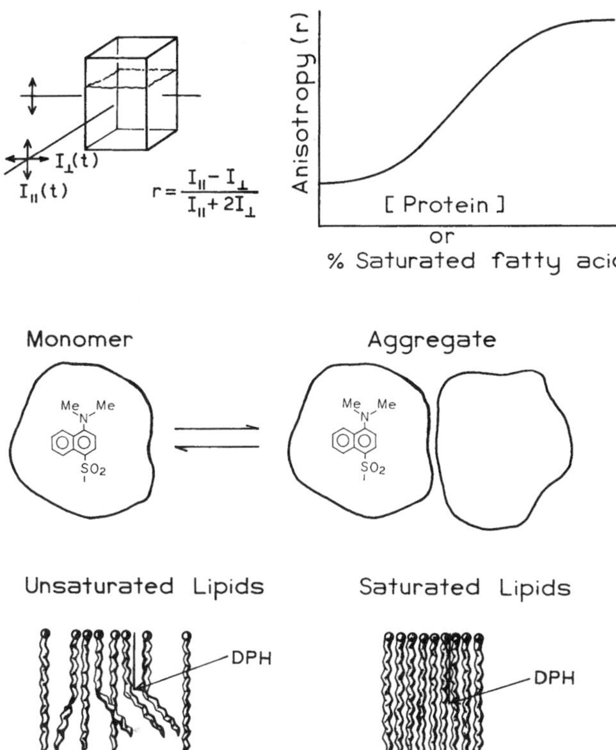


Figure 3. Fluorescence polarization, protein association, and membrane microviscosity.

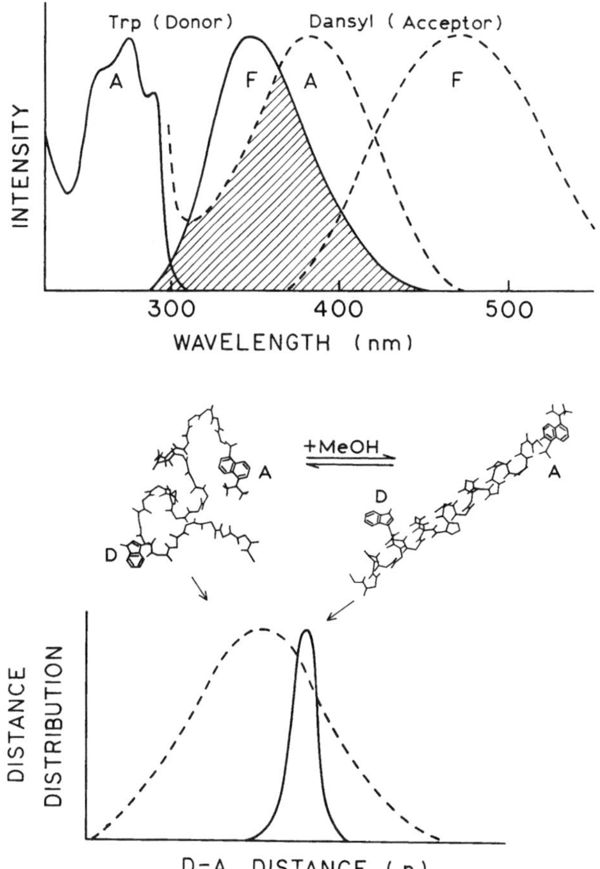


Figure 4. Top: Spectra indicating Förster distances (F). Bottom: Distance distribution in a random coil and α-helical protein.

tion in which the excited state energy is transferred from the donor to an acceptor. This transfer does not involve an intermediate photon, but rather is a coupling of dipoles in which the energy is transferred over a characteristic distance, called the Förster distance. These distances range from 15 to 60 A, which is comparable to the diameter of many proteins and comparable to the thicknesses of membranes. Hence, FRET can be used as a molecular ruler.

FRET allows determination of molecular distances because the Förster distances can be reliably calculated from the absorption spectrum of the acceptor and the emission spectrum of the donor (Figure 4). Once the Förster distance is known, the extent of energy transfer can be used to calculate the donor-to-acceptor (D–A) distance.

The use of FRET to study protein conformation is illustrated in Figure 4. In this case the protein can exist in a random coil state or as an α-helix. Transfer occurs from the single tryptophan residue, the donor D, to a dansyl residue, the acceptor A. The dansyl (dimethyamino naphthyl sulfonyl) group is a frequently used extrinsic label for proteins. The use of modern time-resolved data allows not only distance measurements, but resolution of the range of donor-to-acceptor distances. The data can be used to show that the protein exists alternately in a single conformational state, or with a range of donor-to-acceptor distances when in the random coil state.

## 3 PERSPECTIVES

Numerous additional applications of fluorescence in the biomedical sciences are being developed because of the high sensitivity of fluorescence detection and the desire to eliminate the use of ionizing radiation (X-rays and radioactivity) in the laboratory and in clinical practice. Additionally, there is growing recognition of the value of using longer wavelength probes (red or near-infrared), because tissues are nonabsorbing at these wavelengths, and there is less autofluorescence to interfere with the measurements. These facts suggest the possibility of noninvasive diagnostics based on red/near-infrared probes, which can be excited with simple laser diode sources of the type used in everyday CD players. We have all observed the possibility of noninvasive diagnostics, most simply perhaps when seeing the red light of a flashlight transmitted through our fingers. However, only recently has technology enabled the use of this observation for research and clinical purposes.

*See also* Gene Mapping by Fluorescence In Situ Hybridization; Gene Order by FISH and FACS; Whole Chromosome Complementary Probe Fluorescence Staining.

### *Bibliography*

Chen, R., and Edelhoch, H., Eds. (1976) *Biochemical Fluorescence.* Plenum Press, New York.

Dewey, T. G., Ed. (1983) *Biophysical and Biochemical Aspects of Fluorescence Spectroscopy.* Plenum Press, New York.

Lakowicz, J. R. (1983) *Principles of Fluorescence Spectroscopy.* Plenum Press, New York.

Lakowicz, J. R., Ed. (1991) *Topics in Fluorescence Spectroscopy:* Vol. 1, *Techniques;* Vol. 2, *Principles;* Vol. 3, *Biochemical Applications.* Plenum Press, New York.

Steiner, R. F., Ed. (1983) *Excited States of Biopolymers.* Plenum Press, New York.

Stryer, L. (1978) *Annu. Rev. Biochem.* 47:819–846.

Taylor, D. L., Waggoner, A. S., Lanni, F., Murphy, R. F., and Birge, R. R. (1986) *Applications of Fluorescence in the Biomedical Sciences.* Liss, New York.

## Folding of Proteins: *see* Protein Folding; Protein Modeling.

# Food Proteins and Interactions

*Nicholas Parris and Charles Onwulata*

### *Key Words*

**Emulsion** Thermodynamically unstable dispersion of micellar particles or globules in a liquid medium.

**Foaming** The trapping of air at the surface of a liquid resulting from the rapid diffusion of protein to the air–water interface and the subsequent reduction of the interfacial tension.

**Food Protein Interaction** Interaction of protein molecules and other compounds within their domain which affects their behavior in food products.

**Gelation** Formation of an ordered matrix of functional networks by balancing intramolecular and intermolecular forces between protein aggregates.

**Phospholipids** Class of membrane lipids that possesses polar groups attached to the glycerol residue by a phosphodiester bridge.

**Protein Micelles** Protein aggregates that are formed by the reversible interaction of protein monomers.

**Sedimentation Coefficient(s)** Proportionality constant that defines the sedimentation of a protein as a function of its size and shape.

**Syneresis** Spontaneous exudation of liquid and shrinking of a gel.

Proteins are the most abundant macromolecules found in living cells and account for approximate half of the cell's dry weight. They are required in the food of humans, fish, and most higher animals. Historically, food proteins have been selected for their nutritional value and can be obtained from a wide variety of naturally occurring sources. Proteins undergo a wide range of structural and conformational changes through a variety of complex interactions during processing and storage. Such changes can affect the principal purpose of dietary protein, which is to supply nitrogen and amino acids for the synthesis of proteins in the body. It is through an understanding of these interactions and their effects on functionality that food proteins have played a major role in the food supply.

## 1 PROTEIN STRUCTURE AND CONFORMATION

Proteins are natural compounds composed of amino acids organized at four different levels of structure: primary, secondary, tertiary, and quarternary. The primary strucure consists of amino acids that are sequenced in a linear polypeptide chain and constitute the basic building blocks. The secondary structure describes the conformational arrangement of the proteins in space, which is due primarily to hydrogen bonding of the polypeptide chain resulting in stable conformations. The secondary structure of the polypeptide of the protein is composed of $\alpha$-helices, $\beta$-sheets, and random coils. The tertiary structure refers to the overall architecture of the polypeptide chain whose folding brings into proximity parts of the molecule otherwise widely separated along the backbone. The schematic tertiary structure of carbonic anhydrase is shown in Figure 1 with helices and $\beta$-sheets. Noncovalent association of protein subunits describes their quarternary structures, which are stabilized by hydrogen bonds and van de Waals forces. Systematic denaturation of the organized structures and forces changes the functionality of food proteins.

### 1.1 Food Proteins

Food proteins are derived from a number of sources: plants, meat and fish, milk, eggs, and microbial proteins from unicellular or multicellular organisms (e.g., bacteria, yeast, molds, algae). Plant proteins are broadly classified according to solubility, shape, prosthetic group, and regulatory properties, as well as biological activity. They were first classified on the basis of solubility as albumins, globulins, glutenins, and prolamines. Albumins are the most water-soluble globular proteins. Soybean globular proteins contain globulins such as conglycinin (7S) and glycinin (11S), that are soluble in dilute salt solutions at neutral pH. Glutenins include wheat

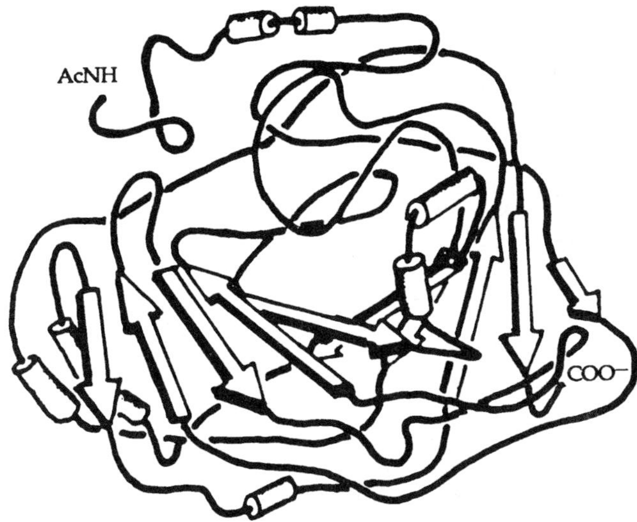


**Figure 1.** Schematic tertiary structure of carbonic anhydrase. Helices are drawn as cylinders; β-sheet strands as arrows with N at the tail and C at the point of each arrow. (From O. R. Fenema, *Food Chemistry,* 2nd ed., Dekker, New York, 1985; reproduced by permission of the author.)

glutenins and are soluble in dilute acid or alkali. Prolamines are corn or wheat storage proteins, are soluble in 70% ethanol.

Muscle or contractile proteins are derived directly from animal tissue and are the most conspicuous food protein in the human diet. Muscle proteins are generally classified into three groups based on their solubility in water: sarcoplasmic proteins, which are soluble in water or dilute salt; myofibrillar proteins, which are soluble in salt solutions >0.6m; and stromal proteins, which are the least soluble class of muscle proteins. Fibrous proteins, actin and myosin, have polypeptide chains arranged in long strands, are generally insoluble, and serve a structural or protective role. Collagen and elastin, the major proteins of connective tissue, occur in several polymorphic forms consisting of three polypeptide chains.

Egg proteins are primarily globular proteins found in the albumen; they contain ovalbumin, conalbumin, ovotransferrin, ovomucoid, and lysozymes. Milk proteins consist of a colloidal dispersion of casein micelles and soluble whey proteins; their stability is of tremendous practical importance in the dairy field. Casein micelles are extremely sensitive to changes in ionic environment and readily aggregate with increased concentration of calcium and magnesium ion.

Other proteins classified according to their prosthetic groups (tightly associated non–amino acid portion) are lipoproteins, glycoproteins, caseins, hemoglobin, and myoglobins. Proteins derived from unicellular organisms are grown on food processing by-products from which the protein is harvested and subsequently purified.

### 1.2 Enzymes

Enzymes are globular proteins that function as specific biological catalysts for chemical reactions in living systems. Given the restricted conditions of temperature and pH in which they operate, enzymes are much more efficient and specific than other catalysts. Enzymes are frequently used in the food industry to modify the functional behavior of food proteins. For instance, proteases from plant sources, such as papain, are used as commercial meat tenderizers. It is important, however, that the nutritive values of the product not be significantly affected by the hydrolysis of peptide bonds and the subsequent loss of essential amino acids.

### 1.3 Energy and Nutritive Value

The nutritional quality of a protein is determined by its amino acid composition. Nine of the 20 identified amino acids are classified as essential because they cannot be synthesized in the human body. The food value of protein is evaluated by quantifying the ratio of essential to nonessential amino acids and calculating protein efficiency ratio (PER), which is determined by the weight gain in the animal divided by the protein intake. Nutrient availability and food protein quality depend on the nature of the cross-linking and denaturation. Protein requirements for maintaining good health and growth are listed as the recommended dietary allowance (RDA) set by the U.S. National Academy of Sciences.

## 2 PROTEIN INTERACTIONS

Molecular forces governing protein interactions determine the relationship of the structure of individual proteins to their functional properties as well as the association of a protein with other compounds in the cell. Forces involved include covalent bonds, ionic interaction, hydrogen bonds, hydrophobic interactions, hydration, and steric repulsion. Covalent bonds include the peptide linkage of the primary structure of the protein as well as disulfide bonds, which are formed between cysteine residues and depend on the conformation of the peptide chain. Bonds or interactions that determine secondary and tertiary structure of proteins are shown in Figure 2. Protein denaturation results in changes in the secondary and tertiary structures. Breaking the peptide bonds is achieved by means of proteolytic enzymes or hydrolysis with strong acid or base. Of the covalent forces, disulfide bonding is the most important in protein interactions. Noncovalent molecular forces, which are one to three orders of magnitude smaller than covalent bonding energy, include hydrogen bonds, hydrophobic interaction, and repulsion forces. Association and disassociation of the molecular forces maintain the integrity of food proteins.

### 2.1 Protein–Protein

Protein–protein interactions often occur as a result of food processing designed to improve the functional properties of proteins for new product application. These interactions occur in two-stage processes consisting of unfolding of the native protein and exposing active sites, followed by association of the polypeptide chain by covalent and noncovalent forces. Gelation is an association of aggregated proteins existing in a three-dimensional network with trapped water molecules. Protein cross-links are formed through sulfhydryl groups or through hydrophobic interactions. To maintain a stable gel, there must be a balance between attractive forces required to form the network and the repulsive forces needed to prevent its collapse or syneresis.

### 2.2 Carbohydrates

Protein and carbohydrates form irreversible complexes in nonoxidation reactions. Food protein heated in the presence of a reducing sugar results in the reaction of the carbonyl groups of the carbohy-

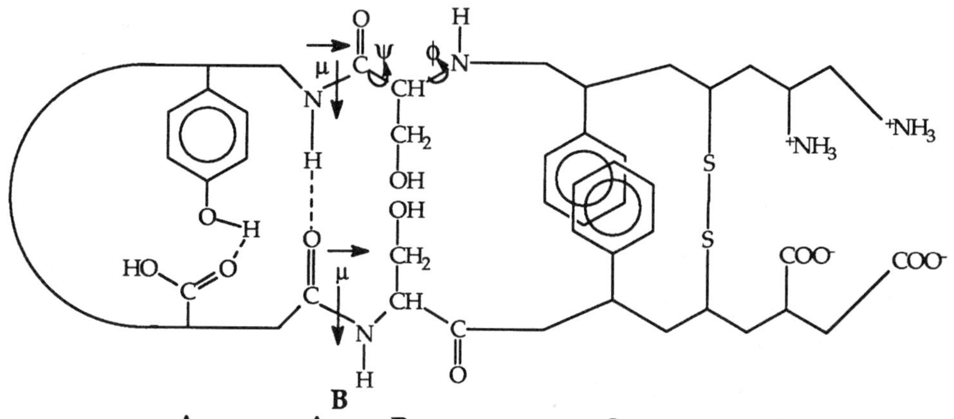

**Figure 2.** Bonds or interactions that determine secondary and tertiary structure of proteins: (A) hydrogen bond, (B) dipolar interaction, (C) hydrophobic interaction, (D) disulfide linkage, and (E) ionic interaction. (From O. R. Fenema, *Food Chemistry,* 2nd ed., Dekker, New York, 1985, reproduced by permission of the author.)

drate with the free amino group of protein followed by a cascade of reactions leading to brown polymers. These maillard reactions occur in preference to other types of protein damage in the drying of milk at moderate temperatures. However, more severe heating, required in the preparation of toasted breakfast cereals, bread, and biscuits, results in late Maillard-type and protein–protein damage.

### 2.3 Lipids

Protein–lipid interactions in nature result in lipoproteins in food systems such as milk and eggs. Proteins stabilize emulsions by assuming the form of lowest free energy at the interface between the two immiscible substances. Interactions between milk proteins and lipids occur through their polar groups; there is no significant modification of the conformation of the proteins or organization of the lipid monolayer at the interface between the water and the oil droplets. Stronger interactions occur in flour–water mixtures, since lipids can bind with gluten proteins to form highly stable lipoglutenin complexes.

### 2.4 Hydration

The native structure and normal functionality of protein is facilitated by the presence of water. Water in the vicinity of protein determines its intrinsic properties in such functions as solubility, swelling, dispersibility, and wettability. Water absorption is considered the most important step to imparting desired functional properties to proteins. Proteins interact with water through their peptide bonds or their amino groups, and their solubility depends on conformational forces (e.g., hydrogen bond, dipole–dipole and ionic interactions, pH, and temperature). The ability of proteins to absorb and retain water plays a major role in the textural stability of food systems that employ swelling of protein matrices. Texturized proteins employing the protein–water balance are designed as analogues that resemble meat and fish patties.

### 2.5 Soluble Ions

Zinc, magnesium, and sulfur are some of the ionic components associated with proteins that increase the proteins' activity and stability. The reactive nature of proteins allows for the manipulation of the ions to accomplish food preservation by controlled denaturation. Proteins are stable within a defined range of pH and have either positive, negative, or neutral charges, depending on the reactive group of amino acids that determine the stability, conformation, and function of the protein in the medium. At extreme pH, proteins aggregate. Small amounts of polyvalent cations, such as calcium and magnesium, are very effective in destabilizing the casein micelle, particularly in conjunction with heat and reduced pH. Polyvalent cations also act as cofactors for enzymes to be catalytically active. Salts used in food processing can denature proteins by charge neutralization or by changing the isoelectric point of the protein.

## 3 FUNCTIONAL PROTEIN INTERACTIONS

Physicochemical properties that enable proteins to affect the characteristics of foods during processing, storage, preparation, and consumption define the functional properties of proteins. Protein functions that affect food utilization are water absorption (viscosity and gelation), surface effects (gelation and foaming), and chemical reactions (textural properties). Functionality of a food product is determined experimentally, since the study of protein structure provides information on physicochemical properties but is not an accurate method of predicting functionality.

### 3.1 Gelation

Gelation is the formation of an extended network of denatured protein molecules in which the intramolecular and intermolecular forces are in an ordered matrix. Gel matrices hold water and other ingredients, which improves the swelling properties of the gel. Strong gels hold water and other ingredients, producing such foods as yogurt, gelatins, and doughs. Gels degrade over time, leading to syneresis or shrinkage of the gel matrix. Gel formation results from controlled protein denaturation and unfolding, formation, and realignment of protein matrices and balance in protein interactions. The cross-linking of proteins in the matrix via disulfide bonds

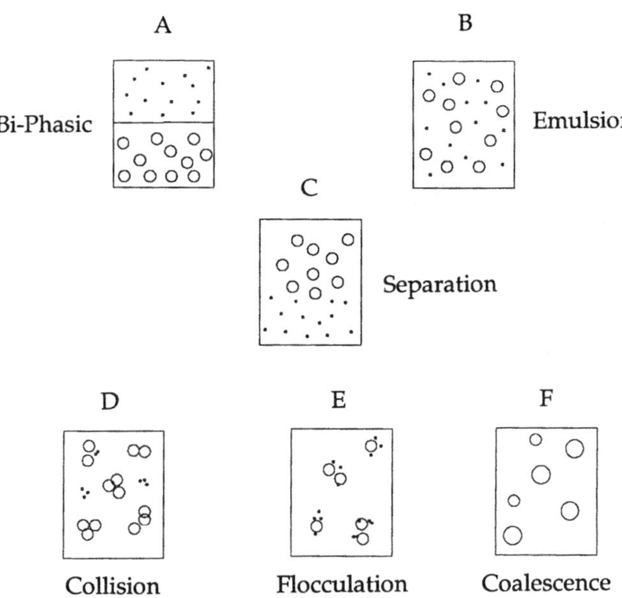


**Figure 3.** Process of emulsion formation, stabilization, and breakdown: (A) biphasic medium of liquid and semisolid emulsion, (B) thermodynamically unstable emulsion, (C) gradual migration of emulsion based on charge separation, (D) particle collision with increasing electrostatic charges, (E) flocculation (i.e., clear aggregation of collided particles), and (F) coalescence and syneresis of emulsion.

results in heat-irreversible gels stabilized by hydrogen bonds. Higher molecular weight proteins tend to form stronger gels.

### 3.2 Emulsification

A stable protein emulsion is produced when proteins in a biphasic medium unfold slightly at the interface and align their nonpolar regions toward an oil phase with their hydrophilic regions toward the aqueous phase, at a minimum of free energy. Force imbalances cause emulsions to separate over time. The breakdown is a result of large positive free energy at the emulsion interface, leading to flocculation, followed by coalescence of the native protein as depicted in Figure 3. Globular proteins with a highly ordered and stable tertiary structure (e.g., β-lactoglobulin, bovine serum albumin, lysozyme) are more likely to unfold and are considered good emulsifiers. Emulsification activity continues to increase with increasing protein denaturation, provided solubility is not compromised. The balance of the hydrophilic and hydrophobic forces in the protein maintains emulsion stability.

### 3.3 Foaming

Foams are emulsions in which gas is dispersed in a continuous aqueous semisolid phase of protein. Foods such as ice cream and whipped toppings are stabilized protein foams. Foam expansion is determined by the volume of foam formed after a known quantity of gas has been incorporated into a food system without breakdown of matrix. Factors that affect foam stability include electric potential, protein concentration, energy input, and presence of salts, sugars, lipids, and metal ions.

### 3.4 Food Processing

Food proteins are thermally processed either to enhance functionality, to improve textural properties, or to minimize natural deterioration. Processing may lead to slight loss in nutritional quality; however, most processes improve quality by destroying antinutritional factors through inactivation of enzymes such as peroxidase. The interactions of proteins are enhanced, leading to formation of new complexes. Factors such as temperature and the presence of salts and oxidizing–reducing agents may be used to produce desirable, high quality foods. The removal or addition of thermal energy can result in denaturation of proteins. Thermal denaturation of food proteins generally occurs between 45 and 85°C, accompanied by exposure of hydrophobic groups to water, resulting in protein aggregation. Low temperature denaturation is mediated by a reduction of hydrophobic interactions in conjunction with enhanced hydrogen bonding, leading to aggregation and precipitation of proteins.

### 3.5 Thermally Induced Mutagens

Mutagens are formed in muscle foods as a result of industrial processing or home cooking (e.g., frying, broiling, boiling, baking). They are generated through decomposition and rearrangement or recombination of endogenous precursors in muscle such as amino acids and sugars, which have individually undergone recombination as in the Maillard browning reaction. Much of the mutagenicity found in meat cooked at moderate temperatures can be attributed to the formation of heterocyclic aromatic amines. *N*-Nitrosamines are readily produced at higher temperatures—for example, in the frying of bacon. These mutagens are consumed daily in beef, pork, poultry, and fish at the low parts-per-billion level. Extensive research efforts to develop methods to reduce or eliminate such compounds from the food supply have been moderately successful.

## *Bibliography*

Bodwell, C. E., and Erdman, J. W., Jr., Eds. *Nutrient Interactions.* Dekker, New York.

Creighton, T. E., Ed. (1983) *Proteins: Structures and Molecular Properties.* Freeman, New York.

El-Nokaly, M., and Cornell, D., Eds. (1991) *Microemulsions and Emulsions in Foods.* American Chemical Society, Washington, DC.

Fennema, O. R. (1985). *Food Chemistry,* 2nd ed. Dekker, New York.

Fieden, C., and Nichol, L. W. Eds. (1981) *Protein–Protein Interactions.* Wiley, New York.

Hudson, B. J. F., Ed. (1987) *Developments in Food Proteins,* Vol. 5. Elsevier Applied Science, London.

Mitchell, J. R., and Ledward, D. A., Eds. (1986) *Functional Properties of Food Macromolecules.* Elsevier Applied Science, London.

Parris, N., and Barford, R. A., Eds. (1991) *Interactions of Food Proteins.* American Chemical Society, Washington, DC.

Phillips, G. O., Wedlock, D. J., and Williams, P. A., Eds. (1984) *Gums and Stabilisers for the Food Industry,* Vol. 2, *Applications of Hydrocolloids.* Pergamon Press, Oxford.

Osborne, T. B. (1924) *The Vegetable Proteins.* Longmans, Green, London.

Waller, G. R., and Feather, M. S., Eds. (1983) *The Maillard Reaction in Foods and Nutrition.* American Chemical Society, Washington, DC.

Whitaker, J. R., and Tannenbaum, S. R., Eds. (1977) *Food Proteins.* AVI Publishing, Westport, CT.

# Fragile X Linked Mental Retardation

*A. J. M. H. Verkerk and B. A. Oostra*

## *Key Words*

**CpG Island** DNA region rich in cytosine and guanine base pairs, usually found in promoter regions of genes.

**Cytogenetics** Study of (stained) metaphase chromosome preparations by means of microscopy.

**Fragile Site** Can be induced in chromosomes by culturing cells in folate-deficient medium.

**Normal Transmitting Male** Phenotypically normal male who is a carrier of a premutation in the *FMR 1* gene.

**Premutation** A mutation in the *FMR1* gene without phenotypic symptoms.

The fragile X or Martin Bell syndrome is an X-linked heritable disease. After trisomy 21, or Down's syndrome, it is the most frequent cause of mental retardation. The syndrome is associated with a (rare) fragile site (referred to as FRAXA) at the end of the long arm of the X chromosome in band Xq27.3. For an X-linked disease, this syndrome shows an unusual pattern of inheritance because some of the carrier females are affected and because of the occurrence of normal male carriers. The prevalence of the fragile X syndrome in the Caucasian population is estimated to be 1 in 1250 males and 1 in 2500 females.

## 1 DESCRIPTION OF THE SYNDROME

### 1.1 Clinical Characteristics

Moderate to severe mental retardation is found in male fragile X patients. Other clinical symptoms are a long face, prominent forehead with relative macrocephaly, large everted ears, and macroorchidism (enlarged testes). Figure 1 shows typical facial characteristics of a male patient. Macroorchidism is found in 65–70% of all fragile X positive adult males and usually develops after puberty. Behavioral abnormalities include hyperactivity, hand-flapping, and poor eye contact. Approximately 30% of female carriers are mildly to moderately mentally retarded, with typical facial characteristics.

### 1.2 Cytogenetic Expression of the Fragile X Site

For a number of years, confirmation of the clinical diagnosis was dependent on detecting the fragile site cytogenetically at Xq27.3 in cultured lymphocytes, fibroblasts, or amniocytes. The appearance of the fragile site, which is visible as a gap or break (see Figure 2), is induced by culturing under specific conditions the cells to be used in chromosome spreads. In mentally retarded males, the fragile site is usually seen in 2–60% of the cells. The site can be detected in only 50% of obligate carrier females.

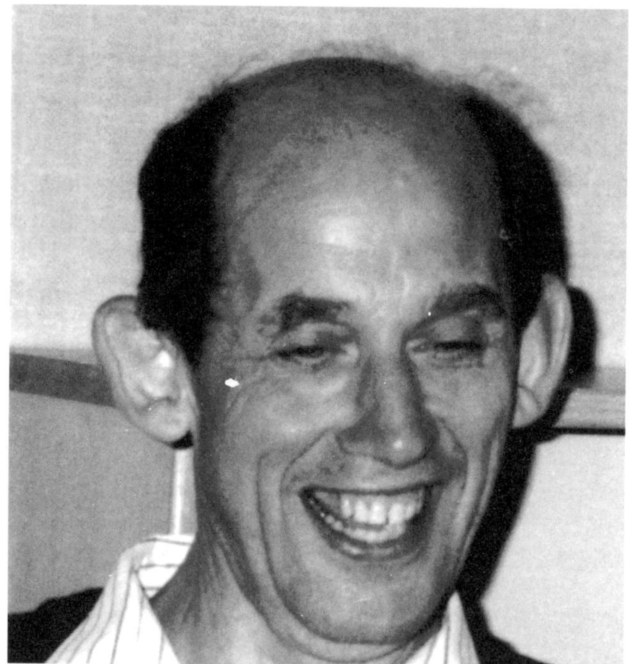

**Figure 1.** Fragile X patient with typical facial characteristics.

### 1.3 Unusual Pattern of Inheritance

In X-linked recessive diseases, females who carry a defective gene on one of their X chromosomes are usually unaffected. Their daughters have a 50% risk of being unaffected carriers, and their sons have a 50% risk of being affected. Fragile X syndrome deviates from this pattern in the following points.

1. Of the carrier females, 30% are mentally retarded, although they are often less severely affected than males. For this reason, the syndrome is also referred to as an X-linked dominant disorder with incomplete penetrance. Sons of affected females have a 50% risk of being affected.

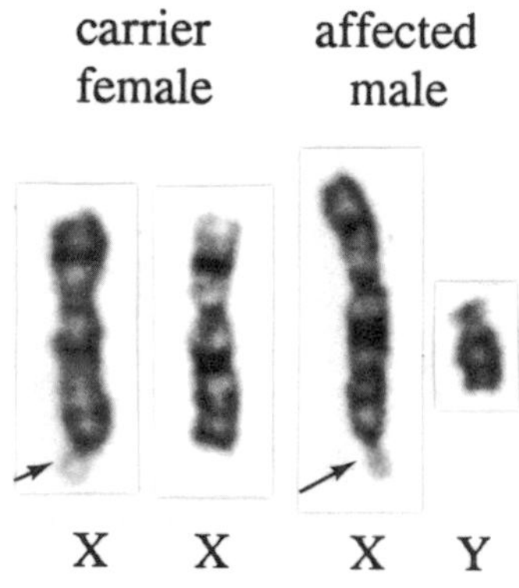


**Figure 2.** Sex chromosomes of a female carrier and an affected male. The fragile site at Xq27.3 is indicated with an arrow. (Chromosomes provided by Dr. J. O. Van Hemel.)

2. Of males who carry the fragile X mutation, 20% are phenotypically normal and are called normal transmitting males (NTMs). They will pass the mutation on to their daughters. These daughters, who are also normal carriers, are again at risk of having affected children. In NTMs and their daughters, no fragile sites at Xq27.3 are detected.
3. The disease phenotype is found in offspring only after transmission of the mutation by a female. Daughters of male carriers are always normal; the daughters of the few male patients that have reproduced are also normal.

## 2 THE FRAGILE X GENE, *FMR 1*

### 2.1 Characterization of *FMR 1*

The fragile X gene, *FMR 1* (fragile X mental retardation—1), has a size of approximately 40 kilobases (kb). The length of the mRNA is about 4.4 kb, with an open reading frame of 1.9 kb. In the 5′ untranslated region of the *FMR1* gene, an unusual CGG repeat is present. The ATG start codon for translation is found after the CGG repeat, indicating that the repeat is not translated into protein. A CpG island is located 250 bp proximal of the repeat. This CpG island is hypermethylated in DNA of fragile X patients. No methylation of the island is found in DNA of normal males and NTMs. CpG islands are found in promoter regions of genes and are C/G rich areas that can be recognized by specific rare-cutting restriction enzymes. Methylation of a CpG island is usually correlated with lack of gene expression. In fragile X patients, no mRNA is transcribed from the *FMR 1* gene. Alternative splicing of the pre-mRNA allows the *FMR1* gene to give rise to several proteins. The longest transcript codes for a protein of 631 amino acids. Homology has been found with RNA binding proteins, but the RNA binding function of *FMR1* protein is still unknown. Various levels of mRNA expression have been found in different organs, with high expression in testis and brain. The *FMR 1* gene has been well conserved during evolution. A gene homologous to *FMR 1* is found in many species, including nematodes, chicken, mice, pigs, and different primates. In the mouse *Fmr 1* gene 97% homology to the human gene is found at the protein level.

### 2.2 Nature of the Fragile X Mutation

The number of CGG repeats is found to be polymorphic in the human, mouse and pig *FMR 1* genes. In normal individuals the number of repeats varies from 6 to 53, with an average of 29. In fragile X syndrome there is an increase in size of this CGG repeat.

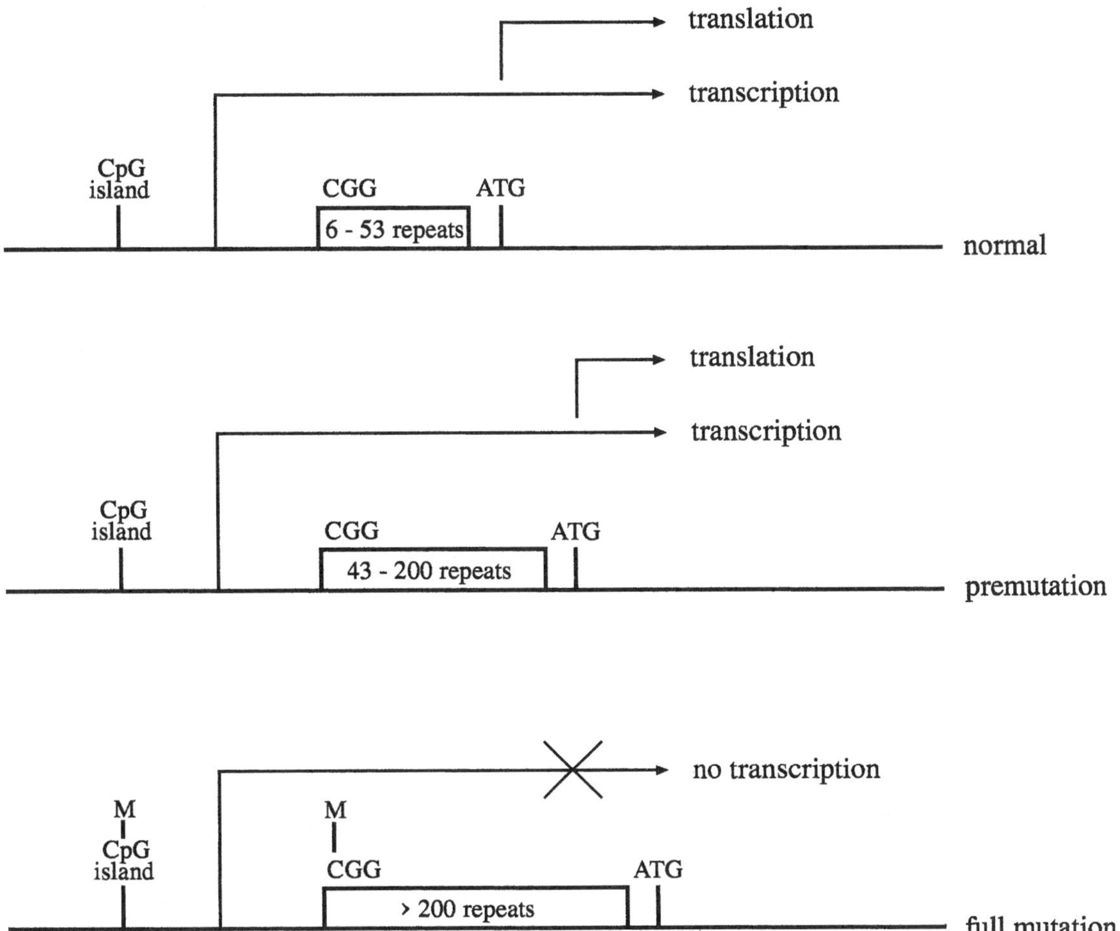


**Figure 3.** Schematic representation of the *FMR1* gene and the molecular basis of the fragile X syndrome (M = methylated). Premutation alleles are found in normal transmitting males and in normal female carriers. Full mutations are found in 100% of affected males. Of the females carrying a full mutation, about 70% are affected.

In general, two classes of mutations are seen in fragile X families: premutations and full mutations. In NTMs and their daughters, the repeat number ranges from 43 to 200. This number of repeats is not associated with the disease phenotype and is called a premutation. In male and female patients and in many carrier females, more than 200 repeats are found, with some expansions exceeding 2000 repeats. This number of repeats is associated with the disease phenotype and is called a full mutation. In the case of a full mutation, the CpG island and the amplified repeat itself are methylated, resulting in loss of transcription of the gene. Figure 3 shows the molecular basis of the fragile X syndrome in schematic form.

Myotonic dystrophy, Kennedy's disease, Huntington's disease, and spinocerebellar ataxia type 1 are four other diseases in which a trinucleotide repeat amplification is found in affected individuals.

### 2.3 Instability and Mosaicism of the CGG Repeat

Normal alleles containing 6–53 repeats are stable. This means that there is no change in repeat size when the gene is passed on to the next generation. Premutation and full mutation alleles are unstable and in general increase in size by passage to the next generation. Sometimes a decrease in size is observed. There is a small overlap between CGG repeat numbers in normal and premutation alleles, which can be distinguished by studying (in)stability of the allele in families. Two kinds of mosaicism are observed in fragile X syndrome patients.

1. Extensive somatic instability is observed in individuals with a full mutation. During mitotic divisions, the repeat increases (or sometimes decreases) in length, resulting in a variable number of repeats in different cells of the same individual. This effect is made visible by Southern blot analysis as a "smear" (see Sections 4.1 and 4.2). The mechanism of repeat amplification is still unknown.
2. In 25% of male patients, in addition to a full mutation, a premutation is found in part of the cells. Despite transcription of the (unmethylated) premutation alleles, these males are mentally retarded. Overall, there is an apparent insufficiency of protein production in the appropriate tissues, resulting in the disease phenotype. This kind of mosaicism is also observed in fragile X females.

## 3 ELUCIDATION OF PAST OBSERVATIONS

Since the nature of the mutation is known, several previously puzzling observations can be explained.

1. In premutation alleles, no methylation of the CpG island of the *FMR 1* gene is found, and transcription of the gene is normal. Therefore NTMs and females who carry a premutation are never mentally retarded.
2. The length of the CGG repeat is correlated with the percentage of cells in which cytogenetic expression is found. Females who carry a premutation and NTMs never show cytogenetic expression. Practically all male patients with full mutations show cytogenetic expression.
3. It was noted by Sherman in 1985 that daughters of NTMs have a risk of 40% instead of 50% of their sons being affected. The finding that the risk for mental impairment in fragile X pedigrees is dependent on the position of individuals in the pedigree has become known as the Sherman paradox. Premutation alleles increase (and sometimes decrease) in size with virtually all transmissions to individuals in the next generation. Premutation alleles transmitted by NTMs always stay in the premutation range in their daughters. Exclusively premutation alleles transmitted by females have a chance of increasing to a full mutation in the offspring (see Section 1.3). The risk of expansion of a premutation to a full mutation in the next generation is correlated with the size of the premutation allele of the mother. A low repeat number has a low risk of expanding to a full mutation, and a high repeat number has a high risk. When a premutation contains more than about 90 repeats, it is almost 100% certain to become a full mutation in the next generation. These risks correspond to the risks calculated by Sherman and explain the non-Mendelian pattern of inheritance in fragile X syndrome.

A premutation can be transmitted through many generations before expanding to a full mutation, resulting in the disease phenotype. Fragile X mutations are predominantly found on a limited number of haplotypes, suggesting founder chromosomes for this mutation.

## 4 DIAGNOSTICS TODAY

### 4.1 Technical Procedure

DNA fragments from individuals are visualized by Southern blot analysis and hybridization with a radioactively labeled specific probe. For DNA diagnosis of the fragile X syndrome, a specific probe is used to demonstrate the presence of a genomic DNA fragment containing the CGG repeat. In normal individuals a 5.2 kb *Eco*RI restriction fragment is found. Premutations are visible as fragments somewhat larger than 5.2 kb (up to 5.7 kb). Full mutations are visible as diffuse broad bands or "smears" due to the presence of different repeat lengths (above 200 repeats) in different cells. By means of the polymerase chain reaction (PCR) technique, it is possible to determine the CGG repeat number of premutation alleles. This technique has limitations, however, and it is still difficult to determine repeat numbers above 200.

### 4.2 Example of DNA Diagnosis in a Fragile X Family

Postnatal and prenatal diagnosis of fragile X syndrome can now be exercised with a high degree of accuracy by detecting the CGG repeat amplification at the DNA level. Figure 4 shows an annotated Southern blot analysis of a fragile X family in which premutations and full mutations are found. In the DNA of the grandmother (1) one normal 5.2 kb band (indicated by an arrow) is seen representing her two normal X chromosomes. The grandfather (2) is a normal transmitting male with a premutation of 70 repeats in the *FMR 1* gene; his DNA shows a band slightly larger than 5.2 kb. He passed his premutation on to his three daughters (4, 8, and 14), who are phenotypically normal carriers. Besides the 5.2 kb band of their normal chromosome, they have a premutation band of increased size (75 and 80 repeats). Daughter 4 has one affected daughter (6), who is mosaic for a premutation and a full mutation. Daughter 8 has two affected sons (12 and 13) with full mutations. Daughter 14 has one affected son (16), who is mosaic for a premutation and a full mutation.

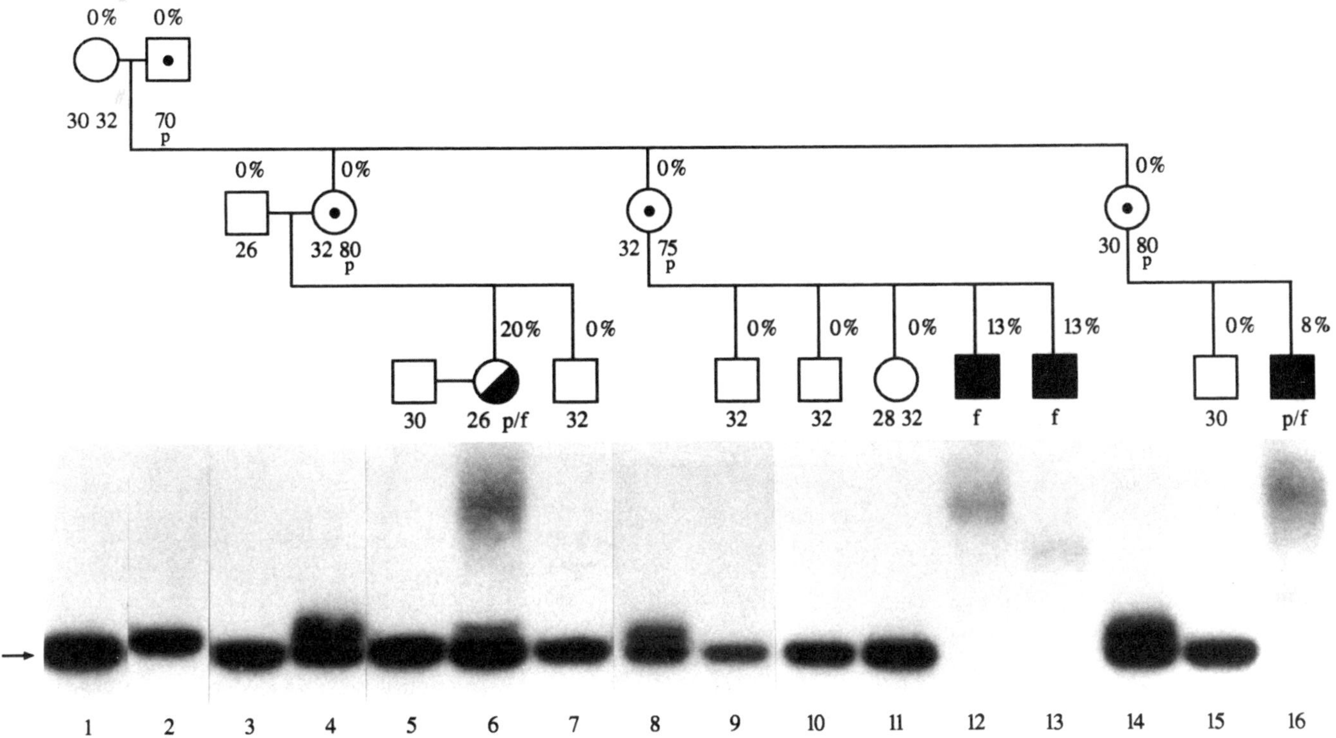


**Figure 4.** Detection of the *FMR1* gene in a fragile X family by means of Southern blot analysis (p = premutation, f = full mutation): ⊡, normal transmitting male; ⊙, normal female carrier; ◑, female carrier with cytogenetic expression; ■, mentally retarded male. Cytogenetic expression indicated by percentage figures; number of CGG repeats for the different X chromosomes (determined by means of PCR) is given beneath round and square symbols. (Example of fragile X DNA diagnosis provided by Drs. D. J. J. Halley and A. M. W. v.d. Ouweland.)

## 5 PATIENTS WITHOUT CGG REPEAT EXPANSION

Since the gene description of the defect has become known, patients with the fragile X phenotype but without CGG repeat expansion or cytogenetic expression have been reported. In several patients the syndrome is caused by deletions of all or part of the *FMR1* gene. In these patients flanking regions are also missing, and thus the involvement of other genes contributing to the disease phenotype could not be excluded. In one patient, however, a de novo point mutation was found in the *FMR 1* gene, resulting in an aberrant protein causing the fragile X phenotype. This indicates that the fragile X syndrome is a single-gene disorder and that loss of function or mutation in the *FMR 1* gene leads to the clinical phenotype of the fragile X syndrome.

*See also* Genetic Testing; Human Genetic Predisposition to Disease; Neuropsychiatric Diseases.

### *Bibliography*


Hagerman, R. J. (1991) Physical and behavioral phenotype. In *Fragile X Syndrome: Diagnosis, Treatment and Research,* R. J. Hagerman and A. C. Silverman, Eds. Johns Hopkins University Press, Baltimore.

Hirst, M. C., Knight, S. J. L., Bell, M. V., Super, M., and Davies, K. E. (1992) The fragile X syndrome. *Clin. Sci.* 83:255–264.

McKusick, V. A. (1992) 309550: Mental retardation, X-linked, associated with marXq28. In *Mendelian Inheritance in Man. Catalogs of Autosomal Dominant, Autosomal Recessive, and X-Linked Phenotypes.* Johns Hopkins University Press, Baltimore and London.

Oostra, B. A., Willems, P. J., and Verkerk, A. J. M. H. (1993) Fragile X syndrome: A growing gene. In *Genome Analysis,* Vol. 6: *Genome Maps and Neurological Disorders,* K. E. Davies and S. M. Tilghman, Eds. Cold Spring Harbor Laboratory Press, Cold Spring Harbor, NY, in press.

X-Linked mental retardation 5, Special issue. *Am. J. Med. Gen.* 43:1–372 (1992).


# Free Radicals in Biochemistry and Medicine

*Barry Halliwell*

### *Key Words*

**Antioxidant** Agent that inhibits oxidative damage to a target molecule.

**Free Radical** Any species containing one or more unpaired electrons.

**Oxidative Damage** Damage produced by attack of reactive oxygen species.

**Oxidative Stress** Imbalance between the generation of reactive oxygen species in vivo and the activity of antioxidant defenses in favor of the former.

**Reactive Oxygen Species** Collective term used by biologists to designate the oxygen radicals (e.g. $O_2^{\cdot-}$, $OH^{\cdot}$ and certain nonradical derivatives of $O_2$, such as $H_2O_2$.

Free radicals and other oxygen-derived species are constantly generated in vivo, both by ''accidents of chemistry'' and for specific metabolic purposes. The reactivity of different free radicals varies, but some can cause severe damage to biological molecules, especially to DNA, lipids, and proteins. Antioxidant defense systems scavenge, and minimize the formation of, oxygen-derived species, but they are not 100% effective. Hence, repair systems exist to deal with molecules that have been oxidatively damaged. Damage to DNA by hydroxyl radicals appears to occur in all aerobic cells and might be a significant contributor to the age-dependent development of cancer.

## 1 WHAT IS A RADICAL?

In the structure of atoms and molecules, electrons usually associate in pairs, each pair moving within a defined region of space (an atomic or molecular *orbital*). One electron in each pair has a spin quantum number of +1/2, the other −1/2. A *free radical* is any species capable of independent existence (hence the term ''free'') that contains one or more *unpaired electrons,* an unpaired electron being one that is alone in an orbital. The simplest free radical is an atom of the element hydrogen, with one proton and a single electron. Table 1 gives some examples of other free radicals. The spectroscopic technique of *electron spin resonance* is used to measure free radicals by recording the energy changes that occur as unpaired electrons align in response to a magnetic field. A superscript dot is used to denote free radical species (Table 1).

## 2 RADICALS IN VIVO

The chemical reactivity of free radicals varies. One of the most reactive is *hydroxyl radical* ($OH^{\cdot}$). Exposure of living organisms to gamma radiation causes homolytic fission of O—H bonds in water (remember that living organisms are greater than 70% water) to give $H^{\cdot}$ and $OH^{\cdot}$. Hydroxyl radical reacts at a diffusion-controlled rate with almost all molecules in living cells. Hence, when $OH^{\cdot}$ is formed in vivo, it damages whatever it is generated next to—it cannot migrate any significant distance within the cell. The harmful effects of excess exposure to gamma radiation on living organisms are thought to be initiated by attack of $OH^{\cdot}$ on proteins, DNA, and lipids.

Whereas $OH^{\cdot}$ is probably always harmful, other (less reactive) free radicals may be useful in vivo. Free radicals are known to be produced metabolically in living organisms. Thus, $NO^{\cdot}$ is synthesized from the amino acid L-arginine by vascular endothelial cells, phagocytes, and many other cell types. Nitric oxide helps to regulate vascular tone and may be involved in the killing of parasites by macrophages. Superoxide radical ($O_2^{\cdot-}$), the one-electron reduction product of oxygen, is produced by phagocytic cells and helps them to kill bacteria. Evidence is accumulating to suggest that smaller amounts of extracellular $O_2^{\cdot-}$ may be generated, perhaps as intercellular signal molecules, by several other cell types, including endothelial cells, lymphocytes, and fibroblasts. In addition to this metabolic production of $O_2^{\cdot-}$, some $O_2^{\cdot-}$ is produced within cells by mitochondria and endoplasmic reticulum: this is usually thought to be an unavoidable consequence of ''leakage'' of electrons onto $O_2$ from their correct paths in electron transfer chains.

Much $O_2^{\cdot-}$ generated in vivo probably undergoes a dismutation reaction, represented by the overall equation

$$2O_2^{\cdot-} + 2H^+ \rightarrow H_2O_2 + O_2 \qquad (1)$$

in which hydrogen peroxide ($H_2O_2$), a nonradical, is formed. $H_2O_2$ resembles water in its molecular structure and is very diffusible within and between cells. As well as arising from $O_2^{\cdot-}$, $H_2O_2$ can be produced by the action of certain oxidase enzymes in cells, including amino acid oxidases and xanthine oxidase. Like $O_2^{\cdot-}$, $H_2O_2$ can have useful metabolic functions. For example, $H_2O_2$ is used by the enzyme *thyroid peroxidase* to help make thyroid hormones. $H_2O_2$ (or products derived from it), can displace the inhibitory subunit from the cytoplasmic transcription factor NF -κB. The active factor migrates to the nucleus and activates genes by binding to specific DNA sequences in enhancer and promoter elements. Thus, $H_2O_2$ can induce expression of genes controlled by NF -κB. This is of particular interest at the moment because NK -κB can induce the expression of genes of the provirus HIV-1, the most common cause of acquired immunodeficiency syndrome.

**Table 1** Examples of Free Radicals

| Name | Formula | Comments |
|---|---|---|
| Hydrogen atom | $H^{\cdot}$ | The simplest free radical known. |
| Trichloromethyl | $CCl_3^{\cdot}$ | A carbon-centered radical (i.e., the unpaired electron resides on carbon). CCl3 is formed during metabolism of $CCl_4$ in the liver and contributes to the toxic effects of this solvent. |
| Superoxide | $O_2^{\cdot-}$ | An oxygen-centered radical. |
| Hydroxyl | $OH^{\cdot}$ | An oxygen-centered radical; the most highly reactive oxygen radical known. |
| Thiyl | $RS^{\cdot}$ | A group of radicals with an unpaired electron residing on sulfur. |
| Peroxyl, alkoxyl | $RO_2^{\cdot}$, $RO^{\cdot}$ | Oxygen-centered radicals formed during the breakdown of organic peroxides. |
| Oxides of nitrogen | $NO^{\cdot}$, $NO_2^{\cdot}$ | Both are free radicals: $NO^{\cdot}$ is formed in vivo from the amino acid L-arginine; $NO_2^{\cdot}$, made when $NO^{\cdot}$ reacts with $O_2$, is found in polluted air and smoke from burning organic materials (e.g., cigarette smoke). |

**Superoxide dismutase**

$$2O_2^{\cdot-} + 2H^+ \rightarrow H_2O_2 + O_2$$

**Catalase**

$$2H_2O_2 \rightarrow 2H_2O + O_2$$

**Glutathione peroxidase**

$$2GSH + H_2O_2 \rightarrow GSSG + 2H_2O$$

**Glutathione reductase**

$$GSSG + NADPH + H^+ \rightarrow 2GSH + NADP^+$$

GSH - reduced glutathione

GSSG - oxidized glutathione

Figure 1. The antioxidant defense enzymes.

## 3 THE ROLE OF TRANSITION METAL IONS

Although at low levels each is often metabolically useful, neither $O_2^{\cdot-}$ nor $H_2O_2$ is very reactive. However, if they come into contact with iron or copper ions, the noxious $OH^{\cdot}$ can be formed:

$$O_2^{\cdot-} + H_2O_2 \xrightarrow{Fe/Cu} OH^{\cdot} + OH^- + O_2 \quad (2)$$

Hence, the presence of such ions in a biological system can cause $O_2^{\cdot-}$ and $H_2O_2$ to be damaging. Because $OH^{\cdot}$ cannot migrate any significant distance, the damage will occur at the sites at which the metal ions are present.

## 4 ANTIOXIDANTS

Living organisms have evolved antioxidant defenses to remove excess $O_2^{\cdot-}$ and $H_2O_2$. *Superoxide dismutase* enzymes (SODs) remove $O_2^{\cdot-}$ by accelerating its conversion to $H_2O_2$ (Equation 1). Human cells have an SOD enzyme containing manganese at its active site (MnSOD) in the mitochondria. An SOD with copper and zinc at the active site (CuZnSOD) is also present, but largely in the cytosol. Enzymes called *catalases* convert $H_2O_2$ to water and $O_2$, but the most important $H_2O_2$-removing enzyme in human cells is *glutathione peroxidase* (GSHPX), the only known human enzyme that requires selenium for its action. GSHPX removes $H_2O_2$ by using it to oxidize reduced glutathione (GSH) to oxidized glutathione (GSSG). Glutathione reductase, a flavoprotein enzyme, regenerates GSH from GSSG, with NADPH as a source of reducing power (Figure 1). Organisms are also careful to keep iron and copper safely protein-bound whenever possible, so that reaction (2) is prevented.

Antioxidant defense enzymes are essential for healthy aerobic life. Thus, in the absence of a rich growth medium, SOD-negative mutants of *Escherichia coli* (obtained by transposon insertions into the cloned structural genes, followed by an exchange between the mutated SOD allele, carried by a plasmid, and the chromosomal wild-type allele) will be impaired in the biosynthesis of certain amino acids and will not grow aerobically. Even when so supplemented, $SOD^-$ *E. coli* cells grow slowly, suffer membrane damage, are abnormally sensitive to damage by $H_2O_2$ (perhaps because of reaction 2), and show a high mutation rate.

In addition to antioxidant defense enzymes, some low molecular mass free radical scavengers exist. GSH can scavenge various free radicals directly, as well as being a substrate for GSHPX. α-Tocopherol is the most important free radical scavenger within membranes. Reactive radicals such as $OH^{\cdot}$ can attack and damage membranes by setting off a free radical chain reaction called *lipid peroxidation.* α-Tocopherol inhibits this sequence by scavenging peroxyl radicals (Table 1), intermediates in the chain reaction.

Antioxidant defenses exist as a balanced coordinated system. Thus, although SOD is important, an excess of SOD in relation to peroxide-metabolizing enzymes can be deleterious. This has been shown by transfecting cells with human cDNAs encoding SOD. The consequences of excess SOD may be relevant to the clinical condition known as *Down's syndrome,* a point being explored by the use of transgenic animals (Figure 2).

## 5 OXIDATIVE STRESS

Normally, the production of $O_2^{\cdot-}$ and $H_2O_2$ is balanced by the antioxidant defense systems; that is, the antioxidants are not present in great excess. One reason for this may be that cells do not scavenge $O_2^{\cdot-}$ and $H_2O_2$ with 100% efficiency because small amounts of these species are useful in vivo. However, if antioxidant levels fall or production of $O_2^{\cdot-}$ and $H_2O_2$ increases, the condition termed *oxidative stress* results.

Cells can tolerate mild oxidative stress, and it often results in upregulation of the synthesis of antioxidant defence enzymes, to restore the balance. Thus, exposure of *E. coli* to increased $O_2^{\cdot-}$ accelerates the biosynthesis of at least 40 different proteins. Nine of these (including MnSOD) belong to the same regulon, controlled by the *sox* locus, which contains two adjacent genes, *sox* R and *sox* S. Expression of *sox* S is increased by $O_2^{\cdot-}$, and it then leads to expression of *sox* R. The excess $O_2^{\cdot-}$ forms an excess of $H_2O_2$, which oxidizes the protein *oxy* R, leading to activation of transcription of another panel of genes, including genes encoding catalase and glutathione reductase. Cells also have repair systems that can deal with moderate free radical damage to DNA, lipids, and proteins.

However, severe oxidative stress can cause cell damage and death. In mammalian cells, oxidative stress appears to cause increases in levels of free $Ca^{2+}$ and free iron within cells. The latter can lead to $OH^{\cdot}$ generation. Some of this $OH^{\cdot}$ generation seems to occur within the nucleus. Hydroxyl radical attacks DNA in a multiplicity of ways. One of the main products of $OH^{\cdot}$ attack on the DNA bases is 8-hydroxyguanine, a miscoding lesion often leading to $G \rightarrow T$ transversions. An excessive rise in intracellular free $Ca^{2+}$ can also activate endonucleases and cause DNA fragmentation.

Oxidative stress can be imposed in several ways. Severe malnutrition can deprive humans of the minerals (e.g., Cu, Mn, Zn, Se) and vitamins (e.g., riboflavin—needed for glutathione reductase, and α-tocopherol) needed for antioxidant defense. More usually, however, the stress is due to production of excess $O_2^{\cdot-}$ and $H_2O_2$.

Several drugs and toxins impose oxidative stress during their metabolism. Carbon tetrachloride is one example (Table 1). Another is paraquat, a herbicide that causes lung damage in humans. Its metabolism within the lung leads to production of large amounts of $O_2^{\cdot-}$ and $H_2O_2$. Cigarette smoke contains lipophilic compounds

(1) Isolate embryos from reproductive tract of female mice

(2) Microinject DNA carrying the gene for human CuZn-SOD and the DNA needed for its expression

(3) Transfer embryos into reproductive tracts of foster mothers

(4) Develop to term

(5) Select and breed progeny expressing human CuZn-SOD in tissues

The resulting mice have elevated levels of CuZn-SOD and

(i) are more resistant than controls to $O_2$ toxicity

(ii) are more resistant than controls to certain toxins

(iii) show abnormal neuromuscular junctions in the tongue

(iv) may show some of the other neurological defects characteristic of Down's syndrome

**Figure 2.** Principles of generating mice transgenic for human CuZn-SOD.

that may enter cells, bind to DNA, and bring about increased production of $O_2^{\cdot-}$ and $H_2O_2$, leading to oxidative DNA damage.

## 6 OXIDATIVE STRESS AND HUMAN DISEASE

Oxidative stress is an inevitable accompaniment of tissue injury during human disease, for the reasons summarized in Figure 3. For example, excess production by phagocytes of $O_2^{\cdot-}$, $H_2O_2$, and other species at sites of chronic inflammation can do severe damage. This seems to happen in the inflamed joints of patients with rheumatoid arthritis and in the gut of patients with inflammatory bowel diseases. Tissue injury can release metal ions from their storage sites within cells, leading to $OH^{\cdot}$ generation. Thus, the main question in human disease is not "Can we demonstrate oxidative

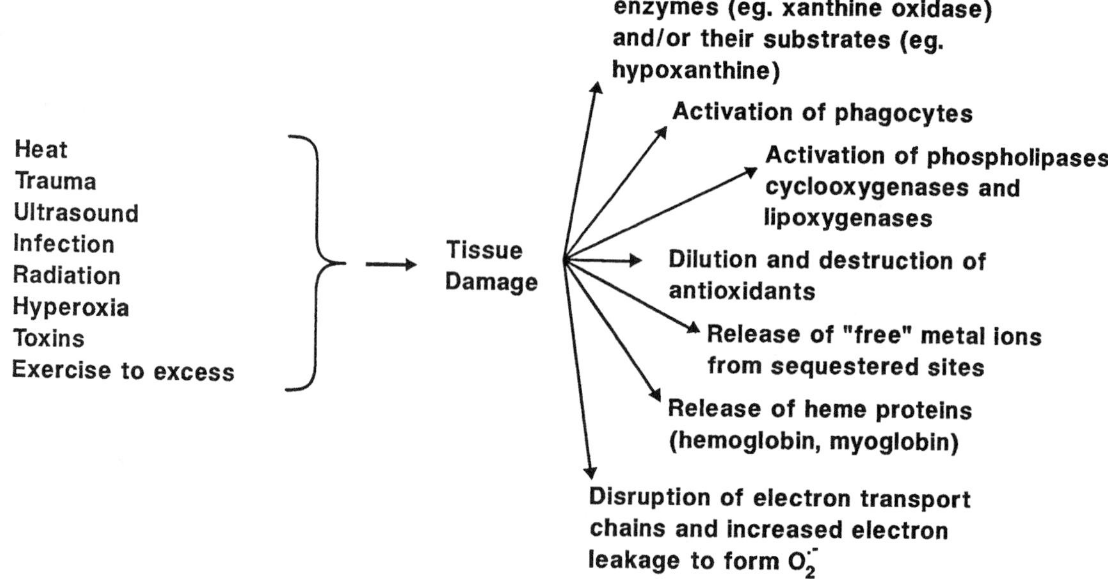


**Figure 3.** How tissue damage can cause oxidative stress. (Reproduced with permission from B. Halliwell, J. M. C. Gutteridge, and C. E. Cross. Free radicals, antioxidants and human disease: where are we now? *J. Lab. Clin. Med.* 119:598–620 (1992).)

stress?'' but rather ''Does the oxidative stress that occurs make a significant contribution to disease activity?'' The answer to the latter question appears to be ''yes'' in at least some cases, including atherosclerosis, and ulcerative colitis. However, it may well be ''no'' in many others. Elucidating the precise role played by free radicals has not been easy because these substances are difficult to measure, but the development of modern assay techniques is helping to solve this problem.

## 7 CONCLUSION

Free radicals are a normal part of human metabolism. Whether they are good or bad is largely a question of amount, location, and chemical nature.

*See also* DNA DAMAGE AND REPAIR; ENVIRONMENTAL STRESS, GENOMIC RESPONSES TO.

### *Bibliography*

Amstad, P., Peskin, A., Shah, G., Mirault, M. E., Moret, R., Zbinden, I., and Cerutti, P. (1991) The balance between Cu, Zn-superoxide dismutase and catalase affects the sensitivity of mouse epidermal cells to oxidative stress. *Biochemistry*, 30:9305–9313.

Halliwell, B., and Gutteridge, J.M.C. (1989) *Free Radicals in Biology and Medicine*, 2nd ed. Clarendon Press, Oxford.

———, Gutteridge, J.M.C., and Cross, C. E. (1992) Free radicals, antioxidants and human disease: Where are we now? *J. Lab. Clin. Med.* 119:598–620.

McBride, T. J., Preston, B. D., and Loeb, L. A. (1991) Mutagenic spectrum resulting from DNA damage by oxygen radicals. *Biochemistry*, 30:207–213.

Scandalios, J. G., Ed. (1992) *Molecular Biology of Free Radical Scavenging Systems*. Cold Spring Harbor Laboratory Press, Cold Spring Harbor, NY.

Schreck, R., and Baeuerle, P. A. (1991) A role for oxygen radicals as second messengers. *Trends Cell Biol.* 1:39–42.

Sies, H., Ed. (1991) *Oxidative Stress, Oxidants and Antioxidants*. Academic Press, London.

Storz, G., and Tartaglia, L. A. (1992) Oxy R: A regulator of antioxidant genes. *J. Nutr.* 122:627–630.

# FUEL PRODUCTION, BIOLOGICAL

*Lee R. Lynd*

### *Key Words*

**Cellulase** Extracellular enzyme that hydrolyzes cellulose.

**Cellulose** $(C_6H_{10}O_5)_n$, a straight-chain polysaccharide that is the main structural material in plants.

**Fermentation** A microbial transformation of organic materials with ATP production via substrate-level phosphorylation.

**Hemicellulose** $(C_6H_{10}O_5)_n$, a group of branched polysaccharides that serve to cement plant fibers together.

**Hexose** A monosaccharide that contains six carbon atoms.

**Pentose** A monosaccharide that contains five carbon atoms.

**Xylose** A wood sugar that contains five carbon atoms.

Biological systems are capable of converting relatively diffuse sources of energy (biomass, sunlight) into more energy-dense fuels such as ethanol, methane, butanol, biodiesel, and hydrogen. Biological fuel production is frequently considered in the context of transportation applications because the energy-concentrating character of biological conversion processes is most valuable in this context, and because of the relatively high price and uncertain supply of transportation fuels. There are radical differences between production of fuels and high value products using biotechnology, with fuel production having higher production volume, lower product value, and a greater premium on efficient processing generally and biological processing in particular. The tools of modern biotechnology are applicable to most schemes for biological fuel production, with recent results showing promise.

## 1 GENERAL FEATURES

The ultimate energy source for most biological fuel options is the sun, and the most frequently considered starting material is biomass in some form. Thus most biologically derived fuels share the desirable features of being renewable and available from indigenous resources, as well as offering the potential for greatly reduced net $CO_2$ emissions. At production levels that can be quite large but are still not commensurate with current fuel utilization in the United States, biological conversion of waste materials is attractive in many instances. For higher production levels, land allocation is required either for biomass production in the form of energy crops or for a more direct conversion of solar energy into fuel. In general, the most important factor impeding the more widespread use of biological fuel production is the cost of production. Both liquid and gaseous fuels can be derived via biological means; Table 1 gives a representative list. This contribution focuses on fuels produced by microorganisms. Biological processes potentially important in fuel applications but not addressed here include production of macrophytes such as woody biomass, grasses, and oil seeds, microbial coal desulfurization, production of enzymes (e.g., cellulases), and mixed acid production via anaerobic fermentation followed by catalytic reduction to yield fuels.

## 2 BIOLOGICALLY PRODUCED FUELS

### 2.1 ETHANOL

Ethanol is the most widely used biologically produced transportation fuel in the world today. Major fuel ethanol industries arose during the 1980s in Brazil and the United States, with government price supports playing a significant role in each case. Ethanol produced in both these countries is not cost-competitive with gasoline, although it could be in the future. With the Clean Air Act Amendments of 1990, corn-derived ethanol does not compete with gasoline in the United States but, rather, with other ''oxygenates'' that facilitate more complete combustion of gasoline. Ethanol has a higher economic value in low level (e.g., 10%) gasoline blends than in neat (unblended) form. However, the fuel properties of neat ethanol are in general excellent, and decreased emissions of ozone precursors are expected for neat ethanol. Ethanol can be produced from sugar- or starch-rich materials, from cellulosic materials (using enzymatic hydrolysis), or from cellulosic materials (using acid hydrolysis). At present three methods are thought to have approximately equal costs of production, although it is noted that there

**Table 1** Summary of Selected Biological Fuel Production Options

| Fuel Option | Status | Limiting Factors | Representative Biocatalysts |
|---|---|---|---|
| Ethanol from starch or sugar crops | Billions of gallons produced annually, mostly in Brazil and the United States | Low process yields (Brazil); cost of substrate (U.S.), especially at a scale exceeding by-product demand | Hexose fermentation: *Saccharomyces cerevisiae,* also *Zymomonas* Amylase: *Bacillus licheniformis* |
| Ethanol from cellulosic materials | | | |
| Enzymatic hydrolysis | Subject of intense R&D effort, entering commercial stage, improvements could allow cost-competitive production of primary fuels from energy crops | Slow rates; less than theoretical yields; pretreatment | Hexose fermentation: *S. cerevisiae* Pentose fermentation: recombinant *E. coli* or yeasts (*Pichia stipitus, Pacchysolen tannophilus*) Cellulase: *Trichoderma reesei* Consolidated processing: *Clostridum* species, *Fusarium oxysporum* |
| Acid hydrolysis | Entering commercial stage; cost-competitive with enzymatic hydrolysis currently but probably not in the long term; likely to find most application for waste materials | Low yields (dilute acid hydrolysis); cost of acid/acid recycling (concentrated acid hydrolysis) | As for starch/sugar crops |
| From synthesis gas | Preliminary studies completed to date by one group; provides a potential means to circumvent limitations associated with hydrolysis processes | Mass transfer limitations in bioreactor design; need to firmly establish control of product selectivity | *Clostridium ljungdahlii* |
| Methane | Widely practiced method of resource recovery from farm and human wastes; gas recovery from solid wastes receiving increasing attention | Slow rates; unattractive economics for dedicated fuel production | Undefined anaerobic consortia |
| Butanol | Formerly an important route for production of industrial solvents from soluble carbohydrates; near-term application possible, especially as a gasoline additive | High-yielding process not yet established (although promising); low product tolerance; no cellulytic organisms | *Clostridum acetobutylicum, C. beijerinckii* |
| Algae-derived diesel | Target for long-term R&D | Nascent status; lack of a transformation system | *Monoraphidium minutum, Cyclotella crypticum* |
| Hydrogen | Target for long-term R&D | Nascent status; decreasing efficiency at high light intensities | *Chlamydomonas reinhardtii, C. moewusii* |

are no full-scale plants processing cellulosic materials in operation today. Processes based on enzymatic hydrolysis of cellulosic biomass are thought likely to be the lowest cost option in the future. The process energy balance associated with converting cellulosic materials into ethanol via enzymatic hydrolysis is clearly favorable, in that the input/output ratio with respect to useful energy is on the order of 5 for current technology and is likely to improve as the technology matures. Production of ethanol from synthesis gas has also been proposed. Although more speculative than other routes at present, this strategy provides an alternative with a distinctive set of features that could prove advantageous.

The existing fuel ethanol industry is based on sugarcane in Brazil and corn in the United States. Sugarcane processing typically involves a dedicated fuel production plant consisting of a cane mill or press and facilities for fermentation and distillation. At present the considerable volume of fibrous bagasse remaining after ethanol production is land-applied or burned, but it could be converted to ethanol. Corn processing can be via either wet or dry milling. Corn steeping associated with wet milling is slow and involves complex chemistry but results in more valuable coproducts. In either case, amylases are used to derive a soluble substrate stream that is fermented and distilled. Coproducts of corn ethanol production include dried process residues that are used as animal feed and have an important beneficial impact on process economics. Technology for corn ethanol production is at a relatively mature state, such that improvements resulting in a 1% selling price reduction are considered significant.

Biologically mediated events in corn ethanol production include starch hydrolysis using amylase enzymes and fermentation of soluble sugars to ethanol using the yeast *Saccharomyces cerevisiae.* Potential improvements provided by biotechnology include development of improved fermentative microorganisms as well as systems for their utilization (e.g., immobilized-cell fermentors). Specific directions for organism development include utilization of the cellulose and hemicellulose (discussed later) present in corn, as well as learning ways to the practical exploitation of the potentially advantageous features of organisms such as *Zymomonas mobilis.*

Processes for converting cellulosic biomass into ethanol involving enzymatic hydrolysis typically consist of pretreatment, biological conversion to ethanol, product recovery, residue processing, and provision of process utilities (steam, electricity, cooling water). Of these stages, biological conversion, utilities, and pretreatment account for the most of the cost of production. The potential for research-driven cost reductions is high for biological conversion and pretreatment, and low for the other process steps. Significant cost reductions are also possible via process engineering aimed at optimizing the integration of the individual process steps. Pretreatment is necessary to render naturally occurring forms of biomass accessible to attack by cellulases. A variety of approaches to pretreatment exist, and several feature a brief exposure to high temperature in combination with acid or ammonia and/or explosive decompression.

Biological conversion of cellulosic biomass, whether to ethanol or to other products, typically involves four elements: production of cellulase enzymes, enzymatic hydrolysis of cellulose, hexose fermentation, and pentose fermentation. Hexoses arise as glucose and cellobiose derived from cellulose hydrolysis; pentoses arise from hydrolysis of the hemicellulose component of biomass. Cellulase enzyme systems, produced by both bacteria and fungi, typically involve multiple proteins carrying out at least three distinct functions (e.g., substrate binding, hydrolysis at the end of a cellulose chain, hydrolysis at the interior of the cellulase chain). Although cellulase is produced commercially for garment and detergent applications, cellulase production is an important focus for applied research because current prices are prohibitively high for fuel applications. Current understanding of the mechanisms and kinetics of cellulose hydrolysis is quite incomplete, with further insights of applied significance likely. While hexose-fermenting yeasts (Table 1) have excellent performance characteristics and have been available since antiquity, available organisms for pentose conversion to ethanol do not perform as well. Development of improved pentose fermenters is an active area of current research. In the most significant application of biotechnology to biological fuel production to date, Dr. Lonnie Ingram of the University of Florida has cloned the pyruvate decarboxylase and acetaldehyde dehydrogenase genes of *Zymomonas mobilis,* which does not utilize xylose, into *E. coli* and *Klebsiella oxytoca,* both of which do ferment pentoses.

Alternative approaches to biological conversion can be differentiated based on the degree of consolidation. Currently, the state-of-the-art process design involves the use of three different biological systems. Cellulase is produced by the aerobic fungus *Trichoderma reesei,* hexose fermentation is carried out by *S. cerevisiae* in conjunction with cellulose hydrolysis, and pentoses are fermented by an organism such as the recombinant *E. coli* just mentioned. If a single organism or system of organisms were to carry out all four elements of ethanol production with high rates and yields, process economics would benefit profoundly, and we would have taken a large and perhaps entirely sufficient step toward making biomass ethanol cost-competitive with conventional liquid transportation fuels on an unsubsidized basis. Thus the biological production of ethanol and other fuels from biomass is very susceptible to biotechnologically driven improvements. Approaches to organism development can be grouped into increasing the substrate-utilizing ability of excellent ethanol producers such as yeast or *Zymomonas,* or improving the ethanol-producing ability of organisms capable of rapidly utilizing cellulose and hemicellulose such as thermophilic bacteria.

## 2.2 Methane

Biologically produced methane is a valuable by-product often available as a result of processing wastes from diverse sources including human and animal excrement, agricultural residues, the food and chemical industries, and municipal solid waste. Methane production has particular advantages in applications that involve relatively small-scale facilities (because the technology is not as capital intensive as that for other biological fuels), dilute waste streams (because product recovery is spontaneous), and/or economic incentives associated with waste treatment. Notwithstanding these advantages, methane production has lower product yields (mass/mass substrate), lower product value per unit mass, and generally a lower formation rate compared to a product such as ethanol. Thus it is unlikely to be competitive when large-scale fuel production is the primary objective.

Methane production involves relatively simple processing technology that does not operate under aseptic conditions, achieves a substantial degree of product separation at no cost (''biogas'' is evolved from digesters via endogenously generated pressure and is typically about 60 mol % methane), and can accept virtually any biodegradable source of organic matter. The dominant piece of equipment used in methane production is the digester (a tank), with additional equipment providing for feed and effluent transfer, and biogas storage. Hundreds if not thousands of different species constitute the microbial population within an anaerobic digester. These microbes may be classified into four trophic groups: the primary fermenters, which convert organic molecules present in the feed into a mixture of organic acids, hydrogen, and $CO_2$; the syntrophic hydrogen-producing bacteria, which convert $> C_2$ organic acids into acetate, $CO_2$, and hydrogen; the acetogens, which convert $H_2$ and $CO_2$ into acetic acid; and the methanogens, which convert acetate and $H_2/CO_2$ into methane. The syntrophic hydrogen-producing bacteria are notable in that their catabolism is nonspontaneous under standard state conditions and can proceed only at very low hydrogen partial pressure, which is normally maintained by methanogens.

Because methane production is carried out by a self-selecting population of microorganisms, the molecular techniques that have revolutionized other areas of biotechnology are of limited applicability and have yet to be applied in a practical context. Improvements in methane production generally fall into the categories of management of the microbial ecosystem and reactor design. Examples of ecosystem management include avoidance and/or detection of organic overloading, provision of adequate moisture (e.g., in a landfill situation) and nutrients (especially micronutrients), and elucidation of conditions that favor formation of anaerobic ''granules'' or biofilms that allow a high rate of processing of soluble compounds. Examples of reactor design approaches are equalization of incoming flows, provisions for retention of biocatalyst (e.g., suitable surfaces for biofilm formation, promotion of gas disengagement from granules), and exploitation of density differences to retain solid substrates relative to the liquid.

## 2.3 Butanol

Butanol is currently produced abiologically from propylene for use as an industrial chemical. However, biological butanol production via carbohydrate fermentation was once widespread and could become so again. Butanol can be used as a fuel either as butanol per se or with the lesser amounts of ethanol and acetone that are coproduced in acetone–butanol–ethanol (ABE) fermentation. As with ethanol, economic factors indicate that initial use of butanol as a fuel is likely to involve low level butanol/gasoline blends. With coproduct credits, the cost of butanol production by corn fermentation has been projected to be competitive with petrochemically based production. Butanol production by both these routes, however, currently costs more than ethanol production on either a mass or an energy basis. Thus price is a significant obstacle to expanded production of butanol to meet fuel demand.

The pathway associated with the ABE fermentation of *Clostridium acetobutylicum,* involving 14 different enzymes, is a model system for branched anaerobic metabolism and has been studied extensively with respect to metabolic control and gene cloning. The solvents acetone, butanol, and ethanol, normally products of secondary metabolism, can be produced in conjunction with growth under conditions of mineral nutrient limitation. Low pH, butyric acid, and ATP nonlimitation have all been associated with the onset of solvent formation, and efforts to promote solvent production via manipulation of the fermentor environment have met with some success.

Genetically based approaches to directing carbon flux toward solvents or butanol in particular also appear promising and are under active investigation. Eleven genes involved in the conversion of acetyl–coenzyme A into acids and solvents have been cloned in *E. coli.* In addition to achieving heterologous gene expression, significant progress has also been made of late in the development of methodologies that allow gene expression in *C. acetobutylicum.* In particular, the restriction endonuclease system has been characterized and protective methylases have been identified, cloned, and used to greatly increase transformation frequencies. Shuttle vectors have also been described suitable for gene transfer between *C. acetobutylicum* and *E. coli* and/or *Bacillus subtilis.* Using these tools, studies are being undertaken aimed at overexpression or modification of genes associated with catabolic product formation. Cloning of cellulase genes into *C. acetobutylicum* is also under investigation.

Over the next few years, development of improved strains with higher butanol yields appears likely, and strains with a broader substrate range may also be developed. The matter of product inhibition is likely to be more difficult to approach because it is probably not determined by any one gene or even a few genes. Finding a way to reduce the cost of butanol recovery is an important component of the goal of improving the competitive position of butanol as a fuel. Such a reduction might be sought through strain development, new insights into mechanisms of inhibition and ways to lessen inhibitory effects, and/or improved product recovery technology.

### 2.4 Biodiesel

A hydrocarbon closely resembling petroleum-derived diesel can be obtained from both oilseeds and algae in the form of esterified fatty acids. Little if any modification of a conventional diesel engine is necessary to utilize such ''biodiesel'' in unblended form, and lower emissions (relative to conventional diesel) are anticipated. The cost of production is a significant obstacle to widespread substitution for petroleum at current market prices. Motivated in part by the prospect of reduced net $CO_2$ emissions, significant expansion of oilseed-derived biodiesel production capacity has begun in Europe and to some extent in the United States.

The potential of biodiesel production using algae is indicated by reports of up to 60% of dry cell mass being lipids. Such high lipid content has been observed under conditions of mineral nutrient limitation, where lipid formation provides a sink for reduced electron carriers that would be used for synthesis of new cells under mineral nutrient sufficiency. Efforts are under way to understand and manipulate the metabolic control of lipid formation, and to develop a transformation system for algae potentially useful in this context.

### 2.5 Hydrogen

Hydrogen is a potentially attractive fuel because of its high energy-to-mass ratio and because its combustion does not produce organic compounds. Moreover, hydrogen production can potentially be coupled to renewable and $CO_2$-neutral energy sources such as solar electricity, or photosynthesis. Technology is at present lacking for the cost-competitive production and utilization of hydrogen, although fuel cells may be coming of age. While these factors relegate hydrogen to the longer term category, the advantages of this gas provide significant incentive for research.

Algae-mediated photosynthetic water splitting resulting in production of molecular hydrogen and coproduction of oxygen has been observed and could form the basis for direct biological conversion into a useful fuel. The mechanism for this process is thought to involve generation of reduced ferredoxin by photosystem I followed by ferredoxin reoxidation with concomitant reduction of protons to hydrogen. The measured efficiency of the process, which is reasonable at low light intensities (on the order of 15% for a process resulting in a chemically stored energy form), is thus far much lower at higher and more practical light intensities. Overcoming this limitation is a key target for applied research.

**Table 2** Comparison of Pharmaceuticals and Fuels[a]

| Feature | Pharmaceuticals | Fuels |
|---|---|---|
| Market potential (U.S.)[b] | | |
| Production, kg product/yr | 10 | $10^{12}$ |
| Sales, billion $/yr | 0.1–1 | 100 |
| Product value | | |
| $/g | $10^4$–$10^5$ | $3 \times 10^{-4}$ |
| $/L unseparated broth | 10–100 | $3 \times 10^{-5}$ |
| Raw material (dry basis) | | |
| To meet market demand, kg | 10,000 | $10^{12}$ |
| Proportion of selling cost | 0.1–1% | 30–70%[c] |
| Cost of production[d] | | |
| Proportion of selling cost | <30% | >75% |

[a]Values given are approximate, often only to ± one order of magnitude. Pharmaceutical values are based on personal communication with David DeLucia (Verax Inc., Lebanon, NH) except where otherwise noted and are representative of mammalian cell production systems in the biopharmaceutical industry.

[b]For a single biopharmaceutical or fuel product.

[c]The upper value is for dedicated energy crops; for waste feedstocks, the percentage is lower.

[d]From Angus Macdonald (Macdonald & Associates, Providence RI). Includes costs for capital and capital-related items (insurance, maintenance, taxes) and operating costs including labor and raw materials; excludes fees associated with licensing, clinical trials, obtaining FDA approval, and recovery of R&D costs.

## 3 FUEL BIOTECHNOLOGY

Production of commodity products, and fuels in particular, is a radically different enterprise from production of the high value products most commonly associated with biotechnology. This difference is illustrated by Table 2, which compares fuel production to production of a hypothetical high value pharmaceutical. The masses of product and raw material to meet U.S. demand differ for the pharmaceutical and the fuel by approximately 11 and 8 orders of magnitude, respectively. Product value differs by approximately 8 orders of magnitude, and the percentage of selling price represented by raw material cost differs by approximately 2 orders of magnitude. Biologically mediated events have among the highest costs of all process steps for representative fuel production processes (e.g., methane, biomass ethanol). Separation typically has the largest cost for pharmaceutical production. Cost-effective and efficient conversion technology is of paramount importance for fuel production, whereas costs associated with R&D recovery, clinical trials, and obtaining approval of the U.S. Food and Drug Administration typically have a larger contribution to the selling price of pharmaceuticals than does the cost of production. The stronger incentive to pursue efficient production technology for fuels is reflected in the distribution of installed capacity: most biological fuel production involves continuous fermentation processes, whereas batch processes are by far the dominant mode of production for pharmaceuticals.

Cells and enzymes involved in fuel production must have both high product yields and high production rates. They must also be robust. In light of the low product value and processing cost margins for fuels, a limited range of measures can be taken to maintain a selectively deleterious phenotype, or to accommodate requirements for exotic nutrients or sensitivity to inhibitors. As indicated here, biotechnology-driven organism development has the potential to significantly improve the practicality of biological fuel production. This enterprise must however be undertaken with cognizance of the distinctive set of constraints operative for biological fuel production applications. Many of the most significant challenges and opportunities for application of biotechnology to fuel production involve modification of organisms that naturally possess characteristics of value in a fuel production context (e.g., high products yields or tolerance, ability to convert low cost substrates). More often than not, current genetic understanding and capabilities with respect to such organisms are rudimentary. In addition, stability of modified organisms is of heightened importance for fuel production in light of the benefits of continuous processing and the limited scope for imposing artificial selective pressure (e.g., with antibiotics).

*See also* Biodegradation of Organic Wastes; Bioprocess Engineering; Carbohydrates, Industrial.

### *Bibliography*

Hobson, P. N., and Whately, A. D. (1993) *Anaerobic Digestion, Modern Theory and Practice.* Elsevier, London.

Klasson, K. T., Ackerson, M. D., Clausen, E. C., and Gaddy, J. L. (1992) Bioconversion of synthesis gas into liquid or gaseous fuels. *Enzyme Microb. Technol.* 4:602–608.

Ingram, L. O., Alterthum, F., Ohta, K., and Beall, D. S. (1990) Genetic engineering of *Escherichia coli* and other enterobacteria for ethanol production. *Dev. Ind. Microbiol.* 31:21–30.

Lembi, C. A., and Waaland, J. R. (1988) *Algae and Human Affairs.* Cambridge University Press, Cambridge.

Lynd, L. R., Cushman, J. H., Nichols, R. J., and Wyman, C. E. (1991) Fuel ethanol from cellulosic biomass. *Science,* 251:1318–1323.

Woods, D. R., Ed. (1993) *Clostridia and Biotechnology.* Butterworths, Oxford.

# Fungal Biotechnology

*Brian McNeil and Linda M. Harvey*

### *Key Words*

**Batch Culture** Closed culture system, to which no major additions of nutrients are made after inoculation of the nutrient medium with the chosen microorganisms.

**Continuous Culture (Chemostat)** Culture system in which the rate of addition of fresh medium balances the rate of removal of spent medium and cells such that, at steady state, the concentration of all reactants is constant.

**Fed Batch Culture** Culture system to which nutrient is added at discrete time intervals or on a continuous basis, with the aim of avoiding either inhibition due to the nutrient or $O_2$ limitation.

**Filamentous Fungi (molds)** Eukaryotic microorganisms, characterized by growth at the apical tip, which form a branched network of vegetative filaments, a mycelium.

**Solid Substrate Fermentation (SSF)** Low water fermentation process in which the fungus grows on or through the solid substrate, often a cereal grain or similar material.

**Submerged Liquid Fermentation (SLF)** Type of liquid fermentation system within which the fungus is aerated and, often, mechanically agitated.

Molds (filamentous fungi) are used in the service of man to produce a wide range of valuable products, to improve feedstuffs, to carry out biotransformations, and to effect bioremediation. The technology used to cultivate and control mold activity either on solid or in liquid media allows the production of high levels (kg/m$^3$) of the desired products from selected microbial strains. Despite this, many problems in cultivating these microorganisms, caused directly by their morphological form, remain to be fully resolved. The biotechnological role of the filamentous fungi appears to be ready to expand, with the development of effective transformation systems and increasing use of organisms such as *Aspergillus niger* as efficient systems for the secretion of heterologous proteins. However, an understanding of the technological means of successfully cultivating these microorganisms will continue to be essential to the exploitation of their biotechnological potential.

## 1 INTRODUCTION

Biotechnology has been defined as the use of whole cells or parts derived therefrom to catalyze the formation of useful products or to treat pollutants.

Within the broad field of biotechnology, the filamentous fungi (molds) have long occupied an important position. This diverse and metabolically versatile group of eukaryotic microorganisms is characterized by the formation of vegetative branching filaments (hyphae), which together form a mycelium. This growth pattern is a result of polarization of growth at the hyphal tip (apex).

Although most groups of molds have economically significant members, biotechnologically, the most important fungi are found within the Fungi Imperfecti, Ascomycetes and Basidiomycetes.

### 1.1 Characteristics of Filamentous Fungi (Molds)

The hyphal mode of growth confers a number of advantages on the filamentous fungi, including the ability to translocate nutrients within the mycelial network, the ability to grow away from the initial growth point, such that hyphae at the edge of a colony constantly encounter fresh nutrients, and the ability to grow into solid materials (penetrative growth), thus allowing the uptake of nutrients whose availability would otherwise be restricted by diffusion.

In the growth process, fungi form branches, and branch frequency is closely related to nutrient availability.

As a consequence of their mode of growth, the filamentous fungi possess a number of metabolic activities that are attractive to the biotechnologist. First among these is the synthesis and excretion of lytic enzymes (proteases, lipases, carbohydrases). Production of such hydrolytic enzymes aids fungal invasiveness. Under appropriate conditions, some fungi can secrete large quantities (≤ 50 g/L) of proteins into their environment.

The filamentous fungi are, as a group, prodigious producers of secondary metabolites. This is a structurally diverse group of products, formed in significant amounts during the stationary growth phase, and includes the penicillins and gibberellins.

## 2 DEVELOPMENT OF FUNGAL BIOTECHNOLOGY

### 2.1 Solid Substrate Fermentation (SSF) Systems

Molds were traditionally used in the production and improvement of foods. Natural colonization of moist grains or cereals led to the discovery that certain molds could improve the flavor, texture, and nutritional value of largely carbohydrate materials.

The next steps involved control of the physicochemical conditions of the process, and, later isolation, selection, and use of particular strains. These processes are the classic examples of solid substrate fermentation (SSF) processes and are characterized by the overgrowth of the substrate, often a grain, by the fungus, which is then used as a source of enzymes to break down substrate components, or, for addition to other substrates. Such processes are often referred to by the Japanese term *koji.* Koji processes are still widely used in the Orient for the formation of a wide range of products (Table 1). Most of these processes are now operated at the industrial scale, and many have been taken up by other countries. Tempeh, for example, in which the action of the fungi produces a high protein material that is flavorsome and useful as a meat supplement, is produced in quantity in both Europe and the United States.

In Europe, traditionally, fungi such as *Penicillium roquefortii* and *P. camembertii* were used to grow on or in cheeses and to enhance flavor by production of lipases and proteases. Further use of SSF came in the industrial production of mushrooms, the most common being *Agaricus bisporus.*

**Table 1** Processes Involving a Koji Stage

| Product | Substrates | Mold | Location |
|---|---|---|---|
| Shoyu (soy sauce) | Soybeans/wheat | *Aspergillus soyae* or *A. oryzae* | Japan (but similar products elsewhere) |
| Miso | Rice/barley soybeans | *A. soyae* | Japan |
| Sake (rice wine) | Rice | *A. oryzae* | Japan |
| Tempeh | Soybeans | *Rhizopus* spp. (*R. oligosporus*) | Indonesia |

All SSF systems are low water systems with relatively low power inputs. Conversely, they are generally recognized as being difficult to monitor, control, and scale up. Oxygen transfer, and the removal of heat and carbon dioxide from such cultures, can be difficult. Several SSF fermenter types are illustrated in Figure 1.

### 2.2 Submerged Liquid Fermentation (SLF) Systems

Citric acid illustrates particularly well the development of the fermentation technology required for convenient formation of mold products in large quantities, at reasonable *rates* in liquid systems. Initially derived from lemon juice, citric acid was first produced industrially by a surface tray process using *Aspergillus niger.* Although this was a relatively lengthy (6–12 days), and labor-intensive process, Pfizer used the method until 1991 in the United States.

From the late 1940s, the use of the stirred tank reactor (STR) for submerged liquid cultivation of *A. niger* resulted in citrate production processes lasting 3–5 days at 25–30°C. A typical STR is shown in Figure 1d. Use of such fermenters allowed high oxygen transfer rates to be maintained, leading to high productivity with low manual labor needs. By comparison with SSF systems, these fermenters were easier to monitor and control. The mold's physical environment could thus be more readily optimized.

The STR is now the industry workhorse, a flexible reactor type used for production of a wide range of products.

Some of the features developed in citrate fermentations that are held in common with other STR mold fermentations include:

1. Use of highly bred (often relatively *genetically unstable*) pure cultures of production strains. Production strains are *usually* stable these days!
2. Use of relatively crude raw materials such as molasses and corn steep liquor on the industrial scale.
3. Continuous aeration with sterile (filtered) air, and continuous stirring.
4. Maintenance of asepsis during fermentation (molds grow more slowly than bacteria and yeasts, thus any contamination by these organisms could lead to displacement of the fungus).

Current annual demand for citric acid, which is around 500,000 tons, is growing at 3–5% per year. Increasingly, the citric acid market is concentrated in the hands of fewer, larger products. Currently, only seven companies produce the vast bulk of citrate used.

#### 2.2.1 Penicillins

At around the time of the development of submerged citrate processes, fermentations in STRs involving pure cultures of *Penicil-*

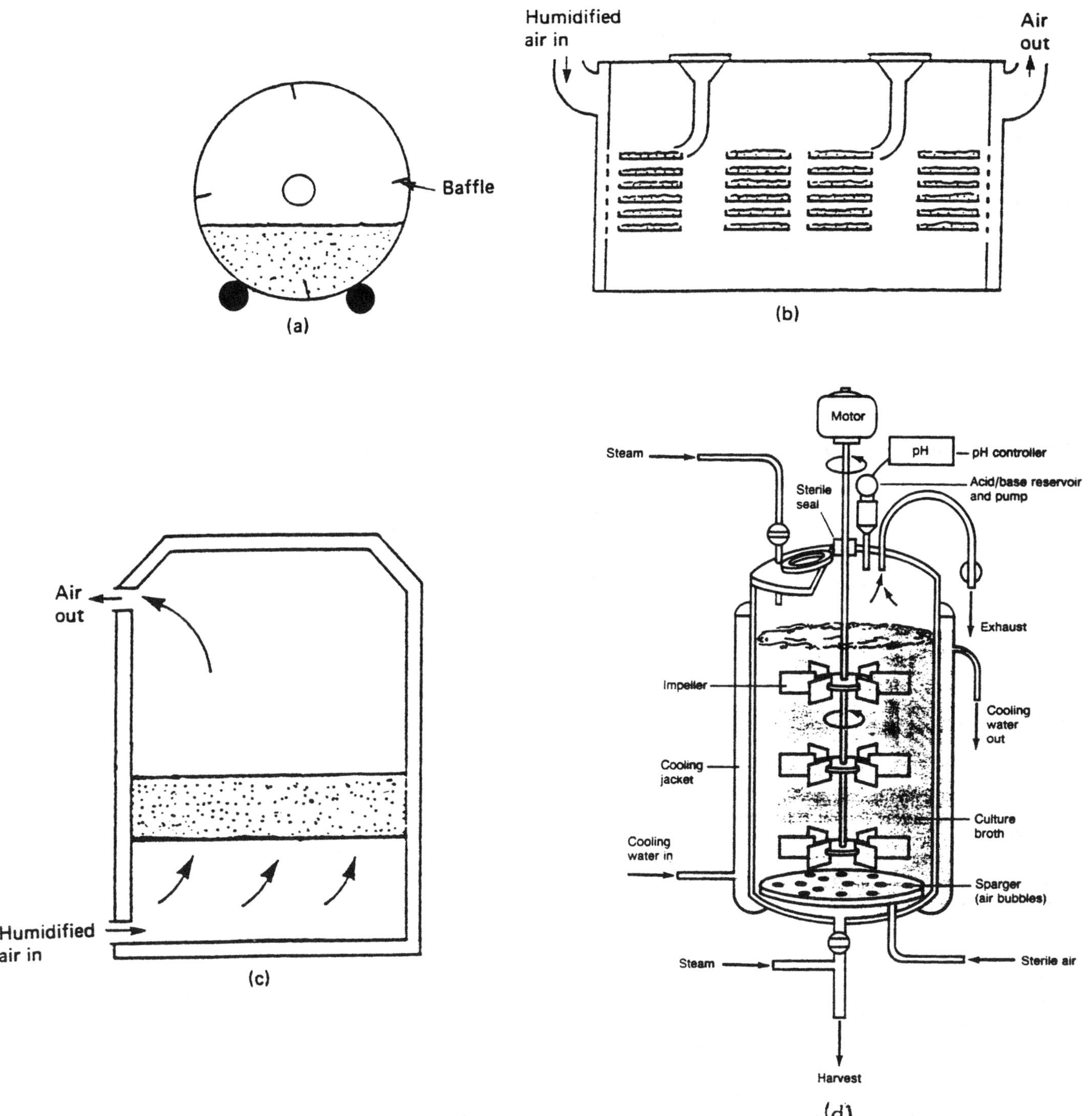


**Figure 1.** SSF and SLF fermenter configurations (a) Rotating drum (SSF), (b) tray system (SSF), (c) forced airflow system, and (d) stirred tank bioreactor (SLF). [(a)–(c) from Kinghorn and Turner (1992); (d) from Crueger and Crueger (1990).]

*lium chrysogenum* for penicillin G production were also being developed. These production processes are conceptually very similar to citrate production.

Initially, the carbon source was lactose, which permitted slow fungal growth and good excretion of the product, the secondary metabolite penicillin G. Later, fed batch processes using glucose feeds were developed. These are the current methods of production for penicillins G and V, which differ chemically only in the attached side chain (phenyl acetic acid in the former, phenoxyacetic acid in the latter). Bulk penicillins G and V are increasingly used as raw materials for conversion into a wide range of semisynthetic penicillins. Penicillins, like citrate, are mature biotechnological products produced in bulk, characterized by ''survival of the fittest'' with respect to producers.

Table 2 Some of the Products of Fungal Biotechnology

| Organism | Product(s) | Use | Mode of Production |
|---|---|---|---|
| *Aspergillus niger* | Citric acid | Acidulant | SLF (usually) |
| *A. oryzae* | Amyloglucosidase and other enzymes | Starch hydrolysis lipolysis, proteolysis, etc. | SLF (fed batch) |
| *Penicillium chrysogenum* | Penicillins G and V | Antibacterial agents | SLF (fed batch) |
| *Fusarium graminearum* | Mycoprotein (Quorn) | Human foodstuff | SLF (continuous culture) |
| *Gibberella fujikuroi* | Gibberellins | Plant growth regulators | SLF |
| *Tolypocladium* spp. | Cyclosporin | Immunosuppressant in transplant surgery | SLF |
| *Sclerotium glucanicum* | Scleroglucan | Tertiary oil recovery; immunostimulant; antitumor agent | SLF (batch) |
| *Agaricus campestris, A. bisporus* | Fruiting bodies (mushrooms) | Human foodstuff | SSF |
| *Lentinus edodes* | Fruiting body or cell extracts | Human foodstuff, animal foodstuff immunostimulant | SLF |

### 2.2.2 Enzyme Production by Fermentation

Fungi are excellent sources of a wide range of lytic enzymes. Most of these are produced using the well-understood STR system described briefly in Section 2.2. In this area the *Aspergilli* are preeminent. The current world marker for industrial enzymes, several of which are from *Aspergilli* by SLF, is around $600 million per year.

### 2.2.3 Other Liquid Fermentation Products

Some additional liquid fermentation products are shown in Table 2. In general, it is much easier to screen mold-derived compounds for antimicrobial activity than for other activities. Indeed, some products such as cyclosporins were originally isolated for their antimicrobial characteristics (antifungal for cyclosporins) before their value as immunosuppressants was realized. It is not unreasonable to state that the use of cyclosporins has revolutionized transplant surgery.

*Mucor* species have been shown to accumulate large quantities of unsaturated fatty acids (e.g., linoleic acid). Although technically successful, SLF-based processes for production are currently not economic.

The production of mycoprotein (Quorn) by ICI/RHM [Imperial Chemical Industries (now Zeneca) and Rank Hovis MacDougall] has been a considerable success. This is one of the few continuous culture SLF processes, if not the only one, operated at the industrial scale. *Fusarium graminearum* is grown in STRs on a defined sugar–mineral salts medium, and the fungal biomass produced is harvested. By virtue of its filamentous nature, the biomass is textured, and this property can be controlled or modified to simulate various meat products. The bland-tasting product readily accepts other flavors. Quorn appears to be an almost perfect foodstuff, having high protein quality, low fat content, zero cholesterol, and a reasonably high fiber content.

Interest is also growing in the role of some fungal polysaccharides to modulate the immune systems of test animals. Such compounds, classed as biological response modifiers (BRMs), include the β-glucan polysaccharide lentinan and scleroglucan. These compounds may stimulate the immune system by enhancement of macrophage activity.

## 3 PRODUCTION MODES

### 3.1 Batch Culture

Batch culture, the traditional mode of cultivation in fermenters, is not suited to formation of products when inhibition by substrate, product, oxygen limitation, or product precursors is a problem. Fungal citrate may be produced in batch culture.

### 3.2 Fed Batch Culture

In a second approach, the fermentation, often after an initial batch phase, is fed regularly or continuously with required nutrients or precursors, such that the level of these compounds is never inhibitory to the culture, or, at such a rate that the consumption of nutrients is just balanced with oxygen consumption, thus avoiding oxygen limitation. Penicillins and amyloglucosidase are produced by the fed batch culture method, as are many other mold secondary metabolites.

### 3.3 Continuous Culture

The chemostat is the most common type of continuous culture system encountered. Here the culture is continuously fed fresh nutrient medium, and spent medium, cells, and product are continuously drawn off. Although not ideally suited to secondary metabolite production, the system is suitable for biomass production, where productivities can be much higher than in batch fermentations.

The continuous culture technique is also widely used on a small (lab) scale to study aspects of cellular biochemistry and physiology.

## 4 PROBLEMS IN CULTIVATION OF FUNGI IN FERMENTERS

Despite a long history of cultivation in both SSF and SLF systems, there are still problems to be overcome. Some of these have been mentioned in relation to SSF systems. In liquid systems, especially the STR, difficulties arise as a result of the particular morphology of the molds.

In the fermenter vessels of liquid systems, formation of a branched mycelium results in a rapid increase in the viscosity of the culture. The rheological character of the fermentation fluid may also change from a low viscosity Newtonian system to highly viscous non-Newtonian flow. The general effect of this trend is to reduce mixing efficiency, oxygen, and heat transfer in the fermenter. The resulting heterogeneity within the fermenter may limit productivity in many cases.

Some producers of fungal metabolites accept these restrictions as inevitable consequences of cultivation of filamentous fungi. Others, however, are actively seeking less ''filamentous'' forms of the producing microorganisms with much shorter hyphae, which will reduce the problems somewhat. It is only reasonable to state, however, that these problems are far from overcome.

## 5 PERSPECTIVE

The use of molds in man's service has a long history. Conversely, their industrial exploitation can be measured in decades. Molds occupy a central role in the fermentation industry and, in the near future, will continue to do so.

However, the future of fungal biotechnology will be heavily dependent on new product development and on the identification of new areas where molds can usefully be applied. There are particularly promising developments in the areas of biotransformations, bioremediation using fungi, and use of filamentous fungi as systems for the expression and production of heterologous proteins.

*See also* BIOPROCESS ENGINEERING; DRUG SYNTHESIS.

### *Bibliography*

Crueger, W., and Crueger, A. (1990) *Biotechnology,* 2nd ed. Sinauer, Sunderland, MA.

Kinghorn, J. R., and Turner, G., Eds. (1992) *Applied Molecular Genetics of Filamentous Fungi.* Blackie, Glasgow.

McNeil, B., and Harvey, L. M. (1990) *Fermentation: A Practical Approach.* Oxford University Press, Oxford.

Ward, I. P. (1989) *Fermentation Biotechnology.* Open University Press, Milton Keynes, U.K.

# Gaucher Disease

*Ernest Beutler*

*Key Words*

**Frameshift** A change in a gene that causes it to be read so that the sets of three nucleotides that comprise a codon are out of synchrony. The wrong triplets are, therefore, read.

**Gaucher Disease** A hereditary disorder characterized by the accumulation of a glycolipid in macrophages throughout the body. Disease manifestations include enlargement of the spleen and liver and bone fractures.

**Glycolipid** Complexes of sugars and fats.

**Hydrophilic** Literally "water loving." Mixes freely with water.

**Hydrophobic** Literally "water hating." Does not mix freely with water.

**Phenotype** The observed effect caused by a mutation in an organism.

**Pseudogene** A copy of a functional gene that has undergone mutational change so that it is no longer functional.

Gaucher disease is the most common of the glycolipid storage disorders. First described in 1882 by P.C.E. Gaucher (1854–1918) in his doctoral thesis as an "epithelioma of the spleen," the condition is now recognized to be the consequence of the deposition of glucosyl ceramide in the macrophages of the body. This glycolipid is a breakdown product of globosides and gangliosides, complex glycolipids that are constituents of the normal cell membrane. Normally the enzyme glucocerebrosidase cleaves glucosyl ceramide into glucose and ceramide, but in Gaucher disease a deficiency of glucocerebrosidase leads to accumulation of glucosyl ceramide.

Clinically, three major types of Gaucher disease have been delineated based on the absence (type 1) or presence and severity (types 2 and 3) of primary central nervous system involvement. Type 1 disease is sometimes designated as the "adult" form of the disease, but since type 1 disease often becomes apparent in early childhood, this term is misleading. Within each type, even within the same ethnic groups, the phenotypes and genotypes can be markedly heterogeneous.

## 1 THE GLUCOCEREBROSIDASE GENE

The glucocerebrosidase gene, located on chromosome 1 at q21, is approximately 7 kb long and contains 11 exons. A 5 kb pseudogene is located about 16 kb downstream. The pseudogene has maintained a high degree of homology with the functional gene. Although it is transcribed, an unusual property for a pseudogene to possess, it does not contain a long open reading frame and has a 55 base pair deletion from what was once the coding region. *Alu* sequences that have been inserted into introns give the functional gene greater length than the pseudogene. The putative TATA- and CAAT-like boxes of the promoter have been identified about 260 bp upstream from the upstream ATG start codon.

The cDNA is about 2 kb long. There are two ATGs at the 5′ end, and both are utilized in translation. The relative importance or function of these two start sites is unknown. Transfer into murine cells of human acid β-glucosidase cDNAs mutated to ablate either one or the other of the start sites has shown that either of the two sites can produce functional enzyme in vivo. The sequence between the upstream and downstream ATG is hydrophilic, while that between the downstream ATG and the codon that represents the amino end of the mature protein is the typical hydrophobic sequence expected in a leader sequence. It has been suggested that alternative splice forms might create one protein product with a hydrophilic and another product with a hydrophobic leader sequence, but such splice forms have not been identified. Messenger RNA of several different lengths has been detected, probably resulting from the existence of alternative polyadenylation sites, alternative splicing, or presence of pseudogene mRNA.

## 2 POLYMORPHISMS AND DEFICIENCY MUTATIONS

Eleven polymorphic sites are known to exist in the introns and flanking regions of the acid β-glucosidase gene, but somewhat surprisingly only four haplotypes have been found. These sites and the haplotypes that have been identified so far are summarized in Table 1. A considerable number of point mutations that cause glucocerebrosidase deficiency, hence Gaucher disease, have been identified. These are summarized in Table 2.

Some genes with the 1448C mutation contain other mutations, each corresponding to the sequence of the pseudogene. In some cases this type of mutation has been shown to represent a crossover between the functional gene and pseudogene with loss of the genetic material between the two. In other cases no mechanism for the formation of abnormal mRNA has been elucidated, and the alleles have merely been referred to as complex alleles, pseudopattern, or "rec" (for recombinant). The mechanism by which these are formed may all be identical to that demonstrated to occur in the XOVR mutation or may represent the result of gene conversion events.

Some of the mutations are relatively mild in their phenotypic effect and have been associated only with type 1 disease. Others are more severe and are observed also in type 2 or type 3, neuronopathic disease. In some instances so few cases have been observed that the severity of the mutation is not known. A severity score has been devised that allows one to express the clinical severity of patients with Gaucher disease. As shown in Figure 1, patients who are homozygous for the 1226G mutation tend to have relatively

**Table 1** Nucleotides at 12 Positions in the Four Glucocerebrosidase Haplotypes

| Haplotype Designation | Frequency | Nucleotide Positions −802 | −725 | −614 | 2128 | 2834 | 3297 | 3747 | 3854 | 3931 | 4644 | 5135 | 6144 |
|---|---|---|---|---|---|---|---|---|---|---|---|---|---|
| 1 + | Common | a | c | c | a | c | g | g | t | g | del | c | g |
| 2 − | Common | g | t | t | g | g | a | g | c | a | a | a | a |
| 3 African | Common African | g | t | t | g | a | a | g | c | a | a | a | a |
| 4 Uncommon | Rare | a | c | c | a | c | g | a | t | g | del | c | g |

**Table 2** Gaucher Disease Mutations

| | | | Point Mutations That Cause Gaucher Disease | | | |
|---|---|---|---|---|---|---|
| cDNA Number[a] | Amino Acid Number[a] | Genomic Number | Nucleotide Substitution | Amino Acid Substitution | Effect | Population Frequency[b] |
| IVS2+1 | | 1067 | G→A[c] | | ? Severe | Uncommon |
| 476 | 120 | 3060 | G→A | Arg→Gln | ? Mild | Rare |
| 481 | 122 | 3065 | C→T | Pro→Ser | Mild | Rare |
| 535[d] | 140 | 3119 | G→C | Asp→His | ? | Rare |
| 1093 | 326 | 5309 | G→A | Glu→Lys | ? | |
| 586 | 157 | 3170 | A→C | Lys→Gln | Severe | Rare |
| 751 | 212 | 3545 | T→C | Tyr→His | ? Mild | Rare |
| 754 | 213 | 3548 | T→A[c] | Phe→Ile | Severe | Uncommon |
| 764 | 216 | 4113 | T→A | Phe→Tyr | Mild | Rare |
| 1043 | 309 | 5259 | C→T | Ala→Val | Mild | Rare |
| 1053 | 312 | 5269 | G→T | Trp→Cys | Mild | Rare |
| 1090 | 325 | 5306 | G→A[c] | Gly→Arg | Severe | Rare |
| 1141 | 342 | 5357 | T→G | Cys→Gly | Severe | Rare |
| 1208 | 364 | 5424 | G→C | Ser→Thr | Mild | Rare |
| 1226 | 370 | 5841 | A→G | Asn→Ser | Mild | Common |
| 1297 | 394 | 5912 | G→T | Val→Leu | Severe | Uncommon |
| 1342 | 409 | 5957 | G→C[c] | Asp→His | Severe | Uncommon |
| 1343 | 409 | 5958 | A→T | Asp→Val | Severe | Rare |
| 1361 | 415 | 5976 | C→G | Pro→Arg | Severe | Rare |
| 1448 | 444 | 6433 | T→C[c] | Leu→Pro | Severe | Common |
| 1504 | 463 | 6489 | C→T | Arg→Cys | Mild | Uncommon |
| 1604 | 496 | 6683 | G→A | Arg→His | Mild | Uncommon |

| Insertions and Deletions that Cause Gaucher Disease | | | | | |
|---|---|---|---|---|---|
| cDNA Number[a] | Genomic Number | Nucleotide Substitution | Amino Acid Substitution | Effect | Population Frequency[b] |
| 84 | 1035 | G→GG | | Severe | Common |
| 1263–1317 del | 5879–5933[c] del | | | ? Severe | Rare |

| Recombination Events That Cause Gaucher Disease | | | |
|---|---|---|---|
| Location of Crossover Event(s)[e] | | Effect | Population Frequency |
| >1343 <1388 | >5957 <6272[f] | Severe | Uncommon |
| >455 <475[d,g] | >3039 <3059[d,g] | ? Severe | Rare |
| >754 | >3548 | | |
| >1317 <1343 | >5932 <5957 | Severe | Uncommon |
| >1343 <1388 | >5957 <6272 | Severe | Uncommon |
| >1225 <1263 | >5588 <5878 | Severe | Rare |

[a]Nucleotide or amino acid position.

[b]Common, high frequency in at least one population; Uncommon, found in a number of unrelated patients; Rare, found in only one or two individuals.

[c]Pseudogene sequence.

[d]Both found in one gene.

[e]Only approximate ranges can be given, since the pseudogene and functional gene contain long identical sequences.

[f]Physical fusion with loss of intergenic segment.

[g]The first range represents crossover from gene to pseudogene; the second from pseudogene to gene. This region contains seven mutations, only six of which are identical with pseudogene sequence. At genomic nt 3474, within the region, the nucleotide conforms to the active gene, not the pseudogene. Thus "conversion" seems imperfect.

*Source:* Beutler, E. (1992) Gaucher disease: New molecular approaches to diagnosis and treatment. *Science,* 256:794–799.

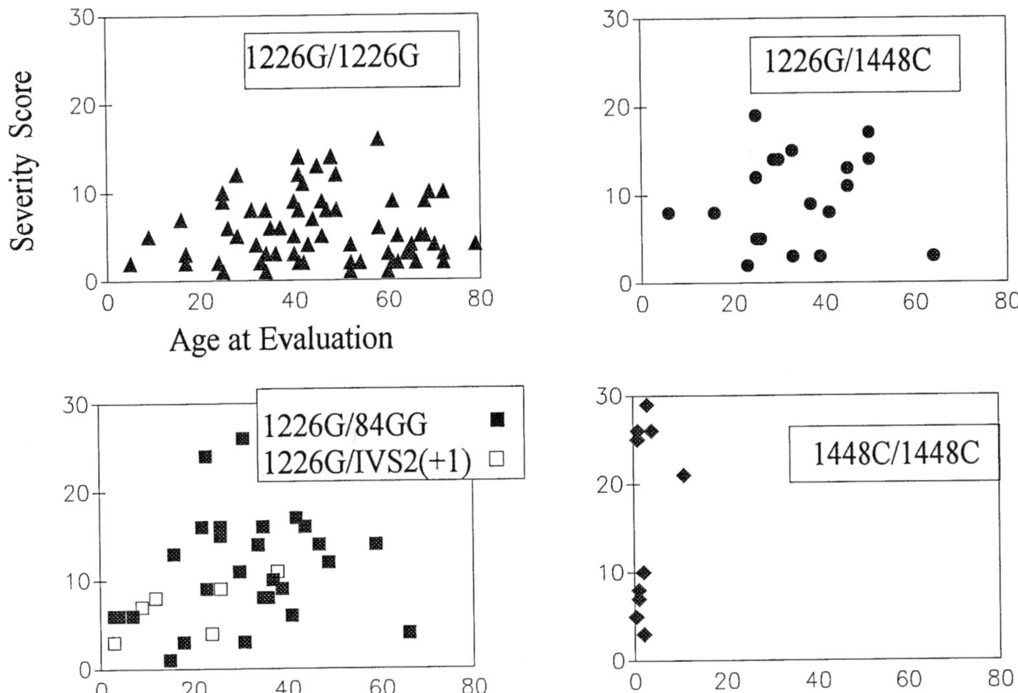


**Figure 1.** The relation between age at the time of evaluation and severity score of patients with four different Gaucher disease genotypes. The most severe disease is seen in the 1448C homozygotes, while many of the patients who were homozygous for the 1226G mutation are elderly and have very mild disease. The severity score is obtained by assigning points for various disease manifestations (e.g., for enlargement of the spleen, abnormally low blood counts, abnormal liver function tests, fractures of the bones, or involvement of the lungs). Thus, patients with a high score have severe multiorgan disease, while those with a score of zero have no disease manifestations at all.

mild disease with a late onset, while patients who also carry one of the more drastic mutations have more severe disease with early onset.

## 3 POPULATION GENETICS

Of the many mutations that have now been documented, only three appear to approach polymorphic frequencies. The most common is the A→G transition, at nt 1226, which produces a protein with an Asn→Ser substitution at amino acid 370 of the mature protein. It is present in about 6% of the Ashkenazi Jewish population. This mutation is the principal cause of the high incidence of Gaucher disease in this ethnic group. An insertion of a guanine at nt 84 of the cDNA is the second common mutation, found in approximately 0.6% of the Jewish population. Because it produces a frameshift even before the amino terminus of the mature protein, this drastic mutation produces no enzyme protein. A T→C transition at nucleotide position 1448, producing a Leu→Pro substitution at amino acid 444 of the mature protein, is present at a relatively high frequency in the Norrbottnian population of northern Sweden. The same mutation also occurs in other populations at low frequencies. It is noteworthy that the homologous position in the pseudogene is occupied by a C, just as in the mutation that produces Gaucher disease.

*See also* Human Genetic Predisposition to Disease.

## *Bibliography*

Beutler, E. (1988) Gaucher disease: New developments. In *Current Hematology and Oncology,* Vol. 6, V. F. Fairbanks, Ed., pp. 1–26. Year Book Medical Publishers, Chicago.

———. (1991) Gaucher's disease. *N. Engl. J. Med.* 325:1354–1360.

———. (1992) Gaucher disease: New molecular approaches to diagnosis and treatment. *Science,* 256:794–799.

Grabowski, G. A., Gautt, S., and Horowitz, M. (1990) Acid β-glucosidase: Enzymology and molecular biology of Gaucher disease. *Crit. Rev. Biochem. Mol. Biol.* 25:385–414.

Kohn, D. B., Nolta, J. A., Weinthal, J., et al. (1991) Toward gene therapy for Gaucher disease. *Hum. Gene Ther.* 2:101–105.

# Gel Electrophoresis of Proteins, Two-Dimensional Polyacrylamide

*Peter J. Wirth*

## *Key Words*

**Amphoteric** Describing a zwitterionic compound containing both positively and negatively charged groups within the same molecule, hence able to function as either acid or base.

**Carrier Ampholytes** Low molecular weight amphoteric polyaminopolycarboxylic acids (pI's from pH 2.5–11), which generate a continuous pH gradient in an electric field.

**Isoelectric Point (pI)** The pH value at which a protein exhibits no net charge (i.e., the molecule has an equal number of positive and negative charges), hence fails to migrate in an electric field.

**Western Blotting** Electrotransfer of separated proteins from gels to a thin support matrix, such as nitrocellulose, to which they bind and are immobilized.

---

The technique of high resolution, two-dimensional polyacrylamide gel electrophoresis (2D-PAGE) of cellular proteins provides an extremely powerful analytical technique for the simultaneous analysis of 2000–3000 individual proteins on a single electrophoretogram. Two-dimensional polyacrylamide gel electrophoresis is the combination of two independent electrophoretic techniques: isoelectric focusing, in which proteins are separated first on the basis of their intrinsic charge or isoelectric point, followed by electrophoresis in the presence of the detergent sodium dodecylsulfate, to separate proteins on the basis of their molecular weight. As a result, this form of electrophoresis has become the methodology of choice for the separation and analysis of complex protein mixtures and has found widespread applications in a broad range of biological systems. Uses include the analysis of phenotypic alterations in protein expression during both normal and abnormal growth and differentiation (including cancer development and other disease states), the identification of specific protein changes induced by mutagens and carcinogens, hormone treatment, mitogen stimulation, and nutrient changes, and the characterization of human and animal tissues and cells. In addition to these analytical applications, 2D-PAGE has recently been used in the isolation of extremely pure proteins for antibody production or for amino acid sequence analysis for subsequent molecular biological applications.

## 1 PRINCIPLE

As illustrated in Figure 1 protein mixtures (intact cells, tissues, or purified protein fractions) are solublized in a denaturing solution containing nonionic detergent(s), urea, a reducing agent (e.g., 2-mercatoethanol, dithiothreitol, etc.), and carrier ampholytes and applied to first-dimension cylindrical tube gels containing an appropriate mixture of carrier ampholytes. In an alternative procedure, commonly known as immobilized pH gradient (IPG) electrophoresis, solublized samples may be applied to ultrathin precast polyacrylamide gel strips containing a performed immobilized pH gradient. Sufficient voltage is applied to the tube gels or strips to induce the individual proteins to migrate to their respective pI values as determined by the protein's content of specific acidic and basic amino acid residues. The polyacrylamide gels are then extruded from the glass tubes using a transfer buffer containing sodium dodecylsulfate (SDS). As a result, the SDS binds to individual proteins and the tube gel is affixed directly onto the top of a second-dimension polyacrylamide slab gel containing SDS. When voltage is applied, the individual proteins, which are negatively charged as a result of their association with SDS, migrate from the cylindrical tube gels (gel strips) and into the slab gel, where they are further separated by molecular sieving according to their size (molecular weight). The resultant electrophoretogram is a two-dimensional map in which individual proteins (polypeptides) are displayed as discrete spots, with the slower migrating, higher molecular weight proteins located toward the top of the gel and the more acidic proteins to the left.

**Two-Dimensional Polyacrylamide Gel Electrophoresis (2D-PAGE)**

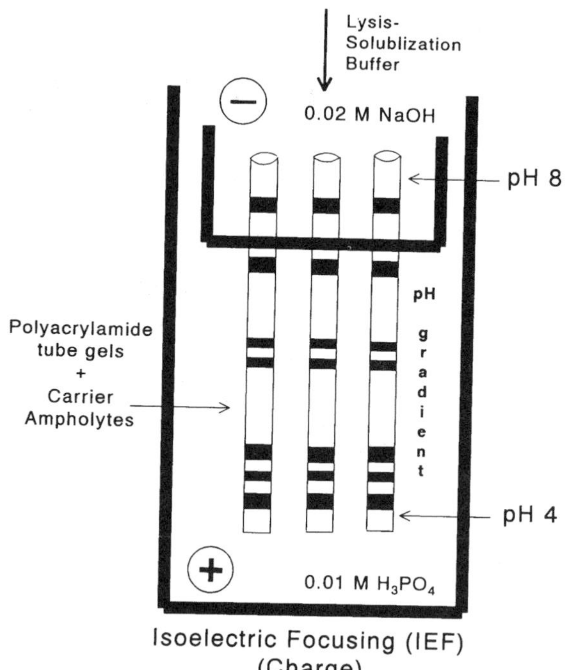


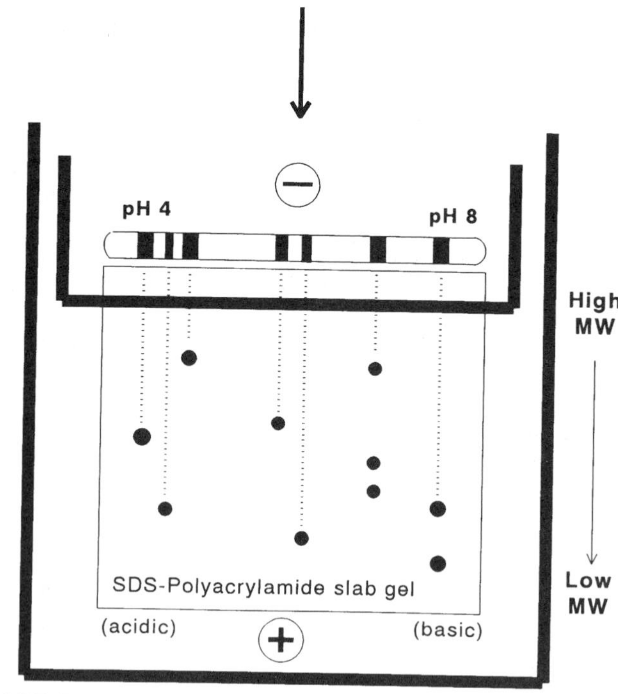


**Figure 1.** Schematic representation of first-dimension (IEF) and second-dimension (SDS-PAGE) procedures of 2D-PAGE.

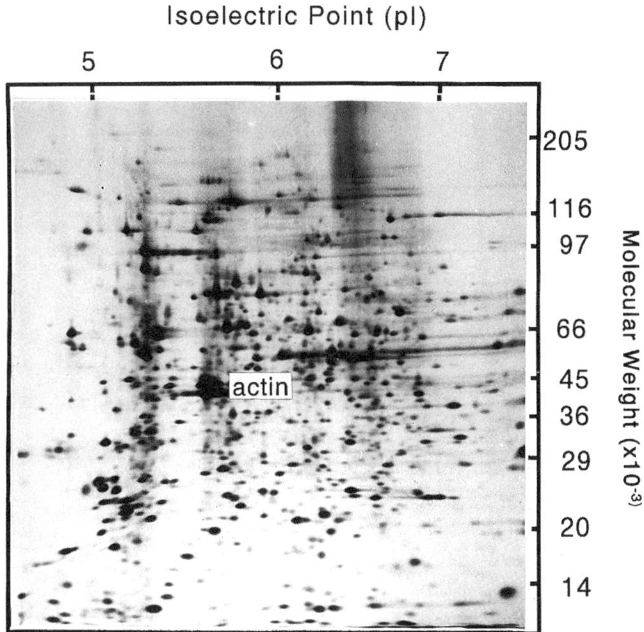


**Figure 2.** Autoradiograph of [$^{35}$S]methionine-labeled polypeptides from cultured rat liver epithelial cells obtained by two-dimensional polyacrylamide gel electrophoresis. Polypeptides were separated in the first dimension by IEF in the pH range of 4.5 (left) to 7.5 (right) and molecular weight range of 14 to 205 kDa.

Figure 2 is the autoradiogram of a typical 2D-PAGE separation of approximately 1000 [$^{35}$S] methionine-labeled whole-cell lysate proteins from cultured rat liver epithelial cells over the pH range of 5–7 and molecular weight of 14,000–205,000. Actin, (pI 5.7, MW 43,000), a ubiquitous structural protein found in a wide variety of cell types, is identified for reference. Although such maps appear extremely complex, the technique is highly reproducible, and routine analysis of the 2D-PAGE patterns can be performed by superimposing the photographic image of one gel over that of another on a light box. For more complex analysis, sophisticated computer-assisted image analysis programs are available.

## 2 TECHNIQUES

### 2.1 First Dimension: Isoelectric Focusing

First-dimension isoelectric focusing (IEF) is typically performed in 4.5% polyacrylamide tube gels (1 mm × 160 mm) containing 2% carrier ampholytes (1.6% pH range 4–8, 0.4% pH 3–10). Protein samples are dissolved in denaturing lysis buffer (9.5 M urea, 1% NP-40, 1% CHAPS, 10 mM dithiothreitol, and 2% carrier ampholytes) to a final protein concentration of 10 μg/μL and 10–100 μg of protein loaded at the basic end of the IEF tube. Samples are electrophoresed at room temperature at constant power (0.02 W/tube) for total of 13.5 kV·h. If IPG strips are utilized in the first dimension, flatbed horizontal electrophoresis is performed, usually for 100–200 kV·h.

### 2.2 Second Dimension: SDS–PAGE

Second-dimension SDS–PAGE electrophoresis is typically performed using vertical slab gels containing 0.1% SDS and a constant percentage of polyacrylamide, although gradient gels containing variable concentrations (e.g. 7–20%) are also used to increase resolution. Following separation in the first dimension, the IEF gel rods are carefully extruded from the glass tubes and affixed directly onto the surface of the second-dimension slab gel, using an SDS-containing transfer/equilibration buffer and held into place with 1% agarose. IPG strips are treated similarly, although a separate equilibration step of the gel strips in an SDS-containing buffer is necessary before transfer to second-dimension slab gels. Electrophoresis is performed at constant current of 20 mA/gel, using a running buffer of 0.192 M glycine, 25 mM Tris base, and 0.1% SDS. Following electrophoresis (4–5 h), individual polypeptides are routinely visualized using a variety of techniques including Coomassie Blue dye staining, silver staining, and autoradiography if cell lysates were metabolically labeled using radioactive amino acids [$^{35}$S, $^{14}$C, $^{3}$H, $^{32}$P, $^{125}$I] prior to electrophoresis, thereby providing the extremely high degree of sensitivity of 2D-PAGE.

## 3 ANCILLARY TECHNIQUES

Practically all analytical techniques applicable for use following 1D-PAGE can similarly be utilized in 2D-PAGE. Two of the most popular procedures employed in combination with 2D-PAGE are Western blotting and immunoblot analysis, using monoclonal and polyclonal antibodies for the identification and analysis of individual proteins. This combination technique has been and continues to be the method of choice to identify specific proteins on the two-dimensional maps. Although the vast majority of applications to date have been for analytical purposes, two new and exciting applications utilizing 2D-PAGE are rapidly emerging. The technique is beginning to be used more and more frequently for the "micropreparative" isolation of extremely pure polypeptides in sufficient quantities for the generation of both monoclonal and polyclonal antibodies. In addition, recent analytical advances in the area of protein microsequencing have now made it possible to obtain N-terminal and internal microsequence information for individual polypeptide spots isolated directly from 2D-PAGE gels. This information can be used for the synthesis of appropriate oligopeptides for antibody preparation, as well as for the synthesis of appropriate oligonucleotide probes for cDNA isolation, gene cloning, and subsequent genetic analysis. These extremely exciting applications provide a valuable link between protein biochemistry and molecular biology.

*See also* HPLC of Biological Macromolecules; Protein Purification; Proteins and Peptides, Isolation for Sequence Analysis of.

## *Bibliography*

Celis, J. E. and Bravo, R. (1984) *Two-Dimensional Electrophoresis of Proteins: Methods and Applications.* Academic Press, New York.

Dunbar, B. S. (1987) *Two-Dimensional Electrophoresis and Immunological Techniques.* Plenum Press, New York.

Hames, B. D., and Rickwood, D. (1990) *Gel Electrophoresis of Proteins: A Practical Approach.* IRL Press, New York.

# Gene Expression, Regulation of

*Göran Akusjärvi*

*Key Words*

**Epigenetic Modification** Denotes changes in the phenotype that are not due to alterations in the genotype (e.g., mutations in the DNA).

**Exon** Eukaryotic genes are encoded in discontinuous segments, where the coding portions are interrupted by noncoding sequences of unknown function (see **introns**). Both exonic and intronic sequences are transcribed into a nuclear precursor RNA. The segments of a eukaryotic gene that are preserved in the mature mRNA are the exon sequences. Prokaryotic genes usually are not split and, thus, are encoded by a continuous DNA sequence.

**Intron** Represents a segment of DNA that is transcribed into RNA in the nucleus of the eukaryotic cell but excised by RNA splicing before the mature mRNA has been exported to the cytoplasm.

**Nucleosome** The basic structural subunit used to condense DNA in a cell. The nucleosome consists of approximately 200 base pairs of DNA wrapped around a protein core made up by histone proteins.

**TATA-Binding Protein (TBP)** As the name indicates, the protein that binds the TATA box.

**TATA Box** A conserved TATAAA sequence found about 25 to 30 base pairs upstream of the transcription initiation site in eukaryotic RNA polymerase II promoters (a similar sequence element is also found in prokaryotic promoters, the −10 element). The TATA box binds the general transcription factor TFIID and helps position the RNA polymerase for correct initiation.

**Transcription Factor D for Polymerase II (TFIID)** A general transcription factor that interacts with the TATA box. It consists of TBP, which makes contact with the TATA box, and several TBP-associated factors (TAFs), which are required for regulation of transcription.

**Uridine-Rich Small Nuclear Ribonucleoprotein Particles (U snRNPs)** The nuclei of eukaryotic cells contain large quantities of several of the so-called U snRNPs. The U1, U2, U4, U5, and U6 snRNPs have all been shown to be involved in RNA splicing. Other U snRNPs serve other functions in other processes.

Genetic information is transmitted between generations of a species in the form of a stable DNA molecule. This molecule is replicated before cell division to ensure that all offspring receive the same genetic constitution. Expression of the genetic information has been summarized in the so-called central dogma, which postulates that the flow of genetic information in a cell goes from DNA to RNA to protein.

Organisms are divided into two major groups, depending on whether their cells possess a nucleus. One group consists of the prokaryotes, which include the bacteria and the blue-green algae. These cells do not have a nucleus. The other group consists of the eukaryotes, which include animals, plants, and fungi. Every eukaryotic cell has a nucleus, which encapsulates the DNA. The mechanisms to regulate gene expression in the two groups of organisms are similar, although eukarotyes generally use more complex regulatory pathways. In prokaryotes, on–off switches of transcription appear to be the key mechanism of control of gene activity, although other mechanisms such as transcriptional attenuation, transcriptional termination, translational control, and mRNA and protein turnover play a significant role in the control of specific genes. In eukaryotes similar mechanisms are in operation. In addition, an increasing number of eukaryotic genes now are shown to be regulated at the level of RNA processing.

## 1 DEFINING A TRANSCRIPTION UNIT

A transcription unit represents the combination of DNA segments that together constitute an expressible unit, whose expression leads to synthesis of a functional gene product(s) that often is a protein but also may be an RNA molecule. In prokaryotes, proteins in a specific metabolic pathway are often encoded by genes that are clustered at the DNA level and transcribed into one polycistronic mRNA that is used to translate the different proteins. In contrast, eukaryotic mRNA are usually functionally monocistronic, encoding only one protein product. Transcription involves synthesis of an RNA chain that is identical in sequence to one of the two DNA strands. Transcription can be subdivided into three stages:

I Initiation, which begins when RNA polymerase binds to the double-stranded DNA molecule and incorporates the first nucleotide(s).

II Elongation, the phase during which the RNA polymerase moves along the DNA template and extends the growing RNA chain by adding one nucleotide at the time.

III Termination, the stage in which RNA synthesis ends and the RNA polymerase complex disassembles from the transcription unit.

This entry covers only RNA synthesis and the maturation of protein encoding messenger RNAs.

### 1.1 Regulation of Transcription in Eukaryotes

Eukaryotic cells contain three RNA polymerases, designated I, II, and III, which are responsible for synthesis of specific RNA molecules within the cell. RNA polymerase I is responsible for synthesis of ribosomal RNA, RNA polymerase II is responsible for synthesis of protein coding mRNA, and RNA polymerase III takes care of transfer RNA and 5S RNA synthesis. The three polymerases are large enzymes consisting of approximately 10 subunits each. Some subunits are shared between the different polymerases, whereas others are unique and probably determine the specificity of the transcription process. Recent data have also shown that the different classes of polymerases use some common transcription factors. For example, the TATA binding protein (TBP), which originally was thought to be specifically used in RNA polymerase II transcription, now has been shown to be required for transcription by all three classes of polymerases, although only RNA polymerase II genes contain a binding site for TBP.

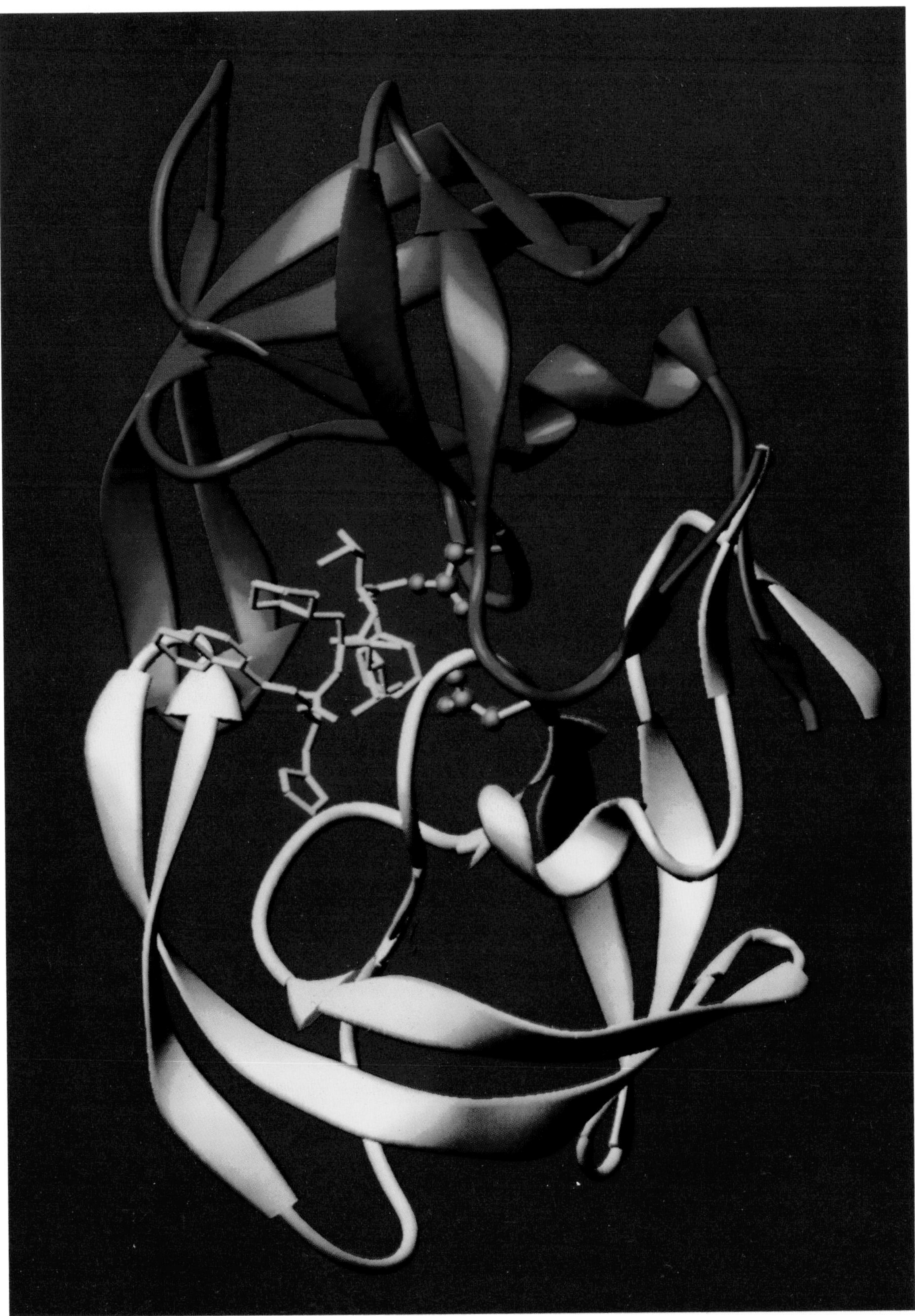

**Plate 1.** Computer graphics image of the HIV protease complexed with an inhibitor, U-75875. The chains of the enzyme are in two different colors, the inhibitor is shown in bond representation, and the active site aspartates are shown in the ball-and-stick format. (See AIDS, Inhibitor Complexes of HIV-1 Protease in, Figure 1.)

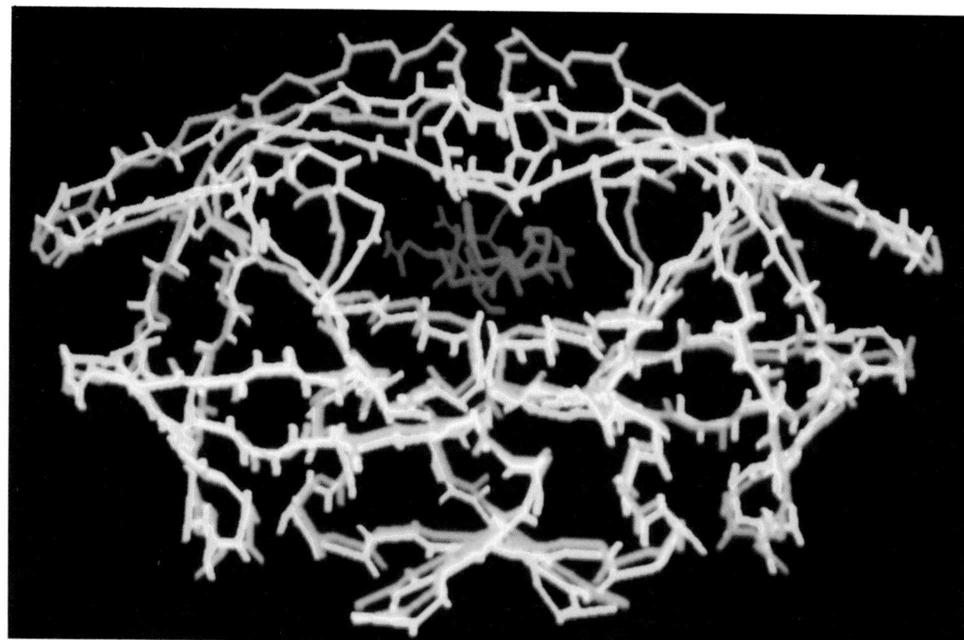

Plate 2. The α-carbon backbone of the homodimer of HIV-1 protease. Apoenzyme is shown in blue. The inhibited enzyme is in yellow complexed with an inhibitor in red. (Courtesy of Dr. K. Appelt.) (SEE AIDS HIV ENZYMES, THREE-DIMENSIONAL STRUCTURE OF, FIGURE 1.)

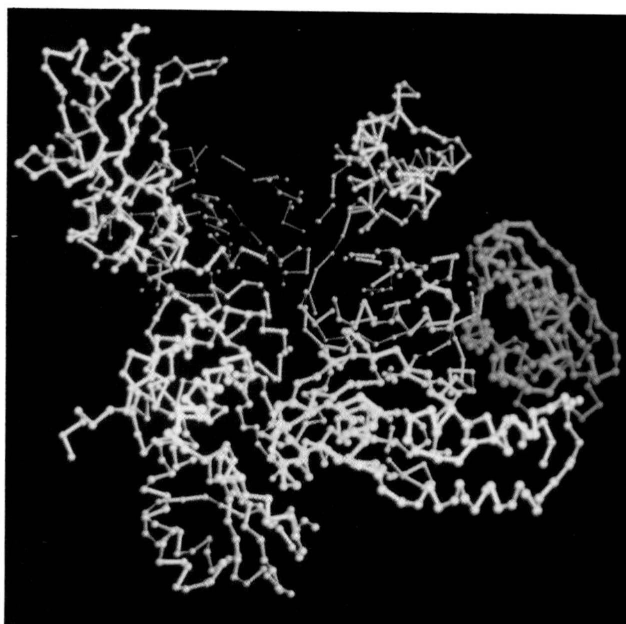

Plate 3. The α-carbon backbone of the heterodimer of HIV-1 reverse transcriptase. Polymerase domains of p66 and p51 are folded differently into four separate subdomains: in blue, terminal subdomain—finger: in purple, catalytic subdomain—palm: in yellow, connecting region between polymerase and RNase H domains: in green, flexible region between connecting and catalytic subdomains—thumb. The RNase H domain is orange. [Reprinted with permission from L. A. Kohlstaedt et al., *Science*, 256:1783–1790 (1992). Copyright 1992 by the American Association for the Advancement of Science.] (SEE AIDS HIV ENZYMES, THREE-DIMENSIONAL STRUCTURE OF, FIGURE 2.)

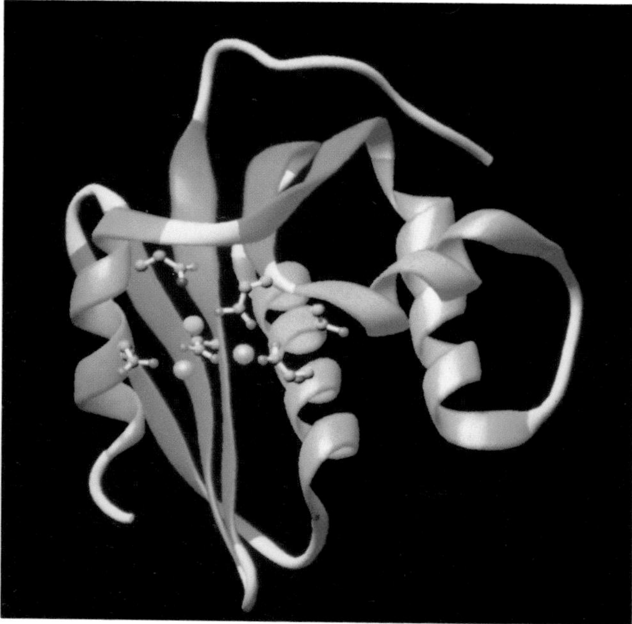

Plate 4. Ribbon representation of the overall fold of the RNase H domain of HIV RT: β-strands are in blue and helices in yellow. Manganese binding sites (silver balls) are shown with respect to the seven invariant residues, which cluster near the catalytic site. Proposed phosphate site is shown as a yellow ball. (Courtesy of Dr. D. Matthews.) (SEE AIDS HIV ENZYMES, THREE-DIMENSIONAL STRUCTURE OF, FIGURE 3.)

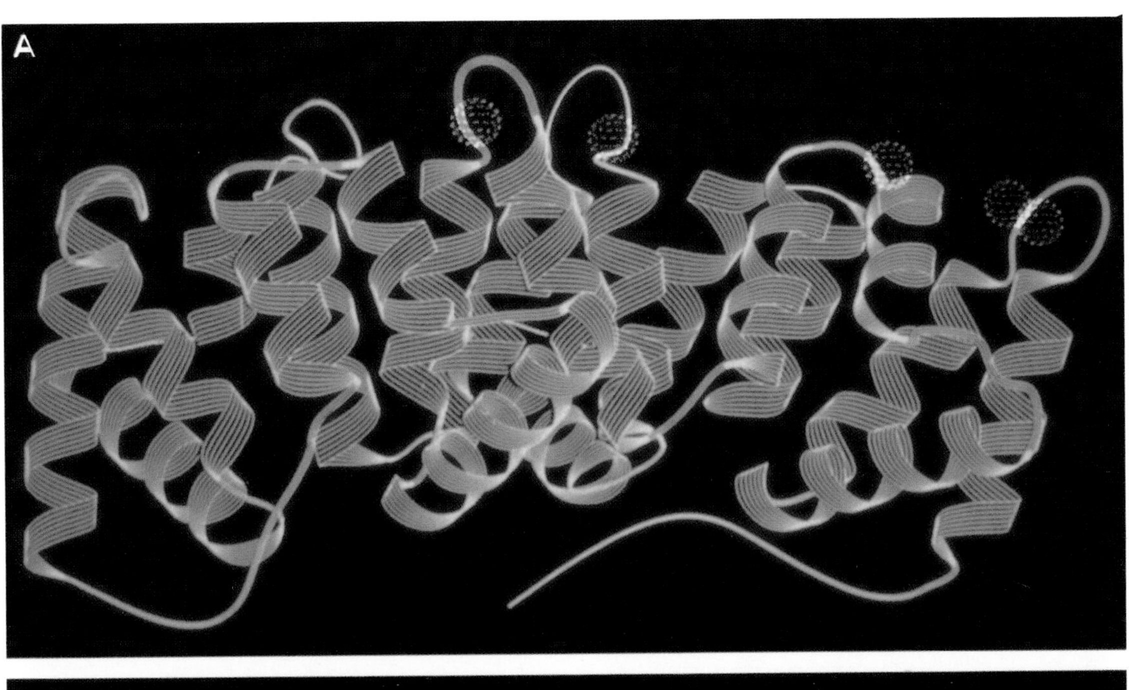


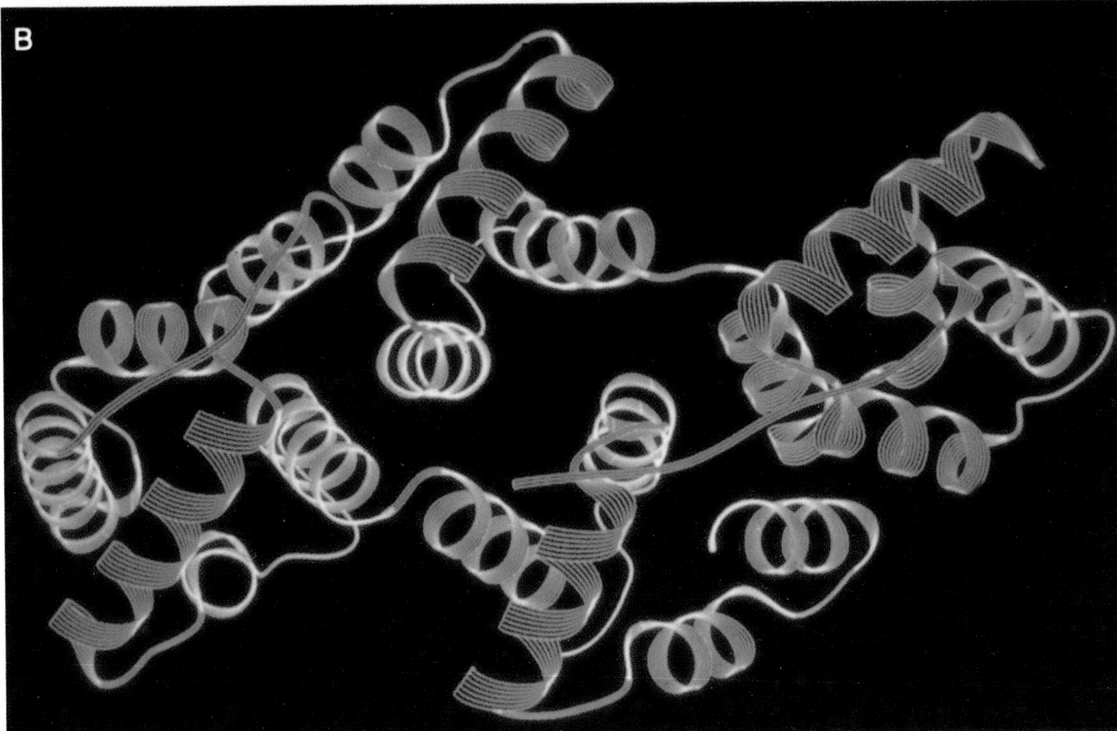


Plate 5. Ribbon plots of the structure of annexin V as determined by X-ray diffraction. (A) "Side" view of the molecule with the calcium-binding sites and the membrane-binding face at the top. Calcium ions are represented by the red spheres. The high affinity sites are represented by the first, second, and fifth spheres from left to right; the third and fourth spheres indicate low affinity ion-binding sites that were identified by lanthanide binding. The extended N-terminus is seen at the bottom. (B) View of the "cytoplasmic" side of the molecule and the N-terminus. The calcium-binding sites are on the opposite face. Note the potential channel structure in the center of the molecule. [Creutz (1992) with permission. Original figure kindly provided by A. Burger and R. Huber.] (SEE ANNEXINS, FIGURE 2.)

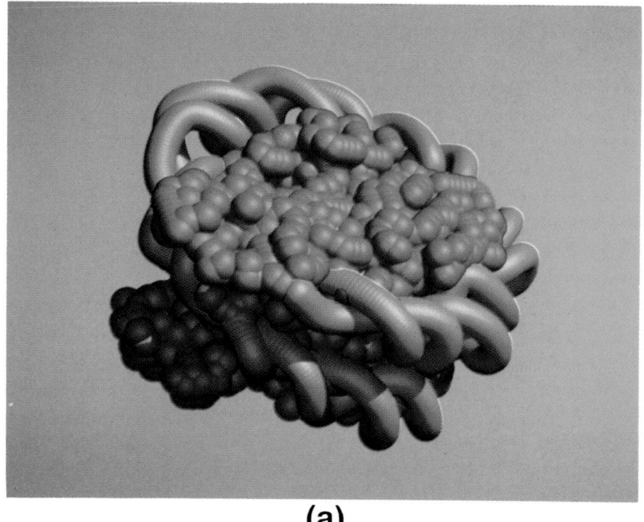

(a)

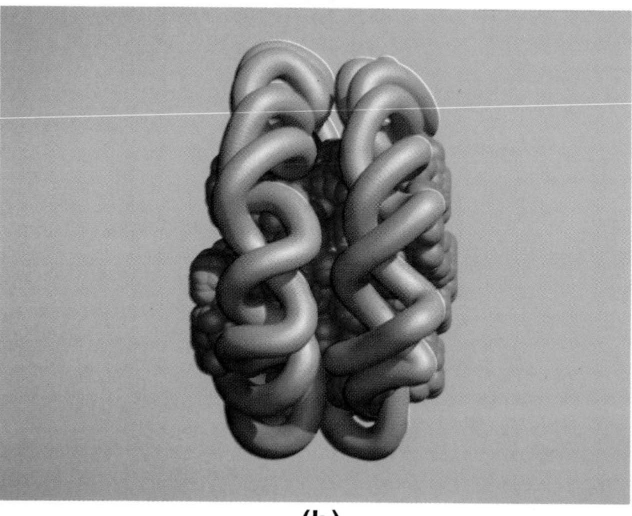

(b)

**Plate 6.** The X-ray crystal structure of the nucleosome core particle at 7.Å resolution showing that the 146 bp of DNA in the particle are wrapped in 1.8 turns of a flat superhelix around the histone octamer. The electron density of the core histone proteins is represented by overlapping green spheres. The paths of the phosphodiester chains of the DNA double helix are shown as intertwined gold tubes. (a) View oblique to both the DNA superhelix axis and the molecular dyad axis. The center of nucleosomal DNA, the binding site of the linker histone, is surrounded above and below by protrusions from the histone core. (b) View down the molecular dyad axis, passing between the gyres of the DNA superhelix. [Adapted from T. J. Richmond, J. T. Finch, B. Rushton, D. Rhodes, and A. Klug, *Nature* 311:532 (1984).] (SEE CHROMATIN FORMATION AND STRUCTURE, FIGURE 1.)

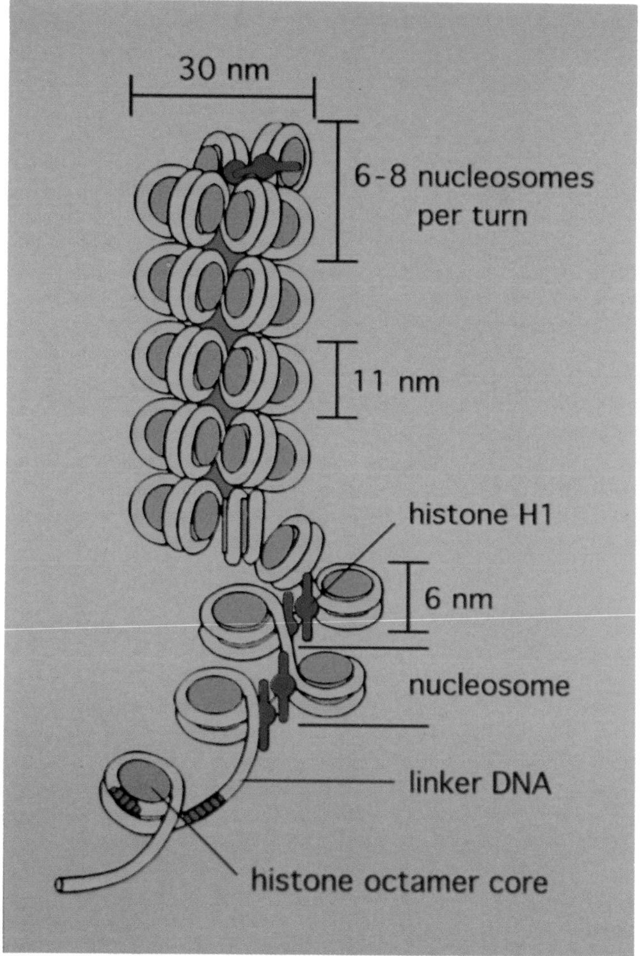


**Plate 7.** Model of the chromatin fiber built from nucleosomes in a higher order helix. The disklike nucleosomes associate to form a higher order helix with 6 to 8 nucleosome units per turn. The colors of the DNA and core histone proteins are the same as in Plate 6. The interactions stabilizing the fiber come from the histone H1, shown in red, thought to be at the center of the helix, and possibly also from the histone tail regions. [Adapted from F. Thoma, T. Koller. and A. Klug, *J. Cell Biol.* 83:403 (1979).] (SEE CHROMATIN FORMATION AND STRUCTURE, FIGURE 2.)

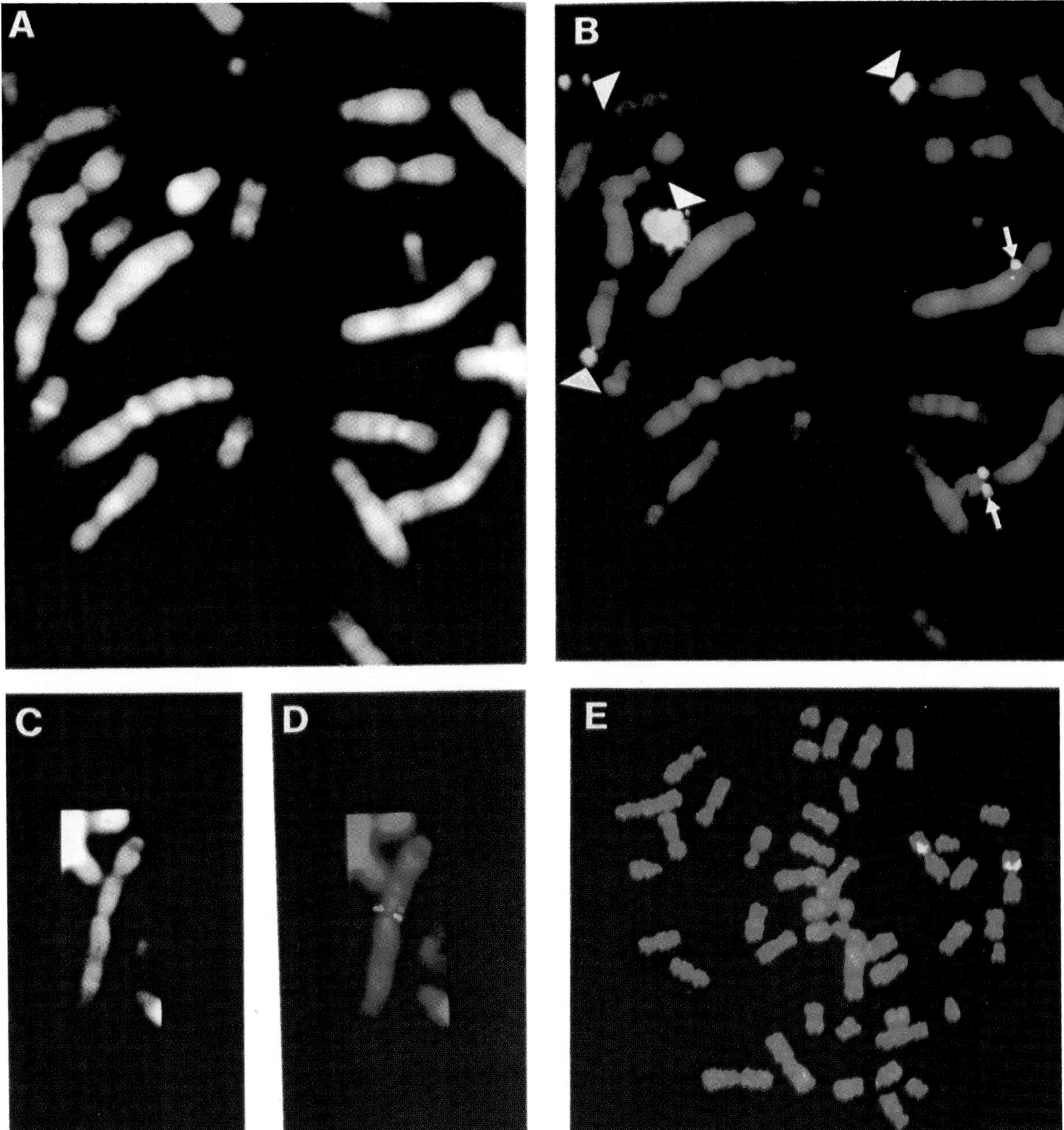


Plate 8. (A) Partial metaphase of Hoechst 33258 stained human chromosomes. This fluorescent dye preferentially binds to AT-rich stretches of DNA creating a G-(Q-) like banding pattern. (B) The same image after computer alignment, merging, and assignment of false colors. The hybridization signal for a small single-copy probe is seen on both homologues (small arrows). The large signal obtained with a repeat sequence probe localized to the acrocentric chromosomes is shown (arrow head) for comparison. Images were obtained using an epifluorescence microscope, RCA-ISIT camera, BioRad MRC 500 workstation, and a Sony video printer. (C) DAPI-stained Q-banded metaphase chromosome. (D) Same image as C, showing two single-copy probes labeled with Fitc (green), and Texas red (red), respectively, localized relative to each other on the DAPI stained (blue) chromosome. Images were obtained using an epifluorescence microscope CCD camera, and BioRad MRC 500 workstation. The photographs were generated by sequential photography from the computer color monitor screen. (E) Fluorescent dye DAPI used in conjunction with PI gives an enhanced R-banding pattern (red). The image is merged and assigned false colors using appropriate computer software. The yellow Fitc hybridization signal is that of a YAC probe. All four sister chromatids are consistently labeled with such large probes. The image was obtained using a confocal microscope, BioRad MRC 600 workstation, and screen photography. (SEE GENE MAPPING BY FLUORESCENCE IN SITU HYBRIDIZATION, FIGURE 3.)

**Plate 9.** Ordering three closely linked cosmid clones using hybridization to the extended chromatin fiber released by alkali treatment of fixed nuclei. The cosmids, each approximately 40 kb long, are differentially labeled by Texas Red (red), FITC (green), and a 50:50 mixture of Texas Red and FITC (giving an orange-yellow color). There is a 5 kb overlap between the red and green labeled cosmids indicated by the yellow signal. The yellow cosmid is separated from the red cosmid by a 50 kb gap. (SEE GENE ORDER BY FISH AND FACS, FIGURE 2.)

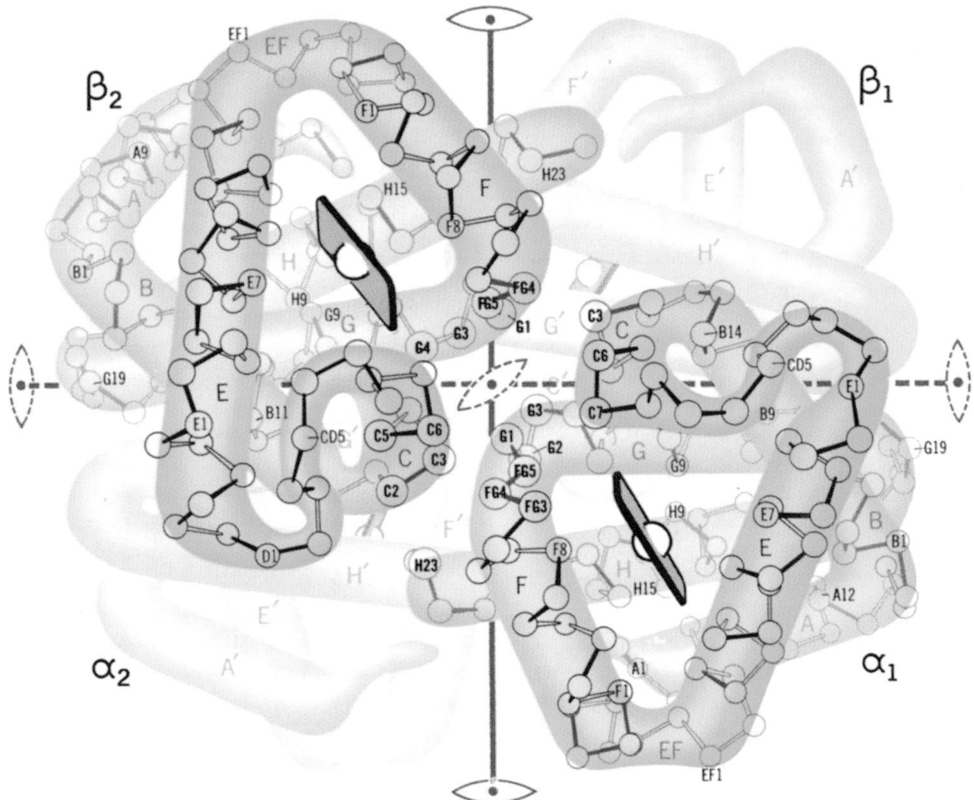


Plate 10. Detailed view of ferric Hb, which is similar to the oxy derivative. Only the α carbons of the main chains are shown. Those whose side chains are involved in contacts between subunits are given boldfaced numbers in large circles. There are two kinds of unlike subunit contact: $\alpha^1\beta^1$ (or $\alpha^2\beta^2$ ) and $\alpha^1\beta^2$ (or $\alpha^2\beta^2$) the first interface being more extensive than the second. Whereas $\alpha^1\beta^1$ contact involves the B, C and H helices and the GH corner, $\alpha^1\beta^2$ contact concerns mainly helices C and G and the FG corner. (Illustration copyright by Irving Geis, from R. E. Dickerson and I. Creis (1983) *Hemoglobin: Structure, Function, Evolution and Pathology*. Benjamin/Cummings, Menlo Park, CA.) (SEE HEMOGLOBIN, FIGURE 1.)

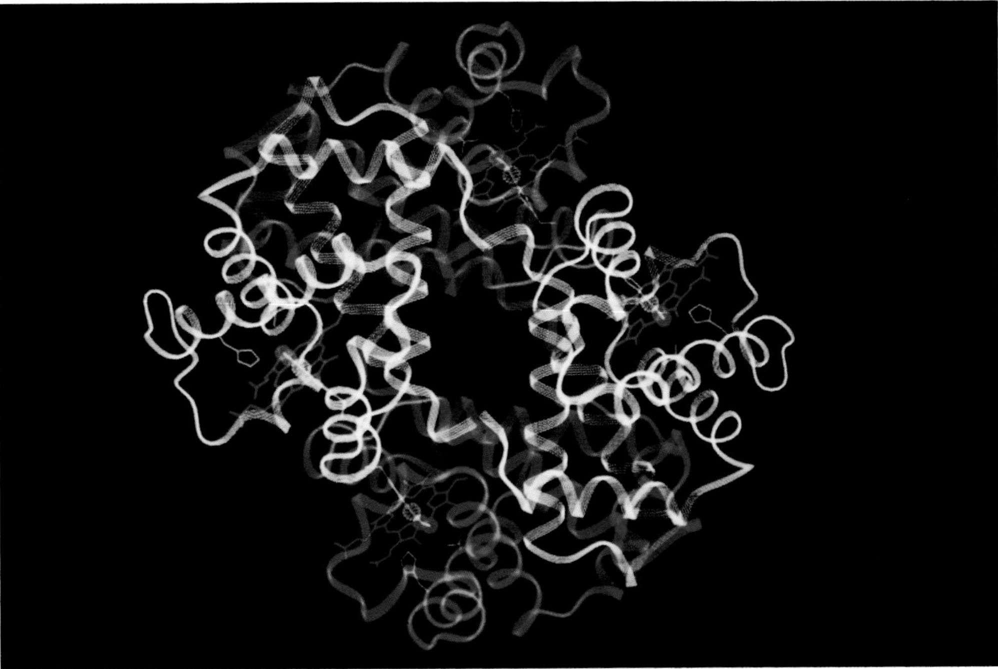

Plate 11. Structure of hemoglobin. The two α-chains are shown as white ribbons, the two β-chains as blue ones. The four heme groups are shown in red, with the central iron atom in green; bound oxygen can be seen as two red spheres close to each heme group. Amino acid residues in vicinity to the heme are shown in green. (SEE HEMOGLOBIN, GENETIC ENGINEERING OF, FIGURE 1.)

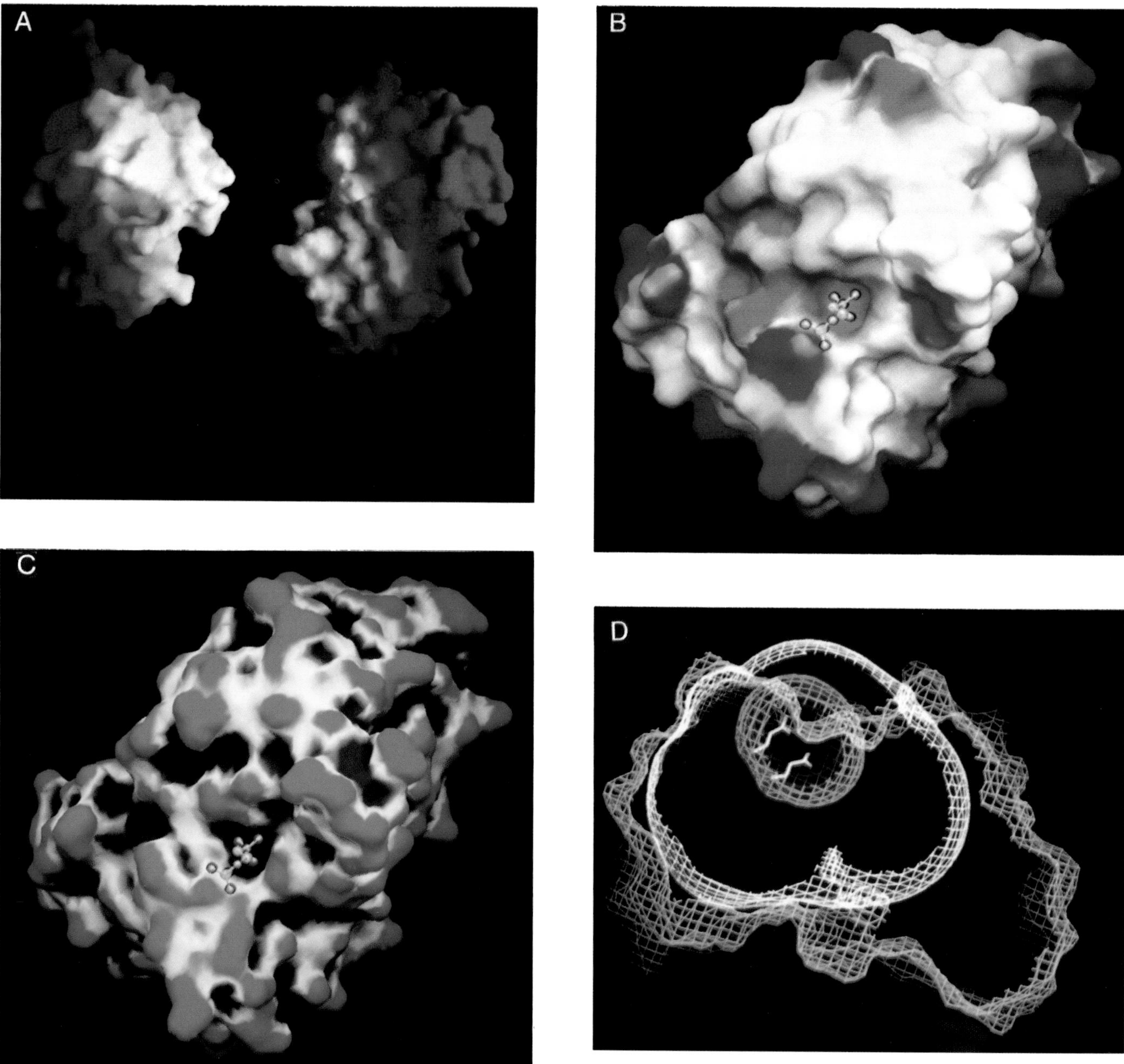


**Plate 12.** (A) Surface representations of the protein antigen (hen egg-white lysozyme, left) and its specific antibody HyHEL-10 Fv fragment (right). The protein surfaces, originally in tight noncovalent interaction, have been pulled apart and color-coded by interprotein atom-atom distances (white to red: close distance to long distance). The picture emphasizes the global convex shape of the antigen surface and its complimentary, valleylike antibody surface. Color graphics in (B) and (C) prepared with the program GRASP (A. Nicholls). (B) Fv fragment of the antiphosphorylcholine antibody McPC 603, color-coded by electrostatic potential (positive to negative: blue to red). The electrostatic field at the binding site is seen to complement well the (opposite) charges on the ligand. (C) Same as (B) except that the Fv fragment surface has been color-coded by surface curvature (convex to concave: green to black). The hapten phosphorylcholine (ball-and-stick representation) is seen in the binding site cavity. (D) Comparison of 1 kT/e electrostatic potential contours generated by a pair of Glu residues H35 and H50 of the HyHEL-5 Fv fragment in the protein and in free solution. In this cross-sectional view, the solvent-excluded surface of the HyHEL -5 Fv fragment is green; the 1 kT/e contour of the aqueous electrostatic potential is red; and the 1 kT/e contour of the potential in the presence of the low dielectric, uncharged protein surrounding the two glutamate side chains is magenta. The figure demonstrates how the low dielectric medium of protein interior, paired with the concave shape of the binding cavity, enhances and modulates the shape of the electrostatic field emanating from the protein. [Color graphics, generated with the program INSIGHT (Biosym Technologies, Inc.), reprinted with permission from J. Novotny and K. Sharp. "Electrostatic fields in antibodies and antibody/antigen complexes." *Prog. Biophys. Mol. Biol.* 58:203–224 (1992).] (SEE IMMUNOLOGY, FIGURE 5.)

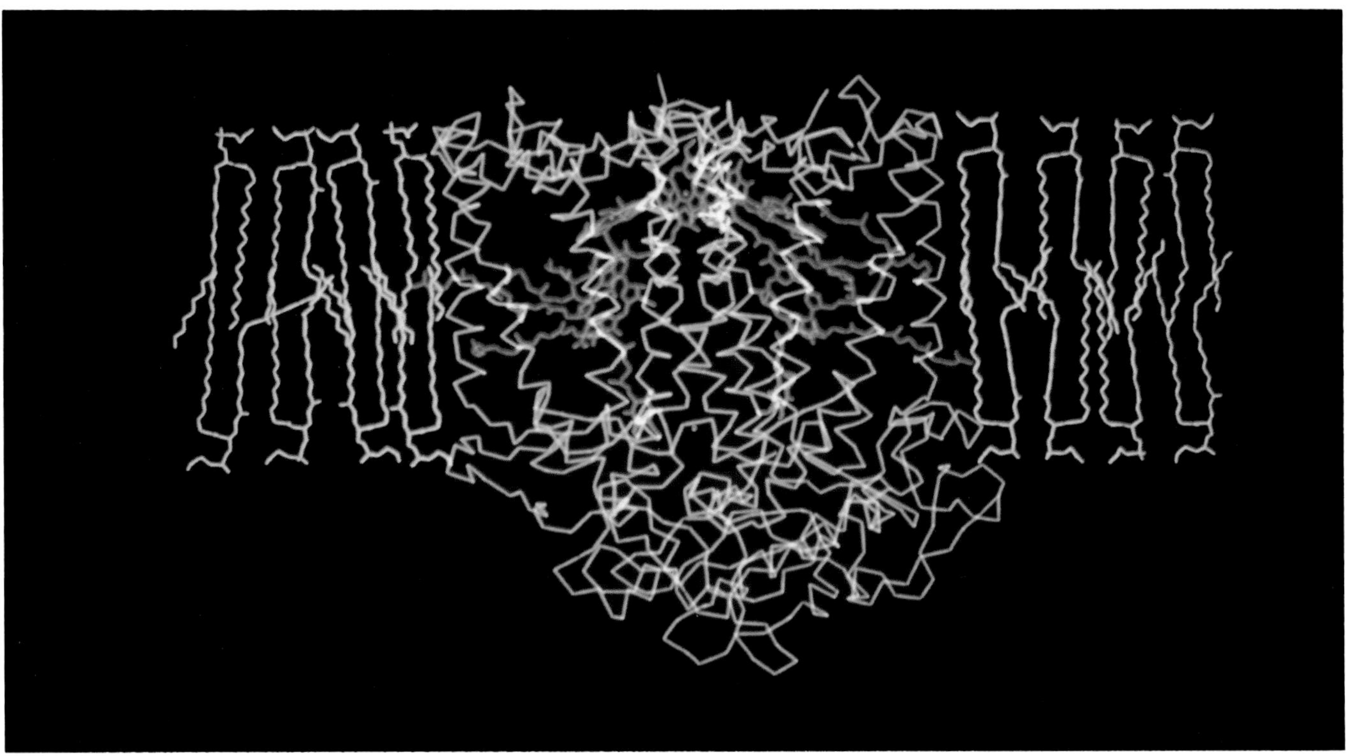

**Plate 13.** Structure of the reaction center of *Rhodobacter sphaeroides* determined from X-ray diffraction analysis. The complex consists of three protein subunits: the L and M core proteins (blue) and the H protein (green). The cofactors involved in excitation energy capture and electron transfer reactions (bacteriochlorophylls, bacteriopheophytins, and quinones) are shown in red. [Reproduced with permission from Yeates et al., *Proc. Natl. Acad. Sci. U.S.A.* 84: 6438-6442 (1987).] (SEE PHOTOSYNTHETIC ENERGY TRANSDUCTION, FIGURE 2.)

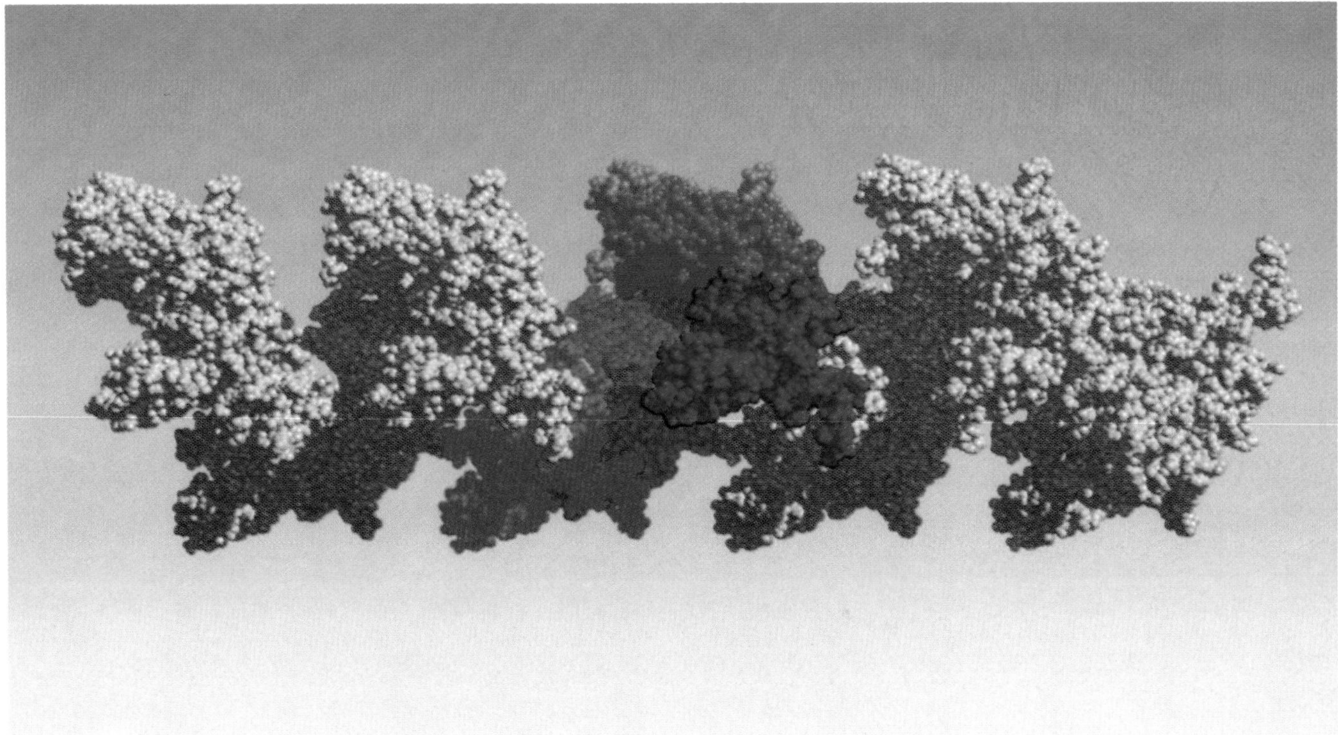

**Plate 14.** Representation of a segment of a helical filament formed by 24 monomers of RecA protein. Six monomers (one helical turn) are colored. [Illustration based on the structure determined by Story el al. (1992) and developed from the coordinates deposited in the Brookhaven Protein Data Bank, using the MidasPlus software obtained from the Computer Graphics Laboratory at the University of California, San Francisco.] (SEE RECA PROTEIN, STRUCTURE AND FUNCTION OF, FIGURE 3.)

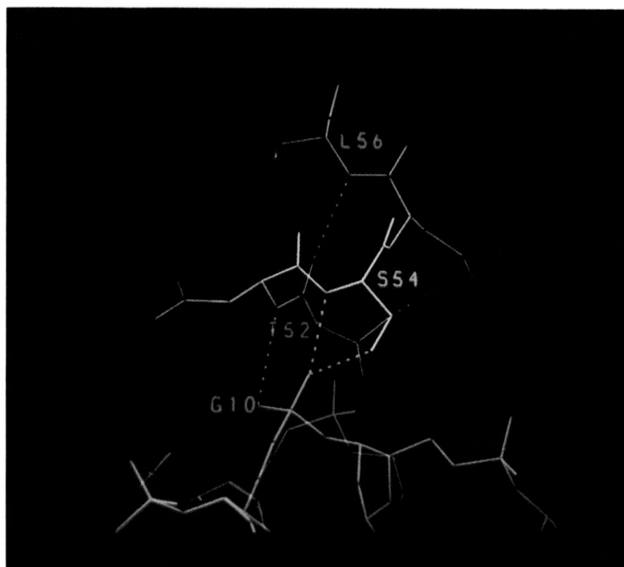


**Plate 15.** Distortion of the DNA phosphate backbone in a MetJ repressor-operator complex. A portion of a protein α-helix (yellow) makes hydrogen bonds (dashed) to oxygens of a DNA phosphate group (blue). The phosphate backbone is distorted, as shown by the superimposed structure of regular B-form DNA (red), where the protein-phosphate contacts could not be made. The relative ease of such distortion of the DNA is sequence-dependent. [From *Nature*, 359: 387–393 (1992).] (SEE REPRESSOR–OPERATOR RECOGNITION, FIGURE 2.)

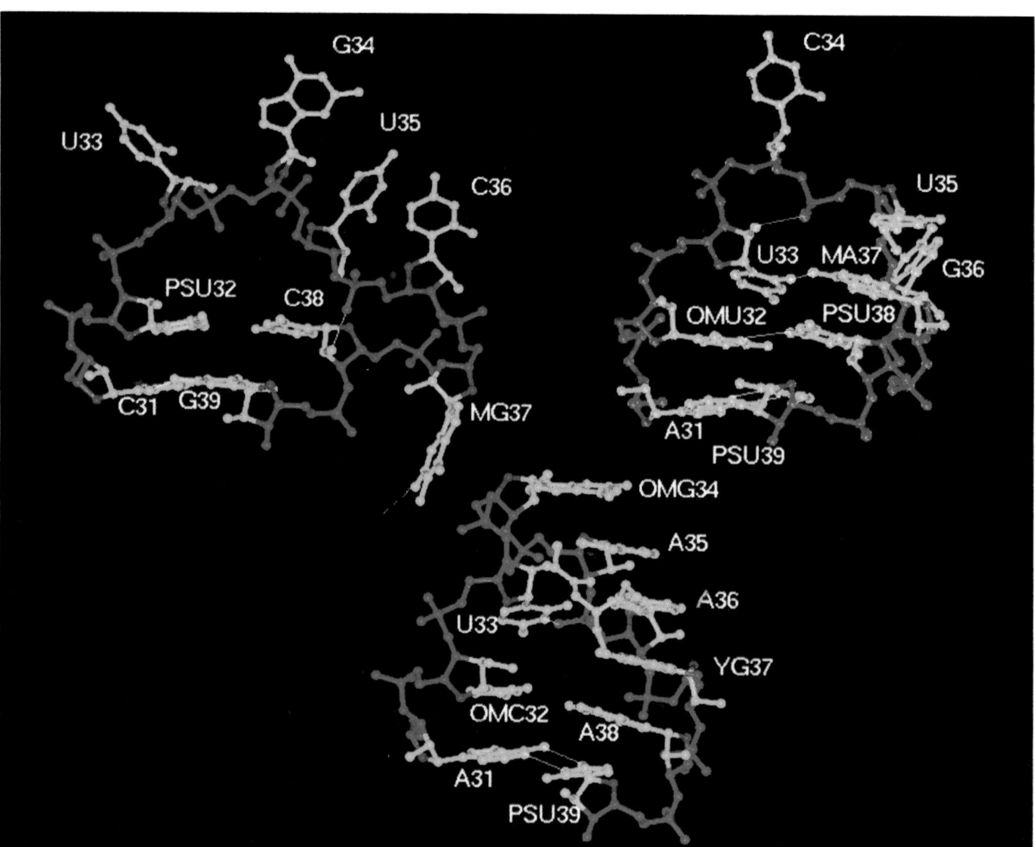


**Plate 16.** Anticodon hairpin loops from three different crystal structures. *Upper left*: the anticodon loop from the $tRNA^{Asp}$–cognate synthetase cocrystal structure; *upper right*, that from the $tRNA^{Gln}$–cognate synthetase cocrystal structure; *bottom*, that from the uncomplexed $tRNA^{Phe}$ structure. For each structure, the last Watson-Crick base pair from the helix is at bottom. Hydrogen bonds are drawn as thin white lines, while the backbone is shown in green and the bases in yellow; 04′ atoms are shown in red to highlight the local direction of the backbone. For example, note the abrupt turns between residues 33 and 34 ($tRNA^{Phe}$; bottom) and between residues 36 and 37 ($tRNA^{Asp}$; upper left). (SEE RNA STRUCTURE, NONHELICAL, FIGURE 2.)

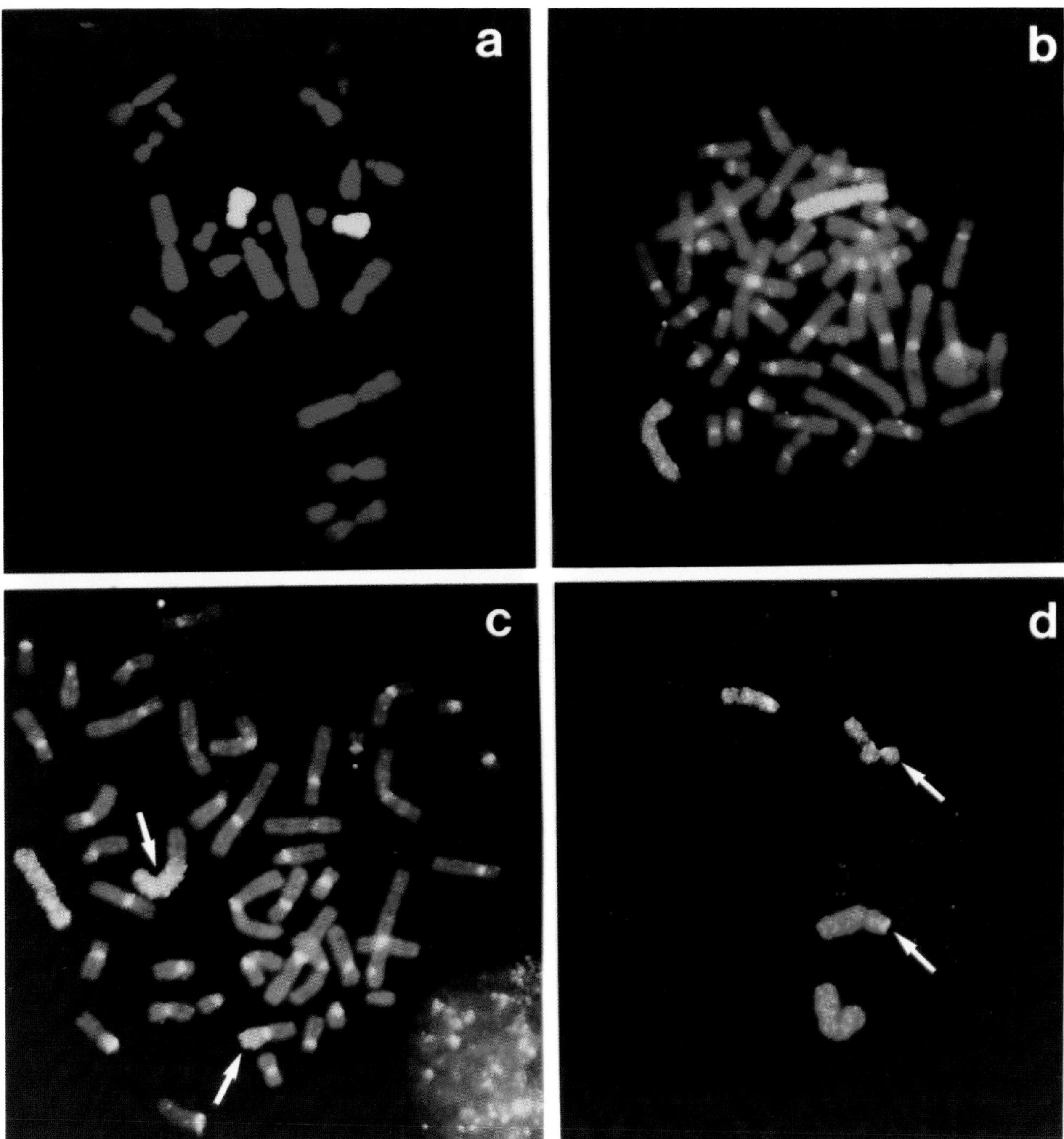


Plate 17. Chromosome-specific staining using FISH. (a) Human chromosomes specifically stained in metaphase spreads from human x hamster hybrid cells by hybridization with human genomic DNA. Hybridization signals appear yellow. Chromosomes to which the probe did not hybridize appear red. (b) Both copies of human chromosome 4 in a normal human metaphase spread stained by hybridizing a WCP for chromosome 4; staining was the same as for (a). Centromeric regions were simultaneously stained with a probe that hybridizes to repeated DNA in the centromeric region of all chromosomes. (c) FISH to a metaphase spread from an irradiated cell population using a WCP to chromosome 4. A radiation-induced exchange between one copy of chromosome 4 and another chromosome is apparent. The derivative chromosomes appear red and yellow (arrows). The centromeres were stained as described in (b) to allow discrimination between translocations (one centromere) and dicentrics (two centromeres) (d) FISH with WCPs to chromosomes 7 (red) and 12 (green) to a metaphase spread prepared from human tumor. Derivative chromosomes resulting from a t(7;12) translocation are visible (arrows) as red and green chromosomes (Photomicrograph courtesy of M. Vooijs.) (See Whole Chromosome Complementary Probe Fluorescence Staining, Figure 1.)

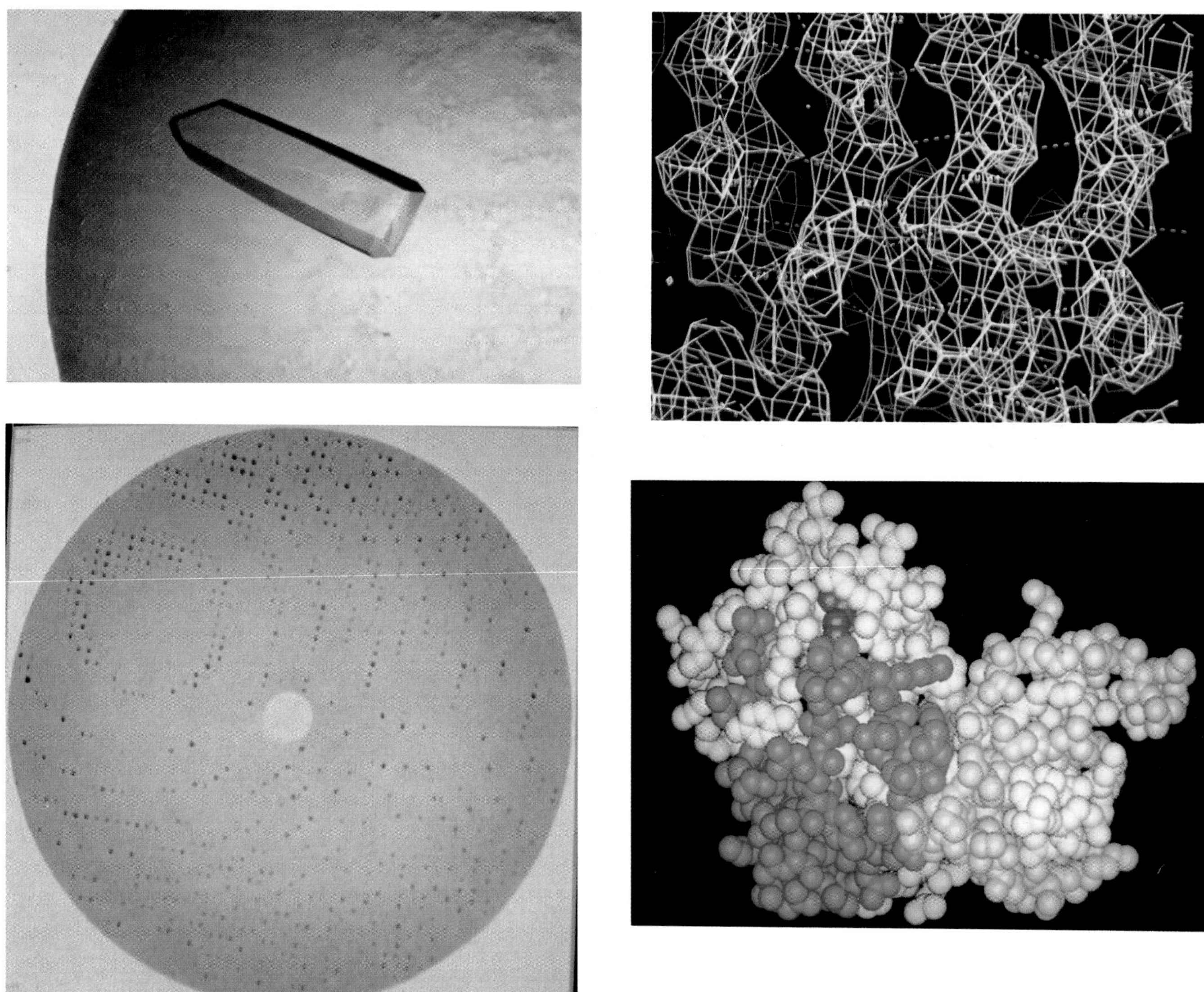

**Plate 18.** Stages involved in the structure determination of a protein molecule [e.g., toxic shock syndrome toxin 1 (TSST-1), a microbial superantigen]. (Top left) Single crystal of TSST-1. (Top right) Diffraction picture of TSST-1 crystal recorded using MAR Research Imaging Plate system (wavelength 1.5418 Å). The data extend to 2.5 Å resolution. (Bottom left) Part of the 2.5 Å resolution electron density map of TSST-1 with atomic positions superimposed. The region displayed shows four strands of a β-sheet. (Bottom right) Space-filling diagram of TSST-1 (drawn using the 2.5 Å crystal structure). (SEE X-RAY DIFFRACTION OF BIOMOLECULES, FIGURE 1.)

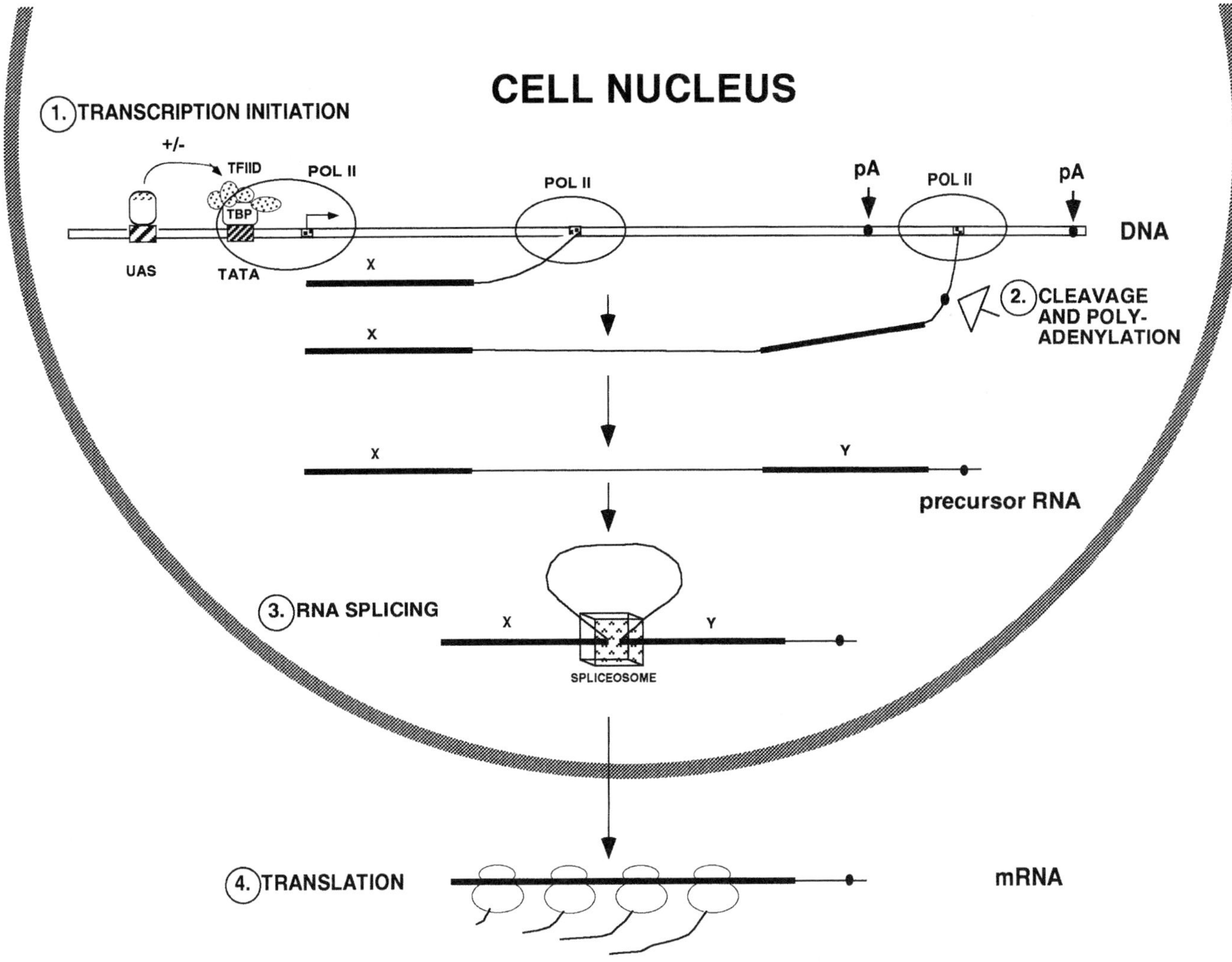


**Figure 1.** Principles of gene regulation in eukaryotes. Three important levels for regulation of gene activity occur within the cell nucleus: 1, transcription initiation; 2, cleavage and polyadenylation; and 3, RNA splicing. Translation (number 4) is physically separated from these events and occurs in the cytoplasm. UAS denotes the binding site for a transcription factor that may have positive or negative effects on transcription (+/−). TFIID consists of the TATA binding factor (TBP), which binds to the TATA box, and TBP-associated factors; POL II, designates RNA polymerase II; X and Y denote exons, and the thin line in the primary transcript designates the position of an intron; pA represents the position of a polyadenylation site. For clarity, the positions of the exons are not marked on the DNA template. The figure is not drawn to scale.

The transcriptional activity of a prototypical RNA polymerase II gene is regulated by a series of DNA elements that can be subdivided into the core promoter element, consisting of the TATA box and the transcription initiation site, and the upstream activating sequences (UAS), which usually are positioned upstream of the core promoter element (Figure 1) but in some cases are found downstream of the transcription start site. Each UAS motif is the binding site for a specific protein, a transcription factor, which may have a positive or negative effect on core promoter activity. Different transcription factors bind to different UAS motifs. Transcription factor D for polymerase II (TFIID) has been shown to play a central role in this process by binding to the TATA box in the core promoter and facilitating the recruitment of the RNA polymerase to the promoter (Figure 1). The assembly of an initiation-competent RNA polymerase at a promoter is a complex process that requires the participation of several additional basal transcription factors. Figure 1 shows only TFIID, since stable binding of this factor to a core promoter appears to be the decisive event committing the promoter for transcription. TFIID is a multiprotein complex consisting of the TATA box binding protein and several tightly associated proteins. Recombinant TBP is as efficient as TFIID in directing transcription from a core promoter element. However, the TBP-associated factors are necessary for regulation of transcription because they allow UAS-binding factors to transmit a signal to the core promoter. UAS-binding transcription factors appears to work by directly interacting with TBP or the TBP-associated factors. When the RNA polymerase leaves the promoter, TFIID remains bound at the TATA box and is ready to help a second RNA polymerase to bind and initiate transcription at the promoter. The activity of TFIID appears to be regulated by inhibitory proteins that interact with TBP. Such TBP-inhibitory protein complexes may serve an important regulatory role by keeping genes that have been removed from inactive chromatin in a repressed but rapidly inducible state.

We are far from understanding how transcription is regulated during development and differentiation. However, some important parameters have been defined. For example, genes that are expressed in specific organs often contain binding sites for cell-specific transcription factors that may vary from cell type to cell type. Thus, tissue-specific transcription is often regulated by the precise arrangement of regulatory UAS motifs in the promoter, the availability of the cognate transcription factors, and the way these transcription factors influence the activity of the promoter. Furthermore, the activity of a specific transcription factor is often regulated by posttranscriptional modification. For example, phosphorylation or dephosphorylation reactions may change the mode by which a UAS-binding transcription factor affects promoter activity. Furthermore, nucleosomes appear to be general repressors of transcription. Thus, there is evidence that either transcription factors or histones, but not both, can reside at a promoter. In activating transcription, UAS-binding transcription factors help TFIID to bind to the core promoter element and, thus, exclude the promoter from being repressed by nucleosome formation. Also epigenetic modification of the promoter may change the activity of a gene without altering the basic nucleotide sequence (e.g., in the methylation of cytosines in the promoter). Increased methylation often correlates with reduced expression of a neighboring gene, whereas a reduction of methylation correlates with high levels of expression.

Although only the efficiency of transcription initiation has been considered, it seems likely that RNA polymerase elongation also is an important parameter for regulation of gene expression. Thus, there are several examples of the RNA polymerase halting at specific pause sites during elongation. To be able to complete a transcript, the polymerase must be capable of overriding this premature transcription stop signal.

### 1.2 Regulation of Transcription in Prokaryotes

Prokaryotic cells contain only one type of RNA polymerase, which is responsible for the synthesis of mRNA, ribosomal RNA, and transfer RNA. The core enzyme is a four-subunit enzyme that is sufficient for transcription elongation. However, the holoenzyme, which is the complete enzyme, also contains the sigma factor (the $\sigma$ polypeptide). The sigma factor is required for proper RNA polymerase binding to a prokaryotic promoter. After the initiation reaction, the sigma factor leaves the polymerase complex and elongation is taken care of by the core polymerase (Figure 2).

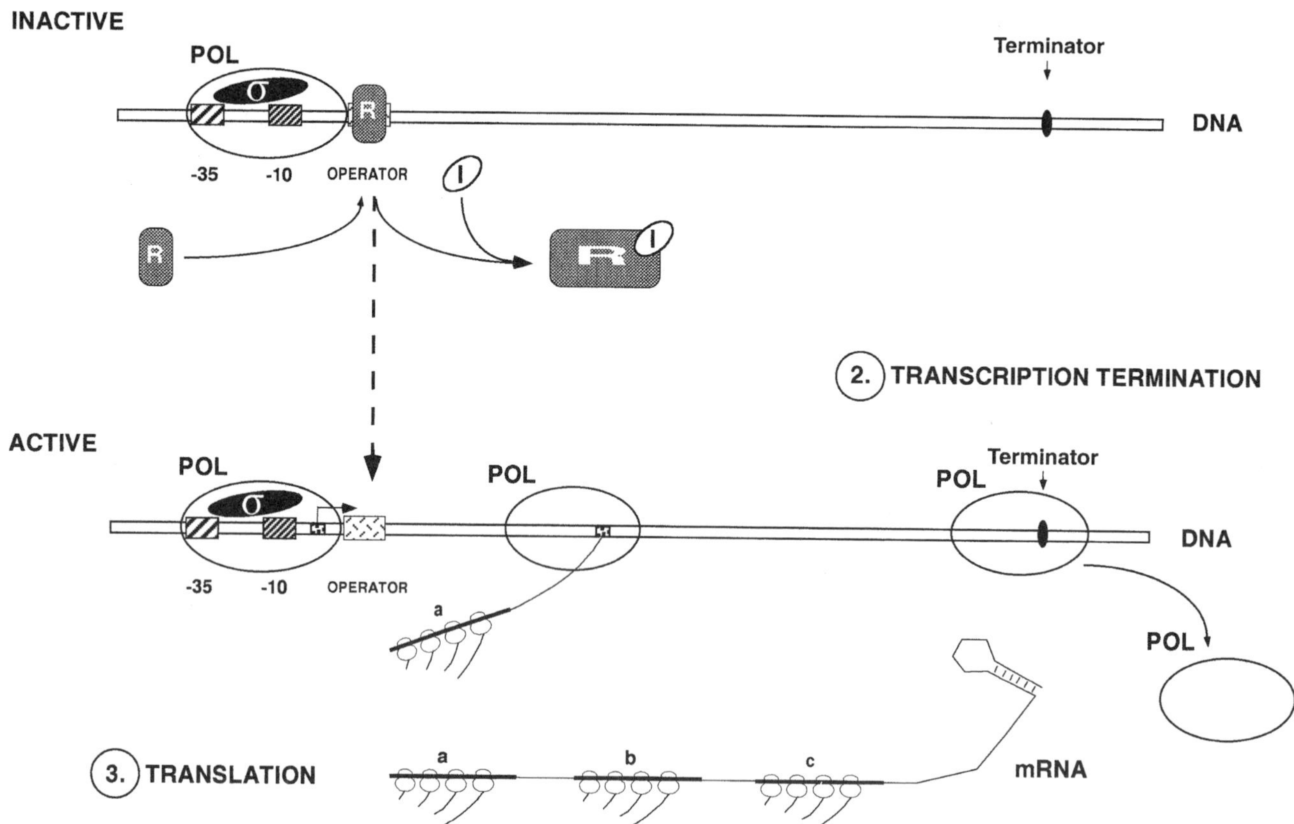


Figure 2. Principles of gene regulation in prokaryotes. The three important levels for regulation of gene activity shown are: 1, transcription initiation; 2, transcription termination; and 3, translation. Here −10 and −35 designate the respective positions of the two conserved sequence elements important for RNA polymerase binding to the promoter; POL, the RNA polymerase; $\sigma$, the sigma factor; R, a repressor protein; and I, an inducer that binds to the repressor protein. Terminator designates a rho-independent transcription terminator signal; a, b, and c, denote three genes cotranscribed into a polycistronic mRNA. For clarity, the positions of genes a, b, and c are not indicated at the DNA level. The figure is not drawn to scale.

A prototypical prokaryotic core promoter contains two conserved sequence motifs, at position −10 and −35 relative to the transcription start site (Figure 2). The −10 consensus sequence, which is TATAAT, resembles the eukaryotic TATA box. The −35 consensus sequence is TTGACA. The spacing between the two elements is of critical importance for the efficiency with which the RNA polymerase binds to the promoter. The exact sequence at the −10 and −35 positions varies slightly between different transcription units. Usually promoters that have a better homology to the consensus sequences also are stronger promoters. An important mechanism to regulate the transcriptional activity of a prokaryotic promoter is to provide the core enzyme with different sigma factors. It has been shown that sigma factors determine promoter specificity by recognizing −10 and −35 elements with different base sequences. An example is sporulation in *Bacillus subtilis,* which uses a cascade of sigma factors to cause the differentiation from a vegetative bacterium to a spore.

Regulation of gene expression can also occur at the level of transcription termination. The 3′ end of a prokaryotic transcript is usually generated by transcription termination rather than by posttranscriptional cleavage. There are two modes of termination of bacterial RNA polymerase transcription: rho-independent and rho-dependent transcription termination. Rho-independent transcription termination is characterized by a hairpin in the secondary structure of the RNA followed by a run of U-residues (Figure 2). RNA synthesis probably slows down or pauses at the hairpin structure. The string of U residues immediately after the hairpin probably then provides the signal that allows the RNA polymerase to dissociate from the DNA template. Rho-dependent transcription termination requires a specific protein, the rho factor. Proper termination does not occur in all termination signals. In some transcription units the polymerase pauses at a hairpin structure but then resumes transcription if the rho factor is not present. There are no U stretches in rho-dependent transcription terminator signals. However, polymerase pausing at secondary RNA structures is an important determinant also for rho-dependent transcription termination. The rho factor binds RNA and is believed to move along the nascent RNA chain after the RNA polymerase. When the rho factor catches up with a stalled RNA polymerase that has paused at a termination signal, it unwinds the DNA–RNA hybrids and causes termination.

Transcriptional repressors are important proteins used to control the activity of prokaryotic promoters (Figure 2). Usually all enzymes used in a specific metabolic pathway are organized into an operon that is transcribed into a polycistronic mRNA. Transcriptional induction or repression then is an important way to control the synthesis of such proteins. In many prokaryotic operons specific repressor proteins control the transcriptional activity of the promoter. Repressors are DNA-binding proteins that block RNA polymerase binding to a promoter, or transcription initiation at the promoter, by associating with an operator sequence that overlaps or is positioned close to the site at which the RNA polymerase binds in the promoter (Figure 2). The paradigm is the *lac* operon in *E. coli.* In this system, synthesis of proteins necessary for usage of lactose as a carbon source is repressed by the *lac* repressor protein if cells are capable of using glucose for growth. This is achieved by the *lac* repressor protein, which binds to an operator sequence immediately downstream of the transcription start site in the *lac* operon (Figure 2) and precludes synthesis of the polycistronic mRNA that would encode the proteins necessary for metabolism of lactose. If cells are grown on lactose as the carbon source, lactose functions as an inducer of *lac* operon transcription by binding to the *lac* repressor protein and converting it to an inactive form that does not bind DNA (Figure 2), and therefore is unable to inhibit synthesis of the enzymes encoded by the *lac* operon.

The *lac* operon represents an example of an inducible system, that is, one in which an inducer activates transcription. However, repressor proteins can also be used to inhibit transcription of an operon. A good example is the *trp* operon in *E. coli.* In this system the repressor protein is normally inactive, and RNA polymerase transcribes the operon encoding the enzymes necessary for biosynthesis of tryptophan. When sufficient quantities of tryptophan have accumulated in the cell, tryptophan binds to the repressor protein, which then becomes activated and binds to the operator and inhibits further synthesis of the enzyme responsible for tryptophan synthesis. Thus, a balanced level of the amino acid is autoregulated through a feedback loop. The *lac* and *trp* operons represent examples of the versatile way specific DNA-binding proteins, repressors, may be used for the negative and positive control of gene expression.

Synthesis of enzymes required for tryptophan production is also regulated at the level of attenuation of transcription. The attenuator region is located between the promoter and the first structural gene in the *trp* operon. It causes RNA polymerase termination in response to high concentrations of tryptophan in the cell. This occurs because the leader region in the *trp* operon is translated into a very short peptide that encodes two tryptophans. The leader mRNA can adopt two alternative conformations depending on whether the amino acid tryptophan is available for incorporation into the leader peptide. If tryptophan is available, the *trp* leader peptide is synthesized and the *trp* leader mRNA adopts a steam–loop structure in which complementary segments are paired, such that a rho-independent transcription termination signal is created. If cells contain low concentrations of tryptophan, translation of the *trp* leader peptide is prematurely terminated and the leader region adopts an alternative conformation that permits transcription to continue to the end of the operon. As a consequence, the *trp* operon is expressed and tryptophan is synthesized in cells. This type of regulation, which requires that transcription and translation be coupled, is unique to the prokaryotes. In eukaryotes the two processes are physically separated by the nuclear membrane.

## 2 REGULATION OF GENE EXPRESSION AT THE LEVEL OF RNA PROCESSING

Virtually all prokaryotic genes are contained as a contiguous DNA segment. In contrast, most eukaryotic genes are discontinuous with the coding sequences (exons) interrupted by stretches of noncoding sequences (introns) (Figure 1). The introns are present at the DNA level and in the primary transcription product of the gene, and they are removed by RNA splicing before the mature mRNA is transported to the cytoplasm. The number of introns varies considerably between genes. For example, the α-interferon gene has no introns, whereas the dystrophin gene has as many as 70. Also the size of introns can vary from fewer than 100 nucleotides to more than 200,000. Short conserved sequence motifs at the beginning (5′ end) and the end (3′ end) of the intron are used as recognition sequences to guide the assembly of a large RNA protein particle, the spliceosome (Figure 1), which catalyzes the cleavage and ligation reactions necessary to mature the final cytoplasmic mRNA. The ends of the introns are, in part, identified by RNA–RNA base pairing between the pre-mRNA and uridine-rich small nuclear ribonucleoprotein particles, the so-called U snRNPs. The conserved sequences at the 5′ and 3′ ends of the intron are surprisingly short, considering the precision with which very large introns are excised during splicing; the best conserved motifs being the GT-AG dinucleotide pair bordering the intron. Since the conserved splice site sequences are short, and not precisely conserved between introns, they occur frequently in the primary sequence of many natural precursor RNAs. This creates conditions that allow the spliceosome to combine different 5′ and 3′ splice sites in a precursor RNA, to produce several alternatively spliced cytoplasmic mRNAs from a single nuclear gene. This of course means that multiple proteins with different primary amino acid sequences and biological activities can be produced from a single eukaryotic gene. Of specific interest, the production of alternatively spliced mRNAs, in many cases, has been shown to be regulated in either a temporal, developmental, or tissue-specific manner. One of the most spectacular examples of an alternative RNA splice site choice being used to regulate expression is the somatic sex determination pathway in *Drosophila.* In this system the sex-lethal and the transformer 1 proteins have been shown to control the maintenance of sex by regulating *Drosophila* gene expression at the level of alternative RNA splicing.

Transcription termination in eukaryotes is an ill-defined process. AT-rich sequences in combination with secondary structures probably determine the end position of RNA synthesis. Usually the 3′ end of a eukaryotic mRNA molecule is specified by an endonucleolytic cleavage of the primary transcript. The RNA polymerase transcribes past the site that specifies the 3′ end of the mature mRNA, and sequences in the RNA are then recognized as targets for an endonucleolytic cleavage, followed by a nontemplated addition of 100 to 200 adenylate residues to the 3′ end; thus forming the poly(A) tail (Figure 1). A highly conserved AAUAAA sequence, 11 to 30 nucleotides upstream of the cleavage site, serves as a key signal for specifying the position of the cleavage–polyadenylation reaction. Since eukaryotic genes often encode multiple potential poly(A) sites, the precise usage of one or another poly(A) site can be used to regulate gene expression. For example, if a poly(A) signal further downstream in a transcription unit is used (Figure 1), a novel exon may be spliced into the mRNA, as is the case in the production of secreted or membrane-bound forms of immunoglobulin M. Alternatively, a new translational reading frame, encoding another protein, may be spliced from such a precursor RNA.

In contrast to eukaryotes, the primary transcripts of prokaryotic protein coding genes usually serve as mRNAs without any modification or processing. However, prokaryotic ribosomal RNAs and transfer RNAs are often matured by RNA cleavage from larger precursor RNAs.

## 3 ADDITIONAL LEVELS FOR REGULATION OF GENE EXPRESSION

Genes are most frequently regulated at the level of RNA production: synthesis or processing. However, gene expression can also be regulated at other levels, such as translational efficiency and mRNA and protein degradation. As is the case for transcriptional regulation, control of gene expression at the level of translation often occurs at the initiation step of the decoding process. Furthermore, gene products that require posttranslational modification or transport to specific cellular compartments may be regulated at each level.

*See also* Genomic Imprinting; Protein Designs for the Specific Recognition of DNA; Repressor-Operator Recognition; Translation of RNA to Protein.

### *Bibliography*

Alberts, B., Bray, D., Lewis, J., Raff, M., Roberts, K., and Watson, J. (1994) *Molecular Biology of the Cell.* Garland, New York.

Lewin, B. (1990) *Genes IV.* Cell Press, Cambridge.

Singer, M., and Berg, P. (1991) *Genes and Genomes.* University Science Press, Mill Valley, CA.

## Gene gun method for introduction of cloned DNA: *see* Plant Cell Transformation, Physical Methods for.

# Gene Mapping by Fluorescence In Situ Hybridization

*Amanda C. Heppell-Parton*

### *Key Words*

**Chromosome Band** Band that is, as a part of a chromosome, clearly distinguishable with available cytogenetic techniques from its adjacent segments (by appearing darker or lighter).

**Haptens** Reporter molecules conjugated to nucleotides, enabling incorporation into nucleic acid probes and subsequent detection of labeled sequence.

**Image Analysis** System for computer analysis of images previously recorded using specialized optical microscope accessories.

**Network** Cross-hybridization of vector DNA fragments with themselves, providing increased signal amplification at the target site.

**Probe** DNA sequence used to detect its homologous location on a chromosome.

**Probe Detection** System for visualizing site of probe nucleic acid after in situ binding of complementary sequences.

**Signal** Fluorescent region indicating the site of bound probe DNA.

**Target (Sequence)** Stretch of DNA on chromosomes with complementary sequence to the DNA probe.

To map a gene is to identify its chromosomal location. This can be achieved using a number of different approaches. However, in situ hybridization (ISH), in which a DNA fragment (probe) representing the gene is seen directly on the chromosome, is one of the most precise. A compilation of these individual localizations enables a gene map to be constructed. In practice, relatively few genes have been isolated. The majority of available probes are anonymous DNA fragments. However, once assigned to a specific position on a chromosome, these marker localizations play an

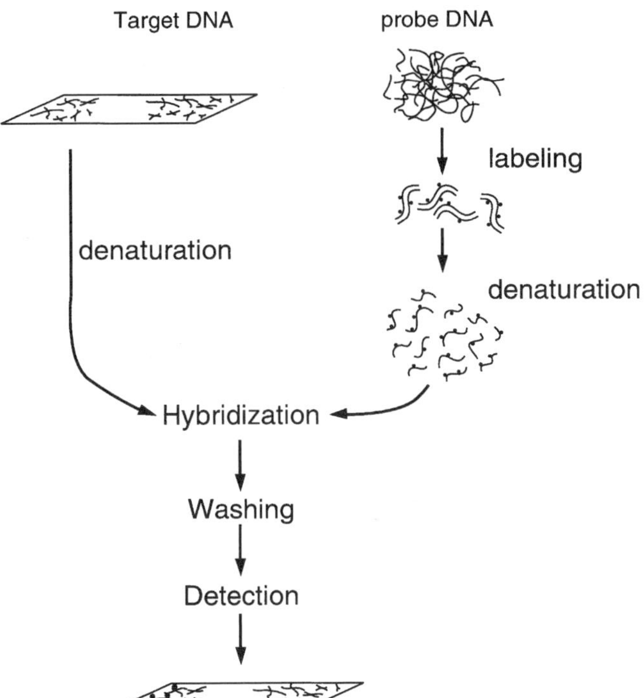


**Figure 1.** Fluorescence in in situ hybridization.

important part in connecting up isolated data to create a complete map.

## 1 INTRODUCTION

In situ hybridization is based on the complementary base pairing of the double-stranded DNA molecule. Increased temperature separates (denatures) the two DNA strands; with decreasing temperature, rejoining (reannealing) occurs. DNA probes recognize and bind (hybridize) to their complementary sequences in the intact chromosomes fixed on a microscope slide. The probe is labeled or tagged with a receptor molecule or hapten. Both probe and chromosomal DNA (template) are denatured separately, then put together on the slide and allowed to reanneal. The probe DNA recognizes its complementary sequence within the intact chromosome and reanneals, forming a hybrid molecule at the target site. Unbound probe is then washed off. In the case of fluorescence ISH (FISH), bound probe is detected either directly, as a result of the fluorescent nature of the hapten, or by the use of binding proteins that recognize the hapten within the probe (Figure 1). These binding proteins either are themselves fluorescently tagged or are subsequently recognized by a second fluorescently tagged binding protein, thus amplifying the signal (Figure 2). Visualization using a fluorescence microscope reveals fluorescently stained chromosomes displaying a chromosome-specific signal in a second fluorescent color at the target site.

Owing to the availability of several fluorochromes, a major advantage of FISH over other mapping techniques is the ability to map probes relative to each other on the same chromosome in multiple colors. This allows the determination of an unambiguous order, which is essential when trying to connect isolated pieces of mapping data.

## 2 HYBRIDIZATION TARGETS

The hybridization target in the majority of cases is the chromosome. Chromosome identification is achieved based on size and on the pattern of the chromosome bands. The probe signal is precisely localized relative to these bands. Banding patterns are reproducible and fall into two main categories: G or Q bands and R bands, as defined by the International System for (Human) Cytogenetic Nomenclature (ISCN). The banded chromosomes are visualized with the aid of fluorescent dyes for chromosomes. The most widely used dyes are *4′,6-diamidino-2-phenylindole* (DAPI) (blue), *Hoechst* 33258 (green), and *propidium iodide* (PI) (red) (Figure 3; see color plate 8). The banding patterns are made visible as a result of the base pair affinity displayed by these dyes.

Less condensed chromosomal material enables the resolution limits of FISH to be increased. When chemical blocking of the cell cycle is removed, the cells are allowed to proceed in synchrony. Careful timing results in a high proportion of more elongated *prophase* and *prometaphase* chromosomes to be harvested.

The resolution limits of the FISH technique on chromosomes allows probe sequences whose complementary target sites on the chromosome are separated by about 1 megabase (1000 kb) to be distinguished and thus ordered relative to each other. To increase the resolution still further, two probes indistinguishable on metaphase chromosomes can be resolved when hybridized to DNA in interphase nuclei. Sequences separated by as little as 50 kb can be distinguished when interphase DNA is used as the target. The DNA in this state is not identifiable as chromosomes, so only the signals are visualized. Two probes can be individually detected if the DNA of each is tagged with a different hapten and subsequently identified with a different fluorochrome (see Section 5). Recently, techniques

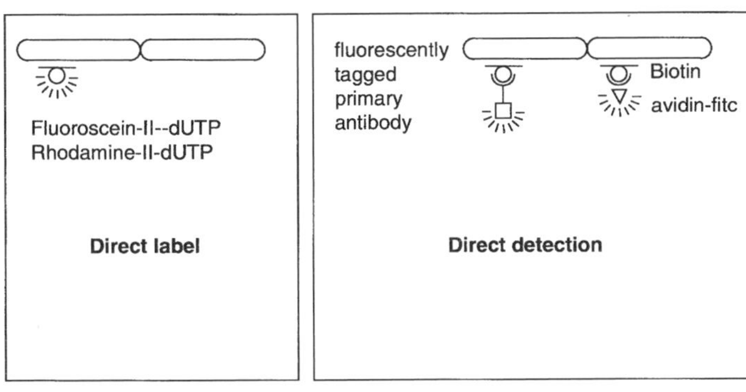


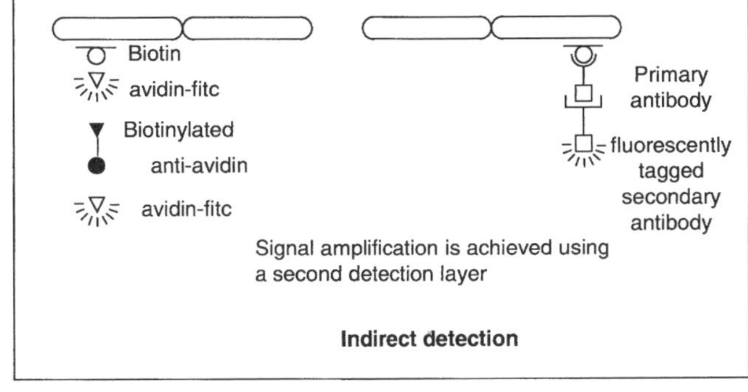


**Figure 2.** Signal detection and amplification.

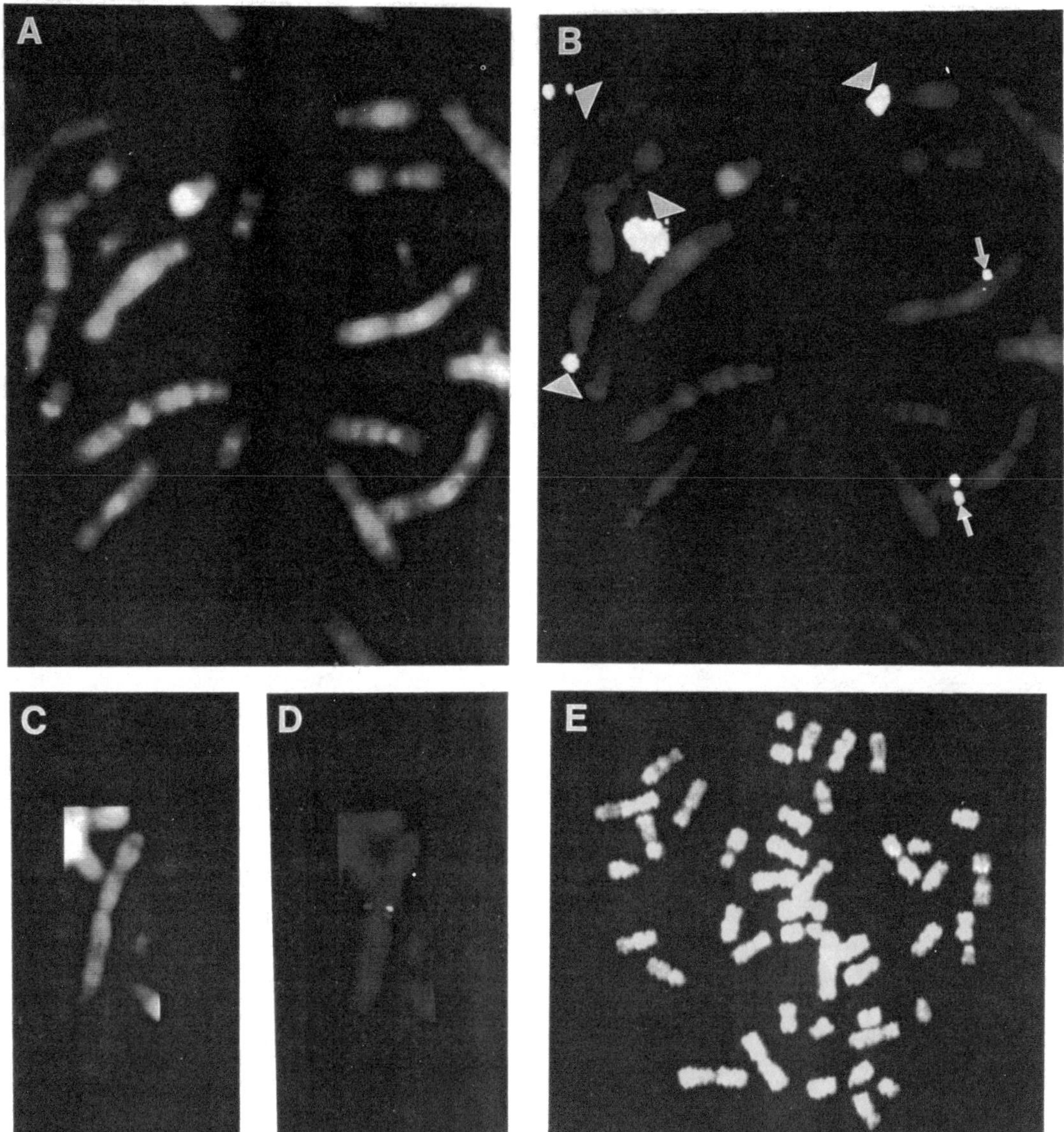


**Figure 3.** (A) Partial metaphase of Hoechst 33258 stained human chromosomes. This fluorescent dye preferentially binds to AT-rich stretches of DNA creating a G-(Q-)like banding pattern. (B) The same image after computer alignment, merging, and assignment of false colors. The hybridization signal for a small single-copy probe is seen on both homologues (small arrows). The large signal obtained with a repeat sequence probe localized to the acrocentric chromosomes is shown (arrow head) for comparison. Images were obtained using an epifluorescence microscope, RCA-ISIT camera, BioRad MRC 500 workstation, and a Sony video printer. (C) DAPI-stained Q-banded metaphase chromosome. (D) Same image as C, showing two single-copy probes labeled with Fitc (green), and Texas red (red), respectively, localized relative to each other on the DAPI stained (blue) chromosome. Images were obtained using an epifluorescence microscope, CCD camera, and BioRad MRC 500 workstation. The photographs were generated by sequential photography from the computer color monitor screen. (E) Fluorescent dye DAPI used in conjunction with PI gives an enhanced R-banding pattern (red). The image is merged and assigned false colors using appropriate computer software. The yellow Fitc hybridization signal is that of a YAC probe. All four sister chromatids are consistently labeled with such large probes. The image was obtained using a confocal microscope, BioRad MRC 600 workstation, and screen photography. (See color plate 8.)

have been described for the preparation of DNA such that in situ hybridization can be used to distinguish sequences separated by only a few kilobases.

## 3 HYBRIDIZATION PROBES

Similar sequences of DNA, constituting approximately 30% are dispersed throughout the genome. Formerly, to avoid cross-hybridization, absence of these repeat sequences was required of any probe to be used for FISH. This limited probe size, since larger sequences, by virtue of their size, carried repeat sequences and required subcloning of smaller unique inserts (500 bp–5 kb) in *plasmid vectors.* Smaller target sites with a smaller signal are detected with decreased efficiency (20–70% of the relevant chromosomes carry the signal), relying on amplification (see Section 4) and network formation for visualization (Figures 3B, D). *Repeat*

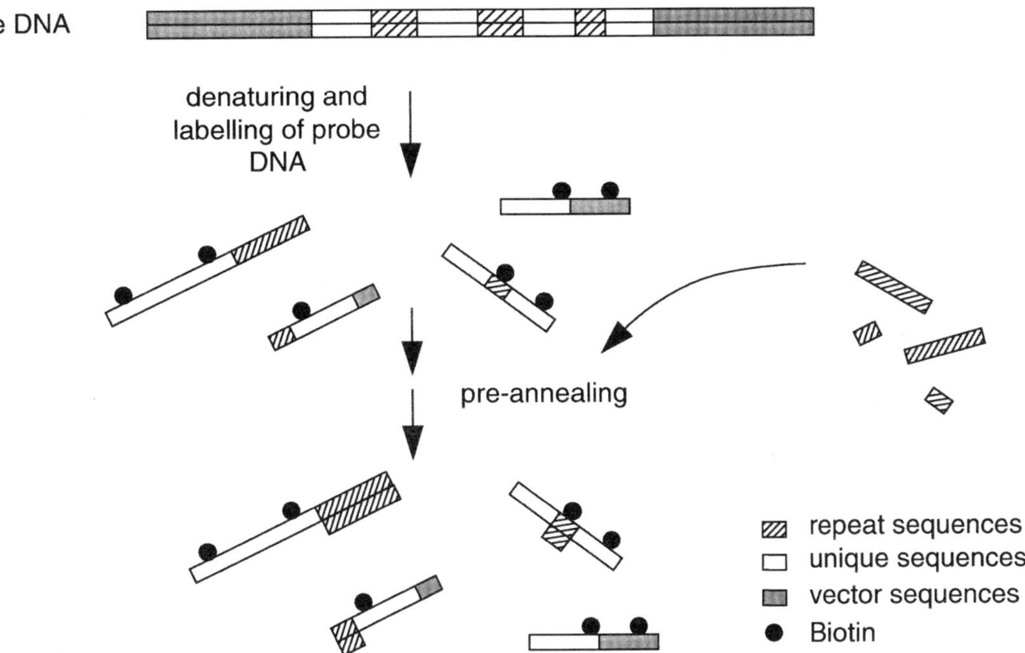


**Figure 4.** CISS hybridization: probe DNA is labeled and denatured. Labeled, single-stranded DNA probe fragments are then allowed to preanneal with an excess of repeat sequence single-stranded fragments. Reassociation of repeat sequences renders such sequences double stranded and thus not available to hybridize to complementary sequences in the target DNA (chromosome). Unique sequences remain single stranded and are subsequently able to bind to their complementary sequences on the chromosome.

*sequence probes* (tandemly repeated stretches of unique sequences) with large target site demonstrated high detection efficiency (90–99%), with larger signal, less reliant on amplification and networking (Figure 3B). Recently, competitive in situ suppression CISS hybridization (Figure 4), in which repeat sequence enriched DNA fragments are allowed to bind to their complementary sequences within the probe, so excluding these sequences from the hybridization, has enabled larger target detection using vectors carrying larger DNA inserts, such as *cosmids, bacteriophage,* and *yeast artificial chromosomes* (YACs) (Figure 3E).

## 4 SIGNAL DETECTION

DNA probes are labeled using haptens, reporter molecules conjugated to nucleotides, which can be incorporated into the DNA sequence of the probe. The techniques most frequently used to achieve this incorporation are nick translation, random primed oligonucleotide labeling and the polymerase chain reaction. The resultant labeled DNA fragments of 200 to 1000 bp are optimal for FISH.

In the case of direct labels, the hapten itself has fluorescent properties and can be visualized immediately after hybridization to the target. However, when detecting smaller target sequences, amplification of the signal is required to allow visualization. To achieve amplification, the hapten is recognized by a binding protein. This binding protein may itself be fluorescently conjugated (direct detection), or it may be subsequently detected by a second fluorescently conjugated binding protein (indirect detection) (Figure 2). The most commonly used haptens are biotin and digoxigenin. Biotin has a high binding affinity for the protein avidin, which is the protein most often used for detection purposes for this hapten; however, antibodies are also frequently used. Recently, the directly fluorescent haptens fluorescein and rhodamine have been used for larger targets. The most commonly used fluorescent labels are *fluoroscein isothiocyanate* (Fitc) (green), *rhodamine* (red), and *Texas red* (red). Other fluorescent tags ranging from blue to infrared are available.

## 5 SIGNAL VISUALIZATION

A standard fluorescence microscope enables fluorescent signals to be visualized. Changing the filter sets allows a number of different fluorochromes to be optimally excited and their emission spectra collected. The separate images then require alignment, achieved using an alignment signal, or avoided, using a dual-band-pass filter. The latter, which produces weaker signals, represents a compromise between the optimal wavelengths.

Smaller probes require greater sensitivity, which is achieved either using a high sensitivity camera or confocal microscopy, plus image analysis. Charge-coupled device (CCD) integrating cameras can capture small stationary images of low luminosity. Low illumination exposure is possible for several seconds without signal fading. Mercury arc lamp light source supplies high peaks on or near excitation wavelengths of the popular fluorochromes. Used in conjunction with a wide range of excitation filters, such equipment will produce multicolored images. Numerous software packages are available for alignment and synthetic color imaging.

In confocal microscopy, light from a point source enables focal illumination and imaging of a single point in the sample, eliminating out-of-focus elements. Coordinated sample scanning allows a complete image to be composed. Laser illumination provides the required bright light source, and high resolution, but it offers a limited range of excitation wavelengths. The standard argon ion laser offers excitation of PI and Fitc simultaneously. Information from both fluorochromes is accumulated from one image, eliminating realignment problems. However, the range of colors that can be analyzed simultaneously is reduced.

## 6 CONCLUSION

The temptations of speed, accuracy, and increased spatial resolution prompted the development of FISH. Since then, this powerful technique has developed rapidly. FISH is particularly important in gene mapping, since it is the only technique to combine accurate gene localization with relative gene order, and both sets of information are required for a complete gene map.

Continuing improvements in probe preparation, detection, and visualization will allow even more information to be accumulated from a single hybridization experiment.

*See also* Human Chromosomes, Physical Maps of; Human Disease Gene Mapping; Whole Chromosome Complementary Probe Fluorescence Staining.

### *Bibliography*

*Review articles*

Gray, J. W., and Pinkel, D. (1992) Molecular cytogenetics in human cancer diagnosis. *Cancer,* 69(6):1536–1542.

Trask, B. J. (1991) Fluoroscence in situ hybridization: Applications in cytogenetics and gene mapping. *Trends Genet.* 7(5):149–154.

———. (1991) Gene mapping by situ hybridisation. *Curr. Opinion Genet. Dev.* 1:82–87.

*Books*

Polak, J. M., and McGee, J. O'D., Eds. (1990) *In Situ Hybridisation—Principles and Practice.* Oxford University Press, Oxford.

Wilkinson, D. G., Ed. (1992) *In Situ Hybridization: A Practical Approach.* Oxford University Press, Oxford.

*Chapter*

Lichter, P., and Cremer, T. (1992) Chromosome analysis by non-isotopic in situ hybridisation. In *Human Cytogenetics—A Practical Approach,* Vol. I, *Constitutional Analysis,* pp. 157–190. Oxford University Press, Oxford.

# Gene Order by FISH and FACS

*Malcolm A. Ferguson-Smith*

### *Key Words*

**Chromosome-Specific Paint** The product of PCR amplification of sorted chromosomes, using random PCR primers. When labeled and hybridized to metaphases by fluorescence in situ hybridization (FISH) techniques, the chromosome paint will give an even distribution of FISH signals along the length of the chromosomes from which the paint was derived.

**Contig** A contiguous series of overlapping cloned DNA sequences.

**Cosmid** A vector capable of cloning inserts of bacterial DNA measuring 20 to 40 kilobases.

**Flow Karyotype** Graphical representation of mitotic chromosomes produced by a dual laser flow cytometer in which the chromosomes are arranged according to size and base pair ratio.

**Gene Locus** The position of a gene on its chromosome.

**Gene Probe** Cloned DNA sequence of part of a gene used to locate the gene on its chromosome and to identify restriction fragments.

**Lymphoblastoid Cell Line** Cell line derived from peripheral lymphocytes transformed and immortalized by Epstein–Barr virus.

**Microsatellite Marker** Polymorphic DNA sequence composed of a variable number of tandemly arranged simple di-, tri-, or tetranucleotide repeats.

**Polymerase Chain Reaction (PCR)** A method for the primer-directed amplification of specific DNA sequences.

**Reciprocal Translocation** The result of chromosome breakage and the exchange of chromosome material between two nonhomologous chromosomes.

**Somatic Cell Hybrid** Cell line derived from the fusion of cells from two different species, most often human and rodent; in the latter case, human chromosomes tend to segregate out of the hybrid.

**Yeast Artificial Chromosome (YAC)** A vector capable of cloning large inserts of DNA measuring 200 to 1000 kb in yeast cells.

Two cytological techniques used in mapping and ordering genes in the Human Genome Project are fluorescence in situ hybridization (FISH) and fluorescence-activated chromosome sorting (FACS).

FISH anneals nonisotopically labeled DNA gene probes to their complementary sequences on chromosomes immobilized on microscope slides. The sites of hybridization, and thus the chromosomal location of the gene, are detected by immunological and other systems in which the antibody to the DNA label is coupled to a fluorochrome. The fluorescent signal is detected by fluorescence microscopy, is collected onto optical disks with the aid of a sensitive digital camera, and is available for detailed examination using image analysis systems. The use of different DNA labels for different probes allows the determination by multicolor FISH of the relative order of several genes along the metaphase chromosome. The order of closely linked DNA sequences may be determined from interphase nuclei or from extended chromatin fibers released from fixed nuclei.

In FACS, mitotic chromosomes in fluid suspension are passed sequentially through a dual laser flow cytometer and sorted into individual groups according to size and base pair ratio. Pure samples of most individual chromosomes can be obtained and used for the chromosome assignment of DNA sequences by the PCR technique. Regional assignment and gene order are achieved by sorting the products of reciprocal translocations.

## 1 INTRODUCTION

The Human Genome Project is an international venture to construct detailed genetic and physical maps of each of the 23 different human chromosomes. The genetic map provides the location of genes and sufficient reference markers to enable new genes to be located rapidly by family (genetic linkage) studies. The physical map consists of a contiguous series of overlapping DNA sequences (contigs) extending throughout the length of each chromosome. Already approximately 80% of the entire human genome is represented in the form of physical maps by such contigs. The final phase of the Human Genome Project aims to sequence the entire DNA of the genome within 10 years. The most interesting parts of the genome, namely, the genes that are transcribed into messenger RNA and translated into proteins, are being sequenced first. The map of these genes is sometimes referred to as the transcriptional or expression map.

The DNA sequence of a particular gene may be determined either from knowledge of the amino acid sequence of its protein product or by a more complex strategy known as positional cloning. Most of the common single-gene disorders that affect human populations (e.g., thalassemia, cystic fibrosis) have now been mapped and sequenced by one or other of these two methods. In positional cloning, families in which members are affected with a recognizable disorder or trait (phenotype) are tested with a series of chromosome-specific DNA reference markers spread evenly across the genome to determine which marker tends to be transmitted through the family in association with the particular phenotype. As members of a chromosome pair assort randomly during the formation of eggs and sperm, linkage of a chromosome marker and the gene locus is the first indication that the gene can be assigned to that chromosome. Genetic recombination due to meiotic crossing over may also separate marker and gene locus if they lie apart from each other on the same chromosome. The linkage analysis is therefore repeated with additional markers from the same chromosome, to assign the gene to the smallest possible chromosomal region. Once this has been achieved, the relevant contig in the physical map can be identified and used to isolate candidates for the gene in question. In the final stage in positional cloning, the various candidate genes are sequenced, whereupon it is possible to look for gene mutations in affected patients and confirm that such anomalies are absent in unaffected controls.

During the construction of genetic maps, new DNA clones are isolated and characterized for their value as genetic markers. Nowadays, the first approach to mapping such markers uses the technique of in situ hybridization, whereby the DNA probe, suitably labeled, is annealed to preparations of chromosomes fixed onto microscope slides. The DNA hybridizes to its complementary sequence on the chromosome, and the site of hybridization is recognized from the label used in making the DNA probe. Radioactive labels were used initially, but these have been replaced by agents such as biotin and digoxigenin, which can be incorporated into the probe DNA and detected by fluorescence microscopy using fluorochromes coupled to appropriate antibodies. Fluorescence in situ hybridization (FISH) has been developed into a very sensitive technique, capable of readily detecting DNA probes containing at least 1 kilobase of DNA sequence. With multiple fluorochromes and appropriate filter systems on the fluorescence microscope, several colors can be used with different probes to map more than one sequence at a time. This capability is particularly valuable when the aim is to determine the order of different DNA markers along the same chromosome.

The most useful markers for genetic mapping are those that show extensive variation (polymorphism) between individuals, but are transmitted without change through families. Microsatellite markers, in which the variation is due to differences in the number of copies of simple di- or trinucleotide repeats, are among the most frequent of these markers and are readily distinguished by use of the polymerase chain reaction (PCR). Using specific DNA primers flanking the microsatellite, PCR amplifies the fragment containing the dinucleotide repeat, which can then be sized by ethidium bromide stained gel electrophoresis.

FISH is not applicable to mapping microsatellite markers, and so alternative strategies are required. One approach uses a series of interspecific somatic cell hybrids formed by fusing human and rodent cells. It is possible to construct hybrid cell lines so that each contains a single human chromosome within a background of rodent chromosomes. PCR amplification of a human microsatellite sequence will occur only if the appropriate chromosome is present. Panels of monospecific hybrids can therefore be assembled for assigning such markers to their respective chromosomes. Occasionally problems arise; a hybrid may be contaminated by part of a second human chromosome that has become incorporated into one of the rodent chromosomes; sometimes the PCR reaction amplifies a rodent sequence that happens to be homologous to the human sequence.

An alternative strategy for assigning microsatellite markers to their respective chromosomes depends on the ability to sort human chromosomes into their respective categories using flow cytometry. In brief, a fluid suspension of chromosomes is prepared from an actively growing cell culture by using colchicine to arrest cells in mitosis. The mitotic cells are disrupted and the released chromosomes are stained with a mixture of two fluorescent dyes, which detect AT-rich and GC-rich DNA, respectively. The stained chromosomes are passed sequentially at a speed of up to 2000 per second through two lasers tuned to excite the fluorescent emission from each dye in a fluorescence-activated cell sorter (FACS). The amount of fluorescence in each chromosome is collected and stored, and this information is used to produce a bivariate flow karyotype (Figure 1) in which large numbers of individual chromosomes form 20 discrete clusters in characteristic relationships. The technique effectively separates all chromosomes except chromosomes 9 through 12, and the distinct pattern of fluorescence in each chromosome enables the FACS to sort each category into separate tubes. Thus a panel of sorted chromosomes can be produced, each tube with 300 to 500 chromosomes, and this is sufficient material to allow a microsatellite probe to be assigned by PCR to its respective chromosome. Chromosomes 9 through 12 can be separated for mapping purposes by exploiting heteromorphisms (for chromosome 9) and translocation derivatives. Similarly, regional assignments may also be made by sorting translocation chromosomes with appropriate breakpoints.

The DNA of chromosomes sorted by FACS can be amplified by PCR and labeled to produce chromosome paints. The amplified DNA is useful in hybridization experiments to assist in the analysis of chromosomes rearrangements, for the chromosome paint will anneal only to the chromosomes from which it is derived. Once again the labeled DNA is detected by FISH. A modification of the method, termed reverse painting, depends on the ability to sort and make paints from rearranged chromosomes. Hybridization onto

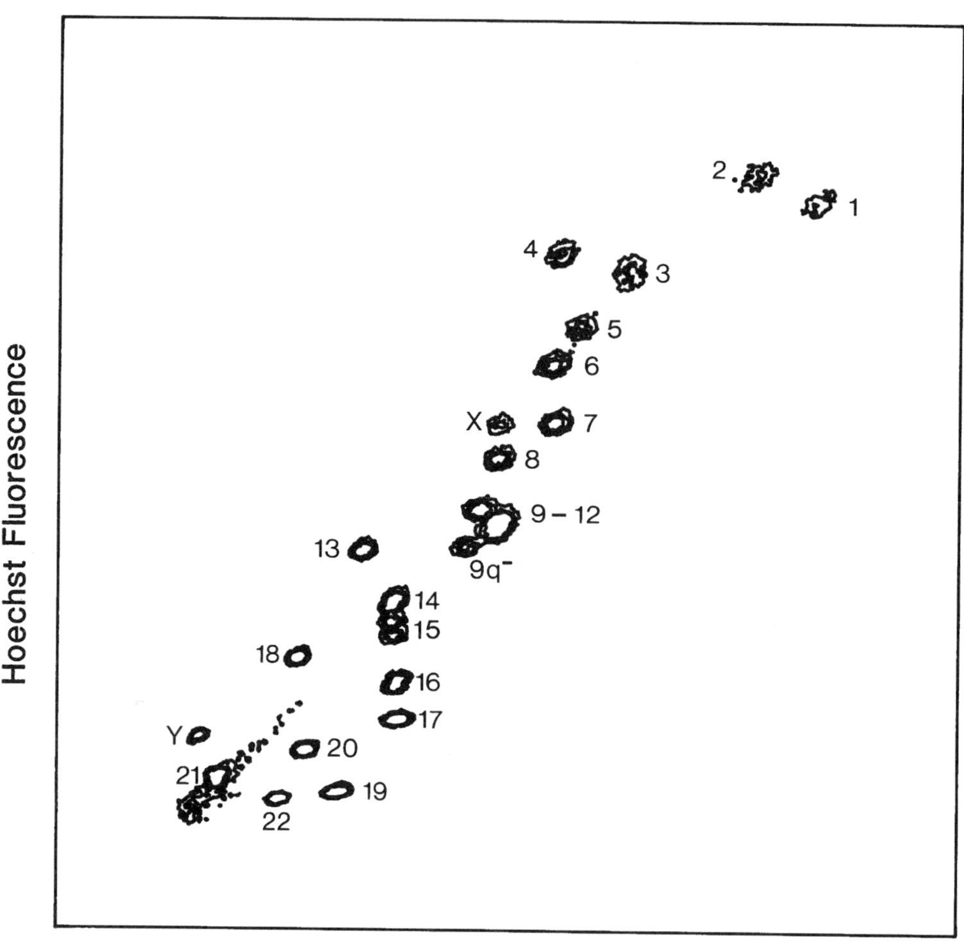


Figure 1. Bivariate flow karyotype showing separation of individual chromosomes according to size and base pair ratio. In this preparation, one homologue of chromosome 9 is separated from the remainder of the 9–12 cluster because it contains substantially less than the average complement of centric heterochromatin.

normal metaphase spreads reveals the origins of the various chromosomes involved in the rearrangement. This is one of the most sensitive techniques for the analysis of complex chromosome arrangements.

## 2 TECHNIQUES

### 2.1 Fluorescence In Situ Hybridization (FISH)

FISH depends on hybridizing labeled nucleic acid probes to cytological preparations of chromosomes and chromatin and detecting the presence of the annealed sequences by fluorochrome-conjugated reagents using a fluorescence microscope.

### 2.2 Fluorescence-Activated Chromosome Sorting (FACS)

High resolution chromosome sorting can be achieved with a commercial fluorescence-activated cell sorter equipped with two argon ion lasers. Preparation of the chromosome suspension, FACS analysis, sorting, and collection of sorted chromosomes are the main steps in the process.

#### 2.2.1 Chromosome Preparation

Chromosome preparations are made by standard methods from short-term peripheral blood cultures, lymphoblastoid cell lines, and a variety of tissue culture cells including skin fibroblasts. Cells are arrested in metaphase by the addition of colcemid (0.1 μg/mL) to the culture for 6 to 12 hours. The cells are then resuspended in low ionic strength buffer (45 m*M* KCl, 10 m*M* $MgSO_4$, 3 m*M* dithiothretol, and 5 m*M* HEPES at pH 8.0) and incubated for 10 minutes at room temperature. Triton X-100 is added to a final concentration of 0.25% and the sample left on ice for 10 minutes.

An alternative technique is to treat the cells in 75 m*M* KCl and resuspend the cell pellet in a buffer containing polyamines and Triton X-100. Chromosomes are released into suspension by 10 seconds of rapid vortexing and stained immediately in Chromomycin A3 (final concentration 40 μg/mL) and Hoechst 33258 (final concentration 2 μg/mL), followed by 2 hours of incubation at 4°C. Fifteen minutes prior to flow analysis, sodium citrate (final concentration 10 m*M*) and sodium sulfate (final concentration 25 m*M*) are added to the sample.

### 2.2.2 Analysis, Sorting, and Collection

Chromosomes are sorted using a dual laser flow cytometer (FACStar Plus or equivalent) equipped with two 5 W argon ion lasers. One laser is tuned to emit 300 mW of light in the UV (351–364 nm) to excite Hoechst fluorescence, and the second laser is tuned to emit 300 mW at 458 nm to excite Chromomycin fluorescence. The chromosome suspension is passed through the two laser beams sequentially, to permit the fluorescence emitted from each chromosome to be collected separately and used to create a flow karyotype (Figure 1) in which large numbers of individual chromosome are represented by discrete clusters in characteristic relationships. Each chromosome cluster has a position in the flow karyotype determined by its relative size and base pair ratio. AT-rich chromosomes tend to sort above a diagonal line drawn through the middle of the chromosome clusters, while GC-rich chromosomes sort below this line. As the chromosomes pass through the two lasers, the fluid system breaks into droplets, some of which will contain one chromosome. Sorting is achieved by giving an electrical charge to the droplets containing the chromosome of interest so that they can be deflected into a container as they pass between two high voltage plates. Highly pure chromosome samples can be collected in this way.

## 3 APPLICATIONS

### 3.1 Gene Order by FISH

Early methods for distinguishing between different human chromosomes depended on their size and centromere position at the metaphase stage of mitosis. Unequivocal identification of every chromosome became possible only when banding techniques of staining were introduced in 1970. At that stage each chromosome could be numbered. Assignment of gene loci to individual chromosome numbers has thus depended on cytological techniques complemented by genetic linkage studies, which are based on the knowledge that if one locus maps to a particular chromosome, all other loci that are linked must also be located on the same chromosome. The relative position of a gene on the chromosome is now most easily determined by FISH, and this requires that at least part of the gene be included in a cloned DNA sequence that is more than one kilobase long. Probes made from cosmid ($\leq$ 40 kb) or yeast artificial chromosome (YAC) clones ($\leq$ one megabase) are most frequently used for FISH mapping because they give strong hybridization signals—characteristically, twin signals in the same position on both chromatids of the chromosome. If two or more probes that map to the same chromosome are hybridized at the same time, it is often possible to determine their order on the chromosome. However, it has been found that closely linked probes cannot be ordered on a metaphase chromosome unless they are at least one megabase apart. This is because the DNA molecule within the chromosome is condensed by several orders of coiling and supercoiling and because the supercoiled chromatin fiber is attached to the chromosome scaffold in a tightly packed series of loops that radiate out from the center in every direction. DNA sequences that are less than one megabase apart on the chromatin fiber may thus appear to be superimposed on one another, or to be in the wrong order relative to the ends of the chromosome.

One strategy for resolving the correct order of two probes that map to the same location at metaphase is to hybridize them to preparations made from a panel of cell lines containing reciprocal translocations that span the region of interest. Suitable translocations can often be obtained from the commercial and private cell banks, which specialize in making such collections available to gene mappers. Clearly, when a translocation breakpoint separates two probes, a single signal will appear on each of the translocation derivatives; if the translocation breakpoint is on either side of the probes, only one translocation derivative will show the signal. The two signals can be distinguished by two-color FISH.

The order of sequences less than one megabase apart may also be determined by FISH, using interphase nuclei in which the chromosomes are extended to 10 times their metaphase length and are not visible as discrete entities. At least three probes are required: one for the sequence whose location is to be determined in relation to the second sequence, and a third whose position in relation to the second is already known. Cytological preparations are fixed and air-dried as for chromosome preparations and cosmid clones are labeled for detection using at least two colors. Nuclei in the G1 phase (i.e., before DNA replication) are chosen for scoring. A typical nucleus will show three signals on each of the two copies of the chromosome in question, and their consensus order along the chromosome usually can be determined from the analysis of 20 to 50 nuclei. Two signals can be resolved if they are more than 50 kb apart, but greater resolution is not possible. As at metaphase; the limitation of resolution is mostly due to the organization of the chromatin fiber into loops radiating from a central scaffold. Measurements made in cytological preparations of the distance between hybridization signals can therefore provide only crude estimates of the relative distance between DNA segments.

The construction of genetic maps has been greatly facilitated by interphase FISH, which has helped to resolve the order of DNA genes and markers too close together to be determined by genetic linkage studies alone. In the construction of physical maps, FISH has played a similar role in the correct ordering of YAC contigs along the chromosome. FISH can readily resolve inconsistencies due to such problems as colligation of two separate DNA sequences within the same YAC. More recently, FISH has been applied to the ordering of cosmid subclones within a single YAC. Interphase FISH does not have sufficient resolution for this purpose, and so alkali treatment and other methods have been developed to release the chromatin fiber from its protein scaffold and fix the extended fiber in the microscope slide in a form suitable for FISH. Cosmid probes hybridized to such extended chromatin fibers typically show a series of interrupted signals in the form of a string of particles along the fiber. By using different combinations of fluorochromes, the order of different cosmids along the fiber can be determined and overlaps and gaps of 5 to 10 kb between cosmids readily detected (Figure 2; see color plate 9) In this particular application, FISH demonstrates its power to bridge the resolution gap between genetic and physical maps.

### 3.2 Gene Order by FACS

Ordering DNA sequences by FISH depends on the availability of a genomic DNA clone that is both large enough and contains at least part of the DNA sequence of interest. For most practical

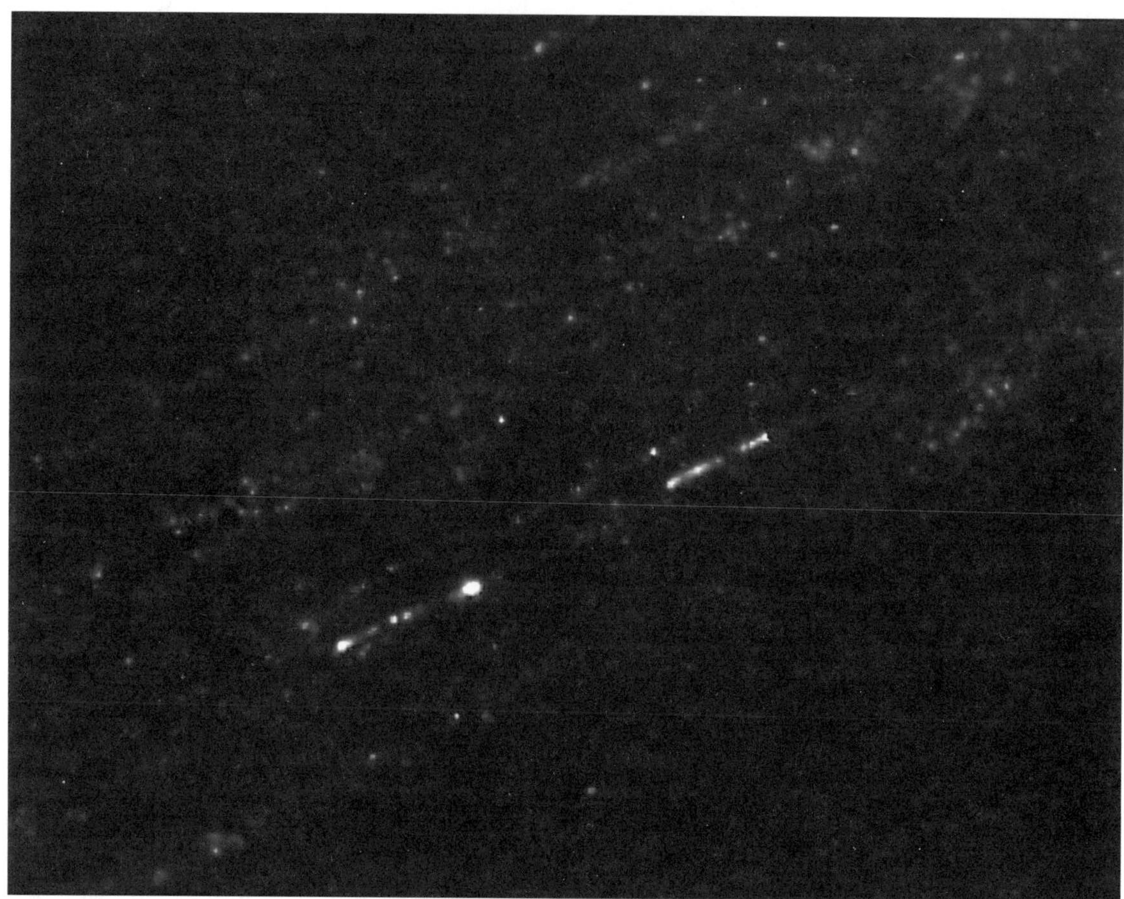

**Figure 2.** Ordering three closely linked cosmid clones using hybridization to the extended chromatin fiber released by alkali treatment of fixed nuclei. The cosmids, each approximately 40 kb long, are differentially labeled by Texas Red (red), FITC (green), and a 50:50 mixture of Texas Red and FITC (giving an orange-yellow color). There is a 5 kb overlap between the red and green labeled cosmids indicated by the yellow signal. The yellow cosmid is separated from the red cosmid by a 50 kb gap. (See color plate 9.)

**Table 1** Flow Dot-Blot Analysis of Chromosome 9 Translocations[a]

| Probe | 9T05 9q12 | 9T10 9q22 | 9T02 9q22.1 | 9T06 9q32 | 9T08 9q33 | 9T14 9q34.1 | 9T03 9q34.1 | 9T04 9q34.1 | 9T12 9q34.2 | 9T01 9q34.3 |
|---|---|---|---|---|---|---|---|---|---|---|
| ALAD | D | D | D | *P* | *P* | *P* | *P* | *P* | *P* | *P* |
| MCOA12 | D | D | D | D | *P* | | *P* | *P* | *P* | *P* |
| ORM | D | D | D | D | *P* | *P* | *P* | *P* | *P* | *P* |
| GSN | D | D | D | D | *P* | *P* | *P* | *P* | *P* | *P* |
| HXB | D | D | D | D | *P* | *P* | *P* | *P* | *P* | *P* |
| CRIP111 | D | D | D | D | D | *P* | *P* | *P* | *P* | *P* |
| AK1 | D | D | D | D | D | D | *P* | *P* | *P* | *P* |
| SPTAN1 | D | D | D | D | D | D | *P* | *P* | *P* | *P* |
| ASSg3 | D | D | D | D | D | D | D | *P* | *P* | *P* |
| T39-2-2 | D | D | D | D | D | D | D | *P* | *P* | *P* |
| ABL3 | D | D | D | D | D | D | D | D | *P* | *P* |
| DBH | D | D | D | D | D | D | D | D | *P* | *P* |
| MCT136 | D | D | D | D | D | D | D | D | *P* | *P* |
| MCT96.1 | D | D | D | D | D | D | D | D | D | *P* |

[a]Ten cell lines from balanced translocation heterozygotes with different translocation breakpoints in chromosome 9q were used to sort the two translocation derivatives and prepare dot blots of approximately 10,000 chromosomes onto nitrocellulose filter disks. DNA probes for each of the loci listed on the left were hybridized to the filter disks: D and *P* indicate the site of hybridization relative to the translocation breakpoint (i.e. on the derivative chromosome carrying either the *distal* or the *proximal* segment of the long arm of chromosome 9). The results give the precise order of many of the loci between 9q12 and 9q34.3.

purposes this means isolating a cosmid clone from an appropriate genomic library. However, many of the most polymorphic DNA markers used in genetic mapping are microsatellite markers tested by PCR. As discussed in Section 1, the markers may be mapped directly, without recourse to FISH, by amplification, using either interspecific somatic cell hybrids or chromosomes sorted by FACS. If the FACS technique is used, gene order is then achieved by sorting translocation derivatives to determine on which side of the translocation breakpoint the microsatellite is located.

Chromosome sorting may also be used for mapping cloned DNA sequences. For this purpose, 10,000 chromosomes of each type are sorted onto nitrocellulose filter disks using mild aspiration from beneath the disk. The chromosomal DNA is denatured and baked onto the disk, which can then be used for filter hybridization with radiolabeled $^{32}$p DNA probes in the usual way. Autoradiography of the filter panel reveals the chromosomal location of the probe. The order of various chromosome-specific probes may be determined using filters prepared from translocation chromosomes (Table 1).

## 4 PERSPECTIVE

For the foreseeable future, FISH will play a very important role in mapping cloned DNA sequences to their positions on the genetic map and in the construction of physical maps. In terms of simplicity and economy, it is the technique of first choice for gene localization. Because quick results are possible, the method is used for confirmation and verification of the identity of DNA clones generated in the molecular biology laboratory. It is the routine test used to exclude coligation during the characterization of clones isolated from YAC libraries. Apart from gene mapping, FISH has an increasing role in the diagnosis of chromosome aberrations, whether constitutional or associated with the pathogenesis of cancer.

The role of FISH in gene mapping is likely to diminish in the more distant future when a fully validated physical map of each chromosome has been achieved based on a contiguous series of overlapping YAC and cosmid clones extending from one end of the chromosome to the other. With such a physical map it will be possible to assign any unknown DNA sequence to a single YAC or cosmid clone in the series, hence to its position on the chromosome. Similarly, the need for chromosome sorting by FACS will diminish in terms of human gene mapping. However, it is likely that gene mapping in many other species will be facilitated by the ease of sorting individual chromosomes. In nonhuman species it is more difficult and usually impossible to construct interspecific cell hybrids containing only one chromosome from one of the hybrid parents. Mapping panels of sorted chromosomes may prove to be an important resource for mapping these genomes.

*See also* DNA Markers, Cloned; Gene Mapping by Fluorescence In Situ Hybridization; Human Chromosomes, Physical Maps of; Hybridization for Sequencing of DNA.

### *Bibliography*


Carter, N. P., Ferguson-Smith, M. A., Perryman, M. T., Telenius, H., Pelmear, A. H., Leversha, M. A., Glancy, M. T., Wood, S. L., Cook, K., Dyson, H. M., Ferguson-Smith, M. E., and Willatt, L. R. (1992) Reverse chromosome painting: A method for the rapid analysis of aberrant chromosomes in clinical cytogenetics. *J. Med. Genet.* 29:299–307.

Fidlerova, H., Senger, G., Kost, M., Sanseau, P., and Sheer, D. (1994) Two simple procedures for releasing chromatin from routinely fixed cells for fluorescence in situ hybridisation. *Cytogenet. Cell Gen.* 65:203–205.

Reed, T., Baldini, A., Rand, T. C., and Ward, D. C. (1992) Simultaneous visualisation of seven different DNA probes by in-situ hybridization using combinational fluorescence and digital imaging microscopy. *Proc. Natl. Acad. Sci. U.S.A.* 89:1388–1392.

Rooney, D. E., and Czepulkowski, B. H. (1992) *Human Cytogenetics: A Practical Approach.* IRL Press, Oxford.

Trask, B. J., and Pinkel, D. (1990) Fluorescence in-situ hybridization with DNA probes. In *Methods of Cell Biology,* Vol. 33. Academic Press, New York.

———, Massa, H., Kenwrick, S., and Gitschier, J. (1991) Mapping of human chromosome Xq28 by two color fluorescence in-situ hybridization of DNA sequences to interphase cell nuclei. *Am. J. Hum. Genet.* 48:1–15.

van den Engh, G., Trask, B., Lansdorp, P., and Gray, J. (1988) Improved resolution of flow cytometric measurements of Hoechst and Chromomycin-A3 stained human chromosomes after addition of citrate and sulphate. *Cytometry* 9:266–270.

Wiegant, J., Kalle, W., Mullenders, L., Brookes, S., Hoovers, J.M.N., Damverse, J. G., van Ommen, G.J.B., and Raap, A. K. (1992) High resolution in-situ hybridisation using DNA halo preparations. *Hum. Mol. Genet.* 1:587–591.


# Genetic Analysis of Populations

*A. Rus Hoelzel*

### *Key Words*

**Demographics** The study of populations, especially their growth rate and age structure.

**DNA Turnover** Contingual gain and loss of DNA regions due to a variety of mechanisms including DNA slippage, gene conversion, transposition, and unequal crossing over.

**Electrophoresis** The movement of molecules on a solid medium through an electric field. Various support media are used, including filter paper, starch gel, agarose, and polyacrylamide.

**Molecular Clock** The rate of molecular genetic change at a given gene, including the assumption of a constant rate.

**Polymorphism** The presence of two or more genetically distinct types in the same interbreeding population.

The molecular genetic analysis of populations involves three main stages: sample collection, analysis, and interpretation. The sample must be representative of the population, and therefore collections need to be random and sufficiently large. The method of analysis and choice of genetic marker will depend on the specific question being addressed. Interpretation of results will depend in part on the method of analysis. DNA analytical methods offer great potential through the analysis of specific regions of the nuclear, mitochondrial, and chloroplast genomes, but numerous mechanisms can affect change, and therefore all interpretation based on time-dependent change must be done cautiously. An understanding of the genetic structure of populations facilitates the understanding of dispersal, reproductive, and social behavior, and provides essential data for the conservation of natural variation.

## 1 INTRODUCTION

It has long been recognized that inherited characters vary and that the quantification of this variation can be used to distinguish local races. However, it was not until the advent of protein gel electrophoresis in the 1960s that the extent of genetic variation in natural populations began to be appreciated. Well over 1000 species have been investigated for protein polymorphisms, and most display extensive variation. Gel electrophoresis became the standard method for the comparison of genetic variation within and between populations, and it is still appropriate for many applications. However, most of the variation present in the genome cannot be detected by this method. The DNA sequence determines, through an RNA intermediary, the amino acid sequence in the polypeptides that combine to form proteins. Only a small proportion of the genome is translated into proteins, and the three-''letter'' codon that determines the sequence of amino acids is degenerate, especially in the third position. Therefore, even within the DNA sequence that encodes the protein, there can be considerable variation that is not expressed. Furthermore, only a small proportion of the changes that do affect the amino acid sequence will also affect the charge properties of the protein, which determine its migration through an electric current. (And consequently, only some changes can be detected by the electrophoretic method.)

Recent innovations have enabled the analysis of DNA directly, providing greater resolution and facilitating the interpretation of variation. This contribution concentrates on the use of these DNA analytical methods to interpret population level genetic variation.

## 2 NATURE OF MUTATIONS

Mutations can occur that affect the sequence of nucleic acids through the conversion of a single base (e.g., a point mutation changing A to T), through the loss by deletion of a string of bases, and through the addition by insertion of a string of bases. Change can also occur through the operation of a variety of DNA turnover mechanisms. This discussion emphasizes two mechanisms that generate variation in repetitive regions of DNA by changing the number of repeated elements: DNA slippage and unequal crossing over (see Figure 1). In each case a misalignment of DNA strands causes the molecule to be repaired in such a way that there is a gain or loss in an array of repeated elements. Repetitive DNA regions are very common in most vertebrate and plant genomes, and change in the copy number of repeats generally occurs at a much higher rate than change due to point mutations.

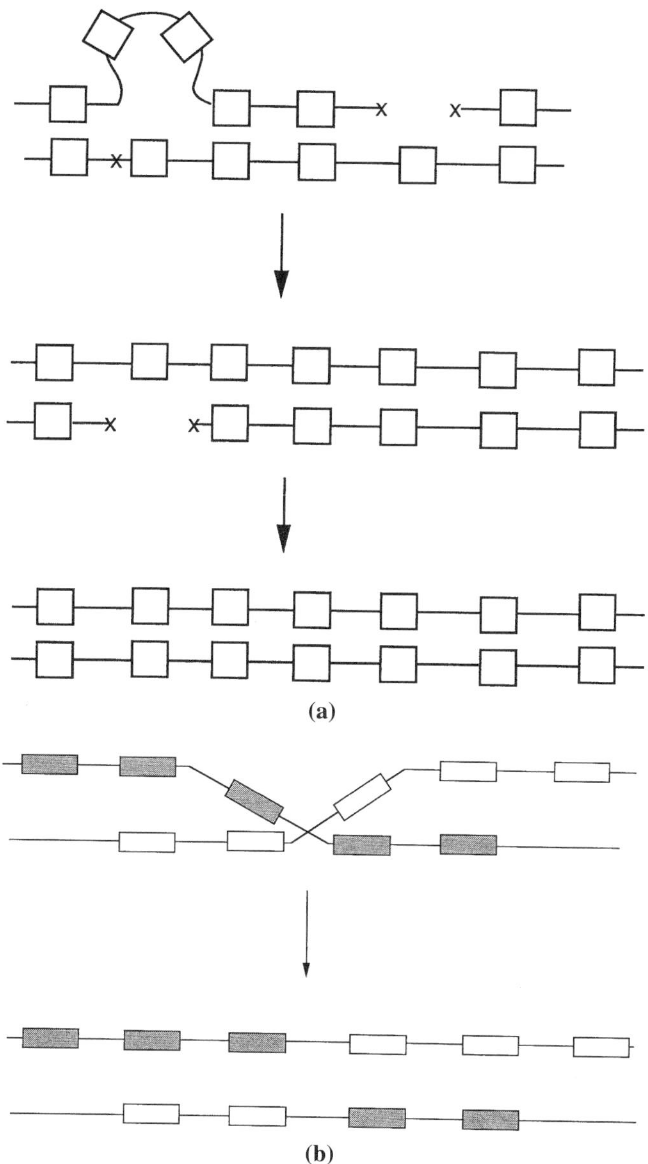


**Figure 1.** (a) One of several DNA slippage mechanisms. In this case the end result is an extention of the array within a single chromatid. Similar mechanisms lead to contraction of the array and slippage during replication. The process illustrated involves endonuclease-mediated breakage, slippage of the upper strand, and polymerase-mediated repair events. The boxes represent repeated elements. (b) Unequal crossing over. Two strands, each with five repetitive elements, undergo an unequal crossing-over event, resulting in the gain of one element for the upper strand and a corresponding loss from the lower strand.

## 3 DISSEMINATION OF GENETIC CHANGE IN NATURAL POPULATIONS

There are three principal mechanisms by which genetic change is spread through a population—natural selection, genetic drift, and molecular drive. The correct interpretation of genetic variation depends on how the DNA region under analysis is affected by each of these mechanisms. Although it is beyond the scope of this review to provide a detailed discussion, the basic tenets should be made clear.

### 3.1 Natural Selection

The morphological, physiological, or behavioral consequences of genetic makeup (genotype) are referred to as the phenotype. Natural selection is the differential reproductive success of different phenotypes. Generally, if a region of DNA is not expressed phenotypically (nor closely linked to a phenotypically expressed gene), it is not exposed to natural selection. On the other hand, some genes are critical to survival or reproductive success and therefore are under strong selective control.

### 3.2 Genetic Drift

Genetic drift is a change in gene frequencies that is a consequence of the random gain and loss of individuals (and genotypes) in a population. According to the neutral theory of evolution, most evolution occurs through the random fixation of selectively neutral mutations. This theory predicts that change accumulating in this way will do so at an approximately constant rate (a molecular clock). When this is true, genetic change can be used to interpret the demographic histories of populations.

(a)

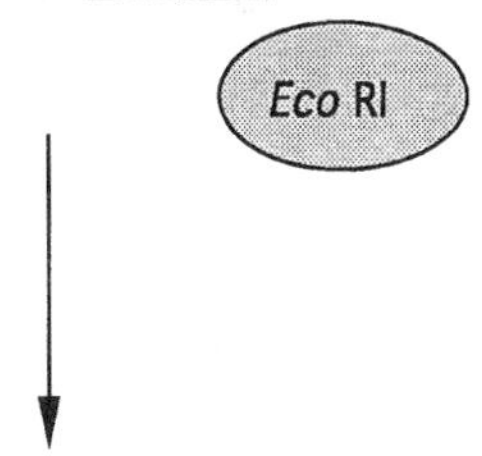




(b)

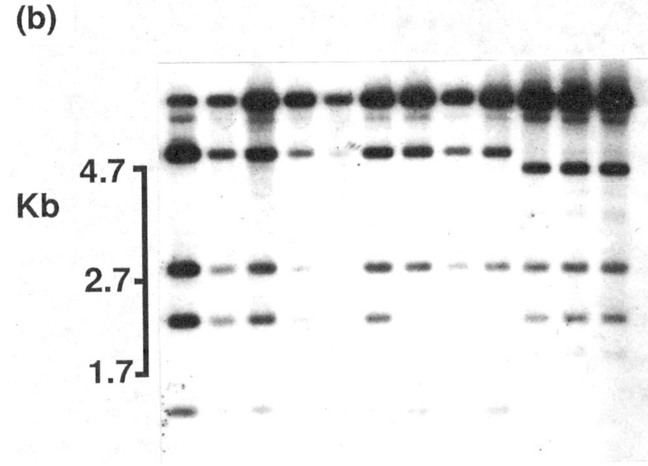

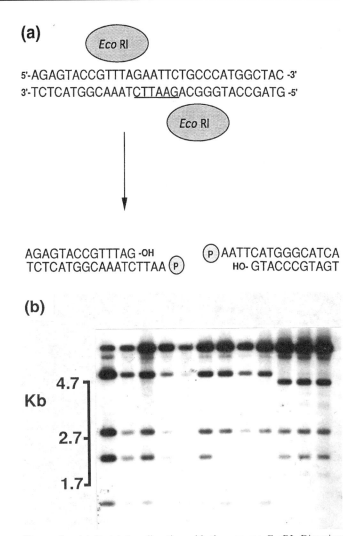

**Figure 2.** (a) Restriction digestion with the enzyme *Eco*RI. Digestion leaves a 3′-hydroxyl group and 5′-phosphate group exposed, and either a 5′ overhang (shown), a 3′ overhang, or blunt ends, depending on the enzyme. *Eco*RI has a 6 bp recognition site; many other common enzymes have 4 bp recognition sites. (b) Panel of 12 lions digested with *Ava*II and hybridized with a heterologous mtDNA probe. (Courtesy of Robyn Hottman and Steve O'Brien).

### 3.3 Molecular Drive

Molecular drive is a term used to describe the way genetic change can be spread through an array of repeated genes, and through individuals in a population, as a consequence of DNA turnover mechanisms. For example, if a mutation occurs in one of the repeats in an array, it can then be spread through the array by unequal crossing over (see Figure 1). Variants homogenized through an array in this way can be homogenized through a population by an analogous but separate process.

## 4 METHODS OF ANALYSIS

There are numerous methods by which genetic variation can be directly or indirectly measured, but just a few have been fundamental to population genetic studies. When genetic variation is being compared between populations or estimated at the population level, it is important to select samples for analysis such that the results are not biased. For example, in sexual species dispersal behavior may vary for males and females, and this could affect kinship among adult males or females sampled from a local area. Furthermore, sampling error will bias small samples. Either a large number of loci or a large number of individuals must be examined.

### 4.1 Enzyme Electrophoresis

When the migration of proteins through an electric field in a gel matrix is measured by enzyme electrophoresis, the charge properties of the protein determine the extent and direction of migration. Tissue homogenized in buffer can be used as the source material, and specific enzymes can then be visualized by immersing the gel in a medium containing the reaction conditions for that enzyme such that a colored dye is incorporated into the end product.

### 4.2 DNA Extraction

DNA is extracted from tissue by removing proteins, fats, and carbohydrates, and then precipitating the DNA out of an aqueous salt solution. This is a straightforward procedure for animal tissues, but it can be difficult with plant material because naturally occurring chemicals may degrade the DNA or inhibit further analysis.

### 4.3 RFLP Analysis

Restriction fragment length polymorphisms (RFLPs) can be analyzed when DNA is digested with a special enzyme called a restriction endonuclease (Figure 2a). These enzymes recognize specific short sequences of DNA (e.g., the enzyme *Eco*RI recognizes the sequence GAATTC) and cleave the DNA at every instance of that sequence. Mutations can occur that create new sites or eliminate old ones. The next step is to separate the DNA fragments by size in a gel matrix. This is done by electrophoresing the samples through an agarose or polyacrylamide gel under buffer. DNA naturally assorts by size (on a roughly logarithmic scale) by this method. The double-stranded DNA in the gel is then made single-stranded (usually with an alkaline solution) and transferred onto a solid membrane that binds DNA (Southern blotting). Then a segment of DNA to be compared between individuals (e.g., a nuclear gene or the mitochondrial genome) is labeled with a radioisotope (or a nonradioactive label) and hybridized to the DNA bound to the membrane. Hybridization is a process whereby complementary single DNA strands combine and reform a double-stranded molecule. Only the fragments of bound DNA that match this DNA probe will be made radioactive. After exposure to X-ray film, these relevant fragments can be visualized as black (exposed) bands on the film (Figure 2b).

### 4.4 DNA Sequencing

The sequencing of the nucleic acid bases that compose a segment of DNA offers the greatest resolution for comparisons of genetic differentiation and permits the greatest scope for interpretation. There are several methods, but the one used most frequently is chain termination. By the chain termination method, a single-stranded DNA template is divided into four reaction mixes, one for each nucleotide base (A, T, C, G). A DNA polymerase (an enzyme that can synthesize DNA along a single-stranded template, starting at a double-stranded/single-stranded interphase), a short (15–25 bases) primer segment of DNA, and all four deoxynucleotide bases are included in each reaction (one of which is labeled either radioactively or with a fluorescent dye). In addition, each reaction will include a smaller quantity of one of the four dideoxynucleotides (either A, T, C, or G). These bases lack the 3′-OH group (Figure 2a) necessary for DNA chain elongation and, therefore, terminate the reaction at every occurrence of that base. A comparison of the length of strands synthesized for each of the four reactions gives the sequence of bases.

### 4.5 PCR Analysis

The polymerase chain reaction (PCR) permits the enzymatic amplification of a specific segment of DNA. This technique has greatly facilitated population level screening of genetic variation by saving time and reducing expense. Two short primers of single-stranded DNA (usually 15–25 bases long) are designed flanking a region of interest. These are often based on published DNA sequences of particular genes or developed from cloned DNA. A programmable heating block is then used to vary the reaction conditions. First the reactions are heated to denature the template DNA, then the temperature is lowered to allow the primers to anneal to the template, then the reaction is heated to an optimal temperature for the polymerase, and the cycle is repeated 20 to 40 times. The key to this technique is the use of a thermostable polymerase (most polymerases would be denatured and rendered inactive by the high temperatures necessary for the DNA denaturation step).

## 5 CHOICE OF GENETIC MARKER AND INTERPRETATION

The nuclear genome is a mosaic of DNA regions under the influence of different mechanisms affecting change, and conserved by selection to different degrees. In addition, there are the mitochondrial and chloroplast genomes with their own unique properties. This variation in the mode and rate of evolution can be used to advantage by the population geneticist. Sections 5.1 to 5.5 describe five types of genetic markers, each with different properties and in some cases, best suited to different levels of analysis.

### 5.1 Enzymes

Enzymes are a functional part of cell physiology and therefore are exposed to natural selection. The apparent annual rate of evolution in these markers varies greatly, from about $10^{-9}$ to $10^{-6}$ change per gene (given and average length of 1000 bp). In some cases—for example, in the histone genes—there is clearly strong selection. This complicates interpretation, but as long as the evolutionary rate is relatively constant over time, these markers are useful for phlyogenetic comparisons. The analytical procedure (enzyme electrophoresis) is relatively fast and inexpensive. The level of resolution possible is relatively low, but this characteristic can be an advantage when the time of divergence between taxa under comparison is great.

### 5.2 Nuclear Genes

Sequence data of nuclear genes, including those encoding enzymes, offer an immediate advantage over enzyme electrophoresis. This is because coding and noncoding regions can be differentiated, and within the coding region of the gene, synonymous and nonsynonymous sites can be distinguished (first vs. second and third codon positions). It also becomes possible to compare sequences and determine the mode of mutation (point mutation, deletion, insertion, rearrangement, etc.), and, in some cases the mechanism. All these factors contribute to the accurate interpretation of the observed variation. For comparisons between populations of the same species, it is necessary to identify both alleles for each individual, and this is typically more difficult by DNA sequencing than it is for enzyme electrophoresis.

### 5.3 Mitochondrial DNA

In multicellular animals, the mitochondrial genome is a double-stranded, closed-circular molecule ranging in length from 15.7 to 19.5 kb. It evolves at a higher rate than comparable nuclear genomic regions (5–10 times faster), and in most cases shows strict maternal transmission (no paternal contribution to the F1 generation). These two factors have made mitochondrial DNA (mtDNA) a very attractive marker for the genetic comparison of populations. A common and very useful method of analysis is to sequence mtDNA regions amplified by PCR. However, interpretation can be complicated by the existence of more than one form in an individual (heteroplasmy), although this is rare. Furthermore, mtDNA is more labile to extreme fluctuations in population size than nuclear DNA, since a single mating pair will pass on four copies of the nuclear genome, but only one haplotype of the mitochondrial genome.

### 5.4 Minisatellite DNA

There is an abundance of repetitive ''satellite'' DNA in the nuclear genome of most complex organisms. These regions are classified according to their composition and structure. Minisatellite arrays are tandem repeats up to 20 to 25 kb long interspersed on most chromosomes. The repeat consists of a highly conserved core sequence, 15 to 40 bp long, and more variable flanking sequences. A high rate of unequal crossing over generates a high level of length variation in these arrays. The ''DNA fingerprinting'' method takes advantage of the conserved minisatellite core sequence and the large number of loci sharing that core repeat. A probe for the core sequence is used in a multilocus RFLP analysis which reveals a ladder of hypervariable bands that can identify an individual or determine paternity with a very small chance of error. While DNA fingerprinting is a powerful tool for the assessment of close kinship, there are various problems with the interpretation of variation in this marker at the population level. For example, variation saturates, and therefore variation within a population can be as great as variation between populations. DNA fingerprinting is more useful to a population geneticist as a tool to reveal reproductive strategy, which in turn has an important effect on the genetic structure of a population.

### 5.5 Microsatellite DNA

Microsatellites are small arrays (typically < 100 bp) of simple di- and trinucleotide repeats; longer repeat elements are less common. These arrays vary in length over time because of DNA slippage. Some loci are hypervariable, but most are less variable than minisatellite loci. In principle, variation at these loci can be interpreted in the same way as for multiple alleles at an enzyme locus. However, there are two main problems. First, little is known about the constancy of DNA slippage rates over time or between loci, and what is known suggests that the rate can vary considerably. This uncertainty will compromise interpretation if the rate is changing over a shorter period than that separating the populations under comparison. Second, since the high level of variation at these loci means that variation in individual reproductive success can have a large effect on overall genotype frequencies, frequencies could change significantly from one generation to the next in the same population, especially in highly polygenous species. Therefore, microsatellites are best applied at the population level only in conjunction with other genetic markers, as well as data on behavior and demographics for the subject species.

## 6 CONCLUSIONS

In general, being able to choose between genetic markers that differ in rates and modes of change has greatly facilitated the interpretation of variation at the population level. However, data still must be interpreted cautiously, and the best approach will be to use several complementary genetic markers.

*See also* DNA Damage and Repair; Genetic Diversity in Microorganisms; Mitochondrial DNA, Evolution of Human; Population-Specific Genetic Markers and Disease.

### *Bibliography*

Hames, B. D., and Rickwood, D. (1990) *Gel Electrophoresis of Proteins, a Practical Approach.* Oxford University Press, Oxford.

Hillis, D. M., and Moritz, C., Eds. (1990) *Molecular Systematics.* Sinauer Associates, Sunderland, MA.

Hoelzel, A. R., and Dover, G. A. (1991) *Molecular Genetic Ecology.* Oxford University Press, Oxford.

———, Ed. (1992) *Molecular Genetic Analysis of Populations, a Practical Approach.* Oxford University Press, Oxford.

Maynard Smith, J. (1989) *Evolutionary Genetics.* University Press, Oxford.

Nei, M. (1987) *Molecular Evolutionary Genetics.* Columbia University Press, New York.

# Genetic Diversity in Microorganisms

*Werner Arber*

### *Key Words*

**Biological Evolution** A nondirected dynamic process of diversification resulting from the steady interplay between spontaneous mutagenesis and natural selection.

**DNA Rearrangement** Results from mostly enzyme-mediated recombination processes, which can be intra- or intermolecular.

**Gene Acquisition** Results from horizontal transfer of genetic information from a donor cell to a receptor cell in transformation, conjugation, or phage-mediated transduction.

**Spontaneous Mutation** Defined here as any alteration of nucleotide sequences occurring to DNA without the intended intervention of an investigator.

**Transposition** DNA rearrangement mediated by a mobile genetic element such as an insertion sequence element or a transposon.

**Variation Generator** Enzyme or enzyme system whose involvement in the generation of genetic variation has been documented.

---

Bacterial genetics took its start about 50 years ago and strongly influenced the development of the molecular genetic strategies and techniques now available to study gene structure and functions of living organisms. Bacterial genetics also revealed natural processes of horizontal gene transfer (transformation, conjugation, phage-mediated transduction) as well as systems (e.g., restriction–modification systems) for holding such gene transfer in tolerably low frequencies to ensure a certain degree of genetic stability. Work with bacterial and bacteriophage systems has also helped to unravel both homologous and nonhomologous enzyme-mediated recombination processes at the molecular level. The acquired knowledge now helps to understand molecular processes contributing to the generation of genetic variation. Among these processes, special attention is given to DNA rearrangements resulting from transposition and from site-specific recombination, which sometimes occurs at secondary crossover sites.

Present knowledge on genetic plasticity of haploid microorganisms offers insights into the molecular basis for the natural interplay between mutagenesis and selection. This approach greatly profits from the short generation times and relatively small genome sizes of haploid microorganisms, which allow one to investigate population genetic questions and to draw conclusions on the mechanisms of evolutionary processes. A general conclusion relates to the existence in the microbial genome of genes with specific evolutionary functions involved in the generation and the limitation of genetic variation.

## 1 INTRODUCTION

Microbial genetics is at the root of molecular genetics and gene technology. Their strategies and methods have become applicable to studies with any kind of living organism. This new advantage has opened interesting possibilities for the investigation and better understanding of the mechanisms of interactions between biological macromolecules supporting life processes. Microbial genetics also offers deeper insights into the process of biological evolution. This contribution concentrates on molecular mechanisms involved in microbial evolution, mainly at the level of spontaneous mutagenesis, and the generation of genetic variation. Instances of genetic diversity can be seen as static pictures of the dynamic process known as biological evolution. At least some of the principles identified in studies with microbial populations are likely to be of general relevance for biological evolution of all forms of life.

## 2 IMPORTANT ROOTS OF MOLECULAR GENETICS

Five discoveries, largely based on work carried out with microorganisms between 1943 and 1953, the decade preceding the publication of the double-helix model for DNA, were essential for the later development of molecular genetics. These are listed briefly as follows.

1. It was realized that bacteria and bacteriophages have genes that can mutate, and that spontaneous mutations normally arise independently of the presence of selective agents. It was also learned that the genetic information of bacteria and of some bacteriophages is carried in DNA molecules rather than in other biological macromolecules such as proteins.
2. The newly discovered phenomena of DNA transformation, bacterial conjugation, and bacteriophage-mediated transduction demonstrated natural means of horizontal gene transfer between different bacterial cells.
3. It was seen that horizontal gene transfer has natural limits, including several systems of host-controlled modification, today known as DNA restriction–modification systems, which were independently described.
4. Mobile genetic elements were identified as sources of genetic instability and seen to represent mediators of genetic rearrangements. Among such rearrangements is the integration of a bacteriophage genome into the genome of its bacterial host strain, which is thereby rendered lysogenic.
5. Structural analysis of DNA molecules led to the double-helix model, offering an understanding of semiconservative DNA replication at the molecular level and thus of information transfer into progeny.

## 3 THE EASE OF MOLECULAR AND POPULATION GENETIC INVESTIGATIONS WITH BACTERIA

Many classical microbial genetic investigations were carried out with *Escherichia coli* K-12. Its genome is a single circular DNA molecule (chromosome) of about $4.7 \times 10^6$ base pairs (bp). It now has an extremely well-established genetics with about 1500 identified genes, which represent, according to estimates, about half of all genes present. The haploid nature of *E. coli* brings about a rapid phenotypic manifestation of mutations. In periods of growth, the rate of spontaneous mutagenesis is about $10^{-9}$ per base pair and per generation. This represents one new mutation in every few hundred cells in each generation. *E. coli* has several well-studied bacteriophages and plasmids. This material facilitates investigations on life processes in these bacteria.

Under good growth conditions the generation time of *E. coli,* measured between one cell division and the next, is very short, in the order of 30 minutes. Upon exponential growth, this leads to a multiplication factor of 1000 every 5 hours. Thus, a population of $10^9$ cells representing 30 generations is reached from an inoculum of a single cell in only 15 hours. This rapid growth rate greatly facilitates population genetic studies and thus investigations on the evolutionary process.

## 4 MECHANISMS AND EFFECTS OF SPONTANEOUS ALTERATIONS OF GENOMIC DNA SEQUENCES

On the filiform DNA molecules of *E. coli* and its bacteriophages and plasmids, the genetic information is stored as linear sequences of nucleotides or base pairs. Genes depend on the presence both of continuous sequences of base pairs (reading frames) that encode specific gene products, usually proteins and of expression control signals that ensure the occurrence of gene expression at the relevant time with the needed efficiency. Mutations can affect reading frames as well as control signals, both of which represent specific DNA sequences. For simplicity, we will use the term ''spontaneous mutation'' to label any type of alteration of a DNA sequence unintended by the investigator.

It is well known that only a relatively small fraction of mutations provide an advantage to the organism and can thus be considered to be useful. Those useful mutations become ''selected'' because they allow the organisms that carry them to succeed by eventually outgrowing other members of a given microbial population. More often, mutations either provide a selective disadvantage or are lethal. In the longer term, such mutations are eliminated from populations. Finally, some mutations are neutral or silent and do not immediately affect the life of the organism. Because of the relatively frequent occurrence of lethal mutations and mutations providing selective disadvantage, a tolerable mutation rate of any haploid organism should be lower than one mutation per genome and per generation. As we have already seen, this criterion is fulfilled by *E. coli.*

Molecular genetic studies have revealed that overall spontaneous mutagenesis is brought about by a large number of specific molecular mechanisms. Studies on spontaneous mutagenesis are difficult because not all these mechanisms act with comparable efficiencies, and these efficiencies may depend on environmental conditions such as temperature. Considerable knowledge has been accumulated, however, and on the basis of available data, we group the mechanisms of spontaneous mutagenesis (i.e., alterations of DNA sequences occurring without intervention of an investigator) into four categories. These are defined in Sections 4.1–4.4.

### 4.1 Infidelity of DNA Replication

An important source of infidelities of DNA replication is likely to depend on tautomeric forms of nucleotides, that is, a structural flexibility inherent to these organic compounds. Base pairing depends on specific structural forms and can result in a long-term mispairing if short-living, unstable tautomeric forms are ''correctly'' used in the synthesis upon DNA replication. For these reasons, we consider mutations resulting in this process not as errors but rather as infidelities. This process is a classical source of nucleotide substitution, and it has an important role in the long-term development of new biological functions.

### 4.2 Environmental Mutagens

In the wide variety of mutagens found in the environment, we include many external and internal chemicals, radiations, and intrinsic metabolic effects. Each of these mutagens and mutagenic conditions contributes in its specific way to the generation of genetic variation.

In part, sequence alterations brought about by replication infidelities and environmental mutagens are efficiently repaired by enzymatic systems. Since, however, the efficiency of such repair is rarely 100%, evolutionarily relevant mutations persist. Some of the repair processes depend on genetic recombination systems, which by themselves can also contribute to spontaneous mutagenesis.

### 4.3 DNA Rearrangements

Indeed, various recombination processes are well known to mediate DNA rearrangements, which often result in new nucleotide sequences. While in haploid organisms general recombination is not essential for propagation, it exerts influence in various ways on the population level as a generator of new sequence varieties. For example, it can bring about sequence duplications and deletions by acting at segments of homology that are carried at different locations in a genome.

Two other, widely spread types of recombination systems are dealt with separately in Sections 5.1 and 5.2: site-specific recombination and transposition. Both are known to contribute to genetic variation. Still other recombination processes, such as the one mediated by DNA gyrase, can perhaps best be grouped as illegitimate recombinations. This group may contain several different molecular mechanisms that act with low efficiency and have remained at least in part unexplained.

### 4.4 DNA Acquisition

While the mutagenesis mechanisms belonging to the three categories explained in Sections 4.1–4.3 are exerted within the microbial genome and can affect any part of the genome, an additional category of spontaneous sequence alterations depends on an external source of genetic information. In DNA acquisition, genetic information indeed originates from an organism other than the one undergoing mutagenesis. This can occur by transformation, by conjugation, or by virus-mediated transduction. In the latter two strategies either a plasmid or a viral genome, respectively, acts as natural gene vector. DNA acquisition represents a particularly interesting source of new genetic information for the receptor bacterium, since the chance that the acquired DNA exerts useful biological functions is quite high in view of the likelihood that it has already assumed the same functions in the donor bacterium. Besides the specific, already mentioned DNA transfer mechanisms, DNA acquisition largely depends on DNA rearrangements belonging to the processes described in Section 4.3. Indeed, some of the mechanisms of horizontal gene transfer depend on a recombinational interaction between the donor genome and the gene vector. Furthermore, the acquired genetic information must find in the receptor cell the possibility of becoming stably inherited. This can be ensured either by recombination into the receptor genome or by its independent maintenance as a plasmid. The recombination mechanism entailed may be general recombination in gene conversion; other possibilities include transposition and site-specific integration.

## 5 ENZYME-MEDIATED DNA REARRANGEMENTS CAN GENERATE GENETIC VARIATIONS

### 5.1 Site-Specific DNA Inversion at Secondary Crossover Sites

Genetic fusions represent the results of joining together segments of two genes (gene fusions) or of two operons (operon fusions) that are not normally together. An operon is a set of often functionally related genes that are copied into messenger RNA (i.e., transcribed) as a single unit. As a result of this organization those genes are coordinately regulated; that is, they are turned on or off at the same time. Therefore, in an operon fusion, one or more genes are put under a different transcription control, but the genes per se remain unchanged. In contrast, gene fusion results in a hybrid gene composed of sequence motifs and often functional domains originating in different genes.

In site-specific DNA inversion, a DNA segment bordered by specific DNA sequences acting as sites of crossing over becomes periodically inverted by the action of the enzyme DNA invertase. Depending on the location of the crossover sites, DNA inversion can give rise to gene fusion or to operon fusion. The underlying flip-flop system can result in microbial populations of two different phenotypic appearances—for example, phage populations with two different host ranges, if the DNA inversion affects the specificity of phage tail fibers, as is the case with phages P1 and μ of *E. coli.*

Occasionally, a DNA sequence that deviates considerably from the efficiently used crossover site and that we call a secondary crossover site, can serve in DNA inversion that thus involves a normal crossover site and a secondary crossover site. This process results in novel DNA arrangements, many of which may not be maintained because of lethal consequences or reduced fitness; but a few new sequences may be beneficial for the life of the organism, hence may be selectively favored. This DNA rearrangement activity can thus be looked at as evolutionarily important. Since many different DNA sequences can serve in this process as secondary crossover sites, although at quite low frequencies, site-specific DNA inversion systems act as variation generators in large populations of microorganisms. I have thus postulated that this evolutionary role of DNA inversion systems may be more important than their much more efficient flip-flop mechanism, which can at most help a microbial population to more readily adapt to two different, frequently encountered environmental conditions. As a matter of fact, other strategies could be used as well for this latter purpose.

### 5.2 Transposition of Mobile Genetic Elements

Already nine different mobile genetic elements have been found to reside, often in several copies, in the chromosome of *E. coli* K-12 derivatives. This adds up to occupation of about 1% of the chromosomal length by such insertion sequences, also called IS elements. At rates on the order of $10^{-6}$ per individual IS element and per cell generation, these mobile genetic elements undergo transpositional DNA rearrangements. These include simple transposition of an element and more complex DNA rearrangments such as DNA inversion, deletion formation, and the cointegration of two DNA molecules. Because of different degrees of specificity in the target selection upon transposition, the IS-mediated DNA rearrangements are neither strictly reproducible nor fully random.

Transposition activities thus also act as variation generators. In addition to DNA rearrangements mediated by the enzyme transposase, which is usually encoded by the mobile DNA element itself, other DNA rearrangements just take advantage of extended segments of DNA homologies, at the sites of residence of identical IS elements, at which general recombination can act. Altogether, IS elements represent a major source of genetic plasticity of microorganisms.

Transposition occurs not only in growing populations of bacteria, but also in prolonged phases of rest. This is readily seen with bacterial cultures stored at room temperature in stabs (little vials containing a small volume of growth medium in agar). Stabs are inoculated with a drop of a bacterial culture taken up with a platinum loop, which is inserted (''stabbed'') from the top to the bottom of the agar. After overnight incubation, the stab is tightly sealed and stored at room temperature. Most strains of *E. coli* keep viable in stabs during several decades of storage. That IS elements exert transpositional activities under these storage conditions is easily seen as follows.

A stab can be opened at any time, a small portion of the bacterial culture removed, and the bacteria well suspended in liquid medium. After appropriate dilution, bacteria are spread on solid medium. Individual colonies grown upon overnight incubation are then isolated. DNA from such subclones is extracted and fragmented with a restriction enzyme. The DNA fragments are separated by gel electrophoresis. Southern hybridization with appropriate hybridization probes can then show whether different subclones reveal restriction fragment length polymorphisms (RFLPs), which are indicative of mutations having occurred during storage.

If this method is applied to subclones isolated from old stab cultures, and if DNA sequences from IS elements serve as hybridization probes, an extensive polymorphism is revealed. No or only little polymorphism is seen with hybridization probes from unique chromosomal genes. Good evidence is available that transposition represents a major source of this genomic plasticity observed in stabs, which at most allow for a very residual growth at the expense of dead cells. One can conclude that the enzymes promoting transposition are steadily present in the stored stabs. Indeed, the IS-related polymorphism increases linearly with time of storage for periods as long as 30 years. For a culture of *E. coli* strain W3110, it turned out to be difficult to guess which genome structure was at the origin of the bacterial population studied after 30 years of storage. On the average, each subclone had suffered about a dozen RFLP changes as identified with hybridization probes from eight different residual IS elements, of which IS*5* was the most active. Nevertheless a very interesting pedigree of the analyzed individual subclones could be drawn, and this offered an amazing insight into the genetic plasticity of *E. coli.*

## 6 PROMOTION AND LIMITATION OF GENE ACQUISITION

Transpositional activities and general recombination acting at IS elements, which are at different chromosomal locations, are a source of associating and dissociating chromosomal genes with natural gene vectors. These mechanisms have been well studied with conjugative plasmids and with bacteriophage genomes serving in specialized transduction. For example, composite transposons, which are defined as two identical IS elements flanking a segment of genomic DNA (often with more than one gene unrelated to the transposition process), are known to occasionally transpose into a natural gene vector and, after their transfer into a receptor cell, to transpose again into the receptor chromosome. Hence, together with other mechanisms, such as site-specific and illegitimate types of recombination, transposition also represents an important promoter of horizontal gene transfer.

One also finds several natural factors that seriously limit gene acquisition. Transformation, conjugation, and transduction depend on surface compatibilities of the bacteria involved. Furthermore, upon penetration of donor DNA into receptor cells, the DNA is very often confronted with restriction endonucleases. These enzymes cause a fragmentation of the invading foreign DNA, which is subsequently completely degraded. Before fragments become degraded, however, they are recombinogenic and may find a chance to incorporate all or part of their genetic information into the host genome. Therefore, we interpret the role of restriction systems as follows: they keep the rate of DNA acquisition low, and at the same time they stimulate the fixation of relatively small segments of acquired DNA to the receptor genome. This strategy of acquisition in small steps can best offer microbial populations the chance to occasionally extend their biological capacities without extensive risk of disturbing the functional harmony of the receptor cell by acquiring too many different functions at once. These considerations have their relevance at the level of selection of hybrids resulting from horizontal DNA transfer. This selection is exerted as one of the last steps in the acquisition process.

## 7 GENETIC DIVERSITY REFLECTS THE PRESENT STATE OF BIOLOGICAL EVOLUTION

Biological evolution is a steady, dynamic process. The four categories of molecular mechanisms of mutagenesis described in Section 4 continuously exert their interactions on microbial populations, so that genetic diversity should steadily increase. However, while some of the newly arising DNA arrangements disappear rapidly by appropriate repair processes, others may at first remain but then submit to selection pressure. Normally, genetic diversity may be kept at a more or less constant level by such selection pressure, together with sampling due to the limited size of the biosphere of our planet, which offers space to only about $10^{30}$ living cells. This balance may hold as long as the overall environmental conditions of life do not undergo drastic alterations, which would of course seriously affect selection. Hence the importance of a relatively large pool of gene functions and life forms, together with some genetic plasticity, to ensure good chances of genetic adaptation in times of changing living conditions.

Gene acquisition can be seen as a very efficient means by which a microbial strain extends its genetic capacities. This function may be of particular relevance if selective pressure undergoes drastic changes—as it has been the case, for example, for enterobacteria since antibiotics became widely used in human and veterinary medicine. Much of our present knowledge of gene acquisition stems indeed from studies on the spreading and selection of drug resistance determinants. In drawing the evolutionary tree of bacteria, DNA acquisition, which we can consider to be a strategy for sharing in the success of others, should be accounted for by adding horizontal shunts between individual branches. Although usually only small DNA segments flow through such shunts at one time or another in horizontal gene transfer, in the vertical flux of genes from one generation to the next in the growing branches of the tree, the entire genome is of course steadily a target of any one of the mechanisms belonging to the first three categories of spontaneous mutagenesis. In comparison with DNA acquisition, and in considering efficiencies per single event, DNA rearrangements internal to a genome may be evolutionarily less efficient and nucleotide substitution still less. However, all these processes make important contributions to the evolutionary process. Nucleotide substitution is a major source of new biological functions, and it also contributes to the amelioration of existing functions. DNA rearrangements can bring about important improvements of existing capacities by, for example, the fusion of functional domains and of DNA sequence motifs. In considering the evolutionary process, we should keep in mind that it largely depends both on the diversity and on the kind of genetic information already available, since time spans needed to develop completely new functions without making use of existing sequences are very long, not in the least because of the extremely large number of specific sequences that are possible in the linear arrangement of nucleotides of a gene and of a genome.

## 8 CONCLUSIONS

Rather than being the result of an accumulation of errors, biological evolution appears to depend on many different specific biological functions. Enzymes acting as variation generators make fundamental contributions to spontaneous mutagenesis, including the rearrangement of existing sequences. In addition, more or less complex enzyme systems and organelles promote horizontal gene transfer and regulate its efficiency. With only a few exceptions, enzymes involved in these processes are not required for life of individuals in populations, and the biological significance of these agents becomes manifested only at the population level. As a matter of fact, spontaneous mutagenesis often does more to hamper the life of an affected individual than it does to convey benefits. Any benefit becomes obvious only at the population level by contributions of occurring mutations to a steady evolution. It is a very interesting concept for our world view to know that among the genes encoded in the genomes of living organisms there are, besides a large number of genes essential for each individual life, also other genes, the products of which can ensure long-term development of life in a wide variety of phenotypic forms. Molecular genetic studies carried out with haploid microorganisms have brought about good evidence for these conclusions. It is quite likely that they might also apply to higher organisms, but these life forms are much less accessible than microorgansims to experimental approaches in search of evidence.

*See also* Bacterial Growth and Division; *E. coli* Genome; Genetic Analysis of Populations; Genetics.

### *Bibliography*

Arber, W. (1991) Elements in microbial evolution. *J. Mol. Evol.* 33:4–12.
———. (1993) Evolution of prokaryotic genomes. *Gene,* 135:49–56.
———. (1994) Bacteriophage transduction, in *Encyclopedia of Virology,* pp. 107–113. Academic Press, London.
———. (1995) The generation of variation in bacterial genomes. *J. Mol. Evol.*
Drake, J. W. (1994) Spontaneous mutation. *Annu. Rev. Genet.* 25:125–146.
Galas, D. J., and Chandler, M. (1989) Bacterial insertion sequences, in *Mobile DNA,* D. E. Berg and M. M. Howe, Eds., pp. 109–162. American Society for Microbiology, Washington, DC.
Glasgow, A. C., Hugues, K. T., and Simon, M. L. (1989) Bacterial DNA inversion systems, in *Mobile DNA,* D. E. Berg and M. M. Howe, Eds., pp. 637–659. American Society for Microbiology, Washington, DC.
Kucherlapati, R., and Smith, G. R., Eds. (1988) *Genetic Recombination.* American Society for Microbiology, Washington, DC.
Moses, R. E., and Summers, W. C., Eds. (1988) *DNA Replication and Mutagenesis.* American Society for Microbiology, Washington, DC.

# Genetic Immunization

*Stephen A. Johnston and De-chu Tang*

### *Key Words*

**Biolistic Device** A device used for delivering genetic material into living cells in a nonlethal manner by inoculating microprojectiles coated with biomolecules directly into either cultured cells or cells in a live organism.

**Cellular Immunity** Activation of cytotoxic T lymphocytes and/or macrophages against specific antigens.

**Humoral Immunity** Antibody production by antigen-activated B lymphocytes.

**Immunization** The elicitation of humoral and/or cellular immunity against specific antigens by presenting the antigen to the immune system in one way or another.

**Viral or Bacterial Vectors** Modified viral or bacterial genomes accommodating foreign genes for the delivery of active genes into cells.

Genetic immunization (GI) is the approach for eliciting immune responses against specific proteins by expressing genes encoding the proteins in an animal's cells. The substantial antigen amplification and immune stimulation resulting from prolonged antigen presentation in vivo would be expected to induce immune responses against the antigen. A variety of biological and physical techniques have been developed for the delivery of active genes into animal cells. Because the manipulation of DNA is simple and versatile, and DNA can be purified to homogeneity with no difficulty, genetic immunization can be a time-, money-, and labor-saving approach compared with the preparation of protein vaccines for eliciting immune responses. Furthermore, because genetic immunization mimics natural infections by expressing the gene in a restricted subset of self-cells for extensive periods of time, it may induce antibody production of higher magnitude and longer duration than conventional means, and it may elicit specific T-cell responses that are normally not achievable through the inoculation of protein immunogens.

## 1 INTRODUCTION

Vaccination represents the most important medical intervention for preventing diseases. It is also an indispensable tool for modern biological research. Although numerous human and animal lives are protected each year by immunization, problems such as expense and refrigeration limit the application of existing technology, and many diseases remain unapproachable with preventive vaccines because the vaccines either are not yet available or are of insufficient efficacy. The past few years, however, have witnessed radical changes in the approach to vaccine development. The conventional way for eliciting immune responses is to inoculate humans or animals with killed or live attenuated pathogens, and more recently, purified protein antigens. Using killed or live attenuated pathogens carries a certain risk of escapes from the killing/inactivation process. Protein purification can be time-consuming, costly, and sometimes difficult or impossible. Recent progress in biotechnology has made it possible to clone genes encoding specific proteins and subsequently to have the genes introduced into the animal for the production of proteins by cells within the body. The manipulation of DNA is easier, less expensive, and more versatile than protein manipulation. The delivery of specific genes into animals for the production of immunogens in vivo has been dubbed ''genetic immunization.'' The first documented vaccine against diseases using vaccinia virus against smallpox is actually a precedent to genetic immunization. The difference is that in genetic immunization the specific genes responsible for immunization are well characterized and portable in various expression vectors. Preparation of nucleic acids for immunization is not only time-saving and more cost-effective than that of proteins, but also it may be safer and more specific because DNA can be purified to homogeneity with ease. Furthermore, genetic immunization speculatively may elicit a more potent immune profile than that resulting from the injection of killed pathogens or protein immunogens (e.g., duration and magnitude of antibody production, T-cell response, etc.) because genetic immunization mimics natural infections by expressing foreign genes in a restricted subset of self-cells for extensive periods of time.

The ideal immunization protocol (1) should have a very low level of side effects and no potential for causing problems in individuals with an impairment of the immune system, (2) should not be tumorigenic or toxic, and (3) should not disseminate within the vaccinee or to other individuals, or contaminate the environment. Effective and long-lasting humoral and cellular immunities are essential. The vaccine should be produced with low cost, and the technique for administration should be simple. While such an ideal vaccine remains an elusive goal, genetic immunization is more ''ideal'' than conventional vaccination for a number of reasons explained throughout this entry. Minimally, it provides an alternative that can be compared with more classical vaccine approaches.

The technology that is essential for the success of genetic immunization is that the genetic material must be delivered into cells in a safe manner, and gene expression must produce sufficient quantities of proteins for the elicitation of immune responses. One type of genetic immunization involves introducing the antigen-producing gene into an infectious, but not pathogenic, carrier. In the other form of genetic immunization the DNA is physically introduced into host cells without using an infectious agent. In all these methods, before gene expression can occur, the delivered DNA must cross subcellular barriers to achieve the correct localization. These techniques are described in the following sections.

## 2 TECHNIQUES

### 2.1 Delivery of Active Genes into Cells by Viral and Bacterial Vectors

Several viral and bacterial live recombinant vaccine vehicles have been developed to produce a new generation of vaccines for eliciting immune responses. These microorganisms have evolved specific mechanisms to accomplish delivery of their genetic material to the host cell. Modern biotechniques have made it possible to stitch foreign genes of interest into viral and bacterial genomes and, as a consequence, the expression of heterologous genes in cells is achieved by exploiting their efficient cellular entry processes. The most advanced vectors as candidates for recombinant vaccines include vaccinia viruses, adenoviruses, retroviruses, BCG (bacillus Calmette–Guerin), *Escherichia coli, Salmonella typhimurium,* and all other poxviruses, herpesviruses, and polioviruses.

Immunity to vectors is a concern for adults who have had prior vaccination or infection. It is not clear how preexisting immunity to any of the adenovirus serotypes would affect primary or secondary immunization with an adenovirus recombinant vector. The efficacy of vaccinia recombinants is also questionable in individuals who have been vaccinated with vaccinia. Existing immunity to these viruses may decrease the immune response to the recombinant protein. Retroviruses are usually slightly immunogenic, and many are immunosuppressive.

### 2.2 Reimplantation of Autologous Cells Expressing the Antigen

The administration to animals of autologous cells that have been manipulated ex vivo is also a viable approach for eliciting immune responses. Target cells are first removed from the body; then protein antigens are produced by gene transfer in vitro, and cells are subsequently reintroduced into the body in a variety of ways, including simple infusion or implantation, and by more complex modes, such as encapsulation within porous membranes. Although more cumbersome in procedure than direct in vivo gene transfer, the ex vivo approach value with respect to safety issues because the specific cell types that express the gene are known and can be controlled within containment.

### 2.3 Intramuscular Injection of DNA

Injection of pure RNA or DNA directly into skeletal muscle leads to significant expression of heterologous genes within the muscle cells. The mechanism of entry of polynucleotides into the muscle cells has not been defined, although it is possible that muscle cells can take up surrounding DNA/RNA when their plasma membranes are transiently disrupted during exercise that stretches muscle cells. It has recently been shown that this approach can also elicit immune responses against a specific protein when the gene encoding the protein is expressed in muscle cells after DNA injection.

### 2.4 Biolistic Delivery of DNA into Skin Cells

We have inoculated gold microprojectiles coated with plasmids expressing human proteins into the skin of mice and have demonstrated humoral immune responses to several antigens. This manipulation was achieved with a handheld biolistic device that can propel DNA directly into cells. The primary response could be augmented with further DNA boosts. Unlike conventional protein-mediated immunization, which usually peaks a few weeks after immunization and then wanes, DNA-mediated antibody levels were sustained for many months after a single inoculation. Although the cellular immunity in response to such a treatment has not been scrutinized, we would be surprised if it had not been developed because the treatment was so much like natural infections in the in vivo setting. Unlike other methods using protein or infectious vectors as the inoculant, plasmid DNA can be purified to homogeneity and is noninfectious; its safety margin over infectious biological vectors, therefore, is obvious. Furthermore, it is possible to restrict inoculations to the outermost skin cells, which are usually shed off in a few days, thus strengthening the safety of the new methods since there will be no long-term persistence of modified cells. In addition, the plasmid DNA approach offers more versatility than other means because the size constraint of biological vectors is obviated when pure plasmid is used as the inoculant; moreover, the immune response may be manipulated more easily, since it is easier to manipulate plasmids by the addition of anchor or signal sequences that move certain proteins to the plasma membrane or secreted into the local tissue environment. The heat stability of plasmid DNA is another advantage because refrigeration would no longer be a process requirement.

## 3 APPLICATIONS

A goal in the development of vaccines is to induce high titer and long-lasting immunity for targeted intervention against diseases with a minimum number of inoculations. An advantage of genetic immunization is the persistence in vivo of antigen-producing cells that could provide continuing immunization for extended time intervals. Genetic immunization has another advantage-namely, that the often difficult steps of protein purification and combination with adjuvant, both routinely required for vaccine development, are eliminated. Since genetic immunization does not require the isolation of proteins, it is especially valuable for antigens that may lose conformational epitopes when extracted and purified biochemically. Genetic immunization is already an efficient way for eliciting immune reactions in laboratory animals, and theoretically it can be expanded to vaccinating farm animals for the development of protective immunities against diseases with little constraint, thus greatly reducing costs. It may also have application to humans, after further testing and development.

## 4 PERSPECTIVES

Genetic immunization is a recent development in vaccine technology. We are now capable of expressing an extrinsic antigen in the authentic tissue environment in a live animal and of presenting the antigen in a native configuration in the tissue context; as a consequence, we can induce both humoral and cellular immune responses. A major barrier to the application of genetic immunization lies in the complications associated with its use. Safety is obviously a major concern if genetic immunization is to be contemplated for human use. Considerable background information has been acquired for some of the viral and bacterial vectors during the extensive use of these agents as vaccines. Advances in biotechnology have made it possible to reduce virulence by alteration of the viral and bacterial genomes. Attenuation may also be obtained through the introduction of lymphokine genes. However, concerns about potential oncogenicity and pathogenicity in infants and immunocompromised individuals have outweighed any need in the general population. With regard to safety, perhaps the most promising way to conduct genetic immunization is the biolistic approach, which operates by inoculating microprojectiles coated with purified plasmid DNA into the outermost skin cells, which have a high turnover rate. Because the DNA is free of any infectious agents, and the modified skin cells are shed during normal processes, biohazard concerns can be minimized.

A current challenge in tumor immunobiology is the elicitation of antitumor immunity by expressing combinations of cytokine genes in tumor cells. Cytokines continuously produced in tumor cells may act as adjuvants for recruiting lymphocytes to tumor deposits and activating effector cells to tumoricidal states. The gene transfer techniques described here are being evaluated for their capabilities to achieve efficient tumor-specific gene expression.

Genetic immunization may be a more economical and potent approach for eliciting immune responses than classical modes of vaccination. It remains to be seen whether it will become a practical method in human vaccination programs.

*See also* HUMAN GENETIC PREDISPOSITION TO DISEASE; MOLECULAR GENETIC MEDICINE; VACCINE BIOTECHNOLOGY.

### *Bibliography*

Johnston, S. A., and Tang, D. (1993) The use of microparticle injection to introduce genes into animal cells in vitro and in vivo. In *Genetic Engineering,* Vol. 15, J. K. Setlow, Ed. pp. 225–236. Plenum Press, New York.

Lanzavecchia, A. (1993) Identifying strategies for immune intervention. *Science,* 260:937–944.

Moss, B. (1991) Vaccinia virus: A tool for research and vaccine development. *Science,* 252:1662–1667.

Mulligan, R. C. (1993) The basic science of gene therapy. *Science,* 260:926–932.



## Genetic Manipulation of Plant Cells: *see* Plant Cells, Genetic Manipulation of.

## Genetic Markers: *see* Population-Specific Genetic Markers and Disease.

## Genetic Medicine: *see* Molecular Genetic Medicine.

## Genetic Mutations and Aging: *see* Aging, Genetic Mutations in.



# GENETICS

*D. Peter Snustad*

### *Key Words*

**Codon** The unit of three contiguous nucleotides in messenger RNA specifying the incorporation of one amino acid in the polypeptide produced by translating that mRNA on the polyribosomes.

**Complementation Test** The introduction of two recessive mutations into the same cell but on different chromosomes (a trans heterozygote) to determine whether the mutations are in the same gene or in two different genes. If both mutations are in the same gene, the $m_1+/+m_2$ heterozygote will exhibit a mutant phenotype, whereas if they are in two different genes, the trans heterozygote will exhibit the wild-type phenotype.

**Gene** The basic unit of genetic information. The unit of function specifying the structure of one primary product, usually one polypeptide chain, and defined operationally by the complementation test.

**Mutation** Heritable change in the structure of the genetic material of an organism. When used in the broad sense, mutations include both "point mutations," involving changes in the structure of individual genes, and gross changes in chromosome structure (chromosome aberrations). When used in the narrow sense, the term "mutation" applies only to "point mutations." Mutation is used to refer both to the *process* by which the change occurs and to the *result* of the process, the *alteration* in the gene or genetic material.

**Recombination** The generation in progeny of combinations of genes that were not present together in either parent, by (1) independent assortment of nonhomologous chromosomes during meiosis or (2) crossing over (breakage and exchange of parts) of homologous chromosomes during meiosis or mitosis.

---

The phenotype of a living organism is controlled by its genotype, the summation of its genetic information, acting within the constraints imposed by the environment in which the organism exists. Much of the genetic material of an organism is organized into basic functional units called genes, with each gene encoding one primary gene product, most commonly one polypeptide. The genetic information of all living organisms, whether viruses, bacteria, corn plants, or humans, is stored in the sequence of bases (purines and pyrimidines) in the deoxyribonucleic acid (DNA) present in their chromosomes. In some viruses, the genetic information is stored in the sequence of bases in ribonucleic acid (RNA). The genetic information is encoded using a four-letter alphabet, namely, the four bases: adenine (A), guanine (G), cytosine (C), and thymine (T). In RNA, uracil (U) replaces the thymine present in DNA. The genetic material of an organism must carry out two essential functions: (1) the genotypic function, transmission of the genetic information from generation to generation, and (2) the phenotypic function, directing the growth and development of the offspring into mature, reproductive adults. The genetic material of an organism is not static; it changes or mutates on occasion to provide new genetic variability, which provides the raw material for evolution. Recombination of genetic material occurs by the independent assortment of nonhomologous chromosomes and by crossing over between homologous chromosomes to provide new combinations of genes and thus new phenotypes on which natural selection can act to allow evolution to progress.

## 1 GENETIC INFORMATION

The genetic information of living organisms is stored in large macromolecules called nucleic acids. These nucleic acids are of two types: DNA, which contains the sugar 2′-deoxyribose, and RNA, which contains the sugar ribose. In all eukaryotic organisms, the genetic information is stored in giant DNA molecules located in from one to many chromosomes, the number depending on the species. In some viruses that contain no DNA, the genetic information is stored in RNA.

### 1.1 Four-Letter Alphabet

The genetic information is stored in nucleic acids using a four-letter alphabet composed of the four bases adenine (A), guanine (G), cytosine (C) and thymine (T) in DNA; uracil (U) replaces thymine in RNA. Although there are only four letters in this genetic alphabet, a vast amount of information can be stored because the nucleic acids utilized for storage are very large. For example, one complete copy of the human genome (all the genetic information in one complete set of the human chromosomes) contains three billion ($3 \times 10^9$) base pairs of DNA. Since the number of possible sequences of four letters used $n$ at a time is $4^n$, one can see that with $n = 3 \times 10^9$, the human genome has the capacity to store a huge amount of information.

### 1.2 The Gene: The Basic Unit of Function

The basic functional unit of genetic information is the gene, defined operationally by the complementation test and most commonly specifying the amino acid sequence of one polypeptide chain. Different forms of a given gene are called alleles. The wild-type alleles of a gene are those that exist at relatively high frequencies in natural populations and yield wild-type or "normal" phenotypes; they are usually symbolized by a + or by a symbol with a + superscript (e.g., $w^+$ for the allele that yields wild-type red eyes in fruit flies). Alleles of a gene that result in abnormal or non-wild-type phenotypes are called mutant alleles; they are usually symbolized by one to three italic letters (e.g., $w$ and $w^{ap}$ for the alleles that cause white and apricot eye color, respectively, in fruit flies). Many eukaryotes, such as corn plants, fruit flies, and humans, contain two copies of their genome in most cells, thus two copies of each of their chromosomes; such eukaryotes are called diploids. Diploid organisms may contain two different alleles of any given gene, in which case they are heterozygous (e.g., $w/w^+$; $w^{ap}/w^+$; $w/w^{ap}$) or two identical copies of a given gene, in which case they are homozygous (e.g., $w^+/w^+$; $w/w$; $w^{ap}/w^{ap}$). A $w^{ap}/w^+$ heterozygous fruit fly has wild-type red eyes. The $w^+$ allele is expressed in this heterozygous fly; $w^+$ is thus called the dominant allele. The $w^{ap}$ allele is not expressed in this heterozygous fly; it is said to be recessive because its effect on the phenotype is masked by the $w^+$ allele.

The complementation test is performed by producing cells or organisms that contain two recessive mutant genes located on two different chromosomes (i.e., trans heterozygotes) and determining whether these cells or organisms have mutant or wild-type phenotypes. If the two mutant genes are allelic—that is, if the defects or mutations are in the same gene—the trans heterozygote will have a mutant phenotype. If the two mutant genes are not allelic—that is, if the mutations are in two different genes—the trans heterozygote will have the wild-type phenotype. The rationale behind the complementation test is illustrated in Figure 1.

## 2 DNA REPLICATION: THE GENOTYPIC FUNCTION

The genetic information of an organism is transmitted from cell to cell during development and from generation to generation during

A. Two mutations in the same gene.

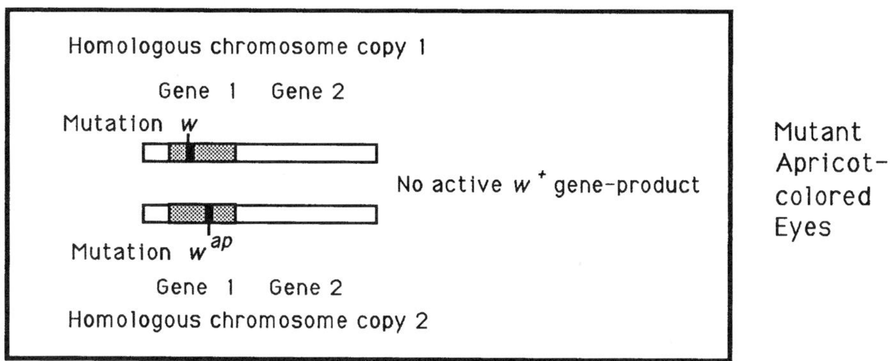


B. Two mutations in two different genes.

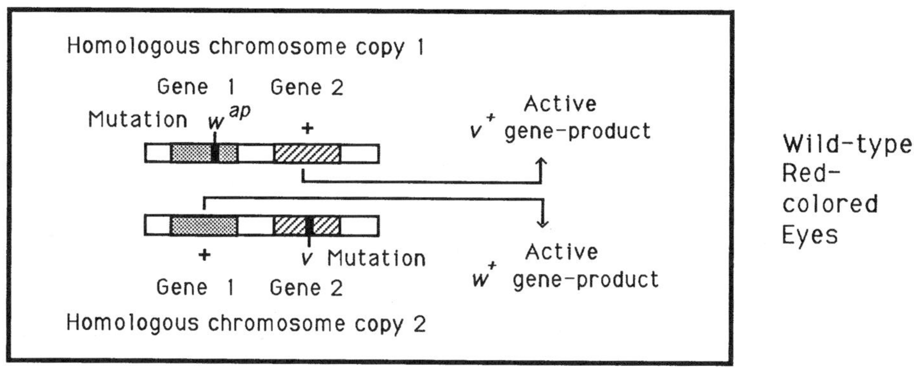


**Figure 1.** Principle of the complementation test used in the operational definition of genes, the basic units of function of genetic material. The operation is to place the two recessive mutations in question in the same cell or cells of a multicellular organism on two separate chromosomes (i.e., to construct a trans heterozygote) and to determine whether this cell or organism has a mutant or a wild-type phenotype. If the phenotype is mutant, the two mutations are in the same gene; this is illustrated for the *w* (white eyes) and $w^{ap}$ (apricot eyes) mutations of *Drosophila* in (A). If the phenotype of the trans heterozygote is wild type, the mutations are in two different genes and are said to complement each other. Complementation between the $w^{ap}$ and *v* (vermilion eye color) mutations of *Drosophila* is illustrated in (B); note that active (wild-type) products of both genes ($w^+$ and $v^+$) are present in the trans heterozygote (B)—thus, the wild-type phenotype.

reproduction by the accurate replication of the sequence of bases in nucleic acids based on the precise base pairing in double-stranded nucleic acids: A with T and G with C. Two properly base-paired strands of DNA are complementary and contain the same genetic information. Thus, when the two strands of a parental double helix of DNA separate, the base sequence of each parental strand will serve as a template for the synthesis of a new complementary strand, as shown in Structure 1.

## 3 GENE EXPRESSION: THE PHENOTYPIC FUNCTION

The genetic information controls the morphogenesis of the organism, be it a virus, a bacterium, a plant, or an animal. This genetic information must be expressed accurately both spatially and temporally to produce the appropriate three-dimensional form of the organism. In multicellular organisms, the genetic information must control the growth and differentiation of the organism from the single-celled zygote to the mature adult. To accomplish this phenotypic function, each gene of an organism must be expressed at the

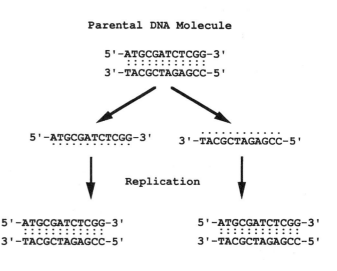

**Parental DNA Molecule**

```
5'-ATGCGATCTCGG-3'
   ::::::::::::
3'-TACGCTAGAGCC-5'

5'-ATGCGATCTCGG-3'      3'-TACGCTAGAGCC-5'

              Replication

5'-ATGCGATCTCGG-3'      5'-ATGCGATCTCGG-3'
   ::::::::::::            ::::::::::::
3'-TACGCTAGAGCC-5'      3'-TACGCTAGAGCC-5'
```

**Two Progeny DNA Molecules**

**Structure 1**

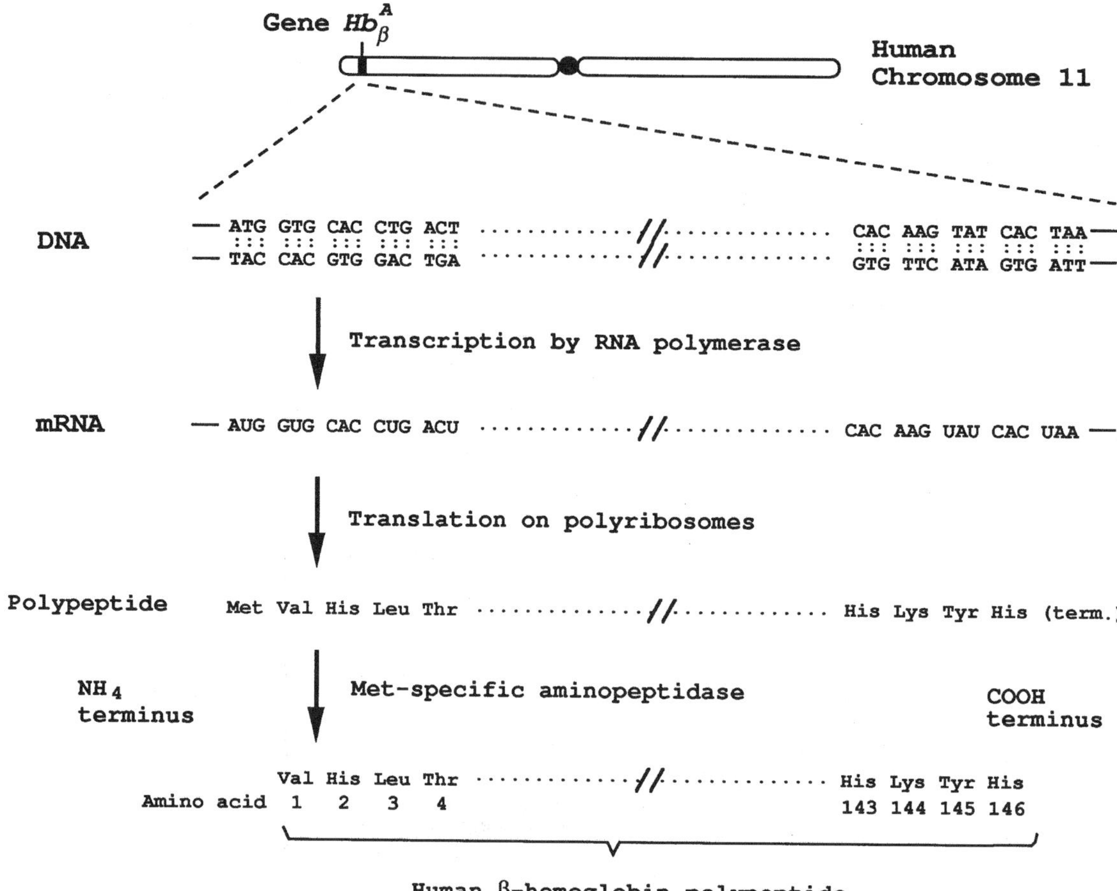


**Figure 2.** Schematic diagram showing the first steps in the expression of $Hb_{\beta}^{A}$, the human gene encoding β-hemoglobin. The steps shown are transcription, translation, and proteolytic removal of the amino-terminal methionine residue from the primary translation product. For simplicity, only the terminal portions of the coding sequence and the polypeptide product are shown.

proper time and in the proper cells during development. The initial steps in the pathways of gene expression, transcription, and translation are quite well elucidated; Figure 2 illustrates these steps for the expression of the human β-hemoglobin gene. In contrast, we are just beginning to understand morphogenesis at the cell, tissue, and organ levels.

### 3.1 Transcription

The first step in gene expression, transcription, involves converting genetic information stored in the form of base pairs in double-stranded DNA into the sequence of bases in a single-stranded molecule of messenger RNA. This process, which is catalyzed by enzymes called RNA polymerases, occurs when one strand of the DNA is used as a template to synthesize a complementary strand of RNA using the same base-pairing rules that apply in DNA replication except that uracil is incorporated into RNA at positions where thymine would be present in DNA (see Figure 2, ''Transcription'' row).

### 3.2 Translation

During translation, the sequence of bases in the mRNA molecule is converted (''translated'') into the specified sequence of amino acids in the polypeptide gene product according to the rules of the genetic code (Figure 2, ''Translation'' row). Each amino acid is specified by one codon, a triplet of three adjacent bases in the mRNA. Translation is a complex process occurring on cytoplasmic macromolecular structures called ribosomes and requiring the participation of many other macromolecules.

### 3.3 Complex Pathways of Gene Action

The pathway through which a gene exerts its effect on the phenotype of the organism is often long and complex, especially in multicellular eukaryotes (Figure 3). The pathways of gene action frequently involve protein–protein and other macromolecular interactions, cell–cell interactions and intercellular communication by hormones and other signal molecules, tissue and organ interactions, and restrictions imposed by environmental factors.

### 3.4 Regulation of Gene Expression

In all organisms, gene expression is highly regulated; thus energy is used to synthesize gene products only when those products are needed for growth and differentiation of the organism. In higher eukaryotes, only a small proportion of the genes in the genome are being expressed in any one cell type. Thus, gene expression is highly programmed to ensure that genes needed to make neurons are turned on only in developing nerve cells, genes needed to make

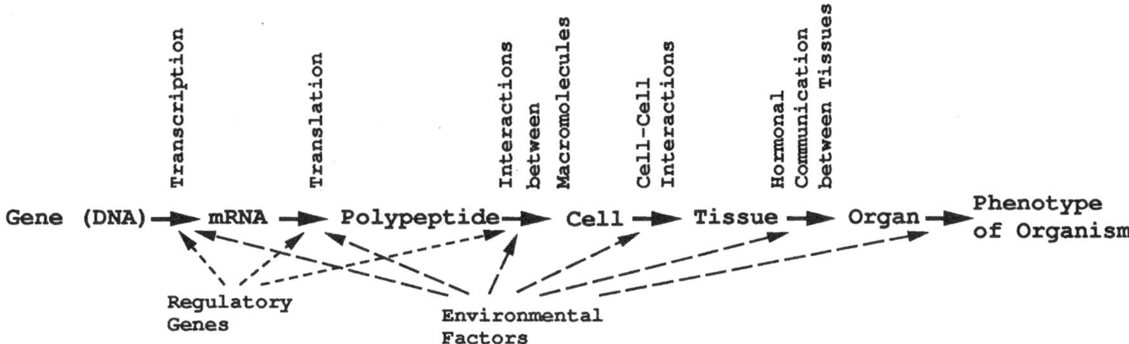


Figure 3. Pathway of gene expression, showing some of the components that may influence the effect a given gene will have on the phenotype of an organism.

red blood cells are expressed only in progenitors of erythrocytes, and so on. The mechanisms by which gene expression is regulated are numerous and beyond the scope of this entry.

## 4 MUTATION: THE ULTIMATE SOURCE OF ALL NEW GENETIC VARIABILITY

Although genetic information must be transmitted from generation to generation with considerable accuracy, it is not static. Rather, it undergoes occasional change or mutation to produce new genetic variability, which provides the raw material for ongoing evolution. The new variant genes produced by mutation are called mutant alleles and often result in abnormal or mutant phenotypes. When used in the narrow sense, mutation refers only to changes in the structures of individual genes. However, in the broad sense, mutation refers to any heritable change in the genetic material and includes gross changes in chromosome structure or chromosome aberrations. There are four types of gross chromosome rearrangement: duplications, deletions, inversions, and translocations. A duplication is the occurrence of a segment of a chromosome in two or more copies per genome. A deletion results from the loss of a segment of a chromosome. An inversion occurs when an internal segment of a chromosome is turned end-for-end relative to its orientation in a normal chromosome. A translocation results when a segment of a chromosome is broken off and becomes attached to another chromosome.

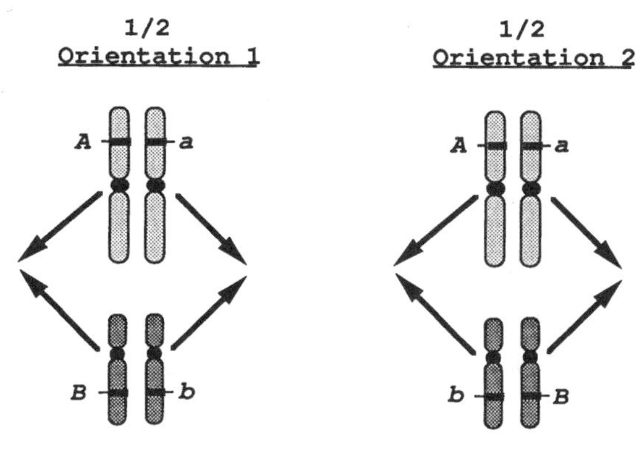


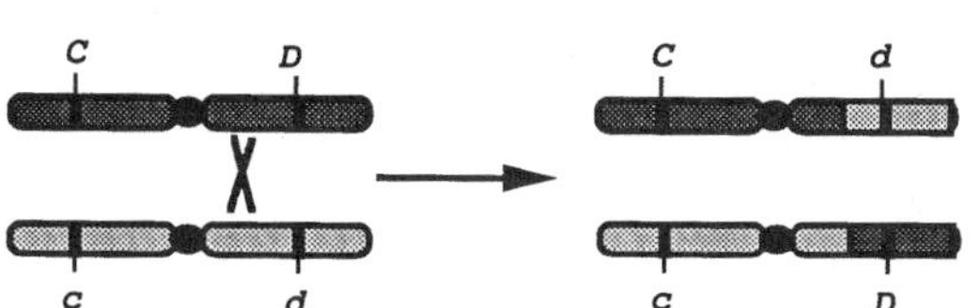


Figure 4. The generation of new combinations of genes by recombination: either (A) the independent assortment of genes on nonhomologous chromosomes during meiosis or (B) crossing over between genes located on homologous chromosomes during meiosis or mitosis.

Point mutations within individual genes may be either base pair substitutions or the insertion or deletion of one or a few contiguous base pairs. The insertion or deletion of one or two base pairs within the coding sequence of a gene alters the codon reading frame in the mRNA; thus, such mutations are referred to as frameshift mutations. Frameshift mutations usually result in totally nonfunctional gene products. In contrast, base pair substitutions usually result in the substitution of a single amino acid in the mutant polypeptide gene product. For example, sickle cell anemia occurs in humans that are homozygous for a β-hemoglobin gene that differs from the normal adult β-hemoglobin gene by a single base pair substitution. This one base pair change in the $Hb_{\beta}^{S}$ gene changes the sixth amino acid of the β-hemoglobin polypeptide from glutamic acid in $Hb_{\beta}^{A}$ homozygotes to valine in $Hb_{\beta}^{S}$ homozygotes, as follows:

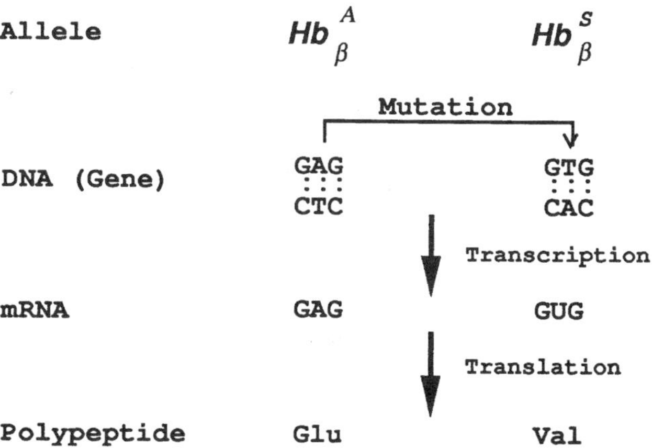

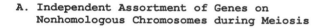

This single amino acid change in the human β-hemoglobin chain results in sickle-shaped red blood cells and in sickle cell anemia in individuals homozygous for the $Hb^s_\beta$ allele. Thus, a single base pair substitution in DNA can have a very large effect on the phenotype of the organism harboring the mutation.

## 5 RECOMBINATION: NEW COMBINATIONS OF GENES TO BE ACTED ON BY NATURAL OR ARTIFICIAL SELECTION

Mutation produces new genetic variability, but the resulting mutant genes must be placed in new combinations with previously existing genes so that natural selection (or, in the case of plant and animal breeding, artificial selection) can preserve the combinations that produce the organisms best adapted to specific environments (or desired by the breeder). These new combinations are produced by recombination mechanisms that are essential to the process of evolution. New combinations of genes on nonhomologous chromosomes are produced by the independent assortment of chromosomes during the first or reductional division of meiosis (Figure 4A). New combinations of genes on the same chromosome are produced by crossing over (breakage and exchange of parts) between homologous chromosomes during meiosis and mitosis (Figure 4B).

*See also* AGING, GENETIC MUTATIONS IN; DNA REPLICATION AND TRANSCRIPTION; GENETIC DIVERSITY IN MICROORGANISMS; HUMAN GENETIC PREDISPOSITION TO DISEASE.

### *Bibliography*

Gardner, E. J., Simmons, M. J., and Snustad, D. P. (1991) *Principles of Genetics,* 8th ed. Wiley, New York.

Kornberg, A. (1980) *DNA Replication.* Freeman, San Francisco.

———. (1982) *Supplement to DNA Replication.* Freeman, San Francisco.

Lewin, B. (1990) *Genes IV.* Oxford University Press, New York.

Singer, M., and Berg, P. (1991) *Genes and Chromosomes.* University Science Books, Mill Valley, CA.

Watson, J. D., Hopkins, N. H., Roberts, J. W., Steitz, J. A., and Weiner A. M. (1987) *Molecular Biology of the Gene,* 4th ed., Vols. I and II. Benjamin/Cummings, Menlo Park, CA.

# GENETIC TESTING

*Frank K. Fujimura*

### *Key Words*

**Autosomal** Related to the nonsex chromosomes.

**Carrier** Individual with one abnormal and one normal copy of a recessive gene; carriers are either male or female for autosomal recessive genes, and female for X-linked recessive genes; carriers generally do not have clinical symptoms of the disease.

**Cystic Fibrosis (CF)** The most common serious genetic disease for Caucasians, with a frequency of one in 2500.

**Denaturing Gradient Gel Electrophoresis (DGGE)** Method for point mutation scanning by electrophoresis in a gel containing a gradient of denaturants (urea and formamide), allowing detection of sequence-dependent differences in denaturation conditions.

**Dominant** Describing a phenotype expressed in heterozygotes.

**Duchenne Muscular Dystrophy (DMD)** An X-linked disease that is the most common of the severe childhood muscular dystrophies.

**Fragile X syndrome** An X-linked disease that is the most common of the heritable forms of mental retardation.

**Gene** A segment of DNA that contains all the information for the regulated biosynthesis of an RNA product.

**Hemoglobinopathy** Disease of one or more of the globin genes whose protein products make up hemoglobin.

**Heteroduplex** A double-stranded DNA molecule; the two strands are derived from two different alleles of a particular locus.

**Heterozygote** Individual carrying two different alleles of a particular locus.

**Homozygote** Individual carrying two copies of the same allele at a particular locus.

**Linkage Analysis** Indirect analysis of a genetic disease gene by tracking polymorphic markers that are physically associated (linked) with the gene of interest.

**Locus** A location on a chromosome or DNA molecule corresponding to a gene or a physical or phenotypic feature.

**Recessive** Describing a phenotype expressed only in homozygotes.

**Short Tandem Repeats (STR)** Polymorphic locus with relatively small (2–4 base pairs) tandemly duplicated sequences, useful as markers for linkage analysis.

**Single-Strand Conformational Polymorphism (SSCP)** Structural variation in single-stranded DNA due to sequence-dependent intrastrand secondary structure. SSCP is detectable by electrophoresis in nondenaturing gels and useful for point mutation scanning.

**Variable Number of Tandem Repeats (VNTR)** Polymorphic locus with relatively large ( 50 base pairs) tandemly duplicated sequences, useful as markers for linkage analysis.

**X-Linked** Describing a genetic condition mapping to the X chromosome; when recessive, the phenotype is expressed only in males.

---

The clinical application of DNA probe technologies for molecular genetic testing is proceeding rapidly. The recent discoveries of the human genes for numerous genetic conditions, including Duchenne muscular dystrophy, cystic fibrosis, fragile X syndrome, and Huntington's disease, have quickly resulted in the transfer of these new technologies into the clinical arena. These new technologies, while very powerful, also have limitations that need to be considered for correct interpretation of test results. The pace of technology transfer challenges the ability of the clinical community and the general public to comprehend and deal with the diverse technical, medical, ethical, legal, and social implications of genetic testing. With the impending expansion of gene discovery from relatively rare genetic conditions such as Huntington's disease to the more common conditions like heritable cancers and heart disease, the potential impact of genetic testing on society is very significant. The results of genetic testing can affect not only the health of the patient, but also many other aspects of the patient's life including reproductive options, lifestyle considerations, employability, and insurability. Furthermore, genetic test results for one individual can have significant clinical and social implications for some of his or her relatives.

These many issues make the delivery of molecular genetic clinical services a challenging but critical task.

## 1 BACKGROUND

The recent interest and activity in the cloning of human genes has resulted in the isolation and characterization of numerous genes relevant to genetic disease. Many of these genes were isolated because the protein product of the gene was known. Clinical testing for these genetic diseases often was available by biochemical methods prior to gene isolation. Although direct genetic analysis may become possible with the isolation of a particular gene, biochemical or other methods, if available, are often preferred over DNA probes at present because of lower costs, shorter turnaround times, and comparable or superior sensitivity and specificity.

However, for many genetic diseases, biochemical tests are unavailable because the nature of the gene products responsible for these diseases has not been determined. The identification of these genes has been the focus of intense activity. The first triumph of ''reverse genetics'' or ''positional cloning''—the use of molecular methods to map, isolate, and characterize genes with unknown gene products—came in 1986 with the cloning of the genes for chronic granulomatous disease, Duchenne muscular dystrophy, and retinoblastoma. Several other genes, notably those responsible for cystic fibrosis, fragile X syndrome, myotonic dystrophy, and Huntington's disease, were isolated subsequently. The importance of identifying genes with unknown gene products for diagnosis is that knowledge of the gene structure often leads to immediate capability for analyzing mutations in the gene or alterations in the expressed gene products.

Genetic diseases can be classified as automosal or X-linked, depending on chromosomal association, and as dominant or recessive, depending on the nature of phenotypic expression. The dominant or recessive nature of genetic disease is not always clearly distinguishable and depends on the phenotype that is being monitored. Molecular methods, by measuring genotype rather than phenotype, provide an analytical approach that is independent of phenotypic variability. There is hope that genotypes can be correlated to phenotypes in the future.

The current applications of DNA testing for genetic diseases include diagnosis or confirmation of diagnosis, carrier detection, and prenatal diagnosis. Future genetic testing very likely will include presymptomatic diagnosis and risk assessment for disease susceptibilities, particularly for heritable cases of cancer, heart disease, and neurological disorders, although these applications are limited in scope at present.

## 2 DIRECT AND INDIRECT GENETIC ANALYSIS

Ideally, DNA analysis for genetic disease involves testing for the specific mutation or mutations relevant for the specific condition. This ideal situation is rarely realized in practice. Direct detection of mutations requires knowledge about the nucleotide sequence of the gene and the specific mutations in the gene that are associated with the disease phenotype. With rare exceptions, the complete characterization of mutations has proven to be a formidable task. Using the example of cystic fibrosis (CF), identification of one mutation, ΔF508, which can account for as much as 80% of the CF mutations in some populations, led to great optimism that a relatively small number of mutations could account for a majority of CF cases. This optimism was soon quenched, however, when it was realized that CF could result from a very large number of different mutations within the CF gene. More than 400 mutations have been identified, some of which are private mutations found only in one family.

Because of the large number of different mutations found for most genetic disease genes, identification of specific mutations is very important for accurate diagnosis. Key to development of a diagnostic test is the characterization of the spectrum, frequency, and clinical significance of mutations within a gene. These mutations can be point mutations, short deletions or duplications, or large deletions or duplications. If a mutation cannot be identified in a patient specimen, the accuracy of the result depends on the frequency of mutations that cannot be ruled out by the analysis. This frequency may not always be known and can vary among different ethnic populations. Therefore, while direct detection of mutations provides very accurate results when the mutations are detectable, the utility of this approach is limited if a significant proportion of mutations within a gene cannot be detected. Consider the case of CF: ΔF508 and six other mutations comprise approximately 80 to 85% of the CF mutations in Caucasians of northern European background. Testing for an additional 15 mutations may increase the detection level to approximately 90% in this ethnic population. To achieve 95% detection of mutations may require analysis for as many as 25 to 50 more mutations.

Although direct detection of mutations is the ideal method for genetic analysis using DNA probes, indirect analysis using linkage can be performed if the gene of interest or the specific mutation in the disease gene has not been characterized. Linkage analysis tracks a disease allele indirectly by following the pattern of inheritance of polymorphic alleles that map very close to the disease gene. This method requires probes for polymorphisms that are linked to the gene of interest. Among the types of polymorphic markers used for linkage analysis are restriction fragment length polymorphisms (RFLPs), variable number of tandem repeat (VNTR) sequences, and dinucleotide or other short tandem repeat (STR) sequences.

Although very powerful, linkage analysis has certain limitations. First, it is necessary to test multiple individuals in a family, which calls for both cooperation and availability of family members. Generally, testing is required of an affected individual who may or may not be available. Because analysis of several family members is necessary, confidentiality of individual patient information may be compromised. The interpretations of a linkage analysis depend on the accuracy of the family pedigree and the clinical diagnosis of affected individuals. Nonpaternity can lead to erroneous interpretations if it is not detected or to social complications if it is. Finally, because linkage analysis is an indirect assay for the gene of interest, there is the possibility of false interpretations due to recombination between the marker being analyzed and the disease gene. The recombination frequency increases proportionally to the distance between the two genetic loci. In spite of these limitations, the method is very useful for carrier analysis and prenatal diagnosis of numerous genetic diseases.

Table 1 lists some of the genetic diseases that are being analyzed by nucleic acid probes, as well as the chromosome locations of these disease genes. Table 1 also indicates whether analysis is performed by linkage (L) or by mutation (M) detection, and whether the gene has been cloned.

## 3 METHODOLOGY

Southern blot analysis is the classic method for RFLP analysis. More recently, nucleic acid amplification, usually using the poly-

Table 1 Genetic Analysis by DNA Probes Using Linkage (L) or Mutation (M) Detection

| Condition | Chromosome | L/M | Cloned |
|---|---|---|---|
| α-1 antitrypsin deficiency | 14 | M | + |
| α-Thallasemia | 16 | M | + |
| Adenomatous polyposis coli | 5 | L, M | + |
| Adult polycystic kidney disease | 16 | L | |
| Breast cancer susceptibility (BRCA1) | 17 | L, M | + |
| Breast cancer susceptibility (BRCA2) | 13 | L | |
| β-Thallasemia | 11 | M | + |
| Charcot–Marie–Tooth disease | 1 | M | + |
| Colon cancer susceptibility (MSH2) | 2 | M | + |
| Colon cancer susceptibility (MLH1) | 3 | M | + |
| Colon cancer susceptibility (PMS1) | 2 | M | + |
| Colon cancer susceptibility (PMS2) | 7 | M | + |
| Congenital adrenal hyperplasia | 6 | M, L | + |
| Cystic fibrosis | 7 | M | + |
| Duchenne/Becker muscular dystrophy | X | M, L | + |
| Fragile X syndrome | X | M, L | + |
| Hemophilia A | X | M, L | + |
| Gaucher's disease | 1 | M | + |
| Hemophilia B | X | M, L | + |
| Huntington's disease | 4 | M, L | + |
| Kennedy's disease | X | M | + |
| Lesch–Nyhan syndrome | X | L, M | + |
| Marfan's syndrome | 15 | M | + |
| Medium chain acyl-coenzyme A dehydrogenase deficiency | 1 | M | + |
| Melanoma susceptibility | 9 | M | + |
| Multiple endocrine neoplasia 1 | 11 | L | |
| Multiple endocrine neoplasia 2A | 10 | L, M | + |
| Myotonic dystrophy | 19 | M, L | + |
| Neurofibromatosis type 1 | 17 | L, M | + |
| Ornithine transcarbamylase deficiency | X | M, L | + |
| Retinoblastoma | 13 | M, L | + |
| Sickle cell anemia | 11 | M | + |
| Steroid sulfatase deficiency | X | L, M | + |
| Tay–Sachs disease | 15 | M | + |
| Werdnig–Hoffman disease | 5 | L | |

merase chain reaction (PCR) coupled with various detection methods, has been replacing Southern blotting. Virtually all RFLPs can be analyzed by PCR with great savings in time and specimen quantity. Furthermore, PCR offers greater flexibility and diversity for analysis of other polymorphisms including VNTRs and STRs, as well as specific mutations. Other amplification methods, including the ligase chain reaction (LCR), also can be used for linkage analysis, as well as for direct mutation detection.

For detection of large deletions, Southern blotting is the general method of choice. Pulsed field electrophoresis is used if the DNA fragments analyzed are very large (> 15–20 kilobase pairs). If specific deletions are of interest, PCR methods can be devised for their detection. Multiplex PCR reactions can be used to scan relatively large portions of the genome in cases of deletions that tend to cluster in "hot spots," effectively replacing multiple Southern blots with a single reaction tube. The clear advantage of multiplex PCR over Southern blotting is the ability to scan mutation "hot spots" using less than 1% the amount of DNA (low specimen size requirement) and taking less than 10% the time (short turnaround time). Furthermore, PCR is much more tolerant of poor DNA specimen quality than is Southern blotting.

Several methods can be used for detection of mutations in PCR or other amplified gene products. The most widespread method uses allele-specific oligonucleotides (ASOs) for the specific mutations of interest. ASO probes usually are about 20 nucleotides long and, upon labeling, are used as hybridization probes for target sequences that are typically immobilized onto a solid support such as a nylon membrane. Labeling of probes can use isotopic ($^{32}P$) or nonisotopic (usually biotinylated or digoxigenin-modified substrate) methods. The target sequences invariably are amplified, and in some cases are separated by gel electrophoresis prior to immobilization. If gel separation is used, the target sequences are transferred onto the support membrane by Southern blotting. Usually, separation is not necessary, and the target sequences are placed directly onto the support as dot or slot blots. An alternative to the dot–slot blot method is the reverse dot blot method in which the ASOs are immobilized onto a membrane or some other support medium, and the membranes are hybridized with target sequences that have been labeled during amplification.

Several other methods have been used for specific mutation detection. One of the first to be employed depends on coincidental restriction site alteration. Certain mutations alter restriction enzyme cleavage sites and can be distinguished from the corresponding wild-type sequence by digestion with the appropriate enzyme. This obviously is not a general method, inasmuch as very few mutations affect restriction enzyme sites. A modification of this method involves the use of a slightly mismatched PCR primer containing sequences that when combined with the mutation sequence of interest, create a restriction enzyme cleavage site at the site of the mutation. This modification extends the utility of the restriction site alteration method of mutation detection, but it is still limited by the need for appropriate sequences at the mutation site that can be altered to create a restriction site.

Allele-specific amplification and allele-specific ligation, utilizing primers complementary to either the wild type or the mutant sequence, provide two alternative means for detection of specific mutations. In addition, some methods are available to screen for the presence of mutations without identifying the specific mutation itself. These methods include single-strand conformational polymorphism (SSCP) analysis, denaturing gradient gel electrophoresis (DGGE), and mismatch cleavage analysis by enzymatic (RNase A) or chemical (piperidine) means. SSCP and DGGE detect differences in electrophoretic mobilities between a wild-type and a mutant sequence that are due to sequence-specific differences in intrastrand structure (SSCP) or in duplex melting properties (DGGE). Mismatch cleavage methods detect unpaired regions in heteroduplex molecules of wild-type and mutant DNA. Of these three methods, SSCP is the most widely utilized, because it is technically less challenging than the other methods. The efficiency of SSCP is about 80 to 90% for detection of point mutations in regions of 200 nucleotides. Dideoxy fingerprinting, a recent modification of SSCP, may increase the efficiency of mutation detection to over 90%. DGGE and chemical mismatch cleavage methods are reported to give greater than 90% detection of point mutations in regions of 500 to 600 nucleotides. These mutation screening methods, although very useful as research tools, have yet to be utilized on a widespread basis in clinical labs.

DNA sequencing provides another option for mutation detection. Sequencing, in theory, should be able to detect every possible mutation within a gene. Practically, DNA sequencing had been used for screening genes for mutations, but its utilization as a clinical diagnostic tool is limited currently by its high cost and labor intensiveness. Should the cost of DNA sequencing drop, genetic analysis by direct sequencing may become feasible. New sequencing strategies coupled with automation will be important factors in future utilization of this method for routine genetic disease testing. Critical to the success of direct sequencing for routine clinical use will be the ability to detect heterozygotes in patient DNA specimens.

## 4 SPECIFIC EXAMPLES

### 4.1 Simple Point Mutation Analysis: Sickle Cell Anemia

The prototype of single mutation analysis is sickle cell anemia, which is an autosomal recessive condition due predominantly to a single mutation in the β-globin gene. This mutation, resulting in the S allele, causes a single amino acid change in β-globin. Individuals homozygous for this mutation have red blood cells that under certain conditions form a characteristic crescent or ''sickle'' shape, resulting in hemolytic anemia and various other clinical complications. Two other mutations, C and E, of the β-globin gene can also result in sickling hemoglobinopathies, usually with less severe phenotype than the S allele. The detection of these mutations is very straightforward by PCR using ASOs or restriction endonuclease digestion, but DNA probe analysis for sickling hemoglobinopathies, in general, is limited to prenatal diagnosis. Routine screening and confirmation of diagnosis is best done by analysis of the protein by other methods (e.g., hemoglobin electrophoresis).

### 4.2 Point Mutation Scanning: Cystic Fibrosis

Sickle cell anemia represents the exception rather than the rule for molecular genetic testing in that a single or limited number of mutations leads to the disease phenotype. For most other monogenic conditions, several, if not many, different mutations are found within the disease gene for different patients and carriers. The example of CF has already been discussed. For CF and many other autosomal recessive genetic diseases, the types of mutation can include missense, nonsense, small deletions, splicing defects, and promoter and other regulatory element changes. The frequencies of any specific mutation may vary for different ethnic groups. Except for special cases, it is impractical to test for every mutation in a particular gene, but information about frequencies of particular mutations for different ethnic groups can allow tailoring of tests to screen for the most likely mutations, given an individual's ethnic background.

CF is the most common severe genetic disease among Caucasians. It is an autosomal recessive condition, and carriers of CF show no symptoms of the disease. Affected individuals carry two mutant CF genes and can exhibit impaired respiratory, digestive, and reproductive functions of varying severity. The frequency of CF in Caucasian populations is about one in 2500 births, meaning that about one in 25 Caucasians is a carrier. More than 400 mutations have been described in the CF gene, which is approximately 250 kilobases (kb) in size and produces an mRNA of about 6.5 kb. Of these mutations, one called ΔF508 accounts for approximately 70% of the CF mutations in the U.S. Caucasian population. None of the other mutations is highly abundant, but particular mutations are very frequent in specific ethnic groups such as the Ashkenazic Jews and the Hutterites.

CF gene analysis generally involves testing for the more common mutations. There is yet no consensus on the number or the set of mutations to be analyzed. Different laboratories are analyzing anywhere from as few as four to six mutations to as many as twenty or more. Detection frequencies are between 80 and 90% in Caucasians, and appreciably lower in most other ethnic groups.

### 4.3 Deletion–Duplication Analysis: Duchenne/Becker Muscular Dystrophies

Duchenne muscular dystrophy (DMD) is an X-linked disease and is the most common of the childhood muscular dystrophies. Affected males usually present with muscle weakness around 2 to 6 years of age, are unable to walk at around 12 years, and seldom survive beyond their twenties. The DMD gene encodes a protein designated dystrophin that is found in muscle and brain cells. A milder form of the disease called Becker muscular dystrophy (BMD) is due to alterations in the same gene as DMD. Approximately 60% of cases of DMD and BMD are due to deletions or duplications that can be detected by Southern blot analysis using the complete 14 kb dystrophin cDNA probe. Up to 80% of the deletions detectable by Southern blot analysis can be detected rapidly by multiplex PCR using up to 18 pairs of primers in deletion ''hot spots'' in the DMD gene. While 60% of DMD mutations can be detected as deletions and duplications, the remaining 40% are presumably due to point mutations or short deletions that lie within the 2500 kb

DMD gene. The extremely large size of this gene makes mutation scanning unfeasible by present methods.

Mutation detection of DMD deletions, when successful, provides a rapid and convenient method for diagnosis of affected males and identification of carrier females (by quantitative PCR or Southern blotting). When mutation detection is unsuccessful, linkage analysis may be an option if appropriate family members are available.

### 4.4 Trinucleotide Repeat Expansion: Fragile X Syndrome

Fragile X syndrome is the most frequent of the heritable forms of mental retardation and exhibits unusual genetic features. Although it is an X-linked genetic disease, females can show symptoms of varying severity. Furthermore, there are clinically normal males that are carriers of fragile X syndrome and transmit the disease through their clinically normal daughters to their grandchildren. The isolation of the fragile X gene provided some clues to the molecular basis of these observations. The gene contains a trinucleotide repeat of CGG. Normal individuals have between 5 and 50 copies of the CGG repeat; affected individuals have more than 200 copies of the repeat. Individuals with 50 to 200 copies of the CGG repeat are said to be carriers of a fragile X premutation. These individuals, very often, do not have symptoms of fragile X but have a high probability of transmitting the disease to their children or grandchildren. Apparently, the CGG repeat becomes destabilized as it increases in size. Thus, alleles with fewer than 50 copies of the CGG repeat appear to be stable and do not change in future generations. Alleles with 50 to 200 CGG repeats are less stable and tend to expand in size when transmitted to offspring. The stability of premutations appears to be size dependent, and alleles with more than 100 repeats have very high probability of expanding to more than 200 repeats. The expansion of trinucleotide repeat sequences in fragile X syndrome appears to occur very early in embryonic development and is restricted to the maternally derived allele. Thus, clinically normal male carriers of fragile X permutations have unaffected daughters, but can have affected grandchildren.

The full fragile X mutation, having more than 200 trinucleotide repeats, shows hypermethylation of the expanded allele in patient DNA samples. Thus, molecular analysis of fragile X syndrome involves detection of expanded DNA using Southern blotting and/or PCR, and determination of the methylation status of the fragile X locus using methylation-sensitive restriction endonucleases.

Expansion of trinucleotide repeats may prove to be a relatively common mechanism for genetic disease. In addition to fragile X syndrome, several other conditions, including Kennedy's disease (spinal–bulbar muscular atrophy), myotonic dystrophy, and Huntington's disease, are due to expansion of trinucleotide repeat sequences. Undoubtedly, other genetic diseases involving trinucleotide repeat expansion will be discovered.

### 4.5 Polygenic Diseases: Heritable Cancer Susceptibilities

Numerous factors, some genetic and some environmental, contribute to the development of cancer. Genetic factors predisposing individuals to certain types of cancer are beginning to be elucidated. Heritable mutations in certain tumor suppressor genes have been shown to predispose individuals carrying these mutations to certain types of cancer. Germ line mutations in the gene for adenomatous polyposis coli (APC) predispose to familial adenomatous polyposis, a rare form of inherited colon cancer. At least four genes, *MSH2*, *MLH1*, *PMS1* and *PMS2*, have been implicated in a more common form of colon cancer, hereditary nonpolyposis colon cancer (HNPCC). Mutations in the *RB1* gene lead to susceptibility for retinoblastoma, an eye tumor, while mutations in the *WT1* gene predispose for Wilms' tumor, a kidney malignancy. Other genes shown to have heritable mutations conferring cancer susceptibility include *NF1* for neurofibromatosis type 1, *TP53* for Li–Fraumeni syndrome, and *RET* for multiple endocrine neoplasia type 2A. A familial breast cancer gene, *BRCA1,* has been mapped to the long arm of chromosome 17. The *BRCA1* gene, after a long, intensive effort by many investigators, has finally been cloned. Recently, a second breast cancer susceptibility gene, *BRCA2*, has been mapped to the long arm of chromosome 13. Together, *BRCA1* and *BRCA2* account for the majority of hereditary breast cancer, but there are other genes that remain to be identified and mapped. A melanoma susceptibility gene, mapped to chromosome 9, has been cloned.

Heritable cancer susceptibility mutations can, in theory, be analyzed using the molecular tools applied to the study of monogenic disease. The spectrum of mutations in these cancer genes appears to be broad, necessitating efficient screening methods for mutation detection. Furthermore, polygenic conditions like cancer introduce into genetic analysis additional complications that usually are not a factor for monogenic disease analysis. One complication is genetic heterogeneity, the condition of several different genes independently conferring susceptibility to a particular type of cancer. When genetic heterogeneity exists, the problem of mutation detection is multiplied by the number of different genes. In the absence of a detectable mutation within a susceptibility gene, linkage analysis generally is not feasible because linkage with one gene cannot be inferred for a particular family without prior analysis of many family members through multiple generations. A second complication is variable penetrance, which can depend on the nature of the mutation and the influence of other genetic and environmental factors. To sort out all the factors influencing expression of a disease phenotype is likely to prove to be a difficult task.

In spite of these limitations and complications, there has been progress in analyzing members of families at risk for some of the rarer conditions with no apparent genetic heterogeneity, such as retinoblastoma and the Li–Fraumeni syndrome. In rare cases of large families, where linkage to a particular locus can be strongly inferred, linkage analysis has been possible. Overall, genetic analysis of familial cancer has been limited. Greater utilization will depend on the isolation and characterization of more cancer susceptibility genes and clinical correlation studies establishing the effectiveness and the clinical significance of detecting mutations within these genes. Coupled with the relatively high frequency of certain malignancies such as breast and colon cancer, the rapid pace of gene discovery suggests that significant progress leading to the routine genetic analysis for heritable cancers will occur in the near future.

## 5 LIMITATIONS AND IMPLICATIONS

Although very powerful, genetic testing does have limitations and pitfalls. As discussed earlier, linkage analysis is dependent on correct pedigree and diagnosis, generally requires a specimen from an affected individual, and can be confounded by recombination. Mutation detection methods utilizing amplification technologies are subject to contamination and can lead to errors if there are sequence polymorphisms at primer or probe sites. Mutation scanning methods must be able to distinguish relevant mutations from benign sequence changes. Because DNA technology is relatively new to the clinical arena, it is essential to educate patients and physicians with respect to the limitations, as well as the benefits, of the technology. In addition to technical issues, a number of social, economic, ethical, and legal issues surround genetic testing. Nonpaternity has been mentioned. Other considerations include reproductive choice, discrimination with respect to insurability and employment, confidentiality of test information, and the psychological impact of test results on individuals. There are anecdotal incidents of misuse of genetic testing information by health maintenance organizations, insurance companies, and employers.

As the range of conditions that can be tested at the DNA level increases, the potential benefits of the technology are immense, and the scope of genetic testing will continue to expand. To become the standard of care, the technology needs improvement in three areas: testing should be less labor intensive, as well as standardized, and accessible at reasonable cost. Furthermore, there is a need to provide individuals with the ability to deal with the complex information that can result from this type of testing. Education is critical to this process, and appropriate mechanisms are necessary to ensure the confidentiality of genetic information and to protect individuals against the misuse or abuse of this information.

*See also* DNA in Neoplastic Disease Diagnosis; Human Genetic Predisposition to Disease; Molecular Genetic Medicine; Population-Specific Genetic Markers and Disease.

### *Bibliography*

Antonarakis, S. E. (1989) Diagnosis of genetic disorders at the DNA level. *New Engl. J. Med.* 320:1153–163.

Caskey, C. T., Pizzuti, A., Fu, Y.-H., Fenwick, R. G., Jr., and Nelson, D. L. (1992) Triplet repeat mutations in human disease. *Science* 256:784–788.

Chirgwin, J. M. (1990) Molecular biology for the nonmolecular biologist. *Diabetes Care* 13:188–197.

Collins, F. S. (1992) Positional cloning: Let's not call it reverse anymore. *Nature Genet.* 1:3–6.

Ostrer, H., and Hejtmancik, J. F. (1988) Prenatal diagnosis and carrier detection of genetic diseases by analysis of deoxyribonucleic acid. *J. Pediatr.* 112:679–687.

Tsui, L.-C. (1992) The spectrum of cystic fibrosis mutations. *Trends Genet.* 8:392–398.

## Genome: *see* specific genome, e.g., *E. coli* Genome; Mouse Genome.

## Genome Research: *see* Automation in Genome Research.

# Genomic Imprinting

*Wolf Reik and Nicholas D. Allen*

### *Key Words*

**Epigenetic Modification** Reversible but heritable alteration of DNA above the level of sequence (e.g., the methylation of C residues in CpG dinucleotides). This constitutes an additional layer of information in the chromosome, which may indicate, for example, the parental origin of the chromosome.

**Fetal Growth Factors** Imprinted genes include those with a strong effect on fetal growth; paternal inheritance of some of these genes increases the size of the fetus, whereas maternal inheritance can decrease size.

**Imprinting and Cancer Syndromes** Because proliferation and differentiation of fetal tissues are affected by imprinting, imprinted genes can be involved in cancer syndromes in the human. This is especially true of childhood tumors, where both monoparental disomy and somatic loss of imprinting have been observed.

**Monoparental Disomy** Inheritance of particular chromosomes in two copies from one parent, with corresponding absence of the chromosome from the other parent. If there are imprinted genes on this chromosome, this will lead to an altered balance of gene product, hence specific phenotypes, or disease in the human.

**Nutrient Transfer** In mammals, there is substantial transfer of nutrients across the placenta from mother to fetus. According to an evolutionary theory based on genetic conflict, genes that have a major influence on this process should be under selective pressure to become imprinted.

**Parental Imprinting** Expression of certain genes from either the maternal or paternal chromosomes, achieved by epigenetic modification of chromosomes.

**Parthenogenesis** Production of offspring from eggs only (no sperm); parthenogenesis is possible in some groups of animals but is absent in mammals because the parental genomes, as a result of imprinting, are functionally inequivalent.

---

Parental imprinting is an important genetic mechanism whereby some genes in an organism are predominantly expressed from either the paternally or the maternally inherited chromosome. Imprinting, which occurs predominantly in mammalian species and in seed plants, leads to the inability to reproduce parthenogenetically because of the functional difference between the parental genomes. Imprinted genes control vital aspects of embryonic development, affecting fetal growth and viability. In mammals, a small number of imprinted genes have been identified, among which are fetal growth factors and receptors. In the human, imprinting is implicated in a number of genetic disorders and cancers. These diseases arise through an imbalance of parental chromosomes, or through somatic imbalance in the expression of imprinted genes. The molecular mechanism of imprinting presumably involves epigenetic modification of chromosomes that are introduced in the gametes and are somatically heritable. Differences in DNA methylation and chromatin structure, and in the timing of replication, have been discovered between the parental alleles of some imprinted genes.

## 1 IMPRINTED GENES

In the mouse, four imprinted genes have been identified so far (Table 1). These are *Igf-2, H-19* and *Snrpn* on chromosome 7 and *Igf-2r* on chromosome 17. Genetic experiments and observations from human disease indicate that more genes are imprinted. Although the total number of imprinted genes in the mouse genome

Table 1 Imprinted Genes in Mouse and Man and Some Genetic Disorders Associated with Imprinting

| Gene | Imprinting | Chromosome | | Disease |
|---|---|---|---|---|
| | | Mouse | Human | |
| *Igf-2* | Maternal | 7 dist | 11p15.5 | Wilms' tumor<br>Rhabdomyosarcoma |
| *H-19* | Paternal | 7 dist | 11p15.5 | Beckwith–Wiedemann syndrome |
| *Igf-2r* | Paternal | 17 prox | 6q | May not be imprinted in the human |
| *Snrpn* | Maternal | 7 | 15q11.2-q12 | Prader–Willi and Angelman syndromes |

is unknown, a number of other chromosomal regions are known to contain imprinted genes. The gene for insulinlike growth factor 2 is maternally imprinted (expressed from the paternal chromosome) except in the leptomeninges and choroid plexus, where both copies are expressed: *Igf-2* is a fetal growth factor with autocrine and short-range paracrine action. Mice that lack *Igf-2* are growth-retarded proportional dwarfs.

The receptor for *Igf-2* is paternally imprinted and possibly identical to the *Tme* locus on proximal chromosome 17; the main function of the receptor, however, is in lysosomal enzyme targeting and binding of mannose 6-phosphate. Its IGF-2 binding capacity may act negatively on ligand-mediated growth stimulation, which is predominantly through the *Igf-1* receptor. Fetuses that lack *Igf-2* receptor are therefore larger than controls. The relationship between *Igf-2* and *Igf-2r* has been demonstrated, since a paternally inherited null mutation in *Igf-2* can rescue the imprinted phenotype of the *Tme* locus, which contains *Igf-2r*.

The *H-19* gene is paternally imprinted. It is very closely linked (within 100 kb) to the *Igf-2* gene on distal chromosome 7, but no protein product of this gene has been identified. The *H-19* transcript possibly down-regulates cellular proliferation.

The *Snrpn* (small nuclear riboprotein N) gene is also on chromosome 7 (in the middle part) and is maternally imprinted. It is predominantly expressed in the brain (starting in the last third of gestation) and is thought to be involved in RNA splicing. Its absence in the human may contribute to some of the symptoms of Prader–Willi syndrome (see Section 3), and it may influence behavioral phenotypes in mouse and man.

## 2 CONSEQUENCES OF IMPRINTING

Studies with mouse chimeras suggest that imprinted gene action can influence the lineage-specific proliferation (and possibly differentiation) of embryonic cells. For example, cells lacking a paternal genome participate poorly in the development of skeletal muscle. These phenotypes have so far not been ascribed to the action of single imprinted genes, and whether they reflect simply an additive action of all the imprinted genes remains to be seen.

The adaptive function of imprinting is unknown at present, but based on the identity of some of the imprinted genes, an evolutionary theory has recently been developed. This theory holds that in organisms in which there is substantial nutritional transfer from one parent to the offspring after conception (e.g., in eutherian mammals, from the mother through the placenta to the fetus), there will be genetic conflict between parental alleles of genes that influence this transfer of resources. In eutherian mammals, paternal alleles will benefit from increasing resource transfer. While maternal alleles will also benefit in principle, they may compromise future reproductive success of the mother (hence their own propagation) by taking too much from the pool of resources. Over evolutionary time, a balance will be reached of the most successful combinations of imprinting maternal and paternal alleles. This theory predicts the imprinting of the *Igf-2* and *Igf-2r* genes very well, but to see how general it is, the isolation of additional imprinted genes must be awaited. However, it will be interesting to see whether in interbreeding natural populations there exist different alleles of imprinted loci. Already, it has been shown that in contrast to the mouse situation, *Igf-2r* may not be imprinted in the human.

## 3 GENETIC DISEASE

Some of the imprinted genes discovered in the mouse, notably *Igf-2* and *H-19*, have also been shown to be imprinted in the human (Table 1). If an imbalance arises therefore between maternal and paternal chromosomes—for example, in the form of a monoparental disomy of particular chromosomes—there will be an increased or decreased gene dosage of imprinted loci. Alternatively, imprints can be lost (or gained) during development by mutational or epigenetic mechanisms resulting in functional disomy. Altered imprinting could thus result in disease phenotypes. Prader–Willi (PWS) and Angelman syndromes (AS) arise from paternal deletion or maternal disomy and from maternal deletion or paternal disomy, respectively, of a region on chromosome 15q11-13. This finding, which indicates that at least two loci are involved, is supported by recent work that reveals nonoverlapping deletions in this region that cause either AS or PWS. At least in AS there are also a number of cases with no deletion or disomy, and these may arise from mutational derepression or repression of an imprinted gene. The genes *DN-34* and *SNRPN* in this region may be imprinted, and *SNRPN* is regarded as a candidate gene for PWS.

The Beckwith–Wiedemann syndrome (BWS) is a fetal overgrowth syndrome associated with a variety of embryonal tumors (Wilms' tumor, rhabdomyosarcoma, etc.). Paternal disomy of chromosome 15p15.5 has been found in this syndrome, and this segment contains the imprinted *IGF-2* and *H-19* genes. Hence overexpression of *IGF-2* (and lack of *H-19*) may contribute to the aberrant growth phenotype. Interestingly, nondisomic cases have now been found in which the maternal *IGF-2* allele is expressed, suggesting that mutational or epigenetic derepression of this imprinted gene can cause the disease.

In Wilms' tumors there is often loss of heterozygosity of chromosome 11p15 with preferential loss of the maternal allele. As in BWS, there are also tumors with no loss of heterozygosity in which both parental *IGF-2* (and sometimes *H-19*) alleles are expressed.

There are also parental transmission effects in monogenic disorders (e.g., Huntington's disease: HD), where early onset cases have predominantly inherited their mutant allele from father. Fragile-X syndrome and myotonic dystrophy, in which onset or severity is dependent on maternal transmission, are also in this category. In these diseases, including HD, there is however expansion of trinucleotide repeats in the mutant genes that generally correlates with expressivity. Thus the influence of imprinting is more difficult to interpret. In other cases of parental transmission effects—for example, in glomus tumor—the genes have not been identified yet and the interpretation awaits molecular analysis.

## 4 MOLECULAR MECHANISM

The molecular mechanism of imprinting must involve epigenetic modifications of DNA. Imprinted genes have therefore been searched extensively for differences in DNA methylation and chromatin structure between the parental alleles (Figure 1). The *Igf-2* gene shows no methylation or chromatin differences in the promoter region of the gene; however, sites within the fourth and fifth introns and a region upstream of the first promoter show parent-specific methylation, with the paternal (expressed) allele being more highly methylated. By contrast, the *H-19* gene shows extensive promoter methylation of the repressed (paternal) copy associated with decreased DNAse I sensitivity. Interestingly, these allelic methylation differences are established after fertilization and may not therefore constitute primary gamete-specific imprints. The tight linkage and reciprocal imprinting of *Igf-2* and *H-19* has led to a hypothesis that the two genes constitute an imprinted domain with a shared regulatory region. If this is correct, then primary gamete-specific imprints may reside outside the immediate vicinity of each gene.

The *Igf-2r* gene also has a promoter methylation imprint (on the repressed paternal allele), but in addition it has an intronic region in which the expressed (maternal) copy is more highly methylated. This intronic methylation imprint is present in the egg before fertilization occurs, and thus may be a primary imprinting signal. Whether the imprinting mechanism is different for different genes remains to be seen. The somatic methylation patterns observed for the three imprinted genes *Igf-2, H-19,* and *Igf-2r* are illustrated in Figure 1.

In addition to methylation and chromatin, differences have been detected in the time of replication of imprinted genes in the cell cycle. In general, the paternal copy of imprinted genes replicates earlier than the maternal one, regardless of whether the genes are maternally or paternally imprinted. In addition, such imprinted domains may be large (in the megabase range) and may contain genes that are not themselves imprinted.

Genetic screens are also being developed to identify genes that may play a role in controlling imprinting, so-called modifier genes. If methylation is indeed involved in imprinting, then the methyltransferase gene itself is of course important. Mice have now been created in which this methylase gene has been mutated, and the effect of the change on imprinted genes is as predicted from the methylation imprints: *Igf-2* and *Igf-2r* expression decreases, whereas *H-19* expression increases. This experiment demonstrates directly that DNA methylation is at least involved in the somatic maintenance of paternal imprints. At present it is necessary to postulate other genes that direct the specificity of methylation imprinting. Work toward isolation of such genes has made use of the observation that transgene methylation can depend on genotype-specific modifier genes, different alleles of which segregate in inbred strains. One such modifier gene has been chromosomally mapped, but none has been cloned so far. Whether there exist in interbreeding populations different alleles of modifier genes that lead to large effects on imprinted genes is a matter of speculation. Modifier genes may also be involved in controlling the specificity of DNA methylation patterns in mammalian cells and may affect penetrance and expressivity in genetic disease.

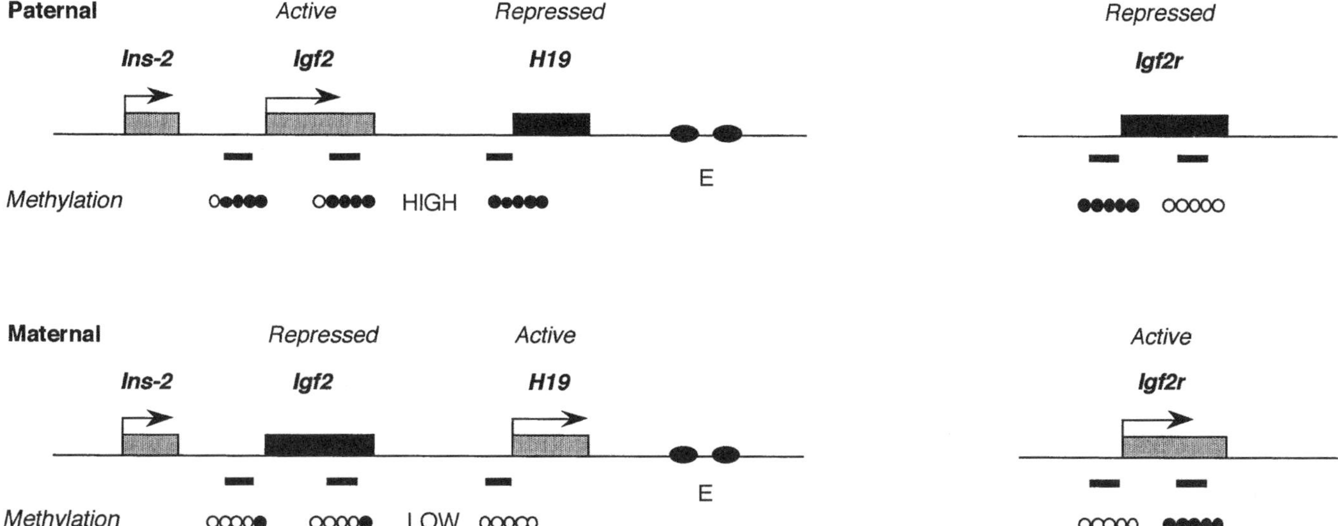


**Figure 1.** Imprinted genes possess differentially methylated parental chromosomes. The tightly linked *Ins-2, Igf-2* and *H-19* genes on mouse chromosome 7 are illustrated together with the downstream *H-19* enhancer region E. The *Igf-2r* gene on mouse chromosome 17 is also shown. Closed circles indicate CpG methylation in the defined regions (solid bars). Gene transcription is illustrated by an arrow. In somatic cells, the paternal alleles of the *Igf-2* and *H-19* genes are more methylated than the maternal alleles. Thus, methylation of *Igf-2* is associated with gene expression, while for *H-19*, methylation is associated with gene repression. Differential methylation is also seen in the *Igf-2r* gene. Maternal methylation in the body of the gene is already present in the unfertilized egg. Paternal methylation in the promoter region arises postzygotically and is associated with gene repression.

*See also* Gene Expression, Regulation of; Human Genetic Predisposition to Disease.

### *Bibliography*

Brandeis, M., Ariel, M., and Cedar, H. (1993) Dynamics of DNA methylation during development. *BioEssays,* 15:709–713.

Feinberg, A. P. (1993) Genomic imprinting and gene activation in cancer. *Nature, Genetics,* 4:110–113.

Haig, D., and Graham, C. (1991) Genomic imprinting and the strange case of the insulin-like growth factor II receptor. *Cell,* 64:1045–1046.

Monk, M., and Surani, A., Eds. (1990) *Genomic Imprinting. Development 1990 Supplement.* Company of Biologists, Cambridge.

Moore, T., and Haig, D. (1991) Genomic imprinting in mammalian development: A parental tug-of-war. *Trends Genet.* 7:45–49.

Reik, W. (1989) Genomic imprinting and genetic disorders in man. *Trends Genet.* 5:331–336.

Sapienza, C. (1990) Parental imprinting of genes. *Sci. Am.* 263:26–32.

Solter, D. (1988) Differential imprinting and expression of maternal and paternal genome. *Annu. Rev. Genet.* 22:127–146.

Surani, A., and Reik, W., Eds. (1992) Genomic imprinting in mouse and man. *Semin. Dev. Biol.* 3:73–160.

## Genomic Responses to Environmental Stress: *see* Environmental Stress, Genomic Responses to

# Glycobiology

*Akira Kobata*

### *Key Words*

**Anomers** Two isomeric structures, formed because the C-1 atom of aldoses becomes asymmetric by making sugar chains. The structures are designated as α- and β-anomers.

**Fc Receptor** Cell surface receptor that binds the hinge region of immunoglobulin G.

**Furanose** The isomeric ring structure formed by four carbons and one oxygen atom of aldoses.

**Glycoconjugates** The generic name of carbohydrates linked to various biomolecules. These conjugates can be classified into glycoproteins, glycolipids, and proteoglycans.

**Glycohormones** Peptide hormones containing sugar chains.

**Lectin** General name for proteins other than antibodies, which bind specifically to particular sugar chain structures.

The development of molecular biology has permitted the elucidation of the roles of nucleic acids and proteins as the molecules containing biological information. Gene technology and protein engineering, which have developed from the knowledge obtained by studies on nucleic acids and proteins, enabled the fermentation industry to utilize living organisms more effectively by directing them to produce a particular protein. Thanks to these biotechnologies, it is possible to handle substantial amounts of bioactive proteins, which are useful but occur in very minute amounts in animal bodies. However, many proteins produced by animal cells contain sugar chains and are called glycoproteins. Because bacteria lack the machinery for glycosylation, recombinant proteins produced by using them as hosts lack sugar chains. Since many of these nonglycosylated proteins do not express the expected biological activities, the functional role of the sugar chains has attracted the interest of molecular biologists. Thus a novel scientific field called *glycobiology* was established on the basis of the use in biology of knowledge obtained from the elucidation of the biological information in the sugar chains of glycoproteins, as well as glycolipids and proteoglycans.

## 1 CHARACTERISTIC FEATURES OF SUGAR CHAINS

Sugar chains (see, e.g., Figure 1) have a distinct characteristic not found in nucleic acids and proteins, which are chains of nucleotides and amino acids, respectively. Let us consider the smallest possible chain unit: A-B. In the case of a ribonucleic acid, only one structure is made when adenylic acid and guanylic acid are assigned to A and B, respectively. In the case of a protein also, only one structure can be formed by assigning alanine to A and threonine to B. The situation of sugar chains is quite different. Suppose that mannose is taken for A and galactose for B. As shown in Figure 1, mannose can be linked at the four positions: C-2, C-3, C-4, and C-6 of galactose residue. Therefore, four isomeric structures can be formed. Now, since a mannose residue can take two anomeric configurations, α and β, there are eight possible isomeric structure of the disaccharide. Furthermore, a mannose residue can occur in the furanose form as well as in the pyranose form shown in Figure 1. Thus 16 different isomeric structures are possible for the disaccharide Man → Gal. When the number of units increases (to three, four etc.), only one structure can be formed in the case of nucleic acids and proteins, because they are linear chain constructs. In contrast, the number of isomeric sugar chains increases in a geometrical progression, because branching can be formed in sugar chains larger than disaccharides. This means that sugar chains but not nucleic acids and proteins have the characteristic feature of being able to form multiple structures with a small number of units.

## 2 SUGAR CHAINS OF GLYCOPROTEINS

The sugar chains found in glycoproteins usually contain more than 10 units. Clearly the structural multiplicity produced by such large sugar chains is theoretically enormous. If it were necessary to handle such large numbers of isomeric structures, the elucidation of the biological information contained in sugar chains might be impossible. Fortunately, studies of the sugar chain structures of various glycoconjugates have revealed a series of structural rules and have indicated that variable regions are limited to a part of their structures. The sugar chains of glycoproteins are classified into two groups: *O*-linked sugar chains and *N*-linked sugar chains. All *O*-linked sugar chains of the mucin type contain *N*-acetylgalactosamine residues at their reducing termini, which are linked to the hydroxyl groups of serine and threonine residues of polypeptide chains. In contrast, *N*-linked sugar chains all contain *N*-acetylglucosamine residues at their reducing termini, which are linked to the amide group of asparagine residues constructing Asn-Xaa-Thr (or Ser) sequence in polypeptide chains.

## 3 STRUCTURAL RULES

By taking the case of *N*-linked sugar chains, the structural rules of sugar chains will be explained below.

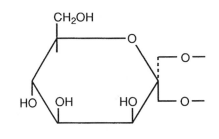

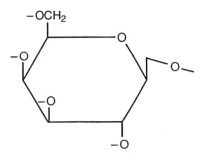

Figure 1. Construction of sugar chains.

All *N*-linked sugar chains contain the pentasaccharide: Manα1→6(Manα1→3) Manβ1→4GlcNAcβ1→4GlcNAc as a common core. By the structures and the location of the sugar residues added to the trimannosyl core, *N*-linked sugar chains are further classified into three subgroups as shown in Figure 2. Sugar chains, which fall into *complex type* contain no other mannose residue than the trimannosyl core. Outer chains with *N*-acetylglucosamine residue at their reducing termini are linked to the two α-mannosyl residues of the trimannosyl core. *High mannose type* sugar chains contain only α-mannosyl residues in addition to the trimannosyl core. A heptasaccharide: Manα1→6 (Manα1→3)Manα1→6(Manα1→3)Manβ1→4GlcNAcβ1→4GlcNAc is commonly included in this type of sugar chain as shown by dotted line in Figure 2. The third type is called *hybrid type,* because the sugar chains have the characteristic features of both complex type and high mannose type sugar chains. One or two α-mannosyl residues are linked to the Manα1→6 arm of the trimannosyl core like in the case of high mannose type, and the outer chains found in complex type sugar chains are linked to the Manα1→3 arm of the core. Presence or absence of the α-fucosyl residue linked to the C-6 position of the proximal *N*-acetylglucosamine residue and the β-*N*-acetylglucosamine residue linked to the C-4 position of the β-mannosyl residue of the trimannosyl core contributes the structural variation of the hybrid type as well as the complex type of sugar chains.

## 4 VARIATION OF COMPLEX-TYPE SUGAR CHAINS

Among the three subgroups of *N*-linked sugar chains, the complex type has the largest structural variation. This variation is formed mainly by the following two structural factors. As shown in Figure 3A, from one to five outer chains are linked to the trimannosyl core by different linkages, resulting in the formation of mono-, bi-, tri-, tetra-, and pentaantennary sugar chains. Various structures are found in the outer chain moieties of complex-type sugar chains as shown in Figure 3B. By combining the antennary and the various outer chains, a large number of different complex-type sugar chains can be formed, and these work as signals of biological recognition important in multicellular organisms.

For example, the galactose-binding receptor on the surface of liver parenchymal cells, which is called hepatic galactose-binding

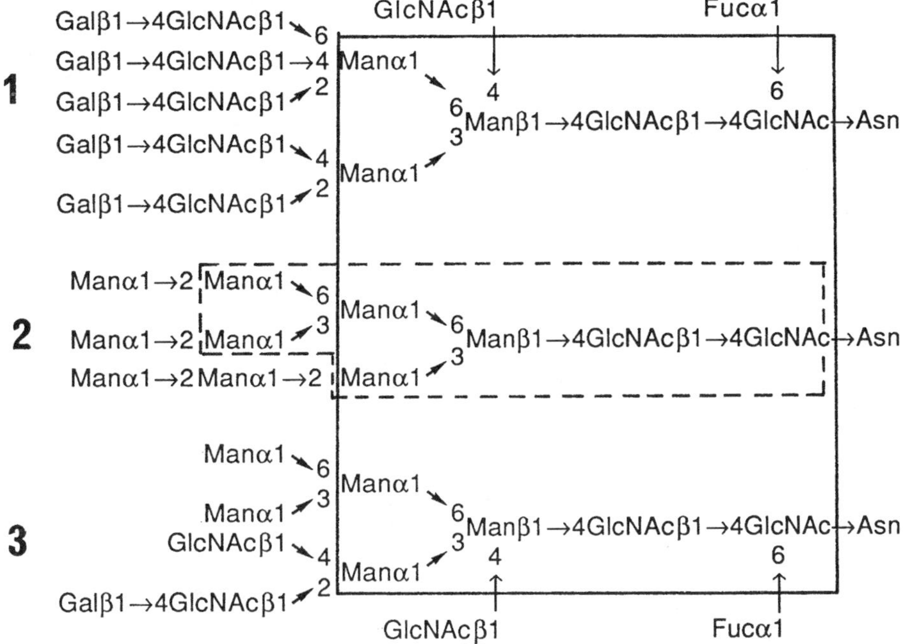


Figure 2. Three subgroups of *N*-linked sugar chains: 1, complex type, 2, high mannose type, 3, hybrid type. Structures within solid box constitute the trimannosyl core common to all *N*-linked sugar chains. The structure enclosed in dashed lines is the common heptasaccharide core of the high mannose type of sugar chains. Structures outside these lines can vary by sugar chains. Gal, galactose; Man, mannose; Fuc, fucose; GlcNAc, *N*-acetylglucosamine.

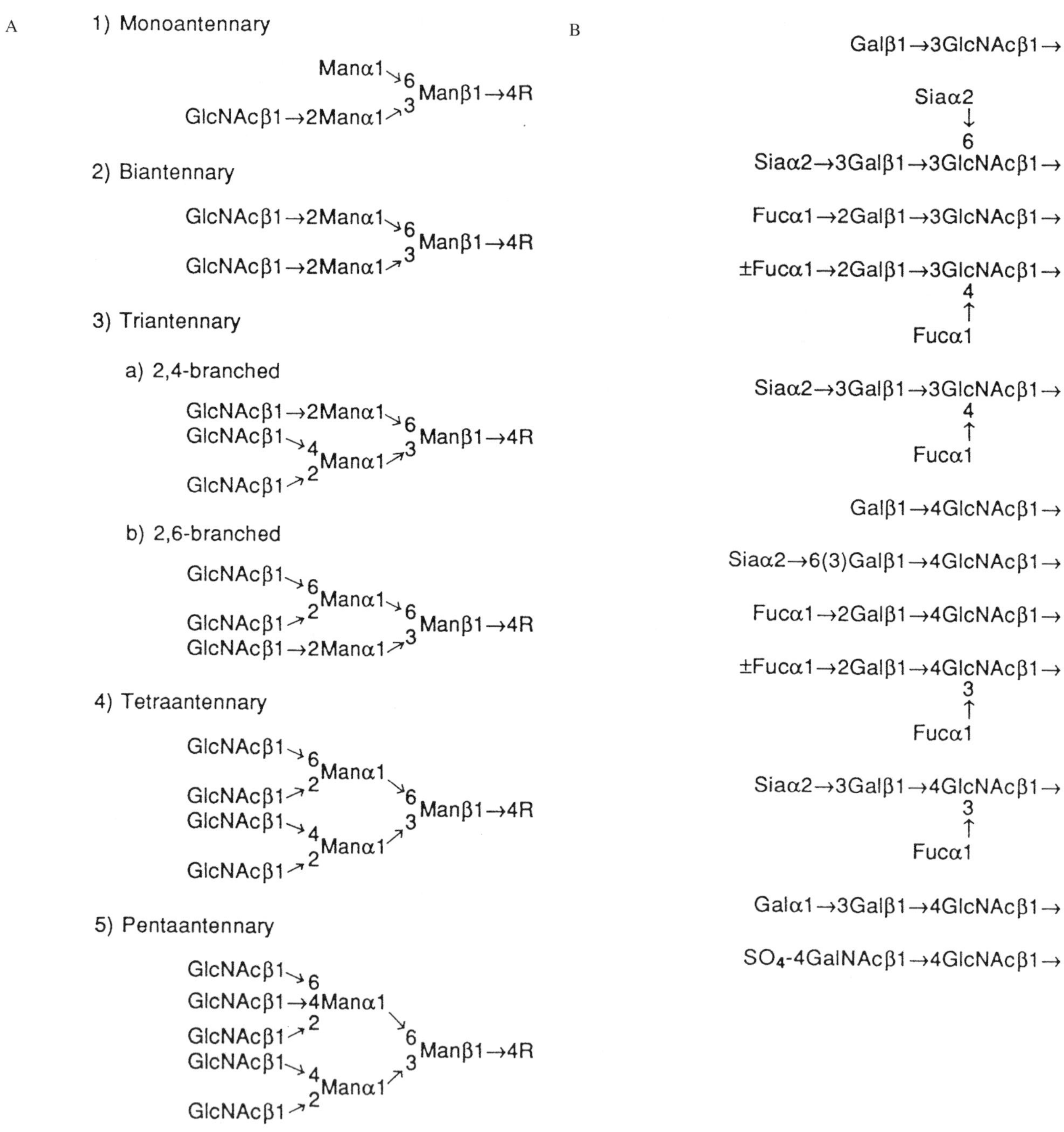


**Figure 3.** Two major elements to form the various structures of complex-type sugar chains: (A) branching of complex-type sugar chains and (B) various outer chain structures found in complex-type sugar chains. Sia, sialic acid.

lectin, binds most strongly with the 2,4-branched triantennary sugar chain, the three outer chains of which are Galβ1→4 GlcNAc. Interestingly, the isomeric 2,6-branched triantennary sugar chain and the biantennary sugar chain with the same disaccharide outer chains bind much more weakly to the lectin, indicating that a particular steric arrangement of the three β-galactosyl residues is important. Accordingly, desialylated ceruloplasmin, which contains the 2,4-branched triantennary chain, is quickly cleared from circulation by being trapped by hepatocytes, while transferrin, which contains only the biantennary sugar chain, is not.

## 5 FUNCTIONAL ROLES OF SUGAR CHAINS

The biological recognition mediated by sugar chains works widely in multicellular organisms, for example, in cell-to-cell recognition phenomena. Hence the information obtained by glycobiology research is essential in understanding the mechanisms of fertilization, morphogenesis, development, and aging. Two examples of the functional role of the sugar chains of glycoproteins are presented.

Four glycohormones have been known to occur in a variety of mammals. Three of them are produced in the same organ: luteinizing hormone and follicle-stimulating hormone are produced by gonadotrophs, and thyroid-stimulating hormone by thyrotrophs in the anterior pituitary. Only chorionic gonadotropin is produced elsewhere: by placental trophoblasts. All these hormones are composed of two noncovalently linked subunits of different sizes, designated α and β. Since α-subunits of all four glycohormones have an identical amino acid sequence within an animal species, the specificity of each hormone to bind to its target cells had been considered to reside in its β-subunit. However, elucidation of the structures of *N*-linked sugar chains of the four glycohormones revealed that the α-subunits of these hormones should not be considered the same. Furthermore, deglycosylated human chorionic gonadotropins obtained by enzymatic or chemical means were found to express no hormonal activity despite binding to the target cells more strongly than their natural counterpart. Therefore, the *N*-linked sugar chains of glycohormones may play crucial roles in their biological functions.

Immunoglobulin G (IgG) has a major role in humoral immunity. This glycoprotein is unique among the serum glycoproteins because it contains more than 30 different biantennary, complex-type sugar chains. That the sugar chains of IgG play an important role in the function of this molecule was shown by successful removal of the galactose residue of the glycoprotein by digestion with *Streptococcus* 6646K β-galactosidase. The degalactosylated IgG binds less effectively to the subcomponent C1q of the first component of complement, and Fc receptor. Furthermore, there are many data indicating the important roles of *N*-linked sugar chains in the construction of a complicated network connecting the immunocompetent cells. Therefore, *N*-linked sugar chains are essential for considering both cellular and humoral immunological systems.

*See also* GLYCOPROTEINS, SECRETORY.

### *Bibliography*

Allen, H. J. and Kisailus, E. C., Eds. (1992) *Glycoconjugates: Composition, Structure, and Function.* Marcel Dekker, Inc., New York.

*Carbohydrate Recognition in Cellular Function* (1989) (Ciba Foundation Symp. 145). John Wiley & Sons Ltd., New York.

Fukuda, M., Ed. (1992) *Cell Surface Carbohydrates and Cell Development.* CRC Press, Ann Arbor.

Fukuda, M. and Kobata, A., Eds. (1993) *Glycobiology: A Practical Approach.* IRL Press, Oxford.

Ginsburg, V., and Robbins, P., Eds. (1984) *Biology of Carbohydrates.* Vol. 2. John Wiley and Sons Ltd., New York.

Welply, J. K. and Jaworski, E., Eds. (1990) *Glycobiology.* Wiley-Liss, New York.

# GLYCOGEN

*Mathieu Bollen and Willy Stalmans*

### *Key Words*

**Glycogen** Polymer of α-D-glucose consisting of linear 1,4-linked chains and 1,6-glucosidic branch points.

**Glycogenin** A tyrosine glucosylated protein that serves as a primer for glycogen synthesis.

**Glycogen Synthase** Enzyme that catalyzes the rate-limiting step of glycogen synthesis (i.e., the transfer of glucosyl units from UDP-glucose to glycogen).

**Phosphorylase** Rate-limiting enzyme of glycogenolysis, catalyzing the conversion of glycogen plus inorganic phosphate to glucose 1-phosphate.

Glycogen is a polysaccharide that is present in most animal cells and in some fungi and bacteria. It is important in the generation of glucose 6-phosphate, which serves as a metabolic fuel for glycolysis. In the liver, glycogen can also be converted to glucose, which contributes to the maintenance of the blood glucose level. At least eight different enzymes are involved in the synthesis and degradation of glycogen, most of them closely associated with glycogen particles.

## 1 STRUCTURE OF GLYCOGEN

Glycogen is a branched homopolymer of α-D-glucose consisting of 1,4-linked linear chains and 1,6-glucosidic branch points. Current knowledge favors a bushlike structure with branches of 10–14 glucosyl units (Figure 1). About half of the glycogen mass is comprised by the external branches. The internal branches carry side chains separated by about four glucosyl units. The origin of the stem is O-glucosidically linked to a tyrosyl residue of the protein primer glycogenin (see Section 2). The bushlike structure of glycogen accounts for the spherical β-particles (30 nm diameter; up to 60,000 glucosyl units) that are present in most cells. In the liver, about 20–40 β-particles are associated into larger complexes known as α-rosettes.

## 2 SYNTHESIS AND BREAKDOWN

The first step in the biogenesis of glycogen is the covalent attachment by an undefined glucosyl transferase of carbon 1 of glucose to a single tyrosine residue of glycogenin (Figure 2). Subsequently, glycogenin autocatalytically extends the glucan chain by six to seven α-1,4-linked residues, using UDP-glucose as the glucosyl donor. This primer is further and similarly elongated by glycogen synthase, which is initially complexed to glycogenin but dissociates during the elongation process. Finally, a branching enzyme transfers a terminal oligoglucan (at least six glucosyl units) from an elongated external chain and attaches its carbon 1 to a carbon 6 in a neighboring chain.

The degradation of glycogen to glucose 1-phosphate requires the concerted action of glycogen phosphorylase and the bifunctional debranching enzyme (Figure 2). In the presence of inorganic phosphate, phosphorylase releases the terminal glucosyl residue of an external chain as glucose 1-phosphate, and continues to do so until the external chains have been shortened to four glucosyl units. Subsequently, the transferase activity of the debranching enzyme removes a maltotriose unit from the α-1,6-linked stub and attaches it through an α-1,4-glucosidic bond to the free carbon 4 of the

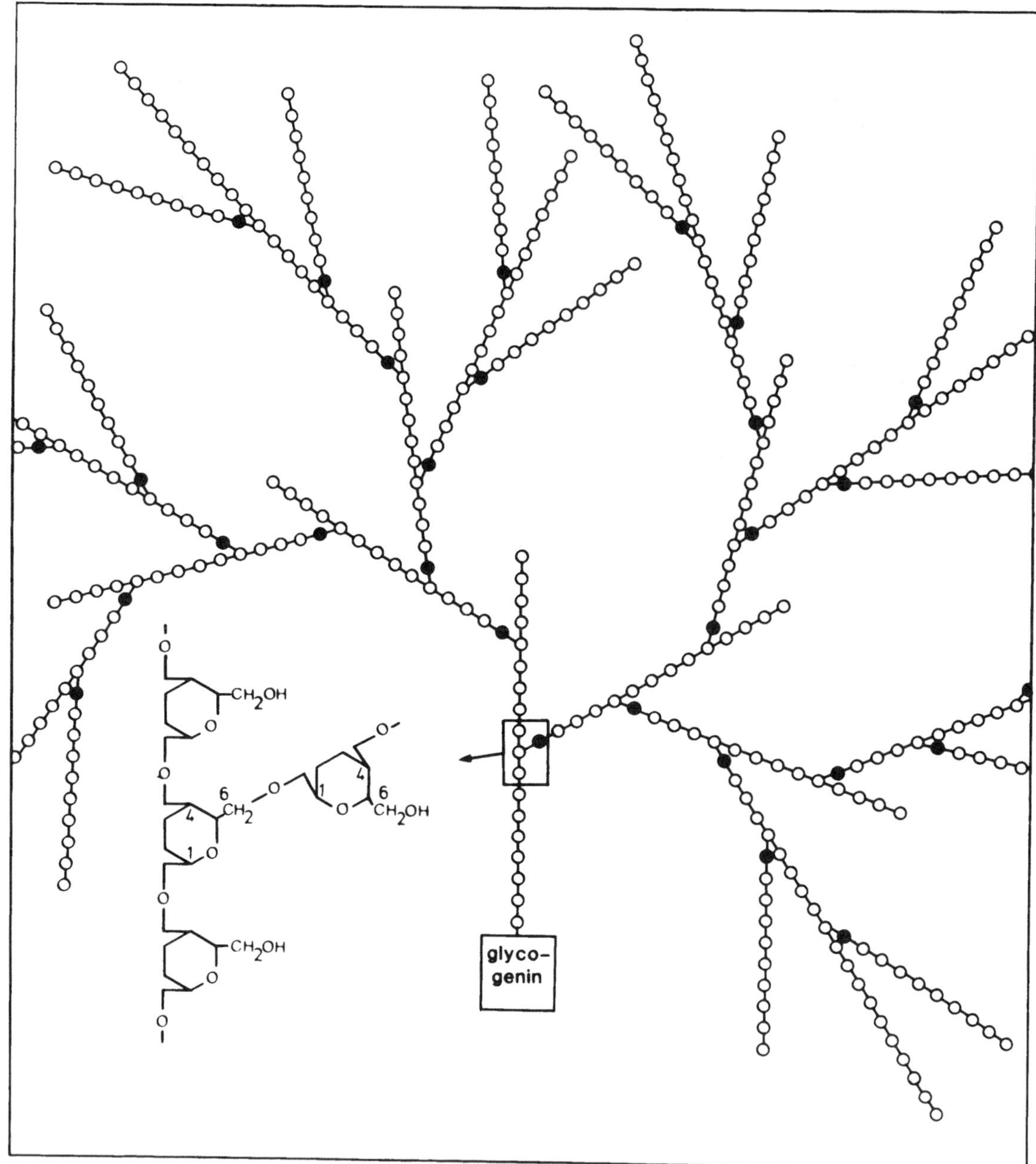


**Figure 1.** The structure of glycogen. ○–○, α-1,4-linked glucosyl units; ○–●, α-1,6-linked glucosyl units.

main chain. The single remaining α-1,6-linked glucosyl unit is then removed as free glucose by the α-glucosidase activity of the debranching enzyme, while additional α-1,4-linked glucosyl residues become available for phosphorylase. In the liver, glucose can be produced from glucose 1-phosphate by the successive actions of phosphoglucomutase and glucose 6-phosphatase.

Apart from the "phosphorolytic" pathway of glycogenolysis, most cells are also equipped with enzymes (e.g., an intralysosomal α-glucosidase) that can hydrolyze glycogen to glucose. Such hydrolytic glycogenolysis is quantitatively insignificant but represents a mechanism for the removal of glycogen from autophagic vacuoles and for glycogenolysis after cell death.

## 3 REGULATORY MECHANISMS

The rate-limiting enzyme of glycogen synthesis is glycogen synthase (Figure 3). This enzyme exists in two interconvertible forms: a physiologically inactive *b*-form and an active *a*-form, which differ in the extent of phosphorylation of multiple serine residues.

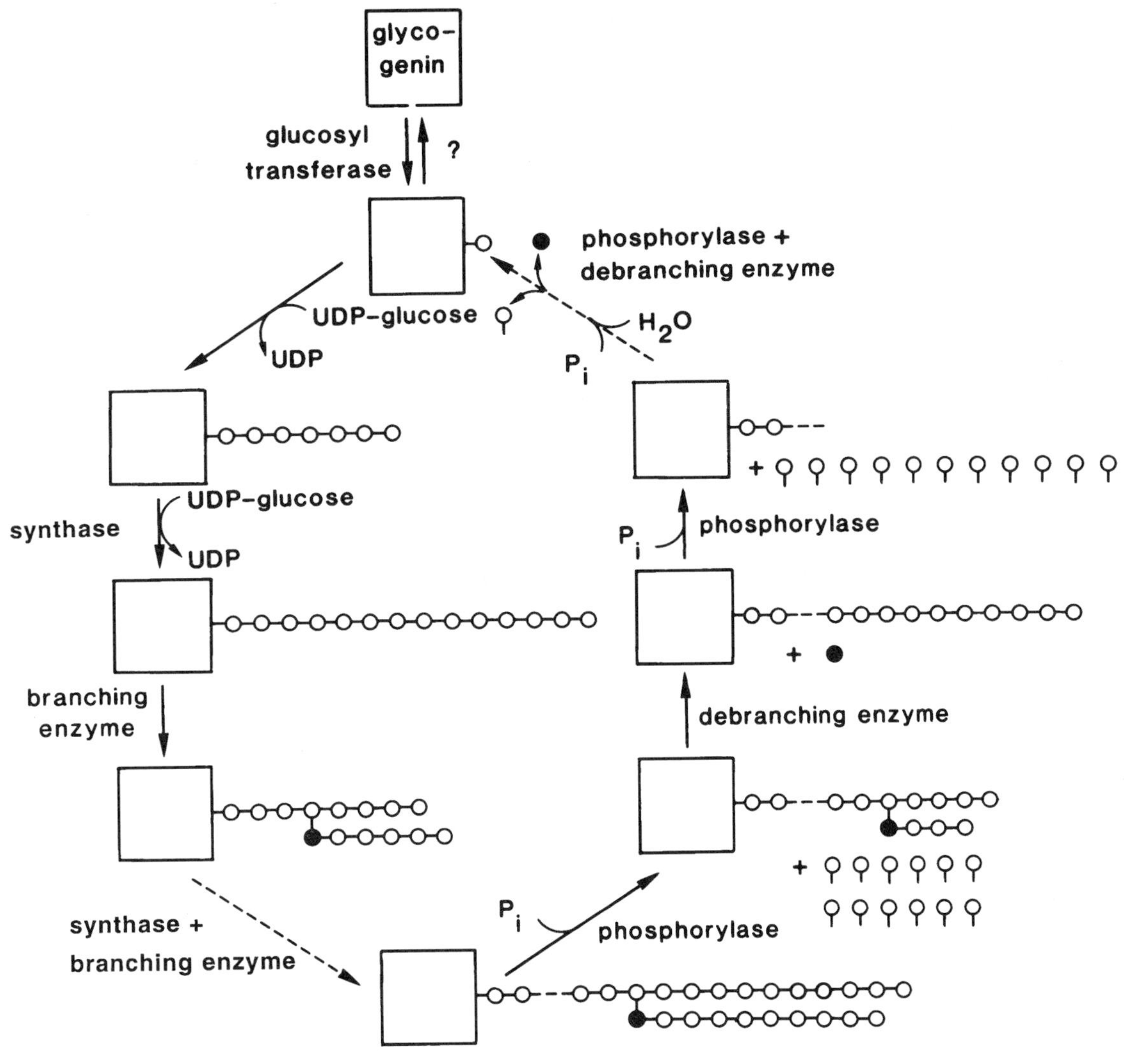


**Figure 2.** Pathway for the synthesis and phosphorolytic degradation of glycogen: ●, glucose; ⚲, glucose 1-phosphate; $P_i$, inorganic phosphate; UDP-glucose, uridine-diphosphoglucose; ○–○, α-1,4-linked glucosyl units; ○–●, α-1,6-linked glucosyl units. In the glycogen particles at the right, dashed lines represent most of the glycogen structure shown in Figure 1.

Glycogen synthase is activated (conversion to the *a*-form) through dephosphorylation by a protein phosphatase, while phosphorylation by various protein kinases (epg., the cAMP-dependent protein kinase) results in a less active enzyme.

Phosphorylase is the rate-limiting enzyme of the phosphorolytic pathway for glycogenolysis. The activity of this enzyme is controlled by the phosphorylation of a single serine residue but, in contrast to glycogen synthase, phosphorylase is converted to the active *a*-form by phosphorylation. Phosphorylase kinase is itself activated through phosphorylation by cAMP-dependent protein kinase, but its activity is also stimulated by the binding of calcium to its calmodulin subunit.

Glycogen metabolism is tightly regulated by hormonal and metabolic signals that alter the phosphorylation state of glycogen synthase and phosphorylase. For example, signaling through the phosphatidylinositol bisphosphate pathway as well as through cyclic AMP will activate protein kinases that inactivate glycogen synthase and activate phosphorylase (Figure 3). A prime regulator of glycogen metabolism in the liver is the circulating glucose concentration. The receptor of glucose is phosphorylase *a,* which then becomes a better substrate for a protein phosphatase. The ensuing inactivation of phosphorylase causes an arrest of glycogenolysis. Since phosphorylase *a* is also a potent allosteric inhibitor of a glycogen-bound synthase phosphatase in the liver, its removal by dephosphorylation allows the activation of glycogen synthase, resulting in the initiation of glycogen synthesis. This mechanism is involved in the postprandial deposition of glucose as hepatic glycogen.

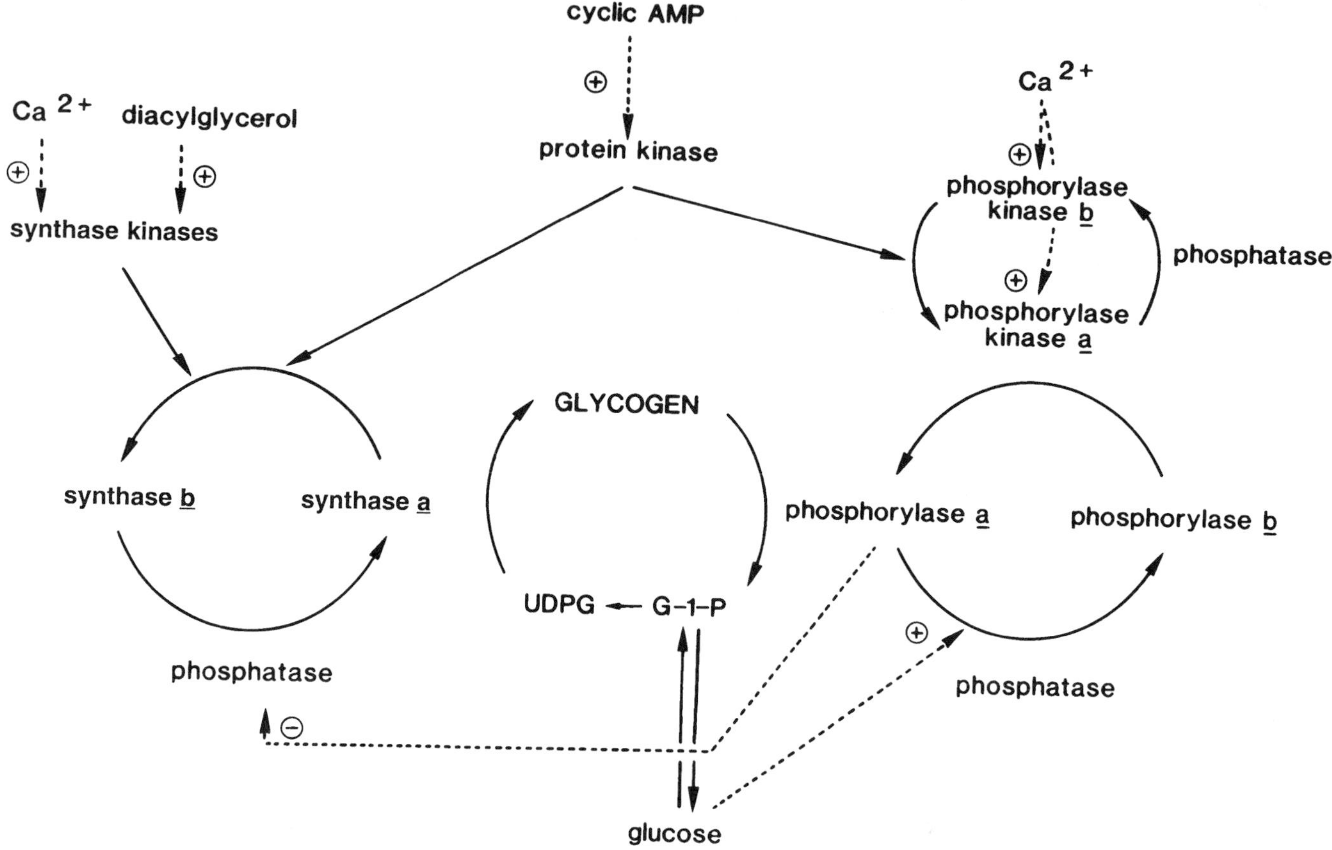


Figure 3. Enzymes involved in the regulation of glycogen synthase and phosphorylase. The control by glucose and the two-step conversion of glucose 1-phosphate to glucose occur only in the liver.

*See also* Carbohydrate Analysis; Carbohydrate Antigens; Glycobiology; Lipopolysaccharides.

### *Bibliography*

Hers, H. G., Van Hoof, F., and de Barsy, T. (1989) Glycogen storage diseases. In *The Metabolic Basis of Inherited Disease,* C. R. Scriver, A. L. Beaudet, W. S. Sly, and D. Valle, Eds., pp. 425–448. McGraw-Hill, New York.

Newgard, C. B., Hwang, P. K., and Fletterick, R. J. (1989) The family of glycogen phosphorylases: Structure and function. *Crit. Rev. Biochem. Mol. Biol.* 24:69–99.

Roach, P. J. (1990) Control of glycogen synthase by hierarchal protein phosphorylation. *FASEB J.* 4:2961–2968.

Smythe, C., and Cohen, P. (1991) The discovery of glycogenin and the priming mechanism for glycogen biogenesis. *Eur. J. Biochem.* 200:625–631.

Stalmans, W., Bollen, M., and Mvumbi, L. (1987) Control of glycogen synthesis in health and disease. *Diabetes/Metab. Rev.* 3:127–161.

## Glycolipids: *see* Lipopolysaccharides.

# Glycoproteins, Secretory

*Alistair G. C. Renwick*

### *Key Words*

**Microheterogeneity** The occurrence of a particular carbohydrate in multiple forms at a glycosylation site. These forms may differ in the structure and composition of their monosaccharide constituents, in the type and position of glycosidic linkages, and in the presence and number of noncarbohydrate substituents.

**Oligosaccharide** Linear or branched carbohydrate that consists of two to twenty monosaccharide units joined by glycosidic bonds.

**Secretory Glycoproteins** Proteins with covalently linked carbohydrate chains, which are synthesized by specialized cells for export to extracellular destinations.

Glycoproteins are a major class of macromolecules, and those secreted by cells, the secretory glycoproteins, are synthesized by specialized cells for export to extracellular destinations: close by, in the case of mucus-secreting cells in the alimentary tract, or distant, as exemplified by the glycoprotein hormones of the anterior pituitary gland, which are released into the vascular system to affect peripheral endocrine tissues.

The chemistry, biochemistry, and molecular biology of glycoproteins are burgeoning areas of intense research that mirror contemporary advances in analytical chemistry and reflect the ubiquity and importance of these macromolecules in biology and medicine. Progress has been facilitated, sustained, and enhanced by concomitant advances in separation science, mass spectrometry, nuclear magnetic resonance spectroscopy, and computing. These are powerful methods that enable rigorous analysis of the most complex biological molecules from minute amounts of starting material, thereby establishing a firm structural foundation for the molecular biologist and the molecular geneticist.

## 1 OLIGOSACCHARIDE STRUCTURES

An oligosaccharide may be defined as a short chain of two to approximately 20 monosaccharide residues linked glycosidically and is synonymous with the terms *glycan* and *sugar chain* in this contribution. Oligosaccharides obtained from secretory glycoproteins form two groups, the N- and O-linked. An N- or asparagine-linked glycan has at its reducing terminus an acetylglucosamine unit that is bonded to the amide nitrogen of the side chain of an asparagine residue in the polypeptide backbone. In contrast, an *O*-glycan contains at its reducing end an *N*-acetylgalactosamine residue that is attached to the hydroxyl groups in the side chains of serine or threonine constituents in the protein. Other glycosyl-amino acids such as β-xylosylserine, β-galactosyl-hydroxylysine, and β-L-arabinosyl-hydroxyproline are commonly found in proteoglycans, collagens, and plant and algal glycoproteins, respectively.

Structural determinations of many *N*- and *O*-glycans, which may coexist in the same glycoprotein molecule, have led to the formulation of rules that permit further classification of both groups. All N-linked glycans share a common pentasaccharide core, also known as the "trimannosyl core": Manα1–6(Manα1–3)Manβ1–4GlcNAcβ1–4GlcNAc, and they fall into three categories labeled high-mannose, complex, and hybrid types (Figure 1). High-mannose oligosaccharides contain only α-mannosyl residues in addition to the core pentasaccharide, and a heptasaccharide structure with two branches is commonly encountered in this type of sugar chain: Manα1–6(Manα1–3)Manα1–6(Manα1–3)Manβ1–4GlcNAcβ1–4GlcNAc. Variations occur in the number and positions of up to four Manα1–2 residues linked to the three nonreducing terminal α-mannosyl components of the heptasaccharide.

In addition to the core pentasaccharide, glycans of the complex variety reveal a plethora of structural changes that stem from outer chain moieties linked to α-mannosyl constituents. Furthermore,

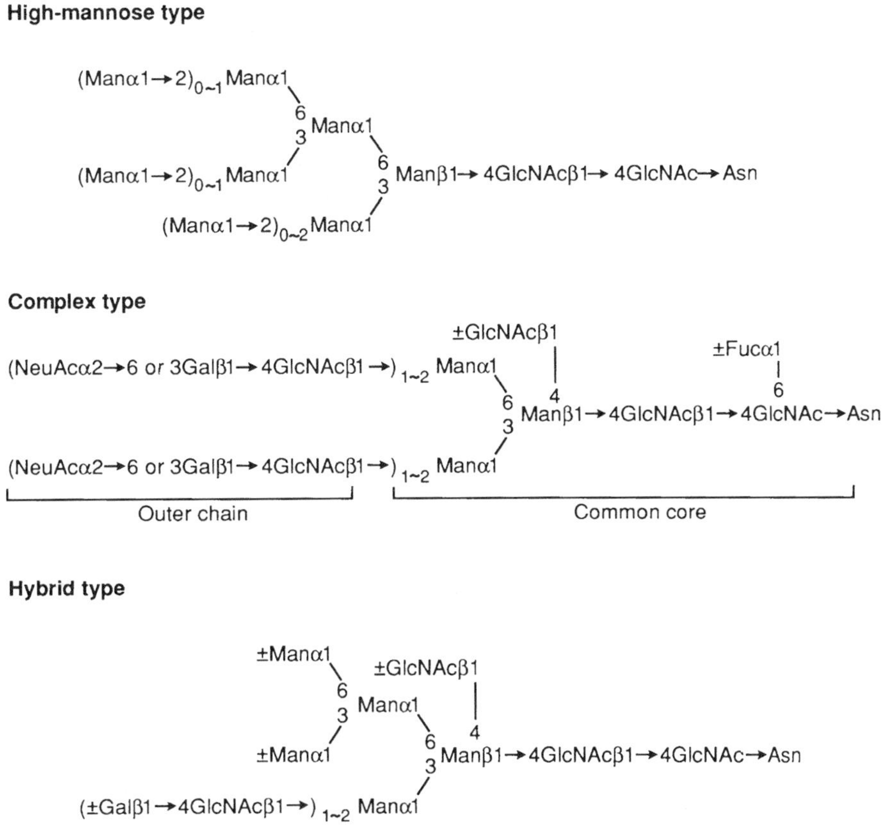

**High-mannose type**

(Manα1→2)$_{0\sim1}$ Manα1
6
3 Manα1
(Manα1→2)$_{0\sim1}$ Manα1
6
3 Manβ1→4GlcNAcβ1→4GlcNAc→Asn
(Manα1→2)$_{0\sim2}$ Manα1

**Complex type**

±GlcNAcβ1
±Fucα1
(NeuAcα2→6 or 3Galβ1→4GlcNAcβ1→)$_{1\sim2}$ Manα1
4
6
6
3 Manβ1→4GlcNAcβ1→4GlcNAc→Asn
(NeuAcα2→6 or 3Galβ1→4GlcNAcβ1→)$_{1\sim2}$ Manα1
Outer chain
Common core

**Hybrid type**

±Manα1
±GlcNAcβ1
6
3 Manα1
4
±Manα1
6
3 Manβ1→4GlcNAcβ1→4GlcNAc→Asn
(±Galβ1→4GlcNAcβ1→)$_{1\sim2}$ Manα1

**Figure 1.** General structures of asparagine- or N-linked sugar chains. The N-linked oligosaccharides derived from glycoproteins can be classified into three subgroups. Glycans of the high-mannose type contain only mannose and *N*-acetylglucosamine residues attached to a common heptasaccharide core. Complex-type sugar chains contain a pentasaccharide core, and structural variations occur through the number and structure of the outer chain components attached to the α-mannosyl residues in that core. Hybrid-type oligosaccharides show structures characteristic of high-mannose and complex-type sugar chains.

**Monoantennary**

Manα1 ↘ 6
Manβ1→ 4GlcNAcβ1→ 4GlcNAc→Asn
GlcNAcβ1→ 2Manα1 ↗ 3

**Diantennary**

GlcNAcβ1→ 2Manα1 ↘ 6
Manβ1 →4GlcNAcβ1 →4GlcNAc →Asn
GlcNAcβ1→ 2Manα1 ↗ 3

**Triantennary**

2, 4 - branched

GlcNAcβ1→ 2Manα1 ↘ 6
GlcNAcβ1 ↘ 4
Manα1 ↗ 3 Manβ1 →4GlcNAcβ1 →4GlcNAc →Asn
GlcNAcβ1 ↗ 2

2, 6 - branched

GlcNAcβ1 ↘ 6
Manα1 ↘ 6
GlcNAcβ1 ↗ 2
Manβ1 →4GlcNAcβ1 →4GlcNAc →Asn
GlcNAcβ1→ 2Manα1 ↗ 3

**Tetraantennary**

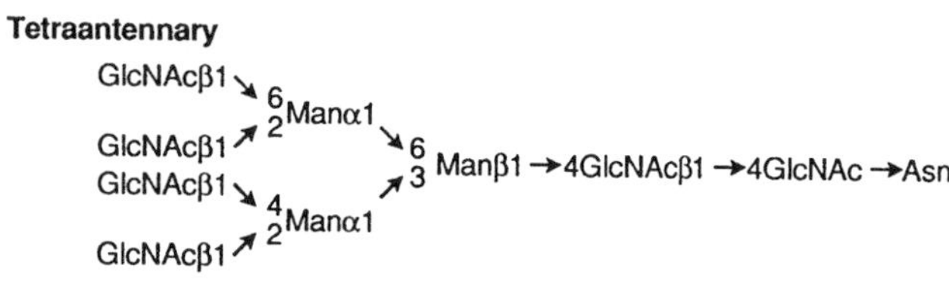


**Pentaantennary**

GlcNAcβ1 → 6
GlcNAcβ1→ 4Manα1
GlcNAcβ1 → 2 ↘ 6
Manβ1 →4GlcNAcβ1 →4GlcNAc →Asn
GlcNAcβ1 ↘ 4 ↗ 3
Manα1
GlcNAcβ1 ↗ 2

Figure 2. Some branched structures of complex-type oligosaccharides. Most of the structural variations that occur in N-linked oligosaccharides are found in the complex type, where one to five outer chains may be linked to the trimannosyl core to form monodi-, tri-, tetra-, and pentaantennary compounds, as shown.

complex-type sugar chains often bear a fucose residue linked to the C-6 position of the proximal *N*-acetylglucosaminyl unit, and a bisecting *N*-acetylglucosamine is frequently encountered at the C-4 position of the core β-mannosyl residue. These *N*-glycans are often highly branched di-, tri-, tetra-, and even pentaantennary structures (Figure 2), and the peripheral components of many complex-type *N*-glycans in animal glycoproteins frequently contain repeating disaccharide units (poly-*N*-acetyllactosamine) in linear or branched arrangements. These repeating units permit a huge range of possible structures to be displayed that include A, B, H, and Lewis blood group antigens and those associated with differentiating and tumor tissues. *N*-Glycans of the complex type may be highly fucosylated, phosphorylated, or sulfated; some are rich in polysialic acid constituents, and more exotic forms have been characterized from plants and primitive organisms.

The third class of *N*-glycans, the hybrid type, contains structural elements of the high-mannose and complex sugar chains. One or two α-mannosyl residues are linked to the mannose α-1–6 component of the trimannosyl core as in high-mannose oligosaccharides. The peripheral structures found in *N*-glycans of the complex type are attached to the mannose α1–3 constituent. Fucosylation of the proximal *N*-acetylglucosamine residue and insertion of a bisecting *N*-acetylglucosamine also occur.

As for the *O*-glycans, there are at least six core classes of the serine (threonine)-*N*-acetylgalactosamine type, which are discussed elsewhere.

## 2 BIOSYNTHESIS

In contrast to the modes of biosynthesis of nucleic acids and proteins, where fidelity of replication and translation is paramount, the *N*- and *O*-glycans are not formed on a template but are fashioned in the endoplasmic reticulum (ER) and in the Golgi complex (Figure 3), which is the most plausible cause of microheterogeneity. While details of the transit of nascent secretory and membrane proteins from rough ER to cell surface have yet to be resolved, many components of the secretory pathways appear to be highly conserved in nature.

It is expedient to consider the formation of *N*-glycans in three stages (early-, middle-, and end-stage processing), but the phenomenon is not discontinuous. The first step occurs when an oligosaccharyltransferase recognizes an asparagine residue in the tripeptide, asparagine-X-serine, or threonine, where X is any amino acid except proline. This enzyme transfers a tetradecasaccharide, common to the formation of all *N*-glycans irrespective of cell type, from a dolichol-linked oligosaccharide to the nascent polypeptide chain, but not at every opportunity; fewer than half of the known consensus sequences in secreted glycoproteins are glycosylated. The newly formed glycopeptide is ''pruned'' by the action of glycosidases in the ER, the transitional ER and the cis-Golgi complex, before midstage processing in the cis- and medial Golgi, whereas glycosyltransferases effect elongation of the glycans. Further modifications that constitute the end stage are carried out mainly in the trans region, which is the exit face of the Golgi stack. The cis and trans faces appear to act as centers for sorting and distribution, and the pathways of secreted glycoproteins, whose ultimate destinations lie in different organelles, diverge in the trans-Golgi network.

*O*-Glycans are made by the stepwise addition of sugars from nucleotide sugars to acceptors, and synthesis is initiated by enzymic transfer of an *N*-acetylgalactosamine to a serine or threonine constituent in the recipient protein. There is no requirement for a consensus sequence of amino acids like that in N-glycosylation, but O-linked sugar chains seem to predominate at β-turns where serine, threonine, and proline residues tend to cluster. While some *O*-glycans are formed early, most are synthesized after the release of nascent peptides from the ribosomes when they have moved to the smooth ER or Golgi.

The transit of glycoproteins and other proteins in the membranous system appears to be unidirectional, facilitated by transport vesicles that bud from each compartment in sequence before fusing with the next. This is not a simple, bulk export process; transit times are variable, with the result that the phenomena of intracellular transport are asynchronous.

## 3 CONTROL OF BIOSYNTHESIS

The absence of a template mechanism in the biosynthesis of oligosaccharides and the nature and complexities of the secretory process make exploration of regulatory events extremely difficult. Existing

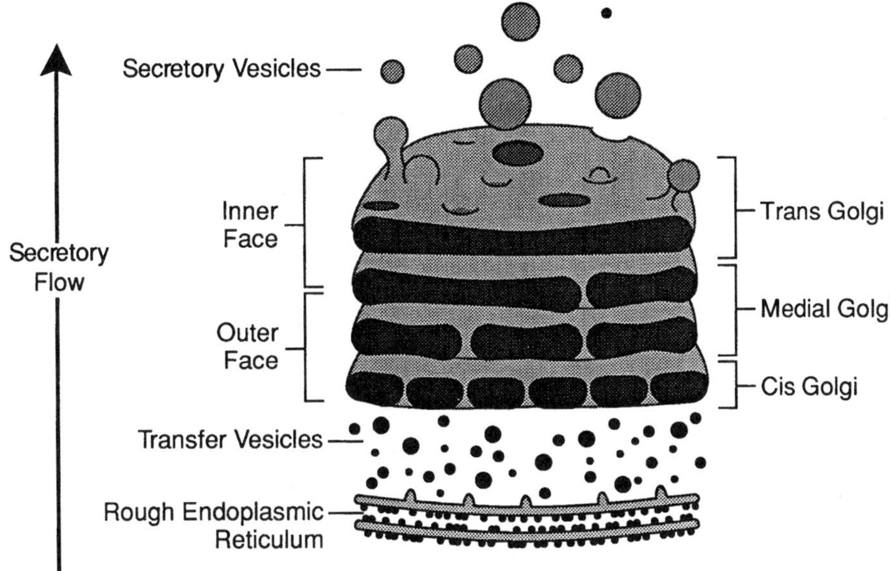


**Figure 3.** Protein secretion in eukaryotes. Newly synthesized proteins and glycoproteins are carried unfolded across the membrane of the endoplasmic reticulum (ER), which contains chaperone proteins that catalyze the folding of nascent polypeptides. The chaperones carry a tetrapeptide sorting signal that allows them to be separated from secretory proteins in the Golgi complex and to be retrieved and returned to the ER. Secretory proteins are transported in small vesicles that bud from each compartment in sequence before fusion with the next.

knowledge may be rudimentary, but there is sufficient to permit delineation of the main features. Most is known about the genes encoding the peptide sequences of many secretory glycoproteins and their regulation, and this area is not considered further. On the other hand, investigation of the control of oligosaccharide synthesis is of recent origin and is fast gathering momentum.

It has been estimated that a hundred or so membrane-bound glycosyltransferases act cooperatively and in competition to synthesize specific *N*- and *O*-glycans. A number have been cloned, and while these enzymes reveal little in the way of sequence homology, they share domains characteristic of the type II membrane-bound proteins. Not only are these enzymes highly specific for substrate and acceptor molecules, but some have strict specificities for branched structures. Local controls exist in their subcellular localization, and catalysis cannot occur if there is inadequate provision of dolichol phosphate and dolichol-linked oligosaccharides, sugar nucleotides, acceptors, and cofactors such as divalent cations. The rate of export to the cell surface is also important in oligosaccharide synthesis because it may exceed that of glycosylation, resulting in the secretion of incomplete structures. The distribution of the glycosyltransferases and glycosidases is of special interest because of the species, tissue, and cell specificities of these enzymes. Their intracellular concentrations fluctuate during differentiation and transformation, in physiological circumstances, and under pathological conditions.

The protein components of most glycoproteins appear to be of significance in local control of N- and O-glycosylation. In some but not all cases, glycosylation at one or more specific sites is obligatory for correct folding of the polypeptide. In many instances improper folding results in rapid degradation and little, if any, secretion. However, a significant number of proteins are secreted without carbohydrate.

Attached oligosaccharides may also affect subsequent glycosylation of the protein backbone. Many make substantial contributions to the size and apparent shapes of intact glycoproteins, and these are well illustrated in the glycoprotein hormones of the human anterior pituitary gland and placenta. The attached glycans affect the association of the two subunits in the formation of the intact, biologically active hormone and its interaction with its receptor. Other examples are found in enzyme–substrate interactions and in associations of glycoproteins with lectins. Such effects most likely result from steric hindrance and/or ionic shielding, but glycans may also affect biological activity by evoking conformational changes in the protein. In addition to local and genetic regulation, one must superimpose nervous and hormonal controls in animals that possess such systems.

## 4 PERSPECTIVES

The emergence of glycobiology and its growing impact on other fields, especially medicine, are among the most significant developments in contemporary science. The opportunities for research are immense, given the breadth, precision, and sensitivity of available analytical methods. Many important questions demand solutions, but the problem central to the biological role of oligosaccharides in glycoproteins is rooted in the lack of three-dimensional structures. Such molecules have resisted crystallization except in a very few instances, and while information gleaned from X-ray diffraction will be exciting, it will be insufficient to define structures in solution. Nuclear magnetic resonance spectroscopy is the method of choice: it is not yet possible to determine the three-dimensional structures of intact glycoproteins by this means, but rapid developments in this form of analysis will surely meet the need.

*See also* GLYCOBIOLOGY.

### *Bibliography*

Allen, H. J., and Kisailus, E. C., Eds. (1992) *Glycoconjugates: Composition, Structure and Function.* Dekker, New York, Basel, Hong Kong.

Drickamer, K., and Carver, J., Eds. (1992) Carbohydrates and glycoconjugates. *Curr. Opinion Struct. Biol.* 2:653–709.

Rothman, J. E., and Orci, L. (1992) Molecular dissection of the secretory pathway. *Nature,* 355:409–415.

Stockell Hartree, A., and Renwick, A. G. C. (1992) Molecular structures of glycoprotein hormones and functions of their carbohydrate components. *Biochem. J.* 287:665–679.

Varki, A. (1993) Biological roles of oligosaccharides: All of the theories are correct. *Glycobiology,* 3:97–130.

## Governmental Regulation: *see* Biotechnology, Governmental Regulation of.

# Growth Factors

*Antony W. Burgess*

*Key Words*

**Cancer** The cell mass that results from the uncontrolled production of genetically altered cells in an animal.

**Cell Production** The process of hierarchical proliferation and maturation of tissue-specific stem cells, which leads to the generation of mature cells in a specific tissue.

**Colony-Stimulating Factors** Proteins that stimulate the proliferation and maturation of myeloid precursor cells.

**Cytokines** Proteins released from one cell which modulate the proliferation, differentiation, and/or function of cells in a specific lineage.

**Differentiation** The alteration in gene expression as precursor cells divide and progress toward their functional forms.

**Growth Factors** Proteins that influence the proliferation and maturation of tissue-specific precursor cells.

**Hormones** Molecular signals released from one cell that modulate the function or production of other cells. In the broadest sense, the term ''hormone'' includes steroids, peptides, and lipids. In traditional physiology, the term was used to describe the molecules released by one organ which acted on the cells of another organ.

**Interleukins** Proteins released from stromal or hemopoietic cells which influence the proliferation, self-renewal, commitment, and/or maturation of hemopoietic cells.

**Lymphokines** Proteins that modulate the proliferation, maturation, or function of lymphoid cells.

**Receptors** Cell surface proteins that are stimulated by exogenous ligands (e.g., growth factors) to transfer a signal to the cytoplasmic compartment of a cell.

**Stem Cells** The cells responsible for the renewal of cells in specific lineages. Stem cells have the capability of dividing to renew themselves or dividing to produce differentiated progeny that are committed to the production of mature cells.

Cell production in all tissues of multicellular organisms is under the control of a network of tissue-specific protein regulators called growth factors. Although the growth factor network may be modulated by circulating hormones, these regulators are usually produced in the tissue in which they function. Growth factors have been identified in the epithelial, neural, lymphoid, myeloid, and hepatic systems; in many cases a number of closely related growth factors are produced in each tissue. As well as controlling the normal production of mature cells, growth factors are important for regulating inflammatory, wound-healing, and antiviral responses. Many diseases appear to be associated with inappropriate growth factor responses. Overproduction of growth factors leads to accumulation of cellular deposits or infiltration of tissues by inflammatory cells. Although cancers arise as the result of numerous genetic lesions, many metastatic cancers appear to be driven by the autocrine action of growth factors.

New therapeutic approaches to cancer involve the use of specific growth factors to improve hemopoietic recovery from cytotoxic or radiation therapy. Furthermore, growth factor antagonists are being developed to suppress the proliferation of the tumor cells. An understanding of the mechanism of action of growth factors and the range of action of these proteins is important for developing effective therapies for autoimmune diseases, cancer, rheumatoid arthritis, skin diseases, and diabetes.

## 1 INTRODUCTION

Multicellular organisms need to control the production, maturation, and function of cells in specific tissues or organs. It is now clear that locally secreted proteins modulate the cell physiology of each tissue. These regulatory proteins have been given a variety of names, including growth factors, cytokines, and interleukins. It is well known that the endocrine systems consists of a network of circulating regulatory molecules (hormones that includes proteins, peptides, and steroids; growth factors can be considered to be an extension of the endocrine system. While there is a direct overlap between the growth factor and endocrine systems, it is often helpful to consider that the major actions of endocrine hormones occur in tissues remote from their tissue of origin, whereas growth factors appear to have a major function within the tissues that produce them.

In the main there is no functional or structural characteristic that distinguishes the different classes of growth factors—most appear to be capable of stimulating multiple biological responses that depend critically on the differentiation state of their target cells. For example, one of the hemopoietic growth factors, granulocyte colony-stimulating factor (G-CSF), stimulates the proliferation of immature bone marrow cells as well as activating bacterial killing by mature neutrophils. For most growth factors, so many distinct biological effects have been observed in vitro that it is not yet possible to define their specific role in particular tissue systems. However, both pharmacological and genetic studies have demonstrated that growth factors can alter cell production, organogenesis, and disease susceptibility in animals. It should be noted that the action of a particular growth factor will usually depend on the presence of other growth factors, the availability of target cells, and the display of the appropriate cell surface receptors (see Figure 1).

Growth factors stimulate cells by binding to specific cell surface proteins called receptors. Upon binding the growth factor, many receptors dimerize and activate intracellular kinase networks. Some receptors, such as the epidermal growth factor receptor, are ligand-dependent tyrosine kinases; others, such as the interleukin-2 receptor, associate with and activate particular membrane-associated tyrosine kinases. Activation of these enzyme systems can have effects on cell metabolism, membrane turnover, cytoskeletal organization, cell movement, and cell division.

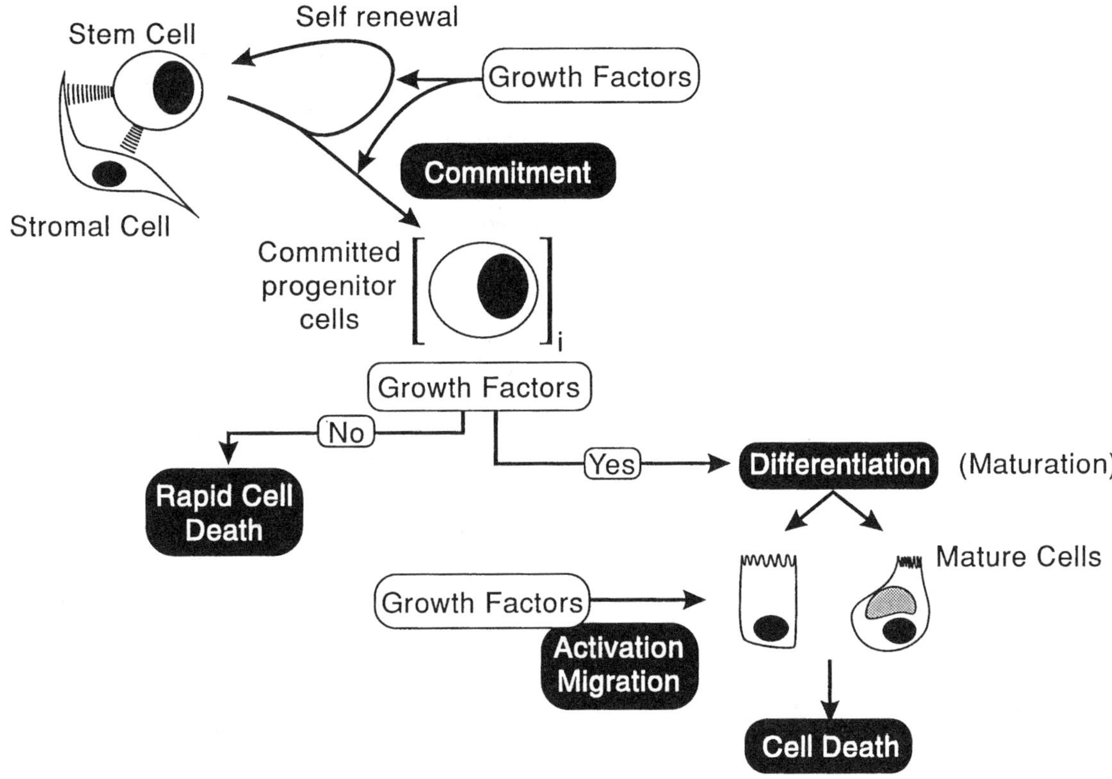


**Figure 1.** There are several sites of growth factor action during the production of cells. The action of a growth factor is likely to be quite different at the different stages of maturation.

## 2 DISCOVERY

The incredible potency and diverse biological activities of each growth factor have led to the multiple, independent discoveries for many of the growth factors. A typical example is the identification of interferon-$\beta_2$ as a member of a family of proteins capable of inducing antiviral responses and its discovery as a growth factor in the lymphoid system, where it was called interleukin-6. Some growth factors, such as fibroblast growth factor, have been identified independently in as many as 20 different biological systems. While these studies emphasize the multiple actions of growth factors, some confusion has developed with regard to their nomenclature and their likely physiological function(s). It should not be assumed that a particular action of a growth factor in vitro is automatically related to its role in the whole animal.

The first growth factor was identified in 1906: erythropoietin (epo). This discovery arose from physiological observations on the humoral control of production of red blood cells. While nerve growth factor (NGF) and epidermal growth factor (EGF) were discovered in the late 1940s, as a result of observations on the innervation of transplanted tumors, most other growth factors have been discovered as a result of their activity in laboratory culture systems. More recently, growth factors such as stem cell factor (also called the c-*kit* ligand) have been identified as a consequence of the characterization of their respective cell surface receptors. Table 1 lists the chromosomal locations on mouse and human of the growth factors mentioned in this entry.

## 3 GROWTH FACTOR FAMILIES

### 3.1 Insulinlike Growth Factors

In 1922 Banting and Best discovered insulin, the regulator of glucose metabolism, in pancreatic extracts. Insulin was the first protein to have its amino acid sequence determined. It is now known that insulin is synthesized as a single precursor protein, which is processed and secreted as a disulfide-linked heterodimer. As well as controlling blood glucose levels, insulin stimulates cell proliferation; however, two other closely related proteins, insulinlike growth factors I and II (IGF-I and IGF-II), are more potent stimulators of cell proliferation. IGF-I was discovered as the protein in plasma responsible for nonsuppressible insulinlike activity. IGF-I, which has been known for some time as a somatomedin, also mediates the action of growth hormone. IGF-II was the first growth factor to be associated with the aberrant growth of tumor cells in the laboratory. While the members of this family circulate in the serum, their availability is controlled by specific binding proteins. The cell surface receptors for insulin and IGF-I are composed of four chains ($\alpha_2$, $\beta_2$). The $\beta$-chains are transmembrane, ligand-dependent tyrosine kinases, and the extracellular $\alpha$-chains bind the ligand. The IGF-II receptor is a large ( 2000 amino acids) single-chain structure with no obvious intracellular catalytic domain. The IGF-II receptor also functions as the receptor for mannose 6-phosphate, which targets acid hydrolases to liposomes. It has been suggested that these two binding specificities might regulate complementary processes during tissue remodeling.

Table 1 Chromosomal Locations of Growth Factors and Their Receptors

| Growth Factor Receptors | Chromosome Location | |
|---|---|---|
| | Human | Mouse |
| Insulin | 11p15 | |
| IGF-I | 12 | 10 |
| IGF-II | 11 | 7 |
| Nerve growth factor | 1p22 | 3 |
| NGF receptor | 17q-12-22 | 11 |
| EGF | 49-25-27 | 3 |
| TGF-α | 2p13 | |
| EGF Receptor | 7p14/12 | 11 |
| *neu* | 17 | 11 |
| TGF-$\beta_1$ | 19q13.1-13.3 | 7 |
| TGF-$\beta_2$ | | 1 |
| TGF-$\beta_3$ | 14q-23-24 | 12 |
| Inhibin α | 2q33 | 1 |
| Inhibin β | 2q33 | 1 |
| Müllerian inhibitory substance | 19 | 10 |
| TNF-α/β | 6p | 17 |
| TNF-receptor (type 1) | 1p36 | 4 distal |
| TNF-receptor (type 2) | 12p13 | 6 distal |
| PDGF A-chain | 7 | |
| PDGF B-chain | 22 | 15 |
| PDGF receptor α | 4q-11-12 | |
| PDGF receptor β | 5q31-32 | 18 |
| a-FGF | 5 | |
| b-FGF | 4 | |
| *wnt-2* | 11 | |
| *hst/ks3* | | 17 |
| Hepatocyte growth factor | 7q11.2-11.21 | |
| HGF receptor (c-*met*) | 7q | 6 |
| IL-1α/β | 2q-13-21 | 2 |
| (α-chains) IL-2 | 4q-26-28 | 3B-C |
| IL-3 | 5q-23-31 | 11A5-B1 |
| IL-4 | 5q31 | 11A5-B1 |
| IL-5 | 5q31 | 11A5-b1 |
| IL-6 | 7P15-P21 | 5 proximal |
| IL-7 | 8q12-13 | 3 |
| IL-9 | 5q22-35 | 13 |
| IL-11 | 19q13.3-13.4 | |
| IL-12A (p35) | 3p12-3q13.2 | |
| IL-12B (p40) | 531-33 | |
| Interleukin 1 (type 1) | 2q12 | 1 centro |
| Interleukin-2α | 10p14-15 | 2A2-A3 |
| Interleukin-2β | 22q11.2-q12 | |
| Interleukin-4 | 16p11.2-12.1 | 7 distal |
| Interleukin-5α | | 6 distal |
| Interleukin-6 | 1 | |
| Interleukin-7 | | 15 proximal |
| M-CSF (also called CSF-1) | 1p13-p21 | 3F3 |
| M-CSF receptor (c-*fms*) | 5q33.2-33.3 | 18D |
| GM-CSF | 5q23-1 | 11A5-B1 |
| GM-CSF receptor α | αxp21-pter,Ypter-p11.2 | |
| GM-CSF receptor β | 229-12.3-13.1 | |
| G-CSF | 17q11.2-21 | 11D-E1 |
| G-CSF receptor | 1p34.2-35.1 | 4 distal |
| Stem cell factor | 12q-22-24 | 10 |
| SCF receptor (c-*kit*) | 4q11-12 | 5 |
| Erythropoietin (epo) | 7q11-22 | 5G |
| epo Receptor | 19pter-q12 | 9 |
| Leukemia inhibitory factor | 22q12.1-12.2 | 11A1-A2 |

### 3.2 Nerve Growth Factors

Nerve growth factor, a humoral substance produced by some tumors, is responsible for innervation of these tumors. The discovery of NGF in the 1940s initiated the search for other tissue-specific growth factors and led quickly to the identification of epidermal growth factor (see Section 3.3). Initially NGF was isolated from the salivary glands of male mice as a complex of three proteins; however, the active component is a single protein of molecular weight 30,000, and the three-dimensional structure of NGF has revealed a novel arrangement of three extended segments of twisted antiparallel β-sheets. Several other neurotropins [brain-derived neurotrophic factor (BDNF), NT-3, and NT4/5] have been discovered. BDNF stimulates peripheral sensory ganglia as well as cholinergic and dopaminergic neurons. More recently it has been reported that there is a glial growth factor, which is a differentially spliced form of the ligand for the *c-neu* receptor.

### 3.3 Epidermal Growth Factor (EGF) Family

It is now 30 years since Stanley Cohen described the discovery of the factor that induces the premature opening of eyelids and tooth eruption in mice. A small protein (53 amino acids) with three disulfide bonds and two distinct folding domains, EGF stimulates lung maturation and the formation of gastrointestinal mucosa, and inhibits the secretion of acid from the gastric mucosa. EGF binds to and stimulates a single-chain tyrosine kinase receptor. There are several growth factors related to EGF—namely, transforming growth factor α (TGFα), amphiregulin, cripto, heparin-binding EGF (HB-EGF), heregulin, and β-cellulin. TGFα was discovered as an autocrine growth factor induced by some tumor viruses. EGF and TGFα are expressed as precursor molecules. The TGFα precursor is expressed on the surface of cells, and the mature protein is released by specific proteolysis. Cripto has been implicated in the growth of colon cancers. Amphiregulin and the c-*neu* ligand bind to distinct receptors within the EGF receptor family. Interestingly, several myxoma and sarcoma viruses also encode proteins homologous to EGF and TGFα.

### 3.4 Transforming Growth Factor (TGF-β) Family

TGF-β was discovered as one of the growth factors released after tumor virus infection of fibroblasts. Indeed, TGF-β, synergizes with TGF-α to induce large colony formation by normal rat kidney (NRK) cells. TGF-β was also isolated independently by its ability to inhibit the proliferation of epithelial cells: it inhibits, as well, the proliferation of both hemopoietic and lymphoid cells. TGF-β induces the differentiation of bronchial epithelial cells and prechondrocytes, but inhibits the differentiation of adipocytes and myocytes. A latent form of TGF-β is released from platelets as a complex with a specific binding protein. Active TGF-β can be released from the complex either by proteolysis or acid treatment. Several other cytokines belong to the TGF-β superfamily: two forms of TGF-β have been isolated from platelets and three other related sequences detected by screening cDNA libraries. Inhibin, the Müllerian inhibitory substances (MIS), the bone morphogenic proteins (BMP), and activin are all related to TGF-β. The availability of the three-dimensional structure for TGF-β will allow models to be developed for the other family members.

TGF-β can induce the synthesis of extracellular matrix components in vivo and in vitro. TGF-β is also a powerful enhancer of monocyte function and a suppressor of lymphopoiesis. Mice have been produced that lack a functional TGF-β1 gene; while they are born as apparently normal animals, within 2 weeks they develop severe autoimmune and hemopoietic defects and die.

### 3.5 Platelet-Derived Growth Factors (PDGF)

For more than 70 years it has been known that serum, but not plasma, stimulates cells to proliferate in culture. However, it was not until the early 1970s that it was discovered that the growth factors in serum were actually released from platelets. It took almost 10 years of protein chemistry to purify and sequence the first PDGF—a disulfide-linked heterodimer of two related polypeptides (A and B), each containing approximately 100 amino acids. At the same time that the amino acid sequence for PDGF was determined, the nucleotide sequence for the oncogene (v-*sis*) encoded by the simian sarcoma virus was characterized. The protein sequence predicted from the v-*sis* gene is closely related to the PDGF-β chain.

Different cell types are known to secrete either PDGF-AA or PDGF-BB homodimers, both of which act as growth factors for fibroblasts, smooth muscle cells, and glial cells. There appear to be two forms of the PDGF receptor monomer (α and β), and these can combine to produce functional receptor dimers: αα, αβ, and ββ. While the αα receptor responds to all forms of PDGF (i.e., AA, AB, and BB), the ββ receptor responds only to PDGF-BB. The PDGF receptors are typical ligand-dependent tyrosine kinases, but the cytoplasmic domain contains an insert that modulates the intracellular signaling processes. PDGF increases the rate of wound healing in rodents, and initial reports indicate that it may be a valuable agent for treating cutaneous ulcers in diabetic patients.

### 3.6 Fibroblast Growth Factors (FGFs)

In the 1930s it was discovered that brain and pituitary extracts stimulated the proliferation of fibroblasts in culture. By the mid-1980s two forms of FGFs had been identified, purified, and sequenced. Both the acidic (a-FGF) and basic (b-FGF) forms are single-chain proteins of approximately 140 amino acids, and 55% of their amino acid sequences are identical. There are several other members of the FGF family: *wnt-2, hst/k53,* and *FGF-5,* and there appears to be a distant homology (20–25%) to interleukin 1α and β. Both a- and b-FGFs stimulate the proliferation of cells, including colonic and breast epithelia, endothelial cells, and muscle cells. Although a-FGF and b-FGF are found associated with the extracellular matrix, neither protein has a classical signal peptide that would direct its secretion. It is still not known how these proteins are released from cells. Although a-FGF and b-FGF bind to the same cell surface receptor, b-FGF has a higher affinity. While high affinity of the FGFs for heparin–sulfate is useful for their purification, this property often interferes with studies of binding to the cell surface. FGFs are potent angiogenic agents and are also capable of accelerating the healing of cutaneous wounds and damaged nerves. b-FGF is mitogenic for oligodendrocytes, astrocytes, and Schwann cells, and it prolongs the survival of neuronal cells in culture. In pharmacological doses, b-FGF can enhance the remyelination of neuronal sheaths and can prevent the death of neurons in the dorsal root ganglion.

### 3.7 Hepatocyte Growth Factor (HGF)

Originally discovered as a factor that increased the motility and spreading of cells (called scatter factor), hepatocyte growth factor (HGF) stimulates the proliferation of primary hepatocytes and increases the invasiveness of endothelial and epithelial cells. HGF appears to be involved in liver regeneration, tumor progression, and several embryological processes. HGF is a 92 kDa disulfide-linked heterodimer, consisting of a light chain (33 kDa) and a heavy chain (62 kDa), which are produced from a single-chain precursor.

The HGF receptor (c-*met*) is a 190 kDa heterodimer in which the α-chain (50 kDa) appears to be extracellular and the β-chain (50 kDa) is a membrane-spanning protein with a cycloplasmic tyrosine kinase domain. It is expressed in a number of epithelial tissues, including liver and colon. Interestingly, a fragment of HGF continuing two Kringle domains binds to c-*met* and stimulates the motility responses and the c-*met* tyrosine kinase activity, but does not stimulate a mitogenic response in primary rat hepatocytes. Mutations in c-*met* have been associated with a number of cancer types.

### 3.8 Hemopoietic Growth Factors

At least 16 cytokines or growth factors have been identified by their action on the production or function of blood cells. The major classes of hemopoietic regulators are interleukins (ILs), lympho-

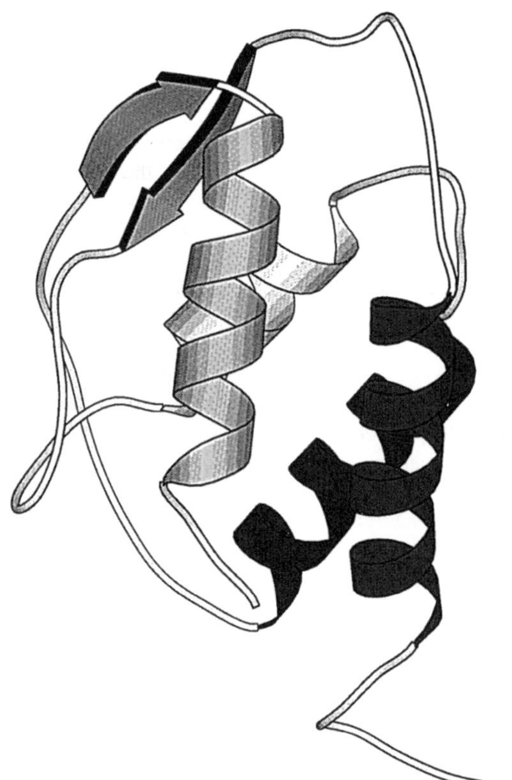

**Figure 2.** The three-dimensional structure of GM-CSF illustrates the typical four-helix bundle found in many of the hemopoietic regulators.

**Table 2** Biological Actions of the Hemopoietic Growth Factors

| Growth Factor | Action |
|---|---|
| Interleukin-1α/β | Stimulates thymocyte proliferation; increases expression of IL-2 receptor on T lymphocytes |
| Interleukin-2 | Stimulates proliferation of T lymphocytes |
| Interleukin-3 (also called multi-CSF) | Stimulates production of cells in all myeloid lineages |
| Interleukin-4 | Stimulates proliferation of B lymphocytes; stimulates production of IgM and potentiates mast cell proliferation in response to IL-3 |
| Interleukin-5 | Induces B-cell proliferation and differentiation in the mouse only; stimulates production and activation of eosinophils |
| Interleukin-6 (also called interferon-$\beta_2$) | Stimulates hemopoietic progenitor cells, B-lymphoid, and myeloma cells and induces acute phase responses |
| Interleukin-7 | Stimulates proliferation of thymocytes, T lymphocytes including cytotoxic T cells, and early B-cell precursors |
| Interleukin-8 | Stimulates neutrophil chemotaxis and acgtivates bacterial killing; inhibits Ig E production by B lymphocytes |
| Interleukin-9 | Stimulates proliferation of T-lymphoid cell lines; potentiates proliferative effect of IL-2 on fetal thymocytes |
| Interleukin-10 | Inhibits cytokine production by T lymphocytes |
| Interleukin-11 | Stimulates production of IgG generating B cells; induces acute phase response; augments ability of IL-3 to stimulate megakaryocyte colonies |
| Interleukin-12 | Stimulates cytokine production by natural killer and T cells |
| M-CSF (also called CSF-1) | Stimulates production, activation, and proliferation of monocytes and macrophages |
| GM-CSF | Stimulates productionand activation of neutrophils, eosinophils, and macrophages |
| G-CSF | Stimulates production and activation of neutrophils |
| Stem cell factor | Stimulates production of all hemopoietic progenitor cells |
| Erythropoietin (epo) | Stimulates production of red blood cells |
| Leukemia inhibitory factor | Inhibits differentiation of embryonal stem cells; induces acute phase responses; stimulates megakaryotic and platelet production; stimulates differentiation of M1 leukemic cell line |
| Thrombopoietin | Stimulates production of platelets |

kines, colony-stimulating factors (CSFs), erythropoietin (epo), and stem cell factor (SCF). Table 2 lists some of the biological actions of these molecules; however, it must be emphasized that as with the other growth factors, each of the factors listed has multiple activities, and their respective actions are not necessarily limited to the hemopoietic system. Several hemopoietic growth factors are being used clinically: in particular, interleukin-2 in conjunction with LAK cells for cancer therapy; granulocyte–macrophage CSF (GM-CSF) and G-CSF to accelerate neutrophil recovery after chemotherapy or bone marrow transplantation; and epo to stimulate red blood cell production in kidney dialysis or cancer patients. Three-dimensional structures have been determined for a number of hemopoietic growth factors—many consist of four helical bundles packed to form a central cylinder (Figure 2). The structure–function relationships of this class of growth factors are being examined in detail, and already several potent antagonists (e.g., for IL-1 and IL-4) have been produced. As might be expected from such a large number of hemopoietic growth factors, several classes of cell surface receptors have been identified: tyrosine kinase (e.g., M-CSF), multichain receptors (epg., IL-2, IL-6, and GM-CSF), and single-chain receptors (epg., G-CSF). Several of the hemopoietic growth factor receptors share adaptor subunits (epg., GM-CSF, IL-3, and IL-5 share a β-subunit, and the leukemia inhibitory factor (LIF), IL-6, the ciliary neurotrophic factor (CNTF), IL-11, and oncostatin M all appear to signal via complexes with the cell surface glycoprotein gp-130).

## 4 PERSPECTIVES

The biological action of the growth factor will be dependent on the differentiation state of the target cell and the presence of other growth factors and/or the existence of cell–cell contacts (see Figure 2). Cell–cell contacts and/or multiple growth factors acting early in many cell production pathways appear to be required to induce self-renewal of the tissue-specific stem cells. In a number of tissue systems the presence of particular growth factors is essential to maintain cell viability at all stages of the differentiation process. If the growth factor concentration decreases, many of the immature cells will die. Mature cells usually have a definite lifetime before the onset of disintegration; however, the functional activity of many mature cells can be activated by growth factors; for example, GM-CSF and G-CSF will prime mature neutrophils to kill bacteria more effectively.

*See also* Cell Death and Aging, Molecular Mechanisms of; Cytokines.

## *Bibliography*

Bradshaw, R. A., Blundell, T. L., Lapatto, R., McDonald, N. Q., and Murray-Rust, J. (1993) Nerve growth factor revisited. *Trends. Biochem. Sci.* 18:48–52.

de Vos, A. M., Ultsch, M., and Kossiakoff, A. A. (1992) Human growth hormone and extracellular domain of its receptor: Crystal structure of the complex. *Science* 255:306–312.

Lieschke, G. J., and Burgess, A. W. (1992) Granulocyte colony-stimulating factor and granulocyte–macrophage colony-stimulating factor. *New Engl. J. Med.* 327:28–35.

Meagher, A. (1990) *Cytokines.* Open University Press, Milton Keynes.

Miyazono, K., Ochijo, H., and Heldin, C.-H. (1993) Enhanced bFGF expression in response to transforming growth factor-β stimulation of AkR-23 cells. *Growth Factors* 8:11–22.

Sporn, M. B., and Roberts, A. B., Eds. (1991) *Peptide Growth Factors and Their Receptors,* Vols. I and II. Springer-Verlag, Berlin.

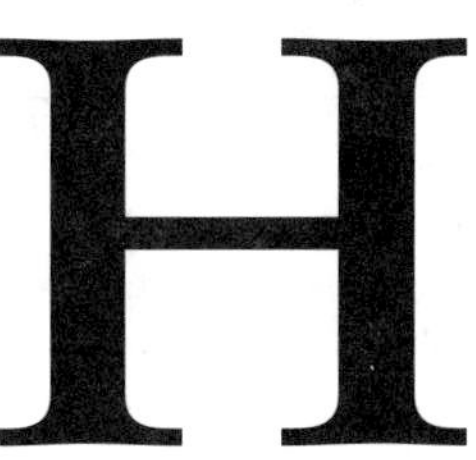

# HEART FAILURE, GENETIC BASIS OF

*Frank Kee and Alun E. Evans*

*Key Words*

**G Proteins** The guanine nucleotide binding proteins: trimeric complexes that are located at the inner aspect of cell membranes and are activated by a range of "signals" (e.g., hormones, growth factors, physical factors) by the substitution of GTP for GDP, initiating a cascade of biochemical events that propagates the "message."

**Photo-Oncogenes** Genes of normal cells, which control proliferation and differentiation. Their abnormal expression, often arising from a mutation at the same locus to form an oncogene, can result in uncontrolled cell growth and malignant transformation.

**Sarcomere** The functional contractile unit of muscle fiber, containing two kinds of interacting protein filament.

Heart failure is a heterogeneous condition, but in most cases it is the result of chronic pressure or volume overload, leading not only to hypertrophy but also manifesting in ultimately harmful more widespread activation of the renin–angiotensin system and the sympathetic nervous system. Superimposed on this global response, quantitative and qualitative modifications of genomic expression take place. Some of these are physiologically linked to the hypertrophic process, and others may ultimately provide avenues for novel therapies.

## 1 EPIDEMIOLOGY OF HEART FAILURE

Heart failure is not a single "disease," and thus the underlying determinants of the condition are more powerfully ascertained through observations on populations. Prospective investigations, such as the Framingham study, suggest an annual incidence of approximately 4.0 and 2.5 per 1000 in men and women, respectively, with a doubling of the rate for each decade of age between 45 and 74 years. The most common antecedents of heart failure are those that result in injury to myocardial cells, or their disorganization or loss, which impairs the ability of the organ to eject blood. Obvious structural derangements of the heart muscle that can cause pump failure include congenital and acquired valvular diseases. Some of these and other distinctly inherited conditions, such as Duchenne muscular dystrophy, have an identifiable hereditary basis. By far the most common risk factors for heart failure, however, are hypertension and ischemic injury, which together underlie some 90% of all cases and have multifactorial etiologies.

## 2 PATHOPHYSIOLOGY OF HEART FAILURE

Heart failure is better regarded as a disorder of the circulation, not merely a disease of the heart. Usually it occurs not at the time of injury to the heart but, rather, when compensatory hemodynamic and neurohormonal mechanisms have been exhausted. First, a decrease in the ability of the ventricles to eject blood activates the sympathetic nervous system. Next, β-adrenergic receptors in the noninjured myocardium increase the force and frequency of contraction. At the same time there is peripheral vasoconstriction from the effects of circulating noradrenaline and angiotensin II, and ultimately, retention of sodium and water, which tends to exacerbate the volume overload of the failing ventricle. The counterbalancing effects of atrial naturetic peptide are also blunted in patients with heart failure.

Prolonged stimulation of the inotropic pathways eventually leads to resistance to their effects, a reduced ability to sequester intracellular calcium, impaired relaxation, and energy deprivation due to depressed levels of high energy phosphates. In response to the increased wall stress, a number of architectural changes develop in the myocardial tissue which exacerbate the imbalance between energy production and expenditure, and affect clinical progression. While the addition of new sarcomeres contributes to myocardial hypertrophy, a reduced density of transverse capillaries and the expression of "slower" fetal isoforms of actin and tropomyosin (tending to reduce oxygen demand and contractility), as well as abnormal interstitial collagen, are also seen in patients with advanced heart failure. Although the underlying determinant of these changes is not known, their genetic basis is slowly being unraveled.

## 3 ALTERED GENETIC EXPRESSION IN HEART FAILURE

### 3.1 CELLULAR HYPERTROPHY

The abnormal expression of a variety of proto-oncogenes appears to contribute to the myocardial hypertrophy seen in heart failure. Some of these, including c-*sis*, encode directly for growth factors. Members of a second class (including *mas*, H-*ras*, and *src*) encode proteins associated with growth factor receptors and second messengers, and a third group (including c-*myc* and c-*fos*) encodes nuclear localizing proteins. The regulation of expression of these genes seems to be intricately linked with the sympathetic and neurohumoral responses to low cardiac output. For instance $\alpha_1$-adrenergic stimulation induces the expression of c-*myc* mRNA and also seems to up-regulate c-*fos* expression. The mitogenic stimulus triggering these changes may be partly attributable to upregulation of platelet derived growth factor gene expression by angiotensin II. Many growth factors, however, are multifunctional and may either inhibit or stimulate growth. Relatively little is known about transcriptional regulation itself, although as for skeletal muscle, there is some evidence of involvement of a family of cell-type-specific HLH (helix–loop–helix) proteins, which dimerize with DNA. However, the mechanism directly linking afterload with the transcriptional and mitogenic response is unclear.

There is also evidence linking genetic susceptibility to the occurrence of particular arrhythmias, and the stretch-induced changes in the expression of genes coding for the synthesis or assembly of the protein constituents of the cardiac potassium channels may contribute to the disposition to arrhythmias in heart failure.

### 3.2 Other Cellular Responses to Low Output

Changes in the intracellular handling of calcium have been clearly demonstrated in the myocardium of patients with heart failure; mRNA levels of the sarcoplasmic reticulum calcium release channel protein (the ryanodine receptor) and the uptake pump ($Ca^{2+}$-ATPase) and its regulatory protein, phospholamban, are all decreased. This probably results in a decrease in the number of excitation–contraction coupling sites, which in turn modifies the force–frequency relationship of myocardial performance.

A decreased expression of the mRNA of inhibitory G proteins (which couple with adenyl cyclate dependent membrane transducers) is also seen in heart failure and probably contributes to deficient production of cyclic AMP and the attenuated inotropic effect of sympathetic stimulation.

## 4 THE GENETIC BASIS OF RISK FACTORS FOR HEART FAILURE

Intriguing interrelationships are beginning to emerge linking the epidemiological risk factors for heart failure, such as myocardial infarction, and the genetic pathophysiology of failure itself. For instance, a well-described mutation in the gene encoding the receptor for low density lipoprotein (LDL) cholesterol underlies familial hypercholesterolemia, a condition that confers a markedly raised risk of coronary thrombosis. It appears however that high levels of LDL could potentiate the pressor neurohumoral response to low cardiac output, as it can inhibit the $G_i$ protein-dependent signal transduction pathways, which mediate endothelial relaxation.

Furthermore, a recently described deletion polymorphism of the gene for angiotensin-converting enzyme (ACE) has been found to increase the risk of infarction, although the same polymorphism does not seem to contribute to a predisposition to hypertension. Whether angiotensin's pressor, trophic, or arrhythmogenic effects on the myocardium are implicated is not known. In view of the improved survival seen in heart failure after ACE inhibition, however, unraveling the association may provide new therapeutic opportunities if drug response turns out to be affected by the underlying genotype.

## 5 IMPLICATIONS FOR THERAPY

The 1970s and 1980s heralded the arrival of technologically advanced and improved methods of diagnosis in cardiology. The dramatic rise in our knowledge of the role of peptide growth factors in cardiac and endothelial remodeling may suggest novel approaches to therapy in the decades ahead. Animal models may already provide a vision of the potential. For example, it is possible to synthesize antisense oligonucleotides that selectively bind to the mRNA of a targeted gene and block the synthesis of the protein. This strategy may one day be used to selectively inhibit the expression of growth factors, adhesion molecules, and cell cycle regulatory genes. An alternative strategy would be to control the expression of the risk factors for heart failure. For instance, successful reduction in LDL levels has been achieved in the Watanabe heritable hyperlipidemic rabbit by transplantation of autologous hepatocytes that had their defective LDL receptor gene "corrected" ex vivo with recombinant retrovirus.

Thus, almost certainly, there will be a move away from the sole use of drugs that modify the circulatory response to low output, to the integration of therapies that readjust the maladaptive myocardial and endothelial growth seen in pathological states.

*See also* Cardiovascular Diseases; Human Genetic Predisposition to Disease; Plasma Lipoproteins.

### *Bibliography*

Dzau, V. J., Gibbons, G. H., Cooke, J. P. and Omoigui, N. (1993) Vascular biology and medicine in the 1990s: Scope, concepts, potentials and perspectives. *Circulation,* 87:705–719.

Katz, A. M. (1990) Cardiomyopathy of overload. A major determinant of prognosis in heart failure. *New Engl. J. Med.* 322:100–109.

———. (1993) Cardiac ion channels. *New Engl. J. Med.* 328:1244–1251.

McFate, W. (1985) Epidemiology of congestive heart failure. *Am. J. Cardiol.* 55:3A–8A.

Packer, M. (1992) Pathophysiology of chronic heart failure. *Lancet,* 340:88–95.

Swales, J. D. (1993) The ACE gene: A cardiovascular risk factor. *J. R. Coll. Phys. London,* 27:106–108.

# Heat Shock Response in *E. coli*

*Takashi Yura and Hirotada Mori*

### *Key Words*

**Heat Shock Gene** Gene that encodes a protein whose synthesis is transiently induced upon shift to a higher temperature.

**Heat Shock Protein** Protein whose synthesis is transiently induced upon shift to a higher temperature.

**Molecular Chaperone** Protein that can assist folding or assembly of other proteins without being a component of the final assembled structures.

**Regulon** Set of genes or operons whose expression is coordinately regulated by a single regulatory protein.

$\sigma^{32}$ Minor sigma factor required for initiating transcription from heat shock promoters located upstream of heat shock genes.

The heat shock response is an adaptive and homeostatic response of bacteria and most other organisms to higher temperature. When cultures of *E. coli* are shifted from low (e.g., 30°C) to high (e.g., 42°C) temperature, the synthesis of heat shock proteins (hsps) is transiently induced at the level of transcription to cope with increased damage in proteins. Immediately upon temperature shift, the cellular level of $\sigma^{32}$ responsible for transcription of *hsp* genes increases rapidly and transiently. This increase in $\sigma^{32}$ results from both increased synthesis (translational induction) and stabilization of $\sigma^{32}$, which is ordinarily very unstable. The primary structure of most hsps is highly conserved during evolution, providing the basis for their important function in normal cell physiology as well as under stress conditions. Some hsps are known as molecular chaperones in that they mediate the correct folding and assembly of polypeptides but are not components of the functional assembled structures.

## 1 INDUCTION OF HEAT SHOCK PROTEINS

When *E. coli* cells are brought from 30°C to 42°C, synthesis of heat shock proteins (hsps) is induced almost immediately, reaches its maximum at about 5 minutes, and declines to a new steady state level in 20–30 minutes; induction occurs coordinately at the level of transcription. Upon exposure to an extreme heat shock (e.g., 50°C), synthesis of most proteins is arrested and those that remain are mostly hsps. Several other stresses, including ethanol and unfolded proteins, also induce hsps.

A minor sigma factor called $\sigma^{32}$, encoded by the *rpoH* (*htpR*) gene, plays a central role in the heat shock response: RNA polymerase holoenzyme containing $\sigma^{32}$ is required for transcription initiated from heat-inductible promoters located upstream of the *hsp* genes. Thus, transcription of a set of *hsp* genes (regulon) is coordinately enhanced under stress conditions, where higher levels of hsps are needed to cope with increased damage in proteins.

## 2 PHYSIOLOGICAL FUNCTION OF HEAT SHOCK PROTEINS

The heat shock regulon under $\sigma^{32}$ control includes genes for about 30 proteins. Some of them (DnaK, DnaJ, GrpE, GroEL, GroES, and HtpG) are molecular chaperones that can assist correct folding and assembly of proteins; consequently, they participate in diverse cellular processes including DNA replication, RNA transcription, protein transport, and protein degradation. These proteins, particularly when overproduced, protect various other proteins from heat inactivation and reactivate denatured proteins. Other hsps (Lon, ClpP, and ClpB) represent either an ATP-dependent protease or a catalytic or regulatory subunit of such a protease. The major functions of hsps appear to be to prevent denaturation of proteins and to reactivate or degrade denatured proteins.

## 3 ROLE OF $\sigma^{32}$ IN TRANSCRIPTION OF *hsp* GENES

The heat shock sigma factor $\sigma^{32}$ is specifically involved in transcription from promoters of the *hsp* genes. Table 1 lists the nucleotide sequences of some representative heat shock promoters. In general, these promoters are recognized by RNA polymerase containing $\sigma^{32}$ but not by polymerases containing $\sigma^{70}$ or other sigma factors.

The factor $\sigma^{32}$ is essential for cell growth at physiological temperatures (20–40°C) primarily because of increasing requirements of GroE (GroEL and GroES) proteins with increasing temperatures. Strains carrying an *rpoH* null mutation grow at or below 20°C in a rich medium, indicating that $\sigma^{32}$ is dispensable under these conditions.

**Table 1** Structure of Some Representative Heat Shock Promoters

| Promoter | −35 Region | Spacing | −10 Region |
|---|---|---|---|
| *groE* | TtTCcCCCTTGAA | 13 | CCCCATtTc |
| *dnaK P1* | TCTCcCCCTTGAt | 14 | CCCCATtTA |
| *grpE* | TgcttCCCTTGAA | 14 | CCCCATaat |
| *lon* | TCTCggCgTTGAA | 14 | CCCCATaTA |
| *clpP* | TgTtatgCTTGAA | 14 | aCCCATaac |
| *clpB* | caataaCCTTGAA | 14 | CCtCATtTA |
| *rpoD Phs* | TgcCaCCCTTGAA | 15 | gaCgATaTA |
| $\sigma^{32}$ consensus | TCTC-CCCTTGAA | 13–17 | CCCCAT-TA |
| $\sigma^{70}$ consensus | TTGACA | 16–18 | TATAAT |

## 4 REGULATORY ROLES OF $\sigma^{32}$ IN THE HEAT SHOCK RESPONSE

### 4.1 Regulation of the $\sigma^{32}$ Level

In addition to the catalytic role in initiating transcription from heat shock promoters, both the level and the activity of $\sigma^{32}$ play an active and dynamic role in regulating expression of the hsp genes. The unusual instability of $\sigma^{32}$ (half-life, 1 min) and translational repression of $\sigma^{32}$ synthesis contribute to the maintenance of $\sigma^{32}$ concentration at a minimum during steady state growth. Following heat shock stress, the levels of $\sigma^{32}$ are rapidly and transiently increased to meet the cellular requirements for higher levels of hsps. This increase in $\sigma^{32}$ level results from increased synthesis and stabilization of $\sigma^{32}$.

### 4.2 Regulation of $\sigma^{32}$ Synthesis

The *rpoH* gene has four promoters: three are transcribed by RNA polymerase-$\sigma^{70}$ and one by RNA polymerase containing a novel σ factor ($\sigma^E = \sigma^{24}$), which is active at very high temperatures (45–50°C). In spite of the complex organization and potential regulatory importance of various promoters, the transient increase in $\sigma^{32}$ synthesis upon mild heat shock (a shift from 30°C to 42°C) results primarily from induction at the level of translation. The 5′ portion of *rpoH* mRNA (ca. 200 nucleotides of coding sequence) is critically involved in controlling thermal induction of $\sigma^{32}$ synthesis, probably mediated by the mRNA secondary structure.

## 5 FEEDBACK REGULATION AND SENSORS IN THE HEAT SHOCK RESPONSE

The rapid accumulation of hsps upon temperature upshift brought about by increased $\sigma^{32}$ level is soon followed by an adaptation period, and the levels of $\sigma^{32}$ and hsps are readjusted to the new steady state growth condition. The readjustment includes a shutoff of heat-induced synthesis of $\sigma^{32}$ and resumed degradation of $\sigma^{32}$ within several minutes after temperature shift. This feedback regu-

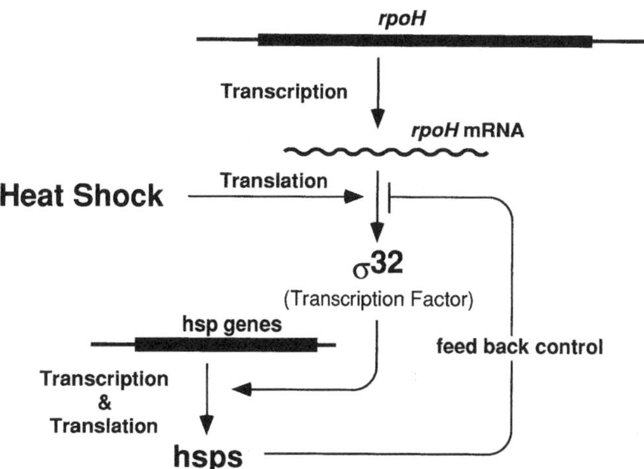


**Figure 1.** Schematic model of the heat shock response in *E. coli.*

lation is mediated by a set of hsps (DnaK, DnaJ, and GrpE) that are involved both in normal shutoff and in destabilization of $\sigma^{32}$.

Although the nature of heat shock sensors is not known, a free pool of DnaK may monitor changes in cellular concentration of unfolded or denatured proteins and negatively regulate the synthesis of hsps; DnaK may also work as a direct thermosensor. Other candidates for sensors include the ribosome, and a hypothetical regulator of translational induction of $\sigma^{32}$ synthesis. A schematic model for regulation of the heat shock response is depicted in Figure 1.

*See also* Bacterial Growth and Division; *E. coli* Genome.

### *Bibliography*

Craig, E. A., and Gross, C. A. (1991) *Trends Biochem. Sci.* 16:135–140.

Drlika, M., and Riley, M., Eds. (1990) *The Bacterial Chromosome.* American Society for Microbiology, Washington, DC.

Ellis, R. L., and van der Vies, S. M. (1991) *Annu. Rev. Biochem.* 60:321–347.

Morimoto, R., Tissieres, A., and Georgopoulos, C., Eds. (1990) *Stress Proteins in Biology and Medicine.* Cold Spring Harbor Laboratory Press, Plainview, NY.

Neidhardt, F. C., Ingraham, J. L., Low, K. B., Magasanik, B., Schaechter, M., and Umbarger, H. E., Eds. (1987) *Escherichia coli* and *Salmonella typhimurium.* American Society for Microbiology, Washington, DC.

Yura, T., Nagai, H., and Mori, H. (1993) *Annu. Rev. Microbiol.* 47:321–350.

# Hemoglobin

*Gino Amiconi and Maurizio Brunori*

### *Key Words*

**Allosteric Protein** A protein whose active site reactivity can be altered by the binding of a ligand (the allosteric effector) at a nonoverlapping site.

**Bohr Effect** The influence of pH on the $O_2$ affinity of hemoglobin.

**Geminate Recombination** Rebinding of a ligand molecule to its original site (the heme iron), after photolysis and before escaping to the solvent to mingle with other ligand molecules. Process kinetics is independent of bulk ligand concentration; process times in water are typically a picosecond to a nanosecond.

Hemoglobin (Hb) is a vital protein, basic to the respiratory function of all vertebrates and some invertebrates, and in plants with nitrogen fixation. In the different species, structural and functional data change widely, according to physiological requirements. The multifaceted behavior of hemoglobin has challenged the interest of scientists from many different backgrounds (biochemists, physiologists, geneticists, biophysicists), who have investigated this protein from extremely diverse points of view (from the quantum chemistry of iron to the regulation of buoyancy in fish by unloading oxygen into the swim bladder). Increasing concern over viral contamination of blood is spurring the development of a blood substitute; solutions of chemically modified hemoglobin represent one option.

Moreover, hemoglobin exhibits biochemical and biophysical phenomena similar to those of many multisubunit enzymes in their interaction with substrates and cofactors. Hemoglobin therefore serves as a prototype to study fundamental problems in biochemistry, such as the determination of mechanisms by which different parts of proteins communicate (i.e., the structural basis of cooperativity and allostery). It is not, therefore, surprising that hemoglobin has been called the hydrogen atom of biochemistry, because understanding its functions is so basic to protein chemistry at large.

## 1 GENERAL ASPECTS

Hemoglobin (Hb), the respiratory protein of erythrocytes, carries oxygen from the lungs to the peripheral tissues and participates in the reverse transport of carbon dioxide. Attention here is focused on human Hb; however, the broad ideas may be successfully transferred to the hemoglobins of other vertebrates and of invertebrates, and in general to more complex respiratory proteins, which are sometimes very large and extracellular. Hemoglobin concentration in normal human erythrocytes is extremely high (340 g/L) and corresponds to a 5.2 mM solution taking a molecular mass of 65,000 g/mol. Even though in the erythrocyte, Hb molecules (which are spheroids measuring 6.4 nm $\times$ 5.5 nm $\times$ 5 nm) are on average only 1 nm apart, they can rotate and flow past one another without hindrance.

## 2 STRUCTURAL FEATURES AND BASIC TERMINOLOGY

Hemoglobin consists of four polypeptide chains (globins). These apoproteins carry heme as their prosthetic group, which is an Fe(II) complex of protoporphyrin IX. In the free heme, Fe(II) is rapidly oxidized to Fe(III) by $O_2$ and water, whereas in Hb Fe(II) combines reversibly with $O_2$ and remains in the ferrous state in both the oxygenated and deoxygenated derivatives. Reversibility in the binding with $O_2$ is achieved in Hb by the folding of the polypeptide chain around the heme group, which is enclosed in a hydrophobic pocket. The subunits of Hb are identical in pairs: the polypeptide chains of two subunits designed $\alpha^1$ and $\alpha^2$ (where indexes denote relative location within the tetramer) each have 141 amino acid residues; those of the $\beta^1$ and $\beta^2$ subunits each have 146 amino acid residues. The tetrameric molecule of adult normal Hb is usually indicated as $\alpha_2\beta_2$. The $\alpha$ and $\beta$ subunits have different sequences of amino acids but fold up spontaneously to a similar three-dimensional structure (Figure 1; see color plate 10), the so-called globin fold. The $\alpha$ and $\beta$ subunits are made up of 7 and 8 helical segments, respectively, interrupted by an equal number of nonhelical sections. The helices are named A to H, and the nonhelical segments lying in between are lettered AB, BC, and so on; the nonhelical portions at the ends of the subunit are named NA (at the amino terminus) and HC (at the carboxyl terminus). Residues within each segment (helical as well as nonhelical) are numbered from the amino end (e.g., A1–A16, EF1–EF8). Therefore, each amino acid residue (usually represented with the three-letter code) is uniquely identified by its topology, its position in the sequence (in parentheses), and the corresponding subunit [e.g., PheCD1(42)$\beta$]. The heme is linked to the globin by the mainly covalent bond between Fe(II) and the imidazole-$N_\varepsilon$ of histidine F8, known as "proximal." The packing of subunits into Hb tetramers is such that there are close

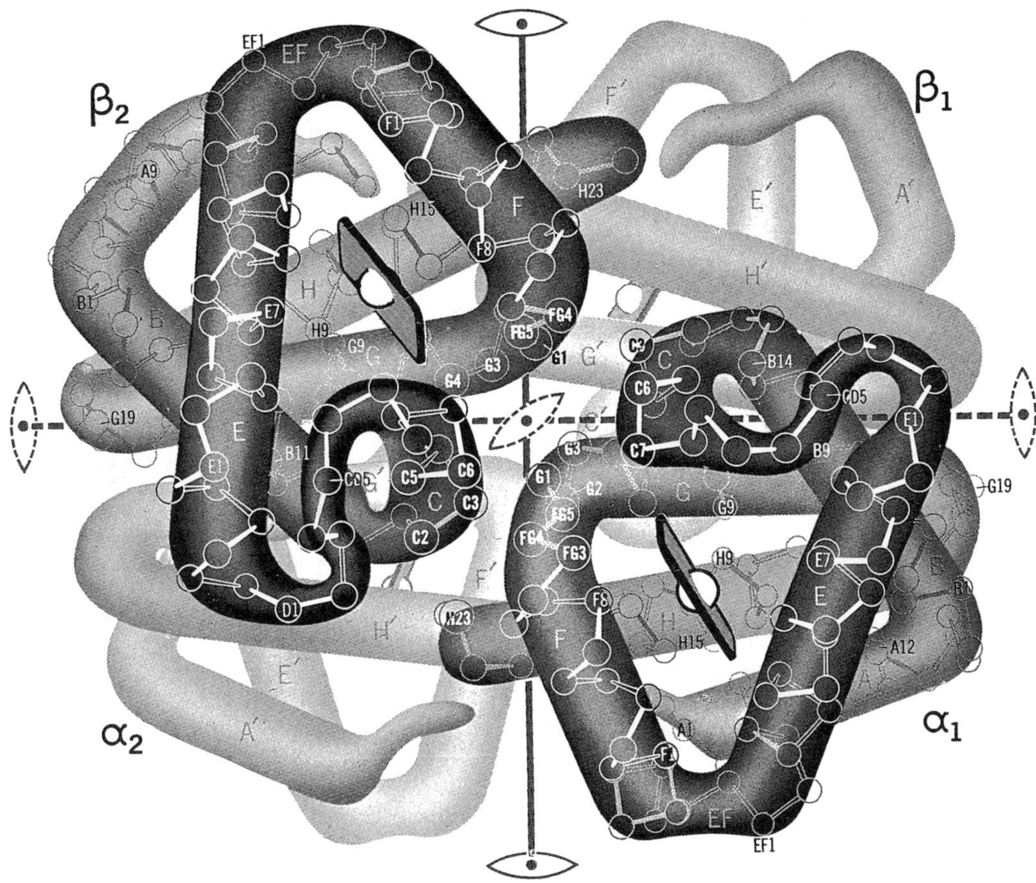


Figure 1. Detailed view of ferric Hb, which is similar to the oxy derivative. Only the α carbons of the main chains are shown. Those whose side chains are involved in contacts between subunits are given boldfaced numbers in large circles. There are two kinds of unlike subunit contact: $\alpha^1\beta^1$ (or $\alpha^2\beta^2$) and $\alpha^1\beta^2$ (or $\alpha^2\beta^2$) the first interface being more extensive than the second. Whereas $\alpha^1\beta^1$ contact involves the B, C and H helices and the GH corner, $\alpha^1\beta^2$ contact concerns mainly helices C and G and the FG corner. (Illustration copyright by Irving Geis, from R. E. Dickerson and I. Geis (1983) *Hemoglobin: Structure, Function, Evolution and Pathology*. Benjamin/Cummings, Menlo Park, CA.) (See color plate 10.)

interlocking connections of side groups between unlike subunits (i.e., α and β), but only little contact between those that are alike (see Figure 1).

## 3 DERIVATIVES WITH HEME LIGANDS

Ferrous Hb with no ligand attached to the Fe is named deoxyhemoglobin or unligated Hb. When a heme ligand of the ferrous protein (e.g., $O_2$, CO, NO, alkylisocyanides, nitroso aromatic compounds) is bound at the Fe(II) (in a 1 : 1 stoichiometry), an electron reorganization occurs, the Fe(II) changing from high spin to low spin. In the presence of $O_2$, the metal ion is subject to a slow spontaneous oxidation, so that ferric Hb or metHb is produced; the water molecule bound to the heme Fe(III) can be replaced by $F^-$, $OCN^-$, $N_3^-$, imidazole, or $CN^-$, where $F^-$ is the weakest and $CN^-$ the strongest ligand.

## 4 THE REACTION WITH HEME LIGANDS

The reaction of Hb with $O_2$ or other heme ligands is associated with large modification of the spectral (see Figure 2) and magnetochemical properties, which change in proportion with the amount of ligand bound to the heme Fe.

### 4.1 Functional Properties at Equilibrium

The fundamental representation of the reversible combination of Hb with $O_2$ is the ligand binding curve, wherein the fraction of heme groups bearing $O_2$ ($Y$) is measured as a function of the partial pressure of $O_2$ ($p$). The sigmoid shape of the curve is in marked contrast to the rectangular hyperbola found for the isolated α or β subunits (and also for myoglobin, see Figure 3A). The sigmoidal curve reflects a phenomenon called cooperativity, a complicated process whereby some $O_2$ binding favors more binding (as would occur in the lungs), or, from the opposite perspective, some $O_2$ release favors more release (as would occur in the peripheral tissues). The shape and the position of the equilibrium curve can be described by the Hill equation in logarithmic terms

$$\log\frac{Y}{1-Y} = \log K + n \log p \qquad (1)$$

Relation (1) has no direct physical meaning but, with a suitable choice of $K$ and $n$, it is able to fit empirically the experimental results with only two parameters: $p_{1/2}$, the value of $O_2$ pressure required to yield half-saturation of hemes ($p$ value at $Y = 0.5$), which indicates the overall affinity of Hb for $O_2$ (the higher the affinity, the lower the $p_{1/2}$), and $n$, the slope in the Hill plot (see

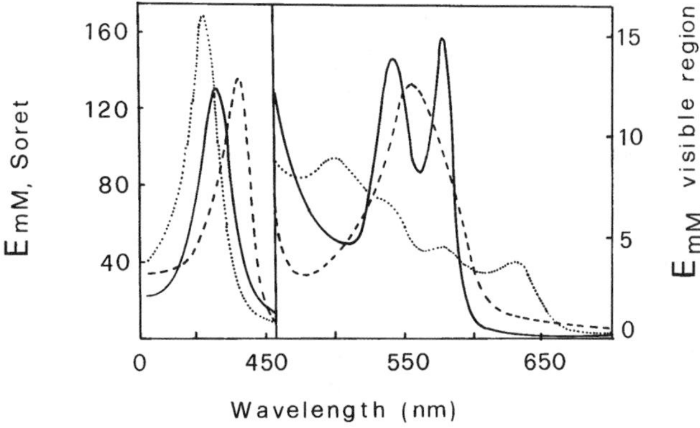


**Figure 2.** Optical absorption spectra of deoxy Hb (dashed curves), oxy Hb (solid curves), and metHb (dotted curves) in the visible and near ultraviolet (the so-called Soret band) regions at pH 7.4: $E_{mM}$, millimolar extinction coefficient in heme (i.e., referred to an equivalent molecular mass of 16,250 g/mol).

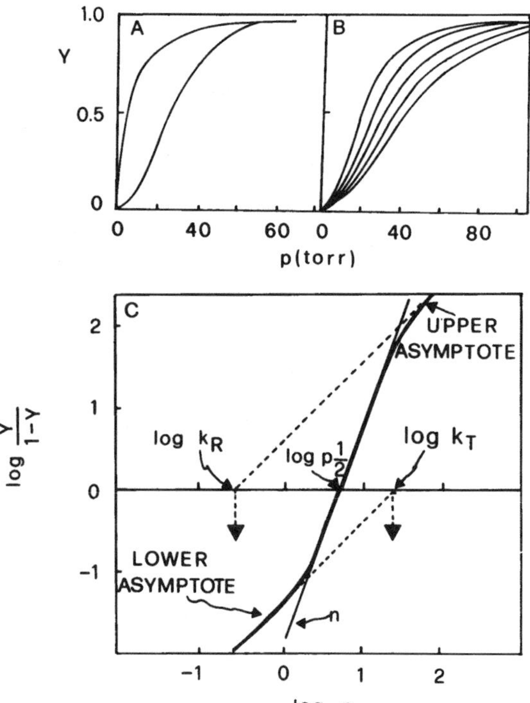


**Figure 3.** Oxygen equilibrium curves of Hb plotted in different ways: in (A) and (B), $Y$, the fractional saturation with $O_2$, is reported as a function of $O_2$ partial pressure $p$; in (C), the Hill plot (see equation 1) is presented. (A) The typical shape of the $O_2$ equilibrium curve for Hb is sigmoid, while the curve for isolated β subunits is hyperbolic. Both proteins are tetrameric ($\alpha_2\beta_2$ and $\beta_4$). In the case of the β subunits, the four $O_2$ binding sites are equivalent and independent; on the other hand, the shape of the Hb oxygenation curve implies that the affinity of the free sites depends on the fraction of hemes already occupied by $O_2$. (B) $O_2$ binding curves for Hb at various pH values. From left to right: pH 7.6, 7.4, 7.2, 7.0, 6.8. The change in $O_2$ affinity with pH is known as the Bohr effect. (C) The Hill plot for Hb oxygenation begins and ends with straight lines, called asymptotes, with positive slope of unity ($n = 1$). The intercepts of these asymptotes with the line parallel to the abscissa when $\log (Y/(1 - Y)) = 0$ (i.e., when the concentration of oxy and deoxy Hb are equal) give the values of the $O_2$ equilibrium constants for the T state ($K_T$) and the R state ($K_R$) (see Section 5.1). Often the asymptotes cannot be determined accurately; in such cases the oxygenation process is described in terms of the maximum slope of the curve, $n$ (taken as a measure of the degree of cooperativity), and the overall oxygen affinity in logarithmic terms (i.e., $\log p_{1/2}$), which corresponds to the value of $\log p$ when $\log Y/(1 - Y) = 0$ (see Table 1).

Figure 3C), which is related to the shape of ligand binding curve and therefore to cooperativity (the higher its value over and above 1, the higher the cooperativity). The $O_2$ affinity of Hb is lowered by $H^+$, $Cl^-$, $CO_2$, 2,3-D-bisphosphoglycerate (BPG), all known as heterotropic ligands. Increasing the concentration of any one of these ligands serves to shift the $O_2$ equilibrium curve to the right (see Figure 3B)—that is, toward lower $O_2$ affinity—thus facilitating $O_2$ release. Table 1 reports the $O_2$ binding parameters at equilibrium of human Hb under different solvent conditions.

### 4.2 Kinetic Aspects

Hemoglobin kinetics with heme ligands approximates second-order reaction in the combination direction and first-order reaction in the dissociation direction. Among the gaseous ligands, at neutral pH and room temperature, NO binds fastest, $O_2$ not quite as fast, and CO significantly more slowly; the dissociation reactions follow a different order (see Table 1). When the rates of binding of the first and fourth gaseous ligands are compared, the kinetic manifestation of cooperativity is observed (see Table 1). In more details, the kinetics of ligation over all chemically meaningful time scales (from picosecond to second) demonstrates the existence of different rate-limiting steps for the various heme ligands and, therefore, the existence of intermediates in the reaction pathway. A four-state scheme is required:

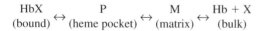

$$\underset{\text{(bound)}}{\text{HbX}} \leftrightarrow \underset{\text{(heme pocket)}}{\text{P}} \leftrightarrow \underset{\text{(matrix)}}{\text{M}} \leftrightarrow \underset{\text{(bulk)}}{\text{Hb + X}}$$

where HbX represents the bound state, Hb + X an unligated molecule (i.e., the two entirely separated free species, Hb and X) and P and M are intermediate states in which the heme Fe has no sixth ligand, but X remains trapped either in the heme pocket (P) or in the protein matrix (M). From the latter states the ligand may either rebind (to yield HbX) in geminate process or escape to the solvent (to yield Hb + X).

**Table 1** Values of Parameters Describing the Functional Properties of Human Hemoglobin[a]

| $O_2$ Equilibria in the Presence and Absence of BPG | | | | | | |
|---|---|---|---|---|---|---|
| Conditions | $p_{1/2}$ (torr) | $n$ | $K_T$ ($\mu M^{-1}$) | $K_R$ ($\mu M^{-1}$) | $L_0$ | $\Delta G^c$ (kJ/mol) |
| | 5.3 | 2.8 | 0.018 | 3.9 | $7.3 \times 10^5$ | 13.0 |
| +2mM BPG | 14.0 | 3.1 | 0.008 | 3.0 | $3.0 \times 10^6$ | 16.3 |

| Kinetics of Binding of Various Ligands | | | | |
|---|---|---|---|---|
| Ligand | $k'_{overall}$ ($M^{-1}s^{-1}$) | $k_{overall}$ ($s^{-1}$) | $k'_4/k'_1$ | $k_1/k_4$ |
| $O_2$ | $4.7 \times 10^6$ | 35 | 5 | 150 |
| CO | $2.0 \times 10^5$ | 0.015 | 40 | 10 |
| NO | $1.5 \times 10^7$ | 0.00005 | 1 | 100 |

[a]In 0.05 M Tris, pH 7.4; $[Cl^-]$ 0.1 M; 20°C.

## 5 MECHANISM OF COOPERATIVE BINDING AND ALLOSTERY

The cooperative binding of $O_2$ by hemoglobin is an example of an allosteric phenomenon. In allosteric binding, the uptake of one ligand influences the affinities of remaining unfilled binding sites.

### 5.1 Concerted Model of Allostery

The structural basis of cooperativity has been (and still is) a subject of contrasting views. Among the various approaches emphasizing the role of subunit–subunit interactions, the concerted model (often referred to as MWC model from the initials of proponents: J. Monod, J. Wyman, and J.P. Changeux) has a very simple algebraic description. Its basic idea is that the intrinsic $O_2$ affinity is determined by the quaternary structure of the protein, rather than by the number of ligand molecules already bound to a tetrameric Hb. This model assumes that there are only two different quaternary structures: one with fewer and weaker interactions among the subunits (fully ligated Hb or R, for relaxed), and the other with more and stronger bonds between the subunits (deoxy Hb or T, for tense). In the R or high affinity state, Hb displays the ability to bind $O_2$ tightly; this functional property is damped but not abolished in the T or low affinity state. On the other hand, the T state interacts more strongly than the R state with heterotropic ligands ($H^+$, $Cl^-$, BPG, $CO_2$). The transition from T to R is a none-or-all process: the symmetry of the molecule is conserved, (i.e., hybrid RT states cannot exist), and the $O_2$ affinity of all subunits in a tetramer would be either equally low or equally high. Small structural changes are linked to oxygenation of the subunits within each quaternary structure, so that intermediate ligation states (i.e., the T state with $O_2$ bound or the R state with less than four $O_2$ bound) will have conformations locally different from those of unligated T and fully ligated R. The assumption of the concerted model is that these local structural differences are not transmitted to neighboring subunits and thus do not to alter their $O_2$ affinity but affect only the relative stability of the R and T quaternary states. Although binding to the four hemes in each of the T and R states is itself noncooperative, the shift in the T $\leftrightarrow$ R equilibrium gives rise to a sigmoid binding curve. The MWC model allows us to describe the ligand binding by Hb at equilibrium with only three independent parameters: the two microscopic association constants of $O_2$ with the protein in the low affinity ($K_T$) and high affinity ($K_R$) states [often reported in terms of a ratio $c(=K_T/K_R)$], and the concentration ratio of the two quaternary states in the absence of ligand, $L_0 = [T]_0/[R]_0$. Adding BPG equimolar with the Hb tetramer shifts the equilibrium in favor of the T structure, with an increase of $L_0$; the same effect is obtained by lowering pH. In turn, $K_T$ decreases with increase in $H^+$, $Cl^-$, $CO_2$, and BPG concentrations, which have no effect on $K_R$. The total free energy of cooperativity can be calculated from the binding of the last and first ligand molecules, approximately equal to $\Delta G^c = 2.303RT \log (K_R/K_T)$ (i.e., the difference in free energy of binding $O_2$ to the R and T states). Table 1 lists values of parameters illustrating the interpretation of the cooperative effects in terms of the MWC model (see also Figure 3C).

### 5.2 Structural Changes Accompanying Binding of $O_2$

The T state is assumed to undergo quaternary interactions between subunits typical of deoxy Hb, while for the R state the subunits are typical of oxy Hb (as determined by single crystal X-ray diffraction data). In deoxy Hb the porphyrins are domed and the Fe is displaced toward the proximal histidine by 0.06 nm from the plane of the porphyrin nitrogens, while in oxy Hb the Fe is "in-plane." On single subunit oxygenation the Fe atom barely moves toward the porphyrin, which remains domed within the T state; this results in strain because the heme is prevented from adopting the geometry optimal with the $O_2$ bound. Relief of strain at the active site requires a movement of the proximal histidine together with the F-helix and the FG-corner (the so-called allosteric core). To accommodate these tertiary modifications, the relative orientation of the $\alpha^1\beta^1$ and $\alpha^2\beta^2$ dimers has to alter; such changes make these dimers a misfit in the quaternary T structure, with consequent promotion of the T $\rightarrow$ R transition, which is identified by major structural changes at the interface $\alpha^1\beta^2$ (and by symmetry $\alpha^2\beta^1$). Adopting the R state requires the Fe to move toward the porphyrin plane with release of strain. All the heterotropic ligands lower the $O_2$ affinity by forming additional hydrogen bonds that specifically stabilize the T structure. BPG binds in the cavity bounded by the amino-termini and helices H of the β subunits, making hydrogen bonds with ValNA1(1), HisNA2(2), LysEF6(82), and HisH21(143). The functionally relevant protons (i.e., those released upon uptake of $O_2$ and at the basis of the Bohr effect; see also Figure 3B) are discharged in deionized solution only from HisHC3(146)β, a residue that forms a hydrogen bond with AspFG1(94)β in the T structure of the same chain. In the presence of $Cl^-$ and/or BPG, other residues [e.g., ValNA1(1)β and LysEF6(82)β] contribute additional Bohr protons.

## 6 EVOLUTIONARY CONSIDERATIONS

Various estimates place the time of duplication of the globin genes from their single ancestral gene during early vertebrate evolution, at approximately 500 million years ago. During the long evolution of the Hb family (based on the amino acid sequence of 32 species of vertebrates) 27 residues in the α subunits and 18 residues in the β subunits have remained invariant: 14 are heme contacts, 16 are subunit contacts essential for allosteric mechanism, 8 are involved in hydrogen bonds or salt bridges within the subunits or at the $\alpha^1\beta^1$ interface, 6 are used for internal nonpolar contacts, and one proline to turn the BC corner in the β subunits. It has been estimated that changes in the Hb genes accumulated at a rate of one residue per α or β subunit every 2 to 3 million years. However, the essential features of the globin fold from very diverse organisms—ranging from mammals to insects to plant root nodules—are preserved even when the sequence homology is very low. Examination of many sequences shows that evolutionary divergence has been constrained primarily by an almost absolute conservation of the hydrophobicity of the residues buried in the helix-to-helix and helix-to-heme contacts. Survival of mutant proteins in the globin family has been restricted to those that maintain the basic globin fold.

## 7 HEMOGLOBIN VARIANTS

Several hundred (647 up to the year 1994) mutant hemoglobins within the human population have been isolated and chemically characterized. In most cases the abnormality consists in the replacement of a single amino acid residue per αβ dimer (variants of α subunits relative to those of β subunits are approximately in a ratio of 1 to 2): all these modifications are consistent with single-base substitution in DNA coding for the globin subunits. Some abnormal hemoglobins have residues deleted or inserted (25 variants); in others the subunits are cut short or elongated (12 variants); others are constituted by fusion of two different subunits (9 variants); and yet others show more than one point mutation in the same polypeptide chain (15 variants). Of the large number of Hb mutants, a significant fraction have deleterious effects on health.

### 7.1 Molecular Basis of Hemoglobin Diseases

Pathological symptoms in carriers have been associated with altered $O_2$ affinity (causing polycythemia if it is raised or cyanosis if it is lowered), decreased stability of the protein (determining hemolysis and clumps of denatured Hb called Heinz bodies), an increased tendency to form metHb (mainly producing cyanosis), and abnormal intracellular polymerization (inducing a contorted banana- or sickle-shaped erythrocyte). In many cases these biochemical abnormalities have been interpreted with the help of three-dimensional structure models. Most of the structural abnormalities giving rise to clinical symptoms are clustered around the heme pockets or in the vicinity of the $\alpha^1\beta^2$ interface, so important in the allosteric transitions.

1. The more common structural pattern causing an increase in $O_2$ affinity is the loss of hydrogen bonds, salt bridges, or nonpolar interactions that stabilize the Hb structure as a whole, and in particular the T state; moreover, some mutations increase $O_2$ affinity because they impair BPG interaction by modifying its binding site.
2. Unstable hemoglobins usually involve an opening of heme pocket or loss of heme and subsequent precipitation of globin. In addition, insertion of a proline into an helix, unfitting substitution of a nonpolar side chain that plugs the interior of protein from the surrounding water, introduction of a charged group inside the molecule, and deletions of single residues or segments of polypeptide chains can all be disruptive for the folded Hb.
3. The M (for metHb) mutants are the class in which the Fe atoms of one type of subunit are permanently oxidized as a result of the replacement of a heme contact residue; the most common mutations are the substitution of the proximal histidine (an invariant residue) or HisE7 (the so-called distal histidine located on the $O_2$ side of the heme) by tyrosine in either the α or the β subunit.
4. In general, the mutations that lead to abnormal Hb behavior are not located on the exterior of the tetramer; yet, the replacement of a surface glutamate GluA3(6)β by a hydrophobic valine in HbS or sickle cell mutant (the most prevalent Hb variant worldwide) has deleterious effects. Upon deoxygenation, in concentrated solutions, HbS molecules aggregate into a solidlike gel, distorting and rigidifying the erythrocytes into a variety of bizarre shapes. As a consequence, the red cells may not be able to traverse the vessels of the microcirculation, with transient to permanent blockage of local oxygenation: resulting organ damage is a major cause of the morbidity and mortality of sickle cell anemia. The key to the erythrocyte sickling is the presence in deoxy Hbs of a hydrophobic pocket for ValA3(6)β [made up of PheF1(85)β and LeuF4(88)β and located between the E and F helices]. Upon oxygenation, in fact, the F helix motion of a β subunit pulls PheF1(85)β of one tetramer away from potential contact with the ValA3(6)β side chain of a second tetramer.

## 8 HEMOGLOBIN SOLUTIONS AS BLOOD SUBSTITUTES

Hemoglobin solutions do not represent suitable blood substitutes for two reasons. First, since tetrameric Hb in dilute solutions easily dissociates into free αβ dimers and thus is rapidly filtered off by the kidney, the half-life of free Hb in the bloodstream is only one hour. Second, $O_2$ release to tissues is poor as a result of the high $O_2$ affinity of the carrier when stripped from BPG, the allosteric effector that modulates the efficiency of oxygenation.

To overcome these drawbacks, several covalent modifications of Hb have been described. The molecular mass of chemically engineered $O_2$ carriers has been increased by (1) intra- and intertetrameric cross-linking with glutaraldehyde or glycolaldehyde, (2) conjugation to dextran and polyethylene glycol derivatives, and (3) protein engineering by fusion of the two α subunits with a glycine residue as a spacer. The $O_2$ affinity has ben reduced by covalent attachment to the subunit of pyridoxal 5′-phosphate (which is known to mimic the functional effect of BPG) or by specific chemical changes induced by site-directed mutagenesis. Other compounds, such as bis(3,5-dibromosalicyl)fumarate and nor-2-formylpyridoxal-5′-phosphate, reduce the $O_2$ affinity and, at the same time, stabilize the tetramer.

*See also* Cytochrome P450; Hemoglobin, Genetic Engineering of; Receptor Biochemistry.

### *Bibliography*

Antonini, E., and Brunori, M. (1971) *Hemoglobin and Myoglobin in Their Reactions with Ligands.* North-Holland, Amsterdam.

———, Rossi Bernardi, L., and Chiancone, E., Eds. (1981) *Methods in Enzymology,* Vol. 76. Academic Press, New York.

Bunn, H. F., and Forget, B. G. (1986) *Hemoglobin: Molecular, Genetic and Clinical Aspects.* Saunders, Philadelphia.

Dickerson, R. E., and Geis, I. (1983) *Hemoglobin: Structure, Function, Evolution and Pathology.* Benjamin/Cummings, Menlo Park, CA.

Eaton, W. A., and Hofrichter, J. (1990) Sickle cell hemoglobin polymerization. *Adv. Protein Chem.* 40:63–279.

Fermi, G., and Perutz, M. F. (1981) *Atlas of Molecular Structures in Biology: Haemoglobin and Myoglobin.* Clarendon Press, Oxford.

International Hemoglobin Information Center. (1994) Hemoglobin Variants List. *Hemoglobin,* 18(2):77–161.

Perutz, M. F. (1990) Mechanisms regulating the reactions of human hemoglobin with oxygen and carbon monoxide. *Annu. Rev. Physiol.* 52:1–25.

Wyman, J., and Gill, S. J. (1990) *Binding and Linkage. Functional Chemistry of Biological Macromolecules.* University Science Books, Mill Valley, CA.

# Hemoglobin, Genetic Engineering of

*Timm-H. Jessen*

### *Key Words*

**Cooperativity** The interaction of subunits of an oligomeric protein, resulting in a sigmoid ligand binding curve.

**Enhancer** DNA sequence upstream or downstream of a gene contributing to an enhanced transcription.

**Promoter** DNA sequence in front of a gene that interacts with the RNA polymerase, hence influences transcription.

**Transgenics** Term usually applied to plants and mammals (mouse, rat) that accommodate a foreign gene in the genome.

**Two-State (MWC) Model** Theory developed by Monod, Wyman, and Changeux explaining cooperativity. An oligomeric protein exists in two conformational states, the T ("tense," usually unligated; low ligand affinity) and the R ("relaxed," usually ligated; high ligand affinity) states, which are in equilibrium.

Hemoglobin, the main protein of the red blood cell, provides organisms with oxygen by transporting this essential energy source from the respective respiratory organ to the tissue. As a result of its comprehensive properties and simple availability, hemoglobin has become a case study in biochemistry. It was therefore not surprising to see that this protein was one of the first subjects genetic engineers dealt with. Nevertheless, it was more than 10 years before fully functional hemoglobin was expressed in bacteria, yeast, and even in mammals. This series of advances has led to new approaches for the treatment of hemoglobin-related diseases.

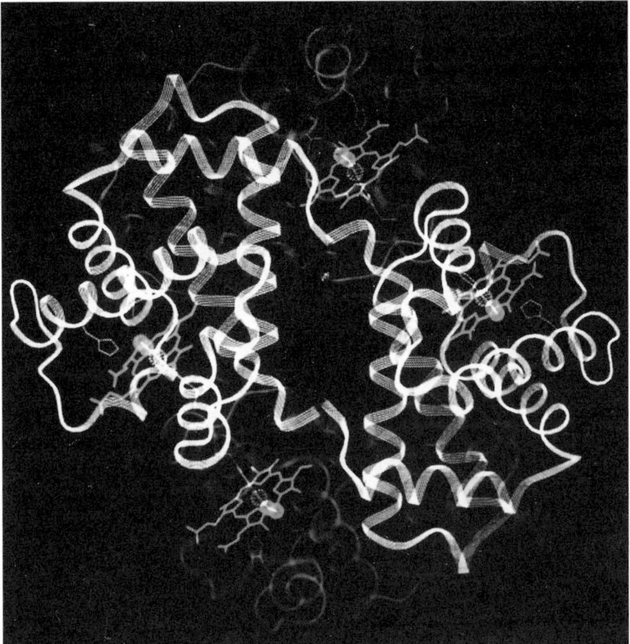

**Figure 1.** Structure of hemoglobin. The two α-chains are shown as white ribbons, the two β-chains as blue ones. The four heme groups are shown in red, with the central iron atom in green; bound oxygen can be seen as two red spheres close to each heme group. Amino acid residues in vicinity to the heme are shown in green. (See color plate 11.)

## 1 STRUCTURE AND PHYSIOLOGY OF THE BODY'S OXYGEN CARRIER

Hemoglobin is a tetrameric protein that appeared some 600 million years ago; it consists of two α- and two β-subunits ($\alpha_2\beta_2$), each carrying an $Fe^{2+}$-containing heme group. Its main task is to supply the body with ample amounts of oxygen. In humans the α-chain comprises 141 amino acids, whereas the β-chain is 5 amino acids longer. Besides a few homogeneous contacts ($\alpha_1\alpha_2$,$\beta_1\beta_2$), it is the vast number of heterogeneous contacts ($\alpha_1$ $\beta_1$,$\alpha_1\beta_2$) that contributes to both the stability of the molecule and its biological function. According to the two-state model of Monod, Wyman, and Changeux (MWC), hemoglobin exists in two conformational states, the T (tense) or deoxy structure and the R (relaxed) or oxy structure, the latter showing a higher oxygen affinity. Upon oxygen uptake, more and more hemoglobin molecules flip from the T state into the R state, in which oxygen binds much more easily to the remaining free heme groups. The result is a sigmoid oxygen binding curve typical of the cooperativity that is inherent in the hemoglobin molecule and many other oligomeric proteins. The subunits are not independent; instead they interact with each other, resulting in an increase in affinity for the ligand (here: oxygen) with increasing saturation. This cooperativity has become the most critical criterion in judging the quality of genetically engineered hemoglobin.

The three-dimensional structure of hemoglobin was determined at low resolution in the late 1950s and at the atomic level (see Figure 1; see color plate 11) 25 years later. In recognition of the enormous impact hemoglobin has had on biology and chemistry, the Noble Prize was awarded to Max Perutz and John Kendrew in 1962. The following decades led to the determination of the molecule's T and R structure; even partially ligated hemoglobin was crystallized and its structure elucidated by X-ray crystallography. This wealth of structural information enabled scientists to understand the physiological behavior of hemoglobin, including its cooperativity, on a molecular level. It also provided insights into the properties of various clinical relevant hemoglobin disorders. At the same time, new questions arose about the importance of certain residues within the molecule and their impact on the function of hemoglobin if replaced by another amino acid. Genetic engineering of hemoglobin was the only way to obtain answers to these questions.

## 2 MOLECULAR BIOLOGY MEETS HEMOGLOBIN

The genes for the α- and β-chains of human hemoglobin were cloned in the late 1970s. However, the first engineered expression of a globin chain was reported in 1980 when the β-globin gene of rabbit was placed under the control of the *lac* UV5 promoter using the β-galactosidase ribosomal binding site and *Escherichia coli* as the appropriate microorganism. The expression was demonstrated by radiolabeling but was far too small to produce β-globin in amounts sufficient for biochemical experimentation. That obstacle was overcome in 1985 when the first recombinant hemoglobin was made by Nagai et al. Human β-globin was produced as a cleavable fusion protein in *E. coli;* the gene was fused to a short coding sequence of the naturally highly expressed lambda bacteriophage *cII* gene. Synthetic DNA was used to introduce between the two proteins a cleavage site for factor $X_a$ (a protease from the blood clotting cascade), to liberate authentic β-globin, which was then reconstituted into functional hemoglobin with heme, and native α-chain was prepared from human blood. This system, slightly modified, worked also for the preparation of functioning hemoglobin from recombinant α-globin. Thanks to the technology of molecular biology, it is now possible to introduce any mutation at any site in human hemoglobin.

Following a thorough purification process, recombinant hemoglobin was proven to have physiological properties identical to

**Table 1** Genetic Engineering of Hemoglobin

| Year | Organism Used for Expression | Protein Expressed in Organism | Ref. |
|---|---|---|---|
| 1980 | *Escherichia coli* | Rabbit β-globin, minute amounts | Guarante et al., *Cell* **20**, 543. |
| 1985 | *Escherichia coli* | Human β-globin, reconstituted to Hb | Nagai et al. *Proc. Natl. Acad. Sci. U.S.A.* **82**, 7252. |
| 1989 | Transgenic mouse | Human hemoglobin | Behringer et al., *Science* **245**, 971. |
| 1990 | *Escherichia coli* | Human hemoglobin | Hoffmann et al., *Proc. Natl. Acad. Sci. U.S.A.* **87**, 8521. |
| 1991 | Yeast | Human hemoglobin | Wagenbach et al., *Bio/Technology* **9**, 57. |
| 1992 | Transgenic swine | Human hemoglobin | Swanson et al., *Bio/Technology* **10**, 557. |

those of native hemoglobin, including full cooperativity. Mutant hemoglobins that were crystallized showed no structural difference from native hemoglobin apart from the introduced mutation and its immediate vicinity. Consequently this system contributed considerably to the understanding of structure–function relationships in hemoglobin, spanning evolutionary aspects as well.

But there was still room for improvement, as can be seen from Table 1. In 1990 scientists succeeded in producing functional hemoglobin directly in *E. coli* by translating α- and β-globin from the same RNA molecule (dicistronic message). The source for the heme was the bacteria's endogenous production. One year later hemoglobin was produced in yeast, thereby adding a further advantage: when a foreign gene is overexpressed in *E. coli,* the methionine (ATG) at the start site of the protein is not efficiently removed, whereas yeast cleaves off the extra residue completely.

Yeast was not the first eukaryotic system used to synthesize recombinant human hemoglobin. The increased understanding of globin gene regulation led to the construction of expression vectors that comprise enhancer elements upstream of the respective globin gene and facilitated an erythroid-specific, high level expression of human hemoglobin in transgenic mice (1989) and in swine (1992); up to 80% of these animals' hemoglobin is the same as that of humans. The advances not only have resulted in a means of producing specially designed hemoglobins but also have made possible the use of animal models for the study of genetically based globin diseases.

## 3 ANIMAL MODELS AND BLOOD SUBSTITUTES

To date, more than 500 different mutations have been identified in human hemoglobin, most of which have little or no clinical relevance. Sickle cell disease remains the most severe genetically based globin disease, claiming some 80,000 children annually. A point mutation causes the glutamic acid (GAG) at position β6 to be replaced by a valine (GTG). Upon deoxygenation, the hemoglobin, termed HbS, polymerizes and forms fibers within the erythrocyte. The red blood cells then tend to assume a sickle shape, which greatly diminishes their high degree of deformability—necessary for traversing the capillaries. As a result, the blood vessels get clogged (see Figure 2). Patients homozygous in HbS suffer from severe organ damage and often die at a young age.

With the tools of molecular biology at hand, the treatment of sickle cell disease has entered a new and promising phase. Transgenic mice expressing human HbS now serve as animal models in the in vivo evaluation of current and new treatments of this fatal disease. The transgenics were bred with mice bearing a deletion in their own α- or β-globin gene, and levels of up to 72% human HbS in the blood of those mice are now obtainable.

Another medical need is a safe and effective blood substitute. A shortfall in supply together with concerns about the transmission of blood-borne infections has enhanced the development of alternatives to blood transfusion. Besides oxygen-carrying perfluorocarbons and liposome-encapsulated hemoglobin, it is the free hemo-

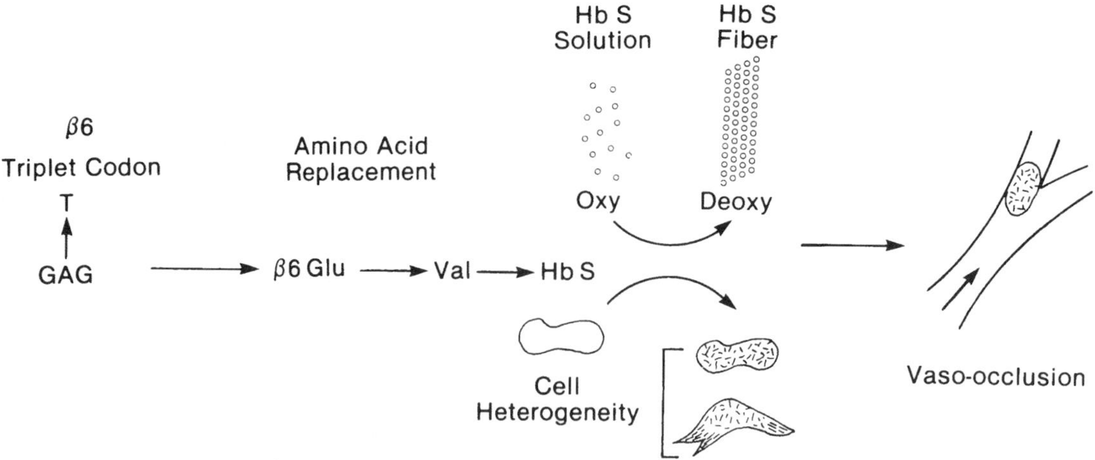


**Figure 2.** Pathogenesis of sickle cell disease: molecular biology offers new approaches for treatment. (From Bunn and Forget, 1986, p. 456.)

globin in the plasma that bears the best prospects of becoming a blood substitute. Modifications are necessary, however, since outside the red blood cell hemoglobin exhibits too high an oxygen affinity, is readily oxidized to methemoglobin, and dissociates into αβ dimers, which allows Hb to escape from the circulation into the urine and can cause severe kidney problems.

So far, highly purified, stroma-free, and chemically polyermized bovine hemoglobin has been the state-of-the-art artificial blood. Now the genetic engineering of hemoglobin has offered a new approach on the way to an appropriate blood substitute. In 1992 work was published reporting a human hemoglobin, produced in *E. coli,* with the two α-chains linked by a glycine and a mutated β-chain leading to a reduced oxygen affinity. The glycine link (α1C-terminus to α2N-terminus) prevents the hemoglobin from dissociating into dimers, whereas the mutation in the β-chain (β108Asn-Lys) is a naturally occurring one (Hb Presbyterian) known to reduce the molecule's oxygen affinity. Phase I clinical trials of this recombinant hemoglobin are under way. Despite a number of questions to be answered (e.g., half-life, oncotic pressure, antigenicity), this approach emphasizes the critical role of biotechnology in modern medicine.

*See also* HEMOGLOBIN.

### *Bibliography*


Bunn, H. F., and Forget, B. G. (1986) *Hemoglobin: Molecular, Genetic and Clinical Aspects.* Saunders, Philadelphia.

Everse, J., Winslow, R. M., and Vandegriff, K. D., Eds. (1994) *Hemoglobins,* Vol. 231 in *Methods of Enzymology.* Academic Press, New York.

Fermi, G., and Perutz, M. F. (1981) *Haemoglobin and Myoglobin,* Vol. 2 in *Atlas of Molecular Structures in Biology,* D. C. Phillips and F. R. Richards, Eds. Clarendon Press, Oxford.

Ogden, J. E. (1992) Recombinant haemoglobin in the development of red-blood-cell substitutes. *TIBTECH,* 10(3):91–96.


# HEMOPHILIA

*Francesco Giannelli*

### *Key Words*

**CpG or HTF Islands** Regions of DNA rich in CpG dinucleotides and with an equal number of GpC dinucleotides. These are usually associated with the start of genes.

**Exon** Part of gene that is found transcribed in mature mRNA.

**Frameshift** Mutation that affects the coding region of a gene by causing alteration in the reading frame used to translate the mRNA.

**Intron** Part of gene flanked by exons and excluded from mRNA.

**Serine Protease** A member of a family of proteolytic enzymes characterised by the presence of a cardinal serine in the catalytic site.

**X-Linked** Attribute of character due to gene on X chromosome (often also referred to as sex-linked).

The two X-linked hemorrhagic diseases hemophilia A and B have made an important contribution to the understanding of human genetics. The first historical record of the inheritance of a disease and of "genetics" motivated advice relates to these diseases, as does the first description of the X-linked pattern of inheritance, the appreciation of equilibria between selection and mutation and the first attempt to estimate mutation rates. In modern times the hemophilias have clearly illustrated the successes and problems of replacement therapy and have provided some insight into the causes of immunological complications associated with the therapeutic administration of the missing gene products. Furthermore, these conditions are members of a large group of diseases characterised by very high mutational heterogeneity. This group is of particular importance because, on the one hand, the wealth of natural mutants that each member possesses allows accumulation of detailed information on the structural features that are important to the functions of the gene and gene product involved in the disease; on the other, this creates the least favorable situation for the development of fully efficient strategies for the provision of the carrier and prenatal diagnoses that are necessary for genetic counseling. In the above respects hemophilia B represents a particularly notable example because it is rapidly becoming the disease with the largest number of mutations characterised, and the one with the most advanced strategy for carrier and prenatal diagnoses and for genetic counseling. Work on hemophilia A is bound to follow the model of hemophilia B, but mutation analysis has revealed that unexpectedly large inversions, splitting the factor VIII gene in intron 22, occur very frequently and cause half of all cases of severe hemophilia A. The inversions can now be easily identified.

## 1 GENETICS OF THE HEMOPHILIAS

Two sex-linked hemophilias are known: hemophilia A, due to deficiency of coagulation factor VIII, and hemophilia B or Christmas disease, due to deficit of factor IX. Coagulation factors VIII and IX exist as inactive molecules in circulation, and upon activation they form a single functional unit, which cleaves and activates coagulation factor X. The genes for factor VIII and factor IX are both on the X chromosome, and both hemophilias show a typical X-linked recessive pattern of inheritance, with affected males and unaffected carrier females transmitting the disease. One in 5000 males suffers from hemophilia A and one in 30,000 from hemophilia B. Such males, until the introduction of modern therapy, had half as much chance of reproducing as normal males, and this reduced efficiency resulted in a loss of hemophilia alleles that was compensated by mutations in the factor VIII or factor IX gene, generating new hemophilia A or B alleles respectively. Thus the rate of renewal of hemophilia alleles was about 1/6 per generation. Clearly both hemophilias show a very high degree of mutational heterogeneity, and most unrelated patients carry mutations of independent origin. Within the population there is therefore a wealth of different hemophilia A and B alleles that can provide information on the features important to the function of the factor VIII and IX genes and proteins.

## 2 THE FACTOR VIII GENE AND ITS MUTATIONS

The factor VIII gene (Figure 1) is located in the most distal region of the long arm of the X chromosome (band Xq28), spans 186 kb

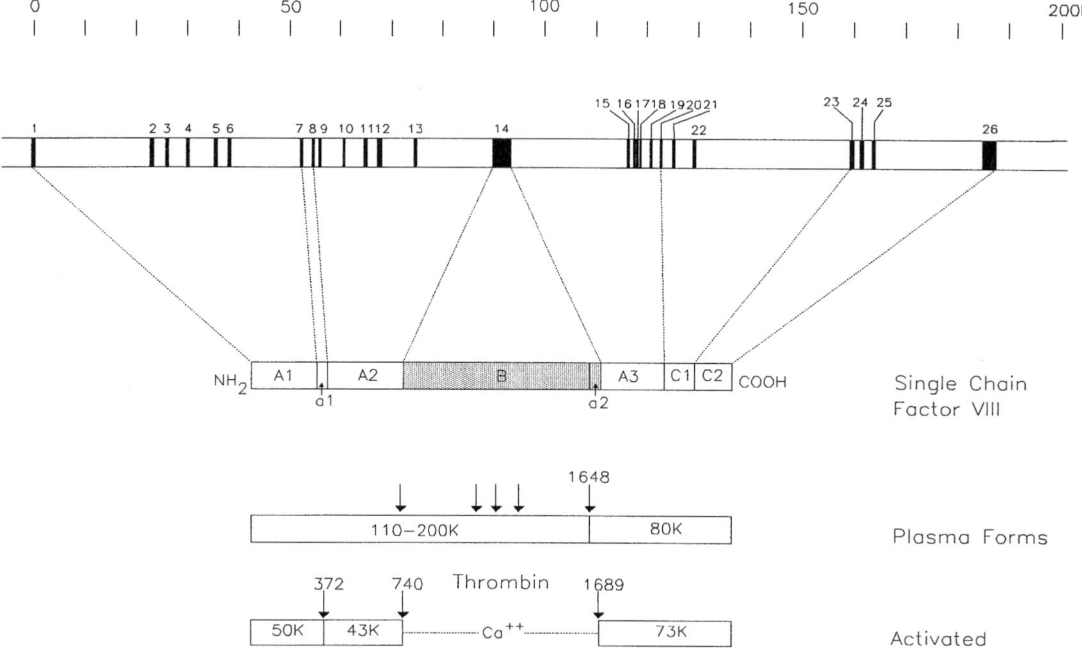


**Figure 1.** Factor VIII gene (upper bar) and protein (lower bars). Exons, indicated by solid boxes, are numbered (1–26). Dotted lines indicate exons coding for the different protein domains (see text). Stippling in single chain factor VIII indicates segment absent from activated factor VIII. Arrows in plasma and activated forms indicate sites of proteolytic cleavage (see text), and numbers above arrows indicate position of amino acid at the amino-terminal end of cleavage sites. Numbers in bars are weight in daltons of individual protein segments.

of DNA, and contains 26 exons. These code for a protein of 2351 amino acids (aa), including a prepeptide of 19 residues needed for the intracellular transport and secretion of factor VIII. This protein (Figure 1) consists of six domains and two small acidic peptides ($a_1$, $a_2$) arranged as follows: $A_1a_1A_2Ba_2A_3C_1C_2$. The A domains, which are homologous to one another and to ceruloplasmin, consist of approximately 350 aa. The B domain is unique and contains 983 amino acids, including 19 glycosylatable Asn residues. The C domains consist of approximately 150 aa and are homologous to one another and to domains of a protein from mammary epithelium. The $C_2$ domain appears to confer affinity for phospholipidic membranes. Factor VIII has a structure similar to factor V, the cofactor of coagulation factor X, and therefore factors V and VIII are evolutionarily related. The former, however, lacks the $a_1$ and $a_2$ acidic peptides, and its B domain is not homologous to that of the latter. Factor VIII is cleaved intracellularly at the boundary between domain B and $a_2$ and also at variable positions within the B domain to form a molecule with a heavy and a light aa chain. In the blood, factor VIII is bound to von Willebrand factor, probably through two sulfated tyrosines in the $a_2$ peptide. Activation of factor VIII requires cleavage at the $a_1A_2$ boundary to form a three-chain molecule and at the $a_2A_3$ boundary to release factor VIII from von Willebrand factor. Activation is also accompanied by cleavage between $A_2$ and B. Activated factor VIII may lose its activity upon cleavage at the $A_1/a_1$ junction or more distally at Arg562 by activated protein C—an important regulator of coagulation.

The large size and complexity of the factor VIII gene has delayed the detection of its mutations, most of which, so far, have been identified by procedures that examine only some of the essential sequences of the gene. However, this research has shown that approximately 4 to 5% of hemophilia A patients have gross deletions or other rare gross rearrangements. The other mutations detected involve small changes, which may cause abnormal RNA splicing, premature termination of protein synthesis (stop codons), miscoding plus abnormal termination of protein synthesis (frameshift mutations), and single aa deletions or substitutions. Splicing abnormalities may result from both single base pair substitutions and small rearrangements (e.g., deletions). Frameshifts result from the insertion or deletion from the coding sequence, usually of a few bases, except 3 or multiples thereof. Stop codons mostly result from single base substitutions. Single aa deletions are due to deletions of three bases from the coding sequences, and single amino acid substitutions commonly result from single base changes. Single aa substitutions have usually been found in patients with mild or moderate disease. Interesting among such substitutions are those that affect residues at activation cleavage sites or the sulfated tyrosines in $a_2$, which are involved in the binding of factor VIII to von Willebrand factor. Many of the reported single base substitutions are transitions of CpG to TpG, a consequence of the proneness of 5-methyl cytosine ($^{Me}C$) to mutate into T. Surprisingly, 40 to 50% of patients with severe disease show gross inversions that result from intrachromosomal homologous recombination between a sequence in intron 22 of the factor VIII gene and either of two further copies of this sequence located 400-500 kb distal to the gene. Such mutations divide the factor VIII gene into two unrelated parts in intron 22.

Intron 22 is the longest in the factor VIII gene and is unusual since it contains a CpG island associated with two transcripts. One of these is transcribed entirely from intron 22 in opposite orientation to the factor VIII message, is entirely contained within intron 22, is 1.8 kb long, and does not contain introns. The other has the same orientation of factor VIII and its coding sequence is identical to that of the terminal segment of the factor VIII mRNA (exons 23–26) except for the first eight codons specified by sequences in intron 22 of the factor VIII gene.

A few nondetrimental amino acid substitutions have been identified in the factor VIII gene. One of these ($Asp_{1241} \rightarrow Glu$) is common, at least in Caucasians.

## 3 THE FACTOR IX GENE AND ITS MUTATIONS

The factor IX gene (Figure 2) is located proximal to that for factor VIII on the long arm of the X chromosome (band Xq27.1); it consists of 34 kb and contains 8 exons. These exons code for a protein of 415 amino acids (Figure 2), consisting mostly of domains common to large families of proteins and preceded by a signal peptide or prepropeptide of 39, 41, or 46 aa. There is a good correspondence between exons and factor IX protein domains. The first exon codes for the prepeptide necessary for intracellular transport and secretion. The second exon codes for the propeptide and Gla region. These two domains are homologous to those of other $Ca^{2+}$-dependent blood coagulation factors and are functionally related, since the propeptide binds the $\gamma$-carboxylase responsible for modifying 12 glutamic acid residues in the amino-terminal domain of circulating factor IX. Such $\gamma$-carboxylated glutamic acids (gla) bind $Ca^{2+}$, and this helps the correct folding of the gla domain. The third exon codes for a short hydrophobic region that joins the Gla domain to the first of two regions homologous to epidermal growth factor (EGF). The EGF domain is shared by large families of proteins, and in factor IX they are coded by the fourth and fifth exons. The first EGF domain contains a high affinity $Ca^{2+}$ binding site while the second does not. The sixth exon codes for a region containing a peptide that is cleaved off during factor IX activation. The last two exons code for the catalytic or serine protease domain, which is common to a large family of proteases. The gene for factor IX and those for three other coagulation proteins (factor VII, factor X, and protein C) are structurally very similar and probably originated from the same ancestral gene by duplication.

Approximately 1200 hemophilia B mutations have been characterized. Only 2 to 3% of these mutations are gross deletions or other gross rearrangements; the rest are single base pair substitutions and small insertions and deletions, usually involving fewer than 20 bp. Such mutations impair the function of the gene in different ways. Changes in the promoter regions have been detected, and they presumably affect gene expression (transcription). These mutations, with only one exception so far, are associated with a disease whose victims improve with age and may become asymptomatic. Other mutations alter the normal RNA splice signals or generate new signals. These presumably alter mRNA processing. The lack of a convenient source of factor IX mRNA prevents experimental confirmation of such inferences. Stop codons and frameshift mutations, which have been found at sites throughout the coding sequence, in general result in the absence of detectable factor IX protein in circulation. Five different single aa deletions and 297 aa substitutions have been detected. These mutations are considered detrimental and reveal important structural features of factor IX, such as a mutation that alters the hydrophobic core of the prepeptide and others that affect the cleavage sites for the release of the pre-, pro-, and activation peptide. In the Gla region, several mutations affect the Gla residues, thus confirming their importance, and in the first EGF domain a number of mutations are found in the region of the high affinity $Ca^{2+}$ binding site. Mutations have been found at several cysteines involved in disulfide bridges, confirming their importance for the stability of factor IX.

Important clusterings of aa substitutions are found in the catalytic region of the protein. Three of these affect the peptides containing the three cardinal amino acids of the active center, $His_{221}$, $Asp_{269}$, and $Ser_{365}$. One further cluster affects the peptide delimited by $Gly_{305}$ and $Val_{313}$, and one the region extending from $Val_{328}$ to $Cys_{350}$. It is possible that these regions are important for the interaction of factor IX with its substrate, factor X, and/or its cofactor, factor VIII, but their function has not been clearly defined. If one classifies

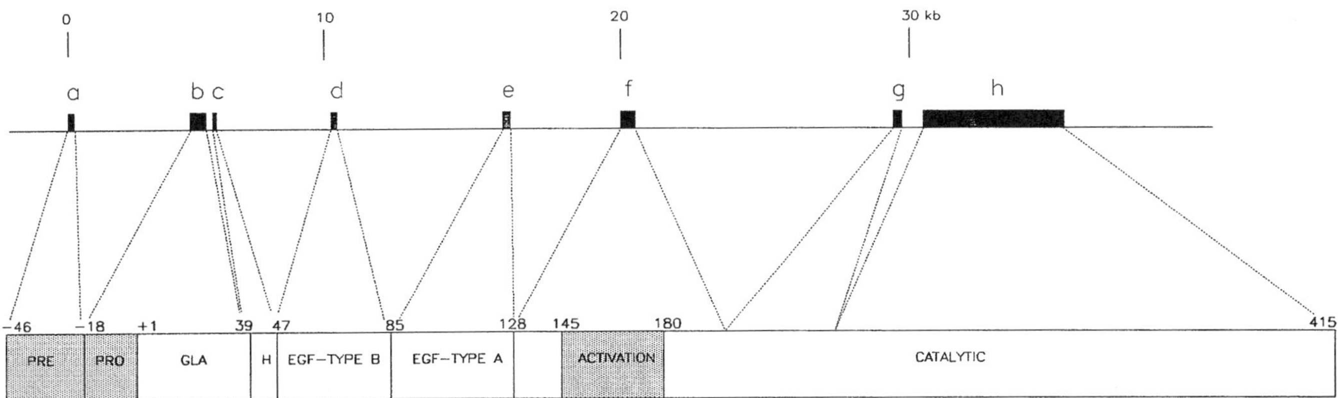


**Figure 2.** Factor IX gene (upper line) and protein (lower bar). Exons are indicated by solid boxes (a–h). Dotted lines show domains coded by individual exons. Stippling indicates protein domains that are cleaved prior to secretion (pre, pro) or at activation. Numbers indicate boundaries of different protein domains: pre, prepeptide; pro, propeptide; gla, gla region; h, hydrophobic stack; EGF—type B, first epidermal growth factor domain with $Ca^{2+}$ binding site; EGF—type A, second EGF domain; activation, activation peptide flanked by carboxyl end of light chain of activated factor IX and amino end of the catalytic region; catalytic, serine protease domain.

the factor IX residues according to how well they are conserved in the homologous factor VII, factor X, and protein C, it is clear that hemophilia B mutations preferentially affect the amino acids that are more conserved.

A number of mutations have been repeatedly observed, and many of these are due to transitions from CpG to TpG, as noted also for the hemophilia A mutations.

The clinical and hematological features of patients with identical mutations are usually similar, suggesting a good correlation between genotype and phenotype. In general one finds that patients with greater reduction in the activity than in the amount of factor IX protein tend to have single aa deletions or substitutions, while patients with either no demonstrable factor IX protein or similar reductions in factor IX activity and amount may have any one of the types of mutation mentioned earlier. A small proportion of patients develop antibodies against therapeutic factor IX that seriously compromise treatment. Such patients usually show gross deletions, stop codons, or frameshifts, indicating that the nature of the hemophilia B mutation is important in predisposing to the complications just named.

## 4 GENETIC COUNSELING IN THE HEMOPHILIAS

Advances in molecular genetics may be used to help the relatives of hemophilic patients through improvement of genetic counseling. This requires the identification of female carriers, who may transmit the disease, and of affected fetuses early in pregnancy. The high mutational heterogeneity of hemophilia A and B has complicated this task. Initially therefore no attempt was made to identify carriers or affected fetuses by direct detection of the gene defect: instead diagnoses were made by examining in each family the transmission of common DNA polymorphisms associated by linkage to hemophilia (Figure 3a). This indirect procedure fails in a large proportion

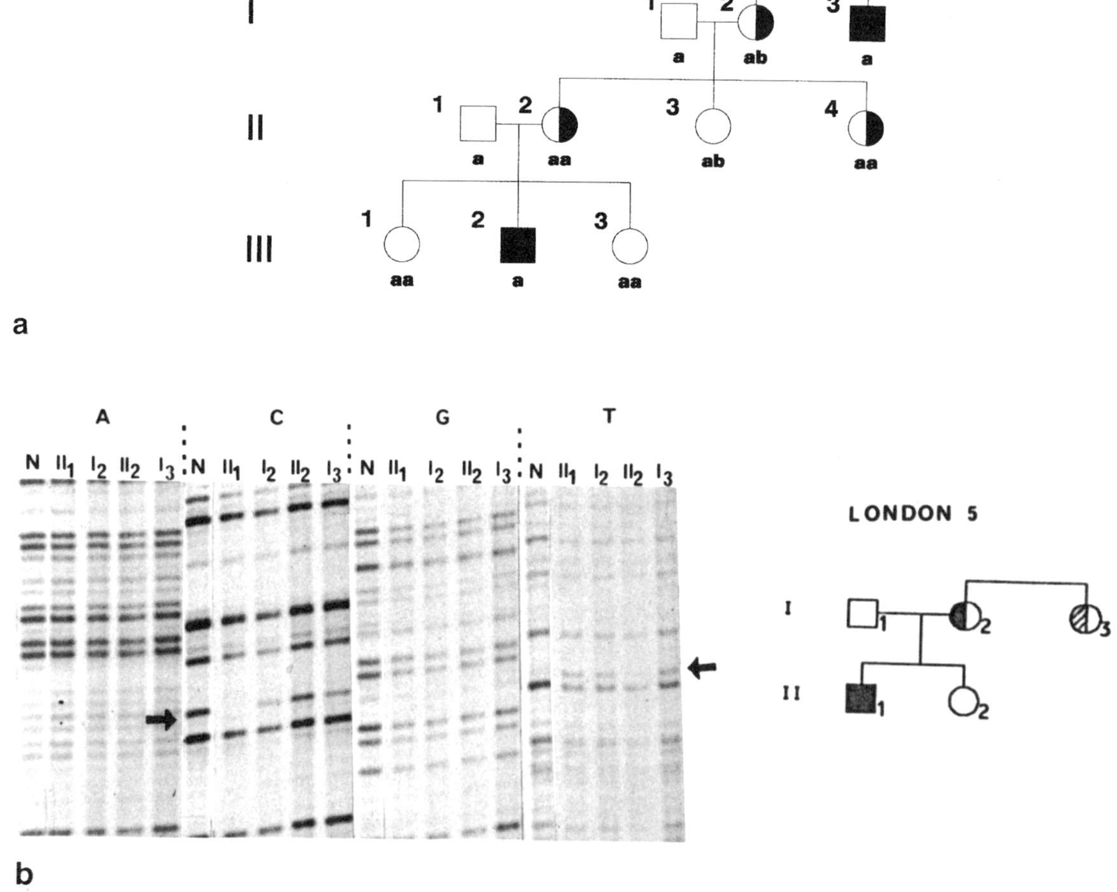


Figure 3. Indirect and direct carrier diagnosis in hemophilia B. (a) Indirect diagnosis, where a and b are two alternative forms of an intragenic DNA marker. The pedigree shows that the hemophilia B gene segregating in the family has the a marker (see $I_3$ and $III_2$). Since $I_2$ is heterozygous for the marker, carrier diagnosis on her daughters is possible. $II_4$ is obviously a carrier because she must have inherited the defective maternal gene (i.e., that with the a marker); $II_3$ appears not to be a carrier because she has apparently inherited the normal maternal gene (i.e., that with the b marker). This conclusion is conditional to paternity being confirmed. Since $II_2$ is homozygous for the marker, this cannot be used for carrier diagnoses on her daughters. (b) Direct diagnosis. Arrows in sequencing gel point to C→T transition in patient $II_1$ (see pedigree on right). This mutation is demonstrated by lack of band in section C track $II_1$, accompanied by abnormal band in same portion of section T track $II_1$. In section T tracks $I_2$ and $I_3$, show the same abnormal band, while $II_2$ has a pattern identical to the normal control (track N). These results indicate that $I_2$ and $I_3$ are carriers and $II_2$ is not. $I_2$ and $I_3$ have a normal band corresponding to the arrow in section C, since they have one normal factor IX gene.

of families either because of lack of informative markers or because of uncertainties about when the hemophilia mutation had arisen in the family.

In hemophilia B, the development of rapid methods for detecting virtually all hemophilia B mutations now allows diagnoses based on the direct detection of the gene defect and ensures success in virtually every family (Figure 3b). In the United Kingdom a national strategy is being implemented for the provision of genetic counseling. This entails the construction of a national confidential database of mutation, hematological, and pedigree information that can be used to provide carrier and prenatal diagnosis to the blood relatives of the patients listed in the database by examination of the region of the gene defective in the index patient. This allows precise, rapid, and economical diagnoses. Similar developments in hemophilia A may occur later, in spite of the size and complexity of the factor VIII gene. The inversion mutations involving intron 22 are now the easiest to identify. Rapid methods begin to be available for the detection of the remaining hemophilia A mutations.

*See also* Genetic Testing; Human Disease Gene Mapping.

### *Bibliography*

Brownlee, G. G. (1989) In *Recent Advances in Haematology,* Vol. 5, pp. 251–264. Churchill Livingstone, Edinburgh.

Giannelli, F., et al. (1992) *J. Med. Genet.* 29:602–607.

———, et al. (1994) *Nucleic Acids Res.* 22:3534–3546.

Green, P. M., et al. (1991) *Blood Coagulation Fibrin.* 2:539–565.

Tuddenham, E. G. D. (1989) *The Molecular Biology of Coagulation,* 2nd ed., pp. 849–877. Bailliere Tindall, London.

———, et al. (1994) *Nucleic Acids Res.* 22:3511–3533.

# Histones

*Gary S. Stein, Janet L. Stein, and Andre J. van Wijnen*

### *Key Words*

**Cell Cycle** The interval between the completion of mitosis in the parent cell and the completion of the next mitosis in one or both progeny cells. The periods of the cell cycle are sequentially defined as mitosis (prophase, metaphase, anaphase, and telophase), $G_1$ (the period between the completion of mitosis and the onset of DNA replication), S phase (the period of the cycle during which DNA replication occurs), and $G_2$ (the period between the completion of DNA replication and the onset of mitosis).

**Histone Proteins** Five principal species of basic chromosomal proteins designated H2a, H2b, H3, H4, and H1, which range in size from 11,000 to 25,000 Da. Histone proteins complex with DNA to form the primary unit of chromatin structure, the nucleosome.

**Nucleosome** The primary unit of chromatin structure in eukaryotic cells, consisting of approximately 200 nucleotide base pairs of DNA and two each of the core histone proteins (H2a, H2b, H3, and H4).

**Posttranscriptional Control** The components of gene expression involving regulation mediated at the level of messenger RNA processing within the nucleus and/or cytoplasm, the translatability and/or stability of mRNA, or the assembly or posttranslational modifications of polypeptides.

**Promoter Regulatory Elements** DNA sequences, generally but not necessarily, 5′ (upstream) from the mRNA transcription initiation site, which modulate the specificity and/or level of transcription.

**Transcriptional Control** The component of gene expression involving the synthesis of RNA, utilizing DNA as a template.

Histones are positively charged nuclear proteins that are ubiquitously represented in eukaryotic cells for packaging DNA into the protein–DNA complex termed chromatin. Histone–DNA complexes form the primary unit of chromatin structure, the nucleosome. Modifications in the interactions of histones with DNA in specific regions of genes occur in association with changes in gene expression. Mammalian and nonmammalian histone genes have been cloned and characterized with respect to the regulation of expression. The histone genes are a multigene family, and most are expressed in proliferating cells at the time in the cell cycle when DNA is replicated, providing histone proteins to package newly replicated DNA into chromatin. Other histone genes are expressed postproliferatively to support structural and transcriptional requirements of specialized cells. Regulatory sequences of histone genes, which determine the specificity of levels of transcription, as well as factors that bind to regulatory elements to mediate histone gene expression, have been identified.

## 1 General Characteristics

### 1.1 The Biological and Structural Properties of Histone Proteins

There are five principal species of histone proteins, designated H2a, H2b, H3, H4, and H1, ranging in size from 11,000 to 25,000 Da. They are positively charged, as a result of high contents of the basic amino acids arginine, lysine, and histidine, which facilitate the interactions of histones with negatively charged DNA molecules. The amino acid sequences of the histone proteins have been highly conserved during evolution, reflecting the conserved role of these proteins in chromatin structure and the apparently stringent requirement to support conservation of the primary unit of chromatin structure, the nucleosome.

The histone proteins are encoded in a multigene family with multiple (e.g., approximately 20 copies in human cells), nonidentical copies of each core (H2a, H2b, H3, and H4) and H1 gene. The histone polypeptides can be separated into the following categories:

1. Those that are represented in most cells and tissues and synthesized only in proliferating cells at the time of DNA synthesis (> 90%).
2. Those that are found in many cells and tissues but are expressed independently of proliferation, either constitutively during the cell cycle or following the completion of proliferation at the onset of tissue-specific gene expression associated with differentiation.
3. Those that are expressed solely in specialized cell types, such as spermatocytes and avian erythrocytes, in which there are

highly specific requirements for modifications in the packaging of DNA into chromatin. In lower eukaryotes, and apparently only in these organisms, there are multiple copies of the histone genes, providing large quantities of "stored" histone mRNA in oocytes that can support histone protein synthesis during the rapid series of initial cell divisions that immediately follows fertilization.

Additional heterogeneity of the histone proteins is reflected by posttranslational modifications that include acetylation, methylation, phosphorylation, and adenosine diphosphate (ADP) ribosylation. Such modifications alter the distribution of charge in specific domains of the histone proteins and, together with hydrophobic bonding, may influence histone–DNA as well as histone–histone interaction. These posttranslational modifications are involved in the incorporation of newly synthesized histones into chromatin and may provide a basis for changes in the interactions of histones with DNA for remodeling chromatin architecture: for example, in condensation of chromatin into discrete chromosomes at the onset of mitosis, and in modification of chromatin structure when the expression of specific genes is activated or repressed. These changes in histone-mediated chromatin structure are rapid and reversible, supporting cellular responsiveness to a broad spectrum of physiological signals that mediate transcription of cell growth, housekeeping, and tissue-specific genes.

### 1.2 The Contribution of Histones to Chromatin Structure

There is a requirement for the ordered packaging of 2.5 yards of DNA within the confines of the mammalian cell nucleus. To accommodate this DNA packaging into nucleosomes, every nucleus contains approximately 300 million histone molecules. Each nucleosome consists of a core particle of approximately 140 base pairs of DNA wound around a complex consisting of two H2a, H2b, H3, and H4 molecules and a linker DNA region of approximately 40–60 base pairs (Figure 1). Under the electron microscope, the nucleosomes appear as a series of beads (protein–DNA complexes) on a string (linker DNA joining the nucleosomes). The H1 histones bind to the linker region and participate in nucleosome–nucleosome interactions. This organization accounts for only a 10 nm chromatin fiber and a packing ratio of 7. The 10 nm beads-on-a-string structures are packed as a 30 nm chromatin fiber, and further packaging results in chromatin fibers of 100 nm. "Nonhistone," sequence-specific DNA binding proteins can mediate DNA conformation and modulate histone–DNA interactions. Thus, it appears that the contributions of histones to transcriptional regulation are through facilitation of conformational properties of DNA that are responsive to gene-specific transcription factors.

## 2 EXPRESSION OF HISTONE GENES

A functional, as well as temporal, relationship between DNA replication and the expression of mammalian core and H1 histone genes was initially indicated by the constant histone–DNA ratio (1:1) observed in a broad spectrum of cells, tissues, and organs, and by the doubling of cellular levels of histone protein during the S phase of the cell cycle. Direct measurements then confirmed that histone protein synthesis is largely confined to S phase and that inhibition of DNA replication results in a rapid cessation of histone protein synthesis. The cellular levels of histone mRNA reflect cellular

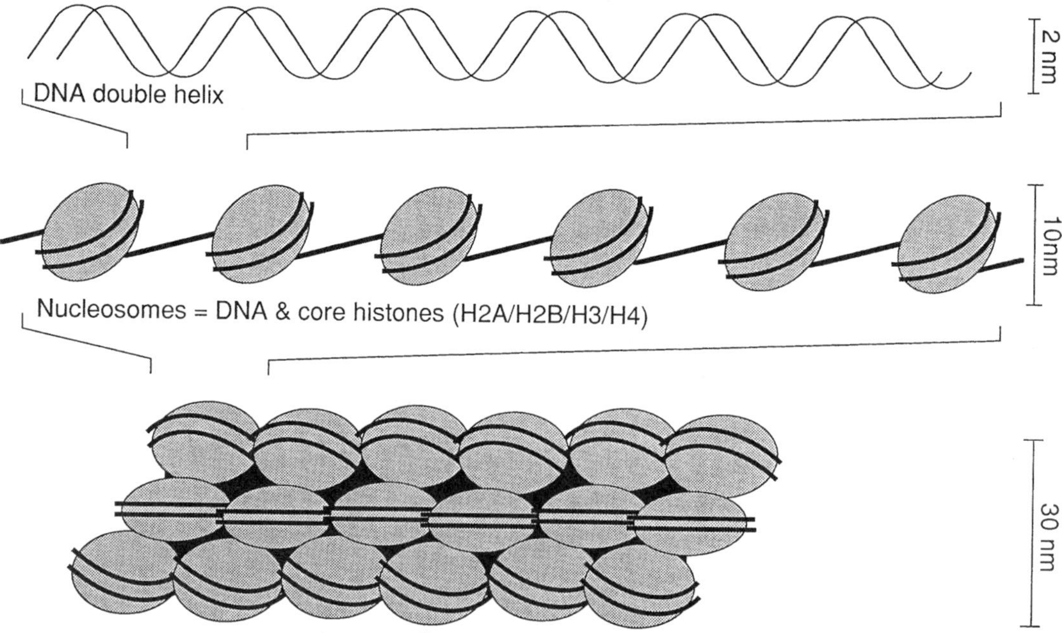

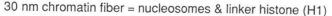

**Figure 1.** Three principal levels of chromatin organization. *Top:* The 2 nm, deproteinized, double-stranded DNA double helix. *Middle:* The organization of DNA into nucleosomes. The beads-on-a-string structure comprises a 10 nm fiber; each bead consists of two each of core histone proteins (H2A, H2B, H3, and H4). The string component of the structure is the DNA. *Bottom:* The higher order organization of chromatin structure mediated by association of nucleosomes through linker histone H1 into a 30 nm chromatin fiber.

levels of both histone protein synthesis and DNA replication. Similarly, inhibition of DNA replication brings about a dose-dependent loss (selective destabilization) of histone mRNAs, which parallels decreases in DNA and histone synthesis. Measurements of histone gene transcription indicate enhanced synthesis of histone mRNAs early during the S phase of the cell cycle.

The increased transcription of histone genes early during S phase and the coordinate accumulation of histone mRNAs for core and H1 histone proteins that closely parallels the initiation of DNA and histone protein synthesis suggest that the onset of histone gene expression is at least in part transcriptionally mediated. Throughout S phase, the synthesis of histone proteins is modulated by the availability of histone mRNAs. The stabilization of histone mRNAs throughout S phase and the destabilization of histone mRNAs when DNA replication is completed or inhibited are highly selective, and largely posttranscriptionally controlled. At the onset of differentiation in mammalian cells, the histone genes that are under cell cycle regulation are down-regulated transcriptionally. When DNA replication is completed during the terminal cell cycle, histone protein synthesis ceases, histone mRNA is degraded, and both basal and enhanced levels of histone gene transcription are abrogated.

## 3 ORGANIZATION AND REGULATION OF HISTONE GENES

### 3.1 Organization of Cell-Cycle-Regulated Histone Genes

In mammalian cells, the cell-cycle-regulated histone genes are organized into clusters of core alone (H2a, H2b, H3, and H4) or core together with H1 histone coding sequences (Figure 2). Within these clusters, which are represented on at least two chromosomes, there is generally a pairing of H2a with H2b genes and H3 with H4 genes. In lower eukaryotes such as sea urchin and *Drosophila,* a similar organization is found for the cell-cycle-regulated genes encoding somatic cell histone proteins. However, the histone genes expressed during oogenesis in these organisms are organized as simple, tandemly repeated clusters that contain one of each of the five types of histone gene.

Despite the clustering of cell-cycle-regulated histone genes, each histone coding sequence is an independent transcription unit with a unique promoter and mRNA coding sequence. All amino acids of the histone protein are encoded in contiguous nucleotides because these genes lack introns. Also noteworthy are the absence of a polyadenylation site and the presence of sequences with hyphenated dyad symmetry that form a stem–loop structure in the 3′ region as well as nontranslated leader and trailer segments of the mRNA that are less than 50 nucleotides long.

### 3.2 Promoter Elements and Transcription Factors That Regulate Cell-Cycle-Dependent Histone Gene Expression

Figure 3 is a schematic representation of the regulatory organization of the initial thousand base pairs of an H4 histone gene promoter. While this region contains the minimal sequences required for regulated expression, the functional limits of the H4 gene appear to extend considerably upstream. Indeed, cis-acting elements up to −6.5 kB may influence developmental expression of the H4 histone gene in vivo in transgenic animals. Two domains of in vivo protein–DNA interactions for the H4 histone gene have been established in the intact cell at single nucleotide (nt) resolution. These have been designated H4-site I (nt −156 to −113) and H4-site II (nt −97 to −47). The proximal promoter domain H4-site I is a bipartite cis-activating element that interacts distally with a member of the ATF family of transcription factors, and proximally with the GC box binding protein (Sp1) HiNF-C. These factors are capable of mediating a fivefold stimulation of transcription. The H4-site II domain represents a mosaic of functional recognition sequences that contribute to H4 gene transcription. H4-site II is a multipartite protein–DNA interaction site for sequence-specific

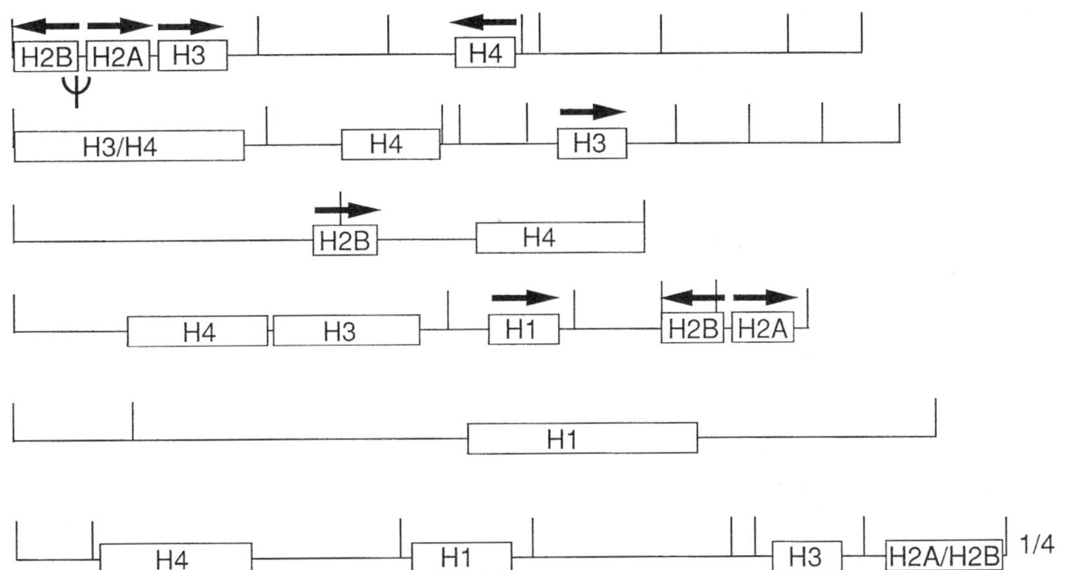


**Figure 2.** The organization of genomic DNA segments containing some of the human histone coding sequences; Arrows designate directions of transcription. H2B and H2A pseudogenes are designated by the symbol Ψ.

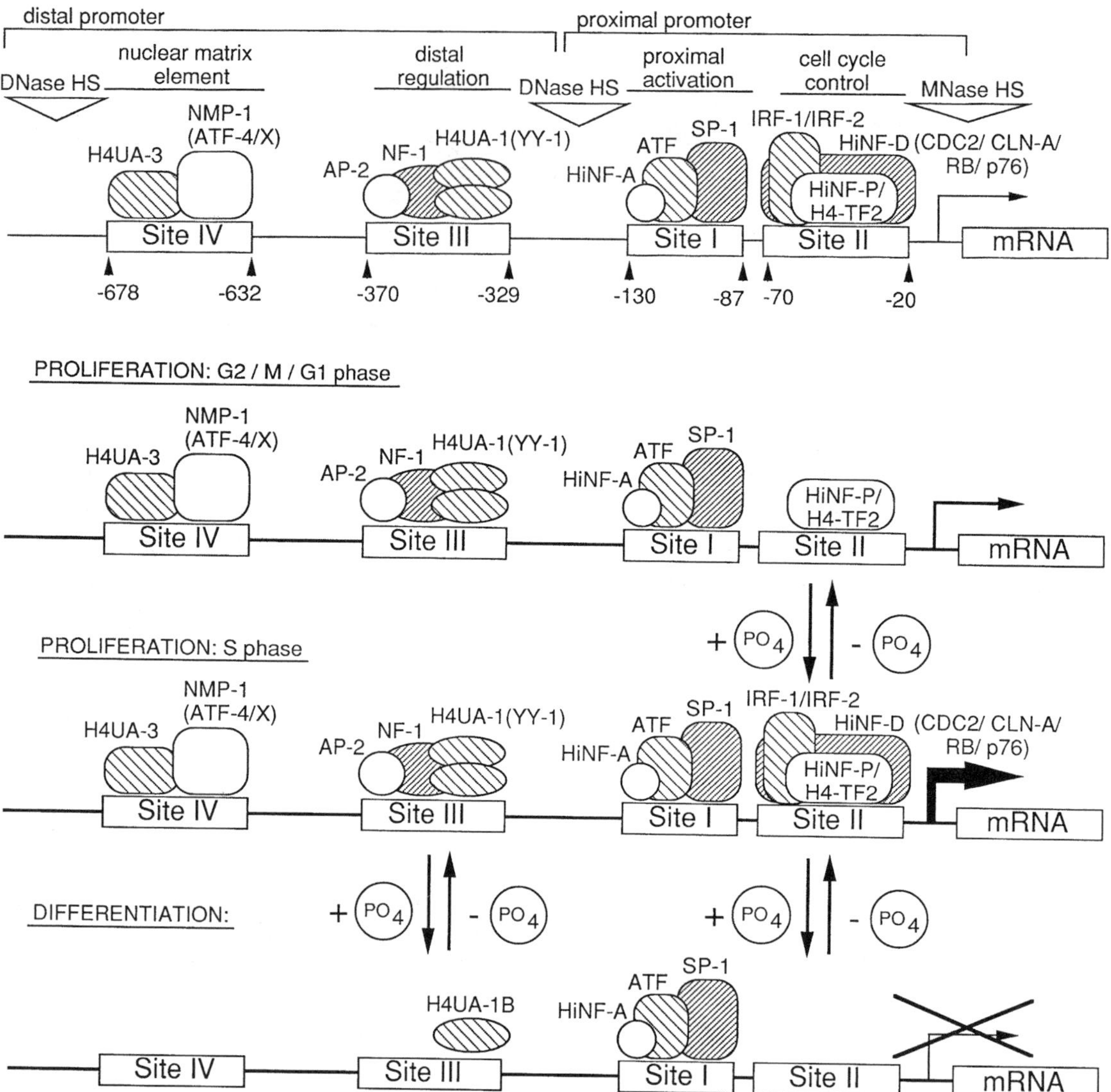


**Figure 3.** Schematic representation of promoter regulatory elements and transcription factors that support histone gene expression. *Top:* The representation and organization of gene regulatory sequences is designated by sites I–IV. The ovals and boxes represent transcription factors. Proximal and distal cell cycle regulatory elements are designated along with nuclease sensitive regions (DNase HS, MNase HS). Also shown are sites of histone gene interactions with the nuclear matrix. *Three lower segments:* Phosphorylation-dependent modifications in interactions of transcription factors with histone gene promoter elements both during the cell cycle and following differentiation. These modificatons in protein–DNA interactions control the extent to which the histone gene is transcribed, which is indicated by the thickness of the horizontal arrows over the mRNA regions of the gene.

factors HiNF-D, HiNF-M, and HiNF-P (H4-TF2). The proximal region of H4-site II spans a TATA motif and is sufficient to mediate accurate transcription initiation in vivo. However, the distal region of H4-site II influences transcriptional competency, as well as the timing and extent of H4 mRNA synthesis in vivo. This site II distal region contains several distinct sequence motifs that either stimulate the basal level of H4 gene transcription (C box) or influence periodic levels of transcription (M box). The distal activating elements H4-sites III and IV encompass regions that stimulate transcription in vivo and interact with the heteromeric nuclear factors H4UA-1 and H4UA-3, respectively. Additionally, H4-site IV overlaps with a putative nuclear matrix attachment site spanning nt −730 to −589. This element interacts with a sequence-specific nuclear matrix protein (NMP-1), and may influence expression of the H4 histone gene promoter by transient anchorage to the nuclear matrix. The integration of mechanisms controlling the coordinately regulated transcription of multiple histone genes may involve several shared promoter-binding activities, including both ubiquitous and histone-gene-specific transcription factors. HiNF-D related protein–DNA interactions are also represented in H3 and H1 histone gene promoters, suggesting the possibility of coordinate transcription factor interactions regulating several histone gene classes.

Insight into transcriptional control of histone gene expression has been provided by identification of modifications in interactions of promoter binding factors within the initial thousand base pairs of a human H4 histone gene promoter at sites I, II, III, and IV, and relating these to the extent of gene transcription. Protein–DNA interactions at these regulatory elements during the cell cycle and with the down-regulation of proliferation during differentiation are schematically shown in Figure 3.

*See also* Chromatin Formation and Structure; Gene Expression, Regulation of; Protein Designs for the Specific Recognition of DNA.

*Bibliography*

Heintz, N. (1991) The regulation of histone gene expression during the cell cycle. *Biochim. Biophys. Acta,* 1088:327–339.

Hnilica, L., Stein, G. S., and Stein, J. L., Eds. (1989) *Histones and Other Basic Nuclear Proteins.* CRC Press, Boca Raton, FL.

Marzluff, W. F., and Pandey, N. B. (1988) Multiple regulatory steps control histone mRNA concentrations. *Trends Biochem. Sci.* 13:49–52.

Osley, M. A. (1991) The regulation of histone synthesis in the cell cycle. *Annu. Rev. Biochem.* 60:827–861.

Stein, G. S., Stein, J. L., van Wijnen, A. J., and Lian, J. B. (1992) Regulation of histone gene expression. *Curr. Opin. Cell Biol.* 4:166–173.

van Wijnen, A. J., van den Ent, F. M. I., Lian, J. B., Stein, J. L., and Stein, G. S. (1992) Overlapping and CpG methylation-sensitive protein/DNA interactions at the histone H4 transcriptional cell cycle domain: Distinctions between two human H4 gene promoters. *Mol. Cell. Biol.* 12:3273–3287.

van Wijnen, A. J. et al (1994) Transcription of histone H4, H3, and H1 cell cycle genes: Promoter factor HiNF-D contains cdc2, Cyclin A and an RB-related protein. *Natl. Acad. Sci. USA,* 91:12882–12886.

## HIV: *see* AIDS articles.

# HPLC of Biological Macromolecules

*Karen M. Gooding*

*Key Words*

**Bonded Phase** Organic coating or layer covering the surface of the solid HPLC support and containing the functional groups responsible for separation.

**Elution** Process of a solute passing through and coming out of a chromatography column.

**Gradient** Systematic variation of the mobile phase composition during an HPLC analysis.

**Packing** Adsorbent, gel or solid support used in the HPLC column.

High performance liquid chromatography (HPLC) is a high resolution separation process using a liquid mobile phase and a column containing microparticulate solid particles coated with a specific functional group. The functional groups, which can be neutral, charged, or hydrophobic, cause separation of components of a mixture by the specific physical interaction. The primary modes of HPLC for biological macromolecules are reversed phase, ion exchange, size exclusion, and hydrophobic interaction chromatography. These rapid and high resolution methods have provided a means of purification, separation, and analysis of peptides and proteins in biotechnology, microbiology, university, and clinical laboratories.

## 1 INTRODUCTION

Liquid chromatography is a separation process in which the components in a mixture migrate in a liquid stream through a packed bed of particles that retard some of the components differentially by a specific physical property. The particles that compose the column have a uniform physical characteristic, such as hydrophobicity, charge, or porosity, which brings about separation by causing molecules or ions to interact or pass through at different rates.

Liquid chromatography has been an important method of separating and purifying proteins and nucleic acids because these substances are soluble and often stable in aqueous buffers. For many years, methods utilized columns containing carbohydrate matrices that achieved good separations in hours or days; flow rates were slow because they were based on gravity. In the mid-1970s, rigid packing materials composed of silica or polymer with diameters of 5 to 10 μm were developed to be used with liquids pressured to several thousand psi. This technique, initially known as high pressure liquid chromatography, is now called high performance liquid chromatography (HPLC). Chemical modification of the surface of the silica or gel support, known as the bonded phase, gives the specific basis for retention.

Columns placed with microparticulate HPLC supports can separate biological macromolecules in minutes with excellent resolution and recovery of biological activity. Since the 1970s, both the technology of producing HPLC columns and the understanding of their operation have improved dramatically, resulting in the widespread use of HPLC for protein and peptide analysis and purification. Although biological macromolecules include polypeptides, polynucleotides, and polysaccharides, this entry primarily discusses polypeptides because of the vast amount of research on the subject. The principles are generally applicable to biomolecules in all three categories.

## 2 INSTRUMENTATION

### 2.1 General

A high performance liquid chromatograph consists of one or more pumps, a sample injector, a column, a detector, and a data recorder. If a single solvent is used as the mobile phase, the method is termed *isocratic.* In many cases, more than one solvent must be used to release the bound molecules and cause them to elute from the column. Multiple solvents are usually combined in a programmed gradient from one composition to another. The time and variation of composition is called the gradient. Figure 1 illustrates the typical configuration of an HPLC with two pumps.

### 2.2 Columns

The column is the key element of the HPLC system. The physical process by which the molecules bind will determine which mobile phase will promote binding and which will release, thereby causing elution. The packing material, or support, in the column is composed of a rigid material, such as silica or a polymer, which can be derivatized or covalently bonded with functional groups; this chemical layer is called the bonded phase. HPLC supports usually have particle diameters of 5 to 10 μm and may be porous or nonporous. For biological macromolecules, pores must be at least 300 A in diameter to allow access, whereas small molecules are typically run on supports with pores of 80 to 100 A diameters.

### 2.3 Detectors

Detectors for HPLC tend to be selective rather than general. Refractive index detectors, which produce a signal for all solutes, are the primary devices used, but their sensitivity is low. Light-scattering or ''mass'' detectors are not very sensitive and have nonlinear

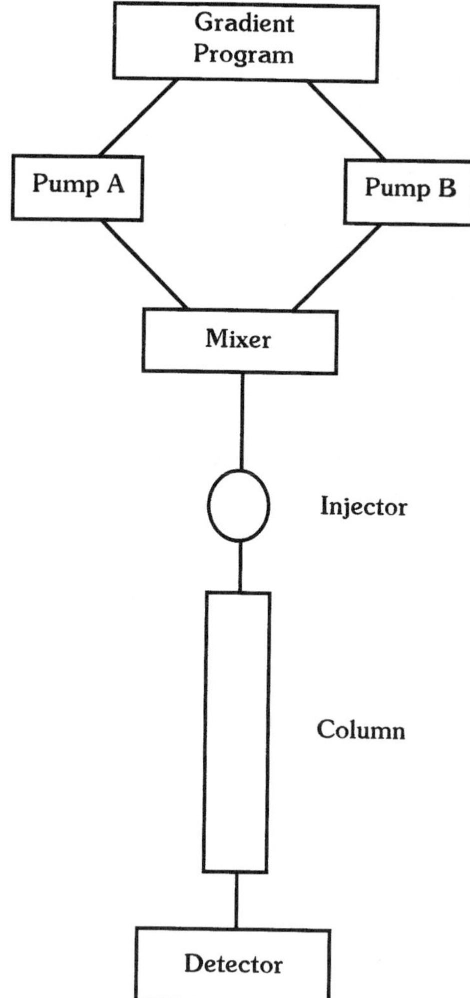


Figure 1. Block diagram of a high performance liquid chromatograph.

calibration curves. Ultraviolet or visible absorption detectors are the most popular. For proteins and peptides, these are usually used at 254 and 210 nm, respectively, with good detectability. Fluorescence, radiochemical, and electrochemical detectors are very selective and sensitive if the molecules of interest have the appropriate chemical properties. On-line mass spectrometry is also used to identify molecules separated by HPLC.

## 3 SIZE EXCLUSION CHROMATOGRAPHY (SEC)

### 3.1 Principles

In concept, SEC is the simplest mode of chromatography. A neutral bonded phase is used with a mobile phase, which eliminates binding. During passage through the column, large molecules, which are excluded from the pores, elute first. Small molecules, which have total access to the pores, elute last. Molecules of intermediate size, which partially penetrate the pores, elute between the two. The position of elution for a range of partially included molecules is linearly related to the logarithm of their molecular weight or hydrodynamic volume. It is essential that there be no interaction between the solutes and the porous support matrix or the relationship to molecular size is no longer valid.

### 3.2 Column Packings

High performance SEC column packings for biomolecules are composed of surface-modified silica or organic gel. It is necessary that the support be rigid to withstand pressure and high flow rates, as well as neutral and hydrophilic to prevent adsorption of the proteins or other macromolecules. Spherical particles of 5 to 10 μm diameter are preferable because they can be packed into columns with high efficiency, thus producing narrow peaks and excellent resolution. Supports for SEC have controlled pores with diameters ranging from 50 to 4000 A. Pore diameter is usually chosen to achieve partial penetration of the solutes. The more homogeneous the pore diameter, the better the potential resolution.

### 3.3 Operation

To achieve good resolution in SEC, operational parameters must be chosen which produce narrow peaks by minimizing diffusion. Slow flow rates (1 mL/min for a 7.8 mm i.d. column) and small samples ($<$ 2–4 mg for a 300 mm $\times$ 7.8 mm i.d. column) are best. To avoid band spreading, sample volumes must be less than 2% of the column volume.

The mobile phase in SEC is generally composed of a buffer such as potassium phosphate at neutral pH and concentration of 0.05 to 0.1 *M* to minimize any ionic effects. The composition of the mobile phase is adjusted to maximize the solubility and biological stability of the macromolecules as well as to eliminate their interaction with the support.

## 4 ION EXCHANGE CHROMATOGRAPHY (IEC)

### 4.1 Principles

High performance IEC separates solute molecules by their surface ionic composition through differential binding to charged groups on the support. Bound molecules are selectively eluted by increasing the salt concentration or changing the pH of the mobile phase. IEC is a popular means of protein separation because it can purify large quantities in mobile phases that yield excellent recoveries of biological activity.

### 4.2 Column Packings

Supports for high performance IEC are generally composed of 5 to 10 μm silica or polymeric gel that has a hydrophilic polymeric pellicular layer containing charged groups. Functional groups for anion exchange include polyethyleneimine (PEI), diethylaminoethanol (DEAE), and quaternized amine (Q). Cation exchange groups are usually carboxymethyl (CM) or sulfopropyl (SP). The packings (functional groups) are designated "weak" or "strong" by the permanency of the charge and their titration curves, similar to acids and bases. For proteins and other macromolecules, large pores must be used to allow access to the internal surface. For proteins with molecular weights under 100,000, pores of 300 A diameter give the best access, and therefore, loading capacity.

### 4.3 Operation

Solutes are bound onto ion exchange columns under conditions of low ionic strength buffer (0.01–0.02 *M*). They are selectively eluted with a gradient of pH or increasing salt. The pH of the initial buffer should maximize charge of both the solute and the functional group on the support. An operational pH of 6 to 8 is common for protein analysis because it is similar to biological conditions. For most ion exchange analyses, a 15 to 30-minute linear gradient to 0.5 to 1.0 *M* salt is satisfactory for separation. Nonionic surfactants or stabilizing agents can be added to the mobile phase to enhance biological recovery. Enzymatic recoveries from IEC are generally high, as are loading capacities.

## 5 HYDROPHOBIC INTERACTION CHROMATOGRAPHY (HIC)

### 5.1 Principles

High performance HIC separates molecules by binding of hydrophobic groups on the surface of a solute to alkyl or aryl ligands on the support. These nonpolar or hydrophobic interactions are enhanced by high concentrations of salt in the mobile phase. Solutes are released by decreasing the salt concentration in a gradient. HIC is a good method for protein purification because the operational conditions are usually nondenaturing and biological recovery is frequently high.

### 5.2 Column Packings

Supports for high performance HIC are generally composed of 300 Å silica of 5 to 10 μm particle diameter having a pellicular hydrophilic polymeric layer into which hydrophobic ligands are incorporated with low ligand density. These ligands are usually short (C-1 to C-3) alkyl chains or benzene rings.

### 5.3 Operation

In HIC, proteins are bound to the packing with high concentrations (1.0–2.0 *M*) of an antichaotropic salt. Antichaotropic salts cause salting-out of proteins and hydrophobic adsorption because they promote the ordering of water molecules at interfaces. Ammonium sulfate is the most popular salt because of its salting-out and solubility characteristics, but sodium sulfate is also commonly used. The mobile phase also contains a suitable buffer to maintain pH. A linear gradient of decreasing salt to the buffer alone selectively elutes the proteins. A 15 to 30 minute linear gradient at neutral pH is usually satisfactory. Many proteins retain biological activity under these analysis conditions.

## 6 REVERSED PHASE CHROMATOGRAPHY (RPC)

### 6.1 Principles

RPC separates molecules by their hydrophobic binding to alkyl chains on the support. This interaction is a noncovalent association of nonpolar moieties in an aqueous solvent. Binding occurs in water that often contains an acidic ion-pairing reagent. Elution results during a gradient to an organic solvent, usually acetonitrile or isopropanol. RPC is an excellent method for peptide analysis and has great potential for resolution, as can be seen in Figure 2. RPC is not universally used for proteins as a result of frequent denaturation by the organic solvent, the acidic solvent, or the highly hydrophobic surface. Unless the proteins can be easily renatured, RPC is more useful for subunit than for intact protein analysis.

### 6.2 Column Packings

Supports for RPC are usually composed of 5 to 10 μm porous silica (80–4000 A) that has alkylsilanes bonded to the surface. For peptide analysis, 300 A is the most popular poré option. Ligands are usually 4, 8, or 18 carbons in length (C-4, C-8, or C-18). There

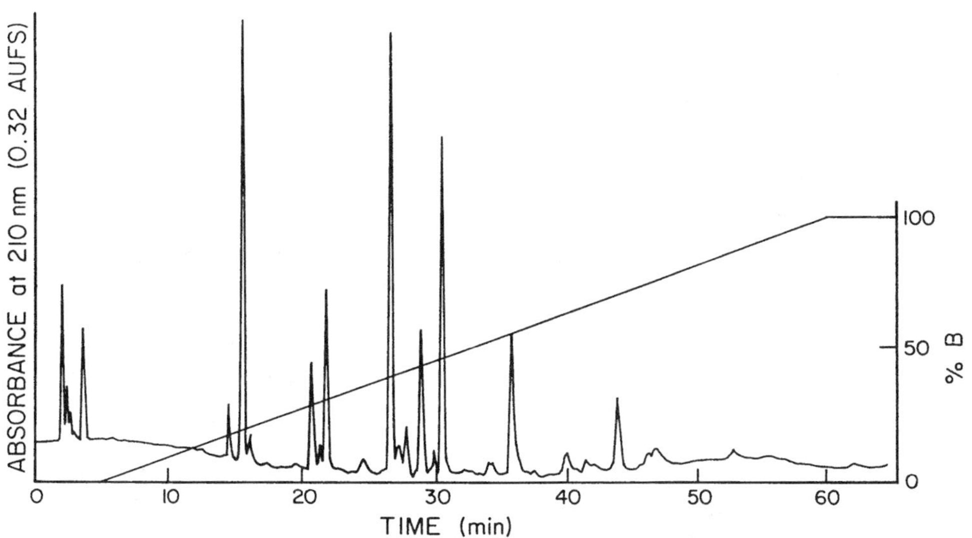


**Figure 2.** Analysis of a tryptic digest of the α-chain of hemoglobin by reversed phase chromatography. (Reproduced with the permission of M & S Instruments Trading Co., Osaka, Japan.)

are slight differences in selectivity for peptides and proteins on different ligand chains. There are also variances in recovery; most notably, certain proteins have low recoveries from C-18 supports. The C-8 and C-18 chains are most popular for peptide analysis and the C-4 and C-8 for proteins.

End capping is a process of reacting residual silanols with a small silane. This is a popular option for RPC supports with 60 to 100 A pores because residual silanols can cause peak tailing for positively charged species. For RPC analysis of polypeptides on sorbents with larger pores, end capping is not universally used, possibly because the presence of some silanols results in a more wettable surface.

### 6.3 Operation

The most common operational conditions for RPC are 50- to 100-minute gradients of 0.1% trifluoroacetic acid (TFA) in water to 0.1% TFA in acetonitrile. The TFA acidifies the eluent and also ion-pairs with positively charged amino acids in the pepties or proteins. Other ion-pairing agents, which vary from TFA in hydrophobicity, can also be used, and alternative organic solvents are sometimes selected. Shallow gradients and slow flow rates result in the best resolution for peptides, but long contact time with a reversed phase column can result in low recoveries for proteins.

*See also* Hemoglobin; Proteins and Peptides, Isolation for Sequence Analysis of.

## *Bibliography*

Gooding, K. M., and Regnier, F. E., Eds. (1990) *HPLC of Biological Macromolecules—Methods and Applications.* Dekker, New York.

Hearn, M.T.W., Ed. (1991) *HPLC of Proteins, Peptides and Polynucleotides.* VCH Publishers, New York.

Heftmann, E., Ed. (1992) *Chromatography: Fundamentals and Applications of Chromatography and Related Differential Migration Methods;* Part A: *Fundamentals and Techniques.* Elsevier Science Publishers, Amsterdam.

———, Ed. (1992) *Chromatography: Fundamentals and Applications of Chromatography and Related Differential Migration Methods;* Part B: *Applications.* Elsevier Science Publishers, Amsterdam.

*J. Chromatogr.,* symposium volumes: (a) *International Symposium of HPLC of Proteins, Peptides and Polynucleotides.* (b) *HPLC '94* or preceding years.

Mant, C. T., and Hodges, R. S., Eds. (1991) *HPLC of Peptides and Proteins: Separation, Analysis and Conformation.* CRC Press, Boca Raton, FL.

# Human Chromosomes, Physical Maps of

*Cassandra L. Smith, Denan Wang, and Dietmar Grothues*

## *Key Words*

***Alu* Repeat** A 300 base pair (bp) DNA interspersed repeat sequence that occurs about once every 3000 bp; the most common repeat element in the human genome. It is GC rich and appears to occur preferentially in light Giemsa bands on condensed metaphase chromosomes.

**Inter-*Alu* PCR** A PCR method that uses primers contained in the *Alu* repeat element to amplify single copy sequences between adjacent *Alu* elements.

***Kpn* Repeat** The second most common repeat sequence in the human genome; occurs on average once every 50,000 bp. This repeat is a LINE repeat, is relatively AT rich, and occurs preferentially in dark Giemsa bands on condensed chromosomes.

**LINE (Long Interspersed Repeats) Elements** Long interspersed repeats that appear to be similar to retroposons.

**PCR (Polymerase Chain Reaction)** A method for amplifying DNA by alternatively denaturing double-stranded DNA, annealing pairs of primers located near each other on complementary strands, and synthesizing the DNA between the primers using DNA polymerase.

**PFG (Pulsed Field Gel Electrophoresis)** A method of electrophoresis that exposes nucleic acids to alternating electrical fields. Fractionation is based on the speed at which the molecules can change directions.

**Polymorphism Link-Up** A mapping approach that establishes continuity between restriction fragments by taking advantage of the naturally occurring polymorphism in different DNAs. In some cell lines hybridized probes appear to identify different fragments, whereas in others they appear to identify the same fragment. The pattern of occurrence of these fragments can be used to determine whether two probes identify the same or adjacent fragments.

**Retroposons** Group of DNA sequence elements that appear to transpose through an RNA intermediate. They do not code for a reverse transcriptase, do not have terminally redundant sequence, and do have a 3′-poly$(A)_n$ stretch and a variable-sized target duplication at the site of integration.

**SINE (Short Interspersed Repeats) Elements** A short interspersed repeat that is too small to function as an independent retroposon.

**STAR (Sequence-Tagged Restriction Site)** A short DNA sequence used to identify the DNA surrounding a restriction nuclease cleavage site.

**STS (Sequence-Tagged Site)** A short DNA sequence used to identify a DNA segment.

**YAC (Yeast Artificial Chromosome)** An artificial yeast chromosome constructed by cloning genomic fragments into vectors that can replicate in yeast and have the following characteristics: a yeast centromere, two telomeric sequences, and a selectable marker.

In the past, most genome studies were limited to organisms having well-developed genetic systems. Now a number of molecular techniques allow the construction of physical maps for virtually any chromosome. This means that genome analysis of an uncharacterized organism can begin with a physical dissection. Knowledge gained from this initial foray should aid the development of the tools necessary to manipulate the DNA in vivo.

This entry reviews the types of physical maps that can be constructed and the methods that are used to construct them. In particular, the focus is on the construction of genomic restriction maps and ordered overlapping libraries, with emphasis on the advantages of top-down mapping approaches.

## 1 INTRODUCTION

Genomic physical mapping experiments were facilitated by the development of pulsed field gel electrophoresis (PFG) [1,2] and accompanying techniques [3,4]. These techniques allow the isolation, characterization, and manipulation of large pieces of DNA, including intact chromosomes, that can range in size up to at least 10 million base pairs, from virtually any organism. For instance, intact megabase (Mb) chromosomes may be analyzed directly; larger chromosomes are analyzed after cleavage into specific Mb pieces, or after cloning as Mb segments.

Bacterial genomes range from approximately 1 Mb to 15 Mb in size, reflecting, in part, the relative ability of these organisms to be free living. Genomes of these sizes are ideal for analysis by PFG, as evidenced by the large number of complete, low resolution, genomic restriction maps that continue to be published for microorganisms. Previously, the study of some lower eukaryotic genomes had been hampered not only by the lack of well-developed genetic systems but by the failure of the chromosomes to not condense during cell division. Thus, the application of PFG to analysis of protozoan genomes, which range in size overlapping that of large bacterial genomes, to about 100 Mb, has been particularly useful.

Large, complex genomes, like the human genome (estimated to be 3000 Mb in size) have, until recently, been particular recalcitrant to molecular dissection. For example, mammalian chromosomes (estimated to be 50–300 Mb in size) condense; hence, they can be seen in the microscope. A method that allows a finer division of these large genomes has been the differential staining of the condensed chromosomes into regions that are estimated to be 5 to 10 Mb in size and are presumed to reflect regional differences in GC content. The division of the genome into chromosomes and chromosomal bands provides convenient pieces to top-down mapping approaches. In top-down mapping approaches, the genome is divided into units that are convenient to study. For instance, the natural division of the genome into chromosomes could represent one such division and chromosomal banding another. Further division would depend on the methods applied for analysis. Molecular approaches to genome analysis allowed the characterization of molecules up to about only 0.05 Mb. The "resolution gap," 0.05 to 10 Mb, was exactly the size range that was most amenable to study by PFG. Thus, the entire size range of DNA molecules can now be analyzed.

The application of PFG analysis to the human genome has been particularly useful for reverse genetic approaches ("positional cloning") of disease genes. Here, genetic analysis allows the region of the genome containing a gene of interest to be narrowed to 1 to 10 Mb. Then PFG analysis can determine the size of the region and the number of putative gene candidates it contains, with the goal of facilitating the isolation of more genetic markers and candidate genes.

## 2 DEFINITION OF TERMS

So many molecular techniques may be applied to characterizing chromosomes that a unifying concept of what constitutes a chromosomal physical map is absent. In some cases this lack has led to ambiguities in what is described in the literature as a genetic map.

Any map will consist of markers or objects, the placement of which suggests that order, and perhaps distance, between pairs of objects is known. (Some might argue that both order and distance are required.) Object order along a chromosome should be maintained irrespective of the method used to construct the map. However, map distances will be method dependent. For instance, the amount of recombination along a chromosome is not constant. Hence, it not surprising that a comparison of the physical and genetic distances along the long arm of chromosome 21 revealed at least a sixfold variation in the recombination frequency. It is quite clear that the ultimate map is the entire sequence of a chromosomal DNA. All maps and objects will be anchored, more or less precisely, to the DNA sequence once it has become available.

Classically, a genetic map was composed of chromosomal loci, and the amount of recombination between the loci was the genetic distance. For eukaryotic organisms, genetic distance, measured in centimorgans (cM), is a measure of the coinheritance of genetic markers. This type of analysis requires examination of several family generations. In contrast to bacteria like *Escherichia coli,* recombination is measured as the time (min) of transfer and integration of DNA from one cell into the chromosome of another cell.

Recently, in some instances, the placement of genes on various physical maps has been referred to as a genetic map (i.e., "gene map"), although only physical distances and locations were known. Here, a physical map is considered to be any map consisting of objects that have been located by physical rather than by genetic methods. Thus, a physical map can consist of objects located along the chromosome, and there are cytogenetic maps, chromosomal breakpoint maps, genomic restriction fragment maps, and overlapping clone libraries. Our only concern is with maps of the last two types of maps, constructed using molecular methods. The first two techniques are covered elsewhere in this volume.

Another concept that introduces some confusion is "locus." Classically this term defined genes ("genetic locus"). More generally, however, a chromosomal locus, representing a location on a chromosome, can consist of objects other than genes. Obviously, locations on physical maps may be defined by a number of different objects (e.g., probe site, restriction enzyme site, clone site, gene site, centromere, telomere). To add to the confusion, new genetic markers based on anonymous DNA sequences now define genetic loci. In this entry, a chromosomal locus is defined as any location on a chromosome that can be identified in a distinctive manner.

## 3 TYPES OF PHYSICAL MAPS

The term "physical maps" has been used to describe both genomic restriction maps and overlapping clone libraries [5–7]. For instance, low resolution genomic restriction maps may be created by hybridization experiments using cloned sequences as probes to order large restriction fragments generated by enzymes that cleave genomic DNA infrequently. Genomic restriction maps may also be created by analyzing restriction sites contained on overlapping clones. The resulting restriction map be of low or high resolution, depending on the restriction enzyme used. In some cases this information is obtained during the ordering of the library, or it may be generated afterward (see Section 5). Furthermore, an overlapping library may in itself represent a map consisting of ordered objects (clones) whose size can be approximated but not stated with certainty.

## 4 DIRECT GENOMIC RESTRICTION MAP CONSTRUCTION

The first step in creating a genomic restriction map is choosing the DNA source. For many organisms or chromosomes this is obvious, since there is a well-characterized isolate or a cell line that may be useful. For the human genome, for instance, some complications associated with the analysis of polymorphic diploid DNA can be avoided by using DNA from hybrid cell lines. The genomic DNA is extracted and purified intact, in agarose, to prevent shear damage. Small chromosomal DNAs may be sized directly by PFG. Other chromosomal DNAs may have to be cleaved with restriction enzymes or other very specific cleavage schemes to pieces that are less than 6 Mb in size before analysis by PFG. This size limit is that of the largest *Schizosaccharomyces pombe* chromosome, the largest molecule of accurately known size that has been fractionated by PFG.

Genomic map construction is similar to putting a puzzle together. It is much easier if all the pieces are known in advance. For small genomes this usually involves finding convenient restriction enzymes that cleave the genome into a reasonable number of fragments that can be resolved by PFG and visualized by simple ethidium bromide staining. The enzymes tested are those with large recognition sequences or those having a recognition site that is likely to occur infrequently in the genome of interest. In some cases, it has been useful to test a battery of enzymes. The usefulness of a particular enzyme may be estimated roughly from the size of the site or from the GC content of the test organism. Somewhat more accurate estimates can be made by comparison with the dinucleotide frequencies in the test organism or with the frequency of occurrence of a particular target sequence in known sequences of the test organism. Even so, these predictions are somewhat inaccurate because the genomic DNA sequences do not occur at random, whereas calculated occurrences assume randomness. Furthermore, the frequency of occurrence in cloned sequences may not be representative, inasmuch as most past molecular studies have focused on genes.

For complex genomes, it is often possible to identify Mb restriction fragments that will be used to construct the restriction map by examining DNA from monosomic hybrid cell lines. Here, human-specific interspersed repetitive hybridization probes are used to identify the human Mb restriction fragments. This approach works as long as the restriction enzyme digestion goes to completion. For the human genome, the restriction enzymes *Not* I and *Sgr*AI have this desired characteristic. In mammalian genomes, many restriction enzymes that cleave infrequently only partially cleave DNA because partial methylation at their recognition sites inhibits total cleavage.

The most commonly occurring interspersed repeat in the human genome is *Alu*. The *Alu* element is a SINE (short interspersed) repeat that is estimated to occur about every 5000 base pairs (bp). The second most commonly occurring interspersed human repeat is the *Kpn* repeat. The *Kpn* element, a LINE (long interspersed) repeat, is estimated to occur at a frequency tenfold less that that of the *Alu* repeat. The use of these two repeats as hybridization probes to hybrid cell line DNA cleaved with *Not* I or *Sgr*AI theoretically should reveal all the human Mb fragments. Furthermore, fragments should be differentially distributed in the human genome in a manner that reflects the differential distribution of these repeat elements into the Giemsa dark and light bands of condensed chromosomes.

The use of repeat sequences as hybridization probes can reveal information about the size and distribution of restriction fragments, but it tells nothing about order. For monosomic hybrid cell line DNA, human-specific repetitive telomeric and centromeric sequences may be used to identify the ends of the physical map as well as to provide an important anchor for regions of condensed chromosomes visualized microscopically. Although these sequences occur on all, or most, human chromosomes, they are limited to the human component of hybrid cell line lines; hence, they act like single-copy probes for these samples. Order information for the remaining interstitial Mb fragments is obtained using single-copy sequences as hybridization probes. Single-copy sequences that have been located on genetic maps can be used for the regional assignment of restriction fragments and can serve as anchors between genetic and the physical maps. The accuracy of fragment location will reflect the accuracy provided by the genetic map.

The construction of a genomic restriction map requires that neighboring fragments be identified. At least one neighbor can be identified unambiguously by hybridizing a single-copy sequence to partially digested genomic DNA. Although more than one partial product may be identified, the difficulty of interpreting the resulting fragments increases dramatically with the number of fragments that need to be ordered. The analysis is simplified if the probe is from a telomere region or is situated on a small fragment located next to a very large fragment. In both cases, the partial digest information reflects the order of fragments in only one direction, as opposed to the bidirectional information obtained from the use of interstitial probes. Many times, the confusion associated with partial digest data can be eliminated by obtaining partial digest data from probes located on adjacent fragments. In this case a different set of partial digest fragments will be identified by each of the probes, but only the interpretation that is consistent with both sets of data will be correct.

There are several other approaches to proving that two restriction fragments are adjacent. By far the most powerful is the use of linking libraries—small insert libraries that contain DNA segments from two adjacent Mb restriction fragments. Hence, a *Not* I linking clone used as a hybridization probe to genomic DNA digested with *Not* I will identify adjacent *Not* I fragments. A complete linking library would consist of all that would be needed to construct, in the most efficient manner, a complete genomic restriction map. In the absence of a complete linking library, partial digest strategies combined with polymorphism link-up and double digestion with a second enzyme can provide information on neighboring fragments.

Polymorphism link-up refers to the results obtained when one probe is used to analyze digested DNA from different cells. For instance, a restriction site may be missing in one cell line, either because there is a mutation at the restriction enzyme site or because methylation is interfering with restriction enzyme cleavage. Hence, when a number of DNA sources are examined, it is often possible to fingerprint the pattern of polymorphism around a restriction site. Thus, probes that are usually located on separate, but adjacent, fragments can, in some DNAs, be found on the same fragment, which is equal to the total size of the two adjacent fragments seen in other cell lines. In such a case, the pattern of occurrence of the two distinct smaller fragments, detected by the different probes, is self-consistent. This is very similar to the more familiar approach of fingerprinting a region using many different restriction enzymes. Megabase fragments then may also be fingerprinted by cutting with a second enzyme. Although the primary data may be different depending on where the probes are located on the fragment, the interpretation of the cleavage sites of the second enzyme will be consistent with data obtained with the two probes.

It is very easy to start maps but very difficult to finish them. Each gap in each map presents a unique problem. It is important to treat each gap individually and to design the best strategy for dealing with the gap. The best strategy will depend on the size of the restriction fragment, the amount of polymorphism in the region, the number of cloned sequences from the region, and whether unassigned Mb fragments exist. In some cases, it is useful to use the Mb restriction fragment themselves as hybridization probes to identify neighboring fragments using partial digests. For the human genome, this can be done by amplifying single-copy human sequences contained between *Alu* elements using inter-*Alu* PCR (polymerase chain reaction) amplification.

## 5 CONSTRUCTION OF OVERLAPPING LIBRARIES AND GENOMIC RESTRICTION MAPS

After the genomic DNA sample has been chosen, the first decision that must be made in constructing an overlapping library is the type of vector to be used. The most distinguishing feature of these vectors is the size of the DNA that can be cloned into them. It is helpful to use the largest genomic pieces possible, so that there are fewer clones to order. In the past, the largest cloning vectors, cosmids, contained cloned sequences about 40 kilobase (kb) in size. Recently, several systems that allow the cloning of larger segments have been developed. For instance, in bacteria, fragments up to 100 kb may be cloned in P1 phage. Less developed is the use of the $F^+$ plasmid, which has the potential to contain very large fragments. For instance, it has been known for some time that *E. coli* can maintain $F^+$ plasmids carrying up to 2.5 Mb of E. coli sequences. Megabase fragments can also be cloned into yeast artificial chromosomes (YAC). Recent progress here has shown that libraries containing clones with more than 1 Mb may be isolated, although not on a routine basis.

The ordering method regulates how the cloning is done as well as which vector is used. For instance, most fingerprinting methods require that libraries be constructed using partial restriction enzyme digests to ensure that overlaps exist. There are a number of ways to fingerprint clones to identify overlaps. Bottom-up library construction strategies usually involve testing individual clones to search for an overlapping restriction fragment or restriction site pattern. This type of fingerprinting is easy to automate. Clones may also be fingerprinted using interspersed repetitive sequences. Here, clones, or restriction fragments of the clones, are hybridized to a set of short oligonucleotides or longer repetitive sequences. A variation on this approach that has proven to be efficient is the use of genomic restriction fragments as hybridization probes. The use of mapped Mb restriction fragments allows the regional assignment of clones. Further division of the clones is possible if pools of smaller restriction fragments are used to fingerprint clones. This method is a multiplex approach because information is gathered from different portions of the genome at the same time. In contrast to the bottom-up strategies inherent in most fingerprinting methods, interspersed repetitive sequences is a top-down mapping approach.

Another ordering approach involves detecting overlaps by DNA sequencing. Here, the sequences may be chosen at random sites (e.g., the sequence-tagged sites: STSs), or the sequences can be collected at specific locations originally called sequence-tagged restriction sites (STARs). The STAR approach may be used with both partial and complete digest libraries. DNA sequence information may be collected at specific restriction sites, including the ends of the cloned sequences. To ensure that overlap information is generated, the STAR approach applied to complete digest libraries must be used in conjunction with another library (e.g., a linking clone library that spans the cloning sites).

Clones may also be hybridized to the entire library to detect overlapping clones. Although this is the slow, conventional walking (or "crawling") approach, the efficiency of the method may be enhanced by taking a multiplex approach—that is, by using intelligent pools of probes to sort overlapping clones from different regions. In fact, most fingerprinting methods have been unable by themselves to lead the construction of a complete map. Eventually, the random approaches had to be abandoned and replaced by focused efforts for filling in the gaps. The usual end game strategy is some form of conventional walking.

## 6 PROSPECTUS

Genomic restriction maps and overlapping libraries are complementary. It is important to have genomic restriction maps because they provide accurate distances and can guard against incorrect information generated from cloning artifacts during library construction. Overlapping libraries are comparatively difficult to construct and maintain; however, they provide genomic DNA inconvenient and useful form.

*See also* DNA SEQUENCING IN ULTRATHIN GELS, HIGH SPEED; GENE ORDER BY FISH AND FACS; NUCLEIC ACID SEQUENCING TECHNIQUES; PARTIAL DENATURATION MAPPING; RESTRICTION ENDONUCLEASES AND DNA MODIFICATION METHYLTRANSFERASES FOR THE MANIPULATION OF DNA; WHOLE CHROMOSOME COMPLEMENTARY PROBE FLUORESCENCE STAINING.

### *Bibliography*

Billings, P. R., Smith, C. L., and Cantor, C. R. (1991) New techniques for the physical mapping of the human genome. *FASEB J.* 5:28–34.

Burmeister, M., and Ulanovsky, L., Eds. (1992) *Pulsed Field Gel Electrophoresis.* Humana Press, Totowa, NJ.

Cantor, C. R., Smith, C. L., and Matthew, M. K. (1988) *Annu. Rev. Biophys. Biophys. Chem.* 17:287–304.

Oliva, R., Lawrence, S. K., Wu, Y., and Smith, C. L. (1991) In *Encyclopedia of Human Biology,* R. Bulbecco et al., Eds., pp. 475–488. Academic Press, San Diego, CA.

Smith, C. L., and Condemine, G. (1990) New approaches for physical mapping in small genomes. *J. Bacteriol.* 172:1167–1172.

———, Klco, S., and Cantor, C. R. (1988) In *Genome Analysis: A Practical Approach,* K. Davies, Ed., pp. 41–72. IRL Press, Oxford.

———, ———, Zhang, T. Y., Fang, H., Oliva, R., Fan, J. B., Bremer, M., and Lawrence, S. (1993) Analysis of megabase DNA using pulsed field gel electrophoresis. In *Methods in Molecular Genetics,* K.W. Adolph, Ed., pp. 155–196. Academic Press, San Diego, CA.

# Human Disease Gene Mapping

*Scott R. Diehl*

*Key Words*

**Association** Correlated occurrence of alleles at two or more loci (including disease alleles or noncoding DNA variants) within individuals in a population due to linkage disequilibrium, population admixture, natural selection, or other processes.

**Candidate Gene** A gene with known or suspected biological functions that are a priori likely to play an important role in the pathophysiological processes of a particular disorder.

**Linkage** Co-occurrence of two or more genes (including disease genes) or other unique DNA fragments within the same region of a chromosome (physical linkage). Or cosegregation or correlated transmission of two or more DNA fragments or phenotypic traits (including a disease) with each other within a family (genetic linkage).

**Microsatellite Marker** A dinucleotide, trinucleotide, tetranucleotide, or pentanucleotide DNA sequence, usually located in a noncoding region of the chromosome, which is repeated in tandem, with the number of repeated $n$-nucleotide units often exhibiting great variation among individuals, thus providing an ideal marker for linkage studies.

Mapping a disease gene involves determining its location on one of the 23 pairs of chromosomes and approximately 3.3 billion DNA bases that comprise the human genome. Increasingly precise determination of the gene's locations from the level of a single chromosome [e.g., 100–300 million base pairs (Mb) of DNA] to a very small subchromosomal region [e.g., < 100 thousand base pairs (kb) of DNA] can lead to ''positional'' cloning of the disease gene and identification of disease mutations. Such knowledge can help us to understand the disease's biochemical and physiological mechanisms, as well as providing insights into the gene's normal functions. These advances may suggest new strategies for diagnosis, treatment, or prevention, and may contribute to an improved understanding of an important biological process in the nondiseased state. Disease gene mapping (with or without disease gene cloning) can lead to prenatal or presymptomatic testing, which may be viewed as very beneficial by some individuals at risk of having a child affected by the disease or at risk of being affected by the disease themselves. However, the increasing ability to provide this knowledge raises new bioethical questions, and a consensus regarding its appropriate use has not been reached within either the biomedical community or society as a whole.

## 1 INTRODUCTION

The past decade has witnessed revolutionary improvements in our ability to map human disease genes. The primary impetus for this change has come from molecular genetics. As is true for most fields of genetics and molecular biology, the development of the polymerase chain reaction has played a central role in this process, opening up new strategies and approaches that were inconceivable prior to its availability. Equally dramatic improvements have been made in molecular cytogenetics, involving techniques such as multicolor fluorescent labeling and interphase in situ hybridization. Likewise, statistical and epidemiological approaches to gene mapping have made major advances. Some changes have been in response to the computational burdens imposed by increasingly large quantities of molecular data, while others are aimed at improving our power to detect disease genes having complex interactions with other genes and the environment.

## 2 GENETIC ARCHITECTURE OF HUMAN DISEASES

Any discussion of disease gene mapping must first consider the many ways that genetic variation can influence the risk of a disease. Human geneticists classify diseases as ''simple'' and ''complex'' genetically (Table 1). It is essential to recognize that these two categories represent extremes of what is really a continuum of features, and many diseases exhibit at least some features of both extremes. Sometimes ''simple'' genetic diseases are referred to as ''single gene disorders'' or ''single major locus only.'' This means that the disease is caused by mutations at one and only one disease gene. Every person who inherits one mutant copy (allele) from either parent of a dominantly transmitted disorder or two mutant alleles (one from both parents) of a recessively transmitted disorder gets the disease without fail. There are very few if any questionable cases clinically, since expression of disease symptoms is usually highly consistent. Furthermore, there are no environmentally induced cases, which might sometimes mimic the symptoms of the ''simple'' genetic disease. Many of the well-known gene cloning success stories of recent years have involved these ''simple'' genetic disorders. Examples include cystic fibrosis, Duchenne's muscular dystrophy, Huntington's disease, and neurofibromatosis.

Progressing to more complicated scenarios, some diseases may be caused by either of two single major genes, located in different places in the genome, which can produce indistinguishable clinical features. Such diseases are said to exhibit locus heterogeneity. A form of hereditary deafness known as Waardenburg syndrome displays locus heterogeneity. There are at least two types of the disorder, one caused by mutation of the *PAX3* gene located on chromosome 2, and others caused by genes at other (not yet determined) chromosomal locations. Allelic heterogeneity, by contrast, occurs when different mutant alleles at the same disease gene lead to significant differences in clinical features. Spinal muscular atrophy is an example of this kind of genetic complexity.

Some disease genes exhibit reduced penetrance. This occurs when individuals carrying the disease allele do not show any disease symptoms. They have escaped the disease's expression as a result of purely stochastic factors or because of protective factors in

**Table 1** Comparison of Simple and Complex Genetic Diseases

| Property | Simple Disease | Complex Disease |
|---|---|---|
| Number of loci involved | One | Two or more |
| Reduced penetrance | No | Yes |
| Phenocopies | No | Yes |
| Locus heterogeneity | No | Yes |
| Allelic heterogeneity | No | Yes |

their environment (e.g., diet, hygiene). For many diseases that are expressed only in older persons, age of onset also needs to be considered. Alzheimer's disease is such a disorder. Young people who are disease gene carriers but have not reached the age when the clinical symptoms of the disease are commonly manifested are called nonpenetrant.

Just as a protective environment may lead to reduced penetrance in disease gene carriers, exposure to a damaging environment can result in phenocopies: the display of classic genetic disease symptoms in individuals without any mutant alleles. Examples include drug-induced psychosis mimicking schizophrenia, cleft lip and palate induced by maternal exposure to teratogens, and deafness induced by maternal rubella.

Other issues concern the number of genetic loci involved in disease risk, the magnitude of each locus's effect, and possible interactions among loci ("epistasis"), which either magnify or decrease the effect of each locus alone. For example, diabetes, many cases of cancer and cardiovascular disease, and some forms of mental illness appear to be "oligogenic" in etiology. This means that there are likely to be at least a few genes (between 2 and 10) that individually contribute a small component of disease risk, although in aggregate their total genetic contribution to disease risk may be very substantial. Highly complex disorders are often strongly influenced by environmental factors such as diet, exercise, and exposure to tobacco. Since the diseases in question constitute the main sources of mortality and morbidity in modern society, major attempts to elucidate their genetic etiology through mapping and other approaches are ongoing. However, their extensive genetic complexity and their interaction with environmental factors present great research challenges.

An examination of two additional common diseases from this perspective will further illustrate these components of genetic complexity. For example, even the relatively "simple" disorder β-thalassemia is characterized by great variation in clinical symptoms because there are so many different β-globin gene mutations (allelic heterogeneity). The impact of this disease can also be influenced by genetic variation at the β-globin or fetal hemoglobin loci (epistasis), as well as by the environment (presence of malaria).

Retinoblastoma occurs both as a result of inherited mutations in a tumor suppressor gene and sporadically in individuals without any hereditary predisposition. The end result in either case is a retinal tumor that lacks both maternal and paternal copies of the tumor suppressor gene. In the inherited form, since the individual starts life already missing one copy of this gene, the chance of somatic loss of both copies is much greater than for nonhereditary forms, in which initially both copies are functional. Consequently, hereditary forms of retinoblastoma often present with tumors in both eyes, while nonhereditary cases usually involve only one eye. In terms of the features of genetic complexity just discussed, this disease involves reduced penetrance (onset of tumors in the hereditary form increases with an individual's age). Nonhereditary forms might be categorized as phenocopies at the level of their germ line (since they do not transmit mutated copies of the disease gene to their offspring), while somatically their tumors derive from the same defective genetic locus as the hereditary forms.

## 3 MAJOR APPROACHES TO DISEASE GENE MAPPING

### 3.1 Linkage Analyses

If two genes are located very close together on the same chromosome (i.e., are "linked"), they will be expected to travel together through a family. In linkage studies, we observe whether a marker locus (DNA variant with known chromosomal location) travels through a family together (formally, "cosegregates") with a disease gene for which the chromosomal location is unknown. If statistically adequate evidence of such cosegregation is observed, we conclude that the disease gene must be located on the same chromosomal region as the marker locus. Such evidence is usually quantified by a statistic called the logarithm of the odds (LOD) score. This is essentially the ratio of the likelihood of an observed pattern of segregation of a disease and a marker locus in a set of families under a hypothesis of linkage to the likelihood under the null hypothesis (viz., that the disease and marker loci are unlinked).

Human linkage analysis has been revolutionized twice during the past decade by technological advances. Prior to this, markers did not exist for most of the genome, the only exceptions being protein-based markers for blood groups, HLA antigens, and a few enzyme polymorphisms. First, restriction fragment length polymorphisms (RFLPs), especially at highly informative variable number of tandem repeat (VNTR) loci provided nearly genome-wide coverage with marker loci. However, many of these loci were not highly informative (i.e., only two allelic variants, often with one variant at a high frequency) and the Southern blotting method necessary to type these markers is slow and costly. More recently, application of polymerase chain reaction (PCR) technology has led to the development of a nearly genome-wide collection of "microsatellite" markers (Figure 1) that are both highly informative and quick and inexpensive to type.

```
left PCR primer - left flanking DNA - tandem repeat - right flanking DNA - right PCR primer
   n = 20 bp          n = 25 bp        variable size        n = 14 bp          n = 19 bp

Maternal Chromosome:   10 repeats: CACACACACACACACACACA      Total PCR Fragment Size =  98 bp
                                         n = 20 bp
Paternal Chromosome:   12 repeats: CACACACACACACACACACACACA  Total PCR Fragment Size = 102 bp
                                         n = 24 bp
```

**Figure 1.** Schematic representation of a hypothetical microsatellite marker locus based on variation in a dinucleotide (CA) tandem repeat. This individual's maternal chromosome contains 10 dinucleotide repeats and following PCR amplification produces a fragment of 98 bp (20 bp + 25 bp + 20 bp + 14 bp + 19 bp). The individual's paternal chromosome produces a fragment that is 4 bp larger (102 bp) due to the presence of two additional repeats (12 dinucleotide repeats).

Real examples and theoretical studies alike show that linkage analysis has considerable power to map genes for fairly complex diseases. Success in mapping three distinct genes for Alzheimer's disease is especially striking. Increasingly sophisticated statistical techniques have been developed to allow for reduced penetrance, phenocopies, and heterogeneity. Theoretical analyses have shown that linkage studies retain substantial power even with these complicating factors. Loci involved in oligogenic disorders can be mapped. Major efforts have been directed at developing ''nonparametric'' methods, which do not require assumptions about the disease's genetic architecture (e.g., dominance or recessivity, penetrance levels, etc.). However, it should be recognized that a price must be paid to overcome disease complexities. For diseases that are highly heterogeneous, where environmental factors play a large role and the effects of individual loci are small, gene mapping studies must have an increasingly larger collection of disease-segregating families to retain statistical power. Thus, the feasibility of linkage analysis may be threatened by the inability to collect and genotype adequate numbers of families and individuals. Especially for rare diseases, it may not be possible to identify and recruit an adequate number of families. Nevertheless, linkage analysis, often combined with other approaches described in Sections 3.2 to 3.5, is likely to continue to be successfully applied for human disease gene mapping.

### 3.2 Association Studies

Association studies constitute a method of gene mapping that has traditionally been applied to collections of unrelated disease cases and matched controls. The expectation is that disease alleles may be correlated (i.e., ''in disequilibrium'') with variable DNA sequences in close proximity on the chromosome. This correlation develops because a new disease mutation always occurs on a chromosome that happens to have a particular set of alleles at the variable DNA sites surrounding it. If mutations are infrequent, many affected persons in the population may have inherited the disease mutation from a common distant ancestor. As the disease chromosome passes through successive generations, meiotic recombination reshuffles variable DNA sites at some distance from the disease mutation. However, the sites very close to the disease gene may continue to be correlated for a long time because crossovers are unlikely to occur in the small region between the disease mutation and very closely flanking variable DNA sites. On a population level, this correlation is observed as a different allele frequency at the closely flanking variable DNA sites in affected (vs. unaffected) individuals. Association tests evaluate whether these allele frequencies differ significantly.

A number of robust examples of disease gene associations have been reported, such as the strong HLA association in diabetes. Since, however, this approach will work only if new disease mutations occur very infrequently, it will not be applicable for all diseases. The breakdown of associations by meiotic recombination also means that for practical purposes the approach can be applied only to candidate genes (see Section 3.4) or to candidate chromosomal regions (identified by linkage analysis or cytogenetic abnormalities), inasmuch as the DNA marker loci will need to be very close to the disease mutation site (e.g., $<< 1$ Mb).

In addition, there are several mechanisms by which associations can occur even when the disease gene is not located near the polymorphic DNA marker, thus potentially leading to a false positive gene mapping result. Great care must be taken in matching disease cases and controls to ensure that they do not differ in race, ethnicity, or geographic origin, as these variables can produce marker allele frequency differences at many locations throughout the genome. A second major concern is population admixture. This means that in relatively recent historical times, mergers have occurred between two or more populations that differ in both disease and marker allele frequencies. This event also results in disease associations with marker alleles that are not necessarily located near the disease gene. To address these problems, statistical geneticists have recently focused attention on developing sampling strategies based on using families rather than unrelated individuals for cases and controls. It is likely that some genetic effects, such as those involving oligogenic disease architectures, may be addressed most powerfully by these family-based association tests.

### 3.3 Chromosomal Abnormalities

Some disease gene mutations involve cytogenetically detectable abnormalities of chromosomes. For disorders such as fragile X disease, chromosome breaks occur in most disease cases, while other genetic disorders may only rarely exhibit mutations visible as cytogenetic rearrangements (vs. point mutations or submicroscopic deletions, insertions, or other small rearrangements). Chromosome breaks may bisect the disease gene itself, thereby preventing synthesis of its full-length mRNA, or the gene's regulation may be pathologically altered due to the ''position effect'' arising from being placed in close proximity to new neighboring genes in the rearranged configuration. Observation of such disease-causing chromosomal aberrations can offer a tremendous short-cut for mapping and cloning a disease gene. Many physical mapping techniques are available which allow researchers to quickly clone DNA surrounding chromosome breakpoints. If the disease gene is bisected by the chromosome break, the initial target for cloning the gene is much smaller than the 10 or 20 Mb region often identified by linkage analysis. This process has been greatly facilitated by major improvements in molecular cytogenetic techniques. However, the approach is unsuitable for mapping diseases that do not exhibit mutations due to chromosomal rearrangements. Furthermore, unless the location of the disease gene near a chromosome breakpoint is confirmed by some other approach (e.g., linkage analysis), there is a risk of pursuing abnormalities unrelated to disease pathology, which may coincidentally occur in one or a few affected individuals. A combined use of techniques was applied to map and clone the type 1 neurofibromatosis gene. Initially, chromosomal rearrangements implicated chromosome 1, 8, 17, 22, and others. When linkage results demonstrated that the gene was on chromosome 17, efforts aimed at cloning the disease gene focused on rearrangements in this chromosome.

### 3.4 Candidate Genes

Although our understanding of a disease's pathological mechanisms may be incomplete, knowledge gained from animal models or from biochemical or physiological studies often allows reasonable guesses about the kinds of genes that might be involved. An example is the amyloid precursor protein located on chromosome 21, which was correctly hypothesized to be involved with amyloid plaques found in the brains of Alzheimer's patients. Linkage studies often focus at least initially on chromosome regions that contain candidate genes before undertaking the larger investment required for a genome-wide search. Sometimes disease genes can be identified directly by scanning for mutations in candidate genes. When disease genes are found by this direct "functional" approach, their location on the human map (if not already known) can be determined subsequently by in situ hybridization or other techniques.

Unfortunately, many diseases either have no good candidates (because very little is known about pathological mechanisms) or too many candidates, such that virtually every chromosomal region contains a likely suspect (e.g., diseases involving the immune system, where immunoglobulins, T-cell receptors, as well as many cytokines and lymphokines could be implicated). In these cases, it may be necessary first to limit the candidate chromosomal region by linkage analysis or cytogenetic abnormalities, and then to focus on reasonable candidates located in this chromosomal region. An example of the convergence of all three approaches occurred for the mapping of Waardenburg syndrome type 1, where a chromosomal rearrangement implicated the distal portion of chromosome 2q. This suspicion was confirmed by linkage analysis, and then mutations were found in a candidate gene (*PAX3*), which produces a similar phenotype in the mouse.

### 3.5 Whole-Genome Molecular Analyses

Two techniques proposed recently offer the potential to greatly facilitate linkage analysis. These are representational difference analysis and genomic mismatch scanning. These methods potentially alleviate the need for typing polymorphic markers in large numbers of extended families. Space limitations prohibit a complete presentation here, but both techniques attempt to clone regions surrounding disease genes based on molecular subtraction and/or hybridization approaches. To date, neither technique has been applied successfully to map a human disease gene, although results of tests in animal models appear promising.

## 4 TRANSCRIPT AND MUTATION IDENTIFICATION

The final step in disease gene mapping is to locate the transcribed region and sites of mutation that cause the disease. If the gene is initially mapped by linkage analysis, which often can only narrow the candidate region to a target region of 20 Mb or more, this last step in the process can be very difficult. An example is Huntington's disease, which required a decade of intensive gene cloning work after the initial mapping by linkage analysis. The first steps used to identify disease gene transcripts often involve association analyses to narrow the region initially defined by linkage (as was done for the cystic fibrosis gene), searching for chromosomal abnormalities in the candidate region, and testing available candidate genes. Additional strategies include (1) identifying gene transcripts by conservation of sequence across different species due to constraints imposed by natural selection (i.e., the "zoo blot" approach), (2) exon trapping molecular techniques designed to locate intron–exon boundaries in genomic DNA, (3) either screening cDNA libraries with genomic DNA from the candidate region or using cDNA hybridization strategies to selectively clone genomic DNA sequences within the region which contain coding sequences, and (4) searching for regions with abundant CG dinucleotide sequences ("CpG Islands"), which are often associated with sites of transcription. Techniques developed for mutation screening once a candidate gene transcript has been found include single-stranded conformational polymorphism, heteroduplex analysis, and direct DNA sequencing approaches.

## 5 FUTURE DIRECTIONS

We are likely to see continued major improvements in the tools available for human disease gene mapping—among them both molecular-based innovations such as automated methods for characterizing very highly polymorphic DNA markers and statistical methods, which allow powerful analyses of highly complex diseases while maintaining robustness and minimizing the chances of false positive findings. Resources that will result from the Human Genome Project will also be invaluable. In the past, whenever a disease gene was localized to a particular chromosomal region, investigators essentially had to carry out a "local" genome project for the previously unexplored region. Additional polymorphic markers had to be generated, genomic DNA in the region had to be cloned by chromosome walking, and transcripts had to be identified and sequenced to determine their likely function. In the near future, all these resources will be made available by the Human Genome Project. Thus, investigators should be able to advance much more quickly from an initial linkage finding to disease gene identification. Furthermore, the number of gene sequences for human, mouse, and other model organisms will increase greatly, along with our capacity to infer the genes' functions using computer algorithms based on their sequence. This knowledge will undoubtedly enhance our ability to identify good candidates for many diseases from this greatly expanded catalogue, either after an initial chromosomal localization or at the start of a gene-mapping study.

*See also* Genetic Testing; Human Genetic Predisposition to Disease; Human Linkage Maps; Population-Specific Genetic Markers and Disease; Whole Chromosome Complementary Probe Fluorescence Staining.

## *Bibliography*

Borgaonkar, D. S. (1994) *Chromosomal Variation in Man.* Wiley-Liss, New York.

Kendler, K. S., and Diehl, S. R. (1993) The genetics of schizophrenia: A current, genetic–epidemiologic perspective. *Schizophrenia Bull.* 19:261–285.

King, R. A., Rotter, J. I., and Motulsky, A. G., Eds. (1992) *The Genetic Basis of Common Diseases.* Oxford University Press, New York.

Lander, E. S., and Schork, N. J. (1994) Genetic dissection of complex traits. *Science,* 265:2037–2048.

McKusick, V. A. (1992) *Mendelian Inheritance in Man.* Johns Hopkins University Press, Baltimore.

Reed, P. W., Davies, J. L., Copeman, J. B., Bennett, S. T., Palmer, S. M., Pritchard, L. E., Gough, S.C.L., Kawaguchi, Y., Cordell, H. J., Balfour, K. M., Jenkins, S. C., Powell, E. E., Vignal, A., and Todd, J. A. (1994) Chromosome-specific microsatellite sets for fluorescence-based, semi-automated genome mapping. *Nature Genet.* 7:390–395.

Thompson, M. W., McInnes, R. R., and Willard, H. F. (1991) *Genetics in Medicine.* Saunders, Philadelphia.

# Human Genetic Predisposition to Disease

*Belinda J. F. Rossiter and C. Thomas Caskey*

*Key Words*

**Codon** Three adjacent nucleotides in messenger RNA that specify which amino acid is to be placed in a protein during translation.

**DNA Polymorphism** Variation in DNA sequence.

**Exon** Portion of a gene that is present in mRNA after RNA processing.

**Gene** Portion of DNA that is required for production of a functional product.

**Genome** Sum of the genetic information contained in one representative of each chromosome pair.

**Intron** Portion of a gene that is excised during RNA processing.

**Mutation** Alteration in a gene.

**Oncogene** Gene that results in uncontrolled cell growth when activated by mutation.

**Phenotype** Observable characteristics of an organism.

**Promoter** DNA sequences controlling the expression of a gene.

**Stop Codon** Triplet nucleotide in mRNA that signals termination of protein synthesis.

**Tumor Suppressor Gene** Gene that normally acts to regulate cell growth, whose inactivation by mutation leads to uncontrolled cell growth.

The human genome is very complex, and alterations causing genetic diseases can occur in a variety of ways. As genetic research advances, more rapid and accurate diagnosis of genetic disease is possible, and genetic methods can even be used as a form of therapy. Some diseases result from the mutation of a single gene, and others have a more complex origin, including multiple genetic and environmental factors.

## 1 THE HUMAN GENOME

The human genome consists of approximately $3 \times 10^9$ nucleotides and includes up to $10^5$ genes. Each gene usually has a coding region that determines the sequence of amino acids in the final protein product. This coding region is split into *exons*, with intervening sequences called *introns*. A *promoter* contains DNA sequences that direct the synthesis of an RNA copy for the entire gene, and splicing signals then control the precise removal of the introns. Each end of the spliced RNA, now called mRNA, is further modified chemically before the molecule is used as a template for protein synthesis (see Figure 1).

## 2 DNA VARIATIONS

Except for identical twins, no two people have exactly the same sequence of DNA. Since less than 5% of the genome consists of protein coding sequence, most person-to-person DNA variations have no apparent effect. Other DNA differences do affect the phenotype, but most of these affect normal (variable) characteristics such as hair and eye color. A small but important fraction of DNA variations result in a phenotype that is not considered "normal," such as a genetic disease. In DNA analysis for diagnosis of a genetic disease, it is important to distinguish between normal DNA variations ("polymorphisms") and the alteration that causes the disorder ("mutation").

### 2.1 Mutations

Several thousand of the genes in the human genome can cause disease if disrupted in some way. A DNA mutation may be a deletion, insertion, duplication, inversion, or other rearrangement (Figure 1), and the extent of a mutation may range from a single nucleotide to a whole chromosome. If the mutation occurs in the protein coding region of a gene, the amino acid sequence may be changed, or even shortened if a premature stop codon is created (Figure 2). A mutation can also affect other portions of a gene, such as the promoter or processing signals (Figure 1). The result of a mutation in a particular gene may be that an altered protein is made, that the protein is not made at all, that the wrong quantity of the protein is made, or that the protein is made at the wrong time in development.

### 2.2 Polymorphisms

Polymorphisms, that is, DNA variations among individuals, are usually one of two types. Single-nucleotide changes, if they happen to fall within the recognition site of a restriction endonuclease, may be detected as restriction fragment length polymorphisms (RFLPs) (Figure 3A). A number of other methods are available for the detection of other single-nucleotide alterations that do not occur within restriction endonuclease recognition sites.

The other major type of DNA polymorphism is a variation in the number of tandemly repeated sequences at a particular location in the genome (Figure 3B). The repeated sequence may be 10–60 nucleotides long, in which case the tandem array is called a "minisatellite." Groups of shorter repeats (2–5 nucleotides each) are called "microsatellites."

## 3 GENETIC THERAPY

Genetic therapy is the supply of a functional gene to cells lacking that function, with the aim of correcting a genetic disorder or acquired disease. Genetic therapy, or "gene therapy," can be broadly divided into two categories. The first is alteration of germ cells, that is, sperm or eggs, which results in a permanent genetic change for the whole organism and subsequent generations. This "germ line gene therapy" is not currently considered an option in humans for technical and ethical reasons. The second type of genetic therapy, "somatic cell gene therapy," is analogous to an organ transplant. In this case, one or more specific tissues is targeted by direct treatment, or by removal of the tissue, addition of the therapeutic gene(s) in the laboratory, and return of the treated cells to the patient. Several clinical trials of somatic cell gene therapy have started, mostly for the treatment of cancers and blood disorders.

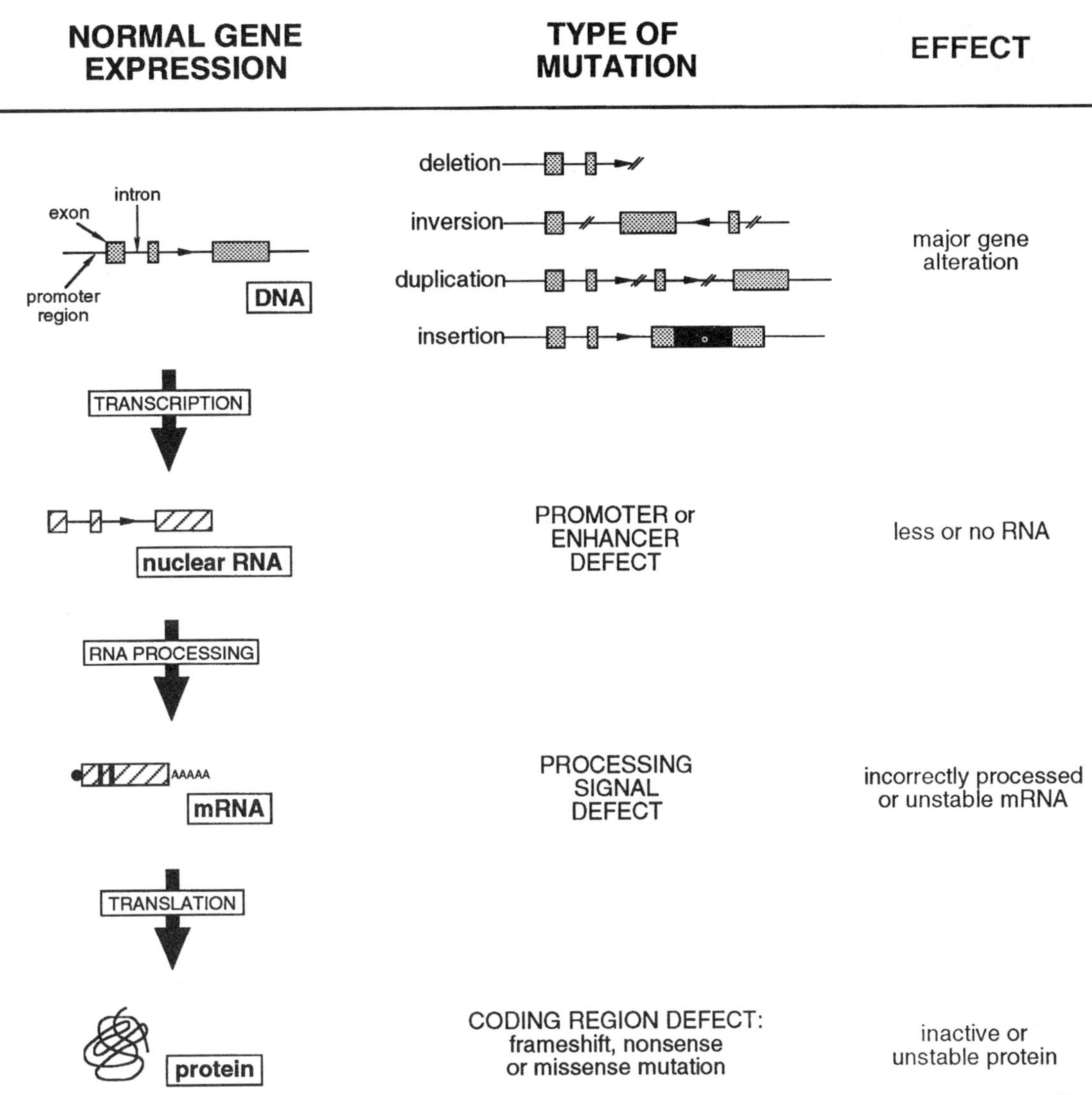


**Figure 1.** Some of the ways a gene can be affected by mutation. Expansion of an unstable repeat region (see Figure 4), also can affect the expression of a gene.

## 4 EXAMPLES OF GENETIC DISEASES

Four examples of genetic diseases are described to illustrate different aspects of human genetic disease.

### 4.1 Symptomless Carriers of a Childhood Disorder: Cystic Fibrosis

Cystic fibrosis has an autosomal recessive inheritance and affects approximately 1 in 2500 Caucasians. The disease is recognized at birth or in early childhood and primarily affects the lungs and pancreas. Children with cystic fibrosis are usually born to parents who do not have the disease, but each carries one copy of a cystic fibrosis mutation. In fact, many cystic fibrosis births occur without any family history of the disease.

Genetic testing methods can now detect the most common cystic fibrosis mutations, even in unaffected carriers of the disorder. It is therefore feasible to test relatives of a cystic fibrosis patient to find other carriers in the same family, or even to offer cystic fibrosis mutation screening to people who do not have a family history of this disorder.

### 4.2 Unstable Genetic Sequences: Fragile X Syndrome and Myotonic Dystrophy

Fragile X syndrome primarily affects males and is the most common cause of inherited mental retardation. Myotonic dystrophy is the most common adult muscle disorder, and symptoms range from being barely detectable to a severe congenital form that includes mental retardation. Both the fragile X gene and the myotonic dystrophy gene contain a polymorphic triplet repeat, that is, a group of three nucleotides repeated a variable number of times (see Section 2.2). Those with a number of repeats within the "normal" range are not affected; but in susceptible individuals, the repeat region can expand during inheritance from parent to child. Once the number of

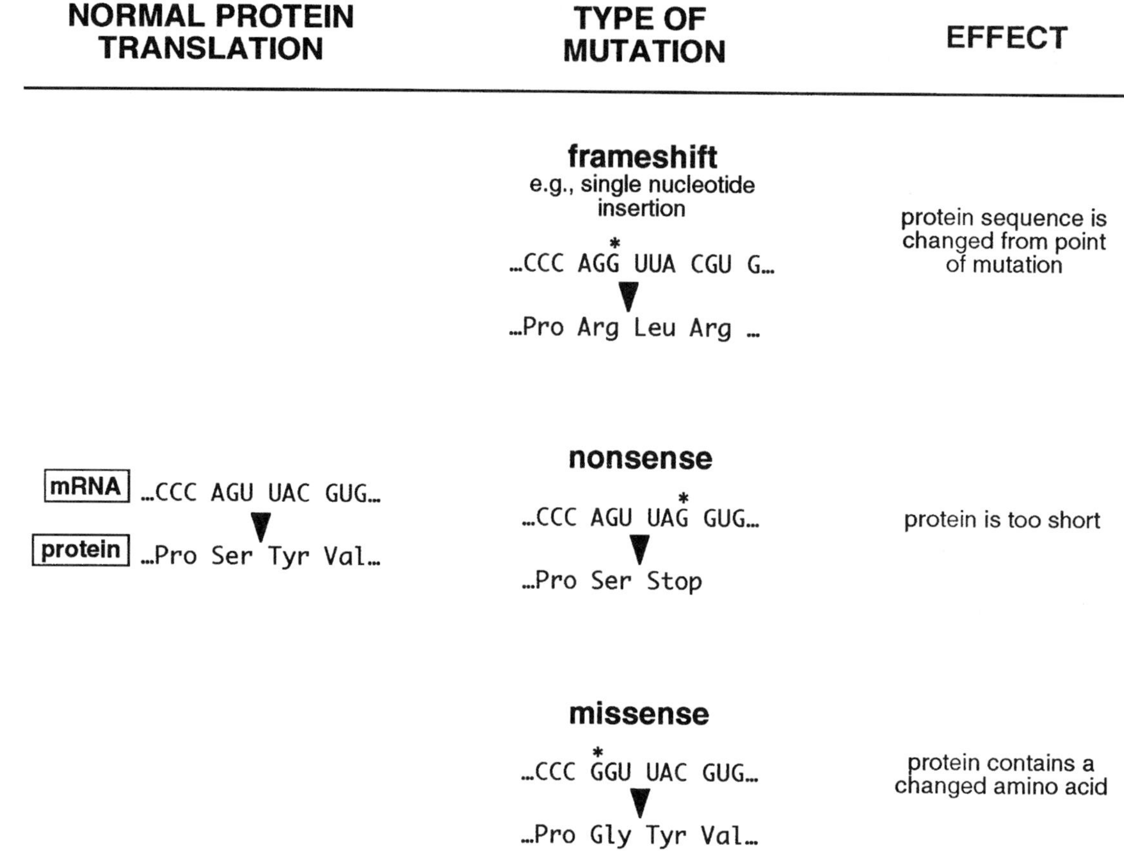

NORMAL PROTEIN TRANSLATION
TYPE OF MUTATION
EFFECT
frameshift
e.g., single nucleotide insertion
...CCC AGG UUA CGU G...
...Pro Arg Leu Arg ...
protein sequence is changed from point of mutation
nonsense
mRNA ...CCC AGU UAC GUG...
protein ...Pro Ser Tyr Val...
...CCC AGU UAG GUG...
...Pro Ser Stop
protein is too short
missense
...CCC GGU UAC GUG...
...Pro Gly Tyr Val...
protein contains a changed amino acid


Figure 2. Some of the mutations that can affect the coding region of a gene.

triplet repeats has exceeded a certain number, the chances of having symptoms of the particular disorder, or of bearing an affected child, increase dramatically (Figure 4).

Genetic testing can reveal expanded repeat regions in fragile X syndrome or myotonic dystrophy families. Such testing can provide an estimate of the likelihood of disease in individuals or their children, although the correlation of repeat expansion and severity

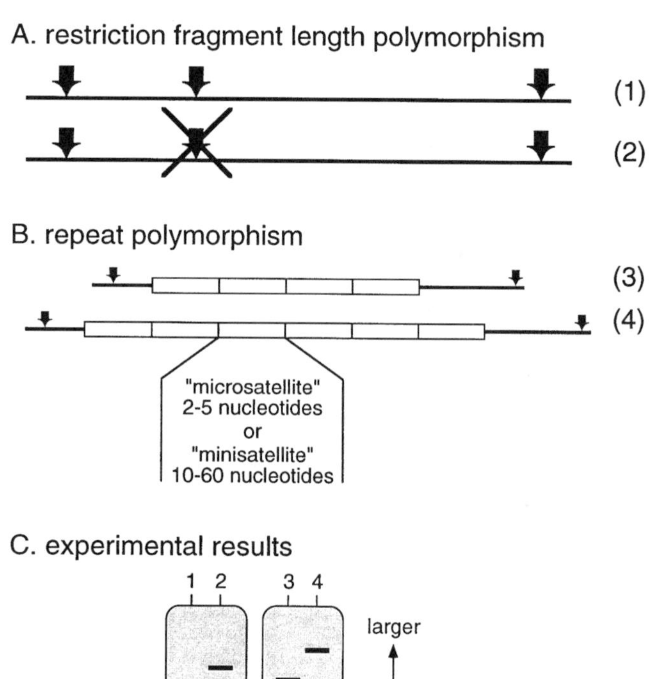


Figure 3. Two major types of DNA polymorphism, and some experimental results. (A) Restriction fragment length polymorphism (RFLP). Heavy arrows indicate restriction endonuclease recognition sites. The presence or absence of a restriction endonuclease recognition site in a specific region of the genome determines the size of the corresponding DNA fragment after digestion by that particular enzyme. The alteration of a single nucleotide in the DNA sequence can determine the presence or absence of a restriction site. (B) Repeat polymorphism. Open boxes represent repeated sequence units that may be short (2–5 nucleotides) or longer (usually 10–60 nucleotides). Repeat polymorphisms differ across individuals in the number of repeated sequence units at a particular location. The length of the fragment containing the sequence repeat indicates the number of repeated units. (C) Experimental results. DNA fragments can be separated by size using gel electrophoresis, and visualized directly or after Southern analysis and autoradiography. The horizontal bars indicate bands that would be seen after such analysis of DNA from (A) and (B). In these examples of RFLP (left-hand panel) and repeat polymorphism (right), the lanes marked 1–4 correspond to the different DNA fragments shown in (A) and (B).

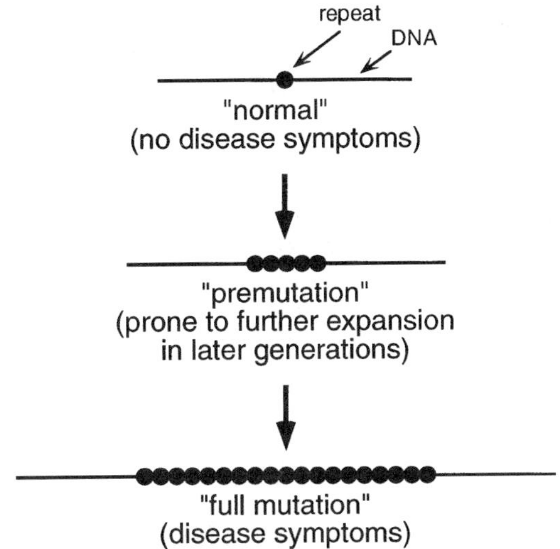


**Figure 4.** Expansion of unstable repeat regions resulting in disease. Some disease genes contain a region of repeated triplet nucleotides that is polymorphic (i.e., the number of repeats is variable), as shown in Figure 3B. People with a so-called *premutation* do not have symptoms of the disorder themselves but have an expanded repeat region that can further increase in size during passage from parent to child. Individuals with a ''full mutation'' are affected with disease symptoms.

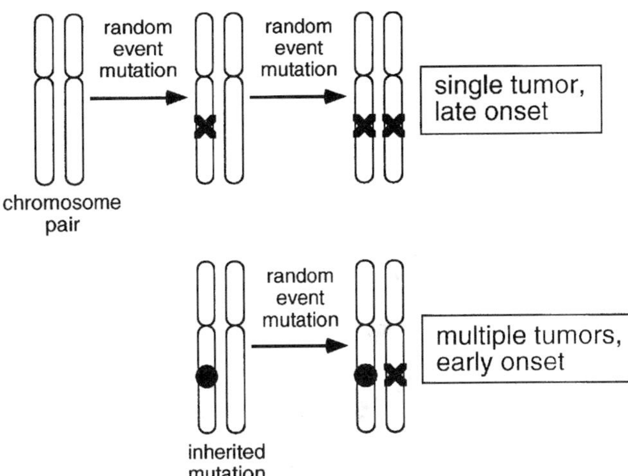


**Figure 5.** The ''two-hit'' model of cancer gene mutation. One of the events in the development of cancer may be a mutation in both copies of an oncogene or a tumor suppressor gene. It is rare for both copies of the gene to be affected by random events, and thus the outcome is usually late onset of a single tumor. If one mutation is inherited, only one further random event is required, and the individual usually experiences earlier onset of multiple tumors.

of disease is not precise. It is also possible to detect a repeat region that has not yet expanded enough to produce symptoms but may do so in subsequent generations.

### 4.3 Presymptomatic Diagnosis of a Late-Onset Disorder: Adult Polycystic Kidney Disease

The first symptoms of adult polycystic kidney disease usually appear at about age 40, and progression of the disease results in chronic renal failure. Renal cysts can be detected by ultrasound before the onset of symptoms (approximately 70% of cases can be detected by age 20), and genetic testing in a family affected by the disease can determine the presence of the adult polycystic kidney disease gene as early as the prenatal period. This is one of many examples of inherited diseases that can be diagnosed by genetic testing before the appearance of symptoms. Most such disorders, however, currently have no treatment or cure.

### 4.4 Predisposition to Genetic Disease: Colorectal Cancer

Colorectal cancer is not the result of disruption of a single gene, but there is strong evidence that a series of genetic events can eventually lead to the disease. A number of genes have been discovered that are frequently altered in colon cancer. Some of these genes are ''oncogenes,'' which can become *activated* by mutation, and some are ''tumor suppressor genes,'' which normally repress cell growth but are *inactivated* by mutation. It is thought that an accumulation of multiple mutations in oncogenes and tumor suppressor genes can eventually lead to colorectal cancer. These mutations may occur spontaneously or they may be inherited (Figure 5). It is not surprising, therefore, that inheritance of one of these mutations increases the risk that an individual will develop cancer. Genetic testing, particularly of someone with a family history of colorectal cancer, may reveal a mutation that is not in itself a *prediction* that cancer will occur, but rather an indication of *predisposition* to that disease.

*See also* Colon Cancer; DNA in Neoplastic Disease Diagnosis; Fragile X Linked Mental Retardation; Genetic Testing; Molecular Genetic Medicine; Neuropsychiatric Diseases; Oncogenes; Population-Specific Genetic Markers and Disease; Tumor Suppressor Genes.

## *Bibliography*

Anderson, W. F. (1992) Human gene therapy. *Science,* 256:808–813.

Caskey, C. T., and Rossiter, B. J. F. (1992) Ninth Ernst Klenk Lecture. Molecular medicine. *Biol. Chem. Hoppe Seyler,* 373:159–170.

Friedmann, T. (1992) A brief history of gene therapy. *Nature Genet.* 2:93–98.

Hall, S. S. (1990) James Watson and the search for biology's ''Holy Grail.'' *Smithsonian,* 20:41–49.

Jordan, E. (1992) The Human Genome Project: Where did it come from, where is it going? *Am. J. Hum. Genet.* 51:1–6.

Rossiter, B. J. F., and Caskey, C. T. (1992) The Human Genome Project and clinical medicine. *Oncology,* 6:61–71.

Thompson, M. W., McInnes, R. R., and Willard, H. F. (1991) *Thompson & Thompson: Genetics in Medicine,* 5th ed. Saunders, Philadelphia.

Verma, I. M. (1990) Gene therapy. *Sci. Am.* 263(5):68–84.

## Human Genome: *see* Body Expression Map of the Human Genome.

# Human Linkage Maps

*Jean Weissenbach*

*Key Words*

**Genotype** The genetic constitution of an organism; usually refers to only one or a few pairs of alleles. Genotyping is the determination of which pair of alleles is present in a given individual.

**Locus (plural: Loci)** The position of a given gene on a chromosome.

**Meiosis** Two successive cell divisions in which four haploid gametes are produced, each with half the number of chromosomes found in the original cell.

**Phase** The combined genotypes of two polymorphic loci on the same parental chromosome (see Figure 2).

**Phenotype** Characteristics produced by the expression of an organism's genes and their interaction with the environment.

**Recombination** The formation of new gene combinations on chromosomes due to crossing over during meiosis.

When allelic forms of genetic traits are transmitted together with a high frequency, they are said to be linked. A genetic linkage map is a map of genetic traits in which distances are based on the measurements of the frequencies of coinheritance. Such distances are related to the physical distances separating the genes for these traits on the chromosome.

In humans it has been difficult to develop genetic linkage maps because of the dearth of distinguishable allelic forms of genes. The observation that DNA sequences can exist with multiple allelic forms has revolutionized human genetic mapping. The new human linkage maps are extremely useful in the localization of disease genes and in prenatal diagnosis and genetic counseling.

## 1 PRINCIPLES

### 1.1 The Concept of Genetic Linkage

The genome of most higher organisms including humans consists of two identical sets of chromosomes, one of which comes from the mother and the other from the father. These organisms, which are said to be diploid, result from the fusion of two haploid gametes, each of which contains only one set of chromosomes from each parent. The gametes are formed during meiosis, a process that results in the formation of haploid egg and sperm cells from diploid germ cells. These basic phenomena have important consequences for genetics which are known as Mendel's laws. These laws were formulated by Gregor Mendel in 1866 (although he was unaware of the existence of chromosomes) and were later reformulated as corollaries of the chromosome theory of heredity.

In diploid organisms, distinct alleles of two different genes residing on nonhomologous chromosomes are distributed to the gametes independently from one another according to Mendel's law of independent assortment, and four types of gamete will be produced with equal probabilities (Figure 1a). Gametes with parental allele combinations will occur at the same frequency as reassorted gametes in which one gene comes from one parent and the second comes from the other parent. Such genes are unlinked.

When two genes are on the same chromosome, alleles residing on the same parental chromosome tend to stay together during gamete formation and are said to be linked. However during meiosis each chromosome replicates, forming two identical sister chromatids, and homologous chromosomes will pair. In mammals two nonsister chromatids in a tetrad resulting from pairing of homologous chromosomes will cross over at least once (Figure 1b). This process is observed as the formation of a chiasma at the chromosome level and involves the breakage and reunion of two of the four chromatids. If crossing over occurs between the two genes mentioned earlier, the parental alleles will be reassorted in the resulting gametes, which are called recombinants (Figure 1b), whereas gametes with the original allele combinations are called parental or nonrecombinant.

The phenomenon of crossing over has two important consequences.

1. The frequency with which a crossover between any two genetic loci occurs is directly related to the chromosomal distance between the loci. The further apart, the greater the chance that a chiasma will occur between them. It is therefore possible to express the distance separating two genes as the percentage of recombination events observed in the total number of individuals in progenies. However more than one crossover can occur between two chromatids. Odd numbers of two-strand crossovers between two loci will produce detectable recombinants, but recombination between two loci will not be observed in the case of even numbers of crossover events. The number of recombinants is therefore not proportional to the distance. For short distances generally inferior to 15%, however, the frequency of double crossovers is extremely low.
2. When three biallelic loci (A/a, B/b, and C/c) are in the indicated order and are fairly close together on the same chromosome, several different combinations of alleles may be found in the gametes (Figure 2). Besides the parental combinations, which will be the most frequent, combinations resulting from one or two crossover events may be observed. Assuming that these crossover events are independent of each other, combinations due to a single crossover between parental chromosomes will be more frequent than those due to two crossovers. The frequency of double crossovers should be the product of the frequencies of the single crossover events. Therefore, the most probable order for the three loci can be determined by enumeration of the phenotypes of the offspring. For relatively short distances, the number of double crossovers is very small, so these distances can be added together up to a value of about 15% recombination frequency.

Because of these two corollaries of the process of meiosis, it is possible in principle to arrange all the genetic markers for a given chromosome in order, and to estimate the distances between them (Figure 2). The principles of genetic mapping are based on these concepts and were developed in the 1920s. They were first used to construct the genetic map of *Drosophila,* and were then applied to genetic mapping of other animal and plant species of scientific or economic interest. As mapping progressed, it was observed that

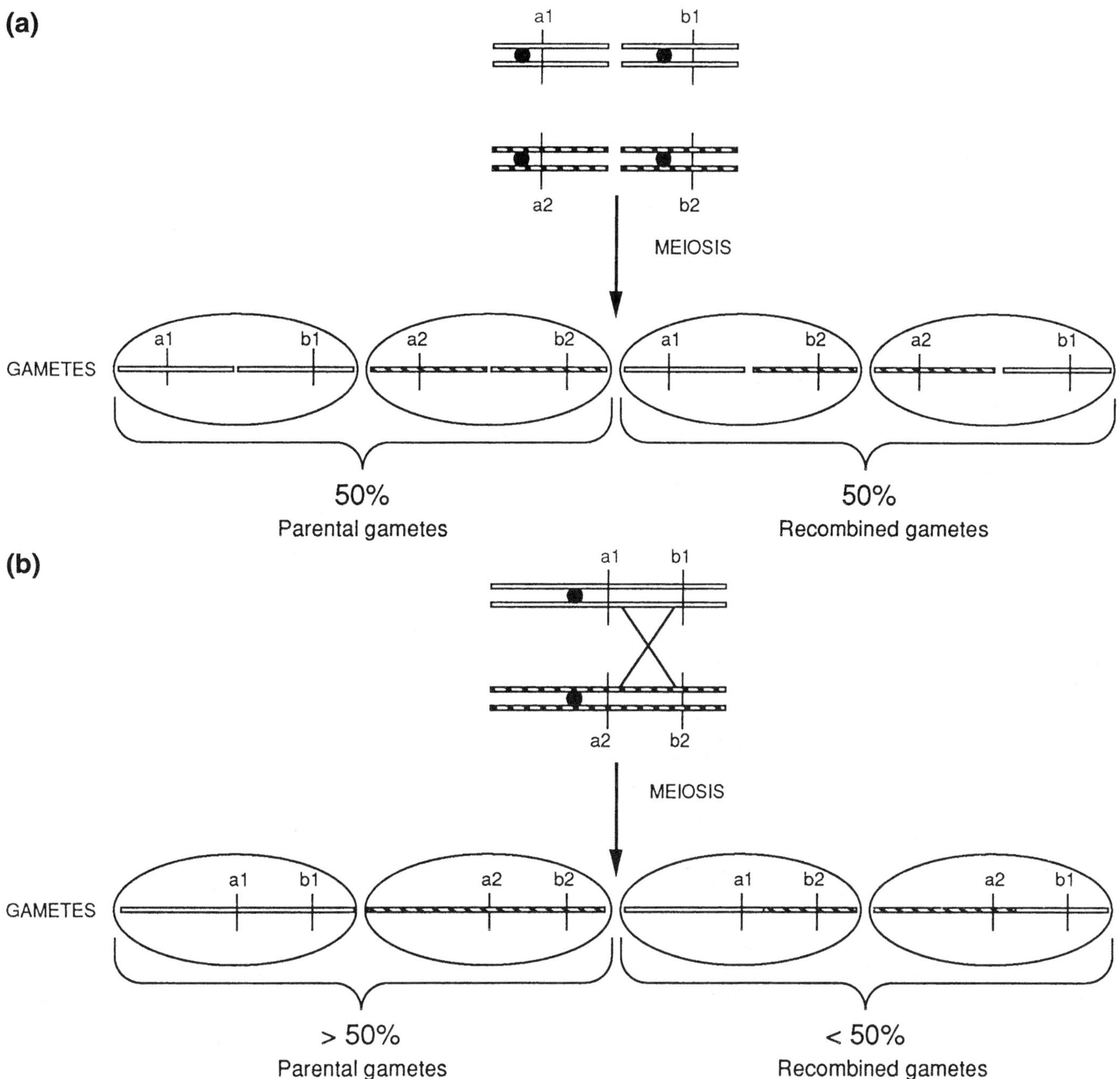


**Figure 1.** (a) Segregation of two pairs of unlinked characteristics. (b) Segregation of two pairs of linked characteristics.

the number of groups of loci that showed genetic linkage was identical to the number of chromosomes in a given species. It has also been possible to verify that the order of the loci on the genetic map could be correlated with the position of cytogenetic anomalies observed along the chromosomes. Because of the nonuniformity in the distribution of crossover events over the length of the chromosome, however, genetic distances exhibit significant distortions of the physical distances.

In humans, mapping progress has been much slower, for two interdependent reasons. First, for obvious ethical reasons, the human species cannot be used for experimentation in classical genetics. Observations were initially limited to studying the transmission of phenotypes and genotypes in families who had given their informed consent to participation in such studies. Second, because of this lack of experimentation, identification of Mendelian phenotypic characteristics has been extremely difficult. To be useful for genetic transmission studies, a characteristic must have at least two allelic forms. For many years such Mendelian characteristics in humans were limited to hereditary diseases and a few proteins, such as those responsible for blood group polymorphisms. For these two reasons, the development of a human genetic map has had a laborious beginning. Since the early 1980s, however, with the application of recombinant DNA technology to human genetics, this situation has improved considerably.

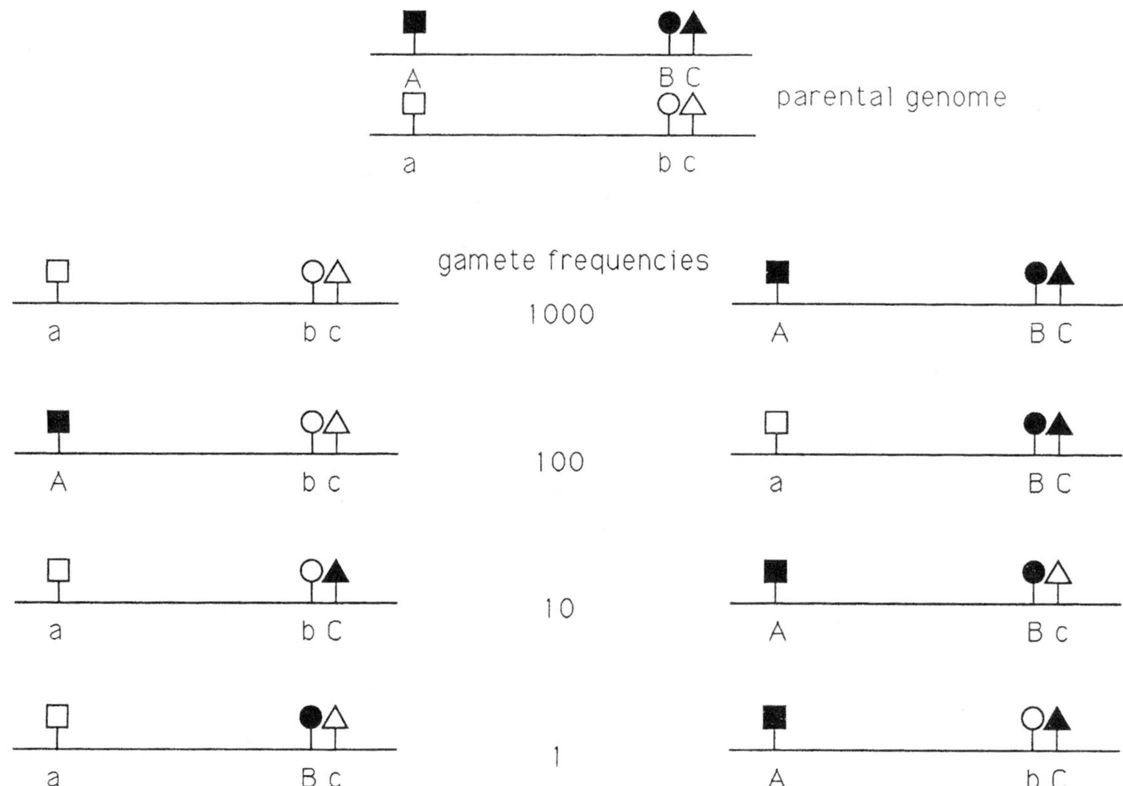


**Figure 2.** Possible allele combinations with three pairs of linked loci. The frequency of the different combination observed in the gametes depends both on distance between the alleles and on their order. The frequency of gametes resulting from two recombination events is equal to the product of the frequency of single recombinations.

## 1.2 DNA Polymorphisms

Several types of variation or polymorphism in DNA have been described since the introduction of cloning techniques. These polymorphisms can be directly observed at the genome level, where the alleles appear codominant. The genotype can therefore be inferred from the phenotype. The usefulness of a polymorphic marker may be estimated from its polymorphic information content (PIC), or more simply by its heterozygosity, which denotes the frequency of individuals heterozygous for a polymorphic system. Markers are chosen progressively, with the goal of obtaining a set of highly informative markers, evenly spaced throughout the genome.

### 1.2.1 Restriction Fragment Length Polymorphisms (RFLPs)

Southern blot analysis of DNA that has been digested with restriction enzymes and hybridized with cloned probes is used to demonstrate differences in length of restriction fragments. These variations in length are due to mutations, which cause either the loss or the gain of a restriction site or, more rarely, insertions or deletions within a restriction fragment (Figure 3). Although RFLPs that result from a gain or loss of a restriction site can be observed only by digesting the DNA with the specific restriction enzyme, insertions and deletions can be observed in a majority of restriction enzyme digestions. These latter RFLPs are frequently biallelic.

### 1.2.2 Variable Number of Tandem Repeats (VNTRs)

A special class of RFLPs due to insertions/deletions has been observed using probes to detect sequences that are repeated in tandem. These tandem repeats are characterized by a large variability in the number of repeated units (Figure 4), as well as variations in the sequences of the repeated motifs. Therefore these polymorphisms have been called VNTRs, for "variable number of tandem repeats." As with other insertion/deletion polymorphisms, they can be observed by using numerous restriction enzymes. Because of their structural resemblance to very long tandem DNA repetitions, which are found in satellite DNA, they are also called minisatellites.

### 1.2.3 Microsatellites

Although theoretically classified as VNTRs, short-sequence nucleotide repetitions (mono-, di-, tri-, and tetranucleotides) form a special group of polymorphisms characterized by the analytic procedures used for genotyping them. These procedures are based on amplification of the polymorphic markers by polymerase chain reaction (PCR) techniques and use of gels to separate the short PCR products. These motifs are very numerous in the genomes of higher eukaryotes, and are at present the most important source of polymorphic markers (Figure 5).

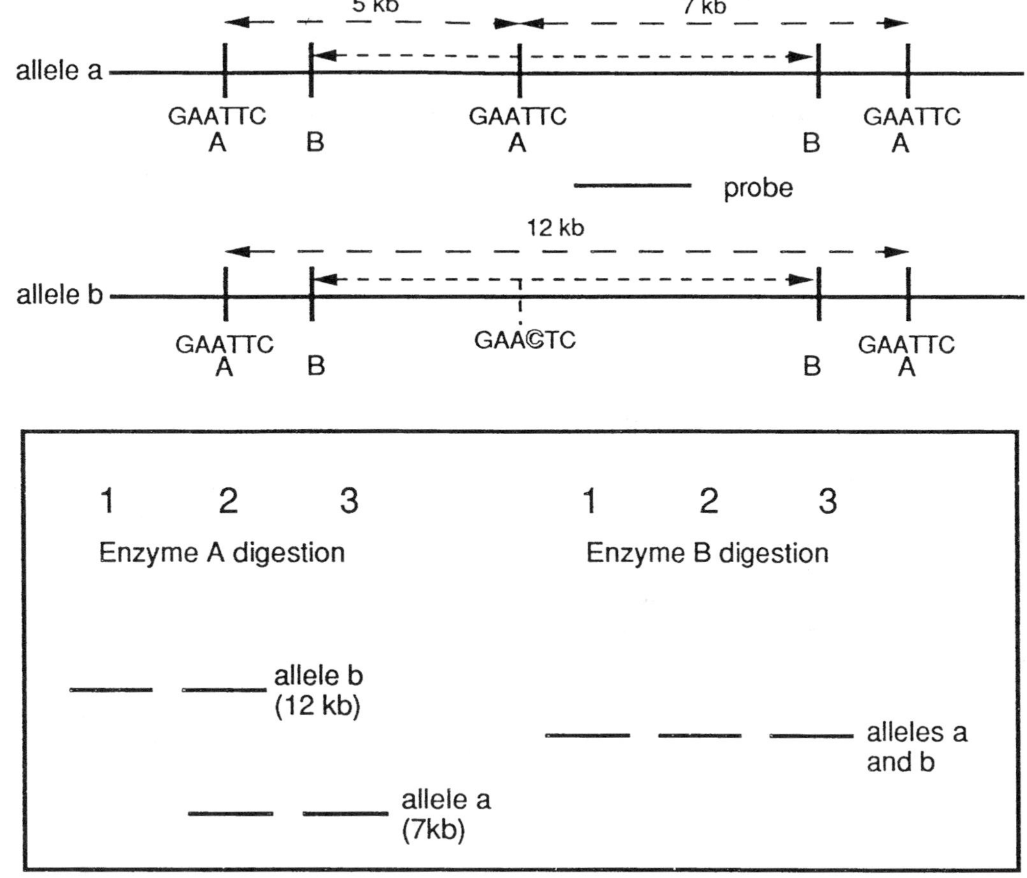


**Figure 3.** Principles and molecular analysis of RFLP. Probing after Southern blotting of DNAs digested with enzyme A (left) and B (right): 1 and 3 are homozygotes and 2 is a heterozygote. The probe detects only the largest enzyme A fragment of allele a.

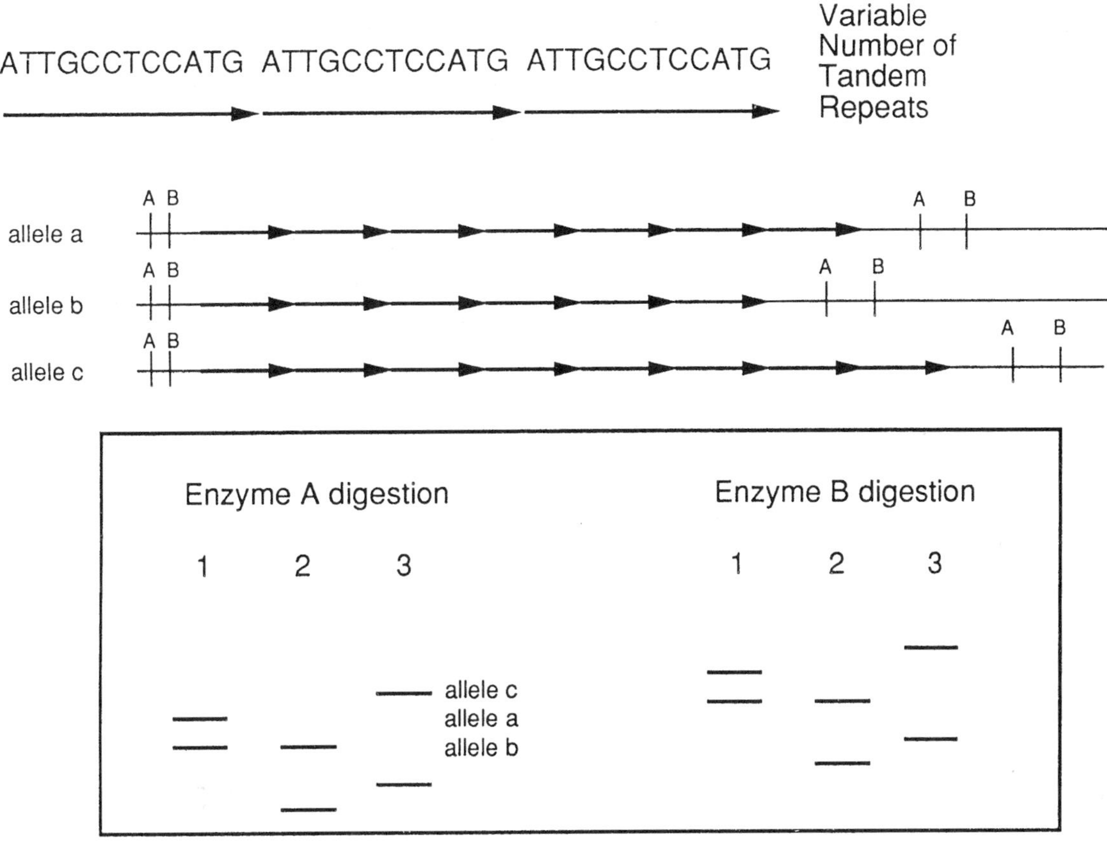


**Figure 4.** VNTRs or minisatellites, observable after digestion of DNA with restriction enzymes. RFLP analysis of VNTRs follows the same principle as in Figure 3.

### 1.3 Establishing Genetic Linkage

In species that are important for experimental or agricultural purposes, the establishment of linkage between two characteristics is based on analysis of numerous offspring of parents of known specific genotype. Segregation of two or more markers is therefore easily observed. The percentage of recombination is usually determined by simply counting the number of recombinant and nonrecombinant offspring. Human families are small, and the phase of the parental generation is frequently unknown. Therefore, a statistical analysis is used to estimate the likelihood of two alternatives, one of which corresponds to linkage and the other to independent segregation of the two genetic markers.

To establish the existence of genetic linkage in humans, it is important to demonstrate that the frequency of recombination is less than 50%. This is done by linkage analysis. The most commonly used methodology for this kind of linkage analysis is a "likelihood method" based on an evaluation of the recombination frequency. The likelihood of linkage is expressed as a recombination fraction $\theta$. In a given family if $n$ is the total number of siblings and $r$ the number of recombinant siblings, the likelihood $L(\theta)$ will be

$$L(\theta) = \theta^r (1 - \theta)^{n-r} \qquad (1)$$

One calculates the ratio of the likelihood of linkage for a given value of $\theta$ versus the likelihood of the absence of linkage for $\theta = 0.5$. Using this hypothesis, $L(0.5) = (0.5)^n$. For practical reasons, one uses the common logarithm ($\log_{10}$) of this value $Z(\theta)$, and this is known as the logarithm of odds (lod) score:

$$Z(\theta) = \log_{10}\left(\frac{L(\theta)}{L(0.5)}\right) \qquad (2)$$

The use of logarithmic values makes it possible to add the lod score values obtained from independent studies. The maximum likelihood method also permits estimations when some data are not available, such as the linkage phase of certain individuals in the family analysis. This is the method presently used for determining the validity of genetic linkage. A lod score greater than 3 indicates that the odds ratio in favor of linkage is 1000:1 and is considered sufficient to permit the conclusion that two loci are genetically linked.

## 2 TECHNIQUES

### 2.1 Analysis of DNA Polymorphisms

The genotypes of DNA polymorphisms are analyzed by classical DNA analysis procedures such as Southern blotting followed by

1) PCR amplification

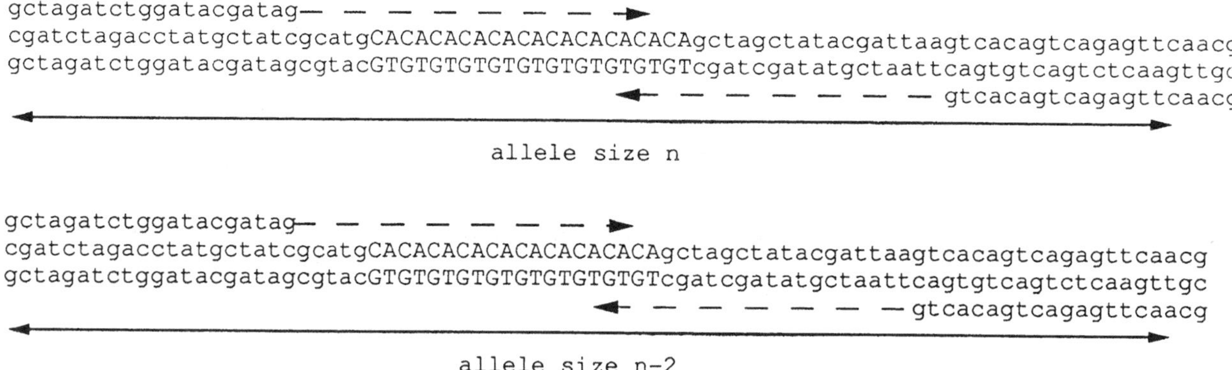


2) Separation of PCR products on denaturing acrylamide gels

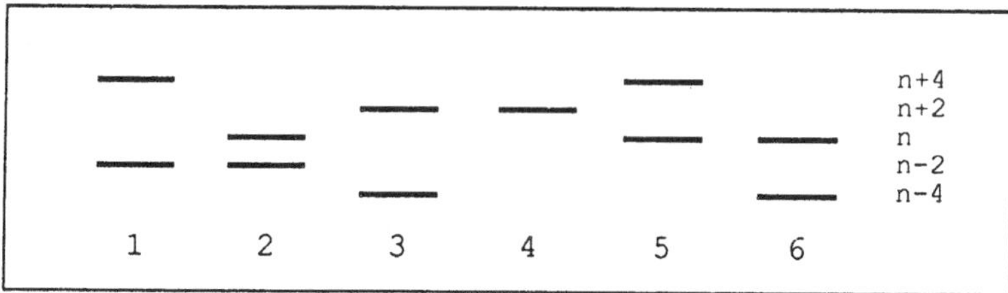

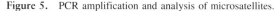

**Figure 5.** PCR amplification and analysis of microsatellites.

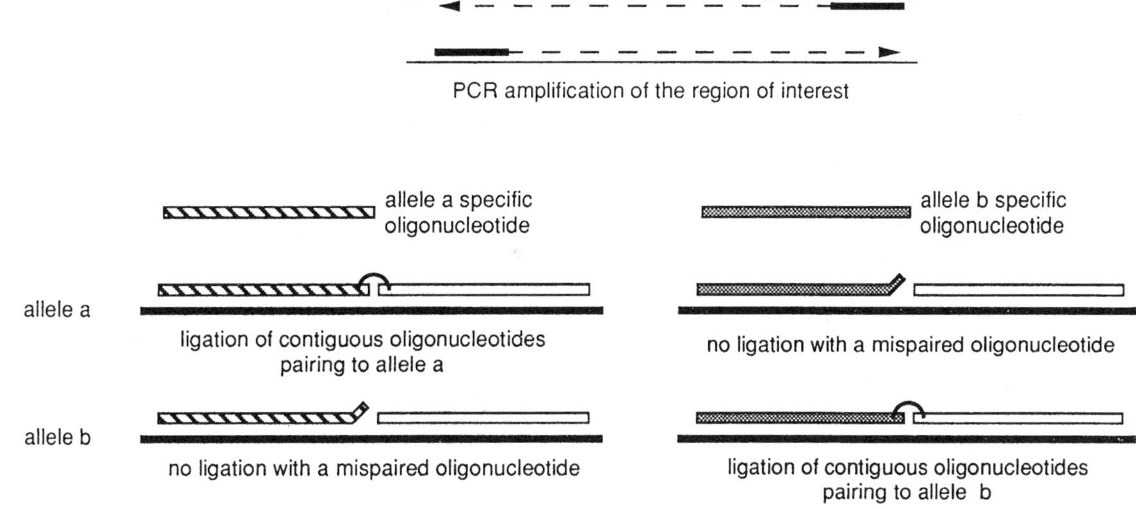


**Figure 6.** Oligonucleotide ligation assay.

DNA hybridization and the PCR (Figures 3–5). Single nucleotide changes can also be observed by an oligonucleotide ligation assay (OLA) using allele-specific oligonucleotide (ASO) annealing (Figure 6).

### 2.2 Reference Maps

The usefulness of polymorphic markers depends on their precise localization on the chromosome. The isolation and mapping of numerous polymorphic markers has led to more and more detailed genetic maps. In fact, the construction of human linkage maps itself depends on segregation analysis of polymorphic markers. The genotypes of all the individuals from several families are studied by the aforementioned methods and then subjected to statistical analysis.

Since the order of the markers depends on the recombination points observed in meiosis, it is highly preferable to study the same meiotic events when placing new markers on a map that has already been constructed. A number of laboratories from several different countries have agreed to pool their efforts by sharing their experimental material and data. This initiative, which was suggested by Jean Dausset and his group, is coordinated by the Centre d'Etude du Polymorphisme Humain (CEPH), which distributes DNA from about 60 large three-generation families to collaborating laboratories. The genotyping results obtained by CEPH contributors are pooled in a common database at CEPH, which regularly updates the information and makes it available to the collaborators and to the scientific community as a whole. About a hundred laboratories have contributed data on more than 7000 genetic markers.

### 2.3 Constructing Linkage Maps

Statistical analysis of the genotypes requires extensive calculations, which are carried out using computer programs that can calculate lod scores as a function of the genetic distances between two loci. A genetic distance value for which the lod score is maximal is obtained in this way. One can also do calculations based on several points, and estimate likelihoods as a function of the order of the loci and their distances apart. Since the number of possible orders for $n$ points is $n!/2$, the number of loci that can be used for multipoint calculations remains limited. Multipoint analysis increases odds values for the relative orders of the loci by several orders of magnitude.

The impossibility of considering the totality of the theoretical possibilities in constructing genetic maps with a sufficient number of loci has led to the development of special computer programs. These programs use strategies that exclude the least probable solutions and begin by evaluating a trial order, which may be called into question as a result of subsequent tests. The final orders so obtained are themselves challenged by testing—for example, by inverting a given order for two consecutive markers. The precision and reliability of such maps depends on the number of meioses analyzed and the informativeness of the markers used.

## 3 APPLICATIONS

### 3.1 Mapping Disease Genes

The identification of genes that are responsible for genetic diseases is based on *positional cloning,* a strategy that begins with the gene's localization. When there are no clues to the gene or the chromosomal region in which it is likely to be found (candidate gene, chromosomal aberrations), it is first necessary to delimit the region in question by scanning the totality of the genome using highly informative, regularly spaced markers. When DNA from a sufficient number of families can be collected, the genotypes of the family members for these markers are analyzed. This procedure serves both to exclude large regions of the genome and also to find one or more polymorphic markers that cosegregate with the disease. The goal is to frame the disease locus with markers that are as close to it as possible. A minimum interval may be defined by precise mapping of the closest recombination points on either side of the abnormal gene. If this interval is small enough (1-2% recombination between delimiting markers), the next step is to look for the gene in cloned segments of DNA covering this interval.

Evidence for genetic linkage can arise from other clues. A combination of alleles from two or more closely linked markers in a gamete is called a haplotype. When alleles of these markers have been redistributed through numerous meioses, the population frequency of each haplotype is given by the product of the frequencies of its constituent alleles. But deviation from random distribution of alleles in haplotypes is sometimes observed and is referred to as *allelic association* or *linkage disequilibrium.* A similar association can be observed between a disease allele of a gene and an allele of a polymorphic marker and is a strong indication for close linkage. Such associations help to narrow linkage intervals and to detect linkage of genes with minor effects in complex genetic diseases like diabetes.

### 3.2 Diagnostic Applications

When the locus for a genetic disease has been framed by several polymorphic markers that are very close to it, prenatal diagnosis of the disease may be offered to families at risk. The diagnosis is based on the same statistical analytic procedure as that for genetic mapping. For this procedure, the disease must be genetically homogeneous, however; that is, it cannot be caused by different genes localized in different regions of the genome. On the other hand, it is sometimes possible to diagnose a genetically heterogeneous disorder that is clinically homogeneous if the family's pedigree is large enough for statistically significant linkage analysis.

## 4 PERSPECTIVES

### 4.1 High Resolution Linkage Maps

The genetic maps currently available are sufficiently extensive and are composed of sufficiently informative markers to enable geneticists to map the majority of monofactorial genetic diseases as long as a large enough number of affected families is available for analysis.

Besides these monofactorial disorders, however, there are numerous etiologically complex diseases involving both genetic and environmental factors. Furthermore, the genetic factors in these diseases may themselves be multiple. In these cases, the effect of the gene may be diminished. A very high resolution genetic map will be required to detect these genes with minor effects.

### 4.2 Sperm Genotyping

Meiotic products can be analyzed directly on isolated haploid cells using PCR techniques. Spermatozoans in particular may be obtained in very large numbers from a single donor and isolated by fluorescent cell sorting. A large number of loci are then amplified simultaneously for genetic linkage analysis. About 30 loci may be genotyped from a single meiosis in this way. Furthermore, a large number of meioses from the same parent may be analyzed, to increase the resolution of the maps of small chromosome regions.

### 4.3 Linkage Maps Based on Radiation Hybrids

Instead of using the recombination points of meiosis, another procedure in which chromosomal breaks are introduced by irradiation can be used to construct high resolution genetic maps of defined regions of the human genome. In a typical experiment, a human–rodent somatic cell hybrid, containing a single human chromosome, is X-irradiated and fused to a recipient rodent cell. The irradiation kills the donor cells and fragments their chromosomes. The fragments are incorporated at random into the new hybrids. By testing a panel of hybrids, it is possible to assay for coretention of markers derived from the human chromosome. Chromosome maps are constructed by mathematical manipulation of the coretention frequencies.

*See also* Human Disease Gene Mapping; Recombination, Molecular Biology of; Restriction Landmark Genomic Scanning Method.

### *Bibliography*

Botstein, D., White, R. L., Skolnick, M. H., and Davies, R. N. (1980) Construction of a genetic linkage map in man using restriction fragment length polymorphisms. *Am. J. Hum. Genet.* 32:314–331.

Ott, J. (1986) A short guide to linkage analysis. In *Human Genetic Diseases: a Practical Approach,* K. E. Davies, Ed., pp. 19–32. IRL Press, Oxford and Washington, DC.

———. (1991) *Analysis of Human Genetic Linkage.* Johns Hopkins University Press, Baltimore and London.

White, R., and Lalouel, J. M. (1988) Sets of linked genetic markers for human chromosomes. *Annu. Rev. Genet.* 22:259–279.

# Human Repetitive Elements

*Jerzy Jurka*

### *Key Words*

**Family of Repeats** Set of similar or homologous repetitive elements in the genome.

**Repetitive Elements** (also: repetitive sequences or repeats) Broad variety of DNA fragments present in multiple copies in the genome.

**Retroposon** Interspersed repeated element amplified through the reverse transcription of an RNA intermediate.

**Retrotransposon** Retroposon containing long terminal repeats similar to those in retroviruses.

**Source Gene** (also: master or founder gene) Common ancestor to family members.

**Transposon** Interspersed repeated element capable of relocating from one chromosomal point to another.

The terms *repetitive elements, repetitive (reiterated) sequences,* and *repeats* describe a broad variety of DNA fragments present in multiple copies in the genome. They can be divided into families of similar or sometimes identical elements. Although this definition includes families of genes with known functions, such as rRNA, tRNA, or histone genes, this entry considers only repetitive elements without clearly assigned biological function, which may represent as much as 60% of the human genome. There are different ways to think about the unresolved question of biological function of the repetitive elements. In some cases, biological function may be attributable to the repetitive family as a whole. For example, tandemly repeated sequences such as satellites may contribute to the nuclear and chromosomal structure. This view may not be applicable to repetitive elements interspersed within unique genomic sequences. Most, if not all, interspersed repetitive families are composed of a limited number of active source genes and huge numbers of pseudogenes (''junk'' DNA), derived from the genes mainly through the reverse flow of information or retroposition. The question about biological function of interspersed repeats applies primarily to the active source genes. The pseudogenes, on the other hand, can best be studied from the point of view of their numerous *effects* on the genome—like modulation of transcription and post-transcriptional processing of mRNA. Such effects taken together represent a major evolutionary force reshaping eukaryotic genomes.

# 1 HISTORICAL BACKGROUND

Historically, the first categorization of repetitive DNA into "highly repetitive" and "middle repetitive" was based on renaturation time and concentration of denatured DNA. The classes of highly and middle-repetitive sequences refer to DNA complexity that cannot be translated satisfactorily into classifications of repeats based on cloning and detailed sequence studies. In general, highly repetitive DNA includes primarily the satellites described in Section 2.1, whereas middle-repetitive DNA includes repetitive elements interspersed with single-copy DNA. However, middle-repetitive DNA may contain sequences with a very high copy number such as the *Alu* family described in Section 4.1. On the other hand, some high copy number interspersed repeats are quite old and diverse, and the acquisition of random mutations over time prevents us from clearly distinguishing them from "unique" DNA based on the sequence complexity alone. Therefore, sequence diversity, patterns or organization, mechanisms of amplification, length, and copy number appear to be increasingly used as specific descriptors of repetitive sequences, and they are basis for the terminology summarized here.

# 2 TANDEM ARRAYS OF REPEATS

## 2.1 Satellites

Satellites are made of arrays of tandemly repeated sequences predominantly located in the centromeric chromatin of different chromosomes. They represent up to 30% of the genomic DNA. Originally, satellites were separated from the bulk DNA using centrifugation in density gradients. The major satellite fractions separated in this way were defined as classical satellites I, II, III, and IV. Classical satellite I was found to be dominated by a major simple component called satellite 1. Satellite 1 contains a basic unit of 42 base pairs originally described as 17-mer (A) and 25-mer (B) arranged in an alternating pattern described as -A-B-A-B-. The major simple components of classical satellites II and III are named satellites 2 and 3, both containing 5′-ATTCC as a basic repeat unit.

Among other well-defined components of classical satellites II and III are so-called α-satellites. First discovered in African green monkeys and subsequently in other primates, α-satellites are composed of basic units (monomers) approximately 171 bp in length. The sequences of alphoid monomers differ 20 to 40% from one another. Diverse monomers can form multimeric units, which then repeat again (Figure 1). This hierarchical organization appears to be chromosome specific. Nonalphoid satellites also show chromosome-specific organization.

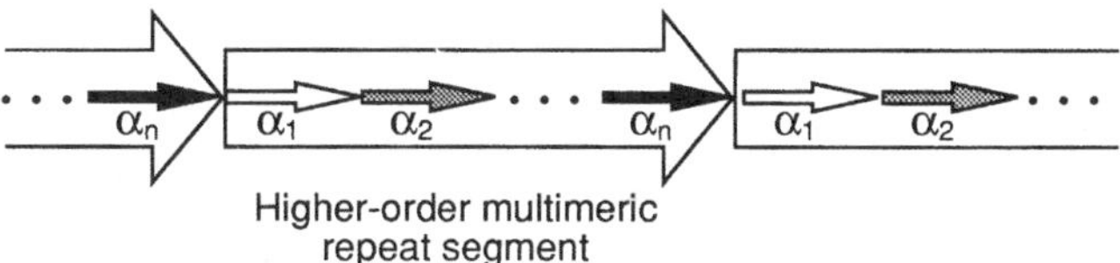


**Figure 1.** A scheme of the hierarchical structure of human α-satellites.

Satellites have been proposed to participate in, for example, chromosome pairing, rearrangements, speciation, and the three-dimensional structure of interphase nucleus, but the evidence for any of these hypotheses remains inconclusive.

## 2.2 Microsatellites and Minisatellites

Microsatellites, also known as simple sequence repeats (SSRs), consist of tandem arrays of mono-, di-, tri-, and tetranucleotides. Unlike satellites, microsatellites are not localized in any particular chromosomal region but are quite randomly interspersed with other genomic DNA. Minisatellites have been defined as arrays of tandemly repeated sequences of at least 10 base pairs interspersed with genomic DNA. They are also known as variable number of tandem repeats (VNTRs). Minisatellites, which are less randomly distributed than microsatellites, show some bias toward chromosome ends. Both micro- and minisatellites are highly variable, hence are useful as polymorphic markers, with a variety of applications from population genetics and genetic mapping of disease loci to forensic studies.

# 3 TELOMERIC AND SUBTELOMERIC REPEATS

The physical ends of human chromosomes, or telomeres, are composed of short tandem repeats up to 30 kb in length that can be described by a general formula $(TTAGGG)_n$. Telomeric repeats are maintained by a telomere-specific reverse transcriptase, called telomerase. Telomeres protect chromosome ends from degradation, fusion, and recombination. In contrast to the highly variable satellites, the TTAGGG motif is conserved throughout vertebrate evolution. Sequences next to telomeric repeats are called subtelomeric repeats. Subtelomeric repeats contain motifs similar to TTAGGG, but, unlike telomeric repeats, they are not conserved in vertebrates.

# 4 RETROPOSONS

## 4.1 Short Interspersed Repeated Elements

Short INterspersed Elements (or SINEs) were originally defined as repeated sequences less than 500 bp long and present in a high copy number (of the order $10^5$ per genome). At least two families match this definition: an old and little-known family of mammalian-wide interspersed repeats (or MIR family), present in at least $10^5$ copies in the human genome, and the thoroughly studied human *Alu* family discussed here.

The human *Alu* family contains around 650,000 retroposed elements (~6% of the total genome), a figure representing a mean of the most extreme estimates. *Alu* repetitive elements are thought to be mostly pseudogenes obtained via the reverse flow of information from a limited number of transcribed source genes. The indirect evidence for this view comes from the fact that *Alu* elements are flanked by short repeats up to 20 bp long and contain 3′ stretches of adenines, called "poly(A) tails," characteristic for retroposed pseudogenes. Furthermore, with the exception of source genes, all *Alu* sequences show rapid loss of CpG (CG) doublets by mutations at the rate about 10 times higher than the neutral rate of mutations. This is also characteristic for CpG-containing pseudogenes.

*Alu* repeats retroposed from different founder genes form subfamilies, which are described by different consensus sequences.

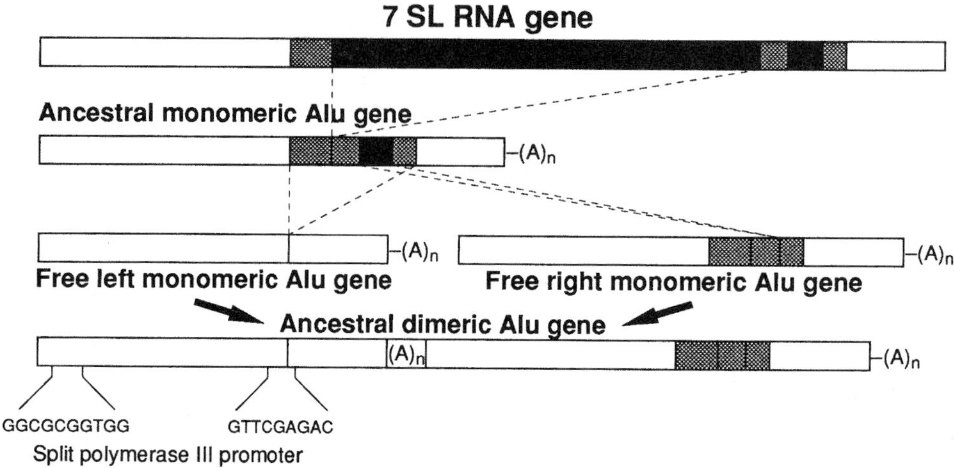


**Figure 2.** A scheme for the origin and early evolution of human *Alu* source genes.

There are two major *Alu* subfamilies known as *Alu*-J and *Alu*-S, which can be split into several sub-subfamilies differing in age and derived predominantly from genes retroposed in the past. However, retroposition of *Alu* elements continues in contemporary human populations. A number of very recently retroposed *Alu* sequences have been identified, primarily in patients with genetic disorders. These sequences were copied from at least two different founder genes.

A typical human *Alu* source gene, as reconstructed from the subfamily consensus sequences, is up to 350 bp long and is composed of two similar but nonidentical monomers originally derived from the 7SL RNA gene as an ancestral molecule (Figure 2). 7SL RNA is normally associated with proteins in a ribonucleoprotein particle that participates in the contranslational secretion of membrane and secretory proteins. *Alu* genes lost some of the 7SL-specific domains and acquired a number of new specific mutations. According to current models, a common ancestor for all human dimeric *Alu* genes was derived from a fusion of independent monomeric genes, of which some were retroposed in the past. At least one monomeric gene, *BC200,* is preserved by natural selection and transcribed in a tissue-specific manner in the human brain, but it is not clear whether it is retropositionally active.

The critical step in retroposition is transcription of *Alu* RNA. Most *Alu* repeats have active internal polymerase III promoters and can be transcribed in vitro by polymerase III. However, very few *Alu* sequences are transcribed in vivo, which indicates that the internal promoter alone is insufficient for *Alu* transcription. Reverse transcription and subsequent integration of *Alu* sequences into the genome are little understood, except that for Alu to be inherited, both steps must occur in the germ line cells.

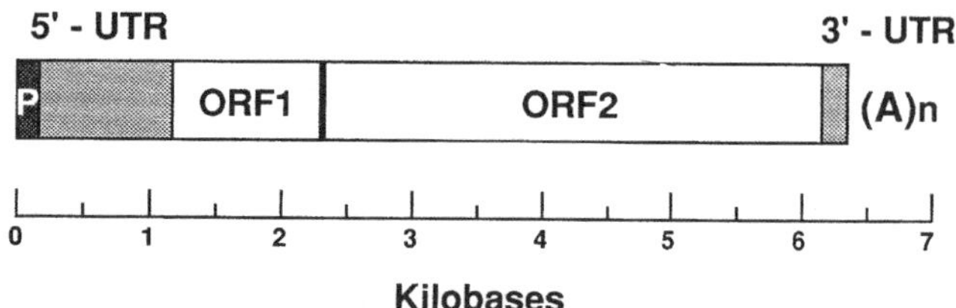


**Figure 3.** A composite scheme of the *L1* source gene.

### 4.2 Long Interspersed Repeated Elements

LINEs (Long INterspersed Elements) were originally defined to include repetitive elements of at least 5 kb, present in at least $10^4$ copies per haploid genome. Only one human LINE family, known as the *LINE-1 (L1), KpnI* or *Kpn* family, has been reported. It contains up to $10^5$ sequences, most of which are 5′-truncated. The most complete *L1* sequences are 6 to 7 kb long. Current models of *L1* transposition postulate formation of full-length RNA transcripts from a limited number of *L1* active source genes under control of the internal polymerase II promoters, followed by the reverse transcription and integration at different chromosomal loci. One of the *L1* source genes has been identified. A composite scheme of a complete *L1* source gene (Figure 3) includes a 5′-untranslated region followed by two open reading frames (ORF1 and ORF2), a 3′-untranslated region, and a poly(A) tail. The 5′-untranslated region contains an RNA polymerase II promoter within the first 155 bp marked by letter "P" in Figure 3, which is distal to the presumed transcriptional start site. This unusual distal location leaves an open possibility of an additional, undiscovered upstream promoter. ORF1 and ORF2 encode polypeptides about 375 and 1300 amino acids long, respectively. ORF2 was shown to be homologous to retroviral and nonretroviral reverse transcriptases. However, unlike retroviruses and viral retrotransposons discussed in Section 6, *L1* lacks long terminal repeats (LTRs).

Like the *Alu* family, the *L1* family includes a number of discrete subfamilies differing in age that have been retroposed from different source genes. The evolutionary history of the *L1* family goes very far back in time, since it is present in all mammals studied. Repetitive elements structurally similar to mammalian *L1* have also been found in insects, plants, and lower eukaryotes.

## 5 MEDIUM AND LOW REITERATION FREQUENCY REPEATS

MERs (MEdium Reiteration frequency repeats) include a variety of repetitive families ranging from hundreds to thousands of elements per haploid genome. To date, around 40 MER families have been reported. Most MER families include elements comparable in size to SINE elements. No internal polymerase promoters have yet been identified in MER elements. Most of them also lack the poly(A) tails and flanking repeats characteristic of retroposons. The origin and mechanism of proliferation of these elements is rather unclear, although, suggestively enough, some MER elements have been identified in defective DNA viruses. A few other MERs represent 3′-ends of old L1 repeats, or homologues of the Transposonlike Human Element 1 (THE1) described in Section 6. The majority, however, are likely to represent independent families. Typically, MER families are relatively old, although they contain subfamilies of different ages. MERs have no detectable open reading frames.

In addition to MER repeats, the human genome contains an undetermined number of repetitive families of 100 or fewer elements. They include processed retropseudogenes of protein-coding mRNAs as well as noncoding DNA of unknown origin.

## 6 ENDOGENOUS RETROVIRUSES AND VIRAL RETROTRANSPOSONS

The human genome contains a number of endogenous proviruses (i.e., DNA intermediates for retroviruses). Typically, they are flanked at both ends by long terminal repeats, and they are often incomplete. Endogenous proviruses are transmitted vertically as Mendelian elements from parents to offspring. They may be one of the potential sources of the reverse transcriptase involved in the retroposition of nonviral RNA. At least one repetitive element, *THE1*, is similar to transposons and mammalian proviruses. *THE1* is about 2.3 kb long and contains long terminal repeats approximately 350 bp in length. It is present in up to 40,000 copies per haploid genome. The *THE1* family appears to be distantly related to the MER10 (*Mst* II) and MER15/MER18 families of repeats.

*See also* Transposons in the Human Genome.

### *Bibliography*


Berg, D. E., and Howe, M. M., Eds. (1989) *Mobile DNA*. American Society for Microbiology, Washington, DC.

Britten, R. J., and Davidson, E. H. (1971) Repetitive and non-repetitive DNA sequences and a speculation on the origins of evolutionary novelty. *Q. Rev. Biol.* 46:111–137.

Brosius, J., and Gould, S. J. (1992) On "genomenclature": A comprehensive (and respectful) taxonomy for pseudogenes and other "junk DNA." *Proc. Natl. Acad. Sci. U.S.A.* 89:10706–10710.

Dombroski, B. A., Mathias, S. L., Nanthakumar, E., Scott, A. F., and Kazazian, H. H., Jr. (1991) Isolation of an active human transposable element. *Science,* 254:1805–1808.

Jurka, J., Walichiewicz, J., and Milosavljevic, A., (1992) Prototypic sequences for human repetitive DNA. *J. Mol. Evol.* 35:286–291.

Korneev, S. A. (1988) Repeat sequences of the human genome. *Genetika,* 24:965–979.

Rogers, J. H. (1985) The origin and evolution of retroposons. *Int. Rev. Cytol.* 93:187–279.

Schmid, C., and Maraia, R. (1992) Transcriptional regulation and transpositional selection of active SINE sequences. *Curr. Opinion Genet. Dev.* 2:874–882.

Shapiro, J. A., Ed. (1983) *Mobile Genetic Elements.* Academic Press, Orlando, FL.

Singer, M. F. (1982) Highly repeated sequences in mammalian genomes. *Int. Rev. Cytol.* 76:67–112.

Weiner, A. M., Deininger, P. L., and Efstratiadis, A. (1986) Nonviral retroposons: Genes, pseudogenes, and transposable elements generated by the reverse flow of genetic information. *Annu. Rev. Biochem.* 55:631–661.

Willard, H. F., and Waye, J. S. (1987) Hierarchical order in chromosome-specific human alpha satellite DNA. *Trends Genet.* 3:192–198.

Zakian, V. A. (1989) Structure and function of telomeres. *Ann. Rev. Genet.* 23:579–604.


# Hybridization for Sequencing of DNA

*William Bains*

### *Key Words*

**Ambiguity** The number of different sequences that could have given rise to a set of hybridization results.

**Sequencing by Hybridization** Using the hybridization of many oligonucleotides to a DNA molecule to determine the base sequence of that DNA.

**Sequencing Chip** A large panel of oligonucleotides, usually a complete set, immobilized onto a surface.

Sequencing by hybridization is the use of the hybridization of large numbers of oligonucleotides to a target DNA to discover the target's base sequence. This gel-free approach to sequencing uses no enzymes or complex chemistry, and so is attractive as a sequencing method suitable for automation. Determining DNA sequence by hybridization alone, however, requires greater control of the hybridization reaction than is presently possible. As well as being a potential method in its own right, hybridization is under development as a complement to conventional sequencing. It can be used as a means of checking conventional sequencing, hence reducing the redundancy (and so increasing the speed) of large-scale, automated sequencing projects, and for comparing newly cloned genes to similar genes that have already been sequenced. In both cases, existing hybridization chemistry can generate large amounts of new sequence data. The large arrays of oligonucleotides can be made as reagents, or synthesized in parallel, immobilized on a "sequencing chip."

## 1 INTRODUCTION

When a short DNA sequence hybridizes to a longer target DNA, we can deduce that a sequence complementary to the probe DNA exists in the target. Usually we use this information to generate specific labeling systems for DNA. However it is also possible to use this knowledge as a way of determining the sequence of the target.

The sequence of a target DNA can be deduced by hybridization data if the hybridization yields a complete list of all the oligonucleotides of a given length that can be found in the target. We can then look for overlaps between the oligonucleotides, and build up the target DNA's sequence as illustrated in Figure 1. This approach to sequencing does not rely on gels or enzymatic synthesis to confer specificity, but rather on the hybridization chemistry. It generates a different sort of data from conventional sequencing, and has the potential for being much easier to automate.

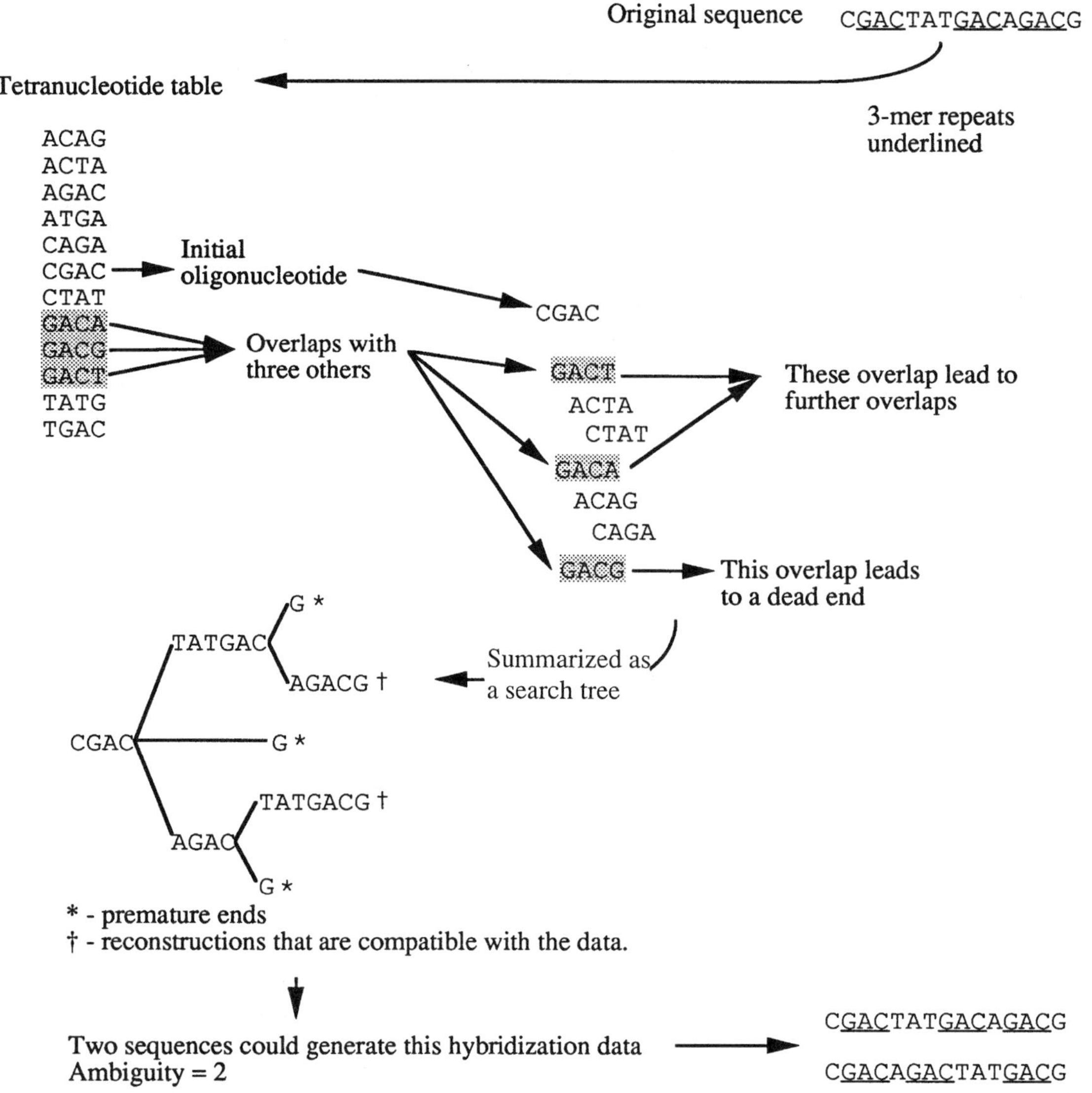


**Figure 1.** Schematic for method of sequencing by hybridization. The *original sequence* is hybridized to tetramers, generating a *table* of all the tetramers that occur in that sequence. From this we search for *overlaps,* constructing a *search tree.* In this case, two reconstructions are found to be compatible with the table: the *ambiguity* is 2. (Note that tetramer "hybridization" is used only to make a simple example; tetramer hybridization is not a realistic practical option.)

## 2 THEORETICAL LIMITATIONS

Theoretical studies assume that hybridization is ideal, with the result that an oligonucleotide hybridizes only to its perfect complement, and does so quantitatively. Even with these restrictions, as Figure 1 illustrates, in some cases multiple possible sequences are compatible with the hybridization data. This is the ambiguity or branch point problem: sequences can be reconstructed to short segments of sequence (underscored in the reconstructions at the bottom of Figure 1), but the order of these subsequences cannot be determined. Their boundaries are repeats whose length is one base less than the length of the probe.

The ambiguity of reconstruction of a sequence depends on the nature of the probes we hybridize to it, and largely on its overall length. Since, as Figure 1 illustrates, repeats of any sequence one base shorter than the probes can cause ambiguity, the longer the probe, the lower the chances of repeats, hence the lower the ambiguity. Hybridizing to a complete set of 65,536 octamers can, in principle, sequence around 170 bases of target DNA; hybridization to 65,536 "mixed" 11-mers can determine the sequence of 700 base targets.

Ambiguity can also be reduced if we include other information in the reconstruction that helps us to order the subsequences. Thus the assumption that the target DNA contains an open reading frame triples the amount of sequence information we can obtain from a hybridization. In addition, hybridization data from DNA fragments that overlap can be used to construct a single, unique sequence.

## 3 PRACTICAL IMPLEMENTATIONS

So massive a hybridization experiment may be imagined in two ways:

Probe Up (format I) experiments immobilize the target DNA and probe it with separate labeled oligonucleotide probes.

Probe Down (format II) immobilizes a large array of oligonucleotides and probes it one at a time with target sequences.

Probe Up, a natural extension of using oligonucleotides to fingerprint clone libraries, generates useful clone mapping data before enough information has been collected to generate a complete sequence. However the number of separate hybridization reactions needed before even one of the clone banks being analyzed yields complete sequence information is very large. Probe Down is an extension of existing DNA diagnostic formats, where polymerase chain reaction (PCR) product is hybridized to an array of probes immobilized on nitrocellulose. It is suited to sequencing short DNAs, such as the PCR-amplified inserts from cDNA clones, and for checking sequences. A single hybridization to the oligonucleotide panel yields the most complete data possible: however the panel must be synthesized first.

A key constraint on the practical realization of either process is the number of probes needed. This is a tradeoff between the number requirement (i.e., to have as many probes as possible) and the problems of making the probes. Successful strategies for Probe Down arrays have built up oligonucleotides in parallel on a surface by synthesizing them in situ. Because of the similarity of both this synthesis strategy and its product to high density integrated circuits, such oligonucleotide arrays are sometimes called ''sequencing chips.''

## 4 EXPERIMENTAL RESULTS

The results so far with this method have been exploratory. Probe Up sequencing has been used to check the sequence of a 100-mer synthetic DNA and to generate sequence from a 150-mer by comparison of closely related, unknown sequences. This shows that Probe Up is feasible, but more work needs to be done to make the hybridization sufficiently robust to be used for sequencing on its own. Probe Down experiments have also used synthetic target DNAs in experiments aimed partly at characterizing the idiosyncrasies of hybridization and partly at characterizing the synthetic chemistry for such arrays.

## 5 REALISTIC HYBRIDIZATION

The differences between the stability of different duplexes is an important limitation on how we could use them to sequence even if characterization has been complete. Short DNA duplexes containing different fractions of A+T will have different melting temperatures from each other. This is because a substantial amount of the binding energy of two single strands is due to the formation of hydrogen bonds between the bases, and G:C base pairs are linked by three hydrogen bonds to the A:T pairs' two. These differences can be reduced by hybridizing in solutions of tetraalkylammonium salts or the amino acid analogue betaine. Since, however, the differences are not eliminated, it will be difficult to use a sequencing chip when all the immobilized oligonucleotides show optimal discrimination. Other ''irregularities'' include stacking interactions between adjacent bases (which are not hydrogen bond dependent) and the formation of internal secondary structure in some probes and targets (which is mediated by base:base hydrogen bonding). If we do not exactly compensate for these effects, the hybridization will appear to generate spurious signals—''errors.''

Stacking interactions have been used imaginatively to generate more data. Short oligonucleotides that cannot hybridize to target DNA on their own can ''stack'' against the end of a longer oligonucleotide:target duplex. This provides a method for extending the duplex by a further 4 or 5 bases, converting an 8-mer ''chip'' into a potential 13-mer ''chip.''

Errors drastically reduce the amount of sequence information that a hybridization yields: a 1% error rate reduces the maximum sequenceable length by at least 10%. Thus if only 1% of 65,536 oligonucleotides gave false positive hybridization signals when hybridizing to a 200-mer DNA target, 75% of the scored ''hybridizations'' would be false, rendering sequence reconstruction impossible. Hybridization must be extremely effective to yield sequence information on its own.

## 6 SEQUENCE CHECKING AND COMPARISON

The problems of errors caused by nonideality of hybridization as well as by true errors in experiments are less important when the hybridization data are used to check a sequence generated by conventional means, or to compare an unknown sequence with a known one. The former could be very useful for reducing the amount of redundant sequencing that must be done in automated sequencing projects to avoid systematic errors in the enzymology of dideoxy sequencing. The latter can be used to sequence genes that are similar to ones already sequenced (e.g., from the same gene family) or to sequence homologous genes from different species. At the moment this type of sequencing is of limited use (except for evolutionary studies), but will become increasingly important in the future.

The errors in hybridization are also less important when the technique is being used to obtain partial sequence information from the sequence (e.g., A+T ratio, coding region potential, repetitive sequence distribution). This information itself can be of value in characterizing clones before more laborious complete sequencing is undertaken. Since, however, current computer methods for processing sequence data cannot handle such partial data, these applications will be limited until a more general definition of ''sequence'' is accepted.

## 7 PROSPECTS

If hybridization is essentially perfect, it is possible in principle, using an array of 65,536 11-mers, to ''sequence'' 700 bases of random DNA or around 2400 bases of DNA that contains an open reading frame. This could be a useful application of the technology, but hybridization of the accuracy needed is not within sight. More immediate applications will be in sequence checking and comparison. Because hybridization generates different types of ''error'' from conventional sequencing, comparing the two is effective at identifying the errors in both, hence reducing the need for highly redundant sequencing in large-scale, semiautomated sequencing projects. This effect will greatly increase the overall speed of large-scale sequencing, by removing the need for multiple resequencing of the same DNA, and by extending the amount of sequence that can be read off a ''conventional'' sequencing gel. Using hybridization information to compare sequences will become increasingly important as the amount of sequence information in public databases increases, and will allow the very rapid determination, by hybridization alone, of the sequence of genes that are similar to known sequences. While issues of fundamental understanding need to be resolved before it is possible to use hybridization alone to sequence entirely uncharacterized DNA, these strategies for using hybridization data to confirm or compare sequence can be implemented now and are consequently the subject of extensive research.

*See also* HUMAN CHROMOSOMES, PHYSICAL MAPS OF; HYDROGEN BONDING IN BIOLOGICAL STRUCTURES; NUCLEIC ACID SEQUENCING TECHNIQUES.

### *Bibliography*


Bains, W. (1991) Hybridization methods for DNA sequencing. *Genomics,* 11:294–301.

Cantor, C. R., Mirzabekov, A., and Southern, E. (1992) Report on the sequencing by hybridization workshop. *Genomics,* 13:1378–1383.

Drmanac, R., Drmanac, S., Jarvis, J., and Labat, I. (1994) Sequencing by hybridization. In *Automatied DNA Sequencing and Analysis Techniques,* J. Craig Ventor, Ed. Academic Press, London.


# HYDROGEN BONDING IN BIOLOGICAL STRUCTURES

*George A. Jeffrey*

### *Key Words*

**Chelated Hydrogen Bond** Bond between two donor hydrogens and one acceptor; also known as a bifurcated bond.

**Hydrogen Bond** Secondary chemical bond between a *donor* and an *acceptor.*

**Hydrogen Bond Acceptor** Electronegative atom with one or more nonbonded electron pairs.

**Hydrogen Bond Cooperativity** The result of an arrangement of hydrogen bonds whereby the energy of the total system is greater than the sum of the individual hydrogen bond energies; also known as the *nonadditivity* property of hydrogen bonds.

**Hydrogen Bond Donor** Hydrogen atom covalently bonded to a more electronegative atom, X—H.

**Polarization** A change in the electronic distribution due to the formation of a hydrogen bond.

**Three-Center Hydrogen Bond** Bond between a donor and two acceptors; also known as a bifurcated bond.

**Two-Center Hydrogen Bond** Bond between a donor and one acceptor.

The components of living systems, proteins, nucleic acids, carbohydrates, and water assemble and disassemble by making and breaking hydrogen bonds. Hydrogen bonds are therefore essential to life processes. This entry describes the principles of hydrogen bond formation and cooperativity and presents some commonly observed hydrogen bonding patterns.

## 1 PRINCIPLES

### 1.1 THE STRENGTH OF HYDROGEN BONDS

It is useful to distinguish *strong, moderate,* and *weak* hydrogen bonds. Although there is no clear demarcation among these categories, they have recognizably different properties, shown in Table 1.

An important property of the hydrogen bonds in biological structures is that compared with the covalent and ionic bonds, they are weak interactions. This weakness permits preferential bonding, which does not impose rigidity. Biological hydrogen bonds can be switched on and off with energies of the same order of magnitude as that of thermal motion at physiological temperatures, thereby permitting the fast intermolecular recognitions and interactions that occur in biological processes.

The donor and acceptor groups that form hydrogen bonds in the three main classes of biological molecules are shown in Table 2.

**Table 1** Distinction Between Strong, Moderate, and Weak Hydrogen Bonds

| Property | Strong Bonds | Moderate Bonds | Weak Bonds |
|---|---|---|---|
| Occurrence | Inorganic structures | Organic and biological structures | |
| Types of bond | _F—H- - -F⁻<br>_O—H- - -O−<br>$\overset{+}{\text{O}}$—H- - -O | _O—H- - -O<br>$\overset{+}{\text{N}}$—H- - -O<br>_N—H- - -O<br>_N—H- - -N | As for moderate bonds, plus<br>C—H- - -O |
| Bond lengths<br>_H- - -A | Narrow range<br>_1.2–1.5<br>_H- - -A ≈ X—H | Broad range<br>_1.5–2.0<br>_H- - -A X—H | Very broad range<br>_2.0–3.2<br>H- - -A ≫ X—H |
| Bond angles<br>_X—H- - -A | Strongly directional<br>_≈ 180° | Weakly directional<br>_150 30° | Not directional<br>_150–90° |
| Bond energy | 50–100 kJ/mol | 50–20 kJ/mol | *20 kJ/mol* |
| IR vibration frequency | *1600 $cm^{-1}$* | _2000–3000 $cm^{-1}$ | *2000 $cm^{-1}$* |
| $^1$H chemical shift | Large downfield: 17 ppm | *17 ppm* | |

**Table 2** Hydrogen Bonding Groups in Biological Structures

| Type of Molecule | Donors | Acceptors |
|---|---|---|
| Amino acids, peptides, and proteins | $-\overset{+}{N}(H_2)-H$ , $>\overset{+}{N}(H)-H$ | $O{=}C(-O^-)<$ , $O{=}C(-OH)<$ , |
| | $\gg\overset{+}{N}-H$ , $-N(H)-H$ , | $O{=}C<$ , $O(H)(C-)$ , |
| | $>N-H$ , $-O-H$ , $-S-H$ , | $-N\ll$ , $O_W(H_2)$ |
| | $H-O_W-H$ , $-C(=O)OH$ | |
| Nucleosides, nucleotides, oligonucleotides, and nucleic acids | $-N(H)-N$ , $\gg N-H$ , | $-N\ll$ , $O{=}C<$ , $O{=}\bar{P}{=}$ , |
| | $\gg\overset{+}{N}-H$ , $P-OH$ , | $O{=}P-$ , $O(C-)(H)$ , $O(C-)(C-)$ , |
| | $-C-O-H$ , $H-O_W-H$ | $O(P-)(H)$ , $S{=}C<$ , $O_W(H_2)$ |
| Carbohydrates, oligo- and polysaccharides | $-C-O-H$ , $-C(=O)OH$ , | $O(C-)(H)$ , $O(C-)(C-)$ , $O_W(H_2)$ , |
| | $H-O_W-H$ | $O{=}C<$ , $O{=}C<$ |

## 1.2 The Electrostatic Origin of Hydrogen Bonds

A moderate or weak hydrogen bond is primarily the consequence of a dipole–dipole interaction. In a covalently bonded hydrogen (or deuterium) atom, the electron density associated with the hydrogen atom participates in the formation of the covalent X—H bond. In consequence, the proton (or deuteron) is unshielded in the direction forward from the covalent bond, relative to the proton in an isolated hydrogen atom. This gives rise to a dipole at the hydrogen end of the $X-\overset{-+}{H}$ bond. The electronegativity of the acceptor atom and its lone pair electron density result in an opposite dipole at the acceptor atom, $\overset{-+}{Y}$. The attraction between these two dipoles is a maximum when X—H- - -Y = 180°, attenuating to zero when the angle is 90°. This electrostatic attraction energy is long-range attenuating as $r^{-3}$, compared with that of van der Waals forces, which attenuate as $r^{-6}$.

## 1.3 Hydrogen Bond Configurations

Since the interaction is primarily electrostatic, a donor may form a hydrogen bond with one or more acceptors. In a *two-center bond,* the hydrogen is bonded to two atoms, one covalently, one by a hydrogen bond. In a *three-center bond,* the hydrogen is bonded to three atoms, one covalently and two by hydrogen bonds. Since these are all attractive forces, the hydrogen should lie close to the plane of the three atoms to which it is bonded: $\alpha + \alpha' + \theta \approx 360°$ in **I**.

In the crystal structures of small biological molecules, except for the zwitterionic amino acids, the proportion of three-center bonds is 20 to 25%. In the zwitterionic amino acids, it is about 75%. Four-center bonds, in which the hydrogen forms three hydrogen bonds, are observed in less than 4% of crystal structures of biological molecules. Chelated hydrogen bonds are formed by one donor and two acceptors, as in **II, III,** or **IV.** Except for **IV,** these are rarely observed in crystal structures.

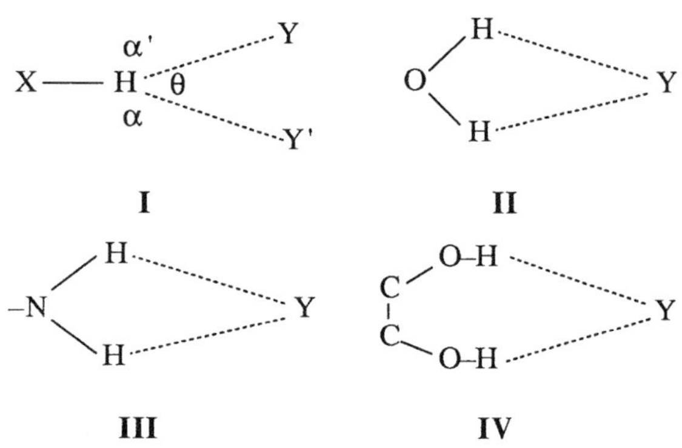

### 1.4 Hydrogen Bond Cooperativity

Hydrogen bonds in crystals form sequential patterns in which the total energy of the pattern is enhanced by *polarization,* that is,

$$E_{n(H\cdots X)} > nE_{H\cdots X}$$

#### 1.4.1 σ-Bond Cooperativity

The O—H group is unique in that it can function as both donor and acceptor to form chains, as in **V,** or homodromic cycles, as in **VI.** This results in a polarization that increases the hydrogen bond energy over that of the sum of individual bond energies.

**V**

**VI**

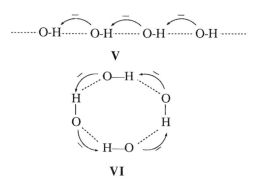

#### 1.4.2 π-Bond Cooperativity

The N—H, O=C, N groups do not have dual donor–acceptor functionality. However, they are covalently bonded to polarizable π-bond systems, as in **VII.** As a result, a system involving a sequence of C⚌N bonds, in proteins or nucleic acids, has enhanced resonance energy as a result of hydrogen bond formation. This effect is also known as *resonance-enhanced hydrogen bonding.*

**VII**

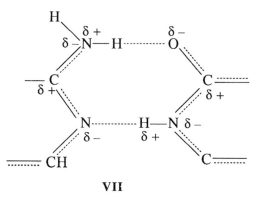

## 2 COMMON HYDROGEN BOND PATTERNS

### 2.1 In Crystal Structures of Proteins

In the *crystal structures of the proteins,* the most common patterns are the α-helix and the β-sheets. The α-helix is formed by twisting the polypeptide chain into a helix so that the donor N—H is hydrogen-bonded to the C=O on the fourth peptide residue along the chain, while keeping the conformation about the peptide C⚌N bond approximately planar, as shown diagramatically in **VIII.** This is also described as a $3.6_{13}$ helix, since there are 3.6 residues per turn, forming a cycle of 13 bonds, including the hydrogen bond.

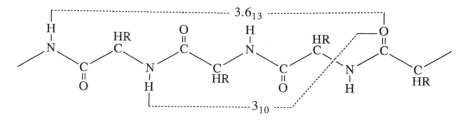

**VIII**

Other helices are possible between the tightest $2_7$ and the widest $4.4_{16}$, but only the α-helix and the $3_{10}$ are observed in globular proteins. The α-helices often extend over long stretches, while the $3_{10}$ occurs only in short segments at the carboxyl termini or at reversals of chain directions.

Other patterns in proteins are the parallel and antiparallel β-sheets **IX** and **X,** of which the latter is the more common. A folded version of the β-sheet is the β-barrel.

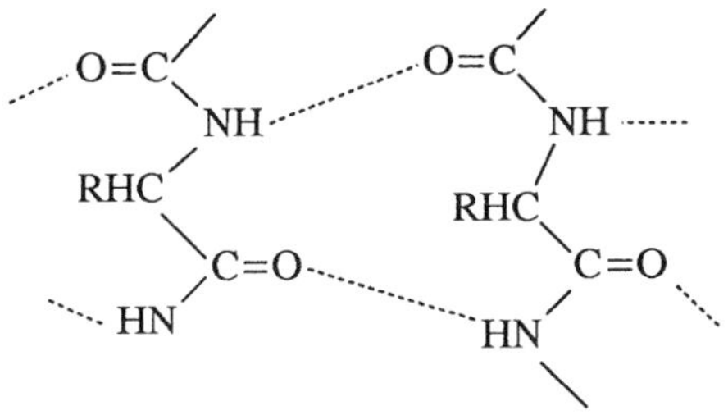

**IX**

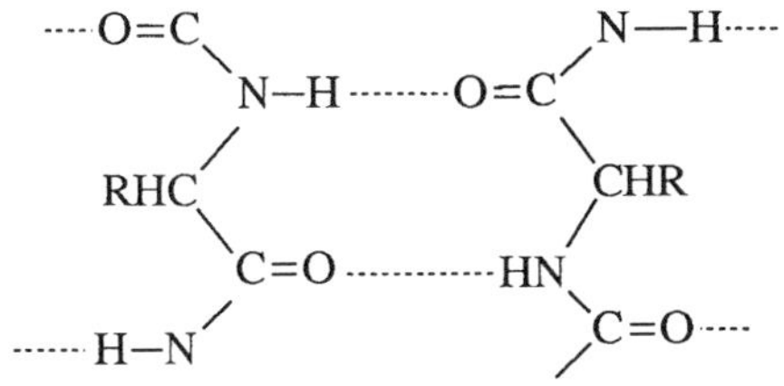

**X**

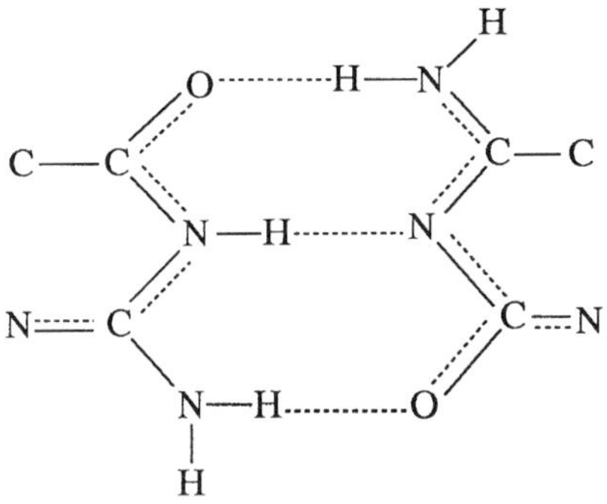

**XI**

## 2.2 In Crystal Structures of Nucleic Acids

In the *nucleic acids,* the most common and famous pattern is the Watson–Crick base pairing between the purine and pyrimidine residues, adenine to uracil, A-U, and guanine to cytosine, G-C, as shown in **VII** and **XI.** Many other base-pairing arrangements are possible, but rarely are such alternates observed.

## 2.3 In Crystal Structures of Carbohydrates

In the *crystal structures of carbohydrates,* the predominant theme is the formation of chains and cycles of hydrogen bonds, which utilize the donor–acceptor property of the —OH groups. Ring or linkage ether oxygens are often incorporated either at the end of finite chains or as the weak components of three-center bonds. Monosaccharides generally form finite or infinite chains, but oligosaccharides, including the cyclodextrins, are generally hydrated, and the hydrogen bonding patterns are irregular networks that include chains and cyclic sequences.

## 2.4 The Water Molecule

The *water molecule* is ubiquitous in biological structures. It is the hydrogen bonding molecule par excellence, being an oxygen atom with dual donor–acceptor functionality. This small molecule can occupy spaces in the packing of irregularly shaped large molecules and enhance the packing energy by its hydrogen bonding capability and versatility. In hydrated crystal structures, water molecules generally donate two hydrogen bonds but may accept either one or two. When they are three-coordinated, the geometry can be planar or pyramidal. However, examples are known of coordination as low as two and as large as seven. Water molecules hydrogen-bond with the functional groups of biological molecules to link the chains into extended networks. The polarizability of these networks provides a mechanism for long-range recognition between biological molecules in an aqueous medium.

*See also* Hybridization for Sequencing of DNA; Sequence Alignment of Proteins and Nucleic Acids.

## *Bibliography*


Hamilton, W. C., and Ibers, J. A. (1968) *Hydrogen Bonding in Solids.* Benjamin, New York.

Jeffrey, G. A., and Saenger, W. (1991) *Hydrogen Bonding in Biological Structures.* Springer-Verlag, Berlin.

Pauling, L. (1939) The hydrogen bond. In *The Nature of the Chemical Bond,* Chapter 12. Cornell University Press, Ithaca, NY.

Pimental, G. C., and McClellan, A. L. (1960) *The Hydrogen Bond.* Freeman, San Francisco.

Schuster, P., Zundel, G., and Sandorfy, C., Eds. (1976) *The Hydrogen Bond. Recent Developments in Theory and Experiment,* Vols. I–III. North Holland, Amsterdam.

Watson, J. D. (1965) The importance of weak chemical interactions. In *Molecular Biology of the Gene,* Chapter 4, pp. 102–140. Benjamin, New York.

# I

## Immunology

*Jiri Novotny and Anthony Nicholls*

### *Key Words*

**Antigenicity** The capability of a molecule to elicit immune response; more specifically, the ability to engage an antibody or T-cell receptor in a noncovalent complex.

**Hapten** A small molecular weight compound (e.g., 2,4-dinitrophenol) covalently attached to a protein carrier. Immunization with a hapten–protein conjugate results in production of hapten-specific antibodies.

**-Topes and -Types: Allotype, Agretope, Clonotype, Epitope, Idiotype, Isotype, Paratope** Various terms are current in experimental immunology to describe immune (serological) phenomena and their assumed molecular basis. Antigenic **epitope** is an operational definition of a surface area, on an antigenic molecule, either directly responsible for, or intimately involved in, specific antibody interaction (synonymous with an ''antigenic determinant''). A **paratope** is the molecular surface of an antibody-combining site, complementary to an antigenic epitope. An **agretope** is that part of an antigenic peptide that mediates its binding to (''presentation by'') major histocompatibility complex molecule. Antigenic determinants that distinguish allelic genes are called **allotypes.** Determinants that distinguish different types of immunoglobulin constant regions are called **isotypes;** those characterizing different immunoglobulin variable regions are called **idiotypes.** A **clonotype** is an antigenic determinant characteristic of a single immune cell clone (i.e., a single variable region such as $V_L$, $V_H$ in antibodies or $V_\alpha$, $V_\beta$ in T-cell receptors, or a combination of those).

The subject of immunology entails understanding immune phenomena in vertebrates, such as (1) production of antibodies specific for individual antigens, (2) mechanisms of cellular immunity enabling specialized cells to recognize and destroy target cells, and (3) definition of somatic ''self''—that is, a capability of the immune system (the antibody and T-cell receptor repertoire) to recognize any chemical structure as either belonging to the body (self) or being foreign. Molecular immunology seeks to explain the foregoing phenomena from the organization and sequences of genes coding for immunoglobulins and related immune molecules, and from three-dimensional structures of these molecules (antibodies, T-cell receptors, antigens). Scientific problems investigated by molecular immunology, which include (i) generation of nucleotide (and, implicitly, amino acid) sequence variability, (ii) structure–function relationships in protein molecules, and (iii) atomic origins of affinity and specificity in protein–ligand complexes, are also central to the rest of molecular biology.

Molecular immunology has been a paradigm of research on structure and stability of biological macromolecules and molecular origins of biological specificity. Antibody engineering is now the most advanced form of protein engineering; and immunological methods, based on the exquisite specificity of antibody–antigen interactions, pervade all modern biology and biochemistry.

## 1 Immune Phenomena and Their Molecular Basis

In vertebrates, objects (cells, molecules, and fragments thereof) recognized as nonself (''antigens'') become targets of immune reactions and are destroyed by processes such as complement fixation, cell lysis, and proteolysis. In a broad sense, all the molecules taking part in these processes belong to molecular immunology: (1) the molecules mediating specific recognition of foreign substances (antibodies, T-cell receptors), (2) antigen-presenting (major histocompatibility complex, MHC) molecules, (3) cell adhesion and (4) other cell surface (''cluster of differentiation,'' CD) molecules, (5) components of complement, and (6) cellular hormones (lymphokines) mediating cell–cell interactions. In a narrower sense, molecular immunology concentrates on antibodies and T-cell receptors and strives to explain the most essential immune phenomena from molecular attributes to these proteins.

The most essential immune phenomena are an exquisite *specificity* of the individual recognition molecules for one of the many possible foreign antigenic structures (the antigenic repertoire); the size, molecular nature, and genetic basis of amino acid sequence *diversity* associated with the immune repertoire; and the ability of antigenic molecules or complexes to elicit specific antigenic response (i.e., their *antigenicity*).

*Specificity* of antibody–antigen association is described here as a special case of that found in all macromolecular interactions (see Section 5). The most important determinants of such specific interactions are the physical characteristics of the two interacting molecular surfaces, such as their shape, complementarity, electrostatic charge, and polarity. Immune repertoire *diversity* (see Section 4) arises in part from gene segment splicing operations that act in and among clusters of immune genes, and in part from targeted somatic mutagenesis of rearranged gene segments in committed cells. Both these genetic phenomena are probably results of actions of specialized endo/exonucleases that remain poorly characterized. Molecular origins of *humoral antigenicity* (involving antibodies, see Section 6.1) and *cellular antigenicity* (involving T-cell receptors, see Section 6.2) are different, reflecting the different pathways through which humoral antigens (typically intact proteins) and cellular antigens (peptides derived from proteins by enzymatic hydrolysis) reach their specific recognition targets. The most conspicuous properties common to all humoral antigenic epitopes are

**Table 1** Polypeptide Chain Structure of Antibody Molecules in Humans

| Immunoglobulin Type[a] | Polypeptide Formula | Molecular Weight | H-Chain Domains | Relative Abundance (%) | Remarks |
|---|---|---|---|---|---|
| IgA | $(\alpha_2 L_2)_2$ | 320,000 | $V_H$-$CH_1$-$CH_2$-$CH_3$ | 13 | Secretory antibody |
| IgD | $\delta_2 L_2$ | 185,000 | $V_H$-$CH_1$-$CH_2$-$CH_3$ | 1 | Rich in sugars |
| IgE | $\varepsilon_2 L_2$ | 200,000 | $V_H$-$CH_1$-$CH_2$-$CH_3$-$CH_4$ | >0.01 | Responsible for allergic reactions |
| IgG | $\gamma_2 L_2$ | 150,000 | $V_H$-$CH_1$-$CH_2$-$CH_3$ | 80 | Activates complement |
| IgM | $(\mu_2 L_2)_5$ | 900,000 | $V_H$-$CH_1$-$CH_2$-$CH_3$-$CH_4$ | 6 | Appears first, activates complement |

[a]All immunoglobulins consist of light chains, L ($\gamma$ or $\kappa$ type, MW $\approx$ 25,000) and heavy chains, H.

their convex shape and the presence of large, polar, and charged amino acid side chains. Peptides with good cellular antigenicity, on the other hand, are characterized by residue types and amino acid sequence patterns that make them to fit well into the peptide-presenting groove on the surface of cell-bound MHC molecules.

## 2 THREE-DIMENSIONAL STRUCTURE OF IMMUNE MOLECULES

### 2.1 Antibodies

Antibodies are soluble serum glycoproteins of the immunoglobulin class with molecular weights ranging from 150,000 for bivalent immunoglobulins G (IgG, see Table 1 and Figure 1) to about 900,000 for pentavalent macroglobulins (IgM, see Table 1). A typical preparation of immune sera consists of a microheterogeneous mixture of molecules of the same antigenic specificity but different amino acid sequence in the antigen-binding parts of the molecule. By contrast, monoclonal antibodies and myeloma proteins are products of a single cell clone—for example, a ''hybridoma'' prepared by polyethylene glycol mediated fusion of spleen-derived, immunocompetent B cells with ''immortal'' myeloma cells. Amino acid sequences of monoclonal antibodies are homogeneous (unique throughout the chain).

The most common antibody IgG, is a molecule consisting of two pairs of identical polypeptide chains, heavy (H, having a molecular weight of 50,000) and light (L, MW 25,000). Interchain disulfide bonds provide covalent bonding among the chains, and a single intrachain disulfide bond characterizes a single domain (i.e., an amino acid sequence region ca. 110 residues long; cf. Figure 1). Amino acid sequences of N-terminal (variable) domains of L and H chains ($V_L$, $V_H$) from different immunoglobulins vary, often by as much as 40%, while sequences of the C-terminal (constant) domains ($C_L$, $C_H1$, $C_H2$, and $C_H3$) are nearly identical for immunoglobulins of the same class. Sequence variability of the V domains is the cause of different antigenic specificities residing in antigen-binding sites of individual antibodies. Each of the $V_L$ and $V_H$ domains contains three hypervariable regions corresponding, in three dimensions, to loops connecting two β-strands of an antiparallel β-hairpin. All six loops from the two domains come into close proximity in the immunoglobulin fold, forming a contiguous surface of the antigen-combining site. Thus, the loops are often termed complementarity-determining regions (CDRs) (Figure 2).

Each sequence domain is an independent, stable folding unit of a characteristic ''immunoglobulin fold,'' an antiparallel β-sandwich formed by two twisted β-sheets, each containing three to five extended β-strands mutually interconnected by hydrogen bonds. In an assembled immunoglobulin molecule, one of the two β-

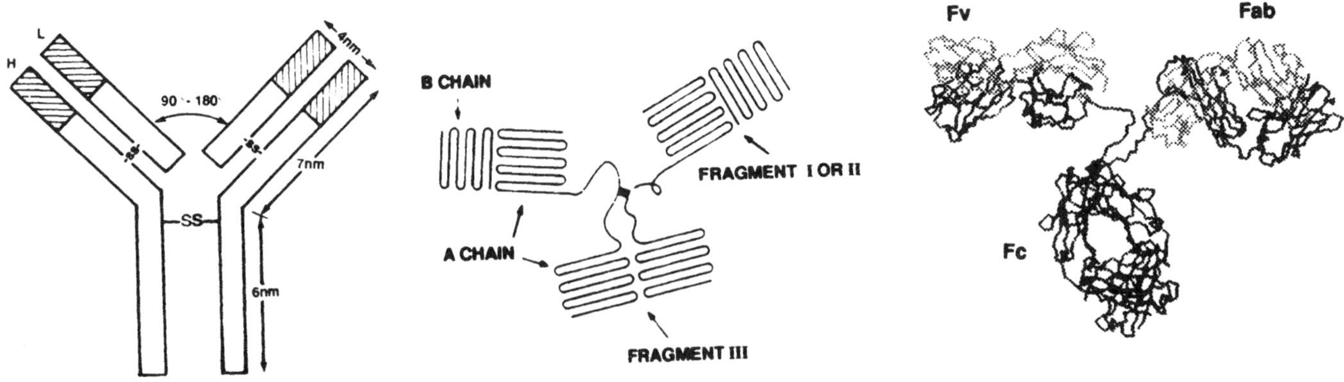


**Figure 1.** Different diagrammatic representations of an IgG molecule. *Left:* chain diagram showing disposition of the four polypeptide chains (light, L, and heavy, H), interchain disulfide bridges (—SS—), and overall dimensions of molecular fragments obtained from X-ray crystallography. *Middle:* schematic picture of the molecule obtained from ultracentrifugation data. [Adapted from M. E. Noelken, C. A. Nelson, C. E. Buckley, and C. Tanford, *J. Biol. Chem.* 240:218–224 (1965).] *Right:* polypeptide backbone trace from the low resolution X-ray crystallographic structure of a mouse IgG2A molecule: light lines, L chains; heavy lines, H chains. Atomic coordinates were obtained by courtesy of Drs. L. J. Harris and A. McPherson, University of California Riverside. Note the compact dimeric modules (Fv, Fab, Fc), and the extended polypeptides (the hinge in the middle of the molecule, the elbows in the midst of Fab fragments). The ''hole'' in the Fc fragment, in between the $CH_2$ domains, is actually filled with oligosaccharides.

sheets of each domain makes a tight noncovalent interaction with a sheet of a domain from a different chain.

Dimeric domain modules are connected by extended polypeptide "linkers" of varying length. The linker connecting the $C_H1$ and $C_H2$ domains, the "hinge," has a particularly conspicuous sequence, rich in Pro and Cys residues. Proteolysis at the hinge produces two identical antigen-binding fragments (Fab) and the crystallizable Fc fragment. Occasionally, a stable $V_L/V_H$ dimeric module, an Fv fragment, can also be obtained. The linker in the middle of the Fab fragment is referred to as an elbow, since the angle between the variable and constant Fab dimeric modules is known to vary in the range of 90–180°. Movement of the $V_L/V_H$ dimer relative to the $C_L/C_H1$ dimer is facilitated by interactions of three $V_H$ and two $C_H1$ amino acids that form the molecular equivalent of a ball-and-socket joint.

Segmental flexibility is an essential property of soluble antibodies important for formation of large antigenic complexes in which polyepitopic antigens (e.g., pathogenic viruses) become cross-linked by bivalent antibodies. A number of physicochemical techniques (electron microscopy of immune complexes; neutron contrast, low angle scattering; fluorescence energy transfer; rotational correlation times) attest to quasi-independent movements of antibody domain modules and to the coexistence in solution of a wide range of antibody conformers.

## 2.2 T-Cell Receptors

T cells display on their surfaces multicomponent receptor complexes noncovalently assembled from distinct groups of subunits, all of which are anchored in the cellular membrane by short (ca. 22 amino acids) transmembrane segments presumed to exist in an $\alpha$-helical conformation. The dimeric, variable, clonotypic subunit (Ti) with polypeptide chains $\alpha$ and $\beta$ (alternatively, $\gamma$ and $\delta$) is the cellular equivalent of the antigen-specific IgG Fab fragment. The remaining CD3 chains consist of $\gamma$- and $\delta$-type glycoproteins (MW ca. 21,000–26,000), the nonglycosylated $\varepsilon$ chain (MW 25,000), and the $\eta$-type chains, which can be any of the $\zeta$ (MW 16,000 forming disulfide-linked dimers), $\eta$ (MW 21,000), or $\gamma$ (MW 7,000) family.

Our only structural information on the T-cell receptor (Ti) is the amino acid sequence of its constituent polypeptide chains, which bears a distant similarity to the Fab fragments of immunoglobulins. The similarity suggests that the Ti polypeptide chains are organized into immunoglobulinlike domains consisting of

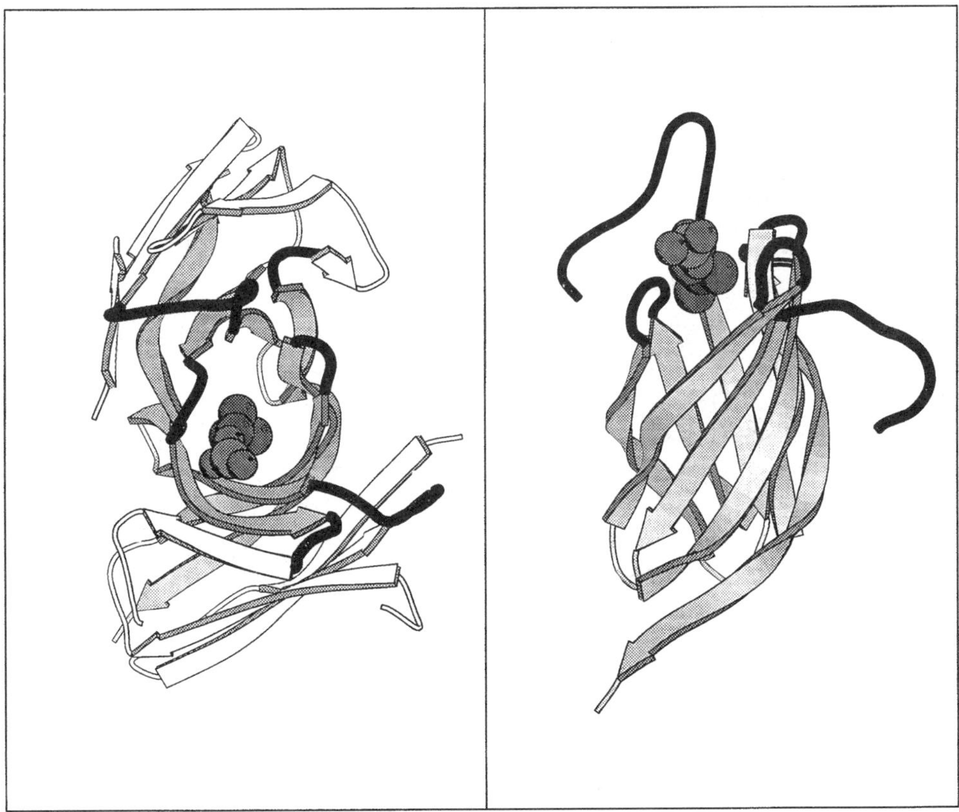

**Figure 2.** Ribbon cartoon of the Fv fragment from a monoclonal antiphosphorylcholine antibody McPC 603: arrows, $\beta$-strands; open ropes, turns and loops; solid ropes, the hypervariable regions. The antigen, phosphorylcholine, is shown in a full space-filling representation. *Left:* the $V_L$ domain is on the top, the $V_H$ domain is on the bottom, and the $V_L/V_H$ interface $\beta$-barrel is clearly visible in the middle, surrounding the antigen. The inner $\beta$-sheets of both domains are shaded, the solvent-facing $\beta$-sheets are unshaded. *Right:* side view of the interface-forming $\beta$-barrel, with the antigen and the hypervariable loops on the top; the outer, solvent-facing $\beta$-sheets of the $V_L$ and $V_H$ domains were deleted for clarity. Picture was generated with the program MOLSCRIPT of P. J. Kraulis [*J. Appl. Crystallogr.* 24:946–950 (1991)] and modified from J. Novotny and E. Haber. [*Proc. Natl. Acad. Sci. U.S.A.* 82:4592–4596 (1985)]. X-Ray crystallographic coordinates, obtained from the Brookhaven Protein Data Bank, were originally described by Y. Satow, G. H. Cohen, E. A. Padlan, and D. R. Davies, [*J. Mol. Biol.* 190:593–604 (1986).]

multistranded antiparallel β-sheet bilayers. Invariant amino acid side chains, which mediate domain–domain interactions and form a constant scaffold for antibody-binding sites, are also conserved in the chains encoded by the Ti genes. It appears that the binding sites of the antigen-specific T-cell αβ chain receptors and of antibodies are very similar in their overall dimensions and geometry: a T-cell receptor molecule probably has an antigen-specific binding site that is fundamentally no different from the conventional binding site of an antibody.

### 2.3 Major Histocompatibility Complex (MHC) Molecules

The MHC proteins are integral membrane proteins with four tightly interacting extracellular domains, one or two transmembrane (possibly α-helical) segments, and intracellular polypeptides. Phenotypically, the most conspicuous feature of the MHC molecules is their allelic variability. In man, the MHC genetic complex has been named HLA (human leukocyte antigen); in mouse, it is called H-2. Two major MHC gene classes, class I and class II, are recognized. Three-dimensional structure of both classes displays the same multidomain motif achieved with surprisingly different oligopeptidic assemblies.

In the HLA class I proteins, the longer α-chain contains the three extracellular domains $\alpha_1$, $\alpha_2$, and $\alpha_3$. On cell surface, the α-chain associates noncovalently with $\beta_2$-microglobulin, whose fold is that of an immunoglobulin C-domain (see Section 2.1). The HLA $\alpha_3$ domain, also having the Ig fold, associates noncovalently with the $\beta_2$ domain in a way different from the antibody constant domain sheet–sheet association. The $\alpha_3/\beta_2$ dimer, being closer to the cell surface, forms a "stem" on which the $\alpha_1/\alpha_2$ domain dimer rests. Both the $\alpha_1$ and the $\alpha_2$ domains fold into an "open sandwich," a flat, up-and-down antiparallel four-stranded β-sheet that supports a long C-terminal α-helix. The $\alpha_1$ and $\alpha_2$ domain helices contain one or two short nonhelical polypeptide segments that interrupt the helix with abrupt kinks. Topology of the $\alpha_1/\alpha_2$ dimer is such that the two β-sheets form a contiguous, antiparallel eight-stranded "bottom" and the two α-helices lie diagonally and antiparallel across the bottom, leaving in between an oblong groove of varying depth, approximately 2 nm wide and 3 nm long. Antigenic peptides are found in the groove, usually in extended conformations (Figure 3). Essentially all the allelic differences cluster inside the groove and, by implication, determine classes of peptide ligands structurally compatible with the groove. Protruding parts of the peptides can then be presented for binding to the T-cell receptor.

## 3 LINEAR STRUCTURE OF GENES CODING FOR ANTIBODIES AND T-CELL RECEPTORS

Modular structure of antibody molecule is paralleled by its gene organization: separate gene segments code for domains and smaller polypeptides. Individual exons are separated by large introns and,

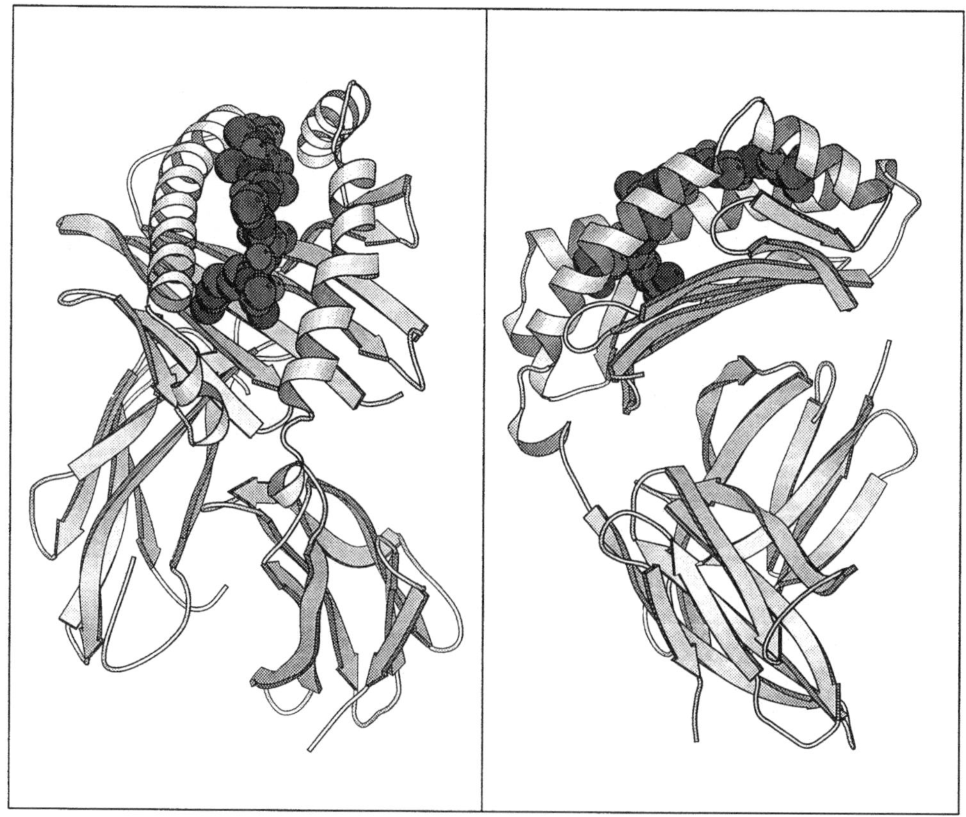

Figure 3. Ribbon diagram of the HLA-B2705 molecule with a peptide "presented." The peptide is rendered in a full space-filling representation: β-strands shown as arrows, α-helices as spiral ribbons. *Left:* "top" view diagonally into the presenting groove in between the $\alpha_1$ and $\alpha_2$ helices. The $\alpha_3$ domain is in front right, the $\beta_2$ microglobulin in back left. *Right:* side view of the HLA-B2705 peptide complex. The picture was generated with the MOLSCRIPT program of P. J. Kraulis. Atomic coordinates, obtained from the Brookhaven Protein Data Bank, were originally described by D. R. Madden, J. C. Gorga, J. L. Strominger, and D. C. Wiley [*Nature,* 353:321–325 (1991)].

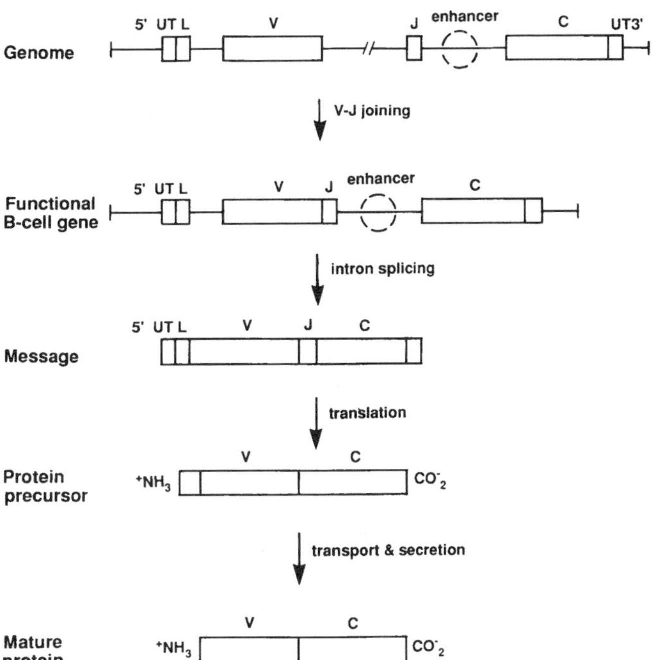


**Figure 4.** Gene rearrangement events leading to immune cell commitment and expression of antibody polypeptide chains, (beginning at the top.) The process of V-J joining in the L-chain genes (V-D-J joining in the H-chain genes) commits the B cell: boxes, exons; lines, introns; UT, untranscribed regions. Next the mRNA is created by further excision of introns. Then the L, V, and C exons are translated on ribosomes into the protein precursor. The mature chain, exported across the cell membrane, lacks the "leader" sequence.

in the process of immune commitment, become translocated and spliced to create a functional antibody gene (Figure 4). For example, in mouse germ line DNA, there are four separate light chain gene segments:

1. The L (leader) gene, which does not appear in the mature antibody sequence, being proteolytically cleaved off the chain in the course of extracellular secretion; the leader peptide directs the L and H chains into, and through, the membrane.
2. The V gene, 93 base pairs downstream from the L gene, coding for all the variable domain sequence except its last β-strand.
3. The short J (joining) gene coding for the C-terminal β strand of the V domain.
4. Some 1200 base pairs downstream from the J gene, the C (constant domain) gene.

Gene loci coding for the heavy chains have a similar structure, however, 10–12 short D (diversity) gene segments, coding for the H chain third hypervariable loop, form an independent gene cluster that is incorporated into the contiguous, rearranged V-D-J gene segment in a separate D-J recombination event.

In their structure and assembly, T-cell receptor genes have many features in common with those just outlined for antibodies. For example, the mouse β chains are assembled from separate V, D, and J segments akin to the immunoglobulin chains. In the α-chain locus, multiple V segments and an unusually large number of J segments have been found.

## 4 MECHANISMS OF ANTIBODY DIVERSITY

Molecular shapes and surface properties of antigens offer a large repertoire of forms, and several levels of combinatorial mechanisms have evolved in immune gene assembly that, together, generate a repertoire of amino acid sequence variants large enough to engage any conceivable antigenic structure.

The following combinatorial levels enter into generation of an antigen-combining site repertoire in man and mouse: (1) L and H chain pairing (chain multiplicity), (2) V-J gene combinations, and (3) V-D-J gene combinations (V, D, and J gene multiplicity). Additional sources of diversity are (4) the so-called N regions of extra nucleotides inserted at the V-D and D-J gene junctions, probably by an action of a 3′-terminal transferase, and (5) somatic hypermutation, which introduces point mutations into fully assembled V genes. Altogether, unrestricted combinations of all these mechanisms can theoretically produce in excess of $10^8$ polypeptide variants.

Little is known about genetic diversification mechanisms that operate in other vertebrate species. In chicken, however, the λ-chain locus contains 25 pseudogenes (i.e., genes truncated in the coding sequence or lacking functional recombination signals) clustered in both the 3′ and 5′ orientations approximately 2400 bases upstream of a single functional V gene. Diversification proceeds via segmental gene conversion in ontology, whereby nucleotide sequences of pseudogene segments are copied, in a seemingly random fashion, into the functional gene. About one gene conversion event is known to occur per 10 cell generations.

## 5 SPECIFICITY OF ANTIBODIES AND T-CELL RECEPTORS

Solution binding studies show antibodies to be capable of recognizing the smallest structural variations of otherwise very similar ("cross-reacting") ligands, such as substitution isomers of small molecular weight haptens. Conceptually, antibody specificity has been discussed in the frameworks of "lock and key" or "induced fit" hypotheses. Crystallographic structures of several free antibodies and antigen–antibody complexes are now available and provide examples of both those binding paradigms.

X-Ray crystallography shows the antibody-binding site to be a concavity surrounding the interface between the light and the heavy chain variable domains. In cases typifying the lock and key (i.e., rigid body interaction) paradigm, a small hapten (phosphorylcholine) enters deeply into the interdomain pocket of the monoclonal antibody McPC 603 (Figure 5B; see color plate 12) or, in the case of the D1.3 antibody–hen egg-white lysozyme complex, the principal antigenic residue (Glu 121) of lysozyme is buried in the interface; the rest of antigen–antibody contract area is rather flat, if irregularly undulated. The antibody 17/9, specific to a nonapeptide antigen from flu hemagglutinin, typifies the induced-fit paradigm. A rearrangement of the H3 hypervariable loop backbone and side chains takes place upon antigen binding, creating a snug binding pocket for the β-turn conformation of the peptide.

It is natural for the curvature of binding site cavities to match that of the antigens: binding sites that accommodate small ligands have high curvatures and appear as pockets (e.g., the antiphosphorylcholine binding site is ca. 1.5 nm wide and 1.2 nm deep, Figure 5B); those directed toward large protein antigens have a low curva-

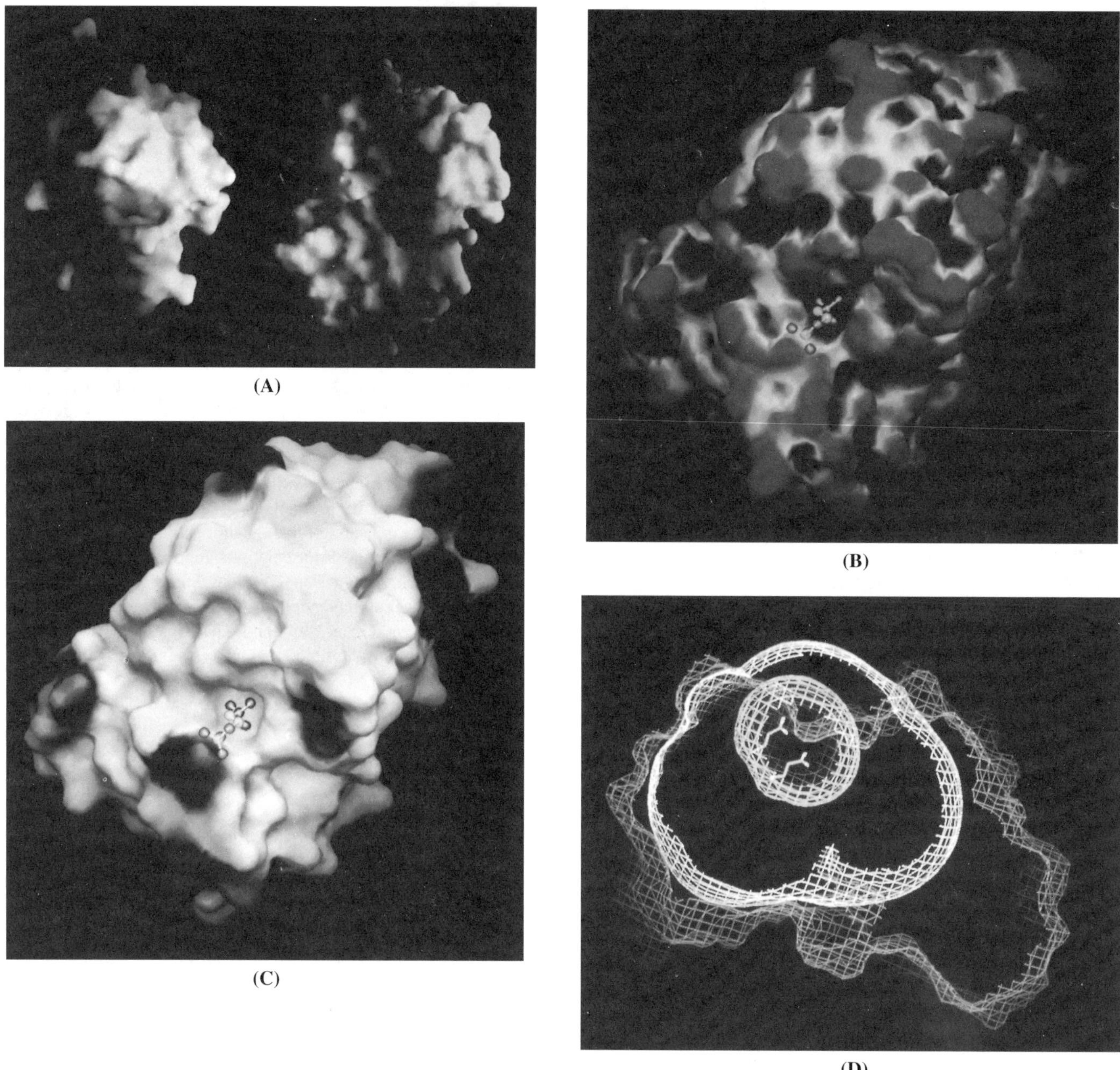


Figure 5. (A) Surface representations of the protein antigen (hen egg-white lysozyme, left) and its specific antibody HyHEL-10 Fv fragment (right). The protein surfaces, originally in tight noncovalent interaction, have been pulled apart and color-coded by interprotein atom–atom distances (white to red: close distance to long distance). The picture emphasizes the global convex shape of the antigen surface and its complementary, valleylike antibody surface. Color graphics in (A) to (C) prepared with the program GRASP (A. Nicholls). (B) Fv fragment of the antiphosphorylcholine antibody McPC 603, color-coded by electrostatic potential (positive to negative: blue to red). The electrostatic field at the binding site is seen to complement well the (opposite) charges on the ligand. (C) Same as (B) except that the Fv fragment surface has been color-coded by surface curvature (convex to concave: green to black). The hapten phosphorylcholine (ball-and-stick representation) is seen in the binding site cavity. (D) Comparison of 1 kT/e electrostatic potential contours generated by a pair of Glu residues H35 and H50 of the HyHEL-5 Fv fragment in the protein and in free solution. In this cross-sectional view, the solvent-excluded surface of the HyHEL-5 Fv fragment is green; the 1 kT/e contour of the aqueous electrostatic potential is red; and the 1 kT/e contour of the potential in the presence of the low dielectric, uncharged protein surrounding the two glutamate side chains is magenta. The figure demonstrates how the low dielectric medium of protein interior, paired with the concave shape of the binding cavity, enhances and modulates the shape of the electrostatic field emanating from the protein. [Color graphics, generated with the program INSIGHT (Biosym Technologies, Inc.), reprinted with permission from J. Novotny and K. Sharp, "Electrostatic fields in antibodies and antibody/antigen complexes," *Prog. Biophys. Mol. Biol.* 58:203–224 (1992).] (See color plate 12.)

ture, being more akin to valleys and grooves than to deep crevices (Figure 5A). Amino acid composition of binding sites is biased toward polar side chains with high dipole moment (Asn, Gln) and aromatic rings—that is, groups well capable of neutralizing the predominantly polar and charged antigenic epitopes (see Section 6.1).

Biophysical importance of an overall concave shape of binding sites relates to the following facts: (1) diffusion away from cavities is significantly slower than from flat surfaces, or through the solvent; (2) (attractive) electrostatic fields are enhanced or focused in cavities, even though solvent quenches the field at other flat or convex parts of protein surface (Figure 5B,D), and (3) hydrophobicity of a surface is a function of its curvature relative to the size and curvature of a water molecule; thus, concave surfaces are more hydrophobic than flat or convex surfaces, providing more "hydrophobic binding energy" (Figure 5C).

## 6 MOLECULAR PROPERTIES OF ANTIGENS: ANTIGENICITY

### 6.1 Humoral (Antibody) Antigenicity

Classical immunochemical experiments have shown that virtually any chemical structure, attached to a protein molecule, can elicit antigenic response. The vast scientific literature focused on antigenic properties of proteins contains several alternative, and often contradictory, theories of antigenicity.

The term *antigenicity* refers to the ability of a protein surface region to be potentially antigenic, while *immunogenicity* refers to the ability of any antigenic site to elicit such a response under particular circumstances (immunization protocol, genetic constellation of the organism, etc.).

Antiprotein antibodies sometimes specifically recognize short peptides (tetra-, penta-, hexapeptides); such antibodies can be elicited by synthetic peptide antigens. The majority of antigenic sites in proteins, however, seem to consist of amino acids that are not contiguous in the amino acid sequence (composite or discontinuous epitopes).

A concept of a "functional (energetic)" epitope, originally formulated in computational analysis of X-ray crystallographic structures, has recently found support in extensive alanine-scanning mutagenesis experiments of antigen–antibody complexes. Thus, only a small part of the total antibody and antigen contact surface (some 30–40% thereof, or three to four amino acid residues) appears to contribute actively to binding, often via hydrogen bonds and buried salt links. This smaller contact area (a functional, or energetic, epitope) belongs to the most protruding parts of the antigenic surface, where side chains such as Arg, Glu, Gln, Lys, and Pro are particularly abundant. For the complex formation to be possible, however, there must be complementarity between additional surface area of the antibody and the antigen. Hence we arrive at different operational definitions of antigenicity, depending on whether we emphasize energetics of complex formation, complementarity of antigen–antibody surfaces, or other phenomena. The conjecture that antigenicity correlates with surface protrusion provides a natural link between the two extreme antigenicity theories ("distinct antigenic epitopes exist" vs. "the whole surface is antigenic") by introducing the concept of antigenic probability that varies along the surface.

### 6.2 Cellular (T-Cell Receptor) Antigenicity

Structural problems related to cellular antigenicity are similar to those of humoral antigenicity: we wish to identify those structural traits of a peptide that make it capable of eliciting a cellular immune response. Molecular mechanisms of T-cell receptor recognition, however, involve antigen presentation by the MHC molecules, and phenomenology of cellular antigenicity is therefore distinctly different from that of the humoral antigenicity.

T-Cell antigens are peptides derived from proteins that became sequestered by presenting cells, after which they were intracellularly cleaved, and then incorporated into the MHC protein structure in the process of its folding. Peptides presented in the MHC peptide-binding groove and specifically recognized by T-cell receptors are often derived from polypeptide segments buried inside foreign protein antigens. The majority of the MHC allelic amino acid replacements modify the shape of the peptide-presenting groove, thereby determining broad classes of peptide sequences that are compatible or incompatible with that allele. Thus, the main determinants of cellular antigenicity are a suitable length (an optimal range for fitting comfortably into the presenting groove appears to be 8–11 residues) and an amino acid sequence compatible with tight binding to an MHC allele.

## 7 PROTEIN ENGINEERING IN IMMUNOLOGY

The modular three-dimensional design of immunoglobulins, T-cell receptors, and MHC antigens lends itself ideally to protein engineering schemes that shuffle, transpose, and reconnect various independent folding units (domains) of various proteins (Figure 6). Protein engineering of immunoglobulins and T-cell receptors has blossomed since middle 1980s, taking advantage of computer-aided structural design based on available X-ray crystallographic coordinates of many antibodies, rapid development of gene cloning technologies, which allowed subcloning of eukaryotic genes into bacterial plasmids, and progress in controlled bacterial expression of proteins from plasmid-inserted genes. Expression vectors such as yeast, baculovirus, and mouse hybridoma cells have been used; large yields of protein chimeras from inexpensive bacterial (*E. coli, Streptomyces* sp.) cultures are becoming common. Proteins are obtained either as secreted, refolded, soluble species or as insoluble, denatured intracellular aggregates—"inclusion bodies" that are easy to isolate but must be solubilized and renatured before a functional protein is obtained.

### 7.1 Stable Fragments, Domain Swaps, Immunotoxins

Some of the first man-made chimeras were mouse antigen-binding domains ($V_L$ and $V_H$) implanted on human constant region domains, $V_H$ domains spliced onto $C_L$ domains giving rise to functional, antigen-specific, $V_H$-$C_L$/$V_L$-$C_L$ light-chain-like dimers, and recombinant mouse antibodies with novel effector functions engineered via H-chain constant domain swaps, deletions (to produce $[Fab]'_2$-like, disulfide-bonded bivalent molecules), and replacements with foreign protein domains (e.g., enzymatically active Fab-staphylococcal nuclease or Fab-*myc* oncogene constructs). Production of large quantities of stable Fv fragments in bacteria has been aided by the design and successful preparation of a single-chain Fv whereby the C-terminus of a $V_H$ (alternatively, $V_L$) domain is connected to the N-terminus of the $V_L$ ($V_H$) domain by a short

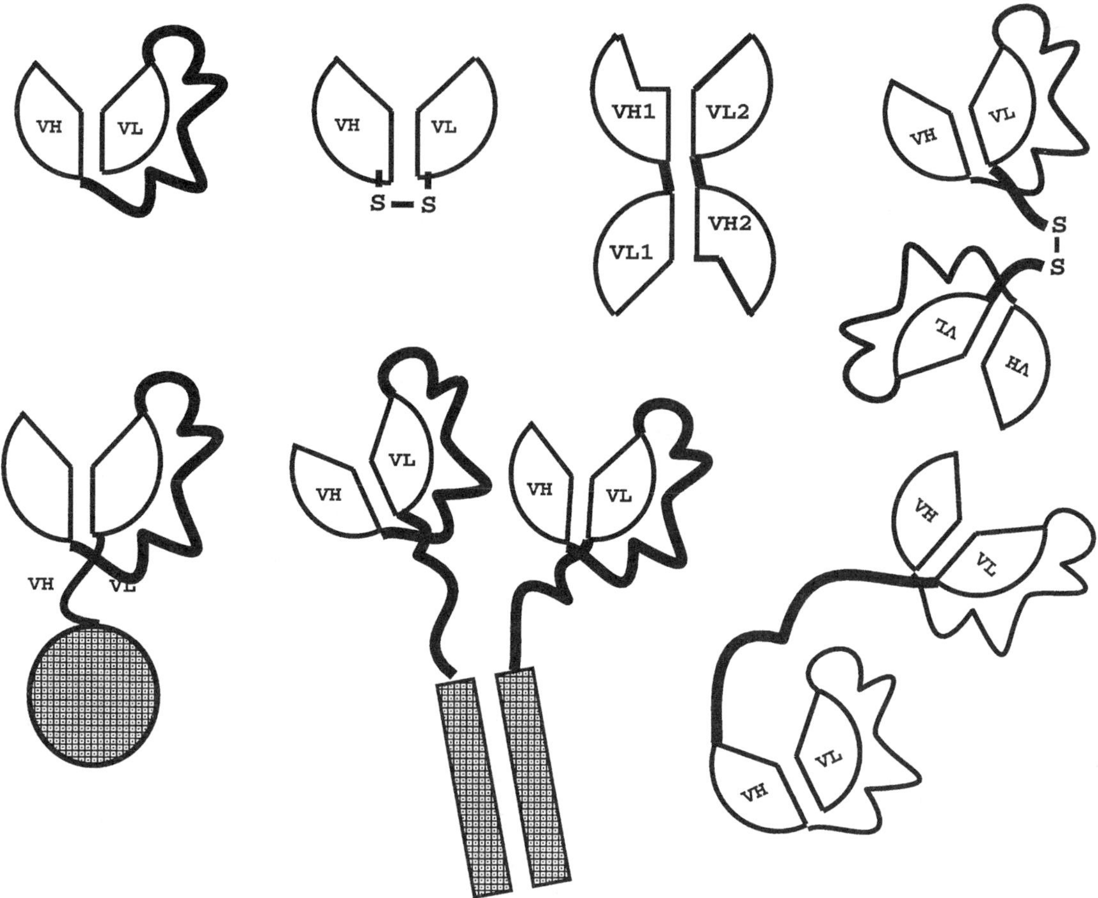


Figure 6. A schematic diagram of protein engineering of antibody binding sites; VH and VL domains are shown as open circular sections, peptide linkers as thick lines, lecine zippers (α-helical dimers) as filled rectangles, cell toxins, enzymes, etc. as a filled circle. Top, from left to right: a single-chain Fv fragment (scFv), a disulfide-bonded Fv fragment (SSFv), a bivalent "diabody", a bivalent, disulfide-bonded scFv dimer. Bottom, from left to right: an scFv attached to an effector molecule (e.g., *Pseudomonas* PE40 cellular toxin), a bivalent scFv-leucine zipper construct, a bivalent scFv chimaera with 2 scFv fragments chained together by a flexible linker. Both homo- and heterodimeric leucine zipper sequences are known allowing, in principle, a noncovalent assembly of homo-bivalent and heterobivalent antibody-like chimaeras. Similarly, homo- or heterobivalent, linker-chained, scFv constructs are possible. Bivalent scFv chimaeras based on 4-helix bundles (the ROP dimer of the α helix-turn-α helix motif) were also described but are not shown. Note also that antigen combining sites (i.e. the three hypervariable loops in the VH and VL domains) can be transferred from one Fv framework to another, and different pairs of VH and VL domains can be recombined to give novel antigen binding sites.

(12–20 amino acid) polypeptidic linker. The single-chain Fv technology has been particularly successful in supporting bacterial production of effective immunotoxins. Thus, single-chain Fv of antitumor specificity with a modified *Pseudomonas* exotoxin (PE40) attached to its C-terminus is a powerful and specific anticancer agent. Other examples of antibody engineering include:

1. CD4 immunoadhesins (proteins composed of IgG constant domains, the Fv modules being replaced with two extracellular domains of the CD4 receptor) for AIDS therapy.
2. Immunoligands such as the interleukin 2 molecule fused to the IgG constant regions.
3. An exogenous peptide epitope implanted in lieu of the H chain third hypervariable loop.
4. Metal coordination sites engineered into an antibody-binding pocket.
5. Chimeric receptors consisting of a single-chain Fv fragment attached to the CD3 $\gamma$ or $\zeta$ transmembrane signal-transducing polypeptides (these chimeras activate T cells, upon stimulation with the original antigen, akin to the intact CD3/Ti receptor complex).
6. Heterofunctional "miniantibodies" consisting of a single-chain Fv fragment, a flexible IgG3 hinge, and an amphiphilic α-helical segment (leucine zipper). In this interesting design, the dimer-forming propensity of the α-helices drives spontaneous generation of a noncovalent, bifunctional, heterospecific chimera.

The single-chain technology has also been applied successfully to T-cell receptors and MHC proteins. For example, Fv-like, soluble, single-chain T-cell receptor $V_\beta V_\alpha$ fragments were expressed in bacteria and shown to possess the same antigen specificity as the parent receptor. Functional, soluble MHC-like proteins were prepared by tethering the three extracellular domains ($\alpha_1$, $\alpha_2$, and $\alpha_3$) of the mouse H-2 class I α chain to the $\beta_2$ domain. Various chimeric constructs consisting of T cell receptor C domains and antibody V domains, or TCR-antibody polypeptide chain heterodimers, were also assembled and shown to be functional.

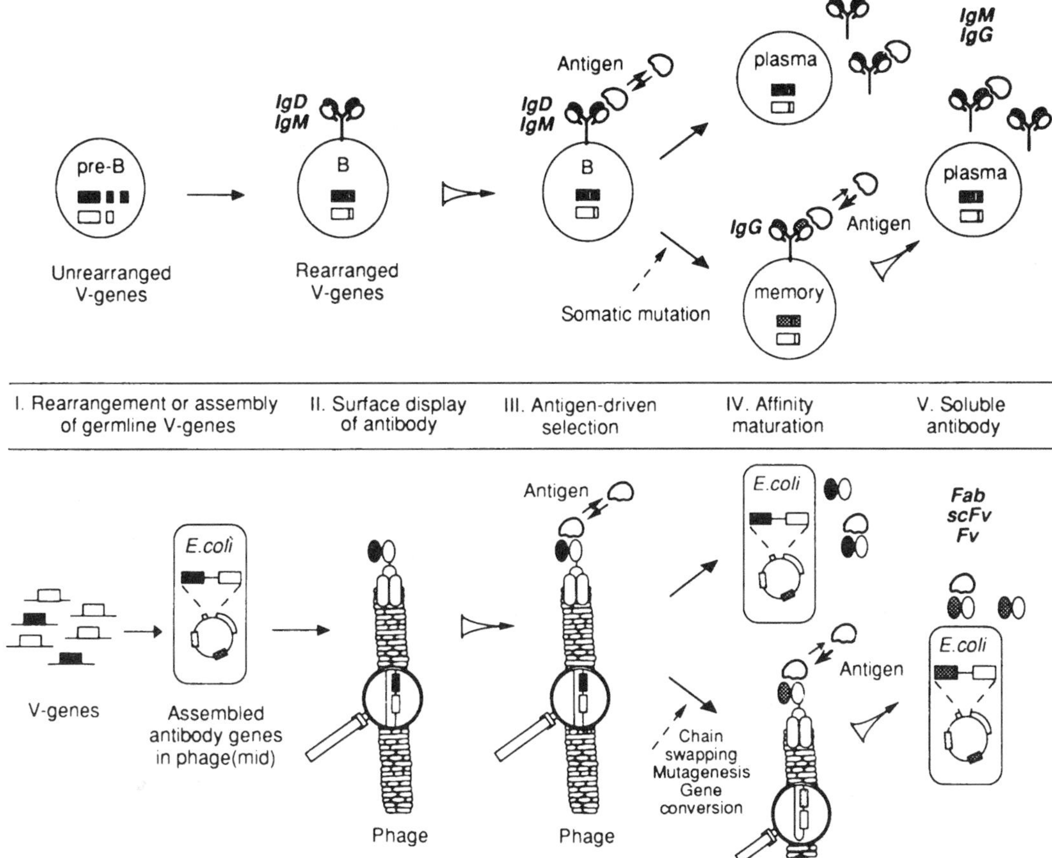


**Figure 7.** The strategy of immune system in vivo and using phage. Antibody heavy ($V_H$, solid units) and light ($V_L$, open units) chain variable domains create the antigen-binding site. The antibody molecules are displayed on the surface of the B cell, where they function as an antigen receptor. Binding of antigen results in proliferation and differentiation of the B cells to short-lived plasma cells making soluble antibody or to long-lived memory cells. Antibody affinity is increased by mutation of the antibody V genes followed by further antigen selection. To mimic the immune system in phage (lower panel), rearranged V genes can be amplified from mRNA using the polymerase chain reaction (PCR), as the ends of rearranged heavy and light V genes are sufficiently conserved to design "universal" primers for PCR amplification of both heavy and light chain V genes from mRNA or cDNA. "Natural" repertoires can be made from unimmunized donors using IgM mRNA, whereas IgG mRNA can be used to tap the postimmunization repertoire. Alternatively, entirely synthetic repertoires of rearranged V genes can be generated. The rearranged heavy and light chain V genes are assembled together randomly to encode either single-chain Fv antibody fragments or Fab fragments and cloned into a phage or phagemid vector to create a library of millions of different phages. For display on gene III, phage vectors give three copies of the antibody-pIII fusion protein while phagemid vectors result in zero to three copies of fusion protein. Binding phage are affinity-purified on antigen and used to infect a nonsuppressor strain of bacteria to produce soluble antibody fragment. The isolated antibody genes can be mutated and the resulting mutant phage subjected to further affinity selection to produce higher affinity antibodies. [Reprinted with permission from J. D. Marks, H. R. Hoogenboom, A. D. Griffiths, and G. Winter, Molecular evolution of proteins on filamentous phage, *J. Biol. Chem.* 267:16007–16010 (1992).]

### 7.2 Immune Libraries; Phage Display Technology

Combinatorial libraries of immune repertoire, or "immune libraries," contain one or more sets of transcripts of antibody L and H chain genes (e.g., cDNA obtained by message transcription using the polymerase chain reaction) cloned into a bacteriophage vector (phage λ, filamentous fd phages). For immune repertoire cloning, a set of oligonucleotide primers was prepared, facilitating amplification of L chain and the N-terminal halves of H chain mRNAs. The primers incorporate restriction sites that allow the cDNA to be force-cloned for sequencing and expression (Figure 7).

*See also* Antibody Molecules, Genetic Engineering of; Autoantibodies and Autoimmunity; Combinatorial Phage Antibody Libraries; Major Histocompatibility Complex; Recognition and Immobilization, Molecular; Superantigens.

## *Bibliography*

Branden, C., and Tooze, J. (1991) Recognition of foreign molecules by the immune system. In *Introduction to Protein Structure,* pp. 161–178. Garland, New York.

Chothia, C., Novotny, J., Bruccoleri, A., and Karplus, M. (1985) Domain association in immunoglobulin molecules. *J. Mol. Biol.* 186:651–663.

Davies, D. R., Padlan, E. A., and Sheriff, S. (1990) Antibody–antigen complexes. *Annu. Rev. Biochem.* 59:439–474.

Harris, L. J., Larson, S. B., Hasel, K. W., Day, J., Greenwood, A., and McPherson, A. (1992) The three-dimensional structure of an intact monoclonal antibody for canine lymphoma. *Nature,* 360:369–372.

Jin, L., Fendly, B. M., and Wells, J. A. (1992) High resolution functional analysis of antibody–antigen interactions. *J. Mol. Biol.* 226:851–865.

Jorgensen, J. L., Reay, P. A., Ehrlich, E. W., and Davis, M. M. (1992) Molecular components of T-cell recognition. *Annu. Rev. Immunol.* 10:835–874.

Novotny, J. (1991) Protein antigenicity: A thermodynamic approach. *Mol. Immunol.* 28:201–207.

# Immuno-PCR

*Takeshi Sano, Cassandra L. Smith, and Charles R. Cantor*

*Key Words*

**Biotin** A water-soluble vitamin (vitamin H) that can be incorporated relatively easily into various biological materials.

**Biotinylation** To incorporate biotin or its analogue into biological material.

**Protein A** A cell wall constituent protein of *Staphylococcus aureus* that specifically binds the Fc domain of an immunoglobulin G molecule without disturbing its antigen-binding ability.

**Streptavidin** A protein, produced by *Streptomyces avidinii* that specifically binds biotin with an extremely high affinity (dissociation constant $K_d \sim 10^{-15}$ M).

Immuno-PCR is an antigen detection system in which the polymerase chain reaction (PCR) is used to amplify a segment of marker DNA that has been attached specifically to antigen–antibody complexes. Because of the enormous amplification capability and specificity of PCR, immuno-PCR allows considerable enhancement in detection sensitivity for a specific antigen over conventional antigen detection systems, such as enzyme-linked immunosorbent assays and radioimmunoassays. A variety of superior characteristics and the versatility of immuno-PCR, including high but controllable sensitivity, simplicity, and easy contamination control, offer great promise for the application of this system in various aspects of biological and biomedical sciences, as well as clinical diagnostics and forensic medicine.

## 1 Principles

### 1.1 Concept of Immuno-PCR

The concept of immuno-PCR, illustrated in Figure 1, is quite simple and is similar to those of conventional antibody-based antigen detection systems, such as enzyme-linked immunosorbent assays (ELISAs) and radioimmunoassays (RIAs).

In immuno-PCR, a molecular linker having a bispecific binding affinity for antibody and DNA is used to attach a marker DNA molecule specifically to an antigen–antibody complex. Using appropriate primers, a segment of the attached marker DNA is amplified by PCR, and the resulting PCR products are analyzed. The presence of specific PCR products demonstrates that the marker DNA molecules are attached specifically to antigen–antibody complexes, indicating the presence of antigen.

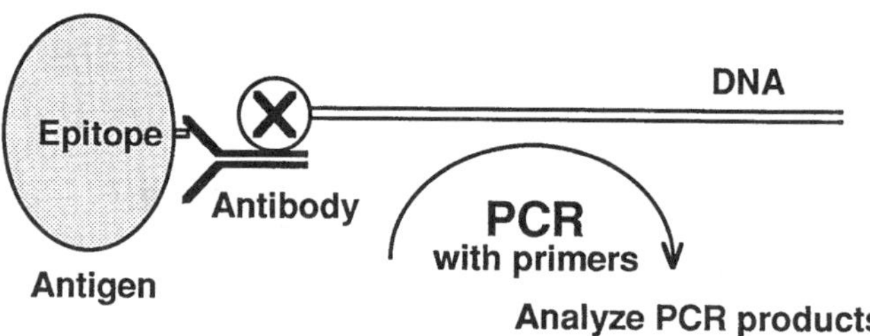


**Figure 1.** Concept of immuno-PCR. See Section 1.1 for explanation. [Reprinted with permission from Sano (1994) *Cell Technology* 13:77–80.]

### 1.2 Molecular Linkers Used in Immuno-PCR

In conventional antigen detection systems, secondary antibodies, directed against a primary antibody, are generally used to attach labels, such as enzymes in ELISA or radioactive labels in RIA, to antigen–primary antibody complexes. In contrast, immuno-PCR uses specific molecular linkers to attach marker DNA specifically to antigen–antibody complexes. Thus, such molecular linkers are key components in immuno-PCR.

To attach marker DNA specifically to antigen–antibody complexes, molecular linkers used in immuno-PCR must possess bispecific affinity for antibody and DNA. One unique and particularly versatile molecular linker for immuno-PCR is a streptavidin–protein A chimera. This chimera has two independent binding abilities: one to biotin, derived from the streptavidin moiety, and the other to the Fc domain of an immunoglobulin G (IgG) molecule, derived from the protein A moiety. This bifunctional binding specificity for biotin and IgG allows the specific conjugation of any biotinylated DNA molecule to antigen–antibody complexes.

Other simple molecular linkers that can be used in immuno-PCR are monospecific multivalent binding molecules, such as streptavidin and avidin, each of which has four biotin-binding sites per molecule. Although these proteins do not bind antibody and biotinylated marker DNA independently, biotinylated marker DNA can be attached to antigen–antibody complexes by the use of biotinylated antibodies. In these methods, biotinylated antibody is first bound to antigen (formation of antigen–biotinylated antibody complexes). Then, streptavidin or avidin is directed to biotin attached to the antibody [formation of antigen–biotinylated antibody–(strept)avidin complexes]. Finally, biotinylated marker DNA is directed to free biotin-binding sites of streptvidin or avidin, followed by amplification of a segment of bound marker DNA by PCR.

## 2 Methods

There are several formats of immuno-PCR. One of the most versatile formats uses microtiter plates as a solid support. In this microtiter plate format, antigen is immobilized on the surface of the wells.

### 2.1 Microtiter Plate Format

The basic immuno-PCR protocol for the detection of antigen immobilized on microtiter plate wells consists of six steps:

1. Immobilization of antigen on the surface of microtiter plate wells.
2. Blocking of free binding sites on well surfaces.
3. Binding of antibody to antigen.

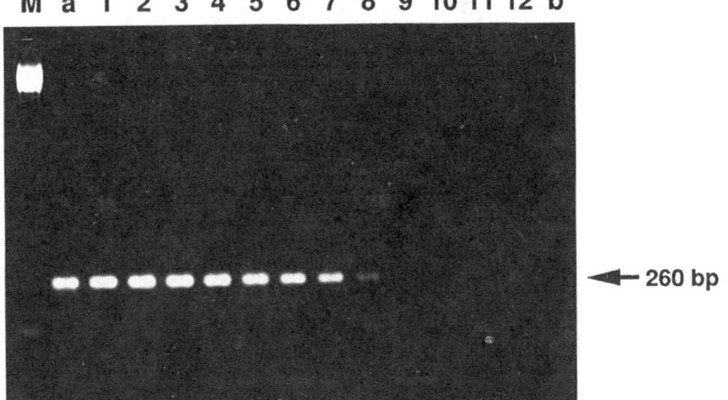


**Figure 2.** Detection of bovine serum albumin (BSA) immobilized on a microtiter plate by immuno-PCR. Immobilized BSA was detected by immuno-PCR using monoclonal anti-BSA and a streptavidin=nprotein A chimera containing a 2.67 kb end-biotinylated linear plasmid as a marker. A 260-bp segment of the marker DNA was amplified by PCR, and the resulting PCr products were analyzied by agarose gel electrophoresis, stained with ethidium bromide. Lanes 1 to 9 contain PCR amplification mixtures with immobilized antigen: lane 1, 96 fmol ($5.8 \times 10^{10}$ molecules); lane 2, 9.6 fmol ($5.8 \times 10^{9}$ molecules); lane 3, 960 amol ($5.8 \times 10^{8}$ molecules); lane 4, 96 amol ($5.8 \times 10^{7}$ molecules); lane 5, 9.6 amol ($5.8 \times 10^{6}$ molecules); lane 6, 0.96 amol ($5.8 \times 10^{5}$ molecules); lane 7, $9.6 \times 10^{-20}$ mol ($5.8 \times 10^{4}$ molecules); lane 8, $9.6 \times 10^{-21}$ mol ($5.8 \times 10^{3}$ molecules); lane 9, $9.6 \times 10^{-22}$ mol ($5.8 \times 10^{2}$ molecules). Lanes 10 to 12 are derived from control wells, where no antigen was immobilized. [Reprinted with permission from Sano et al., *Science* 258:120=n122 (1992).]

4. Binding of the streptavidin–protein A chimera, bound to biotinylated marker DNA, to antigen–antibody complexes.
5. PCR amplification of a segment of the marker DNA with appropriate primers.
6. Analysis of PCR products.

An example of the detection of antigen using the microtiter plate format is shown in Figure 2.

The key factor in this protocol is the efficient and complete removal of unbound and nonspecifically bound antibody and marker DNA, which contribute to background signals. Therefore, extensive washing must be carried out after each step. One of the advantages of the microtiter plate format is that the washing steps can be performed quite easily.

Both monoclonal and polyclonal antibodies can be used in immuno-PCR. However, such antibodies must have the Fc portion that binds to the protein A moiety of the chimera. Because of the extremely high sensitivity of immuno-PCR, very low concentrations of antibody and marker DNA are generally used. For example, the primary antibody can be diluted to concentrations several orders of magnitude below those used in conventional ELISA. Optimization of these concentrations for a given sample is carried out by empirical methods. The use of the microtiter plate format facilitates this optimization step, because many samples can be analyzed simultaneously.

PCR products generated by all immuno-PCR formats can be analyzed by various methods. One simple and easy method, if relatively small numbers of samples are analyzed, is gel electrophoresis. For applications that require the analysis of relatively large numbers of samples, such as in clinical diagnostics, other detection methods are more appropriate. For example, direct incorporation into PCR products of a label, such as fluorochromes and haptens, followed by the detection of the incorporated labels, allows the analysis of large numbers of samples.

### 2.2 Other Formats

Simple modifications of the original protocol just described make possible several other immuno-PCR formats without loss in detection sensitivity. For example, antigen molecules on cell surfaces can be detected by incorporating centrifugation or filtration steps to separate unbound antibody and marker DNA from the cells.

The original protocol is likely to generate high background signals when antigen is to be detected in physiological fluids, such as blood samples, which contain endogenous immunoglobulins. This occurs because of the binding of the protein A moiety of the streptavidin–protein A chimera to sample-derived immunoglobulins. There are two immuno-PCR formats that can be used for the detection of antigen in immunoglobulin-containing samples. One method is the sandwich assay format, in which antigen is first captured by immobilized primary antibody fragments, such as Fab and $F(ab')_2$, thereby removing sample-derived immunoglobulins. Then, captured antigen is detected by the original microtiter plate format. An attractive alternative method is to make a conjugate consisting of antibody, biotinylated marker DNA, and the chimera, in which IgG-binding sites are saturated with antibodies (preconjugation format). Such conjugates should not bind to sample-derived immunoglobulins, because no free IgG-binding site is present.

### 2.3 General Precautions

Immuno-PCR possesses extremely high sensitivity, derived primarily from the enormous amplification capability of PCR. Thus, any nonspecific binding of antibody and marker DNA would cause serious background problems. Extensive washing after the application of antibody and marker DNA is indispensable. Even though some fraction of specifically bound antibody and marker DNA is removed by washing, the overall sensitivity can be recovered by using additional amplification cycles in the final PCR step. In addition, the use of effective blocking agents is of great importance in avoiding nonspecific binding. Both protein blockers, such as nonfat dried milk and bovine serum albumin, and nucleic acid blockers, such as sheared sperm DNA, are used.

Another important factor in avoiding background signals is control of contamination, which is a problem common to all sensitive detection systems. Even though all procedures are performed very carefully, repeated use of the same marker DNA and primers may generate false positive signals. One of the major advantages of immuno-PCR is that marker DNA is purely arbitrary. Therefore, marker DNA molecules and their primers can be changed frequently, as needed, to avoid the generation of false positive signals due to contamination. This characteristic offers easier control of false positive signals than other PCR-based detection methods, in which sample-derived nucleic acids are directly amplified.

## 3 APPLICATIONS AND PERSPECTIVES

Potentially, immuno-PCR can be applied practically and proficiently in many areas. In biological and biomedical sciences, the

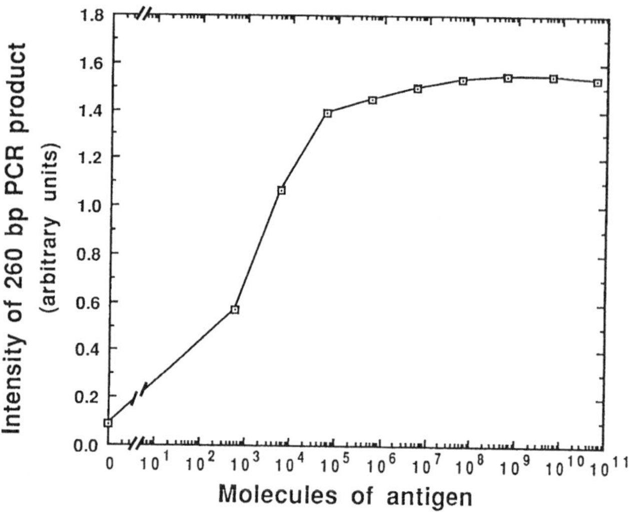


**Figure 3.** Quantitation of PCR product generated by immuno-PCR. The 260 bp PCR fragment shown in Figure 2 was quantitated and plotted as a function of the number of antigen molecules. Note that PCR amplification was not saturated in a range from $10^2$ to $10^5$ molecules of antigen, where the number of antigens can be estimated. [Reprinted with permission from Sano et al., *Science* 258:120=n122 (1992).]

specific detection of biological molecules of interest is one of the most important steps in analysis. The extremely high sensitivity of immuno-PCR should allow the detection of antigens that cannot be detected by conventional systems. Thus, the use of immuno-PCR will allow the analysis of specific antigen at microscopic scales (e.g., at single cell levels). Quantitation of PCR products can provide estimates of the numbers of antigen in given samples, as shown in Figure 3.

One of the most practical applications of immuno-PCR is in clinical diagnostics. The extremely high sensitivity of immuno-PCR will enable the detection of rate antigens, which are present only in very small numbers. This characteristic should allow the diagnosis of pathological conditions at earlier stages of disease or infection development. Another important characteristic of immuno-PCR is its simplicity, which should allow the development of fully automated assay systems. Such automated systems are extraordinarily useful in clinical diagnostics, which call for the repeated analysis of large numbers of samples.

The distinct power and generality of immuno-PCR will allow a tremendous expansion in the applications of antibody-based detection methods, not only to biological but also to nonbiological materials. For example, the extremely high sensitivity of immuno-PCR should enable this technology to be applied to the detection of single antigen molecules, for which no method is currently available.

*See also* IMMUNOLOGY; PCR TECHNOLOGY.

## *Bibliography*

Sano, T., Smith, C. L., and Cantor, C. R. (1994) *Advances in Immunological Methods for Pesticide Analysis.* American Chemical Society, Washington, DC.

# IMMUNOTOXICOLOGICAL MECHANISMS

*Rodney R. Dietert*

## *Key Words*

**B Lymphocytes** The antibody-producing lymphocyte population that matures in the bursa of Fabricius (in birds) or in the mammalian equivalent of this organ.

**Immunomodulation** Alteration in an active immune process or in the functional capacity of the host immune system.

**Macrophages** Mononuclear phagocytes in circulation and tissues that participate in both first-line defense immunosurveillance and in acquired immunity through the presentation of antigen peptides to lymphocytes.

**T Lymphocytes** Lymphocyte population that matures in the thymus and contributes both effector and regulatory functions.

**Xenobiotic Interaction** Interaction of a chemical with an organism apart from the normal metabolism of the organism.

Immunotoxicology is a relatively new interdisciplinary area that has increased in significance concomitant with the recognized importance of immunocompetence to human health. Defined as the identification and analysis of environmental agents or factors that produce immunomodulation, immunotoxicology represents a priority both for the protection of human populations from environmental hazards and for the development of immunotherapeutic pharmaceutical products that can aid in the prevention of human disease. The development and validation of methodologies that provide an accurate assessment of environment–immune system interactions is a cornerstone of immunotoxicology. This entry describes the cellular and molecular approaches used in immunotoxicological assessment and the factors that influence the utility of the resulting data.

## 1 NATURE OF IMMUNOTOXICANTS

Immunotoxicants are environmental factors that produce a significant modulation of the immune system in humans and animals. These factors may be environmental chemicals (e.g., lead), pharmacological agents (e.g., lithium, cyclophosphamide), or physical factors (e.g., ultraviolet B radiation).

Environmentally produced immunosuppression can pose a health risk if the immunosurveillance capacity of the individual is significantly compromised. Not all environmental immunomodulators produce systemic effects. Some factors can influence local immunity. For example, the inhaled pollutant nitrogen dioxide can exert an effect on the local immunity of the lung.

While immunotoxicology has traditionally been recognized as the study of immunosuppressive factors, more recently, chemicals that produce autoimmunity or increased allergic reactions have achieved equal importance within immunotoxicology. Therefore, in considering the environmental modulation of the immune system, any significant deviation from a ''normal functioning'' immune system resulting from environmental exposure must be detectable.

Given the complexity of the immune system, it is not surprising that immunotoxicants have been identified from diverse chemical categories. Host immune defenses comprise numerous cell types and a large number of soluble mediators [e.g., immune cell hormones (cytokines) and metabolites]; therefore, the opportunity exists for chemicals to interact with one or more immune system components. As a result, immunotoxicant identification is problematic without direct examination of the chemical–host interaction. Structural–functional prediction of immunotoxicants can be generally effective within specific xenobiotic chemical categories. For example, the polycyclic aromatic hydrocarbons (PAHs) have a common feature of altering the antigen-presenting capacity of macrophages. Chemically induced immunotoxicity may be widespread, involving many immune system components, or it can be targeted to one susceptible immune cell type. Examples of targeted susceptibility of immune cells to chemicals are B-lymphocyte susceptibility to cyclophosphamide, T-lymphocyte susceptibility to cyclosporin A, and macrophage susceptibility to carrageenan and asbestos.

### 1.1 Chemical Categories

Environmental factors producing immunomodulation are both numerous and diverse. Many immunotoxic agents have been identified through experimental studies, as opposed to direct clinical information. Examples of immunotoxic chemicals are found in many categories, including the following: the polycyclic aromatic hydrocarbons [e.g., 7,12-dimethyl benz(*a*) anthracene], the polyhalogenated aromatic hydrocarbons (PCBs), selected metals (lead), certain pesticides (chlordane), organotins (TBTO), aromatic hydrocarbons (benzene), aromatic amines (acetylaminofluorine), oxidant gases (ozone), naturally occurring fungal products (aflatoxin B1), abused chemicals (cocaine), environmental particles (asbestos), and certain therapeutics (lithium, cyclophosphamide). Several agents have been shown to be important in allergic disease. These include the potential sensitizers nickel, palladium, and toluene diisocyanate (TDI), and ozone and sulfur dioxide, which are modulators of allergic responses.

### 1.2 Indirect Versus Direct Immunotoxic Mechanisms

The modulation of the immune system by chemicals has been shown to occur either through the direct action of the chemical with immune cells or through secondary effects mediated by chemically induced changes in nonlymphoid cells. In addition, some xenobiotics may cause both direct and indirect effects on the immune system. Among these are dioxin [i.e., 2,3,7,8-tetrachlorodibenzo-*p*-dioxin (TCDD)]. Neonatal exposure of mice to TCDD impairs the development of T lymphocytes, probably through an alteration in thymic epithelial cells required to support T-lymphocyte development. In older mice, TCDD appears to exert a direct immunotoxic effect on immune cells, including mouse B lymphocytes.

The interconnections of the immune system with both the neuroendocrine and hepatic systems provide additional mechanistic routes for indirect immunotoxicity. For example, hepatic acute phase responses involve the production of hepatic proteins which, in turn, produce immunomodulation. In addition, immune cells possess cell surface receptors for certain endocrine and neurological mediators; therefore, chemically induced changes in hormone levels or the levels of neuroactive chemicals may have an immunomodulatory effect.

## 2 IMMUNOTOXIC EVALUATIONS

A major challenge in immunotoxicology, as in other areas of toxicology, is to design an evaluation program that will accurately predict the impact of environmental exposure on the human immune system. In most modes of immunotoxicology testing, animal models are used to simulate the possible outcome of human exposure. To the extent that evaluation programs can also provide information on the mechanisms of immunomodulation, the results may help predict the immunotoxic potential of related environmental factors.

Immunotoxicology assessment calls for the balancing of the expense of screening one chemical with the reliability of the information obtained in discerning human and animal health risk. Therefore, the impetus is to utilize fewer immune assays in the evaluation process but to retain the highest level of predictability.

### 2.1 Species Selection

A number of animal models have been employed to evaluate the potential immunotoxicity of environmental factors in humans. The most frequently employed test animal has been the mouse, although in recent years, the rat, the miniature pig, and rhesus and cynomolgus monkeys have been used as models. Other mammals serving as test species include the dog and guinea pig; chickens and fish are the predominant nonmammalian animal models. Human peripheral blood leukocytes have also been used for in vitro tests, although the information obtained from this immune cell source is limited. A novel alternative to this approach is discussed in Section 5.1.

### 2.2 The Tier Approach to Testing

The National Toxicology Program (NTP) of the United States has adopted a two-tiered approach to immunotoxicity testing. Tier I tests cover general immune parameters that provide an indication of pervasive immunomodulation. The tier II assays include specialized immune end points and integrative host–diseases challenges. The second tier tests detect more subtle, yet important changes in the immune system. Recently, the U.S. Food and Drug Administration (FDA) also suggested a tiered approach to immunotoxicology testing. For the FDA, level 1 tests are performed without animal immunizations while level 2 tests include immunizations.

## 3 GENETIC PROBLEMS IN IMMUNOTOXICOLOGY TESTING

One of the problems inherent in immunotoxicity testing is that information obtained in experimental animals usually represents a single genotype, which must ultimately be extrapolated to a more genetically diverse human population. Therefore, the greater the extent to which the original testing can predict the outcome of environment–immune interactions across a broad spectrum of genetic diversity, the more useful the test.

### 3.1 Intraspecies Problems

Immunotoxicity testing is usually performed using one genotype from a given test species. For example, the mouse genotype employed for NTP testing is the F1 female of the cross between the C57B1/6 and C3H inbred strains of mice (B6C3F1). The use of one representative genotype per species is driven by the high cost of testing a single compound and the backlog of candidate compounds awaiting evaluation. However, it is important to recognize that the use of one strain or cross per species means that extrapolation of the results to the entire species (allogeneic extrapolation) may not always be valid. For example, some strains of mice are highly susceptible to immunosuppression induced by ultraviolet B radiation while others are relatively resistant to this environmental factor.

### 3.2 Interspecies Extrapolations

Extrapolation of immunotoxicity data between species (xenogeneic extrapolations) presents certain problems, but it is frequently necessary for the prediction of human health risk. When possible, it is desirable to have information obtained using more than one animal species. If the data produced are in agreement, then extrapolations of the environmental–host immune interaction from animal models to humans can be made with a high level of confidence. The problem arises when conflicting results are obtained from different species. In this case, mechanistic information indicating the basis of the species difference can enhance the predictability of the likely human response to the environmental factor.

## 4 NEW APPROACHES TO IMMUNOTOXICOLOGY TESTING

Since both the NTP and proposed FDA tier testing panels for immunotoxicity testing involve a large number of immune assays, there is considerable interest in the feasibility of using fewer assays for future testing. Recent findings suggest that use of specific combinations of assays in which each assay is dependent on different host protective processes may provide a high degree of predictive reliability. Instead of requiring 20 to 30 different immune assays, it may be possible to use a limited number of assays (e.g., five) and still retain reliability for immunotoxicology testing. Use of a reduced test panel would enable more chemicals to be screened for the same expense and commitment of time presently required to test one chemical. Possible future implementation of a reduced immunotoxicology test panel depends on the outcome of scientific deliberations.

### 4.1 Novel Animal Models

As previously indicated, immunotoxicology information obtained using animal models requires xenogeneic extrapolation to humans, while information obtained directly from humans is inherently limited. To address this problem of direct testing of human immune cells, novel animal models have been used for immunotoxic evaluations that combine the advantages of rodent models with the desirable features of direct examination of human immune cells. Mice with severe combined immunodeficiency syndrome (SCID mice) can be transplanted with human immune stem cells. The resulting xenogeneic mosaic is a mouse containing an intact immune system of human origin. With this modified mouse, in vivo immunotoxicology studies can be conducted on human immune cells. One question raised by this model concerns the extent to which neuroendocrine–immune and hepatic–immune interactions within this mouse are comparable to those interactions in the human.

### 4.2 Predictive Tests

Many immune assays used in immunotoxicity testing employ immune cells exposed in vivo to a test chemical and evaluated in vitro for various characteristics, including functional activity. A pivotal assay for evaluation is the antibody-forming cell (AFC) assay. This assay requires effective cooperation from the three most important immune cell types that function in antigen-specific immune responses. Modulation in the functional capacity of any of the cell populations would result in reduced antibody production in the in vitro culture. When the AFC assay is combined with additional immune end points that reflect first-line host defense processes, the combination of assays can potentially simulate the spectrum of host immune function.

While immunotoxicity testing utilizing cell lines rather than animal-based testing would be desirable, it is problematic for in vitro test systems to model either the complete immune system itself or the interaction of the immune system with hepatic, endocrine, and neurological systems. As a result, in vitro testing, at present, is most useful for the confirmation of direct immunotoxicological processes.

### 4.3 Application of Molecular Biology and Biotechnology to Immunotoxicology

At present, molecular procedures are used to conduct many of the assessments of environmentally determined immunomodulation. Direct examination of gene expression among immune cell populations is a routine procedure that can reflect the functional status of the immune cells. The cytokine genes (e.g., interleukin-1) and the cytokine receptor genes (e.g., interleukin-2 receptor) are frequent candidates for the measurement of transcriptional activity. DNA replication also serves as the basis for certain lymphocyte proliferation assays. In addition, flow cytometry is extensively employed in immunotoxicology for the analysis of cell surface receptors among immune cell populations.

## 5 SUMMARY

The capacity to detect immunomodulation resulting from the exposure to environmental factors is a significant component of our future preventive medicine system. Integrity of the immune system has been recognized as one of the foremost requirements of human health. Yet, the present processes for evaluating potential immunotoxicants have limitations. Among these are the need to conduct many very costly assays to evaluate a single chemical. Additionally, the resulting data usually must be extrapolated from a limited genetic base both within a species and across species to humans. New tests systems such as the SCID mouse transplanted with human cells offer some promise to counter extrapolation problems. In addition, future immunotoxicity testing may employ a limited number of assays; this would permit more chemicals to be evaluated for the same effort. Finally, structural–functional information may be used in some cases for the prediction of the immunotoxic potential of certain environmental chemicals or therapeutic agents.

*See also* CYTOKINES; ENVIRONMENTAL STRESS, GENOMIC RESPONSES TO; IMMUNOLOGY; IMMUNO-PCR; ULTRAVIOLET RADIATION DAMAGE TO DNA.

## *Bibliography*


Burleson, G. R., Dean, J. H., and Munson, A. E., Eds. (1995) *Modern Methods in Immunotoxicology.* Wiley-Liss, New York.

Dean J. H., and Murray, M. J. (1991) Toxic responses of the immune system. In *Casarett and Doull's Toxicology,* 4th ed., M. O. Amdur, J. Doull, and C. D. Klaasen, Eds., pp. 282–333. Pergamon Press, Elmsford, NY.

Hinton, D. (1992) Testing guidelines for evaluation of the immunotoxic potential of direct food additives. *Crit. Rev. Food Sci. Nutr.* 32:-173–190.

Luster, M. I., Munson, A. E., Thomas, P. T., Holsapple, M. P., Fenters, J. D., White, K. L., Jr., Lauer, L. D., Germolec, D. R., and Dean, J. H. (1988) Methods evaluation. Development of a testing battery to assess chemical-induced immunotoxicity. National Toxicology Program's guidelines for immunotoxicity evaluation in mice. *Fundam. Appl. Toxicol.* 10:2–19.

———, Germolec, D. R., and Rosenthal, G. J. (1990) Immunotoxicology: Review of current status. *Ann. Allergy,* 64(5):427–432.

———, Portier, C., Gayla Pait, D., White, K. L., Jr., Gennings, C., Munson, A. E., and Rosenthal, G. J. (1992) Risk assessment in immunotoxicity. *Fundam. Appl. Toxicol.* 18:200–210.

National Research Council. (1992) *Biologic Markers in Immunotoxicology.* National Academy Press, Washington, DC, 206 pp.


# INFECTIOUS DISEASE TESTING BY LIGASE CHAIN REACTION

*Gregor W. Leckie and Helen H. Lee*

## *Key Words*

**Amplicon** The product of nucleic acid amplification which serves as a target in further rounds of thermocycling.

**Anneal** To form double-stranded polynucleotides from two complementary single-stranded polynucleotides by a process of cooling and hybridization.

**Denaturation** Separation of the two hydrogen-bonded, complementary chains of DNA into a pair of single-stranded polynucleotide molecules by a process of heating.

**Extension** Polymerization of nucleotides catalyzed by a DNA polymerase and initiated at the hydroxyl terminus of a hybridized oligonucleotide primer, resulting in replication of a segment of DNA.

**Target** A polynucleotide to which complementary oligonucleotides can anneal.

**Thermocycler** A machine that executes a single, programmable cycle of heating and cooling multiple times.

**Thermostable DNA Ligase** A thermostable enzyme that catalyzes the ligation of two oligonucleotide probes at higher temperatures, thereby reducing the chance of formation of nonspecific ligation products.

**Thermostable DNA Polymerase** A thermostable enzyme that catalyzes the extension of an annealed primer at higher temperatures, thereby reducing the chance of formation of nonspecific extension products.

The ligase chain reaction (LCR) is an oligonucleotide probe directed enzymatic method for the exponential amplification of a nucleic acid target whose sequence is known. LCR requires four oligonucleotide probes, a thermostable DNA ligase, and a thermocycler. The sensitivity of LCR is of the order of 200–300 target molecules, although a modified version termed gap LCR (G-LCR), which also requires thermostable DNA polymerase and a subset of the four types of deoxyribonucleoside triphosphate, can detect fewer than five targets. Neither LCR nor G-LCR efficiently detects RNA. A modified version of gap LCR, termed asymmetric gap LCR (AG-LCR), which also requires reverse transcriptase, can detect as few as 20 RNA molecules. LCR is highly specific, being able to distinguish between target sequences that differ by only a single nucleotide. The sensitivity and specificity with which LCR detects both DNA and RNA targets make it a powerful diagnostic tool with widespread applicability for detection of infectious agents as well as point mutations that are associated with drug resistance of microorganisms, genetic disease, and cancer.

## 1 PRINCIPLES

In the first step of ligase chain reaction (LCR) (Figure 1), a clinical sample containing a specific DNA target sequence is added to a reaction mixture that contains thermostable DNA ligase, nicotinamide adenine dinucleotide, and a vast excess of four oligonucleotide probes in a pH 7.8 buffer. The temperature of the reaction is raised to 94°C to ensure strand separation of both the DNA target and the two pairs of complementary probes. The reaction mixture is then cooled to approximately 55°C, so that probes 1 and 3 anneal at adjacent positions on one strand of the target and probes 2 and 4 bind to the complementary target strand. Thermostable DNA ligase covalently joins the adjacent 3′-hydroxyl group of probe 1 to the 5′ phosphate of probe 3, producing an amplified product, or amplicon. An equivalent process occurs on the complementary target strand between probes 4 and 2. After separation from the original target by heat denaturation, the amplicons themselves serve as targets for unligated probes. Thus consecutive cycles of denaturation and annealing, or thermocycling, result in the exponential amplification of a specific target sequence.

During the annealing step of an amplification thermocycle, unligated LCR probes anneal to their complementary partners as well as to target DNA if it is present. At a low frequency, the blunt-ended LCR probe duplexes can be ligated together, producing complementary LCR amplicons that are further amplified in subsequent cycles of LCR. This process, termed target-independent ligation, limits the sensitivity of LCR to between approximately 200 and 300 target molecules. Target-independent ligation is averted with gap LCR (G-LCR), a modified version of LCR that can detect fewer than five target molecules. G-LCR is more sensitive than LCR because the probes are modified so that the probe duplexes have staggered rather than blunt ends. Target-independent ligation is effectively suppressed because these probe duplexes cannot be ligated together. However, when the modified probes anneal to

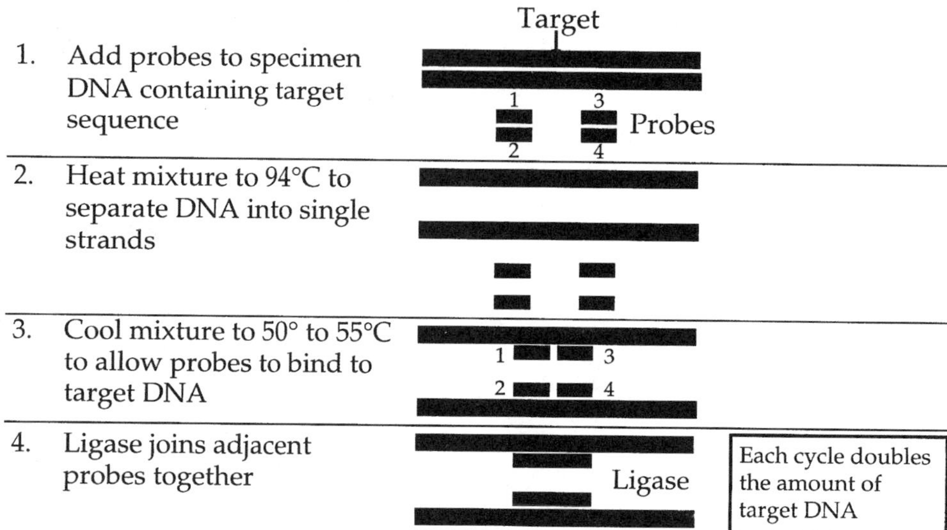
Target
1. Add probes to specimen DNA containing target sequence
1
3
2
4
Probes
2. Heat mixture to 94°C to separate DNA into single strands
3. Cool mixture to 50° to 55°C to allow probes to bind to target DNA
1
3
2
4
4. Ligase joins adjacent probes together
Ligase
Each cycle doubles the amount of target DNA

Figure 1. The mechanism of ligase chain reaction (LCR).

their respective target strands, a gap of one or more bases exists between the ligation termini. G-LCR reactions are therefore supplied with thermostable DNA polymerase, magnesium chloride, and a subset of the four deoxyribonucleoside triphosphate (dNTP) types, along with the other required LCR components, so that these gaps can be filled in by extension. Extension terminates when the nonsupplied dNTP is required. The directly adjacent extended probes are then ligated by thermostable DNA ligase. Repeated rounds of G-LCR also result in exponential amplification of the target sequence.

LCR and G-LCR have limited ability to detect RNA targets because DNA ligase does not efficiently join DNA oligonucleotide probes when they are annealed to an RNA template. A modification of G-LCR, termed asymmetric gap LCR (AG-LCR), can detect as few as 20 RNA target molecules. In AG-LCR (Figure 2), the RNA target is added to a reaction containing probe 4, a heat-sensitive reverse transcriptase, and up to three of the four dNTP types. Reverse transcriptase extends the annealed probe 4 until a nonsupplied dNTP is required (i.e., for 9–15 bases). Extension is then terminated. Amplification cannot proceed unless this target-dependent extension of probe 4 occurs. After heat denaturation, thermostable DNA ligase, thermostable DNA polymerase, and probes 1, 2, and 3 are added to the reaction. After heating and cooling, probes 1 and 3 anneal to the extended probe 4, thus allowing probe 2 to hybridize to probe 1. Thermostable DNA polymerase extends probe 1 using the same dNTP types required for the extension of probe 4. DNA ligase efficiently joins probes 1 and 3 and 2 and 4 because ligation occurs on a DNA not an RNA template. Further rounds of thermocycling result in exponential amplification of the complementary DNA amplicons.

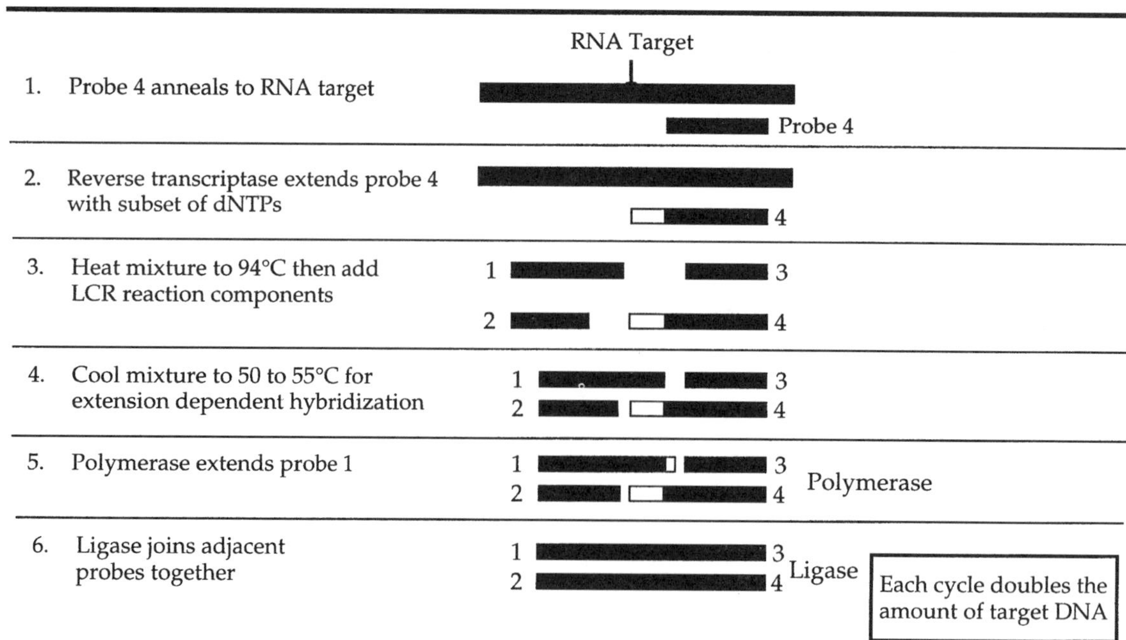
RNA Target
1. Probe 4 anneals to RNA target
Probe 4
2. Reverse transcriptase extends probe 4 with subset of dNTPs
4
3. Heat mixture to 94°C then add LCR reaction components
1
3
2
4
4. Cool mixture to 50 to 55°C for extension dependent hybridization
1
3
2
4
5. Polymerase extends probe 1
1
3
2
4
Polymerase
6. Ligase joins adjacent probes together
1
3
2
4
Ligase
Each cycle doubles the amount of target DNA

Figure 2. The mechanism of asymmetric gap ligase chain reaction (AG-LCR).

LCR is capable of discriminating between two nucleotide sequences that are identical with the exception of one point mutation. The largest discriminatory ability of 10,000-fold has been achieved using G-LCR when the point mutation corresponds to the penultimate nucleotide on the upstream side of the ligation point.

## 2 DETECTION OF INFECTIOUS AGENTS

### 2.1 Sample Preparation

Sample preparation protocols will vary depending on the infectious agent being tested for and the sample being tested. However, all sample preparation protocols should be as simple as possible, to promote high throughput testing and minimize cross-contamination while, at the same time, inactivating and lysing the infectious agent or agents and removing potential inhibitors of LCR. Infectious agents are usually inactivated by heating to 95°C and are lysed by heat treatment alone (if they have weak cell walls) or by a combination of heat treatment and chemical, enzymatic, or mechanical lysis. One advantage of LCR is that it detects a short target sequence of 35–55 bases, and therefore harsh lysis techniques that degrade the released nucleic acid can be used. By comparison, other nucleic acid amplification techniques, such as the polymerase chain reaction (PCR), usually require longer target sequences. Inhibitors of LCR are removed by dilution, centrifugation, or both. More sophisticated techniques now under development concentrate the infectious agent or its nucleic acid while removing LCR inhibitors; such techniques will be useful when the concentration of the infectious agent is low and a high assay sensitivity is required. Ultimately, LCR assays will have internal controls that will monitor for sample inhibition of LCR.

### 2.2 Target and Probe Set Selection

Target sequences comprise 35–55 bases, so that each member of the probe set is sufficiently long for specific hybridization. If possible, targets with significant secondary structure that might reduce LCR efficiency are avoided. Most importantly, the target sequence should be specific to the targeted infectious agent. Specificity is assured by computer searches of available nucleotide sequence databases and by testing the complementary probe set against a wide variety of other infectious agents.

### 2.3 Amplicon Detection

When an LCR reaction is examined by gel electrophoresis, only two bands are usually visible; one band is composed of LCR amplicons while the other consists of unligated probes. In contrast to PCR, LCR rarely generates spurious amplification products of varying size. Consequently, hybridization to a secondary probe is not required to achieve specific detection of LCR amplicons. This is useful when trying to replace a cumbersome and labor-intensive detection procedure, such as gel electrophoresis, with automated techniques. In automated or semiautomated LCR detection procedures, the unligated oligonucleotide probes are labeled with either fluorescent compounds or haptens. Amplicons can be reliably distinguished from unligated probes either by direct measurement of fluorescence or by enzyme immunoassay with colorimetric, chemiluminescent, or fluorogenic substrates. A completely automated microparticle-based detection method performed on an LCx immunoassay analyzer is described in Figure 3. In this method, all four probes are end-labeled at their nonligation termini with either the hapten carbazole or the hapten adamantane, with the result that the resultant LCR amplicons are labeled with both haptens. Following LCR amplification, the analyzer transfers a portion of each LCR reaction mixture into its own individual sample cup, where it is mixed with microparticles that have been coated with polyclonal anticarbazole antibody. The microparticles capture both the amplicons and unligated carbazole-labeled probes. After the capture step, the analyzer transfers each reaction mixture to its own filter, to which the microparticles irreversibly bind. Unligated adamantane-labeled probes are removed by washing with a Tris-acetate-based buffer. The analyzer then adds a polyclonal antiadamantane alkaline phosphatase conjugate, which binds only to amplicon molecules. After a second buffer wash step that removes unbound conjugate, the analyzer adds a substrate, 4-methylumbelliferyl phosphate. This substrate is dephosphorylated by the bound alkaline phosphatase conjugate to produce detectable fluorescence.

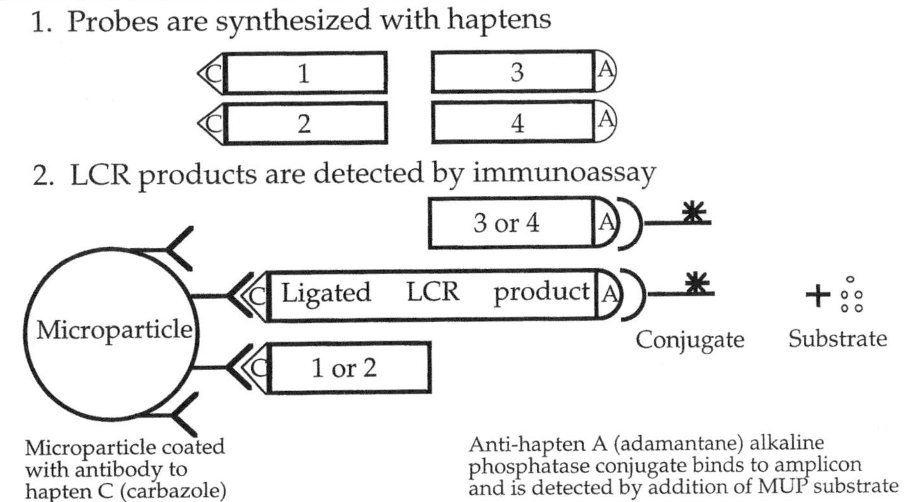


**Figure 3.** Automated microparticle-based enzyme immunoassay detection of LCR amplicon on the LCx Analyzer.

### 2.4 Contamination Control

LCR, like all other techniques capable of nucleic acid amplification, can produce false positive results if samples or reactions become contaminated by small quantities of either target nucleic acid or amplicon. An LCR system that uses the LCx analyzer controls contamination by a combination of (1) physical separation of the different steps (sample preparation, thermocycling, detection) of an LCR assay; (2) use of filter-tip pipets for all pipetting steps to prevent cross-contamination; (3) use of preportioned LCR mixtures, or unit doses, that dispense with the requirement for the tedious and contamination-prone preparation of amplification reactions; (4) automated detection of LCR products, which eliminates operator contact and thereby reduces the physical spread of amplicon; (5) postdetection destruction of amplicon with a binary inactivation reagent (copper phenanthroline and hydrogen peroxide) that is capable of reducing the amount of LCR product by a factor of up to $10^9$; and (6) use of a DNA-destroying solution of 10% household bleach for cleaning contaminated areas.

## 3 APPLICATIONS

LCR, G-LCR, and AG-LCR have been used to detect a wide variety of infectious agents of disease including *Borrelia burgdorferi, Chlamydia trachomatis, Mycobacterium tuberculosis, Neisseria gonorrhoeae,* herpes simplex virus, human immunodeficiency virus (HIV) and provirus, human papilloma virus, hepatitis B virus, and hepatitis C virus. In addition, LCR has been used following PCR to detect *Erwinia stewartii* and *Listeria monocytogenes*. Most of these assays have been qualitative; that is, they can detect only the presence or absence of a particular infectious agent. However, semiquantitative LCR assays for HBV and the HIV provirus have been able to discriminate five- to tenfold differences in the level of DNA target. Semiquantitation is useful when trying to monitor the progression of disease or its response to therapy.

Qualitative LCR assays are capable of highly accurate detection of infectious agents of disease. For example, semiautomated G-LCR assays for detection of the sexually transmitted bacteria *C. trachomatis* and *N. gonorrhoeae* have been developed and tested in clinical trials in the United States. Compared to a combination of culture, the current gold standard technique, and discrepancy analysis, the sensitivity and specificity of the *C. trachomatis* LCR assay were 95 and 99.9%, respectively; the equivalent figures for the *N. gonorrhoeae* LCR assay were 98 and 99.8%. By comparison, overall culture sensitivities for *C. trachomatis* and *N. gonorrhoeae* were 63.1 and 93.5%, respectively. Culture for both microorganisms was 100% specific. These results suggest not only that LCR assays can be highly sensitive, but that contamination as a cause of false positive results can be effectively reduced to insignificance.

The ability to perform multiplex LCR assays is required either if an internal control that monitors sample inhibition is needed or the simultaneous detection of more than one infectious agent is desired. A useful multiplex assay would be one that could simultaneously detect and distinguish between *C. trachomatis* and *N. gonorrhoeae,* since these agents frequently coinfect the urogenital tract, but treatment for the two bacterial infections varies. Finally, if the point mutation discriminatory ability of LCR could be applied in a multiplex format, this reaction could be used to detect such agents as multi-drug-resistant *M. tuberculosis*.

## 4 FUTURE PROSPECTS

Although the basic chemistry of nucleic acid amplifcation by LCR is now well understood, significant work remains to be achieved before the full diagnostic power of this technology is realized. An ideal LCR assay would be automated, homogeneous (incorporating amplification, detection, and, ideally, sample preparation into one reaction vessel), and capable of multiplex detection of nucleic acid targets. Such an assay would also eliminate the potential for generation of false positive results. It is this gap between the current reality and the desired future that researchers will spend the next 10 years trying to bridge.

*See also* Immuno-PCR; Ligation Assays; PCR Technology.

### *Bibliography*

Barany, F. (1991) The ligase chain reaction in a PCR world. *PCR Methods Appl.* 1:5–16.

Birkenmayer, L., and Armstrong, A. S. (1992) Preliminary evaluation of the ligase chain reaction for specific detection of *Neisseria gonorrhoeae*. *J. Clin. Microbiol.* 30:3089–3094.

Chernesky, M. A., Lee, H., Schachter, J., Burczak, J. D., Stamm, W. E., McCormack, W. M., and Quinn, T. C. (1994) Diagnosis of *Chlamydia trachomatis* infection in symptomatic and asymptomatic men by testing first-void urine in a ligase chain reaction assay. *J. Infect. Dis.* 70:1308–1311.

———, Jang, D., Lee, H., Burczak, J. D., Hu, H., Sellors, J., Tomazic-Allen, S. J., and Mahony, J. B. (1994) Diagnosis of *Chlamydia trachomatis* infections in men and women by testing first-void urine by ligase chain reaction. *J. Clin. Microbiol.* 32:2682–2685.

Dille, B. J., Butzen, C. C., and Birkenmeyer, L. G. (1993) Amplification of *Chlamydia trachomatis* DNA by ligase chain reaction. *J. Clin. Microbiol.* 31:729–731.

Landegren, U. (1993) Ligation-based DNA diagnostics. *BioEssays,* 15:761–766.

Lee, H. (1993) Infectious disease testing by ligase chain reaction. *Clin. Chem.* 4:1–3.

Schachter, J., Stamm, W. E., Quinn, T. C., Andrews, W. W., Burczak, J. D., and Lee, H. H. (1994) Ligase chain reaction to detect *Chlamydia trachomatis* infection of the cervix, *J. Clin. Microbiol.* 32:2540–2543.

# Inorganic Solids, Biomolecules in the Synthesis of

*Trevor Douglas and Stephen Mann*

### *Key Words*

**Crystal Lattice** The ordered periodic array of atoms or molecules within a crystalline solid.

**Morphology** The external shape adopted by a solid.

**Supersaturation** The extent to which the equilibrium concentration of solutes ($K_{sp}$) is exceeded.

The use of biological molecules in the synthesis of inorganic solids has grown out of the study of biomineralization. Biological systems are able to tailor the synthesis of inorganic solids using molecular interactions between the organic and inorganic phases. The processes of crystal nucleation and growth can be effectively influenced by using organized molecular assemblies as well as growth modifiers in solution. Langmuir monolayers, vesicles, water-in-oil microemulsions, proteins, gels, and growth additives afford a large degree of control over crystal nucleation and growth as well as the stability and the final shape the crystals will adopt. Well-defined, spatially constrained reaction environments are utilized for nanoscale inorganic material synthesis. Stereochemical, electrostatic, geometric, and spatial interactions between the growing solid and organic molecules have been utilized in controlled crystal formation.

## 1 CRYSTALLIZATION

Crystals can form only if the solution is supersaturated with respect to that material, and their formation involves two stages: nucleation and growth. Many crystalline materials have properties that are anisotropic, and oriented nucleation and growth is therefore important. Recent interest in the properties of small clusters makes the use of spatially constrained reaction environments of particular interest. These clusters exhibit so-called quantum size effects, with their electronic properties having a strong dependence on both the size and shape of the cluster.

### 1.1 Nucleation

Crystals do not form immediately from supersaturated solutions because a kinetic barrier must first be overcome. Crystallization requires the formation of a stable cluster of ions/molecules (critical nucleus) before the energy to form a new surface ($\Delta G_S$) becomes less than the energy released by the formation of new bonds ($\Delta G_B$). The activation energy of nucleation has been formulated as follows:

$$\Delta G_N = \frac{16\pi(\Delta G_S)^2}{3(\Delta G_B)^3} \qquad (1)$$

where $\Delta G_B = kT\ln(S)$ and $S$ is the supersaturation, $k$ is the Boltzmann constant, and $T$ is temperature. Homogeneous nucleation requires that this critical nucleus form spontaneously, as a statistical fluctuation (i.e., the clusters are continually forming and redissolving), from solution. Heterogeneous nucleation comes about through favorable interactions with a substrate, which acts to reduce $\Delta G_S$. The substrate could be an adventitious dust particle contaminating the solution, or it could be an organized molecular assembly designed specifically for the purpose of crystal nucleation.

### 1.2 Growth

Once nucleation has taken place, and provided there is sufficient material, the nucleus eventually grows into a macroscopic crystal. The final morphology of the crystal is affected, for example, by pH, temperature, the degree of supersaturation, and the presence of surface active growth modifiers. Before being incorporated into the lattice, solutes must diffuse to, and be adsorbed onto, the growing crystal surface. Crystal faces that are fast growing will diminish in relative size, while those that grow slowly will dominate the final morphology. The growth rate of a particular face is affected by interactions of molecules adsorbed onto crystal faces.

## 2 ORGANIZED MOLECULAR SYSTEMS

### 2.1 Growth Additives

Additives used in crystal chemistry to affect morphology rely on molecular recognition between an organic and an inorganic phase. The growing face of a crystal has component species in particular geometrical and stereochemical arrangements. When an additive molecule successfully competes with solutes by binding to a face, it slows down the growth of that face because it must be removed before the solute can be incorporated into the growing lattice. The competitive binding of an additive molecule to a particular face requires that it conform to the dictates of lattice geometry, charge density, and stereochemical binding. Additives that bind specifically to certain faces allow crystals to be morphologically tailored. Additives that bind nonspecifically to crystal faces have found use in the inhibition of crystal growth, since they block the growth of all crystal faces. Molecules that affect morphology when free in solution have been found to induce nucleation when they are anchored to a substrate.

### 2.2 Langmuir Monolayers

Surfactant molecules, spread on a water surface, align themselves such that the ionic (or polar) part of the molecule interacts with the water while the hydrocarbon tail is oriented away from the surface. Compressing these molecules together at the air–water interface, in an apparatus with a movable barrier, forms an ordered, two-dimensional array of the surfactants. Compressed Langmuir monolayers above an aqueous subphase, supersaturated with respect to an inorganic solid, induce oriented crystal nucleation. In these systems there is often a high correspondence between the packing of the monolayer and the lattice of the nucleated crystal (geometric factor). Ions adopt unique stereochemical conformations in the crystal lattice that can be mimicked by the monolayer head groups (stereochemical factor). The head group charge results in an accumulation of ions at the interface (electrostatic factor). These three factors are important for oriented nucleation although a degree of mismatch is tolerable.

### 2.3 Vesicles

Small (200–500 Å diameter) synthetic unilamellar vesicles are well-defined reaction environments easily prepared by the sonication of aqueous solutions of phospholipids (or surfactants). When prepared in the presence of metal ions, the vesicle encapsulates those ions and can be separated from the external ions by ion exchange chromatography. Diffusion of membrane-permeable species into the vesicle can be controlled to form the inorganic material of interest as shown in Figure 1. The encapsulated metal cations interact strongly with the charged phosphate head groups of the lipid, and so nucleation is localized at the surface. The slow diffusion of ions across the membrane permits supersaturation to be kept low, and growth of the embryonic nucleus occurs at the expense of further nucleation. Structure, morphology, and particle size are all modified when precipitation is carried out inside a vesicle as compared to bulk solution.

### 2.4 Water-in-Oil Microemulsions

Although only occasionally made from biologically derived materials, oil-in-water microemulsions have proved very useful in the

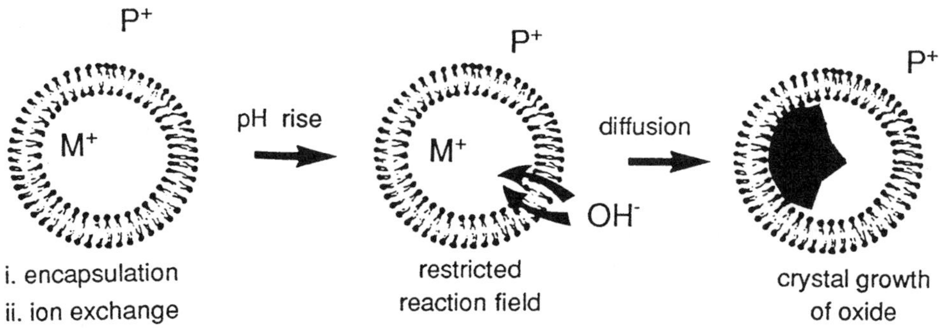


**Figure 1.** Phospholipid vesicles used for the controlled synthesis of inorganic solids.

synthesis and stabilization of semiconductor colloids. These membrane-mimetic structures, encapsulating a pool of water (10–1000 Å in diameter) in an oil phase, have been used in a manner similar to vesicle systems (see Section 2.3). Whereas vesicles are stable, static structures, microemulsions are dynamic, and the water pools, including any dissolved ions, are readily exchanged. Mixture of two microemulsion systems, one containing metal ions ($Cd^{2+}$ or $Zn^{2+}$), the other containing $S^{2-}$ (or $Se^{2-}$) ions, results in the formation of colloidal semiconductor particles stabilized against aggregation. The growth of these particles can be usefully thought of as an inorganic polymerization reaction that stops when the monomers are depleted but continues when they are replenished. Addition of a thiophenolate to the microemulsion system acts to limit the extent of this polymerization and cap the surface of the cadmium sulfide particles.

## 2.5 Ferritin

Ferritin, an iron storage protein, is found in almost all biological systems. It has 24 subunits surrounding an inner cavity 60 to 80 Å in diameter, where the iron is stored as the mineral ferrihydrite ($Fe_2O_3 \cdot xH_2O$—sometimes also containing phosphate). The native ferrihydrite core is easily removed by reduction of the Fe(III) at low pH (4.5) and subsequent chelation and removal of the Fe(II). It is also easily remineralized by the oxidation of Fe(II) at pH exceeding 6, facilitated by specific oxidation–reduction (ferroxidase) and nucleation sites inside the protein cavity, which have been identified by site-directed mutagenesis studies. The intact, demineralized protein (apoferritin) provides a spatially constrained reaction environment for the formation of inorganic particles, which would be rendered stable to aggregation as shown in Figure 2. Oxides of Mn(III), and a uranium oxyhydroxide have been made inside the protein, and an iron sulfide phase prepared by treatment of the ferrihydrite core with $H_2S$. Ferritin is able to withstand quite extreme conditions of pH (4.0–9.0) and temperature (up to 85°C) for limited periods of time, and this property has been used to advantage in the synthesis of magnetite ($Fe_3O_4$) particles within the protein.

## 2.6 Gels

A gel is an extended three-dimensional, loosely cross-linked, polymer permeated by water through interconnecting pores. Gels of molecules such as gelatin, agar-agar, and silica are used as reaction media for crystal growth when especially big, defect-free crystals are desired. Experimentally, reactants (solutes) are allowed to diffuse toward each other from opposite ends of a gel-filled tube. This creates a concentration gradient as the two fronts diffuse through each other, giving rise to conditions of local supersaturation. The gel serves to suppress nucleation, which allows fewer crystals to form, thus reducing the competition between crystallites for solute molecules; the result is larger and more perfect crystals.

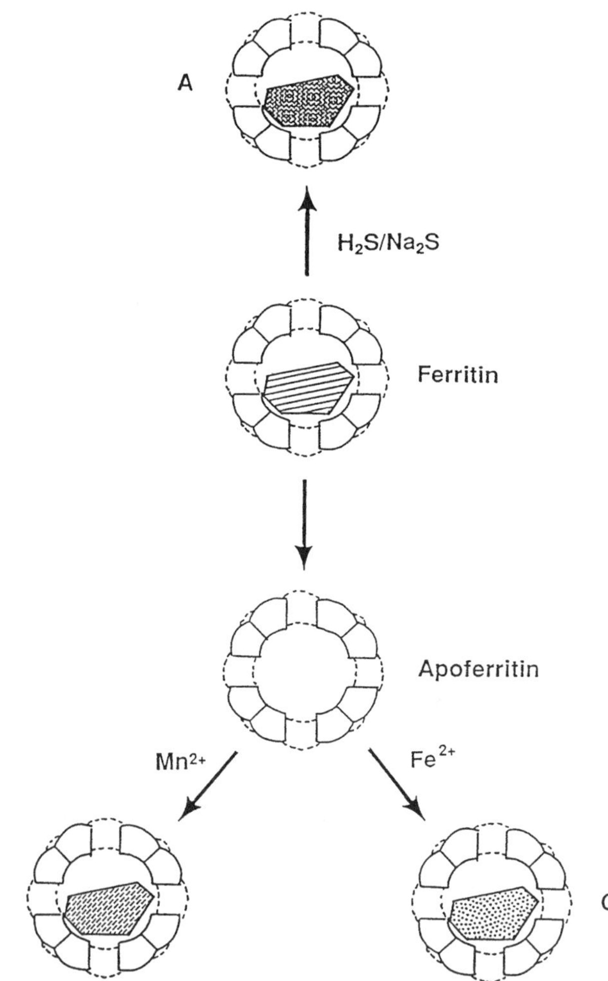


**Figure 2.** Use of ferritin as a reaction vessel: (A) the native ferrihydrite core of ferritin reacts with $S^{2-}$ to yield an iron sulfide; the ferrihydrite core is removed to form apoferritin, which can then be remineralized by the controlled oxidation of (B) Mn(II) and (C) Fe(II).

The gel is easily deformable and exerts little force on the growing crystal. Gelatin is used extensively in the photographic process for the immobilization of silver and silver halide microcrystals.

### 2.7 Dispersants

There is increasing interest in the processing of colloidal particles for use in pigments, ceramic precursors, and solar energy conversion, for example. However, the attractive forces between small colloidal particles can cause irreversible aggregation. Biopolymers such as gelatin, alginate, caseinate, albumin, and dextrin have been used successfully to prevent colloid aggregation. Attachment of these and other biopolymers to the colloid surface (known as steric stabilization) prevents aggregation by creating a repulsive interaction between particles.

### 2.8 Composite Materials

Proteins that have been isolated from biominerals exhibit a number of the properties mentioned in the preceding sections. The production of biocomposite ceramics is a low temperature route to strong, lightweight materials that has not yet been fully exploited. Purified collagen serves as a matrix for calcium phosphate growth, and matrix proteins isolated from bivalves have been shown to mediate nucleation and growth of calcium carbonate. These materials are composites of microscopic crystals held together by a protein "glue" and have the advantages of both the hardness of the inorganic material and the flexibility of the organic matrix. Composite materials such as these often have high fracture toughness, which is thought to arise from interruption, by the protein, of the cleavage planes in the inorganic crystals.

## 3 SUMMARY

Understanding how organized molecular assemblies influence the nucleation and growth of inorganic crystals has allowed some control over those events for the purpose of synthesizing new materials. Molecular recognition at the organic–inorganic interface is key to this understanding. Extensive use of biomolecules has allowed the synthetic chemist to take advantage of unusual properties of shape or size by inducing oriented nucleation (or inhibiting it), controlling the expression of specific crystal faces, and controlling the particle size. Biomolecules have also been useful in stabilizing small particles from aggregation and in providing a crystallization matrix for the production of composite materials.

*See also* Bioinorganic Chemistry; Biomimetic Materials.

### *Bibliography*

Alper, M., Calvert, P., Frankel, R., Rieke, P., and Tirrell, D., Eds. (1991) Materials synthesis based on biological processes. *Mater. Res. Soc. Symp. Proc.* 218.

Fendler, J. H. (1987) Atomic and molecular clusters in membrane mimetic chemistry. *Chem. Rev.* 87:877.

Henisch, H. K. (1988) *Crystals in Gels and Liesegang Rings.* Cambridge University Press, Cambridge.

Heuer, A. H., Fink, D. J., Arais, J. L., Calvert, P. D., Kendall, K., Messing, G. L., Blackwell, J, Rieke, P. C., Thompson, D. H., Wheeler, A. P., Veis, A., and Caplan, A. I. (1992) Innovative materials processing strategies: A biomimetic approach, *Science.* 255:1099.

Lowenstam, H. A., and Weiner, S. (1989) *On Biomineralization.* Oxford University Press, Oxford.

Mann, S., Webb, J., and Williams, R. J. P., Eds. (1989) *Biomineralization: Chemical and Biochemical Perspectives.* VCH, Weinheim.

Rieke, P. C., Calvert, P. D., and Alper, M., Eds. (1990) Materials synthesis utilizing biological processes. *Mater. Res. Soc. Symp. Proc.* 174.

Sikes, C. S., and Wheeler, A. P, Eds. (1991) *Surface Reactive Peptides and Polymers,* ACS Symposium Series No. 444. American Chemical Society, Washington, DC.

Steigerwald, M. L., and Brus, L. E. (1990) Semiconductor crystallites: A class of large molecules. *Acc. Chem. Res.* 23:183–188.

# Insecticides, Recombinant Protein

*James Y. Bradfield and Beverly J. Burden*

### *Key Words*

**Bacillus thuringiensis (Bt)** A spore-forming soil bacterium that produces several orally active insecticidal proteins.

**Baculoviruses** A large group of viruses containing environmentally stable forms that infect only insects and related invertebrates.

**Integrated Pest Management (IPM)** Information-based, long-term maintenance of a pest at acceptable levels, involving chemical and biological agents, and cultural and physical practices.

**Promoter** A segment of DNA that positions RNA polymerase for accurate initiation of gene transcription.

**Ti Plasmid** A bacterial plasmid that transfers DNA sequences into plant chromosomes.

**Transgenic** Endowed with foreign DNA sequences that persist in subsequent generations of the organism.

Insects and related arthropods destroy annually 15% of agricultural produce in the United States and, more seriously, twice that or more in many developing countries. Furthermore, diseases carried by insects pose a serious threat to humans and domesticated animals: for instance, mosquito-borne malaria kills more people worldwide than any other infectious agent.

Devastations by insects continue despite a tenfold increase in the rate of application of synthetic organic insecticides since World War II. It is estimated that today in the United States alone, 250 million pounds of noxious chemicals, some having considerable environmental persistence, are dispensed into nature each year for insect control, at enormous cost to public health, nontarget wildlife, and future environmental quality. It is clear that insect pest management strategies having wholesale reliance on chemical insecticides must be abated.

The emergence of recombinant DNA/genetic engineering methodology 15–20 years ago began a revolution in the biological sciences. Among the many people attracted to the new genetic techniques were (and are) a small number of scientists, consisting chiefly of entomologists, microbiologists, and botanists, who envision that genetic manipulation of organisms associated with pest insects (microbes, host plants, predators, and parasites) will contribute significantly to environmentally safe and effective pest management, mitigating the heavy use of industrial chemicals. This entry summarizes some of the rationale and efforts toward that end. The ideas put forth are not those of the authors by and large, but have been gleaned from primary literature.

## 1 GENETICALLY ENGINEERED INSECT PATHOGENS

Infectious diseases of insects were noted in ancient times, and as early as the 1870s Pasteur suggested that bacteria could be used to control pestiferous species. We now know that all major groups of microorganisms (bacteria, protozoa, yeasts, fungi, rickettsiae, spirochetes, and viruses) are associated with insects. Theoretically, the number of infectious microbes that might be genetically modified to enhance their virulence to pest insects is almost limitless, but as yet very few have begun to be exploited. Among those that have been exploited, viruses of the genus *Baculovirus* and the bacterial genus *Bacillus* (especially species *thuringiensis*) are most prominent.

### 1.1 Baculoviruses

The genus *Baculovirus* (family Baculoviridae) has several qualities that are desirable in a biological insect control agent. Its host range is restricted to insects and a few other invertebrates, it poses no threat to vertebrates or plants, and several naturally occurring strains are already registered for commercial agricultural use. Baculoviruses are especially virulent among caterpillars (order Lepidoptera), agriculturally the most destructive group of insects worldwide. Baculoviruses are readily propagated in cell culture and in insectaries on a large scale. Finally, many baculoviruses are unusual among viruses in that after several rounds of replication in infected (and disintegrating) insects, they are released into the environment embedded in crystals called occlusion bodies or polyhedra. Some 90–95% of the mass of polyhedra consists of multiple units of a protein known as polyhedrin. Polyhedrin is the product of a single gene in the baculovirus genome and is produced massively, late in infection, accounting for 50% or more of all proteins being made by infected insect cells. The occlusion body formed by polyhedrin provides environmental stability to the embedded virions: infectivity can persist several years in the soil and can endure traditional insecticide application equipment. Viral infection occurs upon in-

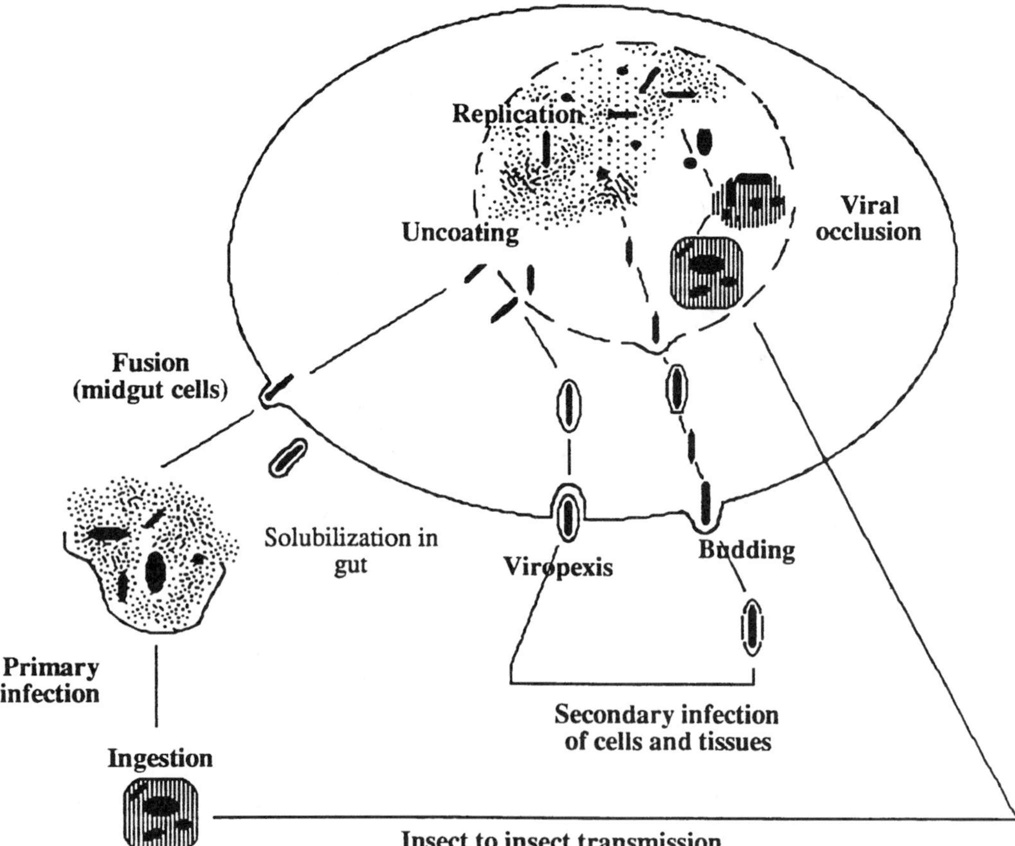


**Figure 1.** Baculovirus life cycle. An insect ingests viral occlusion bodies (polyhedra) that contaminate plant surfaces. Polyhedra dissolve in the midgut of the insect, releasing virus particles that traverse the cell membrane and cytoplasm, to become localized in the nucleus, where viral DNA replication takes place. Within 12 hours postinfection, virus particles begin to bud from the cells for secondary infection of peripheral tissues. About one day postinfection, virus particles become embedded in occlusion bodies that accumulate to large numbers, followed by cell lysis. Eventually, the whole insect disintegrates and liberates millions of infectious occlusion bodies, resulting in insect-to-insect transmission. [From M. D. Summers and G. E. Smith: *A Manual of Methods for Baculovirus Vectors and Insect Cell Culture Procedures.* Texas Agricultural Experiment Station Bulletin 1555 (1988); reproduced by permission of Max D. Summers.]

gestion and dissolution of occlusion bodies in the insect gut. Figure 1 displays the baculovirus life cycle.

While many properties of the baculovirus suggest that naturally occurring forms might be widely useful as insect control agents, deployment of these microbes as commercial insecticides has been insignificant. A major drawback to be addressed by genetic engineering is the slow speed-of-kill by baculoviruses in comparison to synthetic chemicals: insects can survive and continue to feed for several days after viral infection, versus the rapid knockdown delivered by exposure to common insecticides. To increase speed-of-kill by baculoviruses, scientists have begun inserting into the baculovirus genome foreign genes, whose products are toxic to insects, for expression driven by viral promoters. (Since the baculovirus genome consists of double-stranded DNA, addition and substitution of gene sequences are readily accomplished.) Efforts to express genes encoding paralytic toxins from natural predators (spiders, scorpions, and mites) have been successful in the laboratory. The virus–toxin gene recombinants not only reduce feeding time in infected insects compared with effects of the wild-type baculovirus, but they also appear to remain harmless to humans and animals. Other genes that may enhance the virulence of baculoviruses code for insect-specific hormones whose inappropriate expression might disrupt metabolism or water balance. Still other genes and strategies for potentiating viral insecticidal activity, such as insertion of multiple foreign genes in a single recombinant virus, are emerging with discovery of new genes whose products affect insect behavior, and with accruing knowledge of the molecular genetics of the insect–virus relationship. The next several years should result in the commercialization of baculoviruses born of recombinant DNA.

### 1.2 *Bacillus Thuringiensis*

Like many bacteria of the genus *Bacillus, B. thuringiensis (Bt)* is a spore-forming soil inhabitant. A special feature of *Bt* is the formation, during sporulation, of an insecticidal crystalline deposit known as the parasporal crystal. Upon ingestion of spore and its associated crystal by a susceptible insect, the crystal is dissolved in the gut, inducing rapid paralysis of the mouthparts and alimentary canal. Feeding (plant damage) ceases within a few minutes or hours, followed by destruction of the gut and germination of the dormant spores to produce actively multiplying bacteria. Infected insects can linger harmlessly for several days, during which they serve literally as *Bt* factories; eventually they disintegrate with liberation of millions of infectious spores. Dozens of natural isolates of *Bt* have been described having collective activity against some of the most serious insect pests, notably species of Lepidoptera, Diptera (flies, especially mosquitoes and black flies), and Coleoptera (beetles). Formulations for insecticidal use have been registered by the U.S. Environmental Protection Agency for decades and, although the overall impact of *Bt* is small (far less than 1% of insecticides sold are *Bt*-based), it remains the most commercially successful microbial insecticide. Like the baculovirus described earlier, *Bt* is innocuous to vertebrates and most beneficial invertebrates.

Major paralytic component(s) of the *Bt* parasporal crystal are polypeptides called Cry proteins (also known as δ-endotoxins). Cry proteins are typically derived from larger polypeptide precursors. The genes encoding them comprise an evolutionarily related family of molecules that have been classified into four major groups based on molecular structure analysis and the spectrum of insects against which they are active (Table 1). Upwards of two dozen genes distributed among groups *cry* I through *cry* IV are known, and doubtless many more will be discovered. Each *Bt* strain produces a unique set of Cry protein(s). To a significant extent, it is the composition of *cry* genes that accounts for the virulence of a particular *Bt* strain to a particular insect.

**Table 1** *Bacillus thuringiensis* Toxin Genes

| Gene Name | Size (kDa × $10^{-3}$) | | Susceptible Insects |
|---|---|---|---|
| | Toxin | Toxin Precursor | |
| cry IA (a) | 60–70 | 135 | Lepidoptera |
| cry IA (b) | 60–70 | 130 | Lepidoptera |
| cry IA (c) | 60–70 | 135 | Lepidoptera |
| cry IB | 60–70 | 140 | Lepidoptera |
| cry IC | 60–70 | 135 | Lepidoptera |
| cry ID | 60–70 | 130 | Lepidoptera |
| cry IE | 60–70 | 130 | Lepidoptera |
| cry IF | 60–70 | 130 | Lepidoptera |
| cry IIA | 50 | 70 | Lepidoptera and Diptera |
| cry IIB | 50 | 70 | Lepidoptera |
| cry IIIA | 60 | 75 | Coleoptera |
| cry IIIB | Unknown | 75 | Coleoptera |
| cry IIIC | Unknown | 75 | Coleoptera |
| cry IVA | 65–75 | 135 | Diptera |
| cry IVB | 65–75 | 130 | Diptera |
| cry IVC | 60 | 80 | Diptera |
| cry IVD | 40 | 70 | Diptera |

Factors that currently limit the broad acceptance of *Bt* include cost, short half-life on plant surfaces (due to inactivation by sunlight and washing off by rain), slow rate of spread within an insect population and, strain by strain, a narrow range of susceptible insect species compared with conventional industrial chemicals. One strategy being used to address host range and potency limitations is the generation of new isolates carrying novel combinations of *cry* genes. Since the majority of *cry* genes are carried naturally on extrachromosomal plasmids, their transfer into cells for expression can be readily accomplished by the experimenter, using electroporation of other bacterial transformation techniques. This process can also be achieved by conjugation, a means of DNA transfer between bacteria known to occur in nature. Through plasmid transfer, *Bt* strains carrying *cry* genes derived from multiple isolates have already been developed and registered for use against vegetable and forest pests. Development of insecticidally improved *Bt* strains through *cry* gene engineering will accelerate with discovery of new toxin genes and their regulatory mechanisms, and with further elucidation of the interactions of Cry toxins with target sites in the insect gut.

## 2 GENETICALLY ENGINEERED HOST–PLANT SYSTEMS

Delivery of insecticidal chemicals by host plants (or microorganisms closely associated with plants) is attractive for two reasons. First, plant systems that express insect toxicants can reduce the need for repeated foliar spraying with its associated expenses and atmospheric contamination and, second, these systems can deliver the chemicals to insects feeding on plant parts not readily covered by sprays. Through application of genetic engineering technology, novel expression of insect-specific toxins in host–plant systems is beginning to be achieved. It is not surprising that most efforts thus far have focused on expression of the Cry toxins of *Bt,* since these are among the very few gene products (peptides and proteins) known to be both specifically toxic to insects and orally active.

### 2.1 Microorganisms Associated with Plants

A number of microbes naturally colonize or coexist with plants of agricultural importance and are being manipulated to deliver Cry toxins and other gene products. The ubiquitous soil bacterium *Pseudomonas fluorescens* grows nonparasitically on the roots of several plants such as corn and has been engineered with a *cry* I gene. Upon ingestion by insects, this genetically modified *Pseudomonas* has been shown to be lethal, while its wild-type parent is harmless. Some microorganisms are quite intimately associated with plants and thrive harmlessly, or endosymbiotically within tissues of a broad range of hosts. For instance, the bacterium *Clavibacter* multiplies throughout the vascular system and in parts of the reproductive system of a number of plants, including corn and rice. *Clavibacter* and other microbes that systemically inhabit a broad range of agriculturally important crops may be desirable agents for expression of insecticidal genes, since one or a few strains created by genetic engineering could deliver toxicants to several parts of multiple plant types.

### 2.2 Transgenic Plants

Recombinant DNA techniques have been developed for inserting foreign gene sequences into plant chromosomal DNA, providing new phenotypes that persist through ensuing generations. The technology serves as a shortcut to classical and often tedious plant breeding methods and has resulted in stable phenotypes for herbicide tolerance, tolerance of saline soil conditions, plant pathogen immunity, and pest insect determent. With respect to the latter, *Bt* toxin genes have been given greatest attention, although other genes for proteins that deter insect feeding, such as digestive enzyme inhibitors, have been investigated as well.

Methods for integrating new DNA into plant chromosomes vary according to the taxonomic status of the target plant. In the case of dicotyledonous or broad-leafed plants (e.g., cotton, potato, tomato, tobacco), scientists can capitalize on a natural system by which plants acquire foreign genetic material. *Agrobacterium tumifaciens* is a bacterium that infects plants at a wound site, causing a proliferation of cells that results in tumors called crown galls. Crown gall tumors result from transfer of gene sequences from *A. tumifaciens* to the chromosomes of infected plants. As it turns out, the transferred genetic material is derived from an *A. tumifaciens* plasmid—the Ti (tumor-inducing) plasmid. As a bacterial plasmid, the Ti plasmid can be fairly readily engineered with genes of choice for transfer to the plant genome. By applying recombinant DNA and in vitro plant culture methodologies to the naturally occurring *A. tumifasciens*–Ti plasmid system, workers have achieved stable chromosomal integration of *Bt* toxin genes and other genes linked to promoters recognized by plant cells, with varying degrees of expression and protection against insect feeding damage. Examples are listed in Table 2.

Although the *A. tumifaciens*–Ti plasmid system is widely useful for gene transfer into broad-leafed plants, it cannot be applied to monocotyledonous (grassy) plants, a large group that includes some of the most important crops such as corn, wheat, and rice. To generate transgenic monocotyledons, physical means of introducing DNA into cells are being employed. Methods include electroporation, microinjection, lasers that create pores in plant cell walls and membranes and, most effective to date, particle gun bombardment. With the latter, minute metal beads coated with DNA are fired at high velocity into plant cells or seeds. Although some success in plant transformation has been enjoyed using physical techniques (Table 2), biological systems for delivering DNA into plant chromosomes remain attractive, especially those using recombinant plant viruses.

**Table 2** Plants Engineered with *B t* Toxin Genes

| Plant | Gene Transfer Method | Target Insects | Test Status |
|---|---|---|---|
| Tobacco | Ti plasmid | Tobacco hornworm | Field |
| Tomato | Ti plasmid | Tobacco hornworm | Field |
| Cotton | Ti plasmid | Tobacco budworm<br>Cotton bollworm<br>Cotton leafworm | Field |
| Potato | Ti plasmid | Colorado potato beetle<br>Potato tuber moth | Field |
| Corn | Particle gun | European corn borer | Field |
| Poplar | Particle gun | Gypsy moth<br>Tent caterpillar | Laboratory |
| Spruce | Particle gun | Spruce budworm | Laboratory |

## 3 PREDATORS AND PARASITES

Insects have a number of natural enemies in addition to those from the microbial world. Insects are exploited by many parasites, including nematodes, flatworms, tapeworms and, most notably, insects: it has been estimated that half of the 2–3 million species are parasitic hymenopterans (wasps and bees) whose sole targets are other insects. In addition to the large number of parasites, there are thousands of predaceous arthropods for whom insects are the main or only source of sustenance.

Predators and parasites have considerable impact on the management of insect pests. In the United States today it is estimated that natural enemies are responsible for 50% of insect control achieved in agriculture, and improvements in the performance and conservation of these organisms can add to their vast economic and environmental benefits. Future genetic engineering efforts may result in improved phenotypes related to host-seeking behaviors, host range, reproductive performance, and insecticide tolerance in predators and parasites of agriculturally and medically important pest insects. In the meantime, much must be learned of relevant genes and gene pathways and, just as important, efficient means of gene transfer into beneficial arthropods, scarcely achieved to date, need to be developed.

## 4 CONCLUSIONS

Insects are of immense importance in agriculture and medicine, and strategies for their management will be profoundly influenced by recombinant DNA and its associated technologies. This promise is underscored by the recent emergence of microorganisms and plants that have been genetically engineered—successfully so, at the level of the laboratory bench and in some cases field trials—to reduce insect damage while posing no apparent threat to nontarget organisms. The field is in its infancy. Abilities to modify insect activities through applied molecular genetics will increase exponentially with fundamental research on insects and the organisms with which they interact.

Given numerous concerns for the safety of humans, wildlife, the environment, and natural enemies of insect pests, any implementation of genetic technology for insect control should be undertaken only after careful deliberation by experts who can bring to bear fundamental and applied knowledge of many fields of biology, including entomology, botany, microbiology, genetics, and ecology. Naïve and excessive use of synthetic insecticides over the last half-century has resulted, in many cases, in the uselessness of these chemicals as pest control agents. This is because insects are phenomenally adaptive, and populations under heavy selection pressure rapidly develop resistance. It makes little difference whether the pressure is exerted by an industrial chemical delivered from a spray nozzle or a recombinant protein delivered by a plant or microbe. Genetic engineering strategies for insect control must be developed in awareness of the resistance phenomenon and carried out in the context of integrated pest management.

*See also* Bacterial Pathogenesis; Pesticide-Producing Bacteria; Plant Cell Transformation, Physical Methods for.

### *Bibliography*

Baker, R. R., and Dunn, P. E., Eds. (1990) *New Directions in Biological Control: Alternatives for Suppressing Agricultural Pests and Diseases.* Liss, New York.

Croft, B. (1990) *Arthropod Biological Control Agents and Pesticides.* Wiley, New York.

Maramorosch, K., Ed. (1991) *Biotechnology for Biological Control of Pests and Vectors.* CRC Press, Boca Raton, FL.

Pimental, D., and Lehman, H., Eds. (1993) *The Pesticide Question: Environment, Economics, and Ethics.* Chapman & Hall, New York.

Roberts, D. W., and Granados, R. R., Eds. (1989) *Biotechnology, Biological Pesticides and Novel Plant–Pest Resistance for Insect Pest Management.* Boyce Thompson Institute for Plant Research, Ithaca, New York.

Watson, J. D., Gilman, M., Witkowski, J., and Zoller, M. (1992) *Recombinant DNA.* Freeman, New York.

## Intelligence: *see* Memory and Learning, Molecular Basis of.

# Interferons

*Michael J. Clemens*

### *Key Words*

**Antiviral Proteins** Proteins that are induced by the interferons and protect cells from infection by a range of viruses. Examples are the 2′5′-oligoadenylate synthetases, the double-stranded RNA-activated protein kinase (PKR), and the Mx proteins.

**Cytokines** Proteins that are synthesized and secreted by a variety of cell types and regulate the activity of the same or other cell types by interaction with specific cell surface receptors

**Interferons** Proteins that are synthesized and secreted in response to virus infection or antigenic stimulation, and protect other cells from subsequent virus infection. They also have many other important biological effects such as cell growth inhibition and regulation of the immune system.

**Receptors** Proteins, consisting of one or more subunits, on the surface of cells, that bind cytokines in a specific manner and with high affinity and are responsible for transmitting biochemical signals to the interior of cells.

**Signal Transduction** Process whereby the interaction of a cytokine (such as an interferon) with its specific receptor leads to changes in the biochemical activity of a cell.

The interferons (IFNs) are members of the larger group of cytokines, which are factors exerting hormone-like actions on a wide range of target cells and tissues. They share the common property with the hormones of being secreted into the bloodstream and other body fluids by their producer cells and, again like hormones, they exert their effects by interacting with specific receptors on their target cells. Unlike the hormones, however, cytokines, including the IFNs, are not produced by specific endocrine organs but rather are synthesized by a large variety of cell types. They can also exert autocrine and paracrine effects on cells, as well as classical endocrine-type actions.

The receptors for the IFNs transduce their biochemical signals to the interior of the cell by a number of pathways that lead to specific cellular responses. Most if not all of these responses require changes in the pattern of gene expression by the target cells.

## 1 THE INTERFERONS

Although it is now clear that the IFNs can produce a variety of effects on cells, they were originally identified as agents that give rise to an "antiviral state" (i.e., protect cells against infection by viruses). This class of cytokines therefore has a major role to play in host defenses against viral infections, in a manner that is independent of the immune system. Consistent with the antiviral role of the IFNs, major inducers for their production are viruses themselves. However IFN-$\gamma$ is produced in response to antigenic stimulation of T lymphocytes.

Table 1 shows the classification of the human IFNs into three groups. The IFN-$\alpha$ group consists of a large family of closely related proteins, mainly produced by leukocytes. These IFN species can be further divided into two subfamilies: IFN-$\alpha$I and IFN-$\alpha$II (sometimes referred to as IFN-$\omega$). IFN-$\beta$ is a single species, produced by fibroblasts and leukocytes, that is very similar in structure and function to the alpha IFNs. The alpha and beta IFNs share common receptor components on their target cells. In contrast, IFN-$\gamma$ is produced exclusively by T lymphocytes and works through a distinct receptor. Because of its cellular origin, IFN-$\gamma$ can therefore also be classed as a lymphokine. This species is less closely related in structure to the other IFNs.

## 2 INTERFERON RECEPTORS AND SIGNAL TRANSDUCTION

As is the case for many cytokine receptors, the receptors for the IFNs probably consist of more than one subunit, although their full characterization is still awaited.

**Table 1** Properties of the Interferons

| Interferon | Number of Amino Acids | Molecular Mass of Native IFN | Principal Sources |
|---|---|---|---|
| IFN-α family[a] | 166–172 | 16,000 to 27,000 | T and B lymphocytes; monocytes |
| IFN-β | 166 | ca. 20,000 | Fibroblasts; leukocytes |
| IFN-γ | 143 | ca. 50,000 (dimer) | T lymphocytes; natural killer cells |

[a]The human IFN-α family can be divided into two subfamilies: IFN-αI and IFN-αII (also referred to as IFN-ω).

The signal transduction mechanisms used by the IFN system have been the subject of intensive study, and it seems likely that more than one pathway is involved. Activation of one or more tyrosine kinases has been shown to occur when both IFN-α and IFN-γ bind to their respective receptors. This results in the phosphorylation and consequent activation of transcription factors such as ISGF3 (for interferon-stimulated gene factor 3). When phosphorylated, ISGF3 translocates from cytoplasm to nucleus and stimulates the transcription of a number of IFN-α-responsive genes. There is also evidence that protein kinase C and arachidonic acid stimulated pathways may be responsible for some of the biological effects of the IFNs.

## 3 REGULATION OF GENE EXPRESSION BY THE INTERFERONS

Among the genes that undergo transcriptional stimulation following exposure of cells to IFNs are those encoding the 2′5′-oligoadenylate synthetases, the double-stranded RNA activated protein kinase (PKR), and the Mx proteins. These gene products (together with other less well characterized proteins) are involved in protecting cells against a range of infecting viruses.

In addition to these effects, the IFNs bring about many other changes in cellular phenotype, including inhibition of cell proliferation, induction of differentiation, and modulation of functions of cells of the immune system. These actions are summarized in Table 2. The changes in gene expression responsible for these effects are less well understood.

**Table 2** Biological Effects of the Interferons

| |
|---|
| Inhibition of viral replication[a] |
| Protection of cells against other intracellular parasites[a] |
| Inhibition of proliferation of some normal or transformed cells[a] |
| Regulation of cell differentiation[a] |
| Enhancement of class I major histocompatibility antigen expression[a] |
| Induction or enhancement of class II major histocompatibility antigen expression[b] |
| Regulation of expression of receptors for other cytokines and for Fc portion of IgG[b] |
| Induction of cytokines (e.g., interleukin-1, tumor necrosis factor, or colony-stimulating factors)[b] |
| Activation of macrophages[b] |
| Activation of natural killer cells[a] |
| Synergism with interleukin-2 in stimulation of proliferation of B lymphocytes and production of IgG immunoglobulins[b] |

[a]IFNS α, β, and γ.
[b]IFN-γ only.

*See also* Cytokines; Gene Expression, Regulation of.

### *Bibliography*

Balkwill, F. R. (1989) *Cytokines in Cancer Therapy.* Oxford University Press, Oxford.

Clemens, M. J. (1991) *Cytokines.* Bios Scientific Publishers, Oxford.

De Maeyer, E., and De Maeyer-Guignard, J. (1988) *Interferons and Other Regulatory Cytokines.* Wiley, New York.

Meager, A. (1990) *Cytokines.* Open University Press, Milton Keynes.

# Interleukins

*Michael J. Clemens*

### *Key Words*

**Cytokines** Proteins that are synthesized and secreted by a variety of cell types and regulate the activity of the same or other cell types by interaction with specific cell surface receptors.

**Interleukins** Proteins, identified by the abbreviations IL-1 to IL-13, that are synthesized and secreted by a variety of cell types and control many functions of the immune system, as well as numerous other cellular processes.

**Lymphokines** Cytokines that are produced by cells of the immune system (especially activated T lymphocytes).

**Receptors** Proteins, consisting of one or more subunits, on the surface of cells that bind cytokines in a specific manner and with high affinity, and are responsible for transmitting biochemical signals to the interior of cells.

**Signal Transduction** Process whereby the interaction of a cytokine (such as an interleukin) with its specific receptor leads to changes in the biochemical activity of a cell.

The interleukins are proteins that are synthesized and secreted by a variety of mammalian cells in response to a large range of stimuli. They exert their effects on lymphocytes and other target cell types through interaction with specific cell surface receptors and the transcriptional induction of specific genes. The biological responses of cells to interleukins are very varied but particularly include regulation of immune and inflammatory mechanisms. Production of interleukins is also important in mediating the physiological effects of stress.

## 1 INTERLEUKINS AS EXAMPLES OF CYTOKINES

The interleukins constitute a group of small proteins that are produced by, and regulate, lymphocytes and other cells involved in immune defense mechanisms. These proteins are members of the larger group of cytokines which are factors exerting hormonelike actions on a wide range of target cells and tissues. They share the common property with the hormones of being secreted into the bloodstream and other body fluids by their producer cells and, again like hormones, they exert their effects by interacting with specific receptors in their target cells. Unlike the hormones, however, cytokines, including the interleukins, are not produced by specific endocrine organs but rather are synthesized by a large variety of cell types. They can also exert autocrine and paracrine effects on cells, as well as classical endocrine-type actions.

The interleukins are largely, but not exclusively, synthesized by various classes of lymphocytes. These cells also produce a number of other factors which, mainly for historical reasons, are not classified as interleukins.

In common with other cytokines, the interleukins exert their effects mostly by being secreted by their producer cells and then interacting with specific receptors on the surface of their target cells. However some interleukins may remain associated with the membranes of the cells that produce them and exert their effects by direct cell–cell interactions. The receptors for the interleukins transduce biochemical signals to the interior of the cell by a variety of mechanisms, resulting in specific cellular responses. The ability of a cell to respond to a particular interleukin depends on the presence of the relevant receptor on the cell surface, and the nature of the effects produced will depend on the cell type. Many (perhaps all) of these cellular responses require changes in the pattern of gene expression by the target cells.

## 2 THE INTERLEUKINS

The properties of the interleukins described to date are summarized in Table 1. These proteins, which have been characterized most thoroughly in the human and mouse systems, are synthesized in response to a large range of stimuli. In particular, T lymphocytes that are activated by antigenic stimulation produce these molecules. Although the interleukins are characteristically the products of T or B lymphocytes, many can also be produced by other cell types.

## 3 INTERLEUKIN RECEPTORS AND SIGNAL TRANSDUCTION

The receptors for some of the interleukins, such as IL-1 and IL-2, have been well characterized. Other interleukin receptors are still the subject of intensive research. Several appear to consist of multiple protein subunits. Figure 1 shows the structure of the IL-2 receptor as an example. In some cases subunits are shared between the receptors for more than one interleukin and even other cytokines: the receptors for IL-3, IL-5, and granulocyte–macrophage colony stimulating factor, for example, share a common subunit. The presence of all subunits is required for the receptors to exhibit high affinity for the appropriate interleukin.

**Table 1** Properties of the Interleukins

| Interleukin | Molecular Mass of Native Factor ($\times 10^3$) | Principal Sources | Biological Effects |
|---|---|---|---|
| IL-1α | ca. 17.5 | Monocytes and many other cell types | Regulation of differentiation and functions of many cell types in the immune system; control of inflammatory responses |
| IL-1β | ca. 17.3 | Monocytes and many other cell types | Similar effects to those of IL-1α, mediated through binding to the same receptor |
| IL-2 | ca. 15.4 | T lymphocytes | Stimulation of activation and proliferation of T lymphocytes and regulation of a variety of immune responses |
| IL-3 (multicolony stimulating factor) | 15–25 | T lymphocytes | Stimulation of hematopoiesis; promotion of growth of precursors of granulocytes, macrophages, megakaryocytes and erythrocytes |
| IL-4 | 15–19 | T lymphocytes | Regulation of growth and differentiation of T and B lymphocytes; stimulation of Ig E production; effects on macrophages, mast cells, and precursors of megakaryocytes and erythroid cells |
| IL-5 | 45–60 (dimer) | T lymphocytes | Activation and differentiation of eosinophils; activation of B lymphocytes (mouse); regulation of production of cytotoxic T cells |
| IL-6 | ca. 21 | T lymphocytes; monocytes and many other cell types | Proliferation and differentiation of B lymphocytes; induction of acute phase protein synthesis in the liver; enhancement of IL-2 production by T cells; effects on development of hematopoietic precursor cells |
| IL-7 | 20–28 | Bone marrow stromal cells | Proliferation of early B-lymphocyte precursors; stimulation of IL-2 production by T cells |
| IL-8 | ca. 8 | Monocytes and macrophages | Stimulation of chemotaxis by neutrophils and other cell types; degranulation of neutrophils; activation of T cells (IL-8 is a member of a superfamily of related cytokines with chemotactic activity) |
| IL-9 | ca. 14 | Activated T cells | Growth of T-helper cells; activation of mast cells; stimulation of erythropoiesis |
| IL-10 | ca. 18.7 | T-helper cells | Inhibition of cytokine synthesis by another subset of T-helper cells |
| IL-11 | ca. 19 | Mesenchymal adherent cells | Development of megakaryocytes and B lymphocytes; induction of acute phase protein synthesis |
| IL-12 | 35+40 (heterodimer) | B lymphocytes | Proliferation of T cells; activation of killer cells; stimulation of IFN-γ production |
| IL-13 | ca. 100 | T lymphocytes | Activation of B lymphocytes; stimulation of IgG4 and IgE production |

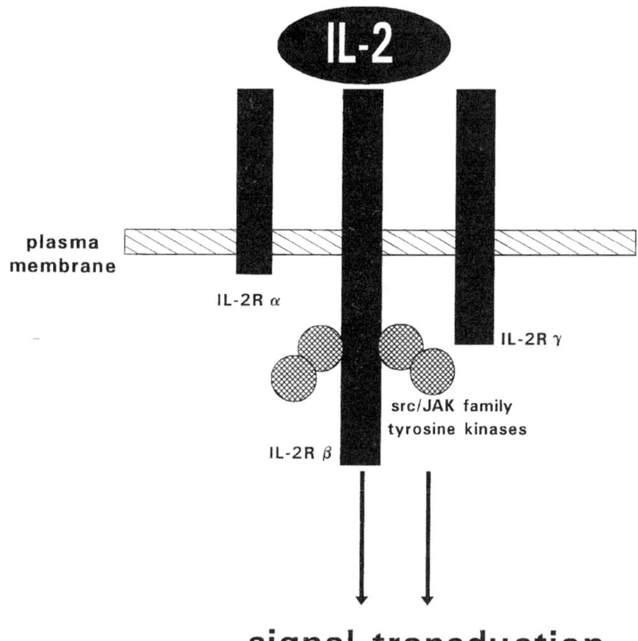


**Figure 1.** The IL-2 receptor (IL-2R), an example of a multisubunit interleukin receptor, consists of three subunits: α, β, and γ (solid bars), which span the plasma membrane (shaded bar) and interact with IL-2 on their extracellular domains. Upon binding of IL-2 to the IL-2R, signal transduction occurs as a result of association of nonreceptor tyrosine kinase enzymes of the *src* and JAK families (shaded circle) with the IL-2R. The biological effects of IL-2 are all mediated by such a signal transduction pathway.

The biochemical basis of the signal transduction mechanisms used by these receptors is still not known with certainty, although stimulation of protein tyrosine kinase activity is a common feature. The receptors themselves do not possess any intrinsic tyrosine kinase activity; rather, nonreceptor tyrosine kinases (e.g., members of the *src* and JAK families) may become associated with interleukin receptors in response to binding of the relevant interleukin (see Figure 1 for the case of the IL-2 receptor).

The biological effects of the interleukins are extremely diverse, as indicated in Table 1. Cellular responses may include stimulation of proliferation (e.g., the effect of IL-2 on T lymphocytes) and/or induction of expression of specific differentiation pathways (e.g., immunoglobulin synthesis and secretion by B lymphocytes in response to IL-6). The range of targets for interleukin-mediated regulation includes not only cells of the immune system but also endothelial cells. Cell types involved in inflammatory responses (monocytes and macrophages) and in the production of blood cells of all lineages also respond to various interleukins.

The biological effects of some interleukins are similar to those of other cytokines. For example, IL-1 exerts a wide range of actions, many of which are also brought about in response to the tumor necrosis factors. IL-1 and IL-6 are of particular interest because of their role in coordinating the synthesis by the liver of acute phase proteins in response to tissue injury or inflammation at distant sites in the body.

*See also* CYTOKINES.

### *Bibliography*

Baggiolini, M., and Sorg, C., Eds. (1992) *Interleukin-8 (NAP-1) and Related Chemotactic Cytokines.* Karger, Basel.

Beadling, C., Guschin, D., Witthuhn, B. A., Ziemiecki, A., Ihle, J. N., Kerr, I. M. and Cantrell, D. A. (1994) *EMBO J.* 13:5605–5615.

Bock, G. R., Marsh, J., and Widdows, K., Eds. (1992) *Polyfunctional Cytokines: IL-6 and LIF,* Ciba Foundation Symposium 167. Wiley, Chichester.

Callard, R. E., Ed. (1990) *Cytokines and B Lymphocytes.* Academic Press, London.

Clemens, M. J. (1991) *Cytokines.* Bios Scientific Publishers, Oxford.

Dawson, M. M. (1991) *Lymphokines and Interleukins.* Open University Press, Milton Keynes.

Meager, A. (1990) *Cytokines.* Open University Press, Milton Keynes.

O'Garra, A. (1989) *Lancet* 1(8644):943–947.

Quesniaux, V. F. (1992) *Res. Immunol.* 143:385–400.

Taniguchi, T., and Minami, Y. (1993) *Cell,* 73:5–8.

# IONIZING RADIATION DAMAGE TO DNA

*Clemens von Sonntag and Heinz-Peter Schuchmann*

### *Key Words*

**Cross-Link** Covalent bond between two macromolecular moieties, usually formed when two macromolecular free radicals combine.

**DNA Strand Breakage** Event induced by certain kinds of chemical damage (most often through free radical processes involving the DNA molecule), or by enzymatic action.

**Free Radical** Chemical species that is highly reactive because it possesses an unpaired electron.

**Heteroatom** Atom in a molecule of an organic compound that is neither a carbon nor a hydrogen atom.

**Ionizing Radiation** Energetic photons (roentgen rays, γ-rays) and energetic particles (β-rays, positrons, α-rays and other fast-moving ions, neutrons, fast electrons from electron accelerators) that cause multiple ionization events along their trajectory through matter.

**Peroxyl Radical** Generated by addition of dioxygen to a free radical.

**Pulse Radiolysis** Technique for studying the kinetics of the chemical reactions of short-lived intermediates: for example, the aqueous solution of a substrate is exposed to a short (nanoseconds-to-microseconds range) burst of energetic electrons from an electron accelerator.

**Radical Ion** Free radical with a positive or negative charge.

Ionizing radiation represents an important risk factor to the living organism. It has always been part of the natural environment, but it is also now a technological phenomenon, lending added urgency to the study of its effects on living matter. It has been established that the most sensitive target in the living cell is the DNA, which undergoes radiation damage through free radical processes. This effect has two practical aspects beyond its purely scientific interest: to assess the radiation risk with a view to minimize it, perhaps by chemical means, and to apply ionizing radiation selectively for therapeutic purposes (e.g., in cancer treatment). These goals will not be optimally achieved without adequate knowledge of the chemical processes that ionizing radiation sets in train in DNA.

## 1 ENERGY ABSORPTION

Absorption of ionizing radiation by matter leads to the formation of radical cations and electrons (reaction 1) and electronically excited molecules (reaction 2).

$$M + \text{ionizing radiation} \longrightarrow M^{\cdot+} + e^- \qquad (1)$$
$$M + \text{ionizing radiation} \longrightarrow M^* \qquad (2)$$

Energy deposition is inhomogeneous. In the case of sparsely ionizing radiation (e.g., γ-rays or high energy electrons) the average distance between two regions of energy deposition is about 200 A in matter of about unit density (i.e., in aqueous media). With densely ionizing radiation (e.g., α-particles) these regions, which are called spurs and are centered on the location of a primary ionization event, strongly overlap to form a continuous ionization track. The rate of energy deposition is proportional to the electron density. In the living cell, about 70% of the energy is absorbed by the water and about 30% by the organic matter and other solutes.

The radiolysis of water leads to the formation of OH radicals, solvated electrons, and H atoms. The water radical cation readily transfers a proton to the neighboring water molecules, thereby yielding an OH radical (reaction 3), and the electron becomes solvated (reaction 4). Electronically excited water may decompose into OH radicals and H atoms (reaction 5).

$$H_2O^{+\cdot} \longrightarrow \cdot OH + H^+ \qquad (3)$$
$$e^- + nH_2O \longrightarrow e^-_{aq} \qquad (4)$$
$$H_2O^* \longrightarrow H^\cdot + {}^\cdot OH \qquad (5)$$

For sparsely ionizing radiation the radiation chemical yields, or $G$ values,* are $G(\cdot OH) = 2.8 \times 10^{-7}$ mol/J, $G(e^-_{aq}) = 2.7 \times 10^{-7}$ mol/J, $G(H^\cdot) = 0.6 \times 10^{-7}$ mol/J. In the spurs, some radical combination occurs yielding the "molecular products" $H_2$ and $H_2O_2$ ($G(H_2) = 0.45 \times 10^{-7}$ mol/J, $G(H_2O_2) = 0.8 \times 10^{-7}$ mol/J). The water-derived radicals are all highly reactive. Hence it is generally accepted that their formation and reactions cause an important part, possibly the major part, of the radiation damage to the living cell.

## 2 RADIATION-INDUCED DNA LESIONS

Ionizing radiation is absorbed by the various cell components with practically equal probability. This is in contrast with UV radiation, which is predominantly absorbed by the cellular nucleic acids. While the mechanism of inactivation in the two cases (UV vs. ionizing radiation) is quite different, DNA again is by far the most sensitive target to damage induced by ionizing radiation. Subionization UV radiation mainly causes base dimerizations (e.g., formation of thymine dimers at 260 nm) with barely any strand breakage. Ionizing radiation produces very little of this type of damage but causes, characteristically, DNA strand breakage. Ionizing radiation base damage is largely due to free radical reactions, including those caused by the OH radicals generated by the radiation absorbed in the aqueous medium surrounding the DNA. In this way a considerable number of different kinds of damage arise, whose products one may group into the following categories:

- Altered bases
- Altered sugar moiety
  - DNA strand break
  - Release of unaltered bases
- Cross-links
  - DNA–DNA
  - DNA–protein

Apart from phosphate–ester bond cleavage, which constitutes a strand break, the phosphate moiety as such is not modified. It is important to realize that the cluster-type energy deposition of the ionizing radiation will cause some of the damaging events also to appear in clusters; that is, there is a considerable likelihood that, especially in double-stranded DNA, one damaged site will be close to another, or even to several others. These "clustered lesions" have also been termed locally multiply damaged sites (LMDS). They may involve damaged bases and one or two single-strand breaks. Two opposite single-strand breaks will result in a DNA double-strand break.

In the living cell, the DNA is surrounded not only by proteins (e.g., histones in eukaryotic cells) but also by a high concentration of low molecular weight organic material, among others the thiol glutathione. These react very efficiently (i.e., at practically diffusion-controlled rates) with the water-derived radicals (${}^\cdot OH$, $e^-_{aq}$, $H^\cdot$) formed in the aqueous surrounding of DNA. Hence these are effectively scavenged, and so the DNA is thus largely protected against, say, OH radical attack. Only OH radicals generated in the close vicinity of DNA stand some chance of reacting with, hence doing damage to, the genetic material.

## 3 REACTION OF DNA RADICALS

Damage by ionizing radiation, induced either by the direct effect or by water radicals, will result in DNA radicals; these are the precursors of the final (nonradical) damage at the product stage. The thiol glutathione (GSH), which is present in living cells at comparatively high concentrations (approaching $10^{-2}$ mol/dm$^3$) can react with the DNA radicals ($R^\cdot$) by H transfer (reaction 6). This reaction is thermodynamically slightly favored because the S–H bond is relatively weak. In competition with this H donation reaction, oxygen can add to the DNA radicals, thereby forming the corresponding peroxyl radicals (reaction 7).

$$R^\cdot + GSH \longrightarrow RH + GS^\cdot \qquad (6)$$
$$R^\cdot + O_2 \longrightarrow RO_2^\cdot \qquad (7)$$

These peroxyl radicals give rise to the final products. Some of the peroxyl radicals are capable of eliminating $HO_2^\cdot/O_2^{\cdot-}$, but the major part will decay bimolecularly (reaction 8).

$$2RO_2^\cdot \longrightarrow \text{products} \qquad (8)$$

Low molecular weight peroxyl radicals often terminate at close to diffusion-controlled rates. In high molecular weight material such

* Expressed in molecules per 100 eV, which corresponds to $1.036 \times 10^{-7}$ mol/J.

as DNA, the diffusion of radical-bearing segments is considerably restricted. Hence they may persist for relatively long times and may undergo reactions other than reaction (8) or the elimination of superoxide.

The reaction of the primary DNA radicals with glutathione is usually termed ''chemical repair.'' However, a repair in the true sense is achieved only if the radical site to be repaired has been created by H abstraction (e.g., reaction 9).

$$RH + {}^{\bullet}OH \longrightarrow R^{\bullet} + H_2O \qquad (9)$$

In DNA, such sites are present only at the sugar moiety or at the methyl group of thymine. For the most part, the OH radical reacts by addition to a C═C or C═N bond (e.g., reaction 10). Subsequent ''repair'' by H donation leads to the formation of a hydrate (reaction 11) rather than the original molecule.

$$R_2C{=}CR_2 + {}^{\bullet}OH \longrightarrow HOCR_2{-}CR_2^{\bullet} \qquad (10)$$

$$HOCR_2{-}CR_2^{\bullet} + GSH \longrightarrow HOCR_2{-}CHR_2 + GS^{\bullet} \qquad (11)$$

Apparently, the cellular enzymatic repair system can cope with this kind of damage much better than the kind resulting from peroxyl radical reactions.

## 4 MODEL STUDIES

Although the chemical nature of some kinds of base damage has been studied in vivo, most of our knowledge of radiation-induced DNA damage is derived from model studies. Electron spin resonance studies for the identification of radicals have mainly been carried out in the solid state at low temperatures. In contrast, practically all product and kinetic, pulse radiolysis, studies have been done in dilute aqueous solutions. Thus they reflect mostly the indirect effect.

In the direct effect, radical cations (and electrons) are the species primarily produced (reaction 1). To mimic this reaction and to study the fate of radical cations in aqueous solutions, investigators have used strongly oxidizing radicals such as the $SO_4^{\bullet-}$ radical, as well as of biphotonic (high laser intensity wavelength, $\lambda = 248$ nm) or monophotonic ($\lambda = 193$ nm) photoionization using laser radiation. Radiation chemically, the $SO_4^{\bullet-}$ radical is generated by reacting the solvated electron (from reaction 1) with peroxodisulfate (reaction 12). The $SO_4^{\bullet-}$ reacts with the nucleobases by electron transfer (reaction 13), yielding nucleobase radical cations.

$$e_{aq}^- + S_2O_8^{2-} \longrightarrow SO_4^{\bullet-} + SO_4^{2-} \qquad (12)$$

$$SO_4^{\bullet-} + \text{nucleobase} \longrightarrow SO_4^{2-} + \text{nucleobase}^{\bullet+} \qquad (13)$$

The nucleobase radical cations are strong acids. Quick deprotonation ensues at a heteroatom, but in thymine, deprotonation can eventually materialize at the exocyclic methyl group carbon, in competition with a nucleophilic addition of water to the carbon-6 position. Although after deprotonation the heteroatom-centered radical predominates, the radical cation in near-neutral media is always present at low ''equilibrium'' concentrations. In contrast to the situation at the heteroatom, deprotonation at the methyl position is practically irreversible under these conditions.

The electrons formed in the ionization process are readily scavenged by the nucleobases (a diffusion-controlled reaction). The radical anions thus formed are strong bases. Hence it is not surprising that those derived from adenine, guanine, and cytosine are protonated by water on the submicrosecond time scale. The thymine radical anion has a $pK_b$ value of about 7.0, hence is much longer-lived than the other radical anions.

Most of these radical anions (and to a lesser extent their heteroatom-protonated forms) have pronounced reducing properties; that is, they are capable of retransferring the electron to an oxidant. However, they are metastable species with respect to ultimate conversion into carbon-protonated intermediates (e.g., in thymine, the final protonation site is C-6). These radicals no longer have reducing properties.

This sequence of events also has a bearing on the reactions of the radical anions with oxygen. For example, the thymine radical anion and its heteroatom-protonated form react with oxygen by forming $O_2^{\bullet-}/HO_2^{\bullet}$, thereby regenerating the nucleobase, while in the reaction of oxygen with the C-6-protonated radical anion, the thymine molecule is destroyed.

Besides the formation of nucleobase–radical cations (on account of the direct effect) and radical anions (by $e_{aq}^-$ attachment), one must consider the reactions of the OH radical as a major contributor to radiation-induced DNA damage. Hydroxyl mostly adds to the double bonds of the nucleobases, but it also abstracts H atoms from the sugar moiety and the methyl group in thymine.

For the investigation of OH radical reactions in isolation, in radiation-chemical experiments it is standard practice to convert the solvated electron (from reaction 4) into OH radicals by saturating the solution to be irradiated with nitrous oxide (cf. reaction 14).

$$e_{aq}^- + N_2O \longrightarrow {}^{\bullet}OH + N_2 + OH^- \qquad (14)$$

As a result of this, the radical species now consist of 90% OH radicals and 10% H atoms; that is, the observed reactions and their products are dominated by the effects of the OH radical. In the pyrimidines, OH addition to the C-5 position yields a reducing radical, while an addition to the C-6 position yields a radical with oxidizing properties. These properties can be defined using suitable redox probes. The pulse radiolysis technique has allowed analysts to characterize these radicals and to determine their yields. The redox titration technique fails in the case of the purines, probably because the reduction potentials of the different radicals that are formed upon OH attack on the purine compound are less far apart than in the case of the pyrimidines. The assignment of the sites of OH addition to this class of compounds is further hampered by rapid ring-opening and water elimination reactions.

Our present knowledge of pyrimidine free radical chemistry (in particular uracil and thymine, their methyl derivatives as well as their nucleosides) is much more extensive than that of the purines. It is obvious from the published data that only a fraction of the primary purine OH adduct radicals have shown up in the form of products. Considerable effort will be required to bring the purine (and cytosine) free radical chemistry to a satisfactory level of understanding. This situation is also reflected in the determination of the base product yields from irradiated DNA, where a considerable deficit in the product yields (related to the primary OH radical yield) is observed.

## 5 DNA STRAND BREAKAGE AND CROSS-LINKING

Solvated electrons do not cause DNA strand breakage, but OH radicals do. In competition wtih addition to the nucleobases, they also abstract H atoms from the sugar moiety. This has two possible consequences: in the subsequent reactions a strand break is induced and an unaltered base is released, or base release occurs without giving rise to strand breakage. Hence base release always predominates somewhat over strand breakage. A number of altered sugars that are related to these processes have been identified, both in the absence and in the presence of oxygen. Based on detailed model studies, the reactions and their kinetics leading to strand breakage in the absence of oxygen are fairly well understood. The primary step is the abstraction of the H atom at C-4′:

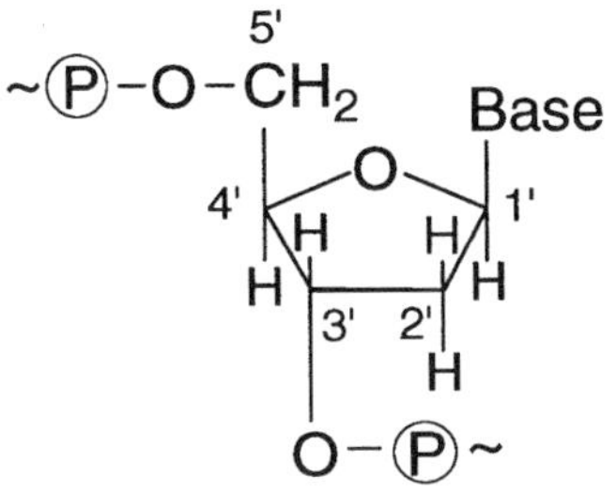

5′
~Ⓟ–O–$CH_2$ Base
O
4′ 1′
H H
H H
3′ 2′
H
O–Ⓟ~

This radical then eliminates a neighboring phosphate (linked to a fragment of the DNA strand), leaving behind a radical cation that can react with water, either at the position that has eliminated the phosphate, or at C-4′. In the former case the other phosphate function may be eliminated by the same mechanism; in double-stranded DNA this sequence of events produces a clean gap in the affected strand (the end groups of the two fragments are phosphate groups), with the loss of some information because of the disappearance of the damaged nucleoside. In the latter case the base is also lost, but additionally at the end of one of the fragment strands a damaged sugar remains linked to the phosphate group—enzymatically speaking, it is a "dirty" end group. Details of the mechanism of DNA strand breakage under conditions of oxygenation are less well understood, but some of the relevant sugar lesions have been detected. Model systems (ribose 5-phosphate) indicate that under these conditions C-5′ should be an additional site of attack and, in analogy, one would expect (no experimental evidence yet) the C-3′ peroxyl radical also to be a potential precursor for strand breakage.

So far, sugar damage has been discussed only in terms of OH radicals attacking this moiety. A contribution of the direct effect (ionization of the sugar moiety and the phosphate groups) must also be considered, but experimental evidence is not yet available. However, there is another interesting aspect. In polynucleotides such as poly(U) and poly(C), there is convincing evidence that base radicals are the major precursors of the sugar radicals that lead to strand breakage and the release of an unmodified base at the site of the damaged sugar. It is less clear whether such a radical transfer from the base to the sugar moiety can also occur in DNA.

In mammalian cells, DNA double-strand breaks are observed alongside single-strand breaks approximately in the ratio of 1 : 25. This poses the question of how these double-strand breaks are formed. It has been argued here that they result from clustered lesions. In the literature, an additional one-hit route has been suggested that involves a radical transfer from the already broken strand to the sugar moiety of the opposite strand, followed by breakage of this strand.

Carbon-centered radicals are known to add to the C═C bonds of nucleobases. Such reactions, as well as radical–radical combination reactions involving macroradicals, in principle allow the formation of DNA–protein and DNA–DNA cross-links. In special cases, such products have been observed with biological material, albeit in yields considerably lower than DNA double-strand breaks.

*See also* DNA Damage and Repair; Free Radicals in Biochemistry and Medicine; Ultraviolet Radiation Damage to DNA.

### *Bibliography*

Dizdaroglu, M. (1991) *Free Radical Biol. Med.* 10:225.

Hüttermann, J. (1991) In *Radical Ionic Systems,* A. Lund, and M. Shiotani, Eds., p. 435. Kluwer, Dordrecht.

Michael, B. D., Held, K. D., and Harrop, H. A. (1983) In *Radioprotectors and Anticarcinogens,* O. Nygaard and M. G. Simic, Eds., p. 325. Academic Press, New York.

Schulte-Frohlinde, D., Simic, M. G., and Görner, H. (1990) *Photochem. Photobiol.* 52:1137.

Steenken, S. (1989) *Chem. Rev.* 89:503.

Teebor, G. W., Boorstein, R. J., and Cadet, J. (1988) *Int. J. Radiat. Biol.* 54:131.

von Sonntag, C. (1987) *The Chemical Basis of Radiation Biology.* Taylor & Francis, London.

von Sonntag, C., Hagen, U., Schön-Bopp, A., and Schulte-Frohlinde, D. (1981). *Adv. Radiat. Biol.* 9:109.

Ward, J. F. (1988) *Prog. Nucleic Acid Res. Mol. Biol.* 35:95.

# Isoenzymes

*Adrian O. Vladutiu and Georgirene D. Vladutiu*

### *Key Words*

**Antigenicity** Property of some substances (mostly proteins) to elicit antibodies after introduction into a foreign (other than self) organism.

**Electrophoresis** Technique used to separate charged particles in solution, by the differences in their rates of migration in an applied electric field.

**Epigenetic** Modification of gene products by events or factors that occur after transcription and translation of the gene.

**Fusion Protein** Protein encoded for by a hybrid gene that has fused part of its original coding sequence with coding sequences of another gene for a different protein (e.g., a readily detected marker).

**Locus** Place on a chromosome occupied by a particular gene or its alleles.

**Michaelis Constant** The experimentally determined substrate concentration at which the enzymatic reaction proceeds at half its maximum velocity.

**Ontogenic Development** The course of the development of an organism from fertilization, through maturity, to death.

**Phenotype** The genetically and environmentally determined characteristics of an organism.

**Polymorphism** Occurrence in the same population of two or more alleles at a locus, with at least one allele having a frequency exceeding 1%.

**Translation** The formation of a peptide chain on the mRNA template from individual amino acids.

Isoenzymes (Greek: *izos,* equal; *zymos,* leaven), also known as isozymes, are enzymes that catalyze the same reaction but are different in structure and often in some of their properties. Isoenzymes are proteins encoded by distinct genes that can be at different loci or can represent different alleles at the same locus. The Commission on Biological Nomenclature of the International Union of Biochemistry recommended that the term "isoenzyme" be restricted to forms of the same enzyme that originate at the level of different genes. The discovery of the heterogeneity of human lactate dehydrogenase (LD) in 1957, as well as the introduction of histochemical techniques for the detection of esterases, led to the proposal of the term "isozyme" by Markert and Møller in 1959 to describe different proteins having similar enzymatic activity.

Differences between various forms of the same enzyme can be genetically determined or can be due to epigenetic changes of the enzyme. Although the latter forms are not isoenzymes, from an operational point of view they have been often called isoenzymes. Newly discovered variants of enzymes cannot be considered isoenzymes until their genetic control is determined. However, this term is sometimes used in a general sense to signify different forms of an enzyme regardless of the mechanisms generating the enzyme diversity.

## 1 DERIVATION OF ISOENZYMES

Many enzymes are coded for by multilocus genes. The genes that code for specific isoenzymes can be on the same chromosome (e.g., chromosome 4 for the four loci of alcohol dehydrogenase in humans) or on different chromosomes (e.g., the hexosaminidase alpha and beta subunits on human chromosomes 15 and 5, respectively). Isoenzymes can also originate from mutations of alternative forms of genes (alleles) at the same locus ("allelozymes": e.g., allelic variants of placental alkaline phosphatase), or they can arise from the modification of the structure or expression of genes in somatic cells—this can result from malignant transformation.

Isoenzymes made up of subunits that are associated in various combinations, producing hybrid molecules when the subunits are different, constitute oligomeric isoenzymes. These molecules may be generated by random associations of units derived from different loci or from multiple alleles at the same locus [e.g., creatine kinase (CK), lactate dehydrogenase (LD), and hexosaminidase]. Lactate dehydrogenase in serum and organs of vertebrates is a tetramer of two different polypeptide subunits, A or H (from heart) and B or M (from muscle), which combine randomly, producing homopolymeric and heteropolymeric molecules. Thus, five different LD isoenzymes are formed (Figure 1), from he most anodal LD-1 (four H subunits), to the most cathodal LD-5 (four M subunits). Lactate dehydrogenase-2, LD-3, and LD-4 are hybrid molecules, with subunit composition $H_3M$, $H_2M_2$, and $HM_3$, respectively. An additional LD isoenzyme, LD-C, composed of four identical subunits (C), which are different from the other two subunits of LD, is characteristic for postpuberal testis (primary spermatocytes). In man, hexosaminidase exists as two main forms: B, with a structure of four beta subunits, and A, with a structure of one alpha and two beta subunits. Because of their different genetic control, the isoenzymes differ in their amino acid structure; however, the amino acid sequence of the active site is usually highly conserved and is the same for different isoenzymes of a single enzyme.

Isoenzymes that are produced as a result of allelic variations (allelozymes) are inherited codominantly; that is, the products of both alleles are expressed. The patterns of these isoenzymes can thus be used to trace their genetic inheritance. On the other hand, isoenzymes that have arisen from genes at multiple loci have been distributed throughout the species population over the course of evolution, resulting in all individuals possessing essentially the same complement of isoenzymes. The existence of multiple gene loci for specific isoenzymes can be explained by two different evolutionary processes. In one process, catalytic activity may have arisen from two or more unrelated events leading to the establishment of two different genes, each with the capacity to code for a different enzyme protein that catalyzes the same reaction. The two

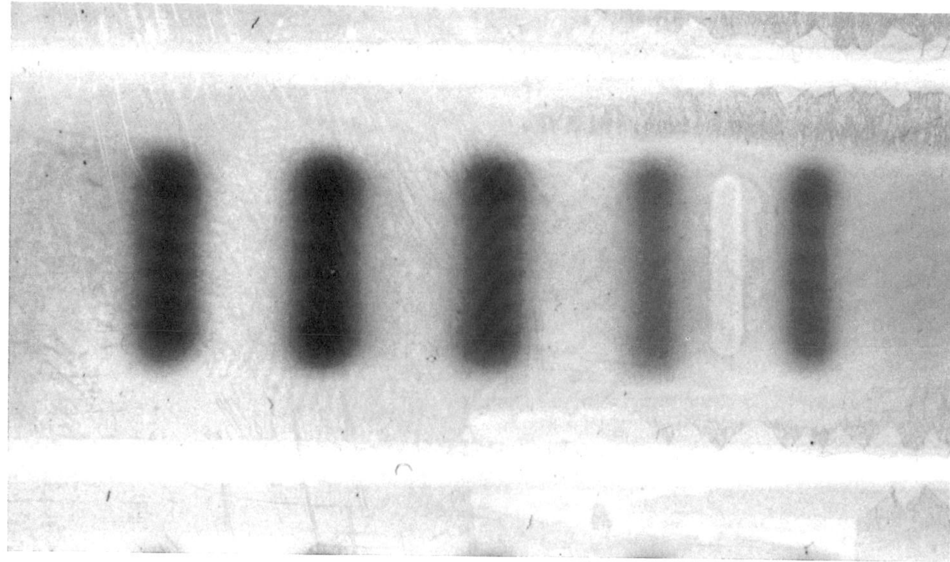

**Figure 1.** Lactate dehydrogenase (LD) isoenzymes from normal human serum. The anode is at the left and the most anodic band is LD-1. The proteins were separated by zone electrophoresis in a thin layer of agarose gel in 0.01 mol barbital buffer, pH 8.6. The gel was overlaid with specific substrate, and color development resulted after incubation at 37°C with tetrazolium salt. Note that all five isoenzymes (LD-1 to LD-5) are equidistant because of the two major subunits of LD, which associate to form tetramers. [Reproduced with permission of Academic Press from R. Dulbecco, Ed. (1991) *Encyclopedia of Human Biology,* Vol. 4, p. 579.]

genes thus formed can undergo mutations and will then code for different enzyme-proteins. In the other, most common process, a single ancestral gene coding for a single enzyme is duplicated. The two resultant genes can mutate and would then be modified favorably for the organism, cell, or organ function that was evolving. This latter process is responsible for the occurrence of isoenzymes controlled by genes at different loci. Although the occurrence of major change in the gene products over time could be predicted, the similarities in the two gene products should be great enough to permit one to ascertain their common ancestry. Such ancestral interspecies homologies exist, for example, in the isoenzymes of LD. For some isoenzymes (e.g., human α-amylase), the two loci occurring after gene duplication remain closely linked.

Genetically determined changes in isoenzymes between individuals can explain abnormalities in metabolism manifested as hereditary metabolic diseases, and can also account for individual sensitivity to various drugs.

## 2 STRUCTURE, FUNCTION, AND DEVELOPMENTAL ASPECTS

By definition, isoenzymes must differ in primary structure to some degree. Full elucidation of these differences requires the application of physical and chemical methods of protein analysis to purified isoenzyme preparations. The complete amino acid sequencing of isoenzymes has been achieved for various isoenzymes (e.g., human aldolases A and B; human carbonic anhydrases I and II, which differ in about 90 of 260 amino acid residues). A sensitive and specific method for analyzing subtle differences in primary structure among isoenzymes is the ''fingerprinting'' technique, which compares two-dimensional maps of the peptides in a partial hydrolysis of the enzyme-protein. This method is especially useful for identifying single amino acid substitutions in products of allelic genes. The elucidation of the secondary and tertiary structures of isoenzymes has been achieved by X-ray crystallography.

Since isoenzymes are coded for by different genes, they are expected to have different properties such as substrate concentration optimum, electrophoretic mobility, resistance to inactivation, sensitivity to inhibitors, Michaelis constant for substrate(s), relative rate of activity with substrate analogues, and sometimes a different extent of substrate specificity. Isoenzymes of LD and carbonic anhydrase, for example, can have different functions because of different properties. The difference in charge distribution over the surface of the molecule can affect the location of the molecule within the cell.

Some isoenzymes show tissue or cell specificity, and individual isoenzymes can be predominantly represented in cells at each stage of their differentiation, from embryo to adult. For example, changes in LD and CK isoenzyme distribution occur during development. The relative proportions of cytoplasmic and mitochondrial forms of certain isoenzymes also change during differentiation. Isoenzymes can differ in their substrate specificity. When substrate specificity is not stringent, isoenzymes can show differences in their activity with substrate analogues; for example, alkaline phosphatases of bone, liver, and kidney differ in this respect from those of placenta and small intestine. Altered substrate specificity has also been shown for isoenzymes of mammalian liver alcohol dehydrogenases and for acid phosphatases (e.g., in rabbit muscle).

The presence of different isoenzymes in certain tissues, cells (e.g., enolase, LD, phosphofructokinase, CK, hexokinase, pyruvate kinase, amylase) or subcellular structures such as lysosomes and mitochondria (e.g., the NAD-dependent malate dehydrogenase, isocitrate dehydrogenase, and adenylate kinase isoenzyme systems, with one isoenzyme in the mitochondrion and another in the cytosol), implies a specialized role for each isoenzyme in cell metabolism. Indeed, isoenzymes allow a fine adaptation of metabolic patterns to ontogenic development or to changes in the environment. One physiological advantage of the isoenzymes is the creation of a compartmentalization of the same catalytic reaction in different metabolic pathways. Compartmentalization allows precision in the maintenance of metabolic patterns unachievable by a single enzyme. Isoenzymes of LD, for example, are represented in different percentages in various organs; this differential distribution has functional significance. A correlation exists between the properties of isoenzymes predominant in certain tissues and the metabolic patterns in these tissues. There are also age-related changes in isoenzymes within certain cells. The proportions of the isoenzymes in a tissue or organ depends on their half-lives (i.e., rates of synthesis and degradation).

Many isoenzymes are named according to their tissue or subcellular structure of origin (e.g., placental and liver alkaline phosphatase); isoenzymes that can be separated by electrophoresis are numbered according to their electrophoretic mobility. Because most enzymes are negatively charged, the most anodic isoenzyme is numbered as the first.

## 3 METHODS OF ANALYSIS

Isoenzymes can be separated by various techniques—zone electrophoresis as well as capillary electrophoresis, elution-convection electrophoresis, isoelectric focusing, or discontinuous electrofocusing. These methods and various types of ion exchange chromatography take advantage mainly of differences in the net electric charge of isoenzyme molecules at a given pH and, to some extent, of differences in molecular size and shape. However, the separation of certain isoenzymes requires other methods such as high performance liquid chromatography (HPLC) or affinity chromatography. Once separated, the isoenzymes can be identified on the basis of some of their properties. For colorimetric detection in serum or tissue extracts, histochemical techniques are used to show enzymatic activity by producing a colored precipitate or a fluorescent substance when fluorogenic substrates are used or when the product of the enzymatic reaction is fluorescent. This technique, known as the zymogram technique, or enzymoelectrophoresis, is commonly performed in clinical laboratories for diagnostic purposes, with the use of scanning densitometers and fluorimeters. Other methods make use of inhibition or denaturation of enzymes by heat, concentrated urea, or other organic compounds; reactivity with different substrates, coenzymes analogues; or measurements of pH optima, Michaelis constants, and so on.

Immunological techniques are also used to separate and to quantitate isoenzymes because most of them differ with respect to antigenicity. Monoclonal antibodies raised against pure isoenzymes are particularly useful to study isoenzyme subunit structure and can detect subtle differences between various forms of enzymes. Immunoinhibition techniques (i.e., measurement of the residual activity after treatment with specific antibodies) can be used for isoenzyme quantitation. Immunological precipitation followed by measurement of the enzyme catalytic activity or of the amount of radioactive label bound to the enzyme allows quantitation and expression of isoenzyme amount either in units of activity or in mass units.

Recently, genetic (protein) engineering techniques, including oligonucleotide-directed, site-specific mutagenesis and chimeric fusion protein formation, have been used to study structure and function of isoenzymes (e.g., human aldolases A and B). With the advent of the transgenic technology, new isoenzyme systems can be made—e.g., by inserting the required genes into the genome, even at specific locations in the genome (by homologous recombination techniques). The genes for the new isoenzymes can be altered to elicit function in specific cells at certain times in development.

## 4 PRACTICAL APPLICATIONS OF ISOENZYME ANALYSIS

Isoenzymes have not only theoretical importance for understanding basic biological processes (e.g., the transformation of one gene into another, the regulation of gene expression, and the significance of metabolic pathways in different tissues), but also their analysis has a broad array of practical applications. These wide-ranging applications include use as markers in human–mouse cell hybrids made for human chromosome mapping and as diagnostic indicators in both hereditary and nonhereditary diseases. Because isoenzymes are valuable markers of gene function, they can be used to study population genetics (e.g., to identify new phenotypes and genetic polymorphisms). With isoenzyme analysis, the measurement of allele frequencies in a population facilitates the study of evolutionary selective pressure.

Isoenzyme analysis has also been used in forensic science, to distinguish species and tissues, and in paternity testing, based on allelic variations as an index for testing.

Isoenzymes can be used for studying cell differentiation and alterations of this process. Because new patterns of isoenzymes commonly occur when fully differentiated cells undergo malignant transformation, isoenzymes are valuable for the diagnosis of malignancy as well as for understanding some mechanisms of malignant transformation. Examples of some ubiquitous isoenzymes that are altered in tumors include aldolase, LD, and β-hexosaminidase.

The organ or tissue specificity of isoenzymes allows them to be used to pinpoint organ damage in various diseases. For example, CK-2 (CK-MB) isoenzyme is increased in the serum of patients with myocardial damage. The diagnostic specificity is further increased by simultaneously measuring several different isoenzymes in serum and other body fluids.

A significant number of relatively rare inborn errors of metabolism result from mutations leading to the deficiency or modification or a specific isoenzyme. The functional consequences of these mutations vary depending on the nature of the alteration. When allelic variation of isoenzymes results in disease, usually the catalytic activity of the mutant enzyme is greatly reduced or absent. The recognition of a specific isoenzyme deficiency not only provides a firm biochemical basis for a clinical diagnosis but also serves as a means to distinguish variants of some metabolic disorders. Ultimately, the biochemical classification of subtypes leads to an understanding of the molecular pathology of individual forms of a disorder.

As a result of mutations at multiple gene loci, a number of disorders arise from alterations of isoenzymes coded for by these loci. The metabolic effects of such alterations consequently are manifested only in the tissues or organs in which the affected isoenzyme is present. For example, in muscle disorders there is a tissue-specific isoenzyme deficiency in either glycogen or fatty acid metabolism leading to symptoms of skeletal muscle injury.

Some lysosomal isoenzyme deficiencies can lead to progressive neurodegenerative disease usually (but not always) manifested in childhood. The best known example is Tay–Sachs disease (deficiency of the alpha subunit of β hexosaminidase A). Examples of other lysosomal storage disorders in which a single genetically distinct isoenzyme is affected include mannosidosis (α-mannosidase A deficiency), metachromatic leukodystrophy (arylsulfatase A deficiency), and Maroteaux–Lamy disease (arylsulfatase B deficiency).

## 5 NONISOENZYMIC MOLECULAR FORMS OF ENZYMES

Posttranslational modification of the enzyme-protein can also result in multiple molecular forms of enzymes. Although these forms are not isoenzymes (i.e., they are not coded for by separate genes), historically and from an operational point of view they have often been referred to as isoenzymes because they have many characteristics similar to isoenzymes—that is, they can exist in different cell organelles and in different metabolic compartments within a single cell (e.g., the cytoplasmic and mitochondrial forms of aspartate aminotransferase). Posttranslational mechanisms of producing enzyme variants include the formation of aggregates of a single unit (homopolymers) or of an enzyme with nonenzymatic proteins (e.g., cholinesterases), and other epigenetic modifications of the initial protein such as combination of the protein with nonprotein molecules (e.g., sialic acid residues). Enzymes with increased mass and altered electrical charge arise when they bind to immunoglobulins (''macro'' enzymes: e.g., macroamylase and macro-CK). Alteration of carbohydrate side chain, acylation, deamidation, sulfhydryl oxidation, and partial cleavage of the polypeptide chain are other mechanisms of producing nonisoenzymic molecular forms of enzymes. The removal of some residues can explain differences in function because, for example, deamidation and phosphorylation may influence secretion and uptake of enzymes by cells. Partial proteolysis of the initial enzyme polypeptide can produce so-called isoforms of enzymes when only one amino acid is removed [e.g., isoforms of CK-2 (CK-MB), CK-$MB_1$, and CK-$MB_2$, which differ by a terminal lysine and can be separated electrophoretically]. It is also conceivable that conformational changes of the polypeptide chains can lead to ''conformational enzymes'' or ''conformers,'' with the same molecular weight and amino acid sequences but different properties, such as changed electrophoretic mobility. The different forms of enzymes that are generated through posttranslational modifications of the enzyme-proteins in general do not differ in their catalytic characteristics but have biological significance and practical applications similar to those of isoenzymes. Once their genetic control has been elucidated, some of the different molecular forms of enzymes may prove to be, in fact, isoenzymes.

*See also* ENZYME ASSAYS; ENZYME MECHANISMS, TRANSIENT STATE KINETICS OF.

### *Bibliography*

Blaten, V., and Van Stirteghem, A., Eds. (1986) Plasma isoenzymes: The current status. In *Advances in Clinical Enzymology.* Karger, Basel.

IUPAC-IUB, Commission on Biochemical Nomenclature. (1977) Nomenclature of multiple forms of enzymes. *J. Biol. Chem.* 252:5939.

Lehman, F. G. (1979) Enzyme and isoenzyme diagnosis of cancer. In *Advances in Clinical Enzymology,* E. Schmidt, R. W. Schmidt, I. Trautschold, and R. Friedel, Eds., pp. 171–195. S. Karger, Basel.

Moss, D. W. (1982) *Isoenzymes.* Chapman & Hall, London.

Ogita, Z.-I., and Markert, C. L., Eds. (1989) *Isoenzymes: Structure, Function, and Use in Biology and Medicine.* Wiley-Liss, New York.

Scandalios, J. G., Whitt, G. S., and Rattazzi, M. C. Eds. (1977–1987) *Isozymes. Current Topics in Biological and Medical Research,* Vols. 1–16. Liss, New York.

# Kallikrein–Kininogen–Kinin System

*Michael E. Rusiniak and Nathan Back*

### *Key Words*

**Crossing Over** The reciprocal exchange of material between homologous regions of chromosomes during meiosis; a DNA strand of one parental molecule base-pairs with a complementary region of a DNA strand from another molecule with which it is being paired.

**Domain** A contiguous stretch of amino acids within a protein sequence related to a functional property of the molecule.

**Gene Conversion** Modification by DNA repair of a mismatched strand of heteroduplex DNA during meiosis; this process results in generation of an extra copy of one of the recombining genes.

**Hageman Factor** The plasma enzyme precursor that undergoes reciprocal activation with prekallikrein at the start of the intrinsic pathway of blood clotting.

**Linkage** The tendency of genes to be inherited together based on proximity within the same chromosome.

**Recombination** Either general (requiring homologous DNA) or site-specific (protein-mediated, not requiring sequence homology) exchange of DNA regions between chromosomes.

**Unequal Crossing Over** A recombination event in which the location of recombining sites is not identical in the parental DNAs.

Mammalian blood contains three major protease cascade systems interrelated through a shared Hageman factor activation mechanism: namely, the blood coagulation, fibrinolysin, and kallikrein–kininogen–kinin (K-K-K) systems. Components of the K-K-K protease cascade system present in blood and a number of different tissues include kinin-forming enzymes, the *kallikreins,* the glycoprotein kinin-containing *kininogen* substrates, biologically active polypeptide *kinins* and kinin-destroying enzymes, the *kininases.* The K-K-K system functions, in part, via formed kinins that regulate tissue local blood flow, functional hyperemia and transmembrane ion transport. *Plasma* kallikrein and kininogen, in collaboration with Hageman factor, help initiate the intrinsic pathway of blood coagulation by contact activation mechanisms. In addition, novel functions of kininogens have been discovered recently; namely, cysteine proteinase inhibition and (in the rat) involvement in the acute phase response. By virtue of the diverse actions of kinins and other system components, the K-K-K protease cascade also is implicated in a wide variety of pathologic conditions.

Recent molecular technologies have enabled the characterization of the molecular biology of K-K-K system components, notably the kallikreins and kininogens, and to a lesser extent, kinin receptors. Whereas plasma kallikrein is encoded by a single gene in the liver, a larger family of glandular kallikrein genes, particularly in rodent species, has undergone concerted evolution to yield a family of enzymes with divergent substrate specificities. Expression of kallikrein genes is tissue specific and developmentally regulated. Kininogens are multidomain, multifunctional proteins. High molecular weight kininogen (HMWK) and low molecular weight kininogen (LMWK) are derived from alternative splicing of a single gene. The region of the modern kininogen gene encoding the heavy chain is proposed to have evolved from the stefin gene progenitor of the mammalian superfamily of cysteine proteinase inhibitors. The present-day kininogen gene also has acquired regions encoding the kinin moiety and a light chain that endows HMWK with a cofactor role in the intrinsic blood coagulation cascade. A third type of kininogen, T-kininogen (found only in the rat), is considered to be derived from an ancestral gene in common with HMWK and LMWK. T-Kininogen is expressed at elevated levels during acute inflammation, whereas the production of HMWK and LMWK remains unchanged. The mechanisms governing the development and regulated expression of K-K-K system components are being elucidated and will enhance our understanding of the role of this protease cascade in health and disease.

## 1 The Kallikrein–Kininogen–Kinin (K-K-K) System

### 1.1 The K-K-K Protease Cascade

Plasma and tissue pathways of kinin formation, the ultimate end product of K-K-K system activation (Figure 1), are embodied in a cascading sequence of limited proteolysis in which the product of one proteolytic action acts as catalyst for the subsequent reaction (Figure 2). The specific kinin-forming plasma or tissue protease kallikrein selectively cleaves kininogen substrate to liberate vasoactive peptide kinins that bear the canonical bradykinin nonapeptide sequence. These vasoactive kinins have localized and short-lived action as they are destroyed rapidly by kinin-inactivating enzyme kininases. Since the discovery of bradykinin in 1951, a family of structurally and pharmacologically similar kinins has been isolated and identified (Figure 3).

Kinins share a high affinity for specific membrane surface receptors present on diverse target cells. The biologic responses elicited by kinin–receptor interaction affect not only local and regional tissue blood flow and ion transport but also smooth muscle and

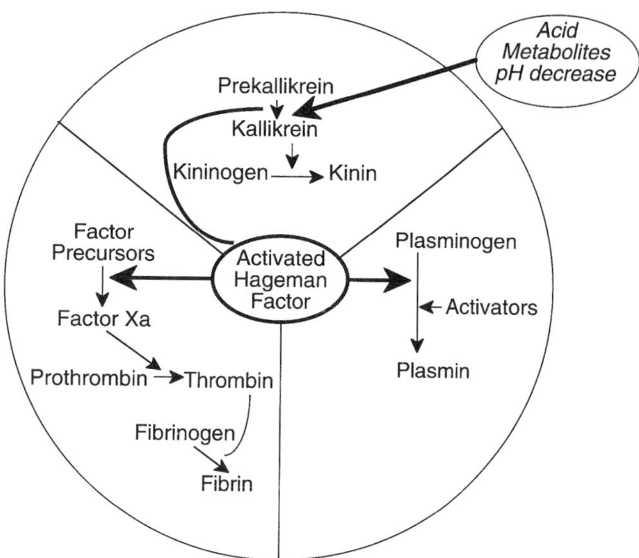


Figure 1. Common Hageman factor activation mechanism of and interrelationship among the blood coagulation, fibrinolysin, and kallikrein–kininogen–kinin protease cascade systems.

vascular tissue, select biochemical and neurochemical pathways, and renal–cardiovascular systems.

## 1.2 Physiological Functions of the K-K-K System

At least four physiological functions have been ascribed to the K-K-K protease cascade system by virtue of the action of formed kinin peptides. They are briefly described as follows.

*Regulation of tissue blood flow.* Kinins regulate local blood flow in exocrine glands (as salivary), and in many vascular tissue beds, as well as in nonexocrine organs as the kidney.

*Regulation of the cardiovascular system.* Kinins dilate and constrict arteries and veins depending on kinin receptor subtypes in various vascular beds. Kinins cause a lowering of blood pressure and are considered to be involved in overall regulation of blood pressure, either by direct action on tissue vasculature or by causing the release of chemical mediators such as histamine, catecholamines, prostaglandins, and/or angiotensins. Kinins also help regulate renal electrolyte and water secretion. Cardiac effects of kinins, which include increased heart rate and stroke volume, serve to elevate cardiac output.

*Effects on nonvascular smooth muscle.* Kinins contract most nonvascular smooth muscle (e.g., bronchioles, intestine, uterus). However, because of species variations, kinins relax rat duodenum and hen rectal cecum but contract guinea pig colon and ileum. The isolated perfused estrus uterus is most sensitive to kinin action and thus is the bioassay model used to quantitate kinin levels. Other methods used to assay kinin levels include radioimmunoassay (RIA) and the enzyme-linked immunosorbent assay (ELISA), which employ polyclonal and monoclonal antibodies. The smooth muscle effects of kinins are of pathologic significance in pain, shock, migraine, asthma, and hypertension.

*Central and peripheral nervous system effects.* Biochemical pathways for kinin formation are present in the central nervous system. Kinins both stimulate and depress the central nervous system depending on administration route and species. Kinin receptors also are at pre- and postsynaptic sites and on ganglia where autonomic nervous system influences are noted.

Apart from the actions just described, kinins stimulate DNA and protein synthesis, promote glucose uptake, initiate neovascularization, and induce cell proliferation.

## 1.3 The K-K-K System in Pathology

The K-K-K system has been implicated in diverse disease states. System component levels change during the course of disease

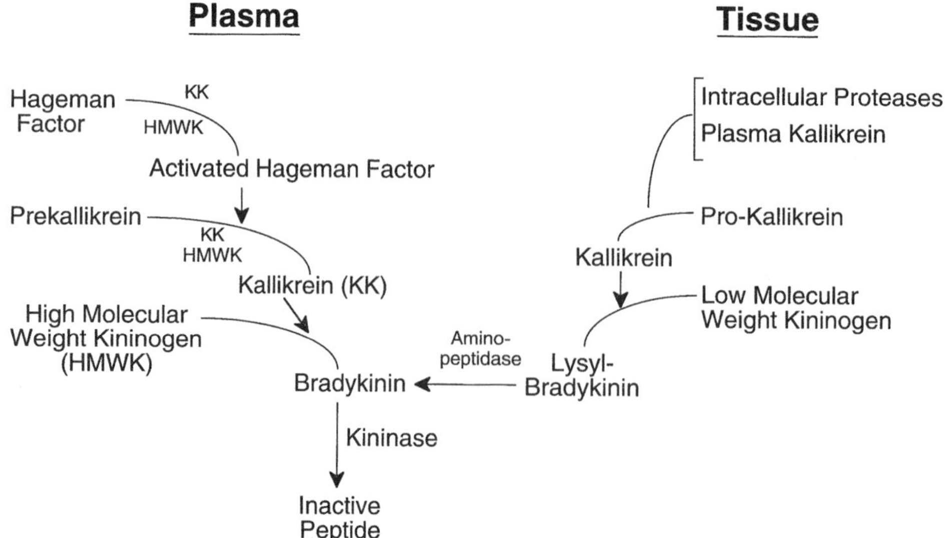


Figure 2. Plasma and tissue (glandular) proteolytic pathways involved in the generation and destruction of bradykinin. In the tissue pathway, the intracellular proteases and plasma kallikrein are not components but rather enzymes that activate the pathway.

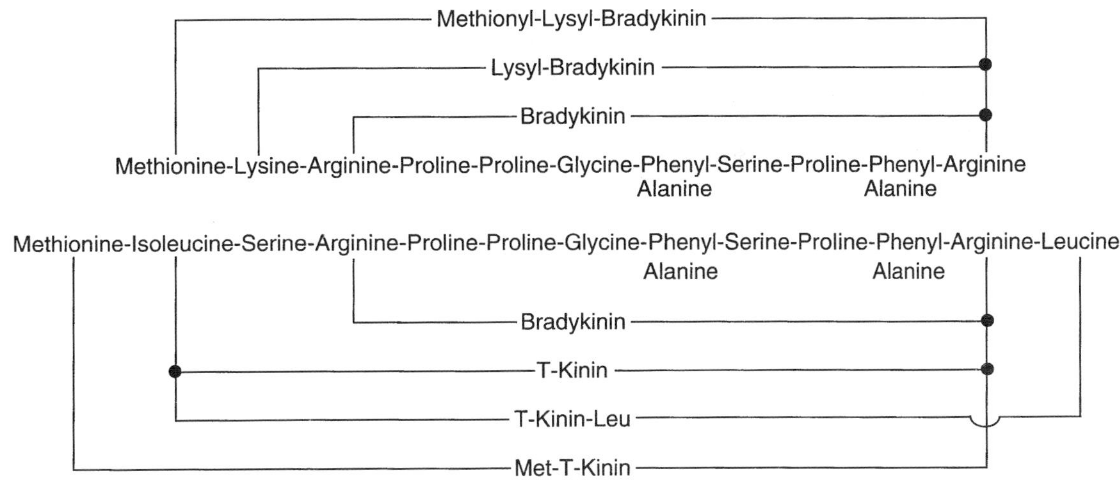


**Figure 3.** Primary structures of bradykinin and other mammalian kinins.

processes. Kallikreins and other proteases digest substrates, releasing kinins and other bioactive mediators. The K-K-K system is purported to be involved in the pathophysiology of inflammation, acute pancreatitis, hypertension, migraine, carcinoid tumor syndrome and the postgastrectomy dumping syndrome, diabetes mellitus, pain, and shock (Table 1).

Kinin peptides, present in inflammatory exudates, are considered to be the mediators for such cardinal symptoms of *inflammation* as edema, heat, redness, and pain. Kinins released during rhinoviral infections and antibody–antigen reactions form the basis for many of the symptoms associated with acute, delayed, and chronic allergic responses.

Symptoms that characterize *acute pancreatitis* include pain, loss of electrolytes and fluid, and hypotension, often leading to shock and mortality. Kinin peptides are released by pancreatic tryptic and kallikrein digestion of kininogen. Protease inhibitor therapy is of benefit in reducing the morbidity and mortality of this life-threatening disease.

Studies of urinary kallikrein levels in essential and experimental *hypertension* strongly support a role for the K-K-K system in the pathogenesis of hypertension. The interrelated K-K-K and renin–angiotensin protease cascades may have concerted roles in blood pressure regulation. In the tissue K-K-K cascade, lysyl-bradykinin is transformed to bradykinin by a plasma aminopeptidase and the bradykinin is quickly broken down to an inactive octapeptide by

**Table 1** Pathologic Conditions in Which the Kallikrein–Kininogen–Kinin System May Be Involved in the Disease Process

| | |
|---|---|
| Acute Pancreatitis | Infection |
| Allergic Phenomena | Inflammation |
| Arthritis | Malignancy |
| Asthma | Migraine |
| Burns | Shock States |
| Carcinoid Syndrome | Anaphylaxis, Burn, Cardiogenic, |
| Cirrhosis of the Liver | Endotoxin, Envenomation, |
| Diabetes | Hemorrhage Hyperfibrinolysis, |
| Dumping Syndrome | Obstetric, Peptone, Surgical |
| Hypertension | Pain |

carboxypeptidase removal of a terminal arginine (Figure 2). This carboxypeptidase, kininase II, also converts inactive angiotensin I to the hypertensive angiotensin II octapeptide.

Kinin is recognized as the specific chemical mediator for pain. Kinins evoke the pain response in experimental animals and migrainelike symptoms in the human. The presence of both kinin-forming proteases and kinins in cerebrospinal fluid during *migraine* attacks suggests a possible etiologic role for kinins in the complex symptomatology associated with migraine.

*Carcinoid tumor and postgastrectomy dumping syndromes* (CTS and PDS) are two distinctly different conditions caused by tumors of the enterochromaffin cells (CTS) and occurring postprandially in individuals with subtotal or total gastrectomies (PDS), respectively. The syndromes share such vasomotor symptoms as cutaneous flushing and such gastrointestinal symptoms as hypermotility and diarrhea. This spectrum of symptoms is presumed to be caused, in part, by kinin release and action. Carcinoid tumor tissue contains a tissue kallikrein shown to cleave human kininogen, thereby releasing lysyl–bradykinin in vitro.

All components of the K-K-K system have been isolated from experimental and human tumor tissues exhibiting many different forms of *malignancy*. K-K-K component levels change during tumor progression and growth; kallikrein levels increase as kininogen substrate concentrations decrease. The contribution of kinin involvement in cell growth and the neoplastic process resides in kinin effects including neovascularization (angiogenesis), increased vascular permeability, alteration of tumor blood flow, and overall maintenance of tumor tissue viability. Protease inhibitors that prevent kinin release were shown to decrease tumor development and growth in experimental animal tumor models.

A vast array of experimental and clinical evidence supports the pivotal role of the K-K-K system in the pathophysiology of *shock*, irregardless of initiating stimulus. The hemodynamic changes leading to the profound and often sustained hypotension occurring in shock, as well as the blood coagulation and electrolyte/fluid biochemical defects, can be traced to massive protease activation and the formation and action of vasopeptide kinin (and other mediators). Shock caused by anaphylaxis, burns, hemorrhage, endotoxins, envenomation, and surgery presents in every case with similar biochemistry and pathophysiology. Once again prophylactic and

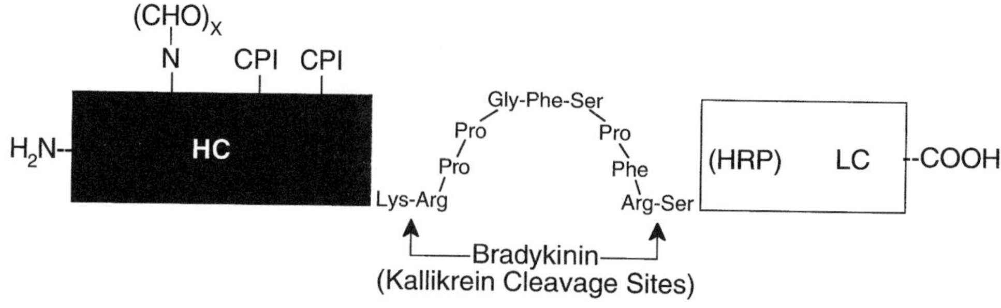


Figure 4. Plasma kallikrein cleavage sites on high molecular weight kininogen at which bradykinin is liberated. Sequential cleavage of the Arg-Ser bond and the Lys-Arg bond releases the light chain (LC) containing the histidine-rich portion (HRP), followed by liberation of the bradykinin moiety from the heavy chain (HC) that contains the cysteine proteinase inhibitor (CPI) domains.

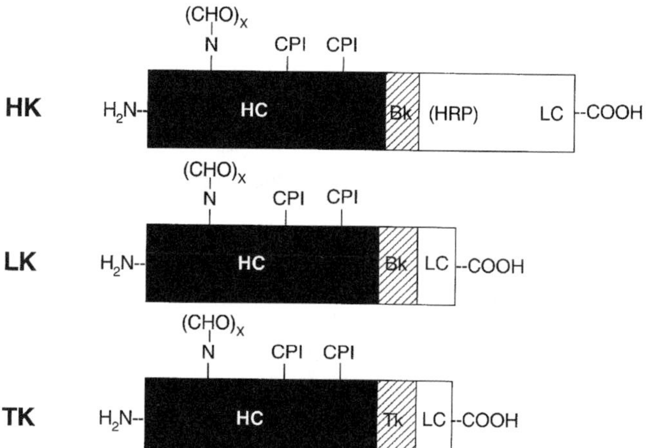


Figure 5. High molecular weight kininogen (HK), low molecular weight kininogen (LK), and T-kininogen (TK). Bradykinin (BK) is included in both HK and LK; however TK contains T-kinin with the Ile-Ser-bradykinin sequence.

therapeutic administration of protease inhibitors has been shown to be efficacious in treating shock.

## 2 STRUCTURE, FUNCTION, AND MOLECULAR BIOLOGY OF K-K-K SYSTEM COMPONENTS

### 2.1 Kininogens

#### 2.1.1 Structure and Function of Kininogens

Kininogens are the natural glycoprotein substrates for the kallikrein family of serine proteases. Kininogens all contain the bradykinin nonapeptide sequence released by kallikrein cleavage at specific sites (Figure 4). All mammalian sera contain two kininogen species, high molecular weight (HMWK) and low molecular weight (LMWK) kininogen forms, with apparent molecular masses of 120 and 68 kDa, respectively. HMWK is cleaved by plasma kallikrein to yield the linear nine amino acid bradykinin peptide. Tissue (glandular) kallikrein cleaves LMWK to form lysyl-bradykinin. A third kininogen, similar in molecular mass to LMWK and found

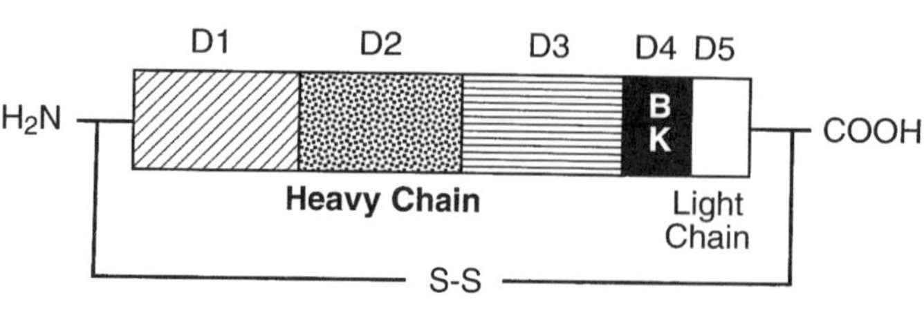

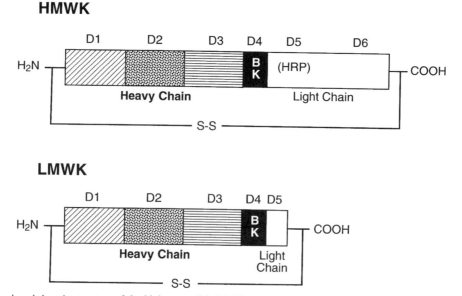

Figure 6. The six-membered domain structure of the kininogens, D1–D6. The common heavy chains of all kininogens have a set of D1–D3 domains; D2 and D3 bear cysteine proteinase inhibitory (CPI) sites. Domain D4 represents the kinin domain, the cleavage site for the kallikreins. The relatively large (45–58 kDa) light chain of HMWK has a histidine-rich domain designated D5 followed by the carboxy-terminal domain D6. LMWK has a very small (4–5 kDa) light chain carboxy-terminal D5 domain.

in rat plasma and kidney, has been termed T-kininogen or thiostatin. T-Kininogen is resistant to cleavage by all known plasma or tissue kallikreins. However, catalytic concentrations of a rodent tumor acid protease and high concentrations of trypsin were found to release from thiostatin a vasoactive kinin peptide with the novel isoleucine–serine–bradykinin sequence.

All three human kininogens share the following structural features generally representative of all mammalian kininogens: a heavy chain (HC) of 362 amino acids, the canonical kinin peptide segment, and a light chain region of 255 amino acid residues for HMWK and 38 amino acid residues for LMWK (Figure 5). The heavy chain of all kininogens has five putative N-linked glycosylation sites. The N-linked oligosaccharides of T-kininogen recently were characterized and were found necessary for cysteine proteinase inhibitor (CPI) function. The light chain of HMWK contains a 91 amino acid ''histidine-rich portion'' (HRP). A single disulfide bond joins the heavy and light chains.

Kininogens appear to be endowed with six distinct domains, as evidenced from structural, functional, and evolutionary studies (see Figure 6 and Section 2.1.2). The heavy chains of all kininogens have a triple set of D1 to D3 domains. D2 and D3 domains function as CPI sites with lengths of 112 to 122 amino acids. D3 includes a sequence that binds thrombospondin, a protein involved in platelet adhesion. Domain D4 is the 19–21 amino acid kinin domain, the cleavage site for the kallikreins. The 45–58 kDa light chain of HMWK includes the histidine-rich portion, D5, followed by the carboxy-terminal domain, D6. The histidine-rich region endows HMWK with a cofactor role in the initiation of blood coagulation via the intrinsic pathway. D6 contains sequences that bind prekallikrein and factor XI. LMWK has a very small (4–5 kDa) light chain, which constitutes its carboxy-terminal D5 domain.

Thus all kininogens are multifunctional. They are kinin donors and potent cysteine proteinase inhibitors. HMWK, by virtue of its ''histidine-rich portion'' and the zymogen binding region in the light chain, is involved with contact activation of blood coagulation. The functional significance of rat T-kininogen remains uncertain. T-kininogen is a major acute phase protein with a dramatically increased expression during the acute inflammatory response. Thus T-kininogen most probably plays a role in the acute phase of the inflammatory response [see Section 2.1.2, ''T-Kininogen (Thiostatin)''].

### 2.1.2 Molecular Biology of the Kininogens

**Evolution of Kininogen Domain Structure** The kininogens comprise a group of multidomain, multifunctional proteins. HMWK has six domains: three cystatinlike sequences (D1, D2, and D3) found in the heavy chain, an internal bradykinin D4 segment, and domains D5 and D6 in the light chain (Figure 6). CPI activity is localized onto the D2 and D3 heavy chain cystatinlike domains. The D5 and D6 light chain domains of HMWK diverge from the LMWK forms in terms of both structure and function. The LMWK light chain has only one domain, D5. The light chain of HMWK thus is larger and more complex; moreover, it contains the histidine-rich portion involved in the intrinsic clotting protease cascade.

The heavy chain, common to all kininogens, is considered to have arisen from the primordial CPI stefin gene (Figure 7). Hypothetically, the stefin gene developed into a cystatin gene encoding two carboxy-terminal disulfide loops. Subsequent gene duplication events led to the heavy chain of the kininogens containing two

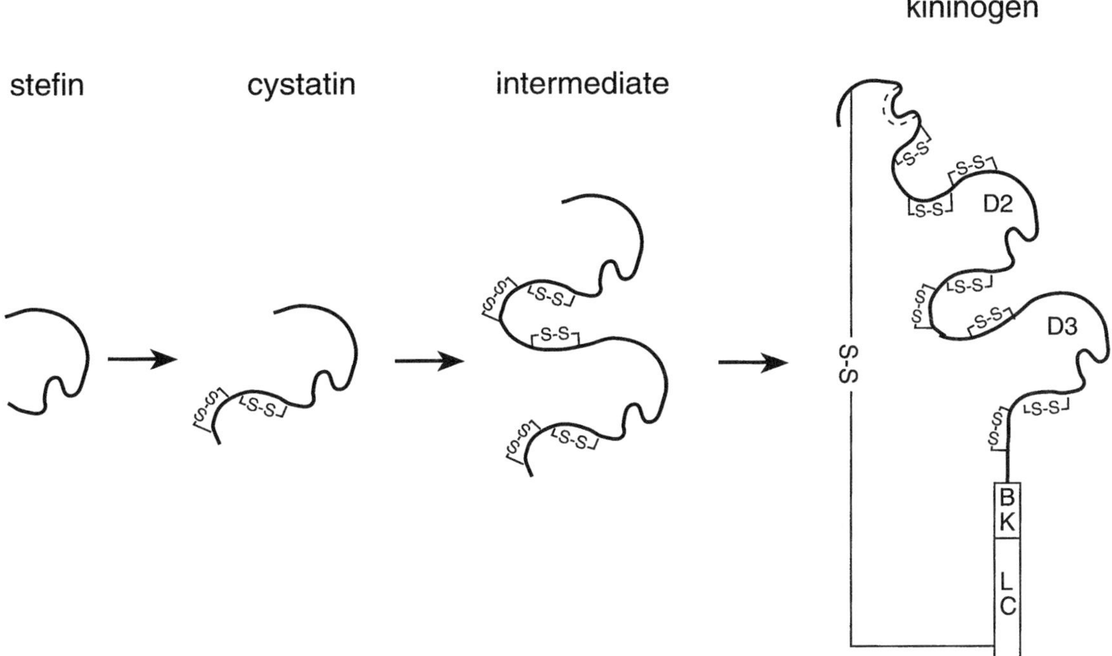


Figure 7. The hypothetical evolution of the kininogen molecule. The primordial stefin gene developed into a gene encoding two carboxy-terminal disulfide loops. Subsequent gene duplication events and acquisition of elements encoding the kinin and light chain domains gave rise to the modern kininogen molecule. The indentations on the domains represent the cysteine proteinase inhibitor active sites (eventually lost on domain 1, dotted lines). Disulfide bonds are as indicated.

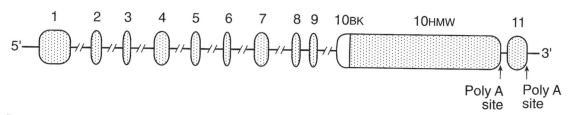

Figure 8. Structure of the human kininogen gene. Exons 1–9 encode the common heavy chain and exon 10 encodes the bradykinin (BK) and high molecular weight kininogen (HMW) light chain regions. Exon 11 is spliced downstream of the BK sequence in exon 10 to encode low molecular weight kininogen (see Figure 9).

cystatinlike domains, which endow kininogens with CPI activity. By virtue of their heavy chain, all kininogens are members of the mammalian superfamily of CPIs. In addition to their CPI function, all kininogens are kinin donors, but only HMWK has acquired the larger, more complex light chain containing the histidine-rich portion.

**Kininogen Gene Structure** The human kininogen gene has been mapped to chromosome 3 (3q26-qter). Kininogens are translated from mRNA, which is identical for all kininogens from the 5′-untranslated region through the bradykinin (BK) segment (up to the twelfth distal amino acid) but divergent downstream from that point. Human kininogens are derived from a single 27 kb gene containing 11 exons (Figure 8). Exons 1 through 9 encode the HMWK and LMWK common sequence. Coding sequences for BK and specific for HMWK are included in exon 10. Exon II contains the coding region specific for LMWK. Figure 9 depicts a proposed mechanism for the generation of kininogen mRNA. Alternative splicing of the gene transcript yields mature mRNA, which then is translated into the HMW and LMW kininogens. For HMWK, exon 10 is transcribed into mature mRNA, whereas for LMWK, exon 11 is spliced downstream of the BK segment in exon 10.

**T-Kininogen (Thiostatin)** Another low molecular weight kininogen, T-kininogen, is the most abundant kininogen in rat plasma and also has been found in rat kidney. This kallikrein-resistant T-kininogen is referred to as thiostatin based on CPI activity resident in the heavy chain D2 and D3 domains. T-Kininogen is the only kininogen that also is an acute phase plasma protein (APPP). The acute phase response is a generalized reaction to a variety of inflammatory stimuli including infection, neoplasia, and disorders of the immune system. The plasma levels of APPPs undergo dramatic alterations during the acute inflammatory response.

T-Kininogen is proposed to have evolved from an ancestral gene which, through gene duplication, gave rise to two gene copies. One gene copy served as the precursor of the so-called K-kininogens, HMWK and LMWK. The other gene copy gave rise to the T-kininogens T1 and TII. It is suggested that in the course of T-kininogen gene development, two mutations occurred. The kallikrein cleavage site (see Figures 3 and 4) was changed by the deletion of six nucleotides. In addition, a mutation was introduced that eliminated the ''read-through'' of the downstream region of exon 10, the transcription of which yields HMWK mRNA. Another gene duplication event is proposed to have led to T-kininogens TI and TII, which have minor structural differences from each other.

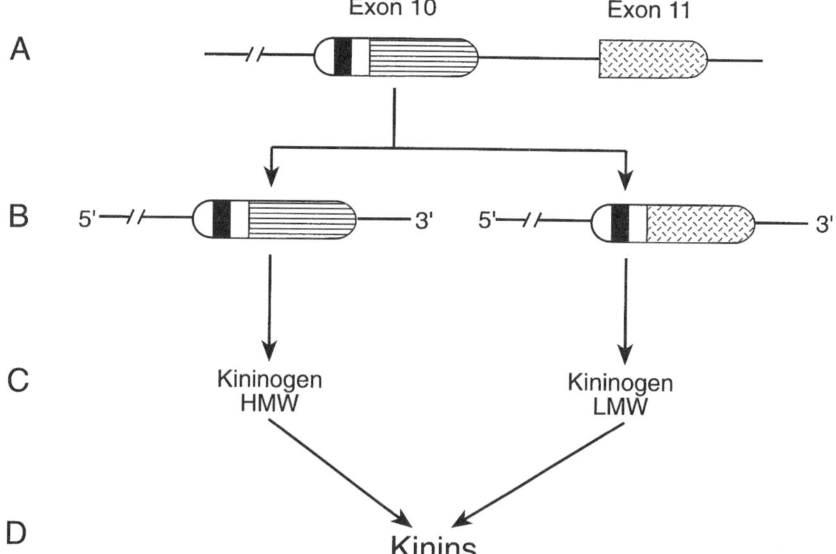


Figure 9. Processing of the kininogen genes into high molecular weight (HMWK) and low molecular weight (LMWK) forms. (A) Kininogen gene as it is found in the genome. The solid box represents the bradykinin (BK) coding element, horizontal lines depict the HMWK encoding region, and stippled areas the LMWK-specific sequence. (B) Transcription of mature mRNA in the liver: exon 10 is transcribed into mRNA to be translated into HMWK (C), exon 11 is spliced downstream of the BK coding region in exon 10, leading to expression of LMWK (C). (D) Kininogens are ultimately acted on by kallikreins to liberate kinin peptides.

Since the new T-kininogen gene product was no longer susceptible to cleavage by kallikrein to yield kinins and also not able to participate as a cofactor in the blood coagulation cascade, it has been proposed that T-kininogen expression became subject to an acute phase promoter. The apparent advantage of this newly evolved system would reside in the increased expression of high levels of a kininogen acute phase protein during inflammation without concomitant acute hypotension. Since T-kininogen possesses CPI activity within its heavy chain domain structure, elevated expression in response to inflammatory stimuli could serve a cytoprotective role by modulation of increased activity of lysosomal cysteine proteinases liberated from damaged cells and tissues. Indeed, during the acute inflammatory response, there is a 10- to 15-fold increase in TI and TII kininogen mRNA expression, whereas levels of K-kininogen mRNA are unaltered. The mechanism of differential expression of K- and T-kininogens during the inflammatory response occurs despite high homology both upstream of and across transcription initiation sites. Kininogen initiation sites are referred to as Cap 1, 2, and 3. In the case of T-kininogen, Cap 2 and 3 are used at greater rates during the acute inflammatory process, whereas the use of Cap 1 is not different from use under normal physiological conditions. In contrast, K-kininogen Cap usage remains unchanged in response to inflammation.

## 2.2 Kallikreins

### 2.2.1 Structure and Function of the Kallikreins

Kallikreins are serine proteases that cleave kininogen at specific sites to yield kinins. Investigators have characterized two types of kallikrein, plasma and tissue (glandular) kallikreins, which differ in their substrate and cleavage site specificities, tissue localization, functions, and molecular features. Plasma kallikrein is an acidic glycoprotein present in blood as an 88 kDa inactive "prekallikrein" proenzyme. Its natural substrate is HMWK, which it cleaves to release the nonapeptide bradykinin (Figure 2). Prekallikrein and factor XI circulate as complexes with HMWK in the blood. These complexes bind to exposed subendothelium through the histidine-rich light chain region of HMWK, where prekallikrein is involved in reciprocal activation with Hageman factor to initiate the blood clotting cascade. Hageman factor is a common activating enzyme to both the fibrinolysin and K-K-K cascade systems as well.

Several distinct tissue kallikreins have been isolated as prekallikrein from various tissues including submaxillary salivary glands, red blood cells, the gastrointestinal tract, pancreas, kidney, and pituitary gland. Molecular weights of tissue kallikreins range from 24 to 45 kDa. Tissue kallikrein cleaves LMWK at a C-terminal arginine–serine bond and N-terminal methionine–lysyl bond to yield lysyl-bradykinin (kallidin), which is transformed to bradykinin by a plasma aminopeptidase.

Kallikreins have been implicated in many physiologic and pathologic processes. Many actions of kallikreins are mediated through the release of bioactive peptide kinins, as noted earlier. The actions of tissue kallikreins may relate to the tissue site of the kallikrein. Thus renal kallikrein may be involved in the regulation of systemic blood pressure and perhaps in the development of hypertension. Salivary kallikrein may affect glandular local blood flow and may even serve a digestive role (in the rat) by hydrolyzing seed capsule gluten. Indeed, *tissue* kallikrein can hydrolyze a number of precursors (epg., *pro*insulin, prorenin, angiotensinogen). Plasma kallikrein functions in such protease cascades as fibrinolysin, blood coagulation, and complement activation and has been found in inflammatory exudates. It also should be noted that thyroid and steroid hormones induce tissue kallikrein gene expression *in a tissue-specific manner*. On the other hand, dopamine and dopamine agonists have been shown to down-regulate tissue kallikrein gene expression in the rat pituitary.

### 2.2.2 Molecular Biology of the Kallikreins

**Kallikrein Gene Families in the Human, Rat, and Mouse** Plasma kallikrein is encoded by a single gene solely in the liver, whereas glandular kallikreins are encoded by multigene families expressed in several tissues. Twenty-four kallikrein genes have been identified in the mouse and 15–20 genes in the rat. A comparatively smaller family of three kallikrein genes has been found in the human. The kallikreins display tissue-specific expression and are found in various tissues, notably the kidney, pancreas, prostate, and salivary gland. The various glandular kallikreins of the human, rat, and mouse species share 60–75% nucleotide sequence homology and are organized with five exons and four introns. However, the protein products of these genes exhibit variations in substrate specificity and amino acid sequence. In fact, varied substrates for rat glandular kallikrein may include atrial natriuretic peptide, proinsulin, vasoactive intestinal polypeptide, low density lipoproteins, and angiotensinogen. Enzymes designated as "true" kallikreins have the capacity to cleave kininogen efficiently.

**Functional Evolution of Kallikrein Gene Families** The amino acid sequence among mouse and rat kallikreins is more highly conserved than between true kallikreins from the mouse compared with true kallikreins from the rat. Products of the human kallikrein gene family show less sequence conservation. Although only 3 human kallikrein genes were identified by genomic blot analysis with a kallikrein cDNA probe, 19 independent clones were characterized by restriction analysis and Southern blot techniques used to screen a human genomic cDNA library with a monkey kallikrein cDNA. These findings suggest that the rodent genes encoding proteins with higher sequence conservation may have remained as tandem repeats, whereas many of the human genes may be spread throughout the genome.

The genetic mechanisms involved in the evolution of the kallikrein family genes and their differential expression in various tissues provide interesting questions for molecular biologists. Human, rat, and mouse genes are clustered on their respective chromosomes. The 15–20 rat kallikrein genes are likely to be closely linked at a single chromosomal locus. Close linkage favors unequal crossing over and gene conversion, leading to concerted evolution, a process whereby kallikrein genes have evolved within a species to share higher sequence identity than exists between true kallikrein genes from two species. In fact, the nucleotide sequences of exonic regions of nine characterized rat genes averages 88%. Extensive sequence identity is found in noncoding regions as well. The concerted evolution of rat kallikrein genes appears to select for sequence homology while allowing for some divergence of function. Indeed, the rat kallikreins display different substrate specificity, which endows them with an array of biological actions including processing of growth factors, peptide hormones, and cleavage of the kininogen substrate.

**Regulated Expression of Kallikrein Gene Families** The human and mouse so-called true kallikrein genes encoding kininogenases share highly homologous 5′-noncoding and 5′-flanking DNA regions, which suggests common regulatory mechanisms. The tissue-specific expression of rat kallikrein genes has been attributed to minor nucleotide differences in upstream flanking regions among different family members. Expression of rat genes is regulated by hormones, salt intake, and dietary protein. Studies of the ontogeny of rat glandular kallikrein mRNA production have revealed developmental regulation in both neonatal kidney and submandibular gland. Although androgen-induced mRNA synthesis has been demonstrated for human prostate cells grown in culture, less is known regarding regulated kallikrein gene expression in the human.

**Human Kallikrein Genes** Three human kallikrein genes, *HKLK1, HKLK2,* and *HKLK3,* have been studied. The *HKLK1* gene, which encodes true kallikrein, is localized on chromosome 19 (19q13). It is expressed in the kidney, pancreas, and salivary gland. The enzyme encoded by *HKLK2* remains unidentified. *HKLK3* encodes prostate-specific antigen (PSA). PSA, a secretion of the human prostate, is used as a marker to detect and monitor prostate cancer. *HKLK2* and *HKLK3* genes are closely linked near the *HKLK1* gene on chromosome 19.

The true kallikrein gene *HKLK1* shares approximately 77% nucleotide sequence identity with *HKLK3.* The identified protein encoded by *HKLK3,* PSA, is approximately 60% homologous with true kallikrein in terms of amino acid sequence. Amino acid residues conferring substrate specificity of true kallikrein have been changed in PSA. The *HKLK2* gene is approximately 86% identical to *HKLK3,* and conservation of a critical catalytic amino acid suggests that the yet-to-be identified gene product may bear kininogenase activity.

Both *HKLK2* and *HKLK3* are expressed in the prostate, whereas true kallikrein is not. The extent of homology of putative regulatory sequences suggests similar tissue-specific mechanisms for controlled expression of both genes. Androgen-dependent upregulation has been demonstrated for these genes in a human prostatic cell line. Nearly the same steroid hormone response element is found 160 base pairs upstream from the Cap site in the 5′-flanking region of both genes. In contrast to true kallikrein, the possible function of *HLK2* and PSA may involve semen liquefaction. PSA has been shown to cleave gel-like substances formed by fibronectin and the high molecular weight protein semenogelin.

### 2.3 Kinin Receptors

Pharmacological evidence suggests as many as five different types of kinin receptor, B1 through B5. The study of kinin receptors has been advanced by the development of novel potent bradykinin receptor antagonists to different receptor subtypes. Notably, the Hoe 140 and NPC 17631 antagonists have enabled further characterization of the B2 receptor, to which the more commonly occurring kinin-induced responses have been attributed. A functional B2 receptor has been cloned from a rat uterus cDNA library. Based on the predicted amino acid sequence, this receptor is homologous with the seven transmembrane G protein coupled receptor family.

*See also* Enzyme Mechanisms, Transient State Kinetics of; Glycoproteins, Secretory; Membrane Fusion, Molecular Mechanism of; Plasma Lipoproteins; Receptor Biochemistry; Recombination, Molecular Biology of.

## *Bibliography*


Abe, K., Moriya, H., and Fujii, S., Eds. (1989) *Advances in Experimental Medicine and Biology Series,* Vol. V, *Kinins.* Plenum Press, New York.

Back, N., and Bedi, G. S. (1991) *Kinins: Chemistry, biology and functions,* in *Encyclopedia of Human Biology,* Vol. 4, pp. 589–601. Academic Press, San Diego, CA.

Bedi, G. S., Balwierczak, J., and Back, N. (1983) A new vasopeptide formed by the action of a Murphy–Sturm lymphosarcoma acid protease on rat plasma kininogen. *Biochem. Biophys. Res. Commun.* 112:621–628.

Carbini, L. A., Scicli, G., and Carretero, O. A. (1993) The molecular biology of the kallikrein–kinin system: III. The human kallikrein gene family and kallikrein substrate. *J. Hypertension,* 11:893–898.

DeLa Cadena, R. A., Wachtfogel, Y. T., and Colman, R. W. (1994) Chapter 11 in *Hemostasis and Thrombosis: Basic Principles and Clinical Practice,* R. W. Colman, J. Hirsh, V. J. Marder, and E. W. Salzman, Eds. (1994) Lippincott, Philadelphia.

El-Dahr, S. S., and Dipp, S. (1993) Molecular aspects of kallikrein and kininogen in the maturing kidney. *Pediatr. Nephrol.* 7:646–651.

MacDonald, R. J., Margolius, H. S., and Erdos, E. G. (1988) Molecular biology of tissue kallikrein. *Biochem. J.* 253:313–321.

Miller, D. H., and Margolius, H. S. (1989) Kallikrein–kininogen–kinin systems, in *Endocrinology,* L. J. DeGroot and G. F. Cahill, Jr., Eds., pp. 2491–2503. Saunders, Philadelphia.

Muller-Esterl, W. (1987) Novel functions of the kininogens. *Semin. Thromb. Hemostasis,* 13:115–126.

———, Iwanaga, S., and Nakanishi, S. (1986) Kininogens revisited. *Trends Biochem. Sci.* 11:336–339.

Regoli, D., Jukic, D., Gobeil, F., and Rhaleb, N.-E. (1993) Receptors for bradykinin and related kinins: A critical analysis. *Can. J. Physiol. Pharmacol.* 71:556–567.

Rusiniak, M. E., Bedi, G. S., and Back, N. (1991) The role of carbohydrate in rat plasma thiostatin: Deglycosylation destroys cysteine proteinase inhibition activity. *Biochem. Biophys. Res. Commun.* 179:927–932.

Schreiber, G., Tsykin, A., Aldred, A. R., Thomas, T., Fung, W.-P., Dickson, P. W., Cole, T., Birch, H., DeJong, F. A., and Milland, J. (1989) The acute phase response in the rodent. *Ann. N.Y. Acad. Sci.* 557:61–86.

Weisel, J. W., Nagaswami, C., Woodhead, J. L., DeLa Cadena, R. A., Page, J. D., and Colman, R. W. (1994) The shape of high molecular weight kininogen: Organization into structural domains, changes with activation, and interactions with prekallikrein, as determined by electron microscopy. *J. Biol. Chem.* 269:10100–10106.

Wines, D. R., Brady, J. M., Southard, E. M., and MacDonald, R. J. (1991) Evolution of the rat kallikrein gene family: Gene conversion leads to functional diversity. *J. Mol. Evol.* 32:476–492.

# L

## Library: *see* type, e.g., Synthetic Peptide Libraries.

## Ligase Chain Reaction: *see* Infectious Disease Testing by Ligase Chain Reaction.

# Ligation Assays

*Claire Delahunty, Pui-Yan Kwok, and Deborah Nickerson*

*Key Words*

**DNA Ligase** Member of a class of enzymes that covalently join DNA fragments by forming a phosphodiester bond between the 5′-phosphoryl and the 3′-hydroxyl groups of juxtaposed DNA strands.

**Ligation Amplification** *In vitro* method to linearly or exponentially amplify DNA segments by ligating pairs of oligonucleotides that are complementary to specific adjacent sequences on a DNA template.

**Ligation Assay** *In vitro* detection method based on the ability of DNA ligases to join juxtaposed strands of DNA or oligonucleotides.

**Polymerase Chain Reaction** Enzymatic method for *in vitro* amplification of specific DNA fragments that uses two oligonucleotides to catalyze synthesis of the DNA strands spanned by the two oligonucleotides.

Ligase-mediated assays allow the discrimination of known sequence variants based on the ability of DNA ligase to covalently join adjacent oligonucleotides only when they perfectly complement a single-stranded DNA template. DNA ligases mediate the formation of a phosphodiester bond between the 5′-phosphoryl and the 3′-hydroxyl groups of juxtaposed DNA strands. In molecular cloning, ligase joins either cohesive or blunt ends between two double-stranded DNA molecules, allowing the cloning of recombinant DNA molecules. In a similar manner DNA ligases will covalently link two juxtaposed synthetic oligonucleotide probes when they are hybridized to a complementary template. The specificity of DNA ligases is such that even a single base mismatch at the probe–target junction will suppress ligation. These reactions are ideal for diagnostic applications involving the detection of any known sequence variations not only because of their powerful specificity but because a single set of assay conditions suffices to distinguish any type of nucleotide substitution or any unique deletion or insertion.

## 1 PRINCIPLES

### 1.1 Oligonucleotide Ligation Assay

One example of a ligase-mediated assay is the oligonucleotide ligation assay (OLA, Figure 1). This assay involves the hybridization of two oligonucleotides to a denatured DNA target. The two oligonucleotide probes lie adjacent to each other, which allows them to be joined by ligase if they correctly base-pair with the target strand. If the probes are mismatched at or near the junction, no bond can be formed. Since oligonucleotide probes are typically 20–25 nucleotides in length, their specific hybridization to a unique part of the genome is assured. After ligation, products can be detected by several methods. In one scheme, the 3′ end of a 5′-biotinylated probe is allowed to ligate with the 5′ end of a 3′ reporter. The double-stranded ligation product is allowed to bind to streptavidin immobilized on a solid support. If there has been a successful ligation event, a newly formed oligonucleotide (40 to 50-mer) containing both the biotin and reporter groups is captured by the support; if ligation has been unsuccessful, no reporter remains after excess reagents have been washed away. The OLA is most effectively used to detect biallelic polymorphisms, including single base substitutions, insertions, or deletions. In a biallelic system three oligonucleotide probes are required; two 5′-biotinylated probes, representing each of the allelic forms of the polymorphism, and one adjoining 3′ reporter probe that is common to both alleles. A separate reaction is assembled for each allele including the corresponding biotinylated probe, the reporter probe, and template DNA. A positive ligation event identifies which sequence variant is present in the template DNA.

The OLA has been improved by advances in methods to exponentially amplify specific DNA or RNA targets. Amplification using the polymerase chain reaction (PCR), the ligation amplification reaction (see Section 1.2), or other techniques generates specific targets with high signal-to-noise ratios. Amplification of the target DNA has several advantages for OLA:

1. Large quantities of specific DNA targets make it possible to use nonisotopic reporter groups.
2. DNA from many different sources can be used, including hair follicles, sperm, blood, epethelial cells, and serum.
3. A variety of different templates can be amplified simultaneously, thus increasing the assay throughput.
4. The entire assay can be performed in a microtiter plate, and a robotic workstation can be employed to semiautomate the procedure.

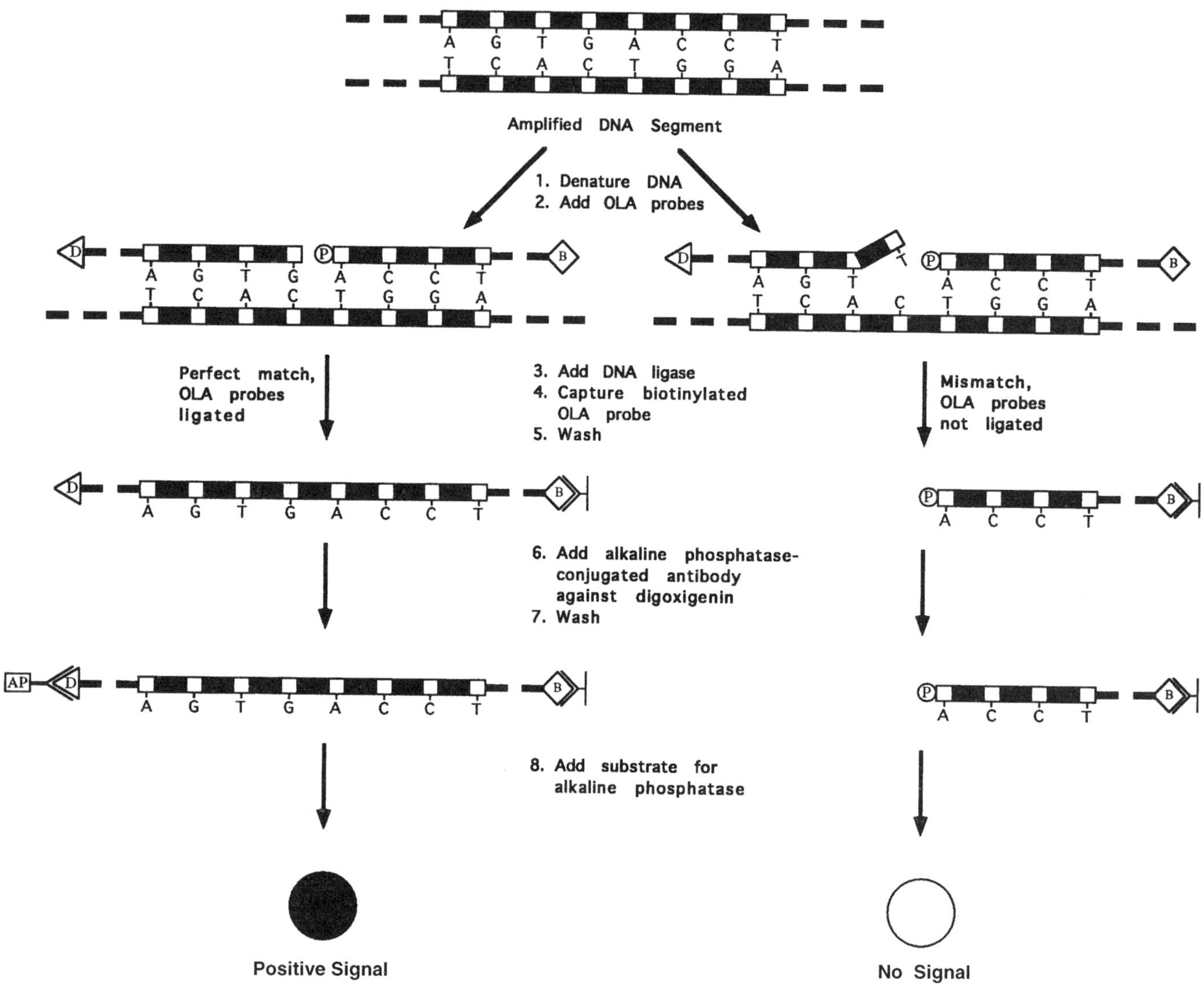


**Figure 1.** Oligonucleotide ligation assay.

## 1.2 Ligation Amplification Reaction

The ligation amplification reaction has also been used to distinguish single base pair mismatches (Figure 2). This method utilizes the ligation of oligonucleotide pairs that are complementary to neighboring sites on a DNA template. The product can be increased either linearly or exponentially during sequential rounds of template-dependent ligation. During linear amplification a single pair of oligonucleotides is hybridized to a template. If they complement the template, oligonucleotides are joined through the action of DNA ligase. The reaction is heated to dissociate the double-stranded ligation product, which can then serve as a substrate for another round of ligation. After multiple rounds of ligation/denaturation, the longer product of ligation is greatly increased over unligated probes and can be detected by size separation using gel electrophoresis. Alternatively, by radioactively or fluorescently labeling the diagnostic oligonucleotide, the allele state of the template can be determined according to whether a tagged ligation product is present or absent. In exponential ligation amplification, two pairs of oligonucleotides are used: one pair complementary to the upper strand of the target DNA and the other complementary to the lower strand. In this way both ligation products can serve as template for subsequent rounds of amplification, resulting in an exponential accumulation of ligated products.

Initially, ligation amplification techniques were limited by the thermal instability of DNA ligase. The heat required to dissociate the duplex DNA also inactivated the enzyme, and therefore fresh enzyme had to be added in each amplification cycle. Recently, however, Barany described a thermostable ligase that remains active through multiple rounds of heating. This enzyme (*Tth* ligase) was purified from either the original *T. thermophilus* strain or a clone in an *E. coli* host, demonstrated ligase activity at temperatures as high as 85°C, and retained activity after repeated exposure to temperatures up to 94°C. The thermal stability of this ligase made

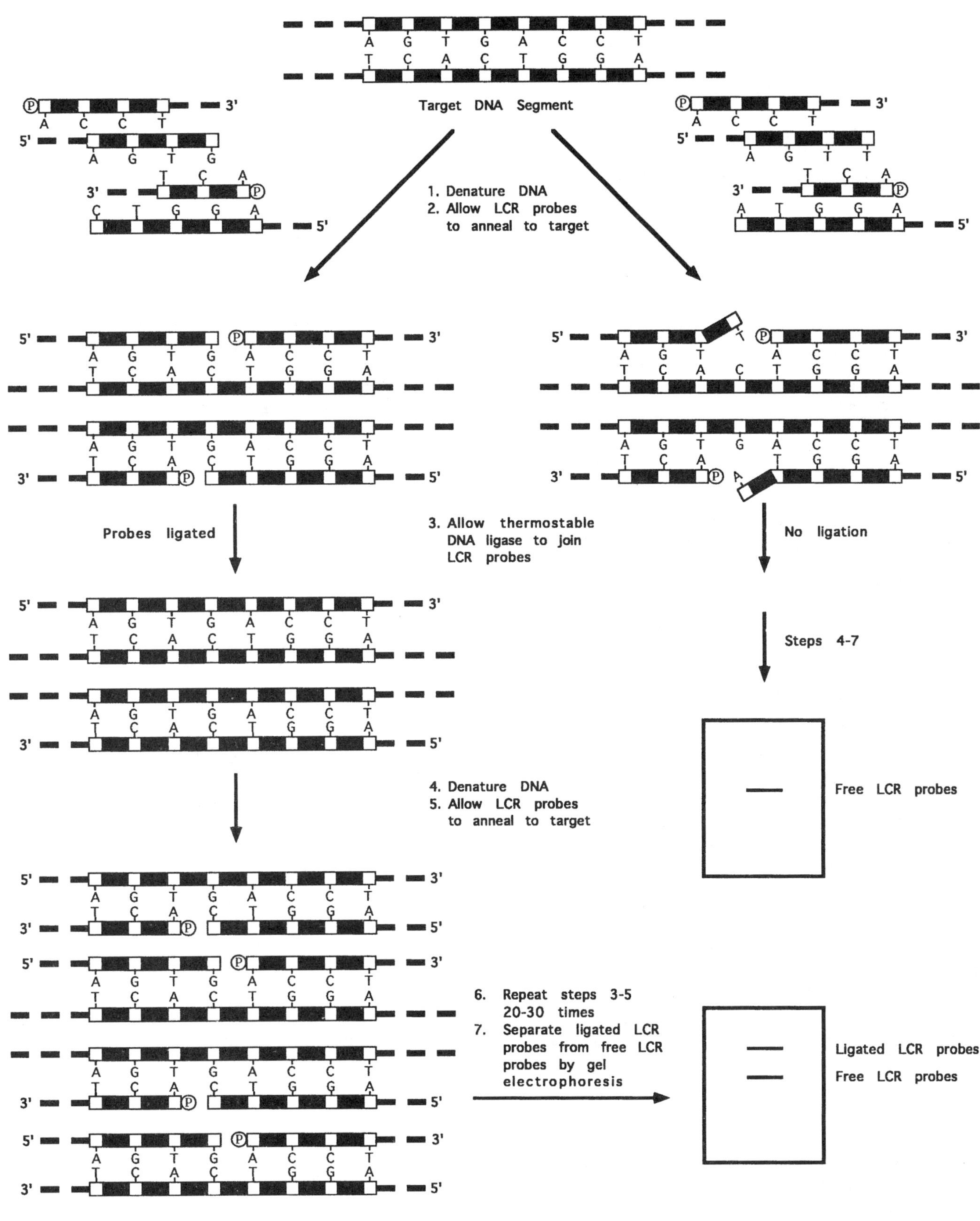


Figure 2. Ligase amplification reaction.

ligation amplification practical as a method for distinguishing single nucleotide mismatches because it eliminated the need to add fresh enzyme in each round of amplification. Additionally, the use of thermostable ligase significantly enhanced the specificity of ligation, since the reaction could now be performed at temperatures near the melting point of the oligonucleotide probes.

### 1.3 Ligase-Mediated Polymerase Chain Reaction

Ligase-mediated PCR combines the specificity of specific oligonucleotide primer hybridization with nonspecific blunt-end linker ligation to amplify and identify sequences when only partial sequence information is available. Target DNA is cleaved, the strands denatured, and a specific primer is hybridized to the part of the target for which the sequence is known. The primer is extended with Sequenase to generate a new blunt end to which a linker primer can be ligated using T4 ligase. The linker primer and a second specific primer are used to amplify the appropriate DNA fragments specifically.

## 2 TECHNIQUES

### 2.1 Automation

The detection of sequence variants via PCR/OLA has been semiautomated, greatly increasing the throughput and potential applications of the method. For each biallelic polymorphism to be screened, two reactions are assembled, each containing one probe or the other (i.e., a diagnostic probe or the reporter probe). The amplified DNA template is denatured and added to each of the two ligation reactions. The probes are allowed to anneal to the template, and DNA ligase is added. After ligation the reactions are transferred to a streptavidin-coated microtiter plate, where the biotinylated oligonucleotide is captured. After the sample has been washed, the presence of the reporter group is detected by addition of an alkaline phosphatase conjugated antibody against digoxigenin followed by substrate for alkaline phosphatase. A Biomek 1000 robotic workstation is used to facilitate the process and increase sample throughput. The workstation can be used for several steps: to assemble reagents for DNA amplification; to mix the oligonucleotide probes with the amplified DNA, distribute ligase to ligation reactions, and transfer the reaction to streptavidin-coated microtiter plates for capture; and to wash away excess reagents for detection. Presently 1200 ligation assays can be processed in a day, and this rate should increase with the development of higher density microtiter plates and better automation.

Recently the ligase amplification reaction assay has been modified for detection of the products with a nonisotopic readout. The incorporation of a chemiluminescent or colorimetric substrate in the ligation product eliminates the need for gel electrophoresis, and the method has the potential for automation.

### 2.2 Combination of Techniques

Both OLA and the ligase amplification reaction are compatible with PCR as an initial amplification technique. In both cases, multiple loci can be amplified simultaneously by PCR and then aliquots of the amplification reaction used for detection of a particular target. Ligation amplification has been combined with a number of other amplification strategies in an effort to improve sensitivity and reduce target-independent ligation of oligonucleotide pairs. Ligation amplification can also be coupled with PCR/OLA. In this powerful technique a DNA target is amplified using PCR and the allele state is determined by the OLA. The signal-to-noise ratio for the ligated product is enhanced by amplifying the ligation product with the LCR prior to capture on streptavidin.

## 3 APPLICATIONS AND FUTURE PROSPECTS

Ligase-mediated DNA assays are finding application in many clinical and research areas. The detection of single nucleotide polymorphisms allows rapid screening for common mutations associated with genetic, malignant, and infectious diseases, as well as a powerful approach to genetic linkage mapping, HLA typing, paternity testing, and forensic testing. PCR/OLA has been used for the detection of polymorphisms responsible for genetic diseases such as sickle cell anemia, cystic fibrosis, and α-antitrypsin deficiency, and for genetic mapping of the T-cell receptor β locus. Ligation amplification methods have been employed to distinguish the normal from the sickle cell allele of the human β-globin gene, to discriminate *Listeria monocytogenes* from other *Listeria* species, and to identify two different point mutations responsible for hyperkalemic periodic paralysis.

Applications of ligase-mediated ligation assays will continue to increase as the technology to perform the assays improves. Advances are being made in the use of higher density formats, which will increase the throughput of the assay, and in the use of multiple nonisotopic reporter groups, which will allow the analysis of several biallelic polymorphisms in a single well of a microtiter plate. The ability to study large numbers of genetic markers in large populations by ligase-mediate DNA assays will open up exciting new ways to identify genes that are responsible for multigenic traits, to follow DNA changes over time as a result of aging or exposure to environmental hazards, and to study genetic diversity and population genetics. Results of such studies will have far reaching benefits to human health and medicine in the twenty-first century.

*See also* Oligonucleotides; PCR Technology.

### *Bibliography*

Barany, F. (1991) The ligase chain reaction in a PCR world. *PCR Methods Appl.* 1:5–16.

Grossman, P. D., Bloch, W., Brinson, E., Chang, C. C., Eggerding, F. A., Fung, S., Iovannisci, D. A., Woo, S., Winn-Deen, E. S. (1994) High-density multiplex detection of nucleic acid sequences: Oligonucleotide ligation assay and sequence-coded separation. *Nucleic Acids Res.*, 22(21):4527–4534.

Nickerson, D. a., Kaiser, R., Lappin, S., Stewart, J., Hood, L., Landegren, U. (1990) Automated DNA diagnostics using an ELISA-based oligonucleotide ligation assay. *Proc. Natl. Acad. Sci. (USA)* 87(22):8923–8927.

# Lipid Metabolism and Transport

*Clive R. Pullinger and John P. Kane*

### *Key Words*

**Apolipoproteins** Proteins associated with the plasma lipoproteins.

**Chylomicrons** A class of triglyceride-rich lipoproteins: density 0.93 g/mL.

**HDL** High density lipoproteins: density 1.063–1.21 g/mL.

**IDL** Intermediate density lipoproteins: density 1.006–1.019 g/mL.

**LDL** Low density lipoproteins: density 1.019–1.063 g/mL.

**Lipase** An enzyme that hydrolyzes triglycerides or phospholipids.

**VLDL** Very low density lipoproteins, a class of triglyceride-rich lipoproteins: density 0.93–1.006 g/mL.

---

Lipoprotein metabolism can be subdivided into transport of exogenous lipids, transport of endogenous lipids from the liver to peripheral tissues, and reverse cholesterol transport. Lipids are transported through the bloodstream as macromolecular complexes called lipoproteins, of which there are a number of classes that can be separated by sequential ultracentrifugation. Lipoprotein particles consist of a hydrophobic lipid core that is stabilized by a monolayer of phospholipid, free cholesterol, and apolipoproteins. A number of the apolipoprotein genes have been cloned. Seven are members of a multigene family. Six of these are clustered at two loci: *apo*AI, *apo*CIII, and *apo*AIV on chromosome 11 and *apo*E, *apo*CI, and *apo*CII on chromosome 19. The apoB gene codes for two major structural proteins, one produced by a novel mRNA editing mechanism. Gene products that play important roles in lipid transport include the high affinity LDL receptor, lipoprotein lipase, hepatic triglyceride lipase, lecithin–cholesterol acyltransferase, and cholesteryl ester transfer protein. The techniques of molecular biology have vastly increased our understanding of lipid metabolism and atherosclerotic heart disease.

## 1 EXOGENOUS TRANSPORT

Prior to absorption, triglycerides are hydrolyzed by lipases to fatty acids and β-monoglycerides. The lipids are absorbed in the small intestine with the aid of emulsifying bile acids. Triglycerides are resynthesized, and much of the cholesterol is esterified in the enterocytes. Exogenous lipids are transported as shown in Figure 1.

### 1.1 Chylomicron Assembly

Hydrophobic triglycerides and cholesteryl esters are packaged in the smooth endoplasmic reticulum (ER) of enterocytes as large stable microemulsion particles called chylomicrons. The nascent chylomicrons (> 1000 Å in diameter) are exocytosed via the Golgi apparatus into the lymphatic system, from which they migrate through the thoracic duct into the circulation. In addition to the structural protein, apoB-48, they contain apoAI and apoAIV. As a result of exchange with HDL, they lose apoAI and apoAIV and gain C apolipoproteins.

Microsomal triglyceride transfer protein (MTP), isolated from intestine and liver, is involved in lipoprotein assembly. The protein and MTP activity were absent in patients with abetalipoproteinemia. This autosomal recessive disease, which is characterized by the absence of apoB-containing lipoproteins in the plasma, occurs when a person is unable to secrete triglyceride-rich lipoproteins from intestine or liver.

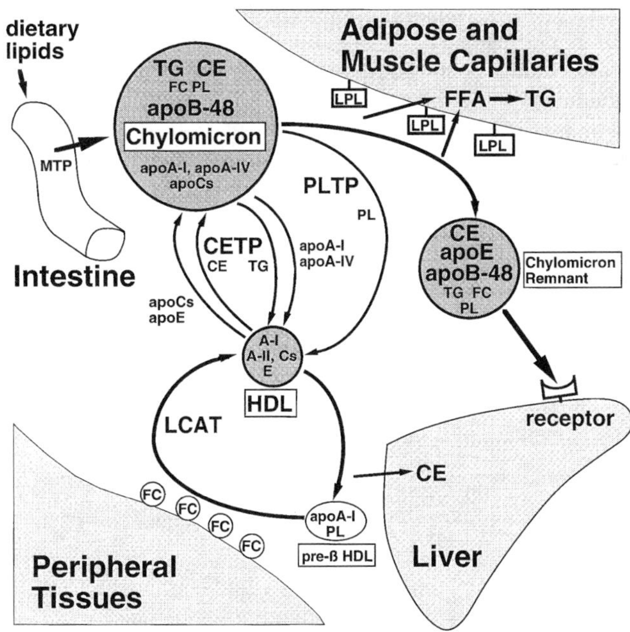


**Figure 1.** Exogenous lipid transport: FC, free cholesterol; CE, cholesteryl ester; PL, phospholipid; TG, triglyceride; FFA, free fatty acid; LPL, lipoprotein lipase; LCAT, lecithin–cholesterol acyltransferase; CETP, cholesteryl ester transfer protein; PLTP, phospholipid transfer protein; MTP, microsomal triglyceride transfer protein.

### 1.2 Chylomicron Catabolism

Most of the triglyceride core of chylomicrons is hydrolyzed by the enzyme lipoprotein lipase (LPL), which is synthesized at a number of sites, notably adipose tissue, skeletal muscle, and the heart. LPL, secreted from the parenchymal cells, is transported to the capillary endothelial surface, where it attaches as a homodimer to a high affinity binding site.

The gene for LPL (chromosomal location, 8p22) spans 30 kb and comprises 10 exons, the first 9 of which code for a 448-residue prepeptide. Two sizes of mRNA are observed: 3.35 and 3.75 kb. The mature enzyme is a 421-residue, 55 kDa, N-linked glycoprotein. Ser 132, Asp 156, and His 241 make up the active site, which is homologous to that in pancreatic and hepatic lipase. Unlike these other two lipases, LPL requires *apo*CII as a cofactor. Mutations in the LPL gene cause the rare recessive disorder, familial LPL deficiency, characterized by the lack, or severely reduced levels, of LPL activity.

The fatty acids released by LPL are taken up by the surrounding tissue, where they are stored as triglyceride or used as fuel. Along with free cholesterol and phospholipid, apoCII is shed as the particle gets progressively smaller. During this process it acquires apoE and cholesteryl esters and loses most of the other apoproteins, except apoB-48, by transfer with HDL. Ultimately chylomicron remnant particles rich in cholesteryl esters are formed. These are rapidly removed by an hepatic chylomicron remnant receptor, which binds apoE with high affinity. Despite much evidence for its existence, this receptor has so far eluded purification and cloning. The $\alpha_2$-macroglobulin receptor (also referred to as the LDL-receptor-like protein or LRP) may play this role. The $\alpha_2$-macroglobulin receptor/LRP is a large multifunctional cell surface receptor consisting of 515 and 85 kDa subunits derived from a 600 kDa precursor. In addition to binding to $\alpha_2$-macroglobulin/protease complexes, the receptor has been shown to bind apoE-enriched βVLDL, *Pseudomonas* exotoxin A, and plasminogen activator–inhibitor complexes. The binding and uptake of these ligands is blocked by a 39 kDa receptor-associated protein (RAP) that copurifies with the receptor. Lipoprotein lipase binds with high affinity to this receptor, which may play a role in the uptake of triglyceride-rich lipoproteins. Its fundamental role is emphasized in mice that are genetically deficient for $\alpha_2$-macroglobulin receptor/LRP. These animals die in utero.

## 2 ENDOGENOUS TRANSPORT

The transport of endogenous lipids is illustrated in Figure 2.

### 2.1 VLDL Assembly

The liver secretes a triglyceride-rich lipoprotein analogous to a chylomicron. These nascent VLDL particles are smaller but have a similar lipid composition. They contain one copy of apoB-100 as the structural protein plus apoAI, apoE, and C proteins. The triglyceride is derived from chylomicron remnants and from fatty acids released by LPL or synthesized de novo in the liver. Unlike chylomicron secretion, which is maintained only throughout the postprandial period, VLDL production continues even during periods of starvation, when it is an important means of providing muscle fuel.

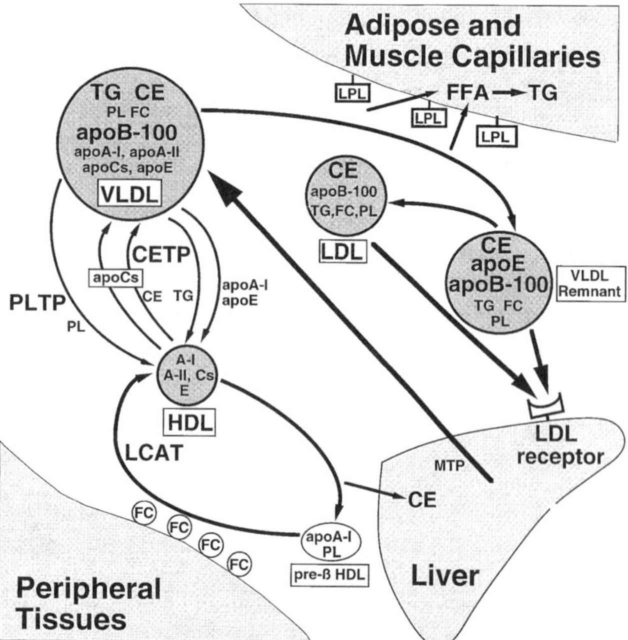


**Figure 2.** Endogenous lipid transport: FC, free cholesterol; CE, cholesteryl ester; PL, phospholipid; TG, triglyceride; FFA, free fatty acid; LPL, lipoprotein lipase; LCAT, lecithin–cholesterol acyltransferase; CETP, cholesteryl ester transfer protein; PLTP, phospholipid transfer protein; MTP, microsomal triglyceride transfer protein.

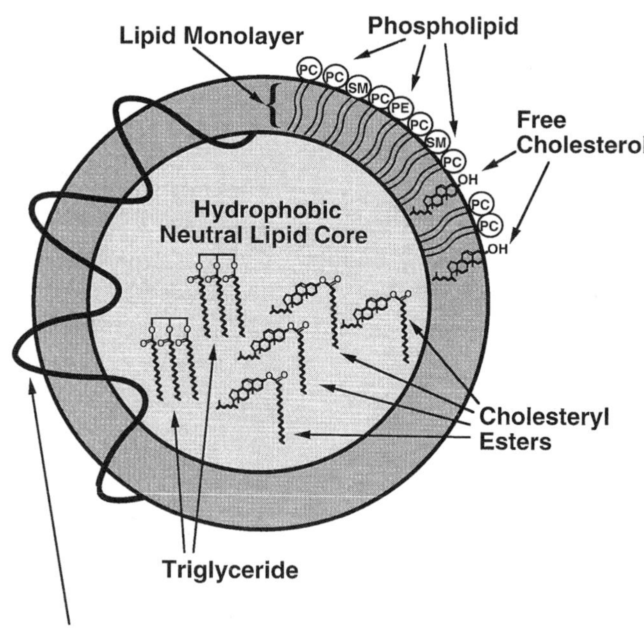


**Figure 3.** Structure of low density lipoprotein: PC, phosphatidylcholine; SM, sphingomyelin; PE, phosphatidylethanolamine.

### 2.2 VLDL Catabolism

The nascent VLDL loses apoAI and apoE and gains more C proteins by exchange with HDL. The triglyceride core of the mature VLDL is hydrolyzed by LPL. Exchange with HDL restores apoE and removes most of the C proteins. The resulting VLDL remnants contain apoB-100 and apoE. About half are removed by the hepatic LDL receptor, which has high affinity binding sites for apoE. The remaining remnants lose more triglyceride, by transfer to HDL and by the action of hepatic triglyceride lipase (HTGL), and are converted to LDL that is rich in cholesteryl esters (Figure 3). During this process the apoE and C proteins are lost, leaving apoB-100 as the sole protein.

HTGL has a similar substrate specificity for triglycerides as LPL but hydrolyzes phospholipid two to three times more effectively. It is not activated by *apo*CII and does not require $Ca^{2+}$. Like LPL, its optimum activity is at pH 8.0–9.0. The gene for HTGL (chromosomal location 15q21) is 35 kb long and is composed of nine exons. The promoter region has tissue-specific glucocorticoid response and cAMP response elements. Mature HTGL is a glycoprotein (60 kDa) of 476 residues. Familial HTGL deficiency, a rare condition, is marked by elevated levels of VLDL remnants and by the presence of triglyceride- and phospholipid-rich LDL and HDL.

A cell surface receptor for VLDL has been cloned in rabbits. It is expressed in the heart and muscle, but not in the liver, in contrast to the LDL receptor, to which it shows a very high degree of homology.

### 2.3 LDL Catabolism

Approximately 70% of LDL particles are removed by the liver—most via the LDL receptor, the remainder by a nonspecific low affinity process. The LDL receptor, a 160 kDa glycoprotein, is present on the surface of most nucleated cells. The binding affinity for LDL is considerably lower than that for VLDL remnants. LDL receptor gene (chromosomal location 19p13.3) is 45 kb long with 18 exons. The 5.3 kb mRNA, which has a long 2.5 kb 3′ untranslated region (UTR) with three 300 bp Alu repeats that may be a recombination hot spot, codes for a 860-residue protein. In addition to restriction fragment length polymorphisms (RFLPs), there is a dinucleotide AT repeat polymorphism in the 3′ UTR. The protein has seven cysteine-rich ligand-binding domains with homology to complement C9, a domain with homology to the epidermal growth factor precursor, an O-linked sugar domain, and a membrane-spanning region. The cytoplasmic C-terminus contains the sequence Asn-Pro-Val-Tyr. This signal targets the receptor to clathrin-coated pits, where it is endocytosed; then it releases the bound LDL in an endosomal compartment and recycles to the cell surface.

The discovery of the LDL receptor and its subsequent cloning has led to an understanding of the molecular basis of familial hypercholesterolemia (FH). FH, an autosomal codominant disorder characterized by elevated LDL levels, the presence of tendon xanthomas, and premature coronary heart disease, is caused by mutations in the LDL receptor gene. The heterozygous frequency is 1 in 500. The mutations result, variously, in null alleles, proteins that are not processed because glycosylation is defective, defects in the ligand-binding domain, receptors that fail to migrate to coated pits, or receptors that do not recycle.

The gene has been targeted in mice using homologous recombination to produce an FH model. The phenotype in these mice can be temporarily reversed using a recombinant LDL receptor adenovirus.

Oxidized or chemically modified LDL is taken up by macrophages via receptor-mediated endocytosis, independent of the LDL receptor. This may be the mechanism, in vivo, for foam cell formation, leading to atherosclerotic lesions. Lipoproteins containing apoB have been isolated from normal and atheromatous arteries. Three "scavenger" receptors, two of which are closely related, have been cloned.

## 3 REVERSE CHOLESTEROL TRANSPORT

The hepatobiliary system is the major means by which cholesterol is excreted. While most of the bile salts and free cholesterol in bile are reabsorbed by the ileum and returned to the liver, a fraction is lost through the feces. Some of this is derived from cholesterol on remnant particles or from hepatic de novo synthesis of cholesterol. The rest is from cholesterol returned to the liver from peripheral tissues. This latter process is termed reverse cholesterol transport. Free cholesterol on plasma membranes is transferred to HDL and is rapidly esterified by lecithin–cholesterol acyltransferase (LCAT), trapping it in the plasma compartment. The initial acceptor is 65 kDa pre-β HDL, a subpopulation of HDL composed of apoAI and phospholipid. The cholesteryl esters are transferred to HDL with α-mobility. HDL are also acceptors of free cholesterol from LDL.

The LCAT gene (chromosomal location 16q22.1) has six exons. Mature LCAT, secreted by the liver, is a 60 kDa, 416-residue glycoprotein with some homology to certain lipases. LCAT is activated by cofactors, apoAI, and apoAIV. Mutations causing LCAT deficiency result in decreased levels of HDL and of LDL cholesteryl esters.

The cholesteryl esters formed by LCAT are moved by cholesteryl ester transfer protein (CETP) from HDL to the apoB-containing lipoproteins in exchange for triglycerides. CETP is a 70 kDa hydrophobic glycoprotein. The gene, located on chromosome 16 (16q13), comprises 16 exons spanning 25 kb and codes for a 493-residue preprotein with a 17-residue signal peptide. It is expressed in the liver, small intestine, adrenal glands, spleen, and adipose tissue. Individuals with familial CETP deficiency have high levels of large, apoE-rich HDL particles and low levels of LDL cholesterol, but do not suffer from premature atherosclerosis. Transgenic mice (a species in which plasma CETP is normally absent) expressing human CETP have a decrease both in the level of HDL cholesterol and in HDL particle size.

Besides the pathway just described, cholesteryl esters can be transferred to liver via endocytosis of apoE-rich HDL particles and by selective uptake by a process involving HTGL.

Although the liver may secrete some nascent discoidal HDL, most HDL particles are formed in plasma, the apoproteins and phospholipids being derived from chylomicrons and VLDL. The formation of HDL particles is mediated by phospholipid transfer protein (PLTP). This protein has an apparent molecular mass of 81 kDa, the cDNA containing an open reading frame that encodes a mature protein of 476 amino acids. PLTP is expressed in many tissues and has homology with CETP, lipopolysaccharide-binding protein, and neutrophil bactericidal permeability increasing protein.

Isoelectric focusing followed by electrophoresis in nondenaturing gels of HDL reveals several subspecies. Some of these contain apoAI but not apoAII and are referred to as LpAI. Others contain both these proteins and are referred to as LpAI/AII. LpAI and LpAI/AII play metabolically different roles, especially with regard to reverse cholesterol transport. Both are able to promote the efflux of cholesterol from the plasma membrane, but LpAI appears to mediate the translocation of intracellular cholesterol to the cell surface. LpAI/AII is the preferred substrate for HTGL. The level of LpAI in plasma is a better measure of the protective potential of HDL than LpAI/AII.

## 4 CHOLESTEROL METABOLISM

Much of the cholesterol in cells is obtained from circulating LDL. However most tissues, in particular the liver, can synthesize cholesterol via a pathway involving the production of isoprenoid intermediates. The enzymes hydroxymethylglutaryl (HMG) CoA-synthetase and -reductase play essential roles in the production of mevalonic acid, the first committed precursor. Several inhibitors of HMG CoA reductase are highly effective in lowering plasma levels of LDL. The genes for HMG CoA synthetase, HMG CoA reductase, farnesyl pyrophosphate synthetase, squalene synthetase (all regulatory enzymes), and the LDL receptor are coordinately regulated by the amount of cholesterol available to the cell. As well as having unique enhancers, they share common cis-acting sterol regulatory elements (SREs) in their promoters. HMG CoA reductase has seven transmembrane domains that anchor it to the smooth ER and are involved in the regulation of degradation of the enzyme. In addition to transcriptional and translational regulation, HMG CoA reductase is regulated posttranslationally by phosphorylation.

Bile acids are synthesized from cholesterol in the liver. The first, rate-controlling step, the formation of 7α-hydroxycholesterol, is catalyzed by cholesterol 7α-hydroxylase, a liver-specific, cytochrome P-450 dependent mixed function oxidase. This enzyme, located in the smooth ER, is regulated at the transcriptional level, with mRNA levels positively correlated to the availability of cholesterol. The gene, which is 12 kb long, lies on chromosome 8 (8q11-q12), contains six exons, and codes for a 504-residue peptide. So far, no disorders of lipid metabolism have been linked to this gene.

Acyl-CoA:cholesterol acyltransferase (ACAT) plays an important role in lipoprotein metabolism and in atherogenesis. This enzyme catalyzes the intracellular esterification of cholesterol with long chain fatty acyl-CoA and is crucial for the regulation of free cholesterol levels within cells. ACAT activity in macrophages and smooth muscle cells of the artery wall is responsible for the accumulation of cholesterol esters in these cells, leading to the formation of foam cells, a typical feature of early lesions in the artery wall. Hepatic ACAT activity helps to maintain the free cholesterol concentration within the liver and plays a role in regulating the secretion of cholesterol into plasma, as a constituent of VLDL. Hence, ACAT is involved in the regulation of apoB secretion from the liver. It is also responsible for regulating the efflux of free cholesterol into bile. ACAT activity in the liver is decreased in patients with cholesterol gallstones, and the decrease contributes to the increased availability of free cholesterol for bile secretion.

ACAT is a membrane-bound protein of the endoplasmic reticulum. It has proved difficult to purify to homogeneity. The ACAT cDNA has an open reading frame that encodes an integral membrane protein of 550 residues. The gene is located on chromosome 1 (1q25).

## 5 TRIGLYCERIDE AND PHOSPHOLIPID METABOLISM

Many triglyceride and phospholipid pathway enzymes are intimately bound to membranes and have proved difficult to clone. Phosphatidylcholine, the major phospholipid, is formed by the action of CDP-choline: 1,2-diacylglycerol choline phosphotransferase, an enzyme intrinsic to the ER and the Golgi apparatus. The gene for this enzyme has been cloned from yeast; it has a molecular mass of 46 kDa and seven membrane-spanning helices.

## 6 THE APOLIPOPROTEINS

### 6.1 The Apolipoprotein Multigene Family

Seven apolipoproteins, all of which are exchangeable among lipoproteins, are members of a multigene family. These are apoAI, apoAII, apoAIV, apoCI, apoCII, apoCIII, and apoE. The genes have a similar structure and evolved from a common ancestor. Each has four exons, except *apo*AIV, in which exons 1 and 2 have fused. Within exon 3 there is, in each case, a conserved region of 33 codons. This amphipathic α-helical lipid-binding domain is a trimer of 11 amino acids. The fourth exons code for variable numbers of consensus, 22-residue, and in some cases 11-residue, lipid-binding tandem repeats.

The apoAI, apoCIII, and apoAIV genes are clustered in a 15 kb region on chromosome 11 (11q23-q24) with the apoAI and apoCIII genes in the opposite orientation. The genes for apoCI, apoCII, apoE, and an apoCI pseudogene are closely linked on chromosome 19 (19q13.2).

### 6.2 ApoAI

The major protein component of HDL, apoAI is an activator of LCAT. The level of apoAI in plasma is approximately 130 mg/dL. The gene (1863 bp) codes for the 267-residue preproapoAI, synthesized in the liver and intestine. The secreted 249-residue propeptide is processed in the plasma to the mature, 28 kDa, 243-residue form. Rare charge variants have been detected by isoelectric focusing. Pro 165→Arg and Arg 173→Cys ($\text{apoAI}_{\text{Milano}}$), and truncated variants, are associated with low levels of HDL. Two mutations, Gly 26→Arg and Leu 60→Arg cause amyloidosis. Overexpression of human *apo*AI in transgenic mice was associated with increased HDL cholesterol levels and prevented early diet-induced atherosclerosis.

### 6.3 ApoAII

A secondary protein component of HDL, apoAII is synthesized and secreted mainly by the liver. The plasma concentration is approximately 40 mg/dL. Like apoAI, it is regulated both transcriptionally and translationally. The gene, 1.3 kb long, is located on chromosome 1 (1q21-q23). The 600 bp mRNA codes for a 100-residue prepropeptide. Following cleavage of the 18-residue signal peptide but prior to secretion, some of the 82-residue propeptide is cleaved, forming the mature 77-residue peptide. The rest is processed in the plasma, where it exists as a homodimer. The physiological function of apoAII is largely unknown. It may be involved in the modulation of LCAT and HTGL activity.

### 6.4 ApoAIV

The 46 kDa glycoprotein known as apoAIV is synthesized mainly in the small intestine. The 2.6 kb gene codes for a 396-residue prepeptide, which after loss of the signal peptide gives rise to the 376-residue mature form. It is secreted on chylomicrons, from which it is rapidly lost to HDL upon entering the blood. Much is found after ultracentrifugation in the lipoprotein-free fraction of the plasma. This is puzzling, since apoAIV has a large number of potentially lipid-binding amphipathic α-helices. The total concentration in plasma is approximately 16 mg/dL. The precise role of apoAIV is unknown, but it activates LCAT and may be involved in reverse cholesterol transport.

### 6.5 ApoB

The level of apoB in plasma is approximately 80 mg/dL: 90% in LDL, 8% in IDL, and 2% in VLDL and chylomicrons. A centile nomenclature is used to describe apoB species based on their molecular weights. The two circulating forms, apoB-48 (264 kDa: 2152 residues) and apoB-100 (549 kDa: 4536 residues), are the products of a single gene on chromosome 2 (2p23-p24). C-terminally truncated apoB-48, found on chylomicrons and their remnants, is produced in the intestine. The editing of apoB mRNA (14 kb) proceeds in the nucleus of enterocytes by an enzyme-mediated process such that codon 2153 becomes UAA (a stop codon) in place of CAA (Glu) in the gene sequence. The enzyme binds both to a species-specific and to a species-nonspecific apoB RNA sequence. In humans the liver makes only apoB-100, found in plasma on VLDL, IDL, and LDL. There is little metabolic regulation of hepatic apoB mRNA levels. The availability of lipid and the extent of degradation of apoB within the cell are important in regulating secretion rates.

Much larger than the exchangeable apolipoproteins, apoB has a relatively high content of amphipathic β-sheet with a large number of short hydrophobic domains distributed throughout its length along with a few amphipathic α-helices. Hence lipid-binding sequences are widely distributed. It is highly glycosylated via 16 N-linked sites, with carbohydrate contributing 35 kDa of the molecular weight of apoB-100. Two regions of apoB (residues 1280–1320 and 3180–3282) are sensitive to cleavage by endoproteases, an indication that they are exposed on the surface of LDL. The region between residues 3000 and 3600 contains sequences responsible for binding to the LDL receptor. Two positively charged sequences in this region (residues 3147–3157 and 3359–3367) bind strongly to heparin. The second of these has close homology to the LDL receptor-binding domain of apoE. Mutations at residues 3500 (Arg→Gln) and 3531 (Arg→Cys) decrease binding to the receptor and cause hypercholesterolemia.

The gene for apoB, which is 43 kb long, is composed of 29 exons. Exons 26 and 29 are the largest—7.6 and 1.9 kb long, respectively. There is a hypervariable region, composed of 15 bp AT-rich tandem repeats, located 200 bp downstream of the gene with 14 common alleles ranging from 25 to 52 repeats. Five single amino acid substitution polymorphisms of apoB, the antigenic group (Ag) series, first detected in antisera from multiply transfused subjects, are in association with apoB RFLPs. A common insertion/deletion polymorphism results in either a 27- or a 24-residue signal sequence. A rare third allele has 29 residues.

A number of individuals with familial hypobetalipoproteinemia (FHB) have apoB mutations that result in the production of C-terminally truncated variants. The length of the truncated protein is monotonically related to the diameter and buoyant density of the lipoprotein formed. Mouse models for FHB have been produced by targeting the apoB gene using homologous recombination.

### 6.6 ApoCI

A minor component of HDL, IDL, and VLDL, apoCI is a 6.6 kDa peptide having a concentration in plasma of approximately 6 mg/dL: 97% in HDL, 1% in IDL, and 2% in VLDL. More than 90% of the circulating apoCI is of hepatic origin, with moderate expression in skin and trace levels of expression in brain. Little is known concerning its physiological function. It was shown to be a highly effective inhibitor of the binding of apoE-enriched β-VLDL to the $\alpha_2$-macroglobulin receptor/LRP. The expression of human apoCI in transgenic mice resulted in high levels of triglyceride-rich lipoproteins, indicating that apoCI may be involved in the regulation of their metabolism. The gene is 4.6 kb long and codes for an 83-residue peptide including a 26-residue signal sequence.

### 6.7 ApoCII

The apolipoprotein apoCII is expressed mainly in the liver with intestinal cells containing only small amounts of mRNA. It is synthesized as a 101-residue preprotein with a 22-residue signal peptide. It is glycosylated within the cell, but is subsequently deglycosylated in the plasma, where it exists mainly as the 8.9 kDa, 79-residue form, although some is further processed by removal of the N-terminal six amino acids to produce a minor isoform. The only known role of apoCII is as the requisite cofactor for LPL. Like the other C apolipoproteins, apoCII is transferred during hydrolysis from chylomicrons and VLDL to HDL. Most of the circulating apoCII (3 mg/dL) is present in HDL; 10% is in IDL, and 30% is in VLDL and chylomicrons.

The first intron of the 3.3 kb gene for apoCII contains 4 Alu type sequences and a $(TG)_n(AG)_m$ microsatellite with at least 15 different alleles. The third intron contains a minisatellite composed of 37 bp tandem repeats found at approximately 60 other loci in the genome. There are two common alleles of six and seven repeats with frequencies of 82 and 18%, respectively. Three charge variants of apoCII have been reported. One is common among African Americans. Homozygous apoCII deficiency causes recessive familial chylomicronemia, often accompanied by the presence of xanthomas and pancreatitis.

### 6.8 ApoCIII

A 8.8 kDa glycoprotein, apoCIII is the most abundant protein in VLDL comprising about 40% of the protein mass. The concentration of apoCIII in plasma is approximately 12 mg/dL, of which 60% is in HDL, 10% in LDL, 10% in IDL, and 20% in VLDL. It exists as apoCIII$_{-0}$, apoCIII$_{-1}$, and apoCIII$_{-2}$, with 0, 1, and 2 residues of sialic acid per molecule, respectively. These isoforms have different isoelectric points and can be separated by isoelectric focusing. The gene (3.1 kb) codes for a 99-residue prepeptide. Removal of the signal sequence gives rise to the mature 79-residue protein. Little is known about the function of *apo*CIII. It inhibits LPL in vitro, and overexpression in transgenic mice causes hypertriglyceridemia.

### 6.9 ApoD

The 29 kDa sialoglycoprotein apoD is related to retinol-binding protein and both are members of the lipocalin superfamily. It shows no homology with the other apolipoproteins. The gene, which is on chromosome 3 (3q26.2-qter), is expressed in a number of tissues, notably the adrenal glands, the kidney, pancreas, and intestine. There is much less expression in the liver. Putative regulatory domains, including steroid hormone regulatory elements, have been reported in the promoter region. The 810-nucleotide mRNA codes for a 189-residue preprotein. Cleavage of the 20-residue signal peptide gives rise to the mature form in which the N-terminal Gln is blocked by cyclization. The level of apoD in the plasma is approximately 10 mg/dL. In addition to the 29 kDa form, much exists as a 38 kDa heterodimer with apoAII. It is found only in the HDL fraction, where it constitutes about 5% of the protein. It possibly plays a role, alongside LCAT and CETP, in reverse cholesterol transport.

### 6.10 ApoE

The 3.6 kb long *apo*E gene produces a 1.1 kb mRNA, which is found in a number of tissues, notably the liver, brain, kidney, adrenal gland, macrophages, spleen, and ovary. It codes for a 317-residue preprotein with a 18-residue signal peptide. The sequence between residues 140 and 160 of the mature 299-residue form is involved in binding to the LDL receptor. Most of the apoE in the circulation is synthesized in the liver. Of the 5mg/dL present, 50%

is in HDL, 10% in LDL, 20% in IDL, and 20% in VLDL. It is a glycoprotein with a single O-linked chain at Thr 194. The carbohydrate contains one or more sialic acid residues. There are three common alleles of the *apoE* gene, ε2, ε3, and ε4, present at frequencies of 9, 77, and 14%, which give rise to E2, E3, and E4 isoforms, respectively. The proteins can be separated by isoelectric focusing, with E4 the most basic and E2 the most acidic. They differ at residues 112 and 158: E2 is Cys 112/Cys 158, E3 is Cys 112/Arg 158, and E4 is Arg 112/Arg 158. The binding affinity of E2 for the LDL receptor is less than 5% that of E3 or E4. The 1% of individuals who are homozygous for the ε2 allele, with the phenotype E2/E2, tend to accumulate β-VLDL in the plasma. Additional unknown genetic or other factors cause 5% of these subjects to develop clinically important dysbetalipoproteinemia with marked accumulation of chylomicron and VLDL remnants.

ApoE deficiency is a rare genetic disorder characterized by less than 1% of normal levels of apoE in plasma. Patients have xanthomas, type III hyperlipidemia, and premature atherosclerosis. There is delayed clearance of chylomicron and VLDL remnant particles, leading to elevated levels of cholesterol-rich VLDL and IDL.

The frequency in the population of the ε4 allele declines with age, but that of ε2 allele increases. There is much evidence for a highly significant genetic association between the ε4 allele and late onset Alzheimer's disease, both in the familial and in the sporadic forms. The effect is proportional to gene dosage, with the ε4/ε4 genotype being a greater risk factor than the ε4/ε3 or ε4/ε2 genotypes. In families with late onset Alzheimer's disease, the presence of the ε4/ε4 genotype alone is associated with the disease in almost all of those over 80 years of age. Additional evidence, supporting the link between *apo*E and Alzheimer's disease, is that the gene locus is in a region of chromosome 19 previously associated with the late onset form. In addition the apoE protein has been found in senile plaques and neurofibrillary tangles along with the β-amyloid protein. In vitro, both apoE4 and apoE3 bind with high specificity, and irreversibly, to β-amyloid, with apoE residues 244–272 being crucial for the interaction. The formation of the complex of β-amyloid with apoE4 is much more rapid than with apoE3.

Overexpression of apoE in mice resulted in increased clearance of triglyceride-rich lipoproteins. ApoE-deficient mice, produced by gene targeting, had substantially elevated levels of plasma cholesterol and rapidly developed atherosclerotic lesions.

### 6.11 Apo(a)

Apo(a), synthesized in the liver, is found on a lipoprotein species Lp(a), where it is covalently attached to apoB. Lp(a) floats in the 1.050–1.080 g/mL density fraction. The level of Lp(a) within the population varies from less than 0.1 mg/dL to as much as 100 mg/dL. This variation is genetically determined at the apo(a) gene locus. High levels are a risk factor for premature atherosclerosis. The function of Lp(a) is largely unknown, although the high degree of homology of apo(a) with plasminogen appears to be responsible for its inhibitory effect on thrombolysis.

The gene for apo(a) is closely linked to that of plasminogen on chromosome 6 (6q26-q27), with the plasminogen gene located approximately 50 kb upstream of the apo(a) gene promoter. The size of the *apo*(a) gene is highly polymorphic and depends on the number of exons that code for a triloop peptide structure held together by three internal disulfide bonds and termed *kringles.* Plasminogen has five kringles, numbered I to V. Apo(a) has a variable number (15–40) of kringles, with high homology to plasminogen kringle IV and a single copy homologous to kringle V. It also has a region that is homologous to the plasminogen protease domain, but it does not have a tissue plasminogen activator cleavage site. The size polymorphism, which is reflected in the molecular weight range (419–838 kDa) of 34 different apo(a) isoforms, can be detected by pulsed-field electrophoresis after digestion with *Kpn* I.

When fed a high fat diet, transgenic mice that express human *apo*(a) had larger areas of lipid-staining atherosclerotic lesions than control animals.

## 7 LIPOPROTEINS AND ATHEROSCLEROSIS

The histological hallmark of atherosclerotic arteries is the foam cell, a monocyte-derived macrophage, which contains droplets of cholesteryl esters. These cells appear to play a central role in the development of the atherosclerotic plaque. Endocytosis of modified lipoproteins via several species of scavenger receptor leads to foam cell formation in macrophages of the subintima. Modification in situ of apoB-containing lipoproteins by oxidation is probably the chief mechanism leading to endocytosis. Metabolic disorders that produce increased levels of atherogenic lipoproteins promote a rapid development of atherosclerosis. A number of metabolic disorders that increase levels of apoB-containing lipoproteins are recognized. Defects in the LDL receptor or of the ligand domain of apoB-100 lead to an accumulation of LDL in plasma and to familial hypercholesterolemia or familial ligand-defective apoB, respectively. The latter defect is usually the less severe of the two, because apoE can serve as an alternative ligand. Defects in the ligand properties of apoE lead to the accumulation of remnants of VLDL and chylomicrons in plasma and to familial dysbetalipoproteinemia. Another genetic disorder (familial combined hyperlipoproteinemia), as yet poorly understood, causes accumulation of VLDL, LDL, or both and is probably due to overproduction of VLDL.

An inverse relationship is recognized between HDL levels and risk of coronary heart disease. Primary hypoalphalipoproteinemia appears to be transmitted in many kindreds as a dominant trait. The molecular basis for this disorder is not yet known in most cases, though defects in the apoAI gene and a case of PLTP deficiency have been reported.

*See also* Cardiovascular Diseases; Heart Failure, Genetic Basis of; Plasma Lipoproteins.

### *Bibliography*

Breslow, J. L. (1988) Apolipoprotein genetic variation and human disease. *Physiol. Rev.* 68:85–132.

Breslow, J. L. (1995) Familial disorders of high density lipoprotein metabolism. In *The Metabolic and Molecular Bases of Inherited Disease,* 7th ed., C. R. Scriver, A. L. Beaudet, W. S. Sly, and D. Valle, Eds., pp. 2031–2052. McGraw-Hill, New York.

Brunzell, J. D. (1995) Familial lipoprotein lipase deficiency and other causes of the chylomicronemia syndrome. In *The Metabolic and Molecular Bases of Inherited Disease,* 7th ed., C. R. Scriver, A. L. Beaudet, W. S. Sly, and D. Valle, Eds., pp. 1913–1932. McGraw-Hill, New York.

Goldstein, J. L., Hobbs, H. H., and Brown, M. S. (1995) Familial hypercholesterolemia. In *The Metabolic and Molecular Bases of Inherited Disease,* 7th ed., C. R. Scriver, A. L. Beaudet, W. S. Sly, and D. Valle, Eds., pp. 1981–2030. McGraw-Hill, New York.

Havel, R. J., and Kane, J. P. (1995) Introduction: Structure and metabolism of plasma lipoproteins. In *The Metabolic and Molecular Bases of Inherited Disease,* 7th ed., C. R. Scriver, A. L. Beaudet, W. S. Sly, and D. Valle, Eds., pp. 1841–1851. McGraw-Hill, New York.

Kane, J. P., and Havel, R. J. (1995) Disorders of the biogenesis and secretion of lipoproteins containing the B-apolipoproteins. In *The Metabolic and Molecular Bases of Inherited Disease,* 7th ed., C. R. Scriver, A. L. Beaudet, W. S. Sly, and D. Valle, Eds., pp. 1853–1885. McGraw-Hill, New York.

Li, W.-H. Tanimura, M., Luo, C.-C., Datta, S., and Chan, L. (1988) The apolipoprotein multigene family: Biosynthesis, structure–function relationships, and evolution. *J. Lipid Res.* 29:245–271.

Lusis, A. J., Rotter, J. I., and Sparks, R. S., Eds. (1992) *Molecular Genetics of Coronary Artery Disease. Candidate Genes and Processes in Atherosclerosis,* Vol. 14: *Monographs in Human Genetics.* Karger, Basel.

Vance, D. E., and Vance, J., Eds. (1991) *New Comprehensive Biochemistry,* Vol. 20: *Biochemistry of Lipids, Lipoproteins and Membranes.* Elsevier, Amsterdam.

# Lipids, Microbial

*Colin Ratledge*

## *Key Words*

**Fatty Acids** Long chain (usually 16 or 18 C atoms) aliphatic carboxylic acids normally found esterified to glycerol (see **Triacylglycerol**).

**Glycolipid** Sugar (usually a mono- or disaccharide) linked to one or more fatty acyl groups; acts as an emulsifying agent by having both water- and lipid-soluble components.

**Lipoglycan** Macromolecular complex in which the glucan or glycan (polysaccharide) backbone molecule has a number of fatty acyl substituents. Overall, the molecule retains its carbohydrate properties.

**Phospholipids** Lipid containing a phospho group, usually refers to lipids based on 1,2-diacylglycero-3-phosphate (also known as phosphatidic acid). A polar substituent (often choline) may also occur on the phospho group, thereby creating a range of phospholipid types. Phospholipids form the major component of most cell membranes.

**Terpenoid Lipids** Lipids based on multiples of the isopentenyl group [$CH_2{:}C(CH_3){\cdot}CH_2{\cdot}CH_2$—] giving rise to sterols, carotenoids, polyprenols, and the side chains of chlorophylls, quinones, etc.

**Triacylglycerol** A lipid in which each of the three alcohol groups of glycerol ($CH_2OH{\cdot}CHOH{\cdot}CH_2OH$) is esterified with a fatty acid. Along with phospholipids (q.v.), triacylglycerols are the predominant lipid types in most eukaryotic cells.

**Unsaturated Fatty Acids** Fatty acids containing one or more double (—CH:CH—)bonds. The position of the double bond is indicated by the first carbon numbering from the carboxylic acid group: $^{1}COOH{\cdot}{}^{2}CH_2{\cdot}{}^{3}CH_2{\cdot}{}^{4}CH_2{\cdot}{}^{5}CH_2{\cdot}{}^{6}CH_2{\cdot}{}^{7}CH_2{\cdot}{}^{8}CH_2{\cdot}{}^{9}CH{:}{}^{10}CH{\cdot}{}^{11}CH_2{\cdot}{}^{12}CH{:}{}^{13}CH{\cdot}{}^{14}CH_2{\cdot}{}^{15}CH_2{\cdot}{}^{16}CH_2{\cdot}{}^{17}CH_2{\cdot}{}^{18}CH_3$ is 9,12-octadecadienoic acid (or linoleic acid). The orientation of the double bond is usually in the Z (or cis) configuration though E (or *trans*) bonds are known. The shorthand nomenclature for linoleic acid is therefore 18:2 (*Z*9,*Z*12).

Microorganisms, like all living cells, contain lipids. Lipids are used in the assembly of membranes, both of the cell itself and of various intracellular organelles. Lipids are also accumulated as storage reserves, which can then be mobilized as sources of carbon, water, and energy. Just as in plants and animals, there is a wide diversity of lipid types: they can be approximately divided as based on either fatty acyls or terpenoid. The former therefore include the triacylglycerols and acylglycerophosphates (phospholipids), and the latter the steroid and carotenoid lipids. Some lipid types, though, are confined to individual groups of microorganisms, whereas others have almost a ubiquitous distribution. The major lipids of Archaea (formerly Archaebacteria), bacteria, yeasts, fungi, and algae are summarized here.

## 1 ARCHAEA (OR ARCHAEBACTERIA)

The lipids of the group of microorganisms called Archaea, which are now regarded as evolutionarily distinct from Eubacteria by virtue of their 16S rRNA sequences and RNA polymerase structures, are composed principally of isoprenoid-derived materials: that is, these lipids are synthesized from mevalonate, not from acetate. The polar lipids making up the cytoplasmic membrane structure are not conventional phospholipids but are diether analogues (**I**).

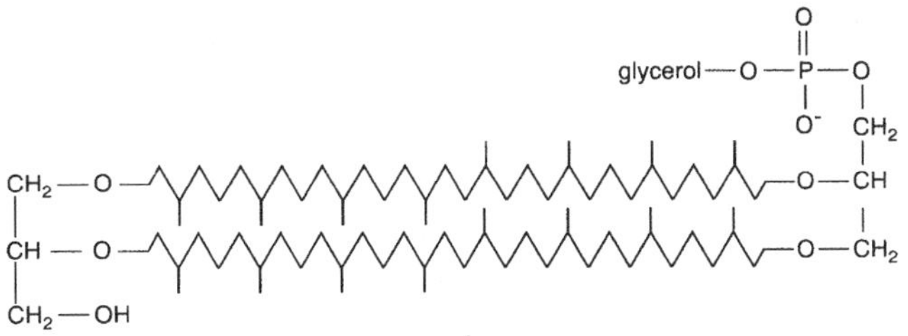

I

Phosphoglycerotetraether lipid from *Methanospirillum hungatei*

Such lipids are found in extreme halophiles (*Halobacterium, Natronobacterium,* etc.), methanotrophs (*Methanospirillum, Methanococcus,* etc.), and the extreme thermophiles (*Thermus, Sulfobolus,* etc.). When glycolipids have been identified, these, too, have been shown to be based on the isoprenoid unit shown in **I;** such glycolipids then can be extended into the macrolipoglycan structures as part of the cell wall. Unusual carotenoids, quinones, pentacyclic diols, and even branched-chain alkylbenzenes occur in many genera. Such lipids are not found in any other gruop of microorganisms.

## 2 EUBACTERIA

With the recognition of the Archaea as an evolutionarily distinct microbial group, the remainder of the "true" bacteria are now referred to as Eubacteria. This group of organisms synthesizes the bulk of its lipid via the usual acetate/malonate, pathway and thus the major acyl groups of the glycerol esters are conventionally C-16 and C-18 fatty acids. However, many bacteria (but not all) synthesize monounsaturated fatty acids by an anaerobic route rather than by oxidation of a saturated fatty acid. By this route, *cis-*

vaccenic acid [18:1(Z,11)] is formed rather than oleic acid [18:1(Z,9)], found in a few bacteria and in all eukaryotic cells. Other fatty acids that are regarded as uniquely bacterial include branched-chain fatty acids, where a methyl group is located at the ω-1 or ω-2 C atom (these are termed iso and anteiso acids). Cyclopropane fatty acids such as lactobacillic acid **(II)**

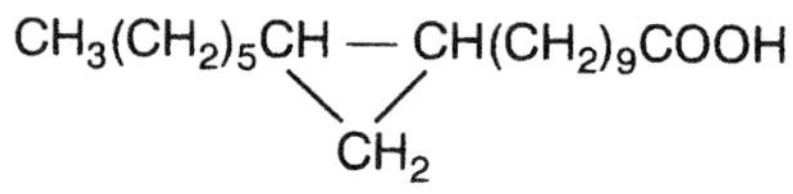

$$CH_3(CH_2)_5CH - CH(CH_2)_9COOH$$
(with a $CH_2$ bridge between the two CH groups)

**II**

Lactobacillic acid

and the corresponding *cyc*-19:0 fatty acid have been found in Gram-positive and Gram-negative bacteria. The specific distribution of fatty acids of different chain lengths and types has permitted classification and identification schemes to be offered based on these characters.

Although it is widely reported that bacteria do not produce di- or polyunsaturated fatty acids, some marine bacteria, including species of *Altermonas, Shewanella,* and *Flexibacteria,* have been isolated from fish that synthesize eicosapentaenoic acid [20:5(5,8,11,14,17)] and other related polyunsaturated fatty acids.

Some bacteria of the *Corynebacterium–Mycobacterium–Nocardia* group synthesize very long chain fatty acids, called mycolic acids, that are up to C-90 in length **(III).**

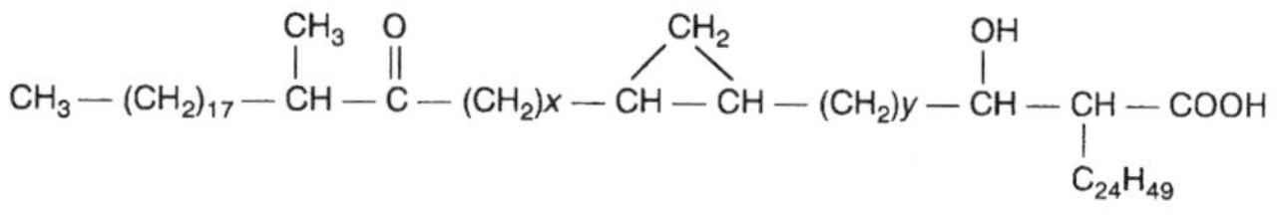

$$CH_3-(CH_2)_{17}-CH(CH_3)-C(=O)-(CH_2)x-CH-CH(CH_2)-(CH_2)y-CH(OH)-CH(C_{24}H_{49})-COOH$$

$x$ = 11 to 21, max. $x + y$ = 39: $C_{90}H_{176}O_4$
$y$ = 28 to 18, min. $x + y$ = 29: $C_{80}H_{156}O_4$

**III**

β-Mycolic acid from *Mycobacterium bovis* BCG

These are associated with the construction of the characteristic lipophilic cell envelope of these bacteria, which include the tubercle and leprosy bacilli (*Mycobacterium tuberculosis* and *M. leprae,* respectively). Mycobacteria also contain a number of complex and antigenic phospholipids.

Phospholipids of bacteria are based on the conventional diacylglycerophosphate structure. However, the commonest types are phosphatidylethanolamine, phosphatidylglycerol, and diphosphatidylglycerol (cardiolipin). Phosphatidylcholine is found only occasionally. Monomethyl- and dimethylethanolamine-based phospholipids occur in the methylotrophs *(Methylococcus, Methylomonas)* and related bacteria.

Although some triacylglycerols are formed in bacteria, many accumulate polyhydroxyalkanoates (PHA) as a major reserve of carbon and energy. The major material is poly-β-hydroxybutyrate (PHB) **(IV),** which in some species, such as *Alcaligenes,* may be up to 85% of the cell biomass.

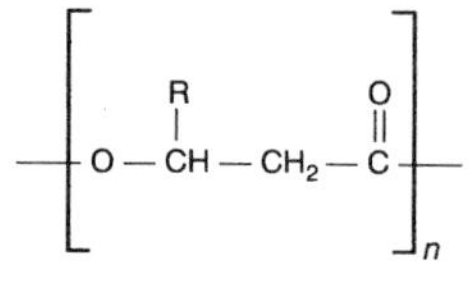

$$\left[ -O-CH(R)-CH_2-C(=O)- \right]_n$$

$n$ = 10,000 – 20,000
($Mr$ up to $2 \times 10^6$ Da)

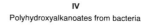

R= — $CH_3$ = poly-β-hydroxybutyrate
= — $CH_2.CH_3$ = poly-β-hydroxyvalerate
= — $(CH_2)_7$— $CH_3$ = poly-β-hydroxyoctanoate

**IV**

Polyhydroxyalkanoates from bacteria

PHB has been found in Gram-positive and -negative bacteria as well as in the Actinomycetes. There is commercial interest in the production of PHB/PHA as a biodegradable plastic. The transfer of the appropriate PHB-synthesizing genes (coding for 3-ketothiolase, acetoacetyl reductase, and PHB synthetase) into plants is being considered as an alternative and cheaper route than large-scale fermentation.

The genetics of bacterial lipid metabolism is under active investigation in several major laboratories around the world, in particular the group led by John E. Cronan (Urbana, Illinois).

## 3 YEASTS AND MOLDS

Yeasts are essentially unicellular fungi and thus share with molds (which are regarded as filamentous fungi) many lipid types as well as routes of biosynthesis and degradation.

The number of different yeast species is about 590, so it has been possible to examine the fatty acids of almost each one and from the findings construct a database for the taxomonic classification of yeasts. Among the interesting and noval discoveries that have been made is that some yeasts are able to synthesize eicosanoid lipids, including a novel metabolite, 3-hydroxyarachidonic acid [3-HO-20:4 ω-6].

Yeasts do not generally show a wide range of fatty acids in their acylglycerol lipids: oleic (18:1), palmitic (16:0), linoleic (18:2), and palmitoleic (16:1) acids are found in most species. The range of lipid types is also not extensive, but yeasts do contain small amounts of sphingolipids (whose function is largely unknown), sterols, and carotenoids, though many species (e.g., all those named *Candida,* which is Latin for "white") are completely devoid of the latter. The predominant phospholipid types are the serine, choline, ethanolanine, and inositol derivatives of phosphatidic acid; diphosphatidylglycerol is also common. *Saccharomyces cerevisiae* (baker's and brewer's yeast) does not synthesize unsaturated fatty acids beyond oleic acid (18:1); reports of the presence of linoleic acid (18:2) and sometimes linolenic acid (18:3) are attributable to the yeast being grown in the presence of animal- or plant-derived materials that contain traces of these acids. Other yeasts, though, synthesize 18:2 and α-linolenic acid [18:3(9,12,15)].

Of the total number of yeasts, only about 30 or 50, have the ability to accumulate lipid (25–75% of biomass) as a major storage material. Such microorganisms are referred to as oleaginous. Details of the underlying biochemistry of lipid accumulation in oleaginous microorganisms have been elucidated: a key role for the enzyme ATP:citrate lyase has been proposed as the principal provider of acetyl CoA units in such species. Commercial interests in using yeasts as potential sources of edible oils, known as single-cell oils, have chiefly centered on the ability of some species such as *Candida curvata* (also referred to *Apiotrichum curvatum*) to produce a facsimile of cocoa butter in which the Δ9-desaturase gene, which converts stearic acid (18:0) to oleic acid (18:1), has been deleted by genetic mutation. Such cells contain up to 50% stearic acid in an overall oil content of about 35 to 40%.

The genetics of lipid synthesis including fatty acid and phospholipid biosynthesis have been studied in detail in *Saccharomyces cerevisiae,* which is not an oleaginous species and therefore does not contain ATP:citrate lyase. Leading investigators include E. Schweizer of Erlangen, Germany, and George M. Carman, Rutgers University, New Jersey.

Filamentous fungi produce a wide range of fatty acids and of lipid types. They may be as diverse as plants in this respect, though many species of fungi (> 60,000 known) have yet to be thoroughly examined for their lipid composition. No systematic survey of fungal lipids has been carried out, and there is still much waiting to be discovered. Little attempt has been made to use lipid components as chemotaxonomic markers, and very little work has been carried out on the genetics or molecular biology of fungal lipids. In some cases, there may be extensive accumulation of triacylglycerols as storage reserves, and oil contents of up to 85% have been recorded. These are also referred to as oleaginous microorganisms.

One of the few differences between the major families of the molds is that members of the lower fungi Phycomycetes (i.e., the orders Mastigomycotina and Zygomycotina, which encompasses 1935 species) produce γ-linolenic acid (GLA) [18:3(6,9,12)] rather than the α-isomer [18:3(9,12,15)] found in all other fungi comprising the slime fungi (Myxomycota) and the higher fungi of the order Ascomycetes (Ascomycotina), Basidiomycetes (Basidiomycotina), and the fungi imperfecti (Deuteromycotina).

The use of fungi to produce GLA and, more recently, dihomo-GLA (20:3), arachidonic acid (20:4), and the highly polyunsaturated fatty acids eicosapentaenoic acid (20:5) (EPA) and docosahexaenoic acid (22:6) (DHA), has been of some commercial interest because of the dietetic importance of these acids. Processes for the production of oils containing each of these fatty acids have been described and fungal oils with GLA have been produced commercially both in the United Kingdom and in Japan in recent years. As the enzyme for the formation of GLA from 18:2 (i.e., Δ6-desaturase) occurs in only a few plants (evening primrose, borage, and *Ribes* spp.), it may be that the molds that produce GLA (*Mucor, Mortierella, Pythium, Rhizopus,* etc.) could be used as a source of the Δ6-desaturase gene. This could be inserted into a plant, such as linseed, which already produces α-linolenic acid, so that if the Δ15-desaturase activity could be removed, the plant would now produce GLA instead of the α-isomer. These fungi appear to be genetically manipulatable, and other aspects of the physiology of *Mucor* spp. have been genetically explored.

## 4 UNICELLULAR ALGAE

Like plants, algae contain a wide variety of lipid types, many of which are intimately associated with photosynthesis. The lipids of the prokaryotic algae, the cyanobacteria (*Anabaena, Anacystis, Oscillatoria, Synechococcus,* etc.), contain a diversity of lipid types and are distinguished from other prokaryotic phototrophs by containing plant chlorophyll rather than bacteriochlorophyll. Unlike their bacterial counterparts, cyanobacteria synthesize polyunsaturated fatty acids, with some producing γ-linolenic acid while other produce the α-isomer. These, and other fatty acyl groups, are associated with complex lipids such as a quinovosyl sulfolipid (= ''plant sulfolipid'') mono- and digalactosyldiacylglycerol and phosphatidylglycerol. Some triacylglycerols may occur as storage material.

Eukaryotic algae can be divided into the freshwater and marine varieties. The former tend to produce fatty acids with a much lower degree of unsaturation than the latter.

Organisms that have been studied for their potential as lipid producers include the halophile *Dunaliella salina,* which accumulates β-carotene and has been grown commercially in the sea and in hypersaline ponds in Western Australia, Isarel, and California. Interest now focuses on species of *Spirulina* and *Porphyridium* (a red alga) as sources of GLA and EPA. Some species can be grown heterophotically in the dark (i.e., using sugar as a carbon and energy source) to produce either EPA or DHA.

Recombinant DNA technologies are now being successfully applied to the green alga *Chlorella ellipsoidea,* to improve production of selected lipids and other products as well. However, this work is very much in its infancy, and a long-term strategy will be needed if success is to be achieved.

*See also* Lipids, Structure and Biochemistry of; Liposomal Vectors; Phospholipids.

### *Bibliography*


Anderson, A. J., and Dawes, E. A. (1990) Occurrence metabolism metabolic role, and industrial uses of bacterial polyhydroxyalkanoates. *Microbiol. Rev.* 54:450–472.

Boom, T. V., and Cronan, J. E. (1989) Genetics and regulation of bacterial lipid metabolism. *Annu. Rev. Microbiol.* 43:317–343.

Carman, G. M., and Henry, S. A. (1989) Phospholipid biosynthesis in yeast. *Annu. Rev. Biochem.* 58:635–669.

Davies, R. J., and Holdsworth, J. E. (1992) Synthesis of lipids in yeasts: Biochemistry, physiology, production. *Adv. Appl. Lipid Res.* 1:119–159.

Kyle, D. L., and Ratledge, C., Eds. (1992) *Industrial Applications of Single Cell Oils.* American Oil Chemists' Society, Champaign, IL.

Ratledge, C., and Evans, C. T. (1989) Lipids and their metabolism. In *The Yeasts,* Vol. 3, A. H. Rose and J. S. Harrison, Eds., pp. 367–455. Academic Press, London and New York.

———, and Wilkinson, S. G., Eds. (1988, 1989) *Microbial Lipids,* Vols. 1 and 2. Academic Press, London and New York.

Zaborsky, O. R., and Attaway, D. H., Eds. (1992) *Advances in Marine Biotechnology: Pharmaceutical and Bioactive Natural Products,* Vol. 1. Plenum Press, New York.


# Lipids, Structure and Biochemistry of

*Donald M. Small*

### *Key Word*

**Lipid** Any molecule of intermediate molecular weight (100–5000) that contains a substantial portion of aliphatic or aromatic hydrocarbon.

The major chemical classes of lipids (Table 1) include substituted hydrocarbons, esters of fatty alcohols and fatty acids, and complex lipids such as acylglycerols, phospholipids, glycoglycerolipids, sphingolipids, steroids, and vitamins. Lipids are found in all living organisms and function primarily as barriers between the cell and the environment and between compartments within the cell. They have specific functions as receptors, sensors, electrical insulators, antigens, biological detergents, and membrane anchors for proteins. Membranes are made of lipids capable of forming a barrier between two aqueous compartments. The membrane-forming lipids include phospholipids, glycolipids, and sphingolipids, whereas oils, fats, and waxes are major sources of stored fatty acids and cellular energy. Most lipids are composed of substituted alkanes, specifically fatty acids, fatty alcohols, or sphingosine, linked to a variety of other chemical groups. For instance, Figure 1 shows the general structure of a glycerol phospholipid. Fatty acids are esterified to the 1 and 2 positions of glycerol and phosphate to the 3 position. The common phospholipids have different bases (choline, ethanolamine, serine, glycerol), which are esterified to the phosphate group.

Table 1 A Chemical Classification of Lipids[a]

- I. Hydrocarbons (normal, branched, saturated, unsaturated, cyclic, aromatic)
- II. Substituted hydrocarbons
  - A. Alcohols
  - B. Aldehydes
  - C. Fatty acids, soaps, acid-soaps
  - D. Amines
- III. Waxes and other simple esters of fatty acids
- IV. Fats and most oils (esters of fatty acid with glycerol)
  - A. Triacylglycerols
  - B. Diacylglycerols
  - C. Monoacylglycerols
- V. Glycerophospholipids (diacyl, O-alkyl, acyl, di-O-alkyl)
  - A. Phosphatidic acid
  - B. Choline glycerophospholipids
  - C. Ethanolamine glycerophospholipids
  - D. Serine glycerophospholipids
  - E. Inositol glycerophospholipids
  - F. Phosphatidylglycerols
  - G. Lysoglycerophospholipids
- VI. Glycoglycerolipids, including sulfates
- VII. Sphingolipids
  - A. Sphingosine
  - B. Ceramide
  - C. Sphingomyelin
  - D. Glycosphingolipids [ceramide monohexosides (cerebrosides), ceramide monohexosides sulfates, ceramide polyhexosides]
  - E. Sialoglycosphingolipids (gangliosides)
- VIII. Steroids (sterols, bile acids, cardiac glycosides, sex and adrenal hormones)
- IX. Other lipids [vitamins (A, D, E, K), eicosanoids, acyl CoA, acyl carnitine, glycosyl phosphatidylinositols, lipopolysaccharides, dolichols]

[a]This chemical classification of lipids is necessarily incomplete and somewhat arbitrary. It progresses from hydrocarbons to more complex chemical structures. Simple esters and glycerol esters yield on hydrolysis the alcohol and/or glycerol and fatty acid. The major membrane lipids, glycerophospholipids, yield fatty acid (or alcohol), glycerol, phosphate, and the appropriate base (choline, ethanolamine, etc.). Sphingolipids yield the base sphingosine and a fatty acid on hydrolysis.

*Source:* D. M. Small and R. A. Zoeller, Lipids. In *Encyclopedia of Human Biology,* Vol. 4. Academic Press, San Diego, CA, 1991.

Phospholipids are amphiphilic (see Figure 2); that is, they have a hydrocarbon part which is soluble in oil (lipophilic) and a water-soluble part (hydrophilic)—for instance, phosphorylcholine. Thus, the molecules can partition between the oil–water interface with the hydrocarbon chains in the oily side of the interface and the water soluble polar groups in the aqueous side. As such, they lower the interfacial tension to form stable interfaces between oil and water which, without these interfacial molecules, would separate from each other. These molecules form the interfaces between globules of nonpolar lipid molecules occurring in cells, such as in adipose tissue cells. In the plasma lipoproteins they form the surface around an oily droplet of triglycerides or cholesterol esters to form a stable interface between plasma and the hydrocarbon core of the lipoprotein. Certain lipids can self-associate so that the hydrophobic parts of the molecules interact and form lamellae, with the polar groups facing toward the water and the hydrocarbon parts interacting with each other. These membranes or bilayers make effective electrical and physical barriers only a few nanometers thick. These lipids are particularly important in cells in that they form the barriers between the external surface of the cell and the cytoplasm (the plasma membrane) and allow the cell to be compartmentalized into its various organelles (mitochondria, microsomes, Golgi apparatus, nuclear membranes, secretory vesicles, etc.). In most membranes the lipids are in the liquid state. That is, the chains are melted like oil. The chains wobble on a nanosecond time scale, move laterally one molecule width (~ 8 A) every 10 to 15 picoseconds, but take days to flip from one side to the other. Under the right circumstances bilayers form closed structures (vesicles), which separate an internal aqueous phase from the surrounding one. Vesicles can be used to trap drugs or DNA and can be targeted to certain cells for specific uptake. When some viruses bud from cells they are encapsulated by a vesicle composed of the plasma membranes of the cell. Of course much transport within cells is carried out by vesicles budding from one organelle, moving to another, and fusing with it. Other lipid molecules with particularly large or highly charged polar groups relative to their hydrocarbon region spontaneously form small spherical, cylindrical, or disklike aggregates in aqueous systems. The molecules of these small aggregates are in rapid equilibrium with a low concentration of monomers in aqueous phase. Such aggregates, called micelles, are capable of solubilizing many other kinds of hydrophobic molecule by dissolving them in the hydrocarbon core of the micelle. The concentration of monomer molecules in the aqueous phase is called the critical micellar concentration (cmc) because when the concentration of lipid increases above the cmc, micelles begin to form.

From the chemical classification of lipids in Table 1, it is impossible to predict what the physical behavior will be like. For instance, both fatty acids and their alkali soaps are classified as substituted hydrocarbons; however fatty acids are largely insoluble in aqueous systems, whereas soaps are quite soluble and form micelles. Thus, to understand how lipids behave within cells and within biological fluids, we have devised a classification based on the behavior of lipids in water and at the aqueous interface (Table 2). The classification applies to lipids with melted chains, since most biological lipids within cells appear to have the liquid chain conformation. The lipids are classified as nonpolar or polar. Nonpolar lipids have extremely low aqueous solubility and do not spread at the air–water or oil–water interface to form a monolayer. That is, they

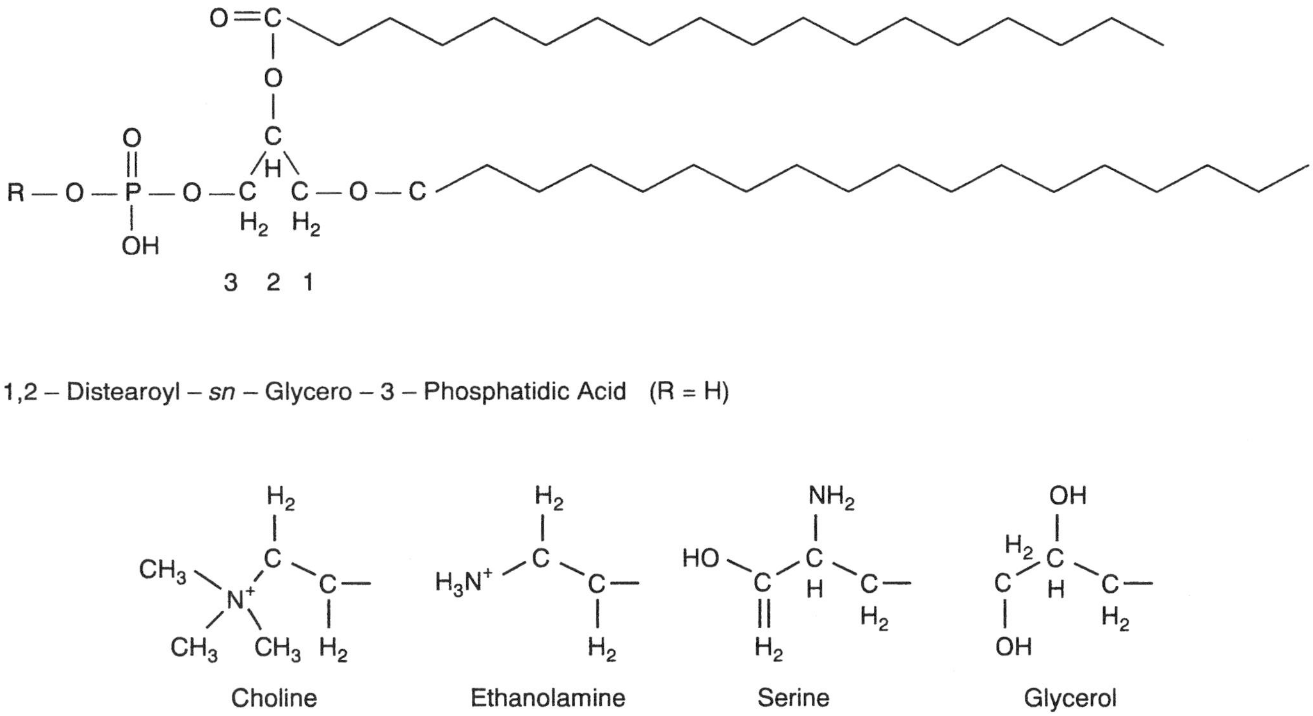

1,2 – Distearoyl – *sn* – Glycero – 3 – Phosphatidic Acid (R = H)

Choline Ethanolamine Serine Glycerol

**Figure 1.** Line models of some phospholipids. If R is hydrogen, the compound is phosphatidic acid as indicated. If the polar moieties shown are substituted at R, the compound is the corresponding phosphatidyl-R (i.e., if R is choline, the compound is phosphatidylcholine. (From D. M. Small and R. A. Zoeller, Lipids. In *Encyclopedia of Human Biology,* Vol. 4, Academic Press, San Diego, CA, 1991.)

do not have bulk or surface solubility. Class I polar lipids have surface solubility but no bulk solubility. They can partition into membranes but cannot form membranes (bilayers) by themselves. Class II lipids (i.e., the phospholipids and glycosphingolipids) are the membrane formers. Class III lipids (i.e., detergents) (are soluble lipids that form micelles. Monolayers of these lipids are unstable. Molecules on the surface and in micelles are in constant rapid equilibrium with molecules in the aqueous phase. Low concentrations of these molecules can penetrate into bilayers and membranes but above a certain concentration will cause disruption, and high concentrations will dissolve membrane lipids into micelles. They also bind to and alter the conformation of proteins. Fortunately, except for a few specific cases (e.g., bile salts in bile and intestinal contents), they are in low concentration, well below their critical micellar concentration, and usually are partitioned within membranes or lipoproteins or bound to specific proteins like albumin.

## 1 SEPARATION, EXTRACTION AND ANALYSIS OF LIPIDS

Lipids are present in a variety of organelles within cells and in extracellular fluids such as blood plasma, bile, milk, and intestinal contents. First, the particular compartment to be characterized (e.g., red blood cells or low density lipoproteins or mitochondria) is isolated. It is then extracted with chloroform/methanol (2:1, v/v). Sufficient chloroform/methanol is required to form a single phase in which the tissue water is dissolved. Insoluble precipitates form, and these materials which include proteins, nucleic acids and polysaccharides, are removed by centrifugation or filtration. Excess water containing acid and salts is then added to form a two-phase system: water/methanol on top and chloroform on the bottom. The lipid partitions into the chloroform phase, which is removed and concentrated by evaporation, and an aliquot is dried to obtain the weight. Lipids are then subjected to high performance liquid chromatography (HPLC) or thin-layer chromatography (TLC) to separate them into the major classes—for example, hydrocarbons, wax esters, sterol esters, triacylglycerols, fatty acids, sterols, and phospholipids. Then each individual lipid class is collected and rechromatographed on a different system to isolate the individual components. For instance, phospholipids would be separated into phosphatidylcholines, sphingomyelins, phosphatidylethanolamines, phosphatidylserines, phosphatidylinositols, and so on. Each individual type of phospholipid—for instance, phosphatidylcholine—can be derivatized and separated on a new HPLC or gas–liquid chromatographic (GLC) column, to discriminate between the different acyl chains on each phosphatidylcholine. These individual peaks are then examined by mass spectrometry or nuclear magnetic resonance techniques to define the chemical structure. The process is long and involved and requires some modifications for each different lipid class. For instance, while the individual molecular species of phosphatidylcholine have been fully characterized for several lipoproteins and even bile, the different molecular species of triacylglycerols are so complex that not even butter fat, which is a highly studied lipid, has been totally dissected into all its individual triacylglycerols. A common short cut is to isolate the individual lipid class (e.g., phosphatidylcholines or phosphatidylserine), then hydrolyze the fatty acids and analyze them by GLC. This procedure gives the overall composition of fatty acids on the specific class but not the molecular species of the individual phospholipids.

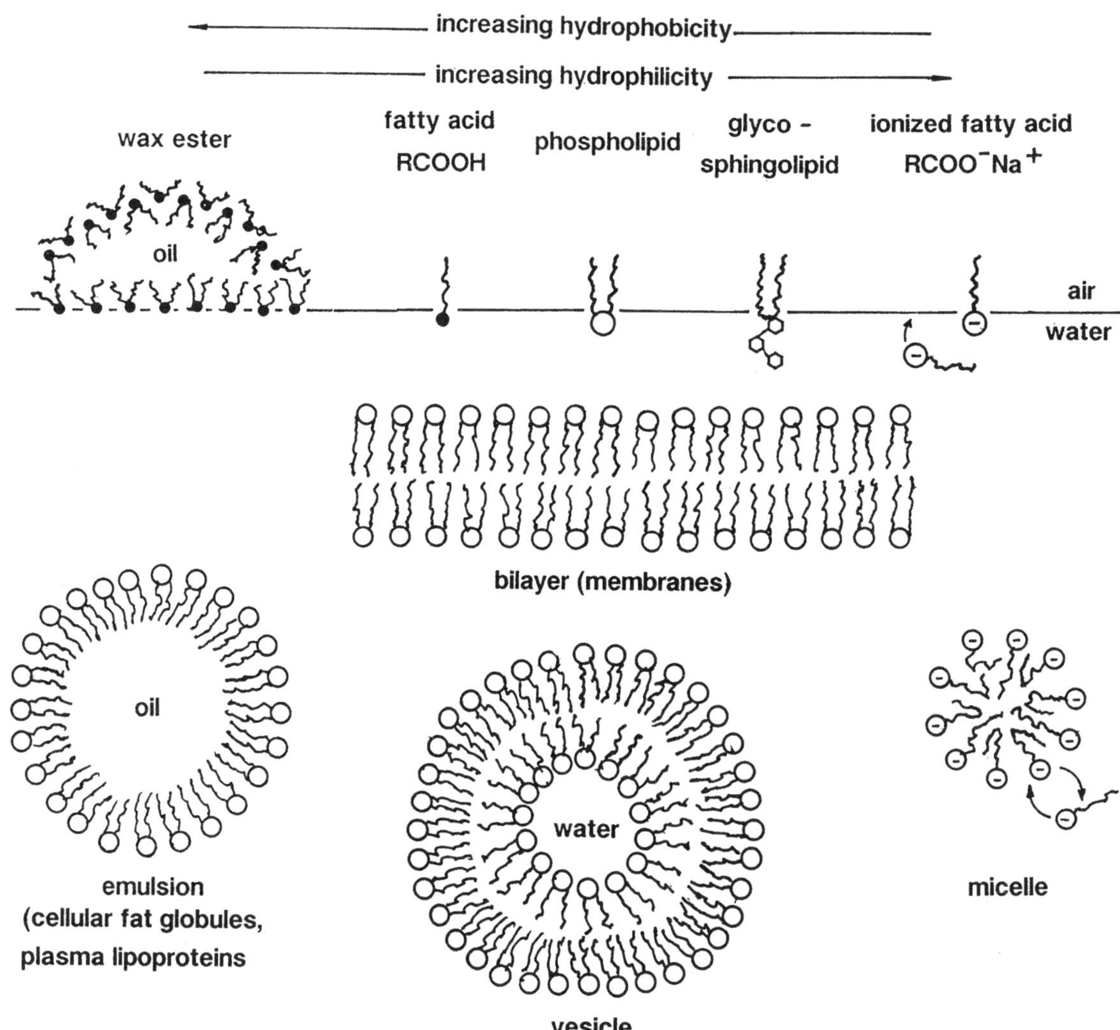


**Figure 2.** Common structures formed by lipids at an air–water interface and in aqueous systems. The upper part of the diagram indicates the behavior of lipids of increasing hydrophilicity (or decreasing hydrophobicity) at an air–water interface. Nonpolar lipids (left) such as wax esters or long chain alkanes form drops of oil that do not spread and simply sit as a lens on the surface of water. They are soluble in organic solvents. The polar lipids (e.g., protonated fatty acids, phospholipids, glycosphingolipids) form stable monolayers. Phospholipids and glycosphingolipids are capable of forming bilayers (membranes) and of stabilizing nonpolar lipids in the center of emulsion droplets by providing the surface with a monolayer of amphiphilic lipid. Ionized fatty acids (epg., sodium or potassium soaps) are detergents and are soluble in water (and insoluble in organic solvents like hexane or benzene); they desorb from the surface rapidly. Above a critical aqueous concentration (the critical micellar concentration, cmc) little aggregates form which are called micelles. The cmc for sodium oleate is about 1.5 mM. High concentrations of detergents can dissolve membranes and solubilize membrane proteins. Fortunately detergents rarely occur physiologically in high enough concentration to damage membranes. Phospholipids and glycosphingolipids membranes can be made to form closed vesicles with an aqueous cavity in the center. Other material such as drugs or DNA may be encapsulated in this space. Ligands for particular receptors may be adsorbed to the surface of it to direct the vesicle to specific cell types, where it is recognized and taken up into the cell.

## 2 OCCURRENCE AND METABOLISM

*Hydrocarbons* contain carbon and hydrogen and can be saturated or unsaturated, branched or cyclic. The source of most hydrocarbons is petroleum. Alkanes can be metabolized in very limited amounts by mammalian cells by converting the alkane to a normal alcohol, which may be further oxidized to fatty acids. Polymeric compounds such as the 40-carbon compound carotene can be split into two 20-carbon vitamin A alcohols in the intestine. The 30-carbon polyisoprenoid squalene, which is synthesized in the liver, can be oxidized by specific enzymes and cyclized to form cholesterol and other steroids. Cholesterol is a precursor of bile acids and adrenal and sex hormones. Other enzyme systems, which are different from those that oxidize alkanes to alcohols, can oxidize and detoxify aromatic hydrocarbons such as benzpyrene.

The predominant class of *substituted hydrocarbons* is the fatty acids (RCOOH), but alcohols (ROH) and aldehydes (RCO) are also found in low concentrations. The major fatty acids (Table 3) consist of saturated, monounsaturated, and polyunsaturated fatty acids that vary in chain length and in double-bond position. Most fatty acids can be synthesized by mammalian cells; however in some mammals linoleic and linolenic acids cannot be synthesized de novo and must be ingested in the diet, hence qualify as essential fatty acids. Linoleic acid is the precursor of arachidonic acid, which is the major source of eicosanoids. Linolenic acid is a precursor for eicosapentenoic and docosahexenoic acids, which form complex

**Table 2** Physical Classification of Biologically Active Lipids

| Class[a] | Surface Properties | Bulk Properties | Examples |
|---|---|---|---|
| Nonpolar | Will not spread to form monolayer | Insoluble | Long chain, saturated or unsaturated, branched or unbranched, aliphatic hydrocarbons with or without aromatic groups (epg., dodecane, octadecane, hexadecane, paraffin oil, phytane, pristane, carotene, lycopene, gadusene, squalene)<br>Large aromatic hydrocarbons (epg., cholestane, benzpyrenes, coprostane, benzphenanthrocenes)<br>Esters and ethers in which both components are large hydrophobic lipids (epg., sterol esters of long chain fatty acids, waxes of long chain fatty acids, and long chain normal monoalcohols, ethers of long chain alcohols, sterol ethers, long chain triethers of glycerol) |
| Polar<br>Class I: Insoluble nonswelling amphiphiles | Spreads to form stable monolayer | Insoluble or solubility very low | Triglycerides, diglycerides, long chain protonated fatty acids, long chain normal alcohols, long chain normal amines, long chain aldehydes, phytols, retinols, vitamin A, vitamin K, vitamin E, cholesterol, desmosterol, sitosterol, vitamin D, unionized phosphatidic acid, sterol esters of very short chain acids, waxes in which either acid or alcohol moiety is less than 4 carbon atoms long (epg., methyl oleate), ceremides |
| Class II: Insoluble swelling amphipathic lipids | Spreads to form stable monolayer | Insoluble but swells in water to form lyotropic liquid crystals | Phosphatidylcholine, phosphatidylethanolamine, phosphatidylinositol, sphingomyelin, cardiolipid, plasmalogens, ionized phosphatidic acid, cerebrosides, phosphatidylserine, monoglycerides, acid-soaps, α-hydroxy fatty acids, monoethers of glycerol, mixtures of phospholipids and glycolipids extracted from cell membranes or cellular organelles (plant glycolipids and sulfolipids) sulfocerebrosides, sphingosine (basic form) |
| Class IIIA: Soluble amphiphiles with lyotropic mesomorphism | Spreads but forms unstable monolayer because of solubility in aqueous substrate | Soluble, forms micelles above a cmc; at low water concentrations forms liquid crystals | Sodium and potassium salts of long chain fatty acids, many of the ordinary anionic, cationic, and nonionic detergents, lysolecithin, palmitoyl and oleyl coenzyme A, and other long chain thioesters of coenzyme A, gangliosides, sphingosine (acid form) |
| Class IIIB: Soluble amphiphiles, no lyotropic mesomorphism | Spreads but forms unstable monolayer because of solubility in aqueous substrate | Forms micelles but not liquid crystals | Conjugated and free bile salts, sulfated bile alcohols, sodium salt of fusidic acid, rosin soaps, saponins, sodium salt of phenanthrenesulfonic acid, penicillins, phenothiazines |

[a]Lyotropic mésomorphism means the formation of liquid crystals (epg., bilayers or other lipid aggregates) on interaction with water.
*Source:* D. M. Small, The physical chemistry of lipids from alkanes to phospholipids. In *Handbook of Lipid Research,* Vol. 4, D. Hanahan, Ed. Plenum Press, New York, 1986.

membrane lipids in the retina and brain. Fatty acids are synthesized de novo from acetate. Acetyl CoA carboxylase and the multifunctional polypeptide *fatty acid synthetase* work together to elongate the chain by two carbons at each step to produce long chain fatty acids, such as palmitic and stearic acids. The chain may be elongated further by fatty acyl CoA elongation system, which adds two carbon fragments. The chain may be desaturated to form a double bond by specific desaturases such as the *Δ9-desaturase,* which forms oleic acid from stearic acid.

Fatty acids are broken down in two carbon units to yield acetyl CoA. The process is called *beta oxidation* and yields energy with each acetyl CoA molecule formed. Very long chain fatty acids are broken down in peroxisomes and short chain fatty acids are broken down in mitochondria. Normal long chain fatty acids like palmitic and stearic acid can be broken down by either organelle. Fatty acids may also be obtained from diet, where they are present as components of dietary fats and oils.

Dietary fatty acids are transported into the intestinal cell, synthesized into *triglycerides* and assembled into chylomicrons, and secreted into the circulation. There the triglyceride of the chylomicron is hydrolyzed by lipoprotein lipase and the fatty acid distributed to tissues in the capillaries and to circulating albumin. The albumin carries fatty acid back to the liver, where it is rapidly taken up and distributed in the cell. Both newly synthesized fatty acid and fatty acid arriving from the plasma either are catabolized through beta oxidation to produce energy or are incorporated into other lipids such as phospholipids or triacylglycerols or other esters. Strictly speaking protonated fatty acid is a class I polar lipid: it binds readily to albumin and in the cells may distribute between membranes and intracellular fatty acid binding proteins. In membranes, however, it is about half-ionized and contributes a net charge to membranes.

When the cellular uptake of fatty acids or its synthesis exceeds the catabolism of fatty acids, the excess must be stored. Plankton and many lower aquatic animals store *wax esters.* The reaction apparently involves an enzyme that utilizes acyl coenzyme A and free alcohol. Waxes are nonpolar molecules and form the cores of storage droplets in both plants and lower animals. In many higher animals triacylglycerols, not wax esters, are the major form of storage lipid.

*Triacylglycerols* are formed by two general pathways. The first utilizes phosphatidic acid. It is converted to 1,2-*sn*-glycerol by removal of the phosphate group and then acylated to triacylglycerol

**Table 3** Names, Formulas, and Selected Properties of Common Fatty Acids

| Fatty Acid | Chemical Name | Δ Formula[a] | ω Formula[b] | Molecular Weight | Melting Point (°C) | Solubility in Water at 25°C (μM) |
|---|---|---|---|---|---|---|
| | *Saturated Fatty Acids* | | | | | |
| Lauric | Dodecanoic acid | 12.0 | | 200.31 | 44.2 | 11.5 |
| Myristic | Tetradecanoic acid | 14.0 | | 228.36 | 54.4 | $7.9 \times 10^{-1}$ |
| Palmitic | Hexadecanoic acid | 16.0 | | 256.42 | 62.9 | $1.2 \times 10^{-1}$ |
| Stearic | Octadecanoic acid | 18.0 | | 284.47 | 69.6 | $1.8 \times 10^{-2}$[c] |
| | *Monounsaturated Fatty Acids* | | | | | |
| Palmitoleic | *cis*-9-Hexadecenoic acid | 16.1 9C | *c*16:1ω7 | 254.40 | 0.5 | $3 \times 10^{-1}$[c] |
| Oleic | *cis*-9-Octodecenoic acid | 18.1 9C | *c*18:1ω9 | 282.45 | 13.4α16.3β | $4.5 \times 10^{-2}$[c] |
| | *Polyunsaturated Fatty Acids* | | | | | |
| Linoleic | *cis*-9,12-Octadecadienoic acid | 18.2 9C12C | *c*18:2ω6 | 280.44 | −5 | |
| Linolenic | *cis*-9,12,15-Octadecatrienoic acid | 18.3 9C12C15C | *c*18:3ω3 | 278.44 | −10(−11.3) | |
| Arachidonic | *cis*-5,8,11,14-Eicosatetraenoic acid | 20.4 5C8C11C14C | *c*20:4ω6 | 304.5 | −49.5 | |
| Eicosapentaenoic[d] | *cis*-5,8,11,14,17-Eicosapentaenoic acid | 20.5 5C8C11C14C17C | *c*20:5ω3 | 302.5 | | |
| Docosahexaenoic[d] | *cis*-4,7,10,13,16,19-Docosahexaenoic acid | 22.6 4C7C10C13C16C19C | *c*22.6ω3 | 328.5 | | |

[a]Formulas are shown as the number of carbon atoms followed by the number of double bonds. The position of each double bond is indicated by the lower of the numbers of the two doubly bonded carbon atoms, counting from the carboxyl carbon and specified as to *cis* (C) or *trans* (T) configuration. Thus, oleic (*cis*-9-octadecenoic) acid is 18.1 9C.

[b]The numbering of the carbon atoms starts from the methyl end, and the location of the first (or only) double bond is indicated by a single number as a suffix preceded by "ω." Oleic acid is therefore designated as *c*18:1ω9, which informs the reader that the most distal double bond is nine carbons from the methyl terminus.

[c]Extrapolated from solubility of shorter chain acids.

[d]Found in marine animals in high concentration and also in retina and brain of mammals.

with an activated fatty acid by *acyl CoA diacylglycerol acyltransferase.* A second pathway involves the reacylation of 2-acyl-*sn*-glycerol, first by *monoacylglycerol acyltransferase,* to form the diglyceride, and then by *diacylglycerol acyltransferase,* to form triacylglycerol. This pathway is prominent in the intestine and is utilized to form the triglyceride, which is transported from the intestine in chylomicrons.

Triacylglycerols are broken down in many places by lipases. In the intestinal lumen, dietary fat is broken down by pancreatic lipases to two moles of fatty acid and one mole of 2-acyl-*sn*-glycerol (monoglyceride). In plasma triglycerides are broken down by *lipoproteins lipase* to the same molecules, which then partition into tissues or are transferred to albumin. In cells the triacylglycerol storage droplets are broken down by *hormone-sensitive lipase,* which produces fatty acid and glycerol in the cell.

The phospholipids shown in Figure 1 are made through several pathways involving either *phosphatidic acid* or *diacylglycerol.* Phosphatidic acid may be produced by the action of phospholipase D on more complex phospholipids or, more commonly, by the acylation of glycerol 3-phosphate with two moles of activated fatty acid to form phosphatidic acid. This key molecule may be hydrolyzed to form diacylglycerol, which can condense with cytosine diphosphate (CDP) choline to form phosphatidylcholine. Phosphatidic acid can interact with cytosine triphosphate to form cytosine diphosphate diacylglycerol (CDP-diglyceride). This intermediate can condense with serine to form phosphatidylserine. Phosphatidylserine may be decarboxylated to form phosphatidylethanolamine, a major membrane lipid.

A second set of reactions sequentially adds methyl groups to phosphatidylethanolamine to produce mono- and dimethylethanolamine and finally phosphatidylcholine. Phospholipids are subject to a variety of phospholipases, which hydrolyze the different ester and ether bonds to break down the phospholipids into various individual groups (fatty acids, glycerol, choline, ethanolamine, etc.), which enter the general metabolic pool. A separate set of reactions forms alkyl (ether) bonds between alcohols and glycerol. Ethers tend to be more stable than esters, and vinyl ether phospholipids such as plasmalogens may play important roles in stabilizing membranes in oxidatively reactive compartments such as lysosomes and peroxisomes.

Glycosphingolipids are formed by the addition of sugars to *N*-acylsphingosine (ceramide). In a complex set of reactions, different sugar moieties are added to ceramide to produce complex glycosphingolipids. This general class of molecules is found in most cells. Some can act as receptors for a variety of toxins and as antigens for antibodies. Others perhaps act as cell-type labels. They are particularly prominent in the brain and nerves, where they form a major portion of the membrane systems. These complex molecules are broken down by glycosyl hydrolases, which catalyze the specific removal of one sugar at a time, ultimately reducing the complex glycosphingolipids to ceramides and sugars. Deficiencies of enzymes along this pathway result in the accumulation of specific glycosphingolipids, and serious neurological disorders accompany these metabolic blocks. For instance the inability to break down gangliosides results in Tay–Sachs disease, and the inability to break down galactocerebroside, which has only one sugar on the ceramide, results in Gaucher disease. The symptoms of these diseases may arise from the physical properties of the lipids. For instance, gangliosides tend to accumulate in Tay–Sachs disease and cause massive swelling and disruption of membrane within the cells. Certainly part of their effect is due to the detergent properties of the accumulating gangliosides. While genetic defects in lipid metabolism or catabolism lead to a wide variety of disease states, many serious diseases involving lipids (e.g., atherosclerosis, obesity, cholesterol gallstone disease) may result from play between the environment and as yet unrecognized genetic factors.

*See also* Lipid Metabolism and Transport; Lipids, Microbial; Lipoprotein Analysis; Liposomal Vectors; Phospholipids; Plasma Lipoproteins; Steroid Hormones and Receptors.

### *Bibliography*


De Loof, H., Harvey, S. C., Segrest, J. P., and Pastor, R. W. (1991) Mean field stochastic boundary molecular dynamics simulation of a phospholipid in a membrane. *Biochemistry,* 30:2099–2113.

Folch, J., Lees, M., and Sloane-Stanley, G. H. (1957) A simple method for the isolation and purification of total lipids from animal tissues. *J. Biol. Chem.* 226:497.

Hanahan, D., Ed. (1978, 1983) *Handbook of Lipid Research* Series, Vols. 1–3, Plenum Press, New York.

Mead, J. F., Alfin-Slater, R. B., Howton, D. R., and Popjak, G. (1986) *Lipids, Chemistry, Biochemistry and Nutrition.* Plenum Press, New York.

Myant, N. B. (1981) *The Biology of Cholesterol and Related Steroids.* Heinemann Medical Books, London.

Paoletti, R., and Kritchevsky, D. (1963–1989); Havel, R. A. and Small, D. M. (1990–present) Eds. *Advances in Lipid Research,* Series, Vols. 1–23, Academic Press, San Diego, CA.

Scriver, C. R., Stanbury, J. B., Wyngaarden, J. B., and Frederickson, D. S., Eds. (1989) *The Metabolic Basis of Inherited Disease,* 6th ed. McGraw-Hill Information Service, New York.

Small, D. M. (1986) The physical chemistry of lipids from alkanes to phospholipids. In *Handbook of Lipid Research,* Vol. 4, D. Hanahan, Ed. pp. 1–672. Plenum Press, New York.

———. (1992) Structure and metabolism of the plasma lipoprotein. In *Plasma Lipoproteins in Coronary Artery Disease,* R. A. Kreisberg and J. P. Segrest, Eds. pp. 57–91. Blackwell, Oxford.

———, and Zoeller, R. A. (1991) Lipids. In *Encyclopedia of Human Biology,* Vol. 4, pp. 725–748. Academic Press, San Diego, CA.

Vance, D. E., and Vance, J. E. (1985) *Biochemistry of Lipids and Membranes.* Benjamin/Cummings, New York.


# Lipopolysaccharides


*Ilkka M. Helander, P. Helena Mäkelä, Otto Westphal, and Ernst T. Rietschel*


### *Key Words*

**Endotoxicity** The acute effects of lipopolysaccharide (endotoxin) on animals.

**Endotoxin** Synonym for lipopolysaccharide, used when referring to the biological activity of LPS; distinct from bacterial exotoxins, which are proteins secreted from cells.

**Gram-Negative Bacteria** Category of bacteria with a cell envelope containing an outer membrane with lipopolysaccharide as an essential component; this property is associated with the negative (red) staining of the cells in the selective Gram staining method.

**Lipid A** The lipid part of a lipopolysaccharide, which is responsible for its endotoxic activities.

**O-Antigen** Synonym for lipopolysaccharide and especially its outermost polysaccharide region showing specific antigenic reactivity.

Lipopolysaccharides (LPS) are unique structural components of the Gram-negative cell envelope. They exist ubiquitously in all Gram-negative bacteria, constituting an essential component of the outer leaflet of the asymmetrical outer membrane (OM) and providing the membrane and thus the bacterial cell with a hydrophilic surface. Therefore the OM can serve as an effective permeability barrier to many agents, such as hydrophobic antibiotics, detergents, dyes, and bile acids. Besides its functional role in the bacterial cell envelope, a multitude of biological, often toxic effects are evoked in mammalian hosts by LPS released from multiplying or disintegrating bacteria. These effects are referred to as endotoxic activities, and the term ''endotoxin'' is therefore often used interchangeably with LPS.

## 1 Structure and Functions of Lipopolysaccharides

### 1.1 General

Figure 1 shows lipopolysaccharides (LPS) in relation to other structural components of a cell envelope, and Figure 2 shows the schematic structure of a complete LPS molecule of *Salmonella.* In general, LPS consists of a complex lipid, termed lipid A, and a covalently linked heteropolysaccharide. The latter is divided into a core oligosaccharide and an O-specific chain, which is built of repeating oligosaccharide units. In certain bacterial groups (e.g. *Neisseria* spp.) the O-specific chain is lacking. Each part of the LPS molecule is endowed with specific biological and physiological functions.

### 1.2 Lipid A

The lipid A component consists most commonly of a β1′-6-linked D-glucosamine (GlcN) disaccharide, carrying at positions 1 and 4′ phosphate groups, which are often partially substituted by charged groups. This lipid A backbone carries four *(R)*-3-hydroxy fatty acids in amide linkage at positions 2 and 2′ and in ester linkage at positions 3 and 3′ (see Figure 2). The hydroxyl groups of some of the fatty acids are further esterified with nonhydroxylated fatty acids, creating the unique 3-acyloxyacyl structure. This structural principle is shared by the lipid A from many bacterial groups. In lipid A of different bacterial origin, however, the type and chain length of fatty acids vary greatly, and in some lipid A types, GlcN is replaced by 2,3-diamino-2,3-dideoxy-D-glucose. Lipid A is responsible for the endotoxic activities attributed to LPS. Lipid A variants with essential deviations from the foregoing structure exhibit reduced endotoxicity. In the bacterium, lipid A anchors the LPS molecule in the OM and is absolutely essential for the survival of the bacterium.

### 1.3 The Polysaccharide

The core oligosaccharide is linked to lipid A via the acidic eight-carbon sugar Kdo (3-deoxyoctulosonic acid), often present as a di- or trisaccharide. The core itself is subdivided into an inner

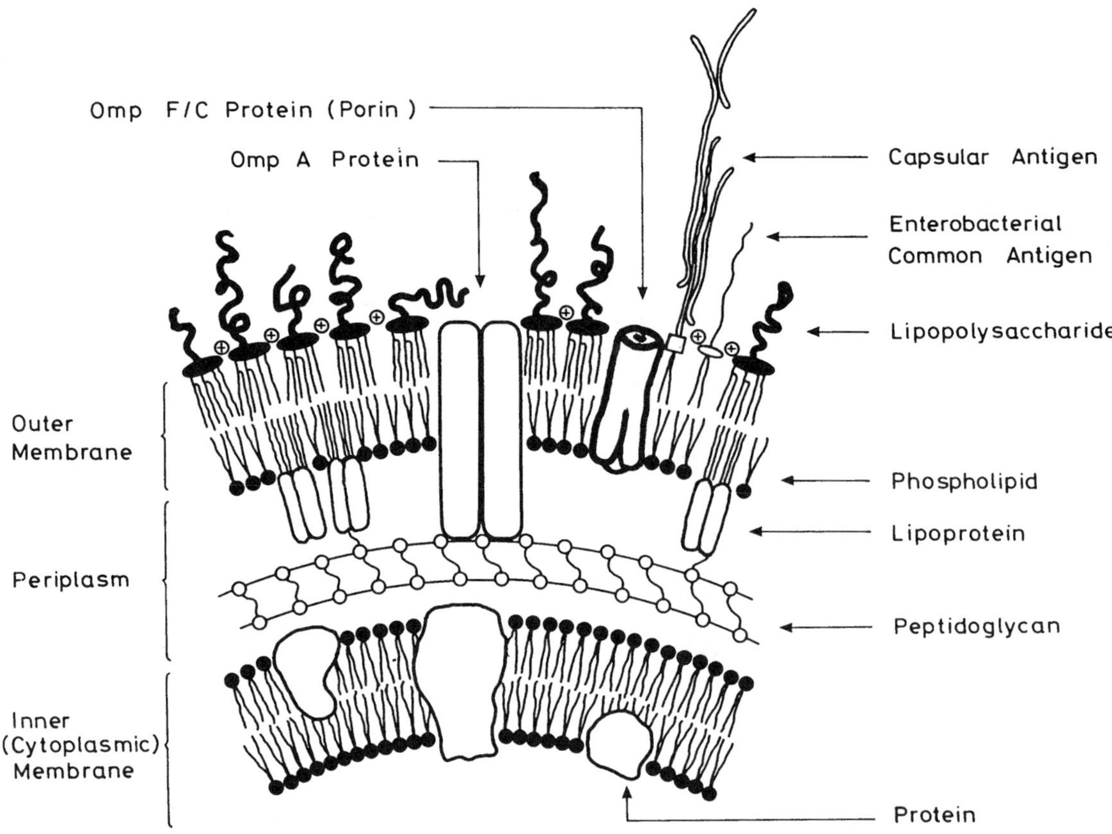


**Figure 1.** The cell envelope of Gram-negative bacteria, showing the location of the lipopolysaccharide (LPS) molecules on the outer leaflet of the outer membrane, which additionally contains phospholipids and proteins, some of which form transmembrane channels (porin proteins). Shown are also other sugar-rich molecules (capsular antigen, enterobacterial common antigen).

and outer core, respectively. In enterobacterial LPS, the inner core contains, in addition to Kdo, neutral seven-carbon sugars (heptoses). Various sites of the inner core are phosphorylated. The outer core is composed of neutral sugars, typically of common hexoses. Only limited variability of the core structure is exhibited by species of one enterobacterial genus. Generally, however, structural variation of core oligosaccharides is larger than that of lipid A. The core functions as receptor for specific bacteriophages, as an antigen, and may modulate the bioactivity of lipid A.

The outermost region of LPS, built of dozens of identical oligosaccharide units, is termed the O-specific chain. It is not essential for the integrity of the OM, and indeed many bacterial groups,

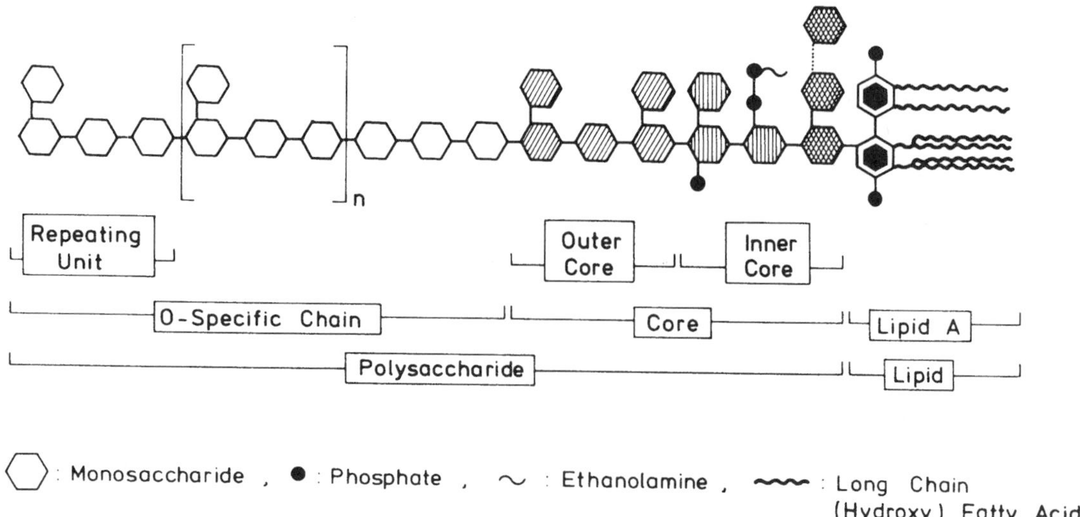


**Figure 2.** Schematic chemical structure of the LPS of *Salmonella enterica* sv. Typhimurium. In distinct LPS molecules, the length of the O-specific chain varies from 0 to approximately 50 repeating units.

including pathogenic ones, are devoid of it. It covers the OM (hence the cell surface), determines the serological specificity of LPS, and protects the bacteria from phagocytosis and complement-mediated killing. In a given bacterial culture, the degree of polymerization of the O-specific repeating units may vary from 0 to approximately 50. This leads to a considerable size heterogeneity of LPS, as revealed by polyacrylamide gel electrophoresis. An immense diversity of constituent sugar components, often including exceptional types of sugar such as 3,6-dideoxyhexoses, is found in the O-antigens.

### 1.4 Macromolecular Diversity of LPS

Bacteria that contain LPS with an O-specific chain are called *S*-form or *S*-type (*S* for smooth colony morphology). Typically, wild-type enterobacteria such as *Salmonella* are of the S type. Some bacterial groups including human pathogens *(Hemophilus, Neisseria, Chlamydia)* naturally lack the polymerized O-specific chain. They produce truncated LPS reminiscent of those of the rough *(R)* mutant bacteria which elaborate incomplete LPS because of genetic loss of the ability to synthesize the O-specific chain or a complete core oligosaccharide. The minimal LPS structure found in a viable bacterium is that of a *Hemophilus influenzae* mutant strain: this LPS consists only of lipid A and one phosphorylated Kdo. Mutants unable to synthesize proper lipid A are nonviable, evidencing an essential role of lipid A for cell viability.

## 2 ENDOTOXIC ACTIVITY OF LPS

In mammals, parenteral administration of endotoxin brings about fever, hypotension, activation of blood coagulation and the serum complement systems and, in larger doses, irreversible shock. Whereas endotoxin-induced complement activation is partly due to the polysaccharide chain, the typical endotoxic effects of LPS are caused by lipid A. Endotoxin stimulates a variety of mammalian cells, including lymphocytes, macrophages, and endothelial cells, via receptor-mediated recognition and signaling to produce powerful endogenous mediator substances. These include proteins such as tumor necrosis factor $\alpha$ (TNF), interferon $\gamma$ and the interleukins IL-1, IL-6, and IL-8, oxygen radicals, and lipids such as prostaglandins and leukotrienes. The mediators act independently or together to cause the systemic effects typically associated with endotoxin. A lethal endotoxic shock may result from massive Gram-negative bacteremic infections. In less severe or localized infections, the array of macrophage products induced by endotoxin in the host is likely to serve as an important defense mechanism.

*See also* Bacterial Growth and Division; Lipids, Microbial.

### *Bibliography*

Morrison, D. C., and Ryan, J. L., Eds. (1992) *Bacterial Endotoxic Lipopolysaccharides,* Vol. 1, *Molecular Biochemistry and Cellular Biology.* CRC Press, Boca Raton, FL.

Nowotny, A., Spitzer, J. J., and Ziegler, E. J., Eds. (1990) *Cellular and Molecular Aspects of Endotoxin Reactions,* Vol. 1. Elsevier Science Publishers, Amsterdam.

Raetz, C.R.H. (1990) Biochemistry of endotoxin. *Ann. Rev. Biochem.* 59:129–170.

Raetz, C.R.H., Ulevitch, R. J., Wright, S. D., Sibley, C. H., Ding, A., and Nathan, C. F. (1991) Gram-negative endotoxin: An extraordinary lipid with profound effects on eukaryotic signal transduction. *FASEB J.* 5:2652–2660.

Rietschel, E. T., and Brade, H. (1992) Bacterial endotoxins. *Sci. Am.* 267:26–33.

Rietschel, E.Th., Kirikae, T., Schade, F. U., Mamat, U., Schmidt, G., Loppnow, H., Ulmer, A. J., Zähringer, U., Seydel, U., Di Padova, F., Schreier, M., and Brade, H. (1994) Bacterial endotoxin: Molecular relationships of structure to activity and function. *FASEB J.* 218:217–225.

Zähringer, U., Lindner, B., and Rietschel E.Th. (1994) Molecular structure of lipid A, the endotoxic center of bacterial lipopolysaccharides. *Adv. Carbohydr. Chem. Biochem.* 50:211–276.

# Lipoprotein Analysis

*E. Roy Skinner*

### *Key Words*

**Apolipoproteins** Specific protein components (apo A, apo B, etc.) associated with lipoprotein particles, having a characteristic distribution and being formed from lipoproteins by removal of their lipid moieties.

**Atherosclerosis** Thickening of artery walls due to the accumulation of cholesterol and fibrous proteins.

**Coronary Heart Disease** The clinical manifestation of atherosclerosis resulting from stoppage of the blood flow to the heart due to the blocking of the constricted coronary artery as a result of thrombosis or embolism.

**Western Blotting** The transfer of resolved proteins from an electrophoretic medium to a nitrocellulose sheet and their identification by binding of antibodies raised against them.

The plasma lipoproteins are macromolecular complexes of lipid and proteins, a major function of which is to transport lipids between different tissues via the vascular system. They form a series of particles of widely differing composition and physical properties, but essentially all contain a hydrophobic core of triacylglycerol and cholesterol, surrounded by a shell of polar lipids (phospholipids and cholesterol) and specific proteins (apolipoproteins). They are classified according to their density into four major classes: chylomicrons, very low density lipoproteins (VLDL), low density lipoproteins (LDL), and high density lipoproteins (HDL) (Table 1). The apolipoproteins serve a variety of functions, which largely determine the role and fate of the particles containing them. They may serve as ligands for specific cell membrane receptors, as activators and modulators of key enzymes in lipoprotein metabolism, and in maintaining the structural integrity of the lipoprotein particles.

The metabolic activities of the different lipoprotein classes are closely integrated. Chylomicrons transport lipid of dietary origin (largely triacylglycerol) from the intestine to the tissues. Upon entering the bloodstream, they receive apolipoproteins from circulating HDL, which acts as a reservoir for these apoproteins. Lipoprotein lipase, located on the vascular endothelium of extrahepatic tissues, catalyzes the hydrolysis of chylomicron triacylglycerol. The fatty acids released enter the tissues, while the chylomicron

Table 1 Properties of the Major Classes of Lipoproteins in Human Plasma

| Lipoprotein Class | Density ($g \cdot cm^{-3}$) | Apoproteins | Role |
|---|---|---|---|
| Chylomicrons | < 0.95 | A-I, A-II, B-48, C-I, C-II, C-III, E | Transport exogenous (dietary) triacylglycerol and cholesterol from intestine to liver and other tissues |
| Very low density lipoproteins (VLDL) | 0.95–1.006 | B-100, C-I, C-II, C-III, E | Released by the liver into the circulation; contain endogenous triacylglycerol and cholesterol |
| Low density lipoproteins (LDL) | 1.006–1.063 | B-100 | Derived from VLDL by lipolysis (and other processes); transport cholesterol to liver and other tissues and regulates de novo cholesterol synthesis |
| High density lipoproteins (HDL) | 1.063–1.21 | A-I, A-II, C-I, C-II, C-III, D, E | Transport endogenous cholesterol from tissues to liver for excretion; provides reservoir for apolipoproteins |

particles, deprived of core triacylglycerol, shrink to form chylomicron remnants after transfer of surplus surface components (cholesterol, phospholipids, and some apolipoproteins) to the HDL pool. The remnant particles also acquire apo E and are taken up by the liver by receptors specific for this apolipoprotein.

VLDL, produced in the liver and rich in endogenous triacylglycerol, undergoes lipolysis in a similar manner to produce intermediate density lipoprotein, part of which is converted to LDL by the action of hepatic lipase. LDL is taken up by the tissues by the B100/E receptor (LDL receptor), the activity of which is regulated according to the cholesterol requirement of each individual cell. LDL also enters cells through a nonsaturable, non-receptor-mediated pathway involving its passage across the cell membrane by pinocytosis. By this means, the organism fulfills the requirement of tissues for cholesterol for cell membrane repair, growth, and steroid hormone synthesis.

The accumulation of cholesterol in the peripheral tissues is countered by the process of reverse cholesterol transport, in which cholesterol that is surplus to the needs of the tissues is transferred back to the liver in association with HDL and excreted in the bile. Aberrations in the normal processes of lipoprotein metabolism, such as changes in receptor activity or rate of apolipoprotein synthesis, may lead to alterations in the distribution of lipids throughout the body. Such changes may be of genetic or environmental origin and are frequently associated with clinical disorders, the most common of which is coronary heart disease. Extensive epidemiological studies have demonstrated a strong relationship between the levels of LDL and HDL and the incidence of this disease and have suggested that measurement of lipoprotein parameters may provide good indicators for identifying people of high coronary risk.

# 1 SEPARATION OF PLASMA LIPOPROTEINS

## 1.1 Flotation Methods

Since lipoproteins have lower hydrated densities than other plasma proteins, they may be separated from the latter by flotation in the preparative ultracentrifuge. This is the method of choice for separating lipoproteins, even though some apolipoproteins become dissociated from the lipoprotein particles and exchange of both lipid and apolipoproteins occurs to some extent during ultracentrifugation. The method is, however, applicable on a small scale for clinical investigation or on large quantities for other studies.

In essence, the method consists of centrifuging plasma (prepared from blood collected in EDTA) at its natural density of 1.006 $g \cdot cm^{-3}$ for 18 hours at 100,000*g* to float the VLDL fraction, which is then removed, and the density of the infranatant adjusted to 1.063 $g \cdot cm^{-3}$ by the addition of NaBr. The sample is recentrifuged to float the LDL, and finally the density of the LDL infranatant is adjusted to 1.21 $g \cdot cm^{-3}$ and the HDL is separated by centrifugation.

## 1.2 Immunological Methods

Lipoproteins are highly heterogeneous with respect to their apolipoprotein composition, and therefore immunoaffinity chromatography provides a valuable method for their separation. The setting up of the method is time-consuming and requires a knowledge of antibody production and protein separation techniques, but immunoaffinity has the advantages of great sensitivity and excellent resolving power. The method entails the production of specific antibody and the covalent linking of the latter to a support medium such as Sepharose. Plasma is incubated with the immobilized antibody, which is packed into a small column and washed to remove nonbound proteins before appropriate solvents are introduced to remove the complexed lipoprotein species containing the specific apolipoprotein antigens.

## 1.3 Precipitation Methods

All precipitation methods are based on the specific interactions of apolipoproteins with precipitating agents and separate lipoproteins according to size and relative surface charge. Heparin/manganese and phosphotungstic acid/magnesium are the most popular agents, so that VLDL and LDL (containing largely apo B) can be separated from HDL (which does not contain apo B). Although the procedure does not separate the lipoprotein fractions from other plasma proteins, it provides a simple and useful tool for the measurement of lipoprotein concentrations and is especially valuable in clinical investigations. The method has been extended for the determination of $HDL_2$ and $HDL_3$ concentrations in plasma by precipitation of $HDL_2$ with dextran sulfate after removal of the apo B-containing lipoproteins.

### 1.4 Separation by Electrophoresis and Isoelectric Focusing

Electrophoresis provides a primary method for the resolution and quantitation of lipoprotein fractions, particularly for subfractions of LDL and HDL. Lipoproteins are applied to a polyacrylamide gradient gel, through which they migrate under the influence of an applied electric field. As the pore size of the gel decreases along its length with increasing polyacrylamide concentration, the migration of each lipoprotein species is progressively reduced according to its particle size, eventually stopping when it reaches its exclusion limit. The distribution of lipoproteins in the plasma may then be evaluated by scanning the gel after either staining or Western blotting.

Isoelectric focusing, by which apolipoproteins are separated according to their differences in isoelectric points, produces greater resolving power and therefore provides a useful method for distinguishing dyslipoproteinemias.

## 2 QUANTITATION AND ANALYSIS OF LIPOPROTEINS

### 2.1 Measurement of Cholesterol and Other Lipids

The concentration of plasma lipoprotein classes is conventionally expressed by reference to their cholesterol concentrations. In turn, concentration is most conveniently measured by enzymatic assays, which are simple, rapid, and accurate and are available commercially in kit form. Chemical and chromatographic methods are also available, but these tend to involve the use of toxic reagents and are time-consuming. Similar tests can be applied for the measurement of triacylglycerols and phospholipids associated with isolated lipoprotein fractions.

### 2.2 Apolipoprotein Concentrations

It is frequently necessary to determine the apolipoprotein composition of lipoprotein fractions, since these are largely responsible for determining the metabolic activities of the lipoprotein particles containing them. In addition, determination of the plasma concentrations of apo B and apo A-I, as measures of the concentrations of LDL and HDL, respectively, is sometimes preferred to measuring the concentration of these lipoproteins by their cholesterol content. Apolipoprotein concentrations are generally determined by immunological methods or by sodium dodecyl sulfate polyacrylamide gel electrophoresis (SDS PAGE). The former include enzyme-linked immunosorbant assay (ELISA), immumonephelometry, radioimmunoassay, and radial immunodiffusion, some of which are available in kit form.

SDS PAGE is best carried out after the lipoproteins have been delipidized using appropriate organic solvents. The apoprotein mixture thereby obtained is submitted to electrophoresis on an SDS-containing gel of uniform polyacrylamide concentration. The gel may be stained, scanned, and quantitated as described in Section 1.4. SDS serves to denature the apolipoproteins, which become uniformly coated with the negatively charged detergent. The migration distance along the gel is therefore proportional to the molecular size of the apolipoprotein, and this can be evaluated by reference to appropriate protein markers. There is considerable intermethod and interlaboratory variation in apolipoprotein determination.

## 3 ANALYSIS OF LIPOPROTEIN SUBFRACTIONS

### 3.1 General Considerations

More recently, a number of laboratories have investigated the separation and characterization of the different types of particles present in each of the major density lipoprotein classes, to determine their respective roles in lipoprotein metabolism. Significant alterations in the distribution of HDL and LDL subfractions are associated with coronary heart disease.

### 3.2 HDL Subfractions

HDL is most conveniently resolved on the basis of particle size into five subpopulations by electrophoresis on polyacrylamide gradient gels. Immunoaffinity and heparin–Sepharose affinity chromatography are also useful methods for isolating HDL subfractions according to apolipoprotein content.

### 3.3 LDL Subfractions

LDL is most effectively separated into three or more subfractions by ultracentrifugation in a density gradient. Gradient gel electrophoresis also resolves LDL subfractions but, on its own, lacks the resolving power for quantitative assessment of LDL subfraction concentrations.

## 4 PERSPECTIVES

The role of the plasma lipoproteins in the development of coronary heart disease is under intensive investigation on a broad front. It is envisaged that studies both at the genetic level and on lipoproteins themselves, including their metabolism, receptor activity, the control of the cholesterol synthesis, and the protective action of HDL, will lead to a fuller understanding of the underlying causes of atherosclerosis. It is hoped that the development of a more reliable means of assessing coronary risk, a greater understanding of the factors that control LDL and HDL levels, and a more rational basis for drug intervention that may result from such studies will reduce the incidence of coronary heart disease substantially within the next decade.

*See also* Plasma Lipoproteins.

## *Bibliography*


Albers, J. J., and Segrest, J. P., Eds. (1986). *Plasma Lipoproteins,* Part B, *Characterization, Cell Biology and Metabolism. Methods in Enzymology,* Vol. 129. Academic Press, London and New York.

Converse, C. A., and Skinner, E. R., Eds. (1992). *Lipoprotein Analysis: A Practical Approach.* Oxford University Press, London and New York.

Segrest, J. P., and Albers, J. J., Eds. (1986) *Plasma Lipoproteins,* Part A, *Preparation, Structure and Molecular Biology. Methods in Enzymology,* Vol. 128. Academic Press, London and New York.

# Liposomal Vectors

*Roger R. C. New*

*Key Words*

**Liposomes** Sealed, usually spherical, vesicles composed of lipid membrane bilayers enclosing a central aqueous compartment.

**Phospholipid** A class of amphiphile; either synthetic or natural, consisting most commonly of two hydrocarbon chains linked to a polar head group via a glycerol bridge. The most widespread example is the neutral phosphatidylcholine; this compound self-associates in aqueous media to give membrane bilayers, which then form liposomes spontaneously.

The use of liposomes for delivery of DNA to mammalian cells to achieve efficient expression is discussed. Liposomes are major candidates for delivery vehicles in human gene therapy.

## 1 SCOPE

In molecular biology, as in other areas of biotechnology, liposomes have played a dual role as models of natural cell membranes, and as agents in their own right for manipulating the behavior and function of cells and organisms. Liposomes are used routinely to introduce DNA into mammalian cells, and basic biophysical studies of the interaction between DNA and phospholipids, or between liposomes and nuclear membranes, shed some light on the way liposomes may be acting to deliver the DNA effectively. Several different methodologies that make use of liposomes are employed, and they all exploit the propensity of phospholipid membranes to become perturbed and fuse together under appropriate conditions (see Figure 1), with the result that DNA is introduced directly into the cell cytoplasm. The mechanism by which DNA passes from the cytoplasm to the nucleus is not understood. In some cases the uncertainty associated with this aspect of delivery may be avoided by making DNA constructs that are capable of cytoplasmic translation, and by combining with RNA polymerase or a gene that produces this polymerase.

## 2 DNA ENTRAPMENT

Liposomes are closed phospholipid vesicles that entrap an aqueous space and may vary in size from 250 A in diameter to several micrometers. Thus they can be constructed to permit entrapment of large macromolecules such as DNA to chromosomes, nuclei, or even whole bacterial or yeast cells. Liposomes are usually prepared by dispersion of phospholipids such as phosphatidylcholine (PC) in water, using mechanical methods that are capable of destroying DNA. However, short lengths of DNA (~ 1000 base pairs) can be incorporated into small sonicated liposomes, and the efficiency of incorporation (and perhaps protection against mechanical destruction) can be markedly increased by the presence of a basic protein such as lysozyme.

The first method employed successfully for the purpose of entrapping large stretches of DNA was a technique in which, instead of being subjected to mechanical means, a process was employed whereby liposome membranes were induced to fuse into large

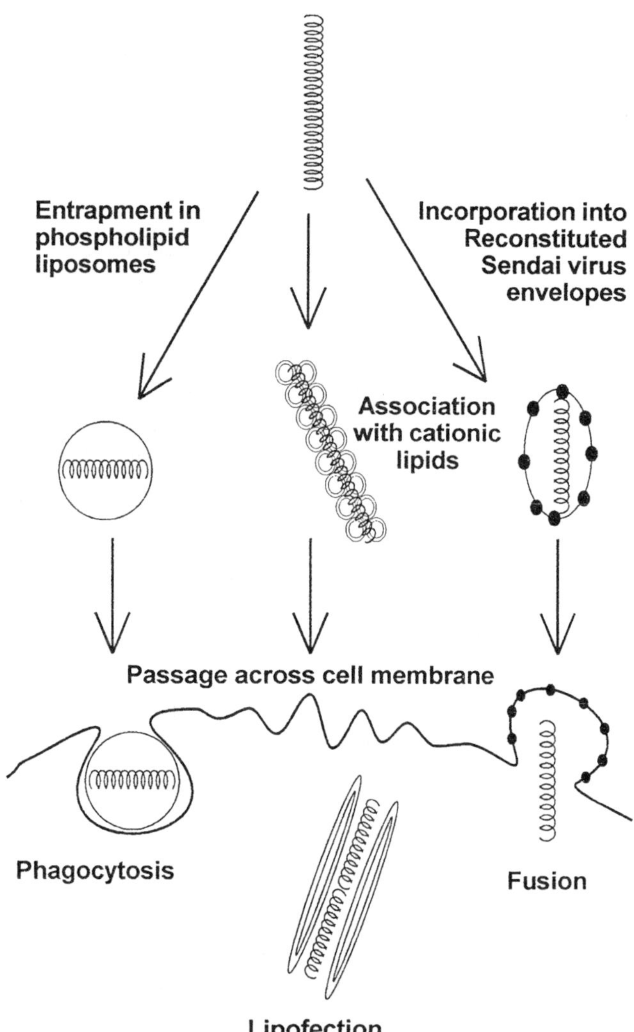

Figure 1. Introduction of DNA into cells by means of liposomes.

sheets by exposure of the anionic lipid phosphatidylserine (PS) to calcium. Unentrapped DNA was separated from liposomal material by flotation of the latter on a density gradient, after high-speed centrifugation. Similar types of large liposome were also produced by two-phase techniques (such as reverse evaporation vesicle and large unilamellar vesicle) in which the liposome membrane is formed at the interface of organic and aqueous solvent phases.

Normally, when liposomes such as those just described come into contact with cells, either in vitro or in vivo, they are taken up into these cells by a process of phagocytosis, and the efficiency of transfection is comparatively low. To deliver DNA into cells in a way that permits its efficient expression, lysosomal degradation must be avoided, and the DNA must be introduced directly into the cytoplasm. The first method devised to achieve delivery of this nature employed liposomes constructed on the same principle as envelope viruses. Thus, the fusion protein of Sendai virus was extracted by detergent and incorporated into vesicles containing SV40 DNA. To bring liposomes into sufficiently close contact for fusion to occur, receptor proteins need to be incorporated into the liposome membrane as well, and in this way specificity of interaction can be introduced, so that the identity of the target cell can be controlled, either in vitro or in vivo. In one refinement of this method, coentrapment of nonhistone chromosomal protein HMG-1 enhances the efficiency of transfection of insulin and *HBV* genes when injected in Sendai virus vesicles directly into rat livers.

A second method adopted for enhanced uptake of liposome-entrapped DNA is to use the endocytic pathway, constructing the liposomes in such a way that they or their contents can escape the endosome into the cytoplasm soon after formation of the endosome, but before the enzymes of the phagosome or secondary lysosome have a chance to act. This is accomplished by employing a liposome with a metastable, pH-sensitive membrane, whose potential to fuse with surrounding membranes is increased as the pH is lowered. Such membranes make use of the fact that phosphatidylethanolamine (PE) is ionizable, and that at low pH, certain forms of PE, such as dioleoyl PE (DOPE), exist as an inverted hexagonal phase, incompatible with a bilayer membrane structure, in preference to a conventional lamellar phase such as is formed by phosphatidylcholine (PC). To control the temperature and pH at which such a phase change takes place, the membrane is generally composed of a mixture of DOPE and PC. Other ionizable lipids such as cholesterol hemisuccinate may be used instead, and the liposomes may be constructed so that they are targetable by incorporating cell-specific antibodies into the membrane.

## 3 DNA–LIPID COMPLEXES

An alternative approach for introducing DNA into cells is to form a DNA–lipid complex by comixing DNA and cationic liposomes; the resultant complex is taken up very efficiently by cells. In this case the DNA is not entrapped inside the liposomes; rather, it associates with them by electrostatic attraction. It is important for the liposomes to be cationic, first to bind well to the negatively charged DNA, second to bind strongly to cells (whose surfaces are also negatively charged), and finally to cause in the cell plasma membrane perturbations that will allow fusion processes to take place and permit entry of the complex into the cell. Once in the cytoplasm of the cell, the complex may still exist intact and have the opportunity to interact with internal organelle membranes (e.g., nuclear, mitochondrial) in the same way. The liposomes commercially available for this purpose (from Gibco-BRL) are composed of a mixture of DOPE and DOTMA (dioleyloxypropyl trimethylammonium chloride) and are small unilamellar vesicles that bind to DNA in optimal proportions of 1:4 wt/wt (DNA/lipid). According to the manufacturers, 1 mg of this material (trademark Lipofectin) is sufficient for 20–30 transfections on 60 mm tissue culture dishes and can result in 5- to 100-fold more efficient transfection than calcium phosphate or DEAE-dextran, depending on the cell type employed.

Other cationic lipids that have been synthesized show the same properties as DOTMA, in some cases with a lower level of generalized toxicity. Thus, in the case of compounds derived from cholesterol linked via a spacer to an amine group, efficacy is greater with tertiary-substituted amine groups than with quaternary, while the converse is the case for toxicity. The chemical nature of the linkage is also important, an ester linkage showing less toxicity than amide or ether. Other systems, such as conjugates of long-chain lipids with spermine or polylysine have been described. It is generally recognized that these compounds need to be presented in combination with DOPE for maximum efficacy. Other lipids (e.g., MOPE, DEPE) do not work as well. DNA–lipid complexes may be found bound to coated pits very soon after addition to cells, and they can then escape from the endosomes very easily if DOPE is present, but not DOPC. As for fusogenic liposomes encapsulating DNA, DOPE is thought to be effective in DNA–lipid complexes because it is fusogenic and easily forms hexagonal phase structures. It also has been suggested that when the complex is formed, the lipids change their phase from lamellar to cubic, this structure being more efficacious for uptake by cells. In a similar manner, chloroquine enhances transfection rates because it prevents the fusion of endosomes with lysosomes, allowing more time for the DNA to escape into the cytoplasm.

In a recently described variation on the DNA–cation–lipid principle, a complex is created by mixing DNA with a cationic peptide such as gramicidin. Gramicidin is well known for its specific interaction with lipid membranes, and this property may be important because other peptides such as polylysine are much less effective in enhancing the uptake of DNA. Addition of DOPE to the complex improves the performance still further and may act to reduce nonspecific toxicity of the preparation. This system has been shown to achieve an efficiency of transfection 10 times that of Lipofectin, which itself displays performance superior to that of pH-sensitive liposomes by an order of magnitude.

Currently, DNA–lipid complexes are taken up mainly via the endocytic pathway into cells, and improvements in performance may be expected as the systems are refined, either to encourage the complex to penetrate through the phagosome membrane after endocytosis (see earlier) or to increase delivery directly through the plasma membrane.

One possibility is to target the complex specifically to a cell surface receptor. While DNA entrapped inside liposomes can be targeted specifically to cells by incorporation of ligand proteins into the liposome membrane, the DNA–lipid complex is completely nonselective in its target cell interactions. In a recently proposed method of inducing targeting capability, however, a biotinylated bisanthracycline structure can be intercalated noncovalently into double-stranded DNA; this structure can be linked to receptor ligands via avidin and used to control the destination of the complex, both extracellularly and inside the cell. Use of such a method will increase the efficiency of transfection and reduce concomitant toxicity still further.

## 4 ADMINISTRATION

Liposomal delivery has become the method of choice for transfection of mammalian cells in vitro, and recently reports of successful utilization in vivo have appeared in the literature. Direct injection into the liver has given rise to incorporation and expression of genetic material in hepatocytes. Inclusion of HMG1 nonhistone protein together with plasmid DNA in the liposome gives improved expression of the β-galactoside marker. Porcine vascular endothelial and smooth muscle cells have been selectively transfected using Lipofectin administered via a double-balloon catheter inserted directly into the artery. Cationic liposomes can also be delivered intravenously or by intratracheal injection into mice to achieve selective expression of marker genes in the lung. No untoward toxicity has been reported at the levels employed. Other workers have demonstrated expression in the rodent liver after intravenous injection of large liposomes. Use of a plasmid containing the rat preproinsulin I gene gave rise to glycemia in treated rats as a result of uncontrolled secretion of insulin. Expression of the human insulin gene in rat liver has also been achieved after injection of liposomes into the hepatic portal vein. Attempts to increase the selectivity of liposomes injected intravenously have made use of ligands located on the outer membrane surface. Transferrin-coated liposomes were found to deliver DNA to erythroblasts in the bone marrow, while inclusion of lactosyl ceramide encourages uptake by hepatocytes (in preference to Kupffer cells), leading to the presence of biologically active DNA detectable in clathrin-coated pits in these cells.

Studies in humans are currently being conducted to evaluate liposomes as delivery vehicles for the cystic fibrosis gene, after administration to nasal or bronchial tissue, and for allogeneic histocompatibility antigens (by direct injection of the gene–lipid complex into melanoma tissue), to stimulate a cytotoxic T-lymphocyte response. Few data are available regarding the stability of incorporation or the persistence of expression in any of the in vivo systems mentioned here.

*See also* Drug Targeting and Delivery, Molecular Principles of; Genetic Immunization; Lipids, Structure and Biochemistry of; Membrane Fusion, Molecular Mechanism of; Phospholipids.


## *Bibliography*

Behr, J. P., Demeneix, B., Loeffler, J. P., and Perez Mutul, J. (1989) Efficient gene transfer into mammalian primary endocrine cells with lipopolyamine-coated DNA. *Proc. Natl. Acad. Sci. U.S.A.* 86:6982.

Brigham, K. L., Meyrick, B., Christman, B., Magnuson, M., King, G., and Berry, L. C. J. (1989) In vivo transfection of murine lungs with a functioning prokaryotic gene using a liposome vehicle. *Am. J. Med. Sci.* 298:278.

Felgner, P. L., Gadek, T. R., Holm, M., Roman, R., Chan, H. W., Wenz, M., Northrop, J. P., Ringold, G. M., and Danielsen, M. (1987) Lipofection: A highly efficient, lipid-mediated DNA-transfection procedure. *Proc. Natl. Acad. Sci. U.S.A.* 84:7413.

Gao, X. A., and Huang, L. (1991) A novel cationic liposome reagent for efficient transfection of mammalian cells. *Biochem. Biophys. Res. Commun.* 179:280.

Haensler, J., and Szoka, F. C. (1993) Synthesis and characterization of a trigalactosylated bisacridine compound to target DNA to hepatocytes. *Bioconjugate Chem.* 4:85.

Kato, K., Nakanishi, M., Kaneda, Y., Uchida, T., and Okada, Y. (1991) Expression of hepatitis B surface antigen in adult rat liver. *J. Biol. Chem.* 266:3361.

Legendre, J.-Y., and Szoka, F. C. (1992) Delivery of plasmid DNA into mammalian cell lines using pH-sensitive liposomes: Comparison with cationic liposomes. *Pharm. Res.* 9:1235.

Loyter, A., Vainstein, A., Graessmann, M., and Graessmann, A. (1983) Introduction of SV40-DNA into tissue culture cells by the use of DNA-loaded reconstituted Sendai virus. *Exp. Cell Res.* 143:415.

Nabel, E. G., Plautz, G., and Nabel, G. J. (1990) Site-specific gene expression in vivo by direct gene transfer into the arterial wall. *Science,* 249:1285.

New, R. R. C. (1990) *Liposomes—A Practical Approach.* Oxford University Press, Oxford.

Nicolau, C., and Cudd, A. (1989) Liposomes as carriers of DNA. *Crit. Rev. Ther. Drug Carrier Syst.* 6:239.

Wang, C. Y., and Huang, L. (1989) Highly efficient DNA delivery mediated by pH-sensitive immunoliposomes. *Biochemistry,* 28:9508.


# Liver Cancer, Molecular Biology of

*Mehmet Ozturk*

## *Key Words*

**Aflatoxins** A family of natural toxins produced by *Aspergillus* species, inducing mutations and liver cancers in animals.

**Hepatitis B Virus (HBV)** A small DNA virus that infects hepatocytes in the liver, causing acute or chronic hepatitis.

**Hepatitis C Virus (HCV)** A small RNA virus infecting hepatocytes and causing acute or chronic hepatitis.

**Loss of Heterozygosity (LOH)** Loss of one of the two heterozygous loci (alleles) on a given chromosome pair.

***p53*** A tumor suppressor gene and its protein product involved in most human cancers.

**X Protein** A transactivating protein encoded by HBV. This protein has transforming activity.

Primary liver cancer, one of the ten most frequent tumors in the world, is closely linked to infection with hepatitis B virus (HBV) or hepatitis C virus (HCV) and intake of aflatoxins and alcohol. HBV may integrate into the genome of the host cells and cause major chromosomal aberrations. In addition, integrated viral sequences may serve as templates for the synthesis of transforming viral proteins in infected hepatocytes. HBV may thus contribute to liver carcinogenesis indirectly, as in chronic liver disease, or specifically, by inducing chromosomal aberrations and/or by giving rise to the synthesis of transforming proteins in hepatocytes. Aflatoxins, which are potent mutagens, appear to induce a "hot spot" mutation at codon 249 of the tumor suppressor *p53* gene. Other mutations affecting *p53* were found in non-aflatoxin-related primary liver cancers. Deletions that affect several chromosomes in primary liver cancers may indicate that other yet unknown tumor suppressor genes are involved in hepatocellular carcinogenesis.

## 1 ETIOLOGY

The great majority of primary liver cancers (PLCs) are hepatocellular carcinomas that derive from hepatocytes. Hepatoblastoma originating from the same cell type, a pediatric form of PLC, occurs rarely. Both tumor types are sometimes called hepatomas to indicate their hepatocyte origin. Primary liver cancer is one of the most frequent tumors worldwide. Between 250,000 and 1,000,000 persons die of this cancer each year in the world. Two-thirds of PLCs occur in southern Africa and southeast Asia.

Chronic infection with hepatitus B virus (HBV) is the major etiology of PLC. This small DNA virus (genomic size 3.2 kbp) may cause chronic hepatitis in patients infected at an early age, and most patients carry infectious viral particles throughout life. It is estimated that there are about 250 million chronic carriers of HBV in the world. Geographic distribution of HBV infection rate correlates well with the geographic distribution of PLCs. In addition, chronic HBV carriers have a very high risk of developing PLC (about a thousandfold compared to noncarriers). Another recently discovered virus, namely, hepatitis C virus (HCV), may also contribute to PLCs in Japan and some other countries. Serum antibodies to this RNA virus have been detected in patients with cirrhosis and PLC. It is not yet known whether HCV contributes to liver cancer nonspecifically by inducing such chronic liver diseases as chronic hepatitis or cirrhosis, or more specifically by inducing malignant changes in infected hepatocytes.

Natural toxins produced by some *Aspergillus* species have also been linked to PLCs in humans. Aflatoxin $B_1$ was found to be the major species of toxins contaminating foodstuffs in certain geographical locations in southern Africa and southeastern Asia, two separate subcontinents that share a high incidence of PLCs. Aflatoxin $B_1$ is a powerful mutagen and a potent liver carcinogen in animals. In western countries, where foods are routinely checked for contamination with aflatoxins and the risk of exposure is almost zero, PLCs are rare, occurring mostly in a setting of chronic viral hepatitis and/or alcoholic cirrhosis. PLCs are rarely linked to a single etiologic factor anywhere in the world, however. In developing countries, patients may be exposed to viruses and aflatoxins, as well as alcohol, whereas in the West viral infection may accompany excessive alcohol consumption. Exposure to multiple carcinogenic factors may complicate the understanding of molecular mechanisms that underlie hepatocellular carcinogenesis. A striking example of this complication has been observed in China, where people who are exposed to both HBV and aflatoxins are extremely vulnerable to PLC development. Elsewhere in continental Asia, people exposed to either HBV or aflatoxins have a moderately increased risk of developing PLCs.

## 2 MOLECULAR GENETICS

Like many tumors found in adults, PLC is a multistep disease in which many genomic changes occur as a result of uncontrolled proliferation of hepatocytes. Under normal physiological conditions, adult hepatocytes are nondividing cells. Indeed, only a minor fraction of hepatocytes appears to undergo cell division as a response to a minor cell loss due to aging or programmed cell death. During liver injury, substantial cell death can occur as a result of exposure to viruses or toxins. Liver injury provokes an increase in the proliferating fraction of hepatocytes. Chronic exposure to viruses or toxins that trigger an abnormal cell death often results in a chronic state of cellular proliferation (i.e., regeneration) in the liver. This chronic state of cell death and regeneration is probably the main source of genetic errors that lead to the malignant transformation of hepatocytes and probably explains the large time lapse (tens of years) between the first exposure to etiologic factors and the clinical manifestation of PLC.

Both epidemiological and molecular studies indicate that five or six independent events are necessary for a normal hepatocyte to develop a fully malignant tumor. It is therefore assumed that several oncogenes and tumor suppressor genes are involved in PLC.

### 2.1 Cellular and Viral Oncogenes

The expression of several proto-oncogenes may be induced during the malignant transformation of hepatocytes. In human PLC, however, no known oncogene has been shown to be consistently activated by a structural change (e.g., point mutations, translocations, amplifications).

Strong epidemiological association between HBV infection and primary liver cancer suggest that the virus plays a role in hepatocellular carcinogenesis. However, it is still unknown whether and how HBV contributes to the oncogenic transformation of hepatocytes. As a free replicating virus, HBV Is rarely found in PLCs. Moreover, patients infected with the virus at birth (vertical transmission from the mother) do not develop PLCs until they reach adulthood. Furthermore, PLCs can occur in patients not infected with HBV.

Thus, one can assume that the viral infection by itself is not sufficient, nor perhaps necessary, for tumor development. However, the persistence of HBV DNA in most PLCs associated with viral infection may indicate that the virus contributes directly to hepatocellular carcinogenesis.

HBV DNA sequences found in PLCs are partial or multiple copies of the virus integrated into the host genome. If the virus can directly contribute to the malignant transformation of hepatocytes, it may use several different mechanisms to accomplish this goal. First, the viral integration itself may disturb the normal expression of cellular genes involved in cell proliferation, differentiation, and apoptosis. In the tumors of woodchucks infected with a virus closely related to HBV, for example, the N-*myc* oncogene was found to be activated by integration of the virus into the host genome. However, there are no examples of such oncogenic activation in humans. Out of hundreds of HBV-related human PLCs studied, only two showed interruption of the integrity of host genes. Both genes (retinoic acid receptor and cyclin A) are involved in cellular proliferation and differentiation. However, such virus-induced genomic alterations appear to be extremely rare in human PLCs.

Another possible mechanism linked to viral integration may entail the expression of transforming viral proteins from integrated viral DNA sequences. Among several HBV proteins, the X protein appears to exhibit features of a transforming protein. Transgenic mice made by the integration of X-protein coding sequences in the germ line DNA often develop multiple liver tumors. Moreover, X protein is a pleiotropic transactivator capable of modifying the expression of several genes in host cells, either directly or indirectly by inducing activation of protein kinase C. In confirmation of such a hypothesis, it can be noted that the HBV DNA integrating into liver cancer cells often carries X-protein coding sequences.

### 2.2 Tumor Suppressor Genes

At least eight chromosomal arms (4q, 5p, 5q, 8q, 13q, 16p, 16q, 17p) appear to show frequent deletions in PLC. This observation suggests that at least eight tumor suppressor genes are involved in the development of these tumors. In one study, the short arm of chromosome 17 was found to be deleted in more than 50% of PLCs. A common region of deletion is 17p3.1. The *p53* gene, also involved in tumors from many other tissues, is located to this region. Initially identified as a protein, p53 is a nuclear protein that forms stable complexes with the transforming large T antigen of SV40. The wild-type *p53* is a tumor suppressor gene that inhibits cell proliferation and induces differentiation and programmed cell death. It has been suggested recently that the primary role of the wild-type *p53* is to check the integrity of the genome and to block

**Table 1** Worldwide Frequency of *p53–249*$^{ser}$ Mutation in Primary Liver Cancers

| Region | Mutants/Total | (%) |
|---|---|---|
| Africa | 10/42 | (24) |
| Asia | 36/186 | (19) |
| Middle East | 1/26 | (4) |
| Europe | 1/104 | (1) |
| Americas | 0/28 | (0) |
| Australia | 0/16 | (0) |
| All countries | 48/402 | (12) |

the survival of cells carrying genetic lesions either by stopping them from dividing or by helping them to die by apoptosis.

In 1991 a ''hot spot'' mutation at codon 249 of the *p53* gene was described in PLCs from China and southern Africa. The majority of mutations identified in both geographical areas were transversions from deoxyguanine to deoxythymine (G:C→T:A). Worldwide, the presence of this codon 249 mutation in PLCs correlates quite well with high risk of exposure to aflatoxins (see Table 1). These observations allowed the establishment of a link between a somatic mutation (G:C→T:A transversion at codon 249 of the *p53* gene) and an endemic form of PLCs. Striking differences between continents and even between neighboring countries strongly associate this mutation with high aflatoxin intake. In addition, a *p53* mutational hot spot at codon 249 appears to be a preferred target for aflatoxin $B_1$, the major aflatoxin species detected in the contaminated foods. Aflatoxin $B_1$ frequently induces G:C→T:A mutations in bacteria and mammalian cells and is a potent hepatocarcinogen in animals. The codon 249 mutation detected in PLCs, which is also a G:C-to-T:A transversion, is now considered to be a rare example of a biological fingerprint of a chemical carcinogen in a human cancer.

Mutations in *p53* also occur in PLCs not related to aflatoxins. However, their spectra (types of mutation) and frequencies vary considerably. The G:C→T:A transversions are rare, and they are scattered on multiple codons on the *p53* gene. In addition, the frequency of *p53* mutations is much lower (5–30%) in PLCs not related to aflatoxin exposure. The presence of *p53* mutations in PLCs suggests that the inactivation of this tumor suppressor gene is a critical step in hepatocellular carcinogenesis. In confirmation of this hypothesis, it has been recently demonstrated that the growth of hepatocellular carcinoma cells was not compatible with the expression of wild-type *p53*. Therefore, *p53* appears to act as a growth suppressor gene in hepatocytes.

At present, *p53* is the only tumor suppressor gene known to be frequently involved in PLCs. Somatic losses of heterozygosity of the retinoblastoma gene *(RB1)* in some PLCs have been described. The *RB1* gene that encodes a 110 kDa nuclear phosphoprotein was discovered initially as a tumor suppressor gene in retinoblastomas. Further studies have shown that the deletions of *RB1* also occur in other primary tumors and/or established malignant cell lines. The frequency of *RB1* deletions in PLCs and the biological significance of the inactivation of this tumor suppressor gene in these tumors remain to be determined.

## 3 PERSPECTIVES

The presence of chromosomal aberrations in PLCs was known for many years. Numerous studies conducted since 1985 have clearly indicated that one of the main sources of genetic changes in PLC is the integration of the hepatitis B virus. Currently, it is well accepted that HBV integration and chromosomal rearrangements related to this integration are random. The absence of common pattern of HBV-related genetic changes in these tumors is the main obstacle to finding a specific role of HBV in hepatocellular carcinogenesis. Since some HBV proteins (X protein in particular) appear to have potential transforming activity, future studies probably will focus on determining molecular pathways of hepatocyte transformation in which viral proteins are involved. However, there are frequent losses in at least eight chromosomal locations in PLCs, none of which appears to be directly linked to HBV integration. The newly discovered HCV appears to be an important etiologic factor in PLCs in certain geographic locations, mainly in Japan. There is however much work ahead in determining whether hepatitis C virus plays any specific role in hepatocellular carcinogenesis. Future research on molecular biology of PLC probably will evolve in the following directions: molecular identification of novel tumor suppressor genes, identification of the biological role(s) of *p53* in hepatocytes, and analysis and identification of the specific roles of hepatitis B and C viruses in hepatocellular carcinogenesis.

*See also* CANCER; ONCOGENES; TUMOR SUPPRESSOR GENES.

### *Bibliography*

Bosch, X., and Munoz, N. (1988) Epidemiology of hepatocellular carcinoma. In *Liver Cell Carcinoma,* P. Bannasch, D. Keppler, and G. Weber, Eds., pp. 3–14. Kluwer Academic Publishers, Dordrecht.

Ganem, D., and Varmus, H. E. (1987) The molecular biology of the hepatitis B viruses. *Annu. Rev. Biochem.* 56:651–693.

Gilman, M. (1993) Pleiotropy and henchman X. *Nature,* 361:687–688.

Marshall, C. J. (1991) tumor suppressor genes. *Cell,* 64:313–326.

Okuda, K. (1992) Hepatocellular carcinoma: Recent progress. *Hepatology,* 15(5):948–963.

Wogan, G. N. (1992) Aflatoxins as risk factors for hepatocellular carcinoma in humans. *Cancer Res.* (Suppl.) 52:2114.

# LUNG CANCER, MOLECULAR BIOLOGY OF

*Jack A. Roth*

### *Key Words*

**Non-Small-Cell Lung Cancer** A group of lung cancers that include adenocarcinoma, squamous carcinoma, and large-cell undifferentiated carcinoma, which together comprise 80% of lung cancers.

**Oncogene** A gene that has a dominant transforming effect on normal host cells.

**Small-Cell Lung Cancer** A specific lung cancer histologic type in which the cells frequently contain neurosecretory granules; the cancers are often disseminated at diagnosis and initially respond well to chemotherapy, but recur rapidly.

**Tumor Suppressor Gene** Homozygous deletion or inactivation of this gene contributes to transformation of the host cell.

The observation that genes responsible for carcinogenesis were altered forms of genes normally present in eukaryotic cells initiated many of the advances in molecular biology that have increased our understanding of lung carcinogenesis. Many of these genes have been implicated in the development of human cancer. An understanding of the molecular basis for lung carcinogenesis is essential for the development of improved methods of diagnosis, staging, treatment, and prevention of lung cancer.

## 1 LUNG MOLECULAR CARCINOGENESIS

Studies in mice with carcinogen-induced lung cancers implicate genes of the *ras* family in the carcinogenesis process (see Table 1). Mouse lung tumors induced by tetranitromethane contained mutated K-*ras* genes, and mice harboring the mutated H-*ras* transgene developed tumors exclusively in the lungs within weeks of birth. Lung tumors can be induced in mice by the tobacco-specific nitrosamine NNK or nitrosodimethylamine (NDMA). Ninety percent of these tumors had a transforming gene by the NIH 3T3 assay, and in all, this gene was K-*ras*. The mutations were generally GC-to-AT transitions, indicating that DNA methylation is the most likely pathway to induction of neoplasia by these carcinogens. This model is of interest because *ras* mutations are commonly found in human lung cancers.

**Table 1** Oncogenes and Tumor Suppressor Genes Altered in Lung Cancer[a]

| SCLC | NSCLC |
|---|---|
| Oncogenes | |
| *c-*myc* | *K-*ras* |
| L-*myc* | N-*ras* |
| N-*myc* | H-*ras* |
| c-*raf* | c-*myc* |
| c-*myb* | c-*raf* |
| c-*erb* B-1 (EGF-R) | *c-*fur* |
| c-*fms* | c-*fes* |
| c-*rlf* | c-*erb* B-1 (EGF-R) |
| | c-*erb* B-2 (Her2, neu) |
| | c-*sis* |
| | *bcl*-1 |
| Tumor Suppressor Genes | |
| **p53* | **p53* |
| **RB* | *RB* |

[a]Asterisks indicate most frequently altered genes in tumors or cell lines evaluated.

*Source:* M. S. Greenblatt, R. R. Reddel, and C. C. Harris, Carcinogenesis, cellular, and molecular biology of lung cancer. In *Thoracic Oncology,* 2nd ed. J. A. Roth, J. C. Ruckdeschel, and T. H. Weisenberger, eds. Saunders, New York, 1995.

### 1.1 Growth Factors and Their Receptors

Autocrine growth factors have been implicated in the stimulation of lung cancer cell growth. The majority of small-cell lung cancer (SCLC) cell lines produce bombesin (gatrin-releasing peptide), and they express a single class of high affinity, saturable binding receptors for bombesin. Bombesin is also a potent stimulator of clonal growth for human SCLC. Non-small-cell lung cancers (NSCLC) express high levels of functional epidermal growth factor receptor (EGFR) and have amplification of the EGFR gene, which suggests that growth factors and their receptors may play an important role in development and maintenance of the malignant phenotype. The expression of EGFR by lung cancer cells suggests that production of a ligand by these cells could mediate an autocrine or paracrine growth stimulation loop. The lung cancer cell lines studied did not produce EGF, but they did produce transforming growth factor type alpha (TGF-α), which binds to the EGFR. The gene *erb* B2 is a member of the EGFR. The *erb* B2 gene product, *p185,* is expressed at higher levels in the tumor than in bronchiolar epithelium. Its expression in adenocarcinomas is independently correlated with diminished survival. Insulinlike growth factor I (IGF-I) is also expressed by lung cancer cells and may participate in autocrine growth stimulation.

### 1.2 *ras* Oncogenes

The K-*ras* oncogene is activated by point mutation in lung cancer cell lines. In a study of fresh human tumors, K-*ras* mutations were confined to adenocarcinomas of the lung, but no mutations were observed in adenocarcinomas from nonsmokers. The presence of K-*ras* mutations is an independent predictor of poor prognosis. Infection of SCLC cell lines with the Harvey murine sarcoma virus altered the phenotype of variant but not classic SCLC. Following infection, the variant SCLC cell line developed features of a large-cell undifferentiated lung carcinoma, including increased expression of carcinoembryonic antigen (CEA) and keratin. Most studies favor the interpretation of *ras* activation as a progression factor in lung cancer. Apparently *ras* is activated in about one-third of adenocarcinomas arising in patients with a history of heavy smoking.

Expression of antisense K-*ras* RNA significantly reduces the growth rate of human lung cancer cells in athymic mice. Antisense constructs can be made that distinguish among members of the *ras* family. Inhibition of K-*ras* reduced the growth rate of H460a human lung cancer cells in an in vivo mouse model but did not alter cell viability or suppress growth in culture. These observations raise the intriguing possibility of specific molecular therapy for cancer. DNA constructs could be inserted to tumor cells that specifically inhibit expression of the oncogenes activated in the cancer cell.

### 1.3 *myc* Oncogenes

A subgroup of SCLC cell lines that have an amplified c-*myc* gene are morphological and biochemcal variants of SCLC (SCLC-V). SCLC-V have a faster doubling time, a higher cloning efficiency, greater tumorigenicity, and greater resistance to radiation than SCLC. In addition, SCLC-V do not express L-dopa decarboxylase or peptide hormones. They do have elevated levels of the bombesin isoenzyme or creatine kinase and neuron-specific enolase, which distinguishes them from NSCLC. The c-*myc* gene was transfected into the H209 classic SCLC cell line. One of the transfectants expressed high levels of c-*myc* and had increases in doubling time and cloning efficiency, but L-dopa decarboxylase levels and bombesinlike immunoreactivity were unchanged.

Analysis of SCLC cell lines for c-*myc* amplification revealed additional *Eco*RI restriction fragments, suggesting *myc*-related genes. Another gene in the *myc* family, L-*myc,* was cloned and showed homology to c-*myc* and N-*myc*. The L-*myc* gene has been cloned and sequenced; it consists of three exons and two introns spanning 6.6 kb of human DNA and has homology with discrete regions of N-*myc* and c-*myc*. The L-*myc* gene encodes a series of nuclear phosphoproteins that arise by alternative mRNA processing. It can cooperate with an activated c-Ha-*ras* to transform primary rat embryo fibroblasts.

Amplification and increased expression of the N-*myc* gene occurs in both SCLC and NSCLC. Increased expression of the gene was associated with poor response to chemotherapy and short survival. Amplification of N-*myc* gene sequences ranging from 5- to 170-fold was observed in SCLC cell lines. Both c-*myc* and N-*myc* were amplified, but only one member of the *myc* family was amplified in any one cell line.

The molecular mechanisms regulating the expression of each of the *myc* family genes are complex. Both c-*myc* and L-*myc* mRNA show loss of transcriptional attenuation that correlated with overexpression in cell lines without gene amplification. Regulation of N-*myc* expression correlates with promoter activity and gene amplification. The SCLC cell lines responsive to bombesin show constitutive expression of L-*myc*; nonresponsive cell lines express N-*myc* or c-*myc*.

The significance of increased expression of *myc* family genes remains uncertain. Initially, c-*myc* amplification was described in SCLC cell lines with variant morphology. This variant morphology is also called small-cell/large-cell carcinoma and is thought to indicate an unfavorable prognosis. Cell lines with the variant morphology have relatively more resistance to chemotherapy and radiation therapy, but a study of pathologic specimens of patients with extensive disease SCLC showed that the variant cell type was rare, occurring in only 4.4% of 550 specimens. There were no significant differences in resopnse rates to chemotherapy or prognosis for patients with ''classic'' and variant morphology SCLC. Amplification of the c-*myc* gene was more frequent in cell lines from SCLC patients with tumor relapse than in untreated patients. Amplification of c-*myc* was associated with shorter survival in relapsed patients. It is likely that increased *myc* expression leads to progression of SCLC, but it is unlikely to be a primary event because it is detected in only a minority of tumors. Clarification of the association of increased *myc* expression with the variant cell type and the significance of this cell type requires additional study. Increased *myc* expression may occur by several mechanisms and is not always associated with gene amplification. Alterations in *myc* expression in NSCLC have not been extensively studied, but in one case several NSCLC showed increased expression of c-*myc*.

### 1.4 Tumor Suppressor Genes

It appears that certain gene products must be present if controlled cell growth is to be maintained. The inactivation or loss of certain genes may thus contribute to tumor growth. Both copies of the gene must be eliminated or inactivated to eradicate the growth suppressive function of the gene in the classic model. Because both copies must be eliminated, the tumor suppressor gene is called ''recessive.'' The retinoblastoma (*Rb*) gene was one of the first tumor suppressor genes to be identified. Patients with the familial predisposition have a germ line inactivation of one copy of the *Rb* gene. The tumor develops when the wild-type allele is either inactivated or deleted. Sporadic retinoblastoma cases have somatic mutations or deletions that eliminate expression of the gene product. This model has stimulated studies searching for consistent chromosomal deletions in human lung tumors.

Deletions in the short arm of chromosome 3 (p14–p23) are often present in SCLC. Cytogenetic studies of fresh tumors confirmed observations made on cell lines. Allelic loss in this region was documented with polymorphic DNA probes. Specific suppressor genes at the 3p locus have not yet been identified. The loss of heterozygosity for alleles on chromosomes 3, 11, 13, and 17 occurs in NSCLC as well. The frequency of 3p deletions in NSCLC is controversial: some have found this deletion in all SCLC and NSCLC specimens, whereas others have found it only in a minority of NSCLC specimens. The high frequency of deletions in both SCLC and NSCLC suggests that loss of specific gene function may be a critical step in the development of lung cancer.

Two genes that are tumor suppressor candidates in lung cancer are the nuclear phosphoproteins *p53* and *Rb*. Loss of heterozygosity on chromosome 13q suggests that the *Rb* locus, located at 13q14, may be deleted. Sixty percent of SCLC and 75% of carcinoid cell lines do not express *Rb* messenger RNA. Of 13 SCLC cell lines reviewed in one study, only three expressed more than a trace amount of *Rb* mRNA. The role of *Rb* in NSCLC appears variable. In one study 90% of NSCLC cell lines expressed *Rb*. In contrast, other studies have shown a significantly higher involvement of altered *Rb* function. For example, another study found that *Rb* protein was absent in 10 of 36 immunostained primary NSCLC tumors.

The *p53* gene encodes a 375 amino acid phosphoprotein that can form complexes with host proteins such as large-T antigen and E1B. Missense mutations of this gene are the most common mutation yet identified in lung cancer. The mechanism of *p53* transformation is complex: wild-type *p53* gene may directly suppress or indirectly activate genes that suppress uncontrolled cell growth. Wild-type *p53* is dominant over the mutant form and thus will suppress the transformed phenotype in human lung cancer cells as well as other tumor types. Absence of or inactivation of wild-type *p53* may, therefore, contribute to transformation. However, some studies indicate that mutant *p53* is necessary for full expression of the transforming potential of the gene. The presence of the mutant *p53* gene can confer a growth advantage to some cells in part by inactivating wild-type *p53* by forming oligomers; the mutant also can cooperate with other genes in the transformation process.

Mutations of *p53* are common in a wide spectrum of tumors. These mutations occur in both NSCLC and SCLC cell lines and fresh tumors. Two types of mutation occur: transitions, in which a purine is substituted for a purine or pyrimidine for a pyrimidine, and transversion, in which a purine is substituted for pyrimidine or vice versa. Transversions have been associated with carcinogens such as benzo[*a*]pyrene. Transitions, which have a predilection for CpG dinucleotides (frequently having 5-methylcytosine residues), are indicative of the spontaneous mutation rate. The majority of mutations in lung cancer are G:C-to-T:A transversions distributed over 10 codons. This suggests that tobacco carcinogens have a strong influence in the etiology of these mutations.

Two other genes may be tumor suppressor genes for lung cancer. Expression of the *nm23* gene is reduced in rodent tumor cells wtih a highly metastatic phenotype. The *nm23* gene is located near the centromere of chromosome 17. Allelic deletion was shown in 5 of 12 informative cases, all of which were adenocarcinomas. The protein–tyrosine phosphatase-$\gamma$ (PTP-$\gamma$) maps to 3p21, a region frequently deleted in lung cancers. Five of ten lung cancers studied had evidence of allelic deletion of this gene. These studies suggest that both these genes may act as tumor suppressors. This will be demonstrated definitely, however, only when insertion of these genes into human lung cancer cells is shown to cause reversal of the malignant phenotype.

*See also* CANCER; HUMAN GENETIC PREDISPOSITION TO DISEASE; ONCOGENES; TUMOR SUPPRESSOR GENES.

## *Bibliography*

Bishop, J. M. (1991) *Molecular themes in oncogenesis. Cell,* 64:235–248.

Cross, M., and Dexter, T. M. (1991) Growth factors in development, transformation, and tumorigenesis. *Cell,* 64:271–280.

Jaggi, R., Hock, W., Ziemiecki, A., Klemenz, R. Friis, R., and Groner, B. (1989) Oncogene mediated repression of glucocorticoid hormone response elements and glucocorticoid receptor levels. *Cancer Res.* 49:2266S.

Roth, J. A. (1992) New approaches to treating early lung cancer. *Cancer Res.* 52:S2652–S2657.

Roth, J. A., Hong, W. K., and Mulshine, J. L., Eds. (1992) *Biology of and Novel Therapeutic Approaches for Epithelial Cancers of the Aerodigestive Tract.* National Cancer Institute Monograph No. 13. NCI, Washington, DC.

Vinocour, M., and Minna, J. D. (1989) Cellular and molecular biology of lung cancer. In J. A. Roth, J. C. Ruckdeschel, and T. H. Weisenburger, Eds. *Thoracic Oncology.* Saunders, Philadelphia.

## Magnetic Field Reception: *see* Electric and Magnetic Field Reception.

# Major Histocompatibility Complex

*James Driscoll*

*Key Words*

**Antigen** A foreign substance such as a virus that elicits an immune response.

**Cellular Immunity** Immunity rendered by intact cells that recognize the nonself antigen. Cellular immunity combats invading pathogens and is mediated through molecules interacting at the cell surface.

**Cytotoxic T Cell (CTL)** A type of T lymphocyte that recognizes cells that display foreign antigens on the surface. Following recognition, the T cell kills the recognized cell, hence the alias, killer T cell.

**Helper T Cell ($T_h$)** A type of T lymphocyte that assists B cells in their antigenic response. Helper T ceils fulfill a requisite function in the stimulation of B cells.

**Humoral Immunity** A form of immunity to combat invading pathogens. Humoral immunity refers to immunity caused by molecules (e.g., antibodies) in solution. Cells of the humoral system (e.g., B cells) secrete antibodies that bind to antigens derived from the pathogen.

**MHC Class I Molecule** A type of MHC-encoded gene product that functions in self-recognition. The class I molecule consists of two polypeptide chains; one is highly polymorphic and the other is a constant polypeptide called β-2 microglobulin. The class I molecule is found on the surface of nearly all cell types.

**MHC Class II Molecule** A type of MHC-encoded gene product that functions in antigen recognition. The class II molecule consists of two polymorphic chains and is found on the surface of B lymphocytes and macrophages.

Higher vertebrates have developed a system to provide a rapid and specific immunological response to counter disease-causing agents. The major histocompatibility complex (MHC) is an immense multigene cluster that occupies nearly 4000 kilobase pairs of DNA. The MHC-encoded proteins fulfill an essential role in the cellular immune response to distinguish between ''self'' and ''nonself'' proteins. Foreign proteins can originate either from tissue transplants or through pathogenic infection. The numerous MHC-encoded proteins are present on nearly all cell types of the body and usually are located at the cell surface. These gene products act in concert with the gene products of the antibody and T-cell genetic loci to provide the selective recognition and efficient removal of infected cells. Recent discoveries in understanding the MHC at the molecular level have had significant impact on unraveling how the body develops cellular immunity. Advances in understanding the structure and function of components of the MHC have profound importance for the disciplines of cell biology, molecular biology, and medicine.

## 1 GENERAL FEATURES OF THE MHC

Tissue transplantation experiments demonstrated that if skin from an animal from one inbred strain were grafted onto the back of an animal from a separate strain, the grafted skin died soon afterward. The graft from the donor animal was said therefore to have been rejected by the recipient. To determine which proteins and genes were responsible for this process of graft rejection, experimenters used inbreeding to systematically reduce the genetic differences between the two strains. This method allowed researchers to define a narrow genomic region (or complex) as being responsible for eliciting the rejection of a tissue or skin graft. Since this gene region was responsible for histocompatibility, the cluster of genes was called the major histocompatibility complex (MHC). Genes encoded within the MHC include some that are responsible for providing a rapid graft response as well as others that are involved in different processes.

## 2 GENES OF THE MAJOR HISTOCOMPATIBILITY COMPLEX

The advent of sophisticated techniques in molecular genetics has greatly facilitated investigation of the MHC, an extraordinarily large gene cluster that occupies approximately 2000–4000 kilobase pairs of DNA within the genome. This multigene complex encodes a vast number of gene products and is considered to be an integral part of the immune response system. Genes of four different types are encoded within the MHC. The first set, discovered more than 50 years ago, consists of the transplantation antigens responsible for self-recognition. This group includes the class I MHC genes, which encode the targets for recognition of ''self'' versus ''nonself.'' These genes are called *H-2D* and *H2-K* in mouse and *HLA* in humans. The resulting protein encoded by the class I MHC gene is a two-chain molecule, one chain being a highly polymorphic or variable polypeptide and the other a low molecular weight constant polypeptide called β-2 microglobulin. The class I MHC molecule works coordinately with a specialized type of T cell (the cytotoxic T cell or CTL). Following the generation of an antigen, the presentation of peptide antigen by the class I MHC molecule allows for MHC class-I-restricted CTL to survey cells for the expression of "nonself" or foreign viral proteins. The CTL destroy such cells to prevent the further growth of either infected or aberrant cells.

The second set of MHC-encoded genes contains the immune response (Ir) genes, which greatly determine the quality or strength of an immune response. The Ir genes encode the class II MHC molecules, which like the class I MHC molecules, assist in distinguishing self from nonself. Class II MHC molecules have two highly polymorphic polypeptide chains and are found mainly on two specialized cell types, macrophages and B lymphocytes. Similar to the class I MHC molecules, the class II MHC molecules also work coordinately with a specialized type of T cell, called the helper T cell ($T_h$). Class II MHC molecules present peptides derived from exogenous antigens to the MHC class-II-restricted $T_h$ cells. Furthermore, $T_h$ cell recognition of an antigen is restricted by the structure of the class II MHC molecule, analogous to the restriction of CTL recognition by the structure of the class I MHC molecule.

The third set of MHC genes, the *Tla* and *Qa* genes, like the other sets of MHC genes, encode cell surface proteins. The *Tla* and *Qa* gene products are structurally similar to the class I transplantation antigens, but at present their function is unknown.

The fourth set of MHC-encoded genes encode the many components of the "complement system," which functions in the elimination of certain antigens. Upon the binding of certain complement proteins to antigen–antibody complexes, the complement proteins effectively destroy the antigen.

## 3 THE ROLE OF THE MHC IN CELLULAR IMMUNITY

The body has developed two distinct mechanisms for immunity: humoral immunity, which is mediated by soluble antibodies that circulate in the body fluids, and cellular immunity, which is accomplished by surface receptor molecules and is mediated by T cells. For cellular immunity to exist, there must be recognition of antigens at the surface of other cells. In fact, the T cell recognizes not only the antigen but also a cell surface protein (the MHC class I or class II molecule) that is presenting the antigen at the cell surface. The cellular response to intracellular or extracellular antigens presents challenges regarding recognition and the appropriate response. Parallel systems have evolved to meet these challenges. Peptides derived from intracellular antigens are normally presented to $CD8^+$ cytotoxic T cells by class I molecules, while peptides from extracellular antigens are presented to $CD4^+$ helper T cells by the MHC class II molecules.

It is now generally accepted that the MHC class I and class II molecules present fragments of antigens, in the form of short peptides, to the T cell. The class I antigens appear to be fragments of proteins generated by endoproteolytic events that originate within the cytoplasm. The resulting peptides are transported subsequently through the endoplasmic reticulum to the cell surface, where they are presented in association with the appropriate MHC-encoded accessory molecules. The MHC class I molecules can be considered to be a cellular device for the presentation of samples of endogenous or foreign proteins for surveillance by the specialized T cells, the CTL. These lymphocytes then function to lyse the infected cell.

Breakthroughs at the molecular level have provided a new understanding of the MHC and, consequently, its role in defending the body against infection. The MHC-encoded proteins all appear to be predominantly highly polymorphic cell surface proteins. Thus a common function of these proteins is to serve as signals for recognition by other cells in the immune system. The MHC, along with the antibody locus and the T-cell locus, comprise the components of the immune response system. Surprisingly, certain MHC-encoded proteins (the class I and class II gene products) bear a striking structural similarity to antibodies. These findings, among others, support the notion that all components of the immune response system have evolved from a single primordial gene to generate an extended immunoglobulin (super)gene family.

*See also* AUTOANTIBODIES AND AUTOIMMUNITY; IMMUNOLOGY; RECOGNITION AND IMMOBILIZATION, MOLECULAR.

### *Bibliography*

Benacerraf, B., and Unanue, E. R. (1979) *Textbook of Immunology.* Williams & Wilkins, Baltimore.

Eisen, H. (1980) *Immunology.* Harper & Row, New York.

Hood, L., Steinmetz, M., and Malissen, B. (1983) Genes of the major histocompatibility complex of the mouse. *Annu. Rev. Immunol.* 1:-529–568.

———, Weissman, I. L., Wood, W. B., and Wilson, J. H. (1984). *Immunology.* Benjamin/Cummings, Menlo Park, CA.

Kimball, J. W. (1983) *Introduction to Molecular Immunology.* Macmillan, New York.

Klein, J. (1986) *Natural History of the Major Histocompatibility Complex.* Wiley, New York.

Nisinoff, A. (1985) *Introduction to Molecular Immunology.* Sinauer, Sunderland, MA.

# MAMMALIAN GENOME

*John Schimenti*

### *Key Words*

**Genome** The DNA contained in all the chromosomes of an organism.

**Genome Project** Scientific undertaking by many countries to ultimately determine the DNA sequence of each chromosome (from humans and other organisms), to find all the genes in the genome, and to determine how they function to form an animal.

**Homology** In the genome context, relatedness between DNA sequences as measured by percentage similarity at the nucleotide level.

**Illegitimate Recombination** Any type of recombination that results in a net gain or loss of DNA, or in which nonallelic sequences recombine.

**Imprinting** The phenomenon in which a gene is active when transmitted by the parent of one sex but not the other.

**Repetitive Sequences** DNA sequences, usually between 150 and 5000 base pairs, which are present in hundreds to thousands of copies throughout the genome and are nonessential to the biological function of that organism.

About 65 million years ago, a cataclysmic event, thought to be an asteroid colliding with the Earth, led to the demise of the dinosaurs and the rise to eminence of the eutherian mammals. The rapid speciation that followed is known as the Mammalian Radiation.

Since 65 million years is a relatively short period of time in evolutionary terms, the salient features of the mammalian genome are shared among the entire class. Nevertheless, the rapid, dramatic diversification of mammals is reflected in tumultuous genetic strategies in which new genes for new jobs are rapidly created, and old genes rapidly recruited to do new things.

Mammals are among the most complicated creatures in existence, and this complexity is reflected by the mammalian genome. It is much larger than that of many other animals, and it contains a relatively large number of genes. These complexities have driven the development of new molecular technologies, such as ways to clone larger pieces of DNA, to generate ''designer'' mutations, and to physically map DNA molecules 100 times larger than just a decade ago. A coordinated effort is now under way to define all the genes in mouse and man.

## 1 CHROMOSOMES OF MAMMALS

Most mammalian species have between 40 and 60 chromosomes (humans 46; mice 40; cows 60; chimps 48; rabbits 44). The gross organization of mammalian chromatin is similar to that of all eukaryotes. That is, each chromosome is composed of an extremely large molecule of DNA, which is packaged by histones into nucleosomes (of about 140 nucleotides each), which are further compressed by higher levels of coiling. The mammalian haploid genome consists of approximately $3 \times 10^9$ bp of DNA.

The gross appearance of chromosomes between species may differ with respect to the centromere location. For example, all the mouse chromosomes are acrocentric (the centromeres are at one end of the chromosome), whereas human chromosomes are metacentric, having distinctive short and long arms. Nevertheless, there is substantial *synteny*—conservation of entire subchromosomal segments having similar/identical gene orders—between chromosomes of different species. The identification of syntenic regions is a major effort in the ongoing Genome Project (Section 7), and comparative efforts help define the genetic rearrangements in the divergence of species.

Sex in mammals is determined by the presence or absence of a Y chromosome. Normal males are XY. Normal females are XX. XO individuals are female. A single gene on the Y chromosome, called the testes-determining factor (Tdf), is alone responsible for making an embryo into a male. Although females (XX) have a double dose of X genes, they generally produce the same amount of gene product as males (XY) as a result of X-chromosome inactivation. In most cells of the female body, one of the X chromosomes is transcriptionally inert and can be visualized cytologically as a compacted piece of chromatin called a Barr body.

## 2 CATEGORIES OF DNA

The DNA of mammals can be grouped simplistically into two classes: functional and nonfunctional sequences. DNA elements of several types are included in each class. The following generalizations must be taken cautiously, since DNA sequences that are now considered nonfunctional may turn out to play roles heretofore unrecognized.

### 2.1 Functional DNA

The most obvious class of functional DNA consists of the polypeptide-encoding genes. Many genes fall into a category of DNA referred to as ''single-copy'' sequences. Strictly interpreted, single-copy DNA is a stretch of sequence that is unique in the genome. Uniqueness is often experimentally defined by technical means such as DNA hybridization (Southern blotting). However, many mammalian genes arose via duplication of an ancestral gene. The products of recent duplications are highly homologous, whereas more ancient duplications display lower sequence similarity owing to evolutionary divergence. In many cases, continuous duplication events have produced families of related genes (the dozen or so globin genes, for example), or ''superfamilies'' (as in the case of the immunoglobulin genes and relatives, which contain hundreds of members). Yet, in general, most molecular biologists do not consider a member of a gene family to be ''repetitive'' DNA (see Section 2.2), a term that implies lack of function. Some sequences, such as satellite DNA sequences, fall into a gray area: while they are present in thousands of copies, and are not genes in a classical sense (they don't encode polypeptides), they may play a role in chromosome structure.

Nontranscribed regulatory sequences, such as promoters and enhancers, are an important class of functional DNA. These sequences are usually closely associated with a gene, often in the 5′ flanking region, such that a discrete fragment containing all the regulatory sequences can be physically isolated and operationally defined by transfection studies or in transgenic animals. However, important regulatory signals can be located at considerable distances from the cognate gene, embedded within nonfunctional intergenic DNA.

Other types of DNA that are more difficult to functionally characterize are those that play roles in chromosome structure (aside from the centromere and telomere) and replication.

### 2.2 Nonfunctional DNA

The most intriguing class of presumably nonfunctional DNA in the mammalian genome contains the repetitive sequences. Repetitive DNA was originally discovered as a significant proportion of total genomic DNA that undergoes rapid reannealing in solution following denaturation. We now know that such sequences are often quite small (150–300 bp) and are present in hundreds of thousands of copies in the genome. In humans, the Alu sequence is the predominant family of repetitive DNA. Alu sequences are about 300 bp in length, and about 500,000 copies exist in the genome. Rodents have a related 150 bp repetitive sequence known as the B1 element. The majority of repetitive sequences appear to have entered and spread into genomes following the Mammalian Radiation, since most mammals have unique classes. These types of small repetitive sequences are called ''SINEs'' (Small Interspersed Elements). Another form, the LINEs (Long Interspersed Elements), range up to several thousand base pairs in length. Because repetitive elements are so prevalent in mammalian genomes, it is thought that they are mobile and have a mechanism of replicating and spreading. A LINE in humans called L1 appears to encode a reverse transcriptase type molecule, which suggests that the LINEs spread in the genome via self-reverse transcription.

Partly through analysis of the ever-growing DNA sequence databases, new repetitive sequences continue to be identified, especially in humans. One such class is known as the MER (medium reiteration). MERs are generally present in a few hundred to few thousand copies in the human genome.

The classes of repetitive elements described here have no known biological function, although they may have played various evolutionary roles or may influence the expression of some genes. A generally accepted distinction between repetitive elements and a gene family is that repetitive elements do not encode a necessary function. Many consider them to be genomic parasites, or "selfish" DNA.

Aside from the protein-coding and repetitive sequences, much of the mammalian genome consists of introns and intergenic DNA. Little study is devoted to the analysis of intergenic DNA, since there is limited interest in sequences that do not have obvious function. However, this DNA may contain sequences important in gene regulation or chromatin structure. Additionally, while particular regions of DNA may not encode any proteins or regulatory signals, they may provide spacing of genes, which is necessary for appropriate expression in the context of nucleosome organization.

The introns of most genes are generally much larger than the exons. Although much of the DNA sequence contained within introns is dispensable, there are several cases of introns containing enhancer elements for gene expression. There is also substantial reason to believe that introns are evolutionarily important, having permitted the segmental construction of genes from composite parts (exons), for example.

## 3 GENOME EVOLUTION

The mammalian genome is remarkably unstable. It is clear that genomes evolve not by a slow, gradual process involving subtle nucleotide mutations, but rather by dramatic, saltatory events. Fossil evidence for this is mirrored at the molecular level.

The gross chromosome complement or karyotype of a species can be altered rapidly by a major chromosome rearrangement such as a translocation or inversion. Reciprocal translocations result in the overall structure of two chromosomes being changed. This exchange does not alter the DNA content of the organism. However, such events are thought to be critical initiators in the divergence of species. If members of a population become physically separated, and one or both fix a translocation event(s), hybrids formed by intermating either will be sterile or will produce aneuploid gametes, a generally lethal outcome. Chromosome fusions make one larger chromosome out of two smaller ones. There is one major chromosomal difference between humans and chimpanzees. Humans have 23 chromosomes in the haploid set, whereas chimpanzees have 24. The difference is due to the fusion of two chromosomes in the human lineage. A dramatic example of tumultuous karyotypic change is revealed by comparison of the mouse and human chromosomes. It is thought that about 150–200 gross chromosomal rearrangements have occurred since their divergence during the Mammalian Radiation.

On a smaller scale, probably the most significant method of rapid genomic change is associated with illegitimate recombination, described in Section 4.3. This activity is responsible for the immediate creation of new genes, which can then undergo further evolution by selection of point changes. When new genes are created by duplication, the extra copies become material for evolutionary "experimentation." Duplications can involve single genes, gene clusters, chromosomal subregions, or entire chromosomes. Comparative analysis of chromosome linkage maps from various mammalian species is beginning to clarify the major genetic changes in chromosome structure that followed divergence.

## 4 RECOMBINATION

### 4.1 Crossing Over

Genetic recombination occurs both in somatic and germ cells of mammals. Crossing over during meiosis generally occurs about once per chromosome; double crossovers during a single meiosis are very infrequent. In mice and humans, the meiotic recombination maps are "longer" in females, meaning that crossing over may be more frequent in the female sex.

There have been several examples of recombination "hot spots" in mice, in which more crossovers occur than would be expected in a given chromosomal interval of known physical size. Conversely, there are regions known as "cold spots," where recombination events occur at a frequency that is lower than expected. The causes for such anomalies are not well understood, although some hot spots have sequence homologies to prokaryotic recombination sequences.

### 4.2 Somatic Recombination

Evidence that somatic recombination occurs in mammals has arisen relatively recently. However, it is now clear that somatic recombination can play an important role in certain diseases. In particular, it can cause a heterozygous cell to become homozygous for a mutant tumor suppressor gene, such as retinoblastoma or *p53*, which can initiate carcinogenesis.

The immune system depends on somatic recombination to generate antibody diversity. Subregions of immunoglobulin genes present in a linked array undergo recombinational rearrangement to generate mature immunoglobulin genes. The randomness of the recombination event, coupled with the large numbers of potential recombinants, underlies the vast diversity our immune systems can generate.

### 4.3 Illegitimate Recombination

"Illegitimate recombination" is a term often used interchangeably with "unequal recombination." It is often perceived as any form of recombination other than typical crossing over. However, illegitimate recombination is a very active process in lower organisms, and emerging data indicate that this category of recombination is highly active in mammals as well.

Homologous but unequal crossing over between chromosome homologues or sister chromatids generates two nonparental chromosomes: one that gains DNA and another that loses DNA. This type of recombination is largely responsible for gene duplication, which is a major molecular mechanism in the evolution of higher organisms. It recruits preexisting genetic material as a substrate for the formation of novel functional units. Extra gene copies created through duplication may ultimately diverge to perform related but specialized developmental and biochemical function. Duplication of subgene fragments, followed by "exon shuffling," is thought to have created the present-day pool of genes from only 1000–7000 exons. Exons often represent functional subunits of proteins.

Gene conversion is a major form of illegitimate recombination. It is the nonreciprocal transfer of genetic information between two related genes or DNA sequences. These sequences can be anywhere in the genome—at allelic positions, adjacent on the chromosome, or on different chromosomes. Gene conversion can influence the evolution of gene families, having the capacity to generate both diversity and homogeneity, depending on the direction of DNA transfer and the number of independent sequences undergoing the exchanges. While gene conversion has long been known to be a frequent event in fungi, only recently have transgenic mouse studies shown that gene conversion is also an active process in the germ line of mammals. Comparative sequence analysis also provides evidence that gene conversion has shaped the evolution of many mammalian gene families. Somatically, gene conversion is another potential mechanism for the reduction to homozygosity of mutant tumor suppressor genes in a heterozygote.

## 5 TECHNIQUES FOR MAMMALIAN GENETIC ANALYSIS

### 5.1 Gene Mapping

Techniques for mammalian genome mapping have progressed dramatically in the past decade. In the first half of the century, visible traits were the only available gene markers. In mice, genes affecting coat color were among the most noticeable traits that were placed onto linkage groups. A major step in the improvement over visible trait mapping was the utilization of molecular markers such as isozyme variants and protein isoforms.

A breakthrough advance was the concept of using random DNA probes to detect genetic heterogeneity in the form of restriction fragment length polymorphisms (RFLPs). RFLPs essentially bypass the requirement for genes as genetic markers. Loci detected by DNA probes could be genetically linked to visible traits. A dramatic example of this technology in human genetics was the discovery of a DNA probe linked to the gene for Huntington's disease. In the past few years, other forms of DNA markers have been employed, which have further simplified genome mapping. Most notable is the application of the polymerase chain reaction (PCR) to identify size differences between loci containing simple sequence repeats (SSRs). SSRs are repeating units of di- or trinucleotides, which are highly unstable and therefore vary in repeat number between individuals or mouse strains. PCR primer pairs corresponding to thousands of SSR-containing mouse and human loci have been identified and used to both map new mutations and to create sophisticated genetic maps of the mouse and human genomes.

The mouse has been, and continues to be, the optimal organism for genetic analysis of mammals. Primary reasons for this are its short breeding time (21-day gestation), the large number of characterized mutations, our ability to introduce new genes and specifically mutate particular genes (Section 6.3), and the existence of numerous inbred (genetically pure) strains of mice. Genetic mapping is simplified by the ability to rapidly generate unlimited numbers of offspring. The *interspecific backcross,* a breeding scheme in which the $F_1$ progeny of two distantly related strains are crossed back to one parent, results in the production of animals that contain crossovers between the chromosomes of the original parents. Since those parents are distantly related (generally of the species *Mus musculus* and *M. spretus*), a high percentage of DNA loci will be polymorphic, that is, will display a molecular difference by either RFLP analysis or with SSR analysis.

### 5.2 Physical Mapping

''Physical mapping'' is the term used to describe molecular characterization and localization of DNA sequences along chromosomes, without respect to visible traits or genes. Such a characterization includes, for example, a restriction site map of a large region, and a determination of the number of nucleotides separating DNA loci. For example, genetic mapping might define close linkage between two DNA markers on an autosome if RFLPs detected by those probes were observed to recombine in heterozygotes less than 1% of the time. Using physical mapping technologies (some techniques are described in Section 7.1), it is determined that these sequences lie 1000 kilobases apart. This information alone would be considered part of a physical map. In mice, 1 centimorgan (cM) corresponds to about 1600 kb.

Techniques for physical mapping have improved rapidly in the past decade. Early methods involved ''chromosome walking,'' or the isolation of overlapping bacteriophage or cosmid clones corresponding to a region of a chromosome. A major advance was the development of pulse field gel electrophoresis, which makes possible the electrophoretic separation of DNA fragments up to about 10,000 kb. This allows the generation of restriction maps spanning large genomic regions. Finally, yeast artificial chromosome (YAC) vectors, which allow cloning of up to 1000 kb of DNA, greatly increased the speed in which ''contigs,'' or contiguous regions of cloned DNA, can be generated.

### 5.3 Positional Cloning

The genetic and physical data obtained by methods such as those just described allows the precise localization of genes responsible for visible traits. Visible traits include human disease genes and mutations that cause physical abnormalities. When such a gene is found to genetically map between two DNA markers, physical mapping can be used to determine exactly how far apart those markers are. YAC cloning can then be applied to isolate all the DNA between those markers. This cloned DNA is then screened for transcriptional units and molecular lesions associated with the mutation. This approach has been responsible for the isolation of numerous disease genes, including cystic fibrosis, muscular dystrophy, and Huntington's disease.

## 6 MAMMALIAN GENOME FUNCTION

### 6.1 Imprinting

The imprinting phenomenon in mammals has to do with the observation that some genes must be inherited intact from a parent of a certain sex. A classic example of this exists for a gene in the mouse known as the *t maternal effect* (*Tme*) gene. If a father is heterozygous for a mutation at *Tme,* and passes on the mutant copy to its offspring, all the pups inheriting the mutation are normal. However, if a female transmits the mutant allele, those embryos die during development. This means that embryos require a functioning copy of this gene from a female gamete to survive; a functioning copy from the father cannot compensate. The explanation for such phenomena is that the paternally inherited gene is transcriptionally inactive, whereas the maternal copy is active. Evidence exists that differential DNA methylation in the female and male germ lines is responsible for imprinting effects.

### 6.2 Transgenic Mice

The mouse is the organism of choice for the examination of mammalian genome function. One very powerful technology has been DNA microinjection into single-celled mouse embryos. A percentage of the injected embryos will contain an integration of the DNA into one of its chromosomes. The resulting ''transgenic'' mice transmit the gene in a Mendelian fashion. This technique allows investigations into gene regulation and the effect of introducing novel genes into particular tissues, and the characterization of the expression patterns of genes during development. Significantly, most human genes are expressed appropriately when introduced into mice.

### 6.3 Targeted Mutagenesis

A revolutionary technology allows scientists to create mutations in any gene in the mouse genome. At the cornerstone of the technique of targeted mutagenesis is the ability to grow, in culture, pluripotential embryonic stem (ES) cells. These cells, which can be injected into early mouse embryos, give rise to all adult tissues, including the germ line. To mutate a gene, ES cells are transfected with a DNA construct containing homology to an endogenous gene. In some of the transfected cells, the introduced DNA recombines with the genomic counterpart, creating a mutation dictated by the construct design. Those cells are injected into embryos to generate mice containing the mutation. Breeding analyses are then performed to study the effect of the mutation in the animal.

## 7 THE GENOME PROJECT

The Genome Project, currently administered by the National Institutes of Health and the U.S. Department of Energy, represents one of the most ambitious scientific endeavors mankind has ever embarked upon. To quote from a Department of Energy program report, ''Acquiring complete knowledge of the organization, structure, and function of the human genome—the master blueprint of each of us—is the broad aim of the Human Genome Project.'' Obviously, this is a massive undertaking. The overall strategy can be broken down into three stages at this point: mapping, sequencing, and functional analysis.

### 7.1 Mapping

The first major goal of the genome project is to generate high resolution genetic and physical maps of the human genome and those of other experimental organisms. Using molecular markers described earlier, such as RFLPs and SSRs, loci have been defined along the length of each chromosome, with no more than a few centimorgans between them, such that any new DNA fragment or gene can be clearly linked to one of these markers. Markers such as the SSR provide what is known as a sequence-tagged site (STS), which means that the locus is associated with a small amount of DNA sequence information. Therefore, researchers around the world can chemically synthesize a probe for these loci, rather than request it from the originator.

### 7.2 Sequencing

Ultimately, it is hoped that the entire DNA sequence of the human and mouse genomes will be determined. However, the vast size of the mammalian genome—$3 \times 10^9$ bp—essentially prohibits the establishment of the complete sequence using standard technology in a reasonable amount of time, even with current automated sequencers.

The major impediments lie in three areas: speed of sequencing, duplicative sequencing, and computer assembly of sequence. Current methods require the use of polyacrylamide gels. In practice, no more than 400–500 bp can be reliably determined from a single template on such a gel. Making and handling these gels is tedious. Therefore, much of project funding is earmarked for the development of new sequencing technologies.

Duplicative sequencing, or the sequencing of a piece of DNA more than once, occurs in ''shotgun'' approaches. In a shotgun approach, DNA from a source (whole genomic DNA, a YAC, or cosmid clone), is randomly cloned and sequenced. To obtain a complete sequence, most pieces of DNA are sequenced repetitively.

Finally, rather sophisticated computing power and software will be required to assemble all the sequence into ''contigs,'' that is, to arrange it in the order in which it actually exists in the chromosomes. This is because sequencing does not proceed unidirectionally from one end of the chromosome to the other; rather, it moves in fits and starts all over the genome. The Genome Project seeks novel technologies to alleviate these problems.

### 7.3 Functional Analysis

It is thought that the mammalian genome contains about 50,000–100,000 genes. The most daunting goal of the Genome Project, and biology in general, is to identify all these genes and determine what they do. This endeavor will likely require at least several decades.

One increasingly popular method of identifying genes is the cDNA approach. This involves sequencing short regions complementary DNA clones isolated from a library of a particular tissue such as brain. This provides an STS for that particular gene, along with enough sequence information to determine whether the cDNA is a ''new'' gene. If this is done for enough tissues, including stages of the developing embryo, it will be possible to determine which genes are specific to certain tissues, and which are ubiquitous. The STS can also be used to map the gene. Ultimately, when the entire genomic DNA sequencing has been determined, mapping of a cDNA STS will be automatic. Another consideration, however, is that computer programs may be able to identify genes from genomic DNA sequence, although it is unlikely that enough information would be made available to characterize the expression pattern of a gene.

The ultimate way to identify the function of a gene is to study the effects of a mutation. Although the technology exists to make specific mutations of genes (Section 6.3), this is currently a labor-intensive process that is clearly impractical for every gene in the genome. For this reason, new methods for more global assignment of gene function must be developed.

*See also* DNA Markers, Cloned; Genomic Imprinting; Mouse Genome; Recombination, Molecular Biology of.

*Bibliography*

Hogan, B., Constantini, F., and E. Lacy. (1986) *Manipulating the Mouse Embryo.* Cold Spring Harbor Laboratory, Cold Spring Harbor, NY.

Kucherlapati, R., and Smith, G. (1988) *Genetic Recombination.* American Society for Microbiology, Washington, DC.

Li, W.-H., and Graur, D. (1991) *Fundamentals of Molecular Evolution.* Sinauer, Sunderland, MA.

O'Brien, S., and Seuanez, H. (1988) Mammalian genome organization: An evolutionary view. *Annu. Rev. Genet.,* 22:323–351.

Suzuki, D., Griffiths, A., Miller, J., and Lewontin, R. (1989) *An Introduction to Genetic Analysis.* Freeman, New York.

## Maps and Mapping: *see type, e.g.,* Body Expression Map of the Human Genome; Gene Mapping by Fluoresence In Situ Hybridization; Human Linkage Maps; etc.

# Mass Spectrometry of Biomolecules

*Raymond E. Kaiser, Jr.*

*Key Words*

**Electrospray Ionization (ESI)** A method for ionization in which a solution is nebulized at high voltage to form multiply protonated molecules (epg., protein/peptides) or deprotonated molecules (epg., oligonucleotides).

**Matrix-Assisted Laser Desorption Ionization (MALDI)** A method for ionization of large molecules, like proteins, using an ultraviolet laser beam. A UV-absorbing matrix, like sinnapinic acid, is mixed with the analyte to aid in vaporization.

**Quadrupole Ion Trap Mass Spectrometer** A simple and inexpensive mass spectrometer constructed of three electrodes (two end cap electrodes and a ring electrode), which are assembled to form a three-dimensional trapping area in which ions are held by radio frequency voltages. This instrument is capable of high sensitivity and high resolution mass analysis in addition to performing tandem experiments within a single mass spectrometer.

**Quadrupole Mass Spectrometer** Instrument for measuring mass-to-charge ratios dependent on an ion's trajectory within a radio frequency and dc field.

**Tandem Mass Spectrometry (MS/MS)** The process by which an ion is isolated by one mass spectrometer followed by collisional dissociation with an inert gas and finally mass analysis in a second mass spectrometer. This method is used for obtaining structural information.

**Time-of-Flight Mass Spectrometer** Instrument for measuring mass-to-charge ratios dependent on the flight time of an ion.

Mass spectrometry is a powerful analytical technique that is often used to identify unknown compounds, to elucidate structural properties of molecules, and to quantify materials. The basic function of a mass spectrometer is to separate and measure ions by their mass-to-charge (*m/z*) ratios. In terms of biomolecular analysis, mass spectrometry offers the ability to provide accurate molecular weight information on subpicomole amounts of peptides and proteins, independent on whether the molecule has been chemically modified. Mass spectrometry is also able to give information on such complex mixtures as proteolytic digests of proteins. Finally, mass spectrometry, when used as a tandem technique, is able to yield partial to complete primary sequence information on peptides.

Mass spectrometry has historically been limited to the analysis of small, volatile molecules. Until recently, this powerful analytical technique has been virtually excluded from the rapidly growing fields of protein chemistry and molecular biology. Limitations in the mass range of mass spectrometers, and difficulties in inducing large biomolecules into the gas phase and in overcoming the great complexity of the instrumentation, have been obstacles to the integration of mass spectrometry into the world of molecular biology. In particular, two ionization techniques, electrospray ionization and matrix-assisted laser desorption ionization, have cocontributed to the fascination in mass spectrometry for biomolecular analysis.

## 1 INSTRUMENTATION

### 1.1 Evolution of the Present State-of-the-Art

Ionization, the first step in obtaining a mass spectrum, is followed by mass analysis and finally data acquisition. The pivotal innovations in mass spectrometry in the recent past have been related to the ionization process, with a few exceptions. For general use, the most common ionization processes are electron ionization and chemical ionization; however, neither is amenable to use with proteins and peptides because of the large amounts of internal energy transferred to the molecule, causing extensive fragmentation, which reduces the amount of useful information obtained for mass analysis of biomolecules. Another difficulty during the ionization of proteins and peptides lies in effectively exciting the molecule into the gas phase, which is difficult due to the low volatility of this class of molecule. Heating the sample is not feasible because of thermal degradation. Clearly, biomolecular mass spectrometry had numerous hurdles to overcome.

Two ionization processes established the groundwork for the recent innovations in biomolecular mass spectrometry: plasma desorption mass spectrometry (PD-MS), where intact proteins are desorbed from a surface like nitrocellulose with a decay product of californium-252, and fast atom bombardment mass spectrometry (FAB-MS), where molecules are dissolved in a liquid matrix and desorbed from a surface with an energetic beam of xenon atoms or cesium ions. PD-MS arrived on the scene in 1974, but the first mass spectrum of a protein, bovine insulin, was not obtained until 1982. FAB-MS arrived in 1983. Both techniques equally helped to demonstrate the applications of mass spectrometry in the area of biomolecular analysis in the 1980s. The problem with both ionization methods was that only a very small family of proteins could be analyzed—those having molecular weights below a few tens of thousands of daltons. Unfortunately, the majority of proteins of interests fall well above this mass range, limiting the usefulness of mass spectrometry during the preceding decade. Also, the sensitivity of the techniques only approached the usefulness for many protein applications, thereby often excluding their use because of the precious nature of many protein samples. For mass spectrometry to be accepted among protein chemists, a mass spectrometric technique with better sensitivity and the ability to analyze larger proteins was needed.

### 1.2 Matrix-Assisted Laser Desorption Mass Spectrometry

Matrix-assisted laser desorption ionization (MALDI) mass spectrometry is one of the techniques that lifted the past restrictions on biomolecular mass spectrometry. With this technique, proteins greater than 300,000 Da have been analyzed, with theoretically no upper mass limit, and femtomole ($\times 10^{-15}$ mol) sensitivities have been reported. In this method, a protein solution is mixed with a matrix, applied to a probe tip, and dried. The matrix is often a low molecular weight organic acid (e.g., sinnapinic acid) that absorbs at the wavelength of the desorbing photon beam (337 nm for a nitrogen laser). The resonant absorption of light allows for efficient energy transfer from the light beam to the matrix. The matrix is believed to dissipate the energy by ''explosive'' vaporization, carrying along the analyte into the gas phase. Normally, the singly protonated (+1) form of the protein is formed, though higher charged states are often observed, especially with certain select matrices.

The mass analyzer most commonly (though not exclusively) used with this powerful ionization technique has been the time-of-flight (TOF) type, which is highly compatible with MALDI. The main advantage of the time-of-flight mass analyzer is that it has no upper mass-to-charge limit, a requirement that has hindered the analysis of singly or doubly charged proteins with extremely high molecular weights. The second advantage of TOF is that both the laser and the mass analyzer are pulsed techniques. The disadvantage of TOF is its poor resolving power relative to quadrupole or sector mass analyzers. Mass spectrometric resolution is defined as $m/\partial m$, where $m$ is the mass of the analyte and $\partial m$ is the width of the measured ion at half its height. Resolution is only a few hundred at 10,000 Da, meaning that the peak would be about 50 Da wide, thereby limiting the analyst's ability to resolve related components (e.g., sodium/potassium adducts, chemical modifications, or microheterogeneity).

One of the most attractive features of MALDI is that the ionization ability is not dependent on components in the matrix, whether the matrix contains other proteins, salts, or other contaminants. Historically, the requirement for a relatively ''clean'' sample often has prevented the use of mass spectrometry as a routine tool for biomolecular analysis, since most biomolecules are isolated and stored in nonvolatile buffers.

### 1.3 Electrospray Ionization Mass Spectrometry

In contrast to MALDI, electrospray ionization (ESI) is termed a ''spray'' method, where the sample (as a solution) is introduced as a spray into the ion source of the mass spectrometer. The liquid sample emerges from a capillary that is maintained at a few kilovolts relative to its surroundings, whereby the resultant field at the capillary tip charges the surface of the liquid dispersing it by Coulomb forces into a spray of charged droplets. This is the process of nebulization by ''electrospray''; if the nebulization is assisted by a flow of inert gas coaxial to the capillary tip, the process is termed ''nebulizer-assisted electrospray'' or Ionspray (trademark of PE Sciex).

In either form, ESI occurs at atmospheric pressure in a region of large electric field and, unlike the generation of singly or doubly charged ions by MALDI, ESI gives extensive charging of the protein or peptide. For example, a protein like human growth hormone (MW 22, 125 Da) will show an approximate Gaussian distribution of ion intensity with the strongest ion current observed for the $(M+16H)^{+16}$ species, as shown in the spectrum in Figure 1. The extent of charging by protonation has been shown, as expected, to correlate quite well with the number of basic amino acid residues in the protein. Because of this extensive protonation, a protein with a high molecular weight will have a mass-to-charge ratio reasonably amenable to analysis by virtually any mass analyzer. The mass analyzer most commonly used for ESI is the quadrupole, which performs much more favorably than TOF analyzers in terms of resolution. The normal mass-to-charge limits of quadrupole instruments extend up to 4000 mass-to-charge units. Just as basic amino acids can be protonated under acidic conditions, acidic amino acids can be deprotonated under basic conditions. To observe deprotonated proteins, the mass spectrometer must be operated under negative ion conditions, a setup that is especially suitable for analysis of oligonucleotides.

The environment of the sample introduced into the ESI source is not nearly as forgiving as for MALDI. Because the ionization process requires charging the sample of interest under high voltage conditions, the presence of nonvolatile salts cause havoc to the analysis. Using samples containing salts above an estimated $10^{-4}$ N usually results in a ''breakdown'' situation, yielding an erratic or unstable electrospray current, thereby limiting analysis. It is often necessary to desalt samples prior to analysis to ensure elimination of nonvolatile salts. If buffering is required, volatile buffers such as ammonium bicarbonate or ammonium acetate can be used with no apparent loss in performance. Though elimination of salts from a sample is sometimes a hindrance and often is regarded as a major disadvantage of ESI by protein chemists, ESI offers the analyst a distinct benefit over MALDI in terms of an ability to interface high performance liquid chromatography (HPLC) for mass spectrometric analysis. This very powerful application is described in more detail in Section 2.2.

## 2 APPLICATIONS

### 2.1 Characterization of Proteins by Molecular Mass Determination

After isolation and purification of a protein, one of the next challenges for characterization is to determine its molecular mass and deduce the number of subunits, if any. Currently, sodium dodecylsulfate (SDS) gel electrophoresis is the most widely used technique for purity and molecular weight determination of proteins. Simplicity of operation, high resolution (especially by two-dimensional gels), and sensitivity are advantages of gel electrophoresis; however, lack of quantification and poor mass accuracy (ca. ±5%) are severe limitations. Hence the value of such molecular mass determinations is very limited in terms of the identification and structural characterization of proteins.

Using mass spectrometry, it is possible to determine the molecular mass of a protein with an accuracy of 0.1 ±0.005% on samples of only a few picomoles or less. At best, this corresponds to a

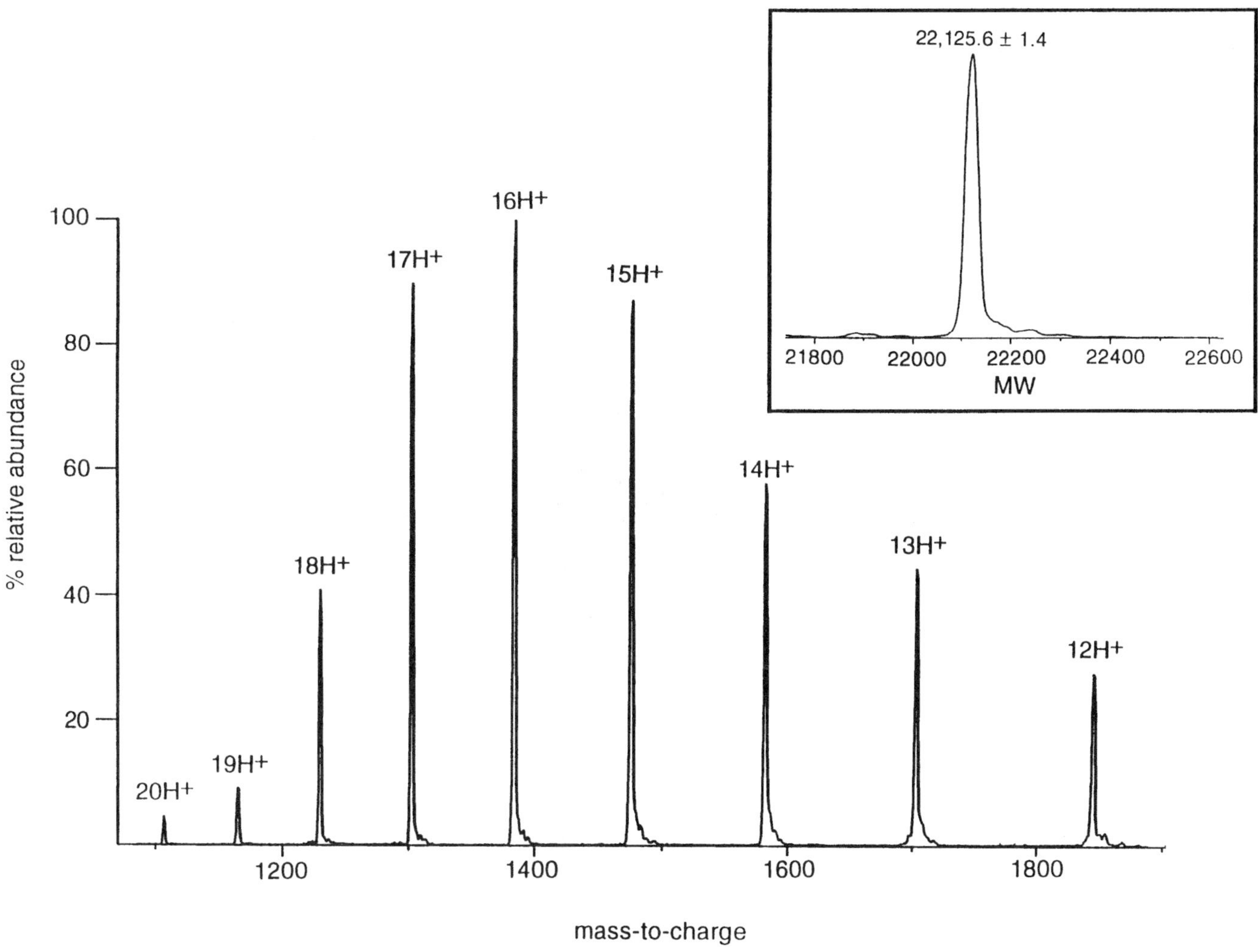


**Figure 1.** ESI mass spectrum of 10 pmol recombinant human growth hormone (calculated average mass is 22,125.1 Da) obtained on a Finnigan triple-quadrupole mass spectrometer. The spectrum shows the multiple protonation phenomenon that is characteristic of electrospray spectra. Insert: computer generated deconvoluted mass spectrum, showing the determined mass to be 22,125.6 ± 1.4 Da.

mass accuracy of 1.5 Da for a 30 kDa protein. This mass accuracy would easily confirm a calculated mass for an expected sequence, which would be instrumental in concluding whether, for example, a protein was cloned correctly, or whether the native protein has undergone posttranslational processing or chemical modification. As illustrated in the insert in Figure 1, the molecular mass of recombinant human growth hormone can be obtained to an accuracy of 22,125.6 ±1.4 Da using an electrospray mass spectrometer (the calculated mass of this protein is 22,125.1 Da). Mutation errors, posttranslational modifications, or chemical modifications can easily be detected with such mass accuracies.

MALDI interfaced to a time-of-flight mass spectrometer is the best alternative for proteins above 80,000 Da. However, the mass accuracy is generally less than ESI by approximately ± 0.05%. Nevertheless, MALDI is 20–100 times more accurate than SDS gel electrophoresis. A large class of glycoproteins are best analyzed by MALDI because of the sometimes extensive amounts of microheterogeneity, which results in a very complex ESI spectrum. With its ability to analyze proteins in complex matrices, MALDI will be one of the most powerful tools available to the protein chemist for quick and simple measurements of molecular mass.

### 2.2 On-Line HPLC-MS for Protein Structure Analysis

Most protein chemists use high performance liquid chromatography for purification and isolation of intact proteins and peptides from complex mixtures. Mass spectrometric analysis of chromatographic fractions has been a central feature in the analytical strategy for characterization. However, there are numerous drawbacks to HPLC fractionation followed by off-line mass spectrometric analysis, such as sample loss due to irreversible adsorption when working at low sample amounts and the time requirements for analyzing numerous HPLC fractions.

Fast atom bombardment and ESI have been used as interfaces for on-line analysis, where the effluent of the HPLC is directly introduced into the ionization source of the mass spectrometer. ESI is the desired technique for on-line analysis of proteins and peptides, and reversed phase chromatography is the desired separation tech-

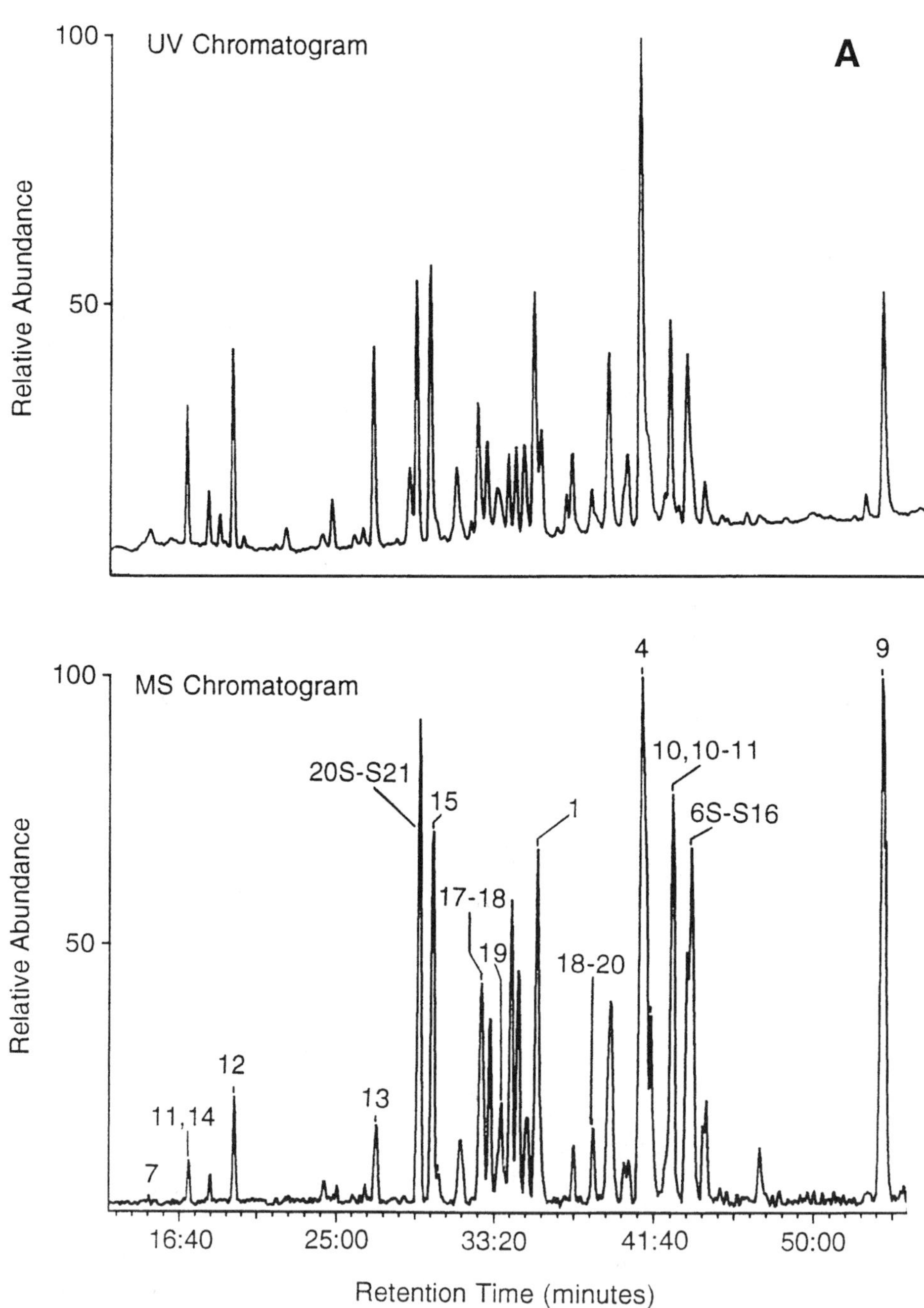


Figure 2. (A) Tryptic digest of 10 μg recombinant human growth hormone. The top HPLC chromatogram is the UV trace and the bottom chromatogram is the ESI LC-MS total ion trace. The numbers on the peaks indicate the tryptic peptides, which were determined by molecular weight. A Vydac C18 (0.46 × 25 cm × 5 μm) column was used with a gradient of 0–50% acetonitrile/0.1% trifluoroacetic acid over 50 minutes at a flow rate of 1 mL/min. The HPLC effluent was split approximately 1:500 to introduce approximately 2 μL/min into the Finnigan electrospray source.

nique. The ideal flow rate introduced into the electrospray source is normally a few microliters per minute; therefore, it is necessary to split the HPLC eluant flow prior to introduction into the source, unless columns are used which are optimized for low flow rates (μL/min) such as fused silica columns. Normally, the eluant not going to the mass spectrometer is directed to a UV detector, which allows correlation of the UV trace with the total ion current trace and analysis of collected fractions by other analytical tests, such as Edman degradation or amino acid analysis.

On-line LC-MS can be used for many applications, such as mapping recombinant proteins. Figure 2 illustrates the UV trace and the total ion current of the tryptic digest of recombinant human growth hormone. When the sequence of a protein is known, an incredible amount of information can be gathered from this type

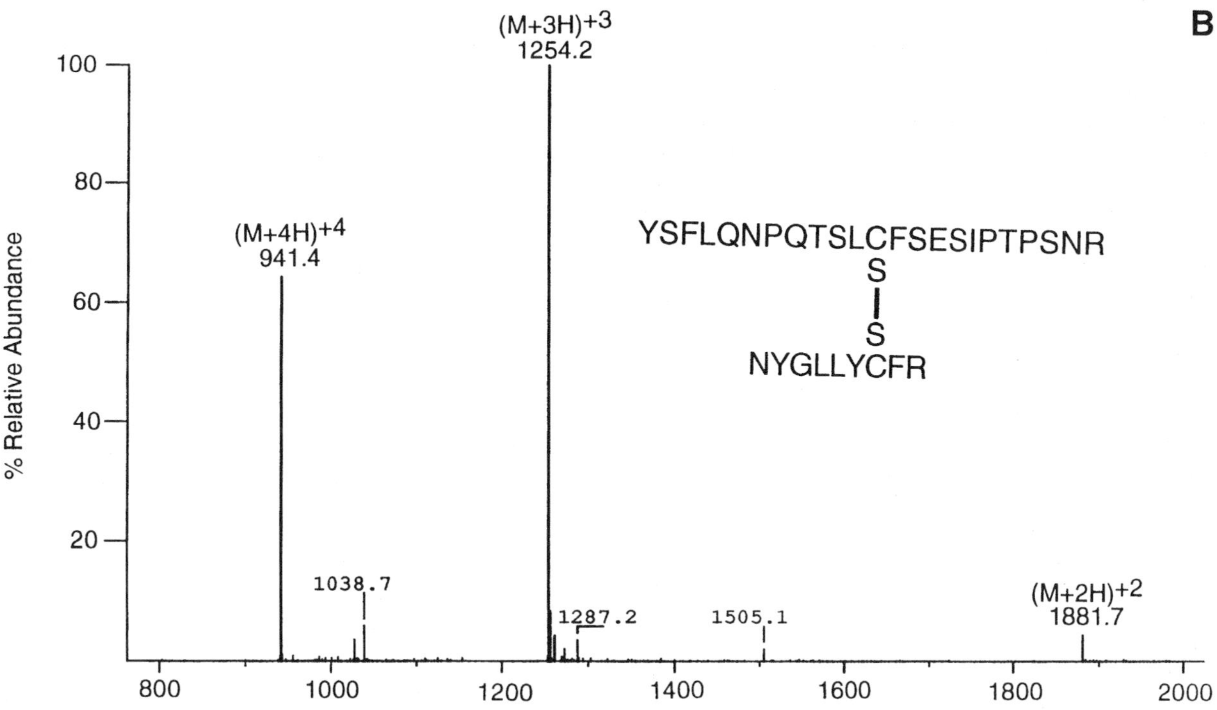


**Figure 2.** (Continued) (B) The electrospray mass spectrum for the T6–S–S-T16 peptide from the LC-MS analysis at a retention time of 2538 seconds.

of analysis. LC-MS data actually are three-dimensional: both HPLC retention and mass analysis data are obtained simultaneously. For example, Figure 2B shows the mass spectrum of the tryptic peptide at retention time 2538 seconds, which corresponds to the expected molecular weight of the disulfide bound T6 and T16 tryptic peptides. In most cases, more than 90% of a protein can be mapped using LC-MS analysis. This technique is especially valuable for characterizing the site of posttranslational or chemical modifications that may result from purification and processing of the recombinant protein. Comparison of the digest of a modified to an unmodified protein will, in most cases, indicate which peptide has been modified because of its change in retention time. The mass spectrum of the shifted peptide will indicate the type of modification; for example, a +16 Da shift in molecular weight from that expected may indicate that a methionine residue has been oxidized. By performing a tandem experiment on the peptide, the actual site of modification can be determined.

## 2.3 Tandem Mass Spectrometry

Both ESI and MALDI give reliable molecular weight information of the intact protein or peptide; however, very little information about individual structures can be obtained. Tandem mass spectrometry (MS/MS) is a means to obtain primary structural information. Numerous mass spectrometer configurations are used for performing tandem experiments; but no matter what type of instrument is used—magnetic, quadrupole or both—all MS/MS equipment operates under the same premise in performing tandem experiments. The first mass spectrometer isolates the ion of interest from a mixture of components and passes this ion to the second sector (or region), which is pressurized with an inert gas like argon or xenon. Depending on the energy of the ion passing through the collision region, varying degrees of fragmentation occur as the ion collides with the neutral gas, in a process referred to as "collision-activated dissociation" (CAD). The resulting fragment ions are mass analyzed by the third sector to yield what is termed an "MS/MS" or "daughter" spectrum. Analysis of the fragment ions yields primary sequence information of the incident ion. Normally, quadrupole mass spectrometers fragment ions by low energy CAD, causing cleavage at predominantly peptide bonds. Magnetic and electrostatic mass spectrometers fragment ions at high CAD energies, often resulting in cleavage of peptide bonds and fragmentation of the amino acid side chain. Because of side chain cleavages in high energy CAD, it is possible to differentiate between isometric ions (epg., isoleucine from leucine, lysine from glutamine) that cannot be directly differentiated by low energy CAD. In any configuration, significant amounts of structural information are obtainable from MS/MS spectra.

Tandem mass spectrometry is uniquely suited for determining the site of posttranslational or chemical modifications, sequence information on N-blocked peptides, disulfide bonding, and C-terminal sequence analysis. Since interpreting MS/MS spectra is sometimes not a trivial task (especially for unknown peptides), advanced mass spectrometric data systems must be developed to expedite interpretation. The potential for this technique is limitless, and it may someday be the primary method for sequencing peptides. The prospect is intriguing because of the speed of MS/MS compared with Edman sequence analysis and the ability of tandem mass spectrometry to simultaneously sequence both the N- and C-termini of peptides.

## 3 PERSPECTIVES

### 3.1 New Developments: Ion Trap Mass Spectrometry

There is mounting evidence that mass spectrometry can allow access to crucial information regarding aspects on the structure of biomolecules. For mass spectrometry to provide the information required for analysis of biomolecules and to keep up with advances in molecular biology, three particular areas of development must progress. The first area has already been discussed, namely, the ability to induce increasingly larger biomolecules to enter the gas phase. The second area is the development of mass spectrometers with higher mass-to-charge capabilities and better performance in terms of resolution and mass-to-charge measurement accuracy. And the third area is sensitivity, which must be increased to permit work on smaller quantities of material.

Recent advances in ion trap mass spectrometer research and development indicate that this compact and inexpensive mass spectrometer may be uniquely capable of fulfilling the requirements for a high performance mass spectrometer suited for biomolecular mass spectrometry. ESI, FAB, and MALDI have been interfaced to an ion trap, and ions with high mass-to-charge ratios (> 50,000 Da) have been analyzed. Another advancement, which has recently been demonstrated, is that this very simple mass analyzer is capable of operating under extremely high resolution conditions (> 15,000 at 1200 Da). In addition, sensitivities for small peptides have been reported to be subfemtomolar for both mass analysis and tandem mass analysis. Continued instrumental development is required to utilize this new and exciting technology for biomolecular mass analysis.

### 3.2 Conclusions

The recent developments of MALDI and ESI have allowed for the mass analysis of almost any protein. Mass spectrometry offers the protein chemist a very complementary tool to protein sequencing, amino acid analysis, and other analytical techniques for determining the identity and structure of proteins and peptides. Though an increasing number of laboratories are integrating mass spectrometry into their protein studies, a number of limitations prevent its use as a general analytical tool. Instrument complexity and cost are the most prohibitive. The development of a simple, versatile, low cost mass spectrometer would greatly facilitate complete integration of mass spectrometry into protein chemistry. Possibly, the quadrupole ion trap will someday fill this role. Either MALDI or ESI will often answer similar questions; however, each method demands different sample preparation procedures, which will dictate the success or failure of mass spectrometric analysis. To fully utilize the incredible potential of mass spectrometry for biomolecular analysis, investigators must understand the limitations and relative advantages of the ionization method and the mass analyzer.

*See also* Protein Analysis by Integrated Sample Preparation, Chemistry, and Mass Spectrometry; Protein Purification; Protein Sequencing Techniques.

### *Bibliography*

Carr, S. A., Hemling, M. E., Bean, M. F., and Roberts, G. D. (1991) *Anal. Chem.* 63:2802–2824.

Cox, K. A., Williams, J. D., Cooks, R. G., and Kaiser, R. E. (1992) *Biol. Mass Spectrom.* 21:226–241.

Fenn, J. B., Mann, M., Meng, C. D., Wong, S. F., and Whitehouse, C. M. (1989) *Science,* 246:64–71.

Hillenkamp, F., Karas, M., Beavis, R. C., and Chait, B. T. (1991) *Anal. Chem.* 63:1193A–1203A.

Huang, E. C., and Henion, J. D., (1990) *J. Am. Soc. Mass Spectrom.* 1:158–165.

Smith, R. D., Loo, J. A., Ogorzalek Loo, R. R., Busman, M., and Udseth, H. R. (1991) *Mass Spectrom. Rev.* 10:359–451.

Van Berkel, G. J., Glish, G. L., and McLuckey, S. A. (1990) *Anal. Chem.* 62:1284–1295.

# Medicinal Chemistry

*David J. Triggle*

### *Key Words*

**Endogenous** Cellular component that interacts with a specific cellular receptor. The term "endogenous ligand" describes cellular molecules that interact with receptors that may have been defined earlier by synthetic or other approaches. A classic example is the discovery of the endogenous opioid peptides for the opiate receptor.

**Ligand** Molecule that binds to receptor to initiate [agonist] or block [antagonist] response. Ligands may be small natural or synthetic molecules [neurotransmitters and their analogues] or may be large proteins or protein–nucleic acid associates [viruses].

**Receptor** Component of cell that interacts specifically with [receives] other molecules and, in appropriate combination, initiates biological response. Receptors may be protein, lipid, nucleic acid, or carbohydrate. They possess the fundamental property, when combined with the appropriate ligand, of expressing the information content of the receptor or the ligand.

*Medicinal chemistry* defines one component of a sequence of events in the drug discovery and development process (Figure 1). The critical steps of lead structure identification and refinement are major contributions of medicinal chemistry to the discovery process. At present, less than one in 5000 candidate molecules achieves clinical status, the cost of new drug introduction is approximately $300 million dollars, and the population is increasingly susceptible to disorders of aging and to newly spreading diseases. Thus there are economic, intellectual, and humanitarian imperatives for developing rapid, efficient, and intelligent processes of drug discovery.

## 1 BACKGROUND

Active principles from natural sources have, until quite recently, represented the major medicines of man. Many of our current major drugs, including antibiotics, opiates, and cardiovascular agents, have their origins and structures firmly rooted in naturally occurring compounds. In its early expression, medicinal chemistry pursued these natural leads with considerable success. In its contemporary expression, medicinal chemistry is a hybrid science at the interface of the several branches of chemistry, structural biology, physiology, pharmacology, biophysics, and clinical medicine. The increasingly interdisciplinary character of medicinal chemistry parallels the progressively more directed approaches to drug discovery that rely increasingly on knowledge of the biochemical pathway to be regulated, the structure of the target protein, and the active conformation and electronic structure of the drug species. The search for active

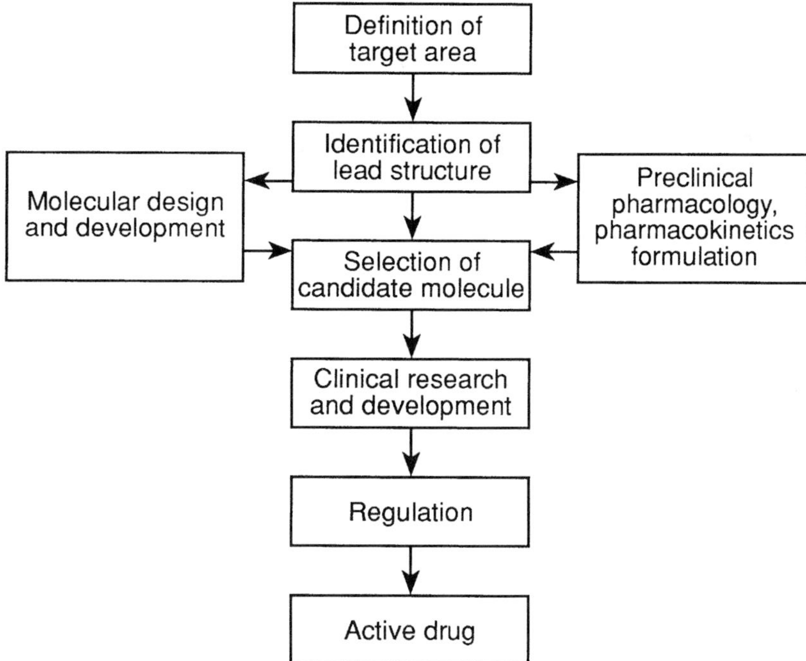


**Figure 1.** The sequence of drug discovery.

structures as leads has been facilitated by rapid and automated screening techniques, including radioligand binding, and most recently by automated approaches for combinatorial chemistry. However, the importance of endogenous structures cannot be ignored. The antitumor activity of taxol (Figure 2) from *Yew* has initiated major synthetic chemical activity; an endogenous ligand anandamide, has been reported for the *Cannabis* (marihuana) receptor; and a new ligand, adenoregulin (Figure 3), active at adenosine and other receptors, has been isolated from the Amazonian frog *Phyllomedusa bicolor* and contributes to the ''hunting magic'' attributed to this species. Similarly, the discovery of the multiple physiological and pathological regulatory roles of nitric oxide in neuronal, cardiovascular, immune, and inflammatory responses provides another example of a significant area of drug discovery made available through the elucidation of endogenous pathways and structures.

## 2 THE RECEPTOR CONCEPT

The concept of the pharmacological receptor, now some 100 years old, has increasingly underwritten drug discovery. The mutual complementarity of drug and receptor binding sites provides the fundamental basis of structure–activity correlations, the definition and interpretation of dose–response relationships, and the elucidation of receptor- and drug-mediated diseases. Attempts to purify receptors commencing in the 1960s and now, joined with molecular biology approaches, have resulted in the isolation, cloning, and expression of more than 100 receptors. These approaches have greatly aided the role of medicinal chemistry in rational drug development by:

1. Generating a molecular description of drug binding sites.
2. Permitting through site-directed mutagenesis a description of site–residue interactions.
3. Providing expressed systems with single receptors and receptor subtypes for drug characterization and screening.
4. Characterizing receptor-mediated diseases in terms of receptor deficiencies in structure or coupling processes.

The isolation and cloning of the human muscarinic receptors $m_1$ to $m_5$ confirmed early studies of drug selectivity, permitted the definition of the differential coupling of these receptor subtypes with the adenylate cyclase and phosphatidylinositol pathways, provided a defined drug characterization and screening system, and permitted construction of models of the receptor binding site.

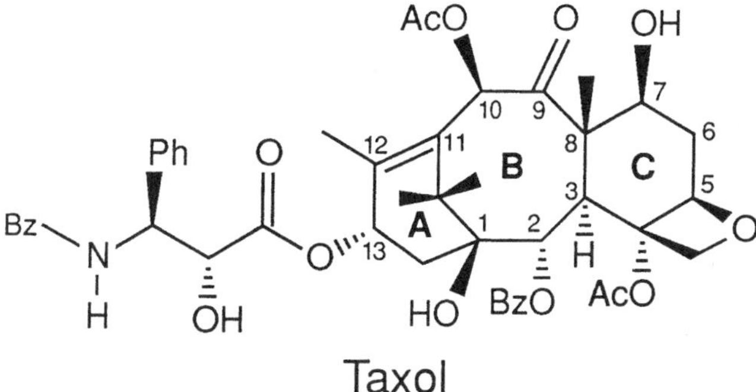

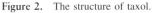

**Figure 2.** The structure of taxol.

## 3 CHEMISTRY

The identification and development of a lead structure are major components of the drug discovery process and represent the major original contributions of medicinal chemistry (Figure 4). The process of lead development is difficult to generalize, but leads may derive from a variety of sources including natural products, existing compounds, biochemical knowledge, unwanted effects of existing agents, and de novo principles.

Lead development usually passes through several stages, including initially semiempirical structure–activity relationships determined by functional group modification. A major objective is to define essential elements of the pharmacophore, as well as areas that offer room for molecular exploitation that may enhance potency, selectivity, metabolic stability, or solubility. Computer-aided mo-

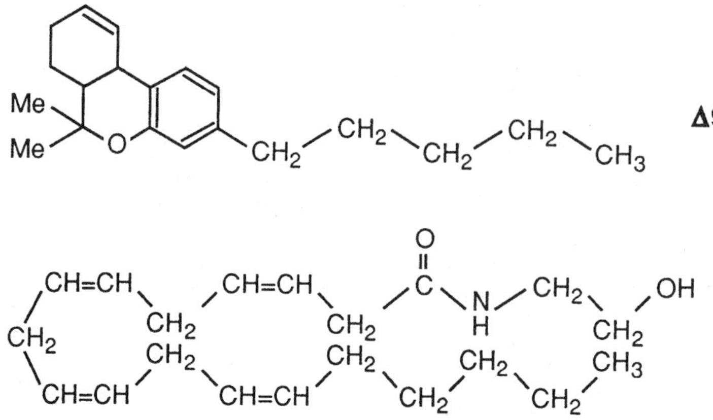

**Δ9 Tetrahydrocannabinol**

**Anandamide**

GLWSKIKEVGKEAAKAAAKAAGKAALGAVSEAV **Adenoregulin**

**Figure 3.** The structures of naturally occurring ligands active at the marihuana (Δ9-tetrahydrocannabinol and anandamide) and adenosine (adenoregulin) receptors.

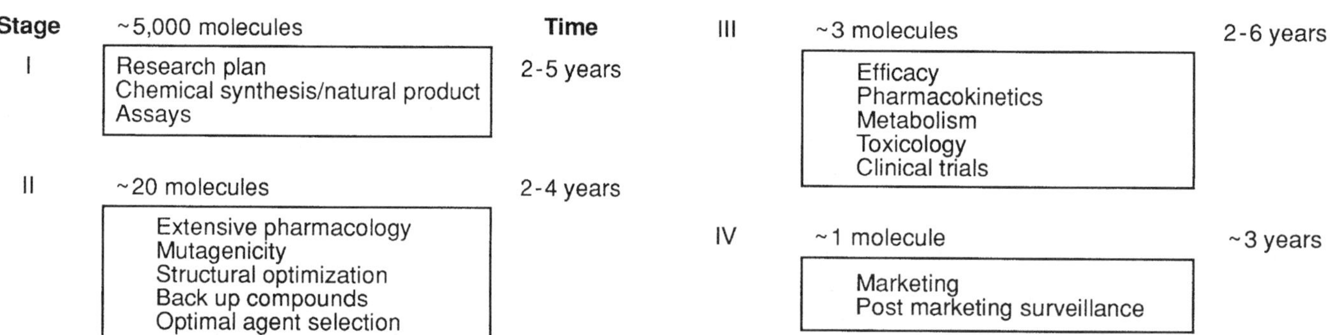


**Figure 4.** The chronology of new drug development.

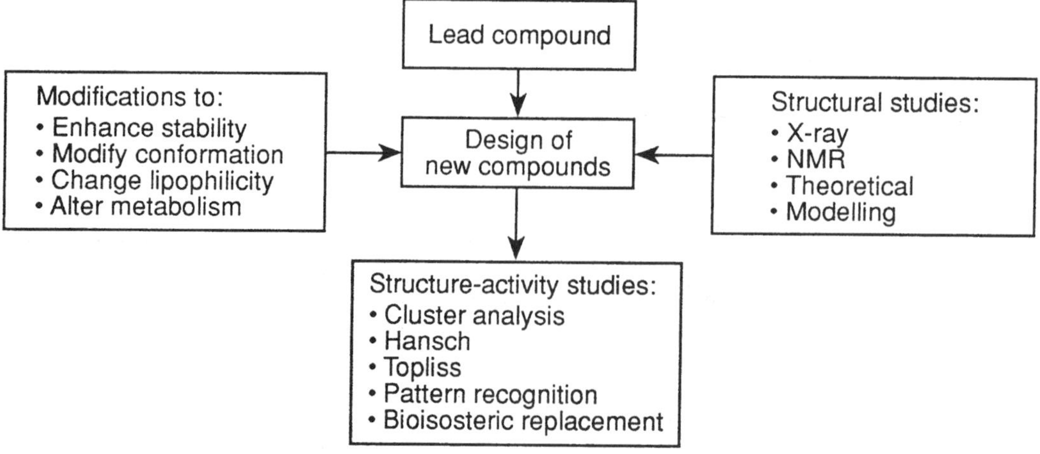


**Figure 5.** The pattern of "lead compound" development.

lecular design techniques, including the definition of quantitative structure–activity relationships (QSAR) and the calculation of electronic distribution, are increasingly employed to refine early empirical and intuitive approaches. Molecular graphic and docking procedures are widely employed, using experimentally determined structural coordinates if available, or calculated minimum energy conformations (Figure 5). Drug–receptor interactions are, almost without exception, stereoselective. The issue of enantioselectivity is of importance to drug development, which increasingly will focus on chiral compounds. Thus the appropriate stereochemistry must be incorporated at an early stage of structure development and synthesis.

Lead structures are frequently peptides or peptide-containing species because these are natural products or endogenous physiological regulators. Peptides present major problems of drug delivery and stability, and peptidomimetic analogue design is a major issue in lead development. The benzodiazepine ring system may represent one such peptidomimetic, since it provides a nucleus for drugs active at several peptide-served receptors including those for cholecystokinin, gastrin, and opiates.

The generation of new structures that serve as leads or develop existing leads is time-consuming and costly. Several methods have been devised to accelerate and automate this process, and these permit rapid construction of peptide and nucleotide libraries. Such combinatorial libraries may then be screened with receptor molecules and the ''active'' sequence identified. These combinatorial products can also be obtained by biological methods in which, for example, a repertoire of random oligonucleotides is inserted into a phage vector and each sequence expressed in one phage. These powerful and sensitive genetic methods provide for rapid enrichment and decoding of active sequences. They are, however, limited to nucleotides and naturally occurring amino acids. New developments will extend this approach to additional building blocks and may ultimately incorporate ''self-designing'' evolutionary chemistry to optimize drug design.

## 4 PERSPECTIVE

Medicinal chemistry is a rapidly changing discipline. Other physical and biological disciplines have contributed to the increasingly ''logical'' or ''rational'' process of drug discovery. Molecular biology has had significant impact by defining target sites and structures and by accelerating, in select fields, the very process of drug synthesis. It is likely that the contributions of medicinal chemistry will become increasingly quantitative in the next decade.

*See also* BIOTRANSFORMATIONS OF DRUGS AND CHEMICALS; DRUG SYNTHESIS; RECEPTOR BIOCHEMISTRY.

### *Bibliography*

Albert, A. (1985) *Selective Toxicity,* 7th ed. Chapman & Hall, London.

Brenner, S., and Lerner, R. A. (1992) Encoded combinatorial chemistry. *Proc. Natl. Acad. Sci. U.S.A.* 89:5381–5383.

Daly, J. W., Caceres, J., Moni, R. W., Gusovsky, F., Moos, M., Seamon, K. B., Milton, K., and Myers, C. W. (1992) *Proc. Natl. Acad. Sci. U.S.A.* 89:10960–10963.

Devane, W. A., Hanus, L., Breuer, A., Pertwee, R. G., Stevenson, L. A., Griffin, G., Gibson, D., Madelbaum, A., Etinger, A., and Mechoulam, R. (1992) Isolation and structure of a brain constituent that binds to the cannabinoid receptor. *Science,* 258:1946–1949.

Hirschmann, R. (1991) Medicinal chemistry in the golden age of biology: Lessons from steroid and peptide research. *Angew. Chem. Int. Ed. Engl.* 30:1278–1301.

Hoffmann, R. (1993) How should chemists think? *Sci. Am.* 268(2):66–73.

Morgan, B. A., and Gainor, J. A. (1989) Approaches to the discovery of non-peptide ligands for peptide receptors and peptidases. *Annu. Rep. Med. Chem.* 24:243–252.

Perun, T. J., and Propst, C. L., Eds. (1989) *Computer-Aided Drug Design. Methods and Applications.* Dekker, New York.

Roberts, S. M., and Price, B. T. (1985) *Medicinal Chemistry. The Role of Organic Chemistry in Drug Research.* Academic Press, Orlando, FL.

Sammes, P. G., and Taylor, J. B., Eds. (1991) *Comprehensive Medicinal Chemistry,* Vols. 1–6. Pergamon Press, Oxford.

Spilker, B. (1988) Multinational drug companies. In *Issues in Drug Discovery and Development.* Raven Press, New York.

Triggle, D. J. (1993) Medicinal chemistry: Through a glass darkly. *Annu. Rep. Med. Chem.* 28:343–350.

Williams, M., and Read, G. L. (1988) *Prog. Drug Res.* 29:329–365.

# Melting: *see* Denaturation of DNA.

# MEMBRANE FUSION, MOLECULAR MECHANISM OF

*Koert N. J. Burger*

### *Key Words*

**Fusogenic Protein** Protein directly responsible for the induction of membrane fusion. Also known as fusion protein.

**Local Point Fusion** Membrane fusion as observed using modern (fast-freeze) electron microscopical techniques: initially restricted to a small area ($< 20$ nm) of the interacting membranes.

**Membrane Fusion** Complete coalescence of two membranes leading to the formation of one membrane-enclosed compartment out of two originally separated compartments (as in cell–cell fusion), or two membrane-enclosed compartments out of one (as in cell division).

**Type II Lipids** Class of fusogenic lipids capable of forming inverted nonbilayer lipid structures.

Membrane fusion is a ubiquitous event in the functioning of a living organism. Life starts as a sperm fuses with an oocyte, leading to its fertilization. Membrane fusion is crucial for the formation of the mature muscle fiber, for exocytosis and endocytosis, and for intracellular membrane traffic. Membrane fusion is used by enveloped viruses to enter and infect cells. Membrane fusion is also an important tool in research and medication. Artificially induced cell fusion is an essential step in the production of monoclonal antibodies, and directed fusion of membrane vesicles may be used to deliver probe molecules and drugs to cells, both in vitro and in vivo.

## 1 METHODS AND MODEL SYSTEMS

Thousands of membrane fusion events occur in any living cell every minute. Considering that even one mistake (e.g., the fusion of an endosome with the nuclear envelope) would be fatal to the cell, it is clear that membrane fusion occurs only under strict control: specific proteins determine exactly where and when membrane fusion takes place. Biomembrane fusion is a multistep process of high complexity, and the actual fusion event (membrane coalescence and pore formation) is extremely fast—in the millisecond range. Only by using appropriate model systems can the molecular mechanism of membrane fusion be unraveled. Biochemical approaches using permeabilized cells and cell-free systems composed of isolated intracellular membranes have identified many of the (protein) factors required for membrane fusion. Molecular genetics has confirmed their importance in the regulation of intracellular membrane traffic. Many of the protein factors are universal, occurring both in yeast and in mammalian cells; and similar proteins are involved in many different membrane fusion events within the cell. However, for none of these factors has a direct role in the induction of membrane fusion been shown. The mechanism by which biomembranes are destabilized and induced to fuse remains largely obscure. Because membrane fusion ultimately requires the coalescence of two lipid bilayers, pure lipid systems have been studied extensively. Our current understanding of the molecular mechanisms of membrane fusion is mainly based on results obtained using these simplest of model systems.

## 2 MORPHOLOGY OF BIOMEMBRANE FUSION

Studies using modern morphological and electrophysiological techniques have provided invaluable information on the process of biomembrane fusion. These techniques, which can be applied to intact biological systems, offer the spatial and temporal resolution so crucial in the analysis of a rapid and local process such as membrane fusion. Biomembrane fusion is consistently observed as a *local point* event that involves only a very small surface area of the interacting membranes. The initial fusion pore has a diameter of only 1 to 3 nm. Obviously, only relatively few lipid molecules (hundred to a few thousand) are directly involved in the fusion process.

## 3 BIOMEMBRANE FUSION DISSECTED

It has proven extremely difficult to discriminate fusogenic factors directly involved in the induction of membrane fusion from secondary factors involved only in processes that precede or follow actual membrane coalescence.

### 3.1 Trigger

Studies on *regulated exocytosis* (i.e., in response to an extracellular stimulus) have revealed a potential role for intracellular free calcium ($Ca_i^{2+}$), ATP, phospholipases, and GTP in the regulation of membrane fusion. A rise in $Ca_i^{2+}$ was long thought to be both necessary and sufficient for the induction of exocytotic membrane fusion. Recent studies, however, indicate that though a rise in $Ca_i^{2+}$ triggers exocytosis in many cell types, calcium is not directly responsible for membrane destabilization and the induction of membrane fusion. In some secretory systems, membrane fusion can even be triggered in the complete absence of calcium. ATP is required for the movement of secretory granules to the plasma membrane. Therefore, sustained release depends on ATP. Like calcium, however, ATP is often not a prerequisite for the fusion reaction itself. Many secretory processes are accompanied by the activation of phospholipases, and substantial amounts of diacylglycerol or unsaturated fatty acids may form. In contrast to calcium and ATP, these lipid products may play a direct role in the induction of membrane fusion by destabilizing the interacting membranes; experimental proof supporting this notion is still scarce, however. Finally, there is more and more evidence stressing the role of GTP-binding proteins in the regulation of membrane fusion during regulated exocytosis and *intracellular membrane traffic*. Extensive studies of cell-free systems reconstituting intracellular membrane fusion events have identified a proteinaceous multisubunit fusion machine that appears to regulate membrane fusion at multiple points in the secretory and endocytotic pathway. The core component is the highly conserved protein NSF (*N*-ethylmaleimide-sensitive factor), and its dephosphorylation may drive membrane fusion. However, a direct involvement of GTP-binding proteins or the intracellular fusion machine in membrane destabilization and the induction of membrane fusion has not (yet) been demonstrated. So far, a direct role for a protein in the induction of biomembrane fusion has been demonstrated only in the case of virus–membrane fusion (see Section 5).

### 3.2 Adhesion

Biomembrane adhesion involves removal of steric barriers, specific recognition, and local close apposition. Cytoskeletal elements and other cytosolic proteins play an important role in the spatial organization of the exocytotic machinery and may coregulate membrane fusion by controlling close apposition of membranes (e.g., in conjunction with calcium). The specificity of biomembrane fusion is guaranteed only if specific receptors exist to control membrane interaction. Evidence for such receptors has been obtained by studying regulated exocytosis in fusion-defective protozoa and in adrenal chromaffin cells. In addition, a sperm surface protein has been identified that interacts specifically with a protein from the zona pellucida of the egg cell. Finally, specific receptors for the NSF-containing intracellular fusion machine have been isolated from bovine brain. Similar proteins are found in other tissues and may have the same function, imposing specificity on an otherwise universal intracellular fusion machine. During membrane fusion, the lipid bilayers of the interacting membranes come into very close contact before they coalesce. Studies on pure lipid bilayers show that at a distance shorter than 30 A two forces dominate: a repulsive hydration force and an attractive hydrophobic force. The repulsive force arises from water tightly bound to the lipid head groups; the attractive force results from hydrophobic attraction between the hydrocarbon interiors of the membranes. Thus, model membrane fusion can be enhanced by incorporating lipids with a low head group hydration, reducing the repulsive hydration force, or by invoking defects in membrane lipid packing, which will expose more of the membrane interior and thereby increase the hydrophobic attraction. Similar principles may underlie the regulation of biomembrane fusion.

### 3.3 Fusion

During membrane fusion the equilibrium bilayer configuration of the membrane is abandoned temporarily (Figure 1a–c). Different molecular models have been proposed for the intermediate or "semi-fusion" stage of membrane fusion, the stage at which the outer lipid monolayers of the two membranes are continuous but the inner lipid monolayers are still separated (b1–b3 in Figure 1). A mechanism has been proposed in which two membranes first

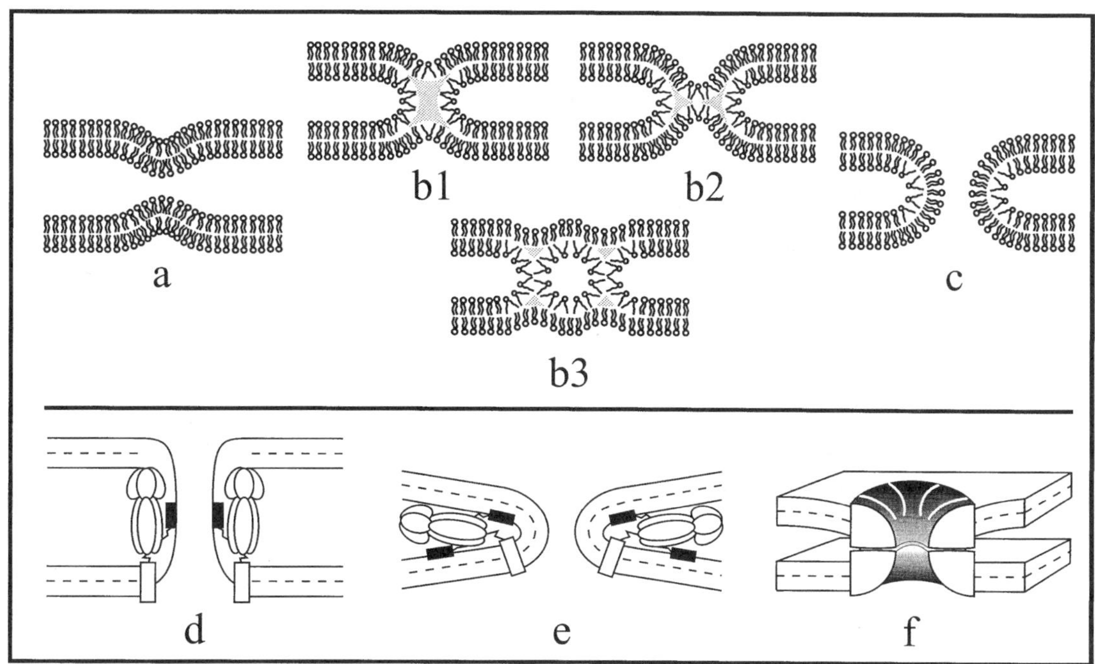


**Figure 1.** Molecular models for membrane fusion. (a=nc) Fusion of pure lipid membranes: (a) adhesion, (b) semifusion, (c) pore formation. Semifusion may involve a bilayer stalk (b1) and trans monolayer contact (b2), or an inverted lipid micelle (b3); hydrophobic interstices are shaded. (d=nf) Models of biomembrane fusion. Fusion pores formed after influenza virus=nmembrane fusion (d, e: viral membrane at the bottom, target membrane at the top); at low pH, HA remains in an upright position and the fusion peptides are extruded sideways (d), or HA tilts and the fusion peptides penetrate target and viral membrane (e). The fusion peptides form amphipathic α-helices (solid rectangles in d and e): the bulky apolar amino acids are present on one side, and the acidic amino acids on the opposite side of the α-helix. A model of the exocytotic fusion pore appears in f. For more details, see text.

join to form a so-called monolayer stalk (b1 in Figure 1). The stalk develops into a trans monolayer contact (b2 in Figure 1), which ruptures, resulting in a small aqueous pore (Figure 1c). A second model predicts the formation of an inverted lipid micelle at the semifusion stage (b3 in Figure 1). These models are mainly based on model membrane studies and on theoretical considerations (energy calculations); the actual involvement of the proposed semifusion intermediates in the act of (bio)membrane fusion has not been directly shown. The possibility that the real semifusion intermediate has yet another geometry, or is simply a local disordering of lipid molecules, cannot yet be excluded. Fusion is not completed before a pore has formed, connecting the two originally separated aqueous compartments. The initial pore probably forms spontaneously, but widening of the pore and release of contents may require influx of water (swelling) or ions.

## 4 SPECIFIC LIPID INVOLVEMENT IN BIOMEMBRANE FUSION

A number of ''fusogenic'' lipids have been proposed to be involved in the induction of membrane fusion. Some of these lipids, such as lysolecithin and monoacylglycerides, and acidic phospholipids, are not likely to be directly involved in the induction of membrane fusion. In contrast, inverted nonbilayer-preferring lipids (''type II lipids'') are present in significant amounts in almost every biomembrane. An example is unsaturated phosphatidylethanolamine, present in all eukaryotic membranes. Type II lipids have a small head group relative to the cross-sectional area of their acyl chains. Upon isolation, they spontaneously form nonbilayer lipid structures with a molecular organization closely related to that of the putative inverted lipid micelle (b3 in Figure 1). Since membrane fusion involves a nonbilayer intermediate, a role for type II lipids in the regulation of biomembrane fusion is very appealing. Moreover, model membrane studies show that physiologically relevant factors such as diacylglycerol, unsaturated fatty acids, and hydrophobic proteins can induce membrane fusion by unmasking the phase preference of type II lipids. Similarly, factors involved in the regulation of biomembrane fusion may (in theory) directly initiate membrane fusion by locally inducing a bilayer-to-nonbilayer lipid structure transition. It should be noted that despite the probable role of type II lipids in biomembrane fusion, the molecular nature of the semifusion intermediate remains controversial. The fusogenic properties of type II lipids can be explained using any of the mentioned models for local point membrane fusion.

## 5 MOLECULAR MODELS OF BIOMEMBRANE FUSION

Current data indicate that proteins are directly involved in the induction of biomembrane fusion. The responsible *fusogenic* proteins, however, have been identified only in the case of the fusion of enveloped animal viruses with their target membrane. The best-characterized viral fusogenic protein is hemagglutinin (HA), a transmembrane glycoprotein of influenza virus. After being taken up by its host cell via endocytosis, influenza virus fuses with the endosomal membrane. The viral nucleocapsid is released into the cytoplasm and the virus starts to multiply. Membrane fusion is triggered by the low pH of the endosomal compartment, and HA is solely responsible for the low pH-induced fusion activity of influenza virus. The HA gene has been cloned, and the molecular structure of the ectodomain of HA has been determined by X-ray crystallography with 0.3 nm resolution. HA is present as a trimeric complex of identical monomers protruding some 13.5 nm from the viral membrane. Each monomer consists of two subunits, HA-1 and HA-2, linked by a disulfide bridge. HA-2 carries at its amino terminus a highly conserved and extremely hydrophobic stretch of about 20 amino acids, the so-called fusion peptide. The fusion peptide plays a crucial role in the induction of membrane fusion, and almost any change in its amino acid sequence by site-directed mutagenesis drastically influences the fusion activity of HA. At neutral pH, the three fusion peptides, one per monomer, are buried in the trimeric ectodomain of HA about 3 nm away from the viral membrane. At low endosomal pH, the interactions between the monomers in the HA trimer are partially lost and the fusion peptides are exposed. In addition, several HA trimers aggregate into a fusion complex. Membrane fusion occurs in the center of the fusion complex and is induced by the fusion peptides (solid rectangles in Figure 1d, e), which dehydrate the intermembrane space or create local defects in lipid packing of the interacting membranes, inducing them to fuse. The exact mechanism of HA-mediated membrane fusion is still under debate: it is not known whether the HA glycoproteins stay in an upright position (Figure 1d) or tilt to allow closer membrane contact (Figure 1e), nor has the molecular nature of the semifusion intermediate been settled yet.

Crucial to both models of HA-mediated membrane fusion is the regulated exposure of a hydrophobic entity, the fusion peptide, in the center of an oligomeric proteinaceous fusion complex. Indeed, this may turn out to be common to many biomembrane fusion processes. The fusogenic hydrophobic entity could be a terminal or an internal stretch of hydrophobic amino acids; alternatively, it could be a hydrophobic element in the tertiary structure, or a multiacylated part of a protein. For example, a membrane protein present in the head of mouse sperm (PH-30) has an internal stretch of amino acids with characteristics typical of viral fusion peptides, and PH-30 could be responsible for sperm–egg fusion. HA-Mediated membrane fusion may even prove to be a valuable model for regulated exocytosis. Morphological data obtained on the protozoan *Paramecium* also suggest the involvement of a multisubunit protein complex in exocytotic membrane fusion, and similar conclusions have been reached on exocytotic membrane fusion in mammalian cells based on electrophysiological measurements. Electrophysiological data indicate that the exocytotic fusion pore has the characteristics of an ion channel, and a model was proposed in which fusion is mediated by a multisubunit (gap junction-like) protein pore that spans both membranes (Figure 1f); the pore expands by incorporating lipids and by lateral dissociation of the pore subunits. However, electrophysiological studies of membrane fusion mediated by influenza HA also indicate the formation of a fusion pore that is very small and has characteristics very similar to those of exocytotic fusion pores. Therefore, in analogy to HA-mediated membrane fusion, exocytotic membrane fusion could involve a fusogenic protein on the vesicle membrane capable of destabilizing the cytoplasmic leaflet of the plasma membrane; and a mixed protein–lipid pore may form via a mechanism very similar to that described for HA-mediated membrane fusion (in Figure 1, compare d and e with f).

## 6 CONCLUSIONS AND PROSPECTS

The willingness of a biomembrane to fuse may be largely determined by a delicate equilibrium between bilayer- and nonbilayer-preferring lipids present in the membrane. During membrane fusion this equilibrium is locally disturbed, leading to a transient loss of the bilayer configuration of the membrane followed by membrane fusion. Proteins play a key role in this process, either by acting themselves as a fusogen or by locally producing a fusogen. Fusogenic proteins have been identified only in virus–membrane fusion, and we are just beginning to understand their mode of action. Well-synchronized biological systems and techniques that offer high temporal and spatial resolution will be required to identify the fusogenic proteins in other biomembrane fusion processes. The vast number of different biomembrane fusion processes may not all operate via the same fusion mechanism. An alternative mechanism may make use of lipid enzymes and a local change in lipid composition or transbilayer lipid asymmetry to induce membrane fusion. The determination of the relative importance of fusogenic proteins and membrane lipids in the regulation of biomembrane fusion will be a major challenge in future research.

*See also* Cell-Cell Interactions; Cytoskeleton-Plasma Membrane Interactions; Lipids, Structure and Biochemistry of.

### *Bibliography*

Almers, W. (1990) Exocytosis. *Annu. Rev. Physiol.* 52:607–624.

Bentz, J., Ed. (1993) *Viral Fusion Mechanisms.* CRC Press, Boca Raton, FL.

Blumenthal, R. (1987) Membrane fusion. In *Current Topics in Membranes and Transport,* Vol. 29. Pp. 203–254. Academic Press, Orlando, FL.

Burger, K.N.J., and Verkleij, A. J. (1990) Membrane fusion. *Experientia* 46:631–644.

Burgoygne, R. D., and Morgan, A. (1993) Regulated exocytosis. *Biochem. J.* 293:305–316.

Düzgüneş, N., and Bronner, F., Eds. (1988) Membrane fusion in fertilization, cellular transport, and viral infection. In *Current Topics in Membranes and Transport,* Vol. 32. Academic Press, San Diego, CA.

Monck, J. R., and Fernandez, J. M. (1992) The exocytotic fusion pore. *J. Cell Biol.* 119:1395–1404.

Rothman, J. E., Ed. (1993) *Reconstitution of Intracellular Transport,* Vol. 219 in *Methods in Enzymology.* Academic Press, San Diego, CA.

Sowers, A. E., Ed. (1987) *Cell Fusion.* Plenum Press, New York.

White, J. M. (1992) Membrane fusion. *Nature* 258:917–924.

# Memory and Learning, Molecular Basis of

*Timothy E. Kennedy and Eric R. Kandel*

### *Key Words*

**Afferent** Describing a neural cell that conducts activity toward the central nervous system. The converse mode of conduction (i.e., toward the periphery) is called **efferent.**

**Associative Learning** The formation of a learned association between two stimuli or between a stimulus and a response. During **classical conditioning,** perhaps the best-studied form of associative learning, an initially neutral stimulus is associated with a highly effective stimulus by being repeatedly paired with it. Through this learned association, the previously neutral stimulus comes to evoke a learned response from the organism. In **operant conditioning,** another form of associative learning, the organism learns to produce a behavioral response in order to receive reinforcement. Unlike classical conditioning, the operant paradigm focuses on behavioral action and the consequences of that action. Much of motivated behavior can be described in these terms.

**Channel** Channels are transmembrane proteins that form pores in cell membranes, making them permeable to ionic species generating ionic currents. An electrical potential is present across all cell membranes. Channels may be sensitive to the electrical potential across the membrane and may be gated, opened or closed, by changes in membrane potential. Alternatively, ligand-gated channels are usually insensitive to membrane voltage and are gated by binding of a ligand such as a neurotransmitter. In some cases, such as the receptor for NMDA (*N*-methyl-D-aspartate), the channel is both ligand and electrically gated.

**Classical Conditioning** See **Associative Learning.**

**Hippocampus** A component of the limbic system implicated in the neural mechanisms of memory.

**Neuron** A nerve cell. Neurons send and receive information to and from other cells via synaptic connections.

**Nonassociative Learning** Learning in which the subject acquires information about the properties of a single stimulus. Thus the stimulus need not be associated with another stimulus or with a response. *Sensitization,* a form of nonassociative learning, describes a form of reflex enhancement due to the exposure of the organism to a stimulus, often a noxious stimulus. Conversely, *habituation* is a form of nonassociative reflex decrement due to the repeated presentation of an innocuous stimulus.

**Operant Conditioning** See **Associative Learning.**

**Postsynaptic** See **Synapse.**

**Presynaptic** See **Synapse.**

**Synapse** A specialized apposition between one neuron and another or between a neuron and an effector cell for the transmission or reception of an intercellular signal via a chemical neurotransmitter. The **presynaptic** side of the synapse initiates the signaling event. The **postsynaptic** side receives the signal. The consequences of synaptic activity can be excitatory or inhibitory for the postsynaptic cell. A presynaptic cell can influence the activity of a postsynaptic cell by opening or closing membrane channels and altering the postsynaptic membrane potential. Alternatively, presynaptic action may be neuromodulatory, activating second-messenger systems within the postsynaptic cell.

---

Learning is the process by which we acquire information about the environment. Memory is the retention of that information over time. An important goal of molecular neurobiology is the elucidation of the molecular mechanisms underlying learning and memory. To characterize these processes, it is necessary to identify the cellular sites within the nervous system responsible for learning and to specify mechanisms that mediate these changes. Significant progress in this regard has come from using invertebrates with relatively simple nervous systems, and from reduced vertebrate preparations such as the hippocampal slice. General principles of the molecular biology of learning and memory are beginning to emerge.

## 1 INSIGHTS FROM STUDIES OF INVERTEBRATES

### 1.1 Short- and Long-Term Sensitization in *Aplysia*

The nervous system of the marine mollusk *Aplysia* is made up of approximately 20,000 neurons. With this simple nervous system, *Aplysia* expresses a behavioral repertoire that includes a number of fundamental forms of learning: nonassociative learning such as sensitization and habituation, and associative learning such as classical and operant conditioning.

Sensitization of a reflex, a nonassociative form of learning, involves the enhancement of the strength of that reflex following the presentation of a noxious stimulus. In *Aplysia,* the gill- and siphon-withdrawal reflexes exhibit both short-term and long-term forms of sensitization. A touch to the siphon, part of *Aplysia's* delicate respiratory apparatus, evokes gill and siphon withdrawal, simple defensive withdrawal reflexes. Such stimulation activates siphon sensory neurons that synapse on interneurons and gill and siphon motor neurons (see Figure 1A). Noxious stimulation of the tail, such as a mild electric shock, activates facilitatory interneurons and produces behavioral sensitization, namely, an increase in the strength and duration of gill- and siphon-withdrawal following a touch to the siphon. The facilitatory interneurons consist of several types. Those best-studied use serotonin as a transmitter. Serotonin strengthens the synaptic connection between the siphon sensory neurons and the gill and siphon motor neurons, a process called presynaptic facilitation. A single shock to the tail produces short-term sensitization, behavioral reflex enhancement lasting seconds to minutes. Figure 1B illustrates a model of the biochemical events underlying short-term sensitization.

Like some forms of learning in vertebrates, repetition of training increases the duration of memory. With repeated stimulation, memory for short-term sensitization grades into memory for long-term sensitization, an enhancement of reflex strength lasting days to weeks. Unlike short-term sensitization, memory for long-term sensitization is dependent on the synthesis of RNA and protein. Inhibitors of protein and RNA synthesis applied during behavioral training, or during serotonin application, block the induction of the long-term process. In addition, long-term sensitization involves the growth of new synapses between the sensory neurons and motor neurons of the reflex. This increase in synaptic connectivity may represent the cellular mechanism underlying the long-term persistence of behavioral reflex enhancement.

As illustrated in Figure 1B, short-term sensitization involves the activation of cytoplasmic second-messenger systems and the covalent modification of existing cellular proteins. In contrast, the dependence on RNA and protein synthesis, and the induction of synaptic growth, suggest that long-term sensitization, in addition to activating cytoplasmic second-messenger systems, is dependent on the activation of a program of changes in gene expression via the activation of nuclear third messengers. Consistent with this idea, the injection into sensory neurons of an oligonucleotide containing the cAMP-responsive element, the binding site for the cAMP-responsive element binding protein transcription factor,

A

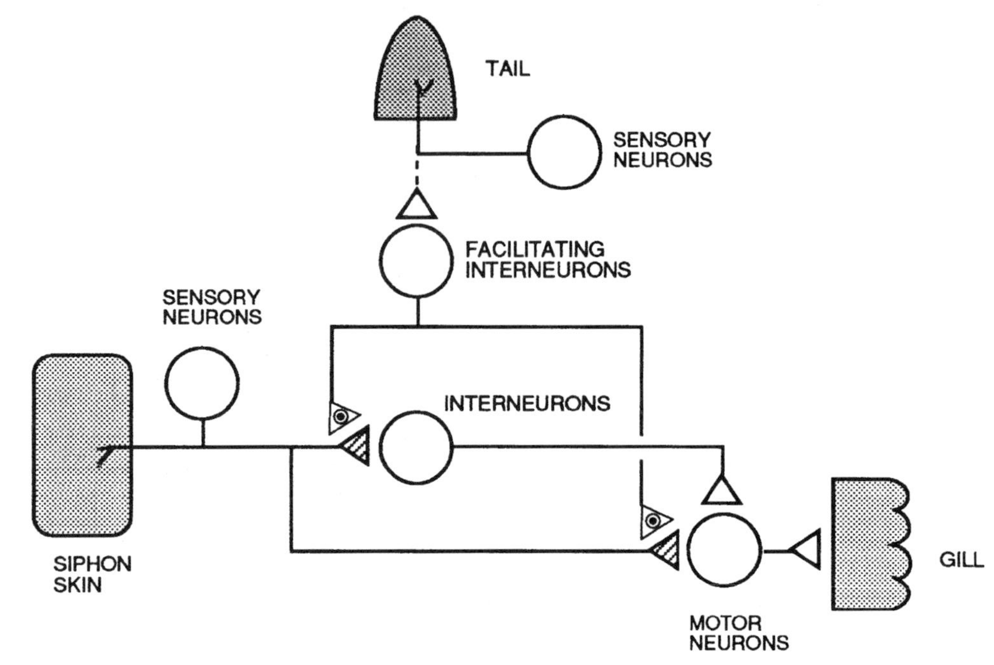
TAIL
SENSORY NEURONS
FACILITATING INTERNEURONS
SENSORY NEURONS
INTERNEURONS
SIPHON SKIN
GILL
MOTOR NEURONS

B

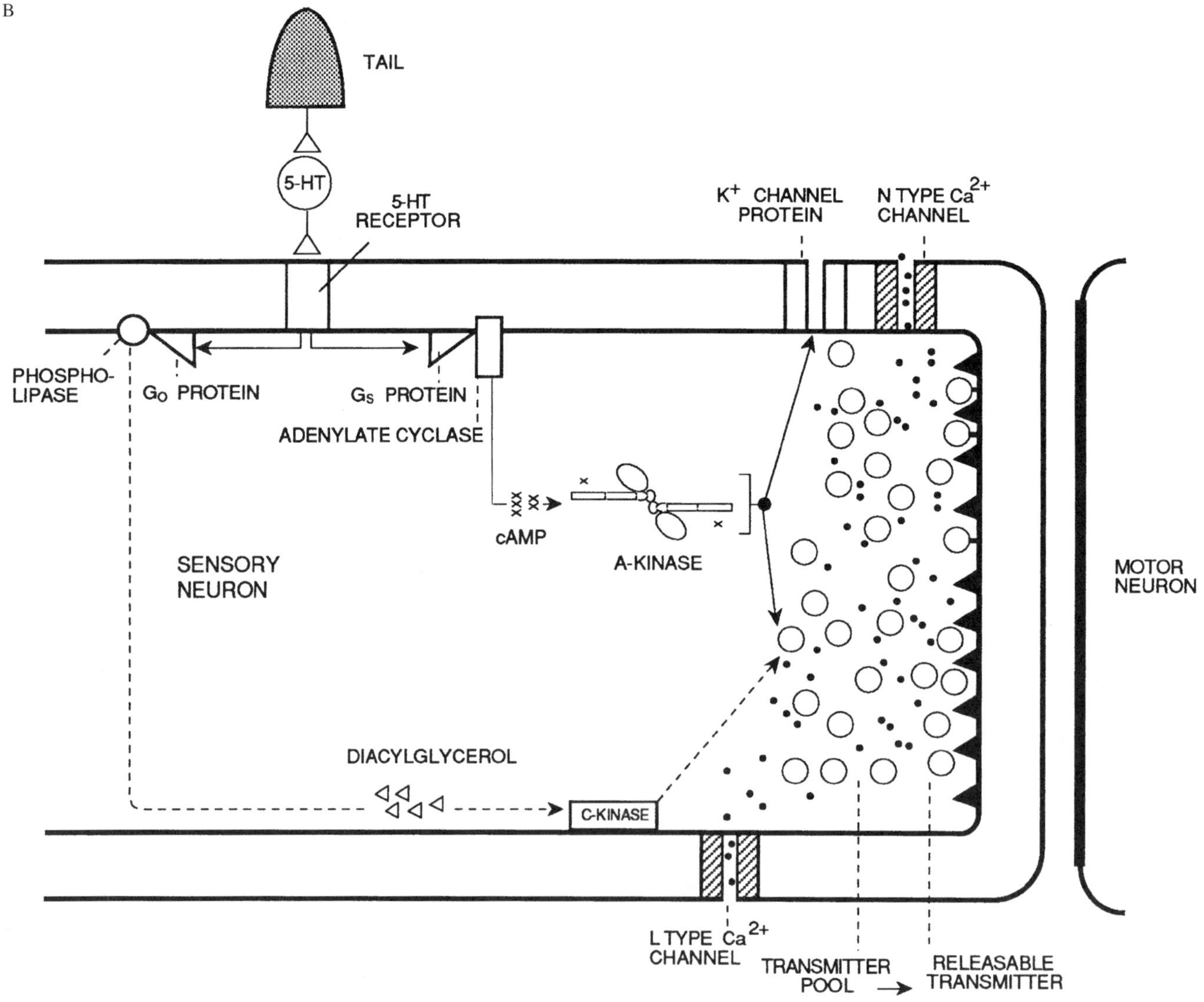

TAIL
5-HT
5-HT RECEPTOR
K+ CHANNEL PROTEIN
N TYPE Ca2+ CHANNEL
PHOSPHO-LIPASE
Go PROTEIN
Gs PROTEIN
ADENYLATE CYCLASE
cAMP
A-KINASE
SENSORY NEURON
MOTOR NEURON
DIACYLGLYCEROL
C-KINASE
L TYPE Ca2+ CHANNEL
TRANSMITTER POOL
RELEASABLE TRANSMITTER

blocks serotonin-induced long-term facilitation of those cells. In addition, specific mRNAs have been shown to increase their steady state level of expression during long-term sensitization in the sensory neuron involved in the behavioral response. These results suggest that the behavioral expression of long-term sensitization in *Aplysia* is dependent on changes in gene expression. Figure 2 illustrates a model of the biochemical events underlying long-term sensitization.

### 1.2 Genetic Analysis of Learning in *Drosophila*

The fruit fly, *Drosophila melanogaster,* exhibits both nonassociative and associative forms of learning in several different behavioral paradigms: olfactory, mechanosensory, and visual. Using a genetic approach, learning and memory mutant phenotypes have been isolated. Two of the best-described learning and memory mutations, *rutabaga,* a calcium-dependent adenylate cyclase mutant, and *dunce,* a cAMP-specific phosphodiesterase mutant, point to the central importance of the cAMP second-messenger system. Mutations in these genes produce deficits in both nonassociative and associative forms of learning. The inducible expression of a pseudo-substrate peptide inhibitor of protein kinase A in transgenic flies similarly interferes with olfactory learning, further implicating the cAMP–protein kinase A pathway. Each of these genes, the *rutabaga* adenylate cyclase, the *dunce* phosphodiesterase, and the catalytic subunit of protein kinase A, is highly expressed in the mushroom bodies (part of the fly brain thought to function as an integrating center for different sensory modalities involved in learning). In addition, in both *dunce* and *rutabaga* mutants, identified mechanosensory neurons in the mushroom bodies have altered neuronal morphologies. This suggests that as in *Aplysia,* the cAMP cascade plays a role in determining synaptic strength and in shaping the synaptic structure, connectivity, and function of the *Drosophila* nervous system.

## 2 INSIGHTS FROM STUDIES OF VERTEBRATES

### 2.1 Long-Term Potentiation may be a Cellular Memory Mechanism

Long-term potentiation (LTP) is one of the best described examples of activity-dependent plasticity in the vertebrate central nervous system. During the induction of LTP, high frequency stimulation of presynaptic afferents produces a long-term increase in the strength of the excitatory postsynaptic potential (EPSP). In the in vitro hippocampal slice preparation, this potentiation of synaptic strength can be shown to last for many hours. In the intact animal, an increase in synaptic strength can be measured weeks after the potentiating stimulation. LTP in the CA1 region of the hippocampus displays a number of interesting properties: cooperativity (more than one afferent fiber must be activated to obtain LTP), associativity (the afferent fibers and the postsynaptic cell must be depolarized together), and specificity (the potentiation is specific to the activated pathway). Having these properties, LTP has attracted a significant amount of attention as a potential mechanism in the vertebrate brain for the storage of information related to learning and memory. A model illustrating the mechanism of induction of LTP in the CA1 region of the hippocampus is shown in Figure 3.

Although LTP has been widely viewed as a likely cellular mechanism for the long-term storage of information in the nervous system, data suggesting that this may in fact be the case have been obtained only recently. Infusion into the lateral ventricle of a rat of 2 Amino 5 phosphonovalerate (APV), a selective inhibitor of the NMDA receptor channel, disrupts both the ability of the animal to perform a spatial memory task (negotiating the Morris water maze) and the ability to induce LTP in the rat's hippocampus. This suggests that an NMDA-dependent process, perhaps LTP, is required to learn this task. In addition, mice in which genes have been removed by homologous recombination have similarly demonstrated that lack of the α-calcium calmodulin dependent protein kinase II, or lack of the tyrosine kinase *fyn,* disrupts both the capacity to learn the Morris water maze task and the ability to induce LTP in the hippocampus. Disruption of the transcription factor CREB (cAMP-responsive element binding protein) does not interfere with learning or short-term memory but selectively blocks the formation of long-term memory. Similarly, LTP induced in the hippocampus of CREB mutant mice decays to baseline levels within 90 minutes, implicating CREB dependent transcription in both long-term memory and LTP.

### 2.2 Neuronal Development and Adult Plasticity

Experience shapes the connectivity of the nervous system during development. For example, in the developing mammalian visual cortex, neuronal activity, driven by the visual experience of the

◄——

**Figure 1.** (A) Schematic illustration of the neuronal circuitry underlying sensitization of the gill-withdrawal response in *Aplysia.* A touch to *Aplysia*'s siphon, part of its external respiratory apparatus, activates siphon sensory neurons, which synapse on both interneurons and motor neurons. Some of these are gill motor neurons, which when activated produce gill withdrawal. Noxious stimulation of the animal's tail activates facilitatory interneurons, which synapse onto the presynaptic terminals of siphon sensory neurons. A subpopulation of these facilitatory interneurons releases the neurotransmitter serotonin. Serotonin strengthens the synaptic connection between the sensory and motor neurons, producing an increase in the vigor and duration of gill withdrawal. This type of behavioral enhancement is a form of nonassociative learning called sensitization. The synaptic strengthening underlying this form of behavioral sensitization is called presynaptic facilitation. (B) Model of the molecular events underlying short-term facilitation within the presynaptic terminal of an *Aplysia* sensory neuron. Following tail stimulation, serotonin released from the presynaptic facilitatory interneuron binds to its receptor on the presynaptic terminal of the gill sensory neuron. This activates at least two second-messenger systems. Activation of the cAMP-dependent protein kinase (the A kinase) leads to the closure of a serotonin-sensitive potassium channel, the S-$K^+$ channel. The removal of this potassium current from the sensory neuron generates a larger influx of presynaptic calcium during the depolarization of the nerve terminal by an action potential. This increase in intraterminal calcium produces an increase in the amount of neurotransmitter released from the sensory neuron, generating a larger response in the postsynaptic motor neuron. In addition to A kinase activation, serotonin activates protein kinase C. The activation of protein kinase C produces an increase in transmitter release which is largely independent of charges in free $Ca^{++}$. Protein kinase C may produce this effect by promoting vesicle mobilization, increasing the population of transmitter vesicles available for release.

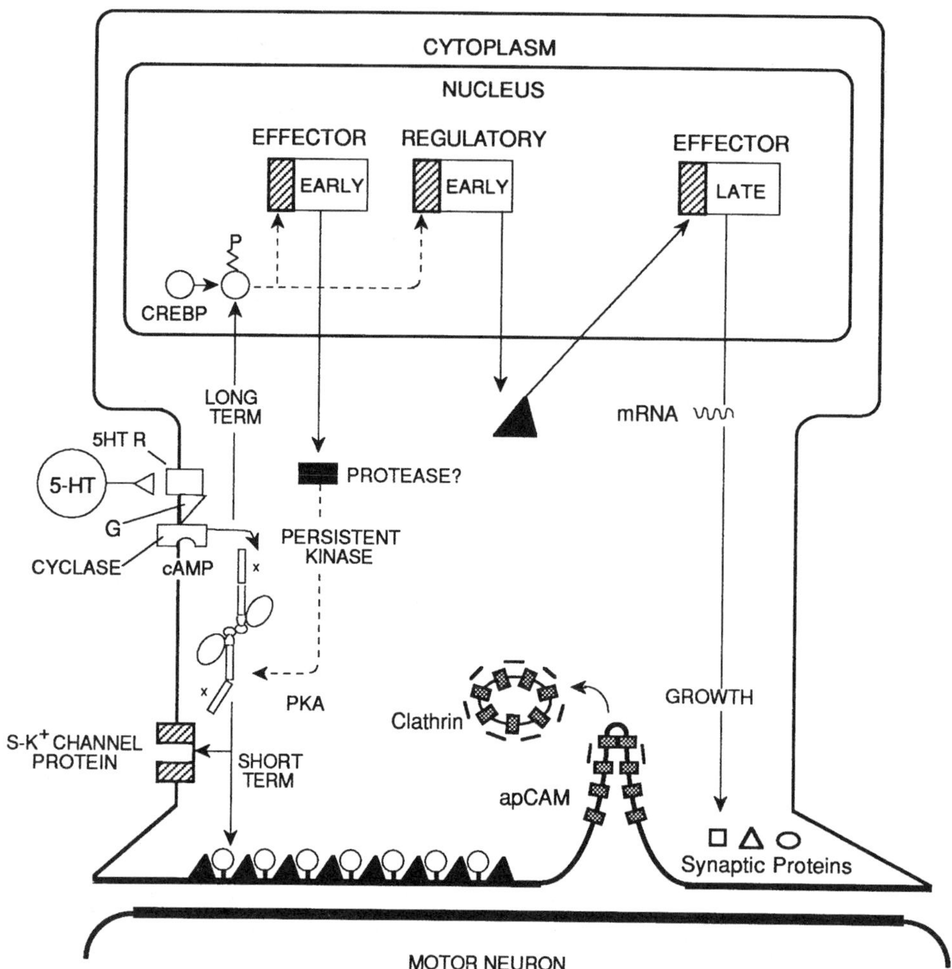


**Figure 2.** Model of the molecular events underlying long-term sensitization in an *Aplysia* sensory neuron. Unlike short-term sensitization, long-term facilitation requires new protein and RNA synthesis. As in short-term sensitization, serotonin released from facilitatory interneurons binds to its receptor on the presynaptic terminal of the sensory neuron and activates adenylate cyclase, generating cAMP, which activates the A kinase. Independent of new protein and RNA synthesis, kinase activation produces short-term effects such as the closure of the S-K$^+$ channel. Unlike short-term sensitization, repeated activation of facilitatory interneurons triggers the induction of biochemical events with a longer time course, producing long-term sensitization. These include the activation of transcription factors such as CREB: cAMP responsive element binding protein initiating a cascade of regulatory and effector gene expression. The transcription factor Aplysia CCAAT enhancer-binding protein (ApC/EBP), an early regulatory gene, is induced with the characteristics of a cAMP-inducible immediate early response gene. Early changes in effector protein expression include a downregulation of apCAM, the *Aplysia* homologue of the neural cell adhesion molecule (NCAM), and an upregulation of clathrin light chains. The downregulation of both apCAM synthesis and cell surface expression may be a permissive step in the process of new synapse formation. The coincident upregulation of clathrin may contribute to the internalization of cell surface apCAM via coated pits. The induction of early regulatory gene expression triggers the expression of later effector genes such as calreticulin, BiP, and other proteins required for the formation of the new synaptic arbors. These long-term changes in synaptic structure likely represent the cellular analogue of long-term behavioral sensitization.

organism, is necessary for the formation of binocularly segregated ocular dominance columns. Similar ocular dominance columns can be induced to form in the frog optic tectum. In the frog, infusion of APV blocks the segregation of the visual inputs, suggesting that the activation of the NMDA receptor is critical for the synaptic rearrangement to occur. A similar mechanism, dependent on calcium and the NMDA receptor, may operate during column formation in the developing mammalian visual cortex. The involvement of the NMDA receptor in the organization of synaptic structure during development is reminiscent of its role in the induction of activity-dependent synaptic strengthening observed during LTP. Results such as these suggest that learning and memory in the adult may retain mechanisms of cellular plasticity utilized during neuronal development. Indeed, recent experiments suggest that synapses in a number of vertebrate brain regions are not fixed structures, but are continually being shaped and maintained by the experience of the adult organism. In vertebrates and invertebrates, these shared cell biological themes suggest that development and learning may be two expressions of the ongoing ontogeny of the organism.

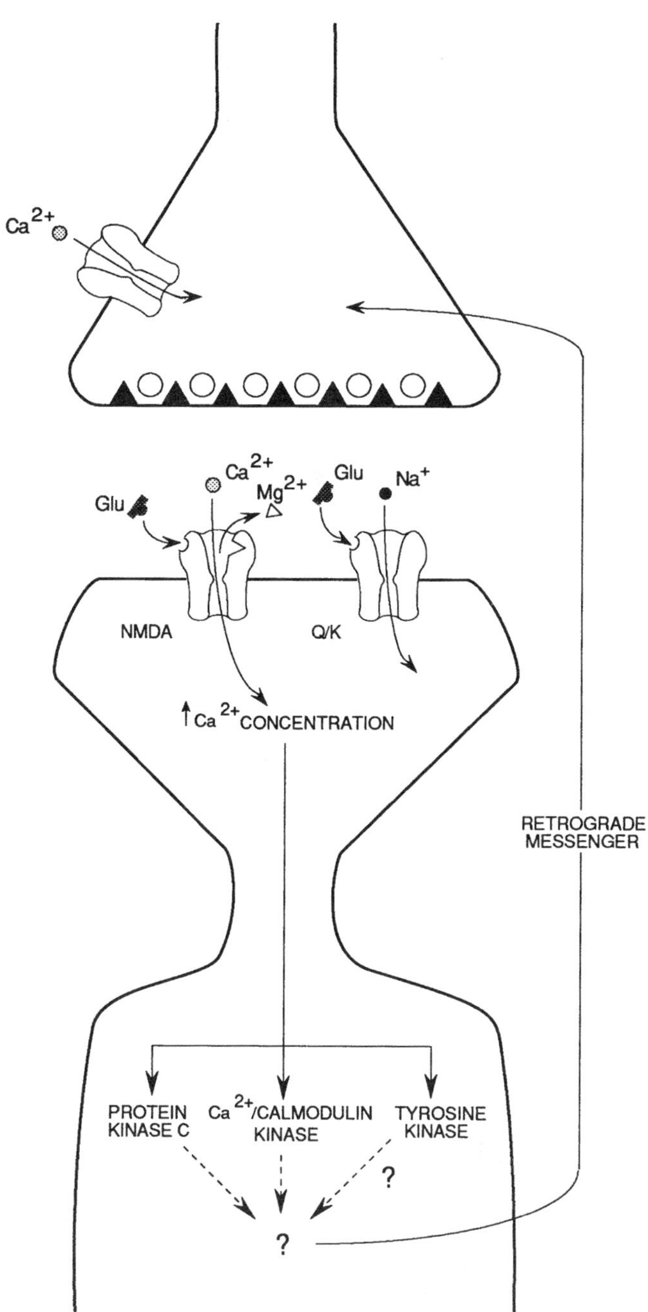


**Figure 3.** Model of the molecular mechanisms of NMDA-dependent long-term potentiation. During low frequency synaptic transmission, the excitatory neurotransmitter glutamate (Glu) is released from the presynaptic terminal and binds to glutamate receptor channels of both the NMDA type and the non-NMDA (Q/K) type. Glutamate binding opens non-NMDA (Q/K) channels, generating a depolarizing $Na^+$ current in the postsynaptic membrane. Atthe resting membrane potential of the postsynaptic neuron, although glutamate may bind to the NMDA receptor channel, the channel is blocked by a $Mg^{2+}$ ion. For current to flow through the NMDA receptor channel, glutamate must activate the non-NMDA receptor channels to the extent that the postsynaptic membrane depolarizes sufficiently to remove the $Mg^{2+}$ block. This dependence on presynaptic activation and postsynaptic depolarization generates the associative quality of LTP induction.

During the induction of LTP, the presynaptic stimulation is sufficient to depolarize the postsynaptic membrane. The NMDA channel opens, generating an NMDA-dependent calcium current in the postsynaptic cell. This influx of $Ca^{2+}$ activates several intracellular second-messenger systems in the postsynaptic neuron. Protein kinase C, the cAMP dependent protein kinase, $Ca^{2+}$/calmodulin-dependent kinase II, and the tyrosine kinase fyn have been implicated in the induction and maintenance of LTP. The induction of LTP produces an increase in the expression of certain immediate early genes, such as the transcription factor zif268. In addition, evidence suggests that the late phases of LTP require protein synthesis and involve changes in synapse morphology.

It has been suggested that these postsynaptic events, initiated by the influx of $Ca^{2+}$ at the NMDA receptor, trigger the release of a cell-permeable retrograde messenger, which travels back across the synapse to the presynaptic terminal, producing a sustained enhancement of transmitter release. Two candidate retrograde messengers have been investigated: arachadonic acid and the gas nitric oxide. The evidence currently favors a physiological role for nitric oxide. The presynaptic action of nitric oxide is enhanced in the presence of presynaptic activity, suggesting that a second associative mechanism is present in the presynaptic maintenance of LTP. Using these pre- and post-synaptic mechanisms, activity converging on a single postsynaptic neuron from multiple inputs generates long-term changes in the strength of the activated synapses.

*See also* CALCIUM BIOCHEMISTRY; SYNAPSES.

## *Bibliography*


Armstrong, R. C., and Montminy, M. R. (1993) Transsynaptic Control of Gene Expression. *Annu. Rev. Neurosci.* 16:17–29.

Bailey, C. H., and Kandel, E. R. (1993) Structural Changes Accompanying Memory Storage. *Annu. Rev. Physiol.* 55:397–426.

Davis, R. L. (1993) Mushroom Bodies and *Drosophila* Learning. *Neuron* 11:1–14.

Dudai, Y. (1989) *The Neurobiology of Memory: Concepts, Findings, Trends.* Oxford University Press, Oxford.

Frank, D. A., and Greenberg, M. E. (1994) CREB: A mediator of long-term memory from mollusks to mammals. *Cell* 79:5–8.

Goelet, P., Castellucci, V. F., Schacher, S., and Kandel, E. R. (1986) The Long and Short of Long-Term Memory—A Molecular Framework. *Nature* 32:419–422.

Kandel, E. R., and Schwartz, J. H. (1982) Molecular Biology of Learning: Modulation of Transmitter Release by Cyclic AMP. *Science* 218: 433–443.

———, ———, and Jessell, T. M., Eds. (1991) *Principles of Neural Science,* 3rd ed. Elsevier, Amsterdam.

Squire, L. R., and Butters, N., eds. (1992) *Neuropsychology of Memory,* 2nd ed. Guilford Press, New York.


# Mental Retardation: *see* Fragile X Linked Mental Retardation.

# Metabolism: *see* Bioenergetics of the Cell.

# METALLOENZYMES

*Walther R. Ellis, Jr. and Gregory M. Raner*

*Key Words*

**Cytochrome P450** One of a family of heme monooxygenases, present in certain pseudomonads and most mammalian cell types, that catalyze the oxidation of a wide variety of structurally diverse compounds.

**Electrophilic Catalysis** The process by which electrostatic stabilization of a negative charge develops in the transition states of certain reactions.

**Oxidase** Enzyme that catalyzes an oxidation using dioxygen as the electron acceptor; O atoms from dioxygen are not incorporated into the product of the oxidation.

**Oxygenase** Enzyme that catalyzes the reaction of dioxygen with an organic substrate in which oxygen atoms (one in the case of monooxygenases; two in the case of dioxygenases) from dioxygen are incorporated into the product.

**Peroxidase** Enzyme that catalyzes the oxidation of an organic or inorganic substrate by hydrogen perioxide.

Metalloenzymes, which comprise approximately one-third of the known enzymes, require stoichiometric quantities of metal ions as cofactors, typically transition metal ions, for their catalytic activities. The roles of metal ions in enzyme active sites (aside from structure maintenance) include electron transfer, oxygen atom transfer, formation of coordinated hydroxide, and electrophilic catalysis, as well as substrate binding. Metalloenzymes catalyze numerous reactions of physiological and environmental importance, including mitochondrial dioxygen reduction, hydrocarbon hydroxylation, nitrogen fixation, photosynthetic evolution of oxygen, and destruction of toxic products of oxygen reduction.

## 1 OCCURRENCE OF METALLOENZYMES

### 1.1 DISCOVERY

The study of metalloenzymes has its roots in investigations that took place nearly a century ago. In 1897 Gabriel Bertrand, working on laccase (a polyphenol oxidase), suggested for the first time that a metal ion was essential for the catalytic activity of an enzyme. At this time, the nature of enzymes was a matter of vociferous debate: Are enzymes protein-based catalysts, or is the catalysis traceable to low level contaminants? In 1926 James Sumner presented a pivotal result—crystallization of jack bean urease, which demonstrated that urease is a protein and that the dissolved crystals catalyze the hydrolysis of urea to form ammonia and carbon dioxide. Sumner failed to detect any metal ions in his urease preparations, and he expressed the view that metal ions are unlikely to play important roles in the enzymatic catalysis of biological reactions. It is ironic that 49 years after Sumner's pioneering crystallization of an enzyme, urease was found to contain nickel ions that are essential for enzyme activity!

Prior to the development of modern methods for trace element analysis and spectroscopic instrumentation, many metalloenzymes (e.g., blue copper oxidases) were isolated simply because they were a colored component of a cell or tissue homogenate. During the last 40 years, numerous nonchromophoric metalloenzymes have been isolated as well. Our current understanding of metalloenzymes is largely based on enzymes containing iron, copper, or zinc. Numerous examples of enzymes containing other metals (Mg, Ca, Mn, Co, Ni, V, Mo, W) are also known; however, the structural and mechanistic information currently available for these substances is less detailed.

### 1.2 BIOLOGICAL IMPORTANCE

Metalloenzymes play key roles in many processes central to human physiology, including the biosynthesis of DNA and certain amino acids, steroid metabolism, destruction of superoxide and hydrogen peroxide, biosynthesis of leukotrienes and prostaglandins, carbon dioxide hydration, neurotransmitter metabolism, digestion, collagen biosynthesis, and, of course, respiration. The latter could be viewed as a process of global bioenergetic importance, one that complements photosynthesis—the dioxygen that is evolved (via the $Mn_4$ oxygen-evolving complex of photosystem II) by photosynthetic organisms is consumed by aerobic microbes and animals (Fe and Cu cytochrome oxidases catalyze the reduction of $O_2$ to $H_2O$).

Within the last two decades it has become widely appreciated that metal ions play important catalytic roles that frequently cannot be matched by protein side chains or organic prosthetic groups. The most significant observation of this type concerns the microbial fixation (equation 1) of molecular nitrogen, a very inert molecule:

$$N_2 + 6H^+ + 6e^- \rightarrow 2NH_3 \tag{1}$$

This reaction is catalyzed by nitrogenase, whose active site consists of a dissociable Mo- and Fe-containing cofactor. The catalysis of such multielectron redox reactions is frequently carried out by metalloenzymes containing multimetal centers (i.e., metal clusters).

### 1.3 METAL ION UTILIZATION

Only a small number of the metallic elements appear to be utilized in biology. Of these, $Na^+$, $Ka^+$, $Ca^{2+}$, and $Mg^{2+}$ are considered macrominerals, or bulk elements; high concentrations of these ions are needed for osmotic homeostasis, neuromuscular transmission, and biomineralization (e.g., bone formation). Other essential metals are present in trace quantities in humans and most other organisms.

The selection of metal ions for incorporation into metalloenzymes is strongly influenced by bioavailability—a given element must be abundant in the environment and must be present in an extractable form. A striking exception to this generalization involves the nearly universal requirement for iron. Organisms have evolved selective uptake mechanisms for this element which, while the most abundant transition metal in the earth's crust, forms insoluble ferric hydroxides in the presence of dioxygen. The incorporation of metal ions into metalloenzymes is also influenced by other, chemically oriented, parameters: ionic radius, charge, preferred coordination geometry, ligand substitution and redox kinetics, aqueous solution chemistry, and thermodynamic stability.

Incorporation of a metal ion, a posttranslational biosynthetic event, requires that the folding of the polypeptide chain be such that several side chains congregate to form an appropriate metal-binding site. The primary metal-binding amino acid side chains are:

imidazole (His), carboxylate (Asp and Glu), thiol (Cys), thioether (Met), and hydroxyl (Ser, Thr, and Tyr). Less frequently, indole (Trp), guanidinium (Arg), and amide (Asn and Gln) groups are used. Backbone carbonyl groups can also participate in metal binding. The side chain functional groups must usually be deprotonated in order for a donor atom (O, N, or S) to form a metal–ligand bond.

## 2 REPRESENTATIVE METALLOENZYMES

### 2.1 Functional Roles of Metal Cofactors

There are two principal types of metal-dependent enzyme. Metalloenzymes contain at least one tightly bound metal ion that is required for activity. Metal-activated enzymes, on the other hand, have rather low affinity for the required metal and generally lose catalytic activity during purification. Enzymes dependent on $Mg^{2+}$, $K^+$, and (for the most part) $Ca^{2+}$ are metal-activated. Dependency on $Na^+$ has yet to be unequivocally demonstrated for an enzyme.

Examples of metalloenzymes are known for each class of enzyme designated by the International Union of Biochemistry: oxidoreductases, transferases, hydrolases, lyases, isomerases, and ligases. The following properties of metal ions make them well suited as catalysts for reactions of these types:

1. Metal ions bind at least three (usually four or more) ligands, thereby promoting the organization of protein structure.
2. With the notable exception of zinc, all the other known transition metal cofactors can exist in two or more oxidation states; metalloenzyme catalysis of oxidation–reduction (redox) reactions is thus quite common.
3. Metal ion cofactors are electrophilic and can serve as effective Lewis acids for binding and activating substrates.

### 2.2 Iron

With few exceptions (certain lactobacilli), all forms of life require iron as a biocatalyst. However, only small amounts are needed—a healthy human adult weighing 70 kg possesses just 4 g of the metal. Most of the body's iron is found in hemoglobin and ferritin; taken together, the body's iron enzymes contain less than 300 mg of iron. The common oxidation states are 2+ and 3+; higher oxidation states are known to be operative during the turnover of some iron enzymes. The widespread occurrence in biology of iron–porphyrin prosthetic groups, or hemes, has prompted the subdivision of iron enzymes into nonheme iron and heme classes.

Table 1 lists some of the iron enzymes that have been found in humans, together with the reactions catalyzed and cofactor contents. Iron can evidently be used to catalyze both redox and acid–base reactions. The active site structures of the intermediates present during turnover are not yet completely understood for any of these enzymes, however.

In the microbial world, nonheme iron enzymes play prominent roles in biodegradations. Perhaps the most exciting reaction (equation 2) is that catalyzed by methane monooxygenase, a binuclear iron enzyme:

$$CH_4 + O_2 + 2e^- + 2H^+ \rightarrow CH_3OH + H_2O \qquad (2)$$

Methane is a cheap compound whose primary industrial use is as a fuel. Functionalization to the corresponding alcohol opens up many alternative uses that have yet to be realized on a large scale because chemists have not been successful in developing an efficient catalyst for this reaction.

Heme proteins figure prominently in biochemistry—their functions include electron transfer, transport and activation of oxygen, and activation of hydrogen peroxide. The latter two functions are associated with several major subclasses of oxidoreductases: oxidases, oxygenases, peroxidases, and catalases. While such activities can also be displayed by enzymes containing nonheme iron, copper, manganese, or even vanadium, it has long been evident that the use of heme cofactors in dioxygen and peroxide metabolism overshadows other alternatives in eukaryotes.

All heme enzymes share one notable feature: during turnover, an oxidized intermediate (oxoferryl iron) is formed from the reaction of peroxide bound to $Fe^{3+}$. Two-electron reduction of dioxygen yields peroxide; not surprisingly, oxoferryl species can also be generated using $O_2$ and additional reducing equivalents. Catalases are unique in that $H_2O_2$ functions as both an electron donor and acceptor.

Cytochromes P450 constitute a family of heme enzymes that comprise as many as 200 members in humans alone. Their roles in drug and steroid metabolism, as well as the activation of potential cancer-causing aromatic hydrocarbons, have led to their present status as the most intensely studied enzyme family. A particularly fascinating aspect of cytochrome P450 biochemistry involves genetic (i.e., transcriptional) regulation—potential substrates can also serve as inducers in vivo.

The mammalian microsomal P450s, although of great clinical interest, have proven difficult to purify and characterize in detail. Studies using a soluble cytochrome P450 from *Pseudomonas putida* have led to the formulation of a detailed mechanism of P450 enzyme

**Table 1** Representative Human Iron Enzymes

| Enzyme | Reaction Catalyzed |
|---|---|
| Aconitase ($Fe_4S_4$) | Citrate $_2$ isocitrate |
| Catalase (heme) | $2H_2O_2 \rightarrow 2H_2O + O_2$ |
| Cytochrome *c* oxidase (2 hemes, 3 Cu) | $O_2 + 4H^+ + 4e^- \rightarrow 2H_2O$; "proton pumping" |
| Cytochromes P450 (heme) | $RH + O_2 + 2H^+ + 2e^- \rightarrow ROH + H_2O$ |
| Lipoxygenase (Fe) | Unsaturated fatty acids → conjugated fatty acid hydroperoxides |
| Myeloperoxidase (heme) | $H_2O_2 + Cl^- \rightarrow OCl^- + H_2O$ |
| Nitric oxide synthetase (heme) | L-$Arg^+ + 2O_2 + 2H^+ + 3e^- \rightarrow$ L-citrullene $+ 2H_2O +$ NO |
| Phenylalanine hydroxylase (Fe) | Phe → Tyr |
| Prostaglandin H synthetase (heme) | Arachidonic acid $+ O_2 \rightarrow PGH_2$ |
| Purple acid phosphatases ($Fe_2$) | Phosphate monoester hydrolysis |

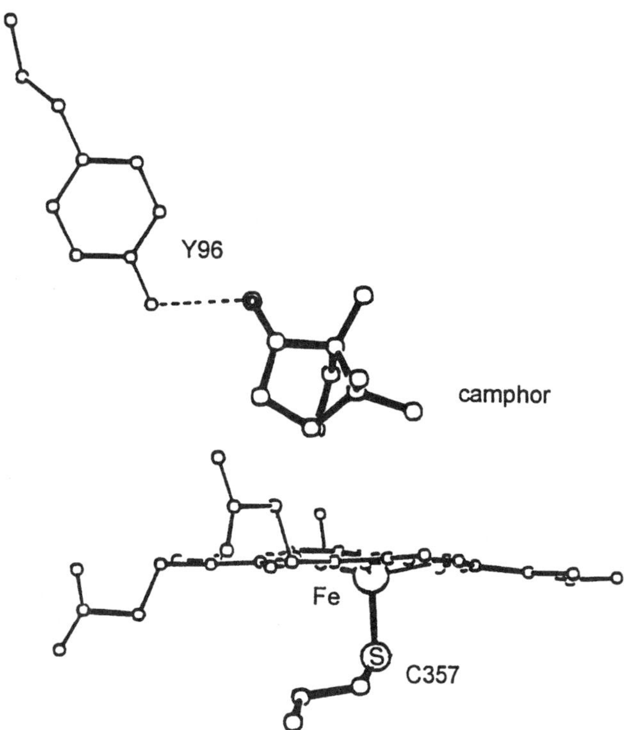


Figure 1. Structure of the *Pseudomonas putida* cytochrome P450$_{cam}$ active site. The substrate, camphor, is shown complexed to the enzyme but does not coordinate to the metal.

action. The availability of a crystal structure of the cytochrome P450$_{cam}$–camphor complex affords an unusual opportunity to inspect the active site of a metalloenzyme–substrate complex (Figure 1).

As indicated in Figure 2, the ferric ''resting'' enzyme is inactive. Reduction of the heme of the ferrous state is required prior to oxygen binding; the necessary reducing equivalents taken up during enzyme turnover are supplied, via a series of electron carriers, by NADH. The key enzyme intermediate contains $Fe^{4+}$ and a porphyrin radical cation—this oxoferryl species oxidizes the bound substrate to form the corresponding alcohol product. The other O atom from dioxygen dissociates as water; hence, all cytochromes P450 are monooxygenases.

### 2.3 Copper

Copper ions have been found in dioxygen carriers, electron carriers, oxidases, superoxide dismutases, and oxygenases. Thus, copper and iron could be viewed as ''cousins'' insofar as their biological roles are concerned. Copper ions in biology evidently use just two oxidation states, 1+ and 2+; there is no compelling evidence for $Cu^{3+}$ in any enzyme. Catalysis of multielectron redox reactions therefore requires more than one copper ion, or an additional redox unit (e.g., tetrahydrobiopterin or 6-hydroxydopa quinone).

Human copper enzymes (Table 2) are receiving increasing attention as a result of a growing awareness of their roles in the metabolism of neurotransmitters and peptide hormones. Also of interest

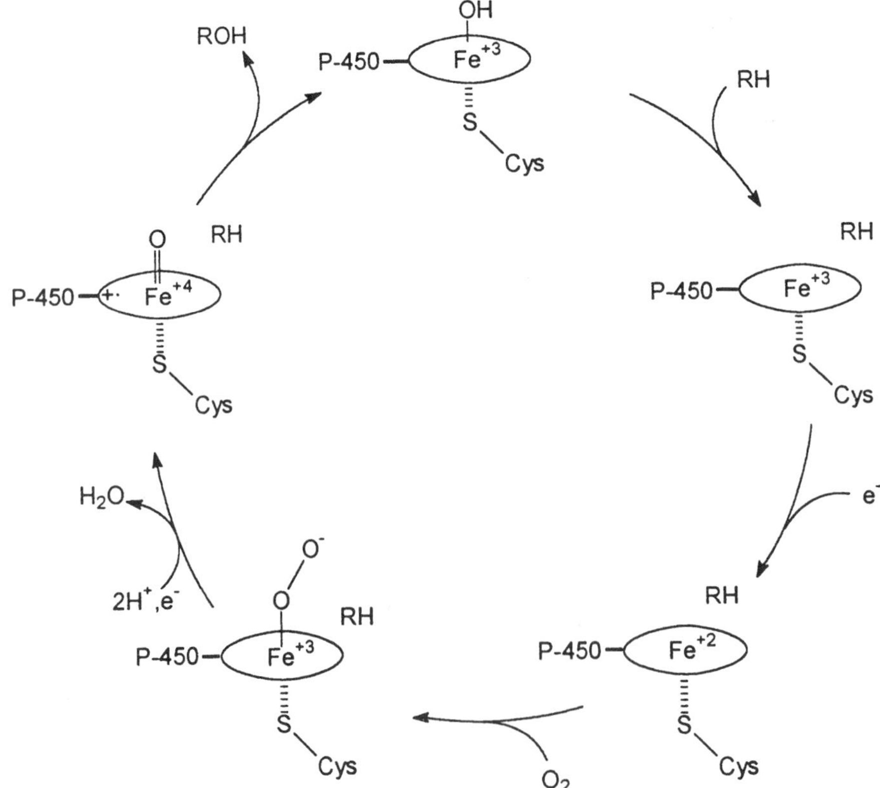


Figure 2. Cytochrome P450$_{cam}$ catalytic cycle. The substrate is designated as RH. Electrons are delivered to the heme in two discrete steps, the second of which is believed to limit the enzyme turnover rate.

Table 2 Representative Human Copper Enzymes

| Enzyme | Reaction Catalyzed |
|---|---|
| Amine oxidases | $RCH_2NH_2 + O_2 + H_2O \rightarrow RCHO + H_2O_2 + NH_3$ |
| Ceruloplasmin | Fe(II) → Fe(III); other oxidations |
| Dopamine β-hydroxylase | Dopamine → norepinephrine |
| Peptidylglycine α-amidating monooxygenase | $PNHCH_2COO^- + O_2 + 2H^+ + 2e^- \rightarrow PNH_2 + H(CO)COO^- + H_2O$ |
| Cu, Zn superoxide dismutase | $2O_2^{\cdot-} + 2H^+ \rightarrow O_2 + H_2O_2$ |
| Tyrosinase | Tyr → dopa |

is copper–zinc superoxide dismutase, a catalyst for the disproportionation (equation 3) of the toxic superoxide anion:

$$2O_2^{\cdot-} + 2H^+ \rightarrow O_2 + H_2O_2 \qquad (3)$$

The X-ray crystal structure of the oxidized bovine erythrocyte enzyme reveals a binuclear active site: four His residues and a water molecule coordinate to the $Cu^{2+}$, while the $Zn^{2+}$ is coordinated to three His and one Asp. A bridging imidazolate, provided by His-61, holds the metal ions 6 Å apart. The role of the zinc appears to be structural: upon reduction of $Cu^{2+}$ to $Cu^+$, the $Cu^+$ bond to the imidazolate bridge is cleaved, while the $Zn^{2+}$ bond to the bridge is retained. These observations have been incorporated into a proposed catalytic mechanism that is illustrated in Figure 3.

### 2.4 Zinc

In 1939 Mann and Keilin discovered zinc in carbonic anhydrase, an enzyme whose function is the catalysis of the reversible hydration of carbon dioxide to the more soluble bicarbonate anion. Since that time more than 300 zinc enzymes have been discovered; X-ray structures for more than a dozen of these are now available. Table 3 lists some of the known human zinc enzymes, together with their activities. Divalent zinc plays a particularly conspicuous role in

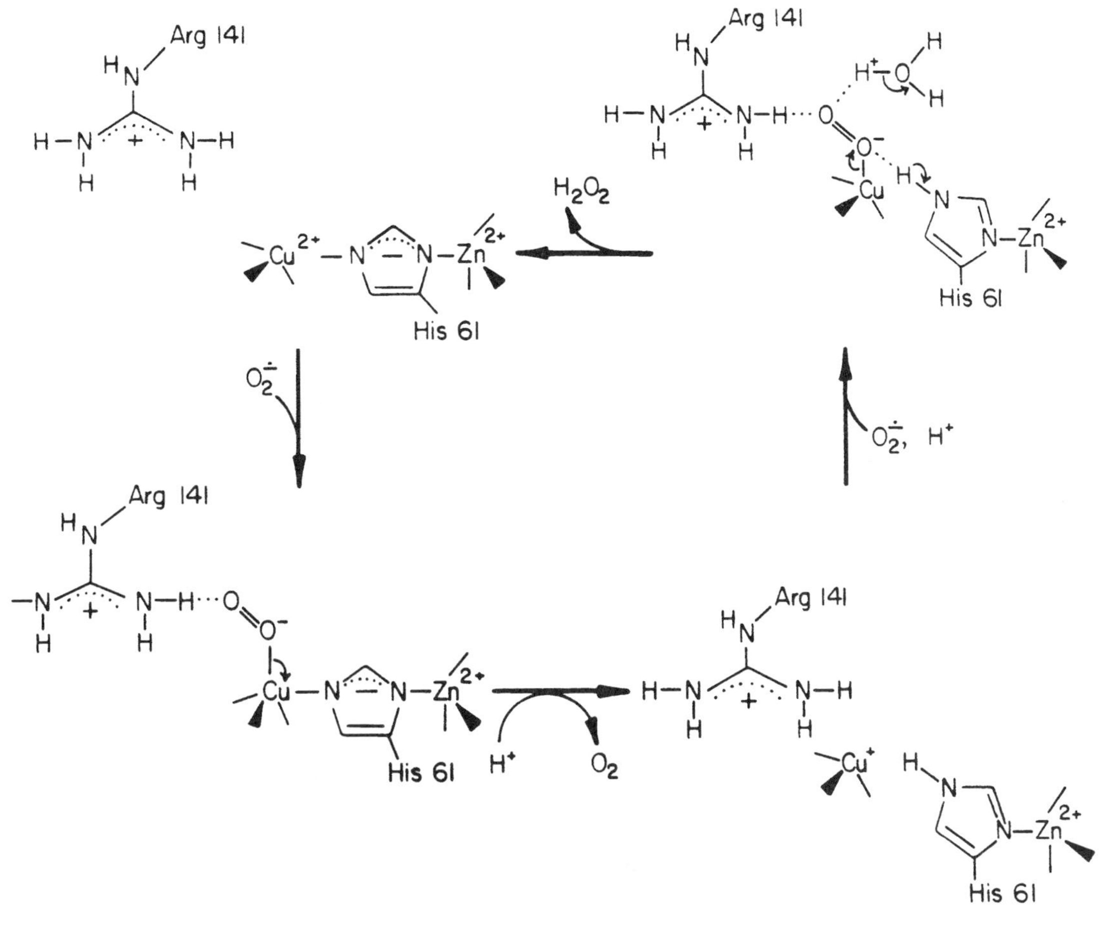

**Figure 3.** Proposed mechanism of action of bovine erythrocyte superoxide dismutase. The copper ion is involved in the catalysis, whereas the zinc plays a purely structural role.

Table 3 Representative Human Zinc Enzymes

| Enzyme | Reaction Catalyzed |
|---|---|
| Alcohol dehydrogenase | RR′CHOH + $NAD^+$ → RR′C=O + NADH + $H^+$ |
| Carbonic anhydrase | $CO_2$ + $H_2O$ → HCO3 + $H^+$ |
| Carboxypeptidases | C-Terminal peptide bond hydrolysis |
| Phospholipase C | Phospholipid hydrolysis |
| Ribonuclease A | RNA digestion (products contain 3′ phosphate termini) |
| RNA polymerase I | RNA biosynthesis (transcriptional elongation) |

hydrolytic and group transfer biochemistry. Zinc is an excellent choice for such tasks because it is redox inactive, easily exchanges ligands, and forms four-, five-, and six-coordinate complexes of comparable stability. Catalytic zinc sites always contain several endogenous (i.e., protein-derived) ligands and coordinated water.

The most important reactivity feature of divalent zinc is its Lewis acidity—it can polarize bound ligands and increase their susceptibility to attack by external nucleophiles or even increase the attacking power of a coordinated base. Water molecules coordinated to $Zn^{2+}$ can deprotonate to form coordinated hydroxide at neutral pH. Alternatively, displacement of the coordinated water by a substrate could lead to electrophilic catalysis.

The kinetics of the carbonic anhydrase reaction increase with increasing pH, and display a sigmoidal pH-activity profile (apparent $pK_a \approx 7.0$). It is now agreed that this reflects the deprotonation of coordinated water to produce a zinc-bound hydroxide ion, as indicated in Figure 4. Further support for the proposed cycle in Figure 4 comes from X-ray crystal structures of bicarbonate complexes of the $Zn^{2+}$- and $Co^{2+}$-substituted enzymes. Bicarbonate is a monodentate ligand in both structures. However, water is also bound to the metal in the $Co^{2+}$ enzyme–substrate complex, indicating that a five-coordinate metal intermediate is possible. Other zinc-catalyzed hydrolytic reactions (e.g., involving carboxypeptidase A) are also believed to proceed via the use of coordinated hydroxide.

## 3 PERSPECTIVES

In addition to the metals mentioned, other metals will undoubtedly be found to have roles as enzyme cofactors in the near future. One fascinating, and perplexing, feature of biological systems concerns the rationale for the selection of a particular metal cofactor for a given biological task. For example, ribonucleotide reductases catalyze the biosynthesis of deoxyribonucleotides from ribonucleotides. At least four different kinds of enzyme, containing either dinuclear iron, dinuclear manganese, iron–sulfur, or cobalt corrin

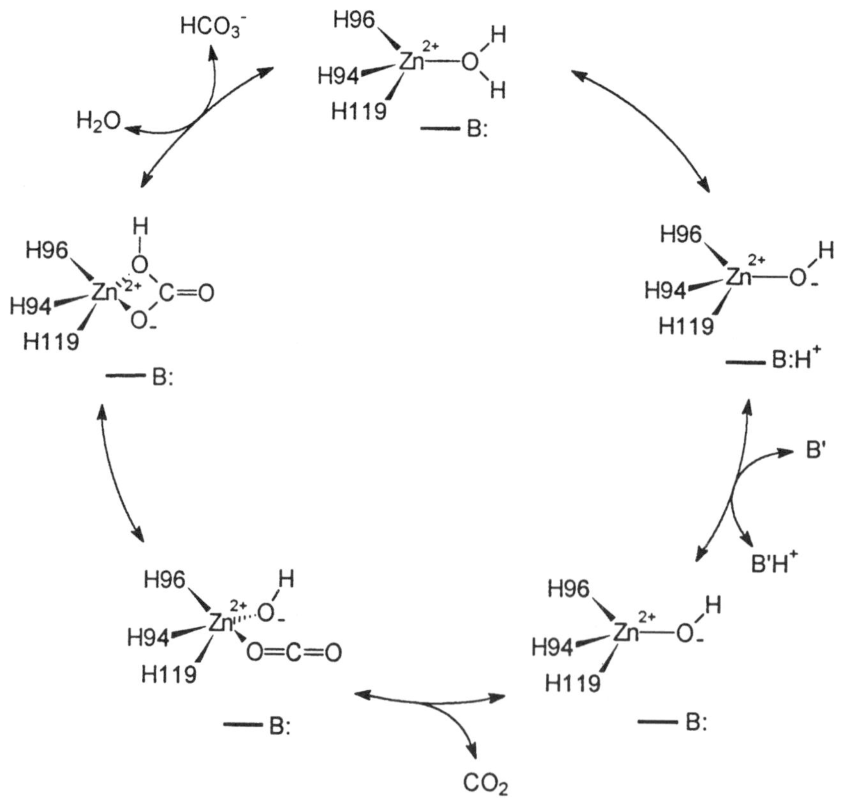

Figure 4. Proposed mechanism for human carbonic anhydrase II. Basic side chains (B and B′) assist the zinc ion in forming coordinated hydroxide, which attacks coordinated $CO_2$ to form an anhydride intermediate.

cofactors, are known. Why are so many different types of cofactor needed?

New, clinically important, roles for metalloenzymes are constantly emerging. For example, nitric oxide, a recently discovered vertebrate hormone, is generated by NO synthases, which are heme enzymes. Furthermore, some target cell responses to NO result from binding of NO to other iron enzymes. Many therapeutic drugs target metalloenzymes. Prominent examples include nonsteroidal anti-inflammatory drugs, such as aspirin or indomethacin, which are the treatment of choice for rheumatoid arthritis. These drugs block prostaglandin production by inhibiting a heme enzyme, prostaglandin H synthase.

Much of the technological interest in metalloenzymes stems from their roles in biotransformations of small molecules (e.g., CO, $CO_2$, $CH_4$, $N_2$, $H_2$, $O_2$, NO). Other metalloenzyme-catalyzed reactions of potential industrial interest include halocarbon and aromatic polycyclic hydrocarbon degradation. Emerging applications for metalloenzymes also involve biosensor fabrication and the synthesis of fine chemicals.

Finally, it is interesting to note that all catalytic RNAs (ribozymes) require divalent metal ions for activity. Ribozymes thus constitute an important new class of enzymes that utilize metal cofactors.

*See also* BIOINORGANIC CHEMISTRY; CYTOCHROME P450; ENZYMES; HEMOGLOBIN.

### *Bibliography*

Bertini, I., Gray, H. B., Lippard, S. J., and Valentine, J. S., Eds. (1994) *Bioinorganic Chemistry.* University Science Press, Mill Valley, CA.

Eichhorn, G., and Marzilli, L., Series Eds. (1979–1993) *Advances in Inorganic Biochemistry,* Vols. 1–10. Elsevier, New York.

Frieden, E., Series Ed. (1980–1993) *Biochemistry of the Elements,* Vols. 1–12. Plenum Press, New York.

Sigel, H., and Sigel, A., Series Eds. (1974–1993). *Metal Ions in Biological Systems,* Vols. 1–29. Dekker, New York.

**Methyltransferases, DNA Modification:** *see* **Restriction Endonucleases and DNA Modification Methyltransferases for the Manipulation of DNA.**

**Microbial Lipids:** *see* **Lipids, Microbial.**

# MITOCHONDRIAL DNA

*Ulf Gyllensten*

### *Key Words*

**Heteroplasmy** The coexistence of multiple forms of mitochondrial DNA in the same individual.

**Mitochondria** Cellular organelles containing the protein machinery for oxidative phosphorylation.

The mitochondria of eukaryotes contain a genome of their own, the mitochondrial DNA. The mitochondrial genome of metazoans is a covalently closed circular molecule, with a size of about 16 kilo nucleotide pairs (kntp). Considerable length differences occur between organismal groups (e.g., ciliata, > 40 kntp; fungi, 17–176 kntp; higher plants, 200–800 kntp). In vertebrates the genome encodes 2 rRNAs, 22 tRNAs, and 13 proteins involved in oxidative phosphorylation. This is a multicopy genome with a predominantly maternal inheritance and, in vertebrates, a high substitution rate. A number of mutations in the mitochondrial DNA of humans have been associated with over 100 diseases, and the accumulation of mutations in the genome has been postulated to be causally related to biological aging.

This contribution is heavily biased toward the vertebrate mtDNA, partly because of the wealth of information available for this group. However, as will be obvious from the comparison of the vertebrate mtDNA to that other groups, a number of unique characteristics set vertebrate mtDNA apart from all other types of mtDNA, partly justifying this bias in the presentation.

## 1 STRUCTURE AND ORGANIZATION

### 1.1 STRUCTURE OF THE MITOCHONDRIAL GENOME

The mitochondria of multicellular animals (metazoa), as well as of other eukaryotes, contain a genome of their own (mtDNA). The metazoan mtDNA genome is a covalently closed circular molecule, with a size that range from about 14 kilonucleotide pairs (kntp) in the nematode *Caenorhabiditis elegans,* to about 42 kntp in scallop, *Placeopecten megellanicus.* Every metazoan mtDNA genome is organized as a single circular molecule, with the exception of that of the cnidarian *Hydra attenuata,* which appears to consist of two 8 kntp circles. In other organismal groups, by contrast, the mtDNA genome is highly variable in size. For instance, that of a ciliata, *Paramecium aurelia,* exceeds 40 kntp; sizes of 17–176 kntp have been found for fungi, and higher plants have a genome size of 200–800 kntp. Particularly in plants, the genome is divided into a number of circles, with the subgenomic circles ranging from 60 to 240 kntp. In protozoans of the order *Kinetoplastida,* the mtDNA is found in certain compartments of the mitochondria (kinetoplasts) and consists of maxicircles (14–39 kntp) and minicircles (0.6–2.5 kntp).

### 1.2 GENE CONTENT

The gene content of metazoan mtDNA is to a large extent conserved: it contains the genes for two RNAs, homologous to the 16*S* and 23*S* ribosomal RNAs (rRNAs) of *E. coli,* 22 tRNAs, and 13 open reading frames. Six of the open reading frames encode enzymes, or components thereof, that are engaged in oxidative phosphorylation: cytochrome *b, COI*–III, ATPase6, and ATPase8. The other seven open reading frames encode subunits of the respiratory chain NADH dehydrogenases (commonly referred to as ND1–6 and ND4L). In addition, a single noncoding region, termed the displacement loop, or D-loop, of about 1–2 kntp, is found between the coding sequences. The term "D-loop" refers to the short three-stranded structure formed by displacement of the H strand by a short, newly synthesized, H-strand segment. Although noncoding, the D-loop contains a number of regulatory signal sequences necessary, for example, for the initiation of H-strand replication and for both H-strand and L-strand transcription. Extensive length variation in the D-loops of different metazoans appears to be due to different types of directly repeated segment. Also, the intraspecific variability in the length of the D-loop is extensive and similarly is caused by variability in the number of directly repeated segments.

In addition to the homologues for the protein-coding genes found in the mtDNAs of all metazoan species, plants carry the genes for several other subunits of the respiratory chain (e.g., ATP synthetase F1 complex, ATP synthetase subunit 9). Also, the mtDNA of some plant species contains the gene for reverse transcriptase. In protozoans, the genome (maxicircle) contains the protein-coding genes of metazoan mtDNA and, in addition, components of complex I (ND7 and ND8) that are found in the nucleus of metazoan species or the chloroplast DNA of plants. Other components that have been found in the protozoan mtDNA genome are the genes for apocytochrome *b* and subunits I, II, and III of *CO*. In fungi, a number of additional open reading frames (ORFs) have been identified, but except for a few, such as the variant 1 protein (VAR1), thought to encode a mitochondrial ribosomal protein, no function has yet been assigned to the products of these ORFs.

### 1.3 Gene Organization

Most vertebrates, as well as invertebrates, have a conserved gene content, but the gene order may vary considerably among groups. The metazoan genomic organization is extremely compact. In contrast to that of nuclear genes, moreover, the metazoan genome lacks introns (excluding the regions of the D-loop for which no functional role has been defined to date). Also, the coding sequences display variable degrees of overlap: either between the 3′ ends of two genes that are encoded in opposite strands of the molecule or, more surprisingly, between genes encoded by the same strand. Examples include the ATPase8 overlap with ATPase6 (by 2–46 ntp in vertebrates and higher invertebrates) and the ND4L overlap with the ND4 (by 7 ntp in vertebrates).

The organization of the genome in higher plants appears to resemble the arrangement of nuclear genes in that coding sequences are intervened by noncoding stretches. The gene order is not strongly conserved among plant species, and genetic exchanges between the chloroplast and the mitochondria appear to occur frequently.

### 1.4 Genetic Code

Using comparisons of the predicted amino acid sequences (based on the nucleotide sequence) and direct protein sequencing, evidence for a unique genetic code of the mitochondrial genome has been obtained. For example, rather than functioning as a stop codon, as it does in nuclear DNA, the codon TGA specifies tryptophan in all mtDNAs, except those of plants. This alteration, thus, is likely to have happened before the divergence of fungi, protista, and metazoa. With the exception of cnidarians, the AGA and AGG codons of mtDNA specify serine, instead of arginine. This alteration appears to have occurred after the cnidarian ancestral lineage diverged from that leading to all other invertebrates. In echinoderms the AUA codon specifies isoleucine and AAA specifies asparagine, instead of lysine. Finally, the AAA codon is thought to specify asparagine also in platyhelminthes. This implies that the specificity of AAA has changed twice from lysine to asparagine on independent evolutionary lineages.

The metazoan transcription machinery also employs unorthodox translation initiation codons. All four ATN triplets are being used as initiation codons in mitochondrial DNA in different species (although the subset of codons used varies considerably between species). Finally, the translation termination of some mammalian mitochondrial proteins ends in either only T or TA, instead of the complete termination codon TAA or TAG. However, the primary transcript of the mtDNA is a multicistronic RNA, and it is believed that precise RNA cleavage and posttranscriptional polyadenylation will generate the complete termination codons.

## 2 THE GENETICS OF MITOCHONDRIAL DNA

### 2.1 Inheritance

The mitochondrial DNA of most metazoan species appears to be predominantly maternally inherited. However, at least in interspecific crosses of mice, the mtDNA in the 5–10 mitochondria of the midpiece of the sperm can enter into the egg and become established as part of the zygote. The contribution in interspecific crosses in mice is on the order of 10 paternal mitochondria in 10,000–100,000 maternal mitochondria (1/10,000). Thus, this type of mechanism may contribute to the generation of individuals with multiple types of mtDNA (heteroplasmy). However, in intraspecific crosses of mice the paternal mtDNA has been shown to be eliminated soon after zygode formation. Low levels of paternal contribution of mitochondrial DNA also have been reported in *Drosophila.* In bivalves (mussels) the contribution appears to be almost equal between sexes, and in forest trees the mitochondria are inherited through the paternal parent. Thus, there is a wide variation between different organismal groups with respect to the inheritance pattern for the organelle.

### 2.2 Heteroplasmy and Segregation

In metazoan species, the mitochondria of single individuals frequently contain a population of molecules differing either by point mutations or in length, a state known as *heteroplasmy*. In higher animals these mutational differences among molecules are to a large extent confined to regions of the D-loop. However, numerous mutations have been found in the coding sequences of human mtDNA, many of them associated with particular mitochondrial diseases (see Section 3). A heteroplasmic state could in principle arise through mutational mechanisms or paternal contribution. The high frequency of some of the variant molecules, however, implies that other forces are needed to increase the frequency of variant molecules: either positive selection for variant molecules or strong segregation during oogenesis. In metazoans a number of reports have described the stability of the heteroplasmic state between generations, suggesting that the segregation mechanisms during oogenesis may not be too dramatic. However, counter to these studies, extremely rapid shifts in the frequency of different mtDNA types have been reported over a single generation in at least one mammalian species (Holstein cows). Also, in studies of disease-associated mtDNA mutations, dramatic changes in the frequency of different mutations have sometimes been seen from one generation to the next. Thus, the rules governing the transmission and segregation of mitochondria between sexual generations are at present unknown. Recently, it has been shown that the mtDNA may spread between organelles within a cell, suggesting that all mitochondria within a cell should be regarded as single unit.

### 2.3 Recombination

In metazoan species there is no evidence for recombination between variant mtDNA molecules. However, recombinational processes seem to be common in plants, where they give rise to the characteristically large number of molecules of different sizes. Particular recombinational sites (repeats) in plant mtDNA appear to function as hot spots for exchange.

### 2.4 Rate of Evolution

The substitution rate for mtDNA, which is the result of the mutation rate times the fixation probability, is measured by calculating the number of differences between the homologous gene segments in different species or individuals with known divergence times. For vertebrate mtDNA, the substitution rate is roughly 5–10 times higher than for nuclear DNA. The causes for this elevated substitution rate are at present unclear. The $\gamma$-polymerase used in the mitochondria for replication has a fidelity similar to or higher than that used for replication of nuclear genes. The higher substitution rate must therefore be due to either a lower efficiency of the repair systems in the mitochondria or a higher mutation rate. A high mutation rate could be the result of the high concentrations of oxidative radicals (superoxide anions, $H_2O_2$), which are a byproduct of oxidative phosphorylation. These compounds react with lipids, proteins, and nucleic acids and form thymine glycols and 8-hydroxyguanosine in DNA, which may inhibit replication and transcription. The reduced replication rate may, in turn, lead to longer exposure of triple-stranded DNA to mutagenic compounds. A similar elevated rate of substitution has not been detected for invertebrate mtDNA, plant mtDNA, or chloroplast mtDNA. Thus, the high substitution rate in vertebrates appears to be unique to this group.

## 3 MALFUNCTION AND DISEASE

The main function of the mitochondrial genome is to supply parts of the protein machinery necessary for oxidative phosphorylation, using a series of five multiple-subunit enzymes located within the mitochondrial inner membrane. Roughly one quarter of these components are encoded by the mtDNA, while the rest are derived from nuclear genes. Thus, a genetic defect on the oxidative phosphorylation could be due to mutations in either of the two sets of genes. Inhibitors to the OXPHOS, such as sodium azide, cyanide, and dinitrophenol, have long been known to result in optic atrophy, deafness, myoclonus, and ataxia. Given the range of symptoms expected from mitochondrial mutations, it is surprising that it was not until 1988 that the first direct observation of a mitochondrial mutation—a deletion, associated with clinical symptoms—was made. This observation, however, was followed by a vast literature on mtDNA mutations in a wide range of symptoms.

### 3.1 Mitochondrial Point Mutations

Point mutations resulting in amino acid replacements have been found to be associated with Leber's hereditary optic neuropathy (LHON), as well as the condition known as NARP: neurogenic muscle weakness, ataxia, and retinitis pigmentosa. A number of different point mutations have been associated with these phenotypes, some (like the one at ntp 11778) account for more than 50% of the LHON cases, while a number of others are private mutations that either alone or through synergistic effects may cause a similar phenotype. The segregation of the main LHON mutation may occur very rapidly, the proportion of mutated genomes rising from less than 50% to almost fixation in a single sexual generation. NARP patients have a very diverse, and diffuse, set of symptoms ranging from ataxia and dementia to proximal neurogenic muscle weakness. However, they have in common a missense mutation at ntp 8993, resulting in a change of amino acid 156 of ATPase6 from a leucine to an arginine.

Since the mitochondria have their own set of tRNAs, mutations in any of these may cause systemic effects by inhibiting translation of mitochondrial mRNAs. A number of different point mutations have been found to be associated with diseases such as MERRF (myoclonus epilepsy with ragged-red fibers) and MELAS (mitochondrial encephalomyopathy, lactic acidosis, and strokelike symptoms). Patients suffering from MERRF (point mutation at ntp 8344) have uncontrolled myoclonic epilepsy and mitochondrial myopathy that may translate into symptoms such as neurosensory hearing loss, dementia, myoclonus, and respiratory failure. The severity of the symptoms appears to be related to the proportion of mutated genomes, and the disease has a progressive development. Thus, an individual with a moderate proportion of mutated mtDNA (20%) and who showed no symptoms at the age of 20, may develop a severe phenotype at the age of 65. This progression is believed to be due to the age-related decline of the oxidate phosphorylation. Similarly, most MELAS patients have a point mutation in the tRNA for leucine (ntp 3243), which also results in inactivation of an overlapping transcriptional termination sequence.

### 3.2 Mitochondrial Deletions/Insertions

Gross alterations (insertions/deletions) of the mtDNA have been found associated with Kearns–Sayre syndrome (KSS), chronic external ophthalmoplegia (CEOP), and Pearson's syndrome. In KSS and CEOP the majority of cases involve the deletion of the region in between, but not including, the origins of replication ($O_H$ and $O_L$). Conceivably, deletions that encompass the two origins of replication inhibit replication and are lost in the competition that occurs during mitochondrial regeneration. More than 95% of the cases carry deletions between $O_H$ and $O_L$: the most frequent deletion (the common deletion, mapping between ntp 8468 and ntp 13446) occurs in 30–50% of the patients. In most of the CEOP, KSS, and Pearson's syndrome cases, only a single type of deletion is found, suggesting that these mutations have been generated by a single event of the alternative mechanisms of replication slippage, homologous recombination, or topoisomerase cleavage. Since most cases are single, spontaneous mutations, with no family history, they are likely to have been generated de novo in a specific cell lineage during development. This suggestion may also be used to explain the large differences in the proportion of deleted genomes found between different tissue types. Again, many KSS and CEOP patients manifest a progression of the severity of the disease with age, concomitant with an increase in the proportion of deleted mtDNA in certain of the muscle fibers. The deleted genomes in these diseases may be postulated to accumulate over time simply as a consequence of their replicative advantage (i.e., their shorter length). In the case of point mutations, however, such as those described earlier, additional factors such as selection must be invoked to explain the rapid shifts observed between generations or tissue types.

Duplications extending from the *CO*II gene to cytochrome *b*, and thus spanning both origins of replication, have been found in the mtDNA of a few patients with Pearson's syndrome. Because of the added number of replication origins, these mutations may have a replicative advantage that contributes to the rapidity of their accumulation.

Cases of respiratory failure, lactic acidosis, and failure of muscle or kidneys have also been found to be associated with extremely low amounts of mtDNA. In a case of lethal mitochondrial myopathy, only 2–12% of the normal amounts of mtDNA were detected. Similar symptoms have been observed in AIDS patients, as a result of zidovudine treatment, which is intended to block viral DNA replication but also exerts a similar effect on mtDNA.

### 3.3 Mitochondrial Mutations and Neurogenerative Diseases

The diseases discussed thus far are but a few of the over 100 in which mitochondrial mutations have been implicated. In addition, some families with diabetes mellitus have a deficient oxidative phosphorylation coupled with an mtDNA deletion, in the heteroplasmic state, which removed $O_L$ and large parts of the mtDNA genome. Also, mtDNA mutations have been implicated in neurogenerative disorders such as Parkinson's, Alzheimer's, and Huntington's diseases. Deficiency of complex I has been reported in the brain, muscle, and mitochondria of patients with Parkinson's disease, and mtDNA deletions have been reported in the brain tissue of these patients. Most cases of Parkinson's disease are sporadic, and only occasionally are the family members affected, as would be expected from a genetically defect oxidative phosphorylation. Similarly, Alzheimer's patients frequently have complex IV defects, and most of the cases appear to be spontaneous (with a fraction being autosomal dominant). The contribution of mtDNA mutations to these diseases are at present unclear.

### 3.4 Mitochondrial Mutations and Aging

A confounding factor in the analysis of mitochondrial diseases is the natural accumulation of mutations with age. A number of different studies have shown that insertion/deletion mutations in skeletal muscle of rats and deletions in human hearts accumulate with age. These studies suggest that there is a progressive accumulation of mtDNA mutations, which lead to the decline of oxidative phosphorylation. Thus, the accumulation of mitochondrial mutations may play a role in biological aging.

*See also* Aging, Genetic Mutations in; Genetics; Mitochondrial DNA, Evolution of Human; Motor Neuron Disease.

### *Bibliography*

Wallace, D. C. (1992) Diseases of the mitochondrial DNA. *Annu. Rev. Biochem.* 61:1175–1212.

———. (1992) Mitochondrial genetics: A paradigm for aging and degenerative diseases? *Science,* 256:628–632.

Wolstenholme, D. R., and Jeon, K. W., Eds. (1992) *Mitochondrial Genomes.* Academic Press, San Diego, CA.

# Mitochondrial DNA, Evolution of Human

*Linda Vigilant*

### *Key Words*

**mtDNA Lineage** A group of closely related mtDNA types of mitochondrial DNA.

**mtDNA Type** A unique mtDNA variant that may be represented by one or several unrelated individuals

**Phylogeny** A depiction of the evolutionary relationships of a group of taxa.

A single small locus of 37 genes, mitochondrial DNA (mtDNA), has provided a disproportionate amount of information about human evolution, diversity, and origins.

## 1 CHARACTERISTICS

### 1.1 A Unique Genome

The evolutionary data is found in the sequence of the mitochondrial DNA, a genetic locus independent of the nuclear genome. Human mtDNA is a compactly organized, circular molecule 16,569 nucleotides in length (Figure 1). The 37 genes encode rRNAs, tRNAs, and proteins involved in energy production by the organelle.

The sheer abundance of mtDNA, present at an average of 500 mtDNA molecules per human cell, facilitates analytical procedures. The high mtDNA copy number is attributed to the fact that DNA replication is not limited to once per cell cycle as it is in the nuclear genome. The replication and transcription of mtDNA in vertebrates has been intensively studied and recently reviewed elsewhere.

A very significant property of mtDNA is the high rate at which mutations accumulate, approximately 2–4% per million years in a wide range of vertebrates. This rate, which is about five to ten times faster than single-copy nuclear DNA, appears to be a consequence of inefficient mismatch repair. The evolutionary changes in mtDNA consist mainly of base substitutions and some length mutations. The rapid accumulation of these mutations provides sufficient data, to allow the comparison of closely related species, or populations within the same species.

Mitochondrial DNA apparently abides by a nearly strictly maternal mode of inheritance in mammals. The elucidation of evolutionary relationships between mtDNAs from different individuals is made possible by both the uniparental inheritance and the lack of recombination of mtDNA molecules.

### 1.2 Analytical Approaches

Both indirect and direct methods have been used to assess the diversity of human mtDNA. The indirect methods consist of clearing the mtDNA with a battery of restriction enzymes and inferring mutations from the alterations in cleavage patterns. A high resolution restriction mapping approach, using 13 enzymes with four-

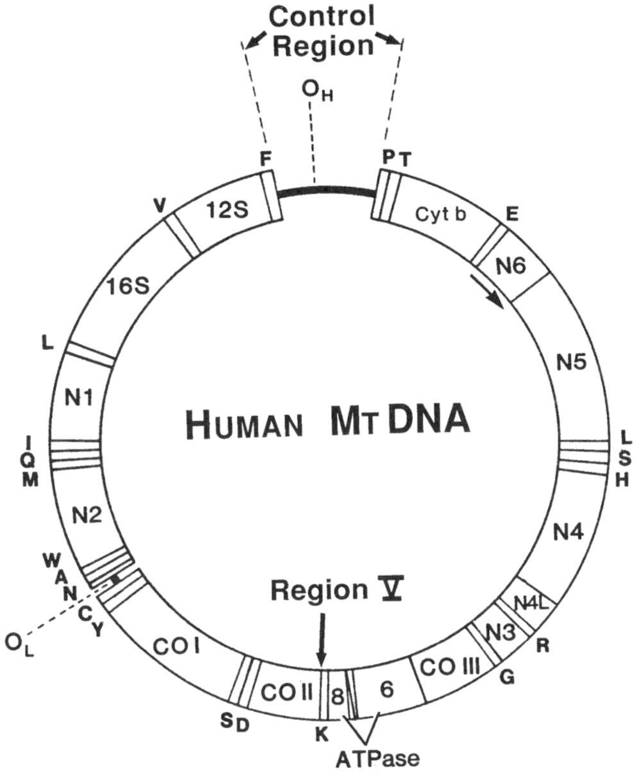


**Figure 1.** The human mitochondrial DNA molecule. Labeled functional regions include two ribosomal RNA genes (12*S* and 16*S*), seven genes for NADH-dehydrogenase subunits (N1–N6), N4L), three cytochrome oxidase subunit genes (CO I–CO III); genes for ATPase subunits 6 and 8 and cytochrome *b*, 22 RNA genes, and the heavy ($O_H$) and light ($O_L$) strand origins of replication. Locations of the control region and region V are indicated.

base recognition sites, is capable of detecting variation in about 1500 nucleotides, or 9% of the genome.

Use of the polymerase chain reaction (PCR) method of DNA amplication has permitted the direct examination of mtDNA sequences. In addition to providing the highest resolution data, PCR-based methods have the considerable advantage of requiring only minute amounts of sample DNA. A noncoding, rapidly evolving segment of mtDNA termed the "control region" is the focus of most direct sequencing studies. The resolution of the data obtained is comparable or superior to indirect data from studies, making PCR the method of choice.

## 2 HUMANS IN HOMINOIDEA

Twenty-five years ago, the suggestion, based on immunological evidence, that humans and African apes shared a common ancestor only 5 million years ago was bitterly contested because it contradicted conventional, fossil-based wisdom), according to which hominoid diversification occurred 20 million years in the past. The more recent date is now generally accepted, but debate continues about both the exact timing and the particular order of events. Difficulty and statistical uncertainty arise from the discovery that the divergence of human, chimpanzee, and gorilla from a common ancestor apparently occurred within a short period of evolutionary time.

Recently, a rigorous examination of mtDNA from five hominoids may have ended this controversy. A sizable portion (almost 5 kb) of mtDNA sequence was determined in several type species (pygmy and common chimpanzees, gorilla, orangutan, and siamang) and compared with the standard human reference sequence. The majority of sequence differences found support a human–chimpanzee clade over a chimpanzee–gorilla clade (Figure 2). This finding agrees with earlier studies based on smaller, less-convincing data sets. Divergence times for mtDNA of gorilla and human are calculated at 7.7 ± 0.7 and 4.7 ± 0.5 million years ago, respectively.

It is probable, but not necessarily true, that the species divergences are identical to the mtDNA divergences. It is primarily the ancestral population size that determines whether a tree based on a single gene actually reflects species relatedness. If present estimates of current and past effective population sizes are correct, the mtDNA tree is likely to represent the species tree.

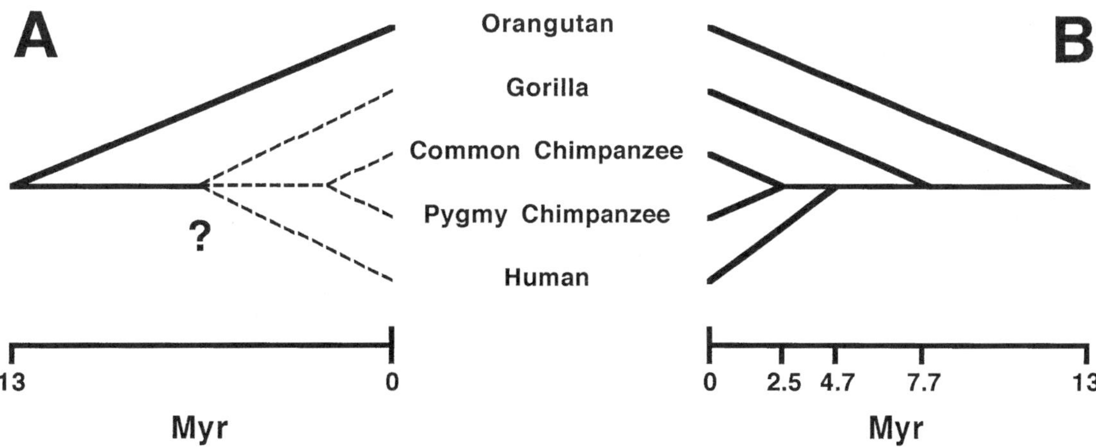


**Figure 2.** Resolution of the branching pattern of the hominoids using mtDNA sequences. (A) The uncertainty surrounding the pattern and timing of the branching events that led to the gorilla, chimpanzee, and human lineages. (B) Branching order and times of divergence (millions of years).

## 3 POPULATION STUDIES

### 3.1 Geographic Specificity

Studies of human mtDNA have been conducted in a variety of populations. The majority of individuals with shared mtDNA types come from relatively small, geographically defined populations. In addition, the spectrum of types seen in a single small population are often closely related. Thus, information from mtDNA can be used to make inferences about the geographic history of the human species.

### 3.2 The Origin of Modern Humans

The uniparental inheritance of mtDNA ensures that all the human mtDNAs in the world today must trace back to a single common ancestor. Much attention has been focused on the identification of this ancient mitochondrial mother of us all. The weight of the data accumulated thus far by several different laboratories indicates that the common mtDNA ancestor most likely lived in Africa about 200,000 years ago.

Two lines of evidence place the ancestor in Africa. First, the highest diversity of mtDNA types is found within the African populations, consistent with an older age for these populations. Second, genealogical reconstructions of the relationships between extant mtDNA types typically place the root in Africa. While the geographic placement of the ancestor has not yet been proven with statistical certainty, an African origin remains the best explanation for all the data.

The time of existence of the ancestor is on firmer statistical ground, with different methods of calculation producing strikingly similar results. The most recent estimates also include an estimate of the standard error and postulate that the ancestor most likely lived about 150,000–200,000 years ago but most certainly no earlier than 500,000 years ago.

This upper limit on the age of the ancestor is significant because of the implications of the mtDNA results for traditional fossil-based theories of modern human origins. A much more ancient age for the ancestor, even as much as 1 million or more years ago, would be needed to support the hypothesis that archaic humans spread throughout the world about a million years ago and evolved into modern humans in several different places at once. Instead, the mtDNA results are consistent with the scenario in which modern humans emerged recently from Africa and displaced the archaic residents of Europe and Asia.

### 3.3 The Region V Mutation

High resolution restriction mapping detects not only base substitutions but also length mutations in the DNA. One particular length mutation, designated the region V mutation, has developed into a useful anthropological marker for people of Asian origin.

The region V of mtDNA is a small noncoding segment flanked by the cytochrome oxidase II and lysyl tRNA genes (see Figure 1). It typically contains two tandemly repeated copies of a 9 bp sequence. Each analysis suggested that a deletion of one of these 9 bp segments arose only once in the evolution of modern human mtDNA types and that it was found solely in types of East Asian origin. More extensive surveys of the deletion in Polynesians, Asians, Amerinds, and others have contributed to an overall picture of the region V deletion in circum-Pacific populations. The data indicate an East Asian origin for the founding population of Polynesia. Both more extensive sampling and high resolution data, such as control region sequences, are needed for phylogenetic analysis and to pinpoint the origin of the region V types.

### 3.4 Amerindian Origins

Data from the region V mutation and other mtDNA markers have been used to link modern Amerindians to their Asian origins. Evidence suggests that only four primary lineages colonized the Americas, with subsequent diversification. This limited number of lineages does not necessarily imply a reduced diversity for these populations, and a high level of mtDNA diversity has indeed been found within a small population of Amerindians of Vancouver Island. It is possible that the Americas were colonized by very few mtDNA lineages, which nevertheless contained substantial ancient diversity. Sampling from more tribal groups is needed to confirm the limited lineage scenario and, in addition, to establish the phylogenetic relationship of Amerindians in the Americas. Analysis of samples of soft tissue and skeletal and fossil remains of pre-Columbian Amerindian individuals also is vital for the determination of population characteristics prior to European contact.

## 4 PROSPECTS

Mitochondrial DNA will continue to exert a tremendous impact on the assessment of the prehistory and current characteristics of our species. Countless populations of anthropological interest have not yet been sampled or studied. Modern molecular analysis requires ever smaller samples and allows us, with ancient DNA, to look directly into our own past.

*See also* Genetic Analysis of Populations; Mitochondrial DNA; Paleontology, Molecular.

### *Bibliography*

Clayton, D. A. (1991) *Annu. Rev. Cell Biol.* 7:453–478.
Ellingboe, J., and Gyllensten, U. B. (1992) *The PCR Technique: DNA Sequencing.* Eaton, Natick, MA.
Horai, S., et al. (1992) *J. Mol. Evol.* 35:32–42.
Wilson, A. C., and Cann, R. L. (1992) *Sci. Am.* 2:68–73.

**Molds: *see* Fungal Biotechnology.**

# Molecular Genetic Medicine

*Bernice E. Morrow and Raju Kucherlapati*

### *Key Words*

**Gene Targeting** Homologous recombination between exogenous DNA and the genome, resulting in modification of the locus within the genome.

**Homologous Recombination (HR)** Recombination or exchange of information between homologous DNA sequences.

**Polymerase Chain Reaction (PCR)** A primer-pair serves in the amplification of a specific DNA sequence present within a template by successive denaturation, annealing, and extension reactions using the enzyme Taq DNA polymerase DNA polymerase.

**Transfection** Introduction into mammalian cells of exogenous DNA in a vector.

---

Molecular studies have revealed a genetic basis for many human diseases, including rare single-gene genetic disorders and more common diseases such as cancer. There is no rational treatment for these diseases. The only way to cure genetic diseases would be by gene therapy—that is, by correcting the genetic defect in affected target cells by the addition or replacement of the mutant gene with its normal counterpart. Gene targeting by homologous recombination (HR), between exogenous sequences and the genome to modify genes, is undergoing rapid technological development with gene therapy as a goal. Furthermore, gene targeting by HR in pluripotent mouse embryonic stem cells has been used to determine the function of genes involved in development and malignant transformation, as well as to develop animal models for human diseases.

## 1 GENETIC BASIS OF HUMAN DISEASE

More than 5000 different human genetic diseases, caused by single-gene mutations, have been identified. In addition to single-gene disorders, which exhibit a Mendelian pattern of inheritance, it is becoming apparent that many common disorders such as cancer, hypertension, atherosclerosis, and mental illness may have genetic components as well.

For most of these disorders, no permanent therapies are available. A rational approach toward the treatment of disorders involving genetic components is (1) to identify the gene(s) that play a key role in the etiology of the disorder, (2) to understand the role of the gene product in health and disease, and (3) to develop an approach for therapy.

Examples of diseases that have single genetic components are cystic fibrosis and several forms of muscular dystrophy including Duchenne muscular dystrophy (DMD). Molecular genetic approaches are now permitting isolation of the genes that are implicated in some of these diseases. Several such genes have already been isolated, including those for dystrophin, mutated in DMD and the cystic fibrosis transmembrane conductance regulator (CFTR) mutated in cystic fibrosis.

Molecular studies have revealed that all cancers have a genetic etiology that occurs within somatic cells. In addition, several forms of cancers are associated with an established familial pattern of inheritance, including retinoblastoma and familial adenomatous polyposis coli. The mutated genes that give rise to malignant cells fall into two categories: oncogenes and tumor suppressor genes, both of which are involved in malignant transformation.

## 2 GENE THERAPY

One way to cure diseases with genetic etiologies would be to use gene therapy—that is, to correct the defective gene in the affected somatic cell type directly by addition or replacement of this gene with one of the normal type. The requisites for gene therapy include availability of the cloned gene and knowledge of the lesion that causes the disease. In the next step, introduction of the normal gene into cells from the affected tissue isolated from the patient, the DNA must integrate within the genome in a stable manner, and its expression must be regulated appropriately.

Much of the effort in gene therapy has been directed toward introducing a normal copy of the affected gene into cells isolated from patients. Retroviral and adenoviral vectors are proving to be useful as vehicles for transfection. The first two patients on whom gene therapy was attempted had adenosine deaminase (ADA) deficiency, a rare metabolic disorder that results in an impaired immune system. The normal *ADA* gene was packaged into a defective retrovirus, where most of the endogenous viral genes were replaced with the *ADA* gene and used to infect lymphocytes isolated from the patients. The infected cells expressing the *ADA* gene were injected back into the patients and *ADA* expression was partially restored, resulting in an improvement in immune function.

Although viral vectors provide an efficient means of transferring a gene into primary cells, they do not permit correction of the original defect, nor do they allow proper regulation of the introduced gene. The ideal mode of gene therapy is to replace the defective gene with its normal counterpart. The strategy that permits such modification utilizes homologous recombination (HR), a normal cellular process. The process of HR, which permits the reciprocal transfer of DNA between paired chromosomes during meiosis, can be used to directly modify a specific gene within the genome. In this way, exogenous DNA introduced into human cells will be targeted to the appropriate locus by utilizing regions of homology between the input DNA and the target. The directed modification of genes by HR is termed gene targeting. Gene targeting has rapidly evolved as a method to precisely modify genes in cultured cells and promises to be the method of choice for gene therapy in the future.

## 3 GENE TARGETING BY HOMOLOGOUS RECOMBINATION

Much of our understanding of the genetic mechanism of HR has been obtained in bacteria and yeast. The enzymatic steps of HR have been characterized using purified bacterial proteins. When DNA is introduced into mitotic yeast cells, HR occurs between homologous sequences within the vector and the genome. Since HR occurs at a very high efficiency in yeast, this organism has served as a model for developing strategies for gene targeting in mammalian cells.

The first demonstration that exogenous DNA could undergo homologous recombination at a specific site within the human genome was in 1985. A plasmid was constructed for modifying the human β-globin locus. This plasmid contained DNA from β-globin locus and a marker gene, neomycin phosphotransferase *(neo)*, which permits selection of transfected cells. Following molecular analysis, a unique DNA fragment that could be formed only by HR was identified.

Following this initial success, more than 100 different mammalian genes were modified by HR. Utilization of HR permits gene disruption or gene inactivation as well as gene correction. Furthermore, subtle alterations in precise locations within the genome can be achieved.

HR can be performed using either an insertion or a replacement vector. In insertion recombination, an insertion vector is linearized

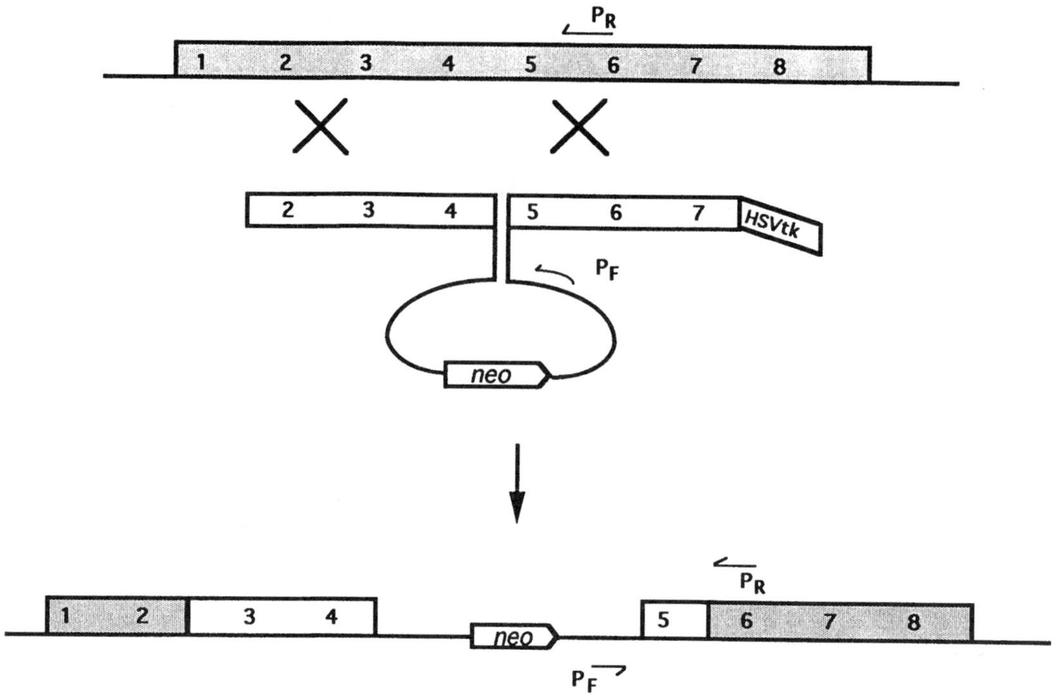


**Figure 1.** Homologous recombination between a replacement vector containing part of the gene to be targeted and the genomic locus. As a result of HR, the neomycin gene and flanking vector sequences disrupt the locus of that particular gene within the genome.

by restriction endonuclease digestion within the sequences to be targeted. Recombination of this vector with its homologous cellular sequences results in the duplication of the homologous sequence. In a replacement event, the two halves of the gene are at opposite ends of the linearized vector and no duplication occurs; however the target sequence is modified such that the vector sequences disrupt the gene.

### 3.1 Selection of HR Events in Mammalian Cells

If gene targeting is to be useful for gene therapy, its efficiency must be high. The efficiency of gene targeting, in turn, depends on two factors, the transfection efficiency and the ratio of homologous to nonhomologous recombination (HR/NHR) events. The stable transfection efficiency depends on the method utilized. Viral vector mediated transfer that does not involve HR confers the highest transfer efficiency of 5 to 100%. Both microinjection and electroporation procedures for transfection have been used in gene targeting. Microinjection, which has a high transfection frequency of $10^{-1}$, is technically difficult. Although the transfection frequency of electroporation is $10^{-3}$, it is the procedure most commonly used because it is the simplest. Although HR/NHR ratios of $10^{-5}$ or less were originally reported, there have been significant improvements in increasing the number of HR events in cultured cells. The use of a linearized target vector, the use of isogenic DNA, and the existence of large regions of homology have all contributed to the improvement of the ratio by several orders of magnitude, to a high of 1 out of 2, and routinely, $10^{-1}$.

When these high frequencies of HR are not achieved, methods to select for targeting events and to rapidly screen for HR are available. One of the most useful enrichment methods has been termed positive negative selection (PNS), a method introduced by M. Capecchi and co-workers. In this strategy (Figure 1) the *neo* gene is inserted between the two regions of homologous sequence that will be used for gene targeting. The presence of the expressed *neo* gene confers resistance to the amino glycoside analogue G418. Cells lacking the *neo* gene will be killed, permitting the growth of transfected cells expressing the *neo* gene. At one or both ends of the targeting sequence, the replacement vector also contains the herpes simplex thymidine kinase (*HSV*tk) gene (Figure 1), which serves as the negative selectable gene. Cells containing a functional copy of this gene are killed by incubation with the drug ganciclovir (GANC). The *HSV*tk gene will be removed following an HR, but not an NHR event, resulting in negative selection of random insertion events. Significant levels of enrichment for homologous recombinants using the PNS technique have been reported.

### 3.2 The In–out Method for Gene Targeting

It is desirable in many circumstances to alter only one or a few nucleotides in a particular gene. Such precise modifications are ideal for gene therapy. Furthermore, these modifications are useful for functional studies of a particular alteration or for producing animal models containing specific alterations that occur in human diseases. One method that permits such modification utilizes a two-step recombination procedure. The first "in" step is a gene-targeting step that uses an insertion vector containing the subtle mutation within the targeting gene. The appropriately targeted cell will have a gene duplication. The second "out" step is a spontaneous HR event between the duplicated region, either by intrachromosomal recombination or via unequal sister chromatid exchange between the homologues. The desired product is a single-copy gene containing the subtle modification, lacking the vector and selection cassettes. This procedure has been tested using the X-linked hypoxanthine guanine phosphoribosyltransferase *(HPRT)* gene—because there is only one copy of the gene, its presence as well as its absence can be easily selected for. An *HPRT*⁻, mouse embryonic stem (ES) cell line was targeted with an insertion vector that corrected the *HPRT* gene and contained a small insertion within the gene. *HPRT+* revertants were selected by exposure to hypoxanthine–aminopterin–thymidine (HAT) medium. In the second step, 6-thioguanine was also used to recover the *HPRT*⁻ revertants. Since the small insertion caused the loss of a restriction site, Southern blot analysis was used to identify the clones containing the appropriately modified gene. The ''in–out'' (also termed the ''hit and run'') procedure was also used to alter the developmental gene, *Hox* 2.6, in ES cells, leaving no vector sequences to interfere with expression of other surrounding genes within the *HOX* gene cluster.

### 3.3 Methods to Analyze HR Events

The polymerase chain reaction (PCR) is a rapid and highly sensitive method to screen for specific gene targeting events. Homologous recombination events usually lead to novel junction fragments, which are not present in the input or target sequences. To detect them, two PCR primers are used, $P_F$ and $P_R$ (Figure 1). One primer ($P_F$) contains sequences present in the exogenous incoming vector DNA. The second primer, in opposite orientation to the first ($P_R$), contains sequences within the HR target. These two primers are used to amplify the template DNA between them. Only when an HR event occurs is the appropriate PCR amplification product formed. Southern hybridization of restriction endonuclease digested DNA is usually performed to confirm the recombination event.

## 4 EMBRYONIC STEM CELLS

A major recent technological achievement in mouse genetics is the ability to culture pluripotent embryonic stem cell lines. The ES cells are established in culture from mouse embryos at the blastocyst stage. They are derived from the inner cell mass of the blastocyst and can proliferate in culture. The ES cells can be manipulated and, when injected into a host blastocyst, they resume normal development and can contribute to all the tissues of the mouse, including the germ line. The combined use of gene targeting and ES cell technology permits the generation of mice with precise genetic modifications.

## 5 GENE TARGETING IN MICE

Figure 2 shows the steps involved in the formation of a mouse homozygous for an altered gene. First an ES cell line is transfected with a gene targeting vector containing selectable markers. The resulting clones are analyzed for gene targeting events by PCR amplification and Southern blot analysis. Positive cells are purified and are microinjected into the blastocoel of a blastocyst from a mouse containing the black coat color gene and surgically implanted into the uterus of a pseudopregnant foster mother, where development of the embyro ensues. A few mice within the litter will be chimeric, with contribution of cells from the donor ES line and the host blastocyst. The mice can be easily distinguished on the basis of their chimeric fur color of agouti (dark and light) black and banded. Most of the chimeric mice are male and can be mated with female mice containing the black coat color gene. If the ES cells contribute to the germ line of the chimeric mice, they produce gametes carrying the agouti coat color, yielding agouti offspring. However, only half these agouti mice will be a heterozygous for the altered gene. Heterozygotes are mated to generate homozygous mice.

### 5.1 Null Alleles Generated by Gene Targeting in the Mouse

Over the past few years, many genes have been altered by HR and introduced into mice to create homozygous null mutations, either to understand the function of certain genes or to create animal models for human disease. In several instances, the mice containing the null mutations have enriched our understanding of how the genes function within the environment of the whole animal. The analyses of genes involved in embryonic development and of proto-oncogenes have been particularly fruitful.

Oncogenes are directly involved in the formation of malignant tumors. Although many of the corresponding cellular proto-oncogenes have been identified, the function of these genes is not known, nor is it known exactly how the mutations give rise to the alterations in cellular phenotype observed in human cancers. Several proto-oncogenes including c-*myb* and c-*src,* have been disrupted in mice to determine the function of these genes.

The *myb* gene product has been implicated in the process of hematopoiesis. Despite normal yolk sac erythropoeisis, mice that lacked the c-*myb* gene died of anemia at day 15 of gestation as a result of a defect in fetal hepatic erythropoiesis. Therefore, the *myb* gene is involved in a process of erythropoiesis specific to the liver. The c-*src* proto-oncogene encodes a plasma membrane tyrosine protein kinase. Little is known about its role in signal transduction pathways or its downstream targets, although its expression patterns are known. Unexpectedly, mice containing a null mutation of the c-*src* gene have abnormal bone development, leading to osteopetrosis, and die a few weeks after birth. However, other tissues in which the c-*src* gene is expressed, such as the brain, show normal development, indicating that other members of the c-*src* gene family of related tyrosine kinases may complement the absence of c-*src* in these tissues.

### 5.2 Animal Models for Human Diseases

Mice have a gene that is homologous to the human cystic fibrosis transmembrane conductance regulator. In human beings, this protein functions as a cAMP-regulated chloride ion channel. It has also been suggested that it may regulate cAMP-dependent endocytosis and exocytosis in secretory epithelial cells. Many different mutations have been identified in the *CFTR* gene, however deletion of three base pairs in exon 10 is responsible for 70% of all cases of cystic fibrosis. Most of the symptoms observed in these patients cannot be readily explained based on the biochemical function of the protein.

Several scientific groups have inactivated the murine *CFTR* gene in ES cells by HR. Mice homozygous for the disruption of the *CFTR* gene have several features common to those found in CF

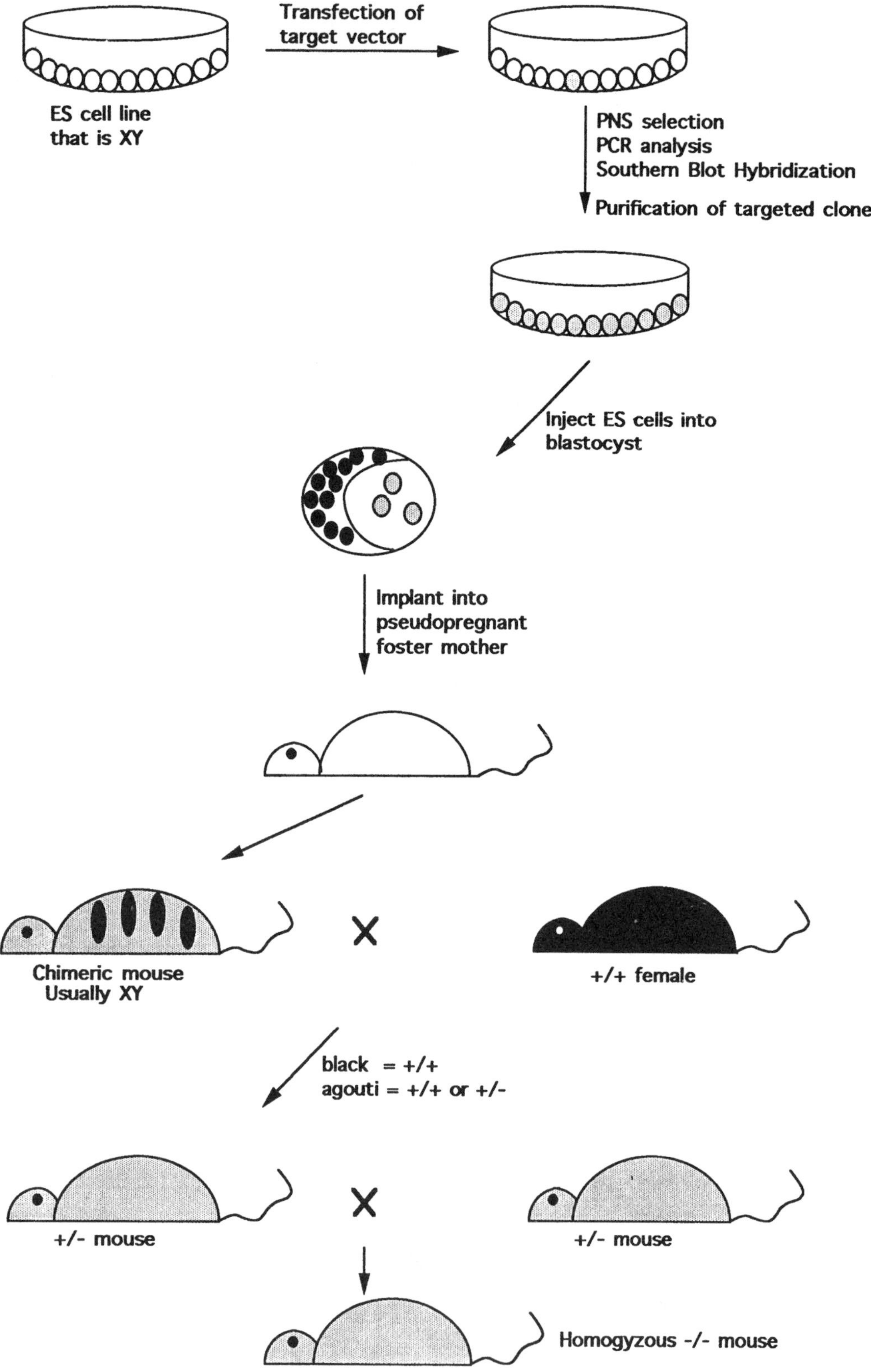


Figure 2. Germ line transmission of a targeted gene introduced into cultured embryonic stem cells, derived from a male embryo and carrying the agouti coat color gene. The targeted cell is selected and amplified. Individual purified targeted ES cells are injected into the blastocyst of a donor mouse carrying the black coat color gene. A mixture of donor and host cells contributes to the formation of the tissues of the chimeric mouse, which is then mated with a normal mouse containing the black coat color gene. If the original ES cell genome is transmitted through the germ line, a heterozygous mouse will be born. The heterozygous mice can be interbred to generate offspring homozygous for the altered gene.

patients. Many of the mice died within 5 days of birth. Most of the deaths were due to intestinal tract blockage and other pathological changes, as observed in human CF. Although the mice do not develop pulmonary infections, the basic complications observed in the mouse resemble the complications in human beings, indicating that this animal model can be used to gain further understanding of the pathology of the disease and may prove useful in development of treatment protocols for human patients.

## 6 CORRECTION OF MUTATED GENES BY HR

The first demonstration of a targeted correction of a mutant gene in mouse ES cells was presented in 1987. An ES cell line containing a spontaneous deletion of part of the *HPRT* gene and its promoter was targeted with an insertion vector containing the missing sequences. Following transfection, *HPRT*+-targeted cells were selected for by culturing in HAT medium. The clones containing the corrected *HPRT* gene were introduced into blastocysts and produced germ line competent chimeric mice.

The autosomal recessive disease sickle cell anemia is caused by a point mutation in the human β-globin gene, resulting in a dysfunctional hemoglobin. There is no known cure for this devastating disease. The goal of gene therapy for sickle cell anemia would be to replace the mutant protein with the correct one by targeting the erythrocyte precursor cells. Then the modified hemoglobin would be injected into patients to repopulate the bone marrow with the normal gene. The human β-globulin gene present in a somatic cell hybrid line expressing the abnormal hemoglobin has been transfected with a replacement vector containing the *neo* gene. The $\beta^S$-globulin allele was replaced with the wild type, $\beta^A$ allele. Furthermore, regulation of $\beta^A$-globulin expression appeared to be normal.

## 7 FUTURE DIRECTIONS

One of the potentials of gene targeting technology is to correct human genes by HR. The advantage of HR is that specific genes can be modified in a precise fashion without affecting other genes. The feasibility of gene targeting has been demonstrated in a number of different cell types for several different genes. One major stumbling block—the low frequencies of HR—seems to be disappearing as the development of efficient methods of gene targeting proceeds. The combination of advanced methods of gene modification by HR, together with our ability to isolate and culture multipotent stem cells from different tissues, promises a future in which many genetic diseases may be amenable to therapy.

*See also* Genetic Immunization; Human Genetic Predisposition to Disease; Motor Neuron Disease; Oncogenes; Retinoblastoma, Molecular Genetics of; Tumor Suppressor Genes.

### *Bibliography*

Bradley, A., Hasty, P., Davis, A., and Ramirez-Solis, R. (1992) Modifying the mouse: Design and desire. *Bio/Technology,* 10:534–540.

Capecchi, M. R. (1989) Altering the genome by homologous recombination. *Science,* 244:1288–1292.

Koller, B. H., and Smithies, O. (1992) Altering genes in animals by gene targeting. *Ann. Rev. Immunol.* 10:705–730.

Singer, M., and Berg, P. (1991) *Genes and Genomes.* University Science Books, Mill Valley, CA.

Thompson, M. W., McInnes, R. R., and Wellard, H. F., Eds. (1991) *Genetics in Medicine,* 5th ed. Saunders, Philadelphia.

## Molecular Motors: *see* Motor Proteins.

# Motor Neuron Disease

*Michael Sendtner*

### *Key Words*

**Amyotrophic Lateral Sclerosis (ALS)** A clinical syndrome defined by degenerative changes of lower and upper motorneurons.

**Motor Neuron Disease (MND)** A variety of sporadic and inherited diseases characterised by muscle weakness due to primary degeneration of motoneurons of the spinal cord, brain stem motor nuclei, and populations of central motoneurons.

**Neurotrophic Factors** Polypeptide molecules that support the survival of specific populations of neuronal cells such as spinal motoneurons in culture and in vivo, via specific signal-transducing receptors.

The term ''motor neuron disease'' (MND) refers to a variety of devastating neurological disorders caused by a generalized degeneration of motor neurons. The common clinical manifestations of such disorders are progressive paralysis, muscle weakness, and very often death by respiratory failure. Inherited and sporadic forms are distinguished, and further clinical distinctions are made according to the onset of the disorder and whether the disease involves only spinal and bulbar motoneurons (spinal muscular atrophy, SMA) or both the lower and upper motoneurons. The most common form of MND is one that attacks adults: sporadic amyotrophic lateral sclerosis (ALS), a noninherited, still incurable fatal disorder with well-known clinical appearance. Progress in the identification of mechanisms leading to the degeneration of motoneurons, and of pharmacological approaches that can interfere with them, has raised hopes that this disorder might be treatable in the future.

## 1 CLINICAL APPEARANCE, PATHOLOGY, AND PATHOPHYSIOLOGY

### 1.1 Inherited Forms of Motor Neuron Disease: Genetic Analysis and Implications for the Pathophysiological Mechanisms Leading to MND

More than 50 different forms of inherited MND have been distinguished on the basis of clinical parameters, and doubts have been raised as to whether such extended subcategorization meets the requirements for rational diagnosis and prospective therapy. The gene defects for several forms of inherited MND have been clarified: a major form of X-linked spinal muscular atrophy is caused by an increase in the size of a polymorphic tandem CAG repeat in the coding region of the androgen receptor gene on the X chromosome. This defect leads to progressive muscle weakness and atrophy, as well as gynecomastia and reduced fertility in affected males.

Another gene defect for a major form of familial amyotrophic lateral sclerosis (ALS) has been identified: an as-yet uncharacterized abnormality of the $Cu^{2+}/Zn^{2+}$-dependent superoxide dismutase gene is associated with a disease characterized not only by the degeneration of both lower and upper motoneurons but also by a marked loss of fibers in other parts of the central nervous system, such as the posterior tract of the spinal cord and the spinocerebellar tract.

It appears likely that some of the yet-to-be-characterized forms of familial MND are allelic for these gene defects, but others might be caused by unrelated mechanisms. This is supported by evidence from animal models. For example, at least three different mouse models exist which show typical signs of degenerative MND: wobbler, pmn (progressive motor neuron disease), and mnd (motor neuron disease). Although the gene defects of the mouse mutants are still not known, gene linkage analysis indicates that the chromosomal locations of the underlying gene defects are different and nonallelic. In addition, several transgenic mice have been generated which develop typical signs of MND: overexpression of neurofilament genes or lack of functional ciliary neurotrophic factor (CNTF) expression leads to a predominant impairment and degeneration of spinal and bulbar motoneurons.

Interestingly, the androgen receptor, the $Cu^{2+}/Zn^{2+}$-dependent superoxide dismutase, and neurofilament genes are widely expressed and are not at all specific for motoneurons or cells that are functionally associated with them. It could be concluded from these results that generalized disorders might manifest themselves as MND, possibly because motoneurons are more vulnerable or sensitive to any of these defects than cells of other types. Moreover, these examples demonstrate that different, unrelated gene defects have a similar clinical appearance resembling typical signs of MND, indicating that MND cannot be considered to be a homogeneous disorder.

## 1.2 Basic Pathological Findings of ALS

Increasing evidence from the last few years has shown that the disease process of ALS, the most common form of MND, might not be limited to motoneurons, but there is little doubt that the motor system is the site of predominant manifestation of the disease. As a characteristic feature of ALS (albeit with significant variation from case to case), loss of motoneurons occurs; the remaining cells appear shrunken and dark, with basophilic cytoplasm and pyknosis of the nuclei. These morphological characteristics very much resemble the changes occurring during apoptosis.

Impairment of axonal transport might underlie many other pathological characteristics. Several types of inclusion bodies have been described in the perikarya of motoneurons, such as basophilic inclusions, which are thought to consist mainly of RNA, eosinophilic Bunina bodies, which resemble autophagosomes, and hyaline inclusions, originally thought to be characteristic for familial ALS. In particular, the latter types of inclusion consist of densely packed phosphorylated neurofilaments, and it is believed that an impairment of slow anterograde axonal transport is responsible for their accumulation in the perikarya of motoneurons. The concept of impaired axonal transport is further supported by the finding that more myelinated nerve fibers are detectable in proximal than in distal parts of motonerves of ALS patients. An early stage in the disease is the failure of motoneurons to maintain distal structures such as the end plates. The disease then proceeds as degeneration of the axons in a dying-back fashion and finally leads to degeneration of the cell bodies in the spinal cord and brain stem. Clinical symptoms are expected to become apparent at the stage when functional deficits occur. This coincides with the stage when motoneuron nerve terminals are affected and not after the cell bodies of the cells have degenerated.

## 1.3 Sporadic MND: The Pathophysiological Mechanisms Are Still Obscure

The majority of cases ($> 90\%$) of MND occur sporadically. It is not yet known whether spontaneous gene defects are responsible or whether epigenetic influences, such as intoxication, infection, autoimmune disorders or uncharacterized combinations of such mechanisms are responsible for the disorder. Intoxication with ingredients of cycad plants is thought to be responsible for the ALS–parkinsonism–dementia complex of Chamorro populations in western Pacific islands. Since, however, the disease has been reported to occur up to several decades after exposure to the toxic agents, it is not clear whether the intoxication is itself responsible for the disease or whether it induces a long-lasting increase in the vulnerability of motoneurons to other agents. It has been suggested that β-*N*-methylamino-L-alanine (L-BMAA) or the increased concentrations of glutamate and abnormal excitatory amino acids detectable in the cerebrospinal fluid (CSF) of ALS patients could induce motoneuron degeneration by *N*-methyl-D-aspartate (NMDA) receptors. For this theory to hold, there would have to be NMDA receptors present on motoneurons of ALS patients. At least in the rat, NMDA receptors seem to be expressed on motoneurons only during the early postnatal period, not in adulthood. It remains to be demonstrated that NMDA receptors are indeed expressed by motoneurons and that this potential mechanism of neuronal degeneration is responsible for the degeneration of motoneurons in human ALS patients.

A great variety of other toxic agents have been suggested to be involved in the pathogenesis of MND, such as heavy and light metals, in particular aluminum. The experimental injection of aluminum into the CSF leads to an accumulation of phosphorylated neurofilaments in the proximal axon region and perikarya of motoneurons, a result that resembles the histopathological findings of transgenic mice overexpressing the H- or L-neurofilament genes. A characteristic of all these experimental conditions is an accumulation of phosphorylated neurofilaments in the motoneuron cell bodies that are associated with impairment of motoneuron function.

All the above-mentioned suggestions on the pathogenesis of MND are well supported by experimental data, but in many patients there are no indications of any kind of intoxication. At present several clinical studies are under way with drugs known to interfere with glutamate or metal intoxications. Their outcomes could be helpful in determining how toxic mechanisms contribute to the pathology of MND. There are many reports of abnormal immune system parameters in MND patients, such as elevated levels of antibodies against gangliosides ($GM_1$), acetylcholinesterase, or $Ca^{2+}$ channels. However, the lack of effect of immunosuppressive therapy of ALS has raised doubts about whether autoimmune mechanisms are primarily responsible for MND. It remains open whether autoimmune mechanisms are primarily involved in the pathogenesis of the disease or are a consequence of the degeneration of motoneurons. Nonetheless, it is conceivable (at least for the anti-$Ca^{2+}$ channel autoantibodies) that such mechanisms could contribute (i.e., by causing acceleration or functional impairment of remaining motoneurons) to the course of the disease.

In summary, many pathophysiological mechanisms have been tentatively identified as responsible for MND, but there is no evidence that interference with any of them results in an effective treatment of the disease. It is premature to deduce that these mechanisms might not be relevant to the pathogenesis of MND, but it must be acknowledged that the basic mechanisms of MND are far from being fully understood.

## 2 IDENTIFICATION OF NEUROTROPHIC FACTORS AND CYTOKINES THAT CAN REGULATE SURVIVAL AND FUNCTION OF MOTONEURONS

Survival of embryonic spinal motoneurons is thought to be dependent on factors obtained from their target tissue, skeletal muscle. It is assumed that such factors are taken up by motoneurons by specific receptors and transported to the perikarya, where they act on specific cellular parameters such as blocking of cell suicide processes, regulation of transmitter synthesis, and other important aspects of motoneuron function. Developing embryos and cultured embryonic spinal motoneurons have been used to identify neurotrophic factors and cytokines capable of supporting the survival of developing motoneurons. Positive effects have been observed for CNTF, brain-derived neurotrophic factor (BDNF), neurotrophins 3 and 4 (NT-3, -4), insulinlike growth factor 1 (IGF-I), and members of the fibroblast growth factor gene family (bFGF, aFGF, and FGF-5). The tissue distribution and temporal expression patterns of these factors differ widely, and it is not yet resolved how the individual factors contribute to the maintenance of motoneuron survival and function, both during development and in the adult.

Thus far, best-characterized neurotrophic factor for motoneurons is CNTF. Interestingly, CNTF is not expressed by the target tissue of lower motoneurons, the skeletal muscle, but by myelinating Schwann cells. The function of this factor seems to be restricted to late postnatal stages, and new evidence suggests that when this factor is absent, there is impairment of motoneuron function in the adult organism. In contrast, many other molecules such as BDNF, IGF-I, and FGF-5 are expressed by skeletal muscle, thus being candidates for target-derived neurotrophic factors for spinal motoneurons. At least some of these factors (CNTF, BDNF, to a lesser degree also NT-3, NT-4, and IGF-I) are highly efficient in rescuing motoneurons both from normal cell death in developing embryos and from lesion-induced cell loss in young postnatal rats. In murine animal models for MND, moreover, CNTF has been shown to be capable of rescuing motoneurons from degeneration. CNTF counteracts the degeneration of motoneuron axons and cell bodies observed in pmn mice, leading to a marked increase in muscle strength. Similar results have been found with CNTF in the genetically unrelated wobbler and mnd mice. These observations raise hope that neurotrophic factors might be used for treatment of human MND. The potential actions of this and other neurotrophic factors are being analyzed, and their effects in patients suffering from ALS and other forms of MND are undergoing tests.

## 3 FUTURE PROSPECTS FOR TREATMENT OF MND

In the past, a large number of clinical trials with a variety of drugs and therapeutical strategies were carried out to find an effective treatment for MND. Among them were plasmapheresis (with the intention of removing toxic or autoaggressive molecules), as well as drugs known to be effective in other autoimmune diseases, such as corticosteroids, ACTH, cyclosporin A, and cyclophosphamide. None were effective. Similar negative effects were reported in trials with growth hormone and thyrotropin releasing factor.

A driving force behind some of the current therapeutic trials for the treatment of MND is the assumption that degeneration of motoneurons might be caused by mechanisms similar to those responsible for the death of other CNS neurons after glutamate overexposure. Therefore considerable attention has focused on clinical trials with glutamate receptor antagonists and glutamate release inhibitors such as dextromethorphan and Riluzole. The recent finding that a form of familial ALS is associated with an alteration of the $Cu^{2+}/Zn^{2+}$-dependent superoxide dismutase gene has drawn attention to the possible role of free radicals in the pathogenesis of the disease. The outcome of therapeutical trials with radical scavengers such as *N*-acetylcysteine is eagerly awaited.

Neurotrophic factors such as CNTF and BDNF are known to be active in rescuing motoneurons from degeneration in vivo under experimental conditions (e.g., the period of naturally occurring motoneuronal cell death in chick embryos; after lesion of motoneuron axons in newborn rats). In addition, CNTF has been shown to be highly effective in preventing motoneuron degeneration in animal models for MND such as the pmn mouse. These results suggest that neurotrophic factors could interact with at least some of the pathophysiological mechanisms responsible for the degeneration of the motoneurons and make the cells resistant to a variety of influences that would otherwise result in functional loss and cell death of the motoneurons. However, the pharmacokinetics of these factors and possible major side effects are problems that could limit the potential usefulness of such agents in the treatment of MND.

Neurotrophic factors such as neurotrophins and CNTF are not secreted into the bloodstream but are released locally to act on their target cells. It is not clear whether the systemic application of these factors would be sufficient to rescue degenerating motoneurons, in particular the upper motoneurons that reside in the central nervous system. The first results from clinical studies with these factors are expected during the next few years, and it is hoped that the outcome of all the above-mentioned trials will result in a better understanding of the most promising ways for treatment of MND.

*See also* Human Genetic Predisposition to Disease; Muscle, Molecular Genetics of Human.

### *Bibliography*

Appel, S. H. (1993) Excitotoxic neuronal cell death in amytrophic lateral sclerosis. *Trends Neurosci.* 16:3–4.

Arakawa, Y., Sendtner, M., and Thoenen, H. (1990) Survival effect of ciliary neurotrophic factor (CNTF) on chick embryonic motoneurons in culture: Comparison with other neurotrophic factors and cytokines. *J. Neurosci.* 10:3507–3515.

Chou, S. M., and Norris, F. H. (1993) Amyotrophic lateral sclerosis: Lower motor neuron disease spreading to upper motor neurons. *Muscle Nerve* 16:864–869.

Côté, F., Collard, J. -F., and Julien, J. -P. (1993) Progressive neuronopathy in transgenic mice expressing the human neurofilament heavy gene: A mouse model of amyotrophic lateral sclerosis. *Cell* 73:35–46.

Delbono, O., García, J., Appel, S. H., and Stefani, E. (1991) IgG from amyotrophic lateral sclerosis affects tubular calcium channels of skeletal muscle. *Am. J. Physiol. Cell. Physiol.* 260:C1347–C1351.

Kalb, R. G., Lidow, M. S., Halsted, M., and Hockfield, S. (1992) N-Methyl-D-aspartate receptors are transiently expressed in the developing spinal cord ventral horn. *Proc. Natl. Acad. Sci. USA.* 89:8502–8506.

La Spada, A. R., Wilson, E. M., Lubahn, D. B., et al. (1991) Androgen receptor gene mutations in X-linked spinal and bulbar muscular atrophy. *Nature* 352:77–79.

Oppenheim, R. W. (1991) Cell death during development of the nervous system. *Annu. Rev. Neurosci.* 14:453–501.

Rosen, D. R., Siddique, T., Patterson, D., et al. (1993) Mutations in Cu/Zn superoxide dismutase gene are associated with familial amyotrophic lateral sclerosis. *Nature* 362:59–62.

Rowland, L. P. (1991) Amyotrophic Lateral Sclerosis and Other Motor Neuron Diseases, Vol. 56 in Advances in Neurology. Raven Press, New York.

Sendtner, M., Schmalbruch, H., Stöckli, K. A., et al. (1992) Ciliary neurotrophic factor prevents degeneration of motor neurons in mouse mutant progressive motor neuronopathy. *Nature* 358:502–504.

Smith, R. A. (1992) Handbook of Amyotrophic Lateral Sclerosis. Dekker, New York.

Xu, Z., Cork, L. C., Griffin, J. W., and Cleveland, D. W. (1993) Increased expression of neurofilament subunit-L produces morphological alterations that resemble the pathology of human motor neuron disease. *Cell* 73:23–33.

# Motor Proteins

*Ravindhra G. Elluru, Janet L. Cyr, and Scott T. Brady*

## *Key Words*

**Cytoskeleton** The structural framework of eukaryotic cells, typically composed of a combination of microtubules, intermediate filaments, and microfilaments.

**Dynein** A family of minus-end-directed, microtubule-based motor proteins widely distributed in microtubule-containing cells. Functions of dyneins may include mitosis, some membrane-bound organelle movements, and ciliary beating.

**Intermediate Filaments** A family of filamentous cytoskeletal structures that are 10 nm in diameter and many micrometers in length in most metazoan cells. They are cell-type specific, but share many characteristics of gene and protein structure. Examples include keratins, vimentin, desmin, neurofilaments, and the nuclear lamins. Intermediate filaments play primarily a structural role in cells and may be important in determining cell morphologies. Intermediate filaments are not present in all cell types, and cytoplasmic intermediate filaments are not found in arthropods.

**Kinesin** The superfamily of microtubule-based motor proteins which includes both plus- and minus-end-directed varieties and is widely distributed in microtubule-containing cells. Functions of kinesins may include membrane-bound organelle movement and mitosis. Also used specifically for the defining member of the superfamily (other members are considered to be kinesin-related proteins).

**Mechanochemical Enzyme** An enzyme that converts chemical energy in the form of nucleoside triphosphates to mechanical energy such as force or motility.

**Membrane-Bound Organelles** Subcellular structures enclosed by a limiting membrane; examples include endoplasmic reticulum, Golgi apparatus, mitochondria, lysosomes, endosomes, and secretory vesicles.

**Microfilaments** A family of polar cytoskeletal structures, 4–6 nm in diameter and highly variable in length, found in all eukaryotic cells. They are composed of actin and a variable set of accessory proteins which are characteristic of cell type and microfilament function. Actins, a small family of highly conserved globular proteins, include both muscle and nonmuscle forms. Microfilaments correspond to the thin filaments in the muscle sarcomere and are a major constituent of the membrane cytoskeleton. They play important roles in cell structure and cellular dynamics. Microfilaments are particularly important in cell locomotion and form the tracks used by myosins.

**Microtubules** A family of polar cytoskeletal structures, 24 nm in diameter and many micrometers long, found in most eukaryotic cells. They are built from heterodimers of α- and β-tubulins plus a variable set of associated proteins (MAPs), which are characteristic of cell type and microtubule function. Microtubules appear as hollow tubules with walls made up of 12–15 protofilaments formed by the highly conserved tubulins. They are usually organized by interaction with structures in the cell center and have a uniform orientation. Microtubules play important roles in cell structure and cellular dynamics. Structural roles include forming the scaffolding of the mitotic spindle, organizing of other cytoplasmic structures, and constituting the structural core of cilia and flagella. Functions in cellular dynamics include changes in cell morphology and serving as tracks for movements mediated by kinesins and dyneins.

**Mitosis** The complex set of processes in cell proliferation by which the genetic material is replicated, divided, and segregated into two equal parts so that each daughter cell receives an identical complement of the mother cell's genome and appropriate cytoplasmic constituents.

**Motor Proteins** Mechanochemical enzymes involved in locomotion or transport.

**Myosin** A family of plus-end-directed, microfilament-based motor proteins found in all eukaryotic cells. Functions of myosins may include muscle contraction, cellular locomotion, and some membrane-bound organelle movement.

**VEC-DIC Microscopy** A type of video light microscopy in which video image processing enhances the contrast of images obtained by differential interference contrast optics. This method allows the real-time detection in living preparations of cytoskeletal and membrane structures of 50 nm or less.

Motor proteins are mechanochemical enzymes that convert energy from hydrolysis of nucleotides to mechanical force. This mechanical force is used to do work moving intracellular components (epg., chromosomes, cytoskeletal elements, synaptic vesicles, mitochondria), cells, tissues, and organisms. Generally, motor proteins bind a structure such as a mitochondrion and move it relative to a cytoskeletal structure—for example, a microtubule—which is tethered to other cytoskeletal elements or to the underlying cell substrate. Motor proteins play roles in events as diverse as chromosome segregation, nuclear migration and fusion, and membrane-bound organelle transport, flagellar beating, cell migration, muscle contraction, and morphogenesis.

Considerable heterogeneity exists in motor proteins. Despite differences at many levels from primary sequence to enzymology and cell physiology, common features allow motor proteins to be grouped into three families: kinesins, dyneins, and myosins. Variations both within and between families generate specificity needed for motor proteins to carry out a diverse set of functions, while similarities provide a common mechanism of action. Study of mechanochemical enzymes provides a basis for understanding cellular processes mediated by motor proteins, including intracellular maintenance and repair, cell division, embryogenesis, and cell migration.

## 1 CELLULAR AND SUBCELLULAR DYNAMICS

Intracellular components must be transported from one site to another because the internal milieu of eukaryotic cells is highly organized, with various cellular domains carrying out specialized functions. For example, DNA replication and transcription, protein synthesis, and energy metabolism are each restricted to specific cellular compartments. Such compartmentalization is essential for biological function, but it imposes certain requirements. Cellular materials must be moved from sites of synthesis to sites of utilization, with the result that the intracellular environment undergoes constant rearrangement.

For example, during mitosis chromosomes must replicate, line up at the metaphase plate, then migrate toward the two poles of the cell to form daughter nuclei. For daughter cells to receive equal genetic complements, such chromosome movements must occur in a precise and reproducible manner. Other specialized functions, such as contraction or locomotion, also require extensive intracellular reorganization. Thus, muscle contraction entails an active rearrangement of the entire cytoskeleton. Protists and some cells in multicellular organisms (embryonic stem cells, lymphocytes, etc.) can migrate through a liquid medium or soft tissue. Such movements require continuous and substantial change in structures like filopodia and cilia.

The movement in all these examples requires generation of force. Motor proteins evolved to perform this role by hydrolyzing nucleotides and converting the resultant energy to mechanical work. Considerable diversity exists in the types of work to be performed, and this functional diversity is reflected in the existence of motor proteins that differ in physical, enzymatic, and physiological properties. The goal here is to identify defining characteristics of motor proteins, and to show how variations in motor proteins allow them to carry out a wide array of functions.

## 2 THE THREE FAMILIES OF MOTOR PROTEINS: KINESIN, DYNEIN, AND MYOSIN

Mechanochemical enzymes can be divided into three families: kinesins, dyneins, and myosins. Categorizations are based on common properties shared by all members of a family. These ''signature'' traits not only delineate the three motor protein families but serve as a framework to understand novel motor proteins. Characteristic properties for each family is outlined in Table 1.

### 2.1 What Makes These Polypeptides Motor Proteins?

Two criteria must be satisfied to qualify a polypeptide as a motor protein: it must be able to obtain energy via nucleotide hydrolysis, and it must use this energy to perform mechanical work. Since many enzymes can hydrolyze nucleotides, the ability to convert such energy to mechanical work is particularly critical. Historically, however, this ability has been difficult to demonstrate directly at the molecular level.

A variety of assays exist to demonstrate whether a polypeptide has nucleotide phosphatase activity. In vitro assays employing purified enzyme are the most rigorous. Typically, nucleoside triphosphate radiolabeled on the $\gamma$-phosphate is added to an incubation mixture, and phosphatase activity is measured by release of radiolabeled orthophosphate. Different radiolabeled nucleotides can be added to determine the specificity of the enzyme's nucleotide phosphatase activity. All known motor proteins are ATPases. Basal ATPase activity is low in all three motor protein families but dramatically increases in the presence of an appropriate cytoskeletal element.

In vitro assays are also used to demonstrate force generation and performance of work by motor proteins. In the most common assay, the investigator attempts to pass cytoskeletal structures across a microscope coverslip coated with the motor protein to be tested. Depending on the motor, either microtubule or microfilament segments are added to the coverslip along with ATP. If the

**Table 1** Properties of the Kinesin, Dynein, and Myosin Families

| Kinesins | Dyneins | Myosins |
|---|---|---|
| Microtubule-based motors | Microtubule-based motors | Microfilament-based motors |
| Binding to microtubules stabilized by AMP-PNP | Released from microtubules by AMP-PNP or ATP | Released from microfilaments by AMP-PNP or ATP |
| Microtubule activated ATPase | Microtubule-activated ATPase | Microfilament-activated ATPase |
| Not sensitive to UV-vanadate cleavage | Sensitive to UV-vanadate cleavage | Less sensitive to UV-vanadate cleavage |
| Most move toward the plus end of microtubules (plus-end-directed motors) | Most thought to move toward the minus end of microtubules (minus-end-directed motors) | Move toward the plus end of microfilaments (plus-end-directed motors) |

polypeptide has mechanochemical activity, the microtubules or microfilament will glide across the coverslip in the presence, but not in the absence of ATP. Movement of cytoskeletal elements can be observed using video-enhanced contrast/differential intereference contrast (VEC-DIC) or video fluorescence microscopy.

These in vitro assays are limited in that they document the potential of a protein to perform work but cannot identify the physiological role of a motor. To this end, alternate approaches must be which combine in vivo analyses with specific functional probes (antibodies, pharmacological agents, etc.) or genetics (mutant phenotypes, antisense oligonucleotides, etc.). Representatives of each family of motor protein have been found to satisfy both criteria, but only a fraction of all putative motor proteins identified to date have been characterized in sufficient detail to permit definition of their biological functions. Since few polypeptides have been rigorously proven to be motor proteins, the focus will be on selected, well-characterized examples from each motor protein family. Other less well-characterized members of each family exist, however, and new members are continually being discovered. The total number of distinct motor proteins in a given cell remains a matter of speculation.

### 2.2 Background and Discovery

Myosin, the first motor protein to be identified, initially was characterized as the mechanochemical enzyme of muscle fibers, where it is the main source of force generation during muscle contraction. The myosins possess a low basal ATPase activity, which is substantially enhanced by the presence of actin microfilaments. All myosins exert their force on actin filaments, not microtubules. Although myosins originally were described in striated muscle, other such proteins have been identified in both muscle and nonmuscle cells. At least seven distinct classes of myosins have been identified, which play a role in a diverse array of actin-based motility events ranging from muscle contraction to organelle motility.

The two best characterized classes of myosins are myosin I and myosin II. Myosin II includes the myosins of smooth and striated muscle, as well as in nonmuscle cells, and dimerizes to form a two-headed rod; these rods associate to form thick, bipolar filaments. Myosin II is thought to play similar roles—enhancing contractility and organizing the cytoskeleton—in all cell types. Myosin I differs from myosin II in both size and function. Myosin I polypeptides are about half the size of myosin II and do not form bipolar filaments. Some myosin I molecules may interact with membranes and play a role in the movement of membrane-bound organelles or the plasma membrane along actin filaments. The distribution and functions of the other five myosin classes are less well understood.

Dyneins were the next motor protein family to be identified. In the 1960s, dyneins were shown to be the motor protein responsible for axoneme or ciliary movements. Axonemal or flagellar dynein forms the outer and inner arms in 9+2 microtubule structure of cilia and flagella. In the presence of ATP, dynein causes flagella to undergo a whiplike motion, which may serve to propel ciliated cells (protists and sperm) through a liquid, clear debris from the laryngeal pharynx, and facilitate migration of specific stem cells during embryogenesis. Multiple dynein isoforms are found within a single flagellum.

Another dynein family member, MAP1C dynein, was discovered in 1986. In contrast to axonemal dynein, which is found only in cilia and flagella, MAP1C dynein has a cytoplasmic localization and is known as cytoplasmic dynein. Cytoplasmic dyneins may play a role in the movement of membrane-bound organelles along microtubules and possibly in the movement of cytoplasmic microtubules.

Kinesin was the last type of motor protein to be discovered. Kinesins were first described during studies of membrane-bound organelle transport in neurons. The axon is an extremely dynamic cellular domain, filled with organelles and structures in a state of continuous movement. Some organelles move in the anterograde direction (from the cell body toward the synaptic terminal) and other move in the retrograde direction (from terminal back to cell body) at rates up to 400 mm/day (2–4 μm/s). Pharmacological manipulations demonstrated a requirement for ATP and microtubules in organelle movements.

Introduction of a nonhydrolyzable analogue of ATP, adenylylimidodiphosphate (AMP-PNP) had a remarkable effect on the motility of organelles, freezing them in place on the microtubules. This observation not only demonstrated the existence of a new type of mechanochemical ATPase but provided a strategy to isolate the ATPase polypeptides. The ATPase bound to microtubules in the presence of AMP-PNP and was released from microtubules by ATP. In 1985 this strategy was used to purify a set of polypeptides that met all criteria for a mechanochemical enzyme involved in organelle movement in axons. This ATPase, kinesin, existed in a wide variety of nonneuronal cells, as well as in neurons. Soon thereafter, a number of proteins containing sequences homologous to the kinesin motor domain, but otherwise distinct, were identified. Discovery of these kinesin-related proteins generated the idea that the kinesin initially discovered in neurons defines a superfamily of related polypeptides. Members of the kinesin superfamily play roles in organelle transport, mitosis and meiosis, and possibly other microtubule-based motility events.

### 2.3 Diversity Among the Motor Proteins

Members of the three motor protein families are characterized by a distinct set of properties, shared by each other member of a given family. These ''signature'' properties can be used to distinguish between the three families and to define the class of motor molecule responsible for a particular form of motility. Variations in composition and sequence confer specificity and allow diversity in the functional roles of motor proteins. In the sections that follow, kinesins, dyneins, and myosins are described and compared in order.

#### 2.3.1 Protein Structure and Composition

As originally described, kinesin is a heterotetramer of 380 kDa with two heavy chains (115–130 kDa each) and two light chains (62–70 kDa each). Electron microscopy shows that kinesin is an 80 nm long rod-shaped molecule, with three structural domains: globular heads, rod-shaped stalk, and fan-shaped tail. Two globular heads can be detected, each of which contains both ATP- and microtubule-binding sites. The rod-shaped stalk has a kink in the middle and separates the globular head domain from the fan-shaped tail. The tail domain is thought to be involved in binding to membrane surfaces (see Figure 1). Kinesin-related proteins are diverse

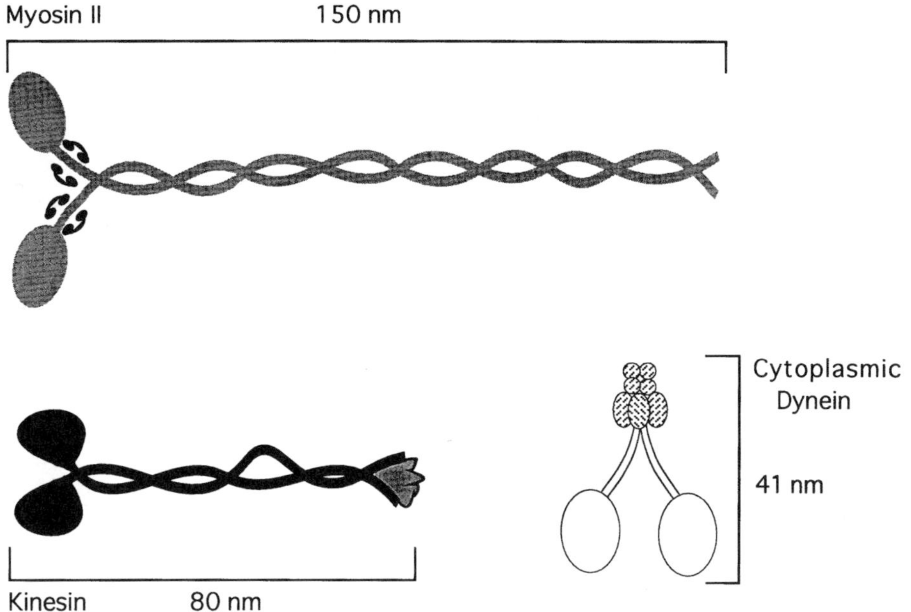


**Figure 1.** Representative members of the myosin, kinesin, and dynein families of motor proteins. All three motor proteins have multiple subunits. Myosin II is comprised of two heavy chain polypeptides each with a globular head and a rod region, which interacts with a second heavy chain to form an α-helical coiled-coil shaft. Each myosin heavy chain has a pair of light chains (regulatory and essential), which bind to their neck regions. The holoenzyme is a heterohexamer. The microtubule-based motor protein kinesin has an organization similar to myosin II. Kinesin also contains two heavy chains oriented in parallel so that both globular head domains are at one end of the molecule. The head domain is followed by a rod domain in which the heavy chains can form an extended α-helical coiled-coil rod with a kink approximately two-thirds down its length. At this point, a kinesin light chain may interact with each heavy chain through an α-helical coiled coil. The opposite end of kinesin exhibits a fan-shaped appearance and is formed by both heavy chain and light chain subunits combined as a heterotetramer. Cytoplasmic and axonemal dyneins are organized differently. They contain a variety of heavy chain, intermediate chain, and light chain subunits. The globular heads of a dynein motor are formed by the heavy chain subunits and are linked by a thin stalk to a basal domain containing variable intermediate and light chains. Figure is not to scale.

in composition, molecular weight, and dimensions, but all appear to include one or more heavy chains with a globular motor domain.

Kinesin globular heads are formed by the heavy chain amino termini, while the heavy chain midregions dimerize to form an α-helical coiled-coil stalk. The fan-shaped tail is formed by the heavy chain carboxy-termini and two light chain subunits. Cells contain multiple isoforms of both heavy and light chains of kinesin, but the purpose of this biochemical diversity is uncertain. Isoforms of the light chain subunits of kinesin may target kinesin to different subclasses of membrane-bound organelles. Kinesin-related proteins may have their motor domains at the amino terminus like kinesin, at the carboxy-terminus, or even in the central region of the polypeptide chain. Little is known about the presence of light chains in kinesin-related proteins, but they are widely assumed to exist.

Like kinesin, dyneins are multisubunit enzymes, but they differ in structural organization. Axonemal dyneins fall into two classes: two-headed enzymes (~1200 kDa) and three-headed enzymes (~2000 kDa). Cytoplasmic dynein or MAP1C dynein is similar in size, morphology, and mass to two-headed axonemal dynein, but it differs in subunit composition. In both two- and three-headed enzymes, heavy chains form globular heads and a thin stalk, which connects to basal domains containing multiple intermediate/light chain subunits. Heavy chain subunits have a mass in excess of 500 kDa and contain both ATP- and microtubule-binding sites. Axonemal dynein and cytoplasmic dynein exhibit considerable homology in the force producing part of the heavy chain. Intermediate and light chain subunits can differ markedly in composition, but homologies may exist between specific subunits. Differences in smaller subunits may reflect adaptation of a basic molecular design to different functions: movement of microtubules relative to each other by axonemal dynein and movement of membrane-bound organelles relative to microtubules by some forms of cytoplasmic dynein.

Myosins are also multisubunit enzymes, and most myosins are elongated, rod-shaped molecules. Myosin II has two structural domains, a globular head domain and a rod-shaped stalk, which is superficially similar to kinesin. The globular amino-terminal head contains functional domains, including one ATP-binding and one actin-binding site per heavy chain. The rod-shaped stalk is formed by dimerization of two heavy chains interacting in a coiled-coil α helix, like kinesin. Unlike kinesin, myosin II tails can interact to form large bipolar myosin thick filaments. The myosin II unit motor contains six polypeptide subunits: two heavy chains (~200 kDa) and two light chain pairs (20 and 16 kDa), with one of each pair of light chains per heavy chain head domain. In contrast, myosin I unit motors have only one heavy chain (~100 kDa) and one light chain that is variable in size. Homologies to myosin II motors are restricted to the head motor domain containing both ATP- and actin-binding sites. Myosin I carboxy-terminal ends are specialized and may contain a nonnucleotide-sensitive actin-binding site, which would allow myosin I to cross-link actin filaments or a membrane-binding site. Myosin light chains are thought to be involved in regulating the mechanochemical activity of myosins. A variety of other myosins have been identified but are less well characterized. These proteins may exist as monomers, dimers, or multimers, but their biology and function remain to be defined.

### 2.3.2 Biochemical and Biophysical Properties

All three motor protein families have an ATPase activity, which is activated by the presence of the appropriate cytoskeletal partner, microtubules (kinesin and dynein) or microfilaments (myosin). In effect, motor proteins utilize ATP only in the presence of appropriate cytoskeletal elements and ATP as the physiologically relevant nucleotide. Maximum rates of nucleotide hydrolysis and translocation appear to be similar for the three families of motor proteins, but specific members may be much less efficient in in vitro assays. Motor proteins have characteristic pharmacological and biochemical properties, which may be used for identification. For example, the three families are similar but not identical in their ability to utilize nucleotides other than ATP.

A major difference between these motor protein families is the effect of AMP-PNP, the nonhydrolyzable analogue of ATP. AMP-PNP stabilizes binding of kinesin to microtubules but weakens binding of dynein to microtubules and myosin to microfilaments, indicating a difference in the enzymatic cycle of three families of motor proteins. In the case of kinesin, hydrolysis of ATP appears to be necessary for the release of kinesin from microtubules. In contrast, binding of ATP to dynein and myosin is sufficient to release them from microtubules or microfilaments, and hydrolysis occurs after release.

Another difference between the three groups is the effect of orthovanadate on ATPase activity. All three ATPases are inhibited by orthovanadate, but effective concentrations vary by a factor of 2 to 10. A more dramatic difference is that dyneins exposed to UV radiation in the presence of orthovanadate and ADP are site-specifically cleaved, a property that has been used to implicate dyneins in cell activities. Myosins are subject to similar cleavage under somewhat harsher conditions, but kinesin does not appear to be subject to UV/vanadate cleavage.

All three types of motor protein mediate directional movements along a cytoskeletal element. Microtubules and microfilaments are polymers formed from tubulin and actin subunits, respectively. Both polymers are polar structures (rapidly growing plus end and slowly growing minus end), and the orientation of each polymer is precisely regulated in cells. Microtubule and microfilament polarity may be determined in a variety of in vitro or in vivo situations, so the direction of translocations can be defined with respect to the polymer polarity. Based on in vitro gliding assays, each motor protein exhibits a characteristic directionality of movement. Many members of the kinesin and myosin families are plus-end-directed motors because they move toward the plus ends of microtubules and actin filaments, respectively. However, some kinesin related proteins are minus-end-directed motors. All dyneins examined to date move toward the minus ends of microtubules, and thus are minus-end-directed motors. Directionality of a specific motor protein is critical for understanding the role it may play in the cell.

## 3 CELLULAR ROLES PLAYED BY MOTOR PROTEINS

How many motor proteins does a cell need for survival and what roles do they serve? This question is particularly intriguing in light of the ever-increasing list of motor proteins in cells. A single cell, such as an intestinal epithelial cell, contains multiple members of all three motor protein families. Although cell-specific motors exist, each cell requires a diverse set of motors for normal function. Individual motor molecules may perform a single cellular function, or the functional roles of these proteins may overlap, with the result that one motor may be able to compensate for loss of another motor if necessary. Two characteristics determine the cellular function of a specific motor protein: the specificity of binding to cargo structures (cytoskeleton, membrane organelles, etc.) and the directionality of movement along a cytoskeletal structure.

The most variable regions of all three motor families are found in the "tail" domains, usually at the carboxyl termini. These domains are thought to specify what a given motor protein can move. Little is known about the nature of this interaction, but evidence exists for specific binding of specific motor molecules to microtubules, microfilaments, various membrane surfaces, and kinetochores. Associated light chains may convey the specificity, but motor heavy chains may also contain relevant targeting information.

The cytoskeleton of cells is highly organized. For example, in Figure 2 microtubules originate near the nucleus of an idealized intestinal epithelial cell and radiate toward the cell periphery. All these microtubules are oriented with minus ends anchored at an organizing center associated with the centrosome, so the plus ends are toward the periphery. In contrast, microfilaments are concentrated in peripheral domains (epg., in microvilli or on the plasma membrane). Membrane microfilaments are anchored at the membrane surface by the plus end, with minus end toward the interior. The situation in real cells may be more complex than shown in Figure 2, but the most common situations are represented.

Organization and polarity of cytoskeletal structures influence the roles played by motor proteins. For example, the kinesin superfamily contains both plus- and minus-end-directed microtubule motors. Kinesin attached to membrane organelles moves them from the Golgi complex near the cell center to the periphery. This feature is important in cells whose membrane organelles must be transported considerable distances—in neurons, for example, organelles are transported down axons that may be a meter or more in length. Kinesin-related proteins with other binding characteristics play different cellular roles, such as mediating chromosome migration from metaphase plate to spindle-pole bodies during mitosis.

Dyneins, like kinesins, are microtubule-based motors, but most dyneins are minus-end-directed motors. Axonemal dyneins produce ciliary beating by attaching stably to one microtubule and exerting a force on an adjacent microtubule in a flagellum. Some cytoplasmic dyneins interact with membrane structures, while others may attach stably to microtubules. Cytoplasmic dyneins appear to move membrane organelles like endosomes and lysosomes from the periphery to the cell center for recycling as well as potentially playing a role in mitosis and meiosis.

Myosins are plus-end-directed, microfilament-based motors. Myosins I and II are thought to be involved in contractile events, such as in muscle or cell locomotion. Actin-rich structures are involved in cell movements, and contraction of the cytoskeleton is necessary for extension and retraction of filopodia by locomoting cells. Other myosin I motors appear to interact with membrane structures and may be important for short-range organelle movements toward the cell periphery. Since regions near the plasma membrane contain few microtubules, myosins may be essential for bringing vesicles to the membrane for fusion events.

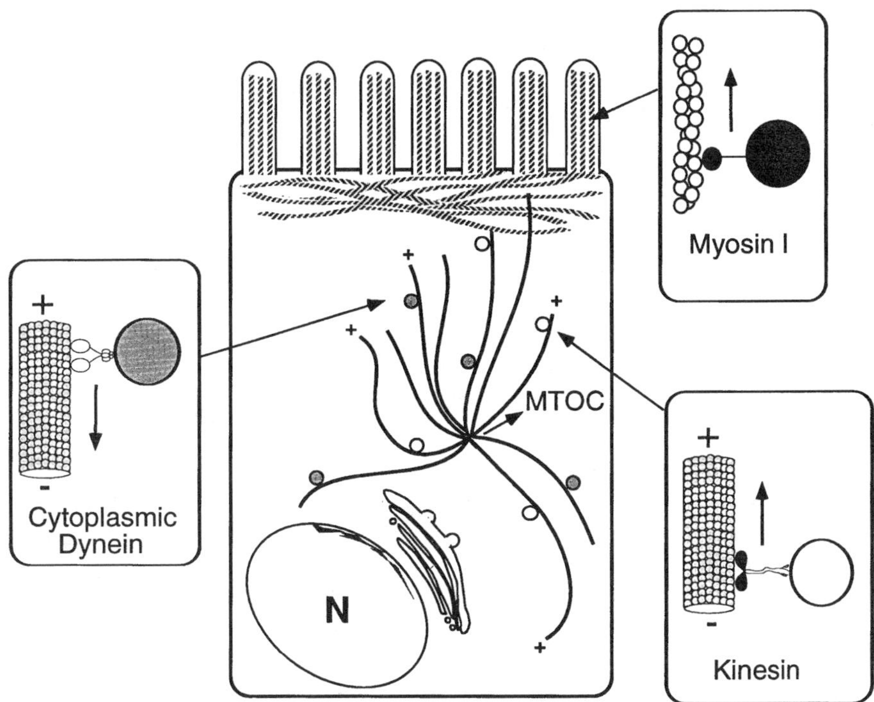


**Figure 2.** Selected cellular roles for the three motor protein families in a specific cell type. Each intestinal epithelial cell contains multiple members of all three motor protein families, which serve a variety of roles in both dividing and interphase cells. Many functions benefit from the inherent polarity of cytoskeletal elements within the cell, which provides a basis for directionality of movements. In most cells, the microtubules originate from a central microtubule organizing center (MTOC) near the centrosome and typically are oriented with the plus ends toward the periphery. In general, kinesins translocate membrane-bound organelles toward the plus ends of microtubules while cytoplasmic dynein moves them toward the minus ends. In most cell types the MTOC from which microtubules radiate is in close proximity to the Golgi apparatus, near the nucleus (N) of the cell. Thus, microtubules serve as "tracks" along which kinesin and dynein translocate Golgi-derived vesicles, mitochondria, and endocytoic organelles. The myosin family members are most prominent in actin-enriched regions of cells such as the cellular cortex near the plasma membrane. In intestinal epithelial cells, the actin corticoskeleton is specialized at the apical surface of the cell to form actin bundles within the microvilli. Myosin I is believed to translocate membrane-bound organelles along the actin bundles of the microvilli to provide essential membrane components for microvilli growth and to change microvilli shape.

## 4 FUTURE PERSPECTIVES

The list of characterized motor proteins continues to grow. Clear criteria for identifying a polypeptide as a motor have been established and used to define three families of motor proteins. Prototypical examples exist for each of three motor protein families, but all have multiple members exhibiting considerable diversity. Most members have been defined by molecular genetic approaches that detect homology with kinesin, dynein, or myosin primary sequence. As a result, cellular roles for many putative motor proteins are uncertain, but a variety of novel roles have been proposed. The near future should produce characterization of additional motors including physiological function, mechanisms of regulation, and the role of motor proteins in disease.

*See also* CELL-CELL INTERACTIONS; CYTOSKELETON-PLASMA MEMBRANE INTERACTIONS; MEMBRANE FUSION, MOLECULAR MECHANISM OF; TRANSPORT PROTEINS.

### *Bibliography*

Brady, S. T. (1991) Molecular motors in the nervous system. *Neuron,* 7:521–533.

Cheney, R. E., Riley, M. A., and Mooseker, M. S. (1993) Phylogenetic analysis of the myosin superfamily. *Cell Motil. Cytoskel.* 24:215–233.

Goldstein, L. S. B. (1993) With apologies to Scheherazade: Tails of 1001 kinesin motors. *Annu. Rev. Genet.* 27:319–351.

McIntosh, J. R., and Pfarr, C. M. (1991) Mitotic motors. *J. Cell Biol.* 115:577–585.

Piperno, G. (1990) Functional diversity of dyneins. *Cell Motil. Cytoskel.* 17:147–149.

Pollard, T. D., Doberstein, S. K., and Zot, H. G. (1991) Myosin—I. *Annu. Rev. Physiol.* 53:653–681.

Porter, M. E., and Johnson, K. A. (1989) Dynein structure and function. *Annu. Rev. Cell Biol.* 5:119–151.

Vallee, R. B. (1993) Molecular analysis of the microtubule motor dynein. *Proc. Natl. Acad. Sci. U.S.A.* 90(19):8769–8772.

———, and Sheptner, H. S. (1990) Motor proteins of cytoplasmic microtubules. *Annu. Rev. Biochem.* 59:909–932.

Wilson, A. K., Pollenz, R. S., Chisolm, R. L., and de Lanerolle, P. (1992) The role of myosins I and II in cell motility. *Cancer Metastasis Rev.* 11:79–91.

# MOUSE GENOME

*Stephen D. M. Brown*

### *Key Words*

**Backcross** A genetic cross involving the intercrossing of two parental strains followed by the backcrossing of progeny (known as $F_1$) to one or other parental strain.

**Genetic Map** A map indicating the linear order of genes along a chromosome as determined by studies of the recombination of gene alleles segregating at meiosis.

**Genome** The sum total of DNA, including all genetic information, carried by a particular organism.

**Mutation** A DNA alteration that may affect the function of a gene, often with deleterious effects on an organism's development, biochemistry, or physiology.

**Physical Map** A map indicating the linear order of DNA fragments along a chromosome.

---

Since the turn of the century, the mouse has been an important organism for the study of mammalian genetics and genetic systems in general. A vast array of mutations have been collected in the mouse, affecting a variety of developmental and physiological processes. Many of these mutations have been genetically mapped on the mouse genome, making the mouse one of the best characterized genetic organisms. The Mouse Genome Project, a worldwide program of mouse genome studies, aims ultimately to provide a dense genetic and physical map of the mouse genome that would be a prelude to acquiring the complete DNA sequence of the mouse. The detailed genetic and physical maps already available have allowed the characterization of some genes involved with the different mutations identified in the mouse, giving researchers unparalleled access to information on the genetic processes of development, differentiation, and cell function. Comparison of the genome maps of mouse and human also allows investigators to relate genetic information obtained in the mouse genome to human biology, including the genetics of disease.

## 1 MOUSE AS A GENETIC TOOL

### 1.1 History

In 1939 some 31 mutations were known in the mouse and only seven genetic linkage groups had been identified. However, the mouse, with its short breeding cycle of around 8 weeks (gestation time, 21 days; 4–6 weeks to sexual maturity) represents an ideal tool for the study of mammalian genetics and genome mapping. A large number of genetic crosses carrying various mouse mutations have been carried out over the last few decades to permit the construction of genetic maps covering the entire 20 chromosomes of the mouse.

### 1.2 The Present Day

Today, more than a thousand mutations are known in the mouse, and the bulk of these have been assigned to one of the animal's 20 chromosomes. In many cases the relative order of mutations on the chromosome is known, forming genetic maps across all the mouse chromosomes. The range of mutations includes gene defects affecting coat color, skin texture, the skeleton, the tail and other appendages, the eye, the inner ear, neurology and neuromusculature, behavior, and reproduction. There are more than 100 mutations affecting the auditory system and more than 100 with effects on the skeleton. In addition, a large number of mutations have been identified that affect the function of a variety of proteins, including enzymes and cell surface antigens. However, for most mutations the underlying gene carrying the mutated DNA sequence is unidentified. The only information available on such a mutation is its genetic position on a chromosome. The Mouse Genome Project studies, involving genetic and physical mapping of mouse chromosomes, aim to provide the resources that will allow access to the relevant genetic locus. This will enable investigators to make use of the myriad mutations in the mouse genome in terms of relating mutational changes to the observed effects on development, differentiation, and physiology.

## 2 GENETIC MAPPING OF THE MOUSE GENOME

### 2.1 Genetic Crosses

To access the gene sequences mutated in the mouse genome, we need to make maps of the DNA sequences on each chromosome. The ultimate goal is to determine the entire DNA sequence of the mouse genome. Nevertheless, a starting point is to construct across the entire mouse genome genetic maps of DNA fragments spaced at sufficiently frequent intervals to allow the eventual linkage of these fragments into a physical map.

One of the pivotal genetic techniques that has markedly improved our ability to construct genetic maps of DNA markers around the mouse genome is the interspecific backcross (see Figure 1). A laboratory strain of mouse (which may be carrying interesting mutations) is crossed to a wild species of mouse—*Mus spretus*. Fertile female progeny can be produced which are subsequently backcrossed to either the parental laboratory strain or the *Mus spretus* strain. Backcross progeny are produced segregating DNA sequence variation from the laboratory or *Mus spretus* strains. DNA from the backcross progeny can be analyzed for a number of types of sequence variation including:

Restriction (enzyme) fragment length variation (RFLV)
Simple sequence length polymorphisms (SSLPs) at short tandem repeats, known as microsatellites

Each DNA marker is called a sequence-tagged site (STS), since at least a portion of its sequence is usually known. The sequence data are readily transferred to another laboratory, and the DNA marker can be reproduced by the polymerase chain reaction (PCR).

Laboratory strains of mice and the wild species *Mus spretus* separated some 2–3 million years ago. Thus, the DNA sequence of the two parental strains is highly diverged, making it relatively easy to identify sequence variation for any DNA marker between the parental genomes. It follows that interspecific backcross progeny can be analyzed for all available DNA markers. The ability to analyze many DNA markers in one backcross in a multipoint fashion makes it possible to order the markers by minimizing the number of observed recombination events (see Figure 1).

### 2.2 Current Genetic Maps of the Mouse Genome

At least 5000 STSs have been mapped across the mouse genome, though they are not evenly spread in each chromosome. Nevertheless, given that the genetic length of the mouse genome is approximately 1800 cM, the genetic map is approaching one STS per centimorgan (cM).

Interspecific backcrosses have the potential to provide very fine genetic resolution: 1000 backcross progeny can provide genetic

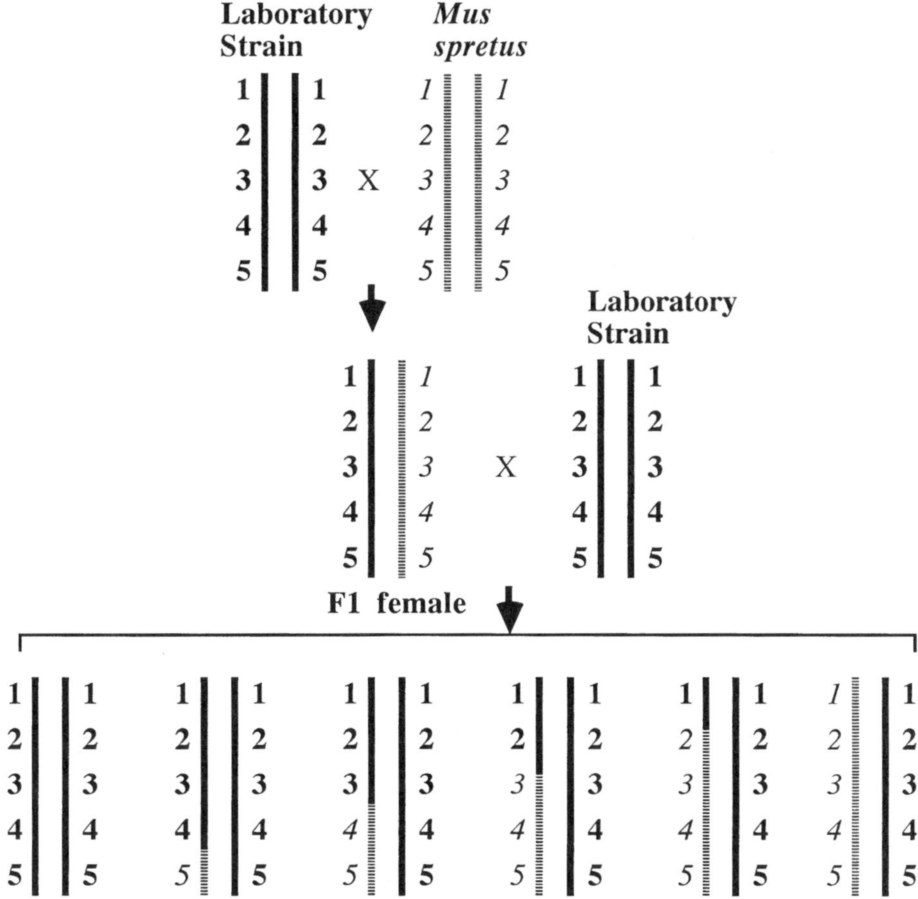


**BACKCROSS PROGENY**

**Figure 1.** The mouse interspecific backcross. Laboratory strains and the wild mouse species *Mus spretus* show extensive DNA sequence variation—most DNA markers show sequence variation between laboratory strains and *Mus spretus*. Laboratory strain sequence variants are shown in bold type and *Mus spretus* strain sequence variants are shown in italics. The backcross illustrates a multipoint analysis of five markers on one mouse chromosome. A number of backcross progeny are recovered derived from recombination between the laboratory strain chromosome and the *Mus spretus* chromosome in the $F_1$ female. The genetic order of markers on the chromosomes is determined by minimizing the number of observed recombination events. Changing the order of markers would necessitate an overall increase in recombination events observed and the appearance of triple recombinants, which usually is very rare.

resolution at the 0.3 cM level with 95% confidence. The DNA content of the mouse haploid genome is $3 \times 10^9$ base pairs. Thus 0.3 cM represents approximately 0.5 Mb of DNA. This is a level of resolution approachable by physical mapping techniques (see Section 3). Indeed, some interspecific backcrosses carrying specific mutations have identified STSs nonrecombinant with the mutation, indicating that the STS lies within 0.5 Mb of the mutated gene. An STS genetic map of the mouse genome approaching 0.5 Mb resolution will be completed by 1996, providing the template for the global physical mapping of the mouse genome. Over 6000 STSs will be used in total to provide a genetic map with sufficient coverage.

## 3 PHYSICAL MAPPING OF THE MOUSE GENOME

### 3.1 Yeast Artificial Chromosome Clones

Yeast artificial chromosome (YAC) vectors carrying centromere, telomere, and ARS (autonomously replicating sequence) sequences are able to replicate in yeast and to carry large inserts of DNA—at least up to the size of the largest yeast chromosome. YAC vectors have been used for the production of large insert mammalian clone libraries from both the mouse and human genomes. Mouse libraries carrying genomic inserts up to 1 Mb have been constructed.

### 3.2 Converting the Genetic Map of the Mouse Genome into a Physical Map

To access all the sequences in the mouse genome, it is necessary to link adjacent STSs with YAC clones. Screening of YAC libraries using PCR allows us to identify YAC clones covering adjacent STSs on the genetic map (see Figure 2). YAC clones carrying an adjacent series of STSs can be linked into an overlapping series of clones called a YAC contig, in a process known as STS YAC contiging. YAC contigs provide all the relevant DNA sequences from a particular region. Some extensive regions of the mouse X chromosome and mouse chromosome 17 have been linked using STS YAC contig analysis. YAC contigs constructed using STSs

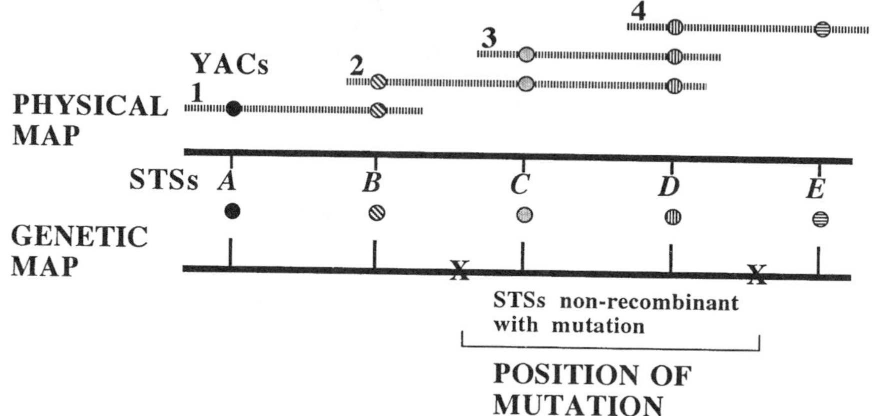


**Figure 2.** STS YAC contiging to create physical maps across genomes. Five STSs (A–E) that have been shown by genetic mapping to lie close together on a chromosome are used to screen for overlying YAC clones. STSs shared by the four YACs allow the assembly of a YAC contig—an overlapping series of YAC clones that provides access to all the DNA sequences in that region. The STSs may be analyzed in an interspecific backcross segregating for a mutation in that region. Some STSs (A, B, and E) may be recombinant with the mutation in backcross progeny, therefore representing STSs that flank the region of the mutation; other STSs may be nonrecombinant (C and D). The genetic analysis defines a region on the overlying YAC physical map in which the mutation is likely to lie. The YAC contig in this region provides the necessary sequences to test for and identify the relevant gene sequences carrying the mutation.

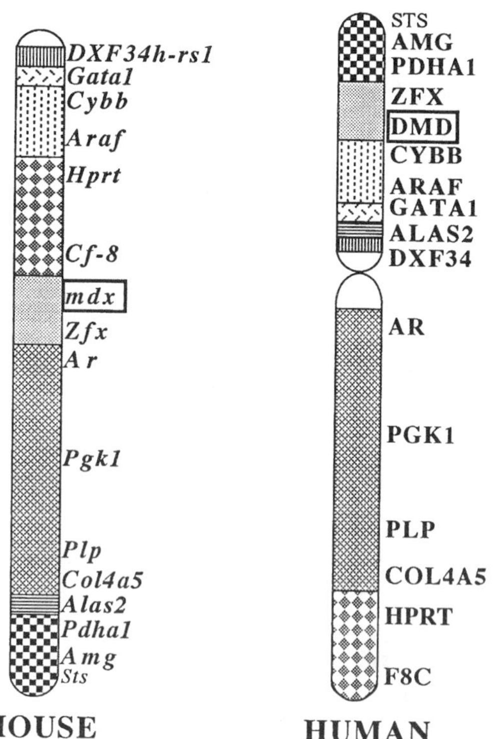


**Figure 3.** Comparative genetic maps of the mouse and human X chromosomes. Comparison of the genetic order of loci between the mouse and human X chromosome reveals a number of regions in which gene content and gene order are conserved (i.e., conserved linkage groups). Each conserved linkage group is indicated by a different shading. In one conserved linkage group mutations causing muscular dystrophy are found in both human (Duchenne muscular dystrophy: DMD) and mouse (*mdx*), suggesting DMD and *mdx* are homologous mutations at the same genetic locus. This possibility has been confirmed: both DMD and *mdx* mutants carry mutations in a gene encoding a muscle protein called dystrophin.

that closely flank a mutation as identified by genetic analysis will contain the relevant gene and are an important starting point for identification of the relevant locus (see Figure 2).

## 4 PERSPECTIVES

### 4.1 Accessing the Physical Map of the Mouse Genome Through Sequencing

YAC clones must be further analyzed to identify the incumbent gene sequences. Techniques include:

- Exon trapping
- Screening of YACs against libraries of cDNA clones
- Identification of sequences conserved between species (sequences that are likely to represent genes)

However, rapid automated DNA sequencing methods, which up till now have been applied in a significant way only for large-scale megabase genome sequencing in lower organisms (such as *Caenorhabditis elegans*) may be increasingly applied to the analysis of large stretches of higher organisms. Computer algorithms are available to search sequence data produced by automated sequencing in the process of identifying appropriate motifs indicative of exons such as splice junction consensus sites.

### 4.2 Further Applications of the Mouse Genome Map

The genetic and physical maps are a powerful resource for comparison with the genome maps of other organisms, especially the human. When the genetic maps of mouse and human are compared, it is found that there are many areas in the two genomes that conserve gene content and gene order. These regions are called conserved linkage groups. Given the density of genes mapped in common between mouse and human, it is believed that the bulk of conserved linkage groups between the two organisms have been identified. Conserved linkage groups allow the identification of:

Putative homologous mutations between the mouse and human genomes (see Figure 3)

Genes in the mouse genome that may be candidates for mutations in the human genome

Genes mapped in the human genome that are potential candidates for mutations in the mouse genome

*See also* Mammalian Genome; Nucleic Acid Sequencing Techniques; Restriction Landmark Genomic Scanning Method.

## *Bibliography*


Anand, R. (1992) Yeast artificial chromosomes (YACs) and the analysis of complex genomes. *Trends Biotechnol.* 10:35–40.

Avner, P. R., Amar, L., Dandalo, L., and Guenet, J. L. (1988) Genetic analysis of the mouse using interspecific crosses. *Trends Genet.* 4:18–23.

Brown, S.D.M. (1992) The physical mapping of the mouse genome. In *Techniques for the Genetic Analysis of Brain and Behaviour: Focus on the Mouse,* D. Goldowitz, D. Wahlstein, and R. E. Wimer, Eds., pp. 215–228. Elsevier, New York.

Burmeister, M., and Ulanovsky, L., Eds. (1992) *Methods in Molecular Biology,* Vol. 12, *Pulsed-Field Gel Electrophoresis.* Humana Press, Towata, NJ.

Copeland, N. G., and Jenkins, N. A. (1991) Development and applications of a molecular genetic linkage map of the mouse genome. *Trends Genet.* 7:113–118.

Lyon, M. F., and Searle, A. G., Eds. (1989) *Genetic Variants and Strains of the Laboratory Mouse,* 2nd ed. Oxford University Press, New York.


# Muscle, Molecular Genetics of Human

*Prem Mohini Sharma*

## *Key Words*

**Dalton** A unit of molecular weight; one dalton equals one-twelfth the mass of carbon-12.

**Exon** Part of the gene that is transcribed to mature mRNA after the intervening sequences (introns) have been spliced out.

**Gene** A hereditary unit specifying the production of a distinct protein (e.g., an enzyme) or RNA.

**Intron** Intervening sequence of DNA, located within a gene, that is not included in the mature mRNA.

**Isoform** One of the several forms in which a protein may exist in various tissues.

**Splicing** Precise excision of the intervening sequences from an RNA primary transcript, followed by ligation of the message to produce a functional molecule.

The study of the molecular biology of *muscle* is one of the classic examples of the intimate association between structure and function and the way in which chemical energy is transformed into mechanical work. The functional unit is the myofibril, which may be either striated or smooth. The molecular machinery involved in muscle contraction comprises force-generating proteins, regulatory proteins, and structural proteins. Isoforms of these proteins confer different regulatory or contractile properties to muscle cells of different types.

## 1 MUSCLE STRUCTURE AND FUNCTION

In vertebrates, there are three classes of muscles: smooth, cardiac, and striated. Typically, smooth muscles are under involuntary (unconscious) control of the central nervous system. They surround internal organs such as the large intestines, the gall bladder, and the large blood vessels. Contraction and relaxation of smooth muscles control the diameter of blood vessels and also propel food along the gastrointestinal tract. Smooth muscle cells can create and maintain tension for long periods.

Muscles under voluntary control have a striated appearance under the light microscope. Striated muscles, which connect the bones in the arms, legs, and spine, are used in complex coordinated activities (e.g., walking, positioning of the head) and can generate rapid movements by sudden bursts of contraction. Cardiac (heart) muscle resembles striated muscle in many respects, but it is specialized for executing the continuous, involuntary contractions needed in the pumping of blood. Approximately 40% of the body is skeletal muscle and almost another 10% consists of smooth or cardiac muscle.

### 1.1 Physiologic Anatomy of Skeletal Muscle

Figure 1 illustrates the organization of skeletal muscle, showing that all skeletal muscles are composed of numerous multinucleate cylindrical fibers ranging between 10 and 100 mm in diameter and several millimeters or centimeters long. Each of these fibers in turn is made up of successively smaller subunits. The entire fiber is surrounded by an electrically polarized membrane with an electrical potential of about −0.1 V, the inner surface of which is negative with respect to the outer surface. This membrane, called the sarcolemma, becomes depolarized physiologically each time a nerve impulse that reaches the motor innervation of the muscle (end plate) activates the membrane. Three cytoplasmic components are highly differentiated in a muscle fiber: myofibrils, sarcoplasma, and the sarcoplasmic reticulum (SR).

#### 1.1.1 Myofibrils

The macromolecular contractile apparatus is made up of actin and myosin filaments. Each muscle fiber contains several hundred to several thousand myofibrils, which are represented by the many small dots in the cross-sectional view of Figure 1C. Each myofibril (Figure 1D) in turn has, lying side by side, about 1500 myosin filaments and 3000 actin filaments, which are large polymerized proteins that are responsible for muscle contraction (see Figure 1E). The thick myofilaments are about 1.5 mm long and 10 nm wide and are separated by a 40 nm space. The thin myofilaments are about 1.0 mm long and 5 nm in diameter. Thick myofilaments are made of myosin, and thin myofilaments are composed of a more complex structure containing several proteins: actin, tropomyosin (TM), and troponin (TN).

Myosin and actin filaments partially interdigitate and thus cause the myofibrils to have alternate light and dark bands. The light bands, which contain only actin filaments, are called I bands be-

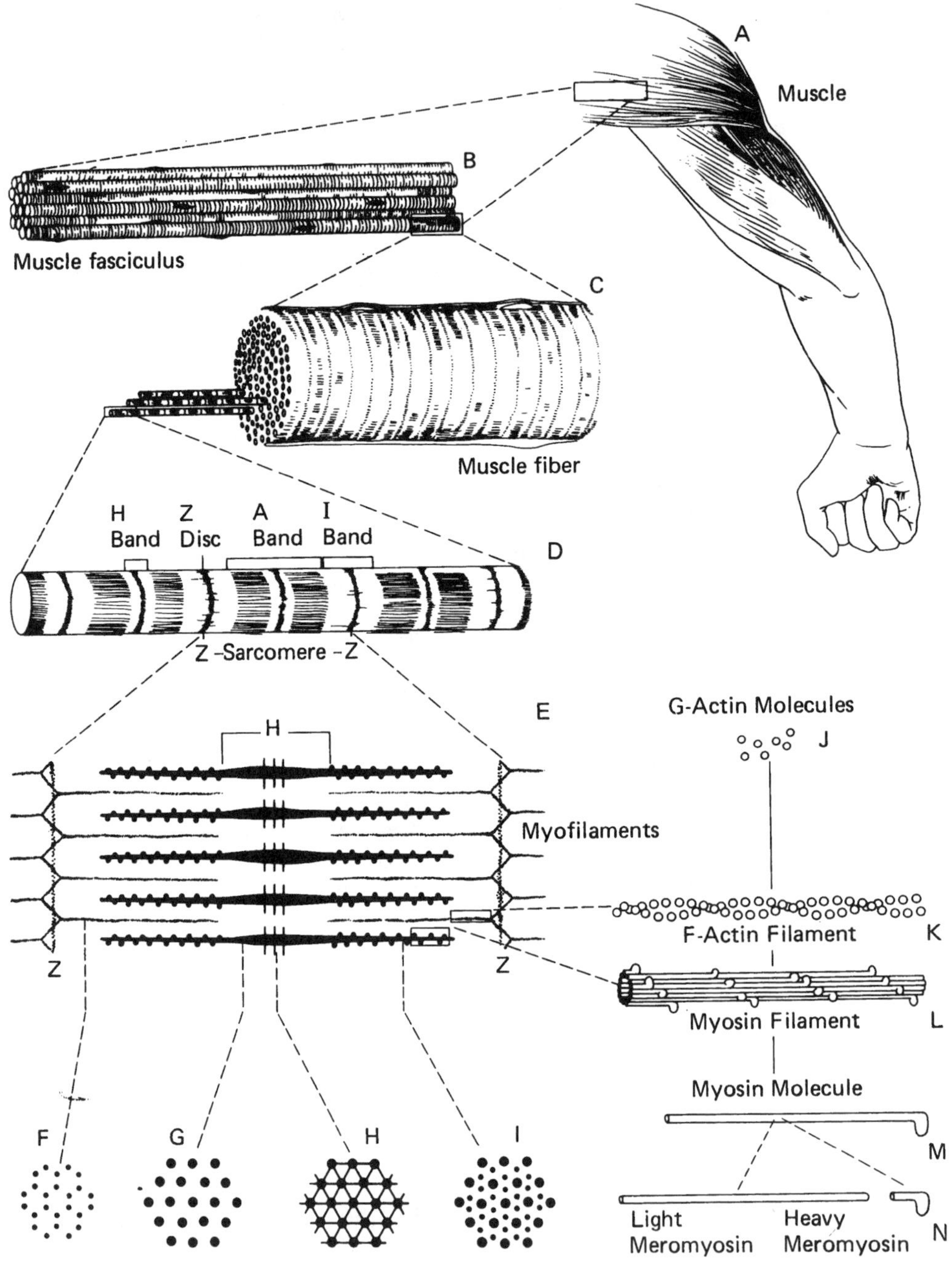


**Figure 1.** Organization of skeletal muscle, from the gross to the molecular level; (F)–(I) are cross-sectional views at the levels indicated. (Drawing by Sylvia Colard Keene. From Bloom and Fawcett; *A Textbook of Histology*, Saunders, Philadelphia, 1975.)

cause they are mainly isotropic and show no directional differences to polarized light (Figure 1D). The dark bands, which contain the myosin filaments as well as the end of the actin filaments that overlap myosin, are called A bands because they are anisotropic (nonisotistic) to polarized light. The two sets of filaments, linked by a system of cross-bridges, protrude from the surfaces of the myosin filaments along the entire extent of the filament, except in the very center (Figure 2B). It is interaction between these cross-bridges and the actin filaments that causes contraction (see Section 3). The light and dark bands are perpendicular to the long axis of the muscle cell along which the muscle contracts.

The actin filaments are attached to the so-called *Z disk* and the filaments extend on either side of the Z disk to interdigitate with the myosin filament (Figure 1E). Localization of the main protein constituents of Z disks locates α-actin in the central domain of the disk, with actin and an 85 kDa protein. Desmin, vimentin, and synemin are located at the periphery. The Z disk passes from myofibril to myofibril, attaching the myofibrils to each other all the way across the muscle fiber. Therefore, the entire muscle fiber has light and dark bands, as do the individual myofibrils. These bands give skeletal and cardiac muscle their striated appearance. The portion of a myofibril (or of the whole muscle fiber) that

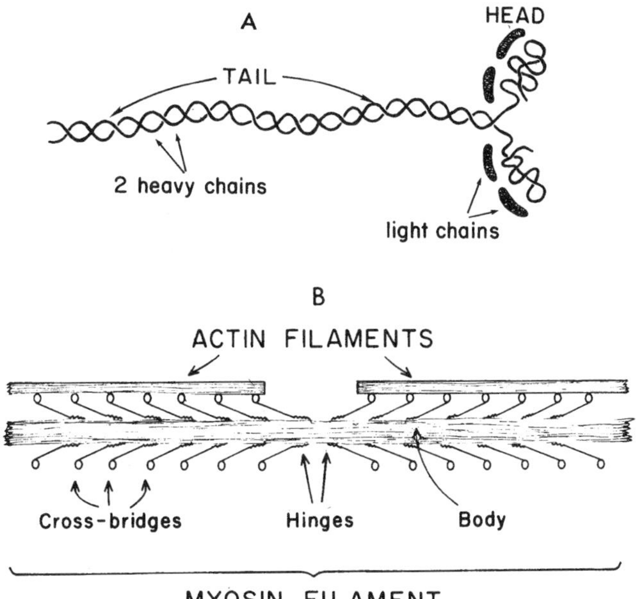


**Figure 2.** (A) The myosin molecule. (B) Combination of many myosin molecules to form a myosin filament. Also shown are the cross-bridges and the interaction between the heads of the cross-bridges and adjacent actin filaments.

lies between two successive Z disks is called a *sarcomere.* The sarcomere is the unit of contraction; movements of actin and myosin within each sarcomere lead to shortening of the sarcomeres and thus, to contraction of the muscle as a whole. When the muscle fiber is at its normal, fully stretched resting length, the length of the sarcomere is about 2.0 mm. At this length, the actin filaments completely overlap the myosin filaments and are just beginning to overlap each other. When a muscle fiber is stretched beyond its resting length, as it is in Figure 1, the ends of the actin filaments pull apart, leaving a light area in the center of the A band, called the *H zone.* In the middle of the H zone an M line can be observed. The H zone rarely occurs in the normally functioning muscle because the normal sarcomere contracts when its length is between 2.0 and 1.6 mm. In this range the ends of the actin filaments overlap not only the myosin filaments but also each other.

### 1.1.2 Sarcoplasm

The myofibrils are suspended inside the muscle fiber in a matrix called sarcomere, which is composed of high concentrations of potassium, magnesium, phosphate, and protein enzymes. Numerous mitochondria lie between and parallel to the myofibrils. These mitochondria may be related to the constancy with which the muscle contracts. A greater number of mitochondria are found in steadily active muscles (e.g., the heart), reflecting the need of the contracting myofibrils for large amounts of ATP formed by the mitochondria.

### 1.1.3 Sarcoplasmic Reticulum

The sarcoplasm also contains an extensive endoplasmic reticulum called the sarcoplasmic reticulum. SR is involved with conduction inside the fiber and with coordination of the contractions and relaxation of different myofibrils.

## 1.2 Sliding Mechanism of Muscle Contraction

In a living muscle fiber, changes with contraction can be observed with use of a phase contrast or interference microscope. Figure 3 illustrates the basic mechanism of muscle contraction. In the relaxed state of the sarcomere, the width of the A band of the myosin fibers remains constant. The ends of the actin filaments derived from two successive Z disks, although barely overlapping each other, completely overlap the myosin filaments. In the contracted state, these actin filaments are pulled inward among the myosin filaments so that they overlap each other to a major extent without any change in the lengths of individual myosin fibers or actin filaments. What does change is the width of the I band, the part of the actin myofilament not covered by myosin. Also, the Z disks are pulled by the actin filaments up to the ends of the myosin filaments. The actin filaments can be pulled together so tightly that the ends of the myosin filaments actually buckle during intense contraction. Thus, muscle contraction occurs by a sliding filament

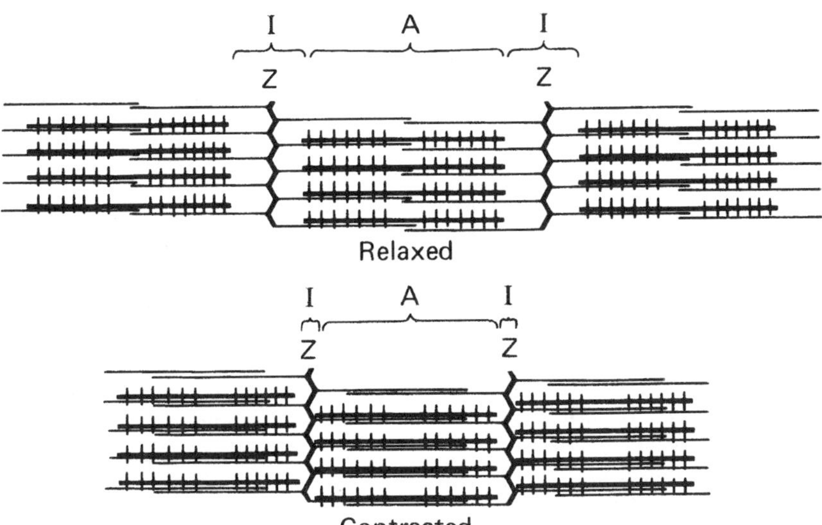


**Figure 3.** The relaxed and contracted states of a myofibril, showing sliding of the actin filaments into the channels between the myosin filaments.

**Table 1** Major Protein Components of Vertebrate Skeletal Myofibrils[a]

| Protein | Percent Total Protein | MW (kDa) | Subunits (kDa) | Function |
|---|---|---|---|---|
| Force-Generating Proteins | | | | |
| Myosin | 45 | 500 | 2 × 200 (heavy chain) | Major component of thick filaments; interacts with actin filaments with hydrolysis of ATP to develop mechanical force |
| Actin | 25 | 42 | | Major component of muscle thin filaments, against which muscle thick filaments slide during muscle contraction |
| Regulatory Proteins | | | | |
| Tropomyosin | 5 | 64 | 2 × 32 | Rodlike protein that binds along the length of actin filaments |
| Troponin | 5 | 78 | 37–42 (Tn-T)<br>22–24 (Tn-I)<br>17–18 (Tn-C) | Complex of three muscle proteins positioned at regular intervals along actin filaments and involved in the $Ca^{2+}$ regulation of muscle contraction |
| Structural Proteins | | | | |
| Titin | 9 | About 2500 | | Large flexible protein that forms an elastic network linking thick filaments to Z disks |
| Nebulin | 3 | 600 | | Elongated, inextensible protein attached to Z disks oriented parallel to actin filaments |
| α-Actinin | 1 | 94–103 | 2 × 95 | Actin-binding protein that links actin filaments together in the region of the Z disk |
| M-line protein | 1 | 165 | | Myosin-binding protein present at the central "M line" of the muscle thick filament |
| C protein | 1 | 140 | | Myosin-binding protein found in distinct stripes on either side of the thick filament M line |

[a]The vertebrate striated myofibril also contains at least 20 other proteins not included in this table.

mechanism. The degree of contraction achieved can be measured by determining the length of the sarcomere (i.e., the distance between Z disks), at rest and when it has shortened. Note that insect muscle, in general, shortens only slightly (ca. 12%), whereas the shortening in vertebrate muscle may be much greater (ca. 43%).

### 1.3 Macromolecular Organization of Smooth Muscles

Smooth muscles have a varied macromolecular organization. They contain thin and thick myofilaments, as do striated muscles, but they lack periodicity and Z disks. In smooth muscles, the contraction is slow, but an extreme degree of shortening may be achieved.

The mechanical, chemical, or electrostatic forces generated by the interaction of the myosin filament cross-bridges with the actin filaments serve to trigger the inwarding sliding of actin filaments toward the myosin filaments.

## 2 MOLECULAR ORGANIZATION AND GENETICS OF THE CONTRACTILE SYSTEM

The molecular machinery involved in muscle contraction comprises force-generating proteins (myosin and actin), regulatory proteins (TM and Tn), and structural proteins (α-actinin in the Z disk, M-disk proteins, and C proteins). The contractile elements of muscle consist of many different proteins (Table 1). Sections 2.1 to 2.6 detail their molecular characteristics and genomic organization. The two most abundant proteins in the contractile element are myosin and actin. About 60–70% of the muscle protein is myosin and about 20–25% is actin. The thick filament is mostly myosin and the thin filament is mostly actin. However, several other proteins listed in Table 1 are also associated with the thin and thick filaments.

### 2.1 Myosin: Its Structure, Function, and Genomic Organization

#### 2.1.1 Overview of Myosin Structure and Function

Myosin is an unusual protein—it is both a globular enzyme and a fibrous structural protein that plays a central role in contractile processes of eukaryotes. Since the discovery of myosin about 50 years ago, biochemical studies have provided a detailed understanding of the structure and organization of this protein in muscle and nonmuscle cells and its structural and enzymatic functions in contractile processes. Myosin has slightly different properties in smooth muscle and in striated muscle; hence their different contractile regulatory mechanisms. Biological and physiological studies reveal that myosin in muscles of vertebrate and invertebrate species has a complex molecular structure that specifies ATPase activity, intramolecular conformational changes, intermolecular interactions with actin during contraction, and assembly into the thick filaments of sarcomeric muscles. A single monomeric myosin molecule contains two identical myosin heavy chain (MHC) subunits of about 200,000 Da and two pairs of myosin light chain (MLC) units of two different types of about 20,000 Da.

The two heavy chains coil around each other to form a double helix. However, one end of each of these chains is folded into globular protein mass called the *myosin head* (Figure 2A). Thus, there are two free heads lying side by side at one end of the double helix myosin molecule; the other end of the coiled helix is called the *tail.* Bound to each myosin head are two MLCs, one of each type. These MLCs help control the function of the head during the process of muscle contraction. The joints between the heads and the tails of the molecule are flexible. Movements of the head are important in generating the force of contraction in muscle.

In muscle, myosin monomers form a specific bipolar aggregate called the *thick filament,* which contains 300–400 myosin molecules. The central portion of one of these filaments is illustrated in Figure 2B, showing the tails of the myosin molecules bundled together to form the body of the filament while many heads of the molecules hang outward to the sides of the body. Also, part of the helix portion of each myosin extends to the side along with the head, thus providing an arm that extends the head outward from the body, as shown. The protruding arms and heads together are called the *cross-bridges,* and each of these is believed to be flexible at two points called *hinges.* There is one hinge where the arm leaves the body of the myosin filament and another where the two heads attach to the arm. The hinged arms allow the heads either to be extended far outward from the body of the myosin filament or to be brought closer to the body. The heads are believed to participate in the actual contraction process, as discussed later (Sections 3.1–3.5).

The total length of the myosin filament is 1.6 mm. The center of the myosin filament, however, is devoid of cross-bridge heads for a distance of about 0.2 mm because the hinged arms extend toward both ends of the myosin filament away from the center. Therefore, in the center there are no heads, only tails of the myosin molecules. Also, the myosin filament itself is twisted so that each successive set of cross-bridges is axially displaced from the preceding set by 120°, thus ensuring that the cross-bridges extend in all directions around the filament.

Another feature of the myosin head, essential for muscle contraction, is the ability to function as an ATPase enzyme. In the absence of actin, this activity is almost undetectable, but whenever pure actin filaments are added, the rate of ATP hydrolysis is increased 200-fold so that each myosin molecule hydrolyzes 5–10 ATP molecules per second, thus providing the ATP's high energy phosphate bond to energize the contraction process.

### 2.1.2 Myosin Genes

The cDNAs for myosin from many different species and types of muscle have been cloned and the amino acid sequences for the different myosin molecules inferred. Numerous isoforms of MLC and MHC proteins are generated from multiple genes or by posttranscriptional mechanisms, resulting in the diversification of the various skeletal and cardiac muscle types. Some of the contractile proteins are expressed at specific developmental stages or at distinct physiological states. There is sufficient similarity between these molecules of myosin to permit the use of examples of any one to illustrate essential points about myosin structure and function. Myosin has evolved slowly, with a high degree of homology, particularly within the head, or globular, region. Even though there is somewhat less absolute amino acid homology within the tail region, functional homology still exists to an extraordinarily high degree in spite of the varying lengths, which range in different species from about 86 nm to about 150 nm. Myosin head varies in amino acid number from about 839 in the rat skeletal muscle to about 850 in the nematode and 849 in chicken gizzard smooth muscle heavy chain. The myosin heavy chain of mammalian species has about 1950 amino acids, with nearly half of them in the globular head piece.

## 2.2 Actin Proteins

Actin is the second major protein and represents about one quarter (20–25%) of the proteins of the myofibril (Table 1). The molecular weight of actin is 42,000 Da. The actin myofilament is made up of two helical strands that cross over every 36–37 nm (Figure 1K). Each of these crossover repeats contains between 13 and 14 globular monomers of G-actin about 5 nm in diameter (Figure 1J). In muscle, actin is present mainly as fibrous or F-actin, which is the polymerized form of G-actin.

Actin is a highly conserved protein that participates in a wide variety of cellular functions in eukaryotes including muscle contraction, ameboid movement, cytokinesis, and mitotic division. Amino acid sequencing data have demonstrated that the present actin isotypes evolved from two major classes of actin, "cytoplasmic" and "muscle." All organisms thus far examined express a cytoplasmic actin form, which is usually used to construct the cellular microfilaments. In simple eukaryotes such as *Dictyostelium, Physarum,* and *Saccharomyces cerevisiae,* only one type of cytoplasmic actin is expressed. In mammals, there are two cytoplasmic actins (β and γ), and in amphibians there are at least three. The second class of actin proteins, the muscle (or "α-like") actins, is found in birds and mammals. It follows that original α-like actin gene evolved from the cytoplasmic type some time before the divergence of birds and mammals. In mammals, four different tissue-specific muscle isotypes have been found: skeletal actin (or α) and cardiac actin, and aortic actin and stomach actin, which are smooth muscle types.

## 2.3 Actin Genes

Considerable information on the structure and organization of actin genes in a wide variety of eukaryotes has been accumulated. The available data reflect the comparative degrees of sequence conservation between species in genes encoding the muscle-specific and the nonsarcomeric isoforms of actin. Even though there may be six or more genes that code for actin, it is a protein with a very highly conserved amino acid sequence. The amino acid sequence of rabbit skeletal muscle actin has 374 amino acid residues.

This gene is transiently coexpressed with a cardiac isoform in fetal cardiac and skeletal muscle tissues. The tissue-specific isoforms then progressively predominate throughout the development of the corresponding tissue. These striated muscle isoforms and the nonmuscle isoforms are encoded by genes localized on different chromosomes: 3, 5, and 17 in the mouse. In lower organisms, in contrast, there is a close physical linkage between several actin genes.

Sequence information on the actin genes from a wide variety of sources indicates several interesting features. The coding sequences are highly conserved, whereas the noncoding sequences are widely divergent and vary to an extent that reflects the putative time course of evolutionary divergence. In fact, the primary amino acid sequences of more than 30 different actin isotypes, with the longest being 375 amino acids, reveal that a maximum of only 32 residues in any of them had been substituted.

Actin as first synthesized is called G-actin. The three-dimensional structure of actin indicates that it is not strictly globular, however. Actin has two distinct domains of approximately equal size that historically have been designated as the large (left) and small (right) domains. Each one consists of two subdomains. Both the N-terminal and the C-terminal are located within the small domain. The molecule has polarity, and when it aggregates to form F-actin, or fibrous actin, it does so with a specific directionality.

The process of formation of F-actin is interesting, but it is not fully understood. G-Actin contains a specific binding site for ATP and a high affinity binding site for divalent metal ions. The $Mg^{2+}$ ion is most likely the physiologically important cation, but $Ca^{2+}$ also binds tightly and competes with $Mg^{2+}$ for the same binding site. It is the G-actin/ATP/$Mg^{2+}$ complex that aggregates to form the F-actin polymer.

A number of proteins in the cytosol can bind to actin. It is not clear what the roles of many of these proteins are, but it does appear that in the sarcomere, β-actin binds to F-actin and plays a major role in limiting the length of the thin filament. α-Actinin, a protein with two subunits of 90–11/kDa each, also binds actin, but its main role appears to be to anchor the actin filament to the Z line of the sarcomere. There are two other major proteins associated with the thin filament, tropomyosin (TM) and troponin (Tn).

## 2.4 Regulatory Proteins

In addition to the force-generating proteins actin and myosin, the coordinated action of the regulatory proteins such as TM and the various Tns is required for the contraction of muscle.

### 2.4.1 Tropomyosins and Tropomyosin Genes

Tropomyosins are components of the contractile systems of the skeletal, cardiac, and smooth muscles and the cytoskeleton of non-muscle cells. Although they are present in all cells, different forms of the protein are characteristic of specific cell types. TM is a rodlike protein composed of two highly α-helical subunits wrapped around each other to form a coiled-coil structure (Figure 4). Skeletal muscle contains two forms of TM termed α and β, and their proportions vary with the fiber type. Both these subunits are 284 amino acids long, and they differ only slightly in amino acid sequence. In cardiac muscle of small mammals such as rodents, only α-TM is expressed. In striated muscle, TM is localized to the thin filaments, where it is found along both grooves of the actin filament. The function of TM in skeletal and cardiac muscle is in association with the Tn complex (Tn I, T, and C: see Section 2.4.2) to regulate the calcium-sensitive interaction of actin and myosin as well as to stabilize and strengthen the filament.

Until recently it was thought that the nonmuscle TMs were structurally distinct from the smooth and skeletal muscle isoforms. Isoforms of TM from platelets were found to be 247 amino acids in length, whereas isoforms from smooth and skeletal muscle are 284 amino acids long. However, molecular cloning has shown that the heterogeneity of nonmuscle TM is due not only to the expression of different isoforms of the classic 247 amino acid nonmuscle protein but also to the expression of different isoforms of smooth muscle TMs. Several cDNA and genomic clones representing the

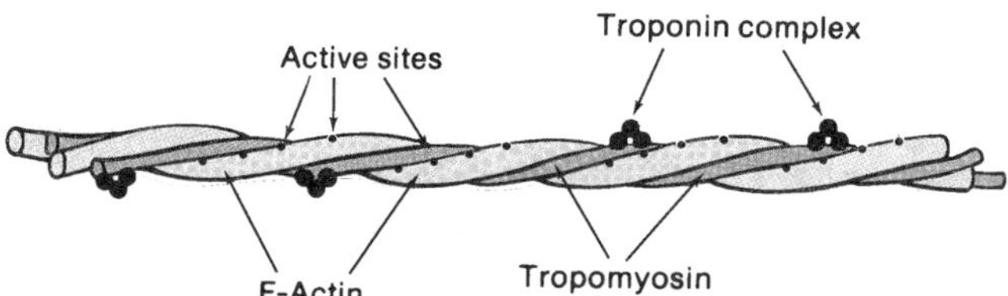


**Figure 4.** The actin filament, composed of two helical strands of F-actin and two tropomyosin strands that lie in the grooves between the actin strands. Attaching the tropomyosin to the actin are several troponin complexes.

TM isoforms from skeletal and smooth muscles and nonmuscle cells (fibroblasts) have been identified and characterized from rodents (rat and mouse) and human species.

### 2.4.2 Troponins, Troponin Proteins, and Genes Encoding the Various Isoforms

Together with TM, the Tns are a family of three small proteins that form a complex of about 80,000 Da (Table 1). Both proteins constitute a $Ca^{2+}$-sensitive switch within the thin filament that regulates the contraction. Each of the Tn complexes binds to TM every 40 nm (i.e., every seven active monomers). The three components present in equimolar proportions are:

Tn-C (MW 17,000–18,000), a small protein that specifically binds $Ca^{2+}$

Tn-I (MW 22,000–24,000), a protein that binds to both actin and Tn-C and inhibits the ATPase of myosin

Tn-T (MW 37,000–42,000), the largest subunit that binds to TM and also interacts with the other Tns (Figure 4).

Troponin has been considered a localized trigger that controls the motion of TM in the thin filament. Observations by electron microscopy have revealed that the Tn complex has a tadpole shape with a globular domain composed mainly of Tn-C and Tn-I, and a long tail that interacts with TM and corresponds mainly to Tn-I.

Troponin is a family of three muscle-specific myofibrillar proteins involved in the calcium regulation of contraction in cardiac and in skeletal muscle. Tn-I proteins have multiple functional domains that are distinct and bind with high affinity to actin and Tn-C. The interactions of these domains regulate actomyosin ATPase activity in resting and contracting muscle. Tn-I also interacts functionally with other muscle proteins, including Tn-T.

**Troponin-C (Tn-C)** The calcium-binding component of the troponin complex, Tn-C, triggers the contraction of skeletal and heart muscle in response to increasing calcium levels in the sarcoplasm. There are at least two similar but biochemically and functionally distinct Tn-C isoforms in striated muscles of higher vertebrates. In nonvertebrates, Tn-C has been found in smooth muscle. In adult vertebrates, the relative concentrations of TnC isoform proteins is muscle-type specific.

From cDNA studies, it is known that human slow (Tn-C) and fast (Tn-C2) skeletal muscle troponin C comprises 162 and 161 amino acids, respectively. The coding regions of both cDNAs are about 74% similar. Only the amino-terminal 39 bp of the coding sequences are dissimilar. The amino acid sequence of human cardiac Tn-C is identical with the amino acid sequence of the slow

Tn-C protein. Southern blot analyses suggests that in the human, both fast Tn-C and slow Tn-C are encoded by single-copy genes. Expression of Tn-C mRNA is developmentally regulated and is tissue specific. Tn-C is structurally and evolutionarily closely related to a number of calcium-binding proteins referred to as members of Tn-C superfamily.

**Troponin-I** The complete gene structure for quail fast skeletal muscle Tn-1 shows that the 4.5 kb gene has eight exons encoding for 186 amino acids from a 830 bp mRNA. Tn-I proteins have two functional domains: the actin-binding domain and the Tn-C binding domain. Interestingly, the actin-binding domain of Tn-I is encoded by a single exon, whereas the Tn-C-binding domain is split into at least two exons. The gene for Tn-I with several muscle-specific contractile protein genes (e.g., chick α-actin, rat α-actin, chick MLC3, and rat cardiac MHC genes) showed the presence of homologous sequences in their 5′ flanking regions, and large introns of similar size that separate the promoter and the first exon coding for the 5′ nontranslated sequences from the protein-coding segment of the second exon. The presence of these common structural features suggests that they may act as regulatory elements in the coordinate control of muscle gene expression during myogenesis.

**Troponin-T** The structures of rat fast skeletal and the chicken cardiac Tn-T genes have been defined. Several salient features are conserved between them. Each contains 18 exons, although the rat skeletal gene is 16.5 kb, whereas the chicken cardiac gene spans 9 kb. The first exon in both genes is a spliced leader. All the translated exons that are spliced constitutively in the rat (e.g., 2, 3, 9–15, and 18) and exons 4 and 16 show a high degree of conservation for size and coding sequence. In contrast, the chicken cardiac exons 5–8 are significantly larger, and exon 17 considerably smaller, than their corresponding rat exons, with little sequence homology.

All the regulatory protein genes can encode multiple isoforms of the proteins by alternative splicing of the primary transcripts. Such alternative splicing is regulated in a tissue-specific and developmental manner.

## 2.5 Structural Proteins

The structural proteins comprise α-actinin, 85K protein, C protein, M-line proteins, and the elastic protein, connection or titin (Table 1).

### 2.5.1 α-Actinin and Its Structure

Actinin is an actin-binding and cross-linking protein found in both muscle and nonmuscle cells at points where actin is anchored to a variety of intracellular structures. It is a homodimer with a subunit molecular weight of 94–103 kDa, the subunit being antiparallel in orientation. It is visualized as a long rod-shaped molecule in the electron microscope, 3–4 nm wide and 30–40 nm long. A number of distinct isoforms of α-actinin have been characterized, including the skeletal and smooth muscle isoforms isolated from brain and cultured fibroblasts. The only clear functional difference between these isoforms is that binding of the various nonmuscle α-actinins to actin is calcium sensitive, whereas binding of the muscle isoforms is not.

α-actinin can be divided into three domains: an N-terminal actin-binding domain, four internal 122 amino acid repeats, and a C-terminal region containing calcium-binding motifs. In chicks, isoforms of α-actinin with different calcium-sensitive actin-binding characteristics appear to be generated by alternative splicing.

**Actin-Binding Domain** The actin-binding domain of α-actinin is contained within the N-terminal region of the molecule and is homologous to residues 1–238 of human dystrophin, implying that dystrophin is also an actin-binding protein. Interestingly, the extreme N-terminal region of dystrophin differs between the brain and muscle isoforms, and is also quite different from that found in α-actinin isoforms. These regions of difference may prove informative in attempts at further defining residues involved in actin binding.

**Central Repeats** The central region of α-actinin contains four internal repeats of about 122 amino acids, homologous to the 100-residue repeats found in spectrin and the approximately 110-residue repeats found in dystrophin. Repeat 1 displays a greater level of identity (90%) than repeats 2, 3, and 4 (80, 76, 80%). The level of identity between a given repeat across species is greater than that between the four repeats within the same species. These repeats in α-actinin appear to be responsible for the formation of the antiparallel dimer.

**The Calcium Binding Domain** The C-terminal region of α-actinin contains two calcium-binding motifs. The smooth and fibroblast isoforms of the protein are 98% identical in nucleotide sequence. Close similarity between the smooth and nonmuscle isoforms of the protein indicates that they may be encoded by the same gene and may arise by alternative splicing of a single primary transcript. In contrast, skeletal isoform is encoded by a separate gene. It is still not clear how many additional (perhaps tissue-specific) isoforms of skeletal, smooth, and nonmuscle α-actinins exist.

### 2.5.2 85K Protein

The 85K protein is present in the central region of the Z disk. As the name implies, it has a molecular weight of 85,000 Da.

### 2.5.3 C Protein

The C protein is present in the A band along the middle portion of the myosin filament. With electron microscopy the anti-C protein antibody is observed to form between seven and nine transverse stripes in each half of the A band.

### 2.5.4 M-Line Proteins

The center of the H band contains a creatine kinase and a structural protein called *myomesin*. Myomesin has a molecular weight of 165,000, can be detected in myoblasts, and has a high affinity for myosin. It acts as a specific marker during the differentiation of myoblasts into myotubes and muscle fibers. Myomesin, which is present in skeletal and heart muscle but not in smooth muscle, plays a key role in the molecular organization of the thick myofilaments.

### 2.5.5 An Elastic Protein of Muscle

A high molecular weight protein isolated from the myofibril, named *connectin* or *titin,* represents 8–12% of the myofibrillar mass. Under the electron microscope, this protein appears as an extremely thin and long ($\geq$ 1 μm) strand. It is extensible and flexible and shows an axial periodicity. This protein interacts with both myosin and actin and forms nets that are concentrated at the A-I junction of the myofibril.

### 2.6 Intermediate Filament Proteins

Intermediate filaments (IFs) are unique cytoskeletal structures 7–12 nm in diameter, which occur in the cytoplasm of most eukaryotic cells. Five subclasses are found in mammalian cells: two complex groups of *acidic* and *basic keratins* (40–70 kDa), a single protein, *desmin* (52 kDa), a single protein, vimentin (55 kDa), the glial fibrillary acidic protein GFAP (50 kDa), and three *neurofilament proteins* (65, 100, and 135 kDa). Cytokeratins A and B, which represent the largest and most diverse classes of IF protein, are found in epithelial cells; desmin is synthesized in skeletal visceral and certain vascular smooth cells; GFAP is found in astroglia and the triplet of neurofilament proteins in neuronal cells; vimentin has a widespread distribution and is found in all mesenchymal derivatives as well as in the progenitors of muscle and neural tissues.

Structural analyses indicates that all IF subunits are built according to a common plan: each has a central helical rod domain flanked by end domains of variable size and chemical characteristics. The different properties of the IFs are due to their end domains; their structural uniformity is attributable to the conservation of the central helical rod domains.

The mammalian vimentin and desmin genes are remarkably conserved between various species. In each species, the two genes contain eight introns in identical positions, suggesting that these two genes arose from a common ancestor. In mammalian skeletal muscle, the replicative presumptive myoblasts synthesize predominantly vimentin. Upon fusion, however, myotubes synthesize high levels of desmin, and as the myotubes mature, the activities of expressing vimentin and desmin genes become mutually exclusive, such that in mature muscle only desmin is found. During the process of maturation of the myotubes, desmin, which initially is homogeneously distributed throughout the myotubes, assembles in the Z bands of the then-forming sarcomeres. In mature muscle fibers, the desmin IFs interconnect myofibrils at the level of the Z bands and attach them to the sarcolemma. In addition, desmin probably serves to attach actin filaments to the Z bands and to fasciae adhering to the intercalated disks, thus helping to maintain the normal myofibrillar architecture. The control of the differential expression of desmin and vimentin genes that determines their transcriptional activity during human myogenesis is not clearly understood.

## 3 REGULATION AND ENERGETICS OF MUSCLE CONTRACTION

The detailed ultrastructural and biochemical information available permits an interpretation of the macromolecular mechanisms involved in muscle contraction.

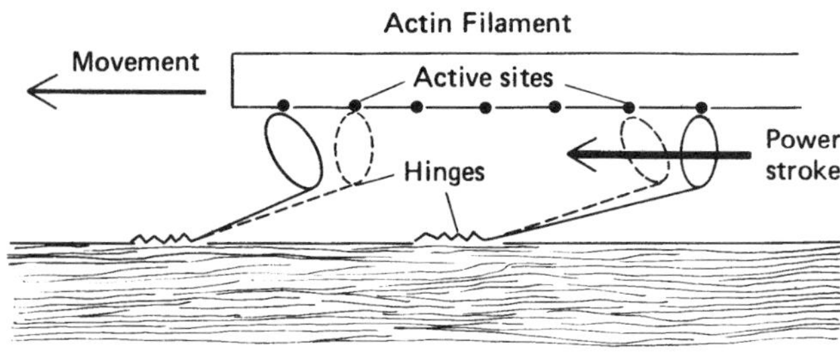


**Figure 5.** The ''walk-along'' mechanism for contraction of the muscle.

### 3.1 The "Walk-Along" Theory of Contraction

It is believed that as soon as the actin filament becomes activated by calcium ions, the heads of the cross-bridges from the myosin filaments become attracted to the active sites of the actin filament. According to the ''walk-along'' theory of contraction, this interaction in some way causes contraction to occur.

This walk-along mechanism is illustrated in Figure 5, which shows the heads of two cross-bridges attaching to and disengaging from the active sites of an actin filament. It is postulated that when the head attaches to an active site, this attachment simultaneously causes profound changes in the intramolecular forces in the head and arm of the cross-bridge. The new alignment of forces causes the head to tilt toward the arm and to drag the actin filament along with it. This tilt of the head of the cross-bridge is called the *power stroke.* Immediately after tilting, the head automatically breaks away from the active site and returns to its normal perpendicular direction. In this position it combines with an active site farther down along the actin filament; then a similar tilt takes place again to cause a new power stroke, and the actin filament moves another step. Thus, the heads of the cross-bridges bend back and forth and step by step walk along the actin filament, pulling the actin toward the center of the myosin filament. Each one of the cross-bridges is believed to operate independently of all others, each attaching and pulling in a continuous but random cycle. Therefore, the greater the number of cross-bridges in contact with the actin filament at any given time, the greater, theoretically, it the force of contraction.

### 3.2 ATP as a Source of Energy for Contraction

When a muscle contracts against a load, energy is required which comes from ATP hydrolysis. No major difference in ATP levels is detectable between a resting and an actively contracting muscle, however, because a muscle cell has an efficient backup system for regenerating ATP. The concentration of ATP present in the muscle fiber, about 4 mM, is sufficient to maintain full contraction for only a few seconds at most. After the ATP has been broken down into ADP, the ADP is rephosphorylated to form new ATP within a fraction of a second.

There are several sources of energy for this rephosphorylation. The first source is phosphocreatine, a compound that carries a high energy phosphate bond similar to those of ATP. The enzyme phosphocreatine kinase catalyzes a reaction between phosphocreatine and ADP to form creatine and ATP. It is the intracellular level of phosphocreatine that drops after a short burst of muscle activity, even though the contractile machinery itself consumes ATP. The pool of phosphocreatine acts like a battery storing ATP energy and recharging itself from the new ATP generated by cellular oxidations when the muscle is resting.

The next source of energy is released from foodstuffs (i.e., carbohydrates, fats, and proteins). The sarcoplasmic matrix contains the glycolytic enzymes as well as other globular proteins such as myoglobin, salts, and high phosphate compounds. Glycogen is present in the matrix as small granules of glycosomes observed under the electron microscope. Glycogen disappears with contraction through glycolysis, and lactic acid is formed, which can be transformed into pyruvic acid to enter the Krebs cycle. Oxidative phosphorylation is the last and most important source of ATP.

## 4 GENETICS OF MUSCLE-RELATED MYOPATHIES

### 4.1 Metabolic Myopathies

Metabolic myopathies are usually classified according to the area of metabolism affected. The only category relevant to this subject is the disorders of carbohydrate metabolism because a deficiency in the enzyme(s) of this pathway leads to the depletion of energy source for muscle contraction. Sections 4.1.1 and 4.1.2 discuss two such genetic disorders.

#### 4.1.1 Glycogen Storage Disease Type VII (Tarui's Disease)

Inherited deficiency of phosphofructokinase (PEK) is characterized by the coexistence of muscle disease and hemolysis. Typically, the disease begins in early childhood and consists of easy fatiguability, transient weakness, muscle cramps, and myoglobinuria after vigorous exercise.

The gene for human PFK is under the control of three structural loci that encode muscle (M), liver (L), and platelet (P) subunits residing on chromosome 1, 10, and 21, respectively, and these isozymes are variably expressed in different tissues. Mature skeletal muscle expresses only the M subunit and contains a single isozyme species, the homotetramer M4. Liver and kidney express only L and P subunits and are not affected in muscle PFK deficiency.

Homozygous deficiency of the M subunit in most cases results from one or more mutations affecting the muscle PFK structure. The catalytically inactive immunoreactive M subunit is present. Genes encoding the human muscle and liver subunit of PFK have been cloned. Tissues expressing the muscle or the liver PFK contain a specific mRNA of 3.0 kb, encoding a protein of about 780 amino acids. Alternative splicing of the transcript encoding the human muscle isoenzyme of PFK has also been observed.

#### 4.1.2 Molecular Mechanisms of McArdle's Disease (Muscle Glycogen Phosphorylase Deficiency)

McArdle's disease is a rare form of metabolic myopathy, genetically transmitted as a recessive autosomal trait. Clinically this disease is characterized by asthenia, cramps, stiffness, and attacks of myoglobinuria following physical effort. These manifestations of the disorder originate from a depletion in the ATP stores, induced by the absence of glycogen phosphorylase activity. This enzyme catalyzes the first step of the glycogenolytic pathway that provides most of the readily available energy for intensive muscle contraction.

In human tissues and those of other mammals, glycogen phosphorylase exists in three isoenzymatic forms: liver, muscle, and brain. The muscle form is the only isoenzyme expressed in skeletal muscle and is also found in heart and brain, along with the cerebral form. This explains why the defect in McArdle's disease is confined to muscle tissue.

Tissues expressing muscle phosphorylase contain a specific mRNA of 3.4 kb, encoding a protein of about 841 amino acids. In four patients with muscle phosphorylase deficiency, the gene was present and apparently not rearranged. However, a clear heterogeneity was found at the mRNA level. Phosphorylase mRNA was undetectable in five patients; in three patients it was in decreased amounts but with a normal length.

### 4.2 Duchenne Muscular Dystrophy

Duchenne muscular dystrophy (DMD), an X-linked recessive disorder resulting in progressive degeneration of the muscle, affects about one in 3500 male children. Affected boys are typically diagnosed around 2 years of age, are no longer ambulatory by 12 years of age, and rarely live beyond the second decade in life. In the last few years, remarkable progress has been made toward the isolation and characterization of the 2 million base pair DMD gene, possibly the largest in the human genome, and recently a 14 kb human DMD cDNA corresponding to a complete representation of the fetal skeletal muscle transcript was cloned. The DMD transcript is formed by at least 60 exons, which have been mapped relative to various reference points within Xp21. The first half of the DMD transcript is formed by a minimum of 33 exons spanning nearly 1000 kb, and the remaining portion has at least 27 exons that may spread over a similar distance.

The gene product, dystrophin, a protein of molecular weight 427,000, appears to be completely absent from DMD muscle. DMD mRNA is most abundant in skeletal and cardiac muscle and less in smooth muscle. DMD selectively affects a subset of skeletal muscle fibers specialized for fast contraction.

Baker's muscular dystrophy (BMD), a less severe and less frequent disease allelic to DMD, is also localized to the same region in the X chromosome. Several other dystrophin-related muscular dystrophies include limb-girdle muscular dystrophy and spinal muscular atrophy.

## 5 MYOGENESIS: MUSCLE-SPECIFIC GENE TRANSCRIPTION

Muscle cells develop from the mesoderm by the determination of multipotent cells to develop into muscle precursor cells, called *myoblasts.* Myoblasts then fuse to form multinucleated myofibers that express muscle-specific genes, including those encoding cardiac and skeletal muscle actins, MHCs, TM, Tn-I and Tn-T, acetylcholine receptor, and muscle-specific creatine kinase. Although we are on our way to understanding how each of these steps is controlled at the molecular level, a number of important pieces to the puzzle are still missing.

Our understanding of the initial steps in myogenesis originated from work on tissue culture cells. Treatment of undifferentiated mouse embryonic fibroblast cells in culture with the chemical 5-azacytidine causes some of the fibroblasts to differentiate into myoblasts, which can fuse to form myotubes. 5-Azacytidine is known to alter the pattern of methylation of cytidine residues in DNA, which may in turn alter the pattern of gene expression. Examination of myogenic lineages led to the discovery that expression of the gene *MyoD* (for myoblast determination) is sufficient to cause conversion of the fibroblasts into muscle cells. Transfection of *MyoD1* DNA into a variety of fibroblast cell lines is sufficient to convert them into a myogenic phenotype. These myogenic cells can be monitored by using immunofluorescence to stain for the presence of muscle-specific proteins in transfected cells. Alternatively, transfection with an RNA virus containing a *MyoD1* gene that is strongly expressed in the infected cells causes dramatic changes in cell morphology. Infected rat fibroblasts become elongated and fuse to become large multinucleated myotubes. In addition to fibroblasts, a number of diverse differentiated cell lines (e.g., melanoma, neuroblastoma, and fat cells) can be caused to express muscle-specific genes.

Recently, additional genes involved in myogenic differentiation have been identified. They are *myd, myf-5, mrf-4, myogenin, herculin,* and *CMD1. CMD1* is apparently the avian homologue of *MyoD1.* The myogenin, myf-5, mrf-4, and herculin proteins are intriguing because each contains a region that is strikingly similar to a 68 amino acid long region of *MyoD1* that is required to initiate myogenesis in transfection experiments. Thus, these proteins apparently belong to a family of skeletal-muscle-specific transcription factors. The region of homology contains two functionally important domains. These are the helix–loop–helix (HLH) domain, which is essential for dimerization of these so-called HLH proteins, and an adjacent basic region, which binds DNA specifically. In addition, a number of other proteins that are not specific to muscle are also HLH proteins.

The advantages of myogenesis for studying gene control of development have stemmed primarily from the ability to monitor the sequence of events of myogenesis in vitro and abundance of muscle-specific proteins. We have now entered a new era, in which the discoveries made with the in vitro systems are being exploited to learn how muscle cell differentiation is regulated at the molecular level during embryonic development. One challenge will be to learn whether interactions between *MyoD1* and other HLH proteins are involved in regulation of myogenesis and—if so—how the synthesis and activity of these proteins are regulated by developmental events, such as embryonic induction.

## 6 MYOBLASTS: VEHICLE FOR GENE THERAPY

A novel approach to drug delivery involves use of cells to introduce genes that continue to express therapeutic proteins. Myoblasts appear to be well suited to this purpose because of their unusual biological properties. In contrast to other cell types, myoblasts become an integral part of the muscle into which they are injected. As a result, myoblasts are currently being used as cellular vehicles for drug delivery of several different endogenous and recombinant gene products for the treatment of both muscle and nonmuscle disorders.

Several features of human muscle make it attractive for use in cell-mediated gene therapy. Human primary myoblasts can be readily isolated from biopsy or autopsy material, enriched to 99%, grown to large numbers ($10^{12}$–$10^{18}$ cells per original isolated cell, depending on donor age), and genetically engineered without losing their potential to differentiate. Transplanted primary human myogenic cells are unlikely to be tumorigenic for two reasons: (1) the malignant transformation of human muscle cells in vitro has never been reported and (2) the cells invariably senesce after a given number of doublings. Upon fusion into myofibers, implanted myoblast nuclei remain viable for at least 6 months. Thus, transplanted myoblasts persist and show stable gene expression.

*See also* MOLECULAR GENETIC MEDICINE; MOTOR NEURON DISEASE; MOTOR PROTEINS.

### *Bibliography*


Berne, R. M., and Levy, M. N. (1993) Contractile mechanism of muscle cells. In *Physiology,* Mosby/Year Book, St. Louis.

Breitbart, R. E., Andreadis, A., and Nadal-Ginard, B. (1987) Alternative splicing: A ubiquitous mechanism for the generation of multiple protein isoforms from single genes. *Annu. Rev. Biochem.* 56:467–495.

Darnell, J., Lodish, H., and Baltimore, D. (1990) Actin, myosin, and intermediate filaments: Cell movements and cell shape. In *Molecular Cell Biology,* Scientific American Books, New York.

Di Mauro, S., Miranda, A. F., Sakoda, S., Schon, E. A., Servidei, S., Shanske, S., and Zeviani, M. (1986) Metabolic myopathies. *Am. J. Med. Genet.* 25:635–651.

Emerson, C. P., Jr., and Bernstein, S. I. (1987) Molecular genetics of myosin. *Annu. Rev. Biochem.* 56:695–726.

Guyton, A. (1991) Contraction of skeletal muscle. In *Textbook of Medical Physiology,* M. J. Wonsiewicz, Ed. Saunders, Philadelphia.

Helfman, D. M., Ricci, W. M., and Finn, L. A. (1988) Alternative splicing of tropomyosin pre-mRNAs in vitro and in vivo. *Genes & Dev.* 2:1627–1638.

Koenig, M., Hoffman, E. P., Bertelson, C. J., Monaco, A. P., Feener, C., and Kunkel, L. M. (1987) Complete cloning of the Duchenne muscular dystrophy (DMD) cDNA and preliminary genomic organization of the DMD gene in normal and affected individuals. *Cell,* 50:509–517.

Li, Z., Lilienbaum, A., Butler-Browne, G., and Paulin, D. (1989) Human desmin-coding gene: Complete nucleotide sequence, characterization and regulation of expression during myogenesis and development. *Gene,* 78:243–254.

Nadal-Ginard, B., Breitbart, R. E., Strehler, E. E., Ruiz-Opazo, N., Periasamy, M., and Mahdavi, V. (1986) Alternative splicing: A common mechanism for the generation of contractile protein diversity from single genes. In *Molecular Biology of Muscle Development,* pp. 387–410. Liss, New York.

Sharma, P. M. (1991) Muscle, molecular genetics. In *Encyclopedia of Human Biology,* Vol. 5, R. Dulbecco, Ed., pp. 205–228. Academic Press, San Diego, CA.

———, Reddy, G. R., Vora, S., Babior, B. M., and McLachlan, A. (1989) Cloning and expression of a human muscle phosphofructokinase cDNA. *Gene,* 77:177–183.

———, ———, Babior, B. M., and McLachlan, A. (1990) Alternative splicing of the transcript encoding the human muscle isoenzyme of phosphofructokinase. *J. Biol. Chem.* 265(16):9006–9010.

Vaisanen, P. A., Reddy, G. R., Sharma, P. M., Kohani, R., Johnson, J. L., Raney, A. K., Babior, B. M., and McLachlan, A. (1992) Cloning and characterization of the human phosphofructokinase gene. *DNA Cell Biol.* 11(6):461–470.


# MYCOBACTERIA

*Johnjoe McFadden and Neil Stoker*

### *Key Words*

**Acid-Fastness** The property of mycobacteria of retaining carbolfuchsin stain even after decolorization in acid–alcohol.

**Adjuvant** Substance that is administered with an antigen to elicit an immune response to the antigen.

**Contig** A series of overlapping (contiguous) recombinant clones.

**Epitope** The portion of an antigen that is recognized by a specific antibody or specific lymphocytes.

**Insertion Sequence** A simple transposon that encodes only functions required for its own transposition.

**Transposon** A mobile genetic element.

---

Mycobacterial genetics is one of the most rapidly advancing fields in bacterial genetics. However, mycobacteria are more than an interesting genetic system. Mycobacterial disease remains one of the most intractable world health problems. Tuberculosis caused by *Mycobacterium tuberculosis* kills between 2 and 3 million people each year, and leprosy caused by *Mycobacterium leprae* continues to afflict millions. In addition, AIDS, with concomitant mycobacterial infections, is bringing tuberculosis back into Western cities and seriously threatens the health services of the developing world.

## 1 INTRODUCTION TO MYCOBACTERIA

Mycobacteria belong in the actinomycete branch of Gram-positive bacteria, a group that includes *Nocardia*, *Corynebacterium* and *Streptomyces* species. These bacteria share a number of common properties including a tendency for mycelial growth, predominantly aerobic metabolism, and DNA with a high (55–70%) GC content. Large amounts of lipid are present in mycobacterial cell walls—approximately 40% of total cell dry weight—causing mycobacteria to grow as rough hydrophobic colonies on solid media and as (funguslike) pellicles in liquid media. Because of their high lipid content mycobacterial cells are difficult to stain by conventional stains such as the Gram stain; once stained, however, they resist decolorization by either 95% ethanol or 3% hydrochloric acid. This property of *acid-fastness* is the basis of the Ziehl–Neelsen staining technique for identification of mycobacteria.

### 1.1 Tuberculosis

Infection with *Mycobacterium tuberculosis* most commonly occurs by inhalation of small droplets containing only a few live tubercle bacilli. In the lung, the bacilli are readily phagocytosed by alveolar macrophages, but pathogenic mycobacteria are able to resist killing by macrophages and can replicate efficiently within macrophages at this site to form the primary focus of infection. This stage of disease is usually clinically silent, and in most cases immunity develops within a few weeks. However about 5% of those initially infected will develop primary tuberculosis soon after infection. Of the remainder, the majority will remain free of tuberculosis throughout their lives; a minority (about 5%) will at some time, usually many years later, develop postprimary tuberculosis as a result of loss of host immunity (associated with, e.g., illness or malnutrition). HIV infection is now the most important factor leading to progression to this stage of disease.

### 1.2 Leprosy

Leprosy exhibits an astonishing spectrum of symptoms from the tuberculoid leprosy form to borderline leprosy to lepromatous leprosy, with many intermediates. The type of disease manifested depends critically on the immune response of the host. Tuberculoid leprosy is characterized by a strong cellular immune response, low levels of circulating antibody, and few bacilli present in lesions. The clinical symptoms are thought to result from a strong cellular immune response to bacterial components in nerves, which results in immunopathology to nerve cells and surrounding tissue. Lepromatous leprosy is characterized by a specific cellular anergy to *M. leprae,* causing massive replication of the organism in tissue but little inflammatory tissue reaction and high levels of circulating antibody. Pathology is due to massive numbers of bacilli in tissue, producing nerve damage and characteristic skin lesions.

### 1.3 Mycobacteria and AIDS

Mycobacterial infections commonly occur in HIV-infected patients. The pathogen most commonly involved in Western AIDS patients is *M. avium*, whereas *M. tuberculosis* commonly infects AIDS patients in the developing world, where tuberculosis is prevalent. In the West, however, *M. tuberculosis* infections also occur frequently among AIDS risk groups that are associated with a high rate of tuberculosis (e.g., intravenous drug abusers).

## 2 GENOME STRUCTURE AND ORGANIZATION

### 2.1 Gene Mapping

The guanidine-plus-cytosine content of mycobacterial DNA is high, ranging from 58 to 69%. Genome sizes have been determined by solution hybridization kinetics and range from 3.3 to 6.8 megabases (Mb) of DNA. Both *M. tuberculosis* and *M. leprae* are currently subjects of genome sequencing projects. A recently produced ordered cosmid library for *M. leprae* contains four contigs of overlapping clones that together account for 2.8 Mb of DNA, close to the expected size of 3.3 Mb for the *M. leprae* genome obtained from hybridization kinetics. Gaps between the contigs are likely to correspond to unclonable segments. Several key housekeeping genes as well as genes encoding major antigens have been identified and mapped.

### 2.2 Mycobacteriophages and Plasmids

Both lytic and lysogenic bacteriophages have been isolated from mycobacteria that are morphologically similar to those found in other genera. Recently the entire DNA sequence of the temperate mycobacteriophage L5 has been determined. The phage is morphologically similar to phage lambda from *Escherichia coli* and contains a linear 52,297 bp genome with cohesive termini. Genes encoding head and tail subunits, DNA polymerase, a putative integrase gene (*int*), tRNAs, and immunity functions have been identified. The L5 lysogens contain a prophage integrated site specifically into a chromosomal attachment site (*att B*) that contains a 43 bp core region. Integration is mediated by the product of the phage *int* gene.

Typical bacterial plasmids have been found in most mycobacteria that have been examined, although not so far in the major human pathogens *M. tuberculosis* or *M. leprae.* These plasmids are cryptic, having no known function. The entire DNA sequence (4837 bp) of the *M. fortuitum* plasmid, pAL5000, has been determined. Five open reading frames have been identified, and the minimal region necessary for replication (1.8 kb) has been located.

### 2.3 Transposable Elements and Repetitive DNA

A number of mobile genetic elements have been discovered in mycobacteria. For example, IS*6110* (also previously termed IS*986*), found in *M. tuberculosis,* is 1358 bp long with 30 bp inverted repeat ends and is a member of family of insertion sequences related to the *E. coli* IS*3*. IS*6110* is present in multiple copies (10–20) scattered throughout the genome of *M. tuberculosis*—in different locations in different strains. It is this property that has made IS*6110* extremely useful for epidemiological analysis. IS*900* was isolated from *M. paratuberculosis* and the related IS*901* from *M. avium* (approximately 1.45 kb). Both belong to a family of atypical insertion sequences that lack terminal inverted repeats, do not produce a target site duplication upon insertion, and have a degree of target site specificity. The transposon Tn*610* was found in the genome of *M. fortuitum* and is a member of the Tn*21* "integron" family of transposable elements that encode sulfonamide resistance. Tn*610* is flanked by two copies of the insertion sequence IS*6100,* which is related to the *E. coli* IS*6* family. Many additional insertion sequences have been identified in mycobacteria.

In addition to functional or putative transposons, a number of repetitive DNA elements of unknown function have been identified in mycobacteria. RLEP is a *M. leprae*-specific repetitive sequence with a 545 bp core sequence. Recently 29 copies of RLEP were mapped to the *M. leprae* contig map. A polymorphic tandem repeat that resembles the repetitive extragenic palindromic (REP) sequences, found in *E. coli,* has been identified in *M. tuberculosis* and other mycobacteria.

## 3 MYCOBACTERIAL GENES AND ANTIGENS

Since 1985, many mycobacterial genes have been characterized. Only a subset of mycobacterial genes contain promoters that function in *E. coli,* although ribosome-binding sites are more conserved. Therefore approaches that involve expression in *E. coli* have usually used *E. coli* promoters. The desired clones are then identified—for example, by screening with antibodies or by complementation of *E. coli* mutations. Genes have also been isolated using DNA probes using mycobacterial protein sequence data or data from other bacterial genera. Cloning of genes directly in mycobacteria has recently been achieved, selecting for either complementation of mycobacterial mutations or acquisition of a new phenotype. Since mycobacterial genome projects are under way, increasingly genes will also be identified purely from analysis of sequence data.

Many genes have been cloned by virtue of the antigenicity of their products, using reactivity to monoclonal antibodies, patient sera, or in certain cases, T-cell reactivity as screening criteria. In certain cases, B- and T-cell epitopes have been mapped in some detail. It is striking that many genes isolated using immunological reactivity have turned out to be heat shock proteins. This finding is attributed to the inherent immunogenicity of the antigens, their high expression in the host environment, and earlier priming to homologues from other organisms that are highly conserved. There is evidence that these proteins can play a role in the development of autoimmune diseases, presumably because of homology with host proteins.

## 4 CLONING OF GENES IN MYCOBACTERIA

Cloning systems have only recently been established in mycobacteria. Plasmid vectors based on the origin of replication from the *M. fortuitum* plasmid pAL5000 have been widely used. Aminoglycoside phosphotransferase genes from Tn*5* and Tn*903,* which give resistance to kanamycin, function well in both *M. smegmatis* and *M. tuberculosis.* Electroporation is an effective method for introducing DNA into mycobacterial cells. In the case of *M. smegmatis,* the efficiency of electroporation was very low, but the rare transformants obtained were found upon curing to be highly electrocompetent, which implies that they were mutants. One of these mutants ($mc^2155$) is now widely used for molecular genetic studies because of the relatively high levels of transformation efficiency that are readily attained ($> 10^5/\mu g$ input DNA). On the other hand, BCG was found to be electrocompetent at the outset.

Shuttle plasmid vectors have also been constructed based on origins of replication from mycobacteriophage D29 and the *Corynebacterium diphtheriae* plasmid pNG2. The broad host range of plasmid RSF1010 also replicates stably in *M. smegmatis* and can be introduced by conjugation. Vectors have been constructed containing *int* gene and *att* site of mycobacteriophage L5; these genetic materials integrate stably into the chromosome and have great potential, for example, in producing recombinant vaccine strains. A variation of this uses homologous recombination with the *pyrF* gene of *M. smegmatis* for integration. Insertion sequences, such as IS*900,* have also been used to integrate foreign genes. Plasmid vectors, which act as phages in mycobacteria and plasmids in *E. coli,* have also been described, based on mycobacteriophages TM4 and L1. Most *E. coli* plasmid origins of replication, such as pMB1, do not function in mycobacteria.

Markers in addition to kanamycin resistance that have been used in vectors are resistance to sulfonamide, streptomycin, viomycin, hygromycin, and fusidic acid, and markers that confer immunity to mycobacteriophage D29 infection. The construction of *pyrF* plasmids allows selection for the plasmid in uracil mutants of *M. smegmatis* and against the plasmid by growth in 5-fluoroorotic acid.

With the inherent adjuvanticity of mycobacteria and the widespread use of BCG as a live vaccine, these genetic tools allow the construction of mycobacteria as live recombinant polyvalent vaccines. Vectors have been developed that express protein antigens from bacterial and viral pathogens, intracellularly, extracellularly, or as lipoproteins attached to the cell surface. Immunization of experimental animals with recombinant BCG has been demonstrated to induce both cellular and humoral immune responses to the cloned foreign antigens.

Chemical mutagenesis has been successfully used with mycobacteria, although the tendency of the cells to clump can be problematical when screening for recessive phenotypes. Also, multiple mutations are likely to occur, and the lack of a general transduction system makes it difficult to transfer individual mutations between strains as a partial remedy for this problem. Transposon mutagenesis, with its promise of single selectable mutations in each cell, has recently become possible. A shuttle plasmid with thermosensitive replication has been constructed and used as an efficient transposon delivery vehicle in *M. smegmatis.* It cannot be used in *M. tuberculosis* because of the narrow temperature range for growth. Some success has been achieved in BCG and *M. tuberculosis* using IS*1096.*

Reverse genetics, in which defined mutations in a gene are incorporated into the chromosome by homologous recombination, is increasingly desirable. Such results have been achieved in *M. smegmatis* but not in *M. tuberculosis,* where illegitimate recombination seems to be more efficient than homologous recombination. The *recA* gene in both *M. tuberculosis* and *M. leprae* is interrupted by an open reading frame encoding a protein that is excised from the RecA protein by posttranslational splicing. It is not known whether this insertion affects the level of homologous recombination in the tubercle bacillus.

## 5 DNA PROBES FOR MYCOBACTERIA

### 5.1 Detection

The slow growth of mycobacteria makes these organisms ideal candidates for the development of rapid DNA probes for diagnosis, and DNA probes for detection of mycobacteria have been developed by Syngene Company (Molecular Biosystems Inc., San Diego, CA) and Gen-Probe (Gen-Probe Inc., San Diego, CA). In the Gen-Probe test, a mycobacterial colony is suspended in lysis buffer and sonicated to release the bacterial ribosomal RNA; then a labeled ribosomal RNA probe is added and hybridized to the sample. Initially, $^{125}$I-labeled probes were used; however, fluorescent probes are now available. Hybridized probe is removed by binding to hydroxyapatite and detected by gamma-counter or fluorimeter. The tests take only a few hours, compared to several weeks for conventional species identification by biochemical tests. Gen-Probe test kits are available for the tuberculosis complex, the *M. avium* complex, and *M. gordonae.* The tests have sensitivity and specificity close to 100% and represent a saving of 15 to 25 days for specific identification of cultures, compared to traditional culture.

### 5.2 Genetic Fingerprinting

Genetic fingerprinting of mycobacterial DNA using DNA probes identifying restriction fragment length polymorphisms has been very successfully applied to subdividing mycobacteria. The insertion sequence IS*6110,* when hybridized to DNA from strains of *M. tuberculosis,* produces complex, highly variable banding patterns that provide excellent markers for contact tracing, identification of the source of outbreaks, and epidemiological studies.

### 5.3 Polymerase Chain Reaction

For rapid detection of mycobacteria in clinical specimens, the polymerase chain reaction has been used to amplify mycobacterial DNA in sputum and tissue samples from patients with leprosy, tuberculosis, and AIDS mycobacterial infections, and from cattle infected with paratuberculosis. However, the use of PCR for laboratory diagnosis of human disease has yet to be fully evaluated in clinical laboratories.

## *Bibliography*


Eiglmeler, K., Honore, N., Woods, S. A., Caudron, B., and Cole, S. T. (1993) Use of an ordered cosmid library to deduce the genomic organization of *Mycobacterium leprae. Mol. Microbiol.* 7:197–206.

Hatfull, G. F., and Sarkis, G. J. (1993) DNA sequence, structure and gene expression of mycobacteriophage L5: A phage system for mycobacterial genetics. *Mol. Microbiol.* 7:395–405.

Jacobs, W. R., Kalpana, G. V., Cirillo, J. D., et al. (1991) Genetic systems for mycobacteria. *Methods Enzymol.* 204:537–555.

McFadden, J. J., Ed., (1990) *Molecular Biology of the Mycobacteria.* Surrey University Press, London.

Stover, C. K., de la Cruz, V. F., Fuerst, T. R., et al. (1991) New use of BCG for recombinant vaccines. *Nature* 351:456–460.

Young, D. B., and Cole, S. T. (1993) Leprosy, tuberculosis and the new genetics. *J. Bacteriol.* 175:1–6.

———, Kaufmann, S. H. E., Hermans, P. W. M., and Thole, J. E. R. (1992) Mycobacterial protein antigens: A compilation. *Mol. Microbiol.* 6:133–145.

# Nematodes, Neurobiology and Development of

*Darryl MacGregor, Ian A. Hope, and R. Elwyn Isaac*

*Key Words*

**ACEDB** The *Caenorhabditis elegans* database; a computer program that contains the physical genome and the genetic maps in graphical form. The locations of sequences, ESTs, and mutations of interest are described. The database is continually updated.

**Expressed Sequence Tags (ESTs)** Clones from a *C. elegans* cDNA library are purified and partially sequenced in an automated sequencer. The resultant sequence is then checked against databases for similarity to previously characterized sequences.

**Genetic Map** All the known mutations of *C. elegans* arranged in the order along the chromosome suggested by genetic experiments.

**Neuropeptides** Usually peptides of between 3 and 30 amino acids long, synthesized in nerve cells for use as neurotransmitters at synapses or as neurohormones.

**Physical Genome Map** A set of ordered cosmids and yeast artificial chromosomes that between them contain all the chromosomal DNA from *C. elegans*. The physical genome map forms the basis of the genome sequencing project.

**Transposon** A short piece of discrete DNA bounded by repeat sequences, which has the ability to insert itself into other DNA sequences. Insertion of a transposon into a gene often causes disruption of the gene, and thus transposons are often used as mutagenic agents.

Nematodes have adapted to an extremely wide variety of habitats ranging from freshwater hot springs to whale placenta.

Many species are parasitic, causing immense economic damage to crops and livestock. Around a third of the world's population suffers from nematode infections, which can result in blindness, malnutrition, and general debilitation. Parasitic nematodes often display complex behavior and development, which determine their interactions with the host animal or plant. A better understanding of nematode neurobiology and development will provide insight into these processes and may help in the rational design of new antinematode drugs.

The nematode *Caenorhabditis elegans* is a powerful model for the study of animal development and behavior. The worm is very amenable to genetic analysis, and a large number of genes required for normal development and behavior have been identified by generating mutations. The entire genome is being sequenced, and sequence information is readily accessible on the ACEDB, European Molecular Biology Laboratory, and U.S. Genome Sequences Databank and (Genbank) databases.

## 1 THE NEMATODA

### 1.1 Nematodes and Their Body Plan

It is estimated that at least 40,000 nematode species exist, of which about 16,000 have been described. About 30% of the described species are parasites of vertebrates; the remainder are either plant or invertebrate parasites or free-living organisms found in soil and mud. With one notable exception, the parasitic species are the most heavily studied because of their economic and medical importance.

The typical nematode has a cylindrical body, tapered at both ends, with a pointed tail and blunt head. The body wall consists of a collagenous cuticle, beneath which lies a cellular or syncytial hypodermis and a layer of longitudinal muscle cells. The muscle cells are divided into four fields by dorsal, ventral, and two lateral hypodermal cords. Respiratory and circulatory organs are absent. The alimentary system consists of a terminal mouth, buccal cavity, muscular pharynx, intestine, rectum, and a ventral, subterminal anus. The body cavity is a pseudocoelom and contains the pharynx, intestine, and gonads.

Nematodes follow a six-stage developmental process: the embryo, four larval stages, and the adult organism. The cuticle is shed after each larval molt.

### 1.2 Nematode Nervous Systems

A nerve ring comprised of neural cell bodies surrounds the pharynx, with sensory and motor neurons extending forward into the head and nerve tracts, which in turn extend alongside the hypodermis to the rear of the animal.

Despite the relative simplicity of the nematode nervous system (most species contain fewer than 250 neurons), many of the neurotransmitters and cellular processes within these neurons are either identical or strikingly similar to the molecules and processes within the nervous systems of higher animals. Many nematode researchers study the neurobiology of these organisms to more readily understand complex mammalian neural systems and their emergent behaviors.

Other researchers are interested in the design of antinematode drugs, and the nervous system has been the target of many of these drugs.

### 1.3 *Caenorhabditis elegans*

*C. elegans* is a hermaphroditic, microscopic, free-living soil nematode, which feeds mainly on bacteria. In the early 1960s Sydney

Brenner suggested that this organism be used as a model for metazoan development.

Many features of *C. elegans* make it an ideal model system. Its size allows it to be handled like a standard laboratory microorganism. The life cycle is relatively short (≈ 3.5 days at 20°C), and the brood size large (> 300). The species is hermaphroditic, and self-fertilization automatically gives rise to homozygotes among the progeny with no need for backcrosses. The male of the species is vital in genetic experiments, as it allows independently isolated mutant alleles to be brought together.

*C. elegans* is transparent at all stages of its development, allowing continual observation of individual cells. This has led to a comprehensive map of all the cell lineages in the adult animal.

### 1.3.1 The Nervous System of *C. elegans*

The nervous system of *C. elegans* has also been fully described at the cellular level. The adult hermaphrodite contains 302 neurons and the male 381. Electron micrographs were used to determine the position and connectivity of every neuron within the animal. This ''wiring diagram'' revealed that the *C. elegans* nervous system contains about 5000 chemical synapses, 600 gap junctions, and 2000 neuromuscular junctions. The neuronal pattern is constant from animal to animal, and for every nerve cell, cell lineage and developmental history are known. Despite its relatively small size, the nervous system can generate and regulate a wide variety of behaviors. The neurons responsible for many of these behaviors have been identified, although the molecular mechanisms underlying them have not been characterized.

### 1.3.2 The Molecular Genetics of *C. elegans*

The genome of *C. elegans* is relatively small (≈ $10^8$ base pairs), and a complete physical and genetic map is available. The physical map is in the form of ordered contiguous cosmids and yeast artificial chromosomes. This physical map forms the basis for the project to sequence the entire *C. elegans* genome.

The genes underlying mutants of interest may be isolated on the basis of the correspondence between the position of the mutation on the genetic map and the position of the gene on the physical map. The vast majority of genes cloned from *C. elegans* have been isolated using some variant of this ''forward genetics'' approach.

New technologies, however, have appeared to facilitate a ''reverse genetics'' in *C. elegans* in which the DNA sequence of a gene is known and the researcher must establish its function in creating a given phenotype. A large library of expressed sequence tags (ESTs) is being built up for *C. elegans.* These ESTs are short (≈ 360 bp) DNA sequences from a cDNA library. The sequence is then searched for homology to known proteins, and the results are made available both in published journals and in the *C. elegans* database, ACEDB. At present 25–30% of the ESTs are similar to known genes from other species, while 60% were unrelated to previously described genes.

*C. elegans* contains a number of transposons, short mobile pieces of DNA, and one of these, *Tcl,* has been used to create a bank of transposon insertion mutants. This library of mutants should contain a strain carrying a *Tcl* insertion in the gene of interest. Using primers to the gene and to *Tcl* sequences, this strain can be identified and the phenotype characterized. The mutant strain can then be restored to wild type by making lines transgenic for the gene of interest. This ''reverse genetics'' approach should enable researchers to characterize a number of genes that would have proven refractory to study using the conventional techniques.

## 2 NEMATODE NEUROBIOLOGY

The microscopic size of *C. elegans* precludes the use of many of the standard techniques favored in neurobiology. Thus, much of the work on nematode neurophysiology and neurochemistry has been carried out on the large porcine parasite *Ascaris suum.* This worm can grow up to 30 cm long, and the large size aids in disjysection of nervous tissues and in the visualization of identified neurons. The neurons of *Ascaris* have also been used in electrophysiological experiments to ascertain the size and frequency of electrical events.

### 2.1 Neurotransmitters in Nematodes

The major neurotransmitter of excitatory motor neurons appears to be acetylcholine, while γ-aminobutyric acid (GABA) is believed to be the main neurotransmitter of inhibitory motor neurons. Other classical neurotransmitters also exist in nematodes, these include dopamine and 5-hydroxytryptamine (5-HT), which have been localized to small subsets of neurons. 5-HT is found in the HSN motor neurons, which control egg-laying in *C. elegans,* and is present in two neurosecretory cells in the pharynx of *C. elegans* and *A. suum.* 5-HT has also been implicated in the regulation of locomotion and carbohydrate metabolism in *A. suum* and in the coordination of male copulatory behavior in free-living nematodes. Immunocytochemistry has revealed richness of peptidergic neurons in nematodes, and several neuropeptides belonging to the FMRF amide peptide family have been isolated from *A. suum, C. elegans,* and *Panagrellus redivivus.*

### 2.2 Molecular Neurobiology of *C. elegans*

A number of the protein molecules involved in the *C. elegans* nervous system have been isolated using the techniques of forward genetics. A mutant has been isolated, and the gene responsible for the mutation characterized. A typical example of this kind of mutation is the *unc*-86 mutation. Strains carrying this mutation were *unc*oordinated, and the gene responsible was found to encode a homeobox-containing protein with homology to mammalian transcription factors. In another approach, drug-resistant mutants are used to characterize genes for the receptor proteins of nematode neurotransmitters (e.g., levamisole-resistant mutants and the acetylcholine receptor).

The simplicity and invariance of the nematode nervous system offers an opportunity to apply the powerful techniques of molecular genetics to the understanding of the development and function of animal nervous systems.

## 3 DEVELOPMENT IN *C. ELEGANS*

### 3.1 Normal Development of *C. elegans*

The embryonic development of *C. elegans* takes place in two stages. In the first the fertilized egg undergoes rapid cell division to generate a 550-cell ovoid with little overt differentiation. The second stage consists of differentiation and morphogenesis with little additional cell division. *C. elegans* has an invariant cell lineage based around six founder cells: AB, C, D, E, MS, and $P_4$. These founder cells give rise to cell lineages within which division occurs at characteristic rates and results in equal-sized daughter cells. The movement and the eventual fate of all the cells are known throughout development.

Postembryonic growth is regular, with the cuticle being shed at the end of each larval stage. During this period the number of somatic nuclei increases to 959 in the hermaphrodite and 1031 in the male. There is little long-range cell movement in development, suggesting that the majority of cells are born where they are required.

### 3.2 Manipulation of Development

Individual cells in a developing animal can be destroyed by a tightly focused microlaser beam and the effect on development ascertained. These experiments revealed that *C. elegans* development is largely cell autonomous, the remaining cells failing to compensate for the structures that would have derived from the destroyed cell(s). However, other experiments have provided examples of nonautonomous development, suggesting that the full extent of intercellular communication in developing *C. elegans* has yet to be elucidated.

Many genes involved in the cell lineages of *C. elegans* have been identified by mutations that affected the development of particular structures such as the vulva. Most of these genes are referred to as homeotic (by analogy with homeotic mutations in *Drosophila*), where the fate of one cell is exchanged with that of another. The gene responsible for the *lin*-12 mutation was isolated and found to encode a transmembrane protein with a sequence similarity to the *Notch* protein, which is involved in cell fate decisions in *Drosophila*. A low stringency hybridization screen of a *C. elegans* gene library with *lin*-12 revealed the presence of *glp*-1, a gene involved in cell–cell interactions distinct from those affected by *lin*-12.

### 3.3 Developmental Cell Death in *C. elegans*

During the development of an adult *C. elegans* hermaphrodite, 131 out of the 1090 somatic nuclei that are born will die. The majority of cells that die would otherwise have been neuronal. These developmental cell deaths occur reproducibly at given times and locations throughout a cell lineage. The pattern of death in these cells is characteristic and conserved. It involves a change in cell morphology, disintegration of the nucleus, and finally, engulfment of the dead cell by its healthy neighbors. A number of genes have been implicated in this process. The most central appear to be *ced*-3 and *ced*-4; in mutants lacking one of these genes, all cells survive. The gene encoding *ced*-4 has been cloned since "reverse genetics" and was shown to encode a novel protein with regions showing sequence similarity to calcium-binding domains.

The activity of the gene *ced*-9 appears to be crucial in preventing cell death in inappropriate cells. It appears that *ced*-9 is a homologue of the human oncogene *bcl*-2. Overexpression of *bcl*-2 in human B and T cells prevents or significantly delays apoptotic cell death. This raises the question, Can *ced*-9 homologues act to prevent cell death in other organisms? Could proteins involved in *C. elegans* programmed cell death cause cell death in other organisms?

These, and many other questions can be asked and, it is hoped, answered by using the nematode *C. elegans* as a model system.

*See also* Cell Death and Aging, Molecular Mechanisms of; Synapses.

### *Bibliography*

Bargmann, C. I. (1993) Genetic and cellular analysis of behaviour in *C. elegans*. *Annu. Rev. Neurosci.* 16:47–71.

Driscoll, M. (1992) Molecular genetics of cell death in the nematode *Caenorhabditis elegans*. *J. Neurobiol.* 23:1327–1351.

Halton, D. W., Ed. (1991) *Parasite Neurobiology*, Vol. 28 in *Symposium of the British Society for Parasitology*. Cambridge University Press, Cambridge.

Wood, W. B., Ed. (1988) *The Nematode Caenorhabditis elegans*, Cold Spring Harbor monograph series 17. Cold Spring Harbor Press, Plainview, NY.

Zuckerman, B. M., Ed. (1980) *Nematodes as Biological Models*, Vols. 1 and 2. Academic Press, New York.

# Neurofibromatosis

*David H. Viskochil and Roger K. Wolff*

### *Key Words*

**Autosomal Dominant** Describing a disorder that does not map to a sex chromosome and is phenotypically expressed in heterozygous individuals.

**Cytogenetic Map** Map depicting the approximate location of chromosomal anomalies based on banding studies. Translocation breakpoints and interstitial insertions and deletions can be mapped by routine chromosome banding studies.

**Dominant Negative Mutation** A mutation in a polypeptide that disrupts the function of the wild-type allele in the same cell.

**Genetic Linkage Map** Map of the relative positions of genetic *loci* on a chromosome, determined on the basis of how often the loci are co-inherited. Distance is measured in centimorgans (cM).

**Genotype** The alleles present at one locus.

**Mutation Rate** The frequency of a mutation at a given locus, expressed as mutations per locus per gamete.

**Penetrance** The proportion of cases in which organisms with a given genotype express the corresponding phenotype.

**Phenotype** The observed biochemical, physiological, and morphological characteristics of an individual.

**Physical Map** Map of the locations of identifiable landmarks on DNA, generally determined by restriction sites and/or overlapping cloned DNA. Distance is measured in base pairs.

**Sporadic** Describing a case of a disease caused by a new mutation.

**Tumor Suppressor Gene** A normal gene involved in the regulation of cell growth that when inactivated can lead to tumor development.

**Variable Expressivity** The extent to which traits are observed in individuals with genetic defects.

---

Neurofibromatosis 1 and neurofibromatosis 2 are distinct clinical entities that provide insight for two important paradigms of molecular genetics. NF1, which is one of the most prevalent heritable conditions worldwide, captures many of the classical principles of human genetics including expressivity, penetrance, pleiotropy, sporadic occurrence, and counseling issues. Even though both conditions carry an increased relative risk for malignancy, NF2 is clearly associated with the development of meningiomas, and its gene is considered to be a classical "tumor suppressor." Both genes were isolated by application of the mapping approach to the cloning of human disease genes, and familiarity with these two conditions bridges the fields of molecular and clinical genetics.

Neurofibromatosis 1 (NF1) and neurofibromatosis 2 (NF2) are distinct genetic disorders. NF1, oftentimes called peripheral NF or Von Recklinghausen disease, is the more common condition, afflicting approximately 1 in 3500 people worldwide. Its typical clinical manifestations tend to involve the skin and peripheral nervous system. NF2, sometimes referred to as central NF, has an incidence of approximately 1 in 40,000 people worldwide; it primarily manifests as tumors involving the central nervous system. Even though NF1 and NF2 share a name, the conditions have distinct clinical signs that usually make diagnosis relatively straightforward. Furthermore, discoveries regarding the molecular biology of the respective conditions have firmly established the separateness of these two distinct clinical entities.

The successful cloning of several genes associated with inherited human diseases has established the utility of map-based approaches for isolating genes that encode proteins of unknown structure or function. According to this paradigm, chromosomal localization for a genetic trait in a family identifies a genomic locus that harbors candidate genes; the identification of mutations carried by patients with the disorder specifies the disease-causing gene. A combination of genetic linkage mapping, cytogenetic mapping, and genomic physical mapping provided an approach to the identification of both *NF1* and *NF2*. After each locus had been precisely mapped, a set of genomic DNA clones was identified that served as probes in both cDNA library screening and mutation analysis of disease family DNA. In this fashion, both genes were genetically mapped in 1987; *NF1* was cloned in 1990, whereas *NF2* was cloned in 1993.

## 1 NEUROFIBROMATOSIS 1

### 1.1 Clinical Aspects

NF1 is inherited as an autosomal dominant condition that is generally diagnosed by physical examination. The clinical features typically seen in this condition include multiple café-au-lait macules, neurofibromas, freckling in the axillae and groin, optic nerve gliomas, Lisch nodules, and specific osseous lesions. NF1 is present in an individual if two or more of these criteria are satisfied (see Table 1). Generally the skin manifestations have minor clinical significance, and approximately two-thirds of affected individuals lead normal lives. The individuals who suffer medically significant problems show any number of the following clinical features: paraspinal and plexiform neurofibromas, learning disabilities, dysplastic scoliosis, intractable headaches, renal artery hamartomas with hypertension, and malignancies such as pheochromocytomas, neurofibrosarcomas, leukemia of myelogenous type, and rhabdomyosarcomas. Although NF1 is fully penetrant, its variability of clinical expression is striking, even within a family whose affected members presumably carry identical *NF1* mutations. NF1 is a progressive condition: affected individuals develop increasing numbers of neurofibromas and NF1-related complications as they age; most, however, have a near-normal life expectancy. Approximately half of NF1 patients have no family history, a circumstance that exemplifies the high mutation rate of this gene. Furthermore, more than 90% of sporadic cases arise from the paternally derived chromosome.

**Table 1** Diagnostic Criteria for NF1: Two or More of the Seven Listed Features are Required for Diagnosis

1. More than five café-au-lait macules exceeding 5 mm in greatest diameter in *pre*pubertal individuals and 15 mm in greatest diameter in *post*pubertal individuals.
2. Two or more neurofibromas of any type or one plexiform neurofibroma.
3. Freckling in the axillary or inguinal regions (Crowe's sign).
4. Optic glioma.
5. Two or more Lisch nodules (iris hamartomas—slit lamp exam).
6. A distinctive osseous lesion such as sphenoid wing dysplasia or thinning of long bone cortex, with or without pseudoarthrosis.
7. A first-degree relative (parent, sibling, or offspring) with NF1 by criteria 1–6.

### 1.2 Gene Identification and Genomic Structure

Genetic linkage mapping provided an approach to the identification of the gene for NF1. Specifically, linkage with DNA markers in NF1 families mapped the locus to the long arm of chromosome 17, near the centromere at band q11.2. The identification and molecular application of two independent balanced translocations enabled investigators to physically map the *NF1* locus and identify four candidate cDNAs that mapped to the region encompassing the translocation breakpoints. Point mutation analysis of DNA from NF1 patients identified the disease-causing gene.

*NF1* is relatively large; its processed transcript is estimated at 9 to 11 kb, and it consists of 59 exons distributed over 350 kb of genomic DNA. The open reading frame encodes a 2818 amino acid protein named neurofibromin. Antibodies directed against various peptide domains detect a large protein of approximately 230 to 280 kiloDaltons by sodium dodecyl sulfate–polyacrylamide gel electrophoresis (SDS-PAGE) Western blot analysis. A single, large intron of the NF1 gene encompasses three genes that map between the two translocation breakpoints, *EVI2A, EVI2B,* and *OMGP;* but *NF1* is transcribed from the opposite strand with respect to the three interdigitated genes. This multigene organization at a single locus has no precedent. Moreover, how the expression of these three genes might be affected by mutations in *NF1,* possibly accounting for variability in expression of the disease, is unresolved.

A search of protein sequence databases for similarities with the peptide predicted from the *NF1* transcript sequence demonstrates significant homology in amino acid sequence between a domain in neurofibromin and the catalytic domains of the mammalian *ras*-GTPase-activating protein (GAP) and its yeast counterparts, *IRA1* and *IRA2*. The GAP-related domain of neurofibromin is capable of interacting with mammalian p21ras and yeast Ras; an expressed peptide segment from neurofibromin that encompasses the catalytic domain stimulates the intrinsic GTPase of p21ras *in vitro.* The implications of these observations are profound, inasmuch as this family of genes is known to be involved in the mechanisms that control cell growth and differentiation through their interaction with the p21ras signal transduction pathway. Furthermore, amino acid sequences outside the catalytic domains of the *IRA1* and *IRA2* genes share homology with neurofibromin, which extends some

300 amino acids in the $NH_2$-terminal direction and more than 800 amino acids in the COOH-terminal direction. Stimulation of the intrinsic GTPase of p21ras is the only known function of neurofibromin; the biological function of the remaining 85% of the peptide is not known.

### 1.3 Pathogenesis in NF1

*NF1* is ubiquitously expressed at low levels; however, the highest level of expression appears in the central nervous system. Neurofibromin amino acid sequence is highly conserved between murine and avian species with approximately 98% and 85% homology, respectively, which suggests that the regions outside the GAP-related domain also play a significant role in normal function. The role of neurofibromin in the p21ras signal transduction pathway is not clear; it could function either as an upstream "negative regulator" of p21ras-GTP or as a downstream "effector" of activated p21ras. It may even have a dual role: in certain tissue it may act entirely as a negative regulator, whereas in other tissue, neurofibromin may act more like an effector in the p21ras pathway.

Attempts to better understand neurofibromin function by its correlation with *NF1* mutations have not been successful. Even though fewer than 15% of all tested NF1 patients have a demonstrable mutation, a full range of mutations has been identified: translocations, deletions and insertions of various sizes leading to frameshifts, and substitutions that lead to premature translational stops and amino acid substitutions. The majority of the few mutations thus far characterized fall in the category of "inactivating" or "loss of function" mutations, in contrast to "dominant negative" or "gain of function" mutations. This observation supports the hypothesis that *NF1* is a tumor suppressor gene. Furthermore, some tissue culture cell lines derived from sporadic tumors have decreased levels of neurofibromin with concomitant increased intracellular levels of "activated" p21ras-GTP, which supports the contention that inactivation of *NF1* leads to abnormal cell growth. However, an inability to detect loss of heterozygosity (chromosome deletion of the remaining normal allele) in the benign neurofibromas from NF1 patients, coupled with the absence of a high association of malignancy in NF1, indicates that *NF1* does not behave like a typical tumor suppressor. Finally, the poor correlation between genotype and phenotype in individuals in whom a mutation has been identified raises the prospect that modifying loci play a major role in the clinical expression of NF1.

## 2 NEUROFIBROMATOSIS 2

### 2.1 Clinical Aspects

NF2, or bilateral acoustic neurofibromatosis, is characterized by the occurrence of bilateral vestibular schwannomas (acoustic neuromas), meningiomas, and spinal schwannomas. The hallmark clinical feature of NF2 is the eighth cranial nerve schwannoma, which causes tinnitus, hearing loss, and facial palsies by exerting pressure on both the seventh and eighth cranial nerves. NF2 is an inherited disorder that affects approximately 1 in 40,000 individuals. The disorder segregates in families consistent with its inheritance as an autosomal dominant trait with an age-related penetrance of greater than 95%; mean age of onset is 21 years. Like NF1, this disorder is diagnosed by the clinical application of a set of diagnostic criteria (see Table 2), which include vestibular schwannomas (identified either by computerized tomography or magnetic resonance imaging), meningiomas, schwannomas, gliomas, and posterior subcapsular lenticular opacities. In distinction to NF1, intracranial and spinal tumors are prominent and consistent features of NF2; this aspect of the condition underlies its relatively high morbidity and shortened life expectancy.

**Table 2** Diagnostic Criteria for NF2

1. Bilateral masses of eighth cranial nerve (CT scan or MRI study) or
2. First degree relative with NF2 *and* either
   (a) Unilateral mass of the eighth cranial nerve
   (b) Two of the following:
       Neurofibroma
       Meningioma
       Glioma
       Schwannoma
       Juvenile posterior subcapsular lenticular opacity

### 2.2 Gene Identification and Genomic Structure

The association of meningiomas with NF2, together with the cytogenetic demonstrations of nonrandom loss of material from the long arm of chromosome 22 in sporadic meningiomas, initially suggested the chromosomal location for NF2. This hypothesis was confirmed by molecular characterization of tumor DNA derived from both sporadic and NF2-related vestibular schwannomas and meningiomas where nearly 60% of tumors had chromosome 22 deletions involving band q12. In addition, genetic linkage analysis performed on an extended pedigree with NF2 mapped the disorder to the region on chromosome 22 designated by marker D22S1. Using a combination of multipoint linkage analysis and tumor deletion analysis, *NF2* was localized to a 6 megabase region, which enabled investigators to focus the physical mapping studies and to identify more subtle deletions in NF2 patients. A candidate cDNA that mapped to the subtle deletions was identified as the NF2 gene in early 1993. In contrast to *NF1, NF2* spans less than 80 kilobases of genomic DNA; its nearest known gene neighbor, *NEFH,* lies approximately 50 to 100 kb centromeric of *NF2.*

### 2.3 Pathogenesis in *NF2*

The clinical features of NF2 support the hypothesis that *NF2* is a tumor suppressor. Analogous to *RB* in retinoblastoma, the *NF2* locus likely harbors a recessive tumor suppressor gene, with tumor progression following a two-hit model. By this model, the hereditary NF2 cases are predisposed to develop multiple tumor foci as a result of having a germ line mutation at this locus. This mutation is recessive to the normal allele and remains masked until a rare somatic mutation, such as chromosome 22 loss or deletion of band q12, inactivates the normal allele. Under this model, sporadic unilateral tumors of the types seen in NF2 patients arise by two rare somatic mutations, inactivating both *NF2* alleles in a single cell.

*NF2* is a unique tumor suppressor because it encodes a novel member of a family of proteins that link the outer cell membrane to the interior cytoskeleton. The NF2 gene product, coined merlin, is a 595 amino acid peptide that has sequence homology with the cytoskeletal protein family containing *m*oesin, *e*zrin, and *r*adixin (hence the "mer" prefix in merlin). It has been hypothesized that inactivation of *NF2* may cause tumor formation by disrupting any one of the many processes in which the cytoskeleton affects cell growth.

## 3 SUMMARY

NF1 and NF2 are clearly distinct clinical entities that arise by mutation of separate genes. A feature that is shared between these two conditions is the surgical management of multiple and recurrent tumors, whether they be schwannomas and meningiomas in NF2 or neurofibromas and gliomas in NF1. The genes for both conditions have been cloned, and molecular characterization of the respective gene products offers promise of new approaches to the medical management of NF1 and NF2.

*See also* HUMAN GENETIC PREDISPOSITION TO DISEASE; TUMOR SUPPRESSOR GENES.

### *Bibliography*

Cavanee, W., Hastie, N., and Stanbridge, E., Eds. (1989) *Recessive Oncogenes and Tumor Suppression.* Current Communications in Molecular Biology, Cold Spring Harbor Laboratory Press, Cold Spring Harbor, NY.

National Institutes of Health (1988) NIH Consensus Development Conference Statement: Neurofibromatosis. *Arch. Neurol.* **45:**575–578.

Riccardi, V.M. (1992). *Neurofibromatosis: Phenotype, Natural History, and Pathogenesis,* 2dn ed. Johns Hopkins University Press, Baltimore.

Shaw, D., Ed. (1994) *Molecular Genetics of Human Inherited Disease.* Wiley, London.

# NEUROPEPTIDES, INSECT

*Edward P. Masler*

### *Key Words*

**Bioassay** A controlled experimental system in which an insect, a tissue, or an organ preparation is exposed to a test compound; the specific response indicates the quality and quantity of the test compound. Considered the most critical test of a compound's activity.

**Neurohemal Organ** Specialized organ, analogous to the pituitary, from which neuropeptides are secreted into the insect blood (hemolymph).

**Neuropeptide** Generic term for peptides of as few as 3 to more than 200 amino acids, produced in specialized neurons, secreted from neurohemal organs, and usually acting as hormones.

**Neurosecretion** Release of material (e.g., biogenic amines, peptides) from secretory granules located in the nervous system in response to internal or external stimuli.

Neuropeptides are involved in nearly all life processes in insects. Activities ranging from behavior and reproduction to energy metabolism and metamorphosis are under the control of one or more neuropeptides. The mechanisms by which insect neuropeptides are synthesized, secreted, act, and are catabolized are similar to those described for vertebrates. It was, in fact, work on insect metamorphosis early in the twentieth century that led to the concept of a neural–endocrine interaction and to the description of neurosecretion. Since insect neuropeptides are so critical to normal physiological functions, detailed studies of insect neuropeptides and careful examination of the mechanisms surrounding their production, action, and inactivation will afford a more thorough understanding of life processes. The argument that this approach will lead to the development of novel, effective, and biorational pest control agents is consistent with an increased awareness of environmental concerns.

## 1 CONCEPT OF NEUROSECRETION

### 1.1 WORK IN THE EARLY TWENTIETH CENTURY

The idea that two apparently disparate systems, nervous and endocrine, could function in unison, was born in large part from the work of Stephan Kopec early in this century. Working with larvae of the gypsy moth, *Lymantria dispar,* he showed that the brain was essential for normal metamorphosis. Ligatures tied about the neck prior to a critical period in larval development prevented the animals, which lived for extended periods following the operation, from metamorphosing to the next developmental stage. Similar results were obtained if the brains of nonligated animals were excised. Kopec proposed, rather prophetically, that a factor from the brain was required for metamorphosis. This was the first demonstration, in any animal, that the nervous system had the capacity for endocrine function. Subsequent work by others, done on a variety of insect species, confirmed and expanded these initial observations.

The manner by which a neural organ, such as the brain, gets its message out involves the release of chemical signals, either biogenic amines or neuropeptides, through the process of neurosecretion. The concept of neurosecretion, which implies that nervous and endocrine systems interact functionally and anatomically, was revolutionary. Initial opposition to the idea gave way to experimental proof as neurosecretion was championed by numerous researchers, notably the husband-and-wife team of Ernst and Berta Scharrer. It eventually became clear that neuropeptides were the key messengers of neurosecretion.

### 1.2 EXISTENCE OF NEUROPEPTIDES

A large variety of physiological processes in insects are controlled by neuropeptides. Some of the first insect neuropeptides described, the ecdysiotropins, are produced by specialized neurons in the brain (neurosecretory cells) and accumulate in a neurohemal organ, the corpus allatum/corpus cardiacum. This complex is an appendage of the brain and functions very much like the vertebrate hypothalamus and pituitary, selectively releasing neuropeptide upon receiving the necessary stimulus. Ecdysiotropins are so named because they stimulate target organs to produce ecdysteroids, steroid hormones essential for metamorphosis and molting.

Because the processes of metamorphosis and molting are so dramatic, they caught the attention of early researchers and led to studies of their underlying causes. Kopec's experiments demonstrated that neck ligature or removal of the brain prevented secretion of ecdysiotropins, synthesis of necessary steroid hormones, and

**Table 1** Some Selected Insect Neuropeptides

| Neuropeptide | Function | Structure |
|---|---|---|
| Adipokinetic hormones | Stimulate mobilization and metabolism of lipid and carbohydrate | pGlu-Leu-Asn-Phe-Thr-Pro-Asn-Trp-Gly-Thr-$NH_2$<br>pGlu-Leu-Thr-Phe-Thr-Ser-Ser-Trp-Gly-$NH_2$ |
| Allatostatins | Inhibit the synthesis and secretion of juvenile hormone | Ala-Pro-Ser-Gly-Ala-Gln-Arg-Leu-Tyr-Gly-Phe-Gly-Leu<br>Gly-Asp-Gly-Arg-Leu-Tyr-Ala-Phe-Gly-Leu-$NH_2$<br>Gly-Gly-Ser-Leu-Tyr-Ser-Phe-Gly-Leu-$NH_2$<br>Asp-Arg-Leu-Tyr-Ser-Phe-Gly-Leu-$NH_2$ |
| Allatotropin | Stimulates the synthesis and secretion of juvenile hormone | Gly-Phe-Lys-Asn-Val-Glu-Met-Met-Thr-Ala-Arg-Gly-Ph |
| Diuretic hormones | Regulate water balance | Cys-Leu-Ile-Thr-Asn-Cys-Pro-Arg-Gly-$NH_2$<br>Cys-Leu-Ile-Thr-Asn-Cys-Pro-Arg-Gly-$NH_2$<br><br>Arg-Met-Pro-Ser-Leu-Ser-Ile-Asp-Leu-Pro-Met-Ser-Val-<br>Gln-Lys-Leu-Ser-Leu-Glu-Lys-Glu-Arg-Lys-Val-His-Ala<br>Ala-Ala-Ala-Asn-Arg-Asn-Phe-Leu-Asn-Asp-Ile-$NH_2$ |
| Ecdysiotropins | Stimulate steroid hormone production | Gly-Ile-Val-Asp-Glu-Cys-Cys-Leu-Arg-Pro-Cys-Ser-Val<br>Leu-Leu-Ser-Tyr-Cys: (A chain of small form)<br><br>pGlu-Gln-Pro-Gln-Arg-Val-His-Thr-Tyr-Cys-Gly-Arg-H<br>Arg-Thr-Leu-Ala-Asp-Leu-Cys-Trp-Glu-Ala-Gly-Val-As<br>of small form)<br><br>[Gly-Asn-Ile-Gln-Val-Glu-Asn-Gln-Ala-Ile-Pro-Asp-Pro<br>Thr-Cys-Lys-Tyr-Lys-Lys-Glu-Ile-Glu-Asp-Leu-Gly-Glu<br>Val-Pro-Arg-Phe-Ile-Glu-Thr-Arg-Asn-Cys-Asn-Lys-Thr<br>Pro-Thr-Cys-Arg-Pro-Pro-Tyr-Ile-Cys-Lys-Glu-Ser-Leu-<br>Ile-Thr-Ile-Leu-Lys-Arg-Arg-Glu-Thr-Lys-Ser-Glu-Glu-S<br>Glu-Ile-Pro-Asn-Glu-Leu-Lys-Tyr-Arg-Trp-Val-Ala-Glu<br>Pro-Val-Ser-Val-Ala-Cys-Leu-Cys-Thr-Arg-Asp-Tyr-Gln<br>Tyr-Asn-Asn-Asn)] × 2 |
| Eclosion hormones | Trigger regimented behavior patterns resulting in ecdysis | Ser-Pro-Ala-Ile-Ala-Ser-Ser-Tyr-Asp-Ala-Met-Glu-Ile-C<br>Glu-Asn-Cys-Ala-Gln-Cys-Lys-Lys-Met-Phe-Glu-Pro-Tr<br>Gly-Ser-Leu-Cys-Ala-Glu-Ser-Cys-Ile-Lys-Ala-Arg-Gly-<br>Ile-Pro-Glu-Cys-Glu-Ser-Phe-Ala-Ser-Ile-Ser-Pro-Phe-Le<br>Lys-Leu-OH<br><br>Asn-Pro-Ala-Ile-Ala-Thr-Gly-Tyr-Asp-Pro-Met-Glu-Ile-C<br>Glu-Asn-Cys-Ala-Gln-Cys-Lys-Lys-Met-Leu-Gly-Ala-Tr<br>Gly-Pro-Leu-Cys-Ala-Glu-Ser-Cys-Ile-Lys-Phe-Lys-Gly<br>Ile-Pro-Glu-Cys-Glu-Asp-Phe-Ala-Ser-Ile-Ala-Pro-Phe-L<br>Lys-Leu-OH |
| Myotropins | Stimulate muscle contraction | A diverse group of peptides from 8±12 residues, amidat C-terminus, some blocked with an N-terminal pGlu, so |
| Oostatic hormone | Prevents ovarian maturation through inhibition of proteolytic enzyme production | Tyr-Asp-Pro-Ala-Pro-Pro-Pro-Pro-Pro-Pro-OH |
| Pheromone biosynthesis activating neuropeptides | Stimulate female sex pheromone production | Leu-Ser-Glu-Asp-Met-Pro-Ala-Thr-Pro-Ala-Asp-Gln-Glu<br>Gln-Pro-Asp-Pro-Glu-Glu-Met-Glu-Ser-Arg-Thr-Arg-Ty<br>Pro-Arg-Leu-$NH_2$<br><br>Leu-Ser-Asp-Asp-Met-Pro-Ala-Thr-Pro-Ala-Asp-Gln-Gl<br>Arg Gln Asp Pro Glu Gln Ile Asp Ser Arg Thr Lys Tyr<br>Pro-Arg-Leu-$NH_2$ |

metamorphosis dependent on these steroids. The factor with which Kopec was dealing was subsequently found to be a peptide and, after years of investigation by numerous laboratories, was isolated recently. As is common with insect neuropeptides, multiple molecular forms, eliciting similar physiological responses, were discovered. A low molecular weight heterodimer, with an A chain of 20 residues and a B chain of 28 residues, was isolated, along with a high molecular weight homodimer. This homodimer consists of two identical 109-residue chains, a sequence ultimately deduced from the cDNA gene sequence. Five C-terminal amino acids may be lost during final processing to the mature neuropeptide (Table 1). Each form, small and large, stimulates ecdysteroid production, but the two show different insect species specificities.

## 2 IDENTIFICATION OF NEUROPEPTIDES

### 2.1 Bioassay

A robust bioassay is essential for the study of a neuropeptide. Since most neuropeptides are initially detected because of their effect on a physiological event, bioassays are usually developed as some mimic of the event. For example, inhibition of molting in neck-ligatured larvae can be overcome by injection of brain extract. Postinjected larvae that molt indicate that the brain extract contained ecdysiotropin. Such an in vivo bioassay is often complemented with an in vitro bioassay. Ecdysiotropins are detected in vitro by incubating test extracts with the steroid-producing organs. Steroid production is then assessed, using an immunoassay specific for the steroid. Although bioassays vary enormously in type, design, execution, and complexity, they are always essential for the detection, isolation, and characterization of neuropeptides.

### 2.2 Biochemical, Chemical, and Immunological Assays

In direct support of the bioassay are chemical and immunological methods. These may include varying combinations of chromatography, enzyme assays, radiochemical tracers, enzyme-linked immunosorbent assays, and radioimmunoassays. Such methods measure either the peptide itself or chemical signals resulting from the action of the peptide (e.g., steroid or cyclic nucleotide levels, enzymatic activities, specific protein synthesis, etc.).

### 2.3 Chemical Features and Neuropeptide Families

Insect neuropeptides vary enormously in structure and size. Despite this great variety, a number of features, although not quite universal, are very common. Most insect neuropeptides are amidated at the C-terminus. This appears to be a mechanism for protection from carboxypeptidase degradation. Many neuropeptides are also blocked at the N-terminus by the formation of a pyroglutamate (pGlu) residue. Inhibition of aminopeptidase digestion is the suspected purpose. A third common, and most intriguing, feature is the existence of neuropeptide families. In some families, sequence homologies and similar activities are common. Often, a bioassay detects more than one biologically active peptide and, in many cases, a series of structurally related peptides exists.

The allatostatins (Table 1) are peptides containing 8 to 13 residues; all are C-terminally amidated, share a Phe-Gly-Leu-$NH_2$ C-terminus, and are active in the same bioassay. Adipokinetic hormones, all having the N-terminal pGlu and C-terminal amide, vary in length from 8 to 10 amino acids but are remarkably conserved across a range of species (two are shown in Table 1).

A second type of family includes members that have similar activities but quite different structures. Ecdysiotropins and diuretic hormones are two examples of such families (Table 1).

A third type of family is represented by sequences that are apparently unrelated but contain subsequences that are similar and have similar effects in the bioassay. For example, the 33 amino acid pheromone biosynthesis activating neuropeptide contains the C-terminal pentamer Phe-Ser-Pro-Arg-Leu-$NH_2$, which is essential for pheromone-stimulating activity. The same pentameric structure is featured in some myotropins and is essential for their ability to elicit muscle contraction. Thus, two seemingly unrelated neuropeptides, having different but essential functions in the insect, share a common structural feature essential to the unique activity of each peptide.

## 3 IMPACT OF NEUROPEPTIDES

### 3.1 Processes Controlled

While essentially all physiological processes in insects are under the influence of neuropeptides, a few selected neuropeptides are described in Table 1 to give a sense of the variety of peptides involved.

#### 3.1.1 Developmental Neuropeptides

In addition to their developmental effects on metamorphosis and molting, ecdysiotropins are involved in reproduction. Specific reproductive ecdysiotropins stimulate insect ovaries to synthesize steroids and undergo maturation. A decapeptide hormone that inhibits ovarian maturation has recently been isolated and characterized. This oostatic hormone is not an antiecdysiotropin per se, since it does not affect steroid production; it does, however, interfere with yolk protein production and serves as an endogenous birth control agent. Neuropeptides involved with metamorphosis are the allatotropins and allatostatins, which control the start and stop, respectively, of the synthesis and secretion of juvenile hormones, sesquiterpene methyl esters involved with both metamorphosis and reproduction.

#### 3.1.2 Metabolic Neuropeptides

The adipokinetic hormones were one of the earliest groups of insect neuropeptides to be sequenced. These are octa-, nona-, and decapeptides, amidated at the C-terminus and blocked by an N-terminal pGlu residue, which stimulate the mobilization of energy stores. Numerous molecular variants have been isolated. Two examples are shown in Table 1. Neuropeptides with similar structures are responsible for regulating heart rate. Diuretic hormones, which regulate water balance, are represented by a homodimer of 18 residues with two cysteine cross-linkages, isolated from locust, and a 41-residue monomer isolated from a moth. Each hormone is amidated at the C-terminus.

#### 3.1.3 Behavioral Neuropeptides

Ecdysis, the shedding of the old cuticle at the end of a molting cycle, is characterized by highly specialized behavioral patterns, such as wiggling and twisting, which facilitate cuticle shedding. These behaviors are under the control of eclosion hormone, which acts directly on the central nervous system. The two hormones isolated so far, from two moth species, are 62-residue monomers with 81% homology. Pheromone biosynthesis activating neuropeptides stimulate production of sex pheromones, used by female moths to attract mates. These 33 amino acid peptides, isolated from two species, are amidated and show 79% sequence homology.

#### 3.1.4 Myotropic Neuropeptides

Amidation, pyroglutamate formation, and sulfated residues are present throughout the myotropic neuropeptides. These comprise a rather large family of similar 8 to 11 residue peptides, stimulating muscle contraction.

### 3.2 Modes of Action and Inactivation

Cyclic nucleotides are involved in the activities of a number of insect neuropeptides including adipokinetic, diuretic, eclosion, and ecdysiotropic hormones. There is some evidence that pheromone biosynthesis activating neuropeptide also utilizes cyclic nucleotides. In contrast, the allatostatins do not seem to rely on these second messengers for activity, and alternative mechanisms are indicated. It is likely that more than one mechanism is involved with neuropeptide signal transduction and that these mechanisms may vary among species.

Mode of action studies present a great challenge to current research on insect neuropeptides. Equally challenging are efforts to understand how neuropeptides are inactivated. Membrane-bound endopeptidase and aminopeptidase enzymes have been reported in *Drosophila,* flies, and locusts. It is proposed that these enzymes digest neuropeptides after they become bound to their receptors and elicit the receptors' signals. Various proteases also degrade circulating neuropeptides. Understanding the roles of these proteases in regulating neuropeptide activities is a very active area of research.

### 3.3 Biosynthesis and the Basis for Variety

Evidence from work on adipokinetic hormone in the locust and molecular genetics in *Drosophila* and moths indicates that insect neuropeptides are synthesized as large precursor polypeptides. Specific active sequences are excised from the precursor and modified (e.g., by amidation), through a series of enzymatic processing steps, to yield the mature active peptide. Minor changes in the excision and modification steps have the potential for production of a great number of variations on a peptide scheme.

## 4 APPLICATION OF NEUROPEPTIDE INFORMATION

Neuropeptides contain coded information, which they transport to targets, resulting in specific physiological responses. If the flow of such information can be controlled, the insect, too, will become controllable. The benefit of this approach is that molecular control agents can be designed to be highly target specific and to have a controlled half-life. Thus, nontarget species would be unaffected and environmental persistence could be eliminated. Research laboratories are now addressing these issues.

*See also* Synapses.

### *Bibliography*


Borkovec, A. B., and Masler, E. P., Eds. (1990) *Insect Neurochemistry and Neurophysiology.* Humana Press, Clifton, NJ.

Hagedorn, H. H., Hildebrand, J. G., Kidwell, M. G., and Law, J. H., Eds. (1990) *Molecular Insect Science.* Plenum Press, New York.

Holman G. M., Nachman, R. J., and Wright, M. S. (1990) Insect neuropeptides (review). *Annu. Rev. Entomol.* 35.

Kerkut, G. A., and Gilbert, L. I., Eds. (1985) *Comprehensive Insect Physiology, Biochemistry and Pharmacology.* Pergamon Press, New York. (A highly comprehensive 13-volume reference work.)

Masler, E. P., Kelly, T. J., and Menn, J. J. (1993) Insect neuropeptides: Discovery and application in insect management. *Arch. Insect Biochem. Physiol.,* 21.

Menn, J. J., Kelly, T. J., and Masler, E. P., Eds. (1991) *Insect Neuropeptides: Chemistry, Biology and Action.* American Chemical Society, Washington, DC.


# Neuropsychiatric Diseases

*F. Owen*

### *Key Words*

**Amyloid Plaques** Extracellular proteinaceous deposits.

**Autosomes** All the chromosomes except X and Y.

**Exons** Regions of DNA that are transcribed.

**Missense Variant** Base change resulting in an amino acid substitution.

**Prion** Proteinaceous infectious particle.

**Pseudoautosomal Regions** Regions of the X and Y chromosomes that may undergo exchange.

In recent years the use of molecular biological techniques in studies of neuropsychiatric diseases has produced some significant advances in our understanding of the etiology of Huntington's disease, Alzheimer's disease, and the transmissible dementias. Despite considerable research effort, there has been relatively little success in establishing the genetic loci of the functional psychoses (i.e., schizophrenia and the major affective disorders). The more important findings for each category are outlined in turn.

## 1 HUNTINGTON'S DISEASE

Huntington's disease (HD) is a chronic degenerative disorder of the central nervous system. The characteristics of the disease include chorea, generalized impairment of motor function, and dementia. The disorder is transmitted as an autosomal dominant trait with the onset of symptoms usually occurring in the fourth or fifth decade of life.

HD was one of the first human genetic diseases to be studied by molecular genetic techniques and the first autosomal disorder to be assigned to a chromosomal location using these methods. Genetic linkage studies were greatly facilitated by the discovery of a very large Venezuelan pedigree, containing hundreds of individuals, descended from a single affected ancestor. Using a polymorphic DNA probe, genetic linkage to the disease in this pedigree was established in 1983. This finding was quickly replicated by other groups, and it was established that the disease locus is on the distal tip of the short arm of chromosome 4.

The use of genetic linkage markers made it possible to offer predictive testing for Huntington's disease in affected families, but 10 years elapsed before the gene and its mutations were eventually identified. Recently, a hitherto unknown gene, designated *IT15,* has been described which contains a polymorphic trinucleotide repeat $(CAG)_n$ that is expanded and unstable in individuals with HD. The normal biological function of the gene is unknown.

## 2 ALZHEIMER'S DISEASE

Alzheimer's disease (AD) is a common, dementing disorder whose prevalence rises steeply with age. Neuropathological studies have indicated that about half of all cases of severe dementia are due to AD. A hallmark of the neuropathology associated with AD is the presence of numerous extracellular senile plaques, composed largely of a peptide about 40 amino acids long, called the β-amyloid protein, which is derived from a much larger protein called the β-amyloid precursor protein (APP).

The disease exists both sporadically and in a familial form, where it is transmitted in an autosomal dominant manner. Since AD and Down's syndrome have some similarities in their neuropathology, it seemed plausible that the gene for AD was on the long arm chromosome 21. However, linkage between AD and markers on chromosome 21 has been reported by some but not all research groups. When the gene encoding APP was located on the long arm of chromosome 21, it led to the suggestion that APP itself was a candidate gene for AD. Recently, the sequencing of exon 17 of the APP gene in one large family with AD revealed a C-to-T transition, resulting in a valine-to-isoleucine change in amino acid 717. Other mutations resulting in a change in amino acid 717 were subsequently reported in different families. However, it is important to point out that mutations in exon 17 of the APP gene in AD are rare and have not been found in many sporadic cases or in affected families. Nevertheless, it is now generally accepted that β-amyloid deposition is of major significance in the pathogenesis of AD.

A genetic locus for early onset familial AD has also been reported to reside on chromosome 14, a finding that has been confirmed by other groups now attempting to identify the gene.

The apolipoprotein E (ApoE) gene on the long arm of chromosome 19, within a region previously associated with late onset familial AD, is currently generating considerable interest. It has now been well established that a common genetic variant of ApoE (ApoE4) is significantly more common in patients with late onset familial and sporadic AD. ApoE is enriched in amyloid plaques and may facilitate their deposition. Clearly, much future research on AD will concentrate on ApoE4.

## 3 THE TRANSMISSIBLE DEMENTIAS

Creutzfeldt–Jakob disease (CJD) and Gerstman–Sträussler–Scheinker syndrome (GSS) are rare dementing illnesses that can be transmitted to a variety of animals through intracerebral inoculation of infected brain material. There are similar neurodegenerative diseases in animals, such as scrapie in sheep and goats and bovine spongiform encephalopathy (BSE) in cattle. CJD usually arises sporadically, but a family history is present in about 15% of cases. GSS is thought to be entirely familial. All these diseases, in animals and man, are associated with an accumulation in the brain of an abnormal protease-resistant isoform of the prion protein (PrP). The diagnosis of CJD or GSS is confirmed postmortem by the presence of spongiform changes in the brain, astrocytic proliferation, and sometimes amyloid plaque deposition. The plaques, however, do not contain the β-amyloid protein found in Alzheimer's disease. Instead, they are immunoreactive to antibodies raised against the PrP, a property that led to the proposal that the PrP was a candidate gene for familial CJD and GSS. Subsequently genetic linkage was established between a prion protein missense variant and GSS, and several point mutation and insertions in the PrP associated with CJD or GSS have been reported. Mutations in the PrP gene have not been detected in sporadic cases of CJD, and the normal function of the gene (located on the short arm of chromosome 20) is still unknown. It is clear, however, that the abnormal isoform of the PrP gene is intimately involved in the disease process.

## 4 THE FUNCTIONAL PSYCHOSES

### 4.1 Schizophrenia

There is ample evidence to suggest that genetic factors operate in etiology of schizophrenia, although the mode of inheritance is unclear. Thus family, twin, and adoption studies have shown that there is a higher rate of schizophrenia in biological relatives of schizophrenics than in the general population. However, every form of model of inheritance from polygenic to a single gene has been proposed to fit the observed data without achieving a consensus. This lack of an acceptable genetic model of inheritance of schizophrenia has no doubt contributed to the conflicting results of genetic linkage studies of the illness.

Significant linkage on the long arm of chromosome 5 between schizophrenia and polymorphic DNA markers in five Icelandic and two English pedigrees has been reported by one group, but several subsequent studies failed to replicate this finding.

Balanced autosomal translocations involving the long arm of chromosome 11, which have tracked with schizophrenia in extended pedigrees, have led to genetic linkage studies on this chromosome. These results too have, so far, been inconclusive.

It has also been suggested that a major susceptibility locus for schizophrenia lies in the pseudoautosomal region of the Y chromosome, and it has been reported that sibling pairs affected with schizophrenia share pseudoautosomal alleles more often than predicted by chance. Other workers have failed to confirm this finding, although research is continuing.

### 4.2 Manic Depression

The pattern of inheritance of manic depression has some similarities with that of schizophrenia, and the mode of transmission is equally obscure. Thus genetic linkage to polymorphic markers in multiply affected families has yet to be convincingly established.

Females have a greater risk than males for the unipolar form of the illness, which has led to the suggestion that an X-linked gene was involved in the etiology of the disease. However, there have been several reports of significant linkage of affective disorders to loci on the X chromosome, while others have reported no evidence of X linkage. This remains an active area of research.

A particularly promising report of significant genetic linkage between polymorphic DNA markers on the short arm of chromosome 11, in a large Amish pedigree with multiple cases of manic depression, was not confirmed by the subsequent research of several groups. Moreover, further work on the Amish pedigree led to a decrease in significance of the original finding, casting more doubt on its validity.

Overall, molecular genetic studies of the functional psychoses have not yet been rewarding. However with the rapid and continuing increase in the availability of polymorphic markers suitable for genetic linkage studies and the refinement of analytical techniques, this may not remain the case.

*See also* ALZHEIMER'S DISEASE; HUMAN GENETIC PREDISPOSITION TO DISEASE.

### *Bibliography*

Harper, P.S. (1991) *Huntington's Disease.* Saunders, London.

McGuffin, P., and Murray, R., Eds. (1991) *The New Genetics of Mental Illness.* Butterworth-Heinemann, Oxford.

Owen, F., and Itzhaki, I., Eds. (1993) *Molecular and Cell Biology of Neuropsychiatric Diseases.* Chapman & Hall, London.

Prusiner, S.B., and McKinley, M.P., EDs. (1987) *Novel Infectious Pathogens Causing Scrapie and Creutzfeldt–Jakob Disease.* Academic Press, San Diego, CA.

Read, T., Potter, M., and Gurling, H.M.D. (1992) The genetics of schizophrenia. In *Schizophrenia: An Overview and Practical Handbook,* pp. 109–125. Chapman & Hall, London.

# NITRIC OXIDE IN BIOCHEMISTRY AND DRUG DESIGN

*Larry K. Keefer*

### *Key Words*

**Endothelium-Derived Relaxing Factor (EDRF)** A substance, now known to be nitric oxide or a closely related molecule, which is synthesized in the endothelial cells and signals the vascular smooth muscle to relax, thereby dilating the blood vessels.

**Nitric Oxide** Compound of molecular formula NO, long known as a toxic gas but recently discovered to play a diversity of critical bioregulatory roles.

**Nitrovasodilators** The traditional name for drugs whose mechanism of action in relaxing vascular smooth muscle involves release of nitric oxide, now preferably known as ''NO donor drugs'' in view of the broad-spectrum pharmacological activity they typically display.

**NO Synthases** The family of enzymes that oxidize L-arginine to nitric oxide.

**Soluble Guanylyl Cyclase** Among the principal receptors for nitric oxide, this enzyme is stimulated by NO to synthesize guanosine 3′,5′-monophosphate (cyclic GMP) as a second messenger in mediating many (but not all) of nitric oxide's biological effects.

Nitric oxide, an easily diffusible molecule of formula NO, is formed in many cell types. It mediates a most impressive variety of bioeffector functions, including vasodilation, inhibition of platelet aggregation, immunologically induced cytostasis, uterine relaxation, intercellular communication in the nervous system, penile erection, gastric accommodation, bronchodilation, and, on the negative side, septic shock, DNA damage, and neurotoxicity. NO is formed by the action of NO synthase on L-arginine, the five-electron oxidation of which yields one molecule each of NO and L-citrulline. The various known isoforms may be either constitutive or inducible, cytosolic or particulate; each contains a reductase as well as a heme domain. Cofactors include two flavins, a pterin, $Ca^{2+}$, and calmodulin, with NADPH supplying the reducing equivalents and $O_2$ as the ultimate electron acceptor. NO's biological effects are expressed by activating some enzymes (e.g., guanylyl cyclase) and inhibiting others (e.g., aconitase or ribonucleotide reductase), or by alternate mechanisms, such as damaging nucleic acids. Disorders caused by either an insufficiency or an excess of endogenously formed NO can be treated with NO donor drugs (e.g., nitroglycerin for angina) or inhibitors of NO synthase (e.g., as an adjunct to immunotherapy of cancer by cytokines), respectively.

## 1 CHEMISTRY

Nitric oxide is a gas of boiling point −152°C. Its maximum solubility in water is only about 1 mM at 37°C. It is easily formed in high energy processes, such as the recombination of reactive nitrogen and oxygen species formed during electrical storms or the incomplete combustion of organic nitrogen compounds during, for example, smoking of tobacco. It is produced by reduction of nitrate (e.g., in the action of nitric acid on certain metals) as well as in the reduction or acid-catalyzed disproportionation of nitrite. It is formed in abundance in the ecosphere as an intermediate in bacterial nitrification and denitrification. This known metabolite of chemotherapeutic as well as carcinogenic *N*-nitroso compounds is produced in stoichiometric amounts during the reaction of ascorbate with nitrosating agents. Its production as a metabolite of the nitrovasodilators and in the enzymatic oxidation of arginine is described in detail in this contribution.

The reactivity of NO is summarized in Figure 1. Its facile coordination to metal centers governs the activation and inhibition of various metalloenzymes. Oxidation by $O_2$ in the gas phase produces the brown product, nitrogen dioxide ($NO_2$), while oxygen converts NO in aqueous solution preferentially to nitrite ($NO_2^-$); both types of oxidation are kinetically second order in nitric oxide and first order in oxygen. NO can also be reduced to $NO^-$, which dimerizes in aqueous solution to form $N_2O$ with loss of water. Though nitric oxide is a radical, it has very little tendency to dimerize; however, it does couple with certain other radicals very readily—for example, at diffusion-controlled rates with superoxide to produce peroxynitrite ion, $ONOO^-$. Its electrophilicity can be harnessed to prepare nucleophile/NO complexes useful for the controlled biological release of nitric oxide (see Section 4).

Common analytical methods for nitric oxide include: its reaction in the gas phase with ozone to produce a chemiluminescent $NO_2$ molecule; its very rapid reaction with oxyhemoglobin to give methemoglobin, which can be detected with great sensitivity by spectrophotometry; and gas chromatography–mass spectrometry. Promising NO-selective electrodes have been described, as has electron spin resonance characterization of metal nitrosyl complexes formed

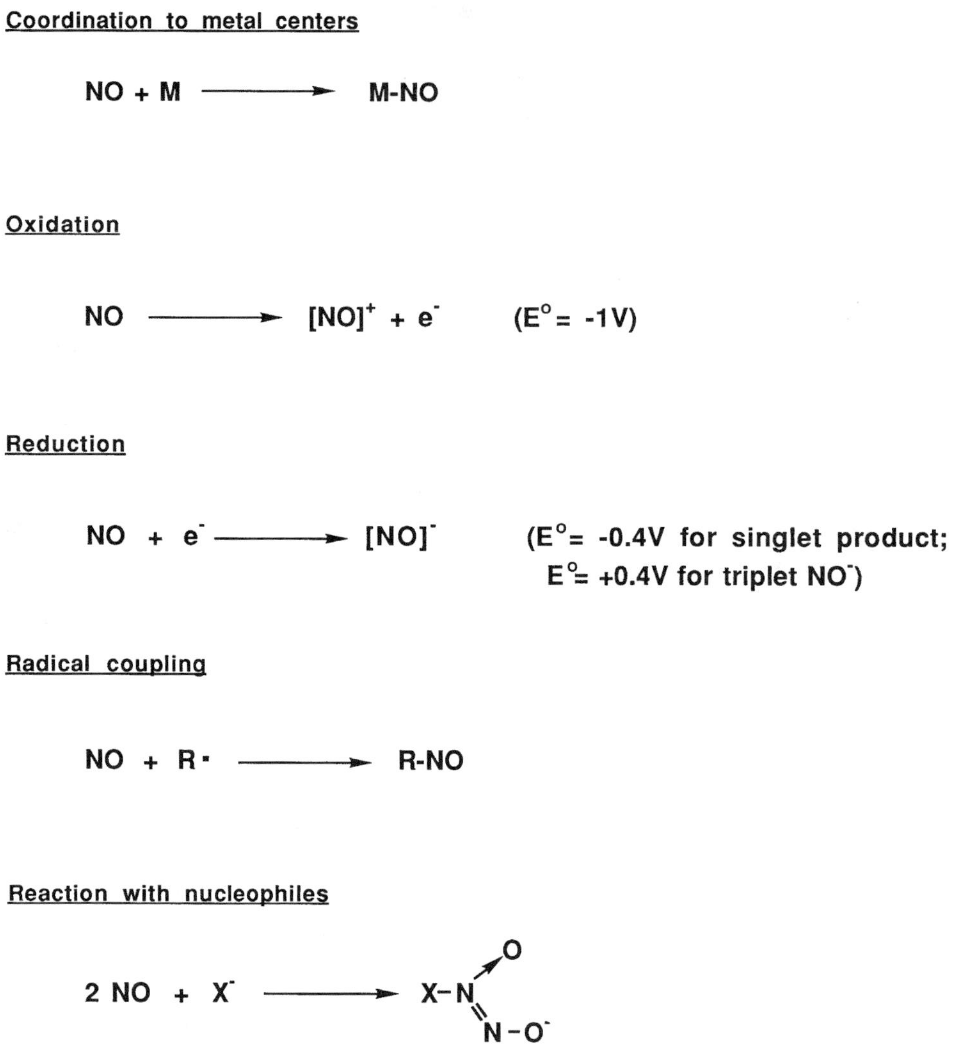

**Coordination to metal centers**

$$\mathrm{NO + M \longrightarrow M\text{-}NO}$$

**Oxidation**

$$\mathrm{NO \longrightarrow [NO]^{+} + e^{-}} \qquad (E^{\circ} = -1\,\mathrm{V})$$

**Reduction**

$$\mathrm{NO + e^{-} \longrightarrow [NO]^{-}} \qquad (E^{\circ} = -0.4\,\mathrm{V\ for\ singlet\ product};\ E^{\circ} = +0.4\,\mathrm{V\ for\ triplet\ NO^{-}})$$

**Radical coupling**

$$\mathrm{NO + R\cdot \longrightarrow R\text{-}NO}$$

**Reaction with nucleophiles**

$$\mathrm{2\,NO + X^{-} \longrightarrow X\text{-}N(\rightarrow O)\text{=}N\text{-}O^{-}}$$

**Figure 1.** Nitric oxide: chemical reactivity. A listing of nitric oxide's important reactivity patterns.

in the interaction of NO with metalloproteins. Indirect methods for nitric oxide analysis include the Griess reaction (a colorimetric assay for nitrite, the oxidation product of nitric oxide in aqueous solution), measurement of urinary nitrate as a metabolic end product of nitric oxide produced in vivo, and analysis for citrulline, the stoichiometric by-product of mammalian NO biosynthesis.

## 2 BIOSYNTHESIS

The principal biosynthetic source of NO in mammals is the enzymatic oxidation of L-arginine. This is accomplished by a family of enzymes known as the nitric oxide synthases, proteins produced by at least three different genes and ranging in molecular weight from about 130 kDa to 160 kDa. A look at the three isoforms that were the first to be well characterized will illustrate both their similarities and their diversity. The predominant form in the bovine endothelium is a constitutive, particulate enzyme, while that in the rat brain is constitutive but cytosolic. The NO synthase found in murine macrophages is cytosolic; though not present constitutively, it becomes a major source of endogenous NO when induced. All contain a reductase domain that passes electrons from reduced nicotinamide-adenine dinucleotide phosphate (NADPH) to flavine-adenine dinucleotide (FAD) to flavine mononucleotide (FMN), and finally to a heme domain. The resulting ferrous protoporphyrin IX moiety coordinates and reduces molecular $O_2$, with tetrahydrobiopterin as an apparent cofactor.

The enzyme is highly substrate specific, converting L- but not D-arginine to $N^{G}$-hydroxyarginine, which it then further oxidizes in a remarkable three-electron process to citrulline plus nitric oxide (Figure 2). In the absence of adequate pterin, uncoupling produces reactive oxygen intermediates, including hydrogen peroxide. Estimates of endogenous NO synthesis in humans range from a basal level of about 1 mmol/day to at least an order of magnitude more than that in people whose immune systems have been appropriately stimulated (e.g., by infection or by cytokine treatment).

## 3 MECHANISMS OF BIOLOGICAL ACTION

Nitric oxide plays a critical role in an astonishing variety of bioregulatory functions, which are partially listed in Table 1. Among the first such actions to be established was its role in vasodilation. There is currently a consensus that the endothelium-derived re-

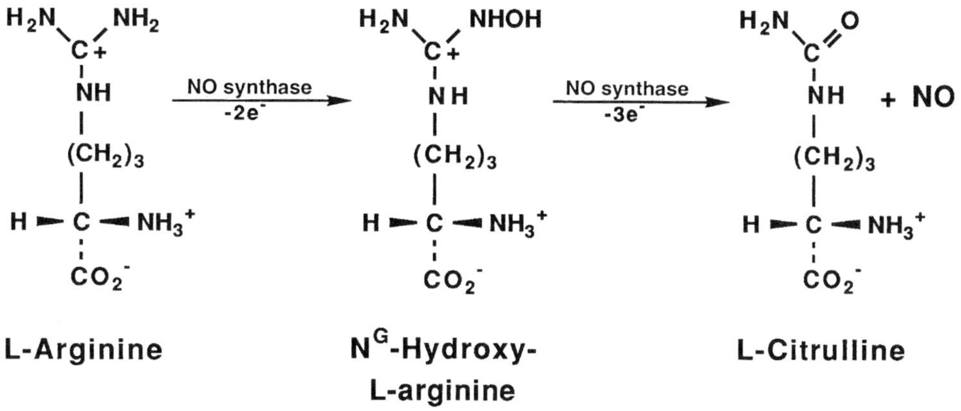

NO synthase $-2e^-$ NO synthase $-3e^-$ + NO

L-Arginine $N^G$-Hydroxy-L-arginine L-Citrulline

**Figure 2.** Nitric oxide: mammalian biosynthesis.

laxing factor (EDRF) is nitric oxide or at least a closely related compound, with very similar properties. Action of an agonist such as bradykinin on the endothelium leads to increases in intracellular $Ca^{2+}$ concentration, allowing the constitutive enzyme to complex calmodulin and convert arginine to nitric oxide. NO synthesized in the endothelium makes its way to the underlying vascular smooth muscle and coordinates to the active-site heme iron of guanylyl cyclase in the muscle cell, changing the enzyme's conformation and switching it on to make cyclic GMP as the ultimate muscle relaxant/vasodilator. Endothelium-derived NO moving in the opposite direction, toward the lumen of the blood vessel, exerts a different effect, also mediated by cyclic GMP—inhibition of platelet aggregation and adhesion.

In the murine immunological system, lipopolysaccharides, interferon γ, and other cytokines induce the synthesis of NO synthase in macrophages and related cells. Since calmodulin is already strongly bound to the mouse macrophage isoform and only very small levels of intracellular $Ca^{2+}$ are required for activity, this isoform is functionally calcium/calmodulin-independent; therefore, synthesis of abundant NO begins as soon as adequate enzyme is available, usually within a few hours. This nitric oxide arrests the growth of invading pathogens and tumor cells by several mechanisms: inhibition of the citric acid cycle enzyme, aconitase, as well as complex I and complex II of the mitochondrial electron transport system (by coordinating to the Fe centers of their active-site iron–sulfur clusters, thus interfering with cellular respiration); quenching radicals involved in ribonucleotide reductase action, thereby preventing DNA synthesis; and combining with the superoxide ion that is also produced by activated neutrophils and macrophages, producing $ONOO^-$ ion, which can inflict oxidative damage.

NO acts as a neurotransmitter via a variety of mechanisms. In the brain, a presynaptic neuron releases glutamate, which activates the *N*-methyl-D-aspartate (NMDA) receptors (one type of glutamate receptor) in the postsynaptic neuron. This in turn opens calcium ion channels and allows calmodulin to bind to the constitutive NO synthase, producing NO as a signal for initiating synthesis of cyclic GMP. NO is suspected to be the retrograde neurotransmitter, produced in a postsynaptic neuron as an easily diffusible signal that causes the cell's presynaptic counterpart to reinforce a memory trace. It is also the agent that acts on certain peripheral nonadrenergic, noncholinergic (NANC) neurons; in the male genitourinary tract, for example, NO signals the corpus cavernosum to relax and allows engorgement of the penis during an erection. Stimulation of NANC neurons also mediates gastric accommodation and the relaxation phase of intestinal motility. Nitric oxide may help prevent contractions of the gravid uterus, and it may be involved in vision, olfaction and nociception. It acts in the lung not only to dilate the pulmonary vasculature but also to increase the caliber of the airways.

Nitric oxide also has deleterious effects. In combination with $O_2$ or superoxide ion, NO can damage DNA and induce mutations; it has been suggested that this genotoxic potential may be responsible for initiating various genetic disorders, including some cancers. In addition, NO-induced hypotension can lead to cardiovascular complications in septic shock patients as well as during cytokine-based immunotherapy.

**Table 1** Nitric Oxide's Bioregulatory Roles: Selected Examples

| |
|---|
| Vasorelaxation |
| Antithrombotic action |
| Immune system function |
| Neurotransmission |
| Penile erection |
| Gastrointestinal motility |
| Uterine relaxation |
| Bronchodilation |

## 4 NITRIC OXIDE AND DRUG DESIGN

Drug development efforts capitalizing on the wealth of mechanistic insight into nitric oxide's multifaceted biological effects are proceeding along two lines. To counteract the negative consequences of endogenous overproduction of NO, as in endotoxic shock, inhibitors of NO synthase are of value. Arginine analogues that reversibly and/or irreversibly inhibit the enzyme are generally effective in this regard. Two frequently used examples are shown in Figure 3. Of these, $N^G$-methyl-L-arginine (NMA) is receiving special attention as an adjunct to the immunotherapy of cancer; since the limiting toxicity is normally hypotension caused by cytokine-induced overproduction of NO, coadministration of inhibitors can reduce this unwanted effect and permit larger doses of cytokine to be given. Efforts are under way to improve the selectivity of inhibition so that desirable functions can proceed unabated while targeted effects are suppressed; examples include the search for drugs that inhibit the inducible but not constitutive isoforms of NO synthase and the

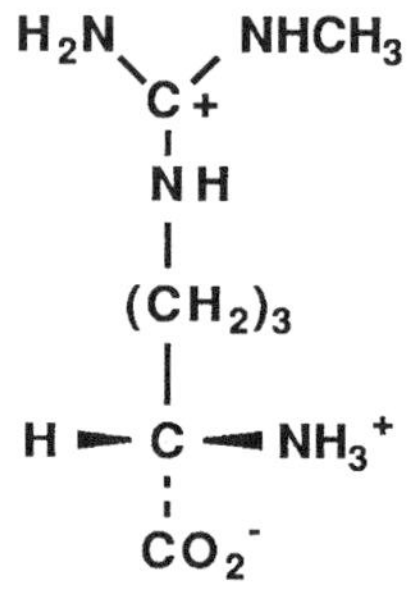

**$N^G$-Methyl-L-arginine (NMA)**

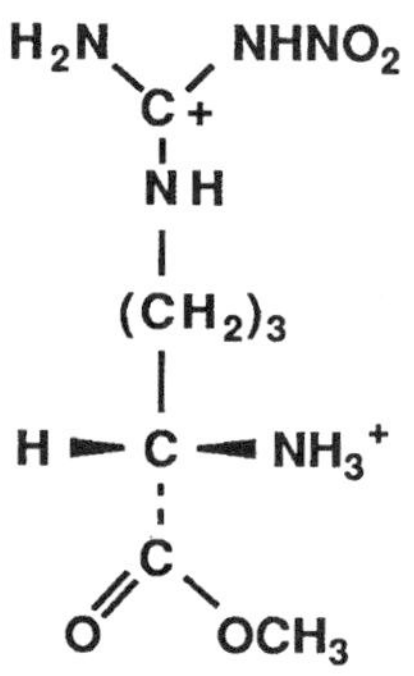

**$N^G$-Nitro-L-arginine methyl ester (L-NAME)**

**Figure 3.** NO synthase inhibitors: selected examples.

use of methylene blue as a guanylyl cyclase inhibitor affecting only processes mediated by cyclic GMP, not other responses to NO.

Drugs have long been available for treating disorders arising from an insufficiency of NO. Historically known as nitrovasodilators because of their effect on the vasculature, these compounds generate NO in the biological milieu and generally display the diversity of pharmacological effects expected of materials having this capacity. Some common examples are shown in Figure 4. Nitroglycerin and other nitrate esters have been used widely as antianginal agents for more than a century; the three-electron reduction of a nitrate ester group required for NO production can be effected by several enzymes, but the thiol cofactors that supply the electrons are easily depleted, leading to tolerance (i.e., the more the drug is used, the less potent it becomes). Sodium nitroprusside also requires activation before NO is released, but this is effected so reliably in vivo that intravenous nitroprusside is used clinically to stabilize

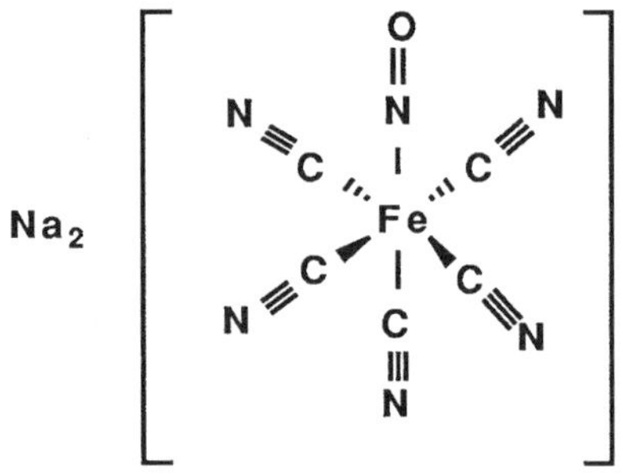

**Sodium nitroprusside (a metal nitrosyl)**

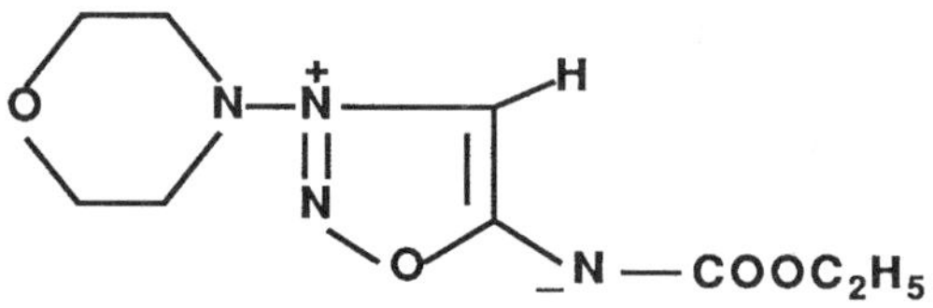

**Molsidomine (a sydnoneimine)**

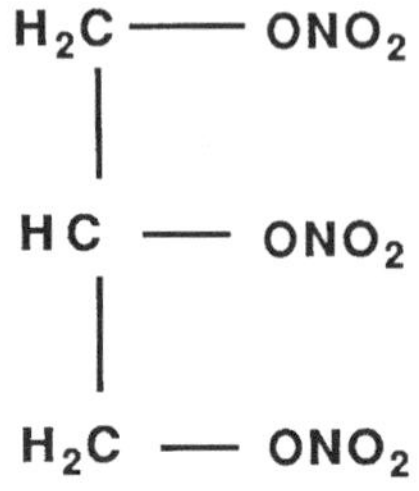

**Nitroglycerin (a nitrate ester)**

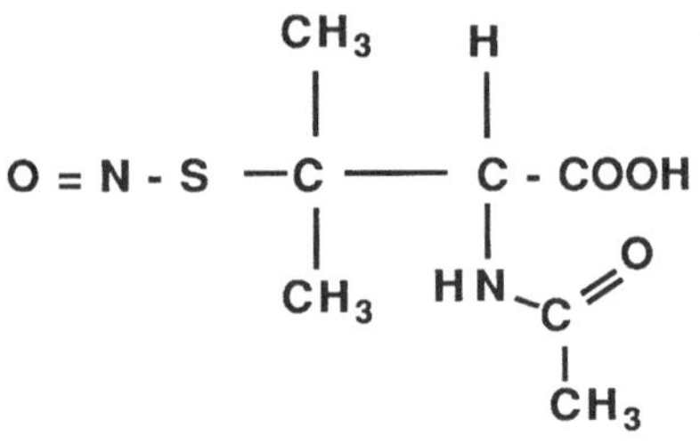

***S*-Nitroso-*N*-acetylpenicillamine (an *S*-nitrosothiol)**

**Figure 4.** Nitrovasodilators: selected examples. Structures of some important NO donor drugs.

patients whose blood pressure would otherwise fluctuate dangerously. Molsidomine is hydrolyzed in vivo to an acyclic form called SIN-1A that is oxidatively activated by $O_2$ to produce $O_2^-$) and an organic radical that fragments to NO; such sydnonimines are used clinically as cardiovascular drugs in some countries. The *S*-nitrosothiols can release NO without electron transfer activation, but the rate of spontaneous NO generation is highly medium dependent and extensive redox metabolism has also been documented. These compounds, which can be formed from NO and thiols under aerobic physiological conditions, are of interest as possible storage and transport forms for the otherwise highly reactive nitric oxide molecule. NO gas itself is useful clinically for treating certain types of respiratory distress. Other nitrovasodilators include the nitrite esters, formerly used as a folk remedy and drug of abuse as well as an important cardiovascular stimulant, and the NONOates, compounds of structure X-[N(O)NO]$^-$ formed by reacting nucleophiles ($X^-$) with nitric oxide, which are useful for the controlled, first-order release of NO in a number of pharmacological research settings.

*See also* SYNAPSES.

## *Bibliography*

Anonymous. (1977) *Oxides of Nitrogen.* World Health Organization (Environmental Health Criteria 4), Geneva.

Moncada, S., Marletta, M. A., Hibbs, J. B., Jr., and Higgs, E. A., Eds. (1992) *The Biology of Nitric Oxide,* 2 vols. Portland Press, London.

———, Palmer, R.M.J., and Higgs, E. A. (1992) Nitric oxide: Physiology, pathophysiology, and pharmacology. *Pharmacol. Rev.* 43:109–142.

Nathan, C. (1992) Nitric oxide as a secretory product of mammalian cells. *FASEB J.* 6:3051–3064.

## Nonaqueous Enzymology: *see* Enzymology, Nonaqueous.

# Nuclear Magnetic Resonance of Biomolecules in Solution

*Betty J. Gaffney and Brendan C. Maguire*

## *Key Words*

**Chemical Shift** Field-dependent parameter in the NMR spectrum which indicates the degree of shielding of the magnetic moment by the microenvironment.

**COSY** Correlated spectroscopy; a two-dimensional NMR technique that depends on through-bond coupling of nuclear spins.

***J* Coupling** Field-independent parameter in the NMR spectrum which indicates the coupling of two or more nuclear magnetic moments through the bonding network.

**NMR** Nuclear magnetic resonance; a spectroscopic technique used to determine structure and dynamics of macromolecules.

**NOESY** Nuclear Overhauser effect spectroscopy; a two-dimensional NMR technique that depends on through-space coupling of nuclear spins.

**Nuclear Overhauser Effect (NOE)** Increase or decrease in intensity of a resonance, which is coupled through space to another resonance whose transition is saturated.

Nuclear magnetic resonance (NMR) spectroscopy provides information directly about the proximity of atoms and about the lifetimes of these spatial relationships. These data can be analyzed to evaluate molecular structures and dynamics. NMR structural analyses are complimentary to those by X-ray crystallography. The NMR approach makes it possible to study molecules in solution, to obtain data from flexible regions of a structure, and to examine selected regions of a structure in samples of appropriate isotopic substitution. At present, structures of proteins and polynucleotides with molecular weights to about 15 kD may be obtained using two-dimensional NMR and samples at millimolar concentrations. Methods to edit the large volume of data from bigger structures extend the size limit to the 30 kD range. These methods include uniform and selective incorporation of $^{15}N$- and $^{13}C$-labeled subunits into biomolecules and display of data in three or more frequency dimensions. The techniques of multidimensional NMR of biomolecules in solution were developed in studies of bovine pancreatic trypsin inhibitor, cytochrome *c,* myoglobin, staphylococcal nuclease, ribonuclease A, hen egg white lysozyme, transfer RNAs, and deoxyoligonucleotides. Larger structures to which NMR techniques recently have been extended include T4 lysozyme; ferredoxin; complexes of the immunosuppressant cyclosporin with cyclophilin; the calcium-binding proteins calmodulin and calcineurin; interleukin-1$\beta$, components of the bacterial sugar phosphotransferase system: HPr and enzyme III; the DNA binding domains of several proteins; electron transport protein complexes; ribosomal RNAs; and DNA tri- and tetraplexes.

## 1 ONE-DIMENSIONAL NMR SPECTROSCOPY

One-dimensional NMR is now used routinely, with high performance liquid chromatography (HPLC) and mass spectral analysis, for the structure determination of small molecules. The principles of one-dimensional NMR serve as an introduction to the sections that follow on the more recent development of multidimensional NMR for studies of biological macromolecules.

### 1.1 Nuclear Moments

The NMR experiment requires nuclei with magnetic moments in the structure of interest. Naturally occurring magnetic isotopes that are abundant include $^{1}H$, $^{14}N$, $^{19}F$, and $^{31}P$. Biomolecules enriched in the nonradioactive isotopes $^{2}H$ (deuterium), $^{13}C$, and $^{15}N$ are very important in one- and multidimensional NMR applications. Although NMR spectrometers can be tuned to allow deuterium nuclei to be observed directly in biomolecules, this hydrogen isotope is used more often to eliminate signals from selected hydrogens when the spectrometer is tuned for $^{1}H$ detection. Tritium ($^{3}H$) NMR has a number of advantages when selectivity is needed, despite the obvious additional difficulty associated with safely handling high levels of radioactivity. The inherent sensitivity of the NMR experiment depends, in part, on the size of the nuclear magnetic moment. Thus the order of sensitivity of the isotopes mentioned, by this criterion, is $^{3}H > {}^{1}H > {}^{19}F > {}^{31}P > {}^{13}C > {}^{2}H > {}^{15}N >$

$^{14}N$. Detection of $^{2}H$- and $^{14}N$- NMR is slightly different because these nuclei also have quadrupole moments.

## 1.2 Basic Features of the One-Dimensional Spectrum

Although all the hydrogens in a biomolecule have the same nuclear moment, each is shielded from an external magnetic field by the immediate structural environment. As a result, there is a spectral dispersion: different nuclei absorb energy at varied fields. Since NMR line widths are very narrow for rapidly tumbling molecules of moderate size, excellent resolution of these lines can be achieved. The position of an NMR line relative to a reference is called a *chemical shift.* Since the magnitudes of chemical shifts scale with magnetic field strength, the dispersion of resonances is greater at higher magnetic fields. For structure determination of biomolecules, it is important to have the highest field magnets available. Magnets with 14 tesla (T) fields (600 MHz for protons) are now widely used, 18 T magnets are becoming commercially available, and a 21 T magnet is under development. Chemical shifts are usually measured in parts per million (ppm) of the external field, a field-independent unit.

Two magnetically different nuclear moments may interact over one to five bonds, splitting the resonance lines of each nucleus. The magnitude of the splitting is given by the coupling constant, *J*. *J couplings* are independent of magnetic field and are reported in hertz. Three-bond *J* couplings are particularly important in structure determination. For example, in proteins, the magnitude of the *J* coupling between α-protons and the NH proton of the same amino acid is related to the torsion angle about the C—N bond and can therefore provide an indication of secondary structure: small $^{3}J_{HN_{\alpha}}$ couplings are characteristic of α-helices; larger ones are found for β-sheets. *J* couplings also provide information about the number of coupled nuclei. This information is revealed as a complex multiplet of lines and it is useful, for example, for assigning resonance from protein side chains.

Nuclei that are close in space, but not necessarily closely linked by bonds, can also give NMR signals via the *nuclear Overhauser effect* (NOE). This effect results from cross-relaxation caused by the interaction of the magnetic dipole moments of two nuclei: when one nucleus is perturbed, the intensity of the signal from the other changes. The NOE intensity is proportional to the inverse sixth power of the distance between the interacting nuclei ($\propto 1/r^6$). This interaction is an exchange of magnetization and takes periods on the order of tens to hundreds of milliseconds to reach maximum intensity. Since secondary magnetization exchange (spin diffusion) can occur, primary NOEs are identified by demonstrating that the NOE intensity as a function of time (the "NOE buildup") extrapolates to zero time. The term *long-range NOE* is applied to effects between nuclei close in space but not closely connected through bonds. Long-range NOEs are essential to determining tertiary structure of biomolecules by NMR. Long-range NOE intensities are usually classified as strong, medium, and weak, corresponding to distances from 1.8 to 2.7 A, 1.8 to 3.5 A, and 1.8 to 5.0 A, respectively.

## 1.3 A Tripeptide Example

The one-dimensional $^{1}H$ NMR spectrum of a tripeptide in $D_2O$, shown in Figure 1, illustrates chemical shifts and magnetization transfer by *J* couplings. The tripeptide is the acetate salt of Lys-

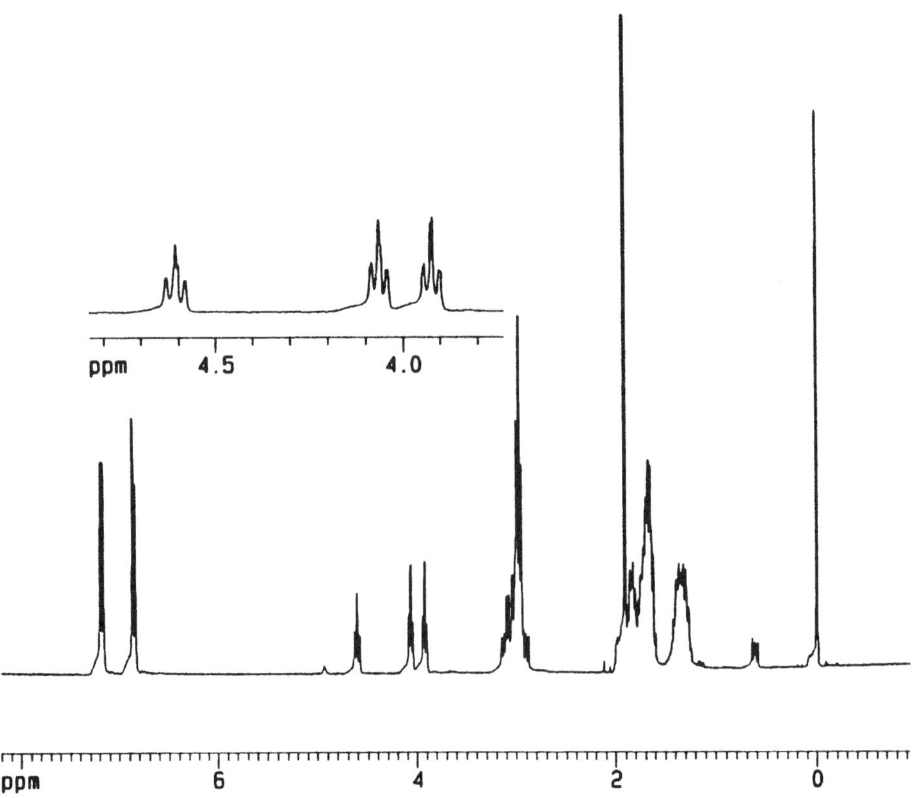


**Figure 1.** One-dimensional proton NMR spectrum of the tripeptide Lys-Tyr-Lys in $D_2O$. This $^{1}H$ NMR spectrum results from 10 mM lysyl-tyrosyl-lysine diacetate in deuterium oxide (pD = 6.9) at 20°C. It was recorded at 300 MHz.

Tyr-Lys. Chemical shifts on the abscissa are relative to the internal standard DSS (2,2-dimethyl-2-silapentane-5-sulfonate), which gives the sharp peak at 0 ppm. In deuterated water, the exchangeable —$NH_2$ and —OH protons are substituted by deuterium and thus do not contribute to the $^1$H NMR spectrum. The peaks above 6 ppm correspond to the ring protons of the tyrosine residue. There are two separate peaks because the pairs of 2,6- and 3,5-protons are in magnetically different microenvironments. In addition, since each pair of protons splits the other by $J$ couplings, each line is a closely spaced doublet. (Other aromatic ring protons from protein side chains or the bases of nucleic acids also exhibit chemical shifts in the 6–9 ppm region.) The three triplets of 1-2-1 intensity seen from 3.9–4.6 ppm are due to the $C_\alpha$—H protons of the three amino acid residues. They are triplets because they are $J$-coupled to two $C_\beta$—H protons. An important part of the assignment strategy is to assign one multiplet in the $C_\alpha$—H region to each amino acid. Multiple peaks are seen in the 2.9–3.2 ppm region. These can be assigned to the $C_\beta$—H protons of Tyr and the ε protons of the Lys residues. The large, sharp singlet at 1.9 ppm is due to acetate counterion. The remaining peaks from 1.6–2.0 ppm arise from the β and δ protons of the Lys residues. The final peak at 1.3 ppm is from the γ protons of Lys. Exact assignments of the β, γ, and δ protons of Lys require two-dimensional methods.

## 2 MULTIDIMENSIONAL NMR SPECTROSCOPY

The frequency of each of the peaks in Figure 1 is the resonant frequency of a nucleus. After an ensemble of nuclei have been prepared for detection by applying a series of pulses to a sample in a magnetic field, the pulses are turned off and the time dependence of magnetization is detected (free induction decay). The data in the time domain $t_2$ is oscillatory and is converted by Fourier transformation to give the individual frequencies contributing to the oscillations. The frequency dimension $\omega_2$ results. An evolution time $t_1$ between the first and later pulses is incremented in sequential experiments. It is followed by a mixing period where magnetization transfer occurs, *via* $J$ coupling, NOE, or other interaction. A second Fourier transform over $t_1$ gives the $\omega_1$ dimension in two-dimensional NMR. A display of $\omega_1$ versus $\omega_2$ is a two-dimensional NMR spectrum with peaks off the diagonal for interacting spins. Multidimensional NMR spectra are usually shown as contour maps so that each peak appears as an oval, or pattern of ovals, and background noise is below the level of the contour. When the correlated spins are coupled to third and fourth nuclei, additional evolution times, $t_3$ and $t_4$, can be introduced to edit the two-dimensional spectrum by frequencies of the third and fourth nuclei, giving spectra of three and four dimensions.

### 2.1 Basic Techniques: COSY and NOESY

The two most widely used two-dimensional NMR methods are correlated spectroscopy (COSY) and nuclear Overhauser effect spectroscopy (NOESY). COSY spectra reflect through-bond coupling and NOESY spectra represent through-space coupling. Both techniques are used in the assignment of resonances, and the NOESY spectra also determine distance constraints around which the structure is built.

Sequential assignment relies on the identification of intraresidue $C_\alpha$H—NH cross peaks through the analysis of COSY-type spectra. Then NOESY spectra of the same sample are analyzed, and $C_\alpha$H—NH cross peaks that were not present in the COSY spectra yield sequential information, which leads to unambiguous assignments of all $C_\alpha$H—NH cross peaks. See Figure 2 for an explanation of the process.

### 2.2 Heteronuclear Experiments and Isotope Editing

It is now possible to synthesize biomolecules uniformly labeled with $^{13}$C and $^{15}$N amino acids. One application is two-dimensional heteronuclear $^{15}$N-$^1$H spectroscopy. The isotope $^{15}$N has a large chemical shift dispersion, permitting the observation of excellent resolution of amide peaks in proteins and amino and imino protons of nucleic acids; $^{15}$N-$^1$H NOEs can be used to differentiate mobile and static sites in polypeptides. The additional dimensions also are provided by $^{15}$N and $^{13}$C in multidimensional NMR. Proteins can be selectively labeled with isotopes in a single amino acid, or in pairs of amino acids, to aid in assigning individual resonances.

### 2.3 Resonance Assignment Strategy and Structure

The sequence of experiments used in determining the NMR structure of a protein can be outlined as follows:

1. Make sequential assignments by first assigning peaks in a COSY-type spectrum to types of residue and then use the NOESY spectra to make the sequential assignments as in Figure 2.
2. Identify secondary structure by determining torsion angles and characteristic NOEs.
3. Determine the intensities of long-range NOEs between assigned nuclei to determine distances.
4. Use ''distance geometry'' and ''simulated annealing'' calculations to assemble the internuclear distances and torsion angles into a structure.

The result of the calculations is a family of closely related structures that satisfy the distance constraints.

## 3 APPLICATIONS

A single multidimensional NMR experiment on a protein having more than 100 residues can take from one day to one week (depending on the dimensionality) for each spectrum. Data analysis takes considerably longer. Solving a structure requires multiple experiments and analyses, and determining a new structure calls for a substantial investment of time. When all the resonances have been assigned, however, a number of further applications can result much more rapidly. Once the assignments and structure of cyclophilin had been obtained, the structure of the drug cyclosporin bound to cyclophilin was determined, using a series of experiments in which either the protein or drug was isotopically labeled. The interactions of other immunologically active molecules, including calcineurin and the interleukins, can be approached by NMR now that assignments and structures have been done for them. Another application is to study the transfer of energy-rich phosphate between components of bacterial sugar transport proteins. Additionally, the structure of staphylococcal nuclease has been widely examined in solution and crystals by NMR and X-ray methods. It is possible

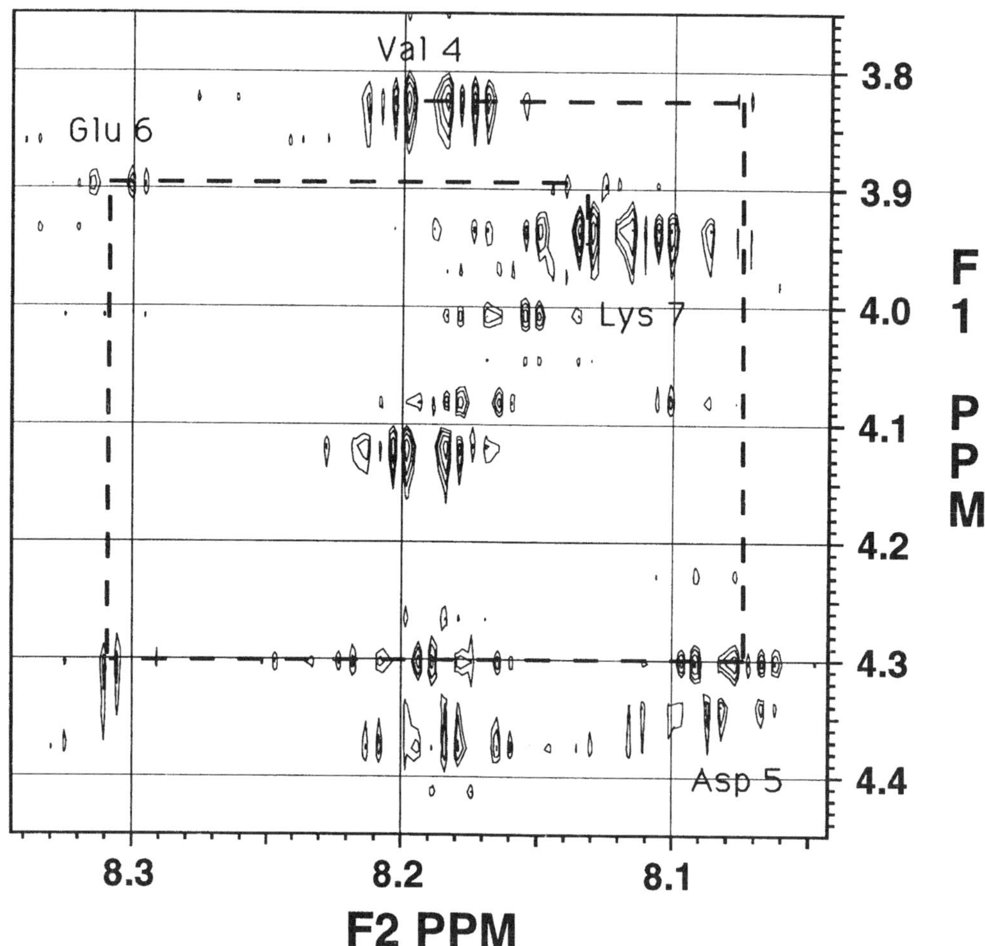


**Figure 2.** An expanded region of a NOESY spectrum of a synthetic tridecapeptide from a 500 MHz $^1H$-$^1H$ NOESY spectrum recorded at 18.5°C. The solution was 9.5 mM in the synthetic tridecapeptide (Leu-Pro-Ala-Val-Asp-Glu-Lys-Leu-Arg-Asp-Leu-Tyr-Ser), 10 mM in KCl, and 90 mM in acetate buffer, (pH = 5.05). The data were processed and the figure generated using the programs nmrPipe and nmrDraw written by Frank Delaglio at the National Institutes of Health. This region illustrates how sequential assignments can be made. The process begins at the peak located at (8.19 ppm, 3.82 ppm). This has been identified as the NOESY cross peak corresponding to $C_\alpha H$—NH of Val4. A line can be drawn along 3.82 ppm, the chemical shift of the Val 4 $C_\alpha H$. A peak at (8.07 ppm, 3.82 ppm) that was not present in the COSY-type spectra defines the NH chemical shift of Asp 5 as 8.07 ppm. The $C_\alpha H$—NH cross peak for Asp 5 is found at (8.07 ppm, 4.31 ppm). The NOESY cross peak between $C_\alpha H$ (Asp 5) and NH (Glu 6) is found at (8.31 ppm, 4.31 ppm) in the same manner, defining the chemical shift of the Glu 6 NH as 8.31 ppm. The $C_\alpha H$—NH peak for Glu 6 is found at (8.31 ppm, 3.89 ppm). The final connectivity is that between $C_\alpha H$ (Glu 6) and NH (Lys 7), giving a cross peak at (8.13 ppm, 3.89 ppm), which leads to the assignment of the $C_\alpha H$—NH peak for Lys 7 to the peak at (8.13 ppm, 3.94 ppm).

to combine information from both techniques to ''dock'' substrates and inhibitors into the binding site of this enzyme.

NMR is uniquely useful in the area of biomolecular dynamics. The rate of exchange of deuterium for protons provides information about the lifetimes of hydrogen bonds in protein and nucleic acid structures; these lifetimes vary from milliseconds to years for exposed and buried hydrogen bonds. In another example, rotation of tyrosine rings in proteins was established by NMR techniques.

For nucleic acids, NMR is most effective as a tool for determining general secondary structure (i.e., A, B, or Z DNA) and base pairing. Peaks are assigned in a sequential manner similar to that used for polypeptides. Determination of the tertiary structure of the nucleic acid polymers is difficult due to the lack of NOE cross peaks for use in determining distances. This lack of known distances can lead to cases of more than one structure fitting the distance constraints imposed by the NOE peaks. NMR has recently been used successfully to prove the existence of triple helical structures and tetraplexes, and to characterize hairpin loop structures.

*See also* Hydrogen Bonding in Biological Structures; Peptides, Synthetic; Protein Modeling.

## *Bibliography*

Ernst, R. R., Bodenhausen, G., and Wokaun, A. (1987) *Principles of Nu lear Magnetic Resonance in One and Two Dimensions.* Oxford University Press, Oxford.

Farrar, T. C., and Harriman, J. E. (1992) *Density Matrix Theory and Its Applications in NMR Spectroscopy,* 2nd ed. Farragut Press, Madison, WI.

Wemmer, W. E. (1991) The applicability of NMR methods to solution structure of nucleic acids. *Curr. Opinion Struct. Biol.,* (I):452–458.

Wüthrich, K. (1986) *NMR of Proteins and Nucleic Acids.* Wiley-Interscience, New York.

*Methods in Enzymology.* (1989) Volumes 176 and 177 are entirely devoted to NMR studies of biological macromolecules.

*Quarterly Review of Biophysics.* (1990) Volume 23 is entirely devoted to NMR studies of biological macromolecules.

# Nucleic Acid Hybrids, Formation and Structure of

*James G. Wetmur*

## Key Words

**Blot** A hybridization experiment in which the target nucleic acid is immobilized on a membrane (or filter). With electrophoretic separation of the target prior to immobilization, the blot is called a Southern blot if the target is DNA and a northern blot if the target is RNA.

**Dissociation Temperature** The temperature at which two strands of a hybrid are 50% dissociated after a specified time.

**Hybrid** A partially or completely double-stranded (duplex) nucleic acid formed by interaction of two single-stranded nucleic acids, either DNA or RNA. Reassociation or renaturation may be used to describe hybridization of completely complementary DNA or RNA strands.

**Melting Temperature** The temperature at which 50% of the base pairs of a polynucleotide:polynucleotide hybrid or 50% of the strands of oligonucleotide:polynucleotide or oligonucleotide:oligonucleotide hybrids have dissociated.

**Oligonucleotide** A short polynucleotide (typically 8–50 nucleotides), which may hybridize to a complementary target in an all-or-none fashion such that the rates of hybridization and dissociation are equal at the melting temperature.

**Polymerase Chain Reaction (PCR)** A method for geometric amplification of nucleic acids employing two sequence-specific oligonucleotides complementary to the DNA strands and a DNA polymerase. Each cycle includes a denaturation step, a hybridization (annealing) step, and a polymerization step.

**Strand Separation Temperature** The temperature at which the two strands of a polynucleotide duplex are dissociated.

Hybridization is the formation of partially or completely double-stranded (duplex) nucleic acid (DNA:DNA, DNA:RNA or RNA:RNA) by sequence-specific interaction of two complementary single-stranded nucleic acids. ''Reassociation'' and ''renaturation'' are terms often used to describe hybridization between completely complementary DNA or RNA strands. Solution hybridization is an integral part of the polymerase chain reaction. Hybridization is often carried out using a labeled probe in an attempt to detect the existence and quantity of complementary sequence in a complex nucleic acid mixture. The target is often immobilized, as in Southern and other blotting techniques. In this discussion of the physical-chemical aspects of hybridization, in particular thermodynamics and kinetics, melting temperatures, hybridization rates, and dissociation rates and temperatures are interrelated. Details are provided for calculation of these useful properties, as well as examples of calculations.

## 1 INTRODUCTION

DNA:DNA, DNA:RNA, and RNA:RNA hybrid duplexes are the products of hybridization reactions between complementary single-stranded molecules. Hybrid formation between completely complementary DNA or RNA strands often is described as reassociation or renaturation. The reverse reaction is called strand separation or dissociation. The structural differences between deoxyribose and ribose in DNA and RNA strands are reflected in different forms of the antiparallel double helices and in quantitative differences in hybrid stability and the rate of hybrid formation. Nevertheless, they are qualitatively the same, and the basic concepts will be treated together.

## 2 STABILITY OF DNA:DNA, DNA:RNA AND RNA:RNA HYBRIDS

### 2.1 Melting and Dissociation Temperatures

The four *different* temperatures defined here are associated with nucleic acid melting. The corresponding physical phenomena are illustrated in Figure 1.

1. $t_m^\infty$, the melting temperature of a polynucleotide duplex, such as a long DNA molecule.
2. $t_{ss}$, the strand separation temperature of a polynucleotide duplex, also called the denaturation temperature in PCR.
3. $t_m$, the melting temperature at which at least one strand of the duplex is an oligonucleotide.
4. $t_d$, the dissociation temperature for an oligonucleotide bound to a polynucleotide on a solid support, such as a blot.

### 2.2 Stability of Polynucleotide Duplexes (DNA:DNA, DNA:RNA, and RNA:RNA)

As indicated in Figure 1, $t_m^\infty$ is the point where the fraction of single-stranded base pairs, $f_{ss}$, is 0.5, leading to molecules containing alternating duplex and denatured (loops or ends) regions. Because the reaction is intramolecular and equilibrium is achieved, $t_m^\infty$ is independent of polynucleotide concentration and time, depending only on the composition of the polynucleotide duplex and the composition of the solvent. Strand separation occurs a few degrees above $t_m^\infty$, at $t_{ss}$, and is not a rapidly reversible reaction. For polynucleotides of length $L$ containing only G:C and A:T or A:U base pairs, $t_m^\infty$ (°C) may be determined from the empirical equation:

$$t_m^\infty = A^* + 16.6 \cdot \log_{10}\left\{\frac{[\text{SALT}]}{1.0 + 0.7 \cdot [\text{SALT}]}\right\} + B^*\,(\%\ \text{G+C}) - \frac{500}{L}$$

where $[\text{SALT}] = [\text{NA}^+] + 4[\text{Mg}^{2+}]^{0.5}$ and potassium may be substituted for sodium, as in the polymerase chain reaction (PCR); $A^*$ is 81.5, 67, or 78°C, and $B^*$ is 0.41, 0.8, or 0.78 °C/(% G+C) for DNA:DNA, DNA:RNA, and RNA:RNA hybrids, respectively. For example, $t_m^\infty$ for a 50% G+C DNA:DNA hybrid of length 250 in 0.05 $M$ KCl, 0.002 $M$ $\text{MgCl}_2$ ([SALT] = 0.23), $t_m^\infty = 81.5 - 11.7 + 0.41(50) - 2 = 88.3$°C.

For imperfectly matched hybrids, $t_m^\infty$ is reduced approximately 1°C for each percent mismatching; $t_{ss}$ is proportionally reduced. Stringency in washing immobilized targets such as filters or blots (Southern, northern, slot) containing polynucleotide probes, a measure of selective dissociation of imperfectly matched hybrids, is

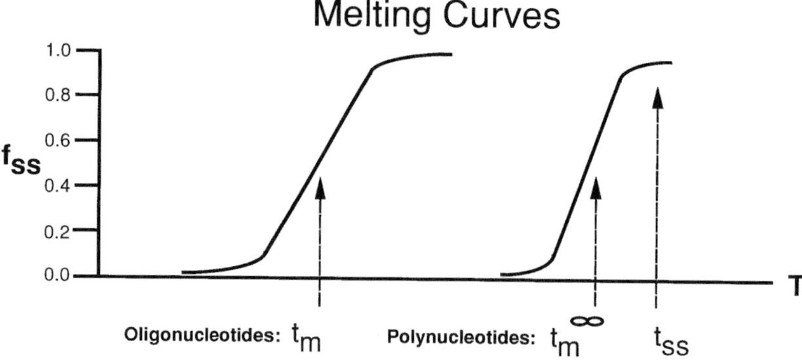


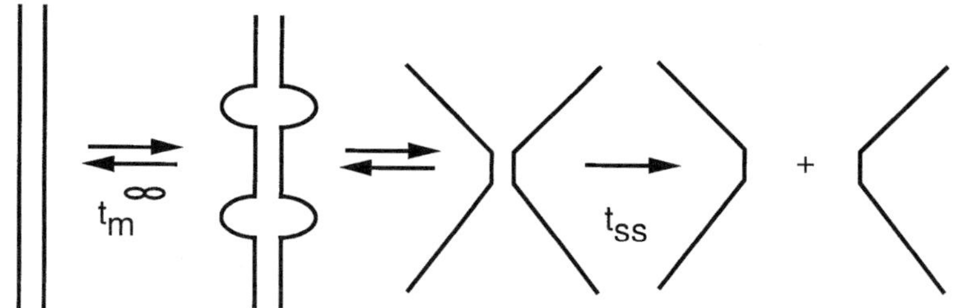


Oligonucleotide Denaturation

$k_r$

$k_2$

+

Forward Rate = Reverse Rate at $t_m$

Only the dissociation rate ($k_r$) is important for washing blots ($t_d$)

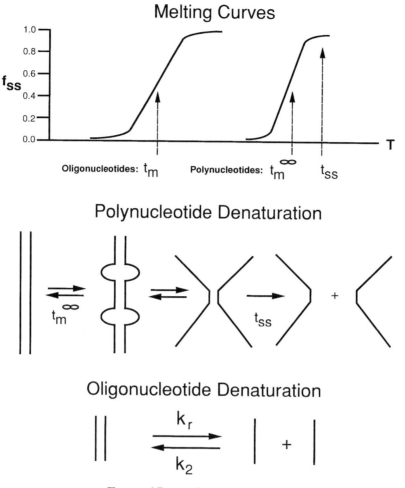

**Figure 1.** DNA melting curves, steps in polynucleotide denaturation, and steps in oligonucleotide denaturation: $T$, temperature; $f_{ss}$, fraction of single-stranded DNA; $t_m$, melting temperature ($f_{ss} = 0.5$) for oligonucleotides; $t_m^\infty$, melting temperature for polynucleotides; $t_{ss}$, strand separation temperature; $k_2$, second-order rate constant for hybridization; $k_r$, rate constant for the reverse reaction; $t_d$, dissociation temperature.

increased as the temperature approaches $t_m^\infty$ in the wash solvent. Use of different solvents (or introduction of modified bases) may affect $A^*$ and/or $B^*$, with different effects for different polynucleotides. For example, base composition may be eliminated as a variable for DNA:DNA hybrids ($B^* = 0$) at 3.2 $M$ $Me_4NCl$ or 2.4 $M$ $Et_4NCl$. In another example, R-loop formation is possible because $t_m^\infty$ is decreased by 0.63°C/% formamide for DNA:DNA, but significantly less for DNA:RNA hybrids.

### 2.3 Stability of Oligonucleotide:Oligonucleotide and Polynucleotide:Oligonucleotide Duplexes

Figure 1 shows the all-or-none melting of complementary oligonucleotides. The more common experiment developed in this section involves an oligonucleotide, in excess at molar concentration $C$, binding to a complementary region on a polynucleotide. Either way, $t_m$, the temperature at which 50% of the duplexes have strand-separated, is the temperature of importance for hybridization to template in PCR and to immobilized target in filter hybridization. Note that $t_m$ is not $t_d$, the dissociation temperature important for washing of filters; $t_d$ is discussed in Section 4.

The $t_m$ for hybridization of excess oligonucleotide probe to a polynucleotide target is given by:

$$t_m = \frac{T\Delta H^\circ}{(\Delta H^\circ - \Delta G^\circ + RT\ln[C])} + 16.6 \cdot \log_{10}\left\{\frac{[\mathrm{SALT}]}{1.0 + 0.7\,[\mathrm{SALT}]}\right\} - 269.3$$

where enthalpy $\Delta H^\circ = \Sigma_{nn}(N_{nn}\,\Delta H^\circ_{nn}) + \Delta H^\circ_e$

free energy $\Delta G^\circ = \Sigma_{nn}(N_{nn}\,\Delta G^\circ_{nn}) + \Delta G^\circ_e + \Delta G^\circ_i$

$T = 298.2$ K and $R = 0.00199$ kcal/mol · K

The first term in the $t_m$ equation is $T_m$ (K) in 1 $M$ NaCl. The symbols $i$, nn, and $e$ indicate contributions from initiation of base pair formation, nearest neighbors, and dangling ends. Complete nearest-neighbor data sets are available for DNA:DNA and RNA:RNA but not DNA:RNA. Estimated average contributions for each of two possible dangling ends are $\Delta H^\circ_e = -5.0$ kcal/mol and $\Delta G^\circ_e =$

$-1.0$ kcal/mol. For the DNA:DNA nearest neighbor values in the following tabulation, $\Delta G_i^\circ = +2.2$ kcal/mol and

| | AA | AT | TA | CA | GT | CT | GA | CG | GC | GG |
|---|---|---|---|---|---|---|---|---|---|---|
| or | TT | | | TG | AC | AG | TC | | | CC |
| $-\Delta H_{nn}^\circ$ | 9.1 | 8.6 | 6.0 | 5.8 | 6.5 | 7.8 | 5.6 | 11.9 | 11.1 | 11.0 |
| $-\Delta G_{nn}^\circ$ | 1.55 | 1.25 | 0.85 | 1.15 | 1.40 | 1.45 | 1.15 | 3.05 | 2.70 | 2.30 |

For example, consider 1 $\mu M$ $(dG)_6$ binding to $(dC)_6$ in a long DNA at 1 $M$ NaCl. $-\Delta H^\circ = 11.0(5) + 5(2) = 65.0$ kcal; $-\Delta G^\circ = 2.3(5) + 1.0(2) - 2.2 = 11.3$ kcal; $RT \ln[C] = -8.2$ kcal; $t_m = 39.9$°C.

## 3 FORMATION OF DNA:DNA, DNA:RNA, AND RNA:RNA HYBRIDS

### 3.1 Hybridization with Polynucleotides

The concentration term used for polynucleotide hybridization is molar nucleotide concentration, $C_0$, not strand concentration, $C$. The second-order rate constant $k_2$ ($M^{-1}s^{-1}$), is given by:

$$k_2 = \frac{k_N' \sqrt{L_s}}{N}$$

where $L_s$ is the length of the shortest strand participating in duplex formation, $N$ is the complexity or the total number of base pairs present in nonrepeating sequences, and $k_N'$ is the nucleation rate constant, discussed shortly. The interpretation of $N$ and $L$ is illustrated in Figure 2. The reaction rates are characterized by $k_2$, whether in solution or immobilized on a solid support such as a Southern blot. When a dilute probe is employed with an immobilized target, stirring may be necessary to maintain homogeneity. When using a labeled probe and an unlabeled target, the concentration governing the reaction is that of the molecule in excess, the driver. When immobilized DNA is the driver, $C_0$ is difficult to determine, being a measure of available target concentration. Hybridization reactions are pseudo–first order if the driver is single stranded.

The time course ($f_{ss}$ of the potential duplex target versus time, $t$) is given as follows:

Second order:

$$\frac{1}{f_{ss}} = k_2 \left(\frac{C_0}{2}\right) t + 1$$

Half-time (seconds):

$$t_{1/2} = \frac{2}{k_2 C_0}$$

Pseudo–first order:

$$f_{ss} = \exp\,[-k_2\,(C_0)\,t]$$

Half-time:

$$t_{1/2} = \frac{\ln 2}{k_2 C_0}$$

For the second-order reaction, a plot of $1/f_{ss}$ versus $t$ gives a slope equal to $k_2$ $(C_0/2)$. For the first-order reaction, a plot of $\ln(1/f_{ss})$ versus $t$ gives a slope equal to $k_2 C_0$.

The rate of hybridization reactions, hence $k_N'$, is zero at $t_m^\infty$. The value of $k_2$ increases to a broad maximum at about 25°C below $t_m^\infty$. At lower temperatures, hybridization rates fall off precipitously as a result of decreased availability of nucleation sites. At optimal temperatures, $k_N'$ for DNA:RNA hybrid formation is 20 to 50% slower than DNA:DNA hybrid formation, depending on the degree of RNA secondary structure. RNA:RNA hybridization rates are similarly reduced. Mismatches up to 10% have little effect on hybridization rates. Solvent viscosity and ionic strength also influence $k_N'$. In solvents with glycerol, ethylene glycol, or formamide, $k_N'$ is reduced because of increased solvent viscosity compared to 1.0 $M$ NaCl at 70°C. The dependence of $k_N'$ on salt concentration is given by

$$k_N' = \{4.35 \cdot \log_{10}\,[\text{SALT}] + 3.5\}\,10^5 \qquad \text{for } 0.2 \le [\text{SALT}] \le 4.0$$

$$k_N' = \{4.35 \cdot \log_{10}\,(1 + 34\,[\text{SALT}]^3)\}\,10^5 \qquad \text{for } [\text{SALT}] < 0.2$$

Thus in typical hybridization buffers approximating 1 $M$ NaCl, $k_N' = 3.5 \cdot 10^5\ M^{-1}s^{-1}$.

### 3.2 Hybridization with Oligonucleotides

Hybridization with oligonucleotides usually takes place far below $t_m^\infty$ (at or below $t_m$), so the decrease in $k_2$ near $t_m^\infty$ is not a factor, and the temperature dependence of $k_2$ for oligonucleotides is small unless the target has interfering secondary structure. Oligonucleo-

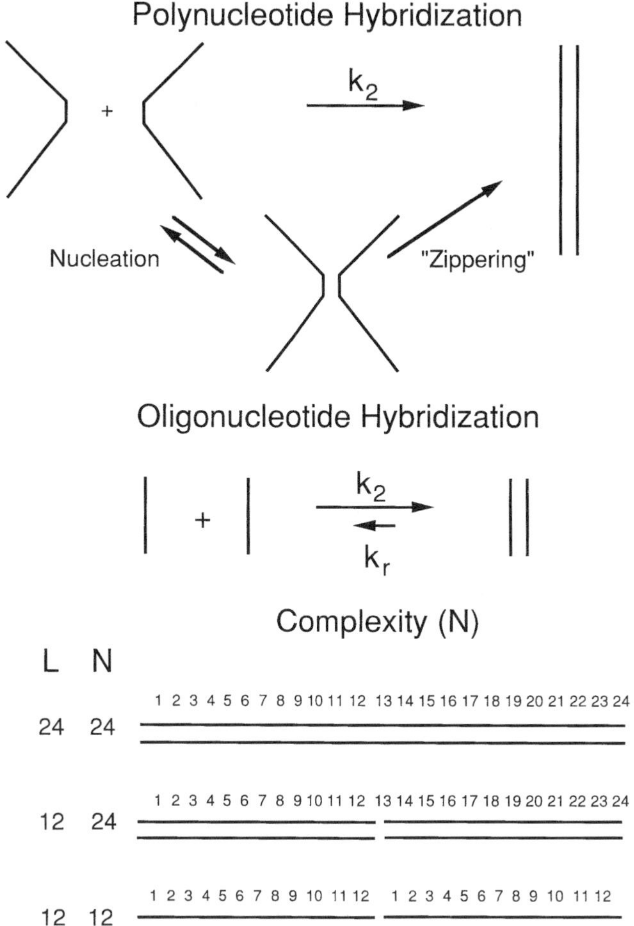


**Figure 2.** Steps in polynucleotide hybridization, steps in oligonucleotide hybridization, and complexity: $k_2$, second-order rate constant for hybridization; $k_r$, rate constant for the reverse reaction; $N$, complexity (= number of nucleotide pairs of unique DNA), $L$, length (= average number of nucleotides in DNA single strands).

tide hybridization has the same salt dependence as polynucleotide hybridization. In fact, oligonucleotide hybridization may be described by the same equations. For oligonucleotide–polynucleotide reactions, the oligonucleotide is usually the driver leading to a pseudo–first-order reaction characterized by a rate constant with $N = L$ of the oligonucleotide. Oligonucleotide probe reactions are very fast: $k_2 = k_N' / \sqrt{L} = 7 \cdot 10^4\ M^{-1}s^{-1}$ for a 25-mer in 1 $M$ NaCl at $t_m^\infty - 25°C$. At 1 n$M$ ($C_0$ = 25 n$M$), $t_{1/2} = \ln 2/(7 \cdot 10^4 \cdot 25 \cdot 10^{-9}) = 396$ seconds (6.6 min). For a 1 $\mu M$ primer in PCR buffer ($k_N' = 7 \cdot 10^4\ M^{-1}s^{-1}$), $t_{1/2}$ is 3 seconds. For every additional 10°C decrease, $k_2$ is reduced 1.2-fold. For 1 $\mu M$ $(dG)_6$ in 1 $M$ NaCl at $t_m$, $t_{1/2}$ = 1.3 seconds. Note that $t_{1/2}$ increases proportional to the degeneracy of mixed oligonucleotide probes or primers.

## 4 OLIGONUCLEOTIDE DISSOCIATION

When hybridized oligonucleotides on solid supports are washed to eliminate unhybridized oligonucleotides, both $t_m$ and the rate characterized by $k_2$ decrease as $C$ decreases, and the hybrid stability depends entirely on the rate constant of dissociation, $k_r$. In Section 2.3, we determined the enthalpy of melting, $-\Delta H°$, and $t_m$, which is a measure of the equilibrium constant for oligonucleotide hybridization, or the ratio of the hybridization and dissociation rates. In Sections 3.1 and 3.2, we determined the $k_2$, which has a small activation energy. With this information, we may determine $k_r$. The activation energy for $k_r$, $E_a$, is approximately 4.0 kcal/mol plus $-\Delta H°$. The dissociation temperature $t_d$ may be defined as the temperature at which 50% of correctly matched hybrid is released in a specified length of time, $t_{wash}$. Because $t_d$ depends on time and not concentration, unlike $t_{1/2}$ for hybridization and $t_m$, $t_d$ is the same whether unique or degenerate probes are being used. It is $t_d$ that may be estimated for extended washing of 14- to 20-mers as 2°C for each dA:dT pair plus 4°C for each dG:dC base pair. The value of $t_d$ may be calculated using:

$$t_d = \frac{T_m}{1 + RT_m \dfrac{[\ln (t_{wash}/t_{1/2})]}{E_a}} - 273.2$$

For example, for 1 $\mu M$ $(dG)_6$ in 1 $M$ NaCl, $E_a = -\Delta H° + 4000 = 69{,}000$ cal/mol, $T_m$ = 313°K (Section 2.3), and $t_{1/2}$ = 1.3 seconds (Section 3.2). Thus $t_d$ = 19°C for 50% probe released with an hour (3600 s) washing. This calculation depends on all the elements of the physical chemistry of hybridization.

*See also* ANTISENSE OLIGONUCLEOTIDES, STRUCTURE AND FUNCTION OF; DENATURATION OF DNA; HYDROGEN BONDING IN BIOLOGICAL STRUCTURES.

### *Bibliography*

Hames, B. D., and Higgins, S. J. (1985) *Nucleic Acid Hybridisation, A Practical Approach.* IRL Press, Oxford.

Wetmur, J. G. (1991) DNA probes: Applications of the principles of nucleic acid hybridization. *Crit. Rev. Biochem. Mol. Biol.* 26:227–259.

# NUCLEIC ACID SEQUENCING TECHNIQUES

*Christopher J. Howe*

### *Key Words*

**Bacteriophage (Phage)** A virus infecting bacteria.

**Oligonucleotide** A short nucleic acid molecule, often artificially synthesized DNA.

**Primer** A short segment of DNA annealing to a DNA or RNA molecule and initiating synthesis of DNA.

**Template** DNA or RNA molecule used to direct synthesis by polymerase.

Nucleic acid sequencing, the determination of the order of nucleotides in a molecule of DNA or RNA, may also include the identification of modifications, such as methylation. Nucleic acid sequencing is not usually an end in itself, but it is an important step toward understanding features such as the function, structure, or evolution of the molecule. For example, determining the sequence of a gene may allow one to predict the amino acid sequence of the protein encoded (although it does not directly provide information on post-transcriptional or posttranslational modifications) and to identify regions of possible secondary structure or sequence motifs that may be recognition sites for proteins involved in control. Comparison of the sequence with that from other organisms or related genes from the same organism may allow evolutionary relationships to be inferred. It is useful to be able to determine the sequence either of cloned DNA or of molecules isolated directly from a host organism. Sequencing of RNA is also important. Sequence determination for both DNA and RNA can be done using enzymatic synthesis or cleavage reactions. Most sequencing is of DNA, using enzymatic synthesis. Computer databases of all published (and many unpublished) nucleic acid sequences are maintained by a number of organizations.

## 1 ENZYMATIC SEQUENCING OF DNA

### 1.1 THE PRINCIPLE

A short oligonucleotide, typically 15 to 20 nucleotides, is mixed with the target DNA. To ensure the formation of a duplex by hydrogen-bonded base pairing, the oligonucleotide must be complementary in sequence to the target. Addition of DNA polymerase and 2′-deoxynucleoside triphosphates (dNTPs) allows DNA synthesis by extension of the oligonucleotide (which is referred to as a "primer"). This would simply extend the primer in the 5′–3′ direction, were it not for the inclusion of 2′,3′-dideoxynucleoside triphosphates (ddNTPs). Although the ddNTPs can be added to a growing DNA chain by means of their 5′-triphosphate, they will terminate the synthesis of that chain because they carry an—H on the 3′ carbon rather than the—OH, which is needed for addition of further nucleotides. Four separate primer extension reactions are done, each containing a different ddNTP: ddATP, ddCTP, ddGTP and ddTTP. Each reaction will generate a "nested" series

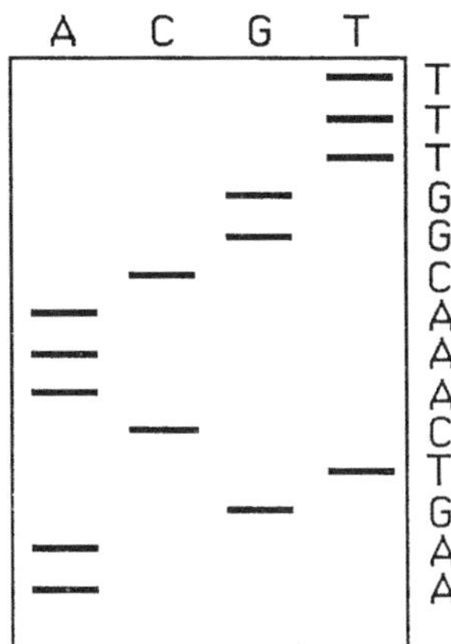


**Figure 1.** Schematic representation of part of an autoradiograph of a sequencing gel; the sequence is indicated up the side.

of newly synthesized molecules, each beginning with the primer and ending at an A (opposite a T) in the tube containing ddATP, at a C (opposite G) in the tube containing ddCTP, and so on. Note that termination does not necessarily take place opposite the first T in the reaction containing ddATP, because dATP is also present. Some of the extending molecules will incorporate the deoxy form and will progress to the next T, where termination likewise may or may not take place. The size distribution of the molecules produced will depend on the ddNTP/dNTP ratio: the higher the ratio, the lower the average length of molecules generated.

Following the synthesis reactions, the products are separated electrophoretically according to size in adjacent lanes of a polyacrylamide gel. To prevent the formation of secondary structures in the DNA, which would cause anomalies in the mobility, the electrophoresis is carried out under denaturing conditions (at high temperature and in the presence of urea). Thin gels are used, allowing greater resolution, and molecules differing in length by a single nucleotide can be separated. The positions of the molecules in the gel are then detected. Usually one of the dNTPs used in the reactions is radioactively labeled (other possibilities for labeling are radioactive end labeling of the primer, and fluorescence labeling, discussed in Section 1.7), and the gel is autoradiographed. A schematic representation of a result is shown in Figure 1. Reading up from the bottom allows the sequence to be deduced one nucleotide at a time.

### 1.2 Generation of Template

It is often convenient if the template for sequencing can be obtained in single-stranded form, and a number of vectors are available for this, based on the filamentous bacteriophage M13. This phage attacks only ''male'' *E. coli* cells, adsorbing to the sex pilus. The infectious particles contain single-stranded DNA, but upon entering the bacterial cell this is converted to a double-stranded replicative form (RF). This material is used to synthesize either more RF DNA or single-stranded DNA, which can then be packaged into phage heads. These are extruded from the bacterial cell without lysing it. In the lab, DNA is cloned into the RF molecule and introduced artificially into a suitable *E. coli* host, after which the phage particles containing ssDNA are collected. The DNA is then purified and used in the sequencing reactions. The phage vectors contain a ''multiple cloning site'' with recognition sequences for a range of restriction enzymes allowing convenient insertion of blunt- or sticky-ended fragments. This multiple cloning site is located within a part of β-galactosidase gene (*lacZ*) engineered into the phage. This ''minigene'' produces a polypeptide that can assemble with a fragment encoded by the host to produce functional β-galactosidase enzyme. Insertion of DNA into the multiple cloning site therefore abolishes the production of β-galactosidase in infected hosts (which are unable to produce the complete enzyme themselves), and this result can be detected on indicator plates.

Other vectors are available that are hybrids between plasmids and phage. They are propagated as plasmids (and contain a similar multiple cloning site to the M13 vectors). However, they also contain an origin for single-stranded DNA synthesis derived from a filamentous phage. When cells carrying these plasmids are superinfected with a helper phage, this origin is activated and single-stranded DNA is synthesized.

Single-stranded DNA can also be obtained by polymerase chain reaction (PCR) techniques. A convenient method is to have an excess of one of the primers, resulting in preferential amplification of the corresponding strand. Sequencing of double-stranded DNA is also possible. The two strands are first separated by heating or treatment with alkali, and the primer allowed to anneal. Since it will anneal to only one strand, sequencing can proceed, provided sufficient primer–template complex can be formed without reannealing of the full-length strands.

### 1.3 Primers

It is clear that to construct a primer that will anneal to the target molecule, some sequence information is necessary at the start. For material cloned into a standard vector, this is not a problem, inasmuch as the sequence flanking the cloning site will be independent of the insert. A single artificially synthesized ''universal primer'' that anneals close to the cloning site can therefore be used. Alternatively, if the sequence of part of the target DNA is already known, a primer complementary to that can be synthesized.

### 1.4 Sequencing Strategies

The amount of information generated in any one sequencing run is limited to a few hundred nucleotides. Most sequencing projects require the determination of larger molecules than this, and the overall sequence is built up in small pieces. This can be done in any of the following ways (and often a combination is used):

1. The larger molecule is fragmented by sonication or restriction enzyme degradation, and the pieces cloned and sequenced.
2. The larger molecule is systematically shortened by exonucleolytic degradation.
3. The larger molecule is cloned intact, and a series of internal primers is used.

In general, when determining any new piece of sequence it is desirable to ensure that both DNA strands have been properly sequenced and that the individual readings for one strand form a continuous overlapping series. Errors caused by secondary structure in the template may not be detected on one strand but will become apparent when it is compared with the other.

## 1.5 Enzymes and Nucleotides

Initially, sequencing was carried out using the Klenow fragment of DNA polymerase. This enzyme had the disadvantage of giving bands of rather uneven intensity, particularly in certain sequence contexts. Subsequently, other enzymes were used which gave more even band intensities, and perhaps the most popular is the polymerase encoded by bacteriophage T7, marketed commercially under the name Sequenase. The use of thermostable polymerases allows the sequencing reactions to be carried out at higher temperatures, reducing the likelihood of artifactual termination due to secondary structure in the template. This procedure has also led to the development of "cycle" sequencing: repeated rounds of synthesis (in the presence of ddNTPs) carried out by cycles of heating and cooling, resembling PCR (although amplification is linear rather than geometric, inasmuch as only a single strand is synthesized in each cycle). Cycle sequencing is particularly useful when only small amounts of template are available. Nonstandard nucleotides, such as dITP, can be used in all cases to reduce problems caused by secondary structure.

## 1.6 Data Collection and Analysis

There are three distinct requirements. The first is the acquisition of data from individual electrophoresis runs. The second is the assembly of these datasets to produce the overall sequence of the DNA under consideration, and the third is the analysis of this sequence (e.g., to deduce coding function). Acquisition can be done fully manually, by simply reading the gel from the bottom up, according to whether successive bands occur in the A, C, G, or T track, and entering the information into a computer file either through a keyboard or a modified mouse. This method has minimum equipment requirements but may be subject to user error (depending on the user!).

Acquisition can be done in a semiautomated way, using a sonic digitizer. In this device, a digitizing pen is pressed onto each band in turn. The signal emitted upon pressing is picked up by two detectors. This allows the calculation of the position of the pen, and therefore the lane it was indicating. The information thus acquired is fed directly into a computer file, allowing the reader to exercise judgment in reading the gel (assessing the significance of possible artifact bands, etc.) while reducing data transcription errors. Acquisition can also be fully automated, by electronic scanning (see Section 1.7). A large number of computer packages are available for assembly of individual gel readings into an overall sequence and its subsequent analysis.

## 1.7 Automation

The manipulations involved in DNA sequencing are essentially independent of the particular sequence to be determined, and the procedure is therefore particularly suitable for automation. This is important for larger scale projects, such as the sequencing of complete genomes. Automation is currently least widespread in template preparation. Procedures being developed include the use of magnetic particle labeling of DNA for purification. Degrees of

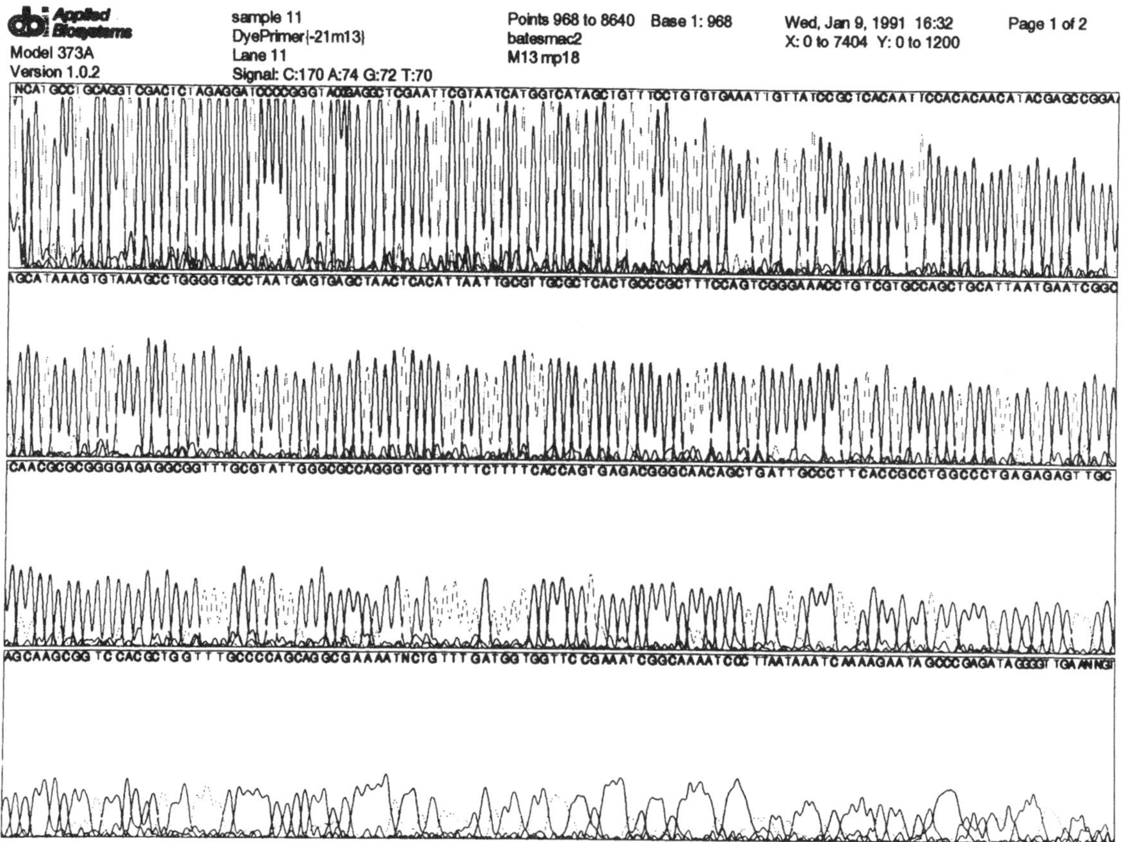


**Figure 2.** Example of the output from an Applied Biosystems 373A automated sequencer. (Original kindly supplied by Dr. D. Watts, Applied Biosystems Ltd., U.K.)

''automation'' of the actual reactions vary from the use of predispensed reagents in microtiter trays and hand-operated multiple pipetters to robot pipetters. Gel reading is quite widely automated. Many systems rely on fluorescent labeling of either the primers or the ddNTPs themselves. A different label is used for each reaction, giving each set of products a different fluorescence pattern. The products of each reaction are mixed (if the fluorescence tags are on the ddNTPs, all four reactions can be done in the same tube) and electrophoresed in a single gel lane. Fluorescence detectors pick up individual bands as they pass through the gel, and determine which nucleotide was present. The information is recorded directly onto computer disk and displayed visually in a series of traces of different colors. A black and white example is shown in Figure 2. By allowing data capture in real time, this approach removes the need for autoradiography.

## 2 CHEMICAL SEQUENCING OF DNA

Sequencing is possible by base-specific cleavage, an approach developed by Maxam and Gilbert in the late 1970s. A DNA molecule is labeled at one end and treated with limiting amounts of reagents, which react specifically or preferentially as follows:

| | |
|---|---|
| G | dimethyl sulfate |
| A + G | piperidine formate |
| C + T | hydrazine |
| C | hydrazine and 1.5–2 M NaCl |
| A + C (stronger at A) | 1.2 M NaOH at 90°C |

Subsequent treatments bring about cleavage at the modified bases, generating nested sets of molecules as with enzymatic sequencing. They are likewise electrophoresed and their positions determined by autoradiography.

Because this technique does not require cloning of the target DNA, it is particularly useful for the sequencing of artificially synthesized oligonucleotides. It is also useful for studies of secondary structure and DNA–protein interactions, both of which will influence the reactivity of individual nucleotide positions. It can be used to probe modifications such as methylation in the same way. Analysis of genomic DNA is possible by this method. For example, the total genomic DNA is digested with a suitable restriction enzyme and then subjected to cleavage reactions. The fragments are separated by electrophoresis, transferred to a membrane, and hybridized with a labeled probe for the region to be sequenced. The fragments picked out will have one end in common (generated by the restriction enzyme), and the other will be sequence specific.

## 3 RNA SEQUENCING

Like DNA, RNA can be sequenced by enzymatic synthesis or degradative methods. In the former, a suitable primer (which therefore requires some preliminary sequence information to be available) is annealed at the 5′ end of the RNA, and sequencing reactions are carried out as with DNA sequencing except that reverse transcriptase is used instead of DNA polymerase. The products of the four reactions are analyzed electrophoretically exactly as before. Degradative methods employ a range of chemical and enzymatic cleavage reagents followed by electrophoretic or chromatographic analysis.

*See also* Gene Mapping by Fluorescence In Situ Hybridization; Human Chromosomes, Physical Maps of; Human Linkage Maps; Partial Denaturation Mapping; Restriction Landmark Genomic Scanning Method; Whole Chromosome Complementary Probe Fluorescence Staining.

### *Bibliography*

Howe, C. J., and Ward, E. S. (1989) *Nucleic Acids Sequencing: A Practical Approach.* IRL Press, Oxford.

Sambrook, J., Fritsch, E. F., and Maniatis, T. (1989) *Molecular Cloning: A Laboratory Manual,* 2nd ed. Cold Spring Harbor Laboratory Press, Cold Spring Harbor, NY.

Watson, A., Smaldon, N., Lucke, R., and Hawkins, T. (1993) *Nature* 362:569–570.

# Nutrition

*C. J. K. Henry*

### *Key Words*

**Basal Metabolic Rate (BMR)** The energy expanded by an individual in a rested, relaxed state and in a thermoneutral environment, 12 to 14 hours after a meal.

**Metabolism** The chemical processes and reactions that take place, every minute of the day, in the cells of the body.

**Minerals** A group of inorganic ions in their elemental form. Minerals may be classified as *macroelements,* which are essential for normal development and functioning of the body and are required at levels of 100 mg or more per day, and *microelements,* essential for the body at levels of 0.01 mg or less per day.

**Nutrition** Nutrition is the study of the biological processes involved in the provision of nutrients to the cells of the body for adequate growth, maintenance, activity, and reproduction. Nutrition is about the ingestion, digestion, absorption, and finally assimilation of the nutrients (or chemical elements) in foods.

**Vitamins** Organic micronutrients that are vital for such basic bodily functions as metabolism, growth, reproduction, and the overall maintenance of health. Vitamins were initially called *accessory factors* because they are required in such minute quantities.

''You are what you eat'' is a statement frequently used to emphasize that the composition of our bodies depends ultimately, to a large measure, on the food we have consumed. The proximate composition of a human body is shown in Table 1. All the constituents listed in Table 1 are obtained from the food we eat. All foods can be broken down into basic chemical constituents. Table 2 gives a simple, but practical, classification of the food constituents.

## 1 FOOD CONSTITUENTS AND THEIR FUNCTIONS

The human body requires energy to sustain life. Carbohydrates, fat, and proteins are used as energy-producing nutrients. A major process of metabolism is the oxidation of foods to provide energy. In most situations, our energy requirements are met from the foods

Table 1 Proximate Composition of the Human Body

| Constituent | Percentage of Body Weight |
|---|---|
| Water | 62 |
| Protein | 18 |
| Fat | 12 |
| Minerals | 7 |
| Carbohydrates | 0.1 |

we eat. When food supplies are limited or absent (e.g., in famine/starvation), the body obtains energy from the breakdown of its own tissues.

## 2 ENERGY METABOLISM

Since all forms of energy can be converted to heat, food energy can be expressed as heat energy. The traditional unit for measuring energy is the *kilocalorie* (kcal or cal). The Calorie, one thousand times larger than the small calorie, is defined as the heat required to raise the temperature of one liter of water from 14.5°C to 15.5°C. More recently the international system of units has come into wide use. The SI unit of energy is the joule (J). The conversion factor for changing kilocalories to kilojoules is:

$$1 \text{ kcal} = 4.184 \text{ kJ}$$
$$1000 \text{ kJ} = 1 \text{ MJ (megajoule)}$$

When oxidized, food constituents that contain carbon and hydrogen (e.g., carbohydrates and fat) release energy: $CO_2 + H_2O$.

The amount of heat produced when a food is burned completely is termed *gross energy* (Figure 1). Since all the food eaten is not digested (some of it being lost in the feces), only food that is broken down and ultimately absorbed can be used by the body as a source of energy. This absorbed energy is called *digestible energy* (DE). Further losses of energy occur in the urine as a result of incomplete oxidation of protein, leading to nitrogenous waste (usually urea) in the urine. When the digestible energy is corrected for urine losses, the resultant factor is called *metabolizable energy* (ME). When any food is consumed, there is an inevitable loss of some energy called *heat increment of feeding*. When metabolizable energy is corrected for this loss, it is called *net energy* (NE). This is ultimately the energy available to the body for maintenance and production. The metabolizable energy values of major nutrients are as follows: protein and carbohydrate, 4 kcal/g; fat, 9 kcal/g, and alcohol, 7 kcal/g.

Table 2 Food Constituents and Their Function

| Food Constituents | Function |
|---|---|
| Carbohydrates | Source of energy for the body |
| Fat | Source of energy; essential fatty acids |
| Protein | For tissue development, growth, and repair |
| Vitamins | For all metabolic processes |
| Minerals | For skeletal growth and metabolic processes |
| Water | For temperature regulation and as a body fluid |
| Fiber | To provide bulk for foods and facilitate bowel movement |

**DIETARY ENERGY**

GROSS ENERGY

Fecal Energy Loss **Digestible Energy**

URINARY ENERGY LOSS **Metabolizable Energy**

**Net Energy** Heat Increment of feeding

**Energy for Maintenance and Growth**

Figure 1. Energy content of food and routes of energy loss.

Table 3 Equations for Calculating Basal Metabolic Rates

| | Males[a] | Females[a] |
|---|---|---|
| *In kcalories per day* | | |
| **Adolescents** | | |
| Age range (years) | | |
| 10–17+ | 17.5*W* + 651 | 12.2*W* + 746 |
| **Adults** | | |
| 18–29+ | 15.3*W* + 679 | 14.7*W* + 496 |
| 30–59+ | 11.6*W* + 879 | 8.7*W* + 829 |
| 60 | 13.5*W* + 487 | 10.5*W* + 596 |
| *In megajoules per day* | | |
| **Adolescents** | | |
| 10–17+ | 0.0732*W* + 2.72 | 0.0510*W* + 3.12 |
| **Adults** | | |
| 18–29+ | 0.0640*W* + 2.84 | 0.0615*W* + 2.08 |
| 30–59+ | 0.0485*W* + 3.67 | 0.0364*W* + 3.47 |
| 60 | 0.0565*W* + 2.04 | 0.0439*W* + 2.49 |

[a]*W* is the average weight (kg).

Table 4 Multiples of BMR to Be Used to Calculate Total Energy Requirement in Subjects with Light, Moderate, and Heavy Occupational Levels

| | Light | Moderate | Heavy |
|---|---|---|---|
| Men | 1.55 | 1.78 | 2.10 |
| Women | 1.56 | 1.64 | 1.82 |

*Source:* Modified from FAO/WHO/UNU Technical Report Series 724, Geneva (1985).

**Table 5** Energy Requirement for a Male Engaged in Heavy Work[a]

| | Hours | kcal | kJ |
|---|---|---|---|
| Resting in bed, at 1.0 × BMR | 8 | 545 | 2,280 |
| Occupational activities at 3.8 × BMR | 8 | 2070 | 8,660 |
| Discretionary activities at 3.0 × BMR | 1 | 205 | 860 |
| For residual time, maintenance energy needs at 1.4 × BMR | 7 | 670 | 2,800 |
| **Total** | | 3490 | 14,580 |
| = 2.14 × BMR | | | |

[a]Description of individual considered: age 35 years, weight 65 kg, height 1.72 m, BMI 22 estimated basal metabolic rate: 68 kcal (284 kJ) per hour.

**Table 6** List of Essential Amino Acids

| |
|---|
| Histidine |
| Isoleucine |
| Leucine |
| Lysine |
| Methionine + cystine |
| Phenylalanine + tyrosine |
| Threonine |
| Tryptophan |
| Valine |

**Table 7** Function and Requirements of Vitamins

| Recommended Daily Allowances of Vitamins | | Major Sources | Physiological Roles | Deficiency Symptoms |
|---|---|---|---|---|
| Vitamin A (retinol) pro-vitamin A (carotene) | | Liver, green and yellow fruits and vegetables, whole milk, cheese, butter, fortified margarine | Maintains integrity of epithelium; healthy functioning of eyes, skin, mucous membrane | Night blindness, faulty bone and tooth development, mucous membrane infections, xerophthalmia, blindness |
| Adults | 1000 μg | | | |
| Pregnancy | 800 μg | | | |
| Lactation | 1300 μg | | | |
| Infants | 375 μg | | | |
| Children | 700 μg | | | |
| Vitamin D (calciferol) cholecalciferol | | Milk, fish oil, liver, butter, egg yolk, fortified margarine; also synthesized in skin, by UV light | Absorption of calcium and phosphorus | Rickets; soft and fragile bones; deformed chest, spine, and pelvis; bowed legs in infants; tetany (in infants); osteomalacia (in adults) |
| Adults | 5–10 μg | | | |
| Pregnancy | 10 μg | | | |
| Lactation | 10 μg | | | |
| Infants | 10 μg | | | |
| Children | 10 μg | | | |
| Vitamin E (tocopherol) | | Vegetable oils, wheat germ, whole grains, leaf vegetables, egg yolk, legumes | Intracellular antioxidant; protects fat from abnormal breakdown; stability of cell membranes | Hemolysis of red blood cells; abnormal fat deposits |
| Adults | 10 mg | | | |
| Pregnancy | 10 mg | | | |
| Lactation | 12 mg | | | |
| Infants | 3 mg | | | |
| Children | 7 mg | | | |
| Vitamin K (phylloquinone) | | Lettuce, spinach, cauliflower, cabbage, kale, egg yolk, soybeans | Required for normal blood clotting | Prolonged blood clotting; hemorrhage |
| Adults | 45–80 μg | | | |
| Pregnancy | 65 μg | | | |
| Lactation | 65 μg | | | |
| Infants | 10 μg | | | |
| Children | 30 μg | | | |
| Vitamin C (ascorbic acid) | | Citrus fruits, strawberries, cantaloupe, tomatoes, kale, parsley, turnip greens, broccoli | Maintains integrity of capillaries; promotes healing of wounds and fractures; aids tooth and bone formation; increases iron absorption and protects folic acid; helps form collagen for healthy connective tissue | Bruise and hemorrhage easily; scurvy and anemia |
| Adults | 60 mg | | | |
| Pregnancy | 70 mg | | | |
| Lactation | 95 mg | | | |
| Infants | 35 mg | | | |
| Children | 45 mg | | | |
| Thiamine ($B_1$) | | Pork, liver, chicken, fish, beef, whole grains, wheat germ, fortified cereal products, nuts, lentils | Required for energy transformation and integrity of central nervous system | Anorexia, fatigue, depression, apathy |
| Adults | 1.3 mg | | | |
| Pregnancy | 1.5 mg | | | |
| Lactation | 1.6 mg | | | |
| Infants | 0.3 mg | | | |
| Children | 1.0 mg | | | |

Table 7 Function and Requirements of Vitamins (Continued)

| Recommended Daily Allowances of Vitamins | | Major Sources | Physiological Roles | Deficiency Symptoms |
|---|---|---|---|---|
| Riboflavin ($B_2$) | | Milk, liver, meat, fish, eggs | Protein metabolism and healthy eyes | Dermatitis, photophobia |
| Adults | 1.3–1.7 mg | | | |
| Pregnancy | 1.6 mg | | | |
| Lactation | 1.6 mg | | | |
| Infants | 0.3 mg | | | |
| Children | 1.0 mg | | | |
| Niacin | | Liver, poultry, meat, fish, eggs, whole grains, enriched cereal products, legumes (epg., peanuts), mushrooms | Required for energy transformation and integrity of central nervous system | Dermatitis, decreased energy, weakness |
| Adults | 15–20 mg | | | |
| Pregnancy | 17 mg | | | |
| Lactation | 20 mg | | | |
| Infants | 5–6 mg | | | |
| Children | 9–13 mg | | | |
| Pyridoxine ($B_6$) | | Pork, organ meats, meat, poultry, fish, corn, legumes, seeds, grains, wheat, potatoes, bananas, green beans | Required for protein metabolism and integrity of central nervous system; facilitates conversion of tryptophan to niacin and hemoglobin synthesis | Nervous irritability, convulsions in infants; weakness; anemias |
| Adults | 2.0 mg | | | |
| Pregnancy | 2.2 mg | | | |
| Lactation | 2.1 mg | | | |
| Infants | 0.4 mg | | | |
| Children | 1–1.4 mg | | | |
| Folic acid, folate, folacin, and tetrahydrofolic acid | | Green leafy vegetables, liver, beef, fish, dry beans, lentils, whole grains, asparagus, broccoli | Red blood cell maturation; folic acid is synthesized in intestinal tract | Anemias |
| Adults | 200 μg | | | |
| Pregnancy | 400 μg | | | |
| Lactation | 280 μg | | | |
| Infants | 30 μg | | | |
| Children | 50–100 μg | | | |
| Vitamin $B_{12}$ | | Only in animal foods, liver, meat, saltwater fish, oysters, milk, eggs | Red blood cell maturation, essential for normal function of all body cells | Neurological degeneration and anemia |
| Adults | 2.0 μg | | | |
| Pregnancy | 2.2 μg | | | |
| Lactation | 2.6 μg | | | |
| Infants | 0.3–0.5 μg | | | |
| Children | 0.7–1.4 μg | | | |

*Source:* Henry, C. J. K. (lecture notes).

## 3 ESTIMATION OF ENERGY REQUIREMENTS

The routes of energy expenditure in all humans may be broadly divided into the following components:

Basal metabolic rate (BMR)
Diet-induced thermogenesis (DIT)
Physical activity

The largest component, basal metabolic rate, represents the energy expended by all the tissues of the body at rest. Since BMR may represent up to 80% of the total energy expenditure, in adults, its determination is the starting point in computing an individual's energy requirements. BMR may be determined by indirect calorimetry or by using predictive equations. The predictive equations for calculating BMR are shown in Table 3.

Once BMR has been predicted, this value must be "adjusted," or increased, to take into account diet-induced thermogenesis (which is the increase in metabolic rate when we consume food) and physical activity. Since physical activity can vary in both intensity and duration, we can express all forms of physical activity as multiples of BMR. Thus a simple approach to estimating total energy requirements in humans is to multiply the BMR by any one of the multiples of BMR, depending on the level of physical activity (light, moderate, or heavy); see Table 4. Table 5 shows an example of such a calculation.

In contrast to adults, the energy requirements of infants and children below the age of 10 can be estimated by calculating the energy intake.

## 4 PROTEINS AND ESSENTIAL AMINO ACIDS

Proteins are composed of amino acids. Most protein foods consumed by man are made up of 20-odd amino acids. Although all protein foods can act as a source of energy, some proteins (e.g., egg protein) are more useful and better utilized for forming body proteins. Such differences result from the different amino acid compositions of the various proteins, and they give rise to the concept of protein quality. Nine of the amino acids (Table 6) cannot be synthesized in our bodies and therefore must be obtained intact from food. These are called *essential amino acids*. Protein synthesis cannot effectively take place in the body if there is a lack of essential amino acids in the diet.

## 5 PROTEIN QUALITY AND CHEMICAL SCORES

Protein sources that have the full complement of essential amino acids are sometimes called *complete proteins*. Proteins that are devoid or deficient in one or more essential amino acids are called *incomplete proteins*. In the most widely used and simplest method of estimating protein quality, the chemical score, we compare the test protein with whole egg protein, which is assigned a value of 100, implying complete utilization. Thus

$$\text{chemical score} = \frac{\text{amount of essential amino acid in test protein (mg/g)}}{\text{amount of same essential amino acid in reference protein or egg protein}}$$

The amino acid that gives the lowest score in relation to the reference protein will dictate the final chemical score and is also called the first limiting amino acid.

## 6 VITAMINS

Table 7 shows the classification, functions, and requirements of vitamins.

## 7 MINERALS

Although minerals are not a source of energy, they play several essential roles in the body. Calcium is an integral part of the skeleton, and many minerals act as catalysts and regulators of the metabolism of enzymes. Minerals are also important for muscular contraction, for nerve excitability, and for acid–base balance. Figure 2 shows a simple classification of minerals.

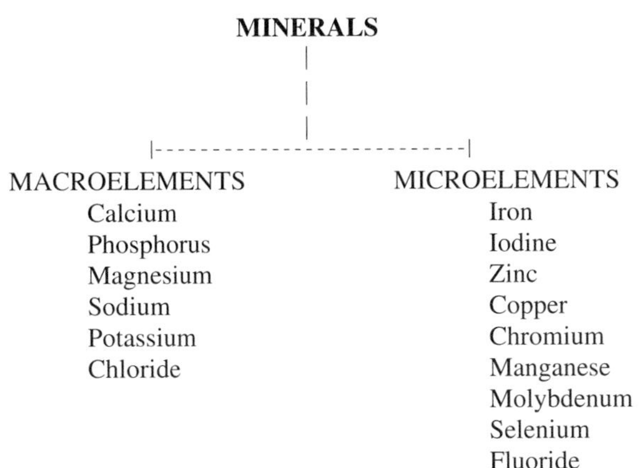


**Figure 2.** A simple classification of minerals.

*See also* BIOENERGETICS OF THE CELL; FOOD PROTEINS AND INTERACTIONS; TRACE ELEMENT MICRONUTRIENTS; VITAMINS, STRUCTURE AND FUNCTION OF.

### *Bibliography*

Eastwood, M. (1990) *Human Nutrition,* Chapman and Hall, London.

Hunt, S. (1990) *Advanced Nutrition and Human Metabolism,* West Pub, St Paul.

ILSI. (1990) *Present Knowledge of Nutrition,* Springer, New York.

Shils, M and Young, V. (1988) *Modern Nutrition in Health and Disease,* Lee and Fobiger, Philadelphia.

Whitney, E. N. and Hamilton, E. M. (1990) *Understanding Nutrition,* West Pub, St Paul.

# Oligonucleotide Analogues, Phosphate-Free

*Rajender S. Varma*

*Key Words*

**Achiral Compounds** Nonchiral molecules that possess the plane of symmetry and consequently are superimposable on their mirror images.

**Internucleotide or Internucleoside Linkage (Backbone)** Linkage connecting the 4′ carbon (ribose moiety) of one nucleoside with 3′ carbon of adjacent ribose unit in an oligomer.

**Modified Phosphate Backbone** The phosphate group present in the natural internucleotidyl phosphodiester linkage, modified by substitutions.

**Nonionic** Neutral or devoid of ionic charge, negative or positive.

**Phosphate-Free (Dephospho) Backbone** Internucleosidyl four- to five-atom linkage that is devoid of phosphate groups.

**Polyamide** Compound consisting of repeating amide ($CO—NH_2$) units, usually formed by condensation of amine and carboxylic acid functional groups.

Oligonucleotides, widely used as diagnostic reagents and as tools in molecular genetics, have found widespread application in biotechnology and, more recently, in medicinal chemistry to block the translation and also the transcription of genes. However, the practical utility of these nucleic acid derivatives is limited because of their instability to nucleases and their inability to penetrate the cell membrane. Also, the expensive phosphate chemistry adds to the cost of the compounds with a natural phosphodiester backbone, and large-scale synthesis has not been achieved. The first generation of analogues contains modified phosphate linkages, especially nonionic oligonucleotides. However, in most cases the resulting linkage consists of a pair of diastereoisomers that are not easily separable. The presence of such diastereoisomeric mixtures weakens considerably the hybridization capabilities of the oligonucleotides with the target sequences, and this limitation has stimulated research activity in the general area of oligonucleotide analogues with phosphate-free backbones (dephospho analogues).

The general criteria for the design of oligomers with altered backbone—namely, charge, hydration, and conformational freedom of the internucleosidyl linkage—are being established. Some of the synthesized phosphate-free internucleosidyl linkages suffer from poor solubility in aqueous media and consequently are of limited practical utility. Recently introduced polyamide linkages exploiting established peptide chemistry and the availability of a wide range of amino acids, however, are best suited to adapt to the practical solid phase synthetic methodologies, while at the same time satisfying the physical requirements for optimum binding to form a duplex and/or triplex.

## 1 INTRODUCTION

As a result of their capacity to base-pair and in view of the goal of achieving sequence-specific recognition of nucleic acids by binding to complementary nucleic acids, oligonucleotide reagents have found widespread application in biotechnology and more recently in medicinal chemistry, to the extent that "antisense" oligonucleotides have been shown to block the translation and also the transcription of genes. Consequently, these compounds are now being actively investigated as a new generation of pharmaceuticals. However, the practical utility of these nucleic acid derivatives is limited because of their inadequate stability in the presence of nucleases and their inability to penetrate the cell membrane. Besides being expensive, the phosphate chemistry involved in the preparation of compounds with a natural phosphodiester backbone evidently is not adaptable to large-scale synthesis of these molecules.

## 2 MODIFIED OLIGONUCLEOTIDE ANALOGUES WITH PHOSPHATE BONDS

The most extensively studied analogues are the first generation of backbone-modified oligonucleotides that contain modified phosphate linkages, especially nonionic oligonucleotides. These include oligonucleotide alkylphosphotriesters 1, oligonucleotide methylphosphonates 2, the related oligonucleotide phosphorothioates 3, and oligonucleotide alkylphosphoramidates 4 (Figure 1). They have attracted attention in view of their resistance to nucleases and because they are readily synthesized by slightly modifying the existing protocols for the synthesis of oligonucleotides with natural phosphodiester backbones.

In most of these cases, the resulting linkage consists of a pair of diastereoisomers that have not been separated. Consequently, an oligomer having $n$ nonionic, internucleotide linkages will consist of a mixture of $2^n$ isomers. The presence of such diastereoisomeric mixtures weakens considerably the hybridization capabilities of the oligonucleotides with target sequences. Moreover, few methods are available for the stereoselective synthesis on automated machines of oligomers with modified phosphate linkages. Nor has it been possible to isolate and separate pure stereoisomers on a large scale, these operations being limited to shorter sequences. Even more important, the large-scale synthesis of modified oligonucleotides, although addressed in recent years, is still inadequate for producing the candidate active compounds on a scale appropriate for clinical, toxicological, and eventual human applications.

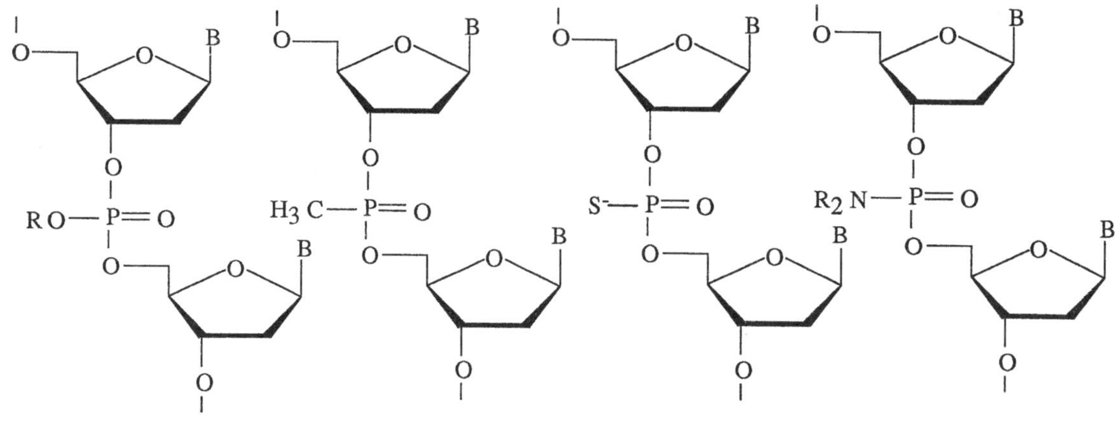

Figure 1. Oligonucleotides with modified phosphate internucleotide linkages.

## 3 PHOSPHATE-FREE BACKBONES

To circumvent the limitations of cost-effective large-scale synthesis, as well as chirality problems associated with modified phosphate oligonucleotides, intense efforts have been made recently to replace the anionic, four-atom, phosphate backbone with a neutral and achiral phosphate-free backbone.

The first-generation approaches to backbone modification have involved replacement of the bridging phosphate group with other functional groups, which include carbonate, oxyacetamide, carbamate, silyl, and ribonucleoside-derived morpholino subunits linked by carbamate groups (Figure 2). The relative instability of these linkages with respect to hydrolysis by base and the poor solubility of the compounds in water are two of the most frequently encountered problems. In some cases, too, the duplex formation appears to differ from that of the natural oligonucleotides as discussed later for peptide nucleic acids (PNA).

### 3.1 Carbonate Linkages

The carbonate-linked oligonucleotides 5 are resistant to hydrolysis by acid but are rather easily cleaved with base. Such base instability severely limits the utility of these compounds in a physiological context.

### 3.2 Carboxymethyl Linkages

Carboxymethyl-bridged oligonucleotide analogues **6** contain one more atom than the natural phosphodiester linkage; the rationale entails the flexibility of the carboxymethyl ester unit. Stable duplexes have been obtained between poly(dA) analogues (15–25 long) containing carboxymethyl internucleotide linkages oligoA′ and oligoU. The usual limitation of poor solubility in water persists with these compounds. Their stability is dependent on the pH, with a half-life of only 7 hours at pH 7.5.

### 3.3 Acetamidate Linkages

Attempts have been made to circumvent the low solubility and the unstable character of the carboxymethyl analogues just described by introduction of an amide unit in place of the 5′-ester oxygen; that is, the—$OCH_2CONHCH_2$—linkage. In addition to poor yields of polymer formation, the product had poor hybridization characteristics, was unstable in water, and tended to be adsorbed onto plastic and glass surfaces.

### 3.4 Carbamate Linkages

Although the carbamate-linked oligomers **7** (Figure 2) are most often prepared using solution chemistry, they can be prepared using

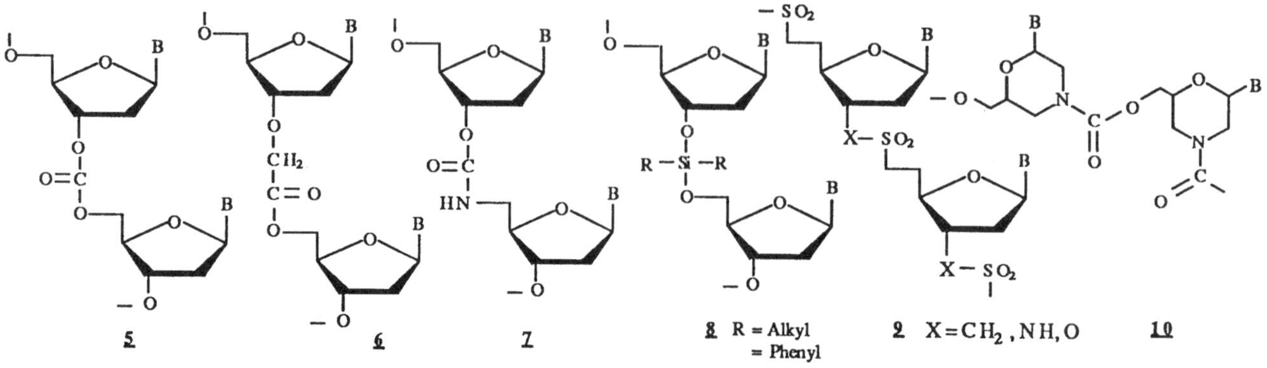

Figure 2. Oligonucleotides with phosphate-free internucleotide linkages.

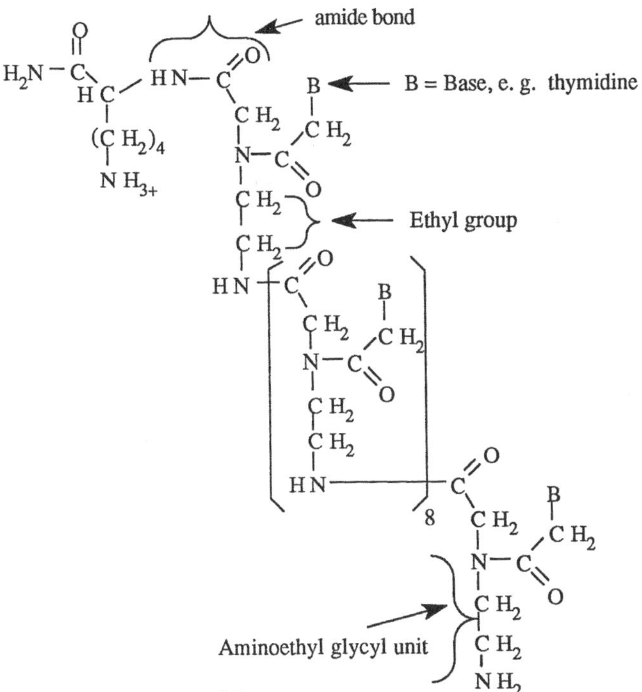

Figure 3. A polyamide nucleic acid oligomer.

polymer support based automated machines. The linkage is stable under physiological conditions and is also resistant to nuclease hydrolysis. The carbamate linkage is stable to hydrolysis by base, a property that simplifies the syntheses of oligomers and favors their use in a biological context.

The simplest carbamate linkage is calculated to be shorter than a phosphodiester linkage by 0.32 A. However, molecular models have suggested that the linkage might allow the oligomer to assume a backbone conformation compatible with duplex formation. Carbamate-linked cytosine containing hexanucleoside formed stable complexes with $d(pG)_6$ and poly(dG) with a melting temperature of 70° and 79°C, respectively, which is indicative of a duplex with *more* thermal stability than that found in the compound formed by the corresponding phosphodiester.

Subsequently, novel oligonucleotide analogues from ribonucleotide-derived morpholino subunits linked by carbamate groups **10** (Figure 2) were prepared. The poor aqueous solubility characteristics of the resulting oligomer were improved by terminal conjugation with polyethylene glycol. A strategy has recently been de-veloped for the synthesis of such hexameric neutral carbamate oligomers on solid support using a combination of the selectively cleavable anchor and the ionizable base-protecting group.

### 3.5 Polyamide Linkages (Peptide Nucleic Acids)

Molecular modeling techniques have been used extensively to examine the ability of the acyclic nucleic acid analogues to adopt low energy conformations that conform to helix formation. Of the several backbone types examined, a polypeptide backbone derived from amino acids indicates preference for binding to DNA.

A model of a neutral and relatively unstructured "oligo-T" polynucleotide amino acid bearing an achiral polyamide linkage instead of the deoxyribose phosphate backbone of DNA has been designed with the aid of a computer program. The successful preparation of such oligomers, consisting of thymine-linked aminoethylglycyl units (**11,** Figure 3), and their ability to recognize their complementary target in double-stranded DNA by strand displacement, indicate that the backbone of DNA can be replaced by a polyamide with retention of base-specific hybridization properties. The use of either 2-aminoethyl-β-alanine or a 3-aminopropylglycine backbone instead of original 2-aminoethylglycine has demonstrated that such extensions using an additional methylene unit, while maintaining the specificity, result in poorer binding affinity. Thus perhaps such extended backbones can be used to fine-tune the binding affinity between PNA (peptide or polyamide nucleic acids) and DNA.

Furthermore, the binding of PNA to a restriction enzyme target completely inhibits the action of the corresponding enzyme. This effect has been demonstrated using plasmids containing double-stranded 10-mer PNA targets proximally flanked by two restriction enzyme sites and challenging them with the complementary PNA or PNAs containing one or two mismatches. Assays of the effect on the restriction enzyme cleavage of the flanking sites revealed complete inhibition with the complementary PNA, thereby suggesting that PNAs can be used as sequence-specific blockers of proteins that recognize DNA.

The strand invasion of duplex DNA by PNA oligomers via formation of a D-loop has been found to be salt-dependent; it can be quantitated by measuring inhibition of *Hin*dIII cleavage at a site overlapping the target sequence ($dA_{10}{\cdot}T_{10}$) for PNA ($T_{10}$ Lys) binding. In binding of PNAs to RNA, a site-specific termination of both reverse transcription and in vitro translation exactly at the position of the PNA·RNA heteroduplex is observed. The cells constitutively expressing a Sendai virus (SV40) large T antigen (T Ag) upon microinjection with either a 15-mer or a 20-mer PNA targeted to the T Ag messenger RNA suppressed the expression of T Ag.

The pioneering Danish work just described reinforces the general importance of the PNA approach and paves the way toward significant additional changes that may occur to improve these polymers for use as diagnostic and pharmaceutical reagents. To address the inherent limitations of preparing the corresponding nucleoamino acid building units in enantiomerically pure form, novel PNA surrogates are designed that incorporates easily accessible α-amino acids such as serine.

The synthesis of semirigid version of neutral or charged PNA molecules is also accessible *via* direct modification of nucleotide building blocks, thus maintaining both the ribose unit and the amide or amine–amide linkage with built-in flexibility of their efficient preparation either by polymerization or by a solid phase stepwise approach. This mode of preparation both contrasts with and complements the earlier approach involving neutral, flexible PNAs, which are synthesized "ab initio."

Since nucleic acid–amino acid conjugates can be converted into oligomers by means of solution phase or solid phase coupling chemistry similar to that used for polypeptides, the distinguishing feature of the amino acid linker motif as the internucleosidyl linkage is chosen based on the application of rational drug design: computer-assisted molecular modeling; solubility and chemical stability criteria.

### 3.6 Sulfur-Based Linkages

A variety of sulfur-based functionalities have been introduced in the internucleosidyl linkages during the last few years, but in some cases experimental details of the preparation are lacking, whereas in others base-specific hybridization cannot be achieved. Almost in all cases described, the chemistry is not adaptable for synthesis of oligomers on solid supports.

The sulfur-based moieties that bear isoelectronic and isosteric resemblance to the natural phosphate linkages include achiral and nonionic 3′,5′-dialkyl sulfone (Figure 2, **9**, X = $CH_2$), 3′- or 5′-sulfonate esters (Figure 2, **9**, X = O), or the corresponding sulfonamides (Figure 2, **9**, X = NH). Potential antisense analogues bearing polar yet nonionic sulfonic acid ester and sulfonic acid amide internucleoside linkages have been prepared. An achiral nuclease-resistant sulfamate ($OSO_2NHCH_2$) linkage that displays duplex formation has also been formed recently.

The preparative and coupling chemistry of the nucleoside sulfonates for the assembly of short oligomers, which has also been detailed, may find application in the synthesis of sulfonyl-containing RNA. Appropriately functionalized building blocks and the corresponding dinucleotide analogues bearing dimethylene sulfide, sulfoxide, and sulfone moieties in the internucleosidyl backbone as potential antisense reagents are described, along with the corresponding analogues bearing a carbocyclic sugar moiety.

### 3.7 Alkylsilyl Linkages

The internucleotide linkage of dialkyl- and diphenylsilyl oligomers (**8**, Figure 2) closely resembles the tetrahedral geometry of the phosphodiester internucleotide bond. These oligomers have been synthesized in solution. Efforts to detect the complexes formed between oligo$(dA)_5$ or oligo$(dT)_5$ analogues containing diisopropylsilyl linkages and complementary oligoribonucleotides were unsuccessful, presumably because of low solubility.

### 3.8 Miscellaneous Linkages

Research activity has intensified during the past few years in the general area of dephospho internucleoside linkages, presumably as a result of the prospective therapeutic value of oligonucleotide analogues. Some of the recent additions are illustrated in Figure 4 and discussed in Sections 3.8.1–3.8.3.

#### 3.8.1 Formacetal and Thioformacetal Linkages

Neutral deoxyoligonucleotide analogues bearing achiral formacetal (**12**) and thioformacetal (**13**) linkages have been synthesized (Figure 4). However, the sulfur analogue (**13**) displays poor binding characteristics when compared to the formacetal (**12**) in the sequence-specific triple-helix formation, inasmuch as substitution at the 5′ position with the larger and less polar sulfur disturbs the usual close contact between the 5′ oxygen and the 6 hydrogen of the thymine ring, as observed by molecular modeling of the corresponding formacetal analogue.

#### 3.8.2 Methylhydroxylamine Linkage

A stereoselective synthesis of a thymidine (T) nucleoside dimer bearing a methylhydroxylamine linkage, (**14**, Figure 4) has been achieved *via* an intermolecular radical reaction. This convergent approach shows that other linkages that bear 3′-C—$CH_2$ moieties in the backbone are nuclease resistant. The $^1$H-NMR studies on T*T dimers containing such linkages exhibit a typical DNA-type stacking interaction of the two thymine bases and a preference for an S-pucker (2′-endo). Also, the reactivity of the amino group present in the internucleosidyl linkage to attach linkers, cleavers, probes, intercalators, and so on can be exploited to improve the antisense properties of these molecules.

#### 3.8.3 *N*-Cyanoguanidine Linkage

The synthesis of *N*-cyanoguanidine linkage (**15**, Figure 4) has been reported to proceed from the conversion of 3′-aminonucleosides to isothiourea followed by reaction with 5′-amino-2′,5′-dideoxynucleosides.

### 3.9 General Criteria for Altered Oligomer Design

As a result of these initial investigations, it appears that the following parameters may serve as a guide toward newer backbone link-

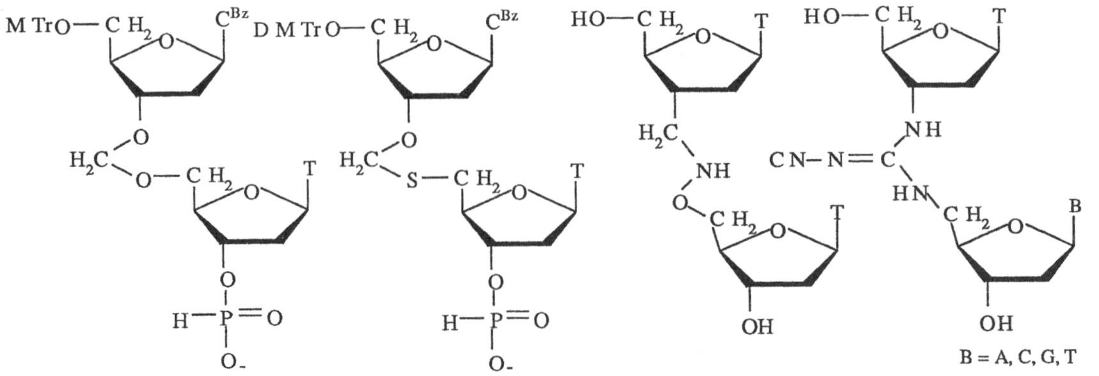

12. Formacetal Linkage

13. Thioformacetal Linkage

14. Methylhydroxylamine Linkage

15. N-Cyanoguanidine Linkage

B = A, C, G, T

Figure 4. Structures of achiral formacetal, thioformacetal, methylhydroxylamine, and *N*-cyanoguanidine linkages.

ages that can give rise to novel oligonucleotide analogues having enhanced stability with respect to nucleases and the capacity to form stable double and triple helices.

1. *Conformational flexibility:* four to six rotatable bonds.
2. *Charge:* negative charge may be inherently unfavorable, and thought should be given to neutral or positively charged alternatives.
3. *Hydration:* the internucleoside linkage should be strongly hydrated to ensure that a hydrophilic face is presented to the aqueous solution.
4. *Chemical stability:* the internucleoside linkages should be stable with respect to general base catalysis.
5. *Synthesis:* the synthetic chemistry should be readily adaptable to modern solid phase synthetic methods.

## 4 CONCLUSION

It may be premature to conclude that a natural phosphodiester chemical linkage is somehow unique with respect to duplex and triplex formation. On the other hand, it may be useful to individually evaluate the physical properties of the phosphodiester linkage, toward the end of establishing essential elements and including them in an alternative linkage that is phosphate free.

The introduction of achiral, nonionic, phosphate-free backbone analogues of oligonucleotides will probably circumvent the cost and stability limitations associated with the phosphate backbone analogues. A large number of the novel linkages reported to date are not adaptable for synthesis on solid support, however. In addition, attempts to synthesize dimers, to investigate their positional incorporation at sites in the oligomers, and to study their stability to nucleases and their binding characteristics with natural nucleic acids only partially address the main question. In this category the amino acid derivatives, oligonucleosides with semirigid structure or flexible polyamides with appropriate charge and hydration characteristics, which exploit in their synthesis the well-established solid phase peptide chemistry, stand apart from other candidate compounds. The pioneering studies by Peter Nielsen and colleagues in Denmark reinforce the general importance of the neutral polyamide nucleic acid approach and point the way toward significant additional changes that may be required to improve these polymers for use as diagnostic and pharmaceutical reagents. The cell uptake behavior and the stability to peptidases still need to be addressed for this promising class of compounds. Such phosphate-free backbone analogues with favorable charge structure for better binding properties would probably replace existing reagents bearing the natural phosphodiester linkages not only for therapeutic purposes but even for diagnostic applications.

*See also* ANTISENSE OLIGONUCLEOTIDES, STRUCTURE AND FUNCTION OF; PEPTIDES, SYNTHETIC.

### *Bibliography*


Sanghvi, Y. S. and Cook, P. D. (1993) In *Nucleosides and Nucleotides as Antitumor and Antiviral Agents,* C. K. Chu, and D. C. Baker, Eds., pp. 309–322. Plenum Press, New York.

Uhlmann, E., and Peyman, A. (1993) Chapter 16 in *Methods in Molecular Biology,* S. Agrawal, Ed., pp. 355–390. Humana Press, Totowa, NJ.

Varma, R. S. (1993) *Synlett,* pp. 621–637.


# OLIGONUCLEOTIDES

*Fritz Eckstein, Olaf Heidenreich, Mabel Ng, and Tom Tuschl*

### *Key Words*

**Annealing** Formation of double stranded DNA or oligonucleotides by formation of Watson-Crick base pairs.

**Antisense** DNA strand in opposite direction to the original, normally the sense strand.

**Conformation** Forms in which the atoms in an oligonucleotide can be arranged without breaking any bonds.

**Diastereomers** Nucleotides which differ in the chirality or handedness at the phosphorus atom.

**H-Phosphonates** Monomeric units of nucleotides for the chemical synthesis of oligonucleotides, bearing an H-phosphonate group at the 3′-position.

**Modification** Chemical derivatization of an oligonucleotide.

**Nucleases** Enzymes which degrade nucleic acids.

**Phosphoramidites** Monomeric units of nucleotides for the chemical synthesis of oligonucleotides, bearing a phosphoamidite group at the 3′-position.

**Polymerases** Enzymes which synthesize nucleic acids from nucleoside 5′-triphosphates.

**Primer** Oligonucleotide which is used by the polymerase to start the polymerisation reaction.

**Recognition** Complementarity according to Watson-Crick base pairing of nucleotides.

**Ribozyme** Catalytic RNA.

**Stability** Term to describe the robustness of nucleic acids against denaturation by temperature or degradation by nucleases.

Oligonucleotides are fascinating because they represent small segments of DNA or RNA. Such short pieces of nucleic acids, commonly 15 to 20 nucleotides long, are more amenable to analysis at the atomic level than their polymeric counterparts. They lend themselves to the search for detailed answers to some fundamental questions in our understanding of the structure and function of nucleic acids. The availability of oligodeoxyribonucleotides and, more recently, of oligoribonucleotides through automated chemical synthesis, has revolutionized research in the nucleic acid field.

For example, the availability of short fragments of DNA in the form of oligodeoxyribonucleotides has permitted crystallization for X-ray structural analysis, thus offering insight into the atomic detail of the A, B, and Z conformations of DNA. The use of oligonucleotides as primers is indispensable in the sequencing of DNA by the Sanger method. These fragments also offer the unique opportunity to interfere, by the so-called antisense methodology, with the expression of genetic information on the basis of their complementarity with a chosen RNA or DNA target sequence. This application is specifically covered under Antisense Nucleotides, Structure and Function, in this volume. The genetic information can also be manipulated in a predictable manner by oligonucleotide-directed mutagenesis, where oligodeoxyribonucleotides serve as mismatch primers. The recently developed polymerase chain reaction has extended the scope of application for oligonucleotides (see PCR Technology, in this volume). Space permits only a brief description here of the more general applications of oligonucleotides. An excellent introduction into nucleic acids, with an emphasis on the chemical aspects, is provided in a textbook. More specialized reviews on oligonucleotides are also available.

# 1 OLIGODEOXYRIBONUCLEOTIDES

## 1.1 Synthesis

The preparation of oligodeoxyribonucleotides is nowadays carried out almost exclusively by automated chemical synthesis on polymer support, mainly using the phosphoramidite approach (Figure 1). This versatile method is also most often used for the synthesis of oligodeoxyribonucleotides containing modified nucleotidic units. Experimental details for such syntheses are provided in the IRL Press's *Practical Approach* series. An alternative but less often used route of synthesis is that by the H-phosphonate method. In a more restricted way, the DNA-dependent reaction, which is catalyzed by DNA polymerase, can be used to specifically extend the sequence of an oligodeoxynucleotide primer annealed to a template.

## 1.2 Structural Variations

### 1.2.1 Base Modifications

Structural changes of the nucleobases have the special appeal of permitting the identification of important functional groups in their interactions with other molecules, such as proteins, or with other nucleic acids, to form higher ordered structures. Many of these syntheses require special protecting groups for the bases. Examples are the syntheses of oligonucleotides containing 6-thiodeoxyguanosine, $O^6$-alkyl-deoxyguanosine, 4-thiothymidine, 2-thiothymidine, $O^4$-alkylthymidine, 2-pyrimidinone, 5-methyl-2-pyrimidinone and 5-methyl-4-pyrimidinone deoxynucleosides, purine deoxynucleoside, and 2,6-diaminopurine deoxynucleoside. Oligonucleotides containing these or other bases have been used to investigate the functional groups essential for the recognition by proteins. This can be done by systematic changes in the possible hydrogen bonding pattern by deletion, modification, functional group reversal, and functional group displacement. A typical example is the study with the restriction enzyme *Eco*RV, which investigated the influence of dA and dT analogues on the catalytic activity of the enzyme.

Base analogues may also be incorporated as a precursor nucleotide bearing a functional group, which permits reaction with alkylamines or aminoalkylthiols after assembly of the oligonucleotide. Two examples of such applications are the chemical cross-linking of two strands of DNA and the coupling of DNA to polymer support for affinity chromatography of DNA-binding proteins.

The 5 position of pyrimidines is traditionally favored for substitution. Examples of oligonucleotides containing such substituents include those carrying a spin label, fluorescein, pyrene, or biotin as a probe for structural and dynamic studies. Illustrative examples for the application of such analogues entail the use of template–primer oligonucleotides bearing a fluorescent label for distance measurements by fluorescence energy transfer in the complex with *Echerichia coli* DNA polymerase I and of an oligonucleotide carrying an azido group for photo-cross-linking with the same enzyme.

### 1.2.2 Phosphate Modifications

**Phosphorothioates** Considerable effort has been expended in modifying the phosphate groups of the internucleotidic linkage in oligodeoxyribonucleotides. The general aims were to increase the stability of this group against hydrolysis by nucleases, to nullify the negative charge for improved entry into cells, and to enhance the reactivity of this group toward further chemical reactions. One of the most versatile modifications has been the change of the phosphate to a phosphorothioate group. Although first introduced into DNA enzymatically, the phosphoramidite method has facilitated the introduction of phosphorothioates into oligodeoxyribonucleotides by reaction of the intermediate phosphite triester with sulfurizing reagents (step 5 in Figure 1). These syntheses result in the formation of a pair of diastereomers, so that for $n$ phosphorothioates, $2^n$ diastereomers will result. Phosphorothioate oligonucleotides have been invaluable in determining the stereochemical course of the enzyme-catalyzed cleavage of internucleotidic bonds. Representative examples for this application are the studies with the restriction enzymes *Eco*RI and more recently, *Eco*RV, which show that no enzyme intermediate is involved in these reactions.

Oligonucleotides containing phosphorothioates show considerable reduction in the rate of hydrolysis by a variety of nucleases. This stability has made them particularly attractive in the application as antisense oligonucleotides (see Antisense Nucleotides, Structure and Function, this volume). The chemical reactivity of sulfur in phosphorothioates has been exploited for alkylation reactions. Reaction with iodoethanol results in the cleavage of the phosphorothioate diester to a small extent, providing a means of identifying positions bearing this particular group. This reaction can be exploited for sequencing such DNA and for site-specific labeling of oligonucleotides.

In an effort to circumvent the problem of diastereomers, the second nonbridging oxygen of the internucleotidic phosphate group has also been replaced by sulfur, resulting in phosphorodithioates which, of course, do not retain a chiral phosphorus. These analogues show promise for application in the antisense field.

**Other Phosphate Analogues** Methylphosphonates are among the oldest analogues used. Their lack of charge has attracted much interest in the antisense technology and under that entry in this volume. Phosphoramidate internucleotidic linkages in oligonucleotides have also attracted much attention, particularly inasmuch as they offer the opportunity for further derivatization by way of *N*-alkylphosphoramidates.

Much imagination has gone into the design of internucleotidic linkages devoid of phosphorus, examples of which are dealt with by Uhlmann and Peyman (1990).

### 1.2.3 Sugar Modifications

Only a few of the great number of sugar-modified oligonucleotides that have been prepared are cited here. Oligonucleotides, in which the glycosidic linkage has the α- rather than the usual β-configuration, are stable against nucleases.

Modifications at the 3′ and 5′ positions of the ribose are of particular interest because they become part of the sugar phosphate

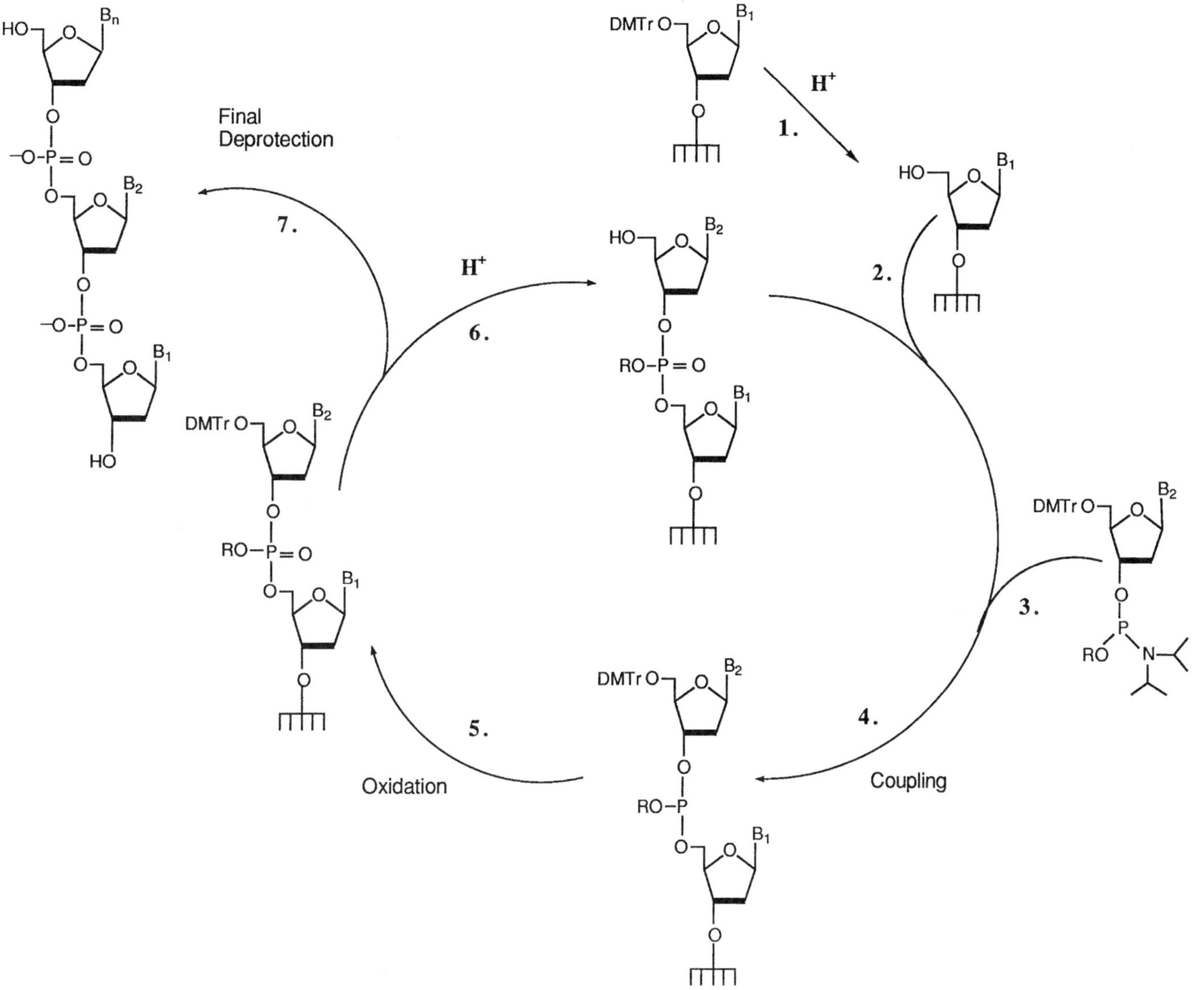

Figure 1. Scheme for oligonucleotide synthesis. The cycle is depicted for the synthesis of oligodeoxyribonucleotides but is essentially the same for the synthesis of oligoribonucleotides, where a protected 2′-hydroxy group is present. Steps 1 and 2 initiate the synthesis; steps 3 to 6 represent the cycle; step 7 is the final deprotection step after completion of the desired number of cycles. B, protected nucleobase; —R, —$CH_2CH_2CN$; DMTr-, dimethoxytrityl-; the first nucleoside, $B_1$, is attached to a polymer support.

backbone. Thus groups or atoms can be introduced here to render an oligonucleotide susceptible to cleavage under mild conditions. Both 3′- and 5′-aminothymidine-containing oligonucleotides have been prepared and have been shown to be acid labile, permitting site-specific cleavage of such internucleotidic linkages. 3′-Thiothymidine in oligonucleotides has been shown to be cleavable by silver ions. However, when incorporated in the *Eco*RV recognition site, the oligonucleotide is resistant to enzyme-catalyzed cleavage.

The 2′ position has attracted much interest for modification because it offers a way to probe the role of this position in many enzyme-catalyzed reactions. 2′-Fluorothymidine has been introduced into oligonucleotides encompassing the restriction site for the restriction endonuclease *Eco*RV. This modification proved to have little effect on the cleavage by this enzyme. There is also little effect in the restriction enzyme catalyzed cleavage of an oligonucleotide that contains 2′,2′-difluoro-2′-deoxycytidine. These results are significant in formulating a mechanism for these restriction enzymes. Further examples for 2′-modified oligonucleotides are cited in Section 3.

## 2 OLIGORIBONUCLEOTIDES

### 2.1 General Aspects

The recent discovery that RNAs can possess catalytic activities has spurred great interest in oligoribonucleotides. A special issue of the *FASEB Journal* is devoted to this subject. A recent review contains additional material.

The specificity of interaction of RNAs with proteins such as the complex of tRNAs with aminoacyl synthetases or the transcriptional complex of the TAT/TAR system has also compounded the interest in oligoribonucleotides. Arising from these interests, questions of secondary structures such as loops and pseudoknots that can be adopted by RNA have come to the fore and have added to the excitement of oligoribonucleotide research (see RNA Secondary Structures, this volume). Furthermore, oligoribonucleotide libraries have recently been used to screen for species with novel biochemical properties.

### 2.2 Synthesis

#### 2.2.1 Chemical Synthesis

The procedures for oligoribonucleotide syntheses are also based on phosphoramidite chemistry and follow the general scheme depicted in Figure 1. However, they are not as well refined as those for oligodeoxyribonucleotides. The favored method involves monomers with a trialkylsilyl-protected 2′-hydroxy group and acylated groups for the exocyclic groups on the bases. However, coupling efficiencies are lower than those for the synthesis of oligodeoxyribonucleotides. Also, during removal of these acyl groups with ammonia, the silyl groups seem also to be at least partially cleaved. This facilitates base-catalyzed hydrolysis and chain cleavage, resulting in lower yields. At present, it seems that purification of the final product by gel electrophoresis is mandatory, although alternatives employing high performance liquid chromatography may eventually replace this step.

#### 2.2.2 Enzymatic Synthesis

In contrast to the synthesis of oligodeoxyribonucleotides, an alternative to the chemical synthesis exists for the preparation of oligoribonucleotides. This approach relies on the transcription of a defined DNA template by DNA-dependant RNA polymerases such as T7 and SP6 polymerases. This is the method of choice for RNAs more than 50 nucleotides long and is routinely used for the preparation of tRNA transcripts and ribozymes other than the hammerhead ribozyme.

### 2.3 Structural Variations

#### 2.3.1 Base Modifications

The hammerhead ribozyme has probably attracted most interest from synthetic chemists, since it can be chemically synthesized, thus permitting the incorporation of modified nucleotides at predefined positions. To understand the structure–function relationship of this catalytically active RNA, modifications have been placed, in particular, at the invariant regions. Many of these base modifications have caused a large decrease in efficiency of the cleavage reaction, indicating a crucial role of certain exocyclic groups for catalysis.

#### 2.3.2 Sugar Modifications

The incorporation of sugar-modified nucleotides into oligoribonucleotides has so far been restricted to 2′ modifications. The simplest change is obviously that from a ribo- to a 2′-deoxyribonucleotide. Indeed, mixed oligonucleotides containing these two sugars have been prepared for hammerhead ribozymes. Hammerhead ribozymes with 2′-fluoro, 2′-aminonucleotides and 2′-*O*-alkyl substituents have also been prepared. These modifications were mainly aimed at determining the necessity of the presence of 2′-hydroxyl groups at particular positions for catalytic activity. As expected, the presence of such substituents also resulted in increased stability against degradation by nucleases.

#### 2.3.3 Phosphate Modifications

Phosphate modifications have so far been limited to the phosphorothioates. These have been incorporated chemically as well as enzymatically at the position of cleavage, to determine the stereochemical course of the reactions catalyzed by the hammerhead and the *Tetrahymena* ribozymes. They have also been incorporated at the 3′ terminus to increase stability against degradation by nucleases.

## 3 CONCLUSION

This brief overview of oligonucleotides illustrates the potential underlying the use of oligonucleotides in yielding valuable information, not otherwise available, on structure–function details of DNA and RNA.

*See also* Antisense Oligonucleotides, Structure and Function of; DNA Structure; Oligonucleotide Analogues, Phosphate-Free.

### *Bibliography*

Allen, D. J., and Benkovic, S. J. Resonance energy transfer measurements between substrate binding sites within the large (Klenow) fragment of *E. coli* DNA polymerase I. *Biochemistry,* 28:9586–9593 (1989).

Blackburn, G. M., and Gait, M. J. *Nucleic Acids in Chemistry and Biology.* IRL Press at Oxford University Press, Oxford, New York, and Tokyo. 1990.

Caruthers, M. H., Beaton, G., Wu, J. V., and Wiesler, W. *Chemical synthesis of deoxynucleotides and deoxynucleotide analogs. Methods Enzymol.* 211:3–20 (1992).

Catalano, C. E., Dwayne, J. A., and Benkovic, S. J. Interaction of *E. coli* DNA polymerase I with azidoDNA and fluorescent DNA probes: Identification of protein–DNA contacts. *Biochemistry,* 29:3612–3621 (1990).

Cowart, M., and Benkovic, S. J. A novel combined chemical–enzymatic synthesis of cross-linked DNA using a nucleoside triphosphate analogue. *Biochemistry,* 30:788–796 (1991).

Eckstein, F. Nucleoside phosphorothioates. *Annu. Rev. Biochem.* 54:367–402 (1985).

Eckstein, F., Ed. *Oligonucleotides and Analogues, A Practical Approach.* IRL Press at Oxford University Press, Oxford, New York, and Tokyo, 1991.

Eckstein, F., and Gish, G. Phosphorothioates in molecular biology. *Trends Biochem. Sci.* 97–100 (1989).

Englisch, U., and Gauss, D. H. Chemically modified oligonucleotides as probes and inhibitors. *Angew. Chem. Int. Engl. Ed.* 30:613–629 (1991).

*FASEB J.* The new age of RNA. Vol. 7 (1993).

Grasby, J. A., and Connolly, B. A. Stereochemical course of the hydrolysis reaction catalyzed by the *Eco*RV restriction endonuclease. *Biochemistry,* 31:7855–7861. (1992).

Gryanov, S. M., and Letsinger, R. L. Synthesis and properties of oligonucleotides containing aminodeoxythymidine units. *Nucleic Acids Res.* 20:3403–3409 (1992).

Larson, C. J., and Verdine, G. L. A high-capacity column for affinity purification of sequence-specific DNA-binding proteins. *Nucleic Acids Res.* 20:3525 (1992).

McPherson, M. J., Ed. *Directed Mutagenesis, A Practical Approach.* IRL Press at Oxford University Press, Oxford, New York, and Tokyo, 1991.

Newman, P. C., Williams, D. M., Cosstick, R., Seela, F., and Connolly, B. A. Interaction of the *Eco*RV restriction endonuclease with deoxyadenosine and thymidine bases in its recognition hexamer d(GATATC). *Biochemistry,* 29:9902–9910 (1990).

Ozaki, H., and McLaughlin, L. W. The estimation of distances between specific backbone-labeled sites in DNA using fluorescence resonance energy transfer. *Nucleic Acids Res.* 20:5205–5214 (1992).
Shaw, J.-P., Milligan, J. F., Krawczyk, S. H., and Matteuci, M. Specific, high-efficiency, triple-helix-mediated cross-linking to duplex DNA. *J. Am. Chem. Soc.* 113:7765–7766 (1991).
Szostak, J. W. In vitro genetics. *Trends Biochem. Sci.* 17:89–93 (1992).
Uhlmann, E., and Peyman, A. Antisense oligonucleotides: A new therapeutic principle. *Chem. Rev.* 90:543–584 (1990).
Usman, N., and Cedergren, R. Exploiting the chemical synthesis of RNA. *Trends Biochem. Sci.* 17:334–339 (1992).
Vyle, J. S., Connolly, B. A., Kemp, D., and Cosstick, R. Sequence- and strand-specific cleavage in oligonucleotides and DNA containing 3′-thiothymidine. *Biochemistry,* 31:3012–3018 (1992).
Williams, D. M., Benseler, F., and Eckstein, F. Properties of 2′-fluorothymidine-containing oligonucleotides: Interaction with restriction endonuclease *Eco*RV. *Biochemistry,* 30:4001–4009 (1991).

## Oligonucleotides, Antisense: *see* Antisense Oligonucleotides, Structure and Function of.

# Oncogenes

*Anthony Byrne and Desmond N. Carney*

*Key Words*

**Oncogene** A gene capable of inducing one or more characteristics of cancer cells.

**Transduction** Genetic recombination occurring in bacteria in which DNA from a donor cell is transferred to a recipient cell by means of a virus.

**Transformation** Conversion of a normal cell in culture to a cell displaying one or more of the characteristics of a tumor cell.

**Tumor Suppressor Gene** A gene that inhibits tumor development.

Oncogenes, by definition, are genes capable of achieving the conversion of a normal cell to a malignant cell, such as that resulting from infection of normal cells by oncogenic viruses. This concept was first established to explain the phenomenon of acute tumor induction by retroviruses. Modern research has expanded our understanding beyond this transformation, and techniques such as gene amplification, gene transfer, and gene knockout have revealed other potential modes. Thus the term "oncogene" now is applied to any gene contributing to cell transformation, regardless of the presence or absence of a comparable sequence in an oncogenic retrovirus.

## 1 INTRODUCTION

Oncogenes, by definition, are genes capable of achieving cell transformation. The concept was first established to explain the phenomenon of acute tumor induction in avian and rodent models by retroviruses. The term "viral oncogene" thus described genomic material in RNA viruses which effected the production and/or maintenance of a malignant state in animals. Normally RNA viruses encode for only three genes—*gag, pol, env*—with the acquisition of a fourth endowing it with transforming abilities (see Figure 1). Such additions have been shown to share common sequences in many instances with cellular genes. Further exploration of this phenomenon has shown that at some point in evolution the "v-*onc*" genetic material actually originated in the eukaryotic cells of the host being hijacked by, and incorporated into, the infectious agent.

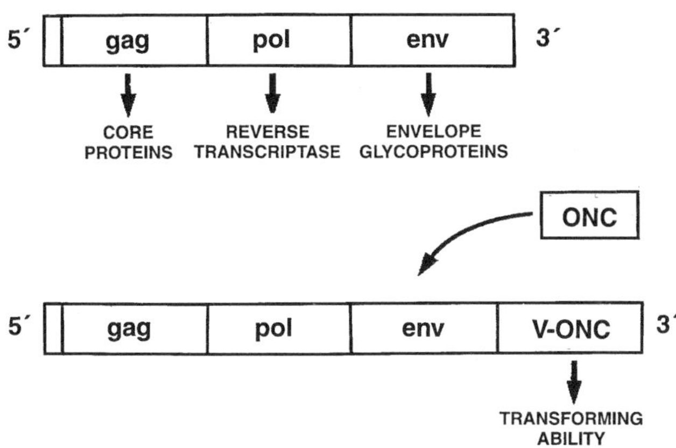


**Figure 1.** Alteration of the retroviral genome.

The origin of viral oncogenes just outlined suggests, ab facto, that there is an innate potential in normal cells for uncontrolled growth in appropriate circumstances. Such an origin defines the presence of precursors—or proto-oncogenes—which, if deregulated, can cause transformation. It also implies that under routine circumstances, such elements play a key role in the intricate cycle of normal cellular proliferation and differentiation. In fact, such material is phyllogenetically ancient, being found in all vertebrate cells examined to date. To clearly differentiate it from the viral counterpart, standard nomenclature designates viral oncogenes as "v-*onc*" and cellular proto-oncogenes "c-*onc*."

Transduction and transformation by retroviruses therefore unveiled for the first time powerful molecular forces capable of cell distortion and proliferation. However, diverse lines of inquiry have expanded our understanding beyond the stereotypical "v-*onc*" template. Elegant developments such as gene amplification, gene transfer, and gene knockout have revealed other potential targets. With this knowledge has come a broadening of the definition of "oncogene." The term now is applied to any gene contributing to cell transformation, regardless of the presence or absence of a comparable sequence in an oncogenic retrovirus.

## 2 ACTIVATION

There are several pathways by which normally functioning proto-oncogenes can be mutated into oncogenic alleles (see Table 1).

### 2.1 Structural Alteration

Structural alteration, leading to an abnormal product, can occur in a number of ways. A *point mutation* in codon 12 of the c-H-*ras* gene of a human bladder cancer cell line was one of the first molecular changes defined as causing oncogenic transformation. Several other point mutations (e.g., at positions 13,61) of the *ras* gene family have also been linked to carcinogenesis. The same

Table 1 Mechanisms of Oncogene Activation with Selected Examples

| Mechanism | Examples |
|---|---|
| Structural alteration | |
| Point mutations | H-*ras*; c-*fms* |
| Translocations | c-*abl*; *bcl*-2 |
| Truncations | pp60*src* |
| Amplification | L-*myc*; *hst* |
| Dysregulation | c-*myc*; *bcl*-2 |

mechanism has been invoked for transformation of other proto-oncogenes such as *gip* and *gsp*—which are associated with carcinoma of the ovary and thyroid, respectively.

Fusion of genes, with loss of part of one, can also result in protein products that have altered function or kinetics. This may occur through *translocation* of a gene from one chromosome to another or to another position on the same chromosome. The classic example is that of the Philadelphia chromosome in chronic myeloid leukemia. Here, breakpoints through the c-*abl* locus on chromosome 9 (q24) and break cluster region (bcr) on chromosome 22 allow splicing of the c-*abl* gene onto the *bcr* gene on 22q11. This results in loss of the first two or three exons of c-*abl,* with an *abl/bcr* transcript read out as a fusion protein (of 210 kDa) with altered tyrosine kinase activity. A variant involving a more 5′ portion of *bcr* and resulting in a 195 kDa protein is associated with acute lymphoblastic activity. Many other translocations have been described. For example, juxtaposition of c-*myc* (from 8q24) with the immunoglobulin enhancer gene on 14q32 results in inappropriate *myc* expression and a proliferative advantage causing Burkitt's lymphoma. More than 80% of follicular low grade lymphomas have a t(14-18) (q32-q21) translocation, which causes abnormal expression of *Bcl*-2, resulting in suppression of normal programmed cell death.

A third event that may cause structural change in oncoproteins is deletion of a coding sequence causing truncation of the gene product. An example is pp60src. Normally phosphorylation of a tyrosine residue (tyr527) near the carboxy terminus regulates the activity of this protein tyrosine kinase. However, truncation of part of the carboxy terminus with loss of the tyrosine increases the enzymatic and transforming activity of the mutant.

### 2.2 Amplification and Dysregulation

An effect comparable to alteration of product structure can be achieved by simply increasing the amount of structurally normal protein. This process of *amplification* is seen, for example, with overexpression of c-*myc* in carcinoma of the lung and breast, n-*myc* in neuroblastoma, and *ErbB2* in adenocarcinoma of stomach, ovary and breast.

Perversion of regulatory factors that normally exert a positive or negative influence on proto-oncogene expression leads to similar results. Such *dysregulation* was hinted at earlier in the cases of c-*myc* and *Bcl*-2, which gain novel transcriptional promoters following translocation close to immunoglobulin genes by coming under the control of the latter's enhancers.

## 3 CLASSIFICATION

As understanding of the processes resulting in, and arising from, oncogenesis has expanded, the task of classifying oncogenes has become more complex. For example, they may be rostered according to location (cytoplasmic vs. nuclear) or on the basis of cooperativity. The latter system arose from the observation that in many in vitro systems, combinations of oncogenes could achieve transformation whereas individual components were impotent alone. For instance, rat embryo fibroblast experiments showed that cytoplasmic *ras* or nuclear *myc* oncogenes were incapable of inducing full transformation if acting alone. However, coincubation of the two achieved the desired effect. Many other examples of such cooperation exist, enhancing the theory that nuclear oncoproteins may immortalize cells but need the assistance of cytoplasmic counterparts to lead to ligand- and anchorage-independent growth, and vice versa.

However, there are exceptions to the possibilities just noted, and perhaps the most self-explanatory classification of oncogenes is based on categorization according to activity of their encoded proteins. Four main classes may be identified: growth factors, transmembrane (growth factor) receptors, transduction factors, and nuclear transcription factors (see Table 2). This system facilitates the identification of the nature and role of oncogenes in initiating, collaborating during, and achieving full cell transformation.

## 4 FUNCTION OF ONCOGENE PRODUCTS

Thus altered alleles may produce proteins that differ from standards in terms of structure, quantity, or temporal expression. Examples of oncoprotein function in Table 2 indicate how individual oncogene activity at the membrane, cytoplasmic, or nuclear level may disrupt normal cell function. This information also implies how cooperation may be important, with prevention of terminal differentiation by one oncogene, for example, allowing the proliferative action of another. Much of such theorizing describes "gain of function," with dominance not simply in a Mendelian sense but with respect to the mechanics of the oncogenes' molecular action. However, loss of function is equally important. Part of the inherent elegance of the cellular genome and part of its defense against disordered regulation is the presence of genes that suppress abnormal activity. Such *antioncogenes* or *tumor suppressor genes* are essential in allowing cells to correct genomic damage prior to progressing in their cycle. Their presence was first indicated in experiments that showed how somatic hybridization of normal with malignant cells inhibits proliferation. Genes such as the retinoblastoma (*Rb,*) p53, and multiple tumor suppressor 1 (*MTS1*) genes have been shown to be of key importance in many tumors, with loss or mutation of the wild type allowing unrestrained growth.

An example of how gain of function, temporally inappropriate expression, and deletion of inhibitory influences may lead to oncogenic transformation is reflected by changes in core genes which intimately affect cell dynamism (see Figure 2). Cyclins are a group of nuclear proteins that show cyclical variation (hence the name) in concentration during the cell cycle. This temporal pattern is essential in allowing appropriate activation of cyclin-dependent kinases (Cdks), crucial stimulants for propulsion of the cell through its cycle. Recent evidence shows that at least some of the cyclin genes are mutable proto-oncogenes. For example, the cyclin D1 oncogene is located on 11q13. Overexpression of its oncoproteins leads to the overstimulation of Cdk and the transition of cells from $G_1$ to S phase, hence to growth-factor-independent cell division. Cyclin D1 has been implicated in parathyroid adenomas, breast and esophageal carcinomas, and B-cell lymphomas. Cyclin E has

**Table 2** Classification of Oncogenes According to Function of the Product[a]

| Class | Oncogene | Product Function | Associated Tumors |
|---|---|---|---|
| Class I Growth Factors | | | |
| | *sis* | PDGFB chain growth factor | Astrocytoma |
| | *hst* | FGF-related growth factor | Kaposi's sarcoma |
| Class II Transmembrane Receptors | | | |
| Receptor protein tyrosine kinases (PTK) | *erb B* | Truncated EGF receptor PTK | Breast/bladder/ovary |
| | *fms* | Mutant CSF-1 receptor PTK | Acute leukemia (ANLL) |
| Receptor without PTK activity | *mas* | Angiotensinlike receptor | ? |
| Class III Transduction Factors | | | |
| Membrane linked | *abl* | Nonreceptor PTK | Leukemia (CML) |
| | *ras* | GTP binding GTPase | Lung/colon |
| Cytoplasmic | *mos* | Protein serine kinase | Mixed parotid tumor |
| Class IV Nuclear Transcription Factors | | | |
| | *myc* | DNA binding protein | Burkitt's/lung/breast |
| | *fos* | Combine with jun to form AP-1 | ? |
| | *jun* | Combine with fos to form AP-1 | ? |

[a]*Abbreviations:* PDFGB, platelet-derived growth factor β; FGF, fibroblast growth factor; EGF, epidermal growth factor; ANLL, acute non-lymphocytic leukemia; CML, chronic myeloid leukemia; AP-1, activating protein-1.

been similarly implicated in breast cancer, and indeed D1 has now been identified as being identical to the previously described *bcl-1* oncogene.

The presence of such gain-of-function mutations at the core of cell cycle regulation suggests that Cdk inhibitors or the kinases themselves may also be likely targets. Such inhibitors have been identified. Expression of one, *WAF1* (also termed *Cip1* and *sdi1*), gives the clearest indication to date of how p53 arrests cell growth and how its absence may allow cell transformation. *WAF1* (wild-type p53 activated fragment 1) encodes a 21 kDa protein that inhibits Cdk2, among other enzymes. Critically, *WAF1* expression is induced by a p53 product, so that normal p53 function is required for control of some crucial kinases.

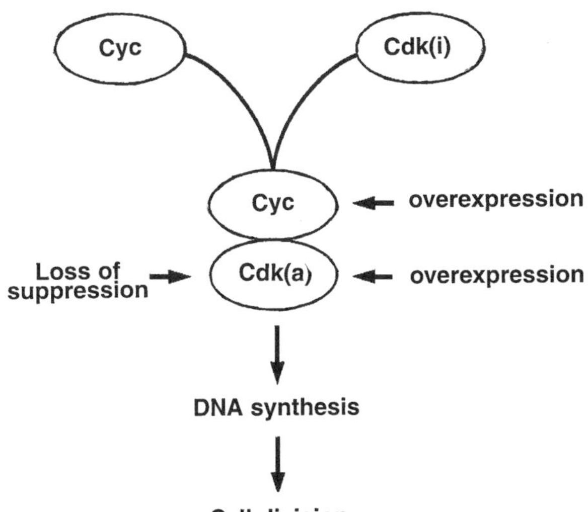


**Figure 2.** Cyclins (cyc) activate cyclin dependent kinases [Cdk(i) = inactive, Cdk(a) = active], which promote DNA synthesis and commit the cell to division. Arrows indicate sites of oncogene activity.

Deletions of a Cdk4 inhibitor gene—*MTS1* (on 9p21)—have been linked to several human cancers including lung cancer, gliomas, leukemias, and malignant melanoma. Indeed, amplification on chromosome 12 of the *Cdk4* gene itself can occur with constitutive kinase expression free of the normal inhibitory actions of transforming growth factor β.

## 5 THERAPEUTIC POTENTIAL

The burgeoning knowledge of oncogenes and proto-oncogenes inevitably opens therapeutic windows. The use of antisense oligonucleotides complementary to base pairs of functional targets is alluring, and promises greater specificity and higher response rates than conventional modalities. In practice, producing molecules that are stable, easily synthesized, and predictable in their behavior has proved difficult, but phase 1 trials commenced in the early 1990s.

## 6 SUMMARY AND FUTURE PERSPECTIVES

Alteration or loss of genetic material playing a key role in the normal cell cycle is fundamental to carcinogenesis. Such mutated molecules, or oncogenes, have been identified at many levels of growth control, even at the nuclear hub of cell cycle control; they include stimulatory ligands, their receptors, and intracellular messengers. Cooperation between these variants allows sufficient autonomy for tumorigenesis or at least provides the basis for secondary changes to achieve transformation. Loss of inhibitors that normally "brake" the cell cycle is of equal importance. Again, loss of a suppressor gene combined with gain of an oncogene has ominous consequences.

Accrual of this knowledge has broad implications for the identification of groups at risk of cancer, for early diagnosis, and for prognostication. Equally important, it provides the template with which to design new therapeutic approaches against such an insidious range of fatal diseases.

*See also* CANCER; COLON CANCER, MOLECULAR BIOLOGY OF; LIVER CANCER, MOLECULAR BIOLOGY OF; LUNG CANCER, MOLECULAR BIOLOGY OF; RETINOBLASTOMA, MOLECULAR GENETICS OF; TUMOR SUPPRESSOR GENES.

### *Bibliography*

Bayever, E., Iversen, P., Smith, L., et al. (1992) Guest editorial: Systemic human antisense therapy begins. *Antisense Res. Rev.* 2:109–110.

Bignami, M., Rosa, S., La Rocca, S. A., et al. (1988) Differential influence of adjacent normal cells on the proliferation of mammalian cells transformed by the viral oncogenes *myc, ras* and *src. Oncogene,* 2:509–514.

Bishop, J. M. (1987) The molecular genetics of cancer. *Science,* 235:305–311.

Bishop, J. M. (1991) Molecular themes in oncogenesis (review). *Cell,* 64:235–248.

Bos, J. L. (1988) The *Ras* family and human carcinogenesis. *Mutat. Res.* 195:244–271.

Calabretta, B., Skorski, T., Szoczylik, C., et al. (1993) Prospects for gene-directed therapy with anti-sense oligonucleotides. *Cancer Treat. Rev.* 19:169–179.

Cartwright, C. A., Eckhart, W., Simon, S., et al. (1987) Cell transformation of pp60src mutated in the carboxy-terminal regulatory domain. *Cell,* 49:83–91.

Cleary, M. L., Smith, S. D., and Sklar, J. (1986) Cloning and structural analysis of cDNAs for *bcl-2* and hybrid *bcl-2*/immunoglobulin transcript resulting from the t(14:18) translocation. *Cell,* 47:19–28.

Ewen, M. E., Sluss, H. K., Whitehouse, L. L., et al. (1993) TGFB inhibition of Cdk4 synthesis is linked to cell cycle arrest. *Cell,* 74:1009–1020.

Harper, J. W., Adami, G. R., Wei, N., et al. (1993) The p21 cdk-interacting protein Cip 1 is a potent inhibitor of G-1 cyclin-dependent kinases. *Cell,* 75:805–816.

Herman, S. A., Heisterkamp, N., von Lindern, M., et al. (1986) Unique fusion of *bcr* and c-*abl* genes in Philadelphia chromosome positive acute lymphoblastic leukemia. *Cell,* 51:33–40.

Hinds, P. W., Dowdy, S. F., Eaton, E. N., et al. (1994) Function of a human cyclin gene as an oncogene. *Proc. Natl. Acad. Sci. U.S.A.* 91:709–713.

Hunter, T. (1991) Cooperation between oncogenes (review). *Cell,* 64:249–270.

Keyomarsi, K., O'Leary, N., Molnar, G., et al. (1994) Cyclin E, a potential prognostic marker for breast cancer. *Cancer Res.* 54:380–385.

Klein, G. (1993) Genes that can antagonise tumour development (overview). *FASEB J.* 7:821–825.

Kurzrick, R., Gutterman, J. V., and Talpez, M. (1988) The molecular genetics of Philadelphia chromosome positive leukemias. *New Engl. J. Med.* 319:990–998.

Land, H., Parada, L. F., and Weinberg, R. A. (1983) Tumorigenic conversion of primary embryo fibroblasts requires at least two cooperating oncogenes. *Nature,* 304:596–602.

Motokura, T., Bloom, T., Kim, H. G., et al. (1991) A novel cyclin encoded by *bel 1*-linked candidate oncogene. *Nature,* 350:512–515.

Nobori, T., Miura, K., Wu, D. J., et al. (1994) Deletions of the cyclin-dependent-kinase-4 inhibitor gene in multiple human cancers. *Nature,* 368:753–756.

Reddy, E. P., Reynolds, R. K., Santos, E., et al. (1982) A point mutation is responsible for the acquisition of transforming properties by the T24 human bladder carcinoma oncogene. *Nature,* 300:149–152.

Rous, P. (1911) A sarcoma of the fowl transmissible by an agent separable from the tumour cells. *J. Exp. Med.* 13:397–411.

Stein, C. A., and Chang, Y. C. (1993) Antisense oligonucleotides as therapeutic agents—Is the bullet really magical? *Science,* 261:1004–1012.

Taparowsky, E., Suard, Y., Frassaro, O., et al. (1982) Activation of T24 bladder carcinoma transforming gene is linked to a single amino acid change. *Nature,* 300:762–765.

Tsujimoto, Y., Finger, L. R., Yunis, J., et al. (1984) Cloning of the chromosome breakpoint of neoplastic B cells with the t(4:18) chromosome translocation. *Science,* 226:1097–1099.

# ORIGINS OF LIFE, MOLECULAR BIOLOGY OF

*James P. Ferris*

### *Key Words*

**Heterocyclic Compounds** Cyclic organic compounds that contain nitrogen, sulfur, and other heteroatoms in place of some of the carbon atoms.

**Interstellar Dust Clouds** Giant clouds, formed by the explosion of stars, which contain organic and inorganic dust and gas.

**Oligomer** A short polymer, usually two to ten monomer units.

**Template-Directed Synthesis** The synthesis of the complementary RNA or DNA chain by condensation of mononucleotide on a preformed RNA or DNA template. The sequence of the synthesized chain is determined by Watson–Crick hydrogen bonding between the mononucleotide and the complementary base on the template.

Scientific research on the origins of life is an investigation of the chemical pathways leading from simple organic compounds to an organization of catalytic and self-replicating polymers. The understanding of how this process may have proceeded on the early earth requires input from astronomy, biology, chemistry, geology and physics to provide information on the formation and environment of the early earth, the chemical processes that can lead to the formation of biological polymers, and the minimal requirements for the formation of a living system.

## 1 INTRODUCTION

### 1.1 NATURE OF THE EARLIEST LIFE

Contemporary life is the result of about 4 billion years of evolutionary development and inevitably differs appreciably from the first life on earth. There is little hope of detecting the original life forms now because the environment of the earth has changed and because the inefficient early life could not compete with the highly evolved life present on earth today. There is a small probability that life exists elsewhere in our solar system. No life was detected on Mars by the *Viking* landers, and there is only faint hope that life or remnants of earlier life forms may be detected by a more exhaustive search of that planet.

Fossils of microorganisms that are believed to have been present on the earth 3.5 billion years ago (bya) have been found in Africa and Australia. Exhaustive radioactive dating provides firm evidence for the age of the rock in which these structures are found. Analyses of thin sections of the rock in which these microfossils lie has revealed that they have filamentous structures similar to those of cyanobacteria (blue-green algae). The morphology does not establish the biochemistry of these life forms, but it does strongly suggest they were photosynthetic.

### 1.2 Environment of the Early Earth

Since relatively complex forms of life were present on earth 3.5 bya, life itself must have originated before that time. An origin of life 4 ± 0.2 bya is consistent with the accretion model for the formation of the solar system. In this model it is assumed that the solar system formed as the result of the condensation of an interstellar dust cloud. The latter stages of the formation of the planets involved impacts of very large bodies, as shown by the presence of extensive cratering on the moon, Mars, and Mercury. Analysis of the size and number of lunar craters suggests that the frequency of the impacts decreased exponentially in the 4.3–4.0 bya time period. The earth's crust would have cooled sufficiently at the end of the accretion period to permit the processes leading to the origins of life to take place.

The early rock record on the earth is fragmentary, but what is known is consistent with this scenario. The Isua supercrustal system of Greenland, which was formed about 3.8 bya, contains sedimentary rock, a finding consistent with the presence on the planet of liquid water at that time. The presence of liquid water means that the earth had cooled sufficiently for the origin and maintenance of life. In fact, controversial evidence for the presence of microfossils in the Isua formation suggests that life was present as early as 3.8 bya.

The geology of the Isua formation suggests that the temperature of early earth was generally in the range of 0–100°C. Less is known concerning the chemical composition of the atmosphere. Initially it was thought to be highly reducing, with large amounts of methane, ammonia, and even molecular hydrogen. Then it was recognized that methane and ammonia would have undergone rapid photochemical destruction by ultraviolet light. The current model for the primitive atmosphere has large amounts of molecular nitrogen and carbon dioxide and much smaller amounts of more reduced carbon and nitrogen compounds. This view is consistent with the present, largely carbon dioxide containing, atmospheres of Venus and Mars, the two planets closest to the earth in the solar system. The molecular oxygen in the earth's atmosphere now is a direct result of photosynthesis, hence the presence of life.

No compelling chemical data exist for the reaction pathways leading to the origins of life. Since biological evolution proceeds from simpler to more complex life, it is assumed that the chemical evolution leading to the first life also proceeded from simpler molecules to more complex biological polymers.

## 2 POSSIBLE SOURCES OF SIMPLE ORGANIC COMPOUNDS

### 2.1 The Earth's Atmosphere

Stanley Miller's production of amino acids by passing an electric discharge through a mixture of methane, ammonia, hydrogen, and water vapor in 1953 was a dramatic demonstration that it is possible to carry out laboratory studies of the origins of life. As noted in Section 1.2, the large proportions of methane, ammonia, and hydrogen present in the gas mixture are not consistent with photochemical models of the primitive atmosphere. Hydrogen cyanide (HCN) is the major product resulting from the action of an electric discharge on gas mixtures containing nitrogen gas and smaller amounts of methane or carbon monoxide. The lower yields of organic materials in these discharge experiments prompted the consideration of other sources of organic compounds.

### 2.2 Comets and Meteorites

Analysis of carbonaceous meteorites falling on the earth revealed the presence of biological molecules including amino acids, carboxylic acids, hydroxy acids, and some of the purine and pyrimidine bases present in nucleic acids. Carbonaceous material constitutes about 20% of the composition of Comet Halley. Consequently, it is believed that comets and meteorites may have delivered to the surface of the earth the organic materials needed for the origins of life.

The problem with this scenario is that the relative velocities of comets and meteors with respect to the earth is usually very high, and many of these bodies are pyrolyzed upon impact with the atmosphere. This is not a total loss, since the carbon and nitrogen compounds formed (e.g., acetylene, carbon monoxide, HCN) may have served as precursors for the biomolecules of life.

### 2.3 Hydrothermal Systems

Marine hydrothermal systems, with their spectacular ''black smokers,'' ''white smokers,'' and unique ecosystems, may have provided yet another environment in which the biomolecules required for the origins of life were formed. These vast systems of seawater circulate through the earth's crust, driven by the internal heat energy of the earth. Temperature ranges from 2°C to 350°C, pressures up to 20–30 mPa (200–300 bar); percolation through potential mineral catalysts and reducing conditions provide an environment that may have been conducive to the synthesis of the reduced compounds utilized for the origins of life. There have been no rigorous experimental tests of this proposal, and it will remain merely a theory until meaningful experimental studies have been performed.

### 2.4 Formation of Nucleic Acid Bases

Nucleic acid monomers (mononucleotides) are complex structures that do not form directly in electric discharges and have not been found in meteorites. The bases have been found in meteorites, and some of them are formed in HCN condensation reactions.

The condensation of HCN (Figure 1) proceeds in a stepwise fashion in aqueous solution to generate a tetramer, diaminomaleonitrile (DAMN). Further reaction of DAMN with HCN generates a mixture of low molecular weight oligomers. Hydrolytic decomposition of the oligomer yields adenine, orotic acid, and uracil. Alternatively, photolysis of DAMN generates 4-aminoimidazole-5-carbonitrile (AICN). Reaction with simple nitrile derivatives of AICN or its corresponding amide (AICA) results in the formation of an array of purine bases.

The facile formation of these heterocyclic compounds from HCN and their presence in meteorites suggests that these bases were available on the primitive earth. The facile synthesis of the heterocycles suggests that mononucleotide building blocks were also available on the primitive earth, but the steps involved in their formation are not as well understood.

### 2.5 Chirality

The molecules present in living systems consist of one or two possible mirror image forms called *enantiomers*. The L-amino acids and D-nucleotides are the basic building blocks of proteins

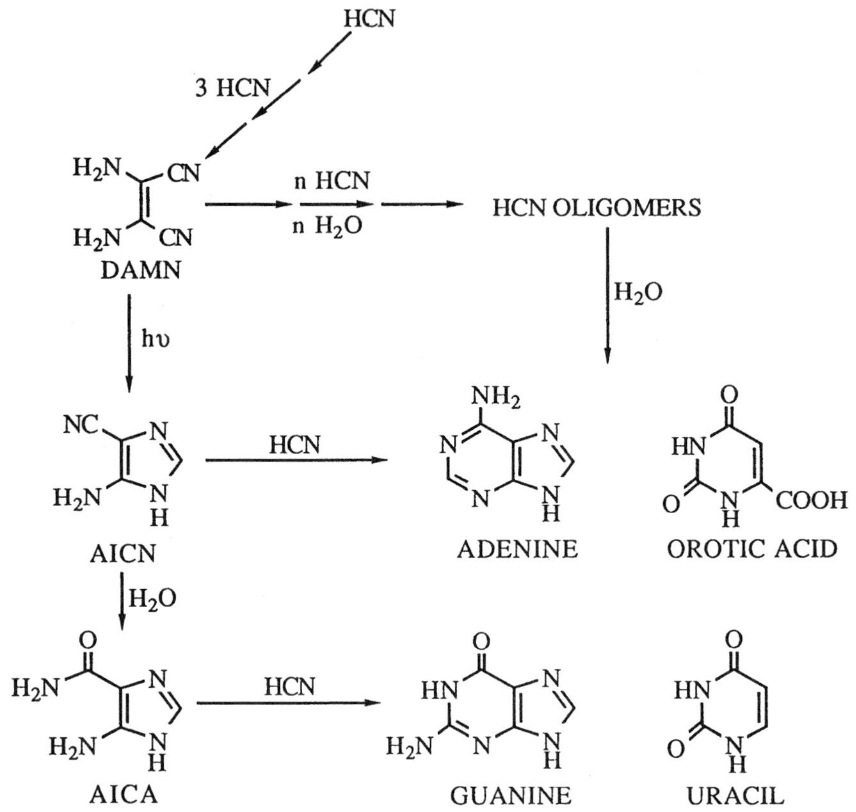

**Figure 1.** Steps in the proposed prebiotic syntheses of purines and pyrimidines from HCN.

and nucleic acids respectively. The mirror image D-amino acids and L-nucleotides rarely appear in contemporary life. It is not known why the enantiomers observed in living systems were selected initially.

## 3 PREBIOTIC FORMATION OF BIOPOLYMERS

The interactions of biopolymers with themselves and with small molecules are an essential part of the biochemistry of living systems. It is likely that biopolymers formed in the presence of water on the early earth, since water was the principal reaction solvent. Proteins and nucleic acids break down in water. Therefore, either special activation conditions existed, to permit these polymerization reactions to take place, or the reactions occurred in the absence of water.

### 3.1 Polypeptides

Initial studies on prebiotic polymerization focused on the formation of polypeptides from amino acids. Most of these early studies invoked the use of heat to drive the reaction under anhydrous reaction conditions. The product mixtures from these reactions contain thermal polymers with molecular weights as high at 4000. Amide (peptide) and other bonds linked these amino acids together. These materials were shown to possess some weak catalytic activity.

An alternative synthesis of polypeptides involves the condensation of aminoacyladenylates on montmorillonite to form the polypeptide adducts of 5′-adenylic acid [equation (1)]. This reaction proceeds in aqueous solution to form oligomers of up to 40 amino acids long. Although it is difficult to envisage the formation of appreciable amounts of such a high energy acyl anhydride in the

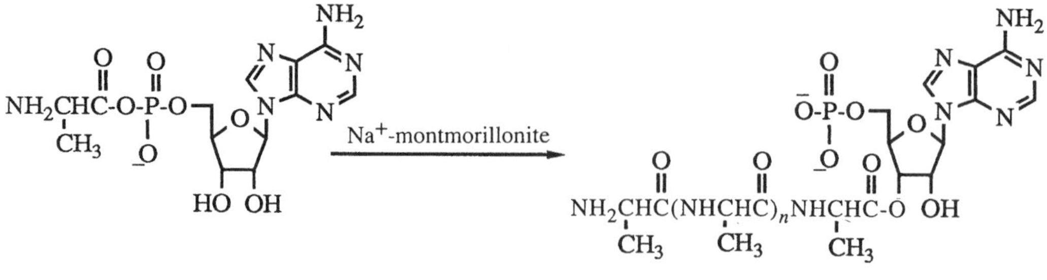

(1)

$n = -2$ to 16

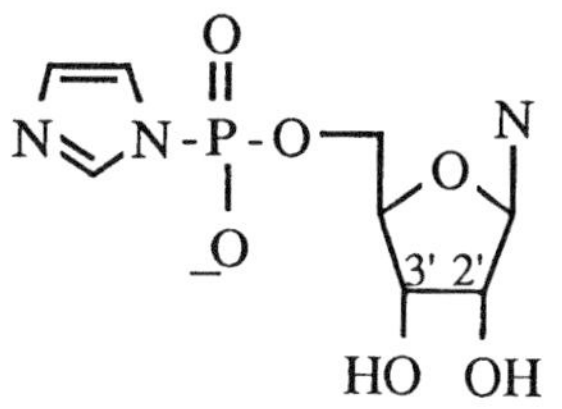

**Figure 2.** The 5′-phosphorimidazolide of nucleosides. N = adenine, guanine, cytosine, or uracil.

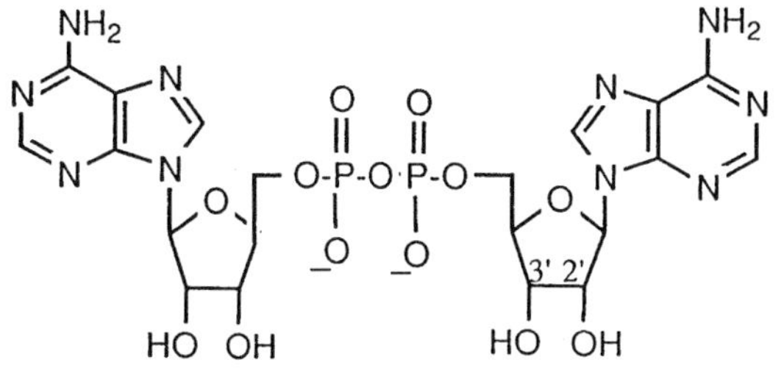

**Figure 3.** The dinucleoside pyrophosphate A$^{5'}$ppA.

presence of water, the experiment does suggest that other, less reactive amino acid derivatives may condense to polypeptides on mineral surfaces.

### 3.2 Polynucleotides

The polymerization of the 5′-phosphorimidazolide nucleosides (5′-ImpN) (Figure 2) proceeds in aqueous solution in the presence of catalysts. Oligomers as long as the hexamer linked by 2′,5′-phosphodiester bonds were observed in reactions of 5′-ImpN in aqueous buffer catalyzed by $Pb^{2+}$. A major synthetic advance was made using uranyl ion ($UO_2^{2+}$) as the catalyst. 2′,5′-Linked oligomers as long as the hexadecamer were formed at room temperature in water.

The synthesis of mainly 3′,5′-linked oligomers were observed in the reaction of 5-ImpA in aqueous solution in the presence of the clay mineral montmorillonite. Oligomers as long as the 10-mer were obtained which have a 2:1 ratio of 3′,5′ to 2′,5′ links. This ratio is increased to 4:1 when the dinucleoside pyrophosphate (A$^{5'}$ppA) (Figure 3) is added to the reaction mixture. The oligomers formed in this reaction contain A$^{5'}$ppA groupings.

The RNA oligomers formed in these reactions are probably not long enough to initiate the processes of replication and catalysis expected in simple living systems, but the research just mentioned does constitute a first step in that direction. In addition, these reactions underline the central role of catalysis in prebiotic reactions. Since protein enzyme and ribozyme catalysis is essential for the maintenance of contemporary life, it is likely that catalysis by metal ions and minerals was required for the formation of the biomolecules essential for the formation of the first life on earth.

## 4 POLYMER REACTIONS

The ultimate goal of origins of the research is the formation of polymers, which replicate and catalyze a sufficient number of reactions to maintain the viability of the reaction system. One model for such a system is the "RNA world" in which RNA oligomers catalyze their own formation and replication from simpler precursors in the absence of proteins. The replication process provides the means of information storage, which is not possible with proteins. The discovery of the limited catalytic action of RNA suggests that protein catalysis is not required. The absence of a requirement for protein makes the postulated "RNA world" much simpler than one requiring both RNA and protein.

### 4.1 Ribozymes

Ribozymes, RNAs that catalyze chemical reactions, are central to a scenario for primitive life on earth based solely on RNA. The possible role of ribozymes in the "RNA world" was demonstrated by the catalytic elongation of a pentamer of cytidylic acid to a series of oligomers as long as the 30-mer by a ribozyme obtained from the protozoan *Tetrahymena.* This experimental finding suggests that oligomers produced in prebiotic processes described in Section 3.2 may have been elongated by the ribozymes of the "RNA world."

### 4.2 Template-Directed Synthesis

The process of the nonenzymatic template-directed synthesis of complementary RNA chains is an important process of information storage and RNA proliferation in the proposed "RNA world." This process has been studied extensively by Leslie Orgel and coworkers, who found that the synthesis of oligomers of guanylic acid [oligo(G)s] proceed very efficiently on poly(C) templates using ImpG (Figure 3) and 2-methyl-ImpG as the starting materials. The reaction proceeds because the hydrogen-bonding and stacking interactions of the guanine bases on the poly(C) template orient the monomer units for reaction. Oligo(A)s are formed less effectively in the reaction of ImpA on poly(U). No reaction is observed in the reactions of the phosphorimidazolides of pyrimidines on purine templates. The failure to discover conditions for the template-directed synthesis of oligomers containing both purine and pyrimidine bases is an important deficiency in the theory of the "RNA world."

*See also* Chirality in Biology; Paleontology, Molecular; Ribozyme Chemistry.

## *Bibliography*

Bonner, W. A. (1991) The origin and amplification of biomolecular chirality. *Origins Life Evol. Biosphere,* 21:59–112.

Cech, T. R. (1986) RNA as an enzyme. *Sci. Am.* 255(5):64.

Day, W. (1984) *Genesis on Planet Earth,* 2nd ed. Yale University Press, New Haven, CT.

Ferris, J. P. (1984). The chemistry of life's origins. *Chem. Eng. News,* 62(35):22.

———, and Usher, D. A. (1988) Origins of life. In *Biochemistry,* 2nd ed., G. Zubay, Ed. Macmillan, New York.

Holland, H. D. (1984) *The Chemical Evolution of the Atmosphere and Oceans.* Princeton University Press, Princeton, NJ.

Joyce, G. F. (1989) RNA evolution and the origins of life. *Nature,* 338:217.

Schopf, J. W., Ed. (1983) *Earth's Earliest Biosphere. Its Origin and Evolution.* Princeton University Press, Princeton, NJ.

Stevenson, D. J. (1990) In *Origin of the Earth,* H. E. Newsom and J. H. Jones, Eds. Oxford University Press, New York.

## Paclitaxel: *see* Taxol® and Taxane Production by Cell Culture.

Paclitaxel: *see* Taxol® and Taxane Production by Cell Culture.

# PALEONTOLOGY, MOLECULAR

*Rob DeSalle*

*Key Words*

**Ancient DNA** Deoxyribonucleic acid isolated from ancient tissue sources; the polymerase chain reaction is the most common method used for isolating DNA from such sources.

**Biomarkers** Chemical or biochemical structures that were part of the biomass of a living ancient organism; also known as molecular fossils.

**Fossilization** The preservation of deceased organisms via mineralization, mummification, compression, or amberization.

**Mineralization, Amberization, and Mummification** Processes of fossilization that preserve materials for examination at the molecular level.

**Paleontology** The study of the structure, composition, and behavior of dead organisms.

Recent developments in modern molecular biology have facilitated the expansion of the new field of molecular paleontology, which is concerned with the isolation and characterization of proteins and nucleic acids from fossil material. This contribution outlines and discusses the various kinds of fossil preservation with respect to the types of preservation most appropriate for producing fossils that will yield direct sequence information. The molecular techniques used to isolate nucleic acids and proteins and to characterize the sequence information from these molecules are described. The biological importance of data obtained from fossils is also discussed.

## 1 DEFINITIONS

### 1.1 PALEONTOLOGY

Paleontology is defined as the study of ancient organisms. Paleontological information was recognized as an important aspect of evolutionary biology before Darwin's time. Indeed, an entire chapter of *On the Origin of Species* was dedicated to an examination of the fossil record. Darwin was preoccupied with the incompleteness of the fossil record and was concerned with the degree to which the fossil record supported his theory of natural selection. Several other examples of the influence of fossils on modern biology can be cited, such as the discovery of entire groups of unusual organisms that have become extinct (dinosaurs), the characterization of mass extinctions, and the use of the fossil record to calibrate divergence times in species comparisons. The utility of fossils for stratigraphic and chronologic correlations has also been repeatedly demonstrated.

### 1.2 MOLECULAR PALEONTOLOGY

The term ''molecular paleontology,'' coined by M. Calvin in 1968 includes the study of the molecular structure, behavior, and composition of ancient organisms. Molecular fossils or biological marker compounds (biomarkers) are types of organic molecules that were originally part of the biomass of an ancient organism. ''Chemical fossil'' refers to the broader term covering atomic and molecular fossils including isotopic features. A subbranch of molecular paleontology that concerns nonfossilized material from deceased organisms is known as molecular archeology. Examination of recently deceased organisms at the molecular level is considered a part of molecular forensics.

In the past, paleontological studies provided information on ancient organisms only at the morphological level. More recently, with the development of gas chromatography, mass spectrometry, scanning and transmission electron microscopy, and modern molecular biology, the molecular aspects of fossils and subfossils have been explored. Three major kinds of molecular information can be recovered from fossils. The first consists of visual structural information using various forms of microscopy. Thin sections of microorganisms observed by means of microscopy have demonstrated the feasibility of this approach, and several fossil microscopic molecular structures have been observed this way. The second kind of molecular data consists of compositional information. Ancient atmospheric conditions, as well as the chemistry of the ancient oceans, can be examined using chemical composition information from fossils. Chemical and biochemical fossils have been examined for molecular structure using combined gas chromatography and mass spectrometry. Some information on the amino acid composition of tissues has been obtained from fossil and subfossil materials. Radioimmunoassay (RIA) techniques have also been used to examine the immunological distances of extinct organisms with their living relatives. The newest and perhaps most exciting class of molecular information obtained from fossils is the analysis of the actual genetic material of fossils, subfossils, and recently extinct organisms.

## 2 FOSSILIZATION

### 2.1 TYPES OF FOSSILIZATION

The condition of the fossilized material is critical to the recovery of molecular information from any fossil. The depositional environment and the rock matrix in which the fossil is deposited, as well as the degree of decomposition of the fossil, are critical factors in determining the state of the fossilized material. Fossilization is an extremely rare event; if it is going to occur, it begins immediately after the death of the organism. In general, only a small portion of the organism is preserved. An organism's body after death experiences three types of decomposition. The first is biological and is accomplished by the activity of bacteria and by autolytic enzymatic processes. Since bacteria are ubiquitous in environments with oxygen, only organisms that are deposited into oxygen-free environments will escape bacterial decomposition and have their soft body parts preserved in any detail. The second type of decomposition is mechanical. Such factors as waves, wind, and other forms of abrasion will facilitate this form of breakdown. The final type of decomposition is chemical, and it can proceed well after the fossilization process has begun. Ancient organisms can be divided into three categories: organisms that have been preserved through mineralization (mineralized fossils), organisms that have been preserved in amber (amber fossils), and organisms that have been preserved but not mineralized or amberized (subfossils and compression fossils).

### 2.2 Mineralization

Several types of mineralization can occur in the fossilization process, and each results in differences in tissue preservation and in the degrees of decomposition that occur. Mineralization usually produces fossils in which the molecules of the preserved organism have been replaced by minerals. Such fossils will be entirely devoid of organic molecules because all the organism's molecules will have been replaced or displaced by mineral matter. Permineralization is the impregnation of an organism with mineral materials. Calcium carbonate and silica are most frequently involved in the permineralization process, which produces a structure that is resistant to chemical change or chemical decomposition. Metasomatism is produced from the exchange of existing minerals by new minerals. This process produces a fossil that is like the original organism in its physical appearance, although the chemical composition of the fossil is drastically different. In plant remains, the oxygen, hydrogen, and nitrogen molecules are sometimes leached away, leaving a carbon residue that retains the original structures of the plant. Incrustation is another process that produces fossilization, most frequently in more recent organismal material. Or, calcium-rich water may flow over the remains of organisms, covering them with a mineral film. The organic material under the film then decomposes, leaving only a negative impression on the mineral layer. In the distillation process, water and gases are distilled away from the organism, leaving on rock a thin film of carbon that can sometimes give rich information about the structure of the distilled organism.

Some tissues are more resistant to the mineralization processes than others. Certain compounds found in organisms, such as chitin and keratin, are rather resistant to chemical decomposition and mineralization. Minerals such as apatite (teeth and bones of vertebrates) and calcium carbonate (shells of invertebrates) often form very resistant structures and parts of organisms. Chitonous exoskeletons of arthropods can be resistant to chemical decomposition by mineralization. Other proteins and nucleic acids (DNA and RNA) are much more susceptible to decomposition.

Some depositional conditions produce fossils remarkably resistant to chemical decomposition and complete mineralization. The Solenhofen deposits in Bavaria and the Clarkia fossil beds of northern Idaho are excellent examples of this type of fossilization. In these areas, organisms were deposited in an extremely low oxygen setting and rapidly buried in fine sediment, which prevented the chemical and biological decomposition of the organic material. During this process, organisms are deposited in a very fine laminated sediment and preserved between two extremely thin sedimentary layers. The layers are usually found as shale deposits, and the organic material is usually preserved in extreme detail. In such deposits it is actually possible to remove the remains of organisms from between the shale layers, and considerable ultrastructural preservation can be observed from these organisms.

### 2.3 Amber, Mummification, and Museum Collections

Another form of fossilization that leaves remarkably preserved organismal remains is amberization, which is produced by the flow of resinous material from coniferous plants. The common factor in the quality of the material found in amber is the thorough dehydration and protection of encased organic material. When the chemical in amber (isoprene and diterpenoids) link to each other, trapped organic material dehydrates and becomes inert. The ability of the flowing resin to completely encapsulate an organism and the bacteriocidal activity of terpenes in the resin suggest that amberization will produce highly protected fossil specimens. The most famous of these deposits are found from the Oligocene epoch of Rumania and the Baltic region and from the Miocene epoch of the Dominican Republic, Mexico, Sicily, and the Appenines of Italy.

Natural desiccation (mummification) is the yet another form of fossilization that can preserve ancient material in a form suitable for molecular analysis. This type of fossilization is extremely rare and involves almost complete preservation of the encased organism. An example of this kind of preservation from the Cretaceous period is the anatosaurus, a dinosaur that was preserved complete with wrinkles in its skin. The preservation of this specimen occurred through extreme dehydration of the animal. Such preservation is most likely to occur if the organism is not deposited in a submarine sedimentary basin, but rather in unusual terrestrial environments. More recent examples of mummification are the ground sloth *(Mylodon)* and other smaller mammals of about 13,000 years ago from the Ultima Esperanza Cave in Chile, bog-preserved human remains of about 8000 years ago from the Windover Pond region of Florida and animals preserved in the La Brea tar pit, about 10,000 years ago. Permafrost-preserved organisms have also been used as a source of ancient material. The most famous are the woolly mammoth found in 30,000-year-old permafrost and the Neolithic human remains (named Oetzi) found in the Alps and believed to be 4000–5000 years old. Other examples of more recently preserved remains have been described.

Human cultural practices also result in mummification or preservation of organisms. The detailed preservation of humans and some domestic animals by the ancient Egyptians is well known. Museum-preserved specimens of an incredibly diverse and large number of organisms would also fit into this category. Teeth, bones, pieces of skins, feathers, alcohol-preserved specimens, and some pinned insects have all been used as sources of material for molecular studies. The continued importance of museum-derived collections should be emphasized as a source of ancient specimens for study. Not all museum material is suitable for molecular analysis, however, since some of the preservation procedures used by museum curators are rather hydrophilic and produce massive damage to the specimen at the molecular level.

### 2.4 Quality of Molecules from Ancient Sources

Typically, in proteins that have been fossilized, the peptide bonds between amino acids will be rapidly degraded. Under special circumstances, especially in shells and skeletal remains, peptide bonds will remain intact and short peptides will be preserved. The position of proteins between (inter-) and within (intra-) crystalline structures in a fossil will effect the longevity of fossil molecules. Intracrystalline molecules would be expected to have greater survival potential because of the protective aspects of the surrounding crystal structure. The decay of amino acid abundance in fossils has been examined in several marine organisms with shell structures. There is a rapid decline in the abundance of amino acids in the first few hundred thousand years after death. The degree of decomposition usually levels off at 2% surviving amino acids. Soluble and insoluble organic residues are also important factors in determining the degree of preservation of amino acids and peptide fragments in fossils. Nucleic acids and DNA in particular would be expected to be much less well preserved than some proteins in fossil materials. DNA is primarily concentrated in soft tissues, which are highly susceptible and vulnerable to degradation during fossilization. In addition, DNA is a highly reactive compound that can easily be broken down into its constituent parts.

## 3 STUDIES IN MOLECULAR PALEONTOLOGY

### 3.1 Microscopy Studies

The gross molecular structure of certain cellular organelles has been examined in several fossils. These studies are accomplished by thin-sectioning rocks containing putative microorganisms. The thin sections are then mounted on slides and observed under microscopy. Perhaps the oldest example consists of bacterial filaments from 3.5-billion-year-old formations in Western Australia, which resemble filaments from contemporary cyanobacteria. Other interesting ancient molecular level structures include nuclear membranes in billion-year old fossils, and the presence of colonial algal characteristics has been noted in microfossils from the Gunflint chert in Michigan. Microscopic examination of amber-preserved fossils has revealed molecular level information for these specimens. Transmission electron microscopy of amber-encased insects reveals astonishing detail at the cellular level. Nuclear membranes, endoplasmic reticulum, and ribosomes are all visible in these ancient materials. Scanning electron microscopy of amber-preserved insects also provides information at the molecular level, although at a much lower level of resolution.

### 3.2 Chemical and Amino Acid Studies

Abelson first showed in the 1950s that amino acids could be isolated from fossilized material. These fossils included representatives of mollusks of various ages, and dinosaurs, fish, and mammals all from the Phanerozoic eon. Studies of chemical fossils involve the characterization of biomarkers from a variety of microorganisms. Such biomarkers as pentacyclic triterpenoids, hopanes, sterols, pigment derivatives, and acyclic isoprenoids are used as signatures for unique fossil molecules and sediments. Isotopic composition of these biomarkers is a characteristic that is used to determine signature. For instance, *n*-alkanes can have even or odd numbers of carbon atoms. Even-numbered long chain *n*-alkanes indicate an origin from plant waxes or algae. Odd-numbered *n*-alkanes indicate a high salinity environment. This approach is immediately applicable to the petroleum industry, since *n*-alkanes and other biomarkers are the major components of petroleum.

### 3.3 Protein and Nucleic Acid Studies

The actual isolation and characterization of proteins and nucleic acids from ancient tissues in phylogenetic studies began with the pioneering work in Allan Wilson's lab at the University of California at Berkeley. Wilson and his colleagues, most notably Russel Higuchi, Ellen Prager, and Svante Pääbo, worked on a variety of ancient tissue sources. Jerold Lowenstein, working at the University of California at San Francisco, used the immunodiffusion and RIA techniques to examine the serum albumin of several extinct organisms. These techniques result in immunological distances between taxa that can be used to infer relationships among organisms. The organisms examined in these studies were mastodon *(Mammut americanum),* woolly mammoth *(Mammuthus primigenius),* Steller's sea cow *(Hydrodamalis gigas),* Tasmanian wolf *(Thylacinus cynocephalus),* and quagga *(Equus quagga).*

The first report of DNA isolated and characterized from ancient tissue appeared in 1984; it concerned the isolation and cloning of a short stretch of mitochondrial DNA (mtDNA) from the quagga, an extinct zebra like mammal. Publications from the Wilson lab followed concerning the isolation and characterization of nucleic acids from woolly mammoth and Egyptian mummies. Several studies since have reported the isolation and characterization of nucleic acids from both plants and animals. Plant remains on the order of hundreds to thousands of years old have been examined. The primary source for most of the animal studies has been bone. Several studies have characterized DNA isolated from bone ranging from tens of years to thousands of years old.

More recently, much older material has been examined. The isolation and characterization of chloroplast DNA from fossilized magnolia leaves (17 million years old), and the isolation and characterization of DNA from Dominican amber-preserved (30 million years old) and Lebanese amber-preserved (120 million years old) insects, are examples of the increase, by several orders of magnitude, of the limits of the age of ancient DNAs. Reports of attempts to isolate DNA from Baltic amber (60–80 million years old), 200-million-year-old fish fossils, and 400-million-year-old marine invertebrates have also appeared.

### 3.4 Ancient DNA Isolation and Verification

The characterization of ancient nucleic acids currently relies on the polymerase chain reaction (PCR). Since, however, this technique is highly susceptible to the production of false positives, hence spurious information, extreme care must be taken in every study involving ancient DNA templates. Several technical precautions are taken in the DNA isolation and PCR amplification stages. Positive displacement pipette tips or plugged, coated pipette tips are used in all manipulations of DNA, buffers, and enzymes. To ensure the accuracy of the ancient DNA reaction, several control reactions are run side by side with the experimental reactions.

Two problems specific to the PCR must also be considered when examining ancient DNA. Since the Taq polymerase used in most PCR reactions preferentially inserts dATP into damaged regions of template DNA, ancient DNA sequences must be scrutinized for the unusual occurrence or frequency of deoxyadenosines in amplified sequences. The second problem concerns the ability of the PCR reaction to form chimeric molecules via PCR jumping. This process occurs when template DNA is damaged and is degraded to sizes in which only one of the priming sites is present on the template, the other having been separated by damage. In these cases, the short templates can serve as primers on other pieces of template DNA, and ''jumping'' occurs across the damaged ends of the original templates. This process results in PCR products from template DNA that can differ in size from the intended product. In addition, if some outside contaminant DNA is present in the PCR reaction, it is possible for a jump from the intended target template to the contaminant to occur. This scenario would produce a chimeric PCR product with sequences from the intended target on one end of the product and sequences from the contaminant on the other.

In addition, extreme care is taken when analyzing sequences obtained from ancient tissues. The verification of ancient DNA sequences is accomplished by detailed phylogenetic analysis. The discovery of a genealogical relationship of DNA sequences from ancient sources with an extant, close relative is usually taken as verification of authenticity. Thus, it is necessary to have as many exemplars or taxa as possible available for the phylogenetic analysis.

## 4 UTILITY OF ANCIENT DNA

Ancient DNA studies can be very useful in many aspects of modern biology. Systematics is perhaps the best example of how these data can be useful. An amber insect study is a useful example of how nucleic acid information from fossils can be important. Information on nearly 400 bases of the small ribosomal subunit (18*S* rDNA) was collected for a number of termites and their close relatives, mantids and cockroaches, and several outgroups. When a phylogenetic analysis using only the extant organisms is performed, the position of cockroaches in the phylogenetic tree is ambiguous. When the extinct amber termite sequences are added to the analysis, however, the tree becomes resolved with respect to the placement of the cockroach. This result supports a well-known observation in paleontology that suggests that fossil information is absolutely essential to understand phylogeny in many groups.

Two other examples involve the use of ancient DNA information to calibrate molecular clocks and the use of ancient DNA in population biology and conservation. Information about the sequences of extinct organisms is essential to the more accurate determination of absolute rates of divergence. If a fossil can be accurately dated and sequences from the fossil can be directly compared to sequences from living organisms, more accurate rates for molecular clock estimates can be obtained. Population level studies have already demonstrated the value of ancient DNA sequences in population genetic studies. The ability of researchers to obtain information from extinct taxa and populations is essential in assessing relatedness among endangered species, and such information should become more prominent in conservation studies.

*See also* MITOCHONDRIAL DNA, EVOLUTION OF HUMAN; ORIGINS OF LIFE, MOLECULAR BIOLOGY OF; PCR TECHNOLOGY.

### *Bibliography*

Abelson, P. H. Organic constituents of fossils. *Carnegie Instit. Washington Year Book,* 53:97–101 (1954).

Bahn, P. G., and Everett, K. *Nature* 362:11–12 (1993).

Calvin, M. *Transa. Leicester Lit. Philos. Soc.* 62:45–69 (1968).

Cano, J., Poinar, H., Pienazek, N. J., Acra, A., and Poinar, G. O. *Nature,* 363:536–538 (1993).

Catteneo, C., Gelsthorpe, K., Philips, P., and Sokal, R. J. *Nature,* 347:339 (1990).

Cooper, A., Mourer-Chauvire, C., Chambers, G. K., von Haeseler, A., Wilson, A. C., and Paabo, S., *Proc. Natl. Acad. Sci. U.S.A.* 89:8741–8744 (1992).

Curry, G. B. In *Molecular Evolution and the Fossil Record,* B. Runnegar and J. W. Schopf, Eds. Paleontological Society, Knoxville, TN, 1988, pp. 20–33.

(a) DeSalle, R., Gatesy, J., Wheeler, W. C., and Grimaldi, D. A. *Science,* 257:1933–1936 (1992). (b) Cano, J., Poinar, H., and Poinar, G. O. *Med. Sci. Res.* 20:249–251, 619–623 (1992).

Gauthier, J., Kluge, A., and Rowe, T. *Cladistics,* 4:105–117 (1990).

Golenberg, E. M., Giannasi, D. E., Clegg, M. T., Smiley, S. J., Durbin, M., Henderson, D., and Zurawski, G. *Nature,* 344:656–658 (1990).

(a) Hagelberg, E., Gray, I. C., and Jeffreys, A. J. *Nature,* 352:427–429 (1991). (b) Hagelberg, E., Sykes, B., and Hedges, R. *Nature* 342:485 (1989).

Higuchi, R., Bowman, B., Frieberger, M., Ryder, O., and Wilson, A. C. *Nature,* 312:282–284 (1984).

Higuchi, R., and Wilson, A. C. *Fed. Proc. Am. Soc. Exp. Biol.* 43:1557–1564 (1984).

Janczewski, D. J., Yuhki, N., Gilbert, D. A., Jefferson, G. T., and O'Brien, S. J. *Proc. Natl. Acad. Sci. U.S.A.* 89:9767–9773 (1992).

Lawlor, D. A., Dicker, C. D., Haiswirth, W. W., and Parham, P. *Nature* 349:785–787 (1991).

Lindhal, T. *Nature,* 362:709–715 (1993).

(a) Lowenstein, J. M., Sarich, V. M., and Richardson, B. J. *Nature,* 291:409–411 (1981). (b) Lowenstein, J. M., and Ryder, O. A., *Experientia,* 41:1192–1193. (1985). (c) Lowenstein, J. M. *Philos. Trans. R. Soc. London,* 292B:143–149. (1980). (d) Prager, E. M., Wilson, A. C., Lowenstein, J. M., and Sarich, V. M. *Science,* 209:287–289 (1980).

Morell, V. *Science,* 257:1902–1904 (1992).

Niklas, K. J., Brown, R. M., Jr. and Santos, R. In *Late Cenozoic History of the Pacific Northwest,* C. J. Smiley, Ed. American Association for the Advancement of Science Press, San Francisco, 1985, pp. 143–159.

Pääbo, S., Irwin, D. M., and Wilson, A. C. *J. Biol. Chem.* 265:4718–4721 (1990).

Pääbo, S. *Proc. Natl. Acad. Sci. U.S.A.* 86:1939–1943 (1989).

Pinna, G. *The Illustrated Encyclopedia of Fossils.* Facts on File, New York, 1985.

Poinar, G. O., and Hess, R. *Science,* 215:1241–1243 (1982).

(a) Rollo, F., Amici, A., Salvi, R., and Garbuglia, A. *Nature,* 335:774 (1988). (b) Rollo, F., Amici, A., and Salvi, R. *Nucleic Acids Res.* 16:3105–3106 (1988). (c) Venanzi, F. M., and Rollo, F. *Nature, 343:25–26 (1990).* (d) Golubinoff, P., Paabo, S., and Wilson, A. C. *Proc. Natl. Acad. Sci. U.S.A.* 90:1997–2001 (1993). (e) Golubinoff, P., Paabo, S., and Wilson, A. C. In *Corn and Culture in the Prehistoric New World,* S. Johannessen and C. A. Hastorf, eds. Westview Press, Boulder CO 1993.

Schopf, W. *Earth's Earliest Biosphere.* Princeton University Press, Princeton, NJ, 1983.

(a) Schopf, W., and Packer, B. *Science,* 237:70–73 (1988). (b) Loomis, W. F., *Four Billion Years.* Sinauer, Sunderland, MA, 1988.

Smiley, C. J., Ed. *Late Cenozoic History of the Pacific Northwest.* American Association for the Advancement of Science Press, San Francisco, 1985.

Summons, R. E. In *Molecular Evolution and the Fossil Record,* B. Runnegar and J. W. Schopf, Eds. Paleontological Society, Knoxville, TN, 1988, pp. 98–113.

Thomas, W. K., Paabo, S., Vilablanca, F., and Wilson, A. C. *J. Mol. Evol.* 31:101–112 (1990).

# Partial Denaturation Mapping

*Ross B. Inman*

*Key Words*

**Cross-Linking** Creation of covalent linkage between various groups in protein molecules. These linkages help stabilize against surface tension forces during preparation for electron microscopy.

**Denaturation of DNA** Breakage of interstrand H bonds in double-stranded DNA with subsequent unwinding and ultimate dissociation of the two strands.

Partial denaturation mapping is an electron microscopic method that allows the visualization of denatured regions within otherwise double-stranded DNA. Denatured regions can be produced either by heat or by high pH treatment in the presence of formaldehyde. Because denatured regions first appear at A+T-rich regions, the pattern of denatured sites is characteristic of the base sequence and can therefore provide identification and polarity information. Additionally, the pattern can be used to determine the position of bound protein or alternative DNA structural alterations (such as hairpin structures) in otherwise normal linear or circular DNA molecules. The procedure can be useful in the investigation of a variety of important biological reactions. The origin and direction of replication has been determined for several bacteriophages using their respective partial denaturation maps as position and polarity markers. Similar information has been obtained for recombinational joints isolated during in vitro and in vivo studies. Another application used the unique denaturation pattern as a landmark to deduce the DNA end first injected during bacteriophage infection.

## 1 INTRODUCTION

Partial denaturation mapping is an electron microscopic method that allows the investigator to deduce the position of some object of interest on a DNA molecule. Because denatured sites first form at the regions in a DNA molecule that are richest in A+T, and because they can be readily observed by electron microscopy, the resulting denaturation pattern can be used as a mapping reference on an otherwise featureless circular or linear DNA molecule. Thus the position of any object or DNA structural alteration that is large enough to be viewed by electron microscopy can be determined with respect to the denaturation map.

The general strategy is as follows. Assume that there is some position of interest on a DNA molecule, such as a bound protein, or an unusual DNA structure, such as a replication branch point. This structure first must be stabilized in some way to ensure that it will remain unchanged during preparation for electron microscopy. Most proteins can be cross-linked with formaldehyde or glutaraldehyde or by consecutive treatment with both fixatives. DNA structures are often quite stable, but the more labile structures can be further stabilized by photo-cross-linking with various psoralen derivatives, which effectively prevent DNA rearrangements and branch migration. The sample is now subjected to conditions that will cause several of the regions richest in A+T to denature. These regions are stabilized with formaldehyde, and the sample is prepared for electron microscopy in a way that permits the denatured sites to be clearly observed. Measurements from such electron micrographs can then be used to establish the position of the object of interest with respect to the denaturation pattern.

## 2 PRINCIPLES OF THE METHOD

Partially denatured samples can be prepared by heating within the melting region of the DNA. The preferred denaturation method, however, is high pH treatment. Any possible reannealing of denatured sites must be prevented by addition of formaldehyde. Two common methods for high pH partial denaturation are as follows.

1. A high pH buffer is prepared containing 67 mM $Na_2CO_3$, 11 mM EDTA, 31% HCHO, and NaOH. Next 3 μL of this buffer is added to 7 μL of a solution containing about 0.04 μg of the DNA and the solution heated. The solution is now added to 10 μL of formamide and 2 μL of 0.1% cytochrome *c* and spread for electron microscopy by a modified protein film method. Usable degrees of denaturation are typically obtained if the buffer is adjusted to 40–50 mM NaOH and the solutions heated to 30°C for 10–60 minutes.
2. An alternative buffer contains 41 mM $Na_2CO_3$, 6 mM EDTA, 19% HCHO, 128 mM NaOH, and 37% formamide; 5 μL of this buffer is added to 7μL of a solution containing about 0.04 μg of DNA. Heating this solution at 50°C for 10–40 minutes produces acceptable degrees of denaturation. After the heating, 2 μL of 0.1% cytochrome *c* is added and the sample spread as indicated.

Once the molecules have been spread and electron micrographs obtained, it is necessary to measure the position of all sites with respect to an end or some arbitrary position on a circular molecule. Furthermore, measurements need to be done on a statistically significant number of molecules (usually 20–40). Because these measurements are time-consuming, every effort has been made to make the process as effortless and as accurate as is reasonably possible. A semiautomatic system has been developed that has proven to be a great aid in obtaining such data.

Figure 1 shows an electron micrograph of a bacteriophage lambda DNA molecule after partial denaturation; in the accompanying map of the positions of the sites in the molecule, the denatured sites are depicted by solid rectangles. Numerous denatured sites can be seen, where the two strands of the double helix have unwound to yield denaturation bubbles. Bacteriophage lambda contains 48.5 kbp of DNA with a contour length of 17.1 μm.

### 2.1 Experimental and Expected Partial Denaturation Maps

It has long been known that the temperature at which DNA denatures depends on base composition. For example, DNA rich in the bases A and T melts at a lower temperature than G+C-rich molecules. It would therefore be expected that the same would hold true at the intramolecular level and that the denatured sites observed after alkali treatment would occur at the regions of the molecule richest in A+T. There are two consequences of this expectation. First, since each phage lambda DNA molecule has exactly the same sequence, the position of the denatured sites should be unique.

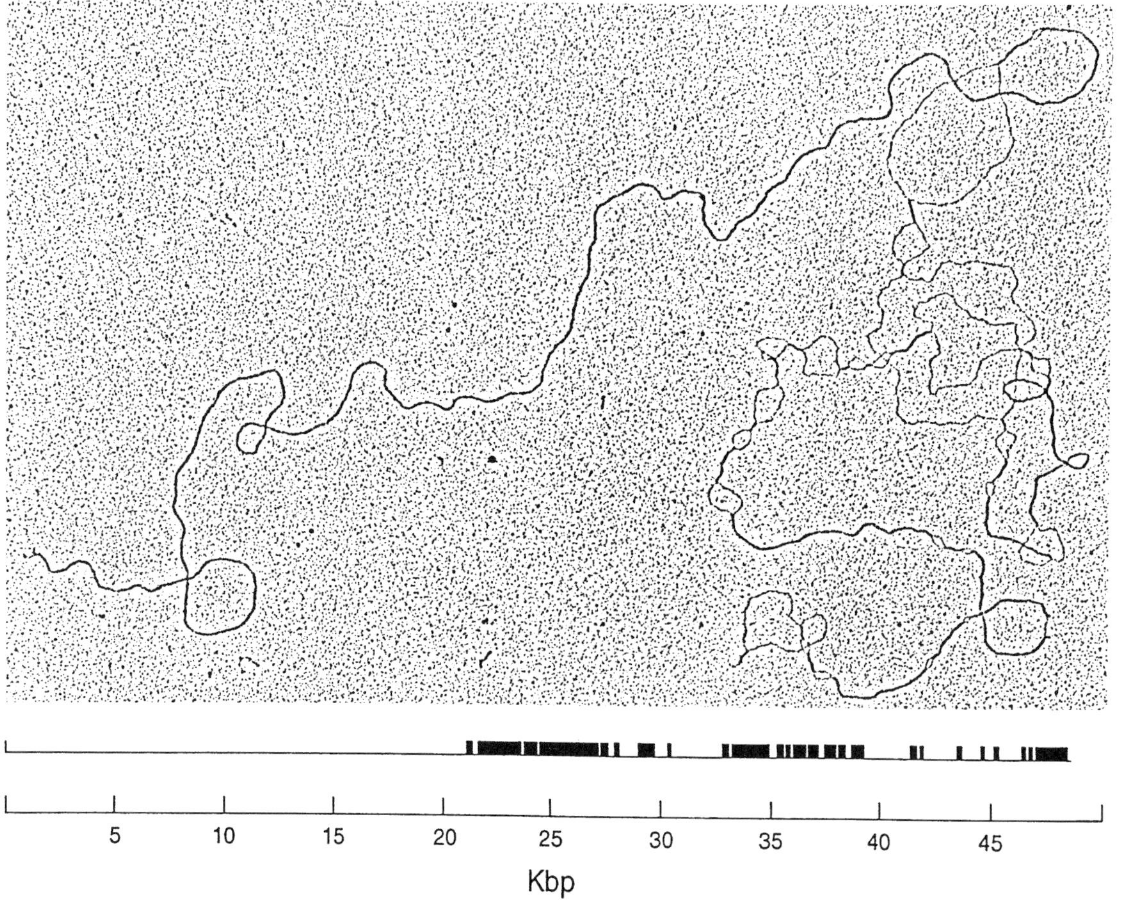


**Figure 1.** Electron micrograph of a partially denatured bacteriophage lambda DNA molecule. DNA from this phage is 48.5 kbp long, and numerous sites can be observed that result from unwinding of the two strands of the double helix because of denaturation produced by high pH treatment. In the map of the molecule, each black rectangle represents the position and size of a denatured site. The DNA image and the map have been aligned so that the leftmost ends correspond to the left-hand end of the lambda genetic map. Lambda DNA is remarkable in that the areas richest in A+T are restricted to the right half of the molecule.

As shown in Figure 2A, this is found to be true; the first denatured sites lie at reasonably unique positions near the center of the molecule. When many molecules are compared, it is obvious that the sites occur in a regular pattern that is unique for the particular DNA sample. Additionally, as the temperature or the concentration of NaOH is increased, new denatured sites should appear, and these also should be located at unique positions. This also has been experimentally confirmed: the degrees of denaturation shown in Figure 2 span the range from 0.6% (Figure 2A), where molecules usually have only two denatured sites, to 23% (Figure 2D). An average denaturation pattern for any given set of denaturation maps can be displayed as a histogram. The average is computed as the fraction of molecules that possess a denatured site at 1000 consecutive sections along the molecule. Figure 3E–H shows examples of such averages for bacteriophage lambda at several increasing degrees of denaturation. From Figure 3, which shows how the pattern develops with increasing degree of denaturation, it can be seen that the region richest in A+T is centered at about 23 kbp and the next richest segments are to be found at 27, 28, 36, and 47 kbp (see Figure 3E).

The experimental maps discussed in connection with Figure 3 have an arbitrary polarity. For instance, if a map is obtained from linear molecules, we will need more information to be able to decide which end of the denaturation map corresponds to the defined leftmost end of the genetic map. Similarly, a partial denaturation pattern obtained from a circular DNA will have an arbitrary starting point and polarity. Section 2.2 describes several methods that make it possible to relate the physical denaturation map to the genetic map.

## 2.2 Relation Between Denaturation and Genetic Maps

There are several ways to relate the physical and genetic maps. The method of heteroduplex mapping results in single-stranded loops on otherwise double-stranded molecules. These loops correspond to the position of deletion, substitution, or addition mutations, which by other methods can be located on a genetic map. Thus the position of these loops can be used to directly relate the physical DNA molecule with the genetic map. If such heteroduplex molecules are partially denatured, the denaturation pattern can be aligned with the same polarity as the genetic map with the aid of the visible heteroduplex loops. This method has been used for alignment of physical and genetic maps of bacteriophage P2 DNA.

Because denatured sites occur at A+T-rich regions, it should be possible to compute and predict the partial denaturation pattern if the DNA sequence is known. The method used to derive predicted

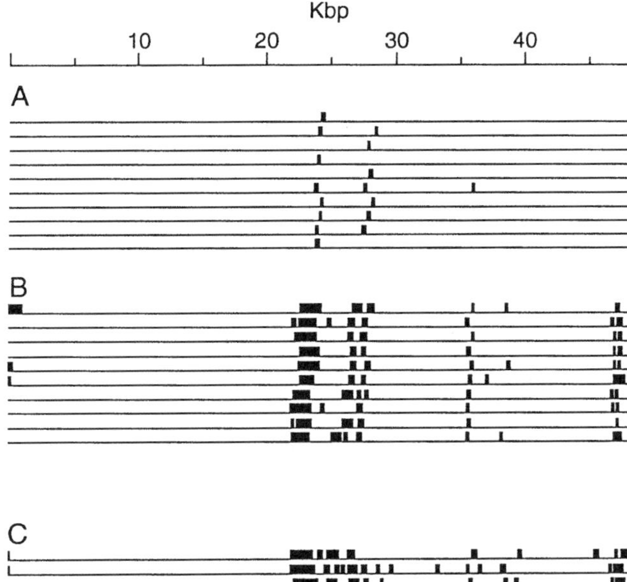

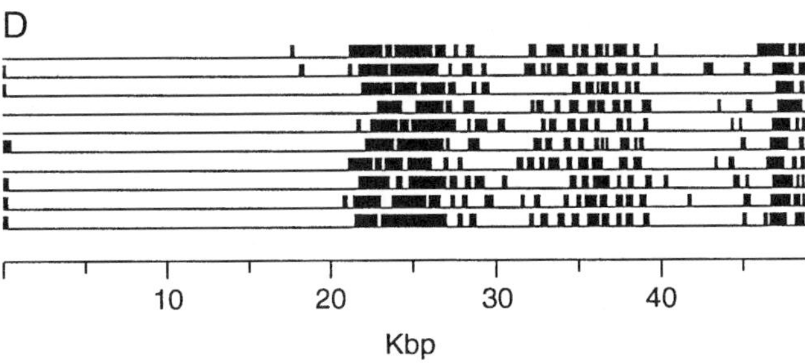


**Figure 2.** Examples of denaturation maps of bacteriophage lambda DNA. Each horizontal line represents a single molecule, and the black rectangles represent the measured position of denatured sites (cf. Figure 1, which shows an electron micrograph of such a DNA molecule and the resulting map). The displayed length of each molecule has been normalized so that the total length is equivalent to 48.5 kbp. The patterns were observed at the following degrees of denaturation: 0.6% (A), 6% (B), 11% (C), and 23% (D).

denaturation maps rests on simply computing a running average A+T content using an averaging segment width equal to the average size of the observed denatured sites. If the running average A+T contents are plotted against the base coordinate midway within each averaging segment, the resulting curve will exhibit peaks and valleys corresponding to A+T- and G+C-rich regions, respectively. If data are plotted only when the A+T content rises above a given value, the resulting curves will show histogramlike peaks, which can be compared with experimentally derived denaturation maps (such as those in Figure 3E–H). Examples of computer-derived denaturation maps of bacteriophage lambda DNA are shown in Figure 3A–D. These predicted maps provide two important pieces of information. First, they give assurance that the experimental maps are in fact a reflection of the A+T-rich segments within a genome. As can be seen, the various regions of high A+T content, as derived from the base sequence, compare quite well with the positions of those found experimentally. We can therefore be sure that within the resolution possible by this technique (40–60 bp), the denatured sites do in fact closely correspond to A+T-rich regions within the DNA molecule. Second, the expected maps show which way the experimental map must be aligned to correspond to the genetic map. In Figures 1, 2, and 3, all maps are aligned to correspond to the defined genetic maps.

The genome shown in the preceding example, bacteriophage lambda, is most unusual in that the left and right halves of the molecule differ greatly in their base composition. As can be seen in Figure 2, the regions richest in A+T are restricted to the right-hand half of the molecule, and in fact, it almost looks as though lambda originally formed from two quite different types of sequence that were joined at about 22 kbp. This is particularly evident in the predicted histogram average (Figure 3D) in which the A+T peaks and the G+C valleys are quite different in the two halves of the molecule. The section to the left, which encompasses the genes responsible for head and tail proteins, has only small fluctuations in base composition, while the genes governing recombination, immunity, replication, and bacterial lysis functions are restricted to the right-hand segment, which has much larger fluctuation in the base composition. Most other bacteriophage genomes studied in this way have a much more scattered distribution of A+T-rich regions, but the A+T-rich regions themselves (as defined by the height of the peaks in the experimental denaturation map histograms) can be of similar magnitude.

## 3 RELATED CONSIDERATIONS

### 3.1 G+C-rich Regions Within DNA Molecules

By inference, the regions in highly denatured samples that are refractive to denaturation will correspond to G+C-rich regions. Experiments that demonstrate these positions are shown in Figure 4. The denaturation patterns are in this case from bacteriophage 186, which is approximately 30.8 kbp long. As well a showing the A+T-rich regions (as observed in the topmost panels), we can see regions that must be quite rich in G+C (bottom panel); these correspond to positions centered at 2.2, 6.5, and 14.8 kbp, and according to the data, the position at 6.5 kbp must be spectacularly rich in G+C.

### 3.2 Complete Denaturation

Under certain conditions it is advantageous to be able to study completely denatured molecules—for instance, when one needs to know whether DNA strands are continuous throughout a molecule or throughout a particular section of a molecule. Rather than completely denature the molecule and visualize the dissociated single strands, intrastrand cross-links can be created by the use of psoralen and UV irradiation. When such molecules are completely denatured, the single strands are held together by the cross-links and the position of strand discontinuities can be observed in a more informative context.

### 3.3 Denaturation at Four- and Three-Stranded Junctions

Studies of recombinational intermediates often require knowledge of strand continuity through the junction. Fortunately DNA regions

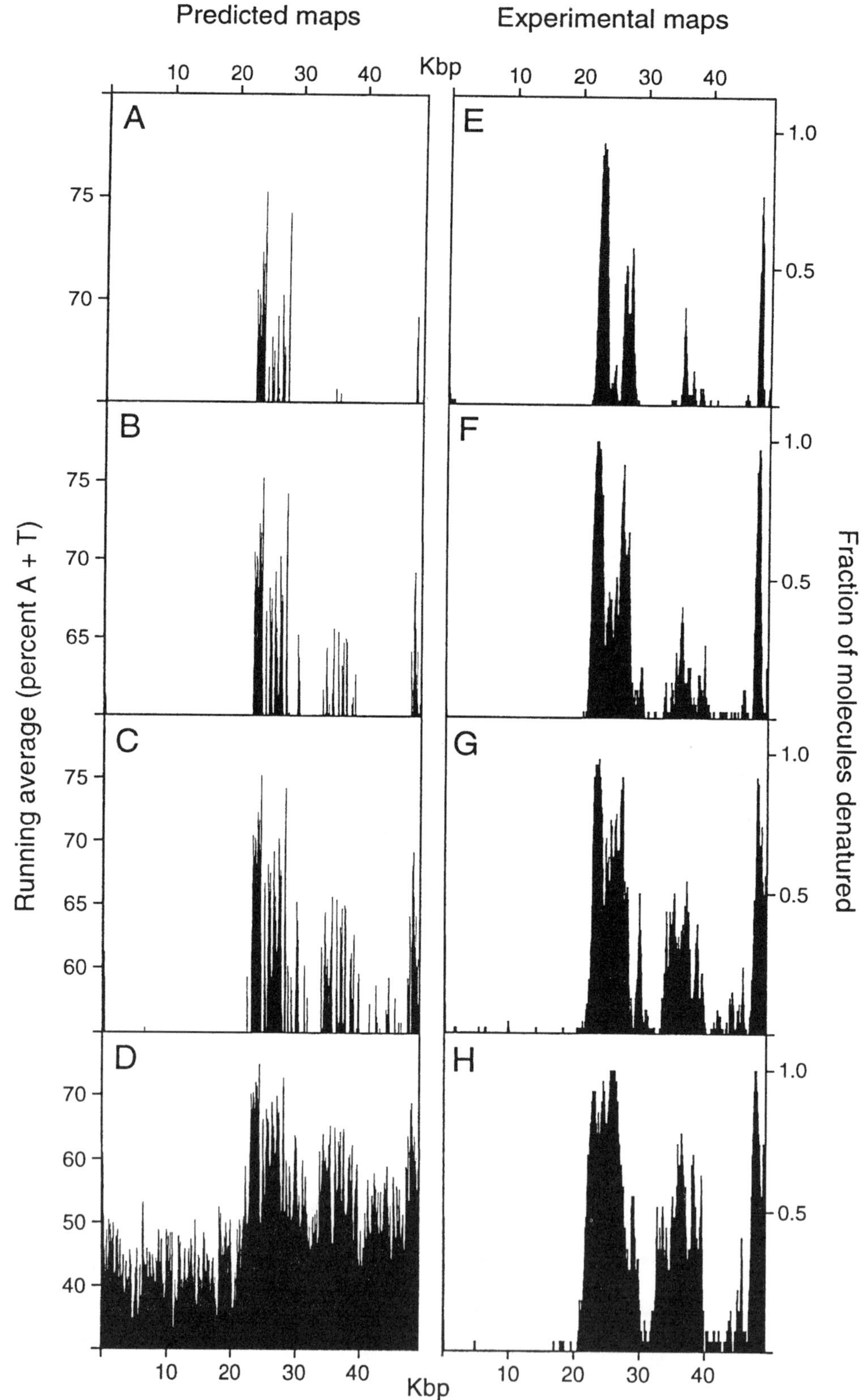


Figure 3. (A–D) Predicted denaturation maps for bacteriophage lambda derived by computing a running average A+T base content with an average segment width of 200 bp. The running average curves were plotted in these examples only when the average rose above 65% (A), 60% (B), 55% (C), and 30% (D). The resulting histogramlike graphs can be compared directly with the experimental results (E–H), which show experimental average denaturation patterns for bacteriophage lambda at several degrees of denaturation. The peaks show the most frequent positions which denature at various degrees of denaturation: (E) average of 47 molecules at 6% denaturation; (F) average of 37 molecules at 11% denaturation; (G) average of 46 molecules at 15% denaturation; (H) average of 27 molecules at 23% denaturation. The averages shown in (E), (F), and (H) correspond to the smaller sets of individual maps in Figure 2B, C, and D, respectively.

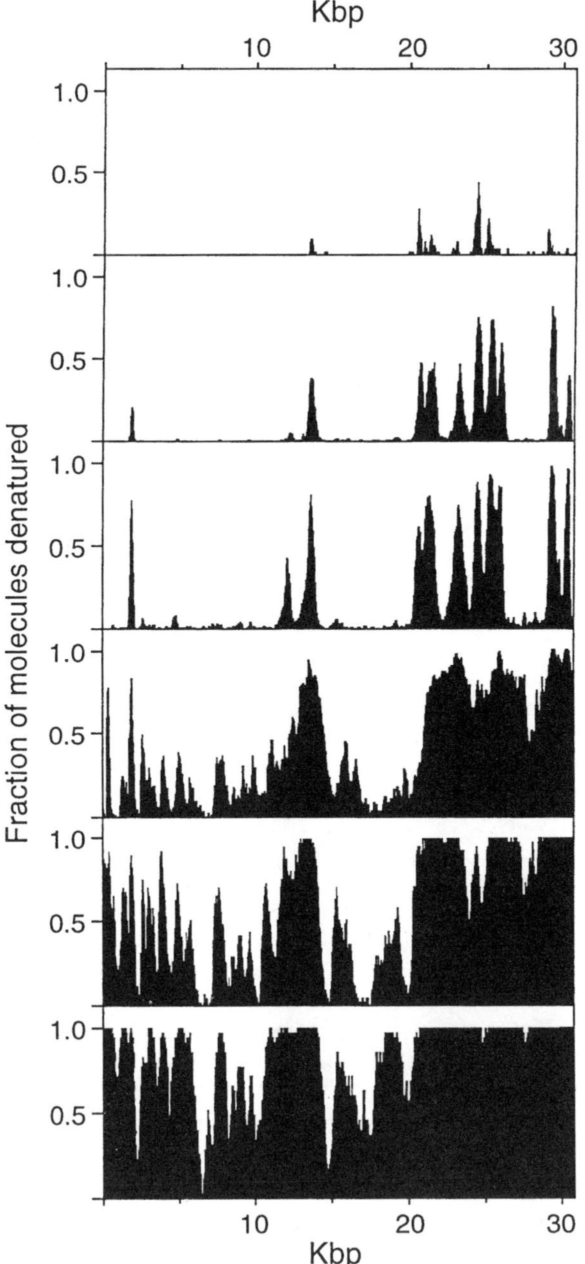


**Figure 4.** Experimental average denaturation patterns for bacteriophage 186 (total length equivalent to 30.8 kbp DNA). Top four panels show the regions richest in A+T as peaks in the histogram. The bottom panels show the regions richest in G+C found at the highest degrees of denaturation (represented by the valleys in these histograms).

around and within such junctions often show a lower helical stability than expected from their A+T content. Thus single-strand continuity within such junctions often can be visualized if the sample is spread under mild denaturation conditions. Holliday junctions in bacteriophage lambda DNA were first observed using such a technique, and more recently the procedure has proved useful for understanding Holliday intermediates and other products in in vitro FLP ("Flip") protein-promoted, site-specific recombination.

*See also* Denaturation of DNA; Electron Microscopy of Biomolecules; Human Chromosomes, Physical Maps of; Nucleic Acid Sequencing Techniques; Scanning Tunneling Microscopy in Sequencing of DNA.

### *Bibliography*

Chattoraj, D. K., Schnos, M., and Inman, R. B. Electron microscopic denaturation map of bacteriophage 186 DNA. *Virology,* 55:439–444 (1973).

Chattoraj, D. K., Younghusband, H. B., and Inman, R. B. Physical mapping of bacteriophage P2 mutations and their relationship to the genetic map. *Mol. Gen. Genet.* 136:139–149 (1975).

Davis, R. W., and Davidson, N. Electron-microscopic visualization of deletion mutations. *Proc. Natl. Acad. Sci. U.S.A.* 60:243–250 (1968).

Funnell, B. E., and Inman, R. B. Comparison of partial denaturation maps with the known sequence of SV40 and ϕX174 RF DNA. *J. Mol. Biol.* 131:331–340 (1979).

Hansen, C. V., Shen, C. J., and Hearst, J. E. Cross-linking of DNA in situ as a probe for chromatin structure. *Science,* 193:62–64 (1976).

Inman, R. B., and Schnos, M. Partial denaturation of thymine- and BU-containing lambda DNA in alkali. *J. Mol. Biol.* 49:93–98 (1970).

Littlewood, R. K., and Inman, R. B. Computer-assisted DNA length measurements from electron micrographs with special reference to partial denaturation mapping. *Nucleic Acids Res.* 10:1691–1706 (1982).

Marmur, J., and Doty, P. Determination of base composition of deoxyribonucleic acid from its thermal denaturation temperature. *J. Mol. Biol.* 5:109–118 (1962).

Meyer-Leon, L., Huang, L.-C., Umlauf, S. W., Cox, M. M., and Inman, R. B. (1988) Holliday intermediates and reaction by-products in FLP protein-promoted site-specific recombination. *Mol. Cell. Biol.* 8:3784–3796 (1988).

Qian, X.-H., Inman, R. B., and Cox, M. M. Reactions between half- and full-FLP recombination target sites. *J. Biol. Chem.* 267:7794–7805 (1992).

Valenzuela, M., and Inman, R. B. Visualization of a novel junction in bacteriophage lambda DNA. *Proc. Natl. Acad. Sci. U.S.A.* 72:3024–3028. (1975).

Westmoreland, B. C., Szybalski, W., and Ris, H. Mapping of deletions and substitutions in heteroduplex DNA molecules of bacteriophage lambda by electron microscopy. *Science,* 163:1343–1348 (1969).

## Patents: *see* Transgenic Animal Patents.

## Pathology: *see* specific disease diagnostic techniques; or organism affected, e.g., Bacterial Pathogenesis; Plant Pathology, Molecular.

# PCR Technology

*Henry A. Erlich*

### *Key Words*

**Anneal** To form double-stranded polynucleotides from two complementary single-stranded polynucleotides by a process of hybridization.

**Denaturation** The separation of the two hydrogen-bonded (complementary) chains of DNA into a pair of single-stranded polynucleotide molecules by a process of heating.

**Extension** The polymerization of nucleotides catalyzed by a DNA polymerase and initiated at the 3′-hydroxy terminus of a hybridized oligonucleotide primer, resulting in replication of a segment of DNA.

**Primer** A short segment of DNA or RNA complementary to the initial portion of a single strand of a DNA template with a free 3′-hydroxy group to which nucleotides are added during DNA synthesis (extension).

***Taq* Polymerase** A thermally stable DNA polymerase from the thermophilic bacterium *Thermus aquaticus,* which resists inactivation during denaturation temperatures and allows primer extension at high temperatures, thereby reducing chances for nonspecific extension products.

---

The polymerase chain reaction (PCR) is an in vitro method for the primer-directed enzymatic synthesis of millions of copies of a specific DNA segment. The reaction is based on the annealing and the extension by a thermostable DNA polymerase of two oligonucleotide primers that flank the target DNA segment. The ability to amplify a specific DNA sequence from a complex template such as human genomic DNA in a 1–2 hour automated reaction has allowed PCR to supplement or even replace many of the traditional molecular cloning applications in basic molecular biology research and has introduced DNA analysis into clinical diagnostics. The PCR-based analysis of allelic sequence variation has been applied in HLA tissue typing for transplantation, in genetic typing for individualization of forensic evidence specimens, and in the diagnosis of genetic disorders such as sickle cell anemia, cystic fibrosis, and the fragile X syndrome. PCR has also been applied to the detection of infectious disease pathogens, such as the human immunodeficiency virus and *Chlamydia,* and, in cancer diagnostics, to the detection of inherited as well as somatic mutations in oncogenes and tumor suppressor genes. The ligation onto any collection of DNA fragments of short DNA sequences that can serve as primer-binding sites confers on that collection the ability to be replicated by PCR, allowing for selective procedures and *in vitro* evolution.

## 1 PRINCIPLES

The polymerase chain reaction (PCR) is an *in vitro* reaction developed by scientists at the former Cetus Corporation in the mid-1980s for the enzymatic synthesis of millions of copies of a specific DNA segment from a complex template. The reaction is based on the annealing and extension of two oligonucleotide primers that flank the target region in duplex DNA; after denaturation of the DNA, achieved by heating the reaction, each primer hybridizes to one of the two separated strands such that extension from each 3′-hydroxy end is directed toward the other. The annealed primers are then extended on the template strand by using a DNA polymerase. These three steps (denaturation, primer binding, and DNA synthesis by primer extension) constitute a single PCR cycle. If the newly synthesized strand extends to or beyond the region complementary to the other primer, it can serve as a primer-binding site and template for subsequent primer extension reactions. Consequently, repeated cycles of denaturation, primer annealing, and primer extension result in the exponential accumulation of a discrete fragment whose termini are defined by the 5′ ends of the primers. This exponential amplification occurs because, under appropriate conditions, the primer extension products synthesized during a given cycle function as templates for the other primer in subsequent cycles. Thus, each cycle of PCR essentially doubles the amount of DNA in the region of interest. The length of the products that accumulate during the PCR is equal to the sum of the lengths of the two primers plus the distance in the target DNA between the primers. The principle of PCR amplification is illustrated in Figure 1. PCR can amplify double-stranded (ds) or single-stranded (ss) DNA and, with the reverse transcription of RNA into a cDNA copy, RNA can also serve as a target. Because the primers become incorporated into the PCR product and some base pair mismatches (away from the 3′ end) between the primer and the original genomic template can be tolerated, new sequence information (e.g., specific mutations, restriction sites, regulatory elements) and labels can be introduced via the primers into the amplified DNA fragment.

## 2 TECHNICAL DEVELOPMENTS

### 2.1 Reagents and Protocols

#### 2.1.1 Primers

The design of oligonucleotide primers to amplify a specific target DNA segment requires some knowledge of the flanking sequence information. The design of the primers should try to minimize the potential for PCR primer artifacts (e.g., the synthesis of template-independent products consisting of the primers and their complementary sequences—sometimes known as "primer-dimer") by avoiding 3′ sequences that are complementary to each other. Secondary structure within the primer and repetitive sequences should also be avoided. Whenever possible, the melting temperatures of the two primers should be very similar, to ensure that a given thermal profile is optimally efficient and specific for both primers. A variety of computer programs are available to aid in PCR primer design. In some cases, the requirement for flanking sequence information can be overcome by ligating a fragment of known sequence to template DNA fragments to serve as a priming site. PCR with random sequence primers can also be useful in amplifying DNA for some purposes.

#### 2.1.2 Polymerases

The initial studies that relied on the PCR to amplify specific targets (e.g., β-globin) from human genomic DNA utilized the Klenow fragment of *Escherichia coli* DNA polymerase I. Although widely used by molecular biologists, this polymerase is not a thermostable enzyme, and its inactivation at the high temperatures necessary for strand separation required the addition of enzyme after the denaturation step of each cycle. This requirement was eliminated by the introduction of a thermostable DNA polymerase, the *Taq* DNA polymerase, isolated from the thermophilic bacterium *Thermus aquaticus.* The use of *Taq* DNA polymerase transformed the PCR by allowing the development of simple automated thermal cycling devices for carrying out the amplification reaction in a single tube containing all the necessary reagents. The availability of thermostable polymerases has not only simplified the procedure for the PCR but has increased the specificity and yield of the amplification reaction. The incorporation of *Taq* DNA polymerase into the PCR protocol allowed the primers to be annealed and

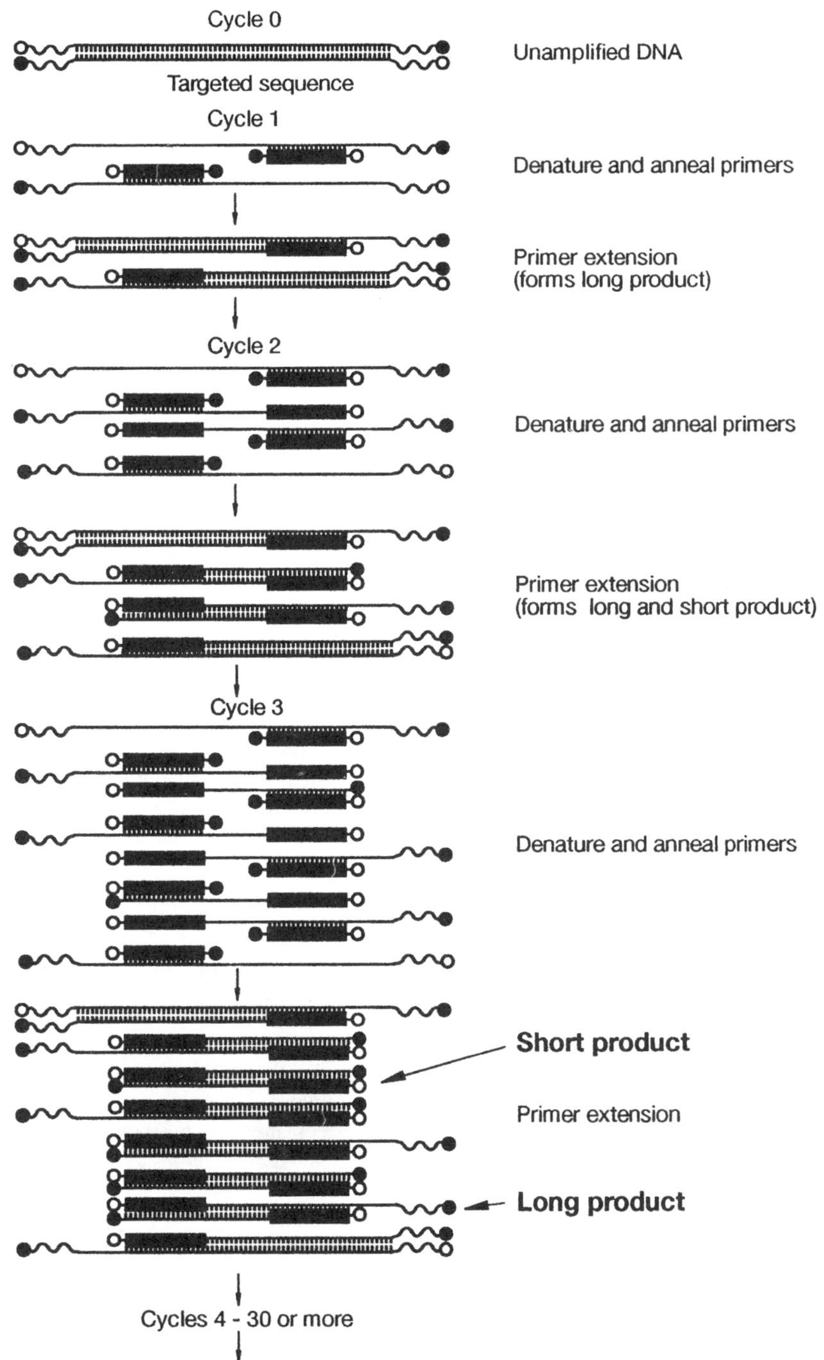


**Figure 1.** Polymerase chain reaction amplification cycles: open circles, 5′ or phosphate end; solid circles, 3′ or hydroxyl end of polynucleotide sequence.

extended at a temperature much higher (e.g., 60°C) than was possible with the Klenow fragment (e.g., 37°C), eliminating most of the nonspecific amplification. Moreover, longer PCR products could be amplified from genomic DNA, probably because there was a reduction in the secondary structure of the template strands at the elevated temperature used for primer extension with a thermostable polymerase.

Since the introduction of *Taq* polymerase, a variety of thermostable DNA polymerases from thermophilic or hyperthermophilic bacteria have been isolated, characterized, and used in the PCR. Some of these polymerases, unlike the *Taq* polymerase, have 3′- to 5′-exonuclease activity (the so-called editing or proofreading function). The cloning of the genes encoding some of these enzymes has permitted the production of genetically engineered variants of these native polymerases, such as a mutant of *Taq* polymerase that lacks the 5′- to 3′-exonuclease activity. In general, polymerases that have or lack the 5′ to -3′- and the 3′- to 5′-exonuclease activities will perform differently in the PCR. For example, polymerases with a 3′- to 5′-exonuclease activity probably will not perform well in the PCR strategy known as allele-specific amplification of sequence-specific priming based on the reduced efficiency of extension for a primer that is mismatched with the template strand at the 3′ end. This limitation of such polymerases is due to the removal of the mismatched base at the 3′ end of the primers, creating a matched 3′ end that can be efficiently extended. On the other hand, these polymerases have a greater fidelity in DNA synthesis as a result of their ability to remove misincorporated and thus mismatched nucleotides.

Many of the new thermostable polymerases have additional useful activities. The thermostable DNA polymerase from *Thermus thermophilus (Tth)* can reverse transcribe RNA efficiently in the presence of $MnCl_2$ at high temperatures. Under appropriate conditions, the DNA polymerase activity can also occur in the presence of $MnCl_2$, allowing both cDNA synthesis and PCR amplification to be carried out in a single-enzyme, single-tube reaction.

### 2.1.3 Protocols

The initial PCR amplifications with the Klenow fragment were not highly specific; although a unique DNA fragment could be amplified approximately 200,000-fold from genomic DNA, only about 1% of the PCR product was, in fact, the targeted sequence. A specific hybridization probe was therefore required to analyze the amplified DNA. Amplification with the *Taq* DNA polymerase, however, greatly increased the specificity of the reaction, so that for many amplifications the PCR products could be detected as a single ethidium bromide stained band on an electrophoretic gel. Conditions that increase the stringency of primer hybridization, such as higher annealing temperatures and lower magnesium chloride concentrations, enhance specific amplification. In addition, the concentrations of enzyme and primers as well as the annealing time, the extension time, and the number of cycles, can affect the specificity of the PCR. The concentration of a specific sequence in a sample can also influence the relative homogeneity of the PCR products. Thus, a single-copy nuclear gene present twice in every diploid cell usually can be detected as a unique band after gel electrophoresis of the PCR products; amplification of a sequence present in only one of 10,000 cells, however, may yield a more heterogeneous gel profile if special procedures to increase the specificity of PCR amplification are not used (see later).

Some recently developed approaches to improve PCR specificity are based on the recognition that the *Taq* DNA polymerase retains considerable enzymatic activity at temperatures well below the optimum for DNA synthesis (72°C). Thus, during sample setup and in the initial heating step of the reaction, primers that have annealed nonspecifically to a partially single-stranded template region can be extended and stabilized before the reaction reaches the targeted 72°C temperature for extension of specifically annealed primers. Some of these nonspecifically annealed and extended primers may be oriented with their 3′-hydroxy termini directed toward each other, resulting in the exponential amplification of a nontarget fragment. If the DNA polymerase is activated only after the reaction has reached high (> 70°C) temperatures, nontarget amplification can be minimized. This can be accomplished by an approach termed ''hot start'': the manual addition to the reaction tube of an essential reagent (e.g., DNA polymerase, magnesium chloride, primers, etc.) at elevated temperatures. Thermolabile physical barriers such as wax beads that separate essential reaction components until the wax melts at high temperatures also increase specific amplification. Thus, the amplification reaction cannot proceed until the temperature in the tube is high enough to melt the wax and allow mixing of the essential reagents.

''Hot start'' has been shown not only to improve specificity but also to minimize the formation of the ''primer-dimer,'' a double-stranded PCR product consisting of the two primers and their complementary sequences. Once formed, this PCR artifact is efficiently amplified; it is often detected in reactions with very few or no specific templates. In rare template reactions, these short amplification products may prevent efficient target amplification by competing with the target fragment for primers and enzyme.

Another approach that can improve PCR specificity is to follow the initial amplification reaction with an additional PCR with internal, single, or double-nested primers. Like the use of oligonucleotide hybridization probes, this approach utilizes sequence information internal to the two ''outer'' primers to identify the subset of amplification products that corresponds to the target fragment. The conventional nesting strategy, however, requires opening the reaction tube, eliminating or decreasing the concentration of the original ''outer'' primers, and then adding the ''inner'' primers. Although this complicated sequence can sometimes be accomplished by simply diluting the first reaction into a second one, the procedure is both inconvenient and risky, in that it provides an opportunity for contaminating the secondary reaction with amplified product from earlier reactions. The problem can be overcome if the outer and inner primers are all present in the initial reaction mix and if the thermal profile is programmed first to allow the outer primers but not the inner primers to amplify the targeted subset of the initial PCR products, and then to allow amplification by the inner but not the outer primers.

## 2.2 Potential Problems

### 2.2.1 Contamination

Given the capacity of PCR to generate trillions of DNA copies from a template sequence, contamination of the amplification reaction for a given sample either with products of an earlier PCR reaction (product carryover) or with exogenous DNA or cellular material can create problems both in research and in diagnostic applications. Valuable precautions to minimize the risk of contamination include attention to careful laboratory procedures, the prealiquoting of reagents, the physical separation of the reaction preparation from the area of reaction product analysis, and the use of dedicated pipets, positive-displacement pipets, or tips with aerosol barriers that prevent contamination of the pipet barrel. Multiple negative controls (no template DNA added to the reaction) are necessary to monitor the process and to reveal any contamination. In genetic typing, as opposed to infectious disease diagnostic testing for the presence or absence of a pathogen sequence, the contamination of a sample reaction often can be detected by a genotyping result with more than two alleles.

Recently, several approaches to minimizing the potential for PCR product carryover have been developed, all based on preventing the amplification products from serving as a template. One such strategy utilizes the principles of the restriction–modification and excision repair systems of bacteria to pretreat amplification reactions and selectively destroy DNA synthesized in an earlier PCR. To distinguish PCR products from sample template DNA, deoxyuridine triphosphate (dUTP) is substituted for deoxythymidine triphosphate (dTTP) in the PCR reagents and is thus incorporated into the amplification products. The presence of this unconventional nucleotide allows products of earlier PCR amplifications to be distinguished from the native DNA of the sample. The enzyme uracil *N*-glycosylase (UNG), added to the reaction prior to thermal cycling, catalyzes the excision of uracil from any potential single- or double-stranded PCR carryover DNA present in the reaction prior to the first PCR cycle. The resulting abasic polynucleotides are susceptible to hydrolysis in slightly alkaline solutions (like PCR buffers) and elevated temperatures. Since UNG itself is inactivated at temperatures used in PCR, the amplification products generated during the thermal cycling are not destroyed and can accumulate.

### 2.2.2 Misincorporation and "Recombinant PCR"

During the course of PCR amplification, a mismatched nucleotide can become incorporated in the newly synthesized strand. The misincorporation rate is dependent on the reaction conditions (e.g., nucleotide concentration, pH, divalent cation concentration) and the thermostable polymerase used in the PCR. For *Taq* polymerase, the misincorporation rate is about $10^{-5}$ nucleotide per cycle.

Another category of very rare PCR artifacts revealed initially by sequencing cloned PCR amplification products consists of the hybrid sequences generated by in vitro recombination or template strand switching. In amplification reactions from heterozygous individuals, hybrid sequences can, under certain conditions (e.g., very high concentrations of PCR products), result from a primer that was partially extended on one allelic template but switches to the other allelic template in a subsequent cycle. Although misincorporation is rare and the appearance of hybrid sequences even rarer, it is nonetheless advisable to distinguish the sequence of the true genomic template from any potential PCR artifact by sequencing multiple clones derived from a single amplification reaction or by cloning the PCR products from several independent amplifications. For most PCR applications, such as direct sequencing or oligonucleotide probe typing, it is the population of amplification products that is analyzed, and therefore rare misincorporated nucleotides are not detected. A specific misincorporation will be present in a significant fraction of the PCR products only if the insertion of an incorrect base occurs during the first cycle of a reaction initiated with very few templates. The determination of individual cloned sequences derived from PCR products, however, can reveal such rare errors.

### 2.2.3 Recent Advances: Long PCR, Kinetic PCR, in situ PCR

About 9 years after the initial report of PCR amplification of a 110 bp β-globin fragment using the Klenow fragment of *E. coli* DNA polymerase I, several modifications of the PCR procedure resulted in the capacity to amplify 30 kb fragments from genomic DNA. Even longer fragments can be amplified from cloned templates of lower complexity, such as insert DNA in λ, cosmid, and yeast artificial chromosome (YAC) clones; for some of these vectors, PCR is able to amplify the whole insert sequence. These developments should have a major impact on cloning and sequencing strategies and are expected to expand even more the critical role played by PCR in molecular biology research. Changes in the DNA extraction protocols and modified buffers prevented single-strand nicks and provided longer and more intact templates for amplifying these very long fragments. Longer extension times (> 15 min) and relatively high denaturation temperatures, as well as the addition of cosolvents such as glycerol and DMSO, were critical in the amplification of these longer PCR products. In addition, specific and efficient amplification required some form of hot start (see Section 2.1.3). Finally, the choice of thermostable polymerase proved to be critical as well in that the long fragments could be produced only by supplementing an efficient polymerase that lacked the 3′- to 5′-exonuclease activity ("proofreading" or "editing") such as *Taq* or *Tth* polymerase with a small amount of another enzyme (e.g., Vent polymerase) that has this function. The rationale for this enzyme blend is that the 3′-exonuclease activity is capable of removing misincorporated bases that otherwise would prevent the newly synthesized strand from being extended beyond the misincorporated (and mismatched) base.

Another technical advance has been the ability to monitor in real time the accumulation of PCR products by the increase in fluorescence of ethidium bromide or other DNA-binding dyes added to the PCR. Video camera monitoring of the fluorescence in the array of reaction tubes in a thermal cycler can provide quantitative data as well as a convenient diagnostic format.

The ability to amplify templates within cells or tissue sections in slides, known as in situ PCR, also represents a significant advance with broad diagnostic as well as basic research potential.

## 2.3 Instruments

In addition to the advances in the reagents and protocols for carrying out efficient and specific amplifications, new instruments for automated thermal cycling and for analyzing the PCR products have been developed. These thermal cyclers have increased rates of heating, cooling, and heat transfer to modified reaction vessels and can accommodate more samples (e.g., the 96-well array). Some of these show a reduced thermal gradient across the heating block, resulting in more precise thermal profiles from well to well. In addition, some new models do not require the use of mineral oil to prevent evaporation and increase the rate of thermal equilibration. Instruments to cycle temperature for amplification reactions on slides rather than in standard plastic microfuge tubes have been developed for in situ PCR. For on-line monitoring of amplification reactions ("kinetic PCR"), instruments that detect the fluorescence accumulating in reaction vessels have also been designed using either fiber-optics or video cameras. In addition, instruments for the gel electrophoretic analysis of PCR products labeled with fluorescent primers have been introduced recently.

## 2.4 Detection Formats

Many different methods for detecting the presence of PCR products or for analyzing their sequence have been developed. Detection based on the identification of electrophoretically resolved, ethidium bromide stained DNA fragments of a specific length has been a widely used method. (Combined with restriction enzyme digestion, this gel-electrophoretic method is capable of distinguishing some sequence polymorphism.) Recently, the analysis of fluorescently labeled PCR products has been facilitated by the availability of laser-scanning instruments. These gel-electrophoretic approaches are valuable for detecting length variation (e.g., the variable number tandem repeat loci used for individual identification and/or genetic mapping) in the PCR products. Alternative separation methods, such as high performance liquid chromatography or capillary electrophoresis, may prove valuable in the future for analyzing length variation. Some gel-electrophoretic techniques, such as denaturing gradient gel electrophoresis or the single-strand conformational polymorphism method, which is based on differential mobility in native gels caused by the secondary structure of single-stranded PCR products, are capable of detecting some of the sequence variation in amplified DNA. Chemicals or enzymes that cleave DNA heteroduplexes with mismatched base pairs represent another approach to analyzing sequence variation in the PCR product.

If the sequence variants to be detected are known, sequence-specific oligonucleotide (SSO) hybridization probes are a valuable and widely used method for the detecting as well as the genetic typing of PCR products. Probe-based detection is critical if the amplification reactions are not completely specific; it also eliminates the need for gel electrophoresis. Immobilizing a panel of probes in a spatial array allows the typing of the PCR product with many probes in a single hybridization reaction; this immobilized probe format has been widely used in diagnostic tests. If the specificity of the PCR is very high, fluorescent labeling of the PCR product followed by detection of fluorescence accumulation represents a very promising approach to developing very simple, rapid, and homogeneous diagnostic tests.

Some detection methods depend on the specificity of primer extension; mismatches between the template strand and the 3′ end of the primer are extended very inefficiently, under the appropriate reaction conditions. Based on this property, genetic typing methods known as allele-specific amplification or sequence-specific priming have been developed in which a PCR product is generated only if the specific allele corresponding to a specific primer pair is present in the sample. These approaches show great potential for the detection of specific sequence variants present in rare cells, such as somatic mutations in an oncogene present in a small proportion of cells within the sample.

## 3 APPLICATIONS

### 3.1 Research

The ability to amplify specific DNA segments from a complex template such as genomic DNA or, following reverse transcription, from RNA, has allowed PCR to replace the construction and screening of recombinant genomic and cDNA libraries for many applications. In other cases, PCR has greatly facilitated subsequent procedures, such as sequencing, oligonucleotide probe hybridization, and restriction site analysis. Given the capacity for automation of the PCR and many of these analytic methods, the genetic characterization of a large number of samples can be carried out within hours. In addition to simplifying many basic molecular biology procedures, thus introducing the capacity for DNA sequence analysis to a wide variety of biological disciplines, PCR has made possible several new experimental approaches to research questions based on the ability to amplify specific segments from very small samples (e.g., individual sperm).

In cell biology, gene expression in specific cell lineages has been analyzed by quantitative PCR methods applied to mRNA templates. These quantitative approaches typically use internal standards preceding the cDNA synthesis step; as noted, now both reverse transcription and PCR can be carried out by a single thermostable enzyme, *Tth,* from *Thermus thermophilus.*

In molecular genetics, PCR has provided a wealth of markers for genetic mapping as well as for the construction of physical maps. In addition to genetic markers based on sequence polymorphism, genetic markers based on length polymorphism—in particular, the frequent and widely distributed simple sequence repeats (e.g., the CA repeats)—have proved very valuable in constructing genetic maps based on recombination within extensive family pedigrees. The ability to type PCR products amplified from individual sperm, which makes possible the detection of recombinant sperm, has introduced a new and very powerful approach to genetic mapping. The analysis of sequences amplified from sperm also provides a novel way of analyzing mutation. The construction of physical maps and the ordering of cloned DNA segments have been facilitated by the identification of unique genomic sequences, known as sequence tagged sites (STSs). The presence in a given cloned fragment of these landmark STSs, each of which is associated with a specific primer pair, can be assayed by simply running a PCR with the appropriate primers.

The analysis of sequence variation, in addition to providing a rich source of genetic markers, has proved invaluable for the identification of specific mutants in critical disease-associated genes—for example, *CFTR* (the cystic fibrosis transmembrane conductance regulator gene)—and has provided significant insights into structure–function relationships. This approach, which offers a relatively easy way to obtain DNA sequence information from a variety of individuals, has revolutionized molecular evolutionary studies; phylogenetic analyses of the sequence database generated by PCR has allowed the reconstruction of evolutionary histories for various species, including our own. The analysis of mitochondrial DNA sequence variation, for example, has suggested an African origin of modern humans and a migration out of Africa about 200,000 years ago. The phylogenetic analysis of HIV sequences derived from patient samples has also proved critical in understanding the natural history of HIV infection and has provided a valuable tool in epidemiological studies and the tracing of HIV transmission.

The ligation of short DNA sequences that can serve as primer-binding sites onto a collection of DNA fragments confers on that entire collection (e.g., genomic DNA fragments) the ability to be replicated by PCR (whole-genome PCR). If the collection of fragments to be replicated contains a random sequence segment, successive rounds of some selective procedures (e.g., binding to a given protein) can be applied and a sequence with the selected property can be isolated. This in vitro evolution approach using PCR has great potential for generating nucleotide sequences with a variety of useful properties.

### 3.2 Clinical Diagnosis

For genetic disorders, such as the autosomal recessive diseases sickle cell anemia, β-thalassemia, and cystic fibrosis, PCR has been used to amplify the gene segments that contain known mutations, and a variety of techniques can be used for direct mutation detection. Such strategies may involve the coamplification of several exons in a single PCR. PCR/SSO probe methods for direct mutation detection are useful for carrier screening as well as for prenatal diagnosis for such diseases, provided the majority of mutations have been characterized. A recently developed format in which the PCR product, labeled with biotinylated primers, is hybridized in a single reaction to an array of probes immobilized on a nylon membrane is used in PCR-based genetic tests for forensic purposes and for human individualization, for HLA tissue typing for transplantation, and for detecting mutations in the *CFTR* gene.

The detection of infectious disease pathogens using PCR amplification of a pathogen-specific DNA segment is a very sensitive and powerful diagnostic test. Many PCR-based infectious disease tests are offered as service tests from clinical laboratories, and commercial test kits are now available. PCR-based tests for the presence of HIV-1 proviral DNA or RNA offer many advantages over conventional diagnostic tests, such as early detection (due to increased sensitivity). Unlike serologic tests, moreover, PCR-based tests are valuable in assessing mother–child transmission. Quantitative PCR tests for HIV RNA, based on reverse transcription of the viral RNA into DNA, are useful not only in clinical diagnosis but in evaluating the efficacy of new drugs. Many other PCR tests for pathogenic fungi and bacteria, as well as DNA and RNA viruses, are currently performed from a variety of clinical samples (blood, urine, sputum, etc.).

There is enormous clinical potential available in the capacity of PCR to detect rare somatic mutations in oncogenes or tumor suppressor genes as well as some of the cancer-associated translocations. Early detection of cancer cells by sampling of sputum, urine, blood, or stool sample is a very exciting possibility. The use of a strategy known as allele-specific amplification, in which the PCR primers amplify the rare somatic mutation selectively, represents an extremely sensitive approach. Already, the ability of PCR to detect specific chromosomal translocations (e.g., the *bcr-abl* gene fusion resulting from the translocation known as the Philadelphia chromosome, associated with chronic myeloid leukemia) has been applied to monitor residual disease in treated patients in many clinical reference laboratories. The identification of inherited mutations, such as in the p53 gene, the retinoblastoma locus, or in the recently identified *HNPCC* (hereditary nonpolyposis colon cancer) locus can also help identify individuals at high risk for cancer.

### 3.3 Forensic Science

Conventional genetic markers, like blood group serologic types or protein electrophoretic variants, cannot be applied to many forms of forensic evidence. In addition, the use of DNA fingerprinting, the analysis of genomic length polymorphisms (variable number tandem repeats, VNTRs) based on Southern blotting, typically requires reasonably large amounts (> 100 ng) of intact, high molecular weight DNA. PCR, however, is capable of analyzing minute quantities of degraded DNA, and this method is now widely used to analyze the genetic patterns of polymorphic DNA sequences amplified from biological evidence found at crime scenes. One PCR-based test, a reverse dot-blot test for polymorphism at the HLA-DQα (*DQA1*) locus (HLA-DQα Amplitype kit), has been applied by many forensic laboratories to the analysis of thousands of cases since *Pennsylvania* vs. *Pestinikis* in 1986, the first case in which such PCR results were admitted as evidence. For most of these evidence samples (typically > 5 evidence specimens/case), DNA typing based on RFLP analysis could not provide a genotyping result.

In general, PCR typing of evidence samples for individual identification can be based either on the analysis of length polymorphism (VNTR regions or di$^-$, tri$^-$, or tetranucleotide repeats, known as short tandem repeat, STR, regions) or sequence polymorphism (see Section 2.4). The amplification of VNTR or STR regions can be carried out by using primers complementary to unique sequences that flank the tandem repeat regions, and allelic variants can be distinguished by determining the length of the PCR products amplified from the sample. In addition, if the two alleles determine PCR products of very different lengths, the smaller fragment can, under conditions of limiting DNA polymerase, be preferentially amplified. In addition, the larger allelic fragment may be preferentially lost when amplifying a VNTR polymorphism from a partially degraded DNA sample. For both VNTR and STR region markers, the definition of alleles in the sample is made using an allelic size ladder, a pool of PCR products comprising the range of allelic size variants in the population. In some cases, however, sequence polymorphism may also be present in VNTR regions, making the definition of allelic length variants by electrophoretic mobility difficult. One recent approach to PCR-based genetic typing, the minisatellite variable repeat (MVR) approach, utilizes *both* the length and sequence polymorphism in some VNTR regions for purposes of individualization.

Sequence polymorphisms can be detected by a variety of methods (see Section 2.4) including the use of oligonucleotide hybridization probes. With the use of the immobilized probe/reverse dot-blot method, in which a panel of oligonucleotide probes is immobilized in a spatial array on a nylon membrane, the complexity of the test is independent of the number of probes because the amplified sample is analyzed with *all* the probes in a single hybridization reaction. As already noted, this format was developed for typing the polymorphism at the *HLA-DQA1* locus; it was the first commercially available PCR-based genetic typing kit. If the genotype of the evidence sample differs from that of a suspect's reference sample, that person is excluded or eliminated from the pool of potential donors of the evidence. If the genotypes of the suspect and the evidence sample match, the possibility that the suspect is in fact the donor cannot be eliminated. The significance of this match for identifying the source of the evidence specimen is a function of the frequency of this specific genotype, as determined by population surveys. Similar dot-blot tests for determining the sequence polymorphism in the D-loop segment of mitochondrial DNA have recently been developed and applied to the forensic analysis of bone fragments.

A recently developed genetic test is based on the coamplification of five additional marker loci (each of which has two or three alleles) and HLA-DQα and analysis of the PCR products by the reverse dot-blot method. If the DNA type of multiple loci of an evidence sample matches that of the suspect, the probability that a random individual in a given population would have the same genetic pattern can be estimated, assuming statistical independence, by multiplying the probabilities of the individual loci. Clearly, as more PCR-based genetic markers are developed and used, the potential for individualization will increase dramatically.

The same genetic tests for individual identity testing in forensic evidence specimens have also been used in paternity testing as well as in the monitoring of bone marrow engraftment by distinguishing donor and recipient cells. Potential mix-ups of cell lines and clinical specimens have also been detected quickly by simple PCR-based assays.

## 4 PERSPECTIVES

In the decade or so since the first report of PCR amplification and its use for genetic diagnosis in 1985, the number of different applications has grown dramatically. The technology of PCR amplification has also evolved over this period as new polymerases and new protocols have increased the efficiency and specificity of PCR as well as the length of amplified DNA fragments. With this recently developed capacity, PCR is likely to replace conventional molecular cloning strategies even more than it has in the past. In addition, new developments such as in situ PCR promise to have a significant impact in both research and diagnostic applications. In general, PCR has made the information embedded in nucleotide sequences more accessible to the basic researcher and clinician alike.

*See also* DNA Fingerprint Analysis; DNA in Neoplastic Disease Diagnosis; Immuno-PCR; Ligation Assays.

### *Bibliography*


Arnheim, N., and Erlich, H. A. (1992) PCR strategy. *Annu. Rev. Biochem.* 61:131–156.

Cheng, S., Fockler, C., Barnes, W. M., and Higuchi, R. (1994) Effective amplification of long targets from cloned inserts and human genomic DNA. *Proc. Natl. Acad. Sci. U.S.A.* 91:5695–5699.

Erlich, H., Ed. (1989) *PCR Technology: Principles and Applications for DNA Amplification.* Stockton Press, New York.

Erlich, H. A., and Arnheim, N. (1992) Genetic analysis using the polymerase chain reaction. *Annu. Rev. Genet.* 26:479–506.

———, Gelfand, D., and Sninsky, J. J. (1991) Recent advances in the polymerase chain reaction. *Science,* 252:1643–1651.

Higuchi, R., Fockler, C., Dollinger, G., and Watson, R. (1993) Kinetic PCR analysis: Real-time monitoring of DNA amplification reactions. *Bio/Technology,* 11:1026–1030.

Innis, M. A., Gelfand, D. H., Sninsky, J. J., and White, T. J., Eds. (1990) *PCR Protocols: A Guide to Methods and Applications.* Academic Press, San Diego, CA.

Mullis, K. B., Ferré, F., and Gibbs, R. A., Eds. (1994) *The Polymerase Chain Reaction.* Birkhäuser, Boston.

Persing, D. H., Smith, T. F., Tenover, F. C., and White, T. J., Eds. (1993) *Diagnostic Molecular Microbiology. Principles and Applications.* American Society for Microbiology, Washington, DC.

White, T., Arnheim, N., and Erlich, H. A. (1989) The polymerase chain reaction. *Trends in Genet.* 5(6):185–189.

———, Madej, R. M., and Persing, D. H. (1992) The polymerase chain reaction: Clinical applications. *Adv. Clin. Chem.* 29:161–196.


# Peptides

*Tomi K. Sawyer, Wayne L. Cody, Daniele M. Leonard, and Mac E. Hadley*

### *Key Words*

**Peptide** A naturally occurring or synthetic (recombinant or chemical) molecule composed essentially of amino acids typically linked together by their amino and carboxy groups. Peptides cover a wide range in structural size and complexity, and differ in the scope of their particular biological properties, which typically may include hormone, neurotransmitter, growth factor, or cytokine activity.

**Peptide Hormone** A subgroup of naturally occurring peptides that are released into the bloodstream from the cell in which they are biosynthesized and delivered to their particular target tissue to stimulate biological effects. Typically, peptide hormones effect their cellular actions by binding to specific receptors and inducing an increase (or decrease) in production of so-called second messengers.

**Peptide Therapeutic** A clinically effective peptide (naturally occurring or synthetic), which was developed with a particular therapeutic utility. Typically, peptide therapeutics are antagonists of the native molecule or are sustained-acting agonists as the result of enhanced stability to peptidases. Peptide therapeutics have been developed for numerous central nervous system, cardiovascular, gastrointestinal, immunological, reproductive, or growth/metabolic disorders.

---

Peptides are a major class of naturally occurring molecules that regulate a variety of biological functions, including central nervous system, cardiovascular, gastrointestinal, immunological, reproductive, or growth/metabolic activities. Peptides are composed essentially of amino acids bonded in a specific manner by backbone amide linkage (peptide bond) and, in some cases, side chain linkage. Over the last 20 years, the development of numerous peptide therapeutics has been successfully advanced for the treatment of diabetes (insulin), osteoporosis (calcitonin), prostate cancer and endometriosis (gonadotropin-releasing hormone), acromegaly and ulcers (somatostatin), diuresis and hypertension (vasopressin), hypoglycemia (glucagon), and hypothyroidism (thyrotropin-releasing hormone).

In recent years the discovery and clinical application of numerous synthetic peptides and their second-generation derivatives (i.e., peptidomimetics and pseudopeptides) have been successfully advanced for yet other life-threatening or disabling afflictions, such as immunological disorders (e.g., AIDS), cancer, neural dysfunction (e.g., Alzheimer's disease), and cardiovascular disease (e.g., hypertension). Finally, the discovery of the peptide sweetener Aspartame™ provides impetus to research and development strategies to identify yet other applications of peptides.

## 1 CHEMICAL PROPERTIES

### 1.1 Peptide Structure

Peptides are composed of two or more amino acids linked together by a peptide bond (—CONH—, Figure 1). Naturally occurring peptides are generally composed of L-stereoisomers (except for Gly) of the 20 or so genetically coded amino acids. Peptides, polypeptides, or proteins may be structurally differentiated based on the number of constitutive amino acids (e.g., "peptide," 2–20 amino acids; "polypeptides," 20–50 amino acids; "proteins," more than 50 amino acids). For our purposes, the term "peptide" will be used generically to designate all such chemical agents or messengers.

The precise linear arrangement, amino→carboxy (N→C) directionality, of amino acids in a peptide is referred to as its primary structure (e.g., substance P: Arg-Pro-Lys-Pro-Gln-Gln-Phe-Phe-Gly-Leu-Met-$NH_2$; also see Table 1). The three-dimensional folding of a peptide through covalent (e.g., S—S, disulfide) or noncovalent (e.g., hydrogen and/or ionic) bonding determines the secondary structure of the peptide. Furthermore, the disulfide linkage between two or more cysteine residues dictates the transformation of linear

*Backbone/Side-Chain Dihedral Angles of Phe in a Peptide*

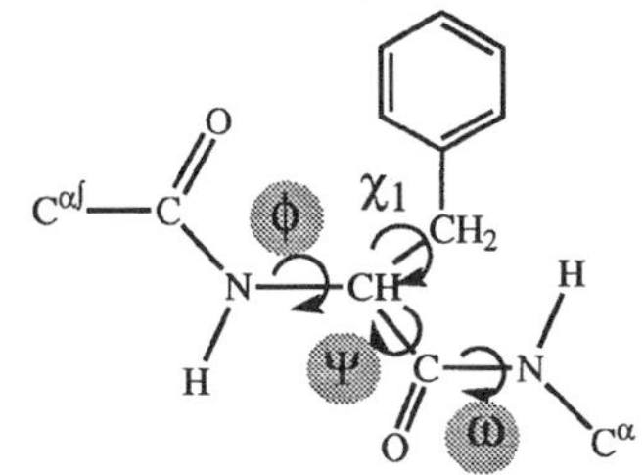


*Cis and Trans Peptide Bond Isomers of Phe-Pro*

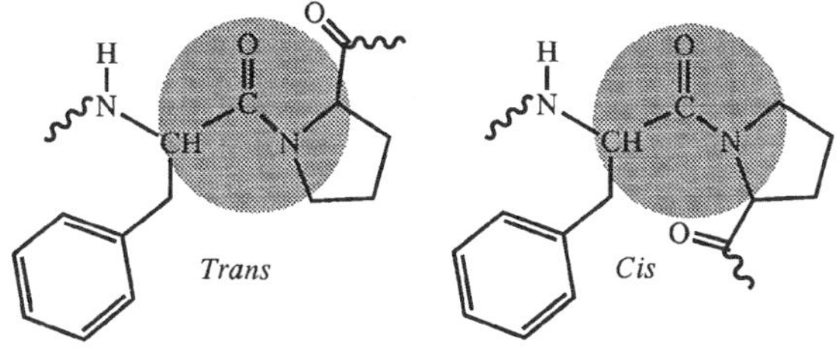


***Naturally-Occurring and/or Synthetic Analogues of Phe***

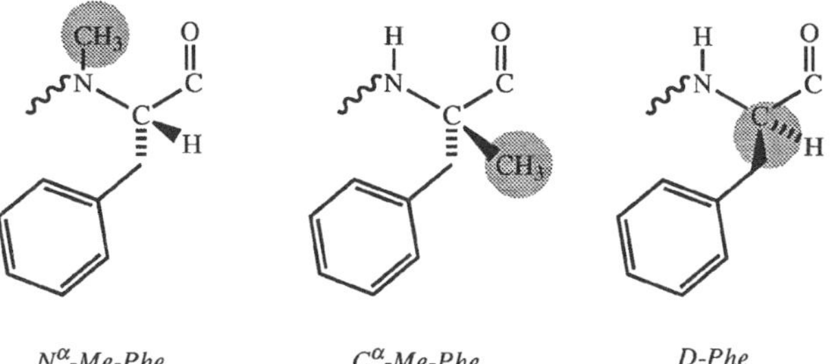


**Figure 1.** Some 3D chemical features of peptides.

peptides to cyclic (or multicyclic) peptides. Such chemically defined restriction of the molecular flexibility of linear peptides leads to higher ordered three-dimensional structural complexity and defined spatial arrangement of constitutive amino acid side chains (i.e., tertiary structure). Contributing to such secondary structure is the tendency of peptides to adopt spatial orientations of both the backbone and side chain functionalities such that hydrophobic amino acids (e.g., Trp, Tyr, Phe, Leu, Ile, Val, Met) form a hydrophobic surface. Similarly, hydrophilic amino acids (e.g., Lys, Arg, Glu, Asp, Thr, Ser, Gln, Asn, His) may form a hydrophilic surface, which provides a high degree of solvation by water. Furthermore, peptides may be composed of two amino acid chains, or subunits, forming a so-called dimeric structure composed of either homologous or heterologous subunits (i.e., quaternary structure). The individual subunits may be either covalently (e.g., disulfide) or noncovalently (e.g., ionic or hydrophobic) bonded together. Both interchain and intrachain disulfide bonds may be present within a peptide. In some cases, such a complex chemical species may be required to establish the biologically active form of a peptide (e.g., nerve growth factor dimer).

The three-dimensional substructure of peptides may be further described in terms of torsion angles between the backbone amine nitrogen ($N^{\alpha}$), backbone carbonyl carbon ($C'$), backbone hydrocarbon ($C^{\alpha}$), and side chain hydrocarbon functionalizations (e.g., $C^{\beta}$, $C^{\gamma}$, $C^{\delta}$, $C^{\varepsilon}$ of Lys) as depending on the amino acid sequence (Figure 1). With respect to the amide bond torsion angle ($\omega$), the trans geometry is typical for most dipeptide substructures; when Pro is the C-terminal partner, however, the cis geometry is possible (Figure 1). It is noted that the three-dimensional structural flexibility is directly related to covalent and/or noncovalent bonding interactions within the amino acid sequence of a particular peptide, and synthetic modifications such as $N^{\alpha}$-alkylation, $C^{\alpha}$-alkylation, and $C^{\beta}$-alkylation effect significant conformational constraints locally as related to backbone and/or side chain torsion angles (Figure 1). Finally, it is noted that backbone amide replacements may be defined in terms of a nomenclature system wherein the functional group substitution of the dipeptide substructure is identified as a $\psi$ [surrogate]; for example, substitution of the reduced bond or aminomethylene surrogate in Leu-Leu would be described as Leu$\psi$[$CH_2$NH]Leu.

### 1.2 Peptide Synthesis

The chemical synthesis of peptides has advanced over the past hundred years, with noteworthy contributions from numerous chemists. In retrospect, the field of peptide chemistry is enriched with scientific achievements related to solution and solid phase synthesis of complex peptides, the purification and analytical characterization of peptides, and, more recently, the development and implementation of specialized technologies allowing the rapid identification of peptide leads (e.g., peptide libraries). The scope of peptide synthesis pervades both naturally occurring peptides and their analogues, which have been developed to exploit conformational modeling hypotheses (local or global three-dimensional structural constraints), molecular recognition and mechanistic properties at biological targets (receptors, enzymes, antibodies, nucleic acids), and/or metabolism by processing or degrading peptidases. The chemical synthesis of peptides has become a rapidly developed scientific art owing to such factors as automated solid phase methods, orthogonal protection/cleavage strategies, new reagents, and documentation of potential side reactions.

Most recently, the chemical synthesis of peptides has been expanded to the preparation of combinatorial peptide libraries for which high volume biological screening (e.g., receptor binding or enzyme inhibition) is performed to identify compounds that possess activity. As opposed to evaluating the native peptide with respect to biological activity, such techniques provide tremendous chemical diversity of potential peptide lead compounds to study molecular recognition and mechanistic aspects of peptide interaction with proteins (receptors, enzymes, antibodies), nucleic acids, or carbohydrate targets. Similarly, phage-based peptide libraries have been developed for such applications.

## 2 BIOCHEMISTRY AND PHYSIOLOGY

### 2.1 Peptide Biosynthesis and Metabolism

Peptides are biosynthesized on ribosomes, where their specific amino acid sequence is determined (translated) by a specific messenger RNA sequence (nucleotide triplet codes for an amino acid). The nucleotide sequences of the RNA are dictated (transcribed) from specific chromosomal deoxyribonucleotide sequences (DNA genes). Cellular biosynthesis of peptides is believed to proceed very specifically, with error rates of less than 1 in $10^4$ amino acids at a rate of 20 amino acids per second. The nascent peptides are subsequently released and transported into the cisternae of the

**Table 1** Primary Structures of Peptides

| Peptide | Primary Structure |
|---|---|
| Aspartame™ | Asp$^{1}$-Phe$^{2}$-OMe |
| Glutathione | γ-Glu$^{1}$-Cys-Gly$^{3}$ |
| Thyrotropin-Releasing Hormone | <Glu$^{1}$-His-Pro$^{3}$-NH$_2$<br>*<Glu, pyroglutamic acid |
| FMRF-amide | Phe$^{1}$-Met-Arg-Phe$^{4}$-NH$_2$ |
| Enkephalin(Met) | Tyr$^{1}$-Gly-Gly-Phe-Met$^{5}$ |
| DPDPE | Tyr$^{1}$-*D*-Pen-Gly-Phe-*D*-Pen$^{5}$<br>*Pen, penicillamine |
| BQ-123 | cyclo(Leu-*D*-Trp-*D*-Asp-Pro-*D*-Val) |
| L-365,209 | cyclo(Pro-*D*-Phe-Ile-*D*-Dhp-Dhp-*D*-MePhe)<br>*Dhp, dehydopiperazyl |
| GHRP | His$^{1}$-*D*-Trp-Ala-Trp-*D*-Phe-Lys$^{6}$-NH$_2$ |
| Ebiratide™ | Met(O$_2$)$^{1}$-Glu-His-Phe-*D*-Lys-Trp$^{6}$-NH-(CH$_2$)$_8$-NH$_2$<br>*Met(O$_2$), methionine sulfone |
| Cholecystokinin-8 | Asp$^{1}$-Tyr[SO$_3$H]-Met-Gly-Trp-Met-Asp-Phe$^{8}$-NH$_2$ |
| Angiotensin II | Asp$^{1}$-Arg-Val-Tyr-Ile-His-Pro-Phe$^{8}$ |
| Sandostatin™ | *D*-Phe-Cys-Phe-*D*-Trp-Lys-Thr-Cys-Thr$^{8}$-ol<br>*Thr-ol, threoninol |
| Oxytocin | Cys$^{1}$-Tyr-Ile-Gln-Asn-Cys-Pro-Leu-Gly$^{9}$-NH$_2$ |
| Vasopressin | Cys$^{1}$-Tyr-Phe-Gln-Asn-Cys-Pro-Arg-Gly$^{9}$-NH$_2$ |
| Desmopressin™ | Mpa$^{1}$-Tyr-Phe-Gln-Asn-Cys-Pro-*D*-Arg-Gly$^{9}$-NH$_2$<br>*Mpa, mercaptoproprionic acid |
| Teprotide | <Glu-Trp-Pro-Arg-Pro-Gln-Ile-Pro-Pro$^{9}$ |
| Bradykinin | Arg$^{1}$-Pro-Pro-Gly-Phe-Ser-Pro-Phe-Arg$^{9}$ |
| HOE-140 | *D*-Arg$^{1}$-Arg-Pro-Hyp-Gly-Thi-Ser-*D*-Tic-Oic-Arg$^{11}$<br>*Hyp, 4-hydroxyproline; Thi, thienylalanine; Tic, tetrahydroisoquinoline-carboxylic acid; Oic, octahydroindole-2-carboxylic acid |
| Gonadotropin-Releasing Hormone | <Glu$^{1}$-His-Trp-Ser-Tyr-Gly-Leu-Arg-Pro-Gly$^{10}$-NH$_2$ |
| Buserelin™ | <Glu$^{1}$-His-Trp-Ser-Tyr-*D*-Ser($^{t}$Bu)-Leu-Arg-Pro$^{9}$-NH-Et |
| Substance P | Arg$^{1}$-Pro-Lys-Pro-Gln-Gln-Phe-Phe-Gly-Leu-Met$^{11}$-NH$_2$ |
| Cyclosporin A | cyclo(MeLeu-MeLeu-Me-Thr[4*R*-4(*E*-2-butenyl)-4-methyl]-Abu-Sar-MeLeu-Leu-MeLeu-Ala-*D*-Ala-MeLeu) |
| α-Melanotropin | Ac-Ser$^{1}$-Tyr-Ser-Met-Glu-His-Phe-Arg-Trp-Gly-Lys-Pro-Val$^{13}$-NH$_2$ |
| Neurotensin | <Glu$^{1}$-Leu-Tyr-Glu-Asn-Lys-Pro-Arg-Arg-Pro-Tyr-Ile-Leu$^{13}$ |
| Somatostatin | Ala$^{1}$-Gly-Cys-Lys-Asn-Phe-Phe-Trp-Lys-Thr-Phe-Thr-Ser-Cys$^{14}$ |
| Endothelin | Cys$^{1}$-Ser-Cys-Ser-Ser-Leu-Met-Asp-Lys-Glu-Cys-Val-Tyr-Phe-Cys-His-Leu-Asp-Ile-Ile-Trp$^{21}$ |
| Glucagon | His$^{1}$-Ser-Gln-Gly-Thr-Phe-Thr-Ser-Asp-Tyr-Ser-Lys-Tyr-Leu-Asp-Ser-Arg-Arg-Ala-Gln-Asp-Phe-Val-Gln-Trp-Leu-Met-Asp-Thr$^{29}$ |
| Galanin | Gly$^{1}$-Trp-Thr-Leu-Asn-Ser-Ala-Gly-Tyr-Leu-Leu-Gly-Pro-His-Ala-Ile-Asp-Asn-His-Arg-Ser-Phe-His-Asp-Lys-Tyr-Gly-Leu-Ala$^{29}$-NH$_2$ |
| Calcitonin | Cys$^{1}$-Gly-Asn-Leu-Ser-Thr-Cys-Met-Leu-Gly-Thr-Tyr-Thr-Gln-Asp-Phe-Asn-Lys-Phe-His-Thr-Phe-Pro-Gln-Thr-Ala-Ile-Gly-Val-Gly-Ala-Pro$^{33}$-NH$_2$ |
| Neuropeptide-Y | Tyr$^{1}$-Pro-Ser-Lys-Pro-Asp-Asn-Pro-Gly-Glu-Asp-Ala-Pro-Ala-Glu-Asp-Met-Ala-Arg-Tyr-Tyr-Ser-Ala-Leu-Arg-His-Tyr-Ile-Asn-Leu-Met-Thr-Arg-Gln-Arg-Tyr$^{36}$-NH$_2$ |
| Adrenocorticotropin | Ser$^{1}$-Tyr-Ser-Met-Glu-His-Phe-Arg-Trp-Gly-Lys-Pro-Val-Gly-Lys-Lys-Arg-Arg-Pro-Val-Lys-Val-Tyr-Pro-Asn-Gly-Ala-Glu-Asp-Glu-Ser-Ala-Glu-Ala-Phe-Pro-Leu-Glu-Phe$^{39}$ |
| Corticotropin-Releasing Factor | Ser$^{1}$-Glu-Glu-Pro-Pro-Ile-Ser-Leu-Asp-Leu-Thr-Phe-His-Leu-Leu-Arg-Glu-Val-Leu-Glu-Met$^{21}$-Ala-Arg-Ala-Glu-Gln-Leu-Ala-Gln-Gln-Ala-His-Ser-Asn-Arg-Lys-Leu-Met-Glu-Ile-Ile$^{41}$-NH$_2$ |
| Growth Hormone-Releasing Factor | Tyr$^{1}$-Ala-Asp-Ala-Ile-Phe-Thr-Asn-Ser-Tyr-Arg-Lys-Val-Leu-Gly$^{15}$-Gln-Leu-Ser-Ala-Arg-Lys-Leu-Leu-Gln-Asp-Ile-Met-Ser-Arg$^{29}$-Gln-Gln-Gly-Glu-Ser-Asn-Gln-Glu-Arg-Gly-Ala-Arg-Ala-Arg-Leu$^{44}$-NH$_2$ |
| Insulin | Gly$^{1}$-Ile-Val-Glu-Gln-Cys-Cys-Thr-Ser-Ile-Cys-Ser-Leu-Tyr-Gln-Leu-Glu-Asn-Tyr-Cys-Asn<br>*(A-Chain, above; B-Chain, below)*<br>Phe$^{1}$-Val-Asn-Gln-His-Leu-Cys-Gly-Ser-His-Leu-Val-Glu-Ala-Leu-Tyr-Leu-Val-Cys-Gly-Glu-Arg-Gly-Phe-Phe-Tyr-Thr-Pro-Lys-Thr$^{31}$ |
| Parathyroid Hormone | Ser$^{1}$-Val-Ser-Glu-Ile-Gln-Leu-Met-His-Asn-Leu-Gly-Lys-His-Leu-Asn-Ser-Met-Glu-Arg-Val-Glu-Trp-Leu-Arg-Lys-Lys-Leu-Gln-Asp-Val-His-Asn-Phe-Val-Ala-Leu-Gly-Ala-Pro-Leu-Ala-Pro-Arg-Asp-Ala-Gly-Ser-Gln-Arg-Pro-Arg-Lys-Lys-Glu-Asp-Asn-Val-Leu-Val-Glu-Ser-His-Glu-Lys-Ser-Leu-Gly-Glu-Ala-Asp-Lys-Ala-Asp-Val-Asp-Val-Leu-Thr-Lys-Ala-Lys-Ser-Gln$^{84}$ |

rough endoplasmic reticulum and then to the Golgi elements, where they may be posttranslationally modified (e.g., sulfated, glycosylated). Vesicles containing the peptide are then pinched off the terminal cisternae of the Golgi apparatus and targeted to intracellular organelles or to the plasma membrane, or are secreted into the extracellular space.

Peptides vary in regard to whether their C-terminus is a free carboxylic acid or carboxamide and whether the N-terminus is a free amine, is acetylated or formylated, or bears some other modified amine group. Some peptides are glycosylated; that is, they are conjugated to one or more carbohydrate groups (e.g., sialic acid). These examples of posttranslational functionalization are often directly linked to the biological activities of the parent peptides as related to the ''bioactive'' conformation and/or ''message'' sequence of the peptide. Such posttranslational ''tags'' also serve as biomolecular zip codes to target the peptide to a particular cellular compartment. Finally, it is important to note that the amino acid sequence itself may differ in varying degrees between species. This information is useful to determine interspecies evolutionary relatedness. Two or more isoforms of a peptide may also exist within an individual species of animal. These isoforms, although chemically similar, may differ either in primary structure length or in other chemical aspects (e.g., site-specific amino acid substitutions and/or side chain modification by glycosylation, sulfation, etc.). Gene duplication followed by single or multiple nucleic acid mutations most likely accounts for such variations in peptide chemical structure and subsequent evolution of various families of peptides.

Some peptides require extensive posttranslational processing. Frequently, N- and/or C-terminally extended amino acid sequences are initially biosynthesized. These propeptides may then be packaged within secretory vesicles along with proteolytic enzymes. These enzymes, endopeptidases or exopeptidases, contribute to appropriate processing of the precursor peptide to yield the active peptide. Thus, such peptides are derived indirectly by way of a propeptide which, itself, may be derived from a prepropeptide. For example, pro-opiomelanocortin is a large inactive peptide (MW 28,500) that is cleaved by specific peptidases into several active peptides, including adrenocorticotropin, β-endorphin, α-melanotropin, and β-melanotropin. Some large plasma proteins also serve as propeptides for peptide production. In contrast, peptidases may also inactivate peptides by splitting the molecules at specific internal peptide bonds. Exopeptidases, both carboxypeptidases and aminopeptidases, cleave off the C-terminal or N-terminal amino acids, respectively. Some peptides may be inactivated by simple deamidation at the C-terminal (if amidated) end of the molecule. Endopeptidase cleavage is generally quite specific, and both exopeptidases and endopeptidases are well characterized in terms of mechanistic properties and substrate specificity. The clinical use of some peptides has been limited significantly because of their short plasma half-life, which may be related to peptidases. Therefore, knowledge of biodegradation mechanisms has been considered to be an important research objective to facilitate the design of synthetic peptide analogues that may exhibit sustained biological activities. Some peptides may also undergo inactivation by other chemical transformations such as that exemplified by insulin, in which its cystine-bridged heterodimeric structure is labile to cleavage by enzymatic reduction of the interchain disulfide bonds. Considerable effort has been devoted to the objective of designing synthetic peptide analogues (including pseudopeptides) and peptidomimetics agents, that are structurally altered to compromise or eliminate biological cleavage–inactivation by peptidases. Alternatively, there has also been considerable effort devoted to directly inhibiting either processing (cleavage–activation) or degrading peptidases to modify the generation or the lifetime of a particular endogenous (or exogenous) peptide, respectively.

### 2.2 Peptide Biological Actions

Peptides serve diverse physiological roles, and such biological properties have been the subject of intensive research to study peptide structure–function relationships. It is also well recognized that a vast number of peptide chemical messengers can manifest their biological activities as regulatory peptide hormones, neurotransmitters, growth factors, and cytokines. Hormones are chemical messengers released by one cell to act on one or more cell types to elicit a physiological response. The largest category of hormones is comprised of peptides, and the biological scope of such peptide hormones is extensive, with examples related to the regulation of the endocrine, cardiovascular, neural, immune, and gastrointestinal systems. In addition, peptide hormones regulate differentiation, growth, reproduction, blood pressure, glucose homeostasis, behavior, and other functions. Peptide hormones normally exist for short times (1–30 min), at very low concentrations ($10^{-12}$–$10^{-9}$ *M*) in bodily fluids or tissues, being rapidly inactivated and/or excreted from the body. The transitory actions of peptide hormones are well suited for continuous regulation (i.e., homeostasis) of various physiological systems.

Peptide hormones, probably without exception, mediate their initial actions at the level of the cell plasma membrane (phospholipid bilayer). Peptide hormones (''first messengers'') interact at the cell surface with protein (or glycoprotein) macromolecular receptors, which bind to the peptide ligand in a very specific manner. This results in activation (signal transduction) of proximally located enzymes or transport proteins, which may be modified in a positive or negative manner to produce a ''second messenger'' within the cell. Alternatively, the peptide–receptor complex may result in kinaselike activities to phosphorylate other cellular signaling proteins, as exemplifed by growth factor receptor-kinases. Many peptide hormones and neurotransmitters effect receptor-mediated stimulation or inhibition of adenylate cyclase via intermediary signaling heterotrimeric proteins known as G proteins. Binding of a peptide to a receptor (or enzyme) results from multiplicity of noncovalent intermolecular interactions, and these events are fundamentally related to structure—activity studies being studied on a plethora of biologically active peptides.

Binding is considered to precede transformation of the receptor (or enzyme) from its inactive to active state by molecular mechanisms often not understood. In many cases it has been determined which substructural features of a peptide are required for binding (i.e., the ''address'' or binding sequence) versus signaling (i.e., the ''message'' or activity sequence). The three-dimensional structural features of a peptide that provide the essential functional group ensemble for molecular recognition (binding) at a target receptor or enzyme may be composed of backbone and/or side chain components, and such a three-dimensional substructure of the peptide has been referred to as the ''pharmacophore.'' Progress in both intuitive and experimentally derived (biophysical and computational chemistry) development of peptide pharmacophore models is providing an opportunity to explore molecular recognition and mechanistic properties of peptide interactions with macromolecular targets, and such studies are of particular significance to the design of peptide, pseudopeptide and peptidomimetic drugs.

## 3 BIOMEDICAL AND PHARMACEUTICAL RESEARCH

A number of pharmaceutical and biotechnology companies are currently designing and producing synthetic or recombinant peptide drugs as well as synthetic peptidomimetic drugs for numerous biomedical applications.

### 3.1 Peptide Pathophysiology

The pathophysiology of regulatory peptides provides many opportunities for therapeutic intervention by recombinant peptides as well as synthetic peptide analogues or peptidomimetics. The pathophysiology of regulatory peptides may be due, for example, to defects in the biosynthesis of the peptide and/or its target receptor (or enzyme). A variety of endocrine disorders result from excessive or deficient peptide hormone levels (Table 2). Within the scope of mimicking or blocking regulatory peptide effects of target receptors, there is a tremendous research effort to identify peptide analogues or peptidomimetic agents that possess potent and selective agonist/antagonist properties at receptors or are effective enzyme inhibitors. Also, such compounds (e.g., receptor-targeted ligands) may be radiolabeled or conjugated to anticancer drugs to yield drugs useful in the diagnosis, localization, or chemotherapy of tumors.

### 3.2 Peptide Drug Delivery

Among the major challenges of developing peptide-based therapeutic agents has been the mode of administration. Peptide drug delivery exists in many forms, including parenteral (intravenous or intramuscular injection), interstitial (subcutaneous implant), oral, nasal, and percutaneous (transdermal) administration. Perhaps, the best-known injectable peptide is insulin, and with respect to implants the recent success of the gonadotropin-releasing hormone (GnRH) analogue Zoladex™ deserves mention. In the case of oral administration, one reason to chemically transform peptides into peptidomimetic or nonpeptide derivatives has been to identify prototypic lead compounds with oral bioavailability. Such efforts have shown success by the development of orally effective angiotensin-converting enzyme inhibitors, renin inhibitors, HIV protease inhibitors, thyrotropin-releasing hormone agonists, fibrinogen antagonists, and so forth. Nevertheless, the possibility of alternative modes of administration remains an intriguing area of research, and some success has been made relative to nasal (oxytocin, Desmopressin™, Buserelin™, and calcitonin), pulmonary, buccal, rectal, and transdermal routes of peptide drug administration. Of particular importance to the oral delivery of peptide-based therapeutics is the intestinal epithelium, which limits passive or active transport of molecules of high molecular weight (typically tetrapeptides and larger). Similarly, stability toward gastric acid and degradative enzymes, including peptidases associated with the epithelial membrane (luminal side) or within such cells, provides another obstacle a peptide-based drug candidate must surpass to access systemic circulation after oral administration.

Table 2 Pathophysiological States of Known Relationship to Regulatory Peptides

| Disease (Symptom) | Relationship to Peptide |
|---|---|
| Addison's disease (abnormal carbohydrate metabolism) | Deficiency in adrenocorticotropin |
| Cushing's disease (increased protein catabolism) | Excess adrenocorticotropin |
| Secondary hypogonadism | Deficiency in gonadotropin |
| Secondary hypergonadism | Excess gonadotropin |
| Hyperglycemia and glucosuria (diabetes mellitus or insulin-dependent diabetes) | Deficiency in insulin |
| Abnormal blood $Ca^{2+}$ (increased), hyperparathyroidism | Excess parathyroid hormone |
| Abnormal growth | |
| Decreased, Laron-type dwarfism | Deficiency in somatomedin |
| Decreased, hypopituitary dwarfism | Deficiency in somatrotropin |
| Increased, in children (giantism) and in adults (acromegly) | Excess somatrotropin |
| Abnormal metabolism (increased), Graves' disease | Excess thyrotropin-mimicking, anti-TSH receptor antibodies |
| Hypovolemia–dehydration (increased), pituitary-type diabetes insipidus | Deficiency in vasopressin |

*Source:* Adapted from Hadley (1995).

### 3.3 Peptide and Peptidomimetic Therapeutics

The first recombinant peptide therapeutic, insulin, was commercialized in 1982. Currently, a number of peptide therapeutics (or preclinical leads) have been advanced using genetic engineering technologies. Based on the success of insulin replacement therapy for the treatment of some types of diabetic mellitus, other structurally complex peptides have been advanced by molecular biology methodologies that permit biosynthesis of peptides not readily obtained by chemical synthesis.

Research into synthetic peptide analogues or peptidomimetic therapeutic agents is one of the major development foci in pharmaceutical and biotechnology companies, and many examples of such compounds have entered into clinical testing or are marketed drugs (Table 3). Synthetic peptides may include native molecules as well as analogues that incorporate side chain and/or backbone modifications. The discovery of many "first generation" peptide antagonists was based on chemically modified analogues. Peptide-based competitive antagonists of bradykinin, oxytocin, phenylthiohydantoin, enkephalin, vasopressin, glucagon, cholecystokinin, gastrin, angiotensin II, GnRH, substance P, and corticotropin releasing factor were reported recently, and such compounds have typically incorporated unusual amino acids (e.g., D-amino acids) and backbone $\psi$[CONH] replacements. The chemical transformation of peptide agonists or antagonists into yet simpler (in terms of molecular weight), metabolically stable, and orally bioavailable "second generation" pseudopeptides or peptidomimetic agents has also met with limited success. In this regard, such research (peptide receptor–targeted) has primarily led to the identification of antagonists. Thus, molecular recognition properties required for the agonist properties of peptides are apparently more complex and, perhaps, uncompromising relative to those effecting antagonist properties. In fact, the identification of many nonpeptide lead compounds as peptide antagonists has been advanced by mass screening assays using existing chemical inventories and various natural product sources. Interestingly, the recent development of peptide and peptidomimetic library technologies has accelerated opportunities to identify novel leads as well as the ability to evaluate analogues

**Table 3** Examples of Recombinant or Synthetic Peptide Therapeutic Agents or Preclinical Leads

| Recombinant Peptides | Proposed/Known Therapeutic Application |
|---|---|
| Atrial natriuretic factor | Potential use in prophylaxsis and/or treatment of acute renal failure |
| Epidermal growth factor | Potential use in skin grafting, eye surgery, and/or treatment of burns or ulcers |
| Somatrotropin | Treatment of growth defects |
| Insulin | Treatment of type-I diabetes |
| Glucagonlike insulinotropic peptide | Potential treatment of insulin-insensitive diabetes |
| Transforming growth factor β | Potential treatment of wound healing and burns |
| Nerve growth factor | Potential treatment of neural plasticity defects as possibly related to Alzheimer's disease |
| Interleukin-1α/β | Potential use in cancer therapy and inflammation |
| Hirudin | Potential use in prevention or treatment of venous blood clots |
| Tissue plasminogen activator | Potential use as an anticoagulant in heart attacks |
| Factor VII | Potential use as a blood clotting factor in hemophiliacs |
| Erythropoietin | Potential uses in treatment of anemia in kidney dialysis patients, AIDS, and cancer |
| Platelet-derived growth factor | Potential uses in promoting growth of fibroblasts and keratinocytes, and formation of new blood vessels |
| Interferon-α | Potential use as an immune stimulant for cancer therapy |
| Interferon-β | Potential use as immune stimulant for treatment of viral diseases and multiple sclerosis |
| Interferon-γ | Potential use as an immune stimulant for treatment of infectious diseases, cancer, and rheumatoid arthritis |
| **Synthetic Peptides** | **Proposed/Known Therapeutic Applications (Relationship to Naturally Occurring Peptide)** |
| Cyclosporin A | Immunosuppressive drug (identical to the natural product peptide) |
| Sandostatin™ | Symptomatic treatment of acromegaly and carcinoid syndrome (hexapeptide agonist analogue of Somatostatin) |
| Desmopressin™ | Treatment of severe diabetes insipidus (nonapeptide agonist analogue of vasopressin) |
| Buserelin™ | Treatment of prostate cancer (nonapeptide agonist analogue of gonadotropin-releasing hormone) |
| HOE-140 | Potential treatment of pain, inflammation, rhinitis, and/or asthma (decapaptide antagonist analogue of bradykinin) |
| Tetracosactide | Diagnostic agent of adrenal function (fragment agonist derivative of corticotropin) |
| BQ-123 | Potential treatment of hypertension, restenosis, and related disorders (cyclic pentapeptidyl, natural product analogue antagonist of endothelin) |
| Calcitonin | Treatment of hypercalcemia, Paget's disease, osteoporosis, and pain affiliated with bone cancer (identical to native peptide) |
| Ebiratide™ | Potential application for central nervous system disorders; cognition treatment (hexapeptide agonist analogue of corticotropin) |
| L-365,209 | Potential use as uterine relaxant for prevention of premature labor (pseudohexapeptidyl, natural product antagonist of oxytocin) |
| DPDPE | Potential analgesic lead; δ-selective opioid agonist (pentapeptide agonist analogue of enkephalin) |
| GHRP-6 | Potential growth hormone secretagogue lead (hexapeptide agonist; native ligand unknown) |

*Source:* Adapted from Sawyer in Amidon, G. and Taylor, M. (1995).

in a structure–activity sense to expedite the overall process of drug candidate selection. The application of biophysical chemistry (e.g., NMR spectroscopy and X-ray crystallography) and computer-assisted molecular design (CAMD) strategies also impacts significantly on the synthetic transformation of peptides to highly chemically modified derivatives and peptidomimetics. In the case of cyclic, conformationally constrained peptides, the synergism of NMR spectroscopy and CAMD-based analysis of three-dimensional structural properties and model building is a promising approach to the rational advancement of peptide-based drug design. In the case of X-ray crystallography (as interfaced with CAMD), the study of peptide or peptidomimetic interaction with proteins is providing high resolution (1.8–3.0 A) molecular maps of the complex to guide further design strategies.

The apparent medical need for peptide or peptidomimetic drugs is already tremendous, and future opportunities will continue to make this area of pharmaceutical research extremely competitive. In the case of the peptide hormones insulin and calcitonin, worldwide sales already surpass one billion dollars. As exemplified by angiotensin-converting enzyme inhibitors (several on the market), the worldwide sales also surpass one billion dollars. Furthermore, the simple dipeptide sweetener Aspartame™ has been credited with being a billion-dollar product in application to a variety of foods and drinks.

*See also* Biotransformations of Drugs and Chemicals; Drug Synthesis; Drug Targeting and Delivery, Molecular Principles of; Medicinal Chemistry; Peptides, Synthetic; Synthetic Peptide Libraries.

## *Bibliography*

Amidon, G., and Taylor, M. Eds. (1995) *Peptide-Based Drug Design: Controlling Transport and Metabolism.* ACS Books, Washington DC.

Fauchere, J.-L. (1986) Elements for the rational design of peptide drugs, in *Advances in Drug Research,* Vol. 15, B. Testa, Ed., pp. 29–69. Academic Press, London.

Giralt, E., and Andreau, D., Eds. (1991) *Peptides 1990, Proceedings of the 21st European Peptide Symposium.* ESCOM Science Publishers, Leiden, The Netherlands.

Gross, E., Meienhofer, J., and Udenfriend, S., Eds. (1979–1987) *The Peptides: Analysis, Synthesis, Biology,* Vols. 1–IX. Academic Press, New York.

Hadley, M. E. (1995) *Endocrinology,* 4th ed. Prentice-Hall, Englewood Cliffs, NJ.

Hider, R. C., and Barlow D., Eds. (1991) *Polypeptide and Protein Drugs—Production, Characterization and Formulation.* Ellis Horwood, Chichester.

Negro-Vilar, A., and Conn, P. M., Eds. (1988) *Peptide Hormones: Effects and Mechanisms of Action,* Vols. I–III. CRC Press, Boca Raton, FL.

Pavia, M. R., Sawyer, T. K., and Moos, W. H., Guest Eds. (1993) *Bioorg. Med. Chem. Lett.*, Vol. 3, Symposium-in-Print, Generation of molecular diversity.

Smith, J. A., and Rivier, J. E., Eds. (1992) *Peptides: Chemistry and Biology, Proceedings of the Twelfth American Peptide Symposium.* ESCOM Science Publishers, Leiden, The Netherlands.

Ward, D. J. (1991) *Peptide Pharmaceuticals—Approaches to the Design of Novel Drugs.* Open University Press, Buckingham, England.

Wieland, T., and Bodanszky, M. (1991) *The World of Peptides—A Brief History of Peptide Chemistry.* Springer-Verlag, Berlin.

Williams, W. V., and Weiner, D. B., Eds. (1993) *Biologically Active Peptides: Design, Synthesis and Utilization,* Technomic, Lancaster, PA.

# Peptides, Synthetic

*Gregory A. Grant*

## *Key Words*

**Peptide** Common usage designates a shortened form of polypeptide. "Peptide" refers to any organic compound composed at least partially of amino acids linked by peptide bonds. The distinction between peptides and proteins is arbitrary, with the borderline often placed at 50 to 100 amino acids.

**Peptide Bond** The amide bond formed between the α-carboxyl group and the α-amino group of two adjacent amino acid residues in a peptide.

**Residue** Amino acid residue is an amino acid in peptide linkage, as opposed to the free amino acid, which contains a free α-amino and α-carboxyl group (see Figure 1).

Synthetic peptides are organic compounds, generated by synthetic chemical techniques, which are composed of two or more amino acids linked together by a peptide bond. They are often intended to resemble or duplicate naturally occurring peptides or segments of naturally occurring proteins. Synthetic peptides are important as tools for research in the biological and biomedical sciences, and a significant number of synthetic peptides have pharmaceutical and commercial uses. For instance, synthetic peptide hormones have useful clinical applications, and the commercial sweetener aspartame, a dipeptide (*N*-L-α-aspartyl-L-phenylalanine-1-methyl ester), is the basis for an entire industry and is perhaps the best known synthetic peptide.

## 1 DESIGN AND STRUCTURE

The design of a particular synthetic peptide is based on its intended use and on considerations of the synthetic chemistry. With modern synthetic techniques it is possible to produce peptides containing in excess of 100 amino acids, although those shorter than 50 are much more common and easier to synthesize. Short peptides tend not to exhibit preferred or stable solution conformations but rather possess a large degree of flexibility. As peptides increase in length, they have a greater tendency to adopt elements of secondary structure, such as helices and β-sheet structures connected by discrete turns, which impart an overall decrease in flexibility. These elements of secondary structure are generally found in naturally occurring proteins and large peptides and are, in part, responsible for their specific biological function. From a knowledge of the features that contribute to the formation of these structures, peptides can be specifically designed to contain them.

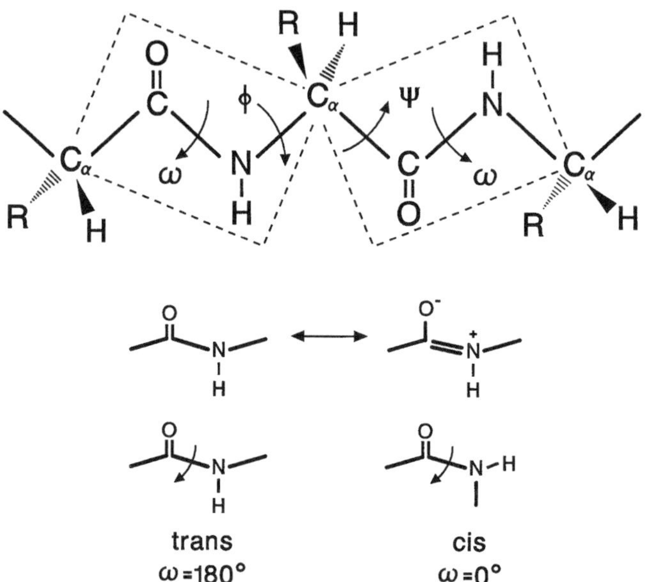


Figure 1. A stretch of peptide demonstrating the planar nature of the peptide bond, the angles that describe the conformation of the peptide backbone, and the chirality of L-amino acids. The resonance structures of the peptide bond and *cis-* and *trans*-peptide bonds are also shown.

Synthetic peptides can be linear, cyclic, or branched and are often composed of, but not limited to, the 20 genetically coded L-amino acids. The synthetic approach allows the introduction of novel or unusual components such as D-amino acids, other unnatural amino acids, ester or alkyl backbone bonds in place of the normal amide bond, *N*- or *C*-alkyl substituents, side chain modifications, and constraints such as disulfide bridges and side chain amide or ester linkages. The result of such changes may be higher biological potency, greater stability resulting in longer biological lifetime, or the ability to specifically interact with or covalently label a biological macromolecule or receptor for localization or structure–function relationship studies.

In its most basic sense, a synthetic peptide consists of at least two amino acids joined by a peptide bond. Peptides composed of more than two amino acids possess a number of peptide bonds in series known as the peptide or polypeptide backbone (Figure 1). This backbone structure is an important contributor to the overall peptide structure. The resonance properties of the carbon–nitrogen peptide bond give it a substantial double-bond character. As a result, the amide bond is flat, with the carbonyl carbon, oxygen, nitrogen, and amide hydrogen all lying in the same plane. Thus, no free rotation occurs around the carbon–nitrogen bond. The torsional angle of that bond, called ω, is defined by the atoms Cα–C(O)–N–Cα, and because of the double-bond character and the large energy barrier to rotation (25 kcal/mol), there are two rotational isomers: trans (ω = 180°) and cis (ω = 0°). The trans isomer possesses the lower energy and is generally found for all peptide bonds not involving proline. The energy of the trans X—Pro bond is somewhat elevated and the barrier to rotation is lowered. Thus, proline-containing peptides will often exhibit cis–trans isomerization.

In addition to ω, the peptide backbone can be completely described with two additional angles, $\phi$ and $\psi$, where $\phi$ is the angle described by C(O)–N–Cα–C(O) and $\psi$ is described by N–Cα–C(O)–N. By convention, the angle $\phi$ is 180° when the two carbonyl carbons are trans to each other and the angle $\psi$ is 180° when the two amide nitrogens are trans to each other.

## 2 NOMENCLATURE

Proper use of nomenclature for expressing peptides is important. The International Union of Pure and Applied Chemists (IUPAC), in conjunction with the International Union of Biochemists (IUB), has formulated a set of guidelines for peptide nomenclature [*J. Biol. Chem.* 260:14–42 (1985)]. These guidelines are too extensive to be described here in their entirety, but some of the more common conventions are presented.

The sequence of amino acids within a peptide is usually designated with either the one-letter or three-letter amino acid code (Table 1). The amino acid symbols denote the L-configuration unless a D- or DL- precedes the symbol. In the case of linear peptides, the convention is to express the peptide left to right from the amino-terminus to the carboxy-terminus. Thus, a pentapeptide might be represented either as V-L-H-A-G or as Val-Leu-His-Ala-Gly. The α-amino group is assumed to be at the left-hand side of the amino acid symbol when using hyphens and at the point of the arrow when using arrows for special cases such as cyclic peptides. Normally, when the termini are free amino or carboxyl groups, additional designation of this condition is unnecessary. To avoid ambiguity, however, the state of the termini is often indicated. Free termini are designated with an H- at the amino end and an OH at the carboxyl end; thus the foregoing peptide would be written H-Val-Leu-His-Ala-Gly-OH. Similarly C-terminal amide and aldehyde groups would be designated by $-NH_2$ and -H, respectively.

Occasionally, the names of small peptides are written out. In this case, the names of all amino acids contributing carboxyl groups to peptide bonds end in -yl, and the amino acid with the terminal carboxyl group retains its original name. The pentapeptide V-L-H-A-G would thus be expressed as valylleucylhistidylalanylglycine. Cyclic peptides are usually expressed on a single line of text, similar to linear peptides, with a line drawn to connect the joined amino acids to indicate the bond leading to the cyclic nature of the peptide. If the amino acids are joined between side chains, such as in disulfide bonds or side chain lactams, the line starts and ends in the middle of the amino acid name or abbreviation, such as

Gly-Val-Glu-Ala-Leu-Ile-Lys-Ala-Thr

If the connection is between the α-amino and α-carboxyl groups of the termini, the line is drawn from the sides of the names, such as

Gly-Val-Glu-Ala-Leu-Ile-Lys-Ala-Thr

In this case, the N→ C direction is read left to right. If the peptide is written out in a circular fashion, as if it were inscribed along the circumference of a circle, the N→ C direction must be designated.

A synthetic analogue of a naturally occurring peptide with which a specific name is associated represents a special case. Examples include insulin, oxytocin, enkephalin, vasopressin, and bradykinin. The peptide hormone oxytocin has the following structure:

$Cys-Tyr-Ile-Gln-Asn-Cys-Pro-Leu-Gly-NH_2$

**Table 1** Common Symbols for Amino Acids in Expressing Peptide Structure[a]

| Residue | Three Letter Symbol | One Letter Symbol |
|---|---|---|
| Alanine | Ala | A |
| Asparagine[b] | Asn | N |
| Aspartic Acid[b] | Asp | D |
| Arginine | Arg | R |
| Cysteine | Cys | C |
| Glutamic Acid[b] | Glu | E |
| Glutamine[b] | Gln | Q |
| Glycine | Gly | G |
| Histidine | His | H |
| Isoleucine | Ile | I |
| Leucine | Leu | L |
| Lysine | Lys | K |
| Methionine | Met | M |
| Phenylalanine | Phe | F |
| Proline | Pro | P |
| Serine | Ser | S |
| Threonine | Thr | T |
| Tryptophan | Trp | W |
| Tyrosine | Tyr | Y |
| Valine | Val | V |
| Hydroxylysine | Hyl | |
| 4-Hydroxyproline | Hyp | |
| α-Aminobutyric acid | Abu | |
| Norvaline | Nva | |
| Norleucine | Nle | |
| Homoserine | Hse | |
| Homocysteine | Hcy | |
| Ornithine | Orn | |
| Citrulline | Cit | |
| α-Aminoisobutyric acid | Aib | |
| γ-Carboxyglutamic acid | Gla | |

[a]Symbols for less common amino acids should be defined in each publication in which they appear.
[b]Asx or B and Glx or Z are used when the distinction between Aspartic acid/Asparagine and Glutamic acid/Glutamine are uncertain. This convention is most commonly used in sequence analysis.

where the Cys residues form a disulfide bridge. If one or more amino acids are replaced in the synthesis of an oxytocin analogue, the new amino acids and their positions are placed in brackets before the name. Thus [$Phe^2$, $Asn^4$]-oxytocin refers to an oxytocin analogue in which the Tyr at position 2 and the Gln at position 4 have been replaced with Phe and Asn, respectively. The prefix *endo* is used to designate the insertion of an extra amino acid into the chain. Thus, *endo*-$Ala^{7a}$-oxytocin denotes an alanine insertion between $Pro^7$ and $Leu^8$. Similarly, deletion of an amino acid is designated by the prefix *des* (e.g., *des*-$Ile^3$-oxytocin denotes that Ile at position 3 has been deleted).

## 3 SYNTHESIS

Peptides can be synthesized either in solution, which is the classical approach, or by solid phase methods. With the exception of specialized applications calling for the use of solution phase chemistry because it is easier and less expensive, or because the necessary

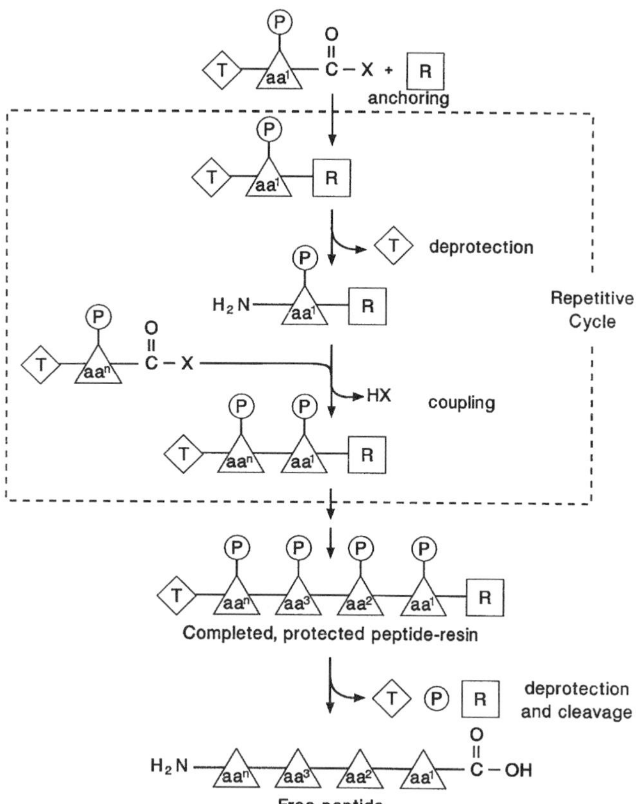


**Figure 2.** Stepwise solid phase synthesis of linear peptides: R, resin or insoluble polymeric support; T, temporary protecting group; P, permanent protecting group; X, activated carboxyl group; $aa^1, \ldots, aa^n$, amino acid residues numbered from the C-terminus. The symbols used simply indicate the general strategy and are not meant to reflect complete chemical accuracy. For instance, the protecting groups are usually chemically altered as they are removed, although the same symbol is used to portray them before and after deprotection.

chemistry is not amenable to the solid phase method, most peptides made for research purposes are synthesized with the solid phase method, introduced by Merrifield in 1963. This technique, for which he was awarded the Nobel Prize in chemistry in 1984, was primarily responsible for opening the way to the widespread use of synthetic peptides in chemical and biomedical investigations.

Solid phase methods are readily automated, and most solid phase syntheses performed today are done with the aid of machines. A variety of peptide synthesizers are commercially available for batchwise and continuous flow operations as well as for the synthesis of multiple peptides within the same run.

### 3.1 The Solid Phase Method

The solid phase approach to peptide synthesis (Figure 2) consists basically of anchoring the growing peptide to an insoluble support or resin. This is accomplished through the use of a chemical handle, which links the support to the first amino acid at the carboxyl terminus of the peptide. Subsequent amino acids are then added, one at a time, in a stepwise manner, until the peptide is fully constructed. The solid phase method allows excess reagents and soluble reaction by-products to be easily removed by filtration and washing, thus doing away with the need to isolate or crystallize the product after each step.

Synthesis of the growing peptide chain is accomplished by formulation of an amide bond, referred to as the "peptide bond," between the free α-amino group of the receiving amino acid (anchored to the support) and the activated α-carboxyl group of the incoming amino acid (in solution). Note that synthesis proceeds from the carboxy end to the amino end. To prevent unwanted side reactions with reactive amino acid side chains and self-coupling of the α-carboxyl group of one incoming amino acid to the α-amino group of another incoming amino acid, these groups must be protected at the appropriate times during synthesis. As depicted in Figure 2, the group protecting the α-amino group of the amino acid being coupled must ultimately be removed, after coupling is complete, to permit the acceptance of the next amino acid during the next cycle. Chemical groups that allow this are referred to as "temporary protecting groups." The reactive amino acid side chains, on the other hand, must remain permanently protected throughout the synthesis, and chemical groups that accomplish this are referred to as "permanent protecting groups." At the end of the synthesis, the completed peptide is cleaved from the resin support, and the permanent protecting groups are removed, usually in a single step process, to yield the desired peptide product.

### 3.2 Chemistry

A number of different chemical approaches for the synthesis of peptides on solid phase supports have been investigated and are in use today. Two general methods have emerged as the most common (Figure 3): "Boc," which uses the *tert*-butyloxycarbonyl group and "Fmoc," which uses the 9-fluorenylmethyloxycarbonyl group for temporary α-amino protection.

A wide variety of permanent protecting groups are available for these two general approaches, and a few of them are depicted in Figure 3. The Boc strategy relies on graded acidolysis to selectively differentiate between temporary and permanent protection. Temporary protecting groups are removed by treatment with trifluoroacetic acid, while cleavage from the resin and removal of permanent protecting groups is usually accomplished with anhydrous hydrofluoric acid or other strong acid. The Fmoc strategy is a milder orthogonal approach in that the two classes of protecting groups are removed by different chemical mechanisms such that either class can be removed while preserving the other class. The temporary Fmoc group is removed by a base-catalyzed β-elimination mechanism, usually employing piperidine. Removal of permanent protecting groups as well as cleavage from the support is accomplished by acidolysis, usually with trifluoroacetic acid.

Formation of the peptide bond between the carboxyl group of the incoming amino acid and the amino group on the peptide–resin requires that the carboxyl group be activated chemically. There are four major coupling strategies that utilize either active esters, preformed symmetrical anhydrides, acid halides, or generation of the active acid in situ with a variety of different reagents.

Peptides modified at the α-amino group can be easily produced by treatment with an acylating agent, such as acetic anhydride, prior to removal of the permanent protecting groups. Resins are also available for both Boc and Fmoc strategies that will directly

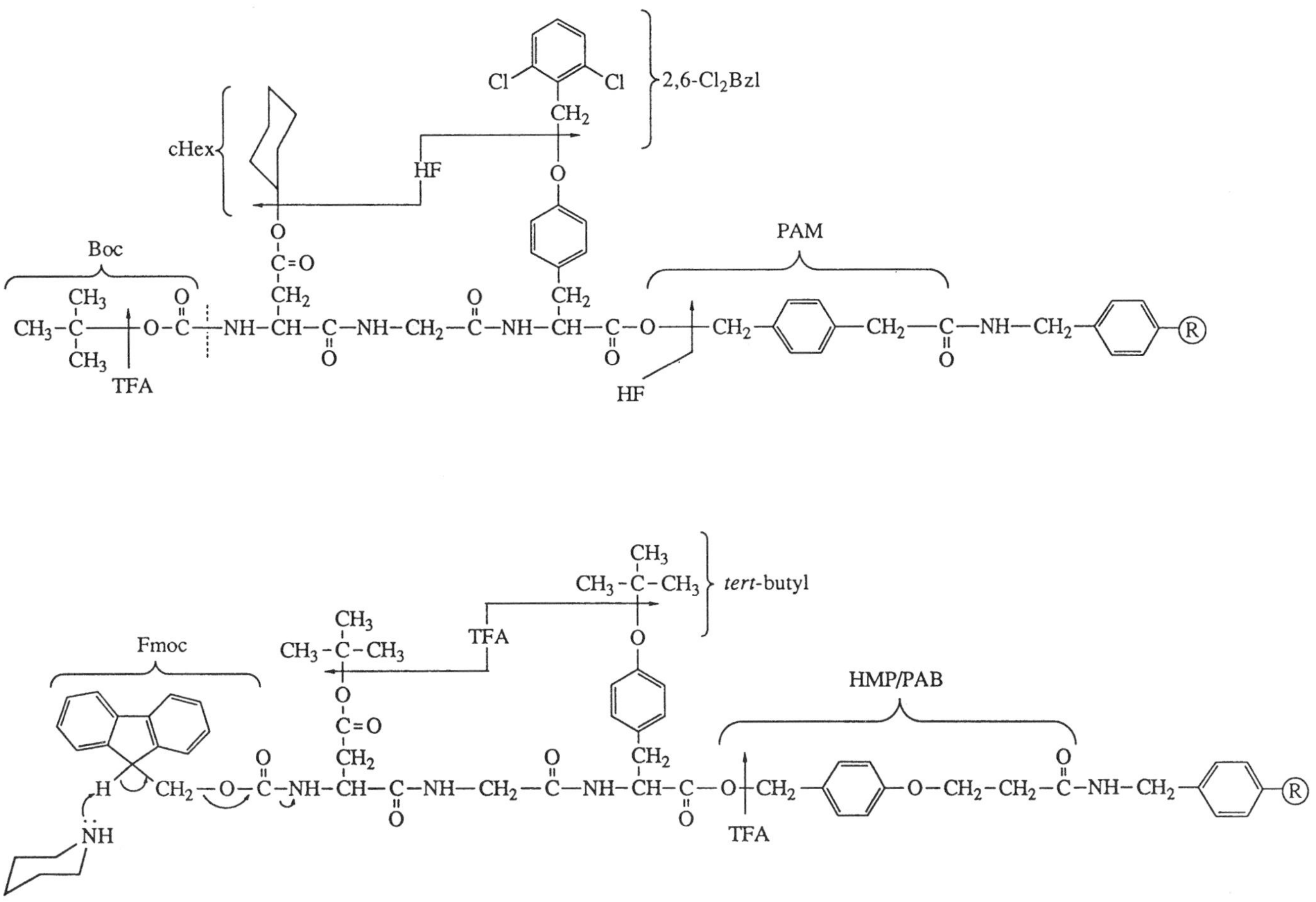

Figure 3. Common protection schemes for solid phase peptide synthesis. *Top:* Protection scheme for solid phase synthesis based on the Boc group utilizing graded acidolysis. The Boc group is removed at each step by trifluoroacetic acid (TFA). Permanent side chain protecting groups and the PAM linkage are cleaved simultaneously by hydrofluoric acid (HF) or other strong acid. *Bottom:* Mild orthogonal protection scheme for solid phase synthesis based on the Fmoc group. The Fmoc group is removed at each step by the base-catalyzed elimination mechanism, usually using piperidine, as shown. Permanent side chain protecting groups and the HMP/PAB linkage are cleaved by treatment with TFA. (From Fields et al., Principles and practice of solid-phase peptide synthesis. In *Synthetic Peptides: A User's Guide,* G. A. Grant, Ed., Freeman, New York, 1992, with permission of the authors and the publisher.)

yield peptides with either a free acid or an amide function at the carboxy terminus.

## 4 EVALUATION

The chemistry used to produce synthetic peptides is complex and can be influenced by a number of factors that cannot always be foreseen or controlled. As a result, one must never assume that the final product is correct until it is proven to be so.

Homogeneity of the product and correct covalent structure are the two main aims in the production of synthetic peptides. Today virtually all peptide purification can be accomplished by high performance liquid chromatography (HPLC) using either reverse phase or ion exchange columns.

Characterization of synthetic peptides is best obtained by a combination of amino acid analysis and mass spectrometry. The positions of modifications and deletions, if present, can be identified by sequencing with either chemical methods (Edman chemistry) or tandem mass spectrometry.

## 5 APPLICATIONS AND PERSPECTIVES

Peptides have become an important class of molecules in biomedical research and medicinal chemistry. Already a number of synthetic hormones or their analogues have found use as therapeutic agents. Synthetic peptide substrates for enzymes such as proteases, kinases, and phosphatases have been used successfully to study enzyme kinetics and mechanism and to develop effective inhibitors.

The ability of synthetic peptides to immunologically mimic regions of whole proteins has resulted in their extensive use as antigens and probes of protein epitopes. Antibodies raised against a native protein bind to relatively small areas on the surface of that protein and can recognize and bind to the same antigenic sequence in the absence of the rest of the protein. Thus, highly antigenic sequences or epitopes can be mapped in proteins by synthesizing overlapping peptides spanning the full sequence of the protein and testing them for binding of the antibody, generated with the intact protein, in an appropriate assay. Epitope mapping of proteins with synthetic peptides has identified regions important for biological activity and has led to the development of vaccines.

Many gene products have been identified and isolated by prediction of antigenic regions in the sequence of the protein deduced from the gene sequence. Synthetic peptides corresponding to these predicted antigenic regions are used to produce antibodies, which in turn are used to specifically interact with the intact protein product for identification and isolation.

Synthetic peptides have been instrumental in receptor localization, characterization, and isolation, and even whole enzymes have been produced by synthetic methods. One striking example of this is the complete chemical synthesis of the human immunodeficiency virus-1 (HIV-1) protease, which possesses full biological activity.

It is evident that there is huge potential in the use of synthetic peptides as therapeutic agents. They show great potential in being effective in regulating such physiological processes as blood pressure, neurotransmission, reproduction, and endocrine functions to name a few. They also show great potential for use in developing new, more effective vaccines and for de novo drug discovery through the development of peptide libraries. Peptide libraries are generated by synthesis of a mixture of a large number of peptides of defined length, where each position, or a defined number of positions, contains a random mixture of amino acids. These heterogeneous mixtures are then screened with some appropriate assay system, and the reactive sequences are isolated and determined.

It seems clear that synthetic peptides will continue to play an increasingly larger role in biomedical research and the treatment of human, animal, and perhaps plant disease.

*See also* AIDS HIV Enzymes, Three-Dimensional Structure of; Amino Acid Synthesis; Peptides; Peptides and Mimics, Design of Conformationally Constrained; Peptide Synthesis, Solid-Phase; Synthetic Peptide Libraries.

## *Bibliography*

Atherton, E., and Sheppard, R. C. (1989) *Solid Phase Peptide Synthesis: A Practical Approach.* IRL Press, Oxford.

Bodanszky, M. (1988) *Peptide Chemistry, A Practical Textbook.* Springer-Verlag, Berlin.

———, and Bodanszky, A. (1984) *The Practice of Peptide Synthesis.* Springer-Verlag, Berlin.

Grant, G. A., Ed. (1992) *Synthetic Peptides: A User's Guide.* Freeman, New York.

Jones, J. (1991) *The Chemical Synthesis of Peptides.* Clarendon Press, Oxford.

Stewart, J. M., and Young, J. D. (1984) *Solid Phase Peptide Synthesis,* 2nd ed. Pierce Chemical Company, Rockford, IL.

Wieland, T., and Bodanszky, M. (1991) *The World of Peptides: A Brief History of Peptide Chemistry.* Springer-Verlag, Berlin.

# Peptides and Mimics, Design of Conformationally Constrained

*Victor J. Hruby and Lakmal W. Boteju*

## *Key Words*

**Peptide Conformations** The various shapes a peptide can attain by rotations about its bonds.

**Peptide Mimetic (Peptide Analogue)** Molecule resulting from structural modification of a (biologically significant) peptide.

**Peptide Topography** The three-dimensional arrangement of the side chain groups of a peptide.

**Receptor** Target on which a biological mediator (e.g., a peptide) acts to elicit a response.

Conformation-constrained peptide mimetics are modified analogues of biologically active peptides that have their natural (flexible) conformation fixed by various constraints. Although important peptide mediators such as hormones and neurotransmitters have great potential for medical applications, their inherent drawbacks, which include lack of receptor selectivity, high flexibility, and biodegradability, complicate their use as drugs. Conformational constraint plays an important part in rational design of peptides to overcome these problems.

## 1 INTRODUCTION

Many important biological mediators in mammalian systems are peptides and proteins. For example, peptide hormones control diverse biological functions from glycogenolysis, to insulin release, to sexual activity, to darkening of the skin. In the central nervous system, peptidic neurotransmitters regulate physiological functions ranging from analgesia and appetite control, to pain and its perception. Thus, these peptides and proteins have great potential for development into useful drugs for the treatment of many human disorders. Peptides and proteins act on receptors to elicit a characteristic biological response. An important aspect in modern drug design involves the rational (chemical) modification of peptide structures to yield molecules that have highly desirable biological properties. Here we discuss the basic principles behind conformational constraint of peptides as a rational design approach to peptide-based drugs.

### 1.1 Peptide Structure and Function

Peptides are formed by a series of amino acids linked together by amide bonds. The topological features of a phenylalanine residue in a peptide are shown in Figure 1. The partial double-bond character in the C—N bond of the amide group makes it relatively rigid, but the other bonds ($\psi$, $\phi$, $\chi$, etc.) generally are not. The conventional notations for the various angles of rotation about bonds in peptides are shown in Figure 1.

The affinity of a peptide for its receptor is due, in part, to the individual affinities of amino acid side chains and, in part, to overall conformational features. It is possible to categorize two types of

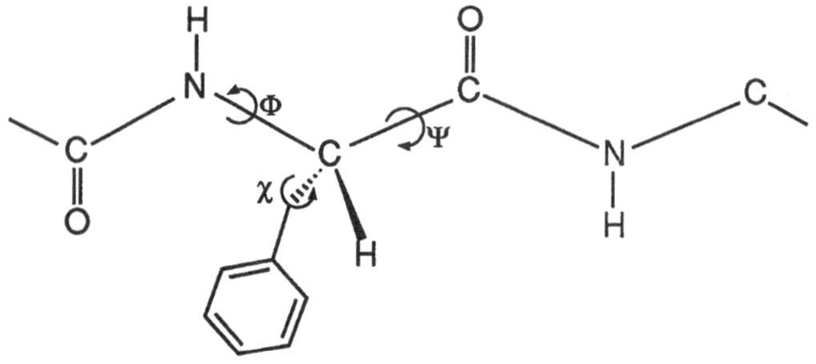

**Figure 1.** The topological features of a phenylalanine residue in a peptide.

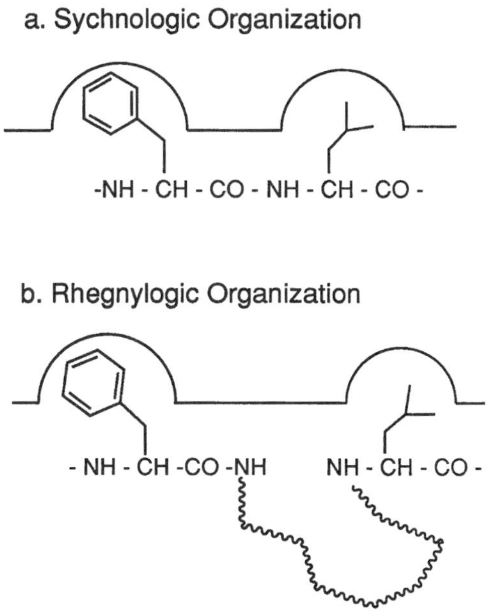

a. Sychnologic Organization

-NH - CH - CO - NH - CH - CO -

b. Rhegnylogic Organization

- NH - CH -CO -NH NH - CH - CO -

Figure 2. Organization of active sites in a peptide.

organization of active regions in a peptide that interacts with the receptor (Figure 2): sychnologic organization, in which the amino acid residues that interact with the receptor are close to one another in the peptide, and rhegnylogic organization, in which the residues that interact with the receptor are far apart, but the conformation of the peptide backbone orients them in a favorable fashion for receptor interaction. In both cases the backbone conformation and topography (relative side chain conformations) are important. The hydrophobicity/hydrophilicity of the amino acid side chain groups and their interactive forces also contribute to the final conformation of the peptide.

### 1.2 Needs for and Uses of Conformational Constraints

Many peptide hormones and neurotransmitters that have potential of being developed into therapeutic drugs are linear peptides that have a high degree of conformational flexibility. For rational drug design, it is desirable to know or have insight into the conformation of the peptide that binds to the receptor. The X-ray crystal structures of peptides can help to overcome this problem, but, to date, only a few peptides suitable for X-ray analysis have been crystallized. In addition, many endogenous peptides can interact with different receptor types, giving rise to different biological responses. To design effective therapeutic drugs, one must overcome this ''multiple-receptor ligand'' problem and modify the peptide to act selectively on only one receptor type.

The design and use of conformationally constrained analogues of peptides helps overcome these problems. In this approach, chemical modification is used to restrict a residue or group of residues in a peptide to a small region of three-dimensional space so that when the peptide interacts with the receptor, the conformation seen in solution will remain as a result of the high activation energy or free energy required to change that conformation. These constrained peptides reveal important insights into the bioactive conformation of the peptide, often are metabolically stable to the action of degradative enzymes, and may themselves be candidates for potent therapeutic drugs.

### 1.3 Peptide Mimetic Agents

The modification of a region or several regions of a peptide molecule provides a structure chemically different from the parent peptide but topographically and functionally similar. The modified structure is termed a ''mimic'' (imitation) of the parent peptide. Section 2 discusses the basic principles used in the design of conformationally constrained peptide mimetic agents.

## 2 METHODS OF CONFORMATIONAL CONSTRAINT

### 2.1 Local Backbone Constraints

Since the amide backbone plays an important role in determining the three-dimensional structure of the peptide, it is a primary target for the design of conformationally constrained peptides. Backbone modifications fall into two main categories, amide bond modifications and modifications of the α-carbon.

#### 2.1.1 Modifications of the Amide Group

**N-Alkylation** Replacement of the amide N—H by an *N*-alkyl group (e.g., *N*-methyl: Figure 3a) greatly restricts the torsion angle ϕ. An important result of N-alkylation is that it facilitates cis–trans isomerization about the CO—N(R) group (Figure 3a), and thus such peptides can have two basic stereochemical structures. N-

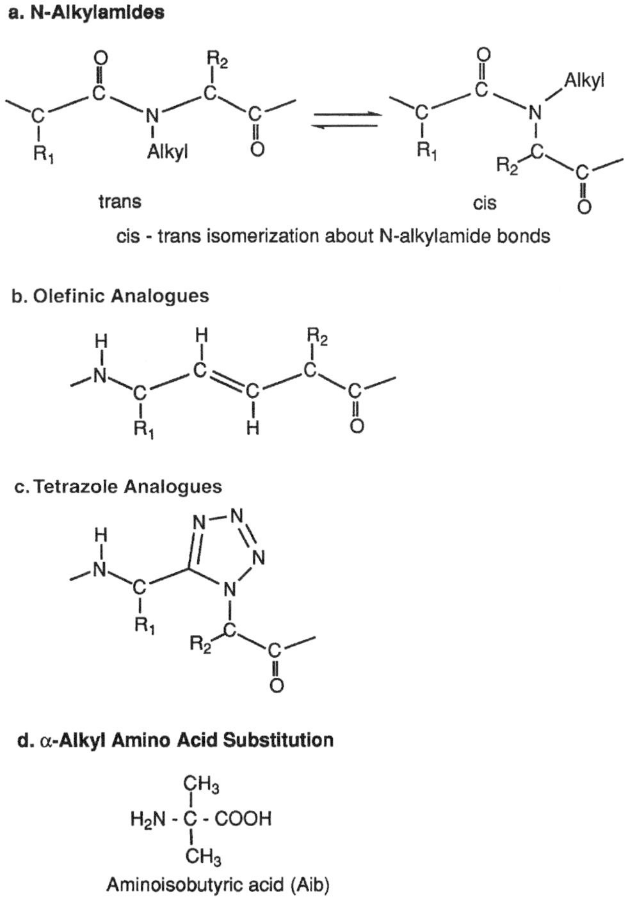

a. N-Alkylamides

trans cis

cis - trans isomerization about N-alkylamide bonds

b. Olefinic Analogues

c. Tetrazole Analogues

d. α-Alkyl Amino Acid Substitution

$H_2N - C(CH_3)_2 - COOH$

Aminoisobutyric acid (Aib)

Figure 3. Modifications of the peptide backbone.

Alkylation also reduces hydrogen-bonding capability of the amide bond, which may further change conformational features of the modified peptide. N-Methylation sometimes is used to resist enzymatic degradation of amide bonds in peptides.

**Olefinic Analogues (ψ[CH=CH])** In this constraint, the entire amide bond is replaced by an olefinic bond (Figure 3b). The bond angles and lengths of the olefinic bond are roughly similar to these properties in the amide bond. The rigidity of the olefinic bond gives a high degree of conformational rigidity to the peptide. Since the electronegative O and N atoms of the amide bond are replaced by carbon atoms, the olefinic analogues have decreased hydrogen-bonding capabilities in the peptide mimetics that contain them. Olefinic peptides also have increased lipophilicity, which may help them in crossing biological membranes.

**Tetrazole Analogues (ψ[$CN_4$])** These analogues have been developed to mimic the *cis*-amide bond conformation. The tetrazole ring twists the two α-centers of the two amino acids into a rigid conformation seen in a *cis*-amide (Figure 3c). Because of the large bulk of the three extra nitrogen atoms, tetrazoles do not seem to be the ideal isosteric replacement for the amide bonds; nevertheless, this type of *cis*-peptide bond mimetic agent is quite easily synthesized, and it has proved valuable in the design of receptor probes.

Numerous other nonrigid modifications are possible.

### 2.1.2 α-Carbon Modification: α-Alkyl Substitution

Substitution of the α-hydrogen atom with an alkyl group restricts the ϕ and ψ torsion angles into two small regions centered around $\phi = -57°$, $\psi = -47°$, and $\phi = 57°$ and $\psi = 47°$. A commonly used α-substituted amino acid is aminoisobutyric acid (Aib) (Figure 3d). The strong tendency to restrict conformational freedom and the added advantage that the α-substituted amino acids are roughly similar in size to the natural amino acids make the former valuable tools in the design of conformationally constrained peptides. It also has been shown that α-substituted amino acids like Aib have a tendency to form helices in peptides.

## 2.2 Local Side Chain Constraints

The conformation of side chain groups in a peptide is an important determinant of its receptor interaction.

### 2.2.1 Dehydro Amino Acid Substitution ($\Delta^E$ or $\Delta^Z$)

The dehydro amino acids have a rigid double bond between the α- and the β-carbon atoms of the side chain (Figure 4a). The rigidity of the double bond restricts the side chain torsion angle $\chi_1$. Also, the steric interaction of the rigid side chain with the peptide serves to restrict the freedom of the ϕ and ψ angles, too. The cis and trans forms of the side chain group, which lead to two possible structures for each dehydro amino acid, are valuable tools in the design of receptor probes.

### 2.2.2 Cyclopropyl Amino Acid Substitution

The cyclopropyl constraint also restricts rotation about the bond between the α- and β-carbons (Figure 4b). The conformational restrictions are similar to those for the dehydro amino acids.

### 2.2.3 D-Amino Acid Substitution

Substitution of a D-amino acid in place of an L-amino acid changes the spatial orientation of the amino acid. Although a conformational effect is not immediately apparent, substitution of D-amino acids can have a significant effect on peptide secondary structure. D-amino acids also may enhance the formation of LD and DL bends to bring about conformational constraints in peptide secondary structure.

### 2.2.4 Increasing the Bulk of the Amino Acid Side Chain

Increasing the steric bulk of the side chain group [e.g., $CH_3$ or $CH_2CH(CH_3)_2$ to $C(CH_3)_3$] may affect the freedom of rotation about ϕ and ψ angles.

## 2.3 Cyclic Constraints

Many naturally occurring peptides occur as cyclic structures. In eukaryotic cells, this is achieved by means of side chain–side chain disulfide bonds between Cys residues. Their cyclic nature imparts to these peptides an increased rigidity. In general, cyclization imposes conformational constraints on linear peptides.

### 2.3.1 N-Terminal to C-Terminal Cyclization

A peptide containing a free COOH at its C-terminus and a free $NH_2$ group at its N-terminus can be cyclized via a CONH linkage (Figure 5a). In addition to imposing a conformational constraint on the peptide, this type of cyclization may increase biological activity for peptides that need their N- and C-termini close together for optimal receptor interaction.

### 2.3.2 Side Chain-to-Backbone Cyclization

A portion of a side chain may be covalently linked to the amide bond backbone (Figure 5b). A linker group [$(CH_2)_n$, etc.] may be used if necessary. In this type of constraint, the relationship of the side chain (R) to the backbone is fixed. Depending on the size and structure of R and the ring size, angles ϕ, ψ, and χ will be fixed or have limited flexibility.

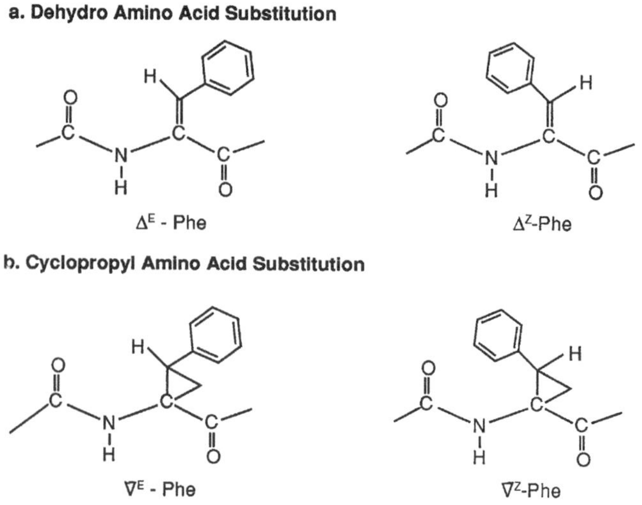

**a. Dehydro Amino Acid Substitution**

$\Delta^E$ - Phe

$\Delta^Z$-Phe

**b. Cyclopropyl Amino Acid Substitution**

$\nabla^E$ - Phe

$\nabla^Z$-Phe

Figure 4. Local side chain constraints.

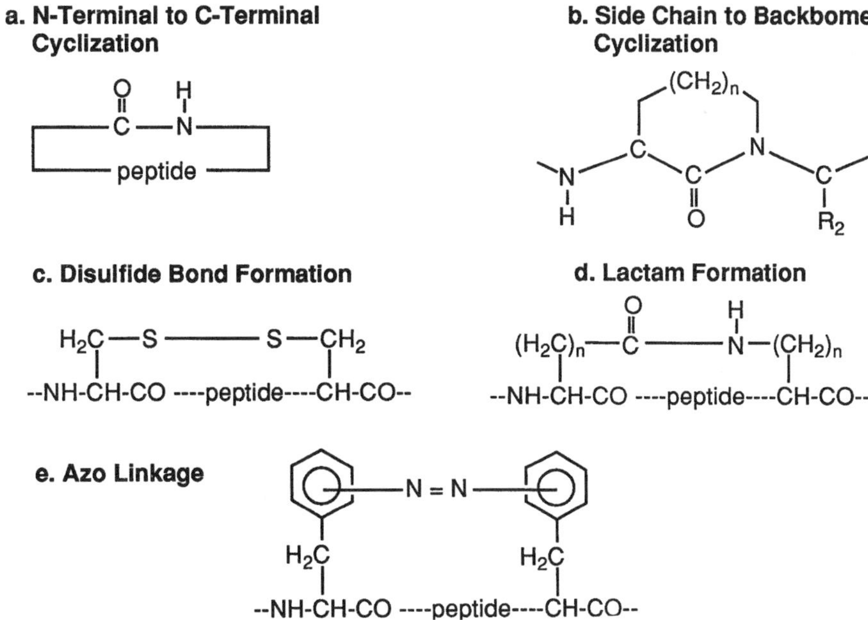


Figure 5. Cyclic constraints.

### 2.3.3 Covalent Side Chain–Side Chain Cyclizations

Side chains that have reactive functional groups offer the possibility of covalent attachment to each other to form cyclic peptides. Thus, ester linkages (lactones), amides (lactams), thioethers, ethers, substituted aromatics, ureas, ketones, and so on, can be formed by reaction of various side chain groups, to form cyclic constrained peptides.

**Disulfide Bond Formation** Two cysteine side chains are linked through covalent modification (Figure 5c). Since disulfide bond formation is easily achieved by chemical means, this method is popular. If the size of the ring formed is relatively small, the $\phi$, $\psi$, and $\chi$ angles can become restricted with this type of cyclization.

**Lactam Formation** An amide bond is formed between a side chain containing an amide group (e.g., Lys, Orn) and a side chain containing a carboxylic group (e.g., Asp, Glu) (Figure 5d). The cyclization can lead to a fixed conformation between the two side chain groups and the backbone. Thus, the angles $\phi$, $\psi$, and $\chi$ have limited freedom.

**Azo-Linked Aromatic Side Chains** Azo linkages have been utilized to join amino acids with aromatic side chains (e.g., Phe or Tyr) (Figure 5e). The conformational restrictions described for the cyclic lactams also apply here.

## 2.4 Designed Secondary Structure

As discussed earlier, hydrogen bonding between carbonyl oxygen and amide NH causes linear peptides to fold into various bends or turns to form secondary structures. Incorporation of constrained functionalities to mimic various bends and turns is an important approach to the rational design of receptor-specific peptide ligands. Some of the commonly utilized constraints to mimic β- or γ-turns are discussed next.

### 2.4.1. δ- and γ-Lactam Constraints

The peptide backbone is modified at a peptide nitrogen that is cyclized with an alkane side chain to form a lactam (Figure 6a). This type of cyclization causes a turn in the basic peptide backbone. According to the size of the lactam ring, the $\psi$ and $\chi$ angles in the peptide mimetic will vary, and thus, turns of different types can be mimicked. Five-membered lactam rings ($n = 0$ in Figure 6a) give analogues that are good mimics of type II′ β-turns, whereas six-membered lactams ($n = 1$) give structures analogous to γ-turns.

### 2.4.2 Bicyclic Thiazolidine Constraints

Bicyclic thiazolidines are a type of β-turn analogue developed to constrain the peptide within a bicyclic thiazolidine system (Figure 6b). The $\phi$ and $\psi$ torsion angles are restricted to values closely resembling those of a type II β-turn.

### 2.4.3 5*H*-6-Oxo-2,3,4,4*a*,7,7*a*-exahydropyrano[2,3-*b*]pyrroles

This bicyclic system (shown in Figure 6c) also mimicks a type II β-turn. Torsion angle restrictions are similar to those described in Section 2.4.2.

### 2.4.4 3-Amino-2-piperidone-6-carboxylic Acid

This monocyclic system (Figure 6d) has the demonstrated ability to fold two ends of a peptide to mimic a turn. NMR studies indicate that the preferred conformation mimics rare types (not type I or II) of β-bends.

### 2.4.5 Spirolactam Analogues

In this novel approach, the lactam analogs described earlier have been further constrained. The spiro-bicyclic systems shown in Fig-

a. Lactam Constraints

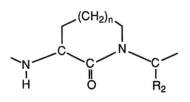

b. Bicyclic Thiazolidines

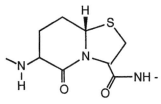

c. Bicyclic oxo pyrroles

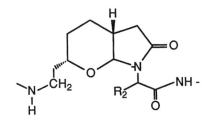

d. 2-Amino-2-piperidone -6-carboxylic Acid

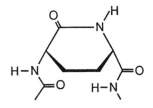

e. Spiro-bicyclic Lactam Analogues

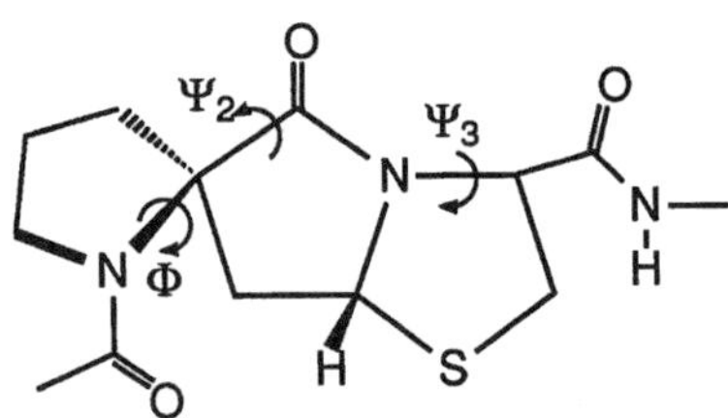

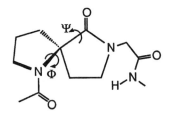

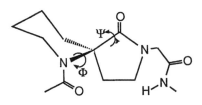

f. Spiro-bicyclic Thiazolidine Analogues

Figure 6. Various types of β- and γ-turn mimetic systems.

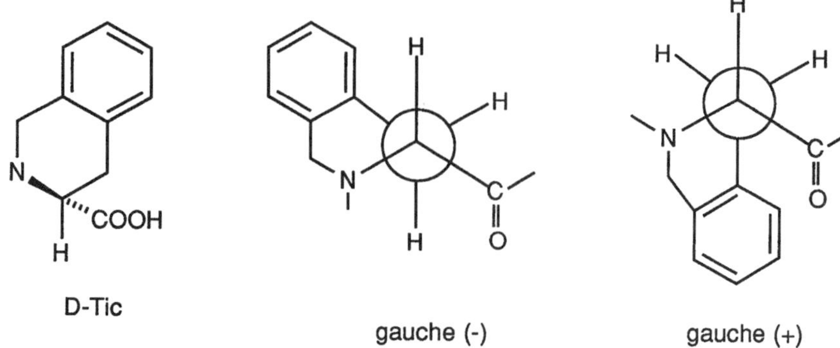


Figure 7. The topographically constrained amino acid D-Tic and its two low energy chairlike conformations.

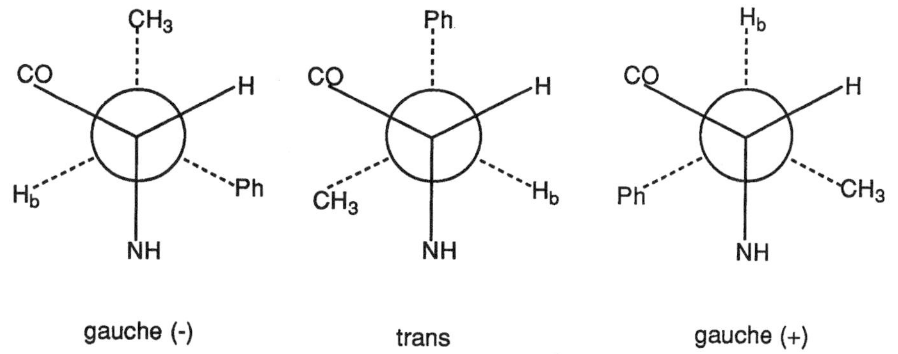

**Figure 8.** The low energy side chain conformations for (2*S*, 3*S*)-β-Methyl-phenylalanine about the bond between the α- and β-carbons. [gauche (−) = − 60°, trans = 180°, gauche (+) = + 60°]

ure 6e have the added constraint of restricting the $\phi_2$ torsion angle as a β-bend mimetic.

### 2.4.6 Spiro-(Bicyclic-Thiazolidine)Lactams

The thiazolidine system is constrained further with a spiro-bicyclic system (Figure 6f). This structure simultaneously restricts three ($\phi_2$, $\psi_2$, and $\psi_3$) torsion angles that are involved in a type II β-turn.

## 2.5 Topographical Constraints

Topography can be defined as *the relative, cooperative three-dimensional arrangements of the side chain groups in a polypeptide.* The topography of the side chains plays a critical role in peptide–receptor interactions. The investigation of topographical aspects of bioactive peptides, which offers great potential in the design of potent and receptor-selective peptides, is an area of intense investigation. We have shown that the preferred topography of side chains can be examined by appropriately constraining, biasing, or fixing side chain conformers in the gauche (−) ($\chi = -60°$), trans ($\chi = \pm 180°$), or gauche (+) ($\chi = +60°$) conformations.

### 2.5.1 Fixing of Side Chain Residues

Linking the α-amino group to a side chain aromatic group in amino acids like Phe, Tyr, or Trp via a methylene bridge can bring about topographical constraint. Thus, 1,2,3,4-tetrahydroisoquinoline-2-carboxylic acid (Tic) (Figure 7) is obtained from Phe in this manner. The $\chi_1$ angle is now restricted to either gauche (−) or gauche (+) and the trans rotamer is excluded. The incorporation of such topographically fixed amino acids into peptides has yielded exciting results in our hands and will continue to be valuable tools for exploring the preferred topography in newly discovered peptides.

### 2.5.2 α,β-Unsaturated Amino Acids

The dehydro amino acids discussed previously also can be used as amino acids containing topographically constrained side chains.

### 2.5.3 β-Alkyl Amino Acids

Topographically constrained amino acids can be obtained by substitution at diastereotopic β-positions. We have developed efficient asymmetric synthetic methods to produce optically pure isomers of β-methylphenylalanine, β-methyltryptophan, and β-methyltyrosine. The side chain conformations of these amino acids are biased as a result of nonbonding interactions between vicinal substituents. As shown in Figure 8, the gauche (−) side chain conformation gives the smallest amount of unfavorable steric interactions for 2*S*,3*S*-β-methylphenylalanine. With similar arguments, it can be shown that the 2*S*,3*R* and 2*R*,3*S* and 2*R*,3*R* conformers all have their own unique side chain topography in a peptide. Incorporation of these isomers into potent peptide ligands such as C[D-Pen$^2$, D-Pen$^5$]Enkephalin (DPDPE) and cholecystokinin analogues has yielded peptides with widely differing potencies and receptor selectivities.

# 3 GENERAL DISCUSSION AND FUTURE PERSPECTIVES

In this brief overview we have outlined some basic principles behind the rational design of peptide ligands utilizing conformational, topographical, stereoelectronic, and dynamic considerations. Using conformational constraints, peptide hormone and neurotransmitter analogues with high potencies and extraordinary selectivities can be designed. State-of-the-art NMR spectroscopy and molecular modeling techniques can help reveal the minimum energy conformation of these peptides. Approaches for conformational constraint will improve as scientists gain a deeper understanding of the molecular forces that determine peptide conformation and topography, and as receptors are ''cloned'' in pure form. With the discovery of new bioactive peptides, conformationally constrained peptide mimetic systems can be expected to undergo exciting developments in the future.

*See also* Medicinal Chemistry; Peptides; Receptor Biochemistry; Recognition and Immobilization, Molecular.

## *Bibliography*

Fauchere, J.-L. (1986) Elements for the rational design of peptide drugs. *Adv. Drug Res.* 15:29–69.

Hruby, V. J. (1981) In *Topics in Molecular Pharmacology,* A.S.V. Burgen and G.C.K. Roberts, Eds., pp. 99–126. Elsevier, Amsterdam.

———. (1982) Conformational restrictions of biologically active peptides via amino acid side chain groups. *Life Sci.* 31:189–199.

———, (1992) Strategies in the development of peptide antagonists. *Prog. Brain Res.* 92:215–224.

———, and Hadley, M. E. (1986) Binding and information transfer in conformationally restricted peptides. In *Design and Synthesis of Organic Molecules Based on Molecular Recognition,* G. Van Binst, Ed., pp. 269–289. Springer-Verlag, Heidelberg.

———, Al-Obedi, F., and Kazmierski, W. (1990) Emerging approaches in the molecular design of receptor specific peptide ligands: Conformational, topographic and dynamic considerations. *Biochem. J.* 268: 249–262.

Kessler, H. (1990) In *Trends in Drug Research,* V. Classen, Ed., pp. 73–84. Elsevier Science Publishers, Amsterdam.

Morgan, B. A., and Gainor, J. A. (1989) Approaches to the discovery of non-peptide ligands for peptide receptors and peptidases, Vol. 24 in *Annual Reports in Medical Chemistry,* R. C. Allen, Ed., pp. 243. Academic Press, San Diego, CA.

Rizo, L., and Gierasch, L. M. (1992) Constrained peptides: Models of bioactive peptides and their protein substructures. *Annu. Rev. Biochem.* 61:387–418.

Spatola, A. F. (1983) In *Chemistry and Biochemistry of Amino Acids, Peptides and Proteins,* Vol. 7, B. Weinstein, Ed., pp. 267–357. Dekker, New York.

Toniolo, C. (1990) Conformationally restrained peptides through short range cyclizations. *Int. J. Peptide Protein Res.* 35:287–399.

# Peptide Synthesis, Solid-Phase

*R. C. Sheppard*

## Key Words

**Coupling Agent** Chemical reagent used to link amino and carboxy groups in peptide synthesis.

**Merrifield Technique** See solid phase synthesis.

**Peptide** A chemical compound formed by the head-to-tail linkage of α-amino-acids through amide (peptide) bonds.

**Protecting Group** Chemical group used to reduce the reactivity of amino, carboxy, or other reactive functions in amino acids and peptides.

**Resin** The insoluble solid support used in solid phase peptide synthesis. Commonly an organic polymer such as polystyrene, polydimethylacrylamide, or polyethylene glycol.

**Solid Phase Peptide Synthesis** Chemical synthesis of peptides by stepwise addition of one protected amino acid residue at a time to an insoluble solid support.

The chemical union of amino acids to form peptides (peptide synthesis) has become an essential tool in modern biology. In addition to their traditional role in the study of structure–function relations in enzymes, hormones, antibodies, toxins, and so on, synthetic peptides have become especially important in molecular biology as immunogens, usually for generating antibodies that cross-react with sequence-related proteins. The chemistry of peptide synthesis is complicated by the polyfunctional nature of the common α-amino acids, and made slow, difficult, and laborious by the large number of consecutive chemical reactions required. A major simplifying and accelerating step was the introduction of solid phase peptide synthesis, in which all the synthetic operations took place on a solid but permeable, gelatinous particle. This process enabled the construction of manual and automatic synthesizers, which have brought synthetic ability to many laboratories that were not chemically oriented.

## 1 PRINCIPLES OF PEPTIDE SYNTHESIS

Selectively cleavable protecting groups are required to mask unwanted reactivities in polyfunctional amino acids and to remove ambiguities in the synthesis. For simple bifunctional amino acids, an amino-protected amino acid (the carboxy component) is chemically linked to a carboxy-protected residue (the amino component):

$$\text{X—NH—CHR}^2\text{—CO}_2\text{H} + \text{H}_2\text{N—CHR}^1\text{—CO}_2\text{Y} \rightarrow$$
$$\text{X—NH—CHR}^2\text{—CO—NH—CHR}^1\text{—CO}_2\text{Y} \; [+ \text{H}_2\text{O}]$$

Joining of the two components (formally, the elimination of water) requires chemical activation of the free carboxyl group, usually by addition of a coupling agent. One of the protecting groups X or Y may then be selectively removed from the product. This liberates a free amino or carboxyl group as in the initial protected amino acid and enables further chain extension. For trifunctional amino acids, additional protecting groups are usually required to mask chemically reactive groups present in side chains.

## 2 SOLID PHASE SYNTHESIS: THE MERRIFIELD TECHNIQUE

The introduction of solid phase methods by Merrifield in 1962 made peptide synthesis both quicker and easier. Merrifield chose as amino-protecting group X the acid-labile *t*-butoxycarbonyl (Boc) function and as carboxy-protecting group Y an *insoluble, gelatinous, styrene–divinylbenzene resin,* which functioned chemically as a more acid-stable benzyl ester (Figure 1). Reactive side chain functions were also usually protected as benzyl derivatives. In the standard Merrifield technique (often known as the Boc–benzyl–polystyrene technique), synthesis proceeds (Figure 1) by attaching the first amino acid to the insoluble resin, cleaving the amino-protecting group with trifluoroacetic acid, neutralizing the amine salt with an organic base, and attaching the incoming Boc–amino acid using the coupling reagent dicyclohexylcarbodiimide (DCCI). Succeeding amino acid residues are then added through repetition of this cycle of deprotection, neutralization, and coupling steps. Synthesis is completed by cleaving the amino, side chain, and carboxy-terminal (resin support) protecting groups in a single treatment with very strong acid, usually liquid hydrogen fluoride. The synthetic peptide is then liberated into solution and may be purified and characterized.

Use of an insoluble carboxy–protecting group confers the following advantages.

1. The growing intermediate peptide is at all times insoluble and is easily and rapidly separated from soluble reagents by simple filtration and washing with solvent.
2. Because soluble reagents can be so easily separated, they can be used in large excess encouraging reactions to proceed rapidly toward completion.
3. The insoluble resin-bound peptide is retained in a single agitated reaction vessel, minimizing mechanical losses and again encouraging high overall efficiency.
4. The reagent addition and separation processes are easily mechanized, permitting the construction of automatic peptide synthesizers.

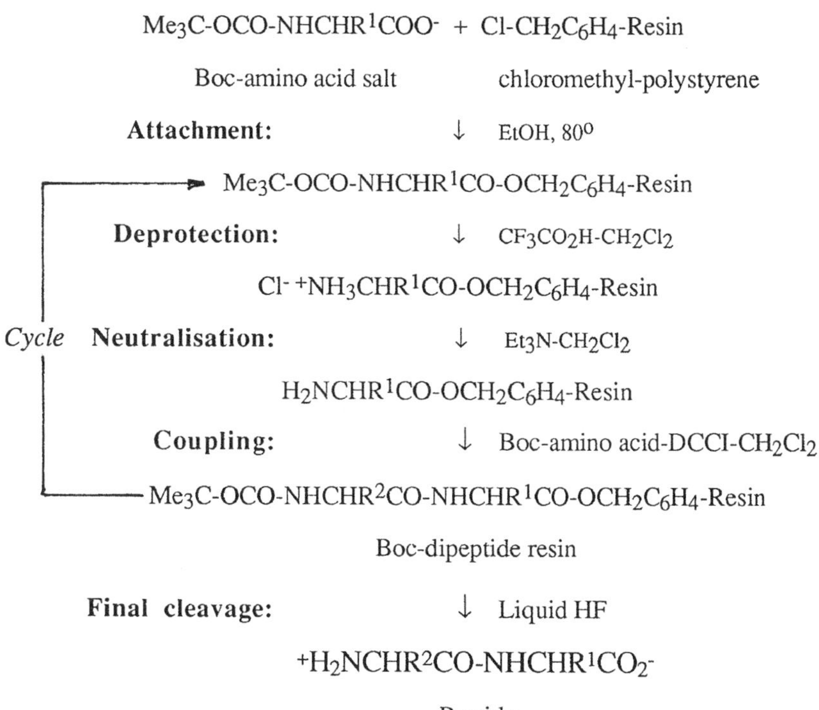


**Figure 1.** Chemistry of the Merrifield technique illustrated for the synthesis of a dipeptide.

Disadvantageous features of the solid phase technique are:

1. No separation of *resin-bound* starting materials and products is possible, and near-quantitative conversions are essential at every stage if serious accumulation of impurities is to be avoided.
2. Because of the solid (gelatinous) nature of the system, analytical control may be difficult and slow.
3. The resin-bound peptide chains may associate together, increasing steric hindrance and slowing chemical reactions.
4. The reagents are chemically vigorous and can cause destructive side reactions, again generating impurities.
5. The usual final cleavage reagent, liquid hydrogen fluoride, is a particularly toxic and hazardous low boiling liquid.

## 3 MODERN VARIANTS OF THE MERRIFIELD TECHNIQUE

The Merrifield technique is still widely practised essentially as illustrated in Figure 1, but a number of modern variants have been introduced which are steadily gaining in popularity. These have affected most of the details of the method, though the solid phase principle has remained unchanged.

### 3.1 The Amino-Protecting Group

The Boc and benzyl protecting groups used in the Merrifield technique have differing degrees of acid lability. It is difficult to cleave Boc groups repetitively through many cycles without slow concomitant cleavage of benzyl-based side chain protecting groups and the peptide resin linkage. This difficulty is now commonly overcome by using the acid-stable, base-labile, 9-fluorenylmethoxycarbonyl (Fmoc) amino-protecting group **(I).**

$CH\text{-}CH_2\text{-}OCO\text{-}$

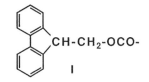

I

The Fmoc group is cleaved rapidly by organic bases such as piperidine (Figure 2) and no neutralization step is required. The strong ultraviolet absorption of the Fmoc group also offers advantages in analytical control of synthesis (q.v.).

### 3.2 Side Chain Protecting Groups

The Fmoc-amino protecting group permits use of very acid-labile protecting groups for amino acid side chains. Commonly these are based on *t*-butyl structures (*t*-butyl esters, ethers, and Boc derivatives) and are cleaved by trifluoroacetic acid. Scavenging reagents (e.g., water, phenol, ethane dithiol, triethyl or triisopropyl silanes) are commonly added to the cleavage reagent. These trap reactive carbonium ions generated from *t*-butyl derivatives, which otherwise might react with the side chains of sensitive residues such as tryptophan or methionine. The Fmoc–*t*-butyl combination of protecting groups avoids altogether the use of very strongly acidic reagents and provides a chemically milder synthesis technique.

### 3.3 The Peptide–Resin Linkage

The flexibility of solid phase synthesis has been enhanced by development of a range of peptide–resin linkages. These may be integral parts of the resin structure (as in the original Merrifield polystyrene resin), or separate linkage agents, which are attached to a more simply functionalized polymer support. For Fmoc-based

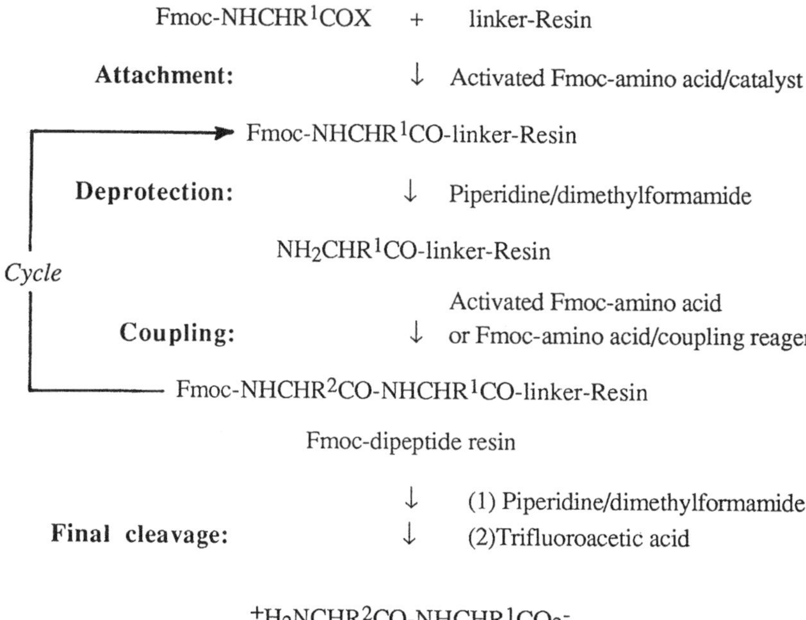


**Figure 2.** Chemistry of the Fmoc technique illustrated for the synthesis of a dipeptide.

solid phase procedures, the *p*-alkoxybenzyl alcohol resin **(II)** forms a benzyl ester with acid lability similar to that of *t*-butyl derivatives. *t*-Butyl-based side chain protecting groups and the peptide–resin linkage may therefore be cleaved simultaneously by mild acid treatment. The separate linkage agent **(III)** used with resins other than polystyrene (see Section 3, 4) produces a similarly labile linkage. Introduction of an additional methoxy group into the aromatic ring enhances the acid lability of derived esters. Peptides linked through the linkage agent **(IV)** may be cleaved under very mildly acidic conditions (1% trifluoroacetic acid in dichloromethane), which leave *t*-butyl derivatives largely unchanged. Side chain protected peptides useful in further synthesis can thus be prepared by the solid phase method. Acid lability can be further enhanced by making the cleavable bond part of a benzhydryl (diphenylmethyl) system rather than a simple benzyl derivative. Peptides attached to the nitrogen atom of the resin-bound linkage agent **(V)** are cleaved at the C—N bond by mild acids, providing a convenient solid phase synthesis of peptide amides.

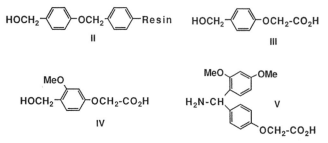

### 3.4 Coupling Reagents and Activated Amino Acids

A number of new coupling reagents have been introduced as alternatives to dicyclohexylcarbodiimide. More soluble carbodiimides [e.g., diisopropylcarbodiimide (DIC)] give more soluble and more easily eliminated coproducts (ureas) and are favored in some techniques. In equimolar amount, carbodiimides are believed to react with carboxy components to generate *O*-acylureas, which react rapidly with the amino component already present or then are added to the reaction mixture. When only 0.5 equivalent of carbodiimide is used in the absence of the amino component, symmetric anhydrides of Boc– or Fmoc–amino acids are formed which may be isolated and combined with the amino component in a separate step. This procedure avoids contact of carbodiimide or its reaction products with the amino–resin and minimizes the risk of side reactions. The BOP **(VI)** and TBTU **(VII)** reagents are now threatening to replace carbodiimides. Both probably generate intermediate esters of hydroxybenzotriazole (HOBt, **VIII**), which react rapidly and cleanly with amino components. Hydroxybenzotriazole itself is often added as a catalyst to other acylation reactions, probably generating the same intermediate esters. Stable preformed activated esters derived from pentafluorophenol **(IX)** or hydroxyoxodihydrobenzotriazine (HODhbt, **X**) and Boc– or Fmoc–amino acids are particularly convenient reagents, especially for automated synthesis. Reactivity of pentafluorophenyl esters is enhanced by added HOBt catalyst.

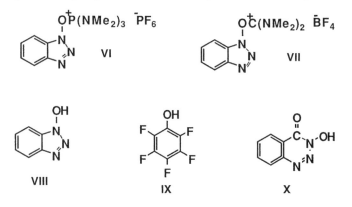

### 3.5 Reaction Media

Dichloromethane (DCM) was selected as acylation reaction medium in the Merrifield technique, because of its good swelling properties for polystyrene gel and the good solubility of Boc–amino acids and their rapid reaction with DCCI in this medium without side reactions. Later it was recognized that inadequate solvation of the growing peptide by nonpolar solvents such as DCM could lead to association of peptide chains within the peptide–resin matrix. In gel systems this may lead to shrinkage of the resin and onset of serious steric hindrance. Dimethylformamide (DMF) and similar dipolar aprotic liquids, which are good solvents for protected peptides, were introduced to reduce this effect. Their use is now widespread even though in DMF, polystyrene resin swells to a gel volume only about half that in DCM. They are especially useful when interchain association effects are observed or expected (i.e., with so-called 'difficult sequences'). Fmoc–amino acids are much more soluble in DMF than in DCM, and acylation reactions involving simple activated ester intermediates (e.g., pentafluorophenyl esters) are also substantially faster.

### 3.6 Solid Supports

A range of solid supports additional to cross-linked polystyrene are now available. Polar polyamide supports, which are particularly well permeated and swollen by solvents such as dimethylformamide, have found wide application in combination with Fmoc–amino acids. Polyamide gels are also available in rigid, physically supported form that enables their use in continuous flow technology (see Section 3.7). More recently, a range of supports based on polyethylene glycol (PEG) have become available which are also suitable for continuous flow techniques. Polyamide and PEG resins are available with a full range of side chain protected Fmoc–amino acids attached through a variety of linkage agents, greatly facilitating the rapid conduct of routine synthesis. Polyamide resins usually also have an internal reference amino acid attached directly to the polymer, facilitating analytical control of the synthesis.

### 3.7 Technology and Instrumentation

Solid phase peptide synthesis can be carried out manually in normal laboratory glassware or automatically in complex synthesizers, usually under computer control. A realistic minimum for the Merrifield and similar techniques is a glass reactor vessel fitted with a lower sintered filter and stopcock. Reagents are added to resin contained in this vessel, which is agitated (usually in a wrist action shaker) and then drained under vacuum or nitrogen pressure. Simple plumbing (e.g., valve-selectable reagent reservoirs and drain connection) is helpful. Various degrees of mechanization may be incorporated, but full automation is difficult because analytical feedback control is cumbersome. In the alternative continuous flow system, the resin is kept stationary in a column reactor through which reagents are pumped in sequence. A minimum manually controlled system requires a low pressure reciprocating or peristaltic pump, a column reactor, and a single valve to switch between recirculate and flow modes. The stationary resin and moving reagent stream render analytical control much easier, and fully automated systems are commercially available.

### 3.8 Analytical Control

Not all peptide coupling or deprotection reactions are straightforward, and there is significant risk of failure in solid phase synthesis. This is minimized by regular analytical control throughout the synthesis, followed by rigorous purification and characterization of the product. A simple color test with ninhydrin is available to assess completeness of coupling reactions in the Merrifield technique. Quantification is possible, though it is time-consuming and is commonly done only after the whole synthesis has been completed. In Fmoc-based continuous flow techniques, release of chromophoric fluorene derivatives into the flowing stream enables deprotection to be monitored photometrically, and indicator or dye-binding techniques enable the progress of coupling reactions to be followed. Because the resin is stationary, direct photometric monitoring of residual resin-bound amino groups is also possible.

### 3.9 Multiple Synthesis

The increasing importance of peptide synthesis is reflected in the apparently ever growing demand for synthetic peptides. Limited multiple synthesis capability is a feature of continuous flow synthesizers, but systems have now been devised permitting simultaneous assembly of many sequences. Generally the efficiency of these techniques is increased if the peptide sequences are related, but this is not necessary. The most popular is probably the T-bag procedure in which resin samples are compartmentalized in porous plastic mesh sachets. They are sorted into groups for addition of individual residues and combined for simultaneous washing and deprotection operations. As a rule, T-bag synthesis is carried out manually. A number of instruments have been devised in which multiple synthesis is carried out using wells drilled or molded in plastic blocks similar to microtiter plates or racks of individual reactor tubes. Activated amino acids and other reagents are added manually or automatically using robotic dispensing systems.

*See also* Amino Acid Synthesis; Peptides.

## *Bibliography*

Atherton, E., and Sheppard, R. C. (1989) *Solid Phase Peptide Synthesis, A Practical Approach,* IRL Press at Oxford University Press, Oxford.

Barany, G., and Merrifield, R. B. (1979) In *The Peptides, Analysis, Synthesis, Biology,* Vol. 2, pp. 3–285, E. Gross and J. Meienhofer, Eds. Academic Press, New York.

Fields, G. B., and Noble, R. L. (1990) Solid phase peptide synthesis utilizing 9-fluorenylmethoxycarbonyl amino acids. *Int. J. Peptide Protein Res.* 35:161.

Jung, G., and Beck-Sicklinger, A. G. (1992) Multiple peptide synthesis. *Angew. Chem. Int. Ed. Engl.,* 31:367.

Stewart, J. M., and Young, J. D. (1984) *Solid Phase Peptide Synthesis,* 2nd ed. Pierce Chemical Company, Rockford, IL.

# Pesticide-Producing Bacteria

*Jeffrey L. Kelly, Michael David, and Peter S. Carlson*


## *Key Words*

**Antibiotic** Substance produced by a microorganism and able in dilute solution to inhibit or kill another microorganism.

***Bacillus thuringiensis* δ-endotoxins** *Bacillus thuringiensis* produces a proteinaceous crystalline inclusion during sporulation that, when ingested by insects, is solubilized, releasing insecticidal proteins called δ-endotoxins. Also called insecticidal crystal proteins (ICPs).

**Pesticide** Agent used to kill or suppress pests; for plant pests, these agents are divided by target into insecticides, fungicides, and herbicides.

**Secondary Metabolites** Compounds produced especially in plants and microorganisms which have no recognized role in the maintenance of fundamental biological processes.

***Streptomyces*** Genus of actinomycetes whose species are mostly soil-inhabiting and form vegetative mycelia; many streptomycetes produce antibiotics as secondary metabolites.


A wide variety of pesticide-producing bacteria have been discovered. Although these microorganisms are active against many pests, the use of bacteria or their products as pesticides is minor in an industry dominated by synthetic chemicals. Both changes in social opinion regarding chemical and biological pesticides and the prospects of improving performance of biologicals contribute to great opportunities for increased use of pesticide-producing bacteria. This entry presents and discusses five well-characterized examples of bacterial pesticides that have or are close to achieving commercial utility. The advantageous characteristics of these example substances, as well as important considerations for future exploitation, are described.

## 1 INTRODUCTION

Bacteria can be used to control plant pests, increase soil fertility, and improve the stress tolerance of crop plants. While many bacteria have been reported to have beneficial effects on crop production, none has gained wide commercial acceptance. The total market share of microbial products for agriculture is limited, amounting to less than 1% of the current $25 billion crop protection market. The cost and performance characteristics of microbial pesticides have never matched their market competition, synthetic chemicals.

Sociopolitical realities, which spotlight the negative aspects of chemical pesticides, and scientific advances in molecular biology, which expand understanding of biological processes, have forced a reevaluation of the potential for biological pesticides. Is it now possible to substitute biology for synthetic chemistry in crop protection? New knowledge and refined methods, including targeted biochemical sites of action, rapid screening, and genetic engineering, have given researchers additional tools with which to identify, produce, and improve bacterial products for agricultural uses. Additionally, the U.S. Environmental Protection Agency, which regulates pesticides, has developed a potentially less stringent evaluation process, Subdivision M, for pesticides of microbial and biochemical origin. The Report on National Biotechnology Policy, prepared in 1991 by the President's Council on Competitiveness, predicts a revolution in industries, including agriculture, that employ biotechnology. The report predicts that in agriculture the new biology will "greatly reduce reliance on toxic pesticides and improve pesticides." The discovery, analysis, and use of pesticide-producing bacteria is the subject of this contribution.

This brief treatment is limited to current understanding of prokaryotic microbes that produce an agent or agents that in turn can kill or suppress a plant pest (i.e., act as a pesticide). Pesticides are conventionally divided into three major categories: insecticides, fungicides, and herbicides. Not all microbial pathogens of plant pests can be considered as pesticide producers, since often no distinct toxic agent or secondary metabolite is involved in producing the disease. On the other hand, nonpathogenic bacteria are often a source of toxins with pesticidal activity. Such toxins are often secondary metabolites—that is, biochemicals that are not directly required for growth and development but serve other functions such as providing chemical defense.

The literature describing pesticide-producing bacteria is of diverse and uneven quality, ranging from anecdotal observations to carefully defined chemical interactions. We focus on a few well-analyzed examples that have achieved or are close to achieving commercial utility. Our discussion includes examples of pesticides comprising the agent alone or the agent along with the producing bacteria, and examples of agents that have provided model compounds for the synthetic chemist.

Bacteria produce two major classes of pesticidal agents: macromolecular proteins (whose synthesis is catalyzed by ribosomes) and low molecular weight secondary metabolites (the products of enzymatic catalysis). In general, proteins have captured the imagination of genetic engineers because they are encoded by discrete, transferable nucleic acid sequences that are easily modified. Secondary metabolites, the products of more complex genetic systems, have been the focus of natural product and synthetic chemists.

## 2 MACROMOLECULAR PESTICIDES: THE δ-ENDOTOXIN PROTEINS OF *BACILLUS THURINGIENSIS*

Perhaps the most thoroughly investigated macromolecular pesticides are the insecticidal δ-endotoxin proteins produced by *Bacillus thuringiensis (Bt)*. *Bt* is a Gram-positive bacterium, known to produce toxic compounds such as chitinases, proteases, phospholipases and α- and β-exotoxins. Most importantly, however, it produces proteinaceous crystalline inclusion bodies during sporulation. Each inclusion body contains from one to five different δ-endotoxin proteins. More than 25 such δ-endotoxins have been identified. Most of these are specifically toxic to immature insects in the orders Lepidoptera, Diptera, or Coleoptera, and this specificity is largely responsible for the popularity of *Bt* as a biological insecticide. However, the recent discovery of strains active against other insect orders and against mites, parasitic nematodes, and protozoa suggest that the potential utility of *Bt* as a biological pesticide is much broader than had been realized. Our discussion is limited to the insecticidal crystal proteins (ICPs) of *Bt,* because they have been most thoroughly characterized.

### 2.1 Genetics and Structure of the δ-Endotoxins

The *Bt* crystalline genes *(cry),* which code for δ-endotoxins, are generally found on large, conjugative plasmids and are associated with transposons and insertion elements. The apparent mobility of *cry* genes has implications for understanding both the high degree of similarity between the crystal protein sequences and the containment of more than one kind of δ-endotoxin in some parasporal inclusion bodies.

With the exception of a crystal protein from *Bacillus thuringiensis israelensis* (CytA), which exhibits cytolytic activity against invertebrate and vertebrate cells and does not share noticeable sequence homology with the other endotoxins, the crystal (Cry) proteins share significant but varying degrees of homology. In 1989 Höfte and Whiteley proposed a classification scheme for *Bt* crystal protein toxins based on the extent of the sequence similarity of their *cry* genes and their range of insecticidal activity. Thus the *cry I* genes encode Cry I proteins (125–140 kDa), which are toxic to insects in the order Lepidoptera; Cry II ICPs (65–75 kDa) are specific to Lepidoptera (Cry IIB) or Lepidoptera and Diptera (Cry IIA); Cry III (65–75 kDa) are toxic to Coleoptera; and Cry IV ICPs are toxic to Diptera and have a molecular mass of 65 to 70 kDa (Cry IVC, Cry IVD) or 125 to 140 kDa (Cry IV A, Cry IVB).

The greater molecular mass of the Cry I and some Cry IV sequences compared to the other ICPs is primarily in the C-terminal end of the protein, which is highly conserved and contains numerous cysteine residues. The C-terminus probably functions to facilitate intermolecular interactions and thus to increase stability of the inclusion body containing Cry I proteins. As described in Section 2.2, most crystal proteins are protoxins that are proteolytically converted into smaller toxic polypeptides in the insect midgut. In addition to C-terminal cleavages, most protoxins undergo limited N-terminal processing, up to 150 amino acids for Cry IIIA. Generally, the protoxins are processed to an active core protein, about 600 to 650 amino acids long. This core protein contains both the toxic domain and the information required for the high specificity of the toxin. Highly conserved regions are found within this core protein. Indeed, Cry proteins that do not undergo C-terminal processing contain at their C-terminus a stretch of about 12 amino acids that is conserved among all Cry proteins. Additional stretches of highly conserved residues are also observed. The N-terminal halves of the active proteins are similar in that they tend to be hydrophobic and to form α-helices and presumed transmembrane domains. Changes made in primary sequence within these regions alter the toxicity of the protein. Changes within the C-terminal half of the active protein, by contrast, affect the range of insects susceptible to the toxin.

### 2.2 δ-Endotoxin Mode of Action

When *Bt* is used as an insecticide, the insect larvae normally feed on a mixture of spores and inclusion bodies. When inclusions containing Cry I are ingested by lepidopterans, the alkaline pH of the insect midgut favors the solubilization of the crystalline bodies, and midgut proteases process the protoxin to an active peptide of some 60 to 70 kDa. It is possible that some of the proteolytic processing is done by spore-associated, *Bt*-encoded proteases. However much of the documented processing is done by insect gut proteases. The specific insecticidal range of each toxin is mediated primarily by the presence of specific receptors on the midgut epithelial cell membranes. However, some of the insect specificity can be conferred by the proteolytic processing itself. For example, the 130 kDa protoxin from the *Bt* subspecies *aizawai* can be processed by gut proteases derived from larvae of the lepidopteran *Pieris brassicae* into a toxin active against both *P. brassicae* and the dipteran *Aedes aegypti;* the larval midgut extract from *A. aegypti,* however, produces a shorter toxin active only against *A. aegypi.* After proteolytic activation, the toxin binds to receptors on the midgut cell membrane. Upon binding, the activated toxin changes its conformation, and the toxic domain is inserted into the midgut cell membrane. According to one proposed model, subsequent oligomerization of a few toxin molecules results in the formation of a nonspecific membrane pore, leading to osmotic imbalance and eventual lysis of the midgut cell.

### 2.3 Genetically Engineered Insecticides Based on δ-Endotoxins

Many of the characteristics that make *Bt* an attractive alternative to chemical insecticides from the standpoint of environmental safety have also limited its commercial success: the toxin has a limited lifetime on plant surfaces after application by spraying; spore release may cause environmental and economic concerns (especially in Japan, where sericulture is a large industry); foliar applications of *Bt* may not reach insects feeding inside plant tissues (e.g., stem borers) or those that suck plant juices (e.g., aphids); and applications must be carefully timed to coincide with the presence of vulnerable stages (usually young larvae feeding on exposed foliage).

Other methods of delivering *Bt* toxins have been devised which address some of the foregoing concerns. For example, Crop Genetics International has engineered and expressed *cry IA(c)* and *cry IA(b)* toxin genes in an endophytic bacterium, *Clavibacter xyli cynodontis.* The engineered bacteria can be inoculated into corn seeds under pressure. When the treated seed germinates, the engineered endophytic bacteria grow along with the host plant. The *Bt* toxin is continuously produced within the plant, providing season-long protection against larvae of the European corn borer. Mycogen Corporation has introduced *Bt* δ-endotoxin genes into *Pseudomonas fluorescens.* When killed and chemically treated (''stabilized''), the engineered bacterial cells encapsulate the crystalline inclusion, resulting in improved persistence of the *Bt* toxin when sprayed onto crops. Ecogen Inc., has created novel combinations of δ-endotoxins in commercial *Bt* strains by a process of selective plasmid curing followed by conjugational transfer of plasmids carrying genes that encode different δ-endotoxin proteins. Thus, *Bt* strains can be obtained that are more potent and toxic to a broader range of pests. *Bt*-derived δ-endotoxin genes have also been introduced directly into the genome of many crop species to confer resistance to insect pests.

## 3 LOW MOLECULAR WEIGHT SECONDARY METABOLITES

### 3.1 Bioherbicide: Bialaphos

One particularly well-studied example of a microbial product with herbicidal activity is bialaphos (Figure 1), a metabolite of the soil

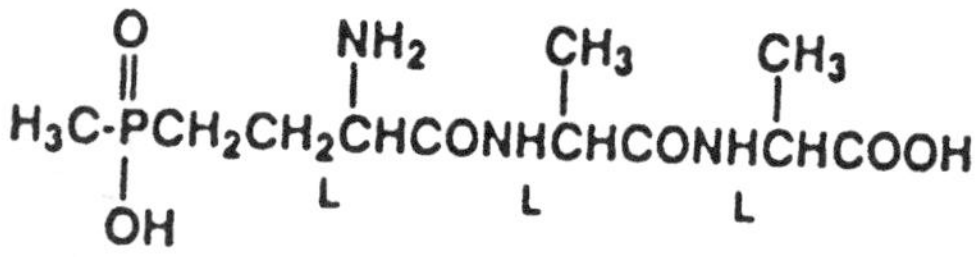

$H_3C-P(=O)(OH)CH_2CH_2CH(NH_2)CONHCH(CH_3)CONHCH(CH_3)COOH$ (L, L, L)

**Bialaphos**

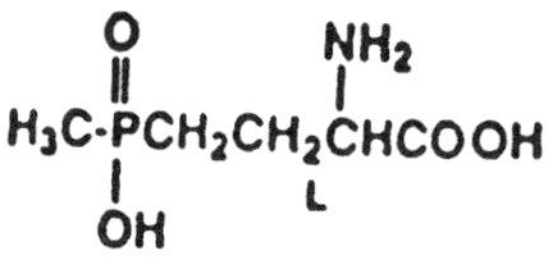

$H_3C-P(=O)(OH)CH_2CH_2CH(NH_2)COOH$ (L)

**Phosphinothricin**

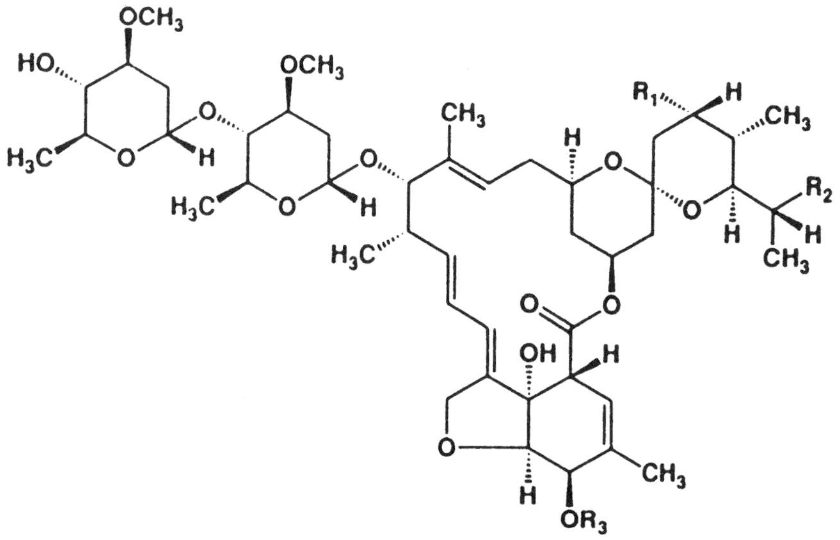

| | $R_1$ | $R_2$ | $R_3$ |
|---|---|---|---|
| Avermectin $A_{1a}$ | | $C_2H_5$ | $CH_3$ |
| Avermectin $A_{1b}$ | | $CH_3$ | $CH_3$ |
| Avermectin $A_{2a}$ | OH | $C_2H_5$ | $CH_3$ |
| Avermectin $A_{2b}$ | OH | $CH_3$ | $CH_3$ |
| Avermectin $B_{1a}$ | | $C_2H_5$ | H |
| Avermectin $B_{1b}$ | | $CH_3$ | H |
| Avermectin $B_{2a}$ | OH | $C_2H_5$ | H |
| Avermectin $B_{2b}$ | OH | $CH_3$ | H |

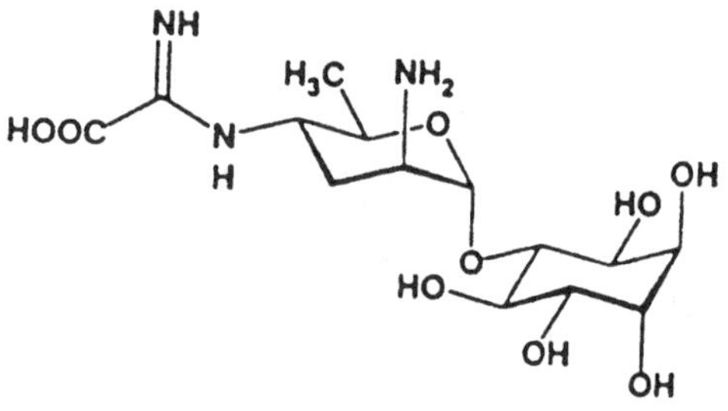

**Kasugamycin**

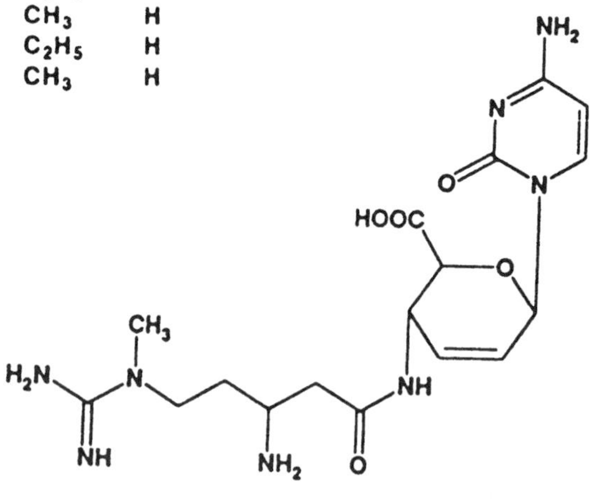

**Blasticidin S**

Figure 1. Structures of examples of low molecular weight secondary metabolites, produced in bacteria and active as pesticides.

microbes *Streptomyces viridochromogenes* and *S. hygroscopicus.* Bialaphos is a broad spectrum herbicide, active against various annual and perennial weeds and producing a phytotoxic response more rapidly than other postemergence, nonselective chemical herbicides such as glyphosate. Peptidases in treated plants metabolize bialaphos to release the active moiety of the molecule, phosphinothricin.

*Streptomyces viridochromogenes* was initially identified as a producer of a metabolite with antibiotic activity. Thereafter this metabolite, termed bialaphos, was demonstrated to be a potent herbicide. Other soil bacteria of the order Actinomycetales have subsequently been shown to produce herbicidal compounds that are analogues of bialaphos or phosphinothricin, such as phosalacine. The biosynthetic pathways of bialaphos and phosphinothricin have been characterized in *Streptomyces hygroscopicus,* and the genes responsible for bialaphos biosynthesis have been cloned. Among these genes, one conferring resistance to bialaphos *(bar)* has been used in various biotechnology applications, including the construction of transgenic plants resistant to the herbicide.

Bialaphos is active as a bactericide against Gram-negative and Gram-positive bacteria and as a fungicide against some important plant pathogens. Bialaphos and phosphinothricin are water soluble; they are readily absorbed by plants and microorganisms, and are translocated rapidly in plants. Commercial herbicides have been produced both by fermentation of *S. viridochromogenes* and by chemical synthesis of phosphinothricin. These herbicides are metabolized in soil after application and have a half-life of 20 to 30 days.

Phosphinothricin inhibits glutamine synthetase (GS) in treated plants, leading to the accumulation of toxic levels of ammonia. GS inhibition also causes glutamine depletion, leading to inhibition of photosynthesis and resulting in severe photodynamic damage.

### 3.2 Bioinsecticide: Avermectins

A well-characterized example of microbial products with insecticidal activity is found in the avermectins (Figure 1), a family of eight closely related macrocyclic lactones from the soil microbe *Streptomyces avermitilis.* They are highly potent against a broad range of insects, nematodes, and mites.

*Streptomyces avermitilis* was discovered by screening fermentation broths for anthelmintic activity. A family of compounds closely related to the avermectins, the milbemycins, has been isolated from the related bacterium *Streptomyces hygroscopicus.* In addition to naturally occurring avermectins formulated commercially from fermentations, more potent derivatives of the parent molecule have been chemically synthesized. In the field, avermectins are degraded in the soil soon after application and are also susceptible to rapid photodegradation on plant surfaces. The avermectins are very potent against a wide range of plant pests and are relatively nontoxic to mammals and plants. They are nearly insoluble in water, but apparently penetrate the leaf lamellae, where they become available as a pesticide reservoir.

Avermectins potentiate the activity of $\gamma$-aminobutyric acid (GABA), an inhibiting neurotransmitter in the nervous system of nematodes and arthropods. Avermectins can cause loss of signal transmission and immobilization. Other modes of action have also been proposed to account for toxicity of these compounds to certain groups of pests.

### 3.3 Biofungicides: Blasticidin S and Kasugamycin

Two microbially produced biofungicides with characterized activities that have achieved commercial use are blasticidin S and kasugamycin.

Blasticidin S is a nucleoside derivative (Figure 1), discovered as a metabolite of *Streptomyces griseochromogenes.* It is a protein synthesis inhibitor that is active in vitro against plant pathogenic fungi and some bacteria. It has been used primarily to control rice blast *(Pyricularia oryzae)* by inhibiting development of mycelia, and, to a lesser extent, also spore germination and spore formation.

Kasugamycin is an amino sugar metabolite of *Streptomyces kasugaensis* and of *S. kasugaspinus* (Figure 1). Like blasticidin S, it is active in vitro against fungi, including *P. oryzae,* and also some bacteria. It has excellent protective activity on rice against *P. oryzae,* suppressing mycelial development by up to 98 to 100%. Like other amino sugar antibiotics, kasugamycin acts by inhibiting protein synthesis.

Kasugamycin has excellent shelf storage stability characteristics and is not toxic to animals. One consequence of repeated use has been the emergence of kasugamycin-resistant strains of *P. oryzae.* Since the resistant strains apparently disappeared after replacement of kasugamycin with alternative fungicides, it has been suggested that resistance is manageable by fungicide rotations.

## 4 THE FUTURE

There are few commercially available pesticidal products from bacteria, but the potential for developing microbial pesticides as control agents appears excellent. Most research is concentrated on identifying novel useful molecules, and many leads are discovered every year. Continued advances in biotechnology, particularly in genetic manipulation, as well as improved culturing systems and advanced formulations, will result in the increased commercial exploitation of pesticide-producing bacteria. To compete successfully in the marketplace, however, microbial pesticides must meet the cost and efficacy criteria set by synthetic chemical agents.

*See also* Bacterial Growth and Division; Plant Cells, Genetic Manipulation of.

## *Bibliography*


Cutler, H. G., Ed. (1988) *Biologically Active Natural Products: Potential Use in Agriculture,* pp. 91–108, 120–128, 182–210. American Chemical Society, Washington, DC.

Dimock, M., Turner, J., and Lampel, J. (1993) Endophytic microorganisms for delivery of genetically engineered microbial pesticides in plants. In *Advanced Engineered Pesticides,* L. Kim, Ed. Dekker, New York.

Feitelson, J. S., Payne, J., and Kim, L. (1992) *Bacillus thuringiensis:* Insects and beyond. *Bio/Technology,* 10:271–275.

Gill, S. S., Cowles, E. A., and Pietrantonio, P. V. (1992) The mode of action of *Bacillus thuringiensis* endotoxins. *Annu. Rev. Entomol.* 37:615–636.

Hoagland, R. E., Ed. (1990) *Microbes and Microbial Products as Herbicides,* pp. 2–52. American Chemical Society, Washington, DC.

Höfte, H., and Whiteley, H. R. (1989) Insecticidal crystal proteins of *Bacillus thuringiensis. Microbiol. Rev.* 53:242–255.

Lambert, B., and Peferoen, M. (1992) Insecticidal promise of *Bacillus thuringiensis. BioScience,* 42:112–122.

Lamjel, J. S., Canter, G. L., Dimock, M. B., Kelly, J. L., Anderson, J. J., Urotoni, B. B., Foulice, J. S. Jr., and Turner, J. T., (1994) Integrative Cloning, Expression, and stability of the *cryIA(c)* gene from *Bacillus thuringiensis* subsp. *kurstaki* in a recombinant strain of *Clavibacter xyli* subsp. *cynodontis*. *Appl. Environ. Microbiol.* 60:501–508.

Omura, S., Ed. (1992) *The Search for Bioactive Compounds from Microorganisms*, pp. 213–262. Springer-Verlag, New York.

Schneider, W. R., Coord. (1989) *Pesticide Assessment Guidelines, Subdivision M: Microbial Pest Control Agents and Biochemical Pest Control Agents*. U.S. Environmental Protection Agency, Office of Pesticide and Toxic Substances, Washington DC.

# PHARMACEUTICAL ANALYSIS, CHROMATOGRAPHY IN

*Satinder Ahuja*

## *Key Words*

**Biopharmaceutics** A study on availability of a compound of interest in a drug product.

**Gas–Liquid Chromatography** Instrumental separation technique in which the mobile phase is gas and the stationary phase is liquid.

**High Performance Liquid Chromatography** Separation method that requires an instrument that can pump the mobile phase under high pressure.

**Impurities** Contaminants; sometimes degradation or transformation products are called impurities.

**Pharmaceuticals (Drugs)** Chemical compounds or formulated products with therapeutic activity.

**Pharmacodynamics** Disposition of a drug in the body.

**Recombinant Products** DNA-derived proteins.

**Thin-Layer Chromatography** Separation carried out on a thin layer of stationary phase applied to a glass plate.

Chromatography comprises a wide variety of separation techniques based on distribution of a sample between a stationary phase, which can be solid or liquid, and a mobile phase, which can be gas or liquid. Since classical methods such as titrimetry or direct spectroscopic analysis do not provide sufficient selectivity and/or detectability, chromatography plays a major role in the analysis of pharmaceutical compounds. The requirements for analysis vary with a given situation. For example, the assay of a new chemical entity (NCE) in the presence of related compounds (by-products, optical isomers, and transformation products) requires mainly selectivity. Detectability generally is not a problem because an adequate amount of the parent compound is almost always available. However, the determination of the related compounds and impurities originating from various sources at trace or ultratrace level calls for methods that are both selective and sensitive. This is where chromatographic methods excel because they can provide unique selectivity and detectability.

Chromatographic methods such as thin-layer chromatography (TLC), gas–liquid chromatography (GLC), and high performance liquid chromatography (HPLC) are commonly used in pharmaceutical analyses. Of these methods, HPLC has most revolutionized the field of separations. The compounds that once were considered too difficult to separate by TLC owing to poor resolution, or too hard to quantitate by GLC because of volatility, polarity, or thermal instability, can be easily separated and quantified by HPLC within a short period of time. HPLC offers selectivity to separate components with slight variations in structure or molecular weight. For compounds with the same molecular weight, the structural differences may involve no more than compounds that are mirror images (i.e., optical isomers resulting from the presence of one or more asymmetric carbon atoms). Chromatography plays a major role in the analysis of pharmaceutical compounds in the following areas:

- Discovery of new compounds
- New drug development
- Transformation products
- Separations of isomers
- Quality control
- Stability studies
- Recombinant products
- Toxicology studies
- Clinical studies
- Biopharmaceutics/pharmacodynamics/clinical pharmacology
- Forensic studies
- Diagnostic studies
- Animal-derived food analyses
- Postmortem toxicology

Most of these topics are discussed in this contribution; the subject of quality control (clinical and stability studies) is discussed briefly in Section 2.

## 1 DISCOVERY OF NEW COMPOUNDS

Most new drugs have been discovered through the synthetic route. It must be recognized that leads for a number of them came from natural products. Isolation and purification of an active principle from a crude mixture extracted from the natural products eventually led to the synthesis of the active principle and other potentially useful related products. Chromatography provides an excellent means of discovery of new compounds, since compounds present even at ultratrace levels can be resolved from related compounds. One approach entails separation of potential compounds resulting as by-products in the synthesis of the target compound, which cannot be resolved by normal inefficient crystallization techniques. Many of the by-products frequently have physicochemical properties and a carbon skeleton similar to the target compound with substituent(s) differing in position or functionality. Alternatively, they may be isomeric compounds that are mirror images, but otherwise not different. Since it is not possible to theorize all by-products, some unusual compounds can be isolated and characterized with this approach.

A more selective approach is based on changes brought about in a chemical entity with reactions such as hydrolysis, oxidation, or photolysis. These reactions are frequently run to evaluate stability. In this case, several theorized new and old compounds are produced. An innovative chromatographer can resolve and characterize the theorized as well as untheorized new compounds.

Another interesting approach depends on characterization of various degradation products produced in the matrices used for pharmaceutical products. The compounds thus produced or those produced during metabolic studies in humans can be resolved by chromatography, and their structure determined by techniques such as elemental analysis, IR, NMR, or mass spectrometry. Two well-known examples of metabolites that became drug products are Tandearil and Pertofrane, which resulted from oxidation and demethylation, respectively, of the parent compound.

## 2 NEW DRUG DEVELOPMENT

Development of new drug products mandates the generation of meaningful and reliable analytical data at various steps of new drug development. To assure the safety of a new pharmaceutical compound or drug, the new drug must meet the established purity standards as a chemical entity and when admixed with animal feeds for toxicology studies or with pharmaceutical excipients for human use for clinical studies. Ideally a high quality drug product should be free of extraneous material and should contain the labeled quantity of each ingredient. Furthermore, it should exhibit excellent stability throughout its shelf life. These requirements demand an analytical methodology sensitive enough to carry out measurements of low levels of drugs and by-products. This need has led to the development of separation methods that are suitable for determination of submicrogram quantities of various chemical entities.

Frequently, trace or ultratrace analysis is required for monitoring transformation products and isomers. Trace analysis can be defined as analysis performed at parts per million (ppm) or a microgram-per-gram (μg/g) level—an analytical landmark that was achieved approximately 30 years ago. Ultratrace analysis can then be defined as analysis performed below the ppm or μg/g level. Meaningful intercomparisons are difficult because of the problems associated with obtaining uniformity of sample, the ease of contamination during sampling and analyses, and the limitations of analytical practices employed. To ensure the reliability of trace and ultratrace data, the methodology must be reproducible, and data must withstand interlaboratory comparisons. The developed methodology can be used for monitoring clinical samples and assessing their stability in the field. Most of these methods are used for quality control in case the product is commercialized.

## 3 TRANSFORMATION PRODUCTS

Transformation products or related products are impurities or by-products that originate during the synthesis of NCE, or they are products arising during storage as a result of the degradation of NCE or its interactions with the excipients, or they are produced as metabolites in the body when the NCE is administered to a patient. These transformation products can be produced during the drug discovery process in a host of other places, as well. Chromatography helps to monitor these transformation products. Isomeric impurities, discussed in the next section, may also be considered transformation products.

## 4 SEPARATIONS OF ISOMERS

The current regulatory position of the U.S. Food and Drug Administration (FDA) is described here with regard to the approval of racemates and pure stereoisomers. Circumstances in which stereochemically sensitive analytical methods are necessary to ensure the safety and efficacy of a drug are noted. Regulatory guidelines for a new drug application (NDA) are interpreted for the approval of a pure enantiomer in which the racemate is marketed, and for clinical investigations to compare the safety and efficacy of a racemate and its enantiomers. The basis for such regulation arose from historical situations (thalidomide, benoxaprofen) as well as currently marketed drugs (arylpropionic acids, disopyramide, indacrinone).

The primary regulatory focus of the Food and Drug Administration is on considerations of clinical efficacy and consumer safety of a potential drug. Because the chiral environment found in vivo affects the biological activity of a drug, the approval of stereoisomeric drugs for marketing can present special challenges. The case of thalidomide is an example of a problem that may have been complicated by ignorance of stereochemical effects. The use of racemates can lead to erroneous models of pharmacokinetic behavior and to the potential for opportunities to manipulate pharmacological activity. It is therefore necessary to design experiments that will unambiguously answer the question of whether a stereochemically pure drug is more effective and/or less toxic than the racemate.

In 1987 the FDA issued a set of guidelines on the submission of NDAs, approaching directly the question of stereochemistry in the manufacture of drug substances. The Federal Food, Drug, and Cosmetic Act requires a full description of methods used in the manufacture of the drug, including testing to demonstrate identity, strength, quality, and purity. Therefore, the NDA should show the applicant's knowledge of the molecular structure of the drug substance. For chiral compounds this includes identification of all chiral centers. The enantiomer ratio, although 50:50 by the definition of racemate, should be defined for any other admixture of stereoisomers. The proof of structure should consider stereochemistry and should provide appropriate descriptions of the molecular structure. The guidelines do not discuss conditions under which a determination of absolute configuration is desirable or essential. Obviously, such information would be appropriate supporting data for the manufacture of optically pure drugs.

U.S. regulatory requirements demand that bioavailability of the drug be demonstrated. When pharmacokinetic models differ between enantiomers, it seems obvious that establishing the bioavailability of the drug from a racemate is a much more complex task, which cannot be accomplished without separation of the enantiomers and investigation of their pharmacokinetics as individual molecular entities.

It is expected that the toxicity of impurities, degradation products, and residues from manufacturing processes will be investigated as the development of a drug is pursued. The same standards should, therefore, be applied to the enantiomeric molecules in a racemate. Whenever a drug can be obtained in a variety of chemically equivalent forms (such as enantiomers), it makes sense to explore the potential in vivo differences between these forms.

## 5 RECOMBINANT PRODUCTS

The following proteins derived from recombinant DNA have been approved for therapeutic use or clinical trials in the United States:

Growth hormone (''Protropin'')
Insulin (''Humulin'')
Interferon-α (''Referon'')
Interferon-γ
Interleukin-2
Tissue plasminogen activator
Tumor necrosis factor

Particular attention must be paid to the detection of DNA in all finished biotechnology products because such DNA might be incorporated into the human genome and become a potential oncogene. It has been claimed that for a biotechnology product to be safe, the absence of DNA at the picogram-per-dose level must be demonstrated.

The isolation and purification of DNA and RNA restriction fragments are of great importance in the area of molecular biology. These fragments are the products of site-specific digestion of larger pieces of DNA and RNA enzymes called restriction endonucleases. The fragments may range in size from a few base pairs to tens of thousands of base pairs. The purification of restriction fragments is key to a number of processes, some of which are cloning of proteins or peptides using engineered vectors (plasmids, phages, cosmids), sequencing of DNA, elucidation of the structure of the genome, and characterization of individual genes and gene effectors.

An ion exchange column can provide DNA and RNA separation within an hour, with resolution equivalent to that obtained using gel electrophoresis. Fragments are visualized using an on-line UV detector, and sample loadings from 500 ng to 50 μg have been applied successfully with recoveries approaching 100%. To date, DNA fragments in the size range of 4 to 23,000 base pairs have been separated effectively. This technique, by allowing fast, high resolution, high recovery purification of DNA and RNA, could enable the molecular biologist to remove one of the most common bottlenecks encountered in a wide variety of experiments.

## 6 TOXICOLOGY STUDIES

To evaluate the toxicology of potential drugs, the candidate compounds are admixed with animal feeds such as rat feed (Purina Laboratory Chow containing a minimum of 23% protein, 4.5% fat, and maximum 6% fiber) with the following composition: meat and bone meal, dried skimmed milk, wheat germ meal, fish meal, animal liver meal, dried beet pulp, ground extruded corn, ground oat groats, soybean meal, dehydrated alfalfa meal, cane molasses, animal fat preserved with BHA (butylated hydroxyanisole), vitamin $B_{12}$ supplement, brewer's dried yeast, thiamine, niacin, vitamin A supplement, D-activated plant sterol, vitamin E supplement, dicalcium phosphate, iodized salt, ferric ammonium citrate, zinc oxide, manganous oxide, cupric oxide, ferric oxide, and cobalt carbonate.

Both GLC and HPLC methods can be used for the analysis of potential drugs in the feed. An electron capture detector can provide detectability down to picograms for some components; however, degradation can occur at the injection port. This means that the developed method not only must be sensitive but should be free of interference from the degradation products and/or impurities; that is, it must be selective. Selective HPLC methods can be developed to circumvent this problem.

## 7 BIOPHARMACEUTICS/ PHARMACODYNAMICS/CLINICAL PHARMACOLOGY

Unless blood level studies of both the extent and rate of absorption are conducted in determining bioequivalence, a generic product's inequivalence with respect to a metabolite may go undetected, and second-pass levels could be dangerously higher than indicated in the labeling.

Interesting investigations have been conducted to determine whether the pharmacodynamics of the central nervous system (CNS) stimulant pentylenetatrazol (PTZ) is altered in renal dysfunction. When PTZ was infused to the onset of maximal seizures, the rats with chemically induced renal dysfunction required higher concentrations, whereas the ureter-ligated rats convulsed at lower concentrations of PTZ than did the corresponding control animals. The effect of experimental renal dysfunction on the convulsant activity of PTZ was examined by a gas chromatographic method for studying pharmacodynamics of PTZ seizure in rats. A nitrogen phosphorus detector was used to provide a detection limit of about 0.5 μg/mL of PTZ.

Catecholamines, biologically active derivatives of the amino acid tyrosine, are produced by cells of the CNS and the adrenal medulla. In the CNS, norepinephrine and dopamine function as neurotransmitters. Even subpopulations of cells within a given area have been shown to contain differences in levels and types of neurotransmitter. Highly sensitive chromatographic methods now available for the analysis of these neurotransmitters will help us better understand brain function.

## 8 FORENSIC STUDIES

Forensic studies usually deal with the analysis of drugs present in powders, natural products, capsules, and tablets picked up by law enforcement personnel. The powders may contain controlled substances such as cocaine, heroin, amphetamine, phencyclidine (PCP), or methaqualone, with adulterants such as quinine, procaine, and lidocaine, and diluents such as starches, sugars, or polyhydric alcohols. Morphine, methadone, methaqualone, and PCP can be analyzed by gas chromatography–mass spectrometry (GC–MS) at the nanogram-per-milliliter level. Liquid chromatography–mass spectrometry (LC–MS) is very useful for methaqualone, cocaine, and a number of other compounds. Gas chromatography with electron capture detector has been found useful for the analysis of heroin samples.

## 9 DIAGNOSTIC STUDIES

Deficiency of xanthine–guanine phosphoribosyltransferase enzyme has been associated with Lesch–Nyhan syndrome as well as with primary gout. The activity of the enzyme is determined by measurement of decrease of the substrate, hypoxanthine, and increase in the product, inosine-5′-monophosphoric acid. A major advantage of using HPLC for enzyme assays is that the simultaneous measurement of both substrate and product reduces the error due to interference from competing enzymes. Similarly, the levels of hypoxanthine and uridine for colorectal cancer and inosine for gastric cancer have been found to be significantly higher in endoplastic mucosa than in normal mucosa.

## 10 ANIMAL-DERIVED FOOD ANALYSES

The use of drugs in feeds necessitates the determination of their content in animals (cows, pigs, etc.) used for human consumption. Drug substances are frequently admixed with feed to confer some useful pharmacological activity such as promotion of growth, and these feeds are referred to as medicated feeds. For convenience, the drugs used in medicated feeds may be divided into the following categories, based on pharmacological activity:

Antibacterials
Coccidiostats
Antidysentery compounds
Anthelmintics
Anabolic agents

HPLC is frequently the method of choice for the analysis of compounds such as ivermectin, zearalenone, tetracyclines, ampicillin, lasalocid, fenbendazole, monesnsin, chloramphenicol, and sedecamycin in animal-derived foods.

## 11 POSTMORTEM TOXICOLOGY

Drugs, including alcohol, are found by medical examiners in more than 50% of cases in postmortem toxicology investigations. Since most drugs have therapeutic levels in the microgram- to nanogram-per-milliliter range, toxicology laboratories must have the capability to detect at these levels in various specimens. Drugs may include narcotics (morphine, codeine, hydromorphone, meperidine, and fentanyl). When death results from a heroin or morphine overdose, one must analyze for morphine since heroin is deacetylated to morphine in blood. GC–MS is a useful technique. Barbiturates and other sedatives/tranquilizers are commonly implicated in drug overdosing, as are some pain killers such as propoxyphene. These compounds can be often analyzed by GC or HPLC.

*See also* MEDICINAL CHEMISTRY; MOLECULAR GENETIC MEDICINE.

### *Bibliography*

Ahuja, S. (1986) *Ultratrace Analysis of Pharmaceuticals and Other Compounds of Interest.* Wiley, New York.

———. (1991) *Chiral Separations by HPLC,* ACS Symposium Series 471. American Chemical Society, Washington, DC.

———. (1992) *Chromatography of Pharmaceuticals: Natural, Synthetic and Recombinant Products,* ACS Symposium Series 512. American Chemical Society, Washington, DC.

Gibaldi, M. (1991) *Biopharmaceutics and Clinical Pharmacokinetics.* Lea & Febiger, Malvern, PA.

Guidelines for Submitting Supporting Documentation in Drug Applications for the Manufacture of Drug Substances. (1987) Office of Drug Evaluation and Research (HFD-100), U.S. Food and Drug Administration, Rockville, MD.

Horvath, C., and Nikelly, J. G. (1990) *Analytical Biotechnology: Capillary Electrophoresis and Chromatography,* ACS Symposium Series 434. American Chemical Society, Washington, DC.

Wainer, I. W., and Drayer, D. E. (1988) *Drug Stereochemistry, Analytical Methods and Pharmacology.* Dekker, New York.

# PHARMACOGENETICS

*Ann K. Daly and Jeffrey R. Idle*

### *Key Words*

**Cytochrome P450** One of a multigene family of enzymes that catalyze oxidative metabolism of a variety of endogenous and xenogenous substrates. All the enzymes contain noncovalently bound heme and use reducing equivalents from NADPH and an atom of oxygen derived from atmospheric oxygen in the oxidation reaction.

**Esterase** Hydrolytic enzymes occurring in blood plasma and other tissues including the liver that metabolize endogenous and exogenous compounds by cleavage of ester bonds.

**Genetic Polymorphism** A Mendelian trait that exists in the population in at least two phenotypes, with each phenotype showing a frequency of at least 1%.

**NADPH** Reduced nicotinamide adenine dinucleotide phosphate.

**Transferases** Enzymes that metabolize endogenous and exogenous compounds by conjugation with various types of small molecule including glucuronic acid, sulfate, glutathione, acetyl groups, and methyl groups. The reactions are often referred to as phase II metabolism and are mainly detoxicating.

Pharmacogenetics is the study of the genetic basis of variation in response to foreign chemicals (xenobiotics), including drugs. In particular, the subject concerns the study of functional genetic polymorphisms in enzymes that metabolize xenobiotics, but it may be broadened to include variation in the transport and binding of xenobiotics. In general, the biochemical and genetic basis of variation in xenobiotic metabolism is now well understood, whereas variation in transport and binding of xenobiotics is, with a few exceptions, still poorly understood. Polymorphisms in xenobiotic metabolism may be responsible for adverse drug reactions, lack of response to drugs, and altered susceptibility to certain diseases associated with exposure to xenobiotics, particularly cancer. A polymorphism is normally defined as a defect that occurs at a frequency of at least 1% in the population; some pharmacogenetic defects, however, occur at much lower frequencies.

## 1 INTRODUCTION

Pharmacogenetic polymorphisms and isolated deficiencies have been detected in a number of different animal species, but this entry covers only those detected in humans. The first pharmacogenetic defect identified, a rare defect in serum cholinesterase, was described by Werner Kalow in 1957. Subsequently, several more common pharmacogenetic polymorphisms were detected and later shown to be due to deficiency in particular enzymes. Table 1 summarizes the polymorphisms and isolated defects of direct relevance to pharmacogenetics that have been identified up to the present. There is evidence for the existence of considerable variation in levels of certain other xenobiotic metabolizing enzymes in populations, but only the enzymes listed in Table 1 show confirmed functional polymorphisms on the basis of bimodal distributions for enzyme activity in vivo. The reason for the high level of polymorphism and variability in xenobiotic-metabolizing enzymes is still unclear but may reflect the nonessential role of these agents in normal metabolism. Many pharmacogenetic polymorphisms show considerable ethnic variation, manifested in differences in frequency and/or the types of mutation present.

Although pharmacogenetic polymorphisms have been known to exist for over 30 years, it is only in the last 5 years that the molecular basis of a number of them has been elucidated, enabling

**Table 1** Enzymes Whose Absence Constitutes a Pharmacogenetic Defect

| |
|---|
| CYP2D6 |
| CYP2C19 |
| Paraoxonase |
| Serum cholinesterase |
| Aldehyde dehydrogenase 2 |
| Glutathione *S*-transferase M1 |
| Glutathione *S*-transferase T1 |
| *N*-Acetyltransferase 2 |
| Alcohol dehydrogenase 2 |
| Thiopurine *S*-methyltransferase |
| Glucose 6-phosphate dehydrogenase |

the establishment of simple genotyping techniques based on analysis by restriction fragment length polymorphism (RFLP) and the polymerase chain reaction (PCR). Recent rapid progress has been due to improvements in techniques for molecular cloning and DNA sequencing that have revolutionized the study of xenobiotic-metabolizing enzymes such as the cytochromes P450, which had been difficult to study. Several types of mutation are associated with pharmacogenetic defects. In some cases no enzyme is synthesized because the coding gene has been completely deleted, or smaller deletions or insertions have given rise to frameshifts. Other defects are due to amino acid substitutions, which alter the catalytic activity of the enzyme and result in altered or complete loss of activity with all or a subset of substrates.

## 2 SERUM CHOLINESTERASE

Certain individuals given the muscle relaxant succinylcholine show prolonged paralysis and apnea upon withdrawal of the drug. Werner Kalow, who demonstrated that the biochemical basis of this adverse drug reaction was a deficiency in the enzyme serum cholinesterase, which metabolizes succinylcholine to succinic acid and choline, termed the defective enzyme the atypical variant. This variant (frequency of homozygote 1:3500) is inactive with succinylcholine as substrate but shows normal activity with neutral esters. Several different enzyme defects have subsequently been identified, including another less common defective variant known as the silent variant (frequency of homozygote 1:100,000) which is inactive with all substrates, and a more common variant (homozygote frequency 1:100) termed variant K, which has little effect on enzyme activity unless it occurs in the heterozygous state with the atypical allele. Recent DNA sequencing studies have shown that the silent allele is due to a frameshift mutation and that the atypical and K alleles have point mutations that give rise to amino acid substitutions. Identification of these and certain even rarer mutant alleles is now possible by use of the polymerase chain reaction.

## 3 DEBRISOQUINE POLYMORPHISM

In 1977 Afaf Mahgoub, Robert Smith, and Jeffrey Idle demonstrated that approximately 3% of a small British population were unable to metabolize the antihypertensive drug debrisoquine and, if given normal therapeutic doses of this compound, developed postural hypotension. Individuals with this deficiency, who comprise ~10% of the population, were termed poor metabolizers. Subsequently the metabolism of debrisoquine was demonstrated to be catalyzed by a cytochrome P450 enzyme now called CYP2D6; metabolism of a number of other drugs was also shown to be impaired in poor metabolizers. There is considerable ethnic variation in the frequency of the poor metabolizer phenotype, with Oriental populations showing a frequency of 0 to 1% for this phenotype compared with ~10% of Europeans.

The gene encoding CYP2D6 has been cloned, sequenced, and localized to chromosome 22q13.1. Analysis of genomic DNA from poor metabolizers has demonstrated that more than 95% of defective *CYP2D6* alleles are of three types, showing either a point mutation at an intron–exon boundary, resulting in introduction of a frameshift during RNA splicing, a single base pair deletion in an exon, or a deletion of approximately 16 kilobases (kb) of DNA, including all of the coding gene. PCR and RFLP analysis of leukocyte DNA (genotyping) allows identification of at least 95% of all poor metabolizers in several European populations.

At least 30 commonly prescribed drugs, mainly cardiovascular agents and psychiatric agents, are now known to be metabolized by CYP2D6, and genotyping prior to drug administration may be useful to allow dose adjustment or the use of an alternative drug in poor metabolizers. A number of epidemiological studies have suggested that poor metabolizers show a reduced susceptibility to lung cancer and bladder cancer but may be at increased risk of developing Parkinson's disease. These findings are still controversial and require confirmation. There are a number of possible explanations for the apparent reduced susceptibility to certain cancers in poor metabolizers. These include linkage of *CYP2D6* to an oncogene, reduced metabolism in poor metabolizers of components of tobacco smoke, which may induce cancer after metabolism, or a role for CYP2D6 in tumor growth. In the case of Parkinson's disease there is evidence that certain environmental and endogenous toxins that appear to induce symptoms of the disease may be neutralized by CYP2D6, and poor metabolizers may therefore have higher levels of these compounds present in the body.

## 4 MEPHENYTOIN POLYMORPHISM

In the early 1980s Adrian Küpfer and colleagues demonstrated that metabolism of the *S*-enantiomer of mephenytoin, an anticonvulsant drug, was polymorphic, with 2 to 6% of Europeans and 15 to 20% of Orientals unable to metabolize this compound. Metabolism of several other drugs has subsequently been demonstrated to cosegregate with that of *S*-mephenytoin. Metabolism of *S*-mephenytoin has recently been demonstrated to be catalyzed by the cytochrome P450 CYP2C19 and the most common genetic defect that gives rise to the poor metabolizer phenotype to be a single base pair mutation in exon 5 of the gene which results in aberrant RNA splicing.

## 5 PARAOXONASE

Serum paraoxonase is an esterase that catalyzes the hydrolysis of a number of compounds, including organophosphate esters found in insecticides and herbicides. Metabolism by paraoxonase of paraoxon, a metabolite of the insecticide parathione, is polymorphic, with 53% of Europeans showing low activity with this compound. The frequency of the polymorphism is lower in certain other ethnic groups (epg., 10% of Japanese show low activity). Not all paraoxonase substrates show polymorphic metabolism. The paraoxonase gene has now been cloned, and the basis of the polymorphism has been demonstrated to be an amino acid substitution at position

192, with arginine associated with high activity and glutamine with low activity. Individuals homozygous for the high activity allele may be protected from some of the toxic effects of organophosphate compounds that show polymorphic metabolism with paraoxonase.

## 6 PHARMACOGENETICS OF ALCOHOL METABOLISM

Two types of dehydrogenase enzyme are responsible for the metabolism of ethanol to acetic acid. Ethanol is first converted to acetaldehyde by alcohol dehydrogenase, and acetaldehyde is then metabolized to acetic acid by aldehyde dehydrogenase. The existence of multiple forms of both dehydrogenases has been demonstrated, and a number of these genes have now been cloned. Amino acid polymorphisms that affect enzyme activity have been described for several alcohol and aldehyde dehydrogenases. The most important of these in terms of ethanol metabolism is a polymorphism in the mitochondrial aldehyde dehydrogenase ALDH2, where substitution of lysine for glutamic acid at position 487 leads to loss of activity. Approximately 50% of Japanese are homozygous for the defective allele and, following ethanol ingestion, show high levels of acetaldehyde, which gives rise to flushing and nausea. Polymorphisms in alcohol dehydrogenases have also been detected, and variant alleles of the enzyme alcohol dehydrogenase 2 that show increased rates of ethanol metabolism have been described. The physiological effects of these polymorphisms are not fully understood, although it has been suggested that the presence of variant alcohol or aldehyde dehydrogenase alleles may be associated with altered susceptibility to alcoholic liver disease.

## 7 *N*-ACETYLTRANSFERASE 2

More than 30 years ago, a polymorphism in the metabolism of isoniazid was described by David Price Evans. This drug was demonstrated to undergo acetylation, with an acetyl group transferred from acetyl coenzyme A by an enzyme known as *N*-acetyltransferase. Approximately 50% of Europeans were unable to carry out this reaction and were termed slow acetylators. Slow acetylators also acetylate certain drugs (e.g., sulfamethazine and caffeine) poorly but carry out acetylation of other compounds (e.g., 4-aminobenzoic acid) normally. It has now been demonstrated that there are two forms of the enzyme—*N*-acetyltransferases 1 and 2 (NAT1 and NAT2)—and that the *NAT2* gene codes for the polymorphic enzyme. Three different allelic variants (M1–M3) associated with the slow acetylator phenotype have been detected. Each contains several point mutations, with some giving rise to amino acid substitutions but others silent. It is believed that the absence of enzyme activity in slow acetylators is due in part to the amino acid substitutions. Effects of silent point mutations on mRNA stability may be important as well, however. Genotyping assays that have been developed using PCR allow identification of in excess of 95% of slow acetylators.

In addition to its role in drug metabolism, NAT2 metabolizes certain carcinogens and procarcinogens. Slow acetylators appear to show an increased incidence of occupationally related bladder cancer as a result of impaired detoxication of compounds such as benzidine, but rapid acetylators have been suggested to show an increased susceptibility to colon cancer, possibly because NAT2 activates procarcinogenic arylamines generated during normal cooking of ingested food.

## 8 GLUTATHIONE *S*-TRANSFERASES

Many hydrophobic and electrophilic compounds are detoxicated by conjugation with glutathione by enzymes from the glutathione *S*-transferase family followed by further metabolism to a mercapturic acid prior to excretion. At least three separate classes of these enzymes occur in humans. Common polymorphisms have been detected in two glutathione *S*-transferases, GSTM1 and GSTT1. In most populations studied, 50% of individuals lack GSTM1 and are unable to conjugate certain substrates including benzo[*a*]pyrene metabolites. Absence of GSTM1 can be detected phenotypically by assay of leukocytes for enzyme activity or by immunoassay. It has been demonstrated that the absence of enzyme activity is due to a large deletion in the coding gene, and a PCR assay in which a product is detected only in those positive for GSTM1 expression has been developed. There have been a number of studies on the association between lack of GSTM1 expression and susceptibility to chemically induced cancer. These studies indicate that GSTM1-deficient individuals may be at increased susceptibility to bladder cancer and possibly to certain forms of lung cancer.

A deficiency in the unrelated glutathione *S*-transferase GSTT1, or theta, has recently been identified. Erythrocytes from 30 to 35% of Europeans do not conjugate certain halomethanes such as dichloromethane with glutathione. Like GSTM1 deficiency, GSTT1 deficiency results from deletion of the gene, and a PCR assay for detecting individuals that express the enzyme has been developed.

## 9 OTHER POLYMORPHISMS OF PHARMACOGENETIC RELEVANCE

Several other polymorphisms of pharmacogenetic interest have been described. A small proportion of several populations lack thiopurine *S*-methyltransferase, an enzyme that metabolizes thiopurine drugs used in cancer chemotherapy and as immunosuppressants. The molecular basis of this deficiency is still unclear. A rare inborn error of metabolism known as fish odor syndrome and characterized by inability to metabolize trimethylamine is due to deficiency of a flavin-containing monooxygenase. Individuals with this deficiency also show impaired metabolism of certain other xenobiotics including nicotine.

Approximately 5% of individuals suffer from a pharmacogenetic defect known as Gilbert's syndrome. Symptoms include reduced glucuronidation of the bile pigment bilirubin and certain xenobiotics, possibly including paracetamol (acetaminophen), due to deficiency in a UDP-glucuronosyltransferase. The exact molecular basis of the syndrome is still unclear, but mutations in the *UGT1* locus of chromosome 2 are likely to be responsible. There is also evidence for considerable interindividual variability in a variety of other xenobiotic-metabolizing enzymes, but it is unclear at present whether these represent genuine pharmacogenetic polymorphisms. However, it is likely that DNA sequencing and in vitro expression studies will resolve this question in the near future.

In addition to defects in the metabolism of drugs, some individuals exhibit metabolic defects that alter their response to drugs; the most common is deficiency of glucose 6-phosphate dehydrogenase. When individuals with this defect are exposed to certain oxidizing drugs, including the antimalarial primaquine and the antibiotic sulfanilamide, red cell lysis may occur as a result of a reduced ability to synthesize reduced glutathione. A large number of variant alleles associated with the deficiency have now been identified. Up to 50% of males in certain racial groups have the deficiency, apparently because heterozygotes are being protected against malaria. Other rarer defects in drug responses, which are also less well understood at the molecular level, include malignant hyperthermia (in which affected individuals suffer serious complications during general anesthesia) and several forms of porphyria (in which individuals suffering from defects in heme biosynthesis suffer toxic responses to cytochrome P450 inducers such as phenobarbitone).

*See also* BIOCHEMICAL GENETICS, HUMAN; CYTOCHROME P450.

### *Bibliography*

Cholerton, S., Daly, A. K., and Idle, J. R. (1992) The role of individual human cytochromes P450 in drug metabolism and clinical response. *Trends Pharmacol. Sci.* 13:434–438.

Evans, D.A.P. (1993) *Genetic Factors in Drug Therapy.* Cambridge University Press, Cambridge.

Kalow, W. (1991) Interethnic variation of drug metabolism. *Trends Pharmacol. Sci.* 12:102–107.

———, Ed. (1992) *Pharmacogenetics of Drug Metabolism.* Pergamon Press, New York.

Nebert, D. W., and Weber, W. W. (1990) Pharmacogenetics. In: *Principles of Drug Action,* 3rd ed., W. B. Pratt, and P. Taylor, Eds., pp. 469–531. Churchill Livingstone, New York.

# PHENYLKETONURIA, MOLECULAR GENETICS OF

*Randy C. Eisensmith and Savio L. C. Woo*

### *Key Words*

**Guthrie Test** A semiquantitative bacterial inhibition assay for the determination of serum phenylalanine levels in newborns.

**Haplotype** The genetic constitution of an individual with respect to one member of a pair of allelic genes.

**Phenylalanine Hydroxylase (PAH)** The enzyme mediating the conversion of L-phenylalanine to L-tyrosine, expressed exclusively in the liver in man.

**Phenylketonuria (PKU)** An inherited metabolic disorder caused by a deficiency of phenylalanine hydroxylase (PAH).

Phenylketonuria (PKU), one of the most common inborn errors of amino acid metabolism, is an autosomal recessive genetic disorder caused by a deficiency of phenylalanine hydroxylase (PAH). PAH deficiency leads to an accumulation of L-phenylalanine in blood, as well as other abnormalities in aromatic amino acid metabolism. The principal clinical feature of PKU is severe and irreversible mental retardation, which can be largely prevented through the administration of a diet low in phenylalanine. More than 75 mutations have been reported in the *PAH* gene. Most are strongly associated with specific haplotypes defined by restriction fragment length polymorphisms (RFLPs) present in or near the *PAH* gene, but they have different distributions in various human populations. The many possible combinations of *PAH* mutations account for most of the metabolic heterogeneity of PKU.

## 1 BIOCHEMICAL BASIS OF PHENYLKETONURIA

### 1.1 IDENTIFICATION OF THE BIOCHEMICAL DEFECT

Phenylketonuria was first recognized as an inborn error of L-phenylalanine metabolism in 1934. By 1952, the biochemical defect in PKU patients was understood as a deficiency in a hepatic enzyme system that mediates the hydroxylation of L-phenylalanine to L-tyrosine. The human phenylalanine hydroxylation system is shown in Figure 1. The primary reaction is catalyzed by PAH, which is a liver-specific, mixed-function oxidase. The reduced cofactor consumed in the hydroxylation reaction, L-erythrotetrahydrobiopterin ($BH_4$), is regenerated by quinonoid dihydrobiopterine reductase (QDPR), formerly referred to as dihydropteridine reductase (DHPR). Deficiencies in PAH cause hyperphenylalaninemia, including the "classical" or "typical" form of PKU. Deficits in QDPR or other enzymes involved in $BH_4$ biosynthesis and regeneration cause the relatively rare "atypical" forms of PKU.

### 1.2 DIETARY TREATMENT AND NEWBORN SCREENING

Restriction of dietary phenylalanine intake reduces serum phenylalanine levels in young PKU patients. Dietary restriction also improves mental development and behavioral performance in treated patients. The benefits derived from dietary therapy may be lessened or lost by its discontinuation, even if this occurs in late adolescence or early adulthood.

Since prompt initiation of dietary therapy is essential to prevent most of the irreversible consequences of PKU, routine newborn screening is now performed in many Western countries, using the Guthrie test. The incidence of PKU is roughly 1 in 10,000 among the general Caucasian population, yielding a carrier frequency of about 1 in 50 for this autosomal recessive disorder.

## 2 THE MOLECULAR GENETICS OF PHENYLKETONURIA

### 2.1 CLONING AND EXPRESSION OF THE HUMAN PAH cDNA

A rat PAH cDNA obtained by immunoprecipitation of rat liver polysomes was used to isolate several PAH cDNA clones from a human liver cDNA library. The longest clone contained an open reading frame encoding a protein of 452 amino acids. Introduction into cultured mammalian cells of an expression vector containing this cDNA clone resulted in the production of mRNA, immunoreactive protein, and pterine-dependent enzymatic activity that were indistinguishable from those obtained from human liver extracts. These studies demonstrated that human PAH is a single-gene product. Since classical PKU is caused by PAH deficiency, this disease has only a single genetic locus.

### 2.2 STRUCTURE AND LOCATION OF THE HUMAN *PAH* GENE

Detailed Southern blot analysis of four overlapping cosmid clones isolated from a human genomic library indicated that the human *PAH* gene is approximately 90 kb long, contains 13 exons separated

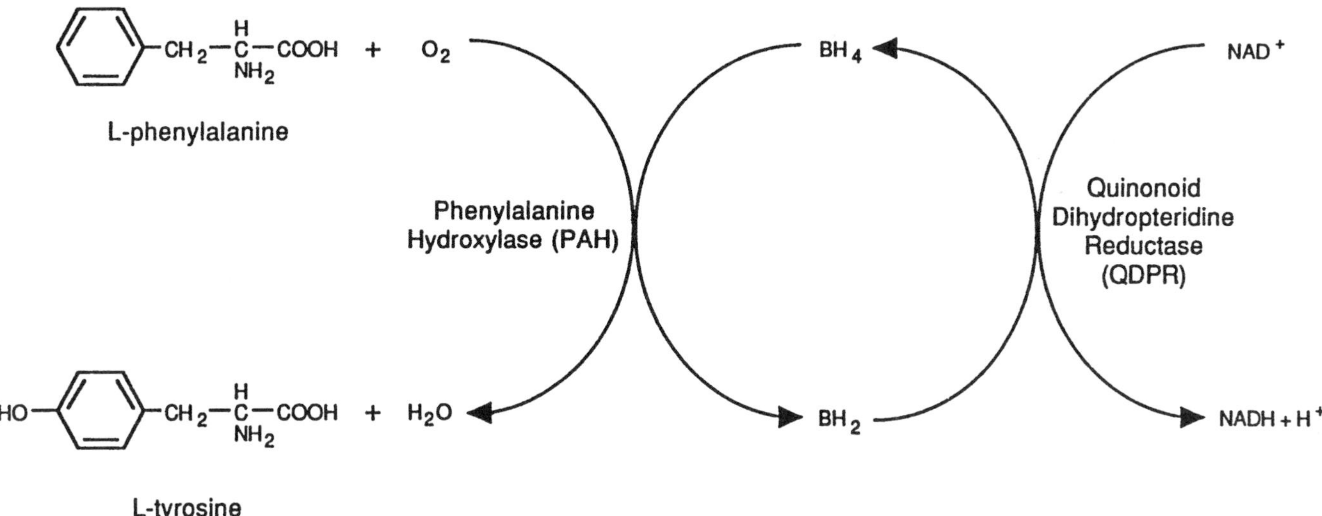


**Figure 1.** The human phenylalanine hydroxylating system.

by introns 1–23 kb long, and encodes a mature messenger RNA of 2.45 kb. Screening of human/rodent cell hybrids containing different combinations of human chromosomes localized the human *PAH* gene/*PKU* locus to chromosome 12. A more precise localization to 12q22–q24.1 was obtained through a combination of deletion mapping and in situ hybridization.

### 2.3 Polymorphisms in the Human *PAH* Gene

An examination of the *PAH/PKU* locus in normal and PKU individuals by Southern blot analysis revealed the presence of several RFLPs in or near the human *PAH* gene. The relative locations of these polymorphic restriction sites are shown in Figure 2. Different combinations of these polymorphic alleles define unique haplotypes that can be used to identify normal or mutant *PAH* alleles, to follow their segregation in PKU families, and to perform prenatal diagnosis and carrier screening in families with a prior history of PKU.

### 2.4 Mutations in the Human *PAH* Gene

Analysis of RFLP-based haplotypes in European PKU families demonstrated close associations between certain haplotypes and mutant chromosomes. These observations suggested that each of these haplotypes may harbor a single predominant mutation. For example, a vast majority of haplotype 2 mutant chromosomes contain a single missense mutation, R408W, while nearly all mutant chromosomes of haplotypes 3 and 6 contain single splicing mutations, IVS12nt1 and IVS10nt546, respectively.

The remaining haplotypes, which are more evenly distributed among normal and mutant chromosomes in most ethnic populations, harbor a larger number of mutations. For example, mutant chromosomes of haplotypes 1 and 4 have each been shown to be associated with more than 20 different mutations, and more than 75 different mutations have been described in the human *PAH* genes of PKU patients from various ethnic groups. Most are single-base substitutions, including at least six nonsense mutations and at least eight splicing mutations, while the remainder are missense mutations. One splicing mutation results in a three amino acid in-frame insertion. A variety of deletions have been reported, including two single-codon deletions, four or five larger deletions, and two single-base deletions and one larger deletion that cause frameshifts. At least a dozen of the missense mutations appear to be caused by methylation and subsequent deamination of highly mutagenic CpG dinucleotides. Although most mutations exhibit strong associations with a single RFLP haplotype, some are present on multiple haplotype backgrounds, presumably because of recurrence.

### 2.5 Molecular Basis of Metabolic Heterogeneity

Deficiencies in PAH result in a broad spectrum of clinical phenotypes, ranging from mild elevations in serum phenylalanine in individuals with normal or near-normal levels of intelligence in

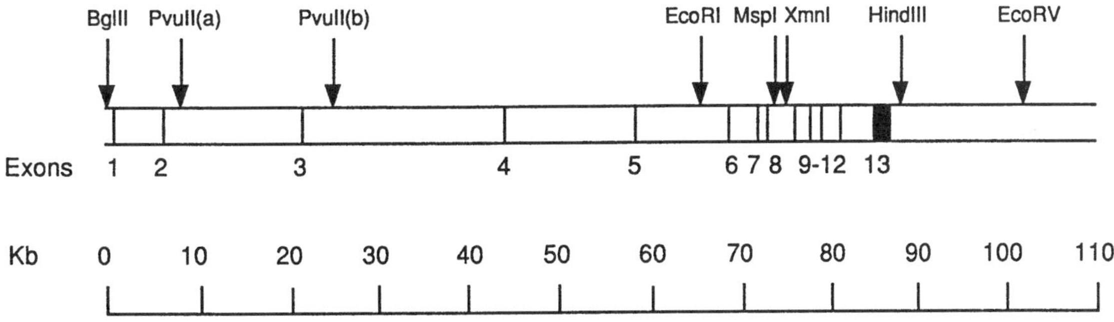

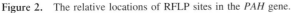

**Figure 2.** The relative locations of RFLP sites in the *PAH* gene.

the absence of treatment to profound increases in serum phenylalanine and urinary phenylalanine metabolites in individuals who will develop severe and irreversible mental retardation unless rigorously treated. One possible cause of this phenotypic heterogeneity lies in the interactions of a number of different *PAH* mutations that exert different effects on the enzymatic activity of the mutant proteins.

Nearly 30 mutations have been examined by in vitro expression analysis. About two-thirds of these mutant proteins are without significant amounts of residual enzyme activity, while the remaining mutations yield mutant proteins with levels of *PAH* activity ranging from 10% to 100% of wild-type levels. There are strong correlations between the amount of enzyme activity predicted from expression studies and the biochemical and clinical phenotypes of patients with PAH deficiencies. Consequently, patients who are homozygous or compound heterozygous for alleles encoding proteins that are largely devoid of enzyme activity have severe clinical and metabolic phenotypes. Patients bearing at least one allele encoding proteins that have significant levels of residual activity exhibit milder forms of hyperphenylalaninemia. Of the mutations not characterized by this type of analysis, nearly all are nonsense or splicing mutations and are presumably associated with severe forms of PKU. Thus, the metabolic heterogeneity in PKU is largely a result of molecular heterogeneity at the *PAH* locus.

*See also* BIOCHEMICAL GENETICS, HUMAN; HUMAN GENETIC PREDISPOSITION TO DISEASE.

### *Bibliography*

Eisensmith, R. C., and Woo, S.L.C. (1991) Phenylketonuria and the phenylalanine hydroxylase gene. *Mol. Biol. Med.* 8:3–18.

———, and Woo, S.L.C. (1992) Molecular basis of phenylketonuria and related hyperphenylalaninemias: Mutations and polymorphisms in the human phenylalanine hydroxylase gene. *Hum. Mutat.* 1:13–23.

Konecki, D. S., and Lichter-Konecki, U. (1991) The phenylketonuria locus: Current knowledge about alleles and mutations of the phenylalanine hydroxylase gene in various populations. *Hum. Genet.* 87:377–388.

Scriver, C. R., Kaufman, S., and Woo, S.L.C. (1988) Mendelian hyperphenylalaninemia. *Annu. Rev. Genet.* 22:301–321.

———, ———, Eisensmith, R. C., and Woo, S.L.C. (1995) The hyperphenylalaninemias. In *The Molecular and Metabolic Basis of Inherited Disease,* Vol. II, 7th ed. C. R. Scriver, A. Beaudet, W. Sly, and D. Valle, Eds. McGraw-Hill, New York.

# PHOSPHOLIPASES

*Suzanne E. Barbour and Edward A. Dennis*

### *Key Words*

**Fatty Acid** A hydrocarbon chain with a terminal carboxylic acid moiety.

**Head Group** The phosphorylated primary alcohol at the *sn*-3 position of a phospholipid molecule.

**Hydrolysis** Decomposition of a chemical using water as a reactant.

**Phospholipid** A glycerol molecule acylated with long chain fatty acids in the *sn*-1 and *sn*-2 positions and containing a polar head group on the *sn*-3 position.

Phospholipases comprise a subclass of lipases that hydrolyze phospholipids. There are many different phospholipases. They are defined by the position they attack on the phospholipid molecule (as shown in Figure 1). Phospholipases are central enzymes in both lipid metabolism and signal transduction. The products of phospholipase hydrolysis are involved in such diverse events as the mobilization of intracellular calcium (inositol trisphosphate), the regulation of protein phosphorylation (diacylglycerol), the disruption of membrane integrity (lysophospholipid), the activation and aggregation of platelets (platelet-activating factor), and the onset of inflammation (arachidonic acid metabolites: prostaglandins and leukotrienes). Many current research efforts are directed at gaining insight into the mechanisms of action of these proteins and the control of their activities.

## 1 PHOSPHOLIPIDS

Phospholipids have long been recognized as the major structural components of biological membranes. More recently, it has become apparent that phospholipids and phospholipid-derived mediators are involved in signal transduction and the control of cell activation and homeostasis.

As shown in Figure 1, the backbone of a phospholipid is a three-carbon molecule called glycerol. The stereospecific numbering positions *sn*-1 and *sn*-2 of the glycerol molecule are occupied by either long chain fatty acids or, in some instances, fatty alcohols. A polar head group is attached to the *sn*-3 position through a phosphate–oxygen bond. The head group can be phosphocholine, -serine, -ethanolamine, -inositol, -glycerol, or hydroxyl (phosphatidic acid).

## 2 PHOSPHOLIPASE SPECIFICITY

Phospholipases are part of a family of enzymes called hydrolases. Hydrolases use water to catalyze the degradation of biological molecules to their component parts. Hence, phospholipases use water to degrade phospholipid molecules. The reaction catalyzed by phospholipase $A_2$ (Figure 2) is an example of this type of catalysis.

Phospholipase attack can occur at a carbon–oxygen or phosphate–oxygen bond. As shown in Figure 1, phospholipases are defined according to the bond that is broken during catalysis. A great deal of information is available about phospholipases $A_2$, C, and D; Sections 3–5 describe these phospholipases in detail.

Less is known about phospholipase $A_1$, which breaks the bond at the *sn*-1 position of phospholipids. Many enzymes are known

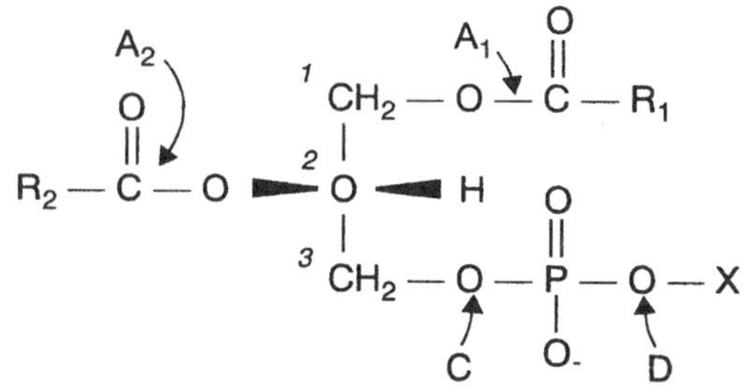

**Figure 1.** Sites of action of the major phospholipases on the glycerol phosphate backbone: phospholipase $A_1$, phospholipase $A_2$, phospholipase C, and phospholipase D. X is any one of a number of polar head groups; $R_1$ and $R_2$ are long chain hydrocarbons. Numbering of the phospholipid backbone is according to the stereospecific nomenclature (*sn*), as indicated.

to have both triglyceride lipase activity and phospholipase $A_1$ activity. A membrane-bound phospholipase $A_1$ has been purified from *Escherichia coli.*

Phospholipase B is a term often used to define phospholipases that catalyze the hydrolysis of both phospholipids and lysophospholipids. Phospholipase B enzymes have been isolated from bacterial and mammalian sources, but little is known about their mechanisms of action.

## 3 PHOSPHOLIPASE $A_2$

REACTION: PHOSPHOLIPID → FATTY ACID + LYSOPHOSPHOLIPID

Phospholipases $A_2$ ($PLA_2$), the most extensively studied of the phospholipases, catalyze the hydrolysis of the *sn*-2 fatty acid from phospholipids. The products of this hydrolysis are free fatty acid and lysophospholipid. The lysophospholipid product is a membrane-lytic agent and must be further catabolized to prevent injury to the cell. Lysophospholipases accomplish this feat by hydrolyzing the remaining fatty acid from lysophospholipids. Alternatively, acyltransferases can esterify another fatty acid onto the *sn*-2 position of the lysophospholipid to regenerate an intact phospholipid molecule. When the phospholipids contain ether linkages in the *sn*-1 position and are acted on by $PLA_2$, the lysophospholipid product is the precursor of platelet-activating factor (PAF), a cellular mediator that is implicated in many disease processes.

The fatty acid product of $PLA_2$ hydrolysis is also a potential cellular mediator. Arachidonic acid is a 20-carbon, unsaturated fatty acid that is primarily esterified in the *sn*-2 position of mammalian phospholipids. After its release by $PLA_2$, free arachidonic acid can be metabolized by the cyclooxygenase or lipoxygenase enzyme system to produce prostaglandins or leukotrienes, respectively. These compounds are potent mediators of inflammation.

### 3.1 SECRETORY PHOSPHOLIPASE $A_2$

Much work in the $PLA_2$ field has focused on secreted enzymes, since these proteins are expressed in relatively large quantities and are readily purified. Secreted forms of $PLA_2$ are found in snake and bee venoms and mammalian pancreatic and synovial exudates. These proteins are believed to be involved in phospholipid digestion and in the extracellular generation of inflammatory mediators.

The secretory forms of $PLA_2$ ($sPLA_2$) are small (~120 amino acids), water-soluble proteins that require calcium as a cofactor for catalysis. Each of these proteins is composed of a single polypeptide chain containing 10–14 cysteines, all in disulfide bonds. Hence, the $sPLA_2$s are very stable molecules. Although the $sPLA_2$s are soluble, they act in or on membranes, micelles, and other lipid–water interfaces rather than on substrates that are free in solution. Thus, the study of these enzymes or even their simple assay for diagnostic purposes requires a special understanding of the role of the interface.

The amino acid sequences of all the known $sPLA_2$s have been compared and aligned. Regions of primary structure conservation among the $sPLA_2$s include the calcium-binding loop and the histidine–aspartic acid pair at the active site. The cysteines constitute the bulk of the sequence conservation between the mammalian, reptile, and insect $sPLA_2$s. Secretory $PLA_2$s can be classified as group I, II, or III based on their disulfide bonding patterns, but there is also a large degree of secondary structure conservation among these proteins. As might be expected, X-ray crystallographic studies indicate that the three-dimensional structures of the secretory $PLA_2$s are conserved as well.

Taken together, these data indicate that the $sPLA_2$s have similar mechanisms of action. However, despite their similarities in sequence and structure, preferences for the head group type and physical form (monomer, micelle, vesicle) of the phospholipid substrate vary among the $sPLA_2$s. Many current research efforts are directed at determining the sequential and structural constraints that contribute to these differences.

### 3.2 CYTOSOLIC PHOSPHOLIPASE $A_2$

Recently, several laboratories have reported the purification, characterization, and cloning of a cytosolic $PLA_2$ ($cPLA_2$). As predicted based on its location in the cytoplasm, this protein shows no sequence homology to the secretory $PLA_2$s. Cytosolic $PLA_2$ is larger than the $sPLA_2$s (~700 amino acids), has no disulfide linkages, and does not require calcium for catalysis. However, $cPLA_2$ translocates from the cytosol to the membrane in response to physiological concentrations (submicromolar) of calcium. This translocation is mediated by a calcium-dependent lipid-binding region near the amino terminus of $cPLA_2$. This region has sequence homology to the isotypes of protein kinase C and phospholipase C that translocate in response to calcium.

There are several functional differences between the $sPLA_2$s and $cPLA_2$. Unlike the $sPLA_2$s, which have no preference for the fatty acid esterified at the *sn*-2 position of the phospholipid, $cPLA_2$ preferentially hydrolyzes phospholipids containing an *sn*-2 arachidonic acid moiety. While the $sPLA_2$s only hydrolyze *sn*-2 fatty acid from intact phospholipid, $cPLA_2$ has lysophospholipase activity as well. Thus, $cPLA_2$ has an inherent mechanism for the catabolism of the membrane-active lysophospholipid products of the $PLA_2$ reaction.

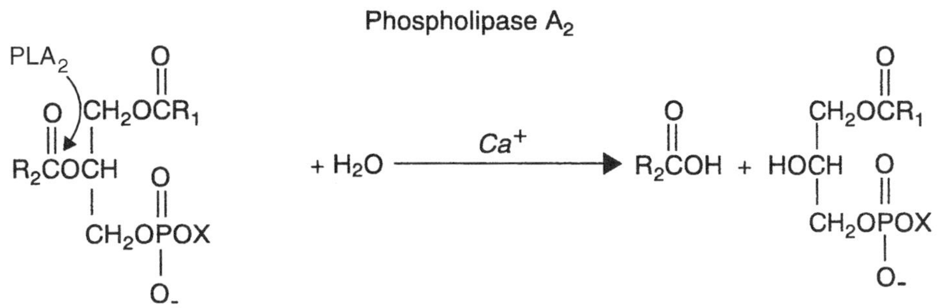

Phospholipase $A_2$

$$\mathrm{CH_2OC(=O)R_1{-}CH(OC(=O)R_2){-}CH_2OP(=O)(O^-)OX} \;(\mathrm{PLA_2}) + H_2O \xrightarrow{Ca^+} \mathrm{R_2C(=O)OH} + \mathrm{CH_2OC(=O)R_1{-}HOCH{-}CH_2OP(=O)(O^-)OX}$$

**Figure 2.** The specific reaction catalyzed by phospholipase $A_2$ in which a phospholipid molecule is hydrolyzed with water to produce a free fatty acid and a lysophospholipid. This enzyme requires calcium for catalysis. Abbreviations as in Figure 1.

## 4 PHOSPHOLIPASE C

REACTION: PHOSPHOLIPID → DIACYLGLYCEROL + PHOSPHOMONOESTER

Phospholipase C (PLC) catalyzes the hydrolytic removal of the head group from phospholipid to produce diacylglycerol (DAG) and a phosphomonoester such as inositol phosphate. While PLCs have little or no preference for the fatty acids esterified in the *sn*-1 and *sn*-2 positions of phospholipids, phosphatidylcholine (PC) preferring and phosphatidylinositol (PI) preferring PLC activities have been described in mammals, insects, and bacteria. In addition, some isoforms of PLC hydrolyze phosphatidylinositol glycans to convert lipid-anchored membrane proteins to their soluble forms.

Three isotypes of PI-specific PLC (beta, gamma, and delta) have been described in detail. Molecular weights and amino acid sequences vary widely among the PLC isoforms, but domains with considerable amino acid sequence homology have been identified and are postulated to be of functional significance. Among these is the carboxy-terminal, calcium-binding region that is believed to bind the calcium required for PLC catalysis. As indicated earlier, this region has homology to the calcium-dependent lipid-binding motifs of $cPLA_2$ and protein kinase C and thus controls PLC translocation as well.

PLC activation has been tied to the signal transduction cascades of many cell surface receptors. The mechanism of activation varies among the receptors and the PLC isoforms. The engagement of hormone receptors by ligand (e.g., vasopressin, thromboxin, angiotensin) results in G protein-mediated activation of PLC-β. These G proteins are members of the $G_q$ family. In contrast, the activities of the gamma isoforms of PLC are regulated by phosphorylation. These proteins are phosphorylated both by growth factor receptor kinases (e.g., EGF, PDGF, NGF) and by the nonreceptor kinases that are involved in the signal transduction cascades of T- and B-cell receptors. The interaction between the receptor kinases and PLC is believed to be mediated by the *src* homology (SH) domains of PLC-γ, which are homologous to the regulatory regions of the *src* kinases.

The products of PLC activation are also involved in signal transduction cascades. When the phospholipid substrate is phosphatidylinositol-4,5 bisphosphate ($PIP_2$), one product of PLC hydrolysis is inositol 1,4,5-trisphosphate ($IP_3$), a mediator of intracellular calcium release. Intracellular calcium fluxes control enzyme activities in many cell activation systems. Diacyglycerol, the other product of PLC hydrolysis, has been linked to the regulation of protein phosphorylation by protein kinase C. This phosphorylation, in turn, alters the activities of cell proteins.

## 5 PHOSPHOLIPASE D

REACTION: PHOSPHOLIPID →PHOSPHATIDIC ACID + ALCOHOL

To date, the phospholipases D (PLD) are the least characterized of the major phospholipases. PLD activities have been described in plants, in trypanosomes, and in a variety of mammalian cell types. Since both PLD and PLC attack the phospholipid head group to produce one water-soluble and one insoluble product, it has been difficult to distinguish between these activities. This problem is alleviated by a second activity that has been ascribed to PLD, the transfer of head groups between intact phospholipids and primary alcohols (transphosphatidylation, or base exchange). Thus, PLD activities can be characterized by the formation of phosphatidylethanol from phosphatidylcholine in the presence of ethanol.

Most of the observed PLD activities are specific for phosphatidylcholine. However, PLDs specific for the inositol phosphate bond of phosphatidylinositol glycans have been described in mammalian serum and trypanosomes. It is believed that PLD hydrolysis of the anchor of the variant surface glycoprotein of trypanosomes allows these organisms to routinely change their major antigenic determinants and thus to avoid host immune defenses.

*See also* ENZYMES; LIPIDS, STRUCTURE AND BIOCHEMISTRY OF; PHOSPHOLIPIDS; PROTEIN PHOSPHORYLATION.

### *Bibliography*

Billah, M. M., and Anthes, J. C. (1990) The regulation and cellular functions of phosphatidylcholine hydrolysis. *Biochem. J.* 269:281–291.

Dennis, E. A. (1983) Phospholipases. In *The Enzymes,* Vol. 16, 3rd ed., P. Boyer, Ed., pp. 307–353. Academic Press, New York.

———, Ed. (1991) *Phospholipases,* Vol. 197 in *Methods in Enzymology.* Academic Press, Orlando FL.

———, Rhee, S. G., Billah, M. M., and Hannun, Y. A. (1991) Role of phospholipases in generating lipid second messengers in signal transduction. *FASEB J.* 5:1294–1300.

Rhee, S. G., and Choi, K. D. (1992) Regulation of inositol phospholipid-specific phospholipase C isozymes. *J. Biol. Chem.* 267:12393–12396.

Waite, M. (1987) *The Phospholipases, Handbook of Lipid Research,* Vol. 5. Plenum Press, New York.

# PHOSPHOLIPIDS

*Dennis E. Vance*

### *Key Words*

**Arachidonic Acid** A 20-carbon fatty acid with four double bonds; the fatty acid precursor of the eicosanoids.

**Diacylglycerol** An intermediate in phospholipid biosynthesis and an activator of protein kinase C.

**Inositol-Trisphosphate** A cyclohexane ring substituted with a hydroxyl group on each carbon and a phosphate on the hydroxyl groups of carbons 1,4, and 5.

**Phosphatidylcholine** Quantitatively the most important phospholipid found in eukaryotes.

**Protein Kinase** An enzyme that phosphorylates a protein on the amino acids serine, threonine, or tyrosine.

**Second Messenger** Compound that is formed within a cell in response to a hormone or other agonist binding to a specific receptor on the cell surface.

Phospholipids are biological compounds that contain phosphorus and have both hydrophobic and hydrophilic moieties. They are ubiquitous components of all biological membranes. Phospholipids provide the basic structure and the permeability barriers of cellular membranes. The absolute requirement for phospholipids in life is underscored by the apparent lack of gene mutations in their metabolism. Such mutations, which would result in severely defec-

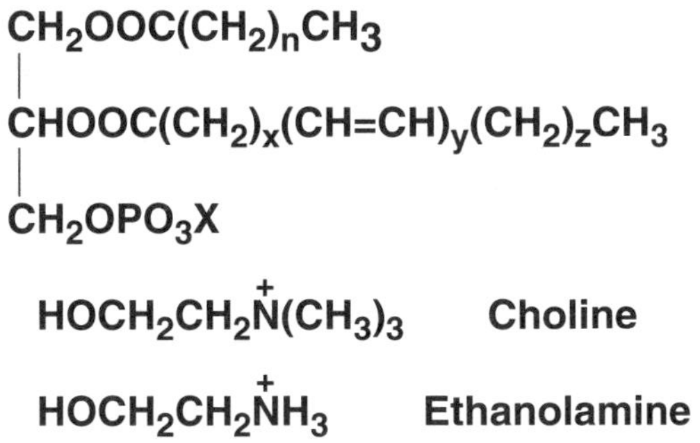

$CH_2OOC(CH_2)_nCH_3$
|
$CHOOC(CH_2)_x(CH{=}CH)_y(CH_2)_zCH_3$
|
$CH_2OPO_3X$

$HOCH_2CH_2\overset{+}{N}(CH_3)_3$ Choline

$HOCH_2CH_2\overset{+}{N}H_3$ Ethanolamine

**Figure 1.** General structure for a phospholipid. The X component can be one of five compounds that contain a hydroxyl group linked to the phosphate residue. The structures of two of these compounds are shown. Usually there is a saturated fatty acid on carbon 1 of the glycerol moiety and an unsaturated fatty acid on carbon 2. In some cases the substituent on carbon 1 can be a long chain alcohol instead of a fatty acid.

tive enzyme activities, probably would be lethal. Phospholipids also act as sources for cellular second messengers such as diacylglycerol and inositol phosphate. Phospholipids are made within each cell from precursors that include fatty acids, glycerol, cytidine 5′-triphosphate (CTP), and ATP, and alcohols such as ethanolamine, choline, glycerol, inositol, and serine. The phospholipids are made by biosynthetic enzymes and degraded by phospholipases. Many of the genes for these enzymes from *Escherichia coli* and yeast, and a few cDNAs from animals, have been cloned and expressed. So far no gene for a phospholipid biosynthetic enzyme from an animal source has been isolated.

## 1 STRUCTURE OF PHOSPHOLIPIDS

The basic structure of all phospholipids is shown in Figure 1. The fatty acid attached to carbon 1 is usually saturated and contains either 16 (palmitate) or 18 (stearate) carbons. Unsaturated fatty acids are usually found on carbon 2. However, there are exceptions: phosphatidylcholine, found in lung surfactant (keeps lungs from collapsing when air is expelled), has palmitate on both carbons 1 and 2. Linked to phosphate on carbon 3 is one of five small compounds (choline, ethanolamine, serine, glycerol, inositol) that have a hydroxyl functional group. The phospholipids are named "phosphatidyl" (the diacylglycerol phosphate moiety) followed by the name of the small compound (e.g., phosphatidylcholine).

## 2 BIOSYNTHESIS OF PHOSPHOLIPIDS

### 2.1 Enzymes and Pathways

The major pathways for the biosynthesis of phospholipids in animals are shown in Figure 2. In the first stage fatty acids are esterified to glycerol 3-phosphate to yield phosphatidic acid. The latter compound either reacts with CTP to yield CDP-diacylglycerol or the phosphate is removed by the enzyme phosphatidic acid phosphohydrolase to yield diacylglycerol, which subsequently reacts with

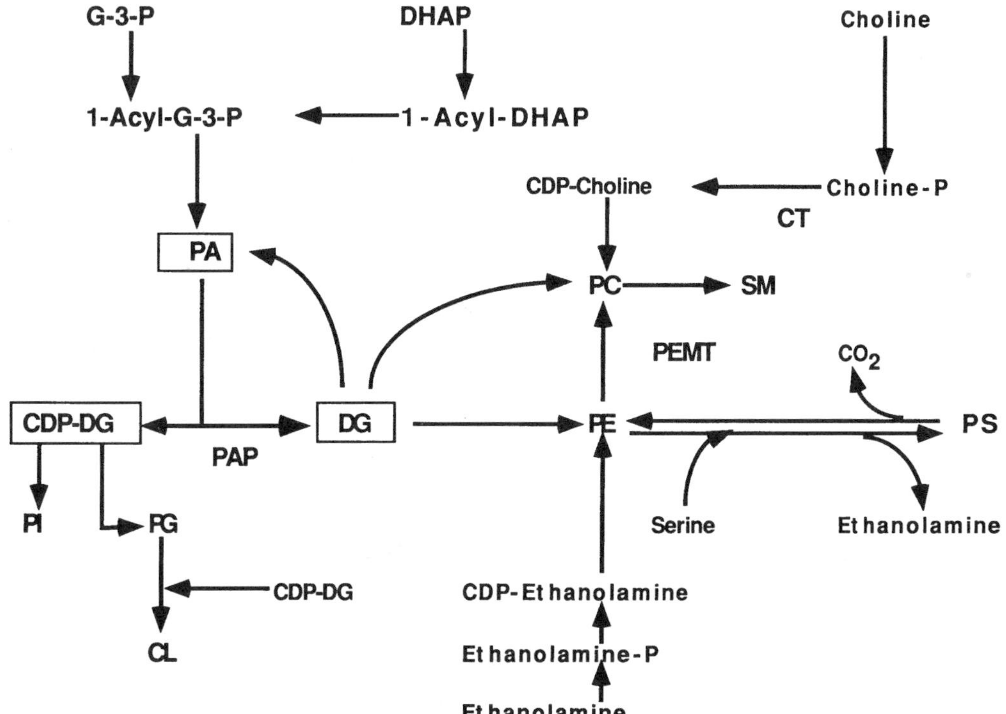


**Figure 2.** Outline for the biosynthesis of the major phospholipids in human cells: DHAP, dihydroxyacetone phosphate; G-3-P, glycerol-3-phosphate; PA, phosphatidic acid; DG, diacylglycerol; CDP-DG, cytidinediphospho diacylglycerol; PI, phosphatidylinositol; PG, phosphatidylglycerol; CL, cardiolipin (also known as diphosphatidylglycerol); PAP, phosphatidic acid phosphohydrolase; PE, phosphatidylethanolamine; PC, phosphatidylcholine; PEMT, phosphatidylethanolamine methyltransferase; CT, CTP:phosphocholine cytidylyltransferase; CDP-choline, cytidinediphosphocholine; choline-P, choline phosphate; CDP-ethanolamine, cytidinediphosphoethanolamine; ethanolamine-P, ethanolamine phosphate; SM, sphingomyelin. In mammals, PEMT is found in significant amounts only in liver.

either CDP-choline to make phosphatidylcholine (PC) or with CDP-ethanolamine to give phosphatidylethanolamine (PE). Note that the formation of a phospholipid de novo always requires CTP.

Phospholipids can be modified in several ways. For example PE (or PC) can react with serine to form phosphatidylserine (PS), which can be decarboxylated to PE (Figure 2). In liver, PE can by methylated to PC by PE methyltransferase. Also, but not shown in Figure 2, the fatty acid moieties of phospholipids can be removed (deacylated) and replaced by another fatty acid (reacylated). Finally, the phosphocholine moiety of PC can be transferred to ceramide (*N*-acylsphingosine) to form sphingomyelin. This particular phospholipid is usually found enriched in the plasma membrane of cells.

The enzymes of phospholipid biosynthesis are found in cytosol, endoplasmic reticulum (ER), Golgi apparatus, mitochondria, and possibly nuclear membrane. Only a few enzymes from animal sources have been purified to homogeneity (choline/ethanolamine kinase, CTP:phosphocholine cytidylyltransferase, CTP:phosphoethanolamine cytidylyltransferase, and PE methyltransferase), despite an enormous effort in many laboratories to purify the other enzymes. More have been purified from *E. coli* and yeast. One difficult problem with membrane-bound enzymes is the need for detergents to accomplish solubilization from the membrane. Second, to keep the enzymes soluble, it is necessary to have detergents in the subsequent purification steps, and this complicates the procedures.

The cDNAs for the kinase, phosphocholine cytidylyltransferase, and the methyltransferase have been cloned. In addition, by using mutants of animal cells, the cDNAs for PS synthetase and PS decarboxylase have been cloned. No particularly unusual features were found for the deduced structures of these enzymes.

### 2.2 Regulation

The regulation of PC biosynthesis is best understood. Generally, the activity of CTP:phosphocholine cytidylyltransferase (CT in Figure 2) is rate limiting and regulated. Cytidylyltransferase exists as a dimer of identical 42 kDa subunits in the cytosol, where it is

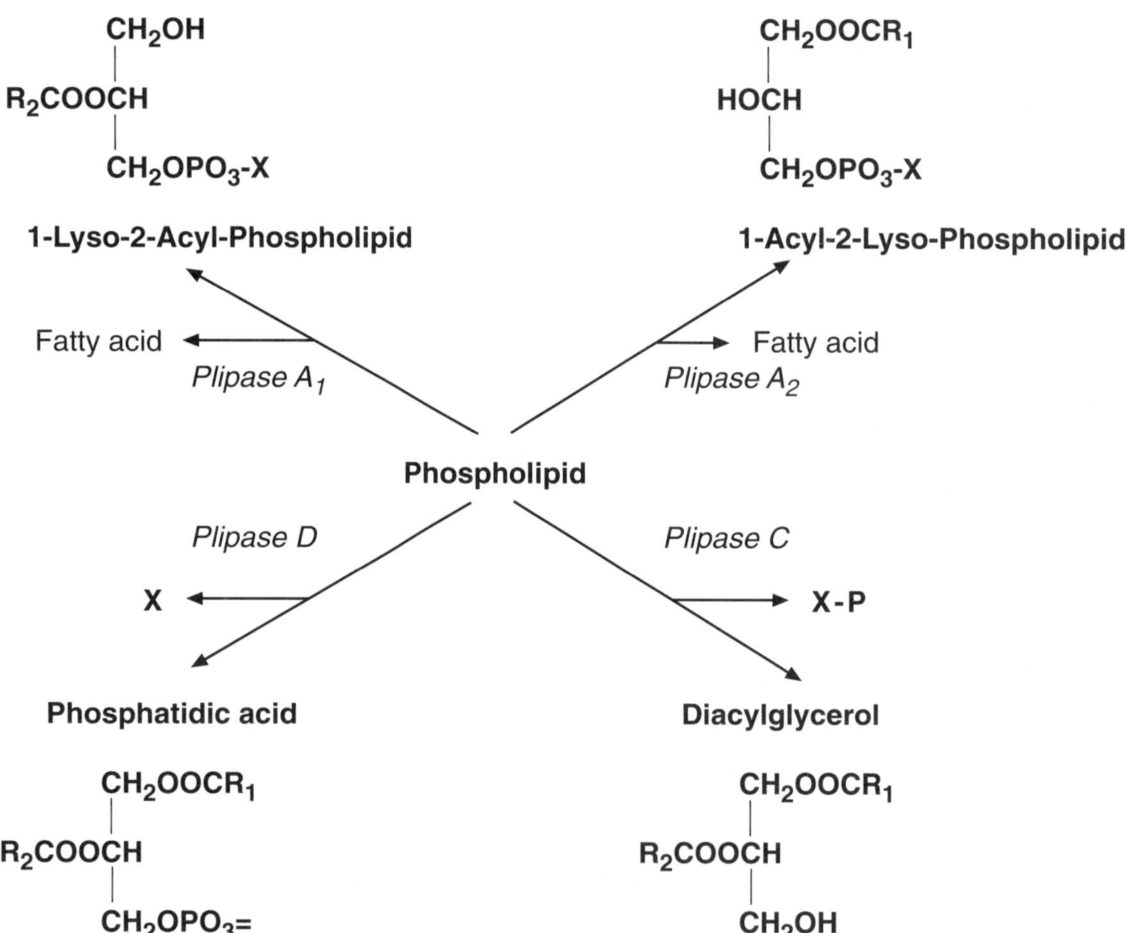


**Figure 3.** Phospholipase activities found in human cells. Lyso indicates the absence of a fatty acid residue at the indicated position on the glycerol backbone. These lyso lipids can be further degraded to fatty acid and glycerol-phospho-X by lysophospholipases. Plipase, phospholipase; X, a compound with a hydroxyl group linked to the phosphate (e.g. choline, ethanolamine).

inactive. Cytidylyltransferase is activated by translocation of the enzyme to membranes (primarily the ER in liver). The binding to the membranes appears to be via an amphipathic helix (the helix is composed of amino acids, which results in a hydrophobic and hydrophilic side). How the binding to the ER activates the enzyme is not known. The binding appears to be mediated by the ratio of bilayer-forming lipids (such as PC, phosphatidylglycerol, phosphatidylinositol) to non-bilayer-forming lipids (such as diacylglycerol, fatty acid, phosphatidylethanolamine). Thus, binding of phosphocholine cytidylyltransferase to the ER is enhanced by an increase in fatty acids or diacylglycerols or by a decrease in the concentration of PC. In contrast, cytidylyltransferase is released into cytosol when the concentrations of fatty acids or diacylglycerols are decreased or an increase in PC occurs. Modulation of cytidylyltransferase activity by PC is a classic example of feedback regulation of a metabolic pathway.

## 3 CATABOLISM OF PHOSPHOLIPIDS

Phospholipids are degraded by enzymes named phospholipases, which occur ubiquitously in nature. The reactions catalyzed and names of the enzymes are shown in Figure 3. Phospholipases can be classified as ''digestive'' or ''regulatory.'' The classic example of a digestive enzyme is phospholipase $A_2$ from the pancreas. This lipase is secreted into the intestinal tract and degrades phospholipids from food. The phospholipase $A_2$ from various snake venoms has also been studied extensively. A third type of digestive phospholipase is found in lysosomes and degrades phospholipids taken into the cell by endocytosis.

There are two types of regulatory phospholipases. One group generates cellular second messengers. Thus, phosphatidylinositol-4,5-bisphosphate is degraded by various phospholipases C to yield inositol trisphosphate. Alternatively, PC can be degraded by a specific phospholipase $A_2$ to release arachidonic acid, which is used for the biosynthesis of eicosanoids. Considerable information is available about the structures and properties of these phospholipases. The other group of phospholipases is involved in the degradation of phospholipids and the deacylation/reacylation of fatty acids on phospholipids. These enzymes are found in virtually all the different compartments of the cell, and very few have been purified and characterized.

## 4 PHOSPHOLIPID TRANSPORT

All organelles of the cell contain phospholipids but not the biosynthetic capacity. Thus, mitochondria contain PC but do not make this lipid. Since phospholipids are virtually insoluble in an aqueous environment, they must be transported by a special mechanism from their site of synthesis (e.g., ER) to another membrane (e.g., mitochondria). How this transport is accomplished is not known. Three possible mechanisms are being considered:

1. There are proteins that bind and transfer phospholipids from one membrane to another in vitro. One such protein in yeast has been shown to be essential for the cell to survive. Whether this protein actually transports phospholipids in yeast remains to be shown.
2. Membranous vesicles are known to transport proteins within cells, and such vesicles may also transport phospholipids.
3. There may be continuity of membranes between adjacent organelles, which would allow the phospholipids to move between the two membranes without being transported in the aqueous cytosolic compartment.

## 5 FUNCTIONAL ASPECTS

### 5.1 Sources of Second Messengers

It has been known for many decades that arachidonic acid in phospholipids is released to form eicosanoids, which are 20-carbon fatty acid derivatives that have hormone-like actions near their sites of synthesis. Among various functions, eicosanoids have been shown to modulate the synthesis of cyclic AMP in some tissues, facilitate ion transport, cause platelets to aggregate, and cause the chemotaxis of polymorphonuclear leukocytes.

In the last decade scientists have recognized that degradation of phosphatidylinositol-4,5-bisphosphate yields two second messengers, diacylglycerol and inositol trisphosphate. Diacylglycerol activates protein kinase C, which is involved in regulation of catecholamine secretion, insulin release, steroid synthesis, and many other cellular responses. Inositol trisphosphate causes the release from the ER of calcium, which has many functions: for example, it activates phosphorylase, which degrades glycogen.

An unexpected discovery in 1979 was that a derivative of PC, 1-alkyl-2-acetylglycerol-3-phosphocholine (platelet-activating factor), caused platelet aggregation and induced a potent antihypertensive response. Various white blood cells, healthy tissues, and tumor cells synthesize and degrade platelet-activating factor.

### 5.2 Anchoring of Proteins to Membranes

Some proteins on the surface cells are covalently linked to phosphatidylinositol in the plasma membrane. The carboxyl terminal amino acid is in amide linkage to ethanolamine, which is linked via a series of carbohydrates to the inositol moiety of phosphatidylinositol. Such proteins can be released from the membrane via hydrolysis of the phosphatidylinositol moiety by a phospholipase C. Proteins linked via phosphatidylinositol on cell surfaces include hydrolases (e.g., alkaline phosphatase), protozoa coat proteins, and proteins involved in cell adhesion. The functional significance of this special lipid anchoring is under study.

*See also* Lipid Metabolism and Transport; Lipids, Structure and Biochemistry of; Liposomes.

## *Bibliography*

Carman, G. M., and Henry S. A. (1989) Phospholipid biosynthesis in yeast. *Annu. Rev. Biochem.* 58:635–639.

Dennis, E. A., and Vance, D. E., Eds. (1992) *Methods in Enzymology,* Vol. 209, *Phospholipid Biosynthesis.* Academic Press, San Diego, CA.

Vance, D. E., Ed. (1989) *Phosphatidylcholine Metabolism.* CRC Press, Boca Raton, FL.

———, and Vance, J. E., Eds. (1991) *Biochemistry of Lipids, Lipoproteins and Membranes.* Elsevier, Amsterdam.

**Photodynamic Inactivation of Viruses:** *see* **Viruses, Photodynamic Inactivation of.**

# Photosynthetic Energy Transduction

*Neil R. Baker*

*Key Words*

**Antenna** Pigment matrices that absorb light energy and transfer it to a photochemical reaction center.

**ATP Synthetase (Coupling Factor)** Proton-translocating ATPase consisting of two multisubunit protein complexes; an intrinsic membrane complex that acts as a proton channel across the thylakoid membrane and an extrinsic complex that has a catalytic site for ATP synthesis.

**Excitation Energy Transfer** Transfer of absorbed light energy between pigment molecules.

**Photophosphorylation** Synthesis of ATP that is coupled to photosynthetic electron transport.

**Photosystem** Multisubunit protein complex that uses light energy to transfer electrons between electron carriers.

**Pigment-Proteins** Proteins that are noncovalently linked to pigment molecules and serve to capture and transfer light energy.

**Reaction Center** Site in photosystem where absorbed light energy is used to drive electron transfer.

**Thylakoid** Vesicle whose membrane contains the photosynthetic apparatus required for light capture, electron transport, and photophosphorylation.

Photosynthesis is the process by which solar energy is used by photosynthetic organisms in the synthesis of organic compounds. An essential feature of photosynthesis in all photosynthetic bacteria and plants is the conversion of solar energy into chemical energy, a phenomenon defined here as photosynthetic energy transduction. Many differences exist among photosynthetic organisms with respect to the components and organization of the photosynthetic apparatus and the end products of photosynthetic energy transduction. For example, the photosynthetic apparatus of oxygen-evolving organisms (eukaryotic plants and cyanobacteria) contains two different photosystems (photosystems I and II), whereas nonoxygenic photosynthesis in the purple and green photosynthetic bacteria involves only one system. However, there are a number of features and principles associated with photosynthetic energy transduction that are common to all organisms. The process is always associated with membranes (thylakoids in oxygen-evolving organisms, chromatophore membranes in photosynthetic bacteria), and it involves multisubunit protein complexes. Initially light energy must be captured and transferred to a photochemical reaction center, where it can be used to drive a reduction–oxidation (redox) reaction. This occurs in complexes, called photosystems in oxygen-evolving organisms, each of which contains a light-harvesting antenna and a photochemical reaction center. The chemical redox energy resulting from the photochemical reaction at the reaction center can then drive a series of redox reactions, commonly termed electron transport reactions, which result in spatial separation of oxidized and reduced chemical species and provide a source of chemical energy to the cell in the form of reducing equivalents. Coupled with electron transport is the pumping of protons across the membrane. The free energy stored in the resulting electrochemical potential difference of protons across the membrane can be used by ATP synthetase to phosphorylate ADP and provide ATP for the cell.

## 1 LIGHT CAPTURE AND EXCITATION ENERGY TRANSFER

Pigments involved in harvesting light energy for photosynthetic energy transduction are associated with proteins, often termed pigment-proteins, that are embedded in or tightly bound to the photosynthetic membrane. Chlorophyll, or bacteriochlorophyll in photosynthetic bacteria, is the major photosynthetic pigment. However accessory pigments, such as carotenoids and bilins, frequently are components of pigment-proteins. An antenna complex consists of groups of pigment-proteins organized to facilitate transfer of absorbed light energy to a photochemical reaction center. Each antenna contains a large number of pigment molecules, often many hundred, and is associated with a single photochemical reaction center. The role of an antenna is to capture light and transfer this excitation energy to the reaction center. If a pigment is directly adjacent to the reaction center, it can transfer its energy directly to the reaction center. However, the majority of pigments in the antenna will not be adjacent to the reaction center and cannot transfer their energy to the reaction center except via other antenna pigments. Excitation energy transfer in the antenna to a reaction center is an extremely rapid and efficient process; it occurs in less than a nanosecond and does not involve any chemical changes. Light absorption and energy transfer within an antenna to a reaction center are illustrated in Figure 1.

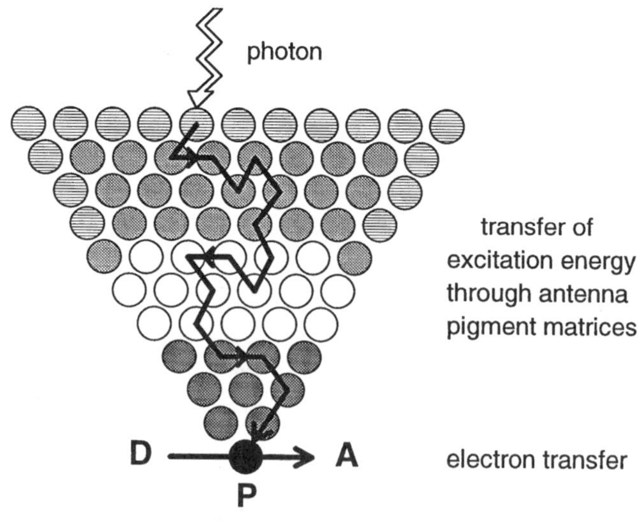


**Figure 1.** Diagrammatic representation of excitation energy transfer to a reaction center within a light-harvesting antenna complex: circles represent pigment molecule components of the pigment-proteins. Groups of pigment proteins are organized into matrices (represented by different shading), which facilitate the transfer of absorbed light energy from peripheral regions of the antenna to the reaction center chlorophyll (P). Upon excitation, P will transfer an electron to a primary acceptor molecule (A) and consequently will receive an electron from a primary donor molecule (D).

Antenna systems can involve more than one multisubunit complex. For example, in higher plant thylakoids a group of pigment-proteins form a distinct multisubunit light-harvesting antenna complex (LHCII) that is weakly associated with a photosystem II complex (see Section 3, Figure 3). LHCII is considered to be a secondary antenna of photosystem II. The pigment proteins of the photosystem II complex constitute the primary antenna. Excitation energy will be transferred from LHCII to the primary antenna of photosystem II, and then to the reaction center chlorophyll, P680.

## 2 REACTION CENTER PHOTOCHEMISTRY

Reaction centers are specific pigment-protein complexes that contain chlorophyll (or bacteriochlorophyll) molecules in a specific microenvironment. The reaction center proteins are associated with a number of other proteins, which together constitute a multisubunit photosystem. An important step forward in understanding the structure and function of reaction centers was the crystallization of reaction centers of purple photosynthetic bacteria, allowing their three-dimensional structures to be resolved using X-ray diffraction analysis. The structure of the reaction center from *Rhodobacter sphaeroides* is shown in Figure 2 (see color plate ?). Features of particular interest are the twofold symmetry of the structure perpendicular to the plane of the membrane, the α-helices of the transmembrane regions of the protein, and the locations of bound cofactors (bacteriochlorophylls, bacteriopheophytins, and quinones). This structure is considered to be similar to that of the reaction center of photosystem II found in oxygen-evolving organisms.

The primary photochemical reactions of photosynthesis take place within the reaction centers of the photosystems: light energy reaching the reaction center chlorophyll (or bacteriochlorophyll), designated P, will generate the excited state of P (P*); P* will then transfer an electron to an electron acceptor (A) within the reaction center and become oxidized to $P^+$. An electron from a donor (D) is then transferred to $P^+$, producing P and $D^+$. This series of events can be summarized as follows:

$$\text{D P A} + \text{light} \rightarrow \text{D P}^* \text{A}$$
$$\text{D P}^* \text{A} \rightarrow \text{D P}^+ \text{A}^-$$
$$\text{D P}^+ \text{A} \rightarrow \text{D}^+ \text{P A}^-$$

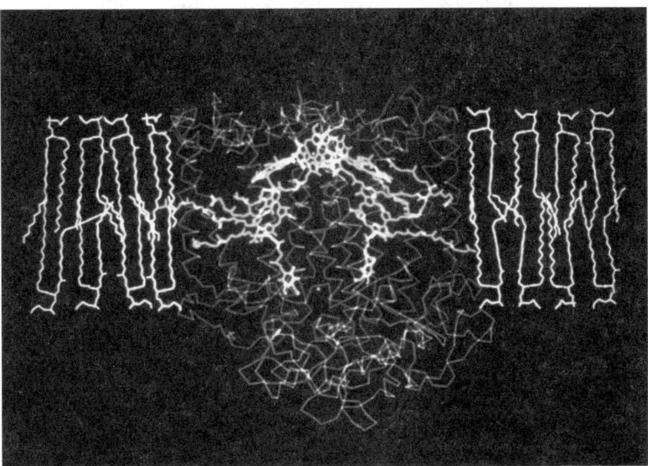

Figure 2. Structure of the reaction center of *Rhodobacter sphaeroides* determined from X-ray diffraction analysis. The complex consists of three protein subunits: the L and M core proteins (blue) and the H protein (green). The cofactors involved in excitation energy capture and electron transfer reactions (bacteriochlorophylls, bacteriopheophytins, and quinones) are shown in red. [Reproduced with permission from Yeates et al., *Proc. Natl. Acad. Sci. U.S.A.* 84: 6438–6442 (1987).] (See color plate 13.)

In almost all cases, A is a chlorophyll or the related pigment pheophytin. The primary photochemical reaction is followed by a series of rapid electron transfers to secondary acceptors (usually quinones). Since the donor and acceptor molecules are located on opposite sides of the complex, these electron transfers result in a spatial separation of electrical charge across the reaction center complex. These sequential electron transfer reactions minimize the possibility of acceptors transferring the electron back to the donor and the energy being dissipated as heat. This is an essential feature of photosynthetic energy storage.

The reaction center chlorophyll of photosystem I of eukaryotic plants and cyanobacteria is designated P700, since when it is photooxidized to $P700^+$, a bleaching (decrease in absorption) of the chlorophyll is observed that is maximal at 700 nm. Similarly the reaction center chlorophyll of photosystem II, termed P680, is bound to two polypeptides known as D1 and D2, which together with at least 20 other polypeptides constitute the photosystem II multisubunit complex.

## 3 ELECTRON TRANSPORT

Although the primary and secondary electron transfer reactions are confined to the photosystem complexes, if the reaction center is to be able to repeatedly perform primary photochemistry, it is essential to remove electrons from the acceptors and to replenish electrons to the donors in the photosystem. This can be achieved by a noncyclic flux of electrons from a reduced substrate through the reaction center to an oxidized substrate, as is the case in oxygenic photosynthesis for electron transfer from water through photosystems I and II to $NADP^+$ (Figure 3). Alternatively, continuous turnover of a reaction center can be maintained by a cyclic electron transfer in which the reduced acceptor transfers electrons back to the oxidized donor via a series of electron carriers in the membrane. Cycling of electrons is found in many photosynthetic bacteria and can occur through photosystem I in oxygen-evolving organisms (Figure 3).

Both noncyclic and cyclic electron transfer processes involve multisubunit protein complexes containing cytochromes and iron–sulfur proteins, often termed Rieske Fe-S centers. In both photosynthetic bacteria and plants, these complexes contain *b* and *c*-type cytochromes (cytochrome *f*, found in higher plant thylakoids, is a *c*-type cytochrome). Transfer of electrons from the photosystems to the cytochrome complexes can involve diffusion of reduced quinones (quinols) through the membrane. A quinone (Q) from a free pool of quinones in the membrane–lipid bilayer binds to a site on the reaction center protein and becomes a bound quinone ($Q_B$). $Q_B$ accepts an electron from the secondary quinone electron acceptor ($Q_A$), which is an integral component of the photosystem. A second electron is then transferred from the primary electron acceptor of the reaction center to $Q_A$ and then onto $Q_B$. The doubly reduced $Q_B$ ($Q_B^{2-}$) accepts two protons from the aqueous phase adjacent to the membrane, thus producing a fully reduced and protonated quinol ($QH_2$). This quinol dissociates from its binding

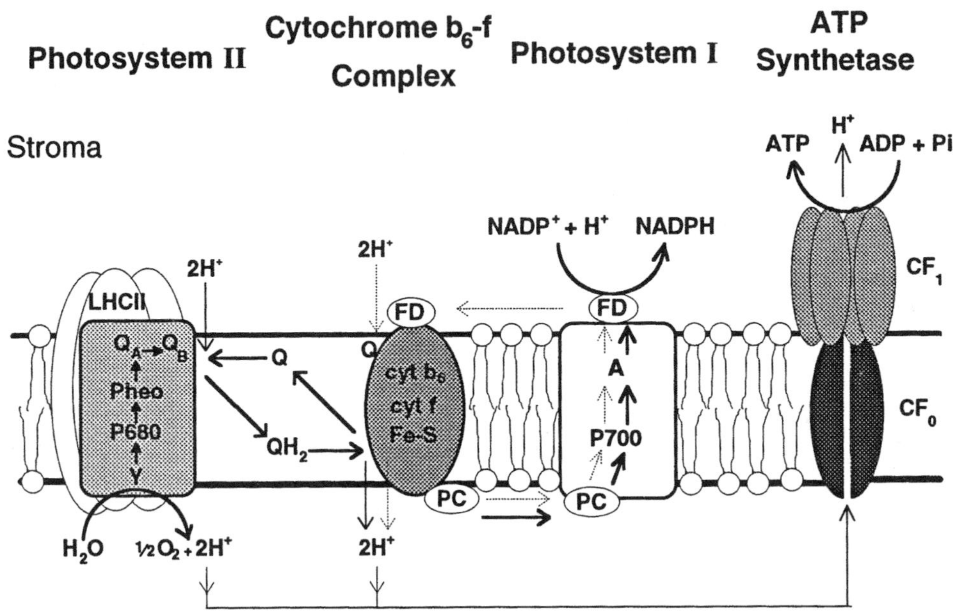


**Figure 3.** The electron and proton transfer events involved in photophosphorylation by the thylakoid membrane of higher plants. The electron transport system comprises of three multisubunit complexes: photosystem II, cytochrome $b_6$–$f$ complex, and photosystem I. A fourth multisubunit complex, the ATP synthetase or coupling factor, is involved in proton transfer and ATP synthesis. Solid arrows indicate events associated with noncyclic electron flow from water through photosystems II and I to $NADP^+$. Dotted arrows represent events involved with cyclic electron flow around photosystem I. In this cyclic process, ferredoxin, a small extrinsic membrane protein, transfers electrons from photosystem I to the cytochrome $b_6$–$f$ complex, where a bound quinone is reduced. Both noncyclic and cyclic fluxes of electrons through the cytochrome $b_6$–$f$ complex are associated with the transfer of protons across the membrane.

Abbreviations: A, primary electron acceptor in reaction center of photosystem I; $CF_0$, intrinsic membrane multisubunit protein complex of ATP synthetase; $CF_1$, extrinsic membrane multisubunit protein complex of ATP synthetase; cyt, cytochrome; FD, ferredoxin; Fe-S, Rieske iron–sulfur center; LHCII, light-harvesting pigment-protein multisubunit complex, which acts as a secondary light-harvesting antenna for photosystem II; P680, reaction center chlorophyll of photosystem II; P700, reaction center chlorophyll of photosystem I; PC, plastocyanin; Pheo, pheophytin (primary electron acceptor of photosystem II reaction center); Q, plastoquinone; $Q_A$, secondary quinone electron acceptor of photosystem II reaction center; $Q_B$, plastoquinone bound to photosystem II reaction center; Y, primary electron donor of the photosystem II reaction center (tyrosine residue located on D1 reaction center protein).

site on the reaction center and diffuses through the membrane to the cytochrome complex, where it transfers its electrons to redox components in the cytochrome complex and releases two protons. The site of proton release at the cytochrome complex is on the side of the membrane opposite the site on the photosystem at which the bound quinone becomes protonated. Consequently, as electrons are transferred from the photosystem to the cytochrome complex, protons are transported across the membrane and give rise to a proton electrochemical potential difference across the membrane. The transfer of electrons from photosystem II to the cytochrome $b_6$–$f$ complex via a quinone and the associated proton pumping across the membrane in oxygenic organisms is summarized in Figure 3. It should be noted that release or consumption of protons in electron transfer reactions besides those involving quinones also can contribute to the transmembrane proton electrochemical potential difference. For instance, in oxygenic photosynthesis protons are released into the thylakoid lumen when water is photooxidized; protons are also consumed on the stromal side of the membrane when $NADP^+$ is reduced to NADPH (Figure 3).

In both noncyclic and cyclic electron transfer, extrinsic membrane proteins, such as plastocyanin in oxygen-evolving organisms (Figure 3) and a *c*-type cytochrome in photosynthetic purple bacteria, are involved in transporting electrons from the cytochrome multisubunit complexes to the oxidized electron donors of the photosystems. These small extrinsic proteins diffuse along the surface of the membrane between the cytochrome and photosystem multisubunit complexes.

## 4 PHOTOPHOSPHORYLATION

It is now generally accepted that as is the case for oxidative phosphorylation in bacteria and mitochondria, photophosphorylation proceeds by the chemiosmotic mechanism. According to the chemiosmotic hypothesis, ion electrochemical potential differences across membranes constitute a source of free energy that can be used by the cell. As discussed in Section 3, photosynthetic electron transport can result in the formation of a proton electrochemical potential difference across the thylakoid membrane. The free energy of this potential difference can be used to drive the phosphorylation of ATP by a membrane-bound ATP synthetase. This multisubunit enzyme, also known as the coupling factor, consists of two discrete parts (Figure 3): a multisubunit integral membrane protein complex (in thylakoids of oxygenic organisms this is called $CF_0$ and consists of four different polypeptides) and a multisubunit extrinsic membrane protein complex (called $CF_1$ and containing five different polypeptides). ATP synthetases of bacterial and chloroplast thylakoids are very similar to the enzyme complex associated with the mitochondrial inner membrane.

The intrinsic complex ($CF_0$) provides a channel for the transport of protons through the membrane, while the extrinsic complex ($CF_1$) contains the catalytic sites for the phosphorylation of ADP (Figure 3). In the presence of a sufficiently large transthylakoid proton electrochemical potential difference, protons will pass through the coupling factor, providing free energy to drive ATP synthesis. In oxygenic photosynthesis, protons accumulate in the lumen of the thylakoid as a result of photosynthetic electron transport, and the extrinsic $CF_1$ is located on the opposite (stromal) side of the membrane, where ADP and phosphate are available in the stroma of the chloroplast (Figure 3). Depending on the thermodynamic conditions, the ATP synthetase can operate in reverse and utilize free energy released in ATP hydrolysis to translocate protons across the membrane from the stroma to the thylakoid lumen. This situation is unlikely to occur in the light, however, when photosynthetic electron transfer reactions are coupled to proton transport across the membrane.

## 5 SUMMARY

Photosynthetic energy transduction in all organisms can be divided into four discrete processes:

1. Capture of light by antennae and the transfer of this excitation energy to reaction centers.
2. Photochemical redox reactions at the reaction centers that result in the separation of electrical charge.
3. Electron and proton transfer reactions that generate a proton electrochemical potential difference across the membrane.
4. The synthesis of ATP, driven by the transport of protons through the ATP synthetase.

*See also* Cytochrome P450; Electron Transfer, Biological.

### *Bibliography*


Deisenhofer, J., and Michel, H. (1989) *Science,* 245:1463–1473.

Golbeck, J. H. (1992) *Annu. Rev. Plant Physiol. Plant Mol. Biol.* 43:293–324.

Hansson, Ö., and Wydrzynski, T. (1990) *Photosynth. Res.* 23:131–162.

Nicholls, D. G., and Ferguson, S. J. (1992) *Bioenergetics 2.* Academic Press, London, and San Diego, CA.

Ort, D. R., and Oxborough, K. (1992) *Annu. Rev. Plant Physiol. Plant Mol. Biol.* 43:269–291.

Staehelin, L. A., and Arntzen, C. J., Eds. (1986) *Encyclopedia of Plant Physiology,* New Series, Vol. 19, *Photosynthesis III.* Springer-Verlag, Berlin and Heidelberg.

Taiz, L., and Zeiger, E., Eds. (1991) The light reactions. Chapter 8 in *Plant Physiology.* Benjamin/Cummings, Redwood City, CA.


# Plant Cells, Genetic Manipulation of

*Roland Bilang and Martin Schrott*

### *Key Words*

**Gene Transfer** Delivery of naked DNA or RNA to a cell.

**Totipotency** Capability of a somatic cell to develop into a differentiated complete organism.

**Transformation** Gene transfer followed by transient or sustained expression of the genetic information.

**Transgenic plants** Plants with transferred genetic information that is stably maintained during mitotic and meiotic cell divisions.

Genetic engineering of plant cells is based on techniques for the delivery into plant cells of functionally defined natural or synthetic nucleic acid sequences. Somatic plant cells can be competent for functional and stable integration of foreign DNA into their own genome as well as for development into differentiated tissue or a complete organism, thus yielding the biological basis for the production of chimeric or fully transgenic plants.

Transfer of nucleic acids to plant cells and production of transgenic plants are routine tools for basic research on plant biology, as well as for plant gene identification and isolation, and the analysis of gene regulation and function. Transgenic crop plants with newly added traits play an important role in the development of concepts for product quality improvement, integrated pest management, and sustainable farming.

## 1 PRINCIPLES

### 1.1 Physiology of the Recipient Cell

Depending on their purposes, experiments require different levels of competence of a cell that has received genetic information. For the analysis of gene regulation and function, competence for uptake and transient expression of the transferred DNA or RNA may be sufficient. For the production of transgenic tissue, a transformed cell must allow the replication of the transferred nucleic acids, either as episomes or as genomic inserts (nonintegrative and integrative transformation, respectively). For the production of fully transgenic plants, transgenic cells must be totipotent, that is, competent for regeneration into a complete, differentiated organism. The competence of a cell for integrative transformation and regeneration depends on its genotype and on its developmental and physiological state. The physiological state may be influenced by plant culture techniques, and it is possible to shift cells from a potentially competent to a competent state by mechanical or enzymatic wounding or by treatment with phytohormones.

### 1.2 Characteristics of ransferable Nucleic Acids

Before being introduced into plant cells, nucleic acids are amenable to rearrangement by in vitro recombination techniques. Strong, constitutive expression signals of plant, bacterial, or viral origin are available. Other promoter sequences confer tissue- or development-specific expression of the introduced genes, or they are inducible by various environmental stimuli. Regulatory and protein-encoding sequences are combined in transcriptional or translational gene fusions. These sequences are delivered to plant cells as parts of plasmid or cosmid vectors, in bacteriophages, or on yeast artificial chromosomes. In self-replicating vectors, components of plant virus genomes provide signals for extrachromosomal replication of the introduced DNA or RNA. Direct gene transfer methods (see Section 2) also allow the introduction of native, noncloned nucleic acids.

**Table 1** Comparison of the Most Widely Used Methods for the Delivery of Nucleic Acids into Plants

| Method | Advantages | Disadvantages |
|---|---|---|
| *Agrobacterium*-mediated gene transfer | High efficiency<br>Insertion of long, nonrearranged DNA stretches<br>Mostly single- or few-copy integration<br>Does not require special equipment | Limited host range<br>Genetic information can be delivered only as T-DNA insert |
| Chemical methods and electroporation | Simple manipulations<br>High efficiency<br>Allows treatment of many samples in one experiment<br>Chemical methods do not require special equipment | High degree of donor DNA rearrangement and multicopy integration events<br>Chemical methods require protoplast culture and regeneration systems for production of transgenic plants<br>Enhanced risk of induction of somaclonal variation |
| Microinjection | Genotype independent<br>High flexibility to reach different target structures<br>Allows visual control of manipulations | Sophisticated and expensive equipment required<br>Much expertise required |
| Ballistic methods | Genotype independent<br>High flexibility to reach different target structures<br>High promise for routine production of transgenic cereals | Sophisticated equipment required<br>High degree of donor DNA rearrangement and multicopy integration events |

### 1.3 Characteristics of the Delivery Systems

Transformation techniques for plants provide the means to let nucleic acids pass the cell wall, plasma membrane, and nuclear envelope without hampering the viability of the target cell. Biological gene transfer systems use the plant pathogens *Agrobacterium tumefaciens* or *A. rhizogenes* as natural vectors. Direct (i.e., vectorless) gene transfer systems include chemical methods, electroporation, microinjection, and biolistics. These most widely and successfully used gene transfer techniques are compared in Table 1, showing their advantages and disadvantages with respect to the production of transgenic plants.

There is no universal gene transfer method suitable for any kind of transformation program. Other gene transfer techniques are still being developed, as well as systems that combine components of more than one of the described methods.

## 2 DNA DELIVERY METHODS

### 2.1 *Agrobacterium*-Mediated Gene Transfer

*A. tumefaciens* and *A. rhizogenes* are capable of transferring a stretch of DNA, the T-DNA, from their large tumor (Ti)- or root (Ri)-inducing plasmids, respectively, to the nuclear genome of host plant cells. Several gene products of the T-DNA interfere with the regulation of the host cell development, thus leading to tumor or enhanced root hair formation. Virulence and host range of a bacterial strain depend on constitutively expressed bacterial chromosomal genes and of *vir* genes located on the Ti or Ri plasmids. Upon induction by plant factors (e.g., acetosyringone), the *vir* system mediates T-DNA excision, transfer to the plant cell, and possibly integration into the nuclear genome. T-DNA transfer occurs to protoplasts or wounded organized tissue. Short border repeats of 24 bp flank the T-DNA; they are the only T-DNA components that need to remain intact for its transfer.

Disarmed vectors have been developed for use in *A. tumefaciens* which, upon release of the T-DNA into the plant cell, do not cause tumor formation, since the native T-DNA genes from *A. tumefaciens* involved in disease symptom development were removed and can be replaced by genes of interest. In binary vectors, a modified T-DNA is located on a separate, *Escherichia coli*–compatible plasmid, and the necessary *vir* functions are provided *in trans* from a Ti-plasmid devoid of T-DNA.

Agroinfection is *Agrobacterium*-mediated transfer to plants of cloned, functional (i.e., replicating) viral DNA.

### 2.2 Chemical Methods and Electroporation

Protoplasts, i.e., cells depleted of the cell wall by enzymatic digestion, can be isolated from callus material, cell suspension cultures, and a wide range of differentiated plant tissues. For genetic transformation, protoplasts can be coincubated with the nucleic acids and chemical substances (e.g., polyethylen glycol, polyvinyl alcohol, calcium phosphate) that lead to reversible permeabilization of the plasma membrane. Nucleic acids are also delivered to protoplasts after encapsidation in liposome vesicles followed by fusion with the plasma membrane or endocytosis.

Changes in membrane permeability for gene transfer can also be induced by high voltage electric fields (electroporation). Electroporation is applicable to protoplasts, isolated cells, and organized tissue.

### 2.3 Microinjection

Microinjection allows the introduction of nucleic acids under microscopic control into subcellular compartments of protoplasts, isolated cells, and cells in multicellular structures such as calli, meristems, and embryos. Recipient cells or tissues are immobilized in sodium alginate, agarose, or poly-L-lysine, or by a holding capillary system. Nucleic acids can be delivered directly to the nucleus with the help of a glass capillary connected to a micromanipulator system. Various culture systems for single cells or small multicellular structures (single cell culture, hanging droplet culture, nurse culture) are used in combination with the microinjection technique.

### 2.4 Ballistic Methods

In ballistic transformation methods nucleic acids associated with gold or tungsten particles in the micrometer range (microcarriers) are transported through cell walls. The microcarriers are accelerated directly or by the help of projectile carriers (macrocarriers). Motive forces are provided by gun powder, by compressed inert gases, or by electric discharge through a small water droplet. Single cells, tissues, or whole organisms are targeted inside a partially evacuated chamber. Various modifications of these basic principles were employed to meet the requirements of many different transformation problems, allowing the fine-tuning of the motive forces and the controlled delivery of the microprojectiles to the target.

## 3 ANALYSIS OF TRANSGENIC PLANT MATERIAL

### 3.1 Analysis of Foreign Gene Products

A number of gene expression vectors have been developed which confer a distinctive phenotype to the recipient cells and therefore allow the positive selection or visual screening of transformed cells, cell lineages and transgenic organisms (Table 2). Activity of the foreign gene products are determined either in situ in transformed tissue or with in vitro enzyme assays of plant extracts. The gene products can be detected directly by immunological methods in vitro (Western blot) or in situ.

### 3.2 Physical Analysis of Foreign DNA

The presence and physical organization of foreign DNA in plant cells is monitored by Southern blot analyses of genomic plant DNA and polymerase chain reaction techniques. Integration of foreign DNA into plant nuclear genomes occurs predominantly at random sites via illegitimate recombination events. Depending on the transformation system used, structural rearrangements of integrated DNA and integration of multiple copies at one or several genetic loci can be frequent. As a rule, foreign DNA remains stably integrated for many generations. However, intrachromosomal homologous recombination between copies of introduced DNA can be a source of physical instability of the locus.

Transcripts of the foreign genes are detected by Northern blot, RNase protection, or $S_1$ protection analysis. Transcription can be affected by neighboring DNA sequences, leading to different expression levels in individual transgenic plants. Initially transcribed foreign DNA can be silenced; that is, the transcription can be inactivated.

## 4 APPLICATIONS

### 4.1 Basic Research

Gene transfer systems have a major impact on the investigation of the regulatory mechanisms of plant development. Mobile genetic elements (transposons) and T-DNA can serve as insertional mutagens, allowing the identification and isolation of coding and regulatory sequences of plant genes. For instance, transgenic plants serve as analytical tools to describe metabolic pathways, to study cis- and trans-acting factors that control gene function, and to understand the mechanisms of plant response to environmental stresses.

### 4.2 Agricultural and Industrial Production

A large number of transgenic mono- and dicotyledonous species have been developed, among them important crop plants such as wheat, rice, maize, soybean, potato, cotton, and alfalfa. Engineered agronomic traits cover tolerance to biotic and abiotic stresses; genetically engineered plants with increased resistance to herbicides, pest damage, and viral, bacterial, and fungal diseases have been produced and are ready to be introduced into agriculture in the next future.

Improvement of crop quality by genetic engineering aims at increasing the nutritional value of food and feed, reducing postharvest losses, and improving suitability and enlarging the spectrum for processing. For these purposes, plants and plant cell cultures are genetically engineered for altered quantity and composition of endogenous products (e.g., storage proteins, carbohydrates, fatty acids) and for the production of new compounds of plant or nonplant origin (e.g., biopolymers or pharmaceutical substances).

**Table 2** Phenotypic Selection and Screening of Transformed Plant Cells

| Marker Gene Product | Organisms of Origin | Phenotype |
|---|---|---|
| Neomycin-phosphotransferase | *E. coli* | Resistance to kanamycin and geneticin |
| Hygromycin-phosphotransferase | *E. coli* | Resistance to hygromycin |
| Streptomycin- and spectinomycin-phosphotransferase | *E. coli* | Resistance to streptomycin and spectinomycin |
| Mutant dihydrofolate-reductase | Mouse *(Mus musculus)* | Resistance to methotrexate |
| Chloramphenicol acetyltransferase | *E. coli* | Resistance to chloramphenicol |
| Phosphinothricin acetyltransferase | *Streptomyces* spp. | Resistance to phosphinotricine |
| Mutant acetolactate synthase | *Arabidopsis thaliana,* Tobacco *(Nicotiana tabacum)* | Resistance to sulfonylureas |
| β-Glucuronidase | *E. coli* | Formation of colored or fluorescent compounds |
| β-Galactosidase | *E. coli* | Formation of colored compounds |
| Luciferase | Firefly *(Photinus pyralis); Vibrio harveyi* | Light emission |
| Diverse regulatory factors involved in anthocyanin biosynthesis | *Zea mays* | Augmented anthocyanin pigmentation |

## 5 PERSPECTIVE

Future successful application of genetically engineered plants relies on the comprehension of the biological processes underlying a desired phenotype. Relevant genes need to be identified and isolated, and the regulatory mechanisms that control their expression must be revealed. Transgenic plants have become a valuable tool to resolve such tasks. The development of simple and efficient techniques for routine and genotype-independent production of transgenic plants from any desired crop species remains another important goal for applied research.

*See also* PLANT CELL TRANSFORMATION, PHYSICAL METHODS FOR; TOMATOES, GENE ALTERATIONS OF.

### *Bibliography*

Christou, P. (1992) Genetic transformation of crop plants using microprojectile bombardment. *Plant J.* 2:275–281.

Fraley, R., and Schell, J., Eds. (1992) Plant biotechnology. *Curr. Opinion Biotechnol.* 3:139–184.

Hooykaas, P.J.J., and Schilperoort, R.A. (1992) *Agrobacterium* and plant genetic engineering. *Plant Mol. Biol.* 19:15–38.

Paszkowski, J., Saul, M.W., and Potrykus I. (1989) Plant gene vectors and genetic transformation: DNA-mediated direct gene transfer to plants. *Cell Cult. Somatic Cell Genet. Plants,* 6:51–68.

Potrykus, I. (1990) Gene transfer to cereals: An assessment. *Biotechnology,* 8:535–542.

# PLANT CELL TRANSFORMATION, PHYSICAL METHODS FOR

*James Oard*

### *Key Words*

**Macrocarrier** Thin, lightweight membrane or disk on which DNA-coated particles are placed for acceleration and penetration into intact cell walls.

**Marker Gene** Cloned gene transferred to cells that confers tolerance to specific antibiotics or herbicides and is used to identify transformed cells.

**Microparticles** Gold spheres or tungsten powder about 0.5 to 3 μm in diameter that are coated with nucleic acids and accelerated under vacuum for delivery to intact cells.

**Particle Bombardment** Physical method of gene transfer by rapid acceleration of DNA-coated microparticles that pierce intact cells and tissues.

The transformation of plant cells by physical methods involves the delivery and expression of nucleic acids in individual cells and organized tissues. This method has been referred to as the particle acceleration method, microprojectile bombardment, the gene gun method, and the biolistic process. The introduction of cloned DNA into living plant cells has allowed significant advances in basic studies of gene structure and function. The physical nature and broad application of this technique have led to the use of several different plant species such as tobacco, maize, and rice in molecular studies of tissue-specific expression and developmental regulation of genes that are difficult or impossible by other approaches.

For certain dicotyledonous plants in the Solanaceae family, the soil bacterial pathogen *Agrobacterium tumefaciens* can serve as a vector for stable gene transfer. However, this strategy has failed with few exceptions for monocotyledonous plants and many species outside the Solanaceae. Physical methods may therefore be particularly useful when gene transfer is unsuccessful by chemical, electrical, or other means. Moreover, the simplicity and facile implementation of this technique allows for rapid assessment of cloned genes over a few days instead of the weeks or months that otherwise might be required.

## 1 PRINCIPLES AND OVERVIEW OF PARTICLE ACCELERATION

The basic approach to physical gene transfer is to accelerate gold or tungsten microparticles coated with nucleic acids for nonlethal penetration of cell walls and membranes. Several different devices have been constructed for particle acceleration. The most common and efficient devices accelerate particles to high velocities by pressurized helium gas or energy released by a water droplet exposed to high voltage. Principal components of the helium pressure system (Figure 1) are gas acceleration tube, rupture disk, macrocarrier with DNA-coated microparticles, stopping screen, and target cells. All components are enclosed in a chamber (not shown) to create a partial vacuum, which helps reduce aerodynamic drag on the accelerated particles.

For operation of this device, a vacuum is first created in the chamber and helium gas under pressure is allowed to enter the gas acceleration tube. Helium is used because it is a light gas that rapidly expands under pressure; moreover, it is readily available and relatively nontoxic to plant cells under short exposure times. Sufficient pressure is created in the tube to break the rupture disk and allow the helium shock wave that is generated to accelerate the macrocarrier and DNA-coated microparticles. The macrocarrier is rapidly accelerated until it is retained by the stopping screen. The microparticles pass through the screen and continue toward the target cells.

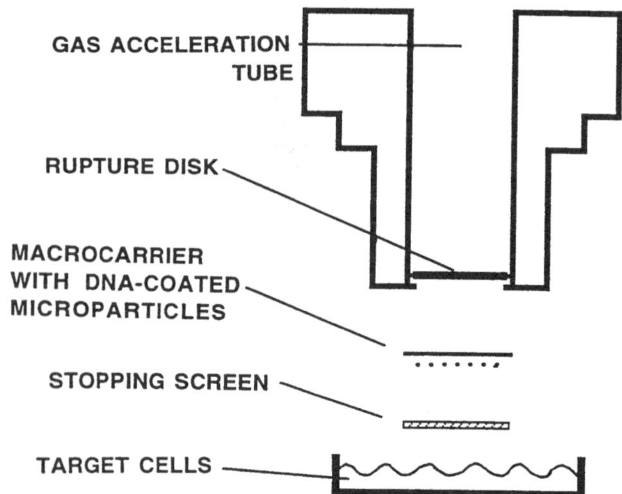


**Figure 1.** Diagram of helium pressure device showing principal components for acceleration of DNA-coated particles. (Redrawn with permission of Bio-Rad Laboratories.)

## 2 TECHNIQUES

### 2.1 Optimizing Vacuum, Helium Pressure, and Velocity

Efficient operation of the helium device as shown in Figure 1 requires removal from the chamber of air, which otherwise would reduce velocity of the DNA-coated particles. In addition, the shock wave of helium appears to inflict less damage to bombarded cells than would have been caused by air molecules in the chamber. A partial vacuum of about 710 to 740 mmHg is the range found to be adequate for most cell types, but the optimum level must be empirically determined for each specific case.

The pressure of helium in the gas acceleration tube is an important factor that helps determine the velocity of the microparticles. Pressures from 450 to 2200 psi are used to accelerate particles, depending on the nature of the material to be treated. For most cells 1000 psi approaches optimal levels, and at higher pressures the helium shock may increase trauma and reduce transformation efficiency. Other variables that can be adjusted to optimize velocity include distance the shock wave travels between rupture disk and macrocarrier, distance between macrocarrier and stopping screen, and distance between stopping screen and target cells.

### 2.2 Macrocarriers and Microparticles

For the helium pressure apparatus, a 2.5 cm diameter plastic (Kapton) membrane, 0.06 mm thick, is used as a macrocarrier. Its small mass permits the membrane to be rapidly accelerated over short distances without creating any debris on impact with the stopping screen, which could damage cells or tissues. The membrane is used only once, being discarded after each bombardment.

Tungsten and gold microparticles coated with DNA are accelerated to penetrate intact cell walls and membranes. The tungsten particles are available in average sizes of 0.5 to 2.0 μm in diameter. The optimum size will depend on the diameter of the target cells, and in most cases, an average size of 1.0 μm has been used for successful transient and stable transformation. The disadvantages of tungsten are irregular particle shape and size, potential toxicity to certain cell types, and surface oxidation, which may reduce precipitation of DNA. Tungsten will also bind together in clumps after addition of DNA, resulting in uneven dispersal of particles over target cells.

In certain cases gold particles are utilized as an alternative to tungsten, to take advantage of the greater uniformity in size (available range is 1–3 μm) and the reduced toxicity. Principal disadvantages of gold are the relatively high cost and the greater variation in efficiency of coating particles, compared with tungsten.

Devising proper methods of coating or precipitating DNA or RNA onto microparticles is one of the most crucial factors for efficient gene transfer. Various protocols have been developed for coating, but virtually all combine a predetermined amount of gold or tungsten (1.25–18 mg) with plasmid DNA (0.5–70 μg), $CaCl_2$ (0.25 M–2.5), and spermidine (0.1 M). This solution is vortexed continually throughout the coating procedure to ensure a uniform reaction mixture. After precipitation of DNA, the microparticles are transferred to the Kapton membranes and allowed to dry just before use.

The distance the microparticles must travel to reach the target cells is another important factor for maximum transfer efficiency. Although each cell type must be tested for optimized conditions, a 10 mm flight distance results in high transformation rates for many species tested. Under certain circumstances, a 100 μm stainless steel mesh can be placed between the stopping screen and the target cells to reduce shock-generated trauma and improve dispersal of microparticles.

### 2.3 Marker Genes for Transformation

The *gus* A (*uid* A) gene from *Escherichia coli* can be used in molecular studies of transient expression, tissue-specific expression, and developmental regulation. The *gus* A gene encodes for the enzyme β-glucuronidase, which cleaves the substrate 5-Br-4-Cl-3-indolyl-β-D-glucuronide (X-Gluc) when taken up by transformed cells. A particular feature of the *gus* A marker system is that the cleaved substrate is converted into an insoluble blue precipitate at the site of enzyme activity and may be visible to the naked eye. An example of *gus* A expression in embryogenic rice callus is shown in Figure 2.

The *bar* gene, an effective selectable marker from *Streptomyces hygroscopicus,* is used primarily for stable transformation experiments. Phosphinothricin is a herbicidal compound that inhibits activity of glutamine synthetase and leads to a lethal accumulation of ammonia in sensitive untransformed cells. The *bar* gene codes for phosphinothricin acetyltransferase, which detoxifies phosphi-

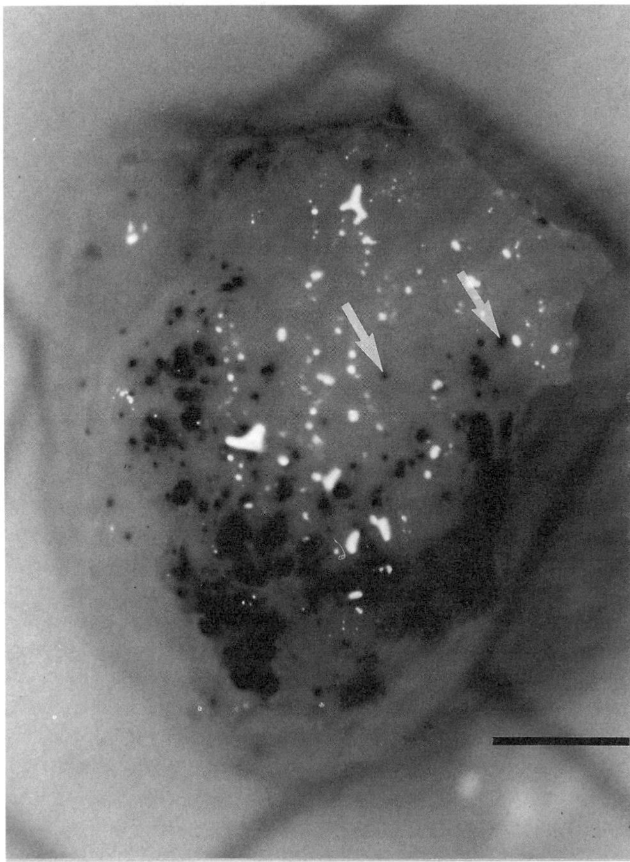

**Figure 2.** Transient expression of *gus*A gene in embryogenic rice callus 48 hours after particle bombardment. Arrows indicate transformed blue cells, which appear black in this print of the original color slide. Scale bar = 0.10 cm. [Reprinted with permission from Oard et al., *Plant Physiol.* 92:334–339 (1990).]

Figure 3. Transgenic rice plant with *bar* gene (left), which is tolerant to spraying of leaves with 500 ppm of the herbicide Basta. Untransformed sensitive control plant (right) is dead after the same herbicide treatment. (Reprinted with permission of Dr. Paul Christou, Agracetus, Inc.)

Figure 4. Growth of stably transformed tobacco cells in presence of the antibiotic kanamycin. Scale bar = 0.30 cm. (Reprinted with permission of Dr. John Sanford, Cornell University.)

nothricin by acetylation. Stable transformants that express the *bar* gene grow at normal or near-normal rates after exposure to lethal doses of phosphinothricin, while sensitive material stops growth, becomes necrotic, and dies within 10 to 21 days. Transgenic rice that expresses the *bar* gene is shown in Figure 3.

Another useful selectable marker from *S. hygroscopicus* is the hygromycin resistance gene that codes for hygromycin phosphotransferase. The enzyme acts by phosphorylation to deactivate the aminocyclitol antibiotic hygromycin B.

For several dicotyledonous plants, the kanamycin gene has been used effectively to transform cell cultures or plants. The gene was obtained from the bacterial Tn5 transposon, which codes for neomycin phosphotransferase. Resistance to the aminoglycoside antibiotics kanamycin, neomycin, and G418 occurs through phosphorylation. Transformed tobacco cells expressing the kanamycin gene are shown in Figure 4. Monocotyledonous plants, on the other hand, are tolerant of high levels of kanamycin and therefore may not be useful in selection of transformed cells.

## 2.4 Cell Manipulation and Selection of Transformants

Several studies in different plant species indicate that for maximum DNA uptake and stable gene expression, cells must be receptive or competent. In most cases actively dividing cells in the early log growth phase produce the highest rates of transformation. Cells at the mid-log phase or later generally result in significantly fewer transformants. The number of times a cell suspension is subcultured just before bombardment also can affect transformation efficiency.

The physical arrangement of cells subjected to bombardment is an important factor in this technique. For liquid cultures, a single layer of cells evenly distributed in a dish is the optimum scheme, providing the greatest number of targets. For liquid tobacco cultures, 5 mL of suspension in early log phase may be placed onto a 7 cm diameter filter paper for bombardment that can result in a stable transformation rate of approximately 1%. This rate can be increased by about 2- to 10-fold by placing tobacco cells on bombardment medium whose osmolarity is adjusted between 400 and 700 mOsm/kg $H_2O$. One day after bombardment, the osmolarity is reduced over a two-day period. This treatment appears to inhibit the growth of untransformed cells and to increase consistency in transformation frequencies over different experiments.

One to two days after bombardment, transient assays for expression of the *gus* A gene may be performed. This type of experiment is often conducted to evaluate efficiency of promoters or other elements that are fused to the *gus* A coding region. The procedure is also useful when testing for expression of promoters in different tissues or at different growth stages.

During transient assays the *gus* A or other reporter genes may be expressed in the nucleus, but these genes may not be integrated into the chromosomes. Therefore, the transient assay is not a reliable indicator for stable transformation. For selection of stably transformed cultured cells expressing the *bar* gene, phosphinothricin can be added to selection media at about 1 to 10 mg/L from zero to 2 weeks after bombardment. In certain formulations, phosphinothricin forms part of the compound bialaphos, which is applied to media at 5 to 20 mg/L, zero to 10 days postbombardment. This

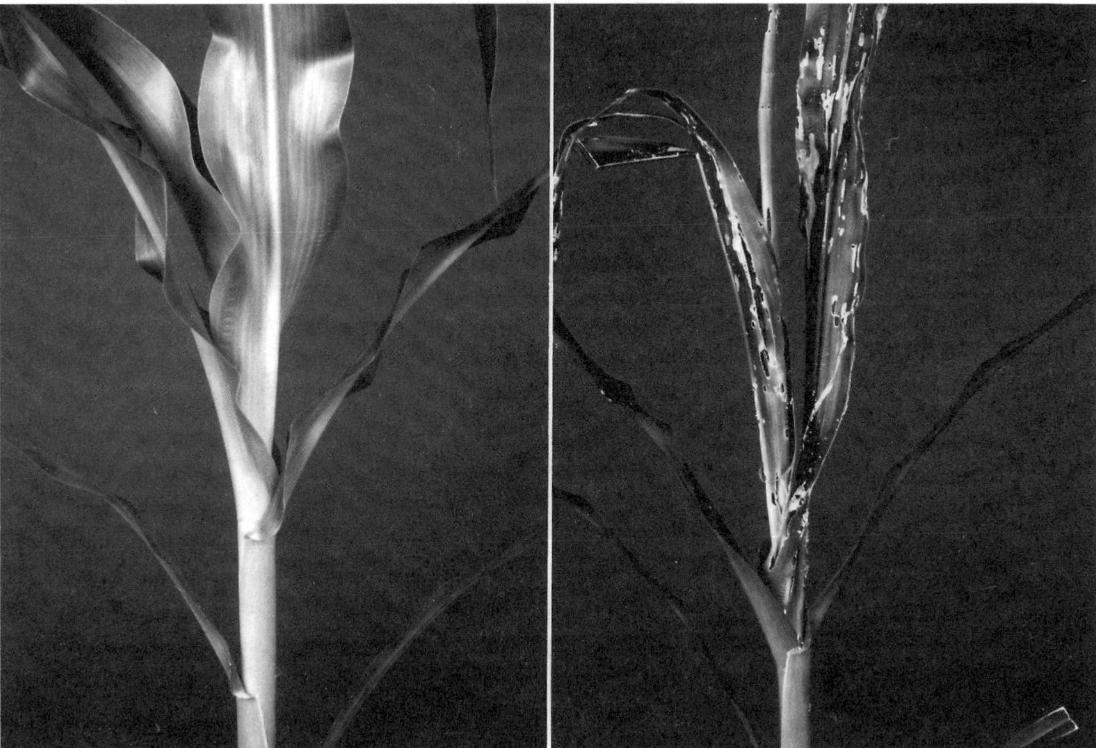

Figure 5. Resistance to European corn borer in transformed maize plant (left) with *Bt* gene. Untransformed sensitive plant (right) showing extensive feeding damage on leaves. (Reprinted with permission of Dr. Michael Koziel, Ciba-Geigy Corp.)

general strategy was used to select transformed rice (Figure 3). The optimum rate and timing of phosphinothricin and bialaphos must be empirically determined when new cell lines or tissues are used.

Hygromycin B is used at about 10 to 100 mg/L to select transformants that carry the hygromycin resistance gene. Kanamycin sulfate is an effective selection agent at about 50 to 350 mg/L for tobacco and certain other dicotyledonous plants, but it should not be used for monocotyledonous species.

## 3 APPLICATIONS IN BASIC AND APPLIED RESEARCH

Physical gene transfer using the biolistic approach is a simple and rapid method to introduce DNA into plant cells that may be unresponsive to other techniques. Tissues or cells that have been transformed by this method include apical meristems, embryogenic calli, cell suspensions, coleoptiles, immature and mature embryos, leaf blades, mature pollen and microspores, and petal, root, and stem sections.

Stable transfer of cloned genes into the nucleus has been achieved in the following plant species: *Arabidopsis thaliana,* cotton, cranberry, maize, papaya, poplar, rice, sorghum, soybean, sugarcane, tobacco, wheat, and white spruce. These plants have also been subjected to physical gene transfer to study the action of structural and regulatory genes, the mechanical infection of viruses, and tissue specificity of promoter elements. In addition, this process has enabled the stable transfer and expression of genes into intact mitochondrial and chloroplast genomes of tobacco and the unicellular *Chlamydomonas.*

Aside from basic genetic and molecular studies, physical gene transfer has proven successful in transfer of genes for practical considerations. Two examples are transgenic maize carrying the *Bt* gene, which confers tolerance to the European corn borer *(Ostrinia nubilalis)* (Figure 5) and herbicide-tolerant rice expressing the *bar* gene (Figure 3).

## 4 PERSPECTIVES FOR IMPROVEMENT OF PHYSICAL GENE TRANSFER

Successful experiments with physical gene transfer methods have clearly demonstrated the utility of those procedures in both basic and applied plant research. In spite of these results, certain areas need refinement for increased transformation efficiency.

The ideal physical gene transfer would involve the delivery of at least one DNA-coated microparticle to each target cell, an improvement that could dramatically enhance transformation rates. Current technology delivers particles to about 15% of the cells, so research to improve dispersal and reduce trauma of microparticles is urgently needed.

A related issue is the method and efficiency of DNA precipitation onto microparticles. The $CaCl_2$/spermadine coating procedure is sensitive to slight changes in conditions, which can lead to inconsistencies over different experiments. Clearly, more work is needed in this area to improve stability in gene delivery. Finally, greater control of velocity and penetration of microparticles in specific cells and tissues would expand the range and finesse of this method. These and other challenges must be addressed before the particle acceleration technique can be truly accepted as a reliable tool for plant biologists.

*See also* Plant Cells, Genetic Manipulation of.

### *Bibliography*

Birch, R. G., and Franks, T. (1991) Development and optimization of microparticle systems for plant genetic transformation. *Aust. J. Plant Physiol.* 18:453–469.

Christou, P. (1992) Genetic transformation of crop plants using microprojectile bombardment. *Plant J.* 2:275–281.

Klein, T. M., Arentzen, R., Lewis, P. A., and Fitzpatrick-McElligott, S. (1992) Transformation of microbes, plants, and animals by particle bombardment. *Bio/Technology,* 10:286–291.

Sanford, J. C., Smith, F. D., and Russell, J. A. (1993) Optimizing the biolistic process for different biological applications. *Methods Enzymol.* 217:483–509.

# Plant Gene Expression Regulation

*Eric Lam*

### *Key Words*

**Antisense RNA** Transcripts that contain the complementary sequence to another RNA species.

**Biolistic Method** Method for the delivery of DNA-coated particles into living cells by mechanical force generated either in a chemical explosion or by a regulated burst of inert gas.

**cis-Acting Element** DNA sequence that can influence the expression of nearby gene(s). Usually this interaction requires a physical linkage between the element and the gene(s) it regulates.

**Cosuppression** The reduction in expression levels, usually at the level of RNA, of a particular gene or gene family by the expression of a transgene that contains the same coding region. The basis of this phenomenon is unclear at present.

**Ti Plasmid** Extrachromosomal DNA in the soil bacterium *Agrobacterium tumefaciens* that contains the functions required for the mobilization and transfer of DNA (T-DNA) into plant cells.

**trans-Acting Factor** Usually refers to protein factors that can influence gene expression by their interaction with cis-acting elements.

**Transgenic Plants** Plants whose nuclear genome has stably incorporated one or more pieces of foreign DNA introduced by an investigator.

The study of plant gene regulation has undergone a tremendous surge since 1988. Promoters and other cis-acting sequences that influence gene expression have been studied extensively in transgenic plants and by rapid transient assays in vitro. The identification of functional cis elements set the stage for the characterization of DNA-binding proteins that may be responsible for their activities. Advances in gene cloning techniques via transposon tagging and direct screening of expression cDNA libraries have resulted in the isolation of numerous genes encoding plant trans-acting factors. These methods bypass the more laborious approach of biochemical purification of regulatory proteins that may be present at very low levels in the nucleus. The isolation of tapetum-specific promoters has allowed the production of transgenic plants that are male sterile through the expression of an RNase gene in the particular reproductive tissues. This type of approach will facilitate the economical production of hybrid seeds in the near future. Alternatively, suppression of an endogenous gene using antisense RNA can be used to create transgenic tomato varieties that are inhibited in the production of the plant hormone ethylene and do not ripen during fruit maturation. These findings serve to illustrate the potential commercial applications that can result from the specific manipulation of gene expression in plants. Detailed studies of the role and regulatory pathways of trans-acting factors on plant development will likely open up more diverse and exciting opportunities for the controlled manipulation of plant morphology and properties.

## 1 PRINCIPLES

### 1.1 Compartments of Gene Expression in Plants

Plant cells contain two structures that are absent in their animal counterparts. The cell wall is a specialized structure that forms a mechanical barrier which prevents cell movement within a plant. The plastids are organelles, which in their many different forms in various plant tissues perform essential functions such as photosynthesis and lipid biosynthesis. Like mitochondria, plastids also contain a set of genetic information that is distinct from that of the nucleus. The plastid genome is consisted of about 120,000 base pairs of DNA and for three higher plant species, the complete sequence of the plastid genomes has been reported. Many of the essential genes involved in housekeeping functions such as protein synthesis are found on the plastid genome. However, the majority of the proteins found in the plastid are encoded in the nuclear genome and their protein products have to be imported. An analogous situation is found with the mitochondria as well. Thus, not only is transcription of genes taking place simultaneously in at least three different compartments within a plant cell, but proteins are also actively being sorted to their different locations from the cytoplasm.

The regulation of mitochondrial gene expression has not been well characterized in plants. This is because this genome apparently exists as a complex array of subgenomic and multimeric forms in plant species that have been examined, presumably the result of active recombination events. In contrast, the promoters and regulation of plastid genes are much better understood. In many respects, the promoters of plastid-encoded genes are very prokaryotic. They usually contain a ''-10''- and a ''-35''-like element upstream from their transcription start site. In addition, RNA polymerase holoenzyme from *E. coli* can utilize many of the plastid promoters effectively in vitro. However, there are also more subtle differences, such as the insensitivity of plastid transcription to the prokaryotic RNA polymerase inhibitor rifampicin and a greater dependence of plastid transcription on superhelicity of template DNA in vitro. From transcription rate assays with isolated plastids, many of the plastid-encoded genes are thought to be regulated at such posttranscriptional levels, as those controlling RNA stability or translational initiation. Thus, the abundance of a particular plastid-encoded pro-

Table 1 Types of Cloned Plant trans-Acting Factor

| DNA-Binding Motif | Gene Names | Species | Function[a,b] | Cloning Method[c] |
|---|---|---|---|---|
| Basic leucine–zipper | TGA1a, TGA1b, TAF1 | Tobacco | N.d. | DS |
| | TGA1, TGA2, TGA3 | *Arabidopsis* | N.d. | HP |
| | CPRF-1, 2, 3 | Parsley | N.d. | DS |
| | HBP-1a, -1b | Wheat | N.d. | DS |
| | EMBP-1 | Wheat | N.d. | DS |
| | OCSBF-1, -2 | Maize | N.d. | DS |
| | *Opaque* 2 | Maize | Seed protein synthesis | TT |
| Basic Helix–loop–helix | R, B | Maize | Anthocyanin synthesis | TT |
| Zinc finger | 3AF1 | Tobacco | N.d. | DS |
| | EPF-1 | Petunia | N.d. | DS |
| Homeodomain | *Knotted* | Maize | Leaf development | TT |
| Homeodomain–leucine–zipper (HD-ZIP) | Athb-1, Athb-2, | *Arabidopsis* | N.d. | HP |
| | HAT-4, -5, -22 | | | |
| | Zmhox1a | Maize | N.d. | DS |
| MADS box | *Agamous* | *Arabidopsis* | Flower development | TI |
| | AGL-1 to AGL-6 | *Arabidopsis* | N.d. | HP |
| | *Deficiens* | *Antirrhinum* | Flower development | TT |
| | *Globosa* | *Antirrhinum* | Flower development | HP |
| *Myb*-like | *Glaborous-1* | *Arabidopsis* | Trichome development | TI |
| | C1, P | Maize | Anthocyanin synthesis | TT |
| | 305, 306, 308, 315, 330, 340 | *Antirrhinum* | N.d. | HP |
| Trihelix | GT-1 | Tobacco | N.d. | DS |
| | GT-2 | Rice | N.d. | DS |

[a]Only well-established functions in plant development are listed.
[b]N.d., not determined.
[c]DS, direct screening with labeled binding sites; HP, nucleic acid screening using homologous probes; TT, transposon tagging; TI, T-DNA tagging.

tein can vary greatly even though its apparent transcription rate is not altered significantly. Much of the field involved in understanding plastid gene regulation is now focused on the mechanisms of RNA turnover and translational initiation.

The nuclear genome of plants is in most respects very much like its counterpart in other eukaryotes. For example, a yeast transcription factor such as GAL4 can function properly in plant cells, while TGA1a, a tobacco transcription factor, can activate transcription in an in vitro system consisting of highly purified mammalian basal factors. The majority of the research in the field of plant nuclear gene regulation has been focused on the class of genes that encode mRNA, those that are transcribed by RNA polymerase II.

### 1.2 Plant Nuclear Transcription Factors

The absence of a facile in vitro transcription system for plant nuclear genes has hampered the biochemical characterization of plant transcription factors for many years. Recently, we have witnessed a tremendous growth in the number of plant DNA-binding proteins whose genes have been cloned. In some cases, the functional properties of the protein have been well studied in vitro with respect to its effect on transcription, while in others, the three-dimensional structure has been resolved. For genes that were first genetically defined by mutant phenotypes, the role of the encoded trans-acting factor in plant development is known, and genes that these factors regulate in vivo are also well characterized. Table 1 summarizes the different types of nuclear trans-acting factor that have been reported so far. It is striking that most of them falls into the same structural families contained as well by DNA-binding proteins from other eukaryotes. These DNA-binding motifs include basic leucine–zipper (bZIP), basic helix–loop–helix (bHLH), zinc fingers, homeobox, MADS box, and *Myb*-like. Interestingly, additional conserved motifs are found in families of plant DNA-binding proteins that have not been discovered in other systems. Two examples are the trihelical motif of the GT-1/GT-2 factors from tobacco and rice, and the HD-ZIP class of factors found in maize and *Arabidopsis thaliana.* Whether factors with DNA-binding domains of these types will be found in other eukaryotes remains to be seen. In contrast to the structural conservation noted in many DNA-binding domains, the activation domains of transcription factors are more difficult to predict based on their primary protein sequences. In plants, as in other systems, the ability of upstream factors to regulate transcription by RNA polymerase is thought to involve protein–protein interactions. It is likely that the activation domains of transcription factors mediate this process. Some of the common characteristics found in known activation domains are abundance of acidic amino acid residues, proline-rich regions, and serine/threonine-rich regions. The three-dimensional structure of one plant trans-acting factor, the TATA-binding protein (TBP) has been resolved recently by X-ray crystallography. A highly conserved factor, TBP is thought to mediate interactions between basal factors and RNA polymerase at the site of transcription initiation. The structural information from this study should facilitate our understanding of this process in the near future.

### 1.3 Promoter Architecture

Promoter analyses have been carried out for many plant genes. These studies usually are based on one of two approaches: characterization of nuclear protein binding sites on the promoter of interest, and definition of functional elements by testing various promoter constructs in an assay system involving living plant cells.

A combination of these two approaches is necessary to identify nuclear factors that may be functionally relevant to the activity of the promoter. A majority of plant promoter analyses that have been reported involve mRNA-encoding genes, and the general promoter architecture that emerges from these studies is one that contains multiple functional elements that interact with a multitude of nuclear factors. Thus, as in animal systems, the specific pattern of transcription activity of a promoter within a plant results from the particular arrangement of binding sites for several different trans-acting factors. Many plant promoters apparently share common motifs that are expected from in vitro studies to bind the same or similar factors. However, their functional properties in the different promoter context may be altered by their interaction with neighboring factors. Thus, a much more diverse pattern of expression can be generated from a relatively limited set of trans-acting factors. This combinatorial model of promoter architecture appears to be conserved among different eukaryotes, and there is still much to be learn about how trans-acting factors interact with each other and their functional consequences on promoter activity.

## 2 TECHNIQUES

### 2.1 In Vivo Assays of Promoter Functions

Promoter function assays in plants can be divided into two groups: those that involve transgenic plants and those that involve plant cells that have recently taken up foreign DNA. The majority of transgenic plant studies have been carried out in easily transformed species such as tobacco and petunia. The transformation procedure usually relies on the transfer of a transgene carrying a particular promoter construct via the soil bacterium *Agrobacterium tumefaciens.* Transformation vectors for this procedure are derived from the T-DNA of *A. tumefaciens,* and transgenes are stably incorporated into the nuclear genome. The activity of the transgenes can then be monitored in the regenerated plants under different conditions. In this way, many promoter elements that are involved in complex regulatory pathways such as light responsiveness or tissue specificity have been defined.

For plants that are difficult to transform, such as monocots, protoplasts isolated from the plants can be obtained by treatment with cell wall degrading enzymes. DNA can be introduced into plant protoplasts by a number of physical techniques including electroporation and polyethylene glycol treatment in the presence of $MgCl_2$. However, since protoplasts behave quite differently from differentiated plant cells, this approach is limited in its use to the definition of physiologically relevant promoter elements. More recently, the biolistic method has become the method of choice for rapid promoter analyses in plants. This technique involves the delivery of the particular DNA construct into plant cells by microprojectiles. As such, it is not limited to any particular plant species or tissue type. Even though only a small percentage of the plant tissues bombarded usually showed expression, this assay can be quantitative if appropriate reference genes are included.

### 2.2 Advances in Cloning Plant Regulatory Genes

Three approaches have been successfully applied toward cloning genes that encode trans-acting factors: (1) gene tagging using transposons or T-DNA insertion mutagenesis, (2) direct screening of cDNA expression libraries with radiolabeled binding sites, and (3) using heterologous genes as probes. The first two approaches have especially been successful in cloning novel DNA-binding proteins. With new advances in transposon tagging in heterologous plant species, it is to be expected that the number of tagged loci and cloned regulatory genes will increase in a rapid pace in the near future.

## 3 APPLICATIONS

### 3.1 Tissue-Specific Expression of Desirable Traits

Techniques such as subtractive hybridization (Figure 1) have been applied successfully in cloning genes that are expressed in a tissue-specific manner. For some of these genes, the promoter elements can confer the same pattern of expression on heterologous coding sequences. One striking example is the anther-specific expression of barnase, a bacterial gene encoding a ribonuclease. This chimeric gene specifically destroys the tapetal cells of the anther, thus resulting in male sterile flowers. This approach, which may facilitate generation of hybrid seeds from crop plants in the future, serves as a good example of the commercial application of a tightly regulated promoter element.

### 3.2 Suppression and Overproduction of Endogenous Genes

Constitutive expression of a gene in the antisense orientation offers an effective way of suppressing the transcript level of the endogenous gene. Although the mechanism for this antisense inhibition is not clear, it has been applied successfully to generate white flowers from red petunia, as well as tomatoes that do not ripen on the vine. Interestingly, overexpression of the sense RNA transcripts for some plant genes also has been found to result in a drastic drop in the level of expression for the particular gene. This phenomenon, called cosuppression, may also be used as a way to down-regulate plant gene expression. More predictable use of this approach, however, awaits the elucidation of the mechanisms involved.

For overproduction of desirable gene products in plants, a frequently used promoter is that of the 35*S* promoter from cauliflower mosaic virus. Many agronomically relevant traits, such as insect and viral resistance, have thus been engineered into commercially important plant species. The ectopic expression of important regulators, such as trans-acting factors that regulate gene expression or key enzymes involved in metabolic pathways, can also be used to study the functional role of these proteins in vivo. However, it became clear from studies in our laboratory, as well as in many others, that high levels of mRNA do not always lead to corresponding increases in the target protein. Thus, there are likely to be unknown feedback pathways by which the level of some key regulators may influence their own translation or turnover. Future studies on these processes will likely uncover novel cellular pathways that play important roles in determining the final level of gene products.

## 4 PERSPECTIVES

With the advent of new techniques for mapping and cloning interesting genes in higher plants, the future looks bright for a rich genetic resource for manipulating plant properties. However, a functional understanding of many of these genes during the life of a plant will likely depend on the ability of researchers to create targeted mutations as well as mosaic analyses with these identified loci in their plant of interest. These studies will ultimately allow

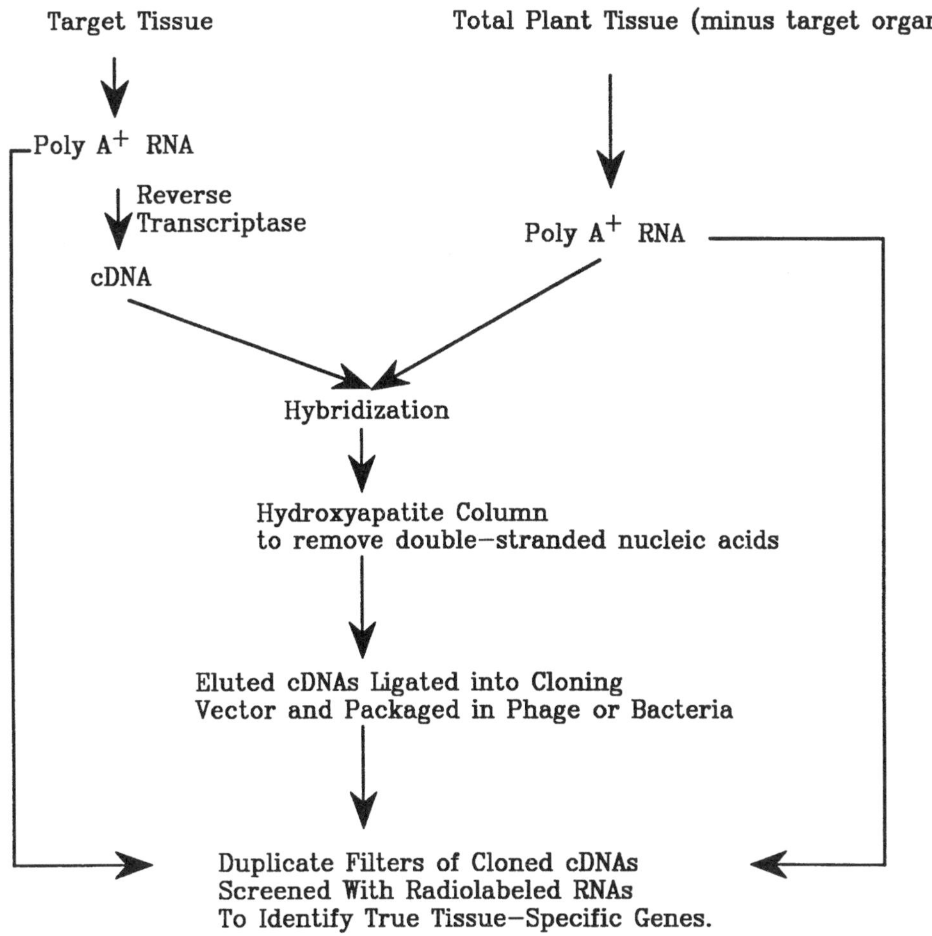


**Figure 1.** Flowchart for the isolation of tissue-specific genes from plant tissues by subtractive hybridization: Poly $A^+$ RNA, transcripts that contain polyadenylated 3′ ends and represent most of the mRNA species.

more predictable alteration of growth and development for important crop plants.

*See also* Plant Cell Transformation, Physical Methods for; Protein Designs for the Specific Recognition of DNA; Repressor-Operator Recognition; Tomatoes, Gene Alterations of; Zinc Finger DNA Binding Motifs.

## *Bibliography*

Benfey, P. N., and Chua, N.-H. (1990) The cauliflower mosaic virus 35*S* promoter: Combinatorial regulation of transcription in plants. *Science,* 250:959–966.

Grierson, D., Ed. (1992) *Developmental Regulation of Plant Gene Expression.* Blackie/Chapman & Hall, New York.

Glick, B. R., and Thompson, J. E., Eds. (1993) *Methods in Plant Molecular Biology and Biotechnology.* CRC Press, Boca Raton, FL.

Grierson, D., and Covey, S. N. (1988) *Plant Molecular Biology,* 2nd ed Blackie, Glasgow.

Katagiri, F., and Chua, N.-H. (1992) Plant Transcription factors: Present knowledge and future challenges. *Trends Biochem.* 8:22–27.

Mariani, C., De Beuckeleer, M., Truettner, J., Leemans, J., and Goldberg, R. B. (1990) Induction of male sterility in plants by a chimaeric ribonuclease gene. *Nature,* 347:737–741.

Theologis, A. (1992) One rotten apple spoils the whole bushel: The role of ethylene in fruit ripening. *Cell,* 70:181–184.

# Plant Pathology, Molecular

*Sarah Jane Gurr*

## *Key Words*

**Elicitors** The active components of cell-free extracts of microbial and plant origin that are capable of inducing defense responses when applied to plant tissues.

**Hypersensitive Reaction** Necrosis and death of an infected plant cell following penetration of the cell wall and establishment of contact with the host protoplast. A common defense reaction in response to incompatible interactions between a plant and a bacterium, virus, fungus, or nematode, where the necrotic tissue isolates the parasite from living plant tissue.

**Monogenic Resistance** Resistance effected by one or a few genes that control a major step in the interaction between a pathogen and host, where incompatibility usually results in a hypersensitive reaction. Also known as vertical or oligogenic resistance, this major form of gene resistance differentiates between races of a pathogen; that is, it is effective against specific races and ineffective against others.

**Polygenic Resistance** Resistance controlled by many genes, each of which alone is ineffective against a pathogen, although together they control the numerous steps involved in biochemical and structural defense responses in plants; also known as multigene or horizontal resistance.

---

Molecular plant pathology is the study of the molecular mechanisms underlying plant–pathogen interactions. Over the past century, great advances have been made toward understanding the biology, genetics, epidemiology, and control of plant disease, but it is the recent advances in molecular biology and the application of such techniques to plant and pathogen that permit us to begin to understand the complexity of such interactions at the molecular level. Control of plant disease in the future will be based on this knowledge and will be afforded by the insertion of disease resistance genes into crops or by engineering microorganisms to outcompete their pathogenic counterparts or to antagonize them.

## 1 INTRODUCTION

Plants have evolved a diversity of mechanisms by which they defend themselves against attack. In a *nonhost interaction,* the pathogen causes little host damage and is unable to replicate; in an *incompatible interaction,* where the plant is resistant and the pathogen is avirulent to the host tissue, damage is localized to a few cells, and this circumstance effectively confines the pathogen to its entry point. Host resistance can be *partial,* known as polygenic resistance (i.e., relying on the synergistic effect of many host genes) or *race specific,* (i.e., monogenic, governed by one or, rarely, a few genes). Race-specific resistance gives a high degree of protection but is unstable because the virulence of the pathogen population changes. In a *compatible interaction* between a susceptible host and a virulent pathogen, the outcome is disease, since the pathogen can multiply and spread throughout the plant. From a genetic perspective, resistance or susceptibility in the plant to avirulent or virulent races of the pathogen is determined by pairs of corresponding genes in the host and pathogen. This concept is known as the gene-for-gene hypothesis of plant–pathogen interactions.

## 2 RECOGNITION

Plant exudates can act as deterrents, or they can stimulate spore germination, cyst hatching, or growth of the pest or pathogen toward the plant surface. Upon arrival at the infection court, attachment to the host can be by structural or biochemical means, such as by the formation of adhesion pads or the production of a mucilaginous sheath. Alternatively, some pathogens, such as certain plant viruses, are directly transferred into the plant cell (e.g., by aphid or nematode vectors). Entry into the host can be through natural openings, such as stomata located by perceiving changes in topography of the leaf surface; through wounds, stimulated by host chemicals released as part of the wound response; or entry may be directly through the plant surface. While a few pathogens use mechanical force to directly penetrate plant tissue, most secrete enzymes that disrupt the structural components of the host cell. So, to penetrate a leaf cell enzymatically, a pathogen must sequentially pass through the host cuticle, which is sometimes overlaid with wax, and through laminated layers of cellulose, pectin, and hemicellulose.

## 3 HOST PHYSICAL AND CHEMICAL BARRIERS

Antipathogen plant defense strategies are based on preexisting or induced structural barriers, or on preexisting or induced toxic or inhibitory compounds.

### 3.1 Structural Defense

Preexisting plant structures such as thick surface hairs or thick cuticles act as effective deterrents to some pathogens. Alternatively, various defense structures can be induced in response to attack. These effectively modify the plant tissue ahead of the invading pathogen in an attempt to limit its spread. For example, the formation of tyloses (outgrowths of the protoplast of a parenchymal cell into an adjacent xylem vessel) ''plugs off'' the xylem of resistant cultivars to spread of the wilt pathogen.

### 3.2 Metabolic Defense

Preformed substances present in plant cells can be inhibitory to pathogens. For example, high concentrations of tannins or phenolics enhance the resistance of young tissue to fungal invasion, while other compounds such as steroidal glycoalkaloids are potent inhibitors of pathogen-derived hydrolytic activity. Inducible plant defenses fall broadly into three responses:

- Accumulation of antimicrobial phytoalexins and cell-wall-bound secondary metabolites
- Accumulation of compounds that modify and reinforce the host cell wall
- Accumulation of proteinase inhibitors and hydrolytic enzymes, which degrade microbial cell walls together with other pathogenesis-related (PR) proteins.

#### 3.2.1 Phytoalexins and Secondary Metabolites

*Phytoalexins* are low molecular weight antimicrobial compounds produced in planta in response to pathogen attack (biotic elicitor) or to chemical or mechanical damage (abiotic inducers). They accumulate in resistant interactions in the healthy plant cells next to damaged tissue, in necrotic cells, and also in certain susceptible host–pathogen interactions. They do not accumulate in biotrophic interactions, that is, in associations the pathogen cannot survive on dead host tissue. Indeed, in certain biotrophic interactions phytoalexin accumulation is thought to be prevented by pathogen-derived suppressors. The key enzymes phenylalanine ammonium lyase *(PAL),* cinnamate 4-hydroxylase *(CH4),* and 4-coumarate:CoA ligase *(4CL)* catalyze the core reactions in *phenylpropanoid metabolism,* leading to a diversity of defense compounds (Figure 1). The enzyme chalcone synthetase *(CHS)* catalyzes the first committed step toward the formation of isoflavonoid phytoalexins, flavonoid pigments, and UV protectants, while 4-coumaroyl-CoA Reductase *(CCR)* catalyzes the initial step toward the biosynthesis of lignin. *PAL* induction has been extensively studied in bean, parsley, and potato; the *PAL* genes respond differently to UV light, to wounding, and to elicitor treatment. In parsley, *4CL* is encoded by two highly homologous single-copy genes which are both activated by UV light or by fungal elicitors. Legume and nonlegume *CHS* genes differ in both number and in function; in parsley the *CHS* gene can be induced by UV light but is insensitive to pathogen attack or to elicitor treatment; in bean, several *CHS* genes are differentially

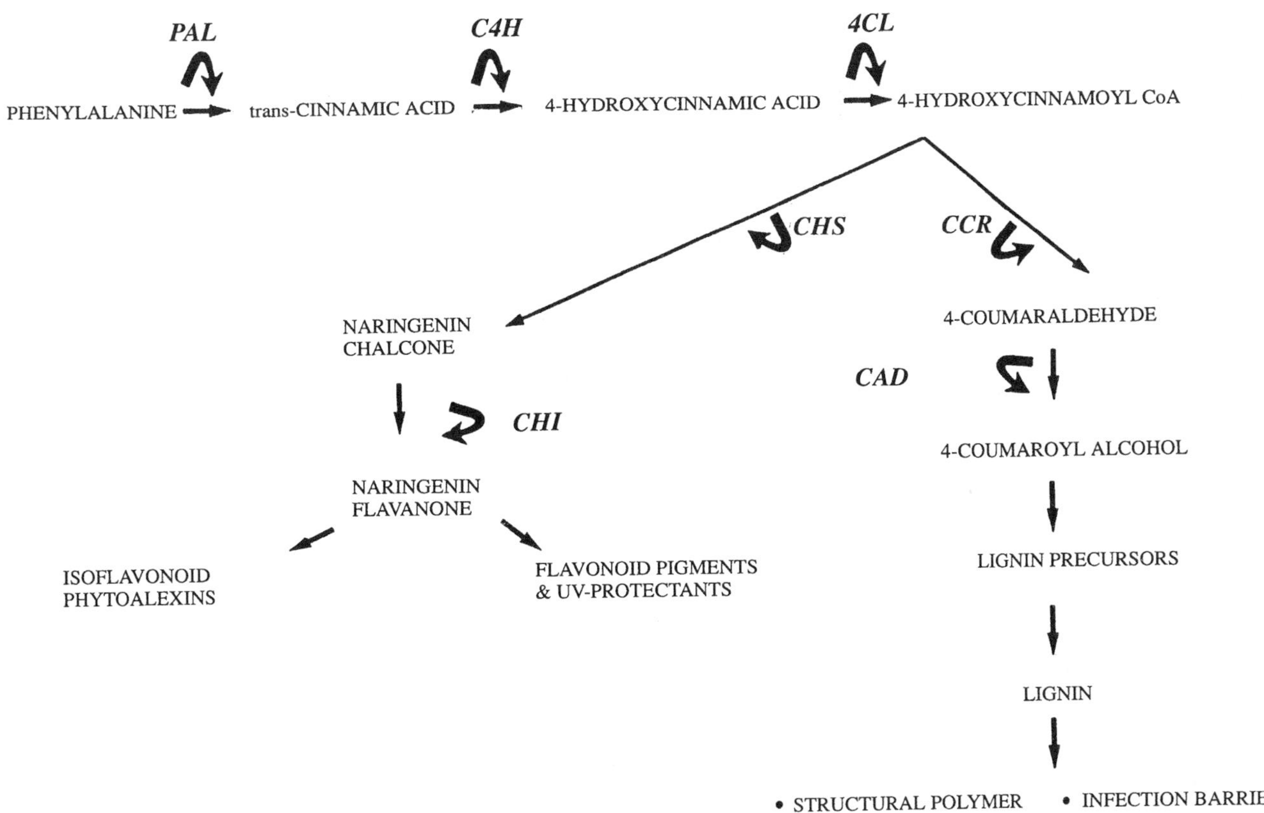


**Figure 1.** The phenylpropanoid pathway elaborated from phenylalanine: *PAL*, phenylalanine ammonia lyase; *C4H*, cinnamate 4-hydroxylase; *4CL*, 4-coumarate CoA ligase; *CHS*, chalcone synthetase; *CHI*, chalcone isomerase; *CCR*, 4-coumaroyl CoA reductase; *CAD*, cinnamyl alcohol dehydrogenase.

expressed in response to elicitors and to wounding, whereas in soybean only one *CHS* gene is activated by UV light or by elicitor treatment.

### 3.2.2. Compounds That Modify and Reinforce Host Cell Wall

*Lignin* molecules, synthesized in response to pathogen attack, can form an effective barrier by confining the pathogen to peripheral host cells. The complete pathway leading to lignin biosynthesis has not been characterized at the molecular level, but cDNA molecules encoding coniferyl alcohol dehydrogenase (*CAD*, see Figure 1) and lignin peroxidase have been isolated. Indeed, the bean *CAD* gene is known to be rapidly transcribed in response to treatment with fungal elicitor.

The *hydroxyproline-rich glycoproteins* (HRGPs) are a group of highly glycosylated and basic proteins encoded by a multigene family. They not only play a central role in the organization of the primary cell wall but accumulate in response to wounding and pathogen attack. The glycine-rich proteins (GRPs) are also cell wall associated, and their gene expression, which is developmentally regulated, is induced in response to wounding.

### 3.2.3. Proteinase Inhibitors and Hydrolytic Enzymes

The expression of two proteinase inhibitor genes, *pin*-1 and *pin*-2, has been extensively studied in tomatoes and potatoes. These developmentally regulated genes are inducible in response to pathogen attack and following wounding. They are expressed both locally at the stimulus site and systemically. Indeed, the systemic accumulation of PI proteins in leaf tissue affords plant protection from grazing pests, since proteinase inhibitors I and II inhibit the activity of the major animal serine endopeptidases trypsin and chymotrypsin. Furthermore, transgenic plants expressing *pin*-1 and *pin*-2 show increased resistance to attack by various pests.

*Thionins,* a family of inducible antifungal proteins rich in cysteine residues, are found in the endosperm of monocotyledonous plants and in the cell walls of mono- and dicotyledons. Their role is thought to be a protective one.

*Zeamatin,* a maize inhibitor protein, and an acidic isoform of a tobacco PR protein show considerable amino acid homology to *thaumatin,* an intensely sweet plant protein. These antimicrobial compounds are known as the *thaumatinlike proteins.*

*PR proteins* can be associated with the cell wall or the vacuole. They accumulate in response to pathogen attack and following treatment with various abiotic elicitors, but their precise role in defense remains unknown. However, a functional role has been ascribed to two PR proteins, the endohydrolases β-1-3-glucanase and -chitinase, which accumulate in response to developmental cues and to defense-related stimuli. Fungal cell walls are mostly made up of β-glucans and chitin which is also present in insect exoskeletons. Therefore the mode of action of such endohydrolases is thought to entail defense gene activation and direct destruction of the pathogen or pest.

### 3.3 Hypersensitivity

Defense can be by the death of an infected plant cell following penetration of the cell wall and the establishment of contact with the host protoplast. This is known as the hypersensitive response (HR) and is characterized by necrotic flecking of the plant tissue, discoloration of the cytoplasm, and compartmentalization of the pathogen. These events are accompanied by the activation of a diversity of inducible defense responses in surrounding tissue, together with the rapid accumulation of high concentrations of antimicrobial phytoalexins around the lesion site. The HR is a common factor in many incompatible bacterial, viral, and fungal–plant interactions. The molecular basis of HR is unknown.

## 4 SIGNALING

The pathways for signals transduced in response to attack, to wounding, or to environmental cues can be considered as local (i.e., the propagation of intracellular and intercellular signals) or systemic, (i.e., long-range signals dispatched around the plant). Various factors have been recognized as signal transducers in animals, such as calcium, phosphoinositides, and cyclic nucleotides, and these are now being considered as possible molecules involved in intracellular signaling in plants. The molecules involved in intercellular signaling also remain obscure, although possible candidates include oligogalacturonide cell wall fragments, ethylene, and glutathione. Several molecules have also been proposed to be involved in systemic signaling. These include salicylic acid, involved in local and distal synthesis of PR proteins; a small polypeptide known as systemin, which is a potent activator of systemic PI gene expression; abscisic acid; oligouronide plant cell wall fragments; chitin and chitosan from fungal cell walls; and methyljasmonate and its free acid jasmonic acid (see Section 11). Changes in redox potential may also play a significant role in the early stages of defense responses, generating superoxide anions or hydroxy radicals before the accumulation of phytoalexin in the HR response. An oxidative burst can be a very rapid manifestation of race-specific resistance and may accelerate cell wall cross-linking.

## 5 PLANT RESISTANCE GENES

The proposed "elicitor–receptor" model describes the interaction between a plant receptor (the product of a constitutively expressed resistance gene) that binds to a race-specific elicitor (a microbial avirulence gene product). Plant resistance genes occur as single specificities at a given locus, or, more usually, as a multiple allelic series, with each allele recognizing a particular pathogen race. Current thinking of the structure of a plant disease resistance gene paints a picture of a complex two-domain unit made up of a recognition element and a hypersensitive response element. Within this unit, rearrangement and recombination events would lead to the emergence of novel resistance gene specificities. While several defense genes induced during host resistance responses have been characterized, plant disease resistance genes have proved recalcitrant to molecular isolation. Toward the goal of cloning resistance genes, research has focused on tagging transposable elements, chromosome walking from linked-marker genes following restriction fragment length polymorphism (RFLP) analysis, using radiolabeled race-specific elicitors as probes for kindred plant gene products, and complementation cloning. Each of these approaches is fraught with technical difficulties, not least of which is the large genome size of plants. The gene-for-gene type of host–bacterium interaction between *Arabidopsis* and *Pseudomonas syringae* pv. *maculicola* (which contains *avrRpm*-1) has led to the mapping of a resistance gene, *Rpm-1.* Since *Arabidopsis* has a relatively small genome (ca. 100 Mb) and yeast artificial chromosome (YAC) libraries are available, the cloning of such a mapped resistance gene by chromosome walking from known markers seems imminent. No resistance genes to fungal pathogens that conform to the gene-for-gene model have been cloned, but transposon tagging of fungal resistance genes has led several groups near to this goal—for example, in the targeting of the *Rp*-1 locus in corn. Transposon tagging has, however, led to the isolation of a plant resistance gene *Hm-1,* which confers resistance to the HC toxin producing corn pathogen *Helminthosporium carbonum.* It encodes a plant-derived pathogen toxin that inactivates reductase but does not conform to a pattern of gene-for-gene interaction. The potential value of being able to manipulate genetically defined resistance genes, particularly in crops, is incalculable. Thus an insight into the structure of disease resistance genes remains a primary goal of molecular plant pathology.

## 6 SYSTEMIC ACQUIRED RESISTANCE

Following infection by a necrotrophic pathogen, plants display a generalized nonspecific "immunity" to a subsequent infection. This induced resistance is effective against a broad spectrum of potential pathogens and lasts for several weeks. Recent work has shown that infection of tobacco plants with tobacco mosaic virus (TMV) or treatment with either salicylic acid or methyl-2,6-dichloroisonicotinic acid leads to the induction of 9-coordinate gene families, the products of which include PR proteins. A knowledge of the molecular basis of induced systemic acquired resistance (SAR) may lead to new strategies for crop protection—that is, by the design of transgenic plants constitutively expressing defense genes or by the exogenous application of novel synthetic chemicals known to induce resistance.

## 7 BIOCHEMICAL WEAPONS OF PATHOGENS

A virulent pathogen can overcome host defenses by using the following biochemical weapons:

- The detoxification of inducible host molecules, such as phytoalexins.
- The use of host-derived molecules as pathogen inducers, such as host cutin monomers, to initiate pathogen cutinase production.
- The disintegration of host molecules, as in the pathogen-derived pectin degrading of cell walls by enzymes, cellulases, hemicellulases, and ligninases.

## 8 TOXINS

Microbial toxins have been implicated in the establishment of certain plant diseases. These compounds kill plant cells at low concentrations either by affecting membrane permeability or by acting as enzyme inhibitors or antimetabolites. Both non-host-specific toxins such as tabtoxin and phaseolotoxin (produced by plant pathogenic bacteria) and fungal-derived host-specific toxins (such as victorin) have roles in pathogenicity.

## 9 GROWTH REGULATORS IN PLANT DISEASE

Many plant pathogens disrupt the normal hormonal balance maintained in a healthy plant by stimulation, mimicry, or retardation of the host growth response or by the production of novel growth regulators. The result can be leaf epinasty, stunting, hyperplasia, hypertrophy, or defoliation.

## 10 AVIRULENCE GENES

The gene-for-gene complementarity hypothesis means that a single host resistance gene is matched by an avirulence gene in the pathogen (see Section 1). Classical genetics has recognized many complementary host–pathogen systems, but few have been characterized by molecular means. The work on genetic determinants of avirulence in plant pathogenic fungi has lagged behind that in bacteria because of poorly defined genetic systems and the lack of DNA-mediated fungal transformation vectors. The first cloned pathogen avirulence gene was obtained by transforming a race of the bacterial soybean pathogen *Pseudomonas syringae* pv. *glycinea* (which gives a susceptible interaction on a given soybean cultivar) with a cosmid library of genomic DNA from a race that induced HR on the same cultivar. The avirulence gene was further defined by transposon mutagenesis and an open reading frame encoding a 100 kDa polypeptide identified as determining the phenotype capable of inducing HR. Similar strategies have been used to clone other phytopathogenic bacterial avirulence genes. The cloned avirulence genes appear to complement specific resistance genes and so support the gene-for-gene theory. Furthermore, it is unlikely, in these examples, that the avirulence gene products are themselves race-specific elicitors that interact directly with a host receptor to trigger resistance, since there is no evidence for the existence of signal peptides involved in the release of the avirulence gene products. The first cloned fungal avirulence gene was *avr*-9 from *Cladosporium fulvum.* Transformation of a fungal isolate lacking *avr*-9 converts it to the avirulent phenotype, which enables it to interact with tomato plants expressing the *Cf*-9 resistance gene, as do wild-type *avr*-9 isolates. The first proven viral avirulence genes were the coat protein genes of tobacco mosaic virus; transgenic tobacco carrying the coat protein genes from either avirulent or virulent strains shows that only plants transformed with the coat protein genes from the avirulent isolates induce HR. Furthermore, a single amino acid change in the coat protein modulates the avirulence phenotype.

## 11 ELICITORS AND SUPPRESSORS

Elicitors, compounds that induce a plant defense response, can be microbial or plant-derived. Such biotic elicitors purified to apparent homogeneity include oligosaccharide fragments of fungal and plant cell wall polysaccharides, fungal cell surface lipids, and certain fungal and bacterial polypeptides. The vast majority of elicitors are not race specific, but good evidence exists that a galactose-mannose-rich glycoprotein of a given race of the fungus *Colletotrichum lindemuthianum* induces phytoalexin in resistant bean cultivars but not in susceptible combinations. The elicitor molecule is thought to interact with a plant membrane receptor, causing the activation of specific genes by an as yet unknown mechanism; the synthesis of a diversity of secondary plant products results. It has recently been proposed that jasmonic acid plays an integral role in the signal transduction pathway (see Section 4), leading to the expression of certain defense genes. Here it is proposed that the elicitor–receptor complex activates a lipase, thereby releasing α-linoleic acid, which is transformed to jasmonic acid and methyljasmonate. In a compatible biotrophic interaction, it has been proposed that fungal-derived suppressor molecules inhibit elicitation by competing for potential elicitor binding sites. However, it should be realized that race-specific suppression would require virulence to be functionally dominant, and this contravenes the genetics of race–cultivar specificity, where avirulence is dominant.

## 12 SUMMARY AND OUTLOOK

Considerable advances have been made in recent years toward an understanding of the events surrounding both pathogen ingress and host defense responses, but no plant resistance gene has been fully characterized. We do not understand the mechanisms of plant resistance gene interaction with the products of avirulence genes, nor do we know how plants perceive and transduce microbial signals. An intimate working knowledge of all such mechanisms associated with disease resistance will lead to the rational design of healthier crops.

*See also* Pesticide-Producing Bacteria; Plant Cells, Genetic Manipulation of.

### Bibliography

Agrios, G. N. (1988) *Plant Pathology.* Academic Press. San Diego, CA, and London.

Bowles, D. J. (1990) Defence-related proteins in higher plants. *Annu. Rev. Biochem.* 59:873–907.

de Wit, P.J.M. (1992) Molecular characterisation of gene-for-gene systems in plant–fungus interactions and the application of avirulence genes in control of plant pathogens. *Annu. Rev. Phytopathol.* 30:391–418.

Gurr, S. J., McPherson, M. J., and Bowles, D. J., Eds. (1992) *Molecular Plant Pathology: A Practical Approach, Vols. I and II.* Oxford University Press, London and New York.

Isaacs, S. (1992) *Fungal–Plant Interactions.* Chapman & Hall, London.

Lamb, C. J., Lawton, M. A., Dron, M., and Dixon, R. A. (1989) Signals and transduction mechanisms for activation of defences against microbial attack. *Cell,* 56:215–224.

Pryor, T. (1987) The origin and structure of fungal disease resistance genes in plants. *Trends Genet.* 3:157–161.

Ryan, C. A. (1992) The search for the proteinase inhibitor-inducing factor, PIIF. *Plant Mol. Biol.* 19:123–133.

# Plasma Lipoproteins

*Thomas L. Innerarity*

### Key Words

**Apolipoprotein (or Apoprotein)** Protein component of a lipoprotein that solubilizes and transports lipids. Certain apolipoproteins also act as ligands for lipoprotein receptors and as cofactors for enzymes.

**Lipoproteins** Spherical particles composed of a nonpolar neutral lipid core surrounded by a shell of polar lipids and apolipoproteins. Lipoproteins transport lipids in the plasma.

**RNA Editing** The alteration of the protein coding sequences of RNA such that the genetic information differs from that encoded by the genome.

**Tissue-Specific Locus Control Element** A tissue-specific cis-acting element that directs the expression of a family of genes in specific organs or tissues.

**Transgenic Mice** Mice into whose chromosomes foreign gene(s) become integrated. This can result in the overexpression of a specific foreign gene or the disruption or alteration of the function of an endogenous gene.

Plasma lipoproteins transport lipids in the plasma. Abnormalities in lipoprotein metabolism are a major cause of atherosclerosis. Most individuals with premature heart disease have either elevated levels of low density lipoproteins (LDL), high levels of remnant lipoproteins known as β-very low density lipoproteins (β-VLDL), low levels of high density lipoproteins (HDL), or high levels of an abnormal lipoprotein called lipoprotein(a) [Lp(a)].

In three areas, the study of lipoproteins has led the way in developing new concepts or has yielded novel information about processes important to biology and especially to molecular biology. First, the use of transgenic animals has furthered our understanding of complex metabolic pathways and the causes of complicated

**Table 1** Characteristics of the Plasma Lipoproteins

| Lipoprotein Class and Density of Flotation (g/mL) | Origin | Major Functions | Major Apoproteins | Principal Lipids |
|---|---|---|---|---|
| Endogenous Lipoproteins | | | | |
| VLDL ($d < 1.006$) | Liver | Transport triglycerides to various tissues. Fatty acids hydrolyzed by lipoprotein lipase are used by the tissues for energy or stored as triglycerides. | B100, E, C-I, C-II, C-III | Triglycerides |
| IDL ($d = 1.006$–$1.019$) | Derived by the lipolysis of VLDL by lipoprotein lipase. | Some are precursors of LDL; some are taken up by the liver and provide cholesterol for the liver. | B100, E | Cholesterol, triglycerides |
| LDL ($d = 1.019$–$1.063$) | Derived from IDL following lipolysis by lipoprotein lipase and hepatic lipase. | Provide cholesterol to various tissues by receptor-mediated endocytosis. Increased levels of LDL cholesterol correlate with cardiovascular disease. | B100 | Cholesterol |
| HDL ($d = 1.063$–$1.21$) | The apoproteins are derived from the hydrolysis of chylomicrons and VLDL and from de novo synthesis by the liver and intestine. | Facilitate removal of cholesterol from extrahepatic tissues for transport to the liver. HDL cholesterol correlates inversely with cardiovascular disease. | A-I, A-II, C-I, C-II, C-III, E | Phospholipid, cholesterol |
| Exogenous Lipoproteins | | | | |
| Chylomicrons ($d < 0.95$) | Small intestine. | Transport triglycerides and cholesterol from the intestine to the plasma. | B48, A-I, A-II, A-IV, E, C-I, C-II, C-III | Triglycerides |
| Chylomicron Remnants ($d < 1.006$) | Derived from chylomicrons following lipolysis by lipoprotein lipase. | Transport cholesterol and triglycerides to the liver. The uptake by the liver is mediated by lipoprotein receptors. | B48, E | Cholesterol, triglycerides |
| Specialized Lipoproteins | | | | |
| β-VLDL ($d < 1.006$) | Two subclasses: I. Intestine II. Liver | Potentially atherogenic lipoproteins. Subclass I represents cholesterol-rich chylomicron remnants; subclass II represents cholesterol-rich VLDL remnants. | B100 or B48, E | Cholesteryl ester |
| Lp(a) ($d = 1.05$–$1.12$) | Liver. | Concentration and phenotypes of Lp(a) correlate with accelerated atherogenesis. | B100, (a) | Cholesterol |

diseases such as atherosclerosis. Second, the mechanism for apo B mRNA editing is now well understood; the hepatic expression of editing activity in man could become a useful and unique mechanism for the lowering of plasma LDL levels. Finally, it has been shown that a single cis-acting element can control the expression of a group of apolipoprotein genes.

## 1 OVERVIEW

Plasma lipoproteins are macromolecular complexes composed of lipids and proteins. Lipoproteins originate mainly in the intestine and the liver and are involved in the transport of lipids to and from cells, tissues, and organs. The plasma lipoproteins are spherical particles composed of a nonpolar, neutral lipid core (triglycerides and cholesteryl esters) surrounded by a shell of polar lipids (phospholipids and unesterified cholesterol) and proteins, known as apolipoproteins. The apolipoproteins enable the insoluble lipids to be transported in an aqueous environment and to influence directly the metabolism of the lipoproteins.

Plasma lipoproteins can be divided into six major classes and two specialized classes. Four of these classes of lipoproteins—very low density lipoproteins (VLDL), intermediate density lipoproteins

**Table 2** Characterization of the Major Apolipoproteins

| Apoprotein | Plasma Concentration (mg/dL) | Molecular Weight (number of amino acids in the mature protein) | Major Sites of Synthesis | Chromosomal Location of the Gene | Functions | Clinical Disorders Due to Genetic Variants or Mutations |
|---|---|---|---|---|---|---|
| A-I | 100–30 | 28,100 (243) | Small intestine, liver | 11q | Structural protein of HDL; activator of lecithin:cholesterol acyltransferase; tissue cholesterol efflux. | Apo A-I-C-III deficiency; apo A-I deficiency; apo A-I multiple mutants; Tangier disease |
| A-II | 30–50 | 17,400 (77) | Small intestine, liver | 1q | Structural protein of HDL. | |
| A-IV | 15 | 43,000 (376) | Small intestine | 11q | Associated with triglyceride transport in chylomicrons. | |
| B100 | 80–120 | 550,000 (4536) | Liver | 2p | Necessary for VLDL biosynthesis and secretion; ligand for the LDL receptor. | A-β-lipoproteinemia; familial hypo-β-lipoproteinemia; familial defective apo B100. |
| B48 | < 5 | 250,000 (2152) | Small intestine | 2p | Necessary for chylomicron biosynthesis and secretion. | A-β-lipoproteinemia; chylomicron retention disease |
| C-I | 5–7 | 6,600 (57) | Liver | 19q | Modulates activation of lecithin:cholesterol acyltransferase. | Hypertriglyceridemia |
| C-II | 3–7 | 8,000 (79) | Liver | 19q | Activates lecithin:cholesterol acyltransferase. | Familial hyperchylomicronemia |
| C-III | 9–13 | 8,750 (79) | Liver | 11q | Modulates receptor uptake of chylomicron remnants. | |
| D | 6–7 | 33,000 (169) | Adrenal, kidney, brain, liver, small intestine | 3q | Unknown. | |
| E | 3–6 | 34,200 (299) | Liver, macrophages in various organs, astrocytes | 19q | Ligand for lipoprotein receptors. | Type III hyperlipoproteinemia; isoforms correlated with high plasma cholesterol levels |
| (a) | 0–100 | 350,000–750,000 (4529) | Liver | 6q | Binds to plasminogen receptors on endothelial cells. | Levels and phenotypes of Lp(a) correlated with cardiovascular disease |

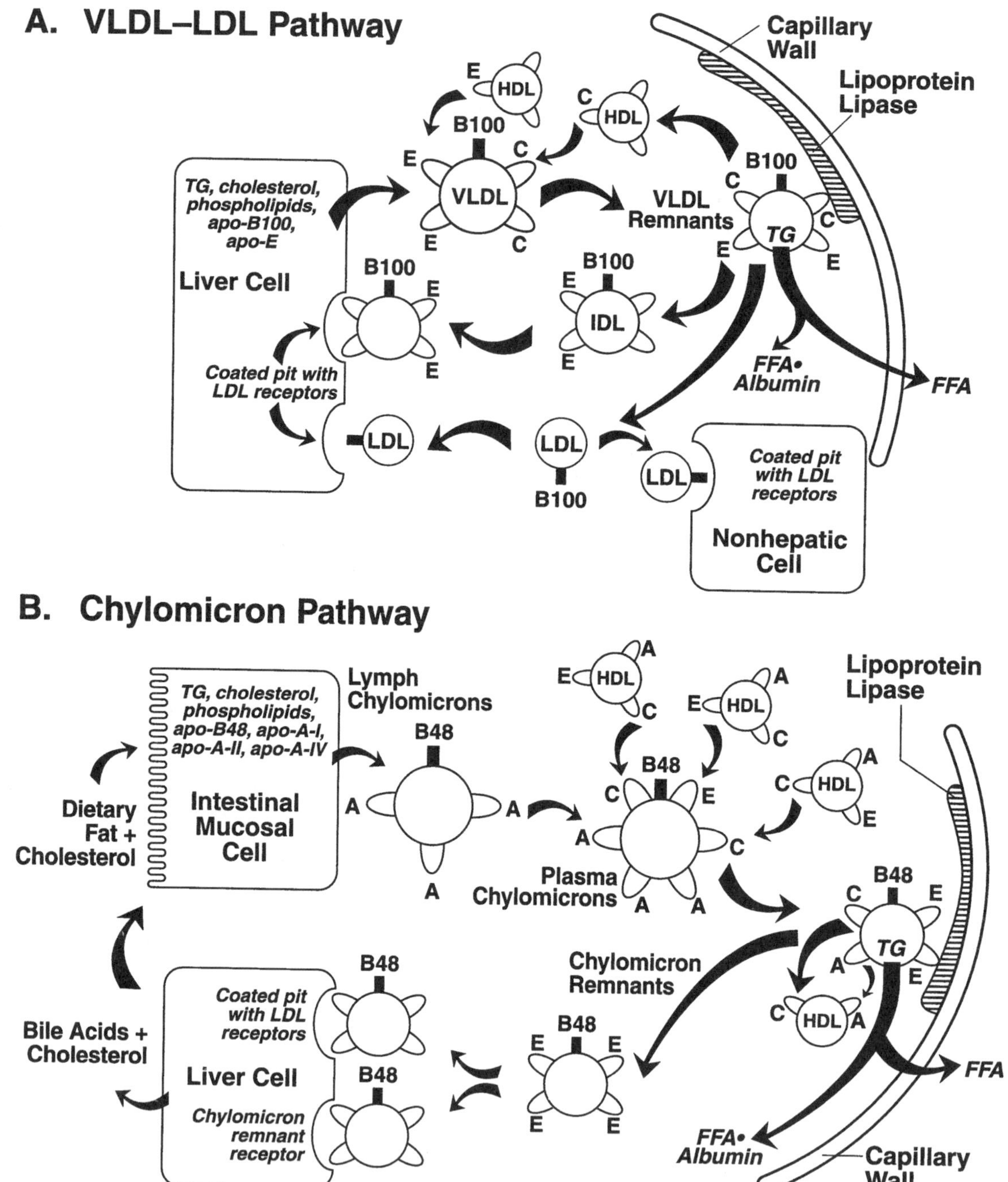


Figure 1. Synthesis and catabolism of lipoproteins derived (A) hepatically and (B) intestinally: FFA, free fatty acids; TG, triglycerides; B100, C, and E are apolipoproteins that direct the fate of lipoproteins.

(IDL), low density lipoproteins (LDL), and high density lipoproteins (HDL)—are derived from the liver and are present in the plasma of fasted subjects (Table 1). Two other major classes of lipoproteins, chylomicrons and chylomicron remnants, are found in the plasma only after a fatty meal (postprandially). The specialized lipoproteins, β-VLDL and lipoprotein(a) [Lp(a)], are believed to be atherogenic. The apolipoproteins have been designated apo(a), apo A-I, apo A-II, apo A-IV, apo B100, apo B48, apo C-I, apo C-II, apo C-III, apo D, and apo E. Table 2 summarizes our current knowledge of the characteristics of these apolipoproteins.

All apolipoproteins except apo(a) share the function of solubilizing and transporting lipids in an aqueous environment. Apolipoproteins have amphipathic α-helical regions that are well suited for this lipid-binding function. In addition, apolipoproteins have a number of other specific functions. Apolipoprotein B100 is necessary for the assembly and secretion of VLDL, while apo B48 is an obligatory component for the biosynthesis and secretion of chylomicrons. Apolipoprotein B100 binds to the LDL receptor, which is the first step in the removal of LDL from the blood. Apolipoprotein E is the ligand that mediates the clearance of rem-

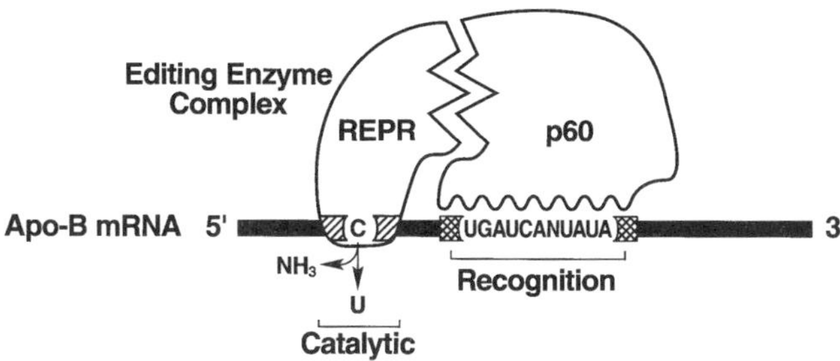


**Figure 2.** Model of the apo-B mRNA editing enzyme complex. The complex is postulated to contain at least two proteins. One, the RNA editing protein (REPR), is the catalytic component that deaminates cytidine 6666. The second component, p60, recognizes and binds to an 11-base sequence 3′ to the edited base.

nants by the LDL and chylomicron remnant receptors. Apolipoprotein C-II is a cofactor for the lipolytic enzyme lipoprotein lipase, and apo A-I is an activator of lecithin:cholesterol acyltransferase (LCAT).

The metabolism of lipoproteins is complex, with many interconnected pathways. For simplification, lipoprotein metabolism can be divided into two main pathways: the VLDL–IDL–LDL pathway and the chylomicron pathway. As shown in Figure 1A, the VLDL are synthesized in the liver. Nascent VLDLs, which contain apo B100 and small amounts of apo E, are released into the liver's space of Disse, from which they enter the plasma. In the plasma, VLDL acquire additional E and C apolipoproteins, primarily from HDL. The mature VLDLs acquire cholesteryl ester from HDL. The VLDLs are converted to VLDL remnants by lipoprotein lipase, which hydrolyzes the triglycerides of the lipoprotein core. Concomitant with the triglyceride hydrolysis is a loss of C apolipoproteins to HDL and a decrease in the size of the lipoprotein particle. Further lipolysis results in the formation of IDLs. The IDLs, which are smaller and richer in cholesterol than the VLDLs, contain apo E and apo B100. Some IDLs are cleared from the plasma via hepatic LDL receptors, and the fraction that escapes clearance by the liver is converted to LDL. The IDLs bind to LDL receptors through the interaction of apo E.

The LDLs that are formed as a result of the VLDL pathway are the major cholesterol-transporting lipoproteins in humans. The apo B100 on the surface of these particles is recognized by hepatic and extrahepatic LDL receptors. This receptor-mediated pathway is responsible for the clearance of about two-thirds to three-fourths of LDLs from the plasma, principally by the liver. The LDLs are believed to be the major atherogenic lipoprotein class in humans.

The chylomicron pathway is responsible for the transport and clearance of dietary lipids that have been absorbed from the small intestine. A distinctive apolipoprotein ''marker'' for intestinally derived lipoproteins is apo B48. In humans, this form of apo B (approximately the amino-terminal 48% of apo B100) is synthesized only in the intestine and is, therefore, unique to chylomicrons and chylomicron remnants. Although both the hepatic form of apo B (apo B100) and apo B48 are encoded by the same gene, apo B48 is not produced by alternate splicing or proteolytic processing but by a novel RNA editing process (see Section 2.1).

Chylomicrons are synthesized when digested dietary fats (fatty acids and cholesterol) are absorbed by intestinal cells (Figure 1B). Nascent chylomicrons appear to be assembled and to accumulate in the Golgi apparatus. The chylomicrons are released into the mesenteric lymph and proceed to the thoracic duct, where they finally enter the general circulation. Newly secreted chylomicrons possess apo B48, apo A-I, apo A-II, and apo A-IV; later, when exposed to HDL and other lipoproteins in the lymph and the plasma, they acquire apo E, apo C-I, apo C-II, and apo C-III.

In the plasma, chylomicrons are acted on by lipoprotein lipase, an enzyme that hydrolyzes much of the core triglyceride, releasing free fatty acids. The depletion of core triglycerides results in the transfer of excess surface components and apolipoproteins from chylomicron remnants to HDL. At the same time, the remnants become enriched in apo E acquired from HDL. The apo E binds to hepatic lipoprotein receptors and mediates the rapid clearance of chylomicron remnants from the circulation.

## 2 MOLECULAR BIOLOGY OF PLASMA LIPOPROTEINS

### 2.1 Apo B mRNA Editing

The investigation of plasma lipoproteins has provided three noteworthy areas of general scientific interest to molecular biologists. The first novel area is apo B mRNA editing. This is a process whereby one nucleotide of the apo B mRNA is altered to form a premature termination codon. The nucleotide cytidine at position 6666 in the apo B mRNA is deaminated to form a uridine. This site-specific deamination converts apo B codon 2153 from a CAA (glutamine) codon to the translational termination codon UAA. The catalytic component (deaminase) of the editing complex has been identified, cloned, and sequenced. As a result of this termination of protein translation, a shortened apo B (apo B48) is formed. This is the first example of RNA editing in mammals, and the process has been extensively investigated. The editing factor complex is composed of the site-specific deaminase and at least one other protein, and unlike many other types of RNA processing such as splicing and polyadenylation, no additional small RNAs or other nucleotide cofactors are involved. However, like other types of mRNA processing, apo B mRNA editing occurs primarily, if not exclusively, in the nucleus. In addition, apo B mRNA editing is linked to another RNA processing event, polyadenylation. The current model of apo B mRNA editing is shown in Figure 2.

### 2.2 Tissue-Specific Locus-Control Elements

A second novel area is the tissue-specific regulation of apolipoprotein gene expression. Transgenic mice have been essential in defining the tissue-specific cis-acting elements that direct the expression of apolipoproteins by the liver and intestine and the expression of apo E in multiple cell types and tissues. With the exception of apo E, the plasma apolipoproteins are expressed primarily in the liver and intestine. Apolipoprotein E is expressed in a wide variety of tissues and organs, including the liver, intestine, brain, testes, kidney, and skin. It has been found in macrophages in all types of tissue. In two different apolipoprotein gene clusters, a single cis-acting element functions as a tissue-specific locus control element, apparently controlling the expression of all the apolipoprotein genes in that cluster.

One of these apolipoprotein gene clusters is located on chromosome 11 at locus q23 and consists of the apo A-I gene closely linked to the apo C-III and apo A-IV genes, in that order. As shown in Figure 3, apo C-III is transcribed in the reverse orientation of the other two genes. In humans, apo A-I and apo C-III are expressed in the liver and intestine, while apo A-IV is expressed primarily

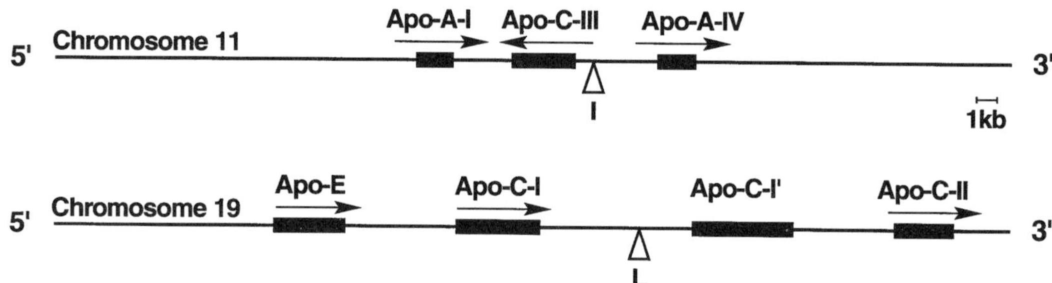


**Figure 3.** Tissue-specific locus-control elements. On chromosome 11, the intestine-specific locus-control element (I) lies between the apo C-III and apo A-IV genes. On chromosome 19, the liver-specific locus-control element (L) is located between the apo C-I gene and the apo C-I pseudogene (apo-C-I′).

**Table 3** Apolipoproteins Expressed in Transgenic Mice

| Gene | Type of Construct | Main Tissues of Expression | Phenotype | Major Finding |
|---|---|---|---|---|
| Human apo A-I, overexpression | Human genomic | Liver, small intestine | ↑ HDL level | Transgenic mice more resistant to atherosclerosis. |
| Mouse apo A-I, deletion | Insertion | | ↓ HDL, cholesterol | Greatly reduced HDL. |
| Human apo A-II, overexpression | Human apo A-I promoter 5′ of apo A-II | Liver | Unique size distribution of HDL | Transgenic mice: none had atherosclerosis on chow diet. |
| Mouse apo A-II, overexpression | Mouse genomic | Liver | Larger HDL | Transgenic mice: atherosclerosis on chow diet. |
| Human apo B, overexpression | Human genomic | Liver | ↑ LDL | Large 80 kb genomic fragment can be used in the generation of transgenic mice. |
| Mouse apo B, deletion | Insertion | | ↓ Cholesterol, apo B | |
| Human apo C-I | Human genomic | Liver | Mild hypertriglyceridemia | |
| Human apo C-III | Human genomic | Liver, intestine | Primary hypertriglyceridemia | Elevated apo C-III causes hypertriglyceridemia. |
| Human cholesterol ester transfer protein | Mouse metallothionine promoter + monkey cDNA | Liver | HDL ↓, cholesterol ↓, apo B ↑ | Cholesterol estertransfer protein activity increases propensity for atherosclerosis. |
| Human apo E, overexpression | Genomic | Liver | ↑ Remnant clearance | ↑ Remnant clearance. |
| Rat apo E, overexpression | Mouse metallothionine promoter | Liver, kidney | ↑ Remnant clearance | Increased apo E lowered cholesterol level in cholesterol-fed transgenic mice. |
| Human apo E, deletion | Insertion | | ↑ VLDL, ↑ chylomicron remnants | Transgenic mice had atherosclerosis on chow diet. |
| Human apo(a) | Transferrin promoter + apo(a) cDNA | Many | Free apo(a), no Lp(a) | Transgenic mice developed atherosclerosis on cholesterol diet. |
| Human LDL receptor, deletion | Insertion | | ↑ LDL, ↑ VLDL | Atherosclerosis on chow-fed diet. |

in the intestine. The expression of gene constructs that contain variable segments of this human gene cluster in transgenic mice has served to identify the elements necessary for the tissue-specific expression of these genes. Hepatic expression requires only 5′ promoter/enhancer elements in the apo A-I and apo C-III genes, whereas the intestinal expression of all three genes requires elements localized to a region between the apo C-III and apo A-IV genes (Figure 3). Although not yet proven, it is likely that the same cis-acting element coordinates the expression of all three genes.

The second apolipoprotein gene cluster that has been studied in transgenic mice is located on chromosome 19 at locus q13 and consists of apo E, apo C-I, an apo C-I pseudogene, and apo C-II, in that order (Figure 3). All three genes are expressed primarily in the liver, although apo E is expressed appreciably in other organs and tissues. The expression of apo E is controlled by a complex set of interacting enhancers and silencers. Of particular interest is the location of the liver-specific element of the apo E/C-I/C-II gene locus in a 154 base pair sequence between the apo C-I gene and the apo C-I pseudogene. Thus, this liver-specific enhancer must act over a distance of more than 15 kb to stimulate the high expression of apo E in the liver. It appears that this element also stimulates the hepatic expression of apo C-I, and it may function as a liver-specific enhancer for all the apolipoprotein genes in this gene cluster. This liver-specific element is similar to the locus-control region described for the human β-globin gene complex. This is a broad region, more than 20 kb 5′ of the ε-globin gene, which is necessary for erythroid-specific expression of the β-globin genes in transgenic mice.

### 2.3 Transgenic Mice

The third key area of research interest is the development of transgenic mice and, through homologous recombination, gene "knockout" mice, both of which have provided animal models for the investigation of altered lipoprotein metabolism and atherosclerosis. The complex interactions involved in lipoprotein metabolism (see Section 1) have made the elucidation of the roles of the individual components in the lipid–lipoprotein pathways very difficult. The use of transgenic mice with overexpression of specific transgenes or an aberration of specific gene function has, in almost every case, provided insights into the function of specific genes in lipoprotein transport. These transgenic mice also have contributed to our understanding of how aberrations in gene function disrupt lipoprotein metabolism and cause atherosclerosis (Table 3). Only a few years ago, mice were thought to be a poor animal model for the investigation of atherosclerosis. In most murine strains, diets rich in cholesterol and fat cause only minimal increases in plasma cholesterol and minimal fatty streaks or no lesions at all. This has now changed with the genetic manipulation of certain genes involved in lipoprotein metabolism, and the genetically altered mice can readily develop atherosclerosis. The overexpression of mouse apo A-II, an abnormal form of human apo E, or human apo(a) causes fatty-streak lesions in transgenic mice. Gene "knockout" by homologous recombination of apo E or of the LDL receptor results in severe atherosclerosis (Table 3).

## 3 PERSPECTIVES

It follows from the work just described that animal models with genetically engineered gene defects that cause specific lipoprotein disorders will be useful for testing therapeutic interventions. Moreover, in the near future, molecular geneticists should be able to delete all the endogenous genes and reconstitute the mouse with human genes to mimic exactly the human lipoprotein transport system.

*See also* Cardiovascular Diseases; Lipoprotein Analysis; Mouse Genome.

### *Bibliography*

Breslow, J.L. (1993) Transgenic mouse models of lipoprotein metabolism and atherosclerosis. *Proc. Natl. Acad. Sci. U.S.A.* 90:8314–8318.

Fazio, S. (1993) Recent insights into the pathogenesis of type III hyperlipoproteinemia. *Trends Cardiovasc. Med.* **3**:191–196.

Innerarity, T.L. (1991) Plasma lipoproteins. In *Encyclopedia of Human Biology*, R. Dulbecco, Ed. Academic Press, San Diego, CA.

Young, S.G. (1994) Transgenic mice expressing human apo-B100 and apo-B48. *Curr. Opin. Lipidol.* **5**:94–101.

Zannis, V.I., Kardassis, D., and Zanni, E.E. (1993) Genetic mutations affecting human lipoproteins, their receptors, and their enzymes. In *Advances in Human Genetics*, H. Harris, and K. Hirschhorn, Eds. Plenum Press, New York.

# Plasmids

*Kimber Hardy*

### *Key Words*

**Conjugation** Transfer of plasmid DNA between cells involving cell–cell or cell–pilus contact. Conjugative plasmids are capable of transferring themselves between cells by conjugation.

**Incompatibility** The inability of different plasmids to coexist stably in the same host cell.

**Mobilization** Conjugative plasmids and conjugative transposons can bring about the transfer between cells of other plasmids that are not themselves capable of conjugation; that is, they are able to transfer certain nonconjugative plasmids by mobilization.

**Pilus** Long thin appendage involved in the transfer of plasmids by conjugation or in the adherence of bacteria to surfaces.

Plasmids are genetic elements that are stably inherited without being a part of the chromosome(s) of their host cells. They are found in bacteria and fungi of many kinds and are not essential to the survival of the host cell, at least not under laboratory conditions. They may be composed of DNA or RNA and may be linear or circular. Plasmids code for molecules that ensure their replication and stable inheritance during cell replication, and they also encode many products of considerable medical, agricultural, and environmental importance. For example, they code for toxins that greatly increase the virulence of pathogenic bacteria. They can also confer resistance to antibiotics, and they enable bacteria belonging to the genus *Rhizobium* to fix atmospheric nitrogen.

Some plasmids can transfer copies of themselves from one cell to another by conjugation and at the same time transfer copies of chromosomal genes. Plasmids are widely used in molecular biology because they provide the basis for many vectors that are used to clone and express genes.

## 1 STRUCTURE

Plasmids have a wide range of structures: they may be composed of DNA or RNA, they may be double- or single-stranded, and they may be circular or linear. The smallest bacterial plasmids are about 1.5 kb and the largest are greater than 1500 kb. The vast majority are circular. However, several very large linear DNA plasmids, up to 500 kb long, have been found in species of *Streptomyces* and *Nocardia.*

Double-stranded DNA plasmids appear to exist as predominantly covalently closed circular molecules in bacterial cells; that is, both the polynucleotide strands forming the double helix are intact. Such molecules are usually twisted upon themselves to form supercoiled molecules. Supercoiling is introduced by DNA gyrase, which uses ATP to drive the negative supercoiling of DNA. Both circular and linear plasmids are found in yeast and other fungi. The best characterized yeast plasmid is the 2μm circle, which does not specify any particular characteristics such as drug resistance and is found in the *Saccharomyces cerevisiae* nucleus. Linear yeast plasmids, composed of either RNA or DNA, can encode protein toxins that inhibit the growth of sensitive yeasts. Yeasts harboring such plasmids are known as "killer strains." Plasmids, both circular and linear DNA molecules, have also been found in the mitochondria of many species of fungi and plants.

## 2 REPLICATION

Since plasmids are, by definition, autonomous replicating elements that are not permanently part of the host chromosome, it follows that the coupling of plasmid replication to the process of cell multiplication is important for stable plasmid maintenance. The mechanisms controlling plasmid replication are understood in greatest detail for plasmids found in *Escherichia coli,* especially the small plasmid ColE1 and the larger drug resistance plasmid R1.

### 2.1 ColE1 Replication

ColE1 is a double-stranded DNA plasmid originally isolated from an *E. coli* strain and present in about 15 copies per cell. ColE1 encodes an antibacterial protein called colicin E1.

From the point of view of the regulation of plasmid replication in relation to bacterial growth, the initiation of plasmid replication is of particular interest. The rate of the initiation of new rounds of replication also controls plasmid copy number. Replication of ColE1 always begins at a particular point on the plasmid—the origin—and is then unidirectional; that is, it proceeds in one direction around the circular molecule. The mechanism that controls ColE1 replication is shown in Figure 1. The two key elements are the RNA molecules, which are transcribed from the ColE1 DNA.

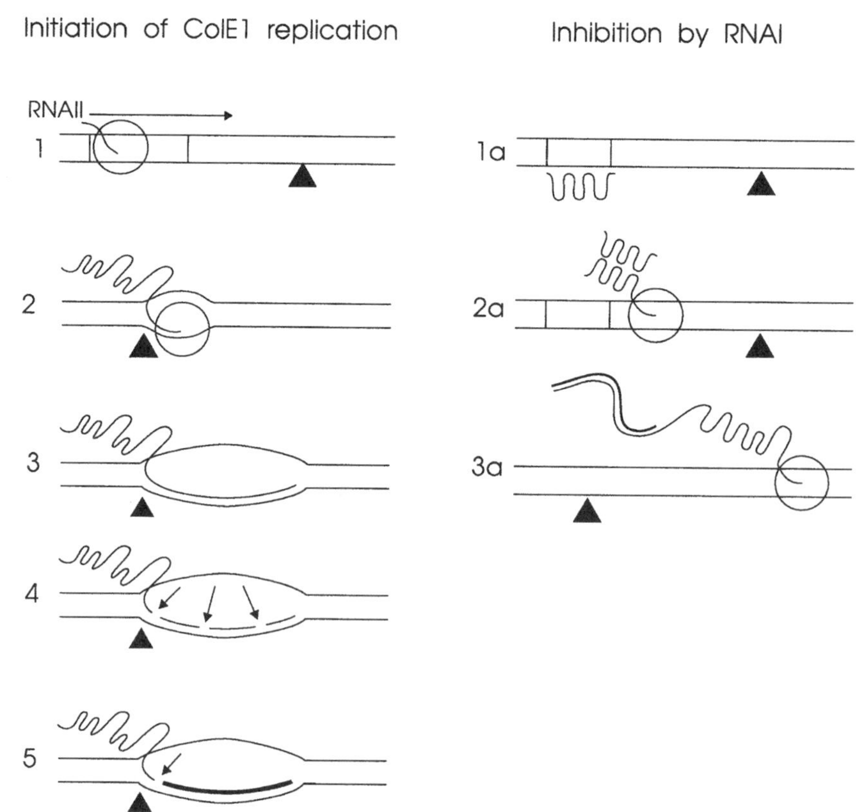


**Figure 1.** Initiation of ColE1 replication and inhibition by RNAI. 1, Synthesis of RNAII (circle represents RNA polymerase); 2, hybridization of RNAII (wavy line) to DNA; 3, elongation of RNAII/DNA hybrid; 4, cleavage of RNAII by RNaseH; 5, synthesis of DNA (bold line) and removal of RNAII; 1a, synthesis of RNAI; 2a, interaction between RNAI and RNAII; 3a, hybridization of RNAI to RNAII inhibits formation of the RNAII/DNA hybrid, and thus inhibits the initiation of ColE1 replication. [Adapted from Eguchi et al., *Annu. Rev. Biochem.* 60:631–652 (1991).]

One of these RNA molecules is called the replication primer (or RNA II); its transcription begins 555 base pairs upstream from the origin of DNA replication. RNA II forms a hybrid with the complementary template DNA near the origin, where it can be cleaved by RNase H, an enzyme that specifically cleaves DNA:RNA hybrids. This leaves an end that has a 3′-hydroxyl group at the origin to act as a primer for DNA synthesis by DNA polymerase I. Hybridization of RNA II with the template DNA is essential for the initiation of plasmid replication. This key step of hybridization is controlled by another RNA molecule, called RNAI, which is encoded by ColE1. RNAI is specified by the same part of the plasmid that encodes the 5′ end of RNAII, but RNAI is encoded by the complementary DNA strand. RNAI can therefore bind to RNAII to prevent it from hybridizing with the template DNA. ColE1 replication is also controlled by Rom (*R*NA *o*ne *m*odulator), a ColE1-encoded polypeptide of 63 amino acid residues. Rom enhances the inhibition brought about by RNAI because it stabilizes the complex formed between RNAI and RNAII, thus preventing RNAII from hybridizing with the template DNA.

Thus ColE1 replication appears to be essentially controlled by a negative feedback mechanism; on this model, the concentration of RNAI varies according to the plasmid copy number. For example, if the plasmid copy number in a cell falls below the norm, the RNAI concentration will be lower, resulting in less inhibition of plasmid replication and a consequent increase in copy number.

Once synthesis has been initiated at the RNA primer and about 500 nucleotides have been synthesized by DNA polymerase I, the synthesis of the opposite strand (lagging strand) begins, and the remainder of the process appears to be similar to the mechanisms of *E. coli* chromosome replication. When replication is completed, the two daughter plasmids become separated, presumably through the action of DNA gyrase, which cuts and rejoins the two strands of the double helix.

ColE1 replication seems to be typical of that of many small, multicopy plasmids that are found in *E. coli* and its close relatives, such as *Shigella* and *Salmonella.* However, several larger, conjugative plasmids, such as the F plasmid, replicate bidirectionally—two replication forks move in opposite directions from the origin. Antisense RNA plays a role, as an inhibitor, in the replication of several types of plasmid. A different replication mechanism is used by small plasmids found in many Gram-positive bacteria, including *Staphylococcus, Streptococcus,* and *Bacillus.* These plasmids replicate by means of an asymmetric rolling circle mechanism, which results in the presence of both double-stranded and single-stranded plasmids in cells.

### 2.2 Plasmid Partioning and Stability

Plasmids are usually stably inherited, at least in the hosts in which they were originally isolated. This implies that daughter cells almost always receive at least one copy of the plasmid at cell division. For plasmids present in 15 or more copies per cell, such as ColE1, it is unnecessary to propose a specific mechanism for partitioning daughter plasmids at cell division to account for such stability because there is a high probability that each daughter cell will have at least one plasmid. However low copy number plasmids would be rapidly lost during cell replication if there were not a specific mechanism to partition copies between cells at cell division. It is clear that such mechanisms do exist, although the way in which they work is not well understood. However, plasmid-encoded proteins that bind to specific regions of plasmid DNA are known to be involved.

The phenomenon of plasmid incompatibility is probably a result of the mechanism of plasmid partitioning, but again the molecular basis is unclear. Some pairs of plasmids are *incompatible;* that is, they are unable to coexist stably in the same host cell. The phenomenon is used to classify bacterial plasmids into incompatibility groups: no two members of the same group can stably coexist. There are more than 25 incompatibility groups among plasmids found in enterobacteria, seven groups of *Staphylococcus* plasmids and 11 groups of *Pseudomonas* plasmids.

### 2.3 Replication of the 2μm *S. cerevisiae* Plasmid

The 2μm circle of the yeast *S. cerevisiae* is a double-stranded DNA plasmid found in the nucleus. Among its 6318 base pairs is a 599bp sequence that is repeated in the plasmid in a reverse orientation. Frequent recombination between these two repeated sequences results in the presence in cells of equal numbers of two different forms of the plasmid. There are between 30 and 100 copies of the plasmid per cell. Replication of the 2μm circle is dependent on host proteins, and it is controlled similarly to chromosomal replication; for example, initiation of replication occurs only once during S phase.

One of the four genes in the plasmid, the *FLP* gene, codes for a site-specific recombinase, which brings about recombination between the inverted repeat sequences. Two genes, *REP*1 and *REP*2, encode proteins that are necessary for efficient partitioning of the plasmids.

The *FLP* gene product plays an important role in the control of plasmid copy number. When the copy number is low, the *FLP* gene product is made. This results in amplification of the plasmid copy number, probably by means of a mechanism that involves the recombination of a newly replicated inverted repeat with an unreplicated repeat during plasmid replication.

## 3 PLASMID TRANSMISSION

Bacterial plasmids can be transferred to other cells by a variety of different mechanisms. Many plasmids are conjugative, or self-transmissible, and encode the means for transferring copies of themselves to other cells. Other transmission mechanisms are *transduction, transformation,* and *electroporation.* In transduction, a plasmid is included in a bacteriophage particle. Transformation, a widely used technique for introducing recombinant DNA into cells, is the uptake of naked DNA by cells. In electroporation, high voltage electrical pulses are used to cause membranes to temporarily lose their semipermeable status, allowing the cells to take up plasmid DNA.

### 3.1 Conjugation

Conjugative plasmids are found in many different genera of bacteria, both Gram-negative and Gram-positive. Some of these plasmids can transfer themselves to a wide range of different bacteria, and some can even transfer DNA to yeast cells. In addition to being able to transfer themselves to other cells, conjugative plasmids sometimes transfer chromosomal genes between cells.

The best studied conjugative plasmid is the F plasmid found in *E. coli* K-12. F is a circular double-stranded DNA plasmid of 100 kb. The functions required for conjugative transfer are encoded by a 33 kb region ("transfer region") of the plasmid. Many of the genes in this region (there are more than 23 *tra* genes) are needed for the synthesis of F-pili, the hollow filaments about 2μm long, which are formed of pilin protein subunits and protrude from the cell.

An early event in the transfer of the F plasmid by conjugation is the cleavage of the F DNA at a site called *ori* T. The F plasmid DNA is then unwound, and a single strand of DNA is transferred, 5′ end first, into the recipient cell. Meanwhile, the single-stranded DNA remaining in the donor forms a template for the synthesis of a complementary strand. Similarly, the single strand transferred to the recipient serves as a template for the synthesis of a double helix.

### 3.2 Surface Exclusion

A cell carrying a conjugative plasmid is a poor recipient for the transfer of the same or closely related plasmids. For example, it is much more difficult to transfer an F plasmid into a cell that already has such a plasmid than to one that lacks it. Two F-plasmid genes, *tra* S and *tra* T, are responsible for surface exclusion. The *tra* S gene product is an inner membrane protein that appears to inhibit DNA transfer, whereas the *tra* T gene product is an outer membrane protein that inhibits the formation of stable mating pairs.

### 3.3 Hfr Strains

Plasmids can, by definition, exist independently of the chromosome of their host cell. However, many plasmids can also recombine with chromosomal DNA, to which they become physically attached. The classical and best studied example is again the F plasmid, which forms so-called *Hfr* (for *H*igh *fr*equency of transfer) when it is integrated into the bacterial chromosome. Since the *tra* genes of the F plasmid are fully functional in such a strain, the F plasmid can transfer not only itself but also chromosomal DNA to a recipient cell. This has proved to be a particularly useful phenomenon for mapping the positions of genes on the *E. coli* chromosome. The *E. coli* chromosome is 4750 kb long, and the linkage map obtained from time-of-entry data is 100 minutes. That is, it takes 100 minutes to transfer a copy the entire *E. coli* chromosome from donor to recipient, and the position of any particular gene can be mapped according to its time of entry. The formation of an *Hfr* involves the presence of insertion elements. Many *Hfr* strains appear to be formed by recombination between identical insertion elements, such as IS2, present on the F plasmid and on the chromosome.

### 3.4 Transfer of Nonconjugative Plasmids by Mobilization

Many conjugative plasmids can transfer nonconjugative plasmids to recipient cells at high frequencies. For instance ColE1 is very efficiently mobilized by the F plasmid. Cotransfer of the F plasmid into the recipient is not necessary for the process. ColE1 has four genes that are essential for mobilization. The proteins specified by three of them form a complex at the ColE1 origin of transfer, *ori*T. Under certain conditions they can act as endonucleases, producing a break in one of the plasmid DNA strands so that the molecule assumes an open circular or relaxed form, a key step in the process of mobilization.

### 3.5 Broad Host Range Plasmids

Broad host range plasmids belonging to the IncP group are able to transfer themselves by conjugation between almost all species of Gram-negative bacteria. Examples include the very closely related, and perhaps identical, R plasmids RK2, RP1, and RP4, which were found in bacteria causing infections in burns patients in Birmingham, England. These plasmids have a series of coregulated operons (called the *kil-kor* regulon) that are involved in replication and conjugation. Several of the genes in these operons are called *kil* genes because their unregulated expression is lethal. The *kor* genes encode repressors that control *kil* gene synthesis. More recently it has been shown that plasmids such as RK2 can also transfer plasmid DNA by mobilization into several species of Gram-positive bacteria, and to yeast cells. Since these plasmids can also transfer chromosomal genes from their hosts, they may well play an important role in evolution by introducing new traits into cells of many different kinds.

## 4 MAJOR TYPES OF PLASMID

### 4.1 R Plasmids

R plasmids, or drug resistance plasmids, were discovered in 1957 in *Shigella* strains that cause bacillary dysentery. R plasmids have since been found in most of the pathogenic bacterial species and, furthermore, plasmid-determined drug resistance has been found for almost all the antibacterial drugs used for therapy. R plasmids are therefore of great clinical importance. Many R plasmids are conjugative, or can be readily mobilized by conjugative plasmids, and the drug resistance genes themselves are often on transposons. These properties have no doubt contributed greatly to the rapid and widespread distribution and evolution of R plasmids in the presence of the selection pressure provided by antibiotics.

There are four principal mechanisms of plasmid-determined drug resistance, described in Sections 4.1.1 to 4.1.4.

#### 4.1.1 Modification of the Antibiotic to Render It Inactive

Two examples of mechanisms that deactivate antibiotics are resistance to chloramphenicol and resistance to penicillins and cephalosporins. Plasmids conferring resistance to penicillins and cephalosporins code for β-lactamases, which hydrolyze the β-lactam ring. R plasmids conferring aminoglycoside resistance (e.g., resistance to streptomycin) encode enzymes that modify the antibiotic through acetylation, nucleotidylation, or phosphorylation. As a result, the antibiotics no longer bind to their target, the ribosome.

#### 4.1.2 Prevention of Drug Uptake

The most common mechanism for plasmid-mediated tetracycline resistance is prevention of drug uptake. The plasmids encode a 43 kDa cytoplasmic membrane protein, which pumps tetracycline out of the cell to prevent the drug from accumulating.

#### 4.1.3 Modification of the Antibiotic Target

Resistance to macrolide (e.g., erythromycin), lincosamide, and streptogramin B antibiotics can be brought about by plasmid-encoded methylases, which methylate specific adenines in 23S ribosomal RNA, the target of the antibiotics.

#### 4.1.4 Substitution of a Resistant Enzyme

Two examples of plasmid-determined resistance by means of enzyme substitution are the plasmid-encoded dihydrofolate reductase and dihydropteroate reductase, which are far more resistant to the effects of trimethoprim and sulfonamide than are the bacterial (chromosome-specified) enzymes.

### 4.2 Virulence Plasmids

The virulence of many species is often crucially dependent on the presence of plasmids. *E. coli* strains that produce enterotoxins are a major cause of diarrhea. Plasmids may encode two kinds of protein toxin, heat-labile toxin (LT) and heat-stable toxin (ST). The heat-labile toxin closely resembles cholera toxin (which is encoded by chromosomal genes of *Vibrio cholerae*). LT is composed of five A subunits (25.5 kDa), which associate with five B subunits (11 kDa). The A protein enters the cell and activates adenylate cyclase, which increases cyclic AMP levels, with the result that electrolytes and water are secreted into the intestinal lumen. The B subunit is required for receptor recognition. The ability of enterotoxigenic strains to colonize the small intestine plays an important part in their pathogenicity. The factors needed for adhesion and colonization are also often plasmid-encoded. The colonization factor antigens are long thin adhesive appendages, called fimbria or pili, and they are often specific for particular animal species. Thus toxigenic *E. coli* strains isolated from man frequently carry plasmids specifying the colonization factor antigens CFAI, CFAII, CFAIII, or CFAIV, whereas strains isolated from calves or lambs have plasmids specifying different adhesive pili (K88, K99, K41, or P987 antigen).

In addition to causing a diarrheal disease, strains of *E. coli* can cause generalized infections in man and animals, growing in blood (bacteremia) and in organs of the body. Plasmid-specified genes that contribute to the virulence of such strains include the *iss* (*i*ncreased *s*urvival in *s*erum gene), which increases bacterial resistance to the bacterial effects of complement, and genes specifying an iron uptake system (there is very little free iron in blood: it is tightly bound to transferrin and to other proteins).

Plasmids also contribute substantially to the virulence of two important enteric pathogens, *Salmonella* and *Shigella,* as well as *Yersinia* species (e.g., *Yersinia pestis,* the cause of plague), *Staphylococcus aureus, Bacillus anthracis* (the cause of anthrax), and *Clostridium tetani* (tetanus toxin is encoded by a plasmid).

*Agrobacterium tumefaciens* causes crown gall, a disease of dicotyledonous plants; the bacteria infect wounded plants and cause the formation of tumors. A plasmid (Ti) in *A. tumefaciens* is essential for virulence, and a part of it (the T-DNA) is transferred to the plant cells, where it becomes integrated into the plant chromosomal DNA. Enzymes encoded by the T-DNA increase the synthesis of auxins, which bring out cell proliferation and the formation of tumors. Wounded plant cells produce phenolic compounds that induce expression of *vir* genes on the Ti plasmid, resulting in the transfer of a single-strand copy of the T-DNA into the plant cell in a process that resembles conjugation. Ti plasmids also have genes for the production and catabolism of metabolites called opines—the genes for opine synthesis are on the T-DNA. Thus it appears that the T-DNA stimulates the plants to produce metabolites that can be specifically used by the bacteria harboring the corresponding Ti plasmid.

### 4.3 Col Plasmids

Bacteriocins are antibacterial proteins synthesized by bacteria. Most of them have a narrow spectrum of action and kill species closely related to the producer strain. Thus colicins, for example, are produced by *E. coli* and closely related genera such as *Shigella* and are lethal only for closely related bacteria.

Colicins kill cells by first binding to protein receptors in the outer membrane. They then penetrate the cell envelope to interact with their specific targets. Colicins kill cells in one of several ways. Many of them, including colicins A and E1, form voltage-dependent ion channels in the cytoplasmic membrane, which dissipates the membrane potential. The target of colicin E3 is the 16S ribosomal RNA, which is cleaved 49 bases from 3′ end, thus activating the ribosome. Cells carrying a ColE3 plasmid are protected from the effects of colicin E3 because the Col plasmid encodes an immunity protein, which binds to and inactivates the colicin. Colicin synthesis is controlled by the host's lex A protein. This binds to a DNA sequence called the "SOS box," which is found upstream in the operator region of SOS-responsive genes. The SOS response can be induced by treatments that damage DNA, such as UV light, since these treatments activate the Rec A protease, which cleaves the lex A protein.

Induction also results in the production of a Col plasmid-encoded "lysis-protein"—a lipoprotein that allows the colicin to be exported from the cell.

### 4.4 Metabolic Plasmids

Many species of *Pseudomonas* carry degradative plasmids, which enable them to catabolize complex organic molecules such as toluene, salicylate, naphthalene, and camphor.

Plasmids enable the bacterium *Rhizobium* to form a symbiotic association with members of the Leguminosae plant family (plants such as clover, pea, soybean, and alfalfa). In the root nodules of such plants the bacteria fix atmospheric nitrogen by reduction to ammonia, which the plant then uses. The sugars produced by the plant during the photosynthetic reduction of $CO_2$ are used by the bacteria. The plasmids carrying genes needed for the symbiosis are relatively large, ranging from 200 to 1500 kb. Plasmid genes (*nod* genes) are required for the formation of root nodules, including infection of the plant cells by *Rhizobium,* and for nitrogen fixation (*nif* genes). The *nod* genes code for enzymes needed for the synthesis of Nod factors, which are oligosaccharides. These oligosaccharides induce the synthesis of host proteins that are involved in nodule formation.

*See also* Bacterial Growth and Division; Bacterial Pathogenesis; Mycobacteria.

## *Bibliography*

Eguchi, Y., Itoh, T., and Tomizawa, J. (1991) *Antisense RNA. Annu. Rev. Biochem.* 60:631–652.

Hardy, K. (1986) *Bacterial Plasmids,* 2nd ed. American Society for Microbiology, Washington, DC.

———, Ed. (1993) *Plasmids: A Practical Approach,* 2nd ed. Oxford University Press, London and New York.

Miller, J. H., Ed. (1991) *Bacterial Genetic Systems,* Vol. 204, *Methods in Enzymology.* Academic Press, San Diego, CA.

# Polymers, Genetic Engineering of

*Maurille J. Fournier, Thomas L. Mason, and David A. Tirrell*

*Key Words*

**Codon Use** Pattern of preferential use of a subset of the codons (nucleotide triplets) that specify a particular amino acid.

**Genetic Instability** Loss or rearrangement of genetic information.

**Polymer** Chain molecule comprising many copies of a repeating monomer unit or sequence of monomers.

**Structural Proteins** Repetitive proteins such as silks, elastins, and collagens, which function at least in part as mechanical load-bearing elements in cells, tissues, or organisms.

Polymers are long chain molecules of repetitive sequence that are widely used as plastics, rubbers, fibers, adhesives, biomaterials, and composite materials. Genetic engineering is now being used to create protein-based polymers related to silks, elastins, collagens, adhesive proteins, viral spike proteins, and coiled-coil proteins. At the same time, protein polymers that bear no direct relation to any naturally occurring proteins are being designed and expressed in microbial hosts. The precise control of macromolecular architecture provided by the protein biosynthetic apparatus raises the prospect of new classes of genetically engineered polymers for high performance materials applications.

## 1 POLYMERS

### 1.1 Structures and Properties

The most important of the structural variables that determine the properties of polymeric materials are topology, chain length, sequence, and stereochemistry. Topological variants include linear and branched chains, stars, rings, dendritic structures, and cross-linked networks. Control of topology can result in remarkable changes in the physical and mechanical properties of polymers; in the case of polyethylene, for example, the linear (high density) form melts at approximately 135°C, compared to about 105°C for the branched (low density) form prepared by radical polymerization. The improved crystallizability of high density polyethylene (HDPE) in turn raises the tensile modulus of the material by a factor of about 10 in comparison with low density polyethylene (LDPE).

Chain length is the single most important determinant of polymer properties. In general, characteristics such as melting temperature and glass transition temperature are strongly dependent on chain length up to degrees of polymerization (DP) of approximately 100, and then show little sensitivity at higher DP. Other properties remain dependent on chain length even for very long chains. For example, the viscosity of a molten polymer varies as $DP^{3.4}$, owing to entanglements between neighboring chains in the melt.

The issue of monomer sequence arises in the synthesis of copolymers (i.e., polymer chains containing two or more structurally distinct monomer units). Linear copolymers prepared by conventional means can be classified into three major groups: statistical copolymers, alternating copolymers, and block copolymers. Most technologically successful copolymers are characterized by statistical sequences that are determined by the relative reactivities of the constituent monomers under the conditions of polymerization used for their production. In these materials copolymerization is used to modify properties such as crystallinity, solvent resistance, permeability, and toughness, which generally vary relatively smoothly with copolymer composition. Certain monomer pairs (e.g., olefins and maleic anhydride) copolymerize in strictly alternating fashion owing to extreme differences in reactivity, and as a result some commercial copolymers are characterized by perfectly alternating sequences. Finally, through the use of ionic or organometallic initiating systems, it is possible to prepare copolymers in which the constituent monomers are segregated from one another into long homopolymeric blocks. Because the chemically distinct segments of block copolymers are in most cases mutually immiscible, these materials form multiphase solids in which it is possible to combine the properties of each of the corresponding homopolymers. For example, block copolymers of styrene and isoprene are useful as thermoplastic elastomers, in which the glassy polystyrene phase allows the material to behave as a vulcanized rubber at room temperature, but to be processed as a thermoplastic at temperatures above 100°C (the glass transition temperature of polystyrene).

Stereochemistry determines polymer properties largely through its effects on crystallinity and melting point. The polymerization of propylene via organometallic ''Ziegler–Natta'' catalyst systems, for example, produces an ''isotactic'' chain structure (**1**), in which nearly all successive monomer pairs adopt ''meso'' stereochemical arrangements. Isotactic polypropylene is a semicrystalline solid melting at approximately 165°C, compared to the stereoirregular (''atactic'') form of the polymer, which does not crystallize. The isotactic form, which is among the most important of all commercial polymers, is sold in quantities of millions of tons per year, while atactic polypropylene is of only minor commercial significance.

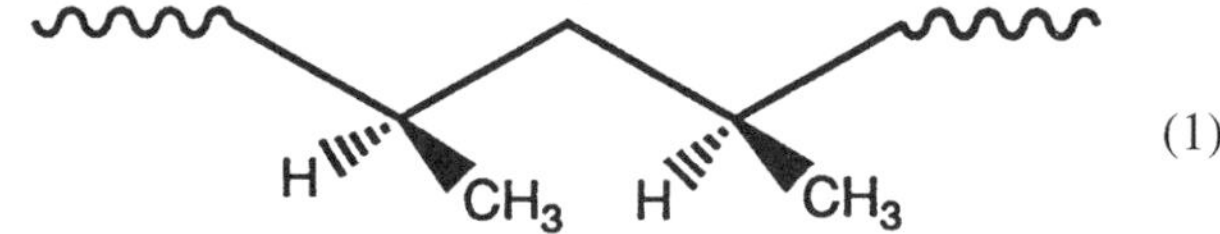


(1)

### 1.2 Synthetic Methods

Polymerization reactions are usefully classified into two kinds of process: step growth polymerizations or polycondensations (e.g., polyamidation or polyesterification) and chain growth polymerizations that involve propagation of reactive radicals, ions, or organometallic species. Both kinds of polymerization are statistical processes, and both produce heterogeneous chain populations characterized by relatively broad distributions of chain length and sequence. Stereochemistry is also subject to control only in a statistical sense. Synthetic polymers are therefore described not in terms of any unique structure, but instead in terms of the average properties of the chain population and the distribution of properties around the average. Such materials are complex mixtures, and this property has had profound impact, both positive and negative, on the technological success of conventional polymers. It has also motivated a continuing search for synthetic methods that provide improved control of macromolecular architecture. Genetic engineering constitutes one such method.

### 1.3 Materials Properties of Natural Polymers

A second motivation for exploring the genetic engineering of polymeric materials arises from the excellent materials properties of certain natural polymers. The dragline silks of spiders, for example, exhibit moduli and tensile strengths comparable to those of high performance synthetic fibers, yet such natural silks require significantly greater elongations to break. As a result, the silks absorb more energy prior to fracture than do the competing synthetic fibers.

## 2 EXPRESSION OF STRUCTURAL PROTEINS

The intriguing properties of the fibrous proteins have stimulated a substantial effort in the development of efficient methods for the synthesis of these polymers and their analogues. Silks, collagens, elastins, adhesive and viral proteins, and proteins forming coiled-coil structures have been expressed successfully in yeast and bacterial hosts.

### 2.1 Silks

The silks illustrate the two major approaches to the microbial expression of structural proteins. The first involves isolation and cloning of the corresponding cDNAs, which has been accomplished successfully for *Bombyx mori* silk fibroins. Although many variants of *B. mori* fibroin have been cloned and characterized, there are as yet no detailed reports of microbial expression of these polymers. Cloning of spider silk cDNAs is also under active investigation.

The second approach to the expression of structural proteins has been used more widely. In this method, the repeating unit of the protein of interest is encoded into a synthetic DNA ''monomer,'' typically 50 to 200 base pairs in length. The monomer is outfitted with nonpalindromic cohesive ends and self-ligated to produce a population of multimers encoding a set of chain length variants of the target polymer. The multimers are then cloned, size-selected, and expressed in the normal fashion.

Several problems with this approach might be anticipated. First, as in other microbial expressions of heterologous proteins, there is the question of codon use by the host organism. While most investigators have adopted coding schemes that reflect the codon use pattern of the host—and in particular, avoid rare codons—there has been no systematic exploration of the relations between codon use and efficiency of expression. The second potential problem is that of genetic instability. Because gene construction via polymerization of oligonucleotides results in extensive sequence homology, repetitive synthetic genes might be expected to be subject to unusually rapid recombination and deletion. Reports of such problems have appeared, but in general genetic instability has not presented insurmountable obstacles to successful expression of uniform, full-length proteins. As in the case of codon use, systematic databases are unavailable, but prudent investigators have adopted coding schemes that reduce the repetitive nature of the artificial gene. This reduction is accomplished by mixing degenerate codons and by avoiding the use of short oligonucleotides in building up the full-length coding sequence.

Many interesting variants of silks have been prepared in this way. Joseph Cappello and co-workers at Protein Polymer Technologies, Inc., have reported the expression in *Escherichia coli* of high molecular weight (70–90,000) proteins that incorporate blocks of the repeating hexapeptide of *B. mori* (Gly-Ala-Gly-Ala-Gly-Ser), and in some polymers have interspersed these sequences with others derived from elastin (Val-Pro-Gly-Val-Gly) or fibronectin (Arg-Gly-Asp-Ser). Copolymerization with elastin reduces the crystallinity of the polymer, while incorporation of the fibronectin cell-binding domain creates a useful substrate for cell culture. The silk–fibronectin polymer can be coated on glass or plastic surfaces and subjected to autoclave sterilization without loss of activity.

Progress has also been made in the cloning and expression of fragments of the silk proteins of the midge (*Chironomus tentans*). Midge larvae spin insoluble fibers consisting in part of a family of very large ($\approx$ 1 MDa) silk proteins designated sp I, which contain approximately 150 copies of an 82-residue core repeat. Steven Case and co-workers, who have prepared synthetic oligonucleotides encoding the core repeat, have reported successful expression in *E. coli* under control of a bacteriophage T7 promoter. The protein has been purified to homogeneity and used in studies of the fiber assembly process.

### 2.2 Collagen

Ina Goldberg and colleagues at Allied Signal, Inc., have prepared a family of synthetic genes encoding tandem repeats of the tripeptide sequence Gly-Pro-Pro, which is of interest by virtue of its analogy to the repeating sequence of collagen. The artificial genes have been cloned in *E. coli* under control of a thermally inducible $\lambda p_L$ promoter. Successful synthesis of a 22 kDa polypeptide has been demonstrated, but the product is labile with respect to proteolysis unless the heat shock response of the host is modified. Use of an *rpo H* 165 mutant is sufficient to stabilize the product protein.

### 2.3 Elastin

The 72 kDa protein tropoelastin forms cross-linked networks that provide the elastic properties of mammalian skin, lungs, arteries, and other tissues. Joel Rosenbloom and co-workers at the University of Pennsylvania have cloned and expressed a full-length tropoelastin cDNA in *E. coli,* under control of a $\lambda p_L$ promoter. The free polypeptide is degraded rapidly in *E. coli,* but expression of tropoelastin fused at its N-terminus to an 81-amino acid fragment of influenza virus NS1 protein yields a stable product. Inclusion of a methionine codon at the junction allows liberation of tropoelastin (which is free of internal methionine residues) by CNBr digestion. After purification, the recombinant tropoelastin exhibits chemotactic activity toward fetal calf ligament fibroblasts, and cross-reactivity with antibodies raised against elastin-derived peptides.

### 2.4 Adhesive Proteins

The adhesive protein of the marine mussel *Mytilus edulis* consists in large part of about 80 repeats of the decapeptide unit Ala-Lys-Pro-Ser-Tyr-Hyp-Hyp-Thr-Dopa-Lys (Dopa = dihydroxyphenylalanine, Hyp = hydroxyproline). The capacity of this material to form strong adhesive bonds in aqueous media has prompted interest in the microbial expression of the protein and several of its analogues.

Robert Strausberg and co-workers at Genex Corporation have isolated a cDNA clone that encodes an adhesive protein consisting of 20 copies of the repeating unit sequence and have expressed this protein (and its dimer, timer, and tetramer) in *Saccharomyces cerevisiae.* The largest of these proteins has a molecular weight of approximately 96,000, and they all segregate with the insoluble proteins of the expression host strain. Use of an expression vector carrying a hybrid promoter composed of fragments of the *S. cerevisiae GAL1* and *MF-α1* promoters produces the adhesive proteins in amounts of roughly 3 to 5% of cellular protein. Treatment of these proteins with a baterial tyrosinase in vitro results in cross-linking and in moisture-resistant adhesion to polystyrene substrates.

Anthony Salerno and Ina Goldberg at Allied Signal have taken the synthetic gene approach to create an analogue of the mussel adhesive protein. The decapeptide repeat was encoded into a 30 base pair DNA monomer consisting of codons that reflect *E. coli* use patterns, and a multimeric DNA encoding 20 copies of the repeat was expressed in *E. coli* under control of a phage T7 promoter. The product was isolated from intracellular inclusion bodies in levels approaching 60% of total cellular protein.

### 2.5 Viral Proteins

The icosohedral capsid of human adenovirus carries fibrillar protein spikes that are believed to mediate the attachment of the virus to the host cell. The spikes are about 200 A long and consist of a dimeric or trimeric assembly of a repetitive 62 kDa protein. The secondary structure of the protein is uncertain, and both cross-β and triple helical models have been proposed.

Corinne Albiges-Rizo and Jadwiga Chroboczek of the European Molecular Biology Laboratory have expressed in *E. coli* the 35 kDa fibrous protein from adenovirus serotype 3, fused at its N-terminus to a short leader sequence derived from the phage T7-based cloning vector. The recombinant protein was isolated as part of an insoluble fraction, and even on denaturing gels was observed to run largely as a trimer. On the basis of the nearly identical gel filtration behavior of the native and recombinant proteins, the authors propose that the native viral fiber also consists of a trimeric protein assembly.

Bacterial expressions of synthetic analogues of viral spike proteins have been reported by John O'Brien and co-workers at E. I. duPont de Nemours and Company. The DuPont group prepared oligonucleotides encoding 15-residue consensus sequences for three such polymers, and after multimerization and cloning in *E. coli,* obtained proteins ranging in molecular weight from 20,000 to 100,000. All the proteins formed inclusion bodies, but all were solubilized in hexafluoroisopropanol (HFIP). Most interesting is that HFIP solutions of several of these polymers were birefringent, indicating the formation of liquid crystalline phases. Fibers spun from the anisotropic solutions were characterized by mechanical properties comparable to those of commercial textile fibers.

### 2.6 Coiled-Coil Proteins

The transcription factors *GCN4, Fos,* and *Jun* form dimeric complexes characterized by coiled-coil structures in which helical chains wrap around one another with a superhelical pitch of approximately 140 A. Kevin McGrath and David Kaplan of the U.S. Army Research, Development, and Engineering Center at Natick, Massachusetts, have reported the design and bacterial expression of a family of polymers modeled on the heptapeptide repeat of the coiled-coil proteins. Their objective is to create protein-based materials capable of intermolecular recognition and self-assembly. Studies of the assembly properties of these materials are at an early stage.

## 3 DE NOVO DESIGN OF ARTIFICIAL STRUCTURAL PROTEINS

Genetic engineering can be used to prepare protein polymers that bear no direct relation to any naturally occurring proteins. Such materials are designed on the basis of principles drawn from polymer chemistry and physics, and have included α-helical and β-sheet proteins, as well as polymers constructed in part from nonnatural amino acids.

### 3.1 α-Helical Proteins

Poly(α,L-glutamic acid) (PLGA) and related polymers have played a central role in investigations of the physical chemistry and materials science of chain molecules. PLGA has been widely used in studies of the helix–coil transition and polyelectrolyte chemistry, and esters of PLGA have been shown to form liquid crystalline solutions, Langmuir monolayers, and Langmuir–Blodgett films.

Monodisperse derivatives of PLGA have been prepared via bacterial expression of artificial genes encoding PLGA fusion proteins. N-Terminal leader sequences have included a 26 kDa fragment of glutathione *S*-transferase (GST, from *Schistosoma japonicum*) and various fragments of the gene 10 protein of bacteriophage T7. The GST fusion allows simple affinity purification on a glutathione column, and the PLGA fragment is liberated by chemical or enzymatic digestion. Alkylation (specifically benzylation) of the glutamate side chains can be accomplished with diazoalkanes, and it has been demonstrated that monodisperse poly(γ-benzyl α,L-glutamate) derivatives form anisotropic (liquid crystalline) solutions in benzyl alcohol.

### 3.2 β-Sheet Proteins

Nearly all flexible, stereoregular polymers crystallize in the form of lamellar aggregates in which the chain lies essentially normal to the lamellar plane and folds regularly at the lamellar surface. With the objective of controlling the thickness and surface chemistry of lamellar polymer crystals, Maurille Fournier, Thomas Mason, David Tirrell, and their co-workers at the University of Massachusetts have prepared several dozen variants of polyalanylglycine containing periodic sequence interruptions designed to dictate the positions and compositions of the chain folds. Expressions in *E. coli* hosts, under control of a phage T7 promoter, have yielded full-length proteins readily purified by simple precipitation procedures. The expected β-sheet structures have been verified in some of these materials, and spectroscopic evidence of adjacent reentry chain folding has been obtained.

### 3.3 Proteins Containing Artificial Amino Acids

The impact of genetic engineering on polymer materials science will be determined in part by the degree to which the methodology can be applied to monomers other than the 20 amino acids normally encoded by messenger RNA templates. Three approaches to the biosynthesis of proteins containing nonnatural amino acids have been described. An intriguing in vitro approach involves chemical charging of a suppressor tRNA that decodes a nonsense codon inserted into the gene of interest at the desired site of substitution. This approach has at least two advantages: it circumvents restrictions on enzymatic charging, and it allows site-specific introduction of the nonnatural monomer. On the other hand, chemical charging of the tRNA is experimentally challenging; moreover, suppression efficiencies are 50% or less, and the practical scale of the method is limited to submilligram quantities of protein. A second *in vitro* method uses a cell-free protein synthesis system supplemented with the amino acid analogue of interest. This method avoids the difficulties of chemical synthesis of the aminoacyl tRNA and of low suppression efficiency, but it is limited to substrates that can be enzymatically activated and charged. The third approach, *in vivo* incorporation of nonnatural amino acids, offers the best prospect for large scale synthesis of proteins containing multiple sites of substitution. This route requires cellular uptake of the analogue as well as enzymatic tRNA charging, but problems of scale and suppression efficiency are avoided. The University of Massachusetts group has used this approach successfully to prepare polymers containing high levels of selenomethionine, trifluoroleucine, *p*-fluorophenylalanine, and 3-thienylalanine.

*See also* BIOMIMETIC MATERIALS.

### *Bibliography*

Aksay, I. A., Baer, E., Sarikaya, M., and Tirrell, D. A., Eds. (1992) *Hierarchically Structured Materials.* Materials Research Society, Pittsburgh.

Alper, M., Bayley, H., Kaplan, D. L., and Navia, M., Eds. (1994) *Biomolecular Materials by Design.* Materials Research Society, Pittsburgh.

Byrom, D., Ed. (1991) *Biomaterials.* Stockton Press, New York.

Vincent, J. F. V., (1990) *Structural Biomaterials.* Princeton University Press, Princeton, NJ.

Viney, C., Case, S. T., and Waite, J. H., Eds. (1993) *Biomolecular Materials.* Materials Research Society, Pittsburgh.

# POLYMERS FOR BIOLOGICAL SYSTEMS

*Robert C. Thomson, Susan L. Ishaug, Antonios G. Mikos, and Robert Langer*

### *Key Words*

**Biodegradable Polymers** Polymers that eventually break down into their smaller molecular weight components, many times down to the monomer.

**Bioresorbable Polymers** A class of biodegradable polymers in which the final lower molecular weight components, such as the monomers, enter metabolic pathways, are absorbed by the body, and eventually are excreted.

**Cross-Linked Polymers** Polymers in which the individual polymer chains are linked by means of covalent bonds.

**Elastomers** Cross-linked polymers which, at room temperature, can be stretched under low stress to at least twice the original length and upon release of stress will return with force to the original length.

**Hydrogels** Hydrophilic cross-linked polymers that swell when in contact with water and biological fluids.

**Monoclonal Antibodies** Antibodies, of the same specificity and having a single amino acid sequence, produced from cloned cells.

**Natural Polymers** Polymers that exist in, and are obtained from, a living organism.

**Synthetic Polymers** Polymers that are manufactured by chemical means.

---

The diversity of mechanical and physical properties that may be achieved using homopolymers and copolymers of different molecular structures has led to their extensive use in a large number of medical devices. These polymeric devices have made it possible to replace certain organs and tissues for which there may be a limited natural supply. Artificial hearts and membrane oxygenators, for example, can prolong human life and improve its quality while donor organs are found. Others, such as polyethylene hip replacements, are designed as permanent implants.

Natural and synthetic biodegradable polymers have found numerous applications in the field of drug delivery. By controlling, and in some cases localizing, the release of pharmaceutical agents, biodegradable polymers have made it possible to treat diseases in a more efficacious manner. Traditionally, medical implants were constructed from commercially available polymers that had been shown to be biocompatible. These materials were essentially invisible to the human body. A second generation of polymers now emerging for tissue engineering applications exhibits specific interactions with cells or tissues and facilitates tissue development and growth. New synthesis schemes and processing techniques are being developed to produce polymers having desired molecular structures and three-dimensional morphologies to respond to technological needs for a variety of pharmaceutical and biomedical applications.

## 1 INTRODUCTION

The biocompatibility of a polymer depends as much on the polymerization process as it does on its chemical structure. The manufacture of any polymer requires the use of a monomer, in many cases an initiator to start the polymerization reaction, and sometimes a catalyst. Inevitably, traces of these substances remain in the polymer even after purification. The toxicity and concentrations of these residues must be taken into account in assessments of polymer biocompatibility. One must consider whether any tissue reaction to the material is due to the polymer or to impurities therein. The material biocompatibility is also dependent on its macroscopic form: sharp edges, rough surfaces, or particles formed by abrasive wear, can negatively influence polymeric implant biocompatibility. Biocompatibility, however, is just one criterion for choosing a particular polymer for a specific application. Equally important in this decision are the physical properties of the material, such as strength, stiffness, and crystallinity.

In recent years, there has been a shift away from the use of commercially available polymers as biomaterials and toward the design of new polymers that possess specialized characteristics. Initially, research focused on designing polymers that exhibited a minimal degree of interaction with biological systems. Such inert polymers are required in orthopedic prostheses and in numerous other applications to prevent fibrous tissue encapsulation, which would impair the performance of the implant. They are also used in blood-contacting devices, such as hemodialysis membranes, in which minimization of thrombosis is of paramount importance. Current research is aimed at the development of polymers that have specific interactions with certain tissues. This area is of particular importance in the design of polymers for organ regeneration, in which cell–polymer interactions are critical in achieving cell growth and in maintaining the cell phenotype. The development of such novel materials, which must then be submitted to the U.S. Food and Drug Administration (FDA), inevitably takes time. Hence we are in a state of transition between the use of conventional polymeric biomaterials and the application of those which have been specifically designed for biomedical use.

## 2 BIODEGRADABLE POLYMERS

We include hydrolysis and enzymatic degradation under the heading of biodegradative processes, but solubilization, which is really a thermodynamic transition to a state of lower free energy, not a degradative process involving chemical reaction, is excluded. When only hydrolysis and enzymatic degradation are considered, a relatively small number of polymers can be classified as truly biodegradable.

Biodegradable polymers are temporary structures that may be utilized to provide mechanical support, deliver drugs at a controlled rate, or provide a scaffold for tissue regeneration. The structures, degradative mechanisms, and biomedical uses of some of the more important biodegradable polymers are outlined next. They are classified according to whether they occur in nature (Section 2.1) or are chemically synthesized (Section 2.2).

### 2.1 Natural Polymers

Several natural polymers are used currently in the biomedical field, and many others are under development. The major advantage of many natural polymers over the synthetics is a highly organized structure, at both molecular and macroscopic levels, which imparts to the polymer such favorable biomaterial characteristics as strength, or the ability to induce tissue ingrowth. Polypeptides with certain amino acid sequences can impart information important to cell adhesion and function, and, therefore, can serve as substrate materials for cell attachment and transplantation. Since, however, these polymers are antigenic, the implantation of such a material into a host stimulates an immune response, leading to possible immune rejection.

#### 2.1.1 Collagen and Gelatin

Collagen is one of the most widely used and best characterized natural biomaterials. There are currently 10 known forms of this hydrophilic polymer (collagens I–X), all differing in chemical and morphological structure. The primary structure of this protein consists of approximately one-third glycyl residues, which impart great rotational mobility to the individual protein chains. The chains form a triple helix called tropocollagen, which is itself coiled. There is a further degree of structural order in which tropocollagen fibers aggregate in a linear arrangement with a precisely defined overlap between fibers. This highly organized structure gives collagen a high degree of structural strength. Indeed, collagen is the major constituent of connective tissues such as bone (type I collagen) and cartilage (type II collagen).

Collagen fibers, bonded by glycosaminoglycans to form a matrix, have been used as a scaffold for the regeneration of damaged skin, to prevent fibrous scar formation during the healing process. In this application, the wound is covered with a microporous form of the matrix, which provides many potential sites for cell attachment. The physical and chemical nature of these sites is such that epithelial cells are induced into the micropores. Over a period of several days, epithelial cells continue to invade the collagen–glycosaminoglycan scaffold and secrete extracellular matrix (ECM), which forms the basis of a new, completely natural layer of skin. As the scaffold is degraded by several enzymes, including collagenase, it is superseded in its supportive role by ECM produced by the epithelial cells. Cell attachment is an important factor in cell transplantation, since it affects greatly both the proliferation and the differentiation of the cells. Collagen–glycosaminoglycan substrates have therefore been widely employed in this method of tissue reconstruction and have been successful in their application to cell-seeded dressings for the treatment of severe burns in humans. They have also been used in animal studies for nerve reconstruction.

Gelatin has the same primary structure as collagen but lacks the higher levels of organization, hence possesses poor mechanical properties. Gelatin, cross-linked with formaldehyde, has been studied in vitro as a drug delivery matrix, an application that seldom requires a high degree of structural strength. Gelatin's lack of organized structure is also not conducive to the induction of tissue into its bulk. This property has been successfully exploited to prevent fibrous tissue ingrowth following canine spinal surgery.

#### 2.1.2 Polysaccharides

Polysaccharides, such as starch, dextrans, chitin, and cellulose, form another interesting class of natural biodegradable polymers. Also known as glycans, these polymers consist of one or more types of monosaccharide unit (glucose, fructose, galactose, etc.) linked by glycosidic bonds.

Starch consists of a mixture of hydrophilic polymers of glucose (glucans), mainly α-amylose and amylopectin. It is a major food reserve in plants and provides many animals, including humans, with an abundant source of nutrients. Amylopectin is a branched glucan, whereas α-amylose is a linear glucan that forms a left-handed helix stabilized by intrachain hydrogen bonds. Neither glucan imparts any significant degree of structural strength to this material, which is not so surprising when one considers that nature did not ''design'' starch with a structural role in mind. Although starch in its natural form has found little use as a biomaterial, it can be cross-linked by chemical treatment to form a hydrogel. This hydrogel, which is enzymatically degradable, was developed for use in triggered drug delivery and has been studied in vitro for the release of naltrexone in response to morphine.

Cellulose is a fibrous material that, unlike starch, plays a structural role in plants. In plant cell walls it is required to withstand large osmotic pressures, and in large plants such as trees it also fulfills a load-bearing function. Cellulose is another glucan, but, unlike α-amylose and amylopectin, the glucose units are linked by β(1–4) bonds.

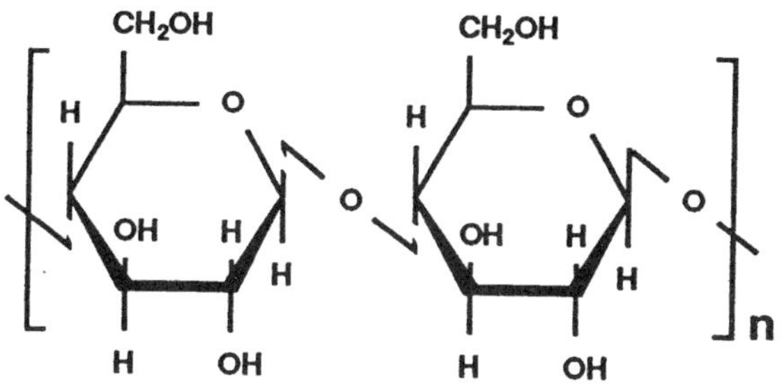

Cellulose

The glucose chains thus formed are hydrogen-bonded with each other and form sheets that impart a great deal of strength to the cellulose fibers. The presence of these hydrogen bonds renders cellulose insoluble in water despite its hydrophilicity. Cellulose has been used for membranes in early hemodialyzers and artificial oxygenators and is still widely employed for hemodialysis membranes today. Furthermore, because of its hydrophilicity, cellulose has been utilized in pharmaceutical formulations to enhance the water uptake and improve drug delivery. Cellulose is not degraded by mammalian enzymatic processes but can be made degradable by chemical treatment. Chemically treated cellulose, in the form of a knitted mesh, has been used as a barrier to prevent the formation of undesirable postoperative tissue adhesions and is currently approved for use in gynecological pelvic surgery.

### 2.1.3 Poly(amino acids)

Poly(amino acids), or proteins, form another group of natural degradable polymers, but for several reasons, and with the notable exception of collagen, they have not been as widely used as one might expect. Each unit of the polymer chain can be any one of the 20 commonly occurring standard amino acids. Hence, over a chain length of hundreds or thousands of amino acid residues, the number of permutations is enormous. Nature, however, through the processes of natural selection, produces only a very small fraction of these permutations based on the requirements of the particular biological system. Nevertheless, there are still many proteins that can be extracted from living systems, and one might consider trying to take advantage of nature's optimization over billions of years of evolution. However, there are some problems associated with this concept, such as the high costs of extraction and purification of the desired protein from a multitude of other compounds, which will undoubtedly be present regardless of the source. Once a given protein has been obtained in a pure form, there is another practical problem to be overcome—that is, since proteins have very poor processing characteristics, it is difficult to manufacture these polymers with the shape and form required for a particular application.

Difficulty of processing is a common problem with many natural polymers, as is their antigenicity, which often causes a chronic immune response and implant rejection. In addition, because the degradation of naturally occurring polymers almost always relies on enzymatic processes, there will inevitably be some patient-to-patient variation in the degradation rate depending on the activity of the specific degradative enzyme in each individual. Proteins in particular, unless they are extracted from the proposed host, tend to stimulate high levels of immune response upon implantation. Hence, with the exception of monoclonal antibodies, proteins are seldom used in their natural form. Monoclonal antibodies offer the exciting possibility of specifically targeting drugs, not just to specific areas of the body, but to specific cell types that exhibit the particular antigen for the antibody. This system of drug delivery may be especially useful for targeting cytotoxic drugs to tumor cells. In this application, it is the specificity of the protein, not its susceptibility to degradation, that is of greatest importance.

In an attempt to overcome some of the problems just outlined, poly(amino acids) that contain only one, two, or at the most three, different amino acids have been synthesized. These polymers are synthetic but mimic some of the positive qualities of their natural counterparts without retaining their antigenic properties. Pseudopoly(amino acids) with chemically modified backbones (nonamide linkages between amino acids) have been produced to improve strength and ease of processing.

## 2.2 Synthetic Polymers

Synthetic biodegradable polymers offer several advantages over natural polymers. Because they are chemically synthesized, it is possible to control (with varying degrees of accuracy depending on, among other factors, the type of polymerization reaction) their molecular weight and molecular weight distribution. These characteristics have a profound effect on the physical properties of the polymer, such as strength and degradation rate. The degradation of synthetic polymers is, in general, brought about by simple hydrolysis, although in some cases enzymatic processes assist in the degradation mechanism. Simple hydrolysis is desirable because the degradation rate does not vary from person to person. In addition, degradation rate data obtained in animal studies can be expected to be reproducible in humans. In the case of biodegradable copolymers, the degradation rate may be controlled by varying the ratio of the monomer units in the polymer. In some instances, it is possible to copolymerize a nondegradable polymer monomer unit with a small proportion of a degradable polymer monomer unit. This technique can be used to make an otherwise nondegradable polymer degradable.

Synthetic polymers also tend to be more easily processed into a finished product. This is particularly important in organ regeneration, since it requires the use of a degradable polymer scaffold with a large surface area and a defined pore morphology. Synthesized polymers are also widely used for drug delivery. Another major advantage of synthetic degradable polymers over their natural counterparts is that they can be produced in bulk and are therefore often much cheaper.

### 2.2.1 Poly(α-esters)

Many polymers may be classed as poly(α-esters), but we consider only the two most widely used, poly(glycolic acid) and poly(lactic acid), which are produced by ring-opening polymerization of glycolide (R = H) and lactide (R = $CH_3$) respectively, schematically represented as follows:

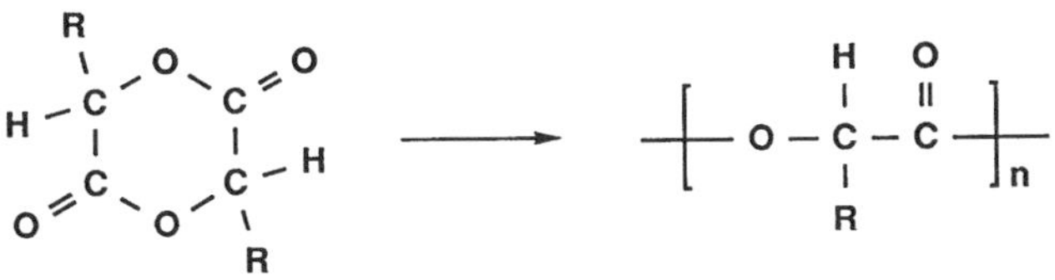

There are three forms of polylactic acid: poly(L-lactic acid), poly(D-lactic) acid, and the stereocopolymer poly(D,L-lactic acid). In this section, we discuss only poly(L-lactic acid), since it is the most widely used of the three.

Poly(glycolic acid) (PGA), a highly crystalline, hydrophilic poly-(α-ester), was one of the first synthetic degradable polymers to find application as a biomaterial. Its degradation in vivo is brought about by simple, nonenzymatic hydrolysis of the polyester bond, which occurs at a rate dependent on the crystallinity of the polymer. Hydrolysis continues to break the polyester bonds at random sites along the polymer chains until, eventually, glycolic acid is produced. As glycolic acid forms, it is processed through normal metabolic pathways and ultimately is eliminated from the body, through the respiratory system, as carbon dioxide. When the degradation products are components of the body's natural metabolic pathways, we classify the polymer as bioresorbable. One of the first applications of PGA was as a biodegradable suture material. PGA has also been used as an artificial scaffold in cell transplantation and organ regeneration. In all its applications, PGA, and its degradation products, have proven to be nontoxic and biocompatible, hence this polymer has gained FDA approval for a number of applications.

Poly(L-lactic acid) (PLLA) is another bioresorbable poly(α-ester). It is less crystalline and more hydrophobic than PGA and therefore degrades at a slower rate. The major degradation mechanism is simple hydrolysis; however, there is a small but nonetheless significant contribution from in vivo enzymatic degradation. The lower degradation rate of PLLA, compared to PGA, is due to the presence of the methyl group, which imparts to PLLA its hydrophobic character and sterically hinders ester bond cleavage. The polymer chains are eventually broken down to individual lactic acid units which, like glycolic acid, are then processed through natural metabolic pathways to produce carbon dioxide, which is exhaled. PLLA is one of the strongest known biodegradable polymers and has therefore found many applications in areas such as orthopedics, where structural strength is an important criterion. A porous form of PLLA has been used as a scaffold for cell transplantation in organ regeneration. The surface chemistry and the pore morphology of this material provide a favorable environment for cell attachment, growth, and differentiation. PLLA has also been used for the controlled delivery of drugs, but copolymers of poly(DL-lactic-*co*-glycolic acid) are more suitable for this application. By altering the copolymer ratio of lactic to glycolic acid, the degradation rate can be tailored to the requirements of the specific application.

### 2.2.2 Poly(ortho-esters)

The ortho-ester polymers degrade by surface erosion and therefore exhibit zero-order degradation kinetics, such that the degradation rate depends only on the wetted surface area of the polymer. These polymers have therefore been exploited as constant rate drug delivery devices. The rate of degradation, hence the rate of drug release, can be controlled by the inclusion of acidic or basic compounds into the polymer matrix.

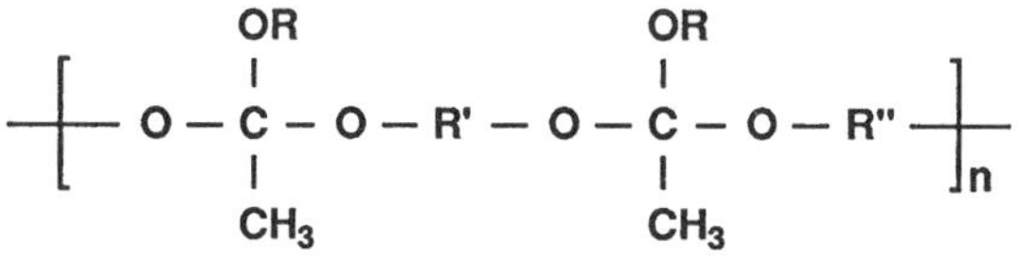

$$\left[ -O-\underset{CH_3}{\overset{OR}{\underset{|}{\overset{|}{C}}}}-O-R'-O-\underset{CH_3}{\overset{OR}{\underset{|}{\overset{|}{C}}}}-O-R''- \right]_n$$

**R, R', R" : Aromatic or Aliphatic Hydrocarbons**

Poly(ortho-esters)

One example of the use of poly(ortho-esters) is in the controlled release of the contraceptive steroid levonorgestrel. The particular polymer used in this case was prepared from the diketene acetal 3,9-bis(ethylidene 2,4,8,10-tetraoxaspiro [5,5]-undecane) and 3-methyl-1,5-pentanediol and was subsequently cross-linked, after the inclusion of the drug, using 1,2,6-hexanetriol. In animal studies, a fairly constant release rate of levonorgestrel could be achieved over a period of about one year. The rate of release could be altered by the inclusion of an acid diol. This is a prime example of one of the advantages of poly(ortho-esters)—the capability of low temperature drug incorporation. Since steroids and many other drugs are vulnerable to thermal degradation, it is essential to avoid high temperatures during the impregnation of the polymer matrix with the drug. The poly(ortho-esters) used in drug delivery applications have low melting points; hence the drug can be mixed into a low temperature polymer melt that cross-links and then forms a solid device with a uniform drug distribution. Uniform drug distribution is an important factor in attaining constant release rates.

### 2.2.3 Poly(anhydrides)

Poly(anhydrides) are formed by a melt condensation reaction and degrade hydrolytically by surface erosion. These polymers are used in controlled drug delivery and there are several excellent reviews on their application in this field.

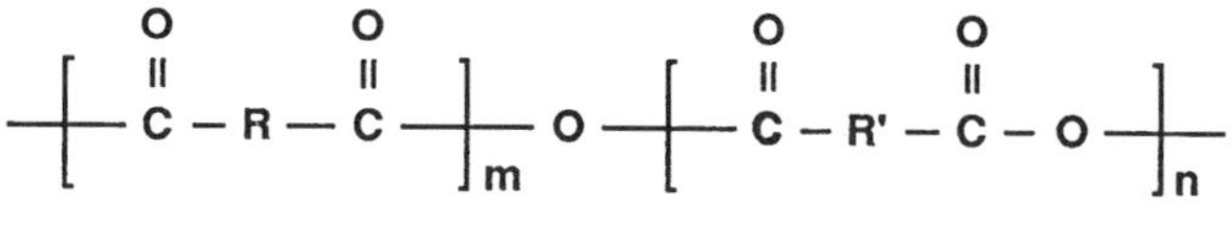

$$\left[ -\overset{O}{\overset{\|}{C}}-R-\overset{O}{\overset{\|}{C}}- \right]_m O \left[ -\overset{O}{\overset{\|}{C}}-R'-\overset{O}{\overset{\|}{C}}-O- \right]_n$$

**R, R' : Aromatic or Aliphatic Hydrocarbons**

**Poly(anhydrides)**

Poly(anhydrides), like poly(ortho-esters), exhibit a constant rate of drug release, and they have been used for the delivery of a number of different drugs. Currently, the most widely used poly(anhydride) is a copolymer of sebacic acid (SA) and 1,3-bis(*p*-carboxyphenoxy)-propane (CPP), which henceforth is referred to as P(CPP-*co*-SA). This copolymer exhibits linear drug release kinetics and has recently been tested in human clinical trials for the controlled delivery of BCNU [carmustine: 1,3-bis(2-chloroethyl)-1-nitrosourea]. BCNU is a chemotherapy agent used in the treatment of glioblastoma multiforme, a particularly severe form of brain cancer with poor prognosis, especially in its advanced stages. The same copolymer has also been studied in animals for the controlled release of insulin in the treatment of hyperglycemia in diabetics.

It has already been mentioned that poly(anhydrides) can exhibit a constant rate of drug release, but it is also possible to customize the release rate. Since altering the comonomer ratio changes the rate of polymer degradation, the rate of drug release may be tailored to that required for a specific application.

## 3 NONDEGRADABLE POLYMERS

The synthetic nondegradable polymers are classed as hydrophobic and hydrophilic. It is the extent of hydrophobicity that affects their interaction with biological systems and determines their medical and pharmaceutical applications.

### 3.1 Hydrophobic Polymers

Hydrophobic polymers possess a wide range of physical and chemical properties and are used in orthopedics and in many diverse applications, including artificial hearts, catheters, and artificial oxygenator membranes.

#### 3.1.1 Polydimethyl Siloxane

Inert sturdy materials are required for many types of biomedical application. Silicone rubber is the most widely accepted artificial polymeric material for surgical use. It is inert, blood compatible, flexible, and elastic at low temperatures. Furthermore, it is a good electrical insulator, does not deteriorate with repeated sterilization, and can be processed into a wide range of shapes. Medical silicone elastomers are predominately poly(dimethyl siloxane) produced by condensation polymerization.

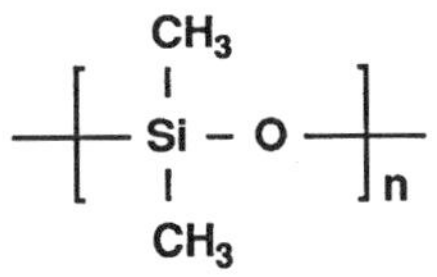

$$\left[ -\underset{CH_3}{\overset{CH_3}{\underset{|}{\overset{|}{Si}}}} - O - \right]_n$$

**Poly(dimethyl siloxane)**

Vinyl or phenyl groups can replace the methyl side groups of the dimethyl siloxane unit, and small amounts are sometimes dispersed throughout the polymer chains. Silicone polymers can have liquid, gel, or rubberlike consistency depending on the degree of cross-linking. There are many silicones available, although we focus only on heat-vulcanized silicone rubbers. Polymer chains of these rubbers are cross-linked at widely separated points when heated with a catalyst, usually dichlorobenzoyl peroxide (DBP), which breaks down into free radicals to start the cross-linking between the side groups of the dimethyl siloxane polymer. Incorporation of the vinyl groups enhances the cross-linking, while phenyl groups contribute to the softness of the rubber. Medical grade silicone is usually composed of polymer chains having 5000 to 9000 monomeric units and 35% silica filler, which is added to enhance the strength. The flexibility of the polymer chain can be attributed to two chemical features: (1) the methyl groups on the silicon atom are extremely mobile and occupy a significant amount of space, increasing the distance between adjacent atoms, and (2) the 160° angle of the Si—O—Si bond results in a highly polar character for the Si—O backbone, wide spacing of the atoms, and a depression of molecular cohesion.

Silicone rubber is most frequently used in reconstructive therapy. The outer shells of breast implants are made of silicone elastomer. Lumina of the first silicone implants were filled with silicone gel, but later, saline-filled implants were used. Although other efforts were made to improve the prosthesis, the saline-filled, silicone-shelled form is the only FDA-approved implant available for use today. Medical tubing and other implants need a firmer silicone, which can be achieved by highly cross-linking the polymer. Cartilage and bone can be replaced with silicone rubber maxillofacial prostheses, which are utilized in nasal supports, jaw and chin augmentation, and orbital floor repair. Silicone rubber has been successful as a prosthesis to replace diseased or mechanically damaged cartilage in finger joints. It has been employed in many cardiovascular applications because its blood compatibility is superior to that of many other materials. Silicone rubber has a high degree of gas permeability to oxygen and has therefore been used for the construction of artificial oxygenator membranes, even though these membranes were inefficient in terms of carbon dioxide elimination.

#### 3.1.2 Polyurethanes

Polyurethanes have the following general repeating chemical structure:

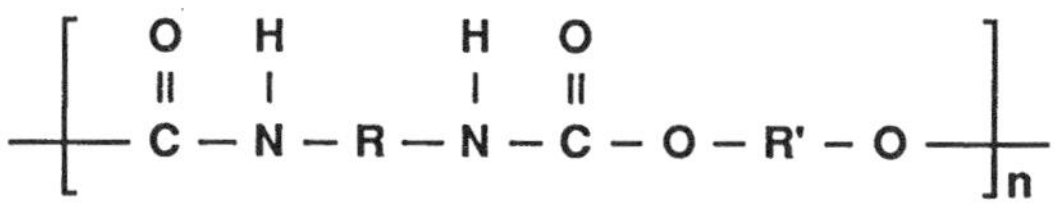

$$\left[ -\overset{O}{\overset{\|}{C}} - \overset{H}{\overset{|}{N}} - R - \overset{H}{\overset{|}{N}} - \overset{O}{\overset{\|}{C}} - O - R' - O - \right]_n$$

**R, R' : Aromatic or Aliphatic Hydrocarbons**

Polyurethanes

The reaction of aliphatic diisocyanates and glycols (also called diols) was the first method used in the production of polyurethanes.

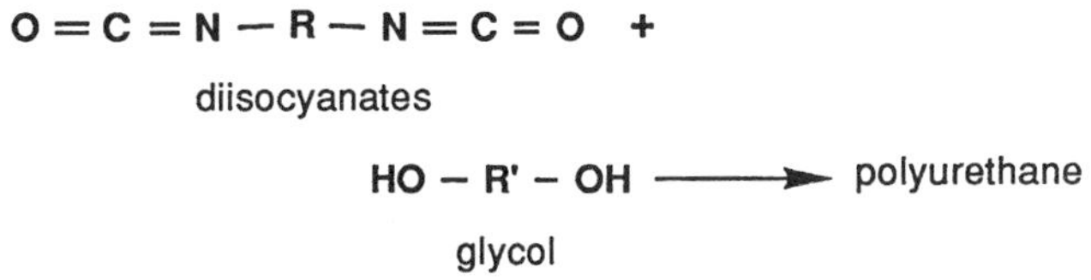

$$\underset{\text{diisocyanates}}{O=C=N-R-N=C=O} + \underset{\text{glycol}}{HO-R'-OH} \longrightarrow \text{polyurethane}$$

Another method involves the reaction of bischloroformates with diamines.

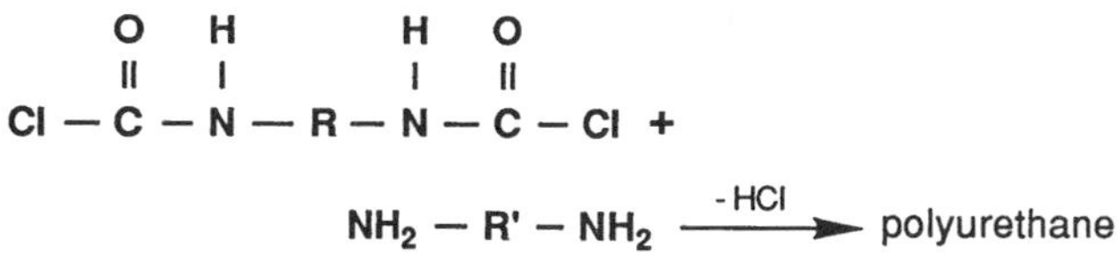

$$Cl - \overset{O}{\overset{\|}{C}} - \overset{H}{\overset{|}{N}} - R - \overset{H}{\overset{|}{N}} - \overset{O}{\overset{\|}{C}} - Cl + NH_2 - R' - NH_2 \xrightarrow{-HCl} \text{polyurethane}$$

Typically polyurethane elastomers consist of alternating rigid and flexible segments. The rigid or hard segments consist of a diisocyanate chain extended with a low molecular weight diol or diamine. The flexible or soft segment is a long chain polyglycol with a molecular weight between 500 and 5000.

Polyurethanes have become popular for biomedical applications because of their relatively good blood compatibility and excellent physical and mechanical properties. Their properties can be altered by making them into copolymers, modifying their surfaces, and fabricating them into various shapes. Probably the most popular commercial biomedical polyurethane is Biomer®, which is a linear, segmented aromatic polyether-based polyurethane. Since polyurethanes have such good blood compatibility compared to other polymers, their applications in medicine tend to focus on blood-contacting devices. Examples include blood filtration membranes, wound dressings, prostheses, catheters, and artificial hearts.

#### 3.1.3 Poly(tetrafluoroethylene)

Of all the perfluorinated polymers that have been used in medicine for in vivo applications, poly(tetrafluoroethylene) (PTFE) is perhaps the most versatile. PTFE has the repeating unit [—$CF_2$—$CF_2$—] and is formed by the addition polymerization of tetrafluoroethylene. It is a strong, tough, crystalline, hydrophobic polymer, which is insoluble in biological/bodily fluids and shows no signs of degradation even after long-term implantation. It may be processed into many forms (solid, porous, composite material, film, etc.), and it is perhaps this diversity of form that has made PTFE the polymer of choice for a number of applications, including vascular prostheses, membrane oxygenators, and hip replacements.

### 3.1.4 Poly(ethylene)

Poly(ethylene) (PE) has the repeating unit [—$CH_2$—$CH_2$—]. It is a crystalline polymer that is formed by the addition polymerization of ethylene gas. Depending on the reaction conditions, three types of PE may be produced: low density poly(ethylene) (LDPE), high density poly(ethylene) (HDPE), and ultrahigh molecular weight poly(ethylene) (UHMWPE). The highly branched nature of LDPE chains prevents their close packing; hence LDPE is significantly less crystalline than either HDPE or UHMWPE. In the past, LDPE has been studied in animals for use in tendon repair and contraceptive devices, but there is some doubt as to its biocompatibility in these applications. A porous form of HDPE has been used to construct total and partial ossicular replacement prostheses. However, UHMWPE, which has a high tensile strength, has overshadowed both LDPE and HDPE in its extensive utilization for orthopedic applications.

### 3.1.5 Poly(vinyl chloride)

Poly(vinyl chloride) (PVC) has the repeating unit [—$CH_2$ —CHCl—]. Its major use is as a material for the medical grade tubing used for indwelling urethal catheters, endotracheal tubes, and so on. Although the tubing is fairly stiff at room temperature, it softens slightly at body temperature; hence, when used internally, it is more comfortable for the patient. Even at body temperature, PVC tubing is more rigid than for example, latex tubing, and this property has two major advantages. First, it is possible to produce tubes with a much smaller wall thickness, without compromising the structural integrity of the tube. When the wall thickness is minimized, the outer diameter is also minimized (without reducing the internal cross-sectional area), and this makes the tube both more comfortable for the patient and more easily positioned by the physician. Second, the rigidity also allows the tubing to be used for purposes that require the application of a vacuum. PVC tubing has a much smoother surface than other tubing types, hence has better flow characteristics and is less vulnerable to buildup of debris on the inner wall.

### 3.1.6 Poly(methyl methacrylate)

Poly(methyl methacrylate) (PMMA) is a tough, strong polymer that is formed by the free radical polymerization of methyl methacrylate (MMA).

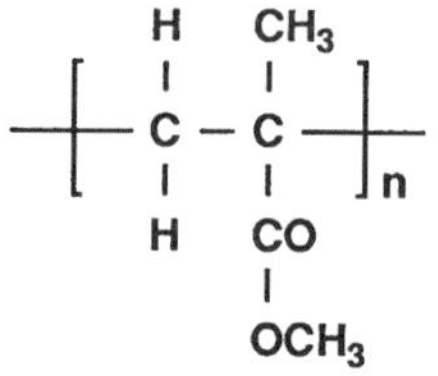

```
      H   CH3
      |    |
  -[- C -- C -]-n
      |    |
      H   CO
           |
          OCH3
```

**Poly(methyl methacrylate)**

Benzoyl peroxide forms free radicals when heated to 60°C and is a commonly used initiator for the reaction. The polymerization reaction can also occur spontaneously while the PMMA is in storage; hence stabilizers, such as hydroquinone, are added. The levels of initiator and stabilizer in the final polymer product are sufficiently low that they do not affect its biocompatibility. PMMA is characterized by high optical clarity and by the ease with which it can be processed. These are desirable properties for ophthalmic application, and indeed this is one of two major areas of use for PMMA. Its other major application is as a bone cement, or resin, which can be cured in situ.

The biocompatibility of PMMA in ocular sites was first observed during World War II, when it was noted that splinters of this material from aircraft canopies provoked surprisingly little tissue reaction when they became lodged in the eyes of airmen. Since that time, high purity PMMA has been used in several ophthalmic applications. Following cataract surgery for the removal of a clouded lens, it is possible to partially restore vision to the affected eye by inserting an intraocular plastic lens made of PMMA. PMMA has also been used to make hard contact lenses, and cosmetic artificial eyes. In addition, it is widely used for bone cements, which are adhesive resins used either to achieve strong bone–prosthesis bonding or to fill bone defects.

## 3.2 Hydrophilic Polymers

Almost all hydrophilic nondegradable polymers are used in their hydrogel form. Hydrogels are hydrophilic cross-linked polymeric networks that swell in aqueous solutions. They are biocompatible and have low interfacial tension, high permeability to small molecules, and a soft and rubbery nature, which renders them good candidates for use as biomaterials. The cross-links can be covalent bonds or secondary bonds. Hydrogels that are covalently cross-linked are insoluble whereas those that are held together by secondary bonds may be dissolved chemically. This section focuses on the nondegradable covalently cross-linked hydrogels. The hydrogel properties are also controlled by the following properties of the monomer(s): hydrophilicity, copolymer ratio for copolymers, cross-linking density, crystallinity, porous structure, and amount of un-cross-linked polymer chains. Since their discovery, hydrogels have been examined for their potential use in replacing soft tissue, as carriers for drugs and bioactive macromolecules, and in other biomedical applications. One limitation of hydrogels is their relatively low mechanical strength, which prevents their use in the replacement of hard tissues.

### 3.2.1 Poly(2-hydroxyethyl methacrylate)

The beneficial biomedical properties of poly(2-hydroxyethyl methacrylate) (PHEMA) were discovered in the 1960s.

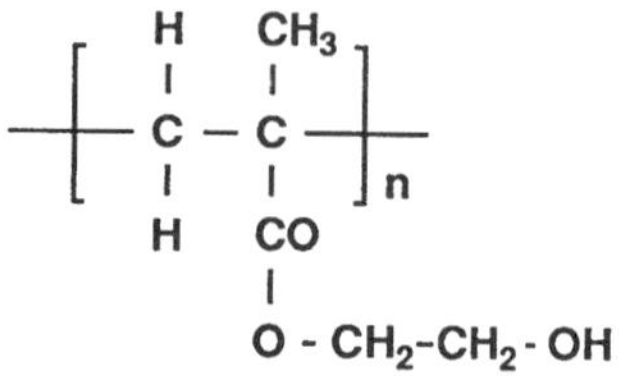

```
      H   CH3
      |    |
  -[- C -- C -]-n
      |    |
      H   CO
           |
          O - CH2-CH2 - OH
```

**Poly(2-hydroxyethyl methacrylate)**

Investigators found that the hydrophilic cross-linked polymer has unique surface properties and will simulate natural tissues, since it can absorb 30 to 90 wt % water or biological fluids. PHEMA implants are biocompatible, do not adhere significantly to the surrounding tissues, and do not promote much tissue ingrowth or calcification.

Soft contact lenses are by far the largest accepted clinical application of PHEMA. In addition, drug delivery has greatly benefited from the discovery of PHEMA. One of the earliest and most studied hydrogel-based wound dressings is composed of PHEMA and poly(ethylene glycol) (PEG). PHEMA can also be used for mammary prostheses and, in orthopedics, as articular cartilage replacements.

### 3.2.2 Poly(vinyl alcohol)

Poly(vinyl alcohol) (PVA) is produced by the hydrolysis of poly(vinyl acetate) and is usually cross-linked by radiation. It has the repeating unit [—$CH_2$—CHOH—]. Hydrogels have inherent biocompatibility properties and therefore have been heparinized in the hope of producing nonthrombogenic materials for implantation devices. Of all hydrogels heparinized, poly(vinyl alcohol) seems to have been studied the most. PVA has potential in controlled drug release applications. It is widely employed in plasmaphoresis and is also used in conjunction with PHEMA and other polymers in the fabrication of certain contact lenses. Limited studies of the use of PVA in hemodialysis membranes, artificial hearts, wound dressings, and blood-contacting devices have also been performed. Since PVA allows the passage of mid-range molecular weight molecules, its potential use in hemodialysis membranes is apparent. In addition, PVA has been used in copolymer forms for arterial applications and mammary prostheses, although problems associated with calcification have appeared.

### 3.2.3 Poly(*N*-vinyl pyrrolidone)

Poly(*N*-vinyl pyrrolidone) (PNVP) is more hydrophilic than PHEMA.

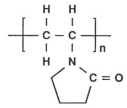

Poly(*N*-vinyl pyrrolidone)

The two polymers can therefore be copolymerized to increase the water uptake of PHEMA. This increase in hydrophilicity is desirable in many applications, such as soft contact lenses and artificial intraocular lenses. Hydrophobic substances such as PMMA, most often used in artificial intraocular lens fabrication, tend to damage the corneal endothelial upon contact. Coating hydrophobic PMMA intraocular lenses with hydrophilic PNVP can hold the loss of corneal endothelial cells to a very acceptable level. The high water uptake that can be achieved with PNVP may also be useful in applications such as hemodialysis membranes and drug release systems, since high water content usually corresponds to easier diffusion of macromolecules.

### 3.2.4 Poly(ethylene oxide)

Poly(ethylene oxide) (PEO) is a relatively inert polymer and has the repeating unit [$CH_2$—$CH_2$—O—]. It is blood compatible and does not seem to bind plasma proteins, such as fibrinogen and thrombin. When PEO is coated onto the surface of materials such as glass, there is a decrease in adhesion of platelets and plasma proteins. It appears that a promising approach to reducing blood–interface interaction is to incorporate PEO into an existing material or apply it as a coating. PEO also has potential as a material for drug delivery devices, wound dressings, contact lenses, or hemodialysis membranes. It has been used as the base for tablets, suppositories, and creams.

### 3.2.5 Poly(acrylamide)

Poly(acrylamide) (PAAm) can be cross-linked by irradiation to form a hydrogel. Since this hydrogel has an ionizable nitrogen atom, the gel's hydrophilicity, hence its swelling, can be seen to vary with changes in temperature, pH, tonicity, and applied electrical fields.

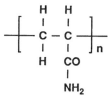

Poly(acrylamide)

For example, ions can be removed from a previously swollen PAAm hydrogel by applying across the gel an electrical field that leaves the charged polymer chains electrically unbalanced. Electrostatic forces cause the charged chains to repel each other, and the polymer swells up to a thousandfold. The unique variable swelling character of this and other ionizable hydrogels gives the materials great potential for use in many applications. One possible use of PAAm is in controlled drug release applications, where the gel could swell and release drugs in specific natural or induced environments. PAAm is also useful for electrophoretic separation of blood plasma proteins, since plasma proteins move through loose PAAm networks, which can be variably ionized. Entrapment of antibodies and antigens for immunoabsorbents has been achieved primarily in PAAm gels. The hydrogel has also been used to entrap cells to permit harvesting of their products.

## 4 CONCLUSIONS

Biomaterials for use in biological systems have been the subject of copious research efforts. Metals are now routinely used in orthopedics. Ceramics also find diverse applications in dentistry. It is polymers, however, that have made the greatest impact in medical and pharmaceutical sciences because of their inertness, ease of fabrication and processing, and high ratio of mechanical strength to density. Polymers are particularly attractive for biological applications because their properties can be tailored in many essential ways by modifying their surface structure and composition, or by copolymerizing different monomers, cross-linking linear chains, and annealing or quenching polymer melts.

Natural and synthetic biodegradable polymers have been utilized in drug delivery and in tissue engineering. Drug delivery systems based on biodegradable polymers facilitate the controlled release of drugs with the concurrent degradation of the polymer. Biodegradable polymers can also provide a temporary scaffold for cell attachment and transplantation to regenerate metabolic organs and repair connective tissues. Many nondegradable polymers have proved to be reasonably biocompatible, and this property was essential to the success a number of applications ranging from soft contact lenses to wound dressings and vascular prostheses.

There have been many successful applications of polymers in biological systems. However, fibrotic scar formation, calcification, thrombosis, abrasive wear, and fatigue of the material are all problems that may occur when a polymer is used in contact with bodily fluids or tissues. Many challenges remain as we seek to improve our understanding of polymer biocompatibility and to develop novel polymers that interact in desired ways with biological systems.

*See also* Polymers, Genetic Engineering of.

### *Bibliography*

Langer, R. (1990) New methods of drug delivery. *Science,* 249:1527–1533.

Peppas, N. A., Ed. (1987) *Hydrogels in Medicine and Pharmacy,* Vol. 2, *Polymers.* CRC Press, Boca Raton, FL.

———, and Langer, R. (1994) New challenges in biomaterials. *Science,* 263:1715–1720.

———, and ———, Eds. (1993) *Biopolymers I,* Vol. 107 in *Advances in Polymer Science.* Springer-Verlag, New York.

Pulapura, S., and Kohn, J. (1992) Trends in the development of bioresorbable polymers for medical applications. *J. Biomater. Appl.* 6:216–250.

Steinbach, B. G., Hardt, N. S., and Abbitt, P. L. (1993) Mammography: Breast implants—types, complications, and adjacent breast pathology. *Curr. Probl. Diagn. Radiol.* 22:39–86.

## Polypeptide: *see* Peptides.

# Population-Specific Genetic Markers and Disease

*L. B. Jorde*

### *Key Words*

**Fitness** An individual's relative success in transmitting his or her genes to the next generation.

**Genetic Drift** Random intergenerational change in gene frequencies due to finite population size.

**Haplotype** The allelic constitution of several loci on a single chromosome (''haploid genotype'').

**Heterozygote Advantage** A fitness differential in which the heterozygote has higher fitness than any of the homozygotes.

**Natural Selection** The evolutionary process in which individuals with a certain gene have higher fitness than those lacking the gene.

**Polymorphism** The occurrence of two or more alleles at a locus in a population, where at least two alleles occur with frequencies greater than 1%.

Variation in the prevalence of genetic diseases results from the action of the evolutionary processes of mutation, natural selection, genetic drift, and gene flow. Genetic markers can be studied in populations to help determine the ways in which these processes interact to influence disease prevalence. Such investigations enhance our understanding of the origins and evolution of genetic diseases. This entry reviews the ways in which genetic markers are used to study disease variation in populations, using specific genetic diseases as examples.

## 1 INTRODUCTION

The prevalence of many genetic diseases varies widely among human populations (see Figure 1 for examples). Cystic fibrosis (CF) is the most common lethal single-gene disorder among Caucasians, affecting approximately 1 in 2500 individuals. Yet it is quite uncommon among Asians, with a prevalence of only 1 in 90,000 births. Sickle cell disease affects 1 in 400 to 600 American blacks, but it is extremely rare among individuals of northern European descent. Tay–Sachs disease, a lethal recessive disorder seen in 1 in 3600 Ashkenazi Jewish births, is uncommon in most other populations.

Variation in the prevalence of genetic disease is but one aspect of human genetic variation. The evolutionary processes affecting ''normal'' human variation (e.g., traits such as height, skin color, blood groups, and DNA polymorphisms) also affect genetic disease variation. Four major evolutionary processes can be identified: mutation, natural selection, genetic drift, and gene flow. The relative contribution of each of these processes to genetic disease variation has been the subject of considerable controversy.

Much of this controversy arises from difficulties in specifying the precise nature of genetic variation. Until fairly recently, population variation in genetic disease was assessed primarily in terms of protein variants. Protein electrophoresis could sometimes establish that different populations had different amino acid substitutions leading to a disease (e.g., for sickle cell disease), but variation at the DNA level was largely unmeasurable. Furthermore, for most disease genes, the gene product, and thus the amino acid sequence, was unknown. Consequently, tracing the origins and evolution of genetic diseases in populations was often a highly speculative endeavor.

The molecular genetic techniques developed during the past 15 years have improved this situation substantially. With thousands of polymorphisms now identified in the human genome, it is usually possible to examine polymorphisms near and within a disease locus. These polymorphisms, or markers, can take the form of restriction fragment length polymorphisms (RFLPs), minisatellite repeat polymorphisms, and microsatellite repeat polymorphisms. If several polymorphisms are measurable in the vicinity of the disease locus, haplotypes can be constructed for chromosomes carrying the disease mutation and for normal chromosomes. For example, a series of four two-allele RFLPs, in which alleles are labeled 1 or 2, could form a haplotype composed of alleles 1, 2, 1, and 1 on the chromosomes bearing a particular disease mutation. Other haplotypes may be observed on normal chromosomes. Because recombination between closely linked polymorphisms is rare, these haplotypes can often be used to trace the evolutionary history of a disease mutation (i.e., under appropriate conditions, all affected individuals with the same chromosome haplotype should be descendants of the same individual). As shown in Sections 2 through 6, comparisons of haplotypes in different populations can reveal whether a genetic disease had independent origins in each population.

As more and more disease genes are cloned, it is becoming possible to analyze not only marker haplotypes but also disease

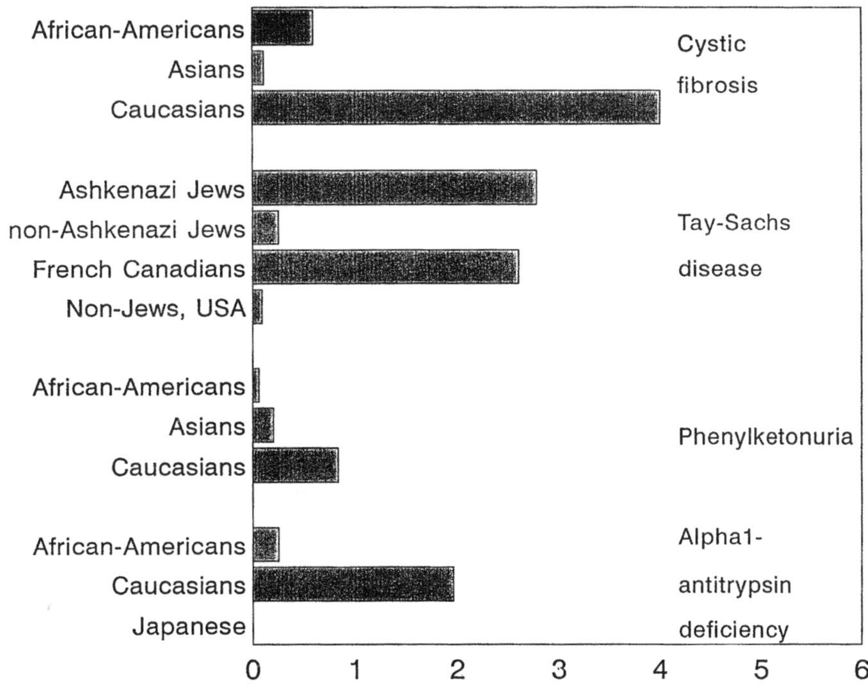


**Figure 1.** Prevalence of selected genetic diseases (per 10,000 births) for a series of populations. The French Canadian population refers only to residents of eastern Quebec.

mutations themselves. This can be accomplished through techniques such as single-strand conformation polymorphism (SSCP) analysis or through direct DNA sequencing. Analysis of mutations sometimes reveals a tremendous diversity of different mutations at a single disease locus. Since the cloning of the CF gene in 1989, more than 400 different mutations have been identified at this locus. These mutations vary in frequency from one population to another.

As a result of advances in molecular technology, our understanding of the role of evolutionary forces in creating and maintaining genetic variation among populations (including variation in genetic diseases) has increased considerably. This entry reviews the evolutionary mechanisms responsible for such variation, illustrating each mechanism with specific disease examples. It also discusses the ways in which molecular genetic markers have contributed to our understanding of these mechanisms.

## 2 MUTATION

Mutation, which is a heritable alteration in DNA sequence, is the ultimate source of genetic variation. While many mutations have no biologic consequence for the organism, some may produce changes in amino acid sequence that result in genetic disease. Mutations can occur as single base substitutions (missense mutations that alter a single amino acid or nonsense mutations that produce a stop codon), deletions and duplications (some of which may produce frameshifts), alterations of splice sites (resulting in inappropriate inclusion of introns or exclusion of exons in the mature messenger RNA), promoter alterations (resulting in decreased transcription of messenger RNA), or insertions of mobile elements.

There is little evidence that mutation rates vary substantially among human populations. Thus, mutation itself does not account for much interpopulation variation in genetic disease. However, mutation rates do vary substantially from one disease gene to another and thus help to explain why some diseases are relatively common and others are relatively rare.

### 2.1 Gene Size and Duchenne Muscular Dystrophy

Duchenne muscular dystrophy (DMD) is one of the more common lethal genetic diseases, affecting approximately 1 in 3500 males in most surveyed populations. Since it is an X-linked recessive disorder, DMD seldom affects females. Affected males nearly always die of cardiac or respiratory failure before age 20 and virtually never reproduce. It is estimated that the DMD gene has one of the highest known mutation rates: approximately $10^{-4}$ mutation per gene per generation. The high mutation rate helps to indicate why DMD is quite common in spite of its lethality in males.

Cloning the DMD gene has helped to explain why the mutation rate is so high. The DMD gene, which spans 2.3 million DNA bases and includes 14,000 base pairs (14 kilobases, or kb) of coding DNA, is the largest gene known in the human. Its large size, quite simply, presents a large ''target'' for mutation. A similar example is given by neurofibromatosis type 1 (NF1), an autosomal dominant disorder that produces benign tumors of peripheral nerves. This disorder also has a high mutation rate, estimated as approximately $10^{-4}$. The NF1 gene is again very large, spanning over 350 kb. The size of the coding DNA, 13 kb, is similar to that of the DMD gene. In general, mutation rate appears to correlate positively with gene size.

### 2.2 Mutation Hot Spots

A second factor influencing mutation rates involves DNA base pair composition. It is known that the CG dinucleotide is highly susceptible to mutation. The cytosine base is usually methylated when it occurs next to a guanine, and methylated cytosine is likely to spontaneously lose an amino group. This sequence produces a transition from cytosine to thymine. Surveys have shown that an inordinately large proportion of disease gene mutations occur at CG dinucleotides. Thus, loci rich in such mutational ''hot spots'' are likely to have higher mutation rates. Examples of genetic diseases that have such mutational hot spots include phenylketonuria (PKU) and hemophilia A and B.

## 3 NATURAL SELECTION

Natural selection can be thought of as a screening process to which genetic variation is subjected. Advantageous variants are positively selected: individuals carrying the variants are more likely to survive and reproduce, thus increasing the frequency of the gene in the population. Disadvantageous variants will be selected against, reducing their frequency in the population. For most genetic diseases, an equilibrium state will eventually be reached in which mutation introduces new copies of a disease gene into the population and selection removes them at the same rate.

### 3.1 Heterozygote Advantage and Sickle Cell Disease

Certain recessive diseases present a slightly more complicated picture. In some cases, the homozygous recessive genotype may undergo negative selection, while the heterozygote may have a selective advantage. In this situation, the disease allele will be maintained at a higher frequency than if the heterozygote had no advantage. A good example of this is found in sickle cell disease. This autosomal recessive disorder, in which the erythrocytes assume a characteristic sickle shape and thus cannot move easily through the capillaries, is often fatal in homozygotes. The heterozygote has no health problems and, in a malarial environment, has decided survival advantage: sickle cell heterozygotes are resistant to infection by the *Plasmodium falciparum* malaria parasite. Consequently, heterozygotes have a survival advantage over both types of homozygote. The frequency of the sickle cell gene is quite high in parts of the world in which malaria has been endemic. It reaches its highest frequency in west-central Africa, where up to one in four individuals is a heterozygous carrier and one in 45 is affected with the disorder. Sickle cell disease is also seen in parts of the Middle East, India, and the Mediterranean region. It is interesting that the frequency of the sickle cell mutation among African Americans is substantially lower than in residents of western Africa. This partly reflects admixture with Caucasians (see Section 6.1), but it is likely also a result of a lack of natural selection in the malaria-free North American environment.

### 3.2 The Hemoglobin S Variant and Sickle Cell Disease

The best-studied of the sickle cell mutations is a single-base substitution (thymine instead of adenine) at the DNA triplet that specifies the sixth amino acid of the β-globin chain. This mutation results in a substitution of valine for glutamic acid known as the hemoglobin S (HbS) variant. At one time, it was thought that the presence of HbS in India and the Mideast was due to migration of Africans who carried the HbS gene. However, more recent investigations using RFLPs show that this is probably not the case. Marker haplotypes that have been examined using RFLPs in and around the β-globin locus show that the disease haplotypes occurring in Africa differ substantially from those occurring in the Mideast and India. Some haplotype variation can be expected as a result of gene conversion or recombination subsequent to the origin of a mutation, but it is often possible to distinguish these events from multiple mutations. The RFLP haplotypes show that the HbS mutation arose at least twice, once in Africa and once in Saudi Arabia or India. Natural selection then operated in each environment independently to raise the frequency of the mutant gene. Molecular studies have shown that Ψβ, a pseudogene (i.e., a nontranscribed gene highly similar to β-globin in DNA sequence), does *not* increase in frequency among populations in malarial environments.

## 4 GENETIC DRIFT AND FOUNDER EFFECT

Often, small, isolated populations exhibit high frequencies of genetic diseases that are rare in other populations. For example, Ellis van Creveld syndrome, a very rare disorder that involves reduced stature, is seen with increased frequency among the genetically isolated Old Order Amish population of Lancaster County, Pennsylvania. In fact, nearly as many cases have been observed in a single Amish kindred as have been observed in the rest of the world. In the Finnish population, which has also been quite isolated until recently, more than two dozen otherwise rare autosomal recessive diseases occur with substantially elevated frequencies. Such variation in disease prevalence is usually the result of the related processes of genetic drift and founder effect. Genetic drift refers to the random fluctuation in gene frequencies that occurs from one generation to the next as a result of sampling a limited number of gametes. In small populations, the degree of fluctuation in gene frequency increases (just as a coin-tossing experiment using only 10 coins is quite likely to result in a large deviation from the expected proportion of 50% heads and 50% tails). Genetic drift can thus lead to high frequencies of genetic diseases in small populations. Founder effect occurs when a small number of individuals group together to form a new population. Because these individuals form a small sample of an original population, their genes are unlikely to constitute a representative sample of that original population. This can again result in a highly skewed distribution of disease genes among the descendants of the founders of the new population. The Old Order Amish population, for example, was founded by only about 50 couples. The founder effect is further exaggerated when certain individuals in the breakaway population have a disproportionately large number of descendants.

Traditionally, demographic history and genealogies have been used to demonstrate drift and founder effect. Genealogical analysis demonstrated, for example, that all 30,000 cases of porphyria variegata in the South African Afrikaner population could be traced to a single couple who emigrated from Holland in the 1680s. This approach is obviously limited by the availability of extensive genealogies.

### 4.1 Haplotype Analysis

More recently, molecular genetic markers have made it possible to assess founder effect at the DNA level. Haplotype analysis is particularly helpful in this regard. It is reasonable to assume that each different haplotype background associated with a disease represents a different founder. Thus, the number of distinct disease haplotypes can be counted to infer the number of founders who introduced a disease into a population. In northern Finland, it was shown that all individuals affected with X-linked choroideremia, a disease that causes blindness, had the same RFLP haplotype (using markers closely linked to the choroideremia gene). This provided evidence that each of these individuals was descended from the same common ancestor. Genealogical analysis proved this to be the case: all affected individuals shared a common ancestor who lived 12 generations ago. A similar exercise showed that the high frequency of PKU among Yemeni Jews can be attributed to the descendants of a single founding couple who lived in the seventeenth century. RFLP haplotype analysis of markers closely linked to the CF locus demonstrated that all studied cases of CF in the Hutterite population could be traced to three founders.

This type of analysis can also show that relatively large numbers of founders have introduced multiple copies of a disease gene into a population. Von Willebrand disease, an autosomal dominant bleeder syndrome, is very common in parts of the Finnish Aland Islands, reaching an incidence of more than 10%. Analysis of haplotypes in and near the von Willebrand locus shows that at least 12 different haplotypes are associated with the disease gene in this population. Since these markers are within 150 kb of one another and since the population was founded only about a thousand years ago, recombination among them is unlikely. Thus, at least 12 founders must have introduced the disease gene into this population. This conclusion is also consistent with genealogical analyses, which show that no single founder produced enough descendants to account for the high prevalence of the disorder.

In addition to its application in isolated populations, haplotype analysis has been used in large continental populations. Analysis of haplotypes near the factor IX locus on the X chromosome shows that approximately 25% of all hemophilia B cases in the United States can be attributed to three individual founders.

### 4.2 Mutation Analysis

The cloning of disease genes often permits direct analysis of population variation in mutations. One of the best examples is a mutation in the *CF* locus characterized by deletion of a phenylalanine residue at position 508 and termed ΔF508; this three-base deletion accounts for approximately 70% of the *CF* mutations found in Caucasians. Extensive population analysis has shown that the proportion of *CF* mutations accounted for by ΔF508 has a north–south cline, with a maximum of about 70 to 90% in northern Europe (where the disease itself is most common) and a minimum of about 50% in southern Europe. The proportion of ΔF508 mutations is even lower among other surveyed populations, such as African Americans (37%) and Ashkenazi Jews (30%). The elevated frequency of CF among Europeans can be traced primarily to the ΔF508 mutation, since the frequencies of non-ΔF508 mutations are quite similar among Caucasians, Asians, and Africans. Among Europeans, the great majority ( 90%) of ΔF508 mutations occur on a single haplotype background, indicating that most current copies of the mutation may have descended from the same common ancestor. Then, as a result either of selection or drift, the frequency of the disease increased substantially in northern European populations.

## 5 DRIFT OR SELECTION?

When a genetic disease is seen with high frequency in a population, it is often unclear whether it is caused by a selective advantage or genetic drift/founder effect. In a very few cases (such as sickle cell disease), a demonstrable selective agent is known. But in most cases, the evidence is far more equivocal. Here again, molecular markers have been helpful.

### 5.1 PKU in Europe

PKU, which is usually the result of mutations at the phenylalanine hydroxylase (PAH) locus, is seen in about 1 in 12,000 Caucasian births. Although more than 60 different mutations have been identified at the *PAH* locus, there are five that are much more common than the others. Analysis of RFLP haplotypes (based on eight polymorphisms) at the *PAH* locus shows that each of these mutations is strongly associated with only one haplotype. Each of these mutation–haplotype combinations probably represents a specific founding event. Among northern Europeans, who have an especially high frequency of the disorder, only two mutation–haplotype combinations account for 40 to 50% of observed cases. A critical point is that *each* of these mutation–haplotype combinations occurs with high frequency. While genetic drift could conceivably elevate the frequency of one mutation in a population (as in the case of ΔF508), it seems unlikely that multiple independent mutations would increase in frequency by chance. Thus, a selective advantage for heterozygotes, as yet unspecified, seems probable.

### 5.2 Tay-Sachs Disease Among Ashkenazi Jews

A similar example is given by Tay–Sachs disease (hexosaminidase A deficiency) in the Ashkenazi Jewish population. The lethal infantile form of this autosomal recessive lysosomal storage disorder is seen in approximately 1 in 3600 live births in this population. As a result of carrier screening and prenatal diagnosis now being performed, however, this number has been declining. Recent molecular analysis has shown that approximately 80% of Tay–Sachs mutations in the Ashkenazi population are due to a four-base insertion, while about 15% are caused by a splice site mutation. As with PKU, it seems unlikely that genetic drift could elevate the frequencies of both these mutations just by chance. Mathematical analysis indicates that it is highly improbable that genetic drift could have elevated even the rarer mutation to its present frequency. Again, it seems more likely that natural selection, in the form of a heterozygote advantage, is responsible. The argument for selection is further reinforced by the observation that three other recessive lysosomal storage disorders (Gaucher disease, Niemann–Pick disease, and mucolipidosis type IV) are also reported with elevated frequency in this population. It is extremely improbable that genetic drift would elevate the frequency of four otherwise rare lysosomal storage disorders in one population. It has been speculated that heterozygous carriers of these disease mutations may have enjoyed increased resistance to infectious diseases in crowded ghetto environments. However, there is little solid evidence regarding the nature of the putative selective agent(s).

### 5.3 Cystic Fibrosis

As noted earlier, the elevated frequency of CF among northern Europeans can be ascribed mostly to a single mutation, ΔF508. The high frequency of ΔF508 could be due either to selection or drift. Case-control studies of the fertility of *CF* heterozygotes have excluded the possibility of a fertility advantage. Heterozygotes may have had a survival advantage in the past, however. One interesting suggestion follows from the role of the *CF* gene in chloride ion transport in epithelial cells. Heterozygotes may have a reduction in transport capacity, protecting them from the ill effects of chloride-secreting diarrhea, a disease that was sometimes lethal in the past. Recent studies of mouse CF models indicate that CF heterozygotes may be partially resistant to the chloride-secreting effects of cholera toxin. As the function of the CF gene becomes better understood, it is likely that we will come to an increased understanding of the possible roles of selection and drift in maintaining a high frequency of this gene in northern Europeans.

## 6 GENE FLOW

Gene flow reduces genetic differences among populations. When two populations manifest the same genetic disease, the question often arises whether the disease mutation arose independently in the two populations or was communicated from one to another via gene flow. When only a disease phenotype can be observed, it is difficult to answer this question. However, more precise molecular characterization of disease loci often illuminates the issue.

### 6.1 Cystic Fibrosis in African Americans

An example is given by CF in the African-American population. This disease affects approximately 1 in 17,000 African Americans, a prevalence much lower than that of Caucasians. One hypothesis is that the disorder was introduced into the African-American population through gene flow from the American Caucasian population. However, analysis of *CF* mutations shows that ΔF508 is quite rare among African Americans and that two-thirds of the observed *CF* mutations are unique to this population. Thus, *CF* almost certainly had independent origins in the African-American population and is not principally derived from Caucasians.

### 6.2 Tay–Sachs Disease in French Canadians

Studies of Tay–Sachs disease in Quebec provide another example. As one would expect, the Ashkenazi Jewish population of this region has an elevated prevalence of this disorder, and the frequencies of the two major mutations closely match those of other Ashkenazi Jewish populations. Tay–Sachs disease is also seen among the French Canadian population of Quebec. In fact, in the eastern portion of the province, the disease gene frequency is as high as among Ashkenazi Jews. Does this reflect gene flow from the Jewish population, or was there an independent origin? Direct examination of the Tay–Sachs mutation shows that the French Canadians usually carry a 7.6 kb deletion that is not found among Jews. Thus, an independent origin is established. Further investigations may ultimately establish whether the high frequency of the disorder among French Canadians is the result of founder effect or natural selection.

## 7 PERSPECTIVES

This review has highlighted some ways in which the study of disease variation in populations can be related to evolutionary processes. Such studies are significant in understanding evolutionary biology. They have other implications as well. An awareness of population variation in disease can aid in clinical diagnosis. For example, a physician evaluating a child who presents with recurrent infections and failure to thrive should include ethnic background in considering a differential diagnosis. If the patient is of African descent, sickle cell disease is a distinct possibility. If the patient is of northern European descent, sickle cell disease would be quite unlikely, but cystic fibrosis should be considered. In some cases, knowledge of the potential adaptive significance of a genetic disease (e.g., the heterozygote advantage for sickle cell disease) may lead to increased understanding of gene function and pathophysiology.

Advances in molecular genetics have greatly improved our understanding of the evolutionary processes leading to population variation in genetic disease. Yet many important questions remain unanswered. As more polymorphic markers are developed, more DNA sequenced, and more disease genes cloned, new and more profound insights are certain to be gained.

*See also* Genetic Analysis of Populations; Human Genetic Predisposition to Disease; Mitochondrial DNA, Evolution of Human.

### *Bibliography*

Bonné-Tamir, B., and Adam, A., Eds. (1992) *Genetic Diversity Among Jews: Diseases and Markers at the DNA Level.* Oxford University Press, New York.

Diamond, J. M., and Rotter, J. I. (1987) Observing the founder effect in human evolution. *Nature* 329:105–106.

Eriksson, A. W., Forsius, H., Nevanlinna, H. R., Workman, P. L., and Norio, R., Eds. (1980) *Population Structure and Genetic Disorders.* Academic Press, London.

Hartl, D. L., and Clark, A. G. (1989) *Principles of Population Genetics,* 2nd ed. Sinauer Associates, Sunderland, MA.

Hill, A.V.S. (1989) Molecular markers of ethnic groups. In *Ethnic Factors in Health and Disease,* J. K Cruickshank and D. G. Beevers, Eds., pp. 25–31. Butterworths, London.

Jorde, L. B., Carey, J. C., and White, R. L. (1995) *Medical Genetics.* Mosby, St. Louis.

McKusick, V. A., Ed. (1978) *Medical Genetic Studies of the Amish.* Johns Hopkins University Press, Baltimore.

Weiss, K. M. (1993) *Genetic Variation and Human Disease: Principles and Approaches.* Cambridge University Press, Cambridge.

# Protein Aggregation

*Jeannine M. Yon*

### *Key Words*

**Aggregates** Association of non-native protein molecules through inter-molecular interactions.

**Inclusion Bodies** Insoluble amorphous aggregates which accumulate in cells, when recombinant proteins are overexpressed in foreign hosts.

**Molecular Chaperones** Proteins which act as helpers of protein folding in cells.

**Protein Folding** The way by which a polypeptide chain acquires its three-dimensional structure.

Protein denaturation and aggregation are frequently occurring phenomena, long recognized as common processes. The aggregation under consideration is very different from precipitation of a native protein at the isoelectric point or upon salting out, which can be reversed by dissolution under appropriate conditions. The aggregates are formed from the nonnative protein upon *inter*molecular interactions, which compete with *intra*molecular interactions. These aggregated species are not in equilibrium with soluble species, complicating the experimentation. Aggregation has been reported to occur during the in vitro refolding for monomeric as well as oligomeric proteins, lowering the refolding efficiency.

Aggregates can also accumulate during the in vivo folding of nascent polypeptide chains synthesized in prokaryotic and eukaryotic cells. The overexpression of genes introduced in foreign hosts often results in aggregated nonnative proteins termed ''inclusion bodies,'' introducing a serious limitation in the production of recombinant proteins. Such a problem remains critical in the development of biotechnology. In any case, protein aggregation is an off-pathway event associated with protein folding or protein association, which results from kinetic competition between proper folding and a side reaction. Such competition, which arises from an early folding intermediate producing misfolded species, can lead to the formation of aggregates.

## 1 PROTEIN AGGREGATION IN IN VITRO REFOLDING

The refolding of an unfolded protein has been recognized to be a spontaneous process under in vitro conditions. Although it was shown by Anfinsen in the 1960s that the three-dimensional structure of a protein is determined by its amino acid sequence, the mechanisms involved in the folding process are not well understood. Solving the folding problem, this ''second translation'' of the genetic code, is of great importance, not only for understanding the mechanisms by which a polypeptide chain acquires its three-dimensional and functional structure, but also for progress in biotechnology.

### 1.1 Monomeric Proteins

The formation of aggregates has been reported for proteins with disulfide bonds as well as for proteins devoid of disulfide bonds. Thermal unfolding of proteins is frequently accompanied by the formation of aggregates and therefore behaves as an irreversible process. Thermal unfolding occurs at temperatures that vary widely according to the protein, since the temperature of optimum stability depends on the balance between hydrogen bonds and hydrophobic interactions. Generally, the products of thermal denaturation are not completely unfolded and retain some structured regions. The unfolding of small proteins induced by urea or guanidine hydrochloride is generally reversible under appropriate conditions (i.e., pH, ionic strength, temperature, and low protein concentration).

The presence of covalent cross-links, such as disulfide bridges, often requires the addition of small amounts of a reducing reagent or a redox mixture to prevent or to correct a wrong pairing of the half-cystines, and further aggregation. The ''reshuffling'' of disulfide bridges during the refolding of a protein takes place through a series of redox equilibria according to either an intramolecular exchange or an intermolecular exchange. It is possible to prevent misfolding and aggregation for proteins with disulfide bonds by monitoring the refolding in rigorously determined redox regeneration mixtures.

The formation of aggregates has also been observed during the refolding of monomeric proteins devoid of disulfide bridges whose polypeptide chain is folded in several structural domains. At a critical concentration of denaturant, which often is very close to that of the end of the unfolding transition, the denaturant induces an apparent irreversibility in the unfolding–refolding transitions. This phenomenon is strongly dependent on the protein concentration and on the time of exposure to the denaturant, reflecting an aggregation process. This apparent irreversibility is also temperature dependent. It has been attributed to side reactions occurring through intermediates on the folding path from which incorrectly folded species are formed. These latter species sometimes, but not always, associate into aggregates through hydrophobic interactions. The nature of the intermediate that generates the incorrectly folded species and/or the aggregates remains a matter of discussion. The outcome of many investigations on protein folding indicates a great variety of early transient intermediates, formed in a fraction of a second. However, the emergence of the final tertiary and functional structure is a late event and is the rate-limiting step in the process. One of these early intermediates, named the *molten globule,* was characterized as having a native like secondary structure with a high flexibility and a tertiary structure not yet formed. In this molten globule, hydrophobic surfaces remain exposed to the solvent, favoring the formation of aggregates. Although this intermediate has been suggested to be a good candidate for the formation of aggregates, there is no conclusive proof that the aggregates arise from the molten globule state. To overcome this complication in the refolding of a protein, one must have very precise conditions of pH, protein concentration, temperature, ionic strength, and redox mixture, when necessary. These conditions differ from one protein to another.

### 1.2 Oligomeric Proteins

With oligomeric proteins, a new degree of complexity appears in the hierarchy of protein structure and folding. It is generally accepted that the early steps in the acquisition of functional structure by oligomeric proteins are practically identical to those occurring in the folding of monomeric proteins. Within a subunit, it is reasonable to assume that first some stretches of structure are formed in several parts of the molecule; these interact to generate subdomains, which then associate to produce structural domains, and the folded monomer is formed upon domain pairing with a nativelike structure. In a last step, subunit association and subsequent conformational readjustments yield the native and functional oligomeric protein. For the process to be completed, the subunit interfaces must be correctly recognized. The refolding of several oligomeric proteins, dimers and tetramers, has been extensively studied by R. Jaenicke and his group, who have described the overall process of formation of an oligomeric protein as a sequence of first-order (folding) and second-order (association) reactions. For oligomeric as well as for monomeric proteins, side reactions could occur in an off-pathway, leading to the formation of aggregates from misfolded species. Once again, there is kinetic competition between folding/association and aggregation, depending on the experimental conditions.

### 1.3 Methods to Detect and Characterize Protein Aggregates

The detection and characterization of aggregates represent important aspects in folding studies. The most usual methods to determine the association state of a protein, available in any laboratory of biochemistry, are gel permeation and sodium dodecyl sulfate polyacrylamide gel electrophoresis with and without cross-linking. The detection of aggregates also can be monitored by other hydrodynamic methods such as analytical ultracentrifugation and viscosity measurements. Classical light scattering is a convenient method to detect and characterize aggregates and to give information on their size. Quasi-elastic light scattering is a dynamic method that can be used to determine macromolecule diffusion coefficients as a function of time (i.e., to follow the kinetics of aggregation). Electron microscopy is often used in the characterization of aggregates in the in vivo folding. In particular, the shape and dimensions of inclusion bodies have been determined by this method.

### 1.4 Rules of Protein Folding Deduced from in Vitro Refolding Studies

Although the mechanism of protein folding is not completely understood, the great number of in vitro folding studies has provided a significant corpus of information allowing the deduction of some main rules.

1. The folding of a polypeptide chain is determined by the amino acid sequence in a given environment. Under appropriate conditions, it is a spontaneous process.
2. The structure of a native protein is thermodynamically controlled; its stability is not very high, as estimated by the variation in free energy $\Delta G$ from the unfolded to the folded state, whose value varies between $-5$ and $-15$ kcal/mol according to the protein.
3. The folding pathway involves partially folded intermediates, which are more or less populated according to the protein and the conditions. Their formation is kinetically controlled. Some of these prefolded intermediates can evolve into misfolded species, which may or may not result in aggregation depending on the conditions. There is kinetic competition between proper folding and misfolding and subsequent aggregation.

## 2 PROTEIN AGGREGATION DURING IN VIVO FOLDING

### 2.1 Folding in the Cellular Environment

The main concepts concerning protein folding were developed from in vitro studies, and it is accepted that the in vitro refolding process is a good model for understanding the mechanisms by which a polypeptide chain reaches its native state in the cellular environment. However, the recent discovery of molecular chaperones has led to a reevaluation of this hypothesis. Indeed, the environmental conditions within cells are markedly different from those used in the in vitro refolding studies. Different factors can affect the in vivo folding pathway, the synthesis, the co- and posttranslational covalent modifications, and the localization of the mature protein. Protein aggregation can occur during the in vivo folding of a nascent polypeptide chain. The phenomenon is often associated with pathological states and can occur in cells either carrying certain genetic lesions or grown under stressful conditions. The overexpression of active recombinant proteins in foreign hosts frequently results in the accumulation of insoluble aggregates, called ''inclusion bodies.'' In biotechnology, the solubilization and proper folding of these aggregates often represents a crucial step in the production of the active product.

### 2.2 The Formation of Inclusion Bodies

#### 2.2.1 Occurrence of Inclusion Bodies

In cells, inclusion bodies appear as amorphous aggregates clearly separated from the rest of the cytoplasm; they form a highly refractile area when observed microscopically. A great variety of experimental approaches indicates that the formation of inclusion bodies results from partially folded intermediates in the intracellular folding pathway of proteins, not from the totally unfolded or native proteins. This is particularly well illustrated by the folding pathway of tailspike *endo*-rhamnosidase from *Salmonella* phage 22, studied by the group of J. King, one of the few systems for which the in vivo pathway has been compared with the in vitro refolding pathway. In the production of engineered proteins, the formation of inclusion bodies generally appears to be a disadvantage, since it requires the aggregates to be dissolved in denaturant and subsequent refolding of the protein. But when the active product can be recovered in a sufficient yield, certain advantages may accrue. Indeed, aggregation generally prevents proteolytic attack, with the exception of pyruvate kinase and bovine pancreatic trypsin inhibitor, expressed in *Escherichia, coli* which coaggregate with a protease in inclusion bodies. The formation of inclusion bodies is also an advantage for the production of proteins that are toxic for the host cells. Furthermore, inclusion bodies contain a great quantity of the expressed protein.

#### 2.2.2 Characteristics of Inclusion Bodies

The characteristics of the aggregates depend on how the protein is expressed. Different sizes and morphologies have been observed. Generally, inclusion bodies appear as dense isomorphous aggregates of nonnative protein separated from the rest of the cytoplasm, but not surrounded by a membrane. They look like refractile inclusions and when large enough can be easily recognized by phase contrast microscopy.

#### 2.2.3 Mechanisms of Inclusion Body Formation

The conformation of the polypeptide chains in inclusion bodies is not known. The particulate character of the aggregates and their properties of light scattering prevent the use of spectroscopic methods. For proteins containing disulfide bridges such as prochymosin, the formation of incorrect intermolecular disulfide bridges can partially account for the formation of aggregates. But inclusion bodies have been also observed for proteins devoid of disulfide bridges. In general, the aggregates result from noncovalent intermolecular interactions, mostly hydrophobic interactions, and they can be dissolved using high concentrations of denaturant. Nevertheless, comparison of in vitro and in vivo studies of the aggregation phenomenon has led to the conclusion that aggregates do not result from the association of completely unfolded molecules, but are generated from early intermediates of the folding pathway.

#### 2.2.4 Strategies for Refolding Inclusion Body Proteins

The main problem in the production of recombinant proteins is the low refolding yield of the biologically active product. The recovery of the active protein involves a succession of several steps, including cell disruption and separation of the insoluble material, solubilization of inclusion bodies in the presence of strong denaturants, refolding, and purification. To maximize refolding yields, stabilizing reagents such as sugars, alcohols, and polyols (including sucrose, glycerol, polyethylene glycol, and isopropanol) may be added. Conversely, the use of solubilizing reagents may be helpful in dissolving the aggregates. Another important factor in the refolding process is the rate of removal of the denaturant. Since there is kinetic competition between correct refolding and the formation of aggregates from an intermediate, conditions that favor folding over the accumulation of aggregates must be found.

### 2.3 Role of Molecular Chaperones in Protein Folding

The recent discovery of molecular chaperones, a ubiquitous class of proteins mediating the correct folding of polypeptide chains in the cellular environment, led to a reconsideration of the mechanisms of in vivo protein folding. Ellis (1987) extended the concept of the molecular chaperone to define a class of proteins whose function is to ensure the correct folding and assembly of newly synthesized proteins through a transient association with the nascent polypeptide chain. Studies of heat shock proteins have widely contributed to the development of this concept. These proteins, induced by high temperatures, associate with proteins that are partially unfolded as a consequence of the thermal stress and thus are prevented from aggregating. Heat shock proteins reportedly have participated in the assembly of oligomeric proteins under normal conditions. These proteins have been classified according to molecular weight into three families, hsp 60 or GroEL, hsp 70, and hsp 90.

#### 2.3.1 The hsp 60 Chaperonins

The chaperonin family is a class of related proteins found in all bacteria and mitochondria and also in chloroplasts. There are two types of chaperonins, cpn 60 and cpn 10 (in *E. coli,* GroEL and GroES, respectively), with subunit molecular weight of 60 and 10 kDa, respectively. These chaperonins are constitutive proteins whose amount increases under stress conditions, such as heat shock, bacterial infection, or an increase in the cellular amount of unfolded proteins. Although chaperonins efficiently facilitate the correct folding and assembly of the polypeptide chain, preventing the formation of aggregates, they are unable to reverse the aggregation process. The dissociation of the complex is dependent on the hydrolysis of ATP for several proteins. ATP hydrolysis is not required for the release of certain proteins however, since a nonhydrolyzable analogue is capable of promoting the release of the protein. These results clearly indicate that the binding, folding, and release of the folded proteins are distinct steps.

#### 2.3.2 The hsp 70 Chaperonins

The heat shock protein family hsp 70 consists of several structurally related proteins, all binding ATP with high affinity. Representatives have been found in bacteria, mitochondria, and chloroplasts, and also in eukaryotic cells. Several are strictly inducible by heat shock or other form of cell stress. Others, referred to as cognate heat shock proteins, are constitutive. It is likely that the constitutive heat shock proteins play a role in posttranslational transmembrane targeting of proteins to cellular organelles, and also in the disassembly of clathrin cages. In yeast cells, the cytosolic hsp 70 proteins are required for membrane translocation, and it was suggested that they are responsible for ATP-dependent unfolding of the preprotein before translocation.

#### 2.3.3 The hsp 90 Chaperonins

Members of the hsp 90 class are also considered as molecular chaperones; indeed the chaperonin function of hsp 90 has been proven by in vitro experiments and has a role in steroid receptors biosynthesis.

#### 2.3.4 The Rules of Protein Folding and the Concept of Molecular Chaperones

By their transient association with nascent, stress-destabilized, or translocated proteins, molecular chaperonins have a role in preventing improper folding and subsequent aggregation. However, they are unable to interact with native proteins, or to refold aggregates once formed. Furthermore, they do not carry information capable of directing a protein to assume a structure different from the one dictated by the polypeptide amino acid sequence. Therefore, they assist the in vivo folding without violating the main rules of protein folding as deduced from in vitro experiments. The molecular details of chaperone–protein interactions are not yet known. The development of in vitro folding and assembly physicochemical studies with highly purified proteins and molecular chaperones will allow substantial progress in the understanding of in vivo protein folding.

#### 2.3.5 Other Proteins That Assist in vivo Folding

Other such proteins that assist in vivo in the correct folding and assembly of proteins have been described. The two enzymes, protein disulfide isomerase and peptidylprolyl-*cis-trans*-isomerase, cannot be regarded as molecular chaperones.

*See also* Chaperones, Molecular; Protein Folding; Protein Modeling.

### *Bibliography*


Elis, R. J. (1987) Proteins as molecular chaperones. *Nature,* 328:378–379.

———, and van der Vries, S. M. (1991) Molecular chaperones. *Annu. Rev. Biochem.* 60:321–347.

Georgiou, G., and de Bernadez-Clark, E. (1991) Protein refolding. ACS Symposium Series 470. American Chemical Society, Washington, DC.

Ghelis, C., and Yon, J. M. (1982) Protein Folding. Academic Press, New York.

Jaenicke, R. (1987) Folding and association of proteins. *Prog. Biophys. Mol. Biol.* 49:117–237.

Riefhaber, T., Rudolph, R., Kohler, H., and Buchner, J. (1991) Protein aggregation in vitro and in vivo: A quantitative model of the kinetic competition between folding and aggregation. *Biotechnology,* 9:825–829.


# Protein Analysis by Integrated Sample Preparation, Chemistry, and Mass Spectrometry

*Kenneth R. Williams and Steven A. Carr*

### *Key Words*

**Edman Degradation** Two-step sequential process by which the $NH_2$-terminal amino acid residue in a peptide or protein is reacted with phenylisothiocyanate and then cleaved off so that it can be identified by HPLC. By repetition of this process, it is possible to determine the amino acid sequence of the first 30 or so residues in a protein or peptide that does not have an acetylated $NH_2$-terminus.

**Electrospray Ionization (ES)** Ionization technique in which analytes are ionized at atmospheric pressure directly from a flowing liquid stream. The ions produced are sampled into the high vacuum analyzer region of the mass spectrometer for mass analysis and detection. Electrospray ionization is most often used on quadrupole mass analyzers.

**Fast Atom Bombardment (FAB) Ionization** Ionization technique in which a solution of the analyte is mixed on a metal sample target with a large molar excess of a water-soluble, viscous liquid matrix such as glycerol or monothioglycerol. Bombardment of this sample–matrix mixture in the ionization source with an energetic beam of atoms or ions (typically Xe or $Cs^+$ at 8–40 keV) forms both sample- and matrix-related ions that are sputtered from the surface of the target, accelerated, and then mass-analyzed in the spectrometer. Most commonly used with magnetic sector based instruments, FAB can also be used with quadrupole mass analyzers.

**High Performance Liquid Chromatography (HPLC)** The high resolution of HPLC derives from the small particle size (typically 5–10 μm for a 0.46 × 25 $cm^2$ column) of the packing material, which permits a high density of the ''stationary phase.'' The latter consists of an inert support (usually silica based) to which is chemically linked the ligand, which binds the components in the mixture to be separated. The ''stationary phase'' used most often for peptide mapping is C-18, which is —$(CH_2)_{17}CH_3$, and the bound peptides are eluted by a ''mobile phase'' consisting of 0.05–0.10% trifluoroacetic acid and increasing concentrations of acetonitrile. This method of chromatography is called ''reversed phase'' because, unlike ''normal phase'' chromatography, the starting mobile phase is more polar than the stationary phase.

**Mass Spectrometry** Branch of science dealing with the formation and characterization of ions produced from intact, neutral molecules and/or fragments formed by decomposition of the original molecules. A mass spectrometer is employed to produce, analyze, and detect the molecular and fragment ions produced. All mass spectrometers have three essential components: an ion source, a mass analyzer, and a detector. Ions are produced from the sample in the ion source using a specific ionization method (see definitions for ES, MALD, FAB, and PD), separated based on their mass-to-charge (*m/z*) ratios in the mass analyzer, and detected, usually by an electron multiplier, which releases and amplifies a cascade of electrons upon impact of a sample-derived ion. ''Soft'' ionization techniques, which primarily produce intact, protonated molecular ions, are essential for the analysis of peptides and proteins. All three components are usually inside a high vacuum chamber (pressure between $10^{-8}$ and $10^{-4}$ torr), although in electrospray ionization the ion source is at atmospheric pressure. A mass spectrum is a plot of the relative abundance of each ion versus*m/z*.

**Matrix-Assisted Laser Desorption (MALD) Ionization** Ionization technique in which a solution of the analyte is first mixed with a larger molar excess of a water-miscible matrix such as sinipinic acid ($10^2$:$10^6$ matrix-to-analyte ratio). A small volume (1–2 μL) of this mixture, dried on a metal sample target, is introduced into the ion source and irradiated with laser light (337 nm if a nitrogen laser is employed). Sample- and matrix-related ions are sputtered from the surface and analyzed, usually in a time-of-flight mass analyzer.

**Plasma Desorption Ionization (PD)** Ionization technique in which energetic particles produced by spontaneous fission of californium-252 pass through a thin metal foil, on the other side of which a sample has been deposited and dried. Sample ions produced are accelerated and mass analyzed by a time-of-flight analyzer.

---

Vast improvements in the sensitivity of detection of mass spectrometry (MS) and in its ability to determine masses for proteins in the 3.5–350 kDa range help to account for the increasing use of this technique in determining the primary structures of peptides and proteins and in detecting and identifying the myriad array of post-translational modifications that may occur on proteins. In addition to being able to determine the molecular masses of peptides and proteins, mass spectrometry has the potential for high sensitivity and rapid primary sequence analysis. Since the technological improvements just mentioned have been coupled with the increased inclusion of mass spectrometers within core facilities, this essential technique is being brought to bear on an ever-increasing array of biomedical problems by investigators who often have little training in this specialty. This contribution offers a realistic appraisal and appreciation of current uses of this rapidly evolving technique and outlines sample preparation procedures that can be used to take maximum advantage of MS analysis of proteins.

## 1 WHAT KIND OF INFORMATION CAN MASS SPECTROMETRY PROVIDE ABOUT PROTEINS AND WHY IS IT USEFUL?

Mass spectrometry can make several critical contributions along the typical pathway to the structural characterization of a ''new'' protein. Since most eukaryotic proteins have ''blocked'' $NH_2$- termini, which preclude direct amino acid sequencing, this pathway often begins with isolating 100–500 pmol of protein, which is cleaved with an enzyme such as trypsin to give several peptides. To permit efficient proteolysis, disulfide linkages are usually opened by either oxidation, which converts each cysteine to cysteic acid, or reduction and alkylation. Following reversed-phase fractionation by high performance liquid chromatography (HPLC) a few of these peptides can be sequenced so that oligonucleotide probes or primers can be synthesized to allow a corresponding cDNA to be isolated. The DNA sequence of this cDNA clone can then be used to predict the amino acid sequence of the protein. Overexpression of this cDNA permits isolation of the micromole quantities of proteins that are required for determining three dimensional structures.

Analysis of the reversed-phase HPLC-separated tryptic digest by on-line HPLC/MS or by off-line MS analysis of collected fractions can greatly assist in selecting chromatographic peaks suitable for amino acid sequencing. That is, MS can help to identify artifact (i.e., nonpeptide) peaks and can qualitatively detect peptide mixtures that would not be suitable for Edman degradation. In addition, the masses of the peptide peaks selected for sequencing can be used to quickly confirm the sequences derived by Edman degradation, and to establish how much of the peptide remains unsequenced based on the difference between the observed molecular mass and that calculated on the basis of the Edman sequence data.

Similarly, once an entire cDNA sequence has been established, the mass of the intact protein and that of several of its tryptic peptides can be used to rapidly confirm the predicted protein sequence and thus eliminate frameshift and other errors that may occur during DNA sequencing. When the mass of the protein and that of one or more of its peptides are larger than predicted by the cDNA sequence, the reason may well be a posttranslational modification such as the addition of a lipid moiety or carbohydrate. The difference between the expected mass of the protein or a peptide and that observed in the MS analysis is often useful for establishing the chemical nature of the posttranslational modification present. For example, a difference of +42 Da, together with lack of a free $NH_2$-terminus, is indicative of the presence of an $\alpha$-*N*-acetyl group. Similarly, a difference of +80 Da suggests the presence of a phosphate or sulfate moiety, while a cluster of molecular ion peaks separated by 162, 203, 291, or 146 Da indicates glycoforms differing by hexose, *N*-acetylhexosamine, *N*-acetylneuraminic acid, or deoxyhexose. Another approach that may be applicable for detecting and localizing posttranslational modifications in proteins involves comparative HPLC tryptic peptide mapping of the protein that has been overexpressed in bacteria (where relatively few posttranslational modifications occur) versus that of the naturally occurring protein. Peaks in the digest of the naturally occurring protein that are absent from the digest of the protein expressed in bacteria are subjected to sequence and mass spectrometric analysis.

Mass spectrometry also provides an attractive alternative to Edman degradation for determining amino acid sequences of peptides less than about 25 residues long. While both Edman degradation and mass spectrometric sequencing usually require at least 5–10 pmol of peptide, mass spectrometry has the potential for greatly increasing the speed of sequence analysis because all the sequence information is obtained in a single experiment rather than in a multistep sequential fashion as with the Edman degradation. In addition, MS sequencing can be used to identify posttranslational modifications and to sequence peptides in mixtures. However, unlike Edman degradation, mass spectrometry cannot beused to determine the $NH_2$-terminal sequence of proteins directly; moreover, ambiguities in sequence assignments can arise, and frequently only partial peptide sequences can be obtained. Moreover, MS sequencing requires considerable expertise and a capital equipment investment exceeding $300,000. It is not surprising, therefore, that few laboratories and even fewer facilities are equippedfor routine MS sequencing. Because access to this promisingapproach is presently limited, it will not be covered in this article.

## 2 WHAT KIND OF MASS SPECTROMETRY IS APPROPRIATE? WHAT ARE THE SAMPLE REQUIREMENTS? HOW ACCURATE ARE THE RESULTING DATA?

Table 1 summarizes the sample requirements and capabilities of the four ionization methods that are most useful for MS of peptides and proteins. There are clear differences in the mass ranges that are accessible with these different methods, but fast atom bombardment (FAB), plasma desorption (PD), and electrospray (ES) all usually require at least 0.5–5 pmol of peptide or protein. Matrix-assisted laser desorption/ionization (MALDI) appears to offer 5–10-fold greater sensitivity than these other methods. The 100-fold sensitivity ranges given in Table 1 reflect differences in the extent of ionization of different peptides and proteins, which largely accounts for the failure of MS to provide more than a qualitative assessment of sample composition and purity.

While it is generally agreed that low levels of residual involatile buffers, salts, and detergents in the protein sample can, to varying degrees, mask the signals or suppress ionization of the peptide or protein analyte in FAB, PD, ES, and MALDIMS, no systematic studies have been reported. Each MS method has certain unique characteristics (see notes to Table 1), and some of these features may be sample specific. In general, MALDIMS is far more tolerant of excipients than the other three MS methods, being able to handle many common sample additives such as alkali metal salts, Tris, urea, phosphate, and guanidine at 100 mM or above. However, it is possible in ESMS to obtain useful data from solutions that are 10 $\mu$M in peptide or protein and up to 50 mM in certain salts such as sodium chloride. More importantly, ESMS is ideally suited for on-line coupling to high performance liquid chromatography which can be used to both desalt and separate mixtures of peptides and proteins (see below). Signals from the analyte would not be observed reliably in FAB at these relative concentrations of analyte to salt. In contrast, FAB has no difficulty with glycerol in the sample (in fact, glycerol is often used as the liquid matrix in FAB), whereas it can severely interfere in ESMS and MALDIMS. Detergents such as Triton, Tween, and especially sodium dodecyl sulfate (SDS), can be a problem for all MS methods at almost any level.

While all these techniques are useful for relatively low molecular weight peptides ($\leq$ 5000 Da), only ES and MALDIMS are useful for analyzing intact proteins. In the case of a 25 kDa protein, both ES and MALDI can routinely measure the mass with an accuracy of 0.01–0.02%. Hence, in this case MS could detect a discrepancy of 2.5 Da between the experimentally determined mass and that predicted from the cDNA sequence. While this level of accuracy is not sufficient to routinely detect deamidation (+1 Da) or the formation of a disulfide bond (−2 Da), it would easily detect an acetylated $NH_2$-terminus (+42 Da) and many other posttranslational modifications such as phosphorylation or the addition of carbohydrate. It is important to note that there is a significant practical distinction between ESMS and MALDI with respect to how the mass scale is calibrated in order to achieve the stated mass accuracy. In MALDI, with a linear instrument, reference standards of known molecular weight that bracket the mass of the unknown must be mixed with the sample being analyzed. The determined masses of these reference compounds are used to adjust the calibration in order to compensate for changes in the initial velocity distribution of ions produced in the ion source. Signals for the reference compounds (often peptides or proteins and one of the matrix peaks) must be observed in the same spectrum as the unknown. The amount of reference compound to add is a matter of trial and error, and achieving the necessary balance can be difficult. In contrast, calibration of the mass scale in ESMS and FABMS only requires an external calibration in order to achieve the desired mass accuracy. External calibration may also be effectively used on MALDI instruments equipped with a reflectron (which compensates for some of the inhomogeneity in the ion energies) to obtain

Table 1 Characteristics of Different Mass Spectrometric Ionization Methods

| Ionization Method | Tolerance of Buffers, Salts, Detergents | Amount (pmol)[a] and Concentration (M) | MW Range (kDa) | Mass Assignment Accuracy | Compatible with LC/CE-MS |
|---|---|---|---|---|---|
| Fast Atom bombardment (FAB) | Volatile buffers to ≤50 mM; volatile acids, bases, solvents; low tolerance (≤20 mM) to involatile buffers, alkali metal salts, detergents | 1–1000<br>$\geq 10^{-6}$–$10^{-4}$[b] | ≤10[c] | ±0.01% up to 5 kDa, lower at higher mass<br>±0.0005% possible using double focusing sector instruments | Yes |
| Plasma desorption (PD) | Similar to FAB, but samples may be washed to remove contaminants after adsorption to nitrocellulose | 1–100<br>$\geq 10^{-6}$–$10^{-5}$ | ≤15 | ± 0.1% | No |
| Electrospray (ES) | Similar to FAB, on-line LC-MS used for contaminant removal | 0.5–50<br>$10^{-6}$–$10^{-5}$[b] | ≤150[d] | ± 0.005–0.01% to 50 kDa<br>± 0.02–0.03% to 150 kDa | Yes |
| Matrix-assisted laser desorption/ionization (MALDI) | High tolerance[e] of alkali metal salts, phosphate, urea, guanidine; tolerant of some detergents to 0.1%; intolerant of SDS | 0.1–10<br>$10^{-7}$–$10^{-5}$ | ≤350 | ± 0.01–0.05% to 25 kDa[f]<br>± 0.05–0.3% to 300 kDa[f,g] | No |

[a]Represents amount required to be delivered to ionization region of MS; two to three times these amounts may be required to deliver the required amount to the instrument.
[b]In the case of on-line liquid chromatography–mass spectrometry sample concentration may be considerably lower because the column concentrates the sample prior to analysis.
[c]May be extended to 20 kDa if unlimited amounts of sample are available.
[d]A few examples at higher mass have been reported.
[e]Most buffers, including phosphate and Tris, are tolerated at ≥ 100 mM; alkali metals, guanidine, and urea at ≥ 100 mM; glycerol and sodium azide at ≥ 1%. The extent of spectral degradation caused by an additive is sample dependent and may relate to the ability of the additive to interfere with incorporation of the peptide or protein into the growing matrix crystals.
[f]Requires the use of internal calibration mass standards or a reflecting time-of-flight instrument.
[g]A large decrease in accuracy above 25 kDa is due to the inability of the time-of-flight analyzer used in MALDIMS to resolve the intense matrix adduct peaks from the molecular ion peak of the protein.

a mass accuracy of 0.01–0.02% for molecular masses up to around 10 kDa. Reflecting time-of-flight instruments also provide significantly higher resolution in this mass range (see below).

Protein samples frequently contain variants of the desired material that may result from incomplete proteolytic processing or from events occurring during production or purification. Oxidation and formation of ''ragged'' $NH_2$- or COOH-termini are common examples. The ability to separate and mass-measure signals for proteins of similar, but not identical, molecular mass is affected by the resolving power of the mass analyzer. In practice, the mass analyzer is not able to resolve the individual isotope peaks of the elements present. Instead, an unresolved peak envelope is obtained for a protein, the centroid of which is the abundance-weighted sum (chemical average mass) of the isotopes of the elements present. Mass resolution is often expressed as the ratio $m/\Delta m$, where $m$ and $m + \Delta m$ are the masses (in atomic mass units or Da) of two adjacent peaks of approximately equal intensity in the mass spectrum. The height of the ''valley'' between the two adjacent peaks (measured from the baseline), expressed as a percentage of the height of peak $m$, is a measure of the degree of overlap. There are presently two operable definitions of the overlap in widespread use, and it is essential to know which is being used when resolution figures are quoted. Historically, the first of these is the so-called 10% valley definition in which the two adjacent peaks each contribute 5% to the valley between them. In practice, it is often necessary to determine the resolution of the mass analyzer using a single peak. This is accomplished by simply measuring the peak width (in Da) at the 5% level. This definition is still frequently used for molecules with masses below 5000 Da where it is possible on certain analyzers to still obtain resolution of the isotope peaks.

For proteins, resolving the isotopes in the molecular ion envelope is generally not possible (except for Fourier transform MS of electrospray-produced molecular ions), nor is it useful in practice to be able to resolve the isotopes of large molecules like proteins. For unresolved isotopic peak envelopes the ''full-width, half-maximum'' (FWHM) definition has come into popular use. The resolution of a peak using this definition is equal to the chemical average mass (in Da) of the sample divided by the width (in Da) of the unresolved isotopic envelope measured at the half-height of the peak. A useful rule of thumb is that the value for the resolution determined using the FWHM definition is approximately twice that obtained using the 10% valley definition. For example, a resolution of 1000 using the 10% valley definition is approximately equivalent to resolution of 2000 using the FWHM definition: clearly, while the resolution value in the latter case is larger, the measured degree of separation of the peaks in Da is identical. It should also be noted that at a resolution of 1000 (FWHM) a peak at mass 1001 will essentially be *unresolved* from the peak at mass 1000. For proteins, another important factor that limits the apparent resolution is the natural width of the isotopic envelope. For a 25 kDa protein, the envelope of isotopes is ca. 11 Da wide at FWHM, corresponding to an apparent resolution of 2300. Having an *instrumental* resolution higher than 2300 will not improve the apparent resolution any further until the isotopes that make up the cluster begin to be resolved. This would begin to occur at a resolution higher than 35,000 (FWHM).

Thus, distinguishing protein variants of similar mass depends on a number of factors in addition to the stated resolution of the mass analyzer, including the absolute mass difference between the components, the width and shape of the protein molecular ion "envelopes," and the mass at which the measurement is being made. For example, detection of an N-terminal formyl group (difference = 28 Da) on a 25 kDa protein requires a minimum instrument resolution of about 2000 (FWHM), which is easily achievable by ESMS on quadrupole mass analyzers. The resolution in MALDIMS on a time-of-flight analyzer is typically less than 1000 (FWHM) for proteins of this mass, which is insufficient to detect such small modifications. Signals due to photochemical or gas-phase adducts of the matrix with the protein are also commonly observed in MALDI. These adducts can give rise to intense signals that, as the molecular weight of the protein increases, fold into the $(M+H)^+$ peak of the protein and skew the observed mass to an artificially high value. These adducts also can interfere with detection of variants.

Commercial availability of the instrumentation and simplicity of operation are two important factors not presented in Table 1. Instruments equipped for FAB, ES, and MALDI are widely available, but presently there are no commercial suppliers of PD instruments. FAB is best accomplished on relatively expensive, complex magnetic sector instruments. ES and MALDI are most often used in conjunction with quadrupole and time-of-flight analysis systems, respectively, both of which are relatively inexpensive and simple to operate.

## 3 STRATEGIES FOR PURIFYING AND PREPARING PROTEINS FOR MASS SPECTROMETRY

The ideal milieu for presenting a sample to the mass spectrometer, regardless of ionization method, is pure water or mixtures of water with highly pure, volatile organic solvents and acids, like those provided by reversed phase high performance liquid chromatography. Direct coupling of HPLC to ESMS (with in-line UV detection) is an ideal combination that accomplishes desalting, molecular mass determination, and preparative fractionation in a single experiment. Approximately 1–5 μL/min of the column flow is diverted to the mass spectrometer for analysis, with the remainder (> 95%) being fraction-collected. Capillary electrophoresis (CE) has also been successfully interfaced to ESMS.

Many samples cannot be isolated under the foregoing conditions, however. For instance, following gel filtration, ion exchange, or hydrophobic interaction chromatography, the purified protein might well end up in 50 mL of a nonvolatile buffer containing 0.5 M salt and 10% glycerol. Assuming that 2 mg was isolated and that the protein has a molecular weight of 50,000, the amount of protein (40 nmol) would be sufficient for any MS technique listed in Table 1, but the protein concentration ($< 10^{-6}$ M) would be too low and/or the nonvolatile salt and buffer concentration too high for any technique other than MALDIMS. Hence, this sample would need to be both desalted and concentrated. Assuming that relatively large losses could be tolerated (because adsorption of dilute protein solutions onto membrane surfaces is usually nonspecific and irreversible), one of the easiest ways of preparing this sample would be via successive concentration and dilution with 50–100 mM ammonium bicarbonate or ammonium acetate in an ultrafiltration device, which is typically capable of rapidly concentrating 500 μL or more down to 25–400 μL. Even if only 5% of this sample (i.e., 2.5 mL) were desalted and concentrated (to a final volume of 50 μL) and the recovery were only 10%, there would still be enough material (~ 200 pmol) at a sufficiently high concentration ($\sim 4 \times 10^{-6}$ M) to have a reasonable chance of succeeding with any of the MS techniques described in Table 1.

Other approaches that might be considered for preparing this sample include vacuum dialysis or dialysis followed by evaporation or lyophilization. Although precipitation with trichloroacetic acid (TCA) can bring about rapid removal of almost all salts and buffers (and some detergents such as SDS), TCA- precipitated proteins are often difficult to solubilize in aqueous buffers. In this instance, 10–50% acetonitrile or hexafluoroisopropanol can often greatly increase solubility.

If the protein can be purified only in the presence of detergent, additional care must be taken during sample preparation. In this instance, PDMS may offer some advantage in that following adsorption onto nitrocellulose, most detergents (and salts, buffers, etc.) may be washed off with water, leaving the protein behind. Another alternative is MALDI, which can tolerate up to 0.1% of nonionic detergents such as octyl glucoside, Tween, and Triton. Whether the detergent can be removed by dialysis depends on its micelle size. Hence, while deoxycholate (4200), CHAPS (6150), and octyl glycoside (8000) may slowly dialyze with a 10,000 Da cutoff membrane, SDS (18,000), Tween 80 (76,000), and Triton X-100 (90,000) probably would not. In the latter three instances, dialysis (to remove high concentrations of nonvolatile salts and buffers) followed by acetone precipitation (to remove the detergent) may represent the best alternative. Since SDS is strongly denaturing, proteins that are acetone precipitated from SDS-containing solutions are often difficult to solubilize. An alternative approach for removing Triton or Tween is via detergent exchange on a gel filtration or DEAE-Sephadex column equilibrated with an exchange detergent (e.g., octyl glucoside).

One approach to dealing with proteins purified by SDS–polyacrylamide gel electrophoresis (PAGE) is to electroelute them after staining a guide strip. Although electroblotting SDS polyacrylamide gels onto polyvinyl difluoride (PVDF) membranes provides an effective means of removing SDS (which may then be washed off the membrane with water), protein recoveries after SDS-PAGE and PVDF blotting are often only ~35–45% and additional losses will be incurred in eluting the protein from the PVDF membrane. Elution can be accomplished in 20–100% yield (average efficiency ~50%) by extraction at 37°C with 40% acetonitrile.

## 4 STRATEGIES FOR PURIFYING AND PREPARING PEPTIDES FROM PROTEINS

Although numerous chemical and enzymatic approaches may be used to cleave proteins, cyanogen bromide (which cleaves after methionine), and trypsin (which cleaves after lysine and arginine) are used most commonly. The advantage of trypsin is that it produces peptides that are in a size range (average length ~9 residues) that separate well via reversed phase HPLC, which is the method of choice for preparing peptide samples for MS. In contrast, cyanogen bromide peptides, which have an average length of ~45 residues, often elute very slowly from reversed phase supports, hence give very broad peaks. In this instance, SDS-PAGE followed by blotting onto PVDF (and subsequent elution) may provide a better approach. Optimal tryptic digestion requires that the substrate be irreversibly denatured, be free of SDS, and be present at a protein concentration exceeding 25 μg/mL. Unlike *Staphylococcus aureus* protease, which cleaves after aspartic and glutamic acid, and lysyl-endopeptidase, which cleaves after lysine, trypsin works well on insoluble substrates. The latter is important because protein denaturation is usually accompanied by precipitation.

Since tryptic cleavages can be carried out in the presence of at least 1 M NaCl and 20% glycerol, it is sometimes possible to simply dry the sample (in a Speedvac) prior to digestion. Alternatively, the sample may be dialyzed and then dried or TCA precipitated and then extracted with acetone (to remove residual TCA and detergents). The latter procedure works extremely well with samples that are in Triton X-100, which coprecipitates the protein and can then be extracted with acetone. After precipitation or drying, the sample may be dissolved in 25 μL of 8 M urea, 0.4 M $NH_4HCO_3$, pH 8, prior to reduction with dithiothreitol (to break disulfide bonds) and carboxymethylation with iodoacetamide (to irreversibly denature the protein). After dilution with water to a final urea concentration of 2 M, the sample may be digested for 24 hours at 37°C with a 1:25 (weight:weight) ratio of trypsin to protein.

If the final purification step involves SDS-PAGE, several excellent approaches may be taken to prepare tryptic peptides in a form that is amenable to reversed-phase HPLC. One of the easiest approaches is to electroelute the (preferably nonstained) protein and then use acetone precipitation to remove the SDS. Alternatively, the protein, stained with Coomassie Blue, may be digested in the gel following extensive washing to remove residual SDS. In this case, the resulting peptides are then recovered in high yield by diffusion. A variation of this protocol takes advantage of the activity of lysyl-*endo*-peptidase in SDS (i.e., at least up to 0.1%). In this instance the Coomassie Blue stained protein is digested with lysyl-*endo*-peptidase in the gel (in the presence of 0.1% SDS) and then an anion exchange precolumn is used to remove the SDS (and much of the Coomassie Blue) prior to reversed-phase HPLC.

Following SDS-PAGE, the protein may also be blotted onto PVDF membranes and then stained with Coomassie Blue. In this instance, the protein may be enzymatically digested on the membrane (following blocking of residual protein binding sites with polyvinylpyrrolidone). Alternatively, the stained protein may be cleaved in situ with cyanogen bromide. The resulting peptides can then be eluted in high yield with 70% formic acid followed by 40% acetonitrile. After removal of the formic acid and acetonitrile by lyophilizing in a Speedvac, these peptides may then be digested with trypsin prior to HPLC separation.

Regardless of which of the foregoing approaches is taken, routine success usually requires a minimum of 100 pmol protein either in solution, in the gel, or on the PVDF membrane. Since our experience is that dye binding and colorimetric protein assays often overestimate the amount of protein by 5–10-fold, it is essential that a representative aliquot of the protein solution, gel band, or PVDF membrane be subjected to acid hydrolysis and ion exchange amino acid analysis to accurately quantitate the amount of protein before the digest is performed. Keep in mind, too, that based on the amount of protein *that was actually digested,* the average recovery of peptides from any procedure that follows SDS-PAGE is ~30%. Hence, if the combined cyanogen bromide cleavage/trypsin digestion procedure were used on an "average" protein that had a 35% blotting efficiency, the overall yield of an individual peptide would be only ~10% (i.e., 30% × 35%) of the amount of protein *originally applied to the gel.* If the same amount of protein were obtained by conventional chromatography and digested in solution, the recovery of an average peptide might well be three- to fivefold higher.

Compared to the variety of options that may be used to digest a protein that is isolated by SDS-PAGE, the separation of the resulting tryptic or lysyl-*endo*-peptidase peptides is straightforward, with reversed phase chromatography representing the clear method of choice. Typically, a 2-hour, linear gradient extending from 0% to 80% acetonitrile in 0.05% trifluoroacetic acid is employed. Optimum separation is usually obtained on a 5 μm particle size, 300 Å pore size, 25 cm long column packed with a C-18 support. Since MS can provide meaningful data on peptide mixtures, gradient times can be shortened to one hour or less, and shorter columns can be used if only mass spectrometric analyses are planned. As the amount of sample decreases, the flow rate of the mobile phase and the internal diameter of the column employed should also be decreased, to maintain or increase the effective concentration of analyte on the column and reaching the detector (UV, MS, or both) and fraction collector.

In general, amounts of proteolytic digests exceeding 250 pmol are best separated at 0.5–0.7 mL/min on a 4.0–4.6 mm (i.d.) column, whereas amounts ranging down to ~25 pmol may be routinely separated at 0.15–0.2 mL/min on 2.0–2.1 mm (i.d.) columns. Amounts of peptides below ~25 pmol are best separated on 1 mm (i.d.) or narrower columns, which typically are eluted with flow rates in the range of 25–50 μL/min. For on-line coupling to the mass spectrometer, it is recommended to use the smallest ID column to provide a high concentration of sample to the instrument. For example, loading 1–5 nmol of a digest on a 2.1 mm (ID) column provides useful data for the large and small peaks in the chromatogram. Similarly, loadings of 200–1000 pmol are used on 1 mm (ID), and 25–250 pmol on a 0.3 mm (ID) packed capillary column. The latter requires the use of specialized equipment to generate reproducible gradients at such low flow rates. In general, 2.1 and 1 mm columns have somewhat lower resolution than standard 4.6 mm (i.d.) columns as a result of "wall effects" and the increased difficulty in packing narrow bore columns.

Unless precautions are taken to minimize dead volumes, significant problems in terms of automated peak detection/collection and postcolumn mixing will be encountered as the flow rates are lowered much below 0.15 mL/min. In this regard, fused silica tubing, which typically has inner diameters ranging from 50 μm to 100 μm, is used between the detector and the fraction collector and a low volume flow cell ($\leq$ 1μL) is substituted for the standard flow cell in the UV detector. The HPLC fractions may be conveniently collected in 1.5 mL Eppendorf tubes, which are then capped to prevent evaporation of the acetonitrile, which in turn tends to prevent loss of the peptide due to absorption and/or precipitation. In most instances these fractions (provided they are tightly capped) can be stored for as long as 1–2 years at 5°C with good recovery.

## 5 CONCLUSIONS

While many protein sequences have been derived by Edman degradation and some by mass spectrometric sequencing, currently, most protein sequences are derived via DNA sequencing. Although limited protein/peptide sequencing is often used to generate the oligonucleotide probes or primers necessary for cloning, mass spectrometry offers a particularly elegant and economical means of rapidly confirming the predicted protein sequence. Frameshift and other errors that may occur during DNA sequencing can be rapidly detected by matching the predicted versus the actual masses for several tryptic peptides. Most importantly, mass spectrometry is currently the only facile means to rapidly detect and, in many instances, identify the plethora of posttranslational modifications that are found on proteins.

*See also* HPLC of Biological Macromolecules; Mass Spectrometry of Biomolecules; Proteins and Peptides, Isolation for Sequence Analysis of.

## *Bibliography*

Angeletti, R. H., Ed. (1993) *Techniques in Protein Chemistry,* Vol. IV. Academic Press, San Diego, CA.

Burlingame, A. L., and Carr, S. A., Eds. (1995) *Biological Mass Spectrometry.* Humana Press, New Jersey.

Crabb, J. W. Ed. (1994) *Techniques in Protein Chemistry,* Vol. I. Academic Press, San Diego, CA.

Findlay, J.B.C., and Geisow, M. J., Eds. (1989) *Protein Sequencing, A Practical Approach.* IRL Press, Oxford.

Mant, C. T., and Hodges, R. S., Eds. (1991) *High-Performance Liquid Chromatography of Peptides and Proteins.* CRC Press, Boca Raton, FL.

Matsudaira, P., Ed. (1993) *A Practical Guide to Protein and Peptide Purification for Microsequencing.* Academic Press, San Diego, CA.

McCloskey, J. A., Ed. (1990) *Mass Spectrometry,* Vol. 193 in *Methods in Enzymology.* Academic Press, San Diego, CA.

McEwen, C. N., and Larsen, B. S., Eds. (1990) *Mass Spectrometry of Biological Materials.* Dekker, New York.

# Protein Analysis by Raman Spectroscopy

*George J. Thomas, Jr.*

## *Key Words*

**Amide Band** A Raman band that originates from a normal mode of vibration localized largely in the peptidyl group(s) of a protein. The so-called Raman amide I and amide III bands, due, respectively, to peptidyl C═O stretching and to a combination of C—N stretching and N—H in-plane deformation, exhibit positions and intensities in the Raman spectrum that are sensitive to the protein main chain conformation and peptidyl hydrogen bonding interactions. The Raman amide bands are thus useful indicators of protein secondary structures.

**Raman Effect** The molecular inelastic scattering of monochromatic radiation, whereby the quanta transferred from the incident photons excite vibrational energy levels in the target molecules. Normal (off-resonance) Raman transitions require that the energy of the incident photon be much greater than the molecular vibrational quanta but much less than the difference in energy between ground and first excited molecular electronic states. This is ordinarily achieved by use of a visible wavelength laser to excite the Raman spectrum. The effect bears the name of its 1928 discoverer, Sir Chandrasekhara V. Raman.

**Resonance Raman Effect** A variant of the normal Raman effect in which the energy of the incident photon is selected in resonance with a molecular electronic transition. The cross section for the resonance Raman effect is several orders of magnitude greater than for the off-resonance Raman effect. However, the resonance Raman mechanism yields vibrational information restricted exclusively to the locus of the chromophoric group.

**Vibrational Normal Mode** A macromolecule of $n$ atoms has $3n - 6$ vibrational degrees of freedom or normal modes. Each Raman band corresponds to one or more vibrational normal modes.

Laser Raman spectroscopy, like infrared spectroscopy, is a method for determining molecular structure by measuring the energies or frequencies of molecular vibrations. Although, the two methods differ fundamentally in the mechanisms of interaction between radiation and matter, one obtains in both cases a *vibrational spectrum* consisting of a number of discrete bands, the frequencies and intensities of which are determined by the nuclear masses in motion, the equilibrium molecular geometry, and the molecular force field. An important advantage of Raman over infrared spectroscopy for biological applications is the virtual transparency of water (both $H_2O$ and $D_2O$) in the Raman effect, which greatly simplifies the analysis of aqueous solutions and facilitates the investigation of hydrogen isotope exchange phenomena. Changes in molecular geometry—particularly the conformational changes characteristic of biological macromolecules—can produce large shifts in Raman band positions, often referred to as ''frequency shifts,'' empowering the technique in the diagnosis of protein secondary structure, side chain configurations, and side chain interactions. Since molecular geometry and force field may be sensitive to interactions between molecules, the Raman method also has the potential for investigating intermolecular interactions, including the formation of biologically important protein complexes. Raman spectroscopy is gaining wide use as a method for probing protein structure, dynamics, assembly, and recognition.

## 1 INTRODUCTION

Raman spectroscopy of proteins comprises a diverse and rapidly expanding field. Reviews of Raman and resonance Raman (RR) applications to a variety of proteins, metalloproteins, retinal-binding proteins, nucleoproteins, and lipoproteins, covering the literature up to about 1986, are contained in the volumes edited by Spiro. A current review (by Austin et al., 1993) of the more specialized area of ultraviolet resonance Raman (UVRR) spectroscopy of proteins is also available. An account of principles and practical considerations is given in the text by Carey and in several reviews. This entry indicates the scope and limitations of the Raman method in molecular biology, illustrates the nature of the data of Raman spectra of proteins, and demonstrates the value of the technique for analysis of protein structures and interactions through examination of representative data. Consideration is given to both *static* and *dynamic* Raman experiments.

### 1.1 Advantages and Disadvantages

The principal advantages of Raman spectroscopy vis à vis other structural methods employed in molecular biology are as follows:

1. Applicability to virtually any sample morphology, including solutions, suspensions, gels, crystals, and amorphous solids.
2. Favorable spectroscopic properties of the aqueous solvent (both $H_2O$ and $D_2O$).
3. High spectral sensitivity for electron-rich chemical groups (e.g., peptide C=O, aromatic rings, C—S, S—H, S—S).
4. Small sample volume ($\approx$ 1 μL).
5. Versatility and flexibility in sample illumination.
6. Nondestructive sampling conditions.
7. Rapid data accumulation.
8. Existence of a large reference database on model compounds.

The principal disadvantages of the method are:

1. Inherent weakness of the Raman scattering process, requiring sophisticated and relatively costly instrumentation.
2. Relatively high solute concentration ($\approx$ 10–100 μg/μL).
3. High sample purity and optical homogeneity.
4. Limited resolution of overlapping bands from structurally equivalent or quasi-equivalent groups in the macromolecule.
5. Inadequacy of molecular force fields for calculating spectra a priori.

Various tactics for optimizing the advantages and circumventing the disadvantages of the Raman method have been developed in recent years. Several examples are illustrated in the applications considered.

### 1.2 Nature of the Data

The complete Raman spectrum of a typical monomeric protein, *E. coli* thioredoxin in aqueous solution, is shown in Figure 1. The sample concentration is 100 μg/μL. The abscissa is the Raman frequency in wavenumber units ($cm^{-1}$); the ordinate is the Raman intensity in arbitrary units. The majority of Raman bands, clustered between 500 and 1800 $cm^{-1}$, originate from vibrational modes of

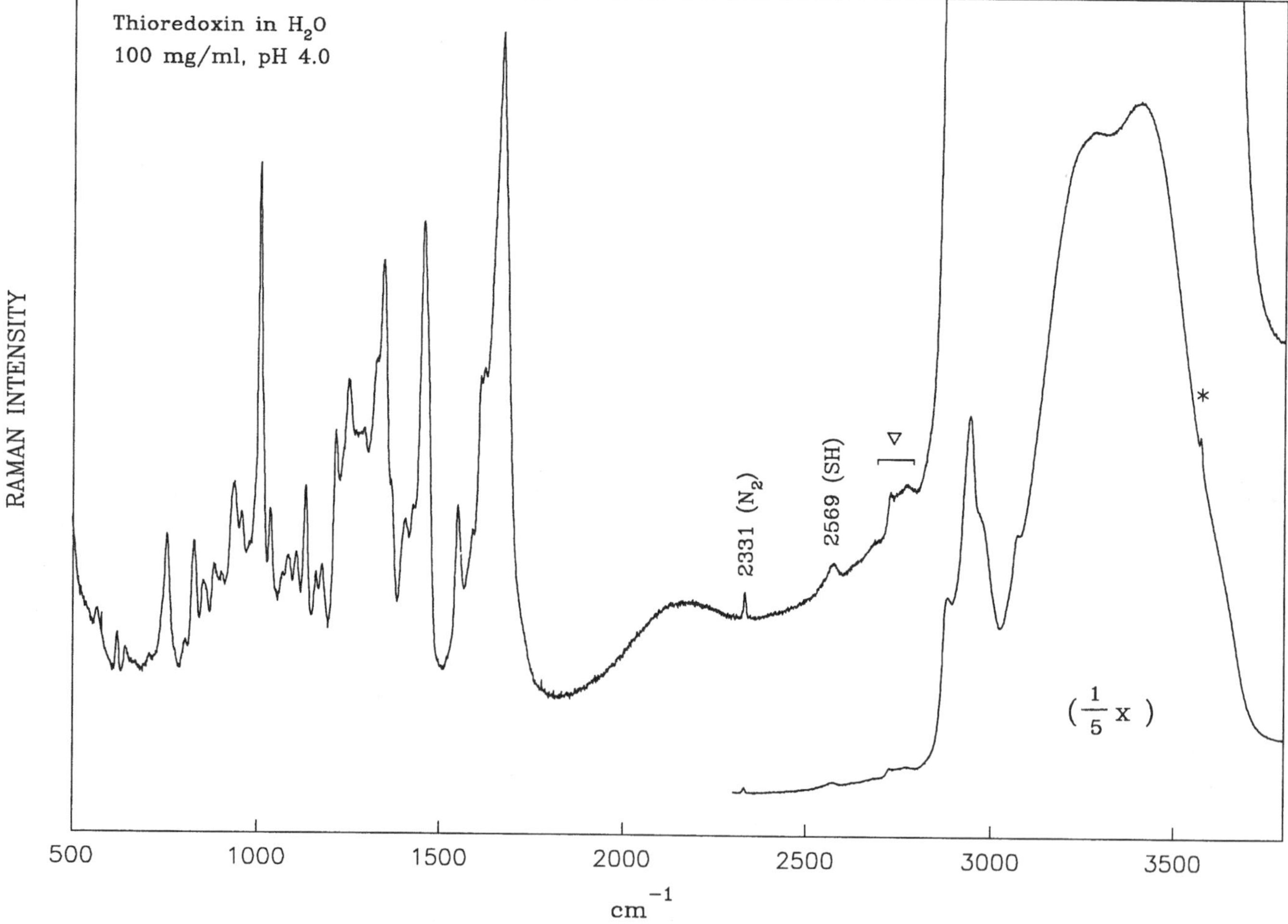


**Figure 1.** Raman spectrum in the 500–3700 $cm^{-1}$ region of reduced thioredoxin at pH 4 and 6°C (sequence: [1]SDKIIHLTDD [11]SFDTDVLKAD [21]GAILVDF-WAE [31]WCGPCKMIAP [41]ILDEIADEYQ [51]GKLTVAKLNI [61]DQNPGTAPLY [71]GIRGIPTLLL [81]FKNGEVAATK [91]VGALSKGQLK [101]EFLDANLA). Protein concentration $\approx$ 100 μg/μL. The Raman band of $N_2$ at 2331 $cm^{-1}$ serves as an internal frequency and intensity reference. The complex Raman band at 2569 $cm^{-1}$ is due to Cys32 and Cys35 sulfhydryl stretching modes. Raman bands just below 3000 $cm^{-1}$, which are due to aliphatic CH stretching modes of most side chains and the very broad, complex band of $H_2O$ solvent ca. 3100–3600 $cm^{-1}$, are evident in the segment recorded at one-fifth amplification. Probable overtone/combination bands (∇) and a laser emission line (*) are also indicated. [Data adapted from Li et al., *Biochemistry* 32:5800–5808 (1993).]

the protein main chain and various side chains. Detailed assignments can be made by analogy with model compound data. The band at 2331 $cm^{-1}$ is due to ambient nitrogen and serves as both a frequency ($cm^{-1}$) and a relative intensity standard. The broad band at 2569 $cm^{-1}$ is due to overlapping contributions from the two cysteinyl SH groups at the thioredoxin active site. (Further discussion of this complex Raman band is given in Section 2.2.2.) The bands near 2800–3000 $cm^{-1}$ are due to the multitude of C—H stretching vibrations of all aliphatic groups (CH, $CH_2$, and $CH_3$) in the protein. The very intense and complex Raman bands between 3100 and 3600 $cm^{-1}$ are due mainly to O—H stretching vibrations of solvent water molecules.

The spectrum shown in Figure 1 represents the *isotropic* Raman scattering, that is, the electric vector of the scattered light is collected for all ($2\pi$) orientations perpendicular to the direction of beam propagation. Additionally, the protein molecules are randomly oriented in solution. The data of Raman spectroscopy, and its potential structural value, may be significantly increased in certain applications by measurement of the *anisotropy* of Raman scattering. With use of oriented samples (e.g., single crystals or fibers), the polarized Raman spectrum can in principle be exploited to determine the orientations of specific molecular subgroups in the macromolecule.

# 2 APPLICATIONS

## 2.1 Quantitative Analysis and Primary Structure

The appearance in a protein Raman spectrum of bands assigned to side chains of specific types constitutes a basis for determining the number of such side chains in the protein. For example, the integrated intensity of the Raman SH band (2569 $cm^{-1}$) in the thioredoxin spectrum of Figure 1 confirms the presence of two sulfhydryls in the protein at the conditions of this experiment (pH 4). Similar quantitative correlations exist for other, but not all, types of amino acid residue. Figure 1 shows further that the spectrum of an aqueous protein contains bands of both the solute and solvent, thus constituting a basis for determination of protein concentration. A roughly linear relationship exists between the protein concentration and the ratio ($I_{2800-3000}/I_{3100-3600}$) of Raman intensities of protein C—H and water O—H bands, over the approximate range of 0.1 to 10 wt % protein.

## 2.2 Analyis of Secondary and Tertiary Structures

### 2.2.1 Main Chain Conformations

The sensitivities of Raman amide I and amide III bands to protein main chain conformation have been extensively documented and discussed. Many investigators have described quantitative or semi-quantitative methods for extracting from the Raman amide band profiles the percentages of $\alpha$-helix, $\beta$-strand, and possibly other secondary structure types present in the protein. However, none of these methods is universally accepted, and use of Raman spectroscopy for this purpose requires considerable caution. When a highly uniform secondary structure is present, such as in the highly $\alpha$-helical coat proteins of filamentous viruses or in the highly $\beta$-stranded structures of the tailspike trimer of P22 virus, Raman amide band analysis tends to be reliable. Changes of secondary structure induced within a protein by environmental changes may also be reliably measured by difference Raman methods, as shown for two different assembly states of the P22 viral capsid in Figure 2.

### 2.2.2 Side Chain Configurations and Interactions

Many protein Raman bands have been definitively assigned to vibrational modes of specific side chains and have been demonstrated to exhibit dependence of frequency, intensity, or both, upon the precise configuration of the side chain and/or its interactions with other protein groups. When such bands are confidently measured in a protein Raman spectrum, they can serve as indicators of detailed molecular geometry or hydrogen bonding interaction of the corresponding side chain. Particularly useful in this respect are bands of tyrosine, tryptophan, cystine, and cysteine. Thus, the complex 2569 $cm^{-1}$ band of thioredoxin (Figure 1) is diagnostic of relatively strong S—H hydrogen bonding for both active site cysteine residues, and detailed analysis of the band as a function of pH (Figure 3) permits determination of the $pK_a$ values governing thiol/thiolate equilibria in the protein.

## 2.3 Analysis of Quaternary Structure

Figure 2 illustrates one kind of Raman band change that can accompany a change in protein assembly state. Another well-studied example is that of the N-terminal domain of lambda repressor. Generally, changes in protein quaternary structure lead to significant changes in bands originating from side chain vibrations and little or no changes in bands originating from the protein main chain. This is in accordance with the relatively large changes expected in surface contacts, without significant intramolecular conformational change, attendant on protein–protein associations.

## 2.4 Analysis of Nucleoprotein Complexes

### 2.4.1 Viruses

Raman spectroscopy has been applied extensively in structural studies of viruses and viral precursors. The methodology is unique in its ability to provide detailed structural information about both protein and nucleic acid constituents of native virions over a broad range of sampling conditions. A comprehensive review to about 1986 has been given, and several more recent applications have been surveyed in a current review (Thomas and Tsuboi, 1993) of nucleic acid complexes. A particular virtue of the Raman method in studies of viruses is the extraordinary spectral detail it affords. This information permits the detection of specific residue interactions, including those underlying protein recognition.

### 2.4.2 Gene Regulatory Complexes

The structural basis of repressor–operator recognition in solution has also been investigated by Raman spectroscopy. This application area requires spectra of exceptional signal-to-noise quality for the detection by digital difference methods of relatively small changes in the diagnostic Raman bands.

## 2.5 Raman Microscopy

Recent developments in confocal light microscopy and improvements in detectors for Raman spectroscopy have permitted implementation of the Raman microscope for analysis of biomolecules

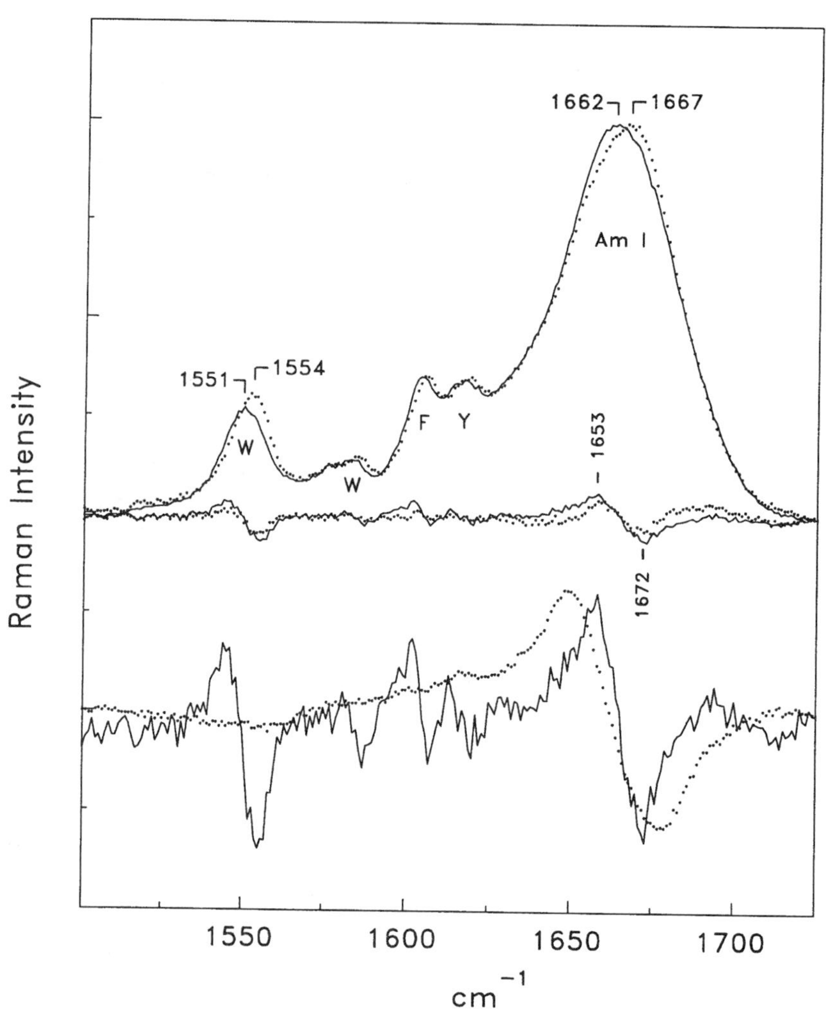


Figure 2. *Top:* Raman amide I spectra of P22 empty procapsid shells (solid line) and mature capsids (dots). The amide I envelopes are centered within the interval (1660–1680 $cm^{-1}$) expected for a protein rich in β-sheet, consistent with subunit folds of many viral capsids. Labels identify Raman band centers and residues to which they are assigned (AmI, amide I; W, tryptophan; Y, tyrosine; F, phenylalanine). *Middle:* Difference spectrum computed with empty procapsid shell as minuend and mature capsid as subtrahend (solid line). Included is the difference spectrum (dots) obtained after applying abscissa shifts of +0.5 $cm^{-1}$ to the empty procapsid and −0.5 $cm^{-1}$ to the mature capsid. Similar results were obtained for oppositely signed shifts (data not shown), demonstrating that observed differences are not due to misalignments within ±1.0 $cm^{-1}$ of either minuend or subtrahend. Difference spectra were computed using established procedures, and assuming a symmetrical amide I band shape compensated for contributions from the 95 side chain carbonyls per subunit. The integrated difference peak at 1653 $cm^{-1}$ and trough at 1672 $cm^{-1}$ each represent 2.0 ±0.4% of total amide I intensity, interpreted as a conversion of 2.0% α-helix in empty procapsids to β-sheet in mature capsids, or a net change involving 4.0% of subunit peptides. Because of inherent broadness and overlap of helix and sheet components of amide I, the net intensity shift may be underestimated by 0.5 to 1.0%. *Bottom:* Fivefold amplification of the observed difference spectrum (solid line) superposed with a model helix-minus-sheet (Pf1 minus β-poly-L-serine) difference spectrum (dots). Other details of data collection are scanning increment, 1 $cm^{-1}$; photon-counting integration time, 1.5 seconds; laser excitation wavelength, 514.5 nm; laser power, 200 mW; sample temperature, 20°C; volume, 5 μL; concentration, 100 mg/mL; pH, 7.5; buffer, 10 mM Tris + 25 mM NaCl + 1 mM EDTA. (Data adapted from Prevelige et al., 1993.)

at the microgram (off-resonance) or nanogram (resonance) levels. Significant further development of this area appears likely in the near future.

### 2.6 Dynamic Processes

The favorability of both $H_2O$ and $D_2O$ as solvents for Raman spectroscopy enables the study of hydrogen isotope exchange processes in proteins and their complexes. Recent applications include the use of Raman flow cell methods to monitor in real time the exchange profiles of protein subunits in viral capsids and mature virions. In principle, the deuteration occurring at peptide NH sites, as well as at exchangeable side chain sites (cysteinyl SH, tryptophanyl NH, tyrosyl OH, etc.), can be monitored independently by Raman dynamic methods.

## 3 PERSPECTIVES

Methods of Raman and resonance Raman spectroscopy provide versatile and sensitive probes of proteins, their biological assemblies and aggregates, and their complexes with other macromolecules. Often, the structural information obtained from Raman spectra of proteins is not obtainable with alternative structural probes. Protein applications of Raman spectroscopy have increased rapidly

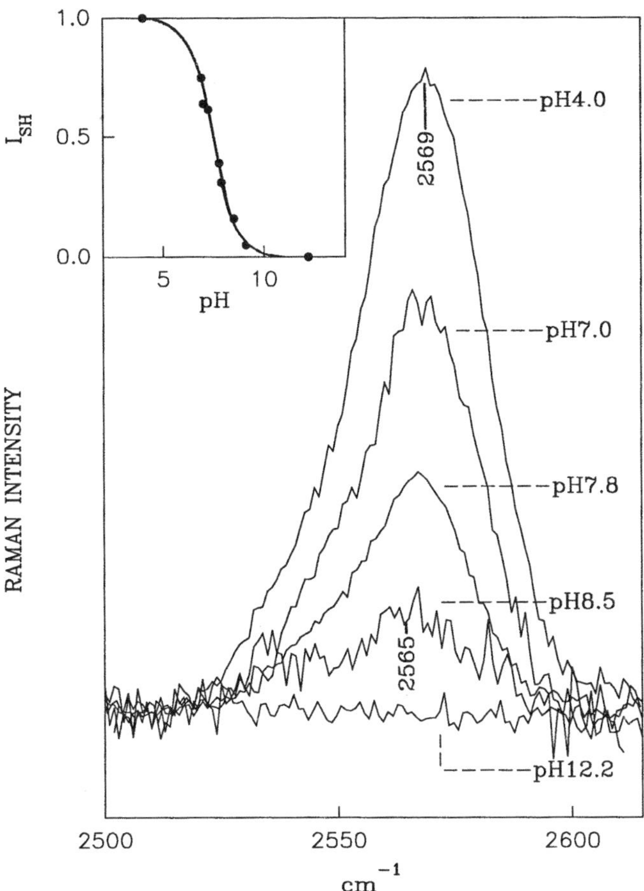


**Figure 3.** Raman spectra in the 2500–2625 $cm^{-1}$ region of reduced thioredoxin at the indicated pH values. The bottom trace, obtained at pH 12.2 (for which all sulfhydryls are titrated to thiolate ion), indicates the level of background noise in this region of the spectrum. The spectra obtained at pH 4.0 and 7.8 represent averages of 300 scans. Other spectra represent averages of 55 scans. Inset (upper left) shows a Raman–pH titration curve obtained from these spectra and additional data not shown. The ordinate in the titration curve is the normalized, integrated intensity of the total Raman SH band. Least-squares fits of the pH 4.0 and 7.8 spectra to two Gauss–Lorentz components show that the two active site sulfhydryls exhibit $pK_{C32} = 7.1$ and $pK_{C35} = 7.9$. [Data adapted from Li et al., *Biochemistry* 32:5800–5808 (1993).]

in number during the past decade and are expected to continue to increase with improvements in the sensitivity and reliability of Raman instrumentation.

The practice of Raman spectroscopy is technically more complex than that of many other spectroscopic probes applied to proteins (e.g., fluorescence, circular dichroism, and Fourier transform infrared spectroscopy). Additionally, the requisite laser irradiation of proteins must be exercised carefully. *The biochemist is well advised to establish protein purity, homogeneity, and integrity both before and after Raman spectra are recorded.* Finally, the Raman spectra of proteins must be interpreted in conjunction with well-established assignments and reliable correlations based on analogous studies of appropriate model compounds of known structure.

*See also* Protein Analysis by X-Ray Crystallography.

### *Bibliography*

Austin, J. C., Jordan, T., and Spiro, T. G. In *Advances in Spectroscopy*, Vol. 20A, R. J. H. Clark, and R. E. Hester, Eds. Wiley, London, 1993.

Carey, P. R. *Biochemical Applications of Raman and Resonance Raman Spectroscopies.* Academic Press, London, 1982.

Greve, J., and Puppels, G. J. In *Advances in Spectroscopy,* Vol. 20A, R. J. H. Clark, and R. E. Hester, Eds. Wiley, London, 1993.

Harada, I. and Takeuchi, H. In *Advances in Spectroscopy,* Vol. 13, R. J. H. Clark, and R. E. Hester, Eds. Wiley, New York, 1986.

Prevelige, P. E., Thomas, D., Aubrey, K. L., Towse, S. A., and Thomas, G. J., Jr. *Biochemistry,* 32:537–543 (1993).

Spiro, T. G., Ed. *Biological Applications of Raman Spectroscopy.* Vol. 1, *Raman Spectra and the Conformations of Biological Macromolecules* (1987); Vol. 2, *Resonance Raman Spectra of Polyenes and Aromatics* (1987). Vol. 3, *Resonance Raman Spectra of Heme and Metalloproteins* (1988). Wiley-Interscience, London.

Thomas, G. J., Jr. In *Biological Applications of Raman Spectroscopy,* Vol. 1, *Raman Spectra and the Conformations of Biological Macromolecules,* T. G. Spiro, Ed. Wiley-Interscience, London, 1987, pp. 135–201.

Thomas, G. J., Jr., and Kyogoku, Y. In *Infrared and Raman Spectroscopy,* Part C (Practical Spectroscopy Series, Vol. 1), E. G. Brame, Jr., and J. G. Grasselli, Eds. Dekker, New York, 1977, pp. 717–872.

Thomas, G. J., Jr., and Tsuboi, M. In *Advances in Biophysical Chemistry,* Vol. 3, C. A. Bush, Ed. JAI Press, Greenwich, CT, 1993.

# Protein Analysis by X-Ray Crystallography

*Gordon V. Louie*

### *Key Words*

**Atomic Model** A description of the three-dimensional arrangement of atoms of a protein molecule (and also of associated ligands and solvent molecules). For each atom in the model, the positional coordinate *(x, y, z)* and the vibrational motion are specified.

**Crystal** A solid composed of a structural motif repeated translationally on a three-dimensional lattice.

**Structure Factor** A description of the total X-irradiation diffracted in a particular direction by an object. The structure factor has two components, an amplitude and a phase, which depend on the spatial arrangement of electrons in the object.

**X-Ray Diffraction** The interference of X-rays scattered by individual electrons in an object. The interference is sensitive to the atomic structure of the object, because the wavelength of X-rays is roughly the same as the typical closest interatomic distance.

X-ray crystallography is one of two techniques (the other being multidimensional nuclear magnetic resonance) for directly determining the three-dimensional arrangement of atoms in protein molecules. This structural information is of great importance because it provides a firm basis for elucidating the molecular mechanism of the biological reactions mediated by proteins. The protein to be studied must be obtained in a crystalline form. From the X-ray diffraction pattern of the protein crystal, an electron density map of the protein can be derived, which in turn can be interpreted in terms of the positions of the atoms in the protein molecule.

## 1 PHYSICAL BASIS OF STRUCTURE DETERMINATION BY X-RAY CRYSTALLOGRAPHY

### 1.1 X-Ray Diffraction

When X-rays irradiate an object (i.e., a collection of atoms), each electron in the object is induced to scatter X-rays in all directions. The total wave diffracted in a particular direction is the result of the superposition of the waves scattered in that direction by each of the electrons in the object; therefore it is dependent on the spatial distribution of electrons in the object. This concept, which is the foundation of structure determination by X-ray diffraction methods, is illustrated in Figure 1 and expressed in the following general X-ray scattering equation:

$$\mathbf{F}(\mathbf{S}) = |F(\mathbf{S})| \exp[i\phi(\mathbf{S})] = \int_{\text{all objects}} \rho(\mathbf{r})\exp(2\pi i\mathbf{r}\cdot\mathbf{S})d\mathbf{r} \qquad (1)$$

The total wave $\mathbf{F}(\mathbf{S})$ is called the structure factor. It contains two pieces of information: the amplitude $|F|$, which can be experimentally measured (as the square root of the intensity), and the phase $\phi$, which is unobservable. The diffraction pattern, or the set of all structure factors, represents the Fourier transform of the object's electron density. Because electron density is localized at atomic positions, the integral in equation (1) can be replaced by a summation over the $N$ atoms constituting the object:

$$\mathbf{F}(\mathbf{S}) = \sum_{n=1}^{N} \rho_n \exp(2\pi i\mathbf{r}_n\cdot\mathbf{S}) \qquad (2)$$

Each atomic electron density distribution $\rho_n$ has two components. The first, the atomic scattering factor, is dependent only on the identity of the atom (i.e., the number and spatial arrangement of electrons). The second, the atomic temperature factor, accounts for thermal motion of the atom and is a measure of the mean-square displacement of the atom from its average position.

The object's electron density distribution (from which atomic positions can be determined) can be retrieved by the following inversion of the Fourier transform (equation 1):

$$\rho(\mathbf{r}) = \int_{\text{all } S} \mathbf{F}(\mathbf{S})\exp(-2\pi i\mathbf{r}\cdot\mathbf{S})d\mathbf{S} \qquad (3)$$

Before equation (3), the electron density equation, can be applied, however, the phase of each structure factor must be determined.

### 1.2 Crystals

A crystal is composed of an object, or motif, repeated translationally on a three-dimensional lattice. The period of repeat in each of the three lattice directions is described by the unit cell, a parallelepiped containing (in the simplest case) a single motif. The motif usually possesses internal symmetry, which interrelates the (integral number of) asymmetric units that constitute the motif. The combination of symmetry elements present is described by the space group, of which 65 are permitted for proteins. The objective of the X-ray crystallographic analysis is to determine the atomic structure of the asymmetric unit, which usually consists of one protein molecule.

### 1.3 X-Ray Diffraction by a Crystal

With a crystal, the contents of each unit cell diffract X-rays identically, and the waves diffracted in specific directions with respect to the crystal lattice axes are constructively reinforced. However, the waves diffracted in all other directions are destructively canceled. The diffraction pattern thus appears—for example, on X-ray film—as an array of discrete spots. In addition, symmetry

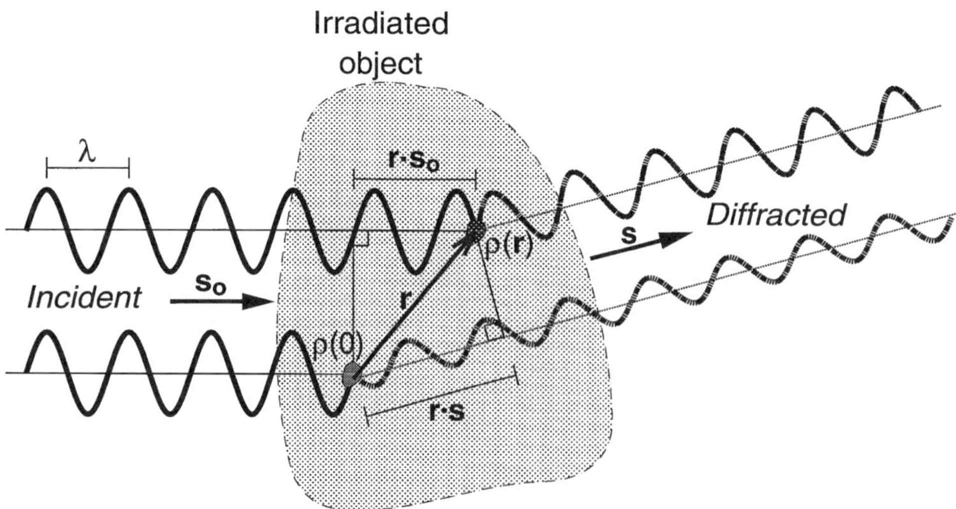


Figure 1. X-Ray scattering in the direction $\mathbf{s}$ by two volumes of electron density. The unit vectors $\mathbf{s}_0$ and $\mathbf{s}$ define the directions of travel of the incident and diffracted X-rays, respectively. The phase difference between the waves scattered by $\rho(0)$, located at the origin, and $\rho(\mathbf{r})$ is $2\pi(\mathbf{r}\cdot\mathbf{s} - \mathbf{r}\cdot\mathbf{s}_0)/\lambda$, or $2\pi\mathbf{r}\cdot\mathbf{S}$, where $\mathbf{S} = (\mathbf{s} - \mathbf{s}_0)/\lambda$. A wave having amplitude $A$ and phase $\phi$ is conveniently represented as a complex exponential $A \exp(i\phi)$. Then, the total wave, diffracted by the entire object in the direction specified by $\mathbf{S}$, can be represented by the integral $\int \rho(\mathbf{r})\exp(2\pi i\mathbf{r}\cdot\mathbf{S}d\mathbf{r})$, where the integration is over all infinitesimal volumes of electron density in the object.

within the unit cell of the crystal gives rise to symmetry in the diffraction pattern.

X-ray diffraction by a crystal can be conveniently represented by saying that for the crystal's real lattice, there is a reciprocal lattice, indexed with the integers *h, k,* and *l,* and this lattice defines the permissible directions of the diffracted X-rays. Each diffracted wave is sometimes regarded as a reflection of the incident X-rays off a specific set of equally spaced lattice planes in the crystal. The resolution of a crystal structure refers to the minimum interplanar spacing of the reflections included in the structure determination. If a crystal is poorly ordered, the high resolution reflections from sets of closely spaced planes cannot be observed.

For X-ray diffraction by a crystal, equation (2) becomes the following structure factor equation:

$$\mathbf{F}(hkl) = |F(hkl)| \exp[i\phi(hkl)] = \sum_{n=1}^{N} \rho_n \exp[2\pi i(hx_n + ky_n + lz_n)] \qquad (4)$$

which allows all structure factor amplitudes and phases to be calculated given the atomic structure in the unit cell. Equation (3), the electron density equation, can also be rewritten:

$$\rho(xyz) = \sum_{\substack{\text{all}\\ hkl}} |F(hkl)| \exp[i\phi(hkl)]\exp[-2\pi i(hx + ky + lz)] \qquad (5)$$

### 1.4 Patterson Map

A Fourier summation calculated without the phases, but with only the (experimentally measurable) squared amplitudes, is called a Patterson function:

$$P(uvw) = \int_{\substack{\text{all}\\ hkl}} |F(hkl)|^2 \exp[-2\pi i(hu + kv + lw)] \qquad (6)$$

In contrast to the electron density equation, which yields a map of all atomic positions, the Patterson yields a map of all interatomic vectors. Although the Patterson map cannot be interpreted to locate directly the atomic positions in a protein molecule, it is extremely useful in phase determination (see Section 2.5).

## 2 OVERVIEW OF THE CRYSTALLOGRAPHIC ANALYSIS OF A PROTEIN

The X-ray crystallographic structure determination of a protein proceeds through the series of steps outlined in Figure 2. These steps are briefly described in Sections 2.1–2.7.

### 2.1 Preparation of Protein Sample

A protein crystallographic analysis begins with at least several milligrams of homogeneous material, free from contaminants that would hamper crystallization. Cloning and expression methods have been invaluable in providing sufficient quantities of many proteins. Large, multidomain proteins, which can be difficult to crystallize, are sometimes more easily studied after dissection into smaller fragments.

### 2.2 Growth of Protein Crystals

Crystals are grown from a supersaturated solution of the protein. The supersaturation state is usually achieved by lowering the protein solubility through addition of a precipitating agent, which is most commonly a salt (e.g., ammonium sulfate, sodium chloride, sodium citrate), an organic solvent (e.g., ethanol, methylpentanediol, acetone), or a polyethylene glycol. Other factors that affect protein solubility include protein concentration, pH, temperature, and the presence of specific ligands and metal ions. The identification of conditions appropriate for protein crystal growth is largely a multidimensional search by trial and error, and often it represents the major obstacle in a crystallographic analysis.

Several methods are available for setting up crystallization trials. Solutions of the protein and the precipitant can be equilibrated by dialysis or by vapor phase or liquid–liquid diffusion. Or, a mixture of the two components can be left to stand undisturbed, or perturbed by slow evaporation, addition of further increments of precipitant, or change in temperature. When small crystals have been obtained, they can be used as seeds in further crystallization setups. Protein crystals at least ~0.1 mm in diameter are required to produce observable diffraction patterns.

Protein crystals typically contain ~50% solvent, which fills extended channels between proteins molecules. These channels allow small ligands (such as substrates, inhibitors, or heavy atom reagents) to be diffused into the crystal.

### 2.3 Characterization of Crystals

A crystal form of a protein is initially characterized by recording a portion of its diffraction pattern (typically a reciprocal lattice plane in which one of the *hkl* indices is zero). These patterns will reveal the extent of observable diffraction from the crystal. Most importantly, the unit cell dimensions can be calculated from the interspot spacings, and the space group can be deduced from the symmetry of the pattern and possibly the systematic absence of certain reflections. The external morphology of the crystal may also provide clues on the internal symmetry.

The ratio of the volume of the unit cell to the molecular mass of protein contained in the cell ($V_M$) gives a measure of the solvent content of the crystal. For most protein crystals, $V_M$ falls in a fairly narrow range, 1.7–3.5 $A^3$/Da. Therefore, if the molecular weight of a crystallized protein is known, the number of protein molecules in the unit cell usually can be deduced from the unit cell dimensions and space group. Alternatively, if the molecular weight of a protein is unknown, it can be estimated from experimentally measured values of the crystal density and solvent content.

The crystallographic space group can sometimes provide information on the symmetry present in the quaternary structure of an oligomeric protein. If the number of oligomers in the unit cell is found to be less than the number asymmetric units expected from the space group, the protomers that make up one oligomer must be related by a crystallographic symmetry element (usually a rotation axis).

### 2.4 Data Collection

For the collection of diffraction data, the protein crystal is irradiated and then, for each diffracted X-ray, the *hkl* indices are assigned and the intensity is measured. Usually, the orientation of the crystal

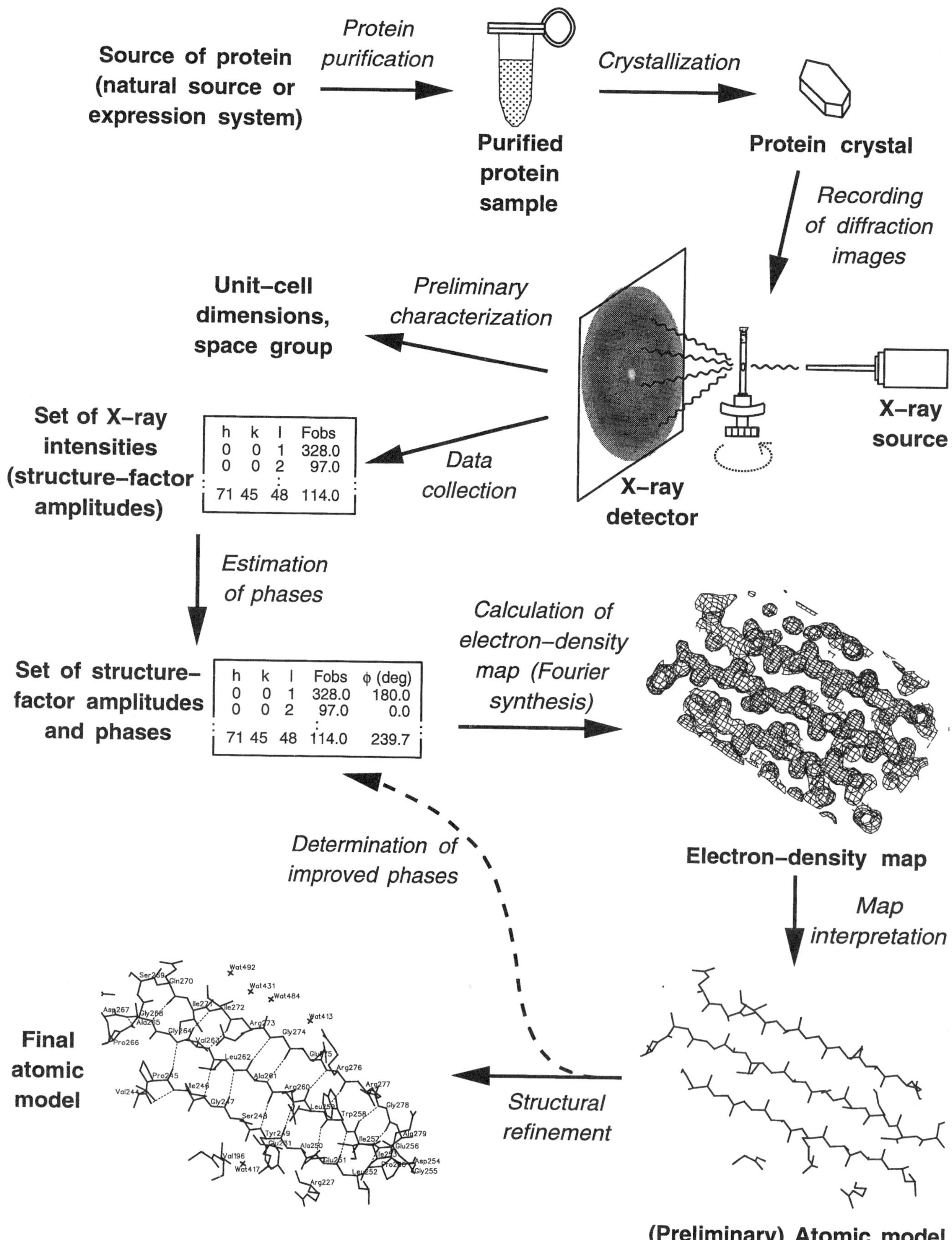


Figure 2. Materials (bold) and methods (italics) involved in the structure determination of a protein by X-ray crystallography.

relative to the incident X-ray beam must be manipulated (e.g., by rotation about an axis perpendicular to the beam, or by precession around the beam) to bring each unique point in the reciprocal lattice into a diffracting position. A complete data set from a protein crystal usually requires several tens of hours of collection time. During data collection, protein crystals are maintained in a hydrated environment within a sealed capillary. Crystals suffer radiation damage with X-ray exposure, resulting in degradation of the diffraction pattern.

The X-rays (of wavelength 1.54 A) emitted when high energy electrons strike a copper anode are the conventional X-ray source in the laboratory. Much more intense X-rays of selectable wavelength are available from electron storage rings at synchrotron facilities. X-ray detectors are of two classes. Diffractometers, which measure one X-ray intensity at a time, are useful for crystals with a small unit cell. Position-sensitive area detectors, which efficiently record an entire region of the diffraction pattern, are of several types, including photographic film, phosphor screen television detectors, multiwire proportional counters, and image plates.

### 2.5 Determination of Phases

In the majority of protein structure determinations, the structure factor phases have been assigned by one of two methods. In the first, isomorphous replacement, the crystal is reacted with a heavy metal reagent, which is bound within the crystal (ideally at a few specific sites on th surface of the protein molecules). Each diffracted X-ray for the heavy atom derivative $\mathbf{F}_{PH}$ will be the sum of contributions from the heavy atom $\mathbf{F}_H$ and the protein alone $\mathbf{F}_P$. The positions of the heavy atom binding sites can be deduced from a $(|F_{PH}| - |F_P|)^2$ Patterson difference (equation 6) and used to calculate both $|F_H|$ and $\phi_H$. These quantities together with the structure factor amplitudes $|F_P|$ and $|F_{PH}|$ then define two possible values for the phase $\phi_P$. The twofold ambiguity in the phase determination can be resolved from the analysis of (at least) one other heavy atom derivative—thus the term multiple isomorphous replacement.

The second method, molecular replacement, can be used if the structure is known for another protein (e.g., the homologous protein from another species) that is closely similar to the unknown protein. The known structure must first be correctly oriented and positioned in the unit cell of the unknown protein, usually through a systematic search involving the Patterson map and the interatomic vectors calculated from the known structure. It can then be used to calculate directly initial phase estimates for the structure factor amplitudes of the unknown protein.

### 2.6 Interpretation of the Electron Density Map

Once phase estimates have been obtained for the measured structure factor amplitudes, an electron density map can be calculated (equation 5). This map is contoured to create envelopes that surround points having an electron density greater than some defined value (usually a multiple of the background level of the map). The atoms of the protein molecule are then fit into the electron density envelopes. For a new protein, the polypeptide chain must be traced from scratch, whereas a molecular replacement solution or in the refinement stages of a structure determination, the existing atomic model is adjusted against the map. Map interpretation is usually performed on a computer graphics display, which allows the map and atomic skeleton to be superimposed and rotated and viewed in any orientation.

The amount of structural detail visible in a map is dependent on resolution. At low resolution (8–5 A), the overall shape of the protein molecule can be seen; in favorable cases, α-helices and β-sheets can be distinguished as continuous rods and slabs of density. At ~3 A resolution, the polypeptide chain backbone can be traced and amino acid side chains identified and fit with a fair degree of certainty. The shapes of the electron density envelopes become increasingly better defined at higher resolution. At 1.5–1.0 A (obtainable only with exceptionally well-ordered protein crystals), individual atoms can be resolved. The building of the initial protein model is greatly aided by a knowledge of the primary structure of the protein. However, fairly accurate amino acid assignments can be derived from a 2.5 A (or better) resolution map for a protein of unknown sequence.

If a native protein and a slightly modified variant (e.g., the protein with a bound ligand, or a site-directed mutant) have the same (isomorphous) crystal form, a difference Fourier map is useful for identifying the structural differences between the two:

$$\Delta\rho(xyz) = \sum_{\substack{\text{all}\\hkl}} [|F_{native}(hkl)| - |F_{variant}(hkl)|] \exp[i\phi(hkl)]\exp[-2\pi i(hx + ky + lz)] \tag{7}$$

### 2.7 Structural Refinement

The initial atomic model will invariably contain considerable inaccuracy, which arises primarily from errors in the phase estimates used in the initial map calculation, and also from inaccurate placement of atoms in their electron density peaks. Structural refinement seeks to maximize the accuracy of the atomic model. Refinement relies ultimately on minimizing the overall discrepancy between the structure factor amplitudes from experimental measurement and those calculated for the atomic model. The most commonly used measure of the overall discrepancy is the crystallographic $R$ factor:

$$R = \frac{\sum_{\substack{\text{all}\\hkl}} \| F_{obs}(hkl)| - |F_{calc}(hkl)\|}{\sum_{\substack{\text{all}\\hkl}} | F_{obs}(hkl) |} \tag{8}$$

which is typically below 0.20 for a well-defined protein structure. An equally important refinement condition consists of having standard bonding geometry and favorable noncovalent contacts for the atomic model.

The adjustments to the positional $(x,y,z)$ and temperature factor parameters of the atomic model that will simultaneously minimize the crystallographic $R$ factor and impose proper interatomic geometry can be derived from least-squares analysis or from simulated annealing with molecular dynamics. Although these computational methods have to a large extent automated the refinement process, they cannot correct gross errors. These are usually corrected manually through periodic inspection of the atomic model against difference electron density maps. In a difference Fourier map,

$$\Delta\rho(xyz) = \sum_{\substack{all\\hkl}} [|F_{obs}(hkl)| - |F_{calc}(hkl)|]\exp[i\phi(hkl)]\exp[-2\pi i(hx + ky + lz)] \tag{9}$$

incorrectly placed atoms will appear in negative density and omitted atoms in positive density (e.g., solvent molecules). Coefficients of the type $2F_{obs} - F_{calc}$, which combine features of both the direct Fourier ($F_{obs}$) and difference ($F_{obs} - F_{calc}$) maps, are also commonly used. At the refinement stage of the structure determination, electron density maps are usually generated using the phases calculated from the atomic model, which will usually be more accurate than the original, experimentally derived phases.

## 3 STRUCTURAL RESULTS

Infrequently, a reported protein structure will contain segments of polypeptide chain with erroneous connectivity, amino-to-carboxy orientation, or amino acid sequence assignment. In well-refined protein structures, such gross errors are rare; the overall accuracy in atomic positions is typically 0.1–0.4 A, depending primarily on resolution. Accuracy is worst for groups located at the surface of the molecule, which are sometimes disordered and therefore lack well-defined electron density.

The crystallographically determined structure for a protein molecule represents an average over all molecules in the protein crystal and also over the time taken to record the X-ray diffraction data. This consideration emphasizes that the crystal structure should be regarded not as static, with atoms fixed in space, but instead as plastic, with atoms oscillating about their average positions. The atomic temperature factors provide a measure of the relative mobilities of different regions of the protein molecule.

The crystal structures of several hundred distinct proteins have been determined, and for most the atomic coordinates are archived in the Protein Data Bank at the Brookhaven National Laboratory. These structures reveal not only the detailed architecture of polypeptide chain folds, but also how proteins have evolved, how enzymes catalyze chemical reactions, and how proteins recognize and bind other molecules. In addition, they provide a basis for the design of altered or entirely new proteins, the design of drugs against specific protein targets, and ultimately the prediction of tertiary structure from amino acid sequence.

*See also* HYDROGEN BONDING IN BIOLOGICAL STRUCTURES; PROTEIN PURIFICATION; X-RAY DIFFRACTION OF BIOMOLECULES.

## *Bibliography*

Blundell, T. L., and Johnson, L. N. (1976) *Protein Crystallography.* Academic Press, London.

Branden, C., and Tooze J. (1991) *Introduction to Protein Structure.* Garland, New York.

Cantor, C. R., and Schimmel, P. R. (1980) *Biophysical Chemistry,* Part II: *Techniques for the Study of Biological Structure and Function.* Freeman, San Francisco.

Ducruix, A., and Giege, R. (1992) *Crystallization of Nucleic Acids and Proteins.* Oxford University Press, New York.

McPherson, A. (1982) *Preparation and Analysis of Protein Crystals.* Wiley, New York.

Perutz, M. F. (1992) *Protein Structure: New Approaches to Disease and Therapy.* Freeman, New York.

Sherwood, D. (1976) *Crystals, X-Rays and Proteins.* Longman, London.

Stout, G. H., and Jensen, L. H. (1989) *X-ray Structure Determination,* 2nd ed. Wiley, New York.

Wyckoff, H. W., Hirs, C.H.W., and Timasheff, S. N., Eds. (1985) *Diffraction Methods for Biological Molecules,* in *Methods in Enzymology,* Vols. 114 and 115. Academic Press, Orlando, FL.

# PROTEIN DESIGNS FOR THE SPECIFIC RECOGNITION OF DNA

*Aaron Klug*

## *Key Words*

**α-Helix and β-Sheet** Elements of secondary structure in proteins. In the α-helix, the polypeptide backbone chain of the protein traces a right-handed helical path; the β-sheet consists of two or more straight extended polypeptide chains adjacent to one another.

**Basic Leucine Zipper** A motif, about 60 amino acids long, found in a large number of eukaryotic genes, consisting of a region of basic amino acids that binds DNA, followed by a leucine heptad repeat that folds into an α-helix, which is used to form a dimer interface (the "zipper") between two protein monomers.

**Coding and Control Region** The coding region is that part of a gene which is copied into RNA; the control region is that part which is not copied but is involved in the regulation of gene activity.

**Helix–Turn–Helix** A structural motif found in some bacterial regulatory proteins, consisting of two α-helices packed at an angle with a tight turn of the polypeptide chain between them.

**Homeobox/Homeodomain** The homeobox is a sequence motif, about 60 amino acids long, originally discovered as a conserved region (box) in homeotic genes, which determine the character of segments in the fruit fly *Drosophila* but now are found more generally in eukaryotes; its three-dimensional structure is known as a homeodomain and contains a helix–turn–helix motif.

**Hydrogen Bond** A noncovalent, weak chemical interaction used in the formation of folded three-dimensional structures of proteins and nucleic acids, and in the interaction between them.

**Protein Domain** A region of a protein that can fold into a stable three-dimensional stretch independently of the rest of the protein.

**Response Element** Term used by geneticists and biochemists to indicate a short piece of the control region of a gene with which the protein product of another gene interacts: in structural terms it corresponds to a DNA-binding site.

**Transcription Factor** A protein that contributes to the regulation of the transcription of the DNA of a gene into RNA, usually by binding to a section of the control region of the gene.

During cell growth and differentiation, various nuclear factors actively interact with the DNA to allow expression of a selected set of genes. The regulation of gene expression is achieved by the binding of such regulatory proteins (or transcription factors) to the recognition sequence of the gene on which they act. Many such proteins achieve binding by having embedded in their structures a domain, or motif, that serves for binding to DNA. Examples are the helix–turn–helix motif found in many bacterial regulatory proteins and in the homeobox region of several eukaryotic proteins, the ''basic leucine zipper'' motif, which forms dimers binding to symmetrical DNA recognition sites, and various classes of ''zinc finger'' domains, which consist of various folded loops of protein stabilized by a zinc ion. Of these, the clasical zinc finger represents a truly modular design, since multiple repeated domains, all chemically distinct but having the same structural framework, can be linked together to recognize longer DNA sequences.

To gain an understanding of how sequence-specific recognition by these different proteins is achieved, it is necessary to elucidate, by X-ray crystallography or two-dimensional NMR spectroscopy, the three-dimensional structures of complexes of these proteins with their cognate DNA binding sites. This entry summarizes and compares the different modes of recognition by different designs, and describes some general principles.

## 1 REGULATION OF GENE EXPRESSION IN HIGHER ORGANISMS: A COMBINATORIAL PRINCIPLE

At any one time most of the genes of a higher organism are silent: only a minority of the genes in a given cell is expressed—that is, being transcribed into messenger RNA, which in turn acts as a template for protein synthesis. The selective expression of any one gene is accomplished primarily through the interaction of protein transcription factors with characteristic DNA sequences located in the control region of the gene, which is most commonly located near to the actual coding region. The binding of a set of such factors, or regulatory proteins, acts as a molecular switch for the activation of RNA polymerase and other components of the transcriptional machinery which are common to all genes. The supply of a particular combination of such transcription factors ensures that a gene is switched on at the right place and at the right time.

The essential idea that gene expression is controlled by binding of a protein to a regulatory region, or promoter, goes back to François Jacob and Jacques Monod, who worked on the bacterium *Escherichia coli* more than 30 years ago. However, whereas transcription control mechanisms in prokaryotes are relatively simple, involving only one or two such factors, it has emerged in the last 10 years or so that more complex organisms (eukaryotes), from yeast to man, need, and have, more complex mechanisms for regulating the expression of their genes. Usually, a combination of several (as many as six) transcription factors is necessary to form a transcription complex that can harness and activate the RNA polymerase to initiate transcription at the right starting point.

A little reflection will convince one why a set of such protein factors is required, rather than a single protein for each gene. If the latter were the case, that protein would have to be coded for by the expression of another gene, which would in turn require another protein transcription factor, and so on, leading to an infinite recurrence. However, if a set of proteins is involved, different combinations can be used for different genes. Thus a smaller number of regulatory proteins can control a large number of genes. In addition, having a set of proteins provides the means for multiple control at the level of the gene, which has the advantage of permitting transcription to be regulated in a quantitative rather than in an all-or-none manner. Having a set of proteins also makes it possible to produce a network of interacting genes, since the protein product of one gene can affect the expression of another. So one has a combinatorial principle at work here operating at the level of a combination of proteins. As we shall see later (e.g., in the case of zinc finger proteins), the principle also operates within individual proteins, where different subdomains can be combined to give greater variety or precision of recognition—a microcosm, as it were, of the macroscopic picture.

In this entry we are concerned not with the actual process of transcription or the wider problems of networking, but rather with the way in which a regulatory protein achieves recognition of the right segment of DNA. How does a transcription factor recognize a specific target site against the vast background of DNA in the nucleus?

## 2 RECOGNITION OF SPECIFIC DNA SEQUENCES BY PROTEINS

Eukaryotic DNA-binding transcription factors, like their prokaryotic counterparts, achieve recognition by having embedded in them a discrete structure or domain that serves for binding to DNA. Most of the structures identified so far fall into a number of different types, each type having a characteristic amino acid sequence and three-dimensional structure. The first such structure to be identified was the so-called helix–turn–helix motif, discovered about 10 years ago by X-ray crystallographic studies of certain bacterial regulatory proteins. Sequences suggestive of similar structured motifs were also subsequently found in proteins involved in the establishment of cellular patterns in developing embryos of *Drosophila* and mice—the so-called homeodomains. However, it was only in 1987 that the first high resolution structure of a complex with DNA of one of these prokaryotic proteins was determined. X-ray crystallographic analyses of a number of other helix–turn–helix motif DNA complexes have provided a precise insight into how this structural motif recognizes specific DNA sequences.

Remarkable progress has been made toward understanding the nature of specific protein–DNA interactions. A number of types of structural motif for DNA recognition have been identified on the basis of amino acid sequence comparisons associated with biochemical studies, and the three-dimensional structures have been determined for members of most of these types. Each motif represents a different solution to the problem of designing a piece of protein surface to fit a particular segment of a DNA double helix.

At first glance the DNA double helix looks like a very uniform structure, with the two strands, the phosphate–sugar backbones, spiraling around each other and held together by the base pairs to form the DNA double helix. The bases on each strand form complementary pairs with the bases on the other strand, using hydrogen bonds to recognize each other. The double helical structure results in major and minor grooves between the chains, each kind with a specific constellation of chemical groupings on the surface. The local structure of DNA varies in detail along its length and is determined by the particular sequence of base pairs. The width and depth of the major and minor grooves, the pattern of chemical groupings in the grooves, and the flexibility of the double helix are all determined by the base sequences, so that a specific sequence will result in a double helix with distinct characteristics. Every DNA sequence is thus recognizable from the shape of the double helix and the chemical identity of the bases. DNA-binding domains of proteins have surfaces complementary to DNA, hence a shape—often containing an α-helix—that fits well into the major (or minor) groove of DNA. This is often achieved by having an α-helix protruding from the bulk of the whole protein, but presented in different ways in different types of protein. When the pattern of a particular amino acid on the surface of the protein matches the pattern of groups on the surface of the DNA double helix, binding takes place through the formation of hydrogen bonds and van der Waals contacts. One might think of this recognition as a small piece of protein "reading" (i.e., interacting with) a short sequence of base pairs.

A protein reading head could thus recognize a short DNA sequence of perhaps three or four base pairs, but such a short sequence occurs too frequently to be unique and uniquely recognizable. High specificity in the selective control of gene expression requires the recognition of a reasonable length of DNA, and nature has found more than one way of putting together reading heads to achieve this. One design exemplified by several classes of DNA-binding proteins, such as prokaryotic helix–turn–helix proteins, basic leucine zipper proteins, and the hormone receptors, described in Sections 3, 4, and 8, consists of combining two monomers to bind as a dimer. The use of dimer does not merely extend the length of DNA sequence recognized, but also brings into play the possibility of a variation in spacing between the two half-sites, so increasing the rarity of occurrence of the particular run of bases to be recognized, hence its degree of uniqueness. Another distinctly different design for sequence-specific recognition is one in which reading heads are directly repeated in tandem, as found in the zinc finger proteins.

## 3 THE HELIX–TURN–HELIX AND HOMEODOMAIN

The prokaryote helix–turn–helix group of DNA-binding proteins is to date the most thoroughly studied. The helix–turn–helix motif consists of two α-helices packed together at an angle of about 120° with a tight turn at the elbow between them. The structure is not stably folded on its own, and there is usually a third helix protruding from the rest of the protein which stabilizes it. The second helix in the helix–turn–helix lies in the major groove of the DNA and carries the main amino acid residues responsible for specific binding (Figure 1A). Hence it is called the recognition helix. Binding involves primarily direct or water-mediated interactions of the side chains on the recognition helix with the exposed chemical groups on the edges of the base pairs, but the whole motif is also fixed in a particular orientation by hydrogen bonds to phosphates on the DNA backbone.

The prokaryotic proteins all bind to DNA as symmetrical dimers, with their twofold axes coincident with a twofold axis of the DNA centered on the palindromic sequence of the DNA binding site. The central base pairs of the latter are not necessarily in contact with the protein, but they can influence the conformation of the whole site by virtue of its detailed geometry and/or by its deformability, allowing a better fit between the protein and DNA. The last point is indeed a general one, namely, that DNA is not a passive participant in the recognition process. The sequence-dependent features in the double helix constitute another, physical level of information in DNA sequences, secondary to the chemical code.

The helix–turn–helix motif in eukaryotes was first identified by the sequence similarity with prokaryotic helix–turn–helix of a highly conserved 60 amino acid sequence now called the homeodomain, which is found in many homeotic genes of *Drosophila.* The eukaryotic motif is however different in two respects. First, the recognition helix is substantially longer, containing at its C-terminal end more basic residues that interact with the phosphates, leading to a somewhat different orientation in the major groove of the DNA (Figure 1B). Second, the parent protein of the homeodomain binds as a monomer, though some proteins also contain other DNA-binding regions (e.g., the POU domain). This originates in

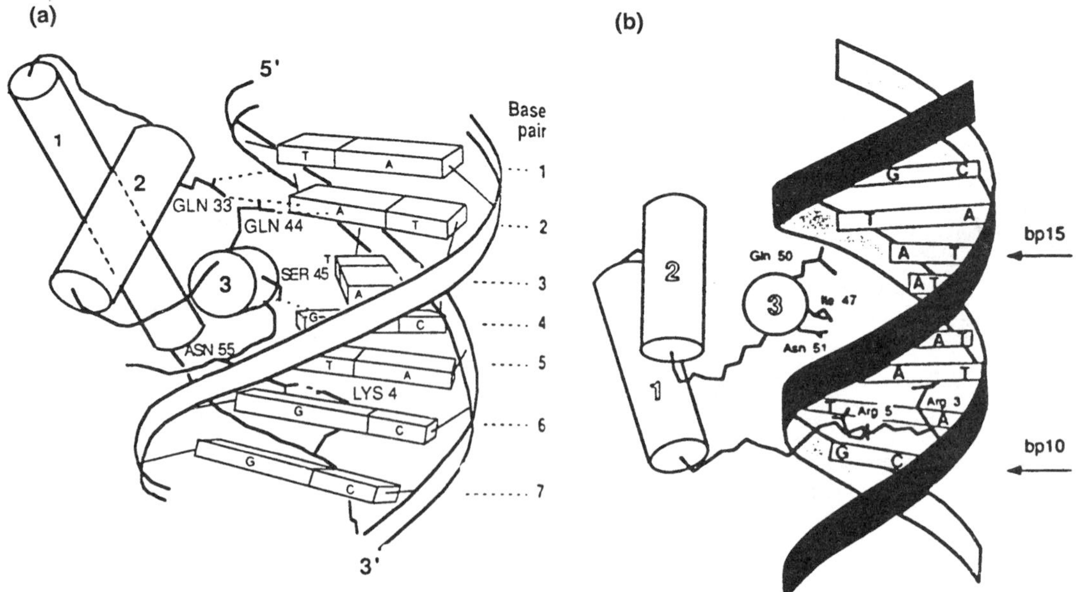


**Figure 1.** Helix–turn–helix protein–DNA interactions. (a) Part of the structure of a complex of the DNA-binding domain of the phage λ repressor and a λ operator site, showing the recognition α-helix, marked 3, inserted into the major groove of the DNA half-site. Some of the hydrogen bond interactions are shown. Helices 2 and 3 constitute the helix–turn–helix motif. [From Jordan and Pabo (1988).] (b) Structure of an engrailed homeodomain–DNA complex, showing the critical contacts made by the recognition helix in the major groove. Note also the additional contacts made by the N-terminal basic region in the minor groove. [From Kissinger et al. (1990).]

a conserved sequence that is found in certain homeotic genes but always appears to be accompanied by a homeodomain.

## 4 THE BASIC LEUCINE ZIPPER

In 1988 it was observed that a number of transcription factors contained a common pattern of repeating Leu residues spaced seven residues apart. A basic region adjacent to the so-called leucine zipper directs sequence-specific binding. The zipper sequence is in fact the motif responsible for dimerization of the parent protein, by forming an α-helical coiled coil of the kind described by Francis Crick 35 years ago. The prototype of the basic leucine zipper family was the yeast transcriptional activator CGN4, which binds specifically as a dimer to a DNA segment 9 base pairs long, containing two palindromic 4 base pair half-sites separated by a single base pair. In the binding site for the CREB proteins, the same half-sites are separated by 2 base pairs.

The crystal structure of the GCN4 basic leucine zipper–DNA complex was determined recently (Figure 2). Each of the two basic leucine zipper monomers, 56 amino acids long, forms a smoothly curved continuous α-helix; the two pack together at their C-terminal ends as a coiled coil, which gradually diverges to allow the residues in the basic region to follow the major groove of each DNA half-site. The DNA, which is straight, is thus grasped, as it were, by a pair of splayed chopsticks. It is noteworthy that the basic regions that are disordered in the free proteins become ordered into helices upon contacting the DNA.

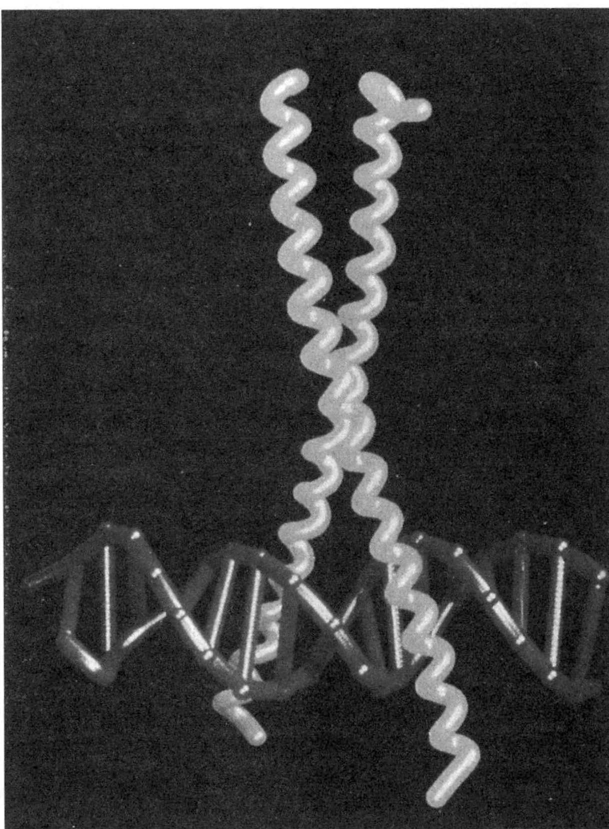

**Figure 2.** Structure of the GCN4 basic leucine zipper–DNA complex, showing a smoothed Cα backbone for the protein. The view is at right angles to the dyad of the complex formed by the protein dimer and the palindromic DNA-binding site. The C-termini of the monomers (at the top) pack together as a coiled coil, which gradually diverges to allow the basic region residues at the N-terminal ends to insert into the major groove of each DNA half-site. [From Ellenberger et al. (1992).]

Although GCN4 functions as a homodimer, it is an important feature of the basic leucine zipper class of proteins that they form both homo- and heterodimers. Thus the homodimer of the proto-oncogene c-*Jun* is unstable and binds DNA inefficiently, whereas the *Jun-Fos* heterodimer is stable and binds Ap-1 DNA sites with much higher affinity. Thus *Fos* is a positive regulator of *Jun*. The use of heterodimers thus allows a combinatorial mode of action and increases the repertoire of regulatory functions for this family of proteins.

Another group of proteins having similarities to the basic leucine zipper group display the helix–loop–helix motif, which so far is characterized only by sequence regularities. A basic region of about 15 amino acids lies immediately N-terminal to a 15 amino acid segment with helical characteristics, which is separated from a second such putative helix by a region of variable length. Members of this family also function as dimers, and in almost all cases a heterodimer is the active DNA-binding species—thus the protein Myc needs the protein Max to be effective.

## 5 β-RIBBON MOTIFS

Other families of DNA binding proteins use α-helices to bind in the major groove—for example, the two main classes of zinc finger proteins (see Sections 7 and 8) and the dimeric DNA-binding domain of the papilloma virus transcriptional activator, whose structure complexed with DNA has recently been determined. It therefore came as a surprise to learn that the prokaryotic MetJ repressor interacts with DNA in a quite different fashion. Here a pair of antiparallel β-strands (a "β-ribbon"), formed between a symmetrical dimer of the protein, interacts with DNA in the major groove (Figure 3). Two other helices from each monomer are important for the stability of the whole dimer structure, which forms a core from which the β-ribbon protrudes so as to be able to lie in the major groove. The other parts of the protein make extensive phosphate contacts on the backbones, and so help fix the whole complex.

The β-ribbon recognition element here requires dimerization to produce its folded structure, but in an earlier example a β-ribbon formed by a continuous loop in a monomer appears to be the main recognition element. This is the bacterial HU protein. Its structure has been determined crystallographically, but no complex with DNA has yet been solved. Chemical protection studies show interactions in the minor groove.

The foregoing examples show that a β-sheet can be used in recognition, but nevertheless it seems clear that an α-helix, with its greater intrinsic stability, is a more suitable element for probing the characteristics of a particular segment of DNA, and this accounts for its more widespread use.

## 6 THE TATA BOX BINDING PROTEIN

In another recent case a large β-sheet is used rather than a ribbon, namely, the TFIID (transcription factor D for polymerase II) TATA box binding protein required for the initiation of transcription for all three classes of eukaryotic polymerase. The first step was the solution of the structure of the isolated protein, and with data from chemical and mutagenesis studies, some of the residues involved

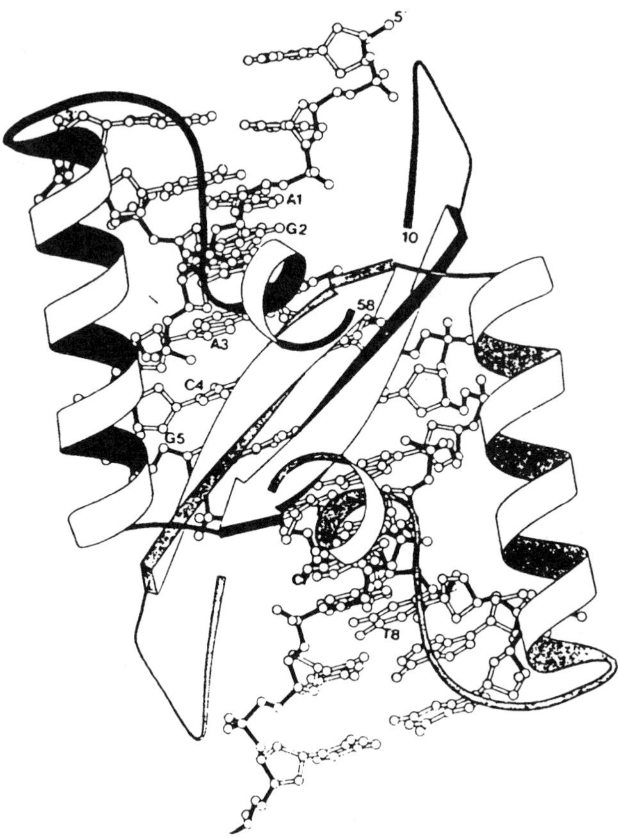


**Figure 3.** Structure of the MetJ repressor β-ribbon–helix motif bound to DNA. Each monomer of the dimer contributes a strand to the β-ribbon. The two helices that support the ribbon are truncated to make it visible. The view is parallel to the dyad axis. [From Somers and Phillips (1992).]

in DNA binding could be located. The whole molecule is saddle shaped, being formed of two almost identical halves related by a pseudodyad; a β-loop is appended from each half, so that the two form as it were a pair of stirrups, which could wrap around grooves in the DNA. Thus it seemed that the saddle could sit astride the helix.

Subsequently, two crystal structures of TATA box binding proteins bound to DNA were determined. It was found that rather than sitting astride the DNA double helix, the TATA box binding protein sits almost parallel to the axis of the DNA, which arches under the concave surface of the protein. This interaction distorts the minor groove so drastically that the double helix is unwound and also bent through about 80°, suggesting how the melting of the double helix necessary for the initiation of transcription may begin. This result also illustrates the point made earlier that the physical properties of special DNA sequences can carry another level of information in their interaction with protein. In this case TATA box binding protein exploits the deformability of the TATA sequence to initiate the unwinding of DNA as a prelude to transcription.

## 7 ZINC FINGER PROTEINS: THE FIRST CLASS

The "zinc finger" motif was first identified by Miller et al. in the *Xenopus* transcription factor IIIA (TFIIIA), where it is repeated consecutively nine times. It consists of a 30 amino acid sequence containing two His, two Cys, and three hydrophobic amino acids, all at conserved positions. On this basis, together with earlier measurement of zinc content and partial proteolysis data, it was proposed that each zinc finger motif forms a small, independently folded, zinc-containing minidomain, used repeatedly in a modular fashion, to achieve sequence-specific recognition of DNA. Since that time, zinc finger motifs of the TFIIIA type have turned up in hundreds of proteins, and they appear to be the most widely used of all types of DNA-binding domains. Indeed, they are estimated to constitute as much as 1% of the human genome. This design may truly be called modular, since all the multiply repeated domains have the same structural framework but can achieve chemical distinctiveness through variations in certain key amino acid residues.

NMR studies first showed that the structured region of the zinc finger motif comprises about 25 amino acids forming a compact unit with a 12 amino acid helix packed against a β-hairpin, confirming an earlier proposal for the structure of the finger by J. Berg. NMR studies of other such peptides showed isomorphous structures (Figure 4A), while those of a two-zinc-finger peptide showed that the two structured finger minidomains folded independently, with a short, flexible linker sequence between them. The mode of binding to DNA was first determined from the crystallographic analysis of a three-zinc-finger peptide from the early response protein Zif 268 bound to its cognate DNA. The three fingers bind in an equivalent manner to a segment DNA 9 base pairs long, so that each finger contacts 3 base pairs, with no gaps between the three sites (Figure 4B). The main interaction for each finger comes from its helical region, which uses amino acid residues three (or six) positions apart in the sequence, and thus facing the same way, to interact with successive (or next nearest) bases on the DNA. These contacts are achieved by the angle at which the recognition helix sits in the major groove. The linkers between fingers are largely extended and make no significant DNA contacts. Hence the linkers seem to play no part in determining the manner in which the fingers interact with DNA, but they do seem to play a wider, secondary role in the binding in solution. There are of course other interactions between the body of each finger and the backbone phosphates, but the specific recognition is effected by the one-to-one interactions between amino acids and base pairs from the recognition helix to a triplet of base pairs. This simple pattern suggested a potential for devising a recognition code for the design of zinc finger modules that would recognize specific DNA sequences, and indeed the specificity of some fingers has already been changed using a small number of site-directed mutations.

This idea gains credence from a statistical study by Jacobs of more than a thousand zinc finger motifs. It was found that amino acids in three positions are highly variable, and these positions are precisely those used to make contact in the Zif 268 complex, namely, those falling on the first, second, and third turns of the recognition α-helix. However, the crystal structure of a second zinc finger–DNA complex has been solved in this laboratory, indicating that zinc finger–DNA recognition may be more complex than at first perceived. The DNA-binding domain is that of the *Drosophila* regulatory protein Tramtrack, which has two fingers and binds with relatively high affinity ($5 \times 10^7$ $M^{-1}$) to a DNA binding site 6 base pairs long. The second finger shows a pattern of contacts similar to those observed in the Zif 268 complex, but the first finger uses a well-conserved Ser residue, which is not in one of the three canonical positions, to make a hydrogen bond with a thymine, at a distorted point of the DNA helix. Clearly one must await further structures of complexes to clarify whether there might be additional variations from the simple pattern shown by Zif 268.

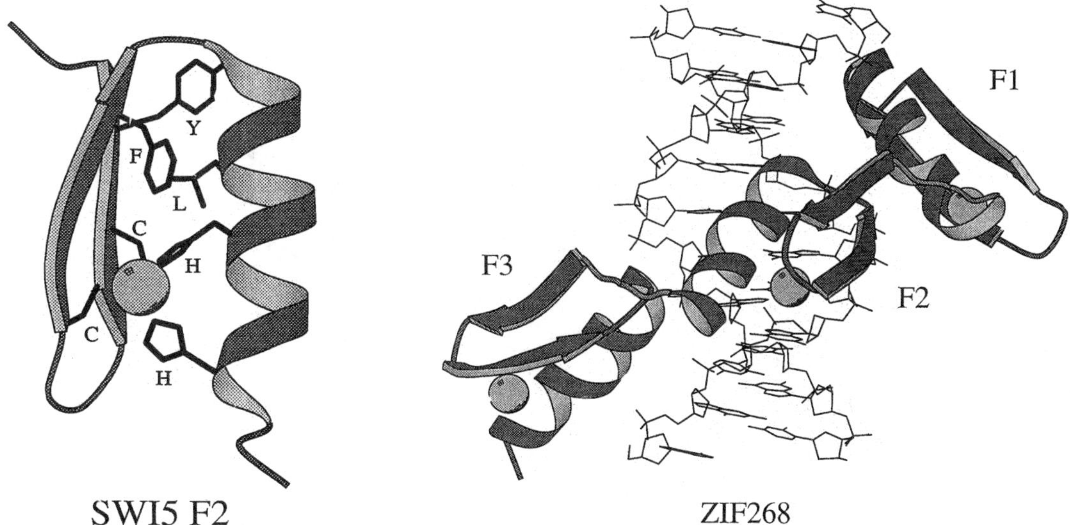


**Figure 4.** Zinc fingers of the first class (TFIIIA type) and their interaction with DNA. (a) Structure of one of the zinc fingers of the yeast protein SWI5 from a two-dimensional NMR study in solution. The two substructures forming the finger, the β-sheet, and the helix are pinned together by the Zn ion (shaded ball) ligated to a pair of Cys and a pair of His. The structure is further stabilized by the cluster formed by the three invariant hydrophobic residues, Phe(F), Tyr(Y), and Leu(L), packed together at the top of the domain. [From Neuhaus et al. (1992).] (b) The structure of the complex formed by the three zinc fingers of Zif 268 and its binding site. Each finger binds in a similar manner to three base pairs in the major groove. The linkers between the three fingers are extended, and the fingers bind consecutively with no gaps in the DNA. Zn ions are represented by balls, as (a). [Adapted from Pavletich and Pabo (1991).]

## 8 A SECOND CLASS OF ZINC FINGERS; HORMONE RECEPTOR DNA-BINDING DOMAINS

Shortly after zinc fingers were discovered in TFIIIA, sequence motifs that appeared to be related were found in several other protein or cDNA sequences of molecules that bind DNA. It was therefore at first thought that these might have a rather similar structure to the TFIIIA-type finger domains. The most important and widespread examples were those from members of the superfamily of hormone-activated nuclear receptors, which play a central role in the control of eukaryotic gene expression and are indeed transcription factors. The DNA-binding domains of such receptors all include two motifs in tandem, each about 30 amino acids long, but each motif contains two pairs of Cys rather than a pair of Cys and a pair of His, as in the first class. They do indeed bind $Zn^{2+}$, but the three-dimensional structure of two such DNA-binding domains, determined in solution using two-dimensional NMR spectroscopy, showed that the receptor DNA-binding domain is structurally distinct from the TFIIIA type of zinc finger. Each of the two motifs in each domain folds up into an irregular loop followed by an α-helix, but the two together form a single structural unit with their helices crossing at right angles, so that the DNA recognition helix (from the first motif) is supported by the helix from the second motif (Figure 5A).

Hormone receptors bind to palindromic sites (response elements, RE) on the DNA as dimers, and the DNA-binding domains alone also form dimers, the dimer interface arising from a region of the loop of the second motif of each receptor. These different roles for the two motifs within one structural unit—namely, helix recognition and dimerization—were deduced by mapping onto the three-dimensional structure data on site-directed mutagenesis from a number of laboratories, particularly those of P. Chambon, R. Evans, and G. Ringold. This combination of structural analysis and biochemical and genetic experiments pointed toward a mechanism of interaction with DNA and a general model was proposed by Härd et al. and Schwabe et al.

The detailed chemistry of the interactions at the protein–DNA interface has, however, had to await a crystal structure of a complex. The first to be determined was that at 2.9 Å resolution by Luisi et al. of the DNA-binding domain of the glucocorticoid receptor (GR) complexed with a DNA segment 18 base pairs long and containing the glucocorticoid response element (GRE). However, the particular DNA segment ($GRE_{s4}$) used was composed of two half-sites (each of 6 base pairs) separated by a nonnative spacing, with four (rather than three) intervening base pairs. As a consequence, the two DNA-binding domains do not bind equivalently to DNA. One monomer of the GR.DNA-binding domain dimer appears to bind specifically to the DNA, with its recognition helix deep in the major groove, while the other binds in the same general way but forms less close, nonspecific interactions.

An intriguing feature of the structure of the DNA-binding domain–$GRE_{s4}$ complex is that the protein–protein contacts at the dimer interface can override the potential specific protein–DNA contacts at the second site in the GRE, yet they are not strong enough to hold the protein as a dimer in solution. The explanation for this must lie in the cooperative nature of the interaction of the DNA-binding domain with the DNA: that is, the protein–protein dimer interface might be induced and stabilized by interactions with the DNA backbone.

The specific–nonspecific character of this complex arises because of the presence of an extra base pair between the two half-sites in the GRE. To obtain specific binding at both half-sites, the native spacing of three base pairs is required. The information contained in the spacing of the two half-sites is an integral part of the binding site "code," which is "recognized" through the

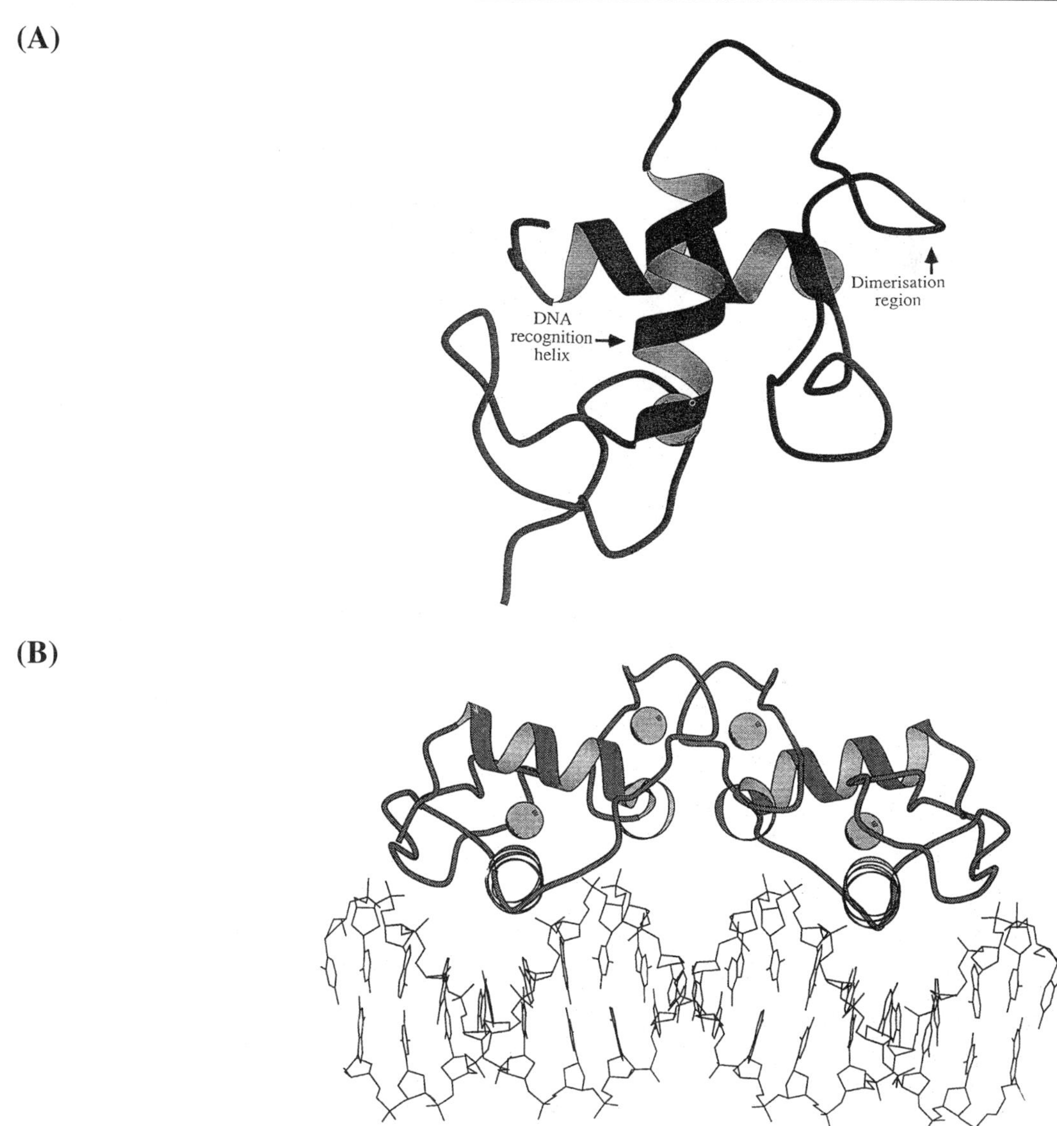


**Figure 5.** Estrogen receptor DNA-binding domain and its interaction with DNA. (A) Structure of the DNA-binding domain of the estrogen receptor as determined by a two-dimensional NMR study in solution. A single structural unit is made by two approximately similarly folded motifs, each consisting of an irregular loop followed by a helix. The Zn ion in each motif is shared between the loop and helix. The conformation of the loop in the dimerization region changes when the dimer is formed on binding to DNA: [From Schwabe et al. (1990).] (B) Structure of the complex formed between the DNA-binding domain of the estrogen receptor and its binding site. The protein binds as a dimer, making identical contacts to the two half-sites on the DNA. The recognition helices bind deep in the major groove, and there are also contacts from basic residues at the dimer interface to the minor groove lying between the half-sites. The view is perpendicular to the dyad of the whole complex. [From Schwabe et al. (1993).]

formation of a protein–dimer interface. This is especially important for members of the nuclear receptor family that can recognize the same half-site sequence, but with different orientations and/or spacings. For these receptors, orientation and spacing are the only means of discrimination.

Recently the crystal structure of a second hormone-receptor-DNA complex has been solved (at 2.4 Å resolution)—the DNA-binding domain of the estrogen receptor (ER), which recognizes a different half-site from GR but with the same native separation of three base pairs between half-sites ($ERE_{s3}$). The protein binds as a symmetrical dimer in an equivalent manner to both half sites (Figure 5B), and now rather more interactions can be seen than were visible in the crystal structure of the DNA-binding domain–$GRE_{s4}$ complex, which probably represents a compromise in the binding of a dimer to a site with an unfavorable spacing between its two halves. The interactions seen in the ER.DNA-binding domain–$ERE_{s3}$ complex are characteristic in number and type of those found in specific interactions in other families of protein–DNA complexes.

## 9 OTHER ZINC-BINDING DOMAINS

Clearly the second class of zinc finger DNA-binding domains is a not a simple variant of the first TFIIIA class. They differ both in their structure and in the way in which they interact with DNA. Above all, the glucocorticoid and estrogen receptors operate as dimers that bind to palindromic DNA sites, whereas the binding of zinc finger of the first class makes no use of the symmetry of the DNA structure nor of base sequence. The latter function as independent modules that can be strung together in a directly repeating (tandem) fashion with no restriction on their number.

It should however be added that some members of the nuclear receptor family (e.g., thyroid, vitamin D, retinoic acid) also bind as dimers, but as nonsymmetrical dimers, to a DNA-binding site made of two directly repeated identical "half-sites." Here clearly another interface is brought into play, but again the discrimination depends entirely on the separation between the repeats in the DNA sequence.

The structure of a member of a third class of zinc-binding domain of a distinct structural type has emerged recently. This is of the GAL4 transcriptional activator, representative of a small family so far found only in yeast. NMR results show that the two zinc ions and six Cys in the DNA-binding domain form a binuclear cluster, with each $Zn^{2+}$ coordinated by four Cys, so that two of the Cys are shared. This cluster holds together two short helices, related by a quasidyad, with one helix inserting into the major groove of the DNA as in the hormone receptors; we thus see the recognition helix supported by a second helix. Again, the molecule binds as a dimer to a palindromic DNA-binding site, with two short half-sites separated by approximately 1.5 turns of the DNA helix so that the DNA-binding domains bind on opposite faces of the DNA.

This GAL4 class of zinc-binding proteins is by no means the final example of proteins in which $Zn^{2+}$ is used structurally to help fold a polypeptide chain into a compact domain that serves for binding nucleic acids. A fourth class is constituted by the Cys-$X_2$-Cys-$X_4$-His-$X_4$-Cys sequences found in the nucleocapsid proteins of retroviruses, which form "stubby" fingers. This general use of zinc was foreshadowed some time ago, but the wide extent to which it happens is becoming increasingly clear, even though the structural information is still limited. A diverse set of families of proteins that bind $Zn^{2+}$ and interact with nucleic acids are becoming uncovered and are loosely grouped under the name of zinc finger proteins. The name is not inappropriate, since the zinc-binding domains in all cases known do grip or grasp the double helix.

## 10 SOME GENERAL PRINCIPLES

Undoubtedly there is still more variety to be discovered in the design of proteins for recognizing DNA, but some principles have emerged. In most cases so far the basic element of specific recognition is that between a short stretch of base pairs (three or four) and a small piece of protein, usually part of an α-helix, but sometimes a β-sheet. These specific base contacts are reinforced by a relatively large number of contacts to the phosphates on the DNA backbone(s), which help fix the orientation of the whole structural element. Even where there are differences in detail with respect to the way in which the recognition helix inserts into the major groove of the DNA, there are nevertheless common features among some of the different eukaryotic families.

One such single interaction element is rarely enough to give sufficient binding energy and a high degree of discrimination between DNA target sites. Thus, very often, two such interaction elements are used when the protein binds as a symmetrical dimer. This also brings into play the spacing between the two half-sites, which must be set accurately to give proper binding. Several examples of the importance of spacing have been given. The dimer design lends itself to still greater versatility by the use of heterodimers.

Another way of putting together interaction elements is by direct repetition, linking them in tandem and, unlike the dimers, making no use of the symmetry of the DNA. This design is exemplified by the zinc finger proteins of the first class (TFIIIA type). They take advantage of the modular features, since there is no restriction on the number of domains that can be deployed (unlike the limit of two in the dimer case), nor on the distances between them. This modular design clearly offers a large number of combinatorial possibilities for sequence-specific recognition of DNA, and it is no wonder that zinc finger proteins of this class are so widely found in nature.

*See also* Chromatin Formation and Structure; Gene Expression, Regulation of; Hydrogen Bonding in Biological Structures; Zinc Finger DNA Binding Motifs.

### *Bibliography*

Ellenberger, T. E., Brandl, C. J., Struhl, K., and Harrison, S. C. (1992) The GCN4 basic-region leucine zipper binds DNA as a dimer of uninterrupted α-helices: Crystal structure of the protein–DNA complex. *Cell,* 71:1223–1237.

Harrison, S. C. (1991) A structural taxonomy of DNA-binding domains. *Nature,* 353:715–719.

Jordan, S. R., and Pabo, C. O. (1988) Structure of the λ complex at 2.5 Å resolution: Details of the repressor operator interactions. *Science,* 242:893–897.

Kissinger, C.R.B., Liu, B., Martin-Blanco, E., Kornberg, T. B., and Pabo, C. O. (1990) Crystal structure of an engrailed homeodomain–DNA complex at 2.8 Å resolution: A framework for understanding homeodomain–DNA interactions. *Cell,* 63:579–590.

Klug, A. (1993) Transcription: Opening the gateway. *Nature,* 365:486–487.

Neuhaus, D., Nakesoko, Y., Schwabe, J.W.R., and Klug, A. (1992) Solution structures of two zinc finger domains from SW15, obtained using two-dimensional $^1$H NMR spectroscopy: A zinc finger structure with a third strand of β-sheet. *J. Mol. Biol.* 228:637–651.

Pabo, C. O., and Sauer, R. T. (1992) Transcription factors: Structural families and principles of DNA recognition. *Annu. Rev. Biochem.* 61:1053–1095.

Pavletich, N. P., and Pabo, C. O. (1991) Zinc finger–DNA recognition: Crystal structure of a Zif268–DNA complex of TFIIIA. *Nucleic Acids Res.* 21:809–817.

Schwabe, J.W.R., Neuhaus, D., and Rhodes, D. (1990) Solution structure of the DNA-binding domain of the oestrogen receptor. *Nature,* 348:458–461.

———, Chapman, L., Finch, J. T., and Rhodes, D. (1993) The crystal structure of the complex between the oestrogen receptor DNA-binding domain and DNA at 2.4 Å: How receptors discriminate between their response elements. *Cell,* 75:567–578.

Somers, W. S., and Phillips, S.E.V. (1992) Crystal structure of the *met* repressor–operator complex at 2.8 Å resolution reveals DNA recognition by β-strands. *Nature,* 359:387–393.

# Protein Folding

*Thomas E. Creighton*

*Key Words*

**Conformation** A three-dimensional structure of a large molecule that differs significantly from other such structures solely by rotations about covalent bonds, but including differences in disulfide bonds.

**Cooperativity** The physical phenomenon by which the occurrence of one event in a protein molecule increases or decreases the probability of further events.

**Denaturant** A reagent that when added to a protein solution, causes the protein to unfold.

**Primary Structure** The covalent structure of the polypeptide chain, excluding disulfide bonds between cysteine residues.

**Secondary Structure** Regular conformations of the polypeptide backbone, in particular the α-helix, β-strands, and certain well-defined reverse turns.

**Tertiary Structure** All aspects of the conformation of a single domain.

To be biologically active, proteins must adopt specific folded three-dimensional, tertiary structures. The genetic information for the protein specifies only the primary structure, yet the process of folding occurs rapidly after biosynthesis of the polypeptide chain. Many purified proteins can be unfolded reversibly in vitro; they will spontaneously refold, so the three-dimensional structure is determined by the primary structure. The folding process has been studied intensively in vitro and is of practical importance for the new discipline of protein engineering, which makes it possible to produce any protein in large amounts, but often in an insoluble, unfolded, inactive and useless form that must be folded correctly.

## 1 THE PROTEIN FOLDING PROBLEM

The protein folding problem can be broken down into four related aspects: (1) the kinetic process or pathway by which the protein adopts its native and biologically active folded conformation, (2) the physical basis of the stability of folded conformations, (3) why the amino acid sequence determines one particular folding process and resultant three-dimensional structure, instead of some other, and (4) the challenge of predicting the three-dimensional structure, given only the amino acid sequence. The last two topics are too uncertain to be described here.

## 2 THE STABLE CONFORMATIONAL STATES OF PROTEINS

Different conformations of a protein occur because of rotations about the covalent bonds of the polypeptide backbone and of the amino acid side chains (Figure 1). Only three conformational states of small proteins are usually stable at equilibrium. Which predominates depends on the conditions. Many proteins exist in membranes or as part of large insoluble structures, but this discussion is limited to those that normally exist in aqueous solution.

### 2.1 The Unfolded State (U)

The ideal unfolded protein is the random coil, in which the rotation angle about each bond of the backbone and side chains (Figure 1) is independent of that of all bonds distant in the sequence, and all conformations have similar free energies. There is only one restriction: atoms of the polypeptide chain cannot overlap in space. Each amino acid residue can adopt a number of different conformations, so a very large number of conformations is possible for an entire protein. For example, with an average of eight conformations per residue, a relatively small protein of 100 amino acid residues would be expected to have $8^{100}$, or $10^{90}$, distinct conformations. Even if only two conformations per residue were possible, $10^{30}$ polypeptide conformations would be expected. This large conformational entropy stabilizes the unfolded state. With a typical sample of an unfolded protein, say 1 μmol (about $10^{18}$ molecules), each molecule will have a unique conformation at each instant of time, and another conformation some $10^{-10}$ second later.

The unfolded state is most stable under unfolding conditions, when its conformational entropy and interactions with the solvent are more favorable than those of more folded conformations with intramolecular stabilizing interactions. Because of their conformational heterogeneity, unfolded proteins are difficult to characterize, but they approximate random coils under strongly unfolding conditions. Under less unfolding conditions, they can have varying degrees of local interactions between groups close in the covalent structure and some tendency to adopt α-helices, but they do not generally contain cooperative folded structures. Unfolded proteins are often insoluble under non-denaturing conditions.

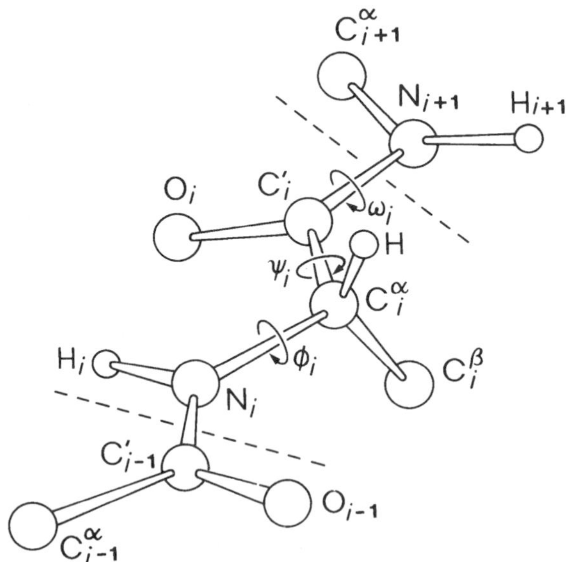


**Figure 1.** Perspective drawing of a segment of polypeptide chain. Only one atom of each amino acid side chain is shown. The limits of a single residue (number $i$ of the chain) are indicated by the dashed lines. There is relatively free rotation about the single bonds indicated by the torsion angles $\phi$ and $\psi$; the peptide bond has partial double-bond character, so its torsion angle, $\omega$, is limited to values of 0° or ±180°, which correspond to the cis and trans conformations, respectively. The polypeptide chain is shown in the fully extended conformation, with $\phi = \psi = \omega = 180°$.

### 2.2 The Fully Folded, Native State (N)

The folded, tertiary structures of many proteins have been determined in great detail by X-ray crystallography and by NMR spectroscopy. They are remarkably compact, consisting primarily of segments of secondary structure packed together; virtually all the interior space is filled with protein atoms. The native structure is essentially a single three-dimensional conformation in which the positions of all atoms, except for a few at the surface, are closely defined. There are, however, varying degrees of flexibility throughout the structure. Proteins containing more than about 200 residues generally contain two or more relatively independent structural units, known as domains, or multiple polypeptide chains, which are often identical.

Proteins with related amino acid sequences are evolutionarily related and always have very similar folded conformations. Otherwise, unrelated primary structures generally produce different folded states. Exceptions are either the result of very distant evolutionary relatedness or of convergence to just a few possible folded conformations. Analysis of the known structures suggests that the number of folding patterns that are possible is in the region of $10^3$.

### 2.3 The So-Called Molten Globule State

Under certain conditions, a number of proteins adopt conformations that are neither fully folded nor fully unfolded and have come to be known as the ''molten globule'' state. The molecule is generally very compact in this state, and with considerable secondary structure, but is otherwise disordered to varying extents; this conformation is understood only poorly.

## 3 GENERAL PROPERTIES OF PROTEIN FOLDING TRANSITIONS

The native states of small proteins may usually be unfolded reversibly by adding denaturants, increasing or decreasing the temperature, varying the pH, applying high pressure, or cleaving disulfide bonds. Unfolding transitions of single-domain proteins are usually two-state at equilibrium, with only the N and U states populated substantially. Partially folded conformations are energetically unstable relative to either U or N under all conditions, except for the molten globule state in some cases. Multidomain proteins often unfold stepwise, with the domains unfolding individually, although there may be varying degrees of interactions between them. Multisubunit proteins usually dissociate first, then the subunits unfold.

### 3.1 Stability of the Folded State

The folded states of proteins are only marginally more stable (10–60 kJ/mol) than the fully unfolded state, even under optimal conditions. The heat capacity of the unfolded state is significantly greater than that of the folded state, so the enthalpies ($\Delta H$) and entropies ($\Delta S$) of unfolding are very temperature dependent, and there is a temperature at which stability of the folded state is at a maximum (Figure 2); consequently, proteins can unfold at low temperatures, although this phenomenon can be observed only under favorable circumstances. The two most important stabilizing interactions in the folded state are believed to be hydrogen bonds and the van der Waals interactions between nonpolar atoms, although this is a very complex subject.

### 3.2 Cooperativity of Folding

The interactions that stabilize folded proteins are individually weak; they stabilize folded conformations only because many of them occur simultaneously, when they can produce a positively cooperative system. This is one of the reasons for the instability of partly folded proteins and the cooperativity of folding transitions.

## 4 KINETIC ASPECTS OF FOLDING

### 4.1 Kinetic Determination of Folding

To find a single conformation by random searching of $10^{90}$ or $10^{30}$ conformations would require, on average, $10^{66}$ and $10^7$ years, respectively. Proteins are observed to fold on the minute time scale, so protein folding pathways probably guide the process. In this case, the N state that results may not be the most stable conformation possible, but that most accessible kinetically. Only a few examples are known: substantial kinetic blocks occur in folding in certain bacterial proteases, and folding of serpin protease inhibitors generates meta-stable conformations, in preference to more stable forms.

### 4.2 Models of Folding

Each unfolded molecule might be imagined to follow a different folding pathway, as in putting together a large, complex jigsaw puzzle. Most models, however, envisage a sequential pathway of folding through a limited number of intermediates. The appearance by random fluctuations of a sufficiently small ''nucleus'' of structure could be imagined to serve as a template for folding of the remainder of the polypeptide chain. Individual elements of nativelike structure, or ''microdomains,'' are unstable in the unfolded protein but could be imagined to interact and stabilize each other. Two or more ''clusters'' or ''embryos'' of local structure might grow until they merged with each other to form the entire structure. The unfolded polypeptide chain under refolding conditions might undergo rapid hydrophobic collapse, at which point the native state might be found relatively rapidly.

## 5 EXPERIMENTAL OBSERVATIONS OF PROTEIN UNFOLDING AND REFOLDING

Protein folding usually occurs on a time scale of seconds to minutes—much more rapidly than expected for a random search. The unstable intermediates in folding are populated only transiently, if at all. Folding pathways are determined most readily if folding is coupled to disulfide bond formation, for then the unstable intermediates can be trapped in a stable form and characterized.

### 5.1 Kinetic Analysis of Complex Reactions

An unusual aspect of the kinetic analysis of protein folding is that the reaction is initiated by changing the conditions, often drastically, from favoring unfolding to favoring refolding. Moreover, each molecule in a sample of a fully unfolded protein will probably have a unique conformation at any instant of time. Kinetic complexities can arise either from the refolding at different rates of subpopulations of the unfolded state or from the accumulation of intermediates (I), when there must be two or more slow steps:

$$U \xrightarrow{slow} I \xrightarrow{slow} N \tag{1}$$

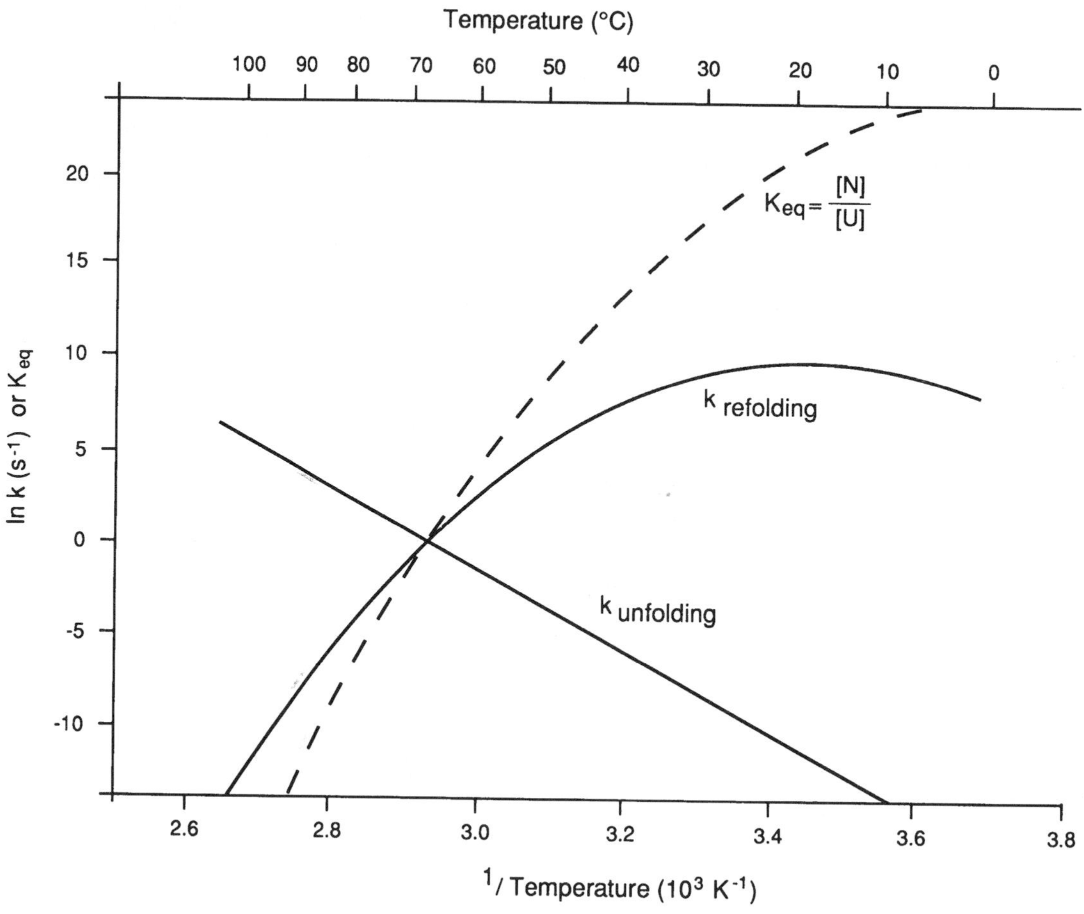


**Figure 2.** Typical temperature dependence of the rates and equilibria of protein folding transitions not involving intrinsically slow isomerizations. The natural logarithms of the rate constants for unfolding and refolding are plotted as a function of reciprocal temperature, in an Arrhenius plot. The similar plot of the equilibrium constant $K_{eq}$ between the native (N) and unfolded (U) states is a Van't Hoff plot. The curvature of the Van't Hoff plot is due to the greater apparent heat capacity of U than of N. The linear Arrhenius plot for the rate of unfolding indicates that the transition state has the same heat capacity as N. The greater heat capacity of U is reflected entirely in the curvature of the Arrhenius plot for the rate of refolding, because $\ln K_{eq} = \ln k_{refolding} - \ln k_{unfolding}$.

The data used to construct this diagram are for hen egg white lysozyme at pH 3, extrapolated to the absence of denaturant. Although the curves for the rates must intersect at $K_{eq} = 1$, it is a coincidence that they then have the value 1 $s^{-1}$. [Reprinted with permission from Creighton, *Biochem. J.* 270: 1–16 (1990).]

Even if an intermediate is detected, it is very difficult to determine its kinetic role.

## 5.2 Protein Unfolding

Upon placement of a native, covalently homogeneous protein into unfolding conditions, unfolding almost always occurs with a single kinetic phase and a single rate constant; it is an all-or-none process in which no partly unfolded intermediates are populated. There is a single rate-limiting step, and all the folded molecules have the same probability of unfolding. The rate usually changes uniformly with variation of the unfolding conditions, suggesting that the mechanism remains unchanged (Figure 2).

## 5.3 Protein Refolding

### 5.3.1 Kinetic Observations

Kinetic complexities in protein refolding usually result from conformational heterogeneity of the unfolded state, with slow- and fast-refolding molecules:

$$U_S \xleftrightarrow{slow} U_F \xleftrightarrow{fast} N \qquad (2)$$

In most cases, this pattern arises from intrinsically slow isomerizations, especially cis–trans of peptide bonds preceding proline residues. In the absence of such slow isomerizations, the fully refolded protein generally appears without a significant lag period and with a single rate constant, suggesting that all the molecules fold via the same rate-determining step and that all preceding and subsequent steps are rapid and reversible. Since the rate of refolding depends not on the initial unfolding conditions but on the final folding conditions, it is determined by the properties the unfolded protein rapidly adopts when placed under the final folding conditions (the ''refolding'' protein). There probably is rapid conformational equilibration prior to the rate-limiting step, so all the molecules converge to follow a common subsequent pathway and the same rate-limiting step.

At low temperatures, the rate of refolding generally increases with increasing temperature, but this trend diminishes, and the rate reaches a maximum and then decreases dramatically at high temperatures (Figure 2).

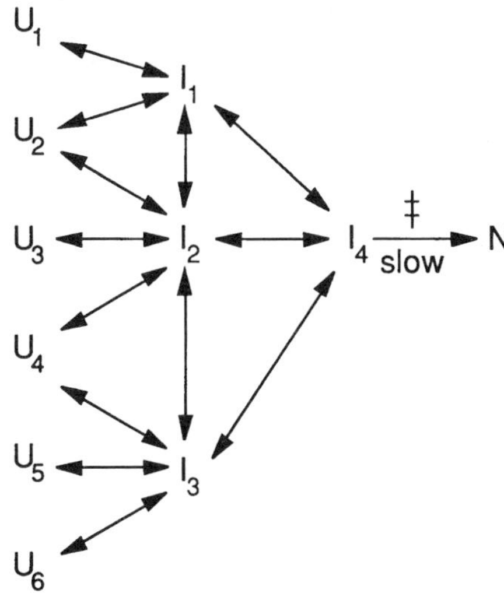


**Figure 3.** A general kinetic model indicated by experimental data for protein folding in the absence of intrinsically slow isomerizations: $U_i$ are various unfolded molecules with different conformations in the initial unfolding conditions, and $I_i$ are partly folded molecules. All kinetic steps indicated by arrows are rapid, except for that labeled "slow"; ‡ indicates the occurrence of the overall transition state. Single-headed arrows represent steps that effectively occur only in the indicated direction under conditions strongly favoring folding. All the unfolded molecules rapidly equilibrate under folding conditions with a few partially folded, marginally stable intermediates, which are also in rapid equilibrium. All the molecules pass through a slow step, which involves a transition state that is a distorted form of the nativelike conformation. Any partly folded intermediates that occur after the rate-limiting step are probably very unstable relative to N. [Reprinted with permission from Creighton, *Biochem. J.* 270: 1–16 (1990).]

### 5.3.2 The Refolding Protein

The refolding protein often, but not always, adopts significant amounts of nonrandom conformation and often appears similar to the molten globule state. Where characterized in detail, however, it consists of subdomains of nativelike secondary structure interacting and stabilizing each other.

### 5.3.3 The Transition State for Folding

Folding transition states are being characterized in detail by altering the folding conditions or the covalent structure of the protein and determining the effects on the rates of unfolding and refolding. The transition state usually appears to be a distorted, high energy form of the native conformation.

### 5.3.4 A General Scheme for Protein Folding in Vitro

A general kinetic scheme indicated by the available experimental data for protein folding in vitro of small single-domain proteins is illustrated in Figure 3.

## 6 PROTEIN FOLDING IN VIVO DURING BIOSYNTHESIS

Protein folding occurs rapidly in the cell after biosynthesis. Individual domains can fold as soon as they are completed, while the remainder of the polypeptide chain is being synthesized. The nascent chain interacts with various cellular factors, or chaperones. These factors do not actively participate in folding but serve to keep the protein unfolded or to inhibit aggregation. There are enzyme catalysts of cis–trans isomerization of prolyl peptide bonds and of formation and rearrangement of protein disulfide bonds.

*See also* Chaperones, Molecular; Peptides and Mimics, Design of Conformationally Constrained; Protein Aggregation; Protein Modeling.

### *Bibliography*

Creighton, T. E. (1990) Protein folding. *Biochem. J.* 270:1–16.
———, Ed. (1992) *Protein Folding.* Freeman, New York.
———. (1993) *Proteins: Structures and Molecular Properties,* 2nd ed. Freeman, New York.
Gierasch, L. M., and King, J., Eds. (1990) *Protein Folding: Deciphering the Second Half of the Genetic Code.* American Association for the Advancement of Science, Washington, DC.
Jaenicke, R. (1991) Protein folding: Local structures, domains, subunits and assemblies. *Biochemistry* 30:3147–3161.
Kim, P. S., and Baldwin, R. L. (1990) Intermediates in the folding reactions of small proteins. *Annu. Rev. Biochem.* 59:631–660.
Nall, B. T., and Dill, K. A., Eds. (1991) *Conformations and Forces in Protein Folding.* American Association for the Advancement of Science, Washington, DC.
Pain, R. H., Ed. (1994) *Mechanisms of Protein Folding.* IRL Press, Oxford.
Privalov, P. L. (1989) Thermodynamic problems of protein structure. *Annu. Rev. Biophys. Biophys. Chem.* 18:47–69.

# Protein Modeling

*Thomas R. Defay and Fred E. Cohen*

### *Key Words*

**α-Helix** Secondary structural element characterized by a right-handed helical conformation with 3.6 residues per turn, and a regular pattern of hydrogen bonds between the carbonyl oxygen of a residue $i$ and the amide hydrogen of residue $i + 4$.

**β-Sheet** Collection of β-strands aligned in a parallel or antiparallel fashion such that the carbonyl oxygens of one strand form hydrogen bonds with the amide hydrogen of the adjacent strands.

**β-Strand** Secondary structural element characterized by an extended conformation for the amino acid chain.

**Denaturation** The loss of protein three-dimensional structure and activity, usually induced by a change in the pH, temperature, or solvent conditions.

**Macromolecules** Polypeptide, polysaccharide, or nucleic acid polymers with a molecular weight typically greater than 2 kilodaltons (1 kDa = 1000 times the mass of a hydrogen atom).

**Secondary Structure** Conformation of the polypeptide chain with a set of repeating backbone dihedral angles. Examples are α-helices and β-sheets.

**Tertiary Structure** The overall three-dimensional structure of the protein. Typically, these structures can be built by packing secondary structure elements together.

Protein modeling is a collection of computational methods used for the description, analysis, and prediction of protein structures and the interaction of proteins with other molecules. Protein structures can provide insight into enzyme mechanisms and are increasingly useful for the design and optimization of novel pharmaceuticals. A major goal of protein modeling is the "protein folding problem," which calls for the ability to accurately predict the structure of a protein from its sequence. The protein folding problem is still considered intractable for most proteins when attempted without additional information (de novo), but headway is being made with restricted systems such as all-helical proteins. Modeling by homology takes advantage of an experimentally determined structure with a sequence similar (homologous) to the protein structure under consideration.

Most representations of protein structures are static, but in solution, proteins are quite flexible and are constantly changing form. The static structure is an average of the most frequently observed positions of the atoms that constitute a protein. Increasingly, dynamic models have been developed to follow the conformational plasticity of a protein over time.

## 1 IMPORTANCE OF PROTEIN STRUCTURE TO PROTEIN FUNCTION/INHIBITOR DESIGN

### 1.1 Origin of Structures: X-ray Crystallography and NMR Spectroscopy

Most protein structures have been determined by the technique of X-ray crystallography. For this method to succeed, suitable conditions must be found to induce the protein into a highly ordered crystalline array capable of coherently scattering X-rays. The intensities of the diffracted X-rays are measured, but phase information is lost. The phase information can be deduced by a variety of methods, including the introduction of heavy atoms into the crystal (multiple isomorphous replacement and multiple anomalous dispersion) or by relation to previously solved proteins of similar structure (molecular replacement). The coordinates of many structures are stored in the Brookhaven Protein Data Bank (PDB).

More recently, multidimensional nuclear magnetic resonance (NMR) methods have been used to determine the structures of many proteins with an accuracy comparable to X-ray crystallography. NMR structures are determined in solution, eliminating the problem of obtaining crystals. Technical limitations make it difficult to determine the structure of proteins larger than 20 kDa by NMR methods.

### 1.2 Biological Lessons Learned: Mechanism/Mutagenesis

The structures produced by X-ray crystallography and NMR spectroscopy have been used to enhance our understanding of biological processes. For instance, the mechanisms of enzyme action have been refined considerably by linking structure analysis with mutagenesis and other experimental techniques.

### 1.3 Recent Applications to Drug Design

Enzymes are proteins that catalyze chemical reactions in clefts on the molecular surface known as active sites. Drugs have been designed to bind or alter the active site, with limited success. More headway has been made by identifying the basic shape of an active site and searching a large database of small molecules to find some that should fit. Crystal structures of complexes of a drug bound to an enzyme have been used to direct a medicinal chemistry effort aimed at developing potent analogues of the lead compound.

## 2 RELATIONSHIP OF SEQUENCE TO STRUCTURE

### 2.1 The Thermodynamic Hypothesis

Most of protein modeling is based on the notion that the final structure of a protein is uniquely determined by its amino acid sequence. A series of experiments by Anfinsen and co-workers demonstrated that a denatured protein (ribonuclease) will spontaneously refold in solution to adopt its active structure. Ribonuclease will refold even if the native disulfide pairings have been scrambled. This behavior has been taken as evidence that proteins fold to reach their thermodynamically optimum state.

### 2.2 Relevance of Chaperonins

Intracellular, molecular chaperones that assist in vivo protein folding have been found. These chaperones most likely function by preventing folding intermediates from aggregating into nonproductive forms.

### 2.3 Kinetic Problems

Catalysts for the protein folding process may take other forms. α-Lytic protease requires an N-terminal Pro segment to fold correctly. Without it, the protein folds to a nonnative-like structure. Under normal folding conditions the pro region is autocatalytically removed after the protein folds. The pro region reduces the activation energy necessary to fold the protein (see Figure 1). In this case, although the minimum energy protease structure is known, the isolated protease sequence cannot overcome the ~126 kJ (30 kcal) activation barrier to folding on a kinetically relevant time scale (< hours).

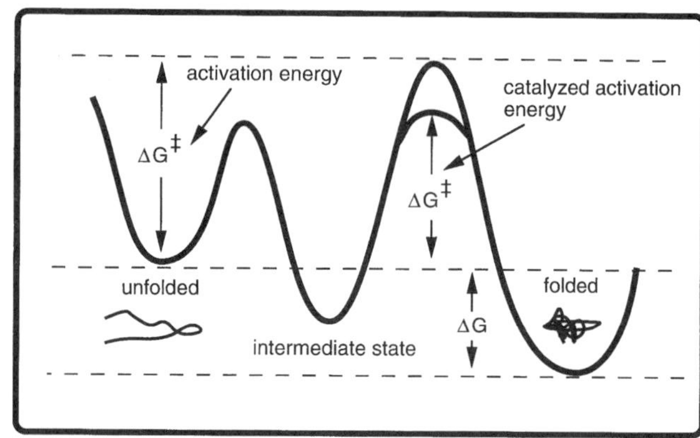


**Figure 1.** Hypothetical activation energy profile for protein folding with a pro region acting as a catalyst. $\Delta G‡$ is the activation energy of folding for the protein with and without the presence of a catalyst; $\Delta G$ is the free energy of folding the protein. The activation energy of the folding process is reduced by the catalyst, thus increasing the rate of the action. Note that the free energy of folding remains constant.

## 3 OVERVIEW OF MODELING PROTEINS

### 3.1 De Novo Versus by Homology

Two general approaches to protein modeling are used: de novo and homology-based methods. Homology modeling requires knowledge of the structure of a sequence that is recognizably similar to that of the desired protein. The known structure is used as a template on which the new sequence is engrafted. For sequences that share identical residues at more than 30% of their aligned positions, model-built structures can be quite accurate and have been used to design novel pharmaceuticals. De novo modeling does not require the initial protein structure. Instead, a model is constructed from an analysis of the sequence, in an attempt to produce a structure that is optimally suited to that sequence. Although de novo methods are unlikely to approach the accuracy of homology-based strategies, in principle they are applicable to a broader range of problems.

### 3.2 Static Versus Dynamic Structures

Rigid models of protein structures capture only one facet of the conformational properties of these macromolecules. A catalytically important residue, when viewed in a static model, may appear to be inaccessible to the ligand molecule. However, this residue may be accessible during a significant part of the molecular trajectory calculated in a dynamic simulation. X-ray crystallography determines an average set of atomic positions best represented by a static structure with a B factor associated with the static disorder and dynamic motion of atoms. Intramolecular distance constraints derived by NMR spectroscopy are used to determine a family of structures consistent with the experimental data. The molecular envelope defined by the family of structures provides a more dynamic view of macromolecular conformation.

## 4 DE NOVO METHODS

### 4.1 Secondary Structure Prediction

Most de novo modeling strategies attempt to identify the secondary structural elements of the protein from the protein sequence and then assemble these structural elements into one or more plausible tertiary structures. The tertiary structures are evaluated for the compatibility with the experimental properties of the protein under study or theoretical properties of proteins in general.

Secondary structure prediction has evolved from early work by Schiffer and Edmunson, who observed that helical sequences tended to segregate hydrophilic residues from their hydrophobic counterparts. When a sequence is displayed as a wheel with consecutive spokes every 100 degrees, helical sequences display a periodicity that is absent from regions destined to form turns. Obviously, this offers little insight into the formation of β-structure. Chou and Fasman compiled the frequency with which each amino acid appears in a specific secondary structural element. The structure of a sequence was then predicted, using the probabilities generated from these frequencies. Garnier et al. improved on this approach by calculating the secondary structure preferences of each amino acid subject to modifications exerted by sequentially proximal residues. Recent recalibration of the Garnier technique has resulted in a secondary structure prediction algorithm that is ~65% accurate for the three-state model (helix, strand, loop).

Neural networks, a computational tool from the machine learning community, have been designed to identify automatically patterns that predict secondary structure. These networks generate relationships among the different amino acids from an analysis of protein sequences and their structures. Neural networks have achieved a 64% success rate. When trained on a set of exclusively helical proteins, networks correctly predict 80% of the conformational preferences of individual residues.

### 4.2 Tertiary Structure Prediction

The tertiary structure of a protein can be approximated by packing secondary structural elements together. The number of plausible tertiary structures is limited by constraints on secondary structure packing. Other tertiary structure constraints are the globular shape of proteins and the tendency of proteins to form a well-packed hydrophobic core. Once a general tertiary structure has been assigned, in principle it can be refined by detailed energy calculations.

### 4.3 Worked Examples and Results: IL-4

The structure of interleukin-4 (IL-4) was calculated by the de novo approach and compared with the structure subsequently determined by NMR methods. Circular dichroism spectroscopy was used to prove that IL-4 is dominated by α-helical structure. Secondary structure prediction methods were used to assign the location of α-helices and loops. A combinatorial algorithm generated all possible juxtapositions of the four helices, subject to the constraints that a hydrophobic core had to be formed and that the interhelical loops had to be able to join neighboring helices. Of the 90,403 structures generated that did not violate steric constraints or disrupt the connectivity of the chain, 311 were consistent with distance constraints imposed by the three disulfide bridges. Solvent-accessible surface area calculations were used to select the energetically most sensible structures. When the three-dimensional structure was solved by NMR spectroscopy, it was clear that the secondary structure was predicted accurately (~90%). Unfortunately, the best predicted structure was the topological mirror image of the NMR structure. The eighth structure on the list resembled the correct structure (rms deviation = 4.8 A).

## 5 MODELING BY HOMOLOGY

### 5.1 Sequence Alignment

Modeling by homology follows a four-step recipe: sequence alignment, framework construction, loop construction, and side chain placement. These steps are detailed in Figure 2. In the first step, the sequence of the protein is aligned with the sequence(s) of a protein(s) of known structure. A sequence alignment with greater than 50% residue identity is desirable, but 20 to 30% is acceptable if several aligned sequences and structures are available.

### 5.2 Framework Construction

After the alignment has been made, a portion of the new protein structure is constructed from the core elements of the known structure. If the structures of several homologous sequences are available, either a representative framework or an aggregate framework structure can be used. This framework may reveal errors in the

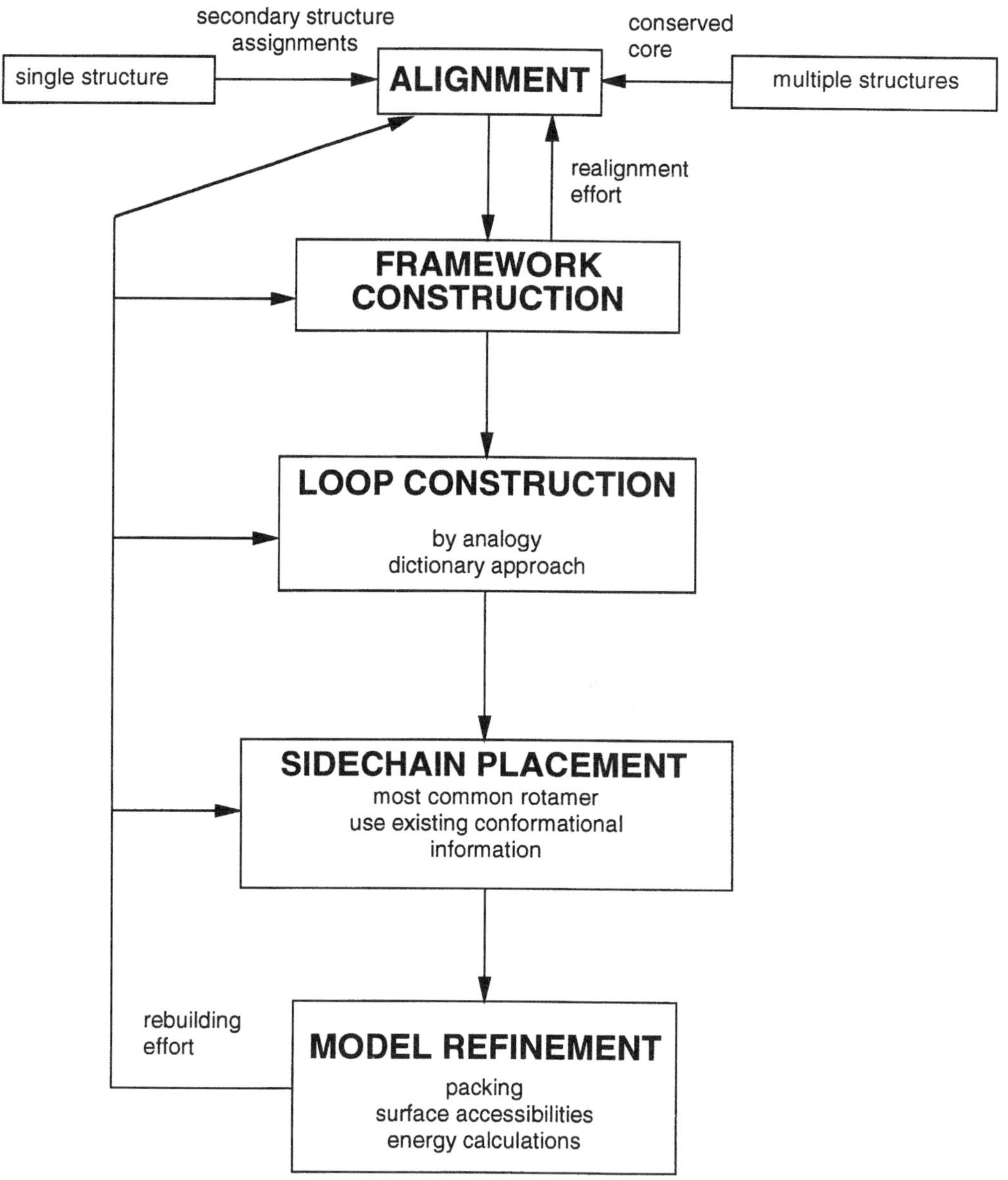


**Figure 2.** Flow chart for modeling by homology. If an error in the structure is detected, the structure is rebuilt from an earlier phase to correct the error. Most errors are found in the framework construction or model refinement phase.

original sequence alignment, which can be corrected. A new framework is then constructed.

### 5.3 Loop Modeling

Helices and sheets contain repeating geometries that are comparatively easy to model because their local conformation and their location relative to other secondary structures is conserved across evolutionarily related proteins. This is not true for the aperiodic loop regions that join secondary structure units. Thus, several methods have been devised for modeling loop structure. Sequence alignment is the simplest and most effective method for anticipating the conformation of a loop with a sequence similar to that of a corresponding loop from the family of known homologous structures. A dictionary of loops, derived from all proteins of known structure assembled in the PDB, is used when it is not possible to identify a loop geometry by the alignment approach. The loop fragment from the PDB provides a plausible conformation for the region of structure under study. This is most effective with short loops (2–10 amino acids), since longer loops are underrepresented in current versions of the protein database. The ab initio method is used to enumerate all possible loop structures and then select the lowest energy conformation. Because of the vastness of conformational space, this approach is computationally difficult and intractable for loops that contain more than seven amino acids.

### 5.4 Side Chain Placement

Usually, protein side chains are placed in conformations that are reminiscent of the side chain geometries adopted by their homologous residues. When no conformational information is available, the residues are placed in their statistically most likely conformation.

### 5.5 Model Refinement and Structure Validation

Energy calculations, including energy minimization and molecular dynamics, are used to locate a structure with sensible steric and electrostatic interactions. Unfortunately, energy calculations are not accurate enough to confirm that the model-built structure is correct. As a result, models are analyzed by a series of empirical measures, such as efficient packing, integration of the hydrophobic core, appropriate residue solvent accessibility profiles, appropriate spatial distribution of charged groups, and agreement of backbone dihedral angles with known dipeptide preferences (the Ramachandran plot). If part of a model appears to be inconsistent with these empirical measures, a rebuilding effort is required (see Figure 2).

### 5.6 Application to Drug Design

Homology model-built structures have been shown to be accurate enough to aid in a drug discovery program.

## 6 MOLECULAR DYNAMICS

### 6.1 Energy Functions and Molecular Mechanics

The attraction or repulsion each atom in a protein exerts on every other atom can be approximated as a sum of interaction energies. The functional form of this expression is derived from structural studies of proteins and small molecules as well as from theoretical and thermodynamic studies. A typical molecular mechanics potential takes the following form:

$$E(\mathbf{x}_1, \ldots, \mathbf{x}_m) = \varepsilon_{bond} + \varepsilon_{ang} + \varepsilon_{tor} + \varepsilon_{vdw} + \varepsilon_{el}$$

$$E = \sum_{i=1}^{m} K_b(r_i - r_b)^2 + \sum_{i=1}^{m} K_a(\theta_i - \theta_a)^2 + \sum_{dihedrals} \frac{K_d}{2}[1 + (\cos n\phi - \chi)] + \sum_i \sum_{j>i} (B_{ij}r_{ij}^{-12} - A_{ij}r_{ij}^{-6}) + \sum_i \sum_{j>i} \frac{q_i q_j}{\varepsilon r_{ij}} \quad (1)$$

where $E$ = total energy of the system
$\mathbf{x}_1, \ldots, \mathbf{x}_m$ = vectors representing the spatial coordinates of all the atoms in the system
$\varepsilon$ = dielectric associated with the molecular environment
$\varepsilon_{bond}, \varepsilon_{ang}, \varepsilon_{tor}, \varepsilon_{vdw}, \varepsilon_{el}$ = components of the total energy representing bond, angular torsional, van der Waals, and electrostatic energy, respectively
$K_b, K_a, K_d$ = force constants associated with bond, angular, and torsional energies, respectively
$r, r_b$ = bond distance and equilibrium bond distance, respectively
$\theta, \theta_a$ = bond angle and equilibrium bond angle, respectively
$n$ = periodicity of rotation
$\phi$ = dihedral angle
$\chi$ = phase angle of the dihedral angle
$i, j$ = different atoms in the protein
$r_{ij}$ = distance between atoms $i, j$
$A_{ij}, B_{ij}$ = nonbonded (Lennard–Jones) repulsion and attraction coefficients for the interacting atoms $i, j$
$q_i, q_j$ = point charges of the atoms $i, j$

### 6.2 Equations of Motion

Molecular dynamics are used to study how the potential equation changes with time given an initial set of atomic positions and velocities (temperature). The motion of the atoms in a protein can be described by Newton's equation of motion:

$$F = -\Delta E(\mathbf{x}_1, \ldots, \mathbf{x}_m) \quad (2)$$
$$F = \frac{m\partial^2 \mathbf{x}(t)}{\partial t^2}$$

where $F$ = force on an atom
$E$ = energy of the system
$\mathbf{x}_1, \ldots, \mathbf{x}_m$ = vectors representing the spatial coordinates of all the atoms in the system
$m$ = appropriate atomic mass
$t$ = time
$x_j(t)$ = position of one atom in the system at a given time

With an integration time step of 1 to 2 femtoseconds, these equations are well behaved over at least 1 to 2 nanoseconds after sufficient equilibration of the system.

### 6.3 What Can You Learn?

Molecular dynamics can reveal protein interactions that static models obscure. Molecular dynamics has been used to follow the movement of molecular oxygen toward the heme iron of myoglobin and to create samples of conformational space accurate enough to permit the calculation of the change in the free energy of binding between two closely related ligands. Computational constraints limit dynamic simulations to the nanosecond time frame. Obviously, this is too short for many biologically important conformational changes (e.g., protein folding).

*See also* Circular Dichroism in Protein Analysis; Nuclear Magnetic Resonance of Biomolecules in Solution; Protein Folding.

## *Bibliography*


Allen, M. P., and Tildesley, D. J. (1989) *Computer Simulations of Liquids.* Clarendon Press, Oxford.

Anfinsen, C. B. (1973) Principles that govern the folding of protein chains. *Science,* 181:223–230.

Branden, C., and Tooze, J. (1991) *Introduction to Protein Structure.* Garland, New York.

Brooks, C. L., Karplus, M., and Pettitt, M. (1988) *Proteins—A Theoretical Perspective of Dynamics, Structure, and Thermodynamics.* Wiley, New York.

Chou, P. Y., and Fasman, G. D. (1974) Prediction of protein conformation. *Biochemistry,* 13:222–245.

Fasman, G. D. (1989) *Prediction of Protein Structure and the Principles of Protein Conformation.* Plenum Press, New York.

———. (1989) Protein conformational prediction. *Trends Biol. Sci.* 14:295–299.

Garnier, J. (1990) Protein structure prediction. *Biochimie,* 72:513–524.

Kuntz, I. D. (1992) Structure-based strategies for drug design and discovery. *Science,* 257:1078–1082.

Lesk, A. M. (1991) *Protein Architecture: A Practical Approach.* IRL Press, Oxford.

Richardson, J. S. (1981) The anatomy and taxonomy of protein structure. *Adv. Protein Chem.* 34:167–339.

# Protein Phosphorylation

*Clay W Scott*

## *Key Words*

**Phosphoprotein Phosphatase** An enzyme that removes phosphate groups from the amino acid side chains of phosphoproteins.

**Protein Kinase** An enzyme that catalyzes the transfer of a phosphate group from a nucleoside triphosphate (usually ATP) to an amino acid side chain of a substrate protein.

**Second Messenger** An intracellular molecule generated by the cell in response to an extracellular signal (first messenger), which triggers a biochemical cascade leading to a change in the behavior of the cell.

The biological activity of many proteins is modulated by the phosphorylation of specific sites within the protein. Protein kinases catalyze the transfer of a phosphate group to the protein, while phosphoprotein phosphatases remove the phosphate group. This reversible modification is utilized by the cell to regulate proteins that are involved in most all cellular functions. Protein phosphorylation plays a particularly prominent role in the transduction of extracellular signals. A hormone, neurotransmitter, or growth factor binds to its cell surface receptor and directly or indirectly activates a protein kinase (or phosphatase), which leads to changes in the phosphorylation states of key regulatory proteins. This cascade of events ultimately results in a change in the behavior of the cell (e.g., contraction, release of neurotransmitter, changes in the gene expression or metabolism).

## 1 IMPORTANCE OF PROTEIN PHOSPHORYLATION IN BIOLOGICAL REGULATION

Protein phosphorylation is one of the most common posttranslational modifications to occur in eukaryotic cells. Phosphorylation is used to control the activity of a variety of cellular proteins, which in turn regulate a vast array of cellular functions. The types of proteins known to undergo phosphorylation and dephosphorylation include metabolic enzymes, cytoskeletal proteins, ion channels, and cell surface receptors. Protein phosphorylation is particularly prominent in signal transduction processes. A cell can rapidly respond to a biological signal by phosphorylating key molecules, thereby initiating a cascade of events resulting in a change in a physiological property such as cell motility, release of neurotransmitters, or modulation of ion fluxes.

The phosphorylation state of a particular protein is regulated by specific protein kinases and phosphoprotein phosphatases (Figure 1). Protein kinases are enzymes that transfer the γ-phosphate from a nucleoside triphosphate (usually ATP) to an amino acid side chain of the substrate protein. Structural studies have shown that phosphorylation can occur within the active site of an enzyme, directly affecting substrate binding. When phosphorylation occurs at sites distal to the catalytic site, it can regulate enzyme function by inducing long-range conformational changes. Phosphorylation has been shown to affect the biological properties of nonenzymic proteins by altering their intracellular location, enhancing their susceptibility to proteolysis, and modulating their ability to interact with other proteins. Removal of the phosphate group is catalyzed by phosphoprotein phosphatases. Dephosphorylation allows the protein to return to its earlier functional state. Both protein kinases and phosphoprotein phosphatases are under stringent regulatory control. The activation state of the relevant kinase and phosphatase will generally determine when a protein becomes phosphorylated and how long it remains phosphorylated.

## 2 CLASSIFICATION AND PROPERTIES OF PROTEIN KINASES

Protein kinases are classified by the amino acid residue they phosphorylate. The vast majority of phosphorylation events in eukaryotic cells occur on serine and threonine residues (by protein–serine/threonine kinases) or tyrosine residues (by protein–tyrosine kinases). A few protein kinases appear to phosphorylate tyrosine as well as serine/threonine residues. It has been suggested that these protein kinases be classified as dual-specific protein kinases if such activity is shown to be physiologically significant.

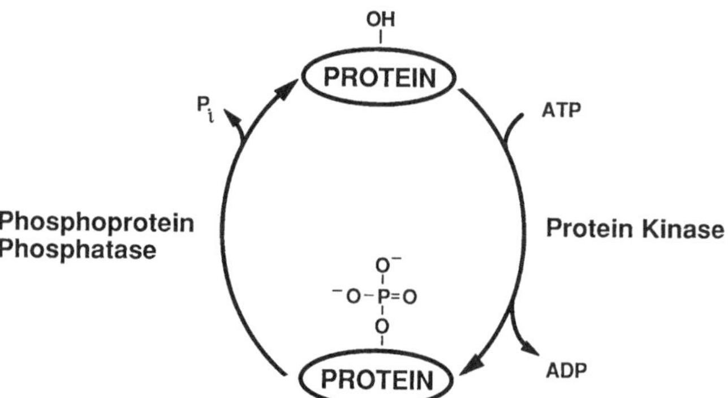


Figure 1. Protein kinases catalyze the transfer of a phosphate group onto a specific amino acid side chain of a substrate protein. ATP usually serves as the phosphate donor, and serine, threonine, and tyrosine residues serve as phosphate acceptors. Phosphoprotein phosphatases remove the phosphate group, allowing the protein to return to its basal state. The relative activities of the protein kinase and phosphoprotein phosphatase generally determine the phosphorylation state of the protein, and thus its functional state.

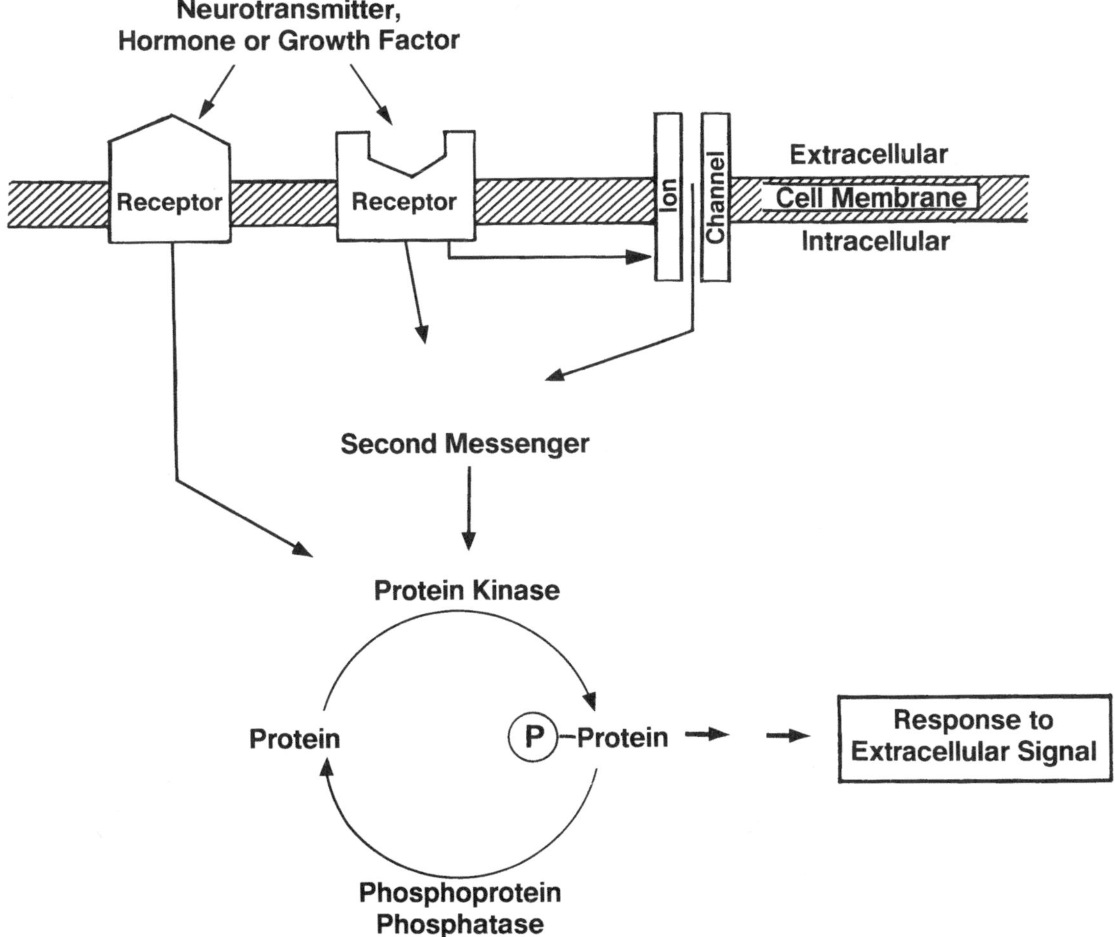


**Figure 2.** Protein phosphorylation is a common mechanism used to transduce extracellular stimuli into intracellular responses. Various extracellular molecules bind to their cell surface receptors and activate protein kinases. Some protein kinases are intrinsic components of receptors and are directly activated by ligand binding. Other protein kinases are indirectly activated by changes in the levels of particular second messengers, including cAMP, cGMP, diacylglycerol, and $Ca^{2+}$. Activation of a specific protein kinase produces an increase in the phosphorylation of selective proteins, thereby propagating the response to the agonist. Phosphoprotein phosphatases are also regulated through the actions of second messengers, both directly and indirectly. (From Scott and Patel, in *Encyclopedia of Human Biology,* Vol. 6, pp 201–211; reproduced by permission of Academic Press.)

Protein kinases recognize phosphorylation sites within particular amino acid sequences, called consensus sequences or recognition motifs. Studies using peptide substrates containing altered amino acid sequences have revealed the importance of primary sequence in distinguishing phosphorylation sites. For example, cAMP-dependent protein kinase will phosphorylate a serine or threonine residue in the sequence-Arg-Arg-X-Ser/Thr-X-, where X represents any amino acid. Eliminating or displacing either arginine residue can dramatically alter the kinetics of phosphorylation. Although amino acid sequence is an essential determinant for substrate recognition, higher orders of structure can also affect phosphorylation. Conformational states that mask or expose phosphorylation sites can affect phosphorylation of the substrate protein.

Many protein–serine/threonine kinases are activated by second messengers, intracellular molecules that are generated in response to extracellular signals (Figure 2). These protein kinases are grouped into classes based on the second messenger that stimulates their activity. These groups include protein kinases activated by cyclic nucleotides (cAMP- and cGMP-dependent protein kinases), calcium plus calmodulin (including $Ca^{2+}$/calmodulin-dependent protein kinase II and myosin light chain kinase), and diacylglycerol (protein kinase C). Other protein kinases have no known activator. These kinases may be regulated by the availability of substrate, by phosphorylation by other protein kinases, or by mechanisms that have not yet been identified.

Several protein–serine/threonine kinases have low basal activities because an inhibitory domain interacts with the catalytic site. The inhibitory domain contains a consensus sequence for the kinase (although in some cases the phosphate acceptor residue is missing). Activation of these kinases is achieved by a ligand- or phosphorylation-induced conformational change that removes the inhibitory domain from the catalytic site, thereby allowing access to substrates.

The protein–tyrosine kinases can be divided into two groups: the cytosolic nonreceptor kinases and the transmembrane receptor kinases. The nonreceptor tyrosine kinases were identified first as the products of viral oncogenes. These viral protein–tyrosine kinases, which are structural variants of normal cellular protein–tyrosine kinases, induce the malignant transformation of cells as a result of their unregulated tyrosine kinase activity. Because of the transforming properties of these viral oncogenes, their normal cellular homologues are felt to play critical roles in controlling cell growth and metabolism.

The transmembrane receptor kinases are cell surface receptors that contain an intrinsic tyrosine protein kinase within their cytoplasmic domains. Examples of receptor tyrosine kinases include the receptors for epidermal growth factor and platelet-derived growth factor. These receptors undergo dimerization following ligand binding. Receptor dimerization leads to activation of the receptor tyrosine kinase and autophosphorylation of the receptor on multiple tyrosine residues. These phosphorylated domains serve as binding sites for signal-transducing proteins, including phospholipase C-$\gamma$, phosphatidylinositol 3′-kinase, and *ras* GTPase-activating protein (GAP). These proteins control different signal transduction pathways. Their binding to the receptor and subsequent phosphorylation by the receptor kinase are important steps in the activation of these pathways which lead to the biological response of the cell.

## 3 CLASSIFICATION AND PROPERTIES OF PHOSPHOPROTEIN PHOSPHATASES

The phosphoprotein phosphatases are categorized by their selectivity for phosphoserine/phosphothreonine or phosphotyrosine residues. Recent evidence suggests that some phosphoprotein phosphatases may have dual specificity, as is the case with the protein kinases discussed earlier. In contrast to the protein kinases, the phosphoprotein phosphatases generally have broad and overlapping substrate specificities. The protein–serine/threonine phosphatases are divided into two groups; type 1, which dephosphorylate the $\beta$ subunit of phosphorylase kinase and are inhibited by protein inhibitor 1 and inhibitor 2, and type 2, which dephosphorylate the $\alpha$ subunit of phosphorylase kinase and are insensitive to inhibitor 1 and inhibitor 2. The type 2 enzymes can be further divided into classes A, B, and C, based on their requirements for divalent cations. Okadaic acid, a potent tumor promoter isolated from marine plankton, is a specific and potent inhibitor of phosphatase-1 and -2A. This agent has been used in intact cells to identify substrates for these two phosphatases and to demonstrate the involvement of protein–serine/threonine phosphatases in signal transduction processes.

The physical characteristics and substrate specificities of the protein–tyrosine phosphatases have not been clearly defined. At least seven different phosphatase activities have been identified based on differences in chromatography properties and sensitivities to various inhibitor molecules.

A family of transmembrane proteins expressing protein–tyrosine phosphatase activity has been described recently. These proteins have receptorlike configurations, suggesting the existence of a class of receptor-linked proteins that utilize the dephosphorylation of tyrosine residues for signal transduction.

## 4 PHOSPHORYLATION CASCADES AND MULTISITE PHOSPHORYLATIONS

A common underlying theme in many signal transduction pathways is the use of protein kinase cascades. A classical example is the regulation by epinephrine of glycogenolysis in skeletal muscle. Stimulation of the $\beta$-adrenergic receptor causes an increase in cAMP production and activation of cAMP-dependent protein kinase. This enzyme phosphorylates and activates phosphorylase kinase, which in turn phosphorylates and activates glycogen phosphorylase. In addition, cAMP-dependent protein kinase phosphorylates and inactivates phosphoprotein phosphatase-1. Thus, protein kinase cascades provide sequential steps for amplification of an extracellular signal and divergence for coordinated regulation of different cellular functions.

Many proteins appear to be phosphorylated at multiple sites, with varying consequences on their biological activity. Examples of multisite phosphorylation have revealed that (1) different protein kinases can phosphorylate a common site on a protein, (2) different protein kinases can phosphorylate distinct sites on a protein yet produce similar changes in biological activity, (3) different protein kinases can phosphorylate distinct sites on a protein, resulting in opposite changes in biological activity, and (4) phosphorylation of one site by a protein kinase can generate a consensus sequence for a second protein kinase. In many instances the multiply phosphorylated protein represents a critical point in a biochemical cascade. Thus, multisite phosphorylation is one approach by which different biological signals, acting through different transduction pathways, can converge to regulate the function of a single protein. This convergence would allow for coordinated regulation of a single cellular process. Reversible phosphorylation of proteins represents a unifying mechanism by which different signaling systems can be integrated to produce the appropriate biological response.

*See also* Bioenergetics of the Cell; Growth Factors.

### *Bibliography*


Boyer, P. D., and Krebs, E. G., Eds. (1986) *The Enzymes,* Vol. 16. Academic Press, Orlando, FL.

———, and ———, Eds. (1987) *The Enzymes,* Vol. 17. Academic Press, Orlando, FL.

Cobb, M. H., Boulton, T. G., and Robbins, D. J. (1991) Extracellular signal-regulated kinases: ERKs in progress. *Cell Regul.* 2:965–978.

Cohen, P. (1992) Signal integration at the level of protein kinases, protein phosphatases and their substrates. *Trends Biochem. Sci.* 17:408–413.

Roach, P. (1991) Multisite and hierarchal protein phosphorylation. *J. Biol. Chem.* 266:14139–14142.

Schlessinger, J. and Ullrich, A. (1992) Growth factor signaling by receptor tyrosine kinases. *Neuron* 9:383–391.

Walaas, S. I., and Greengard, P. (1991) Protein phosphorylation and neuronal function. *Pharmacol. Rev.* 43:299–349.


# Protein Purification

*Murray P. Deutscher*

### *Key Words*

**Chromatography** Separation on solid supports on the basis of charge, size, affinity, or adsorptive properties.

**Electrophoresis** Separation on the basis of movement in an electric field.

Cells contain thousands of different proteins, which participate in essentially every aspect of their structure and function. Included in this arsenal are the enzymes, which catalyze all the reactions of a cell's metabolic pathways, the structural proteins, which determine a cell's shape and internal architecture, and the signaling and regulatory proteins, which are the controllers of cell function. It is the proteins of a cell that determine its unique identity. Understanding how a cell is put together and how it carries out its functions at the molecular level requires a complete and detailed knowledge of its proteins. Such information can be obtained only by separating a protein of interest from its many cohorts, studying it in isolation to determine its individual properties, and examining it in the context of the cellular milieu. The key operation in this protocol is first identifying and then purifying the protein to be studied.

## 1 WHY PROTEINS ARE PURIFIED

On a more detailed level, proteins are purified for many different reasons. Thus, if the protein of interest were an enzyme, we would want to purify it to determine its catalytic properties in the absence of other possible interfering activities. Only in this manner could we be certain that the reaction observed was due solely to the protein under study. Likewise, if we wished to determine how the enzyme carried out its reaction, we would need to undertake studies of its active site and overall structure, which would involve having a pure protein to examine. Furthermore, we might want to know whether the protein is regulated. Such an investigation might include determining whether it is covalently modified or whether it associates with other cellular components. Again, studies of these types are carried out most unambiguously with purified proteins. Similar reasoning, with regard to interactions of proteins with other molecules, would also apply to structural proteins and signaling proteins, which clearly must interact with other components to carry out their functions.

At the investigative level, proteins are purified for other purposes as well. Purified proteins are often needed to generate antibodies that can be used for additional examination of protein structure and cellular localization. Purified proteins also serve as substrates for proteases and modifying enzymes, and in this capacity they help to clarify the mechanism of action of these enzymes. Proteins must also be purified for any detailed physical–chemical analysis of their structure (e.g., crystallization for X-ray analysis).

Proteins are also purified for other reasons. For example, they are used as reagents to carry out metabolic interconversions and to generate products in vitro. Purified proteins also find uses in laundry detergents, in meat tenderizers, in contact lens cleaners, and in other products. Numerous proteins are purified from natural sources or prepared by recombinant DNA techniques for use as therapeutic agents. Insulin, growth hormone, erythropoietin, interferon, the enzyme lactase, and many other proteins are prepared and purified on a large scale to treat various ailments. In all these situations, a highly purified protein is a necessity. Accordingly, purification schemes must be worked out to separate the protein of interest from the thousands of contaminants with which it is mixed.

## 2 STRATEGIES FOR PURIFYING PROTEINS

First and foremost, any purification procedure requires an assay for the protein being isolated. This may be an activity assay when the protein to be isolated is an enzyme, or the protein may be determined directly by the use of a specific antibody, when available, or simply by following a protein of a specific size. In every case, however, a means of detection is essential to follow the course of the purification.

A second important point to be considered in planning a protein purification is the biological source. Obviously, if a protein from a particular organism is the object of interest, no leeway is available for choosing a more suitable starting material. In many situations, however, much more flexibility may be possible, and in these instances, one needs to consider the availability and/or cost of the biological material. In addition, the richness of the source with regard to the protein to be isolated is a prime consideration. The richer the source, the more likely that the protein can be purified in large amounts, and the fewer the steps that may be needed to attain complete purification. Also to be considered is whether the protein can be isolated from a specific organelle and whether it might be overexpressed from a cloned gene. A positive answer in either case would simplify the purification scheme because such a protein has already been enriched relative to contaminants. Another consideration is the use to which the purified protein will be put. This point will determine how much of the protein needs to be purified, as well as how pure the protein needs to be.

In developing the purification scheme itself, a number of strategies generally turn out to be useful. If the protein is being purified for the first time, it usually is more efficient to test each prospective purification step with small trial experiments rather than using all the material in a procedure that might not work or could cause loss or inactivation of the protein. Likewise, it usually makes sense to test the stability of small amounts of the material prior to exposing the entire sample to some uncertain condition. The purification scheme itself generally should begin with higher capacity techniques, proceeding to those of lower capacity as extraneous proteins are removed. Often, the judicious arrangement of the order of purification steps can simplify the overall purification scheme by avoiding unnecessary concentration procedures or steps that require buffer changes. Finally, if the protein of interest has some unique property that distinguishes it from other proteins, that property should be exploited during the purification scheme. In the extreme, such a unique property could result in a one-step purification!

## 3 THE PURIFICATION SCOREBOARD

The goal of any purification procedure is to increase the relative concentration of the protein of interest while maintaining its presence in the highest amounts possible. Thus, at every step in the purification it is essential to evaluate the amount of total protein present and the amount of the protein being purified (expressed in some units). These numbers can be used to evaluate the effectiveness of individual purification steps and to assess the ongoing purification. This is the purification scoreboard; a representative one for hypothetical protein X is shown in Table 1. At every step, one can then determine the degree of purification attained by that particular procedure and the yield or recovery of the desired protein. Relative purity is indicated by the specific activity of the protein of interest (specific protein ÷ total protein). The overall purification, then, is the specific activity at a particular step compared to the specific activity of the starting material. Knowing the specific activity at a particular step in comparison to that of the pure protein would give the degree of purity. In the example shown in Table 1, if the protein is homogeneous after step 5, the material after step 4 is 25% pure. Specific activity measurements are also helpful in deciding what fractions are to be combined after a particular step before proceeding to the next fractionation procedure. One must balance the desire to use the fractions with highest specific

**Table 1** Purification of Protein X

| Step Description | Total Protein (mg) | Protein X (units) | Specific Activity (units/mg) | Relative Purification (-fold) | Recovery (%) |
|---|---|---|---|---|---|
| 1. Crude extract | 1000 | 100 | 0.10 | 1.0 | 100 |
| 2. High capacity | 300 | 75 | 0.25 | 2.5 | 75 |
| 3. Chromatography 1 | 30 | 66 | 2.2 | 22 | 66 |
| 4. Chromatography 2 | 1 | 25 | 25 | 250 | 25 |
| 5. Specific affinity | 0.2 | 20 | 100 | 1000 | 20 |

activity against the undesirability of sacrificing too much material with regard to recovery.

The purification scoreboard of Table 1 shows that protein X was purified 1000-fold with a yield of 20%. In assessing the individual steps, one can see that each gave a reasonable degree of purification, and in all cases except step 4, a reasonably good recovery. Since step 4 resulted in a high degree of purification (>10-fold), one would probably tolerate the high loss of protein X at this step, especially since the overall recovery is not too bad. From this example, it is clear that the purification scoreboard is a very important tool for assessing the effectiveness of a purification scheme and for deciding possible sites for improvement in future attempts.

## 4 PURIFICATION METHODS

A wide variety of protein fractionation methods are available that can be combined to generate a suitable purification scheme. Generally, a large amount of trial and error is involved in identifying just the methods that will be most effective in purifying and maintaining the stability of the desired protein. On the other hand, if only very small amounts of a denatured protein are needed, one may be able to go directly to a high-resolution technique, such as two-dimensional gel electrophoresis, to obtain the protein in one step. Usually, however, one executes a series of purification steps, combining early ones of high capacity and low resolution (when large amounts of protein are present) with lower capacity and higher resolution ones (when less protein is present) at later stages of the purification scheme. It is often most effective to combine purification procedures that take advantage of different properties of the protein of interest (e.g., precipitation properties, charge, size, adsorptive properties, affinity properties). In this manner, one can maximize the overall differences of the desired protein from all the others with which it is mixed. A brief description of widely used purification methods follows and is summarized in Table 2.

Initially, the cell or tissue that has been chosen as starting material must be disrupted to prepare a cell-free extract that allows further fractionation. This is usually accomplished by mechanical means (with homogenizers, blenders, grinders, etc.) or by lysis procedures using detergents or other disruptive agents. At this point, some type of subcellular fractionation is often undertaken to remove cell debris and organelles if the desired protein is cytosolic, or to isolate the appropriate subcellular fraction when this is necessary. In some cases, when the protein of interest is membrane-bound or associated with other cellular structures, detergents or high salt conditions may be necessary to solubilize the protein.

At this stage a bulk method may be used which usually involves some type of fractional precipitation, followed by centrifugation to separate fractions. These procedures may include salting out (generally with ammonium sulfate), precipitation with organic solvents or by decreased pH, or even heat denaturation. The protein of interest should not be damaged by the procedure, and it should be separated sufficiently from the majority of the proteins to give a small purification (a few-fold) with a reasonable recovery (> 70%). These low resolution procedures often serve to concentrate large volumes of extract and to remove nonprotein material, such as nucleic acids. Losses at this stage, although not necessarily great in percentage terms, actually represent a large amount of potentially purified protein.

From this point on, higher resolution procedures are usually employed, generally involving some type of column chromatography. A large number of commercial resins are available for these purposes. Inasmuch as proteins contain large numbers of charged amino acids, which can vary greatly among proteins with regard to number and sign, charged ion exchange resins are often the material of choice at this stage. These materials have high capacities, and by judicious choice of conditions (pH and salt) can serve to afford a high degree of purification. In many situations the use of consecutive ion exchange columns, employing resins of different charges or strengths, can prove very effective. Even the same resin used under different conditions can give additional purification. Ion exchange columns may be used in a stepwise elution mode, or for even better resolution, with salt or pH gradients. Ion exchange chromatography may also serve to concentrate proteins.

Proteins may also be fractionated based on their adsorptive properties or their size. In the former case, hydroxyapatite (a type of calcium phosphate) is usually the medium of choice. Size fractionation can be accomplished using any one of a number of available gel filtration materials that separate proteins in specific size ranges. Nowadays, materials are available that have low affinities for binding proteins, removing the need for concern about interference due

**Table 2** Summary of Protein Purification Methods

1. Extract preparation
   - Mechanical disruption
   - Cell lysis
2. Subcellar fractionation
3. Solubilization of bound proteins
4. Bulk techniques
   - Salting out
   - Precipitation with organic solvents
   - Precipitation by decreased pH or heat
5. Chromatographic procedures
   - Ion exchange
   - Adsorption
   - Gel filtration
   - Affinity resins
6. Electrophoretic procedures
   - Native or SDS-PAGE
   - Isoelectric focusing
   - Two-dimensional gel electrophoresis

to this artifact. Since gel filtration does not involve protein binding to the resin, it is a relatively gentle procedure. Gel filtration can also be used to change buffer conditions. One drawback to the procedure is that it leads to increases in sample volumes.

More specific fractionation of proteins may be obtained by the use of chromatographic resins containing bound ligands with high affinity for particular classes of proteins, or even for specific proteins. For example, resins with certain immobilized dye molecules specifically bind proteins with nucleotide-binding folds, and resins with bound lectins have high affinity for certain carbohydrate-containing proteins. Likewise, proteins with hydrophobic domains can be bound to supports containing alkyl or aryl groups. Specific affinity resins can also be prepared by attachment of a substrate or a competitive inhibitor of a particular protein. In this manner, one can isolate general DNA- or RNA-binding proteins, or even proteins that bind to specific nucleotide sequences. Resins have also been made that contain a bound antibody with high affinity for a certain protein as a means to specifically remove it from a complex mixture. Generally, these procedures are used in later stages of a purification scheme when there may be less interference from potentially destructive or nonspecific binding contaminants. Once bound to an affinity column, specific proteins may be eluted with a mobile form of the affinity ligand or by agents that disrupt the interaction between the protein and the ligand (e.g., salt or detergent). Affinity chromatography is a particularly powerful fractionation tool because it takes advantage of the unique functions of individual proteins to separate them from others of potentially similar structure.

Another class of protein fractionation methods relies on the movement of proteins in an electric field. Different proteins migrate differently in the field based on their net charge, their size and shape, and the properties of the supporting medium. These electrophoretic procedures are used not only for purification purposes, but also analytically, to monitor the effectiveness of a purification scheme and to assess purity. Proteins are generally separated in a supporting medium of acrylamide, whose sieving properties can be adjusted by altering the cross-linking density of the gel matrix. Proteins may be fractionated in their native state, in which their size, shape, and charge will play an important role. Alternatively, proteins may be fractionated after treatment with sodium dodecyl sulfate, which results in denaturation of proteins and allows their subsequent separation based primarily on size.

Proteins are also fractionated electrophoretically in supporting media containing a pH gradient. In this procedure, termed isoelectric focusing, a protein will migrate until it reaches the pH corresponding to its isoelectric point. At this position its net charge is zero, and its movement will cease. Inasmuch as even proteins with similar isoelectric points may be separated by this procedure, the method has a high resolving power. Gel electrophoresis and isoelectric focusing may be combined in two-dimensional gel electrophoresis, a procedure affording very high resolution, in which proteins are separated in one dimension based on isoelectric point, and in the second dimension based on size. Generally, this method is used for analytical purposes.

Through a combination of the various methods mentioned here it is often possible to separate a protein of interest completely from contaminating material. Success in such an endeavor is limited only by the amount of protein available and its stability, and by the patience of the investigator for testing as many procedures as needed to ultimately afford a pure protein.

## 5 CRITERIA OF PURITY

Upon completion of a purification procedure, it is important to assess the purity of the protein of interest. Several criteria may be applied to judge the protein's degree of purity. In theory, the protein should be pure when subsequent purification steps do not lead to any increase in specific activity. Likewise, a pure protein should display a constant specific activity across a chromatographic peak. Most commonly, purity is assessed based on the number of stained protein bands after polyacrylamide gel electrophoresis in buffers containing sodium dodecyl sulfate (SDS-PAGE). Ideally, there should be a method to detect the protein of interest to ensure that the major, or even, the single protein band (or multiple bands if the protein contains different-sized subunits) is the desired one. More stringent criteria to assess purity may be applied, as necessary, such as the presence of a single species upon sedimentation equilibrium or a single amino-terminal sequence upon a sequence analysis.

*See also* Enzyme Assays; Gel Electrophoresis of Proteins, Two-Dimensional Polyacrylamide; Pharmaceutical Analysis, Chromatography in; Protein Analysis by Integrated Sample Preparation, Chemistry, and Mass Spectrometry; Proteins and Peptides, Isolation for Sequence Analysis of.

### *Bibliography*

Deutscher, M. P., Ed. (1990) *Guide to Protein Purification.* Academic Press, Orlando, FL.

Harris, E. L. V., and Angal, S., Eds. (1989) *Protein Purification Methods: A Practical Approach.* IRL Press, Oxford.

——— and ———, Eds. (1990) *Protein Purification Applications: A Practical Approach.* IRL Press, Oxford.

Scopes, R. (1987) *Protein Purification, Principles and Practice,* 2nd ed. Springer-Verlag, New York.

# Proteins and Peptides, Isolation for Sequence Analysis of

*Larry D. Ward and Richard J. Simpson*

### *Key Words*

**Edman Degradation** A chemical procedure for cleaving amino acids sequentially, one at a time, from the N-terminus of proteins and peptides.

**Reversed-Phase High Performance Liquid Chromatography** A high-resolution chromatographic method for separating proteins or peptides based on their hydrophobic properties.

**SDS-PAGE** Sodium dodecyl sulfate polyacrylamide gel electrophoresis: a high-resolution electrophoretic method for separating proteins or peptides based on relative molecular weight.

An essential and often rate-limiting step in many gene cloning strategies is the ability to obtain amino acid sequence information from the protein of interest. Such sequence information allows the design of oligonucleotide primers for use in cloning strategies.

Automated protein-sequencing instruments and the methodology for obtaining such information have evolved considerably in the past decade, with dramatic improvements in sensitivity. It is now possible to obtain sequence data with less than 5 pmol of protein or peptide. Many microsequencing projects, however, are limited not by the sensitivity of the protein sequencing instruments per se but by the ability of investigators to obtain protein/peptide samples in a form suitable for sequence analysis. The Edman chemistry employed in the current generation of protein-sequencing instruments imposes a number of constraints on sample composition which must be taken into consideration when designing a purification strategy. Unlike DNA, however, proteins have physicochemical properties that vary widely, and there is not a universal protocol that can be used to purify all proteins. However, two high-resolution purification procedures, reversed-phase high performance liquid chromatography and sodium dodecyl sulfate–polyacrylamide gel electrophoresis, can be used in most cases. These two protocols can be tailored for purifying various proteins/peptides ranging from growth factors low in relative molecular mass ($M_r$) to high $M_r$ membrane-associated proteins (e.g., receptors) in a form suitable for sequence analysis.

## 1 PRINCIPLES

### 1.1 Edman Degradation

Proteins/peptides are sequenced by degradation from the N-terminus using the Edman reagent, phenylisothiocyanate (PITC). The process is divided into three steps: coupling, cleavage, and conversion. In the coupling step PITC modifies the N-terminal residue of a protein/peptide. Acid cleavage removes the N-terminal amino acid as an unstable anilinothiazolinone (ATZ) derivative and leaves the shortened ($n - 1$) protein/peptide with a reactive N-terminus. The ATZ derivative is converted in the last step to a stable phenylthiohydantoin (PTH) amino acid. Thus, during one cycle of this reaction, the N-terminal residue is removed from a protein/peptide as a PTH derivative and identified by reversed-phase high performance liquid chromatography (RP-HPLC). The shortened protein/peptide is left with a free N-terminus that can undergo another cycle of the reaction. Today, most laboratories perform the Edman degradation with fully automated instruments that are coupled directly to RP-HPLC systems for PTH amino acid detection. When gas phase or pulsed liquid instruments are used, the lowest amount of readable sequence is in the range of 0.2–10 pmol, while the initial yield (percentage of the sample loaded onto the instrument that is sequenceable) can vary from 30 to 80% (provided the N-terminus is not blocked). The repetitive yield from one cycle to the next varies markedly (85–96%) depending on the inherent sequence—for instance, proline residues in a polypeptide can be notoriously difficult to sequence through—as well as on the size of the protein/peptide. Taking these latter limitations together, it is reasonable to expect 10–20 cycles of sequence from 1–5 μg of a 30 kDa protein.

For successful microsequence analysis using gas-phase or pulsed-liquid instruments, investigators should avoid sample contaminants such as aldehydes, oxidants, buffer salts, primary amines, glycerol, sucrose, nonionic detergents, and sodium dodecyl sulfate (SDS). When working at subnanomole levels of sample, the use of classical methods for changing the composition of the sample solvent (e.g., dialysis or gel filtration) and reduction of volume (e.g., lyophilization or ultrafiltration) are fraught with danger. At these levels, proteins/peptides are more susceptible to absorptive losses during handling. Ideally, samples should be in a volatile solvent or buffer and in a volume of 30–100 μl or, alternatively, electroblotted onto an immobilizing matrix. Sample volume and composition is less of a problem with the recently introduced adsorptive biphasic column instrument.

## 2 TECHNIQUES

RP-HPLC and SDS-PAGE (polyacrylamide gel electrophoresis) are currently the only two well-understood and tested high resolution techniques for isolating low microgram amounts of protein/peptides for microsequence analysis. While RP-HPLC is usually considered to be applicable at any stage in a purification protocol, electrophoresis is usually reserved for the last purification step.

### 2.1 Reversed-Phase High Performance Liquid Chromatography

RP-HPLC remains the method of choice for many separation strategies and is especially useful for hydrophilic proteins having a relative molecular mass $M_r$ of less than 40 kDa and most peptides. Separation of proteins/peptides by reversed-phase supports is based predominantly on hydrophobic interactions between amino acid side chains on the protein/peptide and functional groups on the column support. Typical supports are microparticulate (5–10 μm), porous silica-based materials derivatized with varying length alkylsilanes (e.g., octadecyl, C18, or octylsilane C8). For the separation of proteins, large-pore (300 A) "macroporous" packings are recommended, while for the fractionation of peptides, small-pore (60–120 A) "mesoporous" packings are preferred.

Typical chromatographic conditions that one could apply, as an initial approach, for the separation of proteins/peptides would be as follows: column, Brownlee RP-300 (300 A pore size, 10 μm particle size, dimethyloctyloctyl silica); linear 60-minute gradient from 0.1% aqueous trifluoroacetic acid to 60% acetonitrile. The several options for improving RP-HPLC separations include changing the mobile phase solvent, changing the pH, and—especially with peptide separations—switching to a different column packing.

For the purification of low microgram amounts of protein/peptide, consideration needs to be given to the internal diameter (i.d.) of the chromatographic column. There is an inverse relationship between column diameter, sample mass load, and sensitivity. In practice, the optimal column performance for 1.0, 2.1, and 4.6 mm i.d. columns is achieved with protein levels up to 2.5–5.0, 5.0–20, and 30–100 μg, respectively. The peak recovery volumes (< 100 μL) of microbore columns (< 2.1 mm i.d.) are ideally suited for loading samples onto gas phase sequencing instruments. Provided samples are loaded onto these columns at a low organic solvent concentration (below the critical secondary solvent condition required to elute the sample), proteins/peptides can be concentrated from large volumes, simultaneously buffer-exchanged, and recovered in a small volume (< 100 μL) of volatile solvents suitable for sequence analysis.

### 2.2 SDS-Polyacrylamide Gel Electrophoresis/Electroblotting

SDS-PAGE followed by electroblotting of protein/peptide from the gel onto an immobilizing matrix such as polyvinylidene difluoride (PVDF) provides an alternative approach to RP-HPLC for purifying proteins for sequence analysis. This approach is especially useful

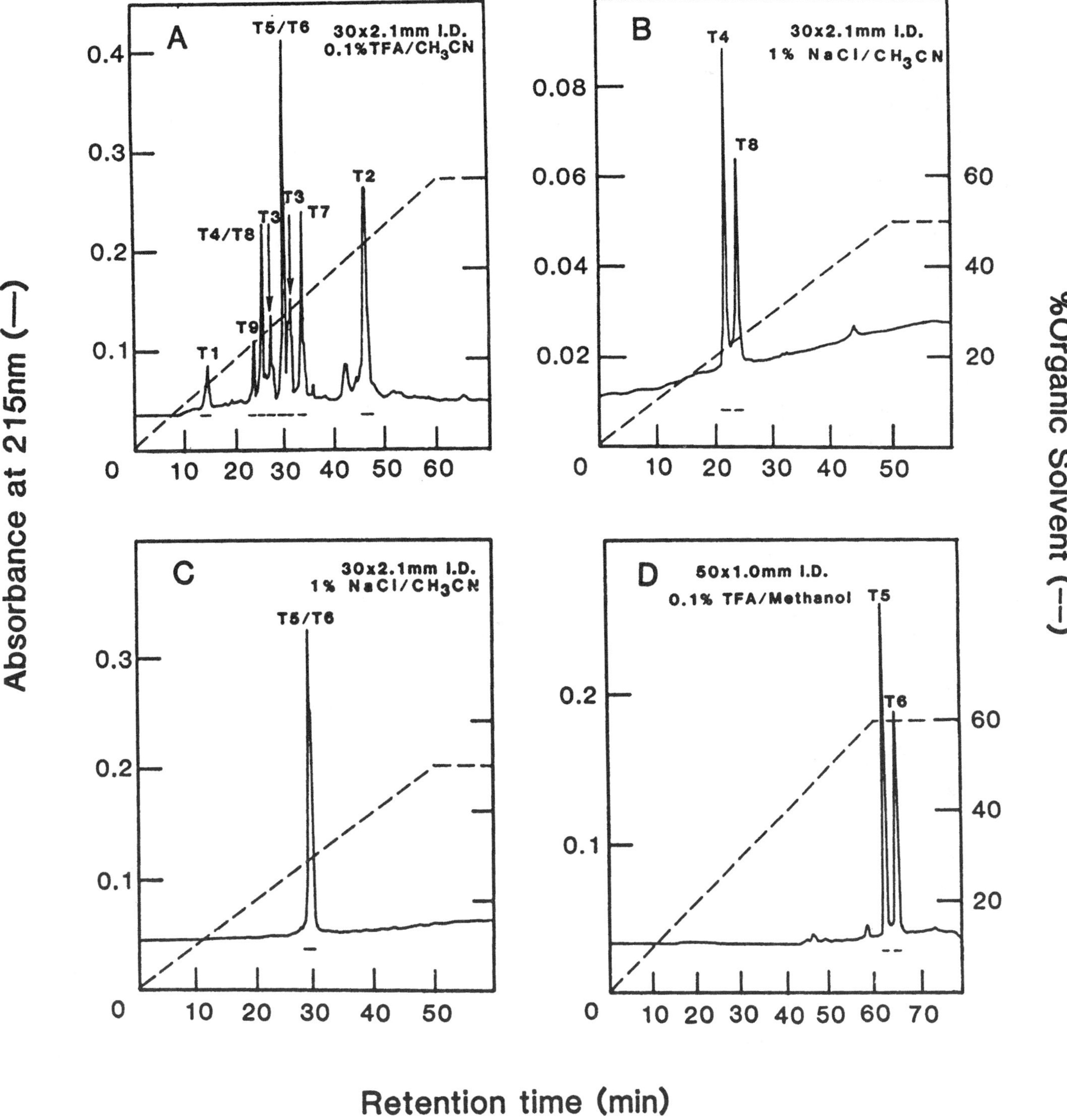


**Figure 1.** Microbore RP-HPLC peptide mapping of murine IL-6. (A) Separation of tryptic peptides of reduced and alkylated murine IL-6 on Brownlee RP-300 (30 mm × 2.1 mm i.d.); linear 60-minute gradient from 0.1% trifluoroacetic acid (TFA) to 0.1% TFA/60% acetonitrile; flow rate, 100 μL/min. (B) Rechromatography of peptides T4 and T8. Same conditions as in (A) but TFA replaced with 1% NaCl. (C) Purification of peptides T5 and T6, which were not resolved using conditions described in (D). Separation of T5/T6 on 5 μm ODS-Hypersil (50 mm × 1 mm i.d.); linear 60-minute gradient from 0.1% TFA to 0.1% TFA/60% methanol.

for proteins that are not conducive to RP-HPLC purification protocols (e.g., high $M_r$ hydrophobic proteins such as membrane proteins), as well as for laboratories not well versed in or equipped with RP-HPLC methodology. After the protein band has been visualized by staining (e.g., Coomassie blue), it can be excised and loaded directly onto the sample cartridge of the protein sequencer. Since this approach requires no specialized equipment, it can be carried out in almost any biochemical laboratory. There are, however, a number of precautions that need to be observed to minimize potential reaction of primary amino groups (i.e., the N-terminus of the sample) during electrophoresis. For example, high quality reagents must be used, gel polymerization times must be prolonged

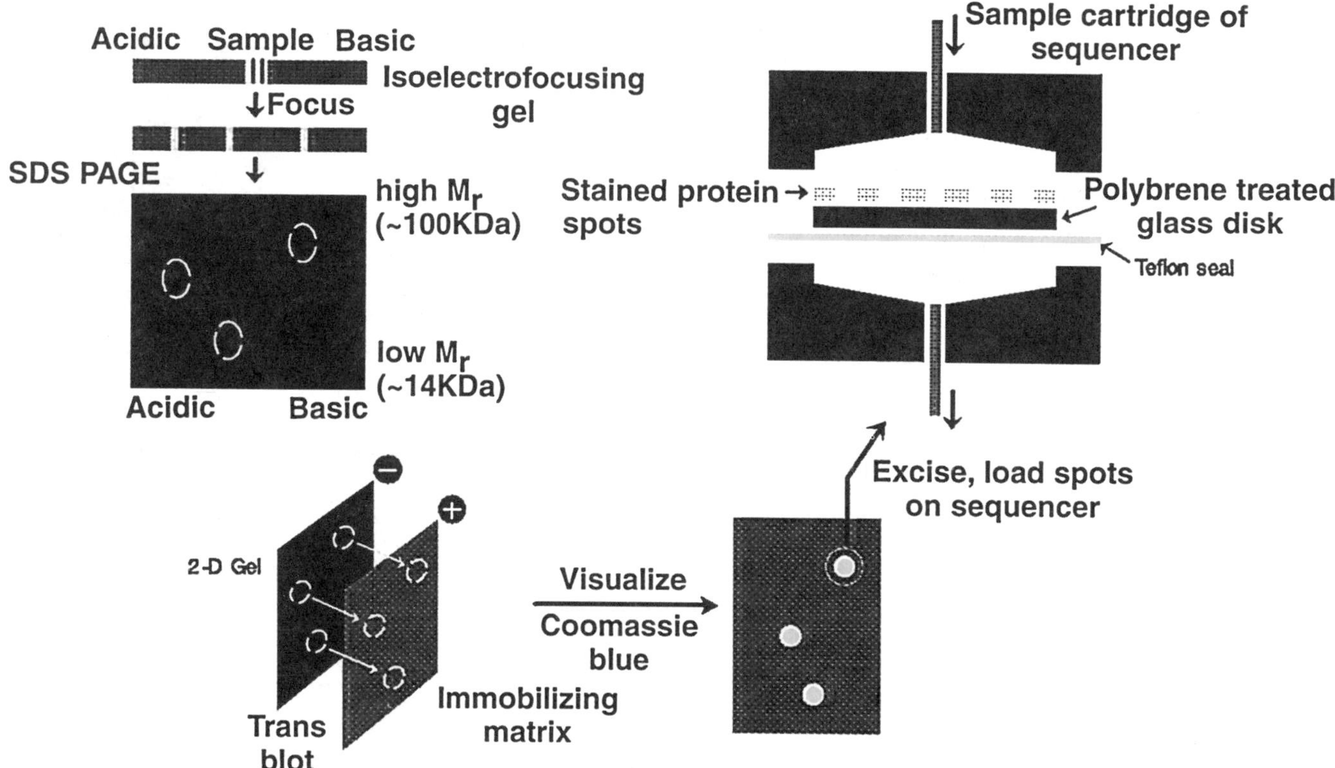


Figure 2. Schematic representation of steps involved in isolating a protein from a complex mixture by 2D gel electrophoresis for N-terminal sequence analysis.

(to minimize the concentration of acrylamide monomer, which can react with the α-amino group), and scavengers such as thioglycollic acid must be included in the running buffer to mop up potential blockers of the N-terminus.

## 3 APPLICATIONS

### 3.1 Complete Sequence Determination of Murine Interleukin-6

A prime example of the utility of microbore column RP-HPLC for protein structural analysis performed in our laboratory is illustrated for murine interleukin-6 (IL-6), a pleiotropic 22 kDa cytokine involved in the regulation and growth of various tissues. Using less than 2 nmol (40 μg) of starting material, the complete amino acid sequence of IL-6 was obtained in a time span comparable to that for recombinant DNA methodologies. After reduction and alkylation of the cysteine residues of IL-6 with iodoacetic acid and desalting by microbore RP-HPLC, the polypeptide was fragmented by various means, including chemically with cyanogen bromide and enzymatically with trypsin, chymotrypsin, and *Staphylococcus* V8 protease. Peptide mixtures were fractionated by microbore RP-HPLC using 2.1 mm i.d. columns. Usually, obtaining homogeneous peptides with complex peptide maps requires a number of chromatographic steps (dimensions), employing various column and solvent-mediated selective effects, as demonstrated in Figure 1. Owing to the different cleavage methods utilized, many of the peptide amino acid sequences obtained overlap, allowing the whole sequence to be pieced together.

### 3.2 Identification of Proteins Separated on 2D Gels

A great advantage of PAGE is the ability to rapidly fractionate complex mixtures of proteins. Indeed, using 2D gel electrophoresis it is possible to resolve up to 3000 discrete proteins on one gel: in the first dimension (1D) proteins are separated on the basis of charge differences using isoelectrofocusing, while in the second dimension (2D) separation is based on molecular weight differences using SDS-PAGE. The ability to identify separated proteins by microsequencing (after electroblotting onto an immobilizing matrix) has greatly increased the utility of this approach. A schematic representation of the steps involved in preparing electrophoretically separated proteins for N-terminal sequence analysis is given in Figure 2. Because some proteins are N-terminally blocked, there is often a need for internal sequence data (also particularly useful for molecular cloning strategies). Thus a variety of methods have been developed for obtaining internal sequence information from electrophoretically separated proteins. The most successful of these strategies involves digesting the protein in situ in the acrylamide gel. Alternatively, electroblotted proteins can be proteolytically digested in situ on the immobilizing membrane (PVDF or nitrocellulose).

In both cases, the generated peptides are extracted and then purified by RP-HPLC.

## 4 FUTURE PERSPECTIVES

Much effort in recent years has been devoted to improving the sensitivity of protein sequence analysis by using new, highly sensitive methods for detecting PTH amino acids. In this regard, both RP-HPLC utilizing microcolumns (< 0.3 mm i.d.) and capillary electrophoresis are highly promising. Using capillary electrophoresis coupled with laser-induced fluorescence detection, subattomole ($10^{-18}$ mole) levels of fluorescently labeled amino acids can be detected. The small loading volumes of these techniques, however, preclude the ready interfacing of these methods to conventional protein-sequencing instruments. Their application will require the development of a new generation of instruments designed for precision submicroliter flow reagent/solvent flows.

Mass spectrometry presents an exciting alternative for obtaining amino acid sequence information. Using electrospray mass spectrometry, sequence information can be obtained on low picomole amounts of sample. The mass spectrometry and Edman degradation procedures are highly complementary, since the two approaches are based on entirely different chemical and physical principles. Mass spectrometry, unlike the Edman degradation procedure, is particularly useful for identifying posttranslational modifications such as phosphorylation and glycosylation sites on proteins.

*See also* HPLC of Biological Macromolecules; Protein Purification; Protein Sequencing Techniques.

### *Bibliography*


Findley, J.B.C., and Geisow, M. J., Eds. (1989) *Protein Sequencing: A Practical Approach.* IRL Press, Oxford.

Kellner, R., Lottspeich, F., and Meyer, H. E., Eds. (1994) *Microcharacterization of Proteins.* VCH, New York.

Matsudaira, P. T., Ed. (1989) *A Practical Guide to Protein and Peptide Purification for Microsequencing.* Academic Press, San Diego, CA.

Shively, J. E., Paxton, R. J., and Lee, T. D. (1989) *Trends Biochem. Sci.* 14:246–255.

Simpson, R. J., Moritz, R. L., Begg G. S., Rubira, M. R., and Nice, E. C. (1989) *Anal. Biochem.* 177:221–236.


# Protein Sequencing Techniques

*Jeffrey N. Keen and John B. C. Findlay*

### *Key Words*

**Edman Degradation** The series of chemical reactions used in protein sequencing, comprising coupling, washing, and cleavage steps.

**N-Terminal Modification (Blockage)** The blocking of the N-terminal amino group of a protein, preventing Edman degradation.

Protein sequencing provides a means of obtaining information about the primary structure of polypeptides. The general methodology has changed relatively little since Pehr Edman introduced his degradative chemistry in the 1950s, despite many attempts to develop alternative chemical methods (e.g., fluorescent isothiocyanates).

The use of automated equipment to perform multiple cycles of the Edman chemistry, thereby eliminating much labor-intensive sample manipulation, has greatly improved the efficiency of protein sequencing. Optimization has allowed the determination of extended sequences of very low abundance proteins (only a few picomoles available).

It was suggested that DNA sequencing would remove the need for routine protein sequencing, yet the latter remains central to modern molecular research. Rather than sequencing entire proteins, however, today's requirements are generally more to supplement and complement other approaches. For instance, protein sequence information is a prerequisite for DNA cloning, providing the information for making oligonucleotide probes and polymerase chain reaction (PCR) primers. It allows the synthesis of peptides for antibody production, provides identification of proteins of interest, and helps in the characterization of recombinant products. A major use of protein sequencing is in the study of posttranslational modifications.

Only mass spectrometry has offered a feasible alternative to routine Edman degradation, allowing very accurate mass determination to be used in the interpretation of sequence data, particularly with respect to posttranslational modifications.

## 1 PROTEIN SEQUENCE ANALYSIS

All modern automated protein sequencers use the Edman degradative chemistry (Figure 1), but the protocols used vary. Table 1 compares the two technologies of adsorptive and covalent sequencing. In gas phase or liquid pulse (adsorptive) equipment (e.g., Applied Biosystems; Beckman/Porton), the sample is noncovalently adsorbed onto a membrane. This is either a glass fiber disk, usually treated with a polycationic carrier (Polybrene) to aid in the retention of the protein, or a piece of polyvinyl difluoride (PVDF) membrane, to which the sample is generally applied, most often by electrotransfer following sodium dodecyl sulfate–polyacrylamide gel electrophoresis (SDS-PAGE). To avoid washout of the noncovalently bound sample, these adsorptive machines use small pulses of liquid reagents [phenylisothiocyanate (PITC) and trifluoroacetic acid (TFA)] and/or gaseous delivery and extraction. Solvent washes are kept to a minimum to reduce washout but still clear away excess reagents and by-products.

In solid phase machines (e.g., MilliGen) the sample is covalently bound to a supporting matrix, usually a chemically modified PVDF membrane. For this reason the delivery of reagents and wash solvents can be increased to allow efficient reaction and cleaning with no sample washout. The main drawback of this technique is that the sample must be free of interfering contaminants prior to coupling to the membrane. Otherwise attachment may be severely compromised.

All current machines use an on-line system of high performance liquid chromatography (HPLC) for the identification of the phenylthiohydantoin (PTH) amino acids recovered at each cycle. The derivative is injected onto a reverse-phase C-18 resin equilibrated with acetate buffer and eluted with an increasing gradient of acetonitrile. The PTH-amino acids are generally detected at 269 nm, with sensitivity around 1 pmol. This result may be enhanced by the use of a diode array detector (e.g., ratio of 269/293 nm; Beckman/Porton), or prior chromatogram subtraction (MilliGen; Applied Biosystems).

## 2 SAMPLE PREPARATION

For efficient sequence analysis, protein samples must be extremely pure with respect to nonprotein contaminants. Compounds that

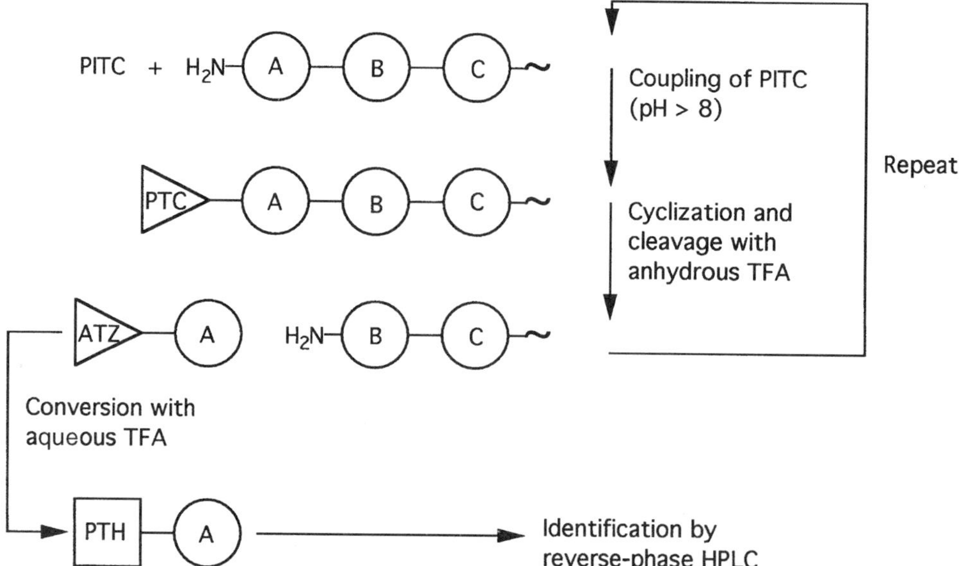


**Figure 1.** Schematic diagram of the Edman degradation procedure. The protein sample is made alkaline with a volatile amine, then exposed to the Edman reagent phenylisothiocyanate (PITC), which reacts with the N-terminal amino group (and some side chains). Excess reagents are washed from the sample with organic solvents, and the modified N-terminal residue is cyclized and cleaved from the peptide chain with anhydrous trifluoroacetic acid (TFA). The N-terminal residue [an anilinothiazolinone (ATZ) amino acid] is washed into a second reaction chamber for conversion to a stable phenylthiohydantoin (PTH) derivative which later can be identified by reverse-phase HPLC. The shortened peptide remains in the original reaction chamber, where it can undergo further rounds of Edman degradation, eventually generating a sequence of residues. (Modified from L. Stryer (1988), *Biochemistry*, W. H. Freeman and Co., San Francisco.)

interfere with the attachment of protein to phenylene diisothiocyanate (DITC) membranes (e.g., primary amines and thiols), or arylamine membranes (e.g., acids and some detergents) for solid phase sequencing, or with the Edman chemistry itself (e.g., primary amines such as ammonium bicarbonate or glycine), must be avoided. Other seemingly unreactive molecules may cause machine problems: for example, SDS may froth and block lines.

Two main approaches are used for sample purification, namely HPLC and SDS-PAGE.

**Table 1** Comparison of Adsorptive and Solid Phase Sequencing

| Adsorptive | Solid Phase |
|---|---|
| Sample loosely held | Sample covalently bound |
| Glass fiber ± Polybrene | DITC-membrane→Lys; arylamine membrane→Asp/Glu |
| PVDF | PVDF/SequeNet |
| Tolerant of small amounts of contaminants | Coupling intolerant of amine/SH (SequeNet; DITC), acids (AA); low yield of attached residues |
| Sample washout problem | No washout; stringent washing allows long runs and clean samples |
| Loss of charged residues in matrix | Charged residues (e.g., phosphorylated amino acids) recovered |
| Long cycle times | Reduced cycle times, due to high flow rates |
| SDS a problem in line blockage and sample washout | SDS solubilization of samples possible |

## 2.1 HPLC

A reverse phase HPLC purification step offers a simple approach to the production of a sequence grade sample. The type of resin used and the column size (analytical, capillary, etc.) will depend on the amount and properties of the protein or peptide to be purified. Small peptides should be purified on C-18 resins, proteins on C-4 resins; large peptides (> 30 residues) and small proteins are more suited to C-8 resins.

The most important consideration for sample cleanup is the choice of mobile phase. A gradient elution system based on water and acetonitrile and incorporating a volatile ion-pairing agent (e.g., TFA) is usually successful. For very hydrophobic samples, the organic phase can be supplemented with 2-propanol to reduce the interaction between sample and column and to allow elution.

## 2.2 SDS-PAGE

Gel electrophoresis provides a relatively simple approach to the purification of specific proteins from rather impure samples. As long as the protein of interest can be identified among the background of bands, electroelution/dialysis or electroblotting can be used to produce a pure sample for sequence analysis.

Electroblotting is the method of choice because it involves direct transfer of protein from the gel to a support matrix, which can be placed directly into the sequencer. Various types of membrane are available specifically for this purpose, based on the original PVDF membrane (Millipore), but comprising double or triple layers with reduced pore size to eliminate transfer through the membrane. Products include ProBlott (Applied Biosystems), Immobilon-PSQ (Millipore), and Fluorotrans (Pall). Since all these membranes retain protein by hydrophobic interaction, they must be manipulated in the sequencer to avoid washout of sample. Although ideally suited to the mild washing conditions of the adsorptive sequencers, they also can be used in solid phase machines following cross-linking of the protein to an overlying polyamino compound (Seque-Net entrapment procedure; MilliGen).

## 3 SEQUENCING PROBLEMS

The commonest problem encountered in protein sequencing is N-terminal modification, which prevents attachment of PITC. In prokaryotes, the formyl group on the initiating methionine residue often remains, but it can be removed with HCl in methanol, thereby allowing sequencing to commence. The commonest modification in eukaryotes is acylation, particularly acetylation, occurring in an estimated 80–90% of proteins. Routine removal of these groups is not possible. Partial success has been reported for acetyl-Ser/Thr, by heating for extended periods in TFA. A hydrolase that has been isolated can remove the acetylamino acid from short peptides. Generally, however, it is necessary to resort to chemical or enzymic proteolysis and peptide purification for internal sequence determination. Another, infrequent, blocking mechanism is the cyclization of N-terminal glutamine to pyroglutamic acid. This derivative can be enzymically removed to generate a free N-terminus for interaction with PITC.

Often it is necessary to retrieve electroblotted proteins for further analysis. Proteolysis on membranes, using cyanogen bromide, BNPS-skatole, or trypsin, generally produces only limited cleavage. The products may elute from the membrane for HPLC purification or remain bound to the membrane for direct sequencing. Enzymes are not particularly suited for such in situ digestions because of steric problems. Therefore it is often preferable to elute the sample for digestion in solution. Generally, quite harsh conditions are required (e.g., mixtures of TFA, detergents, and 2-propanol at high temperatures for extended periods).

The conditions used for proteolysis can be varied greatly (ratio of chemical or enzyme to protein, time, temperature, presence of denaturants, etc.), to produce partial digests for SDS-PAGE purification of large peptides for long solid phase runs, or complete digestion to short peptides for HPLC purification and adsorptive sequencing or mass spectrometry.

## 4 MASS SPECTROMETRY

Extremely accurate mass determination can help considerably in protein sequencing. It can confirm the accuracy of peptide synthesis or completion of peptide sequencing. It can indicate the presence of posttranslational modifications and provide information about blocked peptides and novel covalent cross-links within proteins. It has a role in peptide fingerprinting, rapidly providing data that can be used to identify a protein, with the help of a fragment databank.

The use of mass spectrometry has increased rapidly recently. The types of sample analyzed, amounts required, accuracy, and other factors have enabled its routine use in protein analysis.

Ionization of the protein sample can be achieved in various ways. At first, fast atom bombardment was employed, utilizing a beam of argon or xenon atoms to generate $(M + H)^+$ ions from the sample. These ions fragment spontaneously, or in collision with a gas, to produce smaller ions, thereby providing mass information that may be interpreted in terms of structure. In recent years other methods of sample ionization have been introduced. Plasma or laser desorption of the sample from a solid matrix, or electrospray ionization of a liquid sample, have now become the most frequently used methods.

The method of mass analysis of the ions also varies. Time-of-flight instruments relate the time taken for an ion to reach a detector to its size. The double-focusing magnetic deflection and quadrupole instruments are more complex, and particular fragment ions of interest can be selected according to size and charge by magnetic fields and thence transferred to a detector.

For actual sequencing, another level of complexity must be introduced by means of a second mass analyzer (tandem mass spectrometry), so that after initial ionization and ion selection, further fragmentation can be induced by collision with gas molecules to generate subfragments for mass analysis. The data produced can be converted into overlapping fragment maps to generate sequence information.

Recently, considerable progress has been made in the use of laser desorption time-of-flight mass spectrometry in the analysis of peptides. Refinement of procedures for peptide fingerprinting (including improvement of data banks such as MOWSE), is allowing rapid identification of proteins separated by two-dimensional electrophoresis, complementing the various genome projects underway in the study of patterns of gene expression.

The increased resolution and sensitivity of current instruments and improvements in sample handling have enabled sub-Dalton accuracy of peptide masses, allowing discrimination of all amino acids (except the leucine/isoleucine pair), and thus permitting the use of this technique in direct amino acid sequencing, either by the use of volatile Edman-like chemistry, or by the control of post-source decay of ions, or by the involvement of collision induced dissociation.

## 5 COMPUTER ANALYSIS

When sequence data are available, various computer methods can be used to provide additional information. Protein sequence databanks can be searched to provide an identity to the protein (if not already known), as well as details of homologous proteins. Motif databanks can provide structural and functional information about the protein. Secondary structure can be predicted, providing information on how the protein may fold. Computer modeling of tertiary structure may also be possible, based on related structures. Although the degree of accuracy is still a limitation, the information thus predicted from the sequence can be used as a working model for subsequent studies.

*See also* Proteins and Peptides, Isolation for Sequence Analysis of; Sequence Analysis.

### *Bibliography*

Findlay, J.B.C., and Geisow, M. J., Eds. (1989) *Protein Sequencing: A Practical Approach.* IRL Press, Oxford.

Matsudaira, P. T., Ed. (1989) *A Practical Guide to Protein and Peptide Purification for Microsequencing.* Academic Press, San Diego, CA.

# Protein Targeting

*Brian Austen and David Stephens*

## *Key Words*

**Cotranslational Processing** Occurs when the incomplete, or nascent, protein is still elongating from the ribosome.

**Endocytosis** The taking up of a molecule bound to a receptor protein at the cell surface, after which it is ingested into the cell and delivered to an endosome.

**Endomembrane System** The rough endoplasmic reticulum, the Golgi apparatus, endosomes, and lysosomes, which although physically discontinuous and functionally distinct, operate as a topographically related unit.

**Endoplasmic Reticulum** A subcellular compartment consisting of long, tubular membrane structures. Proteins targeted to the endomembrane system for secretion start their life here on ribosomes bound to the cytoplasmic side of the membranes; from there they are vectorially discharged from the cytoplasm, across into the lumen. The lumen is the major site of protein folding and disulfide formation in the cell.

**Exocytosis** The fusion of the membranes of granules with the plasma membrane, releasing contents into the extracellular space.

**Lysosome** A subcellular single-membrane-bound subcompartment in the endomembrane system; it is responsible for degradation of proteins and other macromolecules, some of which are endocytosed from the exterior of the cell. The pH in the lysosome is maintained at 5.5 by proton-pumping ATPases.

**Molecular Chaperone** A protein that transiently associates with a newly synthesized protein before it has fully folded up, thereby preventing it from aggregating or interacting with other proteins.

**Posttranslational Processing** Occur to the protein shortly after its biosynthesis has been completed.

**Protein Translocation** Occurs when a protein, or part of a protein, moves across a membrane from the cytoplasm to a subcellular compartment, from the space outside the cell into the cell, or from one subcompartment to another.

**Signal Recognition Particle** A ribonucleoprotein consisting of 7*S* RNA and protein subunits of 72, 68, 54, 19, 14, and 9 kDa molecular weight. The 54 kDa subunit contains a binding site for signal sequences and a GTP-binding site. The particle is required for targeting a subset of secretory and membrane proteins to the endoplasmic reticulum.

**Signal Sequence** An N-terminal sequence consisting of eight or more consecutive uncharged, hydrophobic residues that target newly synthesized proteins to the endoplasmic reticulum membrane, to the bacterial periplasm, or to the mitochondrial intermembrane space.

**Subcellular Organelle** A subdivision of the cell which is divided off and bounded by a membranes; eukaryotic organelles include the nucleus, mitochondrion, chloroplast (in plants only), rough endoplasmic reticulum, Golgi apparatus, peroxisome, endosome, and lysosome. Mitochondria and chloroplasts are subdivided by a further membrane(s) into subcompartments.

**Targeting Patch** A surface of a protein, formed by discontinuous parts of its structure brought together by the three-dimensional folding, which determines the protein's targeting.

**Targeting Sequence** A small part of a protein's structure, usually a short part of the primary sequence, which contains all the necessary information to target that protein to a specified location in a particular subcellular compartment.

Protein targeting is the process by which proteins reach their destinations in or out of cells. Newly synthesized proteins are targeted to subcellular organelles (e.g., nucleus, mitochondria, peroxisomes, rough endoplasmic reticulum). Proteins are targeted into the regulated secretory pathway, diverted to lysosomes, or retained in specific subcellular locations against the bulk flow of proteins to the cell surface through the constitutive secretory pathway. Some circulating proteins are taken up from the extracellular medium into cells by endocytosis, and targeted to endosomes or lysosomes. Bacteria also target some proteins for export. Molecular biology has been used to elucidate targeting pathways, identify targeting sequences, and discover targeting and translocation mechanisms.

## 1 TARGETING SIGNALS

Targeting sequences are listed in Table 1. The N-terminal signal sequences that translocate proteins into the lumen of the endoplasmic reticulum (ER), the bacterial periplasm, or the mitochondrial intermembrane space are similar in that they contain a core of uncharged hydrophobic residues, positively charged residues at the N-terminus, and small aliphatic residues adjacent on the N-terminal side of the site at which signal peptidase cleaves. The N-terminal sequences that target proteins from cytoplasm into mitochondria or chloroplasts are amphipathic in structure, with groups of basic residues interspersed with groups of hydrophobic residues. Many targeting sequences are cleaved off after the protein has arrived at its destination, but nuclear targeting sequences, which are small groups of basic residues centrally placed in the proteins, and peroxisomal targeting sequences, which are C-terminal, are not cleaved. Sequences in the cytoplasmic tails or transmembrane regions which act to retain proteins in the membranes of subcompartments, are not cleaved. Fully folded proteins are targeted by a signal consisting of a patch of different parts of the primary sequence brought together by the tertiary folding of the protein. The targeting signal for soluble lysosomal enzymes is composed of mannose 6-phosphate on covalently attached oligosaccharides.

## 2 TARGETING PATHWAYS

The targeting pathways that newly synthesized proteins may follow in mammalian cells are shown by arrows in Figure 1. They may be divided into two groups. The first group is followed by proteins that are initially synthesized with signal sequences, which bring about targeting to the rough ER, and translocation of part or all

**Table 1** Examples of Targeting Sequences[a]

| | |
|---|---|
| Translocation into ER lumen: | MKWVTFLLLLFISGSAFS^RGVF |
| ER stop-transfer sequence: | −KSSIASFFFIIGLIIGLFLVLRVGIH− |
| ER retention sequence (soluble protein) | −KDEL |
| ER retention sequence (membrane protein) | −RRSFIDEKKMP |
| Golgi retention sequence (membrane protein) | −KKWVLIVAFLLFVLIFLSFKKK− |
| Translocation into bacterial periplasm | MKANAKTIIAGMIALAISHTAMA^DDIK |
| Nuclear targeting | −PKKARED− |
| Chloroplast targeting | MAPAVMASSATTVAPFQGLKSTAGLPVSRRSGSLGSVSNGGRIRC^M− |
| Thylakoid targeting | −IKASLKDVGVVVAATAAAGILAGNAMA− |
| Mitochondrial targeting | MLSALARPVGAALRRSFSTSAQNN^AKVAVL− |
| Mitochondrial inter-membrane space targeting | −KLTQKLVTAGVAAAGITASTLLYAD− |
| Peroxisomal targeting | −GSKL |
| Endocytosis | −NPVY− |
| Lysosomal targeting (membrane) | −GY− |
| Yeast vacuole | −QRPL− |
| Lysosomal degradation | −KFERQ− |

[a]Boldface type indicates hydrophobic residues. Other symbols: +, basic residue K, R, H; −, acidic residue D, E; °, hydroxyl side chain S, T; ^, residue

of the protein across its membrane into the lumen. Except for a few small proteins, translocation is cotranslational.

Proteins that are targeted initially to the rough ER (pathway 1) include proteins that become resident in the ER, those that move to the cis-Golgi (pathway 2), and then to the trans-Golgi network, where they may be diverted to endosomes (pathway 3) and lysosomes (pathway 4), or into secretory granules. Proteins in granules move to the cell surface and are secreted into the extracellular medium either by a regulated pathway (5) in which exocytosis is controlled by external stimuli, or by the constitutive pathway (6).

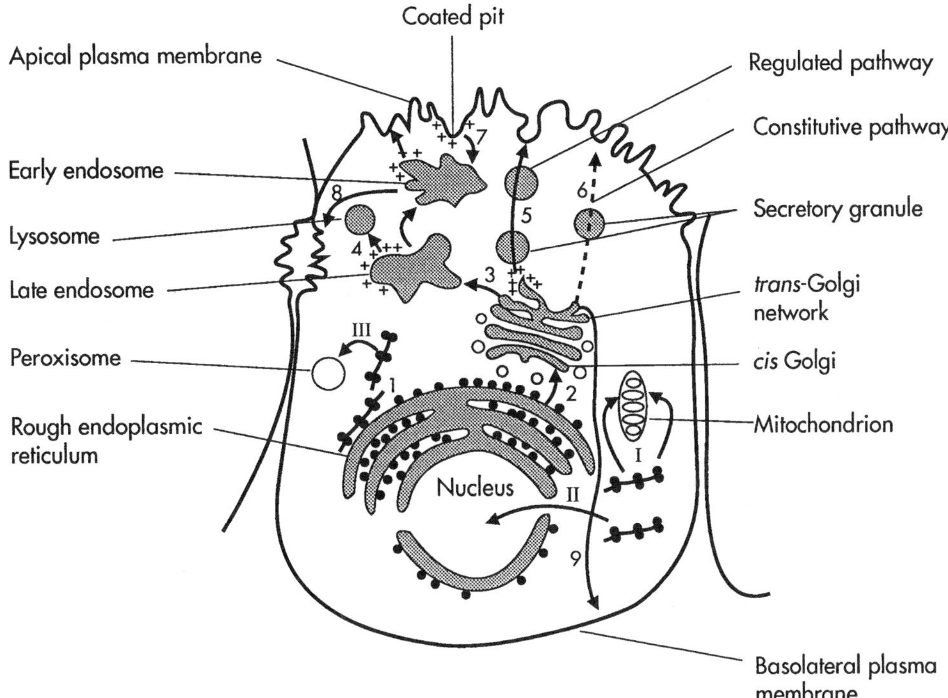


**Figure 1.** Protein targeting pathways in the mammalian cell. The pathways followed by proteins are denoted by arrows. Arabic numerals, which are decoded in the text, show targeting of proteins that are initially biosynthesized on ribosomes attached to the rough ER. The endomembrane system is stippled, and membranes that are coated with clathrin during the formation of transport vesicles are marked with crosses. Roman numerals depict targeting of newly synthesized proteins from the cytoplasm into mitochondria, the nucleus, or the peroxisomes.

For example, parathyroid hormone passes through the regulated pathway, whereas immunoglobulins are secreted from B cells constitutively.

New membrane proteins of these subcellular structures, known as the endomembrane system (shaded in Figure 1), are first targeted to the ER membrane by signal sequences, then integrated into the phospholipid bilayer. These membrane proteins are either moved to the cell surface through the constitutive secretory pathway, retained in the ER or the membranes of the Golgi cisternae by retention sequences in their cytoplasmic tails (Table 1), or transported to endosomes or lysosomes. Protein transport occurs in vesicles that bud off from one compartment and fuse with the next, so that the topography of the polypeptide with respect to the bilayer, which is established at the ER membrane, is retained at the final destination.

Some proteins are endocytosed from the cell surface by recognition of specific cytoplasmic tail sequences (pathway 7), and then targeted to endosomes or lysosomes. In polarized cells, proteins may be targeted to specific parts of the cell surface (e.g., pathway 8 or 9).

The second group of pathways start in the cytoplasm, where proteins are targeted posttranslationally into the subcompartments or membranes of mitochondria (pathway I), into the nucleus (pathway II), or into peroxisomes (pathway III: glyoxysomes in plants).

## 3 EXPERIMENTAL APPROACHES

Targeting sequences are identified by fusing the encoding DNA to a gene coding for a cytoplasmic protein, such as dihydrofolate reductase or globin. The engineered DNA is then inserted downstream of a strong promoter sequence in an expression plasmid. After expression, the location of the fusion protein is ascertained to see if it has been imported into the appropriate subcellular organelle.

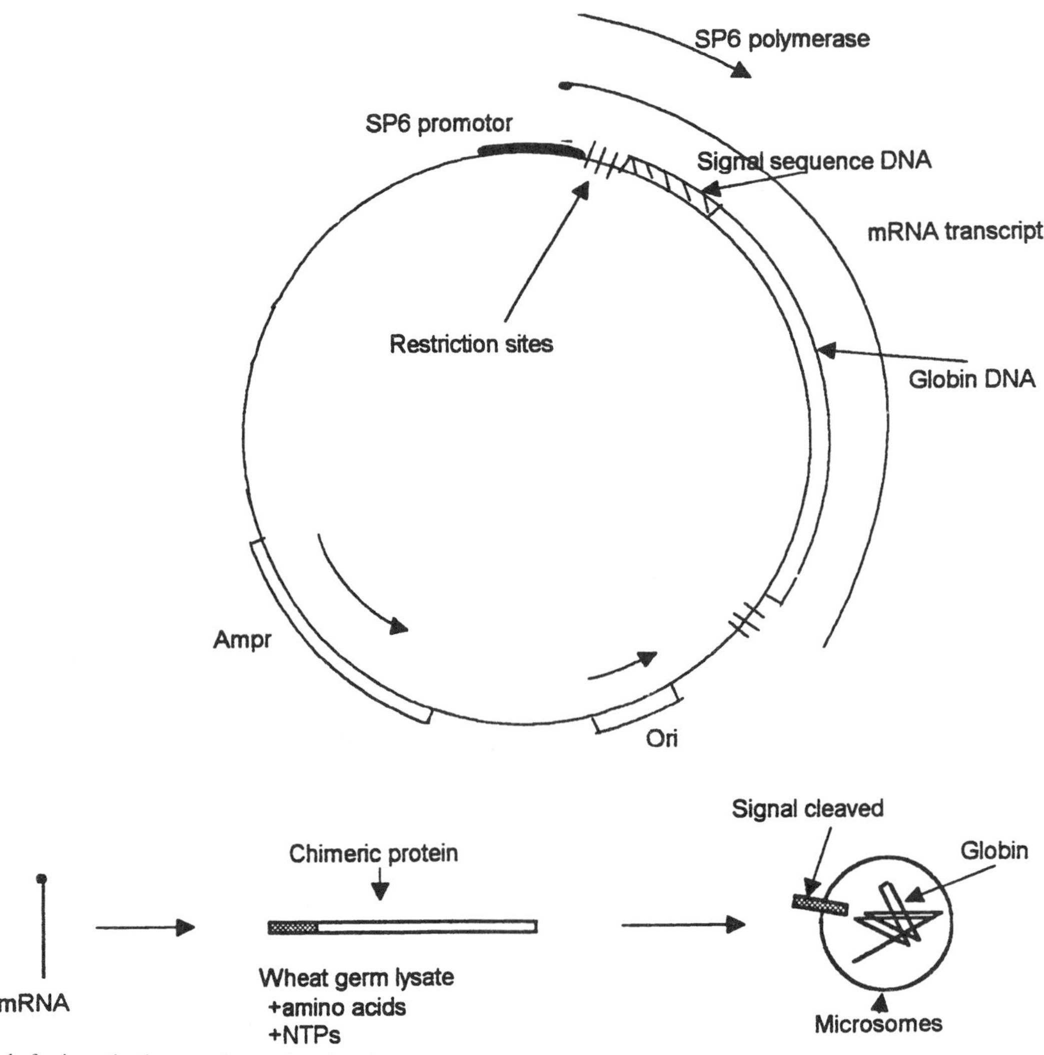


**Figure 2.** Methods for investigating protein translocation. To test the efficacy of a presumed signal sequence, its coding DNA ligated to DNA coding for a cytoplasmic protein, globin, was placed downstream of a strong promoter. The plasmid DNA is amplified in *E. coli* cultured with amphicillin, purified, and transcribed in vitro using SP6 polymerase. The generated mRNA is translated in a wheat germ lysate using radioactive amino acids in the presence of pancreatic microsomes. The translocation of the signal globin chimeric protein is shown by the cleavage of the signal sequence by signal peptidase in the lumen of the microsomes, and by the globin's resistance to proteases added at the finish of translation.

The plasmid may be transcribed in vitro into mRNA, which is then translated using radioactive amino acids and energy sources in cell lysates in the presence of an added subcellular compartment, previously isolated from tissue homogenates by sedimentation (Figure 2). Import of the radioactive translated protein is measured by the shift in mobility determined by gel electrophoresis (SDS-PAGE) caused by modification inside the subcellular compartment or by resistance of the imported protein to added proteases. Reconstituted proteoliposomes containing function proteins can be added instead of subcellular compartments. A number of in vitro assays have been devised to study, for example, vesicle movement between successive compartments of the Golgi, or the packaging of proteins into immature secretory granules.

Alternatively, coding DNA is transfected or infected into intact cells, and the expressed proteins localized by observing whether a compartment-specific modification to the protein has occurred, or by using a microscope to see if the compartment stains with antibodies to the protein. The precise sequence requirement for targeting is explored further by localizing an expressed protein with a targeting sequence that has been altered, either by site-specific mutagenesis or by exposing the whole cell to mutagenic chemicals, then selecting the clones that display a changed targeting phenotype.

Yeast genetics has proved to be a powerful way of identifying components involved in the targeting process. Mutants are isolated and are defective in moving a protein from one compartment to another. Since the mutants are often temperature sensitive, the yeast is grown at a permissive temperature and the defect examined at the nonpermissive temperature. The altered gene is then identified and sequenced. Its role in vivo can be studied by observing the phenotype produced by deletion of the gene, or overexpression of the gene product. Sometimes two mutations exhibit synthetic lethality, exaggerating the defect shown by each mutant individually, and thus demonstrating that the two encoded proteins form a complex and act together. Many of the yeast genes identified have mammalian counterparts.

Proteins functional in protein targeting in bacteria have been identified. The *sec* genes have been identified in mutants which correct for disabling mutations in signal sequences of exported proteins. The *sec* genes have been overexpressed, and the functions of the isolated proteins examined after reconstitution into proteoliposomes.

## 4 TRANSLOCATION INTO THE ER

Proteins destined for the lumen of the endoplasmic reticulum are initially synthesized with N-terminal signal sequences, transient extensions of 15–30 hydrophobic residues. After emerging from the ribosome, the signal sequence attracts binding of a ribonucleoprotein, the signal recognition particle (SRP), which freezes elongation until contact is made with the SRP receptor in the membrane. Elongation then resumes and the nascent chain passes through a translocation complex consisting of sec6lp, a membrane protein with 10 transmembrane domains, and other proteins. The signal sequence is cleaved off by signal peptidase in the ER lumen, and core oligosaccharides are covalently attached during translocation. Disulfide bond formation is catalyzed in the ER lumen by protein disulfide isomerase (PDI), and oligomerization is aided by the chaperone, heavy chain binding protein (BiP).

## 5 RETENTION SEQUENCES

Resident proteins such as PDI are retained in the ER by a C-terminal sequence: KDEL. These proteins are retrieved from a salvage compartment, intermediate between the ER and the Golgi apparatus by a KDEL receptor (*erd2*), which recycles the proteins back to the ER. ER membrane proteins are retained by double lysine residues in the cytoplasmic tails. Membrane proteins are retained in the Golgi by transmembrane sequences, which respond to the cholesterol gradients in intracellular membranes.

## 6 TRANSPORT IN VESICLES

Transport of proteins through the endomembrane system is in vesicles that bud off from a donor compartment and fuse with an acceptor compartment. Secretory proteins, for example, pass from the ER to the cis-Golgi, then through the intermediate compartment to the trans-Golgi and the trans-Golgi network (TGN). In the TGN, they are packaged into granules, from which they are released at the cell surface by exocytosis. Further processing of oligosaccharides occurs in the Golgi, and prohormone cleavage in the TGN.

Vesicles are formed at regions of membranes that become coated by a cage of clathrinlike components, which aid the budding process. A small GTP-binding protein, ARF, is also required. The coats are removed from the vesicles by an ATP-hydrolyzing process, and fusion is then brought about by receptor-mediated processes regulated by proteins known as NSF, SNAPs, and SNAREs.

## 7 LYSOSOMAL TARGETING

At the TGN, some proteins are diverted to endosomes, and then to lysosomes, by interaction with a receptor of mannose 6-phosphate sugars on the protein's oligosaccharide moieties. The complex is transported in vesicles to endosomes, where the low pH causes dissociation of the complex. The receptor recycles back to the TGN, while the lysosomal enzymes move to the lysosome. Lysosomal membrane proteins, in contrast, are targeted by tyrosine-containing sequences in their cytoplasmic tails.

## 8 ENDOCYTOSIS

Proteins are brought into cells by receptor-mediated endocytosis. Specific cytoplasmic tail sequences in receptors such as the low density lipoprotein receptor (LDL) are recognized by adaptins, which bind clathrin in coated pits.

## 9 MITOCHONDRIAL IMPORT

Precursors of matrix proteins synthesized with amphipathic N-terminal transit sequences are complexed in an unfolded state by cytoplasmic chaperone Hsp70 and delivered to outer membrane receptors, which migrate to a translocation complex where the outer and inner membrane are apposed. In the matrix, folding involves mitochondrial Hsp70 and Hsc60, and the transit sequence is cleaved off by a cation-dependent protease. This cleavage exposes a second signal type of targeting sequence in intermembrane proteins, which retracks those proteins back into the intermembrane space.

## 10 NUCLEAR IMPORT

Nucleoplasmin and other large nuclear proteins are imported from the cytoplasm through a pore complex by targeting sequences consisting of one or two short sequences rich in basic residues.

## 11 PEROXISOMAL IMPORT

Peroxisomal enzymes involved in oxidation are imported across the single peroxisomal membrane through dimeric membrane ATPases. AC-terminal targeting sequence Ser-Lys-Leu has been identified in luciferase.

## 12 BACTERIAL TARGETING

About 25% of proteins synthesized in the bacterial cytoplasm either are excreted or are targeted to the inner membrane, periplasm, or outer membrane. Many of these proteins are synthesized with N-terminal signal sequences similar in structure to eukaryotic signals. They are cleaved off in the periplasm by leader peptidase I, a 36 kDa protease similar in function but not in structure to its eukaryotic counterpart.

Many proteins are translocated by mechanisms that involve components mutated in bacterial *sec* mutants. The precursor of an outer membrane protein, pro-OmpA, is stabilized in an unfolded conformation first by binding to the product of *secB*, suggested to be a bacterial counterpart of SRP. The complex then binds to a peripheral membrane protein ATPase secA. Translocation across the membrane is dependent on two inner membrane proteins: secY, which has structural homology to yeast sec61, and secE.

*See also* Cytoskeleton-Plasma Membrane Interactions; Membrane Fusion, Molecular Mechanism of; Mitochondrial DNA; Plasmids; Transport Proteins.

### *Bibliography*


Alberts, B., Bray, D., Lewis, J., Raff, M., Roberts, K., and Watson, J. D. (1989) *Molecular Biology of the Cell,* 2nd ed., pp. 334–335. Garland, New York.

Arvan, P., and Castle, D. (1992) Protein sorting and secretion granule formation in regulated secretory cells. *Trends Cell Biol.* 2:327–331.

Austen, B. M., and Westwood, Q.M.R. (1991) *Protein Targeting and Secretion.* In *Focus Series,* D. Rickwood, Ed. IRL Press, Oxford and Tokyo.

Duden, R., Allan, V., and Kreis, T. (1991) Involvement of β-COP in membrane traffic through the Golgi complex. *Trends Cell Biol.* 1:14–19.

Gething, M.-J., and Sambrook, J. (1992) Protein folding in the cell. *Nature,* 355:33–36.

High, G., and Stirling, C. T. (1993) Protein translocation across membranes: Common themes in divergent organisms. *Trends Cell Biol.* 3:335–339.

Magee, A. I., and Wileman, T. (19) In *Protein Targeting; A Practical Approach,* D. Rickwood and B. D. Hames, Eds. Oxford University Press, Oxford.

Pelham, H. R. B., and Munro, S. (1993) Sorting of membrane proteins in the secretory pathway. *Cell,* 75:603–605.

Rothman, J. E., and Orci, L. (1992) Molecular dissection of the secretory pathway. *Nature,* 355:409–412.

Schatz, G. (1993) The protein import machinery of mitochondria. *Protein Sci.* 2:141–146.

# Radioisotopes in Molecular Biology

*Robert James Slater*

*Key Words*

**Autoradiography** The detection of radioactivity using photographic emulsions.

**Fluorography** Sensitive form of autoradiography in which radioactivity is detected by light emanating from a scintillator (or "fluor") in close contact with the sample.

**Half-life** The time taken for the activity of a radionuclide to lose half its value by decay.

**Quenching** Any process that reduces the efficiency of detection of radioactivity.

**Radioactivity** The emission of ionizing radiations by matter.

**Radioisotopes** Isotopes—that is, atoms with the same atomic number (protons) but differing mass number (protons plus neutrons)—that have an unstable nucleus and decay to a stable state by the emission of ionizing radiations (e.g., $^{14}C$ is a radioisotope of $^{12}C$).

**Radionuclide** An atomic species that is radioactive (e.g., $^{3}H$, $^{14}C$).

**Specific Activity** The rate of decay per unit mass (e.g., dpm/g, Ci/mol, Bq/mmol).

Radioisotopes are used extensively in molecular biology. They can be incorporated into DNA, RNA, and protein molecules both in vivo and in vitro. As a consequence, the metabolism of macromolecules can be investigated, or "traced." Incorporation of radioisotopes allows detection of minute quantities, thereby facilitating such sensitive techniques as Southern blotting and DNA sequencing. Labeling in vitro is much more efficient than labeling in vivo; specific activities regularly reach $10^8$ to $10^9$ dpm/μg for the $^{32}P$-labeling of DNA, for example. The radionuclides $^{3}H$, $^{14}C$, $^{35}S$, and $^{125}I$ are most commonly used for protein labeling. Incorporation into macromolecules of low energy emitters, such as $^{3}H$ and $^{35}S$ has been coupled with autoradiography to provide images with very high resolution (e.g., in chromosome analysis by in situ hybridization or in DNA sequencing). Incorporation of high energy, high specific activity emitters such as $^{32}P$ for nucleic acid labeling and $^{125}I$ for protein labeling has provided great sensitivity in techniques such as Western or Southern blotting, facilitating, for example, the detection of specific genes.

## 1 PRINCIPLES

### 1.1 Radioactive Decay

Isotopes are elements that have the same atomic number but differing mass numbers: for example, $^{12}C$ and $^{14}C$ are different isotopes of carbon. Both have atomic number 6, but their differing mass numbers (12 and 14, respectively) indicate that $^{12}C$ has six and $^{14}C$ has eight neutrons. Atoms that have the incorrect number of neutrons for nuclear stability are referred to as radionuclides or, more commonly, radioisotopes. Radioactivity is the means by which unstable atoms reach a stable state. The radioisotopes most commonly used in molecular biology emit β-radiation (electrons) although some emit electromagnetic radiation (γ or X-rays; see Section 3).

### 1.2 Kinetics of Decay

It is impossible to predict when an individual unstable nucleus will decay. When one is observing the decay of a large number of unstable atoms, however, there is a statistical probability that a certain number will have decayed by a given time. The number that decay per unit time is proportional to the number of unstable atoms present. Mathematically this is represented as

$$\frac{-dN}{dt} \propto N \quad \text{or} \quad \frac{-dN}{dt} = \lambda N \tag{1}$$

where $N$ = number of unstable nuclei,
$t$ = time
$\lambda$ = decay constant ($s^{-1}$)

Therefore, the amount of radioactivity $A$ (the count rate) is

$$A = \frac{-dN}{dt} = \lambda N \tag{2}$$

Integration of equation (1) with respect to time (letting $N = N_0$ at $t$ = o) gives the following expression:

$$\ln \text{count rate (time, } t_2) = \ln \text{count rate (time, } t_1) - \lambda t \tag{3}$$

Thus, if the decay constant is known, the count rate at any time can be calculated from a known count rate and time. In practice, decay constants are rarely quoted, and half-life ($t_{1/2}$) is in many ways a more convenient way to express decay kinetics. The half-life is the time taken for the count rate to drop to half a given value. From equation (3) we can write

$$\ln 2 = \lambda\, t_{1/2} \tag{4}$$

or

$$\lambda = \frac{0.693}{t_{1/2}} \tag{5}$$

so if the half-life is known, a combination of equations (5) and (3) can be used to determine future or past levels of radioactivity. Decay tables based on this mathematics are available in the reference literature and suppliers' catalogues.

### 1.3 Units

The International System of Units (SI system) has the becquerel (Bq) as the unit of radioactivity; 1 Bq is one distintegration per second (1 dps). However, a frequently used unit is the curie (Ci). This unit originates from the number of disintegrations from a gram of pure radium: 1 Ci is therefore $3.7 \times 10^{10}$ dps (or 37 GBq). This is a very large amount of radioactivity, and most molecular biologists use quantities measured in microcuries (μCi); 1 μCi is $2.22 \times 10^6$ dpm (disintegrations per minute) or 37 kBq.

The term "specific activity," which is important in relation to experiments with radioactivity, refers to the amount of radioactivity per unit mass. Thus dpm/g, cpm/g, Ci/mol, Bq/mol, and so on are all units of specific activity. Sometimes it is necessary or desirable to change the specific activity by addition of cold carrier. This can be achieved by applying the following formula:

$$W = Ma\left[\frac{1}{A'} - \frac{1}{A}\right] \quad (6)$$

where $W$ = mass of cold carrier required (mg)
$a$ = amount of radioactivity present (MBq)
$M$ = molecular weight of the compound
$A$ = original specific activity (MBq/mmol)
$A'$ = required specific activity (MBq/mmol)

Units and definitions are summarized in Table 1.

## 2 RADIATION PROTECTION

### 2.1 General Principles

Toxicity is the greatest practical disadvantage associated with the use of radioisotopes. These isotopes produce ionizing radiations which, when absorbed by cells, cause ionization and the production of free radicals, which in turn cause mutation of DNA, hydrolysis of proteins, and ultimately cell death. The unit of dose used in radiobiology is the sievert (see Tables 1 and 2), frequently expressed as a dose rate (e.g., Sv/h). Radiation doses received are related to distance by an inverse square relationship. To calculate dose rates at any distance, use the formula:

$$\text{dose rate}_1 \times \text{distance}^2 = \text{dose rate}_2 \times \text{distance}^2 \quad (7)$$

The various radioisotopes used in molecular biology fall into two broad categories of hazard: those that emit penetrating radiations and present an external radiation hazard ($^{32}$P and $^{125}$I) and those that emit relatively weak radiations and present a low ($^{14}$C, $^{35}$S, $^{33}$P) or zero ($^{3}$H) external radiation risk. Clearly therefore it is incumbent on the laboratory worker to avoid $^{32}$P or $^{125}$I where possible. All radioisotopes present an internal radiation risk, particularly when handled as liquids or gases. But again, the penetrating, higher energy radiations such as from $^{32}$P and $^{125}$I present the greatest risk. The figures for annual limit on intake shown in Table 3 give an indication of the relative risk from ingestion.

Shielding other than that routinely used for storage is not usually required for $^{14}$C, $^{35}$S, $^{33}$P, or $^{3}$H when these isotopes are used at the levels appropriate for most molecular biology experiments. Here the potential risk is through ingestion, which can be avoided by adherence to good laboratory practices applicable to work with all radionuclides: wearing of laboratory coat, gloves, and safety spectacles; working in spill trays lined with absorbent paper; double

**Table 1** Summary of Units and Their Definitions

| Unit | Abbreviation | Definition |
|---|---|---|
| Counts per minute or second | cpm, cps | The *recorded* rate of decay |
| Disintegrations per minute or second | dpm, dps | The *actual* rate of decay |
| Curie | Ci | The number of disintegrations per second equivalent to 1 g of radium ($3.7 \times 10^{10}$ dps) |
| Millicurie | mCi | Ci $\times 10^{-3}$ or $2.22 \times 10^9$ dpm |
| Microcurie | μCi | Ci $\times 10^{-6}$ or $2.22 \times 10^6$ dpm |
| Becquerel (SI unit) | Bq | 1 dps |
| Gigabecquerel (SI unit) | GBq | $10^9$ Bq or 27.027 mCi |
| Megabecquerel (SI unit) | MBq | $10^6$ Bq or 27.027 μCi |
| Electron volt | eV | The energy attained by an electron accelerated through a potential difference of 1 volt; equivalent to $1.6 \times 10^{-19}$ joule |
| Roentgen | R | The amount of radiation that produces $1.61 \times 10^{15}$ ion pairs per kilogram of air [$2.58 \times 10^{-4}$ coulomb/kg (C/kg)] |
| Rad | rad | The dose that gives an energy absorption of 0.01 joule/kg (J/kg) |
| Gray (SI unit) | Gy | The dose that gives an energy absorption of 1 J/kg; thus 1 Gy = 100 rads |
| Rem | rem | The amount of radiation that gives a dose in man equivalent to 1 rad of X-rays |
| Sievert | Sv | The amount of radiation that gives a dose in man equivalent to 1 Gy of X-rays; thus 1 Sv = 100 rem |

**Table 2** Dose Limits Recommended in ICRP Publication 60

| Application | Dose Limit: Occupational | Dose Limit: Public |
|---|---|---|
| Effective dose | 100 mSv in 5 years<br>50 mSv in any 1 year | 1 mSv in a year, averaged over any 5 consecutive years in exceptional circumstances |
| Annual dose in | | |
| Lens of the eye | 150 mSv | 15 mSv |
| Skin (1 cm$^2$) | 500 mSv | 50 mSv |
| Hands and feet | 500 mSv | |
| Dose to surface of abdomen of pregnant woman | 2 mSv to end of pregnancy and intake of radionuclide restricted to 1/20 of an annual limit on intake | |

*Source:* International Commission on Radiological Protection, London, 1991.

containment of stocks for storage and transport; regular and routine monitoring; no eating and drinking (or sucking pencils, etc.) in the laboratory; no mouth pipetting; clear delineation of areas used for radioisotope work; and washing hands when leaving the laboratory.

Dose rates from $\gamma$ sources are quoted as the specific dose rate constant. For $^{125}$I this figure is $1.9 \times 10^{-2}$ $\mu$Sv/h/MBq at one meter.

For $^{32}$P the following formula can be used:

$$\text{dose rate } (\mu\text{Sv/h}) \text{ at 30 cm} = 54 \times \text{activity (MBq)} \qquad (8)$$

Using equations (7) and (8), a 1 MBq source (ca. 27 $\mu$Ci) has a dose rate of 0.9 $\mu$Sv/min at 30 cm and 0.81 Sv/min at 1 cm.

Clearly, shielding is required for work with $^{32}$P and $^{125}$I; the most practical is 1 cm Perspex for $^{32}$P and 1 cm lead-impregnated Perspex for $^{32}$P or $^{125}$I. High energy $\beta$-emitters generate secondary X-rays from absorbers, known as bremsstrahlung radiation. The generation of this radiation increases with the atomic weight of the absorber: hence Perspex is a good shield in this respect. Lead-impregnated Perspex is, however, still suitable for the quantities of $^{32}$P used in a radiobiology laboratory. Ordinary Perspex is not suitable for $^{125}$I.

Vials or Eppendorf tubes containing $^{32}$P or $^{125}$I must not be handled directly even when using a body shield, as doses can potentially by very high. Such receptacles must be manipulated with forceps or other appropriate device or kept in Perspex holders such as those specifically marketed by radioisotope and radiation protection equipment suppliers. Automatic pipets should carry small Perspex shields. Workers using more than 0.8 MBq $^{32}$P should wear fingertip and body dosimeters. In all cases, the member of staff responsible for radiation protection (e.g., radiation protection supervisor or radiation protection adviser) should be consulted. He or she will advise on the laboratory requirements and the type of laboratory required. When dealing with penetrating radiations, remember the basic rules of radiation protection:

**Maximize distance.**
**Minimize time of exposure.**
**Use shielding.**

Do dose estimates on your experiments using equations (7) and (8) and the data in Table 3. Monitor regularly. A sensible precaution when using radioactivity in a protocol for the first time is to do a dummy run with ink in place of the radioisotope solution. A further, simple precaution, is to add radioisotopes to mixtures last where possible; this reduces the time of exposure and risk of contamination.

### 2.2 Unpacking and Dispensing Radioactive Solutions

A protocol for dispensing solutions containing $^{32}$P or $^{125}$I can be summarized as follows: study the data sheet accompanying the radioisotope; do a dose assessment; prepare as much as possible prior to removing the radioisotope from the store; maximize shielding and your distance from the source; do all work in a lined spill tray; return the stock radioisotope to the store as soon as possible; monitor at all times; and apply good laboratory practices as described in Section 2.1.

Shielding is not required for $^3$H but the principle of minimizing contamination still applies. Remember that $^3$H cannot be detected by bench monitors.

## 3 THE CHOICE OF RADIONUCLIDE

The relative order of increasing radiation hazard is $^3$H, $^{14}$C, $^{35}$S, $^{33}$P, $^{32}$P, $^{125}$I, and $^{131}$I. Choose the radioisotope of lowest toxicity where possible. The relative merits of radionuclides used in molecular biology are provided in Table 4.

## 4 NUCLEIC ACID LABELING

Radiolabels are most commonly introduced into nucleic acids by addition of single or multiple labeled nucleotides using enzyme-catalyzed reactions (Table 5). When a labeling reaction is complete, it is necessary to estimate the extent of incorporation. It is not usually necessary to remove free, unincorporated labeled nucleotides from labeled probes; but if it is felt necessary, free label can be removed by precipitation or by use of a "spin column" (Sephadex G25 swollen in buffer in a disposable syringe). Free label from an applied sample remains in the column following centrifugation.

## 5 PROTEIN LABELING

Proteins are most frequently radiolabeled with $^3$H, $^{14}$C, $^{35}$S, or $^{125}$I. Tritium and $^{14}$C are used for reasons of safety and practicality (i.e., because of their long half-life) whenever it is not essential to have proteins labeled to high specific activity. Proteins can be labeled in vivo by incubating cells with labeled amino acids. Alternatively they can be labeled in vitro with *N*-succinimidyl-2,3-[$^3$H]propionate or by reductive methylation with [$^{14}$C]formaldehyde. Both techniques normally result in proteins that retain their biological activity. Methionine labeled with $^{35}$S is used for in vivo labeling and

Table 3 Properties and Radiation Protection Data of Radioisotopes Commonly Used in Molecular Biology

| Property/Radiation Protection Criteria | Isotopes (presented in order of increasing toxicity) | | | | | | |
|---|---|---|---|---|---|---|---|
| | $^3H$ | $^{14}C$ | $^{35}S$ | $^{33}P$ | $^{32}P$ | $^{125}I$ | $^{131}I$ |
| Half-life | 12.3 years | 5730 years | 87.4 days | 25.4 days | 14.3 days | 60.0 days | 8.04 days |
| Mode of decay | $\beta^-$ | $\beta^-$ | $\beta^-$ | $\beta^-$ | $\beta^-$ | $\gamma$ (EC) | $\beta^- + \gamma$ |
| β-Energy (MeV) $E_{max}$ | 0.019 | 0.156 | 0.167 | 0.249 | 1.709 | Auger electrons | 0.806 |
| Monitor | Swabs counted by liquid scintillation | β-counter | β-counter | β-counter | β-counter | γ-probe | β-counter or γ probe |
| Biological half-life (inorganic form) | 12 days[a] | 12 days | 44 days | 19 days | 14 days | 42 days | 8 days |
| Critical organ | Whole body | Whole body/fat | Whole body/testis | Bone | Bone | Thyroid | Thyroid |
| Maximum range in air | 6 mm | 24 cm | 30 cm | 46 cm | 7.2 m | >10 m | >10 m |
| Shielding required | None | Perspex 1 cm[b] | Perspex 1 cm[b] | Perspex 1 cm[b] | Perspex 1 cm | Lead 0.25 mm | Lead 13 mm |
| Annual limit on intake (ALI) MBq[c] oral | 1000[a] | 40 | 100 | 80 | 8 | 1 | 0.8 |
| γ Dose rate at 1 m ($\mu Gy \cdot h^{-1} GBq^{-1}$) | | | | | | 34 | 51 |
| Quantity for notification of [British] Health and Safety Executive (Bq)[d] | $5 \times 10^6$ | $5 \times 10^5$ | $5 \times 10^6$ | $5 \times 10^5$ | $5 \times 10^5$ | $5 \times 10^4$ | $5 \times 10^4$ |
| *Designation of controlled areas*[d] | | | | | | | |
| Air containment ($Bq/m^{-3}$) | $2 \times 10^5$ | $1 \times 10^4$ | $9 \times 10^4$ | $1 \times 10^4$ | $2 \times 10^3$ | $3 \times 10^2$ | $2 \times 10^2$ |
| Surface containment ($Bq/cm^{-2}$) | $4 \times 10^6$ | $2 \times 10^4$ | $2 \times 10^4$ | $2 \times 10^4$ | $2 \times 10^3$ | $1 \times 10^2$ | $1 \times 10^2$ |
| Total activity (Bq) | $3 \times 10^{10}$ | $9 \times 10^8$ | $4 \times 10^9$ | $1 \times 10^9$ | $1 \times 10^8$ | $1 \times 10^7$ | $1 \times 10^7$ |
| Special considerations | Monitoring difficulties lead to high potential internal hazard. High levels of cleanliness necessary. | | | | Potential high source of external radiation. Lead shielding for quantities of ≥370 MBq (10 mCi). Finger dosimeter recommended. | Iodine sublimes, work in fume hood. Spills should be treated with alkaline sodium thiosulfate solution prior to decontamination. Many iodine compounds penetrate rubber gloves—wear two pairs. | |

[a]Figure given is for $^3H_2O$. For compounds such as [$^3$H]thymidine biological half-life is 140 days and the ALI should be reduced to 60 MBq.
[b]Thin Perspex is adequate to reduce dose but has poor mechanical properties.
[c]Figures taken from Publication 61 of the International Commission on Radiological Protection (1991).
[d]Figures taken from *The Ionising Radiation Regulations*, 1985, and therefore apply to the United Kingdom; for supervised areas divide by 3.

in vitro protein synthesis because it provides a high specific activity and good resolution in autoradiography of polyacrylamide gels used for analysis of products.

Radioiodination is used when proteins with a high specific activity are required—for example, in radioimmunoassay and related techniques. Activities of $10^8$ dpm/μg can be achieved. Many methods rely on either direct oxidative iodination or indirect conjugation iodination. Probably the most commonly used technique is the chloramine T procedure, which is relatively simple and inexpensive (for protocol, see Slater, Ed., 1990).

## 6 DETECTION

### 6.1 Ionization Monitors

An ionization monitor consists of a small gas chamber containing two electrodes, a voltage supply, and a scaler. Ionizing radiation enters through a thin-end window, the gas is ionized, and a pulse of current is recorded on the scaler, usually displayed as counts per second, counts per minute, or becquerels per square centimeter. Gas ionization monitors are suitable for detecting $^{32}$P, and they will pick up $^{14}$C, $^{35}$S, and $^{33}$P if a sensitive instrument with a very thin end window is used. They do not detect $^3$H.

Contamination monitors must be officially calibrated each year, but if $^{32}$P is used, it is also worthwhile doing a simple calibration on the instrument for one's own information. Place a known volume of $^{32}$P in an Eppendorf tube and record the decay rate at various distances. Reading 0.1 MBq $^{32}$P at 10 cm (equivalent to a dose rate of approximately 0.5 μSv/min) should give a reading in excess of 50 cps.

### 6.2 Scintillation Counters

Certain substances (termed scintillators or fluors) emit light upon absorption of ionizing radiations. The scintillators used in counters

Table 4 Relative Merits of Commonly Used Radionuclides

| | Advantages | Disadvantages |
|---|---|---|
| $^{3}H$ | Safety<br>High specific activity<br>Wide choice of labeling position in organic compounds<br>High resolution in autoradiography | Low efficiency of detection<br>$^{3}H$ is large compared with $^{1}H$ and may cause an isotope effect<br>Isotope exchange with the environment |
| $^{14}C$ | Safety<br>Wide choice of labeling position in organic compounds<br>High resolution in autoradiography | Low specific activity significantly reduces value in molecular biology |
| $^{35}S$ | High specific activity<br>Good resolution in autoradiography | Relatively long biological half-life |
| $^{33}P$ | High specific activity<br>Safety relative to $^{32}P$<br>Longer half-life than $^{32}P$<br>Good resolution in autoradiography | Short half-life relative to $^{35}S$<br>Cost |
| $^{32}P$ | Sensitivity of detection (10 fg or 50 dpm/cm$^{2}$ in filter hybridization)<br>Short half-life simplifies disposal<br>High specific activity<br>Cerenkov counting | Short half-life effects cost and experimental design<br>External radiation hazard<br>Poor resolution in autoradiography |
| $^{125}I$ | Ease of detection<br>High specific activity<br>Efficient in vitro labeling of proteins<br>Good resolution in autoradiography | Safety: radiation is penetrating and $^{125}I$ accumulates in thyroid; annual limit on intake is very low |

can be solids (e.g., sodium iodide) or liquids (e.g., toluene). The light is detected by one or more of the photomultiplier tubes, which give an electrical signal to a scaler. The number of electrical signals is related to the number of disintegrations in the sample, and the strength of the signal is related to the energy of the absorbed radiation.

Scintillation probes are small handheld instruments useful for monitoring γ-radiation, say, from $^{125}I$. They either display cps values or, for dosimetry purposes, μSv/h. One of these instruments should be available whenever $^{125}I$ is used. As with ionization monitors, scintillation counters should be calibrated officially once per year and can be roughly calibrated by the user as an aid to monitoring (see Section 6.1).

### 6.3 Cerenkov Counting

High energy β-particles travel through water faster than light; along their path, they cause polarization of molecules, which emit photons

Table 5 Nucleic Acid Labeling Methods

| Method | Template | Enzymes | Labeled Precursor | Labeling | Example Applications | Potential Specific Activity Using $^{32}P$ |
|---|---|---|---|---|---|---|
| Nick translation | ds DNA | DNase1, DNA *pol* 1 | dNTP | Uniform | Filter hybridization | $2 \times 10^{9}$ dpm/μg |
| Random primer labeling | ss DNA | DNA *pol* 1 or Klenow enzyme | dNTP | Uniform | Very high sensitivity filter hybridization | $5 \times 10^{9}$ dpm/μg |
| Single primer labeling | ss DNA | DNA *pol* 1 or Klenow enzyme | dNTP | Uniform | Filter hybridization | $10^{9}$ dpm/μg |
| | | | Labeled oligonucleotide | 5′ End | Restriction mapping | |
| RNA polymerase | ds DNA | SP6, T3, T5 or T7 RNA *pol* | NTP | Uniform | Very high sensitivity, filter hybridization | $5 \times 10^{9}$ dpm/μg |
| | | | γ-Labeled NTP | 5′ End | RNA sequencing | $5 \times 10^{9}$ dpm/pmol of ends |
| End-filling | ds DNA | DNA *pol* 1 or Klenow enzyme, or T4 DNA *pol* | dNTP | 3′ End | Restriction mapping, gel markers, S1 mapping | $2 \times 10^{6}$ dpm/pmol of ends |
| Kinasing | ds DNA<br>ss DNA<br>RNA | T4 polynucleotide kinase | γ-labeled NTP | 5′ End | S1 or primer extension mapping, Maxam and Gilbert sequencing | $8 \times 10^{5}$ dpm/pmol of ends |
| Tailing | ds DNA<br>ss DNA<br>RNA | Terminal deoxynucleotidyl transferase | dNTP | 3′ End | Restriction mapping, S1 mapping, Maxam and Gilbert sequencing | $8 \times 10^{5}$ dpm/pmol of ends |
| Ligation | ds DNA<br>ss DNA<br>or RNA | T4 RNA ligase | [5′-$^{32}$P],3′-cytidine biphosphate | 3′ End | Restriction mapping, RNA sequencing | $8 \times 10^{5}$ dpm/pmol of ends |

Table 6 Use and Sensitivity of Autoradiography and Fluorography

| Isotope | Method | Example Applications | Preflashed Film | Intensifying Screen | Temperature of Exposure | Approximate Detection Limit After 24 h Exposure (dpm/cm$^2$) |
|---|---|---|---|---|---|---|
| $^3H$ | Direct | Whenever very high resolution is required (e.g., in situ hybridization) | No | No | Room temp | $8 \times 10^6$ |
| | Fluorography | Acrylamide gels | Yes | No | −70°C | $8 \times 10^3$ |
| $^{14}C$, $^{35}S$, or $^{33}P$ | Direct | DNA sequencing | No | No | Room temp | $6 \times 10^3$ |
| | Indirect | SDS-polyacrylamide gels | Yes | Yes | −70°C | $4 \times 10^2$ |
| $^{32}P$ | Direct | DNA sequencing | No | No | Room temp | $5 \times 10^2$ |
| | Indirect | Southern analysis | Yes | Yes | −70°C | $5 \times 10^1$ |
| $^{125}I$ | Direct | Subcellular localization | No | No | Room temp | $2 \times 10^3$ |
| | Indirect | Western blotting | Yes | Yes | −70°C | $1 \times 10^2$ |

of light as they return to a ground state. The energy of radiation from $^{32}P$ is high enough to cause this Cerenkov effect. Consequently water can be used in place of scintillation fluid for this radioisotope. The counting efficiency is relatively low (about 30% compared with 90% in a scintillation cocktail), and the counter should be specifically calibrated for it, although an acceptable short cut is to count the samples as if they were $^3H$. The big advantages of Cerenkov counting are that aqueous samples can be counted and recovered for experimental use, and there is no accumulation of organic radioactive waste.

### 6.4 Autoradiography

Many experiments in molecular biology rely on autoradiography, in particular, analysis of nucleic acid samples separated on gels (Southern blotting, Northern blotting, DNA sequencing) and detection of particular clones or plaques by hybridization screening. The basic principle is that light, a β-particle, X-rays, or γ-rays can produce silver atoms in crystals of the silver halide present in photographic emulsions. Fluorography is an adaptation of the technique in which radioactivity is detected by light emanating from a scintillator (the "fluor"), which is either incorporated into the sample (e.g., PPO in gels emitting radiation of low energy) or provided as an intensifying screen in autoradiography cassettes (often referred to as indirect autoradiography). Fluorography increases sensitivity but reduces resolution.

A summary of some of the key factors of autoradiography and fluorography is provided in Table 6.

### 6.5 Personal Dosimeters

Individuals who are at risk from occupational exposure to potentially hazardous levels of ionizing radiation may be issued film badges or thermoluminescent detectors (TLDs). The latter contain a phosphor such as LiF, which becomes and remains excited once exposed to radiation. Heat treatment results in light emission, the intensity of which is related to the dose received. Small TLDs can be worn on the fingers, under gloves, for hand dosimetry.

Personal dosimeters are not appropriate for workers using low energy emitters such as $^{14}C$, $^{35}S$, or $^3H$, but they are of value to users of $^{32}P$ or $^{125}I$. Their issue is at the advice of senior staff with appropriate training and responsibility for radiation protection, following dosimetry calculations and estimates of exposure levels to individuals.

*See also* Labeling, Biophysical.

### *Bibliography*


Ballance, P. E., Day, R. L., and Morgan, J. (1992) *Phosphorus-32: Practical Radiation Protection.* H and H Scientific Consultants, Leeds.

Mundy, C. R., Cunningham, M. W., and Read, C. A. (1991) Nucleic acid labelling and detection. In *Essential Molecular Biology A Practical Approach,* T. A. Brown, Ed. Oxford University Press, Oxford and New York.

International Commission on Radiological Protection. (1991) ICRP60: Recommendations of the International Commission on Radiological Protection. *Annals of the ICRP,* Vol. 21, Nos. 1–3. Pergamon Press, Oxford.

———. (1991) ICRP 61: Annual limits on intake of radionuclides by workers based on 1990 recommendations. *Annals of the ICRP,* Vol. 21, No. 4. Pergamon Press, Oxford.

Slater, R. J., Ed. (1990) *Radioisotopes in Biology: A Practical Approach.* Oxford University Press, Oxford and New York.


## Raman Spectroscopy: *see* Protein Analysis by Raman Spectroscopy.

# RecA Protein, Structure and Function of

*Michael M. Cox*

### *Key Words*

**Genetic Recombination** Rearrangement of genetic information within or between nucleic acids.

**Heteroduplex DNA** Double-stranded DNA containing complementary strands derived from different homologous DNA molecules via recombination.

**Heterologous Sequences** Nucleic acid sequences that are dissimilar and evolutionarily unrelated.

**Holliday Intermediate** DNA structure in which two DNA molecules are linked by the breakage, crossover, and rejoining of one strand from each.

**Homologous Sequences** Nucleic acid sequences that are similar or identical by virtue of an evolutionary relationship.

**SOS** A regulated response to DNA damage and other cellular traumas in bacteria.

**Triplex** A nucleic acid structure in which three nucleic acid strands are interwound and hydrogen-bonded.

---

The RecA protein is a central component of the bacterial system that mediates recombinational DNA repair and homologous genetic recombination. It also plays a key role in the regulation of the SOS response to DNA damage. RecA protein forms long, helical nucleoprotein filaments on the DNA. These filaments are the active species in the DNA pairing and strand exchange reactions in recombinational processes and in the induction of SOS. The continuing study of RecA protein offers insights into bioenergetics, DNA structure and topology, and protein–DNA interaction, as well as a broader understanding of DNA recombination.

## 1 OVERVIEW

The product of the *rec*A gene is a multifunctional protein that occupies an important crossroad in bacterial DNA metabolism, with important roles in DNA repair, recombination, and replication. Biochemical interest is focused on the mechanism(s) by which RecA protein mediates its varied functions in DNA metabolism.

### 1.1 Homologous Genetic Recombination

For more than a century, geneticists have made use of the naturally occurring exchanges of genetic information between homologous chromosomes to study heredity and to map genes. The molecular process underlying these genetic exchanges is known as homologous genetic recombination.

A generic model illustrating the kinds of molecular processes that must occur during recombination is presented in Figure 1. A key requirement is the alignment of homologous sequences in two DNA molecules. DNA strands in both DNA molecules must be broken (either before or after alignment) to initiate a strand exchange. The results in formation of a Holliday intermediate, a structure named for Robin Holliday, who first proposed it in 1964. The heteroduplex DNA formed in this exchange is then extended by branch migration. Finally, the Holliday junction is cleaved and the ends repaired.

The first genes with products involved in recombination were identified in *E. coli* in 1965 by A. J. Clark. Recombination genes now include *rec*A, *rec*B, *rec*C, *rec*D, *rec*F, *rec*G, *rec*J, *rec*N, *rec*O, *rec*Q, *rec*R, *ruv*A, *ruv*B, *ruv*C, and *ssb*. Mutations in the *rec*A gene exhibit the most complex phenotypes. The product of this gene is a multifunctional protein called the RecA protein.

### 1.2 Biological Function of RecA

The immediate selective value of RecA protein to the bacterial cell probably lies in its recombinational repair function. In each cell generation, a bacterial cell growing in an aerobic environment is subjected to thousands of DNA lesions, many of which are potentially lethal. Most of the damage is inflicted by highly reactive forms of oxygen. Accurate DNA repair is made possible to a large extent because DNA is double-stranded. A lesion in one strand can be repaired by removing a segment as a template to guide synthesis of a replacement segment. When both strands are damaged (e.g., a double-strand break or cross-link), or when the lesion is in a single-stranded region of DNA, the information for accurate repair must come from another homologous DNA molecule via recombination. The prototypical phenotype of recA mutant cells is an extreme sensitivity to DNA damaging agents. The molecular

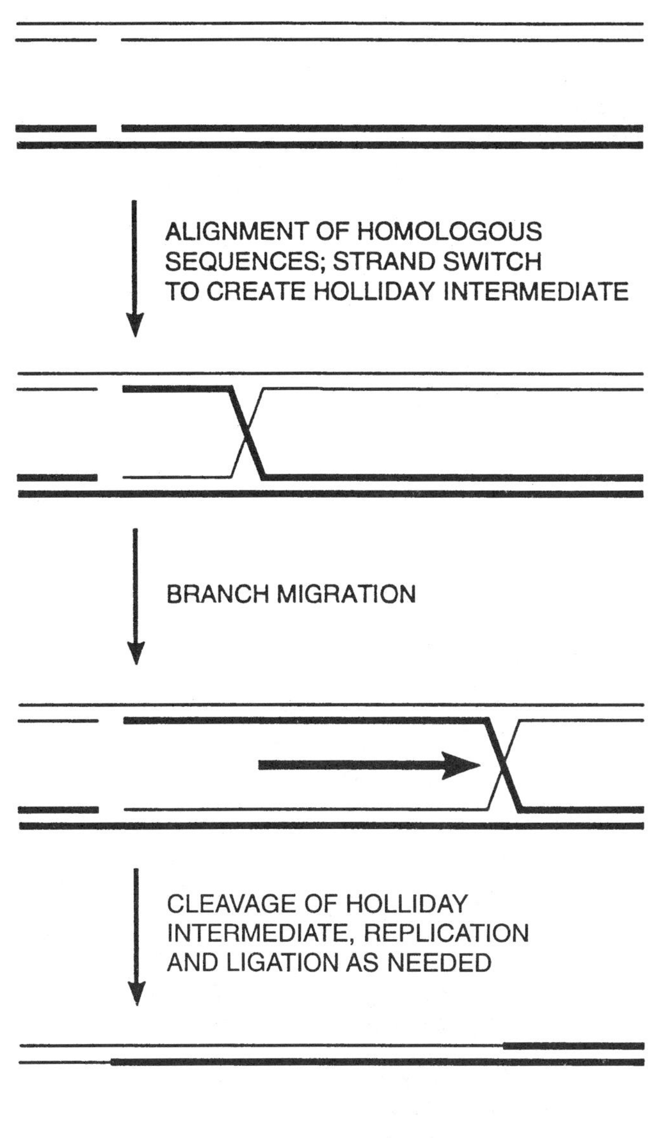


**Figure 1.** One model for homologous genetic recombination between two homologous chromosomes, represented by thick and thin lines. Regions of the DNA intermediates and products in which thick and thin lines are paired are called heteroduplex DNA. There are many variants of this model, and the depiction is designed only to show the kinds of processes that must occur to bring about recombination.

form of RecA is adapted to render recombinational DNA repair maximally efficient.

Homologous genetic recombination and recombinational DNA repair are closely related processes, and RecA protein plays a central role in both. Recombination, even when repair is not the object, has the important result of increasing the genetic diversity, hence viability, of a species.

RecA protein also has a third, and quite different function. It facilitates the autocatalytic cleavage of the *lex*A repressor to induce a response to abnormally extensive DNA damage called the SOS response.

## 2 STRUCTURE

### 2.1 Primary Structure

In *E. coli,* the RecA polypeptide chain contains 352 amino acids with a combined molecular weight of 37,842. The protein occurs in virtually all bacteria, and the *rec*A gene has been sequenced in more than 40 bacterial species. The primary structure is illustrated in Figure 2, with 71 invariant amino acid residues highlighted. The close relationship between bacterial RecA proteins is reinforced by sequence similarities at many other positions. A wide range of RecA mutants are available, and these have facilitated an understanding of structure–function relationships in this polypeptide.

The wide distribution of bacterial *rec*A genes indicates that the protein evolved more than 1.5 billion years ago. The UvsX protein of bacteriophage T4 is a functional analogue of RecA protein with more limited sequence similarity. Eukaryotic genes with significant sequence homology to *rec*A have been found in yeast and plants.

### 2.2 Three-Dimensional Structure

The structure of RecA protein has been determined at 2.3 A resolution by R. Story and T. Steitz. There is a major central domain flanked by two smaller subdomains at the N- and C-termini. The central domain contains a single binding site for ATP or ADP. Monomers in the crystal are packed to form a continuous spiral filament, with six monomers per right-handed helical turn (Figure 3; see color plate 14). These results are consistent with other studies showing that RecA protein forms a helical nucleoprotein filament on DNA. The filament exhibits a deep helical groove. A variety of physical studies indicate that the groove can accommodate up to three DNA strands.

## 3 DNA STRAND EXCHANGE AS A MODEL FOR RECOMBINATION AND RECOMBINATIONAL DNA REPAIR

Typical DNA strand exchange reactions used to study RecA function in vitro are illustrated in Figure 4. The reactions shown are designed for convenience; it is easy to distinguish products from substrates with a wide range of assays. The reaction can involve either three or four DNA strands, as shown, and it nicely mimics several of the putative steps in homologous genetic recombination (cf. Figures 1 and 4). A Holliday intermediate is formed transiently in the four-strand reaction.

### 3.1 Fundamentals

The active species promoting DNA strand exchange is a RecA nucleoprotein filament that completely coats the single-stranded or gapped DNA substrate as the first step in the reaction. The bound single strand is paired to one end of the linear duplex DNA to create the branched molecules shown in Figure 4. The heteroduplex DNA thus created is then extended in a facilitated branch migration reaction that proceeds until products are formed. The branch migration is unidirectional (5′ to 3′ with respect to the single strand initially bound), and proceeds at a rate of three to six base pairs per second. RecA-mediated DNA strand exchange proceeds readily past mismatches, lesions, and even heterologous inserts (up to a few hundred base pairs long) in one or both DNA substrates, a capability that is critical to its function in recombinational DNA repair.

RecA is a DNA-dependent ATPase, and ATP is a hydrolyzed during strand exchange. The apparent efficiency of the reaction is low, with about 100 ATP molecules hydrolyzed per base pair of heteroduplex formed under typical reaction conditions. The manner in which the two DNA molecules are aligned to begin strand

| | | | | |
|---|---|---|---|---|
| AIDENKQKA**L** | AAALGQIEKQ | F**GKG**SI**M**RLG | EDRSMDVETI | 40 |
| STGSLS**L**DI**A** | L**G**AGGLPMGR | IV**E**IY**GPES**S | **GKTT**LTLQVI | 80 |
| AAAQREGKTC | AFIDAEH**A**L**D** | PI**Y**ARK**LG**VD | IDN**L**LCSQ**P**D | 120 |
| TGEQALE**I**CD | ALARSGAV**D**V | IVV**DSVAAL**T | **P**KA**E**IE**G**EIG | 160 |
| DSHM**G**LA**AR**M | M**S**Q**A**M**R**KLAG | NLKQSNTLLI | **F**I**NQ**I**R**M**K**IG | 200 |
| VMF**G**N**PE**T**T**T | **GG**N**ALKF**YA**S** | **VR**LDI**R**RIGA | V**K**EGENVV**G**S | 240 |
| ETRV**KV**V**K**NK | IAA**P**FKQAEF | Q**I**LYGE**G**INF | YGELVDLGVK | 280 |
| EKLIEKA**G**A**W** | YSYKGEKIG**Q** | **G**KANATAWLK | DNPETAKEIE | 320 |
| KKVRELLLSN | PNSTPDFSVD | DSEGVAETNE | DF | |

**Figure 2.** The amino acid sequence of the RecA protein of *E. coli.* The residues shown in bold are invariant among the more than 40 sequenced bacterial *rec*A genes.

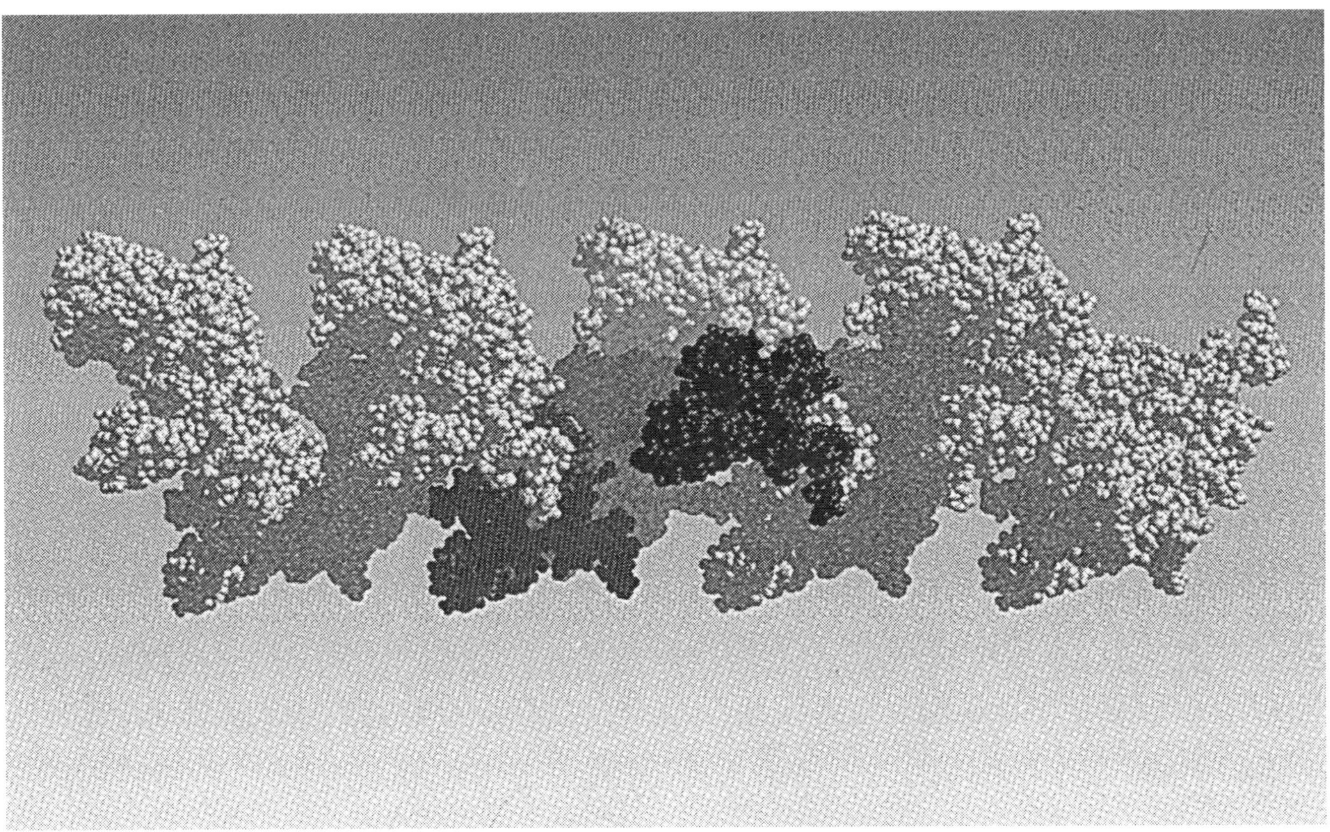

**Figure 3.** Representation of a segment of a helical filament formed by 24 monomers of RecA protein. Six monomers (one helical turn) are colored. [Illustration based on the structure determined by Story et al. (1992) and developed from the coordinates deposited in the Brookhaven Protein Data Bank, using the MidasPlus software obtained from the Computer Graphics Laboratory at the University of California, San Francisco.] (See color plate 14.)

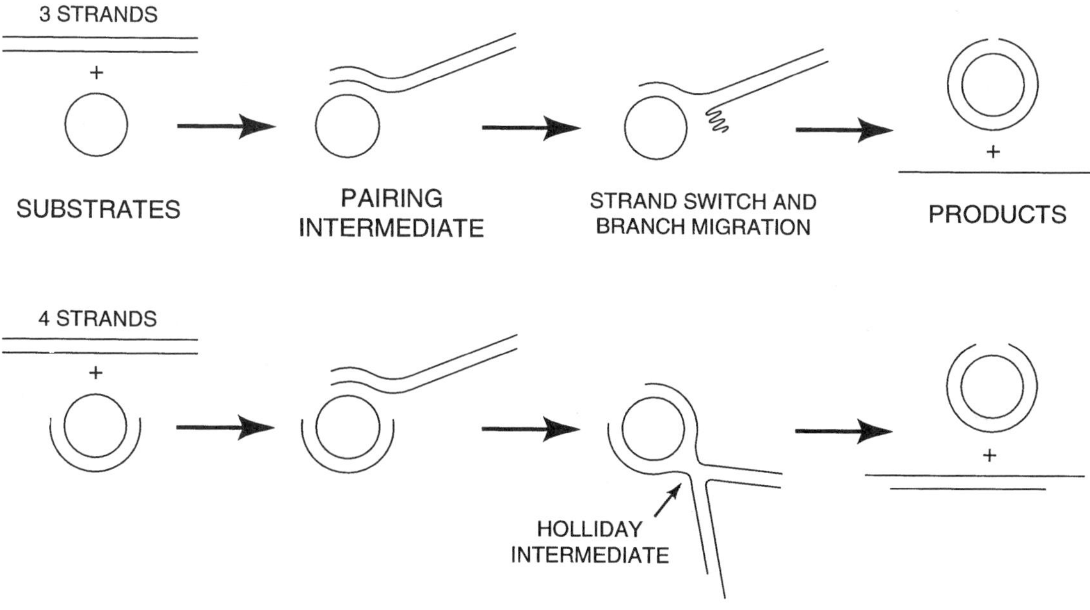


**Figure 4.** DNA strand exchange reactions promoted by RecA protein in vitro; thin lines represent individual DNA strands.

exchange, and the molecular role of ATP hydrolysis, remain as major unresolved mechanistic questions.

### 3.2 DNA Binding and Filament Formation

The filament has a right-handed helical form, with six RecA monomers per turn of the helix. One RecA monomer binds to three bases or base pairs. The bound DNA is extended and underwound, so that the bases are 5.1 A apart, and there are 18 base pairs per turn instead of the usual 10.5. The DNA is bound along the phosphate–deoxyribose backbone.

RecA protein binds much more rapidly to single strands than to duplex DNA. Following nucleation, the filament is extended in the 5′ to 3′ direction along the DNA. If the substrate is a duplex with a single-strand gap, the nucleation occurs in the gap and the filament is extended rapidly to encompass the adjacent duplex. Nucleation on duplex DNA is very slow, although filament extension is rapid once nucleation has occurred. This molecular design tends to restrict RecA filament formation to regions of cellular DNA containing single-strand gaps, which in turn are regions likely to require DNA repair.

Where it occurs, dissociation of RecA monomers from a filament takes place on the end opposite to that at which monomers are added during filament extension.

### 3.3 DNA Pairing

The homologous alignment of two DNA molecules irrespective of sequence is a molecular problem unique to homologous genetic recombination. Possible pairing intermediates in which two duplex DNAs were interwound to form a quadruplex have been discussed for more than 20 years. Most evidence now argues strongly against a four-stranded DNA pairing intermediate in RecA-mediated DNA strand exchange. Even in the four-strand exchange reactions, the initial pairing must occur with three strands within the single-strand gap of the gapped duplex.

Much evidence suggests that a three-stranded DNA pairing intermediate is formed in the course of these reactions. Some evidence has been reported for a stable three-stranded structure that remains intact after RecA removal. This putative triplex DNA should have a structure quite different from known stable DNA triplexes. It is formed in a sequence-independent manner, and like strands must be arranged in parallel to meet the minimal requirements of a viable recombination intermediate.

### 3.4 ATP Hydrolysis

ATP is hydrolyzed uniformly throughout the RecA filament. The Michaelis constant $K_m$ is on the order of 100 μM. The turnover rate for individual monomers approaches 30 per minute for filaments bound to single strands and 20–22 per minute for filaments bound to duplex DNA. ATP is hydrolyzed at similar rates in a DNA-dependent fashion in the presence of very high concentrations (1.5–2.0 M) of salt.

The function of ATP hydrolysis in RecA-mediated reactions remains controversial, largely because of its apparent inefficiency. The strand exchange reactions depicted in Figure 4 are isoenergetic (the products and substrates contain an equivalent number of base pairs), and branch migration is a spontaneous reaction in vitro. Furthermore, substantial DNA strand exchange can occur in the presence of ATP analogues that are not hydrolyzed by RecA. These results show that ATP hydrolysis is not required for DNA strand exchange between homologous DNA substrates.

ATP hydrolysis *is* required for dissociation of RecA monomers from filament ends, although this category accounts for only a small fraction of ATP hydrolytic events. ATP hydrolysis renders the strand exchange reaction unidirectional and is required for the bypass of heterologous inserts and other structural barriers to strand exchange; both these demonstrated functions are particularly relevant to DNA repair. Finally, ATP hydrolysis is required for strand exchange reactions involving four strands. A molecular explanation for these requirements has not been established.

### 3.5 Models

Mechanistic ideas for DNA strand exchange begin with a proposal of P. Howard-Flanders that DNA pairing and strand switching occur within the major helical groove of the RecA filament. Much evidence indicates that DNA pairing and probably the strand switch to form heteroduplex DNA occur within the groove in a process that requires ATP but not its hydrolysis. ATP hydrolysis is required at a later stage, for dissociation of RecA monomers following reaction and/or for some molecular process that renders the reaction unidirectional and forces strand exchange past structural barriers such as heterologous inserts. A model has been proposed that couples ATP hydrolysis to a coordinated rotation of the two DNA substrates to effect a unidirectional strand exchange that could bypass such barriers.

## 4 RecA PROTEIN AS A COPROTEASE

Extensive DNA damage in bacteria triggers a global cellular response mediated by the SOS regulatory system. The key components of the regulatory system are the *lex*A repressor, which regulates the expression of the genes that are induced in SOS, and the RecA protein. Induction of the SOS system requires the inactivation of the *lex*A repressor, which comes about via cleavage at a specific Ala-Gly bond near the center of the repressor. The cleavage is catalyzed by *lex*A itself, and it occurs spontaneously at high pH. At neutral pH the cleavage is facilitated by RecA protein. The signal for induction of the SOS system is the appearance of single-stranded DNA gaps as a result of DNA damage. The RecA nucleoprotein filament that assembles on the DNA in these circumstances is the active form that facilitates the cleavage of *lex*A. Since RecA does not act as a classical protease, but instead facilitates an autocatalytic cleavage of *lex*A, Raymond Devoret introduced the term "coprotease" to describe this RecA activity.

## 5 OTHER RECOMBINATION PROTEINS

More than 20 different proteins play some role in homologous genetic recombination in *E. coli* and other bacteria, including the products of the genes listed in Section 1.1 and a variety of enzymes that also act during DNA replication (such as DNA ligase and topoisomerases). Only a few of the recombination-specific proteins have been studied in any detail. The *rec*B,C, and D genes encode a DNA helicase/nuclease (the recBCD enzyme) that functions primarily during the initiation of recombinational DNA repair at double-strand breaks and recombination during conjugation and transduction. The RecJ protein is an exonuclease that degrades single-stranded DNA 5′ to 3′. The recQ protein is a DNA helicase. The RuvC protein is an endonuclease that specifically recognizes and resolves Holliday intermediates. The activities of many of the other proteins involved in recombination are unknown or incompletely characterized. Reconstitution of a complete recombination system in vitro is one goal of continuing research in this field.

*See also* BACTERIAL GROWTH AND DIVISION; DNA DAMAGE AND REPAIR; *E. COLI* GENOME; RECOMBINATION, MOLECULAR BIOLOGY OF.

### *Bibliography*


Cox, M. M. (1993) Relating biochemistry to biology: How the recombinational repair function of the recA system is manifested in its molecular properties. *BioEssays,* 15:617–623.

Cox, M. M. (1994) Why does RecA protein hydrolyze ATP? *Trends Biochem. Sci.* 19:217–222.

Kowalczykowski, S. C., Dixon, D. A., Eggleston, A. K., Lauders, S. D., and Rehrauer, W. M. (1994) Biochemistry of homologous recombination in *Escherichia coli. Microbiol. Rev.* 58:401–465.

Kucherlapati, R., and Smith, G. R., Eds. (1988) *Genetic Recombination.* American Society for Microbiology, Washington, DC.

Little, J. W. (1991) Mechanism of specific LexA cleavage—Autodigestion and the role of RecA coprotease. *Biochimie,* 73:411–422.

Radding, C. M. (1991) Helical interactions in homologous pairing and strand exchange driven by RecA protein *J. Biol. Chem.* 266:5355–5358.

Roca, A. I., and Cox, M. M. (1990) The RecA protein: Structure and Function. *CRC Crit. Rev. Biochem. Mol. Biol.* 25:415–456.

Smith, G. R. (1989) Homologous recombination in *E. coli:* Multiple pathways for multiple reasons. *Cell,* 58:807–809.

Smith, K. C., and Wang, T. C. (1989) recA-dependent DNA repair processes. *BioEssays,* 10:12–6.

Story, R. M., Weber, I. T., and Steitz, T. A. (1992) The structure of the *E. coli* RecA protein monomer and polymer. *Nature,* 355:318–325.

West, S. C. (1992) Enzymes and molecular mechanisms of recombination. *Annu. Rev. Biochem.* 61:603–640.


# RECEPTOR BIOCHEMISTRY

*Tatsuya Haga*

### *Key Words*

**Agonists** Compounds that bind to and activate receptors (e.g., endogenous ligands such as hormones and neurotransmitters, chemically synthesized compounds, natural products like alkaloids).

**Antagonists** Compounds that bind to but do not activate receptors, hence do inhibit the action of agonists competitively.

**GTP-Binding Regulatory Proteins (G Proteins)** Proteins that are activated by agonist-bound receptors and activate effectors such as adenylate cyclase, phospholipase C, and ion channels.

**Second Messengers** Compounds that are formed in the cells in response to stimulation by agonists on the surface of cells (e.g., cyclic AMP, cyclic GMP, diacylglycerol, inositol triphosphate).

Communications between the cells are performed by cell–cell contact or by chemical substances like hormones, autocoids, or neurotransmitters. These chemical substances are secreted by a cell and recognized by receptors in target cells. The binding of the ligands to their receptors initiates a series of chemical reactions. Some hormones such as steroids and thyroxine pass through cell membranes and interact with their receptors in the cells. Most other hormones and all known neurotransmitters cannot pass through cell membranes and bind to their receptors, which are transmembrane glycoproteins, on the surface of cells. Membrane-bound receptors are classified into three major groups: ion channel receptors, G-protein-linked receptors, and other receptors for growth factors or other proteins. This entry describes the molecular properties of membrane-bound receptors.

## 1 CLASSIFICATION OF RECEPTORS

Receptors have been named and classified by their endogenous ligands and subclassified by specific agonists and antagonists since the introduction of the concept of receptor in the early 1900s by Langley. The molecular entities of receptors were identified in the last decade, enabling us to classify receptors on the basis of structure and function. For example, receptors for acetylcholine have been classified as nicotinic acetylcholine receptors and muscarinic acetylcholine receptors on the basis of their interactions with specific agonists, nicotine and muscarine, respectively. Purification of these receptors and their functional reconstitution into artificial membranes showed that the nicotinic receptor is an acetylcholine-gated ion channel, while the muscarinic receptor is an activator of G proteins, there is no structural or functional similarity between the two. Cloning of complementary DNAs encoding receptor proteins and deduction of their amino acid sequences revealed that a number of receptors have structural characteristics similar to those of the nicotinic or muscarinic receptor and that they constitute two superfamilies of ion channel receptors and G-protein-linked receptors. The structural characteristics common to members of the two superfamilies are numbers of hydrophobic regions that are supposed to be transmembrane segments: subunits of ion channel receptors and G-protein-linked receptors have four and seven such regions, respectively. On the other hand, several groups of receptors for growth factors or other proteins are characterized by a single transmembrane segment.

Cloning of cDNAs encoding receptor proteins has revealed that the number of subtypes for any given receptor is greater than that of subtypes identified by specific ligands. The extreme case is the odorant receptors: more than 100 receptors are reported to exist, and 18 cDNAs encoding odorant receptors have been cloned, but their ligands remain to be identified. One of the challenging projects in the next decade will be to conjecture and synthesize specific ligands for a given receptor with a known amino acid sequence.

## 2 ION CHANNEL RECEPTORS

Endogenous ligands of ion channel receptors are the following neurotransmitters: acetylcholine, glutamate, $\gamma$-aminobutyric acid (GABA), glycine, serotonin, and probably ATP. Ion channel receptors are not known for catecholamines like adrenaline and dopamine, nor for peptides (Tables 1 and 2).

Ion channel receptors are oligomers composed of heterogeneous subunits, although some might be homooligomers, and they incor-

**Table 1** Classification of Membrane Receptors: Endogenous Ligands and Receptor Functions[a]

| Ligands | Subtypes | G Proteins Involved | Function |
|---|---|---|---|
| *Ion Channel Receptors* | | | |
| Acetylcholine | Nicotinic<br>Muscle type<br>Neural type | | Cation channel |
| Glutamate | Non-NMDA<br>AMPA type | | Cation channel |
| | Kallikrein type<br>NMDA | | Cation channel |
| Serotonin | 5-$HT_3$ | | Cation channel |
| GABA | $GABA_A$ | | Chloride channel |
| Glycine | | | Chloride channel |
| *G-Protein-Linked Receptors* | | | |
| **Group 1** | | | |
| (Nor)adrenaline | $\beta_1$, $\beta_2$, $\beta_3$ | Gs | Adenylate cyclase (+)<br>$Ca^{2+}$ channel (L-type) (+) |
| Dopamine | D1, D5 | | |
| Histamine | H2 | | |
| Adenosine | A2 | | |
| CGRP (calcitonin gene-related peptide) | | | |
| Vasopressin | V2 | | |
| Thyrotropin | | | |
| Lutropin-gonadotropin | | | |
| Odorants | 100 | Golf | Adenylate cyclase (+) |
| **Group 2** | | | |
| Acetylcholine | Muscarinic —($m_2$, $m_4$) | Gi1 | Adenylate cyclase (−)<br>$K^+$ channel (+)<br>$Ca^{2+}$ channel (−) |
| Glutamate | Metabotropic (mGluR2–4) | Gi2<br>Gi3 | |
| (Nor)adrenaline | $\alpha_2$ ($\alpha_{2A}$, $\alpha_{2B}$, $\alpha_{2C4}$) | Go | |
| Serotonin | 5HT1A | | |
| Dopamine | D2, D3, D4 | | |
| Adenosine | A1 | | |
| Cannabinoid | | | |
| Somatostatin | SSTR1 SSTR2 | | |
| VIP (vasoactive intestinal peptide) | | | |
| (Light) | Rhodopsin | Gt1 | cGMP phosphodiesterase (+) |
| | Cone opsins (blue, red, green) | Gt2 | |
| Taste (?) | | Gg | cAMP phosphodiesterase (+) |
| (?) | Gz | | (?) |
| **Group 3** | | | |
| Acetylcholine | Muscarinic —($m_1$, $m_3$, $m_5$) | Gq<br>$\alpha_{11}$ | Phospholipase C, $\beta$ type (+) |
| Glutamate | Metabotropic —(mGluR 1, 5) | $\alpha_{14}$<br>$\alpha_{16}$ | |
| (Nor)adrenaline | $\alpha_{1A}$, $\alpha_{1B}$, $\alpha_{1C}$ | | |
| Serotonin | 5HT1C, 5HT2 | | |
| Histamine | H1 | | |
| PAF (platelet-activating factor) | | | |
| Thromboxane | | | |
| Prostaglandin | | | |
| Tachykinin | NK1–3 | | |
| Endothelin | $ET_A$, $ET_B$ | | |
| Vasopressin | V1a | | |
| Bombesin | | | |
| Oxytocin | | | |

**Table 1** Classification of Membrane Receptors: Endogenous Ligands and Receptor Functions[a] (Continued)

| Ligands | Subtypes | G Proteins Involved | Function |
|---|---|---|---|
| (?) | | $\alpha_{15}$ | (?) |
| (?) | | $\alpha_{12}$ | (?) |
| | | $\alpha_{13}$ | (?) |
| *Receptors with a Single Transmembrane Segment* | | | |
| **Group 1** | | | |
| Epidermal growth factor | | | Tyrosine kinase (+) |
| Platelet-derived growth factor | | | |
| Fibroblast-derived growth factor | | | |
| Nerve growth factor | | | |
| Insulin | | | |
| Cytokine | | | |
| **Group 2** | | | |
| Natriuretic peptide | | | Guanylate cyclase (+) |

[a]Table includes receptors that have been cloned, although the list is not exhaustive.

porate the ion channel function into the oligomeric structure. Receptors for acetylcholine, glutamate, and serotonin are cation channels permeable to both $Na^+$ and $K^+$, and receptors for GABA and glycine are chloride channels. The binding of neurotransmitters to these receptors in the postsynaptic membranes leads to the opening of ion channels, followed by the depolarization and excitation of the postsynaptic membranes in the case of cation channels and by hyperpolarization and inhibition of excitation in the case of chloride channels. The time course of these responses is rapid, on the order of milliseconds. Thus ion channel receptors are specified for the purpose of rapid communication in the nervous system.

### 2.1 Nicotinic Receptors

The nicotinic receptor has been studied most extensively by taking advantage of electric organs with a high receptor density and snake venoms that specifically interact with the receptor. A pentameric glycoprotein composed of four distinct subunits ($\alpha_2\beta\gamma\delta$) was purified from electric organs with retention of the acetylcholine-gated cation channel activity. Biochemical studies such as cross-linking of receptors with acetylcholine analogues or channel blockers, molecular biological studies using chemical mutagenesis, and the optical analysis of electron micrographs of receptor microcrystals, put together, indicate three conclusions:

1. Amino-terminal portions of α- and other subunits that lie on outside of cell membranes are involved in the binding of acetylcholine.
2. Four hydrophobic domains in the carboxy-terminal part of each subunit ($M_1$–$M_4$) transverse membranes forming a pore (ion channel) in the middle of 20 transmembrane segments.
3. Five $M_2$ segments line the ion channel, and anionic charges

**Table 2** Classification of Membrane Receptors: Characteristics of Three Groups of Receptors

| Characteristic | Ion Channel Receptors | G-Protein-Linked Receptors | Receptors with a Single Transmembrane Domain |
|---|---|---|---|
| Endogenous ligands | Neurotransmitter | Neurotransmitter<br>Hormone<br>Autoacoid<br>Chemotactic factor<br>Exogenous stimulant | Growth factor hormone<br>Cytokine |
| Structure | Oligomer with a pore | Probably monomer | Monomer or oligomer with (±) catalytic domain |
| Number of transmembrane segments | Four per subunit | Seven | One per subunit |
| Function | Ion channel | Activation of G proteins | Tyrosine kinase<br>Guanylate cyclase (?) |
| Cellular responses | Depolarization or hyperpolarization | Depolarization or hyperpolarization<br>Regulation of function and expression of proteins | Regulation of function and expression of proteins<br>Proliferation or differentiation |

in both sides of $M_2$ segments contribute to the permeation of cations.

The binding of two acetylcholine molecules to the oligomer induces a conformational change, leading to an opening of the channel, followed by a closing of the channel (desensitization). Phosphorylation of nicotinic receptors by cAMP-dependent protein kinase, protein kinase C, or tyrosine kinase is known to facilitate the desensitization.

There are two kinds of nicotinic receptor in skeletal muscle: embryonic type ($\alpha_2\beta\gamma\delta$) and adult type ($\alpha_2\beta\delta\varepsilon$). Nicotinic receptors in the ganglia and brain are similar to but distinct from those in skeletal muscle, and some of the brain nicotinic receptors do not interact with snake venoms.

### 2.2 Glutamate Receptors

Glutamate receptors are classified into ionotropic (ion channel) and metabotropic (G-protein-linked) receptors, and the ion channel receptors are subdivided by specific agonists into NMDA (*N*-methyl-D-aspartate) and non-NMDA types. The non-NMDA types are further classified as AMPA (α-amino-3-hydroxy-5-methylisoxazole-4-propionic acid) or kainate. A number of different subunits have been identified (four for AMPA, five for kainate, and five for NMDA types), but it has not been determined how these subunits are combined to form oligomers in situ.

The molecular properties of the glutamate receptor are less well understood than those of the nicotinic receptor, but the fundamental properties are assumed to be similar. The non-NMDA receptor plays a major role in excitatory synaptic transmission in the central nervous system, whereas synaptic transmission at the neuromuscular junction and in the autonomic ganglia is performed by nicotinic receptors. The NMDA receptor is a cation channel like the non-NMDA receptor but is different from the latter in that the NMDA channel is permeable to $Ca^{2+}$ ions and opens only in the depolarized state because of inhibition by $Mg^{2+}$. The opening of the NMDA channel and entry of $Ca^{2+}$ are known to be necessary for the induction of the long-term potentiation in the hippocampus, which is considered to be related to the formation of memory.

### 2.3 GABA Receptors

$GABA_A$ and $GABA_B$ receptors are ion channel and G-protein-linked receptors, respectively. Fourteen kinds of $GABA_A$ receptor subunit have been identified (six α, five β, four γ, one δ, one ρ) by cDNA cloning, and various oligomers with different combinations of these subunits are thought to exist in situ. The GABA receptor is an anion channel, but its fundamental structure appears to be similar to that of the nicotinic receptor except that more positive amino acids are present in both sides of the transmembrane segments of the $GABA_A$ receptor: these positive charges are considered to contribute to the permeation of chloride ions. $GABA_A$ receptors have a major role in the inhibitory synaptic transmission in the brain. Glycine receptors are also chloride channels and play a role in inhibitory synaptic transmission in the spinal cord.

## 3 G-PROTEIN-LINKED RECEPTORS

The endogenous ligands of G-protein-linked receptors include all the neurotransmitters except glycine, most of the hormones and autacoids, several chemotactic factors, and exogenous stimulants such as odorants. The chemical species of these ligands are diverse and include amines, amino acids, nucleotides, lipids, peptides, and proteins. Rhodopsin, the receptor for light, is also one of the G-protein-linked receptors (Table 1).

Each of the G-protein-linked receptors is composed of a homogeneous protein and most probably exists as a monomer. The function of these receptors is to activate G proteins: that is, to stimulate the dissociation of GDP from G proteins and thereby facilitate the binding of GTP to G proteins. Activated G proteins stimulate or inhibit enzymes that synthesize or break down second messengers. Thus the information that hormones or neurotransmitters are present on the outside of cells is converted to the changes in the concentrations of second messengers, which affect the activities of proteins of different kinds (protein kinases, ion channels, transcription factors, etc.). In some cases G proteins directly regulate ion channels, but the time course of their opening or closing is much slower (10 ms–minutes), compared to the cases of ion channel receptors.

### 3.1 Structure and Function

An amino-terminal tail of the G-protein-linked receptor is considered to be on the outside of cells because consensus sequences for the attachment of sugars are present in this region for almost all receptors of this group. Seven hydrophobic regions (20–25 amino acid residues each) are assumed to traverse the cell membranes forming α-helices. The amino-terminal and carboxy-terminal tails, three intracellular and extracellular loops, are generally short (10–30 amino acid residues) except that some receptors have long tails and/or a long third intracellular loop. Retinal, an acceptor of light, is bound in the transmembrane segments of rhodopsin, and the binding sites of adrenaline and acetylcholine are also reported to be in the transmembrane segments. Receptors for proteins and glutamate have long amino-terminal tails, and these tails are considered to contribute to the ligand binding. The binding sites for G proteins have been reported to be the second and third intracellular loops and the carboxy-terminal tail, where a conformational change is assumed to be induced by the binding of an agonist.

G proteins are composed of three subunits (α,β,γ), and the α-subunits have the binding activity for GTP or GDP and the GTPase activity. G proteins are not transmembrane proteins but are bound to cell membranes, probably through prenyl groups or fatty acids attached to the γ- and α-subunits, respectively. G proteins bound with GTP are dissociated into $\alpha_{GTP}$- and βγ-subunits, and the $\alpha_{GTP}$, and βγ in some cases, activate the effector. The activity is terminated by hydrolysis of GTP followed by formation of the $\alpha_{GDP}\beta\gamma$ trimer. The conversion of $\alpha_{GDP}\beta\gamma$ into $\alpha_{GTP}$ and βγ is facilitated by the action of the receptor. The agonist-bound receptor forms a complex with guanine nucleotide-free G proteins (the aRG complex), which is evidenced by reconstruction of purified receptors and G proteins in artificial membranes. The activation energy of the reaction $G_{GDP} \rightarrow G \rightarrow G_{GTP}$ is considered to be reduced by taking the route of $G_{GDP} \rightarrow aRG \rightarrow G_{GTP}$.

### 3.2 Classification

Seventeen different α-subunit genes have been identified and they are classified into four groups according to the hemology of their amino acid sequences: Gs, Gi/Go, Gq and others (Table 1). Multiple species of four β- and six γ-subunits have been identified, but their functional differences are not yet clear. Receptors may be classified according to the species of α-subunits of G proteins which they activate.

Gs-$\alpha_{GTP}$ activates adenylate cyclase. At least six different adenylate cyclases have been cloned, and all are activated by Gs-$\alpha$. The addition of $\beta\gamma$-subunits in the presence of Gs-$\alpha$ further stimulates adenylate cyclase II and IV and inhibits the activity of adenylate cyclase I. The $\beta\gamma$-subunits alone neither stimulate nor inhibit these adenylate cyclases. Golf-$\alpha$ is an olfactory bulb specific G protein that activates adenylate cyclases like Gs-$\alpha$. Some G-protein $\alpha$-subunits ($\alpha_{i1}$, $\alpha_{i2}$, $\alpha_{i3}$, $\alpha_o$, $\alpha_{t1}$, $\alpha_{t2}$ and $\alpha_g$) are known to be ADP-ribosylated by pertussis toxin and NAD, and the ADP-ribosylated G proteins then lose the ability to interact with receptors. Thus the inhibition by pertussis toxin of receptor-mediated reactions indicates the involvement of these G proteins. In this sense, the inhibition of adenylate cyclase, the opening of inward-rectifier $K^+$ channel, and the inhibition of $Ca^{2+}$ channels are considered to be mediated by one of the three Gi's and Go. Gt (or transducin) is a retina-specific G protein, and Gt-$\alpha_{GTP}$ activates cGMP-phosphodiesterase: $\alpha_{t1}$ and $\alpha_{t2}$ are present in the rod and cone, respectively. Gg (or gustducin), a taste bud specific, transducinlike G protein, is suggested to activate cAMP-phosphodiesterase. Gq-$\alpha$ and related G proteins ($\alpha_{11}$, $\alpha_{14}$, $\alpha_{16}$) activate phosphatidylinositol-specific phospholipase C ($\beta$ types), which catalyzes the formation of inositol 1,4,5-triphosphate and diacylglycerol. The $\beta\gamma$-subunits alone are reported to stimulate the $\beta_2$ type of phospholipase C. The activation by $\beta\gamma$-subunits is inhibited by pertussis toxin when the $\beta\gamma$-subunits are derived from Gi or Go, in contrast with the activation by Gq-$\alpha$. The receptors and effectors for several G proteins such as Gz-$\alpha$, $\alpha_{12}$, $\alpha_{13}$, and $\alpha_{15}$ are not known.

Many receptors are known to activate, Gp and the $\beta$-adrenergic receptor is one of the best-characterized of these receptors. The $\alpha_2$-adrenergic and muscarinic ($m_2$ and $m_4$ subtypes) receptors are typical receptors linked to Gi and Go, and the $\alpha_1$-adrenergic and muscarinic ($m_1$ and $m_3$ subtypes) receptors are linked to Gq. Thus a single kind of neurotransmitter or hormone, noradrenaline, may activate at least three different G proteins through nine different receptors (three kinds each of $\alpha_1$, $\alpha_2$, and $\beta$ receptors). Acetylcholine, glutamate, and serotonin open ion channel receptors and also activate at least two kinds of G protein belonging to the Gi/Go and Gq classes (Table 1). Sensory receptors appear to activate specific G proteins: odorant receptors activate Golf, rhodopsin Gt1, cone pigments Gt2, and taste receptors Gg. The specificity of the interaction between receptors and G proteins is not necessarily absolute. For example, the substance-P receptors expressed in cultured cells or the $\beta$-adrenergic receptors in turkey erythrocytes are reported to activate both adenylate cyclase and phospholipase C.

## 4 RECEPTORS WITH A SINGLE TRANSMEMBRANE SEGMENT

A group of receptors for growth factors, such as epidermal growth factors, platelet-derived growth factors, fibroblast growth factors, and nerve growth factors, are ligand-stimulated tyrosine kinases. These receptors are composed of three domains: an extracellular domain responsible for ligand binding, a single-transmembrane segment, and cytoplasmic domains with both tyrosine kinase activity and tyrosine kinase substrate activity. Tyrosine kinase is activated by the ligand-binding and autophosphorylates the receptors themselves. The phosphorylated receptors bind to several proteins such as phosphorylase C ($\gamma$1 type), GTPase-activating protein (GAP), and phosphatidylinositol 3-kinase through the interaction of specific regions, known as SH2 in these proteins and as phosphorylated tyrosines in the receptors. This binding initiates a series of reactions leading to a regulation of gene expression and then the proliferation or differentiation of cells. The insulin receptor is also a tyrosine kinase.

Receptors for natriuretic peptides are composed of three segments: an extracellular, ligand-binding domain, connected via a single-transmembrane segment to a cytoplasmic domain with guanylate cyclase activity. Other types of guanylate cyclase are soluble proteins and lack transmembrane segments. The soluble guanylate cyclase is activated by nitric oxide through its interaction with a heme group in the enzyme. A group of receptors for cytokines also contain a single transmembrane segment and tyrosine kinase activity.

*See also* ENDOCRINOLOGY, MOLECULAR; STEROID HORMONES AND RECEPTORS.

### *Bibliography*

Barnard, F. A. (1992) Receptor classes and the transmitter-gated ion channels. *Trends Biochem. Sci.* 17:368–374.

Berstein, B., and Haga, T. (1990) Molecular aspects of muscarinic receptors. In *Current Aspects of the Neurosciences,* N. N. Osborne, Ed., pp. 245–284. Macmillan, London.

Galzi, J.-L., Revah, F., Bessis, A., and Changeux, J.-P. (1991) Functional architecture of the nicotinic acetylcholine receptor: From electric organ to brain. *Annu. Rev. Pharmacol. Toxicol.* 31:37–72.

Garbers, D. L. (1992) Guanylyl cyclase receptors and their endocrine, paracrine, and autocrine ligands. *Cell,* 71:1–4.

Hepler, J. R., and Gilman, A. G. (1992) G proteins. *Trends Biochem. Sci.* 17:383–387.

Kobilka, B. (1992) Adrenergic receptors as models for G protein-coupled receptors. *Annu. Rev. Neurosci.* 15:87–114.

Nakanishi, S. (1992) Molecular diversity of glutamate receptors and implications for brain function. *Science,* 258:597–603.

Schlessinger, J., and Ullrich, A. (1992) Growth factor signaling by receptor tyrosine kinases. *Neuron,* 9:383–391.

# RECOGNITION AND IMMOBILIZATION, MOLECULAR

*Irwin M. Chaiken*

### *Key Words*

**Affinity Chromatography** Molecular separation during flow through a solid phase support, by biospecific interaction of a solution phase molecule with counterligand attached to the solid phase.

**Antibody** A protein produced by the immune system which functions as a receptor for an antigen that is a foreign or nonself molecule.

**Biosensor** An instrument that detects molecules by a biological sensing reaction or interaction.

**Biospecificity** Selective molecular function in a biological system, usually due to selective interaction of biomolecules with receptors, antibodies, or other biological macromolecules.

**Design** Creation of a molecule with specific functional properties (e.g., the ability to bind to a receptor or counterligand).

**Diagnostic** An agent that binds to and therein detects the presence of a biomolecule associated with a disease state.

**Interaction Analysis** The measurement of the recognition process between biological molecules (e.g., between a receptor and its counterligand or between antibody and antigen).

**Receptor** A biological macromolecule that binds a complementary biomolecule, or counterligand, resulting in a function (through, e.g., signal transduction or storage and subsequent release).

**Screening** Scanning through a large set of compounds (e.g., natural products or synthetic compounds) to identify those with unique recognition or other activity properties (e.g., the ability to interfere with the interaction of receptor with its counterligand).

**Separation** The process of isolating compounds from mixtures (e.g., by flowing through an affinity chromatographic support to which various molecular components interact with different affinities).

**Therapeutic** An agent that acts to alleviate a disease process (e.g., by recognizing, binding, and therein blocking the deleterious interaction of a receptor and counterligand).

---

Molecular recognition is the selective interaction of molecules, for example, in biology and chemistry. In biology, specific molecular interactions provide the fundamental mechanism for selectivity in every aspect of biological structure and function. Understanding the basic forces that determine molecular recognition helps investigators to understand the mechanisms of biological processes as well as to discover innovative biotechnological methods and materials (e.g., for therapeutics, diagnostics, and separation science).

Molecular immobilization is the attachment of a molecule to a surface. Immobilization that preserves the capability of selective molecular interaction can yield active recognition surfaces. These surfaces can mimic the molecular recognition that occurs at biological surfaces—for example, on and in cells. Immobilized molecules can be used as active surfaces in biotechnology, including analytically to measure interaction processes with heterologous molecules. Hence, molecular immobilization can be used to analyze the mechanisms and utilize the selective power of molecular recognition.

## 1 INTRODUCTION

Macromolecular recognition is pervasive in biology and it provides a paradigm for molecular discovery and design in biotechnology. Essentially all processes in living organisms are underpinned by recognition events. And, increasingly, it is possible to identify the macromolecules responsible for recognition, such as receptors and immunoglobulins in the immune response and adhesion receptors and counterreceptors in cell communication, migration, and development. At the theoretical level, understanding biomolecular recognition can help one to decipher the structural basis of biology. On the practical side, biomolecules provide a powerful repertoire of recognition surfaces that can be used in biotechnology. Recognition

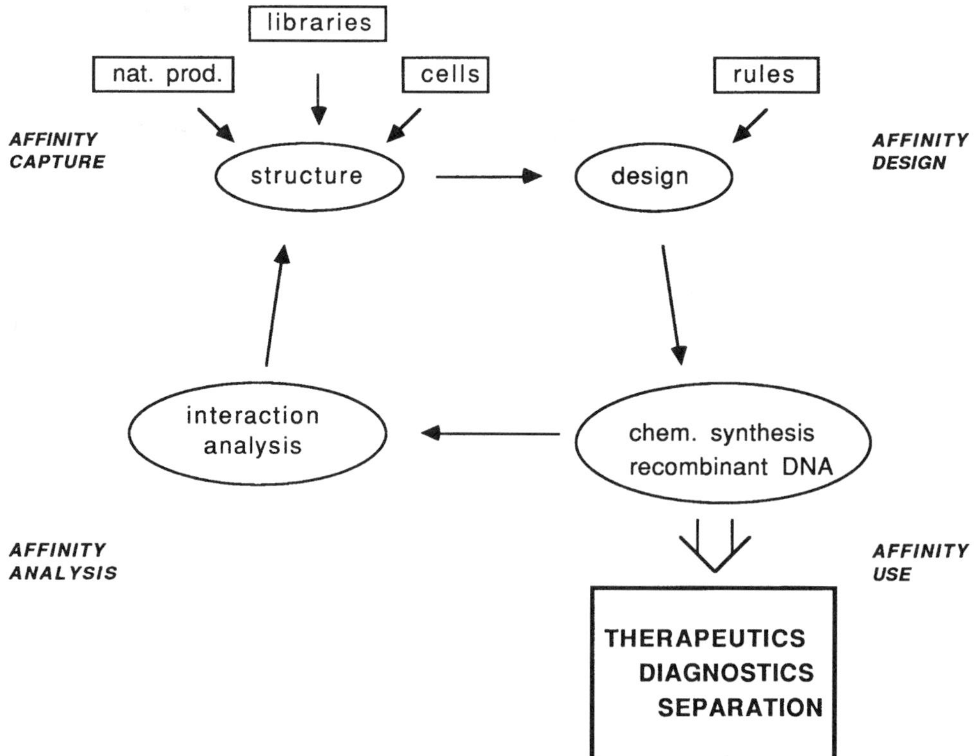


**Figure 1.** The cycle of molecular design in biotechnology and the dominant theme of molecular affinity in this cycle. The scheme shows the linkage of structure determination, redesign, synthesis of redesigned molecules, and their functional analysis as a cyclic process leading from the sources of molecular leads (chemical compounds, natural products, diversity libraries) to molecules of practical use (therapeutics, diagnostics, separation sciences).

elements in natural biomolecules provide starting points for designing molecular agents for therapeutics, diagnostics, and separation sciences by focusing on constructions that either bind to or mimic recognition surfaces of biological macromolecules.

As a culmination of this molecular logic, one might expect that with sufficient understanding of the mechanisms of recognition, interacting molecules could be designed rationally. At present, however, rational design is an ideal and not fully controllable. There is no simple set of rules for designing recognition molecules, and many disciplines must be brought together, as depicted in Figure 1, for the realistic design of molecules of use in biotechnology. This multidisciplinary network is dominated by affinity, including its discovery, design, synthesis, and analysis; hence the network can be seen as an affinity or recognition cycle.

## 2 ANALYZING MOLECULAR RECOGNITION USING IMMOBILIZED LIGANDS

### 2.1 Analytical Affinity Chromatography

Among the central goals in the design of recognition molecules are the analysis of the binding characteristics of native macromolecules and the detection and characterization of interacting counterligands, including other macromolecules. A methodology that has proven increasingly useful for interaction analysis is the use of immobilized ligands. In this methodology, one of the interactors is attached to a solid phase and the other is allowed to interact from free solution. This technique was first devised in a chromatographic format, defined as analytical affinity chromatography. It derived from the observation that affinity chromatography provides a powerful means to purify proteins and other biomolecules with a basic two-step retention–chaotropic elution procedure.

The first demonstration of preparative affinity chromatography was the now-classical demonstration for staphyloccal nuclease of Cuatrecasas et al., in 1969. This and many subsequent cases revealed several key features of immobilized ligand interactions with eluting macromolecules that enable preparative use, namely, *accessibility* of immobilized ligand, *selectivity* of ligand interaction with soluble macromolecule, and *reversibility* of macromolecule binding, which allows its elution without denaturation.

The effectiveness of affinity chromatography as a preparative tool, to purify proteins and other macromolecules selectively from mixtures, predicted the analytical usefulness of affinity chromatography. It was predicted by Dunn and Chaiken in the mid-1970s that isocratic elution of a macromolecule on an immobilized ligand support (i.e., elution with a nonchaotropic buffer at conditions allowing a dynamic equilibrium between association and dissociation) would be directly dependent on the equilibrium constant for the immobilized ligand–macromolecule interaction, hence ought to be reflected in the elution volume. Thus, by measuring the elution volume of a macromolecule on a column with immobilized ligand, one could determine affinity. Furthermore, competitive elution of the macromolecule ought to reflect quantitatively both the matrix and solution interactions of macromolecule and ligand (Figure 2A). The analytical use of affinity chromatography was demonstrated with staphylococcal nuclease (Figure 2B), on the same kind of affinity support that is used preparatively, but under conditions (in particular, decreased concentration of immobilized ligand) that allowed isocratic elution. Similar findings have been reported by now in many other systems (Figure 2C). It is of particular note that interaction analysis on affinity columns can be accomplished over a wide range of affinity and size of both immobilized and mobile interactors and can be achieved on a microscale dependent only on the limits of detectability of the interactor eluting from the affinity column.

### 2.2 Automation and Recognition Biosensors

The analytical use of immobilized ligands has been adapted to methodological configurations that allow for automation and expanded information. An early innovation of analytical affinity chromatography was its adaptation to high performance liquid chromatographs. High performance analytical affinity chromatography provides for more rapid macromolecular recognition analysis, at a more microscale level, and potentially using multiple postcolumn monitoring devices to increase the information gained about eluted molecules. Simultaneous multimolecular analysis also is feasible—for example, by weak affinity chromatography.

Chromatographic recognition analysis with immobilized ligands has led to the evolution of molecular biosensors. A technological breakthrough here was the development by Pharmacia of the surface plasmon resonance (SPR) biosensor called BIAcore. This instrument contains (Figure 3A) the immobilized ligand attached to a dextran layer on a gold sensor chip. It detects the interaction of macromolecules passing over the chip through a flow cell by changes of refractive index at the gold surface using SPR. Some recently obtained analytical binding data on the SPR biosensor are shown in Figure 3B.

The SPR biosensor is similar in concept to the apparatus used in analytical affinity chromatography: both involve interaction analysis of mobile macromolecules flowing over surface-immobilized ligands. The SPR biosensor also provides some unique advantages. These include (1) access to on- and off-rate analysis, thus providing more information for characterization and design, and (2) analysis in real time, a feature with the potential to stimulate an overall acceleration of molecular discovery. An evanescent wave biosensor for molecular recognition analysis called IAsys was introduced recently by Fisons.

Automation in the analytical use of immobilized ligands seems likely to continue to evolve. Analytical affinity chromatography increasingly is being adapted to sophisticated instrumentation and high throughput affinity supports. In addition, new methodological configurations with biosensors are being developed. These advances promise to expand greatly the accessibility of both equilibrium and kinetic data in basic and biotechnological research.

### 2.3 Relevance to Biology

A reasonable concern is that macromolecular interactions with immobilized ligands may not occur in the same way as interactions with ligands in free solution. Hence, properties measured by the former may not be representations of the more conventionally obtained measurements in dilute solution. Nonetheless, with most systems for which solid phase interactions have been quantitated by analytical affinity chromatography and analytical biosensor methods, parameters so measured are actually quite similar to those measured by solution assay methods such as spectroscopy, enzyme inhibition, calorimetry, analytical ultracentrifugation, gel filtration, and equilibrium dialysis. When differences do occur, they usually

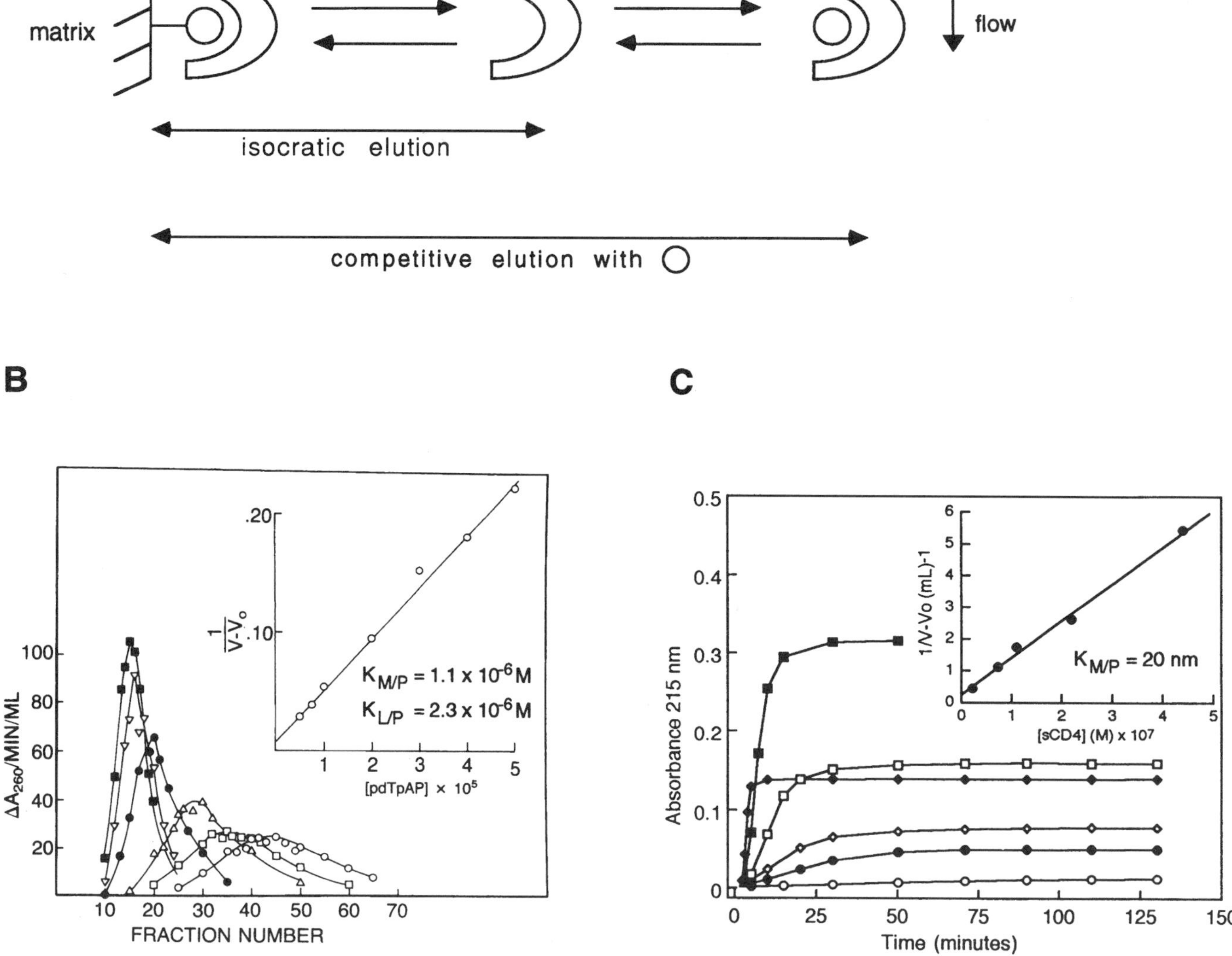


**Figure 2.** Scheme and experimental results of analytical affinity chromatography. (A) Scheme comparing isocratic and competitive elution in affinity chromatography and the consequence of these features in affinity analysis. In isocratic elution, the analyte (here P for protein) associates reversibly with immobilized ligand (M for matrix) and the volume of elution increases as the magnitude of the equilibrium dissociation constant defined as $K_{M/P}$ (for complex of M with P) decreases. In competitive elution, added soluble ligand (L) competes with immobilized ligand for binding to free analyte, so that the volume at which biomolecule elutes depends both on the relative affinities of the immobilized and soluble ligands, reflected in the respective dissociation constants $K_{M/P}$ and $K_{L/P}$ (for complex of L with P), and on the concentrations of these ligand species. (B) Zonal elution affinity chromatographic analysis of staphylococcal nuclease on the immobilized nucleotide ligand thymidine 3′,5′-diphosphate (pdTp). The concentrations (*M*) of competing soluble thymidine diphosphate were: 0.5 × $10^{-5}$; □, 0.75 × $10^{-5}$; △, 1.0 × $10^{-5}$; ●, 2 × $10^{-5}$; ○, 3 × $10^{-5}$; and ■, 4 × $10^{-5}$. *Inset:* linearized plot of competitive elution data and the dissociation constant derived in the analysis. [Adapted from Dunn and Chaiken (1975).] (C) Frontal elution affinity chromatographic analysis of soluble CD4 (soluble domain of T-cell receptor) on immobilized gp120 (envelope protein of HIV). The concentrations (n*M*) of sCD4 in the continuous elution experiment were: ○, 22; ●, 73; ◇, 110; □, 220; and ■, 440. The curve labeled ◆ is void volume elution. *Inset:* linearized variation of isocratic elution data and dissociation constant therein derived. [From Myszka et al. (1992) and Chaiken et al. (1994).]

→

**Figure 3.** Scheme and experimental results of molecular biosensor analysis and example of CD4-gp120. (A) Scheme depicting the flow path of soluble macromolecules over immobilized ligands in the surface plasmon resonance (SPR) biosensor (BIAcore of Pharmacia) and the SPR optical system. (Illustration by Thomas Morton, printed with permission.) (B) Experimental data for affinity analysis of the sCD4-gp120 interaction (Myszka et al., 1992). The sensorgrams are composed of an association phase upon injection of sCD4 at 100 seconds, and a dissociation phase upon return to buffer-alone flow at 580 seconds (after a small shift in response due to a shift in the bulk phase refractive index evident in the first 15 s after shifting from sCD4 in buffer to buffer alone). The sensorgram curves shown were from experiments at 25, 50, 112.5, 225, 550, 1100, 2200, and 4400 n*M* concentrations of sCD4 (increasing response with increasing concentration). (i) Sensorgrams for sCD4 flowing over sensor chip with gp120 immobilized. (ii) and (iii) On- and off-rate analyses from, respectively, sensorgram association (ascending phase) and dissociation (descending phase) processes. The on-rate analysis (ii) used all but the two sensorgrams obtained with the highest two concentrations of sCD4. The off-rate analysis (iii) is for the dissociation phase of the sensorgram obtained at 1100 n*M*.

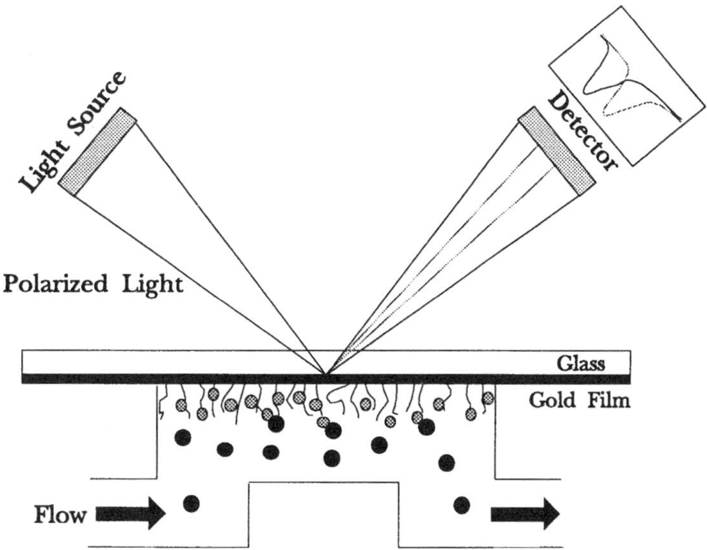
Light Source
Detector
Polarized Light
Glass
Gold Film
Flow

B

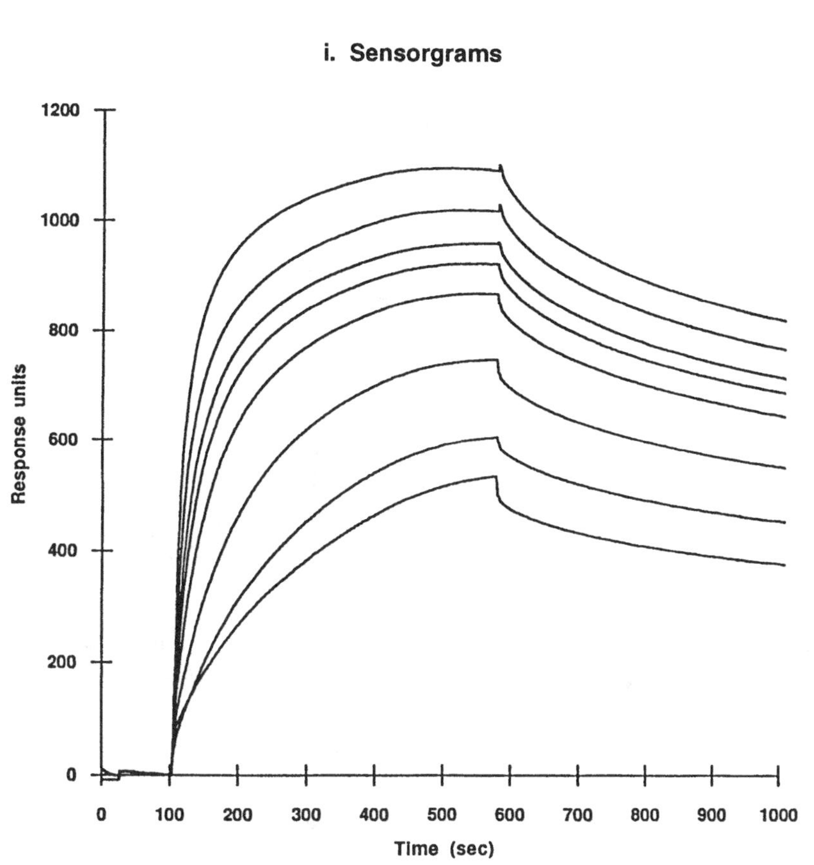
i. Sensorgrams
Response units
Time (sec)

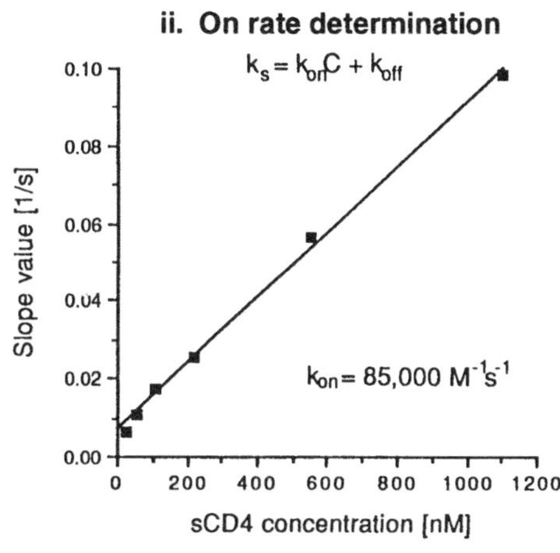
ii. On rate determination
$k_s = k_{on}C + k_{off}$
Slope value [1/s]
$k_{on}$= 85,000 $M^{-1}s^{-1}$
sCD4 concentration [nM]

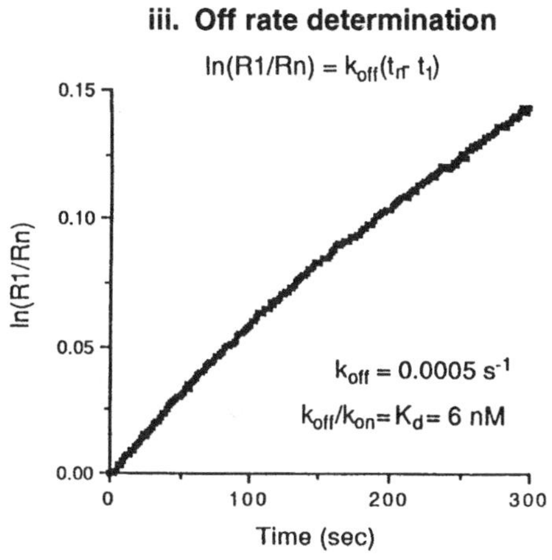
iii. Off rate determination
$\ln(R1/Rn) = k_{off}(t_n - t_1)$
ln(R1/Rn)
$k_{off}$ = 0.0005 $s^{-1}$
$k_{off}/k_{on} = K_d$ = 6 nM
Time (sec)

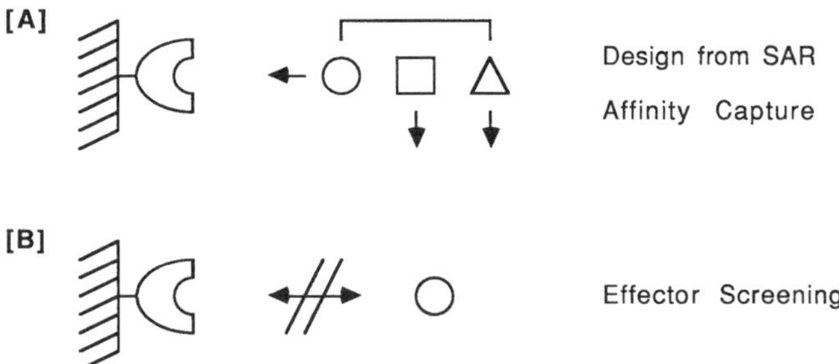


**Figure 4.** Scheme of approaches for using immobilized ligands to discover new recognition molecules. (A) Direct affinity capture and analysis. Direct capture can be used to separate molecular subsets with desired affinity from a diverse mixture or library of compounds. It also can be used to compare interaction properties of a series of compounds—for example, in carrying out a structure–activity relationship (SAR) analysis. (B) Effector screening–sequential testing of selected compounds or pools of compounds for ability to antagonize or enhance interaction of mobile with immobilized interactors. This approach can be used to screen for agonists as well as antagonists of macromolecular self-assembly.

can be explained mechanistically. It was found, for example, that the binding affinity of a bivalent IgA to immobilized phosphorylcholine can be greater than that with phosphorylcholine in solution. This increased affinity measured by analytical affinity chromatography is directly proportional to immobilized ligand density and can be explained by simultaneous bivalent interaction of IgA with the solid phase.

When differences do occur between interaction properties in solution and on immobilized ligands, they may be useful reflections of biological processes. Many macromolecular interactions in biological systems occur on solid phases such as cell surfaces. Hence, increased affinity caused by multivalency, as seen with the IgA/phosphorylcholine system just cited, is likely to be a common theme in biology. For example, multivalency may be a guiding principle in such surface interaction processes as cell adhesion. Furthermore, when heterogeneous samples can be analyzed, as with molecular biosensors, interaction properties can be acquired under solution conditions more closely related to the biological environment than is possible with fully isolated biomolecules in dilute solution. Thus, from several perspectives, measuring interactions with immobilized ligands can provide useful insights into biological recognition not easily obtained by solution methods.

## 3 SCREENING AND DESIGN: THE ROAD TO NEW RECOGNITION MOLECULES

### 3.1 Screening

As described in Section 1, discovering molecular agents in biotechnology in many cases is actually the discovery of recognition molecules (Figure 1). Immobilized ligands can be used as facilitating tools for affinity capture in screening for new recognition molecules. Typically in biotechnology, molecular agents are sought that are either direct binders of macromolecules (e.g., for diagnostics and separation) or antagonists of macromolecular interactions (e.g., for therapeutics). Mass transport affinity methods such as affinity chromatography enable identification of both direct binders and antagonists. Several strategies to do this are shown schematically in Figure 4. Of particular note in screening is the potential (Figure 4A) to use affinity columns for "affinity capture" of interactors from diversity libraries such as phage display libraries. Peptide, oligonucleotide, chemical, and other libraries are increasingly being used as sources of starting materials to identify novel mimetics and antagonists. Affinity chromatography has some potential advantages over the microtiter plate methods that often are used in current library screening practice. These advantages include the ability to select molecules with a specified range of affinity during affinity capture and greater capability to identify affinity constants during or in conjunction with affinity capture. Affinity capture has been used successfully, for example, in screening antibody phage display libraries.

### 3.2 Design

The fast analytical capability of affinity columns and molecular biosensors as already described also has potential use to perform rapid comparative affinity analysis of a series of newly configured molecules both as direct binders to immobilized ligands (Figure 4A) and as competitors of macromolecular interactions with immobilized ligand (Figure 4B). Again, as with screening, other methods such as enzyme-linked immunosorbent assays on microtiter plates often are used as design aids. Once more, however, affinity interactions on immobilized ligands can provide more quantitative information on the interactions, including affinity, stoichiometry, and, in the case of molecular biosensors, rates of association and dissociation. Analytical affinity chromatography has been used as a tool to guide design of "antisense family" peptides as ligands for protein separation and diagnostics. Affinity analysis on immobilized ligands currently is being used to guide both the design of mutant forms of CD4 and the screening and humanization of interleukin 5 monoclonal antibodies.

## 4 FUTURE DIRECTIONS

Molecular recognition is a phenomenon so central to the molecular basis of biology that defining its underlying principles and mechanisms is certain to attract continued scientific investigation. And, discovering and designing recognition molecules also remains an encompassing goal in the biotechnology of therapeutics, diagnostics, and separation science. At the interface of these pursuits is the growing development of enabling instrumentational technologies for detecting and measuring recognition at the molecular level. The merge of pure and applied sciences in the field of biomolecular affinity promises to be an exciting field for some time to come.

*See also* Antisense Oligonucleotides, Structure and Function of; Biosensors; Combinatorial Phage Antibody Libraries; Immunology; Receptor Biochemistry.

## *Bibliography*

Buckle, P. E., Davies, R. J., Kinning, T., Yeung, D., Edwards, P. R., Pollard-Knight, D., and Lowe, C. R. (1993) *Biosensors Bioelectronics,* 8:355–363.

Chaiken, I. M., Ed. (1987) *Analytical Affinity Chromatography.* CRC Press, Boca Raton, FL.

Chaiken, I. M. (1993) *Handbook of Affinity Chromatography,* T. Kline, Ed., pp. 219–227. Dekker, New York.

———, (1993) In *Molecular Interactions in Bioseparation,* T. Ngo, Ed., pp. 169–177. Plenum Press, New York.

———, Rose, S., and Karlsson, R. (1992) *Anal. Biochem.* 201:197–210.
———, Myszka, D., and Morton, T. (1993) In *Advances in Molecular Cell Biology* (*Klaus Mosbach Symposium*). JAI Press, Greenwich, CT.
Cuatrecasas, P., Wilchek, M., and Anfinsen, C. B. (1968) *Proc. Natl. Acad. Sci. U.S.A.* 61:636–643.
Cush, R., Cronin, J. M., Stewart, W. J., Maule, C. H., Molloy, J., and Goddard, N. J. (1993) *Biosensors Bioelectronics,* 8:347–353.
Dunn, B. M., and Chaiken, I. M. (1974) *Proc. Natl. Acad. Sci. U.S.A.* 71:2382–2385.
———, and ———. (1975) *Biochemistry,* 14:2343–2349.
Eilat, D., and Chaiken, I. M. (1978) *Biochemistry,* 18:790–794.
Fagerstam, L. G., Prostell, A., Karlsson, R., Kullman, M., Larson, A., Malmqvist, M., and Butt, H. (1990) *J. Mol. Recognition,* 3:208–214.
Fassina, G., and Cassani, G. (1992) *Biochem. J.* 282:773–779.
———, and Chaiken, I. M. (1987) *Adv. Chromatogr.* 27:248–297.
———, Zamai, M., Brigham-Burke, M., and Chaiken, I. M. (1989) *Biochemistry,* 28:8811–8818.
———, Roller, P. P., Olsen, A. D., Thorgeirsson, S. S., and Omichinski, J. G. (1989) *J. Biol. Chem.* 264:11252–11257.
———, Cassani, G., and Corti, A. (1992) *Arch. Biochem. Biophys.* 196:137–143.
Johnsson, B., Lofas, S., and Lindquist, G. (1991) *Anal. Biochem.* 198:268–277.
Jonsson, U., Fagerstam, L., Iversson, B., Johnsson, B., Karlsson, R., Lundh, K., Lofas, S., Persson, B., Roos, H., and Ronnberg, I. (1991) *Biotechniques, 11:620–627.*
Kasai, K. I., and Ishii, S. I. (1975) *J. Biochem.* 77:261–264.
McCafferty, J., Griffiths, A. D., Winter, G., and Chiswell, D. J. (1990) *Nature,* 348:552–554.
Myszka, D., Granzow, R., and Chaiken, I. M. (1992) *J. Mol. Recognition,* 5:118.
Nichol, L. W., Ogston, A. G., Winzor, D. J., and Sawyer, W. H. (1974) *Biochem. J.* 143:435–443.
Ohlson, S., and Zopf, D. (1993) In *Handbook of Affinity Chromatography,* T. Kline, Ed., pp. 299–314. Dekker, New York.
Swaisgood, H. E., and Chaiken, I. M. (1987) In *Analytical Affinity Chromatography,* I. M. Chaiken, Ed. CRC Press, Boca Raton, FL.

# Recombination, Molecular Biology of

*Hannah L. Klein*

## *Key Words*

**Crossing Over** A recombination that involves physical exchange of DNA between nonsister strands of homologous chromosomes, resulting in a nonparental combination of linked markers.

**Gene Conversion** Nonreciprocal recombination between homologous chromosomes; one allele acts as the donor of information to convert the recipient allele to the donor sequence.

**Heteroduplex** DNA hybrid formed from single strands of two nonsister chromatids; an intermediate in recombination.

**Holliday Junction** Crossed-strand structure consisting of two DNA duplexes joined by a bridge as an intermediate in recombination.

**Mismatch Repair** Repair of mismatched bases in heteroduplex DNA by a dedicated system, resulting in homoduplex DNA.

**Resolution** Cutting of the Holliday junction by a resolvase to disconnect the recombining DNA duplexes, followed by restoration of the parental configuration of markers, resulting in a nonparental configuration of markers termed crossovers.

**Strand Exchange** One of the processes for forming heteroduplex DNA during recombination: a strand from a homoduplex is displaced and a strand from a nonsister duplex is transferred in to form a hybrid region on the chromatid.

**Transformation** Addition of exogenous DNA to a cell to change the genotype of the recipient cell.

---

Homologous recombination is an essential activity of all cells. Meiotic recombination serves two purposes. First, it gives genetic variation to future generations by providing new combinations of alleles in the gametes. Second, meiotic recombination is required for proper segregation of homologous chromosomes at the first meiotic division; in the absence of crossing over, chromosomes segregate randomly. Mitotic recombination, which occurs at significantly reduced rates compared to meiotic recombination, is nonetheless an important aspect of vegetative growth and is one method used by the cell to repair DNA damage that can result from errors in DNA replication or outside agents such as irradiation and toxic chemicals.

## 1 RECOMBINATION IN GENETIC TERMS

Homologous recombination has two different recombinant outcomes. Crossing over, often referred to as reciprocal recombination, involves the physical joining of two parental molecules of homologous chromosomes. Genetically this is detected as recombinant progeny. If one parent has the linked markers AB and the other parent has the alternate alleles ab, the crossover gametes have the genotypes Ab and aB (Figure 1A). Gene conversions, often referred to as nonreciprocal recombination, involve the transfer of information from one parental molecule to another parental molecule. The genotype of the donor molecule remains unchanged, while the genotype of the recipient is converted to the donor genotype (Figure 1B).

Nonreciprocal recombination or gene conversion is best studied in the fungi, where all four products of a single meiosis can be recovered. Gene conversion does not result from mutation. Genetic and molecular studies have shown that the gene conversion product has the same DNA sequence as the donor sequence. Occasionally a meiotic product is recovered that has genetic information from both parents. This mosaic product is called a postmeiotic segregation (PMS) event. In wild-type strains, gene conversion events occur at a much higher frequency than PMS events. PMS events result from the failure to repair a DNA mismatch in heteroduplex DNA. Mutants that are defective for mismatch repair show higher PMS frequencies and reduced gene conversion frequencies, indicating that heteroduplex formation is an intermediate in gene conversion.

Crossing over and gene conversion are related events in meiotic recombination. When a gene conversion occurs in meiotic recombination, about 50% of the time there is an associated crossing over in the flanking region.

A. Crossing-over

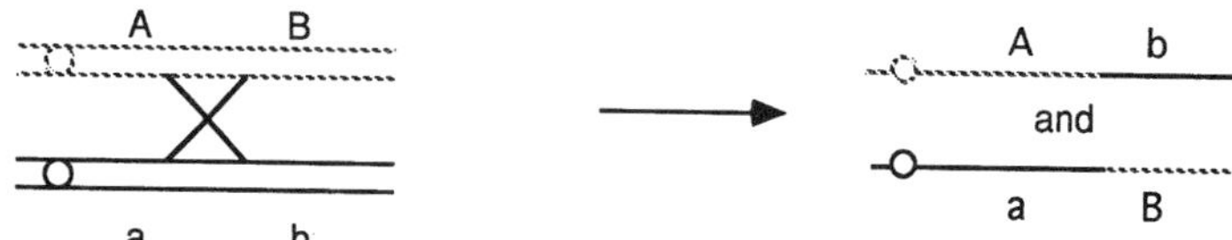


B. Gene conversion

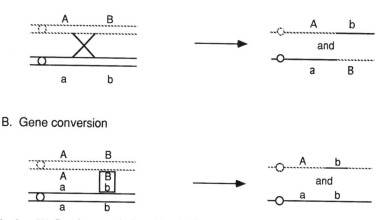


**Figure 1.** Homologous recombination. (A) Crossing over, indicated by the X, generates two recombinant chromosomes. (B) Gene conversion, indicated by the box, yields only one recombinant chromosome (top chromosome on the right). This chromosome has only a small region of substitution, at the B gene, from the donor chromosome. The donor chromosome (shown below the recombinant chromosome on the right) remains unchanged and is of the parental genotype.

The relationships between PMS events and gene conversion events, and between gene conversion events and crossing over, have led to the proposal of several molecular intermediates in recombination. First, a recombination initiating event that breaks the DNA backbone of at least one DNA duplex must occur. Next heteroduplex DNA is formed through the recognition of the homologous chromosome and transfer of the broken strand of DNA into the homologous chromosome. Strand transfer is effected by enzymes that must recognize the homologous partner and promote the formation of heteroduplex DNA. The bacterial enzyme RecA is the prototype strand exchange protein. At some point the strand transfer reaction is stopped and the strand exchange is stabilized by the formation of a crossed-strand structure called the Holliday junction.

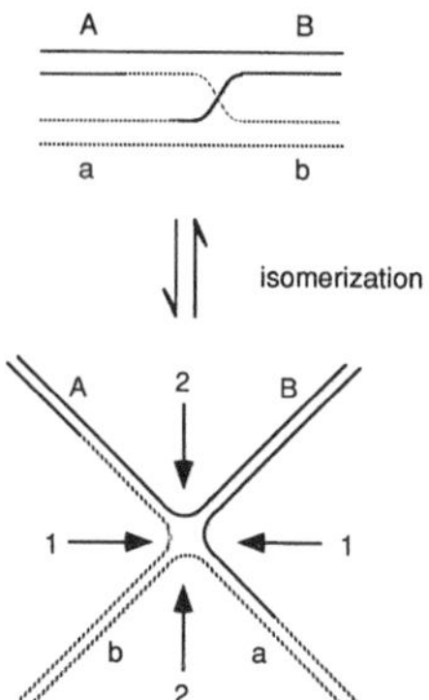


Resolution at sites 1 = noncrossover products A B and a b

Resolution at sites 2 = crossover products A b and a B

**Figure 2.** Resolution of a Holliday junction: in the crossed-strand configuration (*top*), and after isomerization, in an opened configuration, with local regions of heteroduplex (*bottom*). If the junction is resolved by cutting at positions 1 and subsequent ligation to the other cut strand, chromosomes with the parental configuration of A B and a b are formed. If the junction is resolved by cutting at positions 2, chromosomes with the recombinant

Formation of heteroduplex DNA is the first step in gene conversion. Mismatch repair can lead to gene conversion, while failure to repair results in PMS. This is why gene conversion and PMS events are linked and mutants that are defective in mismatch repair show increased PMS and reduced gene conversion.

Once a recombination intermediate with a Holliday junction has been formed, the crossed structure must be resolved by a resolvase, an endonuclease that recognizes the crossed structure. The four-armed structure shows strand equivalence and can be resolved in two alternate modes. Resolution in mode 1 yields the parental configuration of markers, while resolution in mode 2 yields the recombinant configuration (Figure 2). This bimodality explains why gene conversion and crossing over often occur together. The recombination intermediate has heteroduplex DNA and a Holliday junction. Mismatch repair of the heteroduplex results in gene conversion. Resolution of the Holliday junction gives crossing-over products 50% of the time, reconciling the 50% association of crossovers of closely linked markers that flank a gene conversion event.

## 2 RECOMBINATION USING PLASMID SUBSTRATES

DNA transformation experiments in the yeast *Saccharomyces cerevisiae* have led to an alternate model of homologous recombination. When intact plasmid molecules that carry a yeast gene are applied to yeast cells, the cells take up the plasmid and express the yeast gene or marker located on the plasmid after it has been integrated into the yeast genome. The site of integration is not random, instead it occurs by homologous crossing over. Thus transformation in yeast can be viewed as a model for homologous recombination. Although transformation using intact plasmid molecules is relatively efficient, the frequency can be greatly increased if the plasmid molecules are first treated with a restriction enzyme that cuts within the yeast gene located on the plasmid.

When yeast cells are treated with linear plasmid molecules, transformation still occurs by homologous crossing over. If the

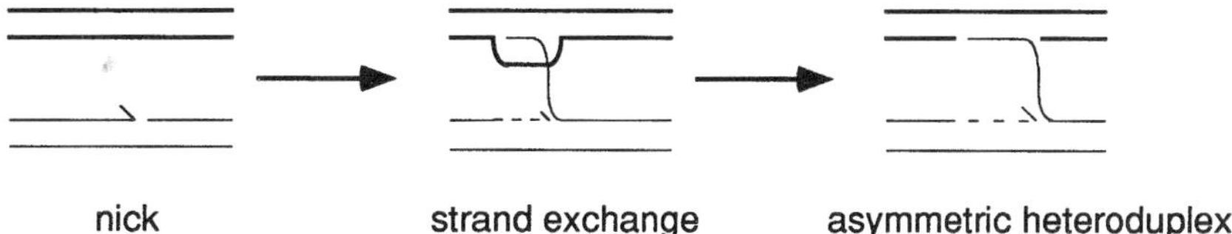


**Figure 3.** Formation of asymmetric heteroduplex in the Meselson–Radding model of homologous recombination. A nick is made in one strand of a chromatid. The 5′ end of the nicked strand is displaced by DNA synthesis at the 3′ end of the nick. The displaced strand then invades the homologous sequence of a nonsister chromatid and by strand exchange forms heteroduplex on one duplex with the displaced single strand, making a structure called a D-loop. The D-loop is then degraded, and the gap on the donor duplex is filled in by repair synthesis.

plasmid molecule contains two unlinked yeast genes A and B, the plasmid molecule will integrate into either the A gene or the B gene with approximately equal efficiency in transformation experiments. However if the plasmid is first treated with a restriction enzyme that cuts only in the A gene, most transformation events will be integrations at the A gene, demonstrating that double-strand breaks are highly recombinogenic.

When the linear transforming plasmid contains a gap in a yeast gene, it is repaired using the chromosomal information. This is a gene conversion event and shows that double-strand breaks are recombinogenic for both gene conversion and crossing-over events. Transformation using gapped molecules has also been performed using molecules that contain a sequence that acts as an origin of DNA replication. Since these molecules can replicate autonomously, they are not obliged to integrate, although the gap must be repaired. Transformation with such molecules is associated with repair (gene conversion) and homologous crossing over (integration) 50% of the time and with simple repair (gene conversion without crossing over). This result gave further support for the double-strand-break repair model as a general model for meiotic recombination.

## 3 MODELS OF RECOMBINATION

The Meselson–Radding model of recombination is based on an earlier model proposed by Holliday. The key features of both models include a nick in a DNA strand to initiate a strand exchange reaction between paired homologous DNA duplexes to form heteroduplex DNA. The Meselson–Radding model proposed that only one of the two recombining duplexes would experience a nick, so heteroduplex DNA would be formed asymmetrically on one duplex (Figure 3). This modification explained the genetic data of the occurrence of PMS events where there was evidence of heteroduplex on only one of the two recombining duplexes. The proposed strand transfer reaction was similar to a reaction that was catalyzed in vitro by the RecA protein.

In this model, strand transfer occurs on one duplex and the gap on the donor strand is filled in by repair synthesis. At some point the concerted strand transfer/formation of heteroduplex/repair synthesis events cease and the free ends are ligated together, forming the Holliday junction (see Figure 2).

Resolution of the Holliday junction by an endonuclease that specifically recognizes the structure releases the two duplexes. Such enzymes have been found in bacteria and their viruses, and similar activities have been reported in eukaryotes. Depending on which of the strands at the Holliday junction are cut, the religated duplexes may have the parental configuration of markers (noncrossover) or the recombinant configuration of markers (crossover) (see Figure 2). The heteroduplex DNA region is recognized by the mismatch repair enzymes and corrected to homoduplex. Depending on which single strand is used as template, the correction can restore normal segregation or give a gene conversion segregation. Failure to correct the mismatch gives a PMS segregation.

The double-strand-break repair model was developed from observations on transformation in yeast that a double-strand break in a homologous DNA sequence stimulated homologous recombination. In this model the initiating event in homologous recombination is a double-strand break (Figure 4). The ends are processed and can invade a homologous duplex in a reaction promoted by a RecA type of protein. The gap is repaired by two rounds of repair synthesis. Heteroduplex DNA flanks the gap, the length of which depends on the initial processing of the break to give long single-strand tails. Heteroduplex DNA may form asymmetrically with respect to the double-strand break. Gene conversions can result in two ways: by repair of the gap or by mismatch repair of the heteroduplex region. PMS events occur only by failure of mismatch repair of the heteroduplex region. Variations include a short gap and long single strand tails flanking the gap, such that most gene conversion events result from mismatch repair.

A particularly novel feature of the model is the double Holliday junction. To reconcile the possible resolution modes of this structure with observations of the occurrence of close crossovers, the model proposes that one Holliday junction is always resolved in the noncrossover mode. The other may be resolved as a crossover or as a noncrossover.

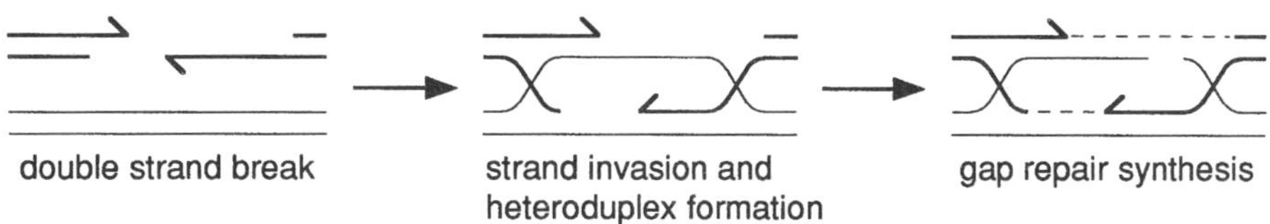


**Figure 4.** Double-strand-break repair with a long 3′ single-strand tail to give a heteroduplex DNA intermediate. When the long 3′ single-strand tail invades the nonsister chromatid, a region of heteroduplex DNA is formed. The gap is eliminated by repair synthesis on both strands.

## 4 MOLECULAR STEPS IN RECOMBINATION

No in vitro system has been established for homologous recombination, but several of the molecular events are deduced from genetic crosses. In a few cases an in vitro assay has been developed for a proposed step, such as the strand exchange reaction, mismatch repair, or the resolution of an artificial Holliday junction. In other cases there are in vivo molecular assays for detection of specific double-strand breaks or formation of heteroduplex DNA. Combined with genetic data, these assays have led to the following scenario in homologous recombination.

Double-strand breaks have been observed at meiotic genetic hot spots of recombination. Many data indicate that this is an early step in recombination in yeast. Whether other initiating substrates such as single-strand nicks and gaps also occur at hot spots is not known because the assays to detect such intermediates in vivo are difficult to perform. Nonhomologous recombination events between chromosomes are rare, however, so there must be a mechanism to determine that the recombining molecules are homologous. Most likely, there exists a system to scan homology, either by side-by-side pairing of homoduplexes or by local melting of the duplexes, with the transient formation of heteroduplexes, which are assessed for the quality of sequence match. Pairing proteins similar to RecA and mismatch repair proteins similar to the bacterial UvrD DNA helicase (DNA helicase II) could be involved in the homology search process. Eukaryotic homologue of both the *RecA* gene and the bacterial mismatch repair genes have been identified and shown to have a role in recombination and repair.

After the recombination initiating lesion has been formed it is processed by exonucleolytic degradation to form a single-strand gap or a break with single-strand tails. In yeast, meiotic double-strand breaks have been shown to have single-strand tails about 300 base pairs in length, with 3′ ends. Pairing and strand exchange is promoted by RecA-like functions that allow the formation of heteroduplex. Continued formation of heteroduplex should require exonucleases, DNA synthesis machinery at the 3′ ends in conjunction with strand displacement (see Figure 3) to repair gaps and breaks (see Figure 4), and DNA ligase to seal the nicks. Single-strand DNA-binding proteins (SSBs) may be involved in the strand displacement reaction.

The recombination intermediate contains both heteroduplex DNA and a crossed-strand Holliday junction. Mismatch repair proteins specifically recognize the DNA mismatches and correct them, using one DNA single strand as template, which can lead to a gene conversion product. The Holliday junction is resolved by specific proteins called resolvases that recognize the unique four-way junction. There is no sequence specificity to the junction, which means that the structure of the Holliday junction must determine the specificity of the cutting reaction to place nicks symmetrically opposed to the junction. Holliday junction cutting will yield two nicked duplexes. Further processing by DNA ligase will restore the molecules to intact duplexes.

## 5 RELATIONSHIP BETWEEN RECOMBINATION AND REPAIR

Several of the molecular intermediates of recombination also occur during DNA repair. There are two types of overlap between repair and recombination. First, repair can occur through a recombinational event. In this situation DNA-damaging agents stimulate recombination. When the recombinational repair pathway misfunctions, an increase in DNA damage (repair deficiency) and a decrease in recombination (recombination deficiency) are seen. Second, there are repair events that are nonrecombinational. These include the recognition and repair of DNA damage, such as a base modification or mismatch and nicks and gaps in a DNA single strand. Enzymes that specifically recognize these lesions bind to them, excise the damaged region, and then fill in the single-strand gap through a repair DNA synthesis reaction. A similar reaction occurs during recombination, and thus there is a functional overlap between recombination and repair enzymes. When this pathway misfunctions, an increase in DNA errors (mutation) and a decrease in mature recombinants (aberrant recombination) is seen, but recombination events are nonetheless initiated.

The recombinational repair enzymes that repair double-strand breaks function in both repair and recombination. In *Escherichia coli* and yeast, such genes when defective show DNA repair defects in vegetative growth and often are also reduced in recombination. The yeast mutants are greatly reduced in meiotic recombination and usually do not yield viable meiotic products.

The second set of enzymes consists of those that function in a repair process that can be independent of recombination but also happens to occur during recombination. Examples of this type of enzyme are the mismatch repair proteins. Mismatch repair occurs whenever DNA mismatches arise. In mitotic growth most mismatches occur from errors during DNA replication; such errors, which are due to the use of an incorrect nucleotide, result in a base pair mismatch. If the mismatch is not corrected by the editing function of the DNA polymerase, it becomes a target for the mismatch repair proteins. Mismatch repair is an essential step in recombination, correcting the heteroduplex that has formed between two recombining DNA strands. In the absence of mismatch repair, mutations will accumulate and the strains will have a mutator phenotype. The strains are able to undergo recombination, but maturation of the heteroduplex is defective. Such mutants will show increased PMS or sectored colonies.

Other common steps involve the repair synthesis that occurs during double-strand-break repair and also during excision repair (repair of UV damage) and mismatch repair. It is not known which DNA polymerases are involved in these repair reactions, although there are data to suggest that excision repair may involve the same DNA polymerases that function in DNA replication. In vitro systems from eukaryotic sources for mismatch repair and recombination will yield answers to this question.

*See also* DNA Damage and Repair; DNA Repair in Aging and Sex; RecA Protein, Structure and Function of.

### *Bibliography*

Haber, J. E. (1992) Exploring the pathways of homologous recombination. *Curr. Opin. Cell Biol.* 4:401–412.

Kucherlapati, R., and Smith, G. R., Eds. (1988) *Genetic Recombination.* American Society for Microbiology, Washington, DC.

Low, K. B., Ed. (1988) *The Recombination of Genetic Material.* Academic Press, Orlando, FL.

Petes, T. D., Malone, R. E., and Symington, L. S. (1991) Recombination in yeast, in *The Molecular Biology of the Yeast Saccharomyces,* J. R. Broach, J. R. Pringle, and E. W. Jones, Eds., pp. 407–521. Cold Spring Harbor Laboratory Press, Plainview, NY.

West, S. C. (1992) Enzymes and molecular mechanisms of genetic recombination. *Annu. Rev. Biochem.* 61:603–640.

## Regulation of Gene Expression: *see* Gene Expression, Regulation of; Plant Gene Expression Regulation.

## Regulations: *see* Biotechnology, Governmental Regulation of.

# Renal System

*Samir S. El-Dahr, R. Ariel Gomez, and L. Gabriel Navar*

### *Key Words*

**Granular Cells** Specialized granular epithelioid cells of the afferent arteriole in the vicinity of the glomerulus. These are the main renin-producing cells of the kidney.

**In Situ Hybridization** Technique that allows the detection and localization of specific messenger RNAs in a tissue section using labeled complementary DNA or RNA probes.

**Juxtaglomerular Apparatus (JGA)** A complex system located at the vascular pole of each glomerulus. It consists of the afferent and efferent arterioles at the glomerular hilus, the granular cells of the afferent arteriole, the extraglomerular mesangium, and the adherent macula densa segment of the ascending limb of Henle's loop.

**Macula Densa Cells** A specialized plaque of epithelial cells located at the distal end of the thick ascending limb approximately 100 to 200 μm upstream from the transition to the distal convoluted tubule and in juxtaposition to the vascular pole of its corresponding glomerulus.

**Reverse Transcription–Polymerase Chain Reaction (RT-PCR)** Two-step technique consisting of the transcription of messenger RNA back into complementary DNA (reverse transcription), and amplification of the target DNA using the standard polymerase chain reaction. The DNA is then resolved on a polyacrylamide or agarose gel for qualitative assessment. This technique can be made quantitative by including in the reaction a competing template as an internal control consisting of genomic or engineered DNA: "competitive RT-PCR." RT-PCR, the most sensitive method used to detect extremely small amounts of specific RNA, is $10^3$ to $10^4$ times more sensitive than traditional Northern blot analysis.

---

The kidney is a complex organ composed of many cell types including vascular smooth muscle, endothelial, epithelial, and fibroblast cells. Each of these cell types is equipped with the molecular mechanisms necessary to produce vasoactive and growth-modulating factors as well as membrane transport proteins and diverse ion channels. The basic functional unit of the kidney is the nephrovascular complex, which consists of the glomerulus and associated afferent and efferent arterioles, the postglomerular capillary system (cortical and medullary), and the nephron. The nephron is further subdivided into the proximal convoluted tubule, the straight proximal tubule, the descending and ascending limbs of Henle, the distal convoluted and connecting tubules, the cortical collecting tubule, and the medullary collecting duct. Renal blood flow, glomerular filtration rate, and urinary excretion are regulated by an intricate interaction among numerous vasoactive and tubular transport mechanisms. Because of the complex interactions between the physically determined processes of glomerular filtration and peritubular reabsorption and the metabolically determined processes that govern tubular transport, it is imperative that the various intrarenal regulatory endocrine, paracrine, transport, and enzymatic systems be subjected to precise control. Much of this control rests on the mechanisms that govern expression of the genes encoding the various components of these systems.

This entry summarizes the current status and the molecular mechanisms that regulate some of these major intrarenal systems. Emphasis is placed on studies that specifically pertain to renal mechanisms and for which data have been obtained from kidney tissue. For reasons of space, each section is limited to selected intrarenal systems; other important systems are not considered. The vasoactive systems discussed are the renin–angiotensin system, the kallikrein–kinin system, the nitric oxide synthase–EDRF system, and endothelin. Vasoactive systems not included in our discussion are the dopaminergic and purinergic systems, prostaglandins–leukotrienes–cytochrome P450 systems, and natriuretic peptides. In Section 2 we discuss platelet-derived growth factor and receptors, insulinlike growth factors and receptors, transforming growth factor β, and epidermal growth factor. Growth factors synthesized and acting in the kidney not discussed in this chapter include fibroblast growth factor and transforming growth factor α. In Section 3 we discuss the various $Na^+$-$H^+$ exchanger isoforms, $Na^+$-$K^+$ ATPases, and water channels, but other transporters recently cloned are not covered.

## 1 VASOACTIVE SYSTEMS

The kidney expresses all the components of many important vasoactive systems. The products of these systems are either vasoconstrictor or vasodilator, and they play important roles in the regulation of renal vascular resistance, glomerular blood flow, and glomerular filtration rate.

The *renin–angiotensin system (RAS)* is vasoconstricting and antinatriuretic. The genes encoding the RAS components (renin, angiotensinogen, angiotensin-converting enzyme, and angiotensin II receptors) are all expressed within the kidney in a cell-specific manner. *Renin,* the key enzyme in this system, is synthesized, stored, and released from the granular juxtaglomerular cells of the afferent arteriole. In the nonstressed animal, renin-expressing cells are limited to the juxtaglomerular portion of the afferent arteriole. Given an appropriate stimulus (e.g., angiotensin-converting enzyme inhibition), renin and its mRNA will extend further up the length of the afferent arteriole, indicating that the preglomerular vascular smooth muscle cells can transform into granular renin-synthesizing cells. The mechanisms underlying the "cell recruitment" phenomenon and the associated phenotypic change remain to be defined.

Once released from granular juxtaglomerular cells, renin cleaves the substrate, *angiotensinogen,* to form the nonactive decapeptide angiotensin I. The carboxy-terminal amino acids histidine and leucine of angiotensin I are subsequently cleaved by circulating or tissue-associated *angiotensin I converting enzyme (ACE),* producing the effector molecule of the RAS angiotensin II.

Following synthesis in the liver, angiotensinogen is released rapidly into the circulation, which accounts for the low levels of stored hepatic substrate. Other tissues expressing the angiotensinogen gene include brain, heart, kidney, adipose tissue, and reproductive organs. Factors up-regulating angiotensinogen gene expression include hormones (glucocorticoids, estrogens, thyroid), sodium depletion, and angiotensin II itself.

Two ACE isoforms are known: somatic and testicular, which are encoded by a single gene using different transcription initiation sites. Somatic ACE mRNA expression is hormonally regulated (steroids, thyroid), activated in resting nonproliferating endothelium, and up-regulated by ACE inhibitors.

Two types of angiotensin II receptor have been identified. Type 1 (AT1) receptors mediate the renal and vascular actions of angiotensin II. The function of type 2 receptors (AT2) remains uncertain at present, although it may be linked to cellular growth. The AT1 receptor gene is expressed at higher levels in the fetal and newborn than in the adult kidneys. AT1 receptor mRNA has been localized in glomeruli, preglomerular vessels, and vasa recta.

The *kallikrein–kinin system* (KKS) is a vasodepressor and natriuretic system and is regulated in concert with the RAS. Kinins, the active product, are formed by cleaving the substrate kininogen by tissue kallikrein. The kidney expresses all the components of the KSS: kallikrein, kininogen, kininases, and bradykinin receptors.

*Tissue kallikrein* is a serine protease that belongs to a multigene family (3 in the human, 24 in the mouse, and 10–17 in the rat). Renal kallikrein is expressed in the distal and connecting tubular cells. Kallikrein gene expression is modulated by hormones (adrenal and gonadal steroids, thyroid), cyclic nucleotides, angiotensin II, and sodium ions. In addition to tissue (renal) kallikrein, the kidney expresses other members of the kallikrein gene family, which display high sequence homology but diverse substrate specificity and functions.

*Kininogen,* the kallikrein substrate, is mainly synthesized in the liver as high and low molecular weight proteins. High and low molecular weight kininogen mRNAs originate from the same gene by differential RNA splicing. Low molecular weight kininogen is believed to be the substrate for tissue kallikrein, whereas the high molecular weight kininogen serves as the substrate for plasma kallikrein. The principal cells of the collecting ducts express low molecular kininogen and its mRNA. The rat expresses an additional unique kininogen molecule, called T kininogen which displays 90% sequence homology with the low molecular weight kininogen. This molecule is a product of an acute phase gene that is stimulated by interleukin-6 in inflammatory conditions.

*Kininase II* is identical with ACE and is highly abundant in the proximal tubular brush border, resulting in the degradation of almost all the filtered kinins. Thus, most if not all intratubular kinins are generated in the distal nephron.

*Bradykinin receptors* are divided into at least two types: B1 and B2. Although both are functionally present in the kidney, the B2 receptor is believed to mediate most of the intrarenal actions of bradykinin. The cDNA encoding the B2 receptor has recently been cloned and it belongs to the seven-transmembrane G-protein receptor gene family. Bradykinin B2 receptor gene is expressed at high levels in the fetus and is down-regulated during development. Bradykinin binding sites are present in the distal nephron and in glomeruli.

*Endothelial-derived relaxing factor/nitric oxide* (EDRF/NO) is a unique endogenously formed small lipophilic molecule that is rapidly diffusible across biological membranes. Nitric oxide is the prototype of a paracrine molecule owing to its extremely short half-life ($<$ 10 seconds in biological tissues). The binding of nitric oxide to the heme group of soluble guanylate cyclase causes an immediate and profound increase in the generation of cyclic guanosine monophosphate. Basal release of endothelial-derived nitric oxide plays a direct role in the regulation of systemic and organ blood flow and in the maintenance of the nonthrombogenic surface of the endothelium.

The amino acid L-arginine is the source of endogenous nitric oxide. The enzyme that catalyzes the conversion of L-arginine to nitric oxide has been cloned from human and bovine endothelium, rat cerebellum, and mouse macrophage. Nitric oxide synthase is expressed constitutively in vascular endothelial cells and cerebellar cells, whereas it is inducible in macrophages, neutrophils, vascular smooth muscle cells, and glomerular mesangial cells. The degree of similarity among the deduced proteins of endothelial, macrophage, and brain nitric oxide synthases (~50%) indicates that nitric oxide synthases are encoded by a gene family. Antibody to rat cerebellar nitric oxide synthase stains rat macula densa cells specifically. This result implicates nitric oxide in modulation of the tubuloglomerular feedback mechanism, which helps control glomerular function. RT-PCR localized the endothelial NO synthetase message in the collecting tubules and glomeruli.

*Endothelin*-1 (ET-1), a vasoconstrictor first purified from the conditioned medium of porcine aortic endothelial cells, is a 21 amino acid molecule derived from a 203 to 212 amino acid precursor (depending on the species), preproendothelin-1 (preproET-1 or big ET-1). The preproET-1 gene consists of five exons spanning over 6.8 kb of genomic DNA on chromosome 6. The mRNA for preproET-1 is expressed in porcine aortic endothelial cells in vivo and porcine aortic endothelial cells, human umbilical vein endothelial cells, and bovine glomerular endothelial cells in vitro.

Levels of preproET-1 mRNA are regulated by physical factors, such as changes in hemodynamic shear stress, and by cytokines, such as transforming growth factor β, as well as by thrombin, cytosolic $Ca^{++}$ and protein kinase C. Intrarenal ET-1 immunoreactivity is greatest in the renal papilla, where staining is predominantly localized to the vasa recta of the distal nephron segments. Cortical immunostaining is localized to the endothelial surfaces of arcuate arteries, veins, arterioles, and peritubular capillaries. Glomerular immunostaining follows predominantly a capillary loop distribution. It has been determined by RT-PCR that ET-1 gene expression is maximal in the glomeruli and inner medullary collecting ducts. ET-1 receptor mRNA is distributed mainly in the glomerulus, vasa recta, arcuate artery, and inner medullary collecting duct. The spatial localization of ET-1 and its receptor within the same nephron segments suggests autocrine/paracrine functions of ET-1 in the kidney.

## 2 GROWTH FACTORS

The kidney synthesizes a number of growth factors and their receptors. These growth factors exert their actions in a paracrine (on adjacent cells) or autocrine (on the synthesizing cell itself) manner. There is evidence that growth factors in the kidney are involved in nephrogenesis and renal cell differentiation, and in the proliferative response to renal injury and regeneration.

### 2.1 Platelet-Derived Growth Factor

*Platelet-derived growth factor (PDGF)* is a covalent dimer of two subunits chains A and B, which exists in three naturally occurring isoforms (PDGF-AA, PDGF-AB, and PDGF-BB). PDGF binds to cells via cell surface receptors, which function as noncovalent dimers of two subunits, α and β. Cells expressing only the receptor PDGFRβ are able to bind only PDGF-BB and cells expressing only PDGFRα are able to bind all forms of PDGF.

PDGF is mitogenic for a variety of cell types, including smooth muscle cells, glomerular mesangial cells, and fibroblasts. PDGF is also a chemotactic factor for monocytes and neutrophils. PDGF increases expression of c-*myc* and c-*fos* proto-oncogenes, which encode proteins involved in the regulation of cell growth and differentiation. PDGF-BB, but not PDGF-AA, induces increased synthesis of both PDGF α- and β-receptor protein, suggesting a positive feedback mechanism that could serve to potentiate autocrine stimulation of growth. Renal injury up-regulates the expression of PDGF and its receptors. In both experimental and human mesangial proliferative glomerulonephritis, the levels of PDGF-B chain and PDGFR increase more than eightfold at a time when mesangial cells are proliferating rapidly. Mesangial proliferation is attenuated by infusion of a blocking antibody against PDGF.

PDGF may also play a role in glomerulogenesis during embryonic life. During the early stages (vesicular, comma-shape, and S-shape) of glomerulogenesis, PDGF B chain is localized to differentiating epithelium of the glomerular vesicle, while PDGFRβ is expressed in the undifferentiated metanephric blastema. As the glomerular tuft forms, both PDGF B chain and PDGFRβ can be detected in an arboreal pattern radiating from the hilus of the glomerular tuft.

### 2.2 Insulinlike Growth Factors

*Insulinlike growth factors (IGFs) I and II* appear to be involved in renal growth and function, both in their capacity as circulating hormones and as local autocrine or paracrine acting tissue factors. IGF-I and II mRNAs are expressed in fetal kidneys; each of these peptides is necessary for growth and development of the renal anlage to take place in vitro, since antibodies to IGF-I or II prevent normal metanephric growth in culture. IGF-I, administered systemically, produces an increase in glomerular filtration rate and renal plasma flow. Chronically elevated circulating IGF-I levels are associated with renal enlargement and glomerular hypertrophy, and increased local IGF-I synthesis has been implicated in the development of compensatory renal hypertrophy.

In situ hybridization studies revealed that IGF-I mRNA is not detected in the human but is abundant in the rat kidney, while IGF-II is abundant in the human but is not detected in the adult rat kidney. IGF-II mRNA is concentrated in the renal vascular system, including afferent arterioles and the medullary interstitium. IGF-I and IGF binding protein-1 mRNAs are colocalized in the rat medullary thick ascending limbs of Henle's loops, but neither is detected in the human kidney. IGF binding protein-2 mRNA is concentrated in glomeruli in both species. The patterns for type I and type II IGF receptor gene expression are identical in both species; however, type I receptor mRNA is more abundant than type II. Both IGF receptor mRNAs are abundant in renal tubular epithelium of the medulla and barely detectable in proximal tubules. The conserved patterns of IGF receptor expression suggests that the role of circulating IGFs in regulating renal function may be similar across species.

### 2.3 Transforming Growth Factors Beta

*Transforming growth factors beta* (TGF-β) are multifunctional polypeptides that have complex effects in development, cell proliferation, and differentiation. In general, these growth factors promote differentiation in many cell types. While they inhibit proliferation in most cells (e.g., fibroblasts, renal epithelial, glomerular and endothelial cells, and lymphocytes), they promote proliferation in other cells (e.g., osteoblasts). TGF-β also stimulate matrix synthesis, inhibit matrix degradation, stimulate the synthesis of receptors for matrix proteins, and induce the synthesis of proteoglycans.

TGF-β and its receptors are located in virtually all organ systems studied, including the kidney, and are up-regulated in the proximal tubules and mesangial cells when exposed to high glucose. Glomeruli from rats with experimental glomerulonephritis express high levels of TGF-β mRNA, and a low protein diet reduces its expression. Under conditions of water deprivation, TGF-β immunoreactivity is seen in the juxtaglomerular cells upstream from the glomerulus and in interlobular arteries, suggesting that TGF-β may play a role as a growth factor or as a phenotypic modulator of JGA and renal arterioles.

### 2.4 Epidermal Growth Factor

The kidney is a major site of epidermal growth factor (EGF) production. This growth factor is synthesized as part of a larger precursor molecule that is an integral membrane protein. The precursor is cleaved extracellularly to produce mature EGF. Within the kidney, EGF has been localized, using immunohistochemistry, to the luminal membranes of cells of the thick ascending limb of Henle's loop and distal tubule, skipping the cells of the macula densa. mRNA for EGF has been colocalized with peptide by in situ hybridization, consistent with renal synthesis of the growth factor occurring at these sites. This location (luminal membrane) poses a dilemma for understanding the potential role of EGF in the kidney because most of the receptors for EGF are located on basolateral membranes of proximal tubule and collecting duct. It is possible that EGF gains access to the basolateral aspects of cells during disruption of tubular integrity, such as during ischemic or toxic acute renal failure.

EGF expression in cells of the distal tubule of kidneys is enhanced following contralateral nephrectomy. Injection of an antibody to EGF in mice following unilateral nephrectomy resulted in a decrease in the labeling index of the renal cortical tubular cells on the second day after uninephrectomy without a reduction in kidney weight, suggesting that EGF may play a role in compensatory renal hyperplasia. Interestingly, reduction of renal mass is accompanied not only by enhanced renal EGF expression, but also by a shift in the distribution of EGF within the thick ascending limb and distal tubules to include the basolateral aspects of the cells.

## 3 TUBULAR TRANSPORT SYSTEMS

Recent advances in molecular biological techniques have permitted the isolation and characterization of the mRNA and genes encoding a number of renal tubular transporters. These results have allowed a better understanding of structure–function relationships and the regulatory mechanisms of tubular reabsorption and secretion of fluids and electrolytes, urinary acidification, and concentration. Progress in this field is very rapid and several major transporters have recently been cloned. Due to space constraints, only a few of these are described.

### 3.1 Renal $Na^+$-$H^+$ Exchangers

The $Na^+$-$H^+$ exchanger is an integral membrane protein that mediates 1:1 electroneutral exchange of extracellular $Na^+$ for intracellular $H^+$. In epithelia, such as the mammalian renal proximal tubule, the $Na^+$-$H^+$ exchanger is responsible for transepithelial flux of $Na^+$, and $HCO_3^-$ and is stimulated by angiotensin II. Alterations in $Na^+$-$H^+$ exchange activity may be involved in the pathogenesis or adaptation to metabolic acidosis, essential hypertension, states of salt-wasting or edema formation, and abnormal growth and development.

cDNAs encoding $Na^+$-$H^+$ exchangers have been cloned from rabbit and rat kidney, and from porcine renal cells (LLC-$PK_1$). The encoded proteins are very similar in primary structure ( 90% homology in nucleotide and amino acid sequences), suggesting a high degree of evolutionary conservation. The cloned cDNAs encode four genes:

1. A renal basolateral $Na^+$-$H^+$ exchanger, *NHE-1*. Transport activity and steady state transcripts levels of *NHE-1* are increased coordinately in LLC-$PK_1$ cells under conditions simulating metabolic acidosis.
2. *NHE-2,* which is expressed in kidney, adrenal, and intestine.
3. *NHE-3,* expressed exclusively in the intestine and kidney, with the kidney cortex having the most abundant message, followed by intestine and kidney medulla.
4. *NHE-4,* most abundant in the stomach, followed by intermediate levels in small intestines and lesser amounts in kidney, brain, uterus, and skeletal muscle.

### 3.2 $Na^+$-$K^+$ ATPase

$Na^+$-$K^+$ ATPases are responsible for maintaining the high $K^+$, low $Na^+$ intracellular environment of higher eukaryotic cells, as well as contributing to the resting membrane potential of most cells. The enzyme is a heterodimer. The α-subunit catalyzes ATP hydrolysis and ouabain binding and is thought to mediate ion translocation. The β-subunit is required for the movement of the α-subunit from the endoplasmic reticulum to the plasma membrane. The β-subunit gene transcript is alternatively spliced into multiple mRNAs. There are at least three α-subunit genes located on three different chromosomes.

The $\alpha_1$ and $\alpha_3$ mRNAs are expressed in the kidney. The $\alpha_2$ mRNA is expressed mainly in the brain. In the kidney, $\alpha_1$ mRNA has been shown to be transcriptionally regulated by thyroxine and hypokalemia, and during postnatal development.

### 3.3 Water Channels

Concentrated urine is produced in response to vasopressin by the transepithelial recovery of water from the lumen of the kidney collecting tubule through highly water-permeable membranes. In this nephron segment, vasopressin regulates water permeability by endo- and exocytosis of water channels from or to the apical membrane.

CHIP28 is a water channel in red blood cells and the kidney proximal tubule but is not expressed in the collecting tubule. The cDNA for the water channel in the collecting duct (WCH-CD) has been cloned and has 42% homology in amino acid sequence to CHIP28. RT-PCR of microdissected nephron segments revealed that WCH-CD transcripts are detected only in the collecting tubule of the kidney. Immunohistochemically, WCH-CD is localized to the apical region of the kidney collecting tubule cells.

Expression of WCH-CD in *Xenopus* oocytes markedly increased osmotic water permeability, suggesting that WCH-CD is the vasopressin-regulated water channel.

*See also* Growth Factors; Kallikrein-Kininogen-Kinen System.

## *Bibliography*


Alper, S. L., and Lodish, H. F. (1991) Molecular biology of renal function. In *The Kidney,* B. M. Brenner and F. C. Rector, Jr., Eds., Chapter 4, pp. 132–163.

El-Dahr, S. S., and Dipp, S. (1993) Molecular aspects of kallikrein and kininogen in the maturing kidney. *Pediatr. Nephrol.* 7:646–651.

Gomez, R. A., Chevalier, R. L., Carey, R. M., and Peach, M. J. (1990) Molecular biology of the renal renin–angiotensin system. *Kidney Int.* 38(Suppl. 30):S18–S23.

Hammerman, M. R., Rogers, S. A., and Ryan, G. (1992) Growth factors and metanephrogenesis. *Am. J. Physiol.* 262:F523–F532.

Mitchell, K. D., and Navar, L. G. (1989) The renin–angiotensin–aldosterone system in volume control. *Bailliere's Clin. Endocrin. Metab.* 3:393–430.


# Repressor–Operator Recognition

*Peter G. Stockley and Simon E.V. Phillips*

### *Key Words*

**Corepressor (Inducer)** Small coeffector molecule that increases (decreases) the affinity of a repressor for its operator site.

**Direct Readout** The situation in which DNA sequence recognition is achieved by amino acid side chains making direct contacts with the edges of DNA base pairs.

**DNA-Binding Motif** Protein structural element primarily responsible for making sequence specific contacts with DNA.

**Indirect Readout** The situation in which DNA sequence recognition involves sensing of sequence-dependent distortions to the DNA conformation.

Transcriptional control is the most important mechanism of differential gene expression in the vast majority of organisms. Repressors control transcription by binding to specific sequences (operator sites) in genomic DNA, hindering the action of RNA polymerase, and thus preventing transcriptional initiation. DNA binding is linked to the physiological state of the cell, either directly by changes in the active repressor concentration or indirectly by sensing of the levels of key metabolites. Operator binding is highly specific and is achieved by a large number of distinct molecular interactions. Repressor–operator interactions provide simple tools to regulate the expression of recombinant proteins in a variety of organisms such as bacteria and yeast.

# 1 PRINCIPLES

## 1.1 The Operon Hypothesis

In the 1950s geneticists observed that bacterial cells were able to regulate the expression of genes for enzymes controlling particular metabolic pathways in response to their physiological state. For instance, the enzymes enabling *E. coli* to metabolize lactose are not constantly present in the cell, but their synthesis can be induced by the addition of lactose to the medium. The extent of induction is dependent on the levels of glucose, the preferred energy source. The results of a series of elegant genetic experiments led to the formulation of the operon hypothesis, which has become the basic model for transcriptional control of gene expression. In general, regulated genes contain two sequence elements at their 5′ ends, a promoter, which defines the starting point for the bacterial RNA polymerase, and an operator, which is the target sequence of a repressor protein. Binding of repressor to the operator interferes with RNA polymerase binding or function, leading to repression of the downstream gene(s). Different enzymes involved in related processes, such as lactose utilization, often are clustered and regulated by a single promoter–operator site; that is, they constitute an operon. It is also possible to have several promoter–operator sites (a regulon) regulated by the same repressor. Repression can occur in response to an environmental stimulus, such as the presence of lactose, by the binding of small coeffector molecules to the repressor proteins. The coeffectors may increase or decrease the affinity of the repressor for its operator site; that is, they can act either as a corepressor or as an inducer of gene expression. In bacteriophage lambda, proteolysis controls the concentration of the active cI repressor whose action controls the switch between lytic and lysogenic growth. The phage has been one of the major model systems for studying the control of gene expression.

## 1.2 The Structures of Repressor Proteins

In the early 1980s the three-dimensional structures of a number of bacterial regulatory proteins were determined by X-ray diffraction techniques. The first three proteins whose structures were available were the bacteriophage lambda repressors cro and cI, and the catabolic activator protein CAP. Comparison of the polypeptide chain fold in all three cases showed that although their overall structures are different, a particular secondary structure element, consisting of two helices, linked by an unusual β-turn, is present in each case. All three proteins are dimers, and the two copies of this helix–turn–helix (hth) motif are positioned on the surface, such that the separation between the two copies of the second helix (helix 2) of each motif is 34 A. This is also the separation of the major grooves of B-form DNA along one face of the DNA duplex, immediately suggesting that the hth is the DNA binding site. This proposal was subsequently confirmed by the so-called helix swap experiment, in which the outward-pointing amino acid residues of helix 2 from two phage repressors, P22 cI and 434 cI, were exchanged using site-specific mutagenesis. The result was that the hybrid proteins showed the operator binding properties of the newly inserted sequences; that is, the hybrid P22 cI protein bound to 434 operators and vice versa. Helix 2 of the hth motif therefore became known as the recognition helix, implying that all operator recognition resided in the residues of such helices. As the three-dimensional structures of repressor–operator complexes have become available, this idea has proved to be an oversimplification (see Section 1.4).

The initial observation of a structural feature conserved between three repressor proteins was unexpected, since there is very little amino acid sequence similarity among them. However, the structural studies allowed key residues within the hth motif to be identified, permitting the construction of an algorithm to search the protein sequence database for potential matches with the hth motif. The search revealed an impressive list of several hundred proteins, all believed to bind DNA, that have the required sequence features of the hth motif. It is now clear that the hth motif defines a family of proteins from both prokaryotic and eukaryotic cells sharing a conserved DNA-binding motif. Although extremely common, the hth is not the only protein structural motif capable of binding to DNA in a sequence-specific fashion. Early model-building studies had suggested that both α-helices and two-stranded antiparallel β-sheets (β-ribbons) had the steric features necessary to interact with duplex DNA. Two distinct families of proteins having β-ribbon DNA-binding motifs have now been characterized. These are based on the *Bacillus stearothermophilus* HU protein, which is a non-sequence-specific DNA-binding protein, and the *E. coli* MetJ protein, the repressor of methionine biosynthesis. It is likely that other families of repressors having novel DNA-binding motifs will be characterized in the near future. Indeed many other DNA-binding motifs have been characterized in eukaryotic transcription factors and DNA-modifying enzymes.

## 1.3 Repressor–Operator Complexes

Operator sites generally contain palindromic sequences, and dimeric repressor proteins bind to them with their twofold axes aligned along the sequence dyad, thus generating two symmetrical sets of protein–DNA interactions. To ensure that such interactions are possible only once or a few times per chromosome, operator sites have an effective minimum length, usually greater than 12 base pairs. Repressors consisting of higher multimers, such as the *E. coli* repressors of lactose (LacR) or arginine (ArgR) biosynthesis, which are a tetramer and a hexamer, respectively, nevertheless seem to be constructed from dimeric units and to retain the symmetry of the interaction with their operator sites. Figure 1 shows the structures of two repressors, one each from the hth and β-ribbon families. In each case the repressor–operator complex is formed by insertion of the DNA-binding motif(s) into the major groove of the DNA duplex. A large number of specific intermolecular contacts are made between the DNA and the repressors, including hydrogen bonds and electrostatic interactions.

Both the TrpR and MetJ repressors regulate the flux through biosynthetic pathways and must therefore alter the levels of repression at their various operons in response to the physiological state of the cell. This is achieved in each case by the binding of a corepressor, which is a product of the biosynthetic pathway, tryptophan for TrpR and *S*-adenosylmethionine (SAM) for MetJ. Each dimeric repressor binds two molecules of corepressor, and the resultant complex, the holorepressor, has a much higher affinity for operator DNA. However, the molecular mechanisms of these increases in affinity are quite different. In the unliganded apo-TrpR, the hth DNA-binding motifs are too close together to fit easily into the adjoining turns of the major groove of B-form DNA. Binding of tryptophan results in a concerted conformational change in the repressor, which increases the spacing between the hth motifs, resulting in greater structural complementarity with the DNA. MetJ, on the other hand, does not appear to alter its conformation signifi-

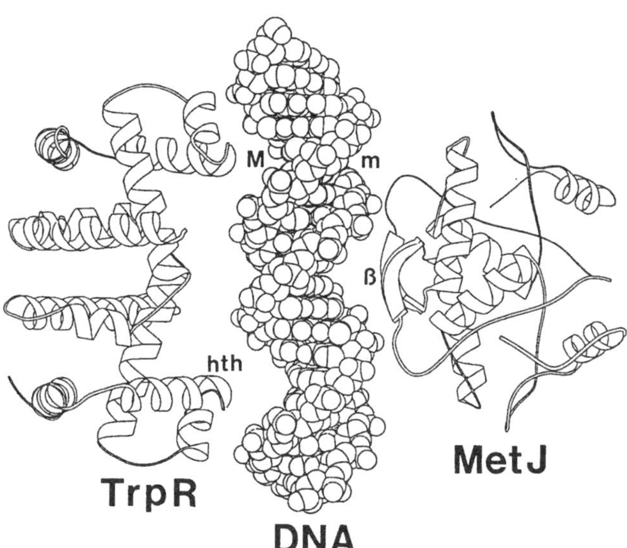


Figure 1. Structures of the active forms of *E. coli* TrpR (tryptophan) and MetJ (methionine) repressors shown alongside two turns of B-DNA; α-helices are indicated by coiled ribbons and β-strands by flat ones. Both repressors are oriented correctly for DNA binding but have been pulled away slightly from the DNA. As TrpR approaches the DNA, its two helix–turn–helix motifs can dock into adjacent turns of the major groove (M). As MetJ approaches, its β-ribbon (β) can dock into the central portion of the major groove. Corepressor molecules have been omitted for clarity; m indicates the position of the minor groove.

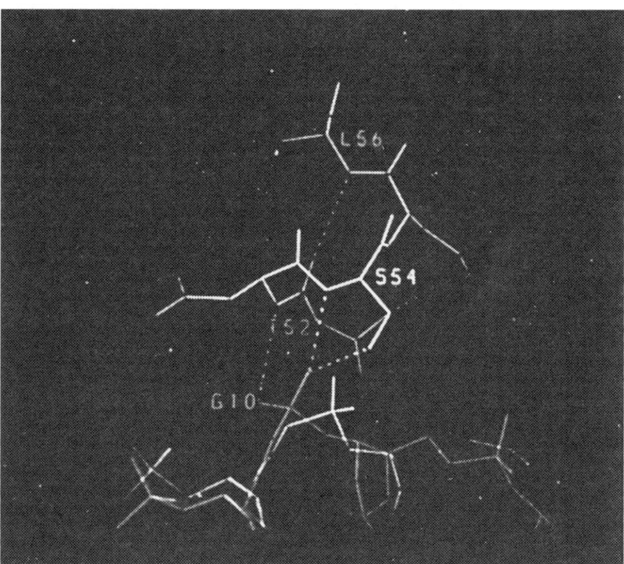


Figure 2. Distortion of the DNA phosphate backbone in a MetJ repressor–operator complex. A portion of a protein α-helix (yellow) makes hydrogen bonds (dashed) to oxygens of a DNA phosphate group (blue). The phosphate backbone is distorted, as shown by the superimposed structure of regular B-form DNA (red), where the protein–phosphate contacts could not be made. The relative ease of such distortion of the DNA is sequence-dependent. [From *Nature*, 359: 387–393 (1992).] (See color plate 15.)

cantly when it binds SAM. In this case it appears that the increase in affinity upon corepressor binding is due to electrostatic effects caused by the positive charge carried by the tertiary sulfur atom of each SAM. These charges increase the positive electrostatic potential on the DNA-binding face of the repressor, helping to neutralize the negative charge of the DNA phosphate backbone. These two cases illustrate that more than one mechanism has evolved to regulate repressor affinity for operator sites.

### 1.4 General Principles of Repressor–Operator Interactions

The sequence specificity of the repressor–operator interactions appears to be achieved in at least two distinct ways. Modeling studies suggested that base pairs could be recognized by formation of a series of hydrogen bond contacts between amino acid side chains and the functional groups on the edges of the base pairs exposed in the grooves of the DNA duplex. Comparison of the different base pairs showed that discrimination would be easier in the major groove because of the number and diversity of possible contacts. Such interactions have now been directly observed in many repressor–operator complexes involving amino acids in hth motifs, β-ribbons, and so on. However, the molecular details proved to be more complex than at first thought. There is no simple code relating amino acid and operator sequences. Instead, complexes contain a constellation of interacting amino acid side chains and DNA bases, and many contacts contribute to the overall specificity. This type of interaction has been termed "direct readout," since the protein makes direct contacts to the edges of operator base pairs.

A second type of interaction, termed "indirect readout," has also been identified. Here complex formation involves sequence-dependent conformational distortion of DNA, resulting in increased intermolecular complementarity or interaction. An example of this type of interaction is shown in Figure 2 (see color plate 15) for MetJ. Two MetJ repressors bind to a minimum operator site (16 bp), each interacting, via direct readout, with the base pairs within each operator half-site. However, the interaction is also sensitive to the sequence at the junction of the two half-sites, where a TA base step is easily distorted away from regular B-form conformation because of the relatively poor stacking energy of pyrimidine bases on purines. In the MetJ–operator complex, this step is overwound (i.e., has a stronger helical twist than expected), resulting in displacement of the phosphate 5′ to the TA step position by about 1.5 A relative to its position in regular B-form DNA. The repressor makes a number of hydrogen bonds to this phosphate, which would be impossible if the phosphate were in a standard conformation.

## 2 TECHNIQUES

The experimental approaches used to study repressor–operator interactions can be divided into in vitro and in vivo techniques.

### 2.1 In Vitro Techniques for Studying Repressor–Operator Interactions

Repressors have many advantages for in vitro studies. For instance, the very high specificity of their interaction with target operators allows experiments to be carried out with unfractionated protein extracts as well as with purified samples. Since affinities for operator sites are often in the nanomolar range, experiments require only tiny amounts of protein and can be performed in conditions of protein excess over operator fragments, the latter usually being radioactively labeled. The gel retardation assay, one of the commonest techniques used to monitor complex formation, works by separating repressor–operator complexes from unliganded operator DNA fragments by electrophoresis in polyacrylamide gels. As a result of changes in the size/shape/charge of the molecule, the repressor–operator complexes migrate more slowly than protein-free DNA, making it possible to determine the amount of complex present by autoradiography, if radioactively labeled DNA is used. The assay has the advantage of revealing the presence of multiple protein–DNA complexes. However retarded species are observed only if the half-life of the complex is greater than the time required to run the gel, which often exceeds several hours.

An alternative assay method involves fractionation of repressor–operator complexes by filtration through nitrocellulose, which binds tightly to proteins and their DNA complexes but allows protein-free DNA to pass through. Since the filtration step is very fast, problems due to rapid dissociation are avoided. However, it is not possible to determine the amounts of multiply bound DNA fragments. For both types of assay, a typical experiment can be used to estimate the affinity of the repressor for the operator site by holding the amount of DNA constant and increasing the repressor concentration across the range at which binding takes place. The protein concentration at which 50% of the DNA is bound is a crude estimate of the apparent equilibrium dissociation constant.

Although useful for comparative quantitation of repressor–operator interactions, these assays provide no information on the precise sequence of the operator that is being recognized. A series of procedures is available to determine the positions of tight interaction between repressors and regions of the DNA duplex. These techniques can be used in a footprinting mode, to analyze the complex itself, or in an interference mode, in which the DNA site is partially modified by a specific reagent prior to complex formation. In the latter case, the molecules still capable of being bound by repressor are fractionated from those in which the modification has interfered with complex formation. In both cases the experiments are performed with radiolabeled DNA, and the products are analyzed on sequencing gels. Examples of footprinting reagents are DNase I, which makes single-stranded nicks along the target DNA; dimethyl sulfate (DMS), which methylates the N-7 position of guanine in the major groove; hydroxyl radicals, which attack the sugar residues along the duplex; and methidium propyl EDTA, which intercalates between bases before cleaving via a hydroxyl radical mechanism. Regions of the DNA buried in the complex can be detected by their reduced reactivity toward these reagents. Binding interference experiments are often performed with ethyl nitrosourea, which ethylates the DNA phosphate groups, hydroxyl radicals, or DMS.

The three-dimensional structures of complexes between repressors and short synthetic DNA duplexes also can be determined by X-ray crystallography or NMR spectroscopy, to provide the context for interpretation of the binding data.

### 2.2 In Vivo Techniques for Studying Repressor–Operator Interactions

Perhaps the most common technique for monitoring repression in vivo involves coupling a promoter–operator site to a reporter gene whose product is easily monitored (e.g., β-galactosidase). The levels of repression at a particular site can then be determined by enzymatic assay after lysis of the cells. Techniques have also been developed to footprint repressor–operator complexes in vivo using reagents (e.g., DMS) that can penetrate living cells. A major problem with both approaches is controlling the levels of both repressor and operator site, since these molecules are usually carried on plasmids of variable copy number.

## 3 APPLICATIONS

The primary application of repressor–operator interactions is in the regulated expression of recombinant proteins in bacteria. The strength of the promoter is also an important consideration. A particularly strong promoter used for such experiments is *tac,* a hybrid between the *trp* and the *lac-uv5* promoters regulated by the LacR repressor. Expression of the regulated gene is induced by the addition of nonhydrolyzable analogues of lactose to the medium. Since the promoter is very strong, however, expression occurs at a low level even in the uninduced state. This effect can be deleterious if the protein product is poisonous to the host cell. A common alternative system involves the lambda promoter–operator site regulated by a temperature sensitive mutant lambda cI repressor. Heating the bacteria to the nonpermissive temperature results in rapid derepression of the regulated gene.

*See also* Gene Expression, Regulation of; Genomic Imprinting; Protein Designs for the Specific Recognition of DNA; Zinc Finger DNA Binding Motifs.

### *Bibliography*

Harrison, S. C., and Aggarwal, A. K. (1990) DNA recognition by proteins with the helix–turn–helix motif. *Annu. Rev. Biochem.* 59:933–970.

Phillips, S. E. V. (1991) Specific β-sheet interaction. *Curr. Opinion Struct. Biol.* 1:89–98.

Ptashne, M. (1992) *A Genetic Switch: Phage Lambda and Higher Organisms,* 2nd ed. Blackwell Scientific Publications, Cambridge, MA.

Travers, A. (1993) *DNA–Protein Interactions.* Chapman & Hall, London.

# Restriction Endonucleases and DNA Modification Methyltransferases for the Manipulation of DNA

*Eric W. Fisher and Richard I. Gumport*

### *Key Words*

**DNA Modification Methyltransferase** An enzyme, commonly called a "methylase," that transfers a methyl group from S-adenosylmethionine onto a specific position of a specific base within a particular recognition sequence in DNA, such that that sequence is rendered refractory to cleavage by the restriction endonuclease recognizing the same sequence.

**Isoschizomers** Restriction endonucleases that recognize identical nucleotide sequences in DNA and hydrolyze the DNA at the same phosphodiester bond. Enzymes recognizing identical sequences but cleaving the DNA at different phosphodiester bonds are termed "neoschizomers" or "heteroschizomers."

**Palindrome** Segment of bases in DNA whose sequence is the same from the 5′ end to the 3′ end in one strand as in its complementary strand. Most type II restriction–modification enzymes recognize palindromic sequences.

**Restriction Endonuclease** Enzyme that recognizes a specific sequence in DNA and hydrolyzes specific phosphodiester bonds to cleave the DNA.

**Restriction–Modification System** A pair of enzymes in numerous species of bacteria consisting of an endonuclease, which cleaves, or restricts, incoming foreign DNA at specific sequences when those sequences are unmethylated, and a modification methyltransferase, which methylates the host DNA at those same sequences to protect it from cleavage by the endonuclease.

**"Star" Activity** The hydrolytic activity arising upon relaxation of the specificity of a restriction endonuclease or methylase. Inaccurate hydrolysis is usually induced by altering the reaction conditions such that cleavage or methylation occurs at sequences not usually recognized.

---

The enzymes of microbial restriction–modification systems are a major technological pillar of modern biotechnology, having provided the capability of dissecting specific DNA fragments from chromosomes. In addition to the utility of these enzymes in technical applications, their exquisite specificity for particular recognition sequences makes them valuable systems for the study of site-specific protein–DNA interactions. To dissect into large pieces very large genomes (e.g., that of humans), a number of groups have sought and found enzymes recognizing sequences of unusual length, whereas others have devised methods to bring about the rare cleavage of large DNA fragments using creative combinations of available restriction endonucleases and methylases. These combinations of enzymes cleave longer, and thus more rare sequences, yielding longer fragments than could have been obtained with the endonuclease, if used singly. Other techniques, such as DNA fingerprinting, have been developed to exploit the specificity of restriction–modification enzymes for analytical and diagnostic purposes; the constantly increasing number of such techniques indicates the indispensability of restriction–modification enzymes in all of molecular biology. Finally, studies of the specificity itself of these remarkable enzymes appear to be auguring the creation of new restriction–modification enzymes that have not evolved naturally.

## 1 THE ENZYMES OF MICROBIAL RESTRICTION–MODIFICATION SYSTEMS

The enzymes of restriction–modification systems in bacteria and other prokaryotes offer their hosts protection from foreign DNA and *in vitro* are essential tools for manipulating DNA as required in modern biotechnology applications. They are also of theoretical interest as models for the study of site-specific protein–DNA interactions, owing to their phenomenal specificity in recognizing particular DNA sequences. A recent incentive to study restriction–modification enzymes is the much publicized Human Genome Project, which suffers from a shortage of tools to precisely dissect into fragments of manageable sizes an essentially continuous three billion base pairs of DNA. The extreme specificity of restriction endonucleases and modification methyltransferases in their site-specific, double-strand cleavage, or nucleotide base methylation, respectively, of a duplex DNA permits the accurate mapping of DNA and its piecewise removal to other vehicles for cloning, sequencing, expression, and further exploitation.

## 2 CHARACTERISTICS OF THE ENZYMES

Restriction endonucleases (commonly "restriction enzymes") are in the hydrolase class of enzymes designated by the International Union of Biochemistry and Molecular Biology (IUBMB). All restriction endonucleases conduct the double-strand, $Mg^{2+}$-catalyzed hydrolysis of the phosphodiester internucleotide linkage of DNA. The action of type II endonucleases, those in widest use as tools of DNA manipulation, as exemplified by the *Eco*RI system, is presented in Figure 1.

In contrast, DNA modification methyltransferases, or "methylases," members of the transferase group in the IUBMB nomenclature, catalyze a fundamentally different reaction using the same DNA sequence as substrate. This reaction deposits a methyl group from the donor *S*-adenosylmethionine (AdoMet) to one of three positions on the target nucleotide residue (Figures 1 and 2), depending on the particular enzyme. The reaction produces the methylated DNA and *S*-adenosylhomocysteine, abbreviated AdoHcy (Figure 1).

The ability of cells, by what was then an unknown mechanism, to "restrict," or limit, the strains of bacteriophage that could infect them was first reported in 1953: bacteriophage grew poorly on some bacterial strains upon initial infection, but reinfection of the same strains by the initial progeny permitted normal viral growth. The reinfecting phage DNA had acquired the "survival" traits of the host chromosome, the heritable modification being an AdoMet-dependent methylation of the phage DNA. Restriction–modification systems presumably exist in bacteria because they provide the host cell defense from foreign DNA, with the methylase protecting the host chromosome from the endonuclease, and the endonuclease degrading incoming DNA that lacks the necessary methylation.

Since the discovery of the *Eco*B and *Eco*K restriction–modification enzymes in 1968, almost 2400 such enzymes, cleaving DNA specifically at 188 different sequences, have been identified. These enzymes are organized into three distinct classes, based on the structures of the proteins themselves, the cleavage and methylation mechanisms, and the nucleotide sequences recognized. The categories are called, simply, types I, II, and III; a fourth, emergent, class, the methyl-dependent systems McrA, McrBC, and Mrr, consists of enzymes whose sequence specificities remain incompletely determined. Type I and III enzymes are similar in that their restriction and methylation components are different subunits of the same holoenzyme, whereas the type II enzymes consist of independent enzymes.

Type II restriction–modification enzymes are the simplest in terms of structure and function. The endonuclease cleaves normally within or adjacent to a palindromic recognition sequence of four to eight base pairs, with the sole requirement for $Mg^{2+}$ as cofactor. Some type II enzymes, designated type IIS, recognize nonpalindromic sequences and cleave at defined positions outside the sequence. A methylase recognizing the same sequence exists for each endonuclease, and representative sequences recognized by members of the two classes are shown in Table 1. While these enzymes are known for their high sequence selectivity, reaction conditions exist for a number of them that produce a relaxation of specificity known as "star" activity. These conditions usually include higher than usual pH, lower ionic strength, and the presence

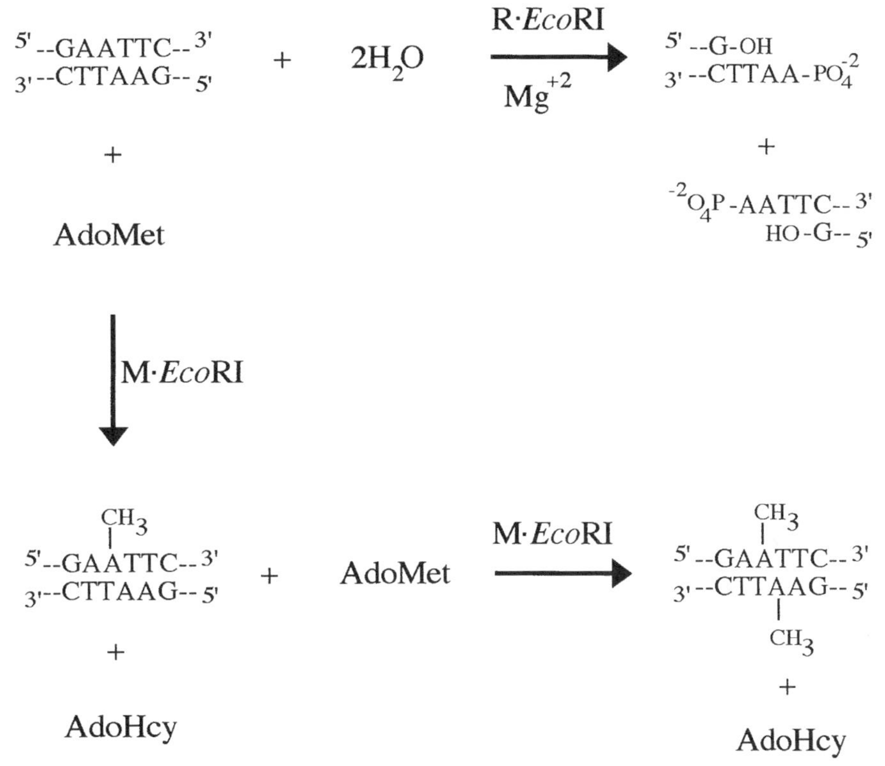

5'--GAATTC--3'
3'--CTTAAG--5' + $2H_2O$ —R·*Eco*RI, $Mg^{+2}$→ 5'--G-OH / 3'--CTTAA-$PO_4^{-2}$ + $^{-2}O_4P$-AATTC--3' / HO-G--5'

+

AdoMet

↓ M·*Eco*RI

$CH_3$
5'--GAATTC--3'
3'--CTTAAG--5' + AdoMet —M·*Eco*RI→ 5'--GAATTC--3' ($CH_3$) / 3'--CTTAAG--5' ($CH_3$) + AdoHcy

+

AdoHcy

**Figure 1.** Microbial restriction and modification as exemplified by the *Eco*RI restriction–modification enzymes. The system consists of an endonuclease (R·*Eco*RI) that hydrolyzes the DNA in both strands at the sequence GAATTC, and a modification methyltransferase (M·*Eco*RI) that methylates the exocyclic amino group of the inner adenosine residue of one strand of the sequence. The methylase can return to modify the other strand. Both the hemimethylated and fully methylated forms of the DNA are refractory to cleavage by the endonuclease. AdoMet represents *S*-adenosylmethionine, the methyl group donor in the methyltransfer reaction, and AdoHcy *S*-adenosylhomocysteine, the by-product.

of $Mn^{2+}$ ion or organic solvents. Researchers should examine products carefully when particular applications require the use of reaction conditions different from those giving high-fidelity cleavage.

## 3 THE SYSTEM OF NOMENCLATURE

Conventionally, the name of any restriction endonuclease or methylase consists of four parts: the letter R· or M·, often dropped for convenience, to designate restriction endonuclease or methyltransferase, respectively; an abbreviation of the genus and species of the organism to three letters, such as *Eco* for *Escherichia coli*; a letter, number, or combination of the two to indicate the strain of the relevant species, such as R for *E. coli* RY 13; and finally a Roman numeral to indicate the order in which different restriction–modification systems were found in the same organism or strain. For example, *Dsa*VI was the sixth restriction–modification system found in *Dactylococcopsis salina.* The history and incompleteness of the discovery process are reflected in this nomenclature, as unidentified bacteria found with these systems receive the genus–species designation *Uba.* The existence of the system *Uba*1453I indicates that, as of July 1993, hundreds (although not as many as 1453) of restriction–modification systems had been identified but their respective host organisms had not. Many of the organisms found earlier have now been identified and the enzymes appropriately renamed.

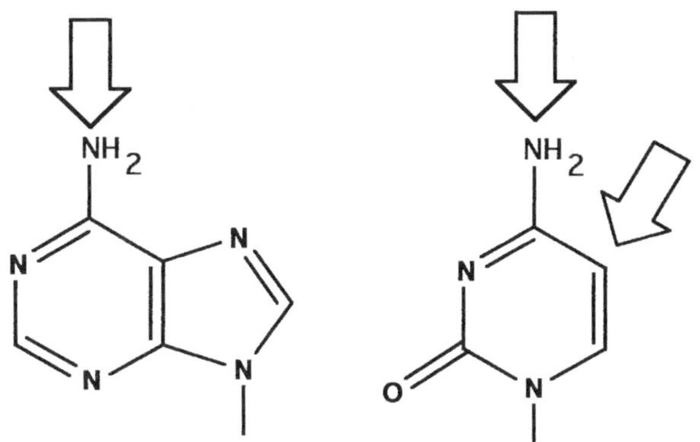


**Figure 2.** Methylation sites of DNA adenine and cytosine methyltransferases. Left: structure of an adenine residue in DNA; right, cytosine. Arrows indicate, from left to right, the atoms to which the methylases transfer methyl groups from AdoMet: The 6-amino group of adenine, the 4-amino group of cytosine, and the 5-carbon of cytosine.

## 4 APPLICATIONS TO PROBLEMS IN MOLECULAR BIOLOGY

Much of the technology of modern molecular biology owes its existence to the advent of restriction–modification enzymes with differing sequence specificities. Most experiments involving DNA manipulation require endonucleases, and DNA methylases are useful in more complicated experiments. Restriction enzymes allow determination of the sizes of unknown DNA fragments of interest, since they can be used to produce fragments of known sizes as reference standards. Restriction enzymes are used to cleave chromosomes into manageable fragments that can be inserted into clon-

Table 1 Representative DNA Sequences[a] Recognized by Microbial Restriction–Modification Enzymes

| **Uninterrupted Palindromes** | | **Interrupted Palindromes** | |
|---|---|---|---|
| GG^CC | *Hae*III | CC^NGG | *Scr*FI |
| G^AATTC | *Eco*RI | GAANN^NNTTC | *Xmn*I |
| GC^GGCCGC | *Not*I | GGCCNNNN^NGGCC | *Sfi*I |
| **Sequences with Points of Two-Base Degeneracy** | | **Sequences with Points of Three-Base Degeneracy** | |
| CMC^CKG | *Msp*A1I | GDGCH^C | *Nsp*II |
| CG^GWCCG | *Rsr*II | RGCB | *Cvi*AIV |
| R^AATTY | *Apo*I | | |
| **Nonpalindromic Sequences[b]** | | **Symbols Used** | |
| CCGC (−3/−1) | *Aci*I | N = A, G, C, or T | W = A or T |
| GGATG (9/13) | *Fok*I | D = A, G, or T | S = G or C |
| GCANNNNNNTCG (12/10) | *Bcg*I | H = A, C, or T | R = A or G |
| | | B = C, G, or T | Y = C or T |
| | | M = A or C | K = G or T |

[a]All sequences are specified as single strands and written 5′ to 3′ with the opposite, complementary strand implied. The point of cleavage by the endonuclease is indicated by ^, and the nucleotide methylated by the methyltransferase is indicated in boldface. Where no point of cleavage or of methylation is specified, none has been determined.
[b]Numbers in parentheses represent the number of nucleotides outside the recognition sequence where cleavage occurs, with negative numbers representing the 5′ direction and positive numbers the 3′ direction from the 3′ end of the sequence. The first number indicates the point of cleavage in the specified strand, and the number after the virgule (/) the point in the unspecified, complementary strand.

ing vehicles for many purposes, even when the fragment of interest contains the restriction site. Such interior sites are protected from cleavage by pretreatment of the target DNA with the cognate methylase, and then the DNA is mechanically sheared into fragments. Oligodeoxyribonucleotide duplex linkers carrying an unmethylated site are ligated to the sheared ends and cleaved with the endonuclease to expose specific ends, and ligation into the vector proceeds normally.

Restriction enzymes are useful in cloning DNA fragments isolated by current methods of DNA amplification using thermostable polymerases. Judicious synthesis of desired restriction sites into the 5′ portions of the primers used in amplification generates desired ends upon cleavage with the appropriate endonuclease.

The Southern hybridization method of characterizing DNA segments requires restriction enzymes to generate discrete target fragments, and the diagnostic method of restriction fragment length polymorphism (RFLP) analysis is often used in paternity and forensic investigations. This method uses a probe (identifier) of the repeated sequence found commonly in members of a population to distinguish those members from one another. The loss or gain of a restriction site in long, repeated sequences will produce a different collection of fragments using this technique and can provide the basis for identifying or excluding members of a population as sources of analyte DNA. If similar restriction fragments from different sources of the same DNA segment are labeled, denatured, and separated by electrophoresis, point mutations often give rise to altered secondary structures and therefore characteristic migration rates during electrophoresis. This method, single-strand conformational polymorphism, identifies point mutations and is useful in the characterization of mutant alleles.

Both sequencing large DNA fragments and the generation of deletion mutations are simplified using restriction enzymes that leave 5′ and 3′ overhanging ends. *E. coli* exonuclease III, which deletes nucleotides at the 3′ end from 3′ recessed or blunt, but not 3′ protruding, termini, allows controlled shortening of a DNA fragment from a single end when the fragment has been cleaved with two restriction enzymes, one leaving the refractory, protruding 3′ end (such as *Pst*I), and the other a susceptible blunt or recessed end (such as *Sma*I or *Eco*RI). This method reduces the number of primers needed in sequencing projects.

Oligonucleotide-directed mutagenesis is simplified for a target sequence abutting a restriction site that can be cleaved to minimize cell transformation by the unmodified vector. The primer is designed such that the restriction site is eliminated at the same time the desired change is made in the DNA segment: products not incorporating the oligomer still possess the restriction site and are linearized by endonuclease treatment prior to transformation, reducing their survival frequency.

Analysis of DNA bending by proteins is expedited by using a plasmid that can be cleaved at unique sites by different restriction enzymes to yield DNA fragments, all of the same size but with the binding site of interest placed at varying distances from the ends of the fragment. The plasmids contain two tandem, parallel repeats of the same sequence of multiple, closely spaced restriction sites, and a unique site between the two repeats for the insertion of the sequence to which the target protein will bind. Gel retardation assays carried out with the resulting restriction fragments indicate the relative extent of bending induced by the bound protein.

The need for endonucleases of specificity exceeding six base pairs, to produce longer fragments from large genomes, has led to novel methods for increasing the effective specificity of the enzymes. The frequency of cleavage by a restriction enzyme is a reciprocal function of the number of nucleotides recognized, that is, $\omega = 4^{-n}$, where $\omega$ is the fraction of phosphodiester bonds, on average, undergoing hydrolysis in DNA of random composition and sequence with equal amounts of the four bases, and $n$ is the number of nucleotides specifically recognized by the enzyme. Frequencies of cleavage of $10^{-5}$ or lower would cleave mega- and gigabase pair chromosomes into appropriately sized fragments. Restriction enzymes are forced to cleave at frequencies lower than their normal sequence specificity allows by creative methylation of the target DNA with the companion methylase. This entails inhibiting methylation at a specific site by reversible obstruction, such that only that site remains cleavable by the endonuclease. This specificity has been achieved using a complementary oligonucleotide that forms a three-strand complex with the desired site, stabilized by the addition of *E. coli* RecA protein, or by adding a DNA-binding protein of high sequence specificity that covers a single restriction site. Inhibition has also been accomplished by first methylating the DNA using an enzyme recognizing another sequence that will, by chance, occasionally overlap the sequence to be cleaved. Because those overlapping sites, upon initial methylation, become refractory to a second methylation step, which uses the methylase corresponding to the endonuclease to be used, those sites are left uniquely susceptible to the endonuclease.

Methylation of a particular restriction site throughout a large fragment of DNA can be used to create a *Dpn*I recognition site where a cryptic *Dpn*I site is formed by the overlap of two methylation sites. *Dpn*I is a type II endonuclease that recognizes G$^{m}$ATC, but not unmethylated or hemimethylated duplex GATC. The use of *Cla*I methylase, which recognizes the sequence ATCGAT and methylates the 3′ A residue, increases the specificity of *Dpn*I from four to ten base pairs where two *Cla*I sites overlap to form the 10 base pair sequence ATC**G$^{m}$ATC**GAT (the *Cla*I site is underlined and the *Dpn*I site is shown in boldface).

In another application, DNA fragments with nicks at defined locations are generated using restriction enzymes. A sample of the fragment of interest is cleaved to completion with the endonuclease, and both the digested DNA and uncleaved copies of the same DNA are mixed, denatured, and renatured. Some renaturation products contain a complete strand hybridized to the two portions of the cut, complementary strand, and these are then purified. This method was used to make a DNA substrate in order to trap a covalent complex between DNA and lambda integrase (Int) by placing a nick three nucleotides downstream from the Int cleavage site. The trinucleotide product diffused away from the complex upon Int cleavage, and the reaction was unable to proceed, allowing the isolation of the Int–DNA intermediate.

## 5 FUTURE DIRECTIONS FOR RESEARCH AND APPLICATIONS

With so many novel and creative applications of restriction–modification enzymes already in use, it seems foolhardy to speculate on the future applicability of these versatile enzymes to molecular biology. Methods for rapidly screening organisms for new enzymes with novel target specificities have greatly increased their numbers, and the application to the design of new restriction and methylation specificities of sophisticated genetic selection techniques will assuredly offer enzymes that have not yet evolved. The identification of structural and recognition domains in several type I endonucleases and type II methylases has resulted in the creation of chimeric enzymes that recognize different sequences and may suggest future protein engineering possibilities. Continued study will likely not only increase our understanding of these exciting enzymes in terms of molecular recognition, but will also result in the introduction of heretofore barely imaginable applications to the repertoire of every biochemist.

*See also* DNA Replication and Transcription; Human Linkage Maps; Restriction Landmark Genomic Scanning Method.

### *Bibliography*


Anderson, J. E. (1993) Restriction endonucleases and modification methylases. *Curr. Opinion Struct. Biol.* 3:24–30.

Bickle, T. A., and Krüger, D. H. (1993) Biology of DNA restriction. *Microbiol. Rev.* 57(2):434–450.

Heitman, J. (1993) On the origins, structures and functions of restriction–modification enzymes. In *Genetic Engineering,* Vol. 15, J. K. Setlow, Ed. pp. 57–108. Plenum Press, New York.

McClelland, M., Nelson, M., and Raschke, E. (1994) Effect of site-specific modification on restriction endonucleases and DNA modification methyltransferases. *Nucleic Acids Res.* 22(17):3640–3659.

Pingoud, A., Alves, J., Geiger, R. (1993) Restriction enzymes. In *Methods in Molecular Biology,* Vol. 16, *Enzymes of Molecular Biology,* M. M. Burrell, Ed. pp. 107–200. Humana Press, Totowa, NJ.

Roberts, R. J., and Halford, S. E. (1993) Type II restriction endonucleases. In *Nucleases,* R. J. Roberts, S. M. Linn, and R. S. Lloyd, Eds., 2nd ed. Cold Spring Harbor, Press, Plainview, NY.

Roberts, R. J., and Macehis, D. (1994) REBASE—Restriction enzymes and methylases. *Nucleic Acids Res.* 22(17):3628–3639.

Szybalski, W., Blumenthal, R. M., Brooks, J. E., Hattman, S., and Raleigh, E. A. (1988) Nomenclature for bacterial genes coding for class-II restriction endonucleases and modification methyltransferases. *Gene,* 74(1):279–280.

Wilson, G. G., and Murray, N. E. (1991) Restriction and modification systems. *Annu. Rev. Genet.* 25:585–627.


# Restriction Landmark Genomic Scanning Method

*Yoshihide Hayashizaki, Shinji Hirotsune, Yasushi Okazaki, Masami Muramatsu, and Jun-ichi Asakawa*

### *Key Words*

**Blocking (in RLGS)** Reduction of background by using enzyme reaction(s) as a pretreatment to interfere with the incorporation of labeling reagents into nonspecific damaged DNA sites.

**(Gel) Electrophoresis (in RLGS)** A method for separating DNA fragments based on their length (molecular weight) and electric charge.

**Genome Scanning** High-speed survey of the presence or absence of landmarks throughout a genome and measurement of their copy number in each locus.

**Restriction Enzyme (Restriction Endonuclease)** Enzyme that cleaves foreign DNA molecules at specific recognition sites (nucleotide sequences).

**Restriction Landmarks** Restriction enzyme sites used as guideposts throughout a genome.

Restriction landmark genomic scanning (RLGS) is a method for the high speed determination of the presence or absence of restriction landmarks throughout a genome and measurement of their copy number, based on the new concept that restriction enzyme sites can serve as landmarks. RLGS employs direct end labeling of genomic DNA digested with a restriction enzyme, and high resolution, two-dimensional electrophoresis. RLGS has the following advantages for genome scanning: (1) high speed scanning ability (thousands of restriction landmarks can be scanned simultaneously), (2) a scanning field that can be extended by the use of different kinds of landmarks in an additional series of electrophoresis, (3) applicability to any organism (because direct labeling of restriction enzyme sites, but not hybridization procedure, is employed as a detection system), (4) copy number of the restriction landmark reflected by spot intensity (thus, haploid and diploid genomic DNAs can be distinguished) and (5) ability to use methylation-sensitive enzyme to screen genomic DNA in the methylated state. Thus, RLGS is a very useful system not only for genome mapping but also for work in a variety of biological and medical fields, including genome mapping and research on cancer, development, and aging.

## 1 CONCEPT OF GENOME SCANNING

"Genome scanning" is a high speed survey of the presence or absence of landmarks throughout a genome and measurement of their copy number in each locus. A landmark is a guidepost on the genome which can be visualized in the form of signals through the assay (scanning) method, representing a one-to-one correspondence of a signal to its locus.

The development of the processes of Southern hybridization and polymerase chain reaction (PCR) has provided quantitative and qualitative methods for visualizing landmarks directly from genomic DNA using a probe or a set of primers. In these systems, a signal specific to each locus can be detected because characteristic molecule recognition of DNA hybridization of probes or primers occurs in a sequence-specific manner.

Recently, our group introduced the "restriction landmark" concept, to suggest that each restriction enzyme recognition site can be used as a landmark. Based on this concept, we developed a restriction landmark genomic scanning (RLGS) method, which employs direct end labeling of genomic DNA digested with a restriction enzyme, and high-resolution, two-dimensional electrophoresis. Because of the strict specificity of sequence recognition of the restriction enzyme, this technique enables us to scan multiple and robust landmarks. Here, we present the concept and method of RLGS, which allows the survey of restriction landmarks throughout a genome.

## 2 PRINCIPLE AND PROCEDURE OF RLGS

### 2.1 Breakthroughs Enabling the Development of RLGS

Direct end labeling followed by electrophoretic separation had been very difficult with mammalian and other highly complex genomic DNAs. However, three significant breakthroughs paved the way to the development of RLGS. First was the recent discovery of various 8 base pair cutters, or rare cutters, which can produce an appropriate number of DNA fragments for analysis. Second was the establishment of a high resolution system of electrophoresis. Methods for one-dimensional thin-layer slab agarose gel or fine disk agarose gel electrophoresis have been developed to enable

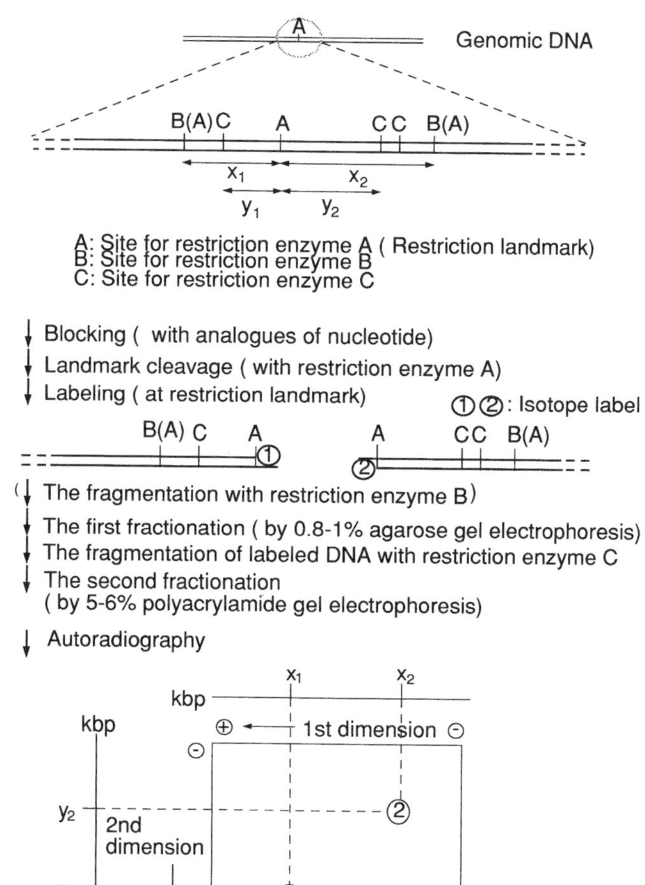


**Figure 1.** Procedure for genome scanning by two-dimensional gel electrophoresis: $x_1$ and $x_2$, respectively, represent the distance from a restriction landmark to the neighboring site for restriction enzyme A or B; $y_1$ and $y_2$ represent the distance from the site of A to C; 1 and 2 indicate the labeled sites at the end of the landmark.

high resolution electrophoretic separation and to subject the one-dimensional agarose samples to high resolution polyacrylamide gel electrophoresis in the second dimension. Third was the development of a technique for blocking the incorporation of nonspecific radioactivity. As the genome size increases, the copy number of DNA fragments generated from the same amount of DNA decreases. However, the background radioactivity produced by the incorporation into nonspecifically damaged sites of genomic DNA does not change. Therefore, the signal-to-noise ratio decreases, depending on the genome size. To overcome this problem, we developed a blocking technique to reduce the nonspecific incorporation of radioactivity based on the principle that the dideoxynucleotide analogues are incorporated into damaged sites, such as nicks, gaps, and/or double-strand breaks.

### 2.2 The Principle and Procedure of RLGS

Figure 1 shows the entire eight-step procedure of the RLGS method. Two different procedures were used, depending on whether the DNA end to be labeled was the 5′ or 3′ protruding end. Sections 2.2.1 to 2.2.8 give the steps followed for the 5′ protruding end of *Not* I.

### 2.2.1 Blocking

To reduce the background generated by the incorporation of radioactivity into the nonspecifically damaged sites, 5 μg of genomic DNA was allowed to react for 30 minutes at 37°C with 8 units of DNA polymerase I in 50 μL of 50 mM Tris-HCl (pH 7.4), 100 mM NaCl, and 10 mM dithiothreitol, 10mM $MgCl_2$; 0.4 μM 2′-deoxyribonucleoside 5′-[α-thio]triphosphate, such as dGCTP[αS], and 40 μM 2′,3′-dideoxyribonucleoside 5′-[α-thio]triphosphate, such as ddATTP[αS], were added for *Not* I digestion. Thereafter, the enzyme was inactivated at 65°C for 30 minutes.

### 2.2.2 Landmark Cleavage (with restriction enzyme A)

To cleave genomic DNA at the restriction landmarks, the treated DNA was digested with 100 units of restriction enzyme A for 1 hour in 100 μL of the reaction buffer appropriate for restriction enzyme A.

### 2.2.3 Labeling

A filling reaction of [α-$^{32}$P]deoxynucleotide with DNA polymerase is used to label the 5′ protruding end. The cleavage ends were filled in with 20 units of Sequenase Ver. 2.0 (USB) in the presence of 0.4 μM 2′-deoxyribonucleoside [α-$^{32}$P]triphosphate, [α-$^{32}$P]dGCTP (6000 Ci/mmol) for 30 minutes at 37°C with 0.16 μM dGCTP[αS] and 40 μM ddATTP[αS] in the same reaction mixture as that used in step 2. To inactivate the enzyme, the sample was incubated at 65°C for 1 hour.

### 2.2.4 Fragmentation of Labeled DNA with Restriction Enzyme B

The fragmentation step achieves further cleavage of the labeled genomic DNA at the restriction landmark sites in a test tube. Because the total length of human or other mammalian genomic DNA is approximately $3 \times 10^9$ bp, more than 3000 DNA fragments with an average length of 1 megabase (Mb) will be produced by digestion with a restriction enzyme, even if *Not* I, the most infrequent 8 bp recognition enzyme, is used. However, such long DNA fragments cannot be prepared from DNA resources such as cancer tissues, and even if these DNA fragments were to be separated by pulsed field gel electrophoresis, the resolution of the technique is not high enough to permit the analysis of so many DNA fragments.

To solve this problem, we employed the digestion reaction of labeled fragments with restriction enzyme B, which appears more frequently in the genome than restriction landmarks. Because the length of the DNA fragments has been reduced, the high resolution thin-layer agarose gel or fine disk agarose gel technique can be used, in which the entire resolution range is expanded to more than 40 cm.

The restriction sites of restriction enzyme A exceed approximately 6000 in the genome (average length of the enzyme is less than 500 kb), and when this restriction enzyme is used, the treatment of restriction enzyme B is sometimes skipped. At present, the technique cannot be used to separate a great many such spots in one gel. However, if fragmentation with restriction enzyme B is omitted, the average length of DNA fragments of almost all 8 bp cutters and rare cutters will be more than 30 kb, and these fragments will be trapped at the top of the agarose gel, resulting in reduction of the spot number.

### 2.2.5 First Fractionation by Agarose Gel Electrophoresis

To obtain a good RLGS profile, the agarose disk gel should be as fine as possible, and the samples should be electrophoresed for as long as possible to precisely separate more than 3000 signals.

First 1.5μg of the DNA from step 4 (Section 2.2.4) was fractionated on an agarose disk gel whose diameter is about 2.4 mm (0.8–1% Seakem GTG agarose; FMC). Then it was electrophoresed in 1 × first-dimension buffer (50 mM Tris-Ac, pH 8.0; 20 mM NaOAc; 18mM NaCl; 2mM EDTA) at 3.3 V/cm for 24 hours.

### 2.2.6 Fragmentation of Labeled DNA with Restriction Enzyme C

To resolve so many signals, the fragments are subjected to further separation. Specifically, the agarose gel cylinder is treated with a reaction mixture containing restriction enzyme C. Usually, an enzyme whose restriction sites occur more frequently than those of restriction enzyme B, such as a 4 bp cutter, is used as restriction enzyme C. This cleavage reaction allows the DNA fragments to be electrophoresed depending on the distance from the site of restriction enzyme A to C ($y_1$ and $y_2$ in Figure 1).

The DNA-containing portion of the gel is excised and soaked for 30 minutes in the reaction buffer appropriate for restriction enzyme C. Thereafter, DNA is digested in the gel with 550 units of restriction enzyme C for 2 hours.

### 2.2.7 Second Fractionation

After connection of the agarose strip to the two-dimensional polyacrylamide gel, the DNA fragments are subjected to further two-dimensional gel electrophoresis. The sizes of the DNA fragments range from 2 kbp to 70 bp.

The gel is fused with a 50 × cm 50 × cm 0.1 cm polyacrylamide gel (5–6% polyacrylamide to acrylamide/bisacrylamide, 29:1) by adding melted agarose to fill up the gap. Two-dimensional electrophoresis was carried out in 1 × TBE buffer at 3 V/cm for 24 hours.

### 2.2.8 Autoradiography

The final gel samples are dried and autoradiographed. After the gel has been dried, a 35 cm × 43 cm area of the original gel is excised and autoradiographed for 3 to 10 days on a film (XAR-5; Kodak) at −70°C using an intensifying screen (Quanta III; DuPont).

### 2.2.9 Steps in RLGS for 3′ Protruding End

The procedures outlined in Sections 2.2.1 to 2.2.8 were used for the 3′ protruding or blunt end labeling, except for the first and third steps.

In the first (blocking) step, DNA was allowed to react with 25 units of terminal deoxynucleotidyl transferase (Toyobo) in the presence of 10 μM 2′,3′-dideoxyadenosine 5′-[α-thio]triphosphate ddATP[αS], for 30 minutes at 37°C. In the third (labeling) step, the cleavage ends were labeled by reaction with 25 units of terminal deoxynucleotidyl transferase in the presence of 0.8 μM [α-$^{32}$P]ddATP (5000 Ci/mmol) for 30 minutes at 37°C. In both steps, the reactions were carried out in 140 mM sodium cacodylate (pH 7.0), 1 mM $CoCl_2$, and 0.1 mM dithiothreitol.

## 3 THE RLGS PATTERN AND ITS ADVANTAGES

Figure 2 shows the RLGS profile that results from using *Not* I, *Eco*RV, and *Mbo* I as restriction enzymes A (restriction landmark), B, and C, respectively. As shown in Figure 1, the $X,Y$ coordinates $(x_1,y_1)$, $(x_2,y_2)$ of each spot indicate the distances from the site of restriction enzymes A through B and from A through C, respectively. In spite of the large size of mammalian or plant genomes, it is difficult to produce two sets of identical valued parameters $(X,Y)$ from different landmarks on the genome. Therefore, one spot on the RLGS profile corresponds to one locus.

RLGS has the following advantages for genome scanning.

1. It has high speed scanning ability. In the RLGS profile of Figure 2, more than 2500 spots appeared in one gel, indicating that the information corresponds to more than 2500 times that obtainable from Southern hybridization or PCR with unique probes (primers).
2. The scanning field can be extended by using landmarks of different kinds in an additional electrophoresis series.
3. The method can be applied to any organism, because direct labeling of restriction enzyme sites, not hybridization procedure, is employed as the detection system.
4. Spot intensity reflects the copy number of the restriction landmark. In Figure 2, several enhanced spots can be seen.

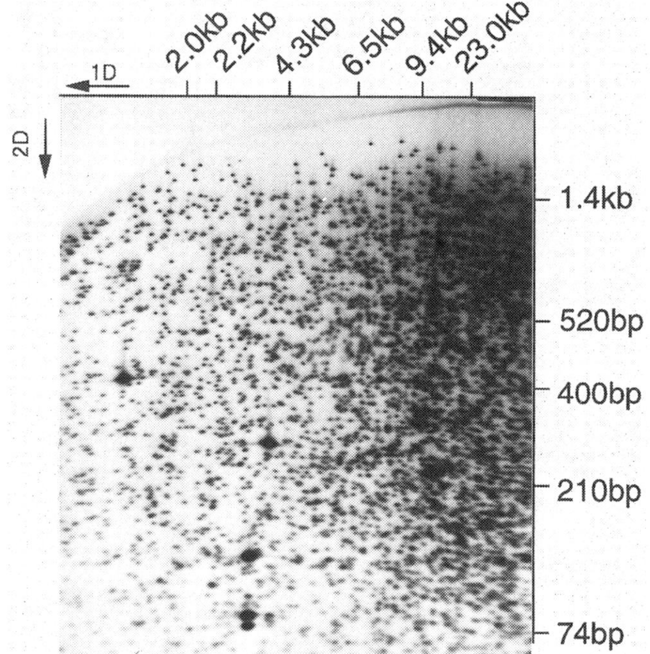


**Figure 2.** RLGS profile of mouse genomic DNA.

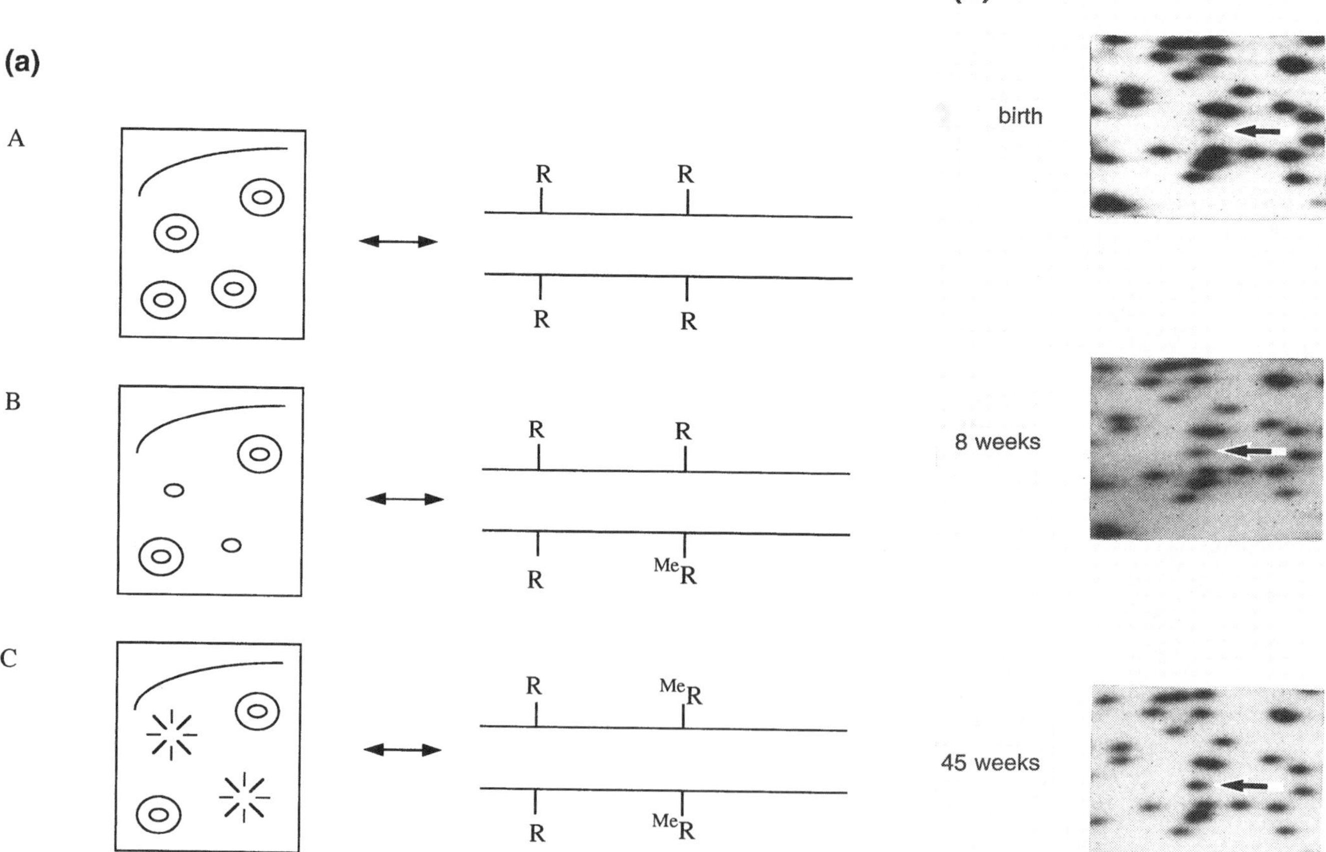


**Figure 3.** Alteration of spot intensity due to the genomic DNA methylation and detection of DNA methylation change depending on aging. (a) Schematic representation: A, when both alleles are demethylated, the spots show diploid intensity; B, when one allele is methylated, the corresponding spot shows half-intensity; C, when both alleles are methylated, the spots disappear: R, restriction landmark; $^{Me}$R, the methylated restriction site. (b) RLGS patterns produced using the mouse liver at birth and at 8 and 45 weeks. Spot indicated by the arrowhead showed increasing intensity with aging.

These spots are derived from mouse ribosomal DNA, which is the typical repetitive sequence in a mammalian genome. The haploid and diploid genomic DNAs can be distinguished in RLGS.

5. When methylation-sensitive enzyme is used, the methylated state of genomic DNA can be screened.
6. When CpG-rich enzymes (*Not* I, *Bss*HII, etc.) are used, CpG islands near the genes are preferentially screened. For example, 89% of all *Not* I spots are located on or near transcripts. Thus, RLGS is very advantageous compared with hybridization-based and PCR-based genome scanning methods.

## 4 APPLICATIONS

The RLGS method can be used in two ways. One is for genome scanning to determine the copy number of each locus corresponding to a spot, using methylation-insensitive enzyme as a restriction landmark. The other use is for the detection of DNA methylation throughout the genome, using methylation-sensitive enzyme.

Generally, the only known site for DNA methylation is the 5-position of the cytosine residue ($^{5Me}$C), which is preferably located on the palindromic dinucleotide 5′CG3′/5′GC3′. In $^{5Me}$CpG methylation-insensitive enzymes such as *Pac* I, *Swa* I, and *Sse*8387I, the cleavage is not affected by DNA methylation in the vertebrate genome. Thus when these enzymes are used on the RLGS profile, the intensity of the spots depends completely on the copy number of the corresponding locus. On the other hand, when $^{5Me}$CpG-sensitive enzyme is used as restriction enzyme A, the intensity of the RLGS spots changes depending on DNA methylation, as shown in Figure 3a.

For example, high speed genetic mapping that allows the identification of the chromosomal location of many RLGS spots has been established. Taking advantage of one feature of RLGS—namely, that spot intensity reflects the copy number of the landmark—amplification in cancer DNA and loss of heterozygosity can be detected. Figure 3b shows an example of systemic detection of DNA methylation changes at the *Not* I site during aging. All three patterns were produced from the liver DNA of the C57BL/6 inbred mouse strain at three stages: birth, 8 weeks, and 45 weeks. The spot represented by the arrowhead is altered in intensity, indicating that DNA methylation state on the restriction landmark *Not* I is variable.

The RLGS method described here can be applied to many biological fields, such as cancer research, high speed genome mapping (large number of analyzable loci), identification of individuals, and phylogenetic analysis of species. Thus, the RLGS method is very useful for scanning not only the change of copy number of each locus, but also the change in DNA methylation.

*See also* AUTOMATION IN GENOME RESEARCH; DNA SEQUENCING IN ULTRATHIN GELS, HIGH SPEED; RESTRICTION ENDONUCLEASES AND DNA MODIFICATION METHYLTRANSFERASES FOR THE MANIPULATION OF DNA.

### *Bibliography*

Dietrich, W., et al. A genetic map of the mouse suitable for typing intraspecific crosses. *Genetics* 131:423–447 (1992).

Hatada, I., et al. A genomic scanning method of higher organisms using restriction sites as landmarks. *Proc. Natl. Acad. Sci. U.S.A.* 88:9523–9527 (1991).

Hayashizaki, Y., et al. Restriction landmark genomic scanning method and its various applications. *Electrophoresis* 14:251 (1993).

Hayashizaki, Y., et al. (1994) A genetic linkage map of the mouse using Restriction Landmark Genome Scanning (RLGS). *Genetics* 138: 1207–1238.

Hayashizaki, Y., et al. Identification of an imprinted U2af binding protein related sequence on mouse chromosome 11 using the RLGS method. *Nature Genet.* 6:33–40 (1994).

Hirotsune, S., et al. New approach for detection of amplification in cancer DNA using restriction landmark genomic scanning. *Cancer Res.* 52:3642–3647 (1992).

Kawai, J., et al. Methylation profiles of genomic DNA of mouse developmental brain detected by restriction landmark genomic scanning (RLGS) method. *Nucleic Acids Res.* 21:5604–5608 (1993).

Lindsay, S., and Bird, A. P. Use of restriction enzymes to detect potential gene sequences in mammalian DNA. *Nature* 327:336–338 (1987).

# RETINOBLASTOMA, MOLECULAR GENETICS OF

*Brenda L. Gallie and John Wu*

### *Key Words*

**Loss of Heterozygosity** Loss of large regions of one chromosome, detected by loss of polymorphic markers on that chromosome: that is, the cells appear to have become homozygous for all markers within the region.

**Retinoma** Benign retinal tumors resembling treated retinoblastoma that do not continue to grow, but spontaneously cease proliferation; caused by mutations in *RB1*.

**Transcription Factor** Proteins that regulate the expression of genes by binding in various protein complexes and DNA.

**Viral Oncoproteins** Proteins of the DNA tumor viruses that induce cellular proliferation and tumors in animals by binding to normal host cellular regulatory proteins, such as the retinoblastoma protein (pRB).

Retinoblastoma, a rare tumor of the infant retina, has become important for understanding cancer in general. Studies of retinoblastoma tumors revealed a previously unsuspected mechanism of cancer initiation, loss of heterozygosity. The retinoblastoma gene (*RB1*) was the first tumor suppressor gene to be defined, and it has turned out to be a transcription factor involved in the regulation of how almost all adult mammalian cells divide. It is not yet known why mutations of *RB1* initiate specific tumors such as retinoblastoma but do not cause a general tumor predisposition.

## 1 THE DISEASE

Retinoblastoma is a rare pediatric tumor that arises from the immature neuroectodermal cells of the developing retina, with an estimated incidence of one in 20,000 births. About half the patients have heritable retinoblastoma and can transmit this tendency as an autosomal dominant trait; these children usually have bilateral and multifocal tumors, diagnosed on average at about one year of age. Some children with a heritable mutation may develop only one

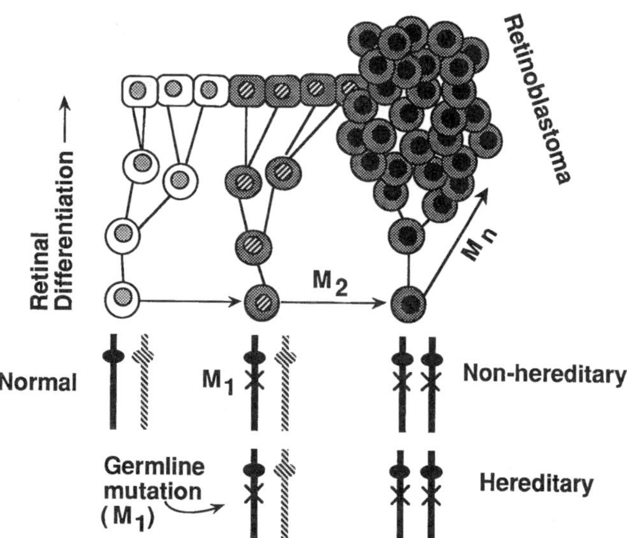


**Figure 1.** Two mutations ($M_1$ and $M_2$) occurring in the same retinal cell are needed to initiate the process of malignant transformation. For malignant progression, however, additional mutations ($M_n$) must accumulate. In nonhereditary retinoblastoma, both $M_1$ and $M_2$ occur in the same somatic retinal cell. In hereditary retinoblastoma, $M_1$ has already occurred in the germ line; the additional mutation ($M_2$) in retinal cells initiates the malignant process.

tumor in one eye, and some mutation carriers develop no tumors at all. All nonhereditary cases develop only one unilateral, solitary retinoblastoma tumor. These manifestations of the disease also can be diagnosed from birth to early childhood, although on average they occur at an older age than the heritable tumors.

Statistical analysis of the different mean ages of diagnosis of hereditary and nonhereditary retinoblastoma led Knudson to hypothesize that the rate-limiting number of mutations for the initiation of this tumor is at least 2. In hereditary retinoblastoma, the first mutation is present in every cell of the patient, and only one somatic mutation is required to initiate the cancer; in nonhereditary tumors, the two mutations must occur in the same somatic retinal cell that eventually is transformed to malignant neoplasm. Comings extended this idea to suggest that the mutations could involve the two allelic homologues at the same locus, illustrated in Figure 1. These hypotheses were later supported by the demonstration of frequent loss of an allele from tumors and mutations in the retained allele(s) at the retinoblastoma locus (*RB1*).

The mechanism of loss of an allele was first recognized to initiate cancer in retinoblastoma. Subsequent studies searched for chromosomal regions in which a similar loss of heterozygosity (LOH) had occurred in tumors of specific tissues, and thus other tumor suppressor genes were identified, such as the Wilms' tumor gene and genes predisposing to adenomatosis of colon and rectum. Several chromosomal regions that show LOH in tumors are suspected of containing tumor suppressor genes not yet discovered. The tumor suppressor gene, p53, also shows LOH in many tumor types.

In heritable retinoblastoma, the first mutation is found heterozygously in germ cells of the individual. The mutation may be inherited, but more commonly it has newly arisen on the father's allele. It is not known why new mutations are more common in fathers' chromosomes than in mothers'. Ninety percent of children with a heterozygous *RB1* constitutional mutation will develop retinoblastoma, a relative risk (RR) of 40,000 compared to the general population. This accounts for the autosomal dominant pattern of transmission of the predisposition to retinoblastoma, although the tumors that arise have mutated both alleles. The same individuals have an RR of 500 to later develop other solid tumors, most commonly osteogenic sarcomas. Radiation, the conventional treatment for treating retinoblastoma in eyes with salvageable vision, increases this risk a further tenfold.

Since only a few patients have a family history of retinoblastoma, most children are diagnosed when one eye contains large tumor(s), causing the white pupil or "cat's eye" that is the classic presenting sign. When such eyes are removed, cure is achieved if no tumor cells have extended outside the eye. Smaller tumors present in an eye with useful vision can often be cured by combinations of chemotherapy, laser therapy, cryotherapy, and focal radiation.

Rare retinal tumors resembling retinoblastomas but not behaving in a malignant fashion arise because of *RB1* mutations and are called retinomas. It is hypothesized that these growths may be initiated by the loss of *RB1* at a late stage of retinal differentiation, when insufficient cell divisions remain before terminal differentiation to permit accumulation of the extra mutations in other cancer genes required for full malignancy. At the other extreme, malignant retinoblastoma tumors that have spread outside the eye have a very poor prognosis, demonstrating mutations in multiple additional growth-related genes. These include, rarely, amplification of the N-*myc* proto-oncogene and, commonly, overexpression of the multidrug resistance gene, p170, which has contributed to the poor success of chemotherapy in retinoblastoma in the past.

Since infant relatives of retinoblastoma patients are also at risk to develop retinoblastoma tumors, they are repeatedly examined, often under anesthetic, to ensure the early discovery of tumors. Identification of the *RB1* mutation(s) in the tumor and constitutional cells of the first affected child makes possible simple molecular screening of relatives to determine which one has the family's mutation. Those unaffected will not require the intense surveillance. New technology to identify each family's mutation based on automated DNA sequencing will soon be available, to permit this testing as a health service.

## 2 THE GENE AND PROTEIN

The *RB1* gene spans 180 kilobases of chromosome 13 band q14.2 and consists of 27 exons. A 4.7 kb *RB1* transcript is expressed in all adult tissues. The *RB1* gene codes for a nuclear phosphoprotein of 928 amino acids (pRB).

There is evidence that *RB1* expression is regulated at a transcriptional level by autoregulatory feedback. First, mutant *RB1* mRNA is undetectable in constitutional cells of individuals heterozygous for germ line *RB1* mutations, in which the normal allele contributes a normal pRB, but mRNA with the same mutation is easily detectable in retinoblastoma cells with no functional protein. Second, transient expression of pRB suppresses expression from an *RB1* promoter fused to a reporter construct. The *RB1* promoter region is TATA-less and contains consensus motifs for transcription factors ATF, Sp1, and E2F.

The apparent molecular mass of pRB (110–116 kDa) depends on the state of phosphorylation of serine and threonine residues, which is tightly regulated in the cell cycle. The protein is hypophosphorylated in $G_0$ and early $G_1$, but hyperphosphorylated in late $G_1$, S, $G_2$, and M phases. The hypophosphorylated form binds the viral transforming proteins, adenovirus E1A, SV40 large T, and papillomavirus E7. The transforming potential of these viral onco-

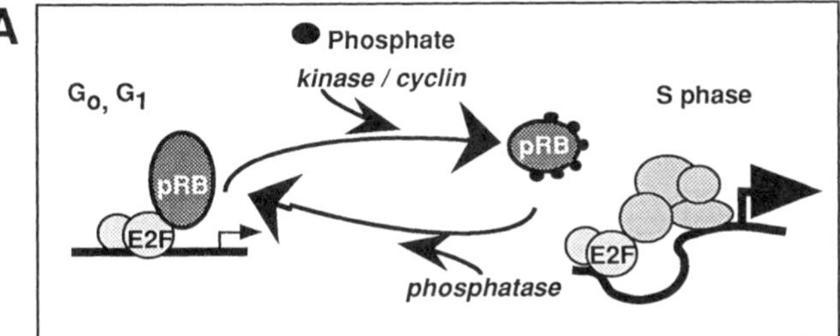


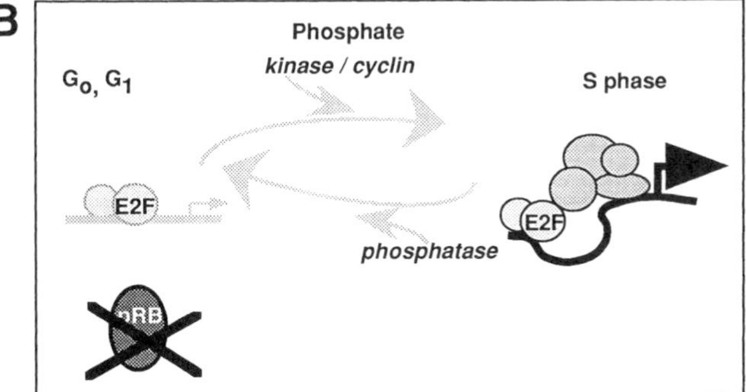


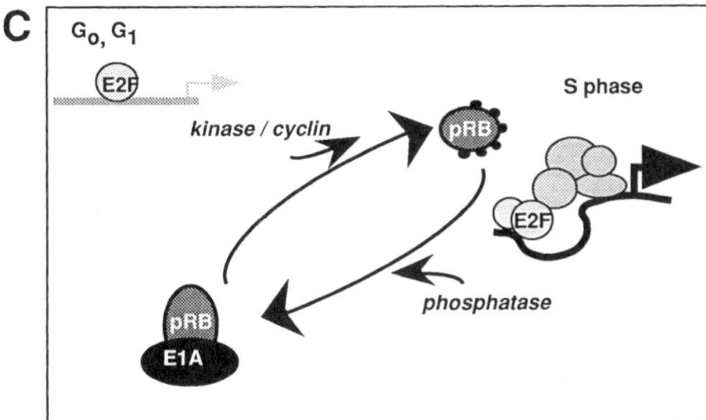


**Figure 2.** (A) The active, hypophosphorylated retinoblastoma protein (pRB) binds and inhibits E2F complexes during $G_0$ and $G_1$. pRB is phosphorylated by cell-cycle-specific kinases, complexed to cyclins, as the cell enters S phase, releasing E2F to activate transcription of responsive genes. After mitosis, pRB is dephosphorylated, making it again available for binding to E2F during $G_1$ phase. (B) Mutations affecting pRB disable it from binding E2F, rendering the cell incapable of achieving the terminal differentiation normal for retina. (C) Viral oncoproteins bind and sequester pRB to achieve malignant transformation.

proteins, which lies in part in their ability to bind and sequester pRB away from its normal role of negative growth regulation, is dependent on critical protein domains. This suggests that hypophosphorylated pRB is the active form, which performs its function in $G_0$ and $G_1$ phases of the cell cycle, as outlined in Figure 2. The binding of pRB to the viral oncoproteins depends on the integrity of two separate domains, which together form a "pocket" for binding. A number of other cellular proteins also bind to the viral oncoproteins in the course of cellular transformation, including the p107 protein, homologous to pRB, and the tumor suppressor gene p53. Thus there is more than one pathway controlling cellular proliferation, and the sum of them all must be neutralized to achieve malignant transformation.

During $G_0$ and $G_1$, hypophosphorylated pRB exerts its negative effect on cellular growth by binding to transcription factors, rendering them unavailable for modification of transcription. The cell is thus blocked from proliferation and not provided an opportunity to differentiate. In late $G_1$, pRB is phosphorylated by cell-cycle-specific kinases related to $p34^{cdc2}$ and $p33^{cdk2}$, becomes inactive, and no longer binds to positive growth regulators, enabling these factors to promote cell proliferation.

Some of the growth response genes regulated by pRB are known. The expression of c-*fos,* itself a transcription factor responsible for early response to growth signals, is repressed by pRB, acting through the "retinoblastoma control element" (RCE) promoter consensus sequence. Hypophosphorylated pRB binds to complexes containing the transcription factor E2F, and as the cell progresses through late $G_1$, hyperphosphorylation of pRB releases E2F to transactivate growth response genes, including c-*myc*.

Introduction of *RB1* cDNA into tumors negative for pRB, including retinoblastoma tumors, shows variable and only partial reversal of the malignant phenotype. Although retinoblastoma initiation requires the loss of both alleles in *RB1,* additional mutations are likely to be involved in the malignant transformation process, probably maintaining malignant growth despite reconstitution with a wild-type *RB1*.

The important role of pRB in development and differentiation has been demonstrated in transgenic mice homozygous for *RB1* mutant alleles. These mice die in utero, manifesting abnormalities in the hematopoietic and nervous systems related to the failure of terminal differentiation. This set of symptoms might reflect a distinct function of *RB1* in promoting differentiation, in contrast to a role of negative growth regulation. Surprisingly, the heterozygous mice do not develop retinoblastoma, but instead develop pituitary malignancy.

Mutations leading to tumors occur throughout *RB1,* with no "hot spots"; most result in truncated proteins that are not detectable in retinoblastoma tumors. A significant proportion of other types of tumor, such as breast cancer, small-cell carcinoma of lung, and bladder cancer, also have mutations in the *RB1* gene. The *RB1* mutations in these tumors likely contribute to tumor progression by providing a selective growth advantage after the tumor is initiated. Individuals with germ line *RB1* mutations do not appear to have increased incidence of all the tumors types that show *RB1* mutations, although they do have increased incidence of specific tumor types, such as osteosarcoma.

## 3 UNANSWERED QUESTIONS

The exact role of *RB1* in the regulation of cellular proliferation and differentiation awaits full elucidation. Since *RB1* is expressed in most normal adult cells, interacts with general transcriptional factors, and plays a role in cell cycle progression at the $G_1$/S transition, it is hard to explain why only a small number of tissues are susceptible to malignant transformation in individuals with a germ line *RB1* mutation. One suggested explanation is that pRB interacts with specific factors in different tissues at critical developmental stages, with the result that the effect of the loss of *RB1* depends on the type of tissue and developmental stage. There may be redundant pathways available for growth control. The presence of pRB is not essential for progression of the cell cycle, since tumor cells that are $RB1^-$ traverse the cell cycle without hindrance. Another explanation might be that *RB1* is essential for the survival of cells of most tissues, in which loss of *RB1* is lethal. Only the few tissues that are susceptible to malignant change have redundant pathways for survival when the loss of *RB1* causes dysregulated cell proliferation.

The results observed in *RB1*⁻ transgenic mice have raised more questions. The important role of *RB1* in development is established, but the tissues affected are unexpected, since there is no increase in hematopoietic or neurological malignancies in individuals with germ line *RB1* mutation. The question is whether these tissues are uniquely sensitive to the loss of *RB1* or whether, perhaps, they are affected just because they are the first to enter terminal differentiation; in the latter case, other developing tissues would also be affected later, if the mice survived long enough. The failure to detect retinoblastoma in heterozygous *RB1*⁻ mice may reflect the requirement for the second allele to be mutated in a developmental window that may be too short in the species studied. This possibility, however, would not account for the occurrence of pituitary tumors.

*See also* CANCER; HUMAN GENETIC PREDISPOSITION TO DISEASE; ONCOGENES; TUMOR SUPPRESSOR GENES.

### *Bibliography*


Gallie, B. L., Dunn, J. M., Chan, H. S., Hamel, P. A., and Phillips, R. A. (1991) *Pediatr. Clin. North. Am.* 38:299–315.

Sopta, M., Gallie, B. L., Gill, R. M., Hamel, P. A., Muncaster, M., Zacksenhaus, E., and Phillips, R. A. (1992) *Semin. Cancer Biol.* 3:107–112.

Knudson, A. G. J., Jr. (1989) *Cancer,* 63:1889–1891.


# RETINOIDS

*Robert R. Rando*

### *Key Words*

**Retinoid** A diterpene polyene derivative related to and including vitamin A.

**Visual Cycle** The sequence of biochemical reactions beginning with the photochemcal cis-to-trans isomerization of the 11-*cis*-retinal Schiff base chromophore of rhodopsin and terminating with the enzymatic resyntheses of 11-*cis*-retinal.

**Visual Transduction** The sequence of biochemical events, initiated by the absorption of light by rhodopsin, which lead to the hyperpolarization of photoreceptors and the initiation of the visual response.

The retinoids are very important in cellular physiology. Their most completely understood roles are found in visual signal transduction in animals and in energy production in halophilic bacterial. The recent discovery of the cell growth, maintenance, and developmental roles of the retinoic acids adds a new and exciting chapter to the biology of the retinoids. All the retinoids are diterpene polyene derivatives, and as such their activities are often controlled by reversible double-bond isomerization reactions involving mono-*cis* and all-*trans* isomers. This entry explores the centrality to vision of double-bond isomerization reactions of retinoids. The possible relationship of insights drawn from this work to the hormonal roles of the retinoic acids is also described.

## 1 PHOTOISOMERIZATION OF RHODOPSIN AND VISUAL TRANSDUCTION

Vision is initiated when the visual pigment rhodopsin absorbs a photon of light, leading to the cis-to-trans isomerization of its protonated 11-*cis*-retinal Schiff base chromophore (Figure 1). 11-*cis*-Retinal, a diterpene retinoid related to vitamin A (all-*trans*-retinol), is covalently linked to the apoprotein opsin at the latter's active site lysine residue. The photochemical isomerization of the chromophore in rhodopsin leads to a series of spectroscopically defined intermediates, resulting in the eventual hydrolysis of the all-*trans*-retinyl Schiff base to produce all-*trans*-retinal and opsin, as shown in Figure 1. The photoisomerization reaction has been reported to be essentially complete in only 200 fs. Interestingly, noise analysis has placed an upper limit on the rate of thermal isomerization of rhodopsin in the neighborhood of centuries, producing in the rods a signal-to-noise ratio of approximately $10^{23}$. One of the spectroscopically defined intermediates, metarhodopsin 2, is the intermediate (R*) that transmits the information that light has been absorbed by interacting with the next molecular entity in the visual cascade, the retinal G protein, transducin.

The interaction of R* with transducin results in catalysis of the exchange of GDP for GTP at the active site of transducin. The transducin–GTP complex activates a rod outer segment phosphodiesterase specific for c-GMP which can hydrolyze c-GMP at the diffusion-controlled limit. This hydrolysis is thought to lower the free concentration of this effector. Since the sodium ion permeability of the rod outer segment sodium channels is gated by c-GMP, hydrolysis of c-GMP results in the hyperpolarization of the rod outer segments. This membrane potential change is signaled to other nerve cells in the retina and communicated to the brain, where the visual image can then be constructed.

The structure of metarhodopsin 2 is of substantial interest because this molecule propagates the light signal. Moreover, rhodopsin represents the best understood member of the homologous family of seven-transmembrane helical receptors, and further molecular insights into its mode of action are likely to be of general significance. Of all the spectroscopically defined bleaching intermediates of rhodopsin, only metarhodopsin 2 possesses a deprotonated Schiff base. Experiments in which a single methyl group is incorporated into the active site lysine residue show that Schiff base deprotonation is obligate in the form of activated rhodopsin capable of fruitfully interacting with transducin. Active site methylated rhodopsin undergoes facile photoisomerization and produces the early photointermediates up to metarhodopsin 1. The long-lived metarhodopsin 1 does not proceed to metarhodopsin 2 and cannot activate transducin, nor can it be phosphorylated by rhodopsin kinase. The idea that Schiff base deprotonation of photolyzed rhodopsin is a key transduction event has gained considerable support from site-specific mutagenic studies. Recent studies in which the active site lysine is removed altogether bring these studies to a satisfying conclusion. Mutant opsins in which the active site lysine residue has been substituted for a glycine or alanine residue are constitutively active with respect to the catalysis of GTP for GDP exchange in transducin. Since these mutants do not bear a positive charge at their active sites; they are able to mimic metarhodopsin 2 in the absence of chromophore and photoisomerization. This suggests that the role of 11-*cis*-retinal is to enable protonated Schiff base formation and the turning off of a potentially

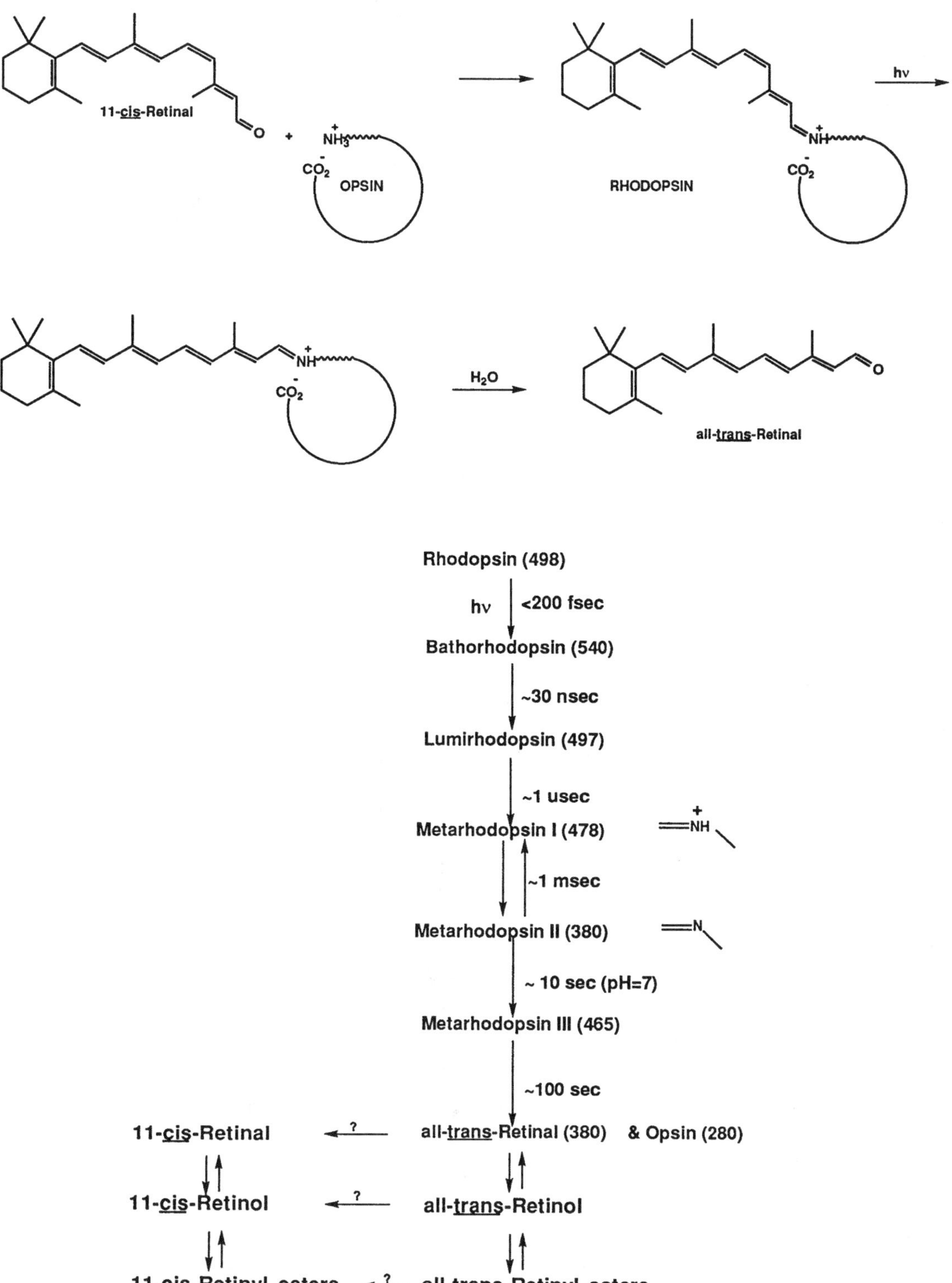


Figure 1. *Top:* the photochemical isomerization of the 11-*cis*-retinal Schiff base of rhodopsin. The various spectroscopically defined intermediates are indicated along with the positions of their maximum wavelengths, in parentheses. *Bottom:* the various retinoid intermediates that occur after all-*trans*-retinal is released from rhodopsin, shown as part of the classical visual cycle.

constitutively activated receptor. Moreover, these results are entirely consistent with the idea that the key molecular event in rhodopsin activation is simply the deprotonation of the active site lysine residue, brought about by the cis-to-trans photoisomerization reaction. It must be assumed that a deprotonation event will be central in the activation of other homologous seven-transmembrane helical signal transducing receptors.

## 2 ENZYMATIC REGENERATION OF 11-*cis*-RETINAL AND THE VISUAL CYCLE

For all-*trans*-retinal to be liberated as a consequence of bleaching, a physiological mechanism must be in place for the regeneration of 11-*cis*-retinal. The all-*trans*-retinal liberated by the bleaching of rhodopsin has been shown to be rapidly reduced enzymatically in the retina by specific nicotinamide-linked retinol dehydrogenases. The vitamin A that is produced is then transported from the rod outer segments to the pigment epithelium, where it is esterified, largely to long chain saturated fatty acid esters, such as those in the palmitate and stearate series. For each of the three all-*trans*-retinoids, there is a corresponding 11-*cis*-retinoid, and the totality of biochemical reactions required to interconvert these six molecules comprises the classical visual cycle, shown in simplified form in Figure 1. Thus, the double-bond isomerization reaction could potentially occur with any one of three substrates and could produce any one of the three corresponding products, making for nine possible isomerization pathways. In addition, the isomerization could occur either in the retinal pigment epithelium (the organ found at the back of the eye in contact with the retina), in the retina proper, or both in the retina and in the pigment epithelium.

In addition to the nature of the substrate for the reaction, the relative stabilities of the *cis*- and *trans*-retinoids also need to be considered. This thermodynamic issue was first pointed out by Pauling. When the retinals and retinyl esters were brought into equilibrium by acid from various isomers, it was found that 11-*cis*-retinoids accounted for approximately 0.1% of the equilibrium mixture. There is a 4.1 kcal/mol difference between 11-*cis*-retinal and its all-*trans* counterpart, with similar differences being observed between the other retinoid congeners. This leads to the important question of where the energy comes from to drive the trans-to-cis isomerization in vivo.

The in vitro biosynthesis of 11-*cis*-retinoids from all-*trans*-retinol was first demonstrated with a membrane preparation from the pigment epithelium, which proved capable of converting all-*trans*-retinol (vitamin A) to a mixture of 11-*cis*-retinol, 11-*cis*-retinal, and 11-*cis*-retinyl palmitate. All-*trans*-retinyl palmitate and all-*trans*-retinal were also generated by these membranes. The isomerase system is highly specific for all-*trans*-retinoids and was found largely, if not entirely, in the pigment epithelium rather than in the retina itself.

Since all-*trans*-retinol is converted into all-*trans*-retinal, all-*trans*-retinyl palmitate, and the three 11-*cis* congeners by the pigment epithelium membrane preparation described, it cannot be assumed that all-*trans*-retinol is the substrate for the isomerase system or that 11-*cis*-retinol is the direct product of the isomerase action. Following the fate of specifically isotopically labeled substrates and using specific ester synthetase inhibitors, such as all-*trans*-retinyl-α-bromoacetate (RBA), allowed for a demonstration of the actual isomerization pathway and showed the obligate intermediacy of retinyl esters in 11-*cis*-retinoid formation. RBA is a potent affinity labeling agent of lecithin retinol acyl transferase (LRAT), the enzyme responsible for retinyl ester biosynthesis in the pigment epithelium. This novel enzyme specifically transfers acyl groups from the *sn*-1 of lecithin to vitamin A and analogues. When pigment epithelial membranes are treated with RBA, neither retinyl esters nor 11-*cis*-retinoids are generated, and all-*trans*-retinyl esters are directly processed to 11-*cis*-retinol. These experiments leave little doubt that all-*trans*-retinyl ester, not all-*trans*-retinol, is processed to 11-*cis*-retinol, and that retinyl ester formation is an obligate part of the isomerase pathway.

Why might the linkage of ester synthesis to isomerization be of interest? As mentioned before, esters are "high energy" compounds and are hydrolyzed with free energies of hydrolysis in the −5 kcal/mol range. If the free energy of hydrolysis of an ester could be coupled to the isomerization process, more than enough energy would be provided to drive the latter process. As applied here, the energies sum as shown in Figure 2A, to yield a mechanism by which ester hydrolysis could be linked to isomerization. Here, the substrate is an all-*trans*-retinyl ester and the product is 11-*cis*-retinol. As predicted by this mechanism, labeling experiments show that the original oxygen of the vitamin A is lost during isomerization. In addition, inversion of the absolute configuration of the prochiral $CH_2OH$ accompanies isomerization. If ester hydrolysis occurs simultaneously with isomerization, then the enzyme catalyzing the isomerization reaction is an isomerohydrolase rather than a simple isomerase, because the substrate and the product of the enzymatic reaction are not isomers of each other.

In the scheme shown in Figure 2A, LRAT has an energy-transducing function and may be a member of a new class of enzymes. Hence it has been studied in some detail. The kinetic mechanism was clearly demonstrated to be ordered as follows: ping pong bi bi. In this case, the lecithin binds to the enzyme first, and an acyl group is transferred from the *sn*-1 position of the phospholipid to an active site residue. Following the desorption of the 2-acyl lysophospholipid, vitamin A binds at the active site and the acyl group is transferred to it. The identity of the active site amino acid to which the acyl group is transferred remains to be determined, but the enzyme is catalytically inactivated by sulfhydryl and serine-directed reagents, suggesting an active site serine or cysteine residue. As mentioned, the enzyme is affinity-labeled by RBA. In experiments using [$^3$H]RBA, a labeled protein of approximately 24 kDa has been identified as a candidate for the holoenzyme or a subunit of it. The finding that the retinyl esters are produced using lecithin as an acyl donor shows that this membrane constituent is being used in a group transfer reaction in much the same way that ATP would be. Therefore the energy to drive the endergonic cis-to-trans isomerization process in the visual cycle originates in the membrane itself. Previously, lecithin was not thought to play a role in the visual cycle, and retinyl esters were thought to be important only as innocuous storage forms of the retinols.

Our current understanding of the visual cycle, which is based on the results already discussed, is shown in Figure 2B. Upon bleaching, the all-*trans*-retinal is reduced by a pro-*R*-specific dehydrogenase in the rod outer segments to produce all-*trans*-retinol. The all-*trans*-retinol travels to the pigment epithelium, probably bound as a complex with the interphotoreceptor retinoid-binding protein (IRBP). The all-*trans*-retinol is esterified via LRAT action in the pigment epithelium-and the retinyl ester is processed to 11-*cis*-retinol. Oxidation of the 11-*cis*-retinol to 11-*cis*-retinal occurs by means of a pro-*S*-specific dehydrogenase followed by the delivery of 11-*cis*-retinal to the photoreceptors, thus completing the visual cycle.

## 3 ISOMERIZATION OF RETINOIC ACIDS

All-*trans*-retinoic acid is known to possess important hormonal and developmental activities which are manifest through nuclear retinoic acid receptors. Recently two groups have observed an isomer of retinoic acid, other than the all-*trans* isomer, that possesses biological activities. Specifically, 9-*cis*-retinoic acid has been shown to be the favored ligand for the retinoid x receptor (RXR) receptor. In fact it appears likely that signaling through the RXR receptors may have much more widespread physiological impact than signaling through the RAR receptors. Whatever the physiological link may be between the two different receptor types, it is of interest to inquire about the possible metabolic relationship between all-*trans*-retinoic acid and its 9-*cis* congener.

It is reasonable to think that what we have learned about isomerization processes in the visual system may be applicable to the problem of 9-*cis*-retinoic acid biosynthesis. The problem of the biosynthesis of 9-*cis*-retinoids may be much simpler than that encountered in the formation of 11-*cis*-retinoids, because 9-*cis*-retinoids are not rendered highly unstable relative to their all-*trans* counterparts by intramolecular steric interactions. Hence, an energy-yielding reaction is not required for the formation of 9-*cis*-retinoids, which comprise approximately 20% of an equilibrium mixture.

A simple mechanism involves the direct isomerization of all-*trans*-retinoic acid to 9-*cis*-retinoic acid. In fact, a process in which all-*trans*-retinoic acid is isomerized to a mixture of 9-*cis*-retinoic acid and 13-*cis*-retinoic acid has already been reported to occur in various cell types. We have demonstrated that bovine liver membranes can nonregiospecifically isomerize all-*trans*-retinoic acid into 9-*cis*-retinoic acid and 13-*cis*-retinoic acid. However, the isomerization process is not saturable. Moreover, the isomerization of all-*trans*-retinoic acid to the congeneric mono-*cis* isomers also occurs in certain buffers in the presence of bovine serum albumin. The spontaneous formation of 9-*cis* isomers in the retinoid series is highly unusual and suggests that isomerization mechanism(s) are possible in the retinoic acid series that are not available in the retinal and retinol series. It is entirely possible that this kind of mechanism operates in cells and is physiologically important. A major drawback of any equilibrium process of this type is that it cannot spatially separate the all-*trans* and 9-*cis*-retinoic acids. It

A

+ LECITHIN →

$EB^-$ = enzyme active-site nucleophile

-OCOR

-EB⁻ -H⁺

Overall Reaction: trans-Retinyl Ester + $H_2O$ → trans-Retinol   -5 kcal/mol (approx.)

trans-Retinol → 11-cis-Retinol   +4 kcal/mol

trans-Retinyl Ester + $H_2O$ → 11-cis-Retinol   -1 kcal/mol (net)

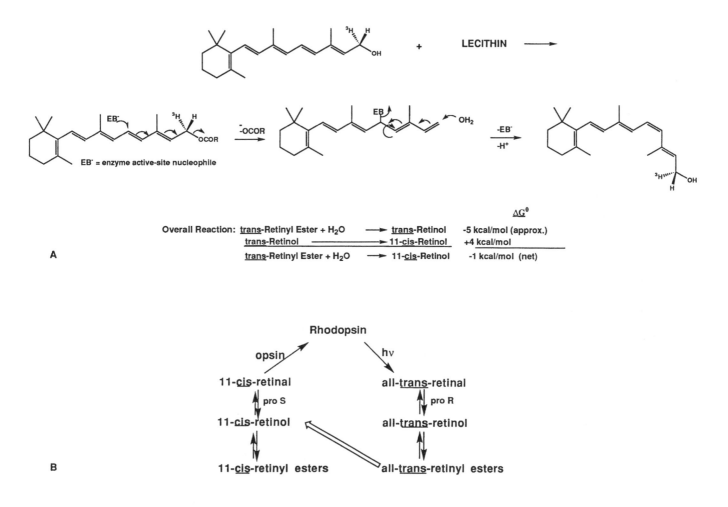

$\Delta G^0$

B

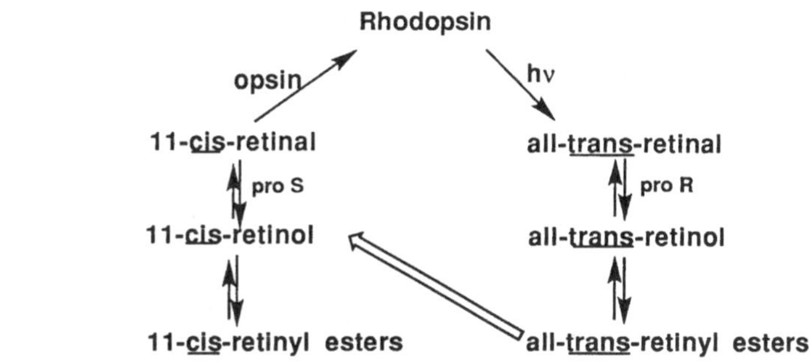


**Figure 2.** (A) The mechanism by which the free energy of hydrolysis of an acyl ester bond is coupled to the endergonic *trans*-11-*cis* retinoid isomerization reaction. (B) Current model of the visual cycle.

is interesting to consider a process analogous to what was described in the biosynthesis of 11-*cis*-retinoids which does not suffer from this drawback. The processing of an all-*trans*-retinyl ester to 9-*cis*-retinol would provide a one-way reaction to the latter (oxidation of 9-*cis*-retinol to 9-*cis*-retinoic acid is feasible via one of the well-described retinol oxidases). An energy-linked mechanism of this type would allow for the physiologically irreversible conversion of an all-*trans*-retinoid to a 9-*cis*-retinoid and for the spa ial separation of the activities of these retinoids.

*See also* VITAMINS, STRUCTURE AND FUNCTION OF.

### *Bibliography*

Mangelsdorf, D. J., Kliewer, S. A., Kakizuka, A., Umesono, K., and Evans, R. M. (1993) Retinoid receptors. *Recent Prog. Horm. Res.* 48:99–121.

Oprian, D. (1992) Molecular determinants of spectral properties and signal transduction in the visual pigments. *Curr. Opinion Neurobiol.*, 2:428–432.

Rando, R. R. (1990) The chemistry of vitamin A and vision. *Angew. Chem. Int. Ed. Engl.*, 29:461–480.

———. (1991) Membrane phospholipids as an energy source in the operation of the visual cycle. *Biochemistry* 30:595–602.

Saari, J. (1990) *Enzymes and Proteins of the Mammalian Visual Cycle.* Vol. 9 in *Progress in Retinal Research.* Pergamon Press, New York, pp. 363–381.

Stryer, L. (1991) Visual excitation and recovery. *J. Biol. Chem.* 266:10711–10714.

# RIBOSOME PREPARATIONS AND PROTEIN SYNTHESIS TECHNIQUES

*Gary Spedding*

### *Key Words*

**Aminoacyl–tRNA Synthetases** Enzymes that activate amino acids and then attach them to their respective tRNAs.

**Cell-Free Translation (Protein-Synthesizing) System** A system of components derived by breaking open cells and releasing their cytoplasmic contents. When cleared of cellular membrane debris and DNA, then supplemented with various ingredients, such a system is capable of synthesizing protein.

**Endoplasmic Reticulum** A system of double membranes in the cytoplasm that is involved in the synthesis of transported proteins. The rough endoplasmic reticulum has ribosomes associated with it; the smooth endoplasmic reticulum does not.

**Microsomes** A fraction of a tissue homogenate enriched in spherical membrane compartments, called vesicles, which are derived from the rough endoplasmic reticulum.

**mRNA 5′ Cap Structure** All eukaryotic cytoplasmic mRNAs and some viral mRNAs have a "cap" at the 5′ end of their sequence; this special methylated residue of guanosine ($m^7G$) is linked via three phosphates to the mRNA. The "cap" structure is important for the efficient translation of these messages.

**Polyribosome (Polysome)** Complex of mRNA and two or more ribosomes actively engaged in protein synthesis.

**Protein Synthesis** The process of translation of an mRNA on the ribosome. Its three important steps, which form the so-called ribosome cycle; are initiation, elongation and termination, and involves the shuttling back and forth of soluble factors to bring in the amino acids and to facilitate movement of the mRNA through the ribosome.

**S30, S45, S100 Supernatants** Cell-free supernatants cleared of membrane fragments, generated by centrifugation of broken cells at 30,000 times the force of gravity (30,000*g*, and 45,000*g*, etc.). Supernatants generated by centrifugation below about 45,000*g* contain all the components: ribosomes, tRNAs, and soluble factors important for the translation of added mRNA. Often such extracts can also transcribe mRNA from exogenously added DNA fragments and form the basis for coupled transcription–translation systems. Supernatants prepared above 100,000*g* have been cleared of polysomes, ribosomes, and microsomes but still contain all the soluble factors required for protein synthesis.

**Translation** The process of reading the information encoded in a messenger RNA for the specified linear amino acid sequence it contains. The sequence of amino acids is dictated by a certain reading frame of triplet codons in the mRNA, and thus translation represents a change from the nucleotide language of mRNA into the amino acid language of protein. The process occurs on ribosomes.

**Translocation** The movement of a protein from one cellular compartment to another. The process involves crossing a membrane (e.g., the endoplasmic reticulum, the bacterial inner membrane, the chloroplast thylakoid membrane, the mitochondrial inner membrane). "Translocation" also refers to the movement of mRNA and tRNA on the ribosome during the events of elongation of protein synthesis.

The field concerning ribosomes and protein synthesis is extremely broad; consequently, only a cursory introduction to the important techniques involved in this arena of research can be presented here. This entry concentrates on the techniques involved in the preparation and uses of ribosomes and soluble components that participate in the process of protein synthesis.

Ribosomes constitute the major piece of molecular machinery of protein synthesis; during the course of translation of mRNA, they interact with many ancillary components (initiation, elongation, and termination factors, GTP, and tRNA). The components of the protein synthetic machinery can all be purified and analyzed to understand their structure and their functioning during the complex events of protein synthesis. Section 1, which details the preparation of ribosomes, begins with a very brief outline concerning ribosome structure, to facilitate the understanding of how, and why, the various ribosomal components are purified.

One of the most powerful approaches in the arsenal of tools available to study the translational apparatus involves the generation of complete cell-free systems for protein synthesis, systems that are capable of faithfully and accurately translating any added mRNA. These systems are used in a variety of ways to elucidate the complex and intricate events involved in the synthesis of proteins, as well as the myriad of control processes that can be involved during translation. Such cell-free systems have been used to study changes in protein synthesis caused by physiological events involved during stress responses, during organismal development or the viral infection of cells, and in the secretion of proteins across membranes. Such studies have a profound and ever-growing influence on the medical field and on the development of biotechnology, in addition to providing a fundamental understanding of cell biochemistry in general.

## 1 RIBOSOME STRUCTURE, AN OUTLINE

Ribosomes are ubiquitous ribonucleoprotein particles present in the cell cytoplasm of all organisms, and also within the organelles, mitochondria, and chloroplasts. As the central piece of the translational machinery, ribosomes are constantly involved in the synthesis of proteins required during the entire life cycle of all organisms. This entry is concerned only with prokaryotic (bacterial) and eukaryotic (animal and plant) cytoplasmic ribosomes, and not with the unique organellar ribosomes, which form the basis of another fascinating subject.

The ribosomes from all sources are composed of two distinct subunit structures, and this is an essential feature for their function. Bacteria possess their own type of ribosomes, which are different from the eukaryotic ribosomes in subtle ways, in particular by being smaller. Bacterial ribosomes are classified according to relative size as 70S, based on their rate of sedimentation in an ultracentrifuge. Their two subunits are similarly classed as 30S and 50S ribosomal subunits. Eukaryotic cytoplasmic ribosomes from plants and animals are larger than their prokaryotic counterparts, having a sedimentation value of 80S. The eukaryotic small and large ribosomal subunits are designated 40S and 60S, respectively.

All ribosomes consist of several (three in bacterial and four in eukaryotes) different RNA molecules and multiple different proteins, all of which can again be individually purified. Bacterial ribosomes consist of more than 50 different ribosomal proteins, and the number is considerably greater in the case of eukaryotic ribosomes. The interested reader may look elsewhere in this volume and in the bibliography for further insights into this vast and fascinating topic.

### 1.1 Initial Steps in the Preparation of Ribosomes: Cell Breakage

Many subtle variations are involved in the preparation of ribosomes from the wide spectrum of organisms. It is first necessary to release the cellular contents by gently, but efficiently, breaking open the cells. This is achieved in any of a number of ways, depending on the cell or tissue source and on the intended use of the components to be purified. Disruption techniques include the grinding of tissues or cells with a coarse abrasive such as alumina, quartz sand, or glass beads, perhaps after freezing in liquid nitrogen, or even via processing in a special tissue grinder; specific lysis techniques that gently dissolve the cellular membranes; forced pressure (e.g., passage of cells under pressure through a small orifice to squeeze the cells open); and the use of homogenization or sonication methods. Special procedures have to be employed during the preparation of ribosomes to prevent unwanted action of contaminating proteases (protein-degrading enzymes) and ribonucleases (RNA-degrading enzymes). These precautionary measures must follow cell disruption because ribosomes, being composed of both protein and RNA molecules, are very susceptible to damage during purification. The elimination of ribonuclease activity is especially important for the preparation of intact polysomes. Once the cells have been efficiently ruptured, and the nuclease and protease problem eliminated, several fairly straightfoward steps are then employed to purify the ribosomes, as described in Sections 1.2 and 1.3.

### 1.2 Bacterial Ribosomes: Preparation

Bacteria are grown in a rich nutrient medium until about the middle of their logarithmic growth phase. Cells are then harvested and washed with a buffered salt solution prior to breaking them open. Often deoxyribonuclease (DNase) is added after cell disruption, to destroy the liberated DNA and thus reduce the viscosity of the extract. All steps in the procedure are performed at temperatures close to freezing and also in the presence of magnesium ions and a reducing agent such as 2-mercaptoethanol or dithiothreitol, to stabilize the ribosomes and the soluble proteins in the extract and maintain them in an intact and functionally active state. Bacterial cell membranes and DNA, together with any components used for grinding open the cells (e.g., alumina or glass beads) are removed by low speed centrifugation. Typically, the broken cell extract is centrifuged at 30,000 times the force of gravity (30,000*g*) for about 45 minutes to yield a cleared supernatant known as an S30 extract (supernatant at 30,000*g*). The S30 supernatant may then be subjected to a high speed ultracentrifugation step at 100,000*g* for about 4 hours to directly pellet the ribosomes from the supernatant.

The ribosomes at this stage are relatively impure, being contaminated by loosely associated protein synthesis factors, and so it is more usual to "wash" them by pelleting the particles, at or above 100,000*g*, through buffers containing a high concentration of sucrose (a so-called sucrose cushion). The ribosomes may then be further purified by careful resuspension in a buffer that contains a high concentration of potassium or ammonium chloride (typically 0.5 *M*), followed by another ultracentrifugation pelleting procedure. It is the "high salt" washing step, which may be repeated as desired, that facilitates the efficient removal of loosely associated nonribosomal proteins and other factors.

After each pelleting step, the highly viscous ribosome-containing material is very gently resuspended in buffer, usually without sucrose, in readiness for further treatment, or for storage as desired. After the very first ultracentrifugation step, the relatively clear ribosomal pellet is usually covered with a gelatinous brownish material that contains residual amounts of DNA and cellular debris. This material must be removed before the ribosomal particles are resuspended. After several salt washings, clean ribosomes, essentially free of all nonribosomal soluble proteins and other contaminants, are obtained. The highly purified ribosomes are finally resuspended in a suitable storage buffer (e.g., 10 m*M* Tris-HCl, 10 m*M* magnesium chloride or acetate, 60 m*M* ammonium chloride, and 3–6 m*M* 2-mercaptoethanol, pH 7.5) and stored at −70°C as small aliquots. The concentrations of ribosome preparations are calculated by measuring the amount of ultraviolet light absorbed by a small sample, at a wavelength of 260 nm and applying appropriate conversion factors. For example, for *E. coli* ribosomes 1 $A_{260}$ unit is equivalent to 23 picomoles (pmol) of particles.

The supernatant generated upon pelleting the crude, non-salt-washed ribosomes, or that obtained after the first salt-washing step, is often kept as a source of unfractionated soluble protein synthesis factors. These supernatants are used as a starting point for further purification of the initiation, elongation, and other factors, or they are used to supplement cell-free protein synthesis systems with the required components in addition to ribosomes (see Sections 1.4 and 2).

As described earlier, the 70S ribosomes are themselves composed of two unequal-sized subparticles designated the 30S and 50S subunits. These subunits may be separated by reducing the magnesium ion concentration (present in buffers to stabilize particles) to very low levels, which causes the 70S ribosomes to dissociate. The free subunits are then separated by centrifugation through sucrose gradients (typically 10–30% w/v sucrose in buffer). The subunits travel through the gradient medium, separating under velocity according to size: the large subunits travel farther down the gradient than the smaller particles. The subunits are recovered by fractionation of the gradients; they are located in appropriate fractions by their property of absorbing ultraviolet light. The fractions containing the respective subunits are pooled and the particles are obtained by pelleting in an ultracentrifuge, after the magnesium ion levels have been reelevated to a concentration compatible with the integrity of intact subunits. They are finally resuspended in a suitable buffer and stored at −70°C. Under appropriate incubation conditions, the subunits may be recombined into intact functional 70S ribosomes free of ancillary protein synthesis factors.

### 1.3 Eukaryotic Ribosomes: Preparation

Many published methods exist for the preparation of eukaryotic ribosomes. All such preparations typically involve the isolation of a postmitochondrial supernatant, followed by the isolation of polysomes or ribosomes.

For the preparation of ribosomes from animal sources, tissues are essentially minced, or chopped up, then homogenized in a buffer composed of potassium and magnesium salts, ethylenediaminetetraacetic acid (a heavy metal chelator), a reducing agent (2-mercaptoethanol or dithiothreitol), and sometimes a ribonuclease inhibitor (heparin, at about 0.2 mg/mL, is commonly used). Occasionally, in addition to the required buffering agent and salts, sucrose is provided at relatively high concentration as an osmoticum in the homogenization buffers to maintain the cellular organelles in an intact state. The resultant homogenate is centrifuged at about 13,000*g* for 30 to 45 minutes at 4°C to remove the mitochondria (which contain their own unique protein synthetic apparatus), nuclei, and cellular debris. The postmitochondrial supernatant is filtered through gauze, and then the ribosomes are isolated by ultracentrifugation through a sucrose layer at or above 100,000*g*. The postribosomal supernatant (the "cell sap") is used as a source of soluble protein synthesis factors.

Not all ribosomes exist free in the cytoplasm, so special detergents are often used during the extraction procedures to "strip off" the ribosomes attached to the endoplasmic reticulum membranes. Then, depending on the precise details of the technique employed and the end product required, the detergent-treated extracts may be applied to continuous or discontinuous sucrose gradients to prepare polysomes, or they may be centrifuged through sucrose cushions to pellet-clean ribosomes, essentially as described earlier.

Following their isolation, eukaryotic ribosomes are "stripped" of endogenous peptidyl–tRNA and some of the soluble factors by incubation in the presence of the antibiotic puromycin (0.5 m*M* final concentration). The puromycin-treated ribosomes may then be reisolated as clean ribosomes by centrifugation through sucrose, or used to isolate ribosomal subunits. A reversible dissociation of ribosomes into subunits occurs in buffers containing potassium chloride (1 *M*) and low levels of magnesium (10 m*M*), following the puromycin treatment.

Eukaryotic ribosomes, unlike their prokaryotic counterparts, do not readily dissociate into subunits by a simple lowering of the magnesium ion concentration. The puromycin-treated preparation is loaded onto sucrose gradients and centrifuged to separate the free 40S and 60S ribosomal subunits. These are then collected and pelleted by ultracentrifugation, carefully resuspended in a storage buffer (e.g., 20 m*M* Tris-HCl, 100 m*M* potassium or ammonium chloride, 5m*M* magnesium chloride, 10 m*M* 2-mercaptoethanol, and 0.25 *M* sucrose, pH 7.6), and stored at −70°C. Following lysis in detergent-containing buffers, ribosomes are prepared from tissue culture cells by methods similar to those already described.

The preparation of ribosomes from plants requires some specialized conditions, but the procedures are essentially similar to those for the preparation of animal ribosomes. With plant cells it is necessary to disrupt the tough and rigid cell walls; this process is followed by differential centrifugation to remove chloroplasts and mitochondria. The cellular organelles are kept intact by use of an osmoticum (sucrose or mannitol) at high concentration, to prevent bursting due to uptake of water, which would result in the contamination of the plant cytosolic 80S ribosomes with organellar ribosomes. Ribonuclease inhibitors usually must be employed to preserve the integrity of the ribosomal RNA. Plant ribosomes and, if desired, their subunits, are prepared from the postmitochondrial/postchloroplastid supernatant in ways similar to those described for animal ribosomes; however detergents are not usually employed in this case.

### 1.4 Ancillary Protein Synthesis Factors

Quite a number of factors and components other than the ribosome are important in the overall events of protein synthesis (see Section 2, Table 1.) All such components can be purified in a functionally intact state. For the purposes of preparing cell-free protein-synthesizing systems, as described in Section 2, relatively simple unfractionated cell extracts often can be used to supply the requisite factors. Occasionally, however, the components must be prepared with a high degree of purity to permit assessment of their individual contributions to protein synthesis, and this is achieved by the use of classical chromatography techniques, or modern methods, such as the high performance and fast protein forms of liquid chromatography (HPLC and FPLC).

In addition to preparing the soluble factors for translation, it is particularly important to prepare high quality intact mRNA to program cell-free translation systems to produce proteins.

### 1.5 Studies of Ribosomes

Purified ribosomes are often used for structural studies of the intact particle, or of the derived subunits. They may also be dissociated completely into their RNA and protein components for detailed structural and functional studies.

Ribosomal proteins can be liberated from the ribosome by extraction in buffered acetic acid and analyzed by highly resolving two-dimensional techniques of gel electrophoresis. The results of such analysis provides comparative information on the number and sizes of proteins from species to species, or changes in the structure of

**Table 1** General Requirements for Polyuridylic Acid=n and Natural mRNA–Dependent Cell-Free Protein Synthesis Assays.

| Component | Function |
|---|---|
| Tris-HCl or HEPES-KOH, (pH 7.0–8.0). | Buffer components to maintain pH in the physiological range. |
| Magnesium chloride or acetate[a] | Magnesium is required to help maintain the structure of the ribosome, and other components. |
| Potassium chloride or acetate, or ammonium salts[a] | Required in correct amounts for efficient translation of mRNA. |
| Dithiothreitol or 2-mercaptoethanol | Reducing agents required to stabilize ribosomes and soluble proteins. |
| ATP and GTP | Energy source to drive reactions. |
| Phosphoenolpyruvate and pyruvate kinase[b] | Part of an energy-regenerating system required to maintain protein synthesis. |
| 19 nonradioactive amino acids[c] | |
| Radioactive amino acid[c] | Used to measure the relative incorporation of amino acids into protein. |
| Polyamines[d] | |
| tRNA[e] | |
| Polyuridylic acid or natural mRNA[f] | Used in optimum amounts. |
| S23, S30 extracts etc.,[g] OR ribosomes, (subunits) polysomes, or microsomes, together with postribosomal supernatants.[g] | All these components are used in carefully optimized amounts to ensure maximal incorporation of the amino acids into protein. |

Other Components

Sometimes increased amounts of, or other soluble components, in addition to the endogenously supplied factors need to be added to the reaction mixtures for optimal protein synthesis.[g] These supplements might include fractionated initiation and/or elongation factors, aminoacyl=ntRNA synthetases, and so on (see also Section 1.5). A special example of an additional component is hemin, which is essential in reticulocyte lysate systems. (See Section 2.2.1.)[g]

[a]Some systems work better with the acetate form of salts than with the chlorides. Sometimes the chloride ions are inhibitory to the reactions involved in protein synthesis. Also, the exact requirement for the potassium and magnesium ions varies considerably with respect to the mRNA being translated.

[b]Phosphoenolpyruvate and pyruvate kinase are added as a nucleoside triphosphate regeneration system in prokaryotic systems. Eukaryotic systems are often supplemented with creatine phosphate and creatine phosphokinase instead.

[c]The radioactive amino acid employed depends on the particular template being translated in the system. When polyuridylic acid is supplied, only radiolabeled phenylalanine can be used, since this is the only amino acid coded for by this synthetic message. In systems programmed with natural message, the radioactive amino acid used is usually methionine, or occasionally leucine. The cold amino acids supplied to the system consist of the other 19 amino acids, which are commonly found in proteins.

[d]Compounds such as spermidine, spermine, and putrescine, which are all polyamines, are sometimes used to stimulate protein synthesis. They apparently stimulate the elongation step and are especially important in the wheat germ system (see Section 2.2.2).

[e]The amounts and composition of the endogenous tRNA species supplied in the cellular extracts are not always geared to the efficient synthesis of exogenously added mRNA. Prokaryotic systems are supplemented witheither phenylalanine-specific tRNA (for polyuridylic acid programmed systems) or ''crudely'' prepared bulk tRNA from *E. coli,* or occasionally with homologous source tRNA when the protein-synthesizing system is derived from a different organism. Eukaryotic systems often incorporate brewer's yeast, calf liver, or wheat germ tRNA.

[f]Carefully prepared, intact, messenger RNAs are added to the system to direct the incorporation of amino acids. Polyuridylic acid is a synthetic template that directs the synthesis of polyphenylalanine. When natural mRNA is used, this can be in the form oftotal cellular RNA, a purified bulk mRNA containing cellular fraction, or even more specifically, a highly purified or synthesized mRNA coding for only one protein species.

[g]The low speed, centrifugally prepared cell-free extracts and the postribosomal supernatants supply the system with the initiation, elongation, and termination factors, aminoacyl=ntRNA synthetases, and other components usually in amounts sufficient to mediate the events of protein synthesis.

proteins due to mutational events that lead to antibiotic resistance or functional perturbation. In the latter case, alterations are judged by changes in the relative electrophoretic mobility compared with the proteins obtained from the wild-type organism. Ribosomal proteins (especially prokaryotic ribosomal proteins) are also individually purified (e.g., employing HPLC techniques) and used for sequence determination, for biophysical (structural) investigations, and in functional studies.

The ribosomal RNAs can be prepared intact from the ribosomes and subunits by, for example, selective extraction methods employing phenol and chloroform. The purified rRNAs are being used to study the secondary and tertiary structure of the molecules; their interaction with specific ribosomal proteins, antibiotics, and other components of the protein synthetic machinery; and the reconstitution of ribosomes along with purified proteins. The reconstitution of prokaryotic ribosomes into functionally intact particles, starting either with all the individually purified components or with subsets of components, is an exquisite and powerful technique that has been extremely useful in determining the roles of the components in the assembly and function of the ribosome.

The ultimate goal of all these studies is the complete structural elucidation of the ribosome, at the finest atomic resolution, a task that is considered to be instrumental for a complete understanding of the precise function of the particle in protein synthesis. Many powerful techniques such as electron microscopy, crystallography, X-ray diffraction, and neutron scattering analysis are all being employed toward this end. Many of these techniques are described elsewhere in this volume.

## 2 CELL-FREE PROTEIN-SYNTHESIZING SYSTEMS

Most of the important details concerning the actual events of protein synthesis have been discovered by using purified ribosomes and appropriately purified components in partial reactions of the overall scheme of protein synthesis, or by the development of reliable cell-free protein-synthesizing systems. These in vitro translation systems, if prepared correctly, will faithfully, accurately, and reproducibly carry out all reactions involved in peptide chain initiation, elongation, and termination, enabling the investigator to translate mRNA at rates that approach those found in vivo. Such cell-free systems, at their simplest, typically employ unfractionated cell-free supernatants (e.g., S30 extracts, prepared from bacteria as described in Section 1.2). These cell-free extracts contain all the important components needed for protein synthesis, including the ribosomes, tRNAs, and factors required for each event in translation.

Alternatively, highly fractionated and therefore better defined systems are used, which employ purified polysomes, ribosomes or microsomes, together with the 100,000*g* postribosomal supernatants. The energy-supplying compounds ATP and GTP, together with a nucleoside triphosphate regenerating system, are supplied to the systems, inasmuch as translational initiation and elongation depend on the availability of these high energy phosphates. Other components are also supplied to the system, as outlined in Table 1.

Cell-free, translationally competent extracts can be rendered free of their own (endogenous) mRNA (see Sections 2.1 and 2.2.1) and then supplemented with exogenously supplied mRNA and a radioactive amino acid. Under appropriate incubation conditions, all the components in a fully active system act correctly and in concert to assemble the amino acids into a fully synthesized protein, as directed by the coding instructions of the supplied mRNA. The radioactive amino acid provided in the mixture permits the synthesis of the protein to be monitored, as described shortly. Further details concerning the components required for cell-free translation are outlined in Table 1.

The messages supplied to the system include synthetic (usually polyuridylic acid) or natural messenger RNAs. Polyuridylic acid, as a unphysiological message, does not allow for the study of initiation events of protein synthesis because it lacks an authentic initiation codon signal. This message also lacks termination (stop) codons. Systems programmed with polyuridylic acid require unphysiologically high magnesium ion concentrations to allow initiation to occur and have been useful in probing the events of elongation of protein synthesis, and in elucidating the mechanism of action of inhibitors of this step in the synthetic cycle of the polypeptide. The best systems are, of course, those programmed with added natural mRNA. These systems should reproducibly exhibit a high rate of protein synthesis over an extended period of time and should initiate translation efficiently either on the endogenous (if not specifically destroyed) mRNA or on exogenously supplied mRNA. Moreover, when presented with a mixture of mRNAs, these systems must be capable of synthesizing a complete spectrum of authentic protein products both faithfully and efficiently.

Following incubation of reaction mixtures, samples are processed to determine whether protein synthesis has indeed taken place in the system. The use of a radioactively labeled amino acid in the assay mixtures allows for the monitoring of protein synthesis over time. The principle behind monitoring protein synthesis is to collect onto filters the acid-insoluble proteins formed during the reaction, then treat the filters to remove free unincorporated (or non-protein-associated) amino acids. Filters are then placed in a scintillation counter for measurement, via radioactive decay, of the amount of amino acid incorporated. Knowing all the other parameters of the assay system allows calculation of the rate and efficiency of protein synthesis.

The method just described, however, measures only the total incorporation of radioactive amino acid into protein; it does not indicate whether authentic, full-length polypeptides have been made. As an alternative, the products that have been synthesized are analyzed by gel electrophoresis and the proteins, which are resolved on the gel, are visualized by exposing the gel to X-ray film. The protein contains the radioactive amino acid used in the assay system, and radioactive decay causes an image to appear on the film. The latter approach is, therefore, routinely employed to test the system for faithful and accurate synthesis of fully completed proteins. Immunological techniques are also often used in conjunction with gel electrophoresis to identify specific proteins synthesized in the reaction.

### 2.1 Bacterial Extracts as Protein-Synthesizing Systems

Bacterial cell-free extracts, together with a standard mixture of components, as outlined in Section 2 and in Table 1, are routinely used for the in vitro synthesis of proteins. Bacterial S30 extracts incorporated into such systems supply all the protein synthetic apparatus required to translate the added natural or synthetic messenger RNA. The S30 lysates can be improved in their capacity to synthesize protein from exogenously added mRNA by preincubating them under carefully defined conditions to remove ribosomes from the endogenous mRNA in the extract. In this process, known as runoff, elongation of polypeptide synthesis occurs without new initiation. The free endogenous mRNA is usually degraded following the removal of ribosomes. The lysate is then often desalted by passage through Sephadex-G25, or preferably by dialysis, which also removes amino acids and nucleotides. These lost, low molecular weight, components are resupplied later.

It is also possible to transcribe exogenously added DNA in well-prepared and supplemented S30 extracts (which also contain the cellular transcription components), and this has led to the development of coupled transcription–translation systems. Such systems offer the efficient transcription of defined sequences of DNA into RNA, and the subsequent, translation of the mRNA into protein. Coupled transcription–translation systems thus provide a very powerful tool for studies of gene expression and its controls.

As an alternative to using simple S30 cell-free extracts, the in vitro translation system can employ purified ribosomes and carefully optimized amounts of the postribosomal supernatant to supply the soluble factors required for the process to occur efficiently. A detailed description of the assembly of all the components required for such cell-free translation systems is provided in Table 1.

No matter which route is taken in generating cell-free systems, the protein-synthesizing reaction mixtures are typically incubated at 30 to 37°C for an appropriate time, then processed to determine the levels and efficiency of protein synthesis, as outlined in Section 2. Efficient systems should show a linear rate of incorporation of amino acids over a period of 30 minutes to an hour, or more.

### 2.2 Eukaryotic Cell-Free Systems

In addition to the bacterial systems that have been developed to study protein synthesis, a number of eukaryotic cell-free systems have been used to study the more complex events involved with these organisms. The two major systems that have been used successfully for many years are those for rabbit reticulocyte lysate and wheat germ. Both these systems are commercially available, and they can also be prepared quite easily in the well-equipped laboratory.

### 2.2.1 Rabbit Reticulocyte Lysate

The preparation of the rabbit reticulocyte lysate system typically involves inducing anemia in New Zealand white rabbits, to elicit stimulation of reticulocyte (red blood cell) production. The blood is then obtained and the reticulocytes, once purified away from contaminating white blood cells, are osmotically lysed by placing in water. The cytoplasmic contents, containing the protein synthetic apparatus, are liberated and the cellular membrane debris is cleared by centrifugation at 20,000*g* for 20 minutes. The supernatant from such preparations naturally contains an abundance of (the reticulocytes' own) endogenous globin mRNA, which would be translated in a fully reconstituted cell-free system, causing problems in the interpretation of the results of translation of any added (heterologous) messenger RNAs. To eliminate this obstacle to efficient analysis, the lysate is treated with an enzyme called micrococcal nuclease. This nuclease, which is calcium ion dependent, degrades the endogenous mRNA. The reticulocyte is a highly specialized cell system and produces an abundance of globin mRNA and only low levels of several other mRNAs, unlike other systems that produce multiple mRNAs and consequently many different and highly abundant proteins. The nuclease, while destroying mRNA, leaves essentially intact the other RNA components of the translational machinery, namely, ribosomal RNA and tRNA. The enzyme is subsequently inactivated by chelating the calcium ions. This is done by sequestering the free calcium in a complex with ethylene glycol bis(aminoethyl ether)tetraacetic acid (EGTA), a compound that is specific for $Ca^{2+}$ ion. When all the calcium has been chelated in this way, the enzyme micrococcal nuclease is effectively silenced. The translational activity of the rabbit reticulocyte lysate is now completely dependent on exogenously added mRNA.

The lysate must, however, be further optimized for mRNA translation by the addition of an energy-generating system, appropriate salts to supply the correct amounts of potassium and magnesium ions, and, sometimes, a mixture of tRNAs to expand the range of mRNAs that can be translated. The abundance of the species of tRNA present in the reticulocyte lysate is primarily geared to the coding capacity of globin mRNA, not to other mRNAs. Hemin must also be added to the system to prevent the inhibition of initiation. If hemin is not added, an inhibitor acts on one of the specific initiation factors, and protein synthesis in extracts ceases after only a few minutes of incubation; this situation is fairly specific to the reticulocyte system.

The lysate, treated and supplemented with micrococcal nuclease, is added to an appropriate cocktail of additional components (see Table 1), which includes a radioactive amino acid and the mRNA to be translated. The mRNA is translated at 30°C, and protein synthesis is monitored by incorporation of the radioactive amino acid into hot, precipitable protein that is insoluble in trichloroacetic acid, or by gel electrophoretic analysis of the protein (synthesized as described in Section 2). The unfractionated rabbit reticulocyte lysate system is very efficient and exhibits a rate of protein synthesis 70 to 100% that of the intact cell.

### 2.2.2 Wheat Germ System

The other major system used to date is that of the wheat germ extract. This system, which is capable of efficiently translating exogenously supplied eukaryotic and prokaryotic mRNAs, is prepared quite simply. Wheat germ is ground with an abrasive, or with liquid nitrogen, and then extracted with a buffer. The resulting extract is cleared of debris by centrifugation at 23,000*g* for 10 minutes, and the pellet that is obtained is discarded. The S23 supernatant is passed through a small Sephadex G-25 column to standardize the ionic conditions, to remove the endogenous amino acids, and also, in this case, inhibitory plant pigments. The translational components are eluted from the column as a turbid fraction, which is collected and stored at −70°C.

Wheat germ extracts are supplemented similarly to the reticulocyte lysate system, but no hemin is required, and spermidine and/or other polyamines are added to stimulate the efficiency of polypeptide chain elongation and to overcome problems of premature termination of protein synthesis that may occur. One major advantage to the wheat germ translation system is that the endogenous translational activity of wheat germ extracts is low compared to untreated rabbit reticulocyte lysates; translation largely depends on added mRNA. The low background incorporation due to any remaining endogenous mRNA can, however, be reduced even further by micrococcal nuclease treatment as described for the reticulocyte system. The wheat germ cell-free translation reactions are carried out optimally at 25°C and processed as described in Section 2.

### 2.2.3 Alternative Cell-Free Systems and Their Uses

While having proved enormously useful over many years, the rabbit reticulocyte, wheat germ, and other earlier established systems have a number of disadvantages. Such systems are, naturally, geared to the synthesis of their own mRNAs, and the correct levels of components and the appropriate control signals to allow for the efficient translation of mRNAs derived from other organisms may not be present. To cite one major example, the reticulocyte is a differentiated anucleate cell that possesses unique regulatory mechanisms (e.g., the control of initiation by hemin, a situation rather specific to the reticulocyte), and also a specific tRNA composition geared for globin synthesis.

As a result of such specific features and potential deficiencies in the two commonly used cell-free translation systems, there has been a recent trend toward the development of newer, homologous cell-free protein-synthesizing systems. In these cases all the components, including appropriate regulatory signals, used in translation, are derived from the same organism, hence are geared to the efficient synthesis of their own mRNAs. These systems are all prepared using variations on the methods already alluded to with respect to the wheat germ and reticulocyte situations. A number of these newer systems are briefly described.

One of the most important alternative cell-free translation systems is that derived from the yeast *Saccharomyces cerevisiae*. Cell-free, translationally competent yeast extracts represent a model system of protein synthesis in a nonspecialized eukaryotic cell and can be used to study the effects on translation of physiological and genetic modifications. The yeast system is expected to deepen our understanding of translation in mammalian cells by taking advantage of the existence of some similarities between yeast and higher eukaryotes. This very important system is not yet commercially available, however.

Translation systems generated from different developmental stages of organisms such as sea urchins, starfish, and the frog *Xenopus laevis* are proving useful in the analysis of translational control during early embryogenesis, and it appears that such control is quite complex in this situation. Apparently more than one level of translational regulation is involved following fertilization with these organisms. Similarly, the characterization at different developmental times of translationally active cell-free systems and translational components from the dormant cysts and the embryos of the brine shrimp *Artemia* has begun to produce insights into the events of protein synthesis involved in the emergence of this organism from dormancy. Such experiments have involved the fractionation of the cell-free lysates, and the crossing over of the ribosomes and soluble factors from one system to another to determine which limiting components are responsible for the differences in rates of overall protein synthesis and/or for the differential specificity of mRNAs that are translated at different developmental stages.

## 3 CELL-FREE TRANSLATION SYSTEMS: APPLICATIONS OF THE TECHNOLOGY

In addition to providing the details of the actual events of protein synthesis, and its regulation, a number of important biological processes and systems have yielded some of their secrets through the use of cell-free translation systems. A few highlights concerning the kinds of information being revealed about biological phenomena by the use of such systems are listed.

1. Details concerning the appropriate cellular response to heat shock and other stress situations, and the translational controls involved in the ensuing important physiological survival mechanism, have come through the use of *Drosophila melanogaster* (fruit fly) cell-free systems.
2. The rabbit reticulocyte system, in particular, has been instrumental in analyzing the specific expression of viral mRNA following virus infection of animal cells, and the concomitant switching off of host cellular mRNA translation.
3. The translational events that occur during the aging process are being unraveled by the use of various cell-free protein-synthesizing systems.
4. Cell-free translation systems provide an important tool in genetic analysis when they are used to confirm the identity and authenticity of protein products encoded by cloned genes.
5. Rabbit reticulocyte lysates, together with other translation systems including cell-free extracts of frog eggs, have also been useful in increasing our understanding of the events involved in a potentially important modern therapeutic approach known as antisense technology. In this technique, specific mRNAs are selected for inactivation by the use of complementary nucleic acid sequences, or by other potentially inhibitory molecules. Antisense technology is potentially useful for combating various diseases and viral infections by means of this specific targeting technique, and the technology is also useful in unraveling the details of protein synthesis.
6. The processing events whereby certain proteins are transported (translocated) into or across membranes, for localization in compartments other than the cytoplasm, can be reproduced and studied in vitro by translation of appropriate mRNAs in cell-free protein-synthesizing extracts supplemented with microsomal membranes. An understanding of these translocation events is considered to be important because incorrectly processed and targeted proteins are characteristic of certain disease states. An understanding of protein targeting also has implications for cancer research.

The foregoing items show how cell-free translation systems can provide a means to answering many specific questions about biological processes. However, it is important to note that many fundamental questions regarding the events and control processes of protein synthesis in prokaryotes and, particularly, in eukaryotes remain to be answered. Cell-free systems, as a complementary approach to in vivo studies and as a confirmation of their results, will continue to be used to provide much more useful information concerning the events and controls of translation. The use of such systems will also undoubtedly have a significant number of additional ramifications with respect to a fundamental understanding of many other aspects of cell and developmental biology, molecular biology, and biochemistry.

*See also* EXPRESSION SYSTEMS FOR DNA PROCESSES; GENE EXPRESSION, REGULATION OF; TRANSLATON OF RNA TO PROTEIN.

## *Bibliography*

Arnstein, H. R. V., and Cox, R. A. (1992) *Protein Biosynthesis: In Focus.* Oxford University Press, Oxford and New York.

Austen, B. A., and Westwood, O. M. R. (1991) *Protein Targeting and Secretion: In Focus.* Oxford University Press, Oxford.

Hames, B. D., and Higgins, S. J., Eds. (1984) *Transcription and Translation: A Practical Approach.* IRL Press, Oxford.

Hill, W. E., Dahlberg, A., Garrett, R. A., Moore, P. B., Schlessinger, D., and Warner, J. R., Eds. (1990) *The Ribosome: Structure, Function, and Evolution.* American Society for Microbiology, Washington, DC.

Liebhaber, S. A. (1989) Use of in vitro translation techniques to study gene expression. In *Molecular Genetics,* E. J. Benz, Jr., Ed. Churchill-Livingstone, Edinburgh.

Merrick, W. C. (1992) Mechanism and regulation of eukaryotic protein synthesis. *Microbiol. Rev.* 56:291–315.

*Methods in Enzymology* Series. Academic Press, San Diego, CA: Noller, H. F., Jr., and Moldave, K., Eds. (1988) *Ribosomes,* Vol. 164; Guthrie, C., and Fink, G. R., Eds. (1991) *Guide to Yeast Genetics and Molecular Biology,* Vol. 194; also, other volumes, particularly Vols. 30 (1974) and 60 (1979), provide extensive methodology on the topics covered in this entry.

Murray, J. A. H., Ed. (1992) *Antisense RNA and DNA.* Wiley-Liss, New York.

Nierhaus, K. N., Franceschi, F., Subramanian, A. R., Erdmann, V. A., and Wittmann-Liebold, B., Eds. (1993) *The Translational Apparatus; Structure, Function, Regulation, Evolution.* Plenum Press, New York.

Richter, J. D. (1991) Translational control during early development. *BioEssays,* 13:179–183.

Schneider, E. L., and Rowe, J. W., Eds. (1990) *Handbook of the Biology of Aging,* 3rd ed. Academic Press, San Diego, CA.

Spedding, G., Ed. (1990) *Ribosomes and Protein Synthesis: A Practical Approach.* Oxford University Press, Oxford.

Thach, R. E., Ed. (1991) *Translationally Regulated Genes in Higher Eukaryotes.* Karger Basel, Switzerland.

Trachsel, H., Ed. (1991) *Translation in Eukaryotes.* CRC Press, Boca Raton, FL.

Wolf, E. C. (1989) The purification and translation of globin mRNA. *Biochem. Educ.* 17:45–47.

# Ribozyme Chemistry

*John M. Burke*

*Key Words*

**Exon** Sequences within a gene or RNA transcript that are represented in the final mature RNA product of a gene. Before RNA splicing, exons are separated by introns.

**Intron** An internal sequence of nucleotides within a gene or RNA transcript that is removed by RNA splicing before the formation of the functional RNA product.

**Ribonucleoprotein** Macromolecular complex comprised of both RNA and protein.

**Ribozyme** An RNA molecule that is endowed with catalytic activity. For *ribo*nucleic acid en*zyme*.

**RNA Processing** Refers to the biochemical reactions the primary RNA transcript of a gene must undergo before it can become a mature, functional RNA.

**RNA Splicing** Removal of intron sequences and joining of exons to generate a mature RNA molecule.

**Self-Cleaving RNA** RNA molecule that is capable of cleaving itself in a site-specific manner.

Ribozymes are RNA molecules that can function to catalyze specific chemical reactions within cells, without the obligatory participation of proteins. Until recently it was believed that all biochemical reactions were catalyzed by protein enzymes. Since 1982, a number of RNA molecules have been found to be endowed with catalytic activity. For example, group I ribozymes take the form of introns, which can mediate their own excision from a self-splicing precursor RNA. Other ribozymes are derived from self-cleaving RNA structures, which are essential for the replication of viral RNA molecules. Like protein enzymes, ribozymes can fold into complex three-dimensional structures that provide specific binding sites for substrates as well as cofactors, such as metal ions. Ribozymes have been developed into highly specific enzymes that are capable of recognizing and catalytically cleaving targeted RNA molecules. A wide variety of applications for targeted ribozymes is under investigation. It is anticipated that ribozymes will become powerful laboratory tools for investigating gene function and may be developed into therapeutic agents useful for the treatment of a variety of viral and genetic diseases.

## 1 RNA Can Encode Genetic Information and Catalyze Biochemical Reactions

RNA is a unique molecule in the biological world because it serves two fundamentally distinct functions. It has long been known that RNA, like DNA, can serve as an informational molecule. For example, messenger RNA transcripts carry the information encoded by genes to cellular protein synthesis machinery. Many viruses, like HIV, use RNA as their genetic information.

Recently, investigators have found that some RNA molecules resemble protein enzymes in that they can catalyze specific biochemical reactions. The discovery that RNA can be a catalyst as well as a storehouse of genetic information has revolutionized our view of RNA and has led to the speculation that catalytic RNA may have played a critical role in the prebiotic evolution of self-replicating systems.

## 2 RNA Structure and Catalytic Function

The catalytic activity of ribozymes results from the folded structure of RNA. Because RNA is a single-stranded molecule, its structure is very complex and is unlike the monotonous structure of the DNA double helix. Like protein enzymes, ribozymes fold into complex three-dimensional structures. Much is known about the secondary structure of ribozymes (Figure 1), which generally consists of a number of short intramolecular helices comprised of Watson–Crick G·C and A·U base pairs, separated by nonhelical segments. It is clear that the nonhelical segments of ribozymes are very important for catalytic activity; these nonhelical nucleotides interact with one another to form tertiary interactions. Because of the virtual absence of high resolution structural data on ribozymes, the details of ribozyme three-dimensional structure are generally unknown.

RNA differs from DNA in the presence of a 2′-hydroxyl group in each nucleotide. This functional group may be important for catalytic activity in three respects. First, it makes RNA much more chemically labile than DNA. Second, it participates in essential tertiary interactions within ribozymes that are essential for catalytic activity. Third, it may participate directly in the chemical reactions that take place at the active site of the ribozyme.

## 3 Self-Splicing RNA

The first catalytic RNA molecules discovered were self-splicing RNA molecules containing group I introns. In the absence of proteins, these molecules can undergo rapid and precise structural rearrangements to excise the intron and covalently join the exons. Group I introns contain characteristic internal nucleotide sequences and have very similar RNA folding patterns (Figure 1); these common structural features are essential for catalytic activity and for recognition of the correct reaction sites in the precursor RNA.

Splicing of precursor RNAs containing group I introns occurs by a two-step reaction pathway (Figure 2). In the first step, GTP attacks the 5′ splice site, becoming covalently linked to the intron and liberating the 5′ exon. In the second step, the 5′ exon attacks the 3′ splice site, joining the exons and liberating the intron RNA. It is clear that the intron RNA itself contains the catalytic center, because it can catalyze additional reactions on internal and external RNA substrates.

A second class of self-splicing introns, the group II introns, is known. These introns share common folded structures and reaction pathways that are distinct from those of group I introns. Interestingly, group II introns appear to be more closely related to those of the nuclear pre-mRNA introns, whose splicing is catalyzed by complex nuclear particles called spliceosomes (see Section 5).

## 4 Self-Cleaving Catalytic RNA

In addition to their importance for RNA splicing, catalytic RNA molecules are essential for RNA replication in some viral systems.

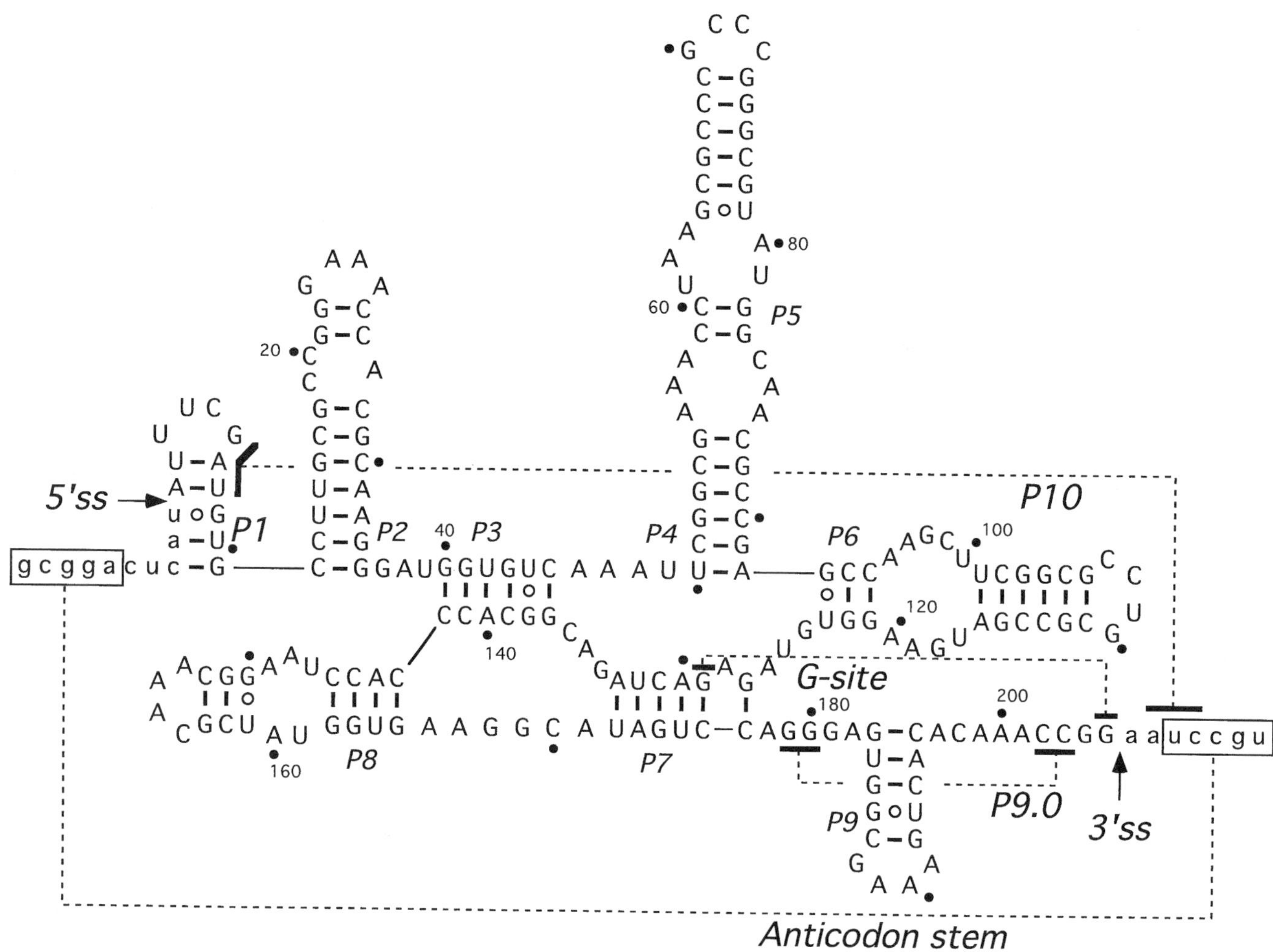


**Figure 1.** Structure of a self-splicing group I intron. The secondary structure of a self-splicing group I intron within a transfer RNA precursor of the bacterium *Azoarcus;* arrows indicate splice sites. Intron nucleotides are indicated in uppercase letters, exon nucleotides in lowercase. All group I introns have very similar RNA folding patterns. The RNA folding is responsible for catalytic activity and for the precise identification of the reaction sites.

For example, a circular RNA molecule associated with tobacco ringspot virus is replicated by a rolling-circle mechanism, as shown in Figure 3. Site-specific RNA cleavage and joining reactions are essential for RNA replication. Instead of using protein ribonucleases and RNA ligases to catalyze these reactions, they are carried out by the RNA molecules themselves. That is, during replication the RNA molecules contain sites at which the cleavage and joining reactions happen, as well as domains that recognize those sites and catalyze the reactions. A similar replication pathway is followed by the RNA genome of an important human pathogen, hepatitis delta virus.

Deleting nonessential RNA sequences serves to isolate the catalytic domains and the sequences that are recognized and cleaved. The catalytic domains function as ribozymes to cleave the substrate in a catalytic manner; that is, each ribozyme molecule can function to cleave a large number of substrate RNA molecules. The hammerhead ribozyme is one such ribozyme derived from self-cleaving RNA molecules. The secondary structure of the ribozyme–substrate complex (Figure 4) consists of one intramolecular duplex within the ribozyme domain and two intermolecular duplexes between ribozyme and substrate. Because the sequence of the intermolecular duplexes can vary in any manner that maintains base pairing, the sequence specificity of the hammerhead ribozyme can be changed by designing and synthesizing ribozymes with altered sequences in the substrate-binding arms. Engineered ribozymes are currently being tested for their ability to cleave targeted RNA molecules inside cells. It is hoped that ribozymes will be useful experimental tools for investigating the function of RNA molecules of interest. In addition, researchers are actively engaged in the development of therapeutic ribozymes for the treatment of serious viral and genetic diseases, for example, AIDS and cancer.

## 5 RIBONUCLEOPROTEIN ENZYMES

Many cellular RNAs are bound to specific proteins, forming ribonucleoprotein (RNP) complexes. These RNPs are responsible for catalyzing a number of critical cell functions. For example, ribosomes are RNP complexes responsible for synthesizing proteins from an mRNA template. The spliceosome is the RNP complex that splices intron-containing mRNA precursors within the nucleus.

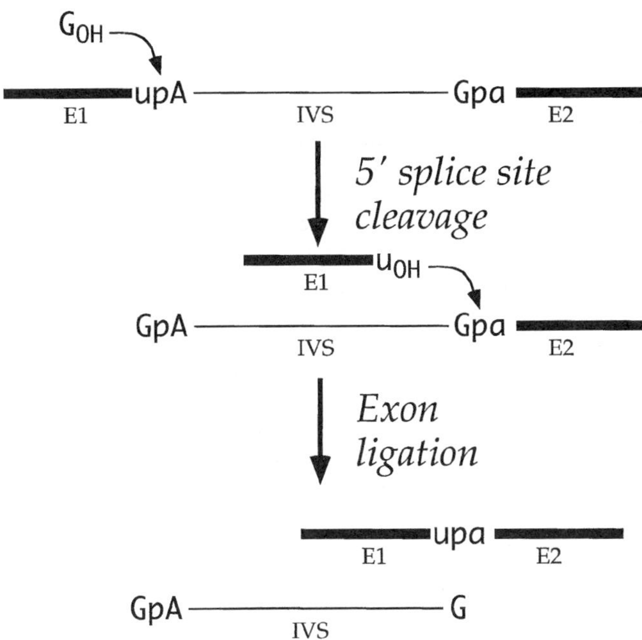


**Figure 2.** RNA splicing pathway for group I introns. First, GTP or another guanosine attacks the 5′ splice site, liberating the 5′ exon and becoming covalently linked to the intron. Second, the 5′ exon attacks the 3′ splice site in a reaction that generates ligated exons and liberates the intron. Each of these transesterification reactions can take place in the absence of proteins.

In most RNP complexes, it is difficult to determine whether the catalytic activity of the complex resides within the protein or RNA components. However, an accumulating body of evidence strongly suggests that RNA may be the catalytic component of the ribosome and the spliceosome.

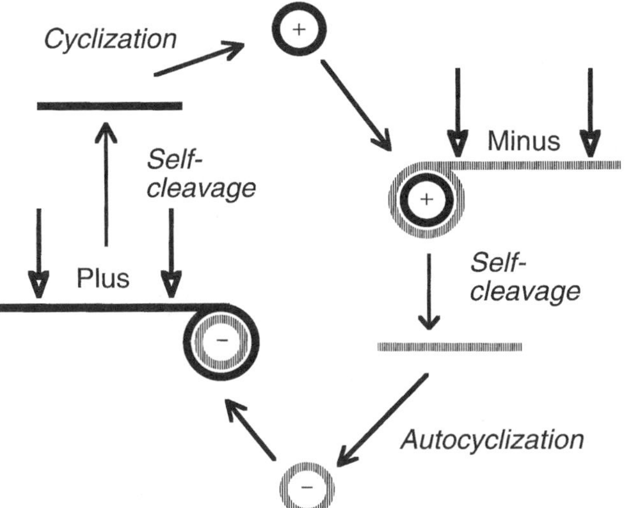


**Figure 3.** Self-cleaving RNA in RNA replication: the replication pathway for the satellite RNA associated with tobacco ringspot virus. Rolling-circle replication gives rise to linear negative- and positive-stranded concatamers that must be cleaved and cyclized at specific sites. The self-cleaving domains in the positive and negative strands are known as the hammerhead ribozyme and the hairpin ribozyme, respectively.

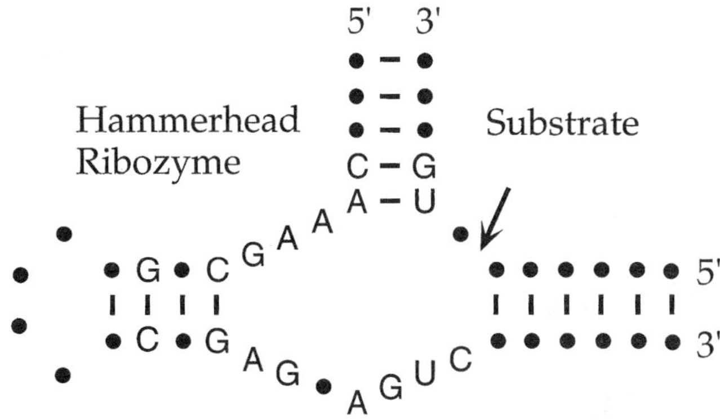


**Figure 4.** The secondary structure of the complex between the hammerhead ribozyme and its substrate; arrow indicates cleavage site. Essential bases are indicated by letters. A dot indicates that any base can function at the specified positions. Two dots joined by a line indicate that any base pair will function. The substrate may be a short RNA oligonucleotide for laboratory biochemical studies, or part of a long RNA molecule (e.g., a cellular mRNA or viral genomic RNA molecule).

*See also* ENZYMES; RNA THREE-DIMENSIONAL STRUCTURES, COMPUTER MODELING OF.

### *Bibliography*


Cech, T. R. (1990) Self-splicing of group I introns. *Annu. Rev. Biochem.* 59:543–568.

Darr, S. C., Brown, J. W., and Pace, N. R. (1992) The varieties of ribonuclease P. *Trends Biochem. Sci.* 17:178–182.

Long, D. M., and Uhlenbeck, O. C. (1993) Self-cleaving catalytic RNA. *FASEB J.* 7:25–30.

Saldanha, R., Mohr, G., Belfort, M., and Lambowitz, A. M. (1993) Group I and group II introns. *FASEB J.* 7:15–24.

Symons, R. H. (1989) Self-cleavage of RNA in the replication of small pathogens of plants and animals. *Trends Biochem. Sci.* 14:445–450.


# RNA REPLICATION

*Claude A. Villee*

### *Key Words*

**Codons and Anticodons** The specific sequence of three nucleotides in messenger RNA (codon) and the complementary sequence of three nucleotides in transfer RNA (anticodon). These nucleotides store and transfer genetic information, which determine the base pairing between specific codons and specific anticodons, and thus determine the order in which amino acids are synthesized into peptide chains.

**Exons and Introns** The initial RNA product of the transcription of DNA, a large molecule called heterogeneous nuclear RNA, is cut and spliced by specific enzymes to yield a shorter messenger RNA composed of a portion of the original heterogeneous nuclear RNA. The pieces of RNA included in the final messenger RNA are termed exons. The other, unused, portions are termed introns, or intervening sequences. It is not clear what function, if any, these intervening sequences may have. For the heterogeneous nuclear RNA to be converted into a messenger RNA, the introns must be deleted from the molecule and the exons must be precisely spliced together to form a continuous message that codes for a specific protein.

**Helix-Destabilizing Proteins** For replication to proceed, the two strands that constitute DNA must be separated physically, the unwinding of the strands is catalyzed by DNA helicases, and the separated strands are then bound by helix-destabilizing proteins, which bind to single-stranded DNA and prevent the reestablishment of the double helix until each strand has been copied.

**Okazaki Fragments** Short DNA chains formed on the lagging strand; each is initiated by a separate primer and then is extended toward the 5′ end of the previously synthesized fragment by DNA polymerase.

**Retroviruses** Retroviruses have a genomic RNA that is a single plus strand, which may serve as a template to direct the formation of a DNA molecule, which, in turn, may act as a template for making messenger RNA. First the virion RNA is copied into a single strand of DNA, which then forms a complementary second DNA strand. This double-stranded DNA is integrated into the chromosomal DNA of the infected cell, and the integrated DNA is transcribed by the cell's own machinery into RNA that either acts as a viral messenger RNA or becomes enclosed in a virus.

**Ribozymes** New class of biological catalysts that, as nucleic acids, can act as enzymes; the RNA that forms the introns may be able to splice itself without the assistance of protein catalysts.

**Sense Strands and Antisense Strands** Of the two strands that comprise the double helix of a DNA molecule, only the sense strand contains a sequence of nucleotides that can be read to form a protein. The complementary strand, termed the antisense strand, has a sequence of nucleotides that if read out, either would give a garbled messenger RNA or would completely lack mRNA.

---

As more has been learned about the process of nucleotide replication it has become evident that DNA replication, RNA replication, and DNA transcription are simply variations on a common theme, that of storing information and then reading out that information as the cell grows and divides. All these processes depend on the formation of a polynucleotide strand, which serves as a template for the production of a complementary new strand of polynucleotide aligned by base pairing of adenine (A) to thymine (T) and guanine (G) to cytidine (C). The replication processes are carried out with great accuracy, with less than one error per thousand base pairs formed.

## 1 REVERSE TRANSCRIPTASE

The central dogma of molecular biology, unchallenged for several decades, was that genetic information always flowed from DNA to RNA to protein. A seminal exception to this rule was discovered in 1964 by Howard Temin. Temin's experiments showed that infection of cells by certain cancer-causing viruses is blocked by inhibitors of DNA synthesis and by inhibitors of DNA transcription. These experiments indicated that the synthesis and transcription of DNA are required for the multiplication of RNA tumor viruses and that in these systems information flows in the opposite direction, from RNA to DNA. Temin suggested that a DNA provirus was essential in the replication of RNA tumor viruses. In these organisms an enzyme was required that would synthesize DNA using RNA as a template. In 1970 Temin and Baltimore identified just such a DNA polymerase, one that uses RNA as a template in the synthesis of DNA. This RNA-dependent DNA polymerase, now termed "reverse transcriptase," has been shown to be present in all RNA tumor viruses. Other RNA viruses that do not form tumors replicate themselves directly, without using DNA as an intermediate.

## 2 STRUCTURE OF DNA AND RNA

Both DNA and RNA are long strands of nucleotides joined together by 3′,5′-phosphodiester bonds (Figure 1). Each nucleotide is composed of a purine or pyrimidine base, a sugar (ribose in RNA and deoxyribose in DNA), and orthophosphate. DNA is composed of four kinds of nucleotide: adenylate (A), guanylate (G), cytidylate (C), and thymidylate (T). RNA contains uridylate (U) instead of T. Uridylate has the same binding properties as thymidylate and base-pairs with adenylate. RNA is usually a single-stranded molecule (Figure 2). Although RNA has only a single-stranded polynucleotide chain, many of the sequences in RNA are mutually complementary, and these sequences can undergo intrachain base pairing to form stem–loop (hairpin) structures (Figure 3).

### 2.1 Structure of DNA

The DNA molecule consists of a very long double helix. The two chains that make up the helix extend in opposite directions and are paired by the Chargaff rules: a nitrogenous base in one chain is paired to a base in the other chain such that an A in one chain always pairs to a T in the other chain and a G pairs with a C. The specific base pairing is determined by the nature of the hydrogen bonds joining the two bases: two bonds join A and T and three hydrogen bonds join G and C. To fit into the double helix, one large base, A or G, must pair with a smaller base, C or T. An A·G pair would be too large and a C·T pair would be too small to fit in the space available in the double helix. In DNA replication the two chains separate momentarily and, with the aid of a number of proteins (some of which are enzymes), a new chain is formed by base pairing to each of the original chains. The product is two DNA molecules, each composed of a double helix.

### 2.2 Information Transfer: DNA to DNA

The Watson–Crick model of DNA suggested a mechanism by which the information in DNA could be copied precisely. Since nucleotides pair with each other in a complementary fashion, A to T and G to C, each of the nucleotide strands in the DNA molecule could serve as a template for the synthesis of its opposite strand. When the hydrogen bonds joining the two strands are broken, the two strands can separate. Each chain can then pair with complementary nucleotides to form the corresponding strand. This results in two DNA double helices, both of which are identical to the original one. Each consists of one original strand from the parent molecule and one newly synthesized complementary strand. The replication of DNA is termed "semiconservative" because each of the original strands is conserved in one of the daughter double strands and constitutes half of the daughter helix.

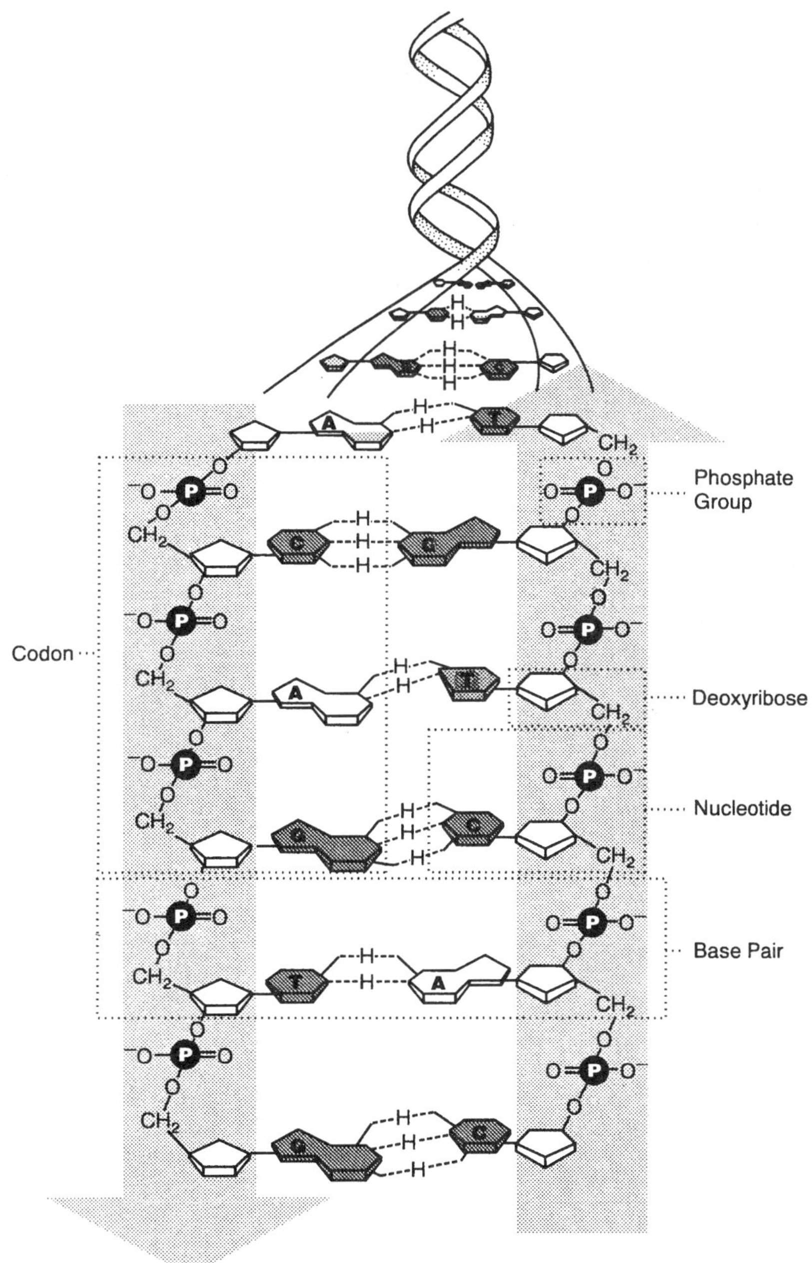


**Figure 1.** A portion of the double helix of DNA, indicating the various portions of the molecule. (From Villee, C. A. (1977) *Biology*. Reprinted by permission of Saunders College Publishing.)

The two strands of DNA in the double helix are physically intertwined; they must be separated for replication to proceed. The separation of the strands has proved to be a complex process. Replication can occur in these very long DNA molecules only if the strain of the unwinding strands is relieved. The unwinding is catalyzed by DNA helicases, enzymes that move along the helix, unwinding the strands as they go. The separated strands are bound by helix-destabilizing proteins that bind to single-stranded DNA, preventing the reestablishment of the double helix until each strand has been copied.

The DNA polymerases that link together the nucleotide subunits in the replication of DNA add nucleotides to the 3′ end of a polynucleotide strand that is paired to the strand being copied. The substrates for the DNA polymerases are dexoyribonucleoside triphosphates. As the nucleotides are joined, two of the phosphates are removed, providing the energy to drive the synthetic reaction. A new polynucleotide chain is elongated by the addition of the 5′ phosphate group of the next nucleotide subunit to the 3′ hydroxyl sugar at the end of the growing strand. Thus, DNA synthesis proceeds in a 5′ → 3′ direction. DNApolymerases can catalyze the addition of nucleotides only at the 3′ end of an existing DNA strand. This leads to the question of how the synthesis of DNA can be initiated when the two strands are separated. This is accomplished by utilizing a short piece of an RNA primer that is synthesized by an aggregate of proteins called a primosome. The RNA primer pairs with a single-stranded DNA template at the point of

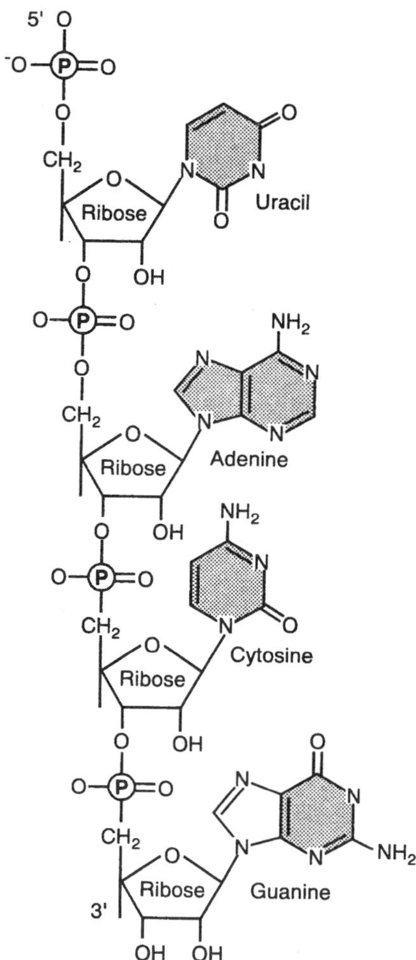


Figure 2. A portion of an RNA strand, with ribonucleoside phosphates linked by 5′–3′ bonds, like the deoxyribonucleosides in DNA. (From Villee, C. A. (1977) *Biology*. Reprinted by permission of Saunders College Publishing.)

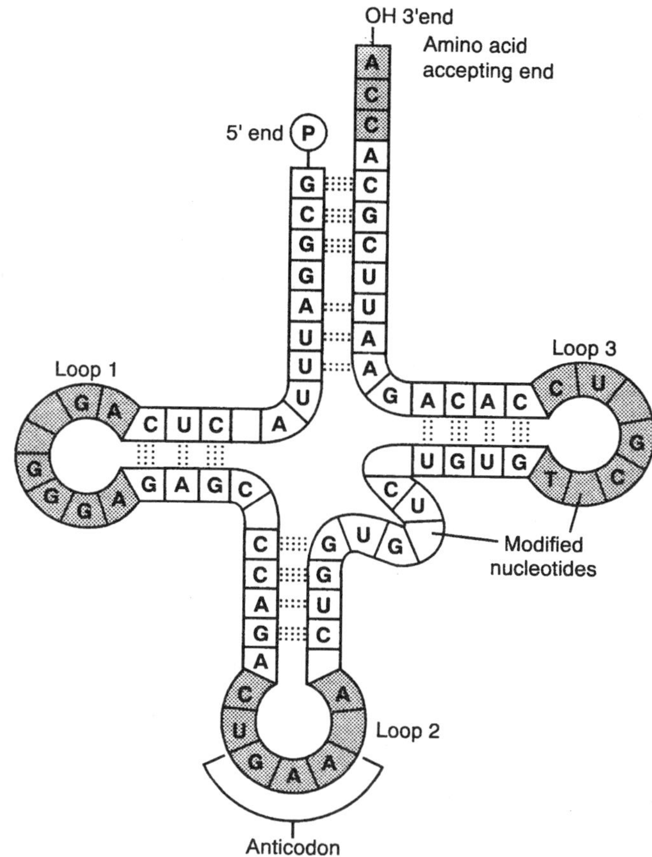


Figure 3. The structure of a transfer RNA molecule, showing the many sites of complementary base pairing (dots indicate hydrogen bonding between A and U and between C and G). Similar complementary base pairing occurs frequently in messenger RNA as well. (From Villee, C. A. (1977) *Biology*. Reprinted by permission of Saunders College Publishing.)

initiation of replication. DNA polymerase can then synthesize the new chain by adding nucleotides to the 3′ end of the RNA primer. When DNA synthesis has proceeded to an appropriate extent, the RNA primer is degraded and removed by specific enzymes and that space is filled in by DNA polymerase.

### 2.3 DNA Replication: Leading and Lagging Strands

Although the complementary DNA strands extend in opposite directions, DNA synthesis can proceed only in the 5′ → 3′ direction. The strand being copied is read in a 3′ → 5′ direction. DNA replication is begun at specific sites on the DNA molecule, called origins of replication, and both strands are replicated at the same time. One, the leading strand, is formed continuously after the process has been initiated. The complementary (lagging) strand is synthesized in short pieces, which are subsequently joined to make a complete DNA chain. DNA is synthesized beginning at a Y-shaped structure called a replication fork. The lagging strand is synthesized in the form of relatively short fragments (Okazaki fragments), each of which is initiated by a separate primer and is then extended toward the 5′ end of the previously synthesized fragment by a DNA polymerase. The complex DNA polymerases serve several functions. As the growing fragment approaches the fragment synthesized previously, one part of the polymerase degrades the RNA primer, permitting other polymerases to fill in the gap between the two fragments. These are then linked by DNA ligase, which joins the 3′ end of one fragment to the 5′ end of another fragment by a phosphodiester bond.

Since similar codes are used in both DNA and RNA, in certain retroviruses, such as the AIDS virus, information is transferred from RNA, which is the genetic material of the retroviruses, to DNA, which becomes an intermediate in the synthesis of the next generation of RNA. Information is transferred directly from RNA to RNA in certain RNA viruses such as the influenza virus or the poliomyelitis virus.

### 2.4 Information transfer: DNA → RNA

The transcription of DNA to form RNA and the replication of RNA are processes that are basically similar to that of DNA replication discussed in Section 2.3. In all these processes one polynucleotide strand serves as a template, and specific polymerases use specific nucleoside triphosphates to add specific nucleotides to the chain (Figure 4). The particular base added in sequence is determined by the complementary pairing of the bases (determined by the hydrogen bonds between the bases) in the initial and forming strands.

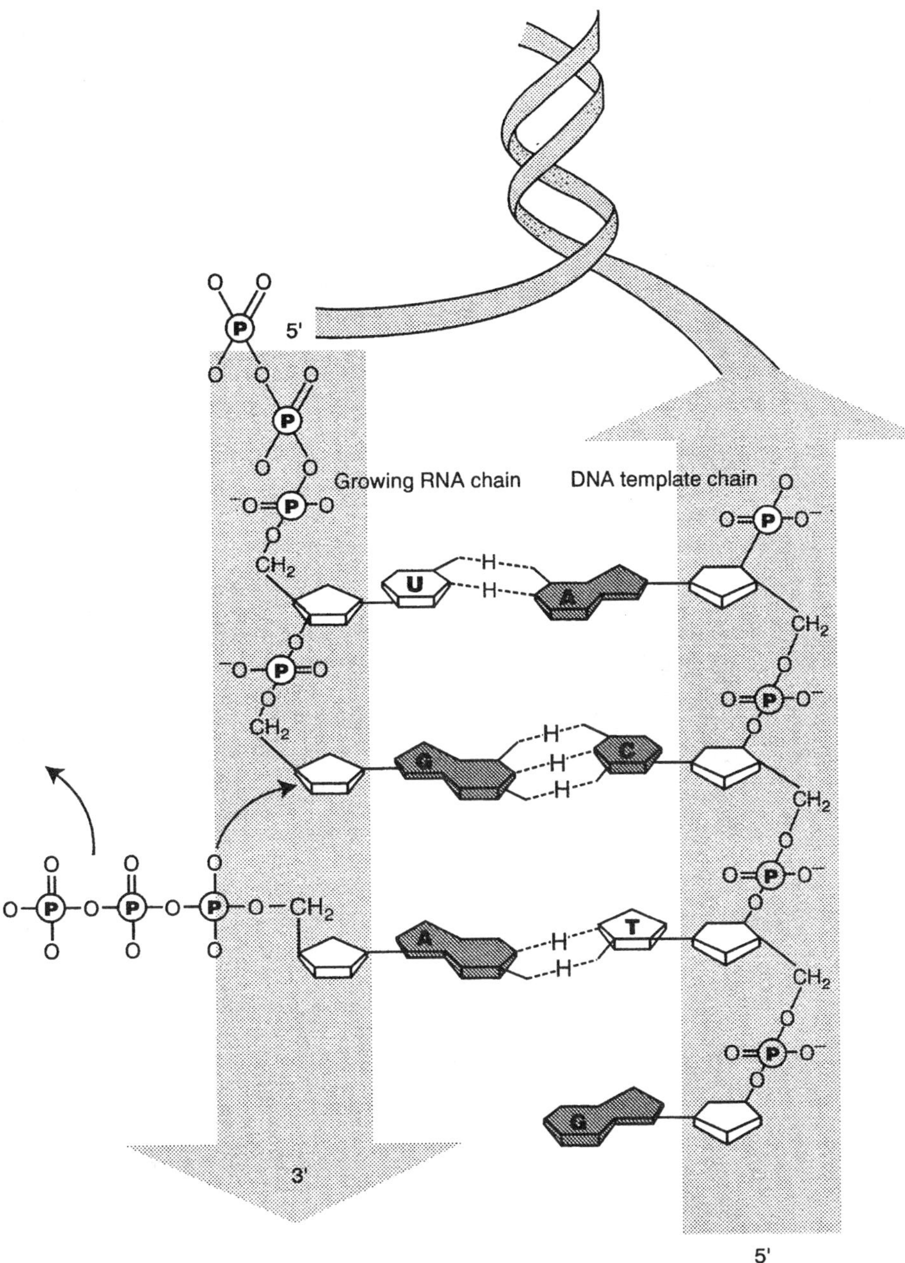


**Figure 4.** The transcription process; the double helix of DNA is unwound from top to bottom. The freed strand of template DNA (on the right) is being copied to yield RNA (on the left). The nucleoside triphosphates being added (lower left) form complementary base pairs with the bases in the DNA template strand. Two phosphates are cleaved from the nucleoside triphosphate. The remaining phosphate is covalently linked to the 3′ end of the growing RNA chain. This process is repeated many times as the RNA strand increases in length. (From Villee, C. A. (1977) *Biology*. Reprinted by permission of Saunders College Publishing.)

Information present in the sequence of nucleotides in one of the DNA strands, the minus strand, is copied, yielding a plus strand of messenger RNA. This transcription process is very similar to that by which DNA is replicated: the plus mRNA strand is formed by complementary base pairing with the minus strand of DNA. Information that has been transcribed into mRNA is used to determine the amino acid sequence in the protein. It is clear that this process involves the conversion of the nucleic acid code of the messenger RNA into an amino acid code of the protein. Hence it is termed a process of translation. The information present in the mRNA is in the form of a genetic code composed of three bases that constitute a codon, the basic unit of genetic information. Each group of three bases in mRNA specifies one of the amino acids in the protein. The codon that specifies the amino acid tryptophan is UGG. Any of six different codons (CGU, CGC, CGA, CGG, AGA, AGG) may specify the amino acid arginine.

The information contained in the codons of mRNA is translated by a very complex process that occurs within the ribosomes in eukaryotic cells. The recognition and decoding of the codons in mRNA is accomplished by transfer RNAs. The anticodons of the tRNA are composed of triplet nucleotides, which recognize a specific codon in mRNA by complementary base pairing. The amino acid specified by the anticodon is attached to the other end of the tRNA molecule. Protein synthesis occurs on the ribosomes, complex subcellular organelles composed of two kinds of subunit. Each contains one or more kinds of ribosomal RNA (rRNA) and a considerable number of proteins. The ribosomes are attached to the 5′ end of the mRNA. The ribosome travels along the mRNA, joining the codon of the mRNA with the anticodon of the tRNA by base pairing. In this way the amino acids are lined up in the proper sequence. The amino acids are then joined enzymatically to form a long polypeptide chain.

### 2.5 Sense and Antisense Strands

Only one of the two strands in DNA contains the correct coding sequence for protein synthesis. This, the minus or sense strand, undergoes transcription to form mRNA. The complementary DNA strand is termed plus, or antisense. The RNA polymerase that catalyzes transcription recognizes a specific promoter sequence of bases at the 5′ end. Messenger RNA is synthesized by the addition of nucleotides, one at a time, to the 3′ end of the growing molecule.

The initiation of RNA replication does not require a primer. The 5′ end of a new mRNA chain begins with a nucleoside triphosphate. The promoter sequences of different genes may differ and this can determine which genes will be transcribed at any given time. Three of the codons in mRNA (UGA, UAA, UAG) serve as ''stop'' signals for the RNA polymerase and bring about the termination of transcription.

The mRNA molecules of eukaryotic organisms undergo posttranscriptional modifications and processing. A molecule of 7-methylguanylate, an unusual nucleotide, is added as a cap at the 5′ end of the mRNA chain. This alteration may be the basis of the greater stability of eukaryotic mRNAs, which have half-lives as long as 24 hours, whereas the half-life of prokaryotic mRNA is about 15 minutes. A long tail of polyadenylic acid, 100–200 adenine nucleotides, is joined to the 3′ end of the mRNA.

### 2.6 Introns and Exons

A third step in the modification of eukaryotic mRNA involves cutting and splicing the mRNA at specific sites. Interrupted coding sequences are present in eukaryotic DNA; these may be quite long, but they do not code for the amino acids present in the final protein product. These noncoding regions within the gene are called intervening sequences or introns, as opposed to exons (which are expressed sequences and part of the protein-coding sequence).

The number of introns present in a gene may vary widely. The β-globin gene, which produces one of the components of hemoglobin, contains two introns. The ovalbumin gene, which codes for one of the proteins in egg white, contains seven introns, and the gene for another egg-white protein, conalbumin, contains sixteen. The combined lengths of the introns may be considerably greater than the combined exon sequences. The ovalbumin gene contains about 7700 base pairs, whereas the sum of its coding sequences (exons) is only 1859 base pairs.

The transcription of a gene containing both introns and exons yields a large RNA transcript termed heterogeneous nuclear RNA. For the hnRNA to be converted into a functional mRNA, the introns must be deleted and the exons must be spliced together to form a continuous message that codes for a specific protein. The splicing reactions are mediated by special base sequences within and to either side of the introns. The splicing reactions may be catalyzed by small nuclear ribonucleoprotein complexes, which bind to the introns and catalyze the cleavage and splicing reactions. The RNA within the intron may have the ability to splice itself without the help of protein catalysts; these nucleic acids act as enzymes. This new class of biological catalysts has been termed ribozymes. The final product, functional mRNA, has had its introns removed and the exons spliced together. It has a 7-methylguanylate cap at the 5′ end and a polyadenylated tail at the 3′ end. It is then ready to pass out of the nucleus into the cytoplasm and, when attached to a ribosome, to serve as the blueprint for the synthesis of a protein with its specific sequence of amino acids.

## 3 VIRAL DNA AND RNA

Much of what we know about genetic machinery in general and about replication and transcription in particular has been derived from studies of viruses. The viruses that infect animal cells exhibit a tremendous variety of shapes, sizes, and genetic strategies. Viruses are classified according to the sequence of reactions by which the mRNA is produced (Table 1). In this classification scheme, a viral mRNA is termed a plus strand and its complementary sequence, which cannot function as a messenger RNA, is termed a minus strand. A strand of DNA complementary to the viral mRNA is termed a minus strand. Production of a plus strand of mRNA requires that a minus strand of RNA or DNA be used as the template. This arrangement permits the characterization of six kinds of virus that infect animal cells. In each of these, the nucleic acid of the virion (the infective particle) ultimately becomes the mRNA of the daughter virus.

The viruses that infect animal cells have genomes composed of DNA or RNA. In some the nucleic acid is single-stranded and in others the nucleic acid is double-stranded.

**Table 1** A Classification of Animal Viruses Based on the Nature of the Nucleic Acid in the Genome: DNA or RNA, Plus Strands or Minus Strands, Double Stranded or Single Stranded

| Class | Genome | Strands | Product | Example |
|---|---|---|---|---|
| I | DNA | Double, ± | Viral mRNA | Adenoviruses |
| II | DNA | Single, + or − | Double-stranded DNA | Parvoviruses |
| III | RNA | Double, ± | +mRNA | Reoviruses |
| IV | RNA | Single, + | Long +mRNA | Poliomyelitis |
| V | RNA | Single, − | +mRNA | Influenza |
| VI | RNA | Single, + | Single DNA | Retrovirus (e.g., HIV) |

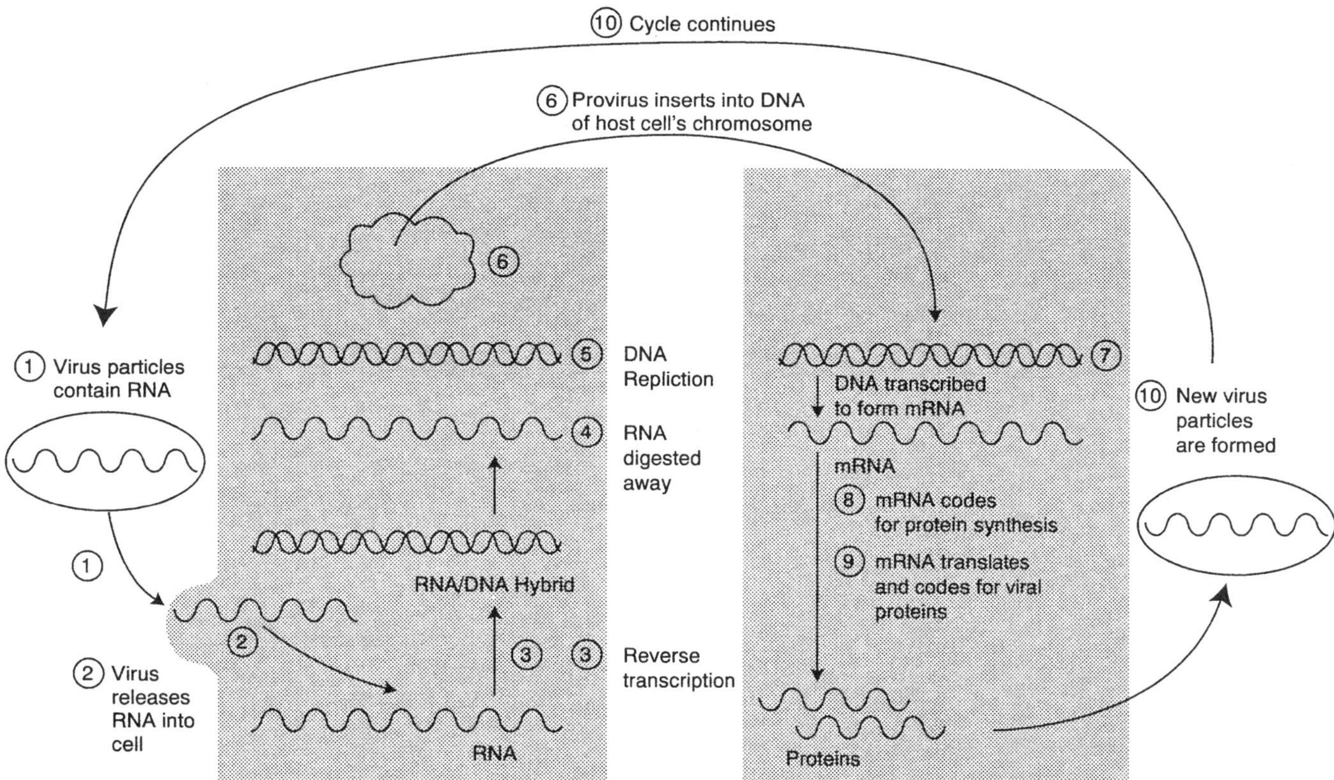


**Figure 5.** The life cycle of an RNA tumor virus, a retrovirus. (1) The virus particle containing RNA attaches to the cell membrane and releases the RNA into the host cell. (2) The reverse transcriptase (3) makes a DNA copy of the RNA, after which the DNA strand is replicated to produce a double-stranded molecule (5). The DNA provirus is inserted into the DNA of the host cell's chromosome (6). The viral DNA uses the cell's transcription apparatus to make RNA copies (7), which are released into the cytoplasm and translated to make essential proteins for the virus particles (8, 9). The RNA and proteins then bud from the cell, forming new virus particles that (10) can infect other cells, completing the life cycle.

The adenoviruses and the SV 40 virus contain double-stranded DNA. The viral DNA enters the nucleus of the host cell. The enzymes that in the uninfected state are responsible for producing cellular mRNAs are diverted to producing viral mRNA. The pox viruses are large viruses, which have their own enzymes for synthesizing mRNA; they undergo replication in the cytoplasm of the host cell.

Parvoviruses are simple viruses containing a single strand of DNA. Some parvoviruses enclose within their capsules both plus and minus strands of DNA, but the strands are present in separate virions. Other viruses enclose only a minus strand within the capsid, and this is copied within the cell into double-stranded DNA, which is then copied to yield messenger RNA (Figure 5).

## 4 INFORMATION TRANSFER RNA TO RNA

Four additional kinds of virus contain RNA rather than DNA genomes. A wide range of animals, including human beings, may be infected by viruses belonging to each of these classes (see Table 1). Viruses with a genome of double-stranded RNA utilize the minus RNA strand as a template for the synthesis of a plus strand of mRNA. The virions of these viruses have segmented genomes containing 8–12 double-stranded RNA fragments, each of which codes for a specific polypeptide. These viruses contain a complete set of enzymes for the synthesis of mRNA.

In viruses with a genome composed of a single plus strand of RNA, the viral genomic RNA is identical to the mRNA and the genomic RNA is infectious by itself. The mRNA is copied into a minus strand of RNA, which then produces more plus strands.

The RNA in the virion of poliomyelitis virus serves as the mRNA to encode all the viral proteins. The proteins are initially synthesized as a single, very long polypeptide chain, which is subsequently cleaved to yield the several functional proteins. The mRNA of these viruses is the same length as the genomic RNA.

The virions of toga viruses are surrounded by a lipid envelope. These viruses synthesize at least two forms of mRNA within the host cell; one is the same length as the RNA in the virion, whereas the other corresponds to the third of the virion RNA at the 3′ end. This is one of a group of rare insect-borne viruses that cause encephalitis in human beings.

A further class of viruses contain a single minus strand of RNA. The RNA in the virion has a base sequence complementary to that of the mRNA. Thus, the virion contains a template for making mRNA, but that template does not itself encode proteins.

Viruses with a single strand of RNA in their genome have in the virion a virus-specific polymerase that synthesizes several different mRNAs from different parts of the single RNA template strand. Each viral mRNA encodes one protein. The influenza virus also has a segmented genome. Each segment is a template for the synthesis of a single mRNA. The minus strands of these viruses alone, in the absence of the virus-specific polymerase, are not infectious. The RNA polymerase of the influenza virus initiates the transcription of each mRNA by a unique mechanism. The polymerase begins the synthesis of mRNA by borrowing 12–15 nucleotides from the 5′ end of the cellular mRNA or the mRNA precursor in the nucleus. This oligonucleotide serves as a primer for the replication of RNA catalyzed by the viral RNA polymerase. The individual mRNAs made by these viruses generally encode single proteins, but some of the mRNAs can be read in different frames to produce two different proteins.

Since some viruses contain RNA but not DNA, in these organisms RNA must be the viral genetic component. This theory was confirmed by the finding that purified RNA preparations from tobacco mosaic virus were infectious even in the absence of any TMV protein. Many other viruses have been shown to contain RNA but not DNA. Thus the potential of RNA for carrying genetic information and undergoing replication cannot be denied.

In most viruses the DNA or RNA codes not only for the coding proteins of the virus but also for the enzymes that are required to replicate the viral nucleic acid. The smaller DNA viruses such as the simian SV-40 virus, and the very small bacteriophage ϕX174, contain very much less genetic information and must rely to a much greater extent on enzymes from the host cells to carry out the synthesis of proteins and DNA. These viruses do contain coding for enzymes that initiate their own DNA synthesis selectively. For a virus to be successful when it invades a cell, it must override the cellular control signals that otherwise would prevent the viral DNA from doubling more than once in each cell cycle.

The discoveries in the 1950s and 1960s that many of the smaller DNA viruses contain only single-stranded DNA can now be appreciated, although at the time it seemed that viruses might have a fundamentally different mode of replication. These DNA viruses begin replication by producing a double-stranded intermediate. This finding reinforced the concept that all DNA replication, all transcription of DNA to RNA, and all RNA replication involve the formation of complementary strands of polynucleotides.

## 5 REPLICASES

The replication of RNA and the multiplication of RNA viruses also involve the formation of complementary chains of nucleotides joined by specific hydrogen bonds between specific purines and pyrimidines. RNA replication is mediated by specific RNA-dependent RNA polymerases (called replicases), which are coded for in the viral RNA chromosome. Some of the RNA viruses, such as poliovirus, have a genome composed of a single-stranded RNA polynucleotide chain. Other viruses, such as the reoviruses, have a double-helical RNA viral chromosome.

## 6 RETROVIRUSES: INFORMATION TRANSFER FROM RNA TO DNA

The genomic RNA of retroviruses is a single plus strand, which serves as a template that directs the formation of a DNA molecule. The process is catalyzed by an RNA-dependent DNA polymerase. The DNA produced in turn serves as the template for the production of mRNA. First the virion RNA is copied into a single strand of DNA, which then forms a complementary second DNA strand. The double-stranded DNA is integrated into the chromosomal DNA of the host cell. Finally the integrated DNA is transcribed by the cell's own metabolic machinery into RNA that either acts as a viral mRNA or becomes enclosed in a virus. This completes the retrovirus cycle (Figure 5). If the retrovirus contains oncogenes (cancer genes), the cell it infects may be transformed into a tumor cell.

## 7 EVOLUTION OF COMPLEMENTARY BASE PAIRING

The evolution of the process of complementary base pairing was a key step in the evolution of life, for it permitted the development of mechanisms for the transfer and storage of genetic information. Most scientists now believe that RNA is the more ancient means of transferring genetic information and that RNA replication preceded DNA replication in the course of evolution. Thus, RNA replication may have been the evolutionary precursor of DNA replication.

The replication of RNA and DNA and the transcription of DNA to RNA and the translation of RNA to form proteins all involve this important basic principle of complementary base pairing, determined by specific hydrogen bonds between the base pairs A·T and G·C. In protein synthesis the lining up of the amino acids is determined by the complementary base pairing of codon and anticodon.

The course of evolution of nucleic acids may have been as follows:

| | | | | |
|---|---|---|---|---|
| | | Virus with single-stranded DNA | → | Virus with double-stranded DNA |
| Virus with single-stranded RNA | → | Virus with double-stranded RNA | | |
| | ↘ | Virus with RNA/DNA hybrid | | |

*See also* Ribozyme Chemistry; RNA Secondary Structures; RNA Structures, Nonhelical; Viruses, DNA Packaging of; Viruses, RNA Packaging of.

### *Bibliography*


Aperion, D., Ed. (1984) *Processing of RNA.* CRC Press, Boca Raton, FL.

Breaker, R., and Joyce, C. (1994) Emergence of a replicating species from an in vitro RNA evolution reaction. *Proc. Natl. Acad. Sci. U.S.A.* 91:6093.

Chamberlin, M. (1982) Bacterial DNA dependent RNA polymerases, in *The Enzymes,* Vol. 15, Part B, P. Boyer, Ed. Academic Press, New York.

Davis, B. D., Dulbecco, R., Eisen, H. N., and Ginsberg, H. S. (1980) *Microbiology,* 3rd ed. Harper & Row, New York.

Hakjiolav, A. A. (1985) *The Nucleolus and Ribosome Biogenesis.* Springer-Verlag, New York.

MacLean, N., Gregory, S. P., and Flavell, R. A., Eds. (1983) *Eukaryotic Genes: Their Structure, Activity, and Regulation.* Butterworths, London.

Stanier, P. Y., Ingraham, J. L., Wheelis, M. L., and Panter, P. R. (1986) *The Bacterial World,* 5th ed. Prentice-Hall, Englewood Cliffs, NJ.

Watson, J., Hopkins, N., Roberts, J., Steitz, J., and Weiner, A. (1982) *Molecular Biology of the Gene,* 4th ed. Benjamin/Cummings, Redwood City, CA.

# RNA Secondary Structures

*John A. Jaeger*

*Key Words*

**Bulge Loop** Regions in which there are unpaired bases on only one side of a helix.

**Hairpin Loop** The unpaired region formed when an RNA folds back upon itself to form a helix.

**Internal Loop** Two or more opposing unpaired bases between two helical segments; internal loops can be symmetric (the same number of unpaired bases on each side of the loop) or asymmetric (a different number of unpaired bases on each side of the loop).

**Multibranch Loop or Junction** Region in which three or more helices join to form a closed loop.

**Pseudoknot** Structure that results when any single-stranded loop forms a helix with another single-stranded region; pseudoknots may be secondary or tertiary structures.

**Tetraloop** Four-base hairpin loop; some common tetraloops in very structured RNAs have the sequence UNCG or GNRA (where N is any base, and R is any purine).

RNA secondary structures can lie flat on a piece of paper with no overlapping strands. Unpaired regions between RNA helices include hairpin loops, bulge loops, internal loops, and multibranched loops or junctions. Secondary structure forms the core for the formation of the overall tertiary structure. RNA secondary structures are important for RNA–RNA and RNA–protein interactions. Chemicals, ribonucleases, and site-directed mutagenesis probe these structures; comparative sequence analysis (phylogeny) and thermodynamic methods predict them.

## 1 TYPES OF RNA STRUCTURE

Primary structure is the sequence of the bases, listed from the 5′ to the 3′ end of the molecule (see Figure 1, top). There are a few definitions of secondary structure, but the simplest is this: the structure lies flat on a sheet of paper with no strands overlapping. Tertiary structure requires these secondary structures to interact out of the plane. Finally, quaternary structures are interactions between two separate molecules (e.g., RNA–RNA or RNA–protein).

## 2 TYPES OF RNA SECONDARY STRUCTURE

Secondary structural elements come in five types: helices, bulges, internal loops, hairpin loops, and multibranched loops or junctions. Pseudoknots can be either secondary or tertiary interactions. The lower portion of Figure 1 annotates the secondary structure.

Helical or base-paired regions contribute most of the stability to RNA secondary structure through hydrogen bonding and base stacking. The Watson–Crick pairs, G·C and A·U, as well as some mismatches, such as G·U, stabilize a helix. Base stacking is such an important stabilizing effect that a single base stacking on the 3′ side of a helix (called a "dangling end") can add as much stability to the structure as a base pair. Helices are on average six base pairs long in secondary structures of small subunit ribosomal RNA. A typical representation of helices in secondary structures is a ladder, where the rungs represent base pairs and the sides represent the sugar–phosphate backbone.

Internal loops form when there are bases that cannot pair on both sides of a helix. Internal loops can be either symmetric (the same number of unpaired bases on each side of the helix) or asymmetric (a different number of unpaired bases on each side of a helix). Two base internal loops are often called mismatches. Originally, studies of homopolymer loops (loops composed of one type of base) suggested that these regions would destabilize secondary structures; recent studies on common small internal loops reveal increased stability due to base stacking and non-Watson–Crick hydrogen bonding. For example, the loop E region of *Xenopus laevis* 5S ribosomal RNA has an unusual structure with many non-Watson–Crick interactions. Internal loops are important sites of RNA–protein interaction in 5S rRNA, and proposed RNA–RNA tertiary and quaternary interactions in group I introns.

Bulges are regions having unpaired bases on only one side of a helix; they can bend RNA backbones. Bulges are important recognition sites for many regulatory and structural proteins. A single-bulged A is a common motif in 5S ribosomal RNA and is the guanosine binding site of group I intron ribozymes (a catalytic RNA). An unusual 2′–5′ linkage at a bulged A is a product of mRNA intron splicing by group II ribozymes, as well as the cellular splicing machinery (small nuclear ribonucleoprotein particles, or snRNPs).

Hairpin loops occur at the end of a helix when the sugar–phosphate backbone folds back on itself to form an open loop; tracing the backbone reveals a hairpinlike structure. Comparisons of small subunit ribosomal RNA structures reveal an uneven distribution of hairpin loop sizes; four base loops are the most common. Many four-base hairpin loops are either GGNRAC or CUNCGG (loop bases underlined; N is any base, and R is either G or A), although other sequences are possible. These special sequence tetraloop hairpins often substitute for one another in ribosomal RNAs. Recent structural studies on the CUUCGG hairpin reveal an additional strained base pair between the first and fourth bases in the loop, and an additional intraloop hydrogen bond. Larger hairpin loops can pair into complex structures involving non-Watson–Crick interactions. Hairpin loops are important for mRNA stability, RNA tertiary interactions, and protein binding sites.

Multibranched loops or junctions occur when three or more helices join to form a closed loop. The crystal structure of $tRNA^{Phe}$ has a four-helix multibranched loop stabilized by helix–helix stacking as well as significant non-Watson–Crick secondary and tertiary interactions. These interactions probably stabilize other multibranched loops. Three-helix multibranched loops are found in 5S rRNA and the hammerhead ribozyme; the former is recognized by the TFIIIA (transcription factor A for polymerase III) ribosomal protein, and the latter is the site of phosphodiester cleavage.

Pseudoknots are special RNA interactions; they result from additional pairing in loop regions of the aforementioned secondary structures. Pseudoknots can form either secondary structures (such as a local folding back on a hairpin structure) or tertiary structures spanning several helical regions. Pseudoknots are important struc-

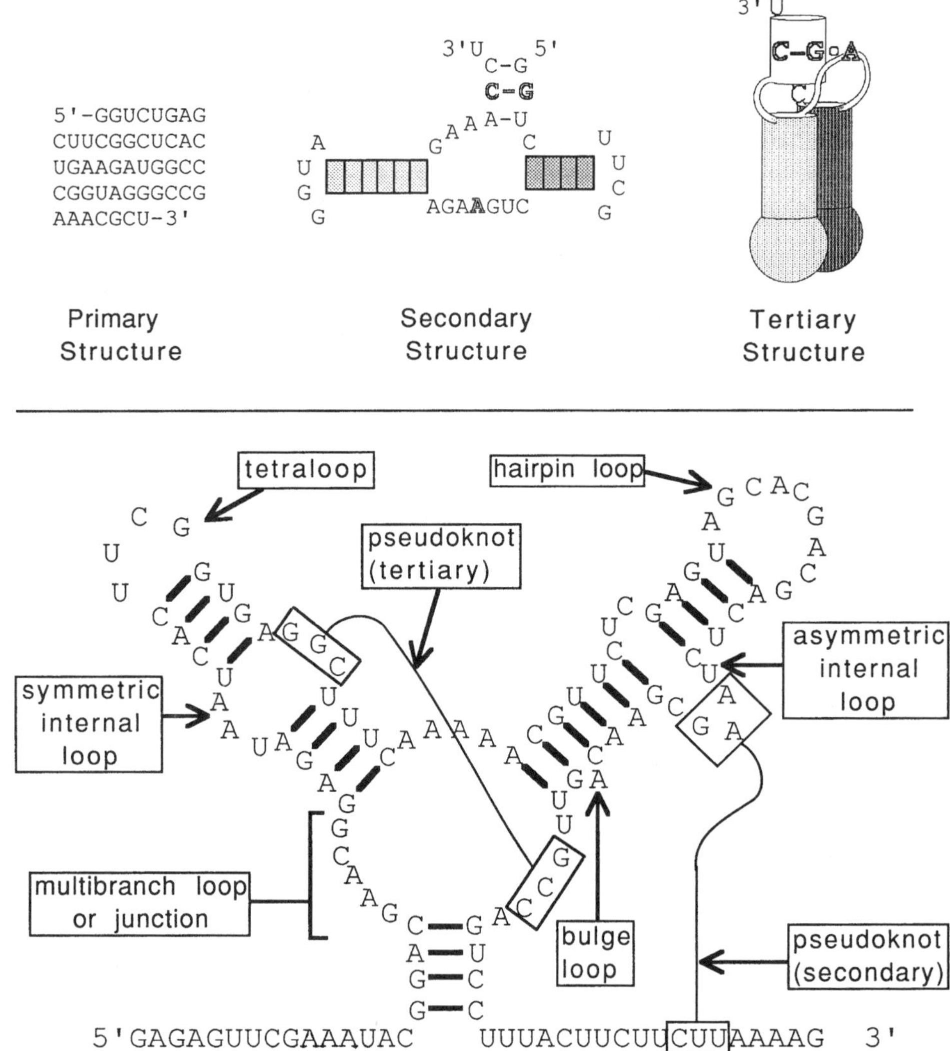


Figure 1. *Top:* The primary, secondary, and a possible tertiary structure of an RNA molecule. The tertiary structure drawing uses cylinders as helices and ribbons as single-stranded regions; it contains a tertiary interaction between a C•G base pair and an A. *Bottom:* RNA secondary structural elements.

tural features in the tRNA-like ends of viroids, 16S rRNA of *Escherichia coli,* and the catalytic core of the group I introns. They are important recognition sites for regulatory proteins, and they cause translational frameshifting in many viruses.

## 3 PREDICTING SECONDARY STRUCTURE

RNA secondary structure is predicted by comparative sequence analysis (phylogeny) or thermodynamics. Comparative sequence analysis finds a common folding pattern of two or more RNAs having a similar function. The existence of compensatory base pairs is proof of helical regions; for example, an A·U pair in one sequence may replace a C·G pair in another sequence. This method cannot predict structure in conserved sequence regions, since there are no compensatory base pairs. Nevertheless, it is the most accurate way of predicting Watson–Crick pairing regions. Thermodynamic predictions assume that the solution structure has the largest equilibrium constant of formation, or, in other words, the most negative Gibbs free energy ($\Delta G°$). RNA secondary structure stability is based on the strong local interactions of base stacking and hydrogen bonding; these local interactions are measured for small oligoribonucleotides and used to predict the structure of large RNAs. Thermodynamic predictions succeed if there is a measured $\Delta G°$ for each interaction.

Even though the stacking and hydrogen bonding schemes in loop regions are not completely understood, a recent thermodynamic model predicts about 70% of the helices of phylogenetically determined secondary structures. Secondary structure prediction programs based on thermodynamics let the user include phylogenetic data as well as constraints from chemical modification and ribonuclease digestion.

## 4 PROBING SECONDARY STRUCTURE

Ribonucleases, chemical modification and cleavage, cross-links, and mutagenesis are four direct probes of RNA secondary structure.

Other methods, such as thermal denaturation, kinetics, and nuclear magnetic resonance spectroscopy, are not discussed here. Experiments must be conducted such that ribonucleases and chemicals modify only once per RNA molecule, since modification or cleavage may alter the secondary structure. Ribonucleases are the simplist RNA structure probes; some enzymes cleave helical regions (such as ribonuclease $V_1$), while others ($T_1$, $T_2$, $S_1$, and $CL_3$) attack only single-strand regions. Base-modifying reagents, like kethoxal or dimethyl sulfate, are important secondary structure probes, whereas chemicals that modify phosphates (e.g., ethylnitrosourea) or cleave the sugar–phosphate backbone [e.g., iron-EDTA (ethylenediaminetetraacetic acid)] are more important as tertiary and quaternary structure probes. Cross-links (unusual covalent bonds) form between two strands of a pairing region with ultraviolet light or special reagents, such as psoralen.

Detection of RNA cleavage or modification is monitored directly using end-labeled RNA or indirectly using reverse transcription. After cleaving 5′- or 3′-$^{32}$P-labeled RNA with ribonucleases or chemicals, the different-sized strands can be separated by denaturing polyacrylamide gel electrophoresis. Each radioactive band on the gel represents a different cleavage site; the longer the fragment, the further the strand cleavage site from the labeled end. Reverse transcription can detect strand scission, but can also detect base modifications and cross-links. Reverse transcriptase synthesizes a complementary DNA strand from the RNA template starting at a labeled DNA primer; it will stop at modified bases and at the end of the RNA. A combination of chemical modification and reverse transcription was used to confirm the secondary structure of the *E. coli* 16S ribosomal RNA, which was originally determined by phylogenetic analysis.

Mutational analysis creates an artificial phylogeny using mutagens or site-directed mutagenesis. Mutagens induce nonspecific changes in an organism's genome; the sequences of viable strains can show compensatory mutations. In contrast, site-directed mutagenesis creates specific mutations in a cloned sequence. The effects of a single base mutation in the helix and of the compensatory base change are observed by a functional assay, such as protein binding or catalytic activity. Mutational analysis is an important tool to determine secondary structure in conserved sequence regions, such as the catalytic core of group I introns.

*See also* RNA Structure, Nonhelical; RNA Three-dimensional Structures, Computer Modeling of.

## *Bibliography*


Chastain, M., and Tinoco, I., Jr. (1991) *Prog. Nucleic Acid Res. Mol. Biol.* 41:131–177.

Jaeger, J. A., SantaLucia, J., Jr., and Tinoco, I., Jr. (1993) *Annu. Rev. Biochem.* 62:255–287.

James, B. D., Olsen, G. J., and Pace, N. R. (1989) *Methods Enzymol.* 180:227–239.

Tinoco, I., Jr., Pugilisi, J. D., and Wyatt, J. (1990) *Nucleic Acids Mol. Biol.* 4:205–226.

Turner, D. H., Sugimoto, N., and Freier, S. M. (1988) *Annu. Rev. Biophys. Biophys. Chem.* 17:167–192.

Westhof, E., and Jaeger, L. (1992) *Curr. Opinion Struc. Biol.* 2:327–333.

Wimberly, B., Varani, G., and Tinoco, I., Jr. (1991) *Curr. Opinion Struc. Biol.* 1:405–409.

Zuker, M. (1989) *Methods Enzymol.* 180:262–288.


# RNA Structure, Nonhelical

*Brian Wimberly*

## *Key Words*

**Conformation** The noncovalent structure of a molecule as described by rotations about its single bonds. The covalent structure is referred to as its configuration.

**Hairpin Loop** An element of RNA secondary structure in which the molecule folds back upon itself to form a Watson–Crick-paired stem, leaving a loop of mismatched and/or unpaired nucleotides.

**Helical RNA or DNA** A segment of nucleic acid with a regular, repeating backbone conformation, as in Watson–Crick-paired helices.

**Mismatch** Any base pair other than the Watson–Crick pairs A·U and G·C. Mismatches, also called noncanonical pairs, are generally less thermodynamically stable than the canonical Watson–Crick pairs.

**Secondary Structure** A picture of the RNA sequence drawn to show the regular, Watson–Crick-paired helices. The elements of secondary structure are Watson–Crick-paired helices, hairpin loops, internal loops, bulges, multistem junctions, and single strands.

**Tertiary Structure** The complete three-dimensional structure of the molecule, derived from the secondary structure by the formation of stabilizing long-range pairings—tertiary interactions—between the elements of secondary structure.

Many functions of biological macromolecules require adoption of a single, three-dimensional structure that is recognized by other biomolecules. While hundreds of three-dimensional structures of proteins are known, only a few RNA structures have been determined. Moreover, the most functionally important parts of RNA—the irregular, nonhelical loops—are also the least well understood in structural terms. Determination of nonhelical RNA structures will improve attempts to predict RNA secondary and tertiary structure, provide insights into the specificity of protein-RNA recognition, and shed light on the mechanism of RNA-mediated catalysis.

## 1 INTRODUCTION

### 1.1 Little Is Known About Nonhelical Nucleic Acid Structure

The elements of regular secondary structure in nucleic acids and proteins—the double helix, the α helix, and the β sheet—were all discovered in the early 1950s. Today, however, knowledge of protein structure has progressed far beyond that of nucleic acid structure. While hundreds of three-dimensional protein structures are known, the number of nucleic acid structures that have been determined to high resolution (most of which have been fully Watson–Crick-paired double helices) is much lower. Thus especially little is known about nonhelical nucleic acid structure, which is unfortunate because it is often these loop structures that are most important to function.

Why have nonhelical DNA and RNA structures been so difficult to determine? For nucleic acids, nonhelical means irregular, and irregular conformations depend much more on sequence than regular, helical ones. The determination of nonhelical nucleic acid structure therefore requires nucleic acids of defined length and sequence, and experimental techniques that can provide conformational details at atomic resolution. Because only a few of the smaller naturally occurring nucleic acids can be isolated from natural sources in sufficient, pure quantities for physical studies, most of the advances in the study of nonhelical nucleic acid structure had to await improved methods for oligonucleotide synthesis. In addition, only X-ray crystallography and very recent NMR techniques allow determination of atomic resolution structures, and only a very few nonhelical nucleic acids have been successfully crystallized. However, improved methods are now available for synthesis, crystallization, and isotopic labeling for NMR studies, which should accelerate the determination of nonhelical nucleic acid structures. As more irregular structures are determined, new rules for their folding will be discovered, which will improve the prediction of nucleic acid secondary and tertiary structure given only the RNA or DNA sequence.

This contribution focuses on high resolution nonhelical RNA structures. While nonhelical conformations are required for some DNA functions, they are relatively rare and often transient structural features. Nonhelical RNA structures, on the other hand, are common, persistent, and crucial to function: typically it is the irregular RNA loops rather than the helices that bind protein, nucleic acid, or ion ligands. The structures of three transfer RNA anticodon hairpin loops are described in detail to illustrate emerging principles of nonhelical RNA structure. Special emphasis is placed on the non-Watson–Crick interactions that stabilize irregular conformations.

Before delving into the details of three-dimensional structures, we briefly review the nomenclature and physical forces that determine RNA conformation.

### 1.2 Nomenclature of RNA Structure

Unlike DNA, which usually consists of two fully Watson–Crick-paired strands, most RNA is found as a single strand, normally a linear chain of the four nucleotide monomers adenine (A), guanine (G), cytosine (C), and uracil (U). Each nucleotide monomer consists of a side chain (a heterocyclic base) bonded to the sugar ring of a sugar–phosphate backbone (Figure 1). Note the presence of the ribose sugar 2′-OH group, which is an H atom in DNA. The nucleotide units are linked head-to-tail, with the O3′ atom of residue i bonded to the phosphorus atom of residue i+1. This configuration of the RNA chain imparts a directionality to it, so that, for example, 5′ AGG 3′ is not the same molecule as 3′ AGG 5′. The three-dimensional structure of an RNA molecule is fully determined by specifying for each nucleotide seven torsion angles describing orientations about rotatable single bonds. Six of these torsion angles ($\alpha$, $\beta$, $\gamma$, $\delta$, $\varepsilon$, and $\zeta$) describe the conformation of the backbone

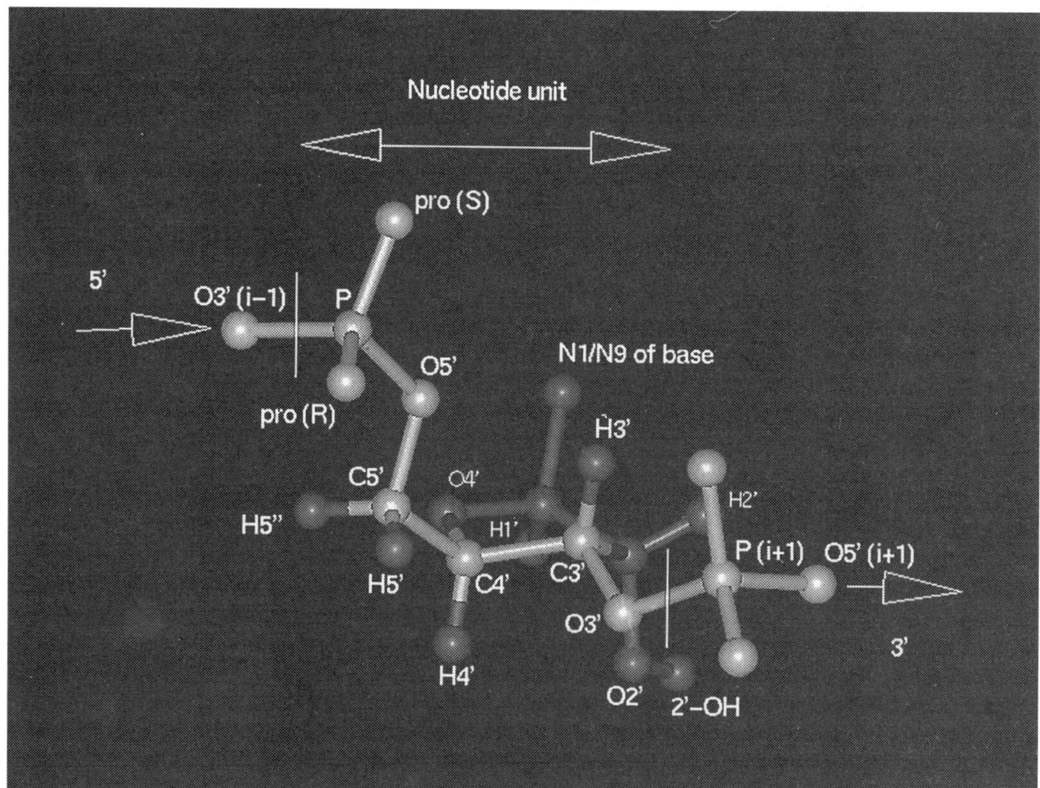


Figure 1. Covalent configuration and nomenclature of RNA. One strand of RNA consists of a linear chain of nucleotide units linked O3′ (residue i-1) to P (residue i), where numbering of residues starts at the left-hand (5′) end. The pro-(*R*) and pro-(*S*) nonesterified phosphate oxygens share a negative charge. The conformation shown is that seen in Watson–Crick-paired double helices. For clarity, only the N1 (pyrimidine)/N9 (purine) atom of the heterocyclic base is shown.

atoms (Figure 1, in white), while one torsion angle ($\chi$) describes the orientation of the base with respect to the ribose sugar. Angle $\alpha$ describes the orientation about the P—O5′ bond, $\beta$ that about the O5′—C5′ bond, and so on, with $\zeta$ (O3′—P) at the end of the nucleotide unit. While this description of RNA structure is very useful for classification of different loop structures, it says nothing about the specific interactions that stabilize the three-dimensional structure.

### 1.3 Physical Forces Determining RNA Conformation

The two most important noncovalent forces affecting nucleic acid conformation are hydrogen bonding and stacking of the aromatic bases. The environment, as determined by interactions with proteins, water, or counterions, is also of great importance, but a detailed description of intermolecular interactions with RNA is beyond the scope of this contribution. Crystallographic studies of oligonucleotides and transfer RNAs reveal that owing to short-range dipole-induced dipole forces, the bases are almost always well stacked. The energy associated with such stacking interactions depends on the identity of the bases and the detailed geometry of the interaction; generally the purine (G and A) bases have a greater tendency to stack than the pyrimidine (C and U) bases. Hydrogen bonds are divided into three categories: base–base, base–backbone, and backbone–backbone. The geometry of Watson–Crick-paired helices allows direct hydrogen bonding between base functional groups (base–base) only. The irregular backbone conformation of loops can allow, in general, close approach of all functional groups, so that all three categories of hydrogen bonds are found. Another kind of backbone–backbone interaction is the close approach of phosphate groups, which may create an ion-binding site. Sections 2.1–2.3 describe the structures of several RNA hairpins in detail, to illustrate specific examples of such interactions and some of the recurrent structural motifs found in nonhelical structures.

## 2 A COMPARISON OF THREE TRANSFER RNA ANTICODON HAIRPIN LOOPS

The differences in conformation between the anticodon loops from three different transfer RNAs (Figure 2: see color plate 16) illustrate

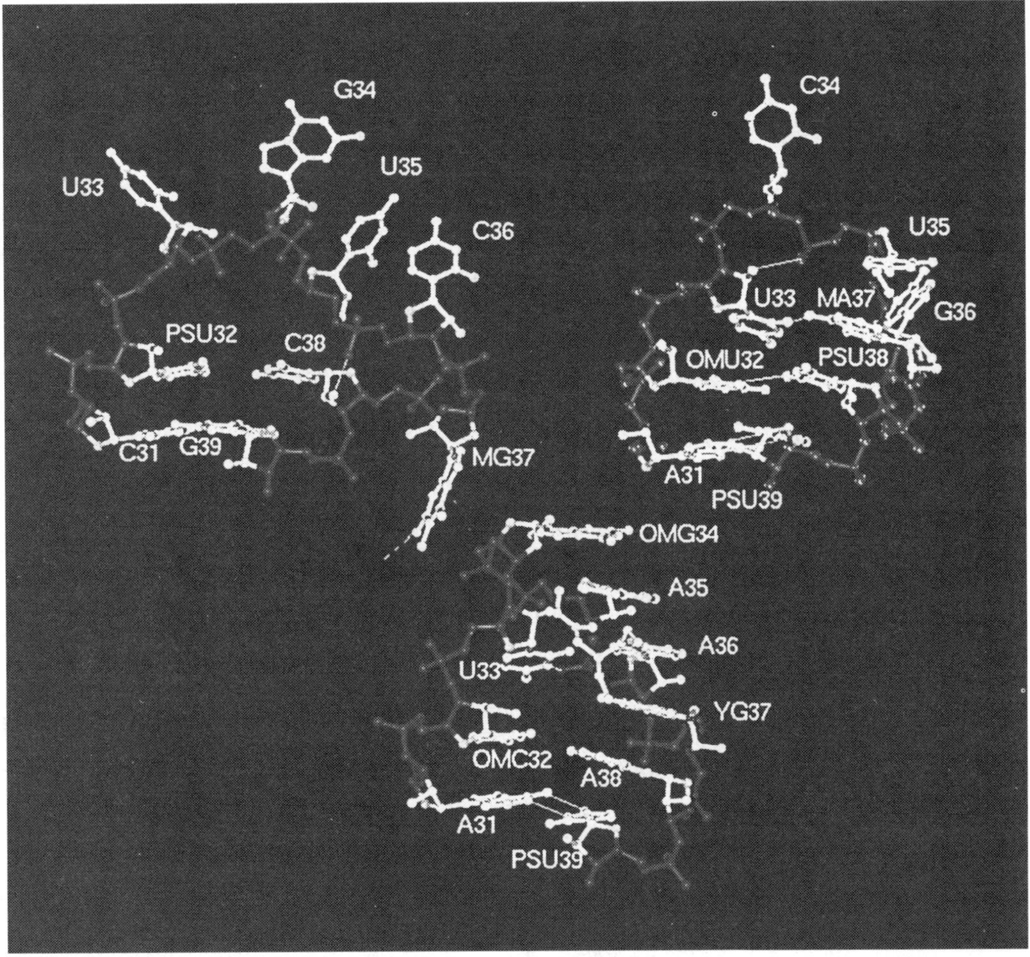


Figure 2. Anticodon hairpin loops from three different crystal structures. *Upper left:* the anticodon loop from the $tRNA^{Asp}$–cognate synthetase cocrystal structure; *upper right,* that from the $tRNA^{Gln}$–cognate synthetase cocrystal structure; *bottom,* that from the uncomplexed $tRNA^{Phe}$ structure. For each structure, the last Watson–Crick base pair from the helix is at bottom. Hydrogen bonds are drawn as thin white lines, while the backbone is shown in green and the bases in yellow; O4′ atoms are shown in red to highlight the local direction of the backbone. For example, note the abrupt turns between residues 33 and 34 ($tRNA^{Phe}$; bottom) and between residues 36 and 37 ($tRNA^{Asp}$; upper left). [See color plate 16.]

the flexibility of large hairpin loops. The loop at upper left is from the tRNA$^{Asp}$–cognate synthetase cocrystal structure, that at upper right is from the tRNA$^{Gln}$–cognate synthetase cocrystal structure, and that at bottom is from the free yeast tRNA$^{Phe}$ crystal structure. The conformations of the free tRNA$^{Asp}$ and tRNA$^{Gln}$ loops are similar to that of the tRNA$^{Phe}$ loop. Therefore, the differences in conformation between the protein-bound loops and the uncomplexed tRNA$^{Phe}$ structure may be interpreted as conformational changes that occur upon binding of the synthetase.

### 2.1 tRNA$^{Phe}$

The tRNA$^{Phe}$ anticodon loop (Figure 2, bottom) has long been considered to be the canonical seven-residue hairpin loop. The structure features helical stacking of the five bases (residues 34–38) at the 3′ side of the loop, and a tight turn motif involving residues 33–36, where the direction of the backbone is reversed [note the difference in orientation of the U33 and OMG34 O4′ atoms (red)]. Because it is in a helical conformation, the anticodon (residues 34–36) is preformed for base pairing to the messenger RNA codon. The tight turn is believed to be a very common structural motif. This ''U-turn'' is stabilized by both base–backbone and backbone–backbone interactions: (1) a hydrogen bond from a U33 base proton to the A36 phosphate, (2) stacking of the A35 phosphate on the U33 base, and (3) a hydrogen bond from the U33 2′ OH to a base nitrogen from A35. Inversion of chain direction is achieved by a rotation of a single torsion angle, α (residue 34), to an unusual value. Finally, note the stacking of both the first and last loop nucleotides (residues 32 and 38) on the adjacent helix.

### 2.2 tRNA$^{Asp}$

As a result of synthetase binding, the other two anticodon loops shown in Figure 2 have very different conformations, with additional tight turns. The tRNA$^{Asp}$ anticodon loop (upper left) could be considered a three-residue hairpin loop (residues 33–35) next to a two-residue bulge (residues 36 and 37). Stacked as they are on the adjacent helix, residues 32 and 38 are in effect a helical extension. The reversal of the chain occurs gradually over residues 33–36, while two tight turns are found in the bulge (i.e., between residues 36 and 37, and between residues 37 and 38). The tight turns in the bulge are stabilized by two hydrogen bonds: the first is between the 2′-OH of residue 38 and the phosphate of residue 36; the second is between a base proton of residue 37 and a phosphate (not shown) from the anticodon stem. Note the close approach of phosphate groups in the bulge.

### 2.3 tRNA$^{Gln}$

Though superficially rather different in conformation, the tRNA$^{Gln}$ anticodon loop has in fact a conceptually similar structure. Unlike the tRNA$^{Asp}$ loop, synthetase binding to the tRNA$^{Gln}$ loop induces the formation of mismatched base pairs between residues 32 and 38, and 33 and 37. These mismatches are stabilized by direct base–base as well as by water-mediated base–base hydrogen bonds. As in the tRNA$^{Asp}$ loop, a backbone–backbone hydrogen bond is found to stabilize a bulged-out conformation, but the location of this feature is different. The hydrogen bond, found between the 2′-OH of residue 33 and the phosphate of residue 35, helps direct a large change in direction of the backbone. Additional tight turns are found among residues 35–37; indeed, residue 36 could be considered to be a bulge (note that the base of residue 35 is stacked on that of residue 37). Thus this hairpin loop could be considered to be a two-residue loop (residues 34 and 35) next to a one-residue bulge (residue 36), much as the tRNA$^{Asp}$ anticodon loop could be considered to be a three-residue loop next to a two-residue bulge. The overall structural similarity of these ''anticodon bulge'' conformations is probably due to the need of the synthetase to recognize the anticodon (residues 34–36), which encourages extrusion of these residues for ease of recognition by the synthetase. Bulged residues are known to be recognized by proteins in several other systems.

The magnitude of the conformational changes induced by synthetase binding suggests that the anticodon loop may be quite flexible. This flexibility may facilitate adoption of conformations other than the ''anticodon bulge'' structure: for example, it is probable that the anticodon loop must adopt at least two rather different conformations during protein synthesis on the ribosome. More generally, large loops not already involved in tertiary interactions may change conformation upon binding other RNA or protein molecules, and bulgelike conformations embedded within larger loops may be a common structural motif.

## 3 EMERGING PRINCIPLES OF NONHELICAL RNA STRUCTURE

Some of the generalizations from these and other structures may be summarized as follows.

*Large loops are flexible.* Binding to other RNA or protein molecules may result in large conformational changes. Bulged-out nucleotides embedded in a larger loop may be a common structural motif.

*Small loops are more stable and less flexible.* Smaller hairpin loops (e.g., the common four-nucleotide hairpin loops) are more thermodynamically stable and less flexible than larger hairpin loops. Because of their relative inflexibility, some small hairpin loops may be preformed for specific interactions with other molecules.

*One or more mismatches stabilize the helix–loop interface.* Loop residues closest to a helix generally stack on the helix and may hydrogen-bond to form a mismatched base pair. In some cases, a pairing geometry very different from Watson–Crick geometry may be used to expose functional groups for tertiary interactions.

*Base–backbone and backbone–backbone hydrogen bonds stabilize turns.* Base–backbone hydrogen bonds are common at very tight turns; a common example is the U-turn motif. Hydrogen bonds between the 2′-OH proton and phosphate oxygens are often found to stabilize a bulged conformation; very often the bulged nucleotides are almost perpendicular to adjacent nonbulged nucleotides.

*Base stacking is extensive but may be sacrificed in special cases.* One or more bulged residues may be completely unstacked from a loop, often in order to engage in interactions with other RNA or protein residues.

*See also* Hydrogen Bonding in Biological Structures; RNA Secondary Structures; RNA Three-Dimensional Structures, Computer Modeling of.

## *Bibliography*

Chastain, M., and Tinoco, I., Jr. (1991) Structural elements in RNA. *Prog. Nucleic Acid Res. Mol. Biol.* 41:131–177.

Gutell, R. R., Larsen, N., and Woese, C. R. (1994) Lessons from an evolving rRNA: 16*S* and 23*S* rRNA structures from a comparative perspective. *Microbiol. Rev.* 58:10–26.

Hill, W. E., Moore, P. B., Dahlberg, A., Schlessinger, D., Garrett, R. A., and Warner, J. R., Eds. (1990) *The Ribosome: Structure, Function, and Evolution.* American Society for Microbiology, Washington, DC.

Jaeger, J. A., SantaLucia, J., Jr., and Tinoco, I., Jr. (1993) Determination of RNA structure and thermodynamics. *Annu. Rev. Biochem.* 62:255–287.

Quigley, G. J., and Rich, A. (1976) Structural domains of transfer RNA molecules. *Science,* 194:796–806.

Saenger, W. (1984) *Principles of Nucleic Acid Structure.* Springer-Verlag, New York.

Varani, G., and Tinoco, I., Jr. (1991) RNA structure and NMR spectroscopy. *Q. Rev. Biophy.* 24:479–532.

# RNA Three-Dimensional Structures, Computer Modeling of

*François Major*

*Key Words*

**Base Pairing** Hydrogen bonding between bases of two polynucleotide chains. In a double helix, the two polynucleotide chains are connected by complementary base pairing.

**Intron** Intervening sequence removed during the maturation of messenger RNA.

**rRNA** Ribosomal RNA.

Since knowledge of the three-dimensional structure of RNA is crucial to the understanding of RNA function, modeling methods are being developed to transform the information derived from experimental data and inferred from sequence analysis into such structures. The predictive nature of these methods is evaluated by their ability to reproduce known structures. Resulting models can then be used to guide investigations of RNA structures by physical and theoretical means. The representation of RNA structural knowledge, and its efficient use in the exploration of the conformational space of RNA, are crucial aspects of computer modeling methods.

## 1 INTRODUCTION

The biology activity of RNA involves atomic interactions in three-dimensional space with other chemical agents, including proteins, DNA, and other RNA molecules. Since structural features reveal chemical function, it is believed that knowledge of the three-dimensional structures of RNA can provide insights into RNA's biological role. The most accurate method for determining such structures at the atomic level is X-ray crystallography. However, this method has been used to determine the three-dimensional structure of only one biologically active RNA, that of transfer RNA (tRNA).

Research into the development of new methods for the determination and prediction of the three-dimensional structure of RNA has intensified since the discovery of unexpected RNA properties (such as self-splicing and enzymatic activity) and the identification of an increasing number of important small RNAs.

The RNA base-pairing pattern, or secondary structure, juxtaposes an important fraction of nucleotides, distant in the sequence, to form regular double helices. Computer programs that encode empirically derived free energy parameters about base pairing and stacking can be implemented to predict the secondary structure of RNA by free energy minimization. At present, though, comparative sequence analysis, or phylogeny, is still the more reliable and productive method. For example, the early consensus secondary structures for tRNA, 5S, 16S, and 23S ribosomal RNA (rRNA), the M1 component of RNAse P, and group I introns have been inferred from phylogeny studies.

## 2 RNA STRUCTURAL INFORMATION

The determination of three-dimensional structures for RNA relies on the identification of base-pairing and tertiary interactions, which stabilize their native conformation. Such structural information can be identified using computer programs that find covariations between distant nucleotides in a set of prealigned, evolutionarily related sequences. A covariation occurs when two nucleotides in two given positions preserve a base-pairing. The probability that the two positions actually form a base pair is high if such covariations occur in many sequences. Prior to the determination of the tRNA crystal structure, secondary and tertiary interactions from 14 different tRNA sequences were inferred and used by many in manual three-dimensional modeling. None of these models reproduced the actual "L" shape of tRNAs, but most predicted interactions were correct. More recently, nucleotide covariations have been studied in tRNA, introns, and rRNA.

Experimental methods are the most important source of RNA structural information, even though sequence databases are increasing so greatly in size that many deduced interactions are fully supported by sequence data alone. Nuclear magnetic resonance data and models have been produced for loops and small domains that are important elements of larger biologically active RNA molecules, such as the common UUCG tetraloop; the 5S rRNA common arm, V-loop, and helix I; the eukaryotic 5S rRNA E-loop; the transactivation response (TAR) RNA-arginine complex; the TYMV RNA pseudoknot; and the rRNA 28S α-sarcin stem-loop. NMR methods also have been used to probe the base pair formation in several hammerhead ribozymes and 5.8S rRNA, and in the study of unconventional base-pairing stability.

Promising techniques for more widespread use of NMR spectroscopy with RNA are being developed. This approach is particularly interesting because it provides Euclidean distances measured between two hydrogen protons [nuclear Overhauser effects (NOE) restraints] that can be used for the deduction of RNA structural features, although the determination of RNA backbone conformation is more difficult since it lacks hydrogen atoms.

Proximity in three-dimensional space can sometimes be determined by cross-linking with ultraviolet irradiation. Ultraviolet irradiation can form covalent bonds where hydrogen bonds (base pairs) occur, which allows for their detection. This kind of information is extremely useful for three-dimensional modeling and has been used in the modeling of the three-dimensional structure of 16S rRNA. Another way to determine the relative position of nucleotides is to evaluate their participation in chemical reactions, using enzymatic or chemical probes, and nucleotide substitution. The translation of data of this type into three-dimensional constraints is complicated by the fact that local modifications could alter structure, lead to further degradation, and therefore disrupt RNA function. Nevertheless, chemical and enzymatic probing data have been derived and used for an investigation of the conformation of domain III of 16S rRNA, for the determination of the secondary structure of *Tetrahymena* rRNA, for the derivation of a three-dimensional model of U1 small nuclear RNA (snRNA), for the analysis of higher order structure of the M1 component of ribonuclease P, and for the secondary structure and a three-dimensional model of 5S rRNA.

Noller et al. have used chemical and enzymatic probes to test their secondary structure model of 23S rRNA and to complete the secondary structure of 16S rRNA initiated by comparative sequence analysis. Studies of the role of 2′-OH groups in catalytic activity of the hammerhead ribozyme and *Tetrahymena* ribozymes, and of the M1 component of RNAse P, have been made by comparing the activity of mixed ribose–deoxyribose analogues.

Electron microscopy has been used for the three-dimensional reconstruction of eukaryotic ribosomes. Verschoor and Frank have recently calculated a three-dimensional model of the 80S ribosome from low dose electron micrographs of negatively stained single particles. The resolution of structures reconstructed from electron micrographs is much lower than those determined by X-ray or NMR methods. Nevertheless, one can still discern the precise orientation and relative positions of subunits and proteins surrounding the RNA.

## 3 COMPUTER MODELING

For nearly three decades, computer programming has been used as a medium for the representation and manipulation of chemical structures. The goal of computer modeling is to develop and implement conceptual models so that RNA structural data, as described earlier, can be organized into three-dimensional structures. From Figure 1, it can be seen that the production of structural information is intimately related to our ability to generate three-dimensional models. Such modeling can be used to test structural hypotheses and suggest future experiments. The most common methods for computer modeling of RNA are presented and compared according to criteria summarized in Table 1.

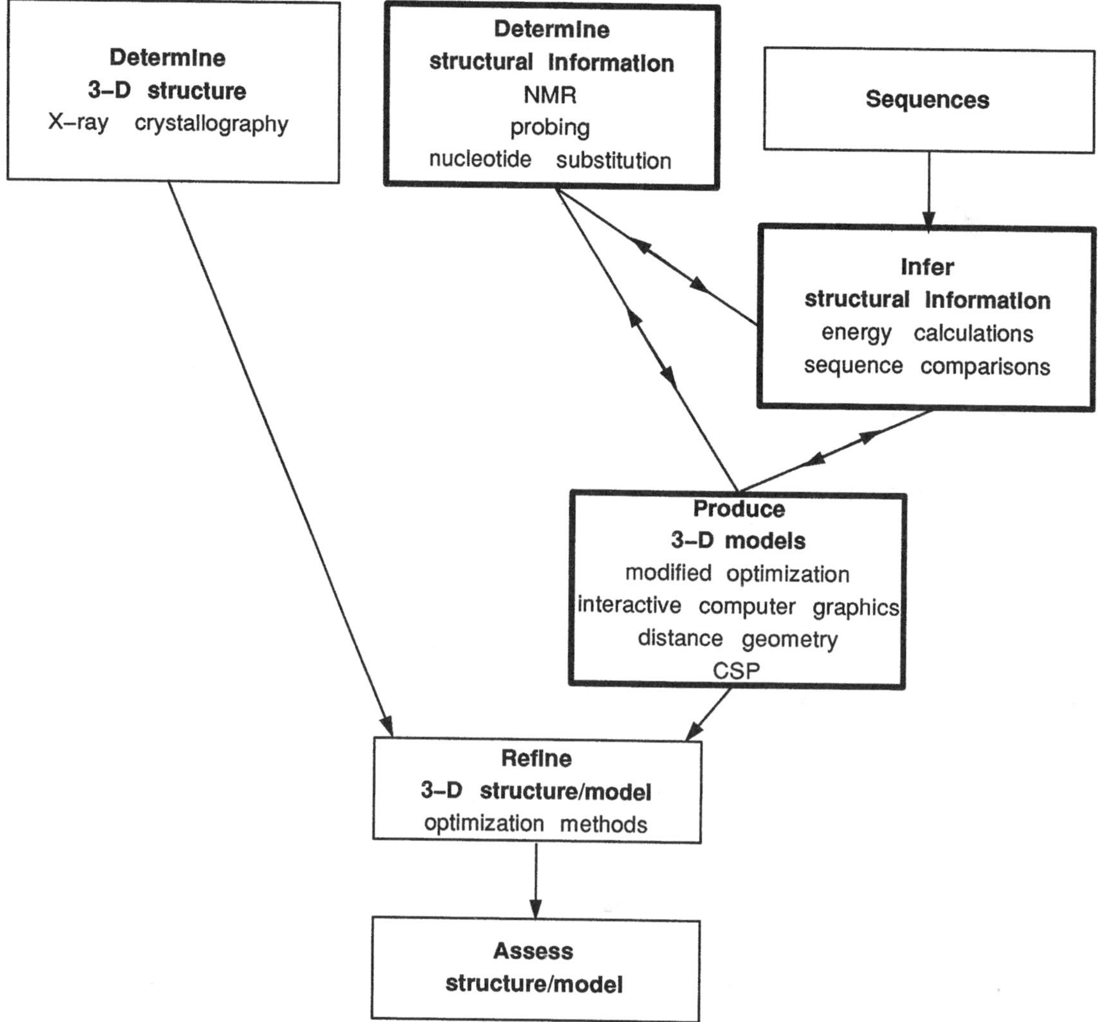


**Figure 1.** An approach to RNA structure determination and prediction. Current computer methods for producing models require structural information determined from experimental results. Future improvements in theoretical methods could permit the determination of RNA structures from sequence only.

**Table 1** Summary of Some Aspects of Computer Approaches to RNA Modeling

| Method[a] | Local Optimization | Constraint | Atomic Model |
|---|---|---|---|
| EM | Yes | Objective function | Yes |
| MD | Yes | Objective function | Yes |
| MC | No | Objective function | Yes |
| SA | Yes | Objective function | Yes |
| ICG | No | User | Yes |
| MMM | No | Objective function | No[b] |
| DG | No | Distance matrix | No[b] |
| CSP | No | Constraint function | Yes |

[a]Models produced by methods that do not perform local optimization are refined using EM or MD. SA is nothing more than MD or MC with heating (for conformational sampling) followed by cooling (for optimization). Only CSP is applying a systematic conformational sampling. All atoms must be grafted to non-all-atom models before local optimization is applied.
[b]Can address all atoms with small RNA molecules.

### 3.1 Numerical Methods

The first category of conceptual models uses an objective function, $f$, to represent the potential energy of a molecule as a function of the relative positions of its constituent atoms. Specific models of $f$ for RNA implement approximations obtained from experimental and quantum mechanical studies. Typically $f$ is a sum of weighted terms for bond lengths and angles, and noncovalently bonded interactions such as repulsion and dispersion. NOE and other available distance constraints can be represented in $f$ by adding terms so that $f$ is penalized when the constraints are not satisfied.

According to classical thermodynamics, a molecular system is most stable at the minimum of potential energy. If $f$ represents the potential energy of the system, then determining the most stable state of a molecular system consists of minimizing $f$. The search for an optimal state is performed by energy minimization (EM), in which the molecule is considered as a mathematical vector, $\boldsymbol{x} = (x_1, x_2, \ldots, x_{3n})$, where $n$ is the number of atoms in the molecule. Since the nature of $f$ is nonlinear (neither convex nor concave) and allows for a large number of local minima, nonlinear programming techniques for such functions will converge to a local minimum that depends on the initial structure. Thus, in practice EM is most useful for local refinement of RNA models (i.e., to solve steric conflicts and to adjust bond lengths and angles) but is impractical for global optimization.

Another method that uses an objective function is molecular dynamics (MD). The difference between MD and EM is that MD considers each atom as a point mass that moves according to the forces exerted on it by all other atoms. To compute the trajectory of the atoms, the equations of motion of classical mechanics are solved numerically. MD at low temperatures has been used as EM. Mei, Kaaret and Bruice have used EM and MD to produce a three-dimensional model of the hammerhead ribozyme.

Theoretically, a systematic grid search over all $3n$ components of $\mathbf{x}$ would overcome the local minima problem. In practice, however, except in the case of very small RNAs, the combinatorial nature of the problem makes it impossible to explore more than a tiny fraction of the conformational space. Nondeterministic methods such as Monte Carlo (MC) and simulated annealing (SA) are used to randomly sample this huge conformational space. Success has been reported only when initial structures are close to the actual structure.

### 3.2 Interactive Computer Graphics

In the absence of automatic methods, global search can be accomplished by interactive computer graphics (ICG). Programs like Insight, Quanta, Macromodel, and FRODO are used to explore interactively the conformational space of molecules. These programs allow for the visualization in three dimensions of the molecule and the modification of any bond length and angle so that the entire conformational space is accessible via the display screen, a mouse, dials, and a keyboard. In this way, all types of structural data and RNA structural rules can be considered and do not have to be explicitly represented in the computer. Further simplifying the task are the large number of nucleotides involved in double helices, which can be considered as rigid objects. Interactive computer graphics is an attempt to mimic physical modeling, which is usually performed using pieces of wire, wood, or plastic. The most important physical model of a macromolecule ever put forward is certainly that of Watson and Crick published in 1953.

The first step of ICG and physical modeling consists of building a sketch model in which the secondary structural elements are roughly positioned to satisfy some constraints. Next, refinements consisting of slight modifications to the position and orientation of the double helices are made to satisfy local constraints and to accommodate the modeling of loop regions. Using a computer, steric conflicts can be eliminated at any time in the process by invoking EM. Using physical modeling, steric conflicts are usually eliminated by visual examination. The process stops when a satisfactory state is reached, that is, when all known constraints are satisfied. So far, this method has produced the most models. Atomic models derived by ICG (sometimes combined with physical modeling) include the tRNA-like 3′ end of turnip yellow mosaic virus, yeast $tRNA^{SER}$, 16S and 5S rRNAs, and group I catalytic introns. Because of its success and great flexibility, the ICG approach is very attractive.

### 3.3 Modified Molecular Mechanics

In optimization procedures, the consideration of structural constraints, such as those used in ICG, introduces a "weighting" problem. If the weight of a constraint term is too large, the term is overemphasized and the procedure moves rapidly to a point at which the constraints are satisfied, even though this point is far from a minimum. On the other hand, if the weight is too small, the constraint is almost not considered. In a modified molecular mechanics (MMM) approach proposed by Malhotra, Tan, and Harvey, a new objective function composed only of constraint terms is created, to emphasize the importance of long-range interactions and to simply the weighting task. Combined with a simplification of nucleotide representation by one to five points (pseudoatoms), this procedure is capable of folding initial random walks to a range of conformations consistent with given experimental data. The method was first tested by producing tRNA models that satisfied all input data, and the investigators then reported 16S rRNA models similar to the one proposed by Stern, Weiser, and Noller using similar input data. Once all the atoms have been grafted to the produced models, a conventional molecular mechanics approach must be applied to obtain low energy models.

### 3.4 Distance Geometry

Distance geometry (DG), a branch of mathematics developed by Blumenthal in the early 1970s, is another approach used to determine positions in three-dimensional space. It concerns the representation or reproduction of structures from internal distances. Distances between $n$ points are represented in a symmetric matrix $M$ of size $n \times n$. A process called embedding reproduces the position of the $n$ points in three-dimensional space.

Crippen in 1977 was the first to relate DG and molecular modeling. The procedure, however, must be slightly modified to accommodate distance intervals produced by the NMR method. An asymmetric matix $M'$ is first defined by introducing the lower bounds of the intervals in the lower triangular part of $M'$ and the upper bounds of the intervals in the upper triangular part. At this point, secondary structure distances also can be introduced in $M'$. Since in practice only a small fraction of $M'$ is filled from experimental and secondary structure data, $M'$ is extended by a process called bound smoothing, which uses the triangle inequality. $M$ is produced by fixing, usually randomly, a value from each distance interval of $M'$.

DG has proved successful for proteins when more than one constraint per degree of freedom is expressed. For large RNA, the number of degrees of freedom must be reduced considerably to avoid the manipulation of extremely large matrices. Thus, nucleotides, and sometimes fragments, are represented by only a few points, or pseudoatoms. In this way, Hubbard and Hearst have obtained good results using secondary structure information and a subset of long-range interactions that were known for tRNA prior to the crystal structure. They then applied the method to the naked 16S rRNA (where no proteins are present) and reported that all models they derived satisfied the known constraints. To obtain low energy models, EM or MD must be applied to the product of DG, since imprecision is introduced during the application of bound smoothing and the selection of random fixed values.

### 3.5 Constraint Programming

Another method is to formulate modeling as a general constraint satisfaction problem (CSP), defined by a set of variables to which are assigned values selected from predetermined value domains, as well as a set of constraints indicating which values can be assigned to the variables. Solving a CSP consists of finding all variable assignments so that all constraints are satisfied. Recently, the RNA modeling problem has been defined as a CSP by Major et al. In their mapping, each nucleotide in an RNA is considered a variable. The value to be assigned to each nucleotide is the Cartesian product of a nucleotide conformation and a spatial transformation. Primary and secondary structure information determines the number of variables and the possible conformations and transformations to be assigned to each nucleotide. Tertiary information is introduced in a ''constraint'' function applied each time a new nucleotide value is assigned. The algorithm performs efficiently by eliminating partial models that contain inconsistencies, and all the models produced are consistent with the input data. Furthermore, this conceptual model of RNA modeling allows for the representation of all types of structural data. The current implementation has been used to reproduce known hairpin loop pseudoknotted structures, and tRNA, demonstrating the feasibility of the approach. The precision of the all-atom (except hydrogen) models produced is in the range of 1 to 4 A of rms deviation with their respective consensus structures. EM is applied to the solutions to obtain low energy structures.

## 4 PERSPECTIVES

All the methods presented here are productive and useful in different contexts, although improvement is necessary in each case since highly underdetermined models almost always result. We expect that the next generation of computer programs will be able to take a simple description of available RNA structural data as input and produce the most likely three-dimensional models. Such programs could automatically choose the appropriate modeling method, or the right combination of methods, according to the input data. The implementation of such programs would be simplified by the establishment of a unified model of RNA structural knowledge. Such a unified model is likely to be the result of multidisciplinary work involving computer scientists, biophysicsts, and molecular biologists. Intuitively, this work could be justified and motivated by observing that RNA modeling by ICG is successfully suggesting ways to explore efficiently the conformational space of RNA and is producing high precision models.

*See also* RNA Secondary Structures; RNA Structure, Nonhelical.

### *Bibliography*


Bazaraa, M. S., and Shetty, C. M. (1979) *Nonlinear Programming: Theory and Algorithms.* Wiley, New York.

Blumenthal, L. M. (1970) *Theory and Applications of Distance Geometry.* Chelsea, New York.

Gautheret, D., and Cedergren, R. (1993) Modeling the three-dimensional structure of RNA. *FASEB J.* 7:97–105.

Kumar, V. (1992) Algorithms for constraint satisfaction problems: A survey. *AI Mag.* 13:32–44.

Major, F., Turcotte, M., Gautheret, D., Lapalme, G., and Cedergren, R. (1991) *Science* 253:1255–1260.

Malhotra, A., Gabb, H. A., and Harvey, S. C. (1993) Modeling large nucleic acids. *Curr. Opinion Struct. Biol.* 3:241–246.

McCammon, J. A., and Harvey, S. C. (1987) *Dynamics of Proteins and Nucleic Acids.* Cambridge University Press, Cambridge.

Saenger, W. (1984) *Principles of Nucleic Acid Structure.* Springer-Verlag, New York.

Varani, G., and Tinoco, I., Jr. (1991) RNA structure and NMR spectroscopy. *Q. Rev. Biophys.* 24:479–532.


# Robotics in Genome Research: *see* Automation in Genome Research.

# S

## Scanning Tunneling Microscopy in Sequencing of DNA

*T. L. Ferrell, D. P. Allison, T. Thundat, and R. J. Warmack*

*Key Words*

**Piezoelectric Crystal** A crystal that changes its volume in response to an electric field applied between electrodes attached to the crystal.

**Plasmids** Small, circular, self-replicating DNA molecules found in bacteria.

**Sequencing** Identifying the order of nucleotide bases along the length of a DNA molecule.

**Tunneling** The phenomenon of penetration of a barrier by a wave under conditions in which a classical particle of the same energy would not penetrate the barrier.

An ongoing international effort to map and sequence the human genome will have a profound effect on future biological research and medical practice. Genes responsible for inherited diseases will be discovered and diagnostic procedures implemented. A better understanding of gene organization and regulation will add to our understanding of human development and aging. However, sequencing the 3 billion nucleotide bases that make up the human genome is a monumental task that will require a significant expenditure of time and money. Therefore the development and evaluation of labor-saving and cost-effective new technologies for sequencing DNA, such as scanning–tunneling microscopy, is an important issue.

### 1 STM Operation

Scanning–tunneling microscopes (STMs) measure the topographical profile of samples and, unlike conventional light or electron microscopes, do not require transmitted waves or lenses for their operation. Imaging is accomplished by detecting changes in the electronic tunneling current between a sharpened tip (mounted on a piezoelectric crystal) and a conducting sample as the tip is serially scanned over the sample surface (Figure 1). A measurable tunneling current is generated when the wave functions of the electrons on the tip have sufficient overlap with the wave functions of the electrons on the sample surface. By applying a difference in electrical potential between the tip and the sample, a tunneling current can be maintained. There is a strong exponential dependence of the tunneling current on the distance between the tip and the sample. When a piezoelectric crystal is used to position the tip within a nanometer of the surface, the current changes by an order of magnitude if the gap between the tip and sample changes by only 0.1 nm. The effect of this strong gap-dependent current is to concentrate the current into a very tiny region only 0.1 to 0.5 nm wide. For electrical conductors and semiconductors, which have abundant charge carriers, atomic resolution is routinely obtained if the surface is flat. Deviations from flatness on a nanometer scale cause switching of the tunneling current to other regions of the tip during a scan. As a result, the tip profile smears local topographical data and degrades the resolution. These effects on the signal set limitations on the topographical relief of samples, a fact that must be recognized in sample selection and image treatment.

#### 1.1 Conductivity

Electrical conductivity of the sample is another limitation that must be considered in the selection of biological samples for scanning–tunneling microscopy. Inevitably, biomolecules tend to be either electrical insulators or poor semiconductors, and this property poses a problem when charged particles are used as probes. If the molecule cannot bleed away excess charge sufficiently, it will distort the results of probing by electrons. Furthermore, electrical insulators typically do not dissipate heat readily, and energy deposited by any probe can cause disintegration. This problem is overcome in conventional scanning electron microscopes by coating samples with a thin metal film; since, however, the highest possible resolution will be necessary to sequence DNA in the electron STM, uncoated samples must be placed directly on a conducting substrate. Electron tunneling can occur through an insulator as thick as 4 to 5 nm, and since the DNA double helix is only about 2 nm thick, a tunneling image can be obtained. However, if the reduction in tunneling current is too great, the electronic feedback system will plunge the tip into the sample. Special precautions must be taken to understand what is happening in a given case.

#### 1.2 DNA Imaging

Because deoxyribonucleic acid molecules contain the genetic code for most living organisms, including man, considerable interest has arisen in the scanning–tunneling microscopy of DNA. One primary objective is to determine the nucleotide base sequence and therefore rapidly and accurately read the genetic code. As a starting point for imaging, the DNA molecules must be placed on a flat conductive surface. Historically, atomically flat highly ordered pyrolytic graphite (HOPG) or gold surfaces were used. However, these surfaces are crystalline and usually chemically inert. Therefore, they lack sites to hold DNA strongly enough to resist removal by the physical and electronic forces exerted by the tunneling tip.

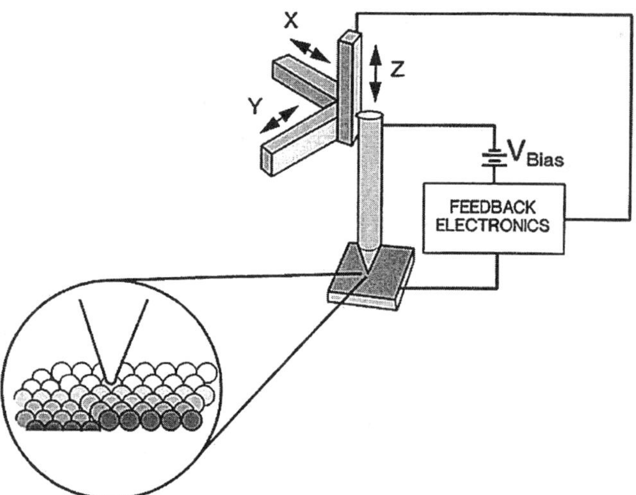


**Figure 1.** Schematic of a scanning=ntunneling microscope. While a constant tunneling current is maintained, an atomically sharp tip is scanned over a sample surface and an electronic image is created.

A secondary problem initially encountered in imaging DNA on these surfaces was that structural defects intrinsic to the substrates could be confused with DNA, since both were seen only occasionally. This was especially true when short fragments of DNA were being imaged, and on surfaces such as HOPG, which bear some irregularities that closely mimic DNA. However, DNA molecules can be positively identified by exploiting differences in conductivity between DNA and mounting surfaces. A dc bias voltage between the tip and the sample is typically applied to maintain tunneling and acquire a topographic image. It is also electronically possible to superimpose an ac bias voltage between the tip and the sample and simultaneously record the derivative of the tunneling current with respect to the applied bias voltage. This procedure yields an image of the local electronic conductance of the material directly beneath the probe tip. There is a large difference noted between DNA and surface artifacts.

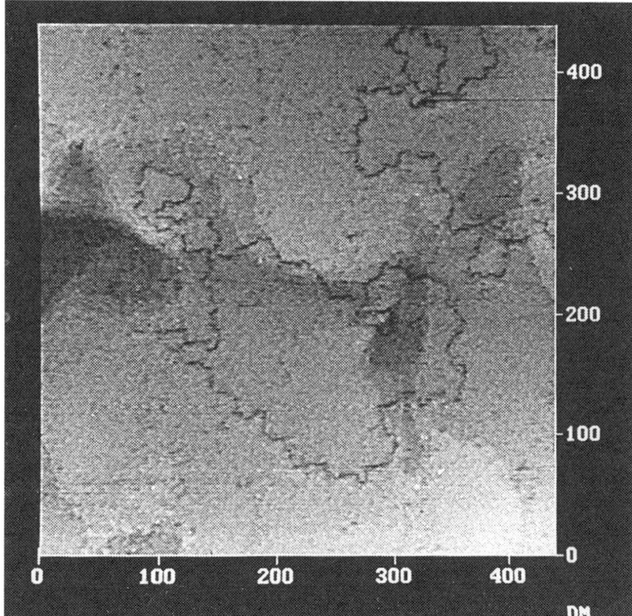


**Figure 2.** High magnification STM image of an open circular pBS$^+$ plasmid DNA molecule chemically bound to a gold surface with 2-dimethylaminoethanethiol. Now that stably bound DNA has been achieved, attention can be focused on improvements in resolution.

### 1.3 STM Techniques

Both chemical and electrochemical techniques have been used to attract and hold DNA to surfaces. One of the more successful techniques, which resulted in the first STM images of entire genetically functional plasmid molecules (Figure 2), involves the coulostatic immobilization of DNA onto chemically modified gold surfaces. To this end, thiol organic compounds are reacted with the gold substrate, and positively charged amino groups on the other end serve as molecular tethers to attract and hold the negatively charged DNA molecules. Such techniques immobilize the biomolecules so well that the DNA is observed at the surface coverage expected, and confusion with rare artifacts is eliminated.

## 2 STM PROGRESS ACHIEVED

### 2.1 Direct Sequencing

Although significant progress has been made in imaging DNA with the scanning–tunneling microscope, sequencing by STM has not been accomplished. Direct sequencing would require denaturation of double-stranded DNA and the imaging at very high resolution of individual bases on one of the resulting single strands. Isolated bases have been imaged by STM with sufficient molecular resolution to identify the double-ring structure of the purine bases (adenine, guanine) and the single-ring pyrimidine bases (thymine, cytosine). Though these results are encouraging, it would require atomic resolution to differentiate topographically between the purine and the pyrimidine bases. For example, separating cytosine from thymine would depend on recognizing the subtle differences of ring positions 5 and 6, since thymine has a methyl group at position 5 and a carbonyl oxygen at position 6, while cytosine has an amino group at position 6. By either physically or chemically tagging one purine and one pyrimidine base, however, all four bases could be identified topographically at molecular resolution.

### 2.2 Secondary Techniques

Alternatively, it may be possible to employ secondary techniques along with topographic imaging to identify nucleotide base sequences. We have already described how topographic and differential conductivity images can be generated simultaneously to differentiate between poorly conductive DNA molecules and surface artifacts. This technique could be refined and extended to the level at which differences in topographic images of the bases could be combined with additional high resolution electronic spectroscopy to perhaps detect conductivity differences in the bases so that the sequence could be visualized through contrast changes.

The incorporation of secondary techniques into scanning probe microscopes, for sequencing DNA, may be extended to the development of hybrid instruments. The photon scanning–tunneling microscope (PSTM), an instrument developed in our laboratory, is analogous to the scanning–tunneling microscope except that electrons are replaced by photons. Instead of detecting the electron tunneling current between an atomically sharp conducting tip and a sample surface, as the tip is scanned over the surface, the PSTM uses a transparent probe tip to collect the frustrated total internal reflection of photons in a sample. The photons are converted to an electrical signal by a photomultiplier tube and processed by the STM control electronic package. Although the spatial resolution of the PSTM is roughly tens of nanometers for visible light, and not as good as the STM, the optical spectroscopic resolution of the PSTM is better than electron spectroscopy by a factor of nearly $10^6$. This instrument demonstrates that the wavelength limit of conventional microscopy is not a fundamental limitation. Therefore, a hybrid instrument combining the high resolution imagery capabilities of STM with sensitive optical spectroscopy of fluorescence to a local optical probe may be possible. This challenging new area of research, combining other analytical techniques to probe microscopies, has attracted the attention of many laboratories. The near-field optical microscope has been improved to follow surface topography and record optical data at a resolution of better than 20 nm. The combination of atomic force microscopy with optical capabilities has already been described, and instruments that can be used to map either fluorescently labeled homologous nucleotide sequences or site-specific proteins to DNA, or even to sequence DNA rapidly and accurately may be developed and perfected.

## 3 CONCLUSION

It is obvious that the visual sequencing of DNA requires the development of new sample preparation methodologies and new instrumentation. There are many promising routes of investigation for such developments, and exciting new discoveries and inventions can be anticipated in the next few years.

*See also* Electron Microscopy of Biomolecules; Nucleic Acid Sequencing Techniques.

### *Bibliography*


Allen, M. J., Balooch, M., Subbrah, S., Tench, R. J., Sickhaus, W., and Balhorn, R. (1991) *Scanning Microsc.* 5:625.

Allison, D. P., Thompson, J. R., Jacobson, K. B., Warmack, R. J., and Ferrell, T. L. (1990) *Scanning Microsc.* 4:517.

———, Bottomley, L. A., Thundat, T., Brown, G. M., Woychik, R. P., Schrick, J. J., Jacboson, K. B., and Warmack, R. J. (1992) *Proc. Natl. Acad. Sci. U.S.A.* 89:10129.

———, Warmack, R. J., Bottomley, L. A., Thundat, T., Brown, G. M., Woychik, R. P., Schrick, J. J., Jacobson, K. B., and Ferrell, T. L. (1992) *Ultramicrosc.* 42–44:1099.

Betzig, E., and Trautman, J. K. (1992) Near-field optics: Microscopy, spectroscopy, and surface modification beyond the diffraction limit. *Science* 257:189–195.

Binnig, G., and Rohrer, H. (1982) *Helv. Phys. Acta* 55:726.

Heckl, W. M., Kallary, K. M. R., Thompson, M., Gerber, C., Hörber, H. J. K., and Binnig, G. (1989) *Langmuir* 5:1433.

Lindsay, S. M., and Barris, B. (1988) *J. Vac. Sci. Technol.* A6:544.

Reddick, R. C., Warmack, R. J., and Ferrell, T. L. (1989) *Phys. Rev. B* 39:767–770.

van Hulst, N. F., Moers, M. H. P., Noordman, O. F. J., Tack, R. G., Segerink, F. B., and Bölger, B. (1993) *Appl. Phys. Lett.* 62:461.


# Scleroderma Diagnosis with Recombinant Protein

*Yoshinao Muro, Kenji Sugimoto, Masaru Ohashi, and Michio Himeno*

### *Key Words*

**Anticentromere Antibodies** In eukaryotic chromosomes, a centromere is the essential domain for proper segregation of chromosomes to daughter cells at mitosis and meiosis. Centromere antibodies are found in about 20% of patients with scleroderma, almost all of whom have limited scleroderma or CREST syndrome.

**CREST Syndrome** A variant of scleroderma symptomatized by *c*alcinosis, *R*aynaud's phenomenon, *e*sophageal dysmotility, *s*clerodactyly, and *t*elangiectasia.

**Recombinant Protein** Large amounts of protein can be expressed by introducing a cloned gene to *Escherichia coli.* Fusion proteins consisting of amino-terminal peptides encoded by a portion of the *E. coli lac* Z gene linked to eukaryotic proteins have been often used.

**Scleroderma** Multisystem disorder of unknown etiology, characterized by abnormal deposition of collagen in the skin and internal organs, and microvascular obliteration. Circulating autoantibodies are a feature of scleroderma.

Highly specific autoantibodies directed to cellular components are detectable in the sera of many patients with systemic autoimmune diseases. Almost all scleroderma patients have been shown to produce autoantibodies to nuclear antigens. Some autoantibodies are specific markers for scleroderma and for subsets of the disease. Anticentromere antibodies (ACAs), for example, are commonly seen in the CREST variant of the disease and are generally associated with a better prognosis. A complementary DNA encoding the antigen can be cloned from an expression library using autoantibodies for selection. Recombinant proteins as chimeras between a protein encoded by the vector and the antigenic protein encoded by the cDNA can be used to investigate the existence of the antibody and the epitope mapping as well as the structure of the functional domain.

## 1 SCLERODERMA AND ANTINUCLEAR ANTIBODIES

For many years, autoantibodies were thought to occur in low frequency and low titer in the sera of patients with scleroderma. However, over the past decade several antinuclear antibodies have been found in patients with this disease, using continuous tissue culture cell lines as the immunofluorescence substrate. These antibodies have been used to clarify the molecular structure of intracellular molecules and their role in the biological function. The identified autoantigens are DNA topoisomerase I (Scl-70), centromere proteins, fibrillarin, RNA polymerase I, PM-Scl, and To. Interestingly, these serological markers have been useful in diagnosis and identifying disease subsets. For example, the majority of patients with antitopoisomerase I antibodies suffer from diffuse scleroderma with progressive internal organ involvement. Patients with the limited form of scleroderma, generally referred to as CREST syndrome, contain in most cases autoantibodies to centromere proteins. This entry emphasizes CENP-B, one of centromere antigens, which is one of the most thoroughly characterized scleroderma autoantigens.

## 2 MOLECULAR CLONING OF CENTROMERE ANTIGEN-CODING GENES

### 2.1 Molecular Cloning of CENP Genes

By immunoblotting analysis with ACA-positive sera, at least three centromere autoantigens have been identified in human cells: CENP-A (17 kDa), CENP-B (80 kDa), and CENP-C (140 kDa). Peptide sequence homologies suggest that CENP-A is a centromere-specific histone H3 variant. CENP-B, which is localized throughout the interior of the centromeric chromatin, has been shown to be a DNA-binding protein that recognizes a specific 17 bp sequence in α-satellite DNA. As for CENP-C, immunoelectron microscopy reveals that the localization of this protein is in the inner kinetochore plate. ACA-positive patient sera were also used to isolate the antigen-coding gene by immunological screening a human cDNA expression library constructed with the λgt11 vector. So far, the cDNA clones for CENP-B and CENP-C have been cloned and sequenced. The cDNA for CENP-B codes for a protein of 599 amino acids (aa), and CENP-C is a polypeptide of 943 aa (Figure 1).

### 2.2 Expression of CENP Genes in *Escherichia coli*

Without further gene manipulation, such cDNA clones isolated from an expression library can be expressed as β-galactosidase fusion proteins, that is, as chimeras between a protein encoded by the vector, β-galactosidase, and the antigenic piece encoded by the cDNA. These fusion proteins can be used for investigating the existence of the autoantibody to CENP-B or -C in patients' sera. However, it is more convenient to reclone all (or part) of the cDNA into other expression vectors under a particular circumstance, for instance, for the characterization of antigenic determinant(s).

For example, we handled CENP-B cDNA as follows. The 2.8 kb CENP-B cDNA fragment of the isolated clone was inserted into plasmids of a commercial bacterial expression system (pGEMEX), so that the whole region of the CENP-B antigen except the initiation codon was located just downstream of T7 phage gene 10 product under control of the T7 promoter. The resulting plasmid was introduced into *Escherichia coli* BL21 (DE3) strain, which contains a single copy of the gene for T7 RNA polymerase in the bacterial chromosome under the control of the inducible *lac*UV5 promoter. Addition of IPTG induces T7 RNA polymerase, which in turn transcribes the insert. As a result, target RNA accumulates to amounts comparable to ribosomal RNA, and target protein constitutes the majority of total cell protein.

The induced cell pellets can be directly used as the source of the recombinant CENP-B. When the expressed CENP-B fusion proteins were analyzed by immunoblotting techniques, recombinant CENP-B proteins produced by the foregoing methods were well recognized by all the sera that had been characterized to be ACA-positive in indirect immunofluorescence studies. These techniques can be applied to determine antigenic regions of recombinant proteins.

## 3 EPITOPE MAPPING AND DOMAIN STRUCTURE OF CENP-B

### 3.1 Epitopes on the N-Terminus

To map the immunoreactive regions, the isolated CENP-B cDNA was treated with exonuclease III for unidirectional deletion from

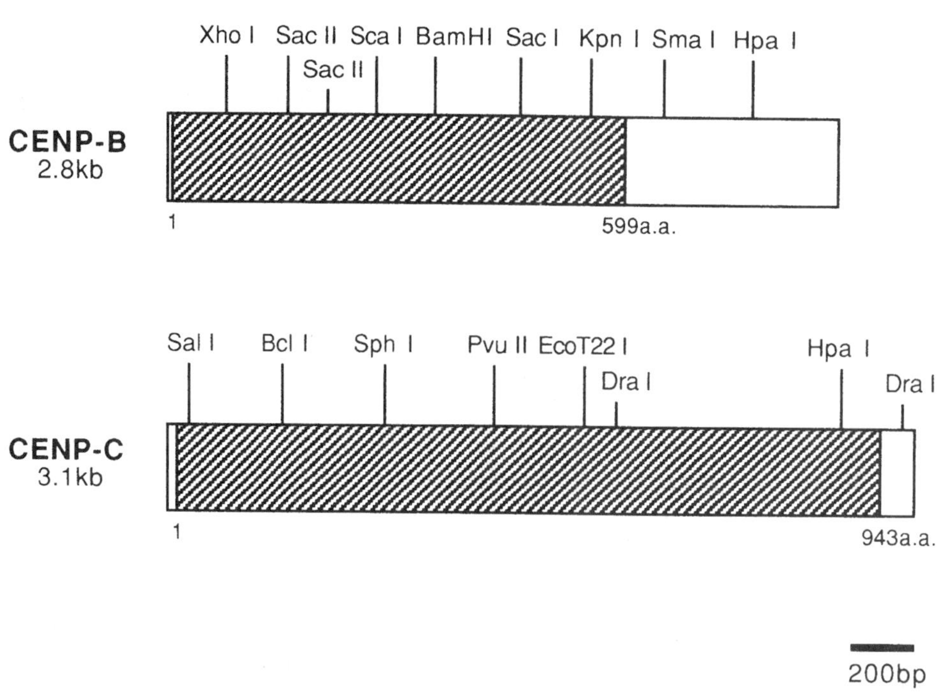


**Figure 1.** Restriction maps of human CENP-B and CENP-C cDNA clones isolated from λgt11 cDNA libraries; shaded boxes indicate open reading frames.

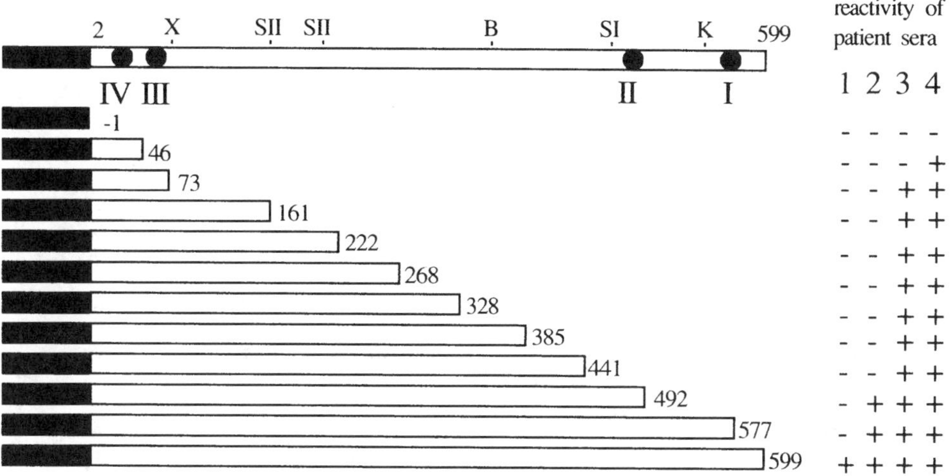


**Figure 2.** Schematic diagram of the truncated CENP-B constructs and the patient groupings. The horizontal bar at the top is the reading frame for CENP-B (residues 2–599). The translated regions are presented as horizontal bars, with the number of terminal amino acid residues indicated on the right in each case. CENP-B truncates are expressed as fusions, with gene 10 proteins, the structural polypeptide of the vector, represented by solid boxes. The reactivity pattern of patient sera to each truncated product is summarized at the far right. The positions of the autoepitopes (epitope I to IV) are also shown in Roman numerals beneath the horizontal bar. Restriction sites are abbreviated as follows: X, *Xho* I; SII, *Sac* II; B, *Bam*HI; SI, *Sac* I; K, *Kpn* I.

the 3′ end. As a result, fusion proteins expressed in *E. coli* were trimmed from the C-terminus (Figure 2). The antigenicity of this set was tested by immunoblotting analysis with 40 ACA-positive sera. According to the reactivity of this truncated series, at least four epitopes from I to IV were estimated to be present on the recombinant CENP-B antigen (Figure 2). Interestingly, the regions of epitopes III (aa 2–73) and IV (aa 2–46) on the N-terminus of CENP-B are within the biologically functional region of CENP-B, because it is shown that the 134 N-terminal residues have specific DNA-binding activity in vitro (Figure 3).

### 3.2 Epitopes on the C-Terminus

To limit the N- and C-terminal boundary of C-terminal epitopes (epitopes I and II), the other plasmid encoding the C-terminal region of CENP-B (residues 438–599) was constructed by inserting the 1.2 kb DNA fragment of another cDNA clone isolated from the human cDNA library into pGEMEX vectors. The 1.2 kb DNA fragment was further dissected by deleting appropriate restriction fragments at the 5′ or 3′ ends or inserting synthetic oligonucleotides. Consequently, the most C-terminal epitope (epitope I) was within the C-terminal 65 amino acid residues. Epitope II was located between positions 438 and 492. Epitope I (residues 535–599) is what is called a ''major epitope''; it reacted to all 40 ACA-positive sera. This is consistent with the data of Verheijen et al. on the antigenicity of the last 60 C-terminal amino acid residues. Epitope II (residues 438–492) comprises the sequence for the casein kinase II site (SDEEE), which was actually phosphorylated in vitro by casein kinase II (Figure 3).

These epitope analyses showed that ACA-positive sera were heterogeneous in reactivity to four epitopes of CENP-B; some of them reacted only to epitope I, and some of them reacted to epitope I and III (or IV). Epitope II was further recognized by a minority of the sera.

In the pioneer study of Earnshaw et al., three epitope regions were predicted on the recombinant CENP-B. The two segments, consisting of aa 6 to 353 and 453 to 556 were found to form epitope regions, which were recognized by minor sera. The major epitope recognized by all sera was limited to the last 147 C-terminal amino acid residues.

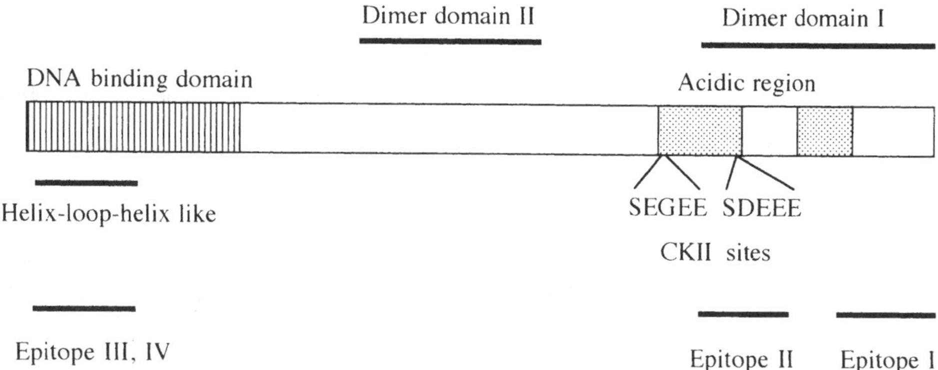


**Figure 3.** Proposed domain structure of CENP-B and four autoepitopes. The DNA-binding domain is hatched and the acidic stretches are shaded. S, E, G, and D show single-letter codes for amino acids; CK II is casein kinase II.

## 4 USE OF ELISA FOR ACA DETECTION

Immunoblotting with cell extracts is not much more sensitive than immunofluorescence for the detection of ACA. Two groups reported the development of an enzyme-linked immunosorbent assay (ELISA) using a cloned CENP-B fragmented as antigen. Earnshaw et al. used the clone encoding 147 amino acid residues from the C-terminal CENP-B fused to 113 kDa of β-galactosidase. Verheijen et al. applied the recombinant protein of glutathione *S*-transferase fused to the last 60 C-terminal amino acid residues to ELISA. In both experiments, nearly all ACA-positive patients' sera gave a positive reaction. In ELISA, the recombinant CENP-B fusion protein is somewhat more sensitive than the immunofluorescence, and the former assay is much easier to use to analyze large numbers of sera. It should be noted that ELISA requires extreme purity of the antigen to avoid detecting reactivity with contaminants.

In the future, synthetic peptides can be also applied to ELISA on the basis of the epitope mapping analysis described here. Judging from reports on other autoantigens, results from the recombinant protein and synthetic peptide approaches may be expected to complement but not necessarily to confirm each other.

## 5 CONCLUDING REMARKS

So far, CENP-A, B, and C, topoisomerase I, fibrillarin, and 75 and 100 kDa proteins of PM-Scl have been cloned with respect to human autoantigens associated with scleroderma. Molecular analysis of antoantigens and autoantibodies using recombinant protein will give us valuable findings for investigating the etiological and pathogenic mechanisms that induce autoimmunity. In our analysis, most ACA-positive patients' sera recognize several antigenic regions of CENP-B and display heterogeneity in their pattern of reactivity to these epitopes. Doubtless, however, recombinant proteins will be of limited value. Posttranslational modifications may contribute to autoantigenic determinants, and the fusion partners of recombinant proteins may interfere with the correct formation of structural (conformational) epitopes. Especially in epitope mapping assays, the use of different vectors or the insertion of different parts of cDNA may influence the level of expression and solubilization of the antigenic peptide. Moreover, it is suggested that a number of autoantigens are subcellular particles, not individual proteins. However, recombinant proteins actually are antigenic in general, and the autoantigens produced in *E. coli* are good substitutes for investigating structures and functions of the real eukaryotic proteins and their clinical applications (e.g., diagnostic tests).

*See also* AUTOANTIBODIES AND AUTOIMMUNITY; IMMUNOLOGY.

### *Bibliography*

Sambrook, J., Fritsch, E. F., and Maniatis, T., Eds. *Molecular Cloning: A Laboratory Manual,* 2nd ed. Cold Spring Harbor Laboratory, Cold Spring Harbor, NY.

Studier, F. W., Rosenberg, A. H., Dunn, J. J., and Dubendorff, J. W. (1990) Use of T7 RNA polymerase to direct expression of cloned genes. *Methods Enzymol.* 185:60–89.

Tan, E. M. (1989) Antinuclear antibodies: Diagnostic markers for autoimmune diseases and probes for cell biology. *Adv. Immunol.* 44:93–151.

Verheijen, R. (1992) B-cell epitopes of scleroderma-specific autoantigens. *Mol. Biol. Rep.* 16:183–189.

**Senescence, Cellular:** *see* **Cell Death and Aging, Molecular Mechanisms of.**

**Senescence, Organismal:** *see* **Aging, Genetic Mutations in.**

# SEQUENCE ALIGNMENT OF PROTEINS AND NUCLEIC ACIDS

*William R. Taylor*

### *Key Words*

**Algorithm** Series of procedural instructions defining a method (commonly a list of coded instructions implemented on a computer).

**Consensus** The reduction of a position in a multiple alignment to a single character representing the majority. Sometimes used more generally to indicate the majority trend (average nature) of the position, as in a *profile*.

**Dynamic Programming** The basic sequence alignment algorithm for optimizing matches.

**Hashing** A method of data storage to enable rapid access to labeled locations. Each location in a *hash table* contains the pointers to all data assigned a common label. For sequences, the labels are tuples and the data are sequence positions.

**Indel** An insertion or deletion (gap) introduced in a sequence to improve an alignment.

**Local Alignment** Region of similarity between two sequences that spans only part of their length.

**Profile** A sequence alignment used to align other sequences or profiles. The amino acids at a position in the profile are used to calculate an average similarity to a single residue or to a position in another profile (as a pairwise sum).

**Regular Expression** A linear sequence pattern that permits the use of "wild cards" (matching any character), set closures (matching any of a group of characters), and gaps (matching a number of characters).

**Relatedness Table** Matrix of similarity measures between sequence elements (amino acids or nucleotides).

**Sequence** String of nucleotide or amino acids codes.

**Tuple** A sequence fragment of fixed length, used in fast matching algorithms.

Alignment, one of the most important and widely used methods of biological sequence analysis, has revealed many unexpected similarities that often give fresh insight into previously uncharacterized systems—allowing the knowledge gained in one area to be transposed directly into the other. Before the end of the century, the sequencing of the genomes of at least three disparate organisms should be complete. These data will provide, for the first time, a view of all the essential proteins needed for life. Analyzing the many similarities, through the alignment of sequences, will give an unrivaled view into the relationships of protein function and the mechanisms of molecular evolution. Many of these comparisons will be difficult, if not impossible, without the aid of three-dimensional structures. Thus the methods of structure-biased sequence comparison are of major importance for achieving the greatest possible extrapolation of the (relatively scarce) structural knowledge into the new wealth of sequence data.

## 1 INTRODUCTION

When a new biological sequence is determined, investigators wish to learn immediately whether the sequence is already known. Taking a broader view, it is then of interest to see whether any other sequence is similar. Any connections to other biological systems can be of great help in elucidating the function, and sometimes structure, of the new protein.

Both these activities involve the comparison of two strings of letters (either nucleotide or amino acids codes), and checking for identity simply requires placing the sequences side by side and counting matching letters. If one sequence is shorter than the other, the process must be repeated for every displacement of one sequence against the other.

When searching for similarity, mismatches must be allowed, and also the possibility of relative insertions and deletions between the two sequences (referred to as *indels* or gaps for short). The latter complication implies that in general, any number of gaps of any size must be allowed at any position in the two sequences. The number of combinations of gaps, even in two short sequences is ''astronomic,'' which means that it is impossible to find the best (the one with most matches) by enumerating all the possibilities. The essence of sequence comparison methods is to solve this problem in a realistic time.

Collected reviews of most of the material described here can be found in the literature. Sankoff and Kruskal cover the more fundamental aspects, while a wide variety of practical methods are reviewed in Doolittle.

## 2 DYNAMIC PROGRAMMING

The problem of gap placement can be solved by a clever, yet simple, class of algorithm called *dynamic programming*. The alignment of one sequence with another can be represented by constructing a grid with a sequence on each axis. Each cell in this matrix links a pair of units (residues or nucleotides) in the two sequences, and an alignment of the two sequences is a path through the matrix that progresses without any backward steps in either sequence. The problem to be solved is to find the path from one end of the matrix (top or left edge) to the other (bottom or right edge) that passes through most matching cells (see Figure 1 for a worked example.)

This algorithm was first applied to biological sequences by Needleman and Wunsch and despite much variation, remains the basic alignment method.

## 3 INCORPORATING EVOLUTIONARY CONSTRAINTS

The basic dynamic programming algorithm will return a best alignment, given a model of how favorable it is to align one amino acid or nucleotide with another. The simple model of Figure 1 may be

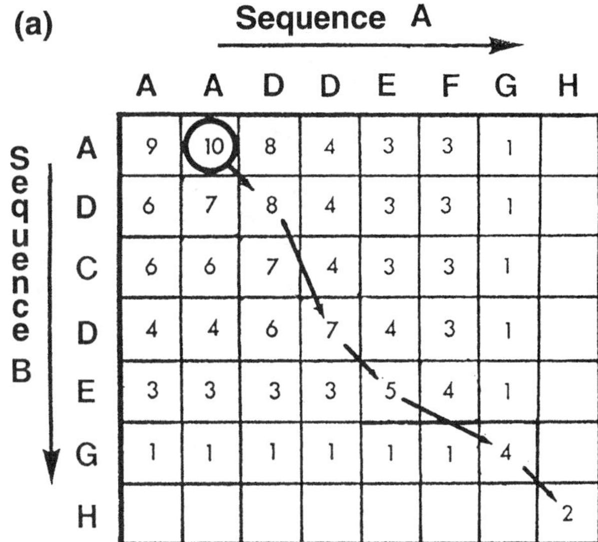


(b) AAD-DEFGH
ADCDE-GH

**Figure 1.** The basic Dynamic Programming algorithm: Steps in this algorithm are illustrated using the alignment of two short sequences: A and B. Matching positions score 2 and the insertion of a gap costs 1. (*a*) Beginning at the ends of the sequences (bottom right of the matrix) the score of 2 is entered in the cell corresponding to the matching Hs. Sums of the scores are accumulated working back towards the start of the sequences. Thus, the cell corresponding to the aligned Gs scores 2 and inherits 2 from the matched Hs, giving 4. None of the other positions on the same row or column have any matches but they still inherit 2 from the matched Hs less 1 for the insertion of a gap (giving rise to lines of 1s). The next diagonal position (EF match) has no score but inherits the preceding score of 4. However an insert of a gap allows Es to match gaining 1 (2 less the gap penalty). This is continued until the matrix is complete. The highest score is then found (circled 10) and its inheritance path is traced back to the opposite corner of the matrix (heavy arrows). (*b*) The resulting alignment. This simple process has the remarkable property that it guarantees to find the best alignment for the given constraints.

adequate for nucleic acids, but it is too simplistic for amino acids that exhibit a wide range of similarity. In addition, the frequency of indels must be balanced against the adopted scoring scheme.

### 3.1 Amino Acid Relatedness

#### 3.1.1 Empirical Estimates

The observation of amino acid substitution between closely related aligned protein sequences allowed the compilation of a score table indicating how likely it is for any two amino acids to be aligned. Although the original table was compiled when relatively few sequences were known, new versions based on more sequences have shown little significant change.

#### 3.1.2 Minimal Base Change

Assuming that the underlying mechanism of amino acid substitution is the substitution of single nucleotides, an amino acid relatedness table can be compiled from the minimal number of base changes needed to convert one codon to the other.

#### 3.1.3 Physicochemical Properties

Direct consideration of the physical and chemical properties of amino acids makes possible the quantification of their similarities and differences. The most important properties are the size and the hydrophobicity of the side chain.

### 3.2 Gap Penalty

A penalty against gaps in a sequence alignment can be incorporated into the standard dynamic programming algorithm simply by subtracting a constant value from each score inherited by any transition other than the diagonally adjacent cell. This step, which imposes a fixed penalty of any size of gap, can be refined to be partly dependent on the gap size. A linear function with positive coefficients is commonly used: $an + b$, where $n$ is the gap length ($a$ = 1 and $b$ = 10 are typical).

Whatever the form of the gap penalty, its value must be in balance with the values found in the relatedness matrix. There is no general rule for the size of the gap penalty, but empirical estimates have been made.

## 4 LOCAL ALIGNMENT METHODS

### 4.1 Smith–Waterman Method

Provided the relatedness matrix contains no negative entries, the dynamic programming method will produce a best alignment over the length of the two sequences considered. If the relatedness matrix has a negative bias, however, the best score will not necessarily be on the top row or the left edge. Finding the best score and following its trail of pointers until a predefined low value (say, zero) is found will define the best alignment fragment or *local alignment*. To find additional good local alignments, it is possible to remove all cells associated with the best and repeat the process.

### 4.2 Sellers' Method

In an elegant method for ''simultaneously'' extracting all local alignments devised by Sellers, the matrix is processed first from bottom right to top left and then in reverse. When the two sets of pointers are compared, those that overlap (point to each other) define the paths of local alignments.

## 5 FAST METHODS

As the sequence databanks have grown, considerable effort has been put into fast alignment methods.

### 5.1 FASTa

One family of methods, typified by the FASTa program, is based on the technique of *hashing;* for a review, see Lipman and Pearson or Pearson. Each peptide or oligonucleotide of a predefined length—referred to as a *tuple*—is recorded in a table, with the positions of all its occurrences kept in the databank. From this table, tuples that are common to two sequences can be found quickly and either displayed in a plot or processed to identify on the plot diagonal tracts that are rich in matches and might constitute a region of alignment.

These programs run about 50 times faster than a standard dynamic programming algorithm but are not as sensitive.

### 5.2 BLAST

The tuples matched by the FASTa-type algorithm must be identical, a requirement that greatly limits the length of the tuples that can be selected for processing. The BLAST program, however, is based on a different algorithm that allows similar amino acids to be matched. Matching fragments found by this algorithm (which cannot include gaps) are limited in length only when the insertion of a gap becomes necessary.

The method is very fast and can search a protein sequence database at the rate of about 500,000 residues per second.

## 6 PATTERN MATCHING METHODS

### 6.1 Regular Expressions

Simple patterns of the *regular expression* class can be matched rapidly against sequence data. They are particularly useful for protein sequences for which groups of amino acids can be defined to reflect physicochemical properties. A large collection of such patterns has been collected from the SWISS-PROT sequence databank.

### 6.2 Complex Expressions

Many derived properties can be associated with protein sequences that do not fit easily into the constraints of a regular expression format. To deal with such properties, more complex methods have been developed which can incorporate data of a probabilistic nature such as secondary structure propensities.

## 7 MULTIPLE SEQUENCE ALIGNMENT

The generalization of the dynamic programming algorithm to multiple sequences, while straightforward, is considered impractical because it calls for a great increase in computation. This limitation has led to the development of a number of methods based on the combination of pairwise alignments, which differ only in the order in which they combine the sequences and in whether some form of abstraction or *consensus* is derived at each stage.

One of the most robust approaches consists of combining the most similar sequence pairs independently into a *profile* and then recalculating the similarity between all single and consensus sequences and progressively bringing together the most similar of these at each stage. This strategy allows the conserved aspects of each subfamily to become apparent before they are aligned at a later stage. The approach retains virtually all the information in each alignment, avoiding the loss that would occur if a more abstract representation were used.

## 8 INCORPORATING STRUCTURAL DATA

In the alignment of protein sequences, the three-dimensional structure of one (or both) of the sequences may be known. This information can greatly influence the evolutionary model of what amino acid substitutions and gaps might be possible at different locations on the sequences. For example, deletions are most common in the loop regions between secondary structures, and the standard alignment method can be modified to incorporate this bias.

In the absence of any structural information, standard sequence alignment uses a single amino acid relatedness table. Where the structure is known, however, different relatedness tables can be used according to the local structure. For example, an alanine-to-leucine change would be more favorable in an α-helix than an alanine-to-valine change, while the reverse would be true in a β-sheet conformation. This approach can be extended to neglect one sequence completely and align amino acids from a sequence of unknown structure with a sequence of structural properties derived directly from a known structure. This has been done using a single property—such as exposure to solvent—or multiple properties (introducing secondary structure states).

The fullest extension of this approach is to evaluate the sequence–structure alignment, not through tables associated with predefined states but directly on the structure itself. This involves difficult computations, since the environment of any position in the model will not be known until the final alignment has been established, yet the assessment of each environment must be used to establish the alignment. This circularity can be broken by a two-stage application of the standard sequence alignment algorithm (known as double dynamic programming), which was developed to compare two protein structures.

*See also* Sequence Analysis; Sequence Divergence Estimation.

### *Bibliography*


Altschul, S. F., Gish, W., Miller, W., Myers, E. W., and Lipman, D. J. Basic local alignment search tool. *J. Mol. Biol.* 215:403–410 (1990).

Bairoch, A., and Boeckmann, B. The SWISS-PROT protein sequence data bank. *Nucleic Acids Res.* 19:2247–2249 (1991).

Barton, G. J., and Sternberg, M.J.E. Evaluation and improvements in the automatic alignment of protein sequences. *Protein Eng.* 1:89–94 (1987).

Bowie, J. U., Lüthy, R., and Eisenberg, D. A method to identify protein sequences that fold into a known three-dimensional structure. *Science* 253:164–170 (1991).

Dayhoff, M. O., Schwartz, R. M., and Orcutt, B. C. A model of evolutionary change in proteins. In *Atlas of Protein Sequence and Structure,* Vol. 5, Suppl. 3, M. O. Dayhoff, Ed. National Biomedical Research Foundation, Washington DC, 1978, pp. 345–352.

Doolittle, R. F., Ed. *Molecular Evolution: Computer Analysis of Protein and Nucleic Acid Sequences,* Vol. 183 in *Methods in Enzymology.* Academic Press, San Diego, CA, 1990.

Jones, D. T., Taylor, W. R., and Thornton, J. M. A new approach to protein fold recognition. *Nature* 358:86–89 (1992).

Lesk, A., Levitt, M., and Chothia, C. Alignment of the amino acid sequences of distantly related proteins using variable gap penalties. *Protein Eng.* 1:77–78 (1986).

Lipman, D. J., and Pearson, W. R. Rapid and sensitive protein similarity searches. *Science* 227:1435–1441 (1985).

Needleman, S. B., and Wunsch, C. D. A general method applicable to the search for similarities in the amino acid sequence of two proteins. *J. Mol. Biol.* 48:443–453, (1970).

Overington, J., Šali, A., Johnson, M. S., and Blundell, T. L. Tertiary structural constraints on protein evolutionary diversity: Templates, key residues and structure prediction. *Proc. R. Soc. London B* 241:132–145 (1990).

Pearson, W. R. Rapid and sensitive sequence comparison with FASTP and FASTA. In *Molecular Evolution: Computer Analysis of Protein and Nucleic Acid Sequences,* Vol. 183 in *Methods in Enzymology,* R. F. Doolittle, Ed. Academic Press, San Diego, CA, 1990, pp. 63–98.

Sankoff, D., and Kruskal, J. B., Eds. *Time Warps, String Edits, and Macromolecules: The Theory and Practice of Sequence Comparison.* Addison-Wesley, Reading, MA, 1983.

Sellers, P. H. On the theory and computation of evolutionary distances. *SIAM J. Appl. Math.* 26:787–793 (1974).

Sibbald, P. R., and Argos, P. Scrutineer: A computer program that flexibly seeks and describes motifs and profiles in protein sequence databases. *CABIOS,* 6:279–288 (1990).

Smith, R. F., and Smith, T. S. Automatic generation of primary sequence patterns from sets of related protein sequences. *Proc. Natl. Acad. Sci. U.S.A.* 87:118–122 (1990).

Smith, T. F., and Waterman, M. S. Identification of common molecular subsequences. *J. Mol. Biol.* 147:195–197 (1981).

Taylor, W. R. Identification of protein sequence homology by consensus template alignment. *J. Mol. Biol.* 188:233–258 (1986).

Taylor, W. R. A flexible method to align large numbers of biological sequences. *J. Mol. Evol.* 28:161–169 (1988).

Taylor, W. R., and Orengo, C. A. Protein structure alignment. *J. Mol. Biol.* 208:1–22 (1989).


# Sequence Analysis

*Martin Bishop*

### *Key Words*

**Alignment** Display of two or more sequences sharing matches, mismatches or gaps at each position.

**Codon Usage** Frequency of the 64 nucleotide triplets for the protein-coding portion of a gene or set of genes.

**Consensus Sequence** Constructed from a multiple alignment of sequences as the nucleotide or amino acid most frequent at each position.

**Dot Plot** Diagram that graphically compares all positions of two sequences by placing diagonals in the matching regions.

**Homology** Hypothesis that sequences are related by descent from a common ancestor.

**Hydropathy** Scale of ordering of amino acids from hydrophilic to hydrophobic.

**Isochores** Regions of a genome in which DNA composition is similar.

***k*-Tuple** Oligonucleotide or peptide subsequence of length *k*.

**Motif** Recognizable subsequence or substructure of nucleic acid or protein, usually of functional significance.

**Neural Network** Computational device capable of automatically learning from a training set of samples. Once trained, the network may be used for recognition of, for example, signal sequences.

**Signal Sequence** Motif that functions by recognizing a second sequence.

**Weight Matrix** Constructed from a multiple alignment of sequences as the frequencies of nucleotides or amino acids at each position.

Sequences of nucleic acids in DNA and RNA and of amino acids in proteins define the primary structure of these molecules. Sequence analysis is carried out using computer programs that implement algorithms to determine sequence properties and to compare sequences. Sequence comparison can indicate whether an RNA or protein molecule or region of DNA is already known (identity) or has some degree of similarity to a known sequence. Sequence similarity may indicate similar structure or function. Sequence analysis can suggest the function of an unknown sequence based on the features it contains. Sequence analysis is a necessary preliminary to detailed experimental studies of structure, function, and interactions of biological macromolecules. Sequences are the information repository of the cell and a natural index to our growing understanding of cellular processes as dynamic systems of interactions between macromolecules.

## 1 PURPOSE OF SEQUENCE ANALYSIS

### 1.1 Prediction of Function

Sequences that are unlike any known sequence may still be made to yield information that can suggest their possible function. The function of nucleic acids and proteins depends on their structure and involves complex interactions in three dimensions. It is not presently understood whether it is possible, in general, to derive structure from sequence. Sequence alone is therefore often inadequate to determine function. Predictions made from sequence analysis need to be experimentally tested. Nevertheless, computer analysis of sequences is valuable in suggesting the most useful experiments to perform.

### 1.2 Revealing Similarity

The first thing to do with a newly determined sequence is to compare it with all known sequences. The outcome may show identity to a known sequence, which may prove disappointing if one is hoping for something new. Similarity to a known sequence may suggest something new that can be characterized with relatively little effort. A totally unknown sequence may be a frustrating result: considerable effort will be needed to understand its function.

Sequence comparison is a nontrivial pursuit, and both statistical and biological considerations are involved. Statistically significant similarities (under some model and at some chosen level of significance) may be biologically meaningless. Sequence motifs that are statistically nonsignificant in similarity may encode the same function (this is likely to occur because the statistical model based on sequence alone is incomplete). In an area fraught with such difficulties, common sense and interpretation based on utility are paramount.

Sequence dissimilarity can range from identity, difference due to sequencing errors, difference due to population polymorphism (individual variants), and differences in multiple copies of a gene in a single individual (multigene families) to wide evolutionary divergence of genes in different organisms. Sequences that are similar due to common function may not share a common ancestral sequence in biological evolution. In general, ideas about the evolutionary relationships of sequences are not experimentally testable. Sequence homology (similarity due to descent from a common ancestor) is a hypothesis, not an observable fact, except in the case of microbial populations with high mutation rates and short generation times, which may be studied experimentally through time.

## 2 ANALYSIS OF SINGLE SEQUENCES

### 2.1 DNA Composition, Isochores, and Codon Usage

Nucleotides in DNA sequences may be counted as singlets, doublets, or triplets in either strand. Doublets or triplets may be counted as overlapping or nonoverlapping in two or three phases, respectively, on either strand. The genomes of various organisms vary considerably in their DNA composition. Warm-blooded vertebrates have a higher G+C content, which correlates with the higher thermal stability of GC over AT base pairs. Composition of regions within a genome can also vary considerably. Mammalian genomes contain relatively GC-rich and AT-rich regions, which are called isochores. Overlapping doublet frequencies are highly characteristic for an organism. CG dinucleotides are less common than expected in vertebrates and angiosperms, probably because spontaneous deamination of 5-methylcytosine to thymine prevents the repair of methylated CpG. In DNA coding for protein, one phase of nonoverlapping triplets will be the phase of translation and the triplets will be codons. In a gene, the possible codons for each amino acid are unevenly used, and the frequency table for the 64 triplets is called codon usage. Codon usage is different between different species and between highly and lowly expressed gene in the same species.

### 2.2 Mapping DNA Sequence Features

Mapping the position of features on a DNA sequence is an important step in investigating its function. It is easy to map sites that can be precisely defined, such as stop codons or restriction enzyme recognition sites. Once DNA has been sequenced, the sizes of the fragments produced with any enzyme can be readily calculated. Features such as promoters, splice junctions, and ribosome binding sites are very difficult to predict because they are hard to specify. Mapping is most simply achieved by comparing the probe sequence with each position of the DNA sequence in turn and noting the hits. More sophisticated algorithms exist for rapid searching in large problems.

### 2.3 Repetitive Sequences

Direct repeats and inverted repeats (sometimes called dyad symmetries) are common in DNA from many sources. Mammalian genomes contain families of long (LINE) and short (SINE) repeats. Repeats of *L1* (*Kpn* I) type are 5000 to 7000 bp long and are present in the genome in $10^3$ to $10^4$ copies. Repeats of *Alu* type are 350 bp long and occur in as many as $9 \times 10^4$ copies. *Alu* repeats make human DNA hard to assemble from gel sequencing reads into the finished sequence. Inverted repeats occur in DNA coding for structural RNA, and these symmetry properties enable the RNA to fold into its secondary structure.

The dot plot is a diagram that reveals the presence of repeats and inverted repeats in sequences. It is also useful for comparing two different nucleic acid or protein sequences to detect regions of similarity. The dot plot is a rectangular array with rows labeled by one sequence and columns labeled by the other. A cell *i, j* can be used to represent the result of comparison of the *j*th residue of sequence A with the *i*th residue of sequence B. The simplest form of dot plot results from placing a diagonal mark in each cell where

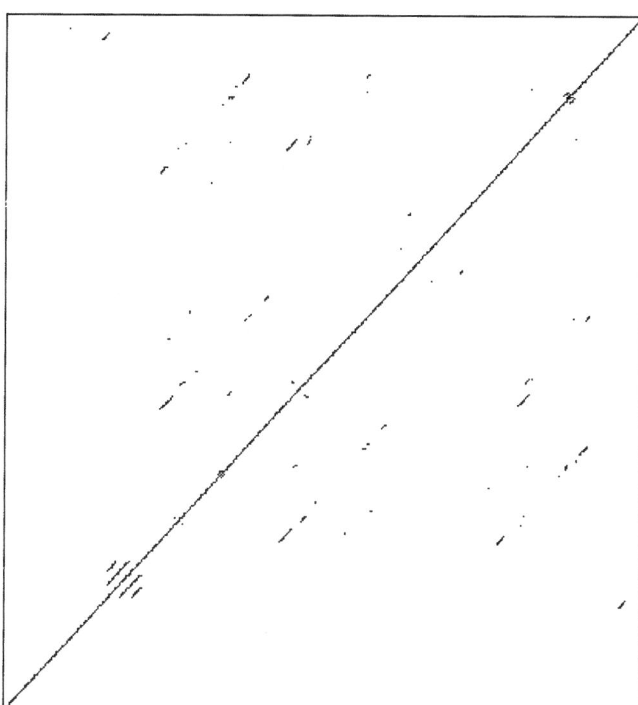

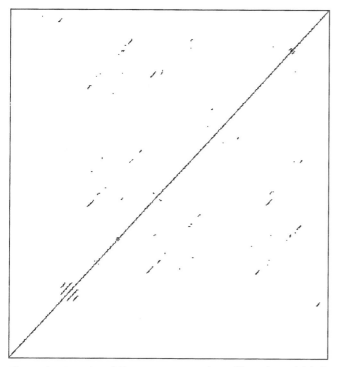

**Figure 1.** Dot plot of Trypanosoma congolense kinetoplast minicircle DNA (GenBank accession M19750) drawn against itself by the Staden program sip. The strong diagonal represents identity while repeat regions are off the diagonal and in mirror image.

the sequence symbols match. The diagram is clarified by drawing only diagonals that reach a threshold length. Partial matches may be displayed by choosing a diagonal of fixed length (e.g., 21) and drawing only diagonals that reach a certain score (e.g., 15) in this span, as shown in Figure 1.

## 2.4 Simple Sequences

Some protein sequences are simple in the sense that they contain stretches with relatively few different amino acids, either in a periodic or nonperiodic arrangement. These have evolved to provide the protein with specific properties. For example, the Atlantic Flounder Antifreeze protein contains many many alanine residues in groups of between 2 and 7 consecutively. Simple sequences of proteins are encoded by DNA, giving rise to simple DNA. There may also be regions of simple DNA that do not code for protein the function of which is poorly understood. Genetic diseases such as myotonic dystrophy and Huntington's chorea may arise by replication slippage in the DNA encoding simple regions of proteins.

## 2.5 Prediction of DNA Sequence Function

### 2.5.1 Coding for RNA

An RNA molecule twists and bends, and the bases interact, folding the chain into local helices separated by single-stranded regions. The cloverleaf folding of tRNA molecules is well documented, while the structure of long RNA molecules is still poorly known. Attempts to predict RNA secondary structure rely on energy models for computing folding stabilities. Predictions are unreliable and need to be confirmed by experimental methods. To predict if whether DNA codes for RNA, one needs to search for the kinds of inverted repeat that are present in RNA folds.

### 2.5.2 Coding for Protein

Coding for protein has an effect on the composition of the DNA, and three factors contribute to this.

1. *Uneven use of amino acids.* Proteins have a restricted range of amino acid composition; for example, alanine is common and tryptophan is rare.
2. *Uneven numbers of codons for amino acids.* Numbers of codons for the different amino acids vary between one and six.
3. *Uneven use of codons.* In a gene, the possible codons for each amino acid are not evenly used.

Coding constraints affect both strands of the DNA, and coding regions tend to have long open reading frames even on the noncoding strand. Indeed, more than one protein may be encoded in a single stretch of DNA on one or both strands. Absence of stop codons is an obvious requirement in a protein coding region. Statistical methods have been developed for predicting protein coding from the properties of the sequence.

Distinction between introns and exons is possible in terms of profiles for observed and expected overlapping oligonucleotide frequencies, particularly pentamers and hexamers. This has been implemented as the contrast words method or the $k$-tuple method.

### 2.5.3 Eukaryotic Gene Structure

Prediction is complicated in eukaryotes by the presence of introns. Most exons are at least 20 codons long and usually at least 50 codons long. Consideration of all the features of eukaryotic gene structure can help to elucidate which are the coding regions. These features are the promoters of transcription; the polyadenylated site at the 3′ end of pre-mRNA; donor, acceptor, and branch point signals of RNA splicing; and the ribosome binding site of initiation of translation.

Control of transcription is much more complex in eukaryotes, with hundreds of differentiated cell types, than in prokaryotes. RNA polymerase II provides transcription of genes coding for proteins, and there are computerized compilations of promoters and transcription factors. The most frequent signal sequences in the promoter region are TATA box, cap site, CCAAT box, and GC box. Searches for these configurations are usually made with consensus sequences or weight matrices. Even the most sophisticated statistical analyses for signals (neural networks) are sometimes ineffective, and false positives (recognition of sites that do not function) and false negatives (failure to recognize sites that do function) occur. Similar remarks apply to the signals of mRNA processing and translation.

The relative spatial arrangement of all the features of eukaryotic genes permits moderate to good prediction. The most successful methods for predicting eukaryotic gene structure proceed as follows:

1. Identify all potential splice sites.
2. Assign coding potential by the hexamer $k$-tuple and codon usage statistics.
3. Combine steps 1 and 2 to suggest the best exons.
4. Enumerate all possible gene structures on the basis of step 3.

5. Attempt to assign initiation context, polyadenylation signal, and promoters.

### 2.6 Prediction of Protein Sequence Function

The very difficult task of predicting protein sequence function is likely to be accomplished only if the target protein is comparable to a protein of known structure and function. Some hints may be obtained from hydropathy plots, secondary structure prediction, and peptide motif searching.

Amino acids may be ordered on the basis of hydropathy: that is, from hydrophilic (seeking contact with water molecules) to hydrophobic (avoiding water and seeking the interior of a protein or a membrane). Plots of hydropathy changes along a protein can suggest sites of B-cell antigenic determinants (hydrophilic) or membrane-spanning regions (hydrophobic).

Proteins are built largely out of α-helices, β-sheets, coils, and turns with the helices and sheets packing together into a variety of more complex motifs. To predict these secondary structures, the most commonly used methods are those of Chou and Fasman or Garnier, Osguthorpe, and Robson. The statistical constribution each residue makes to α-helix, β-sheet, coil, and turn is assessed in a sliding window. The accuracy of prediction is around 50%. In spite of numerous efforts, the accuracy of prediction has not been improved much since these methods were described.

PROSITE is a database of sites and patterns in proteins which are structurally or functionally conserved. Software exists to allow searching of proteins for the targets in PROSITE, either one at a time or in defined spatial relationships.

## 3 COMPARISON OF TWO OR MORE SEQUENCES

### 3.1 Sequence Alignment

A sequence alignment is computed using a score matrix based on the rectangular array described for the construction of dot plots (Section 2.3). In any cell of the matrix, a horizontal move represents a deletion in sequence A, a vertical move represents a deletion in sequence B, and a diagonal move from top left to bottom right represents an identity if $A_j = B_i$ or a substitution if $A_j \neq B_i$. Scoring matrices are used to assign values to all possible changes from one sequence symbol to another. For protein the point-accepted mutation (PAM) matrix of Dayhoff is commonly used. A score is also associated with the introduction of gaps, and there are various models for this. For example, gap score is length independent—higher for the first gap and less for subsequent gaps. Any path through the cells according to the rules of movement has an associated score and can be converted to a sequence alignment. Algorithms are available to determine the best score and the associated path(s) representing alignment(s). In general, there may be more than one path of best score. Suboptimal alignments may be biologically meaningful.

#### 3.1.1 Global Alignment

Global alignment is the arrangement of the whole of sequence A with the whole of sequence B; it is implemented by the Needleman–Wunsch–Sellers algorithm. It is appropriate for the alignment of related genes or proteins with similar content.

#### 3.1.2 Best Location

Sequence A (a short sequence) may be aligned at the best point within a longer sequence B. In ''best location'' alignment every location in B is a potential starting point for the alignment. This method is useful for finding signal sequences in DNA or functional motifs in proteins.

#### 3.1.3 Best Region(s) of Local Similarity

It is also possible to work with the best alignments of a part of sequence A with a part of sequence B. Every location in sequences A and B is a potential starting point for the alignment. The method is implemented by the Smith–Waterman algorithm. This is the most sensitive method for sequence database searching.

### 3.2 Sequence Database Searching

DNA sequences are compiled in GenBank, the Data Library of the European Molecular Biology Laboratory, and the DNA Database of Japan. Protein sequences are compiled in the Protein Identification Resource and in SwissProt. Database searching involves comparing the probe with each sequence in the database in turn. Results above some threshold are collected and reported to the user. The Smith–Waterman search is implemented for parallel-architecture computers such as the Distributed Array Processor and the MasPar by programs such as PROSRCH, BLAZE, and MPSRCH. It takes a few minutes to search all the protein sequences known today with a 500 amino acid probe. This method is impractical on computers of conventional architecture.

For conventional architecture computers, program suites have been written which reduce the run time by taking short cuts. The most popular programs are the BLAST family and the FAST family. BLAST is the fastest and takes only a few minutes to search all known DNA with a 1 kb probe. It does not permit gaps but reports maximal segment pairs (MSPs) between the probe and the database sequence according to parameters that may be set by the user. FAST locates identities of fixed *k*-tuple between the probe and the database sequence in the first step. The best diagonal regions, as they would appear on a dot plot, are then scored by a proportional matching algorithm. Finally there is an attempt to join the regions so found using the Smith–Waterman algorithm. FAST takes about 10 times as long as BLAST.

### 3.3 Multiple Sequence Alignment

When working with a family of related sequences, it is valuable to construct a multiple alignment of the global type. It is not feasible to compute a joint multiple alignment by the Needleman–Wunsch–Sellers algorithm in $n$ dimensions. Instead, various strategies of combining the results of multiple pairwise alignments are used. The best scores of pairwise alignments for the $n$ sequences are first determined. Then the most similar pair is aligned and the next most similar sequence is aligned with the alignment of pairs. This process is continued until all the sequences have been included.

*See also* Sequence Alignment of Proteins and Nucleic Acids; Sequence Divergence Estimation.

## *Bibliography*

Bishop, M. J., Ed. *Guide to Human Genome Computing*. Academic Press, London, 1994.

Bishop, M. J., and Rawlings, C. J., Eds. *Nucleic Acid and Protein Sequence Analysis: A Practical Approach.* IRL Press, Oxford, 1987.

Doolittle, R. F. *Of URFs and ORFs. A Primer on How to Analyze Derived Amino Acid Sequences.* University Science Books, Mill Valley, CA, 1986.

Doolittle, R. F., Ed. *Methods in Enzymology,* Vol. 183, *Molecular Evolution: Computer Analysis of Protein and Nucleic Acid Sequences.* Academic Press, San Diego, CA. 1990.

Gribskov, M., and Devereux, J., Eds. *Sequence Analysis Primer.* Stockton Press, New York, 1991.

Hodgman, T. C., In *Microcomputers in Biochemistry: A Practical Approach,* C.F.A. Bryce, Ed. Oxford University Press, Oxford, 1992.

Lesk, A. M., Ed. *Computational Molecular Biology.* Oxford University Press, Oxford, 1988.

von Heijne, G. *Sequence Analysis in Molecular Biology.* Academic Press, San Diego, CA, 1987.

# SEQUENCE DIVERGENCE ESTIMATION

*Takashi Gojobori and Kazuho Ikeo*

*Key Words*

**Nonsynonymous Substitutions** Nucleotide substitutions that change amino acids.

**Nucleotide (or Amino Acid) Difference** Describing the state of sites where the nucleotides (or amino acids) are different between two homologous sequences that are being compared.

**Nucleotide (or Amino Acid) Substitutions** Nucleotide (or amino acid) changes in the process of evolution, which lead to nucleotide (or amino acid) differences.

**Synonymous Substitutions** Nucleotide substitutions that do not change amino acids.

Methods for estimating sequence divergence are crucial for studies of molecular evolution. In particular, the number of nucleotide (or amino acid) substitutions is a good quantity to use in representing the degree of sequence divergence when two homologous sequences are being compared. Thus, knowledge of the number of nucleotide (or amino acid) substitutions is important for computing evolutionary rate and for constructing phylogenetic trees at the DNA level. This contribution presents statistical methods for estimating the number of nucleotide and amino acid substitutions, putting particular emphasis on their practical usage and usefulness.

## 1 PRINCIPLES

### 1.1 EVOLUTIONARY DIVERGENCE OF NUCLEOTIDE (OR AMINO ACID) SEQUENCES

When two sequences to be compared share a common ancestor, these sequences are called "homologous." Thus, homologous sequences are usually similar, but similar sequences are not necessarily homologous. In the case of homologous sequences, differences in the sequences are, in general, caused by divergence that has taken place during evolution through the accumulation of mutations such as substitution and deletion/insertion.

### 1.2 NUCLEOTIDE (OR AMINO ACID) DIFFERENCES

The degree of nucleotide or amino acid differences can be estimated by the proportion of nucleotide or amino acid sites at which there are differences between the nucleotides or amino acids of two homologous sequences being compared. Let us designate the proportions of nucleotide and amino acid differences by $p_n$ and $p_a$, respectively. For example, let us consider two homologous sequences, both of which have 100 nucleotide sites, when these sequences are aligned with each other. Let us also suppose that nucleotides are different at 10 sites out of 100. For simplicity, any gaps that may exist are excluded from this comparison. (In fact, this is a common practice in studies of molecular evolution, because the evolutionary mechanisms for deletion/insertion have not been clear compared with nucleotide or amino acid substitutions.) Then, $p_n$ is given by $10/100 = 0.1$.

### 1.3 THE NUMBER OF NUCLEOTIDE (OR AMINO ACID) SUBSTITUTIONS

The variables $p_n$ and $p_a$ are important for estimating the number of nucleotide and amino acid substitutions. They do not, however, represent the numbers of nucleotide and amino acid substitutions, because multiple and superimposed nucleotide substitutions at the same site may occur undetected, especially when two homologous sequences have different nucleotides, say A and G, at a particular site, the change of A → T → G may have occurred for one sequence and no change for the other at the site (Figure 1). The observed number of nucleotide differences between two DNA sequences is thus frequently different from the total number of nucleotide substitutions that have actually occurred during their divergence. Thus, comparative studies of DNA sequences call for statistical methods for estimating the number of nucleotide substitutions.

From the perspective of molecular evolutionary studies, both nucleotide and amino acid substitutions can be treated as stochastic processes. Methods for estimating the number of nucleotide substitutions, however, are different from those for estimating the number of amino acid substitutions. Because only four kinds of nucleotide

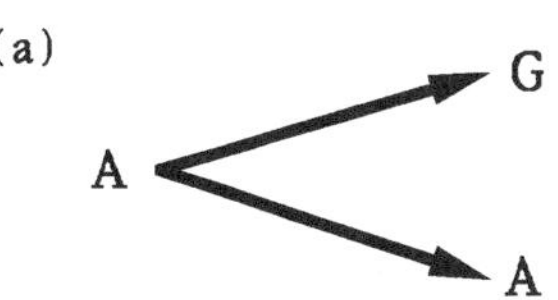


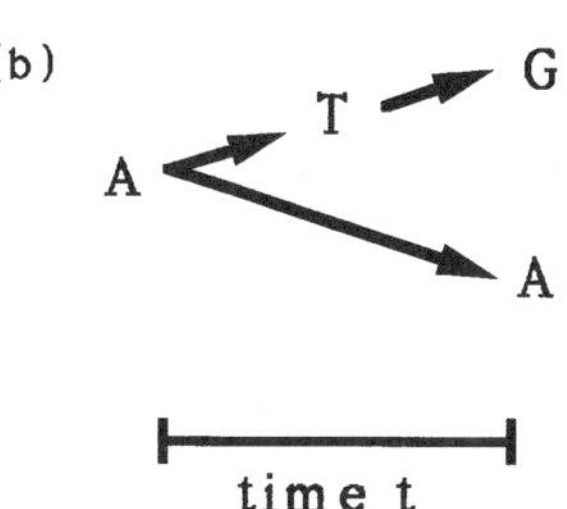


**Figure 1.** Examples of nucleotide changes in the course of evolution: these two descendant DNA sequences have nucleotides A and G at the same site after years. (a) One nucleotide substitution (A → G) has occurred. (b) Two nucleotide substitutions (A → T and T → G) have occurred.

exist, compared with 20 different amino acids, the effect of multiple substitutions on the estimates for the number of nucleotide substitutions is much greater.

### 1.4 Synonymous and Nonsynonymous Substitutions

In the case of protein-coding genes, nucleotide substitutions can be classified according to whether they do or do not cause amino acid changes. Nucleotide substitutions that do not change amino acids are called "synonymous substitutions." On the other hand, nucleotide substitutions that cause amino acid changes are termed "nonsynonymous substitutions." By definition, synonymous substitutions occurring in a gene are exempt from any functional constraints of a gene product. Therefore, the number and pattern of synonymous substitution can, by and large, reflect those of spontaneous mutations. The number and pattern of nonsynonymous substitutions can be a strong indicator of how the gene product is constrained functionally and structurally. Thus, it is important to estimate the numbers of synonymous and nonsynonymous substitutions from the sequence data.

## 2 ESTIMATION METHODS FOR THE NUMBER OF AMINO ACID SUBSTITUTIONS

Let $K_a$ be the average number (per site) of amino acid substitutions between the two homologous sequences being compared. It is assumed that the probabilities of 0, 1, 2, . . . , amino acid substitutions occurring at a particular site are given by $\exp(-K_a)$, $\exp(-K_a)K_a$, $\exp(-K_a)K_a^2/2!$, . . . , in the Poisson process. This assumption is based on the consideration that a substitution at each site is a very rare event for any given period, but extending over an enormous period, the probability of an occurrence is applicable.

In particular, the probability that no substitutions occur at a site is $\exp(-K_a)$. By equating this to the proportion $(1 - p_a)$ of identical amino acids, namely,

$$\exp(-K_a) = 1 - p_a \tag{1}$$

we obtain

$$K_a = -\ln(1 - p_a) \tag{2}$$

where $p_a$ is again the proportion of different amino acids between the two homologous sequences under comparison. The variance of $K_a$ is given by

$$\mathrm{Var}(K_a) = \frac{p_a}{(1 - p_a)n_a} \tag{3}$$

where $n_a$ is the total number of amino acid sites.

## 3 METHODS FOR ESTIMATING THE NUMBER OF NUCLEOTIDE SUBSTITUTIONS

### 3.1 One-Parameter Method

To derive an estimation formula for the number of nucleotide substitutions, we must assume a specific model for the pattern of nucleotide substitutions among the four kinds of nucleotide. In the simplest model, the rate of substitution is assumed to be equal between any pair of nucleotides. This is called the one-parameter model, because the process of substitution is described by a single

**Table 1** Patterns of Nucleotide Substitutions

| | Substituted Nucleotide | | | |
|---|---|---|---|---|
| Original Nucleotide | A | T | C | G |
| One-Parameter Model | | | | |
| A | | $\alpha$ | $\alpha$ | $\alpha$ |
| T | $\alpha$ | | $\alpha$ | $\alpha$ |
| C | $\alpha$ | $\alpha$ | | $\alpha$ |
| G | $\alpha$ | $\alpha$ | $\alpha$ | |
| Two-Parameter Model | | | | |
| A | | $\beta$ | $\beta$ | $\alpha$ |
| T | $\beta$ | | $\alpha$ | $\beta$ |
| C | $\beta$ | $\alpha$ | | $\beta$ |
| G | $\alpha$ | $\beta$ | $\beta$ | |
| Six-Parameter Model | | | | |
| A | | $\alpha_1$ | $\alpha$ | $\alpha$ |
| T | $\beta_1$ | | $\alpha$ | $\alpha$ |
| C | $\beta$ | $\beta$ | | $\alpha_2$ |
| G | $\beta$ | $\beta$ | $\beta_2$ | |

parameter $\alpha$, which stands for the substitution rate of one particular nucleotide such as A for another such as T (Table 1). The parameter is also assumed to be constant over evolutionary time. The estimation formula based on this model is given by

$$K = -\frac{3}{4}\ln(1 - \frac{4}{3}p_n) \tag{4}$$

where $K$ represents the number of nucleotide substitutions and $p_n$ is the proportion of different nucleotides between the nucleotide sequences being compared. This equation is not only simple but also very useful, because it gives a reasonable estimate when sequence divergence is small. Moreover, the variance *Var(K)* of this estimate ($K$) is given by

$$\mathrm{Var}(K) = \frac{\frac{1}{n}p_n(1 - p_n)}{(1 - \frac{4}{3}p_n)^2} \tag{5}$$

where $n$ is the total number of nucleotide sites being compared.

### 3.2 Two-Parameter Method

Nucleotides A and G have similar molecular structures and are called purines. Nucleotides C and T are also similar and are called pyrimidines. Therefore, nucleotide substitutions between purines (i.e., A and G) and between pyrimidines (C and T) may be called transition-type substitutions, while those between purines and pyrimidines are called transversion-type substitutions. For many nuclear genes, the rate of transition-type substitutions is much higher than that of transversion-type ones. Moreover, more than 90% of nucleotide substitutions in the mitochondrial DNA of primates such as humans are reported to be of the transition type. To deal with such a situation, Motoo Kimura developed a two-parameter method for estimating the number of nucleotide substitutions. In this model, $\alpha$ and $\beta$ represent the rates of transition and transversion substitutions (per site per year), respectively. When the proportions of nucleotide sites showing transition-type and transversion-type differences between the two sequences compared are denoted by $P$ and $Q$, respectively, the estimation formula based on this model is given by

$$K = -\frac{1}{2}\ln[(1 - 2P - Q)(1 - 2Q)^{1/2}] \quad (6)$$

The variance of the estimated $K$ obtained by equation (6) is given by

$$\mathrm{Var}(K) = \frac{1}{n}\{(a^2P + b^2Q) - (aP + bQ)^2\} \quad (7)$$

where $a = 1/(1 - 2P - Q)$, $b = [1/(1 - 2P - Q) + 1/(1 - 2Q)]/2$, and $n$ is the total number of nucleotide sites compared.

### 3.3 Six-Parameter Model

Although there are three-parameter and four-parameter methods, we discuss the six-parameter method in this section because when applicable, it generally gives better estimates than the other methods.

The six-parameter model is based on the model originally proposed by Kimura, who called it the 2FC (two-frequency-class) model. Its exact formulation was given by Gojobori and co-workers, who showed that the number of nucleotide substitutions per site between two sequences being compared is given by

$$K = -pq\ln\left(\frac{B_1}{pq}\right) - \frac{2q_Aq_T}{p}\ln\left[\frac{p}{3q_Aq_T}\left(F_{12} - B_1 - \frac{3E_{12}}{B_1}\right)\right] - \frac{2q_Cq_G}{q}\ln\left[\frac{q}{3q_Cq_G}\left(F_{34} - B_1 + \frac{3E_{34}}{B_1}\right)\right] \quad (8)$$

In equation (8), $q_A$, $q_T$, $q_C$, and $q_G$ stand, respectively, for the contents of A, T, C, and G in the nucleotide sequences compared. Furthermore, let us use the following notation:

$$p = q_A + q_T$$
$$q = q_C + q_G$$
$$B_1 = pq - (x_{AC} + x_{AG} + x_{TC} + x_{TG})$$
$$E_{12} = (q_Aq - x_{AC} - x_{AG})(q_Tq - x_{TC} - x_{TG})$$
$$E_{34} = (q_Cp - x_{AC} - x_{TC})(q_Gp - x_{AG} - x_{TG})$$
$$F_{12} = x_{AA} + x_{TT} - x_{AT} - p^2 + 3q_Aq_T$$
$$F_{34} = x_{CC} + x_{GG} - x_{CG} - q^2 + 3q_Cq_G$$

where $x_{ij}$ represents the proportion of sites having the same nucleotide pairs $i$, and $2x_{ij}$ $(i \neq j)$ represents the proportion of sites having different nucleotide pairs $i$ and $j$ $(i, j = \text{A, T, C, G})$.

In this case, the formula for the variance of $K$ is complicated, and a computer program is required for actual use of the model.

### 3.4 Other Methods

There are several other methods. For example, Hasegawa et al. modified Kimura's two-parameter model, taking into account that the nucleotide content varies from one DNA sequence to another. In another example, Lanave and associates developed the 12-parameter method by analyzing the nucleotide substitution patterns of mammalian mitochondrial genes. In these methods, different models of nucleotide substitutions are assumed.

Generally speaking, the method depends on the model of nucleotide substitutions assumed, but the essential way of thinking is the same as that of the methods discussed in Sections 3.1 through 3.3.

## 4 METHODS OF ESTIMATING THE NUMBER OF SYNONYMOUS AND NONSYNONYMOUS SUBSTITUTIONS

When we compare two protein-coding DNA sequences, it is useful to distinguish synonymous substitutions (which do not cause amino acid changes) from nonsynonymous substitutions (which lead to amino acid alterations). If the number of nucleotide differences between two DNA sequences is very small, the number of synonymous and nonsynonymous substitutions can be obtained simply by counting synonymous and nonsynonymous nucleotide differences. However, if there are two or three nucleotide differences between corresponding codons of the two sequences, the distinction between synonymous and nonsynonymous substitutions must be inferred from the sequences by appropriate statistical methods.

### 4.1 Miyata and Yasunaga's Method

Let us compare two nucleotide sequences, codon by codon. When two codons are different at a particular codon site, we can consider minimum pathways leading to difference in the two codons. For example, codons TTT and CCA have six possible pathways (Figure 2). In each of these pathways, we can distinguish between synonymous and nonsynonymous nucleotide differences.

In practice, pathways containing more synonymous nucleotide differences tend to occur with higher probability than those having fewer synonymous nucleotide differences. Moreover, for nonsynonymous nucleotide differences, replacement of similar amino acids tends to occur more frequently than replacement of dissimilar amino acids.

Taking such a situation into account, Miyata and Yasunaga developed a weighted-pathway method (the M-Y method) for estimating the numbers of synonymous and nonsynonymous substitu-

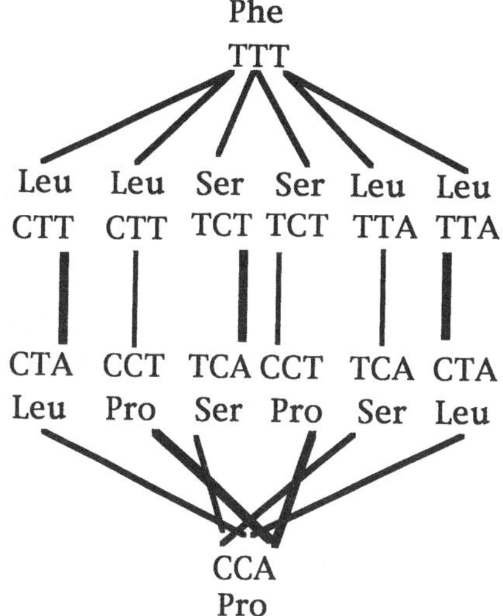


**Figure 2.** Illustration of synonymous and nonsynonymous nucleotide differences in possible pathways when homologous codons (TTT and CCA) are compared.

tions. In this method, the weight given for each nucleotide difference in a pathway depends on the biochemical similarity of the amino acid replacement involved. The estimated numbers of synonymous and nonsynonymous sites may also be obtained by suitably weighting different pathways. Then, the numbers of synonymous and nonsynonymous substitutions can be computed by equation (4), the formula for the one-parameter method, using the estimated proportions of synonymous and nonsynonymous changes. For this method, computer program is needed to perform the calculations.

### 4.2 Nei and Gojobori's Method

Nei and Gojobori's method (the N-G method) is essentially an unweighted version of that of Miyata and Yasunaga. The N-G method is much simpler than the M-Y method but gives a better estimate. In the N-G method, the proportion of synonymous changes at the $i$th position of a codon is denoted by $f_i$ ($i = 1, 2, 3$). The number of synonymous sites for this codon is given by

$$s = \sum_{i=1}^{3} f_i \tag{9}$$

and that of nonsynonymous sites by $n = 3 - s$. For a nucleotide sequence consisting of $r$ codons, the total numbers of synonymous and nonsynonymous sites are, therefore, given by

$$S = \sum_{j=1}^{r} s_j \qquad N = 3r - S, \tag{10}$$

where $s_j$ is the value of $s$ for the $j$th codon.

The total numbers of synonymous and nonsynonymous nucleotide differences can be obtained by

$$S_d = \sum_{j=1}^{r} s_{dj} \tag{11}$$

and

$$N_d = \sum_{j=1}^{r} n_{dj} \tag{12}$$

where $s_{dj}$ and $n_{dj}$ are the numbers of synonymous and nonsynonymous differences, respectively, for the $j$th codon. Thus, the proportions of synonymous ($p_S$) and nonsynonymous ($p_N$) differences are computed by $p_s = S_d / S$ and $p_N = N_d / N$, respectively, where $S$ and $N$ are the average numbers of synonymous and nonsynonymous sites in the two sequences compared.

To estimate the numbers of synonymous ($K_S$) and nonsynonymous ($K_N$) substitutions per site, the one-parameter method can be used by replacing $p_n$ by $p_S$ and $p_N$, respectively, in equation (4). This procedure is necessary to correct for the multiple substitutions.

### 4.3 Other Methods

There are still other methods for estimating the numbers of synonymous and nonsynonymous substitutions. Li et al., who developed another weighted-pathway method, used a weighting scheme based on the expected and observed frequencies of nucleotide substitutions.

A modified version of the method of Li and co-workers is proposed by Pamilo and Bianchi, and Li. The modification was made because Li's method, like the M-Y and N-G methods, tends to overestimate $K_s$ to some extent, particularly when transitional substitutions occur much more frequently than transversional substitutions. Recently, different but similar modifications have been proposed by several researchers.

## 5 PERSPECTIVES

As sequence data accumulate with enormous speed, sequence comparison will play a more important role in conducting the analysis of sequence data. In particular, quantification of sequence divergence is crucial for the prediction of gene function, the search for sequence motifs, the construction of the phylogenetic tree, the evaluation of the functional constraints of genes, and so forth.

We are confident that the methods described here are useful for estimating sequence divergence. However, every effort should be made to develop more useful methods, particularly by incorporating into the model factors such as deletion/insertion information and mutation bias.

*See also* Mitochondrial DNA, Evolution of Human; Paleontology, Molecular; Sequence Alignment of Proteins and Nucleic Acids; Sequence Analysis.

### *Bibliography*

Gojobori, T., Moriyama, E. N., and Kimura, M. (1990) Statistical methods for estimating sequence divergence. In *Methods in Enzymology*, Vol. 183. R. F. Doolittle, Ed. Academic Press, San Diego, CA.

Kimura, M. (1983) *The Neutral Theory of Molecular Evolution.* Cambridge University Press, Cambridge.

Li, W.-H., and Graur, D. (1991) *Fundamentals of Molecular Evolution.* Sinauer Associates, Sunderland, MA.

Nei, M. (1987) *Molecular Evolutionary Genetics.* Columbia University Press, New York.

## Sequencing Techniques: *see* specific method or substrate.

# Sex Determination

*Stephen S. Wachtel*

### *Key Words*

**Gonad** The sex gland, testis or ovary, which produces the haploid gametes and secretes the hormones that determine appearance of the secondary sex traits.

**Sex Determination** The process by which the sexually undifferentiated embryo is induced to become male or female.

**Sex Reversal** Any of several conditions in which gonadal sex is at variance with chromosomal sex or body sex.

**Testis-Determining Gene (*TDF*)** The Y-chromosomal gene that governs male sex determination.

The manner in which the indifferent embryonic gonad is induced to become testis or ovary is a prime question in developmental biology, and may have more general relevance for ontogeny as a whole. Differentiation of the gonad is the key event leading to development of the male or female phenotype. And this key event is influenced by the newly discovered Y-chromosomal testis-determining gene, as well as other, recently assigned, X-linked and autosomal genes. Identification of a pathway of sex-determining genes and their products should provide useful insights into the nature of sex differentiation and the various syndromes of abnormal gonadal development, as well as new approaches to the diagnosis and study of these conditions.

## 1 INTRODUCTION

Sex differentiation can be viewed as a series of orderly processes: establishment of genetic sex at conception, translation of genetic sex into gonadal sex, and translation of gonadal sex into body sex. The genetic factor that sets these processes in motion is located on the Y chromosome. In the presence of the Y, the gonad becomes a testis; the testis secretes testosterone and antimüllerian hormone (AMH), and further development is male. In the absence of the Y, the gonad becomes an ovary; testosterone and AMH are not secreted, and further development is female (Figure 1).

The Y-situated gene is called *TDF* (testis-determining factor). Here we describe the search for *TDF* with emphasis on the molecular methods leading to its characterization in 1990. We continue with a discussion of autosomal and X-linked sex-determining genes, and we conclude with a brief survey of some of the extraordinary conditions that can result when the sex-determining genes are mutated or expressed abnormally.

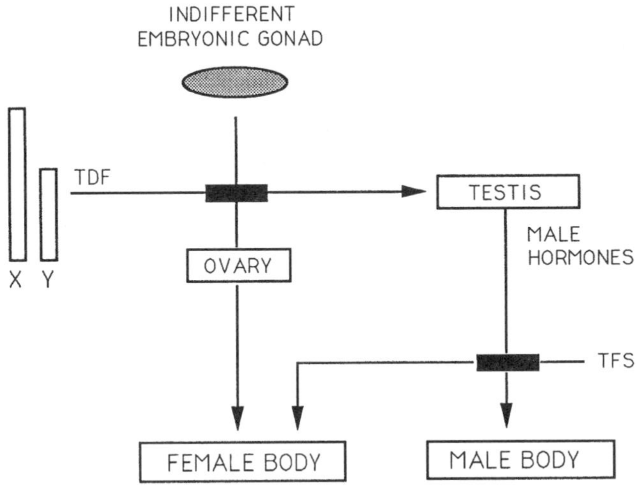


**Figure 1.** Sex determination in the human. In XX embryos, the gonad becomes an ovary and the body develops along female lines. When the *TDF* gene is present, the gonad becomes a testis. The testis secretes testosterone, which induces the wolffian ducts, and antimüllerian hormone (AMH), which blocks formation of the Müllerian ducts; in this case, the end result is the male body type. Response to testosterone and its metabolite, dihydrotestosterone, is mediated by the androgen receptor, encoded by a gene on the X chromosome. When this gene is mutated or absent, the embryo develops as female despite presence of Y chromosome and testes. The resulting condition is called testicular feminization syndrome (TFS). (After S. S. Wachtel and T. R. Tiersch, The search for the male-determining gene. In *Molecular Genetics of Sex Determination*, S. S. Wachtel, Ed., Academic Press, San Diego, CA, 1994, pp. 1–22.)

## 2 THE SEARCH FOR TDF

### 2.1 The H-Y Antigen Gene (*HYA*)

It is a dictum of biology that genes of fundamental significance are conserved and thus relatively unchanged from species to species. On that basis, a number of genes have been proposed as candidates for the Y-linked testis-determining factor. One was the gene for the male-specific H-Y antigen, which causes rejection of male skin grafts by females in inbred strains of the mouse.

When H-Y antigen was found in the heterogametic (XY) sex of diverse species representing a broad spectrum of vertebrate evolution, it was proposed that this cell surface molecule was the product of the testis-determining gene. To test the idea, H-Y was studied in cases of abnormal sex differentiation (e.g., XX males and XY females). If H-Y were the product of the *TDF* gene, it should always be present in association with testicular differentiation, regardless of chromosomal sex or body sex. For nearly a decade, the proposal was supported experimentally, but in 1984, a line of XX male mice lacking the H-Y antigen was discovered, and the search for *TDF* was redirected.

### 2.2 Bkm Satellite DNA

A satellite is a repetitive sequence of chromosomal DNA in which the nucleotide pairs G:C and A:T are represented unequally. Several satellites have been isolated from the sex-determining W chromosome of the banded krait, a poisonous snake. One such satellite—consisting mostly of GATA repeats—was called banded krait minor satellite DNA (Bkm).

The Bkm sequence was found in the W chromosome in a wide variety of snakes. It was also found in the chromosomes of other reptiles, birds, mammals, and even insects. This prompted investigators to ask whether Bkm might be involved in mammalian sex determination. Later studies showed that Bkm is not likely to be the testis-determining gene in man, however, because GATA sequences are found throughout the human genome and there are no concentrations of Bkm in the human Y. Once again the search for *TDF* was redirected, but with the work on Bkm, the study of vertebrate sex determination had passed from the purview of immunogenetics into the realm of molecular biology.

### 2.3 The Zinc Finger Y Gene (*ZFY*)

In 1987 David Page and colleagues at the Whitehead Institute produced a map of the short arm of the human Y chromosome (Yp), containing 13 deletion intervals. One of the intervals—denoted 1A2 and consisting of only 140 kilobases (ca. 0.2% of the Y chromosome)—was present in an XX male who possessed only 300 kb of Y-DNA. It was absent in a woman who had a Y:22 translocation and possessed all but 160 kb of the Y.

One of the probes from the 1A2 interval, designated pDP1007 and consisting of a 1.3 kb *Hin*dIII fragment, hybridized with a single-copy sequence in males from every mammalian species tested, including gorilla, rhesus monkey, owl monkey, rabbit, dog, goat, horse, and cattle. This indicated that the fragment identified by pDP1007 is conserved—that is, that a similar fragment is present in the Y chromosome of each of the species tested. When the amino acid sequence of the corresponding protein was derived, it was found to resemble the sequences of proteins with tandemly repeated "finger" domains. Finger proteins bind DNA and thereby regulate transcription.

Because pDP1007 recognized a tiny fragment of Y-DNA present in an XX male and absent in an "XY" female, because the pDP1007 sequence was phylogenetically conserved and male-specific, and because the corresponding protein was likely to govern transcription, it was suggested that the protein identified by pDP1007 was the product of *TDF*. According to that suggestion, the protein might bind DNA in a sequence-specific manner and thereby regulate transcription of another gene in the pathway leading to testicular differentiation. The gene recognized by pDP1007 was called ''*ZFY*'' for zinc finger Y, because zinc ions establish complexes with finger domains reminiscent of the complexes formed between iron porphyrin and the subunits of hemoglobin.

Yet three findings in particular showed that *ZFY* could not be the *TDF*. First, two ZFY genes were found in the mouse, but neither was expressed in mutant embryonic testes lacking germ cells. Second, in marsupials, *ZFY* was autosomal, hence present in males and females. And last, four human XX males were identified in whom *ZFY* was absent and another part of the Y chromosome was present. This seemed to outline a new region of the Y chromosome in which *TDF* must lie. A search of that region, comprising 35 kb adjacent to the pseudoautosomal boundary, yielded a ''new'' Y-specific gene that is, like *ZFY,* conserved across a broad spectrum of mammalian species. The new gene was called *SRY* (sex-determining region Y).

### 2.4 The Sex-Determining Region Y Gene (*SRY*)

The *SRY* gene was discovered in 1990 by Andrew Sinclair, while working in the laboratory of Peter Goodfellow in London. Initially Sinclair obtained a series of DNA probes from the 35 kb region near the Y-chromosome pseudoautosomal boundary. He used them as probes for corresponding fragments in DNA from males and females of the human, mouse and bovine species. One of the probes identified single-copy, male-specific sequences in all three species. That one was called pY53.3.

Because it hybridized most conspicuously with Y-DNA, a 0.9 kb *Hinc*II subclone of pY53.3 was used to assess the role of the *SRY* gene. Initially, the subclone was used to probe a series of Southern blots containing DNA from males and females of several species, including human, chimpanzee, rabbit, pig, horse, bovine, and tiger. In each case, a unique male-specific sequence was identified (*SRY* was found later on the marsupial Y).

In humans, the pY53.3 subclone identified a 2.1 kb male-specific fragment containing two open reading frames. One of the open reading frames was found to encode a group of amino acids with a ''striking similarity'' to part of the Mc mating-type protein of the fission yeast *Schizosaccharomyces pombe,* and to a conserved DNA-binding motif common among the nuclear high mobility group proteins, HMG1 and HMG2. This was an extraordinary finding, given the phylogenetic distance between yeast and mammals.

Male specificity, extreme phylogenetic conservation, and testis-specific expression (*SRY* transcripts were found in testis, not in other male tissues) seemed to confirm the idea that *SRY* and *TDF* were the same. Yet it remained to be demonstrated unambiguously that *SRY* was the Y-situated gene governing testicular differentiation. This could be done either by showing that *SRY* was absent or mutated in sex-reversed XY women or by demonstrating that transgenic insertion of *SRY* could induce testicular differentiation in genetically female (XX) embryos—in the mouse, for example. Both conditions were met.

The mouse homologue of *SRY* was described by John Gubbay and colleagues in a paper that appeared in July 1990, in *Nature,* back-to-back with the paper of Sinclair et al. When the human pY53.3 was used to probe Southern blots containing DNA from male and female mice, a 3.5 kb male-specific fragment was found. This strongly hybridizing fragment was found not only in normal XY males but also in sex-reversed XX males. The mouse gene was called *Sry.*

As with its human homologue, the mouse *Sry* exhibited a remarkable similarity to the conserved 80 amino acid motif of the fission yeast *S. pombe.* And like its human counterpart, *Sry* was shown to encode a protein with a DNA-binding motif. Moreover, *Sry* RNA transcripts were detected in the urogenital ridge of 10.5-day-old male embryos—just before differentiation of the testis—but not in the corresponding tissue of 10.5-day-old female embryos.

Compelling evidence for a sex-determining role of *Sry* came a year later, when it was found that the gene could induce testicular differentiation if introduced as a ''transgene'' into XX embryos of the mouse. A 14 kb fragment containing *Sry* was injected into fertilized eggs and the eggs were transferred into the oviducts of pseudopregnant females. In one group of experiments, injected embryos were allowed to develop to term. Among the 49 males and 44 females that were born, 5 were transgenic. Two were XY males and two were XX females (later found to be fertile); the last, a sex-reversed XX male mouse, was pictured on the cover of *Nature* in May 1991.

Further evidence for a sex-determining role of *SRY* was provided by studies in sex-reversed XY females of the mouse and human. In the mouse, one group of XY females was originally sired by a chimeric male produced with embryonic cells that had been infected with a retrovirus. Because sex reversal was inherited with the Y chromosome, the mutation (a deletion comprising nearly 11 kb of DNA) was designated $Y^{Tdy.ml}$, or $\not{Y}$ for short. *Sry* was among the sequences found to be deleted in X$\not{Y}$ females.

In sex-reversed human fetuses with XY gonadal dysgenesis, the ovaries degenerate and are represented at around the time of birth by ''streak gonads,'' which are functionally inert. Women with XY gonadal dysgenesis are accordingly sterile and, in general, eunuchoid. By the end of 1993, more than a hundred cases of XY gonadal dysgenesis had been evaluated for mutations within *SRY.* De novo mutations in the *SRY* conserved motif were identified in about 15% of cases. Together with the findings in transgenic XX males, these studies confirmed the testis-determining function of *SRY.*

## 3 X-LINKED SEX-DETERMINING GENES: THE SRVX (*DSS*) GENE

Failure of X-linked or autosomal sex-determining genes could explain why *SRY* mutations are found in only 15% of cases of XY sex reversal (XY gonadal dysgenesis) in the human. In fact, the condition can be transmitted as an X-linked trait in some families, and an X-linked sex-reversing gene was mapped in 1994.

That year, a group of American geneticists headed by Pamela Arn described maternal two half-siblings with ambiguous external genitalia and a small duplication in Xp: in both children, the karyotype was 46,X,dup(X)(p21.2→p22.11),Y, signifying duplication of the region Xp21.2→p22.11. Except for undifferentiated gonads and external genitalia, these children manifested none of the abnormalities found in other sex-reversed patients with duplications in the Xp21 region. This could be explained by the limited size of the duplication in these cases.

In light of these findings, a ''new'' gene, *SRVX* (sex reversal X), was assigned to the region Xp21.2→p22.11, consisting of a few megabases distal to the ornithine *trans*-carbamalase (*OTC*) locus. It was inferred (1) that perturbation of *SRVX,* as in cases of Xp21 duplication, could block male differentiation of the gonad despite the presence of an intact *SRY* sequence; (2) that some cases of XY gonadal dysgenesis could be due to *submicroscopic* duplication involving the *SRVX* locus; and (3) that *SRVX* was part of a pathway of testis-determining genes that include *SRY.*

By evaluation of overlapping duplications, Italian geneticists in the laboratory of Giovanna Camerino narrowed the locus of the new gene to a 160 kb region of Xp21 adjacent to the adrenal hypoplasia congenita locus and within approximately 100 kb of DXS319. They called the gene *DSS,* for ''dosage-sensitive sex reversal,'' and confirmed that having two active copies of the gene blocks male development in XY embryos. Since the gene was absent in certain males, it could not be required for differentiation of testes and thus could not be an essential part of the male-determining pathway.

## 4 AUTOSOMAL SEX-DETERMINING GENES: THE SRA1 GENE

In some familial clusters of XY gonadal dysgenesis, the frequency of affected relatives is consistent with the frequency that would be expected if the condition were inherited as an autosomal trait. At least one autosomal sex-reversing gene has been mapped in the human.

Campomelic dysplasia is a condition characterized by bowing of the long bones, dwarfism, cleft palate, and other congenital anomalies. Because of the severity of the condition, affected children usually die within a few weeks of birth. Patients with campomelic dysplasia can be XX or XY, but two-thirds of the XY patients are sex-reversed. Study of de novo reciprocal translocations of chromosome 17 in three different patients with this condition (two XY females and an XX female) allowed assignment of the genes for campomelic dysplasia (*CMPD1*) and sex reversal (*SRA1*) to chromosome 17q24.3→q25.1.

The association of campomelic dysplasia and sex reversal could be due to a single mutation affecting two developmental pathways or to a deletion involving contiguous genes. There may be other autosomal sex-determining genes. At least five cases of XY sex reversal have been reported in patients with a deletion of chromosome 9p.

## 5 XX SEX REVERSAL

### 5.1 XX Males

Males with a female karyotype are known in several mammalian species, notably human, mouse, goat, and dog. Human XX males, first described in 1964, occur with a frequency of approximately one per 20,000 male births. A chief feature of the XX male syndrome is the small, sterile testis. Femalelike breasts develop in about one-third of cases.

Testicular development in the absence of a Y chromosome could be explained by constitutive activation of downstream testis-determining genes, or alternatively, by Y-X translocation (identified in 70% of cases). These conditions—constitutive mutation and Y-X translocation—would be represented by two classes of XX male, one lacking the *SRY* gene (Y−) and one having it (Y+). So it is remarkable that two such classes have now been identified and that the Y(−) XX males are notable for femalelike breasts (gynecomastia), cryptorchid testes, and abnormal external genitalia—traits uncharacteristic of Y(+) XX males. Occasionally XX males occur in the same kindred as other XX males and sometimes in kindreds with XX true hermaphrodites. These XX males usually are Y(−).

XX males of the mouse were described in 1971. At first, sex reversal in these animals was attributed to mutation of an autosomal gene, *Sxr.* But the XX*Sxr* males were found to be the product of a recurrent nonreciprocal Y-X crossover involving the sex-determining region of the Y (viz., the *Sxr* region). Since XX*Sxr* males are sterile, the condition is transmitted by carrier XY*Sxr* males.

Analysis of the *Sxr* region has revealed the presence of several genes: *Sry,* the testis-determining gene; *Hya,* the gene controlling expression of H-Y antigen; *Spy,* a gene involved in spermatogenesis; *Zfy-1* and *Zfy-2,* homologues of human *ZFY;* and *Ube1-y1* and *Ube1-y2,* genes involved in ubiquitin-mediated protein metabolism.

### 5.2 XX True Hermaphrodites

There are numerous references to hermaphrodites in the ancient literature (e.g., in Plato's *Symposium* and Ovid's *Metamorphoses*). The earliest reference appears in the cuneiform tablets unearthed at the site of the Royal Library at Ninevah (seventh century BCE), and an account of the religious and social obligations of hermaphrodites appears in the Babylonian Talmud (Berachoth 61a), representing an oral tradition committed to writing nearly 2000 years ago.

According to our modern designation, a true hermaphrodite is a person with ovarian and testicular tissue—in particular, ovarian follicles and testicular tubules—in separate gonads or in a single gonad called the ovotestis. Simultaneous occurrence of male and female gonadal tissue is often associated with pronounced ambiguity of the secondary sex traits. The condition may be found in patients with a variety of karyotypes, but 60% of human true hermaphrodites are 46,XX, raising the question, as in XX males, of how testicular differentiation can occur in the absence of the Y chromosome (i.e., in the absence of *SRY*).

Whereas Y-X translocation is common among XX males, it is rare among XX true hermaphrodites. Because this is also true for XX males with ambiguous external genitalia, Y sequences other than *SRY* may be critical in the development of the normal male phenotype. Although cryptic Y-bearing cell lines in the gonad cannot be ruled out in sporadic cases of XX sex reversal, the occurrence of XX males and XX true hermaphrodites in families is often consistent with transmission of an autosomal sex-reversing gene.

## 6 XY SEX REVERSAL

### 6.1 XY Females

When the sex-determining function of the human Y chromosome is compromised, the XY gonad becomes a dysgenetic streak gonad. The resulting condition, called XY gonadal dysgenesis, or Swyer syndrome, is characterized by primary amenorrhea and delayed puberty and occurs in approximately one per 20,000 female births. In addition to having nonfunctional streak gonads, women with XY gonadal dysgenesis have a significant risk of developing germ cell tumors such as gonadoblastoma and dysgerminoma (25–35%).

Mutation in the *SRY* conserved domain in cases of XY gonadal dysgenesis is consistent with the putative sex-determining role of the SRY gene, but the low frequency of de novo mutations (15%) implies other causes of the syndrome and, in particular, other genes in the sex-determining pathway. Mutation outside the *SRY*-conserved box could also account for some cases.

### 6.2 XY True Hermaphrodites

Among the 195 human true hermaphrodites surveyed by Willem van Niekerk, 24 (12%) had the 46,XY karyotype. Of those, 13 had an ovary on one side of the body and a testis on the other; the other 11 had at least one ovotestis. The mosaic phenotype of the true hermaphrodite (testis vs. ovary) implies mosaic expression of the testis inducer, and this is borne out by molecular analysis of the SRY gene.

The subject of one study was a newborn baby with ambiguous external genitalia and 46,XY karyotype. The gonads were described as ovotestes containing tubules, Sertoli cells, spermatogonia, and oocytes. There were no Leydig cells and no intact follicles. DNA was extracted from white blood cells and gonadal tissue and the *SRY*-conserved motif was amplified by means of the polymerase chain reaction (PCR). When the PCR products were sequenced, no mutation was found in *SRY* from the white blood cells, but two mutations were found in gonadal tissue. One, a "silent" mutation, had no effect on the amino acid sequence. The other, a T→A transversion in the second nucleotide of codon 44, induced an amino acid substitution of histidine for leucine. The failure of the two mutations to be identified in all samples of gonadal tissue confirms mosaic expression of the wild-type *SRY* and explains development of the ovotestes in this patient.

## 7 SUMMARY AND CONCLUSIONS

The function of the human Y chromosome was arguable as recently as 1959, when a number of publications appeared showing that the Y is a sex determinant. By 1990, the Y-chromosomal testis-determining gene had been cloned, and soon thereafter a comprehensive Y-deletion map containing 132 loci and 43 ordered intervals was produced; each interval was defined by chromosomal breakpoints and contained fewer than 800 kb of DNA.

By the end of 1994, several sex-determining genes had been identified: the *SRY* testis determinant on the Y chromosome, *SRVX/DSS,* the X-linked sex-reversing gene, and *SRA1,* the autosomal sex-reversing gene on chromosome 17q. Another sex-reversing gene had been indicated on chromosome 9p, and in keeping with its presumptive role as a transcription factor, the SRY protein was found to recognize and bind a particular DNA sequence, 5′-AACAAAG-3′.

These findings were consistent with the notion that maleness is the result of a cascade of events governed by a pathway of sex-determining genes. Yet the role of these genes remained uncertain. For example, *SRY* was believed by some to be the initiator of the cascade. But *Sry* of the mouse was found to be transcribed during days 9.5–12.5 of gestation (the mouse testis appears at ca. day 11.5); this suggested that *Sry* itself is "turned on" by another gene.

If *SRY/Sry* were a positive regulatory element, the testis-determining cascade could be represented in this way:

$$? \rightarrow SRY \rightarrow A \rightarrow B \rightarrow C \rightarrow n \text{ testis}$$

where *SRY* might be activated by another autosomal or X-linked gene, *A, B,* and *C* are other autosomal or X-linked genes downstream of *SRY,* and *n* represents an indeterminate number of genes in the pathway.

Significant advances have accordingly been made in our understanding of the various conditions of sex reversal, and it seems reasonable to look forward to a time, perhaps not far off, when the genetics of sex determination will be thoroughly understood, when many of the human syndromes of sex reversal will be identified and even corrected prenatally, and when the gender of important domestic species will be placed under absolute control.

*See also* Human Repetitive Elements; Recombination, Molecular Biology of; Zinc Finger DNA Binding Motifs.

### *Bibliography*


Gubbay, J., and Lovell-Badge, R. (1994) The mouse Y chromosome. In *Molecular Genetics of Sex Determination,* S. S. Wachtel, Ed., pp. 43–67. Academic Press, San Diego, CA.

Jost, A. (1970) Hormonal factors in the sex differentiation of the mammalian foetus. *Phil. Trans. R. Soc. London B,* 259:119–130.

McLaren, A. (1990) What makes a man a man? *Nature,* 346:216–217.

Sinclair, A. H. (1994) The cloning of SRY. In *Molecular Genetics of Sex Determination,* S. S. Wachtel, Ed., pp. 23–41. Academic Press, San Diego, CA.

Wolf, U., Schempp, W., and Scherer, G. (1992) Molecular biology of the human Y-chromosome. *Rev. Physiol. Biochem. Pharmacol.* 121: 147–213.


# Steroid Hormones and Receptors

*Robin Leake*

### *Key Words*

**Cholesterol** A 27-carbon structure based on four concerted rings, three of which contain six carbons and one, five carbons. It is the parent molecule for synthesis of steroid hormones.

**DNA-Binding Proteins** Nuclear proteins that bind specific sequences of nucleotides within DNA and modulate the rate of transcription of specific structural genes.

**Plasma Carrier Proteins** Molecules present in the plasma that carry steroid hormones and other molecules, partly to protect them from metabolism in the liver and partly to act as a reservoir. Binding is relatively low affinity (except for sex hormone binding globulin) and nonspecific.

**Steroid Hormones** Hormones synthesized from cholesterol and acting through specific receptors, which are found in the soluble compartments of the cell.

**Steroid Receptors** Members of a superfamily of receptors that bind their ligands (hormones) with high affinity and, as a result of binding ligand, acquire a high affinity for specific sequences in the DNA, through which the steroid-regulated genes are controlled.

Many aspects of reproduction, behavior, differentiation, and metabolism depend, in the long term, on the actions of different steroid hormones. Most, if not all, actions of steroid hormones are mediated through specific protein receptors, which are themselves located inside the target cells. This contribution gives a brief introduction to steroid hormones and to the manner in which they achieve their physiological responses.

Steroid hormones are all derived from cholesterol. Cholesterol is an integral part of the cell membrane, and steroid hormones retain this property of being membrane soluble. Consequently, they enter all cells down a concentration gradient. However, target cells can be identified because they are capable of concentrating the particular steroid. This is achieved because steroids act by binding to the intracellular receptors.

## 1 RECEPTORS

Steroid receptors are part of a superfamily of molecules that regulate gene expression by direct interaction with the upstream region of specific structural genes. Other members of the superfamily include the thyroid hormone receptor, the retinol receptor, and the oncogene v-*erb*A. The receptors bind the appropriate steroid with high affinity (dissociation constants around $10^{-10}M$). The ''empty'' steroid receptor is normally located in the nucleus of target cells (empty glucocorticoid receptor can be found in the cytoplasm). In its ''empty'' form, it is part of a multimeric complex made up of the receptor molecule plus a series of other proteins (mainly heat shock proteins) that protect the molecule and prevent it from binding to the DNA until activated. Upon binding steroid, the receptor becomes activated, the heat shock proteins are released, and the DNA-binding domain of the receptor molecule becomes exposed. The activated receptor can now, in conjunction with specific transcription factors, modulate the expression of selected genes. An individual steroid can modulate different structural genes in different target tissues. Much has been learnt about the nature of control of gene expression through the study of the action of activated steroid receptors. For example, each receptor molecule contains two zinc fingers in the DNA-binding domain. Upon activation, the receptor dimerizes and the dimer binds to a consensus site upstream of the structural gene. This binding is achieved through the zinc fingers, which literally reach down into the major groove of the DNA double helix and hold the receptor in the appropriate place on the relevant sequence of nucleotides that define the binding site(s) for that particular receptor. The C- and N-terminal regions of the receptor contain transactivation activities which, in association with other transcription factors, regulate the transcription of the gene.

To understand the action of all steroid hormones, it is necessary to identify the sites and biochemistry of their synthesis; to trace their transport through the bloodstream, uptake into cells, interaction with and activation of receptors; and to discover how the receptors modulate target gene expression.

## 2 CLASSIFICATION

Steroids are classified according to the physiological responses they induce in their target tissues. The different groups to be considered are mineralocorticoids, glucocorticoids, progestins, androgens, and estrogens. In reality, other ligands that bind to members of the superfamily of receptors probably should be included (e.g., vitamins A and D, together with thyroid hormones). However, we limit ourselves here to the classic steroid hormones.

## 3 SYNTHESIS

Overall, steroid hormones are synthesized from cholesterol (which may be of dietary origin or may be synthesized de novo from acetate). Synthesis from cholesterol involves a series of reactions that depend on the cytochrome P450 enzymes. Much of the synthesis, therefore, occurs in microsomes, although some steps take place in the mitochondria. The rate-limiting steps are, initially, the mobilization of the cholesterol (from its stored form as an ester) and the cleavage of the side chain. This initial step is achieved as a result of activation of the cell by a trophic hormone (e.g., corticotropin in the adrenal cortex, or luteinizing hormone, LH, in the ovaries or testis).

The principal *glucocorticoids,* cortisol (humans) and corticosterone (rats and mice), are synthesized mainly in the zona fasciculata of the adrenal cortex. The primary pathway in man involves the 17-hydroxylation of pregnenelone, followed by conversion to 17-hydroxyprogesterone, then 21-hydroxylation to form 11-deoxycortisol.

From a clinical point of view, the output of *adrenal androgens* can be critical in cases of masculinization, hormone-sensitive cancers, and so on. The major androgen produced by the adrenal cortex is dehydroepiandrosterone (DHEA). The adrenal androgens are synthesized in both the zona fasciculata and the zona reticularis. Most of the 17-hydroxypregnenelone follows the pathway to glucocorticoid, but some is oxidized and has the remaining two-carbon side chain removed by the 17,20-desmolase to produce DHEA. Adrenal androgen production increases significantly if any of the hydroxylases on the glucocorticoid pathway are missing (e.g., in congenital adrenal hyperplasia). DHEA is rapidly sulfated to an inactive form, but sulfatases exist to reactivate the molecule. DHEA is only a weak androgen, but androstenedione, produced in small amounts in the adrenal gland by action of the desmolase on 17-hydroxyprogesterone, is much more potent and is important in several clinical conditions. Enzymic reduction of androstenedione at carbon-17 leads to testosterone, the principal circulating biological androgen. Little testosterone is produced in the adrenals, but the enzyme occurs in peripheral tissues, and the reaction is significant in postmenopausal women (hirsutism in postmenopausal women, etc.).

*Mineralocorticoids* are synthesized in the zona glomerulosa of the adrenal gland. When the adrenal cortex is fully developed, corticotropin plays no role in activating mineralocorticoid synthesis and release. This is done, instead, by the renin–angiotensin system, which is itself activated when the kidney becomes aware of abnormal excretion of sodium ions.

The ovaries are the main source of *estrogens* in premenopausal women. The estrogens—estradiol (biologically the most active), estrone, and estriol—are all synthesized both in follicular tissue and in corpora lutea. In the follicular phase of the cycle, the estrogens are synthesized, after activation of the cells by follicle-stimulating hormone, from androgens generated in the thecal cells of the ovary. These are then aromatized to estrogens in the granulosa cells.

Testicular *androgens* are synthesized in the interstitial tissue of the testes by the Leydig cells, after activation by luteinizing hormone. Pregnenelone is converted to progesterone, to 17-hydroxyprogesterone, to androstenedione, and, finally, to testosterone. In

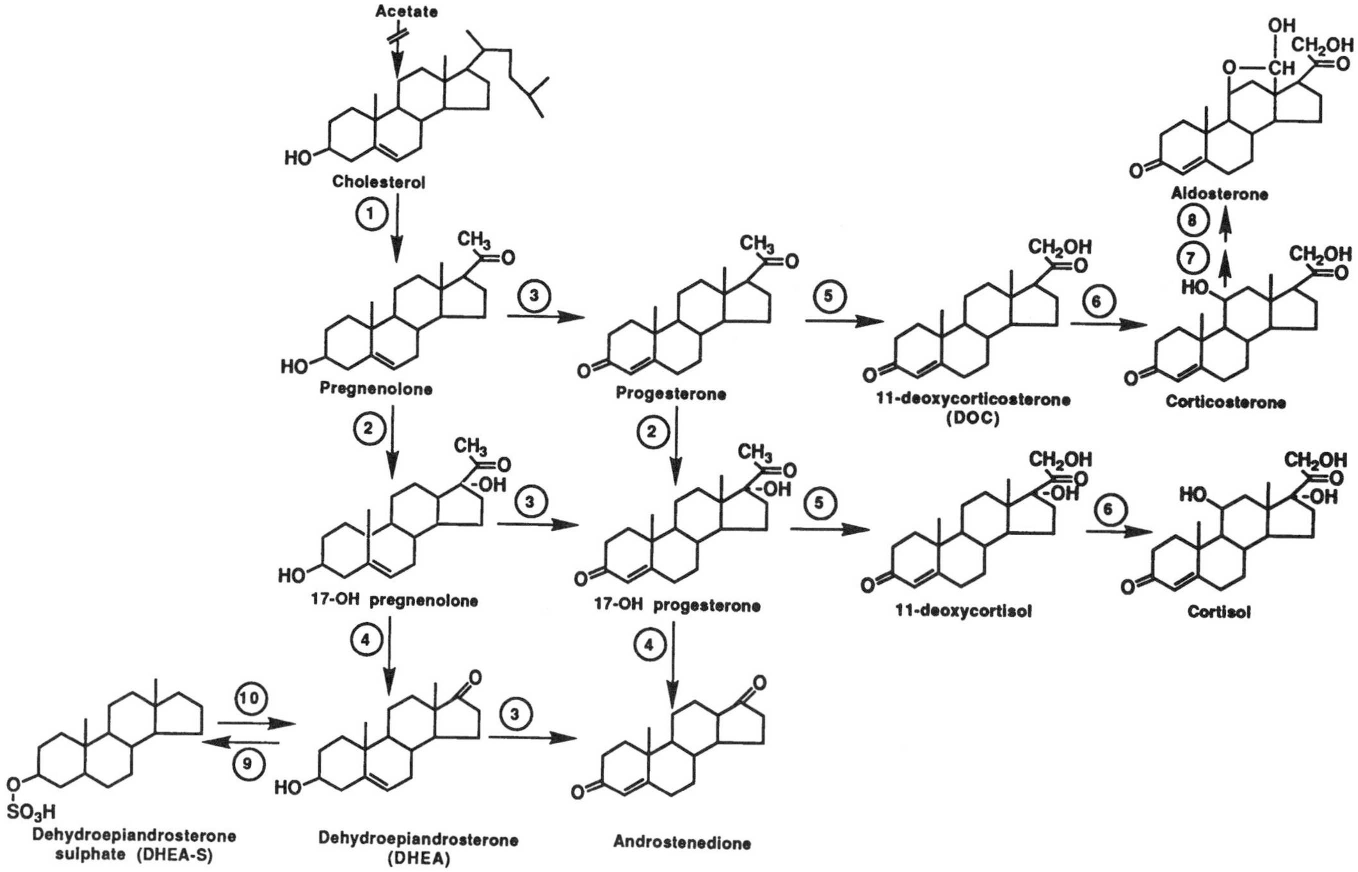

Figure 1. Principal pathways of steroid synthesis in the adrenal cortex. The enzymes are 1, cholesterol side chain cleavage complex; 2, 17-hydroxylase; 3, 3β-hydroxysteroid dehydrogenase; 4, C-17,20-desmolase; 5, 21-hydroxylase; 6, 11β-hydroxylase; 7, 18-hydroxylase; 8, 18-hydroxysteroid dehydrogenase; 9, sulfotransferase; and 10, sulfatase.

some target tissues, testosterone is converted to dihydrotestosterone (DHT) before it becomes fully active.

The pathways of steroid synthesis are shown in Figure 1.

## 4 TRANSPORT

Steroid hormones are carried in the bloodstream by a series of carrier proteins (cortisol binding globulin, sex hormone binding globulin, albumin, etc.). In addition to the role of transporting hydrophobic molecules through the aqueous environment of the bloodstream, these carriers act to protect the steroid from degradation in the liver prior to biological activity at the target tissue. They can also provide a ready reservoir of steroid, at a time of shortage.

## 5 FUNCTION

The biological role of *glucocorticoids* is to stimulate glucose supply at times of hypoglycemia and, correspondingly, to promote release of acetyl CoA units from triglyceride-derived free fatty acids and stimulate protein degradation (on the one hand, to "spare" existing plasma glucose and, on the other, to provide carbon skeletons for gluconeogenesis). Because steroid hormones act by altering gene expression, this response involves synthesis of new enzymes (e.g., liver transaminases to make available carbon skeletons from protein) and so is a long-term response. Glucocorticoids also regulate blood pressure (high levels being associated with hypertension), and adequate levels of glucocorticoid are essential to combat stress; they have anti-inflammatory and antiallergic effects. *Mineralocorticoids* regulate sodium–potassium balance by stimulating the recovery of sodium, chloride, and bicarbonate ions from the bladder back across the glomerular cells into the bloodstream. This activity has corresponding effects on blood volume and blood pressure.

*Estrogens,* at menarche, promote growth of female reproductive tissues and stimulate the laying down of fat in specific sites. In adults, estrogens regulate the female sex tissues and secondary characteristics. Their principal action is to stimulate replenishment of the uterine lining following menses. However, they influence epithelial cells in most reproductive tissues. A major role is to prepare cells for the action of progesterone by inducing synthesis of the progesterone receptor. *Progesterone* promotes the differentiation of the secretory uterine epithelium, preparing it to receive a blastocyst, should fertilization have taken place. A successful pregnancy requires both estrogens and progesterone throughout. Indeed, it is the fall in progesterone levels that signals the onset of labor. *Androgens* (testosterone and DHT) promote male sex character and stimulate maturation of sperm throughout the life of the male.

A variety of synthetic steroids have been developed over recent years to control clinical conditions that arise from abnormal steroid production. For example, much work has been carried out on preparations for the oral contraceptive pill and for hormone replacement therapy. It is important, in designing synthetic steroids, to allow for probable cross-reactivity of the synthetic molecule, given the superfamily nature of the steroid receptors. Antihormones have proved very useful in treating hormone-sensitive cancers (especially prostate and breast cancer). Particularly important is *tamoxifen,* which is a triphenyl-ethylene, rather than a steroid; it binds to the estrogen receptor with lower affinity than the natural estrogens and acts as an antiestrogen in the presence of endogenous estrogens. For this reason, tamoxifen has been probably the most successful drug for treatment of existing breast cancer. There is also a case for giving tamoxifen to otherwise well women who are thought to be at high risk of developing breast cancer. However, tamoxifen is a weak estrogen agonist, rather than being a pure antagonist, and there is a real risk that long-term use of tamoxifen may result in problems of hyperplasia and even cancer in the endometrium.

*See also* CYTOCHROME P450; RECEPTOR BIOCHEMISTRY.

### *Bibliography*

Gower, D. B. (1979) *Steroid Hormones.* Croom Helm, London.

Leake, R. E., and Habib, F. (1987) Steroid hormone receptors: Assay and characterisation, in *Steroid Hormones: A Practical Approach,* B. Green and R. E. Leake, Eds. IRL Division of Oxford University Press, Oxford.

Parker, M. G., Ed. (1991) *Nuclear Hormone Receptors: Molecular Mechanisms, Cellular Functions and Clinical Abnormalities.* Academic Press, New York.

# SUPERANTIGENS

*Monique Lafon*

### *Key Words*

**Enterotoxins** Globular proteins (24–30 kDa) released by bacteria, such as the common human pathogen, *Staphylococcus aureus.* The enterotoxins, which are responsible for food poisoning and occasionally shock, are designated SE (*staphylococcal* enterotoxin) and classified by letters (e.g., SEA, SEB, SEE).

**Major Histocompatibility Class (MHC) Class II Molecules** Antigen-presenting cells bear, on their surface, molecules of the major histocompatibility complex of class II (MHC class II molecules). These molecules bind and present antigen to the T-helper lymphocytes, which cannot recognize soluble antigen. MHC Class II molecules are composed of a variable αβ-complex and an invariant chain, the Ii. The variable αβ-complex is made of two chains, α and β, both of which contain constant and variable (V) elements. Fragments of antigen are bound to portions of the variable α- and β-chains which form a specific groove structure.

**Mouse Mammary Tumor Virus (MMTV)** Retrovirus that causes tumors of the mammary gland in aged female mice. MMTVs encode a superantigen within the 3′ long terminal repeat sequence, which produces a 320 amino acid type II membrane-associated protein.

**T-Cell Receptor** T lymphocytes express on their surface a T-cell receptor (TCR) for the antigen. The TCR is composed of a variable αβ complex and an invariant complex, the CD3. The variable αβ complex is made of two chains, α and β, both of which contain constant and variable (V) elements. Only the variable elements of the α- and β-chains that exhibit an extreme diversity are involved in the antigen recognition. TCRs can be classified according to the nature of their Vα- or Vβ-chains. The β-chain is less diverse than the α-chain. In humans 25 Vβ families have been described so far. The families are numbered (e.g., Vβ 6, Vβ 8).

---

Superantigens are a special class of antigens, defined by the capacity to stimulate a large fraction of T cells (1 in $10^2$ to 1 in $10^3$) and to bind to almost any kind of major histocompatibility class (MHC) class II molecule. Unlike conventional antigens, superantigens do not require processing by antigen-presenting cells, and superantigen activation depends almost exclusively on the variable elements of the β-chain (Vβ) of the T-cell receptor (TCR). These properties are compatible with a model of MHC–superantigen–TCR interaction in which superantigen binds to the outer faces of MHC class II molecules and the Vβ-chain of the TCR.

Superantigens are classified by their action. Thus, the term is applied to a number of distinct molecules from different microorganisms. The best-studied superantigen is the bacterial enterotoxin produced by *Staphylococcus aureus,* one of the most powerful T-cell activators known. Superantigens are also produced by other bacteria and by mycoplasma, and they have been found to be encoded by viruses, such as the mouse mammary tumor retroviruses or the rabies virus. Study of superantigens has provided new insight into the interaction of TCR and MHC class II molecules. Furthermore, evidence has indicated a potential role for superantigens in the pathogenesis of several human diseases.

## 1 INTRODUCTION

The molecular basis of the activation of antigen-specific T-lymphocyte white blood cells is the recognition of a specific peptide antigen by the αβ-chains of the antigen T-cell receptor (TCR) on the surface of the T lymphocytes. To be recognized by TCRs, antigens need to be cleaved proteolytically into small peptides by antigen-presenting cells (APCs) before they can be presented by class I or II major histocompatibility complex (MHC) molecules (upper panel of Figure 1). The TCR–MHC interaction is facilitated by the CD4 and CD8 coreceptors of the T cell. Specific recognition of a peptide is generated by somatic recombination of variable elements of the αβ-chains of both TCR and MHC molecules. As a consequence, the frequency of T-cell response to a particular peptide antigen is usually extremely low (1 in $10^4$ to 1 in $10^6$). Superantigens are a special class of antigens, defined by the capacity to stimulate a large fraction of T cells (1 in $10^2$ to 1 in $10^3$) in the presence of APCs but almost without MHC restrictions. Unlike conventional antigens, superantigens do not require to be processed by APCs, and superantigen activation depends almost exclusively on the Vβ-elements of the TCR. These properties are compatible with a model of MHC–superantigen–TCR interaction in which superantigen

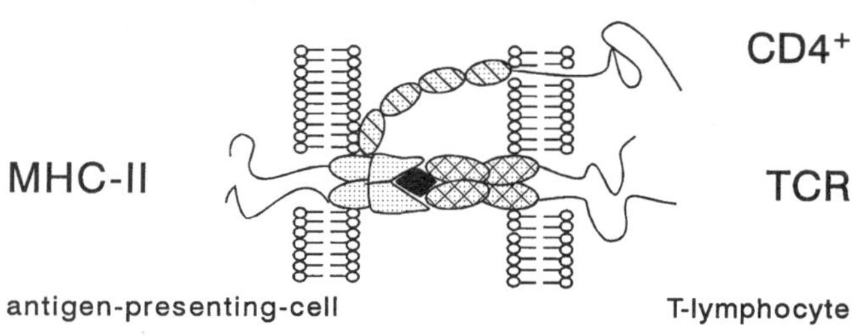


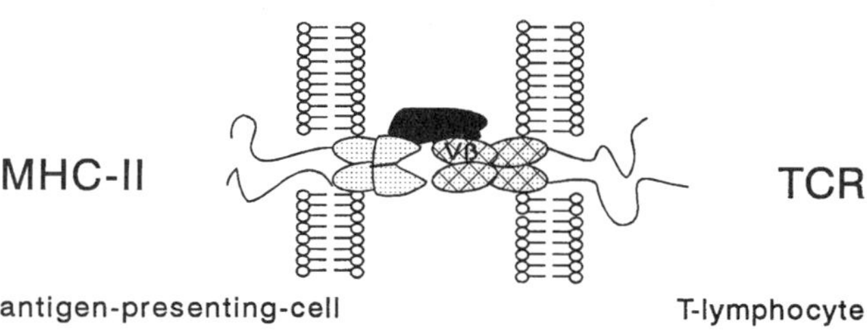


**Figure 1.** Recognition of conventional antigen and superantigen by CD4$^+$ T lymphocyte. *Top:* Peptide antigen (dark) must be processed by the antigen-presenting cells (APCs) and exposed to its surface inside the peptide groove of the MHC II molecule before it can be seen by the T-cell receptor (TCR) of the T lymphocytes. Additional molecules, such as the CD4 molecule, participate in the recognition by strengthening the MHC=nTCR association. Virtually all variable elements of the TCR are involved in these recognitions. As a consequence, very few T cells are stimulated by a specific peptide. *Bottom:* In contrast, the recognition of a superantigen (dark) involves predominantly the variable region of the β-chain of the TCR and virtually any MHC II molecule. As a consequence, superantigens can recruit entire T-cell Vβ families, which can correspond to the stimulation of up to 20% of the T cells. Most of superantigens do not require interaction with the CD4 molecule.

binds to the outer faces of MHC class II molecules and the Vβ-chain of the TCR (lower panel of Figure 1).

Superantigens are classified by their action. Thus, the term is applied to a number of distinct molecules from different microorganisms. The best-studied superantigen is the bacterial enterotoxin produced by *Staphylococcus aureus.* Superantigens are also produced by other bacteria and by mycoplasma, and they have been found to be encoded by retroviruses, such as the mouse mammary tumor retroviruses (MMTVs), which have indicated a new source of potentially severe immunopathological consequences and have spurred the search for new superantigens of viral origin in humans. Although superantigenlike properties have been suggested for the human immunodeficiency type 1 virus (HIV-1), the only viral superantigen described in humans to date has been the rabies virus nucleocapsid.

## 2 SUPERANTIGEN–MHC CLASS II INTERACTION

The superantigens have two well-defined functions: to bind to several alleles and isotypes of MHC class II molecules, and to activate particular T-cell families via the variable portion of the β chain. As a result, a superantigen bridges MHC and the TCR molecules (Figure 2), leading to the activation of T lymphocytes.

### 2.1 Position of the Superantigen-Binding Site on MHC Class II Molecules

MHC class II molecules contain two chains, α and the β, both of which are involved in the binding of superantigens. Superantigens bind to different regions of the variable domains of class II molecules. Bacterial enterotoxins are grouped on the basis of their binding site on MHCs. For example, the *Staphylococcus aureus* enterotoxins A, D, and E (SEA, SED, and SEE, respectively) are in the same group because the binding of one of them to MHC class II molecules abolishes the binding of the other members of the group. The α-chain seems to be required for the binding of toxic shock syndrome toxin 1 (TSST-1) and for the rabies virus nucleocapsid, whereas SEB binds the β-chain. Other superantigens, such as SEA, recognize both chains, suggesting that their binding sites bridge the αβ complex.

The regions of the MHC involved in superantigen recognition probably are located at the end of the peptide-binding groove, however the precise (location) of the site at which superantigens bind to MHC class II molecules has not been identified.

### 2.2 Position of the MHC Class II Binding Site on the Superantigen

Competitive experiments attempting to bind synthetic peptides to SEA indicate that only peptides attached to the very last amino acid of SEA inhibit MHC class II binding. Therefore, it has been proposed that the MHC class II binding site on the SEA is at least the absolute amino terminus of the toxin, the one that contains the zinc ion (large dot in the Figure 2). The zinc ion seems to play an important role for the interaction of some superantigens, like SEA and SEE, with MHC class II molecules.

## 3 SUPERANTIGEN–TCR INTERACTION

The binding of bacterial superantigens to MHC class II molecules is readily detectable by cytofluorimetry or immunoblotting techniques. In contrast, binding to TCR is not detectable and becomes significant only after superantigen has first bound to the MHC class II molecules or has been artificially immobilized. This weakness has led investigators to speculate that binding to MHC class II molecules induces the exposition of superantigen motifs that are otherwise buried in the structure, hence not accessible to the TCR.

### 3.1 Position of the Superantigen-Binding Site on the TCR

The variable domain of the TCR is composed of two chains, Vα and Vβ. The main contributor to T-lymphocyte stimulation by superantigen is clearly the TCR Vβ-chain. The human TCR Vβ repertoire contains at least 50 different gene segments, some of

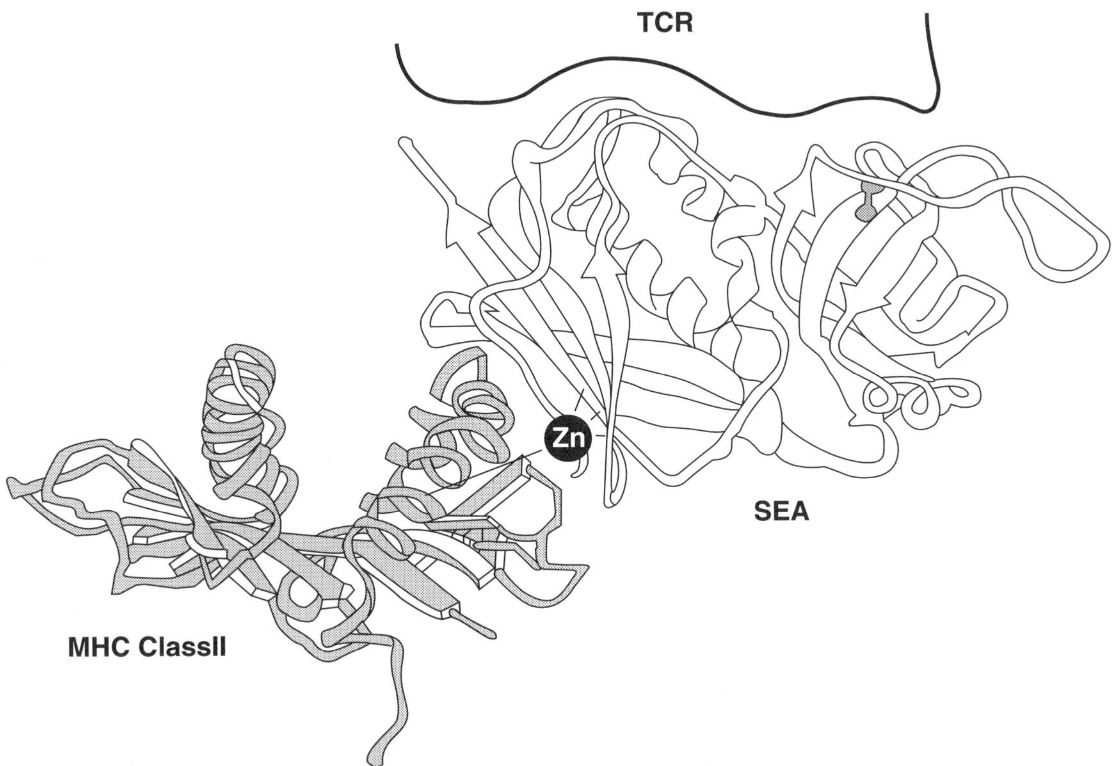


**Figure 2.** Schematic representation of the MHC=nsuperantigen=nTCR spatial interactions. By using the crystal structure of staphylococcal enterotoxin A (SEA) as a model of a superantigen (right) and the crystal structure of an MHC class I molecule, which is similar to an MHC class II molecule (left), the MHC=nsuperantigen interactions can be schematized. Interaction with the VH4 loop of the TCR Vβ is suggested in the top of the drawing; the zinc ion is indicated. (Adapted from Fraser et al., Superantigens: A pathogen's view of the immune system. *Current Communications in Cell and Molecular Biology*. Cold Spring Harbor Laboratory Press, Cold Spring Harbor, NY, 1993, p. 7.)

which are responsive to superantigen. Two methods can be used to determine which Vβs are targeted by a superantigen: the cytofluorimetry method and the polymerase chain reaction (PCR) technique. Cytofluorimetry using monoclonal antibodies specific for different Vβ families permits the characterization of the enrichment of Vβ genes in superantigen-activated lymphocyte samples. Similarly, reverse dot–blot PCR using primers specific for Vβ elements permits the quantification of the percentage of each Vβ families upon superantigen activation. The region of TCR that interacts with superantigens has been shown to be a hypervariable loop in Vβ between residues 70 and 74 (HV4), located to one side of the antigen-binding site.

### 3.2 Position of the TCR-Binding Site on the Superantigen

The locus of the TCR-binding site on SEA was analyzed by preparing hybrids of SEA–SEE molecules containing variable lengths of each superantigen molecule. SEA target T cells were analyzed after stimulation with hybrid molecules that had been first bound to MHC class II molecules. This type of experiment indicates that the MHC class II binding site is located in the final 37 positions of the amino-terminal portion of SEA. Moreover, 206 and 207 residues were found crucial for binding to the TCR β-chain. In the crystal structure of SEB, 206 and 207 residues can be topographically located in a shallow groove on the top of the molecule (Figure 2). This region most likely fits the HV4 loop domain of the TCR.

## 4 SUPERANTIGENS AND DISEASE

Recognition of superantigen by a host is a two-step reaction. First, the superantigen triggers the expansion of certain TCR Vβ subsets; later, the expanded T cells enter a state of unresponsiveness and/or are driven to death. The consequence of massive expansion of superantigen target T cells may be the production of large amounts of unbalanced cytokines, which may have deleterious effects on the host. Similarly the deletion of entire subsets of Vβ families may jeopardize the T-dependent defense capacity of a host infected by microorganisms bearing a superantigen. Therefore, the powerful modulation of the T-cell responses by superantigens is likely to exert drastic effects on the mechanisms of infection and host defense, and therefore on the process and outcome of infection.

### 4.1 Toxic Shock Syndrome

Toxic shock syndrome is a serious disease characterized by rapid onset of fever, rash, and shock. Substances implicated in the pathogenesis of the disease include the bacterial superantigens, the toxin-1 TSST-1, the staphylococcal enterotoxins SEB and SEC, and the streptococcal erythrogenic toxin A (SPE-A), secreted by the group A *Streptococcus.* In humans, these bacterial superantigens stimulate Vβ T cells on a massive scale. These T cells directly release large quantities of cytokines such as interleukin-2 (IL-2), interferon-γ, and tumor necrosis factor-β, or help non-T cells (e.g., macrophages) to release IL-1, all of which could contribute to the induction of shock. It is known that toxic shock syndrome is due to the presence of T cells because mice lacking T cells or treated to suppress T-cell function were protected against the appearance of the disease.

### 4.2 Mouse Mammary Tumor Virus Infection

Mouse mammary tumor virus, a retrovirus that causes mammary tumors in mice, also has been found to exhibit superantigen properties. The viral superantigen corresponds to a type II transmembrane glycoprotein encoded by the open reading frame (ORF) in the 3′ long terminal repeat (LTR). The virus is transmitted maternally via milk. It passes the gut epithelium of newborn mice and elicits a superantigen-mediated immune response in the peritoneal lymph nodes, the Peyer patches. T cells and especially $CD4^+$ T cells are crucial to the infection, since immunosuppressed animals cannot be infected by the virus. B cells are the best presenting cells in the mouse and are easily infected. These infected B cells are caused to multiply by superantigen reactive T cells. Superantigen expression enables the virus to spread infection more efficiently by amplifying the rare infected B cells. Therefore mouse mammary tumor superantigen may facilitate host invasion.

### 4.3 Human Immunodeficiency Virus Infection

Infection by HIV-1 is characterized by a progressive decrease of helper $CD4^+$ T cells. This deletion of $CD4^+$ T cells may have multiple origins, including the destruction of the cells by the cytopathic effect of the virus. However, it has been hypothesized that an HIV superantigen may participate in $CD4^+$ T-cell depletion as follows: HIV-infected $CD4^+$ T cells bearing HIV superantigen first activate T cells expressing the appropriate Vβs and then drive them to death. The nature of the deleted Vβ varies between patients. The type of deleted Vβ could be explained by the extensive mutation capacity of the HIV genes. The variability of HIV-1 encoded superantigen has also been evoked to explain why several Vβs are deleted in the course of the disease in one patient. HIV-1 superantigen could also facilitate the multiplication of the virus by stimulating the virus reservoirs, as has been suggested in the case of mouse mammary tumor retrovirus infection. This point of view found strong support in the observation that T-cell susceptibility to HIV-1 could depend on the TCR Vβ gene expressed by these cells and, in particular, that Vβ12 $CD4^+$ T cells were found both in vivo and in vitro to best replicate HIV-1.

### 4.4 Autoimmune Diseases

Most of the lymphocytes directed to self-motifs (autoreactive lymphocytes) are normally deleted or inactivated during thymic maturation. If autoreactive lymphocytes are reactivated at an inappropriate time, however, a few may escape the tolerance mechanism and initiate autoimmune disorders. The capacity of superantigens to break the barriers of MHC restriction and to activate large numbers of T and B cells has led to the hypothesis that superantigens may activate autoreactive T and B cells to initiate or aggravate autoimmune diseases. Certain autoimmune diseases have characteristics consistent with a role for superantigens in the etiology of the disease. Induction of arthritis in rats by a superantigen encoded by *Mycoplasma arthriditis,* the *M. arthriditis* mitogen (MAM), and the frequent increase of particular Vβ T cells (e.g., Vβ17) in various autoimmune diseases, or T lymphocytes in the synovial tissues and fluids of patients suffering of rheumatoid arthritis, strongly support this point of view.

## 5 PERSPECTIVES

Superantigens represent an entirely new class of antigens which violate the rules of T-cell recognition on several points: (1) superantigens are not processed by antigen-presenting cells into peptides and they associate with MHC class II molecules outside the peptide binding groove; (2) although the presence of antigen-presenting cells is critical for T-lymphocyte activation, the recognition of superantigen is not MHC restricted; and (3) superantigens are specific for a particular TCR Vβ element, regardless of the nature of hypervariable parts of the TCR. These properties make superantigens powerful tools for the study of the mechanism of T-cell activation and deletion in vivo. The powerful modulation of T-cell responses by superantigens is likely to exert a drastic effect on infection and host defense mechanisms, and therefore on the process and outcome of infection. It has been observed in murine autoimmunity diseases that depletion of T cells carrying Vβs implicated in these diseases tended to prevent or reduce pathology. Identifying new viral superantigens should offer important insights into the pathogenesis of human disease and should suggest novel therapeutic approaches as well.

*See also* Immunology; Major Histocompatibility Complex; Recognition and Immobilization, Molecular.

### *Bibliography*

Acha-Orbea, H., and Palmer, E. (1991) Mls-a retrovirus exploits the immune system. *Immunol. Today,* 12:356.

Cole, B. C., and Atkin, C. L. (1991) The *Mycoplasma arthritidis* T-cell mitogen, MAM; a model superantigen. *Immunol. Today* 12:271.

Held, W., Shakhov, A. N., Waanders, G., Izui, A. N., Waanders, G. A., Scarpellino, L., MacDonald, H. R., and Acha-Orbea, H. (1993) Superantigen-reactive $CD4^+$ T cells are required to stimulate B cells after infection with mouse mammary tumor virus. *J. Exp. Med.* 177:359.

Janeway, C. A. (1991) Immune recognition: Mls: Makes a little sense. *Nature* 349:459.

Lafon, M., Lafage, M., Martinez-Arends, A., Ramirez, R., Vuillier, F., Charron, D., Lotteau, V., and Scott-Algara, D. (1992) Evidence of a viral superantigen in humans. *Nature* 358:507.

Pantaleo, G., Graziosi, C., and Fauci, A. (1993) The immunopathogenesis of human immunodeficiency virus infection. *New Engl. J. Med.* 328:327.

Posnett, D. N. (1993) Do superantigens play a role in autoimmunity? *Semin. Immunol.* 5:65.

Swaminathan, S., Furey, W., Pletcher, J., and Sax, M. (1992) Crystal structure of staphylococcal enterotoxin-B, a superantigen. *Nature* 359:801.

# Synapses

*Hui-Quan Han and Robert A. Nichols*

### *Key Words*

**Action Potential** Propagated fluctuation in the membrane electrical potential of the nerve fiber (axon), which travels from the edge of the neuronal cell body to the nerve terminal; also referred to as the nerve impulse.

**Exocytosis** Fusion of neurotransmitter-containing vesicles with the nerve terminal plasma membrane, which results in the release of neurotransmitter into the synaptic cleft.

**Nerve Terminal** Specialized distal ending of a nerve cell fiber (axon); also known as the presynaptic terminal or, in some cases, the terminal bouton.

**Neurotransmitter** Molecule released from the nerve terminal into the synaptic cleft in response to the calcium influx evoked by an action potential (nerve impulse), which acts on discrete target receptor proteins present in or near the synaptic cleft.

---

The term *synapse* (from the Greek, ''to clasp'') was introduced by Sherrington in 1897 to denote the signaling interface between one nerve cell (neuron) and another. The synapse is a discrete, functional entity typically composed of an axonal nerve terminal (presynaptic element) of a nerve cell apposed to the membrane of another cell (postsynaptic element). The postsynaptic cell can be a nerve cell, as would be found within the central and peripheral nervous systems, or cells of other types that are innervated outside the nervous system (e.g., muscle cells and gland cells). Within the nervous system, a nerve cell may make hundreds of synaptic connections with other nerve cells. In spite of such complexity, a number of nerve-terminal-specific proteins have been recently discovered, and the DNA sequences encoding these proteins have been elucidated. Many of the nerve-terminal-specific proteins appear to be common to nearly all chemical synapses, indicating that they play fundamental roles in synaptic signaling.

## 1 GENERAL FEATURES OF SYNAPTIC SIGNALING

### 1.1 Synapse Morphology

There exist three kinds of synapse: chemical, electrical, and mixed. The general features of chemical and electrical synapses are illustrated in Figure 1. Chemical synapses are present throughout the animal kingdom and are by far the major means by which synaptic signaling is mediated. They are present on cell bodies (axosomatic), on the dendrites of nerve cells (axodendritic), and on nerve terminals (axoaxonic). The presynaptic element is usually the bulbous ending of a nerve fiber (typically 0.5–3 μm in diameter), but it may be a swelling, or varicosity, present along the course of a nerve fiber (so-called bouton en passage). Chemical synapses are most readily identified by the presence in the presynaptic nerve terminal of synaptic vesicles—small membranous vesicles (30–60 nm in diameter) that contain neurotransmitter. A majority of the synaptic vesicles appear to be tethered to actin filaments, which are the major components of the nerve terminal cytoskeleton. The synaptic vesicles are largely present in clusters that are focused over so-called active zones, the proposed sites for exocytosis, which may be evident as patches of highly electron-dense material in electron micrographs of some synapses. Voltage-sensitive calcium channels are presumed to be immediately adjacent to the active zones. In addition, numerous mitochondria are clustered in the nerve terminal. In many if not most nerve terminals of chemical synapses, a small number of large ''dense core'' vesicles (~ 100 nm in diameter) are seen in addition to the small synaptic vesicles, and these contain certain peptides as coneurotransmitters. The nerve terminal is separated from the postsynaptic cell by a synaptic cleft (15–24 nm wide), in which is found a filamentous material (not shown), the basal lamina. Neurotransmitter is released into the synaptic cleft, where it interacts with specific neurotransmitter receptors clustered on the surface of the postsynaptic cell, and sometimes on the nerve terminal membrane (not shown).

Electrical synapses have been found in invertebrates and in a few vertebrates. In these synapses the nerve terminal is very closely apposed to the postsynaptic cell (cleft < 3 nm wide), synaptic vesicles are absent, mitochondria are not clustered, and gap junction structures, composed of connexons (hexagonal arrays of membrane proteins), are seen to bridge the synapse. The rare mixed-type synapse contains features of both chemical and electrical synapses.

### 1.2 Synaptic Physiology

There exists an unequal distribution of certain ions between the intra- and extracellular compartments of all cells, including nerve cells. In the resting nerve cell, the membrane displays a significant permeability for $K^+$, for which a high concentration exists within the cell. The outward movement of $K^+$ down its concentration gradient, via K channels, creates a membrane potential wherein the interior of the nerve cell is negative relative to the outside (polarized). The resting membrane potential is predicted by the Nernst equation:

$$E_K = \frac{RT}{zF}\ln\frac{[\mathrm{K}]_o}{[\mathrm{K}]_i}$$

where $E_K$ is the equilibrium potential for $K^+$, while $R$, $T$, $z$, and $F$ are the gas constant, temperature, ion valence, and Faraday constant, respectively. If the membrane potential is reduced (depolarized) to a critical or threshold level, a substantial increase in the permeability for $Na^+$, which is highly concentrated outside the cell, will occur, allowing $Na^+$ to enter via Na channels and trigger an action potential. The action potential is propagated down the axon and invades the nerve terminal. At the chemical synapse, the wave of depolarization accompanying the action potential induces the opening of voltage-sensitive Ca channels residing near the synaptic active zones, allowing $Ca^{2+}$, which is concentrated extracellularly, to enter and trigger the exocytosis of neurotransmitter by fusion of synaptic vesicles with the presynaptic nerve terminal membrane (see Figure 1). The fusion of each vesicle results in the release of a fixed, or quantal, amount of neurotransmitter into the synaptic cleft. The secreted neurotransmitter then binds to a specific receptor on the postsynaptic cell, resulting in either a direct alteration of the postsynaptic membrane potential, which may trigger *or* inhibit the production of an action potential in the postsynaptic cell, or an alteration in the activity of a second-messenger system. Thus, the electrical signal from the axon is converted to a chemical one at the synapse, and this signal transformation creates a significant delay (up to 0.5 ms). At the electrical synapse, the depolarization of the nerve terminal by the action potential is conducted electrotonically through the porous gap junctions to the postsynaptic cell, with virtually no delay.

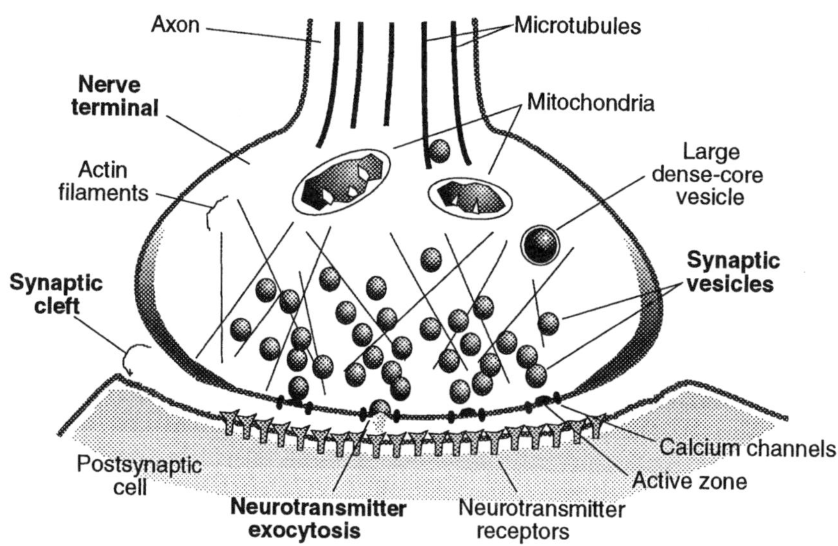


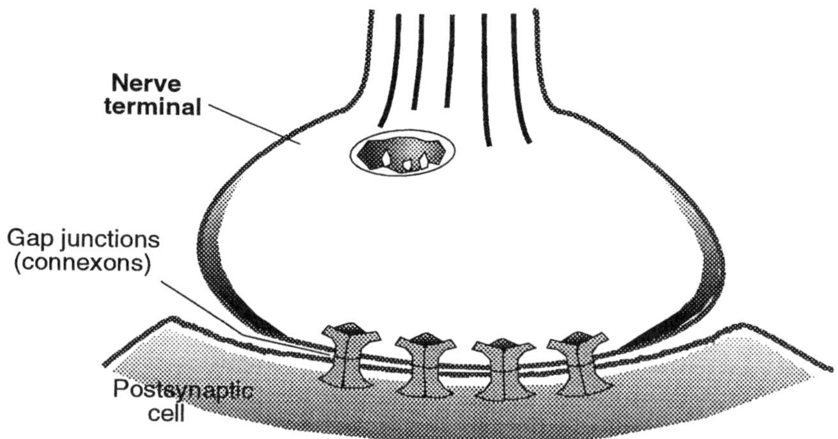


**Figure 1.** Schematic illustrations of the major features of a chemical synapse and an electrical synapse.

## 2 MOLECULAR ELEMENTS OF THE SYNAPSE

### 2.1 Ion Channels

Excitable cells contain plasma transmembrane proteins that respond to changes in the membrane potential by allowing certain ions to pass from one side of the membrane to the other via pores in their internal structures, hence the term "voltage-gated ion channel." Voltage-gated ion channels are classified by the ion for which they are selectively permeable: Na, K, Ca, or Cl. Each class has been found to have numerous members, or subtypes, initially discovered through electrophysiological and pharmacological experimentation, and more recently distinguished using molecular biological approaches. Individual nerve cells have been found to express a unique complement of ion channels from each class, which are differentially distributed in different parts of the nerve cell. Representatives of each ion channel class reside in the nerve terminal. Of particular interest is the strategic localization of clusters of Ca channels near the putative sites for exocytosis, Ca entry being the critical trigger for neurotransmitter secretion.

Members of a second family of ion channels are activated upon binding of neurotransmitter, so-called ligand-gated ion channels. This family mediates the action of neurotransmitter released into the cleft of synapses, where the effect on the postsynaptic membrane potential is rapid and direct but brief. These channels are denoted by the neurotransmitters that bind them, particular examples being glutamate and $\gamma$-aminobutyric acid, which are, respectively, the major excitatory and inhibitory neurotransmitters in mammalian brain. Neurotransmitters also interact with postsynaptic receptors, which are coupled to second-messenger systems, these receptors

**Table 1** Synaptic Vesicle-Associated Proteins[a]

| Protein | Characteristics | Putative Function(s) |
|---|---|---|
| Synapsin I | Amphipathic protein, phosphorylated by CaM-dependent protein kinases (I and II) and cAMP-dependent protein kinase; vesicle and cytoskeleton associated | Regulate availability of vesicles for exocytosis |
| Synapsin II | Vesicle-associated phosphoprotein | Synaptogenic |
| Synaptophysin | Integral vesicle membrane protein; phosphoprotein; pore formation | Fusion process |
| Synaptotagmin | Integrel vesicle protein; cytoplasmic phospholipid-binding domains | Vesicle docking; endocytosis |
| Synaptobrevin/VAMP | Integral vesicle protein; target for tetanus and botulinum toxins; v-SNARE | Vesicle targeting and docking |
| CaM-dependent protein kinase II | Vesicle-associated binding protein (α-subunit) for synapsin; regulation of synapsin via phosphorylation; also phosphorylates synaptophysin | Regulation of neurotransmitter release |
| SNAP-25 | Anchored in vesicle via palmitoylation; v-SNARE | Vesicle targeting and docking |
| Rab 3a | Small GTP-binding protein; vesicle associated | Vesicle cycling; fusogen complex assembly |
| NT Transporters | 12 membrane-spanning domains in vesicle | Neurotransmitter import |
| SV2 | Highly glycosylated integral vesicle protein | Transporter |
| Proton pump | Vacuolar-type pump(?) | Vesicle acidification |
| $pp60^{c\text{-}src}$ | Proto-oncogene, anchored via myristylation; phosphorylates synaptophysin | ? |
| SVAPP-120 | Vesicle-associated phosphoprotein | ? |

[a]CaM, calmodulin; cAMP, cyclic AMP; VAMP, vesicle-associated membrane protein; v-SNARE, soluble NSF (*N*-ethylmaleimide-sensitive fusioin protein) attachment protein receptor (v, vesicle); SNAP-25, synaptosomal associated

typically residing at a distinct class of synapses where their activation yields a more prolonged, neuromodulatory influence. Stimulation of these receptors often results in an indirect modulation of ion channel activity in the postsynaptic cell.

### 2.2 Synaptic Vesicle Proteins

In the last several years, a number of proteins have been found to be specifically associated with synaptic vesicles. Most of these vesicle-associated proteins have been purified and their respective cDNAs have been cloned. One subgroup of these proteins functions in transporting and concentrating neurotransmitter within the vesicle, the neurotransmitter carriers differing across the synapses formed from various neurotransmitter pathways. Another set of vesicle-associated proteins, the synapsins, appear to function as more long-term regulators of neurotransmitter release (see Section 2.4). On the other hand, direct evidence for the functional roles of most of the other vesicle-associated proteins is yet to be obtained. Nonetheless, certain characteristics of these proteins indicate possible actions in the nerve terminal, including vesicle retention in the nerve terminal, vesicle targeting and docking to the presynaptic plasma membrane, pore formation during vesicle fusion in exocytosis, and recycling of vesicle membrane components following exocytosis. The proteins, their distinctive characteristics, and their putative roles in the nerve terminal are summarized in Table 1.

### 2.3 Presynaptic Membrane Proteins

Recently, three proteins have been found which reside specifically in the presynaptic plasma membrane. These proteins were identified by their ability to interact with vesicle-associated proteins, and based on this characteristic, it is proposed that they function as targets for vesicle docking and fusion. These proteins are listed in Table 2. Of particular note is the recent finding that the syntaxins along with the vesicle proteins synaptobrevin and SNAP-25 may correspond to receptors (SNAREs, see Table 1, note *a*) for key fusogen complex proteins (termed NSF and SNAP) found throughout the secretory pathway.

### 2.4 Neurotransmitters and Their Receptors

A neurotransmitter is defined by its localization to the presynaptic terminal, its release upon physiological stimulation, and its postsynaptic action. Neurotransmitters can be categorized into four classes: amino acids or derivatives, biogenic amines, choline esters, and peptides (examples given in Table 3). The nonpeptidergic neurotransmitters are typically synthesized in the nerve terminal, whereas peptides are synthesized in the cell body. Families of several receptor subtypes exist for each neurotransmitter, and the neurotransmitter receptor subtypes may be coupled to different signal transduction systems, yielding an enormously complex array of unique

**Table 2** Nerve Trigeminal Synaptic Membrane Proteins[a]

| Protein | Characteristics | Putative Function(s) |
|---|---|---|
| Syntaxins | Bind synaptotagmin and Ca channel; t-SNARE | Vesicle targeting and docking |
| Neurexins | Ca-dependent binding to synaptotagmin; target for α-latrotoxin | Vesicle docking |
| Physophilin | Binds synaptophysin | Vesicle docking |

[a]t-SNARE, target SNARE (see Table 1, note *a*).

Table 3 Neurotransmitter Classes

| Amino Acids | Biogenic Amines | Choline Esters | Peptides[a] |
|---|---|---|---|
| Glutamate | Dopamine | Acetylcholine | Substance P |
| γ-Aminobutyric acid | Norepinephrine | | Somatostatin |
| Glycine | Serotonin | | Vasoactive intestinal peptide |
| Aspartate | Histamine | | Neurotensin |

[a]Few examples from more than 40 known neuropeptides.

communication pathways within the nervous system. Many of the neurotransmitter receptor subtypes were only recently discovered, using molecular biological techniques. Full characterization of these receptor subtypes will offer new and highly specific therapeutic avenues for treatment of neuropathological disorders arising as the result of dysfunctions in particular neurotransmitter pathways.

## 3 MODULATION OF SYNAPTIC FUNCTION

During the course of ongoing nerve terminal stimulation, the vesicle/cytoskeleton-interacting protein synapsin I undergoes phosphorylation in response to Ca entry via activation of Ca/calmodulin-dependent protein kinases (CaM kinases). Synapsin I appears to be a component in the linkages between synaptic vesicles and the actin cytoskeleton, and phosphorylation by CaM kinase II weakens these links, freeing vesicles and augmenting the release of neurotransmitter during nerve terminal stimulation. Thus, phosphorylation of a key vesicle protein leads to the modulation of synaptic function. The activities of some of the other nerve-terminal-specific proteins may also be modulated during ongoing stimulation of the nerve terminal, providing multiple molecular targets for regulation of the synapse.

*See also* Memory and Learning, Molecular Basis of.

### *Bibliography*

Greengard, P., Valtorta, F., Czernik, A. J., and Benfenati, F. (1993) Synaptic vesicle phosphoproteins and regulation of synaptic function. *Science,* 259:780–785.

Hall, Z. W. (1992) *Molecular Neurobiology.* Sinauer, Sunderland, MA.

Jessel, T. M., Kandel, E. R., Lewin, B., and Reid, L., Eds. (1993) Signaling at the synapse. *Neuron,* 10(suppl).

Kuffler, S. W., Nicholls, J. G., and Martin A. R. (1984) *From Neuron to Brain,* 2nd ed. Sinauer, Sunderland, MA.

Sihra, T. S., and Nichols, R. A. (1993) Mechanisms in the regulation of neurotransmitter release from brain nerve terminals: Current hypotheses. *Neurochem. Res.* 18:47–58.

# Synthetic Peptide Libraries

*Kit S. Lam*

### *Key Words*

**Acceptor Molecule** A molecule, usually a macromolecule, that has the potential of binding to specific peptides, or ligands.

**Amino Acid** The smallest subunit of protein, having an amino, a carboxyl, and a side chain functional group.

**Combinatorial Peptide Library** A large collection of different peptides, with many possible combinations of amino acids joined together.

**Motif** A pattern of amino acid sequences that interact specifically with an acceptor molecule.

**Peptide** A polymeric compound formed by the condensation of two or more amino acids; here we refer to a compound less than 20 amino acids long.

Synthetic peptide library screening is a new method for the rapid discovery of peptide ligands of biological interest. In essence, a library of peptides ($10^4$–$10^8$) is chemically synthesized and subjected to mass screening, whereupon the amino acid sequence of the positive peptide is determined. Unlike the biologic peptide libraries such as the filamentous phage peptide library approach, which can accommodate only the 20 natural amino acids, synthetic peptide libraries have the potential to incorporate D-amino acids and other unnatural amino acids, as well as specific secondary structures or scaffolding structures. The advent of these peptide library methods not only facilitates the drug discovery process, but also provides important information for the fundamental understanding of molecular recognition.

## 1 PRINCIPLES

### 1.1 Random Peptide Libraries

Traditional drug discovery process rely heavily on the random screening of plant and animal extracts, fermentation broths, and individual or mixed synthetic compounds. The process is often laborious. The random peptide library synthesis and screening method is a recent development that shows great promise in speeding up the drug discovery process. The number of peptides present in a library ranges from a thousand to as many as 100 million, depending on the method used. In general, there are two methods of generating a peptide library, namely, the synthetic method and the biologic method. The filamentous phage approach involves the insertion of a stretch of random deoxyoligonucleotide into the pIII gene of filamentous phage, resulting in the expression of a random oligopeptide at the amino terminus of the viral pIII coat protein. The major advantage of this biologic approach is that the size of the grafted peptide is not limited by the constraints of synthetic chemistry, as in the case of the synthetic peptide library. However, the biologic approach suffers from three major disadvantages:

1. Its use is restricted to L-amino acid peptide libraries.
2. Although simple disulfide cyclization is feasible, complicated bicyclic or scaffolding structures or special cyclization chemistries are impossible.
3. In general, screening is restricted to binding assays.

Synthetic peptide libraries, on the other hand, can generally overcome many of these limitations.

### 1.2 Synthetic Peptide Libraries: Synthesis and Screening

Most of the synthetic peptide libraries are synthesized by standard 9-fluorenylmethoxycarbonyl (Fmoc) or *t*-butyloxycarbonyl (Boc) chemistry on solid phase support. The yield of each coupling cycle is in excess of 99%. Cyclization of these synthetic peptide libraries can be accomplished by disulfide formation, β-lactam formation, alkylation, or metal chelation. Since synthetic peptide chemistry is used, unnatural amino acids, such as D-amino acids, *N*-methylated amino acids, glycoamino acids, and γ-carboxyglutamic acid, can be incorporated into the peptide libraries.

When the peptide library has been synthesized, the N-terminal and side chain protecting groups are removed, and the solid phase bound peptides can be screened directly, or cleaved off the solid support and screened by a solution phase assay. Direct binding assay has been commonly used to screen the solid phase bound library. Detection is usually accomplished by enzyme-linked assay or fluorescent assay.

In some peptide library methods, the peptides are released into the solution for screening. The major advantage of this approach is that various standardized receptor, biochemical, or cellular assays can easily be adapted for screening. Purified acceptor molecules are often not needed, and a biochemical or biologic end point may sometimes be measured even if the molecular target is unknown.

## 2 TECHNIQUES

### 2.1 Spatially Addressable Parallel Solid Phase Library

Sometimes a collection of peptides is synthesized on solid phase support in a spatially addressable format. The amino acid sequence of each of these peptides is predetermined. The peptide collection is screened, usually by a binding assay. The positive peptide is then located and its chemical structure determined. The major limitation of this method is that only limited numbers of peptides can be synthesized, and therefore the library is very small. There are three reported techniques based on this strategy:

1. *Multipin technology.* In this method, pioneered by Geysen et al., the solid phase support is a polyethylene pin in a 96-well format. At first, analysis was by enzyme-linked immunosorbent assay (ELISA), but later modifications permitted peptides to be released and analyzed by solution phase assay.
2. *SPOTs-membrane.* Frank et al. synthesized peptides on cellulose membrane or paper by applying the activated amino acids and reagents directly onto the sheet in the form of a spot. Final analysis of the solid phase bound peptide is by enzyme-linked assay.
3. *Light-directed peptide synthesis on chips.* Fodor et al. developed a photolithic masking process in conjunction with an amino acid coupling strategy using photolabile protecting groups to synthesize a minute quantity of a total of 1024 peptides on a single glass slide. Each peptide occupies an area of 50 μm × 50 μm. Screening is accomplished by binding assay with fluorescent-labeled acceptor molecule and fluorescent microscopy. Amino acid coupling efficiency using this photochemistry is reported to be only 85 to 95%. Synthetic yield and purity would therefore be a major problem for the synthesis of longer peptide libraries (e.g., 10- to 15-mers).

### 2.2 Affinity Selection

Zuckermann et al. applied the affinity selection method to the retrieval of specific binding peptides from a mixture of peptides in solution. However, because of high background from nonspecific binding, this method can be applied only to a relatively small peptide library (e.g., <10,000 peptides).

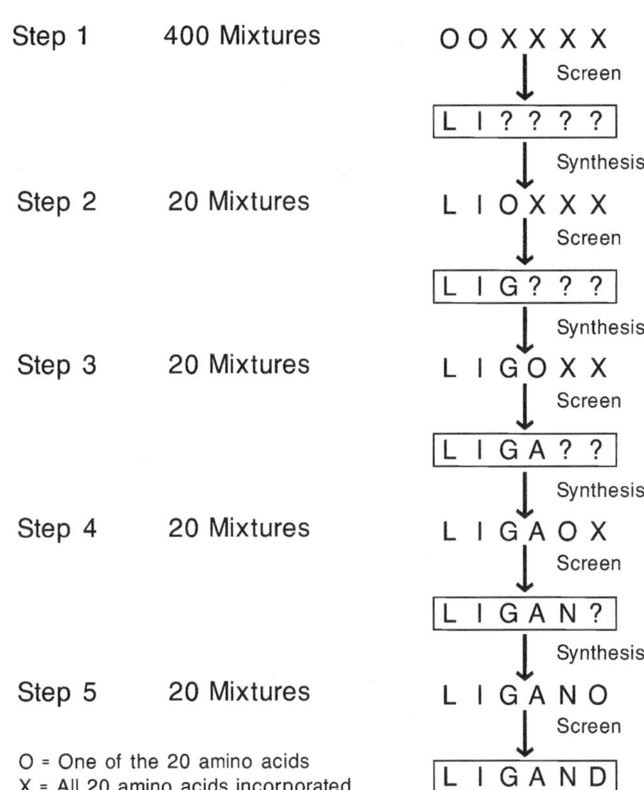


**Figure 1.** Combinatorial library with iterative process.

### 2.3 Synthetic Peptide Combinatorial Library with Iterative Process

In a third approach, the amino acid at each position of the ligand is defined by an iterative process that involves multistep synthesis and analysis (Figure 1). For example, in a combinatorial hexapeptide library, one of the 20 amino acids is incorporated into positions 1 and 2, while all 20 amino acids are incorporated (presumably in an equimolar ratio) into the remaining four residues. This results in the generation of 400 (20 × 20) peptide mixtures (step 1, Figure 1). Each mixture contains 160,000 or ($20^4$) different peptides. Each of the 400 peptide mixtures is then subjected to screening, from which the "active" amino acids for the first two positions can then be defined. In step 2, the first two "active" residues are fixed, and one of the 20 amino acids is incorporated into position 3, while all 20 amino acids are incorporated into the remaining three residues. This results in the generation of 20 peptide mixtures. Each of these 20 mixtures is again screened, and the third "active" residues are then defined. The same iterative process is repeated for three more cycles until the last residue has been defined.

Synthetic combinatorial library constructed by means of an iterative process was first described by Geysen in the multipin system, where a peptide mimotope was defined for a monoclonal antibody. The screening assay used was ELISA. Subsequently, Houghten described a similar method, but the peptides were released into solution prior to screening. More recently, Houghten described the "positional scanning methodology," whereby 120 mixtures of hexapeptides were synthesized with one of the 20 amino acids incorporated into one residue, while all 20 amino acids were incorporated into the remaining five positions. By scanning these 120 mixtures, it is possible to define each of the active residues. However, this approach will only work for acceptor molecules for which

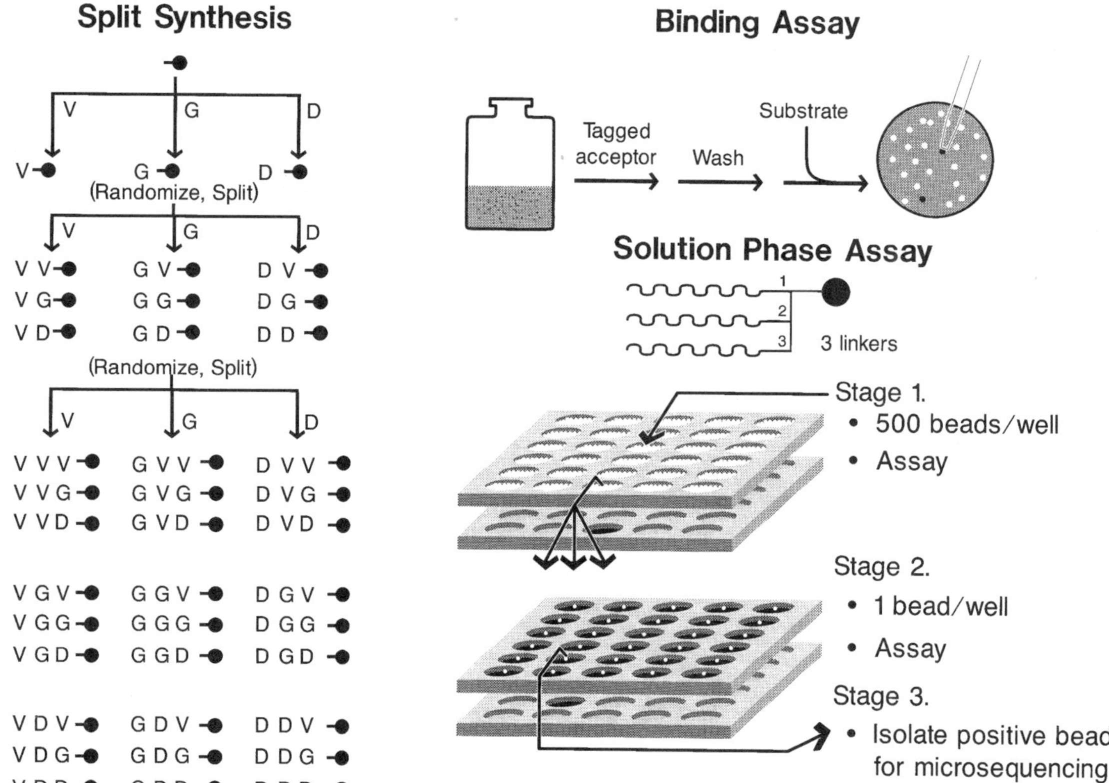


**Figure 2.** Selectide process.

**Table 1** Peptide Library Methods: A Comparison

| Property | Methods | | | | | | |
|---|---|---|---|---|---|---|---|
| | Filamentous Phage | Multipin/ SPOTs | Light-Directed Parallel Synthesis | Affinity Selection | Combinatorial Library (iterative process) Multipin | Combinatorial Library (iterative process) Solution | Selectide Process |
| Number of peptides in a library | $10^7$–$10^9$ | $10^4$ | $10^5$ | $10^4$ | $10^6$–$10^8$ | $10^6$–$10^8$ | $10^6$–$10^8$ |
| Unnatural amino acids | No | Yes | Yes | Yes | Yes | Yes | Yes |
| Constrain structure | Simple disulfide | Yes | Yes | Yes | Yes | Yes | Yes |
| Size of peptide | Small to very large | Small | Small | Small | Small | Small | Small[a] |
| Binding assay | Yes | Yes | Yes | Yes | Yes | No | Yes |
| Solution phase assay | No | Yes | No | No | Probably not | Yes | Yes |
| Screening approach | Parallel | Parallel | Parallel | Parallel | Convergent[b] | Convergent | Parallel |
| Equipment required | | | Complicate equipment, not commercially available | | | | Microsequencer, commercially available |
| Biologic bias | Yes | No | No | No | No | No | No |

[a]We have successfully screened a 15-mer peptide library using the Selectide process.
[b]With the multistep iterative process, usually only one soluton motif may be identified (see Section 2.3).

no more than one motif is possible. If an acceptor has multiple motifs, the result will likely be scrambled and uninterpretable.

### 2.4 Selectide Process

We developed the Selectide approach based on the "one-bead, one-peptide" concept. The peptide is synthesized on solid phase beads using a "split synthesis" method, resulting in a huge library of peptide beads, each of which expresses a single peptide entity (Figure 2). The peptide library is then subjected to screening, and the reactive bead identified, physically isolated, and subjected to microsequencing with Edman degradation chemistry. The main advantage of this library method is that a large number ($10^6$–$10^8$) of peptides can be synthesized and screened rapidly. In addition, since it is a single-step process that does not require iteration, often multiple peptide ligands with completely different motifs can be identified in a single screen. That is, the Selectide process is a parallel approach. This is the opposite of the (convergent) iterative approach, where multistep synthesis and screening result in the emergence of only one, but not necessarily the best, solution motif.

Besides binding assays, the Selectide process has recently been adapted to solution phase assay. Peptide bead libraries with dual orthogonal cleavable linkers are used, and peptides can be released sequentially with a two-step process into solution for screening (Figure 2). Thus, similar to the combinatorial library approach described by Houghten, the screening can be adapted to many existing biological assays. After the bead of origin that tests positive has been recovered, the residual peptide on the bead can be analyzed by a microsequencer.

## 3 APPLICATION

Over the last few years, the synthetic peptide library approach has been applied successfully to various biologic systems. Ligands for antibodies, streptavidin, avidin, and opiate $\mu$ receptors have been identified, as well as ligands for major histocompatibility class molecules of class I, cytokine receptors, adhesion molecules, protein kinases, and $SH_2$ domains. In addition, peptides with protease-inhibitory and antibacterial activities have been discovered. The development of these synthetic peptide library methods not only facilitates the drug discovery process but also provides important information for the understanding of molecular recognition.

## 4 PERSPECTIVES

Table 1 compares seven peptide library methods. In general, there are only three methods of generating and screening huge peptide libraries (i.e., $10^6$–$10^8$ peptides): filamentous phage, combinatorial/iterative, and the Selectide process. Because of the versatility of synthetic chemistry, enormous research efforts are under way to apply the latter two library methods to the generation of totally nonpeptide chemical libraries. This approach may indeed prove to be a revolutionary method in the drug discovery process for the years to come.

*See also* Combinatorial Phage Antibody Libraries; Drug Targeting and Delivery, Molecular Principles of; Receptor Biochemistry.

### *Bibliography*

Jung, G., and Beck-Sickinger, A. (1992) Multiple peptide synthesis methods and their applications. *Angew. Chem. Int. Ed. Engl.* 31:367–383.

Pavia, M. R., Sawyer, T. K., and Moos, W. H. (1993) The generation of molecular diversity. *Bioorg. Med. Chem. Lett.* 3:387–396.

# T

## TAXOL® AND TAXANE PRODUCTION BY CELL CULTURE

*Arthur G. Fett-Neto and Frank DiCosmo*

*Key Words*

**Callus** Mass of disorganized plant tissue, generally with low degree of cell differentiation.

**Explant** Portion of the intact plant used to initiate callus cultures.

**Taxane** Compound related to taxol, bearing an original carbon skeleton. A taxane can be described as a diterpene composed for three pentadecene rings.

**Taxol®** Diterpene amide with antineoplastic properties.

***Taxus*** Genus of gymnosperm trees belonging to the family Taxaceae and commonly known as yew.

Taxol® is an anticancer drug, now commonly called paclitaxel, increasingly used for the approved treatment of ovarian and breast cancers. The supply of this compound was limited because previously the only commercial source was the bark of *Taxus brevifolia,* however, other species of *Taxus* also produce taxol®; *T. brevifolia* plants, relatively slow growing, have low and variable amounts of taxol® and, in many cases, are located in conservation areas. Cell culture of yew species is a viable alternate source of taxol® and related taxanes, which can be used for production and/or semisynthesis of the drug. Cell cultures are readily renewable and yield homogeneous, readily manipulated systems for biosynthetic studies on taxol®, with potential for large-scale commercial production.

### 1 PLANT MATERIAL SELECTION AND PREPARATION

#### 1.1 Type and Source of Explant

The response of yew tissue to in vitro culture is dependent on the type of explant used to start the culture, on the physiological status of the donor plant, and on its genetic makeup as well. Explants that contain meristematic or parenchyma tissue are preferable; because of their low degree of differentiation, these are the types of tissue most likely to display cell proliferation in reaction to the growth regulators in the culture medium. Young stem segments (flexible, but bearing dark green needles) are ideal for rapid initiation of callus to be used in developing cell cultures, whereas mature seed contents (megagametophyte and embryo) are better suited for establishing long-term maintenance callus cultures. The physiological condition of the donor plant must be stable and favorable to growth. Outdoor-grown plants display higher contamination rates and tissue pigmentation when cultures are started in the winter relative to those started in other seasons. Genetic differences between plants of the same species are also a factor to be considered; it may be necessary to test different plants to develop initial in vitro cultures. It may also be beneficial to begin in vitro culture using plants shown to produce high yields of taxol®.

#### 1.2 Sterilization Procedure

Sterilization is accomplished using standard plant tissue culture techniques. The tissue is washed with distilled water to remove dust and solid particles, immersed in 70% (v/v) ethanol for one minute, immersed in NaClO (1.5% v/v) for 30 minutes with a few drops of a commercial surfactant (Tween 20), and rinsed three times with sterile distilled and deionized water. During the sterilization period, the tissue is kept fully immersed and under agitation. Sterilization times may have to be reduced for more sensitive and tender tissues. From our experience, typical contamination rates using outdoor-grown donor plants are under 3% of the total explants. In the winter, however, the incidence of contamination can be 10 times the regular rate.

#### 1.3 Explant Size and Final Preparation

Following sterilization, the exposed cut portions of the tissue that were damaged by the ethanol and NaClO solutions are excised and the healthy tissue is appropriately trimmed and placed in the growth medium. In the case of stem material, segments approximately 1 cm long give good results. Stem segments can be cultured at an approximate angle of 45° with one of the transversely sectioned ends immersed in the medium and the other partly exposed to the air of the culture vessel. Alternatively, longitudinally cut stem segments can be cultured with their cut portions facing the surface of the medium. The contents of longitudinally halved seeds can be placed on the medium after separation from the seed coat with a spatula.

### 2 ESTABLISHMENT OF CALLUS CULTURES

#### 2.1 General Procedure

Initial cultures of yew are relatively difficult to establish owing to the production, release, and oxidation of high levels of phenolics by the explants; this leads to growth inhibition and eventual loss of viability. The inclusion in the medium of phenolic binding compounds such as polyvinylpyrrolidone (PVP) improves the growth response of the explants. In addition, transfers to fresh medium (or rotation of the explants to positions in the culture vessel containing portions of unused medium) at approximately 15-day intervals help prevent growth inhibition due to phenolic oxidation. Following an initial screening of some common medium formulations, the basic medium B5bPVP was established for callus initiation in yew; it consists of B5 medium salts and organic supplements, 3.0% (w/v) sucrose, 1.5% (w/v) insoluble PVP, 0.8% (w/v) agar and 1 mg/L of 2,4 dichlorophenoxyacetic acid (2,4-D). The media are prepared with distilled and deionized water and plant cell culture or reagent grade chemicals, adjusted to pH 5.8 with NaOH and HCl (after the addition of PVP and prior to the addition of agar), and autoclaved (1.1 kg/cm$^2$ at 121°C for 20 min). The cultures can be grown in plastic petri dishes (9 cm diameter) containing 30 mL of medium (3–4 explants per plate) and sealed with Parafilm, or in 30 mL glass vials with polypropylene lids containing 10 mL of medium each (1 explant per vial). Incubation is allowed to proceed at 25°C in darkness.

### 2.2 Medium Optimization

Additional improvement on the basic medium formulation for yew tissue culture can be accomplished by investigating several medium components (e.g., plant growth regulator balance, salt and organic supplement composition) that promote the growth response of related species or seem likely to have a stimulatory effect on the putative biosynthetic pathway to taxol®. The effect of the medium components when included in various amounts in the basic formulation, or when replacing some of its original constituents, was analyzed relative to growth and taxol® yield of developing yew calli. Medium factors with a positive effect on growth and taxol® yield are combined in new formulations to investigate the possibility of synergistic or antagonistic interactions. From such a series of experiments, two improved media formulations were devised: B5bPhe (B5bPVP with 0.1 mM of phenylalanine) and B5Phe8/0GA (B5bPhe salts and organic supplements with 8.0 mg/L of 2,4-D and 0.5 mg/L of gibberellic acid). Gibberellic acid ($GA_3$) is added to the culture medium after autoclaving by filtration through 0.22 μm sterile filters. All other medium components are added before autoclaving.

## 3 ESTABLISHMENT OF CELL CULTURES

Cell cultures can be initiated from approximately two-month-old callus cultures developed from stem segments. Depending on the availability of callus biomass for inoculum and incubator space, cell cultures can be grown in 125, 250, 500, and 1000 mL Erlenmeyer flasks containing one-fifth of their total volume of liquid medium. Basic incubation conditions are 25°C in darkness on gyratory shakers at 110 rpm. Based on the experiments done in callus and some additional optimization for cell culture medium, the following basic formulations were developed for yew cell culture: (1) B5 medium salts, threefold the regular amount of B5 organic supplements, 3.0% sucrose (w/v), 1.5% soluble PVP (approximate MW 40 × 10$^3$), and 1 mg/L of 2,4-D (B5C1), and (2) the same composition of B5C1, but with the following modified balance of plant growth regulators: 4 mg/L of 2,4-D and 1 mg/L of kinetin (B5C2). Medium preparation, pH adjustment, and sterilization procedure are the same as in callus culture medium.

Initially, inoculum weights are preferably high (e.g., one-third the volume of the liquid medium). After 1–2 weeks in culture on a gyratory shaker at 110 rpm, the cells released from the calli are collected and transferred into fresh liquid medium (1:1 or 2:1 proportion of old to fresh medium). The stem debris can be discarded or replated in solid medium to regenerate more callus tissue. During the first months of culture, the media can be replaced at 2-week intervals following 10 minutes of centrifugation of 600 × *g* at 20°C under sterile conditions; in this way, the maximum amount of cell biomass is saved until the density is great enough to start suspension culture propagation in new flasks. Once suspensions have stabilized (generally after two or three cycles of propagation), inoculum amounts can be reduced to one-sixth (w/v) of the liquid medium.

Cell cultures can also be immobilized after the first months of culture to facilitate medium replacement and to aid in the establishment of recalcitrant cell lines. Immobilization is performed by spontaneous adhesion to 5 cm × 5 cm, 10–15 mm thick glass fiber mats with a uniform diameter of 10 μm and fiber density of approximately 61 mg/cm$^2$. Within the first days of contact with the substrate, the cells are spontaneously immobilized. The cells grow on top of and within the mats, forming several layers of cells; cells start to be released into the liquid medium after they have saturated the immobilization substrate. These cells can be used to establish new suspension cultures or new immobilized cultures. Alternatively, cell cultures can be propagated by removing external cell layers with a sterile spatula, or by cutting the mats in two or four portions and placing these on fresh medium flasks, each containing a new glass fiber mat.

## 4 ANALYSES OF TAXOL® AND TAXANE PRODUCTION

### 4.1 Extraction Procedure

At least 0.5 g of fresh tissue or cells is collected in glass test tubes (previously weighed and identified), frozen (−20°C), and homogenized with a glass rod in 1–2 mL of hexane. After 12 hours in hexane at 25°C, the samples are centrifuged at 2500 × *g* for 20 minutes to pellet the cells, and the supernatants are discarded. The pellets are extracted in 1–2 mL of methanol/dichloromethane (1:1) with glass rod rehomogenization, followed by sonication for one hour at room temperature. After centrifugation at 2500 × *g* for 30 minutes, the supernatants are collected; the pellets are used for determining the extracted dry weights at 60°C until constant weight. Alternatively, the tissue can be extracted with 100% methanol. The crude extracts are dried at 25°C, redissolved in 1 mL of dichloromethane, partitioned with 1 mL of water, and centrifuged at 2500 × *g* for 20 minutes. The dichloromethane fraction is collected, dried at 25°C, redissolved in an exact volume of high performance liquid chromatography (HPLC) grade methanol (100–500 μL) and filtered through 0.45 μm filters for HPLC analysis.

### 4.2 Separation, Quantification, and Identification

The taxane analyses are done by HPLC. The samples are passed through a reversed phase microbore column (e.g., Hewlett-Packard ODS Hypersil, 5 μm, 100 mm × 2.1 mm), using a mobile phase containing methanol/water/acetonitrile; the linear gradient elution profile starts with 20:67:13 (solvent A) and ends with 20:27:53 (solvent B) within 50 minutes, and includes a 5-minute wash in both solvents to reestablish the initial condition. The flow rate is 0.2 mL/min, and chromatograms are plotted at 227 nm, using an HPLC system equipped with a photodiode array detector (e.g., Hewlett-Packard 1090A HPLC system). Injection volumes are 10 μL and solvents are all HPLC grade and filtered through 0.45 μm filters prior to use. Quantification is done by the method of external standard; that is, a standard curve is generated with authentic standard and run in the same chromatographic conditions as the samples. Identification of taxol and taxanes is done by retention time, UV spectra, and cochromatography with standard. Further proof of identification can be obtained by collecting the corresponding HPLC peak and performing NMR analyses. Standards for taxol® and related taxanes may be requested for nonhuman research use from the Drug Synthesis and Chemistry Branch, Developmental Therapeutics Program, Division of Cancer Treatment of the U.S. National Cancer Institute. Taxol® is also available from some chemical companies.

## 5 TAXOL® YIELD AND POSSIBILITIES OF MANIPULATION

### 5.1 Characteristics of Taxol® Yield in *Taxus* in vitro Cultures

Experiments with callus and cell cultures of *Taxus cuspidata* have shown that growth and taxol® yield are generally inversely proportional. The reason for the phenomenon, also observed with the production of other secondary metabolites by a number of plant species, is not known with certainty. Possible explanations include the lack of expression of key enzyme genes in cells with low differentiation, depletion of substrates for secondary metabolism into growth processes, unregulated catabolism, and lack of end product transport systems and storage sites.

The analysis of taxol® yields in different in vitro cultures of *T. cuspidata* supports the preceding observation; faster growth rates and/or lower degrees of differentiation are normally associated with lower taxol® amounts. Stem-derived explants of *T. cuspidata* after 2 months of culture in B5bPVP display an average eight-fold increase in fresh weight and taxol® yields of 0.020 ±0.005% of the extracted dry weight (equivalent to the yield of stems of the intact plant). Slow-growing, 6-month-old immobilized cell cultures of *T. cuspidata* grown in B5C2 have variable amounts of taxol®, up to 0.012 ± 0.007% of the dry weight (equivalent to the bark of the intact plant), whereas 6-month-old immobilized cell cultures visually selected for faster growth yield 0.0014 ± 0.0003% of taxol® on a dry weight basis. Year-old cell suspension lines of *T. cuspidata* grown in B5C2 (doubling time of 15 days) have yields of taxol® ranging from 0.0001 to 0.0010% on a dry weight basis; cell suspensions of the same age grown in B5C1 have taxol® yields of 0.0004 ± 0.0002% of the dry weight. These yields may be altered by precursor feeding (see Section 5.2).

Although the yields of other taxanes in yew callus and cell cultures have not yet been investigated in detail, HPLC evidence indicates the presence of baccatin III, 10-deacetyl-baccatin III, 1-dehydroxybaccatin V, and cephalomannine in both types of cultures of *T. cuspidata*. Baccatin III yields in 2-month-old callus of this species grown in B5bPVP are 0.043 ± 0.016% of the extracted dry weight. Taxanes such as 10-deacetyl-baccatin III and baccatin III can be used to produce taxol through semisynthesis.

The media collected from immobilized cell suspensions of *T. cuspidata* and extracted with dichloromethane may contain trace amounts of taxol® and baccatin III as indicated by HPLC analysis. Estimated rates of taxol® accumulation in the media are approximately 4 μg/L/day for both B5C2 and B5C1. These low amounts suggest that taxol® in the media results from cell turnover rather than active secretion; however, these estimates were done with media collected at a single point in time and assuming no taxol® degradation in the medium. In differentiated physiologically active cells, taxol® may be stored inside the cells, possibly in the vacuole. Time course studies on growth and taxol® production in cell suspensions of *T. cuspidata* showed that taxol® concentration in the medium varies during the growth cycle and that taxol® may be degraded after release into the medium. The rates of taxol® accumulation in the medium estimated for immobilized cells may also have been affected by these factors. In a cell suspension line of *T. cuspidata* (FCL1F) grown in B5C2, taxol® accumulation in the medium can be 66% of the total volumetric production of taxol® at stationary phase. Kinetic studies on growth and taxol® accumulation in cell suspensions confirmed the inverse relationship of taxol accumulation and growth in cell cultures (non-growth-linked accumulation pattern).

### 5.2 Precursor Feeding

The problem of overall low taxol® yields, particularly at higher growth rates, can be partly solved by feeding taxol® precursors to the cell cultures. The appropriate combination of type, amount, and feeding time of precursors can significantly increase the yield of secondary metabolites in plant cell cultures. In 2-month-old callus cultures of *T. cuspidata*, phenylalanine supplementation at 0.1 mM in B5bPVP can double the taxol® yield in relation in regular B5bPVP without affecting growth significantly; similar results can be obtained with cell suspensions after a 25-day growth period in medium supplemented with the same amount of this amino acid. Phenylalanine acts as a precursor for the side chain of taxol®. We have also investigated other potential precursors (e.g., benzoic acid) and their impact on taxol® yields. Precursor feeding at the onset of the stationary (taxol®-productive) phase has yielded higher taxol® accumulation in cell suspensions. Results suggest that precursor feeding is a promising approach to increase taxol® yields in yew cell cultures without detrimental effects to biomass generation.

### 5.3 Cell Line Selection

Screening different cell lines for high yields of taxol and/or taxanes may lead to the establishment and propagation of elite cell lines, which can be used for large-scale production purposes. The technique has been successfully applied, for instance, to the production of ajmalicine by *Catharanthus roseus* and shikonin by *Lithospermum erythrorhizon.* Although very laborious and time-consuming, this approach may turn out to be advantageous for long-term potential industrial applications of yew cell cultures as a source of taxol® and taxanes.

## 6 FUTURE PERSPECTIVES

Data suggest that cultured cells of *Taxus* spp., although in early development, represent an alternate source of taxol® and related taxanes, which can be used in the production and semisynthesis of this antineoplastic agent. Scaled-up cell culture studies are being developed with emphasis on nutrient consumption rates, gas composition, cell shearing sensitivity, and precursor feeding schedules. The studies, in combination with the biochemical and growth data accumulated from callus and cell cultures, may provide further insights on the factors that limit the growth and the metabolic pathways leading to taxol® and related taxanes in yew cells.

*See also* BIOPROCESS ENGINEERING; CANCER; DRUG TARGETING AND DELIVERY, MOLECULAR PRINCIPLES OF; MEDICINAL CHEMISTRY.

## *Bibliography*


Charlwood, B. V., and Rhodes, M.J.C., Eds. (1990) Secondary products from plant tissue culture. In *Proceedings of the Phytochemical Society of Europe,* Vol. 30. Clarendon Press, Oxford.

Evans, D. A., Sharp, W. R., Ammirato, P. V., and Yamada, Y., Eds. (1983–1984) *Handbook of Plant Cell Culture,* Vols. 1–6. Macmillan, New York.

Facchini, P. J., DiCosmo, F., Radvanyi, L. G., and Neumann, A. W. (1990) Immobilization of cells by spontaneous adhesion. In *Methods in Molecular Biology,* Vol. 6, *Plant Cell and Tissue Culture,* J. W. Pollard and J. M. Walker, Eds. Humana Press, Clifton, NJ.

Fett-Neto, A. G., DiCosmo, F., Reynolds, W. F., and Sakata, K. (1992) Cell culture of *Taxus* as a source of taxol and related taxanes. *Bio Technology,* 10:1572–1575.

Fett-Neto, A. G., Melanson, S. J., Sakata, K., and DiCosmo, F. (1993) Improved growth and taxol yield in developing calli of *Taxus cuspidata* by media composition modification. *Bio Technology,* 11:731–334.

Nijkamp, H.J.J., Van der Plas, L. H. W., and Van Aartrijk, J., Eds. (1990) Progress in plant cellular and molecular biology. In *Proceedings of the VIIth International Congress on Plant Tissue and Cell Culture,* Amsterdam, June 1990. Kluwer Academic Publishers, Dordrecht.


# THEORETICAL MOLECULAR BIOLOGY

*Andrzej K. Konopka*

## *Key Words*

**Code** A mapping from one symbolic representation (input representation) of a system into another representation (output representation). If the input and output representations can be expressed in the form of elementary symbols chosen from finite alphabets, a code can be seen as a relation between input and output alphabets.

**$k$-Gram ($k$-Tuple)** String of $k$ symbols chosen from a given elementary alphabet. In the case of nucleotide sequences, $k$-grams correspond to oligonucleotides of length $k$. The most frequently used nucleic acids elementary alphabet is {A, C, G, T (or U)}, where the letters stand for adenine, cytosine, guanine, and thymine (or uracil) nucleotides, respectively. Other alphabets used in nucleic acids studies include {K, M}, where K is either guanine or thymine (uracil in RNA) and M is either adenine or cytosine; {R, Y}, where R is a purine (adenine or guanine) and Y is pyrimidine (either cytosine or thymine); {S, W}, where S is either cytosine or guanine and W is either adenine or thymine (or uracil in RNA). If the elementary alphabet contains $n$ symbols and $k$ is fixed, we refer to a set of all $n^k$ $k$-grams as a *$k$-gram* (nonelementary) *alphabet.*

**Motifs and Punctuations** Given a $k$-gram alphabet, we select $k$-grams that we call *motifs* and $k$-grams that we call *punctuations.* Heuristic principles for such choices are complex because they involve knowledge of "extrasequential" facts.

**Pattern** There is no general definition. Pattern can be any logical, geometrical, or (broadly) factual connection between elements of a model that attracts our attention. "Pattern" is usually understood ostensively by experienced explorers of the same model.

**Sequence Pattern** String of motifs and punctuations that begins and ends with a motif. A *contiguous pattern* is a sequence pattern that does not contain punctuations, whereas *noncontiguous pattern* is a sequence pattern that does contain punctuations.

---

The aim of theoretical molecular biology is an extended interpretation of both laboratory and computational experiments in molecular biology proper. The "extended interpretation" consists of several activities, such as the following:

1. Relating the mechanistic and the symbolic (or informational) aspects of biological phenomena through appropriate generalization and modeling. *Example:* the postulation of the triplet nature of the genetic code before it had been determined experimentally.
2. Devising principles for analyzing inexact data (quantitative representations of qualitative descriptions included). *Examples:* database design, annotation, and searches; logic of heuristic reasoning.
3. Relating principles and laws inferred from studies at the molecular level to those derived from studies of organisms, populations, and ecosystems. *Examples:* molecular evolution; molecular evolutionary genetics.
4. Deriving properties of biological systems from principles of physics and chemistry. *Examples:* biological thermodynamics and kinetics; theories of origin of life; molecular mechanics.
5. Modeling (usually mathematical) properties of laboratory tools and designing experiments. *Example:* analysis and interpretation of molecular spectra, X-ray data, and so on.
6. Modeling properties of computational tools and designing computational experiments. *Example:* sequence alignment.

These activities often draw on methods that themselves are distinct research areas with their own theoretical foundations. Some of the methods pertain to physics (e.g., molecular mechanics), some to chemistry (reaction kinetics), and yet others are parts of applied mathematics (combinatorics) or computer science (theory of algorithms). Theoretical molecular biology differs from chemistry, physics, and even biochemistry by systematically considering symbolic interpretations (besides the mechanistic ones) of the biologic role(s) of nucleic acid and protein sequences.

## 1 PRELIMINARIES*

From the early period in molecular biology we know of the three "biological codes": Watson–Crick–Chargaff base-pairing rules, translation code for protein biosynthesis (the genetic code), and

* The focus on symbolic (or informational) aspects of phenomena makes "code" the central, unifying metaphor in the field. Therefore this survey is mostly devoted to biomolecular cryptology, a part of theoretical molecular biology pertaining to the activities 1 and 2 in the introductory list. Theoretical issues pertaining to remaining four activities are scattered throughout other contributions. Because this volume also contains entries devoted to protein and RNA sequence analysis, the present survey explores examples from DNA studies only.

cleavage sites for restriction enzymes. Many biologists assume that more *functional codes* of this kind exist (e.g., we could think of replication code, homologous and illegitimate recombination codes, mutation hot-spots code, protein or RNA folding codes). To specify a functional code, we need to know a great many mechanistic details that contribute to the ''coded'' function. This is a major methodological disadvantage because in most cases we know or surmise that a code exists but we cannot sensibly specify the biological function concerned (the genetic code is a notable exception from this common situation). Moreover, even if a given sequence contains ''encoded'' functions, the corresponding ''messages'' are often superimposed. In this setting, the assumption ''one sequence, one function'' is wrong, whereas the assumptions ''one sequence, many functions'' and ''many sequences, many functions'' are correct.

Alternatives to functional codes are *classification codes.* In principle they are many-to-one mappings from a set of patterns found in (or resulting from) a class of sequences into descriptors (such as a name or a list of features) of this class. We specify a classification code by just listing patterns from its domain. We call each individual pattern from the list a *classification code word* or just a *code word.* Not every pattern found in a given collection of sequences is a classification code word. To qualify as a code word, the pattern must be significant according to predetermined criteria.

The criteria of significance usually involve a complex heuristics called *pragmatic inference.* The name indicates a focus on the ''rules'' of usage for patterns of textual elements. Relation of these rules to syntax and semantics of the alleged ''language'' (in which our text is phrased) is indirect and difficult to demonstrate.

Pragmatic inference explores the entire body of available biological knowledge in addition to statistical analyses. The resulting measure of significance often does not correspond to statistical significance because the rules of pragmatic inference are different from the rules of statistical inference. However, pragmatic inference does involve statistical modeling of $k$-gram and pattern frequency counts.

Simple frequency counts are a vital part of those steps of pragmatic inference in which selection of motifs and punctuations is made. Intuitive judgment resulting from frequency count analyses is also instrumental in defining patterns that are more likely than other patterns to be code words. Although simple frequency counts are insufficient for final determination of classification code words, they allow us to initially compare statistical structures of large collections of sequence data.

## 2 FREQUENCY COUNTS

To determine how ''unusual'' is the occurrence of a given $k$-gram in a long sequence, we compare the occurrence frequency in the data with the frequency expected from a model of chance. Let it be determined from data that the frequency of the $i$th $k$-gram is $F(i)$. Let the expected frequency of this $k$-gram be $F_0(i)$ and the variance (determined from a model of chance) $V(i)$. The ''degree of unusuality'' of the occurrence of the $i$th $k$-gram in the data is a function of $F(i)$, $F_0(i)$, and $V(i)$. The most common such function is a *z-score* defined as

$$z(i) = \frac{F(i) - F_0(i)}{\sqrt{V(i)}} \tag{1}$$

After having calculated the $z$-score $z(i)$, we compare it with a threshold value $z_0$. If $|z(i)|$ exceeds $z_0$, we consider the occurrence of the $i$th $k$-gram unusual.

Models of chance are chosen according to the available knowledge of the situation. A frequently used model is a Bernoulli text in which all 1-grams occur independently of one another and with equal probabilities (i.e., equal to $1/n$). When the elementary alphabet has $n$ letters (1-grams) and $k$ is fixed, the $k$-gram alphabet contains $n^k$ elements. The expected frequency of each of $n^k$ $k$-grams in a (long) sequence of length $L$ is $L/n^k$. If the frequency count is performed ''overlapping,'' the variance of this frequency distribution needs to be evaluated separately for each $k$-gram. The reason for this complication is the dependence of the variance on the self-overlap capacity of individual $k$-grams besides their expected frequency.

The self-overlap capacity of a given $k$-gram $S$ is determined from a polynomial:

$$K_S = \sum_{i=0}^{k-1} a_i x^i \tag{2}$$

Coefficients $a_i$ are 1 if the first and last $(k-i)$-grams in $S$ are identical and 0 otherwise. For example, the polynomial for trinucleotide GGG is $K_{GGG} = 1 + x + x^2$. For trinucleotides GAG and GAT we have $K_{GAG} = 1 + x^2$ and $K_{GAT} = 1$ (no self-overlap), respectively.

The variance of frequency distribution for a $k$-gram $S$ in a Bernoulli text of length $L$ over an elementary alphabet containing $n$ letters is a function of $K_S(1/n)$ given by

$$V_S = \frac{L}{n^k}\left[2K_S\left(\frac{1}{n}\right) - 1 - \frac{2k-1}{n^k}\right] \tag{3}$$

Similar formulas can be derived from nonBernoulli texts.

Failure to consider self-overlap capacity leads to huge errors (hundreds of percent) in variance calculation, particularly in cases of small-size elementary alphabets and large values of $k$.

Examples of 2-gram frequency counts in seven large collections of DNA sequences are shown in Figure 1. Although the ''folklore'' of describing and interpreting $k$-gram frequency counts resembles the descriptive biology of the nineteenth century (see caption to Figure 1), its role in selecting elementary alphabets and $k$-gram candidates for motifs should not be underestimated.

## 3 DETERMINING CLASSIFICATION CODE WORDS

Methods of determining code words explore the symbolic properties of sequences without directly appealing to the meaning of ''encoded'' messages (i.e., function, in molecular biology parlance). They resemble protocols to segment and decipher cryptograms. The linguistic theory behind these protocols draws on the assumption that selected patterns of characters (from an elementary alphabet) should occupy distinct relative positions in every script phrased in a given language. This assumption of the existence of distributional structures seem to be useful in molecular biology, but it is still controversial in linguistics proper.

Distributional structures in nucleotide sequences from large collections of sequences belonging to the same functional class (as introns, exons, tRNAs, 5′- and 3′-UTRs, illegitimate recombination sites, etc.) can be seen on so-called $k$-gram *distance charts.* We

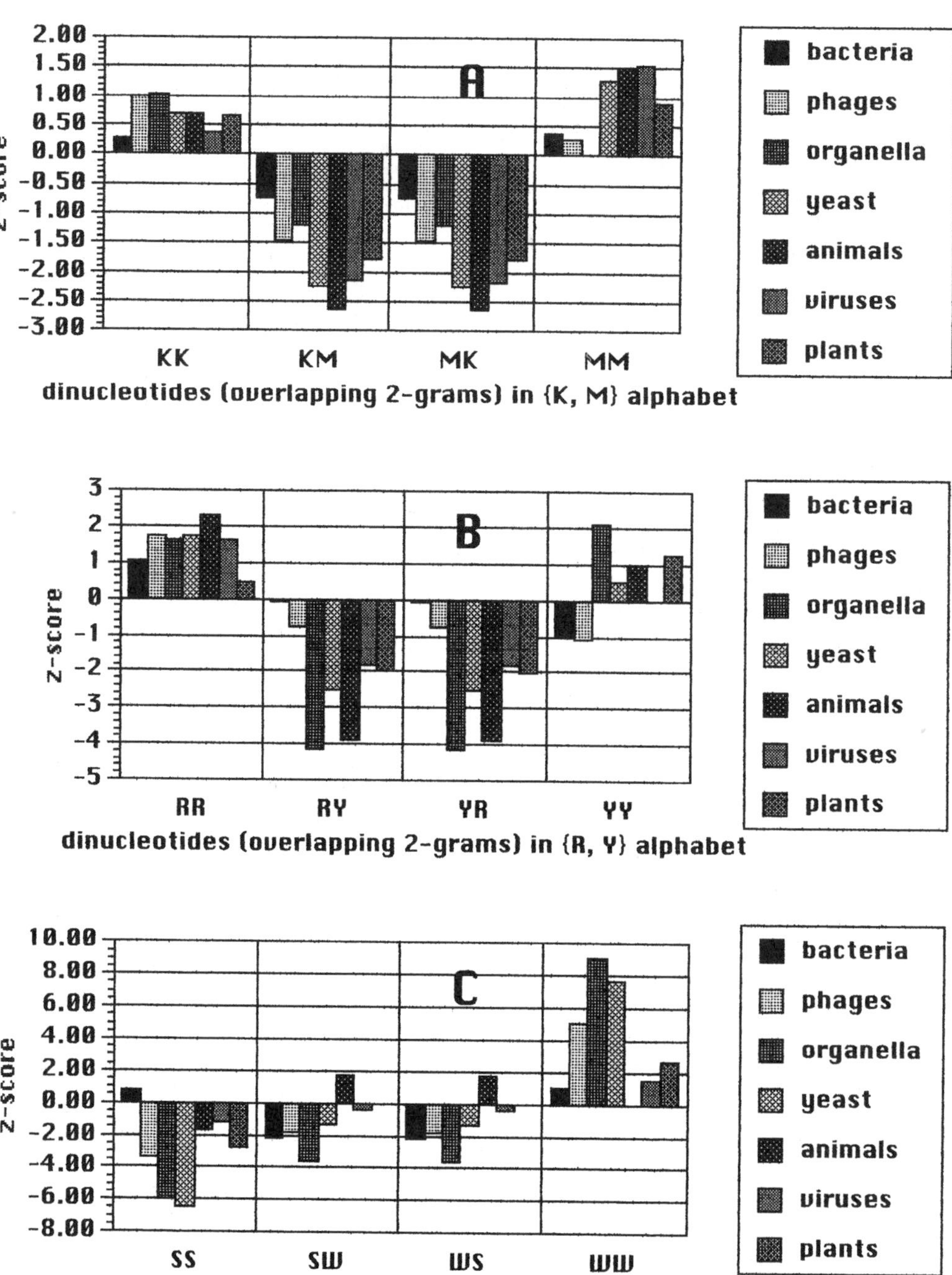


**Figure 1.** Examples of dinucleotide frequency counts in representative (large) samples of DNA sequences from seven different classes of genomes. Three different elementary alphabets were taken into account.

_(A) {K, M} alphabet, where *K* is either guanine or thymine and M is either adenine or cytosine. Assuming that the threshold *z*-score value for unusually high frequency is +3 and for unusually low frequency is −3, none of dinucleotides over this alphabet occurs significantly more often or significantly less often than others in all sequence samples studied. However, ''complex'' dinucleotides KM and MK tend to be slightly but systematically underrepresented.

_(B) {R, Y} alphabet, where R stands for purine (adenine or guanine) and Y stands for pyrimidine (cytosine or thymine). ''Complex'' dinucleotides RY and YR appear to be unusually underrepresented in organella and animal (i.e., primate, rodent, other mammalian, other vertebrate and invertebrate) genomes. This underrepresentation might be due in part to the well-known avoidance of long alternating purine–pyrimidine tracks (sequences with a potential for B-DNA–Z-DNA tertiary structure transition). Note that there is slight underrepresentation of RY and YR in other genomes (bacteria, phages, yeast, viruses, and plants), as well.

_(C) {S, W} alphabet, where S stands for either guanine or cytosine and W stands for either adenine or thymine. It can be seen that dinucleotide SS is underrepresented in phage, organella (mitochondria and chloroplast), and yeast genomes. In contrast, dinucleotide WW is overrepresented in these genomes. Notably, dinucleotides SW and WS are significantly underrepresented in organella (*z*-score −3*) and tend to be underrepresented in other genomes as well.*

select a particular $k$-gram to be a motif and then consider oligonucleotides separating the instances of this $k$-gram to be punctuations. The frequency of occurrence of punctuations plotted against their length is a distance chart. If there were no distributional structure in sequences studied, there would be no particular reason for any punctuation length to occur more frequently than other lengths. In other words, with no distributional structures, punctuation lengths should follow a discrete, uniform frequency distribution. When the distributional structures are present, some punctuation lengths should be much more frequent and some others much more rare than expected from discrete, uniform distribution.

In another kind of distance chart, the motif is not allowed to be a substring of any punctuation. The corresponding distribution of nearest distances between the occurrences of motif is called a *shortest distance chart.* In the absence of distributional structures, the frequency of punctuation lengths would follow geometric distribution. Again, when distributional structures are present, some punctuation lengths will occur more frequently and some others less frequently than expected from geometric distribution.

Distance charts ($z$-scores with regard to discrete, uniform distribution) for motif AC in large collections of nuclear introns and exons from animal genomes (primates, rodents, other mammals, vertebrates, and invertebrates) are shown in Figure 2A; the chart for a mirror-symmetric motif ACA appears in Figure 2B. The corresponding shortest distance charts ($z$-scores with regard to the appropriate geometric distribution) are shown in Figures 2C and 2D. It is clearly seen from this series of charts that exon and intron sequences display $k$-gram distributional structures and that these structures are transparently different for exons and for introns. The former are strongly overrepresentation for punctuation lengths 0, 3, 6, . . . and generally $3k$ ($k$ = 0, 1, 2, 3, . . .) for the trinucleotide ACA, whereas the later favor distances 0, 2, 4, . . . and generally $2k$ ($k$ = 0, 1, 2, . . .) for dinucleotide AC. Inspection of shortest distances (Figures 2C, D) shows that preferred nearest distances between ACs are 0 and 2 in introns. However, exon preferred shortest distances between ACAs (Figure 2D) are the same as distances from Figure 2B (i.e., 0, 3, 6, 9, . . . , and generally $3k$). These findings suggest that AC is involved in ''true'' two-base periodic patterns in introns, whereas ACA is involved in strongly pronounced three-base quasi-periodic patterns in exons.

Other than AC, nonhomopolymeric dinucleotides and the vast majority of nonhomopolymeric (including mirror-symmetric) trinucleotides display distance charts almost identical to those in Figure 2 (data not shown). This finding suggests that predominant patterns in exons and introns are three-base quasi-periodic arrangements of trinucleotides and two-base periodic repeats of dinucleotides, respectively.

Notably, distance charts from the foregoing example allowed us to rediscover the triplet structure of the genetic code from exon sequences alone (i.e., without taking into account our knowledge of the genetic code). In the light of this rediscovery, two-base periodicities observed in introns must reflect meaningful, intron-specific distributional structures. Similarly, we can claim that three-base quasi-periodic patterns involving trinucleotides constitute classification code words for exons.

Many other methods to determine classification code words from corpora of sequences exist. Among the most important are perhaps those based on coincidence indices (string length test and string repetition test) or on Markov analyses (including hidden Markov chains and nonhomogeneous Markov chains). If a given pattern is indicated as a candidate code word by more than one method, its significance is (obviously) high. Two-base periodicities in introns and three-base quasi-periodicities in exons are indeed indicated by several methods other than distance charts from the preceding example. In this sense these periodicities are valid classification code words.

Knowing classification code words has practical consequences. It allows us to develop algorithms for prediction of putative functional domains in unannotated sequences. These algorithms (and the corresponding software) are of paramount importance for all genome sequencing projects, and they are described in other entries in this volume.

## 4 LOCAL COMPOSITIONAL COMPLEXITY: A MAXIMUM ENTROPY PRINCIPLE

Simple frequency counts in nucleotide sequences show that oligonucleotides (i.e., $k$-grams) that do not contain all four nucleotides occur more often than those that do contain all four nucleotides (data not shown). This finding attracted attention of theoretical biologists to the concept and to mathematical models of local compositional complexity. For long stretches of sequences, local compositional complexity can be defined as Shannon entropy over frequencies of mononucleotides. A different measure is needed for local complexity of short oligonucleotides because short strings of symbols do not meet criteria for existence of Shannon entropy.

To define local complexity, we first consider an oligonucleotide of length $L$ and determine its composition $n$(A), $n$(C), $n$(G), and $N$(T). [Obviously $n$(A) + $n$(C) + $n$(G) + $n$(T).] Next we create a vector of sorted nucleotide occurrences without regard to the actual nucleotides participating in the oligonucleotide (only the order counts):

$$\boldsymbol{n} = [n_1, n_2, n_3, n_4] \tag{4}$$

where $n_1 \geq n_2 \geq n_3 \geq n_4$.

Many oligonucleotides of length $L$ can correspond to the same vector (equation 4). The exact number can be determined by counting different rearrangements of nucleotides $W(n)$ and ''colorings'' $F(n)$ of the vector $\boldsymbol{n}$. These quantities have been found to be:

$$W(n) = \frac{L!}{\prod_k n_k!} \tag{5}$$

and

$$F(n) = \frac{4!}{\prod_{i=0}^{L} h_i!} \tag{6}$$

where $h_i$ is the number of components of equation (4) that are exactly equal to $i$.

A definition of local compositional complexity of a given oligonucleotide of length $L$ is

$$C_n = \frac{1}{L} \log_4 W(n) \tag{7}$$

The numerical values of equation (7) for very large $L$ are indistinguishable from values of Shannon entropy.

The theoretical (i.e., a priori) probability of a given vector $\boldsymbol{n}$ has been found to be:

$$P_0 = \frac{F_n}{4^L} \mathrm{W}(n) \tag{8}$$

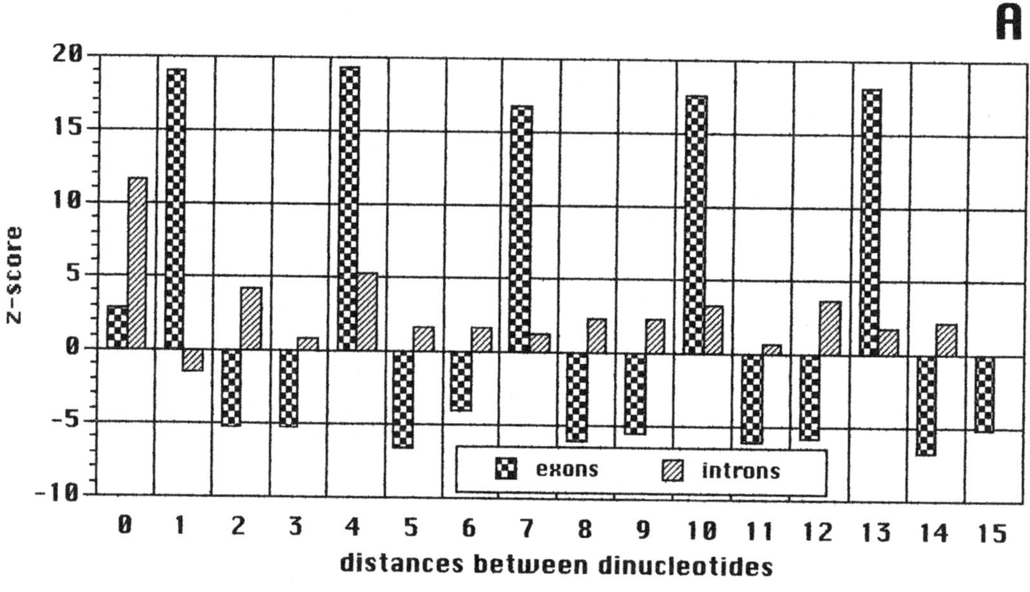


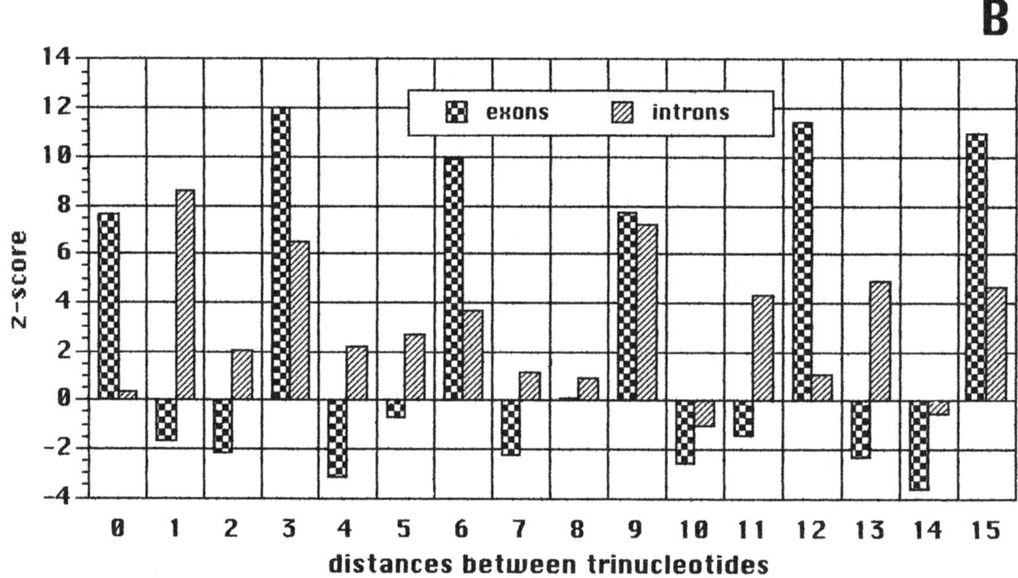


Figure 2. Examples of distance charts for nonhomopolymeric dinucleotides and the corresponding mirror-symmetric trinucleotides in large corpora of animal nuclear exons and introns. [Based on and modified from A. K. Konopka and G. W. Smythers, *Comput. Appl. Biosci.* 3:193–201 (1987).] (A) Dinucleotides tend to be separated by gaps of lengths divisible by 2 in introns. In exons those gaps lengths tend to follow an arithmetic progression with increment 3. This leads to predominant distances of 1, 4, 7, 10, . . . , and generally $1 + 3k$ ($k = 0, 1, 2, \ldots$). (B) Trinucleotides (including the mirror-symmetric ones) tend to be separated by gaps of lengths divisible by 3 in exons.

This theoretical distribution can be used to find distributions of *log-odds ratios* (sometimes called *surprisals*) as a function of complexity of short oligonucleotides in large corpora of functionally equivalent sequences.

Log-odds ratios (suprisals) are defined as

$$S(n) = \log_4 \frac{P_{\text{actual}}}{P_0} \tag{9}$$

All collections of functionally equivalent nucleotide sequences studied thus far display a linear tendency in surprisal versus complexity distributions (see typical examples in Figure 3). The slope of this linear relation is negative in all cases studied. This result is interpreted as a maximum entropy relationship constrained by mean complexity.

It is like that maximum entropy surprisal versus complexity relation (with mean complexity as a constraint) is a general law governing all natural nucleotide sequences. There is preliminary evidence that protein sequences "obey" this principle as well. Besides the generality of this law, sequences belonging to different putative functional domains display systematically different slope values (see examples in Figure 4). This observation suggests that slope values can be used as classification code words as well.

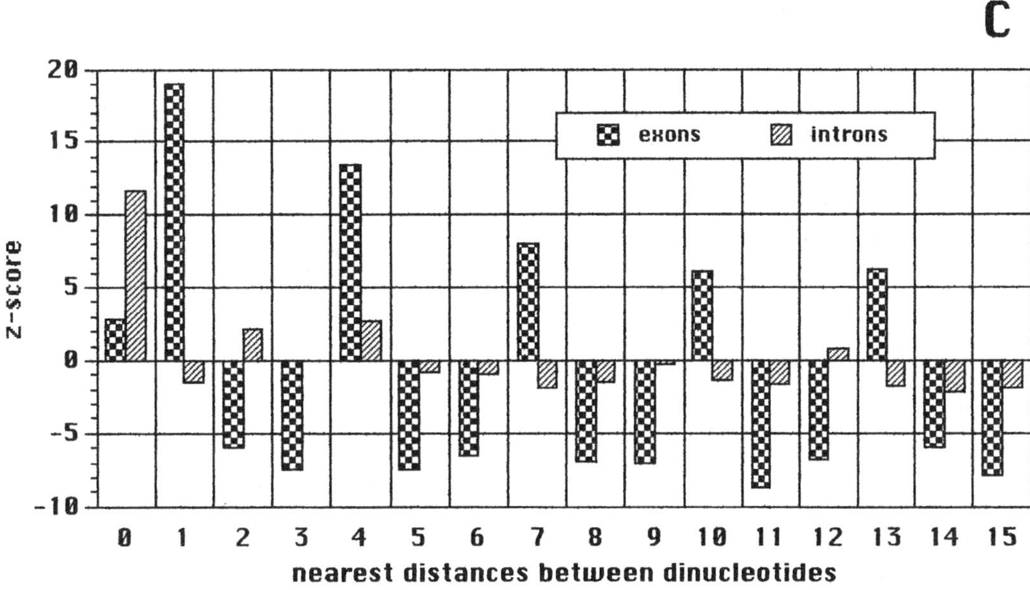


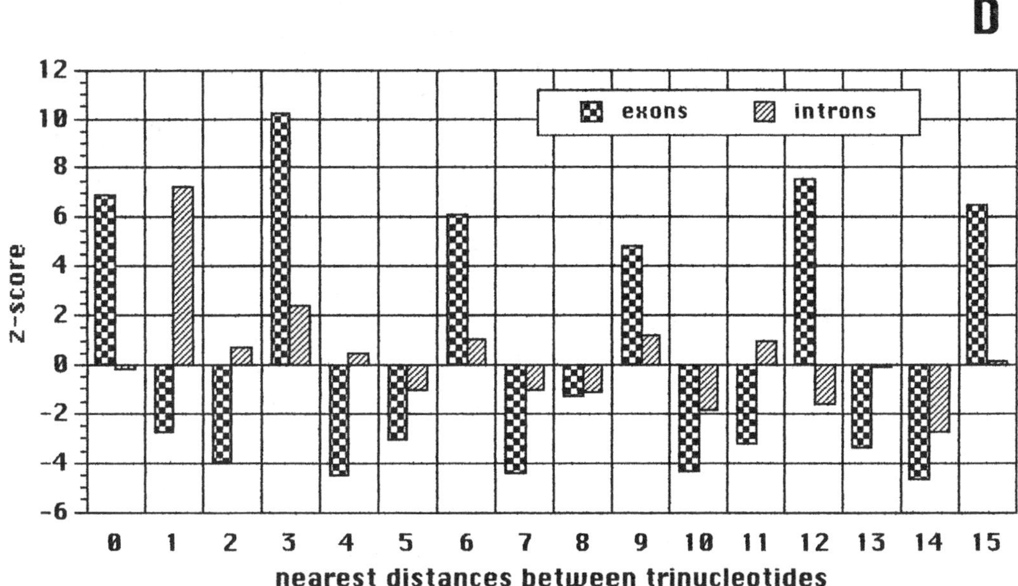


**Figure 2 (continued).** (C) Shortest distance chart displays the same predominant dinucleotide distances in exons as the distance chart (Figure 2A). However, the predominant shortest distance for dinucleotides in introns is 0 nucleotides. This means that the two-base periodicity implied by the distance chart is a "true" periodicity. (D) Shortest distance chart for trinucleotides in exons displays the same, divisible by 3, predominant distances as the distance chart in (B). The three-base periodicity implied by (B) is not a "true" periodicity but a three-base quasi-periodicity.

Search for general principles governing biopolymer sequences (and structures) is the most prevalent trend in theoretical molecular biology today. A maximum entropy principle, as just described, gives us confidence that such general laws indeed exist.

## 5 CONCLUDING REMARKS: A PERSPECTIVE

As far as can be judged in 1993, in the next decade or so theoretical molecular biology will explore two major directions: (1) research on fundamentals of modeling biological phenomena (so to speak, basic research proper) and (2) research to assist molecular biology software development (so to speak, service work demanded by laboratory workers). Solid theoretical foundations—versus a "brute force," mere induction from large number of instances—will be needed in both areas, but here we mention only two general topics pertaining to the first. Descriptions of theoretical issues pertaining to area 2 (software development) are scattered throughout this volume.

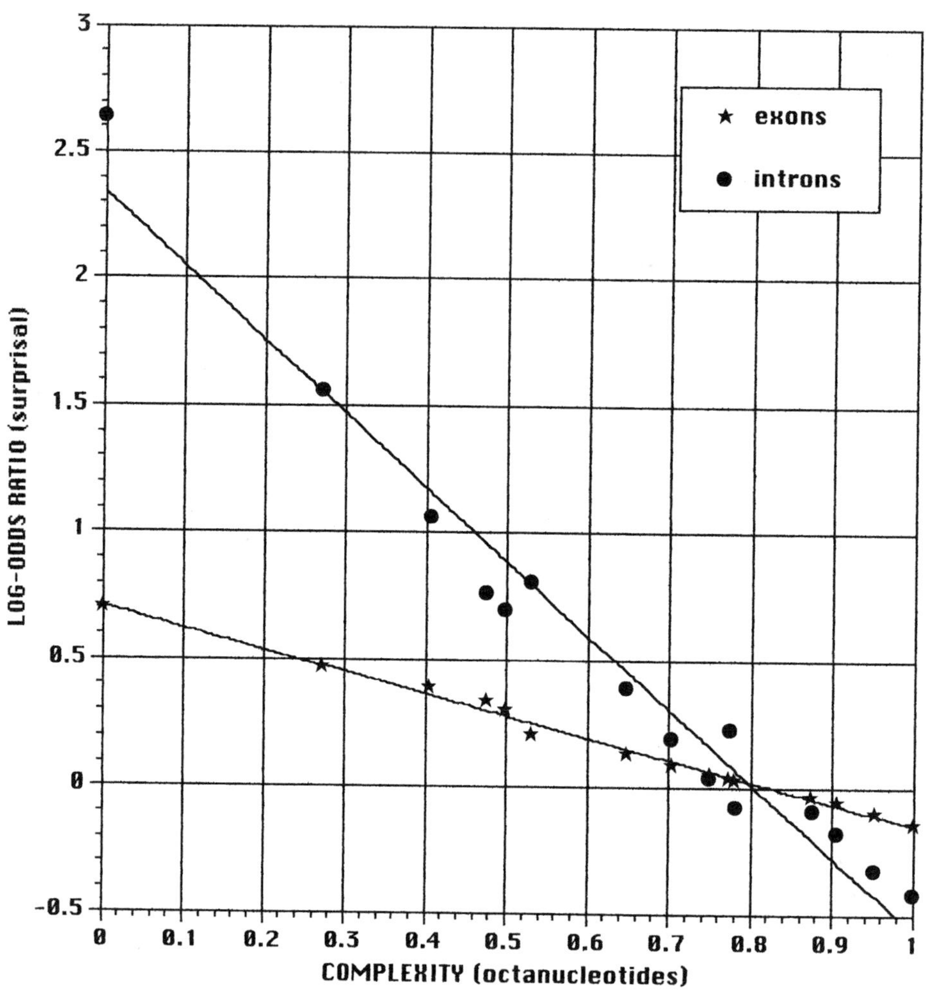


Figure 3. Example of maximum entropy relationship between complexity and surprisal (see text) for octanucleotides in invertebrate introns and exons. All naturally occurring nucleotide sequences studied thus far ( 30 large corpora of sequences) display a linearity (with negative slope) of log-odds complexity distributions for tetra- through octanucleotides. [Based on P. Salamon, and A. K. Konopka, *Comput. Chem.* 16:117–124 (1992).]

### 5.1 Pragmatic and Semantic Theories of Information

The existing (Hartley–Shannon) theory of communication was created for telecommunication systems and digital computers. Despite numerous proposals stating otherwise, it does not seem to provide much new insight into molecular biology (it can be a useful source of statistical methods, though). On the other hand, concepts of ''information,'' ''meaning,'' ''function,'' and ''organization'' seem to be integral parts of biology. If we are to comprehend the phenomenon of life despite our language limitations, we need to develop a theory capable of handling these very general concepts in a ''detached'' way. The need for a theory of information pertinent to biology has been postulated in neurobiology and in developmental biology. Although such a need has not troubled molecular biologists quite yet, it begins to emerge as evidence accumulates for the role of ''epigenetic information'' in controlling gene expression and regulation.

### 5.2 Logic of Heuristic Reasoning

Heuristic reasoning can be conclusive. Yet biologists face the problem of not being able to specify all the rules applied to derive conclusions. Nor are they able to list all assumptions on which those rules ought to operate. It seems that these inabilities are a reflection of the complexity of biological systems themselves. Biological phenomena are often represented by models that are still too complex to be described in a communicable manner. Further and further modeling is required, until our observations can be communicated in a linguistically comprehensive way. The cascade of models gives us the advantage of creating ''communicable reality'' but does not help us to judge the evidence pertinent to ''real'' (i.e., not necessarily communicable) reality. To the contrary, the more advanced a model in a cascade, the greater its ''distance'' (in terms of number of modeling steps) from the modeled system.

From a logical point of view there are two problems here. First, we have no formal system of inference to judge formal correctness of observation sentences that have a variable true value (credibility) and use ill-defined terms. Second, we have no formal system to judge the material adequacy of sentences derived from a ''distant'' model to the properties of modeled phenomena.

The problem of formal correctness seems to be solvable in principle. Progress in dealing with it can be noticed already in the fields of artificial intelligence (AI), pattern recognition (especially development of fuzzy mathematical techniques), and situation logic. In the near future we can expect to have formal tools to derive plausible conclusions from imprecise premises. Or, at least,

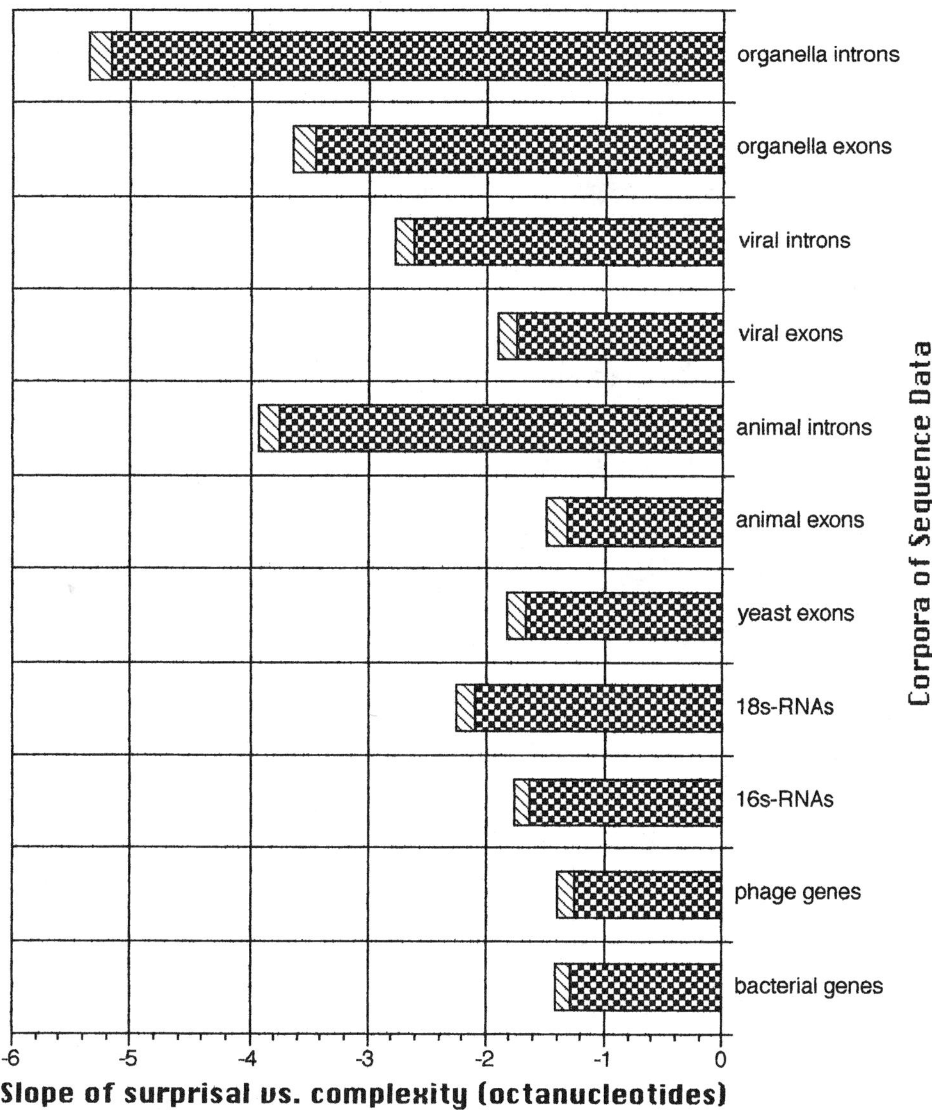


**Figure 4.** Different corpora of sequences generally display different slope values in the maximum entropy relationship between complexity and surprisal (see text). Shaded areas represent standard errors of slope estimate. [Based on P. Salamon, and A. K. Konopka, *Comput. Chem.* 16:117–124 (1992).]

we will have the option of relying on a machine (i.e., reliable AI software) that will perform plausible reasoning for us.

As far as formal judgment of material adequacy is concerned, it is unlikely that the problem is ''addressable'' in its generality. At least within mathematics, the celebrated Gödel theorem precludes such a possibility. However, informal (more or less educated common sense) judgments of material adequacy are possible and, as a matter of fact, all fields of science explore them. This is basically good news for theoretical molecular biologists: biology research will still require human intervention!

*See also* SEQUENCE ANALYSIS; SEQUENCE DIVERGENCE ESTIMATION.

## *Bibliography*


Bell, G. I., and Marr, T., Eds. (1990) *Computers and DNA.* Addison-Wesley Longman, Reading, MA.

Calladine, C. R., and Drew, H. R. (1992) *Understanding DNA.* Academic Press, San Diego, CA.

*Computers and Chemistry.* (1992) Special issue on Open Problems of Computational Molecular Biology, 16(2).

*Computers and Chemistry.* (1993) Special issue on Open Problems of Computational Molecular Biology, 17(2).

*Computers and Chemistry.* (1994) Special issue on Open Problems of Computational Molecular Biology, 18(3).

Doolittle, R. F., Ed. (1991) *Computer Analysis of Protein and Nucleic Acid Sequences,* Vol. 183, *Methods of Enzymology.* Academic Press, San Diego, CA.

Li, W.-H., and Graur, D., (1991) *Fundamentals of Molecular Evolution.* Sinauer, Sunderland, MA.

Rosen, R. (1991) *Life Itself.* Columbia University Press, New York.

Shulman, M. J., Steinberg, C. M., and Westmoreland, N. (1981) The coding function of nucleotide sequences can be discerned by statistical analysis. J. Theor. Biol. 88:409–420.

Smith, D. W., Ed. (1994) *Biocomputing: Informatics and Genome Projects.* Academic Press, San Diego, CA.

Waterman, M. S., Ed. (1989) *Mathematical Methods for DNA Sequences.* CRC Press, Boca Raton, FL.

Wyman, J., and Gill, S. J. (1990) *Binding and Linkage.* University Science Books, Mill Valley, CA.

# Tomatoes, Gene Alterations of

*Belinda Martineau and William R. Hiatt*

*Key Words*

***Agrobacterium tumefaciens*** Soil bacterium carrying a tumor-inducing (Ti) plasmid that allows it to infect and transform plants; products of the Ti plasmid virulence (*vir*) region genes provide for transfer to and insertion of the Ti plasmid T-DNA region into the plant genome.

**Antisense RNA** RNA that is complementary to the mRNA transcribed from an endogenous gene.

**Chimeric Gene** Recombinant gene consisting of the protein-coding region from one gene and the promoter and/or other regulatory regions from another gene(s).

**Promoter** Segment of DNA, located upstream from the coding region of a gene, that regulates spatial and temporal expression.

**Transform** To add exogenous DNA to the genome of an organism (e.g., bacterium or a plant).

Alteration of gene function in tomato is currently achieved using a transformation technology mediated by a harmless version of the plant bacterial pathogen *Agrobacterium tumefaciens.* A cloned gene, modified to effect changes in its spatial and/or temporal pattern of expression or to eliminate its expression, is spliced into a region of *A. tumefaciens* DNA that is subsequently inserted into the tomato genome upon infection of the plant. *Agrobacterium*-mediated gene addition experiments in tomato have increased our understanding of tomato gene expression and gene product function. The technology is now being applied to improve fruit quality and agronomic practices.

## 1 PRINCIPLES

### 1.1 Tomato Transformation and Regeneration

#### 1.1.1 Transformation

*Lycopersicum esculentum,* the cultivated tomato, is most commonly transformed by utilizing the virulence mechanism of the soil bacterium *Agrobacterium tumefaciens.* This bacterium infects tomato and many other dicotyledonous plant species by inserting a segment of *Agrobacterium* DNA into a susceptible individual. Expression of the inserted DNA segment in the infected plant provides the substrates necessary for bacterial sustenance and the "crown gall" of undifferentiated cellular growth by which the disease inflicted by *A. tumefaciens* gets its name.

Genetic engineers have taken advantage of this parasitic relationship between bacterium and plant to introduce their own selected segment of DNA into tomato. The bacterial genes responsible for gall formation and other bacterial functions have been eliminated from the DNA transferred from the bacterium to the plant (T-DNA), thus "disarming" *Agrobacterium* of pathogenicity. Various genes of interest to the genetic engineer are then placed into the bacterial T-DNA and are consequently inserted into the genome of a tomato plant.

#### 1.1.2 Regeneration

Individual cells or a small group of cells from tomato leaves, stems, or other plant parts can be regenerated into an entire, fertile plant. The genetic engineer provides for tomato and bacteria cells a conducive laboratory environment in which to carry out both the DNA transfer process and tomato plant regeneration. The two organisms are incubated together under specific conditions of light, heat, and supporting media, including plant growth regulators in proper proportions. Tomato cells that have permanently acquired T-DNA (i.e., have been stably transformed) are then usually identified by means of a marker gene included in the T-DNA. For most gene alterations of tomato to date, this marker gene has encoded resistance to an antibiotic, kanamycin. By including kanamycin in the regeneration medium, transformed tomato cells expressing the marker gene will be regenerated into whole, fertile tomato plants.

### 1.2 Altering Gene Function Through Gene Addition

The gene alterations possible using *Agrobacterium*-mediated transformation result from the introduction of the T-DNA from the bacterium into the tomato genome. Therefore, tomato gene alterations produced using this system are gene additions. It is the composition of the added genes, customized by the genetic engineer, that dictates the alteration of the function of tomato genes.

## 2 TECHNIQUES

### 2.1 Gene Isolation and Chimeric Gene Construction

Any organism may serve as the source of a gene, especially of a gene's protein-coding region, for addition to tomato. The selectable marker gene that confers resistance in transformed tomato cells to kanamycin, the *kan*$^r$ gene, is a good example of a foreign gene added to tomato. This gene, encoding the enzyme aminoglycoside 3′-phosphotransferase II (APH(3′)II), was originally identified in a bacterial species. Because the genetic code is universal, the specific DNA sequence providing the code for the APH(3′)II protein can be deciphered by the protein translational machinery of a tomato plant as well as by bacteria. However, the DNA sequences that regulate a gene's spatial and temporal function in one organism are not always recognized by other organisms. Therefore, genetic engineers customize a foreign gene to have regulatory DNA sequences that are readily recognized by the tomato plant, in conjunction with the protein-coding region of the foreign gene. The resulting recombinant gene is a molecular chimera, or "chimeric gene," consisting of gene regions of diverse genetic derivation. The *kan*$^r$ gene, for example, contains regulatory sequences that provide for expression of the bacteria-derived APH(3′)II protein encoding gene in tomato cells during the tomato regeneration process. Figure 1 depicts some sample chimeric genes.

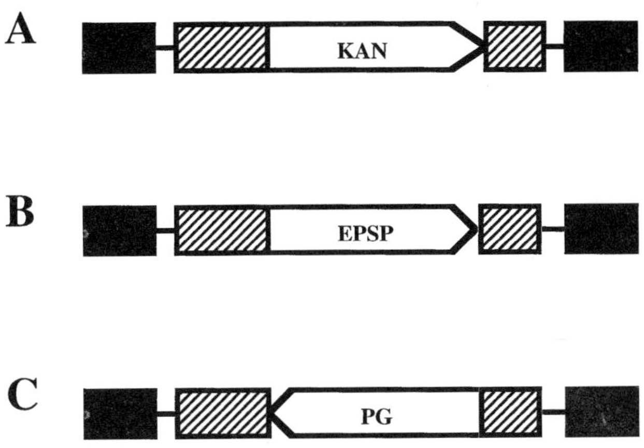


**Figure 1.** Schematic representations of chimeric genes. (A) Sample protein-coding region from a nontomato gene. (B) Sample protein-coding region from tomato (or another plant) gene. (C) Sample protein-coding region from a tomato gene in the antisense orientation. Solid boxes represent the borders of the *Agrobacterium* T-DNA; hatched boxes represent regulatory DNA sequences required for gene expression in plants; arrows represent protein-coding regions. KAN indicates protein-coding region for the *kan*$^r$ gene; EPSP indicates the protein-coding region of an EPSP synthetase gene; PG indicates the protein-coding region for the polygalacturonase gene.

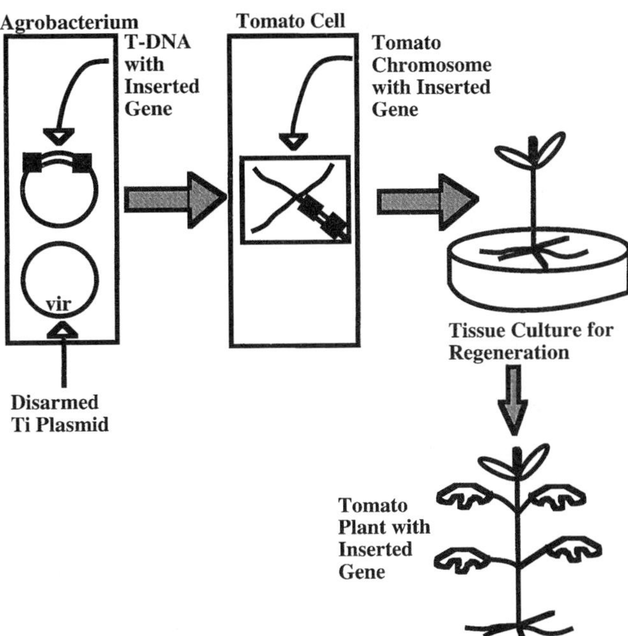


**Figure 2.** Schematic representation of the *Agrobacterium*-mediated transformation and regeneration process for inserting genes into tomato.

### 2.1.1 Altered Regulation of Endogenous Genes

Similarly, construction of chimeric genes can also involve protein-coding regions from genes that are normally present in tomato. The "normal" expression of such an endogenous tomato gene can be altered by eliminating the original regulatory DNA sequences of the cloned gene and replacing them with sequences that determine the gene expression patterns of another (usually a tomato) gene. Since the protein-coding sequences of the gene serving as the source of the regulatory sequences are eliminated in the process, this gene can be of known or unknown function. Typically, gene regulatory sequences, or promoters, are identified by virtue of the desirable level and specificity of their expression in particular plant tissues or organs. This is usually accomplished using a technique referred to as "differential screening" of a cDNA library. The protein-coding region from the cDNA clone with the desired specificity of expression, defined by a start ATG codon and a stop codon, is then eliminated and replaced with that of the endogenous gene. Such constructed chimeric genes, when inserted into the tomato genome via *Agrobacterium*-mediated methods, can augment the spatial and temporal expression of the original endogenous tomato gene.

### 2.1.2 Antisense Genes

Another method of altering tomato gene function is through expression, in transformed tomato plants, of antisense genes. Expression of antisense genes curtails the expression of an endogenous gene. In an antisense gene construction, the protein-coding region (or a portion thereof) from the endogenous gene is placed in reverse orientation with regard to a promoter region, with the result that the complement of the normal mRNA, antisense RNA, is produced during transcription. The presence of antisense RNA interferes with the function of the normal mRNA, and expression of chimeric antisense genes in transformed tomato plants is, therefore, a means of creating specific mutations in selected tomato genes.

## 2.2 *Agrobacterium* Transformation Vectors

Once constructed, a chimeric gene destined for incorporation into the tomato genome must first be placed in the T-DNA of *A. tumefaciens*. The T-DNA region is present on a large tumor-inducing (Ti) plasmid. Since the Ti plasmid is large enough to make conventional cloning techniques cumbersome, the biology of the bacterium has been manipulated to make the procedure more feasible.

While it is the T-DNA, and only the T-DNA, of the Ti plasmid that is transferred to a plant cell during *A. tumefaciens* infection, the proteins responsible for directing that transfer are encoded elsewhere on the Ti plasmid. In fact, this "*vir*" region of the Ti plasmid can function in directing T-DNA transfer without ever being present on the same piece of DNA. As long as the *vir* and the T-DNA regions are present in the same *A. tumefaciens* cell, T-DNA transfer to plant cells can occur. "Binary" vector systems, in which the *vir* and T-DNA regions are located on two separate DNA plasmids, are based on this finding that *vir* gene functions can act "in trans." Especially convenient binary vectors have been developed which allow manipulation of the T-DNA region in *E. coli* as well as in *A. tumefaciens*. As a result of this feature, chimeric genes can be introduced into the T-DNA using standard recombinant DNA techniques. Introduction of the recombinant T-DNA containing binary vector into an *A. tumefaciens* strain carrying the second *vir* region containing vector is subsequently accomplished via standard bacterial conjugation or transformation procedures. The resulting adulterated bacteria, when given the opportunity, will now unwittingly insert the engineered gene in its T-DNA region into the DNA of a tomato cell. An overview of this process is shown in Figure 2.

### 2.3 Analysis of Transformed Plants

The first indication that a tomato plant has been transformed by an *Agrobacterium* strain carrying a selectable marker gene such as *kan*$^r$, as well as a specific chimeric gene of interest, is the plant's growth on media containing kanamycin. Among the transformed plants, there can be variability in the number of sites in the tomato genome into which the T-DNA has been integrated and the number of copies of the T-DNA region inserted at each site. The functioning of the genes within the T-DNA can also vary from one transformed plant to another, an effect believed to be due to local chromosomal influences at the site of insertion. Various standard molecular techniques, designed to identify the specified inserted genes, the products of those genes, and the levels of any biochemical constituents altered through the functioning of those gene products, are therefore used to characterize transformed tomatoes.

Selected transformants are ultimately field tested. Field trials are usually conducted when it is intended that the introduced gene(s) alter biochemical constituents that affect field measured traits. Fruit-related qualities such as total yield, color, and various processing characteristics (viscosity and consistency of paste, yield of tomato solids, etc.) are examples of such traits. A field trial situation is also a good way to ensure that a transformed plant exhibits no unexpected morphological abnormalities that may have resulted from the regeneration process.

## 3 APPLICATIONS

*Agrobacterium*-mediated transformation methods for altering gene function in tomato have been applied to two major areas. One addresses the agronomic practices of pesticide and herbicide use during tomato production. The other deals with fruit quality issues.

### 3.1 Pesticide and Herbicide Applications

In one strategy designed to induce crop plants like tomato to produce their own insecticide, genes encoding the so-called *Bt* toxin proteins have been inserted into plant genomes. These proteins are toxic only to specific lepidopteran and coleopteran insect species (they produce no ill effects in other organisms) and have been used topically for decades to biologically control these major pests. The genes encoding *Bt* toxins have been isolated from the soil bacterium *Bacillus thuringiensis* and used to produce (via *Agrobacterium*-mediated methods) tomato and other crop plants that are unpalatable, even poisonous, for tomato horn worms among other pests.

Genetic engineers also are applying this technology to crop production practices by using it to affect crop plant tolerances to specific herbicides. The herbicide glyphosate serves as a case in point. Glyphosate kills plant cells by inhibiting an important enzyme (5-enolpyruvylshikimate 3-phosphate synthetase; EPSP synthetase) involved in the biosynthesis of aromatic compounds. Tomato and other crop plants have been made more tolerant to glyphosate through the addition of EPSP synthetase genes that have been altered in one of two ways. The plant version of the EPSP synthetase gene has been altered so that significantly more of the enzyme is produced and the plant expressing the altered gene can therefore tolerate higher levels of the herbicide compound. A bacterial gene encoding a glyphosate-resistant version of EPSP synthetase has also been customized for and expressed in tomato. Plants altered in this fashion enable the use of this biodegradable herbicide in place of other more persistent chemicals.

Genetic engineering strategies designed to protect tomatoes and other agriculturally significant crops from viruses, fungi, nematodes, and herbicides such as bromoxynil are also being investigated.

### 3.2 Tomato Fruit Quality Applications

The quality of a fresh tomato is judged primarily on the basis of color, firmness, and general lack of visible damage. To date, the greatest strides in improving tomato fruit quality have been attained by slowing down the process of fruit softening by specifically attenuating the function of the tomato gene encoding polygalacturonase (PG). PG breaks down pectin, a major component of tomato fruit cell walls, and had long been implicated in the fruit softening process. Altering PG gene function as a result of transforming tomato plants with chimeric antisense PG genes resulting in fruit that soften and rot less quickly.

## 4 PERSPECTIVES

The introduction and expression of foreign and endogenous genes in tomato has advanced our understanding of how plant genes and the products they encode function. The resulting knowledge, particularly in areas relating to tomato fruit quality, is being applied to commercial goals. As a convenient model for gene alteration experiments, tomato will continue to play an integral role in deciphering the many remaining obscurities in our understanding of plant molecular biology and physiology. Because of tomato's agricultural relevance, future insights gained can also have wide consequences for the consumer.

*See also* Insecticides, Recombinant Protein; Pesticide-Producing Bacteria; Plant Cells, Genetic Manipulation of.

### *Bibliography*

Gasser, C. S. (1992) Transgenic crops. *Sci. Am.* 266:62–69.

Gray, J., Picton, S., Shabbeer, J., Schuch, W., and Grierson, D. (1992) Molecular biology of fruit ripening and its manipulation with antisense genes. *Plant Mol. Biol.* 19:69–87.

Hofte, H., and Whiteley, H. R. (1989) Insecticidal crystal proteins of *Bacillus thuringiensis*. *Microbiol. Rev.* 53:245–255.

Hooykaas, P. J., and Schilperoot, R. A. (1992) *Agrobacterium* and plant genetic engineering. *Plant Mol. Biol.* 19:15–38.

Mazur, B. J. (1989) The development of herbicide resistant crops. *Annu. Rev. Plant Physiol. Plant Mol. Biol.* 40:441–470.

Nevins, D. J., and Jones, R. A., Eds. (1987) *Tomato Biotechnology*. Liss, New York.

# Trace Element Micronutrients

*Robert J. Cousins*

### *Key Words*

**Ligand** Negatively charged organic molecule that donates electrons to form coordinate bonds with a metal ion.

**Metalloenzyme** Enzyme that requires one or more atoms of a metal for catalytic activity and/or for structure required for activity.

**Metal-Responsive Element (MRE)** Specific sequence of promoter region of gene that enhances transcription through a specific transcription factor (MRE binding protein) that requires metal occupancy for DNA binding.

**RNA Binding Metalloregulatory Protein** Binds in response to metal occupancy to a stem loop of nucleotides called a regulatory element in the 5′ untranslated region of a specific mRNA to influence translation.

**Trace Element Micronutrient** Essential nutrient metal found in the diet in small (microgram or milligram) amounts.

**Zinc Finger Motif** A zinc-binding domain of a protein having a -Cys-Xaa$_2$-Cys-Xaa$_n$-Cys-Xaa$_2$-Cys (or His)- or similar sequence that provides the opportunity for Zn(II) binding.

---

"Trace element micronutrients" is a term that refers to the metals (inorganic elements) in the diet that are essential for specific cellular processes. Required dietary trace elements include copper, iron, manganese, selenium, and zinc. They perform either regulatory, catalytic, or structural roles. Zinc in particular has a major role in maintaining conformations necessary for DNA binding by transcription factors and for protein–protein interactions. Specific trace elements influence transcription through binding proteins that during metal occupancy bind to a specific promoter DNA sequence, a metal-responsive element (MRE), to stimulate transcription. Regulation by trace element micronutrients can also occur through metal-requiring, RNA-binding proteins, which influence translation of specific mRNAs via nucleotide sequences in the untranslated regions. Trace element micronutrients have potential applications in biotechnology, either directly as signaling molecules for regulation of chimeric genes, or indirectly by providing metal-binding sites for targets of therapeutic agents.

## 1 SCOPE OF TRACE ELEMENT MICRONUTRIENTS IN MOLECULAR BIOLOGY AND BIOTECHNOLOGY

### 1.1 Chemical Considerations

"Micronutrients" as a general term describes both essential organic molecules (vitamins) and metals (trace elements). Trace elements required to sustain cellular processes include copper, iron, manganese, selenium, and zinc. Other metals (e.g., nickel) may have nutritional roles, but their roles in molecular biology have not been studied extensively. In humans these substances are required in microgram to milligram amounts per day to maintain balance with endogenous losses. Cellular constituents, particularly proteins, provide ligands (e.g., amine, thiolate, carboxyl) for metal binding. Of the trace element micronutrients, zinc has the most collective influence in molecular biology. Transport systems regulate intracellular trace element concentrations that limit the influence of natural abundance and thermodynamic considerations in determining which elements participate in specific cellular functions.

### 1.2 Classes of Involvement

There are three main classes of involvement of trace element micronutrients in the province of molecular biology. One class is structural, with metals facilitating interaction among various binding groups of specific motifs, which provide altered conformations necessary to achieve unique opportunities for specific interaction. Metalloenzymes, where metals provide a catalytic function, comprise the second class. The third class of involvement is regulatory, as exemplified by the metal-binding, trans-acting proteins that provide signaling to initiate transcription of genes through interaction with specific DNA sequences, the metal-responsive elements (MRE). Metal-responsive RNA-binding proteins are another example. Collectively, these classes of biological involvement of trace elements provide a vital, essential link between the organism and its external nutrient supply.

### 1.3 Cellular Distributions

Concentrations of trace elements in cells differ greatly (usually in the microgram-per-gram range) and are distributed to different organelles. However, metal occupancy at specific binding sites, rather than concentration, determines physiological effects. Cells vary widely in responses to or dependence on the trace element content of the environment. This is particularly evident in unicellular organisms. In contrast, many mammalian cells in situ and in vitro are resistant to metal deprivation, but others are not. Therefore, functional deficiency of the dietary supply may occur only after depletion of a critical pool or compartment of the trace element (e.g., zinc in the nucleus), which affects specific binding sites. Unfortunately, our understanding of the cell biology of trace elements micronutrients is limited.

## 2 STRUCTURAL/CATALYTIC/ REGULATORY ROLES

Characterization of individual metal-binding molecules has shown that stoichiometry is usually one to four metal atoms per binding molecule. There are exceptions: for example, ferritin (4500 atoms of Fe per molecule) or metallothionein (7 atoms of Zn or 12 atoms of Cu per molecule). In addition, abundance of metal-binding molecules ranges from a few to many molecules per cell. Metalloproteins that act as RNA nucleotide transferases, DNA-binding transcription factors, trans-acting regulatory proteins, proteins binding single-stranded nucleotides, and proteins exhibiting metal-dependent interactions with other proteins usually bind metals through thiolate groups from cysteine and/or nitrogen from histidine. Very little is known about the required dietary intake, transfer mechanisms among organelles, or ligand exchange reactions that are necessary for metals to fulfill these regulatory roles. The dietary requirement for the micronutrient consists of the composite cellular and extracellular need for each trace element to satisfy the occupancy needs of these binding sites as balanced against turnover and endogenous losses.

## 3 REGULATION OF GENE EXPRESSION BY TRACE ELEMENT MICRONUTRIENTS

Trace elements provide signals to systems that influence rates of either transcription or translation. They do not act directly. Rather, through recognition, requiring specific coordination chemistry, metals bind to metalloregulatory proteins (sensors or receptors) of low abundance and influence signaling pathways. This activity ultimately augments production of a specific protein. In the classical sense, these micronutrients (metals) act as inducers of the system.

Since the dietary supply provides the source of the inducer, fluctuations in intake and/or physiological factors influencing utilization of the micronutrient are determinants of how effective the trace element will be in activating a metalloregulatory system.

### 3.1 Transcriptional Regulation

Examples of transcriptional regulation by trace elements are found in microbial and animal cells, but detailed descriptions of the cellular apparatus involved have emerged more rapidly from the former. Regulation of mercury resistance factors serves as an example. Other bacterial genes that provide metal regulation are *Dtx*R, *Ans*R, *Cze*R, *fli*C, and *Smt*B. Similarly, the metalloregulatory protein HAP1 of yeast senses heme iron and binds to the apocytochrome *c* gene promoter to increase transcription.

Metallothionein gene expression has been the most widely studied system that is transcriptionally regulated by trace elements. Metallothionein-related copper resistance occurs in *Saccharomyces* and *Candida* organisms. Mammalian metallothionein expression is tissue specific and responds to changes in dietary trace element intake, exposure to toxic metals, and physiological stimuli, with high abundance in liver, intestine, bone marrow, and kidney; expression, however, can be detected in most tissues and cultures of mammalian cells. Dietary zinc intake level provides a proportional stimulus for expression, but the copper intake level over a wide range has little effect. Heme oxygenase, $\alpha_1$ acid glycoprotein, C-reactive protein, cytochrome *c*, superoxide dismutase, apolipoprotein A-I, and creatine kinase genes may be similarly regulated.

### 3.2 Metal Regulatory Elements

A minimum of three components is, at present, considered necessary for transcriptional regulation by a trace element (Figure 1). First of all, the responsive gene must have a metal-responsive element at one or more sites in the promoter upstream from the transcription initiation site. The consensus MRE core sequence for the mammalian metallothionein promoter is CTCTGCRC-NCGGCCC. MREs are frequently located near other core regulatory elements, perhaps for synergistic or differential regulatory purposes.

A trans-acting metalloregulatory protein(s) is the second component needed. Such metal-binding/sensing proteins (depicted in Figure 1 as MRE-BP for metal-responsive element–binding protein, or MeTF for metal transcellular factor) act as transcription factors. Metals bind to these proteins through unique coordination chemistry. Each MRE regulated by a specific metal may have a unique metalloregulatory protein. Tissue specificity of metal regulation requires that genes for the MRE-BP or MeTF be expressed to differing extents requisite to cell function and stage of development and perhaps nutritional or hormonal stimuli. The third component for this metal-responsive unit is the trace element (metal), that is, the inducer. The intracellular metal concentration available to the metalloregulatory protein varies with physiological state and dietary intake.

### 3.3 Translational Regulation

The initial concept of trace element control of translation was developed using as a model the acute induction of ferritin synthesis by Fe(II), which does not require transcription. This translational regulation is controlled by an iron-responsive element binding protein (IRE-BP), which binds iron. With low cellular iron, IRE-BP exhibits high binding affinity for both transferrin receptor mRNA and ferritin mRNA. This feature promotes cellular iron uptake through increased transferrin receptor synthesis and decreased (repressed) ferritin synthesis. When the iron supply is sufficient, receptor mRNA degradation increases and more ferritin mRNA is translated (derepressed). mRNA-binding proteins analogous to IRE-BP may exist which have functional binding domains for other metals. Translational regulation may occur for γ-aminolevulinate synthetase and glutathione peroxidase mRNAs.

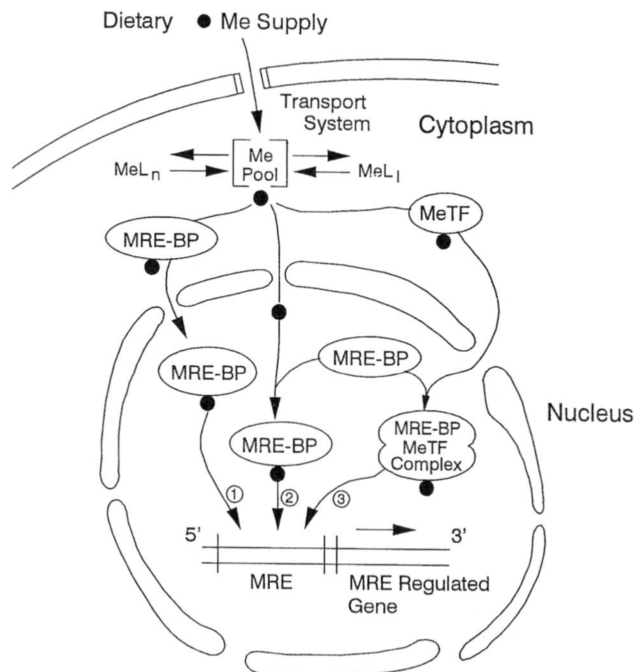


**Figure 1.** Hypothetical modes of transcriptional regulation of gene expression by trace element (metal) micronutrients. After transport into the cell, the metal (Me) interacts rapidly with a variety of ligands (L). The metal may enter the nucleus unbound or via a metal-responsive element binding protein (MRE-BP) or a metal transcellular factor (MeTF). Binding to the MRE of the promoter by the MRE-BP (1 or 2) or a MRE-BP/MeTF complex (3) is required to initiate transcription. The rate of transcription is proportional to the dietary supply of metal and the intracellular Me pool. Metal binding to cellular factors could also remove inhibition of promoter activity. In this mode, metal occupancy would constitute negative regulation. Metallothionein gene expression provides an example of this mode of regulation by a trace element micronutrient.

## 4 APPLICATIONS AND PERSPECTIVES

The application of trace element micronutrients for control of biotechnological systems has been demonstrated. The dramatic growth of transgenic animals producing excess growth hormone through activation of chimeric metallothionein promoter–growth hormone constructs in response to zinc in the drinking water illustrated the potential. Similarly, chimeric constructs using MREs have been used to regulate plant genes by trace elements. Similar experiments in animals and humans will likely occur in the future. The unique physiology and tissue-specific sequestration, as well as nutrition and toxicity of each trace element, need to be considered in these strategies.

The development of pharmaceuticals with actions related to metal chelation or metal donation is of major current interest. Many of these novel compounds will use trace element micronutrients as an active component. Alternatively, naturally occurring cellular components (e.g., a zinc finger transcription factor) could be the target of a specific trace element containing therapeutic or sequestering agent. Dietary intake may need to be adjusted to promote these uses.

*See also* DNA Replication and Transcription; Gene Expression, Regulation of; Zinc Finger DNA Binding Motifs.

### *Bibliography*


Chesters, J. K. (1992) Trace element–gene interactions. *Nutr. Rev.* 50:217–223.

Cousins, R. J. (1994) Metal elements and gene expression. *Annual Review of Nutrition,* R. E. Olson, Ed., pp. 449–469. Annual Reviews Inc., Palo Alto, CA.

Klausner, R. D., Rouault, T. A., and Harford, J. B. (1993) Regulating the fate of mRNA: The control of cellular iron metabolism. *Cell* 72:19–28.

O'Halloran, T. V. (1993) Transition metals in control of gene expression. *Science,* 261:715–725.

Palmiter, R. D., Norstedt, G., Gelinas, R. E., Hammer, R. E., and Brinster, R. L. (1983) Metallothionein–human GH fusion genes stimulate growth of mice. *Science,* 222:809–814.

Pursel, V. G., Pinkert, C. A., Miller, K. F., Bolt, D. J., Campbell, R. G., Palmiter, R. D., Brinster, R. L., and Hammer, R. E. (1989) Genetic engineering of livestock. *Science,* 244:1281–1288.

Rhodes, D., and Klug, A. (1993) Zinc fingers. *Sci. Am.* 268:56–65.

Séguin, C., and Hamer, D. H. (1987) Regulation in vitro of metallothionein gene binding factors. *Science,* 235:1383–1387.

Thiele, D. J. (1992) Metal-regulated transcription in eukaryotes. *Nucleic Acids Res.* 20:1183–1191.


## Transcription: *see* DNA Replication and Transcription.

# Transgenic Animal Modeling

*Carl A. Pinkert, Michael H. Irwin, and R. Jeffrey Moffatt*

### *Key Words*

**DNA Microinjection** A gene transfer technique utilizing micromanipulation to inject DNA constructs (transgenes) directly into pronuclei or nuclei of cells. The foreign DNA integrates into the host cell genome at a random location and usually in concatemers. DNA microinjection is the most commonly used gene transfer technique for creating transgenic mammalian species.

**Embryonic Stem (ES) Cell Transfer** The transfer of genetically manipulable pluripotent embryonic stem cells into a developing embryo by microinjection into blastocysts or coculture of earlier stage ova. The resultant transgenic animal possesses a proportion of cells descended from the ES cell lineage. In mammalian species, this gene transfer technique has been successful in producing only germ line transmission of ES cell lineages in mice.

**Gene Transfer** A set of techniques directed toward manipulating biological function via the introduction of foreign DNA sequences (genes) into living cells.

**Transgenic Animals** Animals either integrating foreign DNA segments into their genome following gene transfer or resulting from the molecular manipulation of endogenous genomic DNA.

**Transgenic Founder** The first-generation transgenic organism directly produced following gene transfer–germplasm manipulation procedures. The term does not imply that the transferred gene (transgene) is heritable to future generations.

**Transgenic Line** A direct familial lineage of organisms derived from one or more transgenic founders, in which the transgene(s) is passed to successive generations as a stable genetic element. The line includes the founder and any subsequent offspring inheriting the specific germ line manipulation.

Transgenic animals and our ability to manipulate the genome of whole animals in vivo are factors that have influenced the life sciences in dramatic fashion. In less than 15 years, practical aspects of germ line manipulation of the genetic composition of animals have allowed researchers to address fundamental questions ranging from medicine and biology, to biotechnology and production agriculture. Studies emanating from basic efforts in gene regulation have led the way to in vivo gene transfer studies focused on the expression of foreign genes and ablation of endogenous genes. Our appreciation of basic biological development in animals and the potential ramifications of novel phenotypes resulting from genetic engineering efforts have been markedly influenced by pioneering efforts in transgenic animal technology.

## 1 INTRODUCTION

### 1.1 Brief History and Definition of Transgenic Animals

The ability to introduce functional genes into animals provides a very powerful tool for dissecting complex biological processes and systems. Hence, transgenic animals represent unique models that are custom-tailored to address specific biological questions. Furthermore, classicai genetic monitoring cannot engineer a specific genetic trait in a directed fashion. For identification of interesting new models, genetic screening and characterization of chance mutations remains a long and arduous task.

In gene transfer, animals receiving new genes (foreign DNA sequences integrated into their genome) are referred to as "transgenic," a term coined by Gordon and Ruddle in 1981. As such, transgenic animals are recognized as specific species variants or strains, following the introduction and integration of new gene(s), or "transgenes," into their genome. More recently, the term "transgenic" has been extended to chimeric or "knockout" mice in which gene(s) have been selectively disrupted or removed from the host genome.

Production of transgenic mice has marked the convergence of earlier advances in the areas of recombinant DNA technology and the manipulation and culture of animal germplasm (Figure 1). Transgenic mice provide powerful models to explore the regulation of gene expression as well as the regulation of cellular and physiological processes. The use of transgenic animals in biomedical, agricultural, and biotechnological fields requires the abilities to target gene incorporation and to control the timing and level of transgene expression. Experimental designs have taken advantage of the ability to direct specific (e.g., cell, tissue, organ specificity) as well as ubiquitous (whole-body) expression in vivo. Furthermore, from embryology to virology, transgenic technology provides unique animal models for studies in various disciplines that otherwise would be all but impossible to develop spontaneously.

### 1.2 Applications of Transgenic Animals

A number of methods exist for gene transfer in mammalian species, including DNA microinjection, embryonic stem (ES) cell transfer, retroviral infection, blastomere–embryo aggregation, teratocarcinoma cell transfer, electrofusion, nuclear transplantation, and perhaps such passive transfer methods as spermatozoa-mediated transfer (although quite controversial).

Transgenic technology has been extended to a variety of animal species in addition to the mouse, including rats, rabbits, swine, ruminants (sheep, goats, and cattle), poultry, and fish. Transgenic amphibians, insects, nematodes, and members of the plant kingdom have also been produced. In the systems explored to date, gene transfer technology is a proven asset in science as a means of dissecting gene regulation and expression in vivo. As such, the primary questions that are addressed concern the roles of individual genes in development or in particular developmental pathways.

There are a number of strategies in the development of transgenic mouse models, including systems designed to study dominant gene expression, homologous recombination/gene targeting and the use of ES cells, efficiency of transformation of eggs or cells, disruption of gene expression by antisense transgene constructs, gene ablation or knockout models, reporter genes, and marking genes for identification of developmental lineages.

## 2 PRODUCTION OF TRANSGENIC LABORATORY ANIMALS

### 2.1 Choice of Animals

The relative importance of using particular strains or breeds of animals in gene transfer experimentation will vary dramatically according to the species under consideration. Probably the most complex system is encountered in the production of transgenic mice, simply because so much work has been done with this species. Here, well-documented differences in reproductive productivity, behavior, related husbandry requirements, and responses to various experimental procedures play major roles in determining the efficiency and degree of effort associated with production of transgenic founder animals.

DNA microinjection, the most direct and reproducible method for producing transgenic animals, necessitates the maintenance of several ''pools'' of animals for specific purposes. Most commonly, donor females are induced to superovulate by using a regimen of injections of pregnant mare serum gonadotropin (PMSG, to stimulate follicular growth and development) followed by human chorionic gonadotropin (hCG, to induce ovulation of mature eggs). These donor females are then mated to fertile males and large numbers of fertilized one-cell eggs (''zygotes'') are obtained surgically. Alternatively, ova may be collected from donor females and then subjected to in vitro fertilization (IVF) to obtain zygotes. In either case, the DNA construct, in a buffer solution, is microinjected into the male pronuclei of the fertilized eggs. Ova that survive microinjection are then surgically transferred to the reproductive tracts of hormonally synchronous female recipients, which carry the embryos to term. In mice, DNA from offspring is isolated from tissue samples (e.g., tail biopsies) obtained at weaning and analyzed by Southern blot analysis and/or methods based on the polymerase chain reaction (PCR), to determine which individuals carry the transgene.

For most other nonmurine species, it will remain important to use donor females that respond well to hormonal synchronization and superovulation, and recipients that are able to carry fetuses to term and care for neonates appropriately. However, methods used for hormonal synchronization, selection of proestrus females, and evaluation of other forms of reproductive behavior may differ significantly from those identified for mice.

### 2.2 DNA Microinjection

The microinjection method involves the use of micromanipulators and an air- or oil-driven apparatus to physically inject the DNA construct solution into the male pronuclei of fertilized one- or two-cell zygotes. As such, the method is also sometimes referred to as ''pronuclear microinjection.''

Virtually any cloned DNA construct can be used, albeit some caveats do apply. Linearized (as opposed to closed, circular) DNA constructs are more readily integrated into the host genome. The presence of extraneous plasmid- or vector-related DNA sequences associated with the construct can adversely affect expression of the integrated construct/transgene. Also, greater care must be taken when handling larger DNA constructs, such as those derived from yeast artificial chromosomes (YACs), to avoid damage by shearing prior to and during microinjection.

With few exceptions, microinjected constructs integrate randomly throughout the host's genome, but usually only in single chromosomal locations. This fact can be exploited to obtain functional linkage of independent transgenes by simultaneous coinjection of more than one DNA construct. In such a case, both constructs tend to integrate in the same randomly located site and may exhibit coexpression. Because integration is spatially random, transgenes may be integrated into either a somatic or a sex chromosome.

The integration process itself is poorly understood, but it apparently does not involve homologous recombination. During integration, multiple copies of a transgene are incorporated into the genomic DNA, predominantly as head-to-tail concatemers. Regulatory elements in the host DNA near the site of integration, and the general availability of this region for transcription, appear to play major roles in affecting the level of transgene expression. This ''positional effect'' is presumed to explain why the levels of expression of the same transgene may vary dramatically between individual founder animals as well as their offspring (or ''lines''). Additionally, the presence of tissue- or cell-type-specific nuclear regulatory factors that act on host genes near the site of integration in *trans* may restrict transgene expression to only a subpopulation of cells even though the transgene is present in virtually all cells of a founder.

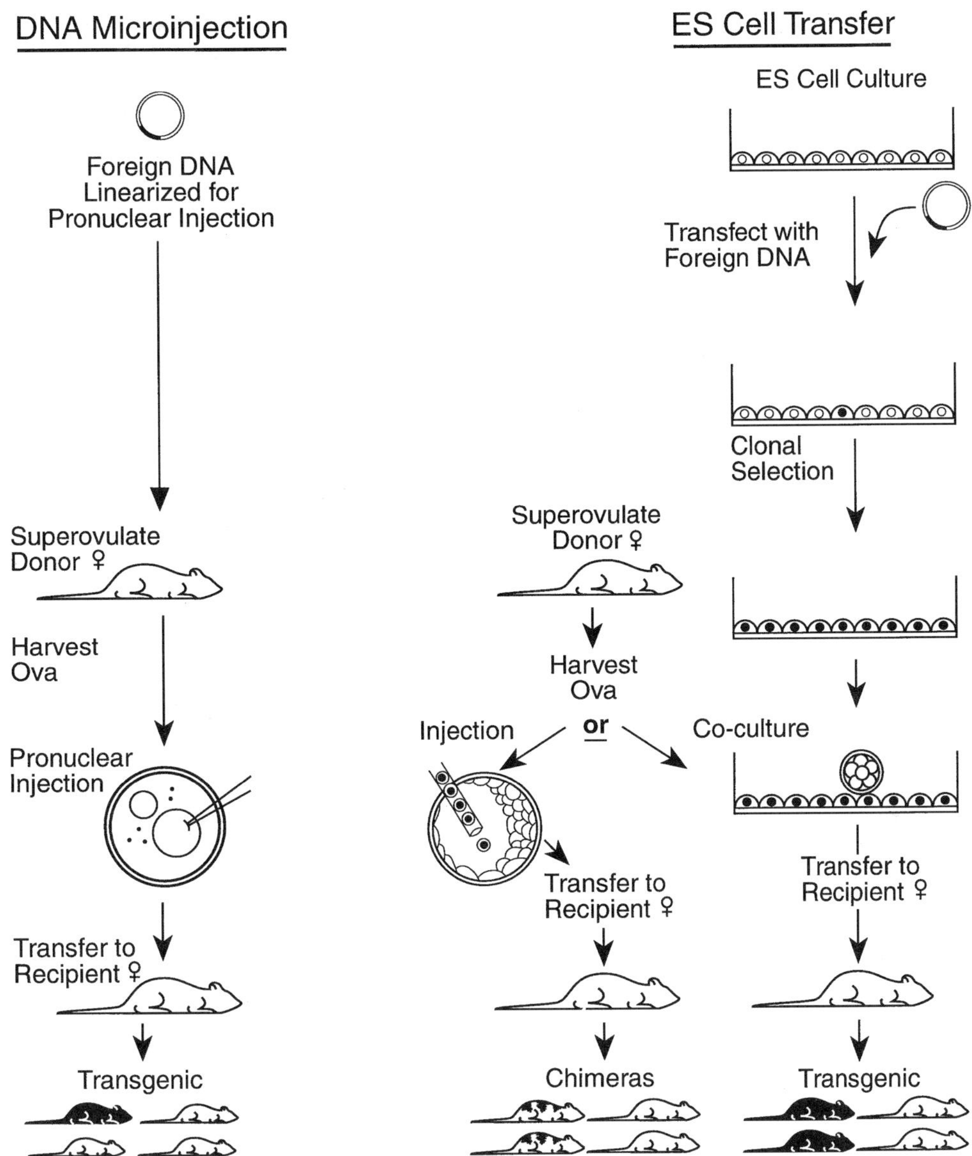


**Figure 1.** DNA microinjection (left) and embryonic stem cell (ES cell) transfer (right), the two most practiced methods for producing transgenic mice. Note that an in vitro culture step is not necessary for DNA microinjection; DNA is injected directly into the male pronucleus of fertilized one-cell ova (pronuclear zygotes). In general, if transgenic mice (represented by black mouse) are derived by DNA microinjection, all their cells contain the new transgene(s).

After clonal selection of transfected ES cells, one of two techniques is used for ES cell transfer. Ova are harvested between the eight-cell and blastocyst stages. ES cells are either injected directly into a host blastocyst (''injection'') or cocultured with eight-cell to morula-stage ova, so that transfected ES cells are preferentially incorporated into the inner cell mass of the developing embryo (''coculture''). With blastocyst injection, transgenic offspring are termed ''chimeric,'' as some of their cells are derived from the host blastocyst and some transfected ES cells (denoted by white mice with black patches). Using coculture and tetraploid embryos, one can obtain founder mice derived completely from the transfected ES cells (denoted as all-black mice).

Host DNA near the site of integration frequently undergoes various forms of sequence duplication, deletion, or rearrangement as a result of transgene incorporation. Such alterations, if sufficiently drastic, may disrupt the function of normally active host genes at the integration site and constitute ''insertional mutagenesis,'' resulting in an aberrant phenotype. Such events cannot be purposefully designed, but they have led to the serendipitous discovery of unsuspected genes and gene functions.

Because gene transfer is accomplished at the one- to two-cell stage, the transgene is incorporated into essentially every cell that contributes to development of the embryo. Thus, transgenic animals (or ''founders'') produced by this method are usually considered ''nonmosaic'' or ''nonchimeric,'' in the sense that the transgene is physically present in the nuclei of all cells of the body. Incorporation of the transgene into cells that will eventually contribute to development of germ cells (sperm or ova) is a common occurrence with this method and makes it possible for offspring of founder animals to inherit the transgene. However, integration of the microinjected DNA construct into the host's genome occasionally is inexplicably delayed. In such a case, if cells of the early embryo (''blastomeres'') undergo mitosis before integration occurs, some but not all of the cells will contain the transgene, and the founder animal, although still considered to be transgenic, will be classified as a mosaic or chimera.

### 2.3 Retroviral Introduction of Transgenes

Transfer of foreign genes into animal genomes has also been accomplished using retroviruses. Although embryos can be infected with retroviruses up to midgestation, early eggs, usually at the 4- to 16-cell stages, are used for infection with one or more recombinant retroviruses containing a foreign gene. Immediately following infection, the retrovirus produces a DNA copy of its RNA genome using the viral enzyme, reverse transcriptase. Completion of this process requires that the host cell undergo the S phase of the cell cycle. Therefore, retroviruses effectively transduce only mitotically active cells. Incomplete infections, in which not all embryonic cells acquire the retrovirus, occur more frequently in work using embryos after the four-cell stage, with resultant chimeric embryos.

The DNA copy of the viral genome, or provirus, integrates randomly into the host cell genome, usually without deletions or rearrangements. However, as is the case for gene transfer by microinjection, because integration is not by way of homologous recombination, this method is not used effectively for site-directed mutagenesis.

### 2.4 Embryonic Stem Cell Technology

Gene transfer has been used to produce both random and targeted insertion or ablation of discrete DNA fragments into the mouse genome. For targeted insertions, where the integration of foreign genes is based on a recombinational gene insertion with a specific homology to cellular sequences (termed homologous recombination), the efficiency of DNA microinjection is extremely low. In contrast, the use of ES cell transfer into mouse embryos has been quite effective in allowing an investigator to preselect a specific genetic modification, via homologous recombination, at a precise chromosomal position.

Technologies involving ES cells, and more recently primordial germ cells, have been used to produce a host of mouse models. Pluripotential ES cells are derived from early preimplantation embryos and maintained in culture long enough to permit the performance of various in vitro manipulations. The cells may then be injected directly into the blastocoel of a host blastocyst or incubated in association with a zona-free morula. The host embryos are transferred into intermediate hosts or surrogate females for continued development. Currently, the efficiency of chimeric mouse production results in about 30% of the live-born animals containing tissue derived from the injected stem cells. This ability to produce ''chimeric'' animals using ES cells has given researchers another tool in their armamentarium to produce transgenic animals. In this set of techniques, the power of gene transfer technology has been catapulted forward because such processes allow for targeted insertions into the genome. Such targeting is extremely important, particularly in areas of gene therapy and correction, for which earlier technologies allowed only for random integration events.

One procedural obstacle in creating ES-cell-derived transgenic animals results from the extreme difficulty of producing and maintaining ES cell lines. This problem lies in a general inability of embryos of most mammalian species, other than mice, to survive in vitro. Additionally, culture conditions and media/sera, as well as the handling of ES cells before and during transfer procedures, play critical roles in experimental efficiencies.

Following identification of true ES cells and subsequent gene transfer procedures, it is necessary to accurately and efficiently identify the ES cells that have integrated foreign DNA. In this regard, the use of marker-assisted selection schemes has greatly simplified mouse experimentation. PCR techniques have also proved to be valuable tools in selecting altered ES cell clones for use in transgenic animal production.

Interestingly, germ line transmission is routinely problematic in some strains of mice but not in others. The underlying biological mechanisms responsible for such differences are not known.

While ES cell lines have been identified for species other than the mouse, production of germ line ES-cell-derived/chimeric animals has not been reported. If ES cells are identified that could give rise to germ-line-competent animals, it is probable that gene transfer into such cells would be successful, based on the wealth of available gene transfer literature.

### 2.5 Other Laboratory Animal Models

Transgenic animal protocols developed in mice have been modified to accommodate production of other transgenic species. In particular, the larger body size of rats and rabbits, their short estrous cycles and gestation periods, their relatively rapid generation times, and their litter-bearing ability have helped make them preferred models for several areas of research employing gene transfer. As with mice, DNA microinjection has been the method used most frequently to introduce foreign DNA. Differences between these species and mice in size of ova, physical response to microinjection, requirements for in vitro ova culture, and numbers of fertilized eggs needed for embryo transfer (to maintain pregnancy), as well as differences in general husbandry practices, are well documented and can be used to optimize production of transgenic rats and rabbits.

Several studies have proven rabbits to be acceptable models for ''gene farming.'' In these experiments, foreign proteins are expressed in transgenic animals preferentially in mammary tissue and are secreted into milk, from which they can be readily purified. To date, transgenic rabbits have been separately produced in which human proteins (including $\alpha_1$-antitrypsin, interleukin 2, growth hormone, and tissue plasminogen activator) and bovine $\alpha$-lactalbumin have been expressed in mammary tissue and harvested from milk.

## 3 PRODUCTION OF TRANSGENIC DOMESTIC ANIMALS

### 3.1 Overview

The success of transgenic mouse experiments led a number of research groups to study the transfer of similar gene constructs into the germ line of domestic animal species. With one exception, these efforts have been directed primarily toward either of two general goals: improvement of the productivity traits of domestic food animal species, or development of transgenic lines for use as ''bioreactors'' (i.e., producers of recoverable quantities of medically or biologically important proteins). These studies have been helpful in identifying basic biological mechanisms as well as a need for precise regulation of gene expression. Since 1985, transgenic farm animals harboring growth-related gene constructs have been created, although ideal growth phenotypes have not been achieved because of an inability to coordinately regulate gene expression and the ensuing cascade of endocrine events. Presently, DNA microinjection is the only method used successfully to produce transgenic livestock.

### 3.2 Traits Affecting Domestic Animal Productivity

Interest in modifying traits that determine the productivity of domestic animals was greatly stimulated by early experiments conducted by Ralph Brinster, Richard Palmiter, and their colleagues, in which body size and growth rates were dramatically affected in transgenic mice expressing growth hormone transgenes driven by a metallothionein (MT) enhancer/promoter. From that starting point, similar attempts followed to enhance growth in swine and sheep by introduction of various growth hormone (GH) gene constructs under control of a number of different regulatory promoters. Use of these constructs was intended to allow for tight regulation of individual transgene expression by dietary supplementation. However, although resulting phenotypes displayed alterations in several areas, including fat composition, feed efficiency, rate of gain, and body composition (lean/fat ratio), they were accompanied by undesirable side effects (e.g., joint pathology, skeletal abnormalities, increased metabolic rate, gastric ulcers, infertility). Such problems, which were attributed to chronic overexpression or aberrant expression of the growth-related transgenes, could be mimicked, in several cases, in normal animals by long-term treatment with elevated doses of GH.

Other productivity traits that are major targets for genetic engineering include altering the properties or proportions of caseins, lactose, or butterfat in milk of transgenic cattle and goats, more efficient wool production, and enhanced resistance to viral and bacterial diseases (including development of ''constitutive immunity'' or germ line transmission of specific, rearranged antibody genes).

### 3.3 Domestic Animals as Bioreactors

The second general area of interest has been the development of lines of transgenic domestic animals for use as bioreactors. Many of these ''gene farming'' efforts have involved attempts to direct expression of transgenes encoding biologically active human proteins. In such a strategy, the goal is to recover, from serum or from the milk of lactating females, large quantities of functional proteins that have therapeutic value. To date, expression of foreign genes encoding $\alpha_1$-antitrypsin, tissue plasminogen activator, clotting factor IX, and protein C were successfully targeted to the mammary glands of goats, sheep, cattle, and/or swine.

Lines of transgenic swine and mice have been created that produce human hemoglobin or specific circulating immunoglobulins. The ultimate goal of these efforts is to harvest proteins from the serum of transgenic animals, for use as important constituents of blood transfusion substitutes, or for use in diagnostic testing.

### 3.4 Examples from Domestic and Miniature Pigs

In contrast to gene transfer in mice, the efficiency associated with DNA microinjection in transgenic livestock, including swine, is quite low. However, two advantages offered by swine over other domestic species are a favorable response to hormonal superovulation protocols (20–30 ova can be collected on average) and a significant uterine capacity, in that swine are litter-bearing species. While the highest transgenic success rate in domestic animals has been achieved with outbred domestic pigs, as might be anticipated, the use of inbred and miniature strains reflects lower overall efficiencies.

A problem encountered initially during the creation of transgenic farm animal species concerned the visualization of the pronuclei or nuclei within the ova. Since swine ova are lipid dense, the cytoplasm is opaque and the nuclear structures are not discernible without some type of manipulation. Overcoming the visualization problem was critical because transgenic pigs were not produced following injection of DNA into the cytoplasm of one- and two-cell ova. Fortunately, it was found that centrifugation of pig ova resulted in stratification of the cytoplasm rendering pronuclei and nuclei visible under differential interference contrast or Nomarski/Smith microscopy. The proportion of transgenic swine that develop from microinjected ova (varying from zero to about 12%) is generally much lower than that observed in mice.

For transgene expression studies targeting pigs, the use of outbred domestic pigs is the most practical way to produce and evaluate potential models. However, miniature or laboratory swine are now used with increasing frequency in biomedical research, where desirable; because of well-characterized background genetics, they are more suitable for human modeling studies (e.g., xenotransplantation research).

In future experiments with domestic animals, the ability to utilize embryonic stem cells would represent a tremendous improvement over the microinjection protocols currently in use. The ability to ''knock out'' or replace endogenous genes has been an extremely valuable tool in mouse genetics and modeling. Yet, characterization and use of embryonic stem cells for gene transfer has not worked to date in any species other than the mouse.

## 4 ANALYSIS OF TRANSGENE INTEGRATION AND EXPRESSION

### 4.1 Transgene Integration

When founder mice are 6–7 weeks of age, biopsies are readily performed to obtain representative tissues for analysis of transgene integration and expression. For integration analysis in mice, typically tail cuts are performed and the tissue is digested with proteinase K followed by DNA extraction using standard protocols. Preliminary determination of transgene integration by the PCR technique can be very useful when the target sequence (the transgene) possesses unique sequences (not endogenous to the genome of the animal). However, other more specific techniques should always be used to confirm the presence of the transgene in PCR-positive samples. These include DNA slot-blotting and Southern blot hybridization, the latter being not only more informative but also less likely to present false positive signals. Additionally, the transgene can be constructed to include a molecular ''tag,'' a unique sequence that is easily detected and has minimal similarity to any endogenous sequence. In some instances phenotypic screening is possible if the expression of the transgene leads to an identifiable change in the appearance of the animal.

Other concerns related to analysis of transgene integration include determinations of copy number of the transgene per cell, orientation of tandemly arranged copies, and the presence of multiple integration sites. These questions can be addressed by Southern blot hybridization following digestion of the genomic DNA with appropriate restriction enzymes. In addition, the functionality of the transgene is an important consideration in the analysis of transgene integration.

### 4.2 Transgene-Encoded mRNA Expression

The analysis of expression of the transgene is essential in determining the utility of the transgenic animals produced. As with integration analysis, the presence or absence of similar or identical endogenous counterparts will determine, to a degree, the strategies that may be most useful. For transgenes that are unique (no endogenous counterpart) or contain some unique sequences, the strategies that can be used are more straightforward. The presence of a novel mRNA transcript or a unique protein (or enzyme activity) is more easily determined than it is when the transcript or protein products are very similar to endogenous transcripts or proteins. As with integration analysis, molecular ''tags'' are also sometimes useful in that the transcripts will contain some unique identifying sequence that can be readily and unequivocally determined.

Once a high quality RNA preparation has been obtained from the appropriate tissue, the presence of mRNA transcripts of the transgene can be determined by RNA slot-blot hybridization. A more informative technique, northern blot hybridization, confirms not only the presence but also the size of the transcript of interest. Additional techniques exist for determining the presence or relative levels of mRNA transcripts from transgenic animals, such as the nuclease protection assay, the reverse transcription-PCR (RT-PCR) assay, and in situ hybridization techniques.

### 4.3 Protein Expression

In addition to the assays for transcription of the transgene mentioned in Section 4.2, various other techniques to determine the translation of the transcript into a protein product are often employed. Most of these depend on the availability of specific antibodies to the protein of interest. They include immunoblotting, radioimmunoprecipitation, and immunohistochemical ''staining'' of tissue sections.

Some experiments involving transgenic animal production focus on measuring the influence of various cis-regulating DNA sequences; the transcript and ultimate protein product are for the most part irrelevant. In these cases, ''reporter genes'' are often employed to simplify determination of expression levels by producing a transcript or protein that is easily and unequivocally detectable.

In analyzing the expression of any transgene, it is always important to show that the expression patterns are consistent and reproducible by evaluating at least two separate lines of transgenic animals. It is not uncommon for the site of integration within a given line of transgenic animals to have profound influences on transgene expression, independent of any transcriptional control sequences in the transgene itself.

## 5 CONCLUSIONS AND FUTURE DIRECTIONS

### 5.1 Gene Transfer Today

The efforts and costs for gene transfer experimentation in animal biology are challenging and limiting. Innovative technologies to enhance experimental gene transfer efficiency in different species are desperately needed. Such enabling techniques would not only bring the cost of individual projects into a reasonable realm but would also increase the likelihood of breakthrough studies in many disciplines. As stated earlier, aside from mouse modeling, multigene targeted (nonspecific) blastomere aggregation, and nuclear transplantation studies, DNA microinjection into preimplantation embryos has been the only reproducible means for heritable gene transfer in the preponderance of mammalian species.

A number of specific achievements would significantly enhance experimental productivity. Enabling technologies would center on the following goals.

1. Development of alternative DNA delivery systems (e.g., embryonic stem cell transfer for stable gene transfer, liposome-mediated gene transfer, spermatogonial transfer, or perhaps targeted somatic cell techniques.
2. Identification of (a) optimal experimental conditions for a given form of gene transfer and (b) animal strains best suited to the individual technologies.
3. Complete animal genome mapping and identification of homologies to human genes.
4. Establishment of routine embryo and germplasm culture systems.
5. Discovery of a means to reduce the number of animals and embryos required (e.g., use of PCR amplification of blastomere DNA, or fluorescence-activated cell sorter analyses and gating of transfected germplasm).
6. Elucidation of mechanisms by which the 129 mouse strain is more suited to the maintenance of germ-line-competent embryonic stem cells with extension to other strains and species.

### 5.2 Future Directions

The use of embryonic stem cells for mammalian gene transfer would be most advantageous. The ability to ''knock out'' or replace endogenous genes has been an extremely valuable tool in mouse genetics and modeling. Yet, characterization and use of ES cells for gene transfer has not worked to date in any species other than the mouse. Numerous laboratories with appropriate tissue and embryo culture background were successful in genetic engineering efforts with mouse models, and ES cell or ''ES-cell-like'' lines were identified for other species. Yet, with hamster, rat, rabbit, and pig stem cell experiments ongoing now for at least 5–7 years, the first bona fide transfected stem cells giving rise to chimeric (and germ-line-competent) founder animals have yet to be documented or reported. Regardless of whether a genetic alteration or deficiency in the mouse genome is responsible for such a dramatic difference when compared to other species, the need for such technology in additional species is great.

For many species, technical drawbacks to both ES cell and gene transfer technologies are manifest in the lack of significant gene characterization and mapping efforts. With the complexity of the mammalian genome, this information is critical for the determination of appropriate genes to engineer and transfer, and in providing preliminary indications of the potential biological consequences of genetic engineering experimentation.

Much has been learned about early development and effects of altering the growth cascade as well as other physiological processes in initial transgenic animal models. These models have had far-reaching effects and have redefined what is possible in the biological, biomedical, and agricultural sectors. While studies of transgene expression and overexpression in mammalian models may not always correlate exactly, nor reveal all the possible effects and consequences when extended to human biology, the utility of transgenic animal models to scientific discovery cannot be overestimated. The use of ES cell or related methodologies to provide efficient and targeted in vivo genetic manipulations offers prospects of creating profoundly useful animal models for biomedical, biological, and agricultural applications.

*See also* Biotechnology, Governmental Regulation of; Genetic Immunization; Mammalian Genome; Transgenic Animal Patients; Transgenic Fish.

### *Bibliography*

Brinster, R. L. (1993) Stem cells and transgenic mice in the study of development. *Int. J. Dev. Biol.* 37:89–99.

Hogan, B., Costantini, F., and Lacy, E. (1994) *Manipulating the Mouse Embryo: A Laboratory Manual.* Cold Spring Harbor Laboratory: Cold Spring Harbor, NY.

Pinkert, C. A. (1994) *Transgenic Animal Technology: A Laboratory Handbook.* Academic Press, San Diego, CA.

# Transgenic Animal Patents

*W. Lesser*

---

### *Key Words*

**Animal Rights** Term indicating the belief that animals have inherent rights beyond protection from cruelty and unnecessary suffering.

**Animal Welfare** A requirement, legal and/or ethical, that the suffering of animals be minimized and, when required, be justified by significant human need.

**Disclosure** The patent law requirement that an invention be described sufficiently to be repeatable. For living organisms, a sample deposit is sometimes required.

**Infringement** The unauthorized use of a patented invention.

**Intellectual Property Rights** Term referring to a group of laws (patent, copyright, trademark, industrial secrets) prohibiting the unauthorized use of the protected creation.

**Nonobviousness (Inventive Step)** Patent law requirement that an invention must not be a trivial extension of what exists.

**Patent (Utility)** Legal document describing an invention and granting the owner the right to prohibit use of the invention. Patents may apply to products, processes, or products-by-process.

**Royalty** Fee charged for the use rights to a patent.

**Scope** The range of similar products, processes, or uses implicitly covered by a patent.

---

The term ''animal patents'' refers to the extension of utility patent protection to higher animals. Patents are intended to provide incentives for private investment in research and development. Such legal protection can be especially important for animals, many of which are self-regenerating. Only a few animals have been patented in the United States, Europe, and Australia, mostly laboratory specimens. Multiple pending applications are largely in the areas of animal disease models, livestock, and animal producers of pharmaceuticals. While the practical effects to date are limited, the field has major potential impacts, raising broad-based but varied concerns about animal welfare and rights, ethics, and economic considerations, especially regarding small farms, and the collection of royalties on progeny. There are no general agreements on how best to proceed, as concerns tend to be individual, but a majority of the public across several countries appears generally supportive of animal biotech and patents. At present, only the United States expressly permits animal patents; 10 countries tacitly allow them, and 54 have exclusionary language, but, as in Europe, interpretation of statutes can be complex.

## 1 INTRODUCTION

The United States Patent and Trademark Office (PTO) created a flap when, on April 7, 1987, the commissioner announced that higher animals were patentable subject matter. The first animal patent, issued on April 12, 1988, was for a cancer-susceptible experimental mouse (Onco-mouse).* Three additional grants, also for lab animals, were made in late 1992. The Onco-mouse was subsequently patented in Europe in October 1991 but remains under review. A number of objections, both moral/ethical and economic, have been raised, and a total of eight bills limiting the practice in the United States have been proposed in Congress. To date no action has been taken, and over 150 applications are pending. The status of animal patents in other countries has not been clarified except in Australia where the first animal patent was issued in March 1993. It is for a pig (other species are claimed as well) with the attribute of controlling the output of supplementary growth hormone which, unregulated, has been found to be crippling or fatal.

This entry reviews the legal, economic and ethical issues associated with animal patents.

## 2 LEGAL CONSIDERATIONS

### 2.1 Legal Status

For an invention to be patented in the United States, it must fit within the class of ''patentable subject matter'' and meet the patent requirements of novelty, utility, and nonobviousness (inventive step) set forth by law (35 USC, Sections 101–103). Patentable subject matter is determined both by statute and by judicial interpretation. In the United States, the key decision explicitly extending patents to living organisms was made in 1980 by the Supreme Court in *Diamond v. Chakrabarty* (447 U.S. 303; 206 *U.S. Pat. Q.* 193) and extended to higher animals in 1987 on internal Patent

* The patent grant was for all ''transgenic non-human mammals,'' but to date only mice have actually been produced.

and Trademark Office appeal [USPTO, B. App. and Int. *Ex parte Allen* U.S.P.Q. 193 (1987)].

The patentability status of higher animals in other countries is less clear. In Article 53(b), the European Patent Convention (EPC) explicitly excludes ''animal varieties'' or essentially biological processes in the production of plants or animals. Yet this clause is subject to interpretation of what is an ''animal variety'' or an ''essentially biological process''; current judgment appears to allow patents for inventions not in a fixed (limited) form of a particular variety. Interpretation of similar wording in laws of more than 50 other countries is even less clear. An additional 10 countries, including Japan, Brazil, Australia, and New Zealand, are mute on protecting animals so that there is no apparent inhibition as the recent grant in Australia indicates.

### 2.2 Patentability Requirements

Animal-based inventions, like all applications, must demonstrate novelty (not previously known), utility (essentially, must serve a purpose and not be detrimental to public interests), and nonobviousness (the invention may not be trivial). In addition, inventions must be disclosed (described) to ensure that they are repeatable. Disclosure for living organisms may require a sample deposit which, for animals, could be in the form of frozen embryos.

Patents provide only negative rights—the right to exclude others; use without permission constitutes infringement. Permission is often granted in exchange for the payment of royalties. For self-reproducing animals, the production of offspring is a complex legal (and economic) matter. When reproduction is implied or necessary, as with breeding stock and milk cows annually for ''freshening,'' it will likely be allowed. Otherwise not. The collection of royalties when owed will, however, be complex (see Section 4). All returns for patents must come from the market, from sales and royalties.

Patent scope refers to the related inventions that are covered. Typically scope is broader for pioneering inventions like patented animals at this early stage.

Eight bills delaying or limiting animal patents were introduced to Congress from 1987 to 1991, but none was enacted. The bills address a series of economic and ethical matters, but activity has declined lately.

## 3 SCIENTIFIC DEVELOPMENTS

Potentially patentable animals are developing along three lines: animal (disease) models, agricultural livestock, and pharmaceutical production. Animal models for cancers, genetic-based diseases (sickle cell anemia), and HIV/AIDS constitute the dominant area of application. Agricultural applications are directed to enhanced growth (more milk, less fat) and changed product traits (lactose-free milk). Such agricultural developments, once commercialized, will dominate lab animals in terms of numbers and could profoundly affect farming (see Section 4). Pharmaceutical production (human factor IX) involves, for example, secretion into cow's or sheep's milk and is a cost-saving approach involving limited numbers of animals. For details on developments, see individual reviews in Fessenden MacDonald.

All these described developments are based on genetic engineering. There is nothing in patent law language preventing the patenting of inventions from traditional technologies (e.g., natural or artificial breeding), but as a practical matter such developments often may not meet patentability requirements.

## 4 ECONOMIC FACTORS

### 4.1 Economic Incentives

Patents are intended to foster private research investments by prohibiting close copying; another justification for patents, not in great favor at this time, is that of ''natural rights''. Inventors cannot compete economically with copiers and, without legal protection, will invest too little or not at all. Patents also provide ''technical teaching'' and encourage advancement by discouraging secrecy.

Theoretically, economists have been largely unsuccessful at defining such factors as optimal scope or duration (see, for example, Reinganum 1982; Harris and Vickers 1987; Loury 1979). Empirically, most evidence comes from Plant Breeders' Rights (PBR), patent-like protection applicable to plant varieties since 1961 (1970 in the United States). PBR operate very similarly to utility patents in terms of concept, requirements and enforcement, but typically are administered by the Department of Agriculture rather than the Patent Office. PBR have been found to increase private investment as expected, but unevenly across crops and not as a complete substitute to public breeding. Experiences with animals should be roughly comparable. Nonetheless, the evidence is limited and contested.

### 4.2 Impacts on Prices and Industry Structure

Many of the concerns with regard to prices and industry structure are applicable to agriculture and parallel the plant debate. Concerns about plants are largely unsubstantiated, primarily because agriculture is a performance-based, low profit sector. A similar situation is expected to apply to animals, even though the breeding sector ranges from very unconcentrated (beef cattle) to highly concentrated (chickens and ''synthetic'' hogs).

### 4.3 Royalty Payments

Breeders of patented animals are potentially liable for royalty payments (see Section 2.1). This issue is especially complex for the large and dispersed agricultural sector. Slow reproduction, especially for cattle, necessitates that potential traits be inheritable, but the large standing herd means that several patented genes could be incorporated in each animal. Lesser (1989) proposed a system of making payments through packing plants, where animals are concentrated. Unless the genes can be readily detected, royalties must be levied on an average basis. Because of the uncertainty and other considerations, breeders are likely to use contracts in conjunction with patents, a not-uncommon process.

### 4.4 Implications for Farm Structure

Critics charge that patented animals will lead to further farm consolidation, even though such inventions are considered to be ''scale neutral.'' However, they require better management and, when combined with other inventions like computerized feeding, are likely to increase concentration slightly. U.S. livestock farm numbers exceed one million, down sharply since the 1930s and still declining, but the bulk of production comes from the 15% of very large operations. Milligan and Lesser (1989) relate the issues to biotechnology, not patents. They argue that the absence of patents would lead to further contracting, ending with large farms only, which would exclude small farms from access no matter how efficient.

### 4.5 Genetic Diversity

The industrialization of agriculture has led to standardization and the shrinking of the gene pool in major commercial breeds. Biotechnology, associated with animal patents, is seen as a means of accelerating that standardization (e.g., Hoyt and Carpenter testimony, Hearings 1987). However, an equally strong argument can be made that biotechnology will promote the introduction of new genes and patenting will encourage it, enhancing diversity. In general, comparisons with the short-lived and rapidly reproducible plants are misplaced.

## 5 ETHICAL ISSUES

### 5.1 Public Attitudes Toward Animal Biotechnology and Patents

Comparable opinion surveys have been conducted in the United States and Japan and New Zealand. For the three countries, respectively, 42, 54, and 58% support animal biotech. Roughly the same percentages support animal patenting. The major reasons for opposition entail the beliefs that the technology is "unethical" and "unnatural."

### 5.2 Animal Welfare

Biotechnology (and patenting) are associated with greater suffering (Fox and Hoyt statements, Hearings 1987), but Adler (1988) points out that there is no scientific connection between the two.

A majority of the public seems to accept some level of animal suffering for promoting basic interests of mankind ("Interest-sensitive speciesism"). The acceptable degree is, according to Singer (1976), difficult to establish on a nonreligious basis, hence is unresolvable in pluralistic societies. Macer (1990), however, argues that the prevailing societal position is to give humans a higher moral status than animals. The European Patent Convention [Article 53(a)] allows the rejection of patents offensive to "ordre public." That article, interpreted as the balance between human benefit and animal suffering, was invoked in rejecting an application for a transgenic mouse for screening hair growth stimulants but not for the Onco-mouse. A systematic delineation has yet to emerge.

### 5.3 Animal Rights

Animal rights at the extreme is species egalitarianism, the position that animals have inalienable rights equal to those of humans. One aspect is that of *species integrity,* the right of every species to exist as a separate, identifiable creature (Rifkin, quoted in Walters statement, Hearings 1987). This concept has no basis in science, which sees species as dynamic, varied, microevolving populations, not as individuals with fixed attributes. Moreover, a species classification is based on a composite of attributes and behavior, factors not alterable by changing, say, 20 of the 50,000 to 100,000 genes in higher animals. Clearly Rifkin is referring to something beyond the ken of patent offices.

### 5.4 Devalue Humanness

The possible introduction of human genes into animals (and vice versa) is seen as clouding the definition of "humanness." Yet there seems to be significant public support for such activities, especially when therapeutic. More broadly, for some observers the bioengineering and patenting of animals reduces animals to mere factories, "forcing upon us a reductionist and materialistic concept of life" (quoted in Nelkin 1992). One component of this position is the use of the term "composition of matter," a U.S. patent law term, when referring to patentable subject matter (see Section 2.1). While many people believe animals to be more than compositions of matter in that they feel pleasure and pain, Brody (1989) argues that they are at least compositions of matter, and it is in that measure they are patented.

These diverse positions describe differing views of life, one holistic, the other able to partition animal life into serving multiple functions, some internal and some in service to mankind. Such major differences in beliefs and perceptions are virtually unresolvable in a pluralistic society. Yet, in democratic systems, the majority appears to support a more utilitarian view of animals, the interest-sensitive speciesism position. Animal patents are compatible with that position.

*See also* Biotechnology, Governmental Regulation of; Transgenic Animal Modeling; Vaccine Biotechnology.

## *Bibliography*

Adler, R. G. Controlling the applications of biotechnology: A critical analysis of the proposed moratorium on animal patenting. *Harvard J. Law Technol.* 1:1–61 (1988).

Beier, F. K. Significance of the patent system for technical, economic and social progress. *Int. Rev. Ind. Prop. Copyright Law,* 11:583 (1980).

Bent, S. A., Schwaab, R. L., Conlin, D. G., and Jeffery, D. D. (1987) *Intellectual Property Rights in Biotechnology Worldwide.* Stockton Press, New York.

Brody, B. Evaluation of the ethical arguments commonly raised against the patenting of transgenic animals. In *Animal Patents: The Legal, Economic and Social Issues,* W. Lesser, Ed. Stockton Press, New York, 1989.

Butler, L. J., and Marion, B. W. *Impacts of Patent Protection in the U.S. Seed Industry and Public Plant Breeding.* University of Wisconsin, NC-117 Monograph 16, September 1985.

Crespi, R. S. What's immoral in patent law? *Tibtech,* 10:375–378 (1992).

Fessenden MacDonald, J., Ed. *Animal Biotechnology: Opportunities and Challenges* (1992). NABC Report 4, National Agricultural Biotechnology Council, 1992. Ithaca, NY.

Foote, R. The technology costs of deposits. In *Animal Patents: The Legal, Economic and Social Issues,* W. Lesser, Ed. Stockton Press, New York, 1989.

Fujiki, N., and Macer, D. R. J., Eds. *Human Genome Research and Society,* Chapter 6. Eubios Ethics Institute, Christchurch, New Zealand, 1992.

Grubb, P. (1986) *Patents in Chemistry and Biotechnology.* Clarendon Press, Oxford.

Harris, C., and Vickers, J. Racing with uncertainty. *Rev. Econ. Stat.* 54:1–22 (1987).

Hoban, T. J., IV, and Kendall, P. A. Consumer attitudes about the use of biotechnology in agriculture and food production. North Carolina State University, July 1992.

House of Representatives, Committee on the Judiciary. Hearings, Patents and the Constitution: Transgenic Animals. U.S. Government Printing Office, Washington, DC, Serial No. 23, 1987.

———. (1989) Report 100-888. U.S. Government Printing Office, Washington, DC.

House of Representatives, Committee on Agriculture. Hearings, Plant Variety Protection Act amendments. U.S. Government Printing Office, Washington, DC, Serial No. 96-CCC, 1980.

Kalter, R. J. Impact of animal growth promotants on the dairy industry. In *Biotechnology and Sustainable Agriculture: Policy Alternatives,* J. Fessenden MacDonald, Ed. NABC Report 1, National Agricultural Biotechnology Council, Ithaca, NY, 1989.

Lesser, W., Ed. (1989) *Animal Patents: The Legal, Economic and Social Issues.* Stockton Press, New York.

Loury, G. Market structure and innovation. *Q. J. Econ.* 194:429–436 (1979).

Macer, D. R. J. *Shaping Genes.* Eubios Ethics Institute, Christchurch, New Zealand, 1990.

Macer, D. R. J. *Attitudes to Genetic Engineering.* Eubios Ethics Institute, Christchurch, New Zealand, 1992.

Machlup, F. An economic review of the patent system. Study 15, Subcommittee on Patents, Trademarks and Copyrights, Committee on the Judiciary, U.S. Senate, 1958.

Milligan, R., and Lesser, W. Implications for agriculture. In *Animal Patents: The Legal, Economic and Social Issues,* W. Lesser, Ed. Stockton Press, New York, 1989.

Nelkin, D. Living inventions: Biotechnology and the public. In MacDonald (ed.), *Animal Biotechnology: Opportunities and Challenges,* J. Fessenden MacDonald, Ed. NABC Report 4, National Agricultural Biotechnology Council, Ithaca, NY, 1992.

Office of Technology Assessment, U.S. Congress. New development in biotechnology: Patenting life—Special Report 5, OTA-BA-370. U.S. Government Printing Office, Washington, DC. April 1989.

———. (1991) *Biotechnology in a global economy,* OTA-BA-494. U.S. Government Printing Office, Washington, DC.

Reinganum, J. A dynamic game of R&D: Patent protection and competitive behavior. *Econometrica* 50:671–688 (1982).

Siebeck, W. E., Ed. "Strengthening protection of intellectual property in developing countries: A survey of the literature," World Bank Discussion Paper 112. The World Bank, Washington, DC, 1990.

Singer, P. *Animal Liberation.* Jonathan Cape, London, 1976.

Thompson, D. B. Concepts of property and the biotechnology debate. In *Ethics and Patenting of Transgenic Organisms,* NABC Occasional Papers No. 1, National Agricultural Biotechnology Council, Ithaca, NY, September 1992.

# TRANSGENIC FISH

*Thomas T. Chen, J. K. Lu, and Kathy Kight*

## *Key Words*

**Electroporation** Process utilizing a series of short electrical pulses to permeate cell membranes, thereby permitting the entry of DNA molecules into animal or plant cells.

**$F_1$ Transgenic Fish** Transgenic progeny derived from a cross between a $P_1$ transgenic individual and a nontransgenic individual or between two $P_1$ transgenic individuals.

**$P_1$ Transgenic Fish** Transgenic fish developed from embryos that received foreign gene transfer.

**Polymerase Chain Reaction (PCR)** An in vitro method for the enzymatic synthesis of a specific DNA sequence, using two oligonucleotide primers hybridizing to opposite strands and flanking the region of interest in the target DNA.

**Transgene** A non-fish-origin gene construct that is introduced into fish for the production of transgenic fish.

**Transgenic Fish** Fish into which a foreign gene (DNA) has been introduced and stably integrated into the genome.

Fish into which a segment of foreign DNA has been introduced and stably integrated into the genome are called transgenic fish. Since 1985, a wide range of transgenic fish species have been produced. These animals play important roles in basic research as well as applied biotechnology. In basic research, transgenic fish provide excellent models for studying molecular genetics of early vertebrate development, actions of oncogenes, and the biological functions of hormones at different stages of development. In applied biotechnology, transgenic fish offer unique opportunities for producing animal models for biomedical research, improving the genetic background of broodfish for aquaculture, and designing bioreactors for producing valuable proteins for pharmaceutical or industrial purposes.

Several important steps are routinely taken to produce a desired transgenic fish. First, an appropriate species must be chosen. The most suitable fish species depends on the nature of the studies and the availability of facilities. Second, a specific gene construct must be prepared. The gene construct contains the structural gene encoding a gene product of interest and the regulatory elements that exert temporal, spatial, and developmental control over the expression of the gene. Third, for the transgene to be integrated stably into the genome of every cell, the gene construct has to be introduced into the developing embryos. Fourth, since not all instances of gene transfer are efficient, a screening method must be adopted for identifying successful transgenic individuals. This contribution discusses these aspects of the transgenic fish technology.

## 1 SELECTION OF MODEL FISH SPECIES

Gene transfer studies have been conducted in several different fish species, including salmon, rainbow trout, common carp, goldfish, loach, channel catfish, tilapia, northern pike, walleye, zebrafish, and Japanese medaka. Depending on the purpose of the transgenic fish studies, the embryos of some species are superior to others as hosts for gene transfer. For example, Japanese medaka *(Oryzias latipes)* and zebrafish *(Barchydanio rerio)* have short life cycles (3 months from eggs to mature adults), produce hundreds of eggs on a regular basis without exhibiting a seasonal breeding cycle, and can be maintained easily in the laboratory for 2–3 years. These eggs are relatively large (0.7–1.5 mm diameter) and possess very thin and semitransparent chorions. These features allow easy microinjection of DNA into the eggs if appropriate glass needles are used. Furthermore, inbred lines and various morphological mutants of both fish species are available. Thus these species are suitable candidates for conducting gene transfer for (1) studying developmental regulation of gene expression and gene action, (2) identifying regulatory elements that regulate the expression of a gene, (3) measuring the activities of promoters, and (4) producing transgenic models for environmental toxicology. However, a major drawback of these two fish species is their small body size, which makes them unsuitable for endocrinological or biochemical analysis.

Rainbow trout, salmon, channel catfish, and common carp are commonly used large body size model species in transgenic fish studies. Since the endocrinology, reproductive biology, and physiology of these species have been well worked out, the fish are well suited for experimental studies on comparative endocrinology as well as the aquaculture application of gene transfer technology. However, the long maturation time of these fish species and single spawning cycle per year will hamper rapid research progress in this field.

Loach, killifish *(Fundulus)*, goldfish, and tilapia are the third group of model fish species suitable for conducting gene transfer studies. Their body sizes are big enough for biochemical and endocrinological studies, and their shorter maturation times, compared to rainbow trout or salmon, allow easier manipulation of transgenic progeny. These fish species are less suitable for conducting gene transfer studies, however, because a well-defined genetic background is lacking, and little is known about the asynchronous reproductive behavior of the animals.

## 2 PREPARATION OF GENE CONSTRUCTS

A transgene used in producing transgenic fish for basic research or biotechnological application is a recombinant gene construct, which will produce a gene product at appropriate levels in the desired tissue(s) at the desired time(s). Figure 1 shows the prototype of a transgene that is usually constructed in a plasmid to contain an appropriate promoter/enhancer element and the structural gene.

Depending on the purpose of the gene transfer studies, transgenes can be grouped into three main types: gain of function, reporter function, and loss of function. The *gain-of-function* transgenes are designed to add new functions to the transgenic individuals or to facilitate the identification of the transgenic individuals if the genes are expressed properly in the transgenic individuals. Transgenes containing the structural genes of mammalian and fish growth hormones (GH, or their cDNAs) fused to functional promoters such as chicken and fish β-actin gene promoters are examples of the gain-of-function transgene constructs. Expression of the GH transgenes in transgenic individuals will result in growth enhancement. Bacterial chloramphenicol acetyltransferase (CAT), β-galactosidase, or luciferase genes fused to functional promoters represent one type of *reporter function* transgene. These reporter genes are commonly used to identify the success of a gene transfer effort.

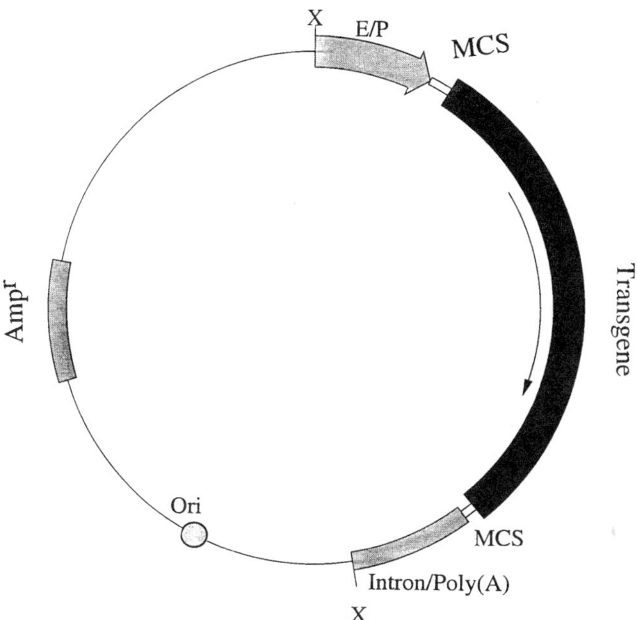


**Figure 1.** Prototype of a transgene construct. A transgene construct has the following elements: the plasmid vector contains a selectable marker gene (e.g., Amp$^r$), an origin of DNA replication (Ori), a restriction enzyme site (X) for removal of the transgene, an enhancer/promoter element (E/P), a transgene, a multiple cloning site (MCS) for inserting the transgene, intron, and a termination sequence, including the poly adenylation cleavage/addition site (Poly(A)/Intron).

A more important function of a reporter gene is to identify and measure the strength of a promoter/enhancer element. In this case, the structural gene of the CAT, β-galactosidase, or luciferase gene is fused to a promoter/enhancer element in question. Following gene transfer, the expression of the reporter gene activity is used to determine the transcriptional regulatory sequence of a gene or the strength of a promoter.

The "*loss-of-function*" transgenes are constructed for interfering with the expression of host genes. These genes might encode an antisense RNA to interfere with the posttranscriptional process or translation of endogenous mRNAs. Alternatively, these genes might encode a catalytic RNA (a ribozyme) that can cleave specific mRNAs and thereby cancel the production of the normal gene product. Although these genes have not yet been introduced into a fish model, they could potentially be employed to produce disease-resistant transgenic broodstocks for aquaculture or transgenic model fish defective in a particular gene product for basic research.

## 3 METHODOLOGY OF GENE TRANSFER

Techniques such as calcium phosphate precipitation, direct microinjection, retrovirus infection, electroporation, and particle gun bombardment have been widely used to introduce foreign DNA into animal and plant cells, as well as germ lines of mammals and other vertebrates. Direct microinjection and electroporation of DNA into newly fertilized eggs have been proven to be the most reliable methods of gene transfer in fish systems.

### 3.1 Microinjection

Microinjection of foreign DNA into newly fertilized eggs was first developed for the production of transgenic mice in the early 1980s. Since 1985, the technique of microinjection has also been adopted for introducing transgenes into Atlantic salmon, common carp, catfish, goldfish, loach, medaka, rainbow trout, tilapia, and zebrafish. The gene constructs that were used in these studies include human or rat growth hormone gene, rainbow trout or salmon GH cDNA, chicken δ-crystalline protein gene, winter flounder antifreeze protein gene, *E. coli* β-galactosidase gene, and *E. coli* hydromycine resistance gene.

In general, gene transfer in fish by direct microinjection is conducted as follows. Eggs and sperm are collected into separate containers. Fertilization is initiated by adding water and sperm to eggs, with gentle stirring to enhance fertilization. Eggs are microinjected within the first few hours after fertilization, using a setup like that shown in Figure 2. The injection apparatus consists of a dissecting stereomicroscope and two micromanipulators, one with an appropriately small glass needle for injection and the other with a micropipette for holding fish embryos in place for injection. Routinely, about $10^6$–$10^8$ molecules of a linearized transgene in about 20 nL is injected into the egg cytoplasm. Following injection, the embryos are incubated in water until hatching. Since natural spawning in zebrafish or medaka can be induced by adjusting photoperiod and water temperature, precisely staged newly fertilized eggs can be collected from the aquaria for gene transfer. If the medaka eggs are maintained at 4°C immediately after fertilization, the micropyle on the fertilized eggs will remain visible for the next 2 hours. The DNA solution can be easily delivered into the embryos by injecting through this opening.

Depending on different fish species, the survival rate of the injected embryos ranges from 35% to 80%, while the rate of DNA

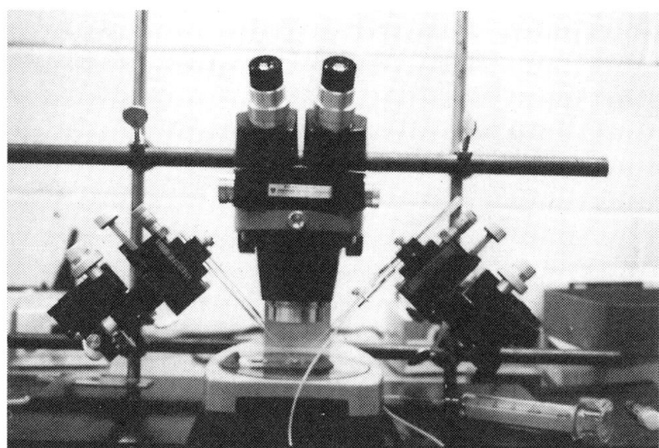

**Figure 2.** A microinjection setup. The apparatus consists of a low power stereo dissecting microscope and two micromanipulators: one for holding the glass injection needle and the other for holding the glass micropipette that keeps the embryo in place during injection.

integration ranges from 10% to 70% of the survivors. The tough chorions of the fertilized eggs in some fish species (epg., rainbow trout and Atlantic salmon) can make insertion of glass needles difficult. This difficulty can be overcome by inserting the injection needles through the micropyle, making an opening on the egg chorions by microsurgery, removing the chorion by mechanical or enzymatic means, preventing chorion hardening by initiating fertilization in a solution containing 1 m*M* glutathione, or injecting the unfertilized eggs directly.

### 3.2 Electroporation

Electroporation is a successful method for transferring foreign DNA into bacteria, yeast, and plant and animal cells in culture. This method has become popular for transferring foreign genes into fish embryos in the past three years. Electroporation utilizes a series of short electrical pulses to permeate cell membranes, thereby permitting the entry of DNA molecules into embryos. Studies conducted in several laboratories showed that the rate of DNA integration in surviving electroporated embryos is about 20% or higher. Although the overall rate of DNA integration in transgenic fish produced by electroporation was equal to or slightly higher than that of microinjection, the actual amount of time required for handling a large number of embryos by electroporation is orders of magnitude shorter than by microinjection. This method is therefore considered to be an efficient massive gene transfer technology.

## 4 FATE OF THE TRANSGENES

### 4.1 Identification of Transgenic Fish

The most time-consuming step in producing transgenic fish is the identification of transgenic individuals. Traditionally, dot blot and Southern blot hybridizations of genomic DNA were commonly used to determine the presence of transgenes in the presumptive transgenic individuals. These methods involve isolation of genomic DNA from tissues of presumptive transgenic individuals, digestion of DNA samples with restriction enzymes, and Southern (or dot) blot hybridization of the digested DNA products. Although this method is expensive, laborious, and insensitive, it tells definitively whether a transgene has been integrated into the host genome. Furthermore, it reveals the pattern of transgene integration if appropriate restriction enzymes are employed in the Southern blot hybridization analysis. To handle a large number of animals efficiently and economically, a polymerase chain reaction (PCR) assay has been adopted. The strategy of the assay is outlined in Figure 3. It involves isolation of genomic DNA samples from a very small piece of fin tissue, PCR amplification of the transgene sequence, and Southern blot analysis of the amplified products. Although this method does not differentiate a transgene that has been integrated in the host genome from one that remains as an extrachromosomal unit, it serves as a rapid and sensitive screening method for identifying individuals that contain the transgene at the time of analysis. In our laboratory, we use this method as a first screening for transgenic individuals from thousands of the presumptive transgenic fish.

### 4.2 Pattern of Transgene Integration

Studies conducted in many fish species showed that following injection of linear or circular transgene constructs into fish embryos, the transgene is maintained as an extrachromosomal unit through many rounds of DNA replication in the early phase of embryonic development. At later stages of embryonic development, some of the transgenes are randomly integrated into the host genome while others degrade, resulting in the production of mosaic transgenic fish. In many fish species studied to date, multiple copies of transgenes were found to integrate in a head-to-head, head-to-tail, or tail-to-tail form, except in transgenic common carp and channel catfish, where single copies of transgenes were integrated at multiple sites on the host chromosomes.

## 5 EXAMPLES OF TRANSGENIC FISH

### 5.1 Growth Hormone Transgenic Fish

Since 1985, many investigators throughout the world have attempted to introduce growth hormone transgenes fused to functional promoters into newly fertilized eggs of channel catfish, common carp, goldfish, loach, medaka, rainbow trout, salmon, and tilapia by microinjection or electroporation. These studies were conducted to produce fast-growing transgenic fish species for aquaculture. While many groups have demonstrated integration or inheritance of transgenes in transgenic fish, Zhang, Du, and Lu and their colleagues have documented that the GH transgene could be integrated into the genome of the target fish species, inherited in subsequent generations, and expressed in transgenic individuals. Furthermore, the expression of the GH transgene resulted in significant growth enhancement in $P_1$ and $F_1$ transgenic fish. In a recent evaluation of the growth performance of seven different $F_1$ transgenic common carp families carrying rainbow trout GH cDNA, Chen and co-workers, observed that the growth performance varied widely among families, ranging from a negative correlation to a positive 60% growth enhancement. Since the transgene is integrated at random locations in the host genome, different growth performance displayed by different $F_1$ transgenic families is expected. The fastest growing transgenic fish can be developed by utilizing a combination of family selection and mass selection of transgenic individuals following insertion of the GH transgene.

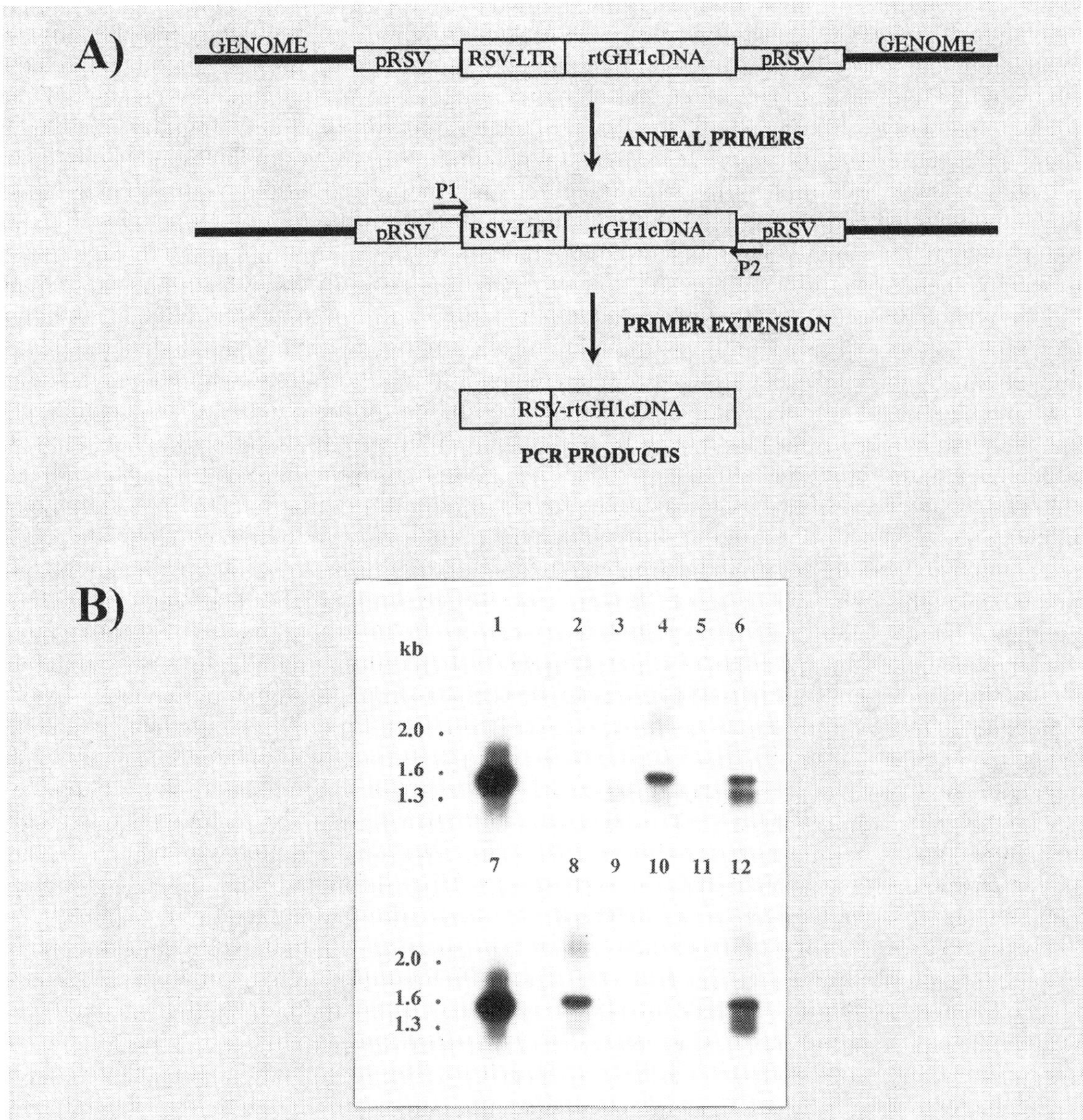


**Figure 3.** Identification of transgene in the presumptive transgenic fish by PCR and Southern blot hybridization. DNA samples were isolated from pectoral fin tissues of presumptive transgenic fish and subjected to PCR amplification. The amplified products were analyzed by electrophoresis on agarose gels and Southern blot hybridization. (A) Strategy of PCR amplification. (B) Southern blot analysis of PCR amplified products: lanes 2–6 and 8–12, PCR products from presumptive transgenic fish; lanes 1 and 7, transgene construct (RSVLTR-rtGH1 cDNA).

### 5.2 Other Transgenic Fish

In addition to the growth hormone gene transfer studies mentioned in Section 5.1, a number of other gene constructs (e.g., antifreeze protein gene, hygromycin resistance gene, neomycin resistance gene, α-globin gene, β-galactosidase gene, chicken δ-crystalline protein gene, bacterial or insect luciferase gene) have been introduced into fertilized eggs of several fish species. A variety of promoters including mammalian or fish metallothionein promoters, simian virus 40 promoter, Rous sarcoma virus long terminal repeat sequence, and fish antifreeze protein promoter were used to drive the expression of the transgenes. In most of those studies, successful gene transfer, expression, and inheritance were reported.

## 6 APPLICATIONS AND PERSPECTIVES

The production of transgenic medaka or zebrafish is a powerful technology for studying the molecular genetics of vertebrate development. By introducing reporter genes linked to promoter/enhancer regions of genes from fish or other organisms, the temporal and spatial expression patterns of these genes can be determined. The same strategy can also be used to define the regulatory sequence of a gene. By making a transgenic fish with a *gain-of-function* gene construct containing a normal fish gene or its homologue under the control of a constitutive or strong promoter/enhancer, the investigator can determine the function of the gene product. For instance, overexpressing a c-*myc* oncogene in a transgenic fish makes possible the determination of the biological effect of over expression of oncogenes. Furthermore, introducing a *loss-of-function* gene construct to inactivate a normal gene in a transgenic fish allows the biological function of this gene product to be determined. For example, transgenic fish carrying an inducible ribozyme transgene for interrupting the production of GH or insulinlike growth factor (IGF) can be produced. By turning on the ribozyme transgene with an appropriate inducer at any developmental stage, the investigator can stop production of GH or IGF in the transgenic fish, thus allowing study of the effect of these hormones on fish growth and development.

Transgenic fish technology has great potential in biotechnological applications as well. Introducing desirable genetic traits into finfish or shellfish can result in the production of superior transgenic strains for aquaculture. These traits may include elevated growth enhancement, improved food conversion efficiency, resistance to some known diseases, tolerance to low oxygen concentrations, and tolerance to subzero temperatures. Recent laboratory progress has shown that transfer, expression, and inheritance of transgenes can be achieved in several finfish species. To realize the full potential of the transgenic fish technology in acquaculture or other biotechnological applications, however, several important scientific breakthroughs are required:

1. Development of more efficient mass gene transfer technologies.
2. Identification of genes of desirable traits for acquaculture and other applications.
3. Development of targeted gene transfer technology (e.g., embryonic stem cell gene transfer method, ribozyme gene inactivation method).
4. Identification of suitable promoters to direct the expression of transgenes at optimal levels during the desired developmental stages.
5. Determination of physiological, nutritional, immunological, and environmental factors that will maximize the performance of the transgenic individuals.
6. Assessment of the safety and environmental impacts of transgenic fish.

When these problems have been resolved, the commercial application of transgenic fish technology will be readily attained.

*See also* PCR TECHNOLOGY.

### *Bibliography*

Buono, R. J., and Linser, P. J. (1992) Transient expression of RSVCAT in transgenic zebrafish made by electroporation. *Mol. Mar. Biol. Biotechnol.* 1:271–275.

Chen, T. T., and Powers, D. A. (1990). Transgenic fish. *Trends Biotechnol.* 8:209–215.

Chen, T. T., Powers, D. A., Lin, C. M., Kight, K., Hayat, M., Chatakondi, N., Ramboux, A. C., Duncan, P. L., and Dunham, R. A. (1993) Expression and inheritance of RSVLTR-rtGH1 cDNA in common carp, *Cyprinus carpio. Mol. Mar. Biol. Biotechnol.* 2:88–95.

Cotten, M., and Jennings, P. (1989) Ribozyme mediated destruction of RNA in vivo. *EMBO J.* 8:3861–3866.

Du, S. J., Gong, G. L., Fletcher, G. L., Shears, M. A., King, M. J., Idler, D. R., and Hew, C. L. (1992) Growth enhancement in transgenic Atlantic salmon by the use of an "all fish" chimeric growth hormone gene construct. *Bio/Technology,* 10:176–181.

Fletcher, G. L., and Davies, P. L. (1991) Transgenic fish for aquaculture, in *Genetic Engineering,* Vol. 13, J. K. Setlow, Ed., pp. 331–370. Plenum Press, New York.

Lu, J. K., Chrisman, C. L., Andrisani, O. M., Dixon, J. E., and Chen, T. T. (1992) Integration expression and germ-line transmission of foreign growth hormone genes in medaka, *Oryzias latipes. Mol. Mar. Biol. Biotechnol.* 1:366–375.

Moav, Boaz, Liu, Z., Groll, Y., and Hackett, P. R. (1992) Selection of promoters for gene transfer into fish. *Mol. Mar. Biol. Biotechnol.* 1:338–345.

Powers, D. A., Hereford, L., Cole, T., Creech, K., Chen, T. T., Lin, C. M., Kight, K., and Dunham, R. A. (1992) Electroporation: A method for transferring genes into gametes of zebrafish *(Brachydanio rerio)* channel catfish *(Ictalurus punctatus),* and common carp *(Cyrinus carpio). Mol. Mar. Biol. Biotechnol.* 1:301–308.

Zhang, P., Hayat, M., Joyce, C., Gonzales-Villasenor, L. I., Lin, C.-M., Dunham, R., Chen, T. T., and Powers, D. A. (1990) Gene transfer, expression and inheritance of pRSV-rainbow trout-GH-cDNA in the carp, *Cyprinus carpio* (Linnaeus). *Mol. Reprod. Dev.* 25:3–13.

# TRANSLATION OF RNA TO PROTEIN

*Robert A. Cox and Henry R. V. Arnstein*

### *Key Words*

**Anticodon** Three consecutive bases in tRNA that bind to a specific mRNA codon by complementary antiparallel base pairing.

**Antiparallel Base Pairing** Pairing through specific hydrogen bonds between base residues of two polynucleotide chains or two segments of a single chain with phosphodiester bonds running in the $5' \rightarrow 3'$ direction in one chain or segment and in the $3' \rightarrow 5'$ direction in the other. In DNA and RNA, the hydrogen bonds usually are formed between complementary base pairs (A with either T or U, and G with C).

**Elongation** The stepwise addition of amino acids to the carboxyl terminus of a growing polypeptide chain.

**Initiation** A multistep reaction between ribosomal subunits, charged initiator transfer RNA, and messenger RNA that results in apposition of the ribosome-bound initiator Met-tRNA with an AUG initiator codon in mRNA. In this position the ribosome is poised to form the first peptide bond.

**Polarity** The asymmetry of a polymer such as a polynucleotide or polypeptide. In DNA the two strands have opposite polarity; that is, they run in opposite directions ($5' \rightarrow 3'$ and $3' \rightarrow 5'$). The polarity of a polypeptide is defined as running from the N-terminus to the C-terminus.

**Reading Frame** One of three possible ways of translating groups of three consecutive nucleotides in mRNA. The appropriate reading frame is determined by the initiation codon.

**Template Strand** The strand of the DNA double helix, also known as the coding strand, that is used as a template for transcription of RNA. It has a base sequence complementary to the RNA transcript.

**Termination** The end of polypeptide synthesis, which is signaled by a codon for which there is no corresponding aminoacyl-tRNA. When the ribosome reaches a termination codon in the mRNA, the polypeptide is released and the ribosome–mRNA–tRNA complex dissociates.

**Translation** The stepwise synthesis of a polypeptide, the amino acid sequence of which is determined by the nucleotide sequence of the mRNA coding region. The genetic code relates each amino acid to a group of three consecutive nucleotides termed a codon. Decoding of mRNA takes place in the $5' \rightarrow 3'$ direction, and the polypeptide is synthesized from the amino to the carboxyl terminus.

**Translocation** The stepwise advance of a ribosome along mRNA, one codon at a time, with simultaneous transfer of peptidyl-tRNA from the A site to the P site of the ribosome.

---

Proteins are essential to the structure and function of living cells. The assembly of polypeptide chains from amino acids and their subsequent modifications, leading to the final three-dimensional protein structure, are exceptionally complex processes; many components are involved and much of the cell's energy is utilized. The linear amino acid sequence of a protein is encoded within the gene as a linear deoxyribonucleotide sequence. Early steps in the biosynthesis of a protein include transcription of the gene and appropriate processing of the transcript leading to the production of mature mRNA. We describe the mechanisms involved in translating mRNA to produce a polypeptide chain which has the amino acid sequence specified by the gene.

## 1 INTRODUCTION

A gene or cistron is defined as the region of DNA that is transcribed into functional RNA. The transcript functions either as such (e.g., tRNA, rRNA) or as a messenger (mRNA), which codes for a single polypeptide chain in the translation process*. A polynucleotide such as RNA is an asymmetrical polymer that is assembled from nucleoside triphosphates by a stepwise mechanism linking the 3′ position of one nucleotide by a phosphate bridge to the 5′ position of the adjacent nucleotide. In the finished polynucleotide chain, the first nucleotide residue has a 5′ position that is not linked to another nucleotide, whereas the last nucleotide has an unlinked 3′ position. Thus, polynucleotide synthesis proceeds from the 5′ to the 3′ terminus and the polymer is said to have a 5′-to-3′ polarity. Usually, linear RNA sequences are written with the 5′ terminus on the left and the 3′ terminus on the right (Figure 1).

The polypeptide chains of proteins are also asymmetrical polymers in which the amino acid residues are linked by peptide bonds between their α-amino and carboxyl groups (Figure 2), leaving a free α-amino group at one end (the amino terminus) of the polymer and a free α-carboxyl group at the opposite end (the carboxyl terminus). The significance of the polarity of RNA and proteins will become evident when the process of protein biosynthesis is explained.

The genetic information stored in DNA is not usable directly for making proteins but must be copied first into mRNA by an enzymatic transcription of segments of DNA containing the genes. Messenger RNA serves as the template for protein synthesis; that is, the linear nucleotide sequence of the mRNA dictates the amino acid sequence of the polypeptide encoded originally by the gene. Conventionally, gene and mRNA nucleotide sequences are written in the 5′-to-3′ direction, which corresponds to the direction in which mRNA is decoded during polypeptide synthesis: the mRNA is read in the 5′-to-3′ direction, and the polypeptide is synthesized from the amino- toward the carboxyl terminus.

The mechanism for translating RNA into protein is complex, and the cell devotes considerable resources to the translational machinery. The components include 20 different amino acids, transfer RNAs, aminoacyl–tRNA synthetases, ribosomes, and a number of protein factors that cycle on and off the ribosomes and facilitate various steps in initiation of translation, elongation of the nascent polypeptide chain, and termination of synthesis with release of the completed polypeptide from the ribosome. The process depends on a supply of energy, which is provided by ATP and GTP. The rate of protein synthesis is approximately 0.15 second per peptide bond at 37°C.

## 2 mRNA STRUCTURE AND THE GENETIC CODE

### 2.1 Structure

The sequence information of a gene is copied (transcribed) into the nucleotide sequence of RNA using one strand of DNA (called the coding strand) as the template. The primary transcript is a single strand of RNA, which is a faithful copy of the *other* strand of DNA (the non-coding strand), with substitution of U residues in place of T residues found in DNA. Sometimes, the primary transcript is altered, as described later, before it functions as mRNA; in these cases the original unmodified transcript is the precursor or pre-mRNA. Usually, mRNAs have nontranslated sequences at the 5′ and 3′ ends in addition to the coding domain. These noncoding sequences sometimes affect the efficiency of translation and the stability of mRNA. Figure 3 shows the structures of typical mRNAs. The decoding process involves base pairing between three bases (i.e., a codon) in the mRNA and the three-base anticodon of a transfer RNA. In a separate reaction, each tRNA is first linked to a particular amino acid, and thus the pairing of mRNA with tRNA determines the sequence of amino acids in the resulting protein.

#### 2.1.1 Prokaryotic mRNA

In organisms that do not have a nucleus (prokaryotes), pre-mRNA usually undergoes little or no modification, with the result that pre-

* The amino acid linked to a charged tRNA is indicated by a prefix and the specificity of the transfer RNA (tRNA) in the aminoacylation reaction is shown as a superscript on the right, e.g. Phe-$tRNA^{Phe}$ indicates phenylalanine-specific tRNA charged with phenylalanine. The anticodon is sometimes indicated as a right subscript or, alternatively, in the superscript after the amino acid, e.g. $tRNA_{UGC}$ or $tRNA^{ALA/UGC.}$ The right hand subscript position is sometimes used to indicate the organism from which the tRNA is derived. The initiator tRNA, which is specific for methionine, is termed 394, $tRNA_f^{Met}$ or $tRNA_f^{Met}$. The elongator methionine-specific RNA, which inserts methionine into internal positions of the growing peptide chain, is termed 396, $tRNA_m^{Met}$ (or $tRNA_m$) when uncharged and Met-$tRNA_m^{Met}$ (or Met-$tRNA_m$) when charged.

A

5'- End

Adenine (A)

Cytosine (C)

Guanine (G)

Uracil (U)

3'- End

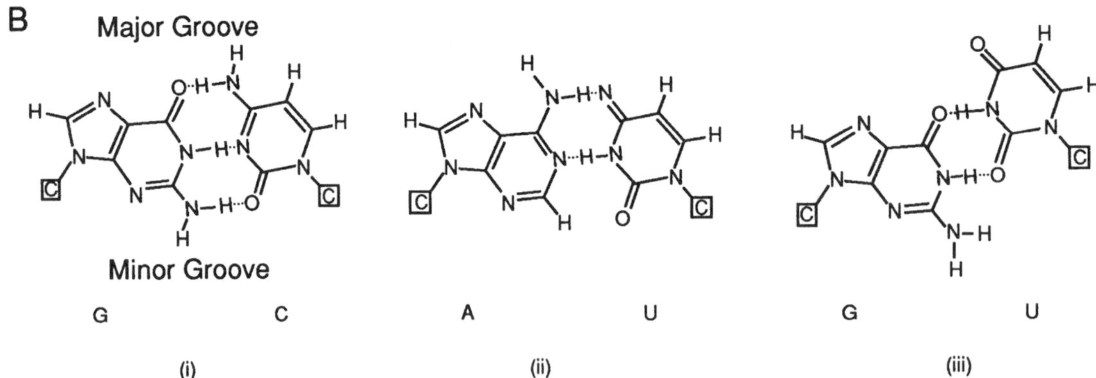

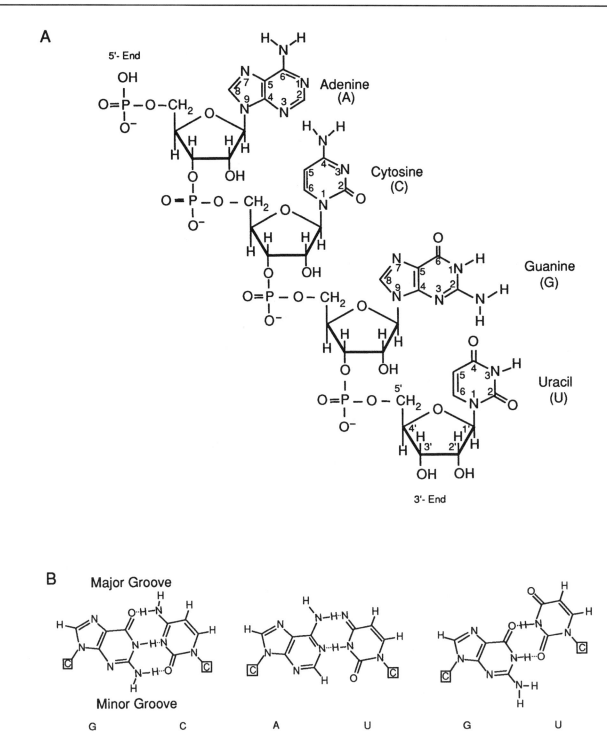

**Figure 1.** Structural elements of ribonucleic acids. (A) Primary structure, indicating the numbering system for purines and pyrimidines. (B) Base-pairing interactions commonly found in RNA: (i) G·C base pair, (ii) A·U base pair, (iii) G·U base pair. Pairs i and ii are Watson–Crick base pairs; iii is a special type of base pairing found in intramolecular bihelical regions. The minor groove of the helix is on the side of the base-pair with the glucosidic bond, of which the carbon atom C1′ of ribose is boxed. Note that DNA contains base pairs of types i and ii only but with thymine (5-methyluracil) in place of uracil. (Reproduced from Arnstein and Cox (1992) with the permission of Oxford University Press.)

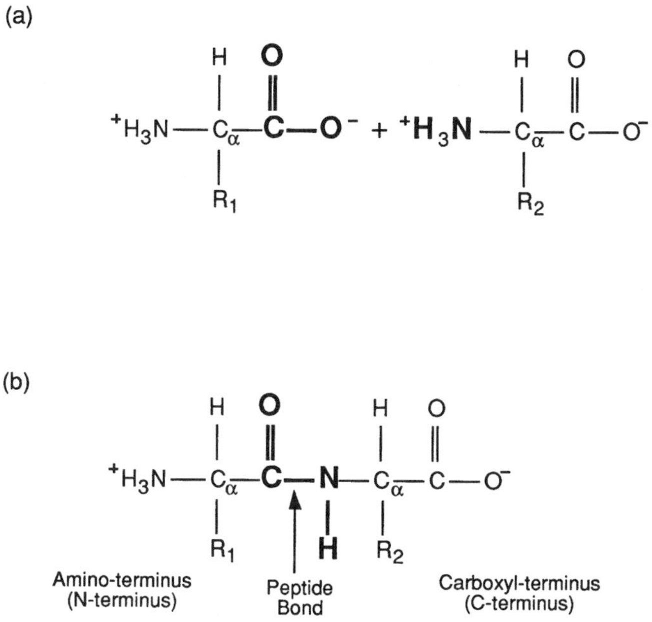

(a)

$$^{+}H_3N-C_\alpha H(R_1)-C(=O)-O^- + {}^{+}H_3N-C_\alpha H(R_2)-C(=O)-O^-$$

(b)

$$^{+}H_3N-C_\alpha H(R_1)-C(=O)-NH-C_\alpha H(R_2)-C(=O)-O^-$$

Amino-terminus (N-terminus)

Peptide Bond

Carboxyl-terminus (C-terminus)

**Figure 2.** Structure of the peptide bond. (a) Two amino acids with different side chains, $R_1$ and $R_2$. (b) A dipeptide formed from the two amino acids shown in (a).

mRNA and mRNA are very similar if not identical. Since pre-mRNA is collinear with DNA, DNA and proteins are usually collinear in these organisms. Gene expression in prokaryotes usually involves the cotranscription of several adjacent genes, and translation of mRNA sequences into polypeptides may begin at the 5′ end of mRNA while transcription is still in progress at the 3′ end.

### 2.1.2 Eukaryotic mRNA

In cells with a nucleus (eukaryotes), the genetic information is stored mainly in the nucleus and to a minor degree in some organelles (mitochondria and chloroplasts). The description that follows pertains only to nuclear genes. Eukaryotic genes are more complicated than prokaryotic genes because the coding region in the former is often discontinuous: the coding sequences or exons are interrupted by intervening sequences (introns). Thus, genes and proteins are usually *not* collinear in eukaryotes. In the nucleus, a complicated set of splicing reactions removes all the introns and fuses the exons into a continuous coding sequence. Other processing steps involve adding a "cap" to the 5′ end of the mRNA adding a polyadenylated tail to the 3′ end. After completion of these nuclear maturation steps the mRNA is transported to the cytoplasm, where it is translated. As with prokaryotic mRNA, the coding region is flanked by 5′ and 3′ nontranslated sequences.

## 2.2 The Genetic Code

With a few minor exceptions (e.g., in mitochondria), the genetic code is universal. It is triplet, comma-less, and nonoverlapping, and, as a consequence, a nucleotide sequence has three possible reading frames (Figure 4).

Since mRNA is normally translated into a unique polypeptide, an essential step in the translation process is the selection of the appropriate reading frame. This is achieved by starting translation at the initiation codon, usually AUG or less frequently GUG, which ensures that the following codons are read in phase within the required reading frame. Of the 64 theoretically possible triplets in the genetic code, 61 correspond to the 20 genetically encoded amino acids found in all, or nearly all, proteins. When GUG is

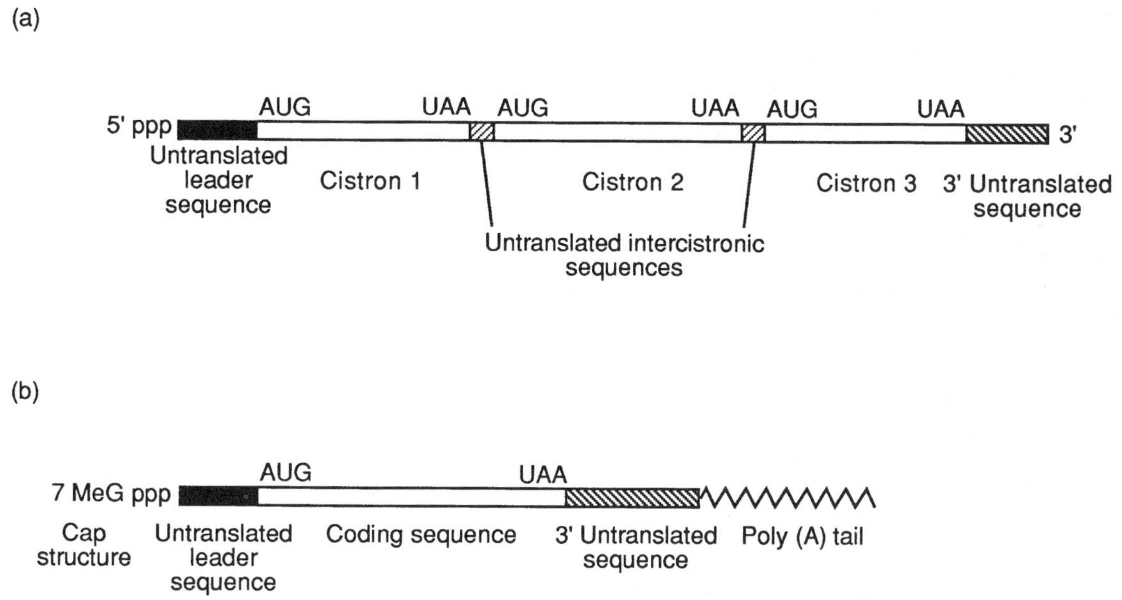

(a)

5′ ppp — AUG — UAA AUG — UAA AUG — UAA — 3′

Untranslated leader sequence | Cistron 1 | Cistron 2 | Cistron 3 | 3′ Untranslated sequence

Untranslated intercistronic sequences

(b)

7 MeG ppp — AUG — UAA — Poly (A) tail

Cap structure | Untranslated leader sequence | Coding sequence | 3′ Untranslated sequence | Poly (A) tail

**Figure 3.** Structure of typical mRNAs: (a) prokaryotic mRNA and (b) eukaryotic mRNA. (Reproduced from Arnstein and Cox (1992) with the permission of Oxford University Press.)

(a) 5′ |AUG|GUA|UUC|AG .|. . . 3′
fMet Val Phe —

(b) 5′ A|UGG|UAU|UCA|G . .|. . 3′
Trp Tyr Ser —

(c) 5′ AU|GGU|AUU|CAG|. . . . 3′
Gly Ile Gln

**Figure 4.** Translation of a polynucleotide sequence into three alternative polypeptides using different reading frames.

used as the initiator codon, it interacts with the anticodon of initiator Met-t $RNA_f^{met}$, but elsewhere it codes for valine. All other codons specify only one amino acid, but many amino acids are specified by two or more (up to six) codons; that is, the code is unambiguous but degenerate. The remaining three codons usually signify termination of synthesis and release of the finished polypeptide chain.

The decoding process involves antiparallel base pairing between the three bases of mRNA codons and the complementary anticodons of tRNA during peptide bond formation (Figure 5). The first and middle bases of the codon form conventional base pairs with the third and middle bases of the anticodon, respectively, but the third base of the codon pairs with the first base of the anticodon by a less stringent interaction. This so-called wobble considerably reduces the number of tRNA species required to decode the 61 sense codons. Thus, the protein synthesis system in the cytosol of eukaryotes contains only a few more than 40 different tRNAs, and in mitochondrial protein synthesis 22 to 24 tRNA species are sufficient.

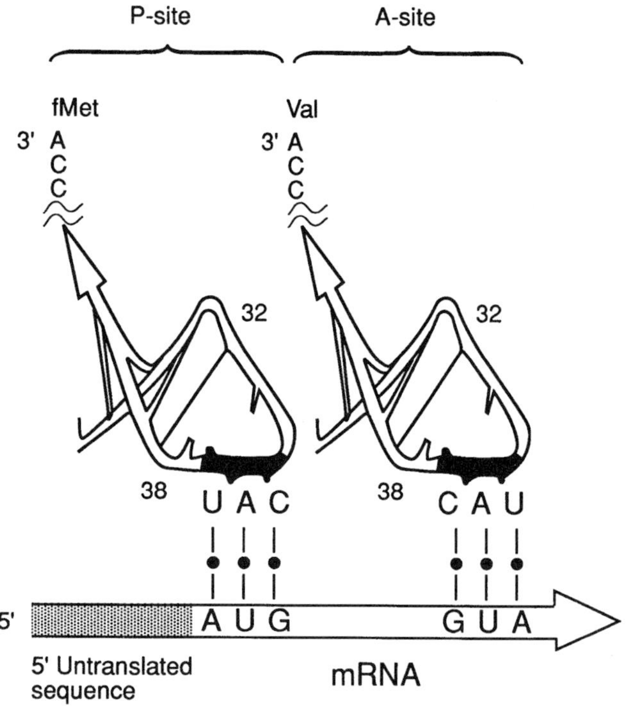


**Figure 5.** Schematic illustration of base pairing between a codon and its anticodon. The diagram shows the interaction at the P site of the ribosome (see Figure 8) between the initiation codon AUG of mRNA and the anticodon CAU of fMet.tRNA fMet and the interaction at the A site between the codon GUA of mRNA and the anticodon UAC of Val.$tRNA^{Val}$. The polarity of the RNA species is revealed by the arrowhead, which points from the 5′ end to the 3′ end. The fragment of tRNA is representative of the general structure (see Figure 6b) placed in the appropriate orientation; the numbers refer to the nucleotide positions measured from the 5′ end. The interaction between the codon and the anticodon is antiparallel, and the three base pairs have a bihelical conformation.

## 3 TRANSFER RNA

By relating individual codons of mRNA to the cognate amino acids, tRNA functions as a key bilingual intermediate in the translation of the genetic code. All transfer RNAs are single-stranded molecules about 80 nucleotides long, with a common 3′-terminal CCA sequence. The secondary structure of tRNA is usually presented in two dimensions as a clover leaf, to highlight the regions of base pairing (Figure 6a). X-Ray crystallography reveals that additional hydrogen bonds give rise to an L-shaped tertiary structure (Figure 6b). The CCA sequence carrying the amino acid is located distal to the anticodon. During translation, the three nucleotides of the anticodon interact, through the formation of antiparallel base pairs, with the three nucleotides of the codon.

The attachment of amino acids to tRNA* involves the formation of an ester bond between the carboxyl group of the amino acid and the 3′-hydroxyl group of the terminal adenosine. It requires specific enzymes, the aminoacyl–tRNA synthetases. There are 20 different synthetases, each specific for one of the 20 amino acids, and each enzyme recognizes something unique in the structure of its cognate tRNA. The structural determinants that ensure the accuracy of this charging reaction vary among the different tRNAs. The anticodon may play a part, but sometimes even a single base elsewhere is sufficient to determine the specificity of the tRNA–synthetase interaction. The accuracy of the synthetase reactions is important because the structure of the amino acid plays no part in its subsequent selection by the mRNA for peptide bond synthesis.

Synthesis of aminoacyl–tRNA **(III)** requires activation of an amino acid (I) through the formation of an intermediate enzyme-bound aminoacyladenylate **(II).**

$$\underset{\mathbf{I}}{\mathrm{E} + \mathrm{R}\cdot\underset{\mathrm{NH_3^+}}{\underset{|}{\mathrm{CH}}}\cdot\mathrm{CO_2^-}} + \mathrm{ATP} \rightarrow \underset{\mathbf{II}}{\mathrm{E}\cdot\mathrm{R}\cdot\underset{\mathrm{NH_3^+}}{\underset{|}{\mathrm{CH}}}\cdot\mathrm{CO}\cdot\mathrm{AMP}} + \mathrm{PP}_i \rightarrow \quad (\mathrm{PP}_i \downarrow 2\mathrm{P}_i)$$

$$\rightarrow \underset{\mathbf{(III)}}{\mathrm{R}\cdot\underset{\mathrm{NH_3^+}}{\underset{|}{\mathrm{CH}}}\cdot\mathrm{CO}\cdot\mathrm{tRNA}} + \mathrm{AMP} + \mathrm{E}$$

where E is aminoacyl–tRNA synthetase.

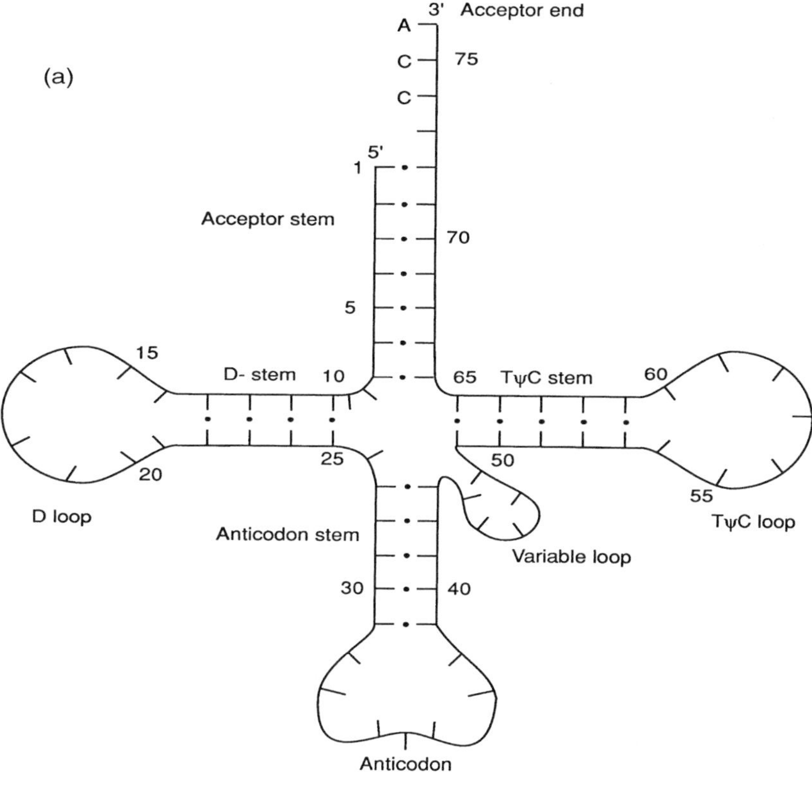


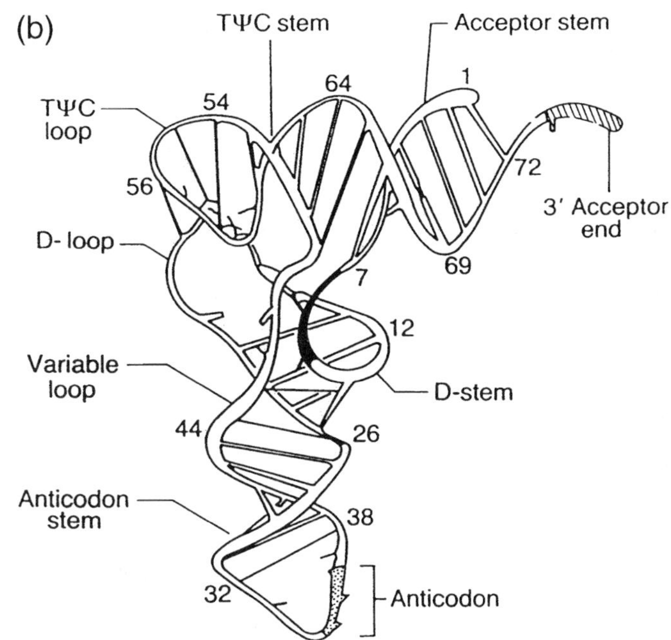


**Figure 6.** Structure of phenylalanyl-transfer RNA: (a) secondary cloverleaf structure and (b) tertiary structure. (Reproduced from Arnstein and Cox (1992) with the permission of Oxford University Press.)

The energy for the reaction is provided by ATP and stored in the ester bond of the aminoacyl–tRNA to be used subsequently for peptide bond synthesis.

## 4 RIBOSOME STRUCTURE AND FUNCTION IN TRANSLATION

### 4.1 Ribosome Structure

Ribosomes are high molecular weight complexes of RNA (rRNA) and proteins. Ribosomes from various sources (prokaryotes, eukaryotic cytoplasm, mitochondria, and chloroplasts) vary in size from 20 to 30 nm in diameter, but all are composed of a large and a small subparticle or subunit (Figure 7) and perform similar functions in protein synthesis.

The small subunit comprises a single rRNA of 0.3 to 0.7 $\times$ $10^6$ Da and single copies of 20 to 30 unique proteins. It has a major function in binding initiator tRNA and mRNA in the initiation of protein synthesis and in decoding the genetic message.

The large subunit comprises a high molecular weight rRNA (0.6–1.7 $\times$ $10^6$ Da) and often one or two smaller rRNAs (0.03–0.05 $\times$ $10^6$ Da) and 30 to 50 different proteins, present, with one exception, as single copies. The large subunit binds aminoacyl-tRNA at the A site, peptidyl-tRNA at the P site, and uncharged tRNA at the E site. The large subunit contains the peptidyl transferase activity, which may reside in the rRNA molecule itself rather than in the associated ribosomal proteins. It is also involved in binding elongation factor G, which is required for translocation (Figure 7).

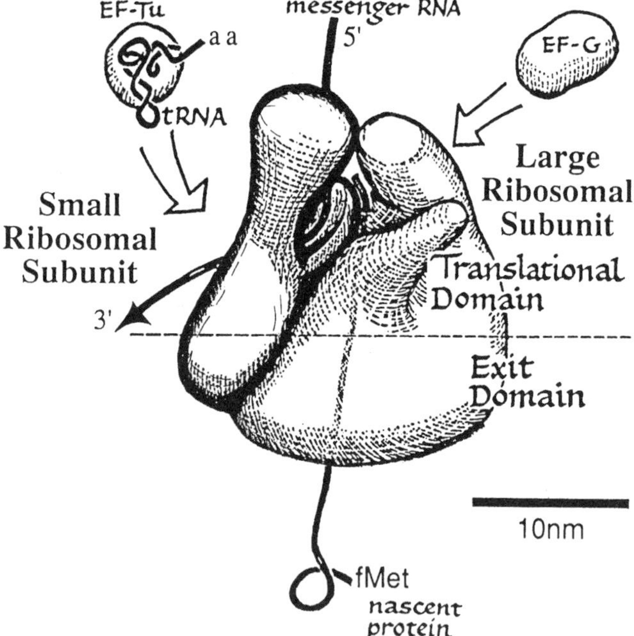


**Figure 7.** Model of the *E. coli* ribosome, based on electron microscopy. Diagram shows the relative orientation of the large and small subunits and other functional components involved in polypeptide chain synthesis. EF-Tu, elongation factor required for binding aminoacyl-tRNA to the ribosomal A site; EF-G, elongation factor required for translocation of the peptidyl-tRNA from the A to the P site. The broken line indicates the boundary between the translational and exit domains of the ribosome. [Based on M. I. Oakes, A. Scheinman, T. Atha, G. Shankweiler, and J. A. Lake in Hill et al. (1990).]

### 4.2 Function of the Ribosome Cycle in Translation

Polypeptide synthesis can be divided into three stages—initiation, elongation, and termination—in a process involving binding of a ribosome to messenger RNA, movement of the ribosome relative to the mRNA in the $5' \rightarrow 3'$ direction, and release of both mRNA and the finished polypeptide from the ribosome (Figure 8).

The ribosome cycle starts with the stepwise formation of an initiation complex. First, charged initiator tRNA (fMet-tRNA$_f$ or Met-tRNA$_f$) binds to the small ribosomal subunit in the P site. Next, mRNA is bound to the small subunit by an interaction involving base pairing between the initiation codon of the mRNA and the anticodon of the initiator tRNA. Last, the large ribosomal subunit is joined to this preinitiation complex (Figure 8a).

After entry of the appropriate aminoacyl-tRNA (as selected by base pairing with the second codon of the mRNA) into the A site of the initiation complex (Figure 8b), the first peptide bond is formed by transfer of the fMet residue from the initiator tRNA to the amino group of the aminoacyl-tRNA in the A site (Figure 8c). This reaction is catalyzed by the peptidyl transferase activity of the large ribosomal subunit.

After formation of the peptide bond, the newly synthesized peptidyl-tRNA is translocated from the A to the P site, together with movement of the mRNA relative to the ribosome through a distance equivalent to one codon. At the same time, the free tRNA in the P site is moved into the exit or E site before leaving the ribosome (Figure 8d and e). Repetition of this process of elongation and translocation results in stepwise growth of the polypeptide chain by one amino acid residue at a time (Figure 8f). Simultaneously, the ribosome moves one codon at a time toward the 3′ end of the mRNA until a termination codon (UAA, UAG, or UGA) enters the A site, whereupon a release factor is bound (Figure 8g) and catalyzes the hydrolysis of the peptidyl–tRNA bond. The finished polypeptide chain is released, and the ribosome dissociates into subunits with the release of both the final tRNA and mRNA (Figure 8h). In the ribosome cycle the same mRNA molecule and the other components needed for protein synthesis can be used repeatedly. The various steps of the translation process are facilitated by protein factors, and usually several ribosomes are attached to one mRNA molecule at the same time, giving rise to polysomes (polyribosomes).

## 5 CONCLUDING REMARKS

The polypeptide chains of all proteins are synthesized by the process just described. This mechanism gives rise to primary polypeptide chains, which are often further modified—for example, by cleavage into smaller peptides, by structural modification of selected amino acid residues, or by the formation of covalent bonds between polypeptide chains. Some of these secondary modifications are related to the folding of polypeptides to the correct secondary and tertiary structure. The production of active enzymes or peptide hormones from inactive precursors (e.g., insulin from proinsulin) also involves modifications of the primary polypeptide. Furthermore, the transport of proteins within the cell or the secretion of extracellular proteins is often linked to structural changes in polypeptide chains either during synthesis or after its completion.

The control of protein synthesis, either by regulation of the amount of mRNA available for translation or by the efficiency

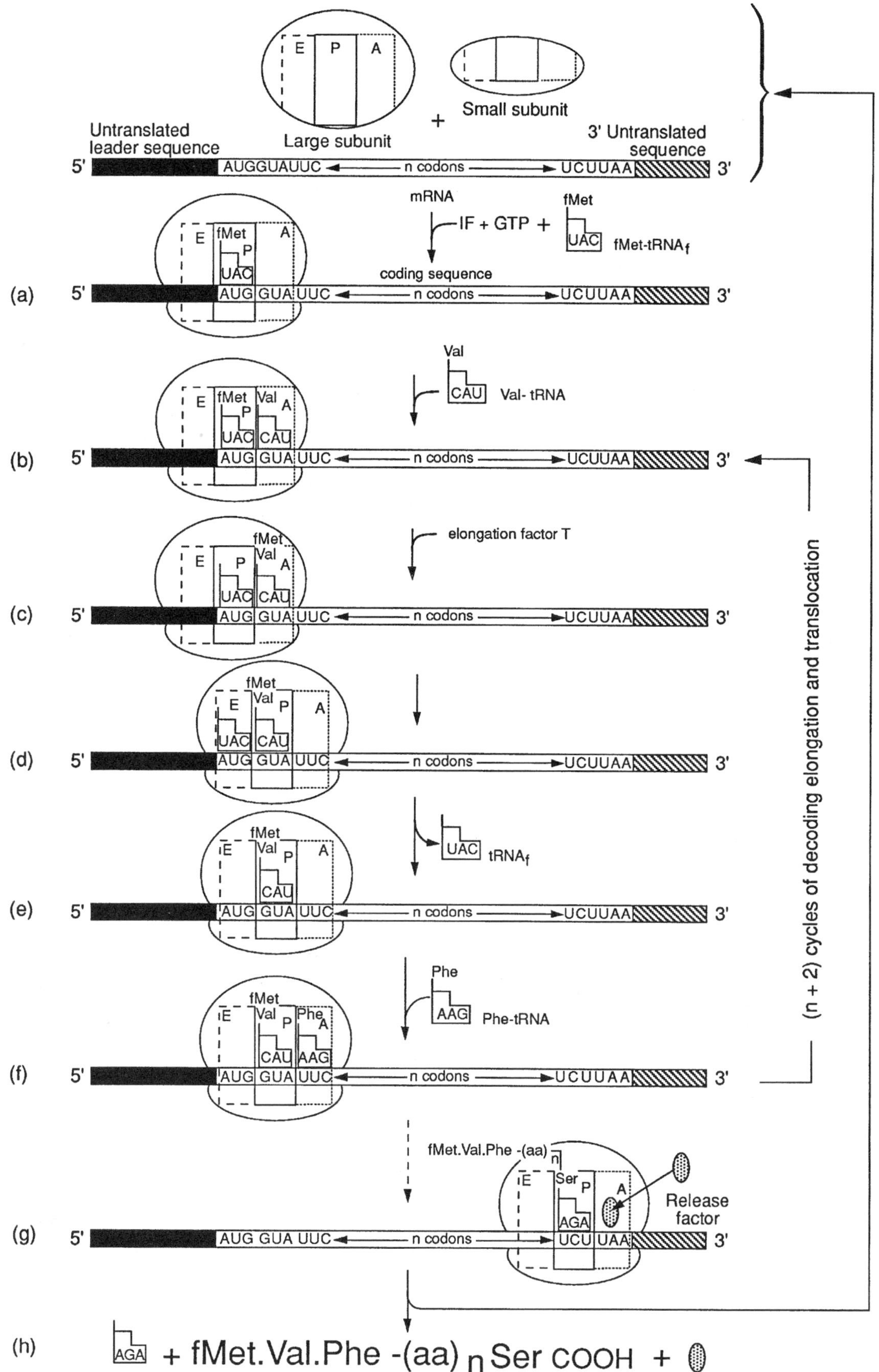


**Figure 8.** Schematic view of the biosynthesis of a polypeptide by translation of prokaryotic mRNA. (a) Formation of the initiation complex from mRNA, ribosomal subunits, initiation factor (IF), GTP, and fMet-tRNA$^{Met}$. The binding sites for aminoacyl-tRNA, peptidyl-tRNA, and uncharged tRNA are designated A, P, and E, respectively. (b) Decoding of the codon GUA with Val-tRNA$^{Val}$. (c) Formation of the peptide bond by transfer of fMet to Val-tRNA$^{Val}$ forming fMet.Val-tRNA$^{Val}$. (d) Translocation of tRNA $^{f}$Met to the E site and of fMet.Val-tRNA$^{Val}$ to the P site with codon UUC aligned with the A site. (e) Ejection of tRNA $^{f}$Met from the ribosome. (f) Decoding of codon UUC with Phe-tRNA$^{Phe}$. (g) Decoding of codon UAA with release factor. Steps b to f constitute the elongation–translocation cycle. The position shown is reached by repeating the cycle $n + 2$ times after formation of fMet.Val-tRNA. (h) Release of the completed polypeptide, fMet.Val.Phe.(aa)$_n$ Ser, ribosomal subunits, release factor, mRNA, and tRNA$^{Ser}$. The cycle may be repeated starting again at (a).

with which it is translated, is important in cell growth and development as a factor determining the level of cellular and extracellular proteins. Subversion of this control occurs in cells infected by viruses: the viral nucleic acid uses the protein-synthesizing machinery of the host cell, thereby changing normal cell metabolism in favor of the synthesis of viral proteins needed for the production of virus progeny.

*See also* Ribosome Preparations and Protein Synthesis Techniques.

### *Bibliography*


Arnstein, H.R.V., and Cox, R. A. (1992) *Protein Biosynthesis.* Oxford University Press, Oxford.

Hill, W. E., Dahlberg, A., Garrett, R. A., Moore, P. B., Schlessinger, D., and Warner, J. R. (1990) *The Ribosome, Structure, Function and Evolution.* American Society for Microbiology, Washington, DC.

Hinnebusch, A. G. and Liebman, S. W. (1991) Protein synthesis and translational control in *Saccharomyces cerevisiae.* In *The Molecular and Cellular Biology of the Yeast: Genome Dynamics, Protein Synthesis and Energetics,* pp. 627–735. E. W. Jones and J. R. Pringle, Eds. Cold Spring Harbor Laboratory Press, Cold Spring Harbor, ME.

Kozak, M. (1983) Comparison of initiation of protein synthesis in procaryotes, eucaryotes and organelles. *Microbiol. Rev.* 47:1–45.

Lee, B. J., and Pirtle, R. M., Eds. (1992) *Transfer RNA in Protein Synthesis.* CRC Press, Boca Raton, FL.

Lewin, B. (1990) *Genes IV.* Oxford University Press, Oxford.

Merrick, W. C. (1992) Mechanism and regulation of eukaryotic protein synthesis. *Microbiol. Rev.* 56:291–315.


# Transport Proteins

*Milton H. Saier, Jr.*

### *Key Words*

**Antiport (Countertransport; Exchange Transport)** Transmembrane transport of two species in opposite directions in a carrier-mediated process.

**Carrier Protein** Transmembrane transport protein that selectively binds a solute and translocates that solute with the binding site across the membrane.

**Channel Protein** Transmembrane transport protein that forms a pore through the membrane, allowing passage of solutes from one side to the other.

**Group Translocation** Transport coupled to chemical modification of the solute transported (e.g., sugar transport and phosphorylation catalyzed by the bacterial phosphotransferase system).

**Symport (Cotransport)** Transmembrane transport of two species, usually a solute and a cation, together in a tightly coupled, carrier-mediated process.

**Transporter (Porter; Permease; Pump)** A transmembrane solute transport system.

**Uniport** Transport of a single species across a membrane by a carrier-type mechanism.

Transport proteins translocate small solutes across biological membranes. Different classes of transport systems are known which function by different mechanisms, couple different forms of energy to transmembrane solute translocation, and comprise distinct families of homologous proteins. Some of these permease classes, many of which extend across the prokaryotic–eukaryotic boundary, share structural and functional characteristics. While some of them probably have a common evolutionary origin, others clearly evolved independently.

## 1 INTRODUCTION

All living organisms probably descended from a common ancestor, but diverged to give rise to three kingdoms: eubacteria, archaebacteria, and eukaryotes. The first two kingdoms comprise the prokaryotes (which lack nuclei), whereas the eukaryotes are typified by eukaryotic organelles such as nuclei, mitochondria, and chloroplasts. Extensive evidence suggests that the emergence of eukaryotes may have occurred some 1.8 billion years ago, although the prokaryotes may have existed more than 2 billion years before that time. Over this enormous time interval of about 4 billion years, proteins, including transmembrane solute transport proteins, were evolving to their present state of complexity.

## 2 CLASSES OF INTEGRAL MEMBRANE TRANSPORT PROTEINS

Recent studies have established that most eukaryotic integral transmembrane solute–transport proteins possess homologous prokaryotic counterparts. For example, the three major types of cation-translocating ATPases, types P, F, and V, are found in both bacteria and eukaryotes (Table 1). Similarly, several mitochondrial electron carrier proteins capable of transmembrane proton translocation possess homologous counterparts in bacteria.

Gram-negative bacteria possess two membranes, an inner and an outer membrane. Outer membrane porin proteins, which transport many solutes nonselectively, have a structurally similar counterpart in the anion channels of the outer mitochondrial membranes of eukaryotes. Other pore-type transporters include the ion channels of nerve and muscle cells, which possess no known homologues in prokaryotes, and the presumed pore-type major intrinsic protein (MIP) family (see Table 1).

In contrast to these channel-type transporters, there exist several classes of carrier-type transporters including several families of homologous uni-, sym-, and antiporters. Most of these families extend across the eukaryotic–prokaryotic division, but a few are restricted to just one of these kingdoms. The ABC-type ATP-driven solute pumps of bacteria have eukaryotic counterparts. By contrast, the permeases of the sugar-transporting bacterial phosphotransferase system, which couples sugar uptake to sugar phosphorylation, do not appear to have a functional equivalent in eukaryotes (Table 1). Sections 3–7 describe some of these families of transport proteins.

## 3 VOLTAGE-SENSITIVE ION CHANNELS

Nerve and muscle cells of higher eukaryotes possess homologous proteins that form either ligand-gated or voltage-sensitive ion channels. The latter group includes cation channels specific for $Na^+$, $Ca^{2+}$, or $K^+$. Proteins that comprise the voltage-sensitive channels that facilitate $Na^+$ or $Ca^{2+}$ transport consist of single polypeptide chains of about 2000 amino acid residues. Each one of these channel proteins contains four homologous internal repeats of about 500 residues, which are similar in sequence. Each of the four repeated

Table 1 Occurrence of Various Homologous Transport Protein Families Across the Prokaryotic–Eukaryotic Boundary

| Protein Family | Prokaryotic | Eukaryotic |
|---|---|---|
| Outer membrane porins | + | + |
| P-type ATPases | + | + |
| F-type ATPases | + | + |
| V-type ATPases | + | + |
| $H^+$-translocating electron carriers | + | + |
| Nerve ion channels | − | + |
| MIP family | + | + |
| Uni-, sym-, and antiporters | + | + |
| ABC-type ATP-driven solute pumps | + | + |
| Phosphotransferase system | + | − |

domains contains six presumed α-helical transmembrane segments. The voltage-sensitive $K^+$ channels are similar in sequence but differ from the $Na^+$ and $Ca^{2+}$ channels in being homotetramers in which each subunit corresponds to one of the four internal repeats of the $Na^+$ or $Ca^{2+}$ channels.

## 4 THE MAJOR INTRINSIC PROTEIN FAMILY

Recent sequence analyses have revealed a family of intrinsic membrane proteins from plants, animals, yeast, and bacteria. These proteins include the major intrinsic protein from the lens of the mammalian eye and the glycerol facilitator from the bacterium *Escherichia coli.* The former protein is probably a channel protein that allows transport of sodium ions across the membrane, while the latter is a channel protein that allows facilitation of glycerol across the membrane. These proteins are similar in size and have six putative membrane-spanning domains. Surprisingly, sequence comparisons revealed that the six transmembrane helical segments in each of these proteins arose by intragenic duplication of a primordial gene encoding just three transmembrane helical segments.

The best characterized of these proteins, MIP, is known to exist in the membrane as a homotetramer in which both the amino- and carboxy-termini of the subunits face the cytoplasm. This arrangement is the same as that established for the voltage-responsive ion channels of nerves and muscle cells discussed in the preceding section.

## 5 SOLUTE UNI-, SYM-, AND ANTIPORTERS OF SIMILAR STRUCTURE

It has long been believed that carrier-type transport systems that allow the transmembrane passage of solutes by facilitation (uniport), countertransport (antiport), or ion cotransport (symport) function by similar mechanisms. Recent sequence analyses have shown that sugar facilitators of eukaryotes are homologous to symporters and antiporters of bacteria. In fact, this large family of proteins extends from enteric bacteria and cyanobacteria to the lower eukaryotes such as yeast, algae, and protozoans, and on up to higher plants and animals including man. These facilitators are specific for sugars, organic acids, organophosphate esters, and drugs. They consist of protein subunits with 12 transmembrane helical segments, all of which are believed to be derived from a common ancestral protein. This family of facilitator proteins is called the major facilitator superfamily (MFS).

In mitochondria, there exist a number of strict antiport carriers, which include an ATP/ADP exchanger and a phosphate porter. Sequence comparisons have shown that these porters are homologous. Each subunit spans the membrane six times, and the dimeric forms of these proteins are presumed to be the structural and functional equivalents of the monomeric forms of carriers that span the membrane 12 times. They nevertheless form a distinct family called the mitochondrial carrier family (MCF), which arose by triplication of a small primordial gene that encoded a protein having only two transmembrane helical segments.

## 6 PERMEASE PROTEINS OF THE BACTERIAL PHOSPHOTRANSFERASE SYSTEM

Permeases of the bacterial phosphotransferase system (PTS) are responsible for the phosphorylation-coupled transport of many sugars into the bacterial cell. They are usually energized by phosphorylation at the expense of phosphoenolpyruvate via two energy-coupling proteins, which are themselves phosphorylated. Each sugar permease of the PTS consists of at least three structurally distinct domains, which either can be fused together in a single polypeptide chain or can exist as two or three interactive polypeptide chains. The permease is itself phosphorylated before the sugar substrate can be transported. In this process, called group translocation, sugar phosphorylation is coupled to transmembrane transport. It is believed that group translocation arose by the superimposition of sugar-phosphorylating enzymes on the transport protein. The transport protein (or protein domain) is probably structurally and functionally equivalent to those of simple facilitators.

## 7 A CLASS OF ATP-DEPENDENT ACTIVE TRANSPORT SYSTEMS

Numerous bacterial permeases as well as several homologous eukaryotic transport systems are similar in organization. Two hydrophilic domains or proteins function to couple ATP hydrolysis to active substrate uptake or efflux, and two hydrophobic domains function as the substrate channel. These proteins constitute what is referred to as the ABC (ATP-binding cassette) superfamily. In bacteria, these transport systems exhibit an exceptionally wide variety of substrate specificities, each transporting one or a few structurally related sugars, amino acids, peptides, anions, intermediary metabolites, drugs, vitamins, proteins, or carbohydrates. Many of them possess extracellular substrate-binding receptors that interact with the transmembrane permease proteins on the external surface of the cytoplasmic membrane and are believed to mediate substrate recognition. Eukaryotic proteins included in this family are the multidrug resistance transporter and the cystic fibrosis chloride transporter of mammalian cells. In most established cases, the transmembrane domains or proteins that catalyze transport consist of six helical membrane-spanning segments with both the carboxy- and amino-terminal residues localized to the cytoplasmic side of the membrane. As for the PTS permeases, the various domains either exist as distinct polypeptide chains or are linked together in various arrangements.

## 8 CONCLUSIONS

Transport proteins translocate their solutes by either a channel-type mechanism or a carrier-type mechanism, and a close relationship between these two kinetically distinguishable transport mechanisms has been postulated. A preformed channel may be a structural requirement for carrier-mediated transport. The hydrophobic domains or proteins of the constituent members of several families of transport proteins exhibit a characteristic pattern of six tightly clustered transmembrane helical segments, with both the amino- and carboxy-termini localized to the cytoplasmic membrane surface. This distinctive structural feature has led investigators to propose that this structural motif may be particularly suitable for the formation of transmembrane channels and carriers.

*See also* Bioinorganic Chemistry; Electron Transfer, Biological; Membrane Fusion, Molecular Mechanisms of.

### *Bibliography*

Higgins, C. F. (1992) ABC transporters: From microorganisms to man. *Annu. Rev. Cell Biol.* 8:67–113.

Marger, M. D., and Saier, M. H., Jr. (1993) A major superfamily of transmembrane facilitators that catalyse uniport, symport and antiport. *Trends Biochem. Sci.* 18:13–20.

Mortlock, R. P., Ed. (1992) *The Evolution of Metabolic Function.* CRC Press, Boca Raton, FL.

Nikaido, H., and Saier, M. H., Jr. (1992) Transport proteins in bacteria: Common themes in their design. *Science,* 258:936–942.

Saier, M. H., Jr. (1985) *Mechanisms and Regulation of Carbohydrate Transport in Bacteria.* Academic Press, Orlando, FL.

Saier, M. H., Jr., Ed. (1990) *Sixth Forum in Microbiology:* Coupling of energy to transmembrane solute translocation in bacteria. *Res. Microbiol.* 141:281–395.

Yeagle, P., Ed. (1992) *The Structure of Biological Membranes.* CRC Press, Boca Raton, FL.

# Transposons in the Human Genome

*Abram Gabriel*

### *Key Words*

***Alu* Sequences** The major class of SINEs in the human genome, named for the presence of a distinctive *Alu* I restriction site.

**Endogenous Retroviruses** Retroviruses that have inserted into the germ line and are inherited in a Mendelian fashion. All known endogenous retroviruses in humans are noninfectious.

**LINEs** The abundant family of long (6–7 kilobase) interspersed nuclear repeats found in all mammalian genomes.

**Orthologous Genes** Genes that are structurally and functionally homologous in different species.

**Retrotransposons** Intracellular mobile genetic elements whose replication requires the reverse transcription of an RNA intermediate.

**SINEs** Highly repeated families of short (100–500 bases) interspersed nuclear sequences found in all mammalian genomes. All SINEs appear to be derived from genes transcribed by RNA polymerase III.

The human genome is home to several classes of multicopy dispersed genetic elements, which in aggregate comprise more than 10% of the entire genome. These elements have many of the hallmarks of a group of mobile genes known as retrotransposons or retroposons, whose replication (i.e., transposition) is thought to involve the following steps: transcription of a genomic copy, reverse transcription of the RNA, and reinsertion of the complementary DNA into a new genomic location, with the creation of flanking target site duplications. Human endogenous retroviruses (HERVs) are found in relatively low numbers and are structurally related to infectious retroviruses, with characteristic long terminal repeats, primer binding sites, and multiple open reading frames (ORFs). In contrast, long interspersed nuclear sequences (LINEs) and short interspersed nuclear sequences (SINEs) have accumulated to remarkably high levels within the genome and are distinct entities lacking long terminal repeats. While phylogenetic analysis suggests that human transposons are stable ancient residents that entered the genome early in primate evolution, recent evidence indicates that subclasses of LINEs and SINEs are still capable of inserting copies into new genomic loci.

The presence of transposable elements in the human genome has many significant implications.

1. Transposons are inherently mutagenic. Insertion of an element into or near a functional gene can alter its expression by causing loss of gene function, by changing the gene's tissue-specific or temporal expression, or by inducing the gene's inappropriate overexpression. Although these changes potentially provide an adaptive advantage, they may also be adaptively neutral or may even result in genetic disease or neoplastic transformation.
2. Recombination between elements can result in deletion, inversion, or duplication of the intervening sequences and can lead to either gene loss or gene duplication. These recombinational events may have contributed to the evolution of the human genome.
3. Transposons can serve as experimental tools. The presence of related elements in other mammalian species has been exploited for phylogenetic studies of mammalian evolution. The polymorphic nature of particular genomic loci resulting from the presence or absence of specific elements has been used as a tool for population genetic studies. The multiple classes of high-copy elements within the genome are useful landmarks for the construction of genetic and physical maps of the human genome.

## 1 LINEs

The human genome contains 50,000 to 100,000 copies of a specific LINE sequences, referred to as L1Hs (L1 of *Homo sapiens*). Full-length, 6 kilobase elements represent less than 10% of the total, with the remainder being variably truncated at their 5′ ends. The vast majority of L1Hs elements have accumulated missense mutations, nonsense mutations, deletions, inversions, and/or frameshifts and therefore lack uninterrupted ORFs. A consensus L1Hs sequence, however, predicts the presence of two long ORFs followed by a poly(dA)-rich 3′ end, all flanked by variable length target site duplications (see Figure 1). The deduced amino acid sequence of the L1 ORF2 has a limited homology to retroviral reverse transcriptase (RT). The sequence of the first full-length, intact L1Hs element (the recently isolated L1.2) confirms the predicted structure.

LINEs have been found dispersed through the genomes of all mammals, suggesting that the ''ancestral'' LINE may have become associated with the genome very early in mammalian evolution. Sequence data, however, indicate that intraspecies variation among

a. Long interspersed nuclear sequence

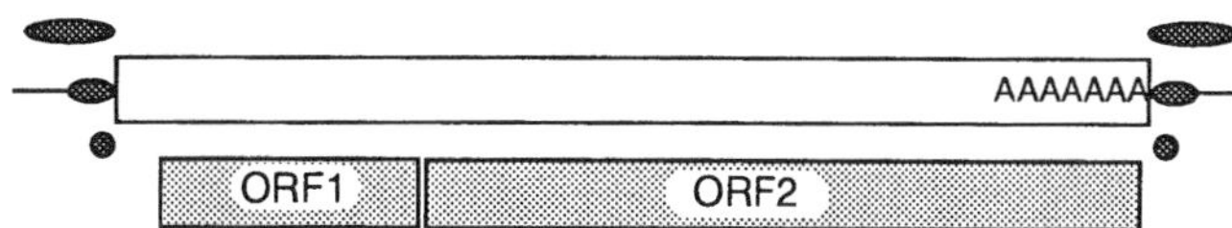


b. Short interspersed nuclear sequence (*Alu* element)

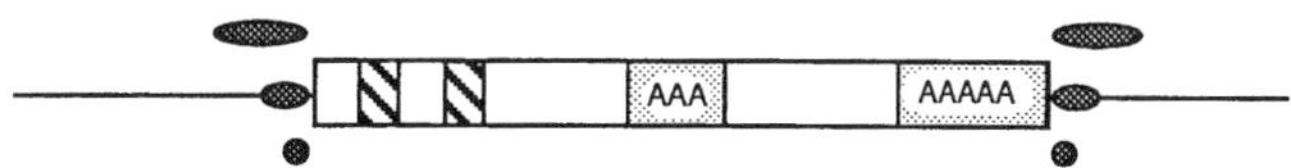


c. Endogenous retrovirus

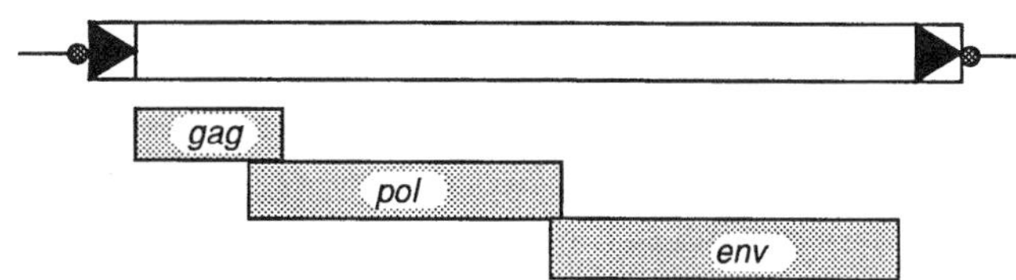


**Figure 1.** General structure of three classes of human transposons. (a) LINEs are 6 to 7 kilobases long and contain two ORFs. Elements are flanked by variable length target site duplications (ovals) and contain a 3′-poly (dA) stretch. (b) *Alu* elements are the most common SINEs in human DNA. They are 300 bases long and form an imperfect dimer, with poly(dA) stretches 3′ to each monomer. Two *pol* III promoter regions are present in the left monomer (striped boxes) and the elements are flanked by variable length target site duplications (ovals). (c) Endogenous retroviruses are 5 to 9 kb long and contain multiple ORFs, corresponding to *gag, pol,* and *env*. Long terminal repeat sequences (solid triangles) flank the coding regions, and characteristic length target site duplications (circles) flank the entire element.

LINEs is much more limited than interspecies variation, a phenomenon known as concerted evolution. These findings have led to the hypothesis that within a given species, either LINEs are homogenized by gene conversion of a few ''master'' copies, or existing LINEs have resulted from the relatively recent amplification of a small number of transpositionally active, full-length ''molecular drivers.''

Although the mechanism by which LINEs transpose is unknown, the key intermediate in the retrotransposon replication model is a full-length RNA, which serves a dual function—it is translated into proteins used in the replication process (e.g., reverse transcriptase) and also is used as the template for reverse transcription. Evidence in support of this model for L1Hs transposition comes from a variety of sources. Discrete, homogeneous, full-length polyadenylated plus-strand LINE RNA is present in the cytoplasm of undifferentiated human teratocarcinoma cell lines, indicating that at least some full-length copies of L1Hs are specifically transcribed and processed in this setting. In more differentiated cell lines, L1Hs RNA is transcribed from both strands, is heterogeneous in size, and is localized to the nucleus. Sequences within approximately the first 100 bases at the 5′ end of L1 can function as promoters, with specificity for teratocarcinoma cells. The observed cell specificity of L1Hs expression is consistent with the conclusion that L1Hs transposition occurs via transcription of a small number of transpositionally competent L1Hs elements early in embryonic development, either within germ cells or in cells destined to become germ cells.

This conclusion is also supported by protein expression studies. Antiserum raised against bacterially expressed L1Hs ORF1 protein recognizes an appropriately sized 38 kDa endogenous polypeptide on Western blots of extracts from teratocarcinoma and choriocarcinoma cell lines. Using the same antiserum, a subset of formalin-fixed, human testicular germ cell tumors can be immunohistochemically stained, implying that these germ cell tumors also express the L1Hs ORF1 gene product. In mouse teratocarcinoma cells, the equivalent ORF1 protein cofractionates with full-length L1 RNA, suggesting the existence of an L1 ribonucleoprotein particle (RNP). Although a function for the ORF1 gene product has not been identified, by analogy to retroviruses it may serve as a capsidlike shell in which reverse transcription occurs. The L1Hs ORF2 from the full-length L1.2 has been cloned into a yeast expression vector and shown to encode a distinct RT activity. Thus L1 encodes a functional enzyme thought to be necessary for its replication. Furthermore, an RT activity identified in human teratocarcinoma cells cofractionates with L1 messenger RNA. While consistent with L1 RNPs, this activity may be due to endogenous retroviruses (see Section 4) and the cofractionation with L1 mRNA fortuitous.

Originally LINEs were thought to be evolutionary relics of a period in mammalian evolution during which these elements were widely amplified and dispersed throughout the genome. According to this model, current LINEs are pseudogenes, independently accumulating random mutations and diverging from one another. While this may be generally true, recent reports of new L1Hs transposition events lend powerful support to the idea that full-length, active elements still reside in the human genome. The best evidence comes from two patients with the X-linked disorder hemophilia A. In each case an insertion of truncated L1 DNA, accompanied by a 3′-poly(dA) tail and perfect flanking target site duplications, was found in exon 14 of the factor VIII gene. In neither case was there a family history of the disease, and in the one case studied in detail, the patient's mother was shown not to have an L1 insertion in either of her factor VIII alleles, indicating that this was a de novo event. Whether the transposition event occurred during maternal gametogenesis or during the early embryonic development of the affected offspring has yet to be determined. The truncated L1 was used as a probe to search the human genome for its progenitor, and L1.2, the first full-length L1Hs element (discussed earlier) was identified. This potentially active element is located on chromosome 22, where it has existed since before the divergence of man, chimpanzee, and gorilla. In addition to these instances of heritable L1 insertions, somatic insertions have been noted within the tumor tissue of two patients with carcinomas. Both an L1 insertion into the proto-oncogene *myc* in a patient with breast cancer as well as an insertion into the tumor suppressor *APC* gene in a patient with colon cancer suggests that transposition events may have played a role in the pathogenesis of these particular neoplasias.

## 2 SINEs

An estimated 500,000 to 1,000,000 copies of various families of short interspersed nuclear sequences are present in the human genome, spaced every 5 to 10 kilobases. These elements are often found inserted within intron sequences and untranslated regions of genes, where they normally do not affect the protein coding capacity of the gene. As with LINEs, the variable length target site duplications at sites of insertion and the presence of 3′-poly(dA) stretches (see Figure 1) are hallmarks of retrotransposition. However, SINEs are quite distinct from LINEs. They average only 300 bases in length and have no protein coding capacity. Therefore they must rely on a nonelement-encoded source of RT for their dispersal, with the leading candidates being either the L1 RT or an RT derived from an endogenous retrovirus.

The *Alu* family is the most widespread class of human SINEs and bears a striking resemblance to the genes encoding the extremely abundant small cytoplasmic 7SL RNA. 7SL RNA is a component of the mammalian signal recognition particle, a structure required for transporting proteins across the membrane of the endoplasmic reticulum. While *Alu* sequences are clearly related to 7SL RNA sequences (~90% identity), the relationship between the two elements is complex. *Alu* elements are arranged as imperfect dimeric structures consisting of two internally truncated tandem copies of 7SL RNA, with each monomer flanked at its 3′ end by adenine-rich stretches that are not present in 7SL RNA (see Figure 1). Two RNA polymerase III promoter regions present in the left half are absent in the right half, yet the right half is 31 bases longer than the left. Thus, while *Alu* sequences are not simple 7SL RNA pseudogenes, they may have resulted from reverse transcription of an aberrantly processed form of this RNA. Interestingly, other families of SINEs in both rodents and primates have strong but independent sequence homology to 7SL RNA or to specific tRNAs, which are also transcribed by RNA polymerase III. These findings have led to the hypothesis that propagation of SINEs requires a member of a multicopy precursor gene family with an internal promoter (e.g., pol III promoter) to be altered so as to become both highly expressed in germ cells and a preferred target sequence for a reverse transcriptase.

Whereas LINEs are present in all mammals examined, *Alu* sequences are specific to primates, suggesting that these elements are of relatively recent origin. The emergence and evolution of the *Alu* family can be inferred from comparisons of their genomic arrangement and sequence in humans and in related primate species. For example, the position of the seven *Alu* sequences within the human α-globin gene cluster is identical to that found in chimpanzees, which diverged from humans approximately 5 million years ago, but is completely different from that found in the prosimian galago, which diverged about 55 million years ago. Individual elements within a species vary by roughly 14% from a derived consensus sequence, as do the orthologous human and chimpanzee sequences examined in the α-globin gene cluster. While these findings are consistent with neutral drift of nonfunctional DNA sequences, *Alu* subfamilies with much higher degrees of sequence identity have been found. This phenomenon is best documented for an *Alu* subfamily (termed HS or PV) that is present in 500 to 2000 copies per haploid genome but is not found at orthologous positions in other primates. Individual members share more than 98% sequence identity. Two de novo *Alu* insertions resulting in genetic disease are members of this subfamily (see later). All these findings indicate recent and ongoing *Alu* amplification and suggest that the members of a small subset of *Alu* elements, present in the genome before the human–ape divergence, are capable of serving as ''master'' copies for transposition.

The retrotransposition model of replication requires that full-length *Alu* sequences be transcribed in appropriate tissues at appropriate times in development. While in vitro transcription experiments using RNA polymerase III demonstrate correct initiation of *Alu* transcripts, in vivo results are less clear-cut. A brain-specific monkey cDNA has been shown to be a 200-nucleotide polyadenylated *Alu* transcript that is also expressed in the human brain. Other short *Alu*-subfamily-specific transcripts have been recovered from HeLa cells, as well as osteosarcoma and thyroid cancer cell lines. A problem that has complicated interpretation of the in vivo transcription data is the ubiquity of *Alu* sequences—because of their presence within introns and in noncoding regions, *Alu* sequences are nonspecifically transcribed in all tissues and cell lines.

*Alu* sequences can cause human genetic disease in a variety of ways. In the most common scenario for *Alu*-mediated gene inactivation, unequal meiotic crossing over between two closely linked elements results in the deletion or duplication of intervening DNA. This genetic mechanism underlies many of the mutant alleles causing familial hypercholesteremia, thalassemia, hereditary persistance of fetal hemoglobin, severe combined immunodeficiency, and hereditary angioedema. As with other mobile elements, a germ line insertion into a functional gene can cause heritable inactivation of that gene. Recently, two case reports of de novo *Alu* insertions have been published. In the first instance, a patient with neurofibromatosis was shown to have an *Alu* insertion into his paternal NF1 allele. Although the insertion was into an intron, it resulted in an aberrant splicing event, presumably as a result of disruption of a normal branch point. In the second case, a person with acholinesterasemia was found to be homozygous for an insertion of an *Alu* sequence within exon 2 of his cholinesterase gene, resulting in a translational frameshift. In both cases, the *Alu* elements had more than 93% sequence identity to the HS family consensus sequence. A more unusual scenario has been proposed for a case of gyrate atrophy. A guanine–cytosine transversion in a preexisting *Alu* element located within intron 3 of the ornithine δ-aminotransferase

*(OAT)* gene created a new donor splice site. Aberrant splicing then resulted in insertion of a fragment of the *Alu* element into the mature *OAT* mRNA.

The ubiquity of *Alu* elements throughout the genome has led to speculations that *Alu* sequences may have broader evolutionary significance in shaping the structure and function of the human genome. Gene duplication by unequal crossing over between *Alu* elements may be involved in generating families of genes with similar structures that serve different functions. Nonhomologous chromosomal rearrangements may be generated by *Alu* recombination, as has been shown for a human XX male with a sex chromosome rearrangement. This process could lead to exon shuffling and the evolution of new genes. On a larger scale, chromosomal translocations might occur between *Alu* elements. The specific arrangement of repeated sequences such as *Alu* sequences on a chromosome may even serve as a cue for homologue pairing during the prophase of meiosis 1, the stage at which meiotic recombination occurs.

## 3 ENDOGENOUS RETROVIRUSES

Retroviruses are infectious agents that behave like transposable elements in that they insert proviral copies of themselves into the host cell's genome. While this is generally a somatic event, endogenous retroviruses are thought to arise by insertion of retroviral proviruses into the host's germ line and their subsequent vertical inheritance. The human genome contains multiple classes of endogenous retroviruses (HERVs), which vary from single-copy elements (HERV-R, S71) to families containing hundreds or thousands of members (HERV-H, THE-1). Most of the families bear strong homologies either to known exogenous retroviruses or to endogenous retroviruses found in other species. In particular, sequences similar to elements RTVL-1 and ERV1 are found at similar loci in closely related primate species, suggesting that germ line insertion of these retroviruses occurred early in primate evolution.

Sequence data from cloned HERVs reveal that these elements retain the general structure of exogenous retroviruses, with long terminal repeats and three ORFs corresponding to *gag, pol,* and *env* (see Figure 1). In all cases so far examined, however, the endogenous elements have accumulated stop codons, deletions, and/or other rearrangements that render them defective. Whether fully functional HERVs exist remains an open question. In mouse, an extensively studied class of endogenous retroviral elements termed intracisternal A-particles (IAPs) are capable of autonomously transposing to new loci within their host genome. IAPs encode a functional *gag* and *pol* gene and are assembled into viruslike particles (VLPs) within the cell, but they lack an *env* gene and are therefore strictly intracellular.

In humans the evidence for functional elements is much more circumstantial and is based on HERV expression studies. Retroviral LTRs contain strong and complex promoters and many HERVs (e.g., HERV-K, RTVL-H) are transcribed in a cell-specific manner from these promoters. Expression of some HERVs (e.g., RRHERV-I, ERV-9, HERV-K) can be induced or repressed by treatment of cells with steroid hormones or retinoic acid. Antigens in some leukemic and choriocarcinoma cell lines, as well as in placenta, cross-react with antibodies raised against various retroviral proteins. Both VLPs and/or reverse transcriptase activity have been reported in placenta, various tumor cell lines, and other human tissues, although their presence has not been definitively linked to specific HERVs. To date, there is no evidence that any HERV can form an infectious viral particle, nor have human mutations caused by de novo HERV germ line insertions been reported.

Instead HERVs have been postulated to cause human disease by more subtle means. For example, HERVs have been implicated in the origin of autoimmune disease based on the finding that sera from some patients with mixed connective tissue disorders cross-react with a 70 kDa U1 small nuclear ribonucleoprotein associated protein that shares antigenic determinants with a p30-*gag* protein from a human endogenous retrovirus. According to this hypothesis, fortuitous expression of the retroviral antigen might initiate an autoantibody response. Furthermore, it has been speculated that somatic transposition of HERVs into tumor suppressor genes or nearby proto-oncogenes could lead to neoplastic transformation, although there is no direct evidence of these occurrences in humans.

The potential for interaction between HERVs and exogenous retroviruses exists at many levels. Recombination between an infectious retroviral provirus and an endogenous sequence could result in the formation of a new retrovirus with altered tissue tropisms or host ranges, a phenomenon clearly observed in mice and resulting in infectious murine leukemia viruses. Recombination need not be limited to DNA. Copackaging of HERV RNA with an exogenous retroviral RNA genome could lead to a chimeric retrovirus via template jumping during reverse transcription. Interestingly, polymerase chain reaction screens of human and other primate genomes with HIV-1 related probes have revealed the existence of endogenous sequences with significant homology to HIV *env*. Exogenous retroviruses could contribute gene products (e.g., reverse transcriptase, *env* glycoprotein) *in trans* that enable a defective endogenous retrovirus to transpose, or possibly even to bud from the cell. *Trans* complementation of endogenous elements has been demonstrated experimentally using an endogenous mouse virus. Intracellular transposition events were observed when HeLa cells were cotransfected with a genetically marked defective virus and a helper construct, which supplied the homologous *gag* and *pol* gene products.

Evolutionary biologists have suggested that HERVs may be involved in such processes as the evolution of gene function and the regulation of tissue-specific and developmentally regulated gene expression. The best documented example of HERV involvement in human gene regulation is the case of the amylase gene cluster. In the human genome, five linked genes encode the differentially expressed enzyme amylase. Two of the genes are expressed only in the pancreas, while the other three are expressed only in the parotid gland. Analysis of the sequences flanking each gene revealed than an identical HERV-E sequence is inserted, in opposite orientation, just upstream of each gene expressed in the parotid gland. Furthermore, when the first 670 bases of the HERV were placed upstream of a reporter gene in transgenic mice, the reporter was expressed specifically in the parotid gland. Thus particular HERV sequences can act as enhancers and confer tissue specificity to nonretroviral host genes. An interesting sidelight to these findings is that rodents (the only other family of mammals that expresses amylase in both the parotid gland and the pancreas) have a completely different arrangement of amylase genes and no nearby endogenous retroviral sequences. This then is an example of convergent evolution.

While most HERVs are clearly related to other retroviruses, the high-copy transposonlike human element (THE-1) family is an exception. Full-length copies are 2.3 kb long and are flanked by 350 bp LTRs. Large numbers of solo LTRs, thought to arise by homologous LTR-LTR recombination, are present throughout the genome. Circular forms, which may be also be products of LTR-LTR recombination, have been observed in HeLa cells. However, the internal sequences of the few copies so far examined contain no long open reading frames and bear no similarity to known retroviruses. Instead, two single-copy genes in the prosimian galago bear strong sequence similarity to THE-1, although they lack LTRs. It has been proposed that these cellular genes were recruited into a transposable element via a retrovirally mediated process. Although no de novo insertions of THE-1 have been reported, two cases of Duchenne muscular dystrophy have been shown to involve deletions in the dystrophin gene, with preexisting THE-1 sequences representing the proximal break point. Thus, THE-1 sequences may serve as sites for illegitimate recombination.

## 4 FUTURE PROSPECTS

While much has been learned about the structure and organization of transposons in the human genome, and there are now specific examples of interactions between these elements and their host, many questions remain unanswered. The mechanism by which LINEs and SINEs transpose is poorly understood. In particular, nothing is known about the process leading to integration into new loci. Likewise, it remains a mystery how these elements became so highly amplified within the genome and whether this amplification is an ongoing process. Do these elements serve a generally useful function or are they simply "selfish DNA"? Is somatic transposition an important component of neoplastic transformation, and do transposons significantly contribute to the generation of mutations leading to human genetic disease? The recent advances in the understanding of human transposons and their potential role in both evolution and disease pathogenesis should make further research in this area an exciting prospect.

*See also* Genetic Analysis of Populations; Genetics; Mammalian Genome; Recombination, Molecular Biology of.

### *Bibliography*

1. Berg, D. E., and Howe, M. M. (1989) *Mobile DNA*. American Society for Microbiology, Washington, DC.
The definitive reference on transposable elements. Particularly relevant chapters are by Varmus and Brown, Hutchison et al., and Deininger (Chapters 3, 26, and 27, respectively).
2. Additional reviews on specific topics relevant to human transposons include:
Deninger, P. L. et al. (1992) *Trends Genet.* 8:307–311. Fanning, T. G., and Singer, M. F. (1987) *Biochim. Biophys. Acta,* 910:203–212.
Larsson, E., Kato, N., and Cohen, M. (1989) *Curr. Top. Microbiol. Immunol.* 148:115–131.
3. Discussions of the phylogeny of retrotransposable elements can be found in:
Doolittle, R. F., Feng, D. F., Johnson, M. S., and McClure, M. A. (1989) *Q. Rev. Biol.* 64:1–30.
Xiong, Y., and Eickbush, T. H. (1990) *EMBO J.* 9:3353–3362.

# Tumor Suppressor Genes

*Prem Mohini Sharma and S. R. Dev Sharma*

### *Key Words*

**Cellular Oncogene** An oncogene formed by mutation or rearrangement of a proto-oncogene in a tumor.

**Chromosome Translocation** Exchange of segments between nonhomologous chromosomes.

**Familial Adenomatous Polyposis** A rare inherited form of colon cancer in which affected individuals develop multiple colon adenomas (polyps).

$G_0$ A quiescent state in which cells are not proliferating.

$G_1$ The stage of the cell cycle between the end of mitosis and the beginning of DNA synthesis.

$G_2$ The stage of the cell cycle between the end of DNA synthesis and the beginning of mitosis.

**Kinase** Enzyme that transfers phosphate groups, usually from ATP, to another molecule.

**Li–Fraumeni Family Cancer Syndrome** A rare hereditary cancer susceptibility, leading to the development of tumors multiple kinds of, which result from inherited mutations of the *p53* tumor suppressor gene.

**MCC** A tumor suppressor gene frequently inactivated in colon and rectum carcinomas.

**Neoplasm** An abnormal growth of cells.

***NF1*** A tumor suppressor gene responsible for inheritance of type 1 (von Recklinghausen) neurofibromatosis.

**Oncogene** A gene capable of inducing one or more characteristics of cancer cells.

**p53** A tumor suppressor gene that is lost or inactivated in a variety of neoplasms, including breast, colon, and lung carcinomas, sarcomas, and leukemias.

**Papilloma** A benign tumor projecting from an epithelial surface.

**Point Mutation** Alteration of a single nucleotide of DNA.

**Preneoplastic** A cell that displays increased proliferative potential and is capable of progressing to the full neoplastic phenotype as a result of further alterations.

**Proto-oncogene** A normal cell gene that can be converted to an oncogene.

***RB*** A tumor suppressor gene identified by genetic studies of retinoblastoma and also frequently inactivated in osteosarcomas, rhabdomyosarcomas, and bladder, breast, lung, and prostate carcinomas.

**Retinoblastoma** A childhood eye tumor.

**Sarcoma** A cancer of connective tissue.

**Transcription Factor** A protein that regulates transcription.

**Tumor Suppressor Gene** A gene that inhibits tumor development.

**Wilms' Tumor** A childhood kidney cancer.

*WT1* A tumor suppressor gene that is inactivated in Wilms' tumor.

---

The oncogenes are one of two distinct classes of genes that contribute to the development of human cancer. They arise from proto-oncogenes as a result of genetic alterations that either increase gene expression or lead to uncontrolled function of the oncogene-encoded proteins. The products of most proto-oncogenes act to stimulate normal cell proliferation, and the unregulated action of the corresponding oncogene proteins leads to the abnormal proliferation characteristic of cancer cells. The other class of genes that are important in carcinogenesis, the tumor suppressor genes, represent the opposite side of the coin of cellular growth control. Whereas oncogenes stimulate cell proliferation and tumor development, the tumor suppressor genes act to inhibit these processes. In many neoplasms, genetic alterations lead to the loss or inactivation of tumor suppressor genes, thereby eliminating inhibitors of cell proliferation. Activation of oncogenes and inactivation of tumor suppressor genes thus are complementary events in the development of cancer, both contributing to increased cell proliferation and loss of normal growth control.

The action of tumor suppressor genes is opposite to that of oncogenes. Tumor suppressor genes are inactivated in cancer cells, and the introduction of functional tumor suppressor genes from normal cells into cancer cells results in loss of tumorigenicity. In contrast, oncogenes are activated in cancer cells, and their introduction into normal cells leads to tumor formation. Not only do oncogenes and tumor suppressor genes both contribute to development of human cancers; the products of some tumor suppressor genes and oncogenes interact directly to regulate each other's expression and function within the cell.

## 1 DEFINITIONS: ONCOGENES, TUMOR SUPPRESSOR GENE

Two classes of genes are implicated in causing or helping to cause cancer. Although greater emphasis has been placed on the study of tumor-promoting oncogenes, the existence of tumor-suppressing genes has long been suspected. Dominant *oncogenes* are altered genes whose presence leads to neoplasia. *Tumor suppressor* genes are genes whose absence allows the malignancy to manifest. Every gene has a matching copy on the paired homologous chromosome. The gene and its copy are each called an *allele.* Oncogenes function in a genetically dominant manner: the presence of an alteration in just one allele is tumorigenic. In the recessively acting tumor suppressor genes, on the other hand, both alleles must be inactivated to cause loss of growth control.

The past several years have brought important discoveries regarding the genetic predisposition to cancer. Antioncogenes/tumor suppressor genes do exist and in fact do play a major role in human carcinogenesis. The current cloned and characterized human tumor suppressor genes include *RB,* p53, *WT1, APC, VHL, NF1,* and *NF2* (Table 1). Several putative genes of this class have been mapped to specific chromosomal bands, and many more are forthcoming. These discoveries have illuminated our understanding of human cancer considerably and have revealed a group of genes that is important in cellular and developmental biology, making this a propitious time to review progress on the subject and to take note of some future prospects for extending this understanding.

## 2 HISTORY

Evidence for the existence of tumor suppressor genes has converged from several distinct lines of work. Early evidence came from somatic cell hybridization studies. Geneticists took cultures of two cell lines, one a tumor cell line and the other normal, and fused them to form cell hybrids that retained the genomes of both parents. If the tumorigenic phenotype is dominant, as oncogene researchers expected, all the hybrids should be tumorigenic because they carry the oncogenes from the tumorigenic parent. But in fact, all the hybrids were normal. Therefore, the normal cells must possess genes that are capable of suppressing the neoplastic phenotype of their tumor cell partners.

Evidence that there were specific tumor suppressor genes in normal cells came from examining revertants of these hybrid cells that regained the tumorigenic phenotype. These hybrid cells had unstable karyotypes and frequently shed chromosomes. By examining hybrids that spontaneously regained the transformed phenotype, investigators could identify the normal chromosome (in human cells, commonly chromosome 11 or 13) that was always lost from such cells. These studies led to the concept that many tumors arise through loss of genetic material and provided the first clue that cancer cells often lose critical growth-regulating information during their progression toward full malignancy.

Human genetics provided a second clue that suggested the existence of tumor suppressor genes. The hereditary and sporadic forms of retinoblastoma led Knudson in 1971 to develop the two-hit hypothesis, in which he postulates that the development of retinoblastoma requires two rare mutations. In the sporadic form of the tumor, both mutations would have to occur within a single retinoblast, an exceedingly infrequent circumstance. In the inherited form, however, Knudson suggested that one of the mutations is already present (inherited) in all retinal cells. Therefore, only a single additional mutation is required for tumorigenesis.

A decade later, analysis of the chromosomes in tumor cells and normal tissues of retinoblastoma patients resoundingly confirmed Knudson's hypothesis. Many of these patients carried deletions in chromosome 13q14, and the mutated gene inherited by afflicted children, termed *RB,* was mapped to this same chromosome. Thus, development of retinoblastoma appeared to require that both copies of the *RB* gene be mutated. The *RB* gene must therefore be a tumor suppressor gene that normally functions to arrest the growth of retinal precursor cells. Even one copy is sufficient to keep growth in check. Loss of both copies of *RB* eliminates the block, and a tumor develops.

A third clue for discovering tumor suppressor genes was suggested by the genetic mechanisms used by evolving tumor cells to eliminate both copies of genes such as *RB.* The first copy of a suppressor gene is inactivated by a somatic (or a germ line) mutation. The chromosomal region carrying the surviving wild-type allele may then be replaced by a duplicated copy of the homologous chromosome region that carries the mutant allele. Most tumors that lack functional copies of a suppressor gene (such as *RB*) display two identically mutated (homozygous) alleles, while the unaffected tissues can be shown to carry one mutant *RB* allele and one normal one. Therefore, the loss of heterozygosity (LOH) at particular chromosome regions in tumor cells (compared with somatic cells from

**Table 1** Cloned Human Tumor Suppressor Gene: Their Chromosomal Location Protein Products and Characteristic Neoplasm Associated with Germ Line Mutations

| Gene | Chromosomal Location | Protein | Neoplasm |
|---|---|---|---|
| *RB1* | 13q14 | 110 kDa transcription modulator of $G_1 \rightarrow S$ transition | Sarcoma, retinoblastoma |
| *WT1* | 11p13 | 45 kDa transcription factor, zinc finger protein | Wilms' tumor |
| p53 | 17p13 | 53 kDa transcription factor, conditional regulator $G_1 \rightarrow S$ | Sarcoma, glioma, breast carcinoma |
| *NF1* | 17q11 | 327 kDa activator of *Ras* GTPase activity | Sarcoma, glioma |
| *NF2* | 22q12 | 66 kDa protein at membrane–cytoskeleton interface | Schwannoma |
| *VHL* | 3p25 | 34 kDa cell membrane protein | Pheochromocytoma, kidney carcinoma |
| *APC* | 5q21 | 310 kDa cytoplasmic protein | Colon carcinoma |
| *DCC* | | 153 kDa cell adhesion molecule | |

the same individual) is generally regarded as evidence for the unmasking of mutations in tumor suppressor genes located in these regions.

The finding of a loss of genetic information at the *RB* locus in association with retinoblastoma formation spurred investigators to examine other malignancies for LOH at *RB* and other chromosomal locations. With polymorphic DNA markers used to survey tumor cell genomes systemically, numerous malignancies show LOH; these include urological neoplasms such as Wilms' tumor, von Hippel–Lindau disease, renal cell carcinoma, breast carcinoma, colon carcinoma, and bladder cancer.

## 3 THE RETINOBLASTOMA GENE

The human retinoblastoma gene is one of the best characterized tumor suppressor genes and serves as a prototype for genes in this category. The locus governs the hereditary susceptibility to many human cancers, most notable of which is retinoblastoma, a malignant tumor arising in eyes of young children. A 180,388 base pair contig encompassing the human *RB* gene has been recently sequenced in its entirety. It is one of the largest contigs of sequenced human DNA reported to date. The gene consists of 27 exons spanning more than 180 kb of genomic DNA within human chromosome 13q14.12 or 13q14.2. The locus produces a 4.7 kb transcript that encodes a nuclear phosphoprotein consisting of 928 amino acids (Figure 1). The *RB* gene is conserved among vertebrates. Human and mouse *RB* share a 91% homology at the amino acid level. This homology extends into the promoter region, which contains potential sites for the transcription factors ATF, Sp1, and E2F.

The *RB* protein ($p110_{RB}$) is a nuclear protein of 110 kDa and is phosphorylated on both serine and threonine residues. The number of $^{32}P$-labeled tryptic peptides indicates that there are at least 10 phosphorylated amino acids. These modifications are thought to be important regulators of $p110_{RB}$ and have been shown to occur in a cell-cycle-dependent manner. Cells in early $G_1$ phase of the cycle contain exclusively unphosphorylated or underphosphorylated p110. At an as yet undefined point later in $G_1$, the protein is hyperphosphorylated ($pp110_{RB}$) and exists in this state until M phase, although the level of phosphorylation does fluctuate slightly. Several studies have implicated the cyclin-dependent kinase (cdk) family of kinases as enzymes responsible for this phosphorylation.

Inactivation of both alleles of this gene has been found in all retinoblastomas examined, and additional studies have implicated *RB* mutations in the development of a variety of tumors (e.g., in sarcomas, in small-cell carcinoma of the lung, and in carcinoma of the breast, bladder, and prostate). Introduction of a wild-type copy of this gene into $RB^-$ tumor cells suppresses their tumorigenicity in nude mice, providing further evidence that the cloned gene has properties consistent with those predicted for a tumor suppressor. The mechanism by which $p110_{RB}$ suppresses tumorigenicity remains unknown. However, a possible explanation emerged with the discovery that $p110_{RB}$ physically associates with oncoproteins of many DNA tumor viruses, namely, the adenovirus E1A protein, the SV40-T antigen, and the E7 protein of human papilloma 16 (HPV16), including the transcription factor E2F. Some $p110_{RB}$ sequences required for these interactions have been mapped to two

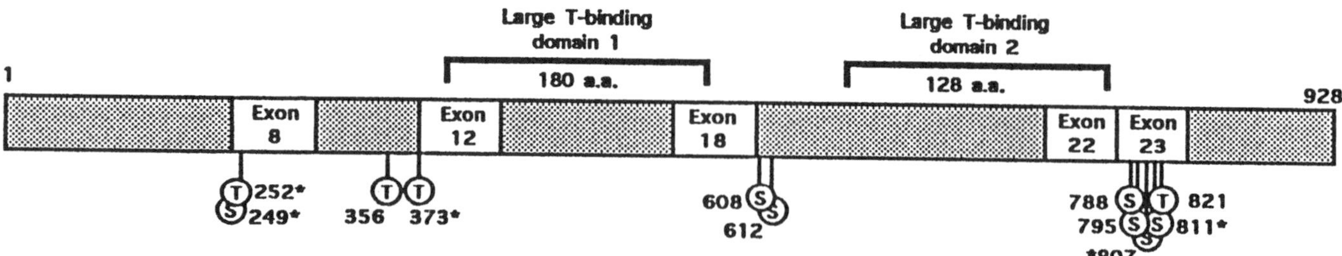


**Figure 1.** The human retinoblastoma protein. The 4.7 kb transcript of *RB* encodes a protein of 928 amino acids; specific exons of *RB* are indicated for reference. Binding of $p110^{RB}$ by viral transforming proteins, $T_{ag}$, or E1A requires two domains (indicated by brackets), one of 180 (large T-binding domain 1; amino acids 393–572) and another of 128 amino acids (large T-binding domain 2; amino acids 646–773). Naturally occurring mutations in *RB* disrupt one of these two domains. Considering conserved sequences between the human and murine proteins, the human protein has ten Thr or Ser residues; seven are preceded by basic or polar residues (Thr 252, Thr 356, Thr 373; Ser 612, Ser 788, Ser 795, Ser 811) and three by nonpolar residues (Ser 608, Ser 807, Thr 821). Five phosphorylation sites have been identified and are indicated by asterisks (*). It is interesting that all these sites are outside the $T_{ag}$/E1A-binding domains.

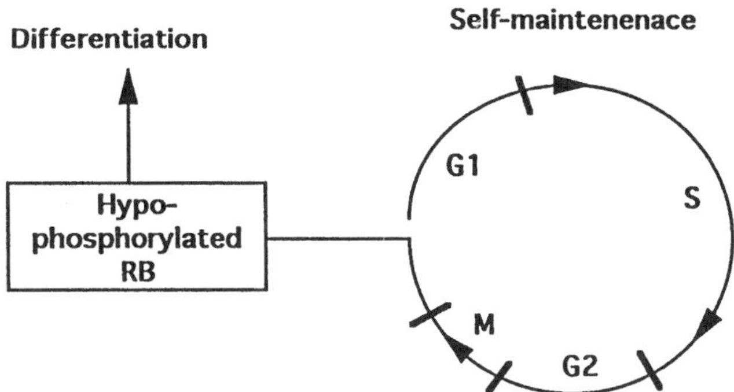


**Figure 2.** Model for the role of *RB* in the regulation of cell differentiation. Phosphorylation of *RB* allows progression through $G_1$ and continuation of a self-maintenance pathway. Un- or hypophosphorylated *RB* blocks cell cycle progression in the $G_1$ phase.

large domains in the carboxy-terminal portion of the protein. These domains are affected in all known naturally occurring mutant *RB* proteins, suggesting an important role for this region of the protein in tumor suppression.

*RB* mRNA is expressed in various tissues during mouse embryonic development, as early as day 10 of gestation. A high level of *RB* expression is detected in megakaryocytes, blood-forming cells of fetal liver, osteoblasts, skeletal muscle, and neurons and glial cells in the central nervous system. Thus, *RB* is expressed by cells undergoing differentiation to produce nondividing cells or giving rise to daughter cells different from the parental cell. *RB* gene knock out mice die in utero around day 14 of gestation. The embryos show disturbed neural and hematopoietic differentiation, indicating that *RB* is vitally important for these processes. This notion is further supported by studies demonstrating that *RB* expression in mouse embryo tissues is highest in cells undergoing differentiation and that *RB* is required for MyoD-induced muscle differentiation.

In summary, the *RB* protein acts as a cell cycle checkpoint at the $G_1$ phase and plays a vital role in the differentiation process, at least, in certain tissues. According to the model depicted in Figure 2, the hypophosphorylated form of *RB* is needed for permanent withdrawal from the cell cycle and initiation of a differentiation pathway.

## 4 THE p53 GENE

The human p53 gene spans about 20 kb of genomic DNA and contains 11 exons. The gene for p53 is localized on the short arm of human chromosome 17p13 and encodes a 393 amino acids (50–55 kDa) protein (Figure 3). The p53 gene is a tumor suppressor gene, negatively regulating the cell cycle by blocking the transit of cells through $G_1$, or from $G_1$ into S. Mutations in the p53 gene are the most common gene mutations identified in carcinomas to date. Human families with Li–Fraumeni syndrome, an inherited predisposition to cancer, have a mutated germ line p53 gene, a rare disorder in which affected patients are at high risk at early age for several different types of malignant lesions. In breast cancer, point mutations and/or small deletions in the p53 gene are found clustered in four ''hot spots'' located in exons 5, 7, and 8 (codons 130–290/domains II–V: Figure 3), which coincide with the four most highly conserved regions of the gene. Three codons (175, 248, 273) stand out as extreme hot spots in colon and breast tumors. The wild-type (wt) p53 can efficiently inhibit the in vitro transformation of primary rodent cells (rat embryo fibroblasts, REF) induced by a variety of oncogenes. In contrast, tumor-derived p53 mutants exhibit a complete loss of this inhibitory capacity. Mutant p53 proteins differ from each other and from the wild-type p53 in conformation, localization, and transforming potential. Mutations in codons 143, 175, and 275 confer strong transforming potential. A serine 135 p53 mutant has an intermediate transforming potential, while the histidine codon 273 allele transforms weakly, if at all. Many mutant p53 proteins are more stable than the normal or ''wild-type'' p53 protein. Because of their higher levels, mutant p53 proteins can be detected more easily than the wild type.

The mRNA transcribed from the p53 gene varies between 1.8 and 3.0 kb, depending on the species. The 3′ untranslated region is long and varies in size from 800 nucleotides for mice to 1800 bp for *Xenopus laevis*. The level of p53 mRNA in normal cells is highest in undifferentiated cells, spleen cells, cells undergoing rapid embryonic development (particularly during midgestation in the mouse), and other rapidly proliferating cell types. In vitro, the level of p53 mRNA appears to be down-regulated in tumor cell lines induced to differentiate and up-regulated in quiescent fibroblasts induced to divide by serum stimulation, suggesting that proliferative state of the cell is correlated with p53 mRNA levels. Overexpression of wild-type p53 in tumor cells can have one of three results: suppression of cell proliferation, induction of apoptosis

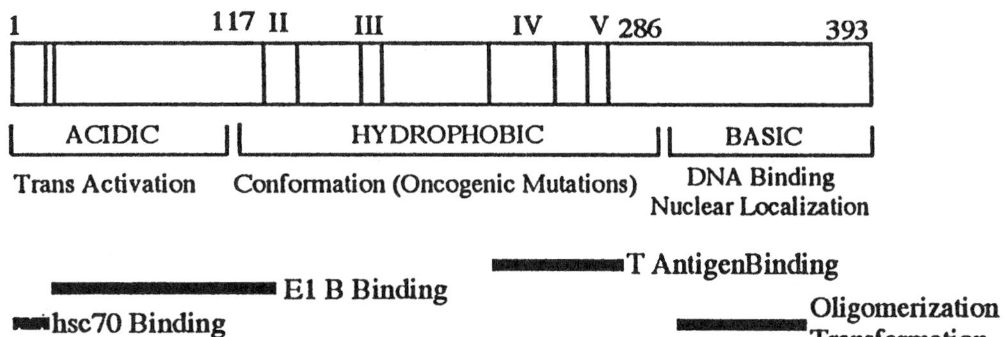


**Figure 3.** Structural features of the murine p53 protein. The 387 amino acid murine p53 protein contains five highly conserved domains, I–V. The protein is also divided into three major regions based on charge. The domains of the protein involved in oligomerization and binding to other viral and cellular proteins are also indicated.

(programmed cell death), or induction of differentiation. The pathway the tumor cell enters probably depends on the tumor cell type and levels of wild-type p53 expressed. Recently, the role for the p53 tumor suppressor gene in execution of apoptosis has been demonstrated in cells produced from animals with homozygous inactivation of the p53 gene.

The mechanisms underlying p53 growth suppression remain undefined. Recent efforts to further discern the function of p53 have centered on the underlying molecular basis for this tumor suppressor. In particular, research has focused on the identification of cellular molecules (specifically DNA and proteins) with which p53 protein associates. Several biochemical features of p53 have been elucidated, and at least two of these are currently of much interest. First, p53 has been shown to repress transcription of *RB*, c-*fos*, β-actin, p53, hsc70, c-*jun*, *MDR*1, and *PCNA* from viral and cellular promoters. This suppression is apparently sequence independent and may involve p53 binding to the TATA-binding protein or to other transcription factors. Second, p53 can bind to DNA in a sequence-specific manner. Kern and co-workers in 1991 demonstrated that p53 binds specifically to DNA fragments containing two or three repeats of TGCCT sequences, and the G residue of a 5 bp repeat is critical for this interaction. Proteins encoded by p53 mutants (e.g., $273Arg_{his}$, $175Arg_{his}$) lose this binding ability. More recently, a 20 bp consensus binding site, consisting of two copies of the 10 bp sequence 5′-PuPuPuC(A/T)(T/A)GPyPyPy-3′, separated by up to 13 bp, has been identified. Both copies of the 10 bp sequence are required for efficient binding by p53. The p53 protein contains a strong transcriptional activation sequence near its amino terminus and can stimulate the expression of genes downstream of its binding site. Such stimulation has been demonstrated in both mammalian and yeast cells, as well as in vitro systems.

The ability of p53 to activate transcription from specific sequences suggests that genes induced by p53 may mediate its biological role as a tumor suppressor. To date, several genes containing p53-binding sites have been identified. These include muscle creatine kinase (MCK), *GADD45, MDM2,* a GLN retroviral element, and *WAF1*. The relation of any of these genes to suppression of cell growth by p53 remains unclear. It has been suggested that *MDM2* may be a feedback regulator of p53 action by being transcriptionally induced and then inhibiting p53 function. In this regard, *MDM2* functions as an oncogene rather than as a tumor suppressor gene. Recently, a gene, highly induced by p53, named wild-type p53-activated fragment 1 *(WAF1)* has been identified. Introduction of *WAF1* cDNA suppressed the growth of human brain, lung, and colon tumor cells in culture. A p53-binding site was identified 2.4 kb upstream of *WAF1* coding sequences. *WAF1* is directly regulated by p53 and can itself suppress tumor cell growth in culture. *WAF1* may thus be an important component of the p53 growth suppression pathway.

Thus, p53 emerges as a DNA-binding protein with an important but undefined role in cell cycle regulation, functioning in the $G_1$–S time frame. It seems likely that new insights into its activities will be published in near future. Additional target genes for which p53 binds and activates or represses transcription will likely develop. Likewise, either by biochemical means or by use of yeast screens, new proteins that bind to p53 are almost sure to be identified. The challenge will then be to determine which of these interactions are physiologically relevant. Some of the more intriguing questions to be illuminated include the following:

What controls the levels of p53 in normal cells?

Why are mutant p53 proteins frequently much more stable in cells than wild-type form?

## 5 THE WILMS' TUMOR GENE, *WT1*

Wilms' tumor or nephroblastoma is the most common solid pediatric malignancy, affecting one in 10,000 young children. In this malignancy, an example of development gone awry, tumors arise from mesenchymal stem cells that normally differentiate into the epithelial components of the nephron. Morphologically, Wilms' tumors are characterized by the disorganized presence of most of the components of normal embryonic kidney: stroma, undifferentiated blastema, dysplastic tubules, and pseudoglomeruloid structures. Additional non-kidney-derived, mesenchymal elements such as muscle or bone are also frequently involved. The genetics and histopathology of this cancer indicate a complicated and heterogeneous etiology involving genes mapping to at least three different locations (11p13, 11p15, and an unmapped locus) in the human genome. A fourth location (16q) contains a gene probably involved in tumor progression, rather than initiation.

One of these genes, the Wilms' tumor *(WT1)* gene, was cloned from chromosomal region 11p13, from the smallest region of overlap in 1990. It was shown to be specifically expressed in the developing kidney. Mapping and isolation of this gene were achieved because deletions of 11p13 were identified in children with the WAGR syndrome, which is characterized by Wilms' tumor, aniridia, genitourinary abnormalities (in at least 50% of boys with the syndrome), and mental retardation. Although *WT*s with gross deletions in chromosome 11p13 were instrumental in localizing and molecular cloning of the *WT1* gene, subsequent work has revealed that the more frequently observed alterations in this locus are missense point mutations and both small and large deletions. Like the p53 tumor suppressor gene, many tumors retain the wild-type allele.

Spanning 50 kb of genomic sequence, the *WT1* gene is organized into 10 exons. The predicted *WT1* polypeptide consists of 449 amino acids and shows several features characteristic of transcription factors, such as nuclear localization, four contiguous Cys2/His2 zinc finger domains in the carboxy-terminal DNA binding domain, and an amino acid terminus rich in proline and glutamine residues (Figure 4). Alternative splicing at two sites—one inserts three amino acids (+/− KTS) between the third and fourth zinc fingers, and the second inserts 17 amino acids of exon 5 (+/− 17 aa)—produces four discrete transcripts. The proximal promoter region of *WT1* is GC-rich and contains CCAAT, Sp1, *NF1,* and *WT1/EGR1* sites. The mRNA is about 3.5 kb; however in some tissues significantly different-sized transcripts have been observed, indicating the presence of *WT1*-related genes.

The expression pattern of *WT1* observed in a number of species both supports its role as a tumor suppressor gene in the kidney and extends its possible functions to differentiation events in many other organs. During embryonic development, *WT1* transcripts are found in the kidney, where the condensing metanephric mesenchyme and primitive renal vesicles are formed, as well as the gonadal ridge mesothelia, and the mesothelial lining of the coelomic cavity and the organs it contains. *WT1* mRNA is also expressed in the spleen and in hematopoietic cell lines. Surprisingly, *WT1* expression is also observed in the spinal cord ventral horn motor neurons and in the area postrema of the rat brain. The latter pattern

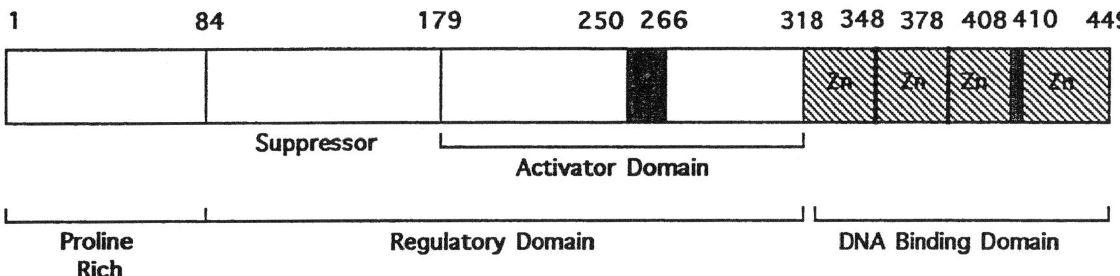


**Figure 4.** Structure of the *WT1* gene showing various functional domains. Four forms of the WT1 protein are produced by alternative splicing. The different forms comprise 429–449 amino acids and range in size from 49 to 54 kDa. Two alternative splices are indicated by solid rectangles: one leads to a 17 amino acid insertion (aa 250–266) upstream of the four zinc fingers (Zn). The other splice inserts three amino acids between the third and fourth zinc fingers. The most common form includes both inserts. The proline/glutamine-rich region, the transcription regulatory domain, and the DNA-binding domain of *WT1* also are indicated.

of expression in the central nervous system persists in the adult animal. The area postrema lacks a blood–brain barrier and may be a chemoreceptor trigger zone for circulation molecules such as angiotensin II. This raises the intriguing possibility that *WT1* participates in the renin–angiotensin cascade by integrating peripheral and central vascular control signals in the area postrema.

The *WT1* transcript undergoes RNA editing in rat in which $U^{839}$ is converted to C, resulting in the replacement of leucine 280 in *WT1* by proline. RNA editing at the same nucleotide is also observed in human testis. In in vitro assay, the *WT1*-leucine$^{280}$ polypeptide represses the *EGR1* promoter 30% more efficiently than the *WT1*-proline$^{280}$. Edited *WT1*-$C^{839}$ mRNA is barely detectable in neonatal kidney, whereas adult rat kidney contain both $U^{839}$ and $C^{839}$-*WT1* mRNA, suggesting a role for the two protein isoforms in growth and differentiation.

The zinc finger domains of *WT1* have sequence similarity to those of the early growth response (EGR) family of transcription factors (e.g., *EGR1* and *EGR2,* and others such as Sp1, krox-2, GLI, and Kruppel) and two of the *WT1* isoforms (−KTS, +/− exons 5) have been shown to bind the GC-rich, *EGR1* consensus sequence CGCCCCCGC. It results in transcriptional repression of the *EGR1* gene and possibly represses the transcription of genes activated by *EGR1.* A second functional *WT1* binding sequence (5′-TCCTCCTCCTCCTCTCC-3′), 3′ to the transcription initiation site of the gene for platelet-derived growth factor A chain (PDGF-A) has been recently identified by DNase I footprinting and gel mobility shift assays. Similar sequences with some divergence have also been identified within the promoters of other growth-related genes, including Ki-*ras,* epithelial growth factor receptor, insulin receptor, c-*myc,* and tumor growth b3.

The evidence of mRNA overexpression of IGF-2, PAX-2, MK, the PDGF, N-*myc,* and the gene for insulinlike growth factor receptor-1 gene (IGF 1-R), has prompted the examination of the notion of *WT1* as a transcriptional regulator of growth-promoting genes in the kidney by several groups. *WT1* protein binds multiple sites in the promoter sequences of IGF-2, PDGF-A, and IGF-1R, to function as a potent repressor and/or activator of transcription in vivo. *WT1* may function in one of two pathways, one of which promotes proliferation, leading to growth of the kidney substructural components during development. Once growth has been completed, the gene may function to limit growth factor production and to enforce a program of differentiation on the proliferating kidney cells. Supporting this line of reasoning is the recent report that maturation of the kidney as well as the urogenital system is seriously defective in recombinant mouse fetuses lacking functional *WT1.* On the other hand, functional loss of *WT1* transcriptional–repressor activity, or a gain of transactivation function by mutation, may result in untimely synthesis of the growth factors in kidney blastemal cells. The unrestrained autocrine growth stimulation could lead to the genesis of Wilms' tumors.

Because of the rarity of Wilms' tumors and their early onset, it has been difficult to decipher the genetic events operative during the initiation and progression of the disease. We have established an animal model system that fulfils the criteria of both the phenotype and genotype of embryonal kidney tumors and would be a valuable tool for furthering the understanding of this type of cancer. Newborn rats administered the direct-acting chemical carcinogen NMU frequently develop kidney tumors with a latency of 4–8 months. Point mutations were found in the *WT1* gene in 7 of 18 tumors. The mRNAs for *WT1,* IGF-2, Pax-2, and MKs, which are all overexpressed in Wilms' tumors, were expressed at levels two- to sevenfold higher in 60% of the tumors compared to newborn kidney. The histopathology of the rat kidney tumors and the presence of genetic alterations in multiple loci are reminiscent of those in Wilms' tumors. These findings establish this as a relevant model system for the human disease. Since the process of tumor formation in these animals mimics tumorigenesis in humans, the rats provide valuable resources for dissecting the multiple steps in carcinogenesis. They also serve as tumor models for testing new treatment strategies such as gene therapy.

In summary, *WT1* serves as a paradigm for how alterations of transcription factor function can lead to oncogenesis. The picture that is emerging suggests that unlike p53 and *RB, WT1* is a highly tissue-specific and developmentally regulated transcription factor that functions as a tumor suppressor gene. *WT1* probably plays a primary role in initiating or maintaining a mesenchymal–epithelial differentiation program in kidney. Loss of *WT1* function leads to no or aberrant differentiation, resulting in unrestrained growth of metanephric blastemal cells in the embryonic kidney, leading to Wilms' tumor. Further work on *WT1* and Wilms' tumor must address the following questions:

Is WT1 involved in metabolic pathway that contains other Wilms' tumor susceptibility genes?

What are the natural downstream target genes of the WT1 protein?

What are the critical protein–protein interactions in the cell that dictate whether *WT1* will activate or repress transcription?

## 6 THE NEUROFIBROMATOSIS TYPE 1 AND 2 (*NF1* AND *NF2*) GENES

Neurofibromatosis is a term used to describe two major human genetic disorders, types 1 and 2 (NF1 and NF2), both of which display autosomal dominant inheritance and involve tumors of the nervous system, but are distinct clinical entities. One of the most common dominantly inherited human diseases, NF1 or von Recklinghausen NF, affects one in 4000 children, half of whom carry new germ line mutants. Neurofibromatosis is characterized by the highly variable expression of an array of features that include neurofibromas, cafe-au-lait macules, Lisch nodules of the iris, and a predisposition to certain malignant tumors such as schwannoma, glioma, pheochromocytoma, and, at low frequency, leukemia, rhabdomyosarcoma, and Wilms' tumor. It is a hereditary disease that is probably caused by the germ line mutations in one allele of the *NF1* gene.

The *NF1* gene mapped to chromosome 17q11 has been cloned and sequenced. The gene is large, encoding a 13 kb transcript and a ubiquitously expressed protein, neurofibromin, of at least 2485 amino acids, with a region of considerable homology to the catalytic domain of $p120_{GAP}$. The gene's function is similar to that of the GAP (GTPase-activating protein) gene product. Loss of activity results in failure of hydrolysis of GTP to GDP by the RAS protein. This loss of neurofibromin function is thought to result in elevated levels of the GTP-bound RAS protein, which transduces certain signals for cell division; the disruption of these signals is believed to lead to tumor formation.

The high germinal mutation rate of *NF1* may be attributed to its heavy methylation. The mutations observed in the gene frequently produce truncated, functionless proteins. In a significant fraction of the characteristic malignant tumors of *NF1,* the second (normal) copy of the gene appears to be mutated or absent. Mutations of the p53 are also observed in some cases of neurofibrosarcomas, as often observed in other sarcomas. Wild-type *NF1* again exemplifies the inhibition of an oncogene by an antioncogene. The tissue-specific existence of both GAP and *NF1* suggests redundancy in the regulation of RAS protein. It is possible that the tissues that are susceptible to tumorigenesis have little GAP function, leaving *NF1* to be sole regulator.

Neurofibromatosis 2 (NF2), by contrast, is a much less common disorder than NF1; it occurs in about one of 40,000 live births and shows a high penetrance. NF2 is a dominantly inherited disorder characterized by the occurrence of bilateral vestibular schwannomas and other central nervous system tumors including multiple meningiomas. Juvenile cataracts are also frequently seen in NF2. Genetic linkages studies and investigations of both sporadic and familial tumors suggest that NF2 is caused by inactivation of a tumor suppressor gene in chromosome 22q12. Sporadic, nonhereditary forms of these tumors are often monosomic for chromosome 22. A candidate NF2 gene has recently been cloned that encodes a 587 amino acid protein. Its protein product is unusual for tumor suppressor genes in that it is homologous with proteins found at the interface between the plasma membrane and the cytoskeleton interface. Mutations in the tumor typically cause truncation of the protein product; *NF2* clearly qualifies as an antioncogene, although its precise mechanism of action is not yet known.

## 7 THE VON HIPPEL–LINDAU (VHL) SYNDROME AND RENAL CELL CARCINOMA

Von Hippel–Lindau disease is a familial cancer syndrome that is dominantly inherited and predisposes affected individuals to a variety of tumors. The most frequent tumors are hemangioblastoma of the central nervous system and retina; less common are bilateral pheochromocytoma and renal cell carcinoma (RCC). The minimum birth incidence of the disease is one in 36,000; penetrance is almost completely by 65 years of age, and median actuarial life expectancy is reduced to 49 years, with RCC being the most common cause of death. The *VHL* gene localized to chromosome 3p25 by linkage analysis has been cloned recently. The *VHL* gene is evolutionarily conserved and encodes two widely expressed transcripts of approximately 6 and 6.5 kb. The partial sequence of the inferred gene product contains an acidic repeat domain found in the procyclic surface membrane glycoprotein of *Trypanosoma brucei.* Its protein product appears to be a cell surface molecule involved in cell adhesion and signal transduction. Abnormalities of the *VHL* gene are observed in a high percentage of nonhereditary renal cell carcinomas, making it the principal gene for RCC. Both these nonhereditary tumors and the tumors found in the syndrome typically reveal mutation or loss of the second copy of the gene, as expected for an antioncogene.

## 8 FAMILIAL ADENOMATOUS POLYPOSIS AND COLON CANCER

The classic syndrome of familial colorectal cancer is that associated with dominant, autosomally inherited premalignant syndrome—namely, familial adenomatous polyposis (FAP), caused by germ line mutation in the adenomatous polyposis coli *(APC)* gene. FAP has a prevalence of around one in 10,000, whereas various lines of evidence suggest that other major genes may be responsible for a significant proportion of colorectal cancers, and familial studies show a threefold increased risk in first-degree relatives of colorectal probands.

FAP, which is transmitted by the *APC* gene, is characterized by the development of hundreds, or carpets of thousands, of colorectal adenomas, some of which become carcinomas. Mapping of the *APC* gene to 5q21-22 by linkage studies in families allowed its identification and the characterization of mutations in affected individuals. In virtually all cases, without prophylactic colonectomy, progression to colorectal cancer occurs and death ensues at an average age of 42 years. In addition to almost universal upper gastrointestinal neoplasia, extracolonic manifestations such as osteomas, epidermoid cysts, desmoid tumors, and congenital hypertrophy of the retinal pigment epithelium (CHRPE) are common. Ocular examination revealed that patients expressing CHRPE tend to cluster within specific families. The extent of CHRPE is dependent on the position of the mutation along the coding sequence. CHRPE lesions are almost always absent if the mutation occurs before exon 9, but systematically present if it occurs after this exon. Thus, the range of phenotypic expression observed among affected patients may result in part from different allelic manifestations of *APC* mutations.

The *APC* mRNA is expressed in normal human and rodent colorectal mucosa and a variety of other tissues. The *APC* gene contains an 8538 bp open reading frame and is predicted to encode a 2843 amino acid polypeptide with a few homologies to other proteins. Its protein product of about 300 kDa is located in the cytoplasm. The *APC* germ line mutations described to date are notable in that they are almost exclusively either single base pair changes leading to termination codons or small deletions, insertions, or splicing mutations causing translational frameshifts and subsequent downstream stops.

A less severe form of familial polyposis, attenuated adenomatous polyposis coli (AAPC), is characterized by a low number of adenomatous polyps (usually fewer than 100), yet patients sustain an elevated risk of colon cancer. The mutant alleles responsible for this attenuated phenotype have been mapped in several families to the adenomatous polyposis coli *(APC)* locus on human chromosome 5q. Four distinct mutations in the *APC* gene have now been identified in seven AAPC families. These mutations, which predict truncation products either by single base pair changes or frameshifts, are similar to mutations identified in families with classical APC. However, they differ in that the four mutated sites are located very close together and nearer the 5′ end of the *APC* gene than any base substitution or small deletions yet discovered in patients with classical APC.

The *MCC* (mutated in colon cancer) gene, a candidate gene located at or near the APC locus and found to be mutant in some colon carcinomas, has been shown to map outside the segments deleted in some APC patients. Three candidate genes—*DP*1, *DP*2, and *DP*3—and the *MCC* genes are located within the deleted region. *MCC* encodes an 829 amino acid protein with a short region of similarity to a G protein (cf. *NF1*). In the absence of confirmation of a constitutional mutation in the *MCC* gene in FAP patients, however, the question of whether the *MCC* gene is indeed the *APC* gene is still open.

Mutation at the *APC* locus is a necessary though not sufficient condition for carcinoma of the colon. Other genetic events are the rule for this tumor, the target being the *KRAS* oncogene, the p53 antioncogene, and the *DCC* (deleted in colon cancer) antioncogene. *KRAS* mutations are featured in larger polyps, where their incidence is 40%. They are rare in very small polyps, and their incidence in carcinomas is about the same as in large polyps. On the other hand, p53 and *DCC* mutations occur uncommonly in polyps but are observed in the majority of carcinomas. It has not been established whether there is a necessary sequence for mutation in these two genes.

The *DCC* gene on chromosome 18q, a region that shows heterozygous loss in more than 80% of colorectal carcinomas, is expressed in most normal tissues but with a greatly reduced or absent expression in colorectal carcinomas. The *DCC* gene encodes a putative translation start site, signal peptide, and hydrophobic transmembrane region dividing the predicted protein into extracellular (1100 amino acid) and intracellular (324 amino acid) domains. Several tissue-specific *DCC* transcripts may be generated through alternative splicing of the primary transcript. The coding region is composed of at least 28 exons and is flanked by lengthy 5′ and 3′ untranslated regions, accounting for a significant portion of the 10–12 kba *DCC*-encoded message. The deduced amino acid sequence of *DCC* shows that it encodes a 190 kDa transmembrane phosphoprotein having attributes of a cell surface receptor.

Mutations are detected in the *DCC* gene, and the predicted amino acid is closely similar to neural cell adhesion molecules (NCAMs) and other cell surface glycoproteins. The characterized mutations usually produce termination codons that in turn lead to truncated, presumably nonfunctional, proteins. No normal copies of *DCC* remain in the tumor cells, as expected for an antioncogene. This is the only antioncogene that has been cloned without the availability of constitutional mutations. So far there is no hereditary condition in which *DCC* is mutant. The gene, for example, is expressed in neural tissues as well as in the gastrointestinal tract. One patient with a constitutional deletion of 18q, the arm that contains the *DCC* gene, developed a brain tumor. Could it be that germ line mutations of *DCC* are associated with brain tumors rather than with colon cancer?

Several lines of evidence suggest that cancers require multiple oncogene mutations to produce a metastatic tumor. For example, six independent mutations have been documented for the development of colon tumors. This number of mutations appears to be excessively high to be accounted for by the normal frequency of spontaneous mutations in cells. A role for the human mutator gene homologue, hMSH2 has been suggested in the development of cancer, with the thought that a defect in the mismatch repair system may lead to an increased rate of accumulation of spontaneous mutations. Recently, a human homologue of the bacterial MutS and *S. cerevisiae* MSH proteins has been identified. This homologue, hMSH2 maps to human chromosome 2p22-21 near a locus implicated in hereditary nonpolyposis colon cancer (HNPCC). A T-to-C transition mutation has been detected in the −6 position of a splice acceptor site in sporadic colon tumors and in affected individuals of two small HNPCC kindreds. The similarity between *S. cerevisiae* msh2 mutants and HNPCC patients with regard to instability of dinucleotide repeat sequences, the correspondence of map locations of HNPCC and the hMSH2 gene, and the detection of a specific mutation in small HNPCC kindreds, as well as sporadic colorectal tumors, suggest that mutations in the hMSH2 gene are responsible for HNPCC.

## 9 SEARCH FOR NEW ANTIONCOGENES

So far eight antioncogenes (*RB1, WT1, VHL, APC, DCC, TP53, NF1,* and *NF2*) have been cloned. All except *DCC* have been found to be mutated in the germ lines of persons predisposed to one or more tumor types. Another seven putative antioncogenes (*BRCA1, RCC, NB1, MLM, MEN1, BCNS,* and *LC1*) have been mapped but not cloned (Table 2). Germinal mutations in all except *LC1* are known to predispose bearers to tumors. Obviously the analysis of hereditary cancer provides excellent entry to the world of antioncogenes.

**Table 2** Uncloned Human Suppressor Gene: Chromosomal Location, Protein Products, and Characteristic Neoplasm

| Gene | Chromosomal Location | Neoplasm |
|---|---|---|
| *NB*1 | 1p36 | Neuroblastoma |
| *MLM* | 9p21 | Melanoma |
| *MEN*1 | 11q13 | Pituitary adenoma |
| *BCNS* | 9q31 | Medulloblastoma, skin carcinoma |
| *RCC* | 3p14 | Kidney carcinoma |
| *BRCA*1 | 17q21 | Breast, ovary carcinoma |

## 10 TRANSGENIC ANIMAL MODELS

The recent development of animal models to knock out tumor suppressor genes further substantiated the idea of cancer suppression. Either p53 or *RB* gene disruption results in mice that are prone to cancer formation. Targeted inactivation of the *WT1* locus in mice leads to death in utero of homozygous mutant *WT1* embryos, which fail to develop kidneys and gonads. However, mice heterozygous for the mutant *WT1* allele undergo normal development and do not develop any malignancies. Obviously, the loss of one allele of *WT1* in the mouse is insufficient for kidney tumor initiation or the urogenital anomalies typical of WAGR patients.

## 11 GENE THERAPY

Somatic gene therapy is defined as the insertion of an exogenous normal gene into somatic cells to correct an abnormality or deficiency of a specific protein. This can be carried out by transfecting the new gene in the presence of an abnormal gene (additional therapy) or by attempting to replace a defective gene by inserting a new one at the same site, using homologous recombination. The genetic information will not be passed on to future generations. Among the formal selection criteria that have been established for gene therapy are the following: (1) the disease is life-threatening; (2) the gene responsible has been cloned; (3) its precise regulation is not required; and (4) a suitable delivery system is available. Considerable effort is going into somatic gene therapy for single gene disorders such as adenosine deaminase immunodeficiency and thalassemia, diseases that might seem to be more amenable to this approach than a complex polygenic disorder such as cancer. Several systems have been explored in malignancy, however, with a view to enhancing the selective destruction of tumor versus normal cells.

The other type of gene therapy, called germ cell transfection or germ cell therapy, poses considerable ethical dilemmas and probably has a less important part to play in cancer therapy. However, at some point in the future it might be used to prevent cancers in individuals carrying defective tumor suppressor genes (e.g., the *RB* gene in retinoblastoma and p53 in Li–Fraumeni syndrome). In addition, gene transfer techniques can be applied to target prodrug activation specifically to tumor cells and also to protect normal tissues against toxic chemotherapy.

## 12 CONCLUDING COMMENTS

It is evident that the term ''tumor suppressor genes'' is applicable to a wide variety of genes involved in normal cellular functioning. The unifying characteristic appears to be the necessity of having one functioning wild-type allele to prevent abnormal proliferation or differentiation at particular stages of cell growth. It is this feature that permits normal activity into a hemizygous state and underlies the association of some of these genes with certain inherited cancer predispositions. To date the genes identified in this category are of diverse function. Some may act within a narrow range of cell types and consequently be involved in the etiology of relatively few tumor types, as with *WT1*. Others, of fundamental importance to the growth and maintenance of a broad range of cells, will be found to be implicated in many tumors as with p53 and *RB*. Since many of these genes appear to act as negative regulators of cellular proliferation, and their presence in a single copy is sufficient for the normal control of proliferation, they may well offer an important approach for the future therapeutic control of abnormal growth.

The continuous efforts in cloning and characterizing new tumor suppressor genes should yield exciting and worthwhile results. In combination with investigation of dominant-acting oncogenes, a much clearer picture of the mechanisms involved in positive and negative signal interactions should emerge. Understanding these mechanisms will provide the basis for development of novel methods of clinical intervention. The hope, of course, is that the use of tumor suppressor genes and/or their products will emerge as an effective clinical strategy in the effort to cure and prevent many forms of human cancer.

*See also* CANCER; ONCOGENES; RETINOBLASTOMA, MOLECULAR GENETICS OF.

### *Bibliography*

Anderson, M. J., and Stanbridge, E. (1993) Tumor suppressor genes studied by cell hybridization and chromosome transfer. *FASEB J.* 7:826–833.

Cooper, G. M. (1992) *Elements of Human Cancer.* Jones and Bartlett, Boston.

Klein, G. (1993) Genes that antagonize tumor development. *FASEB J.* 7:821–825.

Knudson, A. G. (1993) Antioncogenes and human cancer. *Proc. Natl. Acad. Sci. U.S.A.* 90:10914–10921.

Weinberg, R. A. (1991) Tumor suppressor genes. *Science,* 254:1138–1145.

*RB*

Goodrich, D. W., and Lee, W.-H. (1993) Molecular characterization of the retinoblastoma susceptibility gene. *Biochim. Biophys. Acta,* 1155:43–61.

Hamel, P. A., Philips, R. A., Muncaster, M., and Gallie, B. L. (1993) Speculations on the roles of *RB1* in tissue-specific differentiation, tumor initiation, and tumor progression. *FASEB J* 7:846–854.

Hollingsworth, R. E., Jr., Hensey, C. E., and Lee, W.-H. (1993) Retinoblastoma protein and the cell cycle. *Curr. Opin. Genet. Dev.* 3:55–62.

Toguchida, J., McGee, T. L., Paterson, J. C., Eagle, J. R., Tucker, S., Yandell, D. W., and Dryja, T. P. (1993) Complete genomic sequence of the human retinoblastoma susceptibility gene. *Genomics,* 17:535–543.

*p53*

Donehower, L. A., and Bradley, A. (1993) The tumor suppressor p53. *Biochim. Biophys. Acta,* 1155:181–205.

El-Deiry, W. S., Tokino, T., Velculescu, V. E., Levy, D. B., Parsons, R., Trent, J. M., Lin, D., Mercer, E., Kinzler, K. W., and Vogelstein, B. (1993) WAF1, a potential mediator of p53 tumor suppressor. *Cell,* 75:817–825.

Kern, S. E., Kinzler, K. W., Bruskin, A., Jarosz, D., Friedman, P., Prives, C., and Vogelstein, B. V. (1991) Identification of p53 as a sequence-specific DNA-binding protein. *Science,* 252:1708–1711.

Kreidberg, J. A., Sariola, H., Loring, J. M., Madea, M., Pelletier, J., Housman, D., and Jaenisch, R. (1993) *Cell,* 74:679.

Oliner, J. D. (1993) Discerning the function of p53 by examining its molecular interactions. *BioEssays,* 15:703–707.

Prives, C., and Manfredi, J. J. (1993) The p53 tumor suppressor protein: Meeting review. *Genes Dev.* 7:529–534.

Zambetti, G. P., and Levine, A. (1993) A comparison of the biological activities of wild-type and mutant p53. *FASEB J.* 7:855–865.

*WT1*

Campbell, C. E., Huang, A., Gurney, A. L., Kessler, P. M., Hewitt, J. A., and Williams, B. R. G. (1994) Antisense transcripts and protein binding motifs within the Wilms' tumor (WT1) locus. *Oncogene,* 9:583–595.

Hastie, N. D. (1993) Wilms' tumor gene and function. *Curr. Opin. Genet. Dev.* 3:408–413.

Huff, V., and Saunders, G. (1993) Wilms' tumor genes. *Biochim. Biophys. Acta,* 1155:295–306.

Rauscher, F. J., III, (1993) The WT1 Wilms' tumor gene product. A developmentally regulated transcription factor in the kidney that functions as a tumor suppressor. *FASEB J.* 7:896–903.

Sharma, P. M., Yang, X., Bowman, M., Roberts, V., and Sukumar, S. (1992) Molecular cloning of rat Wilms' tumor complementary DNA and a study of messenger RNA expression in the urogenital system and the brain. *Cancer Res.* 52:6407–6412.

———, Bowman, M., Madden, S. L., Rauscher, F. J., and Sukumar, S. (1994) RNA editing in the Wilms' tumor susceptibility gene. WT1. *Genes Dev.* 8:720–731.

———, Bowman, M., Yu, B.-F., and Sukumar, S. (1994) An animal model for human Wilms' tumors: Embryonal kidney neoplasms induced in rats by *N*-Nitroso *N'*-methylurea. *Proc. Natl. Acad. Sci., U.S.A.* (in press).

*NF1 and NF2*

Cawthon, R. M., Weiss, R., Xu, G., Viskochil, D., Culver, M., Stevens, J., Robertson, M., Dunn, D., Gesteland, R., O'Connell, P., and White, R. (1990) A major segment of the neurifibromatosis type 1 gene: cDNA sequence, genomic structure, and point mutations. *Cell,* 62:193–201.

Trofatter, J. A., MacCollin, M. M., Rutter, J. L., Murrell, J. R., Duyao, M. P., Parry, D. M. Eldridge, R., Kley, N., Menon, A. G., Pulaski, K., Haasa, V. H., Ambrose, C. M., Murone, D., Bove, C., Haines, J. L., Martuza, R. L., MacDonald, M. E., Seizinger, B. R., Short, M. P., Buckler, A. J., and Gusella, J. F. (1993) A novel moesin, -ezrin-, rodixin-like gene is a candidate for the neurofibromatosis 2 tumor suppressor. *Cell,* 72:791–800.

Wallace, M. R., Marchuk, D. A., Andersen, L. B., Letcher, R., Odeh, H. M., Saulino, A. M., Fountain, J. W., Brereton, A., Nicholoson, J., Mitchell, A. L., Brownstein, B. H., and Collins, F. S. (1990) Type 1 neurofibromatosis gene: Identification of a large transcript disrupted in three NF1 patients. *Science,* 249:181–186.

Xu, G., O'Connell, P., Viskochil, D., Cawthon, R., Robertson, M., Culver, M., Dunn, D., Stevens, J., Gesteland, R., White, R., and Weiss, M. (1990) The neurofibromatosis type 1 gene encodes a protein related to GAP. *Cell,* 62:599–608.

*VHL*

Latif, F., Troy, K., Gnarra, J., Yao, M., Duh, F.-M., Orcutt, M. L., Stackhouse, T., Kuzamin, I., Modi, W., Geil, L., Schmidt, L., Zhou, F., Li, H., Wei, M. H., Chen, F., Glenn, G., Choyke, P., Walther, M. M., Weng, Y., Duan, D.-S. R., Dean, M., Glavac, D., Richards, F. M., Crossey, P. A., Ferguson-Smith, M. A., Paslier, D. L., Chumakov, I., Cohen, D., Chinault, C., Maher, E. R., Linehan, M., Zbar, B., and Lerman, M. I. (1993) Identification of the von Hippel–Lindau disease tumor suppressor gene. *Science* 260:1317–1320.

*FAP, APC, DCC, and MSH2*

Cleaver, J. E. (1994) It was a very good year for DNA repair. *Cell,* 76:1–4.

Fearon, E. R., Cho, K. R., Nigro, J. M., Kern, S. E., Simons, J. W., Ruppert, J. M., Hamilton, S. R., Preisinger, A. C., Thomas, G., Kinzler, K. W., and Vogelstein, B. (1990) Identification of a chromosome 18q gene that is altered in colorectal cancers. *Science,* 247:49–56.

Fishel, R., Lescoe, M. K., Rao, M. R. S., Copeland, N. G., Jenkins, N. A., Garber, J., Kane, M., and Kolodner, R. (1993) The human mutator gene homolog MSH2 and its association with hereditary nonpolyposis colon cancer. *Cell,* 75:1027–1038.

Gorden, J., Thliveris, A., Samowitz, W., Carlson, M., Gelbert, L., Albertsen, H., Joslyn, G., Stevens, J., Spirio, L., Robertson, M., Sargeant, L., Krapcho, K., Wolff, E., Burt, R., Hughes, J. P., Warringhton, J., McPherson, J., Wasmuth, J. H., Paslier, D. L., Abderrahim, H., Cohen, D., Leppert, M., and White R. (1991) Identification and characterization of the familial adenomatous polyposis coli gene. *Cell,* 66:589–600.

Joslyn, G., Carlson, M., Thliveris, A., Albertsen, H., Gelbert, L., Samowitz, W., Gorden, J., Stevens, J., Spirio, L., Robertson, M., Sargeant, L., Krapcho, K., Wolff, E., Burt, R., Hughes, J. P., Warringhton, J., McPherson, J., Wasmuth, J. H., Paslier, D. L., Abderrahim, H., Cohen, D., Leppert, M., and White, R. (1991) Identification of deletion mutations and three new genes at the familial polyposis locus. *Cell,* 66:601–613.

Kinzler, K. W., Nilbert, M. C., Vogelstein, B., Bryan, T. M., Levy, D. B., Smith, K. J., Preisinger, A. C., Hamilton, S. R., Hedge, P., Markham, A., Carlsom, M., Joslyn, G., Gorden, J., White, R., Miki, Y., Moyoshi, Y., Nishisho, I., and Nakamiru, Y. Identification of a gene located at chromosome 5q21 that is mutated in colorectal cancers. *Science,* 251:1366–1370.

Spirio, L., Olschwang, S., Gorden, J., Robertson, M., Samowitz, W., Joslyn, G., Gelbert, L., Thliveris, A., Carlson, M., Otterud, B., Lynch, H., Watson, P., Lynch, P., Laurent-Puig, P., Burt, R., Hughes, J. P., Thomas, G., Leppert, M., and White, R. (1993) Alleles of the APC gene: An attenuated form of familial polyposis. *Cell,* 75:951–957.

# Ultraviolet Radiation Damage to DNA

*David L. Mitchell*

### *Key Words*

**Abasic** A site in DNA from which a purine or pyrimidine base has been removed, either by the direct action of UVR or during the base excision repair process (i.e., by aglycosylation).

**Dimerization** Covalent bonding of adjacent purine or pyrimidine bases resulting from the resolution of the unstable electronic configuration created by absorption of a photon within the UVR range of light.

**Endonuclease** An enzyme, often associated with the excision repair process, that cleaves a phosphodiester bond to either the 3′ or 5′ side of a base or abasic site in DNA.

**Photohydrate** Type of monobasic photoproduct to which a hydroxyl (—OH) group has been added to the 5 or 6 carbon of a pyrimidine base.

**Photoproduct** A stable change in DNA structure involving a single base or adjacent bases resulting from the dissipation of absorbed UVR.

**Ultraviolet Radiation (UVR)** Wavelengths of the electromagnetic spectrum ranging from 190 to 400 nm; UVR is subdivided into vacuum UV (190–240 nm), UVC (240–280 nm), UVB (280–320 nm), and UVA (320–400 nm).

Air pollution has resulted in global decreases in stratospheric ozone concentrations and an increase in the amount of harmful solar ultraviolet radiation (UVR) reaching the earth's surface. The effects of increased UVR on the human population are complex. The obvious and direct consequences include increased skin cancer and accelerated aging; less obvious and more indirect effects include deterioration of natural ecosystems and world food crops.

## 1 DNA PHOTOPRODUCTS

The sun emits energies at wavelengths that range through 11 orders of magnitude, from 1 nm to 100 m. The vast majority of this energy is biologically irrelevant; ionizing radiations such as high energy particles, X-rays, and γ-rays are expended by atomic collisions in the upper atmosphere, and long wavelength far infrared and microwaves do not have sufficient energy to influence biochemical reactions. The UV spectrum is divided into four regions: vacuum UV (190–240 nm), UVC (240–290 nm), UVB (290–320 nm), and UVA (320–400 nm). Absorption of UVC radiation by stratospheric ozone greatly attenuates these wavelengths, with the result that very little light shorter than 300 nm reaches the earth's surface. Hence, although comprising only a negligible portion of the solar spectrum (i.e., $< 0.0000001\%$), UVB light is responsible for most of the sun's pathological effects.

Because of its maximum absorbance at approximately 260 nm, DNA is considered the major cellular target of UVR; the absorption spectrum of DNA correlates well with photoproduct formation, cell killing, mutation induction, and tumorigenesis. The energy absorbed by DNA is dissipated by various routes, some of which involve single bases (i.e., monomeric damage, single-strand breaks, and abasic sites), others result in interactions between adjacent bases (i.e., dimerizations) as well as between nonadjacent bases (i.e., inter- or intrastrand cross-links), and still others produce interactions between DNA and associated proteins (i.e., DNA–protein cross-links) (Figure 1).

Dimerizations between adjacent thymine and cytosine bases are by far the most prevalent photoreactions resulting from UVC or UVB irradiation of DNA. The relative induction of these photoproducts depends on wavelength, DNA sequence, and protein–DNA interactions. The two major photoproducts are the cyclobutane dimer **(I)** and the (6-4) photoproduct **(II),** named from the types of chemical linkage binding them together. The (6-4) photoproduct undergoes a further UVB-dependent conversion to its valence photoisomer, a Dewar pyrimidinone **(III).** The cyclobutane dimer and (6-4) photoproduct are quite different in their photochemistry (Figure 1), hence their photobiology is quite different as well. In addition to the major dimeric photoproducts, rare dimeric photoproducts may also occur, such as thymine–adenine or adenine–adenine dimers.

The organization of chromatin into nucleosomal particles connected by linker DNA (i.e., "beads on a string") influences sites of dimer induction. Limited flexibility of the DNA helix resulting from histone–DNA interactions inhibits the formation of cyclobutane dimers and modulates their distribution along DNA with a periodicity of 10.3 bases. Unlike the cyclobutane dimer, which forms in equal amounts in core and linker DNA, (6-4) photoproducts occur with sixfold greater frequency in linker DNA. Distribution of (6-4) photoproducts in chromatin is also determined by base modifications associated with metabolic activity. Sequencing analysis has shown that methylation of the 3′-cytosine of a (6-4) photoproduct inhibits its formation.

In contrast to the direct induction of DNA damage by UVB light, UVA light produces much less damage indirectly by generating reactive chemical intermediates. Similar to ionizing radiation effects, UVA radiation generates oxygen and hydroxyl radicals, which in turn react with DNA to form monomeric photoproducts, such as cytosine and thymine photohydrates, as well as strand breaks and DNA–protein crosslinks (Figure 1). In addition, chemical and enzymatic treatments of UV-irradiated DNA have revealed novel sites of photodamage whose structures have yet to be defined.

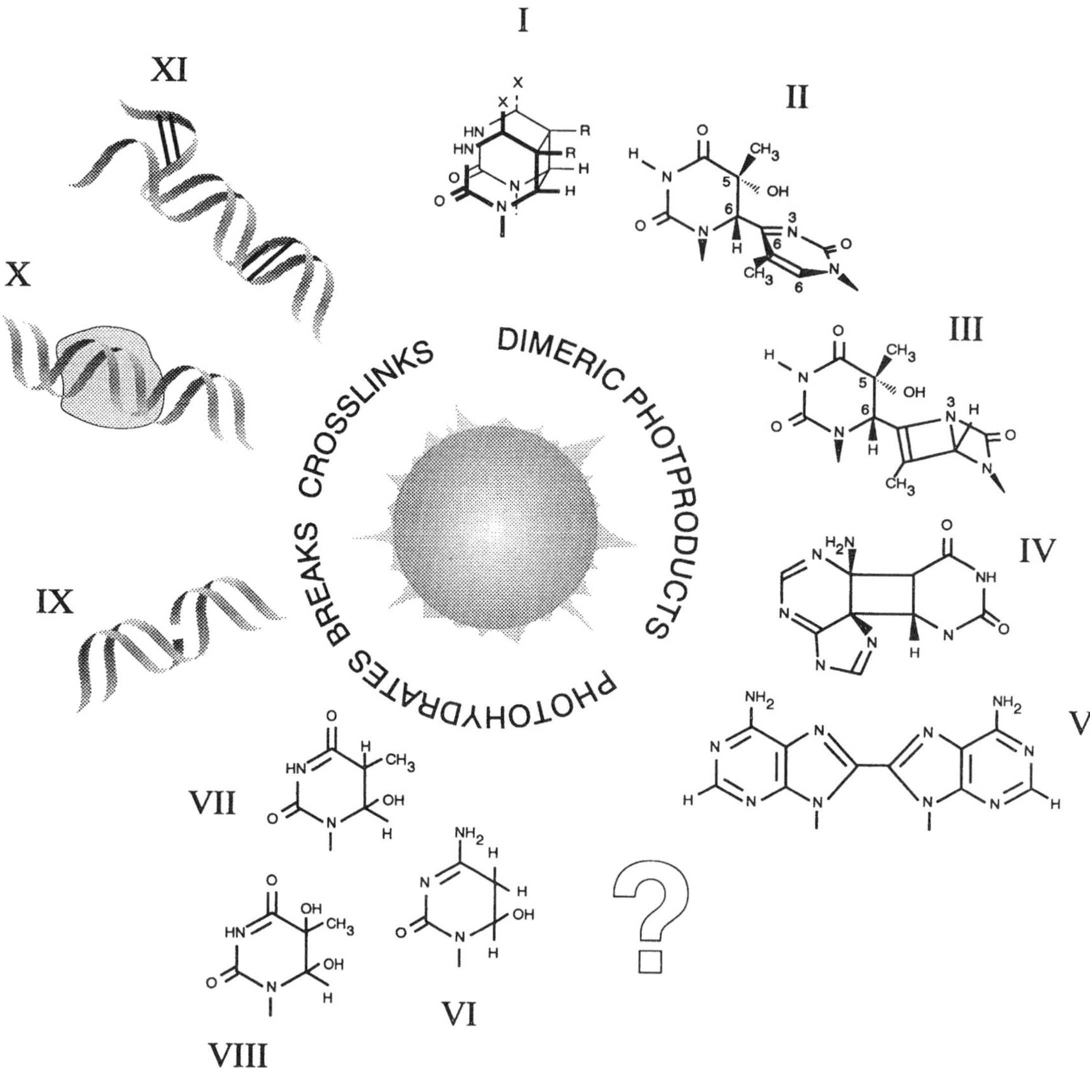


**Figure 1.** Types and structures of known photoproducts induced in DNA by solar UV radiation: **I**, cyclobutane dimer; **II**, (6–4) photoproduct; **III**, Dewar pyrimidinone; **IV**, adenine–thymine dimer; **V**, 8,8-adenine dehydrodimer; **VI**, 6-hydroxy-5,6-dihydrocytosine; **VII**, 6-hydroxy-5,6-dihydrothymine; **VIII**, 5,6-dihydroxy-5,6-dihydrothymine (thymine glycol); **IX**, single-strand break; **X**, DNA–protein cross-link; **XI**, inter- and intrastrand DNA cross-links.

The relationship between the abundance of these photoproducts relative to one another and their relative biological impact depends on the intrinsic effectiveness of the individual photoproduct in cell killing or mutation induction. Hence, the structure and location of a photoproduct that occurs at a low frequency, may elicit a potent biological effect.

## 2 METHODS FOR DETECTING DNA DAMAGE

The sensitive and specific quantitation of photodamage in DNA is essential to understanding the biological effects of UVR. DNA damage can be detected by chromatography, enzymatic or biochemical incision of DNA at sites of photoproducts, or antibody binding to structural damage in DNA. The cyclobutane dimer was the first damage detected in DNA using two-dimensional paper chromatography. Similar techniques, such as thin-layer chromatography and high performance liquid chromatography, have since been adapted to the analysis of other types of base damage. Other procedures measure strand breaks induced directly in DNA by UVR (frank breaks) or resulting from enzymatic or biochemical treatments that cleave DNA at sites of UV photodamage. *Uvr*ABC exinuclease, a partial excision repair complex purified from *Escherichia coli,* cleaves DNA on either side of damage produced by exposure to genotoxic chemicals or UVR (Figure 2, right). Cleavage of specific photoproducts has been achieved by using purified enzymes that combine a glycosylase, which cuts the base from the sugar leaving an abasic site (i.e., an AP site), and an apurinic/apyrimidinic endonuclease, which cleaves the phosphodiester backbone on either side of the AP site (Figure 2, left side). In addition to enzymes, biochemical procedures have been adapted to manifest breaks in DNA at sites of photodamage. For instance, strand breaks can be produced at sites of damage that are alkaline labile, such as AP sites or Dewar pyrimidinones.

Recently, techniques have been developed to explore the fine structure of UV damage induction and repair using strand scission.

Photodamage at the *sequence level* can be mapped as strand breaks in DNA fragments irradiated with high fluences of UVC and UVB and at the *gene level* using Southern blot hybridization of DNA fragments separated by agarose gel electrophoresis. Recently, the distribution of photoproducts at the sequence level in a gene promoter irradiated with low UV fluences in intact cells was analyzed by ligation-mediated polymerase chain reaction.

Antisera raised against UV-irradiated DNA are potent and versatile reagents for the study of various photochemical and photobiological phenomena. Polyclonal and monoclonal antibodies recognize and bind a variety of photoproducts, including the cyclobutane dimer, the (6-4) photoproduct, Dewar pyrimidinone, and thymine glycol. Immunological approaches adapted to the analysis of DNA damage and repair include immunoprecipitation, enzyme-linked immunoassays, radioimmunoassays, quantitative immunofluorescence, and immunoelectron microscopy. Each technique has its own unique attributes and applications. Unlike chromatographic techniques, immunological assays do not require chemical or enzymatic degradation of DNA before analysis; unlike endonucleolytic assays, their sensitivities do not depend on the molecular weight or purity of sample DNA.

## 3 DNA DAMAGE TOLERANCE MECHANISMS

The amount of damage present in genomic DNA at a given point in time depends not only on the amount of UVR absorbed but on the amount of damage repaired. Damage tolerance strategies are complex and vary greatly among organisms. Nearly all organisms have behaviors or natural features that reduce exposure of DNA to solar UVR and reduce the amount of photodamage. Outer coverings, habitat selection, avoidance responses, and physical morphologies may attenuate or eliminate the transmission of UVR to sensitive, internal areas of cells or organisms. Many organisms have evolved biochemical mechanisms to protect themselves against UVR. The UVR-absorbing compounds melanin and anthocyanin are produced in human skin and in plants, respectively. The presence of these compounds reduces UV penetration into cells and tissues and reduces DNA damage. In addition to such pigments,

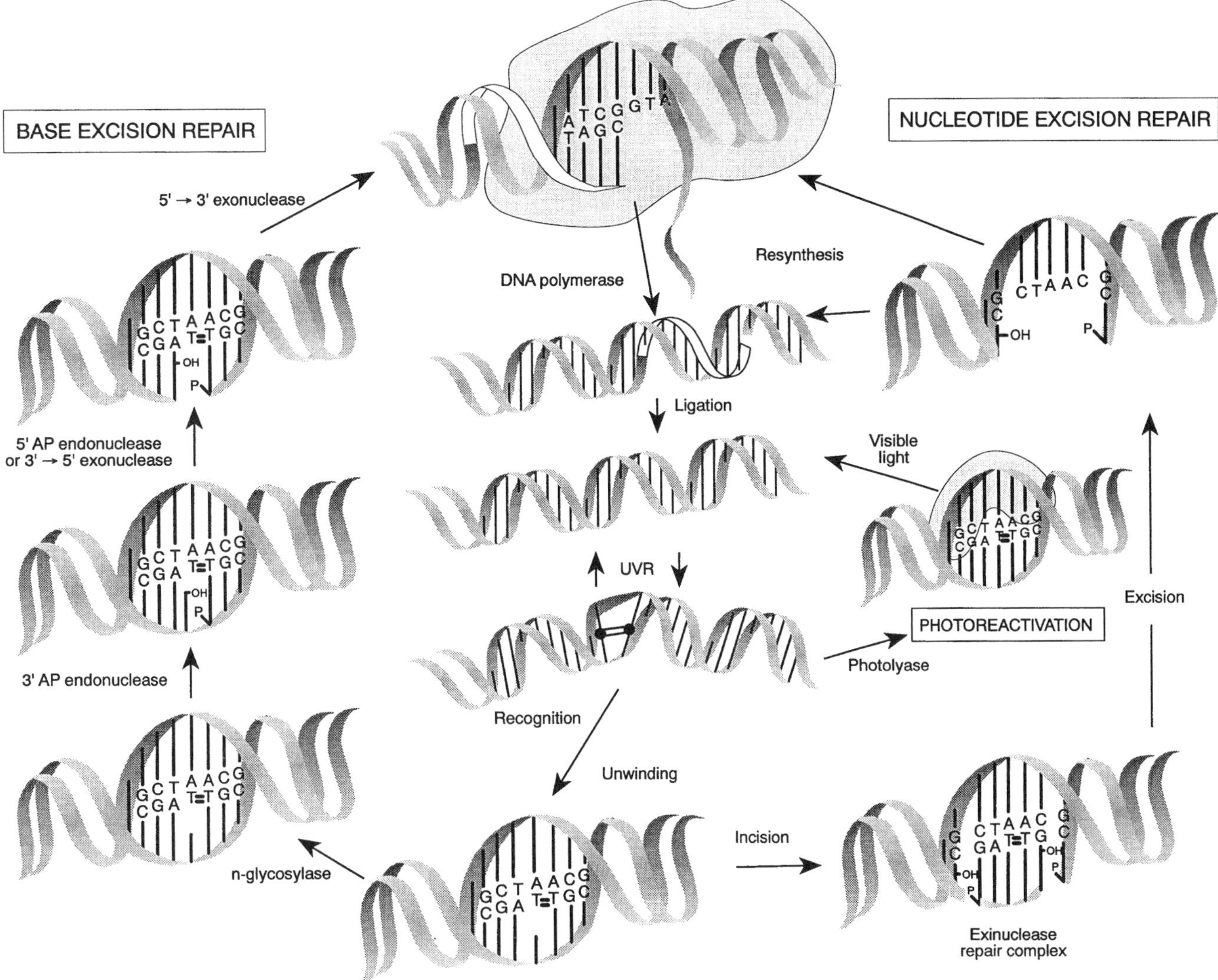


Figure 2. Enzymatic pathways for the repair of UVR damage in DNA: photoenzymatic repair (or photoreactivation), base excision repair, and nucleotide excision repair; T═T, cyclobutane dimer.

colorless UV-absorbing compounds have been identified as possible UV-protective chemicals. Such "natural sunscreens" include flavenoids in terrestrial plants, mycosporine amino acids in fungi, and mycosporinelike compounds in marine organisms.

Once DNA damage has occurred, its removal may proceed in whole or in part by at least two well-studied routes, enzymatic photoreactivation and excision repair (Figure 2). Photoreactivation is the simplest, and probably most ancient repair process. A cyclobutane dimer is specifically recognized and bound by a small enzyme called a photolyase. This enzyme then catalyzes the direct reversal of the dimer upon absorption of a photon within the UVA/visible range of light. Although photoreactivation occurs widely throughout the plant and animal kingdoms, there are several organisms in which it has not been found, including placental mammals.

Organisms that display reduced photoreactivation often have a greater capacity for excision repair. This process varies depending on the type of damage encountered and how that damage is situated in the nucleoprotein complex (i.e., chromatin). The excision repair pathway is thought to proceed by the following steps:

Recognition of the lesion as a structural distortion in the DNA helix

Unwinding or other activity to disassociate the lesion from chromatin proteins and provide accessibility to repair enzymes

Incision of the DNA backbone at or near the site of the lesion

Excision and resynthesis of the DNA around the damaged site

Ligation of the single-strand nick remaining after disengagement of the DNA polymerase complex

This process is very versatile, correcting various classes of chemical damage in addition to those induced by UV radiation.

Excision repair may proceed by one of two routes (Figure 2). *Base excision repair* is initiated by enzymatic recognition of the lesion and scission of the bond between the damaged base and its associated deoxyribose sugar, a process called aglycosylation. Examples of *n*-glycosylases include endonuclease III from *Escherichia coli,* which repairs photohydrates and endonuclease V from T4 phage, which has the unique attribute of recognizing the 5′-pyrimidine of a cyclobutane dimer as a modified base. In the latter case, cleavage of the *n*-glycosyl bond leaves a "dangling" dimer and an abasic site. After removal of the base damage, a 3′-AP endonuclease cleaves the phosphodiester bond 3′ to the abasic site. To prepare the strand for resynthesis, a 5′-AP endonuclease or 3′ → 5′ exonuclease associated with DNA polymerase digests the

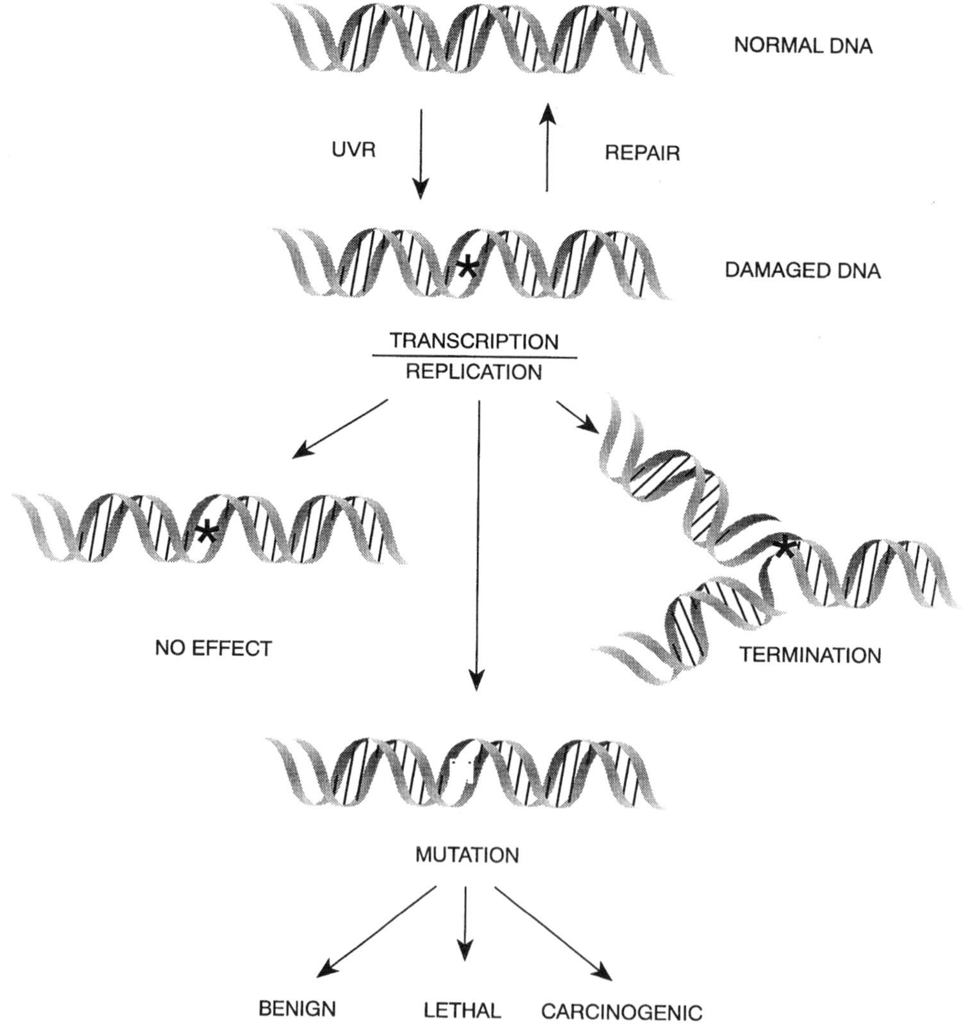

**Figure 3.** Biological consequences of UVR damage in DNA.

remaining abasic site. The damaged strand is removed and resynthesized by a DNA polymerase complex, and the remaining nick is ligated to return the DNA duplex to its original state. The resultant repair ''patch'' of newly incorporated DNA is about 30 bases long in mammalian cells.

The *nucleotide excision repair* pathway recognizes a broad spectrum of UVR-induced photoproducts as well as other ''bulky adducts'' that may be induced by genotoxic agents. This process has been well studied in *E. coli* and is thought to be the primary excision repair pathway in eukaryotes as well. As with base excision repair, the nucleotide excision repair process is initiated by enzymes that recognize and bind the helical distortion created at the damaged site. An exinuclease complex is assembled around the lesion and cleaves the DNA on both sides of the lesion, leaving a gap. This gap is filled by a DNA polymerase, and the strand is ligated to restore the DNA duplex to its original integrity.

As the arrangement of photoproduct formation is influenced by chromatin structure, so the temporal and spatial distribution of photoproduct repair is also determined by its location in the genome. Excision repair in mammalian cells is heterogeneous; different photoproducts are repaired at different rates in the overall genome and are preferentially repaired in transcribing DNA. The (6-4) photoproduct is excised much faster than the cyclobutane dimer, with kinetics that correlate well with DNA repair processes occurring immediately after UV irradiation such as repair incision, repair synthesis, and removal of blocks to DNA replication. In addition, these and other types of UVR and biochemical damage are preferentially repaired in actively transcribing genes and in the transcribed strands of genes.

## 4 BIOLOGICAL CONSEQUENCES OF UVR DAMAGE IN DNA

UVR is a potent and ubiquitous carcinogen responsible for much of the skin cancer in the human population today. Tumor incidence and mortality correlate with exposure: basal and squamous cell carcinomas are most prevalent on the face and trunk in men and on the face and legs in women, areas of the body unprotected from the sun's rays; carcinomas increase with decreasing latitude, corresponding to the level exposure; and tumor incidence is greater among people such as ranchers or fishermen, whose occupations entail high exposure. In addition, the protective action of skin pigmentation results in lower cancer rates in dark-skinned populations as opposed to lighter skinned peoples. The importance of UVR damage and its repair in humans is exemplified by the occurrence of genetic diseases that greatly increase the risk of sunlight-induced skin cancer. In one such disease, xeroderma pigmentosum, a failure in the DNA repair process is associated with a major increase in the rate of onset of squamous and basal cell carcinoma and melanoma.

Photoproducts in cellular DNA can be resolved in several ways, depending on the type of lesion and its genomic location, as well as the type of cell and its developmental state (Figure 3). It is conceivable that some lesions are not perceived as different from normal DNA by the cell. However, it is more probable that unless the damage is repaired, it will disrupt the normal operation of essential cellular processes such as DNA replication or transcription. Cell proliferation may cease or, if the lesion is situated in a gene required for an essential metabolic function, the cell will die. Some types of photodamage are more effective at blocking the progression of DNA or RNA polymerase than others. Bypass by DNA polymerase may result in incorporation of an incorrect complementary base, leading to the production of a mutation. The predominant mutation produced by UVR is a transition mutation resulting in replacement of a cytosine with thymine. A mutation may have several outcomes. First, it may not alter the genetic code, in which case it will not affect normal metabolism. Second, the mutation may produce a truncated or partial RNA transcript and a dysfunctional protein. If this protein is essential, then the mutation is lethal. Finally, it may result in activation of an oncogene or inactivation of a tumor suppressor gene, resulting in the initiation of the carcinogenic process.

Biochemical tolerance to solar UVR must have developed as life emerged on earth. DNA repair mechanisms that originally evolved in response to UVR damage are now faced with a host of newly synthesized genotoxic agents. Understanding the photochemistry and photobiology of UVR damage and repair in various living systems may be important for our continued survival in a changing world.

*See also* DNA Damage and Repair; Environmental Stress, Genomic Responses to; Ionizing Radiation Damage to DNA.

### *Bibliography*

Björn, L. O., Moan, J., and Young, A. R., Eds. (1993) *Environmental UV Photobiology.* Plenum Press, London.

Friedberg, E. C. (1985) *DNA Repair.* Freeman, New York.

———, and Hanawalt, P. C., Eds. 1981 (Vol. I), 1983 (Vol. II), 1988 (Vol. III) *DNA Repair: A Laboratory Manual of Research Procedures,* Vols. 1–3. Dekker, New York.

Hanawalt, P. C. (1989) Preferential repair of damage in actively transcribed DNA sequences in vivo. *Genome,* 31:605–611.

Mitchell, D. L., and Nairn, R. S. (1989) The biology of the (6-4) photoproduct. *Annu. Rev. Photochem. Photobiol.* 49:805–819.

Smith, K. C. (1977) *The Science of Photobiology.* Plenum Press, New York.

Wang, S. Y., Ed. (1976) *Photochemistry and Photobiology of Nucleic Acids,* Vols. I and II. Academic Press, New York.

# Vaccine Biotechnology

*Tilahun D. Yilma*

*Key Words*

**Adjuvant Genes** Genes (often lymphokine genes) that are used for the enhancement of efficacy of recombinant vaccines.

**Baculovirus** A DNA insect virus widely used for abundant and inexpensive generation of recombinant proteins in insect cells or larvae.

**Chimeric** Describing a product of fusion of two or more nucleic acids or polypeptides.

**Recombinant** An organism, nucleic acid, or protein generated by genetic engineering.

**Vaccinia Virus** An orthopox virus of uncertain origin that provides protective immunity against a number of orthopox virus infections, including smallpox.

**Vector** A bacterium, virus, or plasmid that is used for the introduction of DNA molecules or the generation of recombinants.

Using the techniques of molecular biology, effective modern vaccines are being developed by a variety of innovative methods that circumvent some of the problems associated with conventional vaccines. Live recombinant vaccines, which are prepared by isolating an immunogenic gene from a disease organism and inserting it into a vector, are most promising. Such vaccines can provide immunity to agents that cannot be reliably attenuated or inactivated without destroying immunogenicity.

## 1 RECOMBINANT VACCINES

Biotechnological advances have recently led to a new generation of vaccine preparations, developed by methods that diverge completely from traditional efforts. Being readied for use are a variety of subunit, synthetic peptide, and (especially) live recombinant vaccines that may greatly improve our ability to offer protection against infectious disease.

Live recombinant vaccines retain the advantages while overcoming many of the safety hazards and logistical disadvantages inherent in live vaccines. Thus far heterologous genes have been inserted in a variety of expression vectors including simian virus 40, bovine papillomavirus, adenovirus, herpesvirus, and retrovirus. Vaccinia virus (VV) is an especially promising vector for recombinant vaccine preparations that carry immunogenic genes of animal or human disease agents. When the recombinant is introduced into a vaccine, the foreign genes are replicated and expressed along with those of VV, and the vaccinee becomes immune to both. VV's large genome has 19 possible incorporation sites for foreign genes as well as a variety of marker and reporter genes that assist in identifying useful recombinants; the ease of generating VV recombinants has raised the possibility of a single preparation that could immunize against many disease agents.

## 2 VACCINIA VIRUS RECOMBINANTS

The strategy for using VV as a vector involves first constructing a plasmid carrying a chimeric gene. To this end, a VV promoter is fused to a foreign protein coding sequence that is flanked by DNA from a nonessential region of the VV genome. The chimeric gene is incorporated into the VV genome by homologous recombination in tissue culture cells that have been infected with wild-type VV and then transfected with the plasmid. Although any nonessential region of the VV genome can be used as the site of gene insertion, the thymidine kinase (TK) gene locus provides some advantages because recombinants are then $TK^-$. The $TK^-$ phenotype serves to attenuate viral pathogenicity and also can be easily distinguished from wild-type $TK^+$ virus by growth in the presence of 5-bromodeoxyuridine, a thymidine analogue. Rapid screening of recombinants also can be facilitated with the coexpression of reporter genes, such as the *E. coli* β-galactosidase (*LacZ*) gene, along with the foreign gene of interest; *LacZ* gene expression gives the plaque a blue coloration in the presence of 5-bromo-4-chloro-3-indolyl-β-D-galactoside (XGal).

We describe here three of the vaccinia virus recombinants we have constructed and used with varying degrees of success as vaccines: a vaccine for rinderpest (RP) that has induced 100% protection against challenge with more than 1000 times the normally lethal dose, a vaccine for vesicular stomatitis (VS) that has been moderately successful in protecting against the disease, and a vaccine aimed at simian immunodeficiency virus (SIV) that has shown little or no success in inducing protection.

### 2.1 Vaccinia Virus Recombinant Vaccine for Rinderpest

Our most successful effort has been the VV recombinant vaccine produced for rinderpest virus (RPV), the agent of an acute, febrific, highly contagious disease of ruminants (particularly cattle and buffalo in Africa and Asia) with a mortality rate exceeding 90%. Three types of RP vaccine currently used—the caprinized, the lapinized, and the Plowright tissue culture—are all live, attenuated vaccines with the limitations common to such preparations: heat lability, the requirements for tissue culture laboratory facilities and maintenance of a cold chain, and varying degrees of pathogenicity. In contrast, the lyophilized form of VV is heat stable, is easily produced, stored, and transported, and can be administered by personnel with minimal training. Our VV recombinant vaccine for rinderpest is one of two highly effective poxvirus recombinant vaccines—the second being a recombinant vaccine for rabies.

The RPV is an enveloped, single-stranded RNA morbillivirus in the family Paramyxoviridae, along with measles virus of humans, distemper virus of dogs, and peste des petits ruminants virus (PPRV) of goats and sheep. In paramyxoviruses, the hemagglutinin (H) and fusion (F) surface proteins have been shown to provide protective immunity. We propagated the highly virulent Kebete "O" strain of RPV and characterized eight viral proteins. We made cDNA copies of the genes for phosphoprotein (P), H, F, nucleoprotein (N), and the matrix (M), and determined the complete nucleotide sequences. Standard procedures were used to construct VV single recombinants expressing the H gene (v*RVH*) and the F gene (v*RVF*) and a double recombinant that expresses both genes (v*RVFH*). The *TK* and *HA* genes of VV were insertionally inactivated by incorporation of the H and F genes, leading to even greater attenuation.

Protective immune response studies with v*RVH*, v*RVF*, and v*RVFH* were conducted in cattle at the Plum Island Animal Disease Laboratory. Pock lesions developed as early as 4 days in all animals vaccinated with the single recombinants but were limited to the site of inoculation and healed completely by 2 weeks. Cattle vaccinated with the double recombinant showed no pock lesions at the site of inoculation, a strong indication of the level of attenuation. All animals vaccinated with the recombinants produced serum-neutralizing (SN) antibodies to RPV. As expected, control animals had no detectable antibody titer. Two unvaccinated animals housed with each group to assess the transmissibility of VV recombinants from vaccinated to contact animals developed no pock lesions and remained negative to VV by SN and plaque assay reduction assays.

To evaluate protection, all animals were challenged with a heavy dose of RPV subcutaneously in the prescapular lymph node region on day 35 following primary immunization. As low as one $TCID_{50}$ (tissue culture dose that would be infectious to 50% of the subject population) usually induces clinical RP and 100% mortality. Cattle vaccinated with the recombinants were completely protected when challenged, even though the inoculation was greater than 1000 times a normally lethal dose of RPV. Some cattle vaccinated with the F recombinant (v*RVF*) had a significant anamnestic response to RPV after challenge inoculation; this indicates replication of the challenge virus. No anamnestic response at all could be demonstrated in groups vaccinated with a mix of the single recombinants or the double recombinant.

We also tested v*RVFH* and the parental Wyeth strain for pathogenicity by inoculating 12 nude mice intraperitoneally; no clinical disease was detected in the test mice. In addition, we assessed the effects of v*RVFH* and the mixture of single recombinants (v*RVF* + v*RVH*) on the immune systems of rhesus macaques infected with the simian immunodeficiency virus (SIV). These monkeys suffer from an immunodeficiency disease virtually identical to human acquired immunodeficiency syndrome (AIDS). Hence, SIV is a legitimate model for HIV. Four normal macaques and four SIV-infected macaques were vaccinated with v*RVFH*. Both groups developed small pock lesions at the site of inoculation which healed completely by 2 weeks postvaccination. (Apparently, the lack of pock lesions in cattle vaccinated with v*RVFH* is a host-specific phenomenon). However, no clinical disease or dissemination of VV was observed in either normal or SIV-infected macaques for one month following vaccination.

It would seem that RP is an excellent candidate for eradication using the VV recombinant vaccine. A vaccine against one strain will immunize against all, including PPR of sheep and goats. Use of the RP double-recombinant VV in areas of the world where PPRV is endemic would also aid in the control and eradication of PPR.

### 2.2 A Vaccinia Virus Recombinant Vaccine for VSV

One of the earliest VV recombinants was a vaccine for vesicular stomatitis (VS), a contagious disease of horses, cattle, and pigs characterized by vesicular lesions on the tongue and oral mucosa. In a model system the virus can be used to infect mice, causing neuropathy and death in 100% of the subjects. This disease is notable for the confusion of diagnosis in cattle between VS and foot-and-mouth disease, an epizootic with 100% morbidity.

Vaccination for VS is generally prohibited by the U.S. Department of Agriculture because vaccinees cannot be distinguished from naturally infected animals. It is possible to distinguish animals vaccinated with a subunit preparation by serology, however, and this is an essential consideration for epidemiology, disease control, and establishment of quarantine programs. Advantages of subunit and live agent vaccines could be combined in an effective VV recombinant expressing specific VSV proteins. Recombinants of VSV were constructed by the methods described for RPV. DNA copies of mRNA for the external glycoprotein G, shown in mice to be the protective immunogen, were linked to VV promoters and inserted into the genome of VV. The recombinants retained infectivity and synthesized VSV polypeptides similar or identical in size and antigenicity to the natural VSV proteins that were transported normally to the cell surface (perhaps important for antigenicity). Vaccinated mice produced VSV-neutralizing antibodies by day 14 that increased over a 42-day period; booster vaccination on day 28 resulted in a seven- to eightfold increase in SN titer. Upon challenge, all mice that had received primary and booster vaccinations were protected, while 7 of 11 control mice died of encephalitis.

Two successive vaccinations of cattle with the VV recombinant VSV vaccine provided protection, closely correlated with neutralizing antibody titer, for two-thirds of the cattle. Compared with natural infection, this challenge was very heavy.

Still, the lack of complete protection for all vaccinees may indicate a need for cell-mediated immunity that is not induced by this vaccine. It is possible that more effective protection could be provided by constructing recombinants that synthesize considerably more G glycoprotein.

## 3 BACULOVIRUS EXPRESSION VECTORS

The baculovirus recombinant system has the capacity to produce large quantities of proteins that may be used as subunit vaccines. Up to 68% of the total protein of a recombinant baculovirus-infected *Spodoptera exigua* larva is recombinant protein. This protein may also be used in diagnostic tests without purification, since there is no interference or loss of sensitivity noted when plates for enzyme-linked immunosorbent assay are coated with the crude lysate.

A novel approach has also been introduced for production of monoclonal antibodies to one specific protein of a virus, or other agent consisting of several proteins, that does not require purified antigen for either the immunization or the screening phase of the procedure. This system has been effective in producing monoclonal antibodies to uncommon antigens (e.g., Alzheimer's protein) or to antigens that cannot be easily screened for such biological activities as hemagglutination or neutralization (e.g., VSV N protein).

### 3.1 Recombinant VV and Baculovirus Vaccines for SIV

Experimental infection of macaques with SIV provides a valuable animal model for AIDS vaccine research. Whether a live retroviral vaccine could ever meet the safety requirements for human use is highly questionable; recent experimental data have encouraged researchers to explore the potential for using a live recombinant VV vaccine (expressing the glycoproteins of SIV) to prime rhesus macaques. These animals then would be boosted with the same recombinant subunit antigen expressed in mammalian cells, Chinese hamster ovary (CHO) cells, or a baculovirus expression vector. It has been shown that macaques primed with a VV recombinant vaccine, then boosted with a baculovirus subunit product, have greatly enhanced antibody responses compared with animals boosted with the original vaccine.

### 3.2 Prospects for Recombinant VV and Baculovirus Vaccines

Protection has not been afforded against SIV by any recombinant vaccine; however, in vitro suppression of virus was documented and 2 weeks after challenge exposure, the cell-free infectious virus load in plasma (plasma viremia) of macaques primed and boosted with certain preparations of SIV surface glycoprotein was considerably lower than that in unimmunized controls. Results of these studies indicate that immunization with surface glycoprotein may not provide protective immunity against SIV infection, even though the relation between reduction in virus load and delay in the onset of disease remains to be evaluated.

## 4 SAFETY AND EFFICACY OF LIVE AGENT RECOMBINANT VACCINES

The safety of VV recombinant vaccines for general use has been confirmed by the absence of transmission of VV from vaccinated to contact animals. Possibilities of further attenuation of VV recombinant vaccines by insertion of lymphokine genes like interferon gamma (IFN-γ) or interleukin 2 (IL-2) have been investigated. Fusion proteins of IFN-γ (human and murine species) with various immunogens were constructed, and reduced pathogenicity of several VV recombinants expressing these proteins was observed for immunodeficient (athymic, nude) mice.

Development of an adjuvant system for VV recombinant vaccines may be crucial for their general effectiveness. Mixtures of IFN-γ and G glycoprotein of VSV duplicate the adjuvant effect of complete Freund's adjuvant and further increase the anamnestic response on booster administration. Others have documented the adjuvant effects of IL-2. Use of immunoregulatory genes like IFN-γ and IL-2 as adjuvant genes in polyvalent VV recombinant vaccines must be further evaluated. If these lymphokine genes work as adjuvants to augment the immune response, the finding will revolutionize the concept of vaccination. The human IFN-γ gene could be incorporated into vaccinia vectors for human pathogens, and similar vectors could be constructed for other domestic species.

## 5 RECOMBINANT VACCINES FOR HIV

Several laboratories are investigating VV recombinants as vaccines against HIV infection. It has been shown that the HIV antigens expressed by some of these recombinants are correctly processed and transported, and experimental animals as well as human volunteers that were vaccinated developed detectable cell-mediated immunity to the HIV proteins expressed. Innovative solutions to possible safety problems are being investigated. It has been reported that insertional inactivation of its *TK* gene attenuates VV. Also, expression of IL-2 and murine IFN-γ has prevented disseminated VV infection in nude mice.

Chimeric genes coding for IFN-γ and the structural proteins of HIV-1 have been constructed and inserted into VV to develop recombinants that express these genes. These fusion proteins retain the antigenic characteristics of both IFN-γ and HIV. The biologic antiviral activity of IFN-γ could not be documented for all the fusion proteins, but the in vivo attenuating activity of IFN-γ for nude mice was retained at variable rates for all the recombinants; recombinants expressing IFN-γ have reduced pathogenicity for nude mice.

*See also* Expression Systems for DNA Processes; Genetic Immunization; Molecular Genetic Medicine; Plasmids.

### *Bibliography*


Chanock, R. M., Giensburg, H., Lerner, R., and Brown, F., Eds. (1987) *Vaccines 87. Modern Approaches to New Vaccines Including Prevention of AIDS.* Cold Spring Harbor Laboratory, Cold Spring Harbor, NY.

Mahey, B., Ed. (1986) *Virus Vector Vaccines,* AGA: BIOT/86/32. Food and Agriculture Organization of the United Nations, Rome.

National Academy of Sciences and National Research Council (1988) *Science & Technology for Development: Prospects Entering the 21st Century.* NAS/NRC, Washington, DC.

Quinnan, G. V., Ed. (1985) *Vaccinia Viruses as Vectors for Vaccine Antigens.* Elsevier-North Holland, New York.

Yilma, T. (1989) Prospects for the total eradication of rinderpest. *Vaccine,* 7:484–485.

———. (1990) A modern vaccine for an ancient plague: Rinderpest. *Bio-Technology,* 8:1007–1009.

———. (1991) The role of biotechnology in tropical diseases, in J. Williams, K. Kocan, and E. P. Gibbs, Eds., *Tropical Veterinary Medicine: Current Issues and Perspectives,* Vol. 653, *Annals of the New York Academy of Sciences,* pp. 1–6.

———. (1991) A vaccinia virus recombinant vaccine for rinderpest. *World Anim. Rev.* 69:2–6.

———. (1994) Genetically engineered vaccines for animal viral diseases. *J. Am. Vet. Med. Assoc.* 204(10):1606–1615.


**Vaccines, Genetic: *see* Genetic Immunization; Molecular Genetic Medicine.**

# Viral Envelope Assembly and Budding

*Milton J. Schlesinger*

*Key Words*

**Lipid Bilayer** The chemical structure that forms cellular membranes and is composed of phospholipids, sphingolipids, and cholesterol organized in two apposing monolayers.

**Matrix Protein** The viral-encoded polypeptide that links the nucleoprotein with the membrane and binds viral-encoded, membrane-associated glycoproteins.

**Membrane Curvature** The bending of the lipid bilayer that accompanies virus budding and results in an asymmetry of lipid bilayer composition.

**Nucleoproteins** The viral genetic material (i.e., RNA or DNA) and viral-encoded proteins organized in either stable icosahedral or helical structures.

**Transmembranal Glycoproteins** The viral-encoded polypeptides that form spikes on the outer surfaces of enveloped viruses but are anchored to the membrane by hydrophobic sequences that span the lipid bilayer and contain domains that are topologically oriented to the interior of the virus particle or to the cytoplasmic face of the membrane of the infected host cell.

''Assembly and budding'' of enveloped viruses refers to the interactions that occur between viral nucleoprotein complexes and membrane-embedded, viral-specific glycoproteins at infected cell membranes, leading to the release of newly replicated, infectious viruses from host cells.

## 1 GENERAL PROBLEM OF VIRUS RELEASE FROM CELLS AND MODELS OF ASSEMBLY

To continue their cycle of infection, reproduction, and movement among different cells and organisms, newly assembled viruses must be freed from the cells serving as hosts for virus replication. For most ''naked'' viruses (i.e., those containing only a genome packaged in a protein shell and growing in a eukaryotic cell), newly assembled viruses form paracrystalline arrays in the cellular cytoplasm or nucleus; release requires death and disintegration of the host cell. In contrast, viruses containing a lipid bilayer as part of their structure utilize a host cell membrane and a set of biochemical activities that lead to a ''budding'' or membrane extrusion, which frequently occurs in the absence of host cell degradation. Enveloped virus budding has been examined in considerable morphological detail by high resolution electron microscopy, which led to various models that involve basically three distinct events (Figure 1): a nucleation between virus nucleoprotein cores and transmembranal virus-specific polypeptides, followed by additional protein–protein interactions that produce a curvature of the membrane with accompanying asymmetry between lipid bilayers and, finally, a fusion of the virus envelope, which frees the virus from the cell membrane.

The molecular elements in this process were initially identified on the basis of studies of virus mutants (either deleted in portions of virus structural proteins or temperature sensitive for replication), whose phenotypes indicated specific effects on the assembly process, and on the results of chemical cross-linking between proteins and between proteins and membrane lipids. More recently, however, the experimental procedures of cDNA cloning, sequencing, and expression of virus-specific components, of site-directed mutagenesis, and of monoclonal antibody and network antibody techniques have led to a more complete analysis of molecular events involved in enveloped virus assembly. Additional information has come from improved observational techniques such as high resolution cryoelectron microscopy, which allows for examination of viruses in vitrified ice rather than as dehydrated samples.

## 2 COMPONENTS OF ENVELOPED VIRUSES CRITICAL TO ASSEMBLY AND BUDDING

Except for the retroviruses, the packaged viral genome is considered to be completely assembled as either a helical nucleoprotein or an isometric icosahedral shell prior to interaction with host cell membranes. Another important component of the infectious virus, the membrane-embedded glycoprotein, is also considered to be mature in structure and localized to the particular membrane from which the virus will ultimately assemble and bud. In fact, in ''polarized'' cells, where membranes are biochemically and structurally separable, it is the destination of the virus glycoprotein that strictly determines which membrane of the cell will be used for virus assembly and budding. For many enveloped viruses another protein, called the matrix, appears to intervene in this process—perhaps serving as a ''linker'' between glycoprotein and nucleoprotein. Other proteins, both host and virus, may affect the kinetics and efficiency of the assembly and budding (see below).

Another critical requirement in enveloped virus assembly is the rearrangement of lipid components in the bilayer at the site of budding. During assembly and budding, extreme curvature of the bilayer leads to as much as a 25% differential in volume between inner and outer bilayers. It is unlikely that this differential will be attained by a ''flipping'' of phospholipids from inner to outer bilayer, since this activity normally is very slow, although the clustering of transmembrane proteins could perturb the membrane sufficiently to increase the rate of phospholipid movement between bilayers. Cholesterol moves easily across bilayers and could provide for some asymmetry during membrane bending. This sterol binds most tightly to sphingosine-type lipids, which are generally oriented to the extracellular face of the lipid; thus, a clustering of these lipids in domains of virus assembly would promote membrane curvature. Sphingolipid clustering, in turn, could arise by oligosaccharide interactions between virus glycoprotein and glycolipid. Membrane bending may also be ''driven'' by the association of virus–glycoprotein cytoplasmic domains with virus nucleoproteins and/or matrix-type proteins.

## 3 EXPERIMENTAL METHODS FOR STUDYING ASSEMBLY AND BUDDING

Precisely what roles in envelope virus assembly and budding these various molecular components—nucleoproteins, matrix proteins, transmembrane glycoproteins, and lipid-bound polypeptides—ac-

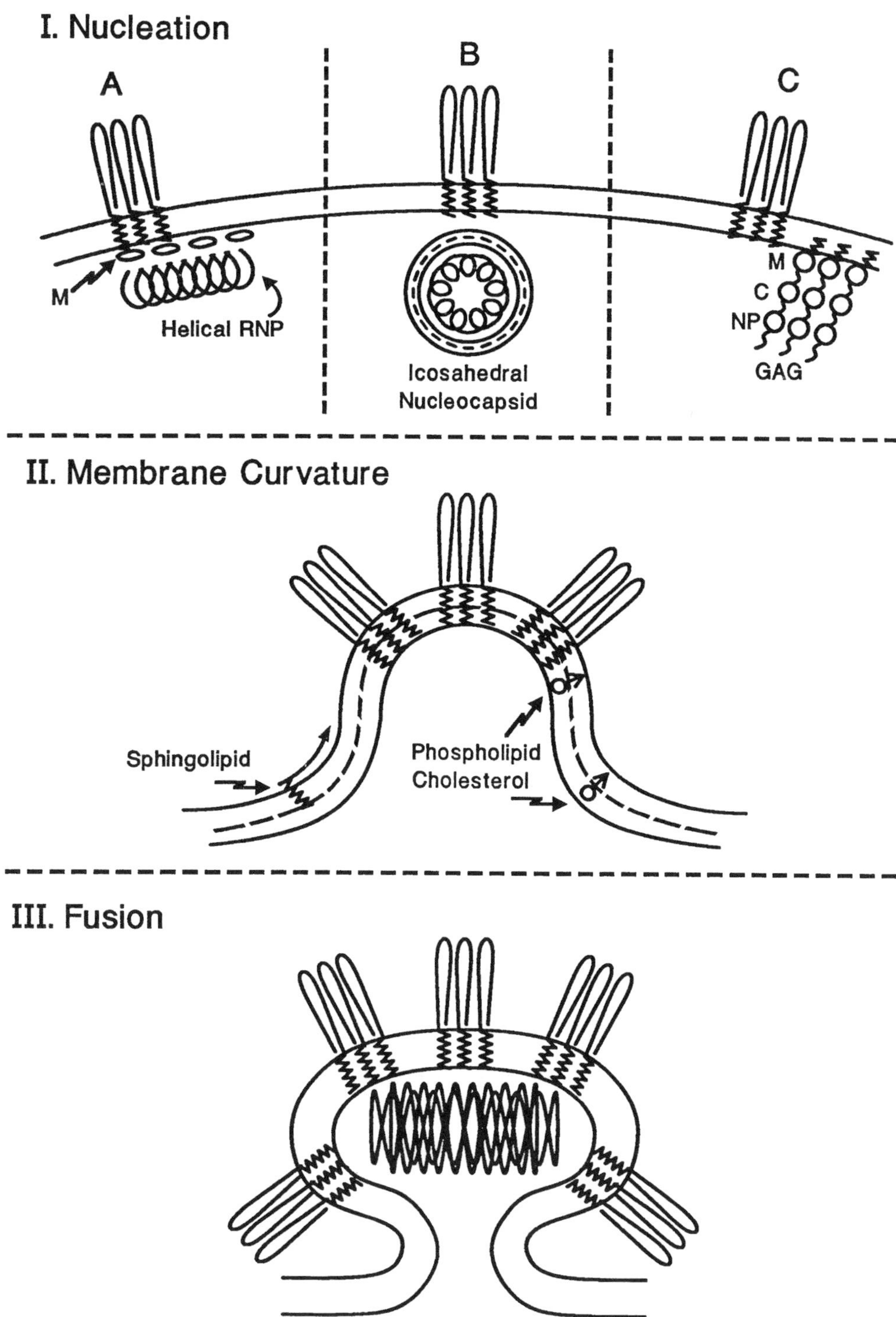


**Figure 1.** Events in the assembly and budding of enveloped viruses, where RNP is the virus–ribonucleoprotein complex; GAG is the complex consisting of the matrix protein M, the capsid protein C, and the nucleoprotein NP.

tually play has been studied with cells in which these proteins could be expressed individually or together with other components. The systems generally employ genetically engineered constructs of plasmid DNA carrying the specific viral-encoded protein in the form of a cDNA and inserted into a permissive host cell. By adding a radioactive amino acid at appropriate times and utilizing specific antibodies, it is possible to measure the virus proteins in both the infected cells and secreted particles. Table 1 is a partial listing of results from experiments of this type carried out for several retroviruses and the alphavirus, Semliki Forest virus. In contrast to the latter, where no budding occurs unless *both* glycoproteins and nucleoprotein cores are present, retrovirus particles are formed

**Table 1** Viral Proteins Identified as Components in Assembly and Budding

| Virus | Components | Type of Data |
|---|---|---|
| HIV-1 | Cytoplasmic domain of gp41 | Mutation–truncation of gene leads to release of noninfectious particle lacking glycoprotein |
| | Glycoprotein | Localization of budding in polarized cells to basal lateral surface |
| | Matrix protein p19 | Small deletions allow for release of particles without glycoprotein |
| Rous sarcoma virus | *GAG* and *env* genes | Particles detected when *GAG* expressed alone; no particles when *env* expressed alone; coexpression forms particles with glycoproteins |
| Semliki Forest virus | Nucleocapsid envelope proteins | Expression of viral structural components with deletions in either nucleocapsid or envelope genes; no budding observed unless both genes are expressed in same cell |

and secreted from infected cells in the absence of the cognate glycoprotein. In fact, for the retroviruses, only the matrix-type protein has been reported to be necessary to form a cell-free particle, suggesting that self-association of this protein is sufficient to form an icosahedral shell.

These various reconstructed systems clearly indicate the minimal components needed for membrane budding; however, they fail to address the problem of "efficiency" and kinetics of virus release. For example, in the cases of the alphaviruses, Semliki Forest and Sindbis, mutants lacking or highly defective in a gene encoding a small hydrophobic, membrane-associated protein form infectious viruses that are indistinguishable from the normal, but these are produced at only 1% of the rate measured for the wild-type virus. For this latter example, the virus appears to have evolved a gene that selectively enhances the efficiency of the assembly and budding process. There are other examples, obtained by selectively truncating a part of the transmembranal envelope gene of the $SIV_{mac}$ retrovirus, which showed that the mutations had different effects in different hosts, with some cells selectively assembling viruses with shorter transmembranal polypeptides. The latter offer strong evidence that host cell, membrane-associated polypeptides are also important components in the assembly and budding process. Their identification and function will be essential in any thorough analysis of the molecular basis for enveloped virus assembly.

*See also* MEMBRANE FUSION, MOLECULAR MECHANISM OF; VIRUSES, DNA PACKAGING OF; VIRUSES, RNA PACKAGING OF.

## *Bibliography*

Caplan, M., and Matlin, K. S. (1989) The sorting of membrane and secretory proteins in polarized epithelial cells, in *Functional Epithelial Cells in Culture,* K. S. Matlin and J. D. Valentich, Eds., pp. 71–127. Liss, New York.

Doms, R. W., Lamb, R. A., Rose, J. K., and Helenius, A. (1993) Folding and Assembly of Viral Membrane Proteins. *Virology,* 193:545.

Dubois-Dalcy, M., Holmes, K. V., and Rentier, B. (1984) *Assembly of Enveloped RNA Viruses.* Springer-Verlag, New York.

Gonzalez, S. A., Affranchino, J. L., Gelderblom, H. R., and Burny, A. (1993) Assembly of the matrix protein of simian immunodeficiency virus into virus-like particles. *Virology,* 194:548.

Harrison, S. C. (1990) Principles of virus structure, in *Fundamental Virology,* 2nd ed. Fields, B. N., Knipe, D. M., Chanock, R. M., Hirsch, M. S., and Melnick, G., Eds., p. 37. Raven Press, New York.

Liljestrom, P., Lusa, S., and Garoff, H. (1991) In vitro mutagenesis of a full-length cDNA clone of Semliki Forest Virus: The small 6,000-molecular-weight membrane protein modulates virus release. *J. Virol.* 65:4107.

Scheele, C. M., and Hanafusa, H. (1971) Proteins of helper-dependent RSV. *Virology,* 45:401.

Suomalainen, M., and Garoff, H. (1992) Alphavirus spike-nucleocapsid interactions and network antibodies. *J. Virol.* 66:5106.

Zingler, K., and Littman, D. R. (1993) Truncation of the cytoplasmic domain of the Simian Immunodeficiency Virus envelope glycoprotein increases *env* incorporation into particles and fusogenicity and infectivity. *J. Virol.* 67:2824.

# VIRUSES, DNA PACKAGING OF

*Philip Serwer*

## *Key Words*

**Bacteriophage** A virus that infects bacteria.

**Capsid, Viral** The protein component of a virus, within which the viral nucleic acid is packaged.

**Concatemer, DNA** An end-to-end aggregate or polymer of DNA.

**Energy, Transduction of** Conversion of energy from one form to another.

**Mutant, Conditional-Lethal** An organism genetically altered so that it is nonviable in some, but not all, circumstances in which the nonmutant organism is viable.

**Procapsid, Viral** The pre-DNA packaging precursor of the mature capsid of a virus.

**Protein, Scaffolding** A protein that is part of and necessary for assembly of a procapsid but is lost during conversion of the procapsid to mature capsid.

The packaging of DNA in preformed viral procapsids is an event of DNA metabolism that is used both as a model for studies of basic science and as a tool for biotechnology. By integrating the disciplines of physical chemistry, genetics, and biochemistry, attempts are made to develop a comprehensive understanding of DNA packaging. By combining the advantages offered by in vitro packaging and the comparatively short bacteriophage life cycle, bacteriophages can be used as tools for studies of both the nucleotide sequence of eukaryotic genomes and the evolution of proteins.

## 1 BASIC PHENOMENOLOGY AND RATIONALE

Of the characteristics that define an organism, purposeful behavior is the most obvious. To determine how purposeful behavior has evolved from interactions of macromolecules that are inanimate when isolated from one another, systems are needed that both mimic purposeful behavior and are simple enough to permit analysis. To provide such systems, researchers have analyzed the processes by which a bacterial virus (also called a bacteriophage or phage) multiplies after it has infected a host cell. One reason for choosing bacteriophage systems was that techniques of genetics were most easily applied to bacteriophages. The science of bacteriophage genetics was assisted by the comparatively short bacteriophage life cycle—13 minutes in the case of the most rapidly growing bacteriophage, T7, and its relatives (at 37°C). The most useful type of bacteriophage mutant has been the conditional-lethal mutant, that is, a mutant that won't grow in one (nonpermissive) condition, but will in another (permissive) condition. Stocks of the conditional-lethal mutant bacteriophage are prepared by growth in the permissive condition; the mutant phenotype is analyzed by growth in the nonpermissive condition. The most often used types of conditional-lethal mutant are the temperature-sensitive mutant (nonpermissive condition: high temperature; permissive condition: low temperature) and the polypeptide chain termination mutant (permissiveness of conditions determined by the host cell used). Isolation of temperature-sensitive mutants is performed by growing bacteriophages in host cells that are spread in a gelled solid support made of agar. For bacteriophage T7 first grown at 30°C and then at 42°C, some temperature-sensitive mutants (arrows in Figure 1) form a cleared zone of killed bacterial cells (plaque), that is smaller than the plaque formed by nonmutant T7. By analysis of the products of its growth in a nonpermissive condition, a conditional-lethal mutant is used to help determine the role of the mutated gene. An example of this analysis is described in Section 3.

In addition to serving in studies of biological processes, some bacteriophages are useful for carrying pieces of cloned DNA from higher organisms. In vitro packaging of bacteriophage DNA joined to a foreign DNA insert is an essential component of some procedures for cloning DNA in double-stranded DNA bacteriophages. The cloned DNA is useful for at least two aspects of applied biotechnology: mapping and sequencing the DNA genome of higher organisms, and causing directed evolution of the protein encoded by the cloned DNA, if the protein is displayed on the outside of the bacteriophage. Thus, bacteriophages have been studied for practical reasons, as well as for reasons of basic science.

## 2 THE LIFE CYCLE OF DOUBLE-STRANDED DNA BACTERIOPHAGES

Initiation of infection occurs by adsorption of a bacteriophage to a host cell. In the case of several double-stranded DNA bacteriophages, the first productive contact appears to be made by fibers that extend from an external projection (tail) from the outer shell of the bacteriophage. The length of the tail varies among the different double-stranded DNA bacteriophages. The outer shell, tail, and tail fiber are all made of protein. The protein component (capsid) of bacteriophage T7 is sketched in Figure 2a; for T7, but not for most other double-stranded DNA bacteriophages, an internal cylinder of protein is coaxial with the external tail. The tail, outer shell, and internal cylinder are joined at a sixfold symmetric connector that is also present in all other double-stranded DNA bacteriophages.

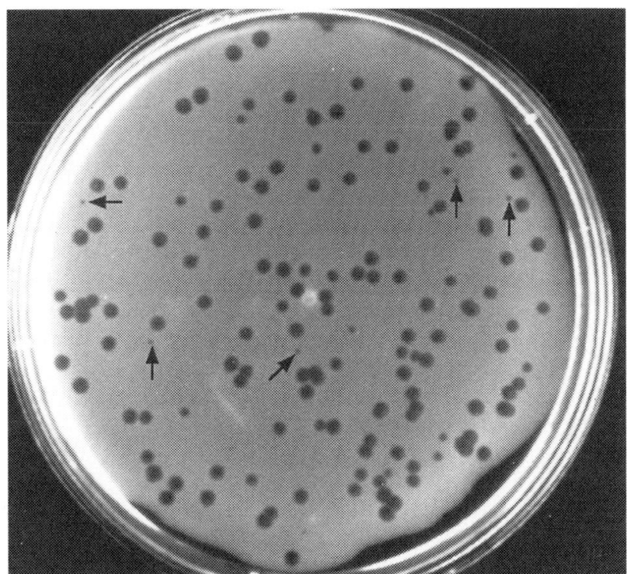

**Figure 1.** Isolation of temperature-sensitive bacteriophage T7 mutants by growth in a host spread in a gelled growth medium.

After adsorption of a bacteriophage to a host cell, the bacteriophage DNA enters the cell, presumably through an axial hole observed in the tail–cylinder complex of T7 (Figure 3). For T7, but not several other bacteriophages, productive infection appears to require injection in a unique direction: the genes that will be expressed first (left end of T7 DNA) must be injected first. The genes injected encode proteins that do the following (in temporal order): inactivate host defenses against infection, transcribe RNA, replicate DNA, and assemble to form the bacteriophage capsid. For all studied double-stranded DNA bacteriophages, both the outer shell and the internal proteins of the capsid are assembled before the genome of DNA is packaged in the outer shell. However, the capsid first assembled (also called a procapsid; illustrated for T7 in Figure 2b) differs from the capsid of the mature bacteriophage in the following ways:

1. The procapsid is rounder, smaller, and more electrically charged (negative at neutral pH) at its surface.

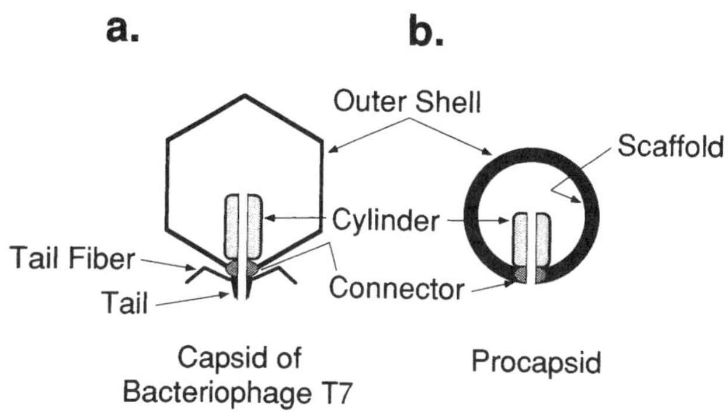


**Figure 2.** Bacteriophage T7 structure: (a) the mature capsid and (b) the procapsid.

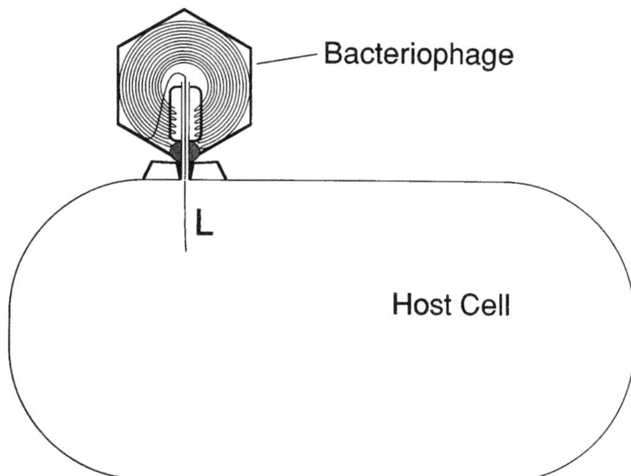


Figure 3. Injection of T7 DNA into a host bacterial cell at the beginning of an infection. The left end of T7 DNA is indicated by the letter, L. In relation to the bacterial host cell, the bacteriophage, 60 nm in outer diameter, is larger in the figure than it is in reality. The host cell is 2000 to 3000 nm long.

2. The procapsid has a scaffolding protein that is necessary for assembly of the outer shell but leaves the capsid during packaging of DNA.
3. The procapsid has neither the tail nor the tail fibers of the mature bacteriophage; these are added after packaging of DNA.

We now discuss the process of the packaging of double-stranded DNA in the procapsids of bacteriophages, using bacteriophage T7 as an example.

## 3 THE DNA PACKAGING PATHWAY

The DNA packaging pathway starts with both the bacteriophage procapsid and its DNA substrate. In the cases of bacteriophage T7 and most (but not all) other double-stranded DNA bacteriophages, the DNA substrate is an end-to-end joined polymer (concatemer) of the mature DNA. During packaging, the concatemer is cleaved to form a mature DNA molecule. In all cases, the mature DNA has the sequence at one end repeated at the other (called terminally repetitious; ABC in Figure 4). Some bacteriophages cleave mature genomes from more than one pair of sites on a concatemer, resulting in the production of mature genomes that are permuted. That is, some genomes begin with B, some with C, and so on. Other bacteriophages cleave from only one pair of sites, thereby producing nonpermuted mature genomes.

The mature T7 DNA has the following properties: nonpermuted genomes, a length of 39.936 kilobase pairs, and a 160 base pair terminal repeat. During packaging, the concatemer is cleaved to mature length. The concatemer might be formed either by multiple cycles of replication around a circular DNA template or by joining of mature genomes by hydrogen bonding of complementary single-stranded terminal repeats. Although the mechanism for concatemer formation is not known, concatemers are both the predominant substrate for DNA packaging in infected cells (in vivo) and the preferred substrate when DNA is packaged in cell-free extracts of T7-infected cells (in vitro).

| Mature Genome | Concatemer |
|---|---|
| ABCDEF...XYZABC | ABCDEF...XYZABCDEF...XYZABCDEF... |

Figure 4. Comparison of a mature bacteriophage genome and a concatemer; letters indicate blocks of nucleotides.

The initial binding of procapsid to concatemeric DNA has not been characterized biochemically. To form this complex, however, all studied bacteriophages are known to require two accessory proteins that are not needed to assemble the procapsid. This information was obtained by infecting a nonpermissive host with bacteriophage that had a conditional-lethal mutation in a gene for an accessory protein. The phenotype observed was normal replication of DNA, normal assembly of procapsids, but no initiation of DNA packaging. For T7, other data indicate that both the nucleotides at a binding site near the right end of a mature genome and the nucleotides at the eventual site of cleavage are necessary for packaging.

To explain this observation, together with the preference for concatemers and also evidence in favor of a blunt-ended substrate at the time DNA enters the T7 capsid, the model of Figure 5 was developed. According to this model (for T7), DNA enters the capsid right end first; this prediction has been found to be correct. However, two possible means of entry exist: (1) entry via a moving right end (illustrated in the top drawing at the right of Figure 5), and (2) entry via a moving fold attached to a fixed right end (illustrated in the bottom drawing at the right of Figure 5). As the DNA enters, eventually the terminal cleavage of the concatemer is made. Initial and terminal cleavages are made by all bacteriophages that package a concatemeric DNA. In the case of a nonpermuted DNA, the terminal cleavage is initiated by filling of the capsid. In at least two cases, the mechanobiochemical transduction for signaling this terminal cleavage requires participation of the connector. In the case of a nonpermuted DNA, one of the accessory proteins is also the nuclease that performs both the initial and the terminal cleavages.

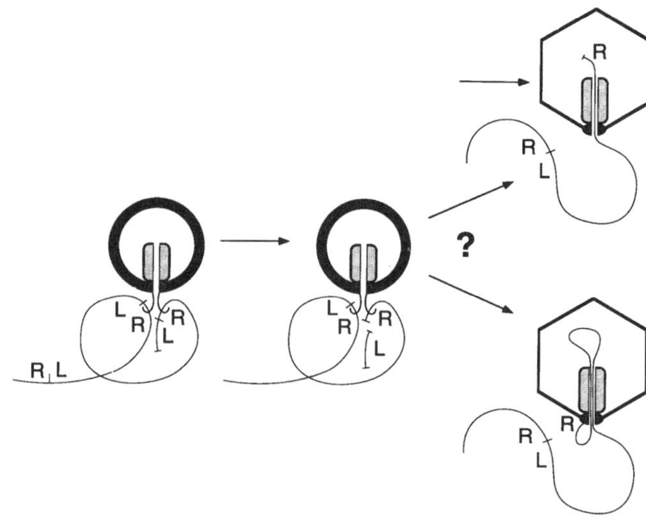


Figure 5. Model for the pathway of packaging bacteriophage T7 DNA. L, left end; R, right end of a T7 genome within a concatemer.

For at least one bacteriophage with nonpermuted DNA (bacteriophage lambda), filling of the capsid improves the efficiency of the terminal cleavage. For T7, the stage at which the procapsid outer shell irreversibly converts to a mature outer shell appears to be very early during DNA entry. However, the timing of this conversion appears to vary among the different bacteriophages.

## 4 MECHANISMS OF ENERGY TRANSDUCTION DURING DNA PACKAGING

Studies of in vitro DNA packaging have revealed that cleavage of ATP is required. Similar release of chemical energy appears to be necessary for all purposeful biological activity. In the case of T7, titration calorimetry revealed that $0.47 \pm 0.05$ kcal/mol of nucleotide pair is the amount of heat released when DNA is expelled from its capsid. Providing this energy is at least one of the possible functions of ATP cleavage. However, neither the mechanism of energy transduction nor all the roles of ATP cleavage have been determined. Mechanisms of energy transduction are not well understood for most biological systems, including muscle. Three types of viable model have been proposed for the transduction of energy during DNA packaging:

1. The connector acts as a worm gear that drives DNA into the capsid during ATP-driven rotation of the connector.
2. The accessory proteins cyclically reach and pull the DNA being packaged.
3. The connector acts as an ATP-dependent gate (i.e., Maxwell's demon) that, possibly with the assistance of an osmotic pressure gradient, fixes inward-directed Brownian motion of the DNA during entry into the capsid.

In vitro systems appear to be necessary for experiments to identify the type of energy transduction that occurs.

## 5 ANIMAL VIRUSES

Many of the double-stranded DNA animal viruses have packaged DNA that is complexed with the histone proteins that also bind the DNA inside a cell. For these viruses, current evidence supports packaging that occurs as the capsid assembles (i.e., no procapsid). However, both herpes viruses and adenovirus have DNA that is packaged without histones. These viruses do form a procapsid. Adenovirus has a packaging signal at one end of a genome. The logical hypothesis is that for both herpes viruses and adenovirus, DNA packaging occurs, as it does for double-stranded DNA bacteriophages, by entry of DNA into a procapsid.

*See also* Bacterial Growth and Division; Bioenergetics of the Cell; Viral Envelope Assembly and Budding.

### *Bibliography*

Black, L. W. (1989) *Annu. Rev. Microbiol.* 43:267–292.

Casjens, S. (1985) In *Virus Structure and Assembly,* S. Casjens, Ed., pp. 75–147. Jones and Bartlett, Boston.

Earnshaw, W., and Casjens, S. (1980) *Cell* 21:319–331.

Hendrix, R. W. (1985) In *Virus Structure and Assembly,* S. Casjens, Ed., pp. 169–203. Jones and Bartlett, Boston.

Serwer, P. (1989) *Chromosomes: Eukaryotic, Prokaryotic and Viral,* K. W. Adolph, pp. 203–223. CRC Press, Boca Raton, FL.

# Viruses, Photodynamic Inactivation of

*James L. Matthews and Millard M. Judy*

### *Key Words*

**Envelope** Lipid-containing membrane that surrounds some viral particles. It is acquired during viral maturation by a budding process through a cellular membrane. Virus-encoded glycoproteins are exposed on the surface of the envelope.

**Photodynamic** Light-mediated damage to biological systems, usually involving the following sequence: light absorption by photosensitizers, energy transfer to oxygen, attack of singlet oxygen on biologic structure.

**Photosensitizer** Dye that reacts to light to form excited states.

**Singlet Oxygen** Oxygen with a paired electric configuration may be produced by reaction of photosensitizer triplet with ground state oxygen.

**Virion** The complete viral particle, which may or may not have an envelope, serves to transfer the viral nucleic acid from one cell to another.

Photosensitizers are typically organic molecular species that participate in chemical reactions following absorption of light. These reactions can involve direct chemical reaction with biological molecules such as covalent bonding of psoralens to bases, especially pyrimidines, of DNA and RNA. Alternatively, they can involve the creation of highly active species of oxygen such as singlet oxygen by direct energy exchange with the triplet state of the light-excited photosensitizer (type II process) or creation of active oxygen species such as the superoxide anion via free radical intermediate species involving a third organic molecular substrate (type I process). These oxygen-dependent processes, types I and II, are called photodynamic processes and the dye molecules that produce them, photodynamic photosensitizers.

Photodynamic photosensitizers such as hematoporphyrin (Figure 1) and benzoporphyrin derivatives (BPD) (Figure 2), merocyanine 540 (Figure 3), and phthalocyanine (Figure 4) have been used to kill tumor cells because of their selective uptake or retention by membranes of these cells. The cells are killed upon irradiation of the photosensitizer-labeled cells with visible light at wavelengths appropriate for maximal absorption by the dye. This type of cytotoxic effect is dependent on the presence of oxygen, which has higher (approximately $10\times$) solubility in membranes than in aqueous media. These short-lived oxygen species created by photodynamic processes damage the membranes of the target cells primarily by lipid oxidation when the photosensitizer is in close proximity to the membrane or is bound within it. Successful use of photosensitizers in malignant tissues has restimulated the examination of their potential use for inactivating viruses, particularly enveloped viruses whose investing membrane might be used to concentrate lipophilic photosensitizers.

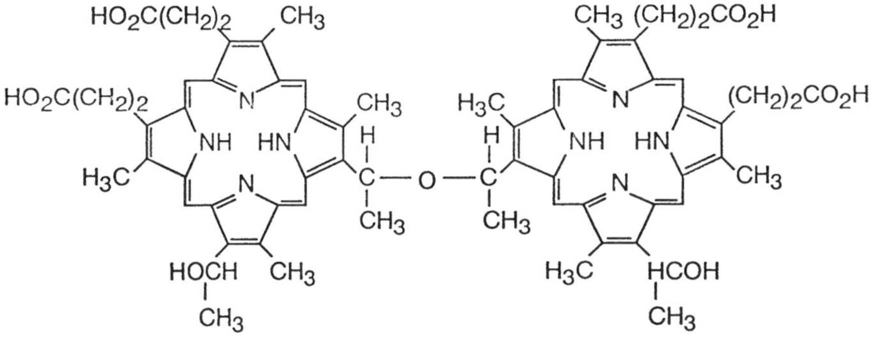

Figure 1. Hematoporphyrin derivative: molecular structure.

## 1 BACKGROUND

The photodynamic susceptibility of microorganisms was recognized as early as 1900 by investigators who used acridine dyes and sunlight. Subsequently, photodynamic therapy was used for herpes simplex virus (HSV) infections employing dyes that act primarily on viral DNA. Methylene blue, neutral red, and proflavine were three of the dyes extensively studied. The use of these dyes and the photodynamic inactivation approach to antiviral therapy was temporarily abandoned because DNA damage due to sensitizer light or UV irradiation exposure disclosed an oncogenic potential of the virus. One photodynamic-treated virus, HSV, was able to transform normal cells in culture, but this capability was not ordinarily expressed because the HSV exerted a lytic action on cells before transformation could occur. Photodynamic treatment with light sensitization possibly resulted in some gene damage, allowing transformation.

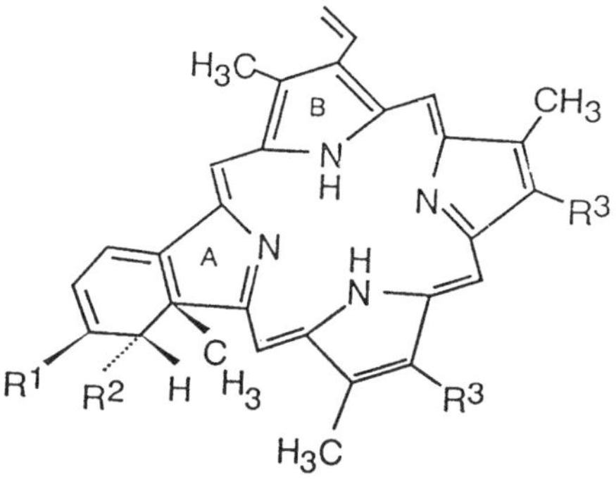

## 2 TYPES OF PHOTOSENSITIZERS

Because some oxygen-dependent photosensitizers with irradiation were successfully employed as antitumor agents, and because they were found to act primarily on organellar and plasma membranes rather than directly on the nucleic acids, these photosensitizers were also investigated for antiviral potential. Examples include hematoporphyrin and benzoporphyrin derivatives, merocyanines, phthalocyanines, carbocyanines, sapphyrins, extended-ring porphyrins, and puerpurins.

This contribution discusses hematoporphyrin derivative (HPD), a complex mixture of ringed tetrapyroles, and dihematoporphyrin ether (DHE) as a prototypic system illustrative of the mechanism of action and utility of these photosensitizers.

The foregoing materials may be activated at 630 nm, a wavelength that falls within the electronic absorption spectrum of the compound and one that is transmitted moderately efficiently in tissues and body fluids. Indeed, activation of hematoporphyrin derivative with laser light of this wavelength has been used to inactivate HSV-1, cytomegalovirus (CMV), and measles virus; naked viruses were not susceptible. The viral suspensions were held static in optically transparent containers or were passed through a transparent tubular flow system through an illuminated region. The virus was suspended in either aqueous buffer or blood and cultured, following no dye and no illumination (control), dye and no illumination (dark control), and dye and illumination (experimental). The

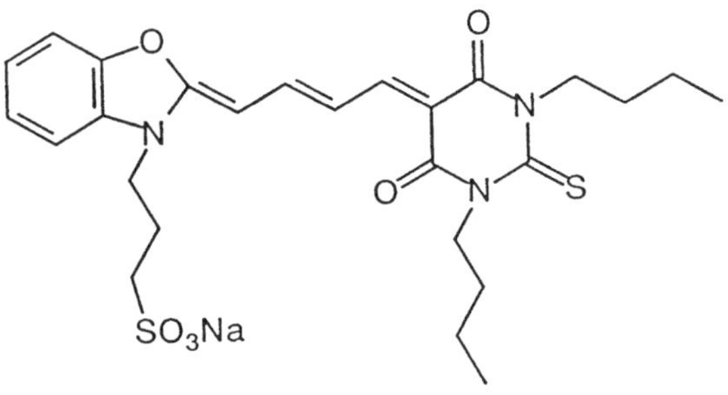

Figure 2. Benzoporphyrin derivative: molecular structure.

Figure 3. Merocyanine 540: molecular structure.

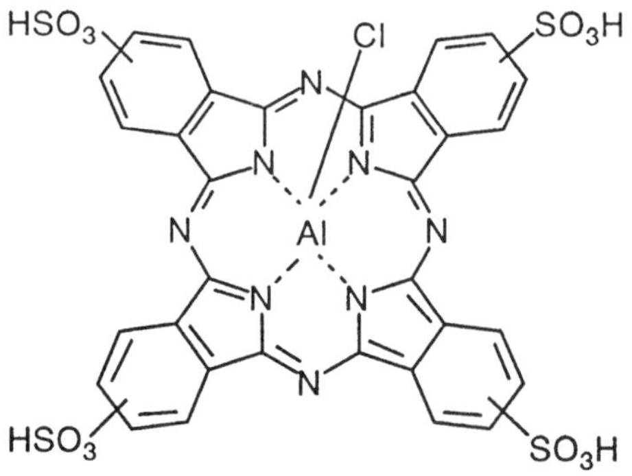

Figure 4. Phthalocyanine: molecular structure.

light source was either an argon-pumped dye laser, a xenon light source equipped with a dichroic heat mirror and a 630 ± 5 nm band-pass filter, or a light-emitting diode (LED) array. Typical intensities of the laser or xenon light source were $1.04 \times 10^{-2}$ W/cm$^2$. HSV-1 was propagated in Vero cells to obtain stock concentrations of $10^6$–$10^7$ (pfu/mL). Viral inactivity was assessed by determining the number of plaque-forming units per milliliter. DHE was used at concentrations of 2.5 μg/mL. Data from this experiment are shown in Table 1.

Photodynamic inactivation of cell-free human immunodeficiency virus (HIV) in cell culture medium also was assessed in the flow cell system. An initial concentration of $4 \times 10^4$ infectious units (iu) was diluted twofold, incubated with 2.5, 10, and 20 μg/mL of DHE in the dark for approximately 30 minutes, and subsequently irradiated in the system at 5 J/cm$^2$. In addition, HIV samples were passed through the system in the presence and absence of light. After photoirradiation, tenfold dilutions of each sample were prepared (neat to $10^{-4}$) and 0.1 mL of each dilution was inoculated into the human CEM T-cell line A3.01. On days 7, 10, 14, and 18, supernatants from the infected cultures were assessed for reverse transcriptase (RT) activity. Figure 5 shows a typical result of this experiment.

The same type of experiment has been used to demonstrate cytotoxicity to viral infected cells and viral suspensions in whole blood. Similar experiments show efficacy of most photosensitizers of this class on enveloped virus.

Even though the presumed primary mechanism of virus eradication is viral envelope damage resulting from singlet oxygen generated upon irradiation, it has been reported that single-stranded DNA breaks occur in nucleocapsids irradiated in the presence of HPD. Virions were readily inactivated and unable to initiate infection. Single-stranded DNA breaks were also observed with adenovirus type 2 treated with HPD, but viral infectivity was not assessed.

A second approach to viral inactivation with photodynamic processes uses the same photosensitizers as in the first approach but the temporal sequence of irradiation is modified: here, the photosensitizer is irradiated prior to exposure to the virus suspension or infected cells. In this case, singlet oxygen is generated but decays long before exposure to the virus. Thus, any virus toxicity or cytocidal activity must be attributed to some other mechanism. Analysis of the postirradiated photosensitizer disclosed a family of new chemical species, which arose from photochemical reactions involving the photosensitizer itself. With some photosensitizers, such as merocyanine 540, the products of irradiation of photosensitizer alone yields a mixture of products that inactivate virus and viral-infected cells. These products maintain viricidal properties for at least 30 days, if refrigerated. Three of these irradiation products have now been isolated (Figure 6). A typical experiment using "preirradiated" merocyanine 540 (PMC540) is shown in Table 2.

The analysis of mechanisms of viral kill using these products is incomplete, but membrane (envelope) damage has been evoked and apoptosis of cells treated in this manner has been noted. The addition of radical quenchers such as glutathione reduces the efficacy of the system.

A third class of photosensitizers has also been developed. These act independently of oxygen and are based on the 3-bromo-4-alkylamino-*N*-alkyl-1,8-naphthalimide skeleton. In fact, some of these new photosensitizers proved successful in neutralizing HIV-1 and inhibiting HIV-1-induced syncytium formation in vitro following light irradiation, while retaining relatively low toxicity toward cells and extremely low toxicities in the absence of light.

**Table 1** Photoinactivation of Herpes Simplex Virus type 1 (HSV-1) in a Flow Cell System[a]

| Concentration of DHE (μg/mL) | Reduction (pfu/mL) In Dark | Reduction (pfu/mL) In Light | $Log_{10}$ Reduction, in Light to Dark |
|---|---|---|---|
| 0 | $3.8 \times 10^6$ | *$2.7 \times 10^6$* | 0.2 |
| 2.5 | $3.1 \times 10^6$ | *$1.0 \times 10^1$* | 5.5 |

[a]HSV-1 photoirradiated in the absence of dye showed a reduction in infectivity of less than 1 $log_{10}$ pfu/mL.

*Source:* J. T. Newman, J. L. Matthews, F. Sogandares-Bernal et. al., Photodynamic inactivation of viruses and its application for blood banking, *Baylor University Medical Center Proceedings* 1(2):8 (1988). Permission granted by *Proceedings*.

The naphthalimide photodynamic effects proceed by forming covalent links with amino acids (e.g., methionine, cysteine, tryptophan, lysine, tyrosine) in intramembranous proteins accessed by the lipophilic photosensitizer. This protein-binding mechanism is supported by bleaching experiments with amino acids in appropriate solvents in the presence and absence of gramocydin, a polypeptide known to form helical channels in membranes. The order of reactivity was found to correlate with the ease with which each reactive compound can act as a single electron donor, suggesting that a single electron transfer (s.e.t.) step is involved in the mechanism. Thus, the mechanism proposed for enveloped virus inactivation is based on the "molecular spring" hypothesis of Marshall et al. In this system, the low energy conformation of the α-helix

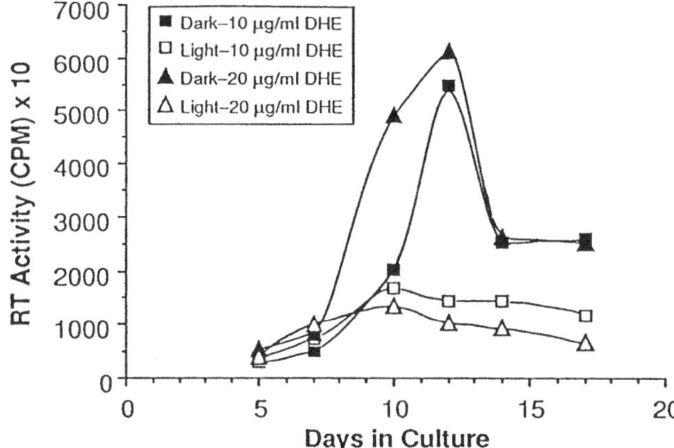


**Figure 5.** Reverse transcriptase activity in counts per minute (cpm) resulting from the inoculation of A3.01 cells with undiluted HIV samples. Samples of HIV at a concentration of $2 \times 10^4$ iu/mL (untreated) were passed through the flow cell in the dark (flow control) or photoirradiated with a light energy density of 5 J/cm$^2$. HIV samples were also photoirradiated at 5 J/cm$^2$ in the presence of either 2.5, 10, or 20 μg/mL of dihematoporphyrin ether (DHE). Tenfold dilutions of each sample were prepared (neat to $10^{-4}$) and 0.1 mL of each dilution was inoculated into the human CEM T-cell line A3.01. Supernatants were obtained from the infected cultures at days 7, 10, 14, and 18 and assessed for reverse transcriptase (RT) activity as described by Chanh et al. RT activity observed with uninfected cell cultures with our system usually ranges from 1000 to 2500 cpm. Each point in the graph is the mean of triplicate assays. [From J. L. Matthews, J. T. Newman, F. Sogandares-Bernal, et al., Photodynamic therapy of viral contaminants with potential for blood banking applications, *Transfusion* 28(1):82 (1988). Permission granted by *Transfusion*.]

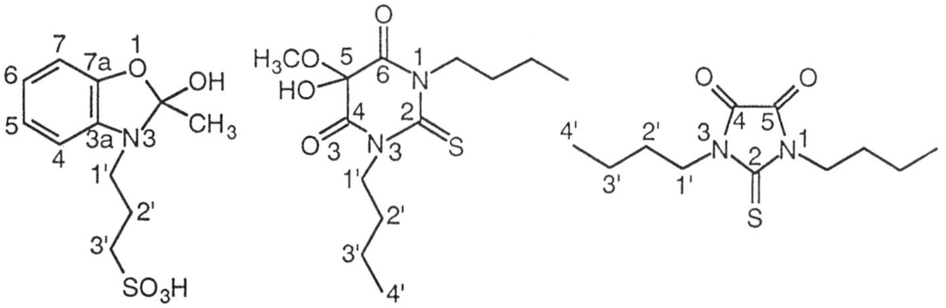

**Figure 6.** Three derivatives of light-activated merocyanine 540: molecular structures.

of transmembrane proteins occurs with a low energy transition to the $3_{10}$ helical conformation. Covalent linking of the photosensitizer to transmembrane protein would prevent the needed conformational change (as occurs during viral invasion of cells at receptor sites), thus precluding delivery of the nucleocapsid or contents to the cell. Gel electrophoresis of photosensitizer-irradiated membrane proteins of HIV indeed show binding of these proteins. Of the compounds tested to date, the monomeric naphthalimides represented by 3-bromo-4-(hexylamino)-*N*-hexyl-1,8-naphthalimide are effective against HSV-1 at concentrations around 250 nM, and the dimeric naphthalimides represented by the bisamide are effective against the same pathogen at concentrations below 100 nM. Both classes of dye are also effective in neutralizing vesicular stomatitis virus (VSV), bovine herpes virus, type 1 (BHV-1), and HIV.

With the increased interest in the potential for employing photodynamic methods for viral eradication, additional photosensitizers that bind to DNA and RNA are being evaluated. In this context, a fourth group of photosensitizers, which can act with or without oxygen, is being recognized. These are the psoralen dyes (Figure 7), substances that have been used extensively in the clinical care of psoriasis and for other medical uses for decades. In the presence of oxygen, these substances are photodynamically active and singlet oxygen is generated. However, these UV-excitable substances can also act in the absence of oxygen as a consequence of their capability to bind DNA and RNA. Indeed, in the presence of UVA, psoralens form covalent adducts with DNA. Presumably this chemistry reflects the fact that psoralens are planar tricyclic furocoumarins, which, in turn, can intercalate between stacked base pairs, especially those derived from pyrimidine bases. Interstrand cross-links are also possible between double-stranded DNA and RNA, between hybrid helices, and between mRNA and DNA. Nucleic acid transcription and replication are thus blocked upon UVA irradiation of psoralen-treated cells.

The foregoing observations suggest that psoralen photosensitizers, when irradiated with UV, could inactivate several types of virus with either RNA and DNA constructs. Not surprisingly, therefore, numerous analogues of psoralen have been synthesized. Many of these have shown efficacy in viral kill, with the antiviral effects of these analogues being dependent on their affinity for single- or double-stranded RNA. Cross-linking of viral nucleic acids with UVA excitation of psoralen photosensitizers is significantly higher than that typically required for virus inactivation. This "overkill" precludes the opportunity for survival or repair of the affected virus. Thus, psoralens have been studied for their potential use in eradicating virus from blood. As with the other sensitizers already described, much of the motive here is to reduce the possibility of transmission of disease via transfusion. HIV, CMV, and other viral pathogens are inactivated with this procedure with only minor effects reported on the red blood cells (RBC) and platelets. The major concern for utilizing this system is the mutagenic potential

**Table 2** Preactivated Merocyanine C540 (PMC-540) Inactivates Cell-Free Human Immunodeficiency Virus, Type 1 (HIV-1) and Simian Immunodeficiency Virus

| MT-4 Cell Infection | | | Viral Antigens (pg/mL)[a] | | | |
|---|---|---|---|---|---|---|
| | | | HIV-1 | | SIV | |
| HIV-1 | SIV | Concentration of P MC-540 (μg/mL) | Day 4 | Day 7 | Day 5 | Day 10 |
| + | ms; | | 735.3 | 956.7 | | |
| + | ms; | 80.0 | 9.5 | 12.4 | | |
| + | ms; | 160.0 | 7.6 | 6.8 | | |
| + | ms; | 200.0 | 8.1 | 8.0 | | |
| ms; | + | | | | 73.0 | 65.9 |
| ms; | + | 80.0 | | | 10.4 | 12.9 |
| ms; | + | 160.0 | | | 4.2 | 5.0 |
| ms; | + | 200.0 | | | 3.5 | 2.8 |

[a]Mean of duplicate determinations using the antigen-capture enzyme-linked immunosorbent assay kits for p24 antigen HIV-1 and core antigen SIV.

*Source:* T. C. Chanh, J. S. Allan, S. Pervaiz, J. L. Matthews, et al., Preactivated merocyanine 540 inactivates HIV-1 and SIV: Potential therapeutic and blood banking applications, *J. AIDS*, 5:191 (1992). Permission granted by Raven Press Publishers.

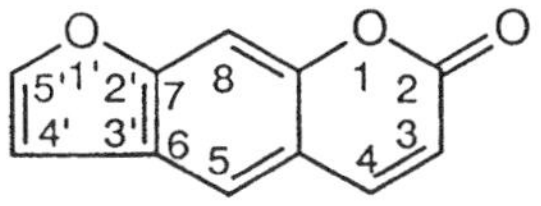

Figure 7. Psoralen: molecular structure.

for photosensitizer/UVA-treated virus. However, since both the RBC and the platelets are anuclear, this mutagenic potential is not as high as it would be in other applications.

## 3 POTENTIAL USES AND PROSPECTS FOR PHOTOSENSITIZERS

Several potential antiviral applications for photosensitizers are being evaluated. The first major approach to application has been the eradication of virus from blood and blood products. Since the photosensitizer can be added to the biological fluid (blood, RBC concentrates, platelet concentrates, etc.) and illuminated in chambers permitting the control and optimization of light penetration, it is possible to activate all photosensitizers within a given volume. This is not possible in the body, where shapes and volumes are irregular and there is light scattering by tissues. The important goal in this blood-cleansing application is to eradicate the virus and viral-infected cells without damaging the normal blood elements. Significant progress has been made where photosensitizers (psoralens, BPD, phthalocyanines, etc.) are effective viral inactivators while causing only minor damage to RBC and platelets. Damage to normal elements has been reduced by using quenchers such as glutathione, mannitol, and various thiols, suggesting that an optimal combination of photosensitizer, light, and quencher will be developed. Fortunately, the viral pathogens of greatest import in blood are the enveloped viruses such as HIV, CMV, HBV, and human T-cell leukemia virus, which are all susceptible to all four of the photosensitizer groups discussed. Other biological fluids of interest include semen, and serum for isolation of various proteins.

Since it is possible to circulate blood outside the body and return it to circulation as is done in dialysis, it is possible that some therapeutic modality might be developed. This extracorporeal approach could permit controlled illumination of blood while in circulation. Interestingly, the preilluminated, photoproduced merocyanine daughter compounds, which do not require direct irradiation of the viral or cell target, may represent a new type of drug derived from light-activated photosensitizers.

The final question to be addressed before these photosensitizers are used for viral eradiation from biological fluids or as therapeutic modalities will be the determination of their efficacy in killing cells containing the provirus forms (i.e., where the viral nucleic acid is integrated within the host cell genome but not yet expressed in terms of an active viron). This is the subject of intense study in ongoing research.

*See also* ULTRAVIOLET RADIATION DAMAGE TO DNA.

### *Bibliography*


Chanh, T. C., Allan, J. S., Matthews, J. L., Sogandares-Bernal, F., Judy, M. M., Skiles, H. G., Leveson, J., Marengo-Rowe, A., and Newman, J. T. Photodynamic inactivation of simian immunodeficiency virus. *J. Virol. Methods,* 26:125–132 (1989).

Chanh, T. C., Allan, J. S., Pervaiz, S., Matthews, J. L., Trevino, S. R., and Gulliya, K. S. Preactivated merocyanine 540 inactivates HIV-1 and SIV: Potential therapeutic and blood banking applications. *J. AIDS* 5:188–195 (1992).

Hanson, C. V., Riggs, J. L., and Lennette, E. H. Photochemical inactivation of DNA and RNA viruses by psoralen derivatives. *J. Gen. Virol.* 40:345–358 (1978).

Horowitz, B., and Valinsky, J. Inactivation of viruses found with cellular components. In *Biotechnology of Blood,* Jack Goldstein, Ed. Butterworth-Heinemann, Stoneham England, 1991, p. 431.

Lytle, C. D., Carney, P. G., Felten, R. P., Bushar, H. F., and Straight, R. C. Inactivation and mutagenesis of Herpes virus by photodynamic treatment with therapeutic dyes. *Photochem. Photobiol.* 50(3):367–371 (1989).

Marshall, G. R., Hodgkin, E. E., Langs, D. A., Smith, G. D., Zabrocki, J., and Leplawy, M. T. Factors governing helical preference of peptides containing multiple α, α-dialkyl amino acids. *Proc. Natl. Acad. Sci. U.S.A.* 87:487–491 (1990).

Matthews, J. L., Newman, J. T., Sogandares-Bernal, F., Judy, M. M., Skiles, H., Marengo-Rowe, A. J., and Chanh, T. C. *Transfusion,* 28:81–83, (1988).

Melnick, J. L., and Wallis, C. L. Photodynamic inactivation of herpes virus. In *The Science of Photomedicine,* J. Regan and J. Parrish, Eds. Plenum Press, New York, 1982, p. 545.

Neyndorff, H. C., Bartel, D. L., Tufaro, F., and Levy, J. G. Development of a model to demonstrate photosensitizer-mediated viral inactivation in blood. *Transfusion,* 30:485–490, (1990).

North, J., Neyndorff, H., King, D., and Levy, J. G. Viral inactivation in blood and red cell concentrates. *Blood Cells,* 18:129–140 (1992).

Raab, O. Über die wirkung fluorescierender Stoffe auf Infusorien. *Z. Biol.* 39:524–530 (1990).

Spikes, J. D. Photodynamic reactions in photomedicine. In *The Science of Photomedicine,* J. Regan and J. Parrish, Eds., Plenum Press, New York, 1982, p. 113.


# VIRUSES, RNA PACKAGING OF

*Dennis H. Bamford*

### *Key Words*

**Packaging, Encapsidation** The process by which the virus genome is translocated and condensed into the virus capsid.

**Virus** An intracellular parasite that utilizes the cellular machinery to synthesize its components, which assemble to a particle containing the genome and protected by protein or protein and lipid layers.

**Virus Capsid** Protective symmetrical protein shell formed by a large number of protein subunits.

**Virus Genome** A nucleic acid molecule. Chemically the genome can be either ribonucleic or deoxyribonucleic acid. These molecules can exist in single- or double-stranded form. Any of these four possible molecules can serve as a virus genome.

A key event in the assembly of a virus is the specific recognition of the genome by the capsid precursor. This interaction must be very selective, to be able to distinguish the viral genome among the number of nucleic acid molecules in the cell. This is particularly true with RNA viruses. The understanding of this event may afford an opportunity to combat viral diseases. RNA and protein interactions are one of the most important macromolecular recognition events in biology, and thus RNA virus packaging offers a good model system to study this largely unknown phenomenon.

## 1 BACKGROUND

Viruses are intracellular parasites that are fully dependent on the metabolic apparatus of the cell. The viral genome contains the information for its multiplication in the cell as well as for directing the synthesis of structural components of the virion. The virus genome can be either DNA or RNA, double stranded (ds) or single stranded (ss). The nucleic acid interacts with the viral structural proteins to form a nucleic acid–protein complex. This event is referred to as nucleic acid packaging or encapsidation. The protein coat thus formed, which protects the nucleic acid, is the virus capsid. The capsid can be organized either helically or icosahedrally. The helical nucleocapsid is an elongated rod-shaped structure in which the coat proteins form a helix around the genome. The icosahedral capsid is a spherical shell-like structure enclosing the condensed viral genome (Figure 1). In some viruses these structures are additionally surrounded by a membrane.

Double-stranded nucleic acids are antiparallel molecules. In the case of RNA, the strand that can serve as the messenger in protein synthesis is known as the (+)-strand and the complementary strand as the (−)-strand. There are RNA viruses that have either positive, negative, or double-stranded dsRNA as the genome. The genome can be a single RNA molecule or segmented. Double-stranded RNA genomes are often segmented. The genome segments can reside in a single capsid or in different capsids (multipartite). The (+)-strand RNA genomes, upon introduction to the cell, can serve directly as messengers, whereas the (−)-strand and dsRNA genomes must be transcribed to (+)-strands in the cell. The (+)-strand synthesis is carried out by virion-associated polymerases either directly or via a DNA intermediate (retroviruses).

One of the key events during virion assembly is the encapsidation (packaging) of the viral genome into a protein capsid. Based on the great variety of genome strategies in RNA viruses, there evidently must also exist variation in the mechanisms used for RNA packaging. It is obvious that only very specific interactions between the RNA and the capsid protein(s) can lead to the very selective packaging and to the formation of a functional metastable virion structure which, upon response to stimuli during the virus entry, leads to virion disassembly and genome activation. Our understanding of the RNA packaging events mostly relies on a few model systems, but information from a great variety of viruses is rapidly accumulating. This entry briefly describes various RNA virus systems about which packaging information is available.

## 2 TOBACCO MOSAIC VIRUS (TMV), THE CLASSIC CASE

TMV is a helical virus composed of a (+)-sense single-stranded, 6395 nucleotide long RNA molecule associated with some 2140 identical capsid protein molecules (Figure 1A). Already in the early 1950s infectious virus particles had been reconstituted from purified RNA and dissociated coat protein subunits. This self-assembly reaction of purified components allowed detailed study of the packaging reaction.

The protein structure initiating the virus assembly is a short (2–3 turn) helix of the coat protein containing 39 to 49 subunits (the so-called 20s particle). A disordered loop in the protein prevents

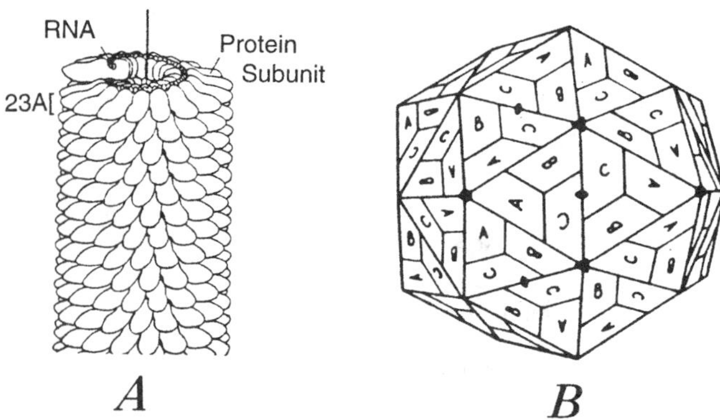


**Figure 1.** Principal structural arrangements of virus capsids. (A) Model of the tobacco mosaic virus: in this helical capsid, the coat proteins are organized helically around the viral RNA. (B) Model of a (+)-strand RNA virus from a plant: in this icosahedral capsid, the coat proteins (A,B,C) form a spherical structure enclosing a condensed viral RNA molecule.

further elongation of this initiation complex. An internal region of the genomic RNA called the "origin of assembly" site (OAS), located in the region between nucleotides 5443 and 5518 from the 5′ terminus, interacts with the nucleating aggregate. The binding of the RNA bases to the protein causes the folding of the disordered loop, leading to helix elongation. The current packaging (assembly) model postulates that the 5′ side of the RNA is encapsidated rapidly and precedes 3′ encapsidation. Furthermore, the upstream RNA passes through the lumen of the growing rod as it is incorporated into the virion. It is not clear how large building blocks are added to the growing rod. The assembly process is greatly dependent on the pH and ionic conditions, as is disassembly upon infection.

## 3 SMALL ICOSAHEDRAL POSITIVE-STRAND RNA VIRUSES

Small (+)-strand RNA viruses have been found infecting bacteria, plants, and animals. These viruses are among the simplest replication systems in biology. High resolution structural data for a number of these viruses are available, and the information has greatly advanced our understanding of the assembly of these simple viruses. The overall structural principles are surprisingly similar in ssRNA viruses infecting plants and animals (Figure 1B), whereas a different arrangement is found in the corresponding bacterial viruses. In spite of the structural similarities of the capsids, the condensed RNA molecule inside the capsid can be organized differently. Either it can be randomly organized (so that it does not contribute significantly to the diffraction pattern), or parts of the RNA can be associated with the capsid proteins in a symmetric fashion (making the nucleic acid visible in the electron density map).

Single-stranded RNA bacteriophage assembly is probably the best understood among viruses of this type. These viruses (MS2, R17, etc.) have one coat protein, which exists as a dimer. Initially coat proteins bind specifically to a 21 nucleotide long asymmetric RNA hairpin loop, which is located in the translation initiation region of the replicase gene in the (+)-strand. This binding leads to translational repression of the replicase synthesis but is most probably also responsible for the initiation of the genome packaging. The initial binding stimulates the binding of additional coat protein dimers to nonspecific sequences, resulting in a highly cooperative assembly of coat proteins that form a shell that encloses the viral RNA. Intact ssRNA bacteriophage capsid contains a single copy of a maturase protein, necessary for infection, which has been shown to bind RNA; capsid proteins, however, can specifically package virus RNA without this protein. ssRNA bacteriophage represents a model in which the RNA genome is condensed to the capsid simultaneously with protein shell assembly. Both picorna virus and certain plant viruses (e.g., southern bean mosaic virus, tomato bushy stunt virus, turnip crinkle virus) must probably assemble via discrete intermediate units. First the different coat proteins form a small aggregate, which associates to a ''pentamer.'' It is not clear whether these particles condense around the viral RNA or whether an empty procapsid is formed which packages the RNA. Anyway, the packaging is very specific for the (+)-sense genomic RNA. The picornavirus genome contains a 5′-terminally linked protein, but it is not the specificity factor for packaging.

Ordered ssRNA has been observed in the crystal structure of bean pod mottle virus, a comovirus. Some 20% of the RNA is icosahedrally oriented with no sequence specificity. Whether this RNA–capsid protein interaction has implications for RNA packaging remains to be seen. These viruses, however, can form stable empty capsids, indicating that correct capsid protein–protein interactions take place without the genomic RNA. A nodavirus (flock house virus) capsid has been shown to bind portions of the genomic ssRNA in a dsRNA form. About 10 bases bind to the coat protein at the icosahedral twofold axis. The dsRNA is in the A form and is formed from the secondary structures of nonunique sequences. This RNA is an integral part of the capsid quarternary structure, suggesting a capsid condensation around the RNA. This model is also supported by the observation that no empty particles have been observed in nodavirus-infected cells.

## 4 DOUBLE-STRANDED RNA VIRUSES

Since cellular polymerases are not capable operating on dsRNA templates, all dsRNA viruses have a virion-associated, RNA-dependent RNA polymerase. Many dsRNA viruses have segmented genomes. These viruses, in addition to the specificity for selecting viral RNA for packaging, must have a mechanism ensuring that one copy of each genome segment is included in the virion. Depending on the virus, the number of segments can vary from a few to more than 10. This section concentrates on two dsRNA viruses whose *in vitro* assembly systems have shed light to the RNA encapsidation reaction.

Bacteriophage $\phi$6 has three genome segments packaged in an icosahedral polymerase complex consisting of four specific virus proteins. This structure is surrounded by an additional protein layer and a membrane. Upon entry to the cell, polymerase complex produces (+)-strands for translation. The largest genome segment encodes for the polymerase complex proteins. These proteins are synthesized early during the life cycle. They assemble, without the action of the viral RNA, into empty icosahedral polymerase complexes. The synthesized (+)-strands serve also as genomic strands, which are packaged to the polymerase complex (procapsid) particles. The packaging signal is located at the 5′ end of each segment (within a few hundred bases in the noncoding region). A procapsid-associated NTPase is proposed to drive the RNA translocation–condensation reaction. The RNA segments are packaged separately and independently (i.e., with no order). However, the next step in the virus life cycle, (−)-strand synthesis inside the polymerase complex particle, does not start until all the genome segments have been packaged. It is very intriguing to address the questions of the specific RNA–protein interactions leading to this high selectivity in the packaging reaction. It remains to be seen whether the Reoviridae follow this general scheme for their RNA packaging.

Yeast dsRNA ''virus'' L-A is a degenerate virus, as are all viruses found in fungi. They only exist intracellularly, no external virus particles having been observed. L-A contains a single dsRNA segment, a major coat protein (gag), and an RNA-dependent RNA polymerase (pol), which is expressed as gag-pol fusion protein. The fusion is achieved by a −1 ribosomal frameshift. A stem–loop structure 400 nucleotides from the 3′ end of the RNA is recognized by the pol domain in the gag-pol fusion protein. The model for L-A assembly involves a gag-pol dimer forming the initiation complex with the (+)-sense RNA. The gag domain of the fusion protein primes the assembly of additional gag proteins to form a shell, condensing the RNA inside. The possibility of packaging of the preformed capsid has not, however, been completely ruled out.

## 5 HEPADNA VIRUSES AND RETROVIRUSES

The packaged nucleic acid is positive-sense RNA in both hepadna- and the retroviruses, but the virus particles contain an RNA-dependent DNA polymerase (reverse transcriptase) that converts the packaged RNA to DNA form. The packaging signal in both cases is located near the 5′ end of the RNA. In hepadnaviruses this signal is 85 to 100 nucleotides long (occurring twice in the genome, but only the 5′ copy is active), whereas in retroviruses the complete signal is several hundred nucleotides long.

The retroviral genome encodes three polyprotein precursors (gag, pol, and env). The gag polyprotein is proteolytically cleaved to several proteins. In the absence of gag cleavage, particles with normal amounts of RNA are produced. This suggests that full length of gag is both required and sufficient for RNA packaging and no polymerase (pol) is needed. One domain in gag (NC) binds RNA, but the binding is unspecific. It is not clear what interactions are responsible for the highly specific genome encapsidation. However, it is clear that for the subsequent steps in the retrovirus life cycle the polymerase is an indispensable element in the virion.

Opposite to the retroviruses, hepadnaviruses that are missing the polymerase are unable to encapsidate genomic RNA, leading to the production of capsids with no RNA. The polymerase contains a domain that is specific for packaging and separate from reverse transcriptase and RNase H domains. That the polymerase packaging domain acts preferentially in cis is shown in experiments in which wild-type genomes are used to complement polymerase-deficient mutant genomes. The packaged genomes are predominantly of wild-type origin. Obviously, the capsid proteins are also necessary for packaging but not for specificity.

## 6 POSITIVE-STRAND ENVELOPED RNA VIRUSES

Sindbis virus, a togavirus, has icosahedral symmetry. The assembly of the virion requires two steps. First the genomic RNA interacts with the capsid protein, leading to the formation of the nucleocapsid. Second, the nucleocapsid specifically interacts with the viral membrane proteins, which are located in the cellular membranes. This event results in the release (budding) of viruses from the plasma membrane. The packaging signal is located internally near the 5′ end of the genome, about 12 kb long. The signal lies between nucleotides 746 and 1226. The actual packaging signal, however, is probably shorter. This binding site lies within the coding region of a nonstructural viral protein nsP1. The crystal structure of the sindbis capsid protein reveals a dimer. However, monomeric capsid proteins bind RNA. The RNA binding site is located between amino acids 76 and 116 of the 264 amino acid long protein. This location is between two domains of the protein and is rather conserved in togaviruses. It has been suggested that the interaction between the sindbis virus RNA and this binding confers specificity on the assembly process.

Mouse hepatitis virus, a coronavirus, has four structural proteins. Three of these are associated with the virus envelope, and the fourth protein interacts with the 31 kb long RNA to form a helical nucleocapsid. An internal 61 nucleotide long packaging signal is located close to the 5′ end of the genome (starting at nucleotide 1381). Implications of a stem–loop formation can be observed in this region. It is possible that the coat protein binding to the packaging site affects also the translation of the virus RNA polymerase. Thus the initiation of assembly would also have a regulatory effect on gene expression, as in ssRNA bacteriophages.

## 7 NEGATIVE-STRAND RNA VIRUSES

The influenza virus, an orthomyxovirus, has eight different genome segments that are associated with four polypeptides forming the ribonucleoprotein core. This core exhibits RNA-dependent RNA polymerase activity and is enclosed in the viral envelope. The terminal untranslated 3′ and 5′ nucleotides of the genome RNAs are involved in forming circles via a panhandle structure. Some 20 nucleotides at the termini of the segments are sufficient signals for transcription, replication, and packaging of the RNAs into the virus particles. However, the orthomyxovirus system is complex, and it remains to be seen whether similarities with other RNA packaging systems exist.

## 8 PERSPECTIVES

Virus RNA packaging is a highly selective event in which almost exclusively viral RNA is encapsidated. The selectivity is obtained by specific RNA–capsid protein interactions. These interactions may also have translational regulatory effects. There is a vast amount of information about DNA–protein interactions, but our understanding of RNA–protein interactions is still relatively limited. RNA viruses surely will be in the forefront in helping us to elucidate the interactions between the two most active cellular components, RNA and protein.

The development of recombinant DNA technologies (including reverse transcription) and the understanding of the packaging signaling in many virus systems offer the opportunity to package foreign RNA into viruses. This new avenue for research has crucial importance in helping us to understand the packaging mechanism. Also it has allowed the construction of a number of vector systems, which are widely used in biological and medical investigations.

*See also* RNA Three-Dimensional Structures, Computer Modeling of; Viral Envelope Assembly and Budding; Viruses, DNA Packaging of.

### *Bibliography*

Bamford, D. H., and Wickner, R. B. (1994) *Semin. Virol.* 5:61–69.

Caspar, D. L. D., and Namba, K. (1990) *Adv. Biophys.* 26:157–185.

Chen, Z., Stauffacher, C., Schmidt, T., Fisher, A., and Johnson, J. E. (1990) In *New Aspects of RNA Positive-Strand Viruses,* M. A. Brinton and F. X. Heinz, Eds., pp. 218–226. American Society for Microbiology, Washington, DC.

Fox, J. M., Johnson, J. E. and Young, J. Y. (1994) *Semin. Virol.* 5:51–60.

Ganem, D. (1991) *Curr. Top. Microbiol. Immunol.* 168:61–63.

Linial, M. L., and Miller, A. D. (1990) *Curr. Top. Microbiol. Immunol.* 157:125–152.

Rossmann, M. G., and Johnson, J. E. (1989) *Annu. Rev. Biochem.* 58:533–573.

Schlesinger, S., Makino, S., and Linial, M. L. (1994) *Semin. Virol.* 5:39–50.

# Vitamins, Structure and Function of

*Donald B. McCormick*

### *Key Words*

**Apoenzyme** Protein moiety of an enzyme that requires a coenzyme.

**Avitaminosis** Disease condition, described as a deficiency syndrome, resulting from lack of a vitamin.

**Coenzyme** Organic molecule, generally derived from a vitamin, that functions catalytically in an enzyme system.

**Holoenzyme** Catalytically active enzyme constituted by coenzyme bound to apoenzyme.

**Hypervitaminosis** Unhealthy condition resulting from excess of a vitamin.

**Hypovitaminosis** Unhealthy condition resulting from too little of a vitamin; interchangeable with avitaminosis.

A vitamin is an organic compound that occurs as a natural component of foods and must be supplied exogenously in small amounts to maintain growth, health, and reproduction of an organism. Hence, a vitamin is essential because it cannot be made in the organism that requires it; yet it functions, usually after metabolic alteration, in indispensable ways. Since only small quantities, usually microgram to milligram amounts per human adult per day, are required to avoid a deficiency disease (avitaminosis), vitamins are considered as micronutrients along with the trace elements.

## 1 CLASSIFICATION

### 1.1 Discovery

Symptoms of diseases that were likely the result of insufficient dietary intake of one or more vitamins are described in early writings. Examples are descriptions of signs of beri-beri (lack of thiamine) given in the ancient Chinese herbals (ca. 2600 B.C.) or scurvy (lack of vitamin C) described in the Ebers papyrus (ca. 1150 B.C.).

That diet could influence such diseases was also suspected in writings from ancient Greek, Roman, and Arab physicians who recognized, for example, that ingestion of liver would both prevent and cure night blindness, which today we know to be an early symptom of hypovitaminosis A. Such empirical relationships set the stage for the experimental phase of vitamin discovery that became scientifically based around the beginning of this century.

### 1.2 Nomenclature

The manner in which vitamins have been named historically is first based on capital Arabic letters (A, B, C, etc.). When it was realized that some (e.g., B) were mixtures more complex than initially perceived, subscript numbers were added to the individually isolated vitamins (e.g., $B_1$ and $B_2$). In some instances, the subscripts were used as it became evident that members of a group were functionally similar. Thus A, which is required for vision, was found to differ in some particular of structure in, for example, its $A_1$ (retinol) and $A_2$ (3-dehydroretinol) forms. Common chemical names, which are receiving greater usage, give a better indication of the types of compound involved. In some cases, these have replaced earlier alphabetical designations (e.g., biotin for vitamin H); in other instances, a letter–subscript combination has given way to a chemical descriptor (e.g., niacin for $B_3$). These common names often reflect the presence of some specific atom; thus *thiamine* denotes sulfur. Chemically functional groups are also indicated: pyridox*amine* with the amine function. Even larger portions of the molecular structure can be reflected: *ribo*flavin with a ribityl chain. Parts of some names reflect a biofunctional property, as with chole*calciferol,* which is an alcohol that "carries" calcium, or reflect source/location, as in *phyllo*quinone from plants.

## 2 INDIVIDUAL VITAMINS

### 2.1 Fat-Soluble Cases

The relative solubilities of vitamins continue to provide a generally useful division. The "fat-soluble" group (i.e., A, D, E, and K) are more soluble in organic solvents, a property that also relates generally to their fatlike absorption, transport, and storage. Hence, they are rather well retained in the body and in some cases become toxic at high dosage. Table 1 provides a list of the fat-soluble vitamins required by the human and many animals.

Among functions of A, only the aldehyde form (retinal), which is metabolically derived from the alcohol (retinol), conjugates with specific proteins to generate pigments involved in vision. However, further oxidation of the *trans*-aldehyde to yield all-*trans*-retinoic acid produces a potent metabolite that can associate with receptors that function as ligand-dependent transcription factors. These affect gene expression, especially as it relates to epithelial differentiation. Also vitamin D is metabolically altered to yield the 1α,25-dihydroxy compound that is hormonelike in acting on nuclear receptors

**Table 1** Fat-Soluble Vitamins

| Group/Vitamers | Adult RDA[a] | Dietary Sources | Physiological Roles | Deficiency | Excess |
|---|---|---|---|---|---|
| A | | | | | |
| Retinol<br>Retinal | 800–1000 μg as retinol equivalents | Provitamins are carotenoids with a β-ionone ring in green and yellow/orange vegetables; retinol in liver, whole milk products | Visual pigments; cellular differentiation and proliferation | Night blindness, xerophthalmia, keratomalacia | Headache, vomiting, diplopia, alopecia, desquamation, bone abnormalities |
| D | | | | | |
| Cholecalciferol<br>Ergocalciferol | 5–10 μg as cholecalciferol (=400 IU of D) | Provitamins are 7-dehydrocholesterol in skin and ergosterol from ergot; fortified milk and dairy products, eggs, margarine | Adsorption of calcium and mineralization of bones and teeth | Rickets (children) and osteomalacia (adults) | Nausea and vomiting; anorexia, diarrhea, and loss of weight; calcification of soft tissue, esp. kidney and heart |
| E | | | | | |
| Tocopherols<br>Tocotrienols | 8–10 mg as α-tocopherol equivalents | Seeds/vegetable oils, margarine, shortening; green leafy vegetables | Antioxidant for cell membrane unsaturated lipids and low density lipoprotein | Anemia (premature infants); nerve and muscle degeneration | Relatively nontoxic, but may antagonize vitamin K in those on anticoagulant therapy |
| K | | | | | |
| Phylloquinones<br>Menaquinones | 55–80 μg | Green leafy vegetables; lesser amounts in milk and dairy products, meats, eggs, cereals, fruits | Formation of blood-clotting proteins that bind calcium ions | Hemorrhagic disease | Relatively nontoxic, but synthetic menadione can lead to jaundice |

[a]The lower range is for women, but needs are modestly higher in pregnancy or lactation.

Table 2 Water-Soluble Vitamins

| Group/Vitamers | Adult RDA[a] | Dietary Sources | Physiological Roles | Deficiency |
|---|---|---|---|---|
| $B_1$ | | | | |
| Thiamine | 1.0–1.5 mg | Pork, organ meats, whole grains, legumes | Coenzyme functions in metabolism or carbohydrates and branched-chain amino acids | Beri-beri, polyneuritis, Wernicke–Korsakoff syndrome |
| $B_2$ | | | | |
| Riboflavin | 1.2–1.8 mg | Milk and dairy products, meats, green vegetables | Coenzyme functions in numerous redox reactions | Cheilosis, angular stomatitis, dermatitis |
| Niacin | | | | |
| Nicotinic acid<br>Nicotinamide | 13–20 mg as niacin equivalents | Liver, lean meats, grains, legumes (can be formed from tryptophan) | Cosubstrates/coenzymes for H-transfer with numerous dehydrogenases | Pellagra with diarrhea, dermatitis, and dementia |
| $B_6$ | | | | |
| Pyridoxine<br>Pyridoxamine<br>Pyridoxal | 1.6–2.0 mg | Meats, vegetables, whole-grain cereals | Coenzyme functions in metabolism of amino acids, glycogen, and sphingoid bases | Nasolateral seborrhea, glossitis, peripheral neuropathy (epileptiform convulsions in infants) |
| Folacin | | | | |
| Folic acid<br>Folyl polyglutamates | 180–200 μg | Liver, yeast, leafy vegetables, legumes, whole-wheat products | Coenzyme functions in single-carbon transfers in metabolism of nucleic and amino acids | Megaloblastic anemia, mental disorder, gastrointestinal disturbances, glossitis, pallor |
| $B_{12}$ | | | | |
| Cyanocobalamin | 2 μg | Meats, eggs, dairy products (not in plant foods) | Coenzyme functions in metabolism of odd-number fatty acids and methyl transfers | Pernicious anemia, neurological disorders, pallor |
| C | | | | |
| L-Ascorbic acid<br>Dehydroascorbic acid | 60 mg | Vegetables and fruits, esp. broccoli and citrus | Formation of collagen and carnitine; important as antioxidant | Scurvy with petechiae, ecchymoses, perifollicular hemorrhage, spongy and bleeding gums |
| Other | | | | |
| Biotin | 30–100 μg (ranges recommended) | Liver, yeast, egg yolk, soy flour, cereals | Coenzyme functions in bicarbonate-dependent carboxylations | Fatigue, depression, nausea, dermatitis, muscular pains |
| Pantothenic acid | 4–7 mg (ranges recommended) | Animal tissues, whole-grain cereals, legumes (widely distributed) | Constituent of CoA and phosphopantetheine involved in fatty acid metabolism | Fatigue, sleep disturbances, impaired coordination, nausea |

[a]The lower range is for women, but needs are modestly higher in pregnancy or lactation.

that exert their effect via gene expression. An important consequence is an increase in the level of $Ca^{2+}$-binding protein in the gut. Alteration of K involves its coenzymelike function in an oxidation–reduction cycle coupled with the carboxylation of glutamyl residues of certain proteins to yield $Ca^{2+}$-binding γ-carboxyglutamyl-containing proteins.

### 2.2 Water-Soluble Cases

Vitamin C and the vitamins in the B-complex group are "water soluble" and as such share the fate of other solutes more compatible with an aqueous, physiological medium; this includes a lesser tendency to be retained for long times in the body and a greater loss by way of urinary excretion. Table 2 lists water-soluble vitamins required by the human and many plants and animals.

The eight vitamin groups that comprise the B complex function after metabolic conversions to coenzymes, which are involved in holoenzymic systems responsible for catalyzing numerous reactions. In the case of vitamin C, a cosubstratelike function is seen, with its oxidation coupled to certain hydroxylations.

## 3 PERSPECTIVES

### 3.1 Quasi-Vitamins

Some compounds not listed in Tables 1 and 2 are legitimate vitamins for certain species, though not the human. For examples, lipoic and *p*-aminobenzoic acids for certain microorganisms or carnitine for the common mealworm represent essential organic micronutrients that cannot be synthesized adequately by the organisms dependent on dietary supply of these compounds. More than microamounts of choline and myoinositol are required at early, growth stages for rodents. There are, furthermore, several compounds that at one time or another have been falsely claimed to be vitamins with reputed benefit to humans. These include such ineffective factors as $B_{15}$ or pangamic acid and $B_{17}$ or laetrile. Only the 13 vitaminic groups listed in Tables 1 and 2 presently account for human needs.

### 3.2 Pharmacological Uses

As mentioned with respect to fat-soluble vitamins, particularly A and D, toxicity (hypervitaminosis) can result from overingestion, especially at chronic rates. Excessive amounts of at least a couple of water-soluble vitamins can also pose a health hazard. Hepatoxicity can result from high amounts of nicotinic acid, and neurological dysfunction has been associated with chronic excess of pyridoxine.

There is growing certainty, however, that amounts of L-ascorbate (C) and α-tocopherol (E) modestly above the present recommended dietary allowance (RDA) levels have beneficial effects as natural antioxidants. The decrease in some specific cancers has been associated epidemiologically with higher intakes of foods rich in C, E, and carotenoids that include the provitamin A β-carotene. Extra folic acid also seems to protect the fetuses in a subset of women who have history of neural tube defects in their newborns. More specific and direct experimentation will be needed to clarify such pharmacological roles of vitamins.

*See also* Nutrition; Retinoids; Trace Element Micronutrients.

## *Bibliography*

Brown, M. L., Ed. (1990) *Present Knowledge in Nutrition,* 6th ed. International Life Science Institute–Nutrition Foundation, Washington, DC.

Combs, G. E., Jr. (1992) *The Vitamins, Fundamental Aspects in Nutrition and Health.* Academic Press, San Diego, CA.

Machlin, L. J. Ed. (1991) *Handbook of Vitamins,* 2nd ed. Dekker, New York.

McCormick, D. B., and Greene, H. L. (1993) Vitamins. In *Tietz Textbook of Clinical Chemistry,* 2nd ed. C. Burtis and E. Ashwood, Eds. Saunders, Philadelphia.

NRC (National Research Council). (1989) *Recommended Dietary Allowances,* 10th ed. National Academy Press, Washington, DC.

Sauberlich, H. E., and Machlin, L. J. (1992) Beyond deficiency. New views on the function and health effects of vitamins. *Ann. N.Y. Acad. Sci.* 669.

Shils, M. E., Shike, M., and Olson, J. A. Eds. (1993) *Modern Nutrition in Health and Disease,* 8th ed. Lea & Febiger, Philadelphia.

# Whole Chromosome Complementary Probe Fluorescence Staining

*Heinz-Ulrich G. Weier, Daniel Pinkel, and Joe W. Gray*

*Key Words*

**In Situ Hybridization** Binding of nucleic acid probes to a target nucleic acid sequence inside a mostly intact cell.

**Label** A reporter molecule attached to a hybridization probe that is normally absent in the hybridization target and can be detected by its fluorescent or antigenic properties.

**Probe** A collection of one or many nucleic acid molecules, each a few hundred base pairs long, carrying covalently bound labels.

**Probe Complexity** A measure of the number of different nucleic acid sequences in a probe.

**Whole-Chromosome Probe (WCP)** A collection of labeled nucleic acid fragments that have sequence homology with regions of the genome distributed over an entire chromosome.

Fluorescence in situ hybridization (FISH) with nucleic acid probes that are complementary to specific chromosomes or parts thereof fluorescently "stains" the region targeted by the probes in interphase and metaphase cells. Probes for FISH typically are labeled by nick translation or random primer extension in the presence of deoxynucleoside triphosphates attached to reporter molecules such as biotin, digoxigenin, or fluorescent molecules (fluorescein, spectrum orange, spectrum green, CY3, Texas Red, etc.). FISH with these probes is accomplished by denaturing the probe and target nucleic acid sequences so that substantial fractions of the probe and target molecules become single stranded. The probe and target are then mixed under conditions that permit the probe molecules to hybridize to complementary sequences in the target cell. Probe molecules that carry nonfluorescent molecules are detected after hybridization by incubation with fluorochrome-conjugated avidin or antibodies that have high affinity for the reporter molecules. Targets as small as 1 kilobase or as large as whole chromosomes can be stained using FISH. The staining of whole chromosomes or substantial portions thereof has proved useful for detection and characterization of human chromosomes in interspecies hybrids, for visualization of chromosomal rearrangements in tumor cells or in cells exposed to clastogenic agents, and for investigation of the chromosomal organization of interphase nuclei.

## 1 PROBES

### 1.1 Species-Specific Hybridization

The first demonstration of whole-chromosome staining was accomplished using genomic DNA from one species to stain chromosomes of that species in mixed-species hybrid cells (e.g., human chromosomes in human × hamster hybrid cells). Species-specific hybridization is accomplished using whole genomic DNA from the target species. Hybridization occurs primarily to interspersed repeated sequences in the chromosomes of the target species, with the result that all DNA from the target species is stained more or less uniformly. The divergence between nucleic acid sequences in humans and rodents is high enough to permit species-specific staining to be easily achieved: that is, interspersed repeats in human genetic material will not bind to interspersed repeats in nucleic acid sequences from rodents. However, the use of genomic DNA as a probe for species-specific staining works less well when the DNA sequence homology is high between the two species (e.g., between human and chimpanzee), and it does not enable staining of specific chromosome types.

### 1.2 Chromosome-Specific Probes

Chromosome-specific staining requires whole-chromosome probes (WCPs), which are composed of a large number of different, labeled DNA sequences that have homology throughout the target region. In general, increasing the probe complexity (i.e., the fraction of the target DNA sequence that has homology to sequences in the probe) will result in more intense and homogeneous staining. Generation of WCPs begins with enrichment of DNA fragments from the target chromosome, as described in the subsections that follow. The enriched DNA fragments are then amplified for use in in situ hybridization. The amplified DNA, however, contains repetitive sequences that are distributed throughout the genome. These sequences will hybridize to all chromosomes. To achieve chromosome-specific hybridization, these repeats must be removed prior to hybridization (quite a labor-intensive process) or inhibited from binding by including unlabeled repetitive DNA in excess during hybridization. Enrichment of chromosome-specific DNA has been accomplished using several different techniques, three of which are described in Sections 1.2.1 through 1.2.3.

#### 1.2.1 Chromosome Sorting

Chromosome sorting was the first technique used for enrichment of chromosome-specific DNA for production of WCPs. In this approach, aqueous suspensions of chromosomes are isolated from metaphase cells and stained with one or more DNA-specific dye. Chromosomes of one type are enriched using fluorescence-activated sorting. Sorter-enriched DNA can be immortalized and

**Table 1** Generation of Whole-Chromosome Probes

| Enrichment Procedure | DNA Immortalization Procedure | Required Number of Chromosomes | Library Amplification | Ref. |
|---|---|---|---|---|
| Flow sorting | Endonuclease digestion and molecular cloning | $10^5$–$10^6$ | Bacterial host | Grayet al. (1987); Collins et al. (1991) |
| Flow sorting | Linker–adaptor PCR | $10^4$–$10^5$ | PCR | Vooijs et al. (1993) |
| Flow sorting | DOP-PCR | $10^2$–$10^3$ | PCR | Telenius et al. (1992); Weier et al. (1994) |
| Cell hybrids | IRS-PCR | $10^2$–$10^3$ | PCR | Liu et al. (1993) |
| Microdissection | DOP-PCR | 10–30 | PCR | Meltzer et al. (1992) |

amplified by cloning restriction enzyme fragments into plasmid, cosmid, or phage vectors, followed by propagation in a bacterial host, or by in vitro DNA amplification using the polymerase chain reaction (PCR). The number of chromosomes required for the establishment of high complexity probes depends on the technique used for immortalization (see Table 1). In general, cloning requires up to $10^5$ to $10^7$ flow-sorted chromosomes, depending on the efficiency of the cloning vector. Far fewer chromosomes are needed if PCR is used to amplify the specific DNA fragments. In fact, high complexity probes have been prepared by PCR amplification of hundreds or thousands of flow sorted chromosomes. Two approaches to PCR amplification have been used. In one, chromosome DNA is digested using a frequently cutting endonuclease, and adaptor oligonucleotides are ligated to both ends of the restriction fragments. The resulting fragments are amplified using PCR with primers to the adaptor oligonucleotides (linker–adaptor PCR). In another approach, chromosomal DNA is PCR-amplified using highly degenerate, mixed-base primers (DOP-PCR). DOP-PCR is particularly simple: it requires almost no manipulation of the chromosomal DNA.

### 1.2.2 Interspersed Repeat Sequence (IRS) DNA Amplification

An alternative technique for the generation of WCPs takes advantage of the presence of repeated sequences mentioned earlier, which are interspersed with single-copy DNA in 10 to 100 kilobase pair intervals throughout primate and rodent genomes. The short and long interspersed repeat DNA sequences (SINEs and LINEs, respectively) provide species-specific primer annealing sites to accommodate the amplification of the intervening single-copy DNA by PCR (IRS-PCR). IRS-PCR amplification of human DNA from human–rodent cells carrying single human chromosomes (or portions thereof) has been used to produce high quality WCPs.

### 1.2.3 Microdissection

Microdissection allows collection of DNA fragments from whole chromosomes or from chromosomal subregions. DOP-PCR amplification of the microdissected DNA from 10 to 20 chromosomes generates WCPs that can be used to stain specific chromosomes or chromosome subregions. These probes are useful for detection and mapping of deletions, gene amplifications, and other structural rearrangements.

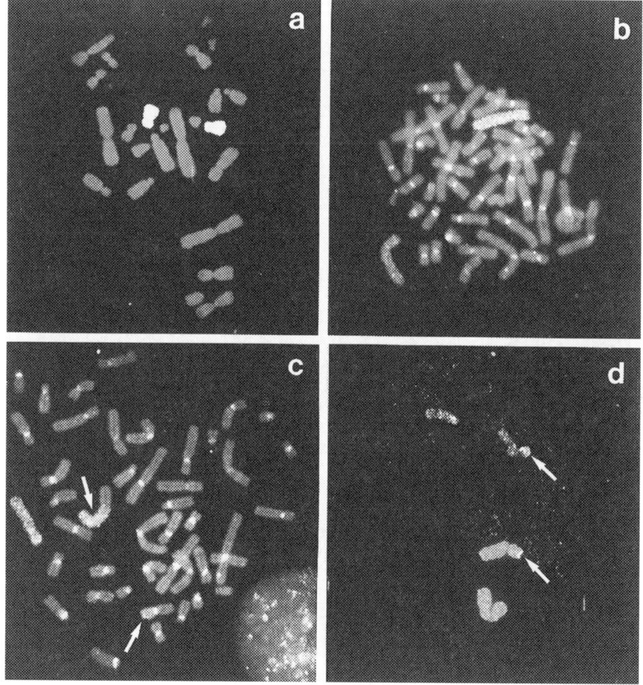


**Figure 1.** Chromosome-specific staining using FISH. (a) Human chromosomes specifically stained in metaphase spreads from human × hamster hybrid cells by hybridization with human genomic DNA. Hybridization signals appear yellow. Chromosomes to which the probe did not hybridize appear red. (b) Both copies of human chromosome 4 in a normal human metaphase spread stained by hybridizing a WCP for chromosome 4; staining was the same as for (a). Centromeric regions were simultaneously stained with a probe that hybridizes to repeated DNA in the centromeric region of all chromosomes. (c) FISH to a metaphase spread from an irradiated cell population using a WCP to chromosome 4. A radiation-induced exchange between one copy of chromosome 4 and another chromosome is apparent. The derivative chromosomes appear red and yellow (arrows). The centromeres were stained as described in (b) to allow discrimination between translocations (one centromere) and dicentrics (two centromeres) (d) FISH with WCPs to chromosomes 7 (red) and 12 (green) to a metaphase spread prepared from human tumor. Derivative chromosomes resulting from a t(7;12) translocation are visible (arrows) as red and green chromosomes (Photomicrograph courtesy of M. Vooijs; see color plate 17)

### 1.2.4 Availability

WCPs have now been produced for each normal human chromosome, for many chromosomal subregions, and for several mouse chromosomes. WCPs for all human chromosomes are available commercially. Table 1 summarizes approaches to WCP production that have been reported and provides references to each.

## 2 APPLICATIONS

### 2.1 Species-Specific Staining of Human DNA in Human × Hamster Hybrid Cells

Figure 1a (see color plate 17) shows the results of fluorescence in situ hybridization (FISH) with human genomic DNA labeled with biotin and hybridized to a metaphase spread prepared from a human × hamster hybrid cell line. After hybridization, the slides were incubated with fluorescein-labeled avidin and counterstained with propidium iodide. The human chromosomes appear yellow because of the superposition of green fluorescence from the hybridized probe (detected with fluorescein) and red fluorescence (propidium iodide), while the hamster chromosomes appear red (propidium iodide only). The technique allows study of the chromosomal composition of human × hamster cell lines, as well as sensitive detection of small rearrangements between human and hamster chromosomes. It provides no information about the identity of the involved chromosomes.

### 2.2 Translocation Detection Using FISH with WCPs

#### 2.2.1 Biological Dosimetry

FISH with WCPs allows analysis of the structural integrity of the target chromosomes in metaphase spreads. Figure 1b, for example, shows FISH to a normal human metaphase spread using a WCP to chromosome 4. The two intact number 4 chromosomes are stained yellow by the hybridization process, while the other chromosomes are stained red by the counterstain propidium iodide. Figure 1c shows FISH with a WCP for chromosome 4 to a metaphase spread prepared from a cell carrying a translocation involving one copy of chromosome 4. The derivative chromosomes resulting from the translocation are visible as red–yellow chromosomes. Translocations and other structural aberrations involving exchanges between nonhomologous chromosomes can be scored rapidly using FISH. As a result, this technique has proved useful for assessment of structural aberration frequencies as a measure of the extent of radiation-induced genetic damage.

#### 2.2.2 Analysis of Subtle or Complex Rearrangements

FISH with WCPs has proved to be a useful adjunct to conventional banding analysis for classification of complex chromosome rearrangements and for detection of subtle chromosomal exchanges. In one approach, FISH with WCPs is used to verify chromosome aberrations suggested by banding analysis and to determine the chromosomal origin(s) of marker chromosomes. Figure 1d, for example, shows dual-color FISH with WCPs for chromosome 7 and 12 to a metaphase spread prepared from cells reported to carry a t(7;12) translocation. The dual-color chromosomes resulting from the translocation are clearly visible.

Analysis of complex rearrangements and marker chromosomes is more difficult, since FISH with several different WCPs may be needed before the identity of the involved chromosomes can be learned. This difficulty can be overcome by preparing WCPs from the aberrant chromosomes (enriched either by chromosome sorting or by microdissection). FISH with these WCPs to normal metaphase spreads immediately shows which chromosomes are involved. FISH with WCPs to the involved chromosomes can then be applied to metaphase spreads from the cells carrying the aberration to determine its structural organization. The major limitation to this analytic procedure is the need to prepare WCPs for each derivative chromosome. However, the development of DOP-PCR has greatly simplified this process.

### 2.3 Chromosomal Organization of Interphase Cell Nuclei

FISH with WCPs to interphase nuclei coupled with optical sectioning microscopy allows investigation of the chromosomal organization of interphase nuclei. Such studies show clearly that chromosomes are organized in discrete, reasonably compact domains in interphase and that these domains are not disrupted during progression through the cell cycle. Investigation of the extent to which the organization is the same in cells of the same type is now being pursued in several laboratories.

### 2.4 Detection of Aneusomy in Metaphase and Interphase Cells

Chromosomal territories marked by chromosomal painting can be scored to detect supernumerary chromosomes in interphase cells. The exact number of a particular chromosome type can be determined easily in metaphase spreads by FISH with WCPs. Hybridization signals in interphase cell nuclei often overlap. This complicates accurate scoring. However, the mean number of chromosomal domains in a population of cells with extra chromosomes will still be higher than values obtained for diploid controls, thus allowing detection of supernumerary chromosomes in mixed populations.

## 3 PERSPECTIVES

Directly labeled whole-chromosome probes are now available for almost all human chromosomes and some rodent chromosomes. Furthermore, WCPs labeled directly with fluorochromes greatly simplify application of whole-chromosome staining by eliminating time-consuming immunocytochemical detection steps. Multiple probes labeled with different fluorochromes can be analyzed simultaneously. WCPs for most of the human chromosomes are now commercially available, and the availability of probes for chromosome subregions is increasing rapidly. Thus, this powerful technique is rapidly approaching the point of being a candidate for routine use in the clinical laboratory.

*See also* Fluorescence Spectroscopy of Biomolecules; Gene Mapping by Fluorescence in Situ Hybridization; Gene Order by FISH and FACS; Partial Denaturation Mapping.

## *Bibliography*

Collins, C., Kuo, W. L., Segraves, R., Fuscoe, J., Pinkel, D., and Gray, J. W. (1991) Construction and characterization of plasmid libraries enriched in sequences from single human chromosomes. *Genomics* 11:997–1006.

Gray, J. W., Dean, P. N., Fuscoe, J. C., Peters, D. C., Trask, B., van den Engh, G., and Van Dilla, M. A. (1987) High-speed chromosome sorting. *Science* 238:323–329.

Liu, P., Siciliano, J., Seong, D., Craig, J., Zhao, Y., de Jong, P. J., and Siciliano, M. J. (1993) Dual *Alu* polymerase chain reaction primers and conditions for isolation of human chromosome painting probes from hybrid cells. *Cancer Genet. Cytogenet.* 65:93–99.

Manuelidis, L. (1990) A view of interphase chromosomes. *Science* 250:1533–1540.

Meltzer, P. S., Guan, X.-Y., Burgess, A., and Trent, J. M. (1992) Rapid generation of region specific probes by chromosome microdissection and their application. *Nature Genet.* 1:24–28.

Pinkel, D., Straume, T., and Gray, J. W. (1986) Cytogenetic analysis using quantitative, high-sensitivity, fluorescence hybridization. *Proc. Natl. Acad. Sci. U.S.A.* 83:2934–2938.

———, Landegent, J., Collins, C., Fuscoe J., Segraves, R., Lucas, J., and Gray, J. W. (1988) Fluorescence in situ hybridization with human chromosome-specific libraries: Detection of trisomy 21 and translocations of chromosome 4. *Proc. Natl. Acad. Sci. U.S.A.* 85:9138–9142.

Telenius, H., Pelmear, A. H., Tunnacliffe, A., Carter, N. P., Behmel, A., Ferguson-Smith, M. A., Nordenskjold, M., Pfragner, R., and Ponder, B. A. (1992) Cytogenetic analysis by chromosome painting using DOP-PCR amplified flow-sorted chromosomes. *Genes, Chromosomes Cancer* 4:257–263.

Trask, B., and Pinkel, D. (1990) Fluorescence in situ hybridization with DNA probes. *Methods Cell Biol.* 33:383–400.

Vooijs, M., Yu, L. C., Tchachuk, D., Pinkel, D., Johnson, D., and Gray, J. W. (1993) Libraries for each human chromosome constructed from sorter-enriched chromosomes using linker–adapter PCR. *Am. J. Hum. Genet.* 52:586–597.

Weier, H.-U., Polikoff, D., Fawcett, J., Lee, K.-H., Cram, S., Chapman, V., and Gray, J. W. (1994) Generation of five high complexity painting probe libraries from flow sorted mouse chromosomes. *Genomics* 24:641–644.

# X-RAY DIFFRACTION OF BIOMOLECULES

*K. Ravi Acharya and Anthony R. Rees*

*Key Words*

**Crystallography** Science of crystal structures; the most widely used method to obtain accurate three-dimensional information of biomolecules.

**Crystals** Three-dimensional array distinguished from an amorphous solid by the regular arrangement of molecules.

**Diffraction** Specific patterns generated by the scattering of X-rays from molecules that are organized in a continuous arrangement in a crystal.

X-Ray diffraction is one the most commonly used techniques in determining the three-dimensional structures of biomolecules. This contribution considers only the biomolecules that form crystals. The detailed study of such crystals using X-ray diffraction, known as macromolecular crystallography, provides a method by which the three-dimensional architecture of the molecule can be established at or near atomic resolution. The resolution of a study critically affects what we can learn of the biomolecule structure. A protein structure at 2 Å resolution, for example, will provide detailed information about atomic positions, specific atomic interactions (intra- and intermolecular hydrogen bonds, ionic interactions, etc.), solvent accessibility, and some indication of flexibility or mobility of the molecule. It is possible that when the macromolecule under study is governed by the crystal packing forces within the crystal, some changes in the conformation of flexible regions will occur. Comparative structural studies in the crystal (using X-ray crystallography) and in solution (using nuclear magnetic resonance spectroscopy) for the same macromolecule, as well as studies of the same molecule in different crystallization conditions, clearly indicate that this ''dynamic'' behavior is not obscured in the solid state. Several experiments reported in the literature have also shown that many macromolecules when crystallized remain active, allowing the experimentalist (crystallographer) to perform time resolved structural studies.

Over the last 10 years a considerable increase in the rate of determining the three-dimensional structures of macromolecules has resulted from developments in many aspects of the X-ray crystallographic method, particularly in the areas of diffraction data collection and computer technology. Macromolecular crystallography has gained momentum through significant advances in protein chemistry and molecular biology, which have led to the availability of complex biological systems for structural studies. The method can now be applied successfully to molecules such as small peptides, proteins, and multi-protein complexes through to large biological systems such as viruses.

## 1 PRINCIPLES

Crystals are three-dimensional arrays with a regular arrangement of molecules. The repeating unit (the unit cell) of a crystal—the basic block from which the whole volume of the crystal is built—may contain multiple copies of the same molecule whose positions are governed by symmetry rules. This format allows compact packing of the molecules within the crystal. In X-ray crystallography, a crystal is bombarded with electromagnetic radiation of a wavelength comparable to the desired resolution (X-rays between 1 and 1.5 Å). The explanation for the diffraction pattern obtained was first described by Bragg and arises mainly because of the scattering of X-rays by the electrons of atoms in the crystal lattice. Since the diffracting power of a single biomolecule is rather weak, three-dimensional crystals that possess billions of molecules are used. The resultant scattered radiation is recombined to form the image of the object, a process that requires the knowledge of ''phases'' (not directly measurable, hence the ''phase problem'' in macromolecular crystallography) and the ''amplitudes'' of the scattered waves (directly measurable) with respect to the others. With reasonable estimates for the phases (through some form of additional information) of all the scattered X-ray beams, it is possible to calculate the ''electron density'' at any point within the crystal. It is important to remember that diffraction occurs simultaneously from all molecules within the crystal lattice; therefore the final three-dimensional structure will be an average picture of the entire volume of the lattice that was illuminated by the X-rays. Crystallography also provides an infinite number of snapshots of the unit cell, since the frequency of X-rays is far higher than the fastest chemical vibrations within a biomolecule.

The information at a particular point within the biomolecule is ''synthesized'' from a complete set of diffraction measurements that obeys Bragg's law of X-ray diffraction,

$$2d \sin \theta = \lambda$$

where $\lambda$ is the wavelength of the radiation, $d$ is the spacing (commonly called resolution in defining the quality of macromolecule structures) within the object being examined, and $\theta$ is the half-angle of scattering. The beams that are deflected by only a small angle (low resolution data, quite often due to the solvent scattering contribution surrounding the protein molecules within the crystal lattice) contribute to the broad features of the molecule under study, while the quality and the extent of the fine detail present in the electron density map depend directly on the higher angle diffraction data (high resolution data). The resolution is an important term in defining the quality of the structure of a macromolecule. The main features of a macromolecule that can be traced at different resolutions are as follows.

| | |
|---|---|
| 6 Å resolution | Outline of the molecule, features such as helices can be identified. |
| 3 Å resolution | Course of polypeptide chain can be traced and topology of folding can be established. With the aid of the amino acid sequence, it is possible to place the side chains within the electron density map. |
| 2 Å resolution | Main chain conformation can be established with high accuracy. Details of side chain conformations can be interpreted easily without amino acid sequence data. Bound water molecules can be identified. |
| 1.5 Å resolution | Individual atoms are almost resolved. Clear water structure. |
| 1.2 Å resolution | Hydrogen atoms may become visible. |

In practice, only a few macromolecules diffract to 1.2 Å resolution. This is because the structure tends to "smear," obscuring the fine details and making the high resolution data too weak to measure. The smearing may arise from packing defects (spatial) or from thermal motion of atoms (temporal). Direct information about the spatial dynamics can also be obtained, using X-ray diffraction, unlike NMR spectroscopy, which probes the temporal dynamics of a biomolecule.

## 2 STAGES OF STRUCTURE DETERMINATION

Several excellent reviews describe in detail the various stages involved in structure determination using X-ray crystallography. We shall highlight only the practical aspects involved in such an approach.

At least a few milligram quantities of pure (usually at least 95% pure), homogeneous material are required. A major boost to protein crystallography has been the recent introduction of cloning and expression techniques that enable the production of large amounts of rare proteins. In addition, advances in site-directed mutagenesis have led to novel protein engineering experiments, which enable the crystallographer to introduce changes that improve crystallization or provide specific sites for the binding of heavy atoms (see Section 2.3.1).

### 2.1 Crystallization

Crystallization is one of the most difficult steps in X-ray crystallography. It is necessary to establish precise conditions under which weak intermolecular forces between biomolecules produce highly ordered crystal packing rather than random aggregation as a precipitate. This procedure depends on achieving supersaturation of the biomolecule in solution, nucleation at a few sites for crystal growth, and sustained growth of single crystals. Despite significant advances in techniques, however, crystallization of a biomolecule is still a black art. One of the most important requirements is that purity and homogeneity be as high as possible.

Standard protocols such as the hanging drop, sitting drop, dialysis, micro and macro seeding, batch, free capillary, and temperature gradient methods are commonly used, depending on the amount of material available. Quite often it is necessary to explore a wide range of conditions, such as sample concentration, pH, concentration of the precipitant (e.g., ammonium sulfate or other salt, polyethylene glycol or other alcohol), added ions or effectors, and their concentrations, temperature, and so on. More recently the use of nonionic detergents has widened the scope of the crystallographic method to include membrane proteins.

### 2.2 Characterization and Data Collection

Once crystals have been obtained, the next step is to characterize them. If they diffract well, it is necessary to determine the crystallographic and noncrystallographic symmetry they exhibit. Crystallographic symmetry is usually detected uniquely from the diffraction pattern, and in general it should not prevent the determination of the structure. However, the presence of noncrystallographic symmetry is sometimes hard to detect and may make structure determination quite difficult. On the other hand, multiple copies of molecules (e.g., protomer structures of a virus capsid) governed by noncrystallographic symmetry can provide an enormous amount of "phasing power," and the resultant electron density map may be of superior quality to that obtained from a simple packing involving crystallographic symmetry operations alone.

In general the intensities of the diffracted beam are recorded by rotating the crystal within the X-ray beam. These intensities provide the amplitudes of the diffracted beam (structure factors) directly. The crystal is usually mounted in a glass or quartz capillary tube of appropriate diameter. Often, crystallographers face two difficulties: (1) since the crystals usually are sensitive to irradiation, the diffraction pattern fades as the crystal is exposed to the X-ray beam, and (2) the diffraction pattern is weak because the diffracted energy is a very small proportion of the total energy in the incident beam and is spread over a large number of reflections stimulated simultaneously.

Crystals with small unit cells (< 50 Å in each dimension) are often suited for a diffractometer, which records one diffraction spot at a time. Unit cells with lengths of 50–150 A in crystals of reasonable size often are suited for routine data collection on an electronic area detector able to record several diffraction spots at once. For molecules with large unit cells (such as viruses) or well-ordered crystals with dimensions $0.1 \times 0.1 \times 0.1$ mm$^3$, the high flux, collimation, and choice of wavelength of synchrotron radiation allows the diffraction spots to be recorded either on a photographic film or on an electronic area detector. Synchrotron radiation has made a major impact by providing an extremely intense and mainly parallel beam of X-rays. Using such intense radiation in combination with electronic area detectors, it is possible to enhance the data collection rate and reduce radiation damage during data collection.

Time-resolved structural studies such as probing enzyme mechanisms require data collection to be fast in comparison to the speed of the reaction. For some systems, standard modes of data collection, possibly aided by using cooling techniques, to lower the working temperature may be sufficiently rapid. Alternatively, very fast data collection using the Laue technique may be employed: here the crystal usually remains stationary, but data are collected using a continuous spectrum of X-ray wavelengths. Depending on the crystal system, if the conditions are favorable it is possible to collect about 70% of the data on one photograph. The potential increase in the rate of data collection with the Laue method is enormous. Most of the recently reported time-resolved experiments using this technique involve data collection in the millisecond time scale.

### 2.3 Phase Determination

After three-dimensional X-ray intensity data have been collected, it is necessary to address the phase problem to determine the structure of the molecule. Quite often it is possible to derive valuable information on the molecular packing from calculations based on native amplitudes. There are three major methods for obtaining phase information: isomorphic replacement, molecular replacement, and anomalous scattering.

#### 2.3.1 Isomorphous Replacement Method

The classical and generally used method described as ''isomorphous replacement'' provides indirect experimental estimates of the protein phase angles by observing the interference effects on the intensities of the scattered beams when heavy atoms (at very low concentrations) are added to the protein. If the heavy atom substitution does not affect the surrounding protein structure (isomorphism), and if more than one pattern of heavy atom binding can be produced, it is possible to obtain estimates of the protein phase angles. Certain chemical groups within the protein are good candidates for combining with heavy atoms; these include free sulfhydryls and single or exposed disulfide bridges. Usually it is necessary to screen a wide range of putative heavy atom derivatives at relatively low resolution (e.g., 3.5 Å). A good single isomorphous derivative can be sufficient for a structure determination, but quite often it is necessary to obtain multiple derivatives to derive phase information.

#### 2.3.2 Molecular Replacement Method

The molecular replacement method makes use of a known three-dimensional structure as an approximate starting model to provide phase angles for the observed structure factor amplitudes from the unknown structure. It is applicable only if there is significant (a minimum of 30% but preferably 50% or higher) amino acid sequence identity to a molecule of known three-dimensional structure. When a family of homologous structures is available, it is preferable to calculate an ''average structure'' for use as a starting model. Quite often, if the relative positions of the copies are not known, the presence of more than one copy of the unknown structure in the asymmetric unit complicates the search. In such cases, phase information must be obtained through the use of an oligomeric search model. Many structural studies now use this technique, either alone or in combination with the method of isomorphous replacement.

#### 2.3.3 Anomalous Scattering

When the biomolecule is rather small, quite often it is rather difficult to prepare isomorphous heavy atom derivatives. In such situations, it is possible to determine the phase angles from the native data. This may be achieved because the scattering properties of atoms change to an absorption edge of the atom. For the usual wavelengths used, this means that quite significant ''anomalous scattering'' effects occur for atoms such as Fe, Cd, and Se, while the effect is much smaller for S and virtually nonexistent for C, N, O, and H. There are two variations in using this method: Hendrickson's method of resolved anomalous scattering, which requires very accurate intensity measurements at one wavelength, and the multiple-wavelength anomalous dispersion method (mainly used for larger molecules), which requires intensity measurements at several wavelengths. A tunable X-ray source such as synchrotron radiation is ideal for performing such experiments. This method is of obvious use for metalloproteins.

Quite often it is possible to improve the accuracy of the initial set of phases when there is a redundancy of information in the asymmetric unit, either arising as a result of several copies of the molecule (noncrystallographic symmetry), or more generally due to the solvent content of the crystal (usually 30–90%). ''Solvent flattening'' is a standard technique for crystals with an average solvent content ($\geq$50%). Where the level density in the solvent region is observed, this can be sufficient to resolve the phase ambiguity for a single isomorphous derivative, or in general to improve the quality of the multiple isomorphous replacement phases to yield an interpretable electron density map. If there are multiple copies of the molecule in the asymmetric unit, or more than one type of crystal, ''molecular averaging'' may also be used to improve the quality of the electron density map.

## 3 MODEL BUILDING AND REFINEMENT

Based on a good set of phases, it is possible to calculate at 3 Å resolution an electron density map that can be used, with the aid of a known amino acid sequence, to build an atomic model. A computer graphics workstation is now a necessity for model-building work. As mentioned earlier, depending on the resolution and quality of the electron density map, it is possible to construct a three-dimensional structure of the biomolecule. When the initial model is ready, it must be refined against the X-ray observations. Such refinement leads in turn to greatly improved phase information, and the quality of the new, refined map should help in resolving the ambiguities in the structure. Thus it is necessary to interleave model building and refinement until all parts of the molecule are well explained in terms of their electron densities and stereochemistry. Assessment of the quality of the final structure is based on two factors: the reliability index ($R$-factor) (which should have values between 0.15 and 0.20 for a well-refined structure) and root-mean-square deviation from covalent bond lengths (typically should range between 0.01 and 0.02 Å).

Figure 1 (see color plate 18) highlights the different stages involved in the structure determination of a protein molecule (e.g., toxic shock syndrome toxin-1, a microbial superantigen).

## 4 PERSPECTIVES

X-Ray crystallography is a vast, highly specialized field, and we have made an attempt to give a bird's-eye view of the technique. Several areas of science are employed, and the technique can be effectively used as a tool to visualize the interactions of biomolecules, an important area in biomedical research. It is absolutely necessary to make a detailed study of molecular structure to reveal clues to how the biomolecule carries out its function. For example, if a protein plays a key role in a particular disease, its three-dimensional structure might enable ''rational design'' to produce therapeutic drugs that act by modifying the protein function. With the aid of detailed knowledge about the structures of proteins in combination with molecular genetics tools, it is now possible to produce ''tailor made'' molecules. The rapid increase in the number

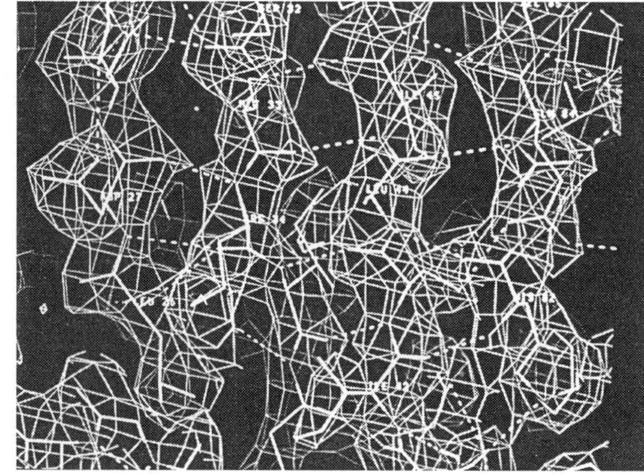

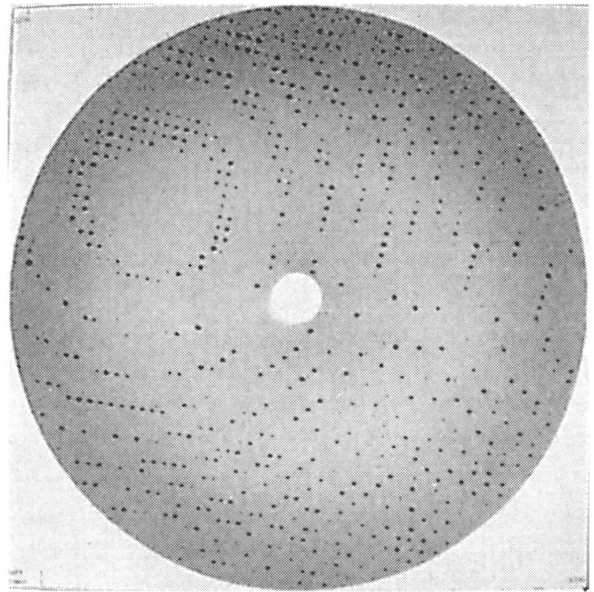

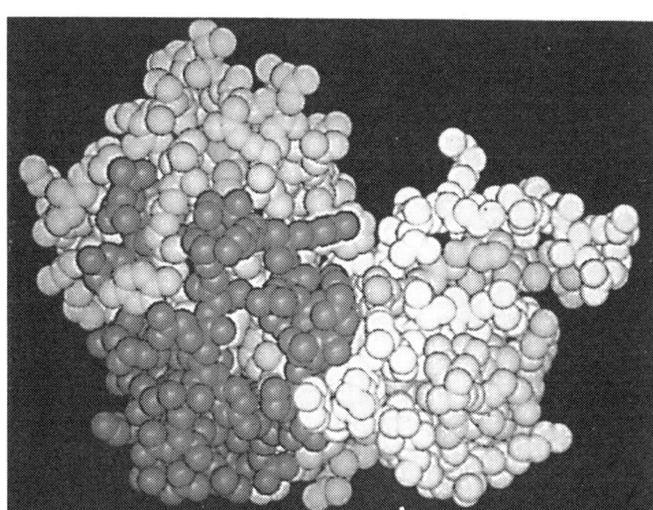

**Figure 1.** Stages involved in the structure determination of a protein molecule [e.g., toxic shock syndrome toxin 1 (TSST-1), a microbial superantigen]. (Top left) Single crystal of TSST-1. (Top right) Diffraction picture of TSST-1 crystal recorded using MAR Research Imaging Plate system (wavelength 1.5418 Å). The data extend to 2.5 Å resolution. (Bottom left) Part of the 2.5 Å resolution electron density map of TSST-1 with atomic positions superimposed. The region displayed shows four strands of a β-sheet. (Bottom right) Space-filling diagram of TSST-1 (drawn using the 2.5 Å crystal structure). [See color plates 18.]

of structures available in the literature will ultimately provide the information for understanding the "folding properties" of molecules and in time will lead to the design of new molecules. X-Ray crystallography has played and will continue to play a central role in modern biochemistry.

*See also* PROTEIN ANALYSIS BY X-RAY CRYSTALLOGRAPHY; PROTEIN MODELING; SUPERANTIGENS.

## *Bibliography*


Acharya, K. R., et al. (1994) *Nature,* 367:94–97.

Blundell, T. L., and Johnson, L. N. (1976) *Protein Crystallography.* Academic Press, London.

Bricogne, G. (1976) *Acta Crystallogr.* A32:832–847.

Brunger, A. T., Kurian, J., and Karplus, M. (1987) *Science,* 235:458–460.

Ducruis, A., and Giege, R., Ed. (1992) *Crystallization of Nucleic Acids and Proteins—A Practical Approach.* IRL Press, Oxford.

Gilliland, G. L., and Davies, D. R. (1984) In *Methods in Enzymology,* Vol. 104, W. B. Jacoby, Ed., pp. 370–381. Academic Press, San Diego, CA.

Hendrickson, W. A. (1991) *Science,* 254:51–59.

Jones, E. Y., and Stuart, D. I. (1992) In *Protein Engineering—A Practical Approach,* A. R. Rees, M. J. Sternberg, and R. Wetzel, Eds., pp. 3–32. IRL Press, Oxford.

McPherson, A. (1982) *Preparation and Analysis of Protein Crystals.* Wiley, New York.

McRee, D. E. (1993) *Practical Protein Crystallography.* Academic Press, San Diego, CA.

Michel, H., Ed. (1991) *Crystallization of Membrane Proteins.* CRC Press, Boca Raton, FL.

Rossmann, M. G., Ed. (1972) *The Molecular Replacement Method,* International Science Review, No. 13. Gordon & Breach, New York.

Schlichting, I. et al. *Nature,* 345:309–315.

Wyckoff, H. W., Hirs, C. H. W., and Timasheff, S. N., Eds. (1985) *Methods in Enzymology,* Vols. 114 and 115. Academic Press, San Diego, CA.

# Yeast Genetics

*Iain Campbell*

*Key Words*

**Diploid** The chromosome state in which each of the chromosomes, except the sex chromosome, is represented twice.

**Haploid** The chromosome state in which each type of chromosome is represented only once.

**Ploidy** The degree of chromosome multiplicity.

**Polyploid** The chromosome state in which each type of chromosome is represented more than twice.

**Protoplast** A cell that has been freed entirely of its cell wall.

**Spheroplast** A cell that is largely, but not entirely, freed of its cell wall.

**Yeast** A lower fungus that reproduces by budding and is characterized by either short or nonexistent mycelia (a multinucleate mass of cytoplasm). The term "yeast" refers particularly to fungi of the genus *Saccharomyces*.

Because of its long association with mankind, baking and brewing yeast, *Saccharomyces cerevisiae,* was an inevitable subject for genetic investigation, along with farmed crops and domesticated animals. However, the additional attraction of *S. cerevisiae* as a subject for genetic research is that as a simple, rapidly growing eukaryotic organism, it can be used as a model system for research on the genetics of eukaryotes in general. Although yeasts of other genera and species have been genetically investigated, this contribution is concerned only with *S. cerevisiae.*

## 1 BASIC PRINCIPLES

Brewing strains of yeast have been isolated in various polyploid states, but *Saccharomyces cerevisiae* is believed to be naturally diploid, with 16 pairs of chromosomes. To avoid the complications of recessive genes upon expression of genetic properties of diploids and higher ploidies, however, haploid cultures are used for much current research work. Although to be expressed, recessive genes must be present on both chromosomes of the pair, a *dominant* gene will be expressed if it is carried on only one chromosome.

Also, whether the genes are *complementary* or *polymeric* is important. The genes controlling the enzymes for biosynthesis of tryptophan, for example, are complementary genes: if any one is missing, the organism can no longer synthesize that amino acid, which must therefore be provided in the culture medium, and the organism is termed *try* .

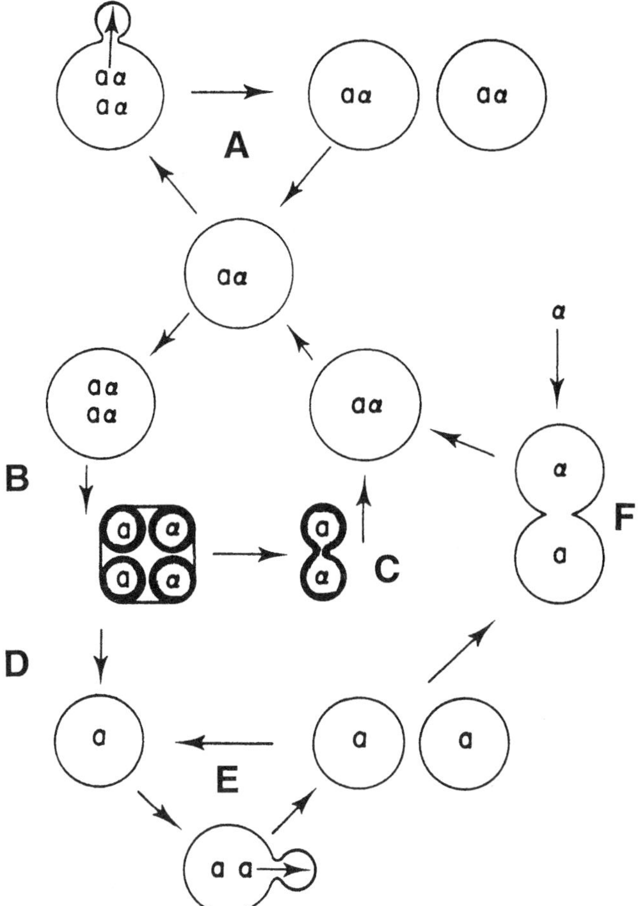


**Figure 1.** Life cycle of *Saccharomyces cerevisiae.* (A) Normal diploid vegetative growth cycle (a and α represent 16 chromosomes apiece): cell multiplication by budding, with concurrent doubling and division of nuclear material a, α. (B) Sporulation: nuclear doubling, and then two successive divisions to form four haploid nuclei, two of a and two of α. but no cell multiplication or division. Sporulation of *S. cerevisiae* occurs only under starvation conditions. (C) Zygote (fusion cell) of a and α spores, which revert, with supply of nutrients, to original diploid vegetative state. (D) Germination of liberated haploid spore, in this case a, in nutrient medium. This occurs only if liberated spores did not encounter the opposite mating type. (E) Haploid vegetative growth cycle. (F) Zygote of a and α haploid cells, restoring original diploid vegetative state.

| precursor → | anthranilic acid → | indole → | tryptophan |
|---|---|---|---|
| *TRY* genes (1) | (2) | (3) | |

Brewing yeasts are able to ferment maltose because they form the enzyme maltase. Several polymeric *MAL* genes are known to con-

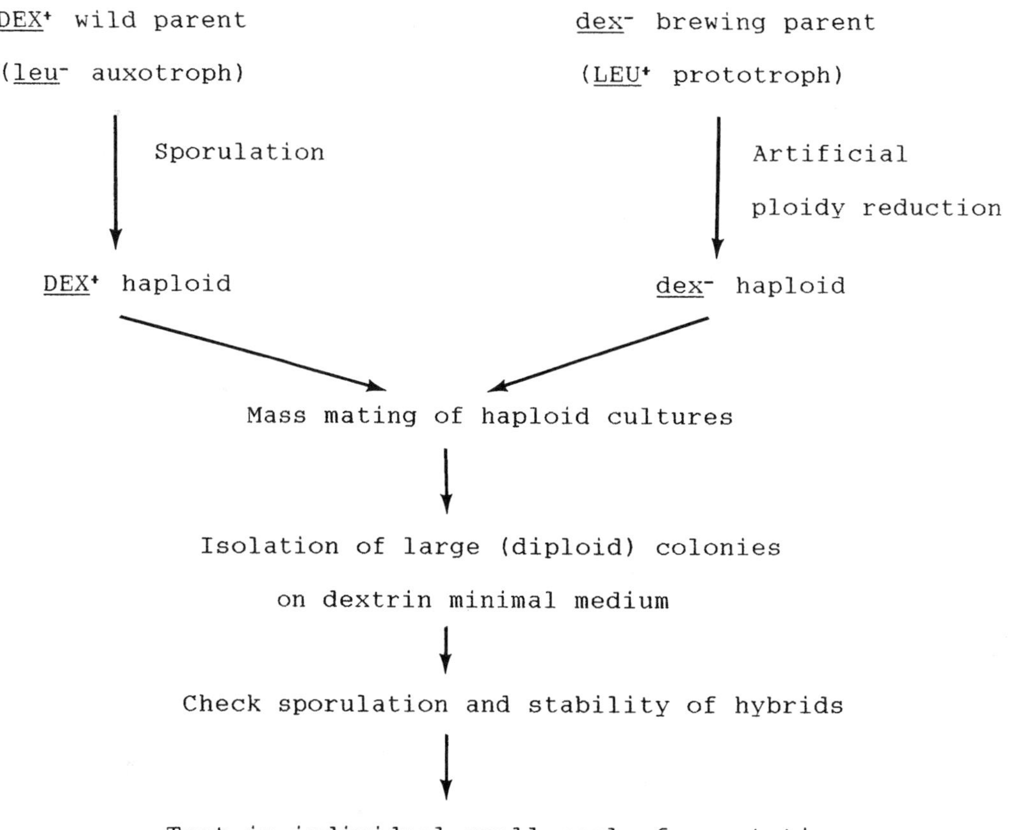


**Figure 2.** Hybridization of diastatic wild yeast and nonsporing brewing yeast. Both cultures were originally grown in malt extract medium; hybridization of haploid cultures and recovery of *DEX*$^+$ *LEU*$^+$ hybrids were performed in defined minimal medium (yeast nitrogen base) without leucine or other amino acids and with starch or dextrin as sole source of carbon. The wild yeast must be auxotrophic (unable to synthesize at least one amino acid or similar essential organic compound). Wild yeast that is prototrophic (self-sufficient), as is usually the case, can be rendered auxotrophic by UV irradiation.

trol maltase production: the more the yeast possesses, the more rapidly it ferments maltose.

## 2 METHODS FOR GENETIC MANIPULATION OF YEASTS

Various methods are available for modification of the genetic properties of a yeast culture:

1. Selection (of one component of a mixture, or of natural mutants).
2. Mutation (by radiation or chemical mutagens).
3. Hybridization of haploid spores or mater cultures (difficult to obtain from brewing yeasts).
4. Rare-mating (one parent is theoretically a nonmater).
5. Spheroplast or protoplast fusion (maters are unnecessary).
6. Transformation (introduction of ''foreign'' genetic elements, e.g., plasmids).

Hybridization, equivalent to a breeding program with higher plants or animals, is based on the life cycle of *S. cerevisiae* (Figure

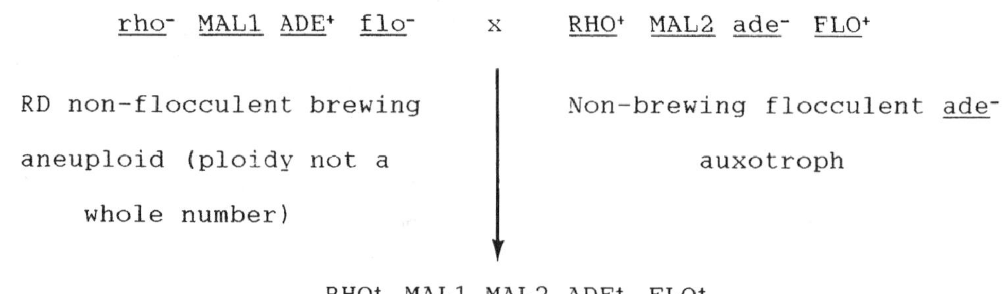


**Figure 3.** Hybridization by spheroplast fusion (in this example, to increase maltose fermentation by combining *MAL* genes 1 and 2 and to introduce flocculence). Respiratory deficiency (RD) and requirement for adenine are the selective markers.

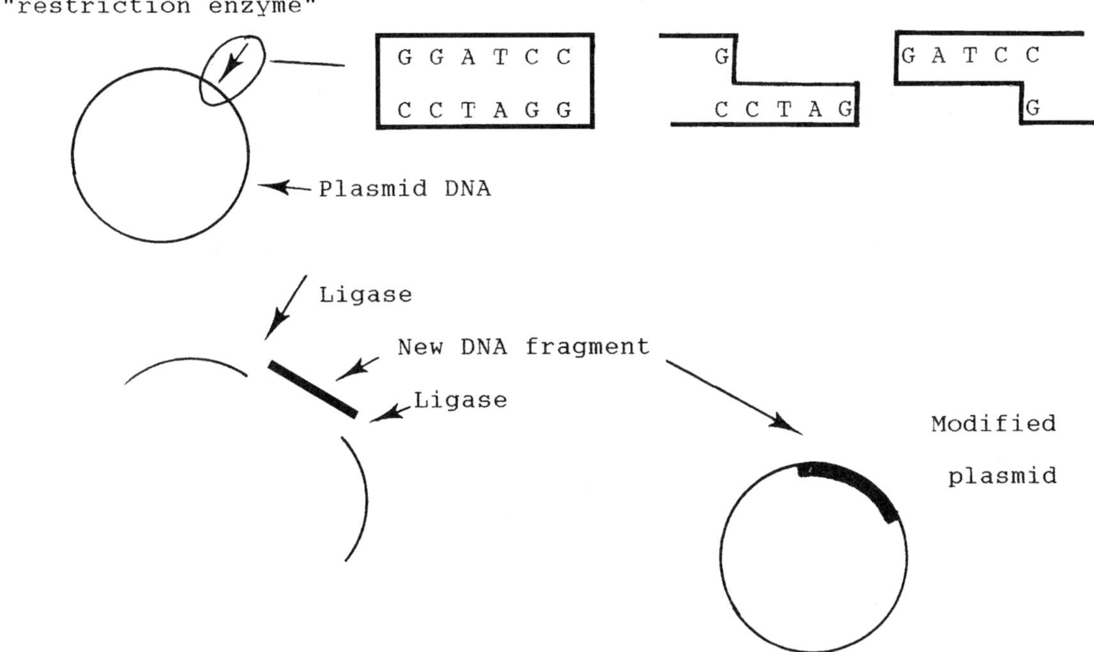

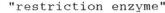

**Figure 4.** Addition of ''foreign'' DNA to plasmid.

1). Spores, or preferably a growing haploid culture of mating type a, will fuse with spores or a mater culture α to form aα diploids. Some selective effect is required after the ''mass-mating'' hybridization to recover the diploids; in practice, if each ''parent'' has specific but different nutritional requirements, only the hybrids are able to grow on a minimal medium (Figure 2). Even though the selective markers themselves are not the genes of interest, if that particular hybridization has taken place, hybridization of the genes of interest in the two ''parents'' may also have occurred. This can be confirmed by tests (e.g., small-scale fermentations) with cultures from individual colonies.

Unfortunately, many strains of brewing yeasts are unable to sporulate, or do so to only a limited extent. Often this is because the yeast strain is not of a suitable ploidy: normally, only diploid or aaαα tetraploid yeasts sporulate. Even if only a few spores are formed, selective recovery of the small production of spores may be possible. Alternatively, artificial haploidization may provide a mater culture: normal nuclear replication is chemically inhibited, so upon successive cell divisions the ploidy is reduced, eventually to haploid. More convenient methods are now available to recover hybrids from cells that theoretically are unable to mate, or cells that mate only rarely under natural conditions. After enzymic removal of the cell wall, spheroplasts (still with some attached cell wall material) or protoplasts (free of wall material) fuse readily to form zygotes (fusion cells), especially in polyethylene glycol (PEG). Originally used simply as a convenient osmotic stabilizer, PEG is now known to stimulate fusion.

Both rare-mating and spheroplast fusion methods require efficient selection to recover hybrids. A brewing yeast may have no convenient nutritional selective markers, and, obviously, genetic manipulation to create such markers may alter valuable brewing properties. One convenient nutritional manipulation that does not alter or damage hereditary properties is the temporary creation of respiratory-deficient (RD) mutants, also known as $\rho^-$ (*rho*$^-$). This is achieved by the inactivation of the mitochondria by ethidium bromide, a manipulation that has no effect on the nuclear genetic system. RD cells grow only on fermentable substrates (e.g., glucose). Ethanol, glycerol, and other compounds that must be utilized oxidatively by yeasts cannot support growth of RD mutants. So, in a genetic cross between RD and RC (respiratory-competent), *RHO*$^+$) mutants, only RC progeny can grow on ethanol- or glycerol-selective media. An example of such a manipulation, to amplify the polymeric genes *FLO,* controlling flocculence, and *MAL,* is shown in Figure 3; clearly neither of these genes themselves can have the required selective effect for recovery of hybrids.

## 3 CYTOPLASMIC GENETIC ELEMENTS

*S. cerevisiae* may have four different and effectively independent types of genetic element: nuclear DNA and, in the cytoplasm, mitochondrial DNA, viral RNA, and the DNA circles of plasmids. RNA for transcription (assembly of amino acids to peptide/protein) is not a genetic system in its own right. The genetic structure of plasmids is easily altered by modern ''genetic engineering'' techniques, and the altered plasmids are readily taken up by spheroplasts under the same conditions required for spheroplast fusion. Plasmids, whose sole natural function seems to be their own replication, can have additional DNA fragments added to specify desired properties of the host organism. The cutting and splicing of plasmids is achieved as shown in Figure 4: a specific restriction endonuclease cuts the DNA chain at a specific site, identified by a unique sequence of bases. A DNA fragment from a different organism

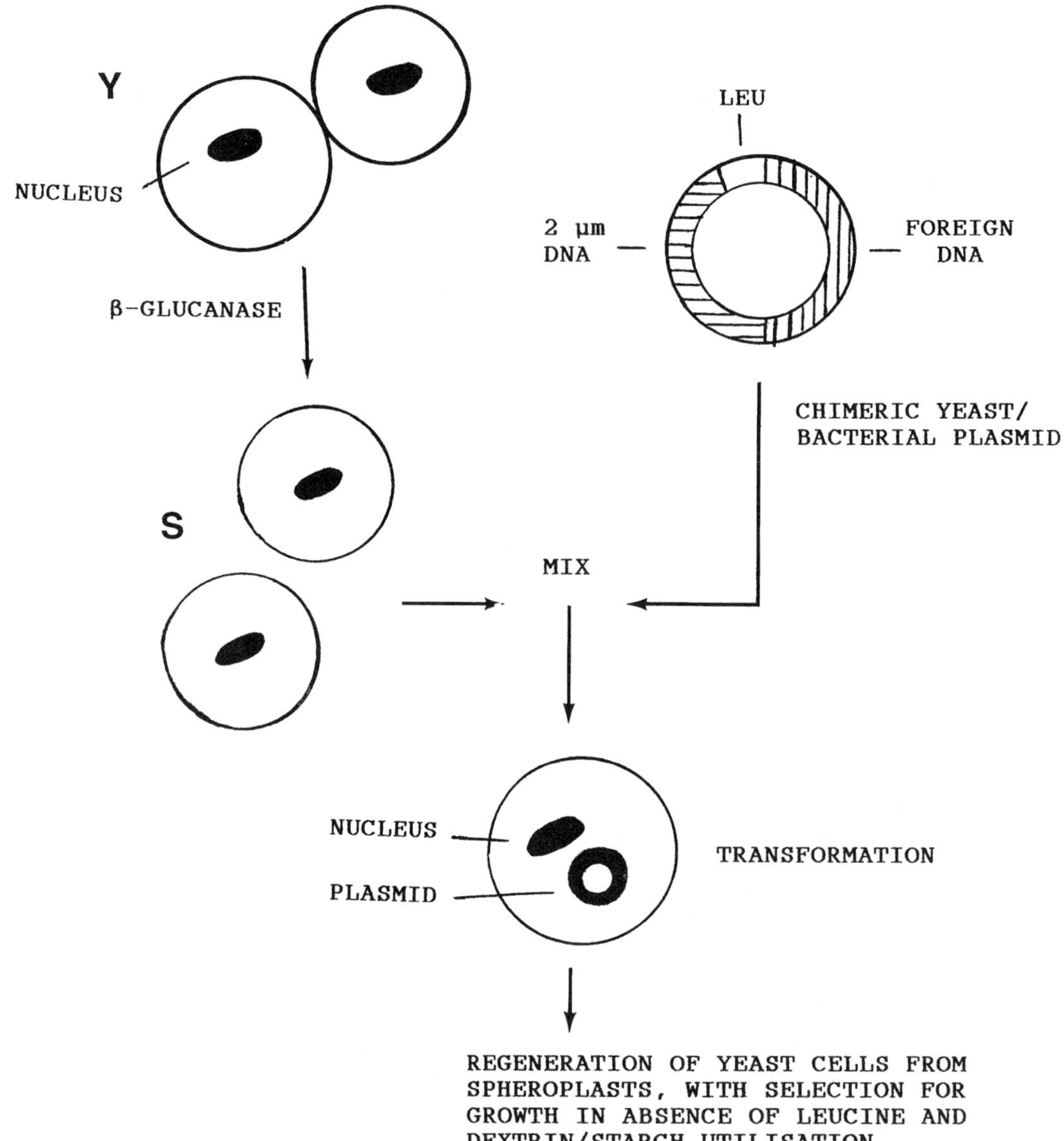


**Figure 5.** Yeast transformation using a plasmid vector. Since the structural strength of yeast cell walls is due to β-glucan, yeast (Y) wall material is removed by β-glucanase to produce spheroplasts (S). ''Chimeric'' yeast/bacterial plasmids (enormously overscale in the diagram) are released by disruption of the bacterial cells in which they are grown and transferred to the host yeast cells by a technique similar to spheroplast fusion. Subsequently, transformed yeasts are selected on medium lacking leucine, and with starch or dextrin as carbon source.

and carrying the required gene, cut by the same endonuclease so that the ends match, can be ''spliced'' to the cut plasmid by a ligase of the same specificity. Ideally, then, the altered plasmid will modify the properties of the yeast as required. Among the various enzymic properties that could be introduced to brewing yeast by this method are amylase/dextrinase for starch utilization (*DEX,* also known as *STA*) and β-glucanase, or more exotic additions (e.g., production of human insulin) for the pharmaceutical industry. The first, and still the simplest, method for obtaining large amounts of plasmid material is to prepare a ''chimeric'' plasmid that is partly bacterial and partly part yeast in its origin. The plasmid, incorporating resistance to an antibacterial antibiotic, is grown in the bacteria in the selective conditions created by that antibiotic; then it is transferred to the yeast, where, by incorporating a gene for the synthesis of, for example, leucine, it is required for growth on a medium lacking amino acids. At the same time, a gene with no selective effect but also incorporated in the plasmid should be expressed. In the example (Figure 5) the gene of interest, *STA,* is itself selective (only *STA*-containing yeasts can grow on starch minimal medium), but that situation is unusual.

Mitochondria and ''viruslike particles'' also influence the properties of brewing yeasts. Many important fermentation properties of yeast (e.g., rate of fermentation, flocculation, or production of flavor compounds) are associated with the genetic system of mitochondria—even though these effects also are under the control of nuclear DNA and, under the anaerobic conditions of fermentation, the mitochondria are not in a functional state. Viruslike particles are associated with the "killer factor" or "zymocin" of yeasts: polypeptide, secreted by "killer" yeasts, which destroys sensitive yeast cells. Although the true situation is more complex, and also involves nuclear DNA, in effect the viral RNA codes for the synthesis and excretion of the killer polypeptide and for the cell wall or membrane structure that confers resistance to zymocins.

## 4 APPLICATION OF GENETIC STRUCTURE TO YEAST TAXONOMY

Since the DNA of the yeast cell specifies its properties, DNA structure is the definitive characteristic of the strain. Several methods are in use, or at a promising stage of development, to apply DNA analysis to identification or classification.

1. *GC ratio* (the proportion of guanine + cytosine in the DNA). The value is useful in classification because it is constant in a particular organism, and related groups of yeasts have similar GC ratios. Since, however, the same overall value can result from many different sequences of the bases on the DNA, the GC ratio is of little value for purposes of identification.
2. *Percentage hybridization.* If the two strands of DNA in a yeast strain are separated and allowed to reassociate, they will combine 100%. If two complementary strands, one each from two *identical* strains, are brought together, they should associate at 100% (although in practice such a high percentage of reassociation is impossible: no two strains are absolutely identical). The relatedness of a known to an unknown organism, assessed as the percentage reassociation, can be used for identification.
3. "*DNA fingerprinting.*" This technique has become well known for its forensic applications, but simpler versions of the same principle can be used for differentiation of yeast strains and are especially useful where standard cultural methods fail to provide a satisfactory differentiation. DNA isolated from yeast is digested by restriction enzymes and the fragments are separated by electrophoresis. The pattern of the stained bands of DNA fragments is a characteristic "fingerprint" of each strain.

*See also* EXPRESSION SYSTEMS FOR DNA PROCESSES.

### *Bibliography*

Broach, J. R., Pringle, J. R., and Jones, E. W., Eds. (1991–1993) *The Molecular and Cellular Biology of the Yeast* Saccharomyces, Vols. 1–3. Cold Spring Harbor Laboratory Press, Plainview, NY.

Campbell, I., and Duffus, J. H., Eds. (1988) *Yeast, a Practical Approach.* IRL Press, Oxford.

Rose, A. H., and Harrison, J. H., Eds. (1987–1991) *The Yeasts,* 2nd ed., Vols. 1–4. Academic Press, London and New York.

# Z

## Zinc Finger DNA Binding Motifs

*John W. R. Schwabe and Louise Fairall*

### *Key Words*

**α-Helix, β-Sheet** Region of regular structure in proteins in which the peptide backbone traces, respectively, a right-handed helical path or a straight extended path with one or more adjacent strands.

**Extended X-Ray Absorption Fine Structure (EXAFS) Analysis** A technique that facilitates the exploration of the environment of a metal ion by comparing the X-ray absorption profile of a sample with that of a model compound.

**Major and Minor Grooves** Spiral ''grooves'' in the DNA that arise through the helical nature of the molecule. At any one position, the two grooves are on opposite faces of the helix. Each base pair is accessible from both grooves. The minor groove is generally the narrower.

**Mutagenesis** The use of genetic engineering to produce proteins whose amino acid sequence differs from that found in vivo.

**Proteolysis** The use of enzymes to digest the flexible regions of a protein, leaving a structured core.

**Transcription Factor** A protein involved in the regulation of the transcription of DNA into messenger RNA (often involves binding to DNA).

***Xenopus laevis*** The African clawed toad, a species used widely in molecular biology research both as a tool and as a model system.

The term ''zinc finger'' was first used to describe a 30-residue, repeated sequence motif that is now known to be extraordinarily widespread in eukaryotic DNA-binding proteins. Consistent with early predictions, it has emerged that these sequence motifs form independent minidomains folded around a central zinc ion. Adjacent zinc-fingers are strung together to make up modular DNA-binding domains—hence the term.

The zinc finger motif was originally identified as a repeated pattern of cysteine and histidine residues with the potential to ligate a zinc ion. Similar patterns of conserved potential metal ligands were found in a number of other families of DNA-binding proteins, and it was suggested that these might be variants or different classes of the zinc finger domain. We now know the structures of the classical fingers, class II and class III zinc-binding motifs, as well as the details of their interaction with DNA. It is clear that there is little structural similarity between these classes of zinc-binding motif, other than their use of a zinc ion to stabilize a small domain. Since the class II and class III zinc-binding motifs do not share the modular nature of the classical zinc fingers, the term ''finger'' is not really applicable and is best avoided.

## 1 INTRODUCTION

For many years biologists have sought to understand the mechanisms by which genes are specifically switched on and off during development and in response to external stimuli. It is now established that the specificity of this regulation is largely brought about by transcription factors that bind, highly selectively, to specific DNA-binding sites associated with their target genes. The selective recognition of these short DNA sequences is often achieved via discrete DNA-binding domains within these transcription factors. To date we have observed comparatively few different classes of DNA-binding motif. It is clear that these motifs have been duplicated during evolution and have subsequently diverged within individual protein environments to bind to different sites on the DNA. This entry focuses on three DNA-binding motifs characterized by the requirement for a zinc ion.

A combination of biochemical, genetic, and biophysical techniques has led to a rather good understanding of how the DNA-binding domains are able to recognize their specific DNA-binding sites. Research into each of the three classes of zinc-binding motif has progressed along similar lines. Genetic and biochemical experiments have identified the minimal DNA target sites and DNA-binding domains. Mutagenesis experiments perturbing these complexes have pinpointed functionally important residues. Two biophysical approaches have also been employed. First nuclear magnetic resonance (NMR) techniques were used to determine the structure of the proteins in the absence of DNA. However, X-ray crystallographic techniques have perhaps been most revealing, yielding the structural details of the interaction between protein and DNA.

## 2 CLASSICAL ZINC FINGERS (TFIIIA TYPE)

The classical ($Cys_2His_2$) zinc finger motif was first observed in 1985 as a repeated 30 amino acid sequence motif in the protein transcription factor TFIIIA from the toad *Xenopus laevis*. This zinc finger motif is defined by four metal ligands (arranged as follows: $CysX_{(2-5)}CysX_{12}HisX_{(2-5)}His$, where X represents any amino acid), and three conserved hydrophobic residues (Figure 1). The observation of this motif, combined with the results of proteolysis studies and extended X-ray absorption fine structure (EXAFS) analysis of metal content and coordination, led to the proposal that this sequence would form an independent minidomain folded around a central zinc ion.

The repeating nature of this motif is a feature common to all proteins containing classical zinc fingers, but the number of repeats varies widely within different proteins, ranging from 2 to 37. These motifs, which have been observed in a very large number of eukaryotic protein sequences, appear to be a ubiquitous structural motif for nucleic acid recognition. To date, more than 200 different cDNA sequences have been found to encode zinc finger motifs, amounting

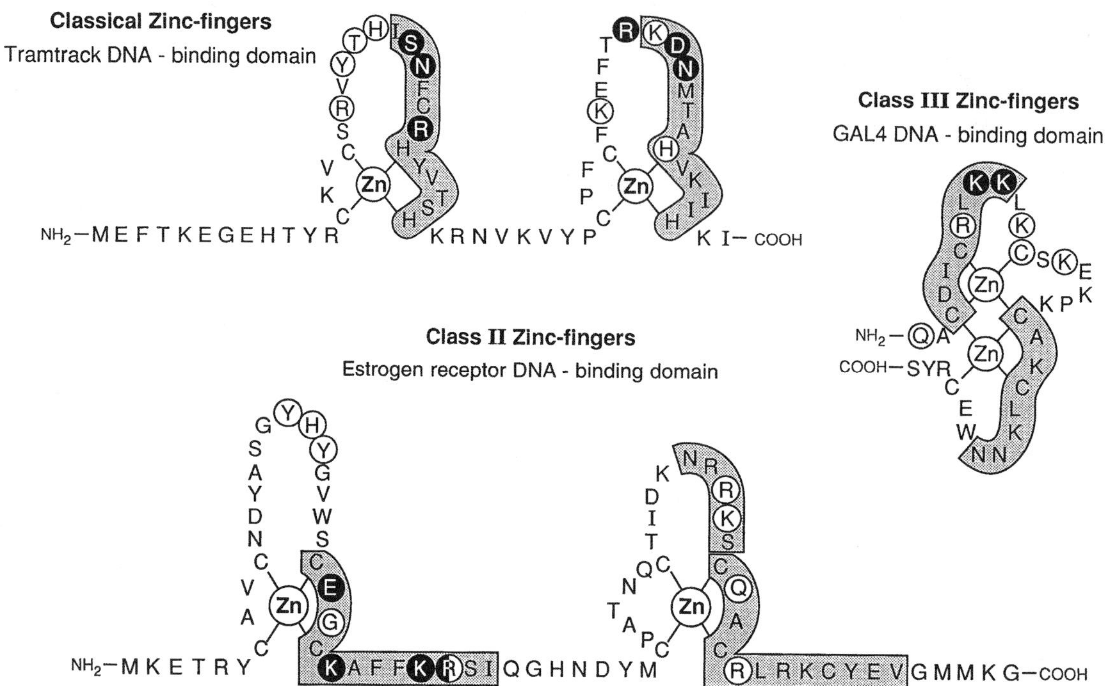


**Figure 1.** Three classes of zinc-containing DNA-binding motif, showing the pattern of zinc coordination. The protein sequence is indicated by the one-letter code. Shaded regions fold to form α-helices. Residues that contact DNA are circled (phosphate contacts in white; base contacts in black). Structures shown in Figures 2 through 4 are generated as cartoons in the program Molscript [From P. J. Kraulis, *J. Appl. Crystallogr.* **24**:946–950 (1991)].

to more than 1200 individual zinc fingers. Indeed it has been estimated that between 300 and 700 human genes encode zinc finger containing proteins.

Prior to the completion of structural studies, model building was used to predict (in hindsight, successfully) the structure of the zinc finger domain based on the structures of other zinc-binding proteins. However, it was not until NMR spectroscopic techniques were used to solve the structure of a single zinc finger from the *Xenopus* protein Xfin that direct evidence revealed that each 30-residue motif folds to form an independent domain, with a single zinc ion tetrahedrally coordinated between an irregular antiparallel two-stranded β-sheet and a short α-helix (Figure 2a).

More recently, NMR studies of constructs with two adjacent zinc finger domains have demonstrated that there is little or no interaction between adjacent domains in solution and that the linker between domains is flexible. Furthermore, there is increasing evidence that the ββα structure in certain zinc fingers (especially the first in a run) may require additional structural elements for stability. For example, in the yeast protein SWI5 an additional N-terminal β-strand is observed in the first of the three zinc fingers. Thus the ββα structure may represent just the minimal core zinc finger domain.

The crystal structures of the zinc finger DNA-binding domains from two proteins have now been solved in complex with their target sites (Zif268 from mice and Tramtrack from fruit flies). These structures illustrate a modular mode of binding: each zinc finger domain binds to DNA, with the N-terminus of the helix oriented directly into the major groove (Figure 2b). Amino acids on the surface of this helix make specific contacts to the base pairs.

The first of these structures (Zif268) suggested that there might be a recognition code that could be generally applicable to all classical zinc finger proteins. In this structure, three amino acids on adjacent turns of the helix have the potential to contact the bases at three adjacent positions on one strand of the binding site. The importance of these three positions was supported by a statistical analysis of many zinc finger sequences. However, the second zinc finger–DNA crystal structure demonstrated that given small distortions of the DNA structure, amino acids at other positions could contact the bases of the binding site. While this implies that the recognition might be more complex than was at first perceived, it seems very likely that a catalog of recognition patterns will be established. This resource will facilitate the design of classical zinc finger proteins to recognize specific target sites of a particular sequence. Both structures illustrate that in contrast to other proteins, water-mediated intermolecular contacts appear to play little or no role in the sequence specificity; in addition, there are rather few contacts made to the sugar–phosphate backbone of the DNA, and each amino acid residue contacts only a single base.

A number of important issues regarding zinc fingers remain unresolved. It is clear that the mode of DNA binding in the two crystal structures would pose a topological problem for proteins containing runs of more than three or four zinc fingers. Since three zinc fingers wrap round a complete turn of the DNA double helix, it seems unlikely that proteins with more than five zinc fingers could achieve such a mode of binding. Consistent with this, biochemical data suggest that the nine fingers of transcription factor A for polymerase III might not bind equivalently and that groups of fingers may bind partially independently.

A final puzzle centers on the ability of TFIIIA to bind both RNA and DNA and the suggestion that proteins such as the *Xenopus* protein Xfin (with 37 zinc fingers) might be used to store RNA (perhaps maternal mRNA transcripts), not to bind DNA. It has been established that the mode of TFIIIA binding to RNA is differ-

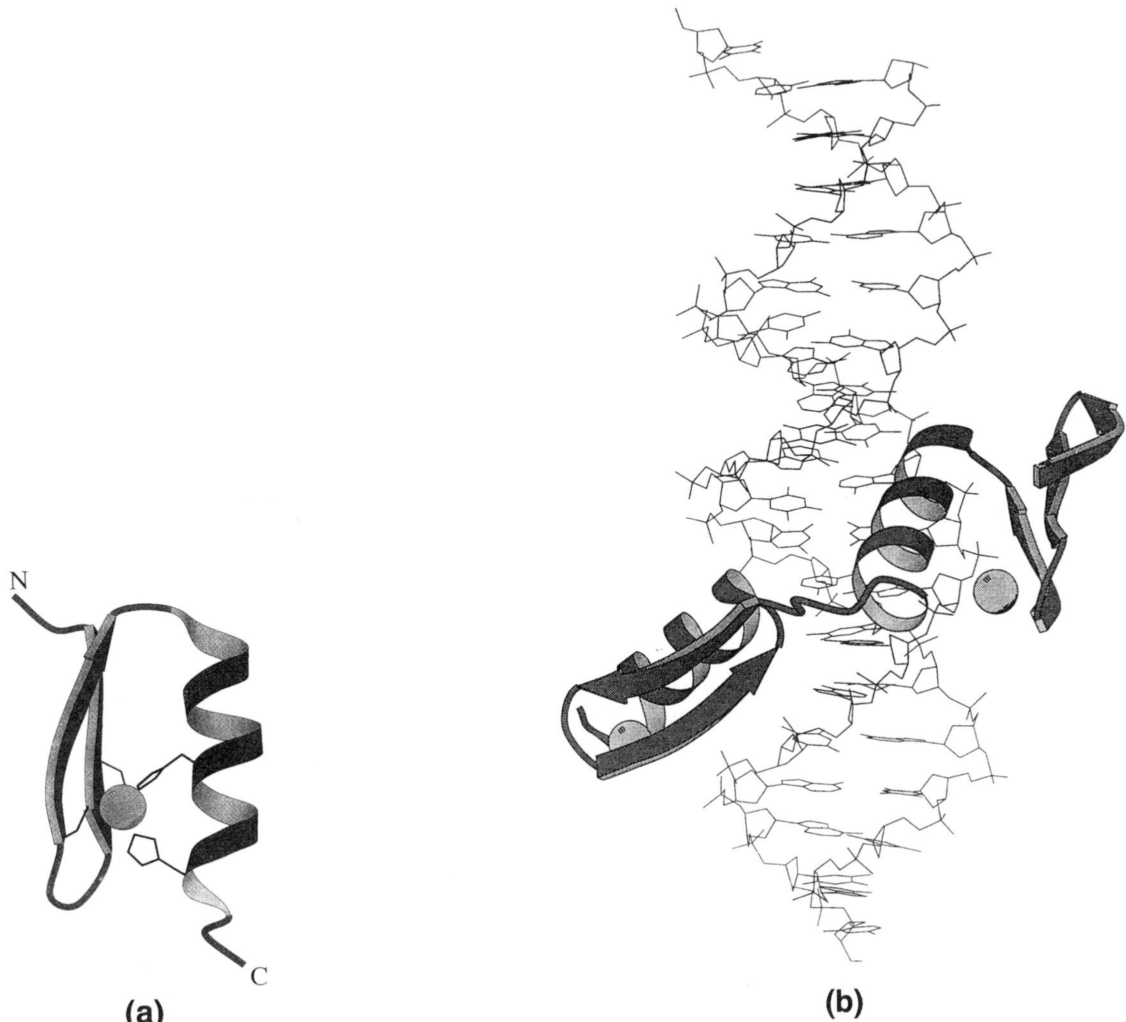


**Figure 2.** (a) The solution structure, determined by NMR spectroscopy, of the second of three classical zinc finger motifs in the yeast transcription factor SWI5. The zinc ion and ligands are shown. [From D. Neuhaus, Y. Nakaseko, J. W. R. Schwabe, and A. Klug, *J. Mol. Biol.* **228:**637–651 (1992).] (b) The crystal structure of the two classical zinc finger motifs of the fruit fly transcription factor Tramtrack in complex with DNA. [From Fairall et al., (1993).]

ent from binding to DNA, but the structural nature of this interaction has yet to be established.

## 3 NUCLEAR HORMONE RECEPTORS (CLASS II ZINC-BINDING MOTIF)

The nuclear hormone receptors form a superfamily of ligand-activated transcription factors. This family includes receptors for the steroid hormones (e.g., estrogens, glucocorticoids), thyroid hormones, retinoids, and many others. Members of this family range from slightly more than 400 to 1000 amino acids, but they all share a highly conserved DNA-binding domain (70–80 residues) that is characterized by eight conserved cysteine residues. These residues have a similar spacing to the cysteine and histidine metal ligands in the classical zinc fingers (Figure 1). Consequently, it was suggested that the DNA-binding domain of the nuclear hormone receptors might contain two variant zinc finger domains.

Like the classical zinc fingers, it was found that the glucocorticoid receptor DNA-binding domain requires two zinc ions for DNA-binding activity. Each of these is bound by a tetrahedral arrangement of cysteine ligands. However, NMR spectroscopic analyses of the DNA-binding domains from the estrogen and glucocorticoid receptors in solution revealed that the two ''zinc finger–like'' motifs fold to form a single structural domain, hence are quite distinct from the classical modular zinc fingers. The structure of the estrogen receptor DNA–binding domain in solution is illustrated in Figure 3a. The dominant feature of this structure is an arrangement of two helices oriented perpendicularly to each other and crossing at their midpoints. At the N-terminus of each of these helices, a zinc-binding pocket is formed from four conserved cysteines, two of which lie in the helix. One region of the protein is comparatively poorly ordered in solution. This region encompasses residues from the first ligands of the second zinc-binding site to the beginning of the second helix. The two zinc-binding motifs have some similarities but are clearly distinct. Whether they derive from an evolutionary duplication of a single motif remains an open question.

The DNA-binding domains of both the estrogen and glucocorticoid receptors are monomeric in solution yet bind highly cooperatively to DNA, forming symmetrical homodimers. Their DNA target sites consist of two conserved six base pair half-sites arranged as a palindrome with three intervening base pairs. The consensus

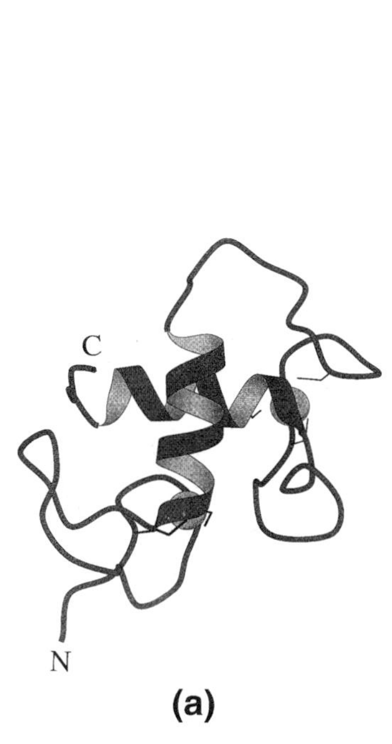


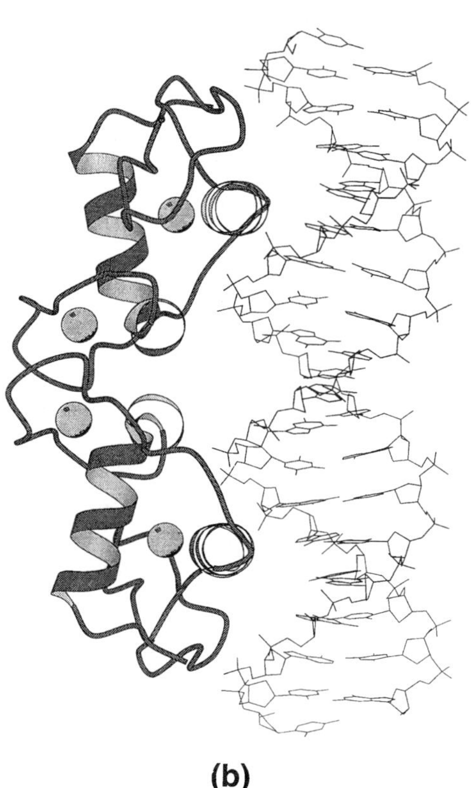


**Figure 3.** (a) The solution structure, determined by NMR spectroscopy, of the DNA-binding domain from the human estrogen receptor containing two class II zinc-binding motifs that are folded to form a single structural unit. The zinc ions and ligands are shown. [From J. W. R. Schwabe, D. Neuhaus, and D. Rhodes, *Nature* **348:**458–461 (1990).] (b) The crystal structure of a dimer of the DNA-binding domain from the human estrogen receptor in complex with DNA. [From Schwabe et al., 1993).]

half-sites recognized by the two receptors differ by only two out of the six base pairs (i.e., estrogen half-site = AG**GT**CA; glucocorticoid half-site = AG**AA**CA). The details of the protein–DNA interaction have been revealed in the crystal structures of the glucocorticoid and estrogen receptor DNA-binding domains in complex with DNA. Figure 3b illustrates the structure of the estrogen receptor complex. Two molecules of the protein bind to adjacent major grooves, from one side of the DNA double helix. In common with many other DNA-binding domains, but in contrast to the classical zinc fingers, each of the proteins in the dimer straddles the major groove of the DNA double helix. The protein makes extensive contacts to the phosphate backbone on either side, orienting the DNA-binding domain such that the recognition helix (''reading head'') enters the major groove, allowing surface side chains to make sequence-specific contacts to the base pairs. Comparison of the complexes of the estrogen and glucocorticoid receptor DNA-binding domains with their target sites reveals differences in the protein–DNA interface that are responsible for the discrimination of the two target sites. These differences are largely consistent with mutagenesis experiments demonstrating that residues at just three positions could switch the specificity of the estrogen receptor to that of the glucocorticoid receptor [glutamate, glycine, and alanine (E, G, and A) at positions 2, 3, and 6 from the N-terminus of the first helix (Figure 1) changed to glycine, serine, and valine]. Interestingly, the region of the protein that was disordered in solution forms the interface between the two proteins of the dimer when bound to DNA, as well as making phosphate contacts. Thus, with these isolated DNA-binding domains, a structured dimer interface is formed only upon binding to DNA.

The DNA targets for members of the nuclear receptor family are not all like those of the estrogen and glucocorticoid receptors. Some receptors bind as monomers to a single half-site, others bind as heterodimers. Those that bind as heterodimers appear to be able to distinguish the spacing between the two half-sites. It is not yet known how this is achieved, but the recent NMR structure of the retinoid X receptor reveals an additional helix C-terminal to the structured domain in Figure 3a. It has been suggested that this may be part of a head-to-tail dimer interface.

## 4 GAL4 AND RELATED PROTEINS (CLASS III ZINC-BINDING MOTIF)

GAL4 is the best-studied member of a group of more than 10 fungal transcription factors with conserved DNA-binding domains containing a 28-residue, cysteine-rich motif with the sequence Cys-$X_2CysX_6CysX_6CysX_2CysX_6Cys$ (Figure 1). This motif ligates two zinc ions that are required for the structural integrity of the protein, hence DNA-binding activity. In addition, however, 26 amino acids C-terminal to the cysteine-rich motif are required for DNA-binding activity. Although this motif has not been observed outside fungi, GAL4 has been found to be a potent activator of transcription in many heterologous systems.

The structure of this DNA-binding domain was solved simultaneously in solution, using NMR spectroscopy, and in the crystal, using X-ray diffraction. These structures show that as had been predicted, the conserved six cysteines of the motif ligate two zinc ions in a single cluster. The first and fourth cysteines are shared between the two zinc ions. In solution, the DNA-binding domain of GAL4 is monomeric, and the residues C-terminal to the cysteine-rich motif are flexible and disordered. In the crystal structure of the protein–DNA complex, a dimer of the GAL4 DNA-binding domain interacts with DNA. The cysteine-rich motif acts as a DNA-binding module and is joined by an extended linker to an α-helical dimerization domain. So like the DNA-binding domain of the steroid hormone receptors, the GAL4 DNA-binding motif is doubled up as a dimer when bound to DNA, such that the protein interacts with a greater number of base pairs as well as measuring the spacing between the two halves of the binding site.

Structurally, the DNA-binding module consists of two short α-helices, each capped at the N-terminus by a pair of cysteine ligands: the first and second cysteines in the first helix and the fourth and fifth in the second helix (Figure 4a). These helices are held in a rigid conformation with respect to each other by the shared zinc ions. Each helix is followed by an extended peptide strand forming two helix–strand motifs with approximate twofold structural symmetry, although the sequence homology is rather low.

Each DNA-binding module of the dimer binds in the major groove of the DNA, 1.5 turns of the double helix apart from its partner (Figure 4b). In each case the C-terminus of the first α-helix is oriented toward the bases of its binding site. This is in contrast to the classical zinc fingers and the steroid hormone receptors, where the N-terminus of the helix is oriented toward the bases. The base-specific contacts are made by two adjacent lysines at the C-terminus of the first helix. Phosphate contacts are made by residues in the first helix–strand motif as well as the extended linker, between the DNA-binding module and the dimerization domain, which follows the path of the DNA backbone.

The GAL4 DNA-binding domain dimerizes, using two amphipathic α-helices near the C-terminus of the DNA-binding domain. The helices pack together in a parallel orientation and are coiled around each other (this structure is termed a coiled coil), burying the hydrophobic residues between them. This coiled-coil structure is oriented perpendicular to the DNA, with the N-termini of the two helices positioned over the minor groove and making additional contacts to the phosphate backbone.

A striking feature of the complex between the GAL4 DNA-binding domain and DNA is that the small, cysteine-rich DNA-binding modules of GAL4 interact in the major groove, 1.5 helical turns apart. This leaves a large stretch of the DNA within the binding site of the dimer exposed, suggesting that other transcription factors (perhaps GAL11) may interact here.

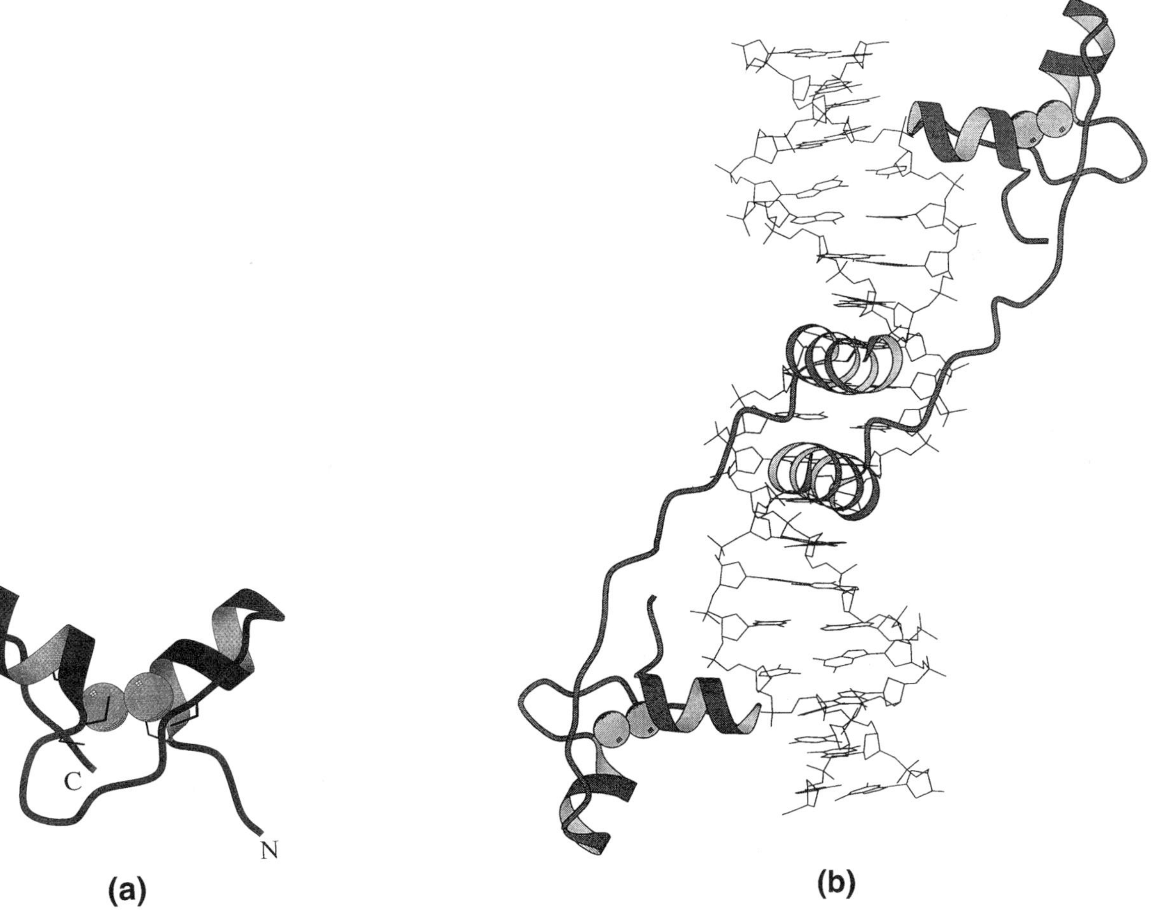


**Figure 4.** The crystal structure of (a) the cysteine-rich DNA-binding module of the yeast transcription factor GAL4 (zinc ions and ligands are shown) and (b) a dimer of the complete DNA-binding domain from the yeast transcription factor GAL4 in complex with DNA. [From Marmorstein et al. (1992).]

## 5 PERSPECTIVES

It is clear that the three classes of zinc-binding motif described represent quite distinct structural solutions to the same biological task of directing specific DNA recognition. However they do share some common features, the first being the incorporation of one or more zinc ions in the structure. The advantage of this is clear. The four-point tetrahedral coordination of the metal ion allows it to serve as a structural buttress joining elements of secondary structure, stabilizing these rather small DNA-binding domains. Why is zinc so well suited to this role? First it should be noted that only the classical zinc finger protein TFIIIA has been shown to ligate zinc in vivo. Both GAL4 and the hormone receptors are found to contain zinc ions when overexpressed in bacteria, but it is known that other metals ions (e.g., $Cd^{2+}$, $Co^{2+}$) can subsitute for the zinc. Nevertheless, it is a fairly safe assumption that zinc is used in vivo, since it is the most abundant metal in cells that is suitable for such a structural role. The wide abundance of zinc within cells can be attributed to its having no redox chemistry under physiological conditions, yet its coordination chemistry is versatile, and it exhibits a wide variation in number and character of coordinating ligands.

These structures also have in common the use of an α-helical element as a ''reading head'' in the major groove of the double helix. However, the orientation of this reading head is markedly different in the three classes of zinc-binding motif. In the classical zinc finger, the N-terminus points down into the major groove. In GAL4 the helix has an almost opposite orientation, with the C-terminus of the helix pointing into the major groove. The recognition helix in the hormone receptors shows an orientation between these two extremes. The variation in recognition helix orientation is a consequence of the different background structures in these protein families. Since this in turn means that the surface amino acids are presented to DNA in different ways in these classes of protein, it is difficult to envisage any recognition code that could be generally applicable across these different classes of domain.

Finally, while these structures have revealed much about the mechanisms of DNA recognition by transcription factors, we have yet to determine the structures of other domains within these proteins. Thus a more complete understanding must await further research.

*See also* CHROMATIN FORMATION AND STRUCTURE; GENE EXPRESSION, REGULATION OF; PROTEIN DESIGNS FOR THE SPECIFIC RECOGNITION OF DNA.

### *Bibliography*

Klug, A., and Rhodes, D. (1987) ''Zinc fingers'': A novel protein motif for nucleic acid recognition. *Trends Biochem. Sci.* 12:464–469.

Pabo, C., and Sauer, R. (1992) Transcription factors: Structural families and principles of DNA recognition. *Annu. Rev. Biochem.* 61:1053–1095.

Rhodes, D., and Klug, A. (1993) Zinc fingers. *Sci. Am.* 268:56–65.

Schwabe, J. W. R., and Rhodes, D. (1991) Beyond zinc fingers: Steroid hormone receptors have a novel structural motif for DNA recognition. *Trends Biochem. Sci.* 16:291–296.

*Original References to the Structures of the Protein–DNA Complexes*

Fairall, L., Schwabe, J. W. R., Chapman, L., Finch, J. T., and Rhodes, D. (1993) The crystal structure of two zinc fingers from Tramtrack bound to DNA. *Nature* 366:483–487.

Luisi, B. F., Xu, W. X., Otwinowski, Z., Freedman, L. P., Yamamoto, K. R., and Sigler, P. B. (1991) Crystallographic analysis of the interaction of the glucocorticoid receptor with DNA. *Nature* 352:497–505.

Marmorstein, R., Carey, M., Ptashne, M., and Harrison, S. C. (1992) DNA recognition by GAL4: Structure of a protein/DNA complex. *Nature* 356:408–414.

Pavletich, N. P., and Pabo, C. O. (1991) Zinc finger–DNA recognition: Crystal structure of a Zif268-DNA complex at 2.1 A. *Science* 252:809–817.

Schwabe, J. W. R., Chapman, L., Finch, J. T., and Rhodes, D. (1993) the crystal structure of the complex between the oestrogen receptor DNA–binding domain and DNA at 2.4 A: How receptors discriminate between their response elements. *Cell* 75:567–578.

# Glossary of Basic Terms

The most basic terms in molecular biology are defined below. These, in combination with the key words listed at the head of each article, provide definitions of all essential terms in this desk reference.

**Alleles** Alternative forms of a given gene, inherited separately from each parent, differing in nucleotide base sequence and located in a specific position on each homologous chromosome, affecting the functioning of a single product (RNA and/or protein).

**Amplification (gene)** The process of replication of specific DNA sequences in disproportionately greater amounts than are present in the parent genetic material, e.g., PCR is an in vitro amplification technique.

**cDNA (complementary DNA)** A DNA copy of an RNA molecule synthesized from an mRNA template in vitro using an enzyme called reverse transcriptase; often used as a probe.

**Chromatin** The complex of nucleic acids (DNA and RNA) and proteins (histones) comprising eukaryotic chromosomes.

**Chromosome** In prokaryotes, the usually circular duplex DNA molecule constituting the genome; in eukaryotes, a threadlike structure consisting of chromatin and carrying genomic information on a DNA double helix molecule. A viral chromosome may be composed of DNA or RNA.

**Cloning vector** See: *vector*.

**Cloning** Asexual reproduction of cells, organisms, genes or segments of DNA identical to the original.

**Complementary base pairing** Nucleic acid sequences on paired polymers with opposing hydrogen bonded bases adenine (designated A) bonded to thymine (T), guanine (G) to cytosine (C) in DNA and adenine to uracil (U) replacing adenine to thymine in RNA.

**DNA (deoxyribonucleic acid)** The molecular basis of the genetic code consisting of a poly-sugar phosphate backbone from which project thymine, adenine, guanine and cytosine bases. Usually found as two complementary chains (duplex) forming a double helix associated by hydrogen bonds between complementary bases.

**DNA polymerase** Enzymes that catalyze the replication of DNA from the deoxyribonucleotide triphosphates using single- or double-stranded DNA as a template.

***E. coli (Escherichia coli)*** A colon bacillus which is the most studied of all forms of life and whose genome is presently the best sequenced and mapped.

**Eukaryote** An organism (or cell) whose cells contain a true nucleus; all living matter except viruses, bacteria and blue-green algae.

**Expression** The process of making the product of a gene, which is either a specific protein giving rise to a specific trait or RNA forms not translated into proteins (e.g., transfer ribosomal RNAs).

**Gene** A DNA sequence, located in a particular position on a particular chromosome, which encodes a specific protein or RNA molecule.

**Homologies** Similarities in DNA or protein sequences between individuals of the same species or among different species.

**Homologous chromosomes** Chromosome pairs, each derived from one parent, containing the same linear sequence of genes, and as a consequence, each gene is present in duplicate (e.g., humans have 23 homologous chromosome pairs but the toad has 11 pairs, the mosquito has three pairs, and so on).

**Heterozygous** Having two different alleles for a given trait in the homologous chromosomes.

**Homozygous** Having two identical alleles for a given trait in the homologous chromosomes.

**Hybridization** The formation of a double-stranded polynucleotide molecule when two complementary strands are brought together at moderate temperature. The strands can be DNA or RNA or one of each; a technique for assessing the extent of sequence homology between single strands of nucleic acids.

**Ligation** The formation of a phosphodiester bond to join adjacent terminal nucleotides (nicks) to form a longer nucleic acid chain (DNA of RNA); catalyzed by ligase.

**Marker** A gene or a restriction enzyme cutting site with a known location on a chromosome and a clear-cut phenotype (expression), or pattern of inheritance, used as a point of reference when mapping a new mutant.

**mRNA (messenger RNA)** RNA used to translate information from DNA to ribosome where the information is used to make one or several proteins.

**Nucleotide** The monomer which, when polymerized, forms DNA or RNA. It is composed of a nitrogenous base bonded to a sugar (ribose or deoxyribose) bonded to a phosphate.

**Oligonucleotide** A polynucleotide 2 to 20 nucleotide units in length.

**Operon** A series of prokaryote genes encoding enzymes of a specific biosynthesis pathway and transcribed into a single RNA molecule.

**Plasmid** An extrachromosomal circular DNA molecule found in a variety of bacteria encoding "dispensable functions," such as a resistance to antibiotics. Often found in multiple copies per cell and reproduces every time the bacterial cell reproduces. May be used as a cloning vector.

**Polymorphism** Difference in DNA sequence among individuals expressed as different forms of a protein in individuals of the same interbreeding population.

**Polynucleotide** The polymer formed by condensation of nucleotides.

**Probe** A radioactively, fluorescent, or immunologically labeled oligonucleotide (RNA or DNA) used to detect complementary sequences in a hybridization experiment (e.g., identify bacterial colonies that contain cloned genes or detect specific nucleic acids following separation by gel electrophoresis).

**Prokaryote** An organism that lacks a true nucleus, a bacterium, virus or blue-green algae.

**Replication** The copying of a DNA molecule duplex yielding two new DNA duplex molecules, each with one strand from the original DNA duplex. Single-stranded DNA replication results in a single-stranded DNA molecule.

**Repressor** A protein that binds to a specific location (operator) on DNA and prevents RNA transcription from a specific gene or operon.

**Restriction mapping** Uses restriction endonuclease enzymes to produce specific cuts (cleavage) in DNA, allowing preparation of a genome map describing the order and distance between cleavage sites.

**Reverse transcription** The synthesis of cDNA from an RNA template as catalyzed by reverse transcriptase.

**RFLP (restriction fragment length polymorphism)** DNA fragment cut by enzymes specific to a base sequence (restriction endonuclease) generating a DNA fragment whose size varies from one individual to another. Used as markers on genome maps and for screening for mutations and genetic diseases.

**Ribosomes** Small cellular components composed of proteins plus ribosomal RNA that translate the genetic code into synthesis of specific proteins.

**RNA polymerase** The enzyme (peptide) that binds at specific nucleotide sequences, called promoters, in front of genes in DNA, catalyzing transcription of DNA to RNA.

**RNA (ribonucleic acid)** A single-stranded polynucleotide with a phosphate oxyribose backbone and four bases that are identical to those in DNA, with the exception that the base uracil is substituted for thymine.

**Transcription** Synthesis of an RNA molecule from a DNA template (gene) catalyzed by RNA polymerase.

**tRNA (transfer RNA)** RNA molecules that transport specific amino acids to ribosomes into position in the correct order during protein synthesis.

**Vector** DNA molecule originating from a virus, a plasmid or cell of a higher organism into which another DNA fragment can be integrated without loss of the vector's capacity for self-replication. Vectors introduce foreign DNA into host cells where it can be reproduced in large quantities.

# Index

Main entries are printed in CAPITAL LETTERS with the page ranges given in **bold face.** The letter f following a page number indicates a figure; the letter t following a page number indicates a table.


Abasic site, defined, 939
Abetalipoproteinemia, 495
Absorbance spectroscopy, for following DNA denaturation, 208–9
Absorption spectra, biomolecules, 318f
Acceptor molecule, defined, 880
Accuracy, of hybridization information, 443
ACEDB, 588
  defined, 587
*Acer saccharum*, 148
Acetamidate linkages, in modified oligonucleotides, 618
Acetaminophen, metabolism of, 120
Acetone-butanol-ethanol (ABE) fermentation, 333–34
Acetylcholine, of nematodes, 588
Acetylcholinesterase, inhibition of, potentiometric enzyme electrodes, 112
*N*-Acetylneuraminic acid, defined, 141
*N*-Acetyltransferase 2, 677
ACHILLES' CLEAVAGE, **1–3**
Achiral compounds
  defined, 617
  with dephospho internucleoside linkages, 620, 620f
  nuclease-resistant sulfamate linkage, 620
Acholinesterasemia, *Alu* insertion in cholinesterase gene, 926–27
Acid-base catalysis, 298–99
Acid-fastness
  defined, 582
  in mycobacteria, 583
Acoustic wave biosensors, 113
Acridinium esters, as chemiluminescent labels, 167
Acrocentric chromosome
  defined, 247
  example, chromosome 21, 249
Actin, 577
  in thin myofilaments, 573
Actin-binding protein (ABP), 205
Actin cytoskeleton, and platelet function, 205
α-Actinin
  and actin stress fiber linkage to plasma membrane, 205
  defined, 204
  structure of, 579
Actinomycin D, effect of, on cell death, 160
Action potential, 877
  defined, 876
Activation, of oncogenes, 625–26, 626t
Activation domain, Tat, 12
Activation energy
  defined, 295
  of nucleation, 467
Active site
  defined, 295
  HIV-1 protease, 16
Active transport, ATP-dependent, 923
Activin, 394
Activity, physical, and energy requirements, 614
Acute lymphoid leukemia (ALL), translocation in, 226, 226t
Acyl-CoA/cholesterol acyltransferase (ACAT), 498
Acyl-CoA diacylglycerol acyltransferase, 508
*N*-Acyltransferases, 119
Adaptor molecule, defined, 277
Addition, mutation by, 360
Adenine (A), 370
Adenocarcinoma, *ErbB2* amplification in, 626
Adenomas
  colon, loss of heterozygosity in chromosome 5q, 188
  defined, 187
Adenomatous polyposis coli (APC), 189–90, 378
  attenuated, 935
    and colon cancer, 189–90
Adenoregulin, 535
Adenosine, 295f
  receptor for, 535
Adenosine deaminase (ADA) deficiency, gene therapy for, experimental, 557
Adenosine triphosphate. *See* ATP
Adenoviruses, 953
  DNA and RNA of, 839
Adenylate charge, 83
Adenylate cyclase, activation by G-protein-linked receptors, 793
Adherence, of pathogens to host cells, 65
Adherin-type junctions, 206
Adhesin, defined, 65
Adhesion
  of cells, 108–9, 154–57
  and membrane fusion, 538
Adipocytes, brown, uncoupled oxidative phosphorylation in, 81
Adipokinetic hormones, 594
Adjuvant
  defined, 582
  mycobacterial, 584
Adjuvant genes, defined, 945
ADP, phosphorylation of, 689
Adrenal androgens, synthesis of, 871
β-Adrenergic, defined, 257
β-Adrenergic receptor, 793
Adrenergic system, defined, 257
Adult polycystic kidney disease, 431
Aerobic organisms
  bacteria, 76–77
  heterotrophs, 81
Afferent conduction, defined, 540
Affinity chromatography
  analytical, 796f
    for molecular recognition, 795
  defined, 793
  screening with, 798
Affinity selection, with synthetic peptide libraries, 881
Affinity sensors, 111
Aflatoxins
  defined, 516
  and liver cancer, 517
  and *p53* mutations, 518
*Agaricus bisporus*, 336
Agarose gel techniques, in restriction landmark genomic scanning, 815
Aggregates
  defined, 728
  protein, detection of, 729–30
Aggregation theory, of cell growth, 58

Aging
cell, during the division cycle, 57-58
and cell death, molecular mechanisms, 158-62
defined, 4, 232
DNA damage theory of, 232-33
and incidence of Alzheimer's disease, 26, 596
maternal, and incidence of Down's syndrome, 248
mitochondrial mutations and, 554
onset and severity of Gaucher disease, 343
organismal, 161-62
AGING, GENETIC MUTATIONS IN, **4-6**
Agonists
β-adrenergic, 258
defined, 257, 789
Agretope, defined, 449
Agricultural applications, of patented transgenic animals, 908
Agriculture, U.S. Department of, 115
*Agrobacterium rhizogenes*, vector for gene transfer, 690
*Agrobacterium tumefaciens*
defined, 896
vector for gene transfer, 690
vector for gene transfer in dicotyledonous plants, 692
*Agrobacterium tumifaciens*, 472
AIDS
defined, 6, 10
HIV ENZYMES, THREE-DIMENSIONAL STRUCTURE OF, **15-18**
INHIBITOR COMPLEXES OF HIV-1 PROTEASE IN, **6-10**
and mycobacteria, 583
THERAPEUTICS, BIOCHEMISTRY OF, **18-24**
TRANSCRIPTIONAL REGULATION OF HIV IN, **10-15**
zidovudine treatment for, 554
Air-water interface, lipid structure at, 506f
Ajmalicine, production of, from cell cultures, 887
Alanine, derivatization of β-carbon, 30
Albuterol, 258, 259
Alcohol
effects on mice, genetic studies, 255-56
and liver cancer, 516
pharmacogenetics of metabolism of, 677
Alcohol dehydrogenases, 118-19
secondary structure, from circular dichroism measurements, 182, 184
stereospecificity of, 172, 172f
Aldehyde dehydrogenases, 118-19
polymorphism in, and alcohol metabolism, 677
Algae, lipids of, 503
Algins, 149
Algorithm, defined, 856
Alignment, defined, 859
Alkaline phosphatase, chemiluminescent detection of label, 167, 167f
Alkanes, bacterial metabolism of, 76, 77f
Alkaptonuria, 74
N-Alkylation, 659-60
Alkylsilyl linkages, in phosphate-free oligonucleotide analogues, 620
Allatostatins, 594
Allatotropins, 594
Alleles
in eukaryotes, 370
in fragile X linked mental retardation, 326
for hemophilia A and hemophilia B, 409
heterogeneity of, 424
modifying oncoprotein function, 627
nonrandom association of, 438
null, from gene targeting, 560
Allele-specific amplification, 646-47
Allele-specific ligation, 377
Allele-specific oligonucleotides (ASOs), 376-77
Allelic loss, defined, 224
Allelotyping
defined, 187
to follow colonic tumor development, 188
Allelozymes, 480-81
Allergic disease, chemicals associated with, 461
Allogeneic extrapolation, of immunotoxicity evaluations, 462
Allogeneicity, defined, 86
Allogeneic transplants, 90
Allosteric protein, defined, 402
Allostery, 300
cooperative binding of oxygen by hemoglobin, 405
defined, 295
and hormone activity, 281
MWC model of, 405, 407
Allotype, defined, 449
Alphabet, genetic, 370
Aluminum, and incidence of Alzheimer's disease, 26
*Alu* repeats, 422, 439-440, 525, 860
defined, 420
*Alu* sequences, 926
defined, 924
elements of, 925f
*Alu* source gene, 440, 440f
ALZHEIMER'S DISEASE, **24-28**, 425, 596
and apoE, 500
in Down's syndrome, 248
mitochondrial mutations associated with, 554
Amberization, 634
defined, 633
Ambiguity in hybridization, 442
defined, 441
Amerindians, mtDNA markers, 556
Amidases, hydrolysis by, 119
Amide band
defined, 737
Raman analysis, 739
Amide bond torsion angle ω, 649, 654
Amination, electrophilic and nucleophilic, α-amino acid preparation, 31-32
3-Amino-2-piperidone-6-carboxylic acid, 662f
Amino acids
β-alkyl, in peptide design, 663
artificial, proteins containing, 716-17
binding of transition metal ions to, 85
chirality of, 28-32, 170
defined, 880
essential, 615
hydrazido-, 31
study of, in molecular paleontology, 635
substitutions, estimating the number of, 864
substitutions in peptides, 660
unsaturated, in peptide design, 663
AMINO ACID SYNTHESIS, **28-32**
Aminoacyl-tRNA synthetases, 918
defined, 824
γ-Aminobutyric acid (GABA)
of nematodes, 588
potentiation by avermectins, 671
Amperometric biosensors, 112
defined, 110
AMP-FLP (amplified fragment length polymorphism), 221
defined, 219
Amphetamine, effects on mice, genetic studies, 255
Amphoteric compound, defined, 343
Amplicon
defined, 463
detection of, LCR, 465-66, 465f
Amplification
allele-specific, 646-47
of c-*myc* in small-cell lung cancer, 520
of genes, in cancers, 626
linear and exponential techniques, 492
PCR, 643f
Amylase
defined, 301
for ethanol production, 332
gene cluster, 927
isoenzymes of, 481
β-Amyloid binding, to apolipoprotein E, 27
Amyloidosis
Dutch-type, 25f, 26
mutations in apoAI associated with, 498
Amyloid peptide, 25f, 26
in Alzheimer's disease, 596
biochemistry of formation, 27-28
circular dichroism measurements, 182
soluble, 28
Amyloid plaques, defined, 595

Amyloid precursor protein (APP), 25, 25f, 26, 27–28
Amylopectin, 149, 718
Amyotrophic lateral sclerosis (ALS)
  defined, 561
  pathology of, 562
Anaerobic metabolism, bacterial, 78
Analgesic, defined, 257
Analyte, defined, 110
Analyte matrix, 110
Analytical methods
  carbohydrate assays, 139t
  for detecting UV damage to DNA, 940–41
  for DNA denaturation assessment, 208–9
  DNA fingerprint analysis, specimen processing and analysis, 220–22
  electron microscopy, 269–73
  ELISA, for ACA detection, 856
  for foreign DNA determination, 691
  for foreign gene product determination, 691
  gel electrophoresis, 345
  for genetic analysis in populations, 361–62
  for genetic testing, 375–77, 377t
  for homologous recombination event detection, 560
  identification of transgene in fish, 913f
  for immuno-PCR product identification, 459
  for isoenzyme separation, 481
  for isoenzyme structure identification, 481
  for LCR amplicon detection, 465–66
  for lipid characterization, 505
  for molecular paleontology, 633, 635–36
  for molecular recognition, with immobilized ligands, 795–98
  for mtDNA assessment, 555
  for neuropeptide identification, 594
  for nitric oxide identification, 597–98
  in PCR, 645–46
  for peptide synthesis control, 667
  in pharmaceutical analyses, 672
  for protein purity assessment, 767
  for protein sequence determination, 771
  reagent requirements in protein synthesis assays, 827t
  for RNA secondary structure assessment, 842–43
  for RNA three-dimensional structure determination, 847–48
  scanning-tunneling microscopy, 851–53
  for following taxol and taxane production, 886–87
  in transgene integration and expression, 905–6
  transient kinetic analysis, 293–94
  in vivo assays of promoter functions, 698
Anandamide, 535, 536f
Ancient sources
  quality of material from, 635
  study of, in molecular paleontology, 635–36
Androgens
  role of, 872
  synthesis of, 871–72
Androstenedione, 871
Aneusomy, detection of, in metaphase and interphase cells, 967
Angelman syndrome (AS), genomic imprinting in, 380–81
Angiogenesis
  defined, 86
  fibroblast growth factors as agents of, 395
Angiosperms
  defined, 45
  DNA organization in, 46
Angiotensin, in essential hypertension, 283
Angiotensin-converting enzyme (ACE), 803–4
  deletion in the gene for, 400
Angiotensinogen, 803, 804
Animal and Plant Health Inspection Service (APHIS), U.S., 115
Animal models
  and blood substitutes, 408–9
  for human disease, 560–61
  for immunotoxic evaluations, 461, 462
  for motor neuron disease, 562
Animal rights, 909
  defined, 907
Animal welfare
  defined, 907
  and transgenic research, 909
Anisole, metabolism of, 118, 118f
Anisotropy, fluorescence, 317
Annealing, defined, 463, 621, 641
ANNEXINS, **33–35**, 133
Anode, defined, 238
Anomalous scattering, for phase determination, X-ray crystallography, 971
Anomers, defined, 382
Antagonistic pleiotropy, 5
  defined, 4
Antagonists
  bradykinin receptor, 490
  defined, 257, 789
Antenna, defined, 686
Anthocyanin, protection against UV damage by, 941–42
Antibiotic resistance, defined, 306
Antibiotics
  defined, 668
  modification of, by plasmids, 712
Antibodies
  biotinylated, in immuno-PCR, 458
  catalytic, 103, 105–6, 300
  chimeric, 37
  defined, 47, 793
  genes coding for, 452–53, 453f
  against HIV-1 envelope, 21
  from immune and nonimmune libraries, 192, 194
  monoclonal, cloning from combinatorial libraries, 191, 191f
  specificity of, 453–55, 455f
  structure of, 36–37, 36f, 450–51, 450f
Antibody-antigen assay, 113
Antibody binding site
  engineering of, 456f
  structure of, 453–55, 455f
Antibody-forming cell (AFC) assay, 462
Antibody libraries, combinatorial phage, 190–94
ANTIBODY MOLECULES
  in electron micrography, 272
  GENETIC ENGINEERING OF, **36–38**
Antibody repertoire, defined, 190
Anticentromere antibodies (ACAs), 853
  defined, 853
Anticodons
  defined, 833, 914
  of tRNA, 915
Antidiuretic hormone (ADH), impairment of, in diabetes insipidus, 282. *See also* Vasopressins
Antigen-binding fragments (Fab), from proteolysis of immunoglobulins, 451
Antigenicity, 455
  cellular, 449, 455
  defined, 449, 479
  humoral, 449
Antigen-presenting cells (APCs), 873
Antigens
  bacterial, carbohydrate, 142–45
  capsular, 142–43
  defined, 47, 523
  human blood group ABO(H), 145–46
  molecular properties of, 455
  mycobacterial, 584
  *O*-Polysaccharide, 144–45
  polysaccharide, capsular, 142
  synthetic peptides corresponding to, 658
  transition state analogues, 105–6, 106f
  tumor-associated carbohydrate, 141
Antinuclear antibodies, in scleroderma, 853–54
Antioncogenes. *See also* Tumor suppressor genes
Antioxidant defense systems, 329f
Antioxidants
  defense mechanisms using, 287, 329–30
  defined, 327
  enzyme defense mechanisms, 287
Antiparallel base pairing, defined, 914
Antipathogen plant defense strategies, 700–702
Antiport, 923
  defined, 922
Antisense blocker, enzymatic, 107

Antisense genes, expression of, tomato plants, 897
Antisense methodology, 621, 622
ANTISENSE OLIGONUCLEOTIDES
analogue of, defined, 39
defined, 38, 621
STRUCTURE AND FUNCTION OF, **38–44**
Antisense RNA, defined, 696, 896
Antisense strands, 838
defined, 834
Antisense therapy, 104
for glioma, 138
Antisocial personality, 254
Antitemplate strategies, for developing antiviral drugs, 24
Antiviral cytokines, list, 203t
Antiviral proteins, defined, 473
*Apis dorsata*, honey production by, 148
*Apis mellifera*, honey production by, 148
ApoE-ε4 allele, gene encoding, 27
Apoenzyme, defined, 185, 960
Apolipoprotein E, 499–500
amyloid binding to, 27
expression of, 707
Apolipoproteins, 150, 151t, 498–500
characterization of, 705t
concentration of, measuring, 513
defined, 494, 511, 703
transport by, 706–7
Apoptosis, 159f, 160f
*Bcl-2* gene involved in, 227
defined, 158
AP-PCR (arbitrary primer-PCR), defined, 219
*ARABIDOPSIS* GENOME, **45–47**
*Arabidopsis thaliana*, defined, 45
Arachidonic acid
defined, 682
metabolism to eicosanoids, 685
metabolism to prostaglandins, 681
Archaea
defined, 301
lipids of, 501
Arginine
binding of, effect on HIV TAR, 11
L-Arginine
oxidation of, to form nitric oxide, 598
8-Arginine vasopressin, 211
Aromatic compounds, bacterial metabolism of, 76, 77f
Aromatic side chains, azo-linked, in peptides, 661
Arrhythmias, genetic susceptibility to, 400
Arylsulfatases, neurodegenerative diseases associated with, 482
Asbestos, macrophage susceptibility to, 461
*Ascaris suum*, neurobiology of, 588
Asilomar meeting, 114
Aspartyl proteases, 7
HIV, 16, 23–24
*Aspergillus niger*, 335
for citric acid production, 336
fructosyl transferase of, 148
Association
allelic, defined, 424
studies of, in gene mapping, 426
Associative learning, defined, 540-41
Asymmetric gap ligase chain reaction (AG-LCR), 463
mechanism of, 464f
Ataxia telangiectasia, excision repair defects, 219
Atenolol, 258
Atherogenesis, and ACAT activity, 498
Atherosclerosis, 150
defined, 149, 511
genetic abnormalities associated with, 152
and lipoproteins, 500, 704
high density, 151
in transgenic mice, 709
Atomic force microscopy, 100, 102, 102f
Atomic model, defined, 741
ATP (adenosine triphosphate)
bacteriorhodopsin role in synthesis of, 70
defined, 194
hydrolysis of, RecA filament, 788
ATPase
motor proteins as, 568
myosin as, 577
$Na^+$, $K^+$, 806
ATP-binding cassette (ABC) superfamily, 923
ATP synthesis, 80
bacteriorhodopsin role in, 68–69
in photosynthetic organisms, 82
via oxidative phosphorylation, 81–82
ATP synthetase, defined, 686
Atropine, 171
AUTOANTIBODIES AND AUTOIMMUNITY, **47–49**
in scleroderma, 853–54
Autoantigens, nuclear, functions of, 49t
Autoimmune disease, 47–49
endogenous retroviruses associated with, 927
insulin-dependent diabetes, 215
and programmed cell death, 161
superantigen in, 876
Autologous cells, modified, for eliciting immune responses, 368
Automated sequencer, defined, 50
AUTOMATION
of amplicon detection, LCR, 465–66, 466f
of DNA fingerprinting, 223
of Edman degradation reactions, 771
IN GENOME RESEARCH, **50–56**
intelligent image processing, 73
of nucleic acid sequencing, 610–11
of oligodeoxyribonucleotide synthesis, 622
of oligonucleotide ligation amplification, 494
in PCR technology, 645
of peptide synthesis, 667
of protein sequencing, 768
and recognition biosensors, 795
use of HUGE, 238
Autoradiography
2D-PAGE separation, 345f
defined, 779
in restriction landmark genomic scanning, 815
use and sensitivity of, 784, 784t
Autosomal dominant trait
adenomatous polyposis coli, 189
Alzheimer's disease, early onset, 26
bilateral acoustic neurofibromatosis, 591–92
defined, 24, 589
Huntington's disease, 595–96
neurofibromatosis 1, 590–91
retinoblastoma, 817–18
Autosomal recessives
cystic fibrosis, 377
phenylketonuria, 678
Autosomes, defined, 247, 374, 595
Autotrophs, energy metabolism in, 80–81, 82
Avermectins, 670f, 671
Avidin, biotin affinity for, 353
Avirulence genes, 703
Avitaminosis, defined, 960
3′-Azido-2′,3′-dideoxythymidine (AZT). *See* AZT
Aziridines, for synthesis of amino acids, 30
AZT (3′-azido-2′,3′-dideoxythymidine), 7, 9, 23
Azurin, reorganization energy, electron transfer, 275

*Bacillus thuringiensis* (Bt), 471
defined, 469
δ-endotoxins of, 668
Backbone, glycerol phosphate, 681f
Backbones
modified phosphate, 617
oligonucleotide, phosphate-free, 618–21, 618f
peptide, modifying local, 659, 659f
phosphate, 40–41
phosphate-free, 617
polypeptide, 654, 654f
Backcross
defined, 569
for mapping the mouse genome, 570–71, 571f
Bacteria
genetic investigations with, 364
gram-negative, cell envelope, 510f
pesticide-producing, 668–71
rough mutant, 511

Bacterial cells
extracts from as protein-synthesizing systems, 828
preparing ribosomes from, 825-26
transmembrane electrochemical potential difference in, 82
BACTERIAL GROWTH AND DIVISION, **57-64**
BACTERIAL PATHOGENESIS, **65-67**
Bacterial targeting, 778
Bacterial vectors, defined, 367
Bacteriocins, 713
Bacteriophage
defined, 36, 306, 608, 950
immunoglobulin gene expression in, 37
λ,
N protein antiterminator, 14
partial denaturation mapping, 637, 640f
λ cI, 1-2
life cycle, double-stranded DNA, 951-52
φ6, 959
φX 174, 5
PM2, DNA of, 271f
T7, 951-52, 952f
BACTERIORHODOPSIN, **68-69**, 94-96, 99, 212f
defined, 68, 98
BACTERIORHODOPSIN-BASED ARTIFICIAL PHOTORECEPTOR, **70-73**, 72f
Baculoviruses, 108
for biological insect control, 470-71
defined, 469, 945
as hosts for glycosylated protein production, 309-10
Baculovirus expression vectors, 946-47
Baker's muscular dystrophy, 581
Balanced growth, in cell replication, 57-58
Banding patterns, FISH gene mapping, 351
Barbiturates, withdrawal from, genetic studies, 255
Barnase, anther-specific expression of, 698
Barr body, 525
Basal metabolic rate (BMR), 612t, 614
defined, 611
Base excision repair, 942-43
Base pairing
between codon and anticodon, 918, 918f
defined, 847
Base pairs
adenine/thymine, 3
defined, 242
Hoogsteen and Watson-Crick, 39, 40f
as a measure of DNA complexity, 173
substitutions in hemophilia B, 411
Watson-Crick, 245f, 916f
Bases
modifying
in oligodeoxyribonucleotides, 622
in oligoribonucleotides, 624
Base-specific cleavage, for nucleotide sequencing, 611
Base stacking, in RNA, 841, 846
Batch culture, 338
defined, 335
B-cell cancers, 227
Becker muscular dystrophy (BMD), 377-78
Beckwith-Wiedemann syndrome (BWS), 380
Becquerel (unit), defined, 780
Behavior, effects of neuropeptides on, 594-95
Benzodiazepines
chlordiazepoxide, 259
inhibition of reverse transcriptase, HIV-1, 23
structure, 22f
withdrawal from, genetic studies, 255-56
Benzoporphyrin, derivatives of, as photosensitizers, 954, 954f
Bernoulli text, 889
Best location alignment, 862
Best region(s) of local similarity, 862
Beta oxidation, 507
*Beta vulgaris*, 147
$BH_4$ ((6*R*)-5,6,7,8-tetrahydrobiopterin), 124-25
Bidirectional replication, 57
Bilateral acoustic neurofibromatosis, 591-92
Binding
actin filaments, by talin, 205
antibody sites for, 36, 36f
catecholamines, by tyrosine hydroxylase, 127
DNA
by RecA protein, 788
sites for Achilles' cleavage, 2
by small molecules, 104-5
enzyme, and solvent characteristics, 305
homophilic, in cell adhesion, 156-57
of inhibitor with enzyme, 105
metal, RNase H, 18
metal ion, 85-86
oxygen, structural changes accompanying, 405
peptide nucleic acid, to RNA and DNA, 43
peptides, 651
promoter/enhancer module, 10
to regulate enzyme reactions, 300
Tat domain, 12-14
*See also* Antibody binding site
Binding affinity
defined, 39
of oligonucleotide analogues, 42-43
Bioassay, defined, 98, 592
Bioavailability, defined, 260
BIOCHEMICAL GENETICS, HUMAN, **73-74**
Biodegradable polymers, 88-89, 89f, 718-20
defined, 717
proposed use in liver regeneration, 91
BIODEGRADATION OF ORGANIC WASTES, **75-79**
Biodiesel production, 334
Bioenergetics, defined, 79
BIOENERGETICS OF THE CELL, **79-83**
Biofungicides, blasticidin S and kasugamycin, 671
Biogas, 333
Bioherbicides, bialaphos, 669-71, 671f
BIOINORGANIC CHEMISTRY, **83-86**
Bioinsecticides, avermectins, 671
Biolistic device, 368-69
defined, 367
transformation using, 691, 692
Biolistic method
defined, 696
for promoter analyses, 698
Biological evolution, defined, 363
Biological response modifiers (BRMs), production of, fungal biotechnology, 338
Bioluminescence, 165-66
defined, 165
tracer for biodegradation reactions, 79
Biomarkers, defined, 633
BIOMATERIALS FOR ORGAN REGENERATION, **86-92**
Biomembrane fusion
models of, 539-40
morphology of, 538
BIOMIMETIC MATERIALS, **93-98**, 98t
Biomineralization, 466-67
Biomolecular cryptology, 888n
BIOMOLECULAR ELECTRONICS AND APPLICATIONS, **98-103**
BIOORGANIC CHEMISTRY, **103-6**
Biopharmaceutics, 674
defined, 672
Biopolymers, prebiotic formation of, 630-31, 630f
BIOPROCESS ENGINEERING, **106-10**
Bioprocessing, defined, 106
Bioreactors
transgenic animals as, 905
whole-cell, 108
Bioremediation, defined, 75
Bioresorbable polymers, 88, 720
defined, 717
BIOSENSORS, **110-13**, 796f
defined, 793
electrochemical, 112
enzymatic, 107
recognition, 795
Bioseparations, 109-10
Biospecificity, defined, 793

Biosynthesis
of glycoproteins, 390
of nitric oxide, 598, 599f
of peptides, 649, 651
of phospholipids, 683-85, 683f
of proteins, folding during, 757
BIOTECHNOLOGY, GOVERNMENTAL REGULATION OF, **114-17**
Biotin
defined, 458
hapten for FISH, 353
Biotinylation, defined, 458
BIOTRANSFORMATIONS OF DRUGS AND CHEMICALS, **117-20**
Biphenyls, polyhalogenated, 117
Biphotonic, defined, 99
Bisheteroarylpiperazines, inhibition of reverse transcriptase, HIV-1, 23
Bislactim ethers, 30
structure, 29
Bladder cancer, 227
Blasticidin S, 670f, 671
BLAST program, for sequence alignment, 858, 862
Bleomycin, 104
Blocking
in immuno-PCR, 459
in RLGS, 813, 815
Blood coagulation system, 483
Blood flow, local, regulation by the kallikrein-kininogen-kinin system, 484
Blood group antigens, 390
ABO(H), 145, 145f
Blood substitutes, 408-9
hemoglobin-based, 406
Bloom's syndrome, excision repair defects, 219
Blot, defined, 605
Blue-white selection, defined, 228
B lymphocytes, defined, 460
Boc method for peptide synthesis, 656-57, 664-65
Body, composition of human, 612t
BODY EXPRESSION MAP OF THE HUMAN GENOME, **120-24**
Body mapping
defined, 121
human genes, adult lung, 123t
Bohr effect, 405
defined, 402
Bombesin, production by lung cancers, 519
*Bombyx mori*, silk fibroins of, 715
Bonded phase, defined, 417
Bond grafts, 91
Bone, organ regeneration, 91-92
Bone morphogenic proteins (BMP), 394
*Bordetella pertussis*, transmission of, 65
Bottom-up design
defined, 50
evolutionary, 52
Boundaries, and nucleosome positioning, 178-79
Bradykinin, 483
generation and destruction of, 484f
liberation from high molecular weight kininogen, 486f
receptors for, 804
antagonists to, 490
structure of, 485f
Bragg's law, 969
Brain
glioma of, 138
mammalian, DNA damage to, 232-33
BRAIN AROMATIC AMINO ACID HYDROXYLASES, **124-27**
BREAST CANCER, GENETIC ANALYSIS OF, **128-30**, 378, 626
*p53* gene, 931
Breast Cancer Linkage Consortium, 129
Breast implants, silicones in, 721
Budding, of enveloped viruses, 948-50, 950f, 960
Bulge loop, 841
defined, 841
Burkitt's lymphoma, translocation in, 227, 626
Butanol
biological fuel production options, 332t
biological production of, 333-34
*N*-Butyldeoxynojirimycin, in vitro inhibition of HIV-1 binding, 21

Cadherins, 156-57
cytoskeletal interactions with, 206
defined, 153
E-, 206
N-, 155f
role in cytoskeleton-plasma membrane interactions, 204-5
*Caenorhabditis elegans*, 587-88
cell death studies, 160
database for genome, 587
delayed senescence in, 6
mtDNA of, 551
Calcitonin (CT), 131
Calcium
binding by annexins, 33-35
binding by selectins, 157
defined, 131
handling of, in heart failure, 400
role in cell adhesion, 156-57
second messenger, 279, 281
in tryptophan hydroxylase regulation, 127
in tyrosine hydroxylase activation, 126
Calcium-binding motifs, of α-actinin, 579
CALCIUM BIOCHEMISTRY, **131-35**
Calcium channels, 877
and exocytosis, 878
Calcium levels, and bacterial virulence, 66
Calcyclin, association of annexin XI with, 35
Calelectrin (annexin VI), 33, 33f
Callus, defined, 885
Calmodulin, 281
calcium-binding sites of, 133
and calcium pump of plasma membranes, 133-34
cell cycle control by, 132
defined, 131
Calpactin (annexin II), 33, 33f
tetramer of, in chromaffin granule aggregation, 35
Campomelic dysplasia, 869
CAMs, L-, 156
CAMs (cell adhesion molecules), 155-56
defined, 153-54
L- and P-, 156
CAMs (neural tissue CAM), N-, 155, 155t
CaMs kinase II ($Ca^{2+}$/calmodulin-dependent protein kinase), 124, 126
Cancer
B-cell, 227
bladder, 227
defined, 136, 392
growth factors in therapy for, 392
hormone-sensitive, steroids for treating, 873
imprinting syndromes, 379
susceptibility to, heritable, 378
taxol for treatment of, 885
two-hit model of gene mutation, 431f
*See also* Colon cancer; Colorectal cancer; Lung cancer
CANCER, MOLECULAR BIOLOGY OF, **136-38**
Cancer family syndrome, 190
Cancer predisposition syndromes, 224, 225t, 378
*Candida curvata*, mutation by deletion in, 503
Candidate genes, 426-27
defined, 424
Capillary gel electrophoresis, 238-39
Capsid
viral
defined, 950, 957
structure of, 958, 958f
Capsular antigens, 142-43
Capsule
defined, 65
mimicry of host markers, by bacteria, 66
Carbamate linkages, in modified oligonucleotides, 618-19
CARBOHYDRATE ANALYSIS, **138-41**
CARBOHYDRATE ANTIGENS, **141-46**
CARBOHYDRATES
chirality of, 170
complexes with protein, 321-22
crystal structures, hydrogen bonds in, 447

INDUSTRIAL, **146–49**
surface, and cell-cell interactions, 157-58
Carbon, metabolic requirement of microbes, 76
Carbonate linkages, in modified oligonucleotides, 618
Carbonic anhydrase
tertiary structure, 321f
zinc in, 549-50, 549f
α-Carbon modification, peptides, 660
Carbon monoxide (CO), as a second messenger, 132
γ-Carboxyglutamic acid, 132
Carboxymethyl linkages, in modified oligonucleotides, 618
Carcinoid tumor syndrome (CTS), 485
Carcinomas
colon, loss of heterozygosity in chromosome 5q, 188
defined, 187
renal cell, 934
*See also* Cancer; Tumors
Cardiac muscles, 573
CARDIOVASCULAR DISEASE, **149–53**
Cardiovascular system, regulation by kinins, 484
Carrageenan, 149
macrophage susceptibility to, 461
Carrier, defined, 374
Carrier ampholytes, defined, 343
Carrier protein, defined, 922
Cartilage, regeneration of, 92
Cascade, of viral production, 14
Catabolic activator protein (CAP), 807
Catabolism, of phospholipids, 685
Catalase (CAT), 287, 329, 329f, 547
Catalysis
by approximation, 298
by engineered antibodies, 38
by enzymes, 295
by metalloenzymes, 546
by RNAs, 623-24
transition states, 297-98, 298f
Catalytic antibodies, defined, 103
Catalytic domain, calcium ion pump, 134
Catalytic sensors, 111
Catecholamines, biosynthesis of, 124, 125f
α-Catenin, homology with vinculin, 206
Cathode, defined, 238
CD4 receptor, and viral envelope binding, 21
cDNA
clone, N-CAM, 155
encoding β-amyloid, 26
randomly primed library, 120
3′-directed library, 120
CELL-CELL INTERACTIONS, **153–58**
adhesion requiring calcium, 206
repulsion, 158
Cell cycle
defined, 413
$G_0$, $G_1$ and $G_2$ stages of, defined, 928
role of *p53* gene in, 189
S phase of, 414–15
Cell death, programmed
in mammals, 160-61
versus pathological, 158, 160
CELL DEATH AND AGING, MOLECULAR MECHANISMS OF, **158–62**
Cell division cycle, 62
Cell division cycle (cdc), roles of calcium in, 132
Cell-free translation (protein-synthesizing) system, defined, 824
Cell membrane, defined, 57
Cell production, defined, 392
Cell repulsion, 154
Cells
counting, with luciferase assay, 168
cultures for producing taxol and taxane, 885-87
distribution of micronutrients in, 899
growth factor action sites, 393f
hypertrophy of, in heart failure, 399-400
structures recognized by autoantibodies, 49t
Cell surface
Fas/APO-1 antigen of, 161
segregation of wall during division, 61, f
synthesis during the division cycle, 60-61, 60f
Cell transplantation, use of collagen-glycosaminoglycan substrates for, 718
Cellular immunity
defined, 367, 523
role of the major histocompatibility complex in, 524
Cellular oncogene, defined, 928
Cellular uptake
defined, 39
of oligonucleotides, 41-42
Cellulase, 333
defined, 331
for ethanol production, 107, 332-33
Cellulose, 97-98, 148
as a biodegradable polymer, 718-19
defined, 331
ethanol production from, 331
Cell wall, plant, 696, 701
Centimorgans (cM), 421
Central diabetes insipidus (CDI), 213
defined, 212
genetic, 211
Central nervous system
genetic diseases affecting, 74
role of kinins in, 484
Centromere, defined, 247
Ceramics
biocomposite, 469
biological, 98
defined, 93
Cerebellum, $IP_3$ in the Purkinje cells of, 134
Cerebral hemorrhage, in Dutch type amyloidosis, 26
Cerenkov counting, 784
Ceruloplasmin, clearance of, 384
Chain of custody (continuity), 220
defined, 219
Chain termination
in genetic analysis of populations, 362
in viral inhibition, 23
Chalcone synthetase, role in plant defense against pathogens, 700-701, 701f
Channel protein, defined, 922
Channel receptors, ion, 278, 279
Channels, 541
calcium, 133
on the endoplasmic reticulum, 281
voltage-dependent, 134
calcium ion release, 134
defined, 131
potassium, 400, 877
for proton transport, 689
regulation of, by nicotinic receptors, 791-92
water, 806
CHAPERONES, MOLECULAR, **162–65**, 391f, 400
defined, 728, 774
hsp, 401
hsp60, 164
hsp70, 163-64
hsp90, 164-65
role in protein folding, 731
SecB, of *E. coli*, 165
Chaperonins, 162, 164, 758
in cell division, 178
hsp70, hsp90, and hsp60, 731
Chargaff rules, 834
Charge displacement, defined, 70
Charge screening, in membranes, 195
Chelate, defined, 83
Chelated hydrogen bond, 445-46
defined, 444
Chemical electric and magnetic field effects, defined, 266
Chemical methods, for gene transfer, 690
Chemical quench-flow analysis, 293–94, 294f
Chemicals, immunomodulating, 461
Chemical score, for protein quality, 615
Chemical shift
defined, 601
one-dimensional NMR, 602
Chemical synapses, 877, 878f
Chemiluminescence, defined, 165
CHEMILUMINESCENCE AND BIOLUMINESCENCE, ANALYSIS BY, **165–68**
Chemotaxis, 109
Chemotherapy, compounds for, 104-5
Chicken ovalbumin upstream promoter (COUP), 11f

Chimera
for destroying anther cells, 698
knockout mice, 901-2
polypeptide or nucleic acid, defined, 945
protein, 455-57
streptavidin-protein A, 458
for studying lipoprotein genes, 152-53
Chimeric compounds
antibodies, 37
peptide nucleic acids, 40
Chimeric gene, 897f
construction of, tomato plant, 896-97
defined, 896
CHIP28 water channel, 806
Chirality
of amino acids, 28-32
center and axis of, 169
circular dichroism spectra associated with, in DNA, 243
defined, 28, 168, 179
in living systems, 629-30
of modified oligonucleotides, 42-43
CHIRALITY IN BIOLOGY, **168-73**
*Chironomus tentans*, silk of, 715
Chitin, 97-98
Chitosan, 149
*CHLAMYDOMONAS*, **173-75**
*Chlorella ellipsoidea*, lipid production by, 503
Chlorophyll, 686-87
electron density, 273
Chloroplasts
of *Chlamydomonas*, 173
DNA, of *Chlamydomonas*, 174
genetic exchanges with mitochondria, 552
transmembrane electrochemical potential difference in, 82, 197
Cholesterol
defined, 870
measurement of, 513
metabolism of, 497-98
cellular, enzymes, 153t
reverse transport of, 497, 512
transport of, 150
Cholesterol 7α-hydroxylase, 498
Cholesteryl ester transfer protein (CETP), 497
Chondrocytes, to maintain bone graft matrix, 92
Choriocarcinoma cells, 925
Chorionic gonadotropin, 385
Choroidermia, X-linked, haplotype analysis of, 727
Christmas disease. *See* Hemophilia, B
Chromaffin granules
role of annexin II tetramer in aggregation of, 35
Chromatin
contribution of histones to the structure of, 414, 414f
electron micrography of, 271f, 272
CHROMATIN, FORMATION AND STRUCTURE, **175-79**
Chromatin fiber, 177-78, 177f, 358f
defined, 175
order of cosmids along, 357
Chromatographic recognition analysis, 795
Chromatography
affinity
for bioseparation, 109
for isolating phage, 37
for bioseparations, 109
defined, 764
perfusion, 109
for protein purification, 766
Chromatosome, defined, 175-79
Chromophore
circular dichroism of, 179
purple, bacteriorhodopsin, 68, 70, 101f
Chromosome band, defined, 350
Chromosomes
1
apoAII gene, 498
glucocerebrosidase gene, 341
plasma membrane calcium pump gene, 134
1 gene for ACAT, 498
20, prion protein on, 596
21
age-dependent abnormalities, 25, 248
in Alzheimer's disease, 427
APP gene on, 26
structure of, 249
22
gene for CYP2D6, 676
L1H element on, 926
in neurofibromatosis 2, 591, 934
2
apoB gene, 498
hereditary nonpolyposis colon cancer, 190, 935
*PAX3* gene of, 424
in Waardenburg syndrome type 1, 427
3
apoD gene, 499
plasma membrane calcium pump gene, 134
in von Hippel-Lindau syndrome, 934
4, gene for alcohol dehydrogenase, 480
5
in APC and Gardner syndrome, 189
*APC* gene, 225
*APC* gene in colon cancer, 934-35
hexosaminidase gene, 480
*MCC* gene in colon cancer, 189
MCC genes in colon cancer, and *fap* genes, 137
6
apo(a) gene, 500
in diabetes mellitus, 215-16, 216f
7
mouse *Igf-2* and *H-19*, 380t
mouse *Igf-2*, *H-19*, and *Snrpm*, 379
8
gene for cholesterol 7α-hydroxylase, 498
gene for lipoprotein lipase, 495
9
deletions linked to cancers, 627
familial melanoma susceptibility gene on, 378
11
*apo*AI, *apo*CIII, *apo*AIV genes, 495, 498
apolipoprotein A-I gene, 707
in Wilms' tumors, 380, 932-34
12
*PAH* gene, 678
plasma membrane calcium pump gene, 134
13, *RBI* gene, 818, 929
14
in Alzheimer's disease, 27
in Alzheimer's disease, early onset, 596
15
in Beckwith-Wiedemann syndrome, 380
gene for hepatic triglyceride lipase, 496
genetic disorders involving, 380-81
hexosaminidase gene, 480
16
gene for CETP, 497
lecithin-cholesterol acyltransferase gene, 497
17
campomelic dysplasia, 869
in colon cancer, 188
gene for breast cancer on, 128-29, 378
gene for neurofibromatosis type 1 on, 426, 934
mouse, *Igf-2r*, 379-80, 380t
in neurofibromatosis 1, 590
p53 gene mutation in colon cancer, 137
*p53* gene on, 931
role in cancer, 225
18
in colon cancer, 188
DCC gene mutation in colon cancer, 137, 189, 935
19
in Alzheimer's disease, 27
*apo*E, *apo*CI, *apo*CII genes, 495, 498
apolipoprotein E gene on, 596
apolipoprotein gene cluster, 709
gene for LDL receptor, 497
abnormalities in, 426
age-dependent abnormalities, 4
changes in liver cancer, 517-18
in colon cancer, loss of heterozygosity, 188

*Escherichia coli*, 263, 264f
human, physical maps, 420-23
hybridization analysis of, 351-52
locations of disease genes, 376t
locations of growth factors and receptors, 394t
locations of identified tumor suppressor genes, 930t
mammalian, 525
morphology of, 250-51
mutant abnormalities, 370
Philadelphia, 136, 226-27, 226t, 626
polytene, 250
X, 253f
rearrangement of, 373
*Drosophila melanogaster*, 252
Robertsonian fusion, 249
sorting, 965-66
translocation t(14;18), in follicular lymphomas, 161, 227
translocation t(15;17), 227
X
and fragile X linked mental retardation, 324-27, 324f
gene for factor IX, 411
gene for factor VIII, 409-10, 410f
mouse and human compared, 572f
and nephrogenic diabetes insipidus, 213t, 214
Y, 867
yeast artificial, 45
*See also* Translocation
Chromosome-specific paint, 355
defined, 354
Chromosome walking, 527
with *Arabidopsis*, 46
defined, 250
Chronic external ophthalmoplegia (CEOP), 553-54
Chronic granulomatous disease, 375
Chronic myelogenous leukemia (CML), translocation in, 226, 226t
Chylomicrons, 150, 507, 707
assembly and catabolism of, 495-96
defined, 494
lipid transport by, 511-12
Ciliary beating, dynein roles in, 568
Cinnamate-4-hydroxylase (CH4), 700, 701f
Circular dichroism (CD)
defined, 179
to measure DNA denaturation, 209
CIRCULAR DICHROISM IN PROTEIN ANALYSIS, **179-84**, 180f
Cis-acting elements
in apolipoprotein expression, 707-9
defined, 696
Citric acid, from *Aspergillus niger* fermentation, 336
Clarkia fossil beds, 634
Classical conditioning vs. operant conditioning, 540-41
Classification, of steroids, by target tissue, 871
*Clavibacter*, in plant hosts, 472
*Clavibacter xylicyndontis*, 669
Clean Air Act, 331
Cleavage
Achilles', 1-3
base-specific, for nucleotide sequencing, 611
landmark, 815
self-, by RNA, 833f
Clonality, 67
defined, 65
Cloning, 228
applications in *Arabidopsis*, 46
of avirulence genes, 703
of cellulase genes, 334
of centromere antigen-coding genes, 854
of DNA markers for a chromosome, 229
of the Duchenne muscular dystrophy gene, 725
of the human phenylalanine hydroxylase cDNA, 678
immune repertoire, 457
of mycobacteria genes, 584-85
of plant regulatory genes, 698
positional, 355, 375, 421, 424, 437, 527
restriction enzymes used in, 812
restriction sites for, 309
stable, to plants, 695
Clonotype, defined, 449
*Clostridium acetobutylicum*, ABE fermentation of, 334
Coagulation factor VIII, in hemophilia, 409-11
Cobalt, in enzymes, 550
Cocaine, effects on mice, genetic studies, 255
Cockayne's syndrome, excision repair defects, 219
Codes
defined, 888
functional and classification, 889
Coding region, defined, 306, 746
Codons
defined, 93, 306, 369, 428, 833
mutant, in Alzheimer's disease, 26-27
Codon usage
defined, 714, 859
in sequence analysis, 860
COENZYMES
BIOCHEMISTRY OF, **185-87**
defined, 960
tetrahydrobiopterin, 124-25
vitamin-derived, 185
Cofactor, defined, 185, 960
Coiled coil
basic leucine zipper, 749
polymers based on proteins with, 716
Coincidence indices, 891
Cold acclimation, 286
Colicins, effects of, on bacteria, 713
Collagen, 97-98, 311-13, 313f
medical application of, 718
synthetic, 715
in tendons, hierarchical structure, 96f
Collagenases, secretion by bacterial pathogens, 66
Collision-activated dissociation (CAD), 533
COLON CANCER, 137, **187-90**, 934-35
Colony picking, automatic, 52-53
Colony-stimulating factors (CSFs), 201, 395
defined, 392
list, 202t
Colorectal cancer
genetic progression in, 226f, 431
*ras* mutation in, 227
Columns, chromatography, 417
Combinatorial library
antibody, defined, 190
peptide, 881-82, 881f
defined, 880
Combinatorial mechanisms, in antibody diversity, 453
COMBINATORIAL PHAGE ANTIBODY LIBRARIES, **190-94**
Cometabolism
defined, 75
of trichloroethylene, 78
Commission of Biochemical Nomenclature, 296
Communication
among cells, 789
Hartley-Shannon theory of, 894
*See also* Information
Compartmentalization, due to isoenzymes, 481
Compatible interaction, host-pathogen, 700
Competitive binding assays, optical sensors for, 113
Competitive elution, in affinity chromatography, 796f
Competitiveness, President's Council on, 116
Complementarity-determining regions (CDRs), 450-51, 451f
antibody, 38
mutagenesis of, 194
Complementary genes, 973
Complementation, defined, 232
Complementation test, 370, 371f
defined, 369-70
Complement system, 524
Complete proteins, 615
Complex expression, 858
Complex-type sugar chains, 383-84, 384f, 389-90, 390f
Composites
defined, 93
hierarchically ordered, 97-98
sensors and actuators embedded in, 196

Computer analysis, in protein sequencing, 773
Computer modeling
 in drug design, 260
 methods, for RNA three-dimensional structures, 849t
 of RNA, numerical methods, 849
 of RNA three-dimensional structures, 847-50
 in sequence analysis, 860-62
COMPUTING, BIOMOLECULAR, **194-97**
Concatemer, DNA, 952, 952f
 defined, 950
Concerted evolution
 in kallikrein genes, 489
 phenomenon of, 925
Concerted model, of allostery, 405
Conditioning, classical versus operant, 540-41
Conductimetry, biosensors based on, 112
Conductivity, sample, effect on STM, 851
Configuration
 at a chiral center, 169
 defined, 168
Conformation
 defined, 754, 843
  DNA, 207
 DNA, and denaturation, 207
 oligonucleotide, defined, 621
 peptide, 658
  constraints on, 659-63
 protein, 754, 754f
  Raman spectroscopic determination, 739
 RNA, physical forces determining, 845
Conformational modeling hypotheses, 649
Congenital disorders. *See* Genetic disorders
Coniferyl alcohol dehydrogenase (CAD), 701
Conjugation
 bacterial, 364, 365
 by broad host range plasmids, 712
 defined, 709
 plasmid, 711-12
Connectin, 580
Consensus binding sequence
 for p53 binding, 932
 Tat, 12-13
Consensus sequences, 309, 357, 858
 for *Alu* repeats, 439-40
 defined, 856, 859
 G-protein-linked receptor, 792
 L1H, 924
 promoter, RCE, 819
 recognition by protein kinases, 763
 in secreted glycoproteins, 390
Constant (C-terminal) domains, of immunoglobulin light and heavy chains, 450
Constitutive immunity, in transgenic animals, 905
Constitutive promoter, 308
Constraint programming, for computer modeling of RNA, 850
Contamination
 in immuno-PCR, 459
 in LCR, 466
 in PCR, 644-45
Contigs
 defined, 50, 354, 583
 Genome Project use of, 528
 in positional cloning, 355, 527
 from the retinoblastoma gene, 930
 YAC, 571-72, 572f
Contiguous pattern, 888
Continuous assays, for enzymes, 291
Continuous culture (Chemostat), 338
 defined, 335
Contractile system
 molecular organization and genetics of, 576-80
 regulation and energetics of, 580-81
Controlled-release systems, for drug administration, 261
Control region, defined, 746
Cooperativity
 defined, 406, 754
 in oxygen binding by hemoglobin, 403, 404, 405
 of protein folding, 755
 σ-bond and π-bond, 446
 state transition, 268
  defined, 266
Coordinated Framework for the Regulation of Biotechnology, 114, 116
Coordination number, of calcium, 132
Copper
 in enzymes, 548-49
 in superoxide dismutase, 329
Coprotease, RecA protein as, 788
Copy number, 307-8
 determining with RLGS, 817
Corepressor
 binding of, 807-8
 defined, 806
Coronary arteries, defined, 149
Coronary heart disease, 511
Corticoskeleton, actin, 569f
Cosegregation, 425
Cosmid library, for *Mycobacterium leprae*, 583
Cosmids, 423
 defined, 173, 354
 probe made from, 357
Cosmid subclones, FISH ordering of, 357
Cosuppression, 698
 defined, 696
COSY (correlated spectroscopy), 603
 defined, 601
Cotranslational processing, defined, 774
4-Coumarate:coA ligase, role in plant defense against pathogens, 700-701
Coupled assays, for enzymes, 291
Coupling agent, 666
 defined, 664
Coupling factor, 688-89
Covalent attachment, to form constrained peptides, 661
Covalent catalysis, enzymatic, 298
Cowden syndrome, 190
C period, bacterial growth, 57
CpG islands (cyclopentadienyl guanosine islands), 229
 association with transcription, 427
 defined, 324, 409
 in fragile X linked mental retardation, 326
 in hemophilia A, 410
CpG methylation, 381, 381f
Creatine kinase, production by lung cancers, 519
CREST syndrome, defined, 853
Creutzfeldt-Jakob disease (CJD), 596
Crops, improvement with genetic engineering, 691
Cross-linked polymers
 defined, 717
 hydrogels, 722-23
Cross links
 defined, 476, 637
 formation of, in ionizing radiation damage, 478-79
 myosin bridges, 577
Crossover, 526, 800f
 defined, 483, 799
 between functional gene and pseudogene, in Gaucher disease, 341
 in meiosis, 432
 nonreciprocal Y-X, in XX*Sxr* males, 869
 secondary site, in DNA inversion, 365
Cryoelectron microscopy, defined, 269
Cry proteins, in *Bt* parasporal crystals, 471
Crystal lattice, defined, 466
Crystallization, 467
 for X-ray crystallography, 970
Crystallography
 defined, 969
 enzyme-inhibitor interactions from, 9
 X-ray, defined, 15
Crystals
 defined, 741, 969
 liquid, defined, 93
 protein, 742
  characterization of, 743
  growth of, 743
Crystal structures, zinc finger binding domains, 980-81
Culture, yew callus, 885-86
Curie (Ci) unit, defined, 780
Cuticle, composite structure of, 97
Cyanobacteria, lipids of, 503
*N*-Cyanoguanidine linkage, 620, 620f
Cycle sequencing, 610

Cyclic electron transfer, 687–88
Cyclic GMP (guanosine-3,5-mono-phosphate)
 and nitric oxide's biological actions, 599
 as a second messenger, 597
Cyclic nucleotides, as second messengers, 279
Cyclic structures, constraints imposed by, peptides, 660
Cyclin-dependent kinases, 627, 627f
Cyclins, 627, 627f
Cyclization
 N- and C- terminal, peptides, 660, 661f
 side chain-to-backbone, peptides, 660, 661f
dimethyl-—Cyclodextrin, absorption enhancement by, 261
Cyclodextrins, 148
Cycloheximide, effect of, on cell death, 160
Cyclooxygenase, role in prostaglandin synthesis, 681
Cyclophosphamide, B-lymphocyte susceptibility to, 461
Cyclopropyl amino acid substitution, in peptides, 660, 660f
Cyclosporins
 from fungi, 338
 T-lymphocyte susceptibility to, 461
Cysteine, Tat, 12
Cysteine proteinase inhibitor, kininogens as, 486f, 487
Cystic fibrosis (CF), 375, 429, 557
 among African Americans, independent origins of, 728
 animal model for, 560-61
 association analysis in identifying, 427
 defined, 374
 among Hutterites, founder effect, 727
 mutation analysis in, 727
 natural selection and, 727-28
 point mutation scanning in, 377
 prevalence of, 724
Cystic fibrosis transmembrane conductance regulator gene, 646
Cytidylyltransferase, 685
Cytochrome
 defined, 194
 transfer of electrons via, 687–88
Cytochrome *c*, 195
 use in electron microscopy, 270
CYTOCHROME P450, **197–200**, 547–48, 548f
 carcinogenic products of catalysis by, 120
 CYP2D6, role in debrisoquine metabolism, 676
 defined, 546, 675
 and hormone synthesis, 871
 monooxygenases, 117–18
 reductase, 118
Cytofluorimetry, for Vβ family characterization, 875
Cytogenetics
 defined, 324
 of Down's syndrome, 249
 of fragile X linked mental retardation, 324
Cytogenic map, defined, 589
Cytokeratins, 580
Cytokine network, 201
CYTOKINES, **200–204**
 defined, 200, 392, 473t, 474
 genes for, for measuring transcriptional activity, 462
 list, 202–3t
 regulation of extracellular matrix molecules by, 310–11, 311t
 in tumors, 369
 in vitro activity of antisense or triplex-forming oligonucleotides, 44t
 *See also* Interferons; Interleukins
Cytology, defined, 224
Cytomegolvirus (CMV), inactivation of, with photodynamic therapy, 954
Cytoplasm
 cell division, segregation during, 61, 61f
 defined, 57, 162
 division cycle
  synthesis and control, 63
  synthesis during, 58–59, 58f
 protein targeting in, 776
Cytoplasmic genetic elements, yeasts, 975-77
Cytosine (C), 370
Cytoskeleton, 195
 defined, 204, 564
 nerve terminal, 877
 organization and polarity in, 568-69
CYTOSKELETON-PLASMA MEMBRANE INTERACTIONS, 204-6, 204t
Cytosol
 calcium in, 131
 defined, 162
 eukaryotic, Tcp 1 protein in, 164
 hsp70 binding in, 163
Cytosolic phospholipase $A_2$, 681
Cytotoxic effect, of photodynamic photosensitizers, 953
Cytotoxic T cell (CTL), defined, 523

DAF, DNA amplification fingerprinting, defined, 219
Dalton (unit), defined, 573
Darvon, 171
Data
 control of flow, automation, 51
 for nucleic acid sequencing, 610, 610f
 from X-ray diffraction, 743, 745
Database searching, sequence, 862
ddC (2′,3′-dideoxycytidine), 7, 9, 23
 structure, 22f
ddI (2′,3′-dideoxyinosine), 7, 9, 23
 structure, 22f
Debrisoquine polymorphism, 676
Degradation, defined, DNA, 207
Dehydroamino acid substitution, in peptides, 660, 660f
Dehydroepiandrosterone (DHEA), 871
Dehydrogenases, 82
 alcohol, 118–19
 aldehyde, 118–19
 defined, 301
Deletion
 in angiotensin-converting enzyme gene, 400
 detection of disease-causing, 376
 of the *FMR 1* gene, 327
 in hemophiiia A, 410
 mitochondrial, 553–54
 mutation by, 360, 373
 oncogene activation associated with, 626
Deletion-duplication analysis, 377–78
Delivery, of nucleic acids into plants, 690, 690t
*Dementia pugilistica*, β-amyloid plaques in, 26
Dementias, transmissible, 596
Demographics, defined, 359
Denaturant, defined, 754
DENATURATION OF DNA, **207–10**
 complete, 639
 defined, 463, 637, 641
 degrees of, 638, 638f
 at four- and three-stranded junctions, 639, 641
Denaturation of proteins, defined, 757
Denaturing gradient gel electrophoresis (DGGE)
 defined, 374
 to measure DNA denaturation, 209
 for mutation screening, 377
De novo modeling, 759
Deoxyadenosine, 295f
Deoxycholate, sodium, absorption enhancement by, 261
Deoxyhemoglobin, 403
 optical absorption spectra, 404f
Deoxyribonucleic acid (DNA). *See* DNA
Deoxyribose-phosphate unit, DNA, 246f
Dependence, defined, 254
Design
 defined, 793
 molecular, cycle of, 794f
 recognition molecules, 798
Desmin, 580
Desmocollins, 206
Desmogleins, 206
Desmoids, in Gardner syndrome, 189
Desmopressin
 effect on blood pressure, 212
 response to, in nephrogenic diabetes insipidus, 214
Desmosomes, 206
*Desulfovibrio desulfuricans*, 108

Detection
in HPLC, 417-18
in PCR, 645-46
*See also* Analytical methods
Devaluation, of humanness, from transgenic animal work, 909
Development
in *Caenorhabditis elegans*, 588-89
cell, autonomous and nonautonomous, 589
changes of lactate dehydrogenase and creatine kinase during, 481
effects of neuropeptides on, 594
Dextrans, 149
Dextran sulfate, in vitro inhibition of HIV-1, 21
Dextrins, 148
DIABETES INSIPIDUS, **210-14**
defined, 210
as an endocrine deficiency, 282
DIABETES MELLITUS, 89, **214-17**
defined, 214
insulin-dependent (IDDM), 215
mitochondrial mutations associated with, 554
Diacylglycerol (DAG), 132
defined, 682
as a second messenger, 685
Diagnosis
antiantibody markers used for, 47-48, 48t
of fragile X genetic defects, 326-27
of genetic disease, 438, 482
of hematologic malignancies, 226-27
of hemophilia B, direct and indirect, 412f
with immuno-PCR techniques, 460
of neoplastic disease, 224-28
of solid tumors, 227-28
Diagnostic, defined, 794
Diastereoisomers, 169, 171
defined, 168
Diastereomer, defined, 28, 621
Diastereoselective reaction, defined, 28
*Dictyostelium*, 155
annexin from, 33
Dideoxy fingerprinting, 377
3,6-Dideoxyhexoses, of *Salmonella* antigens, 144
Diet-induced thermogenesis (DIT), 614
Differential adhesion hypothesis, 154
Differential RNA splicing, defined, 25
Differential scanning calorimetry, for following DNA denaturation, 208
Differentiation, defined, 392
Diffraction, defined, 969
Diffraction patterns, fiber, 243-44
Digestible energy (DI), 612
Digoxigenin, hapten for FISH, 353
Dihydropteridine reductase, 124
Dihydropyridine receptors, 134
3,4-Dihydroxyphenylalanine (dopa), tyrosine hydroxylation to, 124
Diisopropylcarbodiimide (DIC), 666
Dimerization, defined, 939
Dioleoyl phosphatidylethanolamine (DOPE), 515
Dioxin, effect on T lymphocyte, 461
Diphenyl oxazinones, 30
structure, 29
Diploid state
alleles in, 370
defined, 973
Dipole-dipole interaction, in hydrogen bonds, 445
Dipole reactions, 267
Dipyridodiazepinones
inhibition of reverse transcriptase, HIV-1, 23
structure, 22f
Direct assays, for enzymes, 291
3′-Directed cDNA library, 120-21
Directed cDNA library, 122
Directionality, of motor proteins, 568
Direct readout, 808
defined, 806
Direct screening, for cloning plant regulatory genes, 698
Direct sequence analysis, 221
Disaccharides, 147-48
defined, 146
mannose-galactose, 382-83, 383f
Disclosure, defined, 907
Discontinuous assays, for enzymes, 291
Discrimination, by ligase chain reaction analysis, 465
Disease, genetic basis of human, 557
Disease models, animal patents for, 908
Dismutation reaction, of superoxide radicals, 328
Dispersants, biopolymer, 469
Displacement current, bacteriorhodopsin-based photoreceptor, 71
Dissemination, of bacterial pathogens, within the host, 66
Dissociation temperature, defined, 605
Distamycin, DNA binding by, 104-5, 105f
Distance charts, for nucleotides, 891, 892-93t
Distance geometry, for computer modeling of RNA, 850
Disulfide linkages
antibody, 36, 36f
in osteonectin, 132
in peptides, 661
Diuretic hormones, 594
Diversity, of immune phenomena, 449, 453
Division cycle
bacterial, 57-58
variation in, 62
DNA
acquisition of, 365
analysis for, with chemiluminescent acridinium ester label, 166f
ancient, defined, 633
automatic preparation and purification, 54
in biotechnology products, 673-74
categories of, 525-26
complex with HIV-1 reverse transcriptase, 17
enzymatic sequencing of, 608-11
forms of, A,B and Z, 244, 244f
functional, 525
imaging of, STM, 851-52
intramuscular injection of, for genetic immunization, 368
liposomes for delivery of, 514, 514f
methylation of, 187
mitochondrial, 362
nonfunctional, 525-26
phosphorothioate-containing, 103
photoproducts of, 939-40, 940f
radiation-induced lesions, 477
recombinant, for polypeptide production, 96
replication of, as an event in the division cycle, 62
segregation of, during cell division, 61-62, 62f
single-copy, 525
synthesis during the division cycle, 59-60, 59f
DNA, recombinant
defined, 106
for polypeptide production, 95f
DNA-binding motifs, 807
defined, 806
DNA-binding proteins, defined, 870
DNA complexity, defined, 173
DNA DAMAGE AND REPAIR, **217-19**, 232
defined, 217-18, 218f
detecting damage, 940-41
DNA databank, defined, 263
DNA entrapment, 514-15
DNA extraction, 361
DNA FINGERPRINT ANALYSIS, **219-24**, 647
DNA IN NEOPLASTIC DISEASE DIAGNOSIS, **224-28**
DNA-lipid complexes, as liposome vectors, 515
DNA MARKERS, CLONED, **228-32**
DNA microinjection, 902-4, 903f
defined, 901
DNA modification methyltransferase, defined, 809
DNA pairing, models for, 788
DNA polymerase
defined, 234
high temperature, 303
DNA polymorphism, defined, 25, 428
DNA radicals, reactions of, 477-78
DNA rearrangement, 365
defined, 363

DNA REPAIR IN AGING AND SEX, **232–34**
DNA REPLICATION AND TRANSCRIPTION, **234–38**, 364, 836
DNA sequencing, 361–62
 automated, 240, f
 *Escherichia coli*, 263
DNA SEQUENCING, IN ULTRATHIN GELS, HIGH SPEED, **238–41**
DNA sequencing, for mutation detection, 377
DNA slippage, for analysis of mutations, 360, 360f
DNA strand breakage, 478–79
 defined, 476
DNA strand exchange, 786–88, 788f
 models for, 788
DNA STRUCTURE, 209–10, **242–47**, 834
 chromatin packaging, 176
 coding strand, 915
 double helix, 835f
 G+C-rich regions, 639
 packaging, 414
 sequences recognized by restriction-modification enzymes, 812t
DNA turnover, defined, 359
Domains
 actin-binding, 579
 antibody, 36
 antibody recognition of, 49t
 bacteriorhodopsin, 69
 calcium-binding, 133
 catalytic
  calcium ion pump, 134
  tyrosine hydroxylase, 125–26, 126f
 defined, 483
 EGF
  of lectins, 157
  of P-selectin, 155f
 immunoglobulin, 154
 of immunoglobulin light and heavy chains, 450–51
 immunoglobulinlike, 155–56
 integrin-binding, of α-actinin, 205
 kininogen, 486f, 487
 lectin, of P-selectin, 155f
 melting of, in DNA, 209
 polypeptide, defined, 15, 36
 regulatory
  calcium ion pump, 134
  tyrosine hydroxylase, 125, 126f
 structure of CENP-B, 855f
 transcription activation, 277
 transduction, calcium ion pump, 134
 variable, Vα and Vβ, on T-cell receptors, 874–75
Domestic animals, traits affecting productivity, 905
Dominant mutation
 defined, 187, 374
 *Ki-ras*, in colon carcinomas and adenomas, 189
 of oncogenes, 929
Dominant-negative mutation, 189
 in attenuated adenomatous polyposis coli, 190
 defined, 187, 589
Dopamine, activation by addictive drugs, 256
Dose limits, radiation exposure, 781, 781t
Dosimeters, personal, thermoluminescent detectors on, 784
Dot blot analysis, 221
Dot plot, 860–61, 861f, 862
 defined, 859
Double helix
 left-handed, 244
 structure of, 747–48
Double-strand-break repair model, 801, 801f
Double-stranded DNA bacteriophages, 951–52
Double-stranded RNA viruses, 959
DOWN'S SYNDROME
 Alzheimer's disease associated with, 25, 26
 MOLECULAR GENETICS OF, **247–50**
 role of superoxide dismutase in, 329
D period, bacterial reproduction, defined, 57
*DROSOPHILA* GENOME, **250–54**
*Drosophila melanogaster*
 age-dependent changes in proteins, 5
 antagonistic pleiotropy in, 5
 longevity studies, with genetic transformation, 6
DRUG ADDICTION AND ALCOHOLISM, GENETIC BASIS OF, **254–57**
Drugs
 delivery of
  defined, 260
  peptide, 652
 design of, 673
  and protein structure, 758
 developing new, 880
 effects of enantiomers, 171
 metabolism of, defined, 257
 poly(ortho-esters) as delivery devices, 720
 prevention of uptake, by plasmids, 712
 resistance to, 366–67
  plasmid-determined, 712
 therapeutic aspects of metabolism, 120
 tiazofurin, 137–38
 *See also* Rational drug design
DRUG SYNTHESIS, **257–60**
Drug targeting, defined, 260
DRUG TARGETING AND DELIVERY, MOLECULAR PRINCIPLES OF, **260–62**
Duchenne muscular dystrophy (DMD), 375, 557, 581
 defined, 374
 deletion-duplication analysis for, 377–78
 gene size in, 725
 heart failure in, 399
*Dunaliella salina*, lipids of, 503
Duplex, defined, 242
Dyad symmetries, DNA, 860
Dynamic light scattering, to measure DNA denaturation, 209
Dynamic programming
 defined, 856
 for gap placement analysis, 857–58, 857f
Dyneins, 565, 565t, 566
 defined, 564
 structure and function, 567, 567f
Dystrophin, 377, 579
 in Duchenne muscular dystrophy, 581

Ecdysiotropins, 592–94, 594f
Ecdysis, behavioral patterns in, 594–95
Economic incentives, from patents, 908
Economics, of transgenic animal patents, 908–9
Ectoglycosyltransferases, 157
Edge detection, in a bacteriorhodopsin-based photoreceptor, 72–73
Edman degradation, 768, 772f
 defined, 731–32, 767, 771
 mass spectrometry to identify suitable peptides for, 732–33
Effector molecule, 456
Efferent conduction, defined, 540
Efficiency, quantum, of bacteriorhodopsin-based photoreceptors, 71–72
EF-hand proteins, 132–33
 defined, 131
 osteonectin, 132
EGF domain
 of lectins, 157
 of P-selectin, 155f
Egg proteins, 321
Eicosanoids, as hormones, 281
ELAM-1 (E-selectin), 157
Elastases, secretion by bacterial pathogens, 66
Elastin, 97, 313–14, 314f
 in silk polymers, 715
Elastomers, defined, 717
Electrical synapses, 877, 878f
ELECTRIC AND MAGNETIC FIELD RECEPTION, **266–69**
Electric and magnetic field strength, defined, 266
Electric field, intramolecular, in rhodopsin, 195
Electrode, defined, 110
Electron acceptors
 chlorophyll, 687
 metabolic requirement of microbes, 76, 78

Electron density map, interpretation of, 745
Electron donor-acceptor-donor molecule, 101f
Electronic spectra, for DNA structure determination, 243
ELECTRON MICROSCOPY
  for DNA structure determination, 243
  to measure DNA denaturation, 209
  OF BIOMOLECULES, **269–73**
  for partial denaturation mapping, 637
Electron potential drop, 82
Electron spectroscopic imaging (ESI), 272
Electron spin resonance, 328
Electron transfer, biological, 273–76
Electron transport chain, defined, 79
Electron transport system (ETS), 81–82
  in photosynthesis, 99
  photosynthetic energy transduction in, 687–88
Electrophile, defined, 117
Electrophilic catalysis, defined, 546
Electrophoresis
  for bioseparation, 109
  defined, 50, 238, 359, 479, 764
  for protein purification, 767
  for separation of plasma lipoproteins, 513
Electroporation, 558, 584, 690, 912
  defined, 910
  of plasmids, 711
Electrospray ionization (ESI), 530, 734t
  defined, 529, 732
  mass spectrum, 533f
    recombinant human growth hormone, 531f
  for protein measurements, 733
Electrostatic interactions
  in plasma membranes, 195–96
  switching function of, 197
Elements, chemical, essential, 84, 84t
Elicitor-receptor model, of plant resistance, 702, 703
Elicitors, defined, 699
Ellipticity
  defined, for plane-polarized radiation, 179
  molar, 182
Ellis van Creveld syndrome, 726
Elongation
  defined, 914
  prokaryotes in RNA replication, 348
  in RNA replication, 346
Elution, defined, 417
Embryogenesis, translational control during, studying, 830
Embryonic cells, cytokines role in proliferation of, 203t
Embryonic stem (ES) cells
  pluripotent, 528
    mouse lines, 560
Embryonic stem (ES) cell transfer
  defined, 901
  technology of, 904
  transgenic mice from, 903f
Emission spectra, 317, 318f
Emulsification, in foods, 323, 323f
Emulsion, defined, 320
Enantiomeric excess, defined, 28
Enantiomers, 169, 169f, 629
  defined, 168
  physiological responses to, 171
Enantiomorphic compound, defined, 28
Enantioselectivity
  of medical chemicals, 537
  in nonaqueous media, 305
Encapsidation, 958
  of dsRNA viruses, 959
Encephalitis virus infection, tick-borne, antisense oligonucleotide effect on, 43
Endergonic processes, 80
Endocrine disorders
  deficiency, 282
  excess hormones in, 283–84
Endocrine system, control of calcium homeostasis, 131
ENDOCRINOLOGY, MOLECULAR, **277–84**
Endocytosis, 777
  defined, 774
  pathway for DNA entrapment, 515
  transfer of cholesteryl esters to liver by, 497
  uptake of modified LDL by, 497
  uptake of oligonucleotides by, 41–42
Endogenous agents
  defined, 217, 534
  DNA damage from, 218–19
Endogenous genes
  altered regulation of, 897
  suppression and overproduction of, 698
Endogenous ligands, and receptor functions, 790–91t
Endogenous retroviruses, 925f, 927–28
  defined, 924
Endogenous substrates
  defined, 197
  for P450, 199
Endomembrane system, defined, 774
Endonexin (annexin IV), 33f
Endonexin fold sequence, 33f
Endonexin II (annexin V), 33f
  structure, 34f
Endonucleases
  $Ca^{2+}$-dependent, role in apoptosis, 161
  defined, 939
  degradation of end-capped oligonucleotides by, 41
Endopeptidases, 651
Endoplasmic reticulum (ER)
  biosynthesis of glycans in, 390
  calcium ion transport system, 134–35
  chaperoning proteins across the membrane of, 163–64
  cytochrome P450 in, 198
  defined, 162, 774, 824
  protein targeting at, 776
Endothelin-1 (ET-1), 804
Endothelium-derived relaxing factor (EDRF), 598–99
  defined, 597
Endothelium-derived relaxing factor/nitric oxide (ERDF/NO), 804
Endotoxicity
  defined, 509
  of lipopolysaccharides, 511
Endotoxin, defined, 509
δ-Endotoxin proteins, of *Bacillus thuringiensis*, 668–69
  in parasporal crystals, 471
Energetics, chemical, 296–98
Energy
  from foods, 321, 581
  of ionizing radiation, 477
  for muscle contraction, 580–81
  requirements for
    estimating, 614
    by work levels and sex, 612t, 613t
  thermodynamic concepts, 80
Energy function, and molecular mechanics, 761
Energy transduction
  defined, 950
  during DNA packaging, 953
Enhancer/modulatory region, long terminal repeat, HIV, 10, 11f
Enhancers, 525
  defined, 406
Enhancer trap, 253
  defined, 250
Enolates, electrophilic amination of, 31
Enterobacteria, lipopolysaccharide of, 510
Enterotoxins, defined, 873
Enthalpy, 296
Entropy, 296
Envelope, defined, 953
Environment
  of the earth, and origins of life, 629
  and phenotypic expression of genetic disease, 425
  response of isoenzymes to, 481
Environmental policy, and biotechnology, 115
Environmental Protection Agency (EPS), U.S., 116, 471
ENVIRONMENTAL STRESS, GENOMIC RESPONSES TO, **284–89**
Env protein, HIV-1, 24
ENZYME ASSAYS, **289–92**

Enzyme electrophoresis, 361
Enzyme engineering, 107–8. *See also* Genetic engineering
Enzyme inhibitors, 105
 defined, 103
Enzyme-linked immunosorbant assay (ELISA), for apolipoprotein determination, 513
Enzyme-linked immunosorbent assay (ELISA), plate readers for, 291
Enzyme-linked receptors, 278, 278f
ENZYME MECHANISMS, TRANSIENT STATE KINETICS OF, **292–94**
ENZYMES, **295–300**
 α-amino acid synthesis using, 32
 binding of PNA to, 619
 in bioluminescence and chemiluminescence assays, 168
 biosensor electrodes using, 112
 commercial production of, by fermentation, 338
 copper, 549t
 defined, 73
 in foods, 321
 as genetic markers, 362
 iron, 547t
 mediation of DNA rearrangements by, 365–66
 metal-activated, 547
 of microbial restriction-modification systems, 810, 811f
 motor proteins as, 565
 for nucleotide sequencing, 610
 oligoribonucleotide synthesis using, 623–24
 pharmacogenetic defects in absence of, 676t
 phase I, 117–19
 phase II, 119
 polyethylene glycol modified, 304
 posttranslational variation in, 482
 processing of precursor peptides by, 651
 for regeneration of 11-*cis*-retinal, 822
 restriction, *Not* I and *Sgr*AI, 422
 specificity of, and chirality, 171–72
 stereospecific, 172
ENZYMES, HIGH TEMPERATURE, **301–3**
Enzyme substitution, by plasmids, 713
Enzymology, history of, 296t
ENZYMOLOGY, NONAQUEOUS, **303–6**
EPA, 114
Epidemiology
 of heart failure, 399
 of liver cancer, 517
Epidermal growth factor receptor (EGFR), 519
Epidermal growth factors (EGF), 393, 394
 in the kidney, 805
Epigenetic modification
 defined, 346, 379, 479
 of enzymes, 482
 in genomic imprinting, 381
Epinephrine, lead for drug synthesis, 258
Epistasis, 425
Epithelium, infection barrier, 65
Epitopes
 characteristics of, 449–50
 defined, 141, 260, 449, 583
 functional, 455
 mapping of CENP-B, 854–55
Epoxide hydrolases, hydrolysis by, 119
Equilibrium
 defined, 295
 and enzyme reaction rate, 290
 and free energy, 80
Equilibrium constants ($K_{eq}$), 296–97
 defined, 295
 electric and magnetic field-induced changes, 267, 267f
Equine infectious anemia virus (EIAV), activation domain of Tat, 12
Error catastrophe theory, 5
 defined, 4
Erythrocytes, antigens of, 145, 145t
Erythropoietin, 200, 393, 395
 production from recombinant organisms, 108
*ESCHERICHIA COLI*
 Achilles' cleavage of genome, 2
 composition of, 63t
 enzymatic α-amino acid synthesis using, 32
 GENOME, **263–66**
 host cell features, 309
 replication apparatus, 235f
 RNA polymerase, electron micrography, 271f
 RNase H from, 18
Essentiality, 84
Esterases
 defined, 675
 hydrolysis by, 119
Estrogen receptor, DNA binding domain, 752, 752f, 982, 982f
Estrogens
 nuclear receptors for, 278
 role of, 872
 synthesis of, 871
Ethanol
 analysis for, amperometric biosensor, 112
 biological production of, 331–33, 333t
Ethics
 and genetic testing, 379
 of transgenic animal patents, 909
Ethidium bromide staining, 221
Eubacteria, lipids of, 501–2
Euchromatin, 251
Eukaryotes, 370
 cytochrome P450 of, 199
 defined, 234
 DNA replication in, 235f
 gene regulation in, 347f
 gene structure of, 861–62
 origins for replication, 234
 oxidative phosphorylation in, 81
 protein secretion in, 391, 391f
 protein-synthesizing systems extracted from cells of, 828–30
 ribosome preparation from cells of, 826
 transcription in, 237f, 346
Eukaryotic cells
 metabolism in, 80
 senescence in, 162
European Community, policies on biotechnology regulation, 116
European Molecular Biology Library, *Escherichia coli* database, 263
European Patent Convention, 909
Evolution
 and aging theory, 5
 of complementary base pairing, 840
 conservation of cell death, programmed, 161
 constraints for dynamic programming based on, 857–58
 convergent, example of, 927
 and genetic diversity, 366–67
 genome, 526
 and hemoglobin structure, 405
 hominoid, estimate from mtDNA divergence, 555–56, 555f, 646
 of kininogen domain structure, 487–89, 487f
 and mitochondrial DNA, 553
 and origins of life, 628
 and sequence divergence, 863, 863f
 and sequence similarities, 860
 *See also* Genetic drift
Exchanger
 defined, 131
 renal $Na^+$-$H^+$, 806
Excision repair, 219
 of UV-damaged DNA, 942–43, 943f
Excitation energy transfer
 defined, 686
 light capture and, 686–87, 686f
Exclusion, 220f
 defined, 219
Exercise of Federal Oversight Within Scope of Statutory Authority, 115
Exergonic processes, 80
Exocytosis, 878
 of chylomicrons, 495
 defined, 33, 774, 877
 regulated, 538
 sites for, 877

Exogenous agents
  defined, 217
  DNA damaging, 217-18
Exogenous substrates
  for cytochrome P450, 199
  defined, 197
Exons, 838
  defined, 310, 346, 409, 428, 573, 595, 831, 833
  role in illegitimate recombination, 526
Exon skipping, in alternative splicing, 315
Exon trapping, 427
3′-Exonucleases, degradation of oligonucleotides by, 41
Exopeptidases, 651
Explant, defined, 885
Expressed sequence tags (ESTs), 122
  of *Caenorhabditis elegans*, 588
  defined, 587
Expression, of apolipoprotein E, 707
EXPRESSION SYSTEMS FOR DNA PROCESSES, **306–10**
Expression vectors, 307, 307f
Extended X-ray absorption fine structure (EXAFS), defined, 979
Extension, defined, 463, 641
Extracellular fluid (ECF), calcium in, 131
EXTRACELLULAR MATRIX (ECM), 86, 87, 154, **310–15**
  cell adhesion to, 205
Extracellular space, calcium in, 132
Extraction, for bioseparation, 109
Extradimensional bypass, 196

Fab assembly, 192f
Fab fragments, 38
  antibody, complexed with *S. typhimurium* epitope, 144
  binding specificities expressed as, 37
  cloning, 191, 191f
  defined, 36, 154
  from proteolysis of immunoglobulins, 451
  univalent, 155
FAB-MS (fast atom bombardment mass spectrometry), defined, 138
Factor IX gene, 411-12, 411f
Factor VIII, in hemophilia, 409-11
Familial adenomatous polyposis (FAP)
  defined, 928
  screening for colon cancer in, 226
Familial adenomatous polyposis coli (APC), 188, 934-35
Familial amyotrophic lateral sclerosis (ALS), 562
Familial CETP deficiency, 497
Familial chylomicronemia, apoCII deficiency associated with, 499
Familial combined hyperlipoproteinemia, 500
Familial disorders, defined, 187. *See also* Genetic disorders; Hereditary disorders
Familial HTGL deficiency, 496
Familial hypercholesterolemia (FH), 152, 400, 497, 500
  genetic mechanism, 926
Familial hypobetalipoproteinemia (FHB), apoB mutations associated with, 499
Familial hypocalciuric hypercalcemia, 283
Familial juvenile polyposis, 190
Familial ligand-defective apoB, 500
Familial lipoprotein lipase deficiency, 495
Familial male precocious puberty, 283
Familial neurohypophyseal diabetes insipidus, 282
Family of repeats, defined, 438
Fanconi anemia, excision repair defects, 219
Farm structure, impact of biotechnology on, 908
Farnesyl pyrophosphate synthetase, 497
Fast atom bombardment (FAB) ionization, defined, 732
Fast atom bombardment mass spectrometry (FAB-MS), 529, 734t
  for protein measurements, 733
FAST program family, 862
  FASTa, for local alignment, 858
Fat-soluble vitamins, 961-62, 961t
Fatty acids
  defined, 501, 680
  metabolism of, 507
  properties of, 508t
  unsaturated, production from *Mucor* fermentation, 338
Fatty acid synthetase, 507
Fc fragment, defined, 36
Fc receptor, defined, 382
Fed batch culture, 338
  defined, 335
Federal Insecticide, Fungicide, and Rodenticide Act (FIFRA), 116
Feedback, in heat shock response, 401-2, 402f
Fermentation
  for bioprocessing, 106
  defined, 331
  recombinant, 108
Fermi's Golden Rule, 273
Ferredoxin, 302
Ferritin, 468, 468f, 899
  iron in, 547
Fertilization, of fish eggs, 911-12
Fetal growth factors, defined, 379
Fibrillarin, conservation of protein sequence, 48
Fibrinolysin system, 483
Fibroblast growth factors (FGFs), 395
Fibromas, in Gardner syndrome, 189
Fibronectin, 314-15, 314f
  as an adherence molecule, 65, 206
  incorporation in silk polymers, 715
  repeat in N-CAM, 155f, 156
Fibronectinlike receptors, 278-79
Field amplification, by interfacial polarization, 268f
Field reception
  defined, 267
  electric and magnetic, 266
Filamentous fungi (molds)
  characteristics of, 336
  defined, 335
Filtration, for assay of repressor-operator complexes, 809
Fingerprinting
  clone, 423
  clone libraries for, 443
  dideoxy, 377
  DNA, 219-24, 647
  isoenzymes, 481
  of mycobacterial DNA, 585
  of plants, 223
First messengers, peptide hormones as, 651
Fischer projection formulas, 169, 169f
Fish odor syndrome, monooxygenase deficiency in, 677
Fitness, defined, 724
Flagellar system, of *Chlamydomonas*, 175
Flavoprotein monooxygenases, 118
Flexibility, segmental, of soluble antibodies, 451
Flotation, for separation of plasma lipoproteins, 512
Flow cell methods, Raman study in real time, 740
Flow cytometry
  cell cycle analysis using, 64
  chromosome sorting with, 355
Flow karyotype, defined, 354
Flumezanil, 259
Fluorescence, defined, 110, 317
Fluorescence-activated cell sorting (FACS), 229, 359
  gene order by, 357
Fluorescence binding assay, 113
Fluorescence in situ hybridization (FISH), 355, 356, 965
  for chromosome-specific staining, 966f
  gene mapping by, 350-54
  for identifying the Philadelphia chromosome, 226-27
Fluorescence polarization, 317, 319f
Fluorescence resonance energy transfer (FRET), 318-19
FLUORESCENCE SPECTROSCOPY OF BIOMOLECULES, **317–20**

Fluorography
defined, 779
use and sensitivity of, 784t
Fluorophores, 317
Fluoroscein isothiocyanate (Fite), label for FISH, 353
Flybase (computer-based project), 254
Fmoc method for peptide synthesis, 656-57, 665, 666f
Foam cell, 500
Foaming
defined, 320
in foods, 323
Focal adhesions
defined, 204
and integrins, 205
Folded state, stability of, 755
Folding, protein
analysis by circular dichroism, 184
control by glycosylation, 391
rules of, 730, 731
in vivo, aggregation during, 730-31
*See also* PROTEIN FOLDING
Folding transitions, rates of reaction and equilibria, 755-56, 756f
Follicle-stimulating hormone, 385
Food
animal-derived, analysis of, 674-75
constituents and functions, 611-12, 612f
Food and Drug Administration (FDA), 114
approval of antiviral drugs, HIV-1, 23
regulation of products, versus process, 115
regulatory guidelines for drugs, 673
tiered testing in immunotoxicology, 461
Food industry, enzymes used in, 107-8
Food ingredient, defined, 146
Food processing, of proteins, 323
Food protein interaction, defined, 320
FOOD PROTEINS AND INTERACTIONS, **320-23**
Footprinting reagents, 809
Forensic science
identification of illicit drugs, 674
use of isoenzymes, 482
Forks, multiple, 57
Formacetal linkages, in deoxyoligonucleotide analogues, 620, 620f
Formal correctness of biological models, 894
Förster distance, 319, 319f
Forward genetics, isolation of proteins in *C. elegans* nervous system, 588
Fossilization
defined, 633
types of, 633-35
Founder cells, *Caenorhabditis elegans*, 589
Founder effect, and genetic drift, 726-27
Fractionation
by agarose gel electrophoresis, 815
of human plasma proteins, 106-7
Fragile site, defined, 324
FRAGILE X LINKED MENTAL RETARDATION, **324-27**, 375, 426, 429-30
defined, 374
trinucleotide repeat expansion in, 378
Fragmentation, of labeled DNA, in RLGS, 815
Fragments, antibody, fusion proteins from, 38
Frameshifts, 373
in adenomatous syndromes, 189
defined, 187, 341, 409
in hemophilia A, 410
in recessive nephrogenic diabetes insipidus, 214
Framework construction, in homology modeling, 759-60
Framingham study, 399
Franck-Condon factors, 273-74
defined, 273
Free energy of reaction, intraprotein electron transfer, 274-75, 275f. *See also* Gibbs free energy
Free induction decay, in multidimensional NMR, 603
Free radicals, defined, 327, 476
FREE RADICALS IN BIOCHEMISTRY AND MEDICINE, **327-31**
Frequency counts, for *k*-gram sequences, 889, 890f
Frequency shifts, in Raman spectroscopy, 737
Fructose, 147
Fructosyloligosaccharides, 148
Fruit quality, bioengineering to enhance, 898
$F_1$ transgenic fish, defined, 910
Fuel biotechnology, 334t, 335
FUEL PRODUCTION, BIOLOGICAL, **331-35**
Fumarase, stereospecificity of, 172, 172f
Functional analysis, mammalian genome, 528
FUNGAL BIOTECHNOLOGY, **335-39**
products of, 338t
Fungi
biodegradation of organic wastes by, 78-79
cultivation of, 338-39
filamentous, lipids of, 503
as host cells, 309
white rot, biodegradation of organic wastes by, 78-79
Furanose, defined, 382
*Fusarium graminearum*, production of mycoprotein by, 338
Fusion
genetic, 365, 776-77
in oncogene activation, 626
protein-antibody, 38
Fusion peptide, 540
Fusion protein, 309, 853
cytochrome P450, 198
defined, 306, 479
human β-globin, 407
Sendai virus, for DNA entrapment, 514-15
Fusion proteins, 38
Fusogenic lipids, 539
Fusogenic protein, defined, 537
Fv fragments, 38
defined, 36
production of, 455-56

GABA receptors, 792
Gag protein, HIV-1, 24
Gamma radiation, hydroxyl radical formed by, in vivo, 328
$G_1$ and $G_2$ stages of the cell cycle, defined, 928
Gangliosides, 145
$GM_1$, 144
Gangliotetraosylceramide, 143f
Gap ligase chain reaction (G-LCR), sensitivity of, 463
Gardner syndrome, 189, 190
Gas-liquid chromatography, defined, 672
Gastrointestinal tract abnormalities, associated with Down's syndrome, 249
GAUCHER DISEASE, **341-43**, 508
defined, 341
Gelatin, as a drug delivery matrix, 718
Gelation, 321, 322-23
defined, 320
Gel electrophoresis
defined, 813
for genetic analysis of populations, 360
for protein purification, 767, 770, 770f
GEL ELECTROPHORESIS OF PROTEINS, TWO-DIMENSIONAL POLYACRYLAMIDE (2D), **343-45**
Gel filtration, for protein purification, 766-67
Gel retardation assay, 812
for in vitro study of repressor-operator recognition, 809
Gels, 468-69
Geminate recombination, 404
defined, 402
Gene acquisition, promotion and limitation of, 366
Genealogical analysis, 726
Gene constructs, fish, 911, 911f
Gene conversion, 800f
defined, 483, 799
Gene duplication, origin of, 526
Gene expression
defined, 149-50
inhibition of
by antisense enzymes, 107
by antisense oligonucleotides, 44

Gene expression *(Continued)*
lipoprotein system, 153
profile of, 121, 122, 123f
GENE EXPRESSION, REGULATION OF, **346–50**
by trace element micronutrients, 899-900
Gene flow, 728
Gene-for-gene hypothesis of plant-pathogen interactions, 700, 702, 703
Gene locus, defined, 354
Gene mapping, 527
mycobacteria, 583
GENE MAPPING BY FLUORESCENCE IN SITU HYBRIDIZATION, **350–54**
Gene order, by FISH, 357
GENE ORDER BY FISH AND FACS, **354–59**
Gene probe, defined, 354
Gene products, in transformed plant cells, 691, 691t
Gene regulatory complexes, Raman spectroscopic analysis of structure, 739
Genes
actin, 577-78
affecting programmed cell death, conservation of, 161
*Age-1*, *Caenorhabditis elegans*, 6
for antibodies and T-cell receptors, 452-53, 453f
*APC*
in colon carcinoma and adenoma, 188
in familial adenomatous polyposis, 226
*APC* and *MCC*, 934-35
*apo*AI, 152
*apo*CIII, 152
*apo E*, rat, 152
apolipoprotein, 495
*bar*, phosphinothricin acetyltransferase coding by, 693-94, 693f
*Bt* toxin, in plants, 472
*ced*-3, 160, 160f, 161
*ced*-4, 160, 160f 133
*ced*-9, 160, 160f, 161
centromere antigen-coding, 854
*ces*-1 and *ces*-2, 160, 160f
c-*myb* and c-*src*, 560
for coagulation factors, 409
for coagulation factor VIII, 409-11
for collagen, 311-13
*cry*, of *Bacillus thuringiensis*, 471
cystic fibrosis, 377
*DCC*, chromosome 18, 189
defined, 370, 374, 428, 573
*E6*, human papilloma virus, 189
*EIB*, adenovirus, 189
encoding lipoprotein metabolism enzymes, 153t
encoding lipoprotein metabolism proteins, 151t
*erb* B2, 519
for extracellular matrix proteins, 311
for fibronectin, 314-15, 314f
fragile X, *FMR 1*, 325
fusion of, 365
glucocerebrosidase, 341
glucokinase, and diabetes, 215, 216-17
*gus* A, β-glucuronidase coding by, 693
*H-19*, 381
histone, 413
expression of, 414-15
organization of regulation of, 415-16, 415f
*HLA*, 523-24
*hsp*, 401
*hsp70*, 286
human
*c-myc*, *c-fos*, *c-myb*, 15
human LDL receptor, transgenic experiments in, 152
H-Y antigen, 867
hygromycin resistance, hygromycin phosphotransferase coding by, 694
*Igf-2* and *Igf-2r*, 379-80, 381
immunoglobulin, 37
for immunoglobulin domains V, L, J, and C, 453
imprinted, 379-80
kallikrein, 483, 489-90
kanamycin, neomycin phosphotransferase coding by, 694, 694f
kininogen, structure of, 488, 488f
knockout technique for studying, 152-53
*LI* source, 440, 440f
longevity, 6
major histocompatibility complex, 523-24
*MCC*, 189
*mdm2*, in human sarcomas, 189
mitochondrial DNA, 551-52
mouse, genetically imprinted, 379-80
multilocus, coding for isoenzymes, 480
mycobacterial, 584
in myogenic differentiation, 582
myosin, 577
*NFI*, in colonic tumors, 189
*nuc*-1, role in programmed cell death, 160
nuclear, 362
as genetic markers, 362
*p53*, 516, 627, 931-32
in bladder cancer, 227-28
in colon cancer, 188-89
mutations in cancer, 225
phenotypic expression of, 371-73
plant resistance, 702
plasma membrane calcium ion pump, 134
*pol*, 15
*pol* II, direction of RNA polymerase II transcription, 178-79
promoter sequences of, 838
*ras*, in lung cancers, 519
*Rb*, in retinoblastoma and in osteosarcoma, 137
*RBI*, in retinoblastoma, 817, 818
regulation of expression by interferons, 474
regulation of expression of, 372
reporter, for studying transcription, 168
*rpo*H, 401
sarcoplasmic reticulum calcium pump, 135
*sec*, 777
sex-determining region Y (*SRY*), 868
stress-responsive, 285
T-cell receptor, 453
techniques for isolation, tomatoes, 896-98
*Tla* and *Qa*, MHC, 524
*t maternal effect*, 527
tropomyosin, 578
troponin, 578-79
tumor suppressor, 187
vasopressin, mutation in, 213
vasopressin precursor, 211, 211f
vasopressin receptor, mutations in, 213t
*See also* Oncogenes; Proto-oncogenes; Tumor suppressor genes
Genes, heterologous, for cloning plant regulatory genes, 698
Gene signature, 121-22
Gene structure, defined, 310
Gene targeting
defined, 187, 556
by homologous recombination, 557-60
in-out method for, 558-60, 558f
in mice, 559f, 560-61
Gene therapy, 557
myoblasts used in, 582
somatic and germ cell transfection, 936
GENETIC ANALYSIS
of breast cancer, 128-30
*Drosophila melanogaster*, learning in, 543
*Drosophila melanogaster* genome, 253
mammalian, techniques for, 527
OF POPULATIONS, **359-63**
Genetic code, 915, 917-18
mitochondrial DNA, 552
Genetic counseling
DNA profile used in, 223
in hemophilias, 412-13
for risk of breast cancer, 129-30
Genetic disorders
in aging, 162
Alzheimer's disease, 26
apoE deficiency, 500

in atherosclerosis, 152
campomelic dysplasia, 869
central diabetes insipidus, 212
excision repair defects, 219
and genomic imprinting, 380-81
in heart failure, 399-400
hemoglobin mutants, 406
human, 429-31
Li-Fraumeni cancer syndrome, 928
mitochondrial point mutations in, 553
motor neuron disease, 561-62
nephrogenic diabetes insipidus, 211
neurofibromatosis 1, 590-91
neurofibromatosis 2, 591-92
non-insulin-dependent diabetes mellitus, 216-17
PCR for identifying mutations in, 646-47
prevalence of, 724-25, 725f
retinoblastoma, 817-18
simple and complex, 424, 424t
susceptibility to neoplastic disease, 225-26
von Hippel-Lindau syndrome, 934
*See also* BIOCHEMICAL GENETICS, HUMAN; *Familial* entries; Hereditary *entries*
Genetic distance, 437
defined, 128
GENETIC DIVERSITY
and animal patents, 909
IN MICROORGANISMS, **363-67**
Genetic drift, 361
defined, 724
and the founder effect, 726-27
*See also* Evolution
Genetic engineering
of bacteriorhodopsin-based compounds, 95
in biological waste treatment, 79
for improvement of crops, 691
for insect control, 469-73, 669
of mice, 36, 37
plants with *Bt* toxin genes, 472t
of polymers, 714-17
GENETIC IMMUNIZATION, **367-69**
Genetic instability, defined, 714
Genetic linkage map
defined, 589
in Huntington's disease, 595-96
in neurofibromatosis, 590
Genetic manipulation, of yeasts, 974-75
Genetic map
defined, 263, 570
relationship of denaturation to, 638-39, 639f
Genetic marker, defined, 128. *See also* Markers
Genetic noise, and sexual reproduction, 234
Genetic polymorphism, defined, 675
Genetic profiles, 222
in cancer, 224-25
Genetic recombination, defined, 784. *See also* Homologous recombination
GENETICS, **369-74**
of drug addiction and alcoholism, 254-57
of δ-endotoxins, 669
human biochemical, 73-74
of mitochondrial DNA, 552-53
molecular, of liver cancer, 517-18
of muscle-related myopathies, 581
problems in immunotoxicology testing, 461-62
Genetic selection, defined, 228
Genetic Stock Center, *Escherichia coli*, 263
Genetic systems, 173-74, 174t
GENETIC TESTING, **374-79**
Genetic transformation, in aging studies, 6
Gene transfer
*Agrobacterium*-mediated, 690
defined, 689, 901
horizontal, 364
Genogram, 129f
Genome evolution, 526
data from mitochondrial DNA, 554
Genome map, physical, 587
Genome Project, 528
defined, 524
Genomes
aberrations of, in cancer, 137
*Arabidopsis*, 45-47
of *Caenorhabditis elegans*, 588
defined, 1, 45, 263, 284, 428, 524, 570
*Drosophila*, 250-54
*Escherichia coli*, 263-66
HIV-1, 19f
human, 428
body expression map, 120-24
mapping of, 421
mitochondrial, 551
mutation frequency, 5
mycobacterial, 583-84
stress responses within, 287-88
viral, 958
defined, 957
Genome scanning, 814, 814f
defined, 813
Genomic fluidity, 288
defined, 284
GENOMIC IMPRINTING, **379-82**
Genomic library, 228
construction of, 229
Genomic mechanisms, hormone control of, 282
Genomic restriction maps, 421-22
direct construction of, 422-23
Genotype, defined, 254, 432, 589
Gerontogenes, 6
Gerstman-Sträussler-Scheinker syndrome (GSS), 596
Gibbs free energy (Γ*G*), 80, 296-97
defined, 295
Gilbert's syndrome, 677
GLC (gas-liquid chromatography), 138
Glial fibrillary acidic protein (GFAP), 580
Glioma, transplantable, 138
Global alignment, 862
Globin fold, 402
Globins, 402
Globotetraosylceramide, 143f
Glomerulonephritis, poststreptococcal, 66
Glucans, 718
Glucocerebrosidase
deficiency of, in Gaucher disease, 341
haplotypes, 342t
Glucocorticoid receptor (GR), complex with glucocorticoid response element (GRE), structure, 751-52
Glucocorticoid-remediable aldosteronism, 283
Glucocorticoids
nuclear receptor, 277-78
role of, 872
synthesis of, 871
Glucose, 147
biosensor for, 112
carbon source, in penicillin G and V production, 337
Glucose-6-phosphate dehydrogenase, deficiency of, 677-78
Glucose oxidase, 112
Glucosyl ceramide, in Gaucher disease, 341
Glucuronyltransferases, 119
Glutamate receptors, ionotropic and metabotropic, 792
*trans*-Glutaminase, role in cell death, 161
Glutamine synthetase (GS), inhibition by phosphinothricin, 671
Glutathione (GSH), reaction with DNA radicals, 477-78
Glutathione peroxidase, 329, 329f
selenium in, 329
Glutathione reductase, 329, 329f
Glutathione *S*-transferases, polymorphism in, 677
Glutathione transferases, 119
metabolism of methyl bromide via, 120f
*O*-Glycans, 389
Glycans, 718-19
glycoprotein, 140f
*See also* Oligosaccharides
Glycerol phosphate backbone, sites of action of phospholipases, 681f
Glycine, asymmetric derivatization of, 29-30
Glycine-rich proteins (GRP), 701
GLYCOBIOLOGY, **382-85**

Glycocholate, sodium, absorption enhancement by, 261
Glycoconjugates
defined, 141, 382
mammalian, 145-46
GLYCOGEN, **385-88**, 386f
defined, 385
Glycogenin, defined, 385
Glycogenolysis, 386
regulation by epinephrine, 764
Glycogen phosphorylase, muscle, deficiency of, 581
Glycogen storage disease Type VII, 581
Glycogen synthase, 386-87, 388f
defined, 385
Glycohormones, defined, 382
Glycolipids
analysis of, 141
defined, 341, 501
mammalian, 145-46
Glycolysis, 80
Glycoproteins, 146
analysis of, 139-40, 140f
antibodies, 450-51
cell adhesion molecules, 153-54
defined, 154
fibronectins, 314-15
mucins, 146
N-CAM, 156
osteopontin, 132
plasma kallikrein, 489
GLYCOPROTEINS, SECRETORY, **388-91**
Glycosphingolipids, 145
ceramide portion of, 143f
structure at an air-water interface, 506f
synthesis of, 508-9
Glycosylphosphatidylinositol
attachment of N-CAM via, 155f, 156
truncated cadherin attachment via, 156
Glycosyltransferases, defined, 154
GMP140 (P-selectin), 157
Golgi complex, biosynthesis of glycans in, 390
Gonad, defined, 866
Gonadoblastoma, in XY gonadal dysgenesis, 870
G proteins, 279, 280f, 400
defined, 277, 399
receptors linked to, 792-93
transducin, 820
Gradient, defined, 417
Gradualism, in proteins, 196
Graft rejection, 523
*k*-Gram, defined, 888
Gram-negative bacteria
cell envelope, 510f
defined, 509
Granular cells, defined, 803
Granulocyte colony-stimulating factor (G-CSF), 392
Granulocyte-macrophage colony-stimulating factor (GM-CSF), 395, f
Graves' disease, 47
Greiss reaction, nitric oxide assay, 598
Grooves, major and minor, 979
Group translocation, defined, 922
Growth
of crystals, 467
in gels, 468-69
Growth additives, in crystal chemistry, 467
Growth cone collapse, 158
Growth factor pathway, mutation in, 284
GROWTH FACTORS, 200-201, **392-96**
autocrine, in lung cancers, 519
defined, 200, 392
fetal, defined, 379
receptors for, 793
in the renal system, 804-5
Growth hormone
recombinant human, ESI mass spectrum, 531f
in transgenic fish, 912
Growth law, bacterial, 63
Growth rates, and bacterial growth, 63-64
Growth regulators, in plant disease, 703
$G_0$ state, defined, 928
GTP-ase-activating protein (GAP), 590
GTP-binding regulatory proteins (G proteins), defined, 789
Guanine (G), 370
Guanylate cyclase, binding of NO to, 804
Gums
defined, 146
industrial, 149
Guthrie test, for phenylketonuria, defined, 678
Gyrase, DNA, recombination mediated by, 365

Habituation, defined, 541
*Habrobracon juglandis*, radiation studies using, 4-5
*Haemophilus influenzae*, lipopolysaccharides of, 144-45
Hageman factor
activation of the kallikrein-kininogen-kinin systems, 484f, 489
defined, 483
Hairpin loop
defined, 841, 843
RNA anticodon, 845-46, 845f
size and flexibility of, 846
Half-life
defined, 779
of radionuclides, 779-80
*Halobacterium salinarium* (halobirum), bacteriorhodopsin of, 70, 94-96, 99
Halogenated organic compounds
defined, 75
reductive dehalogenation of hydrocarbons, 78
Haloperidol, inhibition of HIV protease by, 16
Halophiles, lipids of, 501
Halo structures, 352
Hamartomatous syndromes, 190
Haploid organisms, 363
defined, 973
Haplotype
allelic association in, 437-38
defined, 678, 724
Haplotype analysis, of the founder effect, 726-27
Haptens, defined, 350, 449
Hartley-Shannon theory of communication, 894
Harvey murine sarcoma virus, 519
Hashing, 858
defined, 856
Head group, defined, 680
Heart disease, associated with Down's syndrome, 248-49
HEART FAILURE, 150
GENETIC BASIS OF, **399-400**
Heat, damage to DNA from, 217
Heat increment of feeding, 612
Heat shock gene
defined, 400
responses to, 285-86
transcription patterns, 15
Heat shock proteins (hsps), 285-86
defined, 400
functions of, 286
*See also* Chaperones
Heat shock regulatory element (HSE), 286
HEAT SHOCK RESPONSE IN *E. COLI*, **400-402**
Heavy metal shadowing, defined, 269
Heinz bodies, 406
α-Helical proteins, polymers based on, 716
Helical RNA, defined, 843
Helicases, DNA, 834, 835
α-Helix
defined, 16, 746, 757, 979
reverse transcriptase, 17
Helix-destabilizing proteins, defined, 834
Helix-loop-helix motif
basic, in plants, 697
basic leucine zipper group, 749
in muscle proteins, 582
*See also* EF-hand proteins
Helix swap experiment, 807
Helix-turn-helix motif, 747, 748-49, 748f, 807
defined, 746
Helper T cell ($T_h$), defined, 523
Hemagglutinin (HA), 539-40
Hematologic malignancies, diagnosis of, 226-27
Hematopoiesis, c-*myb* gene associated with, 560
Hematopoietic growth factors, list, 202t
Hematoporphyrin derivative (HPD), as photosensitizer, 954-55, 954f

Hemes
of cytochrome P450, 198
electron density of, 273
protein, 547
Hemicellulose, defined, 331
HEMOGLOBIN, **402–6**
ferric, 403f
functional properties of, 404t
as an intelligent material, 196
iron in, 547–48
recombinant, 408
structure of, 407, 407f
S variant, 726
HEMOGLOBIN, GENETIC ENGINEERING OF, **406–9**, 408t
β-Hemoglobin, expression of the gene encoding, 372f
Hemoglobinopathy, defined, 374
HEMOPHILIA, **409–13**
A, L1 insertion in, 925–26
B, founder effect, 727
Hemopoietic growth factors, 392, 395
biological actions of, 396t
Hemoproteins
cytochrome P450 monooxygenases, 118
defined, 197
Hemorrhage, cerebral, in Dutch type amyloidosis, 26
Hepadna viruses, packaging of, 959
Hepatic triglyceride lipase (HTGL), 496
Hepatitis B virus (HBV)
defined, 516
duck, effect of phosphorothioates on infection by, 43
Hepatitis C virus (HCV)
defined, 516
and liver cancer, 517
Hepatitis virus, packaging of, 960
Hepatobiliary system, cholesterol secretion via, 497
Hepatocyte growth factors (HGF), 395
Hepatocytes
culture of, 91
malignant transformation of, 517
Hepatomas, 516–17
Heptad repeat
in *APC* gene product, 188
defined, 187
Herbicides
enantiomers of, 171
engineering plants for resistance to, 898
halogenated, biodegradation pathways, 77f
reductive dehalogenation of, 78
Herculin, 582
Hereditary disorders
angioedema, genetic mechanism of, 926
nephrogenic diabetes insipidus, 282
nonpolyposis colon cancer (HNPCC) syndromes, 190, 378
persistance of fetal hemoglobin, 926
Hermaphrodites
XX, 869
XY, 870
Herpes simplex viruses
effect of methylphosphonate oligonucleotides on, in mice, 43
photodynamic therapy for infection by, 954
photoinactivation of, 955, 955t
Herpes viruses, 953
Heteroatom, defined, 476
Heterochromatin, *Drosophila melanogaster*, 251
Heterocyclic compounds, defined, 628
Heteroduplex analysis, 427
Heteroduplex DNA, defined, 374, 784, 799
Heterogeneous nuclear RNA (hnRNA), 838
Heterologous sequences, defined, 784
Heteroplasmy, 552
defined, 551
Heteropolymer, defined, 73
Heterotrophs, 76–77
energy metabolism in, 80–81, 82
Heterozygote, defined, 374
Heterozygote advantage
defined, 724
and sickle cell disease, 726
Heuristic reasoning, 894–95
β-Hexosaminidase A, in Tay-Sachs disease, 482
Hexose, 333
defined, 331
Hierarchical structure, in biological composites, 96f, 97–98
Hierarchy, in biocomputing, 194
High density lipoproteins (HDL), 150–51, 511
analysis of, 513
defined, 495
High performance liquid chromatography (HPLC), 138
for analysis of biomolecules in solution, 601
for bioseparations, 109
defined, 672, 732
isocratic, 417
for peptide purification, 657, 736
sample purification for protein sequencing, 772
*See also* HPLC *entries*
High performance liquid chromatography-mass spectrometry, protein structure analysis, 531–32
High resolution autoradiography (EM ARG), 272
defined, 269
Hill plot, 403–4, 404f
Hinges
immunoglobulin domain, 451
myosin, 577
Hip endoprosthesis, 91
Hippocampus, defined, 541
Histone octamer, defined, 175–76
Histone proteins, defined, 413
HISTONES, **413–16**
in cell division, 178
in chromatin, 176
HIV enzymes, three-dimensional structure, 15–18
HIV protease, defined, 6
Holliday intermediate, 787f, 789
defined, 785
Holliday junctions, 800, 800f, 802
in bacteriophage lambda DNA, 641
defined, 799
double, 801
Holoenzyme
defined, 185, 234, 960
DNA polymerase III, 234
Homeobox, 697
defined, 746
Homeodomains, 747, 748–49
defined, 746
Homeostasis, calcium, 131
Homodimer, protease, 16
Homogeneous assay, defined, 165
Homologous recombination (HR), 557, 785, 785f, 800f
correction of gene mutations by, 561
postmeiotic segregation (PMS) event, 799
Homologous sequences, defined, 785
Homology, 860, 863
defined, 859
genome, defined, 524
Homology modeling, 759–761
Homopolymers, polysaccharide, 142–43
Homozygote, defined, 374
Honey, 148
Hoogsteen base pairs, 39, 40f
Horizontal ultrathin gel electrophoresis (HUGE), 238–39
Hormone receptor DNA-binding domains, zinc finger, 751–52
Hormone response element (HRE)
defined, 277
glucocorticoid, 278
Hormones
classification of, 277
cytokines as, 200
defined, 257, 277, 392
endogenous, leads drug synthesis, 258
glycohormones, 385
and immunotoxic responses, 461
peptide, 648, 651
synthetic, 657
Hormone-sensitive lipase, 508
Host
bacterial invasion of cells, 66
transmission of pathogens to, 65, 952, 952f

Host cell, features of, 309-10
Host-plant systems, genetically engineered, 471-72
Host-vector systems, 307
Hot spots
for mutation, 725-26
of recombination, 802
HPAEC-PAD (high performance anion exchange chromatography with pulsed amperometric detection), 138
H-phosphonate method, 622
H-phosphonates, defined, 621
HPLC. *See* High performance liquid chromatography (HPLC)
HPLC OF BIOLOGICAL MACROMOLECULES, **417–20**
HTF islands, defined, 409
Human antimouse response (HAMA), immunoglobulins, 37
HUMAN CHROMOSOMES, PHYSICAL MAPS OF, **420–23**
HUMAN DISEASE GENE MAPPING, **424–27**
HUMAN GENETIC PREDISPOSITION TO DISEASE, **428–31**
Human genome, 428
Human Genome Project, 50, 249, 355, 427
Human growth hormone, recombinant, analysis of, 532f
Human immunodeficiency virus (HIV), 15, 19-24, 24f
infection by, role of superantigen, 876
photodynamic inactivation of, 955
recombinant vaccines for, 947
*See also* HIV *entries*
Human leukocyte antigen (HLA) class I, structure of, 452
Human leukocyte antigen (HLA) class II
association with insulin-dependent diabetes mellitus, 215
defined, 214
structure of, 452
HUMAN LINKAGE MAPS, **432–38**
HUMAN REPETITIVE ELEMENTS, **438–41**
Humoral immunity, defined, 367, 523
Huntington's disease, 374, 375, 378, 427, 527, 595-96
mitochondrial mutations associated with, 554
H-Y antigene gene (HYA), 867
Hybrid cells, staining of, 967
Hybridization
CISS, 353, 353f
defined, 605
species-specific, 965
by spheroplast fusion, 974f
subtractive, 698, 699f
targets for, 351-52
yeasts, 974f
HYBRIDIZATION FOR SEQUENCING OF DNA, **441–44**
Hybridization protection assay, 166f, 167
Hybridoma
defined, 106
for production of monoclonal antibodies, 108
structure of, 450
Hybrids, formation of, 607-8
Hybrid sequences, artifacts in PCR, 645
*Hydra attenuata*, mtDNA of, 551
Hydration, of proteins, in foods, 322
Hydrocarbons, metabolism of, 506
Hydrocolloids, 149
Hydrogels
from cross-linked polymeric networks, 722-23
from cross-linked starch, 718
defined, 717
Hydrogen
biological fuel production options, 332t
biological production of, 334
Hydrogenases, 302
defined, 301
Hydrogenation, asymmetric, of dehydro amino acid preparation, 32
Hydrogen bond cooperativity, 446
defined, 444
HYDROGEN BONDING IN BIOLOGICAL STRUCTURES, **444–47**, 445t
Hydrogen bonds
acceptors, defined, 444
defined, 444, 746
donors, defined, 444
patterns of, 446-47
in RNA, 846
strength of, 444, 444t
Hydrogenolysis, bacterial, 78
Hydrogen sulfide, from sulfate reduction by anaerobic bacteria, 78
Hydrolysis, defined, 680
Hydropathy, 862
defined, 859
Hydrophilic polymers, nonbiodegradable, 722-23
Hydrophilic substance, defined, 341
Hydrophobic interaction chromatography (HIC), 419
Hydrophobic polymers, nondegradable, 721-22
Hydrophobic substance, defined, 341
Hydrothermal systems, marine, and origin of life, 629
Hydrothermal vent, defined, 301
Hydroxy acids, chirality of, 170-71
Hydroxyapatite, 131
Hydroxyl radicals
chemical reactivity of, 328, 478
from radiolysis of water, 477
Hydroxymethylglutaryl (HMG) CoA-reductase, 497
Hydroxymethylglutaryl (HMG) CoA-synthetase, 497
Hydroxyproline-rich glycoproteins (HRGPs), 701
11β-Hydroxysteroid dehydrogenase deficiency, 283
5-Hydroxytryptamine, of nematodes, 588
Hygromycin B, for selecting transformants with hygromycin resistance genes, 695
Hyperaldosteronism, 283
Hypermutation, somatic, 453
Hypernatremia, in diabetes insipidus, 212
Hyperphenylalaninemia, 678
Hypersensitive reactions
in bacterial infection, 66
defined, 699
in plant defense against pathogens, 702
*See also* Immune response
Hypertension, kallikrein-kininogen-kinin system in, 485
Hyperthermophile, defined, 301
Hyperthermophilic genera, 302t
Hypertriglyceridemia, and apoCIII, 499
Hypervariable regions
defined, 219
of immunoglobulin light and heavy chains, 450
Hypervitaminosis, defined, 960
Hypovitaminosis, defined, 960
Hypoxanthine guanine phosphoribosyltransferase (*HPRT*), for gene targeting, 560, 561
H zone, muscle, 575

ICUMSA (International Commission for Uniform Methods of Sugar Analysis), 138
Identification, from DNA analysis, 223
Idiotype, defined, 449
Illegitimate recombination, defined, 524
Image analysis, 354
defined, 350
software for, 53
Image sensing, by bacteriorhodopsin-based photocells, 70, 72-73
Imidazolidinones, 30
structure, 29
Immobilized ligands, for analyzing molecular recognition, 795-98, 798f
Immune response
cytokine role in, 200
evasion of, bacterial, 66
hepatic acute phase, 461
*See also* Hypersensitive reactions
Immune system, in vivo strategy of, 457f
Immunization, defined, 367
Immunoadhesins, 38
CD4, 456
Immunoaffinity chromatography, for separation of plasma lipoproteins, 512

Immunoassay, defined, 165
Immunoelectron microscopy, defined, 269
Immunogenicity versus antigenicity, 455
Immunoglobulin domain, defined, 154
Immunoglobulin fold, of immunoglobulin light and heavy chains, 450
Immunoglobulins
  IgD, genetic engineering to obtain, 37
  IgG
    structure of, 450-51, 450f
    sugar chains of, 385
  polypetide chain structures, 450t
Immunoglobulin superfamily
  MAG, 155f
  N-CAM, 155-56
IMMUNOLOGY, **449-57**
Immunomodulation, defined, 460
IMMUNO-PCR, **458-60**
  quantitation of PCR product, 460f
  *See also* PCR
Immunosuppression, environmentally produced, 460-62
Immunotoxic evaluations, 461
IMMUNOTOXICOLOGICAL MECHANISMS, **460-62**
Immunotoxins, production of, 456
Imprinting, 527-28
  defined, 524
  genomic, 379-82
  parental, 380-81
Imprinting syndromes, cancer, 379
Impurities, defined, 672
Inactivation, enzyme, 290
Inbred strain, defined, 254
Inclusion, 220f
  defined, 219
Inclusion bodies, 729
  defined, 728
  formation of, 730-31
Incompatibility, plasmid, 308, 709, 711
Incompatible interaction, plant-pathogen, 700
Incrustation, 634
Indel, defined, 856
Indirect assays, for enzymes, 291
Indirect immunofluorescence (IIF), 48-49
  defined, 47
Indirect readout, 808, 808f
  defined, 806
Indium oxide, electrode, bacteriorhodopsin-based photocell, 70, 72
Inducer of short transcripts (IST), 11, 11f
Inducible system, 349
Infectious disease agents, detection by ligase chain reaction, 465-66
INFECTIOUS DISEASE TESTING BY LIGASE CHAIN REACTION, **463-66**
Inflammatory response
  expression of K-kininogens and t-kininogens in, 489
  role of selectins, 157
Influenza virus, 839-40
Information
  chemical flow of, transforming to electrical flow, 110, 111f
  genetic, 370
    RNA encoding of, 831
  pragmatic and semantic theories of, 894
  about protein structure, from circular dichroism, 180
  for RNA three-dimensional modeling, 847-48, 848f
  transfer of
    DNA to RNA, 834-36, 836-838, 838f
    RNA to DNA, 840
    RNA to RNA, 839-40
  visual, processing in bacteriorhodopsin photoreceptors, 72-73
Information processing, with biochips, hypothetical, 102-3
Infrared difference spectra, of intermediates, bacteriorhodopsin photocycle, 69
Infrared spectroscopy, to measure DNA denaturation, 209
Infringement, defined, 907
Inheritance pattern
  of fragile X linked mental retardation, 324-25
  of mitochondrial DNA, 552
  *See also* Genetics
Inhibin, 394
Inhibitors, enzyme, 105
  asymmetric and symmetric peptidic, to HIV protease, 9
  of tyrosine and tryptophan hydroxylases, 127
Initial rate, 289-90
  defined, 289
  estimation of, 291-92
Initiation
  defined, 914
  of DNA replication, 236
    regulation of, 59-60
  of RNA replication, regulation of, 346
Initiator codon, 917-18
INORGANIC SOLIDS, BIOMOLECULES IN THE SYNTHESIS OF, **466-69**
Inosine monophosphate dehydrogenase, in neoplastic cells, 137
Inositol, 134
Inositol triphosphate
  defined, 682
  as a second messenger, 685
Insecticidal crystal proteins (ICPs), 668-69
  defined, 668
INSECTICIDES, RECOMBINANT PROTEIN, **469-73**
Insertional mutagenesis, 903
Insertions, mitochondrial, 553-54
Insertion sequences, 365-66
  defined, 583
In situ hybridization (ISH), 252
  defined, 250, 803, 965
  gene mapping by fluorescence, 350-54
  of insulinlike growth factors, 805
  for mapping candidate genes, 427
In situ PCR, 645
Insulin, 393, 652
  defined, 214
  human gene for, 515-16
  nasal delivery of, 261
  production from recombinant organisms, 108
  receptor for, tyrosine kinase, 793
Insulin-dependent diabetes mellitus (IDMM), 215
Insulinlike growth factors (IGFs), 393, 805
  production by gliomas, 138
Integral membrane transport proteins, 922
Integrase, 7-9
  drugs based on, 24
Integrated pest management (IPM), defined, 469
Integration
  of injected genes, 902-3
    analysis of, 905-6
    in fish, 912
    pattern of, 912
Integrins, 204
  cytoskeletal interactions with, 205
  defined, 204
  role in bacterial invasion of cells, 66
  role in cell adhesion, 109
Intellectual property rights, defined, 907
Intelligent Automation Corporation, 54
Intelligent materials, 196
Interaction analysis, defined, 794
Inter-*Alu* PCR, defined, 420
Inter-American Institute for Cooperation in Agriculture (IICA), recommendations on regulation of biotechnology, 116
Interfacial electric polarization, 268, 268f
  defined, 266
INTERFERONS, 200, **473-74**
  $\beta_2$, 392
  biological effects of, 474t
  defined, 473
  list, 203t
  properties of, 474t
  tumor necrosis factor, 201
Interleukin-1$\beta$ converting enzyme (ICE), homology with CED-3, 161
Interleukin-4 (IL-4), structure of, 759
Interleukin-6 (IL-6), murine
  amino acid sequence, 770
  peptide separation, 769f
INTERLEUKINS, 201, 395, **474-76**
  defined, 200, 392, 474
  properties of, 475t
Intermediate density lipoproteins, defined, 495

Intermediate filaments (IFs), 580
defined, 564
Internal loops, defined, 841
International System of Units, becquerel (Bq) defined in, 780
International Union of Biochemistry
Commission on Biochemical Nomenclature, 296, 479-80
guidelines for peptide nomenclature, 655
International Union of Pure and Applied Chemistry, guidelines for peptide nomenclature, 655
International Unit (IU) of enzyme activity, 289
Internucleoside linkage, defined, 617
Internucleotide linkage, defined, 617
Interphase, chromosomal organization of cell nuclei at, 967
Interphase DNA, target in FISH, 351-52, 357
Interspecific backcross, 527
Interspersed repeat sequence (IRS) DNA amplification, 966
Interstellar dust clouds, defined, 628
Intracellular membrane traffic, 538
Intrinsic membrane proteins, 923
Intron homing, 174
Introns, 526, 838
*Chlamydomonas* Group I, 174
defined, 173, 310, 346, 409, 428, 573, 831, 833, 847
in the factor VIII gene, 410-11
RNA splicing pathway, 833f
of self-splicing RNA, 831, 832f
Inversion
DNA, site-specific, 365
mutation by, 373, 410
Inverted region
electron transfer reactions, 274
in photosynthesis, 276
Inverted repeats, LINE and SINE, 860
In vitro activity, of oligonucleotides, antisense and triplex-forming, 43
In vivo activity, of oligonucleotides, antisense and triplex-forming, 43-44, 44t
Ion channel receptors, 278, 279
Ion channels, 878-79
Ion exchange chromatography (IEC), 418-19
Ionization monitors, for radiation detection, 782
Ionizing radiation
defined, 476
models for effects of, 478
IONIZING RADIATION DAMAGE TO DNA, **476-79**
Ion pump, defined, 68
Ion-selective field effect transistors (ISFETS), 112
Ion trap mass spectrometry, 534
Iron
biocatalyst, 547-48, 548f
in tryptophan hydroxylase, 127
in tyrosine hydroxylase, 125
Iron sulfide, electron transfer in clusters of, 273
Irreversible reactions, in enzyme mechanisms, 293
Isochores
DNA, 860
defined, 859
Isocitrate lyase, circular dichroism spectra, 183f
Isocratic elution, in affinity chromatography, 796f
Isocratic HPLC, 417
Isoelectric focusing
first-dimension, 345
for protein purification, 767
for separation of plasma lipoproteins, 513
Isoelectric point (pI), defined, 343-44
ISOENZYMES, **479-82**
Isoforms
of enzymes, 482
of proteins, 573
tropomyosins, 578
Isomerization, cis-trans, in proline containing peptides, 654
Isomers
defined, 168
separation of, drug analysis, 673
Isomorphous replacement, for protein structure determination, 745, 971
Isoniazid, polymorphism in metabolism of, 677
Isoschizomers, defined, 809
Isotopes, for nuclear magnetic resonance measurements, 601-2
Isotropic Raman scattering, 739
Isotype, defined, 449

Japan, regulation of biotechnology, 116
*J* coupling
defined, 601
in NMR analyses, 602-3
Jumping genes, 285
Jumping libraries, 229, 231f
Juxtaglomerular apparatus (JGA), defined, 803

$K_m$ (Michaelis constant), defined, 289
Kallikrein genes, human, 490
KALLIKREIN-KININOGEN-KININ SYSTEM, **483-90**, 804
Kallikreins, 483, 485, 489-90
tissue, 804
Kanamycin, resistance to, as a gene marker, 896
Kaposi sarcomas, 136
Karyotype, bivariate flow, 355, 356f
Kasugamycin, 670f, 671
Katal, defined, 289
Kearns-Sayre syndrome (KSS), 5, 553-54
Kennedy's disease, 378
Keratins, acidic and basic, 580
Kidney disease, adult polycystic, 431
Kilodalton (kDa), defined, 33
Kinases
cyclin-dependent, 627, 627f
defined, 928
*See also* Protein kinase C (PKC); Tyrosine kinases
Kinesins, 565, 565t
defined, 564
structure and composition, 566-67, 567f
Kinetic energy (KE), 80
Kinetics, 297-98
enzyme, 299-300
hemoglobin, 404
of protein folding, 755, 758, 758f
of protein refolding, 756-57, 757f
of radioactive decay, 779-80
*Kinetoplastida*, mtDNA of, 551
Kininases, 483, 804
Kininogenases, genes encoding, regulated expression of, 490
Kininogens, 804
high molecular weight, 486f
structure and function, 486-87
substrates for, 483
Kinins, 483
chemical mediator for pain, 485
effects on nonvascular smooth muscle, 484
receptors for, 490
*Klebsiella pneumoniae*, encapsulation to evade immune response, 66
*Koji* processes, 336, 336t
*Kpn* repeat, 422
defined, 420
Kringle domains, hepatocyte growth factor, 395

L1Hs, 924-26
Label
for a hybridization probe, defined, 965
in restriction landmark genomic scanning, 815
*See also* Markers
Lactams, formation of, in peptides, 661, 662f
Lactate dehydrogenase (LD)
heterogeneity of, 480
structure of, 480
Lactobacillic acid, 502, 502f
Lactose, 147
carbon source, in penicillin G production, 337
Lactotetraosylceramide, 143f, 145, 145f
Lactulose, 147
Laminins, 91, 315

Landmark cleavage, 815
Langmuir-Blodgett (LB) film
in bacteriorhodopsin-based photocells, 70–72
defined, 70
deposition process, 102–3, 103f
Langmuir monolayers, 467
Latency
defined, 10
and negative transcription control, 14–15
Laue method, 970
lauroyl-1-Lysophosphatidylcholine, absorption enhancement by, 261
Lawrence Berkeley Laboratory, 52–53
Lead
on bioassay, defined, 257
compounds in medicinal chemistry, 535
immunoglobulin peptide, 453
for synthetic targets, drug synthesis, 257–59
Learning, defined, 541
Leber's hereditary optic neuropathy (LHON), 553
LEC-CAM. *See* Selectins
Lecithin-cholesterol acyltransferase (LCAT), 497
by apoAI, 498
Lectins
C-type and S-type, 157–58
defined, 154, 382
domain in P-selectin, 155f
Legal considerations, in animal patenting, 907–8
*Legionella pneumophila*, integrin receptor use by, 66
Lentivirus, HIV, 10
Leprosy, 583
Lesch-Nyhan syndrome, diagnosis of, 674
Lethal mitochondrial myopathy, 554
Leucine zipper
basic, 749, 749f
defined, 746
in plants, 697
*Leuconostoc mesenteroides*, 147, 149
Leucrose, 147
Leukemia
acute lymphoid, gene rearrangements in, 226–27, 226t
acute megakaryoblastic, associated with Down's syndrome, 249
acute promyelocytic, translocation in, 227
chronic myelogenous, 137
gene rearrangements in, 136, 226–27, 226t, 626
human, in SCID mouse, 44
Leukemia inhibitory factor, 201
Leukocytes, cytoskeletal-membrane interactions in, 205
Leukotrienes, production from arachidonic acid, 681
Levallorphan, analgesic, 258
Levonorgestrel, controlled release from poly(ortho-ester) devices, 720
Lewis antigens, of erythrocytes, 145
Library arraying, automatic, 52–53
Life cycle
of automation systems, 52
baculovirus, 470f
HIV-1, 20f
HIV, regulation of, 10
viral, HIV-1, 12–13
Life span
and mechanism of aging, 5
and repair of UV-damage to DNA, 4
Li-Fraumeni cancer syndrome, 378
defined, 928
*p53* mutations in, 189, 931
Ligand-gated ion channels, 878–79
Ligands
and circular dichroism spectra, 184
c-*kit*, 393
defined, 83, 534, 898
heme, 403–4
Ligase
DNA
defined, 491
thermostable, 463, 492, 494
Ligase chain reaction (LCR), 227
infectious disease testing by, 463–66
mechanism of, 464f
Ligation
allele-specific, 377
of calcium ion, 132
Ligation amplification, 492–94, 494f
defined, 491
LIGATION ASSAYS, **491–94**
defined, 491
Light-directed peptide synthesis on chips, synthetic peptide library, 881
Light response element (LRE), 288
Lignin
biodegradation by white rot fungi, 79
synthesis during pathogen attack, 701
LINE (Long Interspersed Repeat) element, 422, 440, 525, 924–26, 926f
defined, 420, 924
Linguistic theory, in deciphering codes, 889, 891
Linkage, 432–33
defined, 128, 224, 424, 483
establishing, 436
Linkage analysis, 425–26, 436
in breast cancer, 128–29
defined, 187, 374
for genetic testing, 375, 378–79
in HNPCC kindreds, 190
for neoplastic disease diagnosis, 225–26
Linkage group
conserved, mouse and human, 572
defined, 173
*uni*, *Chlamydomonas*, 175
Linkage maps
constructing, 437
defined, 589
high resolution, 438
*N*-Linked sugar chains, 389, 389f
Linker
DNA, 178
histone, defined, 176
molecular, in immuno-PCR, 458
polypeptide, in immunoglobulins, 451
Linking libraries, 229, 231f, 422, 423
γ-Linoleic acid (GLA), production from fungi, 503
Lipase, defined, 495
Lipid A
and cell viability, 511
defined, 509
Lipid bilayer, defined, 948
Lipid bilayer membranes, 117
LIPID METABOLISM AND TRANSPORT, **494–500**
LIPIDS
biologically active, classification of, 507t
chemical classification of, 504–5, 504t
defined, 503
fusogenic, 539
interactions with proteins, in foods, 322
MICROBIAL, **501–3**
STRUCTURE AND BIOCHEMISTRY OF, **503–9**
type II, 537
Lipocortin (annexin I), 33f
Lipoglycans, defined, 501
LIPOPOLYSACCHARIDES, **509–11**
antigens of Gram-negative bacteria, 144–45
LIPOPROTEIN ANALYSIS, **511–13**
Lipoprotein lipase (LPL), 495, 508, 707
Lipoproteins, 495
analysis of subfractions, 513
classes of, 151t
defined, 150, 703
low density, 150
metabolism of, 150–51
plasma, 703–9, 709t
synthesis and catabolism of, 706f
LIPOSOMAL VECTORS, **514–16**
Liposomes
biodistribution of, 261
defined, 260, 514
encapsulation of drugs in, 261
Lipoteichoic acids, as adhesins, 65
Lipoxygenase, role in leukotriene synthesis, 681
Lipoyl function, coenzyme, 185
Liquid chromatography, 417
Liquid crystal, defined, 93
Lithium, acquired nephrogenic diabetes insipidus from, 214

Liver
DNA damage in, 232
glucose production in, 386
lipid management in, 150
organ regeneration, 91, 395
synthesis of bile acids in, 498
synthesis of plasma fibronectin in, 315
LIVER CANCER, MOLECULAR BIOLOGY OF, **516–18**
Local alignment
defined, 856
methods for, 858
Local compositional complexity, 891-93
Locally multiply damaged sites (LMDS), from ionizing radiation, 477
Local point fusion, defined, 537
Locus
chromosomal, 421
defined, 374, 432, 479
Locus-control elements, defined, 707-9, 709f
Locus heterogeneity, 424
Logarithm of the odds (LOD), 425, 436
Log-odds ratios, linearity with complexity distributions, 892, 894f
Long terminal repeat (LTR)
defined, 10
nonspecific inhibition of viral, 15
and virus life cycle, 12-13
Long-term potentiation, and cellular memory, 543
Long-term sensitization, *Aplysia*, 544f
Loop modeling, by homology, 760-61
Loss of heterozygosity (LOH), 929-30
in colon tumor formation, 188
defined, 187, 516, 817
Low density lipoproteins (LDL), 496, 496f, 512
analysis of, 513
and atherosclerosis, 150
defined, 495
Luciferase, 165-66
defined, 165
gene for, 168
Luciferin, 165-66
defined, 165
Lumen, defined, 99
Luminescence, cellular, 168
Luminometer, defined, 165
LUNG CANCER
amplification of c-*myc* in, 626
MOLECULAR BIOLOGY OF, **518–21**
*ras* mutation in, 227
Luteinizing hormone (LH), 283, 385
Lymphoblastoid cell line, defined, 354
Lymphocytes
DNA damage in, 232
interleukin interactions with, 474-75
Lymphokines, 201, 395
defined, 200, 392, 474
Lymphomas, 136
B-cell, translocation in, 161, 227
follicular, 227
follicular low grade, translocation in, 626
Lymphopoiesis, suppression of, by transforming growth factor-β, 394-95
Lyon hypothesis, defined, 136
Lyophilized enzyme powder, defined, 303
8-Lysine vasopressin, 211
Lysophospholipid, 281
Lysosomal storage disorders, among Ashkenazi Jews, 727
Lysosomal targeting, 777
Lysosomes
defined, 774
digestive phospholipases in, 685
α-Lytic protease, nonaqueous biocatalysis by, 306

McArdles's disease, 581
McCune-Albright syndrome, mosaicism in, 283
Macrocarriers
defined, 692
for plant cell transformation, 693
Macroencapsulation, defined, 86
$\alpha_2$-Macroglobulin receptor (LDL-receptor-like protein), 495-96
Macromolecular assemblies, electron micrography of, 272
Macromolecular pesticides, 668-69
Macromolecules, defined, 757
Macrophages, defined, 460
Macula densa cells, defined, 803
MADS box, 697
MAG (myelin-associated glycoprotein), 155f
Magnesium, dissociation of ribosome subunits in solutions of, 826
Magnesium ion ($Mg^{2+}$), and ATP reactions, 86
Magnetic induction of electric fields, defined, 266
Magnetosomes, defined, 266
Maize, transgenic, tolerance to corn borers, 695, 695f
Major facilitator superfamily, 923
Major grooves, defined, 979
MAJOR HISTOCOMPATIBILITY COMPLEX (MHC), **523–24**
associated with insulin-dependent diabetes mellitus, 215-16
class I, defined, 523
class II, 524, 874
defined, 523, 873
defined, 215
role in antigenicity, 455
single-chain engineering of fragments, 456
structure of, 452, 452f
Malnutrition, and oxidative stress, 329-30
Maltodextrins, 148
Maltose, 147
MAMMALIAN GENOME, **524–28**
Mammalian-wide interspersed repeats (MIRs), 439
Manganese, in superoxide dismutase, 329-30
Manganese ion ($Mn^{2+}$), binding with RNase H, 18
Manic depression, 596-97
Mannose, in oligosaccharides, 389
α-Mannosidase A deficiency, lysosomal storage disorder associated with, 482
Maple sweeteners, 148
Mapping
of *Caenorhabditis elegans* chromosomes, 587
of chromosome 21, 248f, 249
of a DNA sequence, 860
of *Drosophila melanogaster* chromosomes, 252-53
epitope, 657
of the *Escherichia coli* genome, 263-66, 264f, 265f
genome, 350, 421
jumping and linking libraries for, 231f
Genome Project, 528
of human CENP-B and CENP-C cDNA clones, 854f
of human disease genes, 424-27
mouse genome, 570-71
physical, 421
of the mouse genome, 571-72
of susceptibility to drug abuse, 256-57
of susceptibility to insulin-dependent diabetes mellitus, 215-16
of the Y chromosome, 867
MARCKS protein, 206
Marcus relation, 274, 275
Marker gene, defined, 692
Markers
of alcoholism liability, 255
CpG islands, 229
DNA, in immuno-PCR, 458
for genetic mapping, 355, 421, 425, 646
isoenzymes as, 482
oncogene changes as, 224
for plant cell transformation, 693-94, 896
polymorphic, 434
in neoplastic disease diagnosis, 225-26
population-specific, 724-28
selectable, 308
selection of, for genetic analysis of populations, 362-63
types of, 228
*See also* Microsatellite marker
Markov analyses, 891
Maroteaux-Lamy disease, arylsulfatase B deficiency associated with, 482
Martin Bell syndrome, 324-27
MASS SPECTROMETRY, 732-35
for amino acid sequencing, 771, 773
for analysis of biomolecules in solution, 601

defined, 50, 732
OF BIOMOLECULES, **529–34**
Matrix-assisted laser desorption (MALD) ionization, defined, 732
Matrix-assisted laser desorption ionization (MALDI), 530, 734t
defined, 529
for protein measurements, 733
resolution of, 735
Matrix protein, defined, 948
Maturity onset diabetes of the young (MODY)
defined, 215
genetics of, 215, 216-17
Mean residue weight
in circular dichroism reporting, 181
defined, 179
Measles virus, inactivation of, with photodynamic therapy, 954
Mechanism, of enzyme reactions, transient state kinetic analysis to determine, 292
Mechanochemical enzyme, defined, 564
Medical Research Council (Cambridge, England), 52-53
MEDICINAL CHEMISTRY, **534–36**
applications of collagen fibers, 718
defined, 257
*See also* Drugs; Therapy
Medicine, molecular genetic, 556-61
Medium, for plant tissue culture, 886
Medium reiteration frequency repeats (MERs), 440-41
Meiosis
corollaries of, 432-33
defined, 247, 432
recombination during, 799
Melanin, limitation of UV damage by, 941-42
Melanoma
familial susceptibility gene, 378
ganglioside $GD_3$ of, 144
Melting curve (DNA), 208f, 209
defined, 207
in polynucleotide denaturation, 606, 606f
Melting temperature ($T_m$)
defined, 42, 605
for DNA, 209, 243
Membrane-bound organelles, 568-69
defined, 564
translocation by kinesins, 569f
Membrane curvature, defined, 948
Membrane elution method, 58, 64
MEMBRANE FUSION
defined, 33, 537
MOLECULAR MECHANISM OF, **537–40**
molecular models, 539f
Membrane processes, amplification mode, 268
Membrane proteins, integral ($CF_0$) and extrinsic ($CF_1$), 688-89
Membrane receptors, 278-79, 278f
classification of, 790-91t
Membranes
lipid bilayer, 117
polymer, 90-91
Memory
defined, 541
enzyme and pH, in nonaqueous media, 305
MEMORY AND LEARNING, MOLECULAR BASIS OF, **540–45**
Mendel's laws, 432
Meperidine, 258
Mephenytoin polymorphism, 676
Merlin (protein), 591-92
Merocyanine 540
derivatives of light-activated, 956f
as photosensitizer, 954, 954f
preactivated, effect on HIV-1, 956t
Meselson-Radding model of homologous recombination, 801, 801f
Mesoscopic substrate, plasma membrane, 194-95
Meso structures, 170
Messengers, hormone and neurotransmitter, 277. *See also* mRNA; Primary messengers; Second messengers
Metabolic engineering, 108
Metabolic plant defenses, 700-701
Metabolism, 80
bacterial, and biodegradation of waste, 76-77
defined, 611
of drugs, 117
effects of neuropeptides on, 594
enzyme control of, 74
of food, 612-13
inborn errors of, 74
lipid, 506-9
lipoprotein, 151t, 707
xenobiotic, 117
Metabolites, secondary, 668
Metabolizable energy (ME), 612
Metachromatic leukodystrophy, arylsulfatase A deficiency associated with, 482
Metal ions, in metalloenzymes, 546-47
METALLOENZYMES, **546–51**, 899
defined, 898
Metallothionein, 899
gene for expression of, 900, 900f
Metal regulatory elements, 900
Metal response elements (MRE), 899, 900
Metals, nutritional requirement of microbes, 76
Metamorphosis, insect, 592
Metaphase
images in FACS, 356-57f
images in FISH, 352f
Metasomatism, in mineralization, 634
Metazoans
mitochondrial genome of, 551
organization, 552
Meteorites, carbonaceous, 629
Methane
biological production of, 333
fuel production options, 332t
Methane monooxygenase, 547
Methanogen coenzymes, 187
Methanogenesis, by anaerobic bacteria, 78
Methanotrophs, 77-78
lipids of, 501
Methylase, cognate, 812
Methylation
defined, 187
patterns in genomic imprinting, 381, 381f
Methylene blue, as a guanylyl cyclase inhibitor, 599-600
Methylhydroxylamine linkage, in thymidine nucleoside dimers, 620, 620f
$N^G$-Methyl-L-arginine (NMA), inhibition of NO synthase by, 599
Methyl mercury ($CH_3Hg^+$), 84
Methyltransferases, 119
defined, 1
Mice, genetically engineered, 36, 37
Micelles
casein, 321
lipid, 504, 506f
protein, 320
reversed, 303
Michaelis constant, defined, 479
Michaelis-Menten equation, 292, 299-300
Microdissection
of chromosomes, 966
of chromosomes or chromosome fragments, 966
for genomic library construction, 229
Microemulsions, water-in-oil, 467-68
Microencapsulation, for xenogeneic transplantation, 90-91
Microfilaments
ATPase activity of myosin in the presence of, 568
defined, 564
Microglobulin, β-2, 523
Microheterogeneity, defined, 388
Microinjection, for gene targeting, 558, 690, 911-12, 912f
Microorganisms
asymmetric structures in, 169
DNA fingerprinting of, 223
plant-associated, 472
Microparticles, defined, 692
Microsatellite marker, 355, 363, 425-26, 434
defined, 354, 424
representation of locus, 425f
*See also* Markers

Microsatellites, 228-29, 428, 439
Microscopy, in molecular paleontology, 635
Microsomal triglyceride transfer protein (MTP), 495
Microsomes
  defined, 824
  steroid synthesis in, 871
Microtiter plate wells, for immuno-PCR, 458-59
Microtubule organizing center (MTOC), 569f
Microtubules, 569f
  ATPase of kinesin and dynein in the presence of, 568
  defined, 564
Microviscosity, measuring, 318
Midazolam, 259
Milk proteins, 321
Miller spread, defined, 269
Mimetic agents
  defined, 93
  peptide, 659
Mimicry
  of glycoproteins and glycolipids, by bacteria, 143
  of host surface markers, by bacteria, 66
Mineralization, 634
  defined, 633
Mineralocorticoids
  role of, 872
  synthesis of, 871
Minerals, 615
  defined, 611
Miniantibodies, engineered, 456
Minisatellites, 228-29, 428, 439
  defined, 219
  DNA of, as a genetic marker, 362
  *See also* Variable number of tandem repeats (VNTR)
Minisatellite variable repeat (MVR), 647
Minor grooves, defined, 979
Minus strand, DNA. *See* Sense strand
Mismatch
  base loop, 841
  defined, 242, 843
  in PCR, 645
  and RNA stability, 846
Mismatch cleavage analysis, 377
Mismatch primer, 622
Mismatch repair, 802
  defined, 799
Mispairing, and thermal denaturation, DNA, 210
Missense variant, defined, 595
Mitochondria, 362
  calcium-transporting systems of, 135
  defined, 162, 551
  gene expression in plants, 696
  transmembrane electrochemical potential difference in, 82
Mitochondrial carrier family (MCF), 923
MITOCHONDRIAL DNA, 362, **551-54**
  *Chlamydomonas reinhardtii*, 174
  EVOLUTION OF HUMAN, **554-56**
Mitochondrial encephalomyopathy, lactic acidosis and strokelike symptom (MELAs), 553
Mitochondrial genome, *Drosophila melanogaster*, 252
Mitochondrial import, 777
Mitosis
  defined, 564
  karyotype in, 354, 355
  kinesin-related proteins in, 568
Miyata and Yasunaga method, 865-66
M-line proteins, 579
Mobile elements, 285
Mobilization
  defined, 709
  for transfer of nonconjugative plasmids, 712
Models, from X-ray diffraction data, 971
Modification, oligonucleotide, defined, 621
Modified phosphate backbone, defined, 617
Molecular activity, enzyme, 289
Molecular chaperones, defined, 400, 728, 774. *See also* CHAPERONES, MOLECULAR
Molecular clock, 361
  defined, 359
Molecular computing
  errors in, 196
  Macro-micro (M-m) scheme of, 194
Molecular drive, 361
Molecular dynamics, 761
Molecular engineering, 196
Molecular fossils, 633
MOLECULAR GENETIC MEDICINE, **556-61**
Molecular genetics
  of liver cancer, 517-18
  of phenylketonuria, 678-80
Molecular microscope, 102, 102f
Molecular neurobiology, of *Caenorhabditis elegans*, 588
Molecular recognition, and charged groups, 195
Molecular replacement, for protein structure determination, 745
Molecular systems, organized, 467-69
Molten globule state, 755
Monoacylglycerol acyltransferase, 508
Monoamine oxidase, 118
Monoclonal antibodies (Mabs), 107, 108, 450
  antiphosphorylcholine McPC603, 451f
  defined, 36, 260, 717
  drug targeting with, 719
  linkage of drugs to, 261, 262
Monoclonality
  defined, 136
  of tumors, 136-37
Monogenic resistance, defined, 699
Monokine, 201
  defined, 200
Monomeric proteins, aggregate formation in refolding, 729
Monooxygenase, 77-78
Monooxygenase deficiency, in fish odor syndrome, 677
Monoparental disomy, defined, 379
Monosaccharides, 147
  analysis of, 138
  defined, 146
Morgan, unit of genetic distance, 128
Morphine, 258
Morphogenesis, and cell adhesion, 154
Morpholino nucleoside oligomers, 40-42, 42f
Morphology
  crystalline, additives affecting, 467
  defined, 466
Mortality, due to alcohol or drug abuse, 254
Mosaicism
  in Down's syndrome, 249
  in fragile X linked mental retardation, 326
  in McCune-Albright syndrome, 283
  in transgenic fish, 912
  in XY hermaphrodites, 870
Mosaics, for examining lethal mutants, 253
Mother, mitochondrial ancestral, 556
Motif
  calcium-binding, 579
  defined, 859, 880, 888
  peptide, for predicting protein sequence, 862
  recognition, protein kinase, 763
Motion detection, in a bacteriorhodopsin-based photoreceptor, 72-73
Motoneurons, regulation of, 563
MOTOR NEURON DISEASE, **561-63**
  defined, 561
MOTOR PROTEINS, **564-69**
  defined, 564
MOUSE GENOME, **569-73**
Mouse Genome Project, 570
Mouse mammary tumor virus (MMTV)
  defined, 873
  infection by, 876
M protein, antibodies against, in rheumatic heart disease, 66
mRNA
  binding of antisense oligonucleotides to, 39-44, 39f
  eukaryotic, 917, 917f
  half-life of, 838
  for N-CAMs, 156
  for p53 protein, 931
  poly(A)-tailed, 120
  prokaryotic, 915, 917, 917f
  *RB*, 931
  structure of, and the genetic code, 915-18

Tat-induced levels of, 12–14
for Tat transactivation, 12–14, 13f
transcription of, 236, 346
transgene-encoded expression of, 906
for tyrosine hydroxylase, 125
mRNA 5′ cap structure, defined, 824
MRNA editing, apo B, 707, 707f
mRNA splicing, 310
mtDNA, 551
human, 555f
lineage, defined, 554
region V mutation, 556
type, defined, 554
Mucins, glycoprotein antigens and TACAs from, 146
Müllerian inhibitory substances (MIS), 394
Multibranch junction, defined, 841
Multibranch loop, defined, 841
Multidimensional NMR, 603
Multigene families
hsp60, 164, 164f
hsp70, 163–64, 163f
Multimeric cytoplasmic factor, interferon activation of, 201
Multipin synthetic peptide libraries, 881
Multiple cloning site, 609
Multiple endocrine neoplasia type 2A, 378
Multiple forks, defined, 57
Multiple myeloma, translocation in, 227
Multiple sequence alignment, 862
Multiplex, defined, 50
Multiplexed thermal cyclers, 53–54
Mummification, 634
defined, 633
Murein, recycling of, in cell growth, 60–61
Muscle
as food, 321
mammalian, DNA damage in, 232
mechanism of contraction, 575–76, 575f
MUSCLE, MOLECULAR GENETICS OF HUMAN, **573–82**
Muscular dystrophy, 557
Mutagenesis
chemical, for characterizing mycobacteria, 584
defined, 979
insertional, 903
oligonucleotide-directed, 812
site-directed
of bacteriorhodopsin, 95
p-element, 6
site-specific, 807
to control enzyme activity, 305
in the helix swap experiment, 807
Mutagens
aflatoxin $B_1$, 517
environmental, 364–65
insertional, 691
thermally induced, in foods, 323
Mutants
conditional-lethal, 951
defined, 950
identification of, with PCR, 646
mismatch repair defects in, 799
Mutation, 373–74, 725–26
affecting coding region of genes, 430f
and aging, 4–5, 25f
defined, 187, 217, 232, 370, 428, 570
in DNA, and neoplastic disease diagnosis, 224
and effectiveness of Nivirapine, 17
and effectiveness of nucleoside analogues, 23
effects on genes, 429f
in the gene for amyloid precursor protein, 26–27
in HCHWA-D, 27
of hemoglobin structure, 405–6
identification of, methods for, 74
low frequency, in breast cancer, 128–30
nature of, 360
in pharmacogenetic defects, 676
spontaneous
defined, 363
mechanisms and effects of, 364
TAR loop, effect on Tat transactivation, 11
from ultraviolet radiation, outcomes, 942f
*See also* Chromosomes; Genes; Mutations; Mutations, disorders caused by
Mutation accumulation, 5
defined, 4
Mutation analysis, 727
Mutation rate, defined, 589
Mutations
fragile X, *FMR 1*, 325–26
in the human *PAH* gene, 678
metHb, 406
*p53* gene, 518, 518t
*ras*, in lung cancers, 519
Mutation. *See also* Point mutation
Mutations, disorders caused by, 428
cystic fibrosis, 375
Gaucher disease, 341–43, 343t
hemophilia B, 411–12
neurofibromatosis 1, 590
neurofibromatosis 1 and 2, 591
retinoblastoma, 818, 818f, 819
MVR-PCR (minisatellite variant repeat-PCR), 221, 222f
defined, 219
MWC model, of allostery, 405, 407
Myasthenia gravis, 47
MYCOBACTERIA, **582–85**
lipids of, 502
Mycobacteriophages, 583
*Mycobacterium leprae*, 583
*Mycobacterium tuberculosis*, 583
integrin receptor use by, 66
Mycolic acids, bacterial synthesis of, 502
Mycoprotein, production of, *Fusarium graminearum* fermentation, 338
Myeloma proteins, 450
Myoblasts, 581–82
Myocardial infarction, 150
Myoclonus epilepsy with ragged-red fibers (MERRF), 553
Myofibrils, 573–75, 575f
protein components of, 576t
Myogenesis, 581–82
Myogenin, 582
Myomesin, 579
Myopathies, 581
Myosin head, 577
Myosin heavy chain (MHC), 576–77
Myosin light chain (MLC), 576–77
Myosins, 565, 565t, 566
defined, 564
filament, 575f
structure and function, 567–68, 568f, 576–77
in thick myofilaments, 573–75
Myotonic dystrophy, 375, 378, 429–30
Myotropins, 594
*Mytilus edulis*, adhesive protein of, 715–16

NADH, transfer of electrons to cytochromes P450, 198
NADPH. *See* Nicotinamide adenine dinucleotide phosphate (NADPH)
$Na^+$-$K^+$ ATPases, 806
Naphthalimide, photodynamic effects of, 955–56
National Biotechnology Policy (U.S.), 115
National Toxicology Program (NTP), 461
Native state, fully folded proteins, 755
Natriuretic peptides, receptors for, 793
Natural fields, magnetic and electric, 266
Natural polymers
biodegradable, 718–19
defined, 717
Natural products, as drug synthesis leads, 257–58
Natural selection, 360, 726
defined, 724
Necrosis, 158
Needleman-Wunsch-Sellers algorithm, 862
Negative feedback, in ColE1 replication, 711
Negative staining
defined, 269
of proteins, 270
Negative-strand RNA viruses, 960
Nei and Gojobori method, 866
*Neisseria gonorrhoea*
defense against host immune response, 66
lipopolysaccharides of, 144–45

*Neisseria meningiditis*
antigens, 141f
mimicry of host markers, 66
lipopolysaccharides of, 144–45
polysaccharide antigens of, 142–43
Nematoda, 587–88
NEMATODES, NEUROBIOLOGY AND DEVELOPMENT OF, **587–89**
Neolactotetraosylceramide, 143f
Neonatal severe hyperparathyroidism, 283
Neoplasms
and associated tumor suppressor genes, 930t, 935t
defined, 136, 928
origin of, 224
*See also* Tumors
Nephroblastoma (Wilms' tumor), 932–34, 934f. *See also* Wilms' tumor
Nephrogenic diabetes insipidus (NDI)
cause of, 211
defined, 210
mutations associated with, 213t
Nernst equation, resting membrane potential, 877
Nerve growth factors (NGF), 393–94
Nerve terminal, defined, 877
Nervous system, of *C. elegans*, 588
Net energy (NE), from food, 612
Network, defined, 350
Neural networks
defined, 859
for secondary protein structure prediction, 759
Neuraminic acid, defined, 154
Neurite fasciculation, 156
Neurobiology, nematode, 588
Neuroblastoma, n-*myc* amplification in, 626
Neurodegenerative diseases, 161
NEUROFIBROMATOSIS, **589–92**
*Alu* insertion in, 926
genes for, types 1 and 2, 934
Neurofibromatosis type 1, 378, 590-91
genes for, mapping, 426
mutation rate of, 725
Neurofibromin, 189, 590, 934
Neurofilament proteins, 580
Neurogenerative diseases, mitochondrial mutations in, 554
Neurogenic muscle weakness, ataxia, and retinitis pigmentosa (NARP), genetic disorders, 553
Neurohemal organ, defined, 592
Neuron, defined, 541
Neuronal development, and adult plasticity, 543–44
Neuron-neuron adhesion, 156
Neuron-specific enolase, production by lung cancers, 519
NEUROPEPTIDES
defined, 587, 592
INSECT, **592–95**
insect, 592–94, 594f
myotropic, 595
NEUROPSYCHIATRIC DISEASES, **595–97**
Neurosecretion, 592–94
defined, 592
Neurotransmitters, 674, 879–80, 880t
defined, 877
receptors, inhibition of expression in brain, 43
Neurotrophic factors, defined, 561
Neurotrophins, 161
Neurotropins, 394
Neutrophils, interaction with α-actin and integrin $\beta_1$, 205
Nevirapine, 17
structure, 22f
Nickel, in urease, 546
Nicotinamide adenine dinucleotide phosphate (NADPH), 675
transfer of electrons to cytochrome P450, 198
Nicotine, impaired metabolism of, 677
Nicotinic agonists, heteroregulation of neuronal tyrosine hydroxylase by, 126
Nicotinic receptors, 791–92
NIH Guidelines, for research involving recombinant DNA, 114
Nitrate reduction, by anaerobic bacteria, 78
NITRIC OXIDE (NO), 328, 535, 551, 804
chemical reactivity of, 597–98, 598f
defined, 597
IN BIOCHEMISTRY AND DRUG DESIGN, **597–601**
as a second messenger, 132
Nitric oxide synthases, 281, 598, 804
defined, 597
Nitrogen, fixed, nutritional requirement of microbes, 76
Nitrogenase, 546
Nitrous oxide, withdrawal from, genetic studies, 255
Nitrovasodilators, 600-601, 600f
defined, 597
*N*-linked sugar chains, 382–83, 383f
NMR (nuclear magnetic resonance)
defined, 138
for DNA structure determination, 242–43
NOESY, 603
defined, 601
spectrum of a synthetic tridecapeptide, 604f
Nomenclature
of enzymes, 296
of restriction endonucleases or methylases, 811
of RNA structure, 844–45, 844f
of vitamins, 961
Nonaqueous solvents, enzymes in, 304f
Nonassociative learning, defined, 541
Noncontiguous pattern, 888
Nongenomic effects, of hormones, 281–82
Nonhematopoietic growth factors, list, 203t
Nonhost interaction, 700
Non-insulin-dependent diabetes mellitus (NIDDM), 215
Nonionic, defined, 617
Nonobese diabetic mouse, 216
Nonobviousness (inventive step), defined, 907
Nonpolar lipids, 506f
Nonprocessive transcription
in the absence of Tat, 11, 12
accumulation of TAR RNA in, 15
defined, 10
Non-small-cell lung cancer
defined, 518
epidermal growth factor receptor (EGFR) expression in, 519
Nonsynonymous substitutions, 864
defined, 863
estimating the number of, 865-66, 865f
Noradrenaline
interaction with G proteins, 793
lead for drug synthesis, 258
Normal transmitting male, defined, 324
Normal transmitting males (NTM), for fragile X linked mental retardation, 325
Novrad, 171
N-terminal modification, defined, 771
Nuclear factor kappa B (NF-κB), 10, 11f
inhibition by antisense oligonucleotides, 44
Nuclear factor of activated T cells (NFAT), 10, 11f
Nuclear genes, as genetic markers, 362
Nuclear genome, 251, t
defined, 45
Nuclear hormone receptors, 981–82
Nuclear import, 778
NUCLEAR MAGNETIC RESONANCE (NMR), 601
for DNA structure determination, 242–43
to measure DNA denaturation, 209
OF BIOMOLECULES IN SOLUTION, **601–4**
one-dimensional, 601–3
for protein structure determination, 758
Nuclear moments, 601–2
Nuclear Overhauser Effect (NOE), defined, 601. *See also* NOESY
Nuclear receptors, 277–78, 278f
as transcription factors, 282
Nucleases
defined, 621
resistance of phosphorothiate DNA to, 104
stability defined, 39
stability of oligonucleotides against, 41
Nucleation, in crystallization, 467
Nucleic acid hybridization assay, defined, 165

NUCLEIC ACID HYBRIDS, FORMATION AND STRUCTURE OF, **605–8**
Nucleic acids
binding of metal ions by, 86
crystal structures, hydrogen bonds of, 447
electron microscopy of, 270
formation of, and origin of life, 629
labeling methods, 781, 783t
in molecular paleontology, 635
peptide, 42, 42f
transferable, in plants, 689–90
NUCLEIC ACID SEQUENCING TECHNIQUES, **608–11**
Nucleocytosol, DNA of, *Chlamydomonas reinhardtii*, 173
Nucleophile, defined, 117
Nucleoproteins
defined, 948
"frozen" filaments of, 3
Raman spectroscopic analysis, 739
Nucleosides
analogues to, for HIV treatment, 7, 22f, 23
binding of metallic ions, 85–86
conformation of units, in DNA, 246f
Nucleosome core particle, 177, 177f
defined, 176
Nucleosomes, 176–77
defined, 176, 346, 413
effect on transcription, 348
structure of, 414
ESI study of, 272
and transcription, 178–79
in vitro assembly of, 178
Nucleotide difference, defined, 863
Nucleotide excision repair, of UV damage to DNA, 942–43
Nucleotides
cyclic, as second messengers, 132, 279
interactions of, 103–5
tautomers of, in DNA replication, 364
Nucleotide stains, 104
Nucleotide substitutions
defined, 863
estimating the number of, 864–65, 864t
number of, 863–64
Nucleus, calcium in, 135
Numerical methods, for computer modeling of RNA, 849
Nutrient transfer, defined, 379
NUTRITION, **611–15**
defined, 611

O-Antigen, defined, 509
Occlusion body, of baculoviruses, 470–71
Occupational Safety and Health Administration (OSHA), 114
Odor responses, to enantiomers, 171
Office of Technology Assessment, U.S. Congress, 116
Okazaki fragments, 836
defined, 834
Olefinic analogues, peptides, 660
Oligodeoxyribonucleotides, 622–23
synthetic, pairing with chromosomal DNA, 3
Oligogenic etiology, 425
Oligomer, defined, 628
Oligomeric proteins, aggregation in refolding, 729
OLIGONUCLEOTIDE ANALOGUES, PHOSPHATE-FREE, **617–21**
Oligonucleotide-directed mutagenesis, 622
Oligonucleotide hybridization, 607f, 608
Oligonucleotide hybrids, dissociation of, 608
Oligonucleotide ligation assay (OLA), 437, 437f, 491, 492f
OLIGONUCLEOTIDES, 103–4, **621–24**
allele-specific (ASOs), 376–77
antisense, 38–44
cyclic, 43
defined, 242, 605, 608
dephospho, 40–41, 41f
for producing biomimetic peptides, 96–97
in a sequencing chip, 441
synthesis of, 43, 623f
triple helix forming, 39, 39f, 43
Oligonucleotide synthesis
solid phase, 104f
defined, 103
Oligoribonucleotides, 623–24
Oligosaccharides
analysis of, 138–39
defined, 146, 388
fucosylated, interaction with selectins, 157–58
industrial, 148
of lipopolysaccharides, 509–11
structure of, 389–90, 390f
Oncofetal antigens, 146
ONCOGENES, 284, **625–28**
*bcl*-2, 161
c-Ki-*ras*, mutation in colon cancer, 137
classification by function of product, 626
classification systems, 626–27, 627t
defined, 187, 224, 277, 428, 518, 625, 928
functions of products, 627
mutations in cancer, 560
mutations in lung cancer, 519t
*myc*, in lung cancer, 519–20
N-*myc*, in woodchuck HBV infection, 517
*ras*, activation in lung cancer, 519
viral, nonreceptor tyrosine kinases as products of, 763–64
*See also* Proto-oncogenes
Oncoproteins
cytoplasmic/membrane, in vitro activity of antisense or triplex-forming oligonucleotides, 44t
nuclear, in vitro activity of antisense or triplex-forming oligonucleotides, 44t
Ontogenic development
defined, 479
metabolic patterns in, 481
Open reading frame (ORF)
*APC*, in adenomatous syndromes, 189
chloroplast 23*S* ribosomal RNA gene, 174
defined, 173
*Escherichia coli* chromosome, 265f, 266
*LI* source gene, 440
mitochondrial DNA, 551
*NF1*, in neurofibromatosis 1, 590
*SRY* gene, 868
Operant conditioning, 541
Operon
fusion of, 365
*lac*, 1-2, 349
lux, bioluminescence genes of, 79
R plasmid, 712
*trp*, 349
Operon hypothesis, 807
Opioids, endogenous, 256
Opsonization, 66
Optical activity, defined, 168
Optical biosensors, 113
Optical fibers, in competitive binding assays, 113
Optical purity, defined, 28
Optical switch, 100–102, 102f
Opt(r)odes, 113
Organismal aging, 161–62
Organization of American States (OAS), recommendations on regulation of biotechnology, 116
Organophosphate insecticides, paraoxonase in metabolism of, 676–77
Organophosphorus pesticides, analysis for, 112
Origin of replication, 307–8, 307f
defined, 57, 234
ORIGINS OF LIFE, MOLECULAR BASIS OF, **628–31**
Orthologous genes, defined, 924
Orthomyxovirus system, 960
Osteoblasts, 92
calcium-binding proteins secreted by, 132
Osteocalcin, 132
Osteomas, in Gardner syndrome, 189
Osteonectin, 132
Osteopetrosis, in null mutation of the c-*src* gene, 560
Osteopontin, 132
Osteosarcoma, and *Rb* genes, 137

Overlapping libraries, 421-22
clone, 423
Oxazinones, diphenyl, 30
Oxidants, endogenous, DNA damage from, 218-19
Oxidases
D-amino acid and L-amino acid, 170
defined, 546
Oxidation
in bioluminescence and chemiluminescence, 165-66
of low density lipoprotein, 150
Oxidative damage, defined, 327
Oxidative phosphorylation, 81-82
defined, 79
mitochondrial genome's role in, 553
Oxidative stress, 286-87
defined, 284, 328, 329-30
and human disease, 330-31, 330f
Oxidoreductase
defined, 301
ferredoxin-dependent, 303
Oxygen, binding by hemoglobin, 403-5
Oxygenase, defined, 546
Oxygen binding curve, hemoglobin, 407
Oxygen equilibrium curve, hemoglobin, 404f
Oxygen radical theory, of aging, 161-62
Oxytocin, structure of, 655

Packaging
DNA pathway, 952-53
encapsidation, defined, 957
Packing
defined, 417
for high performance size exclusion chromatography, 418
for hydrophobic interaction chromatography, 419
in ion exchange chromatography, 418-19
for reverse phase chromatography, 419-20
PADGEM (P-selectin), 157
PAGE, 2D- (two-dimensional polyacrylamide gel electrophoresis), 344-45, 344f
Pairing, DNA, 788
Palatinose, 148
PALEONTOLOGY, MOLECULAR, **633-36**
defined, 633
Pallindrome, defined, 810
Pan American Health Organization (PAHO), recommendations on regulation of biotechnology, 116
Pancreas, organ regeneration possibilities, 89-91
Pancreatitis, acute, protease inhibitor therapy for, 485
Panning, 191-92, 191f
combinatorial phage display library, 193f
defined, 190
Papilloma, defined, 928
Papilloma virus
DNA-binding domain of the transcriptional activator, 749
human, 189
Parahormones, 281
Parallel-architecture computers, 862
*Paramecium aurelia*, mtDNA of, 551
Paraoxonase, 676-77
Parasites, insect, 472
Parasporal crystal, insecticide, 471
Parathyroid hormone (PTH), 131, 283
Paratope, defined, 449
Parenchymal cells, regeneration of organ function using, 87
Parental imprinting, 380-81
defined, 379
Parenteral, defined, 260
Parkinson's disease, mitochondrial mutations associated with, 554
Parthenogenesis, defined, 379
PARTIAL DENATURATION MAPPING, **637-41**
Particle acceleration method, 692
Particle bombardment, defined, 692
Partitioning
defined, 306
of plasmids, 711
Parvoviruses, DNA and RNA of, 839
Patent, defined, 907
Patent and Trademark Office (PTO), U.S., 907
Paternity, DNA analysis for establishing, 221f
Pathogenesis
of Down's syndrome, 249-50
in neurofibromatosis 1, 591
in neurofibromatosis 2, 591-92
role of cytokines in, 201, 204
Pathogenicity, defined, 65
Pathogens
identification of, using PCR, 646-47
insect, genetically engineered, 470-71
plant, weapons of, 702
Pathology, kallikrein-kininogen-kinin system in, 484-86
Pathophysiology
of heart failure, 399
of regulatory peptides, 652, 652t
Pathways, of gene action, 372, 373f
Pattern, defined, 888
Pattern matching, for sequence alignment of proteins, 858
Patterson map, 743
Paxillin, focal adhesion protein, 205, 206
PCR. *See* Polymerase chain reaction
PCR TECHNOLOGY, **641-48**
Pearson's syndrome, 553-54
Pectins, 149
Penetrance
defined, 128, 187, 589
of disease genes, 424
of neurofibromatosis 1, 590
variable, in polygenic diseases, 378
Penicillins, fungal biotechnology using, 336-37
*Penicillium camemberti*, 336
*Penicillium chrysogenum*, 336-37
*Penicillium roquefortii*, 336
*Pennsylvania* vs. *Pestinikis*, 647
Pentazocine, 258
Pentose
defined, 331
fermentation of, 333
Peptide analogue. *See* Peptide mimetic
Peptide bond, 917f
defined, 654
Peptide conformations, defined, 658
Peptide hormone, defined, 648
Peptide libraries
comparison of methods, 882t
synthetic, 880-83
Peptide mimetic, defined, 658
Peptide-resin linkage, solid phase peptide synthesis, 665-66
PEPTIDES, **648-53**
biomimetic, 96-97
and recombinant DNA technology, 95f
bonding in, 649f
bulk and conformation of side chains, 660
defined, 648, 654, 664, 880
direct synthesis of, 107
nomenclature of, 655
preparation and purification, 735-37
secondary structure of, 182-84
structure of, 658
SYNTHETIC, **654-58**
PEPTIDES AND MIMICS, DESIGN OF CONFORMATIONALLY CONSTRAINED, **658-63**
PEPTIDE SYNTHESIS, 648, 649, 655-57
SOLID PHASE, **664-67**
Peptide therapeutics, defined, 648
Peptide topography, defined, 658
Peptidoglycan
defined, 57
structure, 61f
synthesis of, 60-61
Peptidyl transferase, 920
Peripheral blood lymphocytes (PBLs), isolation of antibodies from, 37
Periseptal annulus, 62
Permeases, bacterial, 923
Permeation enhancers, 261
defined, 260
Peroxidases
defined, 546
detection of label, chemiluminescent reaction for, 167-68, 167f
oxidation of lignin by, 79
Peroxisomal import, 778

Peroxyl radical, defined, 476
Pertussis toxin, inhibition or receptor-mediated reactions by, 793
PESTICIDE-PRODUCING BACTERIA, **668–71**
Pesticides
defined, 668
engineering plants to produce, 898
reductive dehalogenation of, 78
secondary metabolites as, 669–71, 671f
Peutz-Jegher syndrome, 190
PFG (Pulsed Field Gel Electrophoresis), defined, 420
pH
defined, 83
and DNA denaturation or degradation, 207, 209–10, 637
for protein transfer reactions, bacteriorhodopsin, 69
Phage
antibody isolation using, 37
for antibody library construction, 191, 457
for genomic library construction, 229
immune system mimicking in, 457f
Phage display, defined, 190
Phagemid, replication of pComb3, 193f
Phagocytosis, cellular luminescence for study of, 168
*Phanerocheate* spp., biodegradation of lignin by, 78–79
PHARMACEUTICAL ANALYSIS, CHROMATOGRAPHY IN, **672–75**
Pharmaceutical production, from transgenic animals, patentable, 908
Pharmaceuticals, defined, 672
Pharmacodynamics, defined, 672
PHARMACOGENETICS, **675–78**
Pharmacokinetics, targeting drugs by manipulation of, 261
Pharmacophore, 651
Phase, defined, 432
Phase determination, in X-ray crystallography, 971
Phenotype
defined, 73, 187, 247, 254, 341, 428, 432, 479, 589
Down's syndrome, 248–49
Phenylalanine, as a precursor, taxol production, 887
Phenylalanine ammonium lyase (PAL), in phenylpropanoid metabolism, 700–701, 701f
Phenylalanine hydroxylase (PAH), 124
deficiency of, 678
defined, 678
gene, structure and location of the human gene, 678–80
Phenylalanine hydroxylating system, 679f
Phenylalanine residue, 659, 659f
Phenylalanyl-transfer RNA, 919f
PHENYLKETONURIA, MOLECULAR GENETICS OF, **678–80**
Phenylketonuria (PKU)
defined, 678
founder effect, 727
Phenylpropanoid metabolism, 701f
in plant defense against pathogens, 700-701
Pheophytin, as an electron acceptor, 687
Pheromone biosynthesis activating neuropeptide, 594, 595
Pheromones, insect, chiral, 171
Philadelphia chromosome, 137, 226–27, 226t
defined, 136
fusion protein in, 626
pH memory, 305
Phosphatase, alkaline chemiluminescent detection of label, 167, 167f
Phosphate, nutritional requirement of microbes, 76
Phosphate backbone, modification of, antisense oligonucleotides, 40–41, 41f
Phosphate bonds, modified oligonucleotide analogues with, 617
Phosphate-free backbone, defined, 617
Phosphates
modifying
in oligodeoxyribonucleotides, 622
in oligoribonucleotides, 624
Phosphatidylcholine, defined, 682
Phosphatidylethanolamine (PE), 515
Phosphatidylinositol
derivatives as second messengers, 132
linkage to proteins on cell surfaces, 685
Phosphatidylinositol diphosphate ($PIP_2$), 134
Phosphatidylinositol diphosphate pathway, 387
Phosphatidylserine (PS), for liposome membrane preparation, 514
Phosphinothricin
active moiety of bialaphos, 670f, 671
for selecting transformed rice, with the *bar* gene, 694–95, 694f
Phosphocreatine, muscle contraction and, 580–81
Phosphofructokinase (PEK), deficiency of, in Tarui's disease, 581
Phospholamban, 400
PHOSPHOLIPASES, **680–82**
$A_2$, 681, 685
activities in human cells, 684f
C and D, 682
Phospholipid metabolism, triglyceride and, 498
PHOSPHOLIPIDS, **682–85**
bacterial, 502
defined, 33, 320, 501, 514, 680
for liposome preparation, 514
second messengers arising from, 281
structure of, 504, 505f, 683, 683f
at an air-water interface, 506f
synthesis of, 508
Phospholipid transfer protein (PLTP), 497
Phosphoprotein phosphatases
classification and properties of, 764
defined, 762
Phosphoramidate internucleotidic linkages, 622
Phosphoramidite method, in synthesis of oligonucleotides, 43
Phosphoramidites, defined, 621
Phosphorothioate-containing DNA, 104
defined, 103
Phosphorothioate oligonucleotides, 622
Phosphorylase
defined, 385
in glycogenolysis, 387, 387f
Phosphorylation
activation of tyrosine hydroxylase by, 126
control of transcription factors, 282
defined, 33
protein, multisite, 764
regulation by hormones, 282
Phosphotransferase system (PTS), 923
*Photinus pyralis*, 165–66
Photocells, bacteriorhodopsin-based, 70–72
Photocurrent, wavelength and light intensity dependence, bacteriorhodopsin, 72
Photocycle, bacteriorhodopsin, 69, 70, 71f
Photodynamic damage, defined, 953
Photodynamic protein
defined, 93
mimetics for optical processing, 94–96
Photohydrate, defined, 939
Photoisomerization, 99
defined, 99
of rhodopsin, 820–22, 822f
Photon scanning-tunneling microscope (PSTM), 852–53
Photo-oncogenes, defined, 399
Photophosphorylation, 79
ATP synthetase, 688–89
defined, 686
Photoproduct, defined, 939
Photoprotein, defined, 165
Photoreactivation, reducing DNA damage through, 219, 942
Photoreceptor, defined, 70
Photoresponses, genomic, 288
Photosensitizers
defined, 953
timing of irradiation, and virus toxicity, 955
Photosynthesis, 80, 99
in *Chlamydomonas*, 173
defined, 99
electron transfer in, 273, 276
photosystem I reaction center, 100f
by *Rhodopseudomonas viridis*, 99

PHOTOSYNTHETIC ENERGY TRANSDUCTION, **686–89**
Photosynthetic system, biosynthesis of, 175
Photosystem
chlorophylls of, 687
defined, 99, 686
Phthalocyanine, as photosensitizer, 954, 954f
Phylogeny, defined, 554
Physical mapping, 421–22, 527
of *Caenorhabditis elegans*, 587, 588
defined, 570, 589
Phytoalexins, 700–701
Piezoelectric devices, in biosensors, 113
Piezoelectric material
crystalline, defined, 851
defined, 110
Pigment-proteins
defined, 686
transfer of light energy via, 686–87
Pilus, defined, 709
PKA (cAMP-dependent protein kinase), 124
phosphorylation of tyrosine hydroxylase, 126, 126t
PKC ($Ca^{2+}$/phospholipid-dependent protein kinase), 124
phosphorylation of tyrosine hydroxylase, 126
*Placeopecten megellanicus*, mtDNA of, 551
PLANT CELLS, GENETIC MANIPULATION OF, **689–92**
PLANT CELL TRANSFORMATION, PHYSICAL METHODS FOR, **692–95**
Plant galls, 288
PLANT GENE EXPRESSION REGULATION, **696–99**
PLANT PATHOLOGY, MOLECULAR, **699–703**
Plants
asymmetric structures in, 169
DNA fingerprints of, 223
mitochondrial DNA of, 552
Plasma
formation of HDL particles in, 497
lipoproteins of, 512t
separation of lipoproteins of, 512–13
Plasma carrier proteins, defined, 870
Plasma desorption ionization (PD), 734t
defined, 732
for protein measurements, 733
Plasma desorption mass spectrometry (PD-MS), 529
PLASMA LIPOPROTEINS, **703–9**
Plasma membrane
activation of receptors, 285
calcium transport systems of, 133–34
carbohydrates of, 157–58
cell surface molecules, 155f
as mesoscopic substrate, biomolecular computing, 194–95
myosins' roles in fusion events of, 568
PLASMIDS, **709–13**
adding foreign DNA to, 975f
in bioprocess engineering, 107
Col, 713
*cry* genes of, 471
defined, 851
F, *Hfr* form, 712
metabolic, 713
of mycobacteria, 583
for mycobacteria gene cloning, 584
R, 712
replication of, 57, 61, 710–11f
Ri, 690
structures of, 710
Ti, 469, 472, 690, 696
virulence, 713
Plasmid substrates, recombination using, 800–801
Plasmid vectors
components of, 307–9
in FISH, 352
for gene targeting, 557
for genetic immunization, 369
Plasminogen activator, tissue-type, engineering thrombolytic agent from, 38
*Plasmodium falciparum*, resistance to in sickle cell heterozygotes, 726
Plasmon, 113
Plasticity, cellular, in learning and memory, 544
Plastids, 696
Platelet-activating factor, 685
Platelet-derived growth factors (PDGF), 805
Platelet-derived growth factors (PDGF)fibroblast growth factors (FGFs), 395
Platelets
cytoskeletal-membrane interactions in, 205
tropomyosins from, 578
Pleiotropy
of cytokines, 201
effects of heat shock genes, 286
Ploidy, defined, 973
PNA (peptide or polyamide nucleic acid), 619, 619f
Point-accepted mutation (PAM) matrix, 862
Point mutation, 196, 370, 373
in assembled V genes, 453
defined, 187, 360, 928
in Gaucher disease, 341, 342t
identifying, 210
mitochondrial, 553
oncogenic transformation caused by, 625–26
proto-oncogenes activation by, 225
in sickle cell anemia, 377, 408
Polarity, defined, 914
Polarization
defined, 444
fluorescence, 317, 319f
Polar lipids, structure at an air-water interface, 506f
Poles, formation of, division cycle, 62
*Pol* gene, defined, 15
Policy, U.S. regulatory, 115–16
Poliomyelitis virus, 839
Pol protein, HIV-1, 24
Poly(2-hydroxyethyl methacrylate), 722–23
Polyacrylamide
defined, 238
hydrogel from, 723
Polyadenylation, 707
Polyamide, defined, 617
Polyamide gels, for solid supports, peptide synthesis, 667
Polyamide linkages (peptide nucleic acids), in modified oligonucleotides, 619, 619f
Poly(amino acids), 719
Poly(anhydrides), 720
Poly(A)-tailed mRNA, defined, 120
Polychlorinated biphenyls (PCBs), reductive dehalogenation of, 78
Polycyclic aromatic hydrocarbons (PAHs), effect on macrophages, 461
Polydimethyl siloxane, biomedical applications of, 721
Polydipsia, in diabetes insipidus, 212
Polyesters, for drug delivery, 262
Poly($\alpha$-esters), 719–20
Poly(ethylene), medical applications of, 723
Polyethylene glycol (PEG)
for solid support, peptide synthesis, 667
stimulation of fusion by, 975
Polyethylene glycol modification, defined, 303
Poly(ethylene oxide), 723
Polygenic diseases, testing for, 378
Polygenic resistance, defined, 700
Poly(glycolic acid), 720
Polyhedrin, baculovirus, 470–71
Polyhydroxyalkanoates (PHA), bacterial, 502
Poly(lactic acid), 719–20
Polylinker, defined, 228
Poly(L-lactic acid), 720
Polymer, defined, 93, 714
Polymerase, 642–44, 835–36
active site, HIV-1 reverse transcriptase, 17, 18
defined, 621
DNA, 234
thermostable, 303, 463
*Taq*, 642–43
defined, 642
*See also* Reverse transcriptase; RNA polymerase

Polymerase chain reaction (PCR), 425–26, 436f, 527
in bioprocess engineering, 107
for characterization of ancient nucleic acids, 635–36
in constructing antibody libraries, 191, 191f
cycles of denaturation and synthesis in, 210
defined, 50, 220, 224, 354, 420, 491, 557, 605, 910
for detecting mycobacteria, 585
dot-blot, for identification of Vβ families, 875
in genetic analysis of populations, 362
in genetic testing, 376
in genome research, 53–54
identification of microsatellite markers with, 355
ligase mediated, 494
in ligation assay, 491
multiplex, for muscular dystrophy testing, 377–78
Polymerase complex proteins, 959
γ-Polymerase, in mitochondrial replication, 553
Polymeric genes, 973
POLYMERS
for coating drugs, 261, 262
defined, 260
GENETIC ENGINEERING OF, **714–17**
matrix for organ regeneration, 90, 90f
natural, 715
nondegradable, 720–23
structures and properties, 714–18
POLYMERS FOR BIOLOGICAL SYSTEMS, **717–24**
Poly(methyl methacrylate), medical applications of, 722
Polymorphic information content (PIC), 434
Polymorphism
biallelic, detection of, 491
debrisoquine, 676
and deficiency mutations, 341
defined, 359, 479, 724
disease-causing, 428
DNA, 430f, 434
analysis of, 436–37
defined, 25
genetic, defined, 675
in the human *PAH* gene, 678
sequence, detecting, 647
Polymorphism link-up, 422–23
defined, 420
Polynuclear aromatic hydrocarbons (PAHs), bacterial metabolism of, 76
Polynucleotide duplexes, stability of, 605–6
Polynucleotides
hybridization with, 607–8, 607f
prebiotic formation of, 631, 631f
Poly(*N*-vinyl pyrrolidone), medical applications, 723
Poly(ortho-esters), 720
Polypeptide
calcium ion pump of plasma membranes, 134
motor protein, characteristic of, 565
Polypeptide backbone, 654, 654f
Polypeptide chains
of immunoglobulins, 450–51
gene events controlling, 453f
Ti, structure of, 451–52
Polypeptides, prebiotic formation of, 630–31
Polyphosphoinositide pathway, 281, 281f
Polyploid, defined, 973
Polyprotein precursor, defined, 19
Polyribosome (polysome), defined, 824
*O*-Polysaccharide antigens, 144–46
Polysaccharides
analysis of, 139
biodegradable, 718–19
capsular, of fungi and bacteria, 142–43
defined, 146
industrial, 148–49
of lipopolysaccharides, 509–11
Polysialic acid, in N-CAM, 156
Polytene chromosomes
defined, 250
*Drosophila melanogaster*, 252
genetic analysis, 253
Poly(tetrafluoroethylene), medical applications of, 721
Polyunsaturated fatty acids
bacterial synthesis of, 502
fungal synthesis of, 503
Polyurethanes, biomedical applications, 721
Polyuria, in diabetes insipidus, 212
Poly(vinyl alcohol), 723
Poly(vinyl chloride), medical applications of, 722
Polyvinylpyrrolidone (PVP), for binding phenols, in plant tissue culture, 885–86
Population genetics, in Gaucher disease, 343
Populations genetics, isoenzyme use in study of, 482
POPULATION-SPECIFIC GENETIC MARKERS AND DISEASE, **724–28**
Porphyrin
defined, 99
in optical switches, 100–102, 102f
Positional cloning, 355, 375, 421, 424, 437, 527
Positive negative selection (PNS), 558, 558f
Positive-strand enveloped RNA viruses, 959–60
Postgastrectomy dumping syndrome (PDS), 485
Postmortem toxicology, 675
Postranslational event, defined, 25
Postsynaptic cell, defined, 541
Posttranscriptional control, defined, 413
Posttranslational control, of histone proteins, 414
Posttranslational modification
identifying, 733
protein phosphorylation, 762–64
Posttranslational processes, defined, 774
Posttranslational tags, peptide, 651
Potassium, nutritional requirement of microbes, 76
Potassium channels
cardiac, 400
membrane, 877
Potential energy (PE), 80
Potential energy well, electron transfer, 274f
Potentiometric, defined, 110
Potentiometric biosensors, 112
Prader-Willi syndrome (PWS), 380–81
Pragmatic inference, 889
Precipitation, of plasma lipoproteins, 512–13
Predators, insect, 472
Prediction
of function, from sequence analysis, 860, 861–62
in immunotoxicity evaluations, 462
of protein sequence function, 862
of RNA secondary structure, 842, 847
Pregnancy, cytokines involved in, 203t
Premineralization, 634
Premutation
defined, 324
in fragile X syndrome, 378
Preneoplastic, defined, 928
Presegregation Methocell method, 62
Presynaptic cell, defined, 541
Presynaptic facilitation, 542f
Price, and patent protection, 908
Primary CAMs, 153–54
N-CAM, 156
Primary liver cancers (PLCs), 516–17
Primary messengers, 131
recognition of, by cell surface receptors, 132
Primary structure
of peptides, 648, 650t
of proteins, 320, 754
Raman spectrum analysis of, 739
of RecA protein, 786, 786f
of RNA, 841
Primers
defined, 608, 621, 642
double-nested, 644
for nucleic acid sequencing, 609, 642
Primosome, 835–36

Prion, defined, 595
Prion protein (PrP), accumulation of, in transmissible dementias, 596
Probe complexity, defined, 965
Probe detection, defined, 350
Probe down, 443
Probes, 965-67
autoantibodies as, 48-49
chromosome-specific, 965
for cloning plant regulatory genes, 698
defined, 350, 965
for disease detection, 465
DNA
for detecting mutations, 375
for mycobacteria, 585
fluorescent, 317
hybridization, 352-53, 357, 422
nucleic acid, diseases detected by, 375-76, 375t
for nucleic acid structures, 104
in restriction length fragment polymorphism analysis, 812
Probe up, 443
Procapsid, viral, defined, 950
Processive transcription, defined, 10
Prochirality, 172, 172f
defined, 28, 168
Product inhibition, enzyme reactions, 290
defined, 289
Profile, sequence alignment, defined, 856
Progeria, due to mutation, 162
Progesterone, role of, 872
Prognosis, in neoplastic disease, 227
Programmed cell death, defined, 158
Prokaryotes
defined, 234
DNA replication in, 235f
gene regulation in, 348-49, 348f
origin of replication in, 234
oxidative phosphorylation in, 81
transcription units, 237f
Prokaryotic model
energy metabolism, 80
of RNA-directed elongation regulation, 12-13
Proliferating cell nuclear antigen (PCNA), 49
Promoter, 236, 308, 525, 747
35*S* from cauliflower mosaic virus, 698
alternative use of, collagen genes, 313
bacterial, 238
defined, 234, 306, 407, 428, 469, 896
epigenetic modification of, 348
heat shock, nucleotide sequences of, 401, 401t
in LINEs, 440-41
structure of, 697-98
Promoter/enhancer module, defined, 10
Promoter region
defined, 150, 284
long terminal repeat, HIV, 10-11
Sp1 module, 11f
Promoter regulatory elements
cell-cycle-dependent histone gene expression, 415-16, 416f
defined, 413
Promoter sequences, 838
Propane oxidizers, bacterial, 77-78
Propylene, isotactic polymers of, 714
PROSITE database, 862
Prostaglandin H synthase, 551
Protease
aspartic, 7
defined, 18, 301, 306
effect of inhibitors on tumors, 485
HIV, 6
inhibitors of, 8f, 9, 22f, 23-24
structural and biological properties, 9
structure of, 16, 16f
synthesis of, 658
target for rational design of drugs, 7-9, 7f
inhibitors of, Kunitz type, 27
Protease cascade, kallikrein-kininogen-kinin, 483-84
Protecting groups, in peptide synthesis, 656-57, 656f, 664
Protein, structure, crystal, 446-47
PROTEIN AGGREGATION, **728-31**
PROTEIN ANALYSIS
BY INTEGRATED SAMPLE PREPARATION, CHEMISTRY, AND MASS SPECTROMETRY, **731-37**
BY RAMAN SPECTROSCOPY, **737-41**
BY X-RAY CRYSTALLOGRAPHY, **741-46**
Proteinase inhibitor genes, 701
Protein databank, defined, 263
PROTEIN DESIGNS, FOR THE SPECIFIC RECOGNITION OF DNA, **746-53**
Protein domain, defined, 746
Protein efficiency ratio (PER), in foods, 321
Protein engineering
in immunology, 455-57
for nonaqueous media, 306
PROTEIN FOLDING, **754-57**
analysis with circular dichroism, 184
defined, 162, 728
role of chaperonins in, 164
*See also* Folding, protein
Protein fractionation, 766-67
Protein Identification Resource, 266
Protein interactions, 321-22
functional, in foods, 322-23
Protein kinase
activation by growth factors, 201
$Ca^{2+}$/calmodulin-dependent, 124
cAMP-dependent, 124
$Ca^{2+}$/phospholipid-dependent, 124
classification and properties of, 762-64, 764f
defined, 682, 762
Protein kinase C (PKC), 132, 206, 281, 542f
Protein labeling, 781-82
Protein micelles, defined, 320
PROTEIN MODELING, **757-62**
PROTEIN PHOSPHORYLATION, **762-64**
Protein-protein interactions, in foods, 321
PROTEIN PURIFICATION, **764-67**
Proteins, 105-6
actin-binding, 205
adhesive, 715-16
characterization of, by molecular mass determination, 530-31
chromophores of, 180
coding for, 861
coiled-coil, 716
collagen, 97-98
EF-hand, 131
electron microscopy of, 270-72, 272f
expression in transgenic animals, 906
high mobility group (HMG1 and HMG2), 868
histone, 176
intracellular calcium-binding, 132-33
metal ion binding, 85
M-line, 579
in molecular paleontology, 635
photodynamic, defined, 93
preparation of samples for mass spectrometry, 735
presynaptic membrane, 879, 879t
stability of
engineering, 305-6
monitoring with circular dichroism, 184
strategies for purifying, 765-66
surface, and adhesins, 65
synaptic-vesicle associated, 879, 879t
TAR-binding, 12f
transmembrane, 541
exchanger, 131
pump, 131
viral
components of assembly and budding, 950, 950t
polymers of, 716
*See also* Fusion protein; Proteins, specific
Proteins, specific
85K, 579
apoB-100, apoAI, apoE, C, 496
bacteriorhodopsin, 68-69, 94-96
C, 579
gal4, 982-83
GCN4, 3
IHF, 3
MetJ, 807
p24 *gag*, 14
p53, structure, 931f
p68, 11
pRB (retinoblastoma), hypophosphorylated, 819, 819f

protein A, defined, 458
RB, $p110_{RB}$, 930-31
RecA, 3, 784-89
Sp1, transcriptional activator, 11, 11f
Tat and Rev, 15
Proteins, structural
design of artificial, 716
expression of, 715-16
of muscle, 579-80
scaffolding, defined, 950
PROTEINS AND PEPTIDES, ISOLATION FOR SEQUENCE ANALYSIS, **767-71**
PROTEIN SEQUENCING TECHNIQUES, **771-73**
Protein structure, 320-21
secondary, 182-84
tertiary, from circular dichroism data, 184
*See also* Proteins, structural
Protein synthesis
defined, 824
general requirements for assays of, 827t
Protein synthesis factors, preparation of, from cells, 826
PROTEIN TARGETING, **774-78**
Protein translocation, 776f
defined, 162, 774
Protein-tyrosine kinases, 763-64
Protein-tyrosine phosphatase-y (PTP-y), 520-21
Proteoglycans, 315
analysis of, 139
decorin, 311
Proteolysis, defined, 979
Proton electrochemical gradient, transmembrane, 81
Protonmotive force, 81
defined, 68
Proton pump
defined, 70
light-driven, 68
photosynthetic energy transduction, 688, 688f
Proton release domain (PRD), bacteriorhodopsin, 68f, 69
Proton transfer, in bacteriorhodopsin, 69
Proton uptake domain (PUD), bacteriorhodopsin, 68f, 69
Proto-oncogenes, 225
defined, 224, 928
k-*ras*, 225
in myocardial hypertrophy, 399-400
*rel* family, binding by NF-κB, 10
translocations associated with, 226t
v-*onc* and c-*onc*, 625
*See also* Genes; Oncogenes
Protoplast
defined, 973
fusion of, 975
Proviruses, endogenous, human genome, 441
P-selectin, 155f
Pseudoautosomal regions
association with schizophrenia, 596
defined, 595
Pseudogenes, 438-39f
defined, 341
histone coding, 415f
LINEs as, 925
Pseudoknot, 841-42
defined, 841
*Pseudomonas aeruginosa*, attack on antibodies, 66
*Pseudomonas fluorescens*, 472
1-endotoxin genes in, 669
Psoralen dyes, as photosensitizers, 956-57, 957f
Psychoactive drug, defined, 254
Psychoses, functional, 596-97
$P_1$ transgenic fish, defined, 910
Pulsed field gel electrophoresis (PFG), 421, 527
Pulse radiolysis, defined, 476
Pump
calcium ion, 133-34
sarcoplasmic reticulum, 135-36
transmembrane protein, defined, 131
Punctuations
defined, for a *k*-gram alphabet, 888
frequency of occurrence, 891
Purification scoreboard, proteins, 765-66, 766t
Purines, binding sites for oligonucleotides, 3
Purple membrane, defined, 70
Pyridinone, structure, 22f
Pyrimidine free radicals, 478
*Pyrococcus furiosus*
protease of, 301
sucrose α-glucohydrolase of, 302
*Pyrococcus woesei*, amylase of, 302
Pyrroles, in peptide design, 661, 662f
Pyrroliquinoline quinone (PQQ), coenzyme, 185
Pyruvoyl N-terminus, coenzyme, 185

Quadruplex, oligonucleotide structure, 247f
Quadrupole ion trap mass spectrometer, 530
defined, 529
Quality assurance, in DNA fingerprint analysis, 222-23
Quantitative trait loci (QTL) mapping, 256-57
Quasi-vitamins, 962-63
Quaternary structure
of proteins, Raman spectrum analysis of, 739
proteins, symmetry of, 743
RNA, 841
Quenching
defined, 779
of fluorescence, 317, 319f
Quinones (Q), electron transport by, 687-88
Quinoproteins, 185, 187

Rabbits, reticulocyte lysate from, 829
Racemases, 170
Racemic mixtures, 170
defined, 28, 168
Racemization, defined, 168
Radiation
effect on DNA, 217-18
effect on life span, 4-5
effect on plants, 45
Radiation hybrids, linkage maps based on, 438
Radiation protection, 780-81
Radical, defined, 328
Radical ion, defined, 476
Radioactive decay, 779
Radioactivity, defined, 779
RADIOISOTOPES
defined, 779
IN MOLECULAR BIOLOGY, **779-84**
Radionuclides
choice of, for experimental use, 781
comparison of utility and safety, 783t
defined, 779
Raman effect, defined, 737
Raman microscopy, 739-40
Raman spectra
reduced thioredoxin, 738-39, 738f, 741f
viral capsids, 740f
Raman spectroscopy, to measure DNA denaturation, 209
Randomly primed cDNA library, defined, 120
RAPD (random amplified polymorphic DNA), defined, 220
Rate constants
electric and magnetic field-induced changes, 267, 267f
electron transfer, 273
distance dependence, 275-76, 276f
enzyme reactions, 299-300
first and second order, 293
and free energy, 296-97
photocycle reactions, bacteriorhodopsin, 69
Rational drug design, 795
for AIDS therapy, 7, 9
antisense oligonucleotides for, 39-44
defined, 7
from life cycle information, 21
from structures of protease-inhibitor complexes, 16, 23-24
Reaction center, 686-87, 686f
defined, 99, 686
photochemistry of, 687, 687f

Reaction media, solid-phase peptide synthesis, 667
Reactive oxygen species, defined, 284, 328
Reading frame, 917–18
change in, by alternative use of promoter, 313
defined, 914
*See also* Open reading frame (ORF)
Reading head, 984
Readout, direct and indirect, 806
Rearrangements
chromosomal, 426
analysis of, 967
Reassociation, 605
RECA PROTEIN, STRUCTURE AND FUNCTION OF, **784–89**
RECEPTOR BIOCHEMISTRY, **789–93**
Receptors, 277–79, 279f
angiotensin II, 804
bradykinin, 804
classification of, 789
defined, 257, 392, 473, 658, 794
drug, and drug design, 259–60
fibronectinlike, 278–79
glutamate, 792
G-protein-linked, 792–93
for growth factors, 392
for hemopoietic growth factors, 395
inositol 1,4,5-phosphate, 134
interferon, 473–74
interleukin, 475
interleukin-2, 392, 476f
ion channel, 789–92, 792t
lipoprotein, 151t
low density lipoprotein, 150, 152
membrane surface, interaction with kinins, 483–84
nicotinic, 791–92
pharmacological, 535
serpentine, 279
steroid, defined, 870
steroid hormone, 871
and hsp90, 164–65
T-cell, 451–52
Receptors, cell, 534
and cell adhesion, 109
for cytokines, 200, 474
for site-specific drug delivery, 261
surface, 132
Receptor tyrosine kinases (RTKs), 278
Recessive condition, defined, 187, 374
Recessive mutation, of tumor suppressor genes, 929
Reciprocal translocation
defined, 354
use in mapping with FISH, 357
Recognition
chiral, 171
defined, 621
RECOGNITION AND IMMOBILIZATION, MOLECULAR, **793–98**
Recognition element, β-ribbon, 749
Recognition motifs, for protein kinases, 763
Recognition sequences, for Achilles' cleavage, 2–3
Recombinant DNA
defined, 106
guidelines for regulating, 114
Recombinant PCR, 645
Recombinant products
approved for clinical trials or therapy, 673–74
defined, 672, 945
Recombinant protein, defined, 853
Recombinants, 432
growth hormone, ESI mass spectrum, 531f
peptide, therapeutic uses of, 653
virus-gene, for insect control, 471
Recombinant vaccines, 945
safety and efficacy of, 947
Recombination, 365, 373f, 374
defined, 370, 432, 483
*Drosophila melanogaster* chromosome studies with, 252–53
in Gaucher disease, 342t
geminate, defined, 402
hot spots and cold spots, 526
illegitimate, 524, 526–27, 585
in immunoglobulin assembly, 453
mammalian genome, 526–27
among mtDNA molecules, 553
*See also* Homologous recombination (HR)
RECOMBINATION, MOLECULAR BIOLOGY OF, **799–802**
Redox catalysis, 298–99
Redox centers, modeling, 273
Redox enzyme
in biosensor biomolecule electronic devices, 100
defined, 99
Reductase
HMG CoA, control of cholesterol levels, 153
NADPH cytochrome P450, 198
Reductive dehalogenation, 78–79
defined, 75
of trichloroethylene (TCE), 77f
Redundancy, of cytokines, 201
Reference maps, 437
Refolding
of inclusion body proteins, 730–31
protein aggregation in, 729–30
of proteins, 756–57
Refractive index, changes in, SPR analysis, 113
Regeneration, plant, 896
Regular expression, 858
defined, 856
Regulated exocytosis, 538
Regulation
of bacterial virulence, 66–67, 67t
of cytochrome P450s, 199
of cytoskeleton-membrane interactions, 206
of DNA replication, 236
in enzyme reactions, 295–96, 300
of gene expression, 346–50, 372–73, 747
of glycogenolysis by epinephrine, 764
of heat shock response, 401
of muscle contraction, proteins for, 578–79
of phosphotidylcholine biosynthesis, 684–85
of transcription, 238
Regulation, governmental
of biotechnology, 114–17
of drugs, 673
international, of biotechnology, 116
of waste disposal, 75
Regulatory domain, calcium ion pump, 134
Regulatory proteins, GTP-binding, 789
Regulons, 287, 807
defined, 400
heat shock, 401
Relatedness table, defined, 856
Reliability, in immunotoxicology testing, 462
Renal cell carcinoma, 934
RENAL SYSTEM, **803–6**
Renaturation, 605
Renin-angiotensin system (*RAS*), 803–4
activation of mineralocorticoid synthesis, 871
Reorganization energy, 274
defined, 273
Repair
function of RecA protein, 785–86
relationship with recombination, 802
of UV damage to DNA, 941f
Repair, chemical, of DNA, 478
Repeat polymorphism, 430f
Repeats, amplified, in inherited disorders, 326, 429–31, 431f
Repeat sequence probes, 352–53
Repetitive elements
defined, 438
in mycobacteria, 584
Repetitive sequences
of α-actinin, 579
defined, 524
*LLI (Kpn)* I and *Alu* types, 860
Replica casting, defined, 269
Replicases, 840
Replication
bidirectional, 57
DNA, 370–71
plasmid ColE1, 710–11, 710f
retrotransposition model of, 926

Replication fork
defined, 234
DNA, 178
Replication frequency, 307–8
Replisome, 236
Reporter genes
bioluminescence and chemiluminescence assays using, 168
for in vivo study of repressor-operator interactions, 809
Representational difference analysis, 232
Repressor-operator complexes, 807–8
REPRESSOR-OPERATOR RECOGNITION, **806–9**
Repressors
MetJ, 749, 750f, 807–8, 808f
transcriptional, 349, 807
control by, 308
TrpR, 807–8, 808f
in vitro studies of, 808–9
Repulsion, cell-cell, 158
Residue, defined, 654
Resilin, 97
Resin, defined, 664
Resistance
monogenic, 700
polygenic, 700
Resolution, defined, 799
Resolvase, 800, 802
Resonance-enhanced hydrogen bonding, 446
Resonance Raman effect, defined, 737
Respiration, 82
cellular, rate of, 82–83
electron transfer in, 273
Respiratory-deficient mutants, as markers, 975
Response element, defined, 746
Response profile, bacteriorhodopsin-induced photocurrent, 71, 71f
Restriction endonuclease
cleavage sites, 309
defined, 1, 810, 813
effect on DNA acquisition, 366
RESTRICTION ENDONUCLEASES AND DNA MODIFICATION METHYLTRANSFERASES FOR MANIPULATION OF DNA, **809–13**
Restriction enzymes, 811–12, 812–813
defined, 228, 813
Restriction fragment length polymorphisms, 527
Restriction fragment length polymorphism (RFLP), 221, 228, 361, 425, 428, 430f, 434, 812
for linkage analysis, 375, 376
map of *Arabidopsis*, 46
principles of, 435f
sites in the *PAH* gene, 679f
Restriction fragments, Mb, 422–23
RESTRICTION LANDMARK GENOMIC SCANNING METHOD, **813–17**
Restriction landmark genomic scanning profiles, 816–17, 816f
Restriction landmarks, defined, 813
Restriction maps, of CENP-B and CENP-C cDNA clones, 854f
Restriction-modification systems
defined, 810
DNA, 364
Retention sequences, 777
Reticular systems, intracellular, calcium in, 131
Reticulocytes, lysate from rabbit cells, 829
Reticulum, calcium ion transport systems of, 134–35
Retinal, 68, 68f
binding to rhodopsin, 792
in light-adapted bacteriorhodopsin, 101
11-*cis*-Retinal
regeneration of, 822
Retinal axons, 158
RETINOBLASTOMA, 137, 375, 378, 425
defined, 928
diagnosis of, using PCR, 226
gene for, 930-31, 930f
MOLECULAR GENETICS OF, **817–20**
tumor suppressor gene in, 518, 929–30
Retinoblastoma control element (RCE), 819
RETINOIDS, **820–24**
defined, 820
Retinoma, 818
defined, 817
Retroposition, transcription of *Alu* RNA in, 440
Retroposons, 439–40
defined, 420, 438
Retropseudogenes, of protein-coding mRNAs, 441
Retrotransposons, defined, 438, 924
Retrovir, 23
Retroviruses
defined, 15, 19, 834
information transfer in, 836, 840
for introduction of transgenes, 904
life cycle of, 839f
mouse mammary tumor virus, 873, 876
packaging of, 959
*See also* Endogenous retroviruses
Reversed micelles, defined, 303
Reversed phase chromatography (RPC), 419–20, 419f
Reversed-phase high performance liquid chromatography
defined, 767
for peptide mapping, 769f
for protein purification, 768
Reverse genetics, 375
with *Caenorhabditis elegans*, 588
with mycobacteria, 585
Reverse transcriptase, 834
with AG-LCR, to detect RNA, 463
defined, 19
HIV-1, 17, 17f
inhibition of, 22f, 23
HIV, photodynamic inactivation of, 955f
inhibition of, HIV treatment, 7
telomere specific, 439, 440
Reverse transcription-polymerase chain reaction (RT-PCR), defined, 803
Reversible reactions, 296–97
enzyme, 293
Rev protein, inhibition of, HIV treatment, 24
*R* factor, defined, 242
Rhabdomyosarcoma, genomic imprinting associated with, 380
Rhegnylogic organization, peptide, 659, 659f
Rheumatic diseases, 47
Rheumatic heart disease, M protein antibodies in, 66
*Rhizobium*, plasmids of, 713
Rhodamine, label for FISH, 353
*Rhodobacter sphaeroides*, structure of the reaction center of, 687f
*Rhodopseudomonas viridis*, 99
Rhodopsin, 212f
defined, 194
Rho factor, role in transcription, 349
β-Ribbon motifs, 749
Ribonuclease H
HIV-1, 17–18, 18f
drugs interfering with, 24
Ribonucleoprotein
catalysis by, 832–33
defined, 831
Ribonucleotide reductases, 550–51
Ribosomal RNA, transcription of, 236, 346
Ribosome binding site, 308–9
defined, 306
Ribosome cycle, function in translation, 920
RIBOSOME PREPARATIONS AND PROTEIN SYNTHESIS TECHNIQUES, **824–30**
Ribosomes
structure of, 825–27, 920, 920f
translation in, 837–38
RIBOZYME CHEMISTRY, **831–33**
Ribozymes, 838
defined, 621, 831, 834
in the "RNA world", 631
Ricin, administration with monoclonal antibody linkage, 262
Rieske Fe-S centers, 687–88
Rinderpest, vaccinia virus recombinant vaccine for, 945–46
Risk factors, for heart failure, 400
RNA
7SL, 440
antisense, in gene expression, 39
binding by Tat, 10

RNA *(Continued)*
catalysis by, 831
as a catalyst inhibitor, in bene expression, 107
catalytic molecules, 300
coding for, 861
*nutB* sequence, 13
processing of, defined, 831
self-cleaving, 831-32, 833f
self-splicing, 831
sequencing of, 611
splicing of
defined, 831
differential, 25
strand of, 836f
structure of, 831, 916f
virus, genes encoded by, 625
Ribonucleic acid. *See* RNA
RNA binding metalloregulatory protein
defined, 899
RNA editing, 707f
defined, 704
RNA polymerase, 236, 236t, 372, 839-40
defined, 234
effect of Tat on, 13-14
eukaryotic, 237t
RNA processing, and regulation of gene expression, 349-50
RNA REPLICATION, **833-40**
RNA SECONDARY STRUCTURES, **841-43**
defined, 306
RNase H
active site, HIV-1 reverse transcriptase, 17
cleavage of RNA, DNA-RNA hybrids, 43
defined, 39
RNA STRUCTURE, NONHELICAL, **843-46**
RNA THREE-DIMENSIONAL STRUCTURES, COMPUTER MODELING OF, **847-50**
"RNA world" postulate, 631
Robots, 54-55
Royalties
from breeders of patented animals, 908
defined, 907
rRNA, defined, 847
R state (relaxed state, cooperativity model), 405, 407
Rubredoxin, 302
Ryanodine receptor, 134, 400

S03, S45, S100 supernatants, defined, 824
*Saccharomyces cerevisiae*
2μm plasmid of, 711
Achilles' cleavage of genome, 2, 3
cell-free translation systems from, 829-30
DNA transformation in, 800-801
ethanol production using, 332
hsp70 of, 163-64
life cycle of, 973, 973f
lipid synthesis in, 502-3
*Saccharum officinarum*, 147
Safety considerations
in handling radioactive substances, 781
properties and radiation protection data, 782t
*Salmonella enterica*, lipopolysaccharide of, 510f
Salt bridges, stabilizing reaction complexes, 195
Sample preparation, electron microscopy, 270, 270f
Sandwich assay format, immuno-PCR analysis, 459
Sanger method, 621
SAR (structure-activity relationship)
defined, 257
and drug design, 258
Sarcomas, 188-89
defined, 928
p53 mutants in, 934
Sarcomere, 574-75
defined, 399
Sarcoplasm, 575
Sarcoplasmic reticulum, 575
calcium release channel of, 134
Satellites, 439, 439f
Bkm, 867
α-satellites, 439
Saturation, in enzyme kinetics, 299-300
Sawyer syndrome, 869-70
Scaffold
defined, 86
polymer
for bone regeneration, 91-92
for organ regeneration, 87-88, 87f
Scanning tunneling microscopy, 100, 102
techniques, 852, 852f
SCANNING TUNNELING MICROSCOPY IN SEQUENCING OF DNA, **851-53**
Scavengers, free radical, 329
Schiff base (SB), bacteriorhodopsin, 68, 68f, 69
Schizophrenia, genetic factors in, 596
*Schizosaccharomyces pombe*, chromosome size as standard for PFG, 422-23
Schwannomas, in neurofibromatosis, 591
Scintillation counters, 782-83
SCLERODERMA, 48
defined, 853
DIAGNOSIS WITH RECOMBINANT PROTEIN, **853-56**
Scope, of a patent, defined, 907
Screening
assays for, defined, 257
defined, 794
for recognition molecules, 798
Screening, drug identification, 259
mechanism-based, 21
penicillin inhibitor to HIV protease, 9
SDS-PAGE
defined, 767
for peptide preparation, 736
for protein/peptide preparation, 768-69, 772-73
second dimension, 345
Secondary CAMs, 154
N-CAM, 156
Secondary metabolites
defined, 668
as pesticides, 669-71, 671f
Secondary structure
defined, 843
NMR analysis of, 604
of RecA protein, 786, f
of ribozymes, 831
of ribozyme-substrate complex, 833f
of RNA, 841-43
Secondary structure, peptide
designed, 661
synthetic, 654
Secondary structure, protein, 320, 322f, 754
defined, 757
predicting, 759, 862
Raman spectrum analysis of, 739, 740f
Secondary structure analysis
circular dichroism for, 182-84
circular dichroism utilized in, 179
defined, 179
Second messengers, 279-81
in activation of transcription factors, 283f
calcium, 131
for tryptophan hydroxylase, 127
in tyrosine hydroxylase activation, 126
cyclic nucleotides, 132, 279
defined, 131, 682, 762, 789
generation by regulatory phospholipases, 685
inositol 1,4,5-phosphate, 134
interaction with protein kinases, 763, 763f
phospholipids as sources of, 683
in postsynaptic cells, 541
signal transduction by, 132, 542f
α-Secretase pathway, for processing APP, 27-28
β-Secretase pathway, for APP cleavage, 28
Secretory glycoproteins, defined, 388
Secretory phospholipase $A_2$, 681
Secretory proteins, transport of, 777
Sedimentation, to measure DNA denaturation, 209
Sedimentation coefficients, defined, 320
Sedimentation rate, for classification of ribosomes, 825
Segregation
in cell replication, 57, 61-62

genetic, 433f
defined, 128
and heteroplasmy, 552
Segregation analysis, of breast cancer, 128
Selected line, defined, 254
Selectide process, synthetic peptide library, 882–83, 883f
Selectins, 157–58
E-, 157
L-, 157
P-, 155f
Selection
affinity-based, from phage display libraries, 191–92
genetic, in aging studies, 6
of transformants, plant cells, 694–95
Selective induction of enzymes, 118
Selectivity
adhesive, 154
of phosphoprotein phosphatases, 764
Selenium, in glutathione peroxidase, 329
Self-administration studies, animal models for drug use, 256
Self-cleaving RNA, 831–32
defined, 831
Self-overlap capacity, 889
Self recognition, 523–24
Self-splicing RNA, 831
Sellers method, for local alignment, 858
Semiconductor colloids, 468
Senescence, defined, 158
Sense strands, 838
defined, 834
Sensitivity
of bioluminescence and chemiluminescence assays, 165
of immuno-PCR, 459
of ligase chain reaction, 463
Sensitization
in *Aplysia*, 541, 542f
defined, 541
Separation, defined, 794
Sequence
amino acid
mass spectrometry for determining, 733
and protein structure, 758
defined, 856
Sequence alignment
from dot plots, 862
in homology modeling, 759–61, 761f
multiple, 858
SEQUENCE ALIGNMENT OF PROTEINS AND NUCLEIC ACIDS, **856–59**
SEQUENCE ANALYSIS, **859–62**
Sequence checking, 443
Sequence classes, 251, 251f
Sequence database searching, 862
Sequence-dependent conformational distortion, 808, 808f
SEQUENCE DIVERGENCE ESTIMATION, **863–66**
Sequence organization, *Arabidopsis*, 46
Sequence pattern, defined, 888
Sequence-specific recognition, 748
Sequence-tagged sites (STSs)
chromosomal, 3, 646
defined, 420
expressed, 587
for mapping, 423
mouse genome, 570
Sequencing
DNA
in colon carcinomas and adenomas, 188
defined, 851
denaturation step in, 210
direct, 427
and genome determination, 528
Sequencing by hybridization, 442f
defined, 441
Sequencing chip, 443
defined, 441
Sequencing gel, 609, 609f
Serine, β-lactone derivatives of, 30
Serine proteases
defined, 409
kallikreins as, 489–90
kininogen substrates for, 486–87
nexin-2 inhibitor, 27
Serine-threonine kinases, 279, 281t
Serotonin
biosynthesis of, 124, 125f
as a transmitter, 541, 542f
Serpentine receptors, 278, 278f, 279
activating mutations in, 283
Serum cholinesterase, mutant alleles in deficiency of, 676
Severe combined immunodeficiency syndrome (SCID) mice
genetic mechanism, 926
transplanting human cells to, 462
Severity score, Gaucher disease, 343f
Sex, defined, 232
Sex chromosomes, 525
SEX DETERMINATION, **866–70**
defined, 866
Sex-determining genes
autosomal, SRA1, 869
X-linked, SRVX (*DSS*), 868–69
Sex-determining region, Y (*SRY*) gene, 868
Sex reversal
defined, 866
XX, 869
XY, 869–70
Shannon entropy, 891
β-Sheet
defined, 16, 746, 757, 979
reverse transcriptase, 17
ribonuclease H, HIV-1, 18
β-Sheet proteins, design of polymers based on, 716
Sherman paradox, 326
Shock, role of the kallikrein-kininogen-kinin system in, 485–86
Shortest distance chart, 891
Short tandem repeats (STR)
defined, 374
for genetic testing, 375, 647
Sialic acid, defined, 154
Sickle cell anemia, 373–74, 406, 408, 408f
point mutation analysis, 377
point mutation in, 561
prevalence of, 724
Side chains
peptide
constraint on conformations, 660, 660f
identification with Raman spectroscopy, 739
protecting during synthesis, 665
placement by homology modeling, 760–61
Sigma factors $\sigma^{32}$, 400–401
defined, 400
in prokaryotic transcription, 348–49
Signal
defined, 350
targeting, 774
translation, 308–9
Signal detection, in FISH, 351f, 353
Signaling
by metal-binding proteins, 899
in plant defense against pathogens, 702
synaptic, 877
Signal recognition particle, 777
defined, 774
Signal sequence, defined, 774, 859
Signal transducing factor, calcium ion, 132
Signal transduction
control by phosphorylation, 762
control by vasopressin, 211–12
defined, 473, 474
by interferon systems, 474
by interleukins, 475–76, 476f
by peptide hormones, 651
principles of, 132
protein kinase used in, 764
role of phospholipase C in, 682
Signal transformation, 877
Signal visualization, in FISH, 353–54
Silicone elastomers, medical, 721
Silicone implants, for cartilage replacement, 92
Silks, 715
Simian immunodeficiency virus (SIV)
activation domain of Tat, 12
experimental vaccination for, 946
recombinant vaccinia virus and baculovirus vaccines for, 947
Simian virus SV40, 189

Simple sequence repeats (SSRs), 439, 527
Sindbis virus, packaging of, 959–60
SINE (Short Interspersed Repeat) element, 422, 439–40, 525, 925f
  *Alu*, 924, 926–27
  defined, 420
Single-chain technology, for engineering proteins, 456–57, 456f
Single-gene disorders, 424
Single-strand conformational polymorphism (SSCP)
  analysis of mutations with, 725
  defined, 374
  for screening, mutation identification, 377, 427
Singlet oxygen, defined, 953
Site-specific DNA inversion, 365
Site-specific drug delivery, 261
Site-specific mutagenesis
  to control enzyme activity, 305
  helix swap experiment, 807
*Situs inversus*, 169
Size exclusion chromatography (SEC), 418
Skeletal muscle, 573–75
  organization of, 574f
Skin cancer, from UV damage, 943
Small-cell lung cancer
  bombesin production by, 519
  defined, 518
Smith-Waterman method, for local alignment, 858
Smith-Waterman search, 862
Smooth colony morphology, bacterial, 511
Smooth muscles, 573
  nonvascular, effects of kinins on, 484
  organization of, 576
SNAP-25, 879, 879t
Sodium channel, membrane, 877
Sodium dodecyl sulfate polyacrylamide gel electrophoresis (SDS-PAGE), for apolipoprotein determination, 513
Sodium ion/calcium ion exchanger, 134
Software, architecture of, automated genome research, 56
Solenhofen deposits, 634
Solid phase chemistry, for synthesis of oligonucleotides, 43
Solid phase peptide synthesis, 664–65, 665f
  defined, 664
Solid phase technique
  supports for, synthesis of peptides, 667
  for synthesis of peptides, 656–57, 656f
Solid substrate fermentation (SSF), 336, 337f
  defined, 335
Solid tumors, diagnosis of, 227–28
Soluble guanylyl cyclase, defined, 597
Soluble ions, association with proteins, in food, 322
Solvent engineering versus protein engineering, 304–6, 306t
Somaclonal variation, 288
Somatic cell gene therapy, 428
Somatic cell hybrid
  defined, 354
  for mapping microsatellite markers, 355
Somatic mutation theory
  in aging, 4–5
  defined, 4
Somatic recombination, 526
Somatic self, 449
Somatomedin, insulinlike growth factor, 393
SOS box, 713
SOS regulatory system
  in bacteria, 788
  defined, 785
Source gene, defined, 438
Space group, crystals, 742
Spatially addressable format, synthetic peptide libraries, 881
Species integrity, animal right to, 909
Specific activity
  enzyme, 289
  radioactivity per unit mass, 779, 780
Specificity, 1
  of Achilles' cleavage, 2, 2f
  of antibody binding, expression of, 37
  of autoantibodies, in clinical diagnosis, 48, 48t
  of cadherin binding, 156–57
  chiral, of tRNA synthetases, 170
  of enzyme reactions, 295
  of enzymes
    and chirality, 171–72
    and solvent characteristics, 305
  geographic, of mtDNA patterns, 556
  of immune phenomena, 449, 453–55, 455f
  of PCR amplifications, 644
  phospholipase, 680
  of repressor-operator interactions, 808
  of resins for protein purification, 767
  of RNA reactions with proteins, 623–24
  substrate, of isoenzymes, 481
  of synthetic oligonucleotides, 39–44
Spectra, one-dimensional NMR, 602, 602f
Spectrin, electron micrographs of, 271f
Spectropolarimeter, 179–80
Spectroscopic sensors, 113
Sperm genotyping, 438
Spheroplast, defined, 973
Spheroplast fusion, 975
  hybridization by, 974f
Sphingomyelin, and hormone mediation, 281
Spinal-bulbar muscular atrophy, 378
Spinal muscular atrophy (SMA), 424, 561
Spiro-bicyclic thiazolidine analogues, in peptide design, 663
Spirolactam analogues, in peptide design, 661–63, 663f
Spliceosome, 349–50
Splicing
  abnormal, in hemophilia, 410
  alternative
    of pre-mRNA, 311
    of pre-mRNA in elastin genes, 313–14
    of pre-mRNA in fibronectin genes, 314–15
  *APC* gene, 188
  in the chloroplast of *Chlamydomonas reinhardtii*, 174
  defined, 73, 173, 573
  gene, high and low molecular weight kininogens from, 483
  process of, 838, 975–76, 975f
Splicosome, 832–33
Spontaneous mutation. *See* Mutation, spontaneous
Sporadic disease
  defined, 589
SPOTs membrane synthetic peptide libraries, 881
Squalene synthetase, 497
ssRNA, structure of, 959
Stability
  defined, 621
  of hybrids, 605–7
  of oligonucleotide duplexes, 606–7, 606f
  of proteins, estimating from circular dichroism data, 184
Stability ruler, metal ions, 85, 85t
Stability sequences, metal ions, in biological systems, 84–85
Stacking forces, defined, DNA, 207
Stacking interactions
  in RNA, 845
  in sequencing chips, 443
Staphylococcal nuclease
  in analytical affinity chromatography, 795
  analytical affinity chromatography of, 796f
STAR (Sequence-Tagged Restriction Site), 423
  defined, 420
"Star" activity, defined, 810
Starch, 148–49, 718
  defined, 146
Static model, protein structure, 759
Steady state, enzyme reaction, 290, 299–300
  defined, 289
Steady state kinetic methods, 294
Stem cell factor, 393, 395
Stem cells, defined, 392. *See also* Embryonic stem (ES) cells
Stereoisomers, defined, 168
Stereospecificity, of enzymes, 172
Sterilization, in plant tissue culture, 885
Steroid hormones
  biosynthesis of, role of P450, 198
  defined, 870
STEROID HORMONES AND RECEPTORS, **870–73**

Steroid receptors
defined, 870
hsp90 binding to, 164–65
Sterol regulatory elements (SREs), 497
Sterol response elements (SREs), 153
Stirred tank reactor (STR), 336
Stokes shifts, 317
Stop codons
defined, 428
in hemophilia A, 410
Stopped-flow analysis, 293–94, 294f
STR (short tandem repeat), 221
defined, 220
β-Strand, defined, 757
Strand exchange, defined, 799
Strand separation temperature, defined, 605
Strecker synthesis, asymmetric, synthetic amino acid preparation, 32
Streptavidin
defined, 458
in oligonucleotide ligation assay, 491
*Streptococcus*, group B, capsular polysaccharides, 142
*Streptococcus pneumoniae*, vaccination against, 143
*Streptococcus pyogenes*
fibronectin receptor for lipoteichoic acid of, 65
immune system responses to, 66
mimicry of host markers, 66
*Streptomyces*, defined, 668
*Streptomyces avermitilis*, 671
*Streptomyces avidinii*, 458
*Streptomyces griseochromogenes*, 671
*Streptomyces hygroscopicus*, 671
*Streptomyces kasugaspinus*, 671
*Streptomyces viridochromogenes*, 671
Stress fibers, 205
Striated muscles, 573
Stroma, defined, 99
Structural alteration, in activation of oncogenes, 625–26
Structural proteins, defined, 714
Structural refinement, X-ray crystallography, 745–46
Structure
of antibodies, 36–37, 36f
for defense against plant pathogens, 700
dependence on metallic trace elements, 899
determination of, X-ray diffraction, 970–71
diphenyl oxazinones, 29
of γ-endotoxins, 669
imidazolidinones, 29
incorporating data about, in sequence alignment, 858–59
RNA, 916f
primary, secondary and tertiary, 842f
types of, 841
RNA and DNA, 834–38
*See also* Secondary structure; Tertiary structure
Structure factor, 742–43
defined, 741
equation for, 742
STS. *See* Sequence-tagged sites
Subcellular organelle
defined, 774
Submerged liquid fermentation (SLF), 336–38, 338f
defined, 335
Substituted hydrocarbons, metabolism of, 506–7
Substrate
defined, 295
depletion of, and rate of enzyme reaction, 290
effect of concentration on initial rate of reaction, 299f
Substrate adhesion molecules (SAMs), 154
Subtelomeric repeats, 439
Subtilisin E, nonaqueous biocatalysis by, 306
Subtractive hybridization, for cloning, tissue-specific genes, 698, 699f
Subunit, protein, defined, 15, 295
Sucrose, 147
Sugar chains, 382–85. *See also* Oligosaccharides
Sugars
defined, 146
modifying
in oligodeoxyribonucleotides, 622–23
in oligoribonucleotides, 624
Sulfate reduction, by anaerobic bacteria, 78
Sulfonamides, 259
Sulfotransferases, 119
Sulfur-based linkages, in phosphate-free oligonucleotide analogues, 620
Sulfur-dependent organisms, 301
enzymes from, 302t
SUPERANTIGENS, **873–76**
Supercritical fluid, defined, 303
Superfamily
defined, 197
P450, 199
Superoxide dismutase (SOD), 287, 329, 329f, 549, 549f, 562, 563
Superoxide radical, 328
Supersaturation, defined, 466
Support film, defined, 269
Surface exclusion, 712
Surface mass loading, in optical or acoustic wave biosensors, 113
Surface plasmon resonance (SPR), 113, 796f
Surface plasmon resonance (SPR) biosensor, 795
Surfactants, organized array of, 467
Surprisals, 892
slope of, versus complexity, 895f
Surrogate markers, for disease progression, HIV-1, 24
Sweetener, defined, 146
Switching
through electrostatic interactions, 195–96
by G protein interactions, 279
Sychnologic organization, peptide, 659, 659f
Symbols, for amino acids, 655t
Symmetric inhibitors, $C_2$-, of HIV-1 protease, 22f, 23
Symmetry, bilateral, in animals, 169
Symport, 923
defined, 922
SYNAPSES, **876–80**
defined, 541
Synapsins, 197, 879
Synaptic cleft, 877
Synaptic function, modulation of, 880
Synaptobrevin, 879, 879t
Syneresis, 320
Synexin (annexin VII), 33f
promotion of intermembrane contacts, hypothetical, 35, 35f
Synonymous substitutions, 864
defined, 863
estimating the number of, 865–66, 865f
Syntaxins, 879, 879t
Synteny, 525
Synthesis
biomimetic, advantages of, 96
of glycogen, 385–86
of heat shock proteins, 400–401
of sigma$^{32}$, 401
of polymers, 714–15
of steroids, 871–72, 872f
Synthetase
binding of, effect on tRNA structure, 846
HMG CoA, control of cholesterol levels, 153
SYNTHETIC PEPTIDE LIBRARIES, **880–83**
Synthetic polymers
biodegradable, 719–20
defined, 717
Systemic acquired resistance, in plants, 702
Systemic lupus erythematosus, autoantibodies in, 48–49
System integration, defined, 50
Systems, automation, 55–56

TACA, 144f
Talin, in focal adhesions, 205
Tamoxifen, breast cancer response to, 873
Tandem mass spectrometry (MS/MS), 533
defined, 529
Tandem repeats, 222f, 428, 438–39
defined, 220, 221, 374
in rodent kallikrein gene families, 489
TAR binding protein 1 (TRP-1 for TRP-185), 11

Target-based drug design, 259–60
Targeted mutagenesis, 528
Target-independent ligation, 463–64
Targeting
  antibiotic, modification by plasmids, 712
  bacterial, 778
  infectious disease testing with LCR, 465
  lysosomal, 777
  polynucleotide, defined, 463
  protein, 774–78
Targeting patch, defined, 774
Targeting pathways, 774–75, 775f
Targeting sequence, 775t
  defined, 350, 774
Tarui's disease, 581
Taste, of amino acid enantiomers, 171
Tat, 12
  defined, 10
  HIV-1, inhibitors of, 22f, 23
  HIV, model for, 13f
  transactivator, HIV-1, 12–14
  transcription in the absence of, 11–12
TATA-binding protein (TBP)
  defined, 346
  in RNA polymerase transcription, 346
  three-dimensional structure of, 697
TATA box, 14, 749–50
  defined, 346
  promoter region, HIV-1 LTR, 11, 11f
  promotor sequence, histone octamer binding, 179
Tau protein, in neurofibrillary tangles, 27
Taxane, defined, 885
Taxol, 535
  defined, 885
  production of, 108
  structure of, 535f
TAXOL AND TAXANE PRODUCTION BY CELL CULTURE, **885–87**
Taxonomy, of yeasts, 977
*Taxus*, defined, 885
Tay-Sachs disease, 482, 508
  among Ashkenazi Jews, founder effect, 727
  in French Canadians, independent origin of, 728
  prevalence of, 724
T-bag synthesis, 667
T-cell factor 1a (TCF-1a), 11f
T-cell lymphomas, 136
T-cell receptors, 451–52
  antigenicity of, 455
  binding site on superantigen, 875
  defined, 873
  genes coding for, 452–53, 453f
  interaction with superantigen, 874–75, 875f
  single-chain engineering of fragments, 456–57
  specificity of, 453–55
T cells, apoptosis of, 161
Telomerase, 439
Telomeres, tandem repeats of, 439
Tempeh, production of, 336
Temperature
  Arrhenius-type dependence on, 274
  and bacterial virulence, 66–67
  and copy number, 308
  and DNA degradation, 207
  electron transfer independence of, 273
  food processing, 323
  high, enzymes active at, 301
  normal human, and DNA damage, 218
Template
  defined, 608
  generation of, nucleic acid sequencing, 609
Template-directed synthesis, defined, 628
Template strand, defined, 914
Tenascin, 315
Tendon, structure, 96f
Tensin, focal adhesion protein, 205, 206
Teratocarcinoma cells, 925
Termination
  codon for, 920
  in polypeptide synthesis, defined, 914
  in RNA replication, 346, 349
    eukaryotes, 350
Terminator, 236
Terminus, replication, defined, 57
Terpenes, chirality of, 171
Terpenoid lipids, defined, 501
Tertiary structure
  carbonic anhydrase, 321f
  defined, 843
  determining with NMR, 602
  peptides, 648–49
  RNA, 841
Tertiary structure, protein, 320, 322f, 754
  from circular dichroism data, 184
  defined, 757
  predicting, 759
  spectrum analysis of, 739
Testis-determining factor (TDF), 525
  defined, 866
  search for, 867–68
Tetrahydrobiopterin ($BH_4$), 124–25
1,2,3,4-Tetrahydroisoquinoline-2-carboxylic acid (Tic), in peptides, designed, 663
Tetraloop, defined, 841
Tetrazole analogues, peptides, 660
Texas red, label for FISH, 353
Thalassemia, 926
β-Thalassemia, 425
Thalidomide, 171, 673
THEORETICAL MOLECULAR BIOLOGY, **888–95**
Therapeutic, defined, 794
Therapy
  dietary, for PKU, 678
  gene, 557
  genetic, 428
  in heart failure, 400
  hemopoietic growth factors used in, 395
  for manic-depressive illness, and nephrogenic diabetes insipidus, 214
  and molecular information about cancers, 137–38
  with peptides and peptidomimetics, 652–53, 653t
  retinoic acid, in acute promyelocytic leukemia, 227
  using engineered immunotoxins, 456
  *See also* Drugs; Medicinal chemistry; Peptide therapeutics
Thermal denaturation, of monomeric proteins, 729
Thermal stress, genomic response to, 285–86
*Thermococcus litoralis*, formaldehyde oxidoreductase of, 303
Thermocycler, defined, 463
Thermodynamics
  of biological energy transfer, 80
  of chemical reactions, 296–97
Thermogenesis, in uncoupled oxidative phosphorylation, 81
Thermophiles, lipids of, 501
Thermostable DNA ligase, defined, 463
Thermostable DNA polymerase, defined, 463
*Thermus aquaticus*, source of *Taq* polymerase, 642
Thiazolidines, bicyclic, in peptides, 661, 662f
Thick filament, myosin, 577
Thin-layer chromatography, defined, 672
Thioformacetal linkages, in deoxyoligonucleotide analogues, 620
Thionins, 701
Thiopurine *S*-methyltransferase, deficiency of, 677
Thiostatin (T-kininogen), 487, 488–89
Three-center hydrogen bond, 445
  defined, 444
Three-point attachment hypothesis, chiral recognition, 171
Threonine, derivatives of, 30
Thrombolysis, genetically engineered fusion protein for, 38
Thrombospondin, 315
Thylakoid, defined, 686
Thylakoid lumen, 688f
Thymine (T), 370
Thyroid disease, and Alzheimer's disease, 26
Thyroid hormones, nuclear receptors for, 278
Thyroiditis, autoimmune, 47
Thyroid peroxidase, 328
Thyroid-stimulating hormone, 385
Tiazofurin, inosine monophosphate dehydrogenase inhibition by, 137–38

TIBO, reverse transcriptase inhibitor, structure, 22f
Time constant, for bacteriorhodopsin to state K conversion, 70
Time-of-flight mass spectrometer, 530
 defined, 529
Time-resolved spectroscopy, defined, 68
Tin oxide, electrode, bacteriorhodopsin-based photocell, 70, 72
Ti plasmid, defined, 469, 696
Tissue culture, genomic changes induced by, 288
Tissue plasminogen activator, production from recombinant organisms, 108
Tissue-specific locus control element, defined, 704
Tissue-specific transcription, 348
Titin, in muscle, 580
TLC (thin-layer chromatography), 138
T lymphocytes
 CD4$^{+}$, antigen and superantigen recognition by, 873
 defined, 460
 production of interleukin, 475
 stimulation of IFN-$\gamma$ production, 473
$T_m$, defined, DNA, 207
Tobacco mosaic virus (TMV), 840, 958, 958f
Tocopherol, as a free radical scavenger, 329
Toga viruses, 839
Tolerance, defined, 254
TOMATOES, GENE ALTERATIONS OF, **896–98**
Totipotency, defined, 689
Toxicity
 of antisense oligonucleotides, 44
 of metals, 84
 of nucleoside analogues, 23
 of radioisotopes, 780–81
 of vitamins, overdose, 963
 and xenobiotic metabolism, 120
Toxicology
 postmortem, 675
 studies of potential drugs, 674
Toxic shock, treatment with antibodies to TNF-$\alpha$, 204
Toxic shock syndrome, 875–76
 binding of toxin in, 874
 toxin in, X-ray crystallographic studies, 972f
Toxic Substances Control Act (TSCA), 116
Toxin genes, *Bacillus thuringiensis*, 471t
Toxins
 botulinum, 146
 cholera, 146
 defined, 260
 implication in sporadic motor neuron disease, 562
 in plant diseases, 702
 and primary liver cancer, 517
 secretion by bacterial pathogens, 66
 secretion by plasmids, 713
 tetanus, 146
Trace analysis, of drugs, 673
TRACE ELEMENT MICRONUTRIENTS, **898–901**
 defined, 899
Trans-acting factor
 cloning genes that encode, 698
 defined, 696
 types of, 697, 697t
Transactivation, defined, 19, 23
Transactivation response (TAR) element
 defined, 10
 HIV-1 RNA sequence, 12f
 HIV LTR, 11
Transactivation response element, effects on translation, 14
Transcription, 236, 238, 372
 defined, 73
 factors mediating hormone action, 282t, 283f
 histone, 415f
 HIV, in the absence of Tat, 11–13
 of *hsp* genes, 401
 inactivation by triplex-forming oligonucleotides, 43
 in metazoa, 552
 and nucleosomes, 178–79
 processive and nonprocessive, 10
 process of, 837f
 regulation by trace elements, 900, 900f
Transcription activation domain (TAD), 277
Transcriptional activator, papilloma virus, 749
Transcriptional control, 806
 defined, 413
 operon hypothesis of, 807
Transcription factor
 ATF/AP-1 family of, 415–16
 defined, 234, 746, 817, 928, 979
 DNA-binding, 747
 in eukaryotes, 747
 GALA, yeast, 983, 983f
 in histone gene expression, 416f
 plant nuclear, 697f
Transcription factor D for polymerase II (TFID), 749–50
 defined, 346
Transcription signal, 308–9
Transcription-translation systems, coupled, 828
Transcription unit, defining, 346–49
Transcripts
 *Drosophila melanogaster* genome, 252
 identifying, human disease mapping, 427
Transducer, defined, 110
Transduction
 bacteriophage-mediated, 364
 defined, 625
 photosynthetic energy, 686
 of plasmids, 711
 virus-mediated, 365
 visual, 820, 822
Transduction domain, calcium ion pump, 134
Transfection
 defined, 557
 for protein targeting, 777
 techniques for, 515
Transfer, of genes, methodology, 911–12
Transferases
 defined, 675
 glycosyl, 390–91
 oligosaccharyl, 390
Transferrin, 85
 clearance of, 384
Transfer RNA. *See* tRNA
Transform
 defined, 896
Transformation
 defined, 625, 689, 799
 DNA, 364, 365
 of plasmids, 711
 yeast, using a plasmid vector, 976f
Transformation products, in drug synthesis, 673
Transformation rescue, 253
Transformation vectors, Ti plasmid, 897
Transforming growth factors (TGF-$\alpha$), production by lung cancer cell lines, 519
Transforming growth factors (TGF-$\beta$), 394–95
Transforming growth factors beta (TGF$\beta$), 805
Transgenes
 defined, 910
 gain-of-function, 911, 913
 identification of, in successful transfer, 912
 loss-of-function, 911, 914
 reporter function, 911
Transgenic, defined, 106
TRANSGENIC ANIMAL MODELING, **901–7**
 in tumor suppressor gene studies, 936
TRANSGENIC ANIMAL PATENTS, **907–9**
Transgenic animals
 defined, 901
 production of, 904–5
TRANSGENIC FISH, **910–14**
 defined, 910
Transgenic founder, defined, 901
Transgenic line, defined, 901
Transgenic mice
 apolipoproteins expressed in, 708t
 defined, 704
 from embryonic stem cell transfer, 903
 models for lipoprotein metabolism, 709
 for study of Down's syndrome, 250
 for study of the mammalian genome, 528

Transgenic organisms
as bioreactors, 108
defined, 150, 407, 469
human hemoglobin expression in, 408
plant, 472
for study of lipoproteins, 152–53
Transgenic plants, defined, 689, 696
Transient kinetic analysis, in enzymology, 293–94
Transient state, defined, 292
Transition metal ions, role in hydroxyl radical formation, 329
Transitions, in protein folding, 755
Transition state
defined, 295
for folding, 756–57
stabilization of, in enzymatic catalysis, 298
theory of, 297–98
Transition-state analogues, 105–6, 106f
defined, 103
HIV-1 protease inhibitors, 22f, 23–24
Translation, 372
defined, 306, 479, 824, 915
inactivation of, by antisense oligonucleotides, 43
from mRNA to protein, 837–38
of prokaryotic mRNA, 921f
regulation of, trace element role in, 900
ribosome cycle in, 920
TRANSLATION OF RNA TO PROTEIN, **914–22**
Translation signal, 308–9
Translocation, 373
activation of proto-oncogenes, 225
defined, 224, 824, 915
detection of, with FISH and WCPs, 967
in follicular lymphomas, 161
group, 922
identification with PCR, 646–47
into the endoplasmic reticulum, 777
oncogene activation resulting from, 626
Translocation, chromosome, 928
defined, 928
flow dot-blot analysis, 358t
Philadelphia chromosome creation by, 226, 226t
reciprocal, 227
defined, 354
Y-X, in XX males, 869
Translocation, protein
defined, 162
by hsp70s, 163, 164f
Transmembranal glycoproteins, defined, 948
Transmembrane electron transfer, conditions for, 276
Transmembrane receptor kinases, 764
Transmembrane signaling, and vasopressin receptors, 211–12, 212f
Transmission, of plasmids, 711–12
Transmission electron microscope (TEM), 269–70
Transplantation, 86–87
Transport
in biosensors, 111–12
cell, 108–9
electron, 687–88
endogenous, of lipids, 496–97, 496f
exogenous, of lipids, 495–96, 495f
of lipids, 150–51, 495
mitochondrial, of calcium ion, 135
of phospholipids, 685
of proteins, in vesicles, 777
proton, in the photocycle of bacteriorhodopsin, 69
reticular system, of calcium ion, 134–35
reverse, of cholesterol, 150–51, 497
of steroid hormones, 872
transmembrane, of calcium ion, 133–35
tubular, 805–6
Transporter, defined, 922
TRANSPORT PROTEINS, **922–23**
Transposable elements, 288
defined, 250, 285
in mycobacteria, 584
Transposition
defined, 363
L1H, 925
of mobile genetic elements, 365–66
Transposonlike human element 1 (THE1), 441
Transposonlike human element (THE-1), 927–28
TRANSPOSONS
of *C. elegans*, 588
for cloning plant regulatory genes, 698
composite, 366
defined, 438, 583, 587
IN THE HUMAN GENOME, **924–28**
*Trans*-retinal, defined, 99
*Treponema pallidum*, transmission of, 65
Triacylglycerols
defined, 501
metabolism of, 507–8
Triazolam, 259
Tricarboxylic acid cycle, 81, 82
Trichloroethylene (TCE)
biodegradation of, 77–78, 77f
dehalogenation of, by methanogens, 78
*Trichoderma reesei*, 333
Trigger, for membrane fusion, 538
Triglyceride and phospholipid metabolism, 498
Triglycerides, metabolism of, 507
Trinucleotide repeat expansion, in fragile X syndrome, 378
Tripeptides, NMR analysis of, 602–3
Triplex, defined, 785
Trisomy
16, in mice, 250
defined, 247
*See also* Down's syndrome
tRNA, 837–38, 846, 918
structure of, 836f
transcription of, 236, 346
tRNA synthetases, chiral specificity of, 170
Trophic factors, 161
Tropocollagen, 718
Tropoelastin, 313–14, 314f
Tropomyosin (TM), in thin myofilaments, 573, 578
Troponin (TN), 578–79
C, calcium-binding sites of, 133
in thin myofilaments, 573, 578
Trypsin, for preparing peptides from proteins, 735–36
Tryptophan
fluorescence of, 317
regulation of production, in prokaryotes, 349
Tryptophan hydroxylase, 127
T state (tense state, cooperativity model), 405, 407
Tuberculosis, 583
Tubular transport systems, 805–6
Tubuloglomerular feedback mechanism, role of nitric oxide in, 804
Tumor-associated carbohydrate antigens, defined, 141
Tumor necrosis factors, 200
list, 203t
Tumor progression models, for neoplastic disease diagnosis, 225
Tumors
from activating endocrine mutations, 284
defined, 136
effects on
of antisense oligonucleotides, 44
of hepatocyte growth factor, 395
of photodynamic photosensitizers, 953
embryonal, genomic imprinting associated with, 380
induction by retroviruses, 625
innervation of, 393–94
resistance to therapy, 284
*See also* Neoplasm
TUMOR SUPPRESSOR GENES, 627, **928–36**
*APC*, *p53*, *DCC*, 188
association with chromosomal loss, 228
defined, 187, 224, 428, 518, 589, 625, 928
inactivation of, 225
MCC, 928
mutation in lung cancer, 519t
*NF1*, 591, 928
*NF2*, 591–92
*nm23*, and lung cancer, 520–21

*p53*, 928
and liver cancer, 517-18
and lung cancer, 520
*RB*, defined, 928
*WT1*, 929
Tumor viruses, information flow in, 834
Tunneling, 273, 276
defined, 99, 851
*See also* SCANNING TUNNELING MICROSCOPY IN SEQUENCING OF DNA
Tuple, 858
defined, 856
*k*-Tuple
defined, 859, 888
Turcot syndrome, 189, 190
Turnover
DNA, 359
enzymatic, 298, 305
Turnover number, enzyme, 289
Two-center hydrogen bond, 445
defined, 444
Twofold ($C_2$) symmetry, peptidic inhibitors, HIV protease, 9
Two-hit model
in retinoblastoma, 929
for tumor progression, 591-92
Two-state model (MWC model), defined, 407
Type II lipids, defined, 537
Tyrosine hydroxylase, 124, 125-27
Tyrosine kinases
epidermal growth factor receptor, 392
ligand-stimulated, 793
receptor (RTKs), 278
role in cytoskeleton-membrane interaction, 206
stimulation by interleukin receptor binding, 476

Ubiquinones, 185
UDP-glucuronosyltransferse, deficiency of, 677
Ultraviolet radiation (UVR), defined, 939
ULTRAVIOLET RADIATION DAMAGE TO DNA, **939–43**
Ultraviolet resonance Raman (UVRR) spectroscopy, 737
Unequal crossing over, defined, 483
Unfolded state (U), proteins, 754
Unfolding, rate of reaction, 756, 756f
Unified theory, of aging and sexual reproduction, 232
Uniport, 923
defined, 922
Unit cell, volume of, 743
Units
of DNA transcription, 236-38, 238f
of energy, 612
of enzyme activity, 289
of genetic distance, 421
of molecular weight, 573
of radioactivity, 780, 780t
of RNA transcription, 346-49
Universal ancestor, defined, 301
Unpaired electrons, free radical, 328
Unsaturated fatty acids, defined, 501
Upstream activating sequences (UAS), 347
Upstream stimulatory factor (USF), 11f
Urease
metalloenzyme, 546
in a potentiometric biosensor, 112
Uridine-rich small nuclear ribonucleoprotein particles (UsnRNPs), defined, 346
USDA, 114
Uvomorulin, 156

Vaccination, 367-68
with polysaccharide antigen, 143
VACCINE BIOTECHNOLOGY, **945–47**
Vaccinia virus
defined, 945
for vaccine vector system use, 309-10
Vaccinia virus recombinant, 945-46
Validation, of protein models, 760-61
Variable (N-terminal) domains, of immunoglobulin light and heavy chains, 450
Variable expressivity, defined, 589
Variable number of tandem repeats (VNTR), 425, 434, 435f, 439
defined, 374
for genetic testing, 375
Variation generator, defined, 363
Vasoactive systems, 803-4
Vasodilation, nitric oxide's role in, 598-99
Vasopressin receptors, 211-12
defined, 210
Vasopressins, 211-12
defined, 210
regulation of water permeability, 806
VEC-DIC microscopy, defined, 564
Vectors
adenoviral, for transfection, 557
baculovirus expression, 946-47
defined, 945
delivery of genes by, 584
lambda phage, 229
replacement, in homologous recombination, 558f
retroviral, for transfection, 557
shuttle vector systems, 4
viral or bacterial
defined, 367
delivery of genes by, 368
Vertebrates, substitution rate for mtDNA in, 553
Very low density lipoproteins (VLDL), 150, 512
assembly and catabolism of, 496
defined, 495
Vesicles
phospholipid, for synthesis of inorganic solids, 468f
preparation of synthetic unilamellar, 467
Vesicular stomatitis (VS), vaccinia virus recombinant vaccine for, 946
Veterinary applications, DNA fingerprint analysis, 223
Vibrational normal mode, defined, 737
Vibrational spectra, 737
for DNA structure determination, 243
*Vibrio cholerae*
binding of, to $GM_1$, 146
transmission of, 65
Vimentin, 580
Vinculin
focal adhesion protein, 205
homology with α-catenin, 206
Viral capsid, difference Raman spectra of, 740f
VIRAL ENVELOPE ASSEMBLY AND BUDDING, **948–50**
Viral oncoproteins, defined, 817
Viral vectors, defined, 367
Virion, defined, 953
Virulence
bacterial, 66-67, 67t
defined, 65
Virulence factors, polysaccharide, 142
Viruses
cancer-causing, 136
classification based on nucleic acid in the genome, 838t
complex with antibodies, 451
defined, 307, 957
DNA and RNA of, 838-39
Raman spectroscopic analysis of structure, 739
release from cells, 948
RNA, icosahedral positive-strand, 958-59
in vitro activity of antisense or triplex-forming oligonucleotides, 44t
VIRUSES, DNA PACKAGING OF, **950–53**
VIRUSES, PHOTODYNAMIC INACTIVATION OF, **953–57**
VIRUSES, RNA PACKAGING OF, **957–60**
Viscometry, to measure DNA denaturation, 209
Visual cycle, defined, 820, 822-23, 823f
Visual information processing, defined, 70
Visual transduction, defined, 820
VITAMINS
A, in the visual cycle, 822
B complex, 185
coenzymatic forms and functions, 186t
D, role in calcium homeostasis, 131
defined, 185, 611
fat-soluble, 961-62, 961t

VITAMINS *(Continued)*
functions and requirements for, 613-14t
STRUCTURE AND FUNCTION OF, **960-63**
water-soluble, 962, 962t
$V_{max}$, defined, 295
VNTR (variable number of tandem repeats), 221, 228-29
defined, 220
VNTR analysis, 221-22
Voltage-gated ion channels, 878, 922-23
von Hippel-Lindau (VHL) syndrome, 934
von Recklinghousen neurofibromatosis, 590, 928, 934
von Willebrand factor, 205
binding of factor VIII to, 410
von Willebrand's disease, in Finland, founder effect, 727

Waardenburg syndrome, 424, 427
WAGR syndrome, 932
Walk-along theory of contraction, 580-81, 580f
Waste
biodegradation of, 75-79
organic, examples, 76t
Water channels, kidney, 806
Water molecule, hydrogen bonding in, 447
Water-soluble vitamins, 962, 962t
Watson-Crick base pairs, 39, 40f
Weak-field approximation, 267-68
Weight matrix, defined, 859
Werner's syndrome, 162
Western blotting, defined, 344, 511
Wheat germ, extracts for protein synthesis experiments, 829
WHOLE-CHROMOSOME COMPLEMENTARY PROBE FLUORESCENCE STAINING, **965-67**
Whole-chromosome probe (WCP)
defined, 965
generation of, 966t
Whole genome molecular analysis, 427
Wild-type p53-activated fragment 1 (*WAF1*), 932
Wilms' tumor, 378
defined, 928
gene for, 818, 932-34, 933f
Withdrawal, defined, 254
Wound healing, fibronectins in, 315

Xanthan gum, 149
Xanthin-guanine phosphoribosyltransferase, 674
Xenobiotic interaction, defined, 460
Xenobiotic-metabolizing enzymes, 676
Xenobiotics, defined, 117
Xenogeneic, defined, 86
Xenogeneic extrapolation, of immunotoxicity evaluations, 462
Xenografts, 88-89, 91
*Xenopus laevis*, defined, 979
Xeroderma pigmentosum (XP)
excision repair defects, 219
excision repair defects in, 943
X-linked conditions
defined, 374, 409
Duchenne muscular dystrophy, 377-78
X protein
defined, 516
and liver cancer, 517
X-ray crystallography
defined, 15
protein structure determination by, 744f, 758
X-RAY DIFFRACTION
by a crystal, 742-43
defined, 741
for DNA structure determination, 242, 244f
OF BIOMOLECULES, **969-72**
structure of annexin V, 34-35, 34f
X-rays, scattering by electrons, 742, 742f
XX males, 869
XY females, 869-70
XY gonadal dysgenesis, 868, 869-70
Xylose, defined, 331

Yeast artificial chromosome (YAC), 527, 571
defined, 45, 354, 420
fragments cloned into, 423
libraries of *Arabidopsis* DNA, 46
probe made from, 357
YEAST GENETICS, **973-77**
Yeasts
defined, 973
hemoglobin cloning in, 408
as host cells, 309
lipids of, 502-3
mutant, for protein targeting studies, 777
*Yersinia pestis*, 65
*Yersinia pseudotuberculosis*, integrin receptor use by, 66
Yops proteins, 66

Z disk (muscle), 574
Zeamatin, 701
Zidovudine, 23
effect on mitochondrial DNA, 554
Zinc
in enzymes, 549-50, 550f
in superoxide dismutase, 329
Zinc-binding domains, third class, 753
Zinc finger, 12, 697
ZINC FINGER DNA-BINDING MOTIFS, **979-84**
Zinc finger domains, 747
of the Wilms' tumor gene, 933
Zinc finger motif
classical, 979-81
defined, 899
Zinc finger proteins, 753
first class of, 750, 751f
second class of, 751-52
in steroid receptors, 871
Zinc-finger Y gene (*ZFY*), 867-68
Zoo blot approach, 427
Zwitterionic (molecule), defined, 28
Zymocin, yeast, 977
*Zymomonas mobilis*
ethanol production using, 332
xylose utilization by, 333